1285060   11-21-07 A1285060   Co
UCF
Ref

D1243169

STANDARD
HANDBOOK OF
ELECTRONIC
ENGINEERING

STANDARD HANDBOOK OF ELECTRONIC ENGINEERING

Donald Christiansen   Editor

President and Principal, Informatica; Fellow, The Institute of Electrical and Electronics Engineers; Fellow, World Academy of Art and Science; Eminent Member, Eta Kappa Nu; Editor Emeritus, IEEE Spectrum; *former Staff Director, IEEE; Registered Professional Engineer; formerly Editor in Chief,* Electronics; *Engineering Group Leader, CBS Electronics*

Charles K. Alexander   Editor

Professor of Electrical and Computer Engineering and Dean, Fenn College of Engineering, Cleveland State University; Fellow, The Institute of Electrical and Electronics Engineers; Director, Center for Research in Electronics and Aerospace Technology, CSU; former President, IEEE; Registered Professional Engineer

Ronald K. Jurgen   Associate Editor

Former Senior Editor, IEEE Spectrum; *Life Senior Member, The Institute of Electrical and Electronics Engineers; Editor, the* Automotive Electronics Handbook *and the* Digital Consumer Electronics Handbook

Fifth Edition

McGRAW-HILL

New York   Chicago   San Francisco   Lisbon   London   Madrid
Mexico City   Milan   New Delhi   San Juan   Seoul
Singapore   Sydney   Toronto

The McGraw·Hill Companies

Cataloging-in-Publication Data is on file with the Library of Congress

Copyright © 2005 by The McGraw-Hill Companies, Inc. All rights reserved. Printed in the United States of America. Except as permitted under the United States Copyright Act of 1976, no part of this publication may be reproduced or distributed in any form or by any means, or stored in a data base or retrieval system, without the prior written permission of the publisher.

1 2 3 4 5 6 7 8 9 0   DOC/DOC   0 1 0 9 8 7 6 5 4

P/N 138422-7
PART OF
ISBN 0-07-138421-9

The sponsoring editor for this book was Stephen S. Chapman and the production supervisor was Sherri Souffrance. It was set in Times Roman by International Typesetting and Composition. The art director for the cover was Anthony Landi.

Printed and bound by RR Donnelley.

Previously published as Electronics Engineers' Handbook, copyright © 1997, 1989, 1982, 1975.

This book is printed on acid-free paper.

McGraw-Hill books are available at special quantity discounts to use as premiums and sales promotions, or for use in corporate training programs. For more information, please write to the Director of Special Sales, McGraw-Hill Professional, Two Penn Plaza, New York, NY 10121-2298. Or contact your local bookstore.

Information contained in this work has been obtained by The McGraw-Hill Companies, Inc. ("McGraw-Hill") from sources believed to be reliable. However, neither McGraw-Hill nor its authors guarantee the accuracy or completeness of any information published herein, and neither McGraw-Hill nor its authors shall be responsible for any errors, omissions, or damages arising out of use of this information. This work is published with the understanding that McGraw-Hill and its authors are supplying information but are not attempting to render engineering or other professional services. If such services are required, the assistance of an appropriate professional should be sought.

CONTENTS

Part 3   Circuits and Functions

Section 11. Amplifiers and Oscillators   11.1

Section 12. Modulators, Demodulators, and Converters   12.1

Section 13. Power Electronics   13.1

Section 14. Pulsed Circuits and Waveform Generation   14.1

Section 15. Measurement Systems   15.1

Section 16.   Antennas and Wave Propagation   16.1

Part 4   Systems and Applications

Section 17.   Telecommunications   17.3

Section 18.   Digital Computer Systems   18.1

Section 19.   Control Systems   19.1

Section 20.   Audio Systems   20.1

Section 21.   Video and Facsimile Systems   21.1

Section 22. Broadcast and Cable Systems

Section 23. Navigation and Detection Systems

Section 24. Automotive Electronics

Section 25. Instrumentation and Test Systems

CD-ROM CONTENTS

Subject matter is listed as it relates to Sections and Chapters of the handbook.

Section 20

Ambient Noise and Its Control
Acoustical Environment Control
Mechanical Disc Reproduction Systems
Magnetic-Tape Analog Recording and Reproduction

Section 21

Television Cameras

Section 22

Chapter 22.1: Frequency Channel Assignments for FM Broadcasting
Chapter 22.1: Numerical Designation of Television Channels
Chapter 22.1: Zone Descriptions for Television Broadcasting
Chapter 22.1: Minimum Distance Separation Requirements for FM Broadcast Transmitters in KM (MI)
Chapter 22.1: Medium Wave AM Standard Broadcasting Definitions
Chapter 22.1: Frequency Modulation Broadcasting Definitions
Chapter 22.1: Analog Television (NTSC) Definitions
Chapter 22.1: ATSC Digital Transmission Systems Definitions
Chapter 22.1: Short Wave Broadcasting Definitions
Chapter 22.3: Projections for Digital Cable

Subject matter is listed as it relates to Parts of the handbook.

Part 1   Principles and Techniques

Basic Phenomena
Mathematics, Formulas, Definitions, and Theorems
Circuit Principles

Part 4   Systems and Applications

Electronics in Medicine and Biology

CONTRIBUTORS

Ronald M. Aarts   *Philips Research Laboratories, Eindhoven* (CHAPS. *20.1, 20.2, 20.4*)

Jay A. Alexander   *Agilent Technologies* (CHAP. *25.5*)

Ronald T. Anderson   *IIT Research Institute; Member, IEEE* (CHAPS. *3.1, 3.5*)

Panos Antsaklis   *University of Notre Dame; Fellow, IEEE* (CHAP. *19.1*)

Earl F. Arbuckle, III   *Fox Television Stations, Inc.* (CHAP. *22.1*)

Clarence M. Bailey, Jr.   *Bell Telephone Laboratories; Member, IEEE* (CHAP. *3.3*)

David K. Barton   *Anro Engineering Inc.; Fellow, IEEE* (CHAP. *23.1*)

James F. Bartram   *Raytheon Company; Senior Member, IEEE* (CHAP. *23.4*)

J. C. Baumhauer, Jr.   *AT&T Information Systems* (CHAP. *17.7*)

George Bechtel   *Strategies Unlimited* (CHAP. *7.3*)

Joseph E. Blue   *Naval Research laboratory* (CHAP. *23.4*)

David A. Bosserman   *U.S. Army Electronics Research and Development Command* (CHAP. *9.3*)

A. B. Brown, Jr.   *AT&T Information Systems; Senior Member, IEEE* (CHAP. *17.1*)

Larry V. Caldwell   *U.S. Army Electronics Research and Development Command* (CHAP. *9.5*)

Dwight Caswell   *Consultant* (CHAP. *7.1*)

Rex Chappell   *Agilent Technologies* (CHAP. *25.1*)

Stephen C. Choolfaian   *Pace University* (SEC. *18*)

Joseph L. Chovan   *General Electric Company* (CHAP. *12.5*)

Walter S. Ciciora   *Technology Consultants; Fellow, IEEE* (CHAP. *22.3*)

Munsey E. Crost   *U.S. Army Electronics Research and Development Command* (CHAPS. *5.2, 9.3*)

William F. Croswell   *Harris Corporation; Fellow, IEEE* (CHAPS. *16.1–16.2*)

David E. Cypher   *National Institute of Standards and Technology* (CHAP. *17.4*)

Gordon W. Day   *National Institute of Standards and Technology (Ret.); Fellow, IEEE* (CHAP. *5.6*)

Peter H. N. de With   *University of Technology Eindhoven/CMG Eindhoven B. V.; Senior Member, IEEE* (CHAP. *21.3*)

Sam Di Vita   *U.S. Army Communications Research and Development Command; Fellow, American Chemical Society* (CHAP. *5.1*)

John F. Donlon   *Powerex: Senior Member, IEEE* (CHAP. *5.2*)

Jennifer E. Doyle   *Consultant; Senior Member, IEEE* (CHAPS. *11.3, 11.6*)

Richard L. Doyle   *Doyle and Associates* (CHAPS. *3.1, 3.3, 3.4, 3.5*)

Myron D. Egtvedt   *General Electric Company* (CHAP. *12.4*)

Stanley L. Ehrlich   *Raytheon Company; Senior Member, IEEE* (CHAP. *23.4*)

David L. Evans   *Optoelectronics Division, Hewlett-Packard (CHAP. 9.1)*

George K. Farney   *Consultant; Fellow, IEEE (CHAP. 7.2)*

Brian R. Fast   *Cleveland State University (SEC. 4)*

Joseph Feinstein   *Stanford University; Fellow, IEEE (CHAPS. 7.1, 7.3)*

Thomas Fetter   *Agilent Technologies (CHAP. 25.3)*

Clifton S. Fox   *U.S. Army Electronics Research and Development Command; Member, IEEE (CHAP. 9.1)*

Donald A. Fredenburg   *Raytheon Company (CHAP. 23.4)*

Richard W. French   *Elemek, Inc (CHAP. 11.5)*

Walter R. Fried   *Consultant; Fellow, IEEE (CHAP. 23.3)*

Zhiqiang Gao   *Cleveland State University; Member, IEEE (CHAP. 19.1)*

Glenn B. Gawler   *General Electric Company; Member, IEEE (CHAP. 12.6)*

Joseph M. Giannotti   *U.S. Army Communications Research and Development Command; Senior Member, IEEE (CHAP. 5.1)*

James J. Gibson   *(deceased) formerly with Signal Systems Research; former Fellow, IEEE (CHAP. 21.1)*

Emanuel Gikow   *U.S. Army Electronics Research and Development Command; Member, IEEE (CHAP. 5.1)*

Kurt. E. Gonzenbach   *Martin Marietta Aerospace (CHAP. 3.2)*

Thomas S. Gore, Jr.   *U.S. Army Electronics Research and Development Command; Member, IEEE (CHAP. 5.1)*

Wayne T. Grant   *U.S. Army Electronics Research and Development Command (CHAP. 9.6)*

Carrington H. Greenidge   *Dale Electronics (CHAP. 5.1)*

William A. Gutierrez   *U.S. Army Electronics Research and Development Command (CHAP. 9.4)*

Murray J. Haims   *Consultant; Senior Member, IEEE (SEC. 18)*

Edward B. Hakim   *U.S. Army Electronics Research and Development Command; Member, IEEE (CHAP. 5.2)*

Harry W. Hale   *Iowa State University; Senior Member, IEEE (CHAPS. 10.1–10.5)*

G. Burton Harrold   *General Electric Company; Member, IEEE (CHAPS. 11.1, 11.3)*

George T. Hawley   *Diamond Lane Communications Corp.; Senior Member, IEEE (CHAP. 17.3)*

William M. Hayes   *Agilent Technologies (CHAP. 25.6)*

Jack H. Heimann   *Raytheon Company; Senior Member, IEEE (CHAP. 23.4)*

Joseph P. Hesler   *Eagle Comtronics Incorporated (CHAPS. 11.4, 12.1)*

Mark Hodapp   *Hewlett-Packard (CHAP. 9.1)*

Lee H. Hoke, Jr.   *formerly with Philips Consumer Electronics Company; Fellow, IEEE (CHAPS. 21.4, 22.2)*

Daniel J. Horowitz   *U.S. Army Electronics Research and Development Command (CHAP. 9.2)*

Jerry L. Hudgins   *University of Nebraska, Lincoln (CHAP. 13.1)*

Kevin A. Hughes   *International Telecommunication Union (CHAP. 16.3)*

Paul G. A. Jespers   *Professor of Electrical Engineering, Universite Catholique de Louvain; Fellow, IEEE (CHAPS. 14.1–14.4)*

Amos E. Joel, Jr.   *Executive Consultant; Fellow, IEEE (CHAP. 17.2)*

Virgil I. Johannes   *Virgil I. Johannes Inc.; Fellow, IEEE (CHAP. 17.1)*

Edwin C. Jones, Jr. *Iowa State University; Fellow, IEEE (CHAPS. 10.1–10.5)*

Gareth M. Janney *Hughes Aircraft, Company; Member, IEEE (CHAP. 9.2)*

Howard R. Jory *Communications and Power Industries, Inc.; Fellow, IEEE (CHAP. 7.2)*

Ronald K. Jurgen *formerly with IEEE Spectrum; Life Senior Member, IEEE (CHAPS. 24.1, 24.2, 24.3)*

Samuel Keene *Seagate Technology; Fellow, IEEE (CHAP. 3.6)*

Arthur W. Kelley *North Carolina State University; Member, IEEE (CHAP. 13.2)*

Andrew J. Kennedy *U.S. Army Electronics Research and Development Command; Member, IEEE; Member, American Physical Society (CHAP. 9.5)*

John M. Kikta *Engineering, Grayhill Corporation (CHAP. 5.4)*

Richard C. Kirby *International Radio Consultative Committee and Radiocommunication Bureau, International Telecommunication Union; Fellow, IEEE (CHAP. 16.3)*

Chang S. Kim *Daewoo Crop.; Member, IEEE (CHAP. 11.3)*

Edwin W. Kimball *Martin Marietta Aerospace (CHAP. 3.2)*

John J. Knab *Defense Information Systems Agency; Senior Member, IEEE (CHAP. 17.4)*

Wen H. Ko *Case Western Reserve University; Fellow, IEEE (CHAPS. 6.5, 8.5)*

Samuel M. Korzekwa *General Electric Company; Member, IEEE (CHAPS. 11.2, 11.4)*

Philip T. Krein *University of Illinois; Senior Member, IEEE (CHAP. 13.3)*

Richard A. Kupnicki *Leitch Technology Corp.; Member, IEEE (CHAP. 21.2)*

Stanislaw Kus *IIT Research Institute (CHAPS. 3.1, 3.5)*

Joseph A. Kuzneski *Raytheon Company; Senior Member, IEEE (CHAP. 23.4)*

W. J. Lawless *Paradyne; Member, IEEE (CHAP. 17.7)*

Ramon C. Lebron *NASA Glenn Research Center (CHAP. 15.5)*

W. R. Lehmann *Martin Marietta Aerospace (CHAP. 3.2)*

Y. J. Lin *California Microwave (CHAP. 11.3)*

David Linden *Consultant (CHAP. 5.3)*

C. C. Liu *Case Western Reserve University (CHAP. 8.5)*

Randolph E. Longshore *U.S. Army Electronics Research and Development Command; Member, American Chemical Society (CHAP. 9.6)*

Harold W. Lord *Consulting Engineer; Fellow, IEEE (CHAP. 11.3)*

John W. Lunden *Harris Corporation; Member, IEEE (CHAPS. 11.3, 11.6)*

Gregory J. Malinowski *U.S. Army Electronics Research and Development Command (CHAP. 5.2)*

Patrick R. Manzo *Science Applications, Inc.; Member, American Physical Society (CHAP. 9.2)*

L. A. Marcus *AT&T Information Systems (CHAP. 17.7)*

Daniel W. Martin *(deceased) Acoustical Consultant; former Fellow, IEEE (CHAPS. 20.1–20.4)*

Richard E. Matick *IBM Thomas J. Watson Research Center; Fellow, IEEE (SEC. 18)*

Robert J. McFadyen *General Electric Company (CHAP. 11.2)*

William C. McGee *IBM General Products Division (SEC. 18)*

D. R. Means *Lucent Technologies (CHAP. 17.7)*

M. Mehregany *Case Western Reserve University; Member, IEEE (CHAP. 8.5)*

Tim Mikkelsen *Agilent Technologies (CHAP. 25.7)*

James E. Miller   *U.S. Army Electronics Research and Development Command (CHAP. 9.2)*

Kim W. Mitchell   *Siemens Solar; Member, IEEE (CHAP. 9.7)*

Benton D. Moldow   *Iona College (SEC. 18)*

Robert S. Mroczkowski   *AMP Inc. (CHAP. 5.5)*

Robert A. Myers   *IBM Japan; Senior Member, IEEE (SEC. 18)*

Conrad E. Nelson   *General Electric Company; Senior Member, IEEE (CHAP. 11.3)*

Allen Nikora   *Jet Propulsion Laboratory (CHAP. 3.6)*

Harry N. Norton   *Jet Propulsion Laboratory (CHAPS. 8.1–8.4)*

David O'Brien   *Optoelectronics Division, Hewlett-Packard Company (CHAP. 9.1)*

Neil V. Owen   *Martin Marietta Aerospace (CHAP. 3.2)*

Geoffrey C. Orsak   *Southern Methodist University (SEC. 1)*

George F. Pfeifer   *General Electric Company; Member, IEEE (CHAP. 12.3)*

John H. Pollard   *U.S. Army Electronics Research and Development Command (CHAP. 9.5)*

H. Vincent Poor   *Princeton University; Fellow, IEEE (SEC. 1)*

Nobel R. Powell   *Syracuse Research Corporation; Member, IEEE (CHAP. 12.2)*

Donad H. Preist   *Varian Associates; Fellow, IEEE (CHAP. 7.2)*

Irving Reingold   *Southeastern Center for Electrical Engineering Education; Fellow, IEEE (CHAPS. 5.2, 9.3)*

Glenn Reitmeier   *National Broadcasting Company (NBC); Member, IEEE (CHAP. 21.1)*

Henry C. Rickers   *Reliability Analysis Center RADC (RBRAC); Member, IEEE (CHAPS. 3.1, 3.5)*

Clayton R. Roberts   *Consultant (CHAPS. 11.4–11.5)*

Daniel Rosich   *Pace University (SEC. 18)*

William B. Rouse   *Georgia Institute of Technology (SEC. 2)*

R. M. Sachs   *AT&T Bell Laboratories; Member, IEEE (CHAP. 17.7)*

Matthew N. O. Sadiku   *Prairie View A&M University (CHAPS. 17.5–17.6)*

Andrew P. Sage   *George Mason University; Fellow, IEEE (SEC. 2)*

Edward J. Sharp   *U.S. Army Electronics Research and Development Command (CHAP. 9.2)*

M. B. Shrader   *Varian Associates (CHAP. 7.2)*

Richard R. Shurtz, II   *U.S. Army Electronics Research and Development Command (CHAP. 9.2)*

Paul Skitzki   *Raytheon Company; Member, IEEE (CHAP. 23.4)*

Bernard Smith   *U.S. Army Electronics Research and Development Command (CHAP. 5.2)*

J. Spergel   *General Cable Corporation; Senior Member, IEEE (CHAP. 5.4)*

Scott Stever   *Agilent Technologies (CHAP. 25.2)*

George C. Stierhoff   *IBM Corp. (SEC. 18)*

David H. Su   *National Institute of Standards and Technology (CHAP. 17.4)*

S. Sugihara   *The Aerospace Corp. (CHAP. 3.5)*

Robert S. Symons   *Littton Electron Devices Division; Fellow, IEEE (CHAP. 7.2)*

Fred J. Taylor   *University of Florida; Fellow, IEEE (CHAPS. 10.6–10.7)*

George W. Taylor   *U.S. Army Communications Research and Development Command; Senior Member, IEEE (CHAP. 5.2)*

Stephen W. Tehon   *General Electric Company; Fellow, IEEE (CHAP. 11.3)*

C. A. Tenorio   *Lucent Technologies (CHAP. 17.7)*

S. Tewksbury   *Stevens Institute of Technology (CHAPS. 6.1–6.4)*

John B. Thomas   *Princeton University; Fellow, IEEE (SEC. 1)*

Francis T. Thompson   *Westinghouse Electric Corporation; Fellow, IEEE (CHAPS. 15.1–15.4)*

R. Y. Ting   *Naval Research Laboratories (CHAP. 23.4)*

David A. Torrey   *Advanced Energy Conversion, LLC (CHAP. 13.4)*

E. W. Underhill   *Bell Telephone Laboratories (CHAP. 17.7)*

Stephen J. Urban, Jr.   *Delta Information Systems, Inc.; Member, IEEE (CHAP. 21.5)*

A. L. Van Buren   *Naval Research Laboratories (CHAP. 23.4)*

M. van der Schaar   *University of California Davis; Member, IEEE (CHAP. 21.3)*

P. K. Vasudev   *Sematech; Member, IEEE (CHAPS. 6.1–6.3)*

Bart H. Verbeek   *JDS Uniphase Netherlands BV (CHAP. 9.2)*

John R. Vig   *U.S. Army Electronics Research and Development Command; Fellow, IEEE (CHAP. 5.1)*

Pamela L. Walchli   *Litton Electron Devices (CHAP. 7.2)*

Claude E. Walston   *University of Maryland; Senior Member, IEEE (SEC. 18)*

Wen-Chung Wang   *Polytechnic Institute; Senior Member, IEEE (CHAP. 11.3)*

Harold R. Ward   *Anro Engineering, Inc.; Fellow, IEEE (CHAP. 23.2)*

Steven B. Warntjes   *Agilent Technologies (CHAP. 25.4)*

Gunter K. Wessel   *Syracuse University (CHAP. 11.3)*

James W. Wilbur   *Reliability Analysis Center RADC (RBRAC); Member, IEEE (CHAPS. 3.1, 3.5)*

Arthur B. Williams   *Coherent Communications Systems Corp.; Member, IEEE (CHAPS. 10.6, 10.7)*

Steve Witt   *(CHAP. 25.4)*

Peter Wood   *Westinghouse Research Laboratories (CHAP. 13.5)*

William W. Wu   *Consultare Group Inc.; Fellow, IEEE (CHAP. 17.4)*

P. J. Yankura   *Lucent Technologies (CHAP. 17.7)*

Jason Yorks   *Optical Communications Division, Hewlett-Packard (CHAP. 9.1)*

D. J. Young   *Case Western Reserve University; Member, IEEE (CHAP. 17.8)*

Herbert M. Zydney   *AT&T Bell Laboratories (CHAP. 17.7)*

Note: Affiliations are those effective at the time of authors' original contributions.

CD-ROM

Panos Antsaklis   *University of Notre Dame; Fellow, IEEE*

Earl F. Arbuckle, III   *Fox Television Stations, Inc.*

John C. Belina   *Cornell University; Member, IEEE*

Ilan A. Blech   *Zoran Corporation*

Dudley Childress   *Northwestern University; Member, IEEE*

Donald Christiansen   *Informatica; Fellow, IEEE*

Walter S. Ciciora   *Technology Consultants; Fellow, IEEE*

Wils L. Cooley   *West Virginia University; Senior Member, IEEE*

Nicholas A. Diakedes   *Advanced Concepts Analysis Inc.; Senior Member, IEEE*

B. Dudley   *Consultant; Senior Member, IEEE*

Richard E. Franseen   *U.S. Army Research and Development Command; Member, IEEE*

Zhiqiang Gao   *Cleveland State University; Member, IEEE*

Ronald D. Graft   *U.S. Army Research and Development Command; Member, IEEE*

Kevin A. Hughes   *International Telecommunications Union*

Paul G. A. Jespers   *Universite Catholique de Louvain; Fellow, IEEE*

Clarence A. Johnson   *U.S. Army Research and Development Command; Member, IEEE*

Raymond Kiraly   *Cleveland Clinic Foundation*

Richard C. Kirby   *International Telecommunications Union; Fellow, IEEE*

Wen H. Ko   *Case Western Reserve University; Fellow, IEEE*

Granino A. Korn   *University of Arizona*

Theresa M. Korn   *Consultant*

Paul S. Malchesky   *Steris Corporation*

Daniel W. Martin   *Consultant; Fellow, IEEE*

James O. Meindl   *Georgia Institute of Technology; Fellow, IEEE*

Floro Miraldi   *Case Western Reserve University*

J. Thomas Mortimer   *Case Western Reserve University*

Michael R. Neuman   *Case Western Reserve University*

Harry N. Norton   *Jet Propulsion Laboratory*

P. Hunter Peckham   *Case Western Reserve University; Member, IEEE*

Robert Plonsey   *Duke University; Fellow, IEEE*

Soheyl Pourmehdi   *Neurocontrol Corporation*

Eric Sakk   *Innovative Dynamics*

Edward J. Sharp   *U.S. Army Research and Development Command*

Lawrence J. Thorpe   *Sony Electronics*

PREFACE

Both new and previous contributors have helped add new material to this, the fifth edition of the handbook, quite possibly the most widely used reference book in the fields of electronics and related disciplines.

The book is divided into four major parts: (1) Principles and Techniques, (2) Components and Hardware, (3) Circuits and Functions, and (4) Systems and Applications.

Part 1 covers information and communications theory, systems engineering and management, reliability of electronic components and systems, and computer-aided design.

Part 2 covers electronic and fiber optic components, integrated circuits and microprocessors, UHF and microwave devices, transducers and sensors, and radiant energy sensors and sources.

Part 3 covers filters and attenuators; amplifiers and oscillators; modulators, demodulators, and converters; power electronics; pulsed circuits; measurement circuits; and antennas and wave propagation.

Part 4 covers telecommunications, digital computer systems, control systems, audio systems, video and facsimile systems, broadcast and cable systems, navigation and detection systems, automotive electronics, and instrumentation and test systems.

Revising and updating a work of this magnitude and complexity often seems even more formidable than producing the original. Both editors and contributors face hard decisions in what to delete or compress in order to make room for important new material. Fortunately, we have been able to include a CD-ROM as a companion to this handbook. It enabled us to include material from the fourth edition that the handbook user may need to access, as well as new material that complements various topics in the handbook itself. A separate table of contents for the CD-ROM is provided.

The first edition of the handbook was published in 1975, under the guidance of its founding editor Donald Glen Fink; it was an instant success. While it was not the first handbook published in the field of electronics, it was the first to bring together in one volume the essential principles, data, and design information on the components, circuits, equipment, and systems of electronics engineering as a whole. Earlier handbooks had treated the field primarily from the point of view of its first important application—radio engineering. For example, Keith Henney's *Radio Engineering Handbook* was introduced as a slim volume published by McGraw-Hill in 1933, and had grown to 1750 pages by its fifth edition in 1959.

Don Fink invited me to join him in undertaking the second edition of this handbook, a task I was honored to accept. By the time the third edition was needed, Don had gone on to other projects, but continued as a co-editor and a valued consultant during its preparation; the latter a role he continued into the fourth edition.

I am privileged to have Charles K. Alexander join me as co-editor of this new edition. Dr. Alexander is the Dean of Engineering at Cleveland State University and co-author of the McGraw-Hill textbook *Fundamentals of Electric Circuits*. He was instrumental in developing new material for the chapters on computer-assisted design, telecommunications, and instrumentation. Ronald K. Jurgen again proved invaluable in working with the contributors to the sections on audio, video, and broadcasting systems. He also contributed a completely new section on automotive electronics.

We are grateful to the following new contributors to this work: Ronald Aarts, Jay Alexander, Panos Antsaklis, Rex Chappell, Gordon Day, Peter de With, Richard Doyle, Thomas Fetter, Zhiqiang Gao, William Hayes, David Juedes, Samuel Keene, Ramon Lebron, Tim Mikkelsen, Geoffrey Orsak, Glenn Reitmeier, William Rouse, Matthew Sadiku, Andrew Sage, Scott Stever, Stuart Tewksbury, M. van der Schaar, Bart Verbeek, and Steven Warntjes.

I also wish to thank Steve Chapman at McGraw-Hill for his suggestions on modularizing the material and designing the complementary CD-ROM; Waseem Andrabi for his care in copyediting and production matters; and Nancy T. Hantman for her invaluable typing, research, and record keeping.

DONALD CHRISTIANSEN

"It was with great pleasure that I joined Don Christiansen as a co-editor of this handbook. I want to express my appreciation for his leadership, guidance, and patience in producing this important work. I have worked with Don over the years both as a colleague and a friend. I came to know him well, especially in his role as editor and publisher of *IEEE Spectrum*. At the helm of *Spectrum*, he transformed it into a leading international publication. On a personal note, he helped me launch *IEEE Potentials*, the magazine for IEEE student members.

Don has expressed his thanks to the many new contributors to this edition in his comments above. I want to second that appreciation and also extend a special thank you to Steve Grossman, Agilent Technologies, for the major role he played in developing our new section on instrumentation."

CHARLES K. ALEXANDER

STANDARD
HANDBOOK OF
ELECTRONIC
ENGINEERING

P · A · R · T
1

PRINCIPLES
AND TECHNIQUES

On the CD-ROM

Basic Phenomena
Mathematics, Formulas, Definitions, and Theorems
Circuit Principles

SECTION 1

INFORMATION, COMMUNICATION, NOISE, AND INTERFERENCE

The telephone profoundly changed our methods of communication, thanks to Alexander Graham Bell and other pioneers (Bell, incidentally, declined to have a telephone in his home!). Communication has been at the heart of the information age. Electronic communication deals with transmitters and receivers of electromagnetic waves. Even digital communications systems rely on this phenomenon. This section of the handbook covers information sources, codes and coding, communication channels, error correction, continuous and band-limited channels, digital data transmission and pulse modulation, and noise and interference. C.A.

In This Section:

Section Bibliography:

Of Historical Significance

Davenport, W. B., Jr., and W. L. Root, "An Introduction to the Theory of Random Signals and Noise," McGraw-Hill, 1958. (Reprint edition published by IEEE Press, 1987.)

Middleton, D., "Introduction to Statistical Communication Theory," McGraw-Hill, 1960. (Reprint edition published by IEEE Press, 1996.)

Sloane, N. J. A., and A. D. Wyner (eds.), "Claude Elwood Shannon: Collected Papers," IEEE Press, 1993.

General

Carlson, A. B., et al., "Communications Systems," 4th ed., McGraw-Hill, 2001.

Gibson, J. D., "Principles of Digital and Analog Communications," 2nd ed., Macmillan, 1993.

Haykin, S., "Communication Systems," 4th ed., Wiley, 2000.

Papoulis, A., and S. U. Pillai, "Probability, Random Variables, and Stochastic Processes," 4th ed., McGraw-Hill, 2002.

Thomas, J. B., "An Introduction to Communication Theory and Systems," Springer-Verlag, 1987.

Ziemer, R. E., and W. H. Tranter, "Principles of Communications: Systems, Modulation, and Noise," 5th ed., Wiley, 2001.

Information Theory

Blahut, R. E., "Principles and Practice of Information Theory," Addison-Wesley, 1987.

Cover, T. M., and J. A. Thomas, "Elements of Information Theory," Wiley, 1991.

Gallagher, R., "Information Theory and Reliable Communication," Wiley, 1968.

Coding Theory

Blahut, R. E., "Theory and Practice of Error Control Codes," Addison-Wesley, 1983.

Clark, G. C., Jr., and J. B. Cain, "Error-correction Coding for Digital Communications," Plenum Press, 1981.

Lin, S., and D. J. Costello, "Error Control Coding," Prentice-Hall, 1983.

Digital Data Transmission

Barry, J. R., D. G. Messerschmitt, and E. A. Lee, "Digital Communications," 3rd ed., Kluwer, 2003.

Proakis, J. G., "Digital Communications," 4th ed., McGraw-Hill, 2000.

CHAPTER 1.1
COMMUNICATION SYSTEMS

Geoffrey C. Orsak, H. Vincent Poor, John B. Thomas

CONCEPTS

The principal problem in most communication systems is the transmission of information in the form of messages or data from an originating *information source S* to a *destination* or *receiver D*. The method of transmission is frequently by means of electric signals under the control of the sender. These signals are transmitted via a channel *C*, as shown in Fig. 1.1.1. The set of messages sent by the source will be denoted by $\{U\}$. If the channel were such that each member of *U* were received exactly, there would be no communication problem. However, because of channel limitations and noise, a corrupted version $\{U^*\}$ of $\{U\}$ is received at the information destination. It is generally desired that the distorting effects of channel imperfections and noise be minimized and that the number of messages sent over the channel in a given time be maximized.

These two requirements are interacting, since, in general, increasing the rate of message transmission increases the distortion or error. However, some forms of message are better suited for transmission over a given channel than others, in that they can be transmitted faster or with less error. Thus it may be desirable to modify the message set $\{U\}$ by a suitable *encoder E* to produce a new message set $\{A\}$ more suitable for a given channel. Then a decoder E^{-1} will be required at the destination to recover $\{U^*\}$ from the distorted set $\{A^*\}$. A typical block diagram of the resulting system is shown in Fig. 1.1.2.

SELF-INFORMATION AND ENTROPY

Information theory is concerned with the quantification of the communications process. It is based on probabilistic modeling of the objects involved. In the model communication system given in Fig. 1.1.1, we assume that each member of the message set $\{U\}$ is expressible by means of some combination of a finite set of symbols called an *alphabet*. Let this source alphabet be denoted by the set $\{X\}$ with elements $x_1, x_2, \ldots, x_M$, where *M* is the size of the alphabet. The notation $p(x_i)$, $i = 1, 2, \ldots, M$, will be used for the probability of occurrence of the *i*th symbol x_i. In general the set of numbers $\{p(x_i)\}$ can be assigned arbitrarily provided that

$$p(x_i) \geq 0 \qquad i = 1, 2, \ldots, M \tag{1}$$

and

$$\sum_{i=1}^{M} p(x_i) = 1 \tag{2}$$

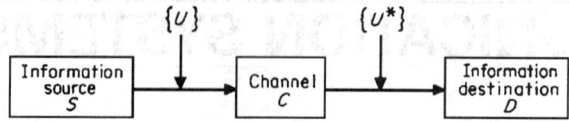

FIGURE 1.1.1   Basic communication system.

A measure of the amount of information contained in the ith symbol x_i can be defined based solely on the probability $p(x_i)$. In particular, the *self-information* $I(x_i)$ of the ith symbol x_i is defined as

$$I(x_i) = \log 1/p(x_i) = -\log p(x_i) \tag{3}$$

This quantity is a decreasing function of $p(x_i)$ with the endpoint values of infinity for the impossible event and zero for the certain event.

It follows directly from Eq. (3) that $I(x_i)$ is a discrete random variable, i.e., a real-valued function defined on the elements x_i of a probability space. Of the various statistical properties of this random variable $I(x_i)$, the most important is the expected value, or mean, given by

$$E\{I(x_i)\} = H(X) = \sum_{i=1}^{M} p(x_i)I(x_i) = -\sum_{i=1}^{M} p(x_i)\log p(x_i) \tag{4}$$

This quantity $H(X)$ is called the *entropy* of the distribution $p(x_i)$. If $p(x_i)$ is interpreted as the probability of the ith state of a system in phase space, then this expression is identical to the entropy of statistical mechanics and thermodynamics. Furthermore, the relationship is more than a mathematical similarity. In statistical mechanics, entropy is a measure of the disorder of a system; in information theory, it is a measure of the uncertainty associated with a message source.

In the definitions of self-information and entropy, the choice of the base for the logarithm is arbitrary, but of course each choice results in a different system of units for the information measures. The most common bases used are base 2, base e (the natural logarithm), and base 10. When base 2 is used, the unit of $I(\cdot)$ is called the *binary digit* or *bit*, which is a very familiar unit of information content. When base e is used, the unit is the *nat*; this base is often used because of its convenient analytical properties in integration, differentiation, and the like. The base 10 is encountered only rarely; the unit is the *Hartley*.

ENTROPY OF DISCRETE RANDOM VARIABLES

The more elementary properties of the entropy of a discrete random variable can be illustrated with a simple example. Consider the binary case, where $M = 2$, so that the alphabet consists of the symbols 0 and 1 with probabilities p and $1 - p$, respectively. It follows from Eq. (4) that

$$H_1(X) = -[p \log_2 p + (1 - p) \log_2 (1 - p)] \text{ (bits)} \tag{5}$$

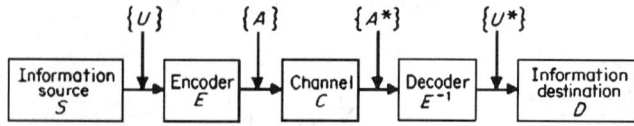

FIGURE 1.1.2   Communication system with encoding and decoding.

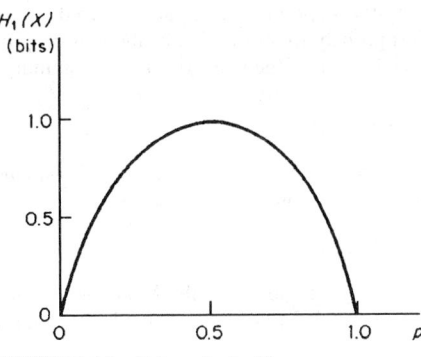

$H_1(X)$
(bits)

1.0

0.5

0

FIGURE 1.1.3 Entropy in the binary case.

Equation (5) can be plotted as a function of p, as shown in Fig. 1.1.3, and has the following interesting properties:

1. $H_1(X) \geq 0$.
2. $H_1(X)$ is zero only for $p = 0$ and $p = 1$.
3. $H_1(X)$ is a maximum at $p = 1 - p = \frac{1}{2}$.

More generally, it can be shown that the entropy $H(X)$ has the following properties for the general case of an alphabet of size M:

1. $H(X) \geq 0$. $\hspace{4cm}$ (6)
2. $H(X) = 0$ if and only if all of the probabilities are zero except for one, which must be unity. $\hspace{1cm}$ (7)
3. $H(X) \leq \log_b M$. $\hspace{4cm}$ (8)

4. $H(X) = \log_b M$ if and only if all the probabilities are equal so that $p(x_i) = 1/M$ for all i. $\hspace{2cm}$ (9)

MUTUAL INFORMATION AND JOINT ENTROPY

The usual communication problem concerns the transfer of information from a source S through a channel C to a destination D, as shown in Fig. 1.1.1. The source has available for forming messages an alphabet X of size M. A particular symbol x_1 is selected from the M possible symbols and is sent over the channel C. It is the limitations of the channel that produce the need for a study of information theory.

The information destination has available an alphabet Y of size N. For each symbol x_i sent from the source, a symbol y_j is selected at the destination. Two probabilities serve to describe the "state of knowledge" at the destination. Prior to the reception of a communication, the state of knowledge of the destination about the symbol x_j is the a priori probability $p(x_i)$ that x_i would be selected for transmission. After reception and selection of the symbol y_j, the state of knowledge concerning x_i is the conditional probability $p(x_i|y_j)$, which will be called the a posteriori probability of x_i. It is the probability that x_i was sent given that y_j was received. Ideally this a posteriori probability for each given y_j should be unity for one x_i and zero for all other x_i. In this case an observer at the destination is able to determine exactly which symbol x_i has been sent after the reception of each symbol y_j. Thus the uncertainty that existed previously and which was expressed by the a priori probability distribution of x_i has been removed completely by reception. In the general case it is not possible to remove all the uncertainty, and the best that can be hoped for is that it has been decreased. Thus the a posteriori probability $p(x_i|y_j)$ is distributed over a number of x_i but should be different from $p(x_i)$. If the two probabilities are the same, then no uncertainty has been removed by transmission or no information has been transferred.

Based on this discussion and on other considerations that will become clearer later, the quantity $I(x_i; y_j)$ is defined as the information gained about x_i by the reception of y_j, where

$$I(x_i; y_j) = \log_b [p(x_i|y_j)/p(x_i)] \hspace{2cm} (10)$$

This measure has a number of reasonable and desirable properties.

Property 1. The information measure $I(x_i; y_j)$ *is symmetric in* x_i *and* y_j; that is,

$$I(x_i; y_j) = I(y_j; x_i) \hspace{2cm} (11)$$

Property 2. The mutual information $I(x_i; y_j)$ is a maximum when $p(x_i|y_j) = 1$, that is, when the reception of y_j completely removes the uncertainty concerning x_i:

$$I(x_i; y_j) \leq -\log p(x_i) = (x_i) \hspace{2cm} (12)$$

Property 3.   If two communications y_j and z_k concerning the same message x_i are received successively, and if the observer at the destination takes the a posteriori probability of the first as the a priori probability of the second, then the total information gained about x_i is the sum of the gains from both communications:

$$I(x_i; y_j, z_k) = I(x_i; y_j) + I(x_i; z_k | y_j) \tag{13}$$

Property 4.   If two communications y_j and y_k concerning two *independent* messages x_i and x_m are received, the total information gain is the sum of the two information gains considered separately:

$$I(x_i, x_m; y_j, y_k) = I(x_i; y_j) + I(x_m; y_k) \tag{14}$$

These four properties of mutual information are intuitively satisfying and desirable. Moreover, if one begins by requiring these properties, it is easily shown that the logarithmic definition of Eq. (10) is the simplest form that can be obtained.

The definition of mutual information given by Eq. (10) suffers from one major disadvantage. When errors are present, an observer will not be able to calculate the information gain even after the reception of all the symbols relating to a given source symbol, since the same series of received symbols may represent several different source symbols. Thus, the observer is unable to say which source symbol has been sent and at best can only compute the information gain with respect to each possible source symbol. In many cases it would be more desirable to have a quantity that is independent of the particular symbols. A number of quantities of this nature will be obtained in the remainder of this section.

The mutual information $I(x_i; y_j)$ is a random variable just as was the self-information $I(x_i)$; however, two probability spaces X and Y are involved now, and several ensemble averages are possible. The *average mutual information* $I(X; Y)$ is defined as a statistical average of $I(x_i; y_j)$ with respect to the joint probability $p(x_i; y_j)$; that is,

$$I(X; Y) = E_{XY}\{I(x_i; y_j)\} = \sum_i \sum_j p(x_i, y_j) \log[p(x_i | y_j)/p(x_i)] \tag{15}$$

This new function $I(X; Y)$ is the first information measure defined that does not depend on the individual symbols x_i or y_j. Thus, it is a property of the whole communication system and will turn out to be only the first in a series of similar quantities used as a basis for the characterization of communication systems. This quantity $I(X; Y)$ has a number of useful properties. It is nonnegative; it is zero if and only if the ensembles X and Y are *statistically independent*; and it is symmetric in X and Y so that $I(X; Y) = I(Y; X)$.

A source entropy $H(X)$ was given by Eq. (4). It is obvious that a similar quantity, the destination entropy $H(Y)$, can be defined analogously by

$$H(Y) = -\sum_{j=1}^{N} p(y_j) \log p(y_j) \tag{16}$$

This quantity will, of course, have all the properties developed for $H(X)$. In the same way the *joint or system entropy* $H(X, Y)$ can be defined by

$$H(X, Y) = -\sum_{i=1}^{M} \sum_{j=1}^{N} p(x_i, y_j) \log p(x_i, y_j) \tag{17}$$

If X and Y are *statistically independent* so that $p(x_i, y_j) = p(x_i)p(y_j)$ for all i and j, then Eq. (17) can be written as

$$H(X, Y) = H(X) + H(Y) \tag{18}$$

On the other hand, if X and Y are not independent, Eq. (17) becomes

$$H(X, Y) = H(X) + H(Y|X) = H(Y) + H(X|Y) \tag{19}$$

where $H(Y|X)$ and $H(X|Y)$ are *conditional entropies* given by

$$H(Y|X) = -\sum_{i=1}^{M}\sum_{j=1}^{N} p(x_i, y_j)\log p(y_j|x_i) \qquad (20)$$

and by

$$H(X|Y) = -\sum_{i=1}^{M}\sum_{j=1}^{N} p(x_i, y_j)\log p(x_i|y_j) \qquad (21)$$

These conditional entropies each satisfies an important inequality

$$0 \le H(Y|H) \le H(Y) \qquad (22)$$

and

$$0 \le H(X|Y) \le H(X) \qquad (23)$$

It follows from these last two expressions that Eq. (15) can be expanded to yield

$$I(X; Y) = -H(X, Y) + H(X) + H(Y) \ge 0 \qquad (24)$$

This equation can be rewritten in the two equivalent forms

$$I(X; Y) = H(Y) - H(Y|X) \ge 0 \qquad (25)$$

or

$$I(X|Y) = H(X) - H(X|Y) \ge 0 \qquad (26)$$

It is also clear, say from Eq. (24), that $H(X, Y)$ satisfies the inequality

$$H(X, Y) \le H(X) + H(Y) \qquad (27)$$

Thus, the joint entropy of two ensembles X and Y is a maximum when the ensembles are independent.

At this point it may be appropriate to comment on the meaning of the two conditional entropies $H(Y|X)$ and $H(X|Y)$. Let us refer first to Eq. (26). This equation expresses the fact that the average information gained about a message, when a communication is completed, is equal to the average source information less the average uncertainty that still remains about the message. From another point of view, the quantity $H(X|Y)$ is the average additional information needed at the destination after reception to completely specify the message sent. Thus, $H(X|Y)$ represents the information lost in the channel. It is frequently called the *equivocation.* Let us now consider Eq. (25). This equation indicates that the information transmitted consists of the difference between the destination entropy and that part of the destination entropy that is not information about the source; thus the term $H(Y|X)$ can be considered a *noise entropy* added in the channel.

CHAPTER 1.2

INFORMATION SOURCES, CODES, AND CHANNELS

Geoffrey C. Orsak, H. Vincent Poor, John B. Thomas

MESSAGE SOURCES

As shown in Fig. 1.1.1, an information source can be considered as emitting a given message u_i from the set $\{U\}$ of possible messages. In general, each message u_i will be represented by a sequence of symbols x_j from the source alphabet $\{X\}$, since the number of possible messages will usually exceed the size M of the source alphabet. Thus sequences of symbols replace the original messages u_i, which need not be considered further. When the source alphabet $\{X\}$ is of finite size M, the source will be called a *finite discrete source*. The problems of concern now are the interrelationships existing between symbols in the generated sequences and the classification of sources according to these interrelationships.

A random or stochastic process x_i, $t \in T$, can be defined as an indexed set of random variables where T is the *parameter set* of the process. If the set T is a sequence, then x_t is a stochastic process with discrete parameter (also called a *random sequence* or *series*). One way to look at the output of a finite discrete source is that it is a discrete-parameter stochastic process with each possible given sequence one of the ensemble members or realizations of the process. Thus the study of information sources can be reduced to a study of random processes.

The simplest case to consider is the *memoryless source*, where the successive symbols obey the same fixed probability law so that the one distribution $p(x_i)$ determines the appearance of each indexed symbol. Such a source is called *stationary*. Let us consider sequences of length n, each member of the sequence being a realization of the random variable x_i with fixed probability distribution $p(x_i)$. Since there are M possible realizations of the random variable and n terms in the sequence, there must be M^n distinct sequences possible of length n. Let the random variable X_i in the jth position be denoted by X_{ij} so that the sequence set (the message set) can be represented by

$$\{U\} = X^n = \{X_{i1}, X_{i2}, \ldots, X_{in}\} \qquad i = 1, 2, \ldots, M \tag{1}$$

The symbol X^n is sometimes used to represent this sequence set and is called the *nth extension of the memoryless source X*. The probability of occurrence of a given message u_i is just the product of the probabilities of occurrence of the individual terms in the sequence so that

$$p\{u_i\} = p(x_{i1})p(x_{i2}) \cdots p\{x_{in}\} \tag{2}$$

Now the entropy for the extended source X^n is

$$H(X^n) = -\sum_{x^n} p\{u_i\} \log p\{u_i\} = nH(X) \tag{3}$$

as expected. Note that, if base 2 logarithms are used, then $H(X)$ has units of bits per symbol, n is symbols per sequence, and $H(X^n)$ is in units of bits per sequence. For a memoryless source, all sequence averages of information measures are obtained by multiplying the corresponding symbol by the number of symbols in the sequence.

MARKOV INFORMATION SOURCE

The memoryless source is not a general enough model in most cases. A constructive way to generalize this model is to assume that the occurrences of a given symbol depends on some number m of immediately preceeding symbols. Thus the information source can be considered to produce an mth-order Markov chain and is called an *mth-order Markov source*.

For an mth-order Markov source, the m symbols preceding a given symbol position are called the *state s_j* of the source at that symbol position. If there are M possible symbols x_i, then the mth-order Markov source will have $M^m = q$ possible states s_j making up the *state set*

$$S = \{s_1, s_2, \ldots, s_q\} \quad q = M^m \tag{4}$$

At a given time corresponding to one symbol position the source will be in a given state s_j. There will exist a probability $p(s_k | s_j) = p_{jk}$ that the source will move into another state s_k with the emission of the next symbol. The set of all such conditional probabilities is expressed by the *transition matrix T*, where

$$T = [p_{jk}] = \begin{bmatrix} p_{11} & p_{12} & \cdots & p_{1q} \\ p_{21} & p_{22} & \cdots & p_{2q} \\ \cdots & \cdots & \cdots & \cdots \\ p_{q1} & p_{q2} & \cdots & p_{qq} \end{bmatrix} \tag{5}$$

A *Markov matrix* or *stochastic matrix* is any square matrix with nonnegative elements such that the row sums are unity. It is clear that T is such a matrix since

$$\sum_{j=1}^{q} p_{ij} = \sum_{j=1}^{q} p(s_j | s_i) = 1 \quad i = 1, 2, \ldots, q \tag{6}$$

Conversely, any stochastic matrix is a possible transition matrix for a Markov source of order m, where $q = M^m$ is equal to the number of rows or columns of the matrix.

A Markov chain is completely specified by its transition matrix T and by an *initial distribution vector π* giving the probability distribution for the first state occurring. For the memoryless source, the transition matrix reduces to a stochastic matrix where all the rows are identical and are each equal to the initial distribution vector π, which is in turn equal to the vector giving the source alphabet a priori probabilities. Thus, in this case, we have

$$p_{jk} = p(s_k | s_j) = p(s_k) = p(x_k) \quad k = 1, 2, \ldots, M \tag{7}$$

For each state s_i of the source an entropy $H(s_i)$ can be defined by

$$H(s_i) = -\sum_{j=1}^{q} p(s_j | s_i) \log p(s_j | s_i) = -\sum_{k=1}^{M} p(x_k | s_i) \log p(x_k | s_i) \tag{8}$$

The source entropy $H(S)$ in information units per symbol is the expected value of $H(s_i)$; that is,

$$H(S) = -\sum_{i=1}^{q} \sum_{j=1}^{q} p(s_i) p(s_j | s_i) \log p(s_j | s_i) = -\sum_{i=1}^{q} \sum_{k=1}^{M} p(s_i) p(x_k | s_i) \log p(x_k | s_i) \tag{9}$$

where $p(s_i) = p_i$ is the *stationary state probability* and is the *i*th element of the vector **P** defined by

$$\mathbf{P} = [p_1 \, p_2 \cdots p_q] \tag{10}$$

It is easy to show, as in Eq. (8), that the source entropy cannot exceed log M, where M is the size of the source alphabet $\{X\}$. For a given source, the ratio of the actual entropy $H(S)$ to the maximum value it can have with the same alphabet is called the *relative entropy* of the source. The *redundancy* η of the source is defined as the positive difference between unity and this relative entropy:

$$\eta = 1 - \frac{H(S)}{\log M} \tag{11}$$

The quantity log M is sometimes called the *capacity* of the alphabet.

NOISELESS CODING

The preceding discussion has emphasized the information source and its properties. We now begin to consider the properties of the communication channel of Fig. 1.1.1. In general, an arbitrary channel will not accept and transmit the sequence of x_i's emitted from an arbitrary source. Instead the channel will accept a sequence of some other elements a_i chosen from a *code alphabet A* of *size D*, where

$$A = \{a_1, a_2, \ldots, a_D\} \tag{12}$$

with D generally smaller than M. The elements a_i of the code alphabet are frequently called *code elements* or *code characters*, while a given sequence of a_i's may be called a *code word*.

The situation is now describable in terms of Fig. 1.1.2, where an encoder E has been added between the source and channel. The process of *coding*, or *encoding*, the source consists of associating with each source symbol x_i a given code word, which is just a given sequence of a_i's. Thus the source emits a sequence of a_i's chosen from the source alphabet A, and the encoder emits a sequence of a_i's chosen from the code alphabet A. It will be assumed in all subsequent discussions that the code words are distinct, i.e., that each code word corresponds to only one source symbol.

Even though each code word is required to be distinct, sequences of code words may not have this property. An example is code A of Table 1.2.1, where a source of size 4 has been encoded in binary code with characters 0 and 1. In code A the code words are distinct, but sequences of code words are not. It is clear that such a code is not *uniquely* decipherable. On the other hand, a given sequence of code words taken from code B will correspond to a distinct sequence of source symbols. An examination of code B shows that in no case is a code word formed by adding characters to another word. In other words, no code word is a *prefix* of another. It is clear that this is a *sufficient* (but not necessary) condition for a code to be uniquely decipherable. That it is not necessary can be seen from an examination of codes C and D of Table 1.2.1. These codes are uniquely decipherable even though many of the code words are prefixes of other words. In these cases any sequence of code words can be decoded by subdividing the sequence of 0s and 1s to the left of every 0 for code C and to the right of every 0 for code D. The character 0 is the first (or last) character of every code word and acts as a comma; therefore this type of code is called a *comma code*.

TABLE 1.2.1   Four Binary Coding Schemes

Source symbol	Code A	Code B	Code C	Code D
x_1	0	0	0	0
x_2	1	10	01	10
x_3	00	110	011	110
x_4	11	111	0111	1110

Note: Code A is not uniquely decipherable; codes B, C, and D are uniquely decipherable; codes B and D are instantaneous codes; and codes C and D are comma codes.

In general the channel will require a finite amount of time to transmit each code character. The code words should be as short as possible in order to maximize information transfer per unit time. The average length L of a code is given by

$$L = \sum_{i=1}^{M} n_i p(x_i) \tag{13}$$

where n_i is the length (number of code characters) of the code word for the source symbol x_i and $p(x_i)$ is the probability of occurrence of x_i. Although the average code length cannot be computed unless the set $\{p(x_i)\}$ is given, it is obvious that codes C and D of Table 1.2.1 will have a greater average length than code B unless $p(x_4) = 0$. Comma codes are not optimal with respect to minimum average length.

Let us encode the sequence $x_3 x_1 x_3 x_2$ into codes B, C, and D of Table 1.2.1 as shown below:

Code B:	110011010
Code C:	011001101
Code D:	110011010

Codes B and D are fundamentally different from code C in that codes B and D can be decoded word by word *without examining subsequent code characters* while code C cannot be so treated. Codes B and D are called *instantaneous codes* while code C is noninstantaneous. The instantaneous codes have the property (previously maintained) that no code word is a prefix of another code word.

The aim of noiseless coding is to produce codes with the two properties of (1) *unique decipherability* and (2) *minimum average length L* for a given source S with alphabet X and probability set $\{p(x_i)\}$. Codes which have both these properties will be called *optimal*. It can be shown that if, for a given source S, a code is optimal among instantaneous codes, then it is optimal among all uniquely decipherable codes. Thus it is sufficient to consider instantaneous codes. A *necessary* property of optimal codes is that source symbols with higher probabilities have shorter code words; i.e.,

$$p(x_i) > p(x_j) \Rightarrow n_i \leq n_j \tag{14}$$

The encoding procedure consists of the assignment of a code word to each of the M source symbols. The code word for the source symbol x_i will be of length n_i; that is, it will consist of n_i code elements chosen from the code alphabet of size D. It can be shown that a necessary and sufficient condition for the construction of a uniquely decipherable code is the *Kraft inequality*

$$\sum_{i=1}^{M} D^{-n_i} \leq 1 \tag{15}$$

NOISELESS-CODING THEOREM

It follows from Eq. (15) that the average code length L, given by Eq. (13), satisfies the inequality

$$L \geq H(X)/\log D \tag{16}$$

Equality (and minimum code length) occurs if and only if the source-symbol probabilities obey

$$p(x_i) = D^{-n_i} \qquad i = 1, 2, \ldots, M \tag{17}$$

A code where this equality applies is called *absolutely optimal*. Since an integer number of code elements must be used for each code word, the equality in Eq. (16) does not usually hold; however, by using one more code element, the average code length L can be bounded from above to give

$$H(X)/\log D \leq L \leq H(X)/\log D + 1 \tag{18}$$

This last relationship is frequently called the *noiseless-coding theorem*.

CONSTRUCTION OF NOISELESS CODES

The easiest case to consider occurs when an absolutely optimal code exists; i.e., when the source-symbol probabilities satisfy Eq. (17). Note that code B of Table 1.2.1 is absolutely optimal if $p(x_1) = 1/2$, $p(x_2) = 1/4$, and $p(x_3) = p(x_4) = 1/8$. In such cases, a procedure for realizing the code for arbitrary code-alphabet size ($D \geq 2$) is easily constructed as follows:

1. Arrange the M source symbols in order of decreasing probability.
2. Arrange the D code elements in an arbitrary but fixed order, i.e., $a_1, a_2, \ldots, a_D$.
3. Divide the set of symbols x_i into D groups with equal probabilities of $1/D$ each. This division is always possible if Eq. (17) is satisfied.
4. Assign the element a_1 as the first digit for symbols in the first group, a_2 for the second, and a_i for the ith group.
5. After the first division each of the resulting groups contains a number of symbols equal to D raised to some integral power if Eq. (17) is satisfied.

Thus, a typical group, say group i, contains D^{k_i} symbols, where k_i is an integer (which may be zero). This group of symbols can be further subdivided k_i times into D parts of equal probabilities. Each division decides one additional code digit in the sequence. A typical symbol x_i is isolated after q divisions. If it belongs to the i_1 group after the first division, the i_2 group after the second division, and so forth, then the code word for x_i will be $a_{i1} a_{i2} \ldots a_{iq}$.

An illustration of the construction of an absolutely optimal code for the case where $D = 3$ is given in Table 1.2.2. This procedure ensures that source symbols with high probabilities will have short code words and vice versa, since a symbol with probability D^{-n_i} will be isolated after n_i divisions and thus will have n_i elements in its code word, as required by Eq. (17).

TABLE 1.2.2 Construction of an Optimal Code; $D = 3$

Source symbols x_i	A priori probabilities $p(x_i)$	Step 1	Step 2	Step 3	Final code		
x_1	$1/3$	1			1		
x_2	$1/9$	0	1		0	1	
x_3	$1/9$	0	0		0	0	
x_4	$1/9$	0	−1		0	−1	
x_5	$1/27$	−1	1	1	−1	1	1
x_6	$1/27$	−1	1	0	−1	1	0
x_7	$1/27$	−1	1	−1	−1	1	−1
x_8	$1/27$	−1	0	1	−1	0	1
x_9	$1/27$	−1	0	0	−1	0	0
x_{10}	$1/27$	−1	0	−1	−1	0	−1
x_{11}	$1/27$	−1	−1	1	−1	−1	1
x_{12}	$1/27$	−1	−1	0	−1	−1	0
x_{13}	$1/27$	−1	−1	−1	−1	−1	−1

Note: Average code length $L = 2$ code elements per symbol: source entropy $H(X) = 2 \log_2 3$ bits per symbol.

$$L = \frac{H(X)}{\log_2 3}$$

TABLE 1.2.3   Construction of Huffman Code; $D = 2$

Source symbols x_i	A priori probabilities $p(x_i)$	Final code	Reduction 1 (Step 5)		Reduction 2 (Step 4)		Reduction 3 (Step 3)		Reduction 4 (Step 2)		Reduction 5 (Step 1)	
x_1	0.40	0	0.40	0	0.40	0	0.40	0	0.40	0	0.60	1
x_2	0.20	111	0.20	111	0.20	111	0.24	10	0.36	11	0.40	0
x_3	0.12	101	0.12	101	0.16	110	0.20	111	0.24	10		
x_4	0.08	1101	0.12	100	0.12	101	0.16	110				
x_5	0.08	1100	0.08	1101	0.12	100						
x_6	0.08	1001	0.08	1100								
x_7	0.04	1000										

Average code length $L = 1(0.40) + 3(0.20) + 3(0.12) + 4(0.08) + 4(0.08) + 4(0.08) + 4(0.04)$
$= 2.48$ code elements/symbol

The code resulting from the process just discussed is sometimes called the *Shannon-Fano* code. It is apparent that the same encoding procedure can be followed whether or not the source probabilities satisfy Eq. (17). The set of symbols x_i is simply divided into D groups with probabilities as nearly equal as possible. The procedure is sometimes ambiguous, however, and more than one Shannon-Fano code may be possible. The ambiguity arises, of course, in the choice of approximately equiprobable subgroups.

For the general case where Eq. (17) is not satisfied, a procedure owing to Huffman guarantees an optimal code, i.e., one with minimum average length. This procedure for code alphabet of arbitrary size D is as follows:

1. As before, arrange the M source symbols in order of decreasing probability.

2. As before, arrange the code elements in an arbitrary but fixed order, that is, $a_1, a_2, \ldots, a_D$.

3. Combine (sum) the probabilities of the D least likely symbols and reorder the resulting $M - (D - 1)$ probabilities; this step will be called *reduction* 1. Repeat as often as necessary until there are D ordered probabilities remaining. *Note*: For the binary case ($D = 2$), it will always be possible to accomplish this reduction in $M - 2$ steps. When the size of the code alphabet is arbitrary, the last reduction will result in exactly D ordered probabilities if and only if

$$M = D + n(D - 1)$$

where n is an integer. If this relationship is not satisfied, *dummy* source symbols with zero probability should be added. The entire encoding procedure is followed as before, and at the end the dummy symbols are thrown away.

4. Start the encoding with the last reduction which consists of exactly D ordered probabilities; assign the element a_1 as the first digit in the code words for all the source symbols associated with the first probability; assign a_2 to the second probability; and a_i to the ith probability.

5. Proceed to the next to the last reduction; this reduction consists of $D + (D - 1)$ ordered probabilities for a net gain of $D - 1$ probabilities. For the D new probabilities, the first code digit has already been assigned and is the same for all of these D probabilities; assign a_1 as the second digit for all source symbols associated with the first of these D new probabilities; assign a_2 as the second digit for the second of these D new probabilities, etc.

6. The encoding procedure terminates after $1 + n(D - 1)$ steps, which is one more than the number of reductions.

As an illustration of the Huffman coding procedure, a binary code is constructed in Table 1.2.3.

CHANNEL CAPACITY

The average mutual information $I(X; Y)$ between an information source and a destination was given by Eqs. (25) and (26) as

$$I(X; Y) = H(Y) - H(Y|X) = H(X) - H(X|Y) \geq 0 \tag{19}$$

The average mutual information depends not only on the statistical characteristics of the channel but also on the distribution $p(x_i)$ of the input alphabet X. If the input distribution is varied until Eq. (19) is a maximum for a given channel, the resulting value of $I(X; Y)$ is called the *channel capacity* C of that channel; i.e.,

$$C = \max_{p(xi)} I(X; Y) \tag{20}$$

In general, $H(X)$, $H(Y)$, $H(X|Y)$, and $H(Y|X)$ all depend on the input distribution $p(x_i)$. Hence, *in the general case*, it is not a simple matter to maximize Eq. (19) with respect to $p(x_i)$.

All the measures of information that have been considered in this treatment have involved only probability distributions on X and Y. Thus, for the model of Fig. 1.1.1, the joint distribution $p(x_i, y_j)$ is sufficient. Suppose the source [and hence the input distribution $p(x_i)$] is known; then it follows from the usual conditional-probability relationship

$$p(x_i, y_j) = p(x_i)p(y_j | x_i) \tag{21}$$

that only the distribution $p(y_j | x_i)$ is needed for $p(x_i | y_j)$ to be determined. This conditional probability $p(y_j | x_i)$ can then be taken as a description of the information channel connecting the source X and the destination Y. Thus, a *discrete memoryless channel* can be defined as the probability distribution

$$p(y_j | x_i) \qquad x_i \in X \text{ and } y_j \in Y \tag{22}$$

or, equivalently, by the *channel matrix D*, where

$$D = [p(y_j | x_i)] = \begin{bmatrix} p(y_1|x_1) & p(y_2|x_2) & \cdots & p(y_N|x_1) \\ p(y_1|x_2) & p(y_2|x_2) & \cdots & p(y_N|x_2) \\ \cdots\cdots\cdots\cdots\cdots\cdots\cdots\cdots\cdots\cdots\cdots\cdots \\ p(y_1|x_M) & \cdots & \cdots & p(y_N|x_M) \end{bmatrix} \tag{23}$$

A number of special types of channels are readily distinguished. Some of the simplest and/or most interesting are listed as follows:

(**a**) *Lossless Channel.* Here $H(X|Y) = 0$ for all input distribution $p(x_i)$, and Eq. (20) becomes

$$C = \max_{p(xi)} H(X) = \log M \tag{24}$$

This maximum is obtained when the x_i are equally likely, so that $p(x_i) = 1/M$ for all i. The channel capacity is equal to the source entropy, and no source information is lost in transmission.

(**b**) *Deterministic Channel.* Here $H(Y|X) = 0$ for all input distributions $p(x_i)$, and Eq. (20) becomes

$$C = \max_{p(xi)} H(Y) = \log N \tag{25}$$

This maximum is obtained when the y_j are equally likely, so that $p(y_j) = 1/N$ for all j. Each member of the X set is uniquely associated with one, and only one, member of the destination alphabet Y.

(**c**) *Symmetric Channel.* Here the rows of the channel matrix D are identical except for permutations, *and* the columns are identical except for permutations. If D is square, rows and columns are identical except for permutations. In the symmetric channel, the conditional entropy $H(Y|X)$ is independent of the input distribution $p(x_i)$ and depends only on the channel matrix D. As a consequence, the determination of channel capacity is greatly simplified and can be written

$$C = \log N + \sum_{j=1}^{N} p(y_j|x_i) \log p(y_j|x_i) \tag{26}$$

This capacity is obtained when the y_i are equally likely, so that $p(y_j) = 1/N$ for all j.

(**d**) *Binary Symmetric Channel* (*BSC*). This is the special case of a symmetric channel where $M = N = 2$. Here the channel matrix can be written

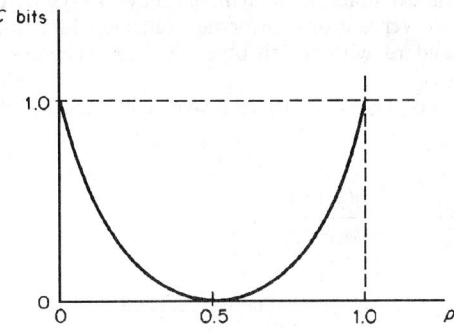

FIGURE 1.2.1 Capacity of the binary symmetric channel.

$$D = \begin{bmatrix} p & 1-p \\ 1-p & p \end{bmatrix} \qquad (27)$$

and the channel capacity is

$$C = \log 2 - G(p) \qquad (28)$$

where the function $G(p)$ is defined as

$$G(p) = -[p \log p + (1 - p) \log (1 - p)] \qquad (29)$$

This expression is mathematically identical to the entropy of a binary source as given in Eq. (5) and is plotted in Fig. 1.1.3 using base 2 logarithms. For the same base, Eq. (28) is shown as a function of p in Fig. 1.2.1. As expected, the channel capacity is large if p, the probability of correct transmission, is either close to unity or to zero. If $p = 1/2$, there is no statistical evidence which symbol was sent and the channel capacity is zero.

DECISION SCHEMES

A decision scheme or decoding scheme B is a partitioning of the Y set into M disjoint and exhaustive sets B_1, $B_2, \ldots, B_M$ such that when a destination symbol y_k falls into set B_i, it is decided that symbol x_i was sent. Implicit in this definition is a *decision rule* $d(y_j)$, which is a function specifying uniquely a source symbol for each destination symbol. Let $p(e \mid y_j)$ be the probability of error when it is decided that y_j has been received. Then the *total error probability* $p(e)$ is

$$p(e) = \sum_{j=1}^{N} p(y_j) p(e \mid y_j) \qquad (30)$$

For a given decision scheme β, the conditional error probability $p(e \mid y_j)$ can be written

$$p(e \mid y_j) = 1 - p[d(y_j) \mid y_j] \qquad (31)$$

where $p[d(y_j) \mid y_j]$ is the conditional probability $p(x_i \mid y_j)$ with x_i assigned by the decision rule; i.e., for a given decision scheme $d(y_j) = x_i$. The probability $p(y_j)$ is determined only by the source a priori probability $p(x_i)$ and by the channel matrix $= D [p(y_j \mid x_i)]$. Hence, only the term $p(e \mid y_j)$ in Eq. (30) is a function of the decision scheme. Since Eq. (30) is a sum of nonnegative terms, the error probability is a minimum when each summand is a minimum. Thus, the term $p(e \mid y_j)$ should be a minimum for each y_j. It follows from Eq. (31) that the minimum-error scheme is that scheme which assigns a decision rule

$$d(y_j) = x^* \qquad j = 1, 2, \ldots, N \qquad (32)$$

where x^* is defined by

$$p(x^* \mid y_j) \geq p(x_i \mid y_j) \qquad i = 1, 2, \ldots, M \qquad (33)$$

In other words, each y_j is decoded as the *a posteriori most likely* x_i. This scheme, which minimizes the probability of error $p(e)$, is usually called the *ideal observer*.

The ideal observer is not always a completely satisfactory decision scheme. It suffers from two major disadvantages: (1) For a given channel D, the scheme is defined only for a given input distribution $p(x_i)$. It might be preferable to have a scheme that was insensitive to input distributions. (2) The scheme minimizes average error but does not bound certain errors. For example, some symbols may always be received incorrectly. Despite these disadvantages, the ideal observer is a straightforward scheme which does minimize average error. It is also widely used as a standard with which other decision schemes may be compared.

Consider the special case where the input distribution is $p(x_i) = 1/M$ for all i, so that all x_i are equally likely. Now the conditional likelihood $p(x_i \mid y_j)$ is

$$p(x_i \mid y_j) = \frac{p(x_i)p(y_j \mid x_i)}{p(y_j)} = \frac{p(y_j \mid x_i)}{Mp(y_j)} \tag{34}$$

For a given y_j, that input x_i is chosen which makes $p(y_j \mid x_i)$ a maximum, and the decision rule is

$$d(y_j) = x^\dagger \qquad j = 1, 2, \ldots, N \tag{35}$$

where $x^\dagger$ is defined by

$$p(y_j \mid x^\dagger) \geq p(y_j \mid x_i) \qquad i = 1, 2, \ldots, M \tag{36}$$

The probability of error becomes

$$p(e) = \sum_{j=1}^{N} p(y_j) \left[1 - \frac{p(y_j \mid x^\dagger)}{Mp(y_j)} \right] \tag{37}$$

This decoder is sometimes called the *maximum-likelihood* decoder or decision scheme.

It would appear that a relationship should exist between the error probability $p(e)$ and the channel capacity C. One such relationship is the *Fano bound*, given by

$$H(X \mid Y) \leq G[p(e)] + p(e) \log (M - 1) \tag{38}$$

and relating error probability to channel capacity through Eq. (20). Here $G(\cdot)$ is the function already defined by Eq. (29). The three terms in Eq. (38) can be interpreted as follows:

$H(X \mid Y)$ is the equivocation. It is the average additional information needed at the destination after reception to completely determine the symbol that was sent.

$G[p(e)]$ is the entropy of the binary system with probabilities $p(e)$ and $1 - p(e)$. In other words, it is the average amount of information needed to determine whether the decision rule resulted in an error.

$\log (M - 1)$ is the maximum amount of information needed to determine which among the remaining $M - 1$ symbols was sent if the decision rule was incorrect; this information is needed with probability $p(e)$.

THE NOISY-CODING THEOREM

The concept of channel capacity was discussed earlier. Capacity is a fundamental property of an information channel in the sense that it is possible to transmit information through the channel at any rate less than the channel capacity with arbitrarily small probability of error. This result is called the *noisy-coding theorem* or *Shannon's fundamental theorem for a noisy channel*.

The noisy-coding theorem can be stated more precisely as follows: Consider a discrete memoryless channel with nonzero capacity C; fix two numbers H and ϵ such that

$$0 < H < C \tag{39}$$

and

$$\epsilon > 0 \tag{40}$$

Let us transmit m messages $u_1, u_2, \ldots, u_m$ by code words each of length n binary digits. The positive integer n can be chosen so that

$$m \geq 2^{nH} \tag{41}$$

In addition, at the destination the m sent messages can be associated with a set $V = \{v_1, v_2, \ldots, v_m\}$ of received messages and with a decision rule $d(v_j) = u_j$ such that

$$p[d(v_j)\,|\,v_j] \geq 1 - \epsilon \tag{42}$$

i.e., decoding can be accomplished with a probability of error that does not exceed ϵ. There is a converse to the noise-coding theorem which states that it is not possible to produce an encoding procedure which allows transmission at a rate greater than channel capacity with arbitrarily small error.

ERROR-CORRECTING CODES

The codes considered earlier were designed for minimum length in the noiseless-transmission case. For noisy channels, the noisy-coding theorem guarantees the existence of a code which will allow transmission at any rate less than channel capacity and with arbitrarily small probability of error; however, the theorem does not provide a constructive procedure to devise such codes. Indeed, it implies that very long sequences of source symbols may have to be considered if reliable transmission at rates near channel capacity are to be obtained. In this section, we consider some of the elementary properties of simple *error-correcting codes*; i.e., codes which can be used to increase reliability in the transmission of information through noisy channels by correcting at least some of the errors that occur so that overall probability of error is reduced.

The discussion will be restricted to the BSC, and the noisy-coding theorem notation will be used. Thus, a source alphabet $X = \{x_1, x_2, \ldots, x_m\}$ of M symbols will be used to form a message set U of m messages u_k, where $U = \{u_1, u_2, \ldots, u_m\}$. Each u_k will consist of a sequence of the x_i's. Each message u_k will be encoded into a sequence of n binary digits for transmission over the BSC. At the destination, there exists a set $V = \{v_1, v_2, \ldots, v_{2n}\}$ of all possible binary sequences of length n. The inequality $m \leq 2^n$ must hold. The problem is to associate with each sent message u_k a received message v_j so that $p(e)$, the overall probability of error, is reduced.

In the discussion of the noisy-coding theorem, a decoding scheme was used that examined the received message v_j and identified it with the sent message u_k, which differed from it in the least number of binary digits. In all the discussions here it will be assumed that this decoder is used. Let us define the *Hamming distance* $d(v_j, v_k)$ between two binary sequences v_j and v_k of length n as the number of digits in which v_j and v_k disagree. Thus, if the distance between two sequences is zero, the two sequences are identical. It is easily seen that this distance measure has the following four elementary properties:

$$d(v_j, v_k) \geq 0 \text{ with equality if and only if } v_j = v_k \tag{43}$$

$$d(v_j, v_k) = d(v_k, v_j) \tag{44}$$

$$d(v_j, v_l) \leq d(v_j, v_k) + d(v_k, v_l) \tag{45}$$

$$d(v_j, v_k) \leq n \tag{46}$$

The decoder we use is a *minimum-distance* decoder. As mentioned earlier, the ideal-observer decoding scheme is a minimum-distance scheme for the BSC.

It is intuitively apparent that the sent messages should be represented by code words that all have the greatest possible distances between them. Let us investigate this matter in more detail by considering all binary sequences of length $n = 3$; there are $2^n = 2^3 = 8$ such sequences, viz.,

000	001	011	111
	010	110	
	100	101	

It is convenient to represent these as the eight corners of a unit cube, as shown in Fig. 1.2.2a, where the x axis corresponds to the first digit, the y axis to the second, and the z axis to the third. Although direct pictorial representation is not possible, it is clear that binary sequences of length n greater than 3 can be considered as the corners of the corresponding n-cube.

Suppose that all eight binary sequences are used as code words to encode a source. If any binary digit is changed in transmission, an error will result at the destination since the sent message will be interpreted incorrectly as one of the three possible messages that differ in one code digit from the sent message. This situation is illustrated in Fig. 1.2.2 for the code words 000 and 111. A change of one digit in each of these code words produces one of three possible other code words.

Figure 1.2.2 suggests that only two code words, say 000 and 111, should be used. The distance between these two words, or any other two words on opposite corners of the cube, is 3. If only one digit is changed in the transmission of each of these two code words, they can be correctly distinguished at the destination by a minimum-distance decoder. If two digits are changed in each word in transmission, it will not be possible to make this distinction.

This reasoning can be extended to sequences containing more than three binary digits. For any $n \geq 3$, single errors in each code word can be corrected. If double errors are to be corrected without fail, there must be at least two code words with a minimum distance between them of 5; thus, for this case, binary code words of length 5 or greater must be used.

Note that the error-correcting properties of a code depend on the distance $d(v_j, v_k)$ between the code words. Specifically, single errors can be corrected if all code words employed are at least a distance of 3 apart, double errors if the words are at a distance of 5 or more from each other, and, in general, q-fold errors can be corrected if

$$d(v_j, v_k) \geq 2q + 1 \qquad j \neq k \tag{47}$$

Errors involving less than q digits per code word can also be corrected if Eq. (63) is satisfied. If the distance between two code words is $2q$, there will always be a group of binary sequences which are in the middle, i.e., a distance q from *each* of the two words. Thus, by the proper choice of code words, q-fold errors can be *detected* but not corrected if

$$d(v_j, v_k) = 2q \qquad j \neq k \tag{48}$$

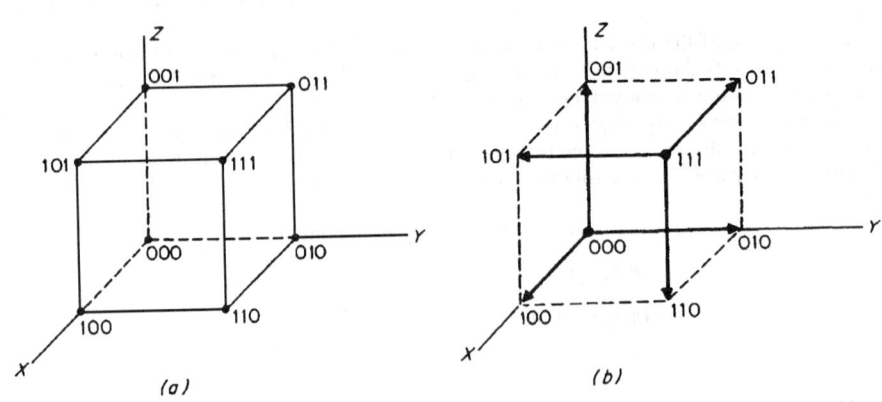

FIGURE 1.2.2   Representation of binary sequences as the corners of an n-cube, $n = 3$; (a) the eight binary sequences of length 3; (b) shift in sequences 000 and 111 from a single error.

TABLE 1.2.4 Parity-Check Code for Single-Error Detection

Message digits	Check digit	Word	Message digit	Check digit	Word
000	0	0000	110	0	1100
100	1	1001	101	0	1010
010	1	0101	011	0	0110
001	1	0011	111	1	1111

Now consider the maximum number of code words r that can be selected from the set of 2^n possible binary sequences of length n to form a code that will correct all single, double, ..., q-fold errors. In the example of Fig. 1.2.2, the number of code words selected was 2. In fact, it can be shown that there is no single-error-correcting code for $n = 3, 4$ containing more than two words. Suppose we consider a given code consisting of the words ..., u_k, u_j, ... All binary sequences of distance q or less from u_k must "belong" to u_k, and to u_k only, if q-fold errors are to be corrected. Thus, associated with u_k are all binary sequences of distance 0, 1, 2, ..., q from u_k. The number of such sequences is given by

$$\binom{n}{0} + \binom{n}{1} + \binom{n}{2} + \cdots + \binom{n}{q} = \sum_{i=0}^{q}\binom{n}{i} \tag{49}$$

Since there are r of the code words, the total number of sequences associated with all the code words is

$$r\sum_{i=0}^{q}\binom{n}{i}$$

This number can be no larger than 2^n, the total number of distinct binary sequences of length n. Therefore the following inequality must hold:

$$r\sum_{i=0}^{q}\binom{n}{i} \le 2^n \quad \text{or} \quad r \le \frac{2^n}{\sum_{i=0}^{q}\binom{n}{i}} \tag{50}$$

This is a *necessary* upper bound on the number of code words that can be used to correct all errors up to and including q-fold errors. It can be shown that it is not *sufficient*.

Consider the eight possible distinct binary sequences of length 3. Suppose we add one binary digit to each sequence in such a way that the total number of 1s in the sequence is *even* (or odd, if you wish). The result is shown in Table 1.2.4. Note that all the word sequences of length 4 differ from each other by a distance of at least 2. In accordance with Eq. (48), it should be possible now to *detect single errors* in all eight sequences. The detection method is straightforward. At the receiver, count the number of 1s in the sequence; if the number is odd, a single error (or, more precisely, an odd number of errors) has occurred; if the number is even, no error (or an even number of errors) has occurred. This particular scheme is a good one if only single errors are likely to occur and if detection only (rather than correction) is desired. Such is often the case, for example, in closed digital systems such as computers. The added digit is called a *parity-check digit*, and the scheme is a very simple example of a *parity-check code*.

PARITY-CHECK CODES

More generally, in parity-check codes, the encoded sequence consists of n binary digits of which only $k < n$ are *information digits* while the remaining $l = n - k$ digits are used for error detection and correction and are called *check digits* or *parity checks*. The example of Table 1.2.4 is a single-error-detecting code, but, in general, q-fold errors can be detected and/or corrected. As the number of errors to be detected and/or corrected increases, the number l of check digits must increase. Thus, for fixed word length n, the number of information digits $k = n - l$ will decrease as more and more errors are to be detected and/or corrected. Also the total number of words in the code cannot exceed the right side of Eq. (50) or the number 2^k.

Parity-check codes are relatively easy to implement. The simple example given of a single-error-detecting code requires only that the number of 1s in each code word be counted. In this light, it is of considerable importance to note that these codes satisfy the noisy-coding theorem. In other words, it is possible to encode a source by parity-check coding for transmission over a BSC at a rate approaching channel capacity and with arbitrarily small probability of error. Then, from Eq. (41), we have

$$2^{nH} = 2^k \tag{51}$$

or H, the rate of transmission, is given by

$$H = k/n \tag{52}$$

As $n \to \infty$, the probability of error $p(e)$ approaches zero. Thus, in a certain sense, it is sufficient to limit a study of error-correcting codes to the parity-check codes.

As an example of a parity-check code, consider the simplest nondegenerate case where l, the number of check digits, is 2 and k, the number of information digits, is 1. This system is capable of single-error detection and correction, as we have already decided from geometric considerations. Since $l + k = 3$, each encoded word will be three digits long. Let us denote this word by $a_1 a_2 a_3$, where each a_i is either 0 or 1. Let a_1 represent the information digit and a_2 and a_3 represent the check digits.

Checking for errors is done by forming two independent equations from the three a_i, each equation being of the form of a modulo-2 sum, i.e., of the form

$$a_i \oplus a_j = \begin{cases} 0 & a_i = a_j \\ 1 & a_i \neq a_j \end{cases}$$

Take the two independent equations to be

$$a_2 \oplus a_3 = 0 \quad \text{and} \quad a_1 \oplus a_3 = 0$$

for an *even-parity* check. For an odd-parity check, let the right sides of both of these equations be unity. If these two equations are to be satisfied, the only possible code words that can be sent are 000 and 111. The other six words of length 3 violate one or both of the equations.

Now suppose that 000 is sent and 100 is received. A solution of the two independent equations gives, for the received word,

$$a_2 \oplus a_3 = 0 \oplus 0 = 0$$
$$a_1 \oplus a_3 = 1 \oplus 0 = 1$$

The check yields the binary check number 1, indicating that the error is in the first digit a_1, as indeed it is. If 111 is sent and 101 received, then

$$a_2 \oplus a_3 = 0 \oplus 1 = 1$$
$$a_1 \oplus a_3 = 1 \oplus 1 = 0$$

and the binary check number is 10, or 2, indicating that the error is in a_2.

In the general case, a set of l independent linear equations is set up in order to derive a binary checking number whose value indicates the position of the error in the binary word. If more than one error is to be detected and corrected, the number l of check digits must increase, as discussed previously.

In the example just treated, the l check digits were used only to check the k information digits immediately preceding them. Such a code is called a *block code*, since all the information digits and all the check digits are contained in the block (code word) of length $n = k + l$. In some encoding procedures, the l check digits may also be used to check information digits appearing in preceding words. Such codes are called *convolutional* or *recurrent codes.* A parity-check code (either block or convolutional) where the word length is n and the number of information digits is k is usually called an (n, k) *code.*

OTHER ERROR-DETECTING AND ERROR-CORRECTING CODES

Unfortunately, a general treatment of error-detecting and error-correcting codes requires that the code structure be cast in a relatively sophisticated mathematical form. The commonest procedure is to identify the code letters with the elements of a finite (algebraic) field. The code words are then taken to form a vector subspace of n-tuples over the field. Such codes are called *linear codes* or, sometimes, *group codes.* Both the block codes and the convolutional codes mentioned in the previous paragraph fall in this category.

An additional constraint often imposed on linear codes is that they be *cyclic.* Let a code word a be represented by

$$a = (a_0, a_1, a_2, \ldots, a_{n-1})$$

Then the ith *cyclic permutation* a^{-i} is given by $a^i = (a_i, a_{i+1}, \ldots, a_{n-1}, a_0, a_1, \ldots, a_{i-1})$. A linear code is cyclic if, and only if, for every word a in the code, there is also a word a^i in the code. The permutations need not be distinct and, in fact, generally will not be. The eight code words

$$0000 \quad 0110 \quad 1001 \quad 1010$$

$$0011 \quad 1100 \quad 0101 \quad 1111$$

constitute a cyclic set. Included in the cyclic codes are some of those most commonly encountered such as the Bose and Ray-Chaudhuri (BCH) codes and shortened Reed-Muller codes.

CONTINUOUS-AMPLITUDE CHANNELS

The preceding discussion has concerned discrete message distributions and channels. Further, it has been assumed, either implicitly or explicitly, that the time parameter is discrete, i.e., that a certain number of messages, symbols, code digits, and so forth, are transmitted per unit time. Thus, we have been concerned with *discrete-amplitude, discrete-time* channels and with messages which can be modeled as *discrete random processes* with *discrete parameter.* There are three other possibilities, depending on whether the process amplitude and the time parameter have discrete or continuous distributions.

We now consider the *continuous-amplitude, discrete-time* channel, where the input messages can be modeled as *continuous random processes* with *discrete parameter.* It will be shown later that continuous-time cases of engineering interest can be treated by techniques which amount to the replacement of the continuous parameter by a discrete parameter. The most straightforward method involves the application of the sampling theorem to band-limited processes. In this case the process is sampled at equispaced intervals of length $1/2W$, where W is the highest frequency of the process. Thus the continuous parameter t is replaced by the discrete parameter $t_k = k/2W$, $k = \cdots, -1, 0, 1, \ldots$.

Let us restrict our attention for the moment to continuous-amplitude, discrete-time situations. The discrete density $p(x_i)$, $i = 1, 2, \ldots, M$, of the source-message set is replaced by the continuous density $f_x(x)$, where, in general, $-\infty < x < \infty$, although the range of x may be restricted in particular cases. In the same way, other discrete densities are replaced by continuous densities. For example, the destination distribution $p(y_j)$, $j = 1, 2, \ldots, N$, becomes $f_y(y)$, and the joint distribution $p(x_i, y_j)$ will be called $f_2(x, y)$.

In analogy with the discrete-amplitude case [Eq. (4)], the entropy of a continuous distribution $f_x(x)$ can be defined as

$$H(X) = -\int_{-\infty}^{\infty} f_x(x) \log f_x(x)\, dx \tag{53}$$

This definition is not completely satisfactory because of some of the properties of this new $H(X)$. For example, it can be negative and it can depend on the coordinate system used to represent the message.

Joint and conditional entropies can also be defined in exact analogy to the discrete case discussed in Chap. 1.1. If the joint density $f_2(x, y)$ exists, then the joint entropy $H(X, Y)$ is given by

$$H(X, Y) = -\int_{-\infty}^{\infty} \int f_2(x, y) \log f_2(x, y)\, dx\, dy \tag{54}$$

and the conditional entropies $H(X|Y)$ and $H(Y|X)$ are

$$H(X|Y) = -\int_{-\infty}^{\infty} \int f_2(x, y) \log \frac{f_2(x, y)}{f_y(y)}\, dx\, dy \tag{55}$$

and

$$H(X|Y) = -\int_{-\infty}^{\infty} \int f_2(x, y) \log \frac{f_2(x, y)}{f_y(y)}\, dx\, dy \tag{56}$$

where

$$f_x(x) = \int_{-\infty}^{\infty} f_2(x, y)\, dy \quad \text{and} \quad f_y(y) = \int_{-\infty}^{\infty} f_2(x, y)\, dx$$

The average mutual information follows from Eq. (15) and is

$$I(X; Y) = -\int_{-\infty}^{\infty} \int f_2(x, y) \log \frac{f_x(x) f_y(y)}{f_2(x, y)}\, dx\, dy \tag{57}$$

Although the entropy of a continuous distribution can be negative, positive, or zero, the average mutual information $I(X; Y) \geq 0$ with equality when x and y are statistically independent, i.e., when $f_2(x, y) = f_x(x) f_y(y)$.

MAXIMIZATION OF ENTROPY OF CONTINUOUS DISTRIBUTIONS

The entropy of a discrete distribution is a maximum when the distribution is uniform, i.e., when all outcomes are equally likely. In the continuous case, the entropy depends on the coordinate system, and it is possible to maximize this entropy subject to various constraints on the associated density function.

The Maximization of $H(X)$ for a Fixed Variance of x. Maximizing $H(X)$ subject to the constraint that

$$\int_{-\infty}^{\infty} x^2 f_x(x)\, dx = \sigma^2 \tag{58}$$

yields the gaussian density

$$f_x(x) = (1/\sqrt{2\pi}\sigma)e^{-x^2/2\sigma^2} \qquad -\infty < x < \infty \tag{59}$$

Thus, for fixed variance, the normal distribution has the largest entropy. The entropy in this case is

$$H(X) = \tfrac{1}{2}\ln 2\pi\sigma^2 + \tfrac{1}{2}\ln e = \tfrac{1}{2}\ln 2\pi e\sigma^2 \tag{60}$$

This last result will be of considerable use later. For convenience, the natural logarithm has been used, and the units of H are nats.

The Maximization of H(X) for a Limited Peak Value of x.   In this case, the single constraint is

$$\int_{-M}^{M} f_x(x)\ dx = 1 \tag{61}$$

One obtains the uniform distribution

$$f_x(x) = \begin{cases} 1/2M & |x| \le M \\ 0 & |x| > M \end{cases}$$

and, the associated entropy is

$$H(X) = -\int_{-M}^{M} \frac{1}{2M} \log \frac{1}{2M} dx = \log 2M \tag{62}$$

The Maximization of H(X) for x Limited to Nonnegative Values and a Given Average Value.   The constraints

$$\int_{0}^{\infty} f_x(x)\ dx = 1$$

and

$$\int_{0}^{\infty} x f_x(x) dx = \mu \tag{63}$$

lead to the *exponential distribution*

$$f_x(x) = \begin{cases} 0 & x < 0 \\ (1/\mu)e^{-(x/\mu)} & x \ge 0 \end{cases}$$

The entropy associated with this distribution is

$$H(X) = \ln \mu + 1 = \ln \mu e \tag{64}$$

GAUSSIAN SIGNALS AND CHANNELS

Let us assume that the source symbol x and the destination symbol y are jointly gaussian, i.e., that the joint density $f_2(x, y)$ is

$$f_2(x,y) = \frac{1}{2\pi\sigma_x\sigma_y\sqrt{1-\rho^2}} \exp\left\{ -\frac{1}{2(1-\rho^2)}\left[\left(\frac{x}{\sigma_x}\right)^2 - 2\rho\frac{xy}{\sigma_x\sigma_y} + \left(\frac{y}{\sigma_y}\right)^2 \right] \right\} \tag{65}$$

where σ_x^2 and σ_y^2 are the variances of x and y, respectively, and ρ is the correlation coefficient given by

$$\rho = \frac{E\{xy\}}{\sigma_x\sigma_y} \tag{66}$$

The univariate densities of x and y are given, of course, by

$$f_x(x) = \frac{1}{\sqrt{2\pi}\sigma_x} \exp\left[-\frac{1}{2}\left(\frac{x}{\sigma_x}\right)^2\right] \quad -\infty < x < \infty \tag{67}$$

and

$$f_y(y) = \frac{1}{\sqrt{2\pi}\sigma_y} \exp\left[-\frac{1}{2}\left(\frac{y}{\sigma_y}\right)^2\right] \quad -\infty < y < \infty \tag{68}$$

In this case we have

$$I(X; Y) = -\tfrac{1}{2} \ln (1 - \rho^2) \tag{69}$$

Thus the average mutual information in two jointly gaussian random variables is a function only of the correlation coefficient ρ and varies from zero to infinity since $-1 \leq \rho \leq 1$.

The noise entropy $H(Y|X)$ can be written

$$H(Y|X) = H(Y) - I(X; Y) = \tfrac{1}{2} \ln 2\pi e \sigma_y^2 (1 - \rho^2) \tag{70}$$

Suppose that x and y are jointly gaussian as a result of independent zero-mean gaussian noise n being added in the channel to the gaussian input x, so that

$$y = x + n \tag{71}$$

In this case the correlation coefficient ρ becomes

$$\rho = \frac{E\{x^2 + nx\}}{\sigma_x \sigma_y} = \frac{\sigma_x^2}{\sigma_x \sigma_y} = \frac{\sigma_x}{\sigma_y} \tag{72}$$

and the noise entropy is

$$H(Y|X) = \tfrac{1}{2} \ln 2\pi e \sigma_n^2 \tag{73}$$

where σ_n^2 is the noise variance given by

$$\sigma_n^2 = E\{n^2\} = \sigma_y^2 - \sigma_x^2 \tag{74}$$

In this situation, Eq. (69) can be rewritten as

$$I(X; Y) = \tfrac{1}{2} \ln (1 + \sigma_x^2 / \sigma_n^2) \tag{75}$$

It is conventional to define the signal power as $S_p = \sigma_x^2$ and the noise power as $N_p = \sigma_n^2$ and to rewrite this last expression as

$$I(X; Y) = \tfrac{1}{2} \ln (1 + S_p / N_p) \tag{76}$$

where S_p/N_p is the signal-to-noise power ratio.

Channel capacity C for the continuous-amplitude, discrete-time channel is

$$C = \max_{fx(x)} I(X;\ Y) = \max_{fx(x)} [H(Y) - H(Y|X)] \tag{77}$$

Suppose the channel consists of an additive noise that is a sequence of independent gaussian random variables n each with zero mean and variance σ_n^2. In this case the conditional probability $f(y/x)$ at each time instant is normal with variance σ_n^2 and mean equal to the particular realization of X. The noise entropy $H(Y|X)$ is given by Eq. (73), and Eq. (77) becomes

$$C = \max_{fx(x)} [H(Y)] - \tfrac{1}{2} \ln 2\pi e \sigma_n^2 \tag{78}$$

If the input power is fixed at σ_x^2 then the output power is fixed at $\sigma_y^2 = \sigma_x^2 + \sigma_n^2$ and $H(Y)$ is a maximum if $Y = X + N$ is a sequence of independent gaussian random variables. The value of $H(Y)$ is

$$H(Y) = \tfrac{1}{2} \ln 2\pi e \left(\sigma_x^2 + \sigma_n^2 \right)$$

and the channel capacity becomes

$$C = \tfrac{1}{2} \ln \left(1 + \sigma_x^2/\sigma_n^2 \right) = \tfrac{1}{2} \ln \left(1 + S_p/N_p \right) \tag{79}$$

where S_p/N_p is the signal-to-noise power ratio. Note that the input X is a sequence of independent gaussian random variables and this last equation is identical to Eq. (76). Thus, for additive independent gaussian noise and an input power limitation, the discrete-time continuous-amplitude channel has a capacity given by Eq. (79). This capacity is realized when the input is an independent sequence of independent, identically distributed gaussian random variables.

BAND-LIMITED TRANSMISSION AND THE SAMPLING THEOREM

In this section, messages will be considered which can be modeled as continuous random processes $x(t)$ with continuous parameter t. The channels which transmit these messages will be called *amplitude-continuous, time-continuous* channels. Specifically attention will be restricted to signals (random processes) $x(t)$, which are *strictly band-limited*.

Suppose a given arbitrary (deterministic) signal $f(t)$ is available for all time. Is it necessary to know the amplitude of the signal for every value of time in order to characterize it uniquely? In other words, can $f(t)$ be represented (and reconstructed) from some set of *sample values* or *samples* $\ldots, f(t), f(t_0), f(t_1), \ldots$? Surprisingly enough, it turns out that, under certain fairly reasonable conditions, a signal can be represented exactly by samples spaced relatively far apart. The reasonable conditions are that the signal be *strictly band-limited*.

A (real) signal $f(t)$ will be called *strictly band-limited* $(-2\pi W, 2\pi W)$ if its Fourier transform $F(\omega)$ has the property

$$F(\omega) = 0 \qquad |\omega| > 2\pi W \tag{80}$$

Such a signal can be represented in terms of its sample taken at the *Nyquist sampling times*, $t_k = \dfrac{k}{2W}$ $k = 0, \pm 1, \ldots$ via the *sampling representation*

$$f(t) = \sum_{k=-\infty}^{\infty} f\left(\frac{k}{2W} \right) \frac{\sin(2\pi Wt - k\pi)}{2\pi Wt - k\pi} \tag{81}$$

This expression is sometimes called the *Cardinal series* or *Shannon's sampling theorem*. It relates the discrete time domain $\{k/2W\}$ with sample values $f(k/2W)$ to the continuous time domain $\{t\}$ of the function $f(t)$.

The interpolation function

$$k(t) = (\sin 2\pi Wt) / 2\pi Wt \tag{82}$$

has a Fourier transform $K(\omega)$ given by

$$K(\omega) = \begin{cases} 1/4\pi W & |\omega| < 2\pi W \\ 0 & |\omega| > 2\pi W \end{cases} \tag{83}$$

Also the shifted functions $k(t - k/2W)$ has the Fourier transform

$$\mathfrak{F}\{k(t - k/2W)\} = K(\omega)e^{jwk/2W} \tag{84}$$

Therefore, each term on the right side of Eq. (81) is a time function which is strictly band-limited $(-2\pi W, 2\pi W)$. Note also that

$$k\left(t - \frac{k}{2W}\right) = \frac{\sin(2Wt - k\pi)}{2\pi Wt - k\pi} = \begin{cases} 1 & t = t_k = k\pi/2W \\ 0 & t = t_n, \quad n \neq k \end{cases} \tag{85}$$

Thus, this sampling function $k(t - k/2W)$ is zero at all Nyquist instants except t_k, where it equals unity.

Suppose that a function $h(t)$ is not strictly band-limited to at least $(-2\pi W, 2\pi W)$ rad/s and an attempt is made to reconstruct the function using Eq. (81) with sample values spaced $1/2W \cdot$ s apart. It is apparent that the reconstructed signal [which is strictly band-limited $(-2\pi W, 2\pi W)$, as already mentioned] will differ from the original. Moreover, a given set of sample values $\{f(k/2W)\}$ could have been obtained from a whole class of different signals. Thus, it should be emphasized that the reconstruction of Eq. (81) is unambiguous only for signals strictly band-limited to at least $(-2\pi W, 2\pi W)$ rad/s. The set of different possible signals with the same set of sample values $\{f(k/2W)\}$ is called the *aliases* of the band-limited signal $f(t)$.

Let us now consider a signal (random process) $X(t)$ with *autocorrelation function* given by

$$R_x(\tau) = E\{X(t)X(t + \tau)\} \tag{86}$$

and *power spectral density*

$$\varphi_x(\omega) = \int_{-\infty}^{\infty} R_x(r)e^{-jwr} dr \tag{87}$$

which is just the Fourier transform of $R_x(\tau)$. The process will be assumed to have zero mean and to be strictly *band-limited* $(-2\pi W, 2\pi W)$ in the sense that the power special density $\phi_x(\omega)$ vanishes outside this interval; i.e.,

$$\varphi_x(\omega) = 0 \qquad |\omega| > 2\pi W \tag{88}$$

It has been noted that a deterministic signal $f(t)$ band-limited $(-2\pi W, 2\pi W)$ admits the *sampling representation* of Eq. (81). It can also be shown that the random process $X(t)$ admits the same expansions; i.e.,

$$X(t) = \sum_{k=-\infty}^{\infty} X\left(\frac{k}{2W}\right) \frac{\sin(2\pi Wt - k\pi)}{2\pi Wt - k\pi} \tag{89}$$

The right side of this expression is a random variable for each value of t. The infinite sum means that

$$\lim_{N \to \infty} E\{|X(t) - X_N(t)|^2\} = 0$$

where

$$X_N(t) = \sum_{k=-N}^{N} X\left(\frac{k}{2W}\right) \frac{\sin(2\pi Wt - k\pi)}{2\pi Wt - k\pi}$$

Thus, the process $X(t)$ with continuous time parameter t can be represented by the process $X(k/2W)$, $k = \dots, -2, -1, 0, 1, 2, \dots$, with discrete time parameter $t_k = k/2W$. For band-limited signals or channels it is sufficient, therefore, to consider the discrete-time case and to relate the results to continuous time through Eq. (89).

Suppose the continuous-time process $X(t)$ has a spectrum $\varphi_x(\omega)$ which is *flat and band-limited* so that

$$\varphi_x(\omega) = \begin{cases} N_0 & |\omega| \leq 2\pi W \\ 0 & |\omega| > 2\pi W \end{cases} \tag{90}$$

Then the autocorrelation function passes through zero at intervals of $1/2W$ so that

$$R_x(k/2W) = 0 \qquad k = \dots, -2, -1, 1, 2, \dots \tag{91}$$

Thus, samples spaced $k/2W$ apart are *uncorrelated if the power spectral density is flat and band-limited* $(-2\pi W, 2\pi W)$. *If the process is gaussian, the samples are independent.* This implies that continuous-time band-limited $(-2\pi W, 2\pi W)$ gaussian channels, where the noise has a flat spectrum, have a capacity C given by Eq. (79) as

$$C = \tfrac{1}{2} \ln (1 + S_p/N_p) \qquad \text{(nats/sample)} \tag{92}$$

Here N_p is the variance of the additive, flat, band-limited gaussian noise and S_p is $R_x(0)$, the fixed variance of the input signal. The units of Eq. (92) are on a per sample basis. Since there are $2W$ samples per unit time, the capacity C' per unit time can be written as

$$C' = W \ln (1 + S_p/N_p) \qquad \text{(nats/s)} \tag{93}$$

The ideas developed thus far in this section have been somewhat abstract notions involving information sources and channels, channel capacity, and the various coding theorems. We now look more closely at conventional channels. Many aspects of these topics fall into the area often called *modulation theory*.

CHAPTER 1.3
MODULATION

Geoffrey C. Orsak, H. Vincent Poor, John B. Thomas

MODULATION THEORY

As discussed in Chap. 1.1 and shown in Fig. 1.1.1, the central problem in most communication systems is the transfer of information originating in some source to a destination by means of a channel. It will be convenient in this section to call the sent message or intelligence $a(t)$ and to denote the received message by $a^*(t)$, a distorted or corrupted version of $a(t)$.

The message signals used in communication and control systems are usually limited in frequency range to some maximum frequency $f_m = \omega_m/2\pi$ Hz. This frequency is typically in the range of a few hertz for control systems and moves upward to a few megahertz for television video signals. In addition the bandwidth of the signal is often of the order of this maximum frequency so that the signal spectrum is approximately low-pass in character. Such signals are often called *video signals* or *baseband signals*. It frequently happens that the transmission of such a spectrum through a given communication channel is inefficient or impossible. In this light, the problem may be looked upon as the one shown in Fig. 1.1.2, where an encoder E has been added between the source and the channel; however, in this case, the encoder acts to *modulate* the signal $a(t)$, producing at its output the *modulated wave* or signal $m(t)$.

Modulation can be defined as the modification of one signal, called the *carrier*, by another, called the *modulating signal*. The result of the modulation process is a modulated wave or signal. In most cases a frequency shift is one of the results. There are a number of reasons for producing modulated waves. The following list gives some of the major ones.

(a) *Frequency Translation for Efficient Antenna Design.* It may be necessary to transmit the modulating signal through space as electromagnetic radiation. If the antenna used is to radiate an appreciable amount of power, it must be large compared with the signal wavelength. Thus translation to higher frequencies (and hence to smaller wavelengths) will permit antenna structures of reasonable size and cost at both transmitter and receiver.

(b) *Frequency Translation for Ease of Signal Processing.* It may be easier to amplify and/or shape a signal in one frequency range than in another. For example, a dc signal may be converted to ac, amplified, and converted back again.

(c) *Frequency Translation to Assigned Location.* A signal may be translated to an assigned frequency band for transmission or radiation, e.g., in commercial radio broadcasting.

(d) *Changing Bandwidth.* The bandwidth of the original message signal may be increased or decreased by the modulation process. In general, decreased bandwidth will result in channel economies at the cost of fidelity. On the other hand, increased bandwidth will be accompanied by increased immunity to channel disturbances, as in wide-band frequency modulation or in spread-spectrum systems, for examples.

(e) *Multiplexing.* It may be necessary or desirable to transmit several signals occupying the same frequency range or the same time range over a single channel. Various modulation techniques allow the signals to share the same channel and yet be recovered separately. Such techniques are given the generic name of

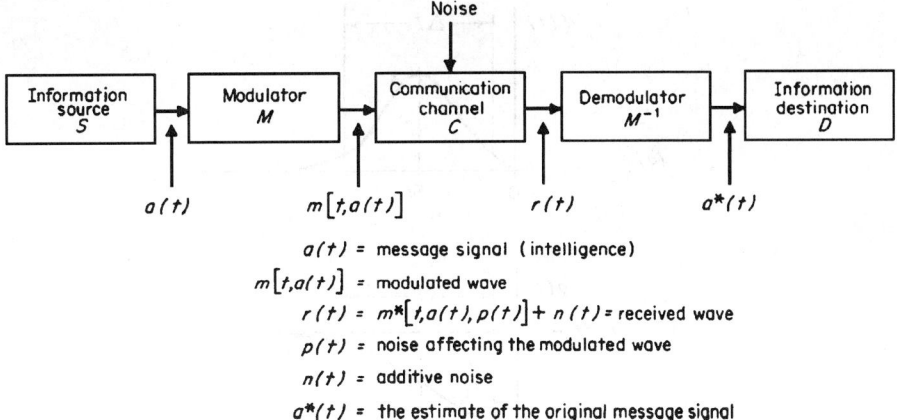

$a(t)$ = message signal (intelligence)

$m[t,a(t)]$ = modulated wave

$r(t)$ = $m^*[t,a(t),p(t)]$ + $n(t)$ = received wave

$p(t)$ = noise affecting the modulated wave

$n(t)$ = additive noise

$a^*(t)$ = the estimate of the original message signal

FIGURE 1.3.1 Communication system involving modulation and demodulation.

multiplexing. As will be discussed later, multiplexing is possible in either the frequency domain (frequency-domain multiplexing FDM) or in the time domain (time-domain multiplexing, TDM). As a simple example, the signals may be translated in frequency so that they occupy separate and distinct frequency ranges as mentioned in item (b).

Thus, the process of modulation can be considered as a form of encoding used to match the message signal arising from the information source to the communication channel. At the same time it is generally true that the channel itself has certain undesirable characteristics resulting in distortion of the signal during transmission. A part of such distortion can frequently be accounted for by postulating noise disturbances in the channel. These noises may be additive and may also affect the modulated wave in a more complicated fashion, although it is usually sufficient (and much simpler) to assume additive noise only. Also, the received signal must be decoded (demodulated) to recover the original signal.

In view of this discussion, it is convenient to change the block diagram of Fig. 1.1.2 to that shown in Fig. 1.3.1. The waveform received at the demodulator (receiver) will be denoted by $r(t)$, where

$$r(t) = m^*[t, a(t), p(t)] + n(t) \qquad (1)$$

where $a(t)$ is the original message signal, $m[t, a(t)]$ is the modulated wave, $m^*[t, a(t), p(t)]$ is a corrupted version of $m[t, a(t)]$, and $p(t)$ and $n(t)$ are noises whose characteristics depend on the channel. Unless it is absolutely necessary for an accurate characterization of the channel, we will assume that $p(t) \equiv 0$ to avoid the otherwise complicated analysis that results.

The aim is to find modulators M and demodulators M^{-1} that make $a^*(t)$ a "good" estimate of the message signal $a(t)$. It should be emphasized that M^{-1} is not uniquely specified by M; for example, it is not intended to imply that $MM^{-1} = 1$. The form of the demodulator, for a given modulator, will depend on the characteristics of the message $a(t)$ and the channel as well as on the criterion of "goodness of estimation" used.

We now take up a study of the various forms of modulation and demodulation, their principal characteristics, their behavior in conjunction with noisy channels, and their advantages and disadvantages. We begin with some preliminary material or signals and their properties.

ELEMENTS OF SIGNAL THEORY

A real time function $f(t)$ and its Fourier transform form a Fourier transform pair given by

$$F(\omega) = \int_{-\infty}^{\infty} f(t)e^{-j\omega t}\, dt \qquad (2)$$

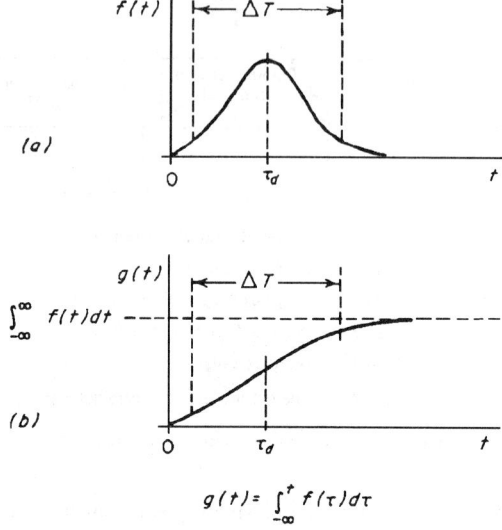

FIGURE 1.3.2   Duration and delay: (*a*) typical pulse; (*b*) integral of pulse.

and

$$f(t) = \frac{1}{2\pi} \int_{-\infty}^{\infty} F(\omega) e^{j\omega t} \, d\omega \tag{3}$$

It follows directly from Eq. (2) that the transform $F(\omega)$ of a real-time function has an even-symmetric real part and an odd-symmetric imaginary part.

Consider the function $f(t)$ shown in Fig. 1.3.2*a*. This might be a pulsed signal or the impulse response of a linear system, for example. The time ΔT over which $f(t)$ is appreciably different from zero is called the *duration* of $f(t)$, and some measure, such as τ_d, of the center of the pulse is called the *delay* of $f(t)$. In system terms, the quantity ΔT is the system *response time* or *rise time*, and τ_d is the system delay. The integral of $f(t)$, shown in Fig. 1.3.2b, corresponds to the step-function response of a system with impulse response $f(t)$.

If the function $f(t)$ of Fig. 1.3.2 is nonnegative, the new function

$$\frac{f(t)}{\int_{-\infty}^{\infty} f(t) \, dt}$$

is nonnegative with unit area. We now seek measures of duration and delay that are both meaningful in terms of communication problems and mathematically tractable. It will be clear that some of the results we obtain will not be universally applicable and, in particular, must be used with care when the function $f(t)$ can be negative for some values of t; however, the results will be useful for wide classes of problems.

Consider now a frequency function $F(\omega)$, which will be assumed to be real. If $F(\omega)$ is not real, either $|F(\omega)|^2 = F(\omega)F(-\omega)$ or $|F(\omega)|$ can be used. Such a function might be similar to that shown in Fig. 1.3.3*a*. The radian frequency range $\Delta\Omega$ (or the frequency range ΔF) over which $F(\omega)$ is appreciably different from zero is called the *bandwidth* of the function. Of course, if the function is a *bandpass* function, such as that shown in Fig. 1.3.3*b*, the bandwidth will usually be taken to be some measure of the width of the positive-frequency

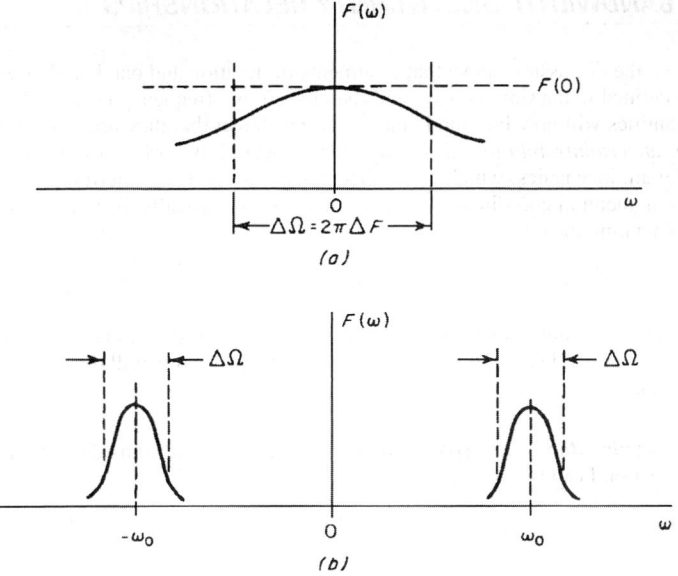

FIGURE 1.3.3 Illustrations of bandwidth: (*a*) typical low-pass frequency function; (*b*) typical bandpass frequency function.

(or negative-frequency) part of the function only. As in the case of the time function previously discussed, we may normalize to unit area and consider

$$\frac{F(\omega)}{\int_{-\infty}^{\infty} F(\omega)\,d\omega}$$

Again this new function is nonnegative with unit area.

Consider now the Fourier pair $f(t)$ and $F(\omega)$ and let us change the time scale by the factor a, replacing $f(t)$ by $af(at)$ so that both the old and the new signal have the same area, i.e.,

$$\int_{-\infty}^{\infty} f(t)\,dt = \int_{-\infty}^{\infty} af(at)\,dt \tag{4}$$

For $a < 1$, the new signal $af(at)$ is stretched in time and reduced in height; its "duration" has been increased. For $a > 1$, $af(at)$ has been compressed in time and increased in height; its "duration" has been decreased. The transform of this new function is

$$\int_{-\infty}^{\infty} af(at)e^{-j\omega t}\,dt = \int_{-\infty}^{\infty} f(x)e^{-j(\omega/a)x}\,dt = F\!\left(\frac{\omega}{a}\right) \tag{5}$$

The effect on the bandwidth of $F(\omega)$ has been the opposite of the effect on the duration of $f(t)$. When the signal duration is increased (decreased), the bandwidth is decreased (increased) in the same proportion. From the discussion, we might suspect that more fundamental relationships hold between properly defined durations and bandwidths of signals.

DURATION AND BANDWIDTH–UNCERTAINTY RELATIONSHIPS

It is apparent from the discussion above that treatments of duration and bandwidth are mathematically similar although one is defined in the time domain and the other in the frequency domain. Several specific measures of these two quantities will now be found, and it will be shown that they are intimately related to each other through various *uncertainty relationships*. The term "uncertainty" arises from the *Heisenberg uncertainty principle* of quantum mechanics, which states that it is not possible to determine simultaneously and exactly the position and momentum coordinates of a particle. More specifically, if Δx and Δp are the uncertainties in position and momentum, then

$$\Delta x \, \Delta p \geq h \tag{6}$$

where h is a constant. A number of inequalities of the form of Eq. (6) can be developed relating the duration ΔT of a signal to its (radian) bandwidth $\Delta \Omega$. The value of the constant h will depend on the definitions of duration and bandwidth.

Equivalent Rectangular Bandwidth DW1 and Duration DT1.   The equivalent rectangular bandwidth DW1 of a frequency function F(w) is defined as

$$\Delta\Omega_1 = \frac{\int_{-\infty}^{\infty} F(\omega) \, d\omega}{F(\omega_0)} \tag{7}$$

where ω_0 is some characteristic center frequency of the function $F(\omega)$. It is clear from this definition that the original function $F(\omega)$ has been replaced by a rectangular function of equal area, width $\Delta\Omega_1$, and height $F(\omega_0)$. For the low-pass case ($\omega_0 \equiv 0$), it follows from Eqs. (2) and (3) that Eq. (7) can be rewritten

$$\Delta\Omega_1 = \frac{2\pi f(0)}{\int_{-\infty}^{\infty} f(t) \, dt} \tag{8}$$

where $f(t)$ is the time function which is the inverse Fourier transform of $F(\omega)$.

The same procedure can be followed in the time domain, and the *equivalent rectangular duration* ΔT_1 of the signal $f(t)$ can be defined by

$$\Delta T_1 = \frac{\int_{-\infty}^{\infty} f(t) \, dt}{f(t_0)} \tag{9}$$

where t_0 is some characteristic time denoting the center of the pulse. For the case where $t_0 \equiv 0$, it is clear, then, from Eqs. (8) and (9) that equivalent rectangular duration and bandwidth are connected by the uncertainty relationship

$$\Delta T_1 \, \Delta\Omega_1 = 2\pi \tag{10}$$

Second-Moment Bandwidth DW2 and Duration DT2.   An alternative uncertainty relationship is based on the second-moment properties of the Fourier pair $F(\omega)$ and $f(t)$.

A *second-moment bandwidth* $\Delta\Omega_2$ can be defined by

$$(\Delta\Omega_2)^2 = \frac{1}{2\pi} \int_{-\infty}^{\infty} (\omega - \bar{\omega})^2 |F(\omega)|^2 \, d\omega / \varepsilon \tag{11}$$

and a *second-moment duration* ΔT_2 by

$$(\Delta T_2)^2 = \int_{-\infty}^{\infty} (t - \bar{t})^2 |f(t)|^2 dt / \varepsilon \tag{12}$$

where the total energy ϵ is given by

$$\epsilon = \frac{1}{2\pi} \int_{-\infty}^{\infty} |F(\omega)|^2 d\omega = \int_{-\infty}^{\infty} |f(t)|^2 dt = 1$$

These quantities are related by the inequality

$$\Delta \Omega_2 \, \Delta T_2 \geq {}^1\!/_2 \tag{13}$$

This expression is a second uncertainty relationship connecting the bandwidth and duration of a signal. Many other such inequalities can be obtained.

CONTINUOUS MODULATION

Modulation can be defined as the modification of one signal, called the *carrier*, by another, called the *modulation, modulating signal,* or *message signal*. In this section we will be concerned with situations where the carrier and the modulation are both continuous functions of time. Later we will treat the cases where the carrier and/or the modulation have the form of pulse trains.

For our analysis, Fig. 1.3.1 can be modified to the system shown in Fig. 1.3.4 in which the message is sent through a modulator (or transmitter) to produce the modulated continuous signal $m[t, a(t)]$. This waveform is corrupted by additive noise $n(t)$ in transmission so that the received (continuous) waveform $r(t)$ can be written

$$r(t) = m[t, a(t)] + n(t) \tag{14}$$

The purpose of the demodulator (or receiver) is to produce some best estimate $a^*(t)$ of the original message signal $a(t)$. As pointed out earlier a more general model of the transmission medium would allow corruption of the modulated waveform itself so that the received signal was of the form of Eq. (1). For example, in wireless systems, multiplicative disturbances can result because of multipath transmission or fading so that the received signal is of the form

$$r_1(t) = p(t)m[t, a(t)] + n(t) \tag{15}$$

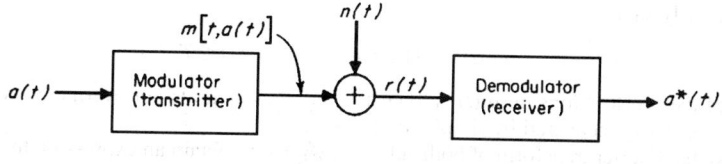

$a(t)$ = message signal (intelligence)

$m[t,a(t)]$ = modulated signal

$r(t) = m[t,a(t)] + n(t)$ = received signal

$n(t)$ = additive noise

$a^*(t)$ = the estimate of the original message signal

FIGURE 1.3.4   Communication-system model for continuous modulation and demodulation.

where both $p(t)$ and $n(t)$ are noises. However, we shall not treat such systems, confining ourselves to the simpler additive-noise model of Fig. 1.3.4.

LINEAR, OR AMPLITUDE, MODULATION

In a general way, *linear (or amplitude) modulation* (AM) can be defined as a system where a *carrier wave* $c(t)$ has its amplitude varied linearly by some *message signal* $a(t)$. More precisely, a waveform is linearly modulated (or amplitude-modulated) by a given message $a(t)$ if the partial derivative of that waveform with respect to $a(t)$ is independent of $a(t)$. In other words, the modulated $m[t, a(t)]$ can be written in the form

$$m[t, a(t)] = a(t)c(t) + d(t) \tag{16}$$

where $c(t)$ and $d(t)$ are independent of $a(t)$. Now we have

$$\frac{\partial m[t, a(t)]}{\partial a(t)} = c(t) \tag{17}$$

and $c(t)$ will be called the *carrier*. In most of the cases we will treat, the waveform $d(t)$ will either be zero or will be linearly related to $c(t)$. It will be more convenient, therefore, to write Eq. (16) as

$$m[t, a(t)] = b(t)c(t) \tag{18}$$

where $b(t)$ will be either

$$b_1(t) \equiv 1 + a(t) \tag{19}$$

or

$$b_2(t) \equiv a(t) \tag{20}$$

Also, at present it will be sufficient to allow the carrier $c(t)$ to be of the form

$$c(t) = C \cos (\omega_0 t + \theta) \tag{21}$$

where C and ω_0 are constants and θ is either a constant or a random variable uniformly distributed on $(0, 2\pi)$.

Whenever Eq. (19) applies, it will be convenient to assume that $b_1(t)$ is nearly always nonnegative. This implies that if $a(t)$ is a deterministic signal, then

$$a(t) \geq -1 \tag{22}$$

It also implies that if $a(t)$ is a random process, the probability that $a(t)$ is less than -1 in any finite interval $(-T, T)$ is arbitrarily small; i.e.,

$$p[a(t) < -1] \leq \epsilon \ll 1 \qquad -T \leq t \leq T \tag{23}$$

The purpose of these restrictions on $b_1(t)$ is to ensure that the carrier is not overmodulated and that the message signal $a(t)$ is easily recovered by simple receivers.

We can take the Fourier transform of both sides of Eq. (18) to obtain an expression for the frequency spectrum $M(\omega)$ of the general form of the linear-modulated waveform $m[t, a(t)]$:

$$M(\omega) = \frac{1}{2\pi} \int_{-\infty}^{\infty} B(\omega - v)C(v) \, dv \tag{24}$$

where $B(\omega)$ and $C(\omega)$ are Fourier transforms of $b(t)$ and $c(t)$, respectively. If the carrier $c(t)$ is the sinusoid of Eq. (21), then

$$C(\omega) = \pi C e^{j\theta}(\omega - \omega_0) + \pi C e^{-j\theta}\delta(\omega + \omega_0) \tag{25}$$

and Eq. (24) becomes

$$M(\omega) = (C/2)e^{j\theta}B(\omega - \omega_0) + (C/2)e^{-j\theta}B(\omega + \omega_0) \tag{26}$$

Thus, for a sinusoidal carrier, linear modulation is essentially a symmetrical frequency translation of the message signal through an amount ω_0 rad/s, and no new signal components are generated. On the other hand, if $c(t)$ is not a sinusoid, so that the spectrum $C(\omega)$ has nonzero width, then $M(\omega)$ will represent a spreading and shaping as well as a translation of $B(\omega)$.

Suppose that the message signal $a(t)$ [and hence $b(t)$] is low-pass and strictly band-limited (0, ω_s) rad/s where $\omega_s < \omega_0$. Then the frequency spectrum of $b(t)$ obeys $B(\omega) = 0$, $|\omega| > \omega_s$, and $B(\omega - \omega_0)$ and $B(\omega + \omega_0)$ do not overlap. The spectrum $B(\omega + \omega_0)$ occupies only a range of positive frequencies while $B(\omega - \omega_0)$ occupies only negative frequencies.

The *envelope* of the modulated waveform $m[t, a(t)]$ is the magnitude $|b(t)|$. If, as mentioned earlier, the function $b(t)$ is restricted to be nonnegative almost always, then this envelope becomes just $b(t)$. In such cases, an *envelope detector* will uniquely recover $b(t)$ and hence $a(t)$. We now consider the common forms of simple amplitude, or linear, modulation.

DOUBLE-SIDEBAND AMPLITUDE MODULATION (DSBAM)

In this case the function $b(t)$ is given by Eq. (13) so that

$$m[t, a(t)] = C[1 + a(t)] \cos (\omega_0 t + \theta) \tag{27}$$

The transform of $b_1(t)$ is just

$$B_1(\omega) = \mathfrak{F}[1 + a(t)] = 2\pi\delta(\omega) + A(\omega) \tag{28}$$

where $A(\omega)$ is the Fourier transform of $a(t)$ and $\delta(\omega)$ is the Dirac delta function. It follows, therefore, from Eq. (26) that the frequency spectrum of $m[t, a(t)]$ is given by

$$M(\omega) = \frac{C}{2}e^{j\theta}[2\pi\delta(\omega - \omega_0) + A(\omega - \omega_0)] + \frac{C}{2}e^{-j\theta}[2\pi\delta(\omega + \omega_0) + A(\omega + \omega_0)] \tag{29}$$

Depending on the form of $a(t)$ a number of special cases can be distinguished, but in any case the spectrum is given by Eq. (26).

The simplest case is where $a(t)$ is the *periodic function* given by

$$a(t) = \eta \cos \omega_s t \qquad \omega_s < \omega_0 \text{ and } 0 \le \eta \le 1 \tag{30}$$

Then the frequency spectrum $A(\omega)$ is

$$A(\omega) = \pi\eta\delta(\omega - \omega_s) + \pi\eta\delta(\omega + \omega_s) \tag{31}$$

For convenience, let us take the phase angle θ to be zero so that

$$M(\omega) = \pi C \left[\delta(\omega - \omega_0) + \delta(\omega + \omega_0) + \frac{\eta}{2}\delta(\omega - \omega_0 - \omega_s) + \frac{\eta}{2}\delta(\omega - \omega_0 + \omega_s) \right.$$
$$\left. + \frac{\eta}{2}\delta(\omega + \omega_0 - \omega_s) + \frac{\eta}{2}\delta(\omega + \omega_0 + \omega_s) \right] \tag{32}$$

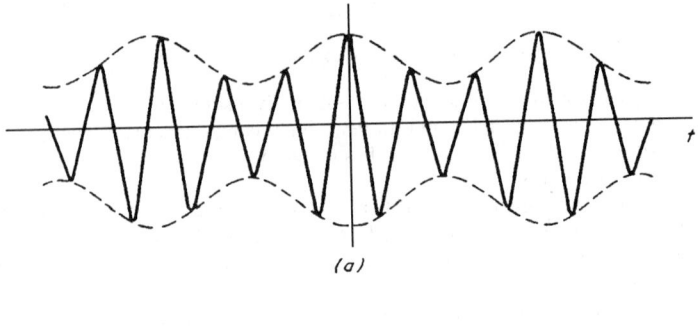

(a)

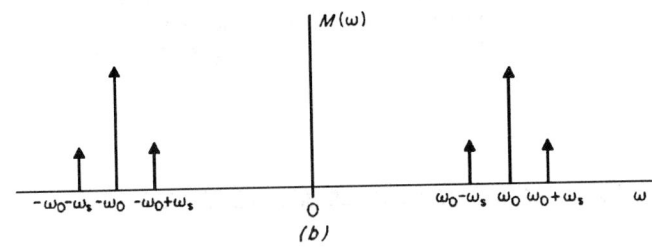

(b)

FIGURE 1.3.5   (a) Double-sideband AM signal and (b) frequency spectrum.

This spectrum is illustrated in Fig. 1.3.5 together with the corresponding modulated waveform

$$m[t, a(t)] = C(1 + \eta \cos \omega_s t) \cos \omega_0 t \tag{33}$$

Note that this last equation can also be written as

$$m[t, a(t)] = \underbrace{C \cos \omega_0 t}_{\text{carrier}} + \underbrace{(C/2)\eta \cos (\omega_0 - \omega_s)t}_{\text{lower sideband}} + \underbrace{(C/2)\eta \cos (\omega_0 + \omega_s)t}_{\text{upper sideband}} \tag{34}$$

In this form it is easy to distinguish the *carrier*, the *lower sideband*, and the *upper sideband*.

Suppose that $a(t)$ is not a single sinusoid but is periodic with period P. Then it may be expanded in a Fourier series and each term of the series treated as in Eq. (32).

DOUBLE-SIDEBAND AMPLITUDE MODULATION, SUPPRESSED CARRIER

If $b(t)$ is given by Eq. (20) so that

$$m[t, a(t)] = Ca(t) \cos (\omega_0 t + \theta) \tag{35}$$

then the carrier is suppressed and Eq. (29) reduces to

$$M(\omega) = (C/2)e^{j\theta} A(\omega - \omega_0) + (C/2)e^{-j\theta} A(\omega + \omega_0) \tag{36}$$

The principal advantage of this technique (DSBAM-SC) over DSBAM is that the carrier is not transmitted in the former case, with a consequent saving in transmittal power. The principal disadvantages relate to problems in generating and demodulating the suppressed-carrier waveform.

VESTIGIAL-SIDEBAND AMPLITUDE MODULATION (VSBAM)

It is apparent from Eq. (29) that the total information regarding the message signal $a(t)$ is contained in either the upper or lower sideband in conventional DSBAM. In principle the other sideband could be eliminated, say by filtering, with a consequent reduction in bandwidth and transmitted power. Such a procedure is actually followed in single-sideband amplitude modulation (SSBAM), discussed next. However, completely filtering out one sideband requires an ideal bandpass filter with infinitely sharp cutoff or an equivalent technique. A technique intermediate between the production of DSBAM and SSBAM is vestigial sideband amplitude modulation (VSBAM), where one sideband is attenuated much more than the other with some saving in bandwidth and transmitted power. The carrier may be present or suppressed.

The principal use of VSBAM has been in commercial television. The video (picture) signal is transmitted by VSBAM with a consequent reduction in the total transmitted signal bandwidth and in the frequency difference that must be allowed between adjacent channels.

SINGLE-SIDEBAND AMPLITUDE MODULATION (SSBAM)

This important type of AM can be considered as a limiting form of VSBAM when the filter for the modulated waveform is an ideal filter with infinitely sharp cutoff so that one sideband, e.g., the lower, is completely eliminated.

BANDWIDTH AND POWER RELATIONSHIPS FOR AM

It is clear that the bandwidth of a given AM signal is related in a simple fashion to the bandwidth of the modulating signal since AM is essentially a frequency translation. If the modulating signal $a(t)$ is assumed to be low-pass and to have a bandwidth of W Hz, then the bandwidth ΔF of the modulated signal $m[t, a(t)]$ must satisfy

$$W \leq \Delta F \leq 2W \tag{37}$$

The upper limit of $2W$ Hz holds for double-sideband modulation and the lower limit of W Hz for single-sideband. For the concept of bandwidth to be meaningful for a single-frequency sinusoidal modulating signal, it is assumed, of course, that a low-frequency sinusoid of frequency W Hz has a bandwidth W. Actually, what is really being assumed is that this sinusoid is the highest-frequency component in the low-pass modulating signal.

The only case where intermediate values of the inequality of Eq. (37) are encountered is in VSBAM. In principle any bandwidth between W and $2W$ Hz is possible. In this respect SSBAM can be considered as a limiting case of VSBAM. From a practical point of view, however, it is difficult (or expensive) to design a filter whose gain magnitude drops off too rapidly.

Power relationships are also simple and straightforward for AM signals. Consider Eq. (27) which is a general expression for the modulated AM waveform $m(t)$. Let the average power in this signal be denoted by P_{av}. The phase angle θ will be fixed if $a(t)$ is deterministic and will be taken to be a random variable uniformly distributed on $(0, 2\pi)$ if $a(t)$ is a random process. Also, it will be assumed that $a(t)_{av}$ is zero. It is clear that P_{av} is given by

$$P_{av} = (C^2/2)[1 + a^2(t)_{av}] \tag{38}$$

where $a^2(t)_{av}$ is the average power in $a(t)$. The first term, $C^2/2$, is the carrier power, and the second term, $C^2/2)a^2(t)_{av}$, is the signal power in the upper and lower sidebands. If $|a(t)| \leq 1$ to prevent overmodulation, then at least half the transmitted power is carrier power and the remainder is divided equally between the upper and lower sideband.

In double-sideband, suppressed-carrier operation, the carrier power is zero, and half the total power exists in each sideband. The fraction of information-bearing power is $1/2$. For single-sideband, suppressed-carrier systems, all the transmitted power is information-bearing, and, in this sense, SSB-SC has maximum transmission efficiency.

The disadvantages of both suppressed-carrier and single-sideband operation lie in the difficulties of generating the signals for transmission and in the more complicated receivers required. Demodulation of suppressed-carrier AM signals involve the reinsertion of the carrier or an equivalent operation. The local generation of sinusoid at the exact frequency and phase of the missing carrier is either difficult or impossible unless a pilot tone of reduced magnitude is transmitted with the modulated signal for synchronization purposes or unless some nonlinear operation is performed on the suppressed-carrier signal to regenerate the carrier term at the receiver. In SSBAM, not only is the receiver more complicated, but transmission is considerably more difficult. It is usually necessary to generate the SSB signal at a low power level and then to amplify with a linear power amplifier to the proper level for transmission. On the other hand, DSB signals are easily generated at high power levels so that inefficient linear power amplifiers need not be used.

ANGLE (FREQUENCY AND PHASE) MODULATION

In angle modulation, the carrier $c(t)$ has either its phase angle or its frequency varied in accordance with the intelligence $a(t)$.

The result is not a simple frequency translation, as with AM, but involves both translation and the production of entirely new frequency components. In general, the new spectrum is much wider than that of the intelligence $a(t)$. The greater bandwidth may be used to improve the signal-to-noise performance of the receiver. This ability to *exchange* bandwidth for signal-to-noise enhancement is one of the outstanding characteristics and advantages of angle modulation.

In the form of angle modulation, which will be called *phase modulation* (PM), the phase of the carrier is varied linearly with the intelligence $a(t)$. Thus, the modulated signal is given by

$$m(t) = C \cos \left[\omega_0 t + \theta + k_p a(t) \right] \tag{39}$$

where k_p is a constant and the *modulation index* $\varnothing_m$ is defined by

$$\varnothing_m = \max \left| k_p a(t) \right| \qquad \text{(rad)} \tag{40}$$

In *frequency modulation*, the instantaneous frequency is made proportional to the intelligence $a(t)$. The modulated signal is given by

$$m(t) = C \cos \left[\omega_0 t + k_f \int_{-\infty}^{t} a(\tau) d\tau \right] \tag{41}$$

where k_f is a constant. The *maximum deviation* $\Delta \omega$ is given by

$$\Delta \omega = \max \left| k_f a(t) \right| \qquad \text{(rad/s)} \tag{42}$$

and, as before, a *modulation index* $\varnothing_m$ by

$$\varnothing_m = \max \left| k_f \int_{-\infty}^{t} a(\tau) d\tau \right| \tag{43}$$

In general, the analysis of angle-modulated signals is difficult even for simple modulating intelligence. We will consider only the case where the modulating intelligence $a(t)$ is a sinusoid. Let $a(t)$ be given by

$$a(t) = \eta \cos \omega_s t \qquad \omega_s < \omega_0 \tag{44}$$

Then the corresponding PM signal is

$$m(t) = C \cos (\omega_0 t + k_p \eta \cos \omega_s t) \tag{45}$$

where the phase angle θ has been set equal to zero. In the same way, the FM signal is

$$m(t) = C \cos [\omega_0 t + (k_f \eta / \omega_s) \sin \omega_s t] \tag{46}$$

Let us now consider the FM signal of this last equation. Essentially the same results will be obtained for PM. The equation can be expanded to yield

$$m(t) = C \cos (\varnothing_m \sin \omega_s t) \cos \omega_0 t - C \sin (\varnothing_m \sin \omega_s t) \sin \omega_0 t \tag{47}$$

where the modulation index $\varnothing_m$ is given by $\varnothing_m = k_f \eta / \omega_s$. Sinusoids with sinusoidal arguments give rise to Bessel functions, and Eq. (47) can be expanded and rearranged to obtain

$$m(t) = \underbrace{C J_0(\varnothing_m) \cos \omega_0 t}_{\text{carrier}} + C \sum_{n=1}^{\infty} J_{2n}(\varnothing_m) [\underbrace{\cos (\omega_0 + 2n\omega_s)t}_{\text{USB}} + \underbrace{\cos (\omega_0 - 2n\omega_s)t]}_{\text{LSB}}$$

$$+ C \sum_{n=1}^{\infty} J_{2n-1}(\varnothing_m) \{ \underbrace{\cos [\omega_0 + (2n-1)\omega_s]t}_{\text{USB}} - \underbrace{\cos [\omega_0 - (2n-1)\omega_s]t}_{\text{LSB}} \} \tag{48}$$

where $J_m(x)$ is the Bessel function of the first kind and order m. This expression for $m(t)$ is relatively complicated even though the modulating intelligence is a simple sinusoid. In addition to the carrier, there is an infinite number of upper and lower sidebands separated from the carrier (and from each other) by integral multiples of the modulating frequency ω_s. Each sideband, and the carrier, has an amplitude determined by the appropriate Bessel function. When there is more than one modulating sinusoid, the complexity increases rapidly.

In the general case, an infinite number of sidebands exist dispersed throughout the whole frequency domain. In this sense, the bandwidth of an FM (or PM) signal is infinite. However, outside of some interval centered on ω_0, the magnitude of the sidebands will be negligible; this interval may be taken as a practical measure of the bandwidth. An approximation can be obtained by noting that $J_n(\varnothing_m)$, considered as a function of n, decreases rapidly when $n < \varnothing_m$. Therefore only the first $\varnothing_m$ sidebands are significant. If the highest-frequency sideband of significance is $\varnothing_m \omega_s$, then the bandwidth is given approximately by

$$BW \approx 2\omega_s \varnothing_m \quad \text{(rad/s)} \tag{49}$$

A more accurate rule of thumb is the slightly revised expression

$$BW \approx 2\omega_s (\varnothing_m + 1) \quad \text{(rad/s)} \tag{50}$$

which may be considered a good approximation when $\varnothing_m \geq 5$.

CHAPTER 1.4
DIGITAL DATA TRANSMISSION AND PULSE MODULATION

Geoffrey C. Orsak, H. Vincent Poor, John B. Thomas

DIGITAL TRANSMISSION

Most of the theoretical foundations for digital data-communication-system design have already been covered in this chapter. The information-theoretic material of Chaps. 1.1 and 1.2 is set in the natural context of digital signals; i.e., it relates directly to discrete-time and discrete-amplitude signals. The sampling theorem of Chap. 1.2 relates continuous-time band-limited signals to discrete-time samples of the signals and forms the natural connection between sampled data and their continuous-time counterpart. There remains the problem of *quantization*, by which a continuum of amplitude values is converted into a finite number of preassigned possible levels. However, we first discuss briefly a form of modulation that stands somewhere between the continuous modulation of Chap. 1.3 and the pulse modulation systems used in digital transmission.

PULSE-AMPLITUDE MODULATION (PAM)

In the previous paragraphs, modulation schemes were considered that operated with sinusoidal carriers. Any other continuous waveform could have been used, although analysis of the resulting modulated signal might have become very complicated. Here systems are considered where the carrier is no longer continuous but consists of a pulse train, some parameter of which is suitably modified by the modulating intelligence. It will be seen that there are natural applications for such modulation schemes and that they may provide striking improvements in noise immunity. One of the simplest of such systems is PAM, where the amplitude of a pulsed carrier $p(t)$ is varied by the modulating signal $a(t)$. Here the modulated signal is given by

$$m(t) = m[t, a(t)] = a(t)p(t) \tag{1}$$

Let us assume that the carrier $p(t)$ is a periodic pulse train with basic pulse shape $f(t)$ and period T. Since the pulse train is periodic, it can be expanded in a Fourier series as

$$p(t) = \sum_{k=-\infty}^{\infty} P_k e^{jk\omega_0 t} \tag{2}$$

where $\omega_0 = 2\pi/T$ and the Fourier coefficient P_k is given by

$$P_k = \frac{1}{T} \int_{-T/2}^{T/2} p(t) e^{-jk\omega_0 t} dt \qquad (3)$$

The PAM signal of Eq. (1) can be rewritten with the help of Eq. (2) as

$$m(t) = a(t) p(t) = \sum_{k=-\infty}^{\infty} P_k a(t) e^{jk\omega_0 t} \qquad (4)$$

with Fourier transform $M(\omega)$ given by

$$M(\omega) = \sum_{k-\infty}^{\infty} P_k \mathfrak{F}\{a(t) e^{jk\omega_0 t}\} = \sum_{k=-\infty}^{\infty} P_k A(\omega - k\omega_0) \qquad (5)$$

where $A(\omega)$ is the Fourier transform of $a(t)$. Suppose now that $a(t)$ is band-limited $(-B/2, B/2)$ rad/s, where $B < \omega_0$. Then it is clear from Eq. (5) that the PAM signal $m(t)$ has a spectrum where the basic spectral shape is repeated periodically throughout the frequency domain; but each spectral pulse $A(\omega - k\omega_0)$ is weighted by the appropriate Fourier coefficient P_k. The value of this coefficient depends on the amplitude, shape, and spacing of the carrier pulse train $p(t)$.

In practice, the width D of the basic pulse $p(t)$ is small compared with the pulse spacing T. The general effect of the modulating process is to simple the intelligence $a(t)$ at a fixed interval of Ts throughout the time domain. If $a(t)$ is strictly band-limited $(-B/2, B/2)$ rad/s, the sampling theorem of Chap. 1.2 ensures that $a(t)$ can be reconstructed exactly from its samples spaced $2\pi/B$ s a part throughout the time domain. Thus a *necessary* condition for exact recovery of $a(t)$ from $a(t)p(t)$ is that

$$T \leq 2\pi/B \qquad \text{or} \qquad BT \leq 2\pi \qquad (6)$$

It is clear from Eq. (5) that this condition is the necessary restriction to prevent overlapping of the repeated spectra of the PAM signal $m(t)$. (Note that ω_0 has been defined as $2\pi/T$.)

In the demodulation of PAM signals, two cases must be distinguished. If the Fourier coefficient P_0 is not zero, the modulating intelligence $a(t)$ can be recovered by low-pass filtering. The resulting output is just the original intelligence $a(t)$ weighted by the constant P_0. In the case where P_0 is zero but P_1 is not, the intelligence $a(t)$ can be recovered by some form of synchronous demodulation.

As a practical matter, the exact waveform given by Eq. (4) is usually not used in PAM. More specifically, the tops of the pulse train are not usually shaped by the modulating signal. Instead they are kept flat, and the pulse height is determined by the value of $a(t)$ at some point in the pulse-length interval D. Thus the PAM signal is not exactly $a(t)p(t)$, and the resulting spectrum $M(\omega)$ does not yield $A(\omega)$ with a constant scale factor. Instead there is some distortion of the spectrum of $A(\omega)$, depending on the exact shape of the pulse train $P(t)$. In many cases this distortion can be kept negligibly small, and in other cases it can be substantially removed by a low-pass equalizing filter in the PAM receiver.

QUANTIZING AND QUANTIZING ERROR

The PAM system just studied involved converting a time-continuous modulating intelligence $a(t)$ into a time-discrete form by *sampling*. The sampling theorem guarantees that $a(t)$ can be reconstructed exactly from its samples $a(k/2W)$, $k = \cdots, -1, 0, 1, \ldots$, speed throughout the time domain, provided $a(t)$ has a spectrum $A(\omega)$ that is strictly band-limited $(-2\pi W, 2\pi W)$ rad/s. In a similar fashion, an amplitude-continuous signal is converted into an *amplitude-discrete* or *digital* signal by quantizing in the amplitude domain into a finite number of fixed distinguishable amplitude levels, each a distance Q apart.

The process of quantizing is irreversible since, regardless of how small the quantization level Q is taken to be, an unresolvable uncertainty of $\pm Q/2$ is associated after quantizing with each amplitude value. Thus a *quantization noise* N_q is inevitably associated with all quantized signals. This noise can be made as small as desired by choosing enough quantization levels or, equivalently, by making each quantization level small enough, but it cannot be eliminated. In the pulsed carrier systems to be discussed subsequently, it turns out that the noise added in transmission can be almost completely eliminated at the receiver. In other words, the type of noise interference that is added externally (say in the channel) is negligible, and the principal source of combination is the quantization noise.

The quantization levels are often taken to be equal; i.e., the spacing between amplitude levels is uniform. Nonuniform quantizing can be used to favor small amplitudes where noise has a greater effect at the expense of large amplitudes. However, for this discussion, it will be assumed that the differences in amplitude levels are all equal to Q. Let us assume that the actual amplitude of the signal being sampled is equally likely to lie anywhere within the particular quantization level. In other words, the quantizing error in an amplitude interval is taken to be a uniformly distributed random variable q with density $p_q(x)$ given by

$$p_q(x) = \begin{cases} 0 & |x| > Q/2 \\ 1/Q & |x| \le Q/2 \end{cases} \tag{7}$$

The quantization noise N_q can be written as

$$N_q = \int_{-Q/2}^{Q/2} x^2 p_q(x)\,dx = \frac{Q^2}{12} \tag{8}$$

The average value of the quantizing error is zero since the distribution of Eq. (7) has zero mean. In terms of 1-V quantization levels the rms error voltage is $1/\sqrt{12}$ V. Suppose that the peak-to-peak value of the signal to be quantized is limited to $\pm A$. Then the quantization level Q can be written as

$$Q = 2A/d \tag{9}$$

where d is the number of quantization levels. The signal-to-quantization-noise power ratio for a sampled signal $a^*(t)$ can then be written as

$$\frac{S}{N_q} = \frac{12[a^*(t)]_{av}^2}{Q^2} = 3d^2 \frac{[a^*(t)]_{av}^2}{A^2} \tag{10}$$

where quantity $[a^*(t)]_{av}^2/A^2$ is the normalized sampled signal power.

SIGNAL ENCODING

The result of amplitude quantizing and time sampling a message signal is a sequence of numbers, i.e., a *digital signal*. We now consider advantageous ways to represent this sequence. Each number could be represented by a pulse of height proportional to that number. The result would be a quantized PAM pulse train. However, there can be great advantages in representing *each* number by a *sequence* of numbers that are allowed to assume fewer values than the original set. Such a system of representation is called a *code*. The principal advantages of encoding a message set are that (1) it is easier to discriminate in the presence of noise between a few possible numbers (pulse heights) than between many and (2) and error in the value of a given code number will affect only part of the information contained in the original sample value. In general, an m-ary code will consist of m possible levels. Then a sequence of n symbols of this code can represent m^n possible message levels since there are $M = m^n$ distinguishable code groups of length n. As an example, suppose that the original message is quantized at eight levels. The message sequence is to be encoded into a *binary* code consisting of two levels or symbols 0 and 1. In this case $m = 2$ and, since M must equal 8, we have

$$M = m^n \quad \text{or} \quad 8 = 2^3$$

Thus each binary code word, representing one of the possible quantization levels, must consist of three binary digits. A logical encoding is

Quantization level	Code word	Quantization level	Code word
0	000	4	100
1	001	5	101
2	010	6	110
3	011	7	111

Of the possible *m*-ary codes that could be used, the binary code ($m = 2$) is the most common, for at least two reasons: (1) the binary code offers the least number of choices, namely, two, in decoding, and (2) binary systems are most easily implemented electronically since the two possible states can be made to correspond to *on-off*, *open-closed*, or *conducting-nonconducting* conditions in circuits and systems.

It is clear from the previous discussion that the process of amplitude quantization, time sampling, and encoding can be used to convert a continuous band-limited message signal into a pulse train consisting of equally spaced pulses of two heights, e.g., zero and unity. The steps by which this is accomplished are as follows: (1) A low-frequency signal $a(t)$ band-limited $(-2\pi W, 2\pi W)$ is quantized into M amplitude levels and sampled at an interval of $1/2W$ s. The result is a sequence of numbers, a digital signal. (2) Each of these numbers is then encoded in a binary code. (3) The corresponding binary sequence becomes a pulse train consisting of pulses of height unity and zero to correspond to 1 and 0, respectively. This binary sequence can now be transmitted directly, in which case it will be called a *baseband* signal, or it can be used as modulating signal in a *pulse-code-modulation* (PCM) system. The steps followed up to this point are as follows:

1. A continuous signal $a(t)$ is sampled and quantized.

2. The quantized sample values are encoded into a pulse train $p(t)$.

3. The pulse train $p(t)$ may be transmitted directly or used as the modulating signal in any appropriate modulation scheme.

The great advantage of this system is its inherent resistance to external noise corruption. If the external noise is not too large, it is clear that the original pulse $p(t)$ can be recovered *exactly* from the received pulse $p^*(t)$ since only a knowledge of the *presence* or *absence* of a pulse and not the pulse *shape* itself is needed at each pulse position to reproduce the original pulse train exactly. Thus, if the signal is large enough compared with corrupting noise, there will be *no error*. Furthermore, if it is desired to transmit the pulse train for a long distance, it can be regenerated exactly as often as desired at repeater stations along the way. In other words, no noise is added in transmission in the large-signal case. This situation is in striking contrast to the analog schemes already studied. Thus, in a properly designed large-signal (encoded) digital-data system, all error that results will arise from the original quantization noise, discussed previously.

The digital-data system can be considered a coded wide-band system that trades complexity and bandwidth for noise immunity. In general, if there are d quantization levels, and if binary pulses are used, each code word must contain n pulses, where

$$d = 2^n \quad \text{or} \quad n = \log_2 d \tag{11}$$

There must be $2nW$ pulses/s in the pulse train $p(t)$ and any two adjacent pulses must be clearly distinguishable. If the bandwidth of the system transmitting the pulse train is made too small, the pulses will be broadened and will interfere with each other, so that errors in identification will result. In fact, the uncertainty relationships previously developed bear directly on this problem. Thus Eq. (13) relates the bandwidth $\Delta\Omega$ and the duration ΔT of a pulse by

$$(\Delta\Omega)(\Delta T) \geq \Delta^1/_2 \tag{12}$$

Since there must be $2nW$ pulses/s, each pulse cannot exceed a width $\Delta T \leq 1/2nW$ s.

It follows directly from Eq. (12) that the system bandwidth $\Delta\Omega$ is bounded from below by

$$\Delta\Omega \geq nW \qquad \text{(Hz)} \tag{13}$$

The digital-data signal requires a transmission bandwidth that is at least n times that of the original intelligence, where n is determined by Eq. (11) and depends on the number of quantization levels desired. If the number of quantization levels is taken to be small (so that n is small), the quantization noise will increase as shown by Eqs. (8) and (9). The output signal-to-noise power ratio owing to quantization noise is given by Eq. (10), where $[a^*(t)]^2_{\text{av}}$ is the power in the quantized signal and must be calculated. For reasonable signal distributions and uniform quantizers it can be shown that the signal-to-noise ratio increases approximately as 2^{2n}.

In many practical situations, equipment inadequacies and/or low-signal conditions may create situations where transmission noise as well as quantization noise is added. In other words, some pulses in the coded pulse train will be incorrectly identified. This problem is most conveniently formulated in terms of a *probability of error*, which gives the average rate at which incorrect pulse identification occurs. This criterion was discussed in Chap. 1.2 and is directly applicable here.

BASEBAND DIGITAL-DATA TRANSMISSIONS

Consider the system of Fig. 1.4.1, where the baseband signal (encoded sequence) is denoted by $p(t)$. To avoid bias (dc level) problems and to increase the effective signal power, it may be convenient to send a sequence of $+1$s and -1s instead of $+1$s and 0s. A noise $n(t)$ is added in the transmission channel and the problem at the receiver is to determine whether a $+1$ or a -1 was transmitted in each time slot of length ΔT.

To obtain a general idea of receiver performance, it will be assumed that the noise $n(t)$ is a zero-mean gaussian process, as discussed in Chap. 1.2 with power spectral density [see Eq. (87)] taken to be the constant $N_0/2$. Such a process is called a *white gaussian noise*. It will also be assumed that the *synchronization* problem has been solved, i.e., that the receiver knows when each pulse in $p(t)$ starts and ends. Such systems are called *coherent*. Any departure from this condition, either from imperfect knowledge or from lack of complete knowledge of the pulse position, will result in performance degradation.

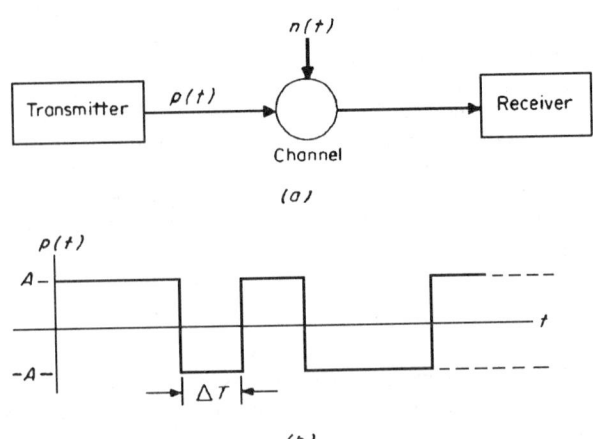

FIGURE 1.4.1 Baseband digital-data transmission; (a) system structure and (b) baseband-encoded sequence.

A general theory of detection and estimation can be used to specify an optimal receiver structure; however, a reasonable (and quasi-optimal) way to proceed is to use an *integrate-and-damp* system, where, at the end of each pulse occurrence time, the random variable

$$V = \int_{\Delta T} [p(t) + n(t)] \, dt \tag{14}$$

is compared with zero. If $V > 0$, it is decided that $+A$ was sent, and if $V < 0$, that $-A$ was sent. It is clear that

$$V = \begin{cases} ADT + N & \text{if } +A \text{ was sent} \\ -ADT + N & \text{if } -A \text{ was sent} \end{cases}$$

where N is a normal random variable (result of a linear operation on a gaussian process) with zero mean and variance σ^2 given by

$$\text{Var } N = \sigma^2 = E\{N^2\} = N_0 \Delta T / 2 \tag{15}$$

An error can occur in either one of two ways: (1) $+A$ is sent and $ADT + N < 0$ or (2) $-A$ is sent and $-ADT + N > 0$. If $+A$ and $-A$ are equally likely a priori, these two errors are equally likely, and the total error probability P_e is

$$P_e = \int_{A\,\Delta T}^{\infty} f_N(x) \, dx \tag{16}$$

where $f_N(x)$ is the normal density function

$$f_N(x) = (1/\sqrt{2\pi}\,\sigma) e^{-x^2/2\sigma^2} \qquad -\infty < x < \infty \tag{17}$$

and $\sigma^2 = N_0 \Delta T / 2$ has already been defined. The expression for P_e can be rewritten, after a change in variable, as

$$P_e = \frac{1}{\sqrt{\pi}} \int_k^{\infty} e^{-y^2} \, dy \tag{18}$$

where $k^2 = A^2 \Delta T / N_0$. Using tables or numerical integration, one can use this last equation to plot P_e versus k, as shown in Fig. 1.4.2. Note that k^2 is a measure of signal-to-noise power ratio. Naturally, as k increases, the probability of error P_e decreases.

This kind of baseband digital signal is often called *amplitude-shift keying* (ASK) when the two signals are $-A$ and $+A$ and asymmetric amplitude-shift keying (AASK) when the two signals are 0 and $+A$.

It is conventional (and convenient) to express the probability of error P_e in terms of the *complementary unit normal cumulative distribution function*

$$Q(x) = \frac{1}{\sqrt{2\pi}} \int_x^{\infty} e^{-y^2/2} \, dy \tag{19}$$

where

$$d^2 = 2k^2 = 2A^2 \Delta T / N_0 \tag{20}$$

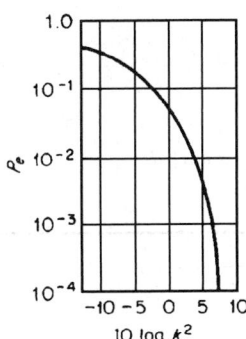

FIGURE 1.4.2 Probability of error plotted against signal-to-noise ratio for an integrate-and-dump receiver.

The preceding calculations are intended to give a general idea of error-probability calculations and receiver design. For voluminous details, see the reference listed at the end of this section under Digital-Data Transmission.

PULSE-CODE MODULATION (PCM)

Thus far, no consideration has been given to the actual transmission of the baseband signal. It is clear that the pulse train cannot be radiated from an antenna of any reasonable size. As in the analog case, some frequency-translation technique must be employed to make radiation practical and/or to allow frequency stacking of more than one baseband signal. A number of systems are used for such *digital signalling.* Some of the commonest follow.

(**a**) *Binary On-Off Keying (BOOK).* This type of binary signalling is almost self-explanatory. A high-frequency sinusoidal signal is switched on an off so that on periods correspond to 1 and off periods to 0 in the PCM wave. In practice the pulsed sinusoids do not start or stop abruptly but exhibit a transient buildup and decay.

(**b**) *Binary Frequency-Shift Keying (BFSK).* In this form of digital signalling, a continuous wave is sent which shifts between two frequencies, one representing the symbol 1 in the PCM wave and the other representing 0.

(**c**) *Binary Phase-Shift Keying (BPSK).* Here the frequency is kept constant, but the phase is shifted 180° whenever the basic signal changes from 0 to 1.

For comparison, Table 1.4.1 lists the error probabilities for these three kinds of PCM signals. Also listed are the error probabilities for baseband ASK and AASK discussed in the previous paragraphs. Note again that k^2 and d^2 are measures of the signal-to-noise power ratio and P_e decreases with increasing k^2 and d^2. In all cases we are dealing with coherent receivers.

TABLE 1.4.1   A Comparison of Binary Data Transmission Systems

Type	(H, K) pair	V	Error probability
AASK	$H: X(t) = N(t)$ $K: X(t) = N(t) + A$	$\int_{\Delta\tau} x(t)\, dt \gtrless l_1$	$P_E = Q(d/2)$
ASK	$H: X(t) = N(t) - A$ $K: X(t) = N(t) + A$	$\int_{\Delta\tau} x(t)\, dt \gtrless l_1$	$P_E = Q(d)$
BOOK	$H: X(t) = N(t)$ $K: X(t) = N(t) + A \cos \omega_0 t$	$\int_{\Delta\tau} x(t) \cos \omega_0 t\, dt \gtrless l_1$	$P_E = Q(d_0/2)$
BPSK	$H: X(t) = N(t) - A \cos \omega_0 t$ $K: X(t) = N(t) + A \cos \omega_0 t$	$\int_{\Delta\tau} X(t) \cos \omega_0 t\, dt \gtrless l_1$	$P_E = Q(d_0)$
BFSK	$H: X(t) = N(t) + A \cos (\omega_0 + \Delta\omega)t$ $K: X(t) = N(t) + A \cos \omega_0 t$ $\Delta\omega_0 = \dfrac{2\pi n}{\Delta T}$	$\int_{\Delta\tau} x(t)\, [\cos \omega_0 t$ $\quad - \cos (\omega_0 + \Delta\omega)t]\, dt$ $\Delta\omega_0 = \dfrac{2\pi n}{\Delta T}$	$P_E = Q(d/2)$

Note: $d^2 = \dfrac{2A^2 \Delta T}{N_0}$ $k^2 = \dfrac{A^2 \Delta T}{N_0} = \dfrac{d^2}{2}$ $d_0^2 = \dfrac{2A^2 I}{N_0}$ $k_0^2 = \dfrac{A^2 I}{N_0} = \dfrac{d_0^2}{2}$

$N(t)$ is WGN of power spectral density $N_0/2$

$A > 0$ $I = \int_{\Delta T} \cos^2 \omega_0 t / dt$

If $I = T/2$, then $d_0^2 = d^2/2$ and $k_0^2 = k^2/2$

SPREAD-SPECTRUM SYSTEMS

In previous paragraphs the treatment of communication systems has been concerned primarily with such things as efficient channel utilization, low probability of error, and (implicitly) the efficient use of signal energy.

Partly because of the increased crowding of the frequency spectrum the partly from the unique needs of mobile communications, the importance of *spread-spectrum communications* has increased significantly in recent years. As the name suggests, such systems tend to be *wide-band*, that is, to require a frequency range substantially greater than that of the information-bearing signal. We have already encountered wide-band systems in FM and PM and in the PCM systems of this section; however, properly speaking, none of these are spread-spectrum systems. This term is generally applied to wide-band systems which, among others, have some or all of the following properties:

1. Low interference to other systems
2. High interference rejection capability
3. Multiple-access capability
4. Antijam capability
5. Covert or secure communications
6. Multipath operating mode

These systems usually achieve the goals listed above by modulation techniques inherently different from those already discussed. The most common techniques are:

1. The impression of random or *pseudorandom sequences* (PN) onto the message sequence before modulation.
2. *Frequency hopping* where a PN sequence is used to switch the carrier frequency.
3. *Time hopping* where a PN sequence is used to switch the position of a message-carrying pulse in a sequence of time frames.
4. *Chirp spread spectrum* where linear frequency modulation of the carrier is used to spread the spectrum. This technique is well-known in radar but is becoming more common in communication systems.

Further details on spread-spectrum systems can be found in the literature cited at the end of this section.

CHAPTER 1.5
NOISE AND INTERFERENCE

Geoffrey C. Orsak, H. Vincent Poor, John B. Thomas

GENERAL

In a general sense, *noise* and *interference* are used to describe any unwanted or undesirable signals in communication channels and systems. Since in many cases these signals are random or unpredictable, some study of random processes is a useful prelude to any consideration of noise and interference.

RANDOM PROCESSES

A random process $X(t)$ is often defined as an indexed family of random variables where the *index* or *parameter* t belongs to some set T; that is, $t \in T$.

The set T is called the *parameter set* or *index set* of the process. It may be finite or infinite, denumerable or nondenumerable; it may be an interval or set of intervals on the real line, or it may be the whole real line $-\infty < t < \infty$. In most applied problems, the index t will represent time, and the underlying intuitive notion will be a random variable developing in time. However, other parameters such as position and temperature may also enter in a natural manner.

There are at least two ways to view an arbitrary random process: (1) as a set of random variables: this viewpoint follows from the definition. For each value of t, the random process reduces to a random variable; and (2) as a set of functions of time. From this viewpoint, there is an underlying random variable each realization of which is a time function with domain T. Each such time function is called a *sample function* or *realization* of the process.

From a physical point of view, it is the sample function that is important, since this is the quantity that will almost always be observed in dealing experimentally with the random process. One of the important practical aspects of the study of random processes is the determination of properties of the random variable $X(t)$ for fixed t on the basis of measurements performed on a single sample function $x(t)$ of the process $X(t)$.

A random process is said to be *stationary* if its statistical properties are invariant to time translation. This invariance implies that the underlying physical mechanism producing the process is not changing with time. Stationary processes are of great importance for two reasons: (1) they are common in practice or approximated to a high degrees of accuracy (actually, from the practical point of view, it is not necessary that a process be stationary for all time but only for some observation interval that is long enough to be suitable for a given problem); (2) many of the important properties of common stationary processes are described by first and second moments. Consequently, it is relatively easy to develop a simple but useful theory (*spectral theory*) to describe these processes. Processes that are not stationary are called *nonstationary*, although they are also sometimes referred to as *evolutionary* processes.

The *mean m(t)* of a random process $X(t)$ is defined by

$$m(t) = E\{X(t)\} \tag{1}$$

where $E\{\cdot\}$ is the mathematical expectation operator defined in Chap. 1.1. In many practical problems, this mean is independent of time. In any case, if it is known, it can be subtracted from $X(t)$ to form a new "centered" process $Y(t) = X(t) - m(t)$ with zero mean.

The *autocorrelation function* $R_x(t_1, t_2)$ of a random process $X(t)$ is defined by

$$R_x(t_1, t_2) = E\{X(t_1)X(t_2)\} \tag{2}$$

In many cases, this function depends only on the time difference $t_2 - t_1$ and not on the absolute times t_1 and t_2. In such cases, the process $X(t)$ is said to be *at least wide-sense stationary*, and by a linear change in variable Eq. (2) can be written

$$R_x(\tau) = E\{X(t)X(t + \tau)\} \tag{3}$$

If $R_x(\tau)$ possesses a Fourier transform $\varphi_x(\omega)$, this transform is called the *power spectral density* of the process and $R_x(\tau)$ and $\varphi_x(\omega)$ form the Fourier-transform pair

$$\varphi_x(\omega) = \int_{-\infty}^{\infty} R_x(\tau)e^{-j\omega\tau}\,d\tau \tag{4}$$

and

$$R_x(\tau) = \frac{1}{2\pi}\int_{-\infty}^{\infty} \varphi_x(\omega)e^{j\omega\tau}\,d\omega \tag{5}$$

For processes that are at least wide-sense stationary, these last two equations afford a direct approach to the analysis of random signals and noises on a power ratio or mean-squared-error basis. When $\tau = 0$, Eq. (5) becomes

$$R_x(0) = E\{X^2(t)\} = \frac{1}{2\pi}\int_{-\infty}^{\infty} \varphi_x(\omega)\,d\omega \tag{6}$$

an expression for the normalized power in the process $X(t)$.

As previously mentioned, in practical problems involving random process, what will generally be available to the observer is not the random process but one of its sample functions or realizations. In such cases, the quantities that are easily measured are various time averages, and an important question to answer is: Under what circumstances can these time averages be related to the statistical properties of the process?

We define the *time average* of a sample function $x(t)$ of the random process $X(t)$ by

$$A\{x(t)\} = \lim_{T \to \infty} \frac{1}{2T}\int_{-T}^{T} x(t)\,dt \tag{7}$$

The *time autocorrelation function* $\mathfrak{R}_x(t)$ is defined by

$$\mathfrak{R}_x(\tau) = A\{x(t)x(t + \tau)\} = \lim_{T \to \infty} \frac{1}{2T}\int_{-T}^{T} x(t)x(t + \tau)\,dt \tag{8}$$

It is intuitively reasonable to suppose that, for stationary processes, time averages should be equal to expectations; e.g.,

$$E\{X(t)X(t + \tau)\} = A\{x(t)x(t + \tau)\} \tag{9}$$

for every sample function $x(t)$. A heuristic argument to support this claim would go as follows. Divide the parameter t of the random process $V(t)$ into long intervals of T length. If these intervals are long enough (compared with the time scale of the underlying physical mechanism), the statistical properties of the process in one interval T should be very similar to those in any other interval. Furthermore, a new random process $X(t)$ could be formed in the interval $(0, T)$ by using as sample functions the segments of length T from a single sample function of the original process. This new process should be statistically indistinguishable from the original process, and its ensemble averages would correspond to time averages of the sample function from the original process.

The foregoing is intended as a very crude justification of the condition of *ergodicity*. A random process is said to be *ergodic* if time averages of sample functions of the process can be used as approximations to the corresponding ensemble averages or expectations. A further discussion of ergodicity is beyond the scope of this treatment, but this condition can often be assumed to exist for stationary processes. In this case, time averages and expectations can be interchanged at will, and, in particular,

$$E\{X(t)\} = A\{x(t)\} = u = \text{a constant} \tag{10}$$

and

$$R_x(\tau) = \Re_x(\tau) \tag{11}$$

CLASSIFICATION OF RANDOM PROCESSES

A central problem in the study of random processes is their classification. From a mathematical point of view, a random process $X(t)$ is defined when all n-dimensional distribution functions of the random variables $X(t_1)$, $X(t_2), \ldots, X(t_n)$ are defined for arbitrary n and arbitrary times $t_1, t_2, \ldots, t_n$. Thus classes of random processes can be defined by imposing suitable restrictions on their n-dimensional distribution functions. In this way, we can define the following (and many others):

(**a**) *Stationary processes*, whose joint distribution functions are invariant to time translation.
(**b**) *Gaussian (or normal) processes*, whose joint distribution functions are multivariate normal.
(**c**) *Markov processes*, where given the value of $X(t_1)$, the value of $X(t_2)$, $t_2 > t_1$, does not depend on the value of $X(t_0)$, $t_0 < t_1$; in other words, the future behavior of the process, given its present state, is not changed by additional knowledge about its past.
(**d**) *White noise*, where the power spectral density given by Eq. (4) is assumed to be a constant N_0. Such a process is not realizable since its mean-squared value (normalized power) is not finite; i.e.,

$$R_x(0) = E\{X^2(t)\} = \frac{N_0}{2\pi} \int_{-\infty}^{\infty} d\omega = \infty \tag{12}$$

On the other hand, this concept is of considerable usefulness in many types of analysis and can often be postulated where the actual process has an approximately constant power spectral density over a frequency range much greater than the system bandwidth.

Another way to classify random processes is on the basis of a model of the particular process. This method has the advantage of providing insights into the physical mechanisms producing the process. The principal disadvantage is the complexity that frequently results. On this basis, we may identify the following (natural) random processes.

(**a**) *Thermal noise* is caused by the random motion of the electrons within a conductor of nonzero resistance. The mean-squared value of the thermal-noise voltage across a resistor of resistance R Ω is given by

$$\bar{v}2 = 4kTR\,\Delta f \quad (\text{volts}^2) \tag{13}$$

where k is Boltzmann's constant, T is the absolute temperature in kelvins, and Δf is the bandwidth of the measuring equipment.

(**b**) *Shot noise* is present in any electronic device (e.g., a transistor) where electrons move across a potential barrier in a random way. Shot noise is usually modeled as

$$X(t) = \sum_{i=-\infty}^{\infty} f(t - t_i) \tag{14}$$

where the t_i are random emission times and $f(t)$ is a basic pulse shape determined by the device geometry and potential distribution.

(**c**) *Defect noise* is a term used to describe a wide variety of related phenomena that manifest themselves as noise voltages across the terminals of various devices when dc currents are passed through them. Such noise is also called current noise, excess noise, flicker noise, contact noise, of $1/f$ noise. The power spectral density of this noise is given by

$$\varphi_x(\omega) = kI^\alpha/\omega^\beta \tag{15}$$

where I is the direct current through the device, ω is radian frequency, and k, α, and β are constants. The constant α is usually close to 2, and β is usually close to 1. At a low enough frequency this noise may predominate because of the $1/\omega$ dependence.

ARTIFICIAL NOISE

The noises just discussed are more or less fundamental and are caused basically by the noncontinuous nature of the electronic charge. In contradistinction to these are a large class of noises and interferences that are more or less artificial in the sense of being generated by electrical or electronic equipment and hence, in principle, are under our control. The number and kinds here are too many to list and the physical mechanisms usually too complicated to describe; however, to some degree, they can be organized into three main classes.

1. *Interchannel Interference.* This includes the interference of one radio or television channel with another, which may be the result of inferior antenna or receiver design, variation in carrier frequency at the transmitter, or unexpectedly long-distance transmission via scatter of ionospheric reflection. It also includes crosstalk between channels in communication links and interference caused by multipath propagation or reflection. These types of noises can be removed, at least in principle, by better equipment design, e.g., by using a receiving antenna with a sufficiently narrow radiation pattern to eliminate reception from more than one transmitter.

2. *Co-channel Interference.* In wireless systems, this refers to interference from other communication systems operating in the same frequency band. It arises, for example, in cellular telephony systems, owing to interference from adjacent cells. It also arises in many types of consumer systems, such as wireless local area networks, that operate in lightly regulated bands.

3. *Multiple-access Interference.* This refers to intranetwork interference that results from the use of nonorthogonal modulation schemes in networks of multiple communicating pairs. Many cellular telephony systems, such as those using code-division multiple-access (CDMA) protocols, involve this kind of interference.

4. *Hum.* This is a periodic and undesirable signal arising from the power lines. Usually it is predictable and can be eliminated by proper filtering and shielding.

5. *Impulse Noise.* Like defect noise, this term describes a wide variety of phenomena. Not all of them are artificial, but the majority probably are. This noise can often be modeled as a low-density shot process or, equivalently, as the superposition of a small number of large impulses. These impulses may occur more or less periodically, as in ignition noise from automobiles or corona noise from high-voltage transmission lines. On the other hand, they may occur randomly, as in switching noise in telephone systems or the atmospheric noise from thunderstorms. The latter type of noise is not necessarily artificial, of course. This impulse noise tends to have an amplitude distribution that is decidedly non-gaussian, and it is frequently highly nonstationary. It is difficult to deal with in a systematic way because it is ill-defined. Signal processors that must handle this type of noise are often preceded by limiters of various kinds or by noise blankers which give zero transmission if a certain amplitude level is surpassed. The design philosophy behind the use of limiters and blankers is fairly clear. If the noise consists of large impulses of relatively low density, the best system performance is obtained if the system is limited or blanked during the noisy periods and behaves normally when the (impulse) noise is not present.

SECTION 2

SYSTEMS ENGINEERING AND SYSTEMS MANAGEMENT

The origins of systems engineering concepts are indeterminate, although the complexity of weapons systems and other military systems during World War II helped formalize the practice. It would be impossible to approach anything near optimization of today's complex systems without the use of systems engineering. It should be no surprise that an important element of systems engineering is the partitioning of a system into smaller components that are more readily managed. But this leads to the necessity for careful, considered communication between and among engineers working on distinctive subsystems, attention to interdisciplinary issues, and an overall management philosophy (and practice) that fosters good communication and good systems engineering at all phases of a system's life cycle.

The two chapters in this section are designed to define the various facets of systems engineering and systems management and to identify some of the techniques and procedures employed in its practice.   D.C.

In This Section:

On the CD-ROM:

- Sage, A. P., and W. B. Rouse, "Discretionary Uses of Systems Engineering."
- Sage, A. P., and W. B. Rouse, "Levels of Understanding, Knowledge Management, and Systems Engineering."

CHAPTER 2.1

INTRODUCTION TO SYSTEMS ENGINEERING

Andrew P. Sage, William B. Rouse

FACETS AND DEFINITIONS

In this chapter, we summarize relevant facets of systems engineering support for management that we believe are essential to the definition, development, and deployment of innovative technologies that feature electronics engineering contributions.

Systems engineering generally involves breaking down a large problem into smaller component problems that are more easily resolved, determining the solution to this collection of subproblems, and then aggregating the solutions such that, hopefully, resolution to the initial larger problem or issue results.

A simple functional definition of systems engineering is that systems engineering is the art and science of producing a product or a service, or a process—based on phased efforts that involve *definition* or the desired end result of the effort followed by *development* of the product or service or process, and culminating with *deployment* of this in an operational setting—that satisfies user needs. The system is functional, reliable, of high quality, and trustworthy, and has been developed within cost and time constraints through use of an appropriate set of methods and tools.

PURPOSES AND APPLICATIONS

The purposes of systems engineering can be defined as follows: *Systems engineering is management technology to assist and support policymaking, planning, decision-making, and associated resource allocation or action deployment.* Systems engineers accomplish this by quantitative and qualitative *formulation, analysis,* and *interpretation* of the impacts of action alternatives on the needs perspectives, the institutional perspectives, and the value perspectives of their clients or customers. Each of these three steps is generally needed in solving systems engineering problems. Issue formulation is an effort to identify the needs to be fulfilled and the requirements associated with these in terms of objectives to be satisfied, constraints and alterables that affect issue resolution, and generation of potential alternate courses of action. Issue analysis enables us to determine the impacts of the identified alternative courses of action, including possible refinement of these alternatives. Issue interpretation helps to rank order the alternatives in terms of need satisfaction and to select one for implementation or additional study.

This particular listing of three systems engineering steps and their descriptions is rather formal. Issues are often resolved this way, especially when there is initially only limited experiential familiarity with the issue under consideration. The steps of formulation, analysis, and interpretation may also be accomplished on an "as-if" basis by application of a variety of often-useful heuristic approaches. These may well be quite appropriate in situations where the problem solver is experientially familiar with the task at hand and the environment into which the task is embedded, so as to enable development of an appropriate context for issue resolution. This requires information. It also requires knowledge, as information embedded into experience-based context and environment, and this may lead to very useful heuristic approaches to use of this knowledge.

The systems approach gains ready acceptance if one or more of these characteristics apply:

1. Complexity cannot be otherwise addressed, and a completely intuitive and unorchestrated approach simply will not work.
2. We are dealing with high-risk situations, and the consequences of errors are too large to be acceptable.
3. There are major investments involved and the potential loss of financial resources because use of inefficient processes to engineer the system is unacceptable.

HUMAN AND SOCIETAL FACTORS

One of the most essential notions is that systems engineering does not deal exclusively with physical products. While the ultimate aim of systems engineering efforts may well be to produce a physical product or service, there are humans and organizations involved, as well as technologies.

Systems engineering and systems management concern more than the specific technical efforts that occur inside a factory. One of the major needs is for a multiple perspective viewpoint across the various issues affecting the engineering of systems. Fundamentally, systems engineers act as brokers and communicators of knowledge across the various stakeholders having interest in a given issue concerned with product or service deployment. These certainly include the enterprise organization for which a system is being engineered. It also includes the technology implementation organization(s) that will be responsible for the detailed internal design efforts needed to bring about a system. This is represented in Fig. 2.1.1. The issues being considered may also

		Information interrogatives				Knowledge interrogatives	
		Entities (what)	Time (when)	Locations (where)	People (who)	Functions/ activities (how)	Purpose/ motivation (why)
Stakeholders	Policy makers						
	Planners						
	Enterprise owners						
	Systems engineers/ architects						
	Builders						
	Impacted publics						

FIGURE 2.1.1   Perspectives of stakeholders to systems engineering and management issues.

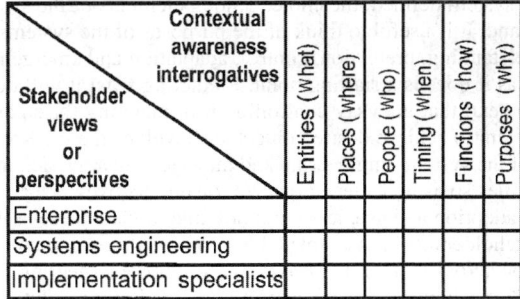

FIGURE 2.1.2 Multiple perspectives framework for engineering a system.

include impacted publics and public sector policy-making bodies, especially when the system being engineered has public sector impacts and implications.

A reduced perspective version of this figure is shown in Fig. 2.1.2. The three rows here correspond to the enterprise perspective, which represents that of the group which desires the system, the systems engineering and management perspective, and the perspective of the group or organization that physically implements the system. The columns contain a total of six interrogatives. The first four of these (what, where, who, and when) generally correspond to structural elements or interrogatives needed to represent the structural issues associated with the phase (definition, development, and deployment) of the system being engineered. The "how" element or interrogative generally corresponds to a functional and activity-based representation of the phase of the system being engineered, and the "why" element generally corresponds to purposeful interrogatives associated with the phase of the system being engineered. It is generally felt that we may define subject matter from a structural, functional, or purposeful perspective. Thus, the representations shown in Figs. 2.1.1 and 2.1.2 are very useful and generally comprehensive ones.

The need for this multiple perspective viewpoint is a major one for successful systems engineering practice. It is a major need in developing an appropriate framework for a system to be engineered, as represented in Fig. 2.1.1.

Systems engineering, applied in its best sense, approaches the problem at three levels: symptoms, institutions and infrastructures, and values. Too often, problems are approached only at the level of symptoms: bad housing, inadequate healthcare delivery to the poor, pollution, hunger, and so forth. "Technological" fixes are developed, and the resulting hardware creates the illusion that solution of real-world problems requires merely the outpouring of huge quantities of funds. Attacking problems at the level of-institutions and infrastructures, such as to better enable humans and organizations to function, would allow the adaptation of existing organizations, as well as the design of new organizations and institutions, to make full and effective use of new technologies and technological innovations and to be fully responsive to potentially reengineered objectives and values. Thus, human and organizational issues are all important in seeking technological solutions. Approaching issues in a satisfactory manner at all three levels will guarantee not only that we resolve issues correctly, but will also guarantee that we resolve the correct issues.

BARRIERS TO ADOPTION

There are significant barriers to adoption of the full armament of systems engineering and management. However, a few central elements of this body of concepts, principles, processes, and methods and tools can be adopted and exploited quite easily. Perhaps most central is the basic question: What's the system? At the very least, the overall enterprise, rather than a particular product, is usually the system that provides the context for formulation, analysis, and interpretation. In many cases, it is important and useful to define the system as the enterprise, competitors, markets, and economy. This broader view enables clearer understanding of the full range of stakeholders and interests.

With the scope of the system defined, the choice among elements of the systems approach depends on the purpose of the effort at hand. It is useful to think of the purposes of the systems engineering and management approach to include understanding, prediction, control, adaptation and emergence, and design. Understanding why the system behaves as it does is a starting point. Predicting what it will do in the future is the next step. Affecting what it will do next, that is to say controlling it or planning for emergence and adaptation, is next. Finally, designing the system to achieve desired objectives involves the full scope of systems engineering and management. Adoption of the systems engineering and management approach involves deciding how to implement the associated activities so as to achieve organizational objectives.

Regarding barriers to adoption and use, it is quite possible to view these as elements of the overall system. This requires adding stakeholders and issues related to power, culture, and so forth. The tasks of formulation, analysis, and interpretation across each of the phases of a systems engineering effort are correspondingly more complicated, but there is, we hope, a greater opportunity for issue resolution by implementing them in a thoughtful manner.

 There are many complex situations where the systems approach is warranted, but, being perceived as discretionary, does not gain wide acceptance. These are discussed on the accompanying CD-ROM under "Discretionary Uses of Systems Engineering" (Sage and Rouse).

THE SYSTEMS VIEWPOINT

The systems viewpoint stresses that there usually is not a single correct answer or solution to a large-scale issue. Instead, there are many different alternatives that can be developed and implemented depending on the objectives the system is to serve and the values and perspectives of the people and organizations with a stake in the solution such that they are rightfully called stakeholders. A system may be defined simply as a group of components that work together in order to achieve a specified purpose. Purposeful action is a basic characteristic of a system. A number of functions must be implemented in order to achieve these purposes. This means that systems have functions. They are designed to do specific tasks. Systems are often classified by their ultimate purpose: service-oriented, product-oriented, or process-oriented systems. We note here that the systems considered by systems engineers may be service systems, product systems, process systems, or management systems. The systems may be systems designed for use by an individual or by groups. These systems may be private sector systems, or they may be government or public sector systems.

SYSTEMS THINKING

"Systems thinking" may be defined in part by the following elements (Senge, 1990):

1. Contemporary and future problems often come about because of what were presumed to be past solutions.
2. For every action, there is a reaction.
3. Short-term improvements often lead to long-term difficulties.
4. The easy solution may be no solution at all.
5. The solution may be worse than the problem.
6. Quick solutions, especially at the level of symptoms, often lead to more problems than existed initially. Thus, quick solutions may be counterproductive solutions.
7. Cause and effect are not necessarily related closely, either in time or in space. Sometimes actions implemented here and now will have impacts at a distance and much later time.
8. The actions that will produce the most effective results are not necessarily obvious at first glance.
9. Low cost and high effectiveness do not necessarily have to be subject to compensatory tradeoffs over all time.
10. The entirety of an issue is often more than the simple aggregation of the components of the issue.
11. The entire system, comprising the organization and its environment, must be considered together.

Systems thinking alone is insufficient for success of an organization. Other necessary disciplines include personal mastery through proficiency and lifelong learning; shared mental models of the organization's markets and competition; shared vision for the future of the organization; and team learning (Argyris and Schon, 1978).

Minimizing Ambiguity

Information and Knowledge.   One of the major tasks of management planning and control is that of minimizing the ambiguity of the information that results from the organization's interaction with its external environment (Weick, 1979).

This task is primarily that of systems management or management control. It is done subject to the constraints imposed by the strategic plan of the organization. In this way, the information presented to those responsible for task control is unequivocal. This suggests that there are appropriate planning and control activities at each of the functional levels in an organization. The nature of these planning and control activities is different across these levels, however, as is the information that flows into them.

Various systems-based approaches to management experience periodic naissance and revival. Characteristically, these revivals are associated with the rediscovery that dynamic phenomena and feedback loops abound in the worlds of organizations. Another common motivation is rediscovery of the importance of getting "out of the box" of looking at microlevel details and looking at the "whole system" from both "inside the box" (system) and "outside the box" from the perspective of all the stakeholders who are impacted by the system.

KNOWLEDGE MANAGEMENT

Efficient Knowledge Conversion

Knowledge management in systems engineering, as in a more general sense, is a very broad concept, and involves many complex interactions between people, process, and technology. Depending on the objectives at hand, knowledge management activities may focus on the explicit or tacit knowledge, or may focus on an integrated approach that considers both explicit and tacit knowledge (Nonaka and Takeuchi, 1995). The Nonaka and Takeuchi theory of knowledge creation comprise four knowledge conversion stages as suggested in Fig. 2.1.3: socialization, externalization, combination, and internalization. The conversion takes place in five

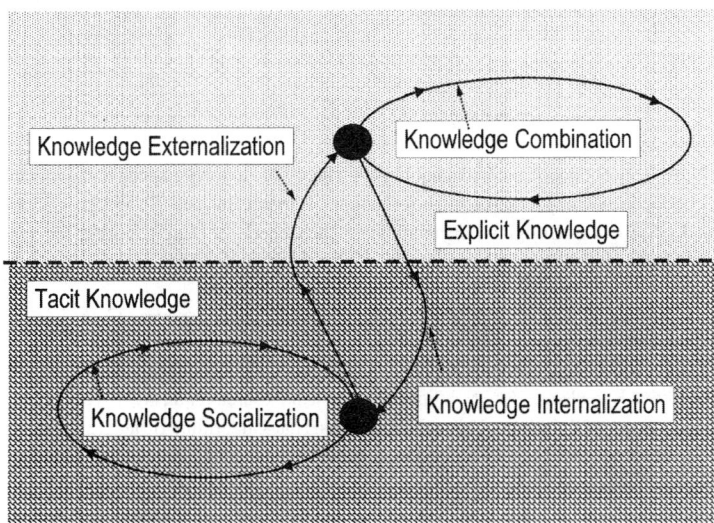

FIGURE 2.1.3   Representation of the four modes of knowledge conversion.

phases: sharing of tacit knowledge, creating concepts, justifying concepts, building an archetype, and cross-level knowledge. Critical to this theory is the concept of levels of knowledge: individual, group, organizational, and interorganization. Sharing primarily occurs during the socialization, externalization, and combination stages. During the socialization stage, sharing occurs primarily at the individual and group levels. In the externalization stage, knowledge is codified and shared at group and organizational levels. In the cross-leveling of knowledge phase, an enterprise shares knowledge both intra- and interorganizationally. This may result in different knowledge content being associated with some emerging issue and thereby enable knowledge creation.

The distinctions between data, information, and knowledge are important. Data represent points in space and time that relate to particular temporal or spatial points. Information is data that are potentially relevant to decision-making. Thus, information relates to description, definition, or outlook. Generally, information is responsive to questions that relate to structural issues of *what, when, where,* or *who*. Knowledge is information embedded in context and may comprise approach, method, practice, or strategy. Generally, knowledge is responsive to questions that relate to functional issues associated with a *how* interrogative, It is sometimes desirable to distinguish wisdom as a higher-level construct that represents insights, prototypes or models, or principles and which would be responsive to questions concerning purpose or *why*. If the distinction is not made, knowledge is needed to respond to the *how* and *why* questions that are generally associated with function and purpose. These six important questions or concerns might be collectively called "contextual awareness interrogatives."

Each of these six interrogatives is important and is needed to be responsive to broad scope inquire regarding modern transdisciplinary endeavors. Systems engineering is one such area.

 Knowledge management involves levels of understanding that may differ for various actors in a systems project. For a more detailed discussion of this topic, see "Levels of Understanding, Knowledge Management, and Systems Engineering" (Sage and Rouse, 1999) on the accompanying CD-ROM.

Organizations are beginning to realize that knowledge is the most valuable asset of employees and the organization. This recognition must be converted into pragmatic action guidelines, plans, and specific approaches. Effective management of knowledge, which is assumed to be equivalent to effective management of the environmental factors that lead to enhanced learning and transfer of information into knowledge, also requires organizational investments in terms of financial capital for technology and human labor to ensure appropriate knowledge work processes. It also requires knowledge managers to facilitate identification, distribution, storage, use, and sharing of knowledge. Other issues include incentive systems and appropriate rewards for active knowledge creators, as well as the legalities and ethics of knowledge management. In each of these efforts, it is critical to regard technology as a potential enabler of human effort, not as a substitute for it.

BIBLIOGRAPHY

Argyris, C., and D. A. Schon, "Organizational Learning: A Theory of Action Perspective," Addison Wesley, 1978.

Brown, J. S., and P. Deguid, "The Social Life of Information," Harvard Business School Press, 2000.

Cook, S. D. N., and J. S. Brown, "Bridging Epistemologies: The Generative Dance Between Organizational Knowledge and Organizational Knowing," *Organization Science*, Vol. 10, No. 4, July–August 1999.

Morgan, G., "Images of Organization," Sage Publications, 1986.

Nonaka, I., and H. Takeuchi, "The Knowledge-Creating Company: How Japanese Companies Create the Dynamics of Innovation," Oxford University Press, 1995.

Polanyi, M., "The Tacit Dimension," Doubleday, 1996.

Popper, K. R., "Objective Knowledge: An Evolutionary Approach," Clarendon Press, 1972.

Rouse, W. B., "Don't Jump to Solutions: Thirteen Delusions That Undermine Strategic Thinking," Jossey-Bass, 1998.

Rouse, W. B., "Essential Challenges of Strategic Management," Wiley, 2001.

Sage, A. P., "Systems Engineering," Wiley, 1992.

Sage, A. P., "Systems Management for Information Technology and Software Engineering," Wiley, 1995.

Sage, A. P., and C. D. Cuppan, "On the Systems Engineering and Management of Systems and Federations of Systems," *Information, Knowledge, and Systems Management*, Vol. 2, No. 4, 2001.

Sage, A. P., and W. B. Rouse (eds.), "Handbook of Systems Engineering and Management," Wiley, 1999.

Sage, A. P., and W. B. Rouse, "Information Systems Frontiers in Knowledge Management," *Information Systems Frontiers*, Vol. 1, No. 3, October 1999.

Senge, P. M., "The Fifth Discipline: The Art and Practice of the Learning Organization," Doubleday/Currency, 1990.

Von Krogh, G., K. Ichijo, and I. Nonaka, "Enabling Knowledge Creation: How to Unlock the Mystery of Tacit Knowledge and Release the Power of Innovation," Oxford University Press, 2000.

Weick, K. E., "The Social Psychology of Organizing," Addison Wesley, 1979.

Zachman, J., "A Framework for Information Systems Architecture," *IBM Systems Journal*, Vol. 26, No. 3, 1987.

Zachman, J., "Enterprise Architecture: The Issue of the Century," and "The Framework for Information Architecture," 2002, available at http://www.zifa.com/articles.htm.

ON THE CD-ROM

Sage, A. P., and W. B. Rouse, "Discretionary Uses of Systems Engineering."

Sage, A. P., and W. B. Rouse, "Levels of Understanding, Knowledge Management, and Systems Engineering."

CHAPTER 2.2
ELEMENTS AND TECHNIQUES OF SYSTEMS ENGINEERING AND MANAGEMENT

Andrew P. Sage, William B. Rouse

SYSTEMS ENGINEERING AS A MANAGEMENT TECHNOLOGY

Systems engineering is a management technology. Technology is organization, application, and delivery of scientific and other forms of knowledge for the betterment of a client group. This is a functional definition of technology as a fundamentally human activity. A technology inherently involves a purposeful human extension of one or more natural processes. For example, the stored program digital computer is a technology in that it enhances the ability of a human to perform computations and, in more advanced forms, to process information.

Management involves the interaction of the organization with the environment. A purpose of management is to enable organizations to better cope with their environments so as to achieve purposeful goals and objectives. Consequently, a management technology involves the interaction of *technology, organizations* that are collections of *humans* concerned with both the evolvement and use of technologies, and the *environment*. Figure 2.2.1 illustrates these conceptual interactions. Information and associated knowledge represents the "glue" that enables the interactions shown in this figure. Information and knowledge are very important quantities that are assumed to be present in the management technology that is systems engineering. This strongly couples notions of systems engineering with those of technical direction or systems management of technological development, rather than exclusively with one or more of the methods of systems engineering, important as they may be for the ultimate success of a systems engineering effort.

Figure 2.2.2 illustrates the view that systems engineering knowledge comprises:

1. *Knowledge principles*—which generally represent formal problem solving approaches to knowledge, generally employed in new situations and/or unstructured environments.

2. *Knowledge practices*—which represent the accumulated wisdom and experiences that have led to the development of standard operating policies for well-structured problems.

3. *Knowledge perspectives*—which represent the view that is held relative to future directions and realities in the knowledge areas under consideration.

Clearly, one form of knowledge leads to another. Knowledge perspectives may create the incentive for research that leads to the discovery of new knowledge principles. As knowledge principles emerge and are

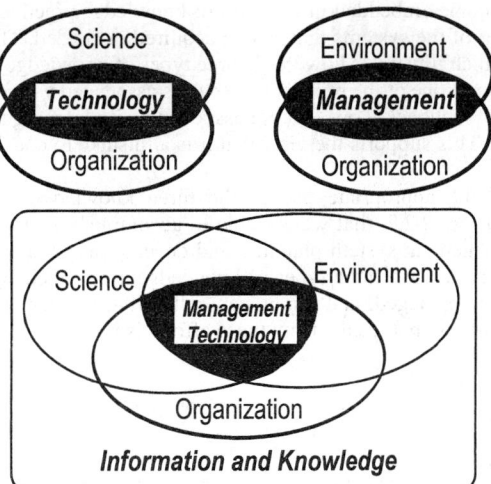

FIGURE 2.2.1 Systems engineering as a management technology.

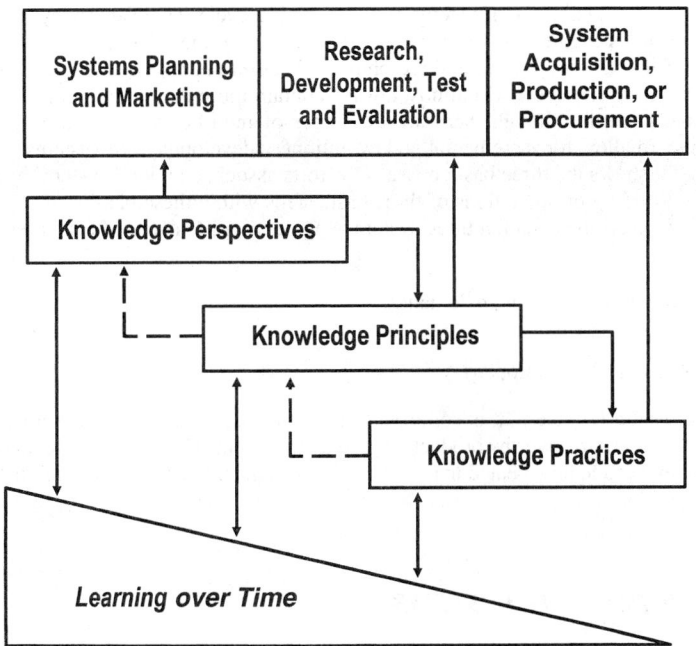

FIGURE 2.2.2 Knowledge types and support for systems engineering efforts.

refined, they generally become embedded in the form of knowledge practices. Knowledge practices are generally the major influences of the systems that can be acquired or fielded. These knowledge types interact as suggested in Fig. 2.2.2, which illustrates how these three types of knowledge support one another. In a nonexclusive way, they each support one of the principal life cycles associated with systems engineering. Figure 2.2.2 also illustrates a number of feedback loops that are associated with learning to enable continual improvement in performance over time. This supports the view that it is a mistake to consider these life cycles in isolation from one another.

It is on the basis of the appropriate use of the three knowledge types—principles, practices, and perspectives—depicted in Fig. 2.2.2, that we are able to accomplish the technological system planning and development and the management system planning and development that lead to new and innovative products, technologies, and services. All three types of knowledge are needed. The environment associated with this knowledge needs to be managed, and this is generally what is intended by use of the term *knowledge management*. Also, the learning that results from these efforts is very much needed, both on an individual and an organizational basis.

LIFE CYCLES

Three Essential Life Cycles. There are three major and quite purposefully different life cycles for technology evolution through systems engineering: system planning and marketing; research, development, test and evaluation (RDT&E); and system acquisition, production, or procurement. These are needed for evolution of trustworthy products and services, and each involves use of one of the three types of knowledge: knowledge perspectives, knowledge principles, and knowledge practices.

Systems engineers are concerned with the appropriate definition, development, and deployment of systems. These comprise a set of phases for a systems engineering life cycle. There are many ways to describe the life-cycle phases of systems engineering life-cycle processes. Each of these basic life-cycle models, and those that are outgrowths of them, comprise these three phases of definition, development, and deployment. For pragmatic reasons, a typical life cycle will almost always contain more than three phases. Generally, they take on a "waterfall" like pattern, although there are a number of modifications of the basic waterfall, or "grand design" life cycle, to allow for incremental and evolutionary development of systems. Figure 2.2.3 represents such a life cycle. It shows the three basic phases of efforts associated with a systems engineering life cycle. It also shows an embedding of the notions of three basic steps within these phases.

Figure 2.2.4 suggests the essential three systems-engineering life cycles. Each is responsive to a particular question:

- Planning and Marketing: What is in demand?
- RDT&E: What is possible?
- Acquisition: What can be developed?

It is only in the response space that is common to all three questions, as suggested in Fig. 2.2.5, that it will truly be feasible to build a trustworthy productive product or system. The life-cycle processes shown in Figs. 2.2.3 and 2.2.4 will often need to be repeated in an evolutionary manner to accommodate successive builds of a technology product.

DEVELOPING TRUSTWORTHY SYSTEMS

Management of systems engineering processes, which we call systems management (Sage, 1995), is necessary for success. Systems engineering may fail at the level of systems management. Often, the purpose, function, and structure of a new system are not identified sufficiently before the system is defined, developed, and

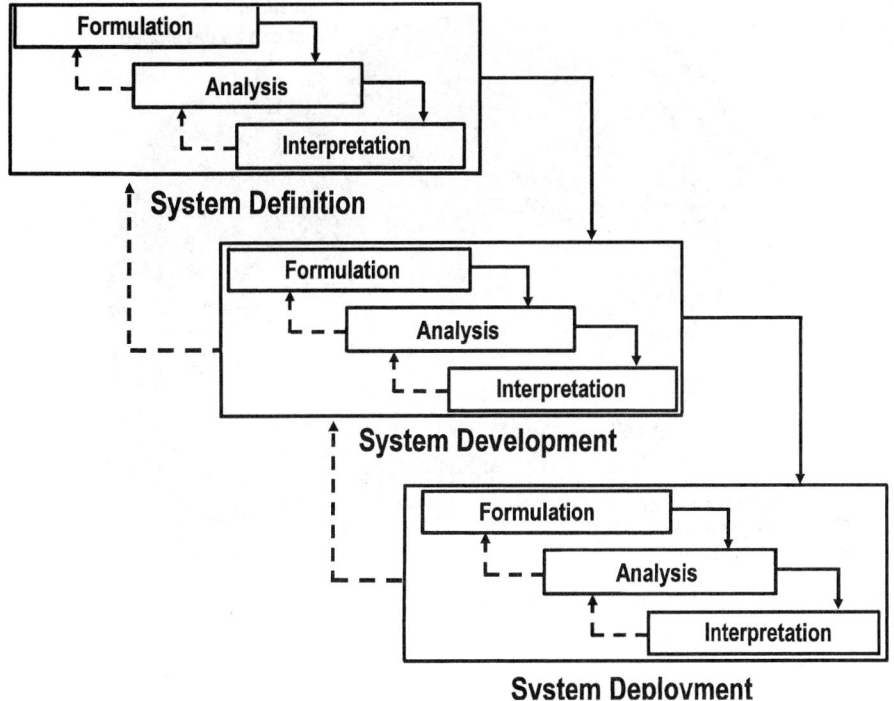

FIGURE 2.2.3   A systems engineering life cycle comprising of three phases and three steps per phase.

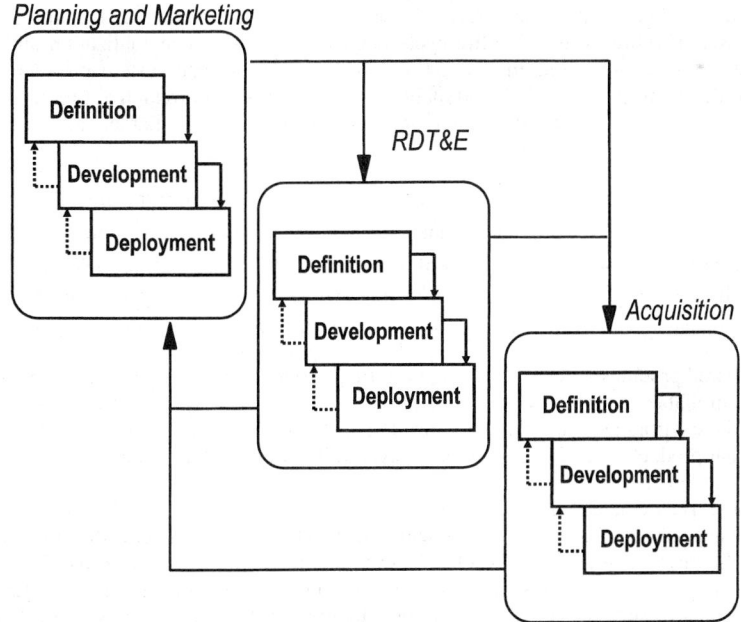

FIGURE 2.2.4   Major systems engineering life cycles and three phases within each.

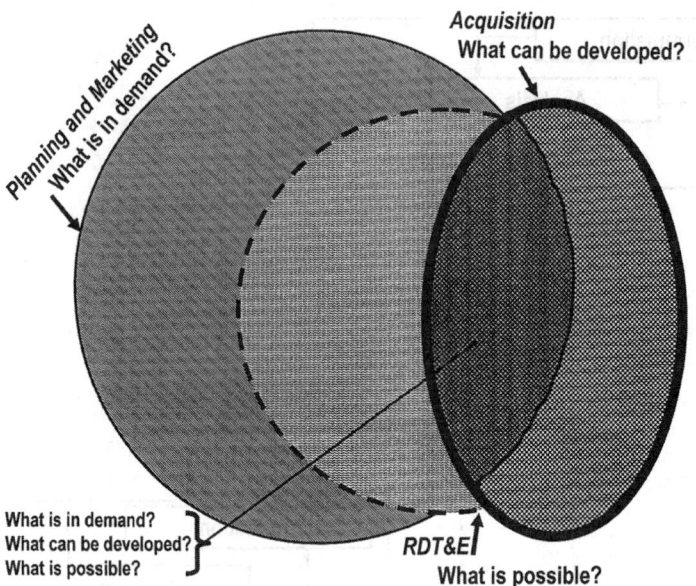

FIGURE 2.2.5 Illustration of the need for coordination and integration across life cycles.

deployed. These failures, generally, are the result of costly mistakes that could have been avoided by more diligent use of systems engineering and systems management.

System management and integration issues are of major importance in determining the effectiveness, efficiency, and overall functionality of system designs. To achieve a high measure of functionality, it must be possible for a system to be efficiently and effectively produced, used, maintained, retrofitted, and modified throughout all phases of a life cycle. This life cycle begins with need conceptualization and identification, through specification of system requirements and architectures, to ultimate system installation, operational implementation or deployment, evaluation, and maintenance throughout a productive lifetime.

There are many difficulties associated with the production of functional, reliable, and trustworthy systems of large scale and scope. Among them are:

1. Inconsistent, incomplete, and otherwise imperfect system requirements and specifications

2. System requirements that do not provide for change as user needs evolve over time

3. Lack of proper concern for satisfying the functional needs of the customer and inappropriate attention to assurance that the overall life cycle of product development and use is conservative of natural resource use

4. Poorly defined management structures for product, or service, development, and deployment

These lead to delivered products and services that are difficult to use, that do not solve the intended problems, that operate in an unreliable fashion, that are not maintainable, that are overly consumptive of natural resources and damaging to the environment, and that, as a result, may become quickly obsolete. Sometimes these failures are so great that products and services are never even fully developed, much less operationally deployed, before they are abruptly canceled.

The major problems associated with the production of trustworthy systems often have more to do with the organization and management of complexity than with direct technological concerns that affect individual subsystems and specific physical science areas. Often more attention should be paid to the definition, development, and use of an appropriate process for production of a product than to the actual product itself. Direct attention to the product or service without appropriate attention to the process leads to the fielding of a low quality and expensive product or service that is unsustainable.

Management and Metrics.   Systems engineering efforts are very concerned with technical direction and management of the process of systems definition, development, and deployment, or systems management. Through adopting and applying the management technology of systems engineering, we attempt to be sure that correct systems are designed, and not just that system products are correct according to some potentially ill-conceived notions of what the system should do. Appropriate metrics to enable efficient and effective error prevention and detection at the level of systems management, and at the process and product level will enhance the production of systems engineering products that are "correct" in the broadest possible meaning of this term. To assure that correct systems are produced requires that considerable emphasis be placed on the front-end of each of the systems engineering life cycles.

ACCURATE DEFINITION OF A SYSTEM

There must be considerable emphasis on the accurate definition of a system, what it should do, and how people should interact with it before one is produced and implemented. In turn, this requires emphasis on conformance to system requirements and specifications, and the development of standards to ensure compatibility, integratibility, and sustainability of system products and services. Such areas as documentation and communication are important. Thus, we see the need for the technical direction and management technology efforts that comprise systems engineering, and the strong role for process and systems management related concerns.

Ingredients associated with the development of trustworthy systems include the following:

- Systems engineering processes, including process development life cycles and process configuration management
- Process risk management, operational level quality assurance and evaluation, and product and process development standards and associated maturity models
- Metrics for quality assurance, to ensure standardization, and for process and product evaluation
- Metrics for cost estimation, and product cost and operational effectiveness evaluation
- Strategic quality assurance and management, or total quality management
- Organizational cultures, leadership, and process maturity
- Reengineering at the levels of systems management, organizational processes and product lines, and products

These ingredients and issues often strongly interact. One of the first efforts in systems management is to identify an appropriate process life cycle that is sustainable and that will lead to production of a trustworthy and sustainable system. This life cycle for the engineering of a system involves a sequence of phases. These phases include identification of client requirements, translation of these requirements into hardware and software requirements specifications, development of system architectures and conceptual designs of the system, detailed design and production of the system, operational implementation and evaluation, and maintenance of the delivered system over time. The precise life cycle that is followed will depend on the client needs. It will also depend on such overall environmental factors as the presence of existing system components, or subsystems, into which a new system must be integrated, natural resource considerations, and the presence of existing software modules that may be retrofitted and reused as part of the new system. These needs generally result in continued evolution and emergence of solutions that are adaptive to organizational needs that are similarly evolving and emergent.

SYSTEM REQUIREMENTS

A necessary first step in the engineering of any system is to determine just what the user wants, the purpose that is to be fulfilled by the resulting system in fulfilling these needs, and the necessary accompanying nonfunctional requirements, such as reliability and quality of the system. The resulting *definition* of user functional needs and associated nonfunctional requirements represent the first and most important phase in the effort to engineer a system. These requirements must be clear, concise, consistent, unambiguous, and represent user goals and purposes for the system to be engineered. A process to identify user needs and requirements must include:

1. Identification of purposes for the system, and associated constraints, functions, and operational environment; and translation of these into a set of functional requirements.

2. Assessment of the allowable risks and the desired quality of the system, and translation of these into a set of nonfunctional requirements

3. Transformation of these functional and nonfunctional requirements into the concise and formal language of a technical specification.

Successful systems engineering is very dependent on the ability to identify user needs in the form of (functional and nonfunctional) requirements and to reflect them in specifications for the delivered system. User needs and requirements that are complete, correct, consistent, and error free, play a major role in ensuring that the delivered system meets the intended purpose.

While there is widespread and general acceptance of the necessity to provide adequate requirements definition, there is often controversy as to the ultimate need, purpose, and cost of performing the requirements and system definition associated activities that are necessary to assure that the engineered system meets user needs. The controversy arises both because detailed system developers often do not visualize that the benefits that may accrue to the final product through implementation of what may be viewed as an expensive total requirements process is comparable to the time and effort involved. System users generally do not understand the complexity associated with transforming their needs to requirements and specifications documents that detailed system developers can use.

Production of well-understood and well-developed requirements information, and the identification and management of risk in the requirements engineering, is a major concern to assure production of systems that meet user needs by engineering a system that is delivered on time and within budget. Many studies suggest that between 50 and 80 percent of the errors found in deployed systems can be traced to errors in identification and interpretation of requirements.

Difficulties related to requirements definition generally revolve around the necessity to elicit or develop requirements information from a variety of sources, including users and/or existing systems. There is usually a lack of definitive information from the user and a lack of agreement on mechanisms to transform these concepts to technical specification documents for use by the detail system developer. These difficulties lie at the interface between the humans involved in the systems engineering process, especially system users and detailed system developers. Once user requirements have been established, these must be transformed to the exact language of technical system specifications; another source of potential problems. There are also potential technical difficulties that relate to hardware, performance, capacity, interfaces, and other issues.

Most often, user originated system level requirements are stated in natural language, and this brings about a high possibility of incorporation of ambiguities, conflicts, inconsistencies, and lack of completeness in the resulting requirements. These problems must be addressed prior to transformation from the informal language of system users' requirements to the formal or semiformal languages of detailed system design. Otherwise, these deficiencies may be incorporated in the final engineered product rather than being resolved at the time of requirements development. If this occurs, these requirements deficiencies must be corrected during later phases of systems development, or following system deployment provided the system is considered to be acceptable at all by the user such as to merit initial deployment.

Strategies for Determining Information Requirements

Davis (1982) has identified four strategies for determining information requirements. Taken together, these yield approaches that may be designed to ameliorate the effect of three human limitations: limited information processing ability, bias in the selection and use of information, and limited knowledge of what is actually needed.

1. The first strategy is to simply ask people for their requirements. The usefulness of this approach will depend on the extent to which the interviewers can define and structure issues and compensate for biases in issue formulation. There are a variety of methods that can be used to assist in this.

2. The second strategy is to elicit information requirements from existing systems that are similar in nature and purpose to the one in question. Examination of existing plans and reports represent one approach of identifying information requirements from an existing, or conceptualized, system.

3. The third strategy consists of synthesizing information requirements from characteristics of the utilizing system. This permits one to obtain a model or structure for the problem to be defined, from which information requirements can be determined. This strategy would be appropriate when the system in question is in a state of change and thus cannot be compared to an existing system.

4. The fourth strategy consists of discovering needed information requirements by experimentation, generally through constructing a prototype. Additional information can be requested as the system is employed in an operational, or simulated setting, and problem areas are encountered. The initial set of requirements for the system provides a base point for the experimentation. This represents an expensive approach, but is often the only alternative when there does not exist the experience base to use one of the other approaches. This approach is therefore equivalent to use of a prototype, either an evolutionary prototype or throwaway prototype.

Each of these four strategies has advantages and disadvantages, and it is desirable to be able to select the best mix of strategies. One's choice will depend on the amount of risk or uncertainty in information requirements that results from each strategy. Here, uncertainty is used in a very general sense to indicate information imperfection.

Information Uncertainties

Five steps are useful in identifying information uncertainties and then selecting appropriate strategies.

1. Identify characteristics of the utilizing system, technology system, users, and system development personnel as they affect information uncertainty.

2. Evaluate the effect of these characteristics on three types of information requirements determination uncertainties:
   a. Availability of a set of requirements
   b. Ability of users to specify requirements
   c. Ability of systems engineers to elicit and specify requirements

3. Evaluate the combined effect of the requirements determination process uncertainties on overall requirements volatility.

4. Select a primary requirements determination strategy.

5. Select a set of specific steps and methods to implement the primary requirements determination strategy.

These steps may be used to identify an appropriate mix of requirements identification strategies. The uncertainty associated with requirements determination, that is to say the amount of information imperfection that exists in the environment for the particular task, influences the selection from among the four basic strategies as indicated in Fig. 2.2.6. This illustrates the effects of experiential familiarity on the part of the system users that are used by a generally experienced requirements engineering team to identify requirements. The factors that influence this information imperfection include:

1. Stability of the environment

2. Stability of organizational management and system users

3. Previous experience of system users with efforts associated with systems engineering and management

4. The extent to which there exists a present system that is appropriate

5. The extent to which a change in requirements will change the usage of present system resources and thereby degrade the functionality of legacy systems and result in this needing to be considered as part of the overall effort

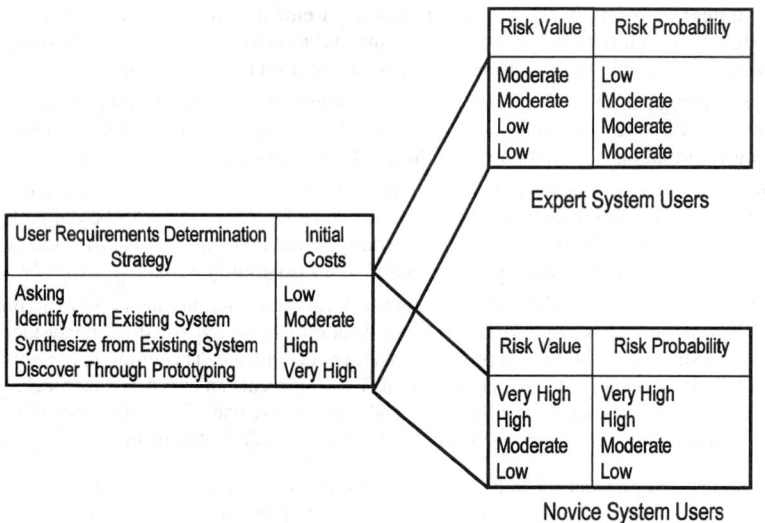

FIGURE 2.2.6   Effects of user familiarity on selecting requirements-determination strategy.

Used appropriately, this effort should enable selection of an appropriate requirements determination approach that can cope with the three essential contingency dependent variables owing to risk elements in requirements determination. To fully enable this it is desirable to also consider the effects of experiential familiarity of the requirements engineering team with both requirements determination in general and for the specific task at hand.

Requirements determination must lead to technological specifications for detailed design and fielding of the system. A purpose of the system definition phase of the life cycle is to determine possibilities of insufficient and/or inappropriate information; that is, information that is sufficiently imperfect such as to make the risk of an unacceptable design too high to be tolerated. The requirements elicitation team should be able to determine the nature of the missing or otherwise imperfect information, and suggest steps to remedy this deficiency. These are truly major efforts in systems and software engineering efforts today and are the subject of many contemporary writings on requirements (Andriole, 1996; Robertson and Robertson, 1999; Sommerville and Sawyer, 1997; Thayer and Dorfman, 1997; Young, 2001).

SYSTEMS ARCHITECTING, DESIGN, AND INTEGRATION

Stakeholder Viewpoints

From a customer point of view, a system is everything that is required to meet the need of the customer to achieve some particular purpose. A system designer may view the system as the product or service described by the design specifications. A logistician may view a system as the system maintenance and logistics efforts. A systems architect may view the system as the structures and interfaces needed to represent the system in a conceptual fashion. Systems engineering is concerned with the total view of the system and encompasses each of these perspectives, and others. Systems engineering must necessarily relate to the enterprise or organization for which the system is being built. It is necessarily concerned with the process for building the system. It is concerned with the systems management and technical direction relative to the implementation agents that construct the system. It is concerned with the technologies and metrics associated with constructing the systems as

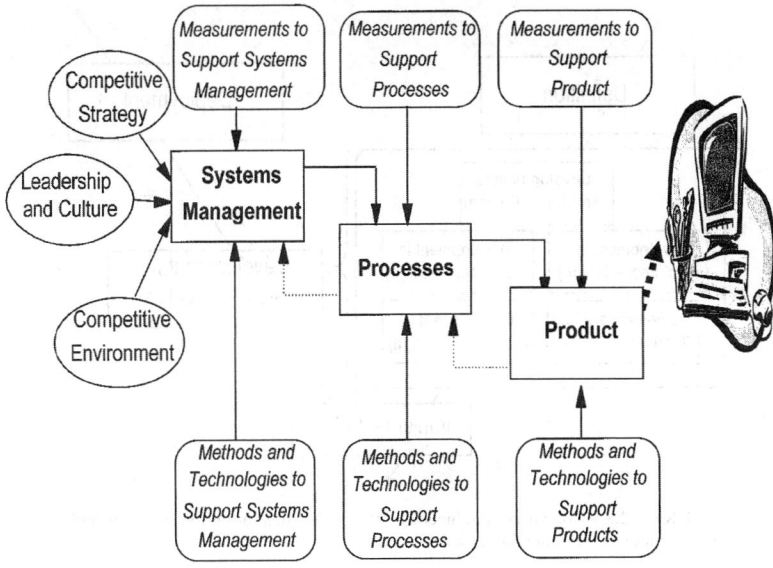

FIGURE 2.2.7   A model for engineering of a system.

well as those associated with the technical direction of this effort. It is concerned with the environment surrounding all of these. Thus, systems engineering is concerned with:

- People
- Processes
- Technologies, in the form of methods and tools
- Metrics
- Systems management
- Environments

All of these considerations are needed in order to engineer a system in the form of a product, service, or process that supports customer needs in an effective manner. Figure 2.2.7 (Sage, 1995) presents a model of system engineering that incorporates these elements. Thus, we see that a multiple perspective view of systems engineering and management is needed. Also needed is a multiple perspective view of systems architecting, design, and integration as an essential ingredient in systems engineering and management.

Integration Issues

Integration is defined in conventional dictionaries as the making of a whole entity by bringing all of the components of that entity together. In other words, integration involves blending, coordinating, or incorporating components into a functioning whole. This definition, while appropriate for systems integration, needs greater specificity in order to illustrate the many facets of integration. We may immediately think of product system integration. This refers to integration of the product or service, which is delivered to the customer. This is the most often used context for the term systems integration. To accomplish external systems integration effectively requires that new system elements be appropriately interfaced with existing, or legacy, systems components. Sometimes these legacy systems are sufficiently "stovepipe" in nature such that the integration with a new system is not possible without reengineering the legacy system. It also requires that it is possible to accomplish integration with later arriving, and potentially unplanned for initially, additional systems elements.

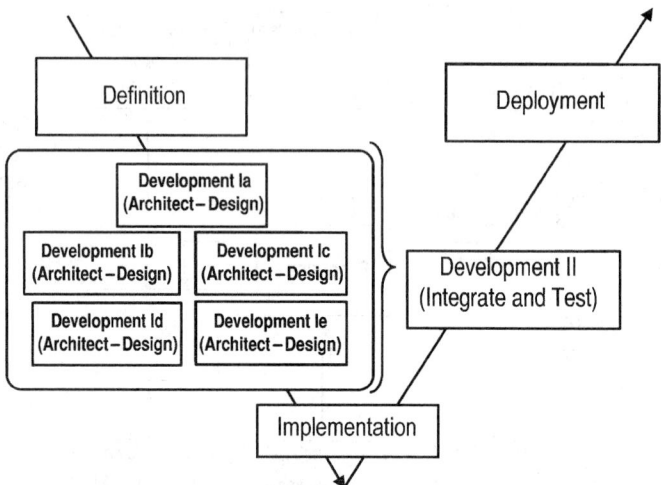

FIGURE 2.2.8   Concurrent engineering "V" model illustrating the need for integration following architecting and design.

This requires much attention to systems level architecting, and design and integration architectures (Sage and Lynch, 1998; Maier and Rechtin, 2001).

This need for system integration brings about a host of systems management and, in many cases, legal and regulatory issues that are much larger in scale and scope than those associated with product development only. In a similar manner, the development of appropriate system level architectures is very important in that efficiency and effectiveness in systems architecting is very influential of the ease with which systems can be integrated and maintained and, therefore, of the extent to which an operational system is viewed as trustworthy in satisfying user functional needs and nonfunctional needs related to quality and sustainability.

Functional, Physical, and Operational Architectures

As we have noted, the life cycle of a system comprises a number of phases that, when completed results in the satisfactory definition, development, and deployment of a system. Three phases are usually not sufficient (without each of them becoming vastly complicated) to represent the plethora of activities needed to properly engineer a system. There are such efforts as the engineering of the functional architecture, followed by the physical architecture, and following this the development of an implementation or operational architecture. This requires detailed design and associated production or manufacturing, operations and support, and other systems deployment efforts that may include disposal or reengineering. These need to be included as phased efforts in any realistic life cycle. Figure 2.2.8 recasts the basic three phase model for engineering a system into a "V" model that more explicitly shows the role of architecting, design and integration in the systems engineering life cycle. Sometimes, the initial efforts in the systems engineering life cycles of Fig. 2.2.8 are called "down-stroke" efforts and those in the latter part of the life cycle are called "up-stroke" efforts.

Not explicitly shown in Fig. 2.2.8 are the drivers for the engineering of a system and how these influence the need for systems integration. It is a rare situation that a new stand-alone system is called for and where this system does not have to be compatible with or interfaced to an existing, or legacy, system. There also exists such drivers for change as:

- A changing technology base because of the emergence of new and innovative technologies and improvements in existing technologies

- Changing organizational needs because of newly emerged competitors, the desire to acquire additional critical core capabilities

- Changing human needs because of the need for better knowledge management and enhanced organizational learning

Thus, we see that systems integration needs and considerations find importance in the:

- Planning effort that results in the very high level definition of a "new" system,
- Classical systems engineering and technical direction efforts that result in management controls for development, and
- Actual implementation engineering efforts that lead to realization of the new system.

It is also very important to determine whether some process or organizational restructuring, or reengineering, is needed to accommodate a new system, or whether a new system should be planned such as to accommodate existing organizational structure and functions. The effort associated with building a system must, if it is to be efficient and effective, consider not only the needs of the system using organization for the new system, but also the capabilities of the teams and organizations that, collectively, are responsible for engineering of the system. In effect, all of these ingredients should be integrated. These create the requirements for systems engineering and management efforts that are enterprise and knowledge management focused. These are major issues in systems and implementation of a system. Each of these views or perspectives needs to be accommodated when determining the architecture of a system to be engineered if the resulting system is to be trustworthy and, ultimately, used to fulfill its intended purpose.

SYSTEMS CONFIGURATION

Following the identification of an appropriate systems engineering process life cycle to use in engineering the system, configuration management plans are identified. This involves defining a specific development process for the set of life-cycle tasks at hand. Metrics of cost analysis or cost estimation are needed to enable this to be done effectively. Today, virtually every new system is software and information technology intensive. Thus, there are major roles for cost and economic estimation for software and information technology based systems. They also include effectiveness analysis or estimation of software productivity indices using various metrics. This couples the notion of development of an information technology or software (or other) product into notions concerning the process needs associated with developing this product. These metrics and indices form a part of a systems management approach for process, and ultimately, product improvement.

Critical aspects of a problem are often a function of how the components interact. Simple aggregation of individual aspects of a problem is intuitively appealing but often wrong. The whole is often not simply the sum of its parts. This does not suggest at all that scientific analysis, in which an issue is disaggregated into a number of component issues and understanding sought of the individual issues, is in any way improper. The formal approach involves three basic efforts:

1. Disaggregation or decomposition of a large issue into smaller, more easily understandable parts
2. Analysis of the resulting large number of individual issues
3. Aggregation of the results to attempt to find a solution to the major issue

APPROACH TO SYSTEMS MATURITY

Problems and issues arise in the course of systems engineering a product or service. Three organizational approaches may be useful in addressing them:

Reactive. Examines a potential issue only after it has developed into a real problem.

Interactive. Examines issues while they are evolving, diagnoses and corrects problems upon detection.

Proactive. Predicts the potential for debilitating issues/problems. Synthesizes a life-cycle process to minimize their likelihood.

While proactive and interactive efforts are associated with greater capability and process maturity, reactive efforts are still generally needed. Of course, another option is the "inactive" approach, favored by an organization that worries little about issues that may well become serious problems if not promptly dealt with.

COMMUNICATIONS AND COMPUTING

The role of communications and computing is ubiquitous in systems engineering and management. Major growth in power of computing and communicating and associated networking is quite fundamental and has changed relationships among people, organizations, and technology. These capabilities allow us to study much more complex issues than was formerly possible. They provide a foundation for dramatic increases in learning and associated increases in both individual and organizational effectiveness. In large part, this is because of the networking capability that enables enhanced coordination and communications among humans in organizations. It is also because of the vastly increased potential availability of knowledge to support individuals and organizations in their efforts.

The need for integration of information technology issues with organizational issues has led to the creation of a field of study, the objectives of which generally include:

- Capturing human information and knowledge needs in the form of system requirements and specifications
- Developing and deploying systems that satisfy these requirements
- Supporting the role of cross-functional teams in work
- Overcoming behavioral and social impediments to the introduction of information technology systems in organizations
- Enhancing human communication and coordination for effective and efficient workflow through knowledge management
- Encouraging human and organizational evolution and adaptation in coping with these challenges.

SPECIFIC APPLICATIONS

There are many specific applications as well as specific issues where systems engineering and management is germane. For example, when a system is to be developed and deployed using concurrent engineering techniques, as opposed to strictly sequential techniques, systems engineering and management is particularly important. Indeed, it may reveal that concurrent engineering is inappropriate to a particular project. Conversely, it may enable successful concurrent engineering of a system that might otherwise fail to meet its expected performance and deadline requirements. See Sage and Rouse (1999). Andrews and Leventhal (1993), Kronlof (1993), and Fiksel (1993).

Another issue on which systems engineering and management has a strong influence is that of physically partitioning a system. The factors bearing on this problem can perhaps best be expressed by these questions:

1. Where ought it to be partitioned from a design standpoint?
2. Where ought it to be partitioned from a development and prototyping standpoint?
3. Where ought it to be partitioned from a fabrication standpoint?
4. Where ought it to be partitioned from a deployment and maintainability standpoint?

The final decision will be a compromise, in which one or more of the factors may have a prevailing influence.

For elaborations and in-depth treatment of topics covered in this chapter, the reader is referred to the "Handbook of Systems Engineering and Management" (Sage and Rouse, 1999).

BIBLIOGRAPHY

Andrews, C. C., and N. S. Leventhal, "Fusion—Integrating IE, CASE, and JAD: A Handbook for Reengineering the Systems Organization," Prentice Hall, 1993.

Andriole, S. J. "Managing Systems Requirements: Methods, Tools and Cases," McGraw-Hill, 1996.

Buede, D. M., "The Engineering Design of Systems," Wiley, 2000.

Davis, G. B., "Strategies for information requirements determination," *IBM Syst. J.*, Vol. 21, No. 1, 1982.

Department of Defense, Directive Number 5000.1–The Defense Acquisition System, Instruction Number 5000.1 and 5000.2–Operation of the Defense Acquisition System, April, 2002. Available at http://www.acq.osd.mil/ar/#5000.

Fiksel, J., "Computer-aided requirements management for environmental excellence," *Proc. Ann. Mtg. Nat. Council Syst. Eng.*, 1993.

Kronlof, K. (ed.), "Method Integration: Concepts and Case Studies," Wiley, 1993.

Maier M. W., and E. Rechtin, "The Art of Systems Architecting," 2nd ed., CRC Press, 2001.

Pascale, R. T., "Surfing the edge of chaos," *Sloan Management Rev.*, Vol. 40, No. 3, Spring 1999.

Pascale, R. T., M. Millemann, and L. Grioja, "Surfing the Edge of Chaos: The Laws of Nature and the New Laws of Business," Crown Business, 2000.

Porter, M., "Strategy and the Internet," *Harvard Business Rev.*, Vol. 79, No. 3, March 2001.

Robertson, S., and J. Robertson, "Mastering the Requirements Process," Addison Wesley, 1999.

Rouse, W. B., "Don't Jump to Solutions: Thirteen Delusions That Undermine Strategic Thinking," Jossey-Bass, 1998.

Rouse, W. B., "Essential Challenges of Strategic Management," Wiley, 2001.

Sage, A. P., "Systems Engineering," Wiley, 1992.

Sage, A. P., "Systems Management for Information Technology and Software Engineering," Wiley, 1995.

Sage, A. P., "Towards systems ecology," *IEEE Computer*, Vol. 31, No. 2, February 1998.

Sage, A. P., and C. L. Lynch, "Systems integration and architecting: An overview of principles, practices, and perspectives," *Syst. Eng.*, Vol. 1, No. 3, 1998.

Sage, A. P., and W. B. Rouse (eds.), "Handbook of Systems Engineering and Management," Wiley, 1999.

Sage, A. P., and C. D. Cuppan, "On the systems engineering and management of systems of systems and federations of systems," *Information, Knowledge, and Systems Management*, Vol. 2, No. 4, 2001.

Sommerville, I., and P. Sawyer, "Requirements Engineering: A Good Practices Guide," Wiley, 1997.

Thayer, R. H., and M. Dorfman (eds.), "Software Requirements Engineering," IEEE Computer Society Press, 1997.

Young, R. H., "Effectiveness Requirements Practices," Addison Wesley, 2001.

SECTION 3

RELIABILITY

Reliability of a component or a system can be considered a design parameter, but, in fact, the intrinsic design reliability may not be realized because of manufacturing defects or misapplication. Thus, the definition of reliability is the probability that a device or system will perform a specified function for a specified period of time under a specific set of conditions. The conditions may also include required maintenance procedures. The first chapter in this section expands on the definitional issues of reliability, including parts and systems modeling, reliability theory and practice, and failure analysis and prediction techniques. Chapter 3.2 discusses very high reliability techniques developed for military and aerospace applications. The special issues involved in semiconductor component reliability are covered in Chap. 3.3.

In a completely new chapter (3.4) the exceptional considerations needed in predicting reliability when using electromechanical and microelectromechanical devices are treated. Although many of the stress-strength relationships applicable to conventional electromechanical components apply to microelectro-mechanical devices, too, differences in the reliability of microelectromechanial systems (MEMS) that may be related to materials, geometries, and failure mechanisms are still under study. Thus far there is a limited amount of parts failure data available; it will be augmented as MEMS technologies mature and as the application of MEMS broadens. Additional material related to this chapter is included on the accompanying CD-ROM.

The design and modeling of electronic systems is covered in greater detail in Chap. 3.5. Finally, in another completely new chapter (3.6), the special concerns of designing software and assuring its reliability are treated. Because military specifications and standards underlie the reliable design and operation of many electronic systems, a summary description of important military reliability documents is included on the CD-ROM.   D.C.

In This Section:

On the CD-ROM:

- *Reliability Standards and Handbooks*. A brief description of 8 MIL handbooks and 12 standards that cover sampling procedures, reliability growth, reliability prediction, reliability/design thermal applications, electrostatic discharge, stress screening, definition of terms, reliability modeling, design qualification, test methods, reliability of space and missile systems, FMECA techniques, failure reporting, analysis, and corrective action.

- *Commonly Used Formulas*. Formulas and equations useful in electromechanical reliability analysis.

- Concepts and Principles useful in probabilistic stress and strength analysis of electromechanical components.

CHAPTER 3.1
RELIABILITY DESIGN AND ENGINEERING

Ronald T. Anderson, Richard L. Doyle, Stanislaw Kus,
Henry C. Rickers, James W. Wilbur

RELIABILITY CONCEPTS AND DEFINITIONS

Intrinsic Reliability

The intrinsic reliability of a system, electronic or otherwise, is based on its fundamental design, but its reliability is often less than its intrinsic level owing to poor or faulty procedures at three subsequent stages: manufacture, operation, or maintenance.

Definitions

The definition of reliability involves four elements: *performance requirements, mission time, use conditions, and probability*. Although reliability has been variously described as "quality in the time dimension" and "system performance in the time dimension," a more specific definition is *the probability that an item will perform satisfactorily for a specified period of time under a stated set of use conditions.*

Failure rate, the measure of the number of malfunctions per unit of time, generally varies as a function of time. It is usually high but decreasing during its *early life*, or *infant-mortality* phase. It is relatively constant during its second phase, the *useful-life period*. In the third, *wear-out* or *end-of-life*, period the failure rate begins to climb because of the deterioration that results from physical or chemical reactions: oxidation, corrosion, wear, fatigue, shrinkage, metallic-ion migration, insulation breakdown, or, in the case of batteries, an inherent chemical reaction that goes to completion.

The failure rate of most interest is that which relates to the useful life period. During this time, reliability is described by the single-parameter exponential distribution

$$R(t) = e^{-\lambda t} \tag{1}$$

where $R(t)$ = probability that item will operate without failure for time t (usually expressed in hours) under stated operating conditions

e = base of natural logarithms = 2.7182

λ = item failure rate (usually expressed in failures per hour) = constant for any given set of stress, temperature, and quality level conditions

It is determined for parts and components from large-scale data-collection and/or test programs. When values of λ and t are inserted in Eq. (1), the *probability of success*, i.e., reliability, is obtained for that period of time.

The reciprocal of the failure rate $1/\lambda$ is defined as the *mean time between failures* (MTBF). The MTBF is a figure of merit by which one hardware item can be compared with another. It is a measure of the failure rate λ during the useful life period.

Reliability Degradation

Manufacturing Effects.   To access the magnitude of the reliability degradation because of manufacturing, the impact of manufacturing processes (process-induced defects, efficiency of conventional manufacturing and quality-control inspection, and effectiveness of reliability screening techniques) must be evaluated. In addition to the latent defects attributable to purchased parts and materials, assembly errors can account for substantial degradation. Assembly errors can be caused by operator learning, motivational, or fatigue factors.

Manufacturing and quality-control inspections and tests are provided to minimize degradation from these sources and to eliminate obvious defects. A certain number of defective items escaping detection will be accepted and placed in field operation. More importantly, the identified defects may be over-shadowed by an unknown number of *latent defects*, which can result in failure under conditions of stress, usually during field operation. Factory screening tests are designed to apply a stress of given magnitude over a specified time to identify these kinds of defects, but screening tests are not 100 percent effective.

Operational Effects.   Degradation in reliability also occurs as a result of system operation. Wear-out, with *aging* as the dominant failure mechanism, can shorten the useful life. Situations also occur in which a system may be called upon to operate beyond its design capabilities because of an unusual mission requirement or to meet a temporary but unforeseen requirement. These situations could have ill-effects on its constituent parts.

Operational abuses, e.g., rough handling, over stressing, extended duty cycles, or neglected maintenance, can contribute materially to reliability degradation, which eventually results in failure. The degradation can be a result of the interaction of personnel, machines, and environment. The translation of the factors which influence operational reliability degradation into corrective procedures requires a complete analysis of functions performed by personnel and machines plus fatigue and/or stress conditions which degrade operator performance.

Maintenance Effects.   Degradation in inherent reliability can also occur as a result of maintenance activities. Excessive handling from frequent preventive maintenance or poorly executed corrective maintenance, e.g., installation errors, degrades system reliability. Several trends in system design have reduced the need to perform adjustments or make continual measurements to verify peak performance. Extensive replacement of analog by digital circuits, inclusion of more built-in test equipment, and use of fault-tolerant circuitry are representative of these trends.

These factors, along with greater awareness of the cost of maintenance, have improved ease of maintenance, bringing also increased system reliability. In spite of these trends, the maintenance technician remains a primary cause of reliability degradation. Reliability is affected by poorly trained, poorly supported, or poorly motivated maintenance technicians where maintenance tasks require careful assessment and quantification.

Reliability Growth

Reliability growth represents the action taken to move a hardware item toward its reliability potential, during development or subsequent manufacturing or operation. During early development, the achieved reliability of a newly fabricated item or an off-the-board prototype is much lower than its predicted reliability because of initial design and engineering deficiencies as well as manufacturing flaws. The reliability growth process, when formalized and applied as an engineering discipline, allows management to exercise control of, allocate resources to, and maintain visibility of, activities designed to achieve a mature system before full production or field use.

Reliability growth is an iterative test-fail-correct process with three essential elements: detection and analysis of hardware failures, feedback and redesign of problem areas, and implementation of corrective action and retest.

Glossary

Availability. The availability of an item, under the combined aspects of its reliability and maintenance, to perform its required function at a stated instant in time.

Burn-in. The operation of items before their ultimate application to stabilize their characteristics and identify early failures.

Defect. A characteristic that does not conform to applicable specification requirements and that adversely affects (or potentially could affect) the quality of a device.

Degradation. A gradual deterioration in performance as a function of time.

Derating. The intentional reduction of stress-strength ratio in the application of an item, usually for the purpose of reducing the occurrence of stress-related failures.

Downtime. The period of time during which an item is not in a condition to perform its intended function.

Effectiveness. The ability of the system or device to perform its function.

Engineering reliability. The science that takes into account those factors in the basic design that will assure a required level of reliability.

Failure. The inability (more precisely termination of the ability) of an item to perform its required function.

Failure analysis. The logical, systematic examination of an item or its diagram(s) to identify and analyze the probability, causes, and consequences of potential and real failures.

Failure, catastrophic. A failure that is both sudden and complete.

Failure mechanism. The physical, chemical, or other process resulting in a failure.

Failure mode. The effect by which a failure is observed, e.g., an open or short circuit.

Failure, random. A failure whose cause and/or mechanism makes its time of occurrence unpredictable but that is predictable in a probabilistic or statistical sense.

Failure rate. The number of failures of an item per unit measure of life (cycles, time, etc.); during the useful life period, the failure rate λ is considered constant.

Failure rate change ($\dot{\lambda}$). The change in failure rate of an item at a given point in its life; $\dot{\lambda}$ is zero for an exponential distribution (constant failure rate), but represents the slope of the failure rate curve for more complex reliability distributions.

Failure, wear-out. A failure that occurs as a result of deterioration processes or mechanical wear and whose probability of occurrence increases with time.

Hazard rate Z(t). At a given time, the rate of change of the number of items that have failed divided by the number of items surviving.

Maintainability. A characteristic of design and installation that is expressed as the probability that an item will be retained in, or restored to, a specified condition within a given time when the maintenance is performed in accordance with prescribed procedures and resources.

Mean maintenance time. The total preventive and corrective maintenance time divided by the number of preventive and corrective maintenance actions accomplished during the maintenance time.

Mean time between failures (MTBF). For a given interval, the total functioning life of a population of an item divided by the total number of failures in the population during the interval.

Mean time between maintenance (MTBM). The mean of the distribution of the time intervals between maintenance actions (preventive, corrective, or both).

Mean time to repair (MTTR). The total corrective-maintenance time divided by the total number of corrective-maintenance actions accomplished during the maintenance time.

Redundancy. In an item, the existence of more than one means of performing its function.

Redundancy, active. Redundancy in which all redundant items are operating simultaneously rather than being switched on when needed.

Redundancy, standby. Redundancy in which alternative means of performing the function are inoperative until needed and are switched on upon failure of the primary means of performing the function.

Reliability. The characteristic of an item expressed by the probability that it will perform a required function under stated conditions for a stated period of time.

Reliability, inherent. The potential reliability of an item present in its design.

Reliability, intrinsic. The probability that a device will perform its specified function, determined on the basis of a statistical analysis of the failure rates and other characteristics of the parts and components that constitute the device.

Screening. The process of performing 100 percent inspection on product lots and removing the defective units from the lots.

Screening test. A test or combination of tests intended to remove unsatisfactory items or those likely to exhibit early failures.

Step stress test. A test consisting of several stress levels applied sequentially for periods of equal duration to a sample. During each period, a stated stress level is applied, and the stress level is increased from one step the next.

Stress, component. The stresses on component parts during testing or use that affect the failure rate and hence the reliability of the parts. Voltage, power, temperature, and thermal environmental stress are included.

Test-to-failure. The practice of inducing increased electrical and mechanical stresses in order to determine the maximum capability. A device in conservative use will increase its life through the derating based on these tests.

Time, down (downtime). See downtime.

Time, mission. The part of uptime during which the item is performing its designated mission.

Time, up (uptime). The element of active time during which an item is alert, reacting, or performing a mission.

Uptime ratio. The quotient determined by dividing uptime by uptime plus downtime.

Wear-out. The process of attrition which results in an increased failure rate with increasing age (cycles, time, miles, events, and so forth, as applicable for the item).

RELIABILITY THEORY AND PRACTICE

Exponential Failure Model

The life-characteristic curve (Fig. 3.1.1) can be defined by three failure components that predominate during the three periods of an item's life. The shape of this curve suggests the usual term *bathtub curve.* The components are illustrated in terms of an *equipment failure rate.* The failure components include:

1. *Early failures* because of design and quality-related manufacturing, which have a decreasing failure rate.
2. *Stress-related failures* because of application stresses, which have a constant failure rate.
3. *Wear-out failures* because of aging and/or deterioration, which have an increasing failure rate.

From Fig. 3.1.1 three conclusions can be drawn: (1) that the *infant-mortality* period is characterized by a high but rapidly decreasing failure rate that comprises a high quality-failure component, a constant-stress-related failure component, and a low wear-out-failure component. (2) The *useful-life* period is characterized by a constant failure rate comprising a low (and decreasing) quality-failure component, a constant stress-related-failure component, and a low (but increasing) wear-out-failure component. The combination of these three components results in a nearly constant failure rate because the decreasing quality failures and increasing wear-out failures tend to offset each other and because the stress-related failures exhibit a relatively larger amplitude. (3) The *wear-out period* is characterized by an increasing failure rate comprising a negligible quality-failure component, a constant stress-related-failure component, and an initially low but rapidly increasing wear-out-failure component.

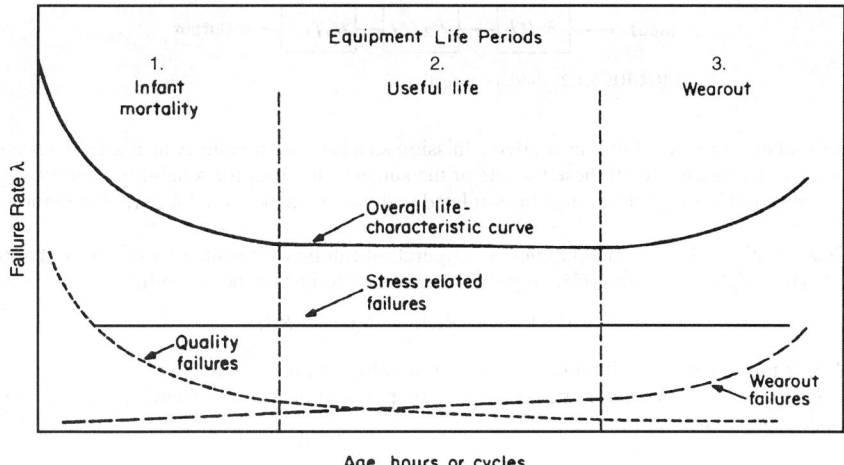

FIGURE 3.1.1   Life-characteristic curve, showing the three components of failure (when superimposed, the three failures provide the top curve).

The general approach to reliability for electronic systems is to minimize early failures by emphasizing factory test and inspection and to prevent wear-out failures by replacing short-lived parts before they fail. Consequently, the useful life period characterized by stress-related failures is the most important period and the one to which design attention is primarily addressed.

Figure 3.1.1 illustrates that during the useful life period the failure rate is constant. A constant failure rate is described by the exponential-failure distribution. Thus, the exponential-failure model reflects the fact that the item must represent a mature design whose failure rate, in general, is primarily because of stress-related failures. The magnitude of this failure rate is directly related to the stress-strength ratio of the item.

The validity of the exponential reliability function, Eq. (1), relates to the fact that the failure rate (or the conditional probability of failure in an interval given at the beginning of the interval) is independent of the accumulated life.

The use of this type of "failure law" for complex systems is judged appropriate because of the many forces that can act on the item and produce failure. The stress-strength relationship and varying environmental conditions result in effectively random failures. However, this "failure law" is not appropriate if parts are used in their wear-out phase (Phase 3 of Fig. 3.1.1).

The approach to randomness is aided by the mixture of part ages that results when failed elements in the system are replaced or repaired. Over time the system failure rate oscillates, but this cyclic movement diminishes in time and approaches a stable state with a constant failure rate.

Another argument for assuming the exponential distribution is that if the failure rate is essentially constant, the exponential represents a good approximation of the true distribution over a particular interval of time. However, if parts are used in their wear-out phase, then a more sophisticated failure distribution must be considered.

System Modeling

To evaluate the reliability of systems and equipment, a method is needed to reflect the *reliability connectivity* of the many part types having different stress-determined failure rates that would normally make up a complex equipment. This is accomplished by establishing a relationship between equipment reliability and individual part or item failure rates.

Before discussing these relationships, it is useful to discuss system reliability objectives. For many systems, reliability must be evaluated from the following three separate but related standpoints: reliability as it

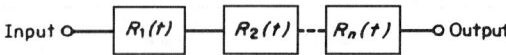

FIGURE 3.1.2   Serial connectivity.

affects personnel safety, reliability as it affects mission success, and reliability as it affects unscheduled maintenance or logistic factors. In all these aspects of the subject, the rules for reliability connectivity are applicable. These rules imply that failures are stress-related and that the exponential failure distribution is applicable.

Serial Connectivity.   The serial equipment configuration can be represented by the block diagram, shown in Fig. 3.1.2. The reliability of the *series configuration* is the product of the reliabilities of the individual blocks

$$R_s(t) = R_1(t)R_2(t) \cdots R_i(t) \cdots R_n(t) \tag{2}$$

where $R_s(t)$ is the series reliability and $R_i(t)$ is the reliability of ith block for time t.

The concept of constant failure rate allows the computation of system reliability as a function of the reliability of parts and components:

$$R(t) = \prod_{i=1}^{n} e^{-\lambda_i^t} = e^{-\lambda_1 t^-} e^{-\lambda_2 t} \cdots e^{-\lambda_n t} \tag{3}$$

This can be simplified to

$$R(t) = e^{-(\lambda_1 t + \lambda_2 t + \cdots + \lambda_n t)} = e^{-(\lambda_1 + \lambda_2 + \cdots + \lambda_n)t} \tag{4}$$

The general form of this expression can be written

$$R(t) = \exp\left[-t \sum_{t=1}^{n} \lambda_i \right] \tag{5}$$

Another important relationship is obtained by considering the jth subsystem failure rate λ_j to be equal to the sum of the individual failure rates of the n independent elements of the subsystems such that

$$\lambda_j = \sum_{i=1}^{n} \lambda_i \tag{6}$$

Revising the MTBF formulas to refer to the system rather than an individual element gives the *mean time between failures* of the system as

$$\text{MTBF} = \frac{1}{\lambda_j} = \frac{1}{\displaystyle\sum_{i=1}^{n} \lambda_i} \tag{7}$$

Successive estimates of the jth subsystem failure rate can be made by combining lower-level failure rates using

$$\lambda_j = \sum_{i=1}^{n} \lambda_{ij} \quad j = 1, \ldots, m \tag{8}$$

where λ_{ij} is the failure rate of the ith component in the jth-level subsystem and λ_j is the failure rate of the jth-level subsystem.

Parallel Connectivity.   The more complex configuration consists of equipment items or parts operating both in series and parallel combinations, together with the various permutations. A parallel configuration accounts for the fact that alternate part or item configurations can be designed to ensure equipment success by redundancy.

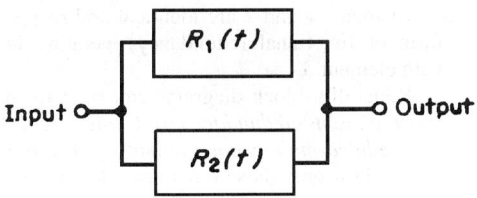

FIGURE 3.1.3 Parallel connectivity.

A two-element parallel reliability configuration is represented by the block diagram in Fig. 3.1.3. To evaluate the reliability of parallel configurations, consider, for the moment, that a reliability value (for any configuration) is synonymous with probability, i.e., probability of successful operation, and can take on values ranging between 0 and 1. If we represent the reliability by the symbol R and its complement $1 - R$, that is, unreliability, by the symbol Q, then from the fundamental notion of probability,

$$R + Q = 1 \quad \text{and} \quad R = 1 - Q \tag{9}$$

From Eq. (9) it can be seen that a probability can be associated with successful operation (reliability) as well as with failure (unreliability). For a single block (on the block diagram) the above relationship is valid. However, in the two-element parallel reliability configuration shown in Fig. 3.1.3, two paths for successful operation exist, and the above relationship becomes

$$(R_1 + Q_1)(R_2 + Q_2) = 1 \tag{10}$$

Assuming that $R_1 = R_2$ and $Q_1 = Q_2$, that is, the blocks are identical, this can be rewritten as

$$(R + Q)^2 = 1 \tag{11}$$

Upon expansion, this becomes

$$R^2 + 2RQ + Q^2 = 1 \tag{12}$$

We recall that reliability represents the probability of successful operation. This condition is represented by the first two terms of Eq. (12). Thus, the *reliability of the parallel configuration* can be represented by

$$R_p = R^2 + 2RQ \tag{13}$$

Note that either both branches are operating successfully (the R^2 term) or one has failed while the other operates successfully (the $2RQ$ term).

Substituting the value of $R = 1 - Q$ into the above expression, we obtain

$$R_p = (1 - Q)^2 + 2(1 - Q)Q = 1 - 2Q + Q^2 + 2Q - 2Q^2 = 1 - Q^2 \tag{14}$$

To obtain an expression in terms of reliability only, the substitution $Q = 1 - R$ can be made, which yields

$$R_p = 1 - (1 - R)(1 - R) \tag{15}$$

The more general case where $R_1 \neq R_2$ can be expressed

$$R_p = 1 - (1 - R_1)(1 - R_2) \tag{16}$$

By similar reasoning it can be shown that for n blocks connected in a parallel reliability configuration, the reliability of the configuration can be expressed by

$$R_p(t) = 1 - (1 - R_1)(1 - R_2) \cdots (1 - R_n) \tag{17}$$

The series and parallel reliability configurations (and combinations of them), as described above in Eqs. (5) and (17), are basic models involved in estimating the reliability of complex equipment.

Redundancy. The serial and parallel reliability models presented in the preceding paragraphs establish the mathematical framework for the reliability connectivity of various elements. Their application can be illustrated to show both the benefits and penalties of redundancy when considering safety, mission, and unscheduled maintenance reliability. Simplified equipment composed of three functional elements (Fig. 3.1.4) can be used to illustrate the technique.

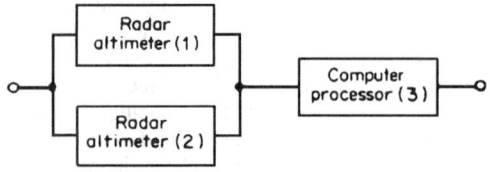

FIGURE 3.1.4   Serial and parallel connectivity.

Elements 1 and 2 are identical and represent one form of functional redundancy operating in series with element 3.

Reliability block diagrams can be defined corresponding to *nonredundant serial, safety, mission,* and *unscheduled maintenance* reliability. The block diagrams show only those functional elements that must operate properly to meet that particular reliability requirement. Figure 3.1.5 depicts the various block diagrams, reliability formulas, and typical values corresponding to these requirements. It indicates that the use of redundancy provides a significant increase in safety and mission reliability above that of a serial or nonredundant configuration; however, it imposes a penalty by adding an additional serial element in the scheduled maintenance chain.

Part-Failure Modeling

The basic concept that underlies reliability prediction and the calculation of reliability numerics is that system failure is a reflection of *part failure.* Therefore, a method for estimating part failure is needed. The most direct approach to estimating part-failure rates involves the use of large-scale data-collection efforts to obtain the relationships, i.e., models, between engineering and reliability variables. The approach uses controlled test

Reliability requirement	Reliability block diagram	Calculated values
1. Serial (nonredundant) reliability	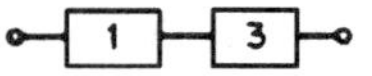 [1]—[3]	$R = R_1 R_3 = 0.842$ MTBF = 575 hr
2. Safety (or mission) reliability	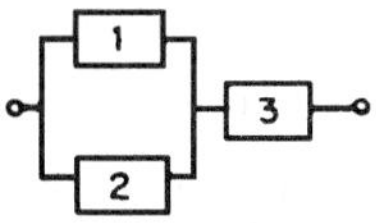 [1] and [2] parallel, then [3]	$R = \left[2R_1 - R_1^2\right] R_3$ $= 0.97$
3. Unscheduled maintenance reliability	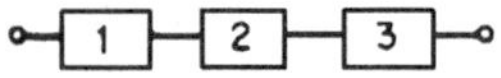 [1]—[2]—[3]	$R = R_1 R_2 R_3 = 0.715$ MTBF = 298 hr

$$R_{\text{Serial}} = R_1 \cdot R_2 \cdots R_n$$

where

$$R_n = e^{-\lambda_n t}$$

$$\text{MTBF} = \frac{1}{\lambda_n}$$

$$R_{\text{Parallel}} = 1 - (1 - R)(1 - R)$$
$$= 2R - R^2$$

$R_1 = 0.85$
$R_2 = 0.85$
$R_3 = 0.99$
$t = 100 \text{ h}$

FIGURE 3.1.5   Calculations for system reliability.

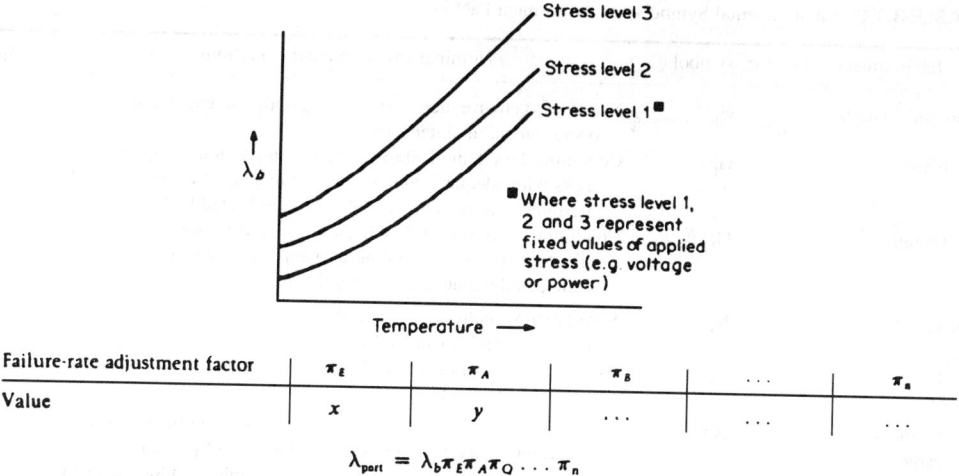

$$\lambda_{\text{part}} = \lambda_b \pi_E \pi_A \pi_Q \cdots \pi_n$$

FIGURE 3.1.6   Conceptual part-failure model.

data to derive relationships between design and generic reliability factors and to develop factors for adjusting the reliability to estimate field reliability when considering application conditions.

These data have been reduced through *physics-of-failure techniques and empirical data* and are included in several different failure rate databases. Some of the more common failure rate databases are: MIL-HDBK-217, Telcordia/Bellcore SR-332, CNET's RDF 2000, British Telecom's database, Reliability Analysis Center's PRISM, and IEEE STD 493. All are suitable for estimating stress-related failure rates. Some even include thermal cycling, on/off switching, and dormant (shelf time) effects. This section will use MIL-HDBK-217 as an example, but all databases provide excellent guidance during design and allow individual part failure rates to be combined within a suitable system reliability model to arrive at an estimate of system reliability.

Although part-failure models (Fig. 3.1.6) vary with different part types, their general form is

$$\lambda_{\text{part}} = \lambda_b \pi_E \pi_A \pi_Q \cdots \pi_n \tag{18}$$

where λ_{part} = total part-failure rate
λ_b = base or generic failure rate
π's = adjustment factors

The value of λ_b is obtained from reduced part-test data for each generic part category, where the data are generally presented in the form of failure rate versus normalized stress and temperature factors. The part's primary-load stress factor and its factor of safety are reflected in this basic failure-rate value. As shown in Fig. 3.1.6, the value of λ_b is generally determined by the anticipated stress level, e.g., power and voltage, at the expected operating temperature. These values of applied stress (relative to the part's rated stress) represent the variables over which design control can be exercised and which influence the item's ultimate reliability.

π_E is the environmental adjustment factor which accounts for the influences of environments other than temperature; it is related to the operating conditions (vibration, humidity, and so forth) under which the item must perform. These environmental classes have been defined in MIL-HDBK-217. Table 3.1.1 defines each class in terms of its nominal environmental conditions. Depending on the specific part type and style, the value of π_E may vary from 0.2 to 120. The missile-launch environment is usually the most severe and generally dictates the highest value of π_E. Values of π_E for monolithic microelectronic devices have been added to Table 3.1.1 to characterize this range for a particular part type.

π_A is the application adjustment factor. It depends on the application of the part and takes into account secondary stress and application factors considered to be reliability-significant.

π_Q is the quality adjustment factor, used to account for the degree of manufacturing control with which the part was fabricated and tested before being shipped in the user. Many parts are covered by specifications that

TABLE 3.1.1 Environmental Symbols and Adjustment Factors

Environment	π_E symbol	Nominal environmental conditions	π_E value[*]
Ground, benign	G_B	Nearly zero environmental stress with optimum engineering operation and maintenance	0.5
Fixed	G_F	Conditions less than ideal to include installation in adequate racks with adequate cooling air, maintenance by military personnel and possible installation in unheated buildings	2.0
Mobile	G_M	Conditions more severe than those for G_1 mostly for vibration and shock; cooling-air supply may also be more limited and maintenance less uniform	4.0
Naval, sheltered	N_S	Surface ship conditions similar to G_F but subject to occasional high shock and vibration	4.0
Unsheltered	N_U	Nominal suface shipborne conditions but with repetitive high levels of shock and vibration	6.0
Airborne, inhabited, cargo	A_IC	Typical conditions in cargo compartments and can be occupied by aircrew without environmental extremes of pressure, temperature, shock, and vibration and installed on long-mission aircraft such as transports and bombers	4.0
Fighter	A_{IF}	Same as A_{IC} but installed on high-performance aircraft such as fighters and interceptors	5.0
Uninhabited, cargo	A_{UC}	Bomb bay, equipment bay, tail, or wing installations where extreme pressure, vibration, and temperature cycling may be aggravated by contamination from oil, hydraulic fluid, and engine exhaust; installed on long-mission aircraft such as transports and bombers	5.0
Fighter	A_{UF}	Same as A_{UC} but installed on high-performance aircraft such as fighters and interceptors	8.0
Space, flight	S_F	Earth orbitral; approaches G_B conditions without access for maintenance; vehicle neither under powered flight nor in atmospheric reentry	0.5
Airborne, rotary winged	A_{RW}	Equipment installed on helicopters	8.0
Missile, flight	M_F	Conditions related to powered flight of air-breathing missiles, cruise missiles, and missiles in unpowered free flight	5.0
Cannon, launch	C_L	Extremely severe conditions related to cannon launching of 155 mm and 5 in guided projectiles; conditions also apply to projectile from launch to target impact	220.0
Missile, launch	M_L	Severe conditions of noise, vibration, and other environments related to missile launch and space-vehicle boost into orbit, vehicle reentry, and landing by parachute; conditions may also apply to installation near main rocket engines during launch operations	12.0

[*]Values for monolithic microelectronic devices.

have several quality levels. Several parts have multilevel quality specifications. Values of π_Q relate to both the generic part and its quality level.

π_N is the symbol for a number of additional adjustment factors that account for cyclic effects, construction class, and other influences on failure rate.

The data used as the basis of MIL-HDBK-217 consisted of both controlled test data and field data. The controlled test data directly related stress-strength variables on a wide variety of parts and were suitable for establishing the base failure rates λ_b.

TABLE 3.1.2 Representative Part-Failure-Rate Calculations

Value	Microcircuits, Gate/Logic Arrays and Microprocessors $\lambda_p = \pi_1\pi_Q(C_1\pi_T + C_2\pi_E)$	Fixed resistor $\lambda_p = \lambda_b\pi_T\pi_P\pi_S\pi_Q\pi_E$	Fixed capacitor $\lambda_p = \lambda_b\pi_T\pi_C\pi_V\pi_{SR}\pi_Q\pi_E$
λ_b	....	0.0017	0.00099
π_E	6.0	4.0	20.0
π_Q	2.0	3.0	1.0
π_L	1.0		
π_T	1.9	1.1	1.9
C_1	0.005		
C_2	0.002		
π_C	....	....	0.81
π_V	....	....	1.6
π_{SR}	....	....	1.0
π_S	....	1.5	
π_P	....	0.44	
$\lambda_p \times 10^{-6}$	0.043	0.015	0.049

MIL-HDBK-217 completely describes failure-rate models, failure-rate data, and adjustment factors to be used in estimating the failure rate for the individual generic part types. Table 3.1.2 presents a tabulation of several models, their base failure rates λ_b, associated π factors, and failure-rate values for several representative part types. The specific procedures for deriving the failure rates differ according to part class and type.

RELIABILITY EVALUATION

Summary

Reliability prediction, failure mode and effects analysis (FMEA), and reliability growth techniques represent prediction and design evaluation methods that provide a quantitative measure of how reliably a design will perform. These techniques help determine where the design can be improved. Since specified reliability goals are often contractual requirements that must be met along with functional performance requirements, these quantitative evaluations must be applied during the design stage to guarantee that the equipment will function as specified for a given duration under the operational and environmental conditions of intended use.

Prediction Techniques

Reliability prediction is the process of quantitatively assessing the reliability of a system or equipment during its development, before large-scale fabrication and field operations. During design and development, predictions serve as quantitative guides by which design alternatives can be judged for reliability. Reliability predictions also provide criteria for reliability growth and demonstration testing, logistics cost studies, and various other development efforts.

Thus, reliability prediction is a key to system development and allows reliability to become an integral part of the design process. To be effective, the prediction technique must relate engineering variables (the language of the designer) to reliability variables (the language of the reliability engineer).

A prediction of reliability is obtained by determining the reliability of the item at the lowest system level and proceeding through intermediate levels until an estimate of system reliability is obtained. The prediction method depends on the availability of accurate *evaluation models* that reflect the reliability connectivity of lower-level items and substantial *failure data* that have been analyzed and reduced to a form suitable for application to low-level items.

Various formal prediction procedures are based on theoretical and statistical concepts that differ in the level data on which the prediction is based. The specific steps for implementing these procedures are described in detail in reliability handbooks. Among the procedures available are parts-count methods and stress-analysis techniques. Failure data for both models are available in most reliability data bases like MIL-HDBK-217.

Parts-Count Method.   The parts-count method provides an estimate of reliability based on a count by part type (resistor, capacitor, integrated circuit, transistor, as so forth). This method is applicable during proposal and early design studies where the degree of design detail is limited. It involves counting the number of parts of each type, multiplying this number by a generic failure rate for each part type, and summing up the products to obtain the failure rate of each functional circuit, subassembly, assembly, and/or block depicted in the system block diagram.

The advantage of this method is that it allows rapid estimates of reliability to determine quickly the feasibility (from the reliability standpoint) of a given design approach. The technique uses information derived from available engineering information and does not require detailed part-by-part stress and design data.

Stress-Analysis Method.   The stress-analysis technique involves the same basic steps as the parts-count technique but requires detailed part models plus calculation of circuit stress values for each part before determining its failure rate. Each part is evaluated in its electric-circuit and mechanical-assembly application based on an electrical and thermal stress analysis. Once part-failure rates have been established, a combined failure rate for each functional block in the reliability diagram can be determined.

To facilitate calculation of part-failure rates, worksheets based on part-failure-rate models are normally prepared to help in the evaluation. These worksheets are prepared for each functional circuit in the system. When completed, these sheets provide a tabulation of circuit part data, including part description, electrical stress factors, thermal stress factors, basic failure rates, the various multiplying or additive environmental and quality adjustment factors, and the final combined part-failure rates. The variation in part stress factors (both electrical and environmental) resulting from changes in circuits and packaging is the means by which reliability is controlled during design. Considerations for, and effects of, reduced stress levels (derating) that result in lower failure rates are treated in the chapter "Derating Factors and Application Guidelines."

Failure Analysis

Failure mode and effects analysis is an iterative documented process performed to identify basic faults at the part level and determine their effects at higher levels of assembly. The analysis can be performed with actual failure modes from field data or hypothesized failure modes derived from design analyses, reliability-prediction activities, and experience of how parts fail. In their most complete form, failure modes are identified at the part level, which is usually the lowest level of direct concern to the equipment designer. In addition to providing insight into failure cause-and-effect relationships, the FMEA provides the discipline method for proceeding part by part through the system to assess failure consequences.

Failure modes are analytically induced into each component, and failure effects are evaluated and noted, including severity and frequency (or probability) of occurrence. As the first mode is listed, the corresponding effect on performance at the next higher level of assembly is determined. The resulting failure effect becomes, in essence, the failure mode that affects the next higher level.

Iteration of this process results in establishing the ultimate effect at the system level. Once the analysis has been performed for all failure modes, each effect or symptom at the system level usually may be caused by several different failure modes at the lowest level. This relationship to the end effect provides the basis for grouping the lower-level failure modes.

Using this approach, probabilities of the occurrence of the system effect can be calculated, based on the probability of occurrence of the lower-level failure modes, i.e., modal failure rate times time. Based on these probabilities and a severity factor assigned to the various system effects, a *criticality number* can be calculated. Criticality numerics provide a method of ranking the system-level effects derived previously and the basis for corrective-action priorities, engineering-change proposals, or field retrofit actions.

Fault-Tree Analysis. Fault-tree analysis (FTA) is a tool that lends itself well to analyzing failure modes found during design, factory test, or field data returns. The fault-tree-analysis procedure can be characterized as an iterative documented process of a systematic nature performed to identify basic faults, determine their causes and effects, and establish their probabilities of occurrence.

The approach involves several steps, among which is the structuring of a highly detailed logic diagram that depicts basic faults and events that can lead to system failure and/or safety hazards. Then follows the collection of basic fault data and failure probabilities for use in computation. The next step is the use of computational techniques to analyze the basic faults, determine failure-mode probabilities, and establish criticalities. The final step involves formulating corrective suggestions that, when implemented, will eliminate or minimize faults considered critical. The steps involved, the diagrammatic elements and symbols, and methods of calculation are shown in Fig. 3.1.7.

This procedure can be applied at any time during a system's life cycle, but it is considered most effective when applied (1) during preliminary design, on the basis of design information and a laboratory or engineering test model, and (2) after final design, before full-scale production, on the basis of manufacturing drawings and an initial production model.

The first of these (in preliminary design) is performed to identify failure modes and formulate general corrective suggestions (primarily in the design area). The second is performed to show that the system, as manufactured, is acceptable with respect to reliability and safety. Corrective actions or measures, if any, resulting from the second analysis would emphasize controls and procedural actions that can be implemented with respect to the "as manufactured" design configuration.

The outputs of the analysis include:

1. A detailed logic diagram depicting all basic faults and conditions that must occur to result in the hazardous condition(s) under study.

2. A probability-of-occurrence numeric value for each hazardous condition under study.

3. A detailed fault matrix that provides a tabulation of all basic faults, their occurrence probabilities and criticalities, and the suggested change or corrective measures involving circuit design, component-part selection, inspection, quality control, and so forth, which, if implemented, would eliminate or minimize the hazardous effect of each basic fault.

Reliability Testing

Reliability Growth. Reliability growth is generally defined as the improvement process during which hardware reliability increases to an acceptable level. The measured reliability of newly fabricated hardware is much less than the potential reliability estimated during design, using standard handbook techniques. This definition encompasses not only the technique used to graph increases in reliability, i.e., "growth plots," but also the management-resource-allocation process that causes hardware reliability to increase.

The purpose of a growth process, especially a reliability-growth test, is to achieve acceptable reliability in field use. Achievement of acceptable reliability depends on the extent to which testing and other improvement techniques have been used during development to "force out" design and fabrication flaws and on the rigor with which these flaws have been analyzed and corrected.

A primary objective of growth testing is to provide methods by which hardware reliability development can be dimensioned, disciplined, and managed as an integral part of overall development. Reliability-growth testing also provides a technique for extrapolating the current reliability status (at any point during the test) to some future result. In addition, it provides methods for assessing the magnitude of the test-fix-retest effort before the start of development, thus making trade-off decisions possible. Many of the models for reliability growth represent the reliability of the system as it progresses during the overall development program. Also, it is commonly assumed that these curves are nondecreasing; i.e., once the system's reliability has reached a certain level, it will not drop below that level during the remainder of the development program. This assumes that any design or engineering changes made during the development program do not decrease the system's reliability.

For complex electronic and electromechanical avionic systems, a model traditionally used for reliability-growth processes, and in particular reliability-growth testing, is one originally published by Duane (1964). It

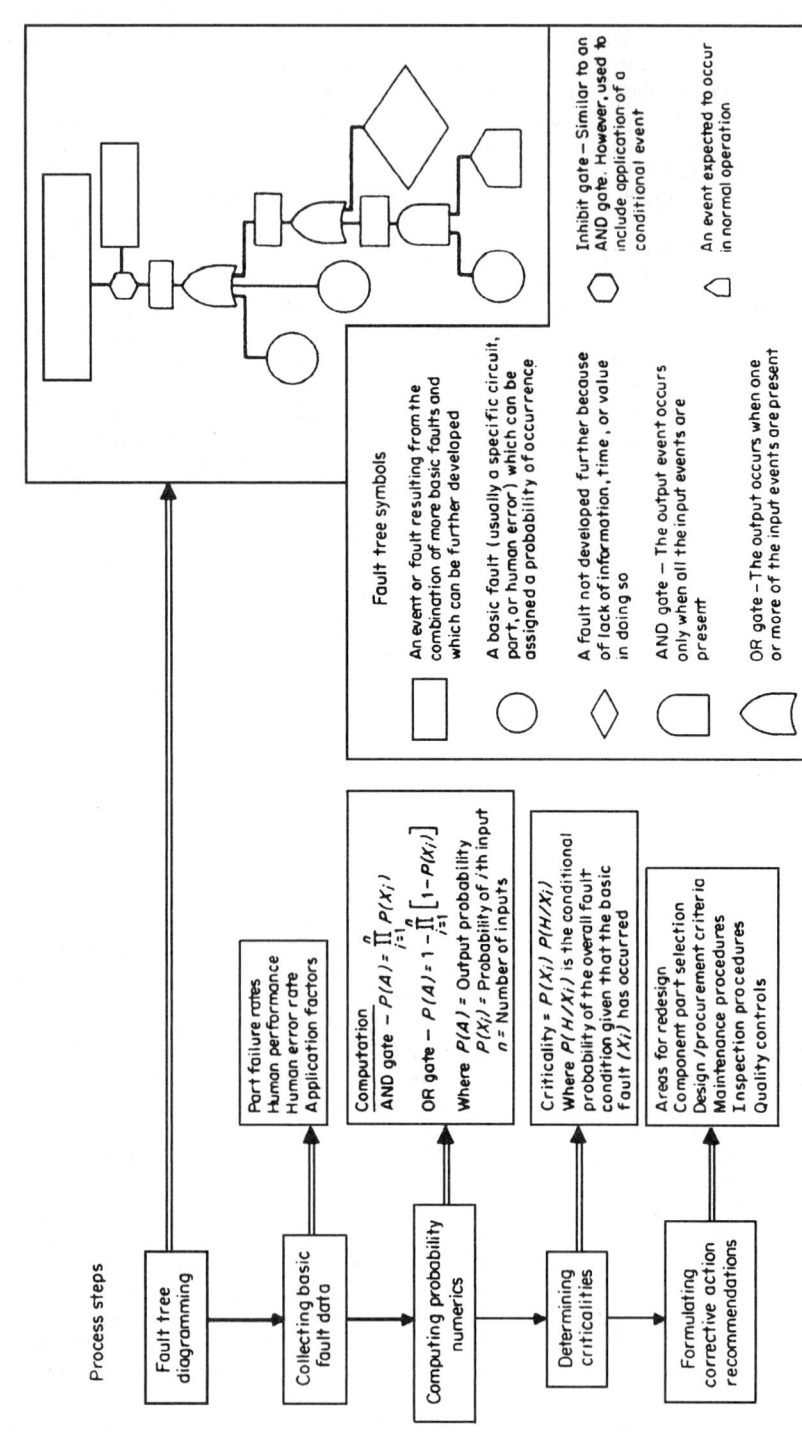

FIGURE 3.1.7 Fault-tree analysis.

Process steps

- Fault tree diagraming
- Collecting basic fault data
- Computing probability numerics
- Determining criticalities
- Formulating corrective action recommendations

Part failure rates
Human performance
Human error rate
Application factors

Computation
AND gate — $P(A) = \prod\limits_{i=1}^{n} P(X_i)$

OR gate — $P(A) = 1 - \prod\limits_{i=1}^{n} \left[1 - P(X_i)\right]$

Where $P(A)$ = Output probability
$P(X_i)$ = Probability of i'th input
n = Number of inputs

Criticality = $P(X_i) \, P(H/X_i)$
Where $P(H/X_i)$ is the conditional probability of the overall fault condition given that the basic fault (X_i) has occurred

Areas for redesign
Component part selection
Design/procurement criteria
Maintenance procedures
Inspection procedures
Quality controls

Fault tree symbols

An event or fault resulting from the combination of more basic faults and which can be further developed

A basic fault (usually a specific circuit, part, or human error) which can be assigned a probability of occurrence.

A fault not developed further because of lack of information, time, or value in doing so

AND gate — The output event occurs only when all the input events are present

OR gate — The output occurs when one or more of the input events are present

Inhibit gate — Similar to an AND gate. However, used to include application of a conditional event

An event expected to occur in normal operation

provides a deterministic approach to reliability growth such that the system MTBF versus operating hours falls along a straight line when plotted on log-log paper. That is, the change in MTBF during development is proportional to T^d, where T is the cumulative operating time and d is the rate of growth corresponding to the rapidity with which faults are found and changes made to permanently eliminate the basic faults observed.

To structure a growth test program (based on the Duane model) for a newly designed system, a detailed test plan is necessary. This plan must describe the test-fix-retest concept and show how it will be applied to the system hardware under development. The plan requires the following:

1. Values for specified and predicted (inherent) reliabilities. Methods for predicting reliability (model, database, and so forth) must also be described.

2. Criteria for reliability starting points, i.e., criteria for estimating the reliability of initially fabricated hardware. For avionics systems, the initial reliability for newly fabricated systems has been found to vary between 10 and 30 percent of their predicted (inherent) values.

3. Reliability-growth rate (or rates). To support the selected growth rate, the rigor with which the test-fix-retest conditions are structured must be completely defined.

4. Calendar-time efficiency factors, which define the relationship of test time, corrective-action time, and repair time to calendar time.

Each of the factors listed above affects the total time (or resources) that must be scheduled to grow reliability to the specified value. Figure 3.1.8 illustrates these concepts and the four elements needed to structure and plan a growth test program.

1. *Inherent reliability* represents the value of design reliability estimated during prediction studies; it may be greater than that specified in procurement documents. Ordinarily, the contract specifies a value of reliability that is somewhat less than the inherent value. The relationship of the inherent (or specified) reliability to the starting point greatly influences the total test time.

2. *Starting point* represents an initial value of reliability for the newly manufactured hardware, usually falling within the range of 10 to 30 percent of the inherent or predicted reliability. Estimates of the starting point can be derived from previous experience or based on percentages of the estimated inherent reliability. Starting points must take into account the amount of reliability control exercised during the design program and the relationship of the system under development to the state of the art. Higher starting points minimize test time.

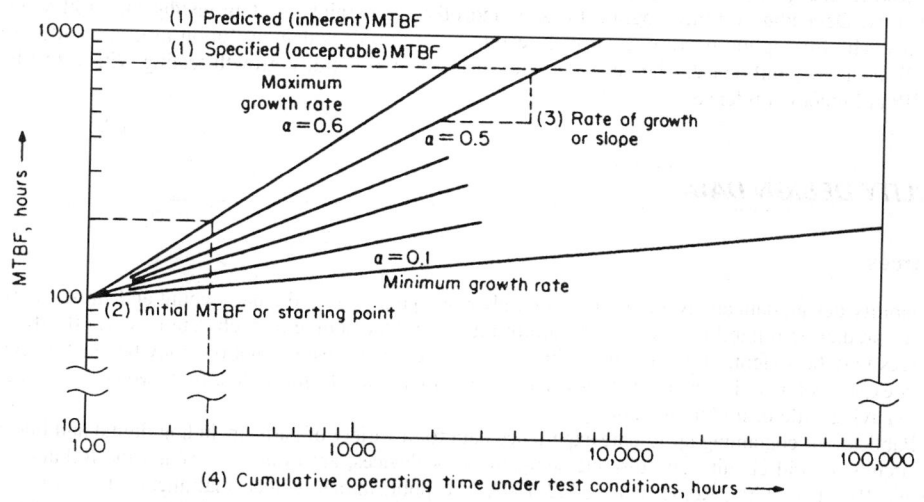

FIGURE 3.1.8   Reliability-growth plot.

3. *Rate of growth* is depicted by the slope of the growth curve, which is, in turn, governed by the amount of control, rigor, and efficiency by which failures are discovered, analyzed, and corrected through design and quality action. Rigorous test programs that foster the discovery of failures, coupled with management-supported analysis and timely corrective action, will result in a faster growth rate and consequently less total test time.

4. The ratio of *calendar time to test time* represents the efficiency factors associated with the growth test program. Efficiency factors include repair time and the ratio of operating and nonoperating time as they relate to calendar time. Lengthy delays for failure analysis, subsequent design changes, implementation of corrective action, or short operating periods will extend the growth test period.

Figure 3.1.8 shows that the value of the growth-rate parameter can vary between 0.1 and 0.6. A growth rate of 0.1 can be expected in programs where no specific consideration is given to reliability. In those cases, growth is largely because of solution of problems affecting production and corrective action taken as a result of user experience. A growth rate of 0.6 can be realized from an aggressive reliability program with strong management support. Such a program must include a formal stress-oriented test program designed to aggravate and force defects and vigorous corrective action.

Figure 3.1.8 also shows the requisite hours of operating and/or test time and continuous effort required for reliability growth. It shows the dramatic effect that the rate of growth α has on the cumulative operating time required to achieve a predetermined reliability level. For example, for a product whose MTBF potential is 1000 h it shows that 100,000 h of cumulative operating time is required to achieve an MTBF of 200 h when the growth rate is $\alpha = 0.1$. A rate of 0.1 is expected when no specific attention is paid to reliability growth. However, if the growth rate can be accelerated to 0.6, only 300 h of cumulative operating time is required to achieve an MTBF of 200 h.

Reliability Demonstration.   Reliability-demonstration tests are designed to prove a specific reliability requirement with a stated statistical confidence, not specifically to detect problems or for reliability growth. The test takes place after the design is frozen and its configuration is not permitted to change. However, in practice, some reliability growth may occur because of the subsequent correction of failures observed during the test.

Reliability demonstration is specified in most military-system procurement contracts and often involves formal testing conducted per MIL-STD-781. This standard defines test plans, environmental exposure levels, cycle times, and documentation required to demonstrate formally that the specified MTBF requirements of the equipment have been achieved. Demonstration tests are normally conducted after growth tests in the development cycles using initial production hardware.

Reliability-demonstration testing carries with it a certain statistical confidence level; the more demonstration testing, the greater the confidence. The more reliability-growth testing performed, the higher the actual reliability. Depending on the program funding and other constraints, system testing may follow one of two options. The first option maximizes growth testing and minimizes demonstration testing, resulting in a high MTBF at a low confidence. The second option minimizes reliability growth testing with a resultant lower MTBF at higher confidence.

RELIABILITY DESIGN DATA

Data Sources

Reliability design data are available from a number of sources. Both the parts-count and stress-analysis methods of predicting reliability rely on part-failure-rate data. One source of such data is MIL-HDBK-217. Other sources may be sought, or estimating techniques using comparative evaluations may be used. Provided similarity exists, comparative evaluations involve the extrapolation of failure data from well-documented parts to those having little or no failure data.

Publications containing up-to-date experience data for a variety of parts, including digital and linear integrated circuits, hybrid circuits, and discrete semiconductor devices, are available through the Reliability Analysis Center, P.O. Box 4700, Rome, NY 13442-4700. The publications include malfunction through distributions, screening fallout, and experienced failure rates. They also publish the PRISM reliability design database.

Physics of Failure

The physical or chemical phenomena leading to the deterioration or failure of electron devices or components in storage under operating conditions is termed *physics of failure* or *reliability physics.* A major source of information on reliability-physics phenomena of electron devices is the *Annual Proceedings of the International Reliability Physics Symposium* (IRPS). This symposium is jointly sponsored by IEEE's Electron Devices Society and IEEE's Reliability Society and continues yearly.

Failure Modes.   A knowledge of the physics of device failure is helpful in predicting and avoiding device failure. Prevalent failure modes are identified in a number of publications, besides the *IRPS Proceedings.* Other sources include MIL-HDBK-217F, and MIL-STD-1547(USAF).

Suspect Devices.   In selecting parts for a particular application and in designing screens to identify potential early-life failures, it is helpful to be aware of failure-aspect device designs. A standard intended for the procurement of "space quality" piece parts for space missions, MIL-STD-1547(USAF), includes an identification of reliability-suspect items. Clearly the identification of such parts *does not suggest their inapplicability for all types of electronic systems.*

BIBLIOGRAPHY

Blischke, W. R., and D. N. Prabhakar Murthy (eds.), "Case Studies in Reliability and Maintenance," Wiley, 2002.

Blischke, W. R., and D. N. Prabhakar Murthy, "Reliability: Modeling, Prediction, and Optimization," Wiley, 2000.

Duane, J. T., "Learning curve approach to reliability monitoring," *IEEE Trans. Aerospace*, Vol. 11, 1964.

IEEE Proc. Annu. Reliability and Maintainability Symp.

IEEE Proc. Int. Reliability Phys. Symp.

Musa, J., A. Iannino, and K. Okumoto, "Software Reliability: Measurement, Prediction, Application," McGraw-Hill, 1987.

O'Connor, P. D. T., "Practical Reliability Engineering," Wiley, 2001.

Ohillon, B. S., "Design Reliability: Fundamentals and Applications," CRC Press, 1999.

Ramakumar, R., "Engineering Reliability: Fundamentals and Applications." Prentice Hall, 1993.

Reliability Standards and Data (some 600 publications on specific topics available from IEC).

CHAPTER 3.2
DERATING FACTORS AND APPLICATION GUIDELINES

W. R. Lehmann, Kurt E. Gonzenbach, Neil V. Owen, Edwin W. Kimball

INTRODUCTION

The following derating guidelines were developed and adopted for use in designing equipment manufactured by Martin Marietta Orlando Aerospace, and it must be recognized that they may be too strict for use by designers of equipment for other markets, such as nonspace or nonmilitary applications. It is also recognized that to achieve higher projected reliability, customers for certain specialized equipment may impose derating factors even more severe than those given here. Nevertheless, the principles underlying the idea of derating are useful in all applications.

Derating is the reduction of electrical, thermal, mechanical, and other environmental stresses on a part to decrease the degradation rate and prolong its expected life. Through derating, the margin of safety between the operating stress level and the permissible stress level for the part is increased, providing added protection from system overstresses unforeseen during design.

The criteria listed in this section indicate maximum application stress values for design. Since safety margins of a given part at failure threshold and under time-dependent stresses are based on statistical probabilities, parts should be derated to the maximum extent possible consistent with good design practice.

When derating, the part environmental capabilities defined by specification should be weighed against the actual environmental and operating conditions of the application. Derating factors should be applied so as not to exceed the maximum recommended stresses.

For derating purposes the *allowable application stress* is defined as the *maximum allowable percentage of the specified part rating at the application environmental and operating condition*. Note that ambient conditions specified by the customer usually do not include temperature rise within a system that results from power dissipation. Thus, a thermal analysis must be performed early in the development phase to be used in the derating process.

Experience has shown that electronic part derating is the single most significant contributor to high reliability. Going from no derating to 50 percent derating can conservatively raise circuit mean time between failure by a factor of two to five times. In addition, important cost benefits can be achieved by optimized electronic part derating. Powerful analytical methodology is presently available for performing trade-off studies aimed at determining the amount of derating most desirable for various product lines.

RESISTOR DERATING AND APPLICATION GUIDELINES

Resistor Types

Variable and fixed resistors are of three types: composition, film, or wire-wound (see Sec. 5). The composition type is made of a mixture of resistive materials and a binder molded to lead wires. The film type is composed of a resistive film deposited on, or inside, an insulating cylinder or filament. The wire-wound type consists of a resistance wire wound on an appropriate structural form.

General Applications.   For ordinary military uses *established-reliability* (ER) part types are contractually required as preferred parts.

1. MIL-R-39005, RBR (fixed, wire-wound, accurate). Higher stability than any composition or film resistors, where high-frequency performance is not critical. Extremely close tolerances of ±1 to ±0.01 percent, with end-of-life tolerance shifts of ±0.02 percent. Operation is satisfactory from dc to 50 kHz. Relatively high cost and large size.

2. MIL-R-39007, RWR (fixed, wire-wound power type). Select for large power dissipation and where high-frequency performance is relatively unimportant. Generally satisfactory for use at frequencies up to 20 kHz, but the reactive characteristics are uncontrolled except for available "noninductive"-type windings at reduced resistance ranges. Wattage and working voltage must not be exceeded. Power derating begins at 25°C ambient. Bodies get very hot at rated power and may affect adjacent components or materials. Also the silicon coating used on some of the RWR resistors can be dissolved by some cleaning solvents. Regardless of purchase tolerance, the design will meet an end-of-life tolerance of ±1 percent.

3. MIL-R-39008, RCR (fixed, composition-insulated). Select for general-purpose resistor applications where initial tolerance need be no closer than ±8 percent and long-term stability no better than ±20 percent at room temperature under fully rated operating conditions. RF characteristics in resistance values higher than about 500 Ω are unpredictable. These resistors generate thermal "noise" that would be objectionable in low-level circuits. They are generally capacitive, are very reliable in catastrophic failure modes, and are also very inexpensive.

4. MIL-R-39009, RER (fixed, wire-wound power type, chassis mounted). Relatively large power dissipation in a given unit size. RF performance is limited. Minimum chassis area for heat dissipation is stated in the specifications and is essential to reach rated wattage. Not as good as RWRs in low-duty-cycle pulsed operation where peaks exceed steady-state rating. End-of-life tolerance of ±1.5 percent.

5. MIL-R-39015, RTR (variable, wire-wound, lead-screw-actuated trimmer). Use for adjusting circuit variables. Requires special consideration in severe environments. These resistors are not hermetically sealed and are susceptible to degraded performance because of ingestion of soldering flux, cleaning solvents, and conformal coating during manufacturing. Should be used with fixed resistors, if possible, in a circuit designed to reduce sensitivity to movable contact shift. Use larger sizes if possible.

6. MIL-R-39017, RLR (fixed, metal film). These resistors (mostly thick film) have semiprecision characteristics and small size. These size and wattage ratings are comparable to those of MIL-R-39008, and stability is between that of MIL-R-39008 and MIL-R-55182. Design-parameter tolerances are looser than those of MIL-R-55182, but good stability makes them desirable in most electronic circuits. RF characteristics in values above 500 Ω are much superior to composition types. Initial tolerances are ±2 percent and ±1 percent, with an end-of-life tolerance of ±5 percent.

7. MIL-R-39035, RJR (variable, non-wire-wound, lead-screw-actuated trimmer). Use for adjusting circuit variables. Use of potentiometers in severe environments requires special consideration. These resistors are not hermetically sealed and are susceptible to degraded performance because of ingestion of soldering flux, cleaning solvents, and conformal coating during equipment manufacturing. Should be used with fixed resistors, if possible, in a circuit designed to reduce sensitivity to movable contact shift.

8. MIL-R-55182, RNR/RNC (fixed, film, high stability). RNR/RNC resistors are available in hermetic and nonhermetic cases. For most applications, where a moderate degree of protection from the environments is provided, the nonhermetic parts have been proven reliable. Use in circuits requiring higher stability than provided by composition resistors or thick-film, insulated resistors and where high-frequency requirements are significant. These thin-film resistors provide the best high-frequency characteristics available unless special shapes are used. Metal films are characterized by low temperature coefficient and are usable for ambient temperatures of 125°C or higher with small degradation. End-of-life tolerance is ±2 percent.

9. MIL-R-55342, RM (fixed, film, chip). Primarily intended for incorporation into hybrid microelectronic circuits. These resistors are uncased, leadless chip devices and have a high degree of stability with respect to time, under severe environmental conditions.

10. MIL-R-83401, RZ (networks, fixed, film). Resistor networks come in dual-in-line, single-in-line, or flat pack configuration. Use in critical circuitry where stability, long life, reliability, and accuracy are of prime importance. They are particularly desirable where miniaturization and ease of assembly are important.

Mounting Guide

Since improper heat dissipation is the predominant contributing cause of wear-out failure for any resistor type, the lowest possible resistor surface temperature should be maintained. The intensity of radiated heat varies inversely with the square of the distance from the resistor. Maintaining maximum distance between heat-generating components serves to reduce cross-radiation heating effects, and promotes better convection by increasing airflow. For optimum cooling without a heat sink, small, power resistors should have large leads of minimum length terminating in tie points of sufficient mass to act as heat sinks. All resistors have a maximum surface temperature that must not be exceeded. Resistors should be mounted so that there are no abnormal hot spots on the resistor surface. Most solid surfaces, including insulators, are better heat conductors than air.

Rating Factors

The permissible power rating of a resistor is another factor that is initially set by the use to which the circuit is put, but it is markedly affected by the other conditions of use. It is based on the hot-spot temperature the resistor will withstand while still meeting other requirements of resistance variation, accuracy, and life.

Self-generated heat in a resistor is equal to I^2R. It is a usual practice to calculate this value and to use the next larger power rating available in conjunction with the derating guides.

Ambient Conditions versus Rating.   The power rating of a resistor is based on a certain temperature rise from a specified ambient temperature. If the ambient temperature is greater than this value, the amount of heat the resistor can dissipate is even less and must be recalculated.

Accuracy versus Rating.   Because all resistors have a temperature coefficient of resistance, a resistor expected to remain near its measured value under conditions of operation must remain relatively cool. For this reason, all resistors designated as "accurate" are very much larger, physically, for a certain power rating than ordinary "nonaccurate" resistors. In general, any resistor, accurate or not, must be derated if it is to remain very near its original measured value when it is being operated.

Life versus Rating.   If especially long life is required of a resistor, particularly when "life" means remaining within a certain limit of resistance drift, it is usually necessary to derate the resistor, even if ambient conditions are moderate and if accuracy by itself is not important. A good rule to follow when choosing a resistor size for equipment that must operate for many thousands of hours is to derate it to one-half of its nominal power rating. Thus, if the self-generated heat in the resistor is $1/10$ W, do not use a $1/8$ W resistor but a $1/4$ W size. This will automatically keep the resistor cooler, will reduce the long-term drift, and will reduce the effect of the temperature coefficient.

In equipment that need not live so long and must be small, this rule may be impractical, and the engineer should adjust his dependence on rules to the circumstances at hand. A "cool" resistor will generally last longer than a "hot" one and can absorb transient overloads that might permanently damage a "hot" resistor.

Pulsed Conditions and Intermittent Loads.   RWR and RER wire-wound resistors can reliably withstand pulse voltages of much greater amplitude than permitted for steady-state operation. When a resistor is used in circuits where power is drawn intermittently or in pulses, the actual power dissipated with safety during the pulses can sometimes be much more than the maximum rating of the resistor. For short pulses the actual heating is determined by the duty factor and the peak power dissipated. Before approving such a resistor application, however, the design engineer should be sure of the following:

1. The maximum voltage applied to the resistor during the pulses is never greater than its permissible maximum voltage.

2. The circuit cannot fail in such a way that continuous excessive power can be drawn through the resistor.

3. The average power being dissipated is well within the rating of the resistor.

4. Continuous steep wavefronts applied to the resistor do not cause malfunctions because of electromechanical effects of high voltage gradients.

Encapsulants.   Resistors embedded in encapsulants require special considerations. Generally, below 1 W all encapsulants raise local temperatures by 20 to 50 percent. Foams increase temperatures in all cases with the effect more pronounced as free air hot-spot temperatures become greater than 100°C. All are highly dependent on the installation's thermal configuration.

Resistor Derating

Resistors that have smaller temperature excursions have a narrower range of resistance shifts because of temperature effects and have slower aging shift rates. In addition, conditions for material and construction failure that may have escaped product testing proceed at a slower rate with lower temperatures and smaller temperature excursions. Derating, therefore, improves stability as well as reliability. For very low power stress, a possible exception is that of carbon composition resistors where a major cause of resistance shift is absorption of atmospheric moisture that can be baked out by moderate self-heating in service.

The resistor derating factors shown in Table 3.2.1 require the application of the principles illustrated in the preceding and following paragraphs. The percentages or ratios are applied to the characteristic or rating that is established, taking into consideration the temperature and duty cycle of actual operation.

Power Derating.   The objective of power derating is to establish the worst-case hot-spot temperature for the resistor. The power dissipated by a resistor causes the temperature to rise above ambient by an amount directly proportional to the amount of power dissipated. The maximum allowable power can vary because of applied voltage and temperature.

Computations of derated power apply to the maximum power permissible under conditions of voltage and ambient temperature. The derating percentage is applied after the permissible power is determined from the specification rating when all conditions and recommendations are observed. For instance, chassis-mounted resistors are designed to conduct most of the heat through the chassis. Thus, power ratings require knowing the thermal resistivity of the mounting surface and its temperature. MIL-STD-1995 defines chassis areas upon which power ratings are based.

TABLE 3.2.1   Resistor Derating Factors

Resistor type	Military specifications (MIL-R-)	Style	Maximum permissible percentage of military specification stress rating		
			Rated power	Voltage[*]	Current
Wire-wound, accurate, fixed 1.0 percent	39005	RBR	50	80	
0.1 percent			25	80	
Wire-wound, power, fixed	39007	RWR	50	80	
Composition, insulated, fixed	39008	RCR	50	80	
Wire-wound, power, chassis-mounted, fixed	39009	RER	50	80	
Wire-wound, lead-screw-actuated	39015	RTR	50	80	70
Film, metal, fixed	39017	RLR	50	80	
Non-wire-wound, lead-screw-actuated	39035	RJR	50	80	70
Film, fixed, high stability	55182	RNR/RNC	50	80	
Chip, film, fixed	55342	RM	50	80	
Networks, film, fixed	83401	RZ	50	80	

[*]Voltage applied should be no more than the smaller of V_{d1} or V_{d2}.

Voltage Derating.   The voltage should be derated to a percentage of the maximum allowable voltage as determined for the specification rating. This voltage may be limited by derated power as well as by the maximum voltage of the resistor. The derated voltage should be the smaller of

$$V_{d1} = C_u V_r \quad \text{and} \quad V_{d2} = \sqrt{P_d R} \tag{1}$$

where V_d = derated voltage
P_d = derated power
C_u = derating constant = (percent derating)/100
V_r = rated voltage
R = resistance value

For ohmic values above the critical value (which depends on the power rating of the device). RCR, RNR/RNC, RLR, and RBR resistors are voltage-limited rather than power-limited. The voltage limitation is related to dielectric breakdown rather than to heat dissipation.

Ratings, Military versus Commercial.   The military ratings of resistors are realistic for long-life performance; commercial resistors with equivalent size, material, and leads which advertise superior ratings should have those ratings subjected to careful scrutiny. Of particular importance is the size (diameter) of the resistance wire used in wire-wound resistors (RWR and RER). The use of smaller wire, of course, allows for higher resistance values in a given size, but the smaller wire is very susceptible to failure. For resistors on cores (both wire-wound and film), the thermal conductivity of the core contributes to both total heat removal and uniformity of the temperature rise along the resistance element. Generally, a size comparison of the commercial resistor with the specified dimensions of a military part with equivalent ratings will indicate whether or not the commercial rating could be acceptable in a military application.

CAPACITOR DERATING FACTORS AND APPLICATION GUIDELINES

Capacitor Types

Electrostatic capacitors, widely used in electronic equipment, include mica, glass, plastic film, paper-plastic, ceramic, air, and vacuum. Electrolytic types are aluminum and tantalum foil and wet or dry tantalum slug. All are affected by three primary stresses: voltage, temperature, and frequency.

Environmental Factors

The characteristic behavior and service life of all capacitors are highly dependent on the environments to which they are exposed. A capacitor may fail when subjected to environmental or operational conditions for which the capacitor was not designed or manufactured. Many perfectly good capacitors are misapplied and will fail when they are used in equipment that subsequently must see environmental conditions that exceed, or were not considered in, the design capabilities of the capacitor. Designers must understand the safety factors built into a given capacitor, the safety factors they add of their own accord, and the numerous effects of circuit and environmental conditions on the parameters. It is not enough to know only the capacitance and the voltage ratings. It is important to know to what extent the characteristics change with age and environment.

Temperature Variations.   Temperature variations have an effect on the capacitance of all types of capacitors. Capacitance change with temperature is directly traceable to the fact that the dielectric constant of the materials changes with temperature.

In general, the lower-dielectric-constant materials tend to change less with temperature.

Capacitance will vary up or down with temperature depending on the dielectric and construction. The temperature can cause two distinct actions to take place that will affect capacitance. Both the dielectric constant of the material and the spacing between the electrodes can be altered. Again, depending on the materials, these two actions tend to either reinforce or offset each other.

The capacitance of polarized dielectrics is a complex function of temperature, voltage, and frequency; nonpolarized dielectrics exhibit less change than polarized materials. Many dielectrics exhibit a very large decrease in capacitance with a relatively small decrease in temperature. The increased power factor at this temperature may raise the dielectric temperature sufficiently to recover lost capacitance. When a capacitor is initially energized at low temperatures, the capacitance will be a small percentage of the nominal value, and if the internal heating is effective, the thermal time constant of the capacitor must be considered.

The *operating temperature* and changes in temperature also affect the mechanical structure in which the dielectric is housed. The terminal seals, using elastomeric materials or gaskets, may leak because of internal pressure buildup. Expansion and contraction of materials with different thermal-expansion coefficients may also cause seal leaks and cracks in internal joints. Electrolysis effects in glass-sealed terminals increase as the temperature increases.

If the capacitor is operated in the vicinity of another component operating at high temperature, the flashpoint of the impregnant should be considered.

Voltage Rating.   Voltage ratings of nonelectrolytic capacitors are uniformly based on some life expectancy before catastrophic failure at some temperature and some voltage stress, since catastrophic failures of capacitors are usually caused by dielectric failure. Dielectric failure is typically a chemical effect and, for hermetically sealed parts where atmospheric contamination of the dielectric does not contribute, is a function of time, temperature, and voltage. The time-temperature relationship is well expressed by assuming that the chemical activity, and therefore degradation, proceeds at a doubled rate for each 10°C rise in temperature; e.g., a capacitor operating at 100°C will have half the life of a similar one operating at 90°C.

Frequency.   This is a capacitor stress most often overlooked by the circuit designer. There are both inductance and capacitance in each capacitor, and obviously, there is a resonant frequency. Depending upon the capacitor type, this resonant frequency may or may not fall in a range troublesome to the designer. In high-frequency applications, NPO ceramic, extended foil-film, mica, and glass capacitors are usually used.

Insulation Resistance.   Increasing temperature usually reduces insulation resistance, increases leakage current and power factor/dissipation factor, and reduces the voltage rating of the part. Conversely, reducing temperature normally improves most characteristics; however, at cold temperature extremes some impregnants and electrolytes may lose their effectiveness. The time of electrification is most critical in the determination of insulation resistance. The effect of the insulation resistance value is also quite critical in many circuit designs and can cause malfunctions if its magnitude and variation with temperature are not considered. The dielectric strength decreases as the temperature increases.

Moisture.   Moisture in the dielectric decreases the dielectric strength, life, and insulation resistance and increases the power factor of the capacitor. Capacitors operated in high humidities should be hermetically sealed.

Aging.   The extent and speed of aging of a capacitor depend on the dielectric materials used in its construction. Aging does not affect glass, mica, or stable ceramic capacitors. The most common capacitors with significant aging factors are the medium-K and hi-K ceramic type (CKR series) and aluminum electrolytic types. Detailed aging and storage-life data are given in MIL-STD-198.

External Pressure.   The altitude at which hermetically sealed capacitors are to be operated will control the voltage rating of the capacitor terminals. As barometric pressure decreases, the ability of the terminals to withstand voltage arcing also decreases. External pressure is not usually a factor to be considered unless it is sufficient to change the physical characteristics of the container housing, the capacitor plates, and the dielectric. Heat transfer by convection is decreased as the altitude is increased. Certain high-density CKR-type capacitors demonstrate piezoelectric effects.

Shock, Vibration, and Acceleration.   A capacitor can be mechanically destroyed or damaged if it is not designed or manufactured to withstand whatever mechanical stresses are present in the application. Movement of the internal assembly inside the container can cause capacitance changes and dielectric or insulation failures because of the physical movement of the electrode and fatigue failures of the terminal connections. The capacitors and mounting brackets, when applicable, must be designed to withstand the shock and vibration requirements of the particular application. Internal capacitor construction must be considered when selecting a capacitor for a highly dynamic environment.

Series Impedance. For solid tantalum electrolytic capacitors, the traditional criterion for their application was to ensure that the circuit provided for a minimum of 3 Ω of impedance for each volt of potential (3 Ω/V) applied to the capacitor. However, advances in the state of the art in recent years (in the control of the purity of tantalum powder and in manufacturing technology) make it possible to use these capacitors in circuits with series impedance as low as 0.1 Ω/V, without a significant impact on equipment reliability. For reliable operation follow these ground rules:

1. Never select the highest voltage rating and capacitance value in a given case size. These parts represent the ultimate in manufacturing capability, are costly, and are a reliability risk.

2. Never select a voltage rating greater than that needed to satisfy the voltage derating criteria herein. The higher voltage ratings require thicker dielectrics, increasing the probability of the inclusion of impurities. Also, the

TABLE 3.2.2 Derating Factors for Capacitors

Capacitor type	Military specifications (MIL-C-)	Style	Maximum permissible percentage of military specification stress rating[a]			
			Voltage[b]	Current[c]	AC ripple	Surge
Fixed, ceramic, temperature-compensating (ER)	20	CCR	50	70		
Fixed, feedthrough	11693	CZR	70	70		
Fixed, paper-plastic (ER), and plastic film (ER)	19978	CQR	70	70	70	70
Fixed, glass (ER)	23269	CYR	75	70	70	70
Fixed, mica (ER)	39001	CMR	80	70	70	70
Fixed, electrolytic tantalum, solid (ER)	39003	CSR[d]	50	70	70	
Fixed, electrolytic tantalum, nonsolid (ER)	39006	CLR	50	70		
Fixed, electrolytic all tantalum nonsolid (ER)	39006/22	CLR 79	80	80[e]		
Fixed, ceramic (ER) (general purpose)	39014	CKR	60	70	70	70
Fixed, electrolytic, aluminum (ER)	39018	CUR	80 min 95 max	75		
Fixed, metallized paper-film (ER)	39022	CHR	50	70	70	70
Chip, fixed, tantalum solid (ER)	55365	CWR	50	70		
Fixed, plastic-film DC or DC-AC, (ER)	55514[f]	CFR	60	70	70	70
Chip, fixed, ceramic (ER)	55681	CDR	60	70		
Fixed, plastic-film DC, AC, or DC-AC	83421	CRH	60	70	70	70

[a]Manufacturer's derating factors must be applied before applying these factors.
[b]Voltage equals instantaneous total of dc, ac, surge, and transient voltage.
[c]Rated current is defined as $I_R = \sqrt{P_{max} / R_{max}}$ and by limiting the current to 0.70 times rated current, power is limited to 0.50 maximum.
[d]Limited to 85°C ambient temperature.
[e]Package for maximum thermal dissipation.
[f]Not hermetically sealed.

lower-voltage-rated parts will typically have smaller slug sizes, which results in a greater internal equivalent series resistance (ESR), which tends to compensate for the reduction of the external series resistance.

3. Always specify the lowest established reliability failure rate available from two or more suppliers. S-level parts are typically manufactured under more rigorous process controls than higher-failure-rate-level parts.

Commercial versus Military-Type Capacitors.   Valid conclusions can be reached concerning life and reliability to be expected from commercial capacitors by comparing their values with those of similar military capacitors of the same dielectric and capacitance. If the commercial capacitor is appreciably smaller than the corresponding military device, it can be safely assumed that the commercial unit has a shorter life and is less reliable.

Capacitor Derating

The capacitor derating factors (Table 3.2.2) should be applied after all derating (stated or implied by the MIL-SPEC or manufacturer) has been applied in the circuit design. The table shows the maximum allowable percentage of voltage and current.

Precautions.   The following checklist will help achieve high reliability:

- Do not exceed the current rating on any capacitor, taking into account the duty cycle. Provide series resistance or other means in charge-discharge circuits to control surge currents.
- Include dc, superimposed peak ac, peak pulse, and peak transients when calculating the voltage impressed on capacitors.
- The MIL-SPEC or manufacturer's recommendations for frequency, ripple voltage, temperature, and so forth, should also be followed for further derating.

SEMICONDUCTOR DERATING FACTORS AND APPLICATION GUIDELINES

General Considerations

Semiconductor device derating should be applied after all deratings stated or implied by the part MIL-SPEC have been used in the circuit design.

For designs using silicon active components, transistors, and diodes, the maximum junction temperature must not exceed 110°C for ground and airborne applications. If the maximum rated junction temperature of the device is ≤150°C, the maximum junction temperature for missile-flight applications must not exceed 110°C. If the maximum rated junction temperature of the device is ≥175°C, the maximum junction temperature for missile-flight applications must not exceed 140°C. (See Table 3.2.3.)

A maximum power rating on any semiconductor device is by itself a meaningless parameter. The parameters of value are maximum operating junction temperature, thermal resistance, and/or thermal derating (reciprocal of thermal resistance). For all semiconductor devices, the mechanism for removal of heat from a junction is usually that of conduction through the leads, not convection. For all silicon transistors and diodes, the maximum operating junction temperature should be 110 or 140°C, respectively.

The method for calculating device junction temperature is

$$T_J = T_A + \theta_{J-A} P_D \tag{2}$$

where T_j = junction temperature
T_A = maximum ambient temperature at component
θ_{J-A} = thermal resistance from junction to air
P_D = power dissipated in device

TABLE 3.2.3   Transistor Derating Factors

Parameter	Derating factor*
Voltage (V_{CEO}, V_{CBO}, V_{EBO})	0.75
Current	0.75
Junction temperature:	
Ground and airborne use	110°C
Missile-flight use ($T_1 \leq 150$°C)	110°C
Missile-flight use ($T_1 \geq 175$°C)	140°C
Allowing for:	
Increase in leakage (l_{CBO} or I_{CEO})	+100 percent
Increase in h_{FE}	+50 percent
Decrease in h_{FE}	−50 percent
Increase in $V_{CE(SAT)}$	+10 percent

*Derating factor (applicable to all transistor types) = Maximum allowable stress/Rated stress

where heat sinks are used, the expression is expanded to

$$T_J = T_A + (\theta_{J-C} + \theta_{C-S} + \theta_{S-A})P_D \qquad (3)$$

where

$$\theta_{J-C} + \theta_{C-S} + \theta_{S-A} = \theta_{J-A}$$

where θ_{C-S} = thermal resistance between case to heat sink (usually includes mica washer and heat-sink compound)

θ_{S-A} = thermal resistance of heat sink
θ_{J-C} = thermal resistance from junction to case

Examples for calculation of junction temperature for various conditions follow.

Thermal Resistance and Power Calculations

Using the types of inputs described above, the following examples demonstrate the ease with which thermal calculations can be performed.

Example 1.   Given: 1N753 reference diode: find; θ_{J-A}. Specifications: $P_D = 400$ mW (max power), $T_J = 175$°C (max junction), $T_A = 25$°C (max ambient).
The manufacturer does not give θ_{J-A}, but it can be calculated using the above data. The maximum power dissipation is calculated from specified maximums at room temperature:

$$T_J = T_A + \theta_{J-A}P_D$$

$$175°C = 25°C + (\theta_{J-A})(0.4)$$

$$\theta_{J-A} = 375°C/W$$

Example 2.   Determine the thermal resistance of a heat sink required for a 2N3716 power transistor that is to dissipate 14 W at an ambient temperature of 70°C.

$$\theta_{J-C} = 1.17°C/W \qquad \text{from 2N3716 specifications}$$

$$\theta_{C-S} = 0.5°C/W$$

TABLE 3.2.4 Contact Thermal Resistance of Insulators

Insulator	Thickness, in	θ_{C-S}, °C/W
No insulation	. . . . .	0.4
Anodized aluminum	0.016	0.4
	0.125	0.5
Mica	0.002	0.5
	0.004	0.65
Mylar	0.003	1.0
Glass cloth (Teflon-coated)	0.003	1.25

($\theta_{C-S} = 0.5$°C/W for mica washer; Table 3.2.4

$$110°C = 70°C + \theta_{J-A}P_D$$

$$\theta_{J-A} = (110 - 70)\frac{1}{P_D} = \frac{40}{14} \approx 2.9°\text{C/W}$$

$$\theta_{J-A} = \theta_{J-C} + \theta_{C-S} + \theta_{S-A}$$

$$2.9 = 1.17 + 0.5 + \theta_{S-A}$$

$$\theta_{S-A} \leq 1.23°\text{C/W}$$

(Heat-sink thermal resistance required; note Table 3.2.5 for thermal resistance of some common commercially available heat sinks. The lowest value of thermal resistance results in the lowest junction temperature of the part.)

Example 3. Determine the maximum power that the 2N3716 can dissipate without a heat sink in an ambient of 70°C;

$$\theta_{J-A} = 35°\text{C/W}$$

(Not given on Motorola data sheets, but for almost all TO-3 devices $\theta_{J-A} = 35$°C/W; see Table 3.2.6.)

$$T_J = T_A + \theta_{J-A}P_D$$

$$P_D = \frac{T_J - T_A}{\theta_{J-A}} = \frac{110 - 70}{35} = 1.14 \text{ W}$$

Example 4. Determine the maximum power that can be dissipated by a 2N2222 transistor (missile-flight use) with an ambient temperature of 70°C. Derating given: 3.33 mW/°C for 2N2222, TO-18. Therefore

$$\theta_{J-A} = \frac{1°C}{3.33 \text{ mW}} = 300°\text{C/W}$$

and

$$P_D = \frac{T_J - T_A}{\theta_{J-A}} = \frac{140 - 70}{300} = 233 \text{ mW}$$

TABLE 3.2.5   Thermal Resistance of Heat Sinks

| Shape | Surface area, in | Volume displacement | | | | | Finish | Thermal resistance, °C/W |
		L, in	W, in	H, in	Vol, in	wt, g		
			Extrusion					
Flat-finned	65	3.0	3.6	1.0	10.8	114	Anod black	2.4
							Bright alum	3.0
							Gray	2.8
	60	3.0	4.0	0.69	8.3	123	Anod black	2.8
	95	3.0	4.0	1.28	15.3	189	Anod black	2.1
	64	3.0	3.8	1.3	15.0	155	Black paint	2.2
	83	3.0	4.0	1.25	15.0	140	Anod black	2.2
	44	1.5	4.0	1.25	7.5	75	Anod black	3.0
	137	3.0	4.0	2.63	31.5	253	Anod black	1.45
	250	5.5	4.0	2.63	58.0	461	Anod black	1.10
	130	6	3.6	1.0	21.5	253	Anod black	1.75
	78	3.0	3.8	1.1	12.5	190	Anod gray	2.9
	62	3.0	3.8	1.3	15.0	170	Anod gray	2.2
	78	3.0	4.5	1.0	13.5	146	Gold alodine	3.0
			Machined casting					
Cylindrical fins, horizontal	30	1.75		0.84	2.0	40	Anod black	8.5
	50	1.75		1.5	3.6	67	Anod black	7.1
	37	1.75		1.5	3.6	48	Anod black	6.65
			Casting					
Cylindrical fins, vertical	7.5	1.5		0.9	4.4	33	Anod black	8.1
	12	1.5		1.4	6.9	51	Anod black	7.0
	25	1.5		2.9	14.2	112	Anod black	5.6
	35	1.5		3.4	16.7	132	Anod black	5.1
	32	2.5		1.5	7.4	94	Anod black	4.5
	20	2.5		0.5	2.45	48	Anod black	6.6
Flat-finned	23	1.86	1.86	1.2	4.15	87	Anod black	5.06
			Sheet-metal					
Vertical fins, square	12	1.7	1.7	1.0	2.9	19	Anod black	7.4
Cylindricals	15	2.31*		0.81	3.35	18	Black	7.1
Horizontal fins, cylindrical	6	1.81*		0.56	1.44	20	Anod black	9.15
	55	2.5		1.1	5.4	115	Gold irridate	7.9

*Diameter.

Semiconductor Derating

Power Derating.   The objective of power derating is to hold the worst-case junction temperature to a value below the normal permissible rating. The typical diode specification for thermal derating expresses the change in junction temperature with power for the worst case. The actual temperature rise per unit of power will be considerably less, but this is not a value that can readily be determined for each unit.

Junction-Temperature Derating.   Junction-temperature derating requires the determination of ambient temperature or case temperature. The worst-case ambient temperature or case temperature for the part is established for the area and for the environmental conditions that will be encountered in service. The ambient temperature for a

TABLE 3.2.6 Thermal Resistance of Packages, °C/W

Package type	Still air	
	θ_{J-A}	θ_{J-C}
TO-3	30–50	1.0–3.0
TO-66	30–50	4.0–7.5
TO-5	100–300	30–90
TO-18	300–500	150–250
TO-99, TO-100	197	60
Flat pack:		
14 lead	200	70
16 lead	195	68
24 lead	170	55
Ceramic dip:		
8 lead	125	50
14 lead	110	50
16 lead	110	50
18 lead	100	40
20 lead	100	40
22 lead	100	40
24 lead	100	40
40 lead	80	30
Leadless chip carrier:		
16 castellation	—	60
20 castellation	—	60
24 castellation	—	60
28 castellation	—	60
44 castellation	—	60
Plastic dip	150	80

device that does not include some means for thermal connection to a mounting surface should include the temperature rise because of the device, adjacent devices, and any heating effect that can be encountered in service.

Voltage Derating.   The voltage rating of a semiconductor device can vary with temperature, frequency, or bias condition. The rated voltage implied by the tabulated rating is the voltage compensated for all factors determined from the manufacturer's data sheet. Derating consists of the application of a percentage figure to the voltage determined from all factors of the rating. Three distinct deratings cover the conditions that can be experienced in any design situation.

1. *Instantaneous peak-voltage derating* is the most important and least understood derating. It is required to protect against the high-voltage transient spike that can occur on power lines as a result of magnetic energy stored in inductors, transformers, or relay coils. Transient spikes also can result from momentary unstable conditions that cause high amplitude during switching turnon or turnoff.

Transient spike or oscillating conditions in test sets or life-test racks or resulting from the discharge of static electricity will cause minute breakdown of surface or the bulk silicon material.

Lightning transients, which enter a circuit along power lines or couple from conducting structural members, are a frequent cause of failure or of damage that increases the probability of failure during service.

2. The *continuous peak voltage* is the voltage at the peak of any signal or continuous condition that is a normal part of the design conditions.

3. The *design maximum voltage* is the highest average voltage. This is essentially the dc voltage as read by a dc meter. The ac signals can be superimposed on the dc voltage to produce a higher peak voltage, providing the continuous peak voltage is not exceeded.

Transistor Guidelines

The major failure modes are the degradation of h_{FE} and increased leakage with prolonged use at elevated temperatures. Depending on the application, this parameter degradation can result in a catastrophic failure or decrease in system performance outside the bound of the worst-case design. It is necessary to maintain the junction operating temperature below 110°C for ground and airborne use (140°C for missile-flight use if device rated $T_J \geq 175$°C). The principle design criteria for silicon transistors are shown in Table 3.2.3. This derating is applicable to all transistor types.

Transistor Application Information.   These general guidelines apply:

- h_{FE} has a positive temperature coefficient. This criterion may not be valid for some power transistors operating at high current levels where h_{FE} decreases with temperature.
- If a maximum leakage is specified at 110°C for ground and airborne use (140°C for missile use if device rated $T_J \geq 175$°C), double the leakage value for end of life condition.
- The ratings of Table 3.2.3 apply for operating junction temperature, not just ambient temperature.
- Typical thermal resistances for common case sizes are described in Table 3.2.6.

Thermal Resistance and Heat Sinks.   Table 3.2.4 lists contact thermal resistance for various insulators, and Table 3.2.5 gives the thermal resistance for various heat sinks. Figure 3.2.1 is used to calculate θ_{S-A} for solid copper or aluminum plates. This is helpful in determining the thermal capabilities of metal chassis. Note that a vertically mounted plate has better thermal properties than a horizontally mounted plate. A list of approximate thermal resistances for various package sizes is given in Table 3.2.6.

Diode Guidelines

The junction-temperature limits specified for semiconductors apply to all diodes. For non-heat-sinked components, a quick calculation can be made to determine the power a given device may dissipate. The calculation described in "Thermal Resistance and Power Calculations," p. 3.28, can be used to determine this parameter. The derating for silicon diodes is given in Table 3.2.7.

For zener diodes, the best worst-case end-of-life tolerance that can be guaranteed is ±1 percent. This places a limitation on the final accuracy of any analog system end of life at some value greater than 1 percent.

Zener-Diode Voltage-Variation Calculation.   The change in zener voltage over a specified operating range is primarily a function of the zener temperature coefficient (TC) and dynamic resistance. The temperature to be used for the TC is the junction temperature of the device in question.

Given: 1N3020B; 10-V zener ±5 percent (add 1 percent end of life)

$$P = 1 \text{ W} \qquad \text{derate } 6.67 \text{ mW/°C}$$

$$Z_z = 7 \text{ } \Omega \qquad \text{at } 25 \text{ mA} = I_{zT}$$

$$\text{TC} = 0.055 \text{ percent/°C} \qquad V_{CC} = 30 \text{ V} \pm 5 \text{ percent (See Fig. 3.2.2.)}$$

The ambient range is –25 to 70°C, V_{CC} (applied through a resistor R) = 30 V ±5 percent, and $R = 800$ ±20 percent.

(*a*) Calculation for the worst-case maximum power of zener and θ_{J-A}

$$P = V_z I_z \qquad \text{neglect TC initially}$$

$$= (10.6 \text{ V}) \frac{31.5 - 10.6}{640}$$

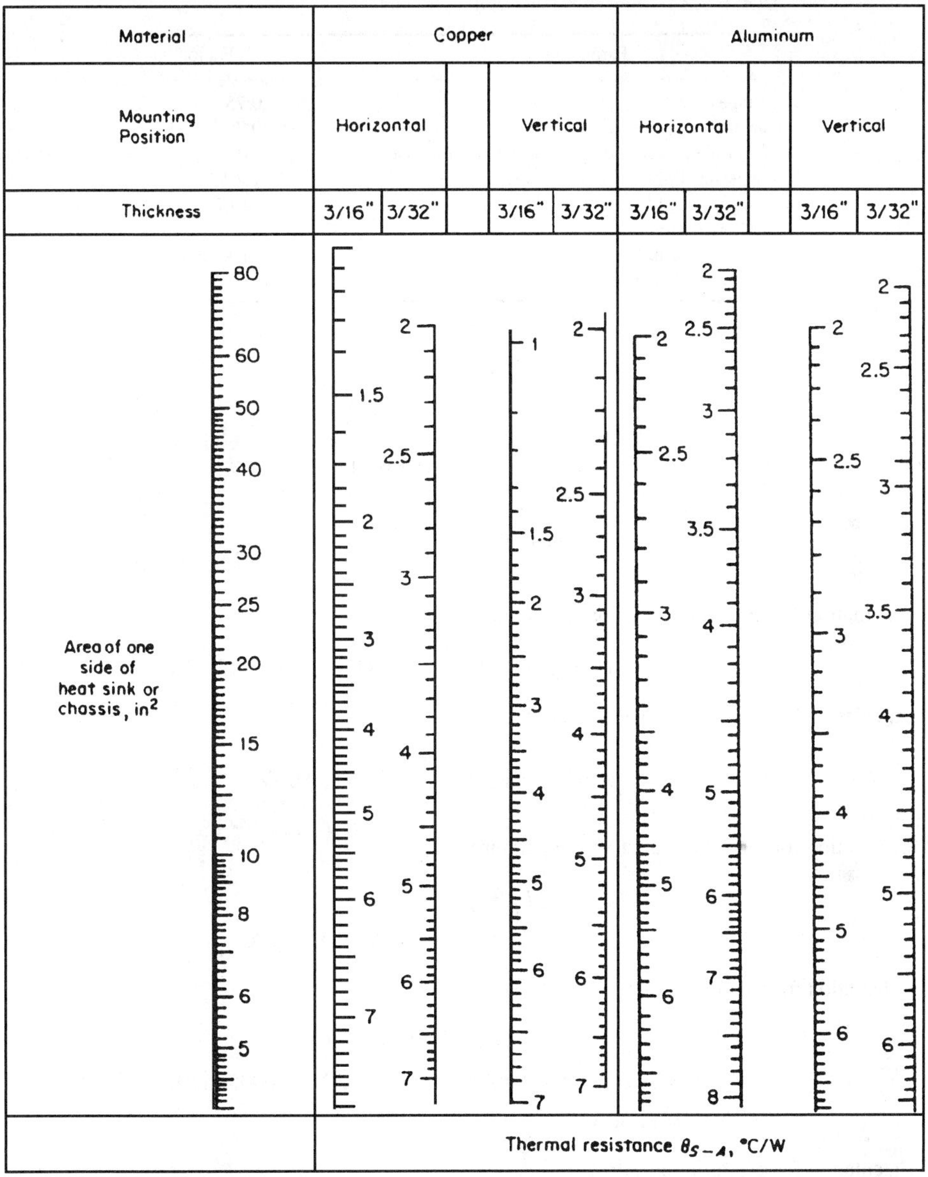

Material	Copper				Aluminum			
Mounting Position	Horizontal		Vertical		Horizontal		Vertical	
Thickness	3/16"	3/32"	3/16"	3/32"	3/16"	3/32"	3/16"	3/32"

Thermal resistance θ_{S-A}, °C/W

Area of one side of heat sink or chassis, in²

Instructions for use: Select the heat sink area at left and draw a horizontal line across the chart from this value. Read the values of θ_{S-A} depending on the thickness of the material, type of material, and mounting position.

FIGURE 3.2.1   Thermal resistance as a function of heat-sink dimensions.

TABLE 3.2.7   Silicon Diode Derating Factors

Parameter	Factor*
Voltage	0.75
Current	0.75
Junction temperature, ground and airborne use	110°C
Missile-flight use ($T_1 \leq 150°C$)	110°C
Missile-flight use ($T_1 \geq 175°C$)	140°C
Allowing for:	
Increase in leakage I_R	+100 percent
Increase in V_F	+10 percent

$$*\text{Derating factor} = \frac{\text{maximum allowable stress}}{\text{rated stress}}$$

$$= (10.6)(32.7 \text{ mA}) = 0.347 \text{ W}$$

$$\theta_{J-A} = \frac{1°C}{6.67 \text{ mW}} = 149°C/W$$

(b) Calculation for maximum junction temperature

$$T_J = T_A + P_D\theta_{J-A} = 70°C + (0.347)(149°C/W) = 121.7°C$$

(c) Calculation for minimum power in zener

$$P = V_z I_z = (9.4)\frac{28.5 - 9.4}{960} = 0.180 \text{ W}$$

(d) Calculation for minimum junction temperature

$$T_J = T_A + P_D\theta_{J-A}$$
$$= -25°C + (0.180)(149°C/W) = 1.8°C$$

(e) Calculation for overall maximum V_z

$$V_z = V_{z.\text{MAX}} + I_z Z_z + (T_J - T_{\text{amb}})(\text{TC})(V_z)$$
$$= 10.6 \text{ V} + (32.7 \text{ mA})(7) + (121.7 - 25)(0.00055)(10)$$
$$= 11.36 \text{ V}$$

(f) Calculation for overall minimum V_z

$$V_z = V_{z.\text{MIN}} - I_z Z_z - (T_{\text{amb}} - T_J)(\text{TC})(V_z)$$
$$= 9.4 - (19.1)(7) - (25 - 1.8)(0.00055)(10) = 9.138 \text{ V}$$

Several iterations may be necessary to determine maximum and minimum zener voltages, because steps (a) and (c) used 10.6 and 9.4 V, respectively, for computing maximum and minimum V_z, while we calculate 11.36 and 9.138. This affects P, which in turn affects the TC term of V_z and maximum and minimum zener currents.

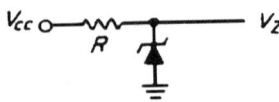

FIGURE 3.2.2   Zener diode V_{CC}.

TABLE 3.2.8 Microcircuit Derating Factors*

Type	Open mounting	Enclosed	Parameter
Digital	0.80	0.70	Fanout/output current
	0.75	0.75	Supply voltage†
	0.75	0.75	Operating frequency
	90°C max	90°C max	C/MOS junction temperature
	100°C max	100°C max	Junction temperature
	110°C max	110°C max	TTL junction temperature
Linear or hybrid	0.80	0.80	Supply voltage†
	0.75	0.65	Output voltage
	0.75	0.75	Operating frequency
	100°C max	100°C max	Junction temperature
	105°C max	105°C max	Hybrid junction temperature

*Derating factor = $\dfrac{\text{maximum allowable stress}}{\text{rated stress}}$

†For devices with dynamic supply voltage ranges only; all others will use manufacturers' recommended supply voltage. Derating below 75 percent of the supply voltage may cause the device to operate below recommended operating voltages.

Integrated-Circuit Guidelines

Derating is a process that improves in-use reliability of a component by reducing the life stress on the component or making numerical allowances for minor degradation in the performance of the component. This technique is applied to integrated circuits in two separate and distinct ways.

The first is to specify a derating factor in the application of the component. Derating factors (Table 3.2.8) are applied to the voltage, current, and power stresses to which the integrated circuit is subjected during operation. Derating factors must be applied knowledgeably and singly; i.e., they must be applied only to a degree that improves reliability, and they must be applied only once throughout the entire cycle that stretches from the design of the integrated circuit to its application in a system.

From the outset, integrated circuits are designed to a set of conservative design-rating criteria. The currents that flow through the conductors and through the wire bonds on a chip, the voltages applied to the semiconductor junctions, and the overall power stress on the entire chip are conservatively defined during the design of an integrated circuit. Therefore it may not be appropriate to derate the integrated circuit further in its application. Derating power consumption of a digital integrated circuit may not be possible, since the circuit must operate at a specified level of power-supply voltage for maximum performance. However, some linear circuits designed to operate over an extended range of power-supply voltages and power dissipations may accept some degree of derating when it is appropriately applied.

Thus, the main area of derating in integrated circuits in not in derating the *stresses* applied to the circuit but in derating the expected and required *performance*. The designer must fully recognize potential performance degradation of integrated circuits over their life. This parametric degradation can require using a digital circuit at less than its full fanout. It can mean designing for an extra noise margin, sacrificing some of it to the degradation of the integrated circuit. It can also mean applying the integrated circuit at performance levels below those guaranteed by the circuit's characterization-specification sheet.

Establishment of derating factors for integrated-circuit parameters must be made after careful analysis of each particular parameter in the circuit. Parameters depending directly on transistor beta, resistor value, or junction leakage are most prone to shift during life. Parameters depending directly on the saturation voltages of junctions and on the ratios of resistors are most likely to remain stable. For digital microcircuits, device fanout should be derated by a factor of 20 percent, logic noise-margin levels should be derated by a factor of 10 percent, and the maximum operating frequency should be derated by a factor of 25 percent. For linear microcircuits, the input offset and input signal voltages should be derated by a factor of 20 percent. The output current and maximum operating frequency should be derated by a factor of 25 percent.

The severity of the application further establishes the degree of proper derating. It is not customary to derate ac parameters, such as delay times or rates, as these parameters do not vary greatly over the life of an integrated circuit. Allowances should be made, however, for unit-to-unit variation within a given integrated-circuit chip. The delay times of separate gates within one integrated-circuit package can vary greatly. These parameters are usually not measured on 100 percent of the units.

Although one may be able to derate an integrated circuit for reliability in specified special cases, one cannot take advantage of the derating designed into the integrated circuit and use it beyond its rating or specified capability.

TRANSFORMER, COIL, AND CHOKE DERATING

General Considerations

The ratings and deratings of transformers, chokes, and coils are covered in the following paragraphs. Transformers are frequently designed for a particular application and can become a major source of heat. Two major considerations result: derating of transformers must include consideration of their heating effects on other parts; and transformer derating requires control of ambient plus winding-temperature rises.

Voltage Derating

Winding voltages are fixed voltages and cannot be derated to any significant degree as a means of improving reliability. The voltages present between any winding and case or between any winding and shield, as specified, should be derated in accordance with the voltage derating factors of Table 3.2.9.

Power Derating

The power dissipated in a transformer should be derated to control the winding temperature to the maximum derated temperature under full load conditions that are normal to the worst-case service conditions.

Temperature rise is determined for service conditions by measurement of winding resistance using the procedure of MIL-T-27B.

The insulation grade of a transformer is rated for a maximum operating temperature. Deratings shown in Table 3.2.9 are allowances of temperature to be subtracted from the rated temperature to determine derated temperature. All considerations of frequency, hot-spot temperature, and other factors included in the manufacturer's data must be allowed for before applying this reliability derating temperature.

TABLE 3.2.9   Coil, Transformer, and Choke Derating Factors

	Maximum permissible percent of manufacturer's stress rating			
	Insulation breakdown voltage		Operating current, A	Allowable winding temp rise, °C
Type	Maximum	Transient		
Coil				
Inductor, saturable reactor	60	90	80	30
General	60	90	80	30
RF, fixed	60	90	80	35
Transformer				
Audio	50	90	80	35
Pulse, low-power	60	90	80	30
RF	60	90	80	30
Saturable-core	60	90	80	30

Current Derating

The maximum current in each winding should be derated in accordance with the percentage deratings shown in Table 3.2.9. The derated current should be considered as the largest current that can flow in the winding under any combination of operating conditions.

In-rush transient currents should be limited to the maximum allowable in-rush or surge rating of the transformer, as shown in Table 3.2.9. The current in all windings should not cause a power dissipation or temperature in excess of the derated temperature requirements.

SWITCHES AND RELAYS

Switch Considerations

Switches are to be applied in circuits with operating current loads and applied voltages well within the specified limits of the type designated. A major problem is contamination, which includes particles and contaminant films on contacts. The storage life on switches may exceed 10 years if they are hermetically sealed. The cycle life for switches can be in excess of 100,000 cycles.

Switch Derating

The contact power (volt-amperes) for general-purpose switches (1 to 15 A) should be derated as shown in Table 3.2.10 from the maximum rated contact power within the maximum current and voltage ratings.

Temperature.   Switches should not be operated above rated temperature. Heat degrades insulation, weakens bonds, increases rate of corrosion and chemical action, and accelerates fatigue and creep in detent springs and moving parts. The derating factor in Table 3.2.10 is defined as the maximum allowable stress divided by the rated stress.

For capacitor loads, capacitive peak in-rush current should not exceed the derated limit. If the relay-switch specification defines inductive, motor, filament (lamp), or capacitive load ratings, they should be derated 75 percent instead of the derating specified in Table 3.2.10.

Relay Considerations

Relays are to be used in circuits with operating current load and applied coil and contact voltage well within specified ratings. The application for each device should be reviewed independently.

Relay cycle life varies from 50,000 to more than 1 million cycles depending on the relay type, electrical loads of the contacts, duty cycle, application, and the extent to which the relay is derated. The storage life of hermetically sealed relays, with proper materials and processes employed to eliminate internal outgassing, is over 10 years.

TABLE 3.2.10   Relay and Switch Derating Factors

Type of load	Percent of rated value
Resistive	75
Inductive	40
Motor	20
Filament	10
Capacitive	75
Contact power[*]	50

[*]Applicable to reed, mercury-wetted, or other loads rated in watt or volt-amperes.

The chief problem in electromechanical relays is contamination. Even if cleaning processes eliminate all particulates, the problem of internal generation of particles, owing wear, is still present.

Relay Derating

The contact power (volt-amperes) should be derated as shown in Table 3.2.10 from the maximum rated stress level for loads of 1 to 15 A.

Film formation on relay contacts, as discussed in Chap. 5.4 can cause excessive resistance. This can be a serious reliability problem with contacts that switch very low voltage or when relays are subjected to long periods of dormancy. Hermetically sealed relays are recommended for military applications, and vendors should be consulted to determine if a particular device is rated for "dry contact" service.

Resistive Loads.   The resistive load is basically what is used to rate relays and what most designers use when calculating the size of the relay needed for a particular circuit. In a resistive load the current flow is practically constant, with the exception of some minor arcing on make or break. As long as the limits of the contacts are not exceeded, the relay should operate reliably throughout its rated life. Unfortunately, this ideal standard is sometimes the most difficult to find in "real world" engineering. When applying derating to a purely resistive load, accepted industry practice would suggest going to 75 percent of the rated value for the device in question.

Capacitive Loads.   The in-rush current in a capacitive load can be very high. This is due to the fact that a capacitor, when charging, acts as a short circuit limited only by whatever resistance may be in the circuit. Therefore, the best way to control this initial surge, and thereby protect the relay contacts, is to add a series limiting resistance. Without the series limiting resistance, contact welding or contact deterioration will shorten the life of the relay. Accepted industry practice would suggest that a derating of 75 percent of the resistive load rating should be applied. Should the relay in question already have a capacitive rating, the same 75 percent of this rating would apply.

Inductive Loads.   Inductors, transformers, coils, chokes, and solenoids are all high inductive loads. In-rush currents may or may not be a problem with such loads, particularly with transformers and solenoids. However, when the circuit is broken, the stored energy in the collapsing magnetic field is dissipated across the opening contacts. The resultant arcing will cause excessive contact deterioration. This situation will be further aggravated when the load circuit has a long time constant. When size and weight preclude the use of a relay with sufficient capacity to handle the transient conditions, arc-suppression circuits should be added. These can be as simple as a diode to a diode-*RC* network. An added benefit to such circuits will be increased contact life. Accepted industry practice would be to derate the contacts to 40 percent of their resistive load rating. If the relay in question has an inductive load rating, then the derating should be 75 percent of this specified value.

Motor Loads.   A motor, in the nonrunning condition, has a very low input impedance and consequently a large in-rush current. As the current starts to flow through the windings, a torque is developed because of the interaction of the magnetic field in the motor and the current in the winding. As the armature of the motor begins to turn, it generates a back emf in the opposite direction of the original applied voltage thus reducing the current flow. However, the greater the load on the motor, the longer the starting time and thus the longer the high starting current will persist in the circuit. Conversely, when the applied voltage is removed, a high inductive spike is produced. Therefore, on turn-on the controlling relay contacts will see very high in-rush currents, for a varying period of time, depending on the motor load: conversely on turnoff, an arc will be produced across the opening contacts. Accepted industry practice for motor loads would suggest a derating to 20 percent of the resistive load rating of the relay. If, however, the relay in question has a motor load rating then the derating should be 75 percent of this specified value.

Incandescent Lamp Loads.   An incandescent lamp filament is considered to be a resistive load. When cold, the resistance of the filament is very low. When power is applied, an in-rush current results that can be as much as 15 times greater than the final steady-state value. The magnitude of these in-rush currents can cause the contacts to either weld shut or degrade to the point where they can no longer be used. Accepted industry practice would be to derate the contacts to 10 percent of their resistive load ratings. If, however, the specific relay in question already has a lamp load rating, then the derating should be 75 percent of this specified value. Good design practice would also dictate that, wherever possible, a series current limiting resistor be used to control these in-rush currents.

Relay Coils.   Coils should not be derated. Any change from nominal voltage, other than ±5 percent tolerance, can have a serious effect on the operation of a relay.

Temperature.   Relays should not be operated above rated temperature because of resulting increased degradation and fatigue. The best practice is to derate 20°C from the maximum rated temperature limit.

CIRCUIT BREAKERS, FUSES, AND LAMPS

Circuit Breakers

Circuit breakers should be sized for each application to protect the circuit adequately from overvoltage or overcurrent. For optimum reliability *thermal circuit breakers* should not be subjected to operation in environments where temperatures vary from specified rating(s) because the current-carrying capability of the sensing element is sensitive to temperature variations. For optimum reliability *magnetic circuit breakers* should not be subjected to dynamic environments exceeding the device limitations.

Derating should depend on the type of circuit breaker or circuit being protected. Normally, circuit breakers of the magnetic type, which are relatively insensitive to temperature, should not be derated except for the interrupting capacity, which should be derated up to 75 percent of maximum rated interrupting capacity. Derating of standard circuit breakers used in high-reactance-type circuits may be required to avoid undesired tripping from high in-rush current.

Thermal-sensitive circuit breakers used outside of specified ratings should be derated to compensate for effects of operating ambient temperatures.

Fuses

Fuses should have ratings that correspond to those of the parts and circuits they protect. These fuse ratings should be compatible with starting and operating currents as well as ambient temperatures. Current-carrying capacity may vary with temperature. An example is given in MIL-F-23419/9.

Fusing should be arranged so that fuses in branch circuits will open before fuses in the main circuit.

Indicator Lamps

Incandescent and gaseous indicator lamps should be protected from voltage or current surges above ratings.

Derating.   Operating incandescent lamps at 94 percent of rated voltage will double the life of the lamp, with only a 16 percent drop in light output. Current is the principal stress parameter for gaseous lamps. Derating current to 94 percent of the lamp's rating will double the life expectancy.

Leds

Temperature and current are the primary stress parameters for light-emitting diodes. The designer should keep maximum junction temperature below 105°C and should maintain average forward current at a level no greater than 65 percent of the rated value.

BIBLIOGRAPHY

Ramakrishnan, A., T. Syrus, and M. Pecht, "Electronic Hardware Reliability," CALCE Electronic Products and Systems Center, University of Maryland.

ASFC PAM 800-27, "Part Derating Guidelines."

U.S. Dept. of the Air Force Military Standard Technical Requirements for Parts, Materials, and Processes for Space and Launch Vehicles, MIL-STD-1547 (USAF).

U.S. Dept. of Defense, "Resistors: Selection and Use of," MIL-STD-199.

U.S. Dept. of Defense, "Capacitors: Selection and Use of," MIL-STD-198.

U.S. Dept. of Defense, "Transformer, Audio Frequency," MIL-T27.

U.S. Dept. of Defense, "Fuse Instrument Type Style FM09 (Non-indicating)," MIL-STD-F23419.

CHAPTER 3.3
SEMICONDUCTOR RELIABILITY

Clarence M. Bailey, Jr., Richard L. Doyle

INTRODUCTION

Electronic systems are designed to operate for a specified period, which is determined by customer requirements, as well as by cost and performance. Since electronic systems consist largely of electronic devices, the system reliability is mostly dependent on the reliability of the individual devices in the system application environment. Semiconductor devices, particularly integrated circuits, considering their fundamental and complex functions, are at the heart of most modern electronic systems.

It is not possible to predict the lifetime or degradation rate (reliability) of any *individual* semiconductor device. However, it is possible to treat populations of such devices probabilistically, and thereby predict average lifetimes. This probabilistic approach is discussed here in terms of different failure rate models, which can be used to describe and predict reliability (on the average) for semiconductor devices. Failure rate λ for all electronic components including semiconductors is considered constant when used in the various reliability databases described in "Part-Failure Modeling," p. 3.10, since these databases use the exponential reliability distribution. However, these databases do not account for failure rate change ($\dot{\lambda}$). $\dot{\lambda}$ is a first-order derivative of failure rate and exhibits a sloped straight line on a linear failure rate chart. When this slope is positive, failure rate is increasing (wearout); when this slope is negative, failure rate is decreasing (infant mortality). The slope is zero when the failure rate is constant, which is generally considered the useful life of the part. However, all three regions are becoming more important as the result of more complex ICs, less than 100 micron trace widths, and lower operating and gate voltages.

This section accounts for all three regions of the failure rate curve by using a more complex reliability distribution such as the Weibull distribution. For the more complex distributions, the hazard rate $\lambda(t)$ is used to represent the instantaneous rate of failure for units of a population that have survived to time t. Electronic device reliability is usually described in terms of a particular hazard rate model, and particular parameter values, given that model. It is noted that the model also applies to electronic passive devices as well; and, in the final analysis, to electronic systems composed of such semiconductor and passive devices. The discussion here is confined, however, to semiconductors, and the model is referred to as the device hazard rate model.

DEVICE HAZARD RATE MODEL

General

Historically, hazard rates have been modeled in terms of the traditional bathtub curve. Such a curve has three regions that are associated with infant mortality, steady-state operation, and wearout. Infant mortality is characterized by an initially high, but rapidly decreasing hazard rate. These early failures come from a small fraction of the population that can be considered weak. The defects in these weak units are usually not immediately fatal

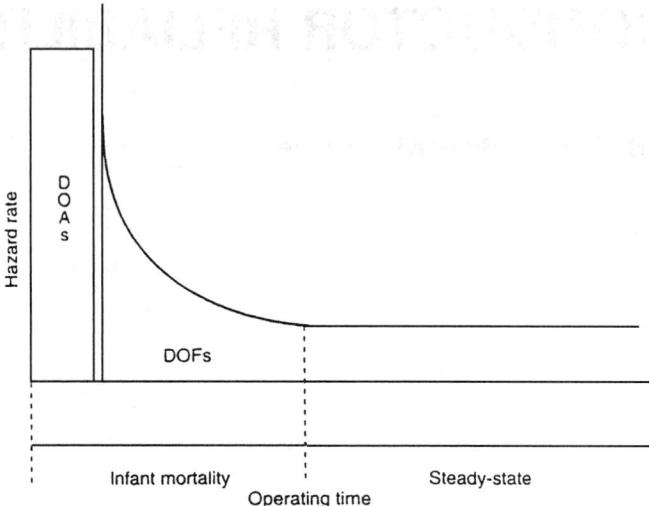

FIGURE 3.3.1   Conceptual reliability model.

but cause failure within a short period of time. After the majority of the weak units fail, operation moves into the steady-state region, in which failures occur at a much slower rate. The steady-state region is therefore characterized by a constant or slowly changing hazard rate. Wearout occurs when the hazard rate rises and the remaining units fail. For most semiconductor devices in normal environments the wearout period is far enough away to have little or no impact on the reliability of the device through normal equipment service life. However, as ICs continue to develop in complexity and physical parameters become smaller, the wearout region is starting to infringe on the useful life of the part. This may have to be addressed in the near future, but at the present time we are only concerned with the infant mortality and steady-state regions of the curve. The wearout region will have to be fit with a separate distribution in the future.

Figure 3.3.1 shows a conceptual reliability model for devices. The failures that occur very early in life are called dead-on-arrivals (DOAs), which are a part of infant mortality and are represented by the vertical box shown in Fig. 3.3.1. Most frequently they occur at first circuit board test. They may have been good when shipped but were found to have failed at various levels of equipment assembly and test. They are sometimes found at first equipment turn-on after shipment to the field. DOAs cannot be related to operating time. A device can test as satisfactory, be assembled into equipment, and then fail to work before the equipment has been in operation. The rate of such failures may in some cases be time-dependent, resulting from the same failure mechanisms as found later in infant mortality. On the other hand, others seem to be event-dependent, owing to handling during equipment manufacture and test. Although their existence is recognized, an accurate quantitative estimate of failure owing to DOAs is not possible; and we do not attempt to include them in our reliability model. Fortunately, most DOAs are found during the equipment manufacturing process. The failures that occur during operation are called "device operating failures" (DOFs). DOFs, with the exception of DOAs, encompass the infant mortality as well as the steady-state failures. Wearout is not included in this conceptual model because, as stated previously, it is not expected to occur during service life.

Failure Time Distributions

The model uses two statistical distributions that are useful in modeling device failures. The *exponential distribution* is characterized by a constant failure rate and is used to describe the steady-state hazard rate beyond the infant mortality region of device life. Pertinent functions for the exponential distribution are listed as follows and are illustrated in Fig. 3.3.2.

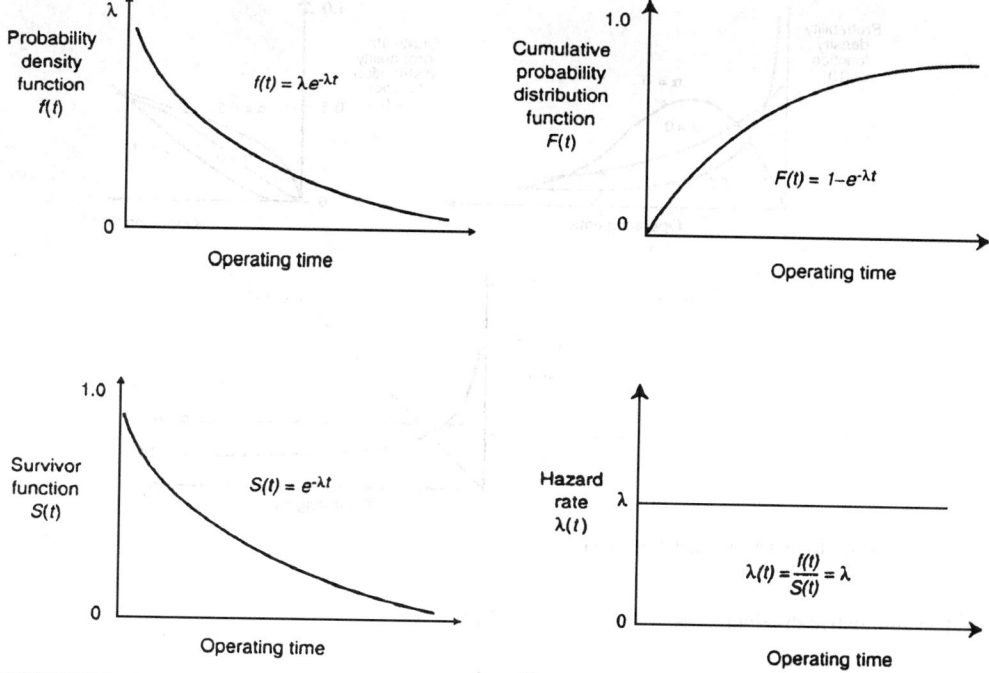

FIGURE 3.3.2 Exponential distribution.

- The probability of failure within some time interval 0 t_0 t the probability density function, is

$$f(t) = \lambda e^{-\lambda t} \qquad (t \geq 0, \lambda > 0)$$

- The probability that a device fails at or before a time t, the cumulative distribution function, is

$$F(t) = 1 - e^{-\lambda t}$$

- The probability of surviving to time t, the survivor function, is

$$S(t) = e^{-\lambda t}$$

- The hazard rate is constant:

$$\lambda(t) = \frac{f(t)}{S(t)} = \lambda$$

The Weibull distribution is especially used for modeling infant mortality failures. For the Weibull distribution, the hazard rate varies as a power of device age. The pertinent functions of the Weibull distribution are listed as follows and are illustrated in Fig. 3.3.3.

- The probability density function is

$$f(t) = \lambda_1 t^{-\alpha} e^{\dfrac{-\lambda_1 t^{1-\alpha}}{1-\alpha}} \qquad (t \geq 0, \lambda_1 > 0, \alpha < 1)$$

where $\lambda_1 > 0$ is the scale parameter of the distribution hazard rate, and α is the shape parameter of the distribution.

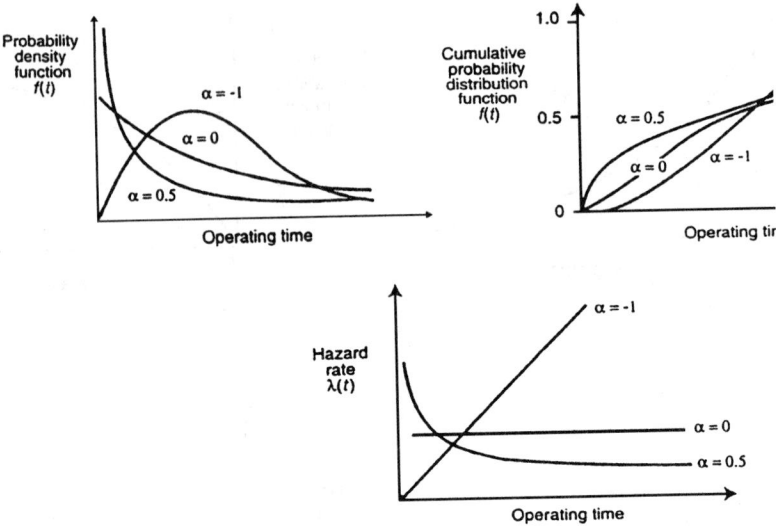

FIGURE 3.3.3 Weibull distribution.

- The cumulative distribution is

$$F(t) = 1 - e^{\dfrac{-\lambda_1 t^{1-\lambda}}{1-\alpha}}$$

- The hazard rate is

$$\lambda(t) = \lambda_1 t^{-\alpha},\ \alpha < 1,\ t > 0$$

- The failure rate change ($\dot{\lambda}$) is

$$\dot{\lambda}(t) = -\alpha\lambda,\ t^{-(\alpha+1)},\ \alpha < 1,\ t > 0$$

- The survivor function is

$$S(t) = e^{\dfrac{-\lambda_1 t^{1-\alpha}}{1-\alpha}}$$

When $0 < \alpha < 1$, the hazard test rate decreases with time ($\dot{\lambda}$ is negative). Therefore a positive α is used to model infant mortality. If $\alpha < 0$ the Weibull increases with device age ($\dot{\lambda}$ is positive) and device wearout may be modeled for this range of the shape parameter. If $\alpha = 0$, the hazard rate is constant ($\dot{\lambda}$ is zero) showing that the exponential distribution is a special case of the Weibull distribution. Since the shape of the Weibull distribution changes with α, it is called the shape parameter.

Specific Hazard Rate Model

Figure 3.3.4 shows the specific hazard rate model used to characterize semiconductor device reliability as well as other electronic devices and systems. The hazard rate in the infant mortality region is modeled by a Weibull hazard rate that decreases with time. In the steady-state region, the reliability is characterized by the exponential distribution where the failure rate is constant. A feature of the Weibull is that the hazard rate is a straight

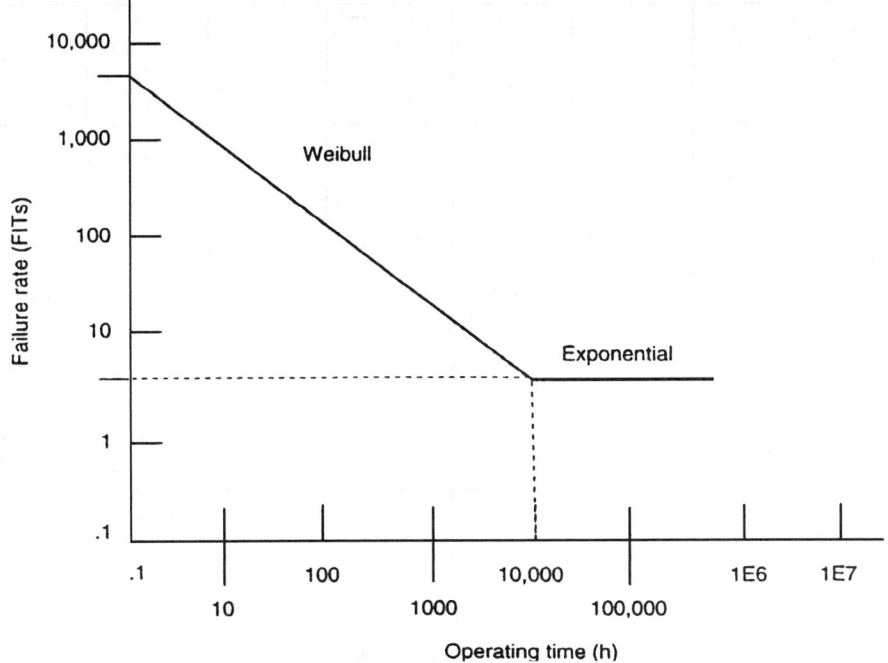

FIGURE 3.3.4 A hazard rate model for electronic components is the basis for the electronic equipment and systems model.

line when plotted on log-log scales as in Fig. 3.3.4. In such a plot, the slope is $-\alpha$, with $0 \leq \alpha < 1$, and the intercept at $t = 1$ h is λ_1, and hazard rate in infant mortality is described by

$$\lambda(t) = \lambda_1 t^{-\alpha}, \ 0 \leq \alpha < 1, \ 0 < t < t_c \tag{1}$$

Beyond some time t_c, assumed to be 10^4 h (slightly over the approximately one year of infant mortality), the hazard rate is assumed to remain constant; that is

$$\lambda(t) = \lambda_L, \quad \text{for } t \geq t_c = 10^4 \text{ h}$$

The hazard rate unit is the FIT (or one failure in 10^9 device hours). The hazard rate shown in Fig. 3.3.4 is typical but does not necessarily correspond to any particular device. As background for this model we note that there are two distinct, but different sources for information on semiconductor device reliability: accelerated life tests, and factory- and field-monitored performance. The former provide information about reliability in the very long term, the latter primarily gives information in the short term. Accelerated test conditions are required to determine the main lifetime distribution of the main population of the devices. For semiconductor devices this is usually well described by the lognormal distribution, and, based on such accelerated tests, we can relate the distribution to normal use conditions. Specifically, the maximum hazard rate, at use conditions, can be determined from accelerated life test data. Figure 3.3.5 shows a possible relationship between accelerated stress test results and the hazard rate model. In contrast, short-term hazard rates can be directly measured from field studies of no more than 2 to 3 years' duration. In those studies, a plot of the logarithm of the observed hazard rate versus the logarithm of time is usually found to fit a straight line and can be modeled by the Weibull distribution. Beyond the infant mortality period, such field studies

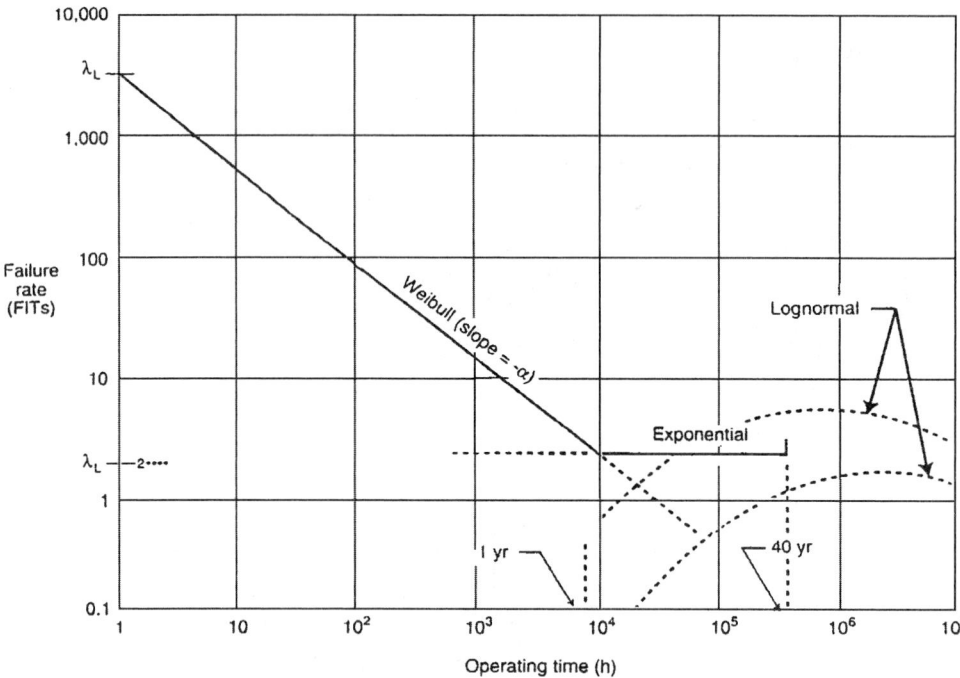

FIGURE 3.3.5   This device failure-rate model is a combination of a Weibull and an exponential distribution. Possible relationships between accelerated-stress results and the model are shown by the two-dashed lognormal curves.

have not contributed much significant information about the hazard rate, except that it is believed that the hazard rate continues to fall or levels off. Hence, we adopt the exponential distribution, with its constant failure rate, beyond 10,000 h—a conservative approach. The 10,000 h crossover point is arbitrary but reasonable. Beyond that point, the hazard rate is changing very little, and the constant failure rate model should be adequate.

Accelerated Life Model

Most modern semiconductor devices are so reliable that at normal use conditions only a very few failures are encountered. Many months of operation of large populations of devices might be needed to acquire statistically significant information on the hazard rate of those devices—at, in many cases, large costs. Another method, requiring less time, quantities of devices, and costs is accelerated testing. In an accelerated life test, a number of devices are subjected to failure-causing stresses that are at levels above what those devices would experience in use conditions. This type of accelerated aging test allows a distribution of failure times to be obtained, albeit at more stressful conditions than use conditions. In order to use this method, we must have a relationship between the distributions of the failure times at accelerated aging conditions to the distribution of failure times at use conditions, called an "accelerated life model." Not only does this accelerated life model allow us to "qualify" or prove-in devices, but it also allows us to develop accelerated stress tests for screening out some infant mortality failures; and, more importantly, relate hazard rate data acquired at one use condition to another different, use condition. Concerning the latter, it is noted that Table 3.3.3 provides possible hazard rate estimates at a reference temperature of 40°C.

We usually characterize an accelerated life model by a linear relationship between failure times at different conditions. If we designate t_1 as the failure time of a device at use conditions, t_2 as the failure time of the device at accelerated (more stressful) conditions, A as an accelerated factor, and $\lambda_1(t)$ and $\lambda_2(t)$ are the hazard rates at use and more stressful conditions, respectively, then

$$\lambda_1(t) = \frac{1}{A}\lambda_2(t/A) \tag{2}$$

The acceleration factor in the preceding equation may be a result of several stress variables. The stress variable of most importance and significance for semiconductor and other electronic devices as well as electronic systems is temperature. For this we use the well-known Arrhenius relationship. Arrhenius fits the temperature dependence of a rate constant k, independent of time, to the general form:

$$k = k_0 e^{-E_n/k_B T} \tag{3}$$

where T = absolute temperature (degree, Kelvin)
$\quad E_a$ = activation energy
$\quad k_B$ = Boltzmann constant
$\quad k_0$ = constant

We can derive from this general form, for two different temperatures T_1 and T_2, an acceleration factor owing temperature:

$$A_T = e^{(E_a/k_B)[(1/T_1)-(1/T_2)]} \tag{4}$$

If the activation energy for the process leading to device failure is known, then the acceleration factor for comparing reliability at two different temperatures can be calculated from the preceding equation. The Boltzmann constant is 8.6×10^{-5} eV/K. (°Kelvin is °Celsius + 273.) The activation energy eV is in electron volts.

INFANT MORTALITY CONSIDERATIONS

Infant Mortality Device Defects

These stem from a variety of device design, manufacturing, handling, and application related defects. Even though the failure mechanisms may vary from product to product and lot to lot, infant mortality can still be modeled adequately. Infant mortality failures are generally caused by manufacturing defects—including such defects as oxide pinholes, photoresist or etching defects, conductive debris on the chip, contaminants, scratches, weak bonds, and partially cracked chips and ceramics. Some infant mortality defects result from surface inversion problems, likely because of gross contamination or gross passivation defects. Many defects result from packaging problems. Some can be attributed to workmanship and manufacturing variations. Such variations can be reduced by changes in design or fabrication techniques, or in handling of the devices during manufacture. One important factor is a result of voltage spikes from *electrostatic discharge* (ESD), particularly at the circuit board manufacturing level. Some devices such as CMOS are more susceptible to this mechanism. Other infant mortality defects are inherent in design rules and constraints, or in the practical limitations of the manufacturing process and material control.

Effect of Temperature on Infant Mortality Hazard Rates

Studies of the infant mortality period show a low activation energy for the failure mechanisms contributing to semiconductor (and other electronic device) failures during infant mortality. These studies indicate an effective activation energy may be in the range of 0.37 to 0.42 eV, indicating that a single energy activation of 0.4 eV is a reasonable estimate for establishing time-temperature tradeoffs in infant mortality. Based on this value of 0.4 eV activation energy, the temperature acceleration factor, A_T, can easily be calculated from Eq. (4). It cannot, however, be used directly as a multiplier for the hazard rate at the desired operating temperature. (That procedure is only correct when the hazard rate is constant as in the exponential distribution.) Instead, it can be shown that when the distribution is Weibull, as in the infant mortality period, the multiplier for the hazard rate is $(A_{IM})^{1-\alpha}$. The hazard rate is then given by

$$\lambda(t) = (A_{IM})^{1-\alpha} \lambda_1 t^{1-\alpha} \tag{5}$$

Effect of Operating Voltage on Infant Mortality Hazard Rates

The dielectric breakdown of an oxide field, in say metal oxide semiconductor (MOS) devices, has been shown to be accelerated by an electric field. Failure analysis of such devices shows that about 30 percent of infant mortality failures are because of oxide-related failures, which are a function of applied voltage and of oxide thickness. Further investigations have established a voltage-dependent acceleration factor, which can be applied to device burn-in, where a voltage stress in excess of operating voltage is applied. Namely, the acceleration factor owing voltage stress is

$$A_V = e^{|C/t_{ax}(V_1 - V_2)|} \tag{6}$$

where C is the voltage acceleration factor in angstroms per volt, t_{ax} is the oxide thickness in angstroms, V_1 is the stress voltage in volts, and V_2 is the operating voltage in volts. A conservative estimate of C is 290 A°/V.

Effect of Temperature Cycling on Infant Mortality Hazard Rates

Mechanical defects such as weak wire bonds, poor pad adhesion, and partially cracked chips on ceramics constitute a significant portion of infant mortality failures. These failure mechanisms involve either plastic deformation or crack propagation, which can be caused by repeated stress in alternate directions resulting in fatigue failure. Fatigue failure can be increased indirectly by increasing the range of ΔT. There is much evidence that temperature cycling of devices results in decreasing hazard rates, and the decrease is a function of the number of cycles. However, there appears to be no general form for an acceleration factor. More than likely, the acceleration factor is a primary function of the materials and geometries of the specific device construction, and would have to be determined by experimentation. Temperature cycling is the major methodology for a technique known as *environmental stress screening* (ESS) or *sampling testing* (EST). The commonly used technique of step-stress testing is the most effective means for determining what kinds and levels of temperature cycling would be most effective for ESS or EST of the particular semiconductor device design type.

Infant Mortality Screening

The term *screening* here refers to the use of some environmental stress as a screen for reducing infant mortality defects. Involved are such accelerated stresses as temperature, temperature cycling, combined temperature and bias, and voltage (as discussed previously for CMOS devices). The selection of the proper stress depends largely on the semiconductor design and the nature and degree of the failure mechanisms contributing to the majority of the infant mortality defects.

Temperature cycling, discussed previously, followed by gross functional and continuity testing is very effective in screening out many mechanical defects. In addition to the defects mentioned before, it will accelerate failures caused by intermetallics in gold-aluminum wirebond systems, if temperature cycling follows a high temperature bake. Temperature cycling is an effective screen for poor seals of hermetic devices; and, in plastic encapsulated devices, temperature cycling accelerates defects caused by fatigue stressing of wire bonds because of plastic material surrounding the lead and bond area.

Screening with temperature alone—called "stabilization bake" or "temperature storage"—is an effective technique for certain failure mechanisms. It involves storing devices at an elevated temperature for a specific period of time. It is used to find diffusion and chemical effects, as well as material deterioration. It accelerates failures caused by oxide and gross contamination defects. It will accelerate the formation of intermetallics in a gold-aluminum wirebond system; if followed by temperature cycling, bond failures are likely to occur sooner. One effective method, referred to commonly as *burn-in*, is discussed in the following section.

Burn-in

Burn-in is an effective means for screening out defects contributing to infant mortality. Burn-in combines electrical stresses with temperature and time, and can be characterized as either static or dynamic. In static burn-in, a dc bias is applied to the device at an elevated temperature. The bias is applied in such a way as to reverse bias as many of the device junctions as possible. In dynamic burn-in the devices are operated so as to exercise the device circuit by simulating actual system operation. Static burn-in is only performed at the device level. Dynamic burn-in offers the option of use at both the device and the system level. When performed at the device level, dynamic system operation is simulated at a high temperature within the capabilities of the device. When performed at the system level, the system is operated at an elevated temperature. Since parts of the system are relatively limited in terms of temperature capability, the system burn-in is generally limited to lower temperatures than device burn-in. Static burn-in appears to be more effective for defects resulting from corrosion and contamination. Dynamic burn-in does not appear to be as effective for these problems. On the other hand, dynamic burn-in provides more access and complete exercise of internal device elements. It appears to be a much more effective screen for more complex, higher scale of integration devices. The choices of the type of burn-in (static or dynamic) and the specific stress conditions therefore depend on the device technologies, complexities, and predominant failure mechanisms, as well as the reliability requirements.

We concentrate here on a dynamic device burn-in based on the assumption of the Weibull distribution for infant mortality. This model assumes that the hazard rates of semiconductor devices are monotonically decreasing. Operation during device or system burn-in produces a certain amount of aging, and will result in a reduced hazard rate during subsequent aging. Subsequent operation begins at the reduced hazard rate and continues to decrease with additional operating time. The effect of burn-in is a function of the burn-in temperature and the subsequent operating temperature. The effective operating time is

$$t_{\text{eff}} = A_T A_V t_{bi}$$

when t_{bi} is the burn-in time at T_{bi}, the ambient burn-in temperature. (A_V is the voltage acceleration factor if performed as a part of the burn-in.) The Arrhenius equation is used to find the acceleration factor, A_{bi}, for burn-in compared to normal operation at the device hazard rate reference temperature. The hazard rate at the reference ambient temperature is

$$\lambda(t) = \lambda_1 (t_{\text{eff}} + t)^{-\alpha}$$

where $t = 0$ corresponds to the start of the device age after the burn-in. Figure 3.3.6 shows that the effect of burn-in is a decrease in the early life hazard rate. It is noted that the modeled burn-in assumes that all infant mortality defects would appear in the equivalent time of burn-in. This is a simplified effect. The stress is typically temperature, but temperature alone may not necessarily eliminate failures that might result from other stresses such as temperature cycling. Experience may well show a higher hazard rate after burn-in than predicted, followed by a more rapid decrease in hazard rate. If the temperature of subsequent operation is not at the device hazard rate reference temperature, then calculation of the hazard rate after burn-in is only slightly more complicated. The hazard rate after burn-in is then

$$\lambda_a(t) = (A_{op})^{1-\alpha} \lambda_1 [t + t_{bi}(A_{bi}/A_{op})]^{-\alpha} \tag{7}$$

where A_{op} is the acceleration factor for the operating temperature relative to the reference temperature, and

$$A_{bi} = A_T A_V$$

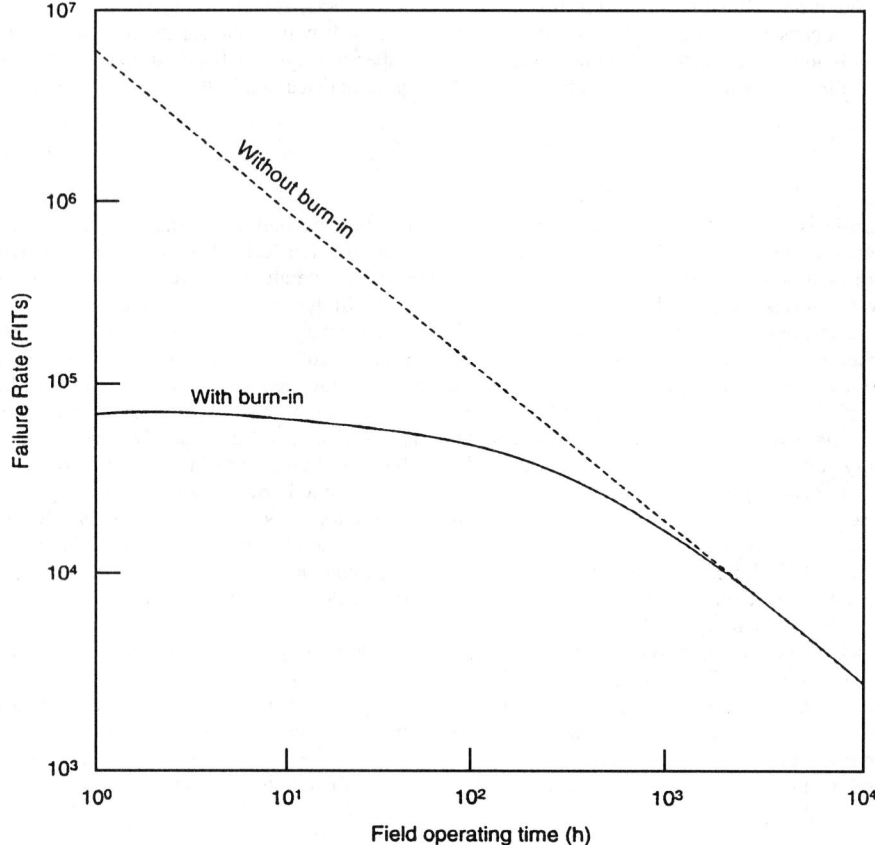

FIGURE 3.3.6   Effect of burn-in.

Infant Mortality Example

Question 1.   If a device has an infant mortality hazard rate, at a reference and use temperature of 40°C, characterized by $\lambda_1 = 15775$ FITs, and $\alpha = 0.7$, and if no burn-in is performed, what percentage of the devices will fail in a year (8760 h)?

Solution.   The expected percent of failures, with no replacement, is given by

$$\frac{\overline{N}}{n} = F(t) = 1 - e^{(\lambda_1/1-\alpha)t^{1-\alpha}} \tag{8}$$

where $\overline{N}$ = expected number of failures
n = number of devices
$F(t)$ = Weibull cumulative distribution function

Then

$$F(t) = 1 - e^{-[10^{-9}(15775)/(1-0.7)](8760)^{1-0.7}} = 0.0008 = 0.08 \text{ percent}$$

Question 2. What percentage of such devices will fail in a year if they have been burned in for 10 h at 125°C?

Solution. The hazard rate after burn-in is

$$\lambda(t) = \lambda_1(t_{eff} + t)^{-\alpha} \tag{9}$$

If $\int_0^t \lambda(t')dt' \ll 1$, then we may use an approximate equation for dropout:

$$\text{percent failing} = 10^{-7}\int_0^{8760} \lambda_1(t_{eff} + t')^{-\alpha}\,dt' = 10^{-7}\lambda_1\int_{t_{eff}}^{t_{eff}+8760} t^{-\alpha}\,d(t)$$

$$\text{percent failing} = 10^{-7}\frac{15775}{1-0.7}\left[(t_{eff}+8760)^{1-0.7} - (t_{eff})^{1-0.7}\right]$$

$$\text{percent failing} = 0.00525\left[(t_{eff}+8760)^{0.3} - (t_{eff})^{0.3}\right]$$

Using $E_a = 0.4$ eV, the acceleration factor, A_T, owing to temperature calculates to be approximately $= 24$. If, in addition, the device is an MOS device with gate oxide thickness of 250 Å, nominally operated at 5.5 V but is burned in at 7.5 V, the equation for voltage acceleration gives a voltage acceleration factor, A_V, approximately $= 10$. So the total burn-in acceleration factor $= (24)(10) = 240$, and the $t_{eff} = 2400$ h. The percentage failing then in 1 year after burn-in is

$$\text{percent failing} = 0.00525[(2400 + 8760)^{0.3} - (2400)^{0.3}] = 0.03 \text{ percent}$$

STEADY-STATE (LONG-TERM) CONSIDERATIONS

Long-term device reliability is defined as the reliability in the postinfant mortality period of device life, also referred to as the steady-state period of life. The hazard rate during this period is modeled as an exponential, constant failure rate. Since experience and theory indicate that, during the steady-state period, the hazard rate is near-constant and slowly changing, the assumption of the constant failure rate is felt to be a reasonable approximation. A knowledge of the contributing failure modes and mechanisms, and how to control them, is necessary to interpret and estimate long-term hazard rates. The discussion of the long-term reliability of semiconductor devices then includes a discussion of the steady-state failure mechanisms and the accelerating stresses.

Steady-State (Long-Term) Failure Mechanisms

Table 3.3.1 is a summary of many of the failure mechanisms in silicon devices. This table lists the processes leading to failure, appropriate accelerating stresses, and the ranges of activation energies. It is important to understand the accelerating stresses, as related to the specific failure mechanism, if one is to use accelerated test data to control and predict long-term hazard rates. Based on the known activation energies for the device failure mechanisms in steady-state, high temperature and other accelerated stress tests can be conducted to determine the effects of the accelerated stress tests on the device failure distribution and lifetime. Hazard rates at lower (use) stress can be extrapolated from the stress test data based on the assumed activation energy. In this activity, the lognormal distribution is found to be extremely valuable. There is much available information on the use of the lognormal distribution in analyzing accelerated testing data. The interested reader is urged to investigate the literature as needed.

Although it is difficult to list all failure mechanisms, it is possible to group some mechanisms together according to the accelerated stresses which affect them:

(*a*) Chemical reaction in contact areas, where contact metals react with the semiconductor material, and growth of intermetallic materials at the bonds of dissimilar metals such as gold and aluminum—elevated temperature, no electrical bias.

TABLE 3.3.1   Time-Dependent Failure Mechanisms in Silicon Devices

Device association	Failure mechanism	Relevant factors	Accelerating factors	Acceleration (apparent E_a)
Silicon-oxide and silicon-silicon oxide interface	Surface charge accumulation	Mobile ions V, T	T	$eV = 1.0 - 1.05$
	Dielectric breakdown	E, T	E	
	Charge injection	E, T, Q_{ss}	E, T	$eV = 1.3$ (slow trapping)
Metallization	Electromigration	T, j, A, gradients of T and j, grain size	T, j	$eV = 0.5 - 1.2$ j to j^4 dependence
	Corrosion (chemical, galvanic, electrolytic)	contamination H, V, T	H, V, T	Strong H effect $eV = 0.3 - 0.6$ (for electrolysis); V may have thresholds
Bonds and other mechanical interfaces	Contact degradation	T, metals, impurities	Varied	
	Intermetallic growth	T, impurities, bond strength	T	A1-Au $eV = 1.0 - 1.05$
	Fatigue	Bond strength temperature cycling	T extremes in cycling	
Metal penetration	Aluminum penetration into silicon	T, j, A	T, j	$eV = 1.4 - 1.6$
Hermeticity	Seal leaks	Pressure, differential atmosphere	Pressure	

V = voltage, T = temperature, E = electric field, j = current density, A = area, H = humidity, Q_{ss} = interfacial fixed charge.

(*b*) Surface inversion, surface charge movement, and dielectric breakdown—elevated temperature and voltage.
(*c*) Weak wire bonds, mismatches of seal materials in hermetic packages, and mismatches of thermal expansion of the chip and its package—temperature cycling.
(*d*) Corrosion of materials, such as electrolytic corrosion of aluminum or gold producing opens (aluminum) or dendritic growth causing shorts (gold)—temperature, humidity, and voltage aided by some contamination.
(*e*) Electromigration of metallization stripes resulting in opens at one end of the stripe—elevated temperature and high current density.
(*f*) Soft errors in MOS DRAM devices, caused by α-particles—the errors are temporary depending largely on the memory cell size or the charge being stored, and the incidence of random α-particles. Accelerated test by a particle source.

Determining Semiconductor Failure Rate

The following is the standard method of determining semiconductor failure rate based on stress testing samples of the integrated circuit of interest.

Example.   Determine the constant failure rate in FITs (failures per billion hours) at 90 percent confidence at a junction temperature of 75°C. Use the following test results: Use an activation energy of 1.0 eV. One lot of 40 pieces tested at 150°C for 1500 with no failures. The other lot of 44 pieces tested at 150°C for

2500 h with no failures. Assume that the 150°C was junction temperature and not a die temperature or environmental chamber temperature.

This section explains how to calculate the failure rate using the summary test data. The failure rate resulting from these data is an average, or estimate, of the typical expected failure rate for this type of product and process technology. This calculation is made for the upper 90 percent confidence limit for the failure rate estimate using Chi-square statistics. The following formula predicts the maximum failure rate or worst-case condition:

$$\lambda(\text{max}) = \frac{\chi^2(1-a)}{2t}$$

(10)

with dof $= 2(r + 1)$

where χ^2 = Chi-square distribution value
   r = number of failures
   dof = degrees of freedom
   t = device hours
   a = statistical error expected in estimate.

For 90 percent confidence, $a = 0.1$ or $1 - a = 0.9$, $(1 - a)$ can be interpreted to mean that we can state with statistical confidence of $1 - a$: (i.e., 90 percent) that the actual failure rate is equal to or less than the calculated maximum $\lambda(\text{max})$ failure rate.

To get the total number of device-hours, multiply the total samples tested (survivors) times the total hours of test.

The test results represent different operating and test conditions, therefore, temperature corrections are required for the data. These data require a correction using the Arrhenius equation for an activation energy of 1.0 eV. The acceleration factor is 369.6 for converting test data from 150 to 75°C.

To find the maximum failure rate at normal operating temperature (30°C) and junction temperature (75°C) with 90 percent confidence, using Eq. (10) we have

$\chi^2 = 4.61^*$
   $r = 0$
dof $= 2$
   $t = (40 \times 1500 + 44 \times 2500 \text{ h})$ unit hours $= 3,240,026$ h
   $a = 0.1$

For 90 percent confidence, $a = 0.1$ for Q failures, since $1 - a = 0.9$ equals 90 percent for P success.

Steady-State (Long-Term) Example

Question.   If a VLSI semiconductor device has a long-term hazard rate, λ_L, of 40 FITs at 40°C, what percent of such devices will fail per year, after infant mortality, when operated at a temperature of 50°C? The activation energy is 0.4 eV.

Solution.   The acceleration factor for operation at 50°C from a reference temperature of 40°C is

$$A_T = e^{(E_a/k_B)[(1/T_1)-(1/T_2)]} = e^{(0.4/8.6\times10^{-5})[(1/273+40)-(1/273+50)]} = 1.58$$

(11)

---

*For a total of 0 failures, $\chi^2 = 4.61$ with 90 percent confidence. Values of χ^2 can be found in a number of statistical tables or using Excel.
   $\lambda(\text{max}) = 4.61/[2 \times (40 \times 1500 + 44 \times 2500 \text{ h})]$
   $\lambda(\text{max}) = 13.559 \times 10 - 6$ failures per hour   or
   $\lambda_{\text{max}}(150°C) = 13.559$ failures per million hours $= 13,559$ FITs
   $\lambda_{\text{max}}(75°C) = 13,559$ FITs/369.6 $= 36.7$ FITs

Since, in the steady-state, constant failure rate period, the acceleration factor is A_T times the hazard rate at the reference temperature, the hazard rate at the operating temperature is approximately = 63 FITs. The percent failure per year is

$$\frac{\overline{N}}{n} = 1 - F(t) = 1 - e^{-(63 \times 10^{-9})8760}$$

$$= 0.00055$$
$$= 0.055 \text{ percent}$$

Crossover Time from Infant Mortality to Steady State

Under the reference operating conditions, the crossover time is that time when infant mortality ends and the long term starts, and they have the same value of hazard rate at 10,000 h. However, for increased temperature, the constant hazard rate of the model increases more than the infant mortality hazard rate, leading to an apparent discontinuity in the model at 10,000 h. But the infant mortality region of the model is only intended to model failures that occur at a rate greater than the long-term rate. Therefore the transition time may be taken to be that time when the infant mortality hazard rate becomes equal to the steady-state hazard rate. Then for operation at accelerated conditions, the transition should occur before 10,000 h.

SEMICONDUCTOR DEVICE HAZARD RATE DATA

General

Semiconductor device hazard rate estimates are presented in this section. These estimates are based on experience, from various sources, with the various devices in communications systems in the field and in factory testing, as well as in accelerated testing. The hazard rates presented are believed to be typical of product from device suppliers who incorporate good quality and reliability programs as a part of their normal manufacturing procedures. The estimates are believed to be well within the current state of the art, and good suppliers should have little or no difficulty in producing devices that meet or exceed these hazard rate estimates.

Reference Operating Temperature

Reference conditions, for the reference operating temperature (case temperature) of 40°C, are similar to those for central office type telephone equipment; that is, the environment is air conditioned, humidity is not a significant factor, and the room ambient temperature is about 25°C. Implicit in this is the assumption that operation of the device in equipment causes its internal ambient to rise 15°C above a 25°C room ambient. If the internal ambient temperature is above 40°C, the hazard rates will be higher than the tabulated numbers.

Hazard Rate Multipliers

The hazard rates of semiconductor devices are affected by the environmental application of the devices, as well as temperature. Table 3.3.2 gives an example of the environmental application factors developed to relate the hazard rates to the environment in which the device is being used. Typically a reliability analysis will be performed using a standard failure rate datebase. Then one should use the failure rate multipliers specified by the database. These reliability databases are described on p. 3.10, "Part-Failure Modeling," and include: MIL-HDBK-217, Telcordia/Bellcore SR-332, CNET's RDF 2000, British Telecom's database, Reliability Analysis Center's PRISM, and IEEE STD 493.

TABLE 3.3.2 Environmental Application Factor (*E*), Example

Environment	*E*
Permanent structures, environmentally controlled	1.0
Ground shelters, not temperature controlled	1.1
Manholes, poles	1.5
Vehicular mounted	8.0

TABLE 3.3.3 Possible Semiconductor Device Hazard Rate Data

Device class	Expected hazard rate (in FITs)	
	λ_L	α
Silicon diode		
A. General purpose	4	0.6
B. Microwave	4	0.6
C. Rectifiers	5	0.6
D. Surge protector	5	0.6
E. Switching	5	0.6
F. Varactors	6	0.75
G. Varistors	2	0.75
Integrated circuits		
A. Digital		
1. Bipolar		
1–100 gates	5	0.4
101–1000 gates	15	0.4
1001–5000 gates	25	0.4
2. CMOS		
1–100 gates	5	0.6
101–1000 gates	12	0.6
1001–10,000 gates	25	0.6
ASIC	40	0.6
3. NMOS		
1–100 gates	2	0.7
101–1000 gates	10	0.7
1001–10K gates	20	0.7
B. Linear		
≤ 100 transistors	10	0.6
101–300 transistors	15	0.6
301–1000 transistors	30	0.6
C. Memory—PROM, EPROM, EEPROM		
1. Bipolar		
16K bits	20	0.75
128K bits	50	0.75
2. NMOS		
8K–256K bits	75	0.6
1M bits	90	0.6

(Continued)

TABLE 3.3.3 Possible Semiconductor Device Hazard Rate Data (*Continued*)

Device class	Expected hazard rate (in FITs)	
	λ_L	α
3. CMOS		
8K–256K bits	70	0.6
512K bits	80	0.6
IM bits	90	0.6
D. Memory—RAM and ROM		
1. Bipolar		
<64K bits	10	0.7
64K bits	20	0.7
256K bits	40	0.7
2. CMOS		
<256K bits	2	0.6
256K bits	4	0.6
1M bits	10	0.6
4M bits	30	0.6
3. NMOS		
<64K bits	6	0.6
64K bits	8	0.6
256K bits	10	0.6
E. Microprocessors		
1. Bipolar		
4 bits	50	0.65
2. CMOS		
4 bits	80	0.7
8 bits	80	0.7
32 bits	80	0.7
3. NMOS		
8 bit	80	0.6
16 bit	80	0.6
F. Digital signal processors		
1. CMOS		
8 bits	25	0.6
16 bits	30	0.6
32 bits	50	0.6
Opto-electronics		
A. Alphanumeric displays		
1. LCD/LED		
1 Character	20	0.6
8 Characters	60	0.6
16 Characters	110	0.6
32 Characters	200	0.6
B. LEDs	10	0.6
C. Opto-isolators	20	0.6
Transistors		
A. FET	20	0.6
B. Microwave	2	0.6
C. NPN, PNP		
up to 6 W	5	0.6
>6 W	20	0.6

Activation Energies

As stated previously an activation energy of 0.4 eV is used for all devices in the infant-mortality period. The acceleration factor, A_{IM}, thus calculated is applied to the hazard rate by the factor, $(A_{IM})^{1-\lambda}$. An activation energy of 0.4 eV may also be used for devices in the steady-state region. This is an "effective activation energy" based on the variety of different failure mechanisms. In the steady state case, the acceleration factor, A_T, thus calculated, is applied directly to the long-term hazard rate.

Hazard Rate Table

If you are using a standard failure rate database, you should use the failure rates of the parts as specified by the database. These reliability databases are described in "Part-Failure Modeling," p. 3.10. You may want to apply the following techniques to the infant mortality portion of the system. Table 3.3.3 presents the hazard rate estimates by device type, and generally by scale of integration, at the reference operating temperature of 40°C. The long-term hazard rate is presented in terms of the parameter λ_L in FITs. The Weibull slope during infant mortality is listed in terms of the parameter α. The infant mortality hazard rate λ_1 is not listed but may be calculated from the relationship

$$\lambda_1 = \lambda_L (10,000)^{\alpha}$$

As stated before, the activation energy which applies during infant mortality and possibly during steady state is 0.4 eV.

BIBLIOGRAPHY

"Reliability Stress Screening of Electronic Hardware," IEC 60300-3-7 (1999–05).

Amerasekera, E. A., and F. N. Najm, "Failure Mechanisms in Semiconductor Devices," 2nd ed., Wiley Europe, 1997.

Chien, W. T-K., and C. H. J. Huang, "Practical 'Building-in' Approaches for Semiconductor Manufacturing," *IEEE Trans. on Reliability*, December, 2002.

CHAPTER 3.4
ELECTROMECHANICAL AND MICROELECTROMECHANICAL SYSTEMS (MEMS) RELIABILITY

Richard L. Doyle

INTRODUCTION

Reliability of electromechanical and microelectromechanical systems (MEMS), like other important design parameters, is established by the design and manufacturing process. Actual reliability (time-to-failure) rarely exceeds desired reliability primarily because of wearout of these systems. The reliability design and analyses techniques discussed in this chapter will aid the designer in using reliability tools to provide a more reliable electromechanical component or system. One must focus attention on design weaknesses so that they may be corrected, protected against, or accepted after consideration. These techniques also provide a means of ensuring that a design will meet specified time-to-failure (reliability to life) requirements.

It is generally true that one can demonstrate elecromechanical reliability by test with a sufficient degree of statistical confidence prior to and during production. This is because of the short life that these electro-mechanical systems have. Their wearout time or cycles is relatively short and the mean-time-between-failures (MTBFs) is measured in hours rather than hundreds of years as is the case with electronic parts. These actual MTBFs can be demonstrated during accelerated life testing.

There are two fundamental types of failures. One results from higher stresses than the part can withstand. The second is caused by the part literally wearing out. Both types of failures are covered in this chapter. However, the second is the most intriguing since wearout comes in many different forms. Wearout can be a result of fatigue, like a spring failing after being subjected to the same load for many years. It can be the result of abrasion, like tire wear on a car. Wear and wearout also are caused by friction, thermal cycling, and so forth. Generally all mechanical parts have a finite life. It can be very long like the pyramids of Egypt or very short like some MEMS motors. We will attempt to quantify the wearout time for all mechanical parts based on stresses and wear in this chapter. In this way, one can predict the MTBF and determine when to perform maintenance.

The classic bathtub curve shown in Chap. 3.1 also applies to mechanical failures. Stage 1 (infant mortality) is very short. It is therefore not necessary to test mechanical components for extended lengths of time to screen out infant mortality failures. The majority of a mechanical component's life is spent in Stages 2 and 3, which show a gradual increase in failure rate during Stage 2 and an accelerated failure rate during Stage 3. This is because mechanical components have continuous wear and fatigue throughout their operating life. An excellent method of selecting an appropriate service time is to determine when (number of operating hours) the failure rate curve begins to accelerate the sharpest, sometimes called the knee of the hazard rate curve. As illustrated in Fig. 3.4.1, higher operating stresses result in shorter life of the part.

Examples of electromechanical devices include motors, sensors, switches, connectors, and so forth. Samples of the MEMS devices and additional applications are presented in Chap. 8.5, "Microsensors and

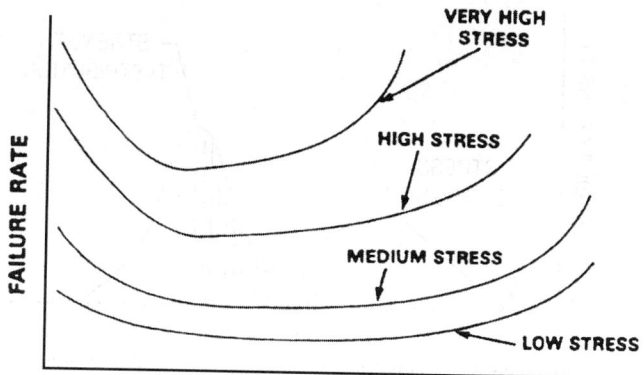

FIGURE 3.4.1   Effect of stress levels on mechanical failure rates.

Actuators." In the design of these systems, and operational requirements must be translated into design requirements, which are then verified by analysis (to provide an assessment of the design) and possibly by testing. The results can be compared with the operational requirements.

MECHANICAL STRESS ANALYSIS

Mechanical stress analysis is an important design procedure by which the maximum actual stresses induced on a material in its application are identified, so that adequate safety margins can be designed in.

Life-cycle phase.   Stress analysis is normally conducted during full-scale development as the detailed design definition progresses.

Purpose.   Stress analysis provides identification of the stresses, and confirmation of compliance with safety margins.

Objectives.   The primary objective of stress analysis conducted on electromechanical systems/ equipment is to identify any life-limiting stresses. This is performed by comparing stress-strength characteristics of parts and materials with internal and/or external loadings so that when the design is exposed to stress levels experienced in service, its life meets the requirements for that equipment. Stress analysis requirements should be invoked as a design technique to reveal system/equipment design deficiencies.

Material Stress-Strength Comparison

A material's strength or capability of handling a given stress varies from lot to lot and manufacturer to manufacturer. This variation for all materials of the same type can be represented by a statistical distribution of material strength. Similarly, the stress applied to a material changes from one point in time to another, with instantaneous changes occurring in temperature, mechanical stresses, transients, vibration, shock, and other deleterious environments.

At a random point in time the environmental effects can combine, reaching stress levels beyond the material's strength, resulting in failure of the material.

This strength-stress relationship can be described graphically by two overlapping statistical probability density distributions as shown in Fig. 3.4.2. For materials rated at the mean distribution used in an environment with an average stress equal to the mean of the stress distribution, the probability of failure is equal to the product of the areas of the two distributions where overlap occurs (i.e., the probability of the stress being greater than the minimum strength times the probability of the strength being lower than the maximum stress). This probability of failure is represented by the solid dark area of Fig. 3.4.2.

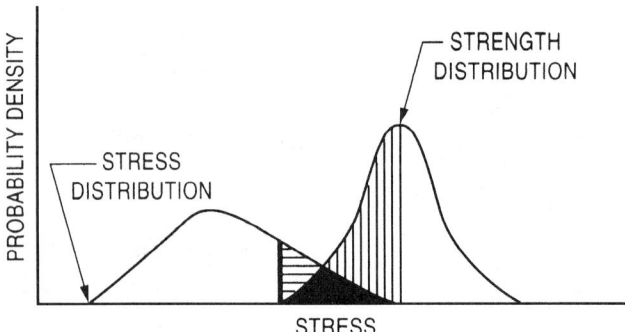

FIGURE 3.4.2 Two overlapping statistical probability-density distributions.

To reduce the probability of failure the potential stress levels can be reduced to a point where there is a very small probability of the stress exceeding the material's strength (stress distribution is shifted to the left), or the material's strength can be increased so that the probability of the combined stresses reaching or exceeding this strength is very small (strength distribution is shifted to the right). In most cases, the strength of the material cannot be increased and the only approach is to decrease the stress on the part. Alternatively, sometimes a different or stronger material can be used that can withstand the higher stresses. Either technique reduces material internal temperatures, decreasing the rate of chemical time-temperature reaction.

Procedure for Mechanical Stress Analysis

This chapter explains the fundamental mechanical stress analysis principles and shows the primary methods and formulas necessary for calculating mechanical stresses, safety factors, and margins of safety associated with electromechanical designs. These calculations are used in the mechanical stress analysis in two ways: (1) They ensure that the structural part has sufficient strength to operate in its service environment and under the loads specified, and (2) to a lesser extent, they ensure that there is no appreciable excess of material or over-design resulting in higher costs and burdening the program with excessive weight, mass, volume, and so forth.

Mechanical stress analysis is a detailed and complex engineering discipline. However, for our purposes simplifying assumptions will be made which provide equations for evaluating the approximate values of the stress, loads, margins of safety, and safety factors.

Several important factors must be considered before performing a mechanical analysis:

1. Important parameters are the loads, the material properties, and the actual calculated stresses.

2. Factors to be considered include the size, material joining techniques, welding, bonding, loads, material properties, moisture, thermal environment, and thermal limitations of the material.

3. Environmental factors such as corrosion and temperature and their resultant stresses must be considered.

4. Actual stress values must be calculated under steady-state loading conditions.

Symbols and Abbreviations

Symbols and abbreviations used in this chapter are defined as per US MIL-HDBK-5E. These symbols are abbreviations will be used throughout this section. with the exception that special statistical symbols are presented as necessary. To go further than this information, the reader should be familiar with the symbols and abbreviations used in mechanical structural analysis.

STANDARD DEFINITIONS, NOMENCLATURE, AND FUNDAMENTAL EQUATIONS

All electromechanical systems, while in operation, are subjected to external loads. These supplied loads must be transmitted through components and between components to the supports, which provide the necessary reactions to maintain equilibrium. In the process, forces and moments are developed in each component of the system to resist the externally applied loads; providing equilibrium for the components. The internal forces and moments at any point within the system are called *stress resultants*.

Normal Force.   If a uniform bar is subjected to a longitudinal force P, the force must be transmitted to its base. Under this loading the bar will increase in length and produce internal stress resultants called *normal forces*, denoted by P. Normal forces are associated with electromechanical components that lengthen under tensile loads and shorten under applied compression loads.

Shear Force.   If a uniform square plate is subjected to a uniformly distributed force f on three edges of the plate, then these forces must be transmitted to the base edge. This form of deformation is called *shear deformation* and is associated with an internal stress resultant called *shear force*, denoted by S.

Bending Moment.   If a uniform beam is subjected to a couple C, the beam is no longer straight, but has become bent. Deformations that bend electromechanical components are associated with internal stress resultants called *bending moments*, denoted by M.

Torque.   If a uniform circular shaft is subjected to a couple, the shaft must twist as the couple is transmitted to the base support. Deformations that twist electromechanical components are associated with internal stress resultants called *torque*, denoted by the vector C.

Fundamental Stress Equations

Internal stress resultants represent the total force or moment acting at any point within a electromechanical system. In reality these stress resultants are distributed over surfaces within the components, and their magnitudes per unit area are called *stress*. Stress is defined by force per unit area. Therefore, bending moments and torques must be described by statically equivalent force systems made up of distributed normal or shear forces in order to evaluate their effect in terms of stress.

Normal and Shear Stress.   Stresses acting perpendicular or normal to a surface are called *normal stresses*, while stresses acting in the plane of the surface are called *shear stresses*. Internal stress resultants do not necessarily act normal or in the plane of the surface in question. For convenience, in mechanical stress analysis, internal force resultants are always represented in component form so that stresses are either normal stresses or shear stresses.

Consider an incremental internal force resultant ΔR acting on an incremental surface area ΔA. The vector sum of ΔP (normal force) and ΔS (shear force) may be used to replace ΔR. The normal force per unit area and the shear force per unit area are then called normal stress f_n and shear stress f_s, respectively. Therefore, average values are defined by

$$f_{n(\text{avg})} = \frac{\Delta P}{\Delta A} \qquad f_{s(\text{avg})} = \frac{\Delta S}{\Delta A} \tag{1}$$

Notice that the correct stress involves more than magnitude and area, but also direction, sense, and location.

Fundamental Strain Equations

All electromechanical components deform under load. They may change size, shape, or both.

Normal Strain.   Normal strain e_n is defined as a change in length per unit of original length. The uniform bar increased in length by δ when subjected to the normal force P. Therefore, average values for normal strain are defined by

$$e_{s(\text{avg})} = \frac{\delta}{L} - \frac{L_{\text{final}} - L_{\text{initial}}}{L_{\text{initial}}} \tag{2}$$

Shear Strain.   Shear strain e_s is defined as change in angle between any two originally perpendicular lengths, measured in radians. A uniform square plate changes in shape when subjected to the shear force S. The square plate becomes a trapezoid under load and forms an interior angle α of less than 90°. Therefore, average values for shear strain are defined in radians by

$$e_{s(\text{avg})} = \frac{\pi}{2} - \alpha \tag{3}$$

When the deformations are small and the angle change is also small, average shear strain can be defined in terms of the shear deformation δ_s,

$$e_{s(\text{avg})} \geq \tan\,(e_{s(\text{avg})}) = \delta_s \frac{}{L} \tag{4}$$

COMMONLY USED FORMULAS

 Formulas and equations related to tensile and compressive action are given in US MIL-HDBK-5E. Formulas extracted from that handbook are provided on the accompanying CD-ROM, under the heading "Commonly Used Formulas in Electromechanical Reliability Studies."

MECHANICAL FAILURE MODES

This section presents mechanical failure modes that might occur and should be evaluated to ensure that the design has sufficient material strength and adequate margins of safety.

Tensile yield strength failure.   This type of failure occurs under pure tension. It occurs when the applied stress exceeds the yield strength of the material. The result is permanent set or permanent deformation in the structural part. This is not generally a catastrophic condition.

Ultimate tensile strength failure.   This type of failure occurs when the applied stress exceeds the ultimate tensile strength and causes total failure of the structural part at this cross-sectional point. This is a catastrophic condition.

Compressive failures.   Compressive failures are similar to the preceding tensile failures only under compressive loads. They result in permanent deformation or total compressive failure causing cracking or rupturing of the material.

Failures due to shear loading.   Yield and ultimate failures occur when the shear stress exceeds the strengths of the material when applying high torsion or shear loads. These failures generally occur on a 45° axis with respect to the principal axis.

Bearing failures.   Bearing failures are similar to compressive failures. However, they are generally caused by a round, cylindrical surface bearing on either a flat or a concave surface like roller bearings in a race. Both yield and ultimate bearing stresses should be determined.

Creep and stress rupture failures.   Long-term loads generally measured in years cause elastic materials to stretch even though they are below the normal yield strength of the material. If the load is maintained continuously and the material stretches (creeps), it will generally terminate in a rupture. Creep accelerates at elevated temperatures. This results in plastic strain because of creep even though the load is in the elastic region. Creep should always be checked under conditions where high loading for long periods of time are anticipated.

Fatigue failures.   Fatigue failures result from repeated loading and unloading of a component. This is a major cause for reducing the design allowable stresses below those listed in the various design tables for static properties. The amount of reduction in the design load is based on complex curves and test data. The amount of load cycling with respect to the maximum load and the number of cycles are the predominant variables. These variables are generally described in Moody diagrams (which relate stress to number of life cycles, *S–N* curves).

Metallurgical failures.   Metallurgical failures are the failures of materials because of extreme oxidation or operation in corrosive environments. Certain environmental conditions accelerate these metallurgical failures. The environmental conditions are: (1) heat, (2) erosion, (3) corrosive media, and (4) nuclear radiation. Therefore, depending on the environment, these conditions should be considered.

Brittle fracture.   Certain materials that have little capacity for plastic flow and are generally brittle, such as glass, ceramics and semiconductor materials, are extremely susceptible to surface flaws and imperfections. The material is elastic until the fracture stress is reached. Then the crack propagates rapidly and completely through the component. These fractures can be analysed using fracture mechanics.

Bending failures.   A bending failure is a combined failure where an outer surface is in tension and the other outer surface is in compression. The failure can be represented by tensile rupture of the outer material.

Failures from stress concentration.   This occurs when there is an uneven stress "flow" through a mechanical design. This stress concentration generally takes place at abrupt transitions from thick gauges to thin gauges, at abrupt changes in loading along a structure, at right angle joints or at various attachment conditions.

Failures from flaws in materials.   This is generally because of poor quality assurance or improper inspection of materials, weld and bond defects, fatigue cracks or small cracks and flaws. These reduce the allowable strength of the material and result in premature failure at the flawed location.

Instability failures.   Instability failures occur in structural members, such as beams and columns particularly those made from thin material and where the loading is generally in compression. The failure may also be caused by torsion or by combined loading including bending and compression. The failure condition is generally a crippling or complete failure of the structural part.

All of the preceding failure modes should be analyzed as necessary to confirm the adequacy of the mechanical design. It should be noted that some of the analysis is very involved, particularly in the case of combined loads or complex structures. These may require the use of a finite-element computer program.

SAFETY FACTORS AND MARGINS OF SAFETY

In order to design a electromechanical component or system to perform a given function in a reliable manner, the designer must understand all the possible ways the component or system could lose its ability to perform that given function. This requires a complete understanding of how the system or component responds to externally applied loads. With the aforementioned modes of failure established, reasonable limits on load can be defined.

The *safety factor* (SF) is a strength design factor defined by the ratio of a critical design strength parameter (tensile, yield, and so forth) to the anticipated operating stress under normal operating conditions. For example, let F be the strength of the material and f_{mw} be the maximum allowable working stress. Then the factor of safety becomes

$$SF = \frac{F}{f_{mw}} \tag{5}$$

The *margin of safety* (MS) is usually expressed as the maximum allowable working stress (f_{mw}) divided by the applied stress (f) minus 1, as shown in the following equation:

$$MS = \frac{f_{mw}}{f} - 1.0 = \frac{F/SF}{f} - 1.0 \tag{6}$$

This is for a simple unidirectional stress. Any negative value indicates that the structural part will fail because the applied stress is in excess of the allowable material strength.

RELIABILITY PREDICTION TECHNIQUES

With electronic hardware, manufacturers produce millions of similar parts that are then incorporated into thousands of different types of electronic equipment. Billions of hours of operation are accumulated which are used to guess at electronic part failure rates. Because of the many different uses, it is possible to accumulate extensive data, with respect to failure rate, under various environmental, temperature, and electrical stress conditions.

With the exception of nuts and bolts type hardware, electromechanical equipment is designed for a specific configuration and use. There is insufficient quantity of any specific device to accumulate the hours necessary to establish a basic failure rate, or variations caused by different stress levels, environments, and temperature extremes.

A "Probabilistic Stress and Strength Analysis" (p. 3.71) provides techniques for determining structural reliability through the computation of the probability of failure. Although this technique provides a good reliability estimate based on stress and strength, it will be found that to use this method for every electromechanical part of a system would be a long and costly effort.

An alternative approach is failure mode, effects and criticality analysis (FEMCA). This procedure does not consider any failure mechanisms that are not critical to equipment performance, or will not otherwise affect the performance of a prescribed mission. Another method is to establish a reasonable safety factor to minimize the risk of part failure.

There are several methods that can be used to determine the reliability of a electromechanical system. These methods may vary from precise to simple estimates. The precise methods require extensive computations for each element that is used in the electromechanical system. The simple approach uses generic data for each of the parts to obtain an estimated reliability, and as such, does not consider stress levels, temperature extremes, and material strengths. Also, one might use accelerated life testing.

Basic Reliability Definition

In order to use standard exponential reliability formulas, it is necessary to assume that the failure rates of all parts within a given unit are constant, that is, they do not increase or decrease with time. Further, those parts that exhibit a wearout mechanism or otherwise have a limited life are replaced at regular intervals prior to any appreciable increase in failure rate.

As an example, an electronic valve has a constant failure rate of 6.5 failures per 1 million cycles. The mean life is 90,000 cycles. The valve is used in equipment that is to have a 10-year life. On an average, the valve will be activated 10 times per hour throughout the equipment's life. Therefore, wearout will occur at 9000 h. In order for the 6.5 failures per 1 million cycles for this device to be valid over the 10-year life of the equipment, this device must be scheduled for replacement every year (8760 h), or more frequently, depending on the consequences of its failure.

Methods of Determining Electromechanical System Reliability

To date, no single method or technique has been accepted as the standard for computing the reliability of electromechanical systems. The following paragraphs briefly describe procedures that are used for computing electromechanical systems reliability.

Generic Failure Rate Data

The easiest and most direct approach to computing the reliability of electromechanical hardware is to use the failure rates of generic parts. As an example, it is possible to find the generic failure rate of an electronically activated valve from several sources. There may be many different valve configurations that fit this general or generic description. However, this valve is probably only one part in a much larger electromechanical system. Therefore, by using generic failure rates for all the parts that make up the system, it is possible to determine which parts will probably be the highest contributors to the system's unreliability. This directs emphasis and design attention to the weak link in the system.

There are several sources of failure rate information with respect to electromechanical hardware. Data sources include:

GIDEP: Government Industry Data Exchange Program

NPRD-95:   "Nonelectronic Parts Reliability Data," 1995, Reliability Analysis Center

MIL-HDBK-217F: "Reliability Prediction of Electronic Equipment," Feb. 1995, U.S. Military Handbook

A description of these data source is as follows:

GIDEP Failure Rate Source.   GIDEP (http://www.gidep.org/gidep.htm) publishes many documents that are continually being updated to provide government and industry with an in-depth perspective of what is currently taking place in practically every technology. Members contribute test data, reports, and other nonrestricted research data and reports.

A GIDEP summary is issued on a yearly basis. The primary drawback to using GIDEP as a source of failure rate data for a part in a specific environment is the lengthy search through computer databases to obtain needed backup data. This is required when environmental conditions differ from the listed environment, or there is an appreciable difference in part application.

NPRD-95.   NPRD-95, "Nonelectronic Parts Reliability Data Notebook" (http://www.rac.iitri.org), published in 1995 by the Reliability Analysis Center, is a report that is organized into four major sections. It presents reliability information based on field operation, dormant state, and test data for more than 250 major nonelectronic part types. The four sections are Background, Generic Data, Detailed Date, and Failure Modes and Mechanisms. Each device type contains reliability information in relation to specific operational environments. The data presented in this reliability publication are intended to be used as failure rate data, not part replacement data (scheduled maintenance). Only verified failures were used in the calculations of the failure rates.

It is noteworthy that: (1) no attempt has been made to develop environmental factors that allow the use of these failure rates for environmental conditions other than those indicated, (2) no attempt has been made to develop temperature history, (3) no attempt has been made to modify these failure rates for various safety factors or materials.

MIL-HDBK-217F.   MIL-HDBK-217F, "Reliability Prediction of Electronic Equipment," Feb. 1995, is a U.S. Military Handbook. It includes 19 sections of various types of electronic parts including seven sections on electromechanical devices. These include: Rotating Devices, Relays, Switches, Connectors, Interconnection Assemblies, Connections, and Meters.

Example Reliability Calculation

Note.   To convert failure rate per million cycles to failure rate per million hours, multiply the number of cycles per hour that the device operates times the cycle failure rate.

A pressure regulator failure rate is 34.0 failures/million cycles. It will operate at 200 cycles per hour. Therefore,

$$3.40 \text{ failures}/1,000,000 \text{ cycles} \times 200 \text{ cycles/hour}$$
$$\text{Failure rate} = 680.0 \text{ failures/million hours}$$
$$\text{MTBF} = 1/0.000680 = 1470 \text{ h}$$

For a mission time of 500 h, the probability of success (P_s) is

$$P_s = \exp(-500 \times 0.000680) = 0.712 = 71.2 \text{ percent}$$

Remember, generic failure rate sources assume that parts display a constant failure rate.

Computing Electromechanical Systems Reliability

The method of computing the reliability of a electromechanical system, either MTBF or probability of success, will depend on the situation. The reliability can normally be determined, in descending order of preference by:

1. Failure rate models (product multipliers)
2. Probabilistic stress and strength analysis for each part
3. Probabilistic stress and strength analysis for each critical part, with generic data for all others
4. Accelerated life testing
5. Parts count techniques for all parts using NPRD-95 data, GIDEP data, and/or MIL-HDBK-217F

Because of costs and time constraints, technique number 2 is rarely imposed. For critical electromechanical equipment, technique 1 should be required. For the majority of electromechanical systems, technique 1 or 3 is usually acceptable.

The use of the Parts Count Technique greatly simplifies the reliability computations of electromechanical systems. This technique, basically, is to list all electromechanical parts used, determine the generic failure rate for each part in a specific environment, add up the failure rates, and then compute the reliability, either MTBF or probability of success.

The computed value may not be within a factor of 10 of the actual test results. But the exercise will provide trend data and show which are the critical parts.

RELIABILITY PREDICTION USING PRODUCT MULTIPLIERS

Introduction

The accuracy of a reliability prediction using the generic failure rate data bank approach cannot be determined because of the wide dispersion of failure rates that occur for apparently similar components. A better method for predicting the reliability of electromechanical equipment is to use the product multiplier approach.

Variations in failure rates for electromechanical equipment are the result of the following:

Multiple Functions. Individual electromechanical components may perform more than one function.

Nonconstant Failure Rate. Failure rates of electromechanical components are not usually described by a constant failure rate because of wear, fatigue, and other stress-related failure mechanisms resulting in equipment degradation.

Stress History. Electromechanical equipment reliability is more sensitive to loading, operating mode, and utilization rate than electronic equipment reliability.

Criticality of Failure. Definition of failure for electromechanical equipment depends on its application. For example, failure owing excessive noise or leakage cannot be universally established.

The above-listed variables associated with acquiring failure rate data for electromechanical components demonstrates the need for reliability prediction models that do not rely solely on existing failure rate data banks.

Models have been developed that consider the operating environment, the effects of wear and other potential causes of degradation. The models are based on failure modes and their causes. Equations were developed for each failure mode from design information and experimental data. Failure rate models were developed from the resulting parameters in the equations and modification factors were compiled for each variable to reflect its effect on the failure rate of individual electromechanical parts.

Failure rate equations for each part, the methods used to generate the models in terms of failures per hour or failures per cycle and the limitations of the models are presented in the *Navy Handbook of Reliability Prediction Procedures for Mechanical Equipment* (NSWC-92/L01, May 1992).

Failure rate models have been developed for the following categories of electromechanical components and parts. The following are the basis for the fundamental part failures. Each of these are relatively simple. The complex equations should be simplified by holding constant the less important parameters. This will provide ease of calculation.

Beam: Based on standard beam stress equations

Shaft: Based on standard shaft stress equations

Plate: Based on standard plate stress equations

Spring: Based on standard spring stress equations

Connector, fitting: Based on standard fitting stress equations

Structural joint: Based on standard fitting stress equations

Bearing: As described on p. 3.69

Gear, spline: Based on standard gear stress equations

Seal, gasket: As described below

Solenoid: Based on standard failure rate data

Motor, pump, compressor, transducer, sensor: Based on standard failure rate data

Failure Rate Model for Seals

A seal is a device placed between two surfaces to restrict the flow of a fluid or gas from one region to another. Static seals, such as gaskets and O-rings, are used to prevent leakage through a mechanical joint when there is no relative motion of mating surfaces other than that induced by environmental changes. The effectiveness of a seal design is determined by the capability of the seal interfaces to maintain good mating over the seal operating environment.

Since the primary failure mode of a seal is leakage, the first step in determining the failure rate of a seal is to isolate each leakage path. In many cases backup seals are used in a valve design and the failure rate is not necessarily proportional to the total number of seals in the value. Each seal design must be evaluated as a potential internal or external leakage path.

A review of failure rate data suggest the following characteristics be included in the failure rate model for seals:

- Leakage requirements
- Material characteristics
- Amount of seal compression
- Surface irregularities
- Extent of pressure pulses
- Fluid viscosity
- Fluid/material compatibility
- Static versus dynamic conditions
- Fluid pressure
- Seal size
- Quality control/manufacturing processes
- Contamination level

The failure rate of a seal or gasket material will be proportional to the ratio of actual leakage to that allowable under conditions of usage. This rate can be expressed as follows:

$$\lambda_1 = \lambda_{b1} \frac{Q_a}{Q_f} \tag{7}$$

where λ_1 = failure rate of seal or gasket, failures per million cycles
λ_{b1} = base failure rate of seal or gasket, failures per million cycles
Q_a = actual leakage rate tendency, in^3/min
Q_f = leakage rate considered to be a device failure, in^3/min

The allowable leakage Q_f is determined from design drawings, specifications or knowledge of component applications. The actual leakage rate Q_a for a flat seal or gasket the leakage can be determined from the following equations:

$$Q_a = \frac{2\pi r_1 (P_s^2 - P_o^2)}{24\, vLP_o} (H^3) \tag{8}$$

where P_s = system pressure, lb/in^2
P_o = standard atmospheric pressure or downstream pressure, lb/in^2
v = absolute fluid viscosity lb—min/in^2
r_1 = inside radius
H = conduction parameter, in
L = contact length

The conduction parameter H is dependent on contact stress, material properties and surface finish. The conduction parameter is computed from the empirically devised formula:

$$H^3 = 10^{-1.1(C/M)} \times f^{1.5} \tag{9}$$

where M = Meyer hardness (or) Young's modulus for rubber resilient materials
C = apparent contact stress (psi)
f = surface finish (in)

The surface finish will deteriorate at a rate dependent on the rate of fluid flow and the number of contaminants according to the following relationship:

$$Z \geq f(\alpha \eta Q dT) \tag{10}$$

where Z = seal degradation
α = contaminant wear coefficient (in^3/particle)^2
η = number of contaminant particles per/in^3
Q = flow rate, in^3/min
d = ratio of time the seal is subjected to contaminants under pressure
t = temperature of operation, °F

The contaminant wear coefficient is a sensitivity factor for the valve assembly based on performance requirements. The number of contaminants includes those produced by wear in upstream components after the filter and those ingested by the system. Combining and simplifying terms provides the following equation for the failure rate of a seal or gasket:

$$\lambda_1 = \lambda_{b1} Xf \left\{ \frac{(P_S^2 - P_O^2) r_1}{Q_f vLP_O} (H^3)(\alpha \eta dt) \right\} \tag{11}$$

where λ_b is the base failure rate of a seal owing to random cuts, installation errors, and so forth, based on field experience data.

$$\lambda_1 = \lambda_{b1} \cdot C_P \cdot C_Q \cdot C_D \cdot C_H \cdot C_f \cdot C_V \cdot C_T \cdot C_N \cdot C_W \tag{12}$$

where λ_1 = failure rate of a seal in failures/million cycles
$\quad\lambda_{b1}$ = base failure rate of seal
$\quad C_P$ = multiplying factor, which considers the effect of fluid pressure on the base failure rate
$\quad C_Q$ = multiplying factor, which considers the effect of allowable leakage on the base failure rate
$\quad C_D$ = multiplying factor, which considers seal size
$\quad C_H$ = multiplying factor, which considers the effect of surface hardness and other conductance parameters on the base failure rate
$\quad C_f$ = multiplying factor, which considers seal smoothness
$\quad C_V$ = multiplying factor, which considers the effect of fluid viscosity on the base failure rate
$\quad C_T$ = multiplying factor, which considers the effect of temperature on the base failure rate
$\quad C_N$ = multiplying factor, which considers the effect of contaminants on the base failure rate
$\quad C_W$ = multiplying factor, which considers the effect of flow rate on base failure rate

Many of the parameters in the failure rate equation can be located on an engineering drawing, by knowledge of design standards or by actual measurement. Other parameters which have a minor effect on reliability are included in the base failure rate as determined from field performance data.

Reliability Prediction of Bearings

From the standpoint of reliability, bearings are by far the most important gear train components, since they are among the few components that are designed for a finite life. Bearing life is usually calculated using the Lundberg-Palmgren method. This method is a statistical technique based on the subsurface initiation of fatigue cracks through hardened air-melt bearing material. Bearing life is generally expressed as the L_{10} life, which is the number of hours at a given load that 90 percent of a set of apparently identical bearings will complete or exceed (10 percent will fail). There are a number of other factors that can be applied to the L_{10} life so that it more accurately correlates with the observed life. These factors include material, processing, lubrication film thickness, and misalignment.

The mean time to failure (MTTF) of a system of bearings can be expressed as

$$\mathrm{MTTF}_B = \frac{L_{10}(5.45)}{N^{0.9}} \tag{13}$$

where MTTF_B = design MTTF for the system of bearings
$\quad L_{10}$ = design life for each individual bearing
$\quad N$ = number of bearings in system

Most handbooks express the L_{10} life of a bearing to be a Weibull distribution as follows (reference "Marks' Handbook," 8th ed., pp. 8–138):

$$L_{10} = \frac{16{,}700}{RPM}\left(\frac{C}{P}\right)^k \tag{14}$$

where L_{10} = rated life in revolutions
$\quad C$ = basic load rating, lb (look up value from handbooks)
$\quad P$ = equivalent radial, load, lb
$\quad K$ = constant, 3 for ball bearings, 10/3 for roller bearings
$\quad RPM$ = shaft rotating velocity, rev/min

The above expression assumes a Weibull failure distribution with the shape parameter equal to 10/9, the value generally used for the number N of rolling element applicable bearings. The life (L_{10}) given in Eq. (14) includes the effects of lubrication, material, and misalignment under design conditions in addition to the common parameters. This value will usually suffice during early design. Typical material processing factors used for vacuum-melt aircraft quality bearings are 6.0 for ball bearings and 4.0 for roller and tapered bearings.

These material processing factors should be incorporated into the basic $MTTF_B$ to determine a representative life of these type bearings. These factors should be multiplied times the $MTTF_B$ to get the expected MTTF.

If the bearings being considered are low-speed (under 100 RPM), the $MTTF_B$ should be multiplied by 2 to account for the fact that the lubrication factor is very low.

MEMS Bearings.   The bearings used in a MEMS application are likely to have metal on metal sliding friction with little or no lubricant. Wear at the bearing surface should be determined by test. These bearings are, however, often used on lightly loaded gear trains. Higher bearing life is easily achieved in these applications if the loads are balanced by using dual gears or a planetary gear train.

The shaft should be a relatively hard material with the gear or armature having a low coefficient of sliding friction with the shaft. The following is a comparison of friction coefficient for different materials (reference "Marks' Handbook," 8th ed. pp. 3–27):

Nickel on nickel	0.53
Zinc on iron	0.21
Lead on iron	0.43
Aluminum on aluminum	1.40

These numbers are reduced by a factor of 3 (approximately) when a lubricant is used.

Reliability Prediction of Other Wearout Parts

Mathematical models provide the capability of predicting the reliability of all types of electromechanical components.

One can also incorporate time-varying parameters such as wear and wearout. To do this it is necessary to quantify the wear rate. This will provide a time to wearout and a probability failure rate about that wearout failure. This same technique works for fatigue life. The only difference is that strength of the material is changing as a function of stress cycles (time and magnitude of load fluctuation). This can be factored into the equation for determining the fatigue life.

Computer Programs for Determining Failure Rate of Parts

Programs for IBM-compatible PCs are available that will generate failure rate data for various electromechanical hardware as described using the product multipliers method.

Mechanical prediction programs are as follows:

MECHREL (mechanical reliability prediction program)

MRP (mechanical reliability prediction program)

RAM commander for Windows

Relex mechanical

MECHREL.   Mechanical reliability prediction program performs a mechanical reliability prediction to the component level using the established reliability parameters of electromechanical parts. Environmental and material factors are used in the calculation of base failure rates for the parts. The component failure rates are then summed to give an overall system reliability prediction.

Hard/Software Re.:   IBM PC

Supplier: Eagle Technology, Inc. 2300 S. Ninth St., Arlington, VA 22204; Phone: (703) 221-2154

Prices: Public Domain

MRP.   Mechanical reliability prediction program addresses the increasing need for predicting failure rates of mechanical equipment. Part of PSI's family of PC-based R&M software. MRP performs reliability prediction analyses in accordance with the "Handbook of Reliability Prediction Producers for Mechanical Equipment," Document DTRC-90/010, prepared by the U.S. Navy David Taylor Research Center under sponsorship of agencies of the DoD. In addition to the mechanical handbook, PSI supports the Reliability Analysis Center document NPRD-95, nonelectronic parts reliability data LIB-91, available as an optional library. The computer program enhances your mechanical reliability predictions by including failure rate data published in NPRD-95.

Hard/Software Req.:   DOS 3.0 or higher, 512 kb free RAM, 3 MB hard disk space, any printer/132 column

Interface Capabilities: Interfaces with RPP, FME, SRP, LSA, and other forms

Supplier: Powertronic Systems, Inc., 13700 Chef Menteur Hwy., New Orleans, LA 70129; Phone: (504) 254-0383

RAM Commander for Windows.   It is a Windows-based reliability, availability, and maintainability (RAM) tool set that implements a wide variety of electronic, electromechanical, and nonelectronic prediction and analysis methods. Robust reliability analysis tools are based on many predictive methods, including graphics engine, large parts library, block diagram modeling, and reliability growth analysis.

Hard/Software Req.:   IBM PC or Comp. Intel Pentium, Microsoft Windows

Interface Capabilities: CAD and other RAM packages

Supplier: Advanced Logistics Developments (ALD) Advanced Automation Corp., P.O. Box 4644, Rome NY 13442-4644; Phone: (800) 292-4519

Relex Mechanical.   Available for Windows, Mac, and MS-DOS, Relex Mechanical performs reliability analyses on MEMS systems per the "Handbook of Reliability Prediction Procedures for Mechanical Equipment." Its main purpose is to aid the user in evaluating and improving his products' reliability. It provides a state-of-the-art user interface with extensive hypertext help, CAD interface, defaults and derating analysis, system modeling and redundancy capabilities, and much more.

Hard/Software Req.:   IBM PC with Windows or Mac

Interface Capabilities: CAD, Text, ASCII, dBase, Lotus, WordPerfect, Paradox

Supplier: Innovating Software Designs, Inc., One Country Drive, Greensburg, PA 15601; Phone: (412) 836-8800

PROBABILISTIC STRESS AND STRENGTH ANALYSIS

Introduction

Probabilistic design can be determined for each type of part. This is determined by taking the basic equation that describes the stresses in the part and performing a Taylor's series expansion on the basic equation. For example, bending stress in a beam would be calculated using the equation MY/I. However, since I and Y are related (not statistically independent) one must use the section modulus (Z). It is very important to used statistically independent variables.

Probabilistic design can be a very important evaluation tool for predicting the probability of failure of an electromechanical system operating in a random loading environment. Ships, airplanes, missiles, and so forth experience random forces that must considered by the designer during the conceptual phases of the design.

The ultimate design goal would be to manufacture a product that never fails. This, however, is usually not achievable because of other considerations, such as cost and weight, which must be integral parts of any design. However, it may be possible to design a product (or system) that has a low probability of failure and not appreciably impact other design constraints.

Conventional stress analysis does not recognize that material properties, design specifications, and applied loads can be characterized as having a nominal value and some variation. Knowledge of the variables, as well as their statistical properties, will enable the designer to estimate the probability that the part will not fail.

The assumption is made in electromechanical stress and strength analysis that all random variables are statistically independent and are normally distributed. Dependent variables and distributions other than normality are complex to analyze and they usually do not enhance the validity of the analysis.

Statistical Concepts and Principles

 Concepts and principles that must be understood to effectively address probabilistic design with respect to the stress-strength relationship are given on the accompanying CD-ROM. See "Concepts and Principles Useful in Electromechanical Stress/Strength Analysis."

BIBLIOGRAPHY

Doyle, R. L., "Handbook of Reliability Engineering and Management," Chapters 18, 19, and 20, McGraw-Hill, 1996.

NSWC-94/L07, "Handbook of Reliability Prediction Procedures for Mechanical Equipment," Naval Surface Warfare Center, 1994.

NPRD-95, "Nonelectronics Parts Reliability Data Notebook and PC Data Disk," Reliability Analysis Center (RAC), 1995.

Reliability Prediction Procedure for Electronic Equipment, TR-332, Telcordia Technologies, 1997. (GR-332-CRE will replace TR-332.)

MIL-HDBK-5E, "Metallic Materials and Elements for Aerospace Vehicle Structures," *MEMS Reliability Assurance Guidelines for Space Applications*, JPI, Publication 99-1, JPL, 1999.

Zintel, J. C., "Applying the Lessons of MIL-SPEC Electromechanical Reliability Predictions to Commercial Products," ECN, Aug. 2002.

Electrical, Electronic, and Electromechanical Parts Selection Criteria for Flight Systems, PD-ED-1236, NASA.

Electrical, Electronics, Electromechanical (EEE) Parts Management and Control Requirements for NASA Space Flight Programs, NHB5300.4(1F), NASA, July 1989.

MIL-STD-975, "NASA Standard Electrical, Electronic, and Electromechanical Parts List," January 1994.

Reh, S., P. Lethbridge, and D. Ostcrgaard, "Quality Based Design and Design for Reliability of Microelectromechanical (MEMS) Systems Using Probabilistic Methods," ANSYS,

Young, W., and R. G. Budynas, "Roark's Formulas for Stress and Strain," 7th ed., McGraw-Hill, 2002.

ON THE CD-ROM:

Commonly Used Formulas. Formulas and equations useful in electromechanical reliability analysis.

Concepts and Principles useful in probabilistic stress and strength analysis of electromechanical components.

CHAPTER 3.5
RELIABLE SYSTEM DESIGN AND MODELING

Ronald T. Anderson, Richard L. Doyle, Stanislaw Kus,
Henry C. Rickers, S. Sugihara, James W. Wilbur

SYSTEM DESIGN

System reliability can be enhanced in several ways. A primary technique is through the application of *redundancy.* Other methods include *design simplification, degradation analysis, worst-case design, overstress analysis,* and *transient analysis.*

Redundancy

Depending on the specific applications a number of approaches are available to improve reliability through redundant design. They can be classified on the basis of how the redundant elements are introduced into the circuit to provide a parallel signal path. There are two major classes of redundancy.

In *active redundancy* external components are not required to perform the function of detection, decision, or switching when an element or path in the structure fails. In *standby redundancy* external elements are required to detect, make a decision, and switch to another element or path as a replacement for a failed element or path.

Techniques related to each of these two classes are depicted in the simplified tree structure of Fig. 3.5.1 Table 3.5.1 further defines each of the eight techniques.

Redundancy does not lend itself to categorization exclusively by element complexity. Although certain of the configurations in Table 3.5.1 are more applicable at the part or circuit level than at the equipment level, this occurs not because of inherent limitations of the particular configuration but because of such factors as cost, weight, and complexity.

Another form of redundancy can exist within normal nonredundant design configurations. Parallel paths within a network often are capable of carrying an added load when elements fail. This can result in a degraded but tolerable output. The allowable degree of degradation depends on the number of alternate paths available. Where a mission can still be accomplished using an equipment whose output is degraded, the definition of failure can be relaxed to accommodate degradation. Limiting values of degradation must be included in the new definition of failure. This slow approach to failure, *graceful degradation*, is exemplified by an array of elements configured to an antenna or an array of detectors configured to a receiver. In either case, individual elements may fail, reducing resolution, but if a minimum number operate, resolution remains good enough to identify a target.

The decision to use redundant design techniques must be based on a careful analysis of the tradeoffs involved. Redundancy may prove the only available method when other techniques of improving reliability have been exhausted or when methods of part improvement are shown to be more costly than duplications. Its use may offer an advantage when preventive maintenance is planned. The existence of a redundant element can

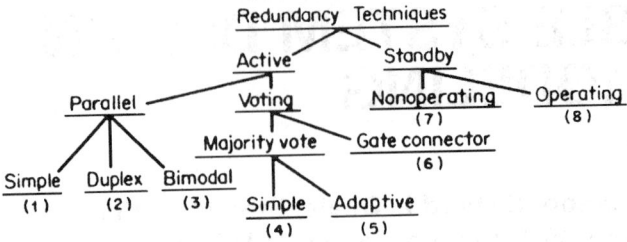

FIGURE 3.5.1 Redundancy techniques.

allow for repair with no system downtime. Occasionally, situations exist in which equipment cannot be maintained, e.g., spacecraft. In such cases, redundant elements may prolong operating time significantly.

The application of redundancy is not without penalties. It increases weight, space, complexity, design, cost, time to design, and maintenance cost. The increase in complexity results in an increase of unscheduled maintenance actions; safety and mission reliability is gained at the expense of logistics mean time between failures (MTBF).

In general, the reliability gain may be minimal for additional redundant elements beyond a few parallel elements. As illustrated in Fig. 3.5.2 for simple parallel redundancy, there is a diminishing gain in reliability and MTBF as the number of redundant elements is increased. As seen for the simple parallel case, the greatest gain, achieved through addition of the first redundant element, is equivalent to a 50 percent or more increase in the system MTBF. In addition to maintenance-cost increases because of the additional elements, reliability of certain redundant configurations may actually be less. This is a result of the serial reliability of switching or other peripheral devices needed to implement the particular redundancy configuration (see Table 3.5.1).

The effectiveness of certain redundancy techniques, especially standby redundancy, can be enhanced by repair. Standby redundancy allows repair of the failed unit (while operation of the unfailed unit continues uninterrupted) by virtue of the switching function built into the standby redundant configuration. The switchover function can also provide an indication that failure has occurred and that operation is continuing on the alternate channel. With a positive failure indication, delays in repair are minimized. A further advantage of switching is related to built-in test (BIT) objectives. Built-in test can be readily incorporated into a sensing and switchover network. Also, the hot swapping of redundant boards or modules minimizes system downtime.

An illustration of the enhancement of redundancy with repair is shown in Fig 3.5.3. The achievement of increased reliability through redundancy depends on effective isolation of redundant elements. Isolation is necessary to prevent failures from affecting other parts of the redundant network. The susceptibility of a particular design to failure propagation can be assessed by application of failure-mode-effects analysis.

Interdependence is most successfully achieved through standby redundancy, as represented by configurations classified as *decision with switching*, where the redundant element is disconnected until a failure is sensed. However, design based on such techniques must provide protection against switching transients and must consider the effects of switching interruptions on system performance.

Furthermore, care must be exercised to ensure that reliability gains from redundancy are not offset by increased failure rates because of switching devices, error detectors, and other peripheral devices needed to implement the redundancy configurations.

Design Simplification

Many complex electronic systems have subsystems or assemblies that operate serially. Many of their parts and circuits are in series so that only one need fail to stop the system. This characteristic, along with the increasing trend of complexity in new designs, tends to add more and more links to the chain, thus greatly increasing the statistical probability of failure.

Therefore, one of the steps in achieving reliability is to simplify the system and its circuits as much as possible without sacrificing performance. However, because of the general tendency to increase the loads on the components that remain, there is a limiting point to circuit simplification. This limit is the value of electrical stress that must not be exceeded for a given type of electrical component. Limit values can be established for

TABLE 3.5.1 Redundancy Techniques

Simple parallel redundancy

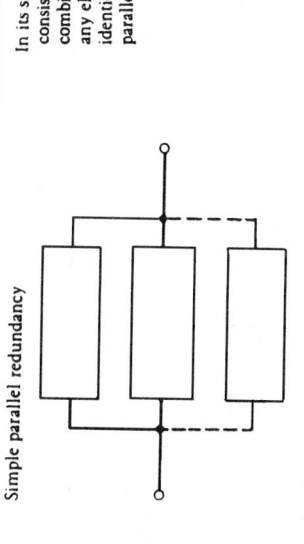

In its simplest form, redundancy consists of a simple parallel combination of elements; if any element fails open, identical paths exist through parallel redundant elements

Duplex redundancy

This technique is applied to redundant logic sections, such as A_1 and A_2 operating in parallel; it is primarily used in computer applications where A_1 and A_2 can be used in duplex or active redundant modes or as a separate element; an error detector at the output of each logic section detects noncoincident outputs and starts a diagnostic routine to determine and disable the faulty element

Majority-voting redundancy

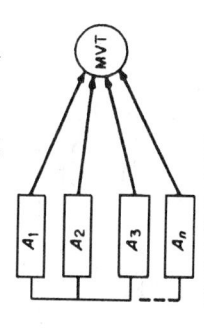

Decision can be built into the basic parallel redundant model by inputing signals from parallel elements into a voter to compare each signal with remaining signals; valid decisions are made only if the number of useful elements exceeds the failed elements

Adaptive-majority logic

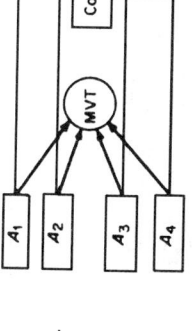

This technique exemplifies the majority-logic configuration with a comparator and switching network to switch out or inhibit failed redundant elements

Gate-connector redundancy

Similar to majority voting; redundant elements are generally binary circuits; outputs of the binary elements are fed to switchlike gates, which perform the voting function; the gates contain no components whose failure would cause the redundant circuit to fail; any failures in the gate connector act as though the binary element were at fault

(Continued)

3.75

TABLE 3.5.1 Redundancy Techniques (*Continued*)

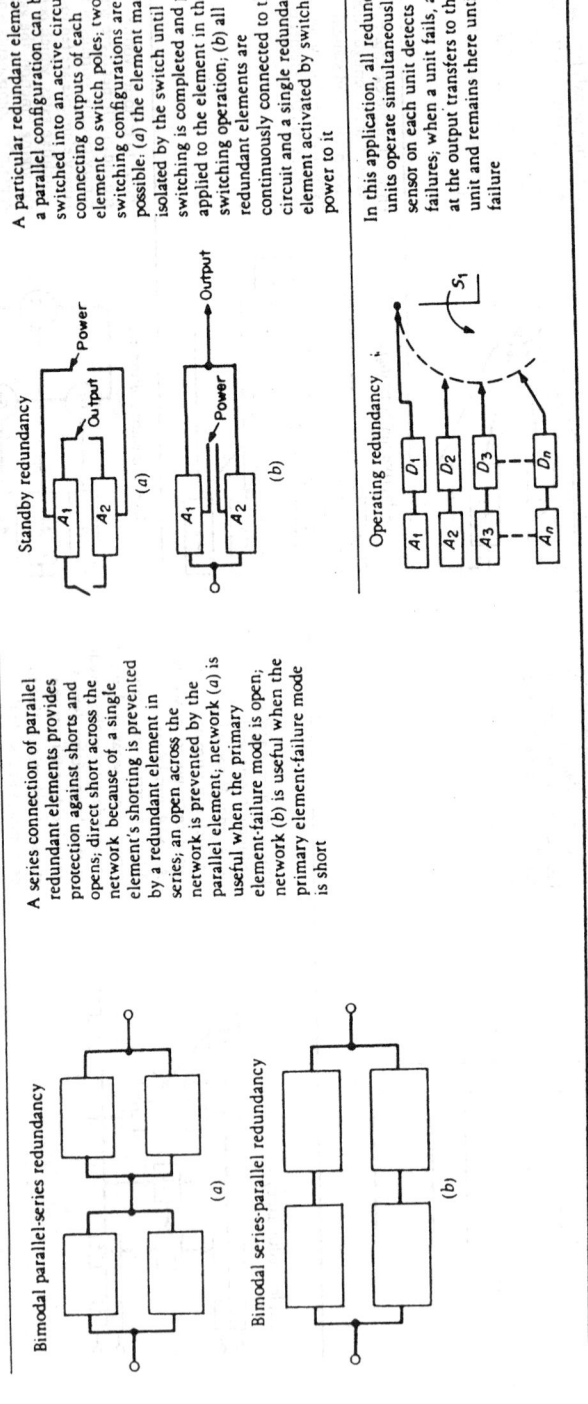

Bimodal parallel-series redundancy

(a)

Bimodal series-parallel redundancy

(b)

A series connection of parallel redundant elements provides protection against shorts and opens; direct short across the network because of a single element's shorting is prevented by a redundant element in series; an open across the network is prevented by the parallel element; network (a) is useful when the primary element-failure mode is open; network (b) is useful when the primary element-failure mode is short

Standby redundancy

(a)

(b)

A particular redundant element of a parallel configuration can be switched into an active circuit by connecting outputs of each element to switch poles; two switching configurations are possible: (a) the element may be isolated by the switch until switching is completed and power applied to the element in the switching operation; (b) all redundant elements are continuously connected to the circuit and a single redundant element activated by switching power to it

Operating redundancy

In this application, all redundant units operate simultaneously; a sensor on each unit detects failures; when a unit fails, a switch at the output transfers to the next unit and remains there until failure

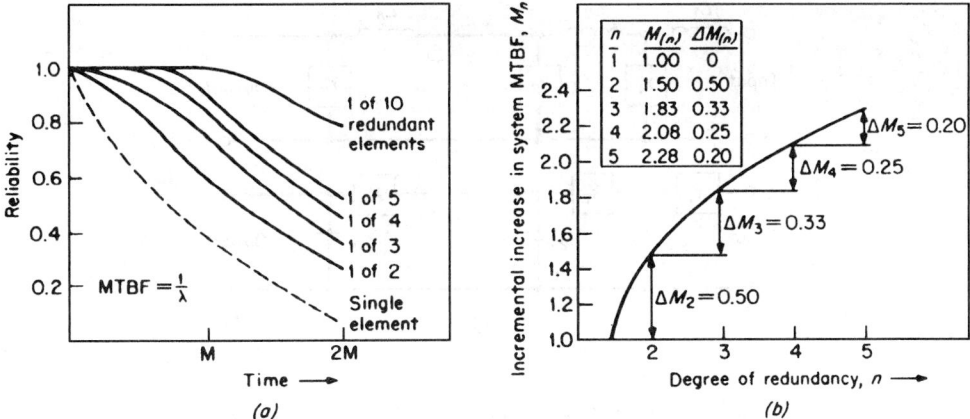

FIGURE 3.5.2   Decreasing gain in reliability as number of active elements increases: (*a*) simple active redundancy for one of *n* elements required and (*b*) incremental increase in system MTBF for *n* active elements.

various types of components as determined by their failure rates. In addition, it is also clear that the simplified circuit must meet performance criteria under application conditions, e.g., worst-case conditions.

Design simplification and substitution involves several techniques: the use of proved circuits with known reliability, the substitution of highly reliable digital circuitry where feasible, the use of high-reliability integrated circuits to replace discrete lumped-constant circuitry, the use of highly reliable components wherever individual discrete components are called for, and the use of designs that minimize the effects of catastrophic failure modes.

The most obvious way to eliminate the failure modes and mechanisms of a part is to eliminate the part itself. For instance, a digital design may incorporate extraneous logic elements. Minimization techniques, e.g., boolean reduction, are well established and can be powerful tools for incorporating reliability in a design through simplification. Simplification can also include the identification and removal of items that have no functional significance.

In addition, efforts should also be directed toward the reduction of the critical effects of component failures. The aim here is to reduce the result of catastrophic failures until a degradation in performance occurs. As an example, consider Fig. 3.5.4, which illustrates the design of filter circuits. A low-pass design, shown in

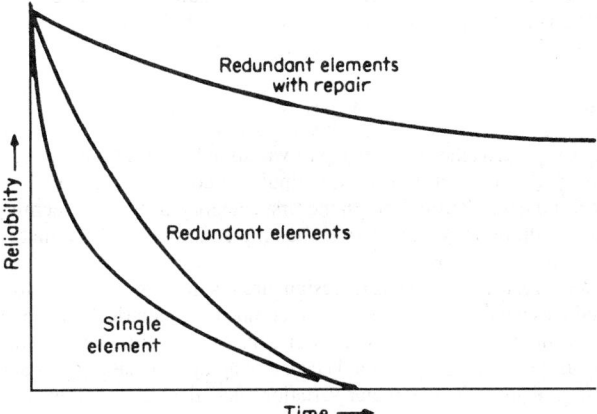

FIGURE 3.5.3   Reliability gain for repair of simple parallel redundant elements on failure.

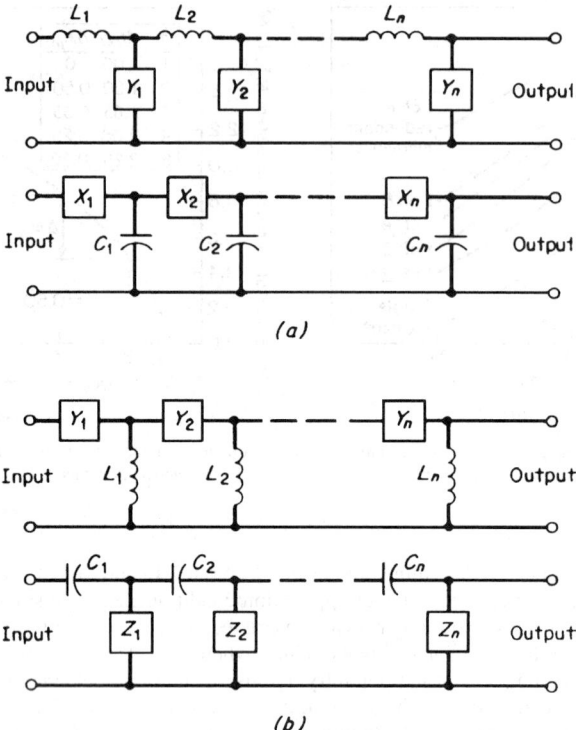

FIGURE 3.5.4   Alternative filter designs: (*a*) low-pass, (*b*) high-pass.

Fig. 3.5.4*a*, can involve either series inductances or shunt capacitances. The latter are to be avoided if shorting is the predominant failure mode peculiar to the applicable capacitor types, e.g., solid tantalum, since a catastrophic failure of the filter could result. Similarly, in the high-pass filter of Fig. 3.5.4*b*, using a shunt inductor is superior to using a series ceramic capacitor, for which an open is the expected failure mode. Here, the solid-tantalum capacitor, if applicable to the electrical design, could be the better risk, since its failure would result only in noise and incorrect frequency, not a complete loss of signal.

Degradation Analysis

There are basically two approaches to reduce part variation because of aging: control of device changes to hold them within limits for a specified time under stipulated conditions and the use of tolerant circuit design to accommodate drifts and degradation time. In the first category, a standard technique is to precondition the component by burn-in. In addition, there is detailed testing and control of the materials going into the part, along with strict control of fabrication processes.

In the second category, the objective is to design circuits that are inherently tolerant to part parameter change. Two different techniques are the use of feedback to compensate electrically for parameter variations and thus provide for performance stability and the design of circuits that provide the minimum required performance, even though the performance may vary somewhat because of aging. The latter approach makes use of analyses procedures such as worst-case analysis, parameter variation, statistical design, transient design, and stability analysis.

In the design of electronic circuits there are two ways to proceed. One is to view the overall circuit specification as a fixed requirement and to determine the allowable limits of each part parameter variation. Each part is then selected accordingly. The alternative approach is to examine the amount of parameter variation expected in each part (including the input) and to determine the output under worst-case combination, or other types of combination, e.g., rms or average deviation. The result can be appraised with regard to determining the probability of surviving degradation for some specified period of time. Optimization programs are helpful in both types of design approach.

Worst-Case Analysis

In worst-case analysis, direct physical dependence must be taken into account. For example, if a voltage bus feeds several different points, the voltages at each of the several points should not be treated as variables independent of each other. Likewise, if temperature coefficients are taken into account, one part of the system should not be assumed to be at the hot limit and the other at the cold limit simultaneously unless it is physically reasonable for this condition to occur. A general boundary condition for the analysis is that the circuit or system should be constructed according to its specifications and that the analysis proceeds from there.

In the *absolute worst-case analysis*, the limits for each independent parameter are set without regard to other parameters or to its importance to the system. The position of the limits is usually set by engineering judgment or circuit analysis. In some cases, the designer may perform several analysis with different limits for each case to assess the result before fixing the limits.

Modified worst-case analyses are less pessimistic than absolute worst-case analysis. Typically, the method for setting the limits is to give limits to critical items as in absolute worst-case analysis and to give the rest of the items limits of the purchase tolerance.

In any worst-case analysis, the values of the parameters are adjusted (within the limits) so that circuit performance is as high as possible and then readjusted so it is as low as possible. The values of the parameters are not necessarily set at the limits; the criterion for their values is to drive the circuit performance to either extreme. The probability of this occurring in practice depends on the limits selected by the engineer at the outset, on the probability functions of the parameters, and on the complexity of the system being considered.

Computer Analyses.   Computer routines are available for performing these analyses on electronic circuits. Generally speaking, the curve of circuit performance versus each independent parameter is assumed to be monotonic, and a numerical differentiation is performed at the nominal values to see in which direction the parameter should be moved to make circuit performance higher or lower. It is also assumed that this direction is independent of the values of any of the other parameters as long as they are within their limits. If these assumptions are not true, a much more detailed analysis of the equations is necessary before worst-case analysis can be performed. This involves generation of a response surface for the circuit performance which accounts for all circuit parameters.

Overstress and Transient Analysis

Semiconductor circuit malfunctions generally arise from two sources: *transient circuit disturbances* and *component burnout.* Transient upsets are generally overriding because they can occur at much lower energy levels.

Transient Malfunctions.   Transients in circuits may prove troublesome in many ways. Flip-flops and Schmitt triggers may be triggered inadvertently, counters may change count, memory may be altered owing to driving current or direct magnetic field effect, one-shot multivibrators may pulse, the transient may be amplified and interpreted as a control signal, switches may change state, and semi-conductors may latch-up in undesired conducting states and require reset. The effect may be caused by transients at the input, output, or supply terminals or combination of these. Transient upset effects can be generally characterized as follows:

1. Circuit threshold regions for upset are very narrow; i.e., there is a very small voltage-amplitude difference between the largest signals which have no probability of causing upset and the smallest signals that will certainly cause upset.

2. The dc threshold for response to a very slow input swing is calculable from the basic circuit schematic. This can establish an accurate bound for transients that exceed the dc threshold for times longer than the circuit propagation delay (a manufacturer's specification).

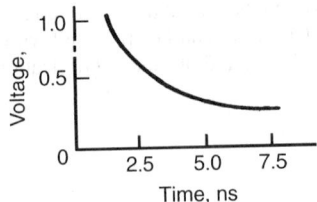

FIGURE 3.5.5   Square-pulse trigger voltage for typical low-level integrated circuit.

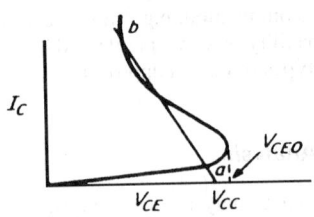

FIGURE 3.5.6   Latch-up response.

3. Transient upsets are remarkably independent in the exact waveshape, depending largely on the peak value of the transient and the length of time over which the transient exceeds the dc threshold. This waveform independence allows relatively easy experimental determination of circuit behavior with simple waveforms such as a square pulse.

4. The input leads or signal reference leads are generally the ones most susceptible to transient upset.

Standard circuit handbook data can often be used to gauge transient upset susceptibility. For example, square-pulse triggering voltage is sometimes given as a function of pulse duration. A typical plot for a low-level integrated circuit is shown in Fig 3.5.5.

Latch-up.   There are various ways in which semiconductors may latch-up in undesired conducting states that require reset (power removal). One common situation is shown in Fig. 3.5.6, which shows an open-base transistor circuit with collector current as a function of collector-emitter voltage and the load line for a particular collector resistance. The collector current is normally low (operating point *a*), but a transient can move the operating level to point *b*, where the circuit becomes latched up at a high current level. The signal required to cause this event can be determined by noting that the collector-emitter voltage must be driven above the V_{CEO} (collector-emitter breakdown) voltage.

Another mode of latch-up can occur when a transistor is grown in a semiconductor substrate, e.g., an *npn* transistor in a doped *p*-substrate. Under usual voltage or gamma-radiation stress, the device can act like an SCR device, latching into conduction.

Transient Overstress.   Although various system components are susceptible to stress damage, the most sensitive tend to be semiconductor components.

Transient Suppression.   There are many techniques available for transient suppression; however, most involve zener diode clamps. Others include some of transistors, SCRs, CMOS, ITL, and diode protection. These techniques are representative of generally applicable methods and are not intended as an exhaustive list.

SYSTEM MODELING

The probability of survival or the reliability function denoted by $R(t)$ is the probability that no failures occur during the time interval $(0, t)$. The most commonly used reliability function is the exponential given by

$$R(t) = e^{-\lambda t} \quad \begin{matrix} \lambda > 0 \\ t \geq 0 \end{matrix} \tag{1}$$

The failure density is the negative of the reliability function. For the exponential case the failure density is

$$f(t) = -R(t) = \lambda e^{-\lambda t} \quad \begin{matrix} \lambda > 0 \\ t \geq 0 \end{matrix} \tag{2}$$

The cumulative distribution $F(t)$ is the probability of failure up to time t and is simply given by $1 - R(T)$. For the exponential case where the density exists we have

$$F(t) = 1 - R(t) = \int_0^t f(z)\,dz = 1 - e^{-\lambda t} \quad \begin{array}{l} \lambda > 0 \\ t \geq 0 \end{array} \tag{3}$$

Thus for the exponential case we have the unique property of having a constant failure function.

In describing the reliability of a complex system the exponential function is rarely appropriate. The reliability function is generally nonexponential because (1) the system is composed of redundant elements (even if each element were itself exponential, the system is no longer exponential), and (2) the system experiences burn-in and wearout properties so that the failure rate is nonconstant. During burn-in the hazard function generally decreases, and during wearout the hazard function generally increases.

For systems formulated as (1), one is given the element reliabilities so that the system reliability can be obtained as a function of them. The particular function obtained depends on the type of redundancy used.

For systems formulated as (2), one is given the hazard function so that reliability is obtained as the solution of the linear differential equation

$$R(t) = -h(t)R(t) \quad \begin{array}{l} R(0) = 1 \\ t \geq 0 \end{array} \tag{4}$$

The solution of Eq. (4) is

$$R(t) = \exp\left(-\int_0^t h(z)\,dz \right) \quad t \geq 0 \tag{5}$$

In general, $h(t)$ need only be piecewise continuous. If failure time is truncated at $t = L$, then $R(t)$ is continuous in the half-open interval $(0, L)$ and the hazard function is not defined at $t = L$.

Reliability for a system is defined here as the product of individual module or branch reliabilities, where each module or branch reliability is made up of reliability functions described in categories (1) and/or (2).

Module-Reliability Models

This section describes various reliability models currently available. The models are developed at the module level, and the system reliability is the product of these module and branch reliabilities. Branch reliabilities are described later. The set of module reliabilities is not exhaustive; others may be developed.

Model 1: Active Exponential Redundant Reliability.   There are n elements in parallel, each element is active with an identical exponential reliability. The module reliability is then

$$R(t) = 1 - (1 - e^{-\lambda t})^n \tag{6}$$

Equation 6 is simply 1 minus the probability that all n elements fail. The symbol λ is the failure rate of the exponential reliability model and is used in the models that follow whenever the exponential reliability model is assumed.

Model 2: Standby Exponential Redundant Reliability.   There are n elements in parallel, each element in standby until called upon to operate. Each element has an identical exponential reliability and does not experience any degradation while on standby. The module reliability is then

$$R(t) = \sum_{x=0}^{n-1} \frac{e^{-\lambda t}(\lambda t)^x}{x!} \tag{7}$$

Model 3: Binomial Exponential (Active) Redundancy Reliability.   There are n elements of which c are required to function properly for the module. Each element is assumed active with an identical exponential reliability. The module reliability is given by the binomial sum

$$R(t) = \sum_{x=c}^{n} \binom{n}{x} (e^{-\lambda t})^x (1 - e^{-\lambda t})^{n-x} \tag{8}$$

In Eq. (8) the variable x is interpreted as the number of successes. If we define N $(N = c, c + 1, \ldots)$ as the number of trials until c successes occur, we have the negative binomial sum for module reliability

$$R(t) = \sum_{N=c}^{n} \binom{N-1}{c-1} (e^{-\lambda t})^c (1 - e^{-\lambda t})^{N-c} \tag{9}$$

If we further define $y = N - c$ $(y = 0, 1, 2, \ldots)$ as the number of failures until c successes occur, we have an alternate form of the negative binomial sum for module reliability

$$R(t) = \sum_{y=0}^{n-c} \binom{c+y-1}{y} (e^{-\lambda t})^c (1 - e^{-\lambda t})^y \tag{10}$$

Model 4: Standby Exponential Redundant Reliability with Exponential Standby Failure.   There are n elements in parallel standby as in model 2, but it is further assumed that the elements in standby have an exponential standby reliability with failure rate λ. Thus, each element has in addition to an identical exponential operational reliability an identical exponential standby reliability. The module reliability is then

$$R(t) = e^{-\lambda t} \sum_{x=1}^{n} \frac{(1 - e^{-\mu t})^{x-1} \Gamma(\beta + x - 1)}{\Gamma(x)\Gamma(\beta)} \tag{11}$$

where $\beta = \lambda/\mu$.
The special case for $\mu = 0$ (no standby failure) given as Eq. (7) is obtained from Eq. (11) by taking its limit as $1/\beta \to 0$.

Model 5: Open-Close Failure-Mode Exponential Redundant Reliability.   There are n elements in series in each of m parallel lines, as shown in Fig 3.5.7. The reliability of the module illustrated is

$$R(t) = (1 - Q_b^n)^m - (1 - R_a^n)^m \tag{12}$$

where $$Q_b = (1 - R_b) = q_b(1 - e^{-\rho t}) \quad Q_a = (1 - R_a) = q_a(1 - e^{-\rho t})$$

and where q_b = conditional probability of failure to close (valve) or failure to open (switch) given system failure = μ/ρ, q_a = conditional probability of failure to open (valve) or failure to close (switch) given system failure = λ/ρ, and $\rho = \lambda + \mu$. There is a total of nm elements in the module A configuration dual to the module illustrated in Fig. 3.5.7 is the "bridged" module, which is the same as that in Fig. 3.5.7 except that the dashed vertical lines are now made solid so that the module units are bridged. For this dual module the reliability is

$$R(t) = (1 - Q_a^m)^n - (1 - R_b^m)^n \qquad R_b \geq Q_a \tag{13}$$

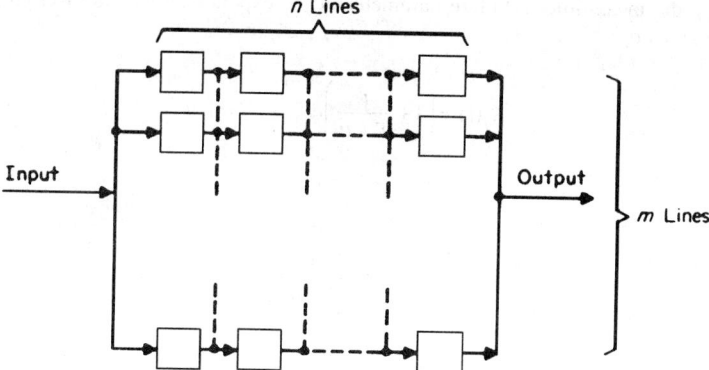

FIGURE 3.5.7   General open-close failure system.

Model 6: Gaussian Redundant Reliability.   There are n identical Gaussian elements in standby redundancy, so that if t_i is the failure time of the ith elements, the module failure time is

$$t = \sum_{i=1}^{n} t_1 \tag{14}$$

The individual elements are independent Gaussian with mean μ_0 and variance σ_0^2. The module then is also Gaussian with mean and variance

$$\mu = n\mu_0 \qquad \sigma^2 = n\sigma_0^2$$

The module reliability is therefore given by

$$R(t) = \frac{c}{\sqrt{2\pi}\sigma} \int_t^{\infty} e^{-(1/2)_\sigma 2(x-\mu)^2} dx \qquad t \geq 0 \tag{15}$$

where

$$c = 1/R(0)$$

Model 7: Bayes Reliability for Inverted Gamma Function.   The module reliability for a single standby element is given by

$$R(t) = R_1(t) + \int_0^t f_1(t_1) R_2(t - t_1) R_{2s}(t_1) dt_1 \tag{16}$$

where $R_i(t)$ = reliability of ith element, $i = 1, 2$
    t_1 = failure time of first element ($i = 1$)
    $f_1(t)$ = failure density of first element $[= -R_1(t)]$
    $R_{2s}(t)$ = reliability of second element in standby mode

Equation (16) is a general formulation of the two-element standby reliability in the sense that the element reliabilities are general. In particular, for the case where the reliabilities are exponential, Eq. (16) reduces to Eq. (11) for $n = 2$. For the case considered here it is assumed that the reliabilities are Bayesian estimates of reliability where the mean-time-to-failure parameter of the exponential has an inverted gamma density. The reliabilities are given by

$$R_j(t) = \left(1 + \frac{t}{T+\mu}\right)^{-(r+v)} \qquad j = 1, 2 \tag{17}$$

and

$$R_{2s}(t) = \left(1 + \frac{t}{T+\mu'}\right)^{-(r+v')} \tag{18}$$

where T = test time
r = number of failures in time T
μ, v, μ', and μ' = inverted gamma parameters

It is noted in Eqs. (17) and (18) that both elements have the same active reliability. In the application of this model the parameters are modeled as follows:

$$\mu = K\theta_A\left(1 + \frac{1}{W^2}\right) \qquad v = \left(2 + \frac{1}{W^2}\right) \quad \text{or} \quad \mu' = \mu/K_1 \qquad v' = v \tag{19}$$

where K and W are the scale factors on the mean and standard deviation of the a priori mean time to failure (inverted-gamma) variable and K_1 is the ratio between the means for the active and standby reliabilities. when $K = 1$, the mean is equal to θ_A, and when $W = 1$, the standard deviation is equal to $K\theta_A$. The relations of Eqs. (17) and (18) define a specific application of Eq. (16).

Model 8: Reliability Model for Weibull Distribution.   The Weibull reliability model is presented in Chap. 3.3 (see "Failure Time Distributions"). Also related to the Weibull distribution is "Specific Hazard Rate Model," also in Chap. 3.3.

Model 9: Accelerated Life Model for all Distribution.   The accelerated life model is presented in Chap. 3.3. This model is a function of temperature and time.

Model 10: Reliability Model for Hazard Function Compound of Three Piecewise Continuous Functions. The general reliability model considered here is for the hazard function described in Chap. 3.3.

$$\lambda(t) = \begin{cases} h_1(t) & 0 \leq t < t_1 \\ h_2(t) & t_1 \leq t < t_2 \\ h_3(t) & t_2 \leq t < \infty \end{cases} \tag{20}$$

The particular case considered here is where h_1 and h_3 are Weibull hazard functions and h_2 is the exponential hazard function. Further, it is assumed that the hazard function is continuous at the transit t_1 and t_2 so that we have

$$\begin{aligned} h_1(t) &= \alpha_1 \lambda_1 t^{\alpha_1 - 1} & 0 \leq t < t_1 \\ h_2(t) &= \lambda & t_1 \leq t < t_2 \\ h_3(t) &= \lambda_2 \lambda_2 t^{\alpha_2 - 1} & t_2 \leq t < \infty \end{aligned} \tag{21}$$

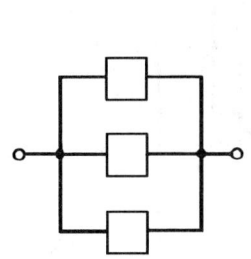

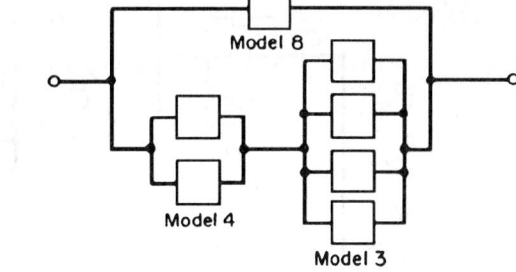

FIGURE 3.5.8 Branch system a consisting of a model 3 module.

FIGURE 3.5.9 Branch system b consisting of models 3, 4, and 8.

where $\qquad\qquad\qquad \alpha_1, \alpha_2, t_1, t_2,$ and λ are input parameters

$$\lambda_i = \frac{\lambda}{\alpha_i t_i^{\alpha_i - 1}} \qquad i = 1, 2$$

Model 11: Tabular Reliability Model. An arbitrary reliability function can be evaluated as an input table of reliability versus time. For each output time, reliability is obtained by interpolation of the input table. This capability is useful as an approximation to a complex reliability model and is also useful in evaluating a reliability function that is available as a plot or a table for which no mathematical model is available.

Model 12: Active Exponential Redundant Reliability with Different Redundant Failure Rate. There are n elements in parallel. The original c elements are exponential each with failure rate λ. The $n - c$ remaining exponential elements in active parallel with the original elements each have failure rate μ. The module reliability is

$$R_s(t) = 1 - (1 - e^{-\lambda t})^c \, (1 - e^{-\mu t})^{n-c} \tag{22}$$

Branch-Reliability Models

A branch-reliability model is a model that evaluates reliability (at each instant in time) for a set of modules arranged in branches. Each branch individually consists of a set of modules. In particular, a branch may be a module itself. Some typical branches are shown in Figs. 3.5.8 to 3.5.10 and are identified as branches a, b, and c, respectively.

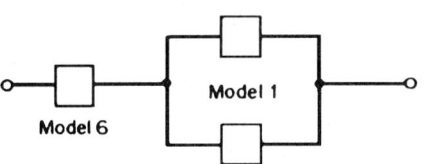

Branch system a is simply a module. Branch system c is a branch consisting of two modules in series and can be evaluated as two separate modules. Branch system b is made up of two subbranches in parallel.

A more complex branch system consisting of subbranch systems a, b, and c is shown in Fig. 3.5.11. Branches a and b are in series in active redundancy to branch system c.

FIGURE 3.5.10 Branch system c consisting of models 1 and 6.

Although the reliability functions for particular branch models are not developed here, clearly it is possible, and a generalized computer program based on both module and branch models can be developed.

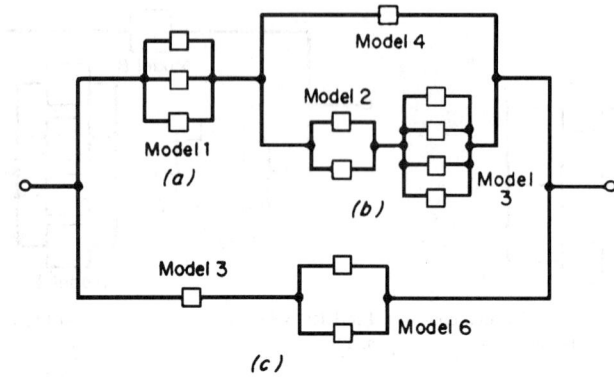

FIGURE 3.5.11   System consisting of branch systems shown in Figs.

The Systems Model

The system reliability model is obtained as the product of the individual branch reliabilities. In addition, the branch reliabilities are multiplied by a system exponential reliability factor and a constant factor; thus

$$R_s(t) = P_0 e^{-\lambda_s t} \prod_{i=1}^{K'} R_1(t) \tag{23}$$

where λ_s = exponential hazard function for system elements not included in branches
  P_0 = system reliability factor
  $R_i(t)$ = ith branch reliability, $i = 1, 2, \ldots, K$
  K' = number of branches in series

BIBLIOGRAPHY

Brombacher, A. C., "Reliability by Design: CAE Techniques for Electronic Components and Systems," Wiley, 1992.

Choi, M., N. Park, and F. Lombardi, "Hardware-Software Co-rehability in Field Reconfigurable Multiprocessor Memory Systems," IPDPS, 2000.

"Fault-Tree Analysis Application Guide," *FTA, RAC.*

Rao, S. S., "Reliability-Based Design," McGraw-Hill, 1992.

"Testability Design and Assessment Tools," *TEST, RAC.*

"Thermal Guide for Reliability Engineers," *RADC-TR-82-172, AD118839, NTIS.*

"Worst-Case Circuit Analysis Guidelines," *WCCA, RAC.*

CHAPTER 3.6
SOFTWARE RELIABILITY

Samuel Keene, Allen Nikora

INTRODUCTION AND BACKGROUND PERSPECTIVE

This chapter describes software reliability, how it can be measured, and how it can be improved. Software is a script written by a programmer dictating how a system will behave under its anticipated operating conditions (called its *main line* function). These conditions are sometimes referred to as "sunny day operations." This part of the software has been tested extensively by the time it is shipped. Problems typically arise in handling program exceptions or branching conditions when these were not properly accounted for in its original design. For example, a planetary probe satellite was programmed to switch over to its backup power system if it did not receive any communications from the ground for a 7-day period. It was assumed that such lack of communication would imply a disruption in the satellite communications power system. The strategy was to switch over to the backup power supply. This switching algorithm was built into the software. In the second decade following launch, the ground crews inadvertently went 7 days without communicating to to the probe. The satellite followed its preprogrammed instructions and switched the system over to backup power, even though the mainline system was still functioning. Unfortunately the standby system was not. So the logical error led to switching the satellite into a failed state. The satellite program was just doing what it had been told to do more than 10 years earlier.

There are many everyday analogies to software faults and failures. Any transaction having residual risks and uncertainties has a counterpart in software reliability. A real estate contract could include a flaw in the title: the house might encroach on an easement, be in a floodplain, or be flawed in its construction. Prudent parties seek to avoid such faults or their likelihood and build defenses against their occurrences. Likewise, a prudent programmer will think proactively to deal with contingent conditions and thereby mitigate any adverse effects.

Managing the programming consistencies problem is exacerbated in developing large programs. This might be considered analogous to achieving a consistent script with 200 people simultaneously writing a novel. There are many opportunities for disconnects or potential faults.

Twenty years ago, software was typically assumed to be flawless. Its reliability was considered to be 1.0. Nothing was thought to break or wear out. Hardware problems dominated systems reliability concerns. Software (and firmware) has now become ubiquitous in our systems and is most often seen as the dominant failure contributor to our systems. For example, the infamous Y2K bug has come and passed. This bug came about because programmers' thought-horizon did not properly extend into the new millenium. When the clock rolled over into the twenty-first century, software that thought the date was 1900 could disastrously fail. Fortunately, its impact was less than forecasted. Problem prevention has cost about $100B worldwide, which is significantly less than many forecasters predicted (http://www.businessweek.com/1999/02/b3611168. htm). Hopefully programmers, or information technology (IT) people, have learned from this traumatic experience. Y2K has sensitized the general public to the greater potential impact of software failures.

Imagine what might have happened if this problem caught us unawares and took its toll. Life as we know it could have been shocked. For example, disruptions could impact supermarket scanners and cash registers. Our commerce could have been foiled by credit cards not functioning.

The use of embedded dates in software is pervasive. There are five exposure areas related to the Y2K or date related type of problem:

1. Operating system embedded dates
2. Application code embedded dates
3. Dates in the embedded software
4. Data (with potential command codes such as 9999 used as a program stop)
5. Dates embedded in interfacing commercial off the shelf (COTS) software

The Y2K software audits had to examine all the above five potential fault areas to ensure that the software was free of the date fault. This type of problem can only be avoided by fully developing the program requirements; this means defining what the program *should not do* as well as what it *should do*. The program requirements should also specify system desired behavior in the face of operational contingencies and potential anomalous program conditions. A good way to accomplish this is by performing failure modes and effects analysis (FMEA) on the high-level design (http://www.fmeainfocentre.com). This will systematically examine the program responses to aberrant input conditions.

Good program logic will execute the desired function without introducing any logical ambiguities and uncertainties. For example, some different C compilers use different order of precedence in mathematical operations. To ensure proper compiling, it is best to use nested parentheses to ensure the proper order of mathematical execution.

We now have an IT layer that surrounds our life and which we depend on. Internet use is pervasive and growing. Our offices, businesses, and personal lives depend on computer communications. We feel stranded when that service is not available, even for short periods. The most notorious outage was the 18-h outage experienced by AOL users in August 1996 (http://www.cnn.com/TECH/9608/08/aol.resumes). This resulted from an undetected operator error occurring during a software upgrade. The cause of failure was thought to be solely because of the upgrade. This confounding problem misled the analysts tracking the problem. Thus, 18 h were required to isolate and correct the problem.

More importantly, many of our common services depend on computer controllers and communications. This includes power distribution, telephone, water supply, sewage disposal, financial transactions, as well as web communications. Shocks to these services disrupt our way of life in our global economy. We are dependent on these services being up and functioning. The most critical and intertwining IT services that envelop our lives, either directly or indirectly are:

1. Nuclear power
2. Medical equipment and devices
3. Traffic control (air, train, drive-by-wire automobiles, traffic control lights)
4. Environmental impact areas (smoke stack filtration)
5. Business enterprise (paperless systems)
6. Financial systems
7. Common services (water, sewer, communications)
8. Missile and weapons systems
9. Remote systems (space, undersea, polar cap, mountain top)
10. Transportation (autos, planes, subways, elevators, trains, and so forth)

SOFTWARE RELIABILITY DEFINITIONS AND BASIC CONCEPTS

Failures can be defined as the termination of the ability of a functional unit to perform its required function. The user sees failures as a deviation of performance from the customer's requirements, expectations, and needs. For example, when a word-processing system locks up and doesn't respond, that is a failure in the eyes of the customer. This is a failure of commission. A second type of failure occurs when the software is incapable of performing a needed function (failure of omission). This usually results from a "breakdown" in defining software requirements, which often is the dominant source of software and system problems. The customer makes the call as to what constitutes a failure. A failure signals the presence of an underlying fault. Faults are a logical construct (omission or commission) that leads to the software failure. Hardware faults can sometimes be corrected or averted by making software changes on a fielded system. The system software changes are made when they can be done faster and cheaper than changing the hardware.

Faults are in correct system states that lead to failure. They result from *latent defects* (these defects are hidden defects, whereas *patent* defects can be directly observed). The first time the newly developed F-16 fighter plane flew south of the equator, the navigational system of the plane caused it to flip over. This part of the navigational software had never been exercised prior to this trial. In the Northern Hemisphere, there was no problem. A fault is merely a susceptibility to failure, it may never be triggered and thus a failure may never occur.

Fault triggers are those program conditions or inputs that activate a fault and instigate a failure. These failures typically result from untested branch, exception, or system conditions. Such triggers often result from unexpected inputs made to the system by the user, operator, or maintainer. For example, The user might inadvertently direct the system to do two opposing things at once such as to go up and go down simultaneously. These aberrant or off-nominal input conditions stress the code robustness. The *exception-handling* aspects of the code are more complex than the operational code since there are a greater number of possible operational scenarios. Properly managing exception conditions is critical to achieving reliable software.

Failures can also arise from a program navigating untested operational paths or sequences that prove stressful for the code. For example, some paths are susceptible to program timing failures. In these instances the desired sequence of program input conditions may violate the operational premise of the software and lead to erroneous output conditions. Timing failures are typically sensitive to the executable path taken. There are so many sequences of paths through a program that they are said to be *combinatory-explosive*. For example, 10 paths taken in random sequence lead to 2^{10} or 1024 combinations of possible paths. Practical testing considerations limit the number of path combinations that can be tested. There will always be a residual failure risk posed by the untested path combinations.

Errors are the inadvertent omissions or commissions of the programmer that allow the software to misbehave, relative to the user's needs and requirements. The relationship between errors, faults, and failures is shown in Fig. 3.6.1.

Ideally, a good software design should correctly handle all inputs under all environmental conditions. It should also correctly handle any errant or off-nominal input conditions. An example of this is can be demonstrated in the following simple arithmetic expression:

$$A + B = C \tag{1}$$

This operation is deterministic when the inputs A and B are real numbers. However, what program response is desired when one of the inputs is unbounded, missing, imaginary, or textural? That is, how should the program handle erroneous inputs? Equation (1) would function best if the program recognized the aberrant condition. Then the software would not attempt to execute the program statement shown above but it would report the errant input condition back to the user. Then this code would be handling the off-nominal condition in a robust manner. It would behave in the manner that the user would desire.

The main line code usually does its job very well. Most often, the main line code is sufficiently refined by the extensive testing it has undergone. The software failures or code breakdowns occur when the software exception code does not properly handle a rarely experienced and untested condition. More frequently the failure driver or trigger is an abnormal input or environmental conditions. An example of a rarely occurring or unanticipated condition was the potential problems posed for programs, failing to gracefully handle the millenium clock rollover. A classic example of improperly handling an off-nominal condition was the infamous

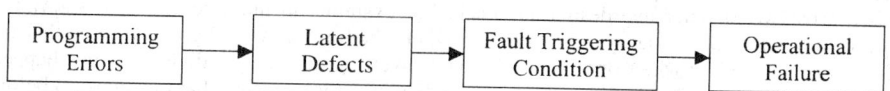

FIGURE 3.6.1   The logical path to software failures.

"Malfunction 54" of the Therac 25 radiation therapy machine. It allowed patients to be exposed to potentially lethal radiation levels because the software controls did not satisfactorily interlock the high-energy operating system mode following a system malfunction. The operator had a work-around that allowed the radiation therapy machine to function but unknown to the operator; the work-around defeated the software interlock.

There is a software development discipline that directly addresses the proper handling of branching and exception conditions. That is the so-called "Clean Room" software design (http://www.dacs.dtic.mil/databases/url/key.hts?keycode=64). It directly ensures validating the branches and exception conditions by routinely checking the exception conditions before it performs the main line operational function. It does these checks at each operational step throughout the program.

System Management Failures

Murphy reports that the biggest cause of problems for modern software stem from either requirements' deficiencies or improperly understood system interfaces [Murp95]. These are communications-based problems, which he labels as *system management* failures. An example of this occurred when the Martian probe failed in the fall of 1999 (http://mars.jpl.nasa.gov/msp98/orbiter/). This failure was because of at least a couple of system management problems. First, there was a reported break down in communications between the propulsion engineers and the navigation scientists, at a critical point in the design cycle. They were working under different assumptions that compromised design reliability. Second, the newspaper headlines publicized another program communication problem. NASA and its contractor had unconsciously been mixing metric and British dimensional units. The resulting tower of Babel situation helped misguide the $125 million dollar space probe.

System management problems are truly communication problems. This author has been a life-long student of system failures (and successes). Today, for the most part, hardware is reliable and capable. Almost all significant system problems are traceable to a communications break down or too limited design perspective on the programmer or designer's part. Also, it needs to be recognized that programmers typically lack the "domain knowledge" of the application area for which they are designing the controlling software. For example, the programmers know the C language or JAVA but they are challenged when it comes to understanding the engineering aspects of a rocket control system. The system development people need to recognize these limitations and be proactive supporting program requirements development.

Failure Modes and Effects Analysis

A FMEA on the new software can also help to refine requirements. This analysis will help focus what the software is expected or desired to do as well as what behaviors are not desired. The software developer is first asked to explain the software's designed functionality. Then the person performing the FMEA will pose a set of abnormal input conditions that the software should handle. The software designer has been focusing on the positive path and now FMEA rounds out the remainder of the conditions that the software must accommodate. Dr. Keene has found design FMEA analysis, nominally takes two hours working directly with the software developer. This analysis has always led to the developer making changes to make the design more robust.

Code Checking and Revision

Designers must consider the possibility of off-nominal conditions that their codes must successfully handle. This is a proactive or defensive design approach. For example, in numeric intensive algorithms, there is always the possibility of divide-by-zero situations where the formula would go unbounded. When many calculations are involved or when one is dealing with inverting matrices, a divide-by-zero happens too frequently. The code can defend itself by checking the denominator to ensure that it is nonzero before dividing. There is a performance cost to do this operation. It is best when the designer is aware of the exposures and

TABLE 3.6.1   Some Typical Fault Tolerance Programming Considerations

---

1. How will you detect a failure in your module? Programmed exception handling or abort with default exception handling?
2. Will you detect out-of-range data?
3. How will you detect invalid commands?
4. How will you detect invalid addresses?
5. What happens if bad data are delivered to your module?
6. What will prevent a divide-by-zero from causing an abort?
7. Will there be sufficient data in a log to trace a problem with the module?
8. Will the module catch all its own faults? (What percentage detection and unambiguous isolation?)
9. Do you have dependencies for fault detection and recovery on other modules or the operating system?
10. What is the probability of isolating faults to the module? Line of code?
11. Does your module handle "exceptions" by providing meaningful messages to allow operator-assisted recovery without total restart?

---

the tradeoffs. A sample list of fault-tolerance considerations that can be applied to the design process is shown in Table 3.6.1.

Code changes are best made at planned intervals to ensure proper review, inspection, integration, and testing to ensure the new code fits seamlessly into the architecture. This practice precludes "rushed" changes being forced into the code. The new code changes will be regression tested through the same waterfall development process (requirements validation, high-level code inspection, low-level code inspection, unit test, software integration and test, and system test) as the base code. This change process preserves the original code development quality and maintains traceability to the design requirements. Using prototype releases and incremental releases helps reveal and refine program requirements. The spiral development model promotes progressive software validation and requirements refinement.

SOFTWARE RELIABILITY DESIGN CONSIDERATIONS

Software reliability is determined by the quality of the development process that produced it. Reliable code is exemplified by its greater understandability. Software understandability impacts the software maintainability (and thus its reliability). The architecture will be more maintainable with reliable, fault-free code requiring minimal revision. The software developer is continually refining (maintaining) the code during development and then the field support programmer maintains the code in the field.

Code reliability is augmented by:

1. Increased cohesion of the code modules
2. Lowered coupling between modules, minimizing their interaction
3. Self describing, longer variable, mnemonic names
4. Uniform conventions, structures, naming conventions and data descriptions
5. Code modularity optimized for understandability while minimizing overall complexity
6. Software documentation formatted for greatest clarity and understanding
7. Adding code commenting to explain any extraordinary programming structures
8. Modular system architecture to provide configuration flexibility and future growth
9. Restricting a single line of code to contain only a single function or operation
10. Avoiding negative logic; especially double negative logic in the design
11. Storing any configuration parameters expected to change in a database to minimize the impacts of any changes on the design
12. Ensuring consistency of the design, source code, notation, terminology, and documentation

13. Maintaining the software documentation (including flow charts) to reflect the current software design level

14. Harmonizing the code after each modification to ensure all the above rules are maintained

There are also aging conditions in software that make it prone to failure. This occurs as its operating environment entropy increases. Consider the case of a distributed architecture air defense system, which had recognizance satellite receivers distributed across the country. These receivers experienced random failures in time. On one occasion the entire countrywide system was powered down and reset at one time. This system synchronization revealed an amazing thing. When the satellite receivers failed, they all failed simultaneously. Synchronizing the computer applications revealed and underlying common fault. A data buffer was tracking some particular rare transactions. It processed them and then passed them along. This register maintained a stack of all its input data, never clearing out the incoming data. So its overflow was inevitable. This overflow was a rare event but the fault was strongly revealed once the system was synchronized.

Computer System Environment

Software reliability is strongly affected by the computer system environment (e.g., the length of buffer queues, memory leaks, and resource contention). There are two lessons learned from software-aging considerations. First, software testing is most effective (likely to reveal bugs) when the distributed software are synchronously aged, i.e., restarted at the same time. To the other extreme, the distributed assets in the operational system will be more reliable if they are asynchronously restarted. Then processors are proactively restarted at advantageous intervals to reset the operating environment and restore the system entropy. This is called "software rejuvenation," or averting system "state accretion failures" (http://www.software-rejuvenation.com). The system assets will naturally age asynchronously as the software assets are restarted following random hardware maintenance activities. The software running asynchronously helps preserve the architectural redundancy of system structure where one system asset can backup others. No redundancy exists if the backup asset concurrently succumbs to the same failure as the primary asset. Diversity of the backup asset helps preclude common-mode failures.

Software and system reliability will be improved by:

1. Strong and systematic focus on requirements development, validation, and traceability with particular emphasis on system management aspects. Full requirements development also requires specifying things the system should not do as well as those desired actions. For example, heat-seeking missiles should not boomerang and return to the installation that fired them. Quality Functional Development (QFD) is one tool that helps ensure requirement completeness, as well as fully documenting the requirements development process (http://www.shef.ac.uk/~ibberson/QFD-Introduction.html).

2. Institutionalizing a "Lessons Learned" database and using it to avoid past problems and mitigating potential failures during the design process, thinking defensively, examining how the code handles off-nominal program inputs, designing to mitigate these conditions.

3. Prototype software releases are most helpful in clarifying the software's requirements. The user can see what the software will do and what it will not do. This operational prototype helps to clarify the user's needs and the developer's understanding of the user's requirements. Prototypes help the user and the developer gather experience and promote better operational and functional definition of the code. Prototypes also help clarify the environmental and system exception conditions that the code must handle. To paraphrase an old saying, "a prototype (picture) is worth a thousand words of a requirements statement," the spiral model advocates a series of prototypes to recursively refine the design requirements (http://www.sei.cmu.edu/cbs/spiral2000/february2000/Boehm).

4. Building in diagnostic capability. Systems' vulnerability is an evidence of omissions in our designs and implementation. Often we fail to include error detection and correction (EDAC) capability as high-level requirements. The U.S. electrical power system, which is managed over a grid system, is able to use its atomic clocks for EDAC. These clocks monitor time to better than a billionth of a second. This timing capability's primary purpose is for proper cost accounting of power consumption. This timing capability

provides a significant EDAC tool to disaggregate failure events that used to appear simultaneous. These precise clocks can also time reflections in long transmission lines to locate the break in the line to within 100 m. The time stamps can also reveal which switch was the first to fail over, helping to isolate the root failure cause.

5. Performing a potential failure modes and effects analysis to harden the system response to properly deal with abnormal input conditions.

6. Failures should always be analyzed down to their root cause for repair and prevention of their reoccurrence. To be the most proactive, the system software should be parsed to see if other instances exist where this same type failure could result. The space shuttle designers did an excellent job of leveraging their failures to remove software faults. In one instance, during testing, they found a problem when system operations were suspended and then the system failed upon restarting the system. Their restart operation, in this instance, somehow violated the original design conception. The system could not handle this restart mode variation and it "crashed." This failure revealed an underlying fault structure. The space shuttle program parsed all 450 KSLOC of operational on board code. Then two other similar fault instances were identified and removed. So, in this instance, a single failure led to the removal of three design faults. This proactive fault avoidance and removal process is called the "Defect Prevention Process (DPP)." DPP was the cornerstone of the highly reliable space shuttle program (http://www.engin.umd.umich.edu/CIS/course.des.cis565/lectures/sep15.html).

7. Every common mode failure needs to be treated as critical, even if its system consequence is somehow not critical. Every thread that breaks the independence of redundant assets needs to be resolved to its root cause the remedied.

8. Studying and profiling the most significant failures. Understanding the constructs that allowed the failure to happen. Y2K is an example of a date defect. One should then ask if other potential date roll over problems existed. The answer is yes. For example, the global positioning system (GPS) clock rolled over August 22, 1999. On February 29, 2000, the Leap Year exception had to be dealt with. Here one failure can result in identifying and removing multiple faults of the same type.

9. Performing fault injection into systems, as part of system development, to speed the maturity of the software diagnostic and fault-handling capability.

Change Management

Operational reliability requires effective change management: Beware small changes can have grave consequence. All too often, small changes are not treated seriously enough. Consequently, there is significant error proneness in making small code changes.

Defect rate: 1 line   50 percent

            5 lines   75 percent

            20 lines   35 percent

The defects here are any code change that results in anomalous code behavior, i.e., changes causing the code to fail. Often, small changes are not given enough respect. They are not sufficiently analyzed or tested. For example, DSC Communications Corp., the Plano Texas Company, signaling systems were at the heart of an unusual cluster of phone outages over a two-month period of time. These disruptions followed a minor software modification. The Wall Street Journal reported, "Three tiny bits of information in a huge program that ran several million lines were set incorrectly, omitting algorithms—computation procedures—that would have stopped the communication system from becoming congested, with messages. . . . Engineers decided that because the change was minor, the massive program would not have to undergo the rigorous 13-week (regression) test that most software is put through before it is shipped to the customer." Mistake!

Post-release program changes increase the software's failure likelihood. Such changes can increase the program complexity and degrade its architectural consistency. Because credit card companies have most of their volume between November 15 and January 15 each year, they have placed restrictions on telephony changes during that time period, to limit interruptions to their business during the holiday period.

EXECUTION-TIME SOFTWARE RELIABILITY MODELS

After the source code for a software system has been written, it undergoes testing to identify and remove defects before it is released to the customer. During this time, the failure history of the system can be used to estimate and forecast its reliability software reliability. The system's failure history as input to one or more statistical models, which produce as their output quantities related to the software's reliability. The failure history takes one of the two following forms:

- Time between subsequent failures
- Number of failures observed during each test interval, and the length of that interval

The models return a probability density function (pdf) for either the time to the next failure, or the number of failures in the next test interval. This pdf is then used to estimate and forecast the reliability of the software being tested. A brief description of three of the more widely used execution time models is given in the following sections. Further details on these and additional models may be found in [Lyu96].

Jelinski-Moranda/Shooman Model

This model, generally regarded as the first software reliability model, was published in 1972 by Jelinski and Moranda (Jeli72). This model takes input in the form of times between successive failures. The model was developed for use on a Navy software development program as well as a number of modules of the *Apollo* program. Working independently of Jelinski and Moranda, Shooman (Shoo72) published an identical model in 1972. The Jelinski-Moranda model makes the following assumptions about the software and the development process:

1. The number of defects in the code is fixed.
2. No new defects are introduced into the code through the defect correction process ("perfect debugging").
3. The number of machine instructions is essentially constant.
4. Detections of defects are independent.
5. During testing, the software is used in a similar manner as the anticipated operational usage.
6. The defect detection rate is proportional to the number of defects remaining in the software.

From the sixth assumption, the hazard rate $z(t)$ can be written as

$$z(t) = KE_r(t)$$

- K is a proportionality constant.
- $E_r(t)$ is the number of defects remaining in the program after a testing interval of length t has elapsed, normalized with respect to the total number of instructions in the code.
  The failure rate $E_r(t)$ in turn is written as

$$E_r(t) = \frac{E_T}{I_T} - E_C(t) \qquad (2)$$

- E_T is the number of defects initially in the program
- I_T is the number of machine instructions in the program
- $E_c(t)$ is the cumulative number of defects repaired in the interval 0 to t, normalized by the number of machine instructions in the program.

The simple form of the hazard rate (exponentially decreasing with time, linearly decreasing with the number of defects discovered) makes reliability estimation and prediction using this model a relatively easy task.

The only unknown parameters are K and E_T; these can be found using maximum likelihood estimation techniques.

Schneidewind Model

The idea behind the Schneidewind model [Schn75, Schn93, Schn97] is that more recent failure rates might be a better predictor of future behavior than failure rates observed in the more distant past. The failure process may be changing over time, so more recent observation of failure behavior may better model the system's present reliability. This idea is implemented as three distinct forms of the model, each form reflecting the analyst's view of the importance of the failure data as a function of time. The data used for this model are the number of observed failures per test interval, where all test intervals are of the same length. If there are n test intervals, the three forms of the model are:

1. Use all of the failure counts from the n intervals. This reflects the analyst's opinion that all of the data points are of equal importance in predicting future failure behavior.

2. Ignore completely the failures from the first s-1 test intervals. This reflects the opinion that the early test intervals contribute little or nothing in predicting future behavior. For instance, this form of the model may be used to eliminate a learning curve effect by ignoring the first few test intervals. This form of the model may also be used to eliminate the effects of changing test methods. For instance, if path coverage testing is used for the first s-1 intervals, and the testing method for the remaining intervals is chosen to maximize the coverage of data definition/usage pairs, the failure counts from the first s-1 intervals might not be relevant. In this case, the analyst might choose to use this form of the model to eliminate the failure counts form the first s-1 test intervals.

3. The cumulative failure counts from the fist s-1 intervals may be used as the first data point, and the individual failure counts from the remaining test intervals are then used for the remaining data points. This approach, intermediate between the first two forms of the model, reflects the analyst's belief that a combination of the first s-1 intervals and the remaining intervals is indicative of the failure rate during the later stages of testing. For instance, this form of the model may be used in the case where the testing method remains the same during all intervals, and failure counts are available for all intervals, but the lengths of the test intervals for the first s-1 intervals are not known as accurately as for the remaining intervals.

The assumptions specific to the Schneidewind model are as follows:

1. The cumulative number of failures by time t, $M(t)$, follows a Poisson process with mean value function $\mu(t)$. This mean value function is such that the expected number of failure occurrences for any time period is proportional to the expected number of undetected failures at that time. It is also assumed to be a bounded, nondecreasing function of time with $\lim_{f \to \infty} \mu(t) = \alpha/\beta < \infty$ where α and β are parameters of the model.

The parameter α represents the failure rate at the beginning of interval s, and the parameter β represents the negative of the derivative of the failure rate, divided by the failure rate.

2. The failure intensity function is assumed to be an exponentially decreasing function of time.

3. The number of failures detected in each of the respective intervals are independent.

4. The defect correction rate is proportional to the number of defects remaining to be corrected.

5. The intervals over which the software is observed are all taken to be the same length. The equations for cumulative number of failures and time to the next N failures are given below.

Cumulative Number of Failures.   Using maximum likelihood estimates for the model parameters a and b, with s being the starting interval for using observed failure data, the cumulative number of failures between and including intervals s and t, Fs,t is given by

$$F_{s,t} = \frac{\alpha}{\beta}\left(1 - e^{-\beta(t-s+1)}\right)$$

(3)

If the cumulative number of failures between intervals 1 and s, X_{s-1}, is added, the cumulative number of failures between intervals 1 and t is

$$F_{s,t} = \frac{\alpha}{\beta}\left(1 - e^{-\beta(t-s+1)}\right)$$

(4)

Time to Next N Failures and Remaining Failures.   At time t, the amount of execution time, $T_F(\Delta r, t)$, required to reduce the number of remaining failures by Δr is given by

$$T_F(\Delta r, t) = \frac{-1}{\beta}\log\left(1 - \left(\frac{\beta\Delta r}{\alpha}\right)e^{\beta(t-s+1)}\right)$$

(5)

Conversely, the reduction in remaining failure, Δr, that would be achieved by executing the software for an additional amount of time T_F is

$$\Delta r(T_F, t) = \frac{\alpha}{\beta}e^{-\beta(t-s+1)}(1 - e^{-\beta T_F})$$

(6)

The Schneidewind model provides a method of determining the optimal value of the starting interval s. The first step in identifying the optimal value of s (s^*) is to estimate the parameters α and β for each value of s in the range $[1, t]$ where convergence can be obtained [Schn93]. Then the *mean square error* (MSE) criterion is used to select s^*, the failure count interval that corresponds to the minimum MSE between predicted and actual failure counts (MSE$_F$), *time to next failure* (MSE$_T$), or *remaining failures* (MSE$_r$), depending on the type of prediction. The first two were reported in [Schn93]—for brevity, they are not shown here. MSE$_r$, developed in [Schn97], is described below. MSE$_r$ is also the criterion for *maximum failures* ($F(\infty)$) and *total test time* (t_t) because the two are functionally related to *remaining failures* ($r(t)$); see Eqs. (9) and (13) in [Schn97] for details. Once α, β, and s are estimated from observed counts of failures, the foregoing predictions can be made. The reason MSE is used to evaluate which triple (α, β, s) is best in the range $[1, t]$ is that research has shown that because the product and process change over the life of the software, old failure data (i.e., $s = 1$) are not as representative of the current state of the product and process as the more recent failure data (i.e., $s > 1$) [Schn93].

Although we can never know whether additional failures may occur, nevertheless we can form the difference between the following two equations for $r(t)$, the number of failures remaining at time t:

$$r(t) = \frac{\alpha}{\beta} - X_{s,t}$$

(7)

where $X_{s,t}$ is the cumulative number of failures observed between the including intervals s and t.

$$r(t_t) = \frac{\alpha}{\beta}e^{-\beta(t_t-s+1)}$$

(8)

where t_t represents the cumulative test time for the system.

We form the difference between Eq. (7), which is a function of predicted *maximum failures* and the observed failures, and Eq. (8), which is a function of *total test time*, and apply the MSE criterion. This yields the following MSE$_r$ criterion for number of *remaining failures*:

$$\text{MSE}_r = \frac{\sum_{i=s}^{t}[F(i) - X_i]^2}{t - s + 1}$$

(9)

The Schneidewind model has been used extensively to evaluate the Primary Avionics Software System (PASS) for the Space Transportation System [Schn92] with very good success, especially employing the procedure for determining the optimal starting interval to obtain better fits to the data.

Musa-Okumoto Logarithmic Poisson Model

The Musa-Okumoto model has been found to be especially applicable when the testing is done according to a nonuniform operational profile. This model takes input in the form of time between successive failures. In this model, early defect corrections have a larger impact on the failure intensity than latter corrections. The failure intensity function tends to be convex with decreasing slope for this situation. The assumptions of this model are:

1. The software is operated in a similar manner as the anticipated operational usage.
2. The detections of defects are independent.
3. The expected number of defects is a logarithmic function of time.
4. The failure intensity decreases exponentially with the expected failures experienced.
5. The number of software failures has no upper bound.

In this model, the failure intensity $\lambda(\tau)$ is an exponentially decreasing function of time:

$$\lambda(\tau) = \lambda_0 e^{-\theta \mu(\tau)} \tag{10}$$

- τ = execution time elapsed since the start of test
- λ_0 = initial failure intensity
- $\lambda(\tau)$ = failure intensity at time τ
- θ = failure intensity decay parameter
- $\mu(\tau)$ = expected number of failures at time τ

The expected cumulative number of failures at time τ, $\mu(\tau)$, can be derived from the expression for failure intensity. Recalling that the failure intensity is the time derivative of the expected number of failures, the following differential equation relating these two quantities can be written as

$$\frac{d\mu(\tau)}{d\tau} = \lambda_0 e^{-\theta \mu(\tau)} \tag{11}$$

Nothing that the mean number of defects at $\tau = 0$ is zero, the solution to this differential equation is

$$\mu(\tau) = \frac{1}{\theta} \ln(\lambda_o \theta \tau + 1) \tag{12}$$

The reliability of the program, $R(\tau_i' | \tau_{i-1})$ is written as

$$R(\tau_i' | \tau_{i-1}) = \left[\frac{\lambda_0 \theta \tau_{i-1} + 1}{\lambda_0 \theta (\tau_{i-1} + \tau_i')} \right]^{1/\theta} \tag{13}$$

- τ_{i-1} is the cumulative time elapsed by the time the $(i-1)$th failure is observed
- τ_i' is the cumulative time by which the ith failure would be observed

The mean time of failure (MTTF) is defined only if the decay parameter is greater than 1. According to [Musa87], this is generally the case for actual development efforts. The MTTF, $\Theta[\tau_{i-1}]$, is given by

$$\Theta[\tau_{i-1}] = \frac{\theta}{1-\theta} (\lambda_0 \theta \tau_{i-1} + 1)^{1-1/\theta} \tag{14}$$

Further details of this model can be found in [Musa87] and [Lyu96].

Benefits of Execution Time Models. There are three major areas in which advantage can be gained by the use of software reliability models. These are planning and scheduling, risk assessment, and technology evaluation. These areas are briefly discussed below.

Planning and Scheduling

- *Determine when a reliability goal has been achieved.* If a reliability requirement has been set earlier in the development process, the outputs of a reliability model can be used to produce an estimate of the system's current reliability. This estimate can then be compared to the reliability requirement to determine whether or not that requirement has been met to within a specified confidence interval. This presupposes that reliability requirements have been set during the design and implementation phases of the development.

- *Control application of test resources.* Since reliability models allow predictions of future reliability as well as estimates of current reliability to be made, practitioners can use the modeling results to determine the amount of time that will be needed to achieve a specific reliability requirement. This is done by determining the difference between the current reliability estimate and the required reliability, and using the selected model to compute the amount of additional testing time required to achieve the requirement. This amount of time can then be translated into the amount of testing resources that will be needed.

- *Determine a release date for the software.* Since reliability models can be used to predict the additional amount of testing time that will be required to achieve a reliability goal, a release date for the system can be easily determined.

- *Evaluate status during the test phase.* Obviously, reliability measurement can be used to determine whether the testing activities are increasing the reliability of the software by monitoring the failure/hazard rates. If the time between failures or failure frequency starts deviating significantly from predictions made by the model after a large enough number of failures have been observed (empirical evidence suggests that this often occurs one-third of the way through the testing effort), this can be used to identify problems in the testing effort. For instance, if the decrease in failure intensity has been continuous over a sustained period of time, and then suddenly decreases in a discontinuous manner, this would indicate that for some reason, the efficiency of the testing staff in detecting defects has decreased. Possible causes would include decreased performance in or unavailability of test equipment, large-scale staff changes, test staff reduction, or unplanned absences of experienced test staff. It would then be up to line and project management to determine the cause(s) of the change in failure behavior and determine proper corrective action.

Likewise, if the failure intensity were to suddenly rise after a period of consistent decrease, this could indicate either an increase in testing efficiency or other problems with the development effort. Possible causes would include large-scale changes to the software after testing had started, replacement of less experienced testing staff with more experienced personnel, higher testing equipment throughput, greater availability of test equipment, or changes in the testing approach. As above, the cause(s) of the change in failure behavior would have to be identified and proper corrective action determined by more detailed investigation. Changes to the failure rate would only indicate that one or more of these causes might be operating.

Risk Assessment. Software reliability models can be used to assess the risk of releasing the system at a chosen time during the test phase. As noted, reliability models can predict the additional testing time required to achieve a reliability requirement. This testing time can be compared to the actual resources (schedule and budget) available. If the available resources are not sufficient to achieve the reliability requirement, the reliability model can be used to determine to what extent the predicted reliability will differ from the reliability requirement if no further resources are allocated. These results can then be used to decide whether further testing resources should be allocated, or whether the system can be released to operational usage with a lower reliability.

Technology Evaluation. Finally, software reliability models can be used to assess the impact of new technologies on the development process. To do this, however, it is first necessary to have a well-documented history of previous projects and their reliability behavior during test. The idea of assessing the impact of new technology is quite simple—a project incorporating new technology is monitored through the testing and operational phases using software reliability modeling techniques. The results of the modeling effort are then compared to the failure behavior of similar historical projects. By comparing the reliability measurements, it is possible to see if the new technology results in higher or lower failure rates, makes it easier or more difficult to detect failures in the software, and requires more or fewer testing resources to achieve the same reliability as the historical projects. This analysis can be performed for different types of development efforts to identify those for which the new technology appears to be particularly well or particularly badly suited. The results of this analysis can then be used to determine whether the technology being evaluated should be incorporated into future projects.

Execution Time Model Limitations

The limitations of execution time models are discussed below. Briefly, these limitations have to do with:

1. Applicability of the model assumptions
2. Availability of required data
3. Determining the most appropriate model
4. The life-cycle phases during which the models can be applied.

Applicability of Assumptions

Generally, the assumptions made by execution time models are made to cast the models into a mathematically tractable form. However, there may be situations in which the assumptions for a particular model or models do not apply to a development effort. In the following paragraphs, specific model assumptions are listed and the effects they may have on the accuracy of reliability estimates are described.

- *During testing, the software is executed in a manner similar to the anticipated operational usage.* This assumption is often made to establish a relationship between the reliability behavior during testing and the operational reliability of the software. In practice, the usage pattern during testing can vary significantly from the operational usage. For instance, functionality that is not expected to be frequently used during operations (e.g., system fault protection) will be extensively tested to ensure that it functions as required when it is invoked.

One way of dealing with this issue is the concept of the testing compression factor [Musa87]. The testing compression factor is simply the ratio of the time it would take to cover the equivalence classes of the input space of a software system in normal operations to the amount of time it would to cover those equivalence classes by testing. If the testing compression factor can be established, it can be used to predict reliability and reliability-related measures during operations. For instance, with a testing compression factor of 10, a failure intensity of one failure per 10 h measured during testing is equivalent to one failure for every 100 h during operations. Since test cases are usually designed to cover the input space as efficiently as possible, it will usually be the case that the testing compression factor is greater than 1. To determine the testing compression factor, of course, it is necessary to have a good estimate of the system's operational profile (the frequency distribution of the different input equivalence classes) from which the expected amount of time to cover the input space during the operational phase can be computed.

- *There is an upper limit to the number of failures that will be observed during testing.* Many of the more widely used execution time models make this assumption. Models making this assumption should not be applied to development efforts during which the software version being tested is simultaneously undergoing significant changes (e.g., 20 percent or more of the existing code is being changed, or the amount of code is increasing by 20 percent or more). However, if the major source of change to the software during test is the correction process, and if the corrections made do not add a significant amount of the new functionality or behavior, it is generally safe to make this assumption. This would tend to limit application of models making this assumption to subsystem-level integration or later testing phases.

- *No new defects are introduced into the code during the correction process.* Although there is always the possibility of introducing new defects during the defect removal process, many models make this assumption to simplify the reliability calculations. In many development efforts, the introduction of new defects during correction tends to be a minor effect, and is often reflected in a small readjustment of the values of the model parameters. In [Lyu91], several models making this assumption performed quite well over the data sets used for model evaluation. If the volume of software, measured in source lines of code, being changed during correction is not a significant fraction of the volume of the entire program, and if the effects of repairs tend to be limited to the areas in which the corrections are made, it is generally safe to make this assumption.

- *Detections of defects are independent of one another.* All execution time models make this assumption, although it is not necessarily valid. Indeed, there is evidence that detections of defects occur in groups, and that there are some dependencies in detecting defects. The reason for this assumption is that it enormously simplifies the estimation of model parameters. Determining the maximum likelihood estimator for a model parameter requires the computation of a joint probability density function (pdf) involving all of the observed events. The assumption of independence allows this joint pdf to be computed as the product of the individual pdfs for each observation, keeping the computational requirements for parameter estimation within practical limits.

Availability of Required Data

Most software reliability models require input in the form of time between successive failures. These data are often difficult to collect accurately. Inaccurate data collection reduces the usefulness of model predictions. For instance, the noise may be great enough that the model predictions do not fit the data well as measured by traditional goodness-of-fit tests. In some cases, the data may be so noisy that it is impossible to obtain estimates for the model's parameters. Although more accurate predictions can be obtained using data in this form [Musa87], many software development efforts do not track these data accurately.

As previously mentioned, some models have been formulated to take input in the form of a sequence of pairs, in which each pair has the form of (number of failures per test interval, test interval length). For the study reported in [Lyu91, Lyu91a, Lyu91b], all of the failure data was available in this form. The authors' experience indicates that more software development efforts would have this type of information readily available, since they have tended to track the following data during testing:

1. Date and time at which a failure was observed.

2. Starting and ending times for each test interval, found in test logs that each tester is required to maintain.

3. Identity of the software component tested during each test interval.

With these three data items, the number of failures per test interval and the length of each test interval can be determined. Using the third data item, the reliability of each software component can be modeled separately, and the overall reliability of the system can be determined by constructing a reliability block diagram. Of these three items, the starting and ending times of test intervals may not be systematically recorded, although there is often a project requirement that such logs be maintained. Under schedule pressures, however, the test staff may not always maintain the test logs, and a project's enforcement of this requirement may not be sufficiently rigorous to ensure accurate test log entries.

Even if a rigorous data collection mechanism is set up to collect the required information, there appear to be two other limitations to failure history data.

- It is not always possible to determine when a failure has occurred. There may be a chain of events such that a particular component of the system fails, causing others to fail at a later time (perhaps hours or even days later), finally resulting in a user's observation that the system is no longer operating as expected. Individuals responsible for the maintenance of the Space Transportation System (STS) and Primary Avionics Software System have reported in private discussions with colleagues of the authors several occurrences of this type of latency. This raises the possibility that even the most carefully collected set of failure history has a noise component of unknown, and possibly large, magnitude.

- Not all failures are observed. Again, discussions with individuals associated with maintaining the STS flight software have included reports of failures that occurred and were not observed because none of the STS crew was looking at the display on which the failure behavior occurred. Only extensive analysis of post-flight telemetry revealed these previously unobserved failures. There is no reason to expect that this would not occur in the operation of other software systems. This describes another possible source of noise in even the most carefully collected set of failure data.

Identifying the Most Appropriate Model

At the time the first software reliability models were published, it was thought that there would be a refinement process leading to at most a few models that would be applicable to most software development efforts [Abde86]. Unfortunately, this has not happened—since that time, several dozen models have appeared in the literature, and we appear to be no closer to producing a generally applicable model. For any given development effort, it does not appear to be possible to determine a priori which model will be the most applicable. Fortunately, this difficulty can be mitigated by applying several different models to a single set of failure history data, and then using the methods described in [Abd86] to choose the most appropriate model results. The methods include:

1. *Determination of how much more likely it is that one model will produce more accurate results than another.* *Prequential likelihood* functions are formed for each model. Unlike likelihood functions, these are not used to find model parameters—rather, the values of the model parameters are substituted into the prequential likelihood functions, as are the observed times between failures, to provide numerical values that can be used to compare two or more models. The larger the value, the more likely it is that the model will provide accurate predictions. Given two models, A and B, the ratio of the prequential likelihood values, PL_A/PL_B, specifies how much more likely it is that model. A will produce accurate predictions than model B.

2. *Determination of whether a model makes biased predictions.* A model may exhibit bias by making predictions of times between failures that are consistently longer than those that are actually observed, or making predictions that are shorter than what's observed. In either case, [Abde86] describes how to draw a *u-plot* that will identify those models making biased predictions. Given two models that have the same prequential likelihood value, the model exhibiting the most bias should be discarded.

3. *Determination of whether a model's bias changes over time.* If a model exhibits bias, that bias may change over time. For instance, early in a testing effort, a model may consistently predict times between failures that are longer than what is actually observed. As testing progresses, the model's bias may change to making predicting shorter times between failures than those actually observed. Such a change in a model's bias may not show up in a u-plot, since the computations involved in producing u-plot does not preserve temporal information associated with the failure history. However, a series of further transformations allow the construction of a *y-plot*, which identifies models whose bias changes over time. Given two models with identical prequential likelihood values and identical u-plots, the model exhibiting a change in bias over time should be rejected over the one that shows no change in its bias.

Applicable Development Phases

Perhaps the greatest limitation of the execution time software reliability models is that they can only be used during the testing phases of a development effort. In addition, they usually cannot be used during unit test, since the number of failures found in each unit will not be large enough to make meaningful estimates for model parameters. These techniques, then, are useful as a management tool for estimating and controlling the resources for the testing phases. However, the models do not include any product or process characteristics that could be used to make tradeoffs between development methods, budget, and reliability.

PREDICTING SOFTWARE RELIABILITY

Software's design reliability is typically measured in terms of faults per thousand lines of source code (KSLOC). The lines of code are the executable lines of code, excluding the commentary lines of source code. Typical fault density levels, as a function of the development process, are shown in Table 3.6.2.

Code is necessarily released with faults in it. These are only estimated faults and not known faults. If they were known faults, the developer would remove them. This author has heard a hardware developer decry the notion of releasing code with faults. This critic claimed that a hardware developer would never do that. The truth is, with software one cannot test or anticipate all the scenarios that the software will have to handle. Some of these bugs

TABLE 3.6.2 Code Defect Rates (Faults/KSLOC)

	Total development	At delivery

Traditional development
  Bottom-up design,
    unstructured code,
    removal through testing
	50–60	15–18

Modern practice
  Top-down design, structured
  code, design and code
  inspections, incremental releases
	20–40	2–4

Future improvements
  Advanced software engineering,
  verification practice, reliability
  measurement
	6–20	<1

Space shuttle code	6–12	0.5*

*The defect density of the space shuttle code is often shown as 0.1 defects/KSLOC. From private correspondence with the software quality lead engineer on the space shuttle, the author has learned that the estimated defect rate is 0.5 at turnover to the NASA customer. The space shuttle code only gets to 0.1 defects/KSLOC after eight additional months of NASA testing.

(another name for faults) will never affect a particular user. Bugs are potential problems in the code that have to be triggered by a set of special conditions. For instance, there could be a calculation in the software having a term of $1/(Z-225)$ that "explodes" or goes to infinity if $Z = 225$. For all other values this term is well behaved.

There was a lot of hype over the thousands of bugs released with Windows 2000 (http://home.rochester.rr.com/seamans). That article states, "Well, it's been 2 months since the release of Windows 2000. Where are we now? After the hype of 63,000 bugs, we've seen a few bugs creep up here and there. Many argue over whether the current bugs out there are major or minor in nature. It seems that a vast amount of the bugs found to date are minor in nature and the ones that may be major (i.e., Windows 2000 doesn't support more than 51 IP addresses) affect only a very few IT operations. In all, I think Windows 2000 has fared rather well considering what has taken place over the last year."

Keene Process Based Reliability Model

It is often required to estimate the software's reliability before the code is even developed. The Keene software reliability model projects fault density based on empirical data correlating fault density to the process capability of the underlying development process. Process capability measurement is most conveniently accomplished using the development organization's Capability Maturity Model (CMM) level, which is explained in the next paragraph. The fault density is projected based on the measured process capability. The Keene model next transforms the latent fault density into an exponential reliability growth curve over time, as failures surface and the underlying faults are found and removed. Thus, failures are traced to the underlying faults and removed. Failures occur and these are the outward manifestation of the underlying fault. The faults are removed over time through software maintenance actions and the corresponding software failure rate diminishes accordingly.

The Software Engineering Institute (SEI) has developed a CMM that is a framework describing the key elements of an effective software development process. It covers key practices for planning, engineering, management, development, assurance, and maintenance that, when followed, should improve the ability of an organization to meet cost, schedule, and quality goals. They also provide five levels of grading which permits good differentiation among process capabilities.

- SEI Level 1 (Initial): Ad hoc process development dependent on individuals. Cost, schedule, and quality are unpredictable. Most companies fall into this category.
- SEI Level 2 (Repeatable): Policies for managing software are in place. Planning is based on prior experience. Planning and tracking can be repeated. Seventy percent of companies are rated at SEI level II or I.

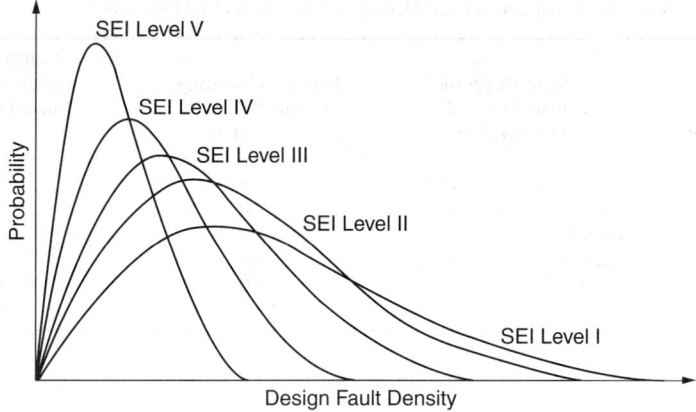

FIGURE 3.6.2   Mapping of the SEI CMM levels versus fault density distributions.

- SEI Level 3 (Defined):   The process for developing the software is documented and integrated coherently. The process is stable and repeatable and the activities of the process are understood organization wide.
- SEI Level 4 (Managed):   The process is measured and operates within measured limits. The organization is able to predict trends and produce products with predictable quality.
- SEI Level 5 (Optimization):   The focus is on continuous improvement. The organization can identify weaknesses in the process and is proactive in preventing faults. The organization is also innovative throughout. It demonstrates a passion to find and eliminate bugs. This organization also is continually improving its development process to reduce the likelihood of bug recurrence. It learns from the errors it finds. It is dedicated to continuous measurable improvement, sometimes referred to as CMI.

Figure 3.6.2 illustrates how the SEI's CMM levels correlate with the design fault density at product shipment as shown below.

The higher capability levels are expected to exhibit lower fault density than the lower capability levels. The higher SEI process levels are indeed more capable. The higher levels are also shown to have less variability, i.e., they are a more consistent predictor of fault density. The figure above depicts that the SEI level is a good predictor (causative factor) of fault density.

The process capability level is a good predictor of the latent fault content shipped with the code. The better the process, the better the process capability ratings and the better the code developed under that process. The projection of fault density according to the corresponding SEI level is now shown in Table 3.6.3. Keene bases these fault density settings on information compiled over a variety of programs. This information is proprietary and is not included in this paper. The SEI Level 5 experience is the most publicized on the space shuttle program of 0.5 defects/KSLOC. The table shows the posited relationship between the development organization's SEI's CMM level and the projected latent fault density of the code it ships.

The software reliability model Eq. (15), shows that the fault content decreasing exponentially over time. The exponential fault reduction model follows from assuming that the number of faults being discovered and removed from the code is proportional to the total number of faults existing in the code, at any point in time. This premise is intuitively appealing and supported by empirical evidence. This relationship expressed mathematically is

$$F(t) = F_0 e^{-kt} \tag{15}$$

where $F(t)$ = current number of faults remaining in the code after calendar time t
F_0 = initial number of faults in the code at delivery, which is a function of the developer's SEI development capability
k = constant of proportionality
t = calendar time in months

TABLE 3.6.3   Software's Projected Latent Fault Density as a Function of CMM Level

SEI'S CMM Level	Maturity profile June 2001 1018 Organizations*	Latent fault density (Faults/KSLOC- all severities)	Defect plateau level for 48 months after initial delivery or 24 months following subsequent deliveries
5	Optimizing: 4.8 percent	0.5	1.5 percent
4	Managed: 5.6 percent	1.0	3.0 percent
3	Defined: 23.4 percent	2.0	5.0 percent
2	Repeatable: 39.1 percent	3.0	7.0 percent
1	Initial: 27.1 percent	5.0	10.0 percent
unrated	The remainder of companies	>5.0	not estimated

*http://www.sei.cmu.edu/sema/pdf/2001aug.pdf

The proportionality constant k is set so the latent fault content drops to a certain percentage of the initial fault discovery rate after a code-stabilizing period of time. One analysis showed field data to stabilize at a 10 percent F_0 level after 48 months [Keen94]. This 10 percent level is appropriate for a CMM Level 1 process. The higher level processes drive the stabilization level lower than 10 percent that is used here. The constant k is determined by solving Eq. (15) for $F(t)$ dropping to 10 percent at 48 months. That analysis shows to be equal 0.048 per months.

Faults are revealed by failures. The failures are then traced to root cause, removed and the fault count goes down correspondingly. So failure the software failure rate goes down in proportion to the number of faults remaining in the code.

For purposes of estimating the operational failure rate of the fielded code, one can apply a linear, piecewise approximation to the exponential fault removal profile. This simplifies calculation of an average failure rate over the time interval. The number of faults found and removed is indicative of the failure rate of the software. That is, it usually takes a failure to reveal an underlying fault in the software. The more failures that occur, the more underlying faults there are uncovered in the code. The software failure rate λ is related to the number of underlying fault removal over time as follows:

$$\lambda(t) \approx \frac{F(t_2) - F(t_1)}{t_2 - t_1} \tag{16}$$

Adding a constant of proportionality ρ,

$$\lambda(t) = \rho \frac{F(t_2) - F(t_1)}{t_2 - t_1} \tag{17}$$

where ρ represents a coverage factor. This is the total number of failure occurrences that a fault manifests before it is removed versus fielded population of systems. The coverage factor is made up of three components:

1. Percentage fault activation is that percent of the population likely to experience a given fault on the average. This factor is a small percentage when there are many copies in a widely distributed system such as an operating system.

2. Fault latency is measured in terms of the number of times that a unit in the failing subpopulation is likely to experience a failure before the failure is analyzed to root cause level and the corrective software patch installed.

3. Percentage critical faults that will cause the system to fail for the user.

The coverage factor is illustrated by considering 10 fielded copies of a software package, where five of the users experience problems. On average these five users experience three failures each before the problem is resolved and the underlying fault removed. The user in this example considers all failures critical. Then the coverage factor would be (5 failures/10 fielded software instances) × (3 repeats) × (100 percent critical failures) over the install base of 10 users. Thus, $\rho = 1.5$. This number is based on the developer's history or by direct estimation of field performance parameters. "Measurement is important, otherwise you are just practicing."

Sometimes a single failure can lead to the removal of several faults. On the space shuttle for instance, a particular problem was experienced when reentering the code, having once exited it. This problem was called the "multi-pass problem." The problem was discovered and resolved. The entire shuttle code was searched looking for other instances of this problem. A single failure led to the correction of three different faults in the code. More often, it takes several failure occurrences before the underlying fault is found and removed.

The failure replication of the same problem will be somewhat offset by failures not being experienced across all systems. That is, the early adopters will first experience the software faults. There is also a subset of all users that stress their systems harder with greater job loads or more diverse applications. Some problems will be found and corrected before the majority of the remaining users have experienced it. This reduces the ρ factor or fault coverage. Chillarege's data show software improvement over time that was slightly steeper than an exponential growth profile (Ram Chillarege, Shriram Biyani, Jeanette Rosenthal, "Measurement of Failure Rate in Widely Distributed Software," Fault Tolerant Computing Symposium, 1995, pp. 424–433). The customer base he reported approached a million users, so there would be plenty of early adopters to find and fix the problems before the majority of users would see them.

Rayleigh Development Model

The defect rates found in the various development phases will successfully predict the latent defects shipped with the product, (from "Metrics and Models in Software Quality Engineering," Chap. 7, by Stephen H. Kan, Addison-Wesley, 1995).

The Rayleigh model is shown below:

- CDF: $F(t) = 1 - e^{(-t/c)2}$
- PDF: $f(t) = (2/t)(t/c)^2 e^{(-t/c)2}$

The Rayleigh model is the best tool to predict software reliability during development, prior to system testing. The Rayleigh model is a member of the Weibull distribution with its shape parameter, $m = 2$. The Rayleigh prediction model forward chains all the defect discovery data, collected throughout the development phases, to predict the software's latent fault density. (Latent errors are those errors that are not apparent at delivery but manifest themselves during subsequent operations. This is contrasted to patent errors, which are obvious by inspection.) The inputs to the Rayleigh model are the defect discovery rates found in the following design phases: high level design, low level design, code and unit test, software integration and test, unit test, and system test.

There are two basic assumptions associated with the Rayleigh model. First is that the defect rate observed during the development process is positively correlated with the defect rate in the field. The higher the Rayleigh curve of defects observed during development, the higher the expected latent defect rate of the released code. This is related to the concept of error injection. Second is that given the same error injection rate, if more defects are discovered and removed earlier, fewer will remain in later stages. The output of the Rayleigh model is the expected latent fault density in the code when it is released.

The following are priority definitions used for quantifying development defects. The Rayleigh analysis can be used to project any or all of these priority defects itemized below:

- Critical (Priority 1): An error which prevents the accomplishment of an operational or mission essential function in accordance with specification requirements (e.g., causes a computer program stop), which interferes with use of the system to the extent that it prevents the accomplishment of a mission essential function, or which jeopardizes personnel safety.
- Major (Priority 2): An error which adversely affects the accomplishment of a mission essential function in accordance with specification requirements so as to degrade performance of the system and for which no alternative is provided. Rebooting or restarting the application is not an acceptable alternative, as it constitutes unacceptable interference with use of the system.

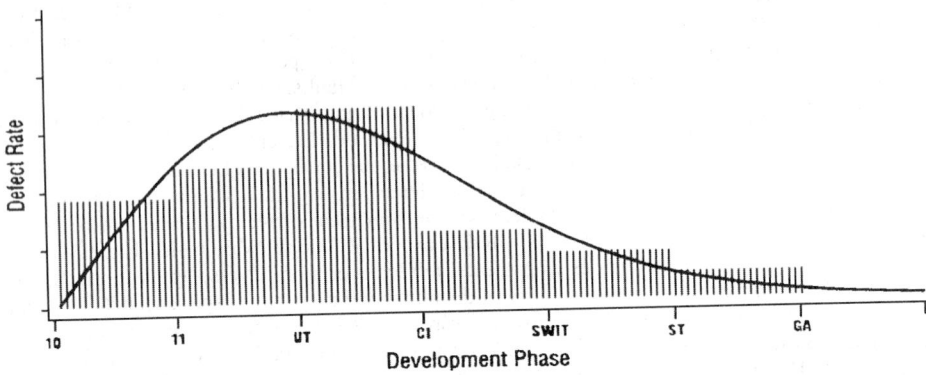

FIGURE 3.6.3   Rayleigh defect removal profile curve.

- Minor (Priority 3): An error which adversely affects the accomplishment of an operational or mission essential function in accordance with specification requirements so as to degrade performance and for which there is a reasonable alternative provided. Rebooting or restarting the application is not an acceptable alternative, as it constitutes unacceptable interference with use of the system.

- Annoyance (Priority 4): An error, which is an operator inconvenience or annoyance and does not affect an operational or mission essential function.

- Other (Priority 5): All others errors.

   The difference between the amount of faults injected into the code and those that escape detection and end up being shipped to the field reflects the goodness of the development process. Defects are removed by a number of development quality processes such as inspections, reviews, and tests. Eight-five percent of all the injected faults were removed during the development phase prior, to the system test, on the highly successful space shuttle program. The better the development process, the sooner the defects are removed during the development process and the fewer defects that escape (detection and removal) and go to the field as latent defects. Once the code is in operational test or the field, the computer aided software reliability estimation (CASRE) tool can be applied to measure its operational reliability.

CASRE

   Computer aided software reliability estimation, developed with funding from the U.S. Air Force Operational Test and Evaluation Center (AFOTEC), is a stand-alone Windows-based tool that implements 10 of the more widely used execution time software reliability models. CASRE's goal was to make it easier for nonspecialists in software reliability to apply models to a set of failure history data, and make predictions of what the software's reliability will be in the future. In 1992, when the first version was released, it was the first software reliability modeling tool that displayed its results as high-resolution plots and employed pull-down menus in its user interface—other tools at that time interacted with the user via command lines or batch files, and displayed their results in tabular form. It also allows users to define combination models [Lyu91a] to increase the predictive accuracy of the models. In addition, it implements the model selection criteria described in [Abde86] to allow user to choose the most appropriate model results from a set of results. Finally, it implements trend tests that can be applied to the failure data to help the user determine whether it's even appropriate to apply reliability models to the data.

   The modeling and model applicability analysis capabilities of CASRE are provided by Statistical Modeling and Estimation of Reliability Functions for Software (SMERFS), a software reliability modeling tool developed by Dr. William H. Farr at the Naval Surface Warfare Center, Dahlgren, Virginia. In implementing CASRE, the original SMERFS user interface and the SMERFS modeling libraries were linked into the user interface developed for CASRE. The combination modeling and trend test capabilities, however, are new for CASRE, as are the model ranking capabilities.

Version 3.0 of CASRE is available from the Open Channel Foundation at http://www.openchannelfoundation.org.

The following steps illustrate a typical application of CASRE to a set of failure data:

1. When a set of failure data is first opened with CASRE, the data are displayed both in text and graphical form. One window displays the text of the failure data, and the other displays a high-resolution plot of the data. CASRE can use two types of failure data—either time between subsequent failures, or the number of failures observed in a test interval of a given length. If the data are in the form of times between subsequent failures, the window displaying text shows the failure number, the elapsed time since the previous failure, and the failure's severity on an arbitrary scale of 0 to 9. The initial plot shows the times between subsequent failures as a function of failure number. For interval data, the text display shows the interval number, the number of failures observed in that test interval, the length of the interval, and the severity of the failures in that interval. The initial plot for this type of data shows the number of failures per test interval as a function of test interval number.

2. Whether the reliability of the software being analyzed is increasing depends on many factors, such as the type of testing being performed, and whether substantial amounts of the software's functionality are being added or modified. To determine the advisability of applying reliability models to the failure data, trend tests should be applied to determine whether the failure data are exhibiting reliability growth. If these trend tests indicate that the failure data exhibit reliability growth, then the user can be confident in applying a software reliability model to the data. However, it may be the case that failures are detected at a nondecreasing, or even an increasing rate, in which case the reliability of the software being analyzed would actually be decreasing. As mentioned above, this could happen if significant amounts of additional functionality are being added while testing is taking place. In this case, application of a reliability model would not be advisable, since all of the models assume that the reliability of the software increases as testing progresses. CASRE implements two trend tests to help users determine whether the data are exhibiting reliability growth—a running arithmetic average, and the Laplace test [Lyu96].

3. If the trend tests indicate it is advisable to apply reliability models to the failure data, the next step is the selection of the models to be used. Because it is generally not possible to determine a priori the most appropriate model for a set of data, it is recommended that several different models be applied to the same set of data and the methods described in [Abde86] be applied to the results to select the most appropriate model. CASRE allows users to apply more than one model to the current set of failure data. Once the models have completed, users can select up to three sets of model results, and plot them in several different ways. Among these are displays of including failure intensity (number of failures per unit time) versus failure number, actual and predicted cumulative number of failures versus time, and software reliability versus time.

4. After the models have been applied to a set of failure data, the next step is determining which model's results are the most appropriate to the data. CASRE provides a facility to help rank the models according to goodness of fit and four criteria described in [Abde86]—prequential likelihood, bias, bias trend, and model noise. The default ordering and weighting of these criteria is based on a study of how to rank software reliability model results [Niko95]. First, CASRE uses the goodness-of-fit tests to screen out those model results that don't fit at a specified significance level—those results are excluded from further consideration in the ranking. Next, CASRE compares the prequential likelihood values for the models—models with higher prequential likelihood values are ranked higher than those having lower values. For those models having the same prequential likelihood value, CASRE uses the extent to which the models are biased to break the tie—models having lower values of bias are ranked higher than models with higher bias values. For those models exhibiting the same amount of bias, CASRE uses the value of the bias-trend criterion to break the tie—ranking with respect to bias trend is the same as it is for bias. There is an additional criterion, "prediction noisiness," or "model noise," which is also available, but its meaning is not as well understood as that of the other three [Abde86]. The default use of this criterion is to break ties among models exhibiting identical values of bias trend—models having lower values of model noise are ranked higher than models having higher values. Model ranks are displayed in a tabular form—the overall rank for each model is displayed, as well as the ranks with respect to each individual criterion.

REFERENCES

[Abde86]   A. A. Abdel-Ghaly, P. Y. Chan, and B. Littlewood, "Evaluation of competing software reliability predictions," *IEEE Transactions on Software Engineering*, Vol. SE-12, pp. 950–967, September 1986.

[Jeli72] Jelinski, Z., and P. Moranda, "Software reliability research," in W. Freiberger (ed.), "Statistical Computer Performance Evaluation," Academic, 1972, pp. 465–484.

[Keen94]   S. Keene, and G. F. Cole, "Reliability growth of fielded software," *Reliability Review*, Vol. 14, No. 1, March 1994.

[Lyu91]   M. Lyu, "Measuring reliability of embedded software: An empirical study with JPL project data," *Proceedings of the International Conference on Probabilistic Safety Assessment and Management*, February 4–6, 1991.

[Lyu91a]   M. R. Lyu, and A. P. Nikora, "A heuristic approach for software reliability prediction: The equally-weighted linear combination model," *Proceedings of the IEEE International Symposium of Software Reliability Engineering*, May 17–18, 1991.

[Lyu91b]   M. R. Lyu, and A. P. Nikora, "Software reliability measurements through combination models: Approaches, results, and a CASE tool," *Proceedings of the 15th Annual International Computer Software and Applications Conference (COMPSAC91)*, September 11–13, 1991.

[Lyu96]   M. Lyu (ed.) "Handbook of Software Reliability Engineering," McGraw-Hill, 1996, pp. 493–504.

[Murp95]   B. Murphy, and T. Gent, "Measuring System and Software Reliability Using an Automated Data Collection Process," Quality and Reliability Engineering International, CCC 0748-8017/95/050341-13pp., 1995.

[Musa87]   J. D. Musa, A. Iannino, and K. Okumoto "Software Reliability: Measurement, Prediction, Application," McGraw-Hill, 1987.

[Niko95]   A. Nikora, and M. R. Lyu, "An experiment in determining software reliability model applicability," *Proceedings of the 6th International Symposium on Software Reliability Engineering.* October, 24–27, 1995.

[Schn75]   N. F. Schneidewind, "Analysis of error processes in computer software," *Proceedings of the International Conference on Reliable Software*, IEEE Computer Society, April 21–23, 1975, pp. 337–346.

[Schn92]   N. F. Schneidewind, and T. W. Keller, "Applying reliability models to the space shuttle," *IEEE Software*, Vol. 9 pp. 28–33, July 1992.

[Schn93]   N. F. Schneidewind, "Software reliability model with optimal selection of failure data," *IEEE Transactions on Software Engineering*, Vol. 19, November 1993, pp. 1095–1104.

[Schn97]   N. F. Schneidewind, "Reliability modeling for safety-critical software," *IEEE Transactions on Reliability*, Vol. 46, March 1997, pp. 88–98.

[Shoo72]   Shooman, M. L., "Probabilistic Models for software reliability prediction," in W. Freiberger (ed.), "Statistical Computer Performance Evaluation," Academic, 1972, pp. 485–502.

BIBLIOGRAPHY

Card, D. N., and R. L. Glass, "Measuring Software Design Quality," Prentice Hall, 1990.

Clapp, J. A., and S. F. Stanten, "A Guide to Total Software Quality Control, RL-TR-92-316," Rome Laboratory, 1992.

Friedman, M. A., P. Y. Tran, and P. L. Goddard, "Reliability Techniques for Combined Hardware and Software Systems," Hughes Aircraft, 1991.

McCall, J. A., W. Randell, J. Dunham, and L. Lauterback, "Software Reliability, Measurement, and Testing RL-TR-92-52", Rome Laboratory, 1992.

Shaw, M., "Prospects for an engineering discipline of software," *IEEE Software*, November 1990.

SECTION 4

COMPUTER-ASSISTED DIGITAL SYSTEM DESIGN

Computer-assisted design continues to become a more powerful tool in digital and analog circuit design, especially using programmable integrated circuits. The different types and manufacturers of such devices fall into two main classes—field programmable gate arrays (FPGAs) and field programmable analog arrays (FPAAs).

We will focus on digital circuits (FPGAs) using the most general process of design. However, the approach works well on all of the rest of the commercial CAD tools and can be applied to the analog devices (FPAAs) as well.   C.A.

In This Section:

CHAPTER 4.1

DESIGN OF DIGITAL SYSTEMS USING CAD TOOLS

Brian R. Fast

OVERVIEW

The use of CAD tools for programming digital circuits has dramatically increased the reusability, processing power, and reliability of digitally designed components. Currently field programmable gate array (FPGA) chips use very high-speed integrated circuits hardware description language (VHDL) to take advantage of these CAD tools in the programming of chips at the lowest gate level. VHDL has become an industry standard language, which is regulated by the Institute of Electrical and Electronic Engineers (IEEE). It is similar, with many conceptual changes described later in this section, to any other programming language (e.g., C, C++, visual basic). The companies that manufacture FPGA devices have software packages that can compile VHDL code that can be simulated or burned to a chip.

One of the potential benefits to VHDL over chip-depended software is the capability to migrate from one chip manufacturer to another. VHDL is an industrial standard language, therefore the software for each chip manufacture should be capable of compiling code designed for another manufacturer's chip. Of course this is dependent on the code not using functions that are proprietary to a specific chip manufacturer.

The increase in processing power is another major advantage of FPGA devices over the standard microprocessor. There are two avenues in which this can be seen. First, the capability of performing concurrent operations is a major advantage of FPGA devices. One chip can be broken into subsystems that perform the processes in parallel. If this was to be implemented with microprocessors there would need to be an independent processor for each subsystem. This would considerably increase the overall design cost and add complexity to system integration. Second, the capability of designing the number of busses to the desired size increases the efficiency and optimizes the design. For example, standard microprocessors have a fixed number of busses and register sizes. FPGA devices are programmed at the gate level, which gives the designer the added freedom to optimize the size and the number of registers throughout the design.

The added reliability of the design can be seen through improved CAD tools and the ease of reusability. CAD tools now provide powerful simulation packages that enable the designer to thoroughly test the components. These simulation packages are typically not as usable in microprocessor packages. For example, the simulation packages for FPGA devices produce waveforms that can be included with testing documentation. Microprocessor-based design can be simulated to show the desired input-output response; however, all of the intermediate details are typically negated. The ease of reusability also increases the reliability of the design, as the component can be used again and again, the confidence level in the component improves. This is also easier to do in an FPGA-based design over a microprocessor design, because integration of the designed component is easier in an FPGA-based design.

There are, however, some disadvantages to the more powerful FPGA-based chip design. Designing with VHDL requires the designer to think in terms of hardware instead of software. This is a major change in programming philosophy, which will take some time for the new designer to become accustomed to. It requires the designer to think of the design as a high-level system, and as signals moving throughout the system. The designer must design all of the overhead tasks that are required (internal interrupts, external interrupts, watchdog timers). The FPGA-based chips do not typically have preconfigured devices on the chip (analog-to-digital converter, pulse-width modulator, communication routines). However, the newer devices are beginning to add a standard microprocessor within FPGA device, which would ease some of the implementation burdens of standard tasks.

ALTERNATIVE PROGRAMMING PHILOSOPHY (BASIC)

One of the major difficulties in working with VHDL/FPGA devices is the alternative programming philosophy. Typically programmers are used to thinking in a linear fashion where variables are thought to contain specific information. The information that is held at a specific moment is dependent on where the program is operating. VHDL requires the programmer to change their method of thinking to become a designer. It requires the designer to think of the variables as wires, which move signals throughout the design rather than moving information. Let us take a look at the design philosophy for a microprocessor and compare it to the VHDL design for a simple problem shown in Eq. (1).

$$F <= (a \text{ AND } b) \text{ AND } (c \text{ AND } d) \tag{1}$$

A typical microprocessor design would begin by loading the contents of a to a register, loading the contents of b to another register, the two registers together, and then store that value in a designated area 1. It would then load the contents of c to a register, load the contents of b to another register, and the two registers together, and then store that value in a designated area 2. It would then load the contents of the designated area 1 to a register, load the contents of the designated area 2 to the register, and the two registers together, and then there is the output. The program would then be repeated over and over to update the changes as a, b, c, or d change. From this process it is seen that programming of a microprocessor occurs, like many programming languages, linearly throughout the appropriate steps.

The implementation of Eq. (1) in VHDL is rather straightforward. However, with this comparison, the difference in the design process is very apparent. For now, the implementation of the VHDL code will be done in a concurrent manner. Later on, the design can be modified to simulate a linear operation. The VHDL designer would first decide if this function is the minimum design needed to generate the appropriate output. This function is of the minimum design, therefore the function will be directly implemented in VHDL. Basically the circuit requires that there be three AND gates. The signals a and b are input to one of the AND gates. The signals c and d are input to the second AND gate. The output of the first and second AND gates are run to the third AND gate that provides the final result. The specific syntax of VHDL will be discussed in the subsequent sections; however, Eq. (1) is also the VHDL code for the desired function. Basically, when sequential statements are not used, in VHDL code the output will change immediately (within a small time delay) following a change in a, b, c, or d. The code would not need a repeating loop to update the output of the function as the values of a, b, c, or d change. The VHDL code would provide the result in an order of magnitude faster than the microprocessor design, and it would also be capable of performing other operations simultaneously. The change from a linear programming to concurrent programming sounds easy; however, it takes a little more effort than would be anticipated.

The most important part of programming in this type of environment is having a good picture of the high-level objective. Having a good high-level diagram is a must for designing VHDL code. An example of the high-level code for Eq. (1) is shown in Fig. 4.1.1. The high-level diagram shown in Fig. 4.1.1 is rather simple for this problem, but as systems become more complicated with multiplexers and other functions, the high-level diagram will make it possible to keep track of what each signal is. This will be seen in the alternative programming section as the high-level diagrams become more complicated, which requires sequential and concurrent processing.

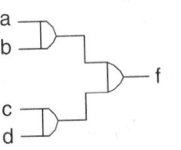

FIGURE 4.1.1   High-level diagram for Eq. (1).

```
-- *****************************************************************
-- Include Libraries
LIBRARY IEEE;
USE IEEE.STD_LOGIC_1164.all;
USE ieee.std_logic_unsigned.ALL;

-- *****************************************************************
```

FIGURE 4.1.2   VHDL code for include libraries.

VHDL/FPGA Program Configuration

This will be the first formal introduction to the syntax of VHDL. There are three major components to the syntax. There are the included libraries, there is the entity that defines the inputs and outputs of the component, and there is the architecture that is the main body of the code.

The included libraries provide standard functions, which are available within the architecture. While deciding which libraries to include, one may want to consider what other chip manufactures support these libraries, in case in the future, a decision is made to use another chip manufacture. The syntax for the libraries is shown in Fig. 4.1.2. Note that two dashes (--) indicates a comment.

The entity section tells the system what inputs and outputs will be coming into and out of the component and how they will be defined. A comparison to a standard programming language would be the type of variable (int, double, float). The syntax for the entity is shown in Fig. 4.1.3.

The main purpose of the entity declaration is to define the signals as input or output and define the size of the signals. For example, number is defined as a vector that is 8 bits wide. In the CAD software the input and output signals will be referenced to a specific I/O port. Take note that an output cannot be feedback to a defined input signal. An input must be set by a signal external to the device. The input and output signals that are declared as STD_LOGIC means that these are only a single bit.

The architecture section defines the actual function of the chip. Within the architecture, there are a few more important declarations. These declarations are known as signals (intermediate locations to store signals), constants, and variables. Variables are only allowed to be used within a sequential statement, which will be covered in subsequent sections. The syntax of the architecture is shown in Fig. 4.1.4.

VHDL/FPGA Concurrent Programming

Very high-speed integrated circuits hardware description language is capable of performing concurrent and sequential operations. This section will focus on concurrent processing; in a later section sequential operations will be introduced. The concurrent processing methodology is the main difference from microprocessor-type design. Concurrent processing means that if one of the inputs changes its value then the output will, nearly, immediately change as well. There isn't a sense of looping or linear programming. Essentially the entire program acts in a parallel manner.

```
-- *****************************************************************************
-- Declaration of INPUT & OUTPUT variables
-- ENTITY must have a reference, which is the same as  the file name
ENTITY example IS
        PORT(         clk           :  IN    STD_LOGIC;
                      button        :  IN    STD_LOGIC_VECTOR(3 DOWNTO 0);
                      flag          :  OUT   STD_LOGIC;
                      number        :  OUT   STD_LOGIC_VECTOR(7 DOWNTO 0));
END example;
-- *****************************************************************************
```

FIGURE 4.1.3   VHDL code for entity.

```
--  ****************************************************************************
--  The actual body of the program
ARCHITECTURE arch OF example IS
          -- Declare constant as needed
          CONSTANT        freq :                    integer := 12587500;
          -- Declare signals and type as needed
          SIGNAL number_now :   std_logic_vector(7 DOWNTO 0);
          SIGNAL number_next:   std_logic_vector(7 DOWNTO 0);
          -- Declare variables
          BEGIN
          -- start of actual body of program
                    -- input desired code here
          -- end of the actual body of program
END arch;
--  ****************************************************************************
```

FIGURE 4.1.4   VHDL code for architecture.

Within this section the designer has access to multiplexers, basic bit operations (AND, OR, NOT, and so forth), and some other general commands. A few of the multiplexer techniques are the WITH and WHEN statement and will be shown in Figs. 4.1.5 and 4.1.6.

In Fig. 4.1.5 the WITH statement is implemented. The actual purpose of the program is not important. It is only shown to illustrate the syntax. Within the figure there is another command that is implemented and that is the "&." The & command enables the designer to append static or dynamic signals within another signal. Another application of the & is a shift operation.

In Fig. 4.1.6 the WHEN statement is implemented. The actual purpose of the program is not important. It is only shown to illustrate the syntax. Another important aspect to all of the multiplexer commands is to remember to include all possible combinations of the select signal. If all possible combinations are not represented, undesirable results could occur.

ALTERNATIVE PROGRAMMING PHILOSOPHY (ADVANCED)

One of the major difficulties of working with FPGA devices is having to design circuits in a concurrent manner. This section will present a method of removing that limitation. There is a method of designing sequential circuits, essentially creating a circuit that functions in a linear fashion. This type of program is called a *process(sel)*. The process function also has a list of signals that are evaluated by the process. There are no fixed number of signals that can be evaluated by the process. The process will be ignored until the *sel* signal changes. Once the signal changes, the process will be entered and the appropriate code will be evaluated.

```
--  ************************************************************
--  The actual body of the program
--  This routine is the meat of the frequency divider
ARCHITECTURE arch OF example IS
BEGIN
          -- shows how to use a WITH statement
          WITH button SELECT
               number <= "1000000" & "0"     WHEN "1000",
                         "010000" & "00"      WHEN "0100",
                         "00100" & "000"      WHEN "0010",
                         "0001" & "0000"      WHEN "0001",
                         "00000000"           WHEN OTHERS;

END arch;
--  ************************************************************
```

FIGURE 4.1.5   VHDL code for WITH.

```
--  **********************************************************************
--  The actual body of the program
--  This routine is the meat of the frequency divider
ARCHITECTURE arch OF example IS
BEGIN
     --  shows how to use a WHEN statement
     number <= "100000000"    WHEN (button = "1000")  ELSE
               "010000000"    WHEN (button = "0100")  ELSE
               "001000000"    WHEN (button = "0010")  ELSE
               "000100000"    WHEN (button = "0001")  ELSE
               "000000000";
END arch;
--  **********************************************************************
```

FIGURE 4.1.6   VHDL code for WHEN.

When designing with a process the designer gains access to a new type of command. This new command is a variable. A variable can only be accessed within a process. This new type of command is needed because of the limitation of working with signals within a process. Within a process the signal will not be updated until the process is left, which means if the signal is changed many times within one process, only the last value of the signal will take effect. The change in the signal will not take place until the process is left. The use of a variable creates difficulty because of the lack of knowledge of how the gate-level design will be configured. This causes the size of the design to grow larger than the use of a typical signal. The added advantage of using a variable will allow the designer to make changes to the variable within the process many times before the process is completed.

There are many advantages to designing a sequential circuit. The most obvious is the capability of taking advantage of a looping statement and an "if" statement, however, the most important addition would be the way the high-level design could be reformulated to create a finite state machine. Essentially being able to break a task into multiple states will enable the designer to systematically solve the problem. During the implementation of a finite state machine a state and a next state need to be known at all times. Another key importance of this type of design is the ability to perform multiple operations on a signal. Remember typically a signal cannot write to itself. However, using a current and next state vector for the signal will enable the designer to get around this limitation. The high-level diagram is shown in Fig. 4.1.7.

Figure 4.1.7 shows how to visualize the state update components. The system begins with two separate processes that depend on two different signals. The first process is dependent on a change in the state_reg register. The second process is dependent on a change in the clock. The first process keeps track of the current state and monitors other signals, not shown, to determine when and what state to post to the state_next register. The second process takes into account a clock signal, which waits for the rising edge of the clock to update the state_reg with the state_next register.

The configuration of the state update component is used to streamline the design in order to reduce the amount of spaghetti code. This configuration also provides a method of performing multiple instances of a function to a signal.

There are a few major drawbacks to this type of design. First, when designing a circuit sequentially the way the circuit is compiled at the gate-level becomes less obvious. This causes the circuit to take up more space within the chip. Second, sequential design runs the possibility of not being able to compile even if the syntax is correct. The resulting sequential design might be too complex.

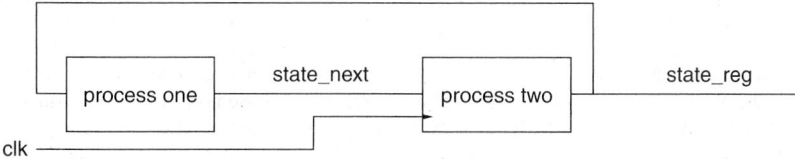

FIGURE 4.1.7   High-level diagram of state update philosophy.

```
-- **************************************************************************
-- The actual body of the program
-- This routine is the meat of the frequency divider
ARCHITECTURE arch OF clkdivider IS
     SIGNAL     count_now:    std_logic_vector(27 DOWNTO 0);
     SIGNAL     count_next:   std_logic_vector(27 DOWNTO 0);
BEGIN
     PROCESS(clk)
       BEGIN
       -- increments count_next each time the input signal
       -- goes high
       if(clk ='1' AND clk'EVENT) THEN
              count_next <= count_now +"0000000000000000000000000001";
       END IF;
     END PROCESS;
     -- resets the counter once the system has gone through the
     -- number of cycles
     PROCESS(count_next)
       BEGIN
       IF(count_next = "1111000000000000000000000000") THEN
              count_now <= "0000000000000000000000000000";
       ELSE
              count_now <= count_next;
       END IF;
     END PROCESS;
END arch;
-- **************************************************************************
```

FIGURE 4.1.8   VHDL code for a process and performing multiple instances of a function on a signal.

VHDL/FPGA Sequential Programming

This section will focus on presenting a sequential programming syntax for VHDL. The main illustrations will show how to set up a process, read the rising edge of the clock signal, and how to perform multiple instances of a function on one signal.

Figure 4.1.8 shows the typical syntax for a process and how to determine the rising edge of the clk signal. The implementation of the architecture shown in Fig. 4.1.8 also shows how to perform multiple instances of a function to a signal. This method is done by creating a current signal (defined as count_now) and by creating a next signal (defined as count_next).

Creating a process to implement the state update component shown in Fig. 4.1.7 would follow the same methodology of the process shown in Fig. 4.1.8. Instead of updating the next state by evaluating a function, the output would be the next state, which is dependent on the current state and the inputs to the component.

Timing Considerations

Timing considerations are the most critical components to any embedded system design. This is true with FPGA-based designs as well. The requirements that are standard to any embedded system still remain; however, there is a new timing requirement that must be considered in order to optimize the system. In the development of the states of the system, the designer must consider whether to have fewer states with more packed into the function or to have more states with less packed into the function. The key trade-off is throughput versus clock speed. Within the process that updates the current state based on the clock frequency, the max clock frequency would be dependent on the longest cycle through the process. Therefore, it is typically more desirable to have more states that implement less than to have less states that implement more. The timing considerations and max clock frequency can be determined using the CAD tools.

Simulation with CAD Tools

A key advantage to using the new CAD tools is the powerful set of simulation packages. Windows can be created, which show the waveform for any signal throughout the system. The simulations can show the final value and any intermediate value throughout the design. This is a major advantage for debugging and system verification.

VHDL/FPGA Component Integration and Configuration

Integrating many VHDL components together into a big design is much easier than implementing functions within a typical microprocessor. The overall design can be broken into components and each component can be independently designed and compiled. The final product can link the components together to form the overall design.

In order to illustrate this concept, a kitchen timer example will be shown. The actual function of the design is not as important as how the system is integrated together. However, for clarity, a brief explanation will be given. The system begins by placing the initial value to begin counting from LSreg (seconds) and MSreg (tens of seconds). There will be a clock divider, which will create a 1-s pulse for the counter circuit to count in order to update the clock every second. The counter component will take the initial values of LSreg and MSreg and decrement the value every second until the timer reaches zero. The updated counter will be output to the seven seg display component, which will display the appropriate number on the display. This process will stop when the counter reaches zero or it will reset when the Reset_Switch is initiated. The high-level design is shown in Fig. 4.1.9.

Within Fig. 4.1.9 it is seen that there are four major subsystems for the kitchen timer example—complete system, clkdivider, counter, and the seven seg display. Each of these subsystems is an individual component, individual file, in the design of the overall system. The inputs and outputs of each block are declared and the function of each block is declared within the architecture. The clkdivider, counter, and seven seg display are the core functions, which dictate what is actually happening in the circuit. The complete system component links all of the subsystems together.

Within the complete system component the design will define the inputs and outputs to the entire system and link the inputs and outputs of each component together. There is one more function of the complete system component, which will complete the design. When there is a signal, which is an output from one component and an input to another component (not an output for the system), there needs to be an intermediate signal defined, which will be used to link the signal. Figure 4.1.10 shows the code that was designed for the complete system component.

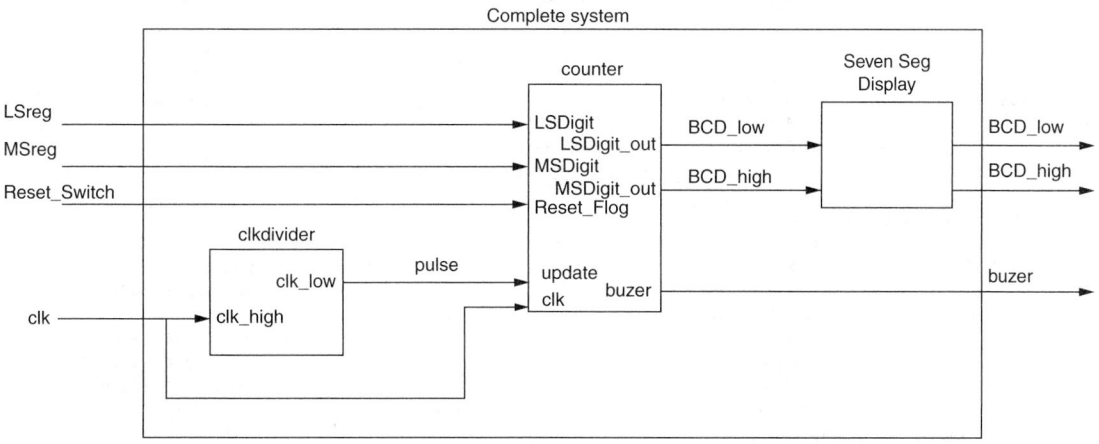

FIGURE 4.1.9   High-level diagram of kitchen timer example.

```
-- ****************************************************************
-- Program Name    :   completesystem.vhd
-- File Name       :   desktop\new folder\altera\lab 2b
-- Programed By     :   Brian Fast
-- Date            :   April 1, 2004
-- Purpose         :   This Program will pull all of the subsystems
--                     together for the kitchen timer program
--                     1) clock divider
--                     (output a rising pulse every second)
--                     2) Binary To Decimal Decoder and Subtractor
--                     3) 7-Seg Display
-- Input    :   clock signal -> 25.175 MHz
--              Reset_Switch -> pull high then low to start
--              LSreg -> binary representation of decimal value
--              input valid only between 0-9 but can handle
--              15-0 but anything above 9 will default to zero
--              MSreg -> binary representation of decimal value
--              input valid only between 0-9 but can handle
--              15-0 but anything above 9 will default to zero
-- Output   :   LS7seg -> seven segment display decimal representation of
--              Least Significant Digit of counter
--              MS7seg -> seven segment display decimal representation of
--              Most Significant Digit of Counter
-- Chip     :   10K70RC240-4
-- Board    :   Altera University Program Development Board UP2
--
-- NOTE     :   Must Compile These Files and have them
--              in the same folder
--              1) counter.vhd
--              2) clkdivider.vhd
--              3) sevensegdisplay.vhd
--
--              This program creates a program that joins all
--              of the subprograms compiled before
--
--              I/O pin numbers are can be predefined in the
--              .acf file
--              I/O pin number verification be seen in the
--              .rpt file
--
-- ****************************************************************

-- Include Libraries
LIBRARY IEEE;
USE IEEE.STD_LOGIC_1164.all;
-- ****************************************************************
-- Declaration of INPUT & OUTPUT variables
-- ENTITY must have a reference which is the same as  the file name

ENTITY completesystem IS
      PORT(   clk             :       IN      std_logic;
              LSreg           :       IN      std_logic_vector(3 DOWNTO 0);
              MSreg           :       IN      std_logic_vector(3 DOWNTO 0);
              Reset_Switch    :       IN      std_logic;
              LS7seg          :       OUT     std_logic_vector(6 DOWNTO 0);
              MS7seg          :       OUT     std_logic_vector(6 DOWNTO 0);
              DP7seg          :       OUT     std_logic_vector(1 DOWNTO 0);
              Buzer           :       OUT     std_logic);
END completesystem;
-- ****************************************************************
```

FIGURE 4.1.10   VHDL code for the complete system component that links all the other components together.

```
-- ****************************************************************
-- The actual body of the program
-ARCHITECTURE arch OF completesystem IS
        -- component declaration
        -- files which are interconnected to make up the entire file
        -- include:
        --              file name
        --              inputs
        --              outputs
        COMPONENT clkdivider
                PORT(
                        clk_high        :       IN      STD_LOGIC;
                        clk_low         :       OUT     STD_LOGIC);
        END COMPONENT;

        COMPONENT counter
                PORT(
                        clk             :       IN      STD_LOGIC;
                        update          :       IN      STD_LOGIC;
                        LSDigit         :       IN      STD_LOGIC_VECTOR(3 DOWNTO 0);
                        MSDigit         :       IN      STD_LOGIC_VECTOR(3 DOWNTO 0);
                        Reset_Flag      :       IN      STD_LOGIC;
                        LSDigit_out     :       OUT     STD_LOGIC_VECTOR(3 DOWNTO 0);
                        MSDigit_out     :       OUT     STD_LOGIC_VECTOR(3 DOWNTO 0);
                        Buzer           :       OUT     STD_LOGIC);
        END COMPONENT;

        COMPONENT sevensegdisplay
                PORT(
                        LSinput         :       IN      STD_LOGIC_VECTOR(3 DOWNTO 0);
                        MSinput         :       IN      STD_LOGIC_VECTOR(3 DOWNTO 0);
                        LSdisplay       :       OUT     STD_LOGIC_VECTOR(6 DOWNTO 0);
                        MSdisplay       :       OUT     STD_LOGIC_VECTOR(6 DOWNTO 0);
                        DPdisplay       :       OUT     STD_LOGIC_VECTOR(1 DOWNTO 0));
        END COMPONENT;

        -- interconnection signals
        -- signals that connect outputs to inputs within the system
        SIGNAL pulse      : STD_LOGIC;
        SIGNAL BCD_low    : STD_LOGIC_VECTOR(3 DOWNTO 0);
        SIGNAL BCD_high   : STD_LOGIC_VECTOR(3 DOWNTO 0);

        BEGIN
        -- mapping the inputs and outputs to and from each subsystem
                clkdivider_unit: clkdivider
                PORT MAP(clk_high=>clk, clk_low=>pulse);

                counter_unit: counter
                PORT MAP(clk=>clk, update=>pulse, Reset_Flag=>Reset_Switch, LSDigit=>LSreg,
                        MSDigit=>MSreg, LSDigit_out=>BCD_low, MSDigit_out=>BCD_high,
                        Buzer=>Buzer);

                sevensegdisplay_unit: sevensegdisplay
                PORT MAP(LSinput=>BCD_low, MSinput=>BCD_high, LSdisplay=>LS7seg,
                                MSdisplay=>MS7seg, DPdisplay=>DP7seg);

END arch;
-- ****************************************************************
```

FIGURE 4.1.10   (*Continued*) VHDL code for the complete system component that links all the other components together.

EXAMPLE DESIGN

A complete example design will be shown in this section. The example will be the remaining components of the kitchen timer problem. The complete system code has already been shown in Fig. 4.1.10. The remainder of the components shall be shown in Figs. 4.1.11 to 4.1.13.

```
-- ****************************************************************
-- Program Name    :    clkdivider.vhd
-- File Name       :    desktop\new folder\altera\kitchen timer
-- Programed By    :    Brian Fast
-- Date            :    April 2, 2004
-- Purpose         :    Clock Divider
--                      take a high frequency clock and output a
--                      lower frequency cycle
-- Input           :    currently configured for system clock (25.175 MHz)
--                      can be changed by changing the constant values
--                      freq_in = input frequency/(2*desired output frequency)
--                      freq_in_switch = (input frequency/(2*desired output frequency))/2
-- Ouput           :    currently configured for clock signal at (60 Hz)
-- Chip            :    10K70RC240-4
-- Board           :    Altera University Program Development Board UP2
-- Software        :    Altera Max+Plus v10.22
--
-- NOTE            :    This file will be used as a black box within
--                      another file so the I/O pins will not be
--                      set within this file.  The I/O signals will
--                      be interconnections set within the main program
--                      which is where the I/O pins will be set
--
--                Num    Description                                        By      Date
-- ------------------------------------------------------------------------------------------
-- Status    : 1.0    Began Initial Program                               BRF     4/2/2004
--                    The program is working correctly
--                    The program takes an input frequency
--                    and outputs a lower frequency
--                    which is dependent on the values set
--                    in the constants
--                    freq_in & freq_in_switch
--                    these constants are dependent on the
--                    input frequency and the desired
--                    output frequency
--                    the input output frequency is not
--                    determined but is dependent on the
--                    input frequency output frequency
--                    size of the registers and the
--                    internal speed of the code
--
--
--                                  -----------
--                                 | freq      |
--                 high frequency ---->| divider   |---> lower frequency
--                 system clock    |           |     new desire frequency
--                 [clk_high]       -----------      [clk_low]
--
--
-- ****************************************************************
```

FIGURE 4.1.11 VHDL code for the clock divider component.

```
-- Include Libraries
LIBRARY IEEE;
USE IEEE.STD_LOGIC_1164.all;
USE ieee.std_logic_unsigned.ALL;
-- ********************************************************
-- Declaration of INPUT & OUTPUT variables
-- ENTITY must have a reference which is the same as  the file name

ENTITY clkdivider IS
        PORT(   clk_high               : IN    STD_LOGIC;
                clk_low                : OUT   STD_LOGIC);
END clkdivider;
-- ********************************************************
-- ********************************************************
-- The actual body of the program
-- This routine is the meat of the frequency divider

ARCHITECTURE arch OF clkdivider IS
-- Set constant by the equation below
-- freq_in = (the input frequency/(2*desired frequency))
-- the max freq_in is dependent on the output frequency
CONSTANT        freq_in        :       integer := 12587500;

-- Set constant by the equation below
-- freq_in_switch = (the input frequency/desired frequency)/2
CONSTANT        freq_in_switch :       integer := 6293750;

-- used for testing
--CONSTANT       freq_in        :       integer := 34215;
--CONSTANT       freq_in_switch :       integer := 17107;
-- temporary registers used to keep track of how many input signals
-- have been input to the system
SIGNAL          count_now:             std_logic_vector(27 DOWNTO 0);
SIGNAL          count_next:            std_logic_vector(27 DOWNTO 0);

BEGIN
        PROCESS(clk_high)
                BEGIN
                -- increments count_next each time the input signal
                -- goes high
                -- keeps tracked of the number of cycles input by the system clock
                -- via clk_high
                if(clk_high ='1' AND clk_high'EVENT) THEN
                        count_next <= count_now
                        +"0000000000000000000000000001";
                END IF;
                -- determines if the output signal should be high or low
                -- by waiting until based on the constants set above
                -- ie. the input frequency and the desire output frequency
                IF(count_now < freq_in_switch) THEN
                        clk_low <= '1';
                ELSE
                        clk_low <= '0';
                END IF;
        END PROCESS;
        -- resets the counter once the system has gone through the
```

FIGURE 4.1.11   (*Continued*) VHDL code for the clock divider component.

```
                -- number of cycles equal to the input frequency
           PROCESS(count_next)
                   BEGIN
                   IF(count_next = freq_in) THEN
                           count_now <= "000000000000000000000000000000";
                   ELSE
                           count_now <= count_next;
                   END IF;
           END PROCESS;
     END arch;
     -- *******************************************************************
     -- END OF PROGRAM
```

FIGURE 4.1.11 (*Continued*) VHDL code for the clock divider component.

```
-- *******************************************************************
-- Program Name      :     counter.vhd
-- File Name         :     desktop\new folder\altera\kitchen timer
-- Programed By      :     Brian Fast
-- Date              :     April 3, 2004
-- Purpose           :     BCD Counter
--           The program will take the 4 bit binary number input to the system
--           on the MSDigit and LSDigit and initialize the counter
--           to that value then the reset is pulled high then low.
--           The inputs are 2 registers - 4 bit which represent a decimal value
--           one for the one's place holder and one for the ten's place holder
--           The system will then decrement the value
--           by one every time that the update pin has a positive
--           transition.  The decremented value will then be
--           put on the output pins.
-- Input    :   LSDigit = connected to four input pins = input binary number on pins
--              MSDigit = connected to four input pins = input binary number on pins
--              the inputs can handle a decimal number from 15-0
--              only an entry between 9-0 will be accepted
--              any number greater then 9 will default back to 9
--              this limitation is due to the fact that the counter is intended
--              to count in decimal with decimal inputs
-- Ouput    :   LSDigit_next
--              MSDigit_next
--              outputs can be decimal number from 15-0 but only an output of
--              the range 0-9 decimal will be output
-- Chip     :   10K70RC240-4
-- Board    :   Altera University Program Development Board UP2
-- Software :   Altera Max+Plus v10.22
--
-- NOTE     :   This file will be used as a black box within
--              another file so the I/O pins will not be
--              set within this file.  The I/O signals will
--              be interconnections set within the main program
--              which is where the I/O pins will be set
--
--              Num   Description                       By        Date
-- -----------------------------------------------------------------------
-- Status :    1.0   Began Initial Program             BRF       4/2/2004
--             1.01  Program Working for BCD Counter   BRF       4/6/2004
```

FIGURE 4.1.12 VHDL code for the counter component.

```
--
--
--                                        ------------
--                                       |            |
--        system clock [clk]  (1 bit ) ---->|  2 seg     |---> [LSDigit_out] (4 bits)
--              [update]  (1 bit ) ---->|    BCD     |---> [MSDigit_out] (4 bits)
--            [LSDigit]  (4 bits) ---->|  counter   |
--            [MSDigit]  (4 bits) ---->|            |
--          [Reset_Flag] (1  bit) ---->|          0 |---> [Buzer] (1 bit)
--                                       |            |
--                                        ------------
--                                 Currently this program is decrementing a 2 seg
--                                 BCD counter from the value loaded down to
--                                 zero. The program begins when the Reset_Flag
--                                 is pulled high and then low. The program
--                                 continues to decrement the value until it
--                                 reaches zero zero. Then the value remains
--                                 zero zero until a new value is loaded.
--
-- ************************************************************
-- Include Libraries
LIBRARY IEEE;
USE IEEE.STD_LOGIC_1164.all;
USE ieee.std_logic_unsigned.ALL;


-- ************************************************************
-- Declaration of INPUT & OUTPUT variables
-- ENTITY must have a reference which is the same as the file name

ENTITY counter IS
        PORT(           clk             : IN     STD_LOGIC;
                        update          : IN     STD_LOGIC;
                        LSDigit         : IN     STD_LOGIC_VECTOR(3 DOWNTO 0);
                        MSDigit         : IN     STD_LOGIC_VECTOR(3 DOWNTO 0);
                        Reset_Flag      : IN     STD_LOGIC;
                        LSDigit_out     : OUT    STD_LOGIC_VECTOR(3 DOWNTO 0);
                        MSDigit_out     : OUT    STD_LOGIC_VECTOR(3 DOWNTO 0);
                        Buzer           : OUT    STD_LOGIC);
END counter;
-- ************************************************************
-- ************************************************************
-- The actual body of the program
-- This routine is the meat of the frequency divider

ARCHITECTURE arch OF counter IS
SIGNAL    LSDreg_now    :       std_logic_vector(3 DOWNTO 0);
SIGNAL    LSDreg_next   :       std_logic_vector(3 DOWNTO 0);
SIGNAL    MSDreg_now    :       std_logic_vector(3 DOWNTO 0);
SIGNAL    MSDreg_next   :       std_logic_vector(3 DOWNTO 0);
SIGNAL    Done_Flag     :       std_logic;
BEGIN

        ------------------------------------------------------------------
        -- Counter Routine
        -- counts down the Least Significant Digit until it gets to zero
        -- then decrements the Most Significant Digit
        -- then decrements resets Least Significant Digit back to nine
        -- until it gets to zero zero
```

FIGURE 4.1.12   (*Continued*) VHDL code for the counter component.

```
-- then the buzzer goes off
-- if the initial value set for the input is greater then 9
-- then the initial value will be set to 9
PROCESS(clk)
BEGIN
        IF(clk = '1' AND clk'EVENT) THEN
                -- checks to see if the value is greater then
                IF(MSDreg_now > 9 OR LSDreg_now > 9) THEN

                        IF(MSDreg_now > 9) THEN
                                MSDreg_next <= "1001";
                        END IF;

                        IF(LSDreg_now > 9) THEN
                                LSDreg_next <= "1001";
                        END IF;

                ELSIF(Done_Flag = '0') THEN
                        IF(LSDreg_now > 0) THEN
                                LSDreg_next <= LSDreg_now - 1;
                                MSDreg_next <= MSDreg_now;
                        ELSE
                                LSDreg_next <= "1001";
                                MSDreg_next <= MSDreg_now - 1;
                        END IF;

                END IF;
        END IF;
END PROCESS;
------------------------------------------------------------------
------------------------------------------------------------------
-- Puts updated output onto LSDigit_out and MSDigit_out
-- when the update pin provides a positive transition which is
-- provided every second.  Therefore the output display is a clock
-- counting in seconds.
PROCESS(update, Reset_Flag)
BEGIN
        IF(Reset_Flag = '0')THEN

                Buzer <= '0';
                Done_Flag <= '0';
                LSDreg_now <= LSDigit;
                MSDreg_now <= MSDigit;

        ELSIF(update ='1' AND update'EVENT) THEN

                LSDreg_now <= LSDreg_next;
                MSDreg_now <= MSDreg_next;

                IF(LSDreg_now = "0001" and MSDreg_now = "0000") THEN
                        Buzer <= '1';
                        Done_Flag <= '1';
                END IF;
        END IF;
END PROCESS;
----------------------------------------------------------------
```

FIGURE 4.1.12 (*Continued*) VHDL code for the counter component.

```
        LSDigit_out <= LSDreg_now;
        MSDigit_out <= MSDreg_now;
END arch;
-- *****************************************************************
-- END OF PROGRAM
```

FIGURE 4.1.12 (*Continued*) VHDL code for the counter component.

```
-- *****************************************************************
-- Program Name    :    sevensegdisplay.vhd
-- File Name       :    desktop\new folder\altera\kitchen timer
-- Programed By    :    Brian Fast
-- Date            :    April 6, 2004
-- Purpose         :    display output in correct form
-- Input   :  Two Four Bit Registers that Represent a unsigned
--            decimal number (0-9)
--            The two regisiters represents the Least Significant Digit
--            and Most SIgnificant Digit
-- Ouput   :  Two seven bit Registers that Represents the decimal
--            number of the the two corresponding binary inputs.
--            the seven bit output will represend the appropriate bit
--            pattern to display the value on a 7 segment LED display
-- Chip    :  10K70RC240-4
-- Board   :  Altera University Program Development Board UP2
-- Software :  Altera Max+Plus v10.22
--
-- NOTE    :  This file will be used as a black box within
--            another file so the I/O pins will not be
--            set within this file.  The I/O signals will
--            be interconnections set within the main program
--            which is where the I/O pins will be set
--            Num              Description                 By        Date
-- -------------------------------------------------------------------------
-- Status :  1.0              Began Initial Program       BRF       4/6/2004
--
--                                      --------------
--                                      |  two        |   input to 7-seg display
--                                      |  seven      |
--            [LSinput] (4 bits)  ----->|  seg        |---> [LSdisplay] (7 bits)
--                                      | Display     |---> [MSdisplay] (7 bits)
--            [MSinput] (4 bits)  ----->|             |---> [DPdisplay] (2 bits)
--                                      |             |
--                                      --------------
--                            Currently the Program is making the appropriate
--                            Decimal to 7 seg display conversion.
--
-- *****************************************************************
-- Include Libraries
LIBRARY IEEE;
USE IEEE.STD_LOGIC_1164.all;
USE ieee.std_logic_unsigned.ALL;
```

FIGURE 4.1.13 VHDL code for the seven segment display component.

```
-- *****************************************************************
-- Declaration of INPUT & OUTPUT variables
-- ENTITY must have a reference which is the same as  the file name

ENTITY sevensegdisplay IS
        PORT(
                        LSinput         :       IN      std_logic_vector(3 DOWNTO 0);
                        MSinput         :       IN      std_logic_vector(3 DOWNTO 0);
                        LSdisplay       :       OUT     std_logic_vector(6 DOWNTO 0);
                        MSdisplay       :       OUT     std_logic_vector(6 DOWNTO 0);
                        DPdisplay       :       OUT     std_logic_vector(1 DOWNTO 0));
END sevensegdisplay;
-- *****************************************************************
-- *****************************************************************
-- The actual body of the program

ARCHITECTURE arch OF sevensegdisplay IS
        -- Look up table for seven segment LCD
        --                      a
        --                     ---
        --              f | _ | b
        --              e | g | c
        --          dp.    ---
        --                      d
        --
        -- 1 equals off
        -- 0 equals on
        --                                      gfedcba
        CONSTANT seg7_0: std_logic_vector(6 DOWNTO 0):= "1000000";
        CONSTANT seg7_1: std_logic_vector(6 DOWNTO 0):= "1111001";
        CONSTANT seg7_2: std_logic_vector(6 DOWNTO 0):= "0100100";
        CONSTANT seg7_3: std_logic_vector(6 DOWNTO 0):= "0110000";
        CONSTANT seg7_4: std_logic_vector(6 DOWNTO 0):= "0011001";
        CONSTANT seg7_5: std_logic_vector(6 DOWNTO 0):= "0010010";
        CONSTANT seg7_6: std_logic_vector(6 DOWNTO 0):= "0000010";
        CONSTANT seg7_7: std_logic_vector(6 DOWNTO 0):= "1111000";
        CONSTANT seg7_8: std_logic_vector(6 DOWNTO 0):= "0000000";
        CONSTANT seg7_9: std_logic_vector(6 DOWNTO 0):= "0010000";
        CONSTANT seg7_n: std_logic_vector(6 DOWNTO 0):= "0111111";
        CONSTANT seg7_x: std_logic_vector(6 DOWNTO 0):= "1111111";
        CONSTANT DP_on : std_logic_vector(1 DOWNTO 0):= "00";
        CONSTANT DP_off: std_logic_vector(1 DOWNTO 0):= "11";
BEGIN
        -- reads the value on the Least Significant Input (4 bits)
        -- and then displays the appropriate Binary to Decimal Conversion
        -- the output is then feed to the LCD display.
        -- that conversion is taken care of in the look up table above
        WITH    MSinput SELECT
                        MSdisplay <=    seg7_0          WHEN "0000",
                                        seg7_1          WHEN "0001",
                                        seg7_2          WHEN "0010",
                                        seg7_3          WHEN "0011",
                                        seg7_4          WHEN "0100",
                                        seg7_5          WHEN "0101",
                                        seg7_6          WHEN "0110",
```

FIGURE 4.1.13   (*Continued*) VHDL code for the seven segment display component.

```
                    seg7_7          WHEN "0111",
                    seg7_8          WHEN "1000",
                    seg7_9          WHEN "1001",
                    seg7_x          WHEN OTHERS;
-- reads the value on the Most Significant Input (4 bits)
-- and then displays the appropriate Binary to Decimal Conversion
-- the output is then feed to the LCD display.
-- that conversion is taken care of in the look up table above
WITH    LSinput SELECT
               LSdisplay <=  seg7_0          WHEN "0000",
                             seg7_1          WHEN "0001",
                             seg7_2          WHEN "0010",
                             seg7_3          WHEN "0011",
                             seg7_4          WHEN "0100",
                             seg7_5          WHEN "0101",
                             seg7_6          WHEN "0110",
                             seg7_7          WHEN "0111",
                             seg7_8          WHEN "1000",
                             seg7_9          WHEN "1001",
                             seg7_x          WHEN OTHERS;
               DPdisplay <= DP_off;
END arch;
-- ************************************************************************
```

FIGURE 4.1.13   *(Continued)* VHDL code for the seven segment display component.

CONCLUSION

In conclusion this chapter has gone through the process for designing integrated digital systems using VHDL. Details were provided for the appropriate syntax and for design philosophies, which will help a designer get started. This chapter was intended for someone who has a familiarity with digital and embedded system design.

INFORMATION RESOURCES

Chip supplier: http://www.altera.com/#

Chip supplier: http://www.latticesemi.com/products/fpga/index.cfm?qsmod=1#xpga

Chip supplier: http://www.atmel.com/products/ULC/

Chip supplier: http://www.xilinx.com/

Article about VHDL component programming:

http://www.circuitcellar.com/library/print/0899/Anderson109/8.htm

Software: http://www.electronicstalk.com/news/qui/qui132.html

Overview: http://www.ecs.umass.edu/ece/labs/vlsicad/ece665/spr04/projects/kesava-fpga.pdf

FPAA:

http://bach.ece.jhu.edu/~tim/research/fpaa/fpaa.html

http://www.designw.com/

http://www.ee.ualberta.ca/~vgaudet/fpaa/index.html

http://howard.engr.siu.edu/~haibo/research/mfpa/

http://www.circuitcellar.com/library/newproducts/158/anadigm.html

http://www.ece.arizona.edu/~cmsl/Publications/FPAA/mwc98.pdf

http://www.imtek.uni-freiburg.de/content/pdf/public/2004/becker.pdf

FPAA manufacturer: http://www.anadigm.com/

http://www.zetex.com/

FPGA & FPAA manufacturers list:

http://www.alexa.com/browse/general?catid=57799&mode=general

FPGA & FPAA manufacturer:

http://www.latticesemi.com/products/index.cfm?CFID=4003705&CFTOKEN=67611870

BIBLIOGRAPHY

Brown, S., and Z. Vranesic, "Fundamentals of Digital Logic with Verilog Design," McGraw-Hill, 2003.

Pierzchala, E., "Field-Programmable Analog Arrays," Kluwer Academic Publishers, 1998.

Razavi, B., "Design of Analog CMOS Integrated Circuits," McGraw-Hill, 2001.

Uyemura, J. P., "A First Course In Digital Systems Design," Brooks/Cole Publishing, 1999.

P · A · R · T
2

COMPONENTS

 On the CD-ROM

Properties of Materials

SECTION 5

ELECTRONIC AND FIBER OPTIC COMPONENTS

This section reviews the basic components in their discrete form that make up electronic circuits and systems. The first two chapters cover fundamental passive and active discrete components, respectively. Another chapter is devoted to batteries and fuel cells. While batteries convert chemical energy into electric energy via an oxidation-reduction electrochemical reaction, the reactants are contained within the cell, whereas in a fuel cell, one or both of the reactants are not embedded, but fed from an external supply when power is needed. A chapter on relays and switches compares the characteristics of solid-state and electromechanical versions. Another on connectors describes various types of connectors along with design, application, and reliability issues. Finally, a completely new chapter explores the characteristics and applications of optical fiber components, ranging from the optical fiber itself to connectors, couplers, fiber gratings, circulators, switches, and amplifiers.

Since the selection of materials to fabricate electronic components is fundamental to their operational characteristics and expected lifetimes, valuable background information covering a broad range of electronic materials is provided on the accompanying CD-ROM.

The treatment of specialized components is covered in separate sections of the handbook: Section 6, Integrated Circuits and Microprocessors; Section 7, UHF and Microwave Devices; Section 8, Transducers and Sensors; and Section 9, Radiant Energy Sources and Sensors.   D.C.

In This Section:

On the CD-ROM:

Blech, I. A., "Properties of Materials," reproduced from the 4th edition of this handbook, is a comprehensive review of conductive and resistive materials, dielectric and insulating materials, magnetic materials, semiconductors, electron-emitting materials, and radiation-emitting materials.

CHAPTER 5.1
PASSIVE COMPONENTS

Carrington H. Greenidge, Thomas S. Gore, Jr., Emanuel Gikow, Joseph M. Giannotti, John R. Vig, Sam Di Vita

RESISTORS

Carrington H. Greenidge

Introduction

The resistive function is a fundamental element of electronics. It results from the characteristic of a material to impede the flow of electrons through that material in a measured or calibrated manner. Resistors are devices that exhibit this primary characteristic along with secondary characteristics. The primary characteristic, which is measured in ohms, is derived from the resistivity (rho) of the material and the geometry of the resistance element. The basic relationships between resistance, voltage (volts), current (amps), and power (watts) are defined by Ohm's law relationships: $I = E/R$ and $P = I * E$.

In general, resistive devices have an equivalent circuit as shown in Fig. 5.1.1, where series inductance (L) and shunt capacitance (C) are secondary (reactance) characteristics that affect device response at high frequencies. The use of different materials and geometries results in a wide range of secondary characteristics and capabilities that provides an expanding portfolio of resistor products to meet new application requirements.

Resistor Terminology

Resistance value. The primary characteristic of the device measured in ohms (Ω).

Nominal resistance value. The ohmic value attributed to a resistor and identified by color coding or some other identification technique.

Resistance tolerance. The allowed variation of resistance value from the nominal value; it is marked on the resistor or identified by some other technique.

Resistance stability. The estimate of the variation of resistance value over time when operated within ratings.

Power rating. The maximum value of steady-state power that can be dissipated by a resistor, over long periods of time without significant change in resistance value.

Voltage rating. The maximum voltage that may be applied across the element without significant change in resistance value.

Temperature rating. The temperature at which the resistor will operate at full rated power. Operation of the resistor at temperatures above the established rating is possible at reduced power levels up to a temperature called the zero power level or zero derating. This temperature usually reflects the hot-spot temperature of the resistor.

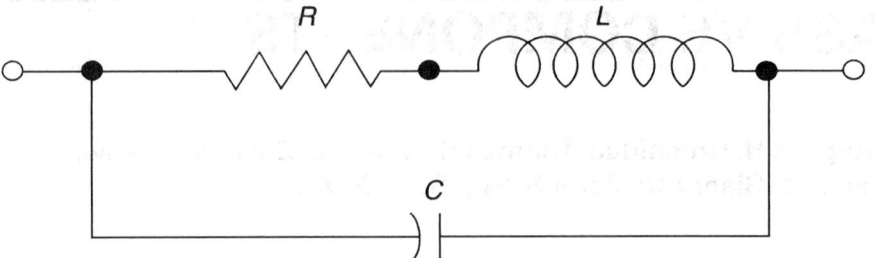

FIGURE 5.1.1   Equivalent circuit of a resistor.

Temperature coefficient of resistance (TCR). The variation of resistance value as a function of temperature expressed in parts per million/degree Celsius or as a percentage per degree Celsius. TCR is usually retraceable.

Voltage coefficient of resistance (VCR). The variation of resistance value as a function of applied voltage expressed in parts per million (PPM) or percent per volt. VCR is usually retraceable.

Noise. The instantaneous voltage instability that occurs in a resistor owing to thermal excitation of the structure (Johnson noise) or current effects (current noise). Most resistors have basal noise index and third harmonic (THI) levels. Significant deviation from these indices suggests the presence of structural abnormalities.

High-frequency effects (see Fig. 5.1.1). The application of ac voltages produces inductive and capacitance reactance in many resistors. Series inductance increases the overall impedance of the resistor while parallel capacitance decreases the impedance.

Fixed resistors. Resistors that in the completed state cannot be adjusted.

Adjustable resistors. Resistors that in the completed state can have the resistance value changed by some external technique that is nonreversible.

Variable resistors. Resistors that in the completed state can be changed by some technique (usually mechanical) that is reversible.

Precision resistors. Precision resistors are used in applications requiring tight tolerance (less than ±1 percent), tight TCR (less than ±100 PPM), long-term resistance stability (less than ±1 percent drift), and low noise.

Semiprecision resistors. Semiprecision resistors are used in applications in which long-term stability (up to ±5 percent) is important but initial tolerance (±1 percent) and TCR (±100 PPM) requirements are not as critical.

General-purpose resistors. These devices are used in applications where wide resistance tolerances and resistance shifts (±5, 10, and 20 percent) are acceptable.

Fixed Resistors for Thru-Hole Applications

Fixed wirewound resistors are constructed using an appropriate length and diameter of resistance wire that is wound around a mandrel or core to achieve a specific resistance. Leads, lugs, or other terminals (including solderable terminations for surface mount applications) are provided at each end of the resistance wire, mechanically secured and protected in an appropriate package. Wirewound resistors provide very tight tolerance, excellent long stability, high temperature capability, and very low noise. They have poor high-frequency response because of inductive properties and high-resistance value limitations because of thin wire fragility. Wirewound resistors are used in precision and power applications.

Fixed carbon composition resistors consist of a slug of inert binder mixed with measured amounts carbon particles to give the desired nominal resistance. Axial lead terminations are embedded in each end of the slug during a formation process. The entire assembly is molded into a plastic housing that provides mechanical stability. Alternatively, a carbon particle/binder slurry is established on the outside wall of a glass tube, terminal leads are mounted at each end using a conductive cement and the assembly is molded into a protective case.

Carbon composition resistors are in the general purpose category and provide a wide resistance range. The slug construction provides exceptional transient overload protection and these devices have exceptional reliability. The high-frequency response of these resistors is diminished because of capacitive reactance.

Fixed film resistors (carbon film, cermet, and metal films) are constructed by establishing the appropriate film on an insulating substrate. Terminations, usually cap and lead assemblies, are provided at each end. The film resistance element is adjusted to value using a mechanical or a laser spiraling technique that cuts a helix path into the film or by the mechanical removal of film from the surface (unlike the unadjusted glass tube carbon comp device described previously). A protective coating is usually established over the entire structure to provide mechanical and environmental protection.

Carbon film resistors are available primarily as general-purpose resistors but can fit into semiprecision applications where ± 1 percent tolerances are needed but looser TCRs are acceptable.

They are available over a wide resistance value range (in some specialty areas going into the teraohm range) and they have a good high-frequency response. Their environmental performance, however, is not equal to the metal film and cermet devices described as follows.

Metal film (thin film, thin metal film, or metal oxide) resistors are used in precision (< ±1 percent) and semiprecision (±1 percent) applications. They are the most stable film devices with TCRs ±25 and ±50 PPM/°C generally available at reasonable cost. Tighter tolerances and TCRs are available at higher cost. Metal films are selected over precision wirewound devices because of lower cost, better high-frequency response and wider resistance range.

Cermet films are used primarily in semiprecision applications, but are also available for some precision (< ±1 percent) and general purpose (> ±1 percent) use. The element structure (different ratios of insulating CERamic and conductive METal particles, fused together on a substrate) provides a wider resistance range than metal film or wirewound resistors. The long-term stability is not as good and tighter tolerances and TCRs are not available.

Resistor networks are structures with multiple fixed elements established on one substrate. The elements are usually Cermet films but precision devices are available in thin film technologies. There are two basic packages, the single in-line package (SIP) and the dual in-line package (DIP) with three standard circuits or interconnections (isolated element, pull-up, and terminator) as shown in Fig. 5.1.2.

Custom circuits, both simple and complex, can be designed for special applications. The termination (pin) count for SIPs ranges from 2 to 20 or more while DIPs are usually 14 or 16 pins. Pin spacing is usually 0.100 in.

Resistor networks in general have a higher per element cost than discrete devices. However, savings in assembly cost, improved circuit board real estate usage and more uniform electrical characteristics (stability, TCR, and high-frequency response) drive the selection of resistor networks.

Fixed Resistors for Surface Mount Applications

Film chip resistors are constructed by establishing a resistive element on the top surface of a rectangular ceramic chip. Solderable terminations are created at each end and short extensions of the terminations are also created on the underside of the chip to enhance the bond strength after solder attachment. A glassy or plastic passivation layer is usually formed over the resistive element for environmental protection. Resistance value is established using laser trim techniques. Film chip resistors have excellent power handling capability because of the thermally efficient mounting technique. This allows for improved board component density compared to axial devices. As new, more accurate placement equipment is developed, the physical size of the parts is shrinking. The size, XXYY, in millimeters, is defined by XX the approximate length and YY the approximate width. The size trend, from 1206 to 0805 to 0603 to 0402, is clear. Larger sizes, 1210, 2012, and 2520 are used when higher power levels are required.

Cermet and metal film chips have similar electrical and performance characteristics as the axial designs from which they were derived. Because of size standardization, the electrical performance is more uniform and assembly operations are simplified. Film chip resistors are now the workhorse resistive component.

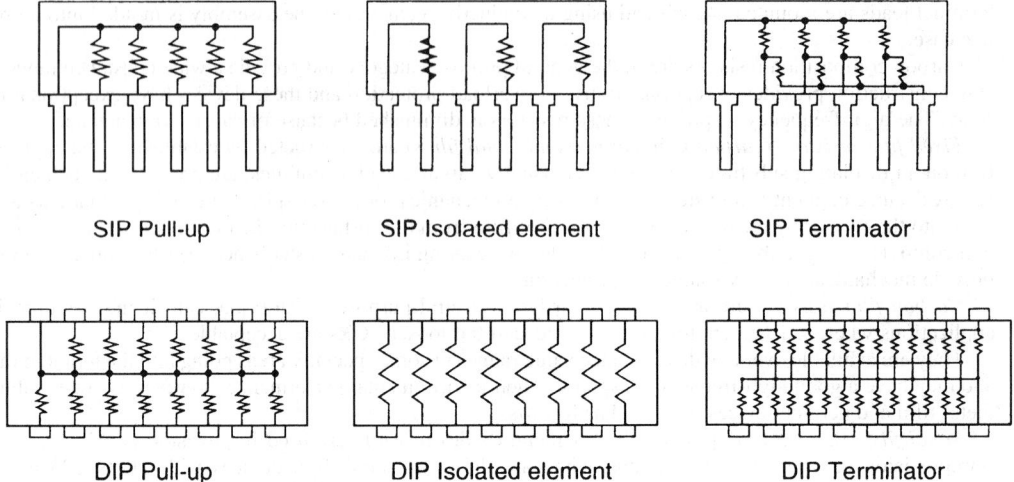

SIP Pull-up   SIP Isolated element   SIP Terminator

DIP Pull-up   DIP Isolated element   DIP Terminator

FIGURE 5.1.2 Standard SIP and DIP circuits.

The metal electrode face-bonding (MELF) resistor is the surface mount technology (SMT) equivalent of the axial designs described previously. Modifications to the protective coating process leave the endcap backwalls exposed, which provides the solderable surfaces required for surface mounting. MELFs are not as popular as the rectangular chip because of more difficult assembly routines but are used because they are the lowest cost tight tolerance, tight TCR SMT device.

Wirewound chips have been developed to satisfy the needs for precision wirewound performance in the 100 percent SMT assembly environment (as opposed to the mixed through-hole and SMT environment). The performance characteristics of wirewound chips and fixed wirewound resistors are similar.

Surface mount networks are multiple element devices derived from the DIP concept with either standard or custom internal circuitry. "Gull"-shaped or "J"-shaped terminations (which sit on the solder pads of an SMT board) are molded into the completed device body. Lead spacing can be 0.100 to 0.50 in. and can be placed on opposing or on all four sides.

Chip arrays are the true SMT derivatives or resistor networks and are essentially sets of chip resistors integrated into a monolithic structure (although internal circuitry can be much more complex, including film capacitors, than just isolated elements). Chip arrays use termination constructions similar to chip resistors with pitch down to 0.025 in.

Variable Resistors

Potentiometers are precision devices designed with either film, wirewound, or conductive plastic elements and are available in single or multiturn and single multisection units. The electrical output in terms of voltage applied across the element is linear or follows a predetermined curve (taper) with respect to the angular position contact arm. *Precision potentiometers* can provide superior resolution (tracking of the actual resistance to the theoretical resistance defined by the taper) where precise electrical and mechanical output and quality performance are required. *General purpose potentiometers* are used as gain or volume controls, voltage dividers or current controls in lower precision circuits. These devices can be ganged together and/or provided with rear-mounted switches. *Trimmer potentiometers* are lead-screw actuated, multiturn devices principally used as "set and forget" devices to control low current or bias voltages or to balance circuit variables.

Rheostats are larger, variable resistors, usually wirewound, for power applications. They are used for controls for motor speed, ovens, heaters, and other applications where higher current levels are present.

CAPACITORS

Thomas S. Gore, Jr.

Introduction

A capacitor consists basically of two conductors separated by a dielectric or vacuum so as to store a large electric charge in a small volume. The capacitance is expressed as a ratio of electric charge to the voltage applied $C = Q/V$, where Q = charge (C), V = voltage (V), C = capacitance (F). A capacitor has a capacitance of one farad when it receives a charge of 1C at a potential of 1V. The electrostatic energy in watt-seconds or joules stored in the capacitor is given by

$$J = \tfrac{1}{2}CV^2$$

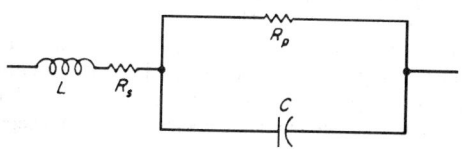

FIGURE 5.1.3 Equivalent circuit of a capacitor; R_s = series resistance owing to wire leads, contact terminations, and electrodes; R_p = shunt resistance owing to resistivity of dielectric and case material, and to dielectric losses; and L = stray inductance owing to leads and electrodes.

Dielectric. Depending on the application, the capacitor dielectric may be air, gas, paper (impregnated), organic film, mica, glass, or ceramic, each having a different dielectric constant, temperature range, and thickness.

Equivalent Circuit. In addition to capacitance, a practical capacitor has an inductance and resistance. An equivalent circuit useful in determining the performance characteristics of the capacitor is shown in Fig. 5.1.3.

Equivalent Series Resistance (ESR). The ESR is the ac resistance of a capacitor reflecting both the series resistance R_s and the parallel resistance R_p at a given frequency so that the loss of these elements can be expressed as a loss in a single resistor R in the equivalent circuit.

Capacitive Reactance. The reactance of a capacitor is given by

$$X_c = 1/2\pi fC = 1/\omega C \quad (\Omega)$$

where $f = \omega/2\pi$ is the frequency in hertz.

Impedance (Z). In practical capacitors operating at high frequency, the inductance of the leads must be considered in calculating in impedance. Specifically,

$$Z = \sqrt{R^2 + (X_L - X_C)^2}$$

where R is the ESR, and X_L reflects the inductive reactance.

The effects illustrated in Fig. 5.1.3 are particularly important at radio frequencies where a capacitor may exhibit spurious behavior because of these equivalent elements. For example, in many high-frequency tuned circuits the inductance of the leads may be sufficient to detune the circuit.

Power Factor (PF). The term PF defines the electrical losses in a capacitor operating under an ac voltage. In an ideal device the current will lead the applied voltage by 90°. A practical capacitor, owing to its dielectric, electrode, and contact termination losses, exhibits a phase angle of less than 90°. The PF is defined as the ratio of the effective series resistance R to the impedance Z of the capacitor and is usually expressed as a percentage.

Dissipation Factor (DF). The DF is the ratio of effective series resistance R to capacitive reactance X_C and is normally expressed as a percentage. The DF and PF are essentially equal when the PF is 10 percent or less.

Quality Factor Q. The Q is a figure of merit and is the reciprocal of the dissipation factor. It usually applies to capacitors used in tuned circuits.

Leakage Current, DC. Leakage current is the current flowing through the capacitor when a dc voltage is applied.

Insulation Resistance.   The insulation resistance is the ratio of the applied voltage to the leakage current and is normally expressed in megohms. For electrolytic capacitors, the maximum leakage current is normally specified.

Ripple Current or Voltage.   The ripple current or voltage is the rms value of the maximum allowable alternating current or voltage (superimposed on any dc level) at a specified frequency at which the capacitor may be operated continuously at a specified temperature.

Surge Voltage.   The surge voltage applicable to electrolytic capacitors is a voltage in excess of the rated voltage, which the capacitor will withstand for a specified limited period at any temperature.

Fixed Capacitors

Precision Capacitors.   Capacitors falling into the precision category are generally those having exceptional capacitance stability with respect to temperature, voltage, frequency, and life. They are available in close capacitance tolerances and have low-loss (high-Q) dielectric properties. These capacitors are generally used at radio frequencies in tuner, rf filter, coupling, bypass, and temperature-compensation applications. Typical capacitor types in this category are mica, ceramic, glass, and polystyrene. The polystyrene capacitor has exceptionally high insulation resistance, low losses, low dielectric absorption, and a controlled temperature coefficient for film capacitors.

Semiprecision Units.   Paper- and plastic-film capacitors with foil or metallized dielectric are nonpolar and generally fall between the low-capacitance precision types, such as mica and ceramic, and the high-capacitance electrolytics.

General Purpose.   Electrolytic aluminum and tantalum capacitors and the large-usage general-purpose (high-K) ceramic capacitors both have broad capacitance tolerances, are temperature-sensitive, and have high volumetric efficiencies (capacitance-volume ratio). They are primarily used as bypass and filter capacitors where high capacitance is needed in small volumes, and with guaranteed minimum values. These applications do not require low dissipation factors, stability, or high insulation resistance found in precision and semiprecision capacitors. On a performance versus cost basis, the general-purpose capacitors are the least expensive of the groups. High-capacitance aluminum electrolytic capacitors have been designed for computer applications featuring low equivalent series resistance and long life.

Suppression Capacitors.   The feed-through capacitors are three-terminal devices designed to minimize effective inductance and to suppress rf interference over a wide frequency range. For heavy feed-through currents, applications in 60- and 400-Hz power supplies, paper or film dielectrics are normally used. For small low-capacitance, low-current units, the ceramic and button-mica feed-through high-frequency styles are used.

Capacitors for Microelectronic Circuits.   Table 5.1.1 lists representative styles of discrete miniature capacitors electrically and physically suitable for microelectronic circuit use (filtering, coupling, tuning, bypass, and so forth). The chip capacitor, widely used in hybrid circuits, is available in single-wafer or multilayer (monolithic) ceramic, or in tantalum constructions, both offering reliable performance, very high volumetric efficiency, and a wide variety of capacitance ranges at moderate cost. Temperature-compensating ceramic chips are used where maximum capacitance stability or predictable changes in capacitance with temperature are required.

General-purpose ceramic and solid electrolytic tantalum chips are used for coupling and bypass applications where very high capacitance and small size are necessary. The ceramic chips are unencapsulated and leadless, with suitable electrodes for soldering in microcircuits. The beam-leaded tantalum chips are attached by pressure bonding. The tantalum chip is also available in a multiple-unit assembly in the dual-in-line package for use on printed-circuit boards.

Transmitter Capacitors.   The principal requirements for transmitter capacitors are high rf power-handling capability, high rf current and voltage rating, high Q, low internal inductance, and very low effective series resistance. Mica, glass, ceramic, and vacuum- or gas-filled capacitors are used primarily for transmitter applications. Glass dielectric transmitter capacitors have a higher self-resonating frequency and current rating than comparable mica styles. The ceramic capacitors offer moderately high rf current ratings, operating temperatures to

TABLE 5.1.1 Characteristics of Typical Capacitors for Microelectronic Circuits

Type of capacitor	Typical capacitance range	Typical working voltage range, V, dc	Temperature range, °C	Dissipation factor, %	Minimum insulation resistance, MΩ at 25°C	Temp. coeff., ppm/°C	Max. capacitance change over temp. range, %	Termination
Chip:								
Ceramic, temperature-compensating	1 pF to 0.027 mF	50–200	–55 to +125	0.1 at 1 MHz	50,000	0 to –750	…	Metallized
Ceramic, general-purpose	390 pF to 0.47 μF	25–200	–55 to +125	3.0 at 1 kHz	10,000	…	±15	Metallized
Tantalum oxide (beam-lead), dry	100–3000 pF	12–50	–55 to +85	0.6 at 1kHz	10,000	+200	…	Beam lead
Tantalum oxide Polar, solid electrolyte	0.1–47 μF	3–35	–55 to +85	4.0 at 120 Hz	…	…	±10	Metallized
Metallized, film (metal case)	0.1–10 μF	50–100	–55 to +85	2.5 at 1 kHz	5,000	…	±10	Axial
Variable ceramic trimmer	Min. 1–3 pF, max. 5–25 pF	25	–55 to +125	0.2 at 1 MHz	10,000	…	±5–±15	Printed circuit

105°C, and high self-resonant frequencies. The gas or vacuum capacitor is available in a wide range of capacitance and power ratings and is used where very-high-power, high-voltage, and high-frequency circuit conditions exist. These units are also smaller than other transmitting types for comparable ratings. The circuit performance of the transmitter capacitor is highly dependent on the method of mounting, lead connections, operating temperatures, air circulation, and cooling, which must be considered for specific applications.

Variable Capacitors

Variable capacitors are used for tuning and for trimming or adjusting circuit capacitance to a desired value.

Tuning Capacitor (Air-, Vacuum-, or Gas-Filled).   The parallel plate (single or multisection style) is used for tuning receivers and transmitters. In receiver applications, one section of a multisection capacitor is normally used as the oscillator section, which must be temperature-stable and follow prescribed capacitance-rotation characteristics. The remaining capacitor sections are mechanically coupled and must track to the oscillator section. The three most commonly used plate shapes are semi-circular (straight-line capacitance with rotation), midline (between straight-line capacity and straight-line frequency), and straight-line frequency (which are logarithmically shaped). For transmitter applications, variable vacuum- or gas-filled capacitors of cylindrical construction are used. These capacitors are available in assorted sizes and mounting methods.

Special Capacitors

High-Energy Storage Capacitors.   Oil-impregnated paper and/or film dielectric capacitors have been designed for voltages of 1000 V or higher of pulse-forming networks. For lower voltages, special electrolytic capacitors can be used which have a low inductance and equivalent-series resistance.

Commutation Capacitors.   The widespread use of SCR devices has led to the development of a family of oil-impregnated paper and film dielectric capacitors for use in triggering circuits. The capacitor is subjected to a very fast rise time (0.1 ms) and high current transients and peak voltages associated with the switching.

Reference Specifications.   Established reliability (ER) specifications are a series of military (MIL) specifications that define the failure-rate levels for selected styles of capacitors. The predicted failure rate is based on qualification-approval life testing and continuous-production life testing by the capacitor manufacturer.

INDUCTORS AND TRANSFORMERS

Emanuel Gikow

Introduction

Inductive components are generally unique in that they must be designed for a specific application. Whereas resistors, capacitors, meters, switches, and so forth, are available as standard or stock items, inductors and transformers have not usually been available as off-the-shelf items with established characteristics. With recent emphasis on miniaturization, however, wide varieties of chokes have become available as stock items. Low inductance values are wound on a nonmagnetic form, powdered iron cores are used for the intermediate values of inductance, and ferrites are used for the higher-inductance chokes. High-value inductors use both a ferrite core and sleeve of the same magnetic material. Wide varieties of inductors are made in tubular, disk, and rectangular shapes. Direct-current ratings are limited by the wire size or magnetic saturation of the core material.

Distributed Capacitance

The distributed capacitance between turns and windings has an important effect on the characteristics of the coil at high frequencies. At a frequency determined by the inductance and distributed capacitance, the coil becomes a parallel resonant circuit. At frequencies above self-resonance, the coil is predominantly capacitive. For large-valued inductors the distributed capacitance can be reduced by winding the coil in several sections so that the distributed capacitances of the several sections are in series.

Toroids

The toroidal inductor using magnetic-core material has a number of advantages. Using ungapped core material, the maximum inductance per unit volume can be obtained with the toroid. It has the additional advantages that leakage flux tends to be minimized and shielding requirements are reduced. The inductance of the toroid and its temperature stability are directly related to the average incremental permeability of the core material. For a single-layer toroid of rectangular cross section without air gap,

$$L = 0.0020N^2b\mu_d \ln (r_2/r_1)$$

where L = inductance (μH)
    b = core width (cm)
    μ_d = average incremental permeability
    r_1 = inside radius (cm)
    r_2 = outside radius (cm)
    N = total number of turns

Incremental permeability is the apparent ac permeability of a magnetic core in the presence of a dc magnetizing field. As the dc magnetization is increased to the point of saturating the magnetic core, the effective permeability decreases. This effect is commonly used to control inductance in electronically variable inductors.

Inductors with magnetic cores are often gapped to control certain characteristics. In magnetic-cored coils the gap can reduce nonlinearity, prevent saturation, lower the temperature coefficient, and increase the Q.

Adjustable Inductors

The variable inductor is used in a number of circuits: timing, tuning, calibration, and so forth. The most common form of adjustment involves the variation of the effective permeability of the magnetic path. In a circuit comprising an inductor and capacitor, the slug-tuned coil may be preferred over capacitive tuning in an environment with high humidity. Whereas the conventional variable capacitor is difficult and expensive to seal against the entry of humidity, the inductor lends itself to sealing without seriously inhibiting its adjustability. For example, a slug-tuned solenoid can be encapsulated, with a cavity left in the center to permit adjustment of the inductance with a magnetic core. Moisture will have little or no effect on the electrical performance of the core. The solenoid inductor using a variable magnetic core is the most common form of adjustable inductor. The closer the magnetic material is to the coil (the thinner the coil form), the greater the adjustment range will be. Simultaneous adjustment of both a magnetic core and magnetic sleeve will increase the tuning range and provide magnetic shielding. In another form, the air gap at the center post of a cup core can be varied to control inductance. Not only is the result a variable inductor, but also the introduction of the air gap provides a means for improving the temperature coefficient of the magnetic core by reducing its effective permeability.

Power Transformers

Electronic power transformers normally operate at a fixed frequency. Popular frequencies are 50, 60, and 400 Hz. Characteristics and designs are usually determined by the voltampere (VA) rating and the load. For example, when used with a rectifier, the peak inverse voltage across the rectifier will dictate the insulation requirements. With capacitor input filters, the secondary must be capable of carrying higher currents than with choke input filters. There are a number of ways by which the size and weight of power transformers can be reduced, as follows:

Operating frequency. For a given voltampere rating, size and weight can be reduced as some inverse function of the operating frequency.

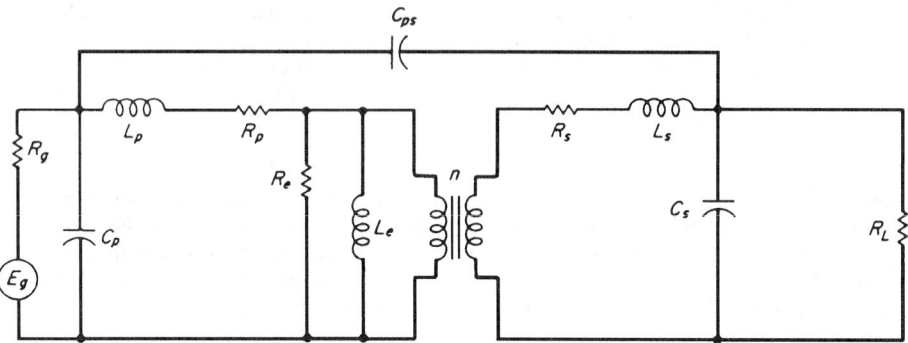

FIGURE 5.1.4   Equivalent circuit of a broadband transformer; E_g = generator voltage, R_g = generator impedance, C_p = primary shunt and distributed capacitance, L_p = primary leakage inductance, L_s = secondary leakage inductance, C_s = secondary shunt and distributed capacitance, R_p = primary winding resistance, R_e = equivalent resistance corresponding to core losses, L_e = equivalent magnetizing (open-circuit) inductance of primary, n = ideal transformer primary-to-secondary turns ratio, C_{ps} = primary-to-secondary capacitance (interwinding capacitance), R_s = secondary winding resistance, R_L = load impedance.

Maximum operating temperature. By the use of high-temperature materials and at a given VA rating and ambient temperature, considerable size reduction can be realized by designing for an increased temperature rise.

Ambient temperature. With a given temperature rise and for a fixed VA rating, transformer size can be reduced if the ambient temperature is lowered.

Regulation. If the regulation requirements are made less stringent, the wire size of the windings can be reduced, with a consequent reduction in the size of the transformer.

Audio Transformers

These transformers are used for voltage, current, and impedance transformation over a nominal frequency range of 20 to 20,000 Hz. The equivalent circuit (Fig. 5.1.4) is the starting point for the basic analysis of the transformer frequency response.

A prime consideration in the design, size, and cost of audio transformers is the span of the frequency response. In wide-band audio transformers the frequency coverage can be separated into three nearly independent ranges for the purpose of analysis. Thus, in the high-frequency range leakage inductance and distributed capacitance are most significant. In the low-frequency region the open-circuit inductance is important. At the medium-frequency range, approximately 1000 Hz, the effect of the transformer on the frequency response can be neglected. In the above discussion, the transformation is assumed to be that of an ideal transformer.

Miniaturized Audio Transformers

Frequency Response.   The high-frequency response is dependent on the magnitude of the leakage inductance and distributed capacitance of the windings. These parameters decrease as the transformer size decreases. Consequently, miniaturized audio transformers generally have an excellent high-frequency response. On the other hand, the small size of these transformers results in increased loss and in a degradation of the low-frequency response, which is dependent on the primary open-circuit inductance.

Air Gap.   The primary open-circuit inductance is proportional to the product of the square of the turns and the core area, and inversely proportional to the width of the gap. As the transformer size is reduced, both the core area and number of turns must be reduced. Consequently, if the open-circuit inductance is to be maintained, the air gap must be reduced. A butt joint has of necessity a finite air gap; as a result the full ungapped inductance is not realized. An interlaced structure of the core, where the butt joints for each lamination are staggered, most closely approximates an ungapped core. A substantially higher inductance, 120 H, is possible.

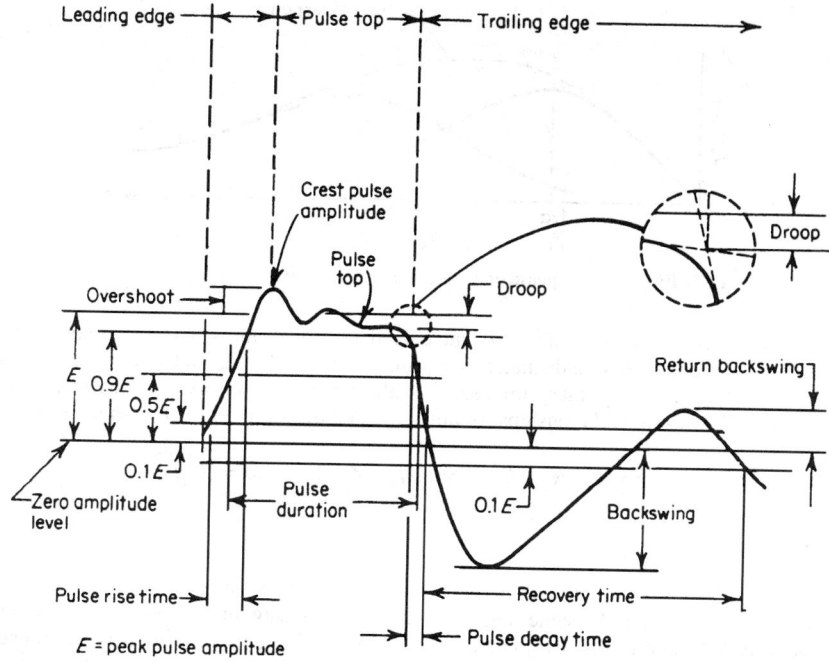

FIGURE 5.1.5 Pulse waveform.

The problem with taking advantage of this effect is that as the air gap is reduced, the allowable amount of unbalanced direct current flowing in the transformer winding must be lowered to prevent core saturation.

Pulse Transformers

Pulse transformers for high-voltage or high-power applications, above 300 W peak, are used in modulators for radar sets. Their function is to provide impedance matching between the pulse-forming network and the magnetron. Prime concern is transformation of the pulse with a minimum of distortion. Lower-power pulse transformers fall into two categories: those used for coupling or impedance matching similar to the high-power pulse transformers, and blocking oscillator transformers used in pulse-generating circuits. Pulse widths for such transformers most commonly range from about 0.1 to 20 μs.

Assuming the pulse transformer is properly matched and the source is delivering an ideal rectangular pulse, a well-designed transformer should have small values of leakage inductance and distributed capacitance. Within limits dictated by pulse decay time the open-circuit inductance should be high. Figure 5.1.5 shows the pulse waveform with the various types of distortions that may be introduced by the transformer.

Broadband RF Transformers

At the higher frequencies, transformers provide a simple, low-cost, and compact means for impedance transformation. Bifilar windings and powdered-iron or ferrite cores provide optimum coupling. The use of cores with high permeabilities at the lower frequencies reduces the number of turns and distributed capacitance. At the upper frequencies the reactance increases, even though the permeability of the core may fall off.

Inductive Coupling

There are a variety of ways to use inductive elements for impedance-matching or coupling one circuit to another. Autotransformers and multiwinding transformers that have no common metallic connections are a common method of inductive coupling. In a unity-coupled transformer, N_1 = number of primary turns, N_2 = number of

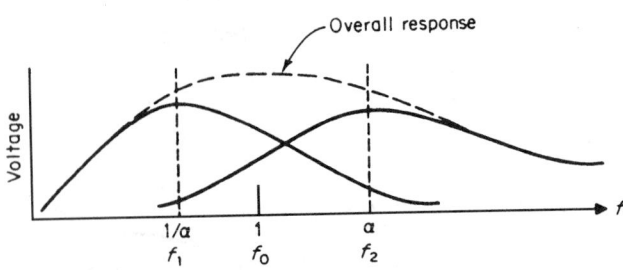

FIGURE 5.1.6   Response of a staggered pair, geometric symmetry.

secondary turns, k = coefficient of coupling, M = mutual inductance, n = turns ratio, L_1 = primary open-circuit inductance, L_2 = secondary open-circuit inductance, I_1 = primary current, I_2 = secondary current, E_1 = primary voltage, E_2 = secondary voltage, Z_1 = primary impedance with matched secondary, Z_2 = secondary impedance with matched primary. Transformer relationships for unity-coupled transformer, $k = 1$, assuming losses are negligible:

$$n = N_2/N_1 = E_2/E_1 = I_2/I_1 = Z_2/Z_1 \qquad M = \sqrt{L_1 L_2}$$

Single-Tuned Circuits

Single-tuned circuits are most commonly used in both wide-band and narrow-band amplifiers. Multiple stages that are cascaded and tuned to the same frequency are synchronously tuned. The result is that the overall bandwidth of the cascaded amplifiers is always narrower than the single-stage bandwidth. The shrinkage of bandwidth can be avoided by stagger tuning. A stagger-tuned system is a grouping of single-tuned circuits where each circuit is tuned to a different frequency. For a flat-topped response the individual stages are geometrically balanced from the center frequency. In Fig. 5.1.6, which illustrates the response of a staggered pair, f_0 = center frequency of overall response, and f_1, f_2 = resonant frequency of each stage. The frequencies are related as follows:

$$f_0/f_1 = f_2/f_0 = \alpha$$

Double-Tuned Transformers

One of the most widely used circuit configurations for i.f. systems in the frequency range of 250 kHz to 50 MHz is the double-tuned transformer. It consists of a primary and secondary tuned to the same frequency and coupled inductively to a degree dependent on the desired shape of the selectivity curve. Figure 5.1.7 shows the variation of secondary current versus frequency.

Bandwidth

A comparison of the relative 3-dB bandwidth of multistage single- and double-tuned circuits is shown in Table 5.1.2. Most significant is the lower skirt ratio of the double-tuned circuit, i.e., relative value of the ratio of the bandwidth at 60 dB (BW_{60}) to the bandwidth at 6 dB (BW_6).

POLED FERROELECTRIC CERAMIC DEVICES

Joseph M. Giannotti

Introduction

The usefulness of ferroelectrics rests on two important characteristics, asymmetry and high dielectric constant. Poled ferroelectric devices are capable of doing electric work when driven mechanically or mechanical work

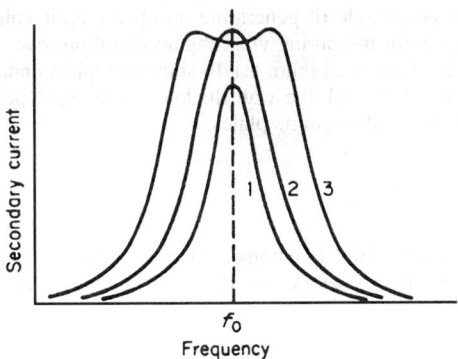

FIGURE 5.1.7 Variation of secondary current and gain with frequency and with degree of coupling: 1 = undercoupled; 2 = critically coupled; 3 = overcoupled.

when driven electrically. In poled ferroelectrics, the piezoelectric effect is particularly strong. From the design standpoint, they are especially versatile because they can be used in a variety of ceramic shapes. Piezoelectricity is the phenomenon of coupling between elastic and dielectric energy. Piezoelectric ceramics have gained wide use in the low-frequency range up to a few megahertz over the strongly piezoelectric nonferroelectric single crystals such as quartz, lithium sulfate, lithium niobate, lithium tantalate, and zinc oxide. High dielectric strength and low manufacturing cost are prime factors for their usefulness.

The magnitude and character of the piezoelectric effect in a ferroelectric material depend on orientation of applied force or electric field with respect to the axis of the material. With piezoelectric ceramics, the polar axis is parallel to the original dc polarizing field. In all cases the deformations are small when amplification by mechanical resonance is not involved. Maximum strains with the best piezoelectric ceramics are in the range of 10^{-3}. Figure 5.1.8 illustrates the basic deformations of piezoelectric ceramics and typical applications.

Transducers

The use of barium titanate as a piezoelectric transducer material has been increasingly replaced by lead titanate zirconate solid-solution ceramics since the latter offer higher piezoelectric coupling, wider operating temperature range, and a choice of useful variations in engineering parameters. Table 5.1.3 gives the characteristics of different compositions.

The high piezoelectric coupling and permittivity of PZT-5H have led to its use in acoustic devices where its high electric and dielectric losses can be tolerated. For hydrophones or instrument applications PZT-5A is a better choice, since its higher Curie point leads to better temperature stability. The low elastic and dielectric losses of PZT-8 composition at high drive level point to its use in high-power sonic or ultrasonic transducers. The very low mechanical Q of lead metaniobate has encouraged its use in ultrasonic flaw detection, where the low Q helps the suppression of ringing. The high acoustic velocity of sodium potassium niobate is of advantage in high-frequency thickness-extensional thickness-shear transducers, since this allows greater thickness and therefore lower capacitance.

Since ceramic materials can be fabricated in a wide range of sizes and shapes, they lend themselves to designs for applications that would be difficult to achieve with single crystals. Figure 5.1.9 illustrates the use

TABLE 5.1.2 Relative Values of 3-dB Bandwidth for Single- and Double-Tuned Circuits

No. of stages	Relative values of 3-dB bandwidth		Relative values of BW_{60}/BW_6	
	Single-tuned	Double-tuned[*]	Single-tuned	Double-tuned
1	1.00	1.00	577	23.9
2	0.64	0.80	33	5.65
3	0.51	0.71	13	3.59
4	0.44	0.66	8.6	2.94
6	0.35	0.59	5.9	2.43
8	0.30	0.55	5.0	
10	0.27	0.52	4.5	

[*]Based on identical primary and secondary circuits critically coupled.

of simple piezoelectric elements in a high-voltage source capable of generating an open-circuit voltage of approximately 40 kV. Piezoelectric accelerometers suitable for measuring vibrating accelerations over a wide frequency range are readily available in numerous shapes and sizes. Figure 5.1.10 shows a typical underwater transducer, which uses a hollow ceramic cylinder polarized through the wall thickness. Flexing-type piezoelectric elements can handle larger motions and smaller forces than single plates.

Resonators and Filters

The development of temperature-stable filter ceramics has spurred the development of ceramic resonators and filters. These devices include simple resonators, multielectrode resonators, and cascaded combinations thereof,

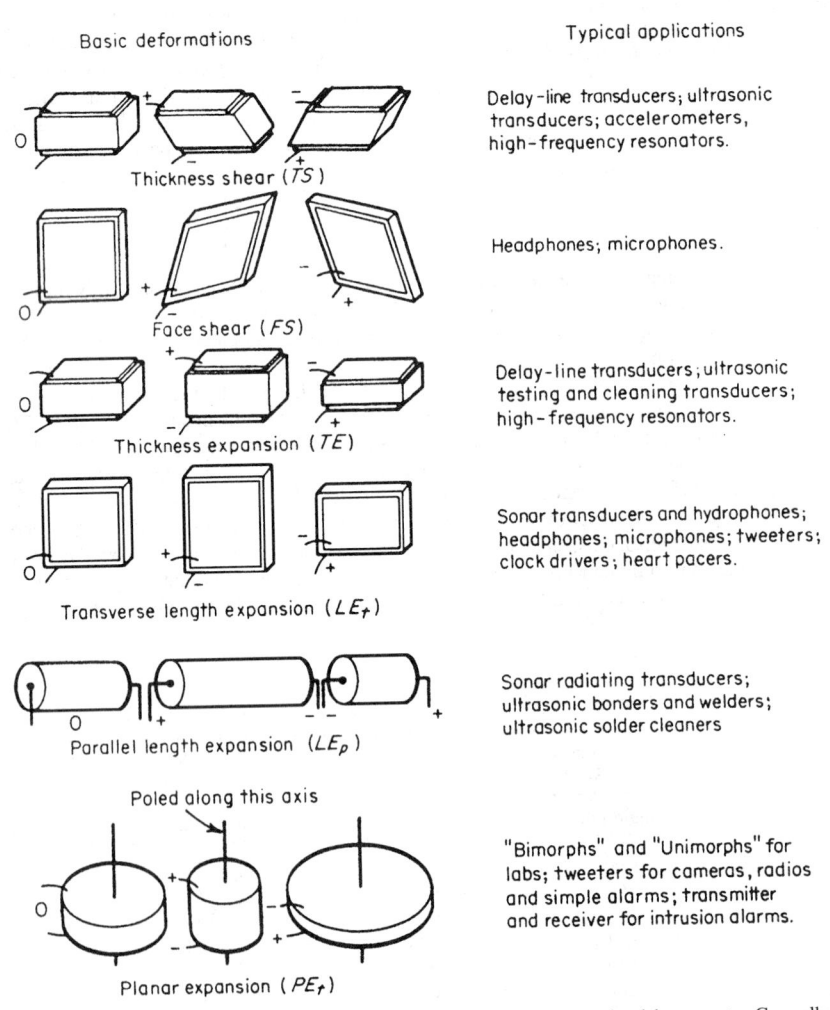

Basic deformations   Typical applications

Thickness shear (*TS*) — Delay-line transducers; ultrasonic transducers; accelerometers, high-frequency resonators.

Face shear (*FS*) — Headphones; microphones.

Thickness expansion (*TE*) — Delay-line transducers; ultrasonic testing and cleaning transducers; high-frequency resonators.

Transverse length expansion (*LE$_t$*) — Sonar transducers and hydrophones; headphones; microphones; tweeters; clock drivers; heart pacers.

Parallel length expansion (*LE$_p$*) — Sonar radiating transducers; ultrasonic bonders and welders; ultrasonic solder cleaners

Poled along this axis

Planar expansion (*PE$_t$*) — "Bimorphs" and "Unimorphs" for labs; tweeters for cameras, radios and simple alarms; transmitter and receiver for intrusion alarms.

FIGURE 5.1.8 Basic piezoelectric action depends on the type of material used and the geometry. Generally, two or more of these actions are present simultaneously. TE and TS are high-frequency (greater than 1 MHz) modes and FS, LE$_t$, LE$_p$, and PE$_t$ are low-frequency (less than 1 MHz) modes. The thickness controls the resonant frequency for the TE and TS modes, the diameter for PE$_t$, and the length for the LE$_t$ and LE$_p$ modes.

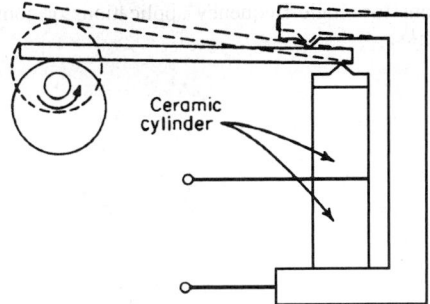

FIGURE 5.1.9 High-voltage generator.

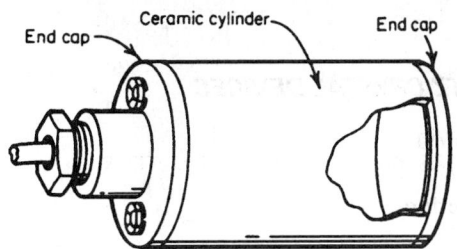

FIGURE 5.1.10 Underwater sound transducer.

mechanically coupled pairs of resonators, and ceramic ladder filters, covering a frequency range from 50 Hz to 10 MHz.

Two lead titanate-zirconate compositions, PZT-6A and PZT-6B ceramics, are most widely used for resonator and filter applications. PZT-6A, having high electromechanical coupling coefficient (45 percent) and moderate mechanical Q (400), is used for medium to wide bandwidth applications, while PZT-6B, with moderate coupling (21 percent) and higher Q (1500), is used for narrow bandwidths. The compositions exhibit a frequency constant stable to within ± 0.1 percent over a temperature range from –40 to +85°C. The frequency characteristics increase slowly with time at less than 0.1 percent per decade of time.

Ceramic Resonators

A thin ceramic disk with fully electroded faces, polarized in its thickness direction, has its lowest excitable resonance in the fundamental radial mode. The impedance response of such a disk and its equivalent circuit are shown in Fig. 5.1.11.

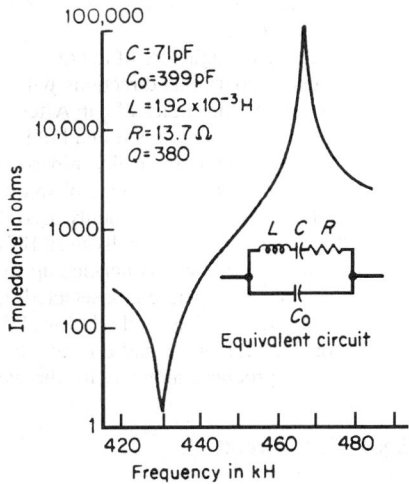

FIGURE 5.1.11 Impedance response of a fundamental radial resonator.

TABLE 5.1.3 Ceramic Compositions

	k_{33}	k_p	$\epsilon_{33}^T/\epsilon_0$	Change in N_1 −60 to +85°C, %	Change in N_1 per time decade, %
PZT-4	0.70	0.58	1,300	4.8	+1.5
PZT-5A	0.705	0.60	1,700	2.6	+0.2
PZT-5H	0.75	0.65	3,400	9.0	+0.25
PZT-6A	0.54	0.42	1,050	<0.2	<0.1
PZT-8	0.62	0.50	1,000	2.0	+1.0
$Na_{0.5}K_{0.5}NbO_3$	0.605	0.46	500	?	?
$PbNb_2O_6$	0.38	0.07	225	3.3	?

k_{33} = coupling constant for longitudinal mode.
k_p = coupling constant for radial mode.
ϵ_{33}^T = permittivity parallel to poling field, stress-free condition.
N_1 = frequency constant (resonance frequency × length).

Ceramic resonators can be used in various configurations for single-frequency applications or combined in basic L-sections to form complete bandpass filter networks.

QUARTZ CRYSTAL DEVICES

John R. Vig

Introduction

Piezoelectric crystal devices are used primarily for frequency control, timing, transducers, and delay lines. The piezoelectric material used for most applications is quartz. A quartz crystal acts as a stable mechanical resonator, which, by its piezoelectric behavior and high Q, determines the frequency generated in an oscillator circuit. Bulk-wave resonators are available in the frequency range from about 1 kHz to 300 MHz. Surface-acoustic-wave (SAW) and shallow-bulk-acoustic-wave devices can be made to operate at well above 1 GHz.

In the manufacture of the different types of quartz resonators, wafers are cut from the mother crystal along precisely controlled directions with respect to the crystallographic axes. The properties of the device depend strongly on the angles of cut. After shaping to required dimensions, metal electrodes are applied to the quartz wafer, which is mounted in a holder structure. The assembly, called a *crystal unit* (or *crystal* or *resonator*) is usually sealed hermetically in partial- to ultrahigh-vacuum.

To cover the wide range of frequencies, different cuts, vibrating in a variety of modes, are used. Above 1 MHz, the AT cut is commonly used. For high-precision applications, the SC cut has important advantages over the AT and the occasionally used BT cuts. AT-, BT-, and SC-cut crystals can be manufactured for fundamental-mode operation at frequencies up to about 100 MHz. Above 40 MHz, *overtone* crystals are generally used. Such crystals operate at a selected odd harmonic mode of vibration. AT-, BT-, and SC-cut crystals vibrate in the thickness shear mode. Below 1 MHz, tuning forks, X-Y and NT bars (flexure mode), +5° X cuts (extensional mode), or CT and DT cuts (face shear mode) can be used. Over 200 million 32.8-kHz tuning-fork crystals are produced annually for the world watch market.

Equivalent Circuit

The circuit designer treats the behavior of a quartz crystal unit by considering its equivalent circuit (Fig. 5.1.12). The mechanical resonance in the crystal is represented by L_1, C_1, and R_1. Because it is a dielectric with electrodes, the device also displays an electrical capacitance C_0. The parallel combination of C_0 and the motional arm, C_1-L_1-R_1, represents the equivalent circuit to a good approximation. As shown in Fig. 5.1.12b, the reactance of this circuit varies with frequency.

The Q values ($Q^{-1} = 2\pi f_s R_1 C_1$) of quartz-crystal units are much higher than those attainable with other circuit elements. In general-purpose units, the Q is usually in the range of 10^4 to 10^6. The intrinsic Q of quartz is limited by internal losses. For AT-cut crystals, the intrinsic Q has been experimentally determined to be 16×10^6 at 1 MHz; it is inversely proportional to frequency. At 5 MHz, for example, the intrinsic Q is 3.2×10^6.

Oscillators

The commonly used crystal oscillator circuits fall into two broad categories. In series-resonance oscillators, the crystal operates at series resonance, i.e., at f_s. In parallel-resonance or antiresonance oscillators, the crystal is used as a positive reactance; i.e., the frequency is between f_s and f_A. In this latter mode of operation the oscillator circuit provides a load capacity to the crystal unit. The oscillator then operates at the frequency where the crystal unit's reactance cancels the reactance of the load capacitor. When the load capacitance is changed, the oscillator frequency changes. An important parameter in this connection is the capacitance ratio $r = C_0/C_1$. Typically, the value of r is a few hundred for fundamental-mode crystals. It is larger by a factor of n^2 for nth-overtone crystals.

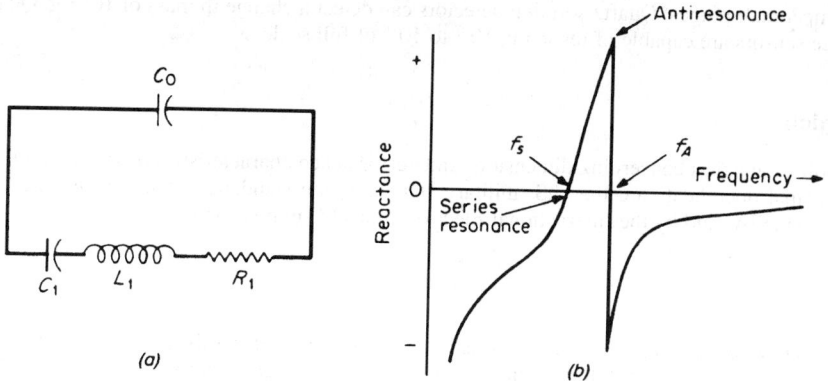

FIGURE 5.1.12   (*a*) Equivalent circuit and (*b*) frequency-reactance relationship.

When a load capacitor C_L is connected in series with a crystal, the series-resonance frequency of the combination is shifted from f_s by Δf, which is related to the other parameters by

$$\frac{\Delta f}{f_s} \approx \frac{C_0}{2r(C_0 + C_L)}$$

For a typical fundamental-mode AT-cut crystal unit with $r = 250$, $C_0 = 5$ pF, and $C_L = 30$ pF, the shift is 286 ppm. If such a crystal unit is to remain stable to, for example, 1×10^{-9}, the load reactance owing to C_L and the other circuit components must remain stable to within 1.2×10^{-4} pF. The frequency can also be "tuned" by intentionally changing C_L. For the above example, a change of 1 pF in C_L shifts the frequency by nearly 10 ppm.

Filters

Quartz crystals are used as selective components in crystal filters. With the constraint imposed by the equivalent circuit of Fig. 5.1.12*a*, filter design techniques can provide bandpass or bandstop filters with prescribed characteristics. Crystal filters exhibit low insertion loss, high selectivity, and excellent temperature stability.

Filter crystals are designed to have only one strong resonance in the region of operation, with all other responses (unwanted modes) attenuated as much as possible. The application of energy-trapping theory can provide such a response. If electrode size and thickness are selected in accordance with that theory, the energy of the main response is trapped between the electrodes, whereas the unwanted modes are untrapped and propagate toward the edge of the crystal resonator, where their energy is dissipated. It is possible to manufacture AT-cut filter crystals with greater than 40-dB attenuation of the unwanted modes relative to the main response.

Transducers

Whereas in frequency control and timing applications of quartz-crystal devices the devices are designed to be as insensitive to the environment as possible, quartz crystals can also be designed intentionally to be highly sensitive to environmental parameters such as temperature, mass changes, pressure, force, and acceleration. Quartz-crystal transducers can exhibit unsurpassed resolution and dynamic range. For example, one commercial "quartz pressure gage" exhibits a 1 ppm resolution, i.e., 60 Pa at 76 MPa (0.01 lb/in^2 at 11,000 lb/in^2), and a 0.025 percent full-scale accuracy. Quartz thermometers can provide millidegrees of absolute accuracy over wide temperature ranges. Quartz sorption detectors can detect a change in mass of 10^{-12} g. Quartz accelerometer/force sensors are capable of resolving 10^{-7} to 10^{-8} of full scale.

Standardization

Standardization exists concerning dimensions and performance characteristics of crystal units and oscillators. The principal documents are the U.S. military standards, the standards issued by the Electronics Industry Association (EIA) and by the International Electrotechnical Commission (IEC).

Temperature

The frequency versus temperature characteristics are determined primarily by the angles of cut of the crystal plates with respect to the crystallographic axes of quartz. Typical characteristics are shown in Figs. 5.1.13 and 5.1.14. The points of zero temperature coefficient, the *turnover points*, can be varied over a wide range by varying the angles of cut.

The frequency-temperature characteristic of AT- and SC-cut crystals follow a third-order law, as shown in Fig. 5.1.13 for the AT cut. A slight change in the orientation angle (7 min in the example shown in Fig. 5.1.13) greatly changes the frequency-temperature characteristic. Curve 1 is optimal for a wide temperature range (–55 to 105°C). Curve 2 gives minimum frequency deviation over a narrow range near the inflection temperature T_i. The frequency versus temperature characteristics of SC-cut crystals is similar to the curves shown in Fig. 5.1.13 except that T_i is shifted to about 95°C. The SC cut is a doubly rotated cut; i.e., two angles must be specified and precisely controlled during the manufacturing process. The SC cut is more difficult to manufacture with predictable frequency versus temperature characteristics than are the singly rotated cuts, such as AT and BT

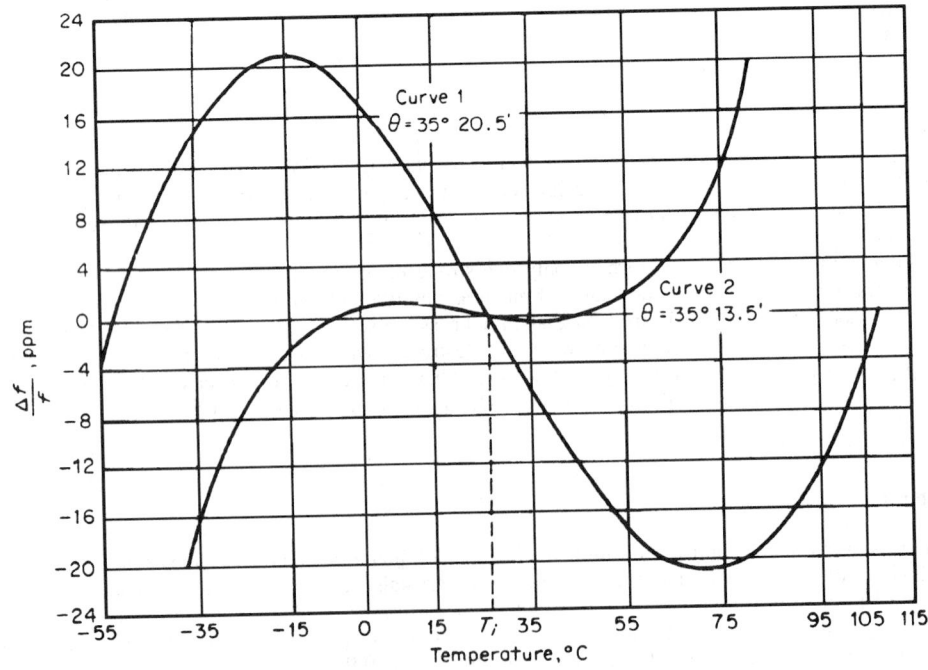

FIGURE 5.1.13   Frequency-temperature characteristics of AT-cut crystals.

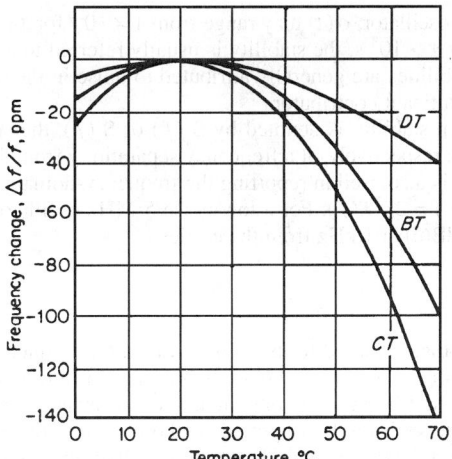

FIGURE 5.1.14 Parabolic frequency-temperature characteristic of some crystal cuts.

cuts. Many of the crystal cuts have a parabolic frequency versus temperature characteristic, as shown in Fig. 5.1.14 for the BT, CT, and DT cuts. The turnover temperatures of these cuts can also be shifted up or down by changing the angles of cut.

To achieve the highest stability, oven-controlled crystal oscillators (OCXO) are used. In such oscillators, the crystal unit and the temperature-sensitive components of the oscillator are placed in a stable oven the temperature of which is set to the crystal's turnover temperature. OCXOs are bulkier and consume more power than other types. In addition, when an oven-controlled oscillator is turned on, one must wait several minutes for the oscillator to stabilize. During warm-up, the thermal stresses in the crystal can produce significant frequency shifts. This thermal-transient effect causes the typical warm-up time of an oscillator to be several minutes longer than the time it takes for the oven to stabilize. The thermal-transient effect is absent in SC-cut crystals. In oscillators which use SC-cut crystals, the warm-up time can be much shorter.

In temperature-compensated crystal oscillators (TCXOs) the crystal's frequency versus temperature behavior is compensated by varying a load capacitor. The output signal from a temperature sensor, e.g., a thermistor network, is used to generate the correction voltage applied to a varactor. Digital techniques are capable of providing better than 1×10^{-7} frequency stability from –55 to +85°C. TCXOs are smaller, consume less power than OCXOs and require no lengthy warm-up times. A major limitation on the stabilities achievable with TCXOs is the thermal hysteresis exhibited by crystal units.

Aging

Aging, the gradual change in a crystal's frequency with time, can be a result of several causes. The main causes are mass transfer to or from the resonator surfaces (owing to adsorption and desorption of contamination) and stress relief within the mounting structure or at the interface between the quartz and the electrodes. The observed aging is the sum of the aging produced by the various mechanisms and may be positive or negative. Aging is also sometimes referred to as *drift* or *long-term stability*.

The aging rate of a crystal unit is highest when it is new. As time elapses, stabilization occurs within the unit and the aging rate decreases. The aging observed at constant temperature usually follows an approximately logarithmic dependance on time. When the temperature of a crystal is changed, a new aging cycle starts.

A major reason for the aging of low-frequency units (below 1 MHz) is that mechanical changes take place in the mounting structure. Properly made units may age several ppm/year, half that aging occurring within the first 30 days. Crystal units for frequencies of 1 MHz and above age primarily because of mass transfer. General-purpose crystal units are usually housed in solder-sealed or resistance-welded material enclosures of the HC-6 or HC-18 configuration that are filled with dry nitrogen. The aging rate of such units is typically specified as 5 ppm for the first month: over a year's time their aging may be from 10 to 60 ppm. Higher-quality general-purpose crystals may age as little as 1 ppm per year. If a lower aging rate is desired, overtone crystals in clean glass, metal, or ceramic enclosures should be used. Advanced surface-cleaning, packaging, and ultrahigh-vacuum fabrication techniques have resulted in units that age less than 5×10^{-11} per day after a few days of stabilization, or $<1 \times 10^{-8}$ per year.

Short-Term Stability

Short-term stability, or "noise," in the time domain $\sigma(\tau)$ is usually expressed as the 2-sample deviation of the fractional frequency fluctuations for a specified averaging time. The averaging times τ over which $\sigma(\tau)$ is specified generally range from 10^{-3} to 10^3 s. For a good oscillator, $\sigma(\tau)$ may range from 1×10^{-9} for $\tau = 10^{-3}$ s to 5×10^{-13} for $\tau = 1$ s to 1×10^{-12} for up to 10^3 s. For $\tau > 10^3$ s, the stability is usually referred to as *long-term stability* or aging. For $\tau < 1$ s, the short-term instabilities are generally attributed to noise in the oscillator circuitry although the crystal itself can also be a significant contributor.

When measured in the frequency domain, short-term stability is denoted by $S_\phi(f)$ or $S_y(f)$, the spectral density of phase fluctuations and frequency fluctuations, respectively, at a frequency separation f from the carrier frequency v. $\pounds(f)$, the single-sideband phase noise, is also used in reporting the frequency-domain stability. The three quantities are related by $f^2 S_\phi(f) = v^2 S_y(f) = 2f^2 \pounds(f)$. For a low-noise 5-MHz oscillator, $\pounds(f)$ may be -115 dB(c) at 1 Hz from the carrier, and -160 dB(c) at 1 kHz from the carrier.

Thermal Hysteresis

When the temperature of a crystal unit is changed and then returned to its original value, the frequency will generally not return to its original value. This phenomenon, called *thermal hysteresis* or lack of *retrace*, can be caused by the same mechanisms as aging, i.e., mass transfer because of contamination and stress relief.

For a given crystal unit, the magnitude of the effect depends on the magnitude and direction of the temperature excursion, on the thermal history of the unit, and on the design and construction of both the crystal unit and the oscillator circuit. The effect tends to be smaller in OXCOs, where the operating temperature of the crystal is always approached from below, than in TCXOs, where the operating temperature can be approached from either direction. Thermal hysteresis can be minimized through the use of clean, ultrahigh-vacuum fabrication techniques (which minimize the mass-transfer contribution) and through the use of properly mounted SC-cut crystals (which minimize the stress-relief contributions). The magnitude of the effect typically ranges from several ppm in general-purpose crystals to less than 1×10^{-9} in high-stability crystals operated in OXCOs.

Drive Level

The amplitude of vibration is proportional to the current through a crystal. Because of the nonlinearities of quartz, the frequency of a crystal and the incidence of interfering modes are functions of the drive level. The drive level must be stated when a crystal's parameters are specified. The oscillator designer's ability to improve signal-to-noise ratios by using a higher drive level is limited. In SC-cut crystals, the drive level effects are significantly below those observed in AT- and BT-cut units. When the drive level is increased, the frequency of AT- and SC-cut crystals increases and the frequency of the BT-cut crystal decreases.

Acceleration, Vibration, and Shock

The frequency of a crystal unit is affected by stresses. Even the acceleration owing to gravity produces measurable effects. When an oscillator using an AT-cut crystal is turned upside down, the frequency typically shifts about 4×10^{-9} because the acceleration sensitivity of the crystal is typically 2×10^{-9} g^{-1}. The sensitivity is the same when the crystal is subjected to vibration; i.e., the time-varying acceleration owing to the vibration modulates the frequency at the vibration frequency with an amplitude of 2×10^{-9} g^{-1}. In the frequency domain, the vibration sensitivity manifests itself as vibration-induced sidebands that appear at plus and minus the vibration frequency away from the carrier frequency. The acceleration sensitivity of SC-cut crystals can be made to be less than that of comparably fabricated AT- or BT-cut crystals.

Shock places a sudden stress on the crystal. During shock, the crystal's frequency changes because of the crystal's acceleration sensitivity. If during shock the elastic limits in the crystal's support structure or in its electrodes are exceeded, the shock can produce a permanent frequency change. Crystal units made with chemically polished crystal plates can withstand shocks in excess of $20{,}000g$ (11 ms, $1/2$ sine). Such crystals have been successfully fired from howitzers.

Radiation

The degree to which high levels of ionizing radiation affect the frequency of a crystal unit depends primarily on the quality of quartz used to fabricate the device and on the dose. When crystals made of natural quartz are subjected to steady-state ionizing radiation, their frequencies are changed permanently by approximately a few parts in 10^{11} per rad at a 1 Mrad dose. To minimize the effect of such radiation, high-purity cultured quartz should be used. The frequency change per rad is higher at lower doses. Pulse irradiation produces a transient frequency shift because of the thermal-transient effect. This effect can be minimized by using SC-cut crystals. Energetic neutrons change a crystal's frequency by about 6×10^{-21} per neutron per cm^2.

INSULATORS

Sam Di Vita

Introduction

Ceramics and plastics are the principal materials for electronics insulation and mounting parts. Ceramic materials are outstanding in their resistance to high temperature, mechanical deformation, abrasion, chemical attack, electrical arc, and fungus attack. Ceramics also possess excellent electrical insulating properties and good thermal conductivity and are impervious to moisture and gases. These properties of ceramics are retained throughout a wide temperature range and are of particular importance in high-power applications such as vacuum-tube envelopes and spacers, rotor and end-plate supports for variable air capacitors, rf coil forms, cores for wire-wound resistors, ceramic-to-metal seals, and feed-through bushings for transformers.

The properties of plastics differ rather markedly from ceramics over a broad range. In a number of properties, plastics are more desirable than ceramics. These include lighter weight; better resistance to impact, shock, and vibration; higher transparency; and easier fabrication with molded-metal inserts (however, glass-bonded-mica ceramic material may be comparable with plastic in this latter respect).

Ceramic Linear Dielectric Insulators

Ceramic insulators are linear dielectrics having low loss characteristics, that are used primarily for coil forms, tube envelopes and bases, and bushings, which all require loss factors less than 0.035 when measured at standard laboratory conditions. Dielectric loss factor is the product of power factor and dielectric constant of a given ceramic. Military Specification MIL-I-10B, Insulating Compound Electrical, Ceramic Class L, covers low-dielectric-constant (12 or under) ceramic electrical insulating materials, for use over the spectrum of radio frequencies used in electronic communications and in allied electronic equipments. In this specification the "grade designators" are identified by three numbers, the first representing dielectric loss factor at 1MHz, the second dielectric strength, and the third flexural strength (modulus of rupture).

Table 5.1.4 lists the various types of ceramics and their grade designators approved for use in the fabrication of military standard ceramic radio insulators specified in MIL-I-23264A, insulators—ceramic, electrical and electronic, general specification for. This specification covers only those insulators characterized by combining the specific designators required for the appropriate military standard insulators used as standoff, feed-through, bushing, bowl, strain, pin, spreader, and other types of insulators.

Currently, Grade L-242 is typical of porcelain, L-422 of steatite, L-523 of glass, L-746 of alumina, L-442 of glass-bonded mica, and L-834 of beryllia.

High-Thermal-Shock-Resistant Ceramics

Lithia porcelain is the best thermal-shock-resistant ceramic because of its low (close to zero) coefficient of thermal expansion. It is followed in order by fused quartz, cordierite, high-silica glass, porcelain, steatite beryllium oxide, alumina, and glass-bonded mica. Those materials find wide use for rf coil forms, cores for wire-wound resistors, stator supports for air dielectric capacitors, coaxial cable insulators, standoff insulators, capacitor trimmer bases, tube sockets, relay spacers, and base plates for printed radio circuits.

High Thermal Conductivity

High-purity beryllium oxide is unique among homogeneous materials in possessing high thermal conductivity comparable with metal, together with excellent electrical insulating properties.

TABLE 5.1.4 Property Chart of Insulating Ceramic Materials Qualified under MIL-I-10

Class L ceramics	Grade designators	Power factor, 1 MHz	Dielectric constant, 1 MHz	Dielectric loss factor, 1 MHz	Dielectric strength, V/mil
Steatite:					
Unglazed	L-523	0.00069–0.0010	6.42+	0.0041–0.0063	230
Glazed	L-543	0.0008–0.0014		0.005–0.008	330
Porcelain:					
Glazed	L-232	0.0076–0.0099	5.42–6.01	0.041–0.059	249
Zircon:					
Unglazed	L-433	0.0012–0.0014	8.14–8.22	0.010–0.012	259
Glazed	L-413	0.00119	8.92	0.011	191
Alumina:					
Unglazed	L-746	0.0001–0.0008	8.14+	0.0009	500
Glass:					
Borosilicate (Pyrex)	L-622	0.00074	4.19	0.0031	226
High silica (Vycor)	L-541	0.0017	3.78	0.0065	363
Glass-bonded mica:					
Unglazed	L-442	0.0017–0.0018	7.08–7.44	0.012–0.013	382
Forsterite:					
Glazed	L-723	0.0003	6.37	0.002	200
Cordierite:					
Unglazed	L-321	0.0049	4.57	0.022	245
Wallastonite:					
Unglazed	L-621	0.0004	6.49	0.003	293
Berylia	L-834	0.00015	6	0.0009	295

Care must be exercised in the use of beryllium oxide because its dust is highly toxic. Although it is completely safe in dense ceramic form, any operation that generates dust, fumes, or vapors is potentially very dangerous.

Some typical uses of beryllium oxide are:

1. Heat sinks for high-power rf amplifying tubes, transistors, and other semiconductors
2. Printed-circuit bases
3. Antenna windows and tube envelopes
4. Substrates for vapor deposition of metals
5. Heat sinks for electronic chassis or subassemblies

Polycrystalline sintered diamond powder is also an excellent dielectric. It has the advantage of a thermal conductivity twice as good as copper and three to four times that of beryllia. It provides a cost-effective solution for critical heat sinks for high-power diodes and similar solid-state devices.

Plastic Insulators

The term *plastics* usually refers to a class of synthetic organic materials (resins) that are sold in finished form but at some stage in their processing are fluid enough to be shaped by application of heat and pressure. The two basic types of plastics are *thermoplastic resins*, which, like wax or tar, can be softened and resoftened repeatedly without undergoing a change in chemical composition, and *thermosetting resins*, which undergo a chemical change with application of heat and pressure and cannot be resoftened.

Choice of Plastics

Some of the differences between plastics that should be considered when defining specific needs are degree of stiffness or flexibility; useful temperature range; tensile, flexural, and impact strength; intensity, frequency, and duration of loads; electrical strength and dielectric losses; color retention under environment; stress-crack resistance over time; wear and scratch resistance; moisture and chemical resistance at high temperature; gas permeability; weather and sunlight resistance over time; odor, taste, and toxicity; and long-term creep properties under critical loads.

Reinforced Plastics

These comprise a distinct family of plastic materials that consist of superimposed layers of synthetic resin-impregnated or resin-coated filler. Fillers such as paper, cotton fabric, glass fabric or fiber, glass mats, nylon fabric—either in the form of sheets or macerated—are impregnated with a thermosetting resin (phenolic, melamine, polyester, epoxy, silicone). Heat and pressure fuse these materials into a dense, insoluble solid and nearly homogeneous mass, which may be fabricated in the form of sheets or rods or in molded form.

CHAPTER 5.2
ACTIVE DISCRETE COMPONENTS

George W. Taylor, Bernard Smith, I. Reingold, Munsey E. Crost, Gregory J. Malinowski, John F. Donlon, Edward B. Hakim

POWER AND RECEIVING TUBES

George W. Taylor, Bernard Smith

Introduction

Power and receiving tubes are active devices using either the flow of free electrons in a vacuum or electrons and ions combined in a gas medium. The vast majority of uses for low-power electron tubes are found in the area of high vacuum with controlled free electrons. The source of free electrons is a heated material that is a thermionic emitter (cathode). The control element regulating the flow of electrons is called a *grid*. The collector element for the electron flow is the *anode*, or *plate*. Power tubes, in contrast to receiving-type tubes, handle relatively large amounts of power, and for reasons of economy, a major emphasis is placed on efficiency. The traditional division between the two tube categories is at the 25-W plate-dissipation rating level.

Receiving tubes traditionally provided essential active-device functions in electronic applications. Most receiving-type tubes produced at the present time are for replacement use in existing equipments. Electronic functions of new equipment designs are being handled by solid-state devices.

Power tubes are widely used as high-power-level generators and converters in radio and television transmitters, radar, sonar, manufacturing-process operations, and medical and industrial x-ray equipment.

Classification of Types

Power and receiving-type tubes can be separated into groups according to their circuit function or by the number of electrodes they contain. Table 5.2.1 illustrates these factors and compares some of the related features. The physical shape and location of the grid relative to the plate and cathode are the main factors that determine the amplification factor μ of the triode. The μ values of triodes generally range from about 5 to over 200. The mathematical relationships between the three important dynamic tube factors are

$$\text{Amplification factor } \mu = \Delta e_b / \Delta e_{c1}$$

$$\text{Dynamic plate resistance } r_p = \Delta e_b / \Delta i_b \qquad e_b, e_{c1}, i_b = \text{const}$$

$$\text{Transconductance } Sm \text{ or } Gm = \Delta i_b / \Delta e_{c1}$$

TABLE 5.2.1 Tube Classification by Construction and Use

Tube type	No. of active electrodes	Typical use	Relative features and advantages
Diode	2	Rectifier	High back resistance
Triode	3	Low- and high-power amplifier, oscillatgor, and pulse modulator	Low cost, circuit simplicity
Tetrode	4	High-power amplifier and pulse modulator	Low drive power, low feedback
Pentode	5	Low-power amplifier	High gain, low drive, low feedback, low anode voltage
Hexode, etc.	6 or more	Special applications	Multiple-input mixers, converters

where e_b = total instantaneous plate voltage
  e_{c1} = total instantaneous control grid voltage
  i_b = total instantaneous plate current

Note that $\mu = Gmr_p$. Figure 5.2.1 shows the curves of plate and grid current as a function of plate voltage at various grid voltages for a typical triode with a μ value of 30.

The tetrode, a four-element tube, is formed when a second grid (screen grid) is mounted between grid 1 (control grid) and the anode (plate). The plate current is almost independent of plate voltage. Figure 5.2.2 shows the curves of plate current as a function of plate voltage at a fixed screen voltage and various grid voltages for a typical power tetrode.

Cooling Methods

Cooling of the tube envelope, seals, and anode, if external, is a major factor affecting tube life. The data sheets provided by tube manufacturers include with the cooling requirements a maximum safe temperature for the various external surfaces. The temperature of these surfaces should be measured in the operating equipment. The temperature can be measured with thermocouples, optical pyrometers, a temperature-sensitive paint such as Tempilaq, or temperature-sensitive tapes.

The envelopes and seals of most tubes are cooled by convection of air around the tube or by using forced air. The four principal methods used for cooling external anodes of tubes are by air, water, vapor, and heat

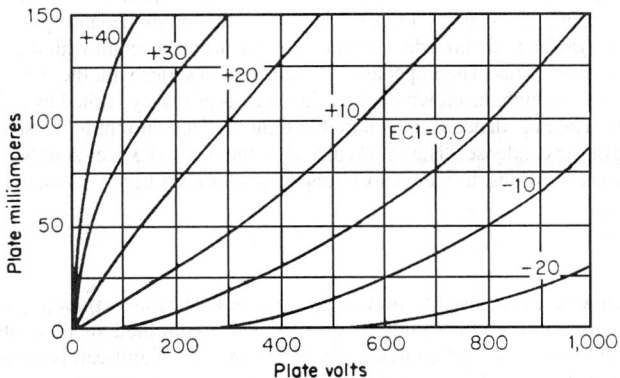

FIGURE 5.2.1 Typical triode plate characteristics.

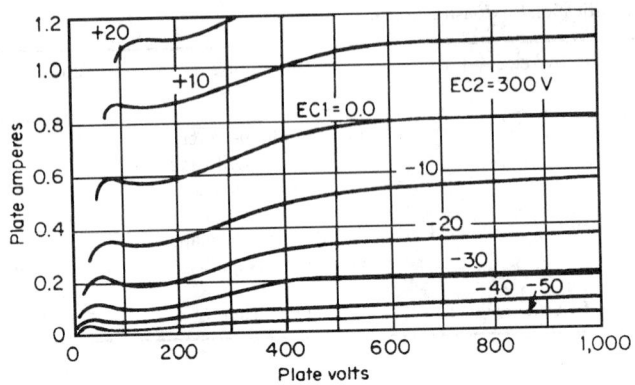

FIGURE 5.2.2 Typical tetrode plate characteristics.

sinks. Other cooling methods occasionally used are oil, heat pipes, refrigerants, such as Freon, and gases, such as sulfahexafluoride.

Protective Circuits

Arcs can damage or destroy electron tubes, which may significantly increase the equipment operating cost. In low- or medium-power operation, the energy stored in the circuit is relatively small and the series impedance is high, which limits the tube damage. In these circuits, the tube will usually continue to work after the fault is removed; however, the life of the tube will be reduced. Since these tubes are normally low in cost, economics dictates that inexpensive slow-acting devices, e.g., circuit breakers and common fuses, be used to protect the circuit components.

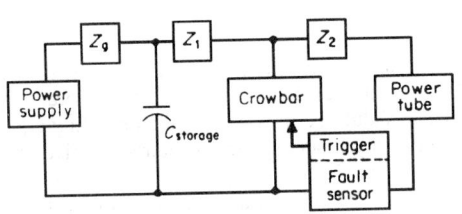

FIGURE 5.2.3 Energy-diverter circuit.

High-power equipment has large stored energy, and the series impedance is usually low. Arcs in this type of equipment will often destroy expensive tubes, and slow-acting protective devices offer insufficient protection. The two basic techniques used to protect tubes are *energy diverters* and special *fast-blow fuses*. The term *crowbar* is commonly used for energy diverters. The typical circuit for a crowbar is shown in Fig. 5.2.3. In the event of a tube arc, the trigger-fault sensor unit "fires" to a crowbar, which is a very-low-impedance gas-discharge device. The firing time can be less than 2 μs. The low impedance in the crowbar arm is in shunt with the Z_2 and tube arm and diverts current from the tube during arcs. The impedance Z_2, which is in series with the tube, is used to ensure that most of the current is diverted through the crowbar. The value of Z_2 is primarily limited by efficiency considerations during normal operation. The impedance Z_1 is required to limit the fault current in the storage condenser to the maximum current rating of the condenser. The impedance Z_g is the internal power-supply impedance. Devices used as crowbars are thyratrons, ignitrons, triggered spark gaps and plasmoid-triggered vacuum gaps.

Fast Fuses

Two types of fast-blow fuses are used to protect power tubes. They are *exploding-wire* and *exothermic* fuses. Exploding-wire fuses require milliseconds to operate and are limited in their ability to protect the tube. Exothermic fuses, although faster-acting than exploding wires, are significantly slower than crowbars in clearing the fault current in the tube. A second disadvantage of fuses is that after each fault the power supply must be turned off and the fuse replaced. For this reason, fuses are limited to applications where the tubes seldom arc. The major advantage of fuses is low cost.

CATHODE-RAY STORAGE AND CONVERSION DEVICES

I. Reingold, Munsey E. Crost

Storage Tubes

Electronic charge-storage tubes are divided into four broad classes: electrical-input-electrical-output types, electrical-input-visual-output types, visual-input-electrical-output types, and visual-input-visual-output types. An example of each class is cited in Table 5.2.2.

Tubes under these classes in which storage is merely incidental, such as camera tubes and image converters, are classed under conversion devices.

Electrical-Input Devices

Electrical-Output Types. The *radechon*, or *barrier-grid storage tube*, is a single-electron-gun storage tube with a fine-mesh screen in contact with a mica storage surface. The metal screen, called the *barrier grid*, acts as a very-close-spaced collector electrode, and essentially confines the secondary electrons to the apertures of the grid in which they were generated. The very thin mica sheet is pressed in contact with a solid metal back-plate. A later model was developed with a concave copper bowl-shaped backplate and a fritted-glass dielectric storage surface. A similarly shaped fine-mesh barrier grid was welded to the partially melted glass layer. The tube is operated with backplate modulation; i.e., the input electrical signal is applied to the backplate while the constant-current electron beam scans the mica storage surface. The capacitively induced voltage on the beam side of the mica dielectric is neutralized to the barrier-grid potential at each scanned elemental area by collection of electrons from the beam and/or by secondary electron emission (equilibrium writing). The current involved in recharging the storage surface generates the output signal.

In a single-electron-gun *image-recording storage tube* (Fig. 5.2.4) information can be recorded and read out later. The intended application is the storage of complete halftone images, such as a television picture, for later readout. In this application, write-in of information is accomplished by electron-beam modulation, in time-sharing sequence with the readout mode. Reading is accomplished nondestructively, i.e., without intentionally changing the stored pattern, by not permitting the beam to impinge upon the storage surface. The electron-gun potentials are readjusted for readout so that none of the electrons in the constant-current beam can reach the storage surface, but divide their current between the storage mesh and the collector in proportion to the stored charge. This process is called *signal division.*

One type of double-ended, multiple-electron-gun *recording storage tube* with nondestructive readout operates by recording halftone information on an insulating coating on the bars of a metal mesh grid, which can be penetrated in both directions by the appropriate electron beams. Very high resolution and long storage time with multiple readouts, including simultaneous writing and reading, are available with this type of tube. Radio-frequency separation or signal cancellation must be used to suppress the writing signal in the readout during simultaneous operation.

Figure 5.2.5 is a representative schematic drawing of a double-ended, multiple-electron-gun *membrane-target storage tube* with nondestructive readout, in which the writing beam and the reading beam are separated by

TABLE 5.2.2   Storage-Tube Classes

Type of input	Type of output	Subclass	Representative example
Electrical	Electrical	Single-gun	Radechon
		Multiple-gun	Graphechon
	Visual	Bistable	Memotron
		Halftone	Tonotron
Visual	Electrical	Nonscanned	Correlatron
	Visual	Time-sharing	Storage image tube

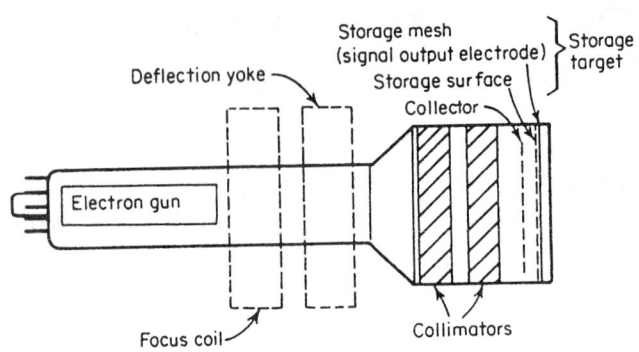

FIGURE 5.2.4   Electrical-signal storage tube, basic structure. (*Hughes Aircraft Corp.*)

the thin insulating film of the storage target. In this case there is a minimal interference of the writing signal with the readout signal in simultaneous operation, except for capacitive feed-through, which can readily be canceled. Writing is accomplished by beam modulation, while reading is accomplished by signal division. Very high resolution and long storage time with multiple readouts are available with this group of tubes.

The *graphechon* differs from the two groups of tubes just described in that its storage target operates by means of electron-bombardment-induced conduction (EBIC). Halftones are not available from tubes of this type in normal operation. The readout is of the destructive type, but since a large quantity of charge is transferred through the storage insulator during writing, a multitude of readout copies can be made before the signal displays noticeable degradation. In simultaneous writing and reading, signal cancellation of the writing signal in the readout is generally accomplished at video frequencies.

Visual-Output Types.   This class comprises the *display storage tubes (DSTs)* or *direct-view storage tubes (DVSTs)*. Figure 5.2.6 shows a schematic diagram of a typical DVST with one electrostatic focus-and-deflection writing gun (other types may have electromagnetic focus and/or deflection) and one flood gun, which is used to provide a continuously bright display and may also establish bistable equilibrium levels. The storage surface is an insulating layer deposited on the bars of the metal-mesh backing electrode. The view screen is an aluminum-film-backed phosphor layer.

The *memotron* is a DST that operates in the bistable mode; i.e., areas of the storage surface may be in either the cutoff condition (flood-gun-cathode potential) or in the transmitting condition (collector potential), either of which is stable. The focusing and deflection are electrostatic. Normally, the phosphor remains in the unexcited

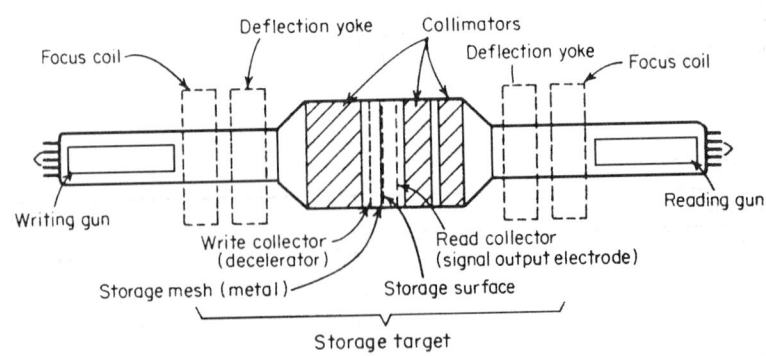

FIGURE 5.2.5   Typical double-ended scan converter, basic structure. (*Hughes Aircraft Corp.*)

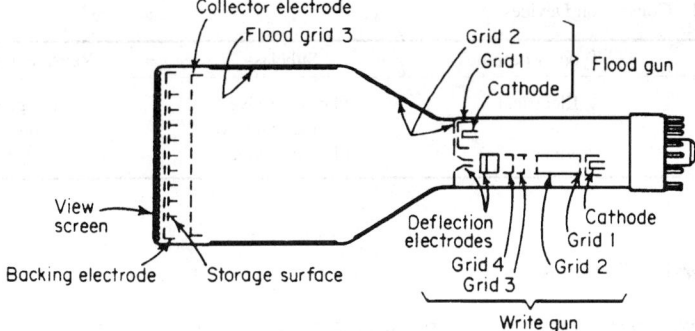

FIGURE 5.2.6 Cross-sectioned view of direct-view storage tube. (*Westinghouse Electric Corp.*)

condition until a trace to be displayed is written into storage; then this trace is displayed continuously until erased, so long as the flood beam is maintained in operation.

The *tonotron* is typical of a large number of halftone DSTs. These operate in the nondestructive readout mode, with the storage surface at or below flood-cathode potential. Writing is accomplished by depositing halftone charge patterns by electron-beam modulation.

Visual-Input Devices

Electrical-Output Types.   The *correlatron* is a storage tube that receives a visual input to a photoemissive film, focuses the photoelectron image upon a transmission-grid type of storage target, where it is stored, and later compares the original image with a similar image. A single total output current is read out for the entire image, with no positional reference. The purpose of this comparison is to ascertain whether the first and second images correlate.

Visual-Output Types.   In the *storage image tube*, a positive electron image can be stored upon the insulated bars of the storage mesh. Then, if the photocathode is uniformly flooded with light to produce a flood electron cloud, a continuously bright image of the stored charge pattern can be obtained on the phosphor screen in the following section of the tube. A high degree of brightness gain can be achieved with this type of tube, or a single snapshot picture of a continuous action can be "frozen" for protracted study.

Conversion Devices

The conversion devices discussed receive images in visible or infrared radiation and convert them by internal electronic processes into a sequence of electrical signals or into a visible output image. Some of these devices may employ an internal storage mechanism, but this is generally not the primary function in their operation. These tubes are characterized by a photosensitive layer at their input ends that converts a certain region of the quantum electromagnetic spectrum into electron-hole pairs. Some of these layers are photoemissive; i.e., they emit electrons into the vacuum if the energy of the incoming quantum is high enough to impart at least enough energy to the electron to overcome the work function at the photosurface-vacuum interface. Others do not emit electrons into the vacuum but conduct current between the opposite surfaces of the layer by means of the electron-hole pairs; i.e., they are photoconductive. The transmission characteristics of the material of the entrance window of the tube can greatly modify the effective characteristics of any photosurface. If the material is photoemissive, the total active area is called a *photocathode.*

The types of conversion devices discussed are divided into visual-input devices, electrical-output and visual-output types. An example of each type is cited in Table 5.2.3.

TABLE 5.2.3   Conversion Devices

Type of input	Type of output	Subclass	Representative example
Visual	Electrical	Photoemissive Photoconductive	Image orthicon Vidicon
	Visual	Photoemissive	Image amplifier

Visual-Input Conversion Devices

Electrical-Output Types.   This class covers the large group of devices designated camera tubes. The defining attributes of the *vidicon* are a photoconductive rather than photoemissive image surface or target, a direct readout from the photosensitive target rather than by means of a return beam, and a much smaller size than the above camera tubes. The original vidicons used coincident electromagnetic deflection and focusing. Many later versions employ either or both electrostatic focusing and deflection.

A very important group of tubes is the *semiconductor diode-array vidicons*. In place of the usual photoconductive target, these tubes include a very thin monolithic wafer of the semiconductor, usually single-crystal silicon. On the beam side of this wafer a dense array of junction photodiodes has been generated by semiconductor diffusion technology. These targets are very sensitive compared with the photoconductors, and they have very low leakage, low image lag, and low blooming from saturation.

Visual-Output Types.   The *image tube*, or *image amplifier*, with input in the visible spectrum is used principally to increase the light level and dynamic range of a very low light-level image to a level and contrast acceptable to a human observer, a photographic plate, or a camera tube. The image tube consists basically of a photoemissive cathode, a focusing and accelerating electron-optical system, and a phosphor screen.

Since the photocathode would be illuminated by light returning from the phosphor screen, the internal surface of the phosphor is covered with a very thin film of aluminum that can be penetrated by the high-energy image electrons. The aluminum film also serves as the tube anode and as a reflector for the light that would otherwise be emitted back into the tube.

SEMICONDUCTOR DIODES AND CONTROLLED RECTIFIERS

Gregory J. Malinowski, John F. Donlon

Semiconductor Materials and Junctions

Diodes and controlled rectifiers are fabricated from semiconductor materials, a form of matter situated between metals and insulators in their ability to conduct electricity. Typical values of electrical resistivity of conductors, semiconductors, and insulators are 10^{-6} to 10^{-5}, 10 to 10^4, and 10^{12} to 10^{16} $\Omega \cdot$ cm, respectively. Silicon is the most widely used semiconductor material. Other semiconductor materials such as germanium, gallium arsenide, selenium, cadmium sulfide, copper oxide, and silicon carbide have electrical properties that make them useful in special applications.

Semiconductor devices develop current flow from the motion of *charge carriers* within a crystalline solid. The conduction process in semiconductors is most easily visualized in terms of silicon. The silicon atoms have four electrons in the outer shell (valence shell). These electrons are normally bound in the crystalline lattice structure. Some of these valence electrons are free at room temperature, and hence can move through the crystal; the higher the temperature, the more electrons are free to move. Each vacancy, or hole, left in the lattice can be filled by an adjacent valence electron. Since a hole moves in a direction opposite to that of an electron, a hole may be considered as a positive-charge carrier. Electrical conduction is a result of the motion of holes and electrons under the influence of an applied field.

Intrinsic (pure) semiconductors exhibit a negative coefficient of resistivity, since the number of carriers increases with temperature. Conduction owing to thermally generated carriers, however, is usually an undesirable effect, because it limits the operating temperature of the semiconductor device.

At a given temperature, the concentration of thermally generated carriers is related to the energy gap of the material. This is the minimum energy (stated in electron volts) required to free a valence electron (1.1 eV for silicon and 0.7 eV for germanium). Silicon devices perform at higher temperatures because of the wider energy gap.

The conductivity of the semiconductor material can be altered radically by doping with minute quantities of *donor* or *acceptor impurities*. Donor (*n*-type) impurity atoms have five valence electrons, whereas only four are accommodated in the lattice structure of the semiconductor. The extra electron is free to conduct at normal operating temperatures. Common donor impurities include phosphorus, arsenic, and antimony. Conversely, acceptor (*p*-type) atoms have three valence electrons; a surplus of holes is created when a semiconductor is doped with them. Typical acceptor dopants include boron, gallium, and indium.

In an *extrinsic (doped) semiconductor*, the current-carrier type introduced by doping predominates. These carriers, electrons in *n*-type material and holes in *p*-type, are called *majority carriers*. Thermally generated carriers of the opposite type are also present in small quantities and are referred to as *minority carriers*. Resistivity is determined by the concentration of majority carriers.

Lifetime is the average time required for excess minority carriers to recombine with majority carriers. Recombination occurs at "traps" caused by impurities and imperfections in the semiconductor crystal. Semiconductor junctions are formed in material grown as a single continuous crystal to obtain the lattice perfection required, and extreme precautions are taken to ensure exclusion of unwanted impurities during processing. However, in some applications the characteristics associated with short lifetime are desirable. In these cases electron and/or proton irradiation and/or heavy metal doping such as gold are used to create recombination sites and lower the life time.

Carrier mobility is the property of a charge carrier that determines its velocity in an electric field. Mobility also determines the velocity of a minority carrier in the diffusion process. High mobility yields a short transit time and good frequency response.

pn Junctions

If *p*- and *n*-type materials are formed together, a unique interaction takes place between the two materials, and a *pn* junction is formed. In the immediate vicinity of this junction (in the *depletion region*), some of the excess electrons of the *n*-type material diffuse into the *p* region, and likewise holes diffuse into the *n*-type region. During this process of recombination of holes and electrons, the *n* material in the depletion region acquires a slightly positive charge and the *p* material becomes slightly negative. The space-charged region thus formed repels further flow of electrons and holes, and the system comes into equilibrium. Figure 5.2.7 shows a typical *pn* junction.

To keep the system in equilibrium, two related phenomena constantly occur. Because of thermal energy, electrons and holes diffuse from one side of the *pn* junction to the other side. This flow of carriers is called *diffusion current*. When a current flows between two points, a potential gradient is produced. This potential gradient across the depletion region causes a flow of charge carriers; drift current, in the opposite direction to the diffusion current. As a result, the two currents cancel at equilibrium, and the net current flow is zero through the region. An energy barrier is erected such that further diffusion of charge carriers becomes impossible without the addition of some external energy source. This energy barrier formed at the interface of the *p*- and *n*-type materials provides the basic characteristics of all junction semiconductor devices.

pn Junction Characteristics

When a dc power source is connected across a *pn* junction, the quantity of current flowing in the circuit is determined by the polarity of the applied voltage and its effect on the depletion layer of the diode. Figure 5.2.8 shows the classical condition for a reversed-biased *pn* junction. The negative terminal of the power supply is connected to the *p*-type material, and the positive terminal to the *n*-type material. When a *pn* junction is reverse-biased, the

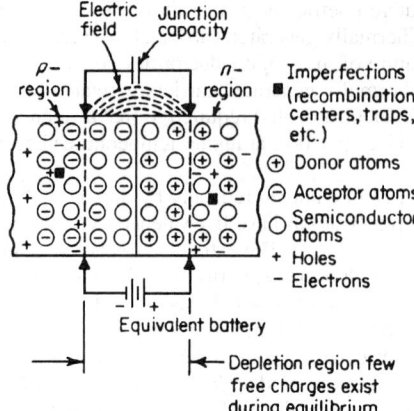

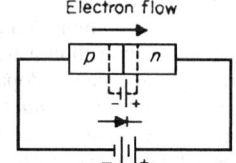

FIGURE 5.2.7   *pn* junction.

FIGURE 5.2.8   Reversed-biased diode.

free elections in the *n*-type material are attracted toward the positive terminal of the power supply and away from the junction. At the same time, holes from the *p*-type material are attracted toward the negative terminal of the supply and away from the junction. As a result, the depletion layer becomes effectively wider, and the potential gradient increases to the value of the supply. Under these conditions the current flow is very small because no electric field exists across either the *p* or *n* region.

Figure 5.2.9 shows the positive terminal of the supply connected to the *p* region and the negative terminal to the *n* region. In this arrangement, electrons in the *p* region near the positive terminal of the supply break their electron-pair bonds and enter the supply, thereby creating new holes. Concurrently, electrons from the negative terminal of the supply enter the *n* region and diffuse toward the junction. This condition effectively decreases the depletion layer, and the energy barrier decreases to a small value. Free electrons from the *n* region can then penetrate the depletion layer, flow across the junction, and move by way of the holes in the *p* region toward the positive terminal of the supply. Under these conditions, the *pn* junction is said to be *forward-biased*.

A general plot of voltage and current for a *pn* junction is shown in Fig. 5.2.10. Here both the forward- and reverse-biased conditions are shown. In the forward-biased region, current rises rapidly as the voltage is increased and is quite high. Current in the reverse-biased region is usually much smaller and remains low until the breakdown voltage of the diode is reached. Thereupon the current increases rapidly. If the current is not limited, it will increase until the device is destroyed.

Junction Capacitance.   Since each side of the depletion layer is at an opposite charge with respect to each other, each side can be viewed as the plate of a capacitor. Therefore a *pn* junction has capacitance. As shown in Fig. 5.2.11, junction capacitance changes with applied voltage.

DC Parameters of Diodes.   The most important of these parameters are as follows:

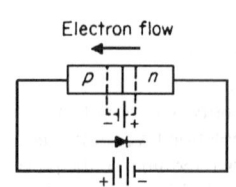

FIGURE 5.2.9   Forward-biased diode.

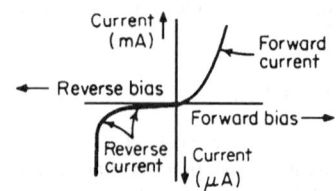

FIGURE 5.2.10   Voltage-current characteristics for a *pn* junction.

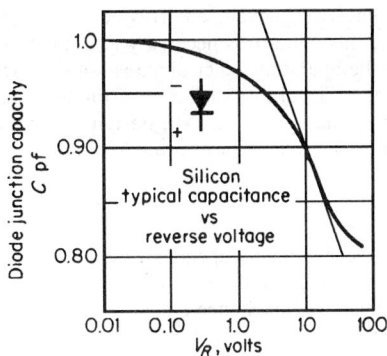

FIGURE 5.2.11   Diode junction capacitance vs. reverse voltage.

Forward voltage V_F is the voltage drop at a particular current level across a diode when it is forward-biased.

Breakdown voltage BV is the voltage drop across the diode at a particular current level when the device is reverse-biased to such an extent that heavy current flows. This is known as *avalanche*.

Reverse current I_R is the leakage current specified at a voltage less than BV when the diode is reverse-biased.

AC Parameters of Diodes.   The most important of these parameters are as follows:

Capacitance C_0 is the total capacitance of the device, which includes junction and package capacitance. It is measured at a particular frequency and bias level.

Rectification efficiency R_E is defined as the ratio of dc output (load) voltage to the peak of the input voltage, in a detector circuit. This provides an indication of the capabilities of the device as a high-frequency detector.

Forward recovery time t_{fr} is the time required for the diode voltage to drop to a specified value after the application of a given forward current.

Reverse recovery time t_{rr} is the time required for the diode principle current to recover to a specified value after switching from the on state to the reverse blocking state. *Reverse Recovery Charge Qrr* and *recovery softness factor RSF* are also characteristics of the diode turn-off process that are important to the proper application of a rectifier diode.

Transient thermal impedance provides data on the instantaneous junction temperature rise above a specified reference point such as the device case as a function of time with constant power applied. This parameter is essential in ensuring reliable operation of diodes in pulse applications.

Small-Signal Diodes

Small-signal diodes are the most widely used discrete semiconductor devices. The capabilities of the general-purpose diode as a switch, demodulator, rectifier, limiter, capacitor, and nonlinear resistor suit it to many low power applications.

The most important characteristics of all small-signal diodes are forward voltage, reverse breakdown voltage, reverse leakage current, junction capacitance, and recovery time.

Silicon Rectifier Diodes

Silicon rectifier diodes are *pn* junction devices that have average forward current carrying capability ratings upward of 8000A and reverse blocking voltage ratings upward of 10,000 V. An ideal rectifier has an infinite reverse resistance, infinite breakdown voltage, and zero forward resistance. The silicon rectifier approaches these ideal specifications in that the forward resistance can be on the order of a few thousandths of an ohm, while the reverse resistance is in the megohm range.

Since silicon rectifiers are primarily used in power supplies, thermal dissipation must be adequate. To avoid excessive heating of the junction, the heat generated must be efficiently transferred to a heat sink. The relative efficiency of this heat transfer is expressed in terms of the thermal resistance of the device. The thermal-resistance range is typically 0.1 to 1°C/W for stud mount diodes and 0.009 to 0.1°C/W for double side cooled disc diodes.

Zener Diodes

Zener diodes are primarily used as voltage reference or regulator elements. This performance is based on the avalanche characteristics of the *pn* junction. When a source of voltage is applied to the diode in the reverse direction (anode negative), a reverse current I_r is observed. As the reverse potential is increased beyond the knee of the current-voltage curve, avalanche-breakdown current becomes well developed. This occurs at the zener voltage V_z. Since the resistance of the device drastically drops at this point, it is necessary to limit the current flow by means of an external resistor. Avalanche breakdown of the operating zener diode is not destructive as long as the rated power dissipation of the junction is not exceeded. The ability of the zener diode to maintain a desired operating voltage is limited by its temperature coefficient and impedance. The design of zener diodes permits them to absorb overload surges and thereby serve as transient voltage protection.

Varactor Diodes

The varactor diode is a *pn* junction device that has useful nonlinear voltage-dependent variable-capacitance characteristics. Varactor diodes are useful in microwave amplifiers and oscillators when employed with the proper filter and impedance-matching circuitry. The voltage-dependent capacitance effect in the diode permits its use as an electrically controlled tuning capacitor in radio and television receivers.

Tunnel Diodes

The tunnel diode is a semiconductor device whose primary use arises from its negative conductance characteristic. In a *pn* junction, a *tunnel* effect is obtained when the depletion layer is made extremely thin. Such a depletion layer is obtained by heavily doping both the *p* and *n* regions of the device. In this situation it is possible for an electron in the conduction band on the *n* side to penetrate, or tunnel, into the valence band of the *p* side. This gives rise to an additional current in the diode at a very small forward bias, which disappears when the bias is increased. It is this additional current that produces the negative resistance of the tunnel diode.

Typical applications of tunnel diodes include oscillators, amplifiers, converters, and detectors.

Schottky Barrier Diodes

This diode (also known as the surface-barrier diode, metal-semiconductor diode, and hot-carrier diode) consists of a rectifying metal-semiconductor junction in which majority carriers carry the current flow. When the diode is forward-biased, the carriers are injected into the metal side of the junction, where they remain majority carriers at some energy greater than the Fermi energy in the metal; this gives rise to the name *hot carriers*. The diode can be switched to the OFF state in an extremely short time (in the order of picoseconds). No stored minority-carrier charge exists.

The reverse dc current-voltage characteristics of the device are very similar to those of conventional *pn*-junction diodes. The reverse leakage current increases with reverse voltage gradually, until avalanche breakdown is reached.

Schottky barrier diodes used in detector applications have several advantages over conventional *pn*-junction diodes. They have a lower noise and better conversion efficiency, and hence have greater overall detection sensitivity.

Light Sensors (Photodiodes)

When a semiconductor junction is exposed to light, photons generate hole-electron pairs. When these charges diffuse across the junction, they constitute a photocurrent. Junction light sensors are normally operated with a load resistance and a battery that reverse-biases the junction. The device acts as a source of current that increases with light intensity.

Silicon sensors are used for sensing light in the visible and near-infrared spectra. They can be fabricated as phototransistors in which the collector-base junction is light-sensitive. Phototransistors are more sensitive than photodiodes because the photon-generated current is amplified by the current gain of the transistor.

Light-Emitting Diodes (LEDs)

These devices have found wide use in visual displays, isolators, and as digital storage elements.

LEDs are principally manufactured from gallium arsenide. When biased in the avalanche-breakdown region, *pn* junctions emit visible light at relatively low power levels. LEDs are capable of providing light of different wavelengths by varying their construction.

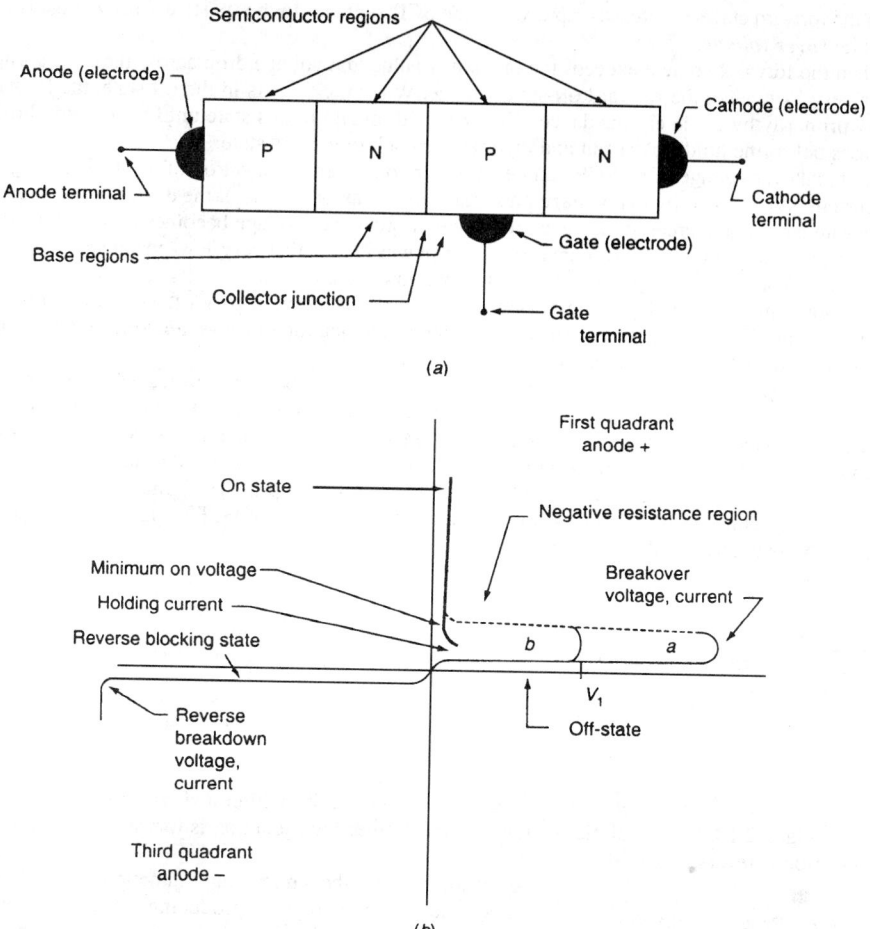

FIGURE 5.2.12 (*a*) SCR junction diagram; (*b*) typical SCR voltage-current characteristics. Curve *a* applies for zero-gate current; curve *b* applies when minimum gate current to trigger is present and the off-state voltage is V_1. (Electronic Industries Association)

Silicon Controlled Rectifiers (SCRs)

A silicon controlled rectifier is basically a four-layer *pnpn* device that has three electrodes (a cathode, an anode, and a control electrode called the *gate*). Figure 5.2.12 shows the junction diagram and voltage-current characteristics for an SCR.

When an SCR is reverse-biased (anode negative with respect to the cathode), it is similar in characteristics to that of a reverse-biased silicon rectifier or other semiconductor diode. In this bias mode, the SCR exhibits a very high internal impedance, and only a very low reverse current flows through the *pnpn* device. This current remains small until the reverse voltage exceeds the reverse breakdown voltage; beyond this point, the reverse current increases rapidly.

During forward-bias operation (anode positive with respect to the cathode), the *pnpn* structure of the SCR is electrically bistable and may exhibit either a very high impedance (OFF *state*) or a very low impedance (ON *state*). In the forward-blocking state (OFF), a small forward current, called the *forward* OFF-*state or leakage current*, flows through the SCR. The magnitude of this current is approximately the same as that of the reverse-blocking current that flows under reverse-bias conditions. As the forward bias is increased, a voltage point is reached at which the forward current increases rapidly, and the SCR switches to the ON state. This voltage is called the *forward breakover voltage*.

When the forward voltage exceeds the breakover value, the voltage drop across the SCR abruptly decreases to a very low value, the forward ON-state voltage. When the SCR is in the ON state, the forward current is limited primarily by the load impedance. The SCR will remain in this state until the current through the SCR decreases below the holding current and then reverts back to the OFF state.

The breakover voltage of an SCR can be varied or controlled by injection of a signal at the gate. When the gate current is zero, the principal voltage must reach the breakover value of the device before breakover occurs. As the gate current is increased, however, the value of breakover voltage becomes less until the device goes to the ON state. This enables an SCR to control a high-power load with a very-low-power signal and makes it suitable for such applications as phase control and high power conversion.

The most important SCR parameters are forward or ON-state voltage V_T, OFF-state and reverse breakover voltage V_{BO} and $V_{(BR)R}$, gate trigger current I_{GT}, rate of application of OFF-state voltage dv/dt, circuit commutated turn-off time t_q, and transient thermal impedance.

Gate-turn-off thyristors (GTOs) are basically SCRs that are specially fabricated so that the gate does not completely lose control when the GTO is in the latched conducting state. It can be restored to the blocking state by the application of reverse bias to the gate. The GTO finds application in dc control, dc-to-dc converters, and self-commutated ac-to-dc, dc-to-ac, and ac-to-ac converters where the fact that it does not need an external force-commutating circuit is an advantage. The GTO finds primary application in very high power circuits where the other turn-off devices (bipolar transistors, IGBTs, MOSFETs, and so forth) do not have the necessary current and/or voltage ratings.

TRANSISTORS

Edward B. Hakim

Introduction

A bipolar transistor consists of two junctions in close proximity within a single crystal. An *npn* transistor is shown in Fig. 5.2.13. In normal bias conditions, the emitter-base junction is forward-biased and the collector-base junction is reversed-biased.

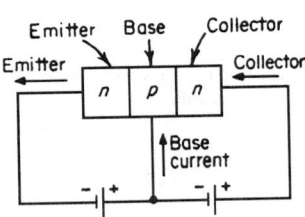

FIGURE 5.2.13   An *npn* junction transistor.

Forward bias of the emitter-base junction causes electrons to be injected into the base region, producing an excess concentration of minority carriers there. These carriers move by diffusion to the collector junction, where they are accelerated into the collector region by the field in the depletion region of the reverse-biased collector junction. Some of the electrons recombine before reaching the collector. Current flows from the base terminal to supply the holes for this recombination process. Another component of current flows in the emitter-base circuit because of the injection of holes from the base into the emitter.

Practical transistors have narrow bases and high lifetimes in the base to minimize recombination. Injection of holes from the base into the emitter is made negligible by doping the emitter much more heavily than the base. Thus the collector current is less than, but almost equal to, the emitter current.

In terms of the emitter current I_E, the collector current I_C is

$$I_C = \alpha I_E + I_{CBO}$$

where α is the fraction of the emitter current that is collected and I_{CBO} is a result of the reverse-current characteristic of the collector-base junction. Increase of I_{CBO} with temperature sets the maximum temperature of operation.

High-frequency transistors are fabricated with very narrow bases to minimize the transit time of minority carriers across the base region. Although germanium has a higher carrier mobility, silicon is the preferred material because of its availability and superior processing characteristics.

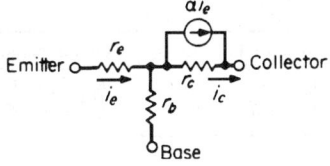

FIGURE 5.2.14   Common-base T-equivalent circuit.

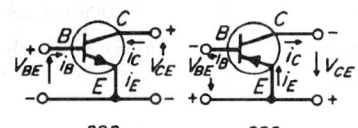

FIGURE 5.2.15   Transistor symbols.

Circuit Models of the Transistor

Performance of the transistor as an active circuit element is analyzed in terms of various small-signal equivalent circuits. The low-frequency T-equivalent circuit (Fig. 5.2.14) is closely related to the physical structure. This circuit model is used here to illustrate the principle of transistor action. Carriers are injected into the base region by forward current through the emitter-base junction. A fraction α (near unity) of this current is collected. The incremental change in collector current is determined essentially by the current generator αi_e, where i_e is the incremental change of emitter current. The collector resistance r_c in parallel with the current generator accounts for the finite resistance of the reverse-biased collector-base junction. The input impedance is a result of the dynamic resistance r_e of the of the forward-biased emitter-base junction and the ohmic resistance r_b of the base region.

The room temperature value of r_e is about $26/I_E$ Ω, where I_E is the dc value of emitter current in milliamperes. Typical ranges of the other parameters are as follows: r_b varies from tens of ohms to several hundred ohms; α varies from 0.9 to 0.999; and r_c ranges from a few hundred ohms to several megohms. The symbolic representations of an *npn* and *pnp* transistor are shown in Fig. 5.2.15. The direction of conventional current flow and terminal voltage for normal operation as an active device are indicated for each. The voltage polarities and current for the *pnp* are reversed from those of the *npn*, since the conductivity types are interchanged.

Transistors may be operated with any one of the three terminals as the common, or grounded, element, i.e., common base, common emitter, or common collector. These configurations are shown in Fig. 5.2.16 for an *npn* transistor.

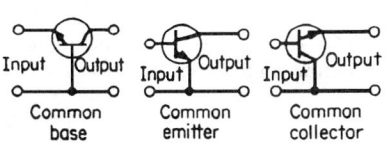

FIGURE 5.2.16   Circuit connections for *npn* transistor.

Common Base.   The transistor action shown at the left in Fig. 5.2.16 is that of the common-base connection whose current gain (approximately equal to α) is slightly less than 1. Even with less than unity current gain, voltage and power amplification can be achieved, since the output impedance is much higher than the input impedance.

Common Emitter.   For the common-emitter connection, only base current is supplied by the source. Base current is the difference between emitter and collector currents and is much smaller than either; hence current gain I_c/I_b is high. Input impedance of the common-emitter state is correspondingly higher than it is in the common-base connection.

Common Collector.   In the common-collector connection, the source voltage and the output voltage are in series and have opposing polarities. This is a negative-feedback arrangement, which gives a high input impedance and approximately unity voltage gain. Current gain is about the same as that of the common-emitter connection. The common-base, common-emitter, and common-collector connections are roughly analogous to the grounded-grid, grounded-cathode, and grounded-plate (cathode-follower) connections, respectively, of the vacuum tube.

h *Parameters.*   Low-frequency performance of transistors is commonly specified in terms of the small-signal *h* parameters listed in Table 5.2.4. In the notation system used, the second subscript designates the circuit connection (*b* for common-base and *e* for common-emitter). The forward-transfer parameters (h_{fb} and h_{fe}) are current gains measured with the output short-circuited. The current gains for practical load conditions are not greatly different. The input parameters h_{ib} and h_{ie}, although measured for short-circuit load, approximate the input impedance of practical circuits. The output parameters h_{ob} and h_{oe} are the output admittances.

TABLE 5.2.4 Transistor Small-Signal h Parameters

	Input parameter	Transfer parameter	Output parameter
Common-base	$h_{ib} \approx r_e + (1 - \alpha)r_b$	$h_{fb} \approx \alpha$	$h_{ob} \approx 1/r_c$
Common-emitter	$h_{ie} \approx r_e/(1 - \alpha) + r_b$	$h_{fe} \approx \alpha/(1 - \alpha)$	$h_{oe} \approx 1/r_c(1 - \alpha)$

The current gain of the common-base stage is slightly less than unity; common-emitter current gains may vary from ten to several hundred. Input impedance and output admittance of the common-emitter stage are higher than those of the common-base circuit by approximately h_{fe}. Nomenclature and units for h parameters are given in Table 5.2.5.

Although matched power gains of the common-base and common-emitter connections are about the same, the higher input impedance and lower output impedance of the common-emitter stage are desirable for most applications. For these reasons, the common-emitter stage is more commonly used. For example, the voltage gain of cascaded common-base stages cannot exceed unity unless transformer coupling is used.

The common-collector circuit has a higher input impedance and lower output impedance than either of the other connections. It is used primarily for impedance transformation.

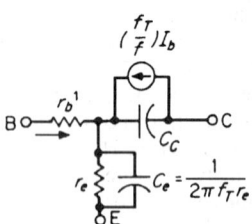

FIGURE 5.2.17 High-frequency common-emitter equivalent circuit.

High-Frequency Limit.   The current gain of a transistor decreases with frequency, principally because of the transit time of minority carriers across the base region. The frequency f_T at which h_{fe} decreases to unity is a measure of high-frequency performance. Parasitic capacitances of junctions and leads also limit high-frequency capabilities. These high-frequency effects are shown in the modified equivalent circuit of Fig. 5.2.17. The maximum frequency f_{max} at which the device can amplify power is limited by f_T and the time constant $\dot{r}_b C_C$, where $\dot{r}_b$ is the ohmic base resistance and C_C is that portion of the collector-base junction capacitance which is under the emitter stripe. Values of f_T greater than 2 GHz and f_{max} exceeding 10 GHz are obtained by maintaining very thin bases (<30 μm) and narrow emitters (<3 μm).

Transistor Voltampere Characteristics

The performance of a transistor over wide ranges of current and voltage is determined from static characteristic curves, e.g., the common-emitter output characteristics of Fig. 5.2.18. Collector current I_C is plotted as a function of collector-to-emitter voltage V_C for constant values of base current I_B. Maximum collector voltage for grounded emitter is limited by either punch-through or avalanche breakdown, whichever is lower, depending on the base resistivity and thickness. When a critical electric field is reached, avalanche occurs because of intensive current multiplication. At this point current increases rapidly with little increase in voltage. The common-emitter breakdown voltage BV_{CEO} is always less than the collector-junction breakdown voltage BV_{CBO}. Another

TABLE 5.2.5 h-Parameter Nomenclature

Parameter	Nomenclature	Unit
h_{ib}	Input impedance (common-base)	Ω
h_{ie}	Input impedance (common-emitter)	Ω
h_{fb}	Forward-current transfer ratio (common-base)	Dimensionless
h_{fe}	Forward-current transfer ratio (common-emitter)	Dimensionless
h_{ob}	Output admittance (common-base)	S
h_{oe}	Output admittance (common-emitter)	S

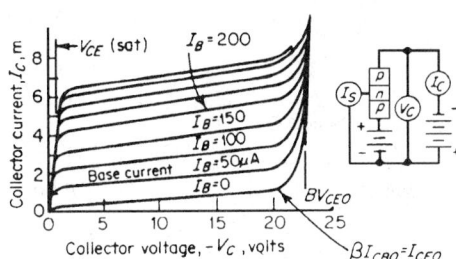

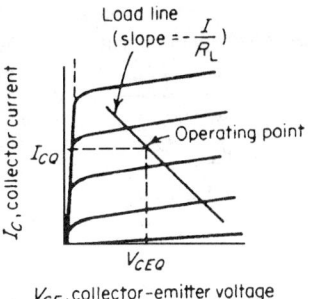

FIGURE 5.2.18 V_C-I_C characteristic for grounded-emitter junction transistor.

FIGURE 5.2.19 Load line for linear-transistor-amplifier circuit.

characteristic evident from Fig. 5.2.18 is the grounded-emitter saturation voltage $V_{CE,\text{sat}}$. This parameter is especially important in grounded-emitter switching applications.

Two additional parameters, both related to the emitter junction, are BV_{EBO} and V_{BE}. The breakdown voltage emitter to base with the collector open-circuited BV_{EBO} is the avalanche-breakdown voltage of the emitter junction. The base-to-emitter forward voltage of the emitter junction V_{BE} is simply the junction voltage necessary to maintain the forward-bias emitter current.

The leakage current I_{CBO} in the common-base connection is the reverse current of the collector-base junction; common-emitter leakage is higher by the factor $1/(1 - \alpha)$ because of transistor amplification. In either case, the leakage current increases exponentially with temperature. Maximum junction temperatures are limited to about 100°C in germanium and 250°C in silicon. The locus of maximum power dissipation is a hyperbola on the voltampere characteristic curve. Power dissipation must be decreased when higher ambient temperatures exist. Large-area devices and physical heat sinks of high thermal dissipation are used to extend power ratings.

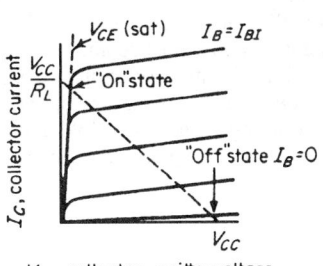

Dynamic variations of voltage and current are analyzed by a load line on the characteristic curves, as in vacuum tubes. For a linear transistor amplifier with load resistance R_L, the output varies along a load line of slope $-1/R_L$ about the dc operating point (Fig. 5.2.19). Since the minimum voltage $V_{CE,\text{sat}}$ is quite low, good efficiencies can be obtained with low values of supply voltage. The operating point on the V_{CE}-I_C coordinates is established by a dc bias current in the input circuit. Transistor circuits should be biased for a fixed emitter current rather than a fixed base current to maintain a stable operating point, since the lines of constant base current are variable between devices of a given type and with temperature.

The common-emitter circuit can also be used as an effective switch, as shown by the load line of Fig. 5.2.20. When the base current is zero, the collector circuit is effectively open-circuited and only leakage current flows in the collector circuit. The device is turned on by applying base current I_{BI},

FIGURE 5.2.20 Switching states for common-emitter circuit.

which decreases the collector voltage to the saturation value.

Field-Effect (Unipolar) Transistors

There are two general types of field-effect transistors (FET): junction (JFET) and insulated-gate (IGFET). The IGFET has a variety of structures, known as the metal-insulator-semiconductor (MISFET) and the metal-oxide-semiconductor (MOSFET).

The cross section of a p-channel JFET is shown in Fig. 5.2.21. Channel current is controlled by reverse-biasing the gate-to-channel junction so that the depletion region reduces the effective channel width. The input

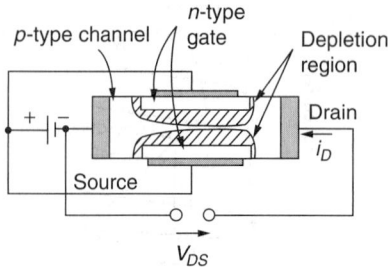

FIGURE 5.2.21   A *p*-channel junction field-effect transistor.

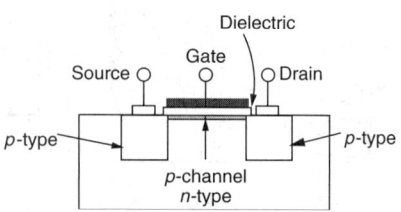

FIGURE 5.2.22   A *p*-channel MOS transistor.

impedance of these devices is high because of the reverse-biased diode in the input channel. In fact, the voltampere characteristics are quite similar to those of a vacuum tube. Another important feature of the junction FET is the excellent low-frequency noise characteristics, which surpass those of either the vacuum tube or conventional (bipolar) transistor.

The cross section of the *p*-channel MOSFET (or MOS transistor) is shown in Fig. 5.2.22. This device operates in the depletion mode. For zero gate voltage, there is no channel, and the drain current is small. A negative voltage on the gate repels the electrons from the surface and produces a *p*-type conduction region under the gate. Compared with the JFET, the MOS transistor has a wider gain-bandwidth product and a higher input impedance (>100 GΩ).

Power MOSFETs offer the major advantage over bipolar transistors in that they do not suffer from second breakdown. Their safe dc operating areas are determined by their rated power dissipation over the entire drain-to-source voltage range up to a rated voltage. This is not the case for bipolar transistors. The superiority in power handling capability of MOSFETs is also true for the pulsed-power operating mode.

Further operational and reliability advantages of FETs over bipolar devices are obtained by the use of gallium arsenide (GaAs) in place of silicon (Si). The advantages include enhanced switching speeds resulting from electron velocity twice that of Si; lower operating voltages and lower ON resistance resulting from a five-fold greater electron mobility; and 350°C maximum operating temperature versus 175°C.

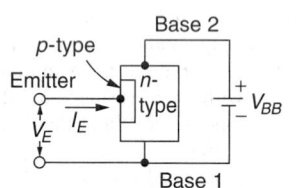

FIGURE 5.2.23   Unijunction transistor.

Unijunction Transistors

A unijunction transistor is shown in Fig. 5.2.23. The input diode is reverse-biased at low voltages owing to *IR* drop in the bulk resistance of the *n*-type region. When V_E exceeds this drop, carriers are injected and the resistance is lowered. As a result, the *IR* drop and V_E decrease abruptly. The negative-resistance characteristic is useful in such applications as oscillators and as trigger devices for silicon controlled rectifiers.

CHAPTER 5.3
BATTERIES AND FUEL CELLS*

David Linden

BATTERIES

A battery is a device that converts the chemical energy contained in its active materials directly into electric energy by means of an oxidation-reduction electrochemical reaction. This type of reaction involves the transfer of electrons from one material to another. In a nonelectrochemical reaction this transfer of electrons occurs directly, and only heat is involved. In a battery (Fig. 5.3.1) the negative electrode or anode is the component capable of giving up electrons, being oxidized during the discharge reaction. It is separated from the oxidizing material, which is the positive electrode or cathode, the component capable of accepting electrons. The transfer of electrons takes place in the external electric circuit connecting the two materials and in the electrolyte, which serves to complete the electric circuit in the battery by providing an ionic medium for the electron transfer.

Components of Batteries

The basic unit of the battery is the cell. A battery consists of one or more cells, connected in series or parallel, depending on the desired output voltage or capacity. The cell consists of three major components: the anode (the reducing material or fuel), the cathode or oxidizing agent, and the electrolyte, which provides the necessary internal ionic conductivity. Electrolytes are usually liquid, but some batteries employ solid electrolytes that are ionic conductors at their operating temperatures. In addition, practical cell design requires a separator material, which serves to separate the anode and cathode electrodes mechanically. Electrically conducting grid structures or materials are often added to each electrode to reduce internal resistance. The containers are of many types, depending on the application and its environment.

Theoretical Cell Voltage and Capacity

The theoretical energy delivered by a battery is determined by its active materials; the maximum electric energy (watt-hours) corresponds to the free-energy change of the reaction. The theoretical voltage and ampere-hour capacities of a number of electrochemical systems are given in Table 5.3.1. The voltage is determined by the active materials selected; the ampere-hour capacity is determined by the amount (weight) of available reactants. One gram-equivalent weight of material will supply 96,500 C, or 26.8 Ah, of electric energy.

---

*Material for this chapter has been abstracted from the "Handbook of Batteries," 3rd ed., D. Linden and T. B. Reddy, (eds.), McGraw-Hill, 2001.

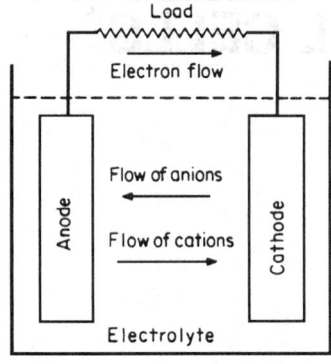

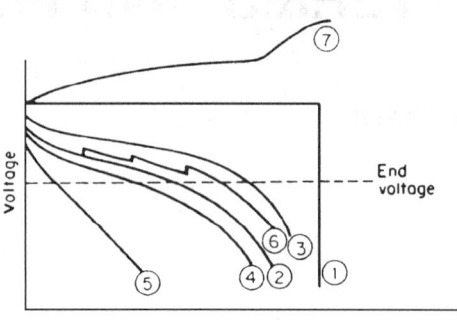

FIGURE 5.3.1 Electrochemical operation of a cell or battery during discharge.

FIGURE 5.3.2 Battery discharge characteristics.

Factors Influencing Battery Voltage and Capacity

In actual practice, only a small fraction of the theoretical capacity is realized. This is a result not only of the presence of nonreactive components (containers, separators, electrolyte) that add to the weight and size of the battery but also of many other factors that prevent the battery from performing at its theoretical level. Factors influencing the voltage and capacity of a battery are as follows.

Voltage Level.   When a battery is discharged in use, its voltage is lower than the theoretical voltage. The difference is caused by *IR* losses because of cell resistance and polarization of the active materials during discharge. This is illustrated in Fig. 5.3.2. The theoretical discharge curve of a battery is shown as curve 1. In this case, the discharge of the battery proceeds at the theoretical voltage until the active materials are consumed and the capacity fully utilized. The voltage then drops to zero. Under load conditions, the discharge curve is similar to curve 2. The initial voltage is lower than theoretical, and it drops off as the discharge progresses.

Current Drain of the Discharge.   As the current drain of the battery is increased, the *IR* loss increases, the discharge is at a lower voltage, and the service life of the battery is reduced (curve 5). At extremely low current drains it is possible to approach the theoretical capacities (in the direction of curve 3). In a very long discharge period, chemical self-deterioration during the discharge becomes a factor and causes a reduction of capacity.

Voltage Regulation.   The voltage regulation required by the equipment is most important. As is apparent by the curves in Fig 5.3.2 design of equipment to operate to the lowest possible end voltage results in the highest capacity and longest service life. Similarly, the upper voltage limit of the equipment should be established to take full advantage of the battery characteristics. In some applications, where only a narrow voltage range can be tolerated, voltage regulators may have to be employed to take full advantage of the battery's capacity. If a storage battery is used in conjunction with another energy source, which is permanently connected in the operating circuit, allowances must be made for the voltage required to charge the battery, as illustrated in curve 7, Fig. 5.3.2. The voltage of the charging source must just exceed the maximum voltage on charge.

Type of Discharge (Continuous, Intermittent, and so Forth).   When a battery stands idle after a discharge, certain chemical and physical changes take place, which can result in voltage recovery. Thus the voltage of a battery that has dropped during a heavy discharge will rise after rest period, giving a sawtooth discharge. Curve 6 of Fig. 5.3.2 shows the characteristic of a battery discharged intermittently at the same drain in curve 2. The improvement resulting from the intermittent discharge depends on the current drain, length of the recovery period, discharge temperature, end voltage, and the particular battery system and design employed. Some battery systems, during inactive stand, develop a protective film on the active-material surface. These batteries, instead of showing a recovery voltage, may momentarily demonstrate a lower voltage after stand until this film is broken by the discharge. This is known as a *voltage delay.*

TABLE 5.3.1 Voltage, Capacity, and Specific Energy of Major Battery and Several Fuel Cell Systems—Theoretical and Practical Values

Battery type	Anode	Cathode	Reaction mechanism	Theoretical values[†]				Practical battery[‡]		
				V	g/Ah	Ah/kg	Specific energy Wh/kg	Nominal voltage V	Specific energy Wh/kg	Energy density Wh/L
Primary batteries										
Leclanché	Zn	MnO_2	$Zn + 2MnO_2 \rightarrow ZnO\cdot Mn_2O_3$	1.6	4.46	224	358	1.5	85[4]	165[4]
Magnesium	Mg	MnO_2	$Mg + 2MnO_2 + H_2O \rightarrow Mn_2O_3 + Mg(OH)_2$	2.8	3.69	271	759	1.7	100[4]	195[4]
Alkaline MnO_2	Zn	MnO_2	$Zn + 2MnO_2 \rightarrow ZnO + Mn_2O_3$	1.5	4.46	224	358	1.5	145[4]	400[4]
Mercury	Zn	HgO	$Zn + HgO \rightarrow ZnO + Hg$	1.34	5.27	190	255	1.35	100[6]	470[6]
Mercad	Cd	HgO	$Cd + HgO + H_2O \rightarrow Cd(OH)_2 + Hg$	0.91	6.15	163	148	0.9	55[6]	230[6]
Silver oxide	Zn	Ag_2O	$Zn + Ag_2O + H_2O \rightarrow Zn(OH)_2 + 2Ag$	1.6	5.55	180	288	1.6	135[6]	525[6]
Zinc/O_2	Zn	O_2	$Zn + \frac{1}{2}O_2 \rightarrow ZnO$	1.65	1.52	658	1085	—	—	—
Zinc/air	Zn	Ambient air	$Zn + (\frac{1}{2}O_2) \rightarrow ZnO$	1.65	1.22	820	1353	1.5	370[6]	1300[6]
Li/$SOCl_2$	Li	$SOCl_2$	$4Li + 2SOCl_2 \rightarrow 4LiCl + S + SO_2$	3.65	3.25	403	1471	3.6	590[4]	1100[4]
Li/SO_2	Li	SO_2	$2Li + 2SO_2 \rightarrow Li_2S_2O_4$	3.1	2.64	379	1175	3.0	260[5]	415[5]
$LiMnO_2$	Li	MnO_2	$Li + Mn^{IV}O_3 \rightarrow Mn^{IV}O_2(Li^+)$	3.5	3.50	286	1001	3.0	230[5]	535[5]
Li/FeS_2	Li	FeS_2	$4Li + FeS_2 \rightarrow 2Li_2S + Fe$	1.8	1.38	726	1307	1.5	260[5]	500[5]
Li/$(CF)_n$	Li	$(CF)_n$	$nLi + (CF)_n \rightarrow nLiF + nC$	3.1	1.42	706	2189	3.0	250[5]	635[5]
$LiI^{(3)}$	Li	$I_2(P2VP)$	$Li + \frac{1}{2}I_2 \rightarrow LiI$	2.8	4.99	200	560	2.8	245	900
Reserve batteries										
Cuprous chloride	Mg	CuCl	$Mg + Cu_2Cl_2 \rightarrow MgCl_2 + 2Cu$	1.6	4.14	241	386	1.3	60[7]	80[7]
Zinc/silver oxide	Zn	AgO	$Zn + AgO + H_2O \rightarrow Zn(OH)_2 + Hg$	1.81	3.53	283	512	1.5	30[8]	75[8]
Thermal	Li	FeS_2		2.1–1.6	1.38	726	1307	2.1–1.6	40[9]	100[9]
Secondary batteries										
Lead-acid	Pb	PbO_2	$Pb + PbO_2 + 2H_2SO_4 \rightarrow 2PbSO_4 + 2H_2O$	2.1	8.32	120	252	2.0	35	70[10]
Edison	Fe	Ni oxide	$Fe + 2NiOOH + 2H_2O \rightarrow 2Ni(OH)_2 + Fe(OH)_2$	1.4	4.46	224	314	1.2	30	55[10]
Nickel-cadmium	Cd	Ni oxide	$Cd + 2NiOOH + 2H_2O \rightarrow 2Ni(OH)_2 + Cd(OH)_2$	1.35	5.52	181	244	1.2	35	100[5]
Nickel-zinc	Zn	Ni oxide	$Zn + 2NiOOH + 2H_2O \rightarrow 2Ni(OH)_2 + Zn(OH)_2$	1.73	4.64	215	372	1.6	60	120
Nickel-hydrogen	H_2	Ni oxide	$H_2 + 2NiOOH \rightarrow 2Ni(OH)_2$	1.5	3.46	289	434	1.2	55	60
Nickel-metal hydride	$MH^{(1)}$	Ni oxide	$MH + NiOOH \rightarrow M + Ni(OH)_2$	1.35	5.63	178	240	1.2	75	240[5]
Silver-zinc	Zn	AgO	$Zn + AgO + H_2O \rightarrow Zn(OH)_2 + Ag$	1.85	3.53	283	524	1.5	105	180[10]
Silver-cadmium	Cd	AgO	$Cd + AgO + H_2O \rightarrow Cd(OH)_2 + Ag$	1.4	4.41	227	318	1.1	70	120[10]

(Continued)

TABLE 5.3.1 Voltage, Capacity, and Specific Energy of Major Battery and Several Full Cell Systems—Theoretical and Practical Values (*Continued*)

Battery type	Anode	Cathode	Reaction mechanism	Theoretical values[†]				Practical battery[‡]		
				V	Ah/kg	g/Ah	Specific energy Wh/kg	Nominal voltage V	Specific energy Wh/kg	Energy density Wh/L
Zinc/chlorine	Zn	Cl_2	$Zn + Cl_2 \rightarrow ZnCl_2$	2.12	394	2.54	835	—	—	—
Zinc/bromine	Zn	Br_2	$Zn + Br_2 \rightarrow ZnBr_2$	1.85	309	4.17	572	1.6	70	60
Lithium-ion	$Li_x C_6$	$Li_{(i-x)} CoO_2$	$Li_x C_6 + Li_{(i-x)} CoO_2 \rightarrow LiCoO_2 + C_6$	4.1	100	9.98	410	4.1	150	400[5]
Lithium/ manganese dioxide	Li	MnO_2	$Li + Mn^{IV}O_2 \rightarrow Mn^{IV}O_2(Li^+)$	3.5	286	3.50	1001	3.0	120	265
Lithium/ iron disulfide[2]	Li(Al)	FeS_2	$2Li(Al) + FeS_2 \rightarrow Li_2FeS_2 + 2Al$	1.73	285	3.50	493	1.7	180[11]	350[11]
Lithium/ iron monosulfide[2]	Li(Al)	FeS	$2Li(Al) + FeS \rightarrow Li_2S + Fe + 2Al$	1.33	345	2.90	459	1.3	130[11]	220[11]
Sodium/ sulfur[2]	Na	S	$2Na + 3S \rightarrow Na_2S_3$	2.1	377	2.65	792	2.0	170[11]	345[11]
Sodium/ nickel chloride[2]	Na	$NiCl_2$	$2Na + NiCl_2 \rightarrow 2NaCl + Ni$	2.58	305	3.28	787	2.6	115[11]	190[11]
Fuel cells										
H_2/O_3	H_2	O_2	$H_2 + {}^1\!/_2 O_2 \rightarrow H_2O$	1.23	2975	0.336	3660	—	—	—
H_2/air	H_2	Ambient air	$H_2 + ({}^1\!/_2 O_2) \rightarrow H_2O$	1.23	26587	0.037	32702	—	—	—
Methanol/O_2	CH_3OH	O_2	$CH_3OH + {}^3\!/_2 O_2 \rightarrow CO_2 + 2H_2O$	1.24	2000	0.50	2480	—	—	—
Methanol/air	CH_3OH	Ambient air	$CH_3OH + ({}^3\!/_2 O_2) \rightarrow CO_2 + 2H_2O$	1.24	5020	0.20	6225	—	—	—

[†]Based on active anode and cathode materials only; including O_2 but not air (electrolyte not included).
[‡]These values are for single cell batteries based on identified design and at discharge rates optimized for energy density, using midpoint voltage.
(1) MH = metal hydride, data based on 1.7 percent hydrogen storage (by weight).
(2) High temperature batteries.
(3) Solid electrolyte battery (Li/I_2 (P2VP)).
(4) Cylindrical bobbin-type batteries.
(5) Cylindrical spiral-wound batteries.
(6) Button type batteries.
(7) Water-activated.
(8) Automatically activated 2- to 10-min rate.
(9) With lithium anodes.
(10) Prismatic batteries.
(11) Value based on cell performance.
Source: D. Linden and T. B. Reddy (eds.). "Handbook of Batteries," 3rd ed. McGraw-Hill, 2001.

Temperature During Discharge. The temperature at which the battery is discharged has a pronounced effect on its service and voltage characteristics. This is because of the reduction in chemical activity and the increase in battery internal resistance at lower temperatures. Curves 3, 2, 4, and 5 of Fig. 5.3.2, can also represent discharges at the same current drawn but at progressively reduced temperatures. The specific characteristics vary for each battery system and discharge rate, but generally best performance is obtained between 20 and 40°C. At higher temperatures, chemical deterioration may be rapid enough during discharge to cause a loss in capacity.

Size. Battery size influences the voltage characteristics by its effect on current density. A given current drain may be a severe load on a small battery, giving a discharge similar to curve 4 or 5 (Fig. 5.3.2), but be a mild load to a larger battery and give a discharge similar to curve 3. It is often possible to obtain more than a proportional increase in the service life by increasing the size of the battery. The absolute value of current, therefore, is not the key influence, although its relation to the size of the battery, i.e., the current density, is important.

Age and Storage Condition. Batteries are a perishable product and deteriorate as a result of chemical action that proceeds during storage. The type of battery, design, temperature, and length of storage period are factors that affect the shelf life of the battery. Since shelf-discharge proceeds at a lower rate at reduced temperatures, refrigerated storage extends the shelf life. Refrigerated batteries should be warmed before discharge to obtain maximum capacity.

General Characteristics

The many and varied requirements for battery power and the multitude of environmental and electrical conditions under which they must operate necessitate the use of a number of different battery types and designs, each having superior performance under certain discharge conditions. In addition to the theoretical values, the practical performance characteristics of the major primary and secondary batteries are listed in Table 5.3.1. A comparison of the performance of some typical batteries is given in Fig. 5.3.3.

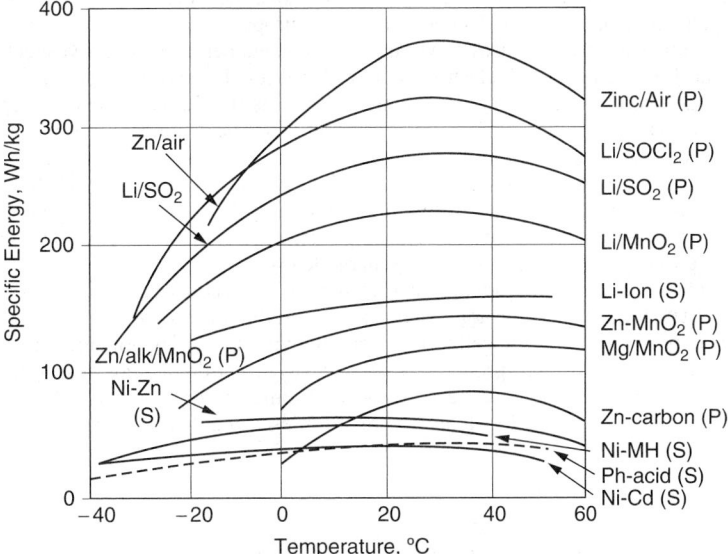

FIGURE 5.3.3 Comparative performance of battery system; primary (P), secondary (S).
Source: D. Linden and T. B. Reddy (eds.), "Handbook of Batteries," 3rd ed. McGraw-Hill, 2001.

TYPES OF BATTERIES

Batteries are generally identified as primary or secondary.

Primary batteries are not capable of being easily recharged electrically and hence are used or discharged a single time and discarded. Many of the primary batteries, in which the electrolyte is contained by absorbent or separator materials (i.e., there is no free or liquid electrolyte), are termed *dry cells.*

Secondary or rechargeable batteries are those that are capable of being recharged electrically, after discharge, to their original condition by passing current through them in the opposite direction to that of the discharge current. They are electric-energy storage devices and are also known as storage batteries or accumulators.

Reserve batteries are primary types in which a key component is separated from the rest of the battery. In the inert condition, the battery is capable of long-term storage. Usually, the electrolyte is the component that is isolated. The battery is activated by introducing this component into the cell structure. In other systems, such as the thermal battery, the battery is inactive until it is heated, melting a solid electrolyte, which then becomes conductive.

PRIMARY BATTERIES

The general advantages of primary batteries are reasonably good shelf life; high energy densities at low to moderate rates; little, if any, maintenance; and ease of use. Typical theoretical and practical characteristics of these batteries are shown in Table 5.3.1 and applications in Table 5.3.2.

Leclanché Battery (Zn/MnO$_2$)

The Leclanché dry cell, known for over 100 years, is still one of the more widely used worldwide but has given way to the popular alkaline cell in the United States and Europe.

The most common construction uses a cylindrical zinc container as the negative electrode and manganese dioxide as the positive element, with electrical connection through a carbon electrode. A paste or paper separator separates the two electrodes. Another common design is the flat cell, used only in multicell batteries, which offers better volume utilization. The Leclanché cell is fabricated in a number of sizes of varying diameter (or cross section) and height.

The capacity varies with different discharge conditions (higher capacities are available at lower current or power densities) to the point where shelf deterioration during the long discharge causes a loss of capacity. Performance is usually better under intermittent discharge, as the cell is given an opportunity to recover. This effect is particularly noticeable at the heavier discharge loads.

Leclanché cells operate best at normal temperatures (20°C). The energy output of the cell increases with higher operating temperatures, but prolonged exposure to high temperatures (50°C) will cause rapid deterioration. The capacity of the Leclanché cell falls off rapidly with decreasing temperature and is essentially zero below −20°C. The effects are more pronounced at heavier current drains. For best operation at low ambient temperatures, the Leclanché cell should be kept warm by some appropriate means. On the other hand there is an advantage in storing batteries at low temperatures for preserving their capacity.

Zinc Chloride Battery

A modification of the Leclanché battery is the zinc chloride battery. The construction is similar to the conventional carbon-zinc cell but the electrolyte contains only zinc chloride, without the saturated solution of ammonium chloride. The zinc chloride battery is a high-performance battery with improved high-rate and low-temperature performance and a reduced incidence of leakage.

TABLE 5.3.2   Major Characteristics and Applications of Primary Batteries

System	Characteristics	Applications
Zinc-carbon (Leclanché) Zinc/MnO_2	Common, low-cost primary battery, available in a variety of sizes	Flashlight, portable radios, toys, novelties, instruments
Magnesium (Mg/MnO_2)	High-capacity primary battery, long shelf life	Military receiver-transmitters, aircraft emergency transmitters
Mercury (Zn/HgO)	Highest capacity (by volume) of conventional types, flat discharge, good shelf life	Hearing aids, medical devices (pacemakers), photography, detectors, military equipment but in limited use due to environmental hazard of mercury
Mercad (Cd/HgO)	Long shelf life, good low- and high-temperature performance, low energy density	Special applications requiring operation under extreme temperature conditions and long life, in limited use
Alkaline (Zn/alkaline/MnO_2)	Most popular general-purpose premium battery, good low-temperature and high-rate performance, moderate cost	Most popular primary-battery, used in a variety of portable battery operated equipments
Silver/zinc (Zn/Ag_2O)	Highest capacity (by weight) of conventional types, flat discharge, good shelf life, costly	Hearing aids, photography, electric watches, missiles, underwater, and space application (larger sizes)
Zinc/air (Zn/O_2)	Highest energy density, low cost, not independent of environmental conditions	Special applications, hearing aids, pagers, medical devices, portable electronics
Lithium/soluble cathode	High energy density, long shelf life, good performance over wide temperature range	Wide range of applications (capacity from 1 to 10,000 Ah) requiring high energy density, long shelf life, e.g., from utility meters to military power applications
Lithium/solid cathode	High energy density, good rate capability and low-temperature performance, long shelf life, competitive cost	Replacement for conventional button and cylindrical cell applications
Lithium/solid electrolyte	Extremely long shelf life, low-power battery	Medical electronics, memory circuits, fusing

Source: D. Linden and T. B. Reddy (eds.) "Handbook of Batteries," 3rd ed., McGraw-Hill, 2001.

Magnesium Primary Batteries (Mg/MnO_2)

The magnesium battery was developed for military use and has two principal advantages over the zinc dry cell: (1) it has twice the capacity or service life of an equivalently sized zinc cell, and (2) it can retain this capacity during storage in an undischarged condition. The construction of the magnesium dry cell is similar to that of the cylindrical zinc cell, except that a magnesium can is used instead of the zinc container. The magnesium cell has a mechanical vent for the escape of hydrogen gas, which forms as a result of a parasitic reaction during the discharge of the battery. Magnesium batteries have not been fabricated successfully in flat-cell designs.

The good shelf life of the magnesium battery results from a film that forms on the inside of the magnesium can, preventing corrosion. This film, however, is responsible for a delay in the battery's ability to deliver full output voltage after it is placed under load. Nevertheless, there has been little commercial interest in this battery because of the generation of hydrogen and the relatively poor storageability of a partially discharged cell.

Zinc/Mercuric Oxide Battery (Zn/HgO)

The zinc/mercuric oxide battery was another important zinc anode primary system. This battery was developed during World War II for military communication applications because of its good shelf life and high volumetric energy density. In the postwar period, it was used in small button, flat, or cylindrical configurations as the power source in electronic watches, calculators, hearing aids, photographic equipment, and similar applications requiring a reliable long-life miniature power source. In the past decade, the use of the mercuric oxide battery has about ended mainly because of environmental problems associated with mercury and with its replacement by other battery systems, such as the zinc/air and lithium batteries, which have superior performance for many applications.

Alkaline-MnO$_2$ Cell (Zn/MnO$_2$)

In the past decade, an increasing portion of the primary battery market has shifted to the Zn/alkaline/MnO$_2$ battery. This system has become the battery of choice because of its superior performance at the higher current drains and low temperatures and its better shelf life. While more expensive than the Leclanché battery on a unit basis, it is more cost-effective for those applications requiring the high-rate or low-temperature capability, where the alkaline battery can outperform the Leclanché battery by a factor of 2 to 10. In addition, because of the advantageous shelf life of the alkaline cell, it is often selected for applications in which the battery is used intermittently and exposed to uncontrolled storage conditions (such as consumer flashlights and smoke alarms), but must perform dependably when required. Most recent advances have been the design of batteries providing improved high rate performance for use in cameras and other consumer electronics requiring this high power capability.

Cadmium/Mercuric Oxide Battery (Cd/HgO)

The substitution of cadmium for the zinc anode (the cadmium/mercuric oxide cell) results in a lower-voltage but very stable system, with a shelf life of up to 10 years as well as performance at high and low temperatures. Because of the lower voltage, the watt-hour capacity of this battery is about 60 percent of the zinc/mercuric oxide battery capacity. Again, because of the hazardous characteristics of mercury and cadmium, the use of this battery is limited.

Zinc/Silver Oxide Battery (Zn/AgO)

The primary zinc/silver oxide battery is similar in design to the small zinc/mercuric oxide button cell, but it has a higher energy density (on a weight basis) and performs better at low temperatures. These characteristics make this battery system desirable for use in hearing aids, photographic applications, and electronic watches. However, because of its high cost and the development of other battery systems, the use of this battery system, as a primary battery, has been limited mainly to small button battery applications where the higher cost is justified.

Lithium Primary Batteries

The lithium battery is the most recent development with a number of advantages over other primary-battery systems. Lithium is an attractive anode because of its reactivity, light weight, and high voltage; cell voltages range between 2 and 3.6 V, depending on the cathode material.

The advantages of the lithium battery include high energy density (see Fig. 5.3.3), high power density, flat discharge characteristics; excellent service over a wide temperature range, down to –40°C or below. Excellent shelf life of 5 to 10 years at normal room temperatures is expected.

Nonaqueous solvents must be used as the electrolyte because of the solubility of lithium in aqueous solutions. Organic solvents, such as acetonitrile and propylene carbonate, and inorganic solvents, such as thionyl chloride, are typical. A compatible solute is added to provide the necessary electrolyte conductivity. A number of different materials (sulfur dioxide, carbon mono-fluoride, vanadium pentoxide, copper sulfide) are used as the active cathode material. Hence the name lithium battery refers to a family of battery systems, ranging in size from 100 mAh to 10,000 Ah; they all use lithium as the anode but differ widely in design and chemical action.

The lithium primary batteries can be classified into three categories. The smallest are the low-power solid-state batteries with excellent shelf life, and are used in applications such as cardiac pacemakers and battery backup for volatile computer memory, where reliability and long shelf life are paramount requirements. In the second category are the solid-cathode batteries, which are designed in coin or small cylindrical configurations. These batteries have replaced the conventional primary batteries in watches, calculators, memory circuits, photographic equipment, communication devices, and other such applications where its high energy density and long shelf life are critical. The soluble-cathode batteries (using gases or liquid cathode materials) constitute the third category. These batteries are typically constructed in a cylindrical configuration, as flat disks, or in prismatic containers using flat plates. These batteries, up to about 35 Ah in size, are used in military and industrial applications, lighting products, and other devices where small size, low weight, and operation over a wide temperature range are important. The larger batteries have been used in special military applications or as standby emergency power sources.

Air Depolarized Batteries

These batteries, particularly the zinc/air battery system, are noted for their high energy density, but they had been used only in large low-power batteries for signaling and navigational-aid applications. With the development of improved air electrodes, the high-rate capability of the system was improved and small button-type batteries are now used widely in hearing aids, electronics, and similar applications. These batteries have a very high energy density as no active cathode material is needed. Wider use of this system and the development of larger batteries have been slow because of some of their performance limitations (sensitivity to extreme temperatures, humidity and other environmental factors, as well as poor activated shelf life and low power density). Nevertheless, because of their attractive energy density, zinc/air and other metal/air batteries are now being seriously considered for a number of applications from portable consumer electronics and eventually for larger devices such as electric vehicles, possibly in a reserve or mechanically rechargeable configuration.

Solid Electrolyte Batteries

Most batteries depend on the ionic conductivity of liquid electrolytes for their operation. The solid electrolyte batteries depend on the ionic conductivity of an electronically nonconductive salt in the solid state, for example, Li^+ ion mobility in lithium iodide. Cells using these solid electrolytes are low-power (microwatt) devices but have an extremely long shelf life and can operate over a wide temperature range. The absence of liquid eliminates corrosion and gassing and permits the use of a hermetically sealed cell. The solid electrolyte batteries are used in medical electronics (in devices such as heart pacemakers), for memory circuits, and other such applications requiring a long-life, low-power battery.

SECONDARY (STORAGE) BATTERIES

Secondary batteries are characterized, in addition to their ability to be recharged, by high-power density, high discharge rate, flat discharge curves, and good low-temperature performance. Their energy densities are usually lower than those of primary batteries. Tables 5.3.1 and 5.3.3 list the characteristics and applications of secondary batteries.

The applications of secondary batteries fall into two major categories: (1) applications where the secondary battery is used essentially as a primary battery but recharged after use; secondary batteries are used in this manner for convenience (as in portable electronic devices), for cost savings (as they can be recharged rather than

TABLE 5.3.3 Major Characteristics and Applications of Secondary Batteries

System	Characteristics	Applications
Lead-acid: Automotive	Popular, low-cost secondary battery, moderate specific-energy, high-rate, and low-temperature performance, maintenance-free designs	Automotive SLI, golf carts, lawn mowers, tractors, aircraft, marine
Traction (motive power)	Designed for deep 6–9 h discharge, cycling service	Industrial trucks, materials handling, electric and hybrid electric vehicles, special types for submarine power
Stationary	Designed for standby float service, long life, VRLA designs	Emergency power, utilities, telephone, UPS, load leveling, energy storage, emergency lighting
Portable	Sealed, maintenance-free, low cost, good float capability, moderate cycle life	Portable tools, small appliances and devices, TV and portable electronic equipment
Nickel-cadmium: Industrial and FNC	Good high-rate, low-temperature capability, flat voltage, excellent cycle life	Aircraft batteries, industrial and emergency power applications, communication equipment
Portable	Sealed, maintenance-free, good high-rate low-temperature performance, excellent cycle life	Railroad equipment, consumer electronics, portable tools, pagers, appliances, and photographic equipment, standby power, memory backup
Nickel-metal hydride	Sealed, maintenance-free, higher capacity than nickel-cadmium batteries	Consumer electronics and other portable applications, electric and hybrid electric vehicles
Nickel-iron	Durable, rugged construction, long life, low specific energy	Materials handling, stationary applications, railroad cars
Nickel-zinc	High specific energy, extended cycle life and rate capability	Bicycles, scooters, trolling motors
Silver-zinc	Highest specific energy, very good high-rate capability, low cycle life, high cost	Lightweight portable electronic and other equipment, training targets, drones, submarines, other military equipment, launch vehicles and space probes
Silver-cadmium	High specific energy, good charge retention, moderate cycle life, high cost	Portable equipment requiring a lightweight, high-capacity battery, space satellites
Nickel-hydrogen	Long cycle life under shallow discharge, long life	Primarily for aerospace applications such as LEO and GEO satellites
Ambient-temperature rechargeable "primary" types [Zn/ MnO_2]	Low cost, good capacity retention, sealed and maintenance-free, limited cycle life and rate capability	Cylindrical cell applications, rechargeable replacement for zinc-carbon and alkaline primary batteries, consumer electronics (ambient-temperature systems)
Lithium ion	High specific energy and energy density, long cycle life	Portable and consumer electronic equipment, electric vehicles, and space applications

Source: D. Linden and T. B. Reddy (eds.), "Handbook of Batteries," 3rd ed., McGraw-Hill, 2001.

replaced), or for power drains beyond the level of primary batteries; (2) applications where the secondary battery is used as an energy-storage device, being charged by primary energy source and delivering its energy to the load on demand. Examples are automotive and aircraft systems, emergency and standby power sources, and hybrid electric vehicles.

Lead-Acid Batteries

This is the most widely used and economical secondary battery. It uses sponge lead for the negative electrode, lead oxide for the positive, and a sulfuric acid solution for the electrolyte. As the cell discharges, the active materials are converted into lead sulfate and the sulfuric acid solution is diluted; i.e., its specific gravity decreases. On charge the reverse actions take place. A key factor is its low cost, good performance, and cycle life.

The lead-acid battery is designed in many configurations, from small sealed cells with a capacity of 1 Ah to large cells, up to 12,000 Ah. The automotive SLI battery is by far the most popular and the one in widest use. Most significant of the advances in SLI battery design are the use of lighter-weight plastic containers, the improvement in shelf life, the "dry-charge" process, and the "maintenance-free" design. The latter, using calcium-lead or low-antimony grids, has greatly reduced water loss during charging (minimizing the need to add water) and has reduced the self-discharge rate so that batteries can be shipped or stored in a wet, charged state for relatively long periods.

The lead-acid industrial storage batteries are generally larger than the SLI batteries, with a stronger, higher-quality construction. Applications of the industrial batteries fall in several categories. The motive power traction types are used in materials-handling trucks, tractors, mining vehicles, and, to a limited extent, golf carts and personnel carriers, although the majority in use are automotive-type batteries. A second category is diesel locomotive engine starting and the rapid-transit batteries, replacing the nickel-iron battery in the latter application. Significant advances are the use of lighter-weight plastic containers in place of the hard-rubber containers, better seals, and changes in the tubular positive-plate designs. Another category is stationary service: telecommunications systems, electric utilities for operating power distribution controls, emergency and standby power systems, uninterruptible power systems (UPS), and in railroads, signaling and car power systems.

The industrial batteries use three different types of positive plates: tubular and pasted flat plates for motive power, diesel engine cranking, and stationary applications, and Planté designs, forming the active materials from pure lead, mainly in the stationary batteries. The flat-plate batteries use either lead-antimony or lead-calcium grid alloys. A relatively recent development for the telephone industry has been the "round cell," designed for trouble-free long-life service. This battery uses plates, conical in shape with pure lead grids, which are stacked one above the other in a cylindrical cell container, rather than the normal prismatic structure with flat, parallel plates.

An important development in lead-acid battery technology is the valve-regulated lead-acid (VRLA) battery. These batteries operate on the principle of oxygen recombination, using a "starved" or immobilized electrolyte. The oxygen generated at the positive electrode during charge can, in these battery designs, diffuse to the negative electrode, where it can react, in the presence of sulfuric acid, with the freshly formed lead. The VRLA design reduces gas emission by over 95 percent as the generation of hydrogen is also suppressed. Oxygen recombination is facilitated by the use of a pressure-relief valve, which is closed during normal operation. When pressure builds up, the valve opens at a predetermined value, venting the gases. The valve reseals before the cell pressure decreases to atmospheric pressure. The VRLA battery is now used in about 70 percent of the telecommunication batteries and in about 80 percent of the uninterrupted power source (UPS) applications.

Smaller sealed lead-acid cells are used in emergency lighting and similar devices requiring backup power in the event of a utility power failure, portable instruments and tools, and various consumer-type applications. These small sealed lead-acid batteries are constructed in two configurations, prismatic cells with parallel plates, ranging in capacity from 1 to 30 Ah, and cylindrical cells similar in appearance to the popular primary alkaline cells and ranging in capacity up to 25 Ah. The acid electrolyte in these cells is either gelled or absorbed in the plates and in highly porous separators so they can be operated virtually in any position without the danger of leakage. The grids generally are of lead-calcium-tin alloy; some use grids of pure lead or a lead-tin alloy. The cells also include the features for oxygen recombination and are considered to be VRLA batteries.

Lead-acid batteries also are used in other types of applications, such as in submarine service, for reserve power in marine applications, and in areas where engine-generators cannot be used, such as indoors and in mining equipment. New applications, to take advantage of the cost-effectiveness of this battery, include load leveling for utilities and solar photovoltaic systems. These applications will require improvements in the energy and power density of the lead-acid battery.

A lead-acid battery can be charged at any rate that does not produce excessive gassing or high temperatures. The most common practice for recharging a fully discharged battery is to start the charge at the $C/5$ rate (amperes, corresponding to one-fifth the rated capacity of the battery) for 5 h, which will return about 80 to 85 percent of the battery's capacity. The charging current is then tapered to about one-half to complete the charge. In an emergency, fast or boost charging is used. In this type of charge, the current should not exceed the C rate or the battery can suffer damage.

The battery can also be *float*- or *trickle*-charged when it is continuously connected to an electrical system. The current should be regulated at a low level to maintain the battery in a charged condition (sufficient just to replace capacity loss because of stand) and to prevent overcharging. The battery manufacturers can supply detailed charging instructions.

Alkaline Secondary Batteries

Most of the other conventional types of secondary batteries use an aqueous alkaline solution (KOH or NaOH) as the electrolyte. Electrode materials are less reactive with alkaline electrolytes than with acid electrolytes. Furthermore, the charge-discharge mechanism in the alkaline electrolyte involves only the transport of oxygen or hydroxyl ions from one electrode to the other; hence the composition or concentration of the electrolyte does not change during charge and discharge.

Nickel-Cadmium Batteries. The nickel-cadmium secondary battery is the most popular alkaline secondary battery and is available in several cell designs and in a wide range of sizes. The original cell design used the pocket-plate construction. The vented pocket-type cells are very rugged and can withstand both electrical and mechanical abuse. They have very long lives and require little maintenance beyond occasional topping with water. This type of battery is used in heavy-duty industrial applications such as materials-handling trucks, mining vehicles, railway signaling, emergency or standby power, and diesel engine starting. The sintered-plate construction is a more recent development, having higher energy density. It gives better performance than the pocket-plate type at high discharge rates and low temperatures but is more expensive. It is used in applications, such as aircraft engine starting and communications and electronics equipment, where the lighter weight and superior performance are required. Higher energy and power densities can be obtained by using nickel foam, nickel fiber, or plastic-bonded (pressed-plate) electrodes. The sealed cell is a third design. It uses an oxygen-recombination feature similar to the one used in sealed lead-acid batteries to prevent the buildup of pressure caused by gassing during charge. Sealed cells are available in prismatic, button, and cylindrical configurations and are used in consumer and small industrial applications.

Nickel-Iron Batteries. The nickel-iron battery was important from its introduction in 1908 until the 1970s, when it lost its market share to the industrial lead-acid battery. It was used in materials-handling trucks, mining and underground vehicles, railroad and rapid-transit cars, and in stationary applications. The main advantages of the nickel-iron battery, with major cell components of nickel-plated steel, are extremely rugged construction, long life, and durability. Its limitations, namely, low specific energy, poor charge retention, and poor low-temperature performance, and its high cost of manufacture compared with the lead-acid battery led to a decline in usage.

Silver Oxide Batteries. The silver-zinc (zinc/silver oxide) battery is noted for its high energy density, low internal resistance desirable for high-rate discharge, and a flat second discharge plateau. This battery system is useful in applications where high energy density is a prime requisite, such as electronic news gathering equipment, submarine and training target propulsion, and other military and space uses. It is not employed for general storage battery applications because its cost is high, its cycle life and activated life are limited, and its performance at low temperatures falls off more markedly than with other secondary battery systems.

The silver-cadmium (cadmium/silver oxide) battery has significantly longer cycle life and better low-temperature performance than the silver-zinc battery but is inferior in these characteristics compared with the nickel-cadmium battery. Its energy density, too, is between that of the nickel-cadmium and the silver-zinc batteries. The battery is also very expensive, using two of the more costly electrode materials. As a result, the silver-cadmium battery was never developed commercially but is used in special applications, such as nonmagnetic batteries and space applications. Other silver battery systems, such as silver-hydrogen and silver-metal hydride couples, have been the subject of development activity but have not reached commercial viability.

Nickel-Zinc Batteries.   The nickel-zinc (zinc/nickel oxide) battery has characteristics midway between those of the nickel-cadmium and the silver-zinc battery systems. Its energy density is about twice that of the nickel-cadmium battery, but the cycle life previously has been limited because of the tendency of the zinc electrode toward shape change, which reduces capacity and dendrite formations, which cause internal short-circuiting.

Recent development work has extended the cycle life of nickel-zinc batteries through the use of additives in the negative electrode in conjunction with the use of a reduced concentration of KOH to repress zinc solubility in the electrolyte. Both of these modifications have extended the cycle life of this system so that it is now being marketed for use in electric bicycles, scooters, and trolling motors in the United States and Asia.

Hydrogen Electrode Batteries.   Another secondary battery system uses hydrogen for the active negative material (with a fuel-cell-type electrode) and a conventional positive electrode such as nickel oxide. These batteries are being used exclusively for the aerospace programs, which require long cycle life at low depth of discharge. The high cost of these batteries is a disadvantage, which limits their application. A further extension is the sealed nickel/metal hydride battery where the hydrogen is absorbed, during charge, by a metal alloy forming a metal hydride. This metal alloy is capable of undergoing a reversible hydrogen absorption-desorption reaction as the battery is charged and discharged, respectively. The advantage of this battery is that its specific energy and energy density are significantly higher than that of the nickel-cadmium battery. The sealed nickel-metal hydride battery, manufactured in small prismatic and cylindrical cells, is being used for portable electronic applications and are being employed for other applications including hybrid electric vehicles. Larger sizes are finding use in electric vehicles.

Zinc/Manganese Dioxide Batteries.   Several of the conventional primary battery systems have been manufactured as rechargeable batteries, but the only one currently being manufactured is the cylindrical cell using the zinc/alkaline-manganese dioxide chemistry. Its major advantage is a higher capacity than the conventional secondary batteries and a lower initial cost, but its cycle life and rate capability are limited.

Advanced Secondary Batteries

A number of battery systems are being studied for such applications as electric vehicles and utility power which require a high specific energy (in the order of 200 Wh/kg), high power density and low cost. These include: (1) high temperature systems using molten salt electrolytes, (2) high temperature systems using ceramic or glass electrolytes, (3) aqueous zinc halogen batteries. Some of this work, particularly for electric vehicles, has been deemphasized in favor of lithium-ion and other conventional batteries operating at ambient temperatures, which are now demonstrating energy and power capabilities approaching those of high temperature batteries.

Rechargeable Lithium Batteries

Rechargeable lithium batteries operating at room temperature offer several advantages compared to conventional aqueous technologies, including: (1) Higher energy density (up to 150 Wh/kg, 300 Wh/L), (2) Higher cell voltage (up to about 4 V per cell), and (3) Longer charge retention or shelf life (up to 5 to 10 years).

These advantageous characteristics result in part from the high standard potential and low electrochemical equivalent weight of lithium.

Ambient-temperature lithium rechargeable batteries, on the other hand, do not have the high-rate capability (because of the relatively poor conductivity of the aprotic organic or inorganic electrolytes that must be used because of the reactivity of lithium in aqueous electrolytes) nor, in some instances, the cycle life of conventional rechargeable batteries. In addition, the rechargeable lithium cells that use lithium metal as the negative electrode present potential safety problems, which are more challenging than those with primary lithium batteries.

This is because of a three- to fivefold excess of lithium that is required for these types of cells in order to obtain a reasonable cycle life and to the reactivity of the porous high surface area lithium that is formed during cycling with the electrolyte. For these reasons, these types of batteries have had little commercial success except for small, low energy coin batteries where an aluminum lithium alloy is generally used.

Another type of rechargeable lithium battery that has been recently developed and is achieving commercial success, is the "lithium-ion" battery. This battery does not use metallic lithium and thus should minimize the safety concerns. The lithium-ion battery uses lithium intercalation compounds as the positive and negative materials. As the battery is cycled, lithium ions exchange between the positive and negative electrodes. The negative electrode material is typically a graphite carbon, a layered material. The positive material is typically a metal oxide, also with a layered structure, such as lithium cobalt oxide or lithium manganese dioxide. In the charge/discharge process, lithium ions are inserted or extracted from the interstitial space between atomic layers within the active material. Liquid electrolytes, using lithium salts in organic solvents, are the most common electrolytes. There is increasing interest and activity with polymer electrolytes as these allow for a greater flexibility in cell design and better safety characteristics.

Lithium-ion batteries are presently manufactured in a variety of designs in sizes as small as 0.1 Ah to large units of about 100 Ah but with the major emphasis on consumer electronic applications such as laptop computers, camcorders, and cell phones. Various cell designs have been developed including cylindrical, wound prismatic, and flat plate prismatic configurations. The lithium-ion batteries have a number of advantages, including a high cell voltage, a high specific energy and energy density, a competitive specific power, long cycle life, and a broad operating temperature range. Lithium-ion batteries should not be overcharged nor overdischarged. Electronic management circuits or mechanical disconnect devices are usually used with lithium-ion batteries to provide protection from overcharge, overdischarge, and over temperature conditions.

It is expected that the market share enjoyed by the lithium rechargeable battery will increase significantly during the next decade because of its high energy density and its ability to meet consumer demands for lighter weight and longer service more adequately than the conventional aqueous rechargeable batteries. Larger cells also are being considered for a number of military and commercial applications, including underwater propulsion and electric vehicles.

RESERVE BATTERIES

Batteries that use highly active component materials to obtain the required high energy, high power, and/or low-temperature performance, are often designed in a reserve construction to withstand deterioration in storage and to eliminate self-discharge prior to use. These batteries are used primarily to deliver high power for relatively short periods of time after activation in such applications as radiosondes, fuzes, missiles, torpedoes, and other weapon systems. The reserve design also is used for batteries required to meet extremely long or environmentally severe storage requirements.

In the reserve structure, one of the key components of the cells is separated from the remainder of the cell until activation. In this inert condition, chemical reaction between the cell components (self-discharge) is prevented, and the battery is capable of long-term storage. The electrolyte is the component that is usually isolated, although in some water-activated batteries the electrolyte solute is contained in the cell and only water is added.

The reserve batteries can be classified by the type of activating medium or mechanism that is involved in the activation:

Water-activated batteries: Activation by fresh or seawater.

Electrolyte-activated batteries: Activation by the complete electrolyte or with the electrolyte solvent. The electrolyte solute is contained in or formed in the cell.

Gas-activated batteries: Activation by introducing a gas into the cell. The gas can be either the active cathode material or part of the electrolyte.

Heat-activated batteries: A solid salt electrolyte is heated to the molten condition and becomes ionically conductive, thus activating the cell. These are known as thermal batteries.

Activation of the reserve battery is accomplished by adding the missing component just prior to use. In the simplest designs, this is done by manually pouring or adding the electrolyte into the cell or placing the battery in the electrolyte (as in the case of seawater-activated batteries). In more sophisticated applications the electrolyte storage and the activation mechanism are contained within the overall battery structure, and the electrolyte is brought automatically to the active electrochemical components by remotely activating the activation mechanism. Thermal batteries are activated by igniting a heat source that is contained within the battery.

FUEL CELLS

Fuel cells, such as batteries, are electrochemical galvanic cells that convert chemical energy directly into electrical energy and are not subject to the Carnot cycle limitations of heat engines. Fuel cells are similar to batteries except that the active materials are not an integral part of the device (as in a battery), but are fed into the fuel cell from an external source when power is desired. The fuel cell differs from a battery in that it has the capability of producing electrical energy as long as the active materials are fed to the electrodes (assuming the electrodes do not fail). The battery will cease to produce electrical energy when the limiting reactant stored within the battery is consumed.

The electrode materials of the fuel cell are inert in that they are not consumed during the cell reaction, but have catalytic properties that enhance the electroreduction or electrooxidation of the reactants (the active materials).

The anode active materials used in fuel cells are generally gaseous or liquid (compared with the metal anodes generally used in most batteries) and are fed into the anode side of the fuel cell. As these materials are more like the conventional fuels used in heat engines, the term "fuel cell" has become popular to describe these devices. Oxygen or air is the predominant oxidant and is fed into the cathode side of the fuel cell.

TABLE 5.3.4 Types of Fuel Cells

---

1. Solid Oxide (SOFC): These cells use a solid oxygen-ion-conducting metal oxide electrolyte. They operate at about 1000°C, with an efficiency of up to 60 percent. They are slow to start up, but once running, provide high grade waste heat that can be used to heat buildings. They may find application in industrial and large-scale applications.

2. Molten Carbonate (MCFC): These cells use a mixed alkali-carbonate molten salt electrolyte and operate at about 600°C. They are being developed for continuously operating facilities, and can use coal-based or marine diesel fuels.

3. Phosphoric Acid (PAFC): This is the most commonly used type of fuel cell for stationary commercial sites such as hospitals, hotels, and office buildings. The electrolyte is concentrated phosphoric acid. The fuel cell operates at about 200°C. It is highly efficient and can generate energy at up to 85 percent (40 percent as electricity and another 45 percent if the heat given off is also used).

4. Alkaline (AFC): These are used by NASA on the manned space missions, and operate well at about 200°C. They use alkaline potassium hydroxide as the electrolyte and can generate electricity up to 70 percent efficiency. A disadvantage of this system is that it is restricted to fuels and oxidants that do not contain carbon dioxide.

5. Proton Exchange Membrane (PEM): These cells use a perfluorinated ionomer polymer membrane electrolyte that passes protons from the anode to the cathode. They operate at a relatively low temperature (70 to 85°C), and are especially notable for their rapid start-up time. These are being developed for use in transportation applications and for portable and small fuel cells.

6. Direct Methanol (DMFC): These fuel cells directly convert liquid methanol (methyl alcohol) in an aqueous solution that is oxidized at the anode. Like PEMs, these also use a membrane electrolyte, and operate at similar temperatures. This fuel cell is still in the development stage.

7. Regenerative (RFC): These are closed-loop generators. A powered electrolyzer separates water into hydrogen and oxygen, which are then used by the fuel cell to produce electricity and exhaust (water). That water can then be recycled into the powered electrolyzer for another cycle.

Sources: (1) Connecticut Academy of Science and Engineering Reports, Vol. 15, No. 2000.
        (2) D. Linden and T. B. Reddy (eds.), "Handbook of Batteries," 3rd ed. McGraw-Hill, 2001.

Fuel cells have been of interest for over 150 years as a potentially more efficient and less polluting means for converting hydrogen and carbonaceous or fossil fuels to electricity compared to conventional engines. A well-known application of the fuel cell has been the use of the hydrogen/oxygen fuel cell, using cryogenic fuels, in space vehicles for over 40 years. Use of the fuel cell in terrestrial applications has been developing slowly, but recent advances has revitalized interest in air-breathing systems for a variety of applications, including utility power, load leveling, dispersed or on-site electric generators and electric vehicles.

Fuel cell technology can be classified into two categories:

1. Direct systems where fuels, such as hydrogen, methanol and hydrazine, can react directly in the fuel cell

2. Indirect systems in which the fuel, such as natural gas or other fossil fuel, is first converted by reforming to a hydrogen-rich gas that is then fed into the fuel cell

Fuel cell systems can take a number of configurations depending on the combinations of fuel and oxidant, the type of electrolyte, the temperature of operation, and the application, and so forth. Table 5.3.4 lists the various types of fuel cells distinguished by the electrolyte and operating temperature.

More recently, fuel cell technology has moved toward portable applications, historically the domain of batteries, with power levels from less than 1 to about 100 W, blurring the distinction between batteries and fuel cells. Metal/air batteries, particularly those in which the metal is periodically replaced, can be considered a "fuel cell" with the metal being the fuel. Similarly, small fuel cells, now under development, which are "refueled" by replacing an ampule of fuel can be considered a "battery."

Table 5.3.1 gives both theoretical and practical values of voltage, capacity, and specific energy for several fuel cell systems.

REFERENCE

D. Linden and T. B. Reddy (eds.), "Handbook of Batteries," 3rd ed., McGraw-Hill, 2001.

BIBLIOGRAPHY

Crompton, T. R. "Battery Reference Book," Butterworths, 1990.

Energy Storage Systems, U.S. DOE (periodical).

Journal of the Electrochemical Society, (periodical).

Kiehne, H. A. "Battery Technology Handbook," Dekker, 1989.

Linden, D. "Handbook of Batteries," 2nd ed., McGraw-Hill, 1995.

Pistoia, G. "Lithium Batteries," Elsevier, 1994.

Proceedings of the Annual Battery Conference on Applications and Advances, California State University.

Proceedings of the International Power Sources Symp, Academic Press, (annual).

Proceedings of the Power Sources Conference, The Electrochemical Society.

CHAPTER 5.4
RELAYS AND SWITCHES

J. Spergel, John M. Kikta

INTRODUCTION

The primary function of switches and relays is the transmission and control of electric current accomplished by mechanical contacting and actuating devices. In recent years, solid-state (nonmechanical) switching devices have come into wide use and their applications are extending rapidly.

ELECTROMAGNETIC RELAYS

The simplified diagram of a relay shown in Fig. 5.4.1a illustrates the basic elements that constitute an electromagnetic relay (EMR). In simplest terms, it operates through the use of a coil which, when energized, attracts the spring-loaded armature; this, in turn, moves a set of contacts to or from a set of stationary contacts. The most common EMR types are as follows:

General-purpose. Design, construction, operational characteristics, and ratings are adaptable to a wide variety of uses.

Latch-in. Contacts lock in either the energized or deenergized position until reset either manually or electrically.

Polarized (or polar). Operation is dependent on the polarity of the energizing current. A permanent magnet provides the magnetic bias.

Differential. Functions when the voltage, current, or power difference between its multiple windings reaches a predetermined value.

Telephone. An armature relay with an end-mounted coil and spring-pickup contacts mounted parallel to the long axis of the relay coil. Ferreeds are also widely used for telephone cross-point switches.

Stepping. Contacts are stepped to successive positions as the coil is energized in pulses: they may be stepped in either direction.

Interlock. Coils, armature, and contact assemblies are arranged so that the movement of one armature is dependent on the position of the other.

Sequence. Operates two or more sets of contacts in a predetermined sequence. (Motor-driven cams are used to open and close the contacts.)

Time-delay. A synchronous motor is used for accurate long time delay in opening and closing contacts. Armature-type relay uses a conducting slug or sleeve on the core to obtain delay.

Marginal. Operation is based on a predetermined value of coil current or voltage.

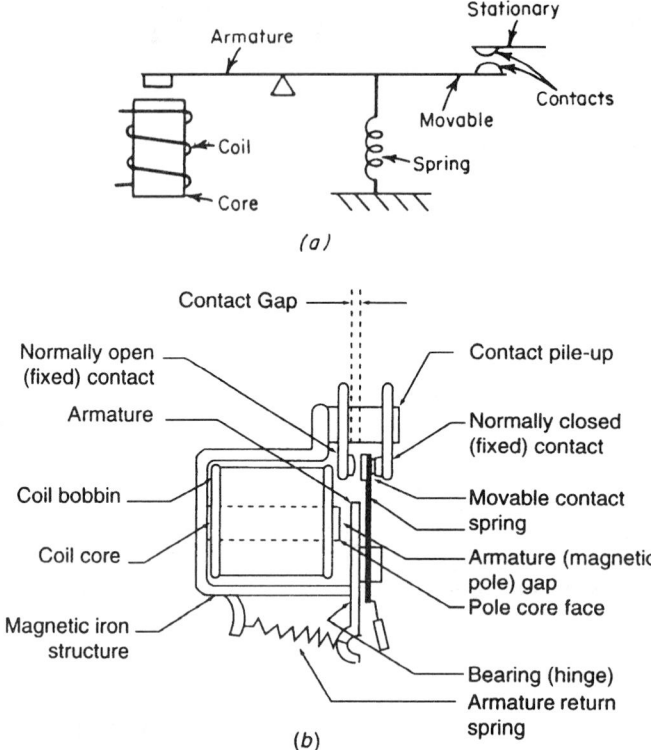

FIGURE 5.4.1   (*a*) Simplified diagram of a single-pole, double-throw, normally open rely; (*b*) structure of conventional relay (from NARM Engineers Relay Handbook, 4th ed.).

Performance Criteria

The design or selection of a relay should be based on the following circuit-performance criteria:

Operating frequency. Electrical operating frequency of relay coil

Rated coil voltage. Nominal operating voltage of relay coil

Rated coil current. Nominal operating current for relay

Nonoperate current (or voltage). Maximum value of coil current (or voltage) at which relay will not operate

Operate voltage (or current). Minimum value of coil voltage (or current) at which switching function is completed

Release voltage (or current). Value of coil voltage (or current) at which contacts return to the deenergized position

Operate time. Time interval between application of power to coil and completion of relay-switching function

Release time. Time interval between removal of power from coil and return of contacts to deenergized position

Contact bounce. Uncontrolled opening and closing of contacts due to forces within the relay

Contact chatter. Uncontrolled opening and closing of contacts due to external forces such as shock or vibration

Contact rating. Electrical load on the contacts in terms of closing surge current, steady-state voltage and current, and induced breaking voltage

Figure 5.4.2. illustrates some of the contacting characteristics during energization and deenergization.

General Design and Application Considerations

The dynamic characteristics of the moving system, i.e., armature and contact assembly, are primarily determined by the mass of the armature and depend on the magnet design and flux linkage. Typical armature configurations are clapper or balanced armature, hinged or pivoted lever about a fixed fulcrum; rotary armature; solenoid armature; and reed armature. Contact and restoring-force springs are attached or linked to the armature to achieve the desired make and/or break characteristics. The types of springs used for the contact assembly and restoring force are generally of the cantilever, coil, or helically wound spring type. Primary characteristics for spring materials are modulus of elasticity, fatigue strength, conductivity, and corrosion resistance. They should also lend themselves to ease of manufacture and low cost. Typical materials for springs are spring brass, phosphor bronze, beryllium copper, nickel silver, and spring steel.

Contacts.   These include stationary and moving conducting surfaces that make and/or break the electric circuit. The materials used depend on the application; the most common are palladium, silver, gold, mercury, and various alloys. Plated and overlaid surfaces of other metals such as nickel or rhodium are used to impart special characteristics such as long wear and arc resistance or to limit corrosion.

The heart of the relay is the contact system that is typically required to make and/or break millions of times and provide a low, stable electrical resistance. The mechanical design of the relay is aimed principally at achieving good contact performance. Because of the numerous operations and arcing often occurring during operation, the contacts are subject to a wide variety of hazards that may cause failure, such as:

Film formation. Effect of inorganic and organic corrosion, causing excessive resistance, particularly at dry-circuit conditions.

Wear erosion. Particles in contact area, which can cause bridging between small contact gaps.

Gap erosion. Metal transfer and welding of contacts.

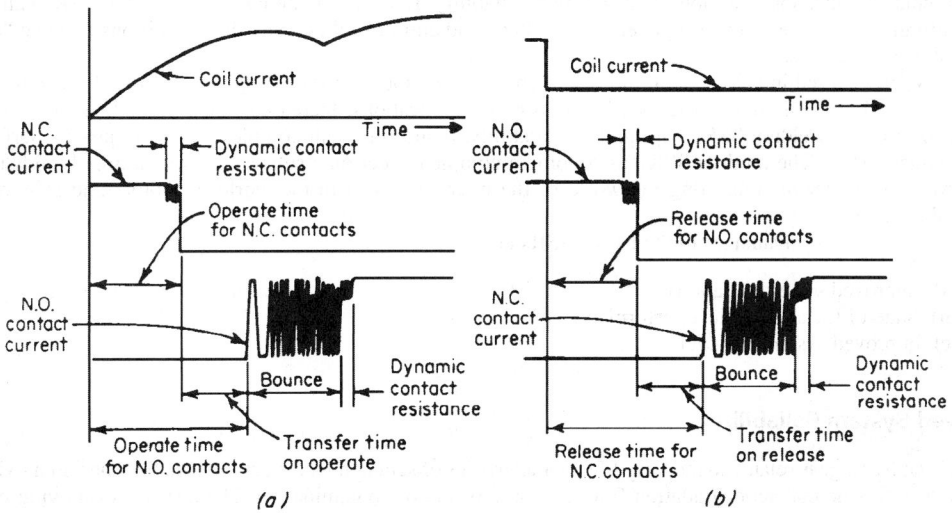

FIGURE 5.4.2 Typical oscillograph depictions of contacting characteristics during (*a*) energization and (*b*) deenergization of a relay. (*Automatic Electric, Northlake, IL.*)

Surface contamination. Dirt and dust particles on contact surfaces can prevent achievement of low resistance between contacts and may actually cause an open circuit.

Cold welding. Clean contacts in a dry environment will self-adhere or cold-weld.

One of the major factors in determining relay-contact life is the arcing that occurs at the contact surface during the period in which the contacts are breaking the circuit. Contact life (and hence relay reliability) can be greatly enhanced by the addition of appropriate contact-protection circuits. Such circuitry can reduce the effects of load transients, which are especially deleterious. A variety of circuits may be employed using bifilar coil windings, *RC* networks, diodes, varistors, and so forth. As a rule of thumb, for compact relays with operating speeds of 5 to 10 ms, the approximate parameters for suitable *RC* networks are approximately *R*, as low as possible but sufficient to limit the capacitor discharge current to the resistive load rating of the contacts, and *C*, a value in microfarads approximately equal to the steady-state load current in amperes.

Details on the effects of various methods of suppression are given in the *19th Relay Conference Proceedings.*

Packaging

Relays are packaged in a wide variety of contact arrangements (see Table 5.4.1) and in many package configurations ranging from T05 cans and crystal cans to relatively large enclosures, as well as plastic encapsulation, and open construction with plastic dust covers. The packaging adopted depends on the environment and reliability requirements. If the environment is controlled, as in a telephone exchange, a relatively inexpensive and simple package can be used. In a military or aerospace environment, hermetically sealed or plastic-encapsulated packages are essential, to prevent corrosion of contacts. In this regard the reed switch has had great impact on relay design since it provides a sealed-contact enclosure in a relatively simple and inexpensive package. It has become widely used in the telephone and electronics industry because it has been able to extend relay life to millions of cycles. A comparison between the reed and typical conventional relays is given in Table 5.4.2. Solid-state relays (SSRs) are also used extensively. Table 5.4.3 compares SSRs with electromagnetic relays.

SOLID-STATE RELAYS

An SSR performs a switching function using semiconductor devices and technologies to provide electrical isolation between the control circuit (input) and the load circuit (output) without electromechanical moving parts or contacts, thus ensuring long life and high reliability. It should be noted, however, that SSRs can be very unforgiving when encountering overloads, voltage and current spikes, and other conditions that can "destroy" solid-state devices.

SSRs are used in commercial, industrial, and military applications throughout the world including sophisticated avionics and microprocessor-based systems applications. Increased functional integration and miniaturization are yielding SSRs with greater reliability, reduced system complexity, and improved performance characteristics. The modern SSR has become a complete electronic subsystem, which greatly aids the electronics engineer in connecting the world of the microprocessor to the world of motors and solenoids. See Table 5.4.3.

The primary advantages to the use of SSRs are

(a) Improved system reliability
(b) State of the art technical performance
(c) Improved system life cycle costs

Improved System Reliability

This advantage is related to the long-life characteristics of semiconductor devices. An SSR has no moving parts and suffers no contact degradation from arcing, erosion, or contamination. The absence of moving parts and contact chatter coupled with the low mass of semiconductor devices and proper construction and packaging techniques results in a virtual immunity to shock and vibration effects.

TABLE 5.4.1 Symbols for Relay Contact Combinations Established by American National Standards Institute (ANSI)

Form	Description	ANSI symbol	Form	Description	ANSI symbol
A	Make or SPSTNO		L	Break, Make, Make, or SPDT (B-M-M)	
B	Break or SPSTNC		M	Single pole, Double throw, Closed Neutral SP DT NC (This is peculiar to MIL-SPECS)	
C	Break, Make, or SPDT (B-M), or Transfer		U	Double make, Contact on Arm. SP ST NO DM	
D	Make, Break or Make-Before-Break, or SPDT (M-B), "Continuity Transfer"		V	Doule break, Contact on Arm. SP ST NC DB	
E	Break, Make, Break, or Break, Make Before Break SPDT (B-M-B)		W	Double break, Double make, Contact on Arm. ST DT NC-NO (DB-DM)	
F	Make, Make SPST (M-M)		X*	Double make or SP ST NO DM	
G	Break, Break or SPST (B-B)		Y**	Double break or SP ST NC DB	
H	Break, Break, Make, or SPDT (B-B-M)		Z	Double break, Double make, SP DT NC-NO (DB-DM)	
I	Make, Break, Make, or SPDT (M-B-M)				
J	Make, Make, Break, or SPDT, (M-M-B)				
K	Single pole, Double throw Center off, or SPDTNO				

★ Not to be confused with preliminary ("X") make
★★ Not to be confused with a late ("Y") break

Special A	Timed close	T.C. OR T.C.
Special B	Timed open	T.O. OR T.O.

Multi-point selector switch   —OR—

The heavy arrow indicates the direction of operation. Contact chatter may cause some electrical discontinuity in forms D and E.

TABLE 5.4.2 Estimated Load-Life Capability of Two-Pole Miniature Relays

	Dry reed. 0.125 A	Conventional crystal can. 5 A	Miniature power, 10 A
λ at 1×10^6 (%/10^4)	0.002	1.2	0.80
$R_{(0.999)}(10^3$ operation)	700	2.8	6.0
$R_{(0.90)}(10^3$ operation)	10,000	60	120.0

λ = failure rate in percent/10,000 h operation; $R_{(0.999)}$ = operating life with 99.9 percent probability; $R_{(0.90)}$ = operating life with 90 percent probability.

The technical characteristics of an SSR can reduce system complexity. Isolation techniques can prevent a series of secondary system failures from rippling through a system when a primary failure occurs. The lower power requirements can also contribute to reduced system power consumption, reduced thermal effects, and downsizing of system power supplies.

State-of-the-Art Performance

Many SSRs can be driven by either CMOS or TTL logic with no need for buffers or relay driver stages. Constant current input circuitry allows a wide control voltage range without need for external current limiting resistors or sources. They also allow a wide range of switching speeds to match user needs. Absence of moving parts provides clean bounce-free switching. This characteristic, coupled with zero-voltage turn-on and zero-current turn-off techniques, results in low EMI/RFI levels that are far below FCC and VDE specifications. The introduction of solid-state power FET relays has provided a quantum leap in switching technology with low "on" resistance with virtually no bipolar offset voltage. This allows them to switch low-level microvolt analog signals that have always been beyond the capability of bipolar solid-state devices, as well as high power levels with significantly reduced power dissipation.

TABLE 5.4.3 Relative Comparison of Electromagnetic Relays (EMRs) vs. Solid-State Relays (SSRs)

Characteristic	EMRs	SSRs	Advantage
Life	From 100,000 to millions of cycles. Reed contacts are outstanding.	No moving parts. When properly designed should last life of equipment.	SSR
Isolation	Infinite dielectric isolation.	Not dielectrically isolated; however, several techniques are available to achieve up to 10 kΩ.	EMR
EMI (RFI)	Can generate EMI by switching of its coil, thereby requiring special isolation (i.e., shielding).	Noise generated is negligible compared with EMR.	SSR
Speed	Order of milliseconds.	Up to nanoseconds.	SSR
Operate power	Uses more power than SSR.	Lower power requirements but requires continuous standby power.	SSR
Contact voltage drop	Relatively low voltage drop because of low contact resistance.	High voltage drop which is dissipated into heat.	EMR
Thermal power dissipation	Primarily concerned with dissipating coil power.	Higher voltage drop develops appreciable heat to be dissipated.	EMR

Improved System Life Cycle Costs. This aspect stems from the SSRs' greater reliability and reduced failure rate, which yields lower production costs, lower system acquisition costs, and reduced field repair costs.

Anatomy of an SSR. Every SSR consists of at least three parts: (1) input or control circuitry, (2) isolation, and (3) output or load circuitry.

Input. The input circuitry is classified as dc, ac, or ac–dc, depending on the control requirements. Input control circuitry may also include such features as TTL/CMOS compatibility, positive or negative logic control, series/parallel digital interface, and so forth.

Isolation. The isolation between the input and output may be accomplished optically, magnetically (transformer), or, in special applications, using a mechanical relay (typically a reed relay) to drive the output device.

Output. Output circuitry is also classified as dc, ac, or ac–dc, and may use a variety of output devices, depending on the application. For ac operation a thyristor (2 SCRs or triac) is generally used. For dc applications a bipolar device or power FET transistor may be used. For ac–dc operation power FETs are used, although bipolar transistors may be used for currents up to 100 mA. Additionally, the output may be single or multiple pole, single or double throw, and may incorporate such features as time delay, latching, break-before-make logic, and so forth.

SWITCHES

Switches are electromechanical devices used to make and break, or select, an electrical circuit by manual or mechanical operation.

An examination of manufacturers' and distributors' catalogs makes it obvious that switches come in a bewildering array of types, shapes, sizes, actuator styles, ratings, and quality levels.

Beside the basic ability of a switch to make, break, and carry the required electrical load (including the inductive or capacitive, low level or logic level effects), switches can be characterized in additional ways, usually dependent on the sensitivities of the application, such as electrical noise, capacitance, frequency, contact make or break speed, contact bounce, and ability to withstand various environmental conditions.

Contact Make/Break Time and Bounce Time. Make time is the time required by a switch contact to travel from an open position to a closed position. Break time is the time required by a switch contact to travel from a closed position to a fully open position. Bounce time is the interval between an initial contact and steady contact during which the moving contact bounces or makes intermittent contact as a result of its impact on a stationary contact.

Electrical Noise and Capacitance. Electromagnetic radiations may occur during make and break, causing noise interference in sensitive circuits or high-gain amplifiers. Arc suppression may be necessary to reduce such noise to acceptable levels.

Selection Considerations. Once the basic switch type, size, and rating has been selected, the exact configuration must be considered and tailored for the application. Each particular switch type, and sometimes each manufacturer, has its own list of options and peculiarities, so manufacturers should be consulted at an early stage.

For rotary switches, the number of decks (sections), poles per deck, positions per pole, shorting and/or non-shorting contacts, and coding need to be considered, as well as, on the mechanical side, factors such as shaft and mounting bushing configuration, push-to-turn, pull-to-turn, and spring return features.

For toggle switches, contact configuration (single pole normally open, single pole normally closed, double throw, multiple poles, and so forth), the desirability of the bathandle configuration, and whether a lock is required, are among the factors to be considered.

Rocker switches are similar to toggles so similar factors apply. They also provide the option of illumination. Keypads and keyboards offer such a wide range of options that manufacturers should be consulted at the outset of the selection process.

For a simple on-off situation, a two-position rotary or two-position toggle, a push-push, or a push-pull switch are good choices. For data input applications, a numerical keypad and/or a keyboard is an obvious choice.

Application Notes.   Switch selection must not overlook the ergonomic factors of the application. When a rotary switch is among the options, its choice will help minimize the panel space required.

Pushbutton switches often require the operator to view them in order to operate them, whereas toggle and rocker switches can be operated by feel. Rotary switches can sometimes be operated by feel, with the practiced operator counting clicks to position the knob correctly. Numerical keypads can be operated by touch, but special keyboards, notably the flush type, require eye contact for correct operation.

Switches designed to carry high electrical loads are physically large in order to accommodate the larger contacts to handle the load and dissipate heat, while switches designed for low and logic level loads can be very small, like subminiature toggles and DIP switches.

Standards.   Switches used in appliances and equipment that handle line power will require a UL (Underwriters Laboratories) listing for use in the United States, a CSA listing for use in Canada, and other listings for use in other countries. These listings require the listing agency to run or observe special tests of the product to be listed, and continuous follow-up testing to prove that they can meet the necessary safety requirements. There are no universal tests or listing agencies although there are efforts being made in that direction.

Many switch standards are available from the Electronic Industries Association, the U.S. Department of Defense, Underwriters Laboratories, and others.

Ratings.   Switch manufacturers publish one or more ratings for their products. Each represents a point on a load versus life curve, usually developed from testing at room conditions unless otherwise noted. Although not always stated, each load-life curve represents a given set of life limiting criteria: typically a certain maximum allowable value of contact resistance, a minimum value of insulation resistance between adjacent contacts or between contacts and ground (mounting bushing, case, or similar reference point), and a minimum value of dielectric strength (breakdown voltage across insulation). Increasing the ambient temperature during the test will generally result in higher values of contact resistance and reduced insulation resistance, which can equate to shorter life. These should be weighed against the application requirements. Typically the use conditions are more lenient than the test conditions for the load-life curve. For instance, many electronic circuits are of the high impedance variety where contact resistances of 20 or 50 milliohms are insignificant, but where insulation resistance can be important.

Arcing.   Typical power loads will cause an arc to be drawn when a circuit is broken, which causes electrical erosion of the contacts. The erosion products are conductive, and if deposited on insulating surfaces, can result in a reduction of insulation resistance and dielectric strength. Higher currents produce hotter arcs and more erosion. Inductive loads also increase arcing. Fast making and breaking of contacts, as in snap-acting switch types, reduce the effect of arcing. Reduced barometric pressure, as in high altitudes, reduces the dielectric strength of air, which enhances the effect of arcing, so a switch should be derated for high-altitude operation.

Life tests for rating purposes are usually run on a continuous basis. This gives the contacts little time to cool between carrying high loads and breaking arcs, whereas typical applications require switches to operate infrequently, giving the contacts plenty of cooling time, and resulting in an application life that is longer than the test life.

Many switch applications in electronics do not switch power loads; they are more likely to handle low level or logic level loads. Typical low level tests are run at 30 mV and 10 mA, while logic level tests are run at 2 to 5 Vdc and 10 mA or less. Low levels can be troublesome because they are run below the melting voltage of the contact material and below the breakdown voltage of contact contaminants.

Contact pairs will be of the butt type, wiping type, or a combination of these basic types. Even small amounts of lateral wiping action between contact pairs can provide significant cleaning of the contact faces, thus minimizing contact resistance but creating some mechanical wear of the contact surfaces. Butt contacts will exhibit less wear but have higher contact resistance. Most rotary switches used in electronics have a considerable amount of wiping action, while sensitive or snap action switches and some types of toggle/rocker switches are closer to pure butt types (although most have a slight amount of wipe).

BIBLIOGRAPHY

Definitions of Relay Terms, Electronic Industries Association.

EIA-448 Series, Standard Test Methods for Electromechanical Switches (ANSI/EIA-448-78) (R83), EIA-448/-IB—2/-3A/-4/-19/-20/-21/-22/-23/-24, Electronic Industries Association.

EIA/NARM Standard for Solid-State Relay Service (ANSI/EIA-443-79), EIA-443.

EIA/NARM Standard: Testing Procedures for Relays for Electrical and Electronic Equipment (ANSI/EIA-407-A-78) (R83), EIA-407-A.

Engineers Relay Handbook, Electronic Industries Association.

Generic Specifications for Keyboard Switches of Certified Quality (ANSI/EIA-4980000-A-89), EIA-4980000-A, Electronic Industries Association.

Relay Conference Proceedings, 1978 to present, National Association of Relay Manufacturers and EIA Relay Division, Electronic Industries Association.

CHAPTER 5.5
CONNECTORS

Robert S. Mroczkowski

INTRODUCTION

This chapter provides an overview of the structure and function of electronic connectors. Because of space limitations a limited number of topics will be discussed. References are given to provide additional detail for interested readers.

Defining a Connector

A connector can be defined in terms of its structure or its function. A functional definition describes what the connector is expected to do, such as carry a signal or distribute power, and the requirements it must meet, such as the number of mating cycles and electrical requirements. A structural definition describes the materials of manufacture and design of the connector. The discussion will begin with a functional definition.

A connector provides a *separable* connection between two functional units of an electronic system without *unacceptable* effects on signal integrity or power transmission.

A few comments on the highlighted elements in the definition are in order. The *separable* interface is the reason for using a connector. Separability may be required for manufacturing, maintenance/upgrading or portability/multifunction capability. The number of mating cycles a connector must support without degradation in performance depends on the reason for the separability requirement. A manufacturing application may require only a few matings while portability/multifunctional capability may require hundreds or thousands of mating cycles. Once in place, however, the connector must not introduce any *unacceptable* effects on the electrical/electronic function. The effects of connectors on signal degradation is becoming increasingly important as pulse rise times fall to a nanosecond and below.

To understand how these requirements are considered in connector design and selection, a short discussion of connector structure follows.

CONNECTOR STRUCTURE

A connector consists of four basic elements as outlined by Mroczkowski (Ref. 1).

1. The contact interface
2. The contact finish
3. The contact spring element
4. The connector housing

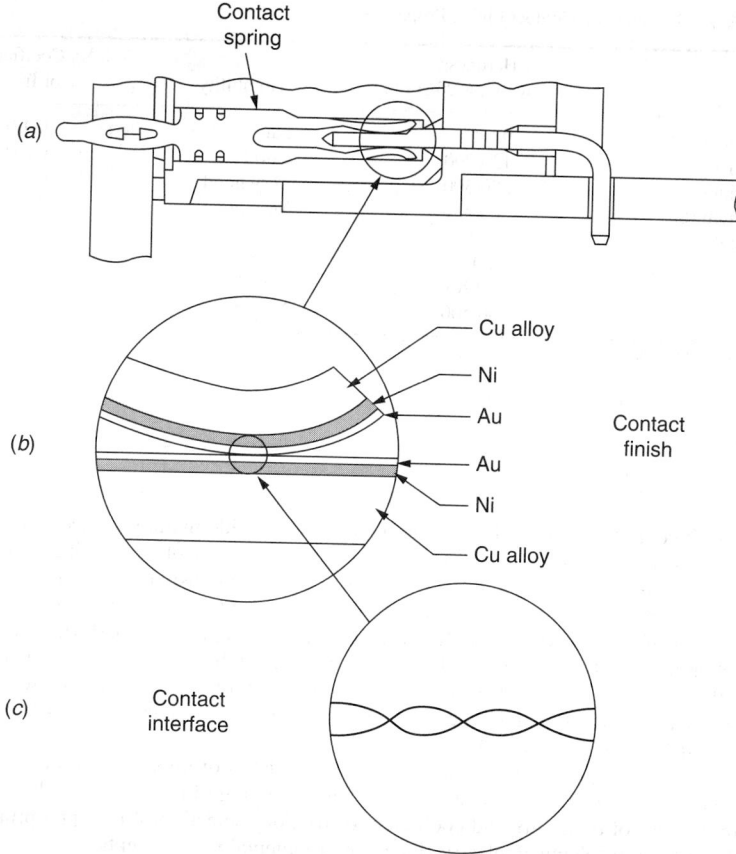

FIGURE 5.5.1   The structural components of a connector: (*a*) the contact spring and connector housing, (*b*) the contact finish, and (*c*) the contact interface.

A schematic illustration of these components is provided in Fig. 5.5.1. Each of these components will be briefly considered.

Contact Interface(s)

There are two contact interfaces of interest. The separable, or mating, interface has already been mentioned as the major reason for using a connector. In addition to the separable interface, it is also necessary to attach the connector to the subsystems that are to be connected. Such connections are generally "permanent," such as solder joints or crimped connections. A wide variety of separable and permanent connection designs are used in connectors as will be discussed in the section on connector applications.

At this point it is sufficient to say that contact interfaces, whether separable or permanent, are established by generating metallic interfaces between the two halves of the connector or between the connector and the subunit to which it is connected. Williamson (Ref. 2) provides an informative discussion of the fundamental structure of contact interfaces. Design considerations for establishing contact interfaces are discussed in Whitley and Mroczkowski (Ref. 3).

TABLE 5.5.1 Selected Contact Finish Properties

	Hardness (Knoop@25g)	Durability	Coefficient of friction
Gold (soft)	90	Fair	0.4/0.7
(hard)	130/200	Good	0.3/0.5
Palladium	200/300	Very good[*]	0.3/0.5
Palladium (80)- Nickel (20)	300/450	Very good[*]	0.2/0.5
Tin	9/20	Poor	0.5/1.0
Silver	80/120	Fair	0.5/0.8
Nickel	300/500	Very good	0.3/0.6

[*]with gold flash.

Contact Finish

The contact finish consists of a metallic coating intended to provide mating surfaces that optimize the ability to establish and maintain a metallic contact interface. Optimization of the mating surfaces can be simply described as meeting the need to avoid the formation of, or facilitate displacement of, surface films at the mating interface. The dominant contact finishes are noble metals and tin. Noble metal finishes include gold, palladium, and alloys of these two metals. Noble metal finishes "optimize" performance by being corrosion resistant which minimizes film formation. Tin, on the other hand, has a surface oxide that is readily disrupted on mating of the connector. Tin finishes are, however, subject to reoxidation during use, a failure mechanism called *fretting corrosion* (Bock and Whitley, Ref. 4). For additional discussion of contact finishes, see Antler (Refs. 5, 6, 7) and Mroczkowski (Ref. 8).

A brief descriptive compilation of contact finish characteristics of importance for connectors is provided in Table 5.5.1. The values realized depend strongly on the processing of the finish and the state of lubrication of the surface in the case of durability and coefficient of friction. Selection of the appropriate finish for a given application depends on mechanical, electrical, and environmental requirements.

Mechanical requirements include the following.

Mating durability. The number of mating cycles the connector can support without degradation of the finish. Mating durability depends on the hardness of finish system, the contact normal force and the contact geometry.

Connector mating force. The force required to mate the connector depends on the contact normal force, the coefficient of friction of the finish system, the contact geometry, and the number of positions in the connector.

The dominant electrical requirement is a low and stable value of connector resistance. From a contact finish viewpoint the contact resistance is dependent on the film formation and displacement characteristics of the finish. It is in the respect that noble and non-noble finishes display significant differences a previously mentioned.

The environmental characteristics of the operating environment, in particular the temperature, humidity, and corrosive species, determine the type and structure of films, which will form on contact interfaces. Films on noble metal finishes generally result from corrosion of exposed copper. Tin shows good environmental stability in most environments because of the self-limiting oxide film. The degradation of tin is primarily related to small motions, called *fretting* of the contact interface leading to a fretting corrosion (Bock and Whitley, Ref. 4). Such motions can arise from mechanical disturbances or differential thermal expansion mismatch stresses.

Noble metal finishes are generally considered as more forgiving and versatile in connector applications. Evaluation of the sensitivity of an application to fretting susceptibility is the major consideration limiting the use of tin finishes.

TABLE 5.5.2   Selected Properties of Commonly Used Contact Spring Materials

	Young's modulus, (E) (10^6 kg/mm^2)	Electrical conductivity (% IACS)	0.2 percent Offset yield strength (10^3 kg/mm^2)	Stress relaxation
Brass (C26000)	11.2	28	40/60	Poor
Phosphor bronze (C51000)	11.2	20	50/70	Good
Beryllium-copper (C17200)	13.3	20/26	55/95	Excellent

Contact Spring System

The contact spring system has two functions, mechanical and electrical. There are two different mechanical considerations, those for the separable interface and those for the permanent connection. In most separable connections the contact spring system consists of a spring member and a supporting surface. As indicated in Fig. 5.5.1 the deflection of the receptacle contact spring as it mates to the post provides the contact normal force, which establishes and maintains the integrity of the separable contact interface. The contact normal force is a major design consideration for connectors as discussed and Whitley and Mroczkowski (Ref. 3). Establishing a mechanical permanent connection, as opposed to a metallurgical connection such as a soldered or welded joint, involves controlled deformation and force considerations as discussed in Ref. 1. A balance of strength, for spring forces, and ductility, for manufacturing and formation of permanent connections, is required from a contact spring material. Most connector contact springs are copper alloys because such alloys have a good combination of strength, ductility, and electrical conductivity.

For a discussion of contact spring material selection issues see Bersett (Ref. 9), Spielgelberg (Ref. 10) and Lowenthal et al. (Ref. 11). Table 5.5.2 provides a limited selection of materials characteristics of importance for three of the more commonly used contact spring materials. A range of yield strength data is shown since different temper alloys may be used. From a user viewpoint, selection of an appropriate spring material for a given application depends primarily on the application temperature and the stress relaxation characteristics of the spring material. Stress relaxation resistance is important because it results in a reduction in the contact normal force as a function of time and temperature.

Electrically, the contact spring sections provide the conducting path between the permanent connection to the subsystems and the separable interface. The conductivity and geometry of the contact spring elements determine the resistance introduced by the connector. Figure 5.5.2 illustrates the sources of resistance in a connector. Three different resistances are indicated, the connection resistances of the permanent connections, R_{conn}, the bulk resistances of the receptacle spring and the post, R_b, and the resistance of the separable interface, R_c. In a typical connector the permanent connection resistances will be of the order of tens or hundreds of microohms, the bulk resistances of the order of a few to a few tens of milliohms, and the separable interface resistance of the order of milliohms. The resistivity of the contact spring material is a factor in all these resistances and is, therefore, a material selection parameter. As indicated in Table 5.5.2 however, the resistivity of the copper alloys used in connectors varies over a limited range. For power applications the bulk resistance becomes increasingly important and low resistivity alloys may be required.

Although the bulk resistance dominates the overall connector resistance, it is the contact and permanent connection resistances that are variable and can degrade during the application lifetime of a connector. An understanding of connector degradation mechanisms and how they can be controlled is a major element of connector design and selection. For an overview of connector design/degradation issues see Mroczkowski (Ref. 12).

Connector Housing

The connector housing also performs electrical and mechanical functions. There are two levels of electrical function, one at low frequencies and another for high frequencies. For low frequencies, the primary electrical function is insulation. The dielectric material of the housing insulates and isolates the individual contacts electrically, to allow

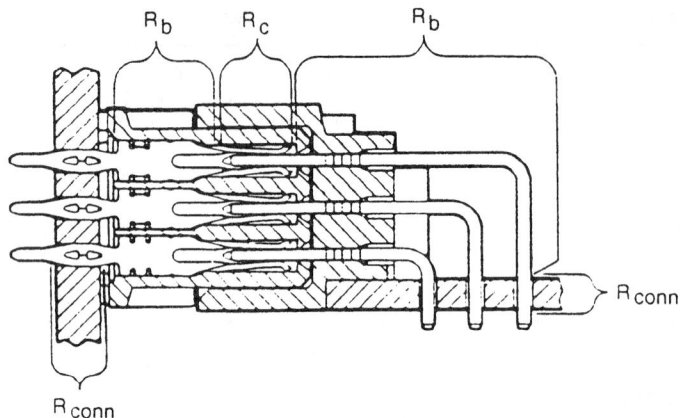

FIGURE 5.5.2 The contribution of various connector components to connector resistance: the connector resistance, R_{conn}; the bulk resistance of the contact spring, R_b; and the interface (contact) resistance R_c.

multiposition connectors. For high-frequency connectors in addition to insulation, the dielectric properties and geometry of the housing significantly impact on the characteristics of the connector as a transmission line. With the increasing processing speed capability of microelectronics, connector transmission line parameters such as characteristic impedance, propagation delay, and cross talk become major system design considerations. For additional discussion of these issues, see Southard (Ref. 13) and Aujia and Lord (Ref. 14).

Mechanically the connector housing latches and supports the contact springs within the housing and locates them mechanically. Control of mechanical location is critical to both mating characteristics of the connector separable interface and to proper permanent connections to the subunits to which the connector is attached.

Most connector housings are made from thermoplastic resins for cost and performance reasons. The electrical and, in most cases, mechanical properties of engineering thermoplastics are generally satisfactory for typical connector applications. Selection of a housing material is typically dependent on a particular manufacturing or application related requirement, such as molding of a long high pin count connector, high-temperature applications (greater than 100°C) or surface mount soldering requirements. For additional discussion on housing functions and material selection criteria see Walezak et al. (Ref. 15).

Table 5.5.3 contains a selection of housing material characteristics of importance in connector housings. The values shown are nominal values indicative of the resin families. For details the manufactures literature should be consulted because a wide variation in properties can be realized within a resin family depending on reinforcement and other additives. In most cases it is the stability of these characteristics through processing and in application that dictates the appropriate housing material for a connector. Surface mounting requirements are arguably the most demanding on connector housings.

One other housing function merits mention. The connector housing provides an environmental shield for the contacts and contact interfaces to decrease corrosion susceptibility in severe operating environments. This function is, of course, dependent on the housing design and independent of the housing material.

CONNECTOR DESIGN

As has been implied, the performance and reliability of electronic connectors depends on the design and material of manufacture of these connector components. Important design parameters for separable interfaces include selection of the contact finish, establishment of an appropriate contact normal force, and attention to

TABLE 5.5.3 Selected Properties of Commonly Used Polymers

	Flexural modulus (10^6 kg/mm^2)	Heat deflection temperature @264 psi(°F)	UL temperature Index (°C)	Dielectric strength (V/M)
Polyamide	0.7	666	130	17.4
Polybutylene Terephthalate	0.8	400	130/140	24.4
Polyethylene Terephthalate	1.0	435	150	26.0
Polycyclohexane Terephthalate	0.9	480	130	25.4
Polyphenylene Sulfide	1.2	500	200/230	18.0
Liquid crystal Polymers	1.5	650	220/240	38.0

the mating interface geometry all of which interact to determine mating mechanics, durability, and susceptibility to degradation of the separable interface. For a permanent interface minimizing the contact resistance and maximizing resistance stability, by control of the permanent connection process, are the dominant parameters of design. Housing performance characteristics of importance include the stability of the housing through the required assembly process, particularly surface mount, and its temperature and environmental stability relative to the application operating environment.

With this overview of connector structure as a context, attention now turns to connector applications and the functional requirements a connector is expected to meet.

CONNECTOR FUNCTION

As stated earlier, functionally, a connector is intended to provide a separable interface between two elements or subsystems of an electronic system without an unacceptable effect on the integrity of the signal, or power, being transmitted.

The requirement on separability is usually defined in terms of the number of mating cycles a connector must provide without an impact on performance, a number that will vary with the application.

For the purposes of this discussion, unacceptable refers to limits on the resistance that the connector introduces into the circuit or system. Both the magnitude of the resistance and the allowed change in resistance depend on the application. For signal applications the most important requirement is often resistance stability. However, the magnitude of the resistance is also important and may be dominant in power distribution applications.

A brief review of classes of connector applications is in order at this point. In this discussion, connector types are considered in terms of the structure of the connector and the medium to which the permanent connection is made. Connector applications are discussed in terms of the system architecture, a "levels of interconnection" approach, and the connector function, signal, or power applications.

CONNECTOR STRUCTURE/TYPE

With respect to connector structure or type, categorization in terms of the circuit elements being connected is useful. There are three such categories:

1. Board to board
2. Wire to board
3. Wire to wire

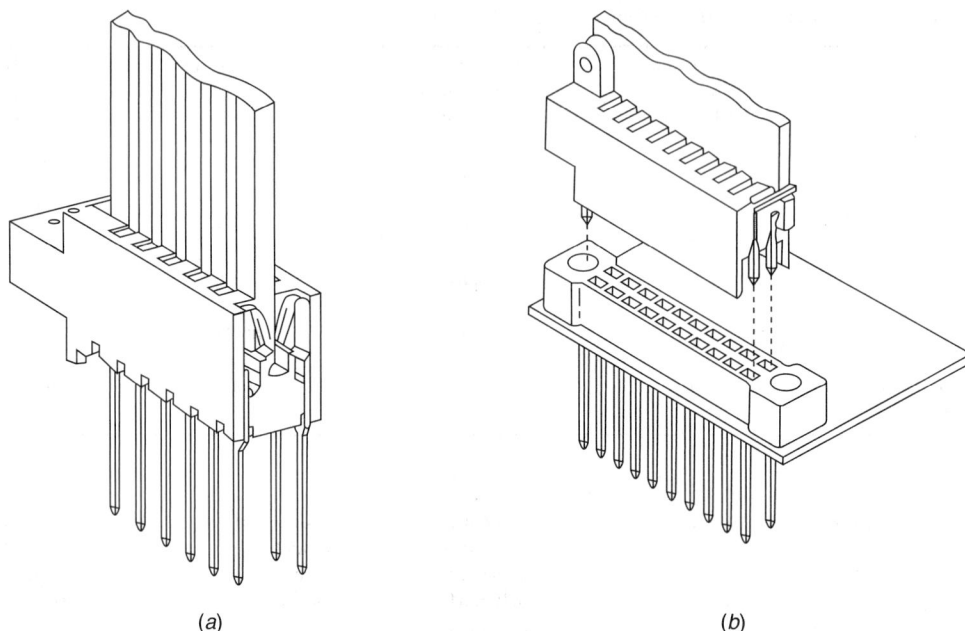

(*a*)   (*b*)

FIGURE 5.5.3   Printed circuit board connectors: (*a*) card edge (one piece) and (*b*) two piece.

It should be mentioned that the term *wire* also includes cables. In addition to the discussion of connector type, this section also includes a discussion of the classes of permanent connections, both mechanical and metallurgical.

Board-to-Board Connectors

It is in board-to-board connectors that the impact of the tremendous advances in speed of microprocessor chip technology are most apparent. These connectors are closest to the chip and, therefore, face the greatest requirements on maintaining the speed capability of the chip. In many cases board-to-board connectors must be considered as transmission lines with the accompanying requirements on controlled impedance, cross talk and shielding (EMI/RFI). Transmission line requirements have led to the development of connectors with intrinsic ground planes to provide improved high-speed performance with minimal effects on density. For a discussion of such connectors, see Sucheski and Glover (Ref. 16) and Aujla and Lord (Ref. 14). In addition to the transmission line requirements the increased functionality of microprocessor technology has led to a need for high input/output (I/O) connectors. Board-to-board connectors in pin counts in excess of 1000 are available.

There are two basic types of board-to-board connectors, one piece, or card edge, which mate to pads on the board, and two piece, which have an independent separable interface. Examples of each are shown in Fig 5.5.3.

Two-piece connectors are available in higher pin counts and densities because of greater multirow capability (up to eight rows), improved tolerance control, and the possibility of integral ground planes. In addition, they can be placed anywhere on the board. Centerline spacings in two-piece connectors include 0.100 and 0.050 in. and 1 mm with smaller spacings coming online.

Wire-to-Board Connectors

In some cases wire-to-board, or cable-to-board, connectors also face high speed requirements, which impacts on both the connector and the cable design. However, many wire-to-board connectors are used in applications where speed is not as important. The I/O densities also tend to be lower, but connectors with over 100 positions are not uncommon. There are a wide variety of connector styles and configurations in wire-to-board connectors. Connectors using 25-square technology, square posts 0.025 in. on a side, are one of the dominant wire-to-board technologies with connector systems using 1 mm and 15-square (0.015 in on a side) posts coming on line. Pin and socket connectors are also common.

Wire-to-Wire Connectors

Wire-to-wire connectors are generally located at a distance from the chip and may not face high-speed requirements, with the exception of coax connectors where controlled impedance and cross talk remain important parameters. Many wire-to-wire connectors are used external to the enclosure or equipment so ruggedness, grounding, and shielding become important design considerations. Wire-to-wire connectors are often variations on wire-to-board design, differing only in the media to which the permanent connections are made, to a wire or cable instead of a board.

Types of Permanent Connections

Because these connector types are discussed in terms of their permanent connections to the circuit elements or subsystems being interconnected, a brief discussion of permanent connection types is in order. There are two basic types of permanent connections: mechanical and metallurgical.

Mechanical Permanent Connections.   The dominant mechanical permanent connection technologies include:

- Crimped connections
- Insulation displacement connections (IDC)
- Press-in or compliant connections
- Wrapped connections

Crimped and ID connections are wire/cable connections. Press-in connections are made to plated through holes in printed wiring boards. Wrapped connections are made to square posts. A brief description of the mechanisms of contact generation and maintenance for each of these technologies is in order. Illustrations of each permanent connection technology are provided in Fig. 5.5.4.

Crimped connections. The permanent interface in crimped connections is established when the striped wire and crimp barrel are deformed in a controlled fashion by the crimp tooling. The deformation process displaces surface films and creates the contact interface. Residual stresses in the wire and wire barrel provide a restraining force that maintains the permanent connection interface integrity.

Insulation displacement connections. Insulation displacement connections are made by insertion of a wire into a properly designed IDC terminal. The wire insulation is displaced and the conductors deformed in a controlled fashion during the insertion process, which disrupts surface films and generates the contact interface. Elastic restoring forces resulting from the deflection of the terminal contact beams maintain the interface integrity. A wide variety of IDC terminal designs are used depending on the application requirements.

Press-in connections. Press-in connections are made by insertion of pins with controlled geometry compliant sections into a plated through hole (PTH) in a printed wiring boards. The contact interface is generated as the pin is inserted into the PTH, disrupting surface films and creating elastic restoring forces in the compliant section of the pin. The pin restoring force maintains the interface integrity. A wide variety of IDC terminal designs are available.

Wrapped connections. Wrapped connections are made to square posts with controlled radii on the post corners. The contact interface is generated by wrapping a stripped wire section around the post several times under controlled tension. The contact interface is generated as the post corners penetrate the wire. Residual tensile forces in the wire after wrapping maintain the interface integrity.

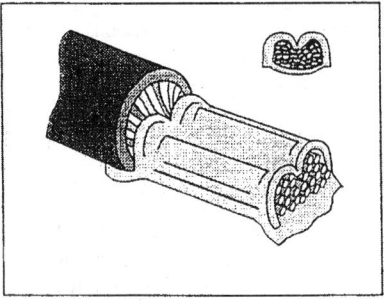
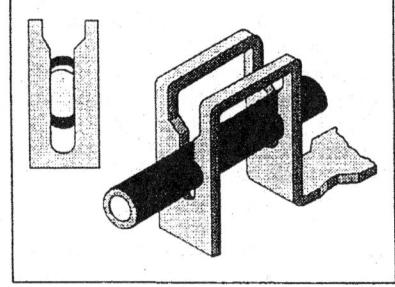
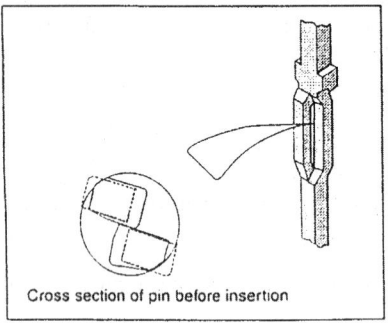
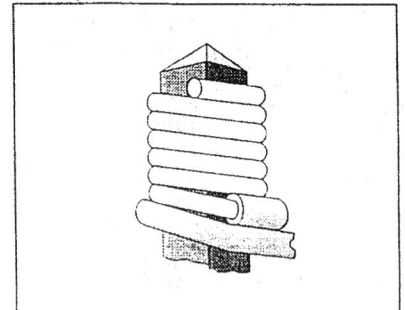

Cross section of pin before insertion

FIGURE 5.5.4   Permanent connection technologies.

For additional discussion of these technologies, see Ref. 1.

Metallurgical Permanent Connections.   Metallurgical permanent connections include soldered and welded connections. Soldered connections are generally to printed wiring boards and welded connections to wire and cable. There are two basic soldering technologies used today, through hole technology (THT) and surface mount technology (SMT). In most cases, THT consists of leads inserted into plated through holes and wave soldered. SMT connections are made to pads on the printed wiring board surface. SMT processing generally involves secondary source of solder applied to the leads or the board and a solder reflow process. Several different reflow processes are used, but the technology appears to be trending toward conduction reflow in inert gas with an infrared assist. For a discussion of soldering technologies, see Manko (Ref. 17).

From this structural discussion of connectors, attention now turns to a more application related categorization.

APPLICATION FACTORS

As mentioned previously, connector applications will be considered from two viewpoints, where the connector is used (levels of interconnection), and how the connector is used (signal/power).

Levels of Interconnection

The level of interconnection is defined by the points in the system that are being connected, not the connector type, as discussed by Granitz (Ref. 18). In brief, the six levels of packaging, and an associated connection or connector type, are:

1. Chip pad to package leads, e.g., wire bonds

2. Component to circuit board, e.g., DIP socket

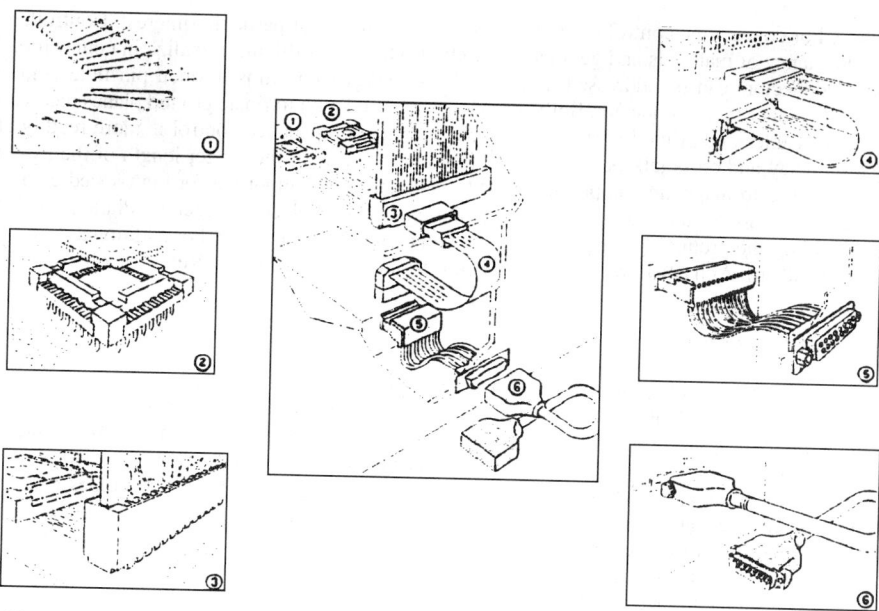

FIGURE 5.5.5   Levels of packaging in a typical electronic system.

3. Circuit board to circuit board, e.g., card edge connector

4. Subassembly to subassembly, e.g., ribbon cable assembly

5. Subassembly to input/output, e.g., D subcable assembly

6. System to system, e.g., coax cable assembly

A schematic illustration of these levels of packaging is provided in Fig. 5.5.5.

Level-1 connections are intended to be permanent connections and will not be discussed. Level-2 connections, generally accomplished through sockets, share many of the design features and requirements of connectors, but generally are subjected to only a few mating cycles. The separability requirement that has been used as a defining characteristic of a connector becomes important at levels 3 and above.

These levels of interconnection differ primarily in the application environment, but also in the durability requirements on the connector. Levels 3 through 5, being inside the enclosure, are expected to be benign with respect to corrosion and mechanical disturbances, but possibly more demanding with respect to temperature than level 6. In general, the number of mating cycles the connector must support increases, and the level of training of the user decreases, from level 3 to level 6. These differences impact on the ruggedness and "user friendliness" required of the connector. Electrically, performance requirements, particularly in terms of signal speed, are more stringent the closer the connection is to the chip level. Cables and cable connectors, however, may face more demanding electrical requirements even at higher levels of interconnection. It should also be noted that many connector types see usage in more than one level of interconnection.

Signal Connectors

Signal connectors are those that carry relatively small currents, milliamps to a few amperes. The functional requirements the connector must meet, however, depend primarily on the frequencies of signals that the connector must conduct. Signal connectors to be used at high frequencies, hundreds of megahertz or rise times less than 1 ns, conditions, which are increasingly common, must be considered as transmission lines. For such connector systems, characteristic impedance, propagation delay, and cross talk become critical performance parameters. For connectors, control of characteristic impedance is of particular interest. Controlled impedance requires constancy of materials and geometries, both of which are difficult to realize in connectors. Impedance control in connectors can be addressed, however, by signal/ground ratios in open pin field connectors or by added ground planes to the connector. Both approaches are used with intrinsic ground planes receiving increased attention. For cables and printed wiring boards, characteristic impedance control is more readily attained, but propagation delay and cross talk become more important because of the longer length of the associated transmission lines. Electromagnetic compatibility (EMC) also becomes important for high-speed connectors. EMC includes consideration of the impact of both external and internal electromagnetic radiation on system performance. Shielding and grounding considerations take on increased importance. For additional discussion of these issues, see Southard; Aujla & Lord; Sucheski & Glover; and Katyl & Simed (Refs. 13, 14, 16, and 19).

Power Contacts/Connectors

Contacts and connectors for power applications face different requirements. For power contacts/connectors two effects become important. The first is the loss of voltage arising from the resistance the connector introduces in the system. Systems have a millivolt drop budget to ensure that sufficient driving voltage is available to the chips. Excessive losses in the connector, whether bulk or interface related, can disrupt system performance. Second, the Joule, or I^2R, heating, which accompanies high current flow becomes a factor in system operating temperature. Increasing system temperature generally results in degraded performance or system life. Reduction and control of connector resistance addresses both these issues. From a bulk resistance perspective, selection of high conductivity spring materials is imperative. From a control perspective, ensuring the stability of connector interface resistance takes on new significance. For additional discussion of power contact/connector considerations see Corman and Mroczkowski (Ref. 20).

There are two approaches to power distribution, discrete power contacts and the use of multiple signal contacts in parallel. Discrete power contacts are generally large, to reduce bulk resistance and allow for higher normal forces to ensure interface stability. The size becomes a limiting factor and leads to the parallel signal contact approach. With the development of high pin count connectors allocation of some of the pins to current distribution becomes a viable option. Parallel signal contact consideration include:

- Ensuring an equal distribution of current through the contacts
- Accounting for thermal interactions between the contacts
- Appropriate derating of the contact current capacity for multiple contact applications

For additional discussion of parallel contact application considerations, see Ref. 2.

REFERENCES

1. Mroczkowski, R. S., Materials considerations in connector design, *Proc. 1st Elec. Mat. Conf. American Soc. Materials*, 1988a.
2. Williamson, J. B. P., The microworld of the contact spot, *Proc. 27th Ann. Holm Conf. Elec. Contacts*, 1981.
3. Whitley, J. H., and Mroczkowski, R. S., Concerning normal force requirements for precious metal plated connectors, *Proc. 20th Ann. Conn. Interconn. Tech. Symp.*, 1987.
4. Bock, E. M., and Whitley, J. H., Fretting corrosion in electrical contacts, *Proc. 20th Ann. Holm Conf. Electrical Contacts*, 1974.
5. Antler, M., The tribology of contact finishes for electronic connectors: Mechanisms of friction and wear, *Plat. Surf. Fin.*, October 1988a, 75, pp. 46–53.
6. Antler, M., The tribology of contact finishes for electronic connectors: The effect of underplate, topography and lubrication, *Plat. Surf. Fin.*, November 1988b, 75, pp. 28–32.
7. Antler, M., Contact materials for electronic connectors: A survey of current practices and technology trends in the U.S.A. *Plat. Surf. Fin.*, June 1991, 78, pp. 55–61.
8. Mroczkowski, R. S., Connector contacts: critical surfaces, *Adv. Mat. Processes*, Vol. 134, December 1988b, pp. 49–54.
9. Bersett, T. E., Back to basics: Properties of copper alloy strip for contacts and terminals, *Proc. 14th Ann. Conn. Interconn. Technology Symposium*, 1981.

10. Spiegelberg, W. D., Elastic resiliance and related properties in electronic connector alloy selection, *Proc. ASM Int. 3d Elec. Mat. Proc. Congress*, 1990.

11. Lowenthal, W. S., Harkness, J. C., and Cribb, W. R., Performance comparison in low deflection contacts, *Proc. Internepcon/UK84*, 1984.

12. Mroczkowski, R. S., Connector design/materials and connector reliability, *AMP Technical Paper P351-93*, 1993.

13. Southard, R. K., High speed signal pathways from board to board, *Electron. Eng.*, September 1981, pp. 25–39.

14. Aujla, S., and Lord, R., Application of high density backplane connector for high signal speeds, *Proc. 26th Ann. Conn. Interconn. Symp. Trade Show*, 1993.

15. Walczak, R. L., Podesta, G. P., and McNamara, P. F., High performance polymers: Addressing electronic needs for the nineties, *Proc. 21st Ann. Conn. Interconn. Tech. Symp.*, 1988.

16. Sucheski, M. M., and Glover, D. W., A high speed, high density board to board stripline connector, *Proc. 40th Elec. Comp. Tech. Conf.*, 1990.

17. Manko, H. H., "Soldering Handbook for Printed Circuits and Surface Mounting," Van Nostrand Reinhold, 1986.

18. Granitz, R. F., Levels of packaging, *Inst. Control Syst.*, August 1992, pp. 73–78.

19. Katyl, R. H., and Simed, J. C., Electrical design concepts in electronic packaging, in D. Seraphim, R. Lasky, and C. Li (eds.), "Principles of Electronic Packaging," McGraw-Hill, 1989, Chapter 3.

20. Corman, N. E., and Mroczakowski, R. S., Fundamentals of power contacts and connectors, *Proc. 23d Ann. Conn. Interconn. Tech. Symposium*, 1990.

21. Mroczkowski, R. S., Conventional versus Insulation Displacement Crimps, *Connection Technology*, July 1986.

CHAPTER 5.6
OPTICAL FIBER COMPONENTS

Gordon W. Day

INTRODUCTION

Optical fibers are thin (~100 μm diameter), normally cylindrical, strands of optically transparent material (usually glass or plastic) that are capable of guiding light over long distances with low loss. The basic principles by which optical fibers guide light were discovered in the nineteenth century, applied to the transmission of images in the first part of the twentieth century, and to telephony and data communications in the last part of the twentieth century.

Optical fiber communications systems provide many advantages over their predecessors, the greatest of which is their ability to transmit information at incredibly high rates. In laboratory demonstrations, rates of over 10 Tb/s, or the equivalent of roughly 150 million simultaneous telephone calls, have been achieved in a single fiber. In theory, much higher rates are possible. Optical fiber communications is the key enabling technology of the Internet.

Bundles of optical fibers with their relative positions fixed (imaging bundles) continue to be used to transmit images, and optical fibers have other applications in medicine, sensing, and illumination.

This section covers some of the most important properties of optical fiber and the components used with fiber, especially in communications systems.

Transparency

Glass, especially with very high silica content, can be highly transparent. Early optical fiber used for imaging bundles typically attenuated visible light by factors in the range of 0.1 dB/m (2.3 percent/m) to 1 dB/m (21 percent/m), low enough to permit useful transmission of images over distances of meters. Analyses performed in the 1960s indicated that much lower levels of attenuation could be achieved, perhaps lower than 10 dB/km, making it possible to consider the use of optical fiber as a communications medium over distances of kilometers. In the 1970s, these predictions were fulfilled, and levels of attenuation continued to be reduced until they now approach what is considered to be the fundamental limit in silica glass, approximately 0.15 dB/km at wavelengths near 1550 nm.

Attenuation mechanisms include intrinsic absorption, which in glass generally occurs in the ultraviolet and mid-infrared regions, absorption by impurities, losses associated with waveguide properties, and Rayleigh scattering.

In the spectral region from the visible through about 1700 nm, absorption by the OH$^-$ ion and Rayleigh scattering dominate. Absorption arises from the overtones of the fundamental 2.7 μm OH$^-$ lines at 1390 and 950 nm. In most fiber intended for telecommunication applications, manufacturing techniques eliminate the absorption at 950 nm, and the absorption at 1390 nm is no more than a few tenths of a dB (Fig. 5.6.1), and, in some cases, is almost completely eliminated.

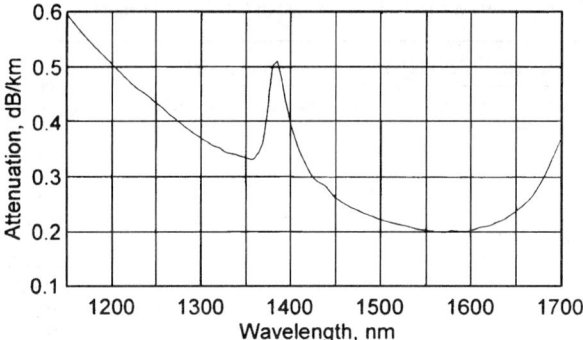

FIGURE 5.6.1   Spectral attenuation of a typical silica optical fiber.
Source: R. B. Kummer, Lucent.

The fundamental mechanism limiting attenuation in glass in the visible and near infrared regions is thus Rayleigh scattering, which varies inversely with the fourth power of wavelength. In this case, the attenuation of power in the fiber is given by

$$P(z) = P(0)e^{-\alpha_R z}$$

where $P(0)$ and $P(z)$ are the power in the fiber at the input and at position z along the direction of propagation, respectively, and α_R is the attenuation coefficient associated with Rayleight scattering. Expressed in dB, the attenuation is given by

$$\text{Attenuation (dB)} = -10 \log \frac{P(z)}{P(0)} \propto \lambda^{-4}$$

Three spectral regions have become important in communications, because fiber properties are favorable and, in some cases, because other components are available for those regions. The first to be exploited was the region around 850 nm, away from the 950 nm OH$^-$ absorption, and where GaAs lasers and LEDs, as well as Si detectors, were readily available. This region continues to be used for short distance systems such as local area and premise networks.

The second is the region around 1300 nm, where Rayleigh scattering is about a factor of 5 lower and, as discussed below, chromatic dispersion is very small. While this region also remains of interest, for high-performance systems it has been superceded by systems operating around 1550 nm, where Rayleigh scattering losses in the fiber are still lower and, more importantly, high-performance optical fiber amplifiers have been developed. These longer wavelength regtions are commonly divided for convenience into "bands." While the terminology has not been formally standardized, common usage is approximately as shown in Table 5.6.1.

TABLE 5.6.1   Optical Communication Bands

Name	Wavelength range, nm
O-band	1260–1360
E-band	1360–1460
S-band	1460–1530
C-band	1530–1565
L-band	1565–1625
U-band	1625–1675

In order to facilitate communications using many wavelengths simultaneously, a technique known as wavelength division multiplexing (WDM), the International Telecommunications Union (ITU) has established a standard frequency grid (Recommendation G.694.1) defined by

$$f_n = 193.1 + n\Delta \quad \text{(THz)}$$

Where n is any integer, including zero, and Δ may be 12.5, 25, 50, or 100 GHz. Using the defined value for the speed of light in vacuum ($c = 2.99792458 \times 10^8$ m/s), the corresponding vacuum wavelengths can be computed, as shown in Table 5.6.2.

TABLE 5.6.2   WDM Channels as Defined in ITU Recommendation G.694.1

Central frequencies (THz) for channel spacing of				Central wavelength
12.5 GHz	25 GHz	50 GHz	100 GHz	
⋮	⋮	⋮	⋮	⋮
⋮	⋮	⋮	⋮	⋮
193.1625				1552.02
193.1500	193.150	193.15		1552.12
193.1375				1552.22
193.1250	193.125			1552.32
193.1125				1552.42
193.1000	193.100	193.10	193.1	1552.52
193.0875				1552.62
193.0750	193.075			1552.73
193.0625				1552.83
193.0500	193.050	193.05		1552.93
⋮	⋮	⋮	⋮	⋮
⋮	⋮	⋮	⋮	⋮

TYPES OF OPTICAL FIBER

A simple optical fiber consists of long thin concentric cylinders of transparent material—an inner *core* having a slightly higher index of refraction than that of the surrounding *cladding*. The difference in index between the two regions is usually 1 percent or less; in glass, both values are typically near 1.5. While glass, and some other materials from which optical fibers are made, is inherently strong, it must be protected from surface abrasion that can cause it to fracture easily. And its performance can be degraded by certain types of bending, especially small quasiperiodic bends, known as microbends.

Most optical fibers are therefore protected with one or more additional polymeric coatings, sometimes called *buffers*, or *buffer layers*.

In a simple, ray-optics, view, a structure of this sort guides light because light incident at high angles (relative to normal) on a region of lower index of refraction is completely reflected. This is the principle of total internal reflection, discovered in the nineteenth century. A more complete electromagnetic analysis allows the structure to be characterized as a waveguide, guiding a finite number of spatial modes, not unlike those of typical microwave waveguides (see Fig. 5.6.2).

Step-Index Multimode Optical Fibers

The optical fibers used in imaging bundles, and those used in the earliest experimental optical communications systems, are known as step-index optical fibers, meaning that the transition from the index of refraction of the core to that of the cladding is abrupt. Generally, the indices of refraction in the core and cladding are fairly uniform.

In a fiber that is used in an imaging bundle, or in illumination, the difference in index of refraction between the core and the cladding is one of the most important parameters, in that it governs the angular field of view for which light can be collected and guided. In an imprecise analogy to microscope systems, the *numerical aperture* of a step-index multimode fiber is given by

$$NA = \sqrt{n_1^2 - n_2^2} = \sin\theta_{max}$$

where n_1 and n_2 are the indices of refraction of the core and cladding, respectively, and θ_{max} is the half-cone acceptance angle for rays that intersect the axis of the fiber.

Numerical aperture, along with the size of the core and the wavelength of the light, also determines how many modes can propagate in the fiber.

Step-index multimode fibers rarely find application in communications, because the different modes propagate at significantly different velocities. Pulses of light that propagate in more than one spatial mode are broadened in time through this process, which is known as intermodal distortion, or sometimes intermodal dispersion. A step-index multimode fiber with a core diameter of 50 μm, and an index of refraction difference $(n_1 - n_2)$ of about 1 percent may have an effective bandwidth of only about 20 MHz in 1-km length.

Graded-Index Multimode Optical Fiber

To reduce intermodal distortion, the refractive index of the core can be tailored so that all modes propagate at approximately the same velocity. The variation in refractive index with radius (the "refractive index profile") that minimizes intermodal distortion (maximizes bandwidth) is roughly parabolic with the highest refractive index on the axis. Sometimes the refractive index profile is specified by fitting index data to an equation of the form

$$n(r) = n_1 \sqrt{1 - 2\Delta \left(\frac{r}{a}\right)^g} \quad \text{for } r \le a$$

and

$$n(r) = n_1 \sqrt{1 - 2\Delta} \equiv n_2 \quad \text{for } r \ge a$$

where $\Delta \equiv \dfrac{n_1^2 - n_2^2}{2n_1^2}$

$n(r)$ = refractive index as a function of radius
n_1 = refractive index on axis
n_2 = refractive index of the cladding
a = core radius

The parameter g defines the shape of the index profile and is used in some specifications (e.g., Telecommunications Industry Association specifications TIA-4920000-B, TIA-492AAAB).

Because the refractive indices of glass vary with wavelength, the maximum bandwidth is achieved at a specific wavelength, and may vary dramatically over the wavelength range in which the fiber may be used.

In fibers that are not optimized for maximum bandwidth at the wavelength of operation, the actual bandwidth will depend on the excitation conditions, that is, the manner in which light is coupled into the fiber. Light coupled into a fiber from a light emitting diode (LED) will typically excite most of the modes of the fiber, and often results in the maximum intermodal distortion (minimum bandwidth) for a given length. This specification of bandwidth

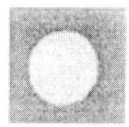
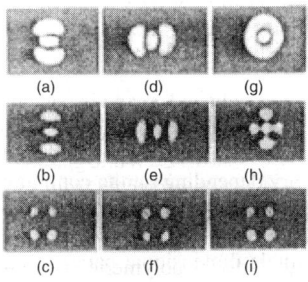
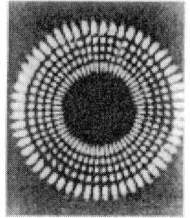

FIGURE 5.6.2 Modes of an optical fiber: Top, the lowest order, HE_{11} or LP_{01} mode; Middle, several low-order modes; Bottom, a high-order mode.
Sources: E. Snitzer and H. Osterberg, *J. Opt. Soc. Am.* 51 499 (1961); W. J. Stewart, Plessey.

is sometimes called the *effective modal bandwidth* or the bandwidth with an "overfilled launch." Light coupled into a fiber from a laser will typically excite fewer modes, and may yield a lower intermodal distortion (higher bandwidth), but the bandwidth, in this case, may depend strongly on the specific range of modes excited. Another specification that may be used to describe these effects is *differential mode delay*, which is generally measured by focusing a laser on the core of a multimode fiber and determining the difference in group delay (the propagation time of a short pulse through a specified length of fiber) as a function of the transverse position of the focused spot. Details of various methods of specifying the bandwidth of graded-index multimode fiber can be found in TIA standards TIA-455-220 and TIA-455-204.

The attenuation of multimode optical fiber may also vary among different modes. This effect is sometimes known as differential mode attenuation. As a result, the specification of attenuation is also dependent on the manner in which light is launched into the fiber and, further, the attenuation coefficient may vary with length as modes with higher attenuation disappear. Differential mode attenuation can also cause the bandwidth of the fiber to vary nonlinearly with length. Commonly, an overfilled launch is used for attenuation specifications, but that is the measurement condition for which these effects are usually the greatest. The measurement of attenuation in multimode fiber is specified in TIA-455-46-A and TIA/EIA-455-50-B.

Graded-index multimode fibers are most commonly used in data networks of under 1 km in length, such as access networks and local area networks. Standards for Gigabit Ethernet and 10 Gigabit Ethernet produced by the Institute of Electrical and Electronics Engineers (IEEE), IEEE 802.3z and IEEE 802.3ae, respectively, are based on these fibers.

Single-Mode Optical Fiber

When, for a given difference in index of refraction between the core and cladding, the ratio of the core radius to wavelength is sufficiently small, the fiber will only support one spatial mode, which is commonly designated the HE_{11} (alternatively the LP_{01}) mode. For a step-index fiber, this occurs when

$$\lambda \geq \frac{2\pi a}{2.405}\sqrt{n_1^2 - n_2^2}$$

where a is the core radius.

The value of λ for which this expression is an equality is usually designated λ_c and is known as the cut-off wavelength. The cut-off wavelength is generally determined by observing the apparent change in attenuation that occurs when the wavelength of light passing through the fiber is scanned through the cut-off wavelength (TIA standard TIA/EIA-455-80B). The cut-off wavelength may vary depending on the configuration (such as bends) of the fiber when measured. Manufacturers thus sometimes specify the change in cut-off wavelength that occurs when a fiber is cabled.

The size of the HE_{11} mode is generally specified through a single-dimensional parameter, known as the mode-field diameter (MFD), which varies with wavelength as well as the physical and optical properties of the fiber. The most commonly used definition of MFD is one known as Petermann II. Several measurement methods for MFD are described in TIA standard TIA/EIA-455-191A. Knowing the MFD permits the calculation of loss when fibers with different MFDs are joined; and, for a given optical power level, the larger the MFD, the lower the irradiance in the fiber, which usually leads to a reduction in nonlinear optical effects. Typical single-mode fibers have mode-field diameters in a range of 10 μm.

For purposes of predicting the magnitude of nonlinear optical effects, it is important to know the effective area A_{eff} of an optical fiber. The TIA (TIA/EIA-455-132A) defines the effective area as

$$A_{eff} = \frac{2\pi\left[\int_0^\infty I(r)r\,dr\right]^2}{\int_0^\infty I(r)^2 r\,dr}$$

which, for a given class of fibers, can usually be estimated by

$$A_{eff} = k\pi\left(\frac{MFD}{2}\right)^2$$

where k is a constant specific to that fiber class.

Pulses of light propagating in a single-mode fiber increase in duration because of several dispersive effects—the variation of the index of refraction of the glass with wavelength, which leads to variation of the group velocity with wavelength (chromatic dispersion), the variation of group velocity resulting from the parameters of the optical waveguide (waveguide dispersion), and differences in group velocity with polarization (polarization mode dispersion).

Generally, chromatic dispersion and waveguide dispersion are not distinguished in specifications, and sometimes the term *chromatic dispersion* is used to include both effects. Though specifying dispersion versus wavelength might be useful, it is more common to specify two dispersion parameters, the *zero dispersion wavelength* λ_0, and the *zero dispersion slope* S_0, which is the slope of the dispersion near λ_0, and has units of ps/(nm^2 km). For a typical step-index single-mode fiber, sometimes called a *conventional* or *nondispersion-shifted fiber*, the dispersion is dominated by chromatic dispersion, and increases with increasing wavelength, exhibiting zero dispersion at a wavelength between 1310 and 1320 nm. For such fibers, S_0 is typically in the range of 0.1 ps/(nm^2 km), and it is possible to estimate the dispersion at nearby wavelengths $D(\lambda)$ using the expression

$$D(\lambda) = \frac{S_0}{4}\left(\lambda - \frac{\lambda_0^4}{\lambda^3}\right)$$

More complex refractive index profiles can provide waveguide dispersion that differs substantially from that of a step-index profile. This can be exploited to shift the zero dispersion wavelength to another, typically longer, wavelength, or to change the variation of dispersion with wavelength. Fibers in which the dispersion has been shifted to the region around 1550 nm are commonly called *dispersion-shifted* fibers. Fibers that are designed to maintain a low level of dispersion over an extended wavelength range are commonly called *dispersion flattened* fibers. Fibers in which dispersion *decreases* with increasing wavelength are called *dispersion compensating* fibers and can be inserted into a system to reduce the total dispersion.

In WDM systems, nonlinear optical effects can lead to cross talk between channels. Most nonlinear effects are strongest when dispersion is near zero and all wavelengths involved in the process propagate at the same velocity. Maintaining a small but nonzero dispersion over the range of wavelengths used can thus be useful in minimizing cross talk. This has led to the development of various types of "nonzero-dispersion-shifted" fibers, which typically have a zero dispersion wavelength just shorter or just longer than the band in which they will be used. Figure 5.6.3 shows dispersion curves for a variety of fibers.

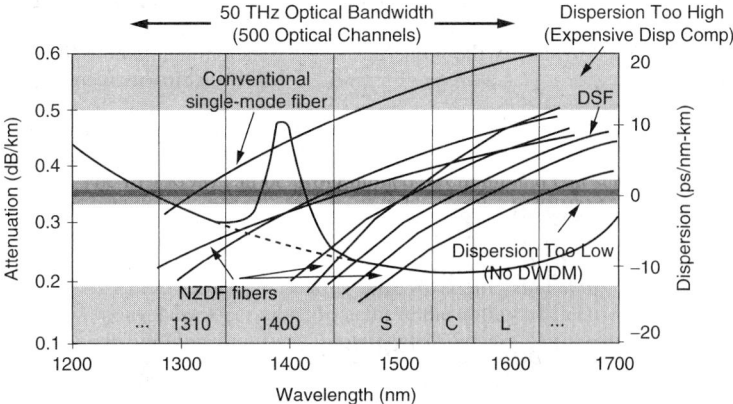

FIGURE 5.6.3   Spectral attenuation of a typical fiber and dispersion characteristics of several types of fiber.
Source: R. B. Kummer, Lucent.

The third source of dispersion in a single-mode fiber arises from the fact that different polarizations within the single spatial mode propagate with slightly different phase and group velocities. These polarization effects arise from stress, either inherent to the fiber or induced mechanically, and imperfections in its cylindrical symmetry. The result is a fiber property known as polarization-mode dispersion (PMD), which, like chromatic dispersion and waveguide dispersion, causes the duration of a pulse to increase as it propagates.

If the polarization properties of the fiber were stable and light did not couple between polarization states, it would be relatively easy to measure and specify PMD. The fiber could be modeled as a combination of a single retarder (a linearly birefringent element) and a single rotator (a circularly birefringent element) and would exhibit an orthogonal pair of stable eigen-polarizations (polarizations for which the output polarization state is the same as the input state). The PMD would be relatively easy to determine and would scale linearly with length.

Instead, the fiber parameters change with the temperature of the fiber and with handling, so PMD can only be specified statistically. Normally, the measurement is of the differential group delay (DGD) between principal polarization states, which are polarization states that, over an appropriate range of wavelengths, represent the fastest and slowest propagation velocities. Theoretically, and well observed in practice, the DGD measured over a range of wavelengths, or over time, follows a Maxwellian distribution, as shown in Fig. 5.6.4. The mean of the distribution is usually called the PMD or mean DGD. Several measurement methods are used, as specified, for example, in TIA standards TIA/EIA-455-113, TIA-455-122-A, TIA/EIA-455-124, and TIA/EIA-455-196.

To further complicate PMD measurements, in long lengths of fiber, light is coupled between polarization states, which tends to decrease the observed DGD. For lengths of fiber long enough for the coupling to be in equilibrium (longer than the "coupling length") the differential group delay between principal states will increase with the square root of length rather than the length. Similar effects are observed when fibers are concatenated. Methods of addressing these effects, which generally lead to the specification of a *link value* or *link design value* of PMD, are discussed in International Electrotechnical Commission (IEC) Technical Report TR 61282-3.

Single-Mode Fibers with Special Polarization Properties

In certain applications, it may be desirable to use an optical fiber with a very large linear birefringence, that is, a fiber in which orthogonal linear polarizations have significantly different phase velocities. One way to achieve this is to design and manufacture the fiber to have an elliptical core. Another is by incorporating, into the cladding, regions that apply a transverse stress to the core. In either case, the resulting, stress-induced, birefringence can be in the range of 10^{-3}.

Though they may have additional applications, fibers with a large linear birefringence are commonly called *polarization-maintaining fibers*. Linearly polarized light, launched into the fiber with its direction of polarization along one of the axes of birefringence, will tend to remain linearly polarized along that axis, even when the fiber is bent or twisted.

Several methods are used to specify the properties of polarization-maintaining fibers. The beat length, L_B is the propagation distance over which a phase difference between light in the two polarizations of 2π radians

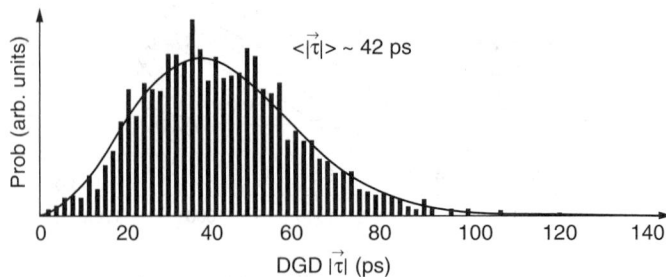

FIGURE 5.6.4   Statistical distribution of differential group delay between principal states showing Maxwellian distribution. PMD is the average DGD.
Source: YAFO Networks.

accumulates. It is given by $L_B = \lambda / \Delta n$, where Δn is the difference in effective (phase) refractive index between the two polarizations and λ is the wavelength. For a typical polarization maintaining fiber, L_B may be in the range of a few millimeters.

Another common specification is *cross talk*, which is generally given for a specific length. This is the fraction of light coupled into the orthogonal polarization when the polarization of the input light is aligned with one of the polarization axes. It is generally given in decibels as

$$\text{Polarization cross talk (dB)} = 10 \log \frac{P_{min}}{P_{max}}$$

where, P_{min} and P_{max} are the minimum and maximum power levels observed as a polarizer is rotated in the output.

Sometimes, a coupling parameter, commonly known as the *h-parameter* and defined, for a given length L, by

$$hL = \tanh^{-1}\left(\frac{P_{min}}{P_{max}}\right)$$
$$\approx \left(\frac{P_{min}}{P_{max}}\right) \quad \text{for} \left(\frac{P_{min}}{P_{max}}\right) \leq 0.1$$

is also specified. A typical value for h in a high-quality polarization-maintaining fiber is between 10^{-5} and 10^{-4}.

Fibers Manufactured from Other Glass Systems

Optical fibers made from various fluoride compounds can transmit light in the 2- to 5-μm region. Mixtures of zirconium, barium, lanthanum, aluminium, and sodium fluoride have been studied, including a combination of all of these compounds known as ZBLAN. Because the fundamental Rayleigh scattering in that wavelength range would be very small, it was once thought that these fibers could provide levels of attenuation much lower than those of silica fiber. That possibility has not been demonstrated, but fluoride fibers have become useful as hosts for a wide variety of rare-earth dopants, enabling optical amplifiers and lasers to be developed for wavelength ranges where they were not previously available.

Glasses with a high lead content (e.g., SF-57) can have very low stress-optical coefficients, that is, the index of refraction in these glasses is less affected by applied stress than in other glasses. Optical fiber manufactured from these glasses exhibits less birefringence when bent or twisted than ordinary fiber. Because the attenuation in such fiber is relatively high, they are not useful for telecommunications applications, but they may be very useful in certain optical fiber sensor applications.

Plastic Optical Fiber

Optical fiber can also be produced from plastic material, most commonly the acrylic PMMA. Typical plastic optical fibers (POF) are much larger in diameter than silica fibers, often as large as 1 mm, and have a thin cladding. Both step-index and graded-index multimode fibers are available. Generally POF is intended for short-distance applications, perhaps up to 100 m, sometimes within buildings or vehicles.

PMMA fibers have a characteristic spectral attenuation shown in Fig. 5.6.5. Commonly, they are used at a wavelength of about 650 nm, where the attenuation is a local minimum and inexpensive lasers are available.

OPTICAL FIBER CONNECTORS

A wide range of connectors is available for connecting optical fiber; some of the more common designs for single fibers are shown in Fig. 5.6.6. The two key specifications for a connector are the insertion loss and the return loss. For most high-quality connectors, insertion loss is often specified as a maximum value, typically in the range of a few tenths of a dB. The return loss is the fraction of light reflected by the connector. It is usually specified as a positive value in dB (a return loss of 40 dB meaning that the reflected signal is four orders

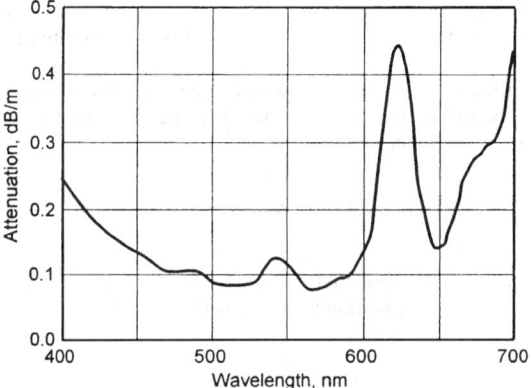

FIGURE 5.6.5   Attenuation of typical PMMA plastic optical fiber.
Source: Mitsubishi Rayon.

of magnitude smaller than the incident signal). Depending on design, connectors are commonly specified to have return loss values between 20 and 60 dB.

Most connectors rely on ceramic ferrules to hold the fibers and align them transversely. In most high-quality connectors, the ferrule and the fiber are polished together and the connectors are designed to bring the two fibers into contact. This is commonly known as a physical contact (PC) design. In some types of connectors the fiber and ferrule is polished at an angle to the axis, which decreases the return loss. This is known as an angled physical contact (APC) design. Connectors are often available factory installed on short lengths of cabled fiber (fiber pigtails) for splicing to a longer length, and on patch cords, with connectors, sometimes with a different connector type on each end.

FIBER COUPLERS

If the cores of two or more fibers can be brought into close enough proximity that the modes overlap to some degree, light will be coupled between or among the fibers. This is most readily achieved by twisting the fibers

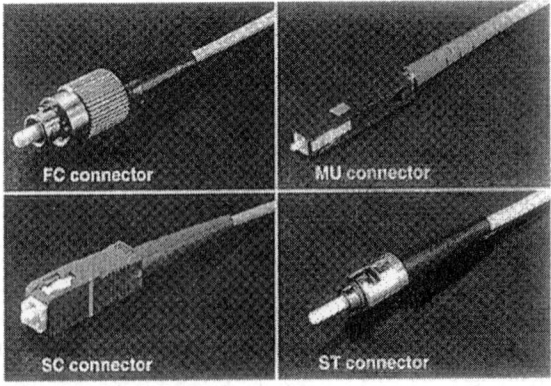

FIGURE 5.6.6   Common types of optical fiber connectors.
Source: Furukawa Electric Co.

together, heating them until they are soft, and pulling until cores reach the desired separation. The degree of coupling will depend on the wavelength, the separation and size of the cores in the interaction region, and the length of the interaction region.

The simplest device based on this principle is a 2 × 2 coupler that can be used to tap a portion of the light propagating in fiber (Fig. 5.6.7). Couplers of this sort are usually designed for a specific wavelength, and are commonly available in tap ratios (ratio of power output from the secondary output port to the total power from both output ports) of 1 (20 dB), 10 (10 dB), and 50 percent (3 dB), among others. Typically the wavelength range over which the tap ratio is maintained within a stated range is specified. Another important parameter is insertion loss, which is typically defined as the total decrease in power in the through-port. For example, a 3-dB coupler may have an *insertion loss* of 3.3 dB, meaning that 50 percent (3 dB) of the light is coupled to the secondary port and 0.3 dB is lost to absorption, scattering, or other effects.

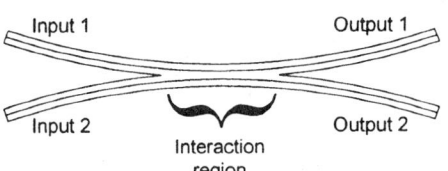

FIGURE 5.6.7   Schematic of a 2 × 2 fiber coupler.

Depending on design and manufacturing processes, the coupling ratio may be strongly wavelength-dependent, and this effect can be used to produce couplers for combining or separating wavelengths. Common applications include separating or combining signals at 1300 and 1550 nm in local networks and coupling pump wavelengths into fiber amplifiers.

FIBER GRATINGS

The refractive index of glass containing germanium can be increased slightly by exposing it to intense ultra-violet light. When this effect is exploited to produce a periodic variation in refractive index along the direction of propagation (a grating), several useful devices can be created. Usually, the manufacturing method involves illuminating the fiber transversely with a laser operating in the 200 to 300 nm spectral range. Periodic variations in the laser intensity can be achieved either by interference of two beams or by using a mask, similar to those used in lithography.

Fiber Bragg Gratings (Short-Period Gratings)

When the spatial periodicity Λ of the variation in refractive index is half the wavelength of light in the fiber, that is, when

$$\lambda_0 = 2n_{co}\Lambda$$

where λ_0 is the vacuum wavelength, and n_{co} is the effective (phase) refractive index of the LP_{01} mode in the fiber, light will be strongly reflected.

Depending on the details of the grating configuration, very high levels of reflection can be achieved over very narrow spectral ranges, in some cases sufficiently narrow to reflect one wavelength of the ITU WDM grid while transmitting adjacent wavelengths. When combined with other components (e.g., Fig. 5.6.12) this permits the removal of a single channel from a WDM system.

If the periodicity of the grating is chirped, that is, the periodicity varies from one end of the grating to the other, the reflection spectrum is broadened. More importantly, chromatic dispersion is created because light reflected from the far end of the grating is delayed relative to light reflected from the near end. Further, by apodizing the refractive index variation, that is, varying the degree of index modulation along the grating, the delay versus wavelength can be a smooth, quasi-linear, function of wavelength suitable for compensating for dispersion in a fiber. These grating characteristics are illustrated in Fig. 5.6.8.

Because their periodicity is affected by temperature and strain, fiber Bragg gratings can also be used as sensors for those measurands, or for other measurands that can lead to changes in strain or temperature. And

Types of Fiber Bragg Gratings

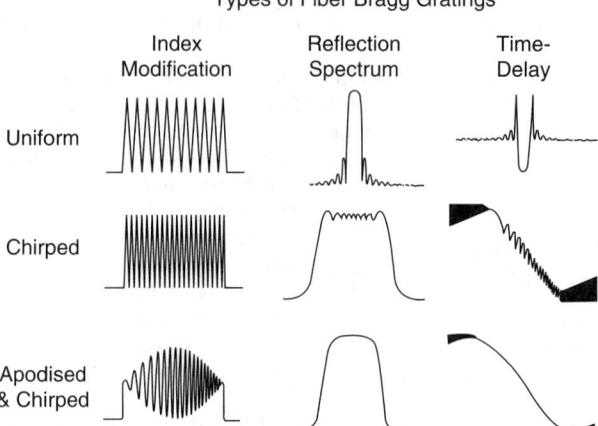

FIGURE 5.6.8   Reflection and dispersion characteristics of several types of fiber Bragg gratings.
Source: Southampton University.

because many fiber Bragg gratings can be incorporated into a fiber or network of fibers, quasi-distributed sensor networks can be developed.

Long-Period Gratings

In addition to coupling light from a forward propagating LP_{01} mode to a reverse propagating LP_{01} mode, gratings can be used to couple light from the LP_{01} mode to either forward or reverse propagating cladding modes. This occurs when the following phase matching condition is met:

$$\lambda_0 = \Lambda(n_{co} \pm n_{cl})$$

where n_{cl} is the effective (phase) refractive index of a cladding mode. The positive sign applies to the coupling between modes propagating in opposite directions, in which the spatial period will be similar to that satisfying the Bragg condition described above. In fact, many fiber Bragg gratings will exhibit transmission minima at wavelengths somewhat shorter than the Bragg condition. The negative sign applies to guided and cladding modes propagating in the same direction. In this case, the spatial period Λ will typically be in the range of several hundred micrometers, hence the term *long-period grating*. Figure 5.6.9 shows the transmission spectrum of a grating with a period of 198 μm, in which the transmission minima correspond to coupling of light into individual cladding modes.

The principal application of long-period gratings is as nonreflective band-rejection filters. They are sometimes used to reject wavelengths in WDM systems, to attenuate amplified spontaneous emission or adjust gain flatness in erbium-doped fiber amplifiers, and to reject Stokes wavelengths in Raman amplifiers.

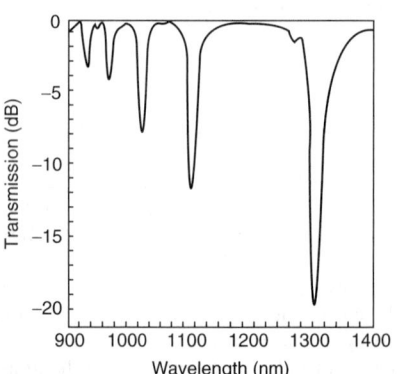

FIGURE 5.6.9   Spectral transmission of a long-period grating.
Source: A. Vengsarkar et al. J. Lightwave Tech., 14, 58–65 (1996).

OPTICAL FIBER ISOLATORS

Isolators are nonreciprocal two-port devices that have a low attenuation in one direction and a high attenuation in the other direction. They are frequently used in optical systems, usually to block light from reflecting back into a laser and damaging it or causing it to become unstable.

Most isolators are based on the Faraday effect, which is a magnetically induced circular birefringence usually observed as a rotation of the plane of polarization of linearly polarized light as it passes through the device. The magnetic field is parallel to the direction of propagation and the plane of polarization is rotated clockwise or counterclockwise depending on the material and whether the light is propagating in the same or opposite direction as the magnetic field.

In a simple isolator, the material, the magnitude of the magnetic field, and the length of propagation in the material are chosen so that the plane of polarization is rotated by exactly 45°, and the rotating element is placed between polarizers that are oriented at 45° to each other. Light entering the device polarized so that it is transmitted through the input polarizer is rotated so that it also passes through the second polarizer. However, light propagating through the device in the opposite direction is rotated so that it is blocked by the input polarizer.

As described, the isolator works well for linearly polarized light, polarized so that it is transmitted by the input polarizer. However, for arbitrary states of polarization, the attenuation in the forward direction is variable and can be high.

A polarization-independent isolator can be built by dividing the input polarization state into orthogonal polarizations, rotating them separately, and recombining them at the output. One implementation of this approach is shown in Fig. 5.6.10. A wedged birefrigent plate causes one linear polarization to be refracted while the other is undeviated. After the plane of polarization of each polarization is rotated 45° by a Faraday rotator, a second wedged birefringent plate, rotated 45° to the first, causes the two polarizations to again propagate in parallel directions, so they can be coupled into fiber. However, for light propagating in the reverse direction, the axial and refracted polarizations are orthogonally polarized, respectively, relative to the forward propagating beam, and are thus refracted at angles that will not be coupled into the fiber.

Optical fiber isolators can usually provide isolation (ratio of the forward to reverse transmittance) in the range of 40 dB, with an insertion loss of less than 1 dB. In multistage, higher performance devices, the isolation can exceed 70 dB.

OPTICAL FIBER CIRCULATORS

Optical fiber circulators, like their microwave counterparts, are multiport devices that use the Faraday effect to direct light from one port to another. In a 3-port circulator, the most common form (Fig. 5.6.11), light entering

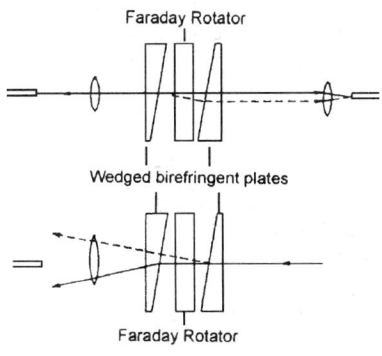

FIGURE 5.6.10   Design of polarization insensitive optical fiber isolator.
Source: M. Shirasaki, U.S. Patent 4,548,478.

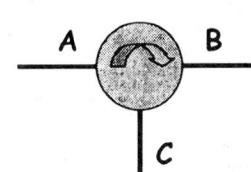

FIGURE 5.6.11   Schematic of a circulator.

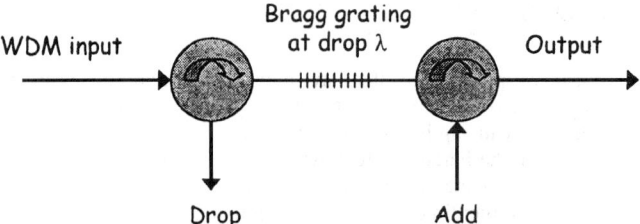

FIGURE 5.6.12 Add-drop multiplexer.

port A is directed to port B; light entering port B is directed to port C, and light entering port C is directed to port A. Most circulators employ Faraday rotation and other technologies similar to those used in isolators.

Circulators are versatile components for manipulating optical signals, as illustrated with the two examples below. Figure 5.6.12 shows how two circulators with a Bragg grating between them can be used to drop and add signals at a particular wavelength. Figure 5.6.13 shows how a single circulator can be used with a chirped fiber Bragg grating to compensate for chromatic dispersion.

WDM MULTIPLEXERS AND WAVELENGTH ROUTERS

In a WDM system, it is necessary to combine and separate (multiplex and demultiplex) different wavelength channels, often large numbers of channels with very small wavelength separations. This can be done with bulk optics and diffraction gratings or thin-film filters, or with combinations of fiber Bragg gratings and wavelength dependent couplers, but most such systems become complex when large numbers of channels are required. An alternative is a planar waveguide device, known (among other names) as an "arrayed waveguide grating (AWG)," and depicted in Fig. 5.6.14.

The AWG is a phased array device, usually produced in silica. Light from each of N input waveguides is divided equally, in an input coupler, among an array of waveguides in which adjacent guides differ in length by a precisely fixed amount. The device is designed so that, for light from a specific input fiber, constructive interference of light from the arrayed waveguides occurs at a specific output port. Thus, one use of the device is as a demultiplexer, as shown in Fig. 5.6.15, which is the superposition of the outputs of an AWG for broadband input light. The AWG is reciprocal, so it can similarly be used as a multiplexer.

The AWG can also be used as a router. If, as illustrated in Fig. 5.6.16, the device is designed so that N wavelengths ($\lambda_1, \ldots \lambda_N$) from input port 3 are directed to output ports N, $N-1$, $\ldots$, 1, respectively, then the same N wavelengths entering input port 2 will be directed to output ports 1, N, $N-1$, $\ldots$, 2, respectively. The device is thus a router, permitting one wavelength channel from each input channel to be directed to each output channel.

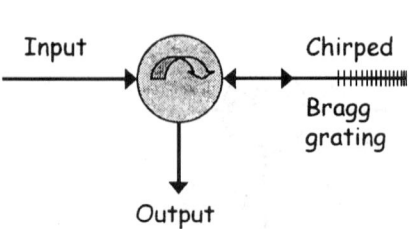

FIGURE 5.6.13 Dispersion compensator.

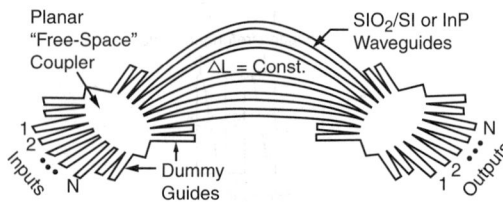

FIGURE 5.6.14 Schematic of an arrayed waveguide grating (AWG).
Source: K. Okamoto, NTT.

SWITCHES

Switches permit networks to be reconfigured, and channels to be rerouted, without conversion from optics to electronics, permitting all-optical networks. For some applications, switching times must be fast, comparable to a data period, but for most applications relatively slow switching is adequate. These slower switches are often called optical cross-connect switches and usually provide the ability to switch the light from any of N input fibers to any of N output fibers. Several switching technologies have been developed—optomechanical designs, including those based on microelectromechanical structures (MEMs); thermo- and electro-optic effects in planar waveguides; and liquid/bubble filled waveguides. Some approaches are wavelength selective and others are wavelength independent.

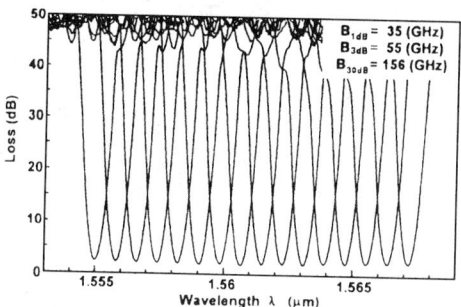

FIGURE 5.6.15 Demultiplexing properties of a 16 channel, 100 GHz spacing AWG.
Source: K. Okamoto. NTT.

One common switch structure is based on 2×2 switches in which the two input ports are either connected to the two corresponding output ports ($1 \rightarrow 1$, $2 \rightarrow 2$) or to the opposite output ports ($1 \rightarrow 2$, $2 \rightarrow 1$). Arrays of these switches can provide $N \times N$ switching, as shown in Fig. 5.6.17. Note, however, that not all switching combinations are possible. For example, in Fig. 5.6.17, if input port 1 connected to output port 1, then input port 2 cannot be connected to output ports 2, 3, or 4. This is known as *blocking*.

Figure 5.6.18 shows two approaches to the construction of $N \times N$ nonblocking switches based on MEMs technology. In Fig. 5.6.18*a*, a two-dimensional array of N^2 MEMs mirrors, which can be independently positioned either in or out of the beam, provide the switching. In Fig. 5.6.18*b*, two arrays of N MEMs mirrors are used. This approach uses fewer mirrors, but the mirrors require greater positioning control.

FIBER AMPLIFIERS

Optical amplifiers permit an optical signal to be amplified, preserving whatever modulation may be present, and allow designers to compensate for attenuation in fiber, losses in other components, and coupling between components. Perhaps most importantly, they can simultaneously amplify many different wavelengths and they are thus the principal enabler of WDM technology.

Three principal types of optical amplifiers are available: semiconductor optical amplifiers (SOAs), which are based on the same compound semiconductor technologies used in semiconductor lasers (SOAs are outside

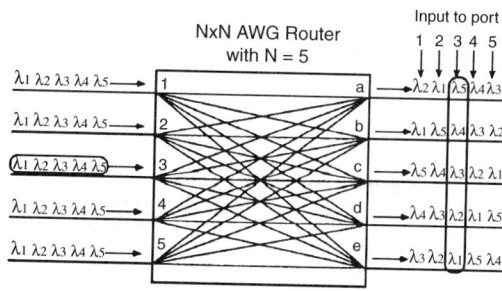

FIGURE 5.6.16 Wavelength routing, as may be accomplished with an AWG.
Source: K. Okamoto, NTT.

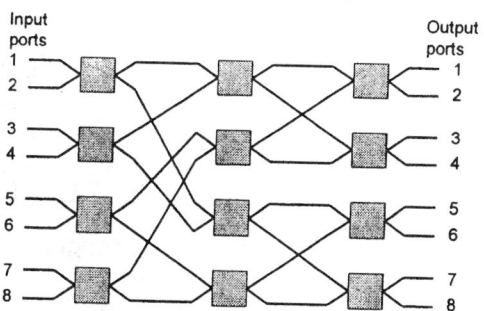

FIGURE 5.6.17 An 8×8 cross-connect switch based on twelve 2×2 switches.

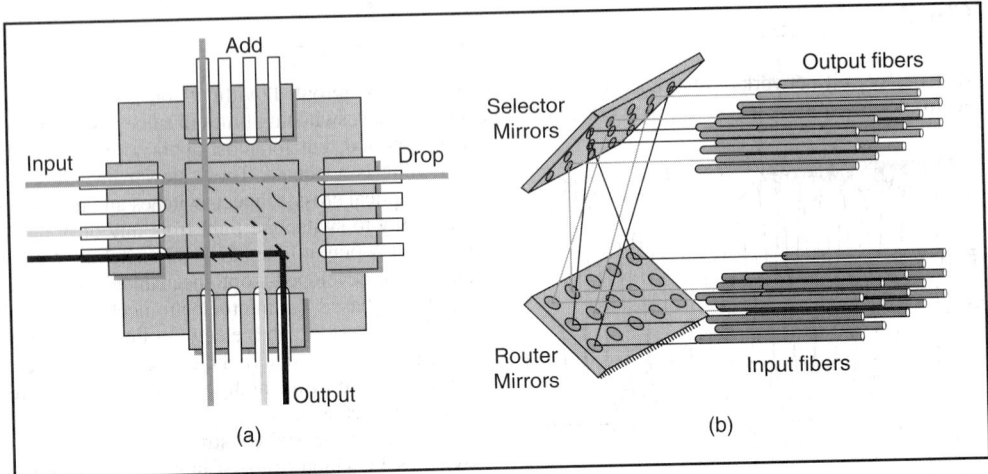

FIGURE 5.6.18 Two configurations nonblocking $N \times N$ cross-connect switches based on MEMs mirrors.
Source: Optical Micromachines.

of the scope of this section); rare-earth-doped fiber amplifiers, in which the gain is provided by any of several rare-earth ions, incorporated into a fiber and pumped by an appropriate laser source; and Raman fiber amplifiers, in which the gain is provided by the Raman effect in ordinary transmission fiber.

Rare-Earth-Doped Fiber Amplifiers

At least eight of the rare-earth lanthanide ions have been shown to provide optical gain when incorporated into silica or other glass hosts—praseodymium (Pr), neodymium (Nd), terbium (Tb), holmium (Ho), erbium (Er), thulium (Tm), ytterbium (Yb)—alone or in combination. In principle, these ions provide the possibility of gain through most of the spectral region from less than 1300 nm to greater than 1600 nm.

Erbium-Doped Fiber Amplifiers. Of the rare-earth-doped fiber amplifiers, the most successful have been those based on erbium-doped silica fiber. The erbium-doped fiber amplifier (EDFA) is important because it provides gain in the 1535- to 1565-nm region, spanning the region of lowest attenuation in a silica fiber, and it has proven effective and reliable. Alternate host glasses, including tellurium-oxide-containing glass can extend the spectral region to wavelengths greater than 1600 nm.

Figure 5.6.19 shows a diagram of the basic elements of an EDFA. At the center is a length of Er-doped fiber, perhaps 10 to 20 m long, which is pumped with the light of one or two pump lasers, usually operating at either

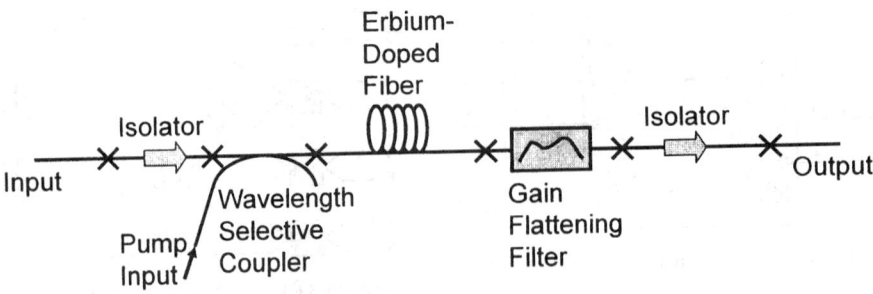

FIGURE 5.6.19 Schematic of an erbium-doped fiber amplifier (EDFA).

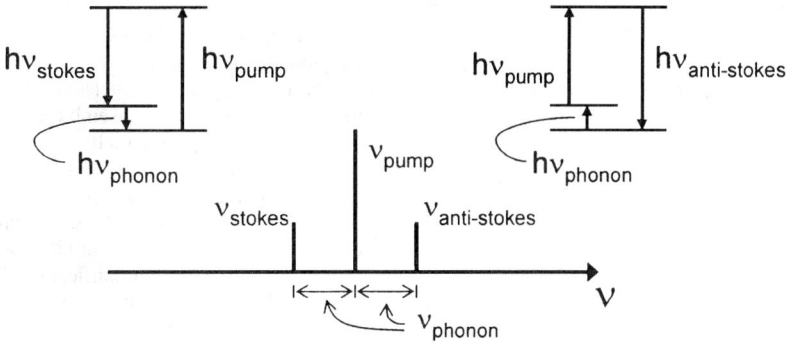

FIGURE 5.6.20   Energy levels and frequencies corresponding to Stokes and anti-Stokes amplification.

980 or 1480 nm, and coupled into the amplifying fiber through WDM couplers. Pumping at 980 nm generally provides lower noise, but may have a lower efficiency. Isolators before and after the amplifier suppress retroreflections.

The small-signal gain typically exhibits a peak at around 1530 nm and a broad shoulder on the longer-wavelength side of the peak. In WDM systems where it is desirable that each channel experience the same gain, filters may be incorporated into the design to flatten the gain spectrum; gain flatness of ±0.5 dB or better is often achieved in this way.

Depending on doping levels, fiber length, and pump power, the amplifier may provide a peak gain of 20 dB to more than 30 dB. When the input power level is such that the upper level of the amplification transition in Er is depleted, the gain saturates. Corresponding output powers may be as high as 100 mW or more.

Noise in EDFAs arises principally from amplified spontaneous emission (ASE), which depends on fiber properties and on pump power and wavelength. Noise figures in the range of 4 to 6 dB are typical.

The ITU, which provides several recommendations on the description of optical fiber amplifiers (Recommendations G.661, G.662, and G.663), categorizes EDFAs into three types: *booster power amplifiers*, which are high saturation power devices intended to be used with a transmitter to increase signal power lever; *preamplifiers*, which are low noise devices intended to be used with an optical receiver to improve sensitivity; and *line amplifiers*, which are used between passive fiber sections to compensate for fiber attenuation and other component losses.

Other Rare-Earth-Doped Fiber Amplifiers.   Praseodymium-doped fiber amplifiers, usually pumped around 1017 nm, provide high gain and high saturation power in the 1280- to 1340-nm region, spanning the wavelength of zero-chromatic dispersion. Unfortunately, amplification in the Praseodymium ion works best when the host is fluoride fiber, which is not as well developed as silica fiber.

Thulium, or thulium-ytterbium co-doping, in fluoride fibers has been shown to provide useful gain at wavelengths generally in the S-band, when pumped with a Nd:YAG laser at 1064 nm, and neodymium-doped fibers can provide amplification in the 1320-nm region.

Raman Fiber Amplifiers

In some respects, optical fiber amplifiers based on the Raman effect provide an attractive alternative to rare-earth-doped fiber amplifiers. For a given gain, they require a higher pump power, but Raman fiber amplifiers can provide amplification in any transmission fiber and, when a suitable pump source is available, provide gain at any wavelength where the fiber is transparent. When the pump and signal propagate in opposite directions, the Raman amplifier can provide very low effective noise.

The Raman effect is based on the coupling of light among an intense pump, a low-level signal, and optical phonons, which are mechanical oscillations that have frequencies characteristic of the material and are strongly damped. As shown in Fig. 5.6.20, gain can be achieved at frequencies corresponding to both the difference and sum of the pump and phonon frequencies. These are known as the "Stokes" and "anti-Stokes"

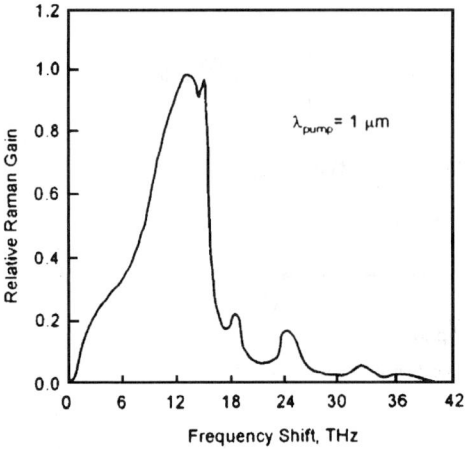

FIGURE 5.6.21   Raman spectrum of a germano-silicate fiber.
Source: R. Stolen, *Proc. IEEE* 68, 1232 (1980).

frequencies, respectively. In silica, the phonon spectra peaks roughly 12 THz (~90 nm) away from the pump frequency (Fig. 5.6.21).

Raman amplifier products are generally pump modules that consist of an ensemble of high power lasers, often four but experimentally as many as 20 or more, with control electronics to control the power of each laser independently. This permits tailoring the gain spectrum to acceptable flatness. Usually the amplifiers are designed to use Stokes amplification, which is more stable than anti-Stokes amplification. Raman amplification is polarization sensitive, so it is common to incorporate polarization diversity into optical systems associated with the pump.

It is also common to insert the Raman pump module into a system so that the pump propagates in the opposite direction from the signal. Thus, the greatest amplification is achieved where the signal would otherwise have been attenuated to its minimum level. This improves overall noise performance.

STANDARDS

Standards cited can be obtained from the following sources:

International Telecommunications Union (ITU)
Place des Nations
1211 Geneva 20
Switzerland
http://www.itu.int

International Electrotechnical Commission (IEC)
3, rue de Varembé
P.O. Box 131
1211 Geneva 20
Switzerland
http://www.iec.ch

Telecommunications Industry Association (TIA)
(Member of Electronic Industries Alliance, EIA)
2500 Wilson Blvd., Suite 300
Arlington, VA 22201 USA
http://www.tiaonline.org

Institute of Electrical and Electronics Engineers (IEEE)
Customer Service
445 Hoes Lane
Piscataway NJ 08855-1331 USA
http://standards.ieee.org

BIBLIOGRAPHY

Agarwal, G. P., "Nonlinear Fiber Optics," 3rd ed., Academic Press, 2001.

Becker, P. C., N. A. Olsson, and J. R. Simpson, "Erbium-Doped Fiber Amplifiers, Fundamentals and Technology," Academic Press, 1999.

Derickson, D. (ed.), "Fiber Optic Test and Measurement," Prentice Hall, 1998.

Kaminow, I. P., and T. L. Koch (eds.), "Optical Fiber Telecommunications," Vols. IIIA and IIIB," Academic Press, 1997.

Kaminow, I., and Tingye Li (eds.) "Optical Fiber Telecommunications," Vols. IVA and IVB, Academic Press, 2002.

Kartalopoulos, S. V., "DWDM Networks, Devices, and Technology," Wiley Interscience/IEEE Press, 2003.

Miller, S. E., and A. G. Chynoweth (eds.), "Optical Fiber Telecommunications," Academic Press, 979.

Miller, S. E., and I. Kaminow (eds.), "Optical Fiber Telecommunications II," Academic Press, 1988.

Okamoto, K., "Fundamentals of Optical Waveguides," Academic Press, 2000.

Ramaswami, R., and K. N. Sivarajan, "Optical Networks: A Practical Perspective," 2nd ed., Academic Press, 2002.

SECTION 6

INTEGRATED CIRCUITS AND MICROPROCESSORS

Digital and analog integrated circuits are the fundamental building blocks of today's electronic systems, digital ICs being dominant. They can emulate many functions for which analog circuitry was previously required. Nevertheless, because the real world is largely analog in nature, the analog IC remains the primary interface.

The extraordinary rate of progress in making dynamic random-access memories (DRAM) smaller and more powerful is reflected in their wide use, and, for example, in the rapid obsolescence of lesser-performing computer models.

The microprocessor, over its 30-some-year life span, has become a powerful tool for the design engineer, who embeds it in a broad range of intelligent digital devices.

Finally, Chap. 6.5 in this section covers nanotechnologies. The rapid developments in nanofabrication techniques have spawned a new era in microcomponent developments, including microelectromechanical components (see Chap. 8.5).   C.A.

In This Section:

CHAPTER 6.1
DIGITAL INTEGRATED CIRCUITS*

P. K. Vasudev, S. Tewksbury

INTRODUCTION

This and the following two sections will discuss the design and application of two major classes of integrated circuits (ICs): digital and analog and their application to both memory and logic circuits, which are shown for both bipolar and complementary MOS (CMOS) technologies.

Digital circuits are the most widespread and commonly used integrated circuits today. They process signals in binary bits. They are distinguished by their function and their performance. Analog circuits are less common and process signals as waveforms.

Analog circuits play a critical role intrinsic to the connection of electronic systems to the physical world and analog electronics have a rich and long history. More recently, digital electronics have become sufficiently sophisticated that computations on numerical values (representing the value of analog signals at regularly spaced intervals in time) can often be used to perform several traditional signal processing functions that previously required analog circuits. This transition of several traditional analog circuit functions into the digital world has accelerated, providing the capability of "programming" the digital computations to perform different traditional analog circuit functions or "reconfiguring" the digital components of an integrated circuit to change the operations performed. However, the analog circuits usually are required for the interface between the physical world (where things are usually analog) and the digital world. The combination of both analog and digital circuitry on the same VLSI circuit has allowed a wide range of applications to develop. Such "mixed signal" ICs are seen in various "embedded systems" such as those used in automobiles (sensors and computers measuring and controlling ignitions, and so forth).

In this section, the basic parameters used to measure the performance of digital circuits are first introduced. This is followed by a description of the different digital integrated circuit technologies, and their function in systems.

Logic Performance Parameters

Although there are a wide variety of parameters of interest, such as environmental, operating speed, voltage range, availability and cost, five key parameters are generally used in comparing digital circuit families.

Speed

This indicates how fast a digital circuit can operate. It is usually specified in terms of gate propagation delay, or as a maximum operating speed such as the maximum clock rate of a shift register or a flip-flop.

Gate propagation delay is defined as the time between the transition of an input signal between two states and the resulting response of the output signal.

---

*The contents of this chapter have been extracted and significantly updated from "Integrated Circuits" by F. F. Mazda, Butterworth, 1983.

Power Dissipation

This gives a measure of the power which the digital circuit draws from the supply. It is measured as the product of the supply voltage and the mean supply current for given operating conditions such as speed and output loading. Low power dissipation is an obvious advantage for portable equipment. However, since the amount of power which can be dissipated from an integrated circuit package is limited, the lower the dissipation, the greater the amount of circuit which can be built into a silicon die.

Speed-Power Product

The "power-delay" product is obtained by multiplying the power dissipation of a component by the delay between the input signal change and the output response. The product represents the energy associated with the functional operation of the component. The concept of power-delay product extends well beyond its use with simple devices or logic functions to larger scale systems. In general, a given component will be characterized by an energy (power-delay product) but allow trade-offs between the delay (or speed—reciprocal of the delay) and the power dissipation with the constraint that the power-delay product remain constant. This conforms to our intuitive expectation that if we are willing to accept higher power dissipation in a component we can achieve higher speed operation. Figure 6.1.1 illustrates the general theme. Here, three different circuits are characterized by different energies (i.e., representing different energies associated with different VLSI technologies) but each of these circuits can be adjusted either for high-speed operation at high-power dissipation or for low-power dissipation operation at lower speeds).

Current Source and Sink

This measures the amount of current which the digital circuit can interchange with an external load. Generally digital circuits interconnect with others of the same family and this parameter is defined as a *fan-out*, which is the number of similar gates which can be driven simultaneously from one output. The *fan-in* of a circuit is the number of its parallel inputs.

Noise Susceptibility and Generation

Digital circuits can misoperate if there is noise in the system. This may be a slowly changing noise, such as a drift in the power supplies, or high-energy spikes of noise caused by switching transients. Some types of logic

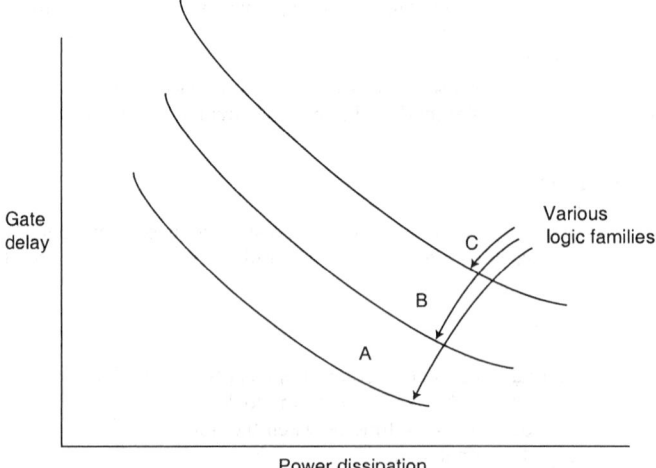

FIGURE 6.1.1 Speed vs. power curves.

families can tolerate more noise than others, and generally the faster the inherent speed of the logic, the more it is likely to be affected by transient noise.

Digital circuits also generate noise when they switch and this can affect adjacent circuits. Noise generation is a direct function of the current being switched in the circuit and the speed with which it is changed.

Comparisons of Various Digital Circuit Families

Today's advanced VLSI circuits draw upon several advantages of silicon CMOS circuits over earlier technologies. During the era of lower density digital circuits, eras of small-scale integration (SSI) and large-scale integration (LSI), bipolar circuit technologies were dominant. Fabricated vertically, small distances were more easily obtained than laterally, an advantage leading to higher performance bipolar circuits than found in MOS circuits. The table below shows the relative advantages of those earlier versions of digital technology. As the number of transistors in an IC increased and as technologies evolved to provide very small lateral distances, MOS circuits became increasingly competitive. The initial NMOS logic (using an NMOS pull-down circuit in combination with a pull-up resistor) was quickly replaced by CMOS technologies, once fabrication techniques had advanced to allow fabrication of both NMOS and PMOS transistors in the same substrate (essentially by creating "wells" deeply doped opposite to the substrate doping to create the other type of MOSFET). Today, CMOS is clearly the dominant technology and each successive generation of microfabrication technology increases the relative advantages of CMOS over the earlier bipolar forms of digital logic. Some of the principles of these earlier bipolar circuits are described later, providing an understanding of the various ways in which basic devices could be arranged to yield the basic digital functions of logic. In contrast to analog circuits which require a very linear response to input analog signals, digital logic circuits are deliberately nonlinear—switching between one state and another in response to the input signal and producing a fundamental regeneration of the binary logic states at the output of the circuit.

Comparison of Logic Families (1 = best, 6 = worst)

Logic family	Speed	Power	Fan out	Noise
DTL	4	4	3	3
TTL	3	4	3	3
ECL	1	6	2	2
NMOS	5	2	2	2
CMOS	3	1	1	1

BIPOLAR LOGIC CIRCUITS

This section discusses digital circuits using bipolar technology.

Saturating Bipolar Logic Circuits

One of the earliest digital circuit families was the resistor-transistor logic (RTL) family. Figure 6.1.2a shows a 3-input NAND gate. When the three transistors are OFF, R4 *pulls up* the output voltage D toward Vcc. If any of the three transistors turns ON (base voltage high), that transistor *pulls down* the output voltage toward 0 V. To provide full output voltage swings, the ON (OFF) state resistance of a transistor must be small (large) compared to R4. Diode-transistor logic (DTL), Fig. 6.1.2b, was another early bipolar logic technology, using pull-down diodes in combination with a pull-up resistor to drive the base of the output transistor. Replacing the multiple input resistors with simpler diodes leads to smaller gate areas. In addition, the dependence of the output resistance of the RTL circuit on inputs A to C is eliminated.

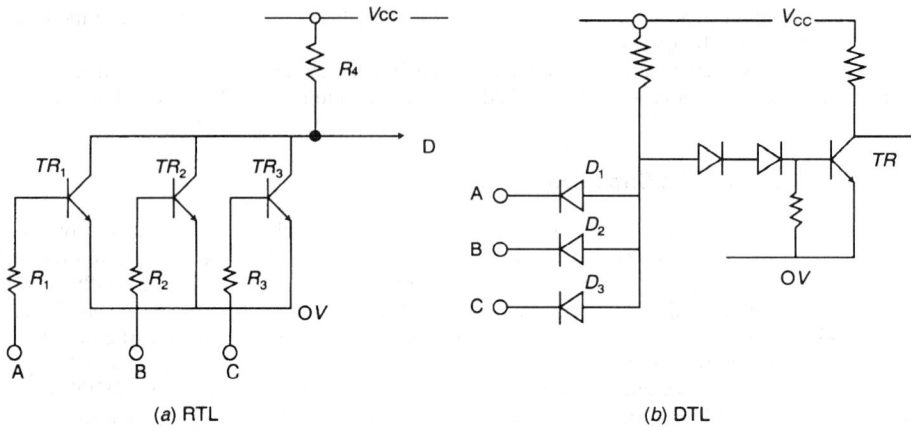

FIGURE 6.1.2   Resistor-transistor logic and diode-transistor logic.

The RTL and DTL logic families were quickly replaced by transistor-transistor logic (TTL) (Fig. 6.1.3), further simplifying the input configuration (here three emitters placed in the same transistor structure) while adding an active pull-up transistor TR1 in combination with the pull-down transistor to achieve faster outputs. The TTL logic family developed into a variety of subfamilies, each optimized for different performance objectives and remained the mainstream digital logic technology through the evolution from SSI through MSI to LSI. The increasing number of gates per IC required a decrease in power dissipation per gate to avoid a concurrent increase in the power dissipation of the IC. This limited the extension of the TTL logic family into VLSI since the TTL gate's power dissipation could not be decreased sufficiently. CMOS technologies have replaced TTL as the mainstream logic technology in VLSI. However, the TTL circuits continue to see use in BiCMOS, a technology that combines CMOS circuits with bipolar circuits on the same IC to achieve speed and current drive advantages as needed.

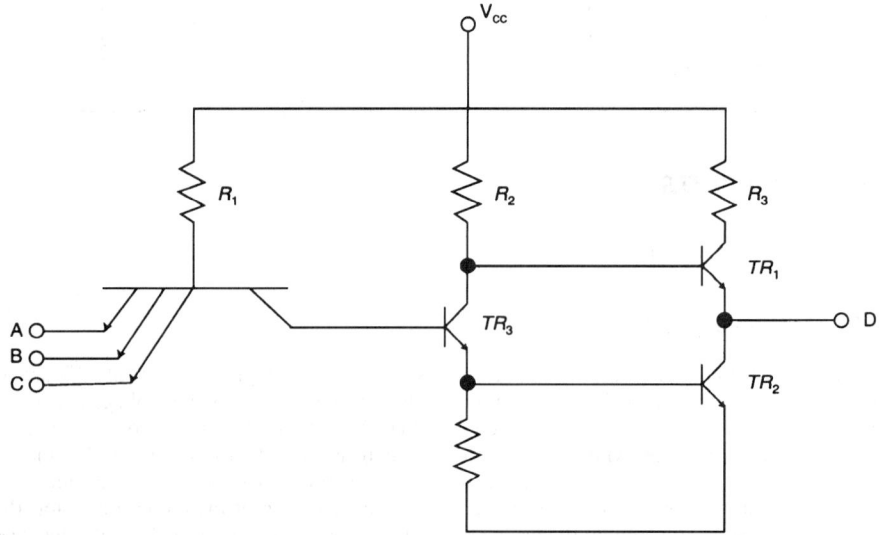

FIGURE 6.1.3   Transistor-transistor logic gate.

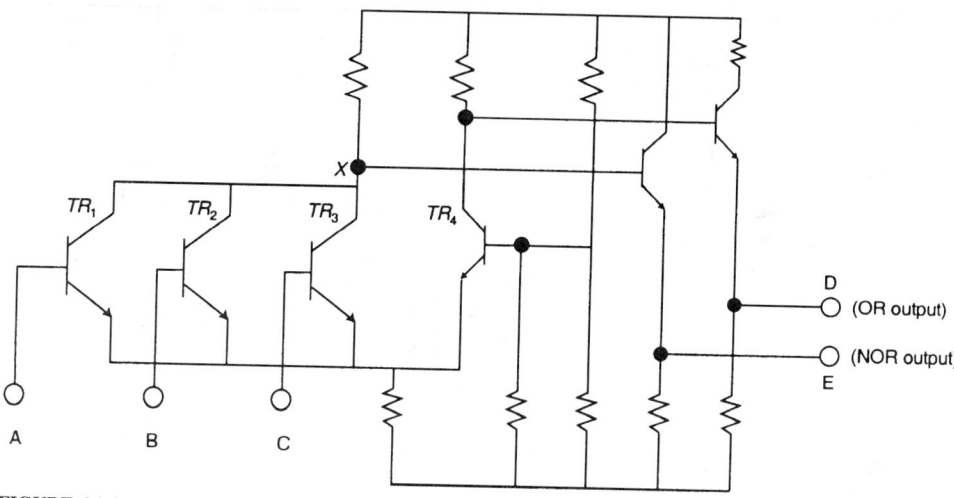

FIGURE 6.1.4 Emitter-coupled logic gate.

Nonsaturating Bipolar Logic Circuits

When in the saturated ON state, an excess charge density appears within the bipolar transistors. Switching the transistor to the OFF state requires that this excess charge density be removed, limiting the speed of the gates. One method to reduce the charge storage delays is to use Schottky transistors in the TTL gate. The Schottky diode prevents the transistor from saturating, leading to faster switching times. Such Schottky TTL logic circuits, either in low-power versions or high-speed versions, generally replaced the saturating TTL circuits discussed earlier.

Another nonsaturating bipolar logic family was the standard when high speed was a priority. This is the emitter-coupled logic (ECL) family (Fig. 6.1.4) in which separate transistors for each input and optimized designs for the highest possible speed are used. The amplifier consisting of TR4 and the input transistors drives the output transistors providing both noninverted and inverted outputs. The output transistors are designed with low resistance, providing high output current drive and high speed. Current flow into the gate from the power supply is nearly constant, avoiding noise problems seen in other bipolar logic circuits. The ECL logic circuits used supply voltages different from TTL logic and also had smaller output voltage swings, making connection of TTL logic and ECL logic difficult to interface (special interface circuits were available).

MOS LOGIC CIRCUITS

MOS technologies emerged early in the history of digital logic but suffered from poor performance because of the device structures being fabricated as lateral structures (with larger distances) in contrast to the bipolar device structures which took advantaged of small distances achieved with vertical structures. As device dimensions decreased, this disadvantage also decreased, with MOS device performance improving with each scaling to smaller device sizes. Initial MOS logic circuits used either NMOS or PMOS transistors (Figs. 6.1.5 and 6.1.6), a fabrication constraint since these different transistors used different type substrates (P-type and N-type, respectively).

Figure 6.1.7 shows a basic three input NMOS gate as the pull-up circuit and a resistor used as the pull-down element. That pull-down resistor is realized as a transistor in the ON-state to achieve a small area resistor. The symbol for the MOS transistor directly illustrates one feature that provides a fundamental advantage over bipolar transistors. In particular, the input is coupled capacitively to the device, with the result that all inputs are to capacitors, rather than to the resistances appearing at bipolar transistor inputs. For this reason, the output current from D need not be a dc current to maintain the output level. Instead, the output current is merely that needed to charge the capacitance of the MOS input being driven, going to zero when that capacitance has been

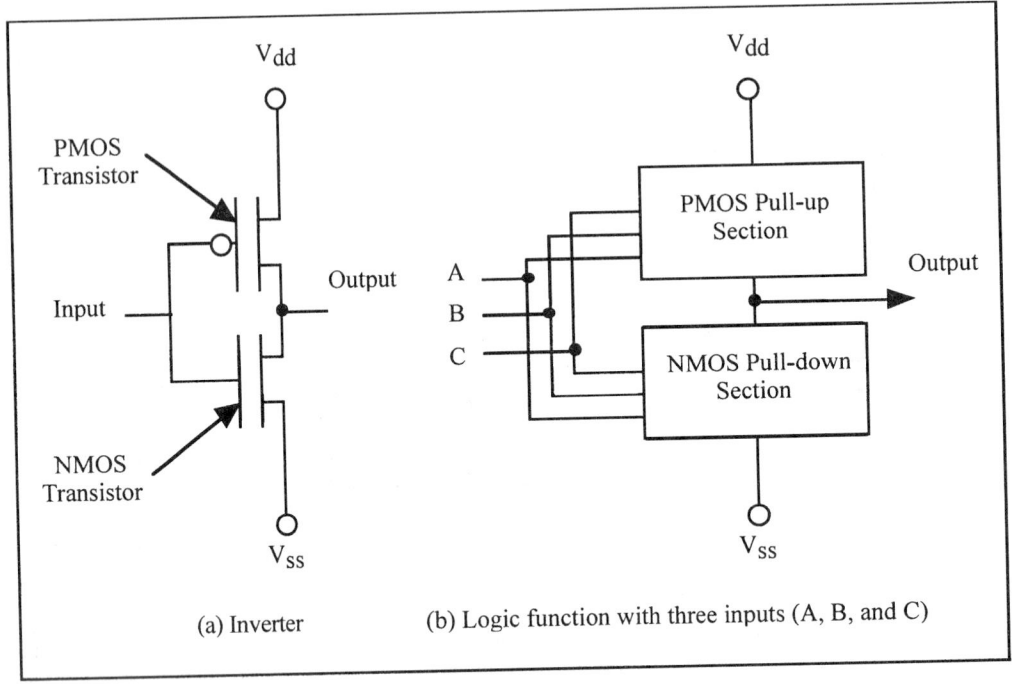

FIGURE 6.1.5   CMOS logic gates.

charged. The result is that, aside from small leakage currents, the dc power dissipation is zero when the pull-up transistors are OFF (output is low). The NMOS circuit does dissipate power when one or more pull-up transistors is in the ON-state since a dc current flows through that transistor through the pull-down resistor.

As technologies advanced, it became possible to efficiently dope selective regions of a substrate in such a manner as to obtain a "well" doped opposite to the substrate. For example, if the substrate is P-type (to create NMOS transistors), a deep well can be doped N-type and PMOS transistors fabricated in that N-type "substrate" region. Being able to fabricate both NMOS and PMOS transistors in the same silicon substrate allows today's CMOS circuits to be created. Figure 6.1.5a shows a basic inverter using a PMOS transistor for the pull up and an MOS

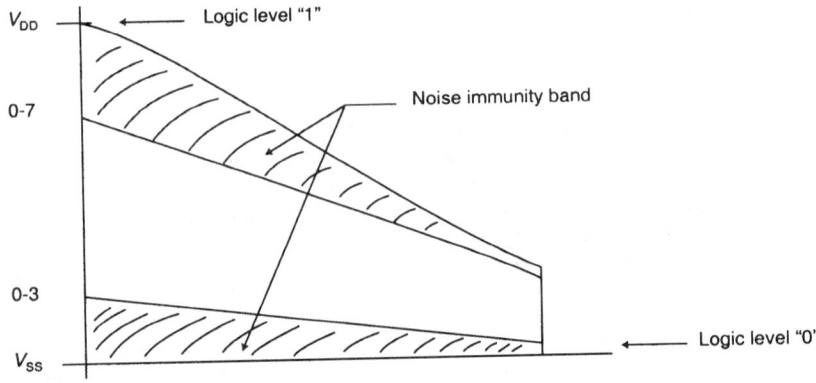

FIGURE 6.1.6   Noise immunity bands for CMOS logic.

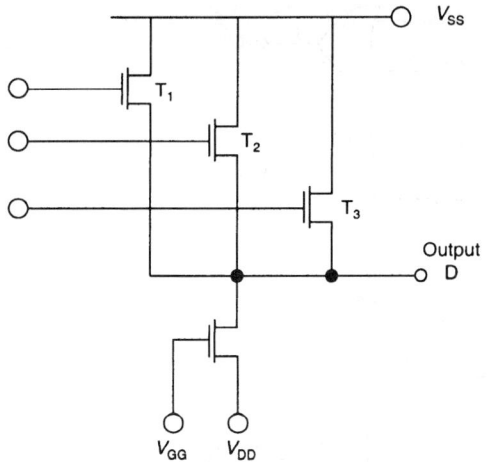

FIGURE 6.1.7   NMOS NOR gate.

transistor for pull down. Figure 6.1.5b shows the general organization of a CMOS circuit, consisting of a pull-up circuit using PMOS transistors and a pull-down circuit using NMOS transistors. Properly designed, the same set of digital inputs to the pull down are also applied to the pull up, as shown. Noise immunity characteristics of CMOS logic circuits are shown in Fig. 6.1.6.

The CMOS logic circuits using pull-up and pull-down sections such as shown in Fig. 6.1.5b require the same number of transistors in the pull-up section as in the pull-down section but do provide static logic functions (the output is held indefinitely so long as the inputs do not change). Dynamic CMOS logic replaces either the pull-up (pull-down) section with a single transistor used to precharge the capacitive load seen by the output (capacitive since outputs drive the gate of subsequent logic circuits and the gate is essentially a capacitance). When this transistor is turned on and the corresponding pull-down (pull-up) section is OFF, the capacitive load is precharged to logic "1" ("0"). Next, the precharge transistor is turned OFF and the pull-down (pull-up) section is activated, either leaving the output unchanged or driving the output capacitance to logic "0" ("1"). Figure 6.1.8 illustrates the general technique for such "dynamic logic," with the clock signal used to control the activations of the precharge pull-up transistor and of the pull-down section in this example.

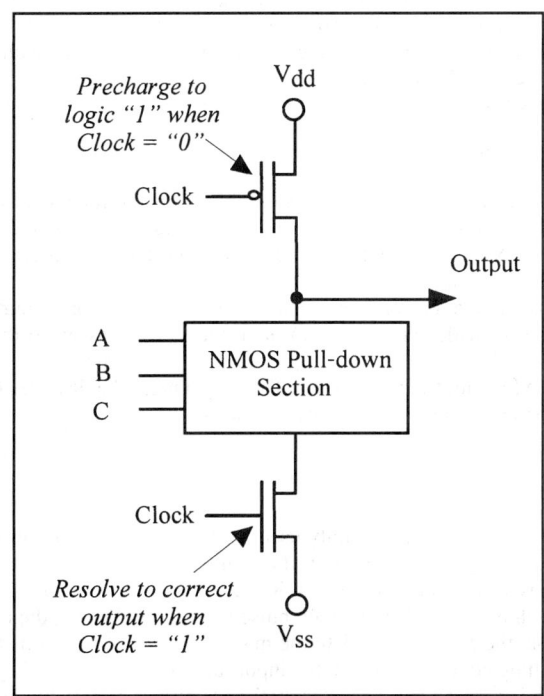

FIGURE 6.1.8   Dynamic CMOS logic gate.

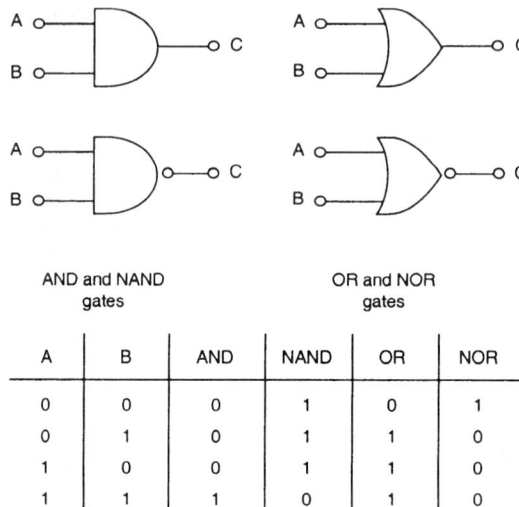

A	B	AND	NAND	OR	NOR
0	0	0	1	0	1
0	1	0	1	1	0
1	0	0	1	1	0
1	1	1	0	1	0

FIGURE 6.1.9   Commonly used gates and truth table.

DIGITAL LOGIC CIRCUITS

The circuits which are described in this and subsequent sections can be fabricated using any of the bipolar or CMOS logic technologies described earlier. The electrical characteristics such as speed and power consumption will be determined by the logic family but the function will be the same in all cases. The circuits described range from simple to more complex types.

Gates

Figure 6.1.9 shows commonly used gates and a truth table gives the functional performance in logic 1 and 0 states. Gates are usually available as hex inverter, quad two input, treble three input, dual four input, and single eight input. AND-OR-INVERT gates are also available and these are used to connect the outputs of two gates together in a wired-OR connection.

In a CMOS transmission gate a p- and an n-channel transistor are connected together. A gate signal turns on both transistors and so provides an ac path through them, whereas when the transistors are off the gate blocks all signals.

Gates are also made in Schmitt trigger versions. These operate as the discrete component circuits and exhibit an hysteresis between the ON and OFF switching positions.

Flip-flops

Flip-flops are bistable circuits which are mainly used to store a bit of information. The simpler types of flip-flops are also called latches. Figure 6.1.10 shows the symbol for some of the more commonly used flip-flops. There are many variations such as master-slave, J–K, edge-triggered, and gate flip-flops.

In the master–slave set–reset flip-flop, a clock pulse is required. During the rising edge of the clock information is transferred from the S and R inputs to the master part of the flip-flop. The outputs are unchanged at this stage. During the falling edge of the clock the inputs are disabled so that they can change their state without affecting the information stored in the master section. However, during this phase the information is transferred from the master to the slave of the flip-flop is disabled when no clock pulse is present.

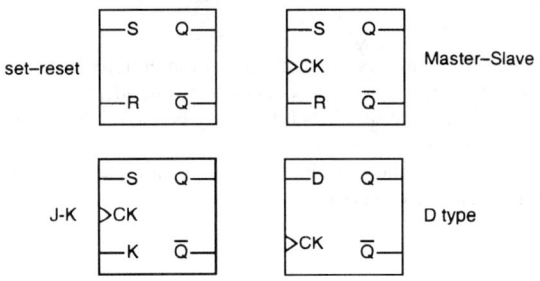

FIGURE 6.1.10   Commonly used flip-flops.

J–K flip-flops are triggered on an edge of the clock waveform. Feedback is used internally within the logic such that the undeterminate state, when both inputs are equal to logic 1, is avoided. Now when the inputs are both at 1, the output will continually change state on each clock pulse. This is also known as *toggling*.

D-type flip-flops have a single input. An internal inverter circuit provides two signals to the J and K inputs so that it operates as a J–K flip-flop, having only two inputs modes.

Counters

Flip-flops can be connected together to form counters in a single package. There are primarily two types, asynchronous and synchronous. The synchronous counters may be subdivided into those with ripple enable and those with parallel or look-ahead carry.

In an asynchronous counter the clock for the next stage is obtained from the output of the preceding stage so that the command signal ripples through the chain of flip-flops. This causes a delay so asynchronous counters are relatively slow, especially for large counts. However, since each stage divides the output frequency of the previous stage by two, the counter is useful for frequency division.

In a synchronous counter, the input line simultaneously clocks all the flip-flops so there is no ripple action of the clock signal from one stage to the next. This gives a faster counter, although it is more complex since internal gating circuitry has to be used to enable only the required flip-flops to change state with a clock pulse. The enable signal may be rippled through between stages or parallel (or look-ahead) techniques may be used, which gives a faster count.

Counters are available commercially, having a binary or BCD count, and which are capable of counting up or down.

Shift Registers

When stored data are moved sequentially along a chain of flip-flops, the system is called a shift register. Commercial devices are available in sizes from four bits to many thousands of bits.

The shift register is serial-in, serial-out, but it is possible to have systems that are parallel data input and output. The only limitation is the number of pins available on the package to accommodate the inputs and outputs. Shift registers can also be designed for left or right shift.

In the register, input data ripple through at the clock pulse rate from the first to the last stage. Sometimes it is advantageous to be able to clock the inputs and outputs at different rates. This is achieved in a first-in first-out (FIFO) register. Each data bit has an associated status and input data are automatically moved along until it reaches the last unused bit in the chain. Therefore the first data to come in will be the first to be clocked out. Data can also be clocked into and out of the register at different speeds, using the two independent clocks.

Data Handling

Several code converter integrated circuits are available commercially. A typical example is a BCD to decimal converter. These converters can also be used as priority encoders. An example may be considered to be a ten-input priority encoder. Several of the lines 0 to 9 may be energized simultaneously but the highest number will generate a BCD output code.

Another example is an eight-channel multiplexer. The channel select lines connect one of the eight data input lines to the output line using BCD code.

Timing

A variety of commercial devices are available to give monostable and astable mutivibrators. Most of these incorporate control gates so that they are more versatile when used in digital systems. A gated monostable will trigger when the voltage at the transistor goes to a logic 1.

External resistors and capacitors are used to vary the duration of the monostable pulse. By feeding the output back to the input, this circuit can also be operated as an astable multivibrator.

Drivers and Receivers

Digital integrated circuits have limited current and voltage drive capability. To interface to power loads, driver and receiver circuits are required, which are also available in an integrated circuit package. The simplest circuit in this category is an array of transistors. Usually the emitters or collectors of the transistors are connected together inside the package to limit the package pin requirements.

For digital transmission systems line drivers and receivers are available. The digital input on the line driver controls an output differential amplifier stage, which can operate into low impedance lines. The line receiver can sense low-level signals via a differential input stage and provide a logic output.

Adders

Adders are the basic integrated circuit units used for arithmetic operations such as addition, subtraction, multiplication, and division. A half-adder adds two bits together and generates a sum and carry bit. A fuller adder has the facility to bring in a carry bit from a previous addition. Figure 6.1.11 shows one bit of the full adder.

The basic single-bit adder can be connected in several ways to add multibit numbers together. In a serial adder the two numbers are stored in shift registers and clocked to the A and B inputs one bit at a time. The carry out is delayed by a clock pulse and fed back to the adder as a carry in.

Serial adders are slow since the numbers are added one bit at a time. A parallel adder is faster. The carry output ripples through from one bit to the next so that the most significant bit cannot show its true value until the carry has rippled right through the system. To overcome this delay, look ahead carry generator may be used. These take in the two numbers in parallel, along with the first carry-in bit, and generate the carry input for all the remaining carry bits.

Adders can be used as subtractors by taking the two's complement of the number being subtracted and then adding. Two's complementing can be obtained by inverting each bit and then adding one to the least significant bit. This can be done within the integrated circuit so that commercial devices are available which can add or subtract, depending on the signal on the control pin.

Adders are used for multiplication by a process of shifting and addition, and division is obtained by subtraction and shifting.

FIGURE 6.1.11   Full adder.

Magnitude Comparators

A magnitude comparator gives an output signal on one of three lines, which indicates which of the two input numbers is larger, or if they are equal. Figure 6.1.12 shows one bit of magnitude comparator. Multibit numbers can be compared by storing them in shift registers and clocking them to the input of the single-bit magnitude comparator, one bit at a time, starting from the most significant bit. Alternatively, parallel comparators may be used where each bit of the two numbers is fed in parallel to a separate comparator bit and the outputs are gated together.

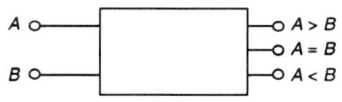

FIGURE 6.1.12   Magnitude comparator.

Rate Multiplier

Rate multipliers can be connected to give a variety of arithmetic functions. A typical example is an adder for adding the numbers X and Y. The clock input in all cases is produced by splitting a single clock into several phases. For the adder $Z = X + Y$ and for the multiplier $Z = XY$.

BIBLIOGRAPHY

Katz, R. H., "Contemporary Logic Design," Benjamin/Cummings Publishing, 1994.

Mano, M., "Digital Design," Prentice Hall, 2002.

Middleton, R. G., "Understanding Digital Logic," Howard W. Sams, 392 pp., 1982.

Oklobdzija, V. G. (ed.), "High-Performance System Design: Circuits and Logic," IEEE Press, 1999.

Pate, R., and W. Berg, "Observe Simple Design Rules When Customizing Gate Arrays," EDN, November 1982.

Proudfoot, J. T., "Programmable logic arrays," *Electronics and Power*, November/December 1980.

Sanchez-Sinencio, E., and A. G. Andreou (eds.), "Low-Voltage/Low Power Integrated Circuits and Systems," IEEE Press, 1999.

Sandige, R. S., "Digital Design Essentials," Prentice Hall, 2002.

Twaddell, W., "Uncommitted IC Logic," EDN, April 1980.

Uyemura, J. P., "Introduction to VLSI Circuits and Systems," John Wiley and Sons, 2002.

Walker, R., "CMOS Logic Arrays: A Design Direction," *Computer Design*, May 1982.

CHAPTER 6.2
ANALOG INTEGRATED CIRCUITS

P. K. Vasudev, S. Tewksbury

INTRODUCTION

Digital logic circuits are designed to preserve the concept of binary states (logic "1" and "0") as voltage levels that are restored to levels defined by the supply voltages at the output of each logic gate. This provides a substantial simplification in the design of complex systems composed of millions of gates—at any point the voltage levels merely reflect the supply voltages at the logic gate driving that point. Analog circuits, on the other hand, must be designed to preserve the concept of a continuum of voltage levels, with the output of an analog component preserving the transformation of an input analog signal into a desired output analog signal. The result is that analog integrated circuits generally do not reach the levels of tens of millions of transistors such as seen in digital logic circuits. The term *analog circuit* has been used somewhat loosely in the past, including circuits such as analog-to-digital converters that convert a continuum of voltage levels into a discrete number corresponding to a set of voltage levels. More recently, the term *mixed signal* VLSI has appeared, representing the placement of both analog circuits and digital circuits on the same integrated circuit. Such mixed signal VLSI merely continues the evolution of the general technology to provide compact and low-cost electronic solutions of practical needs.

One of the developments in digital logic ICs has been the development of *programmable* logic components that can be programmed by the user to implement the logic function desired by the user. Several families of such programmable logic have been developed and present technologies allow not only user programmed logic ICs containing millions of gates but also programmable logic ICs in which the user can embed sophisticated functions such as microprocessors and signal processors. Although not having the efficiency of gate utilization or the speed of fully custom-designed logic ICs, these programmable logic ICs have empowered virtually any user to create low-cost but sophisticated logic circuits through the simple expedient of downloading a configuration file into the IC.

Analog circuits have recently moved also towards such *programmable analog ICs*, providing a set of standard analog circuit functions (operational amplifiers, resistors, and so forth) that can be *programmed* by the user—configuring the interconnections among the analog circuit elements and adjusting the values of some of those elements (e.g., setting the value of a resistor). As these programmable analog IC technologies evolve, programmable VLSI including both digital logic and analog circuit elements will become available, providing the user with powerful components that can be optimized for the specific needs of the user by simply downloading a configuration program to the IC.

There are many distinct types of analog circuit function (operational amplifiers, phase-locked loops, low-noise amplifiers, analog switches, voltage regulators, analog/digital converters, modulators, and so forth). Within each of these analog circuit function types, there are wide ranges of performance specifications. Amplifiers may be intended for lower frequency applications such as audio amplifiers or may be intended for very high frequency applications such as the input amplifier of an RF receiver. Voltage regulators may be required to provide switched local power (generally low current) to subsections of an integrated circuit or may be required to provide high currents to separate electronic components. Because of the vast differences among the various applications of a particular type of analog circuit, there are corresponding vast differences among the design approaches for those various types. In this chapter, basic analog circuit functions that can be routinely manufactured are reviewed. Specialized techniques for special purpose analog circuits with specialized (and high) performance specifications are not considered. The operational amplifier plays a special role in routine analog circuits and is emphasized here since it demonstrates several of the basic concepts in analog circuit design. Analog-to-digital converters also play a special role, serving as the interface between the analog world and the digital world. Basic analog-to-digital conversions are also discussed. Readers interested in the design and use of analog circuits will find a substantial amount of information (product data sheets, application notes, tutorials, and so forth) available at the websites of the analog IC manufacturers. Although the remarkable evolution of digital ICs is well known (e.g., through the rapid introduction of more powerful personal computers), the equally remarkable advances in analog circuits are far less well known. Although hidden from view within automobiles, cellular telephones, audio equipment, and other consumer products, the miniaturization of analog circuits plays a major role in the rapid expansion of electronics into products of all types, including applications requiring miniaturization of sophisticated analog (and mixed analog/digital) circuit functions.

OPERATIONAL AMPLIFIERS

The operational amplifier, or op amp, was originally developed in response to the needs of the analog computer designer. The object of the device is to provide a gain block whose performance is totally predictable from unit to unit and perfectly defined by the characteristics of an external feedback network. This has been achieved by op amps to varying degrees of accuracy, largely governed by unit cost and complexity. Nevertheless, the accuracy of even low cost units has been refined to a point where it is possible to use an op amp in almost any dc to 1 GHz amplifier/signal processor application.

Ideal Operational Amplifier

The *ideal* operational amplifier of Fig. 6.2.1 is a differential input, single-ended output device that is operated from bipolar supplies. As such, it can easily be used as a virtual earth amplifier. Differential output is possible, although not common, and single supply operation will be discussed later. The ideal op amp has infinite gain, infinite bandwidth, zero bias current to generate a functional response at its inputs, zero offset voltage (essentially a perfect match between the input stages) and infinite input impedance. Because of these characteristics, an infinitesimally small input voltage is required at one input with respect to the other to exercise the amplifier output over its full range.

Hence, if one input is held at earth, the other cannot deviate from it under normal operating conditions and becomes a "virtual earth" of the feedback theory definition. The shortfalls against the ideal of practical op amps are now considered with their application consequences.

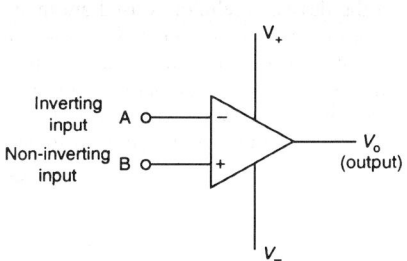

FIGURE 6.2.1   Ideal operational amplifier.

Input Offset Current

Op-amp fabrication uses monolithic integrated circuit construction, which can produce very well-matched devices for input stages and so on by using identical geometry for a pair of devices fabricated at the same time on the same chip. Nevertheless, there is always some mismatch, which gives rise to the input offset current. This is defined as the absolute difference in input bias current, i.e.,

$$I_{\text{diff}} = |I_{\text{A}} - I_{\text{B}}|$$

Effect of Input Bias and Offset Current

The effect of input bias current will be to produce an unwanted input voltage, which can be much reduced by arranging for the bias current to each input to be delivered from an identical source resistance. Figure 6.2.2

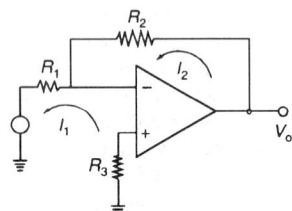

shows a simple inverting amplifier with a gain defined by R_2/R_1, in which R_3 is added to achieve to this effect. In this example the effect of amplifier input impedance is ignored, and signal input impedance is assumed to be zero. In order to balance the bias current source resistance, R_3 is made equal to R_2 in parallel with R_1.

Obviously, the values of the feedback network resistors must be chosen so that, with the typical bias current of the op amp in use, they do not generate voltages that are large in comparison with the supplies and operating output voltage levels. The other sources of error, after balancing source resistances, are the input offset current, and the more insidious effects of the drift of input offset and bias current with temperature, time, and supply voltage, which cannot easily be cor-

FIGURE 6.2.2   Simple inverting amplifier.

rected. Op-amp designers have taken great trouble to minimize the effect of these external factors on the bias and offset current, but, of course, more sophisticated performance is only obtained from progressively more expensive devices. As might be expected, a high precision applications will require an op amp with a high price tag.

Input Offset Voltage and Nulling

As mentioned before a mismatch always exists between the input stages of an op amp, and the input offset voltage (V_{os}) is the magnitude of the voltage that, when applied between the inputs, gives zero output voltage. In bipolar input op amps the major contributor to V_{os} is the bare-emitter voltage mismatch of the differential input stage. General purpose op amps usually have a V_{os} in the region of 1 to 10 mV. This also applies to the modern MOSFET input stage op amps, which achieve an excellent input stage matching with the use of ion implantation. As with the input current parameters, input offset voltage is sensitive to temperature, time, and to a lesser extent input and supply voltages. Offset voltage drift with temperature is often specified as μV per mV of initial offset voltage per °C. As a general rule, the lower the offset voltage of an op amp, the lower is temperature coefficient of V_{os} will be.

Enhanced performance low V_{os} op amps are often produced today by correcting or *trimming* the inherent unbalance of the input stage on chip before or after packaging the device. Techniques used are mainly laser trimming of thin film resistor networks in the input stage and a proprietary process known as *zener zapping*.

Other op amps are available with extra pins connected for a function known as *offset null*. Here a potentiometer is used externally by the user with its slider connected to V_+ or V_- to adjust out the unbalance of the device input stage. A note of caution should be sounded here when using an op amp with an offset null feature in precision circuitry; the temperature coefficient of V_{os} can be changed quite significantly in some op-amp types by the nulling process.

Open Loop Gain

This is one parameter where the practical op-amp approaches the infinite gain ideal very closely and typical gains of 250,000 and higher are quite common at zero frequency. However, it is also common for the gain to start to fall off rapidly at low frequencies, e.g., 10 Hz, with many internally compensated op amps (see Fig. 6.2.3). The commercially available μA741 op amp, for instance, has its gain reduced to unity at around 1 MHz. The closed loop gain is limited by the open loop gain and feedback theory indicates that the greater the difference between open and closed loop gain, the greater the gain accuracy.

Settling Time

Frequently in op-amp applications, such as digital-to-analog converters, the device output is required to acquire a new level within a certain maximum time from a step input change.

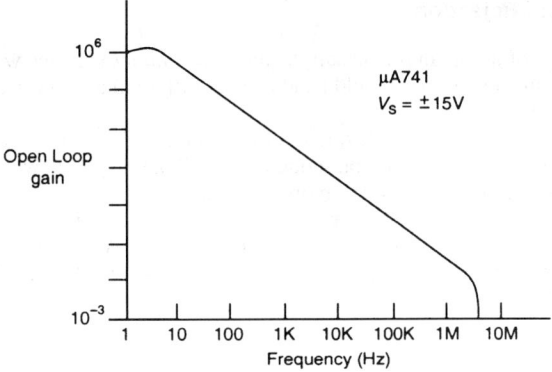

FIGURE 6.2.3   Frequency sensitivity of open loop gain.

The slew rate is obviously a factor in this time, but transient effects will inevitably produce some measure of overshoot and possibly ringing before the final value is achieved. This time, measured to a point where the output voltage is within a specified percentage of the final value, is termed the *settling time*, usually measured in nanoseconds. Careful evaluation of op-amp specifications is required for critical settling time applications, since high-slew-rate op amps may have sufficient ringing to make their settling times worst than medium slew rate devices.

Output Capabilities

As would be expected because of internal design limitations, op amps are unable to achieve an output voltage swing equal to the supply voltages for any significant load resistance. Some devices do achieve a very high output voltage range versus supply, notably those with CMOS FET output stages. All modern op amps have short-circuit protection to ground and either supply built in, with the current limit threshold typically being around 25 μA. It is thus good practice to design for maximum output currents in the 5- to 10-mA region.

Higher current op amps are available but the development of these devices has been difficult because of the heat generated in the output stage affecting input stage drift. Monolithic devices currently approach 1 A and usually have additional circuitry built in to protect against thermal overload and allow output device protection, in addition to normal short-circuit protection.

Power Supply Parameters

The supply consumption will usually be specified at ±5 V, for example, and possibly other supply voltages additionally. Most performance parameters usually deteriorate with reducing supply voltage. Devices are available with especially low power consumption but their performance is usually a trade-off for reduced slew rate and output capability.

Power supply rejection ratio (PSRR) is a measure of the susceptibility of the op amp to variations of the supply voltage. The definition is expressed as the ratio of the change in input offset voltage to the change in supply voltage (μV/V or dB). No op amp ever has a PSRR of less than 70 dB. This should not be taken to indicate that supply bypassing/decoupling is unnecessary. It is good practice to decouple the supplies to general purpose op amps at least every five devices. High speed op amps require careful individual supply decoupling on a by-device basis for maximum stability, usually using tantalum and ceramic capacitors.

Common Mode Range and Rejection

The input voltage range of an op amp is usually equal to its supply voltages without any damage occurring. However, it is required that the device should handle small differential changes at its inputs, superimposed on any voltage level in a linear fashion.

Because of design limitations, input device saturation, and so forth, this operational voltage range is less than the supplies and is termed the common mode input voltage range or swing. As a differential amplifier, the op amp should reject changes in its common mode input voltage completely, but in practice they have an effect on input offset and this is specified as the common mode rejection ratio (CMRR) (μV/V or dB). Note that most general-purpose op amps have a CMRR of at least 70 dB (approximately 300 μV offset change per volt of common mode change) at dc, but this is drastically reduced as the input frequency is raised.

Input Impedance

The differential input resistance of an op amp is usually specified together with its input capacitance. Typical values are 2 MΩ and 1.4 pF for the μA 741.

In practice, the input impedance is usually high enough to be ignored. In inverting amplifiers, the input impedance will be set by the feedback network. In noninverting amplifiers, the input impedance is "bootstrapped" by the op-amp gain, leading to extremely high input impedance for configurations with the feedback case of voltage follower. This bootstrap effect declines as A_{ol} drops off with rising frequency, but it is safe to assume that the circuit input impedance is never worse than that of the op amp itself.

Circuit Stability and Compensation

To simplify design-in and use, many op amps are referred to as "internally" or "fully" compensated. This indicates that they will remain stable for closed loop gains down to unity (100 percent feedback). This almost certainly means that performance has been sacrificed for the sake of convenience from the point of view of users who require higher closed loop gains. To resolve this problem, many op amps are available that require all compensation components to be added externally, according to manufacturers' data and gain requirements. A compromise to this split has been the so-called undercompensated op amps, which are inherently stable at closed loop gains of 5 to 10 and above.

The aim of the standard compensation to shape the open loop response to cross unity gain before the amplifier phase shift exceeds 180°. Thus unconditional stability for all feedback connections is achieved. A method exists for increasing the bandwidth and slew rate of some uncompensated amplifiers known as feedforward compensation. This relies on the fact that the major contributors of phase shift are around the input stage of the op amp and bypassing these to provide a separate high frequency amplifying path will increase the combined frequency response.

OPERATIONAL AMPLIFIER CONFIGURATIONS

Op amps are available ranging from general purpose to ultraprecision. The low-to-medium specification devices are often available in duals and quads. There is a growing trend to standardize on FET input op amps for general-purpose applications, particularly those involving ac amplifiers because of their much enhanced slew rate and lower noise.

Specific devices are available for high speed and high power output requirements. Additionally, some op amps are available as programmable devices. This means their operating characteristics (usually slew rate, bandwidth, and output capability) can be traded off against power supply consumption by an external setting resistor, for instance, to tailor the device to a particular application. Often these amplifiers can be made to operate in the micropower mode, i.e., at low supply voltage and current.

Single-Supply Operational Amplifiers

It is entirely possible to operate any op amp on a single supply. However, the amplifier is then incompatible with bipolar dc signal conditioning. Single-supply operations is entirely suitable for ac amplifiers.

Most later generation op amps have been designed with single supply operation in mind and frequently use pnp differential input arrangements. They have an extended input voltage range, often including ground-in single-supply mode, and are referred to as single-supply op amps.

Chopper and Auto Zero Operational Amplifiers

Many attempts have been made to circumvent the offset and drift problems in op amps for precision amplifier applications. One classic technique is the chopper amplifier, in which the input dc signal is converted to a proportional ac signal by a controlled switch. It is then applied to a high gain accuracy ac amplifier, removing the inherent drift and offset problems. After amplification, dc restoration takes place using a synchronous switching action at the chopper frequency. This is available on a single monolithic device.

Other approaches have used the fact that the output is now required continually, i.e., as in analog to digital converters, and have used the idle period as a self-correction cycle. This again involves the use of a switch, but in this case it is usually used to ground the input of the op amp. The subsequent output is then stored as a replica of actual device error at that moment in time and subtracted from the resultant output in the measurement part of the cycle.

OPERATIONAL AMPLIFIER APPLICATIONS

Op amps are suitable for amplifiers of all types, both inverting and noninverting, dc, and ac With capacitors included in the feedback network, integrators and differentiators may be formed. Active filters form an area where the availability of relatively low cost op amps with precisely defined characteristics has stimulated the development of new circuit design techniques. Filter responses are often produced by op amps configured as gyrators to simulate large inductors in a more practical fashion. Current-to-voltage converters are a common application of op amps in such areas as photodiode amplifiers.

Nonlinear circuit may be formed with diodes or transistors in the feedback loop. Using diodes, precision rectifiers may be constructed, for ac to dc converters, overcoming the normal errors involved with forward voltage drops. Using a transistor in common base configuration within the feedback loop is the basic technique used for generating logarithmic amplifiers. These are extensively used for special analog functions, such as division, multiplication, squaring, square rooting, comparing, and linearization. Linearization may also be approached by the *piecewise* technique, whereby an analog function is *fitted* by a series of different slopes (gain) taking effect progressively from adjustable breakpoints.

Signal generation is also an area where op amps are useful for producing square, triangle, and sine wave functions.

Instrumentation Amplifiers

Many applications in precision measurement require precise, high gain differential amplification with very high common mode rejection for transducer signal conditioning, and so forth. To increase the common mode rejection and gain accuracy available from a single op amp in the differential configuration, a three-op-amp circuit is used, which is capable of much improved performance. The standard instrumentation amplifier format, may be assembled from individual op amps or may be available as a complete (often hybrid) integrated circuit. The inputs are assigned one up amp each which may have gain or act as voltage followers. The outputs of these amplifiers are combined into a single output via the differential op-amp stage. Resistor matching and absolute accuracy is highly important to this arrangement. Suitable resistor networks are available in thin film (hybrid) form.

Comparators

The comparator function can be performed quite easily by an op amp, but not with particularly well-optimized parameters. Specific devices are available, essentially modified op amps, to handle the comparator function in a large range of applications. The major changes to the op amp are to enable the output to be compatible with the logic levels of standard logic families (e.g., TTL, ECL, CMOS) and trade off a linear operating characteristic against speed. A wide common mode input range is also useful. Thus, all the usual op-amp parameters are applicable to comparators, although usually in their role as an interface between analog signals and logic circuits, there is no need for compensation.

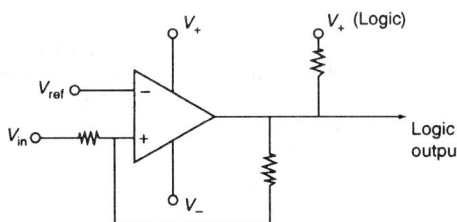

FIGURE 6.2.4   Voltage comparator using op amps.

A typical comparator application shown in Fig. 6.2.4. The circuit has a separate ground terminal to reference the output swing to ground while maintaining bipolar supply operation for the inputs. The output TTL compatibility is achieved by a pull-up resistor to the TTL supply rail V_{CC}, as this is an open collector type comparator. Higher speed comparators will of necessity employ "totem-pole" output structures to maintain fast output transition times. The circuit produces a digital signal dependent on whether the input voltage is above or below a reference threshold. To clean up the switching action and avoid oscillation at the threshold region, it is quite often necessary to apply hysteresis. This is usually relatively easy to achieve with a small amount of positive feedback from the output to input.

General-purpose comparators exhibit response time (time from a step input change to output crossing the logic threshold) in order of 200 ns. Their output currents are limited, compared to op amps, being usually sufficient to drive several logic inputs. Often a strobe function is provided to disable the output from any input related response under logic signal control.

Comparator Applications.   Voltage comparators are useful in Schmitt triggers and pulse height discriminators. Analog to digital converters of various types all require the comparator function and frequently the devices can be used independently for simple analog threshold detection. Line receivers, RC oscillators, zero crossing detectors, and level shifting circuits are all candidates for comparator use. Comparators are available in precision and high speed versions and as quads, duals, and singles of the general-purpose varieties. These latter are often optimized for single-supply operation.

ANALOG SWITCHES

Most FETS have suitable characteristics to operate as analog switches. As voltage controlled, majority carrier devices, they appear as quite linear low value resistors in their "on" state and as high resistance, low leakage path in the "off" state.

Useful analog switches may be produced with JEETs (junction field effect transistors) or MOSFETs (metal oxide semiconductor FETs). As a general rule, JFET switches are lowest "off" leakage. Either technology can be integrated in single- or multichannel functions, complete with drivers in monolithic form.

The switched element is the channel of a FET or channels of a multiple FET array, hence the gate drive signal must be referred to the source terminal to control the switching of the drain-source resistance. It can be seen that the supply voltages of the gate driver circuit in effect must contain the allowable analog signal input range. The FET switched element has its potential defined in both states by the analog input voltage and additionally in the off stage by the output potential. The gate drive circuit takes as input of a ground refereed, logic compatible (usually TTL) nature and converts it into a suitable gate control signal to be applied to the floating switch element. The configuration will be such that the maximum analog input range is achieved while minimizing any static or dynamic interaction of the switch element and its control signal. Figure 6.2.5 shows a typical analog switch in block diagram form. This is a dual SPST function. Because of their close relationship to mechanical switches in use,

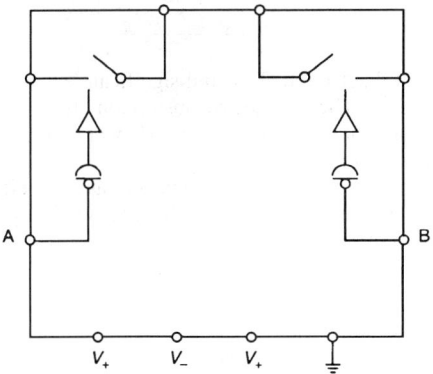

FIGURE 6.2.5 Dual SPST analog switch.

mechanical switch nomenclature is often used, e.g., SPST is single pole, single throw, and DPDT is double pole, double throw, both poles being activated by the same control single. Break-before-make switching is frequently included to avoid shorting analog signal inputs by a pair of switch elements.

The equivalent of single pole multiway switches in monolithic form is referred to as *analog multiplexers*.

Analog Switch Selection

Selection is based on voltage and current handling requirements, maximum on resistance tolerable, minimum off resistance and operating speed. Precautions have to be taken in limiting input overvoltages and some CMOS switches were prone to latch-up, a destructive SCR effect that occurred if the input signal remained present after the supplies have been removed. Later designs have since eliminated this hazard.

SAMPLE AND HOLD CIRCUITS

A sample and hold takes a *snapshot* of an analog signal at a point in time and holds its voltage level for a period by storing it on a capacitor. It can be made up from two op amps connected as voltage followers and an analog switch.

This configuration block diagram is shown in Fig. 6.2.6. In operations, the input follower acts as an impedance buffer and charges the external hold capacitor when the switch is closed, so that it continually tracks the input voltage (the "sample" mode). With the switch turned off, the output voltage is held at the level of the input voltage at the instant of switch off (the "hold" mode). Eventually, the charge on the hold capacitor will be drained away by the input bias current on the follower. Hence the hold voltage *droops* at a rate controlled by the hold capacitor size and the leakage current, expressed as the droop rate in $mVs^{-1} \mu F^{-1}$.

When commanded to sample, the input follower must rapidly achieve a voltage at the hold capacitor equivalent to that present at the input. This action is limited by the slew rate into the hold capacitor and settling time of the input follower and it is referred to as the acquisition time. Dynamic sampling also has other sources of error. The hold capacitor voltage tends to lag behind a moving input voltage, due primarily to the on-chip charge current limiting resistor. Also, there is a logic delay (related to "aperture" time) from the onset of the hold command and the switch actually opening, this being almost constant and independent of hold capacitor value. These two effects tend to be of opposite sign, but rarely will they be completely self-cancelling.

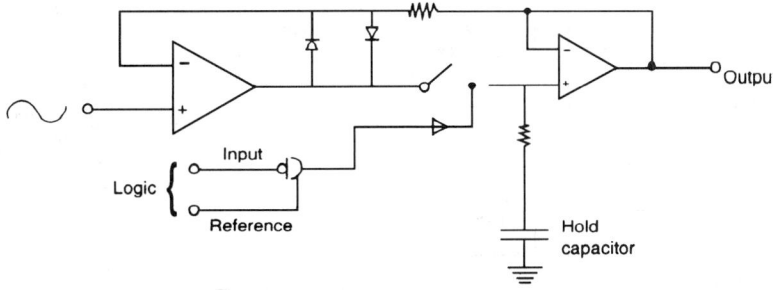

FIGURE 6.2.6 Sample and hold circuit using op amps.

SAMPLE AND HOLD CAPACITOR

For precise applications, the hold capacitor needs more careful selection. To avoid significant errors, capacitors with a dielectric exhibiting very low hysteresis are required. These dielectric absorption effects, are seen as changes in the hold voltage with time after sampling and are not related to leakage. They are much reduced by using capacitors constructed by polystyrene, polypropylene, and PTFE.

The main applications for sample and hold circuits are in analog to digital converters and the effects of dielectric absorption can often be much reduced by performing the digitization rapidly after sampling, i.e., in a period shorter than the dielectric hysteresis relaxation time constant.

DIGITAL-TO-ANALOG CONVERTERS

Digital-to-analog converters (DACs) are an essential interface circuit from the digital world into the analog signal processing area. They are also the key to many analog-to-digital converter techniques that relay on cycling a DAC in some fashion through its operating range until parity is achieved between the DAC output and the analog input signal. All DACs conform to the block diagram of Fig. 6.2.7. The output voltage is a product of the digital input word and an analog reference voltage. The output can only change in discrete steps and the number of steps is immediately defined by the digital inputs. Should this be eight, e.g., an 8-bit DAC, the number of steps will be 256 and the full-scale output will be 256 times voltage increment related to the reference.

Although various techniques can be used to produce the DAC function, the most widely used is variable scaling of the reference by a weighing network switched under digital control. The building blocks of this kind of DAC are (a) reference voltage; (b) weighing network; (c) binary switches; and (d) an output summing amplifier. All may be built in, but usually a minimum functional DAC will combine network and switches.

R-2R Ladders

The weighing network could be binary weighted with voltage switching and summation achieved with a normal inverting op amp. However, even in this simple 4-bit example, there is a wide range of resistor values to implement and speed is likely to suffer due to the charging and discharging of the network input capacitances during conversion. The resistor ladder network employs current switching to develop an output current proportional to the digital word and a reference current. The current switching technique eliminates the transients involving the nodal parasitic capacitances. Significantly, only two resistance values are required and the accuracy of the converter is set by the ratio precision rather than of the absolute value. Analog IC fabrication is capable of accommodating just such resistance value constraints quite conveniently. Monolithic DACs are available in 8- and 12-bit versions and recently in 16 bit. Sixteen-bit and above DACs are also available in hybrid form.

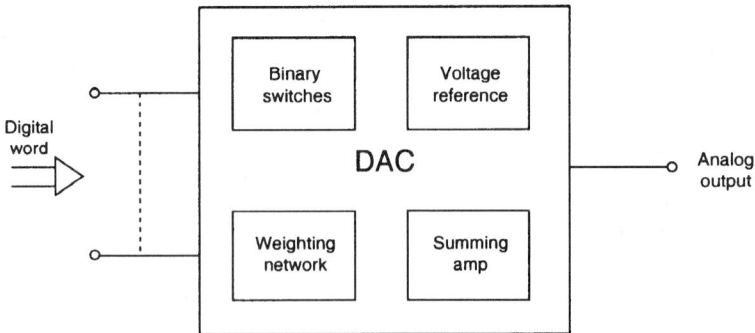

FIGURE 6.2.7 Digital-to-analog converter.

Resolution, Linearity and Monotonicity

Resolution has already been touched upon as being defined by the digital word length and reference voltage increment. However, it is important to note that it is quite possible to have, say, a 12-bit DAC in terms of resolution which is incapable of 12-bit accuracy. This is because of nonlinearities in the transfer function of the DAC. It may well be accurate at full scale but deviate from an ideal straight line response because of variations in step size at other points on the characteristic. Linearity is specified as a worst case percentage of full-scale output over the operating range. To be accurate to n bits, the DAC must have a linearity better than $1/2$ LSB (LSB = least significant bit, equivalent to step size) expressed as a percentage of full scale (full scale = 2 in. × step size).

Differential linearity is the error in step size from ideal between adjacent steps and its worst-case level determines whether the converter will be monotonic. A monotonic DAC is one in which, at any point in the characteristic from zero to full scale, an increase in the digital code results in an increase in the absolute value of the output voltage. Nonmonotonic converters may actually "reverse" in some portion of the characteristic, leading to the same output voltage for two different digital inputs. This obviously is a most undesirable characteristic in many applications.

Settling Time

The speed of a DAC is defined in terms of its settling time, very similar to an op amp. The step output voltage change used is the full-scale swing from zero and the rated accuracy band is usually $+1/2$ LSB. Settling times of 100 = 200 ns are common with 8-bit, 4-2R ladder monolithic DACs. The compensation of the reference op amp will have a bearing on settling time and must be handled with care to maintain a balance of stability and speed.

Other DAC Techniques

There are many other methods that have been proposed and used for the DAC function. There is insufficient space here to make a full coverage. However, mention should be made of the pulse width technique. The digital inputs are assigned time weighing so that in, say, a 6-bit DAC, the LSB corresponds to 1 time unit and the MSB to 32 time units. The time units are derived by variable division from a master clock and usually clocked out at a rate which is some submultiple of the master clock.

The duty cycle will then vary between 0 and 63/64 in its simplest form. This pulse rate is integrated by an averaging filter (usually RC) to produce a smooth dc output. The full scale is obviously dependent on the pulse height and this must be related back to a voltage reference, albeit only a regulated supply. This type of DAC* is often used for generating control voltages for voltage controlled amplifiers used for volume, brightness, color, and so forth (TV application), and it is possible to combine two 6-bit pulse width ratio outputs for 12-bit resolution (not accuracy) to drive voltage controlled oscillators, i.e., varactor tuners. This is one application (tuning) where nonmonotonicity can be acceptable.

ANALOG-TO-DIGITAL CONVERTERS

Analog-to-digital converters (ADCs) fall into three major categories: (a) direct converters; (b) DAC feedback; and (c) integrating.

Direct or "Flash" Converters

Flash converters have limited resolution but are essential for very high-speed applications. They also have the advantage of providing a continuous stream of the digitized value of the input signal, with no *conversion time* waiting period. They consist of a reference voltage which is subdivided by a resistor network and applied to the inputs of a set of comparators. The other inputs of the comparator are common to the input voltage. A digital encoder would then produce, in the case of a 3-bit converter, for example, a weighted 3-bit output from the eight comparator inputs. The method is highly suitable to video digitization and is usually integrated with a high-speed digital logic technology such as ECL or advanced Schottky TTL. Flash converters in monolithic form of up to 9-bit resolution have been built.

Feedback ADCs

Feedback ADCs use a DAC within a self-checking loop, i.e., a DAC is cycled through its operating range until parity is achieved with the input. The digital address of the DAC at that time is the required ADC output. Feedback ADCs are accurate with a reasonably fast conversion time. Their resolution limitations are essentially those of the required DAC cost, performance, and availability.

Feedback ADCs require a special purpose logic function known as a successive approximation register (SAR). The SAR uses an iterative process to arrive at parity with the ADC input voltage in the shortest possible time. It changes the DAC addresses in such a way that the amplitude of the input is checked to determine whether it is greater or smaller than the input on a bit sequential basis, commencing with the MSB. This implies continuous feedback from a comparator on the analog input and DAC output connected to the SAR.

Figure 6.2.8 shows a block diagram of a 10-bit ADC available on a single chip. The 10-bit DAC is controlled by a 10-bit SAR with an internal clock, and the DAC has an on-board reference. The digital output is fed through tristate buffers to ease the problems of moving data onto a bus-structured system. These are high impedance (blank) until the device is commanded to perform a conversion.

After the conversion time, typically 25 μs for this type of device, a data ready flag is enabled and correct data presented at the ADC output. An input offset control is provided so that the device can operate with bipolar inputs. Because of the nature of the conversion process, satisfactory operation with rapidly changing analog inputs may not be achieved unless the ADC is preceded by a simple and hold circuit.

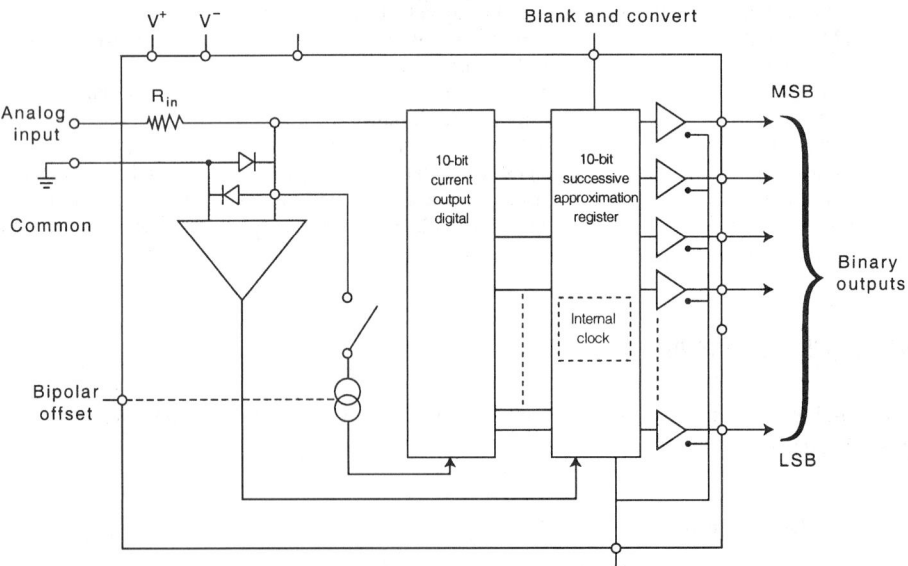

FIGURE 6.2.8   10-bit analog-to-digital converter.

Integrating ADCs

Integrating ADCs use a variety of techniques, but all usually display high resolution and linearity. They also have a much greater ability to reject noise on the analog input than other types of ADC. The penalty that is paid for this is a much longer conversion time which is also variable with actual input voltage. These factors make integrating ADCs very suitable for digital voltmeter, panel meter, and other digital measuring applications.

BIBLIOGRAPHY

Jaeger, R. C., "Microelectronic Circuit Design," Irwin McGraw-Hill, 1997.

Muller, J., and T. Kamins, "Device Electronics for Integrated Circuits," John Wiley and Sons, 2003.

Ng, K. K., "Complete Guide to Semiconductor Devices," Wiley Interscience, 2002.

Terrell, D., "Op Amps, Design, Application and Troubleshooting," Butterworth-Heinemann, 1996.

CHAPTER 6.3
MEMORY CIRCUITS

P. K. Vasudev, S. Tewksbury

DYNAMIC RAM

Dynamic RAM (DRAM) provides the highest density IC memory capacity of the various RAM/ROM technologies. With a market driven by the evolution of computers to higher performance products with increasingly powerful microprocessors (and correspondingly larger and faster main memory), the DRAM technologies are among the most sophisticated of the VLSI technologies. This section reviews the basics of DRAM circuits, drawing on the evolution from the earlier small DRAMs (1 kbit 3-transistor/cell memories to 64 kbit 1-transistor/cell memories) to more advanced DRAMs. Given the sophistication of contemporary DRAM ICs, it is not possible to describe in detail today's advanced technologies. However, it is important to recognize that advances other than the underlying transistor-level design and fabrication technology have played a substantial role in providing fast data transfers between the DRAMs and microprocessors.

DRAMs (like SRAMS and ROMS) are generally organized as an array of cells, each cell holding one bit of data. When an address of a data bit is applied, that address specifies which row (the row address part of the address word) of the array is to be selected and, given the row, which data element of the row is to be selected (according to the column address portion of the address word). Earlier DRAM technologies were designed to access a single data bit cell for a given input address, with a new address applied to access another data bit cell. The data bit is stored as a charge on a capacitor. To read that data bit, the row address is applied and the small output voltage from each cell of the row is connected to individual column lines. Each column line connects through a switch to the corresponding (e.g., Nth) bit of each row and only the addressed row's switch is closed. The small signals from each of the selected row's cells are transmitted to a linear array of amplifiers, converting that small signal into a digital signal depending on whether charge was or was not stored on the cell's capacitor. In this sense, an entire row of the DRAM array is read at the same time and that entire row of data is available when the column address is applied. Earlier DRAM architectures selected one of these data bits as the desired output data, discarding the others. The inefficiency of this approach is clear and architectures have advanced to take advantage of the reading of an entire row of data. One approach was to retain the separate row and column addresses but to allow successive column addresses to be applied (without requiring a new read of the row) in order to select successive data bits already placed in the output stage (avoiding the delay time necessary to complete the reading of the row and amplification of the small signals from the cells of the row). However, the data transfer rates between the DRAM module and the microprocessor remained a substantial bottleneck to increased performance. Microprocessor architectures were advancing rapidly, including addition of cache memories to avoid, when possible, having to access the external DRAM to acquire instructions and data. With the advanced microprocessor architectures, a quite different approach became possible, greatly increasing the net data transfer rates between the DRAM module and the microprocessor. In particular, an entire row can be accessed by the row address and then, rather than reading individual bits of that row by applying a new column address each time, the entire row of data can be transferred to the microprocessor. Although somewhat simplified relative to the actual operation, the basic approach is as follows. A row address is applied and the small signals corresponding to data stored in the cells of that row are passed to the amplifier array, generating digital signals that are loaded into a parallel-load/serial-shift register. Once loaded, the data in the shift register can be clocked out at a high data rate without requiring application of a new address. It is through this process that the very high data rates can be achieved. If the address of the row that will be needed next is known when the current row of data is being shifted out of the DRAM, that address can be applied so that the small signals from that next row are being transferred to the amplifiers and converted to digital signals—ready to be loaded into the register for shifting out as soon as the present data in the register have been shifted out.

This provides a fast transfer of data from the DRAM to the microprocessor. A corresponding approach allows fast data transfer from the microprocessor into the DRAM. To write data, data corresponding to a row are shifted into an input data register. When the register has been filled, its contents are loaded into a parallel data register and are available for writing into the row specified by the write address. While the data in that parallel data register are being written into the DRAM, the next data to be written can be shifted into the serial shift register. The basic approach discussed above has actually been in use for some time, appearing in video RAM. In that case, an entire row of a screen is loaded from memory into a parallel-in/serial-out data register of the read operation to read the row and an entire row of a screen is loaded into the serial-in/parallel-out data register of the write operation. While a row of a data image is being shifted out, another row of the data image can be loaded.

The above example of architectural approaches to relax performance bottlenecks (in this case the data transfers between DRAM and the microprocessor) illustrates an important principle that will increasingly characterize VLSI components. In particular, system-level solutions will increasingly define the functional characteristics of mainstream technologies such as DRAM and microprocessors, providing performance enhancements unachievable through purely technology and circuit design approaches.

Space prohibits detailed discussion of these architectural schemes, or the rather sophisticated techniques that have been used to miniaturize the surface area of the capacitance. However, the underlying principles are similar to earlier memory designs, as summarized below.

DRAM is the lowest-cost, highest-density RAM available. Since the 4k generation DRAM (dynamic RAM) has held a 4 to 1 density advantage over static RAM, its primary competitor. Dynamic RAM also offers low power and a package configuration which easily permits using many devices together to expand memory sizes. Today's computers use DRAM for main memory storage with memory sizes ranging from 16 kbytes to hundreds of megabytes. With these large-size memories, very low failure rates are required. The metal oxide semiconductor (MOS) DRAM has proven itself to be more reliable than previous memory technologies (magnetic core) and capable of meeting the failure rates required to build huge memories. DRAM is the highest volume and highest revenue product in the industry.

Cell Construction

The 1k and early 4k memory devices used the three-transistor (3-T) cell shown in Fig. 6.3.1. The storage of a "1" or a "0" occurred on the parasitic capacitor formed between the gate and source of transistors. Each cell had amplification thus permitting storage on a very small capacitor. Because of junction and other leakage paths the charge on the capacitor had to be replenished at fixed intervals; hence the name dynamic RAM. Typically a *refresh* pulse of a maximum duration was specified.

The next evolution, the 1-T cell, was the breakthrough required to make MOS RAM a major product. The 1-T cell is shown in Fig. 6.3.2. This is the 16k RAM version of the cell. The 4k is similar except that only one level of polysilicon (poly 1) is used. The two-level poly process improves the cell density by about a factor of 2 at the expense of process complexity. The transistor acts as a switch between the capacitor and the bit line. The bit line carries data into and out of the cell. The transistor is enabled by the word line, which is a function of the row address. The row address inputs are decoded such that one out of **N** word lines is enabled. **N** is the number of rows which is a function of density and architecture. The 16k RAM has 128 rows and 128 columns. The storage cell transistor is situated such that its source is connected to the capacitor, its drain is connected to the bit line and the gate is connected to the word line. Higher-density DRAMs today have evolved more sophisticated structures, involving trenches and stacked capacitors to continue meeting the requirements of the 1-T cell. By simultaneously driving down the cell size, while maintaining the required storage capacitance of

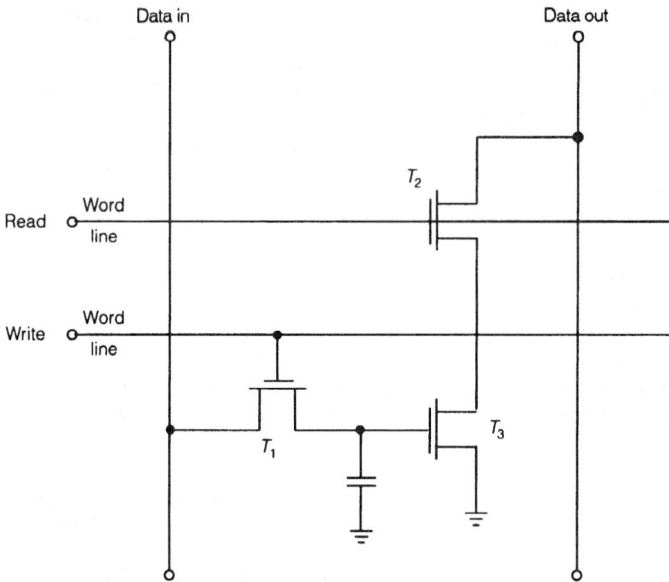

FIGURE 6.3.1 Dynamic RAM cell.

bit line, DRAMs have been able to sustain the steady increase in DRAM capacity, moving beyond the 1 GB level. This steady increase is expected to continue.

Cell Process

The principal process used in early generation DRAMs was the double-level polysilicon gate process. The process uses n-channel devices which rely on electron mobility. The 1k RAM (1103) used a p-channel process which relied on hole mobility. The p-channel process although simpler to manufacture (it is more forgiving to contamination) is by nature slower than the n-channel process (electrons are much faster than holes) and cannot operate a 5-V levels. Therefore once manufacturing technology advanced enough to permit n-channel to yield, it quickly displaced p-channel techniques for memory. Today the DRAM processes are extremely sophisticated, using trenches, stacked capacitors, and high-impedance dielectrics that allow densities in the 100-million gate level to be achieved. They typically use CMOS technology and power supplies as low as 1 V or so.

System Design Consideration Using Dynamic RAMs

Dynamic RAMs provide the advantages of high density, low power, and low cost. These advantages do not come for free; the dynamic RAM is considered to be more difficult to use than static RAMs. This is because dynamic RAMs require periodic refreshing in order to retain the stored data. Furthermore, although not generic to dynamic RAMs, most dynamic RAMs multiplex the address bits which requires more complex edge-activated multi-clock timing relationships. However, once the system techniques to handle refreshing, address multiplexing, and clock timing sequences have been mastered, it becomes obvious that the special care required for dynamic RAMs is only a minor inconvenience.

A synopsis of the primary topics of concern when using dynamic RAMs follows.

Address Multiplexing. The use of address multiplexing reduces the number of address lines and associated address drivers required for memory interfacing by a factor of 2. This scheme, however, requires that the RAM

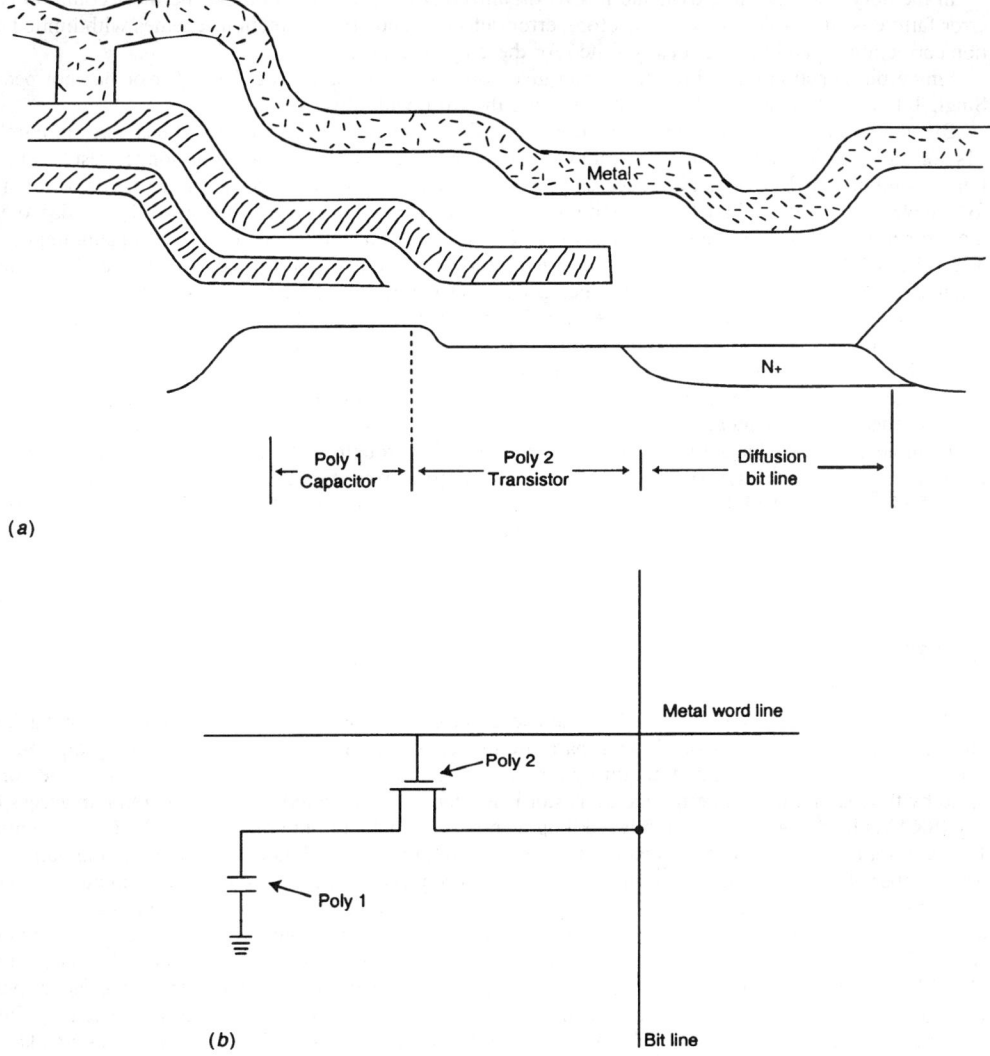

FIGURE 6.3.2 (*a*) DRAM cross section; (*b*) schematic of single-transistor cell DRAM.

address space be partitioned into an *X–Y* matrix with half the addresses selecting *X* or ROW address field and the other half of the address selecting the *Y* or column address field. The ROW addresses must be valid during the RAS clock and must remain in order to satisfy the hold requirements of the on-chip address latches. After the ROW address requirements have been met, the addresses are allowed to switch to the second address field, column addresses, a similar process is repeated using CAN (column address strobe) clock to latch the second (column) address field. Note, for RAS-only refreshing the column address field phase is not required.

Error Detection and Error Correction. In a large memory array there is a statistical probability that a soft error(s) or device failure(s) will occur resulting in erroneous data being accessed from the memory system. The larger the system, the greater the probability of a system error.

In memory systems using dynamic RAMs organized *N* by one configuration, the most common type of error-failure is single-bit oriented. Therefore, error detection and error correction schemes with limited detection/correction capabilities are ideally suited for these applications.

Single-bit detection does little more than give some level of confidence that an error has not occurred. Single-bit detection is accomplished by increasing the word width by 1 bit.

By adding a parity bit and selecting the value of this bit such that the 4-bit word has an even number of "1s" or an odd word number of "1s," complementing any single bit in the word no longer results in a valid binary combination. Notice that it is necessary to complement at least two bit locations across the word in order to achieve new or new valid data word with the proper number of "1s" or "0s" in the word. In order to extract the error information from the stored data word. This concept was introduced by Richard Hamming in 1950. The scheme of defining the number of positions between any two valid word combinations is called the "Hamming distance." The Hamming distance determines the detection/correction capability.

Correction/detection capability can be extended by additional parity bits in such a way that the Hemming distance is increased. Figure 6.3.2 shows how a 16-bit word with five parity bits (commonly referred to as "check bits") can provide single-bit detection and single-bit correction. In this case, a single-bit error in any location across the word, including the check bits, will result in a unique combination of the parity errors when parity is checked during an access.

As indicated above, the ability to tolerate system errors is brought at the expense of using additional memory devices (redundant bits). However, the inclusion of error correction at the system level has been greatly simplified by the availability of error correction chips. These chips provide the logic functions necessary to generate the check bits during a memory write cycle and perform the error detection and correction process during a memory ready cycle.

STATIC RAM

DRAM requires that the charge stored on the capacitance of the data cells be refreshed at a regular interval since the charge slowly leaks off through parasitic diodes. In addition, the signal provided by the cells when accessed is a small analog signal, requiring amplification and noise avoidance. These penalties are compensated by the requirement that only one transistor is needed for each memory cell. Static random access memory (SRAM) holds its data indefinitely, so long as power is applied to the memory. Each cell is essentially a flip-flop, set to logic "1" or "0" when written and providing a full-amplitude logic signal to the output stage. The number of transistors needed for this storage and to connect the flip-flop to the input and output data lines is typically six. With a regular digital signal provided when reading the SRAM, read delays are substantially smaller than in a DRAM, the fast access time justifying the smaller amount of data that can be stored on an SRAM IC (that lower storage capacity resulting from the larger number of cell transistors than needed in the DRAM). Several applications of this faster SRAM appear in a personal computer—including video memory, cache memory, and so forth. Since there is no need for special microfabrication steps to create the minimum area, large capacitance cells of the DRAM ICs, the SRAM can be fabricated on the same IC as standard digital logic. Although there have been demonstrations of the fabrication of microprocessors and DRAM on the same IC, the general need for more DRAM ICs than microprocessor ICs in a personal computer leads to no major advantage of this more complicated technology. On the other hand, the higher capacity of the DRAM memory technology may drive cointegration with logic circuitry for other applications requiring that higher-density RAM.

Just as in the case of the DRAMs discussed earlier, architectural principles are also a significant part of the SRAM IC design. With multiple levels of SRAM cache memory being used in contemporary microprocessor systems, the cache memories connected directly to the DRAM memory can draw upon some of the advantages of reading/writing entire rows of memory in parallel. Cache memories also generally require sophisticated algorithms to place memory from some real DRAM address location into a local RAM accessed with a different address. Translations and other system-level functions related to such needs can be embedded directly within the SRAM IC itself. As in the case of the DRAMs, space here does not permit a discussion of these advanced themes. However, the underlying principles of the basic SRAM cell are captured in the earlier technologies discussed next.

Organization

Static RAMs in the 1k (k = 1024) and 4k generations are organized as a square array (number of rows of cells equals the number of columns of cells). For the 1k device there are 32 rows and 32 columns of memory cells. For a 4096 bit RAM there are 64 rows and 64 columns. The RAMS of the seventies and beyond include on-chip decoders to minimize external timing requirements and number of input/output (I/O) pins. Today, much higher-density SRAMs are being built using the same basic array design, but employing scaled silicon structures.

To select a location uniquely in a 4096-bit RAM, 12 address inputs are required. The lower-order 64 addresses decode to select one column. The intersection of the selected row and the selected column locates the desired memory cell.

Static RAMs of the late seventies and early eighties departed from the square array organization in favor of a rectangle. This occurred for primary reasons: performance and packaging considerations.

Static RAM has developed several application niches. These applications often require different organizations and/or performance ranges. Static RAM is currently available in 1-bit I/O, bout bit I/O, and 8-bit I/O configurations. The 1- and 4-bit configurations are available in a 0.3-in-wide package with densities of 1k, 4k, and 16k bits. The 8-bit configuration is available in a 0.6-in-wide package in densities of 8k and 16k bits. The 8-bit I/O device is pin compatible with ROM (read only memory) and EPROM (electrically programmable ROM) devices. All of the static RAM configurations are offered in two speed ranges. Ten nanoseconds typically are used as the dividing line for part numbering.

Construction

The static RAM uses a six-device storage cell. The cell consists of two cross-coupled transistors, two I/O transistors, and two load devices. There are three techniques that have been successfully employed for implementing the load devices: enhancement transistors, depletion transistors, and high-impedance polysilicon load resistors.

Enhancement transistors are normally off and require a positive gate voltage (relative to the source) to turn the device on. There is a voltage drop across the device equal to the transistors' threshold. Depletion transistors are normally on and require a negative gate voltage (relative to the source) to turn them off. There is no threshold drop across this device. Polysilicon load resistors are passive loads which are always on. Formed in the polylevel of the circuit they are normally undoped and have very high impedances (5000 MΩ typically). They have the advantage of very low current (nanoamperes) and occupy a relatively small area. All present large density, state of the art, static RAMs now use this technique for the cell loads. Further enhancements have been made by manufacturing a two-level polyprocess, thereby further decreasing cell area.

The operation of the polyload SRAM cell is shown in Fig. 6.3.3. The cross-coupled nature of the flip-flop is easily noticed.

Characteristics

Static RAM is typified by ease of use. This comes at the expense of a more complex circuit involving six or more transistors in a cross-coupled array. Thus, the total transistor count or density is four times higher than a DRAM.

The fundamental device is often referred to as a ripple through static RAM. Its design objective is ease of use. This device offers one power dissipation mode "on". Therefore its selected power (active power) is equal to its deselected power (standby power). One need simply present an address to the device and select it (via chip select) to access data. Note that the write signal must be inactive.

The address must be valid before and after read/write active to prevent writing into the wrong cell. Chip select gates the write pulse into the chip and must meet a minimum pulse width which is the same as the read/write minimum pulse width.

Read modify write cycles can be performed by this as well as other static RAMs. Read modify write merely means combining a read and write cycle into one operation. This is accomplished by first performing a read

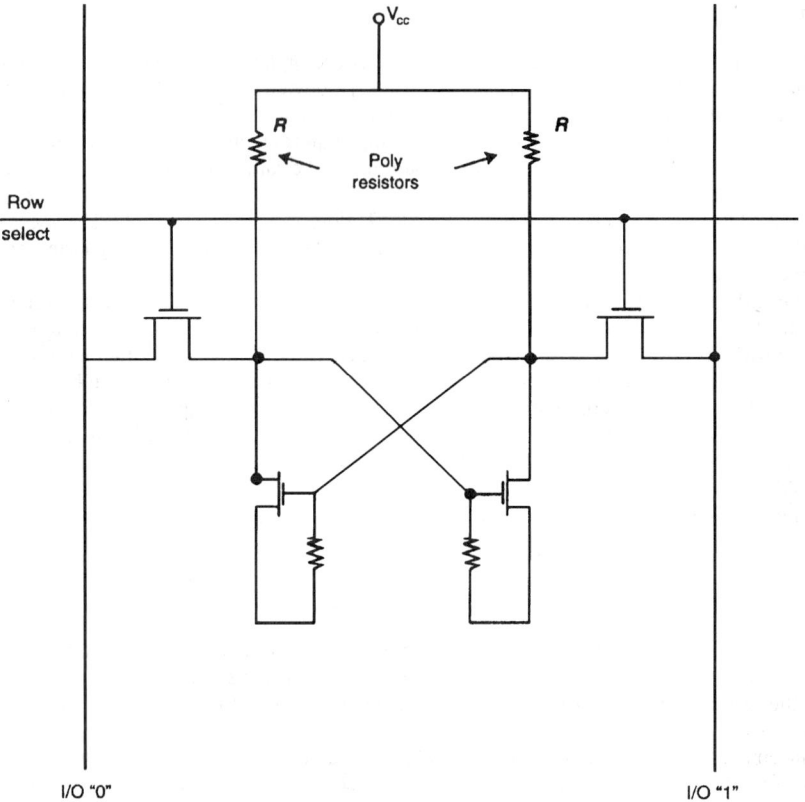

FIGURE 6.3.3 Polyresistor load static RAM cell.

then activating the read write line. The status of the data out pin during the write operation varies among vendors. The product's data sheet therefore be read carefully.

The device achieves its ease of use by using very simplistic internal circuitry. To permit a nonsynchronous interface (no timing sequence required) all circuitry within the part is active at all times. That static "NOR" row decoder is typical of the circuitry used. For a 4k RAM with 64 rows, this circuit is repeated 64 times with the address lines varying to uniquely identify each row.

In the static row decoder, a row is selected when all address inputs to the decoder are low making the output high. The need to keep the power consumption of this decoder at a minimum is in direct conflict with the desire to make it as fast as possible. The only way to make the decoder fast is to make the pull-up resistance small so that the capacitance of the word line can be changed quickly. However, only one row decoder's output is high; while the output of each of the other 63 decoders is low, causing the resistor current to be shunted to ground. This means that the pull-up resistance must be large in order to reduce power consumption.

Application Areas

Static RAM applications are segmented into two major areas based on performance. Slower devices, greater than 40 ns access time, are generally used with microprocessors. These memories are usually small (several thousands of bytes) with a general preference for wide word organization (by 8 bit). When by 1 organized memories are used, the applications tend to be "deeper" (larger memory) and desire lower power or parity.

The by 4-bit product is an old design which is currently being replaced by 1k × 8 and 2k × 8 static RAM devices. Statics are used in these applications because of their ease of use and compatibility with other memory types (RAM, EPROM). The by 4 and by 8 devices use a common pin for data in and data out. In this configuration, one must be sure that the data out is turned off prior to enabling the data in circuitry to avoid a bus conflict (two active devices fighting each other drawing unnecessary current). Bus contention can also occur if two devices connected to the same bus are simultaneously on fighting each other for control. The by 8 devices include an additional function called output enable (OE) to avoid this problem. The output enable function controls the output buffer only and can therefore switch it on and off very quickly. The by 4 devices do not have a spare pin to incorporate this function. Therefore, the chip select (CS) function must be used to control the output buffer. Data out turns off a specified delay after CS turn off. One problem with this approach is that it is possible for bus contention to occur if a very fast device is accessed while a slow device is being turned off. During a read/modify/write cycle the leading edge of the read write line (WE) is normally used to turn off the data out buffer to free the bus for writing.

The second segment is high-performance applications. Speeds required here are typically 10–20 ns. This market is best serviced by static RAM because of their simple timing and faster circuit speeds. The requirement of synchronizing two "reads" as required in clocked part typically cost the system designer 10–20 ns. Therefore, fast devices are always designed with a ripple through interface. The by 1 devices have been available longer, and tend to be faster than available by 8 devices. As a result the by 1 device dominates the fast static RAM market today. Applications, such as cache and writable control store, tend to prefer wide word memories. The current device mix should shift over time as more suppliers manufacture faster devices.

Process

The pin-outs and functions tend to be the same for fast and slow static devices. The speed is achieved by added process complexity and the willingness to use more power, for a given process technology.

The key to speed enhancement is increasing the circuit transistor's gain, minimizing unwanted capacitance, and minimizing circuit interconnect impedance. Advanced process technologies have been developed to achieve these goals. In general, they all reduce geometries (scaling) while pushing the state of the art in manufacturing equipment and process complexity. Figure 6.3.4 shows the key device parameters of MOSFET scaling, one of which is device gain. The speed of the device is proportional to the gain. Because faster switching

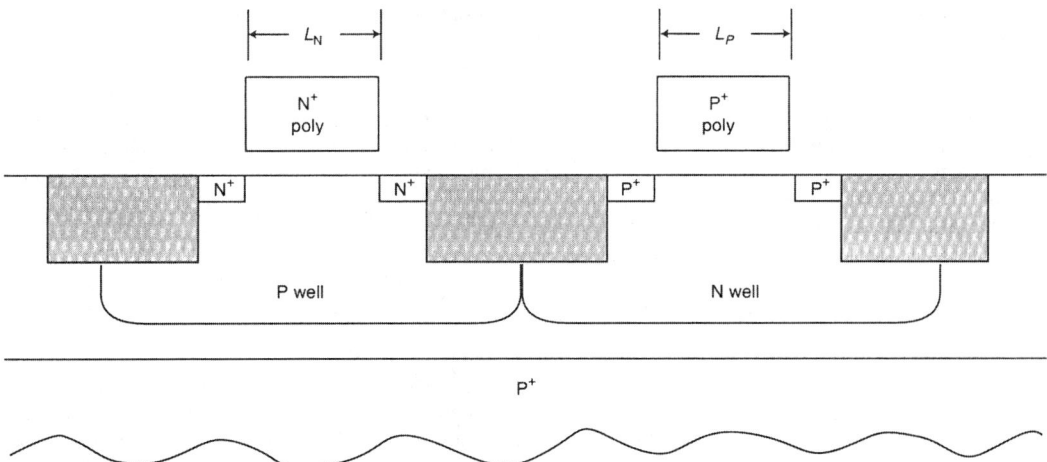

FIGURE 6.3.4   Cross section of submicron scaled CMOS gates for high-density SRAMs.

speeds occur with high gain, the gain is maximized for high speed. Device gain is inversely proportional to the gate oxide thickness and device length. Consequently, scaling these dimensions increases the gain of the device.

Another factor that influences performance is unwanted capacitance, which appears in two forms: diffusion and Miller capacitance. Diffusion capacitance is directly proportional to the overlap length of the gate and the source. Capacitance on the input shunts the high-frequency portion of the input signal so that the device can only respond to low frequencies. Second, capacitance from the drain to the gate forms a feedback path creating an integrator or low-pass filter which degrades the high-frequency performance.

One of the limits on device scaling is punch through voltage, which occurs when the field strength is too high, causing current to flow when the device is *turned off*. Punch through voltage is a function of channel length and doping concentration thus channel shortening can be compensated by increasing the doping concentration. This has the additional advantage of balancing the threshold voltage, which was decreased by scaling the oxide thickness for gain.

NONVOLATILE MEMORIES

DRAM and SRAM memories retain their data so long as power is applied but lose their data when the power is disrupted or turned off. In this sense, they are "volatile" memories. Nonvolatile memories retain their data even when the power is turned off. In the case of personal computers, the hard disk drive serves the essential "nonvolatile memory" function, in combination with a nonvolatile IC memory capable of starting the process of loading information from the hard drive into the computer system. The fundamental importance of nonvolatile IC memory can be seen by the rapid trend toward programmable components within many electronics applications not able to handle the inconvenience of an external hard drive. The term *embedded system* reflects this insertion of microprocessors into a vast range of applications. As one example, the microprocessor used to control a refrigerator requires storage of its *program*, a requirement that can be handled by a relatively simple programmable read only memory (PROM). Other applications of embedded systems require memories that can be written as well as read, with the additional requirement that the information written into the memory be preserved when power is removed. If this writing is very modest, then extensions of simple PROM ICs with slow write times is acceptable, as in the case of an electronic address book. But as the embedded systems have become more sophisticated and migrated closer to a personal computer in their general (not detailed) operation, the equivalent of a hard disk drive in IC form with not only fast reads but also fast writes becomes more important. To support such applications, there have been major advances in nonvolatile memory technologies recently, in several cases drawing on storage mechanisms quite different than those used in traditional LSI/VLSI components.

To illustrate the extent to which small embedded systems mimic larger personal computer systems, the example of the Motorola MC680xx 16/32-bit microprocessor can be used. This microprocessor (introduced in its initial version, the MC68000, in 1979 and requiring only about 68,000 transistors) was the processor that drove the early Apple computers. With today's VLSI able to provide several hundred millions of transistors in a single IC, placing the MC68000 on a VLSI circuit would require only about 0.001 percent of the IC area (assuming 68 million transistors). Equivalently, 1000 MC68000 microprocessors could be placed on a single IC. In this sense, adding a microprocessor to a digital circuit intended for embedded applications is not a major consumer of IC resources. As such embedded systems (particularly those that embed an earlier generation personal computer in its IC) continue to advance into a wider range of applications, the demand for high-performance nonvolatile memories with high-speed read and write capabilities (as well as low-voltage write conditions) will increase. As noted earlier, SRAM (with cells that are basically flip-flops) can be fabricated on the same IC as logic circuits, a major advantage over DRAM in many applications. Nonvolatile memories that can be fabricated on the same IC as logic will also be of major importance as the underlying microfabrication technologies continue to advance, seeking system-on-a-chip or system-on-a-(few) chip (set) capabilities.

The field of advanced technologies for nonvolatile memories is in a state of flux and a variety of quite different technical approaches are under study. Underlying these new technologies and approaches there remains the basic principles of nonvolatile memory seen in the traditional PROM, EPROM, and EEPROM technologies that continue to play a major role in systems designs.

Mask-Programmed ROM

In mask-programmed ROM, the memory bit pattern is produced during fabrication of the chip by the manufacturer using a mask operation. The memory matrix is defined by row (X) and column (Y) bit-selection lines that locate individual memory cell positions.

ROM Process

For many designs, fast manufacturing turnaround time on ROM patterns is essential for fast entry into system production. This is especially true for the consumers' "games" market. Several vendors now advertise turnaround times that vary from 2 to 6 weeks for prototype quantities (typically small quantities) after data verification. Data verification is the time when the user confirms that data have been transferred correctly into ROM in accordance with the input specifications.

Contact programming is one method that allows ROM programming to be accomplished in a shorter period of time than with gate mask programming. In mask programming, most ROMS are programmed with the required data bit pattern by vendors at the first (gate) mask level, which occurs very early in the manufacturing process. In contact programming, actual programming is not done until the fourth (contact) mask step, much later in the manufacturing process. That technique allows wafers to be processed through a significant portion of the manufacturing process, up to "contact mask" and then stored until required for a user pattern. Some vendors go one step further and program at fifth (metal) mask per process. The results in a significantly shorter lead time over the old gate-mask programmable time of 8 to 10 weeks; the net effect is time and cost savings for the end user.

ROM Applications.   Typical ROM applications include code converters, look-up tables, character generators, and nonvolatile storage memories. In addition, ROMs are now playing an increasing role in microprocessor-based systems where a minimum parts configuration is the main design objective. The average amount of ROM in present microprocessor systems is in the 10 to 32 kbyte range. In this application, the ROM is often used to store the control program that directs CPU operation. It may also store data that will eventually be output to some peripheral circuitry through the CPU and the peripheral input/output device.

In a microprocessor system development cycle, several types of memory (RAM, ROM, and EPROM or PROM) are normally used to aid in the system design. After system definition, the designer will begin developing the software control program. At this point, RAM is usually used to store the program, because it allows for fast and easy editing of the data. As portions of the program are debugged, the designer may choose to transfer them to PROM or EPROM while continuing to edit in RAM. Thus, he avoids having to reload fixed portions of the program into RAM each time power is applied to the development system.

Electrically Erasable Programmable ROM

The ideal memory is one that can perform read/write cycles at speeds meeting the needs of microprocessors and store data in the absence of power. The EEPROM (electrically erasable PROM) is a device which meets these requirements. The EEPROM can be programmed in circuit and can selectively change a byte of memory instead of all bytes. The process technology to implement the EEPROM is quite complex and the industry is currently on the verge of mastering production.

EEPROM Theory.   EEPROMs use a floating gate structure, much like the ultraviolet erasable PROM (EPROM), to achieve nonvolatile operation. To achieve the ability to electrically erase the PROM, a principle known as Fowler-Nordheim tunnelling was implemented. Fowler-Nordheim tunnelling predicts that under a field strength of 10 MV cm^{-1} a certain number of electrons can pass a short distance, from a negative electrode, through the forbidden gap of an insulator entering the conduction band and then flow freely toward a positive electrode. In practice the negative electrode is a polysilicon gate, the insulator is a silicon dioxide and the positive electrode is the silicon substrate.

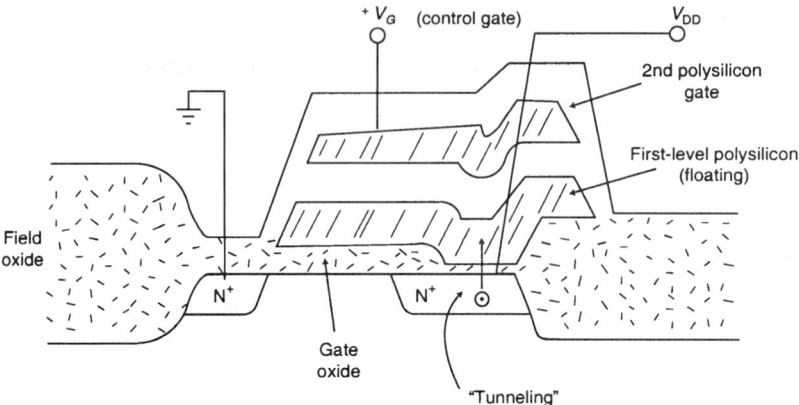

FIGURE 6.3.5   EEPROM cell using Fowler-Nordheim tunnellings.

Fowler-Nordheim tunnelling is bilateral, in nature, and can be used for charging the floating gate as well as discharging it. To permit the phenomenon to work at reasonable voltages (e.g., 20 V), the oxide insulator needs to be less than 200 Å thick. However, the tunnelling area can be made very small to aide the manufacturability aspects (20-nm oxides are typically one-half the previously used thickness).

Intel Corporation produces an EEPROM based on this principle. Intel named the cell structure FLOTOX. A cross section of the FLOTOX device is shown in Fig. 6.3.5.

The FLOTOX structure resembles the original structure used by Intel for EPROM devices. The primary difference is in the additional tunnel-oxide region over the drain. To charge the floating gate of the FLOTOX structure, a voltage V_G is applied to the top gate and with the drain voltage V_D at 0 V, the floating gate is capacitively coupled to a positive potential. Electrons will then flow to the floating gate. If a positive potential is applied to the drain and the gate is grounded, the process is reversed and the floating gate is discharged.

The EEPROM designs introduced today are configured externally to be compatible with ROM and EPROM standards which already exist. Devices typically use the same 24-pin pin-out as the generic 2716 (2k × 8 EPROM) device. A single 5 V supply is all that is needed for read operations. For the write and clear operations, an additional supply (V_{pp}) of 20 V is necessary. The device reads in the same manner as the EPROM it will eventually replace.

EEPROM Applications. The EEPROM has the nonvolatile storage characteristics of core, magnetic tape, floppy, and Winchester disks but is a rugged low-power solid-state device and occupies much less space. Solid-state nonvolatile devices, such as ROM and EPROM, have a significant disadvantage in that they cannot be deprogrammed (ROM) or reprogrammed in place (EPROM). The nonvolatile bipolar PROM blows fuses inside the device to program. Once set the program cannot be changed, greatly limiting their flexibility. The EEPROM, therefore, has the advantages of program flexibility, small size, and semiconductor memory ruggedness (low voltages and no mechanical parts).

The advantages of the EEPROM create many applications that were not feasible before. The low power supports field programming in portable devices for communication encoding, data formatting and conversion, and program storage. The EEPROM in circuit change capability permits computer systems whose programs can be altered remotely, possible by telephone. It can be changed in circuit to quickly provide branch points or alternate programs in interactive systems.

The EEPROMs nonvolatility permits a system to be immune to power interruptions. Simple fault tolerant multiprocessor systems also become feasible. Programs assigned to a processor that fails can be reassigned to the other processors with a minimum interruption of the system. Since a program can be backed up into EEPROM in a short period of time, key data can be transferred from volatile memory during power interruption and saved. The user will no longer need to either scrap parts or make service calls should a program bug be discovered in fixed memory. With EEPROM this could even be corrected remotely. The EEPROM's flexibility will create further applications as they become available in volume and people become familiar with their capabilities.

Erasable Programmable ROM

The EPROM like the EEPROM satisfies two of our three requirements. It is nonvolatile and can be read at speeds comparable with today's microprocessors. However, its write cycle is significantly slower, like the EEPROM. The EPROM has the additional disadvantage of having to be removed from the circuit to be programmed as contrasted to the EEPROM's ability to be programmed in circuit.

EPROM is electrically programmable, then erasable by ultraviolet (UV) light, and programmable again. Erasability is based on the floating gate structure of *n*- or *p*-channel MOSFIT. This gate, situated within the silicon dioxide layer, effectively controls the flow of current between the source and drain of the storage device. During programming, a high positive voltage (negative if *p*-channel) is applied to the source and gate of a selected MOSFET, causing the injection of electrons into the floating silicon gate. After voltage removal, the silicon gate retains its negative charge because it is electrically isolated (within the silicon dioxide layer) with no ground or discharge path. This gate then creates either the presence or absence of a conductive layer in the channel between the source and the drain directly under the gate region. In the case of an *n*-channel circuit, programming with a high positive voltage depletes the channel region of the cell; thus a higher turn-on voltage is required than on an unprogrammed device. The presence or absence of this conductive layer determines whether the binary 1 bit or the 0 bit is stored. The stored bit is erased by illuminating the chip's surface with UV light. The UV light sets up a photocurrent in the silicon dioxide layer which causes the charge on the floating gate to discharge into the substrate. A transparent window over the chip allows the user to perform erasing, after the chip has been packaged and programmed, in the field.

Programmable ROM

This PROM has a memory matrix in which each storage cell contains a transistor or diode with fusible link in series with one of the electrodes. After the programmer specifies which storage cell position should have a 1 bit or 0 bit, the PROM is placed in a programming toll which addresses the locations designated for a 1 bit. A high current is passed through the associated transistor or diode to destroy (open) the fusible link. A closed fusible link may represent a 0 bit, while an open link may represent a 1 bit (depending on the number of data inversions done in the circuit). A disadvantage of the fusible-link PROM is that its programming is permanent; that is, once the links are opened, the bit pattern produced cannot be changed.

Shadow RAM

A recently introduced RAM concept is called the shadow RAM. This approach to memory yields a nonvolatile RAM (data are retained even in the absence of power), by combining a static RAM cell and an electrically erasable cell into a single cell. The EEPROM shadows the status RAM on a bit-by-bit basis. Hence the name shadow RAM. This permits the device to have read/write cycle times comparable to a static RAM, yet offer nonvolatility.

The nonvolatility has a limit in that the number of write cycles is typically 1000 to 1 million maximum. The device can, however, be read indefinitely. Currently two product types are offered. One permits selectively recalling bits stored in the nonvolatile array, while the other product recalls all bits simultaneously.

Typical memory subsystems used in computers employ a complex combination of dynamic RAMs, static RAMs, and nonvolatile ROMs.

BIBLIOGRAPHY

Donnelly, W., "Memories—New Generations Push Technology Forward," *Electronic Industry*, October 1982.

Eaton, S. S., and D. Wooton, "Circuit Advances Propel 64 K RAM Across the 100 ns Barrier," *Electronics*, 24 March 1982.

Threewitt, B., "A VLSI Approach to Cache Memory," *Computer Design*, January 1982.

Whittier, R. J., "Semiconductor Memories," *Mini-Micro Systems*, December 1982.

Wilcock, J. D., "Semiconductor Memories," *New Electronics*, August 17, 1982.

CHAPTER 6.4
MICROPROCESSORS

S. Tewksbury

INTRODUCTION

This chapter reviews microprocessors, ranging in complexity and performance from 8-bit microprocessors widely used in embedded systems applications to very high performance 32- and 64- bit microprocessors used for powerful desktop, server, and higher-level computers. The history of the evolution of microprocessor architectures is rich in the variety of hardware/software environments implemented and the optimization of architectures for general-purpose computation. A variety of simple 8-bit microprocessors introduced over 30 years ago continue to thrive, adding to the basic microprocessor architecture a variety of other analog and digital functions to provide a *system-on-a-chip* solution for basic, intelligent products. These include digital cameras, electronic toys, automotive controls, and many other applications. These applications represent cases in which the programmable capabilities of microprocessors are embedded in products to yield "intelligent modules" capable of performing a rich variety of functions in support of the application. Many of these applications are intended for portable products (such as cameras) where battery operation places a premium on the power dissipated by the electronics.

Starting from the early 8-bit microprocessors, most manufacturers (Intel, Motorola, and so forth) followed a path of evolutionary development driven by the exponentially increasing amount of digital logic and memory that could be placed on a single IC. The class of microprocessor generally called *microcontroller* made use of the additional circuitry available on an IC to integrate earlier external circuitry directly onto the microcontroller IC. For example, 8-bit microcontrollers based on the Motorola 6800 and Intel 8085 8-bit microprocessors have evolved to include substantial nonvolatile memory (able to hold programs without power), significant internal RAM (above to provide storage for data being manipulated), digital ports (e.g., serial ports, parallel ports, and so forth) to the external world, and various analog circuit functions. Among the analog circuit functions are analog-to-digital and digital-to-analog converters, allowing the microcontroller to capture information from sensors and to drive transducers. With increasing circuitry per IC, functions such as the analog-to-digital converters have evolved into more complex components—for example, allowing a multiplicity of different analog input signals to be converted into digital signals using an analog multiplexing circuit and allowing programmable gain to amplify the incoming analog signal to levels minimizing the background noise included in the digital signal.

As the density of VLSI ICs has increased, earlier microprocessors requiring the full IC area have become smaller in area required. The 8- and 16-bit controllers can now be embedded in VLSI ICs to extend many of today's applications to "intelligent" systems. This embedding of microprocessor cores into other digital circuit functions has extended to the area of programmable logic (general-purpose arrays to logic that can be user programmed to perform a custom function). For example, Altera programmable logic ICs includes IP (Intellectual Property) cores such as a 200 MHz 32-bit ARM9 core processor along with embedded RAM cores. Similar capabilities exist for other manufacturers of programmable logic. Today, the combination of programmable logic allows a user to efficiently implement custom logic functions of considerable complexity (i.e., the Xilinx Virtex family of programmable logic provides 8 million gates and can operate at a 300 MHz clock rate) while including the familiar functionality of well-established microprocessors for overall control. As the VLSI technologies continue to advance, the more powerful microprocessors of today will eventually migrate into these embedded applications as even more powerful microprocessors emerge for the desktop/laptop computer market.

Another evolutionary path was to create increasingly powerful microprocessors for applications such as desktop computers. Drawing power directly from a plug, minimizing power dissipation is less a priority (aside from the inevitable side issue of heating due to power dissipation). Instead, raw computing performance is the objective. Having to provide connections among integrated circuits to handle different peripheral functions (floating point arithmetic units, modules managing memory access though cache and other techniques, image processing modules, and so forth) can create a substantial performance barrier because of data transfer rate limitations. Migrating such peripheral functions essential to the basic computer operation can greatly improve the performance of the overall computer (microprocessor and peripherals). As an example, the evolution of the Motorola 68000 family starts with the basic MC68000. It then progressed to the MC68010, to the MC68020, to the MC68030, to the MC68040, and beyond not by changing the core microprocessor architecture greatly but rather by moving "onto the chip" such functions as the floating point accelerator, the memory management unit, virtual memory control, cache memory to reuse data/instructions loaded from RAM, and so forth.

The greater computational power of these early 16/32-bit microprocessors (such as the Motorola MC68000) is attractive in many embedded applications. As a result, microcontrollers have evolved beyond the 8-bit microprocessor cores to include these classical 16/32-bit microprocessor cores. At the same time, the microcontrollers retain the integration of other peripherals that support the applications (such as automotive and entertainment) into which they are embedded.

Returning to the microprocessors used for desktop computers, workstations, servers, and so forth, more recent advances have challenged and overcome many limitations that were once considered fundamental. Microprocessors such as the Pentium introduced in 2001–2002 operate at clock rates well above 1 GHz, a clock rate that was deemed unrealizable not long ago. Figure 6.4.1 shows the evolution of clock rates for Intel microprocessors. A high-end microprocessor recently demonstrated by Intel operates with voltages near 1 V

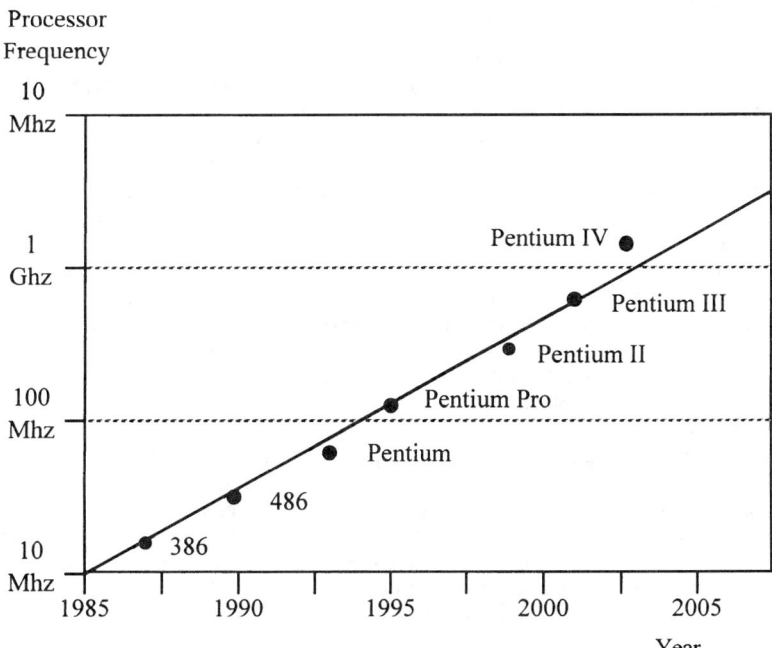

FIGURE 6.4.1 Clock rates of Intel Microprocessors. (From R. Ronen et al., "Coming Challenges in Microarchitectures and Architecture," 2001.)

and at power levels of over 100 W. That corresponds to currents of 100 A or greater. To respect how large this current is, not that most homes have 100 A or 200 A services to operate the entire house. Today's high-end microprocessors employ well over 100,000,000 transistors on a single integrated circuit not much larger than a few square centimeters. Internally, the architectures have become sufficiently sophisticated that the hardware/firmware providing operating system functions are a fundamental part of the microprocessor. The architectures themselves have become highly refined.

The IBM Power4 microprocessor's basic description illustrates this sophistication, With 170 million transistors, the Power4 microprocessor is a 64-bit microprocessor operating at frequencies greater than 1 GHz, using an eight-instruction-wide design with superscalar operation and out-of-order execution of instructions. More than 200 instructions can be pending completion in each of the two on-chip processor cores.

Programming these contemporary microprocessors requires sophisticated programming support tools and generally require that programs be written in higher-level programming language (such as C++). Programming basic programs in assembly language is virtually impossible.

In a very real sense, microprocessors have become ubiquitous throughout our technologically driven society. Dozens appear in an automobile, invisible to the owner but providing the intelligence and control that are at the heart of the advanced automotive systems. Portable video game players contain some of the most remarkable microprocessors available, supporting advanced visualization in the palm of your hand.

CMOS Technology

Digital circuits today are based on complementary MOS (CMOS), CMOS having emerged as a preferred technology for a variety of reasons, perhaps the most important being the low dc power dissipation provided. Despite the dominance of pure CMOS, a hybrid version of CMOS and bipolar devices (called BiCMOS) has been developed to provide the advantages of low power in CMOS with the capabilities of high speed in bipolar circuits. Such BiCMOS technologies are used, for example, in the Intel Pentium microprocessors.

Digital electronics used a standard +5 V supply voltage over several generations, driven in part by the earlier dominance of bipolar circuits. With the migration to a dominance of CMOS, lower voltage operation has been possible and supply voltages have been decreasing significantly over the past decade. Initially, voltages were decreased to 3.3 V, with subsequent advances leading to further reductions to voltages of about 1 V. These reductions in supply voltage provide substantial performance advantages. For example, the lower voltage leads to a lower power consumption per device, an important reduction for holding overall chip power dissipations within acceptable limits as the number of devices per chip has increased. In addition, the steady decrease in feature sizes leads to cases in which the electric fields within the IC increase if the voltages are not reduced. Reduction of supply voltage in combination with reductions in feature sizes has helped prevent excessive electric fields and their associated effects. Beyond the simple reduction of externally provided supply voltage, contemporary microprocessor chips have evolved to include sophisticated internal power management subsystems—generating internal voltages as needed as well as allowing selective turning off of various sections of the system to reduce power consumption. Although voltage levels have decreased over the past several years, today's advanced, high-performance microprocessors require remarkably high current supplies (several tens of amps up to 100 A or higher). These high currents present complications associated with voltage losses along power interconnections within the IC because of the nonzero resistance of those lines and voltage drops due to that resistance. Such complications have led to specialized metalization layers for distribution of power throughout the IC.

The trends toward very high performance microprocessors are quite visible through their impact on successive generations of personal computers. Less visible have been the dramatic advances in microprocessor based systems intended for embedded applications (e.g., microprocessors embedded in automobiles, in audio equipment, in toys, and so forth). Microprocessor architectures from earlier generations of microprocessors provide the core computing power for these embedded applications while the addition of a wide range of peripheral functions (including analog input and output) have provided essentially single-chip solutions for intelligent embedded systems. In contrast to high-performance desktop computers, which receive power from a wall plug, these embedded applications often receive their power from batteries. Minimization of power dissipation is essential to provide long battery life.

These two application areas have led to two distinct roadmaps for future generation technologies—one roadmap for the high-performance microprocessors and another roadmap for the battery-powered, lower-performance microprocessors. The roadmap generated at regular intervals by the Semiconductor Industry Associates (SIA) organization provides not only a perspective on the current state of the technologies but also on the planned progression of those technologies over the next 10 years. The roadmaps can be viewed at the SIA website.

General Overview

Contemporary microprocessors include many of the functions that in earlier generation computers were implemented on separate integrated circuit chips. In addition, their architectural designs have become highly sophisticated, supporting not only routine user programs but also the underlying operating system associated with the computers. To provide a basic overview of computer architectures, the basic architecture of earlier generation systems implementing the computer on one or more PC boards, rather than on a single integrated circuit, is used here. The reader is advised, however, that far more sophisticated systems have replaced these earlier generation systems.

Figure 6.4.2 illustrates a typical multichip computer based on the earlier generation Intel microprocessors and peripheral chips. The role of the various blocks is as follows.

Processor.   The block labeled "processor" was one of the earlier generation microprocessors (e.g., 8088 and higher generations). The internal processor architecture is discussed in the next section. In general, the processor manages the operation of the entire system, communicating with the peripheral functions using a "system bus" shown as the horizontal line. This system bus includes parallel wires (address lines) to select the peripheral and an internal data storage item in the peripheral (or memory), parallel wires (data lines) to transfer data to and from the peripherals and memory, a read/write line to specify whether data are being transferred to or from the microprocessor, and interrupt line(s) allowing external peripherals to "interrupt" the currently

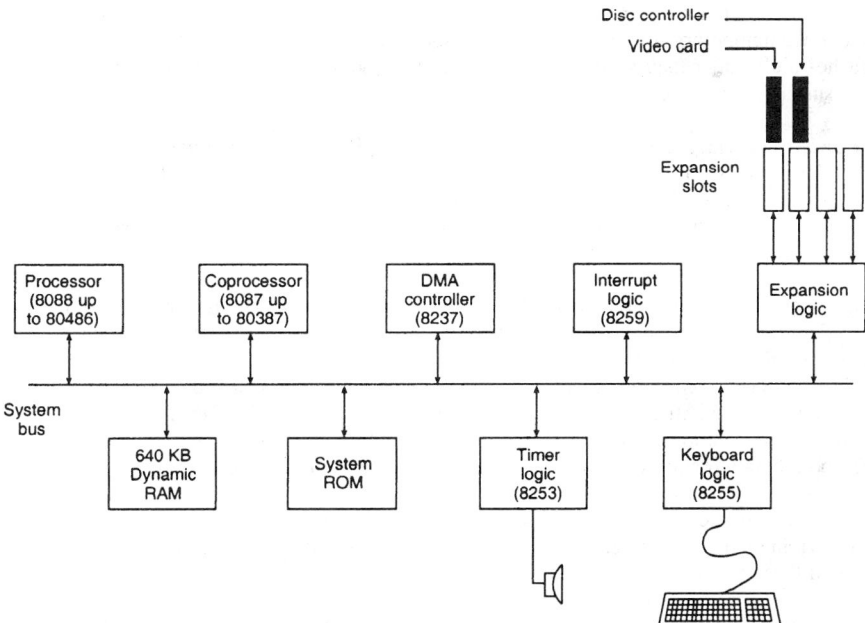

FIGURE 6.4.2   Block diagram of a typical PC motherboard.

executing program to allow the peripheral to communicate at its request with the microprocessor. The bus also includes other lines for functions such as providing resets and clocks.

Dynamic RAM. The dynamic RAM (640k shown, corresponding to the DOS system RAM capacity in those earlier computers) provides storage of both data and instructions. To access a particular item in the RAM, the address lines provide the location of the specific information involved in the data transfer. Today's RAM is far larger than the 640k shown, with standard personal computer memory capacities approaching 1 Gbyte.

System ROM. This nonvolatile memory provides the startup information needed to start the computer in a desired state. During startup, special logic in the processor accesses this startup information.

Disk Controller. The external disk hard drive stores user programs, operating systems, and other information, retained when the power is turned off. The disk controller manages the exchange of data between the microprocessor system (and its RAM) and the disk drive. There has been a substantial evolution in the philosophy and technologies of disk controllers over the past two decades, evolving from manufacturer specific controllers to standardized controller functions and data interfaces. Today's computer systems employ disk drives of substantial capacity (tens to hundreds of Gbyte capacity). The technologies used in contemporary disk drives have advanced dramatically in parallel with the advances in VLSI technologies, providing not only much larger capacities but also smaller and more rugged drives.

Cache Memory. Although the speeds of microprocessors have increased dramatically, memory components have seem a less dramatic increase. Memory access times (DRAMs) are typically several tens of nanoseconds. With microprocessor clock rates having increased to 1GHz and higher, a read or write of external memory can take the microprocessor about 50 to 100 clock cycles. Static RAMS have substantially faster response times, but have lower data storage capacities per IC. This has led to the introduction of "staged" memories, with instructions (and in some cases data) transferred from the slow DRAM into fast SRAM where the instructions (and data) can be retrieved more quickly. If an instruction is reused (and in the case of a loop instruction sequence) or data are reused (as in several computational algorithms), the second use of the instruction (data) is more quickly retrieved from the SRAM. Rather sophisticated algorithms are used to decide when information in a cache can be replaced by new information and to search the cache for instructions (content-addressable memories). Contemporary microprocessors have included one or more stages of cache within the IC. First-level caches are smaller than second-level caches, but significantly faster (two to three cycles for first-level cache versus six to 10 cycles for the second-level cache).

Video Card. The video card provides access to the display of the computer. As in all other components, video displays have seen dramatic improvements over the past 20 years, including not only the transition to high-resolution color displays but also to flat panel displays such as used with laptop computers. The expectations of the display drivers (hardware and software) have increased sufficiently that video cards continue to play a major role in the performance of the overall computer system.

Interrupt Logic. While the processor is in control, it determines to whom it will talk and is oblivious to any needs originating at a peripheral. For example, the serial port of the computer may receive data (e.g., through a modem) and the processor must be made aware that there are data to be read before the next data byte is received by the serial port. Interrupts provide the mechanism through which the peripherals can "interrupt" the processor's current activities and request that it handle requests originating in the peripherals. The "8259" shown for the interrupt logic is an classic integrated circuit included on computer board to interface interrupts from the peripherals for presentation to the processor. The interrupts ask the microprocessor to interrupt its operation and service the peripheral requesting service. The microprocessors allow external inputs to be turned on or off. When an interrupt occurs, the software program branches to a predetermined location where commands determine which peripheral generated the interrupt and then to a location where the program handles the needs of the interrupting peripheral.

Keyboard Logic. The "8255" component shown for interfacing the computer to a keyboard was also an integrated circuit designed specifically for this purpose.

DMA Controller.   The direct memory access controller provides a high-speed means of streaming data to and from the computers RAM. When active, this replaces the normal single word at a time operations provided by the processor across the system bus.

Coprocessor.   The capabilities of operations performed by the microprocessor is limited by the number of transistors available, leading to cases in which some desired operations on data that would normally be provided by the processor must be performed instead by an auxiliary integrated circuit. The classic example was floating point computations, since the standard microprocessor's arithmetic unit was designed to operate only on integers. Addition of the coprocessor allowed floating point instructions to be added to the instruction sets, the data being exchanged between the microprocessor and the floating point coprocessor (bypassing the internal arithmetic unit of the microprocessor) for fast hardware execution. The alternative is a slower sequence of instructions executed within the microprocessor to perform the floating point computations as a program. Like many of the other peripherals shown, such floating point coprocessors have migrated into the microprocessor IC. This has led to contemporary architectures with multiple arithmetic units and an ability to execute more than one arithmetic function at a time. This ability is actually employed, and contributes to the ability of microprocessors to execute (on average) one instruction for each clock period.

Timer.   The timer function (shown as the earlier 8253 integrated circuit) allows a variety of timing functions to be performed. These include measuring time, generating timing pulses for various peripherals, and so forth.

Not Shown.   Not shown in Fig. 6.4.2 are several other peripheral functions such as serial ports for modems and parallel ports for printers. Specialized integrated circuits were also developed to handle these peripheral ports and have evolved over time to include several types of peripheral ports (USB, Firewire, RGB TV, SCSI, and several others). These "standardized" interfaces allow components from different manufacturers to be used in combination with different computers and have played a substantial role in the availability of a rich set of reasonably priced peripherals.

Microprocessor Architecture

Next, the basic internal architecture of the processor shown in Fig. 6.4.2 is discussed. Again, it is necessary to review the general architecture from the perspective of earlier generation microprocessors because of the great complexity of today's high-end microprocessors. Figure 6.4.3 shows the general architecture of an elementary microprocessor. Such architectures are typical of the 8-bit microprocessors used in microcontrollers.

The microprocessor's internal bus is shown as an address bus and a data bus, extending to the external peripherals as shown in Fig. 6.4.2. The data bus "width" (number of parallel wires) defines the size of the data/instruction unit transferred into and out of the microprocessor. Typically, this data bus width also defines the type of microprocessor. An 8-bit microprocessor has an 8-bit data bus, a 16-bit microprocessor has a 16-bit data bus, and so forth. In the case of the microprocessor reading an instruction from the external memory, the data bus width also constrains the number of instructions available. The basic sequence of operations in performing an instruction includes first the reading of the instruction itself, then the reading of any constants associated with the instruction (e.g., numerical constants, address offsets, and so forth) and then, if the data for the instruction is stored in memory, the memory locations where these alterable data items are stored. With an 8-bit data bus, the first instruction item can represent up to 256 instructions, normally setting the limit on the number of instructions in the microprocessor's instruction set. With a 16-bit data bus, the first instruction can represent up to 64,000 instructions (allowing a rich instruction set or an organization of the instruction op code and operand information in a highly efficient manner, as in the Motorola 68000 processor family).

The width of the address bus defines the amount of memory data storage supported. In 8-bit and 16-bit microprocessors, the width of the address bus was twice that of the data bus. For example, an 8-bit microprocessor would have a 16-bit address bus, able to address $2^{16} = 65,536$ (equal to "64K" in computer jargon) memory locations. More memory can be addressed (as in the PC example in Fig. 6.4.2 using "memory banks," essentially an external selection of which set of a multiplicity of 64K memory units is accessible through the address bus. A 16-bit microprocessor with a 32-bit address bus is able to directly address $2^{16} = 4.3$ billion memory locations.

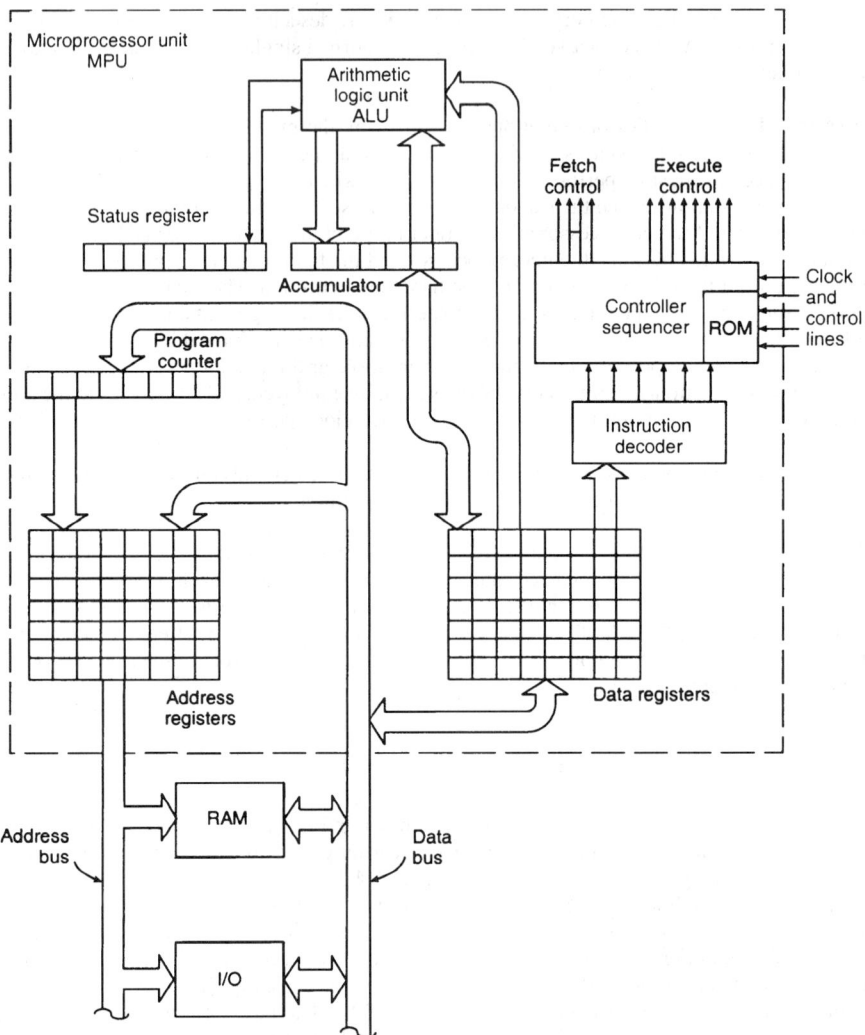

FIGURE 6.4.3 Architecture of an elementary microprocessor.

The internal logic units of the microprocessor shown in Fig. 6.4.3 provide the following functions.

Arithmetic Logic Unit. The arithmetic logic unit (ALU) can perform a variety of arithmetic and logic functions on data inputs (either on two input data items or on a single input data item). Standard arithmetic functions are add, subtract, multiply, and divide. Logic functions include the bit-wise AND, OR, XOR functions. These are also functions allowing the data word to be shifted in position. The specific operation to be performed is determined by input control signals generated by the controller. If a given ALU function requires more than one clock period (e.g., the multiply operation), then the controller/sequencer controls the multiple addition steps to complete the multiplication using a single multiply instruction.

Figure 6.4.3 shows a register receiving the output of the ALU and feeding that output back into the ALU during the next cycle. This provides, for example, an efficient means of adding a long string of numbers—this particular function leading to the name "accumulator" associated with the register.

Microprocessors initially used a single arithmetic unit. However, many programs would allow more than one arithmetic operation to be performed at the same time. By including multiple arithmetic units (perhaps different types of arithmetic units or other functions), more than one execution cycle can be launched at the same time. Such architectures are known as superscalar architectures and have become standard in 32-bit and higher microprocessors. The superscalar architectures, combined with high-performance cache memories and pipelined techniques can execute more than one instruction per clock cycle. For example, the Motorola PowerPC G4 microprocessor supports 1064 MB/s data rates and executes up to 16 simultaneous instructions per clock cycle.

Status Register. The status register is closely associated with the arithmetic logic unit and holds information regarding the data. This information is extracted by analyzing the data, deciding, for example, whether the number is zero, whether the number is negative, whether a carry or borrow state was generated from the high-end of the data in the operation, whether the result is an overflow (larger than the maximum number allowed), and so forth. This information is used most often in establishing the condition in conditional instructions (e.g., the assembly language equivalent of "if the number is negative then do this").

Instruction Decoder and Controller/Sequencer. The instructions are stored as binary data words (e.g., 8-bit data for an 8-bit microprocessor) in memory, using binary codes to represent the specific task to be performed in completing the reading of the instruction (if a multiword instruction) and carrying out the operation to be performed on the instruction. This binary code must be decoded and then provided to a sequential logic circuit that can generate the various electronic signals to control the various components within the microprocessor. For example, if the instruction says to add the number 10 to the number stored in the accumulator, then the first instruction defines the operation (addition) and the data to be used. The number 10 would be stored in memory along with the instruction so a second read of the memory would be required (for an 8-bit microprocessor). Following the reading of the number 10 into the microprocessor, the controller/sequencer would apply the two data items to the arithmetic unit and tell the ALU to perform an operation. Then, the controller/sequencer would transfer the result to the location specified in the instruction. Although sounding like a complex task, the controller/sequence can be efficiently implemented using efficient realizations for the basic set of stored information and feedback through registers to sequence the current state to the next state until the entire instruction has been executed.

Data Registers and Address Registers. Figure 6.4.3 shows two banks of registers, one the set of data registers and the other the set of address registers. The distinction between the data and address notations is largely based on the use of address registers to provide part of a memory address (e.g., the base address of a large table). Otherwise, the registers are similar and provide the function of "scratch pad" storage on the microprocessor chip. By taking data inputs to the ALU from these registers (and storing results from the ALU into these registers for later reuse), the delays associated with transferring data to and from external memory are avoided. As an example of the use of this scratch pad data storage, consider the following three instructions: (a) X = B + C, (b) Y = D + E, and (c) Z = X * Y. Upon completion of the first instruction, the value of X could be stored in data register D0. On completion of the second instruction, the value of Y could be stored in data register D1. Then at some later instruction, instruction (c) could be performed by retrieving the values of X and Y stored in the data registers D0 and D1. An example of using the address registers to generate addresses is illustrated by the example of addressing external memory to obtain successive components of a vector V(j). The address (base address) of the first component V(0) could be stored in address register A0. Then, the ALU/accumulator can be used to generate the sequence 0, 1, 2, ... By using the so-called indexed addressing mode, the address in A0 and the number from the ALU are added without having to perform an explicit addition instruction in the ALU.

Program Counter. The program counter is a counter that is automatically incremented after each instruction has been retrieved from the external RAM so that the number in the counter is the address for the start of the next instruction. This automatic incrementing reflects the typical case of instructions being executed in sequence. In the case when the program jumps to some other place, the offset between the current value of the program counter and the desired value of the program counter is added to the count in the program counter. Since the program counter always points to the next address, this automatically causes program execution to jump to the desired instruction in memory.

Included among the address registers discussed previously (or defined separately) is a specialized register called the "stack." This register performs a critical function when external interrupts occur or when the program calls a subroutine stored in some area of memory. Once the program jumps to the subroutine (or the "subroutine" used to respond to the interrupt), the microprocessor has lost the value of the next instruction that would have been executed (memory address following the current instruction). When such jumps occur, the location of the *next* instruction in sequence, in particular the address given by the program counter, is stored automatically in the memory location whose address is in the stack register and the contents of the stack pointer is either incremented (or decremented depending on the microprocessor family used). The address in the stack pointer at any time is the memory location in which the last item was stored, allowing that address to be used to retrieve that item. On retrieving an item, the address in the stack pointer is decremented (or incremented) so that the address points to the previous memory location. In this manner, the stack pointer operation implements in memory a "last in, first out" data structure. The memory structure is generally called a *stack*, suggesting a stack of addresses from which you retrieve the address at the top of the stack, exposing the address underneath. Return from the subroutine to the point in the program from which the jump was called is achieved simply by placing the address stored in the stack in the program counter.

Instruction Execution Sequence

The overall execution of an instruction involves a basic sequence of three steps, starting with the reading of the instruction from memory. The three steps are (1) the reading of the instruction, (2) the decoding of the instruction within the microprocessor, and (3) execution of the instruction (including the storage of the result of an instruction). Figure 6.4.4 illustrates this sequence of steps in combination with a clock signal driving the progression. The time from the start of the reading of the instruction (called "fetch" in Fig. 6.4.4) to the completion of the instruction including writing of any results into storage is called the instruction cycle time. The example illustrated in Fig. 6.4.4 is merely representative of several different approaches that can be used to arrange the orderly completion of these three basic steps. For example, in some 8-bit microprocessors, a single clock cycle is used to complete a single basic instruction. When the clock is in one state (e.g., high), the instruction information is fetched and when the clock signal is in the other state (e.g., low) the instruction is decoded and executed. There are also important cases in which a multiplicity of clock cycles are needed for some instructions. For example, a multiply operation in a microprocessor with only a addition-based ALU requires that a sequence of multiplies and adds be performed to complete the overall instruction. In this case, each addition operation will consume one or more clock cycles.

Figure 6.4.4 suggests that only one of the three operations can be performed at any given time. However, the fetch operation involves interfacing to the external memory (setting addresses and reading external data into the microprocessor), whereas the other operations are performed within the microprocessor. It is therefore possible to arrange for the fetching of the next instruction to be started while the decoding/execution stages of the current instruction are being completed. It is also possible to separate the decode and execution operations and perform them in parallel. For example, while executing the current instruction the decoder could be decoding the next instruction while at the same time the microprocessor is fetching still one more instruction. The approach is called "pipelining" (reflecting the output from one step being fed into the next step while a new action is being input to that first step). Pipelining provides a faster throughput of instructions (more instructions per second can be executed) but does not intrinsically reduce the time to execute a given instruction (i.e., pass a single instruction through the pipe). Figure 6.4.5 illustrates the basic principle, here taking the basic instruction cycle as having four phases—fetching an instruction (F), decoding the

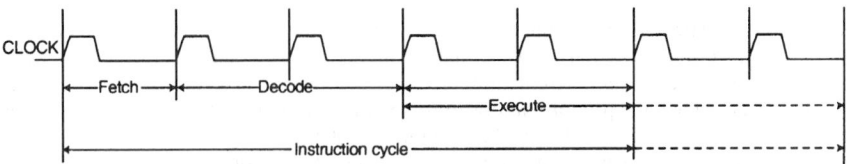

FIGURE 6.4.4 Instruction cycle timing.

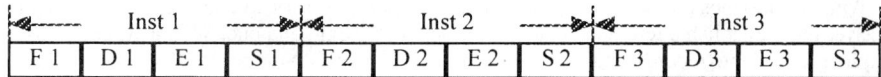

(a) Sequential completion of each instruction in order.

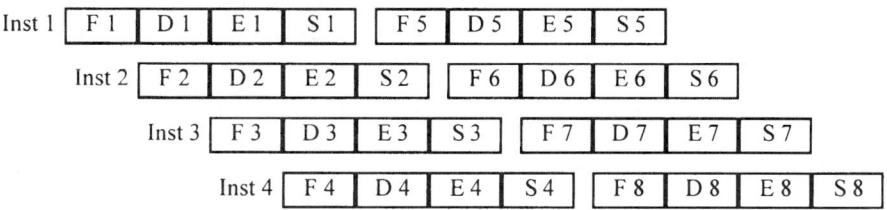

(b) Pipelined execution (fetching next instruction while executing current instruction).

FIGURE 6.4.5   Instruction step sequencing: (a) nonpipelined; (b) pipelined.

instruction (D), executing the instruction (E), and storing the result (S). Figure 6.4.5*a* shows the operation without pipelining, each instruction starting after the previous has completed. In Figure 6.4.5*b*, each of the four steps can be performed in parallel, allowing a higher throughput rate for instructions. Optimizing the performance of a pipeline is complicated by the common use of branch instructions in programs. To handle such cases, sophisticated schemes to manage the pipeline have been developed. Modern high-performance microprocessors are characterized by an increasing number of pipeline stages (e.g., the Pentium Pro has more than 10 pipeline stages).

Serial Data Control

Microprocessors interface with a variety of devices that transmit and receive data in serial streams. The interface logic converts serial data to parallel data and vice versa. A programmable communication interface is used to handle this conversion and samples for data transmission, both synchronous and asynchronous.

Serial protocol involves three possible serial communication techniques. Two of the most widely used are *synchronous* and *asynchronous* serial communication. The third, *isosynchronous*, is a hybrid.

Asynchronous communication is best suited for data transmission between two devices where data are sent at low speed in intermittent, small groupings. A typical application is data entry through a CRT keyboard. Since the receiving device has no way of knowing when data will be sent, the asynchronous format requires *framing information* for each character transmitted. This enables the receiver, such as a universal synchronous-asynchronous receiver-transmitter (USART), to detect a valid data signal properly.

While the receiver waits on a dead line for possible transmission, it constantly samples for the leading edge of the start bit to occur. Sampling is performed by the USART much faster than the transmission baud rate (bit rate). For example, a receiver may sample the signal edge at eight times the baud rate. If the baud rate is 100 (100 bits/s), the USART would sample for the leading edge every $^1/_{800}$ s. Once the edge is detected, however, the receiver must ensure that it is receiving a valid signal and not a transmission caused by line noise. Upon detecting a possible start it is receiving a valid signal and not a transmission caused by line noise. Upon detecting a possible start bit, the receiver steps off a half-bit time to see that the bit is a logical 0. Once the start bit is detected, the receiver times a 1-bit time sampling for the remaining data and stop bits. Through this routine, the receiver synchronizes on each character transmitted.

In a synchronous protocol there is clocking between the two systems. The transmitter generates the clock and transmits data on the leading edge of the clock pulse. The receiver uses this clock to read in the serial data stream.

With synchronous protocol, bit synchronization is not necessary, as data are expected continuously. Still, the receiver must establish a reference indicating where data begin and end. With a synchronous protocol, characters are grouped into records, and *framing characters* are added to the record. These framing characters are SYN (or synch) characters. The USART uses the SYN characters to determine the boundaries of the message. SYN characters consist of a synchronization pattern, a pattern not likely to occur during a normal message. They are generated by the transmitter. The receiver samples these SYN characters bit by bit to establish its reference.

The isosynchronous format of serial communication retains the clock interconnect of the synchronous protocol, but it does not generate SYN characters. As with the asynchronous format, a start bit is generated. As the protocol uses clock synchronization, the repetitive samplings of the asynchronous format are eliminated. The hybrid protocol reduces the amount of MPU time devoted to message recognition, eliminates software overhead required by framing characters, and implements a greatly simplified protocol.

In receiving and transmitting the serial protocol in any form, the USART strips (or inserts) the framing characters and bits from the serial data stream and converts the data into a parallel format to be placed on the data bus. Figure 6.4.6 shows the elementary function of a serial I/O interface.

As far as the MPU is concerned, the USART consists of the data bus buffer, a control, and status registers. The receive data and transmit data buffers lie passively in the path of received and transmitted data and do not need direct access. The USART receives the serial-data-in signal and transmits the serial-data-out control signal. The MPU can service the transmissions on an interrupt-driven basis.

The rate of data transfer in serial data communications is stated in bps (bits per second). Another widely used terminology is baud rate. Digital systems normally are based on signals that can take one of two possible values (one corresponding to logic "1" and the other to logic "0"). However, it is also possible to allow a signal to take on a larger number of discrete values. In such cases, each signal value represents more than a single binary bit. For example, if a signal amplitude can assume one of four (eight) different values, each signal amplitude received represents two (three) bits. The baud rate is the number of signal values transmitted per second. If the baud rate is constant, then the number of bits per second transmitted increases as the number of

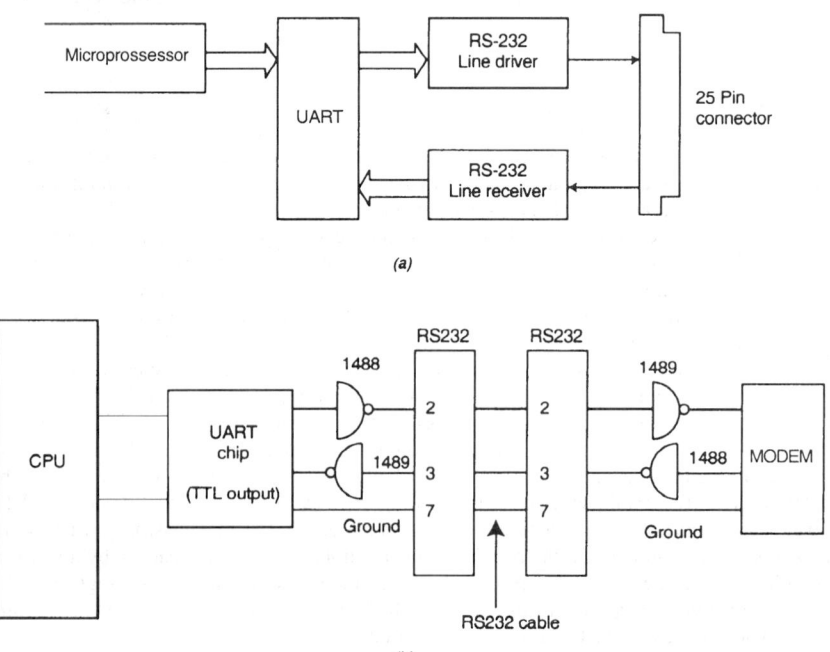

FIGURE 6.4.6 Generalized serial port: (*a*) using RS-232 interface; (*b*) using voltage level converters such as MC1488 and MC1489 chips.

TABLE 6.4.1   RS232 Comparison with RS422 and RS423

	RS232	RS422	RS423
Maximum cable length (ft)	50	4000	4000
Maximum speed (baud)	20K	10M/40 ft	100K/30 ft
		1M/400 ft	10K/300 ft
		100K/4000 ft	1K/4000 ft
Logic 1 voltage level	−3 to −25	A > B	−4 to −6
Logic 0 voltage level	+3 to +25	B > A	+4 to +6

signal levels increases. Serial data communications usually is designed for transmission over as long a distance as possible, with the result that there is significant attenuation (and noise) associated with the received signal level, causing transmission of binary signals to be more robust. This dominance of binary signal transmission has led to the terms *bps* and *baud rate* often being used (though incorrectly) as equivalent.

The serial port of a personal computer is a familiar serial data transmission port. The RS232 standard is common, with characteristics shown in Table 6.4.1. Notice that the voltage levels representing logic 1 and 0 levels are not conventional digital logic levels but rather positive and negative voltages, respectively. These voltage levels are one of the reasons why power supplies for personal computers must provide +/− 12–15 V in addition to the standard digital logic supply voltages. The RS232 standard allows serial data transmission over a modest distance at a modest rate. Table 6.4.1 also shows the characteristics of two extensions of the RS232 standard, both providing substantially higher data rates (or modest data rates over very long distances).

The serial port evolved from peripherals designed to connect a computer to the telephone network through a modem (modulator/demodulator). Table 6.4.2 shows the pins of an RS232 cable using a 25 pin D-type connector. Normally, only a subset of the control signals shown are used, allowing a smaller number of cable pins. Table 6.4.3 shows the pins of a 9-pin serial connector in which the common signals are retained. These tables show that the RS232 connection provides two serial data streams, one transmitted data and the other data being received. In addition, there are a number of control signals that manage the exchange of data between the computer and the peripheral. Because of the origins in connecting computers to telephone networks, terminology related to that connection are still used today. The connectors on opposite ends of an RS232 cable are not equivalent. For example, if one connector is configured according to the connector pins listed in Table 6.4.3, then the other end will require that its transmit data be placed on pin 2 (for reception by the far end) and that its receive data be taken from pin 3 (from transmission by the far end). Control signals also show different pin connections on opposite ends of the cable. The two ends are distinguished by the terminology DTE (Data Terminal Equipment, e.g., the computer) and DCE (Data Communications Equipment, e.g., the connection to the telephone wire). RS232 ports have evolved to provide connections to a wide range of peripherals, well beyond the single example of a modem. Often, the full set of control signals is not required. In cases not requiring end-to-end control signals, only the R × D, T × D, and GND signals shown in Tables 6.4.2 and 6.4.3 are needed.

The RS232 serial port generally transmits ASCI characters. The relationship between the number code (7-bit binary) and the ASCI

TABLE 6.4.2   RS232 Pins

Pin	Description
1	Protective ground
2	Transmitted data (T × D)
3	Received data (R × D)
4	Request to send (RTS)
5	Clear to send (CTS)
6	Data set ready (DSR)
7	Signal ground (GND)
8	Data carrier detect (DCD)
9/10	Reserved for data set testing
11	Unassigned
12	Secondary data carrier detect
13	Secondary clear to send
14	Secondary transmitted data
15	Transmit signal element timing
16	Secondary received data
17	Receive signal element timing
18	Unassigned
19	Secondary request to send
20	Data terminal ready (DTR)
21	Signal quality detector
22	Ring indicator
23	Data signal rate select
24	Transmit signal element timing
25	Unassigned

TABLE 6.4.3 IBM PC 9-Pin Signals

Pin	Description
1	Data carrier detect (DCD)
2	Received data (R × D)
3	Transmitted data (T × D)
4	Data terminal ready (DTR)
5	Signal ground (GND)
6	Data set ready (DSR)
7	Request to send (RTS)
8	Clear to send (CTS)
9	Rind indicator (RI)

characters is shown in Table 6.4.4. There is also an 8-bit code, including several foreign language characters.

The standard printer interface on desktop computers transmits data in parallel, rather than serial. Figure 6.4.7 shows the basic printer cable's connector. There are two data rate links, one an 8-bit data link from the computer to the printer and the other an 8-bit data link from the printer to the computer. Control signals in this case reflect the handshake between the computer and the printer.

Recently, there has been an emphasis on serial data communications technologies capable of handling a far wider range of peripherals than can be handled by the low data rate of RS232 (and its extensions). USB and Firewire are two examples, both allowing connections to not only modems and printers but also to cameras, hard disk drives, and so on. These newer serial data technologies provide high data rates and allow a single serial data port to be connected to a multiplicity of peripherals. These newer

TABLE 6.4.4 The ASCII Code

DECIMAL	HEX	ASCII	DECIMAL	HEX	ASCII	DECIMAL	HEX	ASCII	DECIMAL	HEX	ASCII
0	00	NUL	32	20		64	40	@	96	60	r
1	01	SOH	33	21	!	65	41	A	97	61	a
2	01	STX	34	22	"	66	42	B	98	62	b
3	03	ETX	35	23	#	67	43	C	99	63	c
4	04	EOT	36	24	$	68	44	D	100	64	d
5	05	ENQ	37	25	%	69	45	E	101	65	e
6	06	ACK	38	26	&	70	46	F	102	66	f
7	07	BEL	39	27	'	71	47	G	103	67	g
8	08	BS	40	28	(	72	48	H	104	68	h
9	09	HT	41	29	)	73	49	I	105	69	i
10	0A	LF	42	2A	*	74	4A	J	106	6A	j
11	0B	VT	43	2B	+	75	4B	K	107	6B	k
12	0C	FF	44	2C	,	76	4C	L	108	6C	l
13	0D	CR	45	2D	-	77	4D	M	109	6D	m
14	0E	SO	46	2E	.	78	4E	N	110	6E	n
15	0F	SI	47	2F	/	79	4F	O	111	6F	o
16	10	DLE	48	30	0	80	50	P	112	70	p
17	11	DC1	49	31	1	81	51	Q	113	71	q
18	12	DC2	50	32	2	82	52	R	114	72	r
19	13	DC3	51	33	3	83	53	S	115	73	s
20	14	DC4	52	34	4	84	54	T	116	74	t
21	15	NAK	53	35	5	85	55	U	117	75	u
22	16	SYN	54	36	6	86	56	V	118	76	v
23	17	ETB	55	37	7	87	57	W	119	77	w
24	18	CAN	56	38	8	88	58	X	120	78	x
25	19	EM	57	39	9	89	59	Y	121	79	y
26	1A	SUB	58	3A	:	90	5A	Z	122	7A	z
27	1B	ESC	59	3B	;	91	5B	[	123	7B	{
28	1C	FS	60	3C	<	92	5C	\	124	7C		
29	1D	GS	61	3D	=	93	5D	]	125	7D	}
30	1E	RS	62	3E	>	94	5E	^	126	7E	~
31	1F	US	63	3F	?	95	5F	–	127	7F	DEL
127	7F	DEL									

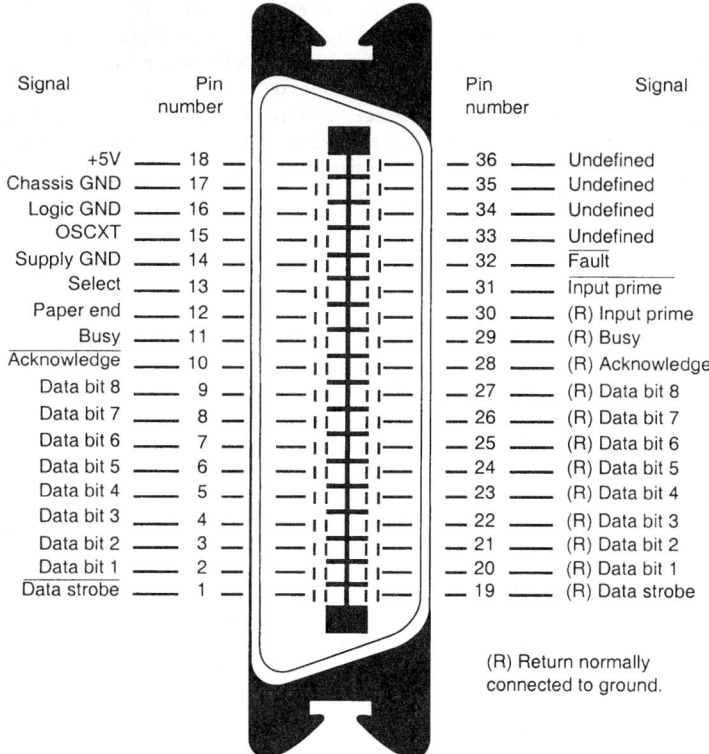

Signal	Pin number			Pin number	Signal
+5V	18			36	Undefined
Chassis GND	17			35	Undefined
Logic GND	16			34	Undefined
OSCXT	15			33	Undefined
Supply GND	14			32	Fault
Select	13			31	Input prime
Paper end	12			30	(R) Input prime
Busy	11			29	(R) Busy
Acknowledge	10			28	(R) Acknowledge
Data bit 8	9			27	(R) Data bit 8
Data bit 7	8			26	(R) Data bit 7
Data bit 6	7			25	(R) Data bit 6
Data bit 5	6			24	(R) Data bit 5
Data bit 4	5			23	(R) Data bit 4
Data bit 3	4			22	(R) Data bit 3
Data bit 2	3			21	(R) Data bit 2
Data bit 1	2			20	(R) Data bit 1
Data strobe	1			19	(R) Data strobe

(R) Return normally connected to ground.

FIGURE 6.4.7   The Centronics parallel printer interface using a 36-pin connector.

serial data communication technologies provide significantly smaller connectors and smaller diameter cables, well suited for today's laptop computers but also providing improved performance for desktop computers.

Disk Interfaces

The evolution of interfaces to hard disk drives illustrates the major impact that standardization plays in the area of computer systems. Early computer manufacturers provided both the computer and the hard disk drive, using a proprietary interface requiring that the disk drives be obtained from the same manufacturer as the computer. In those earlier computers, most of the functional tasks required to read data from and write data to a hard disk drive was managed by a disk controller located in the computer itself, the disk drive having little built-in intelligence. As VLSI advanced, it became possible to place much of the functionality needed to manage the reading and writing of data in the hard disk drive unit itself. Once this responsibility was separated from the computer, hard drives could be purchased from a number of manufacturers, so long as the cable connection between the computer and hard drive was satisfied. This drove the evolution to interface and cable standards to connect hard drives to computers, allowing the disk drive market to emerge and provide drives for virtually any computer with the standardized interface.

Standard disk drive interfaces developed to support the Intel-based PC included the ESDI (Enhanced Small Device Interface) and IDE (Integrated Device Electronics). The SCSI (Small Computer System Interface) standard (pronounced "scuzzy") emerged for general computer use, not tied to any particular computer platform. The standards have had to adapt to the remarkable increases in disk drive capacity and data transfer rate, leading to the highly sophisticated hard disk drives in use today. For example, the SCSI standard has passed through several generations, successive generations providing higher data rates and less bulky connectors. These interfaces are used not only for hard disk drives but also for tape drives, CD drives, and DVD drives.

These disk drive interface standards generally require rather bulky cables, limiting their extension to other peripherals such as digital cameras. The need to support a wide range of external devices has stimulated the development of less bulky but high data rate interfaces such as USB (modest data rate) and Firewire (high data rate), signaling a new generation of standardization of cables and connectors for personal computer use as the variety of peripherals expand.

BIBLIOGRAPHY

Antonakos, J. L., "The Pentium Microprocessor," Prentice Hall, 1997.

Ayala, K., "The 8086 Microprocessor: Programming and Interfacing the PC," West, 1995.

Bakoglu, H., and T. Whiteside, "RISC System/6000 Hardware Overview," IBM RISC Technologies, IBM, 1990. SA23-2619.

Bartee, T., "Computer Architecture and Logic Design," McGraw-Hill, 1991.

Carter, J. W., "Microprocessor Architecture and Microprogramming," Prentice Hall, 1995.

Gibson, V., "Microprocessors: Fundamental Concepts and Applications," Delmar, 1994.

Gilmore, C., "Microprocessor Principles and Applications," McGraw-Hill, 1989.

Goody, R., "Intel Microprocessors," Glencoe/McGraw-Hill, 1992.

Hall, D., "Microprocessors and Interfacing," 2nd ed., Glencoe/McGraw-Hill, 1991.

Hennessy, J. L., and D. A., Patterson, Computer Architecture: A Quantitative Approach, 2nd ed., Morgan Kaufmann, 1996.

Hester, P., "RISC System/6000 Hardware Background and Philosphies," IBM RISC Technologies, IBM, 1990.

Horwath, R., "Introduction to Microprocessors Using the MC6809 or MC6800," McGraw-Hill, 1992.

Kleitz, W., "Digital and Microprocessor Fundamentals," Prentice Hall, 1990.

Leventhal, L., "Microcomputer Experimentation with the Intel SDK-86," BBS College, 1987.

Mazidi, M., and J., "The 80 × 86 IBM PC and Compatible Computers," Vol. I and II, Prentice Hall, 1995.

Mazur, H., "The History of the Microcomputer—Invention and Evolution," *Proc. IEEE*, Vol. 83(12), pp. 1601–1608, 1995.

Mueller, S., "Upgrading and Repairing PCS," 4th ed., Que, 1994.

O'Connor, P., "Digital and Microprocessor Technology," 2nd ed., Prentice Hall, 1989.

Pack, D. J., and S. F., Barrett, "68HC12 Microcontroller," Prentice Hall, 2002.

Putnam, B. W., "Digital and Microprocessor Electronics: Theory, Applications, and Troubleshooting," Prentice Hall, 1986.

Ronen, R. A., Mendelson, K. Lai, S-L. Lu, F. Pollack, and J. P., Shen, "Coming Challenges in Microarchitectures and Architecture," *Proc. IEEE*, Vol. 89 (3), 2001, pp. 325–340.

Smith, A. J., "Cache Memories," ACM Computer Surveys, Vol 14(3), pp. 473–530, 1982.

Thompson, A., "Understanding Microprocessors: A Practical Approach," Delmar, 1994.

Treibal, W., and A. Singh, "The 8088 and 8086 Microprocessors," Prentice Hall, 1992.

Uffenback, J., "The 8086/8088 Family, Design, Programming and Interfacing," Prentice Hall, 1987.

Urganiak, K., "Experiments for the Intel 8088," Delmar, 1994.

Valvano, J. W., "Embedded Microcomputer Systems," Brooks/Cole, 2000.

Yeager, K., "The MIPS R10000 Superscalar Microprocessor," *IEEE Micro*, Vol. 16(4), pp. 28–40, 1996.

NANOTECHNOLOGIES

Wen H. Ko

INTRODUCTION

As integrated circuit technology and microtechnology progress and expand into all phases of human activities with micron-size devices and systems, we find that materials, devices, and even systems at the nanometer scale are interesting for study and may have great potential for new science, engineering, and industries. The national nanotechnology initiative of National Science Foundation, U.S.A., states "Nanotechnology is concerned with materials and systems whose structures and components exhibit novel and significantly improved physical, chemical, and biological properties, phenomena, and processes because of their small nanoscale size. Structural features in the range of about 10-9 to 10-7 m (1 to 100 nm) determine important changes as compared to the behavior of isolated molecules (1 nm) or of bulk materials."

The research and technology on nanoscale materials and some atomic scale biochemical films existed long before the organized initiative. However, only with the recent advances in sciences and technology, it is possible to perform atomic or molecular manipulation and to initiate organized efforts to study and develop all fronts of nanoscale natural world. The flourishing activities under the general topic of nanotechnology now include:

1. Materials with unique or modified properties
   a. Nanostructured metals, semiconductors, and superconductors
   b. Nanoparticles—clusters, crystals, and composites
   c. Hybrid materials—alloys, ceramics, and mono- or multilayer materials
   d. Supermolecular, dendrimers, polymers
   e. Nanotubes, nanorods, molecular wires
   f. Superlattice, quantum wires, and quantum dots
   g. Nanoporus systems
2. Science of nanomaterials—physical, chemical, and biological analysis; characterization and modeling of surfaces; nanomaterials and nanostructures aimed at understanding of them
3. Nanofabrication and processing—chemical vapor deposition (CVD) and other controlled deposition methods, molecular beam epitaxy growth, surface treatment, and programmable self-assembly techniques
4. Nanomanipulation and instrumentation—Nanoprobes, atomic force microscope (ATM), scanning tunnel current microscope, scanning electrochemical microscope, nanoactuators and controllers, and many other instruments
5. Nanoelectronics, nano-optics, and nanomagnetism—device characteristics and potential application
6. Application to electronics, information technology, robotics, fluidics, communication, and energy research
7. Nanobiotechnology and application in medicine and pharmacology
8. Others

The MEMS has served as a pathway for nanotechnology by providing instruments and tools for nanoresearch such as:

1. Microprobe devices, nanometer actuators, and controllers make these tools available for many researchers
2. Microfluidic systems for biomedical studies and self-assembly instruments with fast, multiple unit parallel process capability
3. Microsensors and actuator for nanofabrication and processes
4. Models and methods to understand nanoscale materials, devices, and properties

From engineering aspect of nanotechnology there are two parallel approaches. One is the bottom-up approach. This is to build materials and devices from atoms or molecules up, starting from the modification of the atomic or molecular structures or composition of material to alter their property. This includes research on self-assembly techniques, nanomaterials, composites, nanotubes, quantum dots … as well as biochemical and medical related activities. Using these modified materials and nanoassembly techniques applications in biochemical and medicine can be developed. The other is the top-down approach. That is to shrink the microdevices and systems to nanoscale such as the nanotransistors/switches, and 90 nm and smaller integrated circuits, nanoscale sensors and actuators, and thin film and thick film systems. Both approaches are being pursued with great enthusiasm and efforts. The choice depends on which one is more suited for which research activity.

The field of nanotechnology is so broad and is advancing with tremendous speed over many wide fronts; it is not possible for this section to cover it at any depth. Hopefully these few pages will call attention to this field and that the references, particularly the journals and Internet addresses, will serve as beginning steps to understand the scope and potential of this field.

CARBON NANOTUBES

The carbon nanostructures are the most well-known nanostructures, since the discovery of C_{60} in 1985 by Buckminster Fuller. Among all types of carbon nanostructures the carbon nanotubes may be the most studied material. A brief discussion of carbon nanotubes and applications from the electrical engineer's viewpoint is given here as an introduction to the vast world of nanomaterials and nanotechnology.

There are two main types of carbon nanotubes (CNT)—the single-walled nanotubes (SWNTs) and the multi-walled nanotubes (MWNTs). They can be thought of as single or multiple layers of graphite sheets seamlessly wrapped into cylindrical tubes. Depending on the manner the graphite sheet is rolled from the tube, a pair of integers (n, m) is used to denote the nanotube type that defines the direction of rolling and the diameter of the tube. There are three types of CNT: the armchair ($n = m$), zigzag (n or $m = 0$), and chiral (any other n and m). Armchair SWNTs are metals, those $n - m = 3k$ [k is an integer] are semiconductors with small band gap.

Carbon nanotubes can be produced, in small quantity, by several methods, such as laser ablation of carbon, carbon arc discharge, and chemical vapor deposition of hydrocarbon vapor over a dispersed Fe or other catalysts. One method used decomposition of CH_4 or CO over Al_2O_3 supported Mo or Fe:Mo (9:1), at 700 to 850°C.

SECTION 7

UHF AND MICROWAVE COMPONENTS

The three chapters in this section cover passive microwave components, vacuum electronic devices, and microwave semiconductor devices.

A broad range of passive devices (Chap. 7.1) enable the transmission, sampling, filtering, and impedance matching of UHF and microwave power. Rigid rectangular waveguide is widely used for transmission. Flexible coaxial cable is employed for short runs, in cases where its higher attenuation can be tolerated. In lower power applications microstrip technology provides advantages in miniaturization and conformal configurations.

Chapter 7.2 is devoted principally to vacuum electronic devices, both traditional and advanced. Modern materials and design methods have advanced the RF vacuum device orders of magnitude in performance over that of the earlier glass envelope vacuum tube. While magnetrons and klystrons were used in World War II receivers, postwar funding of particle accelerator development resulted in klystrons having a 10^3 increase in peak power. Later developments designed to overcome bandwidth limitations of klystrons include the clustered cavity technique. The klystrode, an advanced version of the inductive output tube (IOT), offers reduced length and weight as a result of prebunching the beam. The traveling-wave tube (TWT) is a linear-beam device that amplifies microwave signals to high power levels over broad bandwidths. It finds diverse applications in communications, radar, and electronic countermeasure systems, and aerospace and avionics where low weight and volume, and minimizing power consumption are required. A further step to miniaturization of RF systems is the microwave power module (MPM), which exploits both vacuum electronics and solid-state technology. In the MPM, a high-gain monolithic integrated circuit drives a miniaturized helix TWT—both powered by a high-density electronic power conditioner. Gyrotrons (or cyclotron resonance masers) are a class of microwave generators that promise applications in the millimeter-wave region. Gyrotron oscillators have been developed that can produce 1 MW for 1 s at frequencies greater than 100 GHz.

Advances in semiconductor materials, particularly compound semiconductors, have enhanced the performance of microwave and millimeter wave devices (Chap. 7.3). Monolithic microwave integrated circuits (MMICs) enable small, lower cost RF and microwave components like amplifiers, oscillators, and switches for use in high speed/high frequency applications like cell phones and optical fiber communication systems. They are used, for example, in microwave power modules (MPMs).   D.C.

In This Section:

CHAPTER 7.1
PASSIVE MICROWAVE COMPONENTS

Dwight Caswell, Joseph Feinstein

INTRODUCTION

While the physical concepts and mathematical theory underlying electromagnetic-wave propagation in confined structures were developed at the end of the nineteenth century, the practical utilization of wavelengths shorter than 1 m began during World War II. A wide variety of devices is now available for the transmission, sampling, filtering, and impedance matching of UHF and microwave power. Because of the short wavelengths at these frequencies (10 cm at 3 GHz, 1 cm at 30 GHz), most of these components use distributed elements and obtain specific reactances by judicious use of short-circuited lengths of transmission line. However, the trend toward microcircuitry has led recently to the introduction of lumped elements in some low-power, low-Q applications. In addition, the use of strip and microstrip transmission lines marks a return to open-wire media, with the attendant radiation loss kept low by close spacing and the presence of the dielectric filler.

Except for special applications (such as rotating joints for antenna feeds where a cylindrical member is essential), the use of rectangular waveguide, dimensioned to transmit in the dominant (lowest-order) mode, is standard for high-power transmission. Coaxial cable is used for short-distance runs where the advantage of its flexibility outweighs its higher attenuation. Ridged waveguide is useful for designing matching sections and for providing very-wide-bandwidth single-mode transmission. Oversized cylindrical guide operated in the circular electric mode is finding use in millimeter-wave-carrier telephony, where its extremely low attenuation justifies the special precautions which must be taken to avoid mode conversions.

Reciprocal and Nonreciprocal Components.   Of the components which have become standard in this field, perhaps the most unusual are ferrite devices. When biased with the proper magnetic field, ferrites act as nonreciprocal elements with respect to microwave transmission in an appropriate frequency band. This behavior allows isolation of a signal source from reflections and the separation of incident from reflected power along the same transmission line (by means of a ferrite device called a *circulator*). It is also possible to vary the phase of a transmitted wave by adjusting the magnetizing field on the ferrite. Such phase shifters are capable of handling high powers with low loss.

All other types of microwave components, such as hybrid junctions and directional couplers, are reciprocal in their action. The latter are employed for power division and for signal sampling. A wide variety of transmission components such as variable attenuators, matched-load terminations, and slotted lines are used in microwave-measurements and design.

High-Q resonators are formed from completely enclosed short-circuited sections of waveguides with slit or loop coupling. Extremely low loss dielectric cylinders can also be used as high-Q resonators and have the advantages of smaller size and easy coupling to microstrip or other transmission lines. Resonators using lengths of strip, microstrip or coaxial lines are frequently convenient, but have lower Q's. Lumped elements, such as varactor diodes or YIG spheres, provide electrically tunable resonators for microwave oscillators and receivers.

Transmission-Line Relationships

The basic transmission-line equations are derived for distributed parameters R (series resistance), G (shunt conductance), L (series inductance), and C (shunt capacitance), all defined per unit length of line. Some useful relations are shown in Table 7.1.1. For zero losses (R, $G = 0$) one obtains the ideal line expression shown on the right. Table 7.1.2 gives some equations that are useful for relating the measured voltage standing wave ratio (VSWR) to wave transmission and reflection.

Matching

A microwave circuit typically consists of several devices connected together or connected through sections of transmission line. Mismatches between the devices decrease transmitted power, increase reflected energy and make the performance of the circuit more frequency dependent. Match between devices may be achieved by using capacitive or inductive elements suitably positioned along the transmission line. Another approach is to

TABLE 7.1.1   Summary of Transmission-Line Equations

Quantity	General line expression	Ideal line expression
Propagation constant	$\gamma = \alpha + j\beta = \sqrt{(R + j\omega L)(G + j\omega C)}$	$\gamma = j\omega\sqrt{LC}$
Phase constant β	Im γ	$\beta = \omega\sqrt{LC} = 2\pi/\lambda$
Attenuation constant α	Re γ	0
Impedance, characteristic	$Z_0 = \sqrt{\dfrac{R + j\omega L}{G + j\omega C}}$	$Z_0 = \sqrt{\dfrac{L}{C}}$
Input	$Z_{-1} = Z_0 \dfrac{Z_r + Z_0 \tanh \gamma l}{Z_0 + Z_r \tanh \gamma l}$	$Z_{-1} = Z_0 \dfrac{Z_r + jZ_0 \tan \beta l}{Z_0 + jZ_r \tan \beta l}$
Of short-circuited line, $Z_t = 0$	$Z_{oc} = Z_0 \tanh \gamma l$	$Z_{oc} = jZ_0 \tan \beta l$
Of open-circuited line, $Z_r = \infty$	$Z_{oc} = Z_o \coth \gamma l$	$Z_{oc} = -jZ_0 \cot \beta l$
Of line an odd number of quarter wavelengths long	$Z = Z_0 \dfrac{Z_t + Z_0 \coth \alpha l}{Z_0 + Z_r \coth \alpha l}$	$Z = \dfrac{Z_0^2}{Z_r}$
Of line an integral number of half wavelengths long	$Z = Z_0 \dfrac{Z_r + Z_0 \tanh \alpha l}{Z_0 + Z_r \tanh \alpha l}$	$Z = Z_r$
Voltage along line	$V_{-1} = V_i(1 + \Gamma_0 e^{-2\gamma l})$	$V_{-1} = V_t(1 + \Gamma_0 e^{-2\beta l})$
Current along line	$I_{-1} = I_i(1 - \Gamma_0 e^{-2\gamma l})$	$I_{-1} = I_t(1 - \Gamma_0 e^{-2\beta l})$
Voltage reflection coefficient	$\Gamma = \dfrac{Z_r - Z_0}{Z_r + Z_0}$	$\Gamma = \dfrac{Z_r - Z_0}{Z_r + Z_0}$

l = length of transmission line.

TABLE 7.1.2   Some Miscellaneous Relations in Low-Loss Transmission Lines

Equation	Explanation
$r = \dfrac{1 + |\Gamma|}{1 - |\Gamma|}$	r = VSWR
$|\Gamma| = \dfrac{r - 1}{r + 1}$	$|\Gamma|$ = magnitude of reflection coefficient
$\Gamma = \dfrac{R - Z_0}{R + Z_0}$	Γ = reflection coefficient (real) at a point in a line where impedance is real (R)
$r = \dfrac{R}{Z_0}$	$R > Z_0$ (at voltage maximum)
$r = \dfrac{Z_0}{R}$	$V < Z_0$ (at voltage minimum)
$\dfrac{P_r}{P_i} = |\Gamma|^2 = \left(\dfrac{r - 1}{r + 1}\right)^2$	P_r = reflected power P_i = incident power
$\dfrac{P_t}{P_i} = 1 - |\Gamma|^2 = \dfrac{4r}{(r + 1)^2}$	P_t = transmitted power
$\dfrac{\alpha_r}{\alpha_m} = \dfrac{1 + \Gamma^2}{1 - \Gamma^2} = \dfrac{r^2 + 1}{2r}$	α_m = attenuation constant when $r = 1$, matched line α_r = attenuation constant allowing for increased ohmic loss caused by standing waves
$r_{max} = r_1 r_2$	r_{max} = maximum VSWR when r_1 and r_2 combine in worst phase
$r_{min} = \dfrac{r_2}{r_1}\, r_2 > r_1$	r_{min} = minimum VSWR when r_1 and r_2 are in best phase
$|\Gamma| = \dfrac{|X|}{\sqrt{X^2 + 4}}$	Relations for a normalized reactance X in series with resistance Z_0
$|X| = \dfrac{r - 1}{\sqrt{r}}$	
$|\Gamma| = \dfrac{|B|}{\sqrt{B^2 + 4}}$	Relations for a normalized susceptance B in shunt with admittance Y_0
$|B| = \dfrac{r - 1}{\sqrt{r}}$	

use transformers in the transmission line, such transformers having an appropriate length and impedance. Yet another approach is to taper from one impedance to another.

Smith Chart.   The matching elements or transformers can be calculated graphically using a Smith chart which plots complex impedance. The standing wave pattern in a transmission line repeats itself every half-wavelength. The angular position of the radius of a circle represents the phase along a transmission line. Rotating the radius of a circle a complete revolution corresponds to a phase change of $\frac{1}{2}$ wavelength. Circles of constant resistance are tangent to the circumference of the Smith chart at a single point. These two features are illustrated in Fig. 7.1.1a. The center of the Smith chart has been normalized to a value of 1. To change to a 50-ohm transmission line, all of the resistance values would be multiplied by 50. A constant voltage standing wave

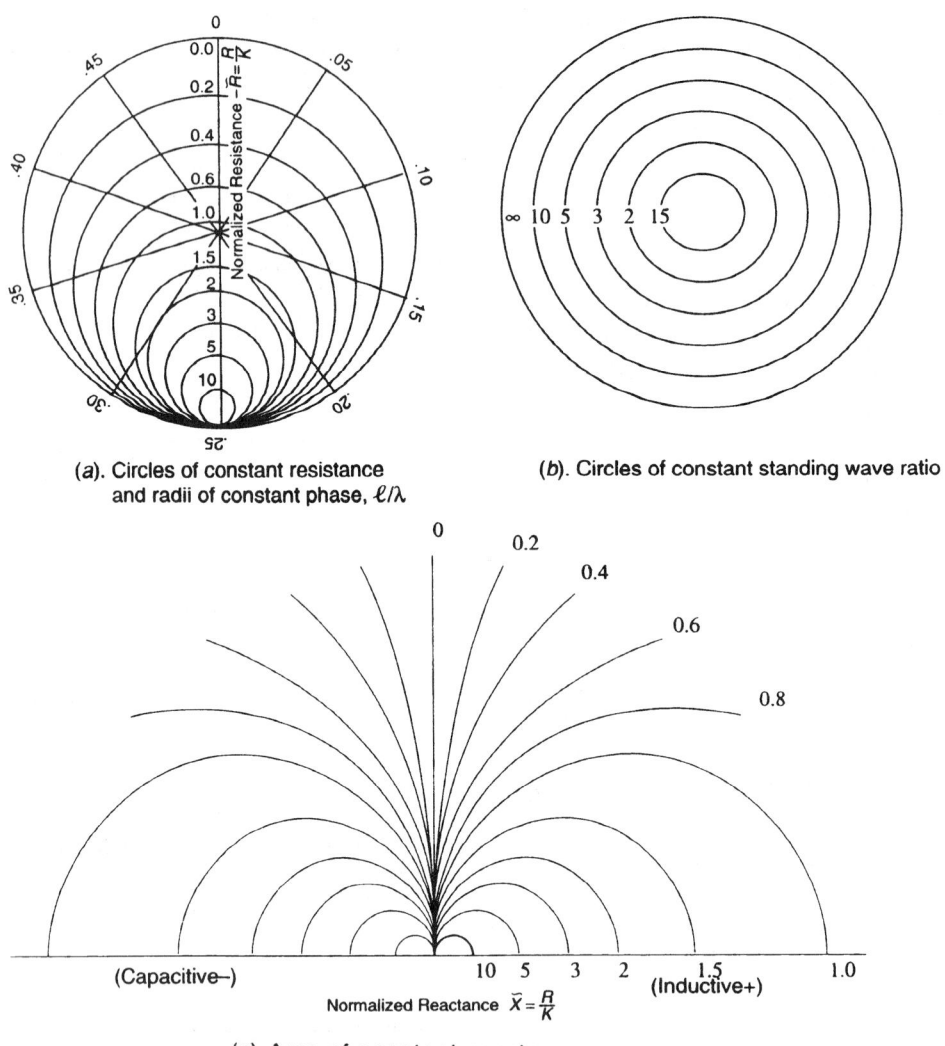

(a). Circles of constant resistance
and radii of constant phase, ℓ/λ

(b). Circles of constant standing wave ratio

(c) Arcs of constant reactance

FIGURE 7.1.1 Elements of a Smith chart.

ratio is represented by a circle concentric with the center of the Smith chart. This is illustrated in Fig. 7.1.1*b*. Circles are not shown on the chart because their value can be determined by measuring the distance from the center of the chart to the point of interest. For the normalized Smith chart, the resistance value along the zero reactance line is numerically equal to the VSWR, varying between 1 and infinity. Normalized reactance is a series of arcs as shown in Fig. 7.1.1*c*. The actual values of reactance can be obtained by multiplying by the normalizing constant. Figures 7.1.1*a* and 7.1.1*c* are superimposed to obtain the Smith chart, Fig. 7.1.2. The impedance of P_1 is 0.6 Ω. The impedance of P_2 is $0.5 - j1.0$. The impedance of P_3 is $0.5 + j0.5$. P_3 can be matched by adding a capacitive reactance $-j1.0$ at a distance of 0.72 λ toward the generator, as shown.

The Smith chart may also be used to represent admittance Y, where $Y = 1/Z$. For a normalized admittance Smith chart ($Y_0 = 1$ at the center) capacitive reactance ($-jX/Z_0$) is replaced by inductive susceptance ($-jB/Y_0$); inductive reactance ($+jX/Z_0$) is replaced by capacitive susceptance ($+jB/Y_0$).

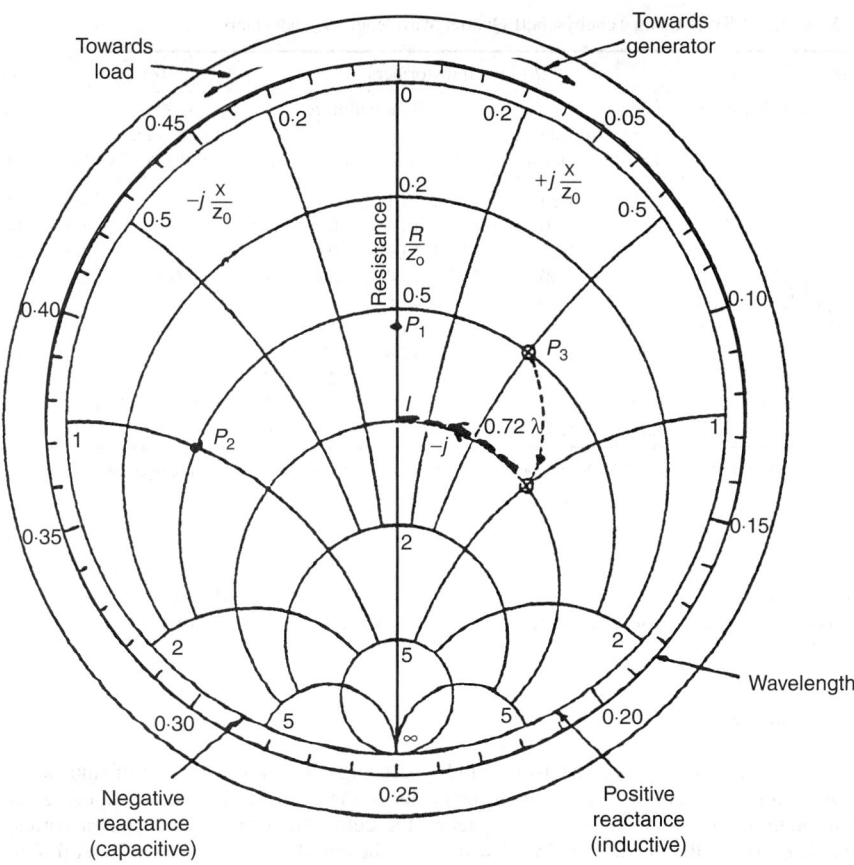

FIGURE 7.1.2 Smith chart representation of complex impedance. Impedance of P_1 is 0.6 ohm, P_2 is $0.5 - j1.0$, P_3 is $0.5 + j0.5$. P_3 is matched with a capacitive impedance $-j1.0$ with a phase of $0.72\ \lambda$ toward the generator.

Quarter Wavelength Transformers. One or more quarter wavelength sections of transmission line may be used to match from one impedance to another. Table 7.1.3 gives the VSWR which will result when one, two, or three transformer sections are used. R is the ratio of the two impedances to be matched. ω_g is the bandwidth over which match is to be achieved. If a single quarter wavelength section is used, the impedance of the transformer is

$$Z \text{ (impendance of matching transformer)} = \sqrt{Z_L \times Z_H}$$

where Z_L and Z_H are the characteristic impedance of the lower and higher impedance transmission lines, respectively. When two quarter wavelength sections are used and when the impedance of the lower impedance transmission line, Z_L, is normalized to unity, the impedance of the first transformer is the same as Z_1 in Table 7.1.4a. The impedance of the second quarter wavelength section, Z_2, is R/Z_1. For three matching transformers, the transformer impedances are

$$Z_1 = Z_1 \qquad Z_2 = \sqrt{R} \qquad Z_3 = R/Z_1$$

TABLE 7.1.3 Maximum VSWR Using Tchebyscheff Quarter Wavelength Transformers

(a) One transformer					(b) Two transformers					(c) Three transformers				
Impedance ratio, R	Bandwidth, ω_q				Impedance ratio, R	Bandwidth, ω_q				Impedance ratio, R	Bandwidth, ω_q			
	0.2	0.4	0.6	0.8		0.2	0.4	0.6	0.8		0.2	0.4	0.6	0.8
1.25	1.03	1.07	1.11	1.14	1.25	1.00	1.01	1.03	1.05	1.25	1.00	1.00	1.01	1.02
1.50	1.06	1.13	1.20	1.27	1.50	1.01	1.02	1.05	1.09	1.50	1.00	1.00	1.01	1.03
1.75	1.09	1.19	1.30	1.39	1.75	1.01	1.03	1.07	1.13	1.75	1.00	1.00	1.02	1.04
2.00	1.12	1.24	1.38	1.51	2.00	1.01	1.04	1.08	1.16	2.00	1.00	1.01	1.02	1.05
2.50	1.16	1.34	1.53	1.73	2.50	1.01	1.05	1.12	1.22	2.50	1.00	1.01	1.03	1.07
3.00	1.20	1.43	1.68	1.95	3.00	1.01	1.06	1.14	1.27	3.00	1.00	1.01	1.03	1.08
4.00	1.26	1.58	1.95	2.35	4.00	1.02	1.08	1.19	1.37	4.00	1.00	1.01	1.04	1.11
5.00	1.32	1.73	2.21	2.74	5.00	1.02	1.09	1.23	1.45	5.00	1.00	1.01	1.05	1.13
6.00	1.37	1.86	2.45	3.12	6.00	1.03	1.11	1.26	1.53	6.00	1.00	1.02	1.06	1.15
8.00	1.47	2.11	2.92	3.86	8.00	1.03	1.13	1.33	1.67	8.00	1.00	1.02	1.07	1.18

Source: IRE Trans. PGMTT-8, pp. 478–482 (September 1960). Leo Young, "Tables for Cascaded Homogeneous Quarter-Wave Transformers."

Matching with a Taper. Tapered transitions may be used to match from one transmission line to another. The VSWR is effected by the shape of the taper, i.e., linear or some other shape, and the length of the taper. In general a better match can be achieved using quarter wavelength sections when compared to a taper of the same length. The voltage reflection coefficient Γ is plotted in Fig. 7.1.3 for an impedance ratio of 2:1 for various taper lengths.

Rectangular Waveguide

The electric field pattern of a dominant-mode (TE_{10}) rectangular waveguide is a half sinusoid across the transverse guide dimension with its maximum at the center of the broad wall. Microwave energy cannot be transmitted through a waveguide in frequencies below the cutoff frequency, which has a corresponding cutoff wavelength λ_c. When the inside broad wall width is a, the cutoff wavelength $\lambda_c = 2a$. A different mode (electric field configuration) can be transmitted through the waveguide when the frequency is greater than $c/2b$, where c is the speed of light and b is the narrow dimension of the waveguide. The normal waveguide operating

TABLE 7.1.4 Z_1 for Tchebyscheff Quarter-Wave Transformers

(a) Two-section transformer						(b) Three-section transformer					
Impedance ratio, R	Bandwidth, ω_q					Impedance ratio, R	Bandwidth, ω_q				
	0.0	0.2	0.4	0.6	0.8		0.0	0.2	0.4	0.6	0.8
1.00	1.0000	1.0000	1.0000	1.0000	1.0000	1.00	1.0000	1.0000	1.0000	1.0000	1.0000
1.25	1.0573	1.0581	1.0603	1.0641	1.0697	1.25	1.0282	1.0288	1.0305	1.0335	1.0383
1.50	1.1066	1.1080	1.1123	1.1197	1.1305	1.50	1.0520	1.0530	1.0561	1.0618	1.0709
1.75	1.1501	1.1521	1.1583	1.1690	1.1846	1.75	1.0725	1.0739	1.0783	1.0864	1.0993
2.00	1.1892	1.1918	1.1997	1.2136	1.2338	2.00	1.0906	1.0924	1.0980	1.1083	1.1246
2.50	1.2574	1.2611	1.2724	1.2921	1.3211	2.50	1.1217	1.1242	1.1319	1.1460	1.1686
3.00	1.3160	1.3207	1.3352	1.3604	1.3976	3.00	1.1479	1.1509	1.1605	1.1779	1.2062
4.00	1.4142	1.4208	1.4410	1.4764	1.5289	4.00	1.1907	1.1947	1.2074	1.2308	1.2689
5.00	1.4953	1.5036	1.5292	1.5740	1.6408	5.00	1.2252	1.2301	1.2455	1.2741	1.3207
6.00	1.5650	1.5750	1.6056	1.6593	1.7397	6.00	1.2543	1.2600	1.2779	1.3110	1.3655
8.00	1.6817	1.6947	1.7347	1.8052	1.9110	8.00	1.3021	1.3091	1.3312	1.3725	1.4409

Source: IRE Trans. PGMTT-8, pp. 478–482 (September 1960). Leo Young, "Tables for Cascaded Homogenous Quarter-Wave Transformers."

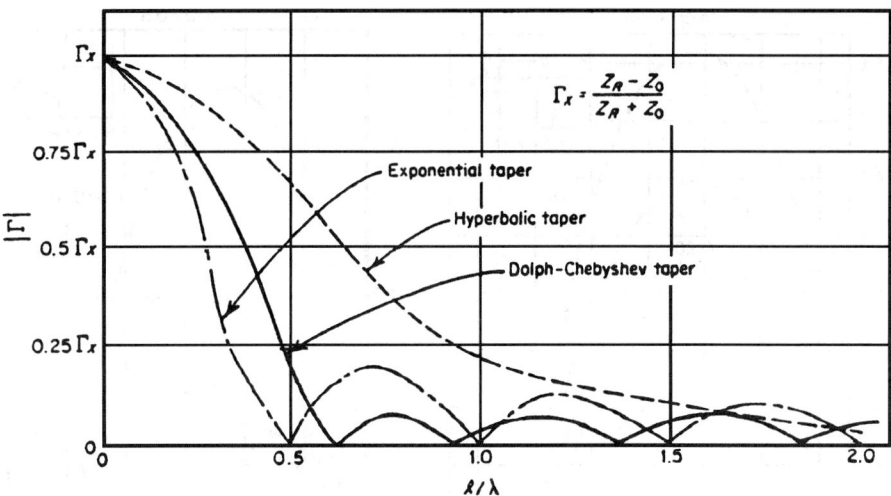

FIGURE 7.1.3 Use of tapers for impedance matching.

frequency range is significantly higher than the cutoff frequency and significantly lower than the frequency where the second mode can be transmitted.

When higher power must be transmitted than is possible because of breakdown in air with these restrictions on dimensions, pressurization of the air or a sulfur hexafluoride gas fill is generally used.

Unlike the case for the transverse electromagnetic (TEM) mode of a coaxial or parallel-wire line, the guide wavelength λ_g in rectangular waveguide departs from the free-space wavelength λ_0 and varies with frequency, producing phase distortion in a wide-band signal. The relationship is

$$\lambda_g = \lambda_0 / \sqrt{1 - (\lambda_0 / \lambda_c)^2}$$

Ridge Waveguide

To obtain broader bandwidth in a single mode, as well as increased flexibility in the choice of impedance, ridge waveguide is generally employed. The variation of cutoff wavelength with dimensions for single and double ridges (always inserted on the broad wall of a rectangular guide to give capacitive loading) is given in Fig. 7.1.4. A large increase in bandwidth is made possible by this type of guide.

Circular-Mode Transmission

The circular electric (TE_{01}) mode is currently employed in long-distance broadband communications at millimeter wavelengths because of its low-loss transmission property. The electric field pattern and the wall currents form concentric rings in this unusual mode. Conversion to lower-order modes occurs if the cross section is even slightly elliptical or if the guide axis is curved. Various forms of mode suppression are employed; a common technique consists in fabricating the cylindrical wall from a tightly wound helix with loss between the turns to damp out modes with axial components of current. Attenuation as low as a few decibels per mile has been achieved.

Coaxial Line

The attenuation and power handling capabilities of rigid coaxial lines of various outer diameters are shown in Fig. 7.1.5. The frequency scale on this figure extends only to 3 GHz because such high-power transmitting coaxial cable is generally not used at higher frequencies.

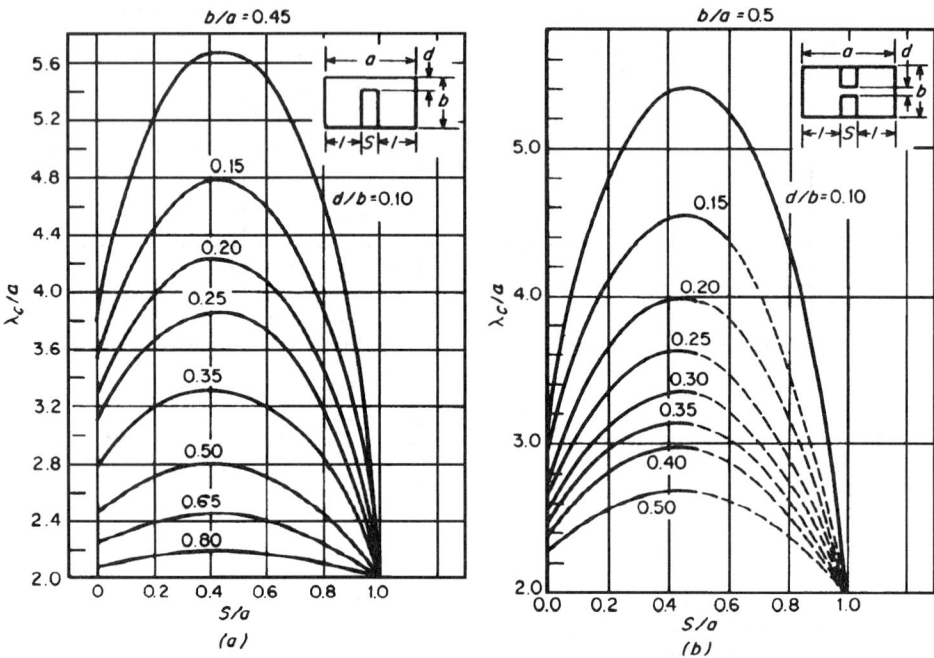

FIGURE 7.1.4   Cutoff wavelength of ridged waveguide: (*a*) single ridge; (*b*) double ridge.

Care must be taken to avoid operation of a coaxial line at wavelengths where it becomes possible for additional modes to propagate. This occurs when the mean circumference between the inner and outer cylinders forming the transmission path equals a full wavelength; a stable standing wave is then possible in a circumferential direction. The higher attenuation and reduced power-carrying capability which accompany the small dimensions necessary to avoid higher-order modes generally lead to the choice of waveguide as a transmission medium at frequencies above 3 GHz for medium and high-power applications.

Strip Transmission Line

Strip transmission lines consist of a flat conductor between two ground planes. The characteristics are similar to coaxial line; however, they are easier to change by varying the width of the circuit, the proximity of adjacent lines or by the addition of stubs or other circuit elements. Impedance is controlled by the width of the strip; coupling is controlled by the spacing between lines. The circuit is flat and is easily fabricated. The material between the ground planes and the strip can be air or a low-loss dielectric material such as those enumerated in Table 7.1.5. Components such as directional couples, filters, switches, and isolators are frequently made in stripline. Connectors are usually provided to join one component to another. Characteristic impedance Z_0 of a dielectrically loaded strip line (ϵ = dielectric constant) is given in Fig. 7.1.6. Figure 7.1.7 illustrates typical attenuation curves. A stripline with 0.125-in. ground plane spacing and teflon fiber glass dielectric will usually handle 25 kW peak power and 500 W average power in C-band. The dielectric material and the ground plane spacing must be chosen so that the line propagates only the TEM mode.

Microstrip Transmission Line

A microstrip transmission line has a conductive circuit on top of a dielectric substrate with a ground plane below the substrate. The advantages of microstrip include low cost production in large volume; the ability

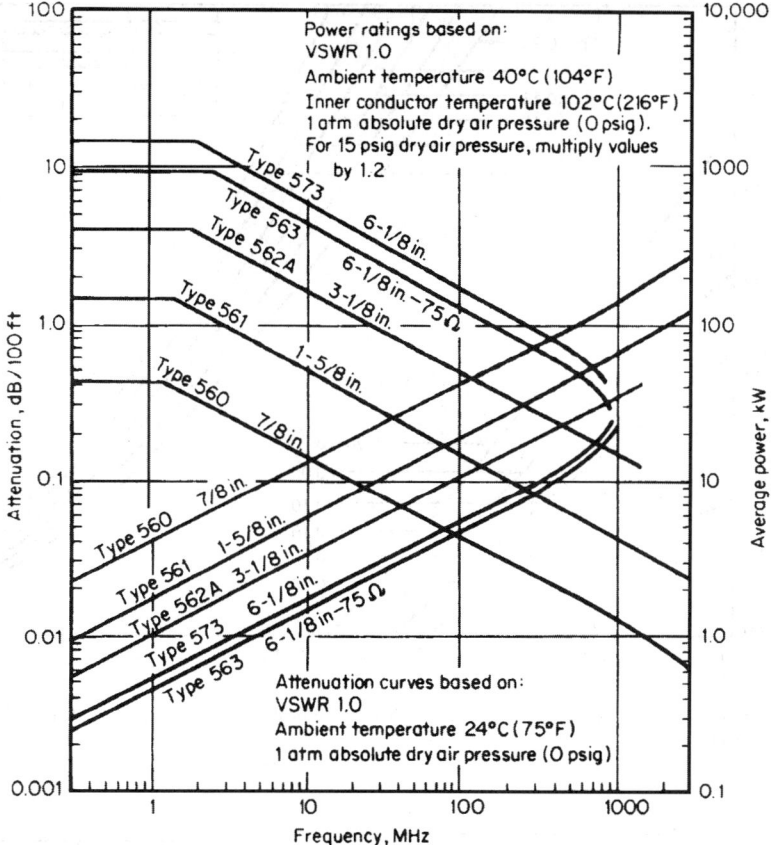

FIGURE 7.1.5 Power rating and attenuation for rigid coaxial line.

to mount components (such as resistors and dielectric cavities) on top of the circuit; and to adjust the circuit which is open so it can be easily worked on. No connectors are required between circuit elements, thereby reducing size. The circuit itself has somewhat higher loss than stripline, but the elimination of connectors and the reduction in size frequently results in surprisingly low loss. Disadvantages include relatively low power capability (satisfactory performance of 100 W average at 2 GHz), tight tolerances, launching difficulty above 15 GHz, and leakage across the open circuit. Absorptive material and grounding screws may be required. Typical substrate materials are shown in Table 7.1.5. Figure 7.1.8 shows characteristic impedance of microstrip. In microstrip, part of the field is above the circuit in air and part below the circuit in dielectric. The resulting effective dielectric constant is less than the dielectric constant of the substrate as shown in Fig. 7.1.9.

Finline

A finline consists of rectangular waveguide with a metallic fin attached along the centerline of the top wall of a waveguide and another fin along the centerline of the bottom wall (Fig. 7.1.10). The fin must have electrical continuity with the waveguide walls. A dielectric substrate is located adjacent to the fin and extends from the bottom to the top wall of the waveguide. This structure has low loss and can operate up to 100 GHz.

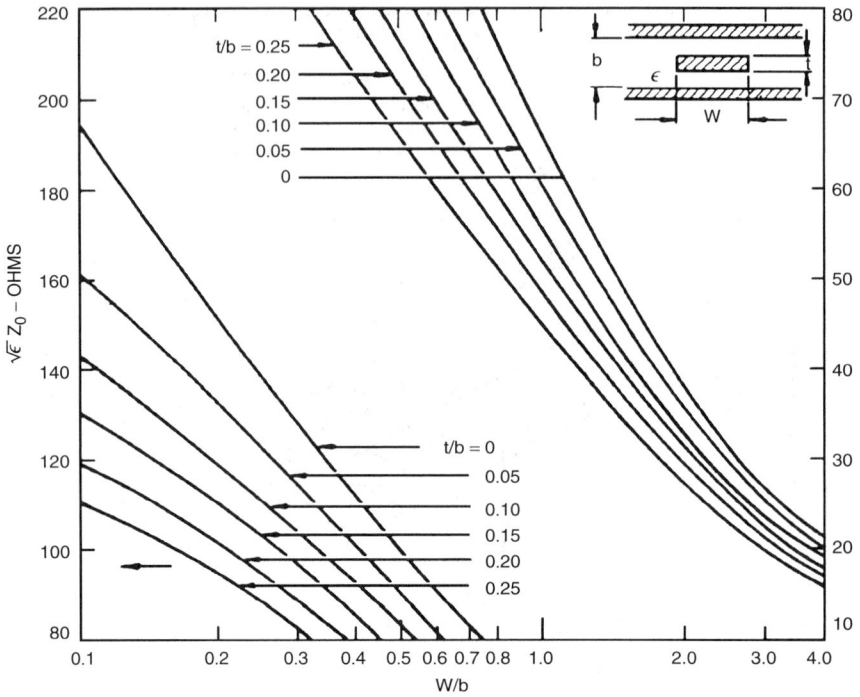

FIGURE 7.1.6   General curves for characteristic impedance of dielectrically loaded stripline (dielectric constant = ϵ).

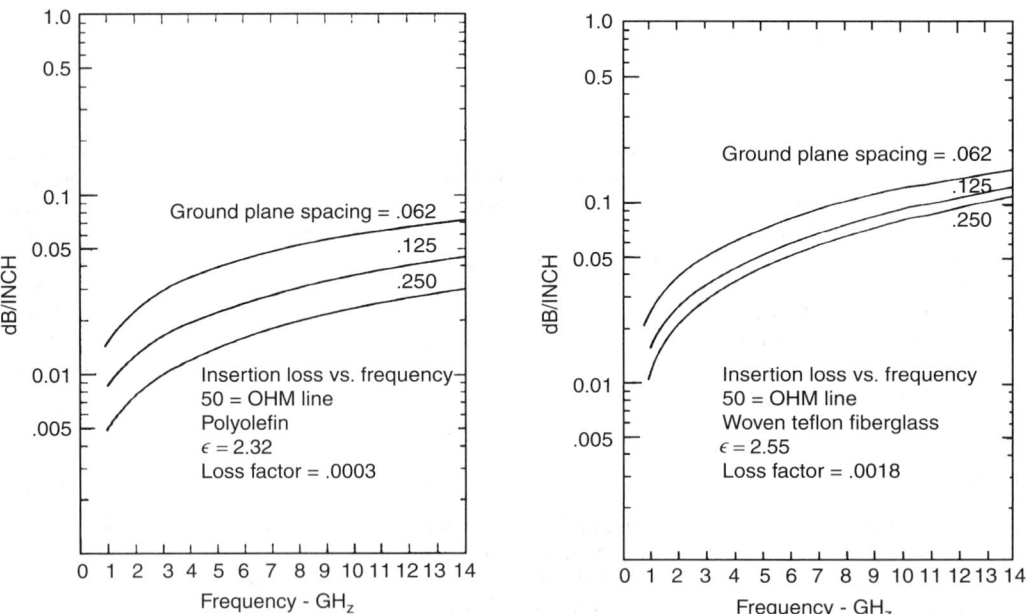

FIGURE 7.1.7   Insertion loss vs. frequency for polyolefin and fiberglass, 50-ohm line.

TABLE 7.1.5 Dielectric Materials

Material	Dielectric constant	Loss factor	Useful temp range, °C	Flexibility	Coeff. of thermal expansion × 10⁶, °C	Surface finish, μ in*
Woven TFG	2.55	0.0015	−60 to +200	Good	18.5	N/A
Microfiber TFG,						
Duroid 5870	2.33	0.0005	−60 to +200	Good	5	N/A
Duroid 5880	2.2	0.0006	−60 to +200	Good	32	N/A
Polystyrene	2.53	0.0003	−60 to +100	Very poor	7	N/A
Reinforced	2.62	0.002	−60 to +100	Poor	5.7	N/A
Polyphenelene oxide (PPO)	2.55	0.0016	−60 to +200	Fair	29	N/A
Polyolefin	2.32	0.00015	−60 to +100	Excellent	4.4	N/A
Quartz Teflon	2.47	0.0006	−60 to +200	Good	18.5	N/A
Polymide, Micaply 5032	4.8	0.01	−60 to +250	Good	9	N/A
Epsilam-10	10.0	0.002	−60 to +150	Good	11–23	N/A
99.5% alumina	9.9	0.00008	Up to 500	Very poor	7.5	<3
Quartz (fused silica)	3.78	0.0001	Up to 500	Very poor	0.55	<1
Sapphire	9.4	0.00008	Up to 500	Very poor	7.7	<1
	11.6				8.3	
99.5% BeO	6.6	0.0004	Up to 500	Very poor	7.8	3–7
Boron nitride	4.4	0.0003	Up to 500	Poor	1–2	N/A

*N/A = Not available.

Transitions Between Transmission Lines

Waveguide to Coaxial Line.   The maximum E field is in the center of the waveguide at a distance $\lambda_g/4$ from a short at the end of the guide. An extension of the center conductor, located at the point of maximum E field, acts like an antenna to couple energy from the coaxial line into the waveguide, Fig. 7.1.11. Typically the diameter of

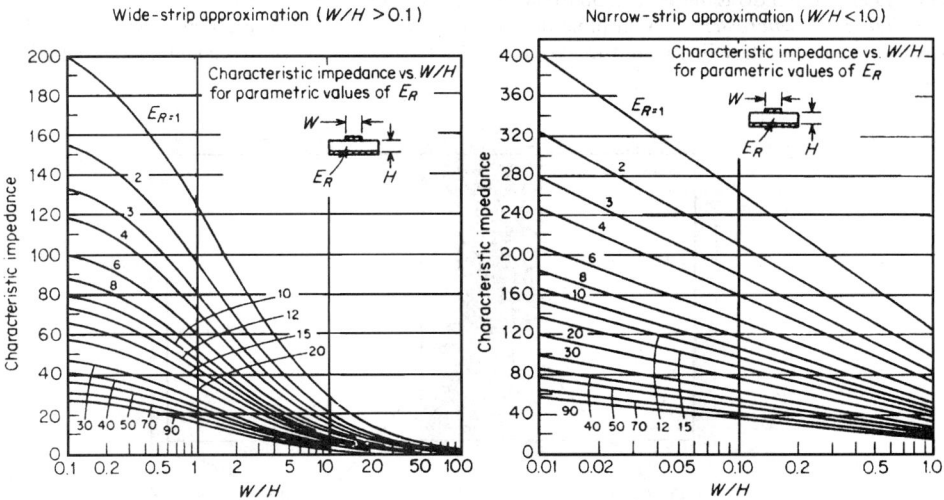

FIGURE 7.1.8 Characteristic impedance of microstrip line: wide-strip approximation (*left*); narrow-strip approximation (*right*).

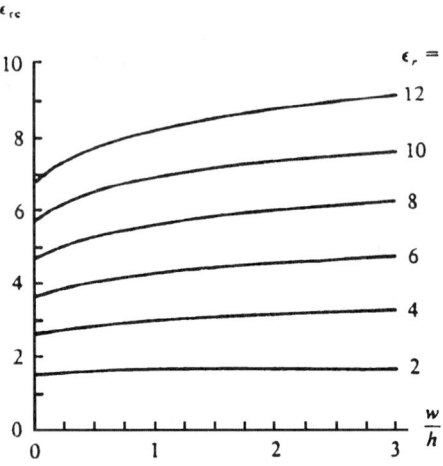

FIGURE 7.1.9   Plot of the effective dielectric constant vs. *w/h* for microstrip transmission lines, with the relative dielectric constant as a parameter.

the probe is enlarged and enclosed in a dielectric cylinder to increase bandwidth. Coupling from coaxial line to waveguide may also be achieved by using a loop, as shown in Fig. 7.1.12 which couples to the magnetic fields.

Waveguide to Stripline or Microstrip.   The center conductor of a stripline may be extended into a waveguide forming a probe. Increasing the width of the center conductor at the end of the probe may improve bandwidth. Similarly the conductor and substrate of a microstrip circuit, but not the ground plane, may be extended directly into the guide, Fig. 7.1.13.

Coaxial Line to Microstrip.   The center conductor of a coaxial line is pressed against or soldered to the strip conductor of microstrip. The outer conductor of the coaxial line is grounded to the microstrip groundplane. At frequencies above 15 GHz, the microstrip substrate thickness may be as little as 0.010 in. which usually requires decreasing the diameter of the coaxial line as illustrated in Fig. 7.1.14. The center conductor must not come so close to the groundplane that it introduces a capacitive discontinuity. It may be necessary to short out unwanted modes with grounded lines or screws near the transition. The exact mechanical configuration and capacitive loading become critical.

Directional Couplers

Waveguide Directional Couplers.   Waveguide directional couplers are employed primarily to sample power for measurements. Two waveguides may be located side by side, one above the other, either parallel to each other or crossing each other. Holes are drilled in the common wall to permit coupling power between guides. In Fig. 7.1.15 the coupling = $10 \log P_1/P_4$, while the directivity = $10 \log P_4/P_4'$ (P_1 = power in at port 1, P_4 = power out at port 4 with power P_1 in a port 1, P_4' is the power at port 4 with power P_1 in at port 2). Coupling values range from 3 dB to more than 30 dB with directivity in excess of 40 dB.

Stripline and Microstrip.   A stripline or microstrip coupler consists of a main transmission line in close proximity to a secondary line as illustrated in Fig. 7.1.16. The definitions of coupling and directivity are the

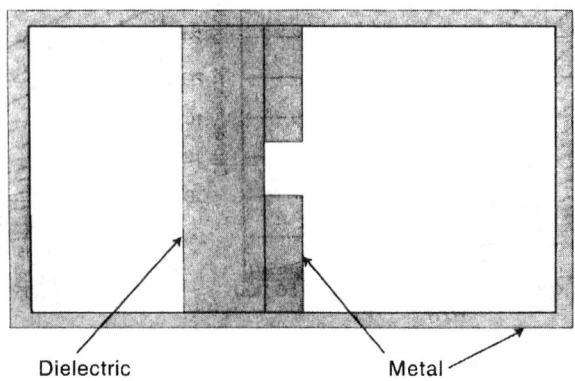

Dielectric   Metal

FIGURE 7.1.10   Finline configuration.

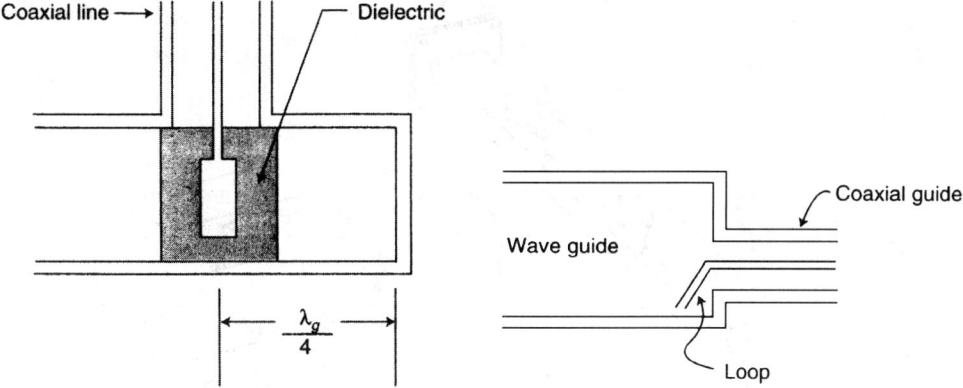

FIGURE 7.1.11 Electric field coupled waveguide to coax adapter.

FIGURE 7.1.12 Magnetic field coupled waveguide to coax adapter.

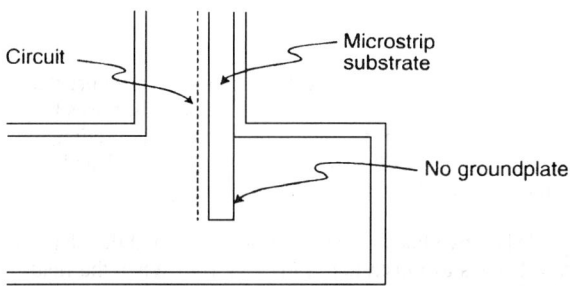

FIGURE 7.1.13 Waveguide to microstrip adapter using circuit probe.

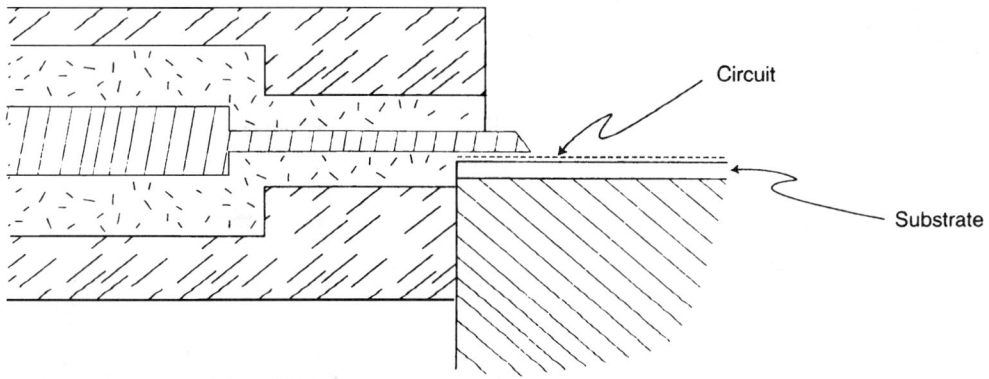

FIGURE 7.1.14 Coaxial line to microstrip adapter using constant impedance transformer in coaxial line.

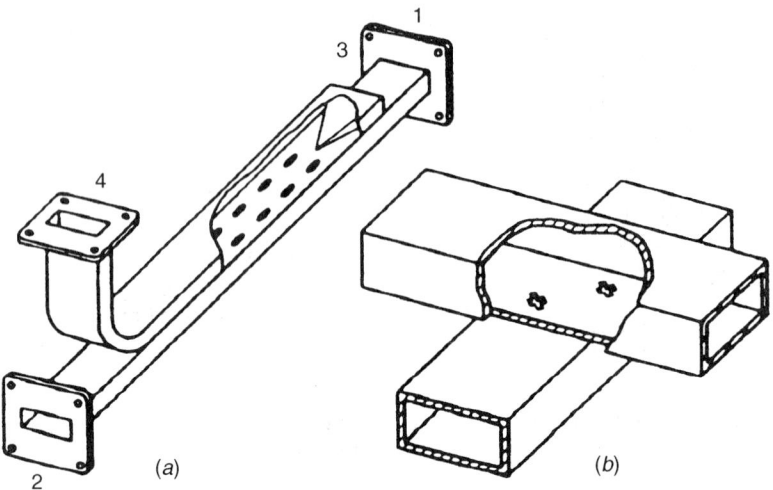

FIGURE 7.1.15 Directional couplers in waveguide: (*a*) with array of openings, (*b*) crossguide coupler.

same as for waveguide, but the coupled part is adjacent to the input rather than at the far end. Adding quarter wavelength coupling sections on either side of the center section increases bandwidth and reduces ripple. The added quarter-wave sections are less tightly coupled than the center section and are equally disposed about it. For microstrip, the velocity of propagation is different for even and odd modes. The addition of localized capacitances along the line, Fig. 7.1.17, improves the directivity.

Even and Odd Modes.   When two lines are in close proximity and the phase of the energy is the same, there is even-mode symmetry of fields as illustrated in Fig. 7.1.18*a*. When the fields are 180° out of phase there is odd mode symmetry, Fig. 7.1.18*b*. The impedance of the two modes are different. In addition, in microstrip, one mode has more field in air than the other, resulting in different propagation velocities.

In-Line Power Dividers.   A Wilkinson power divider consists of a single line separated into two quarter wavelength sections as shown in Fig. 7.1.19. The divider can have a VSWR of 1.25:1 or less over an octave. Power division is frequency independent.

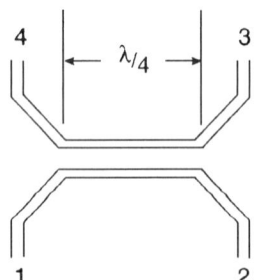

FIGURE 7.1.16 Stripline directional coupler.

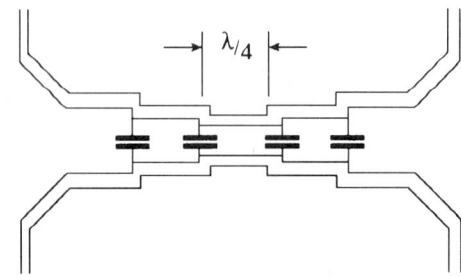

FIGURE 7.1.17 Microstrip coupler with capacitors to improve the directivity.

(a)   (b)

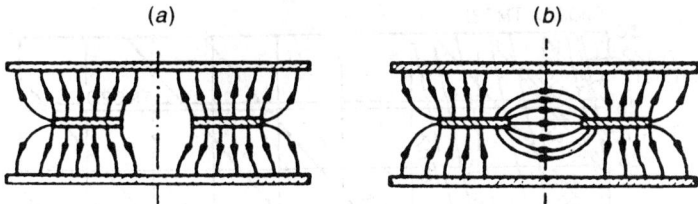

FIGURE 7.1.18 Even and odd modes on a homogeneous coupled line.

Resonant Circuits

Resonant circuits are generally formed from a short-circuited length of transmission line. Many modes can exist within the cavity created by the shorted line, as illustrated in Fig. 7.1.20 for a right circular cylinder. The size of the cavity and the quality of resonance, Q, is determined by the mode selected. A resonator with a Q of 100 has poor selectivity, while a resonator with a Q of 5000 will have good selectivity. The Q is defined as

$$Q = \frac{\text{Total energy stored in electric and magnetic fields}}{\text{Energy dissipated per cycle}}$$

The operating frequency can be changed by mechanically changing the cavity geometry, or by introducing a metal or dielectric post.

Dielectric disks and cylinders can be used as cavities. They are most easily coupled to microstrip, but can be used with coaxial line or waveguide, or other types of transmission lines. Materials are available from 0.7 to 30 GHz, with typical dielectric constants of 36, which reduces the dimensions by a factor of 6 when compared to air. Q ($1/\tan \delta$) is typically 30,000/frequency in GHz, where $\tan \delta$ is frequently used as a measure of loss in materials. Tuning of 5 percent or more is achieved by moving metal or dielectric tuning stubs near the dielectric cylinder.

Mechanically tuned resonant cavities are generally used for frequency determination. For a transmission wavemeter, such a cavity is coupled in series into the transmission path, while for an absorptive indication it is coupled in shunt. A dominant-mode (TE_{111}) cylindrical resonator is most widely used for this purpose, but for highest selectivity a circular electric-mode (TE_{011}) resonator is employed. Dielectric resonators are frequently used to tune or stabilize oscillators.

Filters

Frequency filters are used to separate the components of a composite waveform for signal-processing purposes or to suppress RF interference (RFI), which results from the spurious output of transmitters. The latter problem

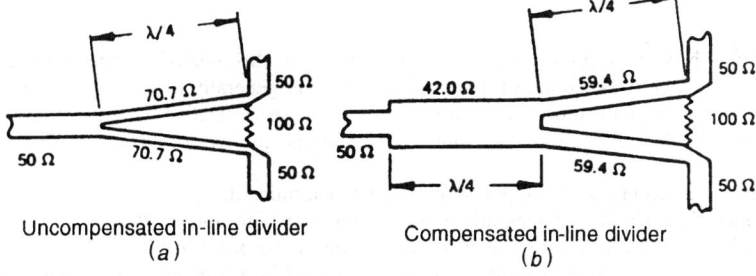

Uncompensated in-line divider (a)   Compensated in-line divider (b)

FIGURE 7.1.19 Compensated and uncompensated in-line three-port power dividers.

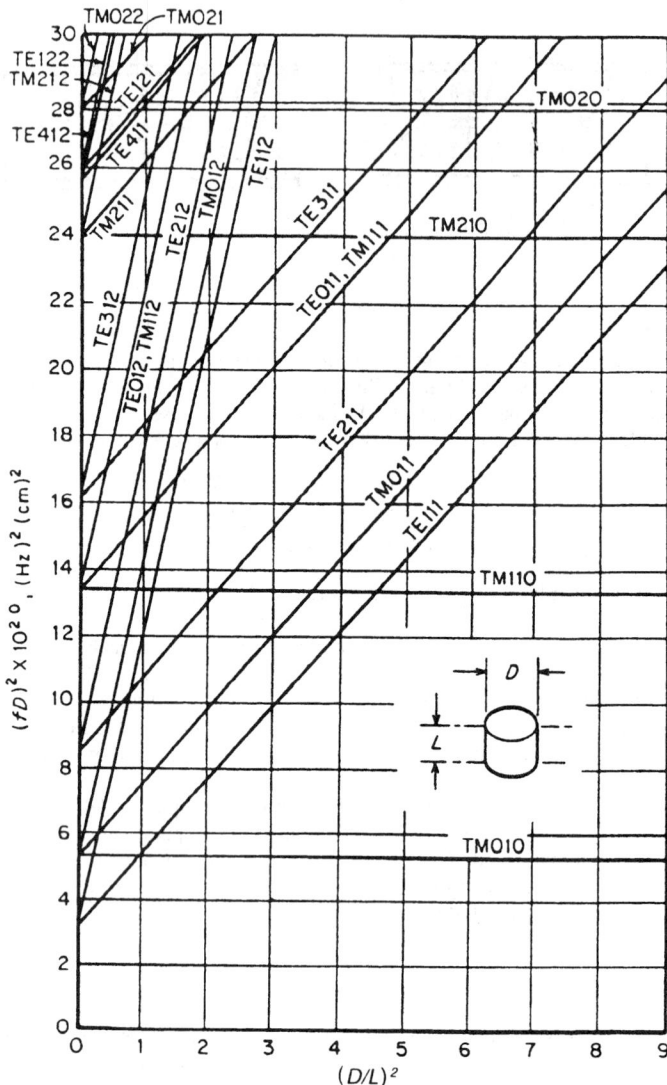

FIGURE 7.1.20   Mode chart for right circular cylinder.

has only recently become serious at microwave frequencies, as this area of the spectrum has become congested. High-power capability is required for such filters, leading generally to the use of waveguide structures. A section of waveguide beyond cutoff constitutes a simple high-pass reflective filter. Loading elements in the form of posts, irises, or stubs are employed to supply the reactances required for conventional lumped-constant-filter design.

The desired skirt steepness and stopband attenuation determine the number of sections, as at lower frequencies. A disk-loaded coaxial line is generally used as a low-pass high-power filter. Insertion loss of reflective filters is typically 0.1 to 0.2 dB, with stopband attenuation of the order of 50 dB. Absorption filters avoid the reflection of unwanted energy by incorporating lossy material in secondary guides which are coupled through leaky walls (typically small sections of guide beyond cutoff in the passband). These filters are effective primarily against harmonics.

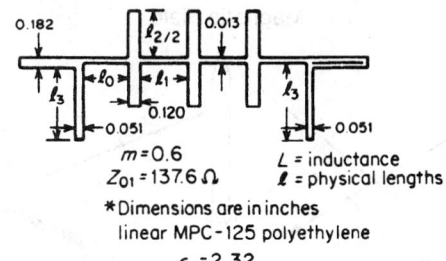

$m = 0.6$
$Z_{01} = 137.6\,\Omega$

L = inductance
l = physical lengths

*Dimensions are in inches
linear MPC-125 polyethylene
$\epsilon_r = 2.32$

Frequency	Dimensions			
f_c	l_0	l_1	$\mathit{l}_{2/2}$	l_3
	$0.084\lambda_{\epsilon_r}$	$0.100\lambda_{\epsilon_r}$	$0.144\lambda_{\epsilon_r}$	$0.200\lambda_{\epsilon_r}$
1 GHz	0.6653	0.7920	1.093	1.548
2 GHz	0.3326	0.3960	0.533	0.756
3 GHz	0.2220	0.2640	0.333	0.493

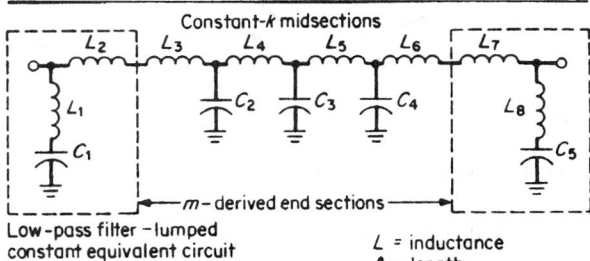

Low-pass filter – lumped
constant equivalent circuit

L = inductance
l = length

FIGURE 7.1.21   Strip-line filter design.

For low-power applications, strip-line filters are widely used because of their compact sizes and low cost. Typical dimensions for a low-pass filter of this type are shown in Fig. 7.1.21.

Other Components

Among the components useful for measurement purposes are wavemeters, attenuators, and matched terminations. In waveguide, variable attenuators take the form of thin absorptive material introduced tangential to the electric field typically through a slot in the broad wall of rectangular guide to produce minimum reflection. Load terminations are tapered attenuators designed to produce at least 40 dB of return loss, while maintaining a good match (maximum VSWR less than 1.2) through the specified band. For high power, water cooling is provided either around a loaded ceramic or dielectric absorber or by introducing a tube of water directly into the guide to act as the absorber. Calorimetric determination of power is possible with such water loads. A wide range of grounded resistive elements are used as terminations.

Ferrite Components

Isolators.   Isolators transmit microwave power in one direction with little attenuation, while power transmitted in the opposite direction is absorbed. Attenuation in the transmit direction may be less than 0.5 dB, while isolation in the opposite direction may exceed 20 dB. Isolators are relatively well matched in both directions, a VWSR less than 1.20:1 is typical. Such an isolator will reduce a mismatch from 2:1 to less than 1.3:1. When used between stages of an amplifier they may eliminate the need for tuning between stages and may improve gain flatness. Isolators are available in waveguide, in stripline with or without coaxial connectors and in microstrip. At 1.0 GHz a differential phase shift type waveguide isolator can handle 4000 kW peak power and

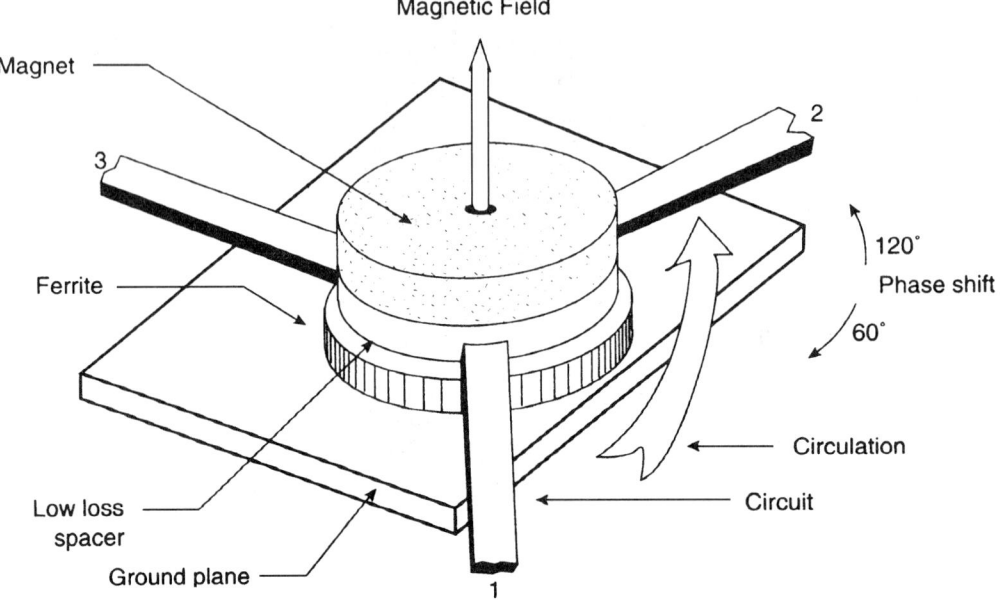

FIGURE 7.1.22 Configuration of microstrip junction circulator.

over 250 kW average power. At higher frequencies, such as Ku-band, a microstrip isolator may handle only a few watts and weigh only a few grams.

Isolators and circulators are based on the interaction of a circularly polarized magnetic component of microwave energy with the magnetic field of a ferromagnetic molecule, usually iron, which has been aligned by the application of a steady dc magnetic field. The ferromagnetic molecules are held in a nonconductive crystalline lattice through which microwave energy can pass with little loss. The interaction between the microwave magnetic field and the magnetic field of the molecule is a vector product interaction which tends to cause the molecule to process like a gyroscope, and is frequently referred to as gyromagnetic interaction. The direction of the applied dc magnetic field determines the direction for transmission and attenuation of the isolator. Reversing the direction of the applied field reverses the direction of the isolator.

Circulators.   The most common type of circulator uses a single junction and can be built in waveguide, stripline or microstrip. Figure 7.1.22 illustrates a single junction microstrip circulator. The stack-up of parts consists of a groundplane (frequently metalized on the ferrite), a ferrite disk, a conductive metal circuit with arms at 120° relative to each other, a spacer to keep the microwave fields out of the magnet, and a magnet to supply the dc magnetic field. Due to the nonreciprocity of the magnetically biased ferrite, the phase shift between ports in the circulation direction is 120° while the phase shift in the opposite direction is 60°. Energy transmitted from port one to port two is shifted 120°, while energy from 1 to 3 is shifted 60° and energy from 2 to 3 is shifted 60°. Energy going either direction is in phase at port 2 and adds together. Energy at port 3 is out of phase and cancels so no energy is transmitted to this port. Due to symmetry, this applies to all ports. Typical characteristics of junction circulators is shown in Table 7.1.6. Placing a termination at port 3 converts the circulator to an isolator.

Differential Phase Shift Circulators.   High-power circulators are generally built in waveguide as shown diagrammatically in Fig. 7.1.23. If the phases are added up as shown, it will be seen that power in at 1 goes to 2, power in at 2 goes to 3, and so on. It should be noted that in a folded hybrid Tee, power in at 1 has the same phase in path A and path B. Power in at the vertical arm of the Tee is 180° out of phase for the two paths. A similar phase shift occurs for power traveling in the opposite direction. There is a 90° phase shift across the short slot hybrid as shown. Placing termination at ports 3 and 4 converts the circulator into an isolator.

TABLE 7.1.6   Characteristics of Typical Junction Circulators

Circulator type	Center frequency, GHz	Bandwidth, %	Isolation, dB	Insertion loss, dB	Power capacity Avg, W	Peak kW
Waveguide	12.4–18	. . . .	20	0.3	20	1
	18–26.5	. . . .	17	0.5	15	0.5
Strip transmission	4–8	. . . .	20	0.4	35	
line	12.4–18	. . . .	18	0.5	25	0.25
Switching	2.9	8.9	26	0.35		15
	35	5	15	0.5		15
Lumped constant	1.2	30	20	0.6		
	0.4–0.5	30	20	0.4		

Source: *Microwave J.,* November 1978.

Circulators may be used to simultaneously connect a transmitter and receiver to a single antenna as shown in Fig. 7.1.24. Transmitted power in at 1 goes to the antenna at 2. Received energy entering the antenna at 2 travels to the receiver at 3. Transmitter energy reflected from the antenna also goes to the receiver. The reflected power must be low enough it does not damage the receiver or a fast-acting switch must be provided to protect the receiver. If the transmitter and the receiver are operating at different frequencies, protection may be provided by a filter.

Phase Shifters.   The simplest reciprocal phase shifter uses a thin ferrite rode centered within a rectangular waveguide. An axially directed magnetic field provides a full 180° phase variation at 9 GHz with a change of a few hundred gauss. A more cost-effective approach is the dual-mode reciprocal phase shifter, shown in Fig. 7.1.25, which uses Faraday rotation. Toroidal nonreciprocal phase shifters, in which the ferrite takes the form of a rectangular closed loop in waveguide, switched by means of an axial current-carrying conductor, are also popular. Digital phase shifting is generally obtained by latching ferrites, since the remanent magnetization states of the hysteresis loop require relatively small bias fields.

Limiters.   The nonlinear behavior of ferrites at high power levels is used in ferrite limiters. Such limiters have replaced TR gas-discharge tubes in radar. Peak powers of tens of kilowatts at X band can be handled, but with considerable spike-energy leakage, which requires a follow-on solid-state (*pin*) diode stage. Typical insertion loss is

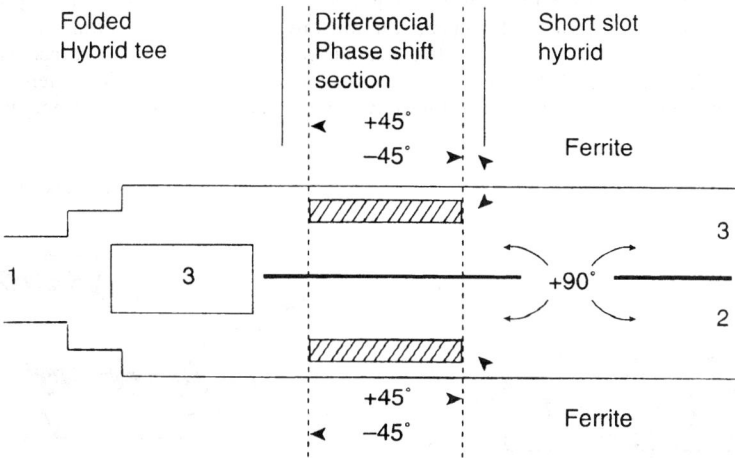

FIGURE 7.1.23   Diagram of differential phase shift waveguide circulator.

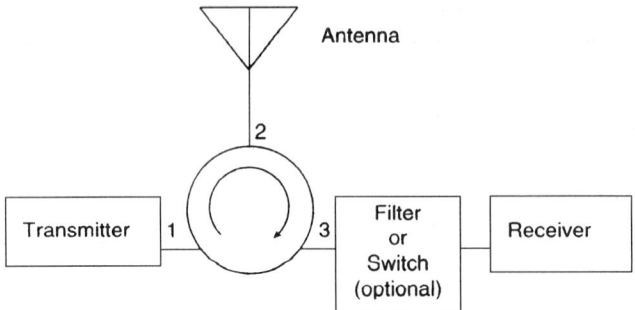

FIGURE 7.1.24 A transmitter and receiver connected to same antenna.

about 0.5 dB, with 30 dB of flat limiting and about 6 dB of spike limiting for the ferrite alone. The diode stage increases the insertion loss to about 1 dB but reduces all leakage an additional 30 dB. The recovery time is very short, typically 100 ns, and is determined primarily by the diode section.

Acoustic-Wave Devices

Acoustic waves owe their usefulness to their relatively low velocity, typically 10^{-5} times the electromagnetic velocity, permitting relatively long electric-signal delay times to be obtained in a physically small space. Both bulk-mode propagation and surface waves have been used, the latter gaining in popularity because of the relative ease of access to intermediate points along the propagation path for structure shaping and taping.

Transducers. The transducers designed for the two types of acoustic modes are physically quite different, but both contain electrodes spaced either a quarter or half an acoustic wavelength in a piezoelectric material. Microwave bulk-wave transducers consist of multiple films of ZnO or CdS, each approximately $\lambda/4$ thick, separated by similar films of gold or other nonpiezoelectric material deposited on the bulk medium to provide electric-to-acoustic wave coupling and impedance transformation.

Surface acoustic waves (SAW) are generally excited by interdigital transducers. An interdigital transducer, illustrated in Fig. 7.1.26, consists of two sets of interleaved metal electrodes, called *fingers*, deposited on the piezoelectric substrate. To generate a wave an RF potential is applied between the adjacent sets of fingers, which are spaced by a distance equal to one-half wavelength at the transducer design frequency. A typical 100-MHz transducer on $LiNbO_3$ has aluminum fingers 0.2 μm thick by 9 μm wide with 9-μm gaps.

The wave excited by the RF potential between a pair of fingers travels at the surface-wave velocity. By the time the wave arrives midway between the next pair of fingers, the RF excitation potential has reversed sign, and the wave excited by the second pair of fingers will be in phase with the wave from the first pair. Thus the

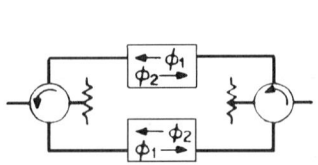

FIGURE 7.1.25 Dual-mode reciprocal phase shifter.

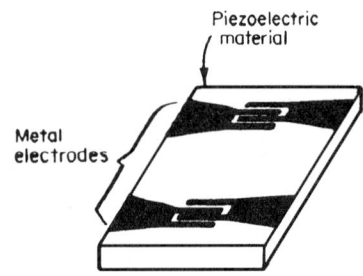

FIGURE 7.1.26 Piezoelectric surface-wave microwave device with interdigital electrodes.

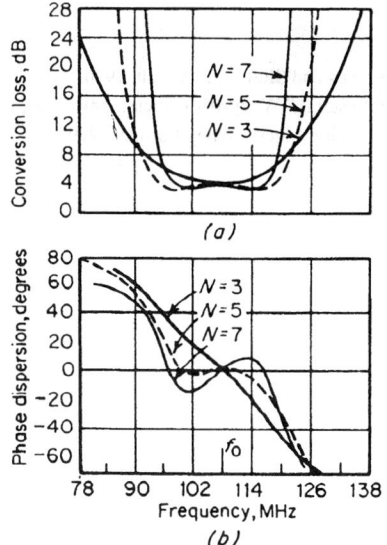

FIGURE 7.1.27 Coupling between electric port and one acoustic port for transducers with three, five, and seven interdigital periods: (*a*) theoretical conversion loss; (*b*) phase dispersion.

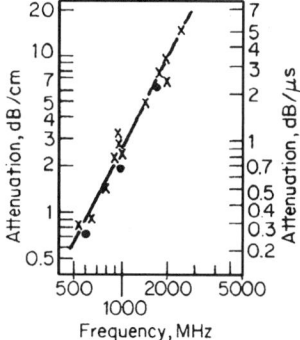

FIGURE 7.1.28 Surface-wave attenuation of Y-cut lithium niobate.

excitation due to the second pair is added to the excitation from the first, and so on. The mechanism is reciprocal, and hence the transducer that excites a wave will also detect it.

The transducer has a fractional bandwidth of $1/N$, where N is the number of finger parts. Electrically, the transducer is represented by a capacitance shunted by a radiation resistance which depends on the choice of finger length.

A surface-wave transducer is a three-port device, i.e., with one electric and two elastic ports. Figure 7.1.27 shows the conversion loss as a function of frequency from a 50-Ω source to one of the two acoustic outputs for three different transducers on a lithium niobate surface. These calculated curves show that for this particular case the use of five finger pairs provides the widest bandwidth and smallest conversion loss. The attenuation of the wave, once it has been launched, is given in Fig. 7.1.28 for lithium niobate.

The lowest operating frequency of acoustic surface-wave devices is limited by the allowable size to typically 10 MHz. At present the upper operating frequency is limited by fabrication difficulties to about 1 GHz with possible operation to 3 GHz.

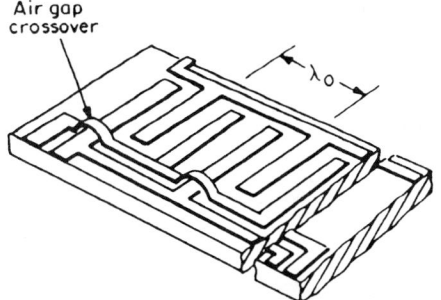

FIGURE 7.1.29 Three-phase unidirectional transducer. Each electrode is 120° out of phase, so that the acoustic waves add in one direction and cancel in the other.

Other Acoustic-Wave Devices. Bandpass filtering is the principal commercial application of SAW technology. Such devices are replacing *LC* filters in television receiver i.f. circuits. Minimum stop-band rejection of the order of 60 dB with in-band response flat to ±0.1 dB and phase deviation from linearity of only a few degrees are typical of these filters. The major drawback of SAW devices, their high insertion loss (of the order of 15 dB), can be reduced to less than 3 dB by using a three-phase unidirectional transducer structure, as shown in Fig. 7.1.29. Other acoustic-wave devices include: (a) dispersive filters used primarily for pulse compression; (b) chirp transformers that can be used to obtain a Fourier transform of an input signal; and (c) compact, low-cost SAW resonators, which are useful below 1 GHz.

BIBLIOGRAPHY

Gupta, K. C., G. Ramesh, B. Inder, and B. Prakash, "Microstrip Lines and Slotlines," Artech, 1996.

Simons, R. N., "Coplanar Waveguide Circuits, Components, and Systems," Wiley-IEEE Press, 2001.

Wong, K., "Design of Nonplanar Microstrip Antennas and Transmission Lines," Wiley Europe, 1999.

CHAPTER 7.2
MICROWAVE GENERATORS AND AMPLIFIERS

**Robert S. Symons, Donald H. Preist, M. B. Shrader,
Pamela L. Walchli, George K. Farney, Howard R. Jory**

KLYSTRONS

Robert S. Symons

Introduction

For high frequencies, linear-beam tubes overcome the transit-time limitations of grid-controlled tubes by accelerating the electron stream to high velocity before it is modulated. Modulation is accomplished by varying the velocity, with consequent drifting of electrons into bunches to produce RF space current. The RF circuit for coupling signals to and from the electron beam are generally integral parts of the tube. Two basic types are important today, klystrons and traveling-wave tubes. Different versions of each are used as oscillators and amplifiers.

In a klystron, the RF circuits are resonant cavities which act as transformers to couple the high-impedance beam to low-impedance transmission lines. The frequency response is limited by the impedance-bandwidth product of the cavities but can be increased by stagger tuning and by multiple-resonance filter-type cavities.

Reflex Klystrons

In the reflex klystron a single resonator is used to modulate the beam and extract RF energy from it, making the tube simple and easy to tune. The beam passes through the cavity and is reflected by a negatively charged electrode to pass through again in the reverse direction. With proper phasing determined by applied voltages, oscillating modes occur for n + three-quarters-cycle transit time between passes through the cavity. The frequency can be modulated by varying voltage on the reflector (which draws no current). Reflex klystrons have been used as test signal sources, receiver local oscillators, pump sources for parametric amplifiers, and low-power transmitters for FM line-of-sight relays. Reflex-tube frequencies cover the entire microwave range from 1 to 140 GHz. In new applications, they have largely been replaced by solid-state devices.

Two-Cavity Klystron Oscillators

In all klystrons except the reflex, the beam goes through each cavity in succession, and so there is no feedback. The tube is a buffered amplifier, with each stage isolated from those upstream. Electromagnetic feedback may be provided to make an oscillator.

The specialized two-cavity oscillator has a coupling iris in the wall between the cavities. This tube is more efficient and more powerful than the reflex klystron. It can be frequency-modulated by varying the cathode voltage around the center of the oscillating mode but requires more modulator power than a reflex klystron.

Two-cavity oscillators have been used where moderate power and stable frequency for low side band noise are needed. Examples are the transmitter source in Doppler navigators, pumps and parametric amplifiers, and master oscillators for cw Doppler radar illuminators.

Again in new applications, most requirements that were met by two-cavity oscillators are now met by solid-state devices unless extremely low noise requirements exist.

Extended-Interaction Oscillators

At millimeter wave frequencies the losses in klystron cavities make it hard to build up the impedance necessary to oscillate with the very small low-current beams required.

If a series of cavities are coupled together and interact sequentially with the beam in the proper phase, the total interaction impedance increases directly with the number of cavities. The circuits of extended-interaction oscillators resemble those of traveling-wave tubes. Since they operate with a complete standing wave (at the cutoff of the traveling-wave passband), the tubes can be classed as klystrons. Various names are used for tubes of this type. The Laddertron uses a ladder-shaped periodic circuit and a flat-ribbon electron beam. Multicavity klystron oscillators use coupled cavities and cylindrical beams. Communication and Power Industries (formerly Varion) makes a number of extended-interaction klystron oscillators with power output ranging from 1 kW at 15 GHz to 1 W at 300 GHz. Most of these oscillators operate with a beam voltage near 10 kV. They provide very low noise characteristics in the order of −120 dB below carrier power in a 1-Hz bandwidth close to the carrier when used with quiet power supplies.

Two-Cavity Amplifiers

In the simplest klystron amplifier, the driving signal is coupled through a transmission line to input cavity. The cavity voltage produces velocity modulation of the beam. After a single drift space, the resultant density modulation induces current in the output resonator, from which power is extracted through another transmission line. As with nearly all klystron amplifiers, for efficiency, the Q of the output cavity is adjusted so the RF voltage almost stops electrons at the center of the bunch. The beam is usually focused electrostatically.

The gain of a two-cavity klystron is about 10 dB. Use is limited because more gain is desired in high-power tubes and solid-state amplifiers are available at low powers.

Multicavity Klystron Amplifiers

Downstream from the input cavity, cascaded intermediate cavities are inserted between the input and output cavities. They usually have no external coupling and are driven by the RF beam current and in turn remodulate the beam velocity. Figure 7.2.1 shows a three-cavity klystron schematically.

Each cavity tuned to the signal frequency adds about 20 dB of gain to the 10 dB of a two-cavity klystron. Net gain of up to about 60 dB is practical. If higher gain is attempted, one must do something to suppress fast secondary electrons traveling in the reverse direction through the tube and creating a feedback path, and one must be careful that one does not create a tube with a high noise figure by exciting a high-power beam with a very low-power drive signal. In klystrons with five or more cavities, they are usually added to increase bandwidth or efficiency rather than gain and consequently are stagger tuned. The characteristics that make the klystron desirable in the various applications in which they are used include fairly linear amplification of the RF input signal and separation of the beam formation, RF interaction and collection functions in different parts of the electron tube. The separation, in turn, allows the designer of

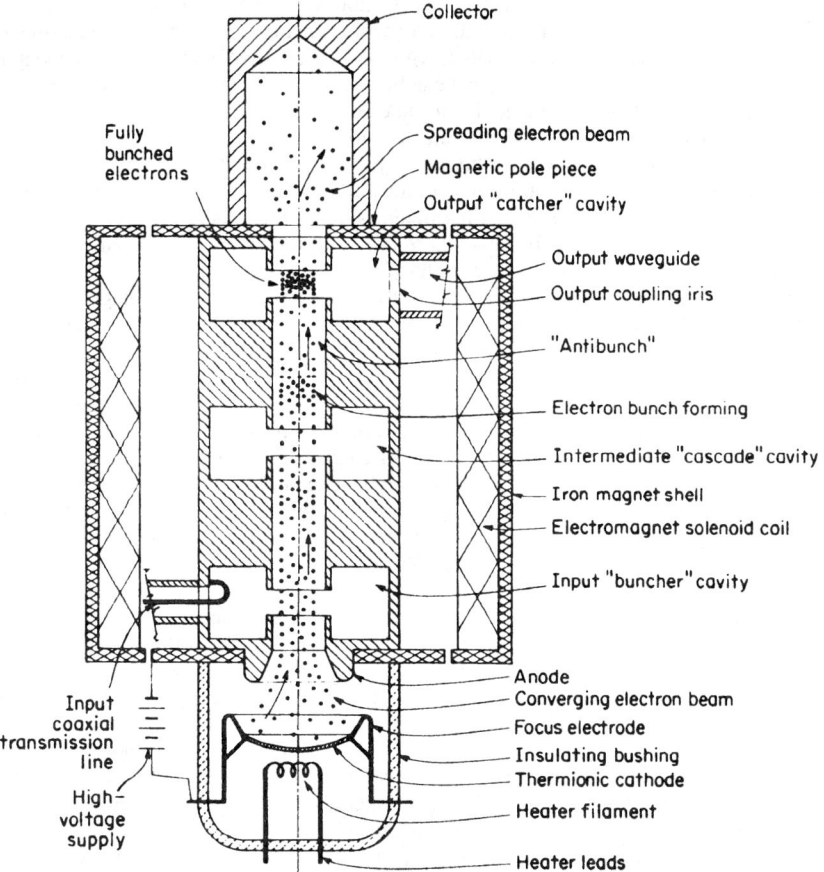

FIGURE 7.2.1 Cross section of cascade klystron amplifier. (*Varian Associates*)

the tube to optimize it for life, gain, bandwidth, low noise, efficiency or power output. Multicavity klystrons provide the economic solution to many requirements. Air cooling is often used for tubes with output power below 5 kW, and boiling water or forced liquid cooling is used for higher power. Magnetic focusing is used to control the beam in most power amplifier klystrons. Permanent magnets are used if size and required field strength permit, and solenoids or other electromagnets are used for larger and higher-power klystrons.

CW multicavity klystron amplifiers are employed in industrial-heating applications, electron-storage rings and superconducting electron linear accelerators used in high-energy nuclear physics experiments, satellite-ground-station tropospheric-scatter and space-communications transmitters, and UHF television broadcast transmitters. Pulsed multicavity klystrons are used in scientific and medical electron linear accelerators and in proton linear accelerators. They are also used in many radars in which pulse compression, Doppler processings, or spectral purity is an important requirement. A discussion of klystron designs optimized for these various applications follows:

CW klystrons for industrial heating and those used to make up the synchrotron losses of electrons in the storage rings used in nuclear physics experiments must have very large output power and high efficiency. A number of klystrons ranging in operating frequency from 200 to 2450 MHz with continuous output power up to one-half or 1 MW have been manufactured. At 200 MHz such a klystron is about 6 m in length, but the length scales inversely as frequency and directly as the square root of beam voltage. In such tubes, efficiency is achieved by using enough cavities so that, with the desired drive power, the fundamental component of current in the beam reaches the maximum that can be achieved with synchronously tuned cavities prior to reaching the output cavity. At this distance down the electron beam, the second harmonic content of the current may be used to excite a cavity tuned near the second harmonic in order to sweep electrons located between the electron bunches into the bunches. Alternatively, the beam is allowed to drift for a fairly long distance while the electrons between bunches continue to drift toward the bunches and the bunches deteriorate to some extent. In either approach, following the second-harmonic cavity or the long drift length, usually two cavities tuned to frequencies somewhat above the fundamental frequency (inductively tuned cavities) provide RF electric fields that push most electrons back toward the centers of the bunches and raise the fundamental component of current to a very high level. Using such techniques, together with high beam voltages (V) and low beam currents (I), efficiencies (η) reaching 70 to 80 percent have been achieved. An empirical relationship that fits the experimental data is $\eta = [90 - 20 \times 10^6 (I/V^{3/2})]$ for $(I/V^{3/2}) \leq 3 \times 10^{-6}$.

Klystrons for communications applications require bandwidth, and so it is customary to use somewhat lower beam voltages and higher beam currents. Bandwidth scales directly as the dc beam conductance [I/V] so it increases slowly as the power output of the tube is increased by raising beam voltage (because [$I/V^{3/2}$], the electron beam "perveance," is a constant of the electron gun design. Bandwidth will increase more rapidly as the perveance, itself, is raised. As discussed above, a perveance increase will reduce efficiency, so there is a trade-off. Nevertheless, the efficiency increasing techniques discussed above remain effective, and bandwidth can be increased further by stagger tuning the input cavity and low-level intermediate cavities. Many tropospheric scatter communications tubes have been designed to operate at frequencies from UHF through 5 GHz at power levels up to several tens of kilowatts, and satellite communications klystrons with power outputs from 800 W to 15 kW have been designed to operate from 5 to 30 GHz. For deep-space communication even higher powers are used ranging from 500 kW at 2450 MHz to 200 kW at X-band. Figure 7.2.2 shows the gain characteristics of a VA-884D, a 14-kW cw broadband klystron designed for ground-to-satellite communication and tunable over the 5.925 to 6.425 GHz band. The klystron is a fairly good linear amplifier from zero signal up to 2 to 3 dB below saturated output, but the curvature of the gain characteristic does introduce some third-order intermodulation products, $(2f_2 - f_1)$ and $(2f_1 - f_2)$. Figure 7.2.3 shows the necessary reduction in the average power of a two-equal-carrier signal relative to the klystron saturation power as a function of the desired third-order intermodulation product level relative to the carrier powers. Another type of distortion in klystrons is am-to-pm conversion. It is caused by the increase in transit angle at the output cavity gap when the tube is operating very efficiently near maximum power and electrons in the bunches are nearly

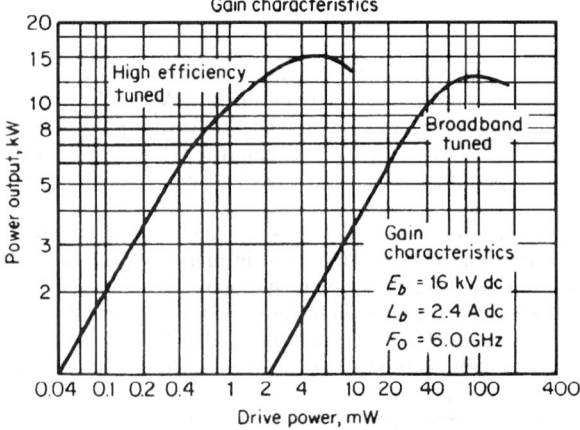

FIGURE 7.2.2   Typical gain, output power and drive power characteristics.

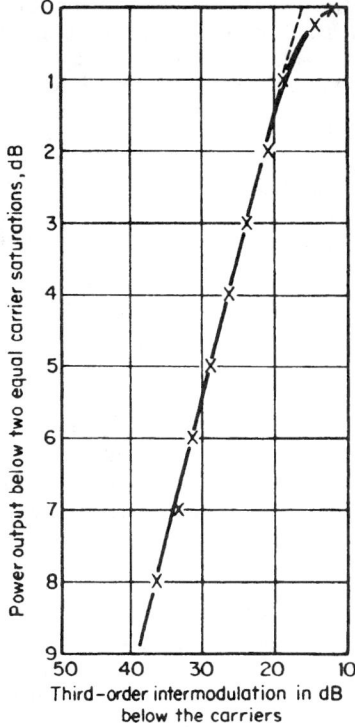

FIGURE 7.2.3   Third-order intermodulation distortion under two equal carrier conditions.

stopped. The variation in phase amounts to about 20°. Lowering the output cavity Q slightly below that value which gives maximum efficiency will substantially reduce the am-to-pm conversion. Variation of the beam voltage also causes phase modulation, so good filtering of the beam voltage supply is important. The percentage change in phase through a klystron is equal to a maximum of one-half the percentage change in beam voltage because of the square root relationship between beam voltage and electron velocity. At very high beam voltages this is somewhat less because of relativistic effects.

Figure 7.2.4 shows bandwidth curves for the klystron with various levels of drive power. Figure 7.2.5 shows the trade-off between gain and bandwidth as the cavity resonant frequencies are stagger tuned, and Fig. 7.2.6 shows fundamental and harmonic power outputs as a function of drive power. The operating bandwidth of many communications klystrons may be located anywhere within an entire communications band by means of tuners in the cavities. These sometimes take the form of a capacitive paddle which may be moved relative to the cavity gap, or sometimes the form of a moveable copper cavity wall with tungsten spring contacts which can vary the volume or "inductance" of the cavity. In either case, motion is transmitted through the vacuum wall of the klystron by means of a metal bellows. "Channel tuners," which move all cavities simultaneously to settings appropriate to one of a fixed number of broadband channels, are common on tubes designed for satellite-communication ground stations. A typical power supply schematic for a cw amplifier is shown in Fig. 7.2.7. Protective devices often required for klystron amplifiers include sensors to monitor cooling air or water flow, collector overtemperature, cathode-heating-time delay, cathode overcurrent, body overcurrent, and output waveguide arcing (photodetector and PIN diode switch).

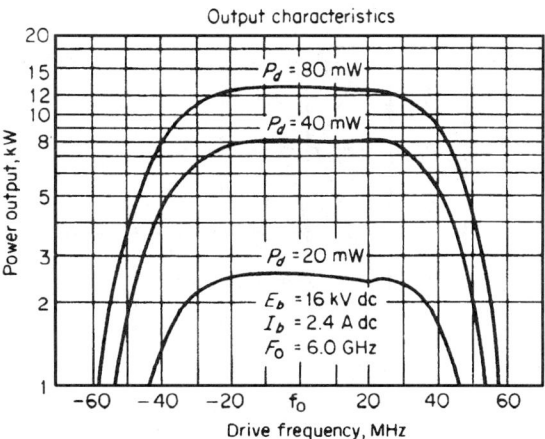

FIGURE 7.2.4   Gain and output vs. frequency characteristics under saturated and unsaturated RF drive conditions.

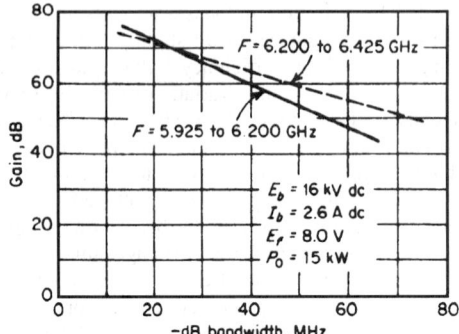

FIGURE 7.2.5 Type VA-884 series klystron gain-bandwidth characteristics.

UHF television broadcast klystrons are like other communications klystrons except that the designs are optimized to amplify the NTSC signal efficiently. In the NTSC signal, the lower sideband is vestigial and the carrier is suppressed. Typical peak output power of television klystrons is 30 to 60 kW. Operating beam voltages range from 20 to 30 kV. The required bandwidth is 6 MHz. Thus, four or five cavity stagger-tuned klystrons are used to provide gain bandwidth and efficiency. In "internal-cavity" klystrons, the cavity wall is the vacuum wall and tuning is accomplished with a capacitive paddle near the interaction gap of the cavity. The paddle is actuated through a bellows. With this type of construction, tubes with three different cavity sizes are required to cover the 470 to 806 MHz UHF TV band. "External-cavity" klystrons have a ceramic vacuum wall inside each cavity, and tuning over the entire 470 to 806 MHz band is accomplished by moving two cavity walls at each end of an air-filled waveguide-like cavity. The synchronizing pulses that are transmitted at the end of each line scan require full output power from the transmitter about 10 percent of the time. White level is 10 percent of the peak RF voltage or 1 percent of the peak power, and black level is about 78 percent of peak voltage or 60 percent of peak power. Thus, for an average gray raster, the average output power is about one-third the peak-of-sync power. Before 1974, when electric power was cheap, it was usual to operate klystrons at constant input power, and the output cavity was heavily loaded to a low Q to improve linearity and reduce am-to-pm conversion at the expense of efficiency. Thus the average efficiency was about 11 or 12 percent. After 1974, as the cost of power rose, it became attractive to operate klystrons at two different beam current levels, a lower current during the visual scan period and a higher current during the synchronizing pulses. To vary the current, a control electrode was incorporated in the electron gun. The Q of the output cavity was also increased. The increased amplitude and phase distortion in the klystron is compensated by predistortion in the exciter at the expense of partially regenerating the lower sideband which is then again eliminated by a high-power filter between the final amplifier and the

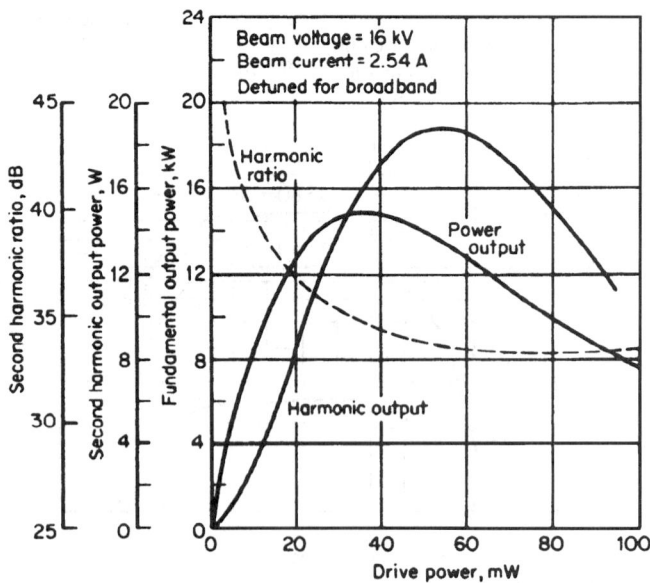

FIGURE 7.2.6 Harmonic output of a typical klystron.

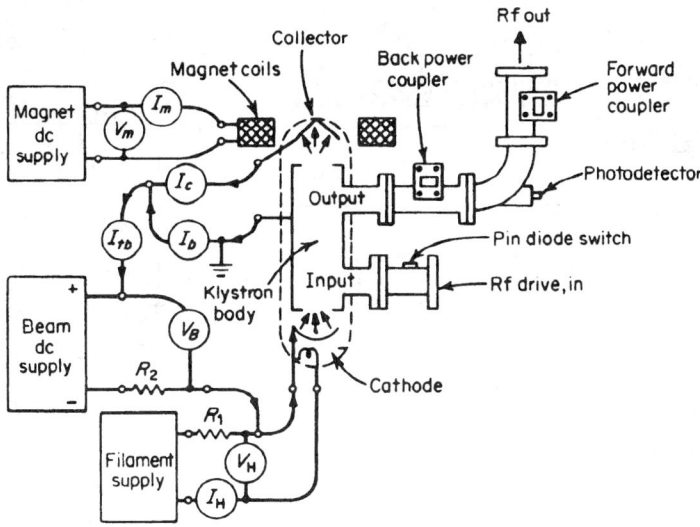

FIGURE 7.2.7   Circuits for a cw klystron amplifier.

antenna. In this way, average efficiency is increased to about 18 percent. More recently, UHF TV klystron efficiency has been increased further by means of multiple electron collector electrodes at reduced potentials, progressively closer to cathode potential. In such "multistage depressed collectors" (MSDC) klystrons, electrons tend to be collected at the lowest possible potential, and the power input goes up and down with drive level and output power in much the same way as it does in a triode or tetrode class-B amplifier. In this way efficiency is further increased by a factor of 2.5 to 3. Figure 7.2.8 shows a simplified schematic diagram of a circuit in which an MSDC klystron might be used.

Klystrons for radar and accelerator applications operate at higher voltages and currents than the cw klystrons described above. Because the electrons are traveling faster, the drift lengths between cavities and the cavity gaps can be longer. Because the gaps can be longer the drift-tube radius can also be larger without excessively lengthening the transit angle for electrons on the axis over that for electrons near the drift-tube wall. For this reason, pulse klystrons can be big enough to handle large average powers. Radar and accelerator klystrons operate at beam potentials and beam currents up to several hundred kilovolts and several hundred amperes. The Stanford Linear Accelerator Center (SLAC) has developed and manufactured many fairly conventional klystrons with peak output power at 2856 MHz ranging from 20 to 60 MW and pulse lengths of several microseconds. A 12 MW peak, 20 kW average power, 115 ms pulse length, 805 MHz klystron is used in the ion injector at the Fermi Laboratory (Fig. 7.2.9). The tube operates at 180 kV and 155 A. Long-pulse (several hundreds of microseconds) klystrons with frequencies between UHF and S-band, peak power of several megawatts and average power of several hundred kilowatts have been used to detect ballistic missiles. The Federal Aviation Agency uses many 2-to-5-MW klystrons in air-route surveillance radars (ARSR) operating in the 1250 to 1350 MHz band, and even more 1-to-2-MW klystrons in airport surveillance radars (ASR) operating in the 2700 to 2900 MHz band. An identical klystron is used in the National Oceanic and Atmospheric Administration's NEXRAD Doppler weather radar which gives superior tornado and clear-air-turbulence warnings. The coherence and high-peak-power, short-pulse performance of klystrons make them ideal for detecting close-in low-radar-cross-section moving targets such as weather disturbances. It is interesting to note that the ASR/NEXRAD klystron will operate over the voltage range from 60 to 80 kV and produce output from 500 kW to 2 MW with efficiency ranging from 45 to 52 percent over the full range of voltages. This is not atypical; most klystrons will operate from one-half to about twice their design power by varying beam voltage and allowing the current to follow. It is also interesting to note that some of the 5-to-7 MW S-band klystrons used to power 16-to-20 MeV cancer-therapy linear accelerators are made using many of the same parts used in ASR/NEXRAD radar klystrons. The ASR/NEXRAD klystron is air

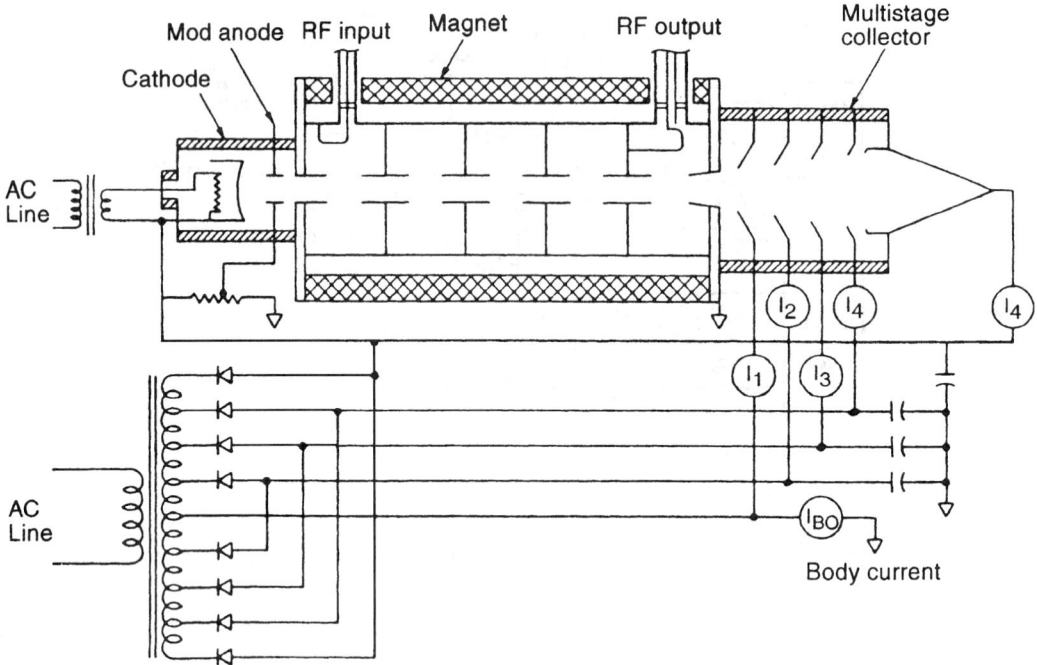

FIGURE 7.2.8 Simplified schematic for high-efficiency, depressed-collector-klystron UHF-TV final amplifier.

cooled; other radar and accelerator klystrons are liquid cooled. Nearly all are solenoid focused. Because the military, for good reason, places a premium on large instantaneous bandwidth, very few tunable klystrons remain in military service. By raising perveance, by full exploitation of stagger tuning, and by using single-gap, double-tuned output circuits, it is possible to achieve about 4 percent bandwidth at the 1-MW level and 10 percent at the 10-MW level in otherwise conventional klystrons. More modern broadband klystrons usually use some form of extended interaction which will be discussed in the next section.

All klystrons are usually operated with the cavities at ground potential. Short-pulse klystrons are usually "cathode pulsed." That is, the cathode is connected to the negative going terminal of the secondary of a pulse transformer. The primary is connected by means of a switch (solid-state or gaseous electron tube) to a charged artificial transmission line (pulse-forming network). Longer pulse lengths are usually produced by incorporating a control electrode such as a floating (or "modulating") anode in the klystron. When such an anode is used, the cathode of the klystron is connected to the negative end of a power supply incorporating a large energy storage capacitor, and electronic switches are connected between the cathode and the modulating anode (the "off switch") and between the modulating anode and ground (the "on switch"). Sometimes control grids are used in front of the cathodes of pulsed klystrons, but they are fragile and are easily damaged by arcs. When dc power supplies are used either with modulating anodes or grids, it is common to use rapid discharge or "crow bar" circuitry across the power supply to protect the tube if an arc is sensed. These usually operate in less than 1 μs and are used in conjunction with an inductor which limits the rate of rise of the fault current.

Extended Interaction Klystrons

Extended interaction klystrons (EIK) use resonant circuits in which a weak electric field interacts with the electron beam over some distance or over several gaps which are so phased that the energy gain or loss of an

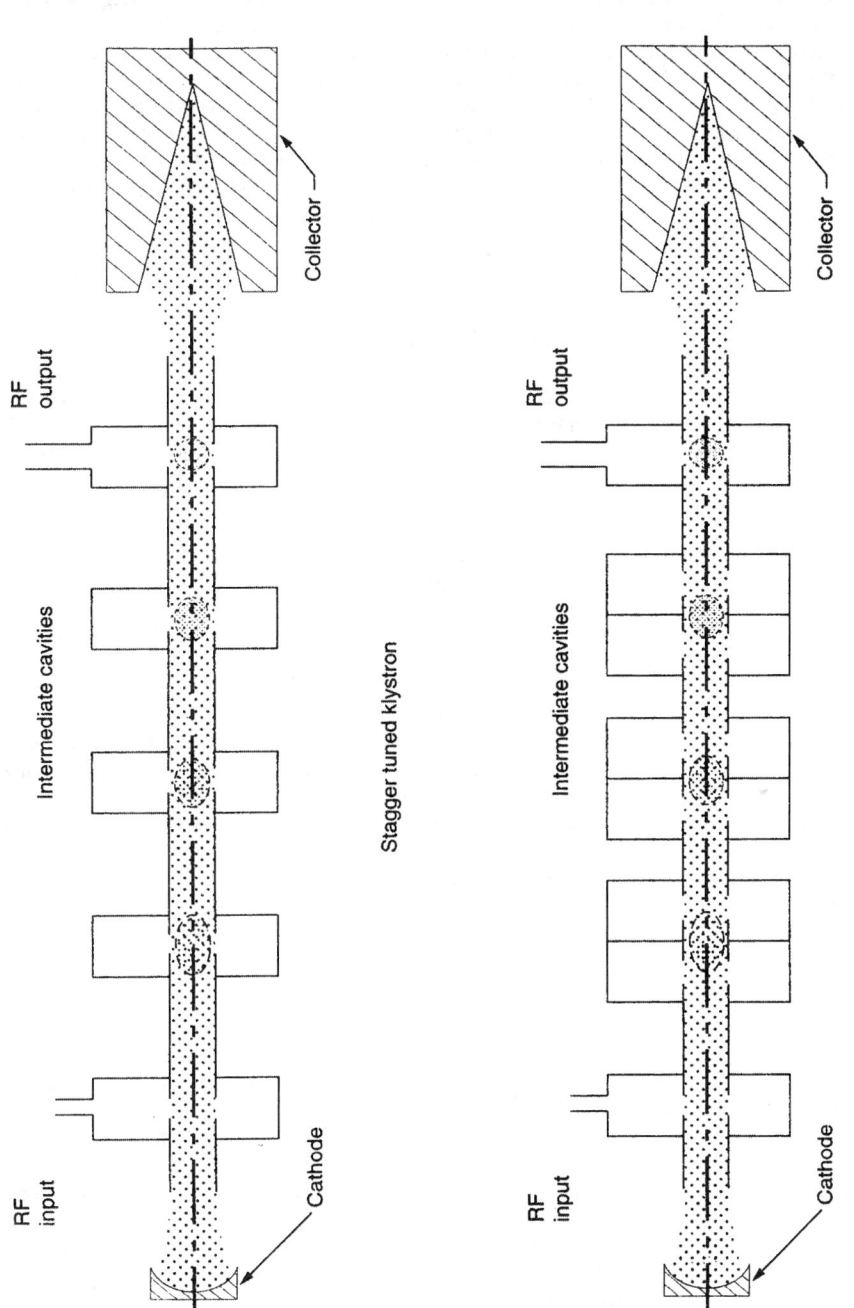

FIGURE 7.2.9 Conventional and clustered—cavity™ klystrons.

individual electron is cumulative. The advantage of such resonant circuits is that the integral of E^2 over the volume or the stored energy of the circuit is always less than it would be if the same voltage were developed at a single gap. This fact offers several possibilities:

Extended interaction output circuits (EIOC) can develop a greater impedance over a larger bandwidth, or in the highest power accelerator klystrons, the fields in the output circuit can be reduced to prevent RF arcing. Because stagger tuning of the cavities of conventional klystrons can achieve bandwidths which exceed those of single-gap output circuits, EIOCs have sometimes been used on such tubes, which have then been referred to as EIKs. At SLAC double-, triple-, and quadruple-coupled cavity circuits are being used in 150 MW, 3 GHz klystrons and 50 MW, 14 GHz klystrons. The circuits usually have been constant impedance filter circuits which support either a standing wave or a traveling wave. In standing wave resonators, the beam interacts with the forward traveling component of the wave which has been designed to have a phase velocity very nearly the same as the average beam velocity. Most traveling wave circuits are really much like traveling-wave tube output circuits in which the fields grow exponentially toward the output end. Such a circuit is used in the Twystron[R], so called because it uses a klystron buncher and a cloverleaf TWT output circuit. A more sophisticated approach to EIOC design makes use of a tapered impedance output circuit in which the filter circuits which couple the output gaps have image impedances which are progressively reduced in an inverse arithmetic taper (i.e., 1, 1/2, 1/3, . . .). Thus, when used with a current saturated electron beam in which the RF current can be made fairly constant over some distance, the current on the circuit builds up linearly (i.e., 1, 2, 3, . . .) and the voltage at all gaps is the same. This results in lower stored energy and reduces arcing for the same output power.

Extended-interaction buncher circuits have been used to a more limited extent. The first extended interaction klystron built by Wessel-Berg at Stanford about 1960, used three short-circuited helix sections as cavities and gave excellent bandwidth. Several millimeter wavelength klystrons, with power output ranging up to 2 kW at 94 GHz, have been built using ladder circuits to minimize circuit losses. Various manufacturers have proposed klystrons using pairs of cavities coupled together to achieve greater bandwidth. In the mid-1980s during computer simulations of such tubes, it was found that the coupling between the cavities of the pair was largely irrelevant because cavities, so closely spaced that there is little gain between them, are already coupled in the proper phase by the exciting current bunch so they cooperate in forming a new bunch farther downstream. This kind of klystron was named a clustered-cavity[TM] klystron, and in a sense, it is a special case of extended interaction which is somewhat easier to design because one does not have to deal with coupling between cavities. Figure 7.2.9 shows a schematic comparison of a conventional klystron and a clustered-cavity klystron of the same power and gain. The clustered-cavity klystron will have twice the bandwidth if the paired cavities are artificially loaded to one-half the Q values of the single cavities they replace. A 13 percent bandwidth, 3 MW, 3 GHz clustered-cavity klystron is being manufactured. Computer simulations indicated that 30 percent bandwidths may be available from megawatt klystrons of this type.

BIBLIOGRAPHY

Granatstein, V. L., R. K. Parker, and C. M. Armstrong, "Scanning the technology: vacuum electronics at the dawn of the twenty-first century," *Proc. IEEE*, Vol. 87, No. 5, May 1999.

Phillips, R. M., and D. W. Sprehn, "High-power klystrons for the next linear collider," *Proc. IEEE*, Vol. 87, No. 5, May 1999.

Symons, R. S., "Scaling laws and power limits for klystrons," *1986 IEEE Int. Electron Devices Meet. Digest*, pp. 156–159.

Symons, R. S., and J. R. M. Vaughan, "The linear theory of the clustered-cavity klystron," *IEEE Trans. Plasma Sci.*, Vol. 22, No. 5, pp. 713–718.

"Twystron" is a registered trademark of Varian Associates Inc.

"Clustered-Cavity" is a registered trademark of Litton Systems, Inc.

Portions of the material on klystrons were adapted from material supplied by Richard B. Nelson.

INDUCTIVE OUTPUT TUBES (KLYSTRODES)

Donald H. Preist, M. B. Shrader

Tube and Cavity Configurations

The name *Klystrode* was coined to indicate that the output cavity, load system, and collector are basically as in a klystron. The input cavity and the input part of the tube resemble a triode. In all the UHF/TV amplifiers so far built the output cavity system is the external cavity type using a cylindrical ceramic window concentric with the electron beam. Many klystrons in service use this arrangement and a coupled secondary cavity connected to the output load, to maximize bandwidth (see Fig. 7.2.10). The input cavity is also external in that the major part of the circuitry is outside the vacuum envelope. Two types are in service.

The older type (Fig. 7.2.10) uses a small amount of controlled positive feedback and gives 1 to 3 dB more power gain than the second type (Fig. 7.2.11), but is hard to adjust. The second type is much easier to adjust and gives adequate gain, typically 22 dB. This is the favored approach in new amplifiers. In the very high power amplifiers built for "Big Science" applications the output cavity is in vacuum and the output window is in the coaxial transmission line to the load (Fig. 7.2.11). The input cavity is external as in the TV amplifiers. Both types have been used successfully, but the second type is preferred.

The most significant difference between the IOT/Klystrode and the klystron is the input cavity/tube system. The grid and cathode are at high negative dc potential with respect to ground. The correct RF voltage must appear between grid and cathode, even though the source of the RF drive power must be at dc ground potential. Also the grid to anode RF impedance must be minimized to prevent feedback; if positive this will tend to cause self-oscillation, and if negative it will tend to reduce power gain.

Transmitter Configurations for UHF/TV

Most of the earlier high-power installations used a pair of 60 kW tubes giving 120 kW P. S. visual and a third giving 13 kW CW aural. Later, transmitters featuring combined aural and visual amplification in the same tube were put into service. The exacting NTSC specifications could be met only by a high degree of amplitude and

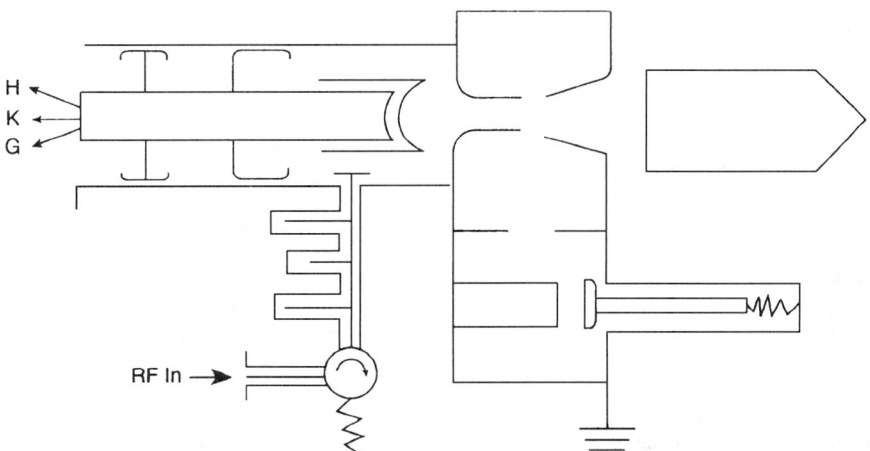

FIGURE 7.2.10   Input cavity provides controllable positive feedback to enhance gain. It is stabilized by stub tuner/Ferrite circulator combination. Resonant cavities iris coupled.   (Loop has been used.) Preferred for 450 MHz and higher. Magnetic field between anode and collector not shown.

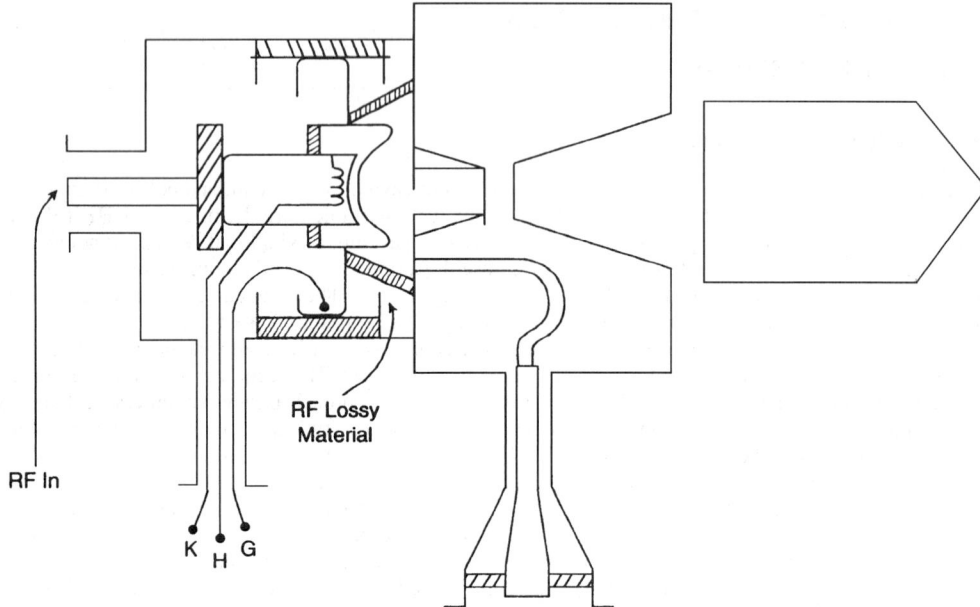

FIGURE 7.2.11 Grid-anode part of input cavity is heavily damped by lossy material for stabilization. Bringing out H, K, G connections radially makes cathode extension available for coupling RF drive through HV dc insulation. Output cavity in vacuum shows loop coupling to load. Fine-tuning "paddle" not shown. Magnetic field not shown.

phase linearity. This was achieved by a combination of careful tube design and a sophisticated precorrection module which allowed the residual nonlinearity of the IOT/Klystrode and the solid-state driver amplifier to be reduced to a very low level.

The IOT itself is more linear than the klystron at black level in TV service, and it resembles a tetrode at low level (Fig. 7.2.12).

Properties of the IOT/Klystrode as an Electron Device

First, it must be emphasized that the tube, unlike the klystron and the TWT is not a velocity-modulated tube. It uses a density-modulated electron beam as in a triode or tetrode. If the electron transit time through the tube is small in terms of the operating frequency, e.g., 1 rad, the efficiency will be high and may approach the theoretical limit of 78.5 percent. The "Big Science" tubes have shown 70 to 74 percent at output powers of several hundred kW. However, as the frequency is raised and the transit time or angle increases, the efficiency will eventually fall, but it will exceed the triode efficiency because the IOT anode is at high dc potential at the time of maximum current flow.

The tube (Fig. 7.2.10) is a five-electrode device with a spherical cathode and grid and an apertured anode at high dc potential. A reentrant cavity is placed between the anode and the next electrode, the tailpipe. Insulated from this is the final electrode, the collector. The electron beam is guided through the cavity by a magnetic field.

Because of the action of the negatively biased grid and the superimposed RF voltage the beam current flows for about half the RF cycle and is zero for the rest of the cycle. It passes through the aperture in the anode and the first part of the cavity without interception, at constant velocity. Next, it passes through the gap in the cavity where the RF field is decelerating, then through the second field-free region in the tailpipe with minimal interception due to the magnetic field, and finally enters the collector which is usually at ground dc potential. The two field-free regions are field-free because they are cylindrical metal pipes long enough to behave as waveguides beyond cut-off at the operating frequency.

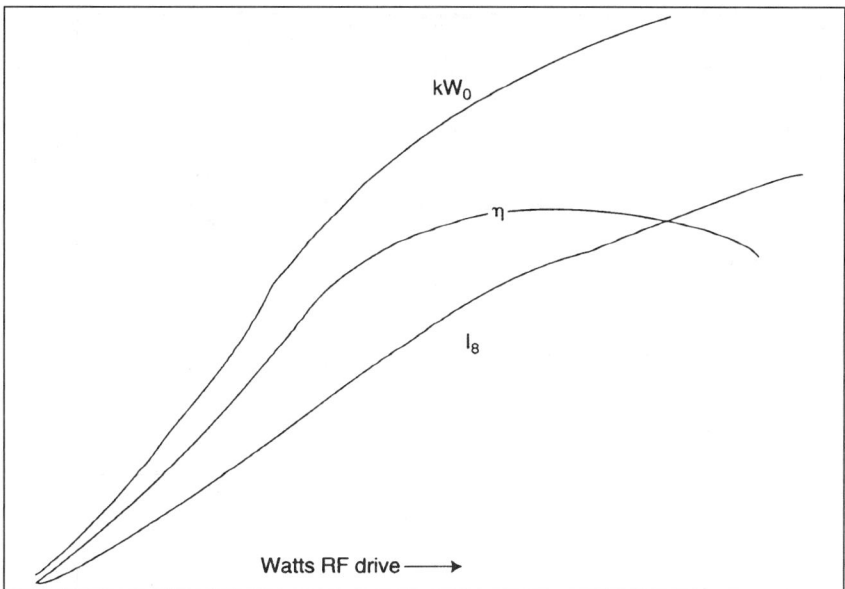

FIGURE 7.2.12 Power output vs. RF drive shows fair linearity. Efficiency stays relatively flat over 2:1 change in drive. Beam current is zero for zero drive.

Reference 2 contains a detailed analysis of the electron field and wave interactions. From the user's viewpoint the outstanding features are: (a) a complete absence of interaction between the output load impedance and the grid-cathode conditions, except when the load impedance approaches infinity; (b) the absence of a "dc blocking capacitor" carrying RF current as in a tetrode; (c) the output cavity system is at dc ground potential; (d) the power dissipated on the tailpipe is small; (e) the power output is not limited by the collector as this can be made indefinitely large and easy to cool, in principle. In practice there may be constructional problems with a very large collector on a super power tube. It should be noted, though, that the collector has to dissipate only the dc beam power minus the RF output power. In a conventional klystron the entire beam power has to be dissipated when the RF drive goes to zero.

Other significant performance characteristics can be seen from Fig. 7.2.12. Because the beam current falls to zero when the RF drive falls to zero, and varies monotonically with drive level, the device resembles the class B linear amplifier well-known to tetrode users. It is this feature that provides high average efficiency with an amplitude-modulated signal such as in TV, compared to a klystron. This is even more important with systems having a high peak-to-average signal ratio such as some digital HDTV systems.

In high power CW or long pulse (10 ms) service the IOT/Klystrode has performed well at 425 MHz (Ref. 5) and at 267 MHz (Ref. 4) as an RF power source for driving proton accelerators (Fig. 7.2.13). Here a set of characteristics become important which do not apply in TV service. Among these are: (1) absence of need for a high level pulser because the RF drive can be pulsed; (2) at VHF the physical size is very much smaller than a Klystron; (3) a high-power ferrite circulator may not be needed between the tube and the accelerator, and (4) the efficiency is higher than that of a klystron or a tetrode (70 to 74 percent measured).

Basis for Output Improvements

There are several reasons why IOT/Klystrodes can produce hundreds of kW compared to the 100 watts of Haeff's developmental tube, described in 1939 (Ref. 1).

1. The evolution of techniques for designing and fabricating electron guns, beam-focusing systems, cooling systems, cavities, and output windows over decades of microwave tube development,

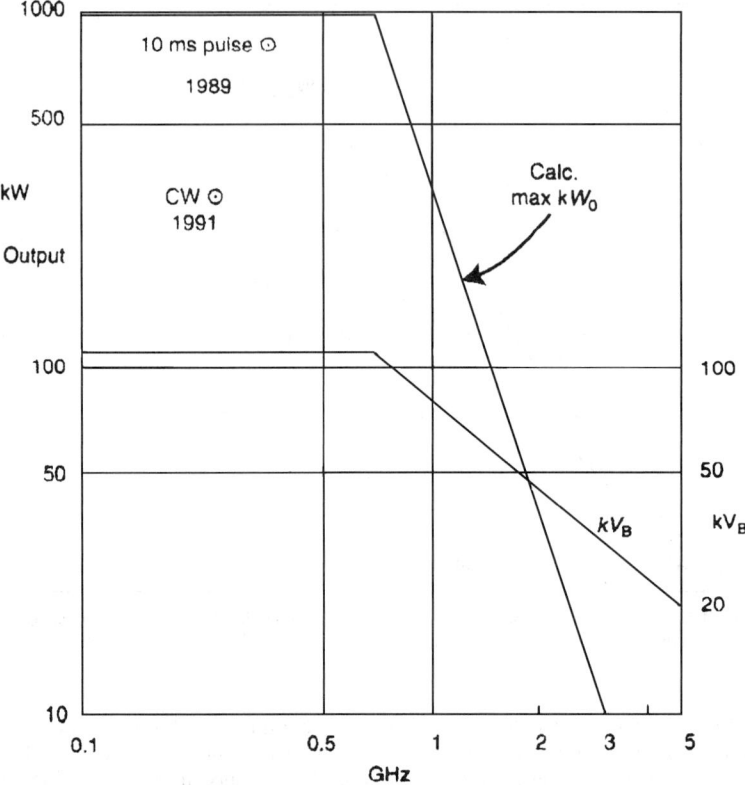

FIGURE 7.2.13 Assumptions made in calculating maximum kW are (*a*) constant dc perveance 0.3 microperv, (*b*) maximum usable gun anode voltage 113 kV up to 750 MHz 20 kV at 3 GHz, (*c*) tube dimensions vary inversely with *f* above 750 MHz, (*d*) efficiency 70 percent.

2. The availability of pyrolytic graphite, an excellent material for making grids,

3. The availability of high purity alumina for output windows and coatings for multipactor suppression, and interelectrode insulators, and

4. The invention of the tungsten matrix impregnated cathode.

REFERENCES

1. Haeff, A. V., "An UHF power amplifier of novel design," *Electronics*, pp. 30–32. February 1939.

2. Preist, D., and M. Shrader, "The Klystrode—an unusual transmitting tube with potential for UHF/TV," *Proc. IEEE*, Vol. 70, No. 11, November 1982.

3. Ostroff, N. S., "UHF transmission technology," *Broadcast Engineering Magazine*, February 1994.

4. Sheik, J. Y. et. al., "Operation of a High Power C. W. Klystrode with the RFQ1 Facility," A.E.C.L. Research, Chalk River Labs, 1992.

5. Preist, D. H., and M. B. Shrader, "A high power Klystrode with potential for space application," *IEEE Trans. ED*, Vol. 38, No. 10, October 1991.

6. Granatstein, V. L., R. K. Parker, and C. M. Armstrong, "Scanning the technology: vacuum electronics at the dawn of the twenty-first century," *Proc. IEEE*, Vol. 87, No. 5, May 1999.

TRAVELING-WAVE TUBES

Pamela L. Walchli

Introduction

The traveling-wave tube (TWT) is a linear-beam device which amplifies microwave signals to high-power levels over broad bandwidths. It was invented by Rudolf Kompfner in the latter part of World War II and developed into a viable device by J. R. Pierce and L. M. Field at Bell Telephone Laboratories in 1945. Today, the TWT finds diverse application in communications, radar guidance, and electronic countermeasure systems.

Basic structure.   All TWTs comprise four basic elements: (1) an electron gun, (2) an RF interaction circuit, (3) an electron-beam magnetic focusing system, and (4) a collector to dissipate the spent beam power.

The major difference between the various types of TWTs lies in the RF interaction structure. A schematic representation of a typical TWT is shown in Fig. 7.2.14. At the left is the electron gun, which forms the beam; at the center are the RF interaction structure (in this case, a helix) and the magnetic beam-focusing system; on the right is the collector that absorbs the spent beam power.

Theory of operation.   The purpose of the interaction structure is to slow the RF signal so that it travels at the same speed as the electron beam. Electrons enter the structure during both positive and negative phases of an RF cycle. Those entering during a positive phase are accelerated; those entering during a negative phase are decelerated. The electrons that experience a velocity increase catch up with the electrons that have been slowed down, forming electron bunches. These bunches produce an alternating current superimposed on the dc beam current. The alternating current induces growth of the RF circuit wave, which, in turn, forms tighter electron bunches and thus a larger component of alternating current.

Growth of the wave on the circuit occurs because the velocity at which the beam is traveling forces the electron bunches to enter a decelerating phase of the RF field. In the decelerating field, the electrons are slowed, transferring their energy to the RF wave. This cycle is limited by one or more severs and ultimately by the extraction of the RF power through the output connector. The sever absorbs the RF power that has been built up on the circuit but does not affect the ac component of current in the beam.

The modulated beam drifts through the sever region, and induces a new RF wave in the next circuit section, where the interaction process begins again. The purpose of the sever is to absorb reflected power, which travels in a backward direction on the circuit. The reflected power arises from an imperfect match between the RF circuit and the output connector. Without the sever, regenerative oscillations would be induced.

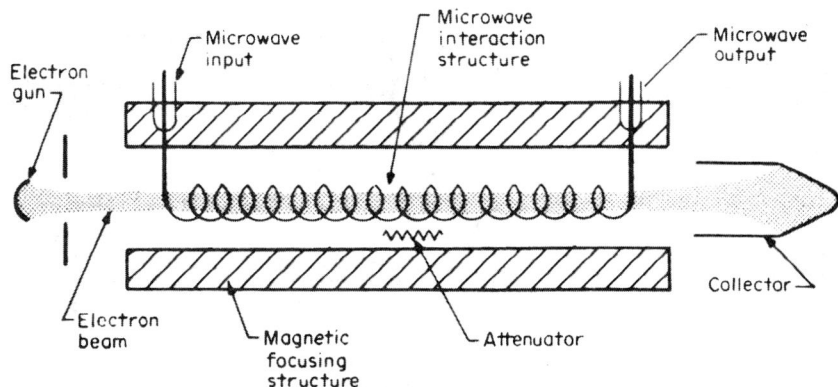

FIGURE 7.2.14   Basic elements of a typical TWT.

At any given frequency, a certain level of drive power will cause the maximum degree of bunching and thus the greatest amount of output power. This condition is known as *saturation*. For small drive signals, typical TWTs have 40 to 70 dB of gain.

The Electron Gun

The electron gun forms a high-current-density pencil beam of electrons, part of whose energy is converted to RF power through interaction with the wave traveling along the RF circuit. In the typical electron gun, electrons are emitted from a spherical cathode and converged to the required beam size by focusing electrodes. The final converged beam size is maintained through the interaction structure by either permanent magnet or electromagnet focusing.

On-off switching modulation of the electron beam is accomplished by applying a pulse to one of four electrodes: (1) cathode, (2) anode, (3) focus electrode, or (4) grid.

Cathode and anode pulsing.   In the first method, the cathode is pulsed negatively with respect to the grounded anode, requiring both the full beam voltage and current to be switched. Alternatively, anode modulation involves switching the full beam voltage between cathode potential and ground, but the current switched is just that intercepted on the anode, usually only a few percent of the full beam current.

Focus-electrode pulsing.   The focus electrode, which normally operates at or near cathode potential, can be biased negatively with respect to the cathode to turn the beam off. The voltage swing required is usually one-third or less of the full cathode voltage. Since the focus electrode draws no current, reduction in power requirements is significant.

Grid pulsing.   Switching power requirements are minimized with grid modulation. A grid structure, to which the modulating voltage is applied, is placed directly in front of the cathode surface. The amount of voltage swing needed is typically only one-twentieth or less of the full beam voltage.

Some common grid structures in use are shown in Fig. 7.2.15. The grid in Fig. 7.2.15a is a simple, single intercepting grid. To turn the beam on, a voltage positive with respect to the cathode is applied to the grid, drawing current from the full cathode surface. The grid webs intercept the current drawn from the cathode area directly behind the grid. This interception limits the duty cycle at which the tube can operate.

Current drawn by the grid can be minimized by schemes such as those shown in Figs. 7.2.15b and 7.2.15c. The structure in Fig. 7.2.15b is composed of two grids, the one nearest the cathode surface operating at cathode potential and the outer one operating at the modulating voltage. The inner grid, identical in pattern to the outer grid, prevents emission from the cathode surface directly behind it, effectively eliminating intercepted current on the control grid. This inner grid is referred to as a *shadow grid*. The shadow grid may be attached directly to the cathode surface, as in Fig. 7.2.15c. This kind of structure has the trade name Unigrid.

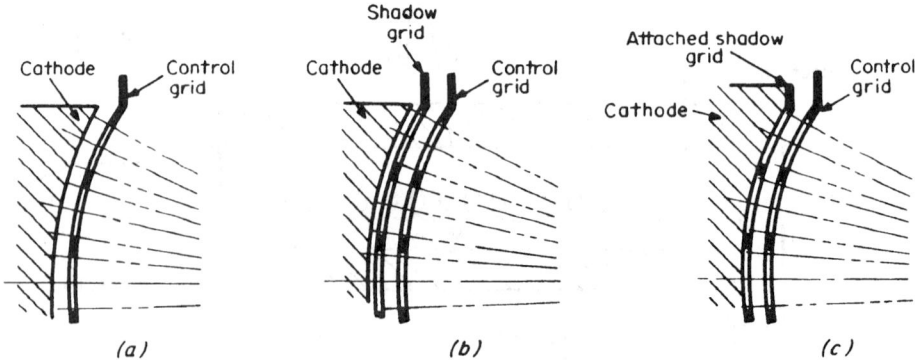

FIGURE 7.2.15   Grid structures used in TWTs: (*a*) simple intercepting grid; (*b*) double grid, with shadow grid operated at cathode potential; (*c*) unigrid type, with shadow grid attached directly to cathode.

A simple grid has application in smaller guns, where high amplification factors are not required. The grid operates at cathode potential and therefore intercepts no current. To turn off the beam, a voltage negative with respect to the cathode is applied to the grid. Amplification factors around 10 are typical for this kind of electron gun.

Magnetic Beam Focusing

Without a focusing system the electron beam would spread due to the mutually repulsive forces on like-charged particles, causing the electrons to strike the RF circuit. A magnetic beam-focusing system is the most widely used and is usually implemented in one of three ways: (1) electromagnetic, (2) permanent magnet, or (3) periodic permanent magnet.

Electromagnet Focusing. Electromagnet focusing is used primarily on very high power coupled-cavity TWTs. Tight beam focusing is required in these tubes because significant interception on the RF circuit is intolerable at the power levels in question. Disadvantages of this kind of focusing are size, weight, and consumption of power, but all can be reduced somewhat by wrapping the windings of the solenoid directly on the tube body. Solenoidal focusing is illustrated in Fig. 7.2.16a.

Permanent-Magnet Focusing. Permanent-magnet focusing is possible where the interaction structure is short, e.g., in low-gain or millimeter-wave tubes. It can be used in place of solenoidal focusing in these kinds of tubes. This focusing system is shown in Fig. 7.2.16b.

Periodic-Permanent-Magnet Focusing. Periodic-permanent-magnet focusing is used on almost all helix TWTs and most coupled-cavity TWTs. A periodic-permanent-magnet (PPM) structure is shown in

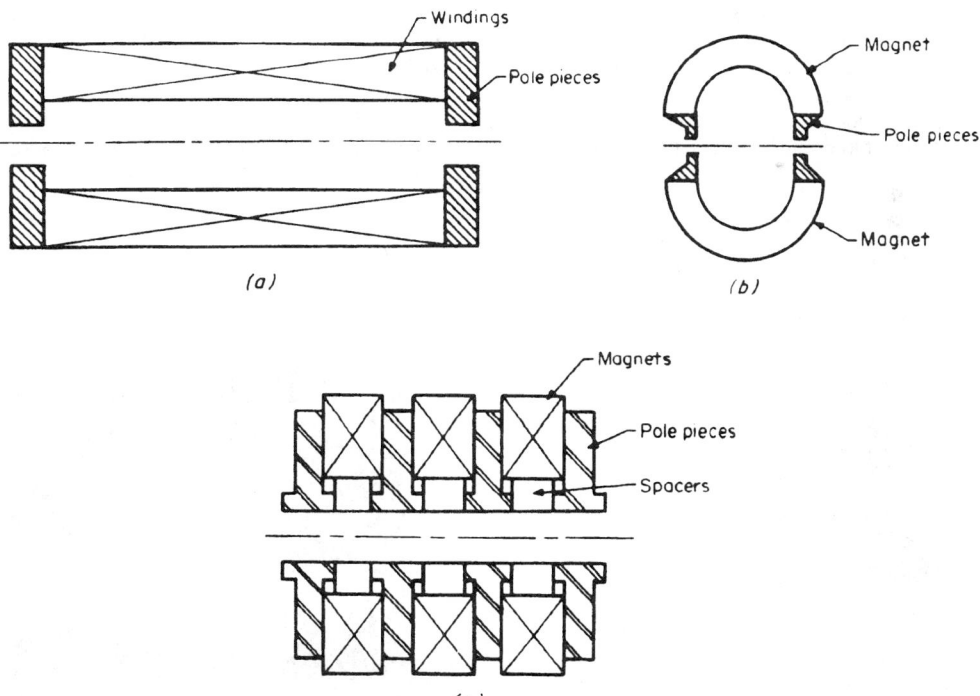

FIGURE 7.2.16 Magnetic focusing arrangements: (*a*) solenoidal type; (*b*) permanent-magnet type; (*c*) periodic permanent-magnet structure.

Fig. 7.2.16c. The magnets are arranged with alternate axial polarity in successive cells. In helix TWTs the pole pieces (with nonmagnetic spacers) may form the tube's vacuum envelope, or the pole pieces and spacers may be slipped over a stainless-steel tube that serves the same purpose. In coupled-cavity TWTs the cavity walls themselves are the pole pieces.

This kind of focusing provides a major reduction of tube size and weight, along with the elimination of the magnet power supply. The drawback of this scheme is that the electron beam ripples with a periodicity of the length of one magnet cell. This increases beam interception on the RF circuit and thus generally limits the use of PPM focusing to lower average-power TWTs.

The Interaction Circuit

The fundamental principle of operation of a TWT is that an electron beam moving at approximately the same velocity as an RF wave traveling along a circuit gives up energy to the RF wave. Since the RF wave travels at the speed of light, a method must be found to slow the forward progress of the wave to roughly the same velocity as that of the electron beam. The beam speed in a TWT is typically between 10 and 50 percent of the velocity of light, corresponding to cathode voltages of 4 to 120 kV. The two structures that accomplish the slowing of the RF wave are the helical and coupled-cavity circuits.

Helix circuits.   The helix (Fig. 7.2.17) is supported inside the vacuum envelope by three or more ceramic support rods, which also conduct heat away from the helix. A helix interaction structure is used where bandwidths of an octave or more are required, since over this range the velocity of the signal carried by the helix is almost constant with frequency. For greater than octave-bandwidth operation, the variation of velocity with frequency can be modified by the introduction of metal loading segments near the helix, causing the phase velocities of a wider range of frequencies to be more nearly in synchronism with the beam velocity.

The helix provides satisfactory performance over the range of frequencies from 500 MHz to over 40 GHz. However, the typical helix circuit is limited in average power-handling capability to a few hundred watts. Peak power levels above several kilowatts cannot, in general, be achieved because of circuit RF instabilities. Higher peak power levels can be obtained by eliminating these oscillations with a special type of helix circuit consisting of the superposition of a helix wound in a right-hand sense on a helix wound in a left-hand sense. Two practical implementations of this configuration are the ring-loop and ring-bar circuits (Fig. 7.2.18). Peak powers of hundreds of kilowatts are attainable, but average power capability is no better than that of the conventional helix

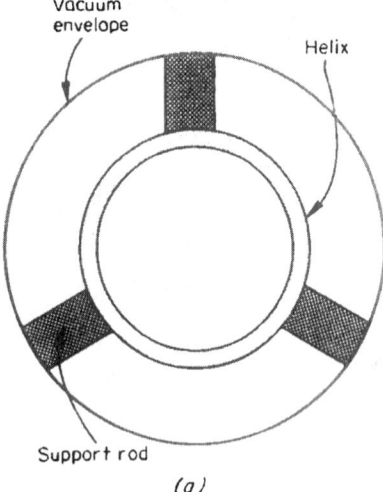

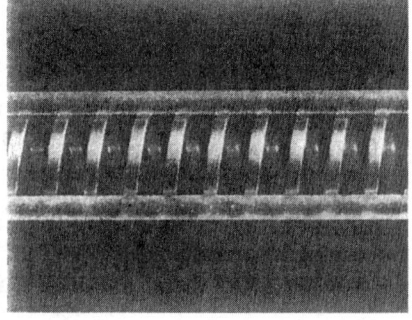

(a)   (b)

FIGURE 7.2.17   Helix circuit: (*a*) end view; (*b*) side view.

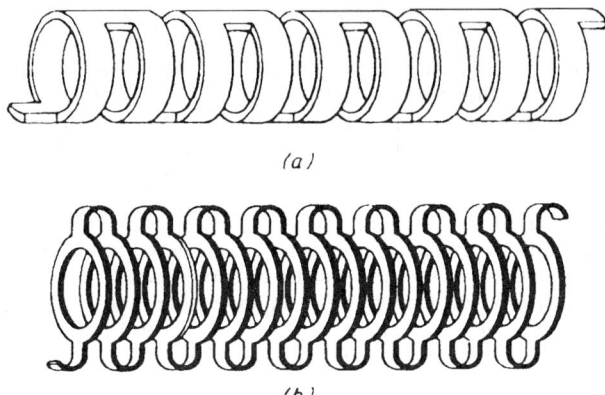

FIGURE 7.2.18 Structures composed of two helixes superimposed in opposite sense of rotation: (*a*) ring-bar circuit; (*b*) ring-loop circuit.

circuits, since the structures are supported in a like manner. Because the ring-bar and ring-loop circuits are dispersive, the maximum bandwidth of a tube using them is typically only one-third octave.

Coupled-Cavity Circuits. Because of its superior ability to dissipate heat, the coupled-cavity structure is capable of both high peak and average power over moderate bandwidths. Coupled-cavity tubes find applications from 2 GHz up to nearly 100 GHz. Bandwidths of 10 percent are typical, although tubes with 40 percent bandwidth have been developed.

The coupled-cavity circuit consists of resonant cavities coupled through slots cut in the cavity end walls, resembling a folded waveguide. This arrangement results in a bandpass filter network that is highly dispersive, limiting the tube bandwidth. The two most common kinds of coupling schemes are illustrated in Fig. 7.2.19. The structure in Fig. 7.2.19*a* is a forward fundamental circuit, also called a cloverleaf circuit from the shape of its cavities. It is used primarily on extremely high peak power coupled-cavity tubes or in the output section of a hybrid klystron TWT, known as a Twystron. Typical performance of a tube with a forward fundamental circuit is 3 MW peak and 5 kW average at S band. Figure 7.2.19*b* illustrates the more commonly used coupled-cavity structure, the single-slot space harmonic circuit. A peak power of 50 kW and an average power of 5 kW at X band are typical for space harmonic TWTs.

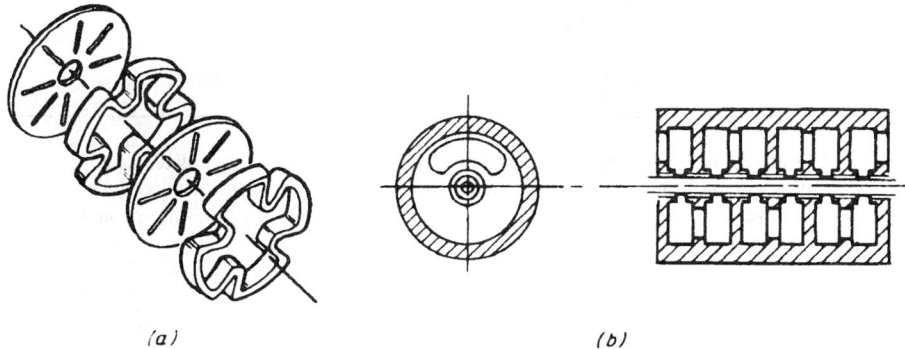

FIGURE 7.2.19 Coupled-cavity circuits: (*a*) forward fundamental circuit ("cloverleaf"); (*b*) single-slot space harmonic circuit.

The Collector

The function of the collector is to collect the electron beam after it has passed through the interaction structure and dissipate the remaining beam energy. During interaction electrons give up various amounts of energy, and some actually gain energy. Typically, the slowest electrons lose no more than 50 percent of their original energy; the fastest gain at most 20 percent with the remainder distributed between these extremes. If a TWT had an interaction efficiency of 20 percent, the average electron would possess 80 percent of its original energy.

Single-Stage Collectors.   The overall efficiency of the TWT can be increased by operating the collector at a voltage lower than the full beam voltage, a practice known as *collector depression.* This introduces a potential difference between the interaction structure and the collector through which the electrons pass. The amount by which a single-stage collector can be depressed is limited by the remaining energy of the slowest electrons; i.e., the potential drop can be no greater than the amount of energy of the slowest electrons or they will be turned around and reenter the interaction structure, causing oscillations.

Multistage Collectors.   Efficiency can be increased still more by introducing multiple depressed-collector stages. This method provides for the collection of the slowest electrons on one stage, while allowing those with more energy to be collected on other stages depressed still further. Figures 7.2.20a and b represent the configuration of power supplies (less the heater supply) to operate a gridded TWT with a single-stage and a multistage depressed collector, respectively. Calculations of the overall efficiency of such TWTs are shown in the following table, assuming a beam power of 5 kW (10 kV, 0.5 A) and an interaction efficiency of 15 percent:

	Voltage, kV	Current, A	Power, W
Single-stage collector			
Helix supply	10	0.025	250
Collector supply	5	0.475	2,375

$$\text{Overall efficiency} = \frac{750 \text{ W}}{250 \text{ W} + 2375 \text{ W}} = 29\%$$

	Voltage, kV	Current, A	Power, W
Multistage collector			
Helix supply	10	0.025	250
Collector stage, 1	5	0.23	1,150
2	2.5	0.15	375
3	1	0.085	85
4	0	0.01	0

$$\text{Overall efficiency} = \frac{750 \text{ W}}{250 \text{ W} + 1150 \text{ W} + 375 \text{ W} + 85 \text{ W}} = 40\%$$

Helix TWTs are cooled conductively by mounting the tube in a metal baseplate, which is in turn attached to an air- or liquid-cooled heat sink. Coupled-cavity tubes below 1 kW average power are cooled convectively by drawing air over the entire tube length. Higher-power coupled-cavity tubes are cooled by circulating liquid over the tube body and collector.

Microwave Power Module

In keeping with the general trend in electronic components toward minimization of system size, the microwave power module (MPM) is a complete microwave power amplifier contained within a volume of less than 20 in^3. This system consists of a micro-miniature helix TWT driven by an MMIC preamplifier, and a

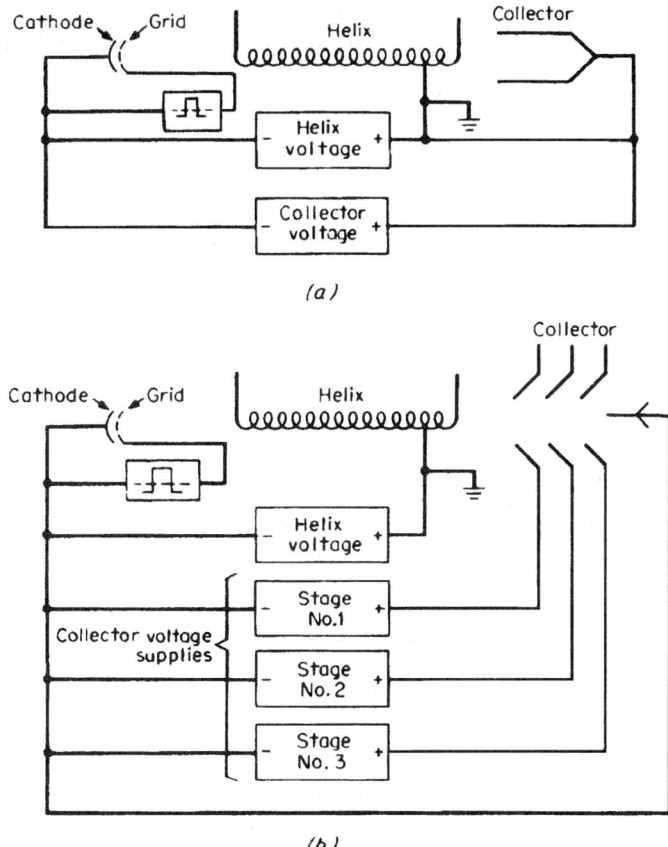

FIGURE 7.2.20   Power supplies for TWTs: (*a*) single collector TWT; (*b*) multistage depressed collector TWT.

high-density electronic power conditioner that powers both the TWT and the MMIC amplifier. Applications include electronic countermeasures, radar and communications transmitters. These modules may also be combined for use in shared aperture phased arrays.

BIBLIOGRAPHY

Abrams, R. H., and R. K. Parker, "Introduction to the MPM: What it is and where it might fit," *IEEE MTT-S Int. Symp. Dig.*, 1993.

Brees, A., G. Dohler, J. Duthie, G. Groshart, G. Pierce, and R. Wayrich, "Microwave power module (MPM) development and results," *IEDM Dig.*, 1993.

Gewartowski, J. W., and H. A. Watson, "Principles of Electron Tubes," Van Nostrand, 1965.

Gilmour, A. S., "Principles of Traveling Wave Tubes," Artech House, 1994.

Gittins, J., "Power Traveling Wave Tubes," Elsevier, 1965.

Granatstein, V. L., R. K. Parker, and C. M. Armstrong, "Scanning the technology: vacuum electronics at the dawn of the twenty-first century," *Proc. IEEE*, Vol. 87, No. 5, May 1999.

Liao, S. Y., "Microwave Devices and Circuits," Prentice Hall, 1980.

Mendel, J., "Helix and coupled-cavity traveling-wave tubes," *Proc. IEEE*, Vol. 61, March 1973.

Pierce, J. R., "Traveling-Wave Tubes," Van Nostrand, 1950.

Smith, T. I., "The microwave power module—a versatile RF building block for high power transmitters," *Proc. IEEE,* Vol. 87, No. 5, May 1999.

Staprans, A., E. E. McCune, and J. A. Ruetz, "High power linear beam tubes," *Proc. IEEE,* March 1973, Vol. 61.

CROSSED-FIELD TUBES

George K. Farney

Crossed-Field Interaction Mechanism

A crossed-field microwave tube is a device that converts dc electric power into microwave power using an electronic energy-conversion process similar to that used in a magnetron oscillator. These devices differ from beam tubes in that they are potential-energy converters rather than kinetic-energy converters. The term *crossed field* is derived from the orthogonality of the dc electric field supplied by the source of dc electric power and the magnetic field required for beam focusing in the interaction region. Typically, the magnetic field is supplied by a permanent-magnet structure. These tubes are sometimes called *M tubes*.

The electronic interaction is illustrated schematically in Fig. 7.2.21. Electrons moving to the right in the figure experience electric field deflection forces ($f_e = -\epsilon E$) toward the electrically positive anode, while the magnetic deflection forces ($f_m = -\epsilon v \times B$) resulting from the motion of the negatively charged electron in the orthogonal magnetic field cause deflection toward the negative electrode. This electrode is also called the *sole*.

The forces are balanced when an electron is traveling in a parallel direction between the electrodes with a velocity numerically equal to the ratio of the dc electric field to the magnetic field ($v_e = E/B$). Any alteration of the electron velocity leads to an unbalanced condition. Reduction of the electron forward motion causes the magnetic deflection force to become less, and the electron trajectory is deviated toward the positive electrode. Conversely, an increase of velocity causes a greater magnetic deflection force, which causes trajectory deviation toward the negative electrode.

Electronic interaction with a traveling wave occurs when the positive electrode is an RF-guiding slow-wave circuit whose phase velocity for the traveling wave is numerically equal to the ratio of the dc electric field to the magnetic field ($v_p = E/B$). Under these conditions synchronous interaction occurs between the RF fields on the slow-wave circuit and the stream of electrons traveling in the interaction region.

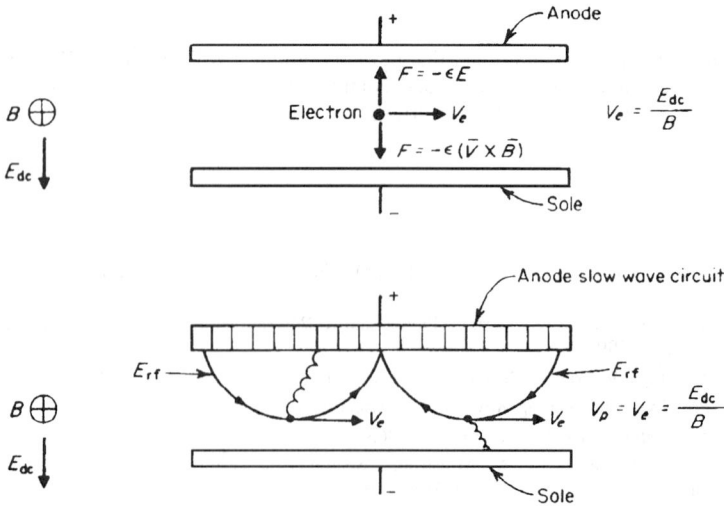

FIGURE 7.2.21   Forces exerted on a moving electron in a crossed-field environment.

Two general kinds of motion result, as illustrated schematically in Fig. 7.2.21, where a moving frame of reference is shown traveling from left to right at a velocity equal to the phase velocity of the circuit wave, so that the instantaneous RF fields are seen as stationary. The electronic motion resulting from interaction with the tangential components of the additional RF electric fields depends on the location of the electron relative to the phase of the RF fields of the slow-wave circuit. Those located so that their forward motion is retarded by the RF electric field are slowed, and the energy they lose is transferred to the RF wave on the circuit. These slower-moving electrons are subsequently accelerated toward the anode by the dc electric field, and their velocity is increased to the synchronous condition. The energy-exchange cycle can then be repeated. Electrons moving in this phase of the RF field pattern transfer energy to the RF wave on the circuit while maintaining nearly constant kinetic energy. The energy transfer results from the loss of potential energy of the electrons as they move to the anode.

Electrons located in the alternate phase of the RF field pattern are accelerated by the RF field and move away from the anode. The intensity of the slow wave decreases exponentially with distance away from the slow-wave circuit so that the magnitude of this interaction decreases. The result is the transfer of dc electric power to microwave power on the slow-wave circuit, with the phase-sorted electrons in the crossed-field interaction region providing the necessary coupling mechanism. Electron current thus flows to the anode only in the region of suitably phased RF electric fields.

The components of the RF field which are perpendicular to the forward motion of the electrons exert forces which phase-lock the electron near the center of the pattern. These regions are called *spokes* because of the similarity, in a magnetron oscillator, to the spokes in a rotating wheel. The phase locking of the sorted space charge relative to the traveling RF wave on the slow-wave circuit reduces the effect of power-supply variations on the electron trajectories. The details of the electron trajectories are extremely complex and have been calculated only approximately, using sophisticated computer techniques.

It is an important fundamental of crossed-field interaction that very high electronic conversion efficiency can be obtained because the kinetic energy of the electrons lost as heat upon ultimate impact with the slow-wave circuit can be designed to be a small fraction of the total potential energy transferred from the power supply. The ideal electronic conversion efficiency is given by $\eta = 1 - V_0/V$, where η is efficiency, V_0 is the synchronous voltage, and V is the cathode-to-anode voltage. Large ratios of V/V_0 lead to high efficiencies.

The crossed-field magnetron oscillator achieved prominence as a source of microwave power for radar applications during World War II. Since that time many kinds of crossed-field devices were investigated. One type of device obtains current from a thermionic cathode, electron gun located external to the crossed-field interaction space similar to that used in electron beam tubes. A second type uses electron current supplied by thermionic and/or secondary emission from a negative electrode facing the slow-wave circuit. This is similar to a cathode in a magnetron. These are called emitting sole tubes. Both types are illustrated schematically in Fig. 7.2.22.

Slow-Wave Circuits for Crossed-Field Tubes

Electron current in crossed-field interaction moves toward the slow-wave circuit rather than through the circuit as in beam tubes. This leads to the use of open circuits that present an RF waveguiding surface to the electron stream. Maximum energy conversion efficiency is usually obtained when the current is intercepted on the slow-wave circuit; so the structures must withstand the thermal stress associated with electron bombardment. Electronic interaction can occur using either forward-wave or backward-wave traveling-wave circuits, as well as with circuits supporting a standing wave. Examples of circuits suitable for use in forward-wave interaction are various *meander lines, helix-derived structures, bar* and *vane structures*, which are capacitively loaded by ground planes, and *capacitively strapped-bar circuits*.

A helix-coupled vane circuit and a ceramic-mounted meander line are shown in Fig. 7.2.23a. The most common backward-wave circuits are derivatives of the interdigital line and strapped-bar and vane circuits. Examples of a choke-supported interdigital line and a strapped-bar circuit are shown in Fig. 7.2.23b. Traveling-wave circuits are used mostly in amplifiers. Standing-wave circuits are resonant and used typically in magnetron oscillators. The most commonly used standing-wave circuits are composed of arrays of quarter-wave resonators that may or may not be strapped for improved oscillating-mode stability.

Variations of these circuits include *hole-and-slot resonators* and *rising-sun* anodes. Examples are shown in Fig. 7.2.23c. Cooling of vane structures for high average power is obtained by heat conduction along the vanes to the back wall of the anode to a heat sink, which may be liquid- or forced-air-cooled. Bar structures are cooled by passage of liquid coolant through the tubular bars of the slow-wave circuit.

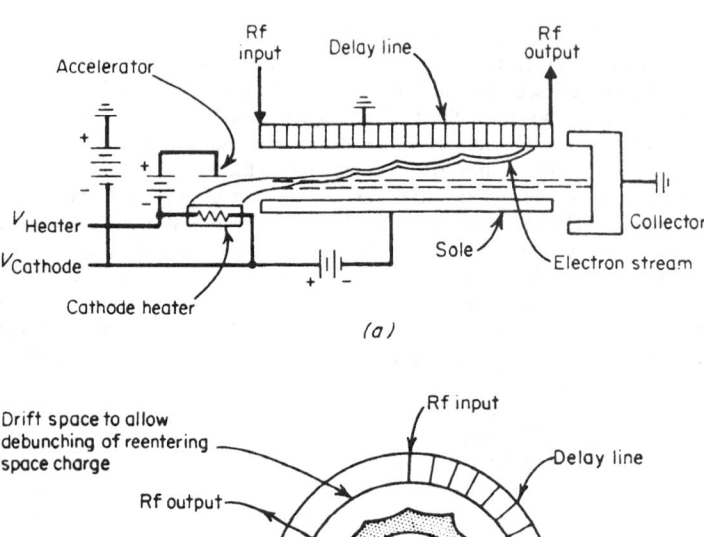

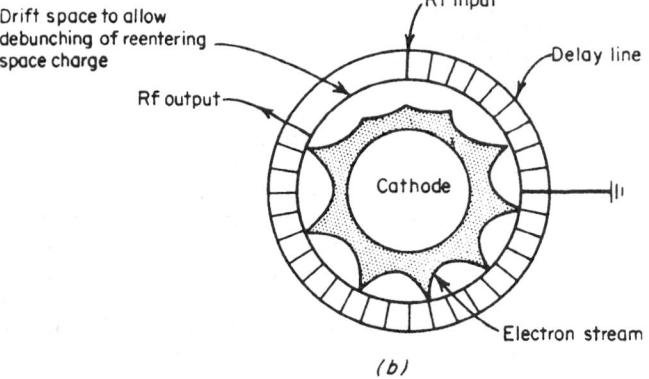

FIGURE 7.2.22   (*a*) Linear injected beam and (*b*) reentrant emitting-sole crossed-field amplifier.

During the time of intense research and development for crossed-field tubes, there were a large number of concepts under investigation. It was common to display a family tree of these to illustrate their similarities and differences. They were separated into two major groupings of injected beam and emitting sole devices. There were subgroups in each dependent on whether the device was an oscillator or amplifier, whether it used reentrant or nonreentrant electron streams and whether it used forward wave, backward wave or standing wave interactions with a slow-wave circuit. Many of these are no longer of general interest since some were not fully developed and others were made obsolete by newer technologies and/or by changes in performance requirements.

Magnetron oscillators are single-port devices. Both the slow-wave circuit and the electron stream are reentrant; i.e., the circular geometry is always used. Traveling-wave crossed-field oscillators are single-port devices but use a nonreentrant electron stream. They use either the linear or circular format.

Injected-beam and emitting-sole amplifiers are two-port devices with RF input and output ports. They are fabricated in both linear and circular format. Linear tubes must use a nonreentrant electron stream. Some circular-format amplifiers use a reentrant electron stream and some do not. Both forward-wave and backward-wave amplifiers have been developed.

Crossed-Field Oscillators

Conventional Magnetrons.   The conventional magnetron is an emitting-sole, circular-format, reentrant-stream device with electronic interaction between the circulating current and a π-mode, RF standing wave on the slow-wave circuit. Oscillation builds up from noise contained initially in thermionic-emission current from

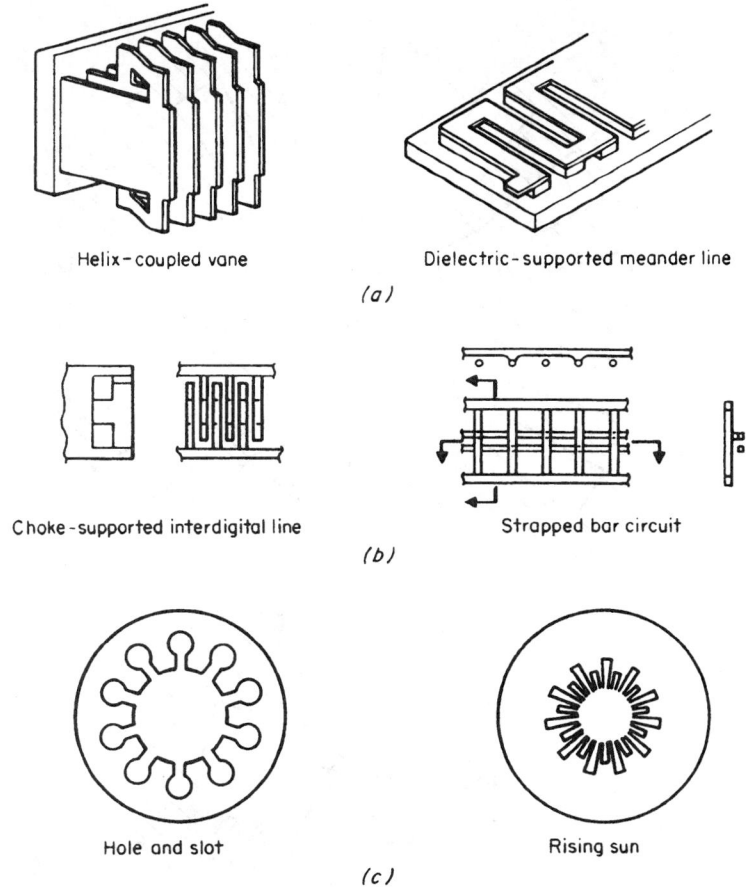

Helix–coupled vane

Dielectric–supported meander line

(a)

Choke–supported interdigital line

Strapped bar circuit

(b)

Hole and slot

Rising sun

(c)

FIGURE 7.2.23 Slow-wave circuits for crossed-field tubes.

a heated cathode. During operation interaction current is obtained from a circulating hub of space charge supplied primarily by secondary electron emission from the cathode surface. This is illustrated in Fig. 7.2.24. Large peak currents are obtainable, permitting the generation of high peak power at lower voltages than are used for beam tubes of comparable peak power.

Pulsed magnetrons have been developed covering frequency ranges from a few hundred megahertz to 100 GHz. Peak power from a few kilowatts to several megawatts has been obtained with typical overall efficiencies of 30 to 40 percent, depending on the power level and frequency range. Continuous-wave magnetrons have also been developed with power levels of a few hundred watts, in tunable tubes, at an efficiency of 30 percent. As much as 25 kW cw has been obtained for a 915-MHz fixed-frequency magnetron at efficiency greater than 70 percent.

Pulsed magnetrons are used primarily in radar applications as sources of high peak power. Low-power pulsed magnetrons find applications as beacons. Magnetrons operate electrically as a simple diode, and pulsed modulation is obtained by applying a negative rectangular voltage pulse to the cathode with the anode at ground potential. Voltage values are less critical than for beam tubes, and line-type modulators are often used to supply pulse electric power. Tunable cw magnetrons are used in electronic countermeasure applications. Fixed-frequency magnetrons are used as microwave heating sources.

Mechanical tuning of conventional magnetrons is accomplished by moving capacitive tuners, near the anode straps or capacitive regions of the quarter-wave resonators, or by inserting symmetrical arrays of

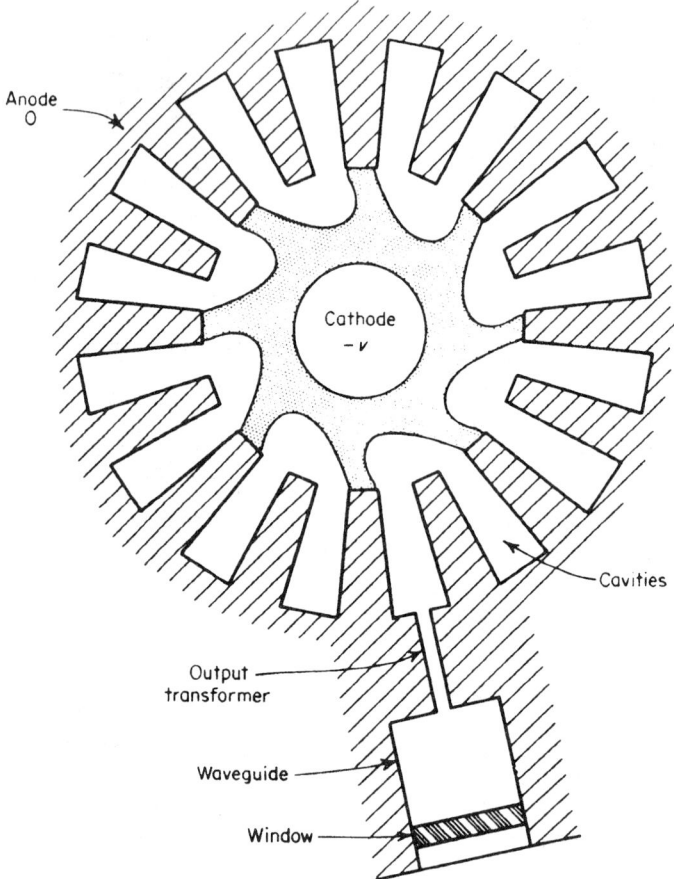

FIGURE 7.2.24   Conventional magnetron structure.

plungers into the inductive portions. Tuner motion is produced by a mechanical connection through flexible bellows in the vacuum wall. Tuning ranges of 10 to 12 percent bandwidth are obtained for pulsed tubes and as much as 20 percent for cw tubes.

Coaxial Magnetrons

The frequency stability of conventional magnetrons is affected by variations in the microwave load impedance (frequency pulling) and by cathode current fluctuations (frequency pushing). When the mode control becomes marginal, the tube may occasionally fail to produce a pulse. The coaxial magnetron minimizes these effects by using the anode geometry shown in Fig. 7.2.25. Alternative cavities are slotted to provide coupling to a surrounding coaxial cavity. π-mode operation of the vane structure provides in-phase currents at the coupling slots which excite the TE_{011} circular electric coaxial mode. The unique RF field pattern of the circular electric mode permits effective damping of all other cavity modes with little effect on the TE_{011} mode, and oscillation in other cavity modes is thereby prevented. Additional resistive damping is used adjacent to the slots but removed from the vanes to prevent oscillation in unwanted modes associated with RF energy stored in the vanes and slots that does not couple to the coaxial cavity.

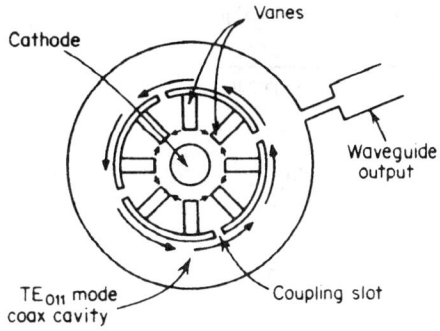

FIGURE 7.2.25   Coaxial magnetron coupling.

The oscillation frequency is controlled by the combined vane system and resonant cavity. Sufficient energy is stored in the TE_{011} cavity to provide a marked frequency-stabilizing effect on the oscillation frequency. Hence the coaxial magnetron is much less subject to frequency pushing and pulling than conventional magnetrons, and it exhibits fewer missed pulses. Tunable versions of this tube type are tuned by a movable end plate in the coaxial cavity similar to a tunable coaxial wavemeter. This is illustrated in Fig. 7.2.26. The larger resonant volume for energy storage leads to a slower buildup time for oscillation than in conventional magnetrons. This causes greater statistical variation in the starting time for oscillation (leading-edge jitter).

These factors are compared in Table 7.2.1 for the SFD-349 coaxial magnetron, which was designed as an improved retrofit for the 7008 conventional magnetron. The operating efficiency of the SFD-349 was deliberately degraded to meet retrofit requirements. Typically, coaxial magnetrons operate with an efficiency of 40 to 50 percent or higher.

Spurious Noise.   The circulating space charge in the hub of both conventional and coaxial magnetrons contains wide-band noise-frequency components that can couple to the output. In conventional magnetrons this spurious noise can couple directly to the output waveguide. Spurious noise power measured in a 1-MHz bandwidth is typically greater than 40 to 50 dB below the carrier. The coaxial cavity in the coaxial magnetron provides some isolation between the spurious noise coupled to the vanes and the output waveguide. The spurious-noise power from coaxial magnetrons is typically 10 to 20 dB lower than conventional magnetrons of comparable peak power level.

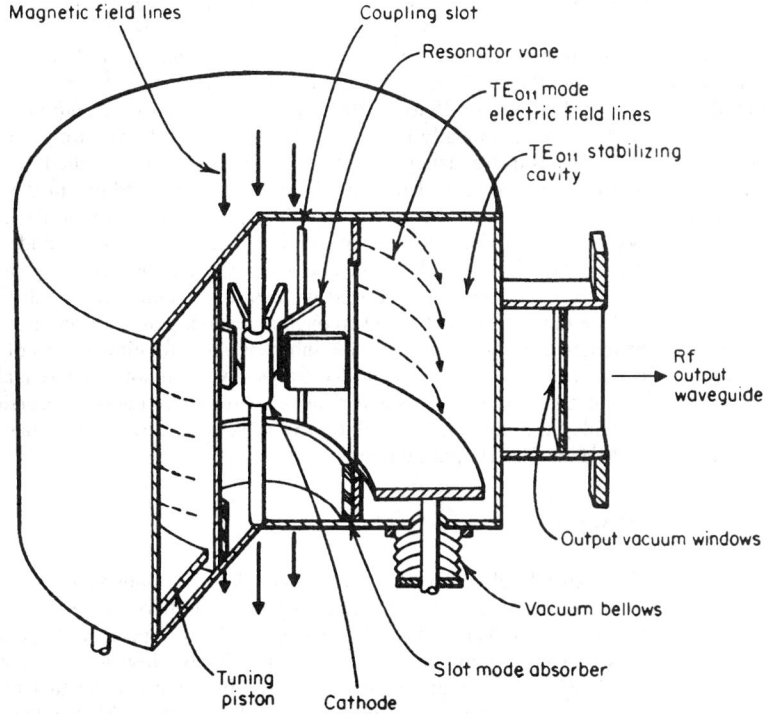

FIGURE 7.2.26   Schematic of coaxial magnetron.

TABLE 7.2.1 Comparison of the 7008 Magnetron and the SFD-349 Coaxial Magnetron

	7008	SFD-349
Efficiency, %	38	38
Leading-edge jitter, rms, ns	1.2	1.5
Pushing factor, kHz/A,	500	100
specified Typical	200	50
Pulling factor (VSWR 1.5),	15	5
MHz		
Spectra side lobes, dB	8–9	12–13
Missing pulses, %	1	0.01
Pulse-frequency jitter, rms, kHz	60	5
Life, h, specified	500	1,250
Typical	700–800	3,000–3,500

Frequency-Agile Magnetrons

To improve radar-signal detection and electronic countermeasures, rapid frequency-changing signal sources have been developed. Frequency-agile conventional magnetrons are available with rapidly rotating capacitive tuners (*spin-tuned magnetrons*) or hydraulic-driven, *mechanically tuned* tubes. The operational advantages of the coaxial magnetron are preserved in frequency-agile dither-tuned and gyro-tuned coaxial magnetrons.

Dither-tuned magnetrons use a mechanically tuned coaxial magnetron with an integral motor and *resolver* to provide high-speed, narrow-band frequency-agile operation. Mechanical linkage between the rotating motor and the tuning plunger provides approximately sinusoidal tuning of the magnetron frequency. Mechanical limitations imposed by acceleration forces determine the attainable tuning range and tuning rates. A voltage output from the resolver is made proportional to the magnetron frequency and is used to adjust the receiver local oscillator to track the rapidly tuned frequency of the magnetron. X-band tubes, 200 kW, with narrow-band dither-tuned frequency ranges of 30 to 50 MHz, are tuned at rates of 200 Hz. Wider-band frequency excursions of 250 to 500 MHz are dithered at rates of 25 to 40 Hz. Some of these tubes are equipped with servo motors for tuning. Frequency can be set electronically to provide rapid changes of fixed-frequency operation or can be dither-tuned with various shapes of frequency-tuning curves. These tubes are called *Accutune magnetrons*.

Gyro-tuned coaxial magnetrons use several rotating dielectric ceramic paddles in the stabilizing coaxial cavity, which cause frequency variation as they are rotated in a plane normal to the RF electric field of the TE_{011} mode. The anode vane system of the tube is surrounded by a ceramic cylinder bonded to the ends of the coaxial cavity to form the vacuum wall for the electronic interaction region. The stabilizing cavity, containing the tuning, is outside of this vacuum wall and is pressurized with sulfur hexafluoride, to inhibit arcing or corona caused by high RF fields. The tuner drive motor and frequency readout generator are also located within the pressurized section of the magnetron. The symmetry and inherently low rotational mass of the dielectric paddles result in a mechanism in which tuning speed and RF frequency excursion are essentially independent. It is therefore possible to attain higher tuning rates and relatively broader frequency excursion simultaneously than is achieved currently with dither tuning. K_u band, 60 kW peak power, gyro-tuned magnetrons obtain frequency excursions of 300 MHz at 200-Hz tuning rates.

Crossed-Field Amplifiers (CFAs)

Injected-Beam Types. Extensive development efforts have been devoted to nonreentrant forward-wave amplifiers (*TPOM*) for both cw and pulsed application. Continuous-wave amplifiers with a few hundred watts output have been developed with gain of 20 to 30 dB and with efficiency of 20 to 35 percent. Proper control of the electron stream at large values of gain requires the beam to be physically close to the anode at synchronism. Operation at low values of V/V_0 (3 to 6), together with increased insertion loss for tubes with greater circuit length for greater gain, leads to lower efficiency values. Like TWTs, the attainable bandwidth is dependent on the electron-beam optics and the dispersiveness of the slow-wave circuit. Half-octave bandwidth has been obtained at constant-voltage settings, and full-octave bandwidth has been obtained with adjustment of the anode-to-sole voltage. Stable operation at gain in excess of 20 dB requires circuit severs or distributed attenuation as in TWTs. Useful gain in excess of 30 dB is difficult because of excessive noise buildup in the electron stream.

Narrow-band (10 percent), pulsed, high-peak-power (5 MW) injected-beam amplifiers, which operate with efficiencies greater than 50 percent, were developed in France for use in radar application. These tubes have gain of 11 to 15 dB. The lower gain values (greater input signal level) permit the use of less critical beam optics, shorter slow-wave circuits, and greater ratios of V/V_0.

Injected beam CFAs were developed in the United States for electronic countermeasure applications, but the tubes were not widely deployed.

Emitting-Sole Crossed-Field Amplifiers

Electron current for emitting-sole crossed-field amplifiers is obtained from the sole electrode in the interaction space by electron-beam back bombardment, as in the magnetron oscillator. Unlike the magnetron, these amplifiers do not require thermionic emission to initiate current flow. Current flow can be started in an emitting-sole CFA by the admission of an RF signal to the input of the slow-wave circuit when the proper magnetic field and anode-cathode voltage are present in the interaction region. Amplifiers with RF-induced current flow are called *cold-cathode* amplifiers, regardless of cathode temperature, provided there is no thermionic emission from the cathode. In the absence of an RF input signal, these amplifiers remain quiescent even with full operational voltage applied. The details of the starting mechanism of RF-induced current flow are not fully understood, but the phenomenon is reliable in a properly designed amplifier.

Radio-frequency-induced current flow permits several modulation techniques for pulsed emitting-sole CFAs. These include cathode-pulsed CFAs, dc-operated CFAs with combination of dc voltage and a pulsed turn-off voltage, and dc-operated CFAs with only dc voltages applied. Cathode-pulsed amplifiers, as do magnetron oscillators, use pulse modulators to supply the required dc electric power during amplification. Direct-current-operated amplifiers obtain electric input power for amplification from a dc power supply. Electron current flow is initiated by an RF input signal and is terminated at the end of the RF input signal either by a voltage pulse or a dc bias voltage applied to a quench electrode.

Cold-cathode starting for cathode-pulsed amplifiers is assured by correct temporal alignment so that the RF input pulse bridges the cathode voltage pulse. The RF signal is present on the slow-wave circuit as the applied voltage pulse increases to synchronous value. Current flow is initiated, and amplification occurs during the voltage pulse and ceases upon removal.

Both forward-and backward-wave cathode-pulsed CFAs are available. They use the circular-formal reentrant-stream geometry illustrated in Fig. 7.2.27. A reentrant electron stream is advantageous because electrons that have delivered only part of their available potential energy to the circuit wave as they leave the interaction region can reenter for further participation, thereby leading to higher electronic conversion efficiency. (Overall amplifier efficiencies of 45 to 50 percent or more are common.) The leaving phase-sorted electrons contain RF modulation. Some reentrant-stream amplifiers use a circuit geometry with the RF input and output ports spatially separated by a sufficiently long circuit-free region (called the *drift space*) for internal space-charge forces to cause dispersal of the electron spokes. This removes the RF modulation from the reentrant stream while preserving the reentrant-stream efficiency advantage. Other reentrant backward-wave amplifiers (*Amplitrons*) use the modulated electron stream to create a regenerative amplifer. A minimal drift space is used, so that the electron spokes reenter the interaction region before the modulation is dispersed. By suitable design of the shorter drift-space dimensions, the modulated stream reenters with positive phase to enhance the interaction.

Greater electronic conversion efficiency (amplifier efficiency in excess of 70 percent has been obtained) can be obtained at the expense of lower gain-bandwidth product than can be obtained with amplifiers which remove the reentrant modulation. The use of regenerative amplification is not feasible with a forward-wave amplifier because, at a fixed operating voltage, the simultaneous regenerative gain at frequencies other than the drive signal could lead to unwanted auxiliary oscillations or selective peaks in spurious-noise output power. This is avoided with a backward-wave amplifier because the dispersive slow-wave circuits require different voltages to obtain adequate amplification of separated frequencies.

Spurious Noise.   Dispersal by space-charge forces of the nonphase-locked electrons in a drifting spoke can lead to a rapid buildup of broadband spurious-noise components in the reentering electron stream. This is

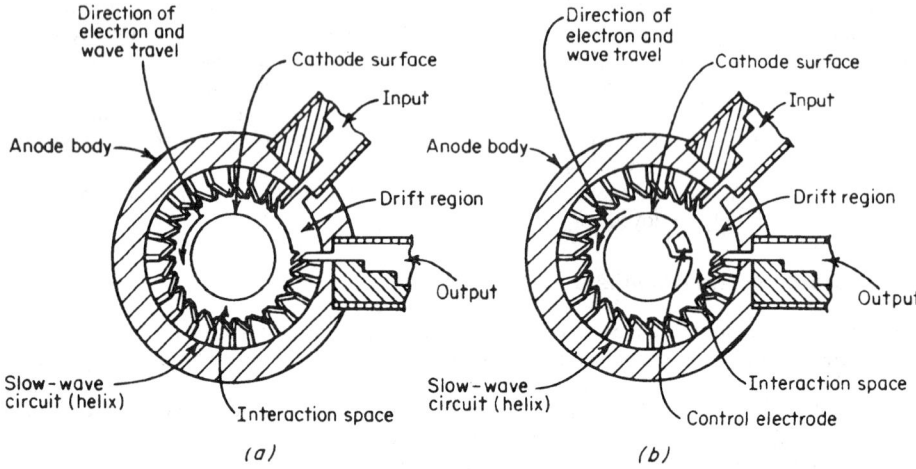

FIGURE 7.2.27   Diagrams of reentrant-stream crossed-field amplifiers: (*a*) cathode-pulsed; (*b*) with control electrode turn-off.

prevented from becoming severe by use of a sufficiently large RF input signal to lock out noise-signal growth. CFA power output as a function of RF drive signal is illustrated in Fig. 7.2.28. Reduction of the RF input signal leads to a reduced amplifier output signal at higher values of gain but is accompanied by an increased relative amount of broadband noise-power output. (Reentrant emitting-sole CFAs with terminated input have been used for high-efficiency broadband noise generators.) Adequate lockout of the noise power is obtained at reduced signal gain when the amplifier is driven well into saturation. The rapid growth of noise at small drive signals precludes the use of distributed attenuation and of circuit severs for emitting-sole amplifiers.

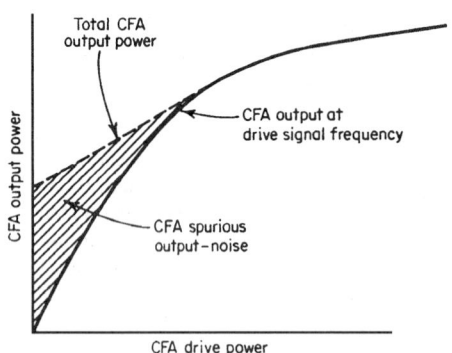

FIGURE 7.2.28   Emitting-sole crossed-field power output.

Relatively short slow-wave circuits are used with the minimum attainable insertion loss between the RF input and output connections. These amplifiers are called *transparent tubes*. Reflected signals from a load mismatch travel backward through the amplifier to the RF input, where they can be reflected, possibly leading to oscillation. Judicious use of ferrite isolators and circulators at the RF input and/or output minimizes this effect. The overall stable gain of emitting-sole CFAs is limited to 20 dB or less (typically 13 to 15 dB) because of the requirement for a sufficiently large RF input signal for adequate lockout of spurious-noise power and the need to limit gain to avoid oscillations caused by multiple reflected signals. Transparency is often used to advantage in radar systems employing the final amplifier in the transmitter chain in a nonoperating feedthrough mode to provide coarse programming of the output power level.

Bandwidth Characteristics.   Cathode-pulsed forward-wave amplifiers offer 10 to 15 percent instantaneous bandwidth at a fixed value of pulsed voltage. Backward-wave amplifiers provide only 1 to 2 percent instantaneous bandwidth under comparable conditions but can accommodate 10 percent bandwidth by adjustment of cathode voltage. The static impedance of both tube types varies as a function of frequency. The constant voltage versus frequency characteristics of forward-wave amplifiers is readily accommodated by a hard-tube cathode modulator, providing nearly constant power output across the frequency band of the amplifier. Constant power output versus frequency for a variable-voltage backward-wave amplifier is nearly achieved with a constant-current modulator. For restricted bandwidth (4 to 6 percent) this condition is approximated by using a line-type modulator.

Modulation Requirements.   A simplification in modulator requirements is obtained with a broadband, dc-operated, RF-triggered CFA. The termination of the RF input signal after RF turn-on leaves uncontrolled circulating space charge that can generate a cw spurious output signals of magnitude as large as half the amplified signal output. To avoid this, a *control electrode* (also called quench electrode) isolated from the cathode is mounted as part of the cathode structure and is located in the drift space. This location minimizes interference with amplifier performance (Fig. 7.2.27*b*). This electrode is pulsed positive with respect to the cathode, coincident in time with removal of the RF input signal. The circulating space charge in the interaction space is collected upon the control electrode, and the cathode current flow is terminated. Modulator requirements are simplified because the pulse voltage required for the turn-off electrode is typically one-quarter to two-thirds of the anode-cathode voltage and the peak collected current is less than one-half of the peak cathode-current flow during amplification. The duration of the collection time for the circulating current is approximately one transit time for electron flow around the interaction region. Typically, this is a small fraction of the time duration of the amplified signal. Consequently, the modulator energy required for the control electrode per amplifier pulse is much less than that required from the modulator for full-cathode-pulsed amplifiers.

Secondary Emission Cathodes.   A variety of materials are used for secondary-electron-emission cathodes in RF-triggered amplifiers. The selection is based on the amplifier drive signal level, peak power output, and the intended operating voltage for the tube. Materials used include pure metals, such as aluminum, beryllium, and platinum, as well as a variety of composite materials, such as dispenser cathodes and cermets. Dispenser cathodes and pure-platinum cathodes are suitable for drive signal levels in excess of 10 kW. Amplifiers with drive signals from a few hundred watts to 10 kW are better accommodated with metals supporting oxide surface layers such as aluminum and beryllium. Oxide layers are susceptible to erosion under electron-beam bombardment, and some CFAs employ a low-level background pressure of pure oxygen supplied from a suitable reservoir to rejuvenate and extend the active cathode life.

Frequency range.   Emitting-sole amplifiers have been developed at frequency ranges extending from VHF to K_u band, with experimental models at lower and higher frequencies. Examples of peak power levels available include 100 kW at K_u band, 1 MW at X band, and 3 MW at S band. Average power levels vary from 200 W at K_u band to several kilowatts at lower frequencies. Laboratory models have demonstrated as much as 400 kW of cw power at S band.

Noise Power.   Noise-power output from reentrant stream emitting role CFA's with a space charge dispersing drift space measured in a 1-MHz bandwidth far from the carrier frequency is typically greater than 35 dB below the carrier power level. Noise levels greater than 45 dB below the carrier are not uncommon. Many radar applications require very low noise power between spectral lines of a pulsed signal. Much effort has been expended in recent years for further reduction in noise power. Signal-to-noise ratios (S/N) close to the carrier frequency are now exceeding 50 dB$_c$/MHz. In some experimental tubes, S/N ratios of greater than 65 dB$_c$/MHz have been measured, but this performance is not now obtainable across a full operating band by production tubes. Phase locking of the space charge by the drive signal also minimizes phase variation due to voltage change and drive signal variation. Saturated amplifiers with 12 to 15 dB gain have output phase variations of 3 to 8° for a 1 percent change in anode/cathode voltage. A comparable phase change occurs with a 1 dB variation in drive-signal level.

Applications.   The primary use for emitting-sole CFAs is for transmitter tubes in coherent radars. CFAs have been used in pulse compression radars and pulse-coded and phased-coded radars, as well as phased-array radars. Lightweight low-voltage cathode-pulsed amplifiers are attractive for airborne applications. High-power dc-operated CFAs are used in ground-based radars.

BIBLIOGRAPHY

Collins, G. B., "Microwave Magnetrons," McGraw-Hill, 1948.

Okress, E., "Crossed-Field Microwave Devices," Vols. 1 and 2, Academic Press, 1961; "Microwave Power Engineering," Vol. 1, Academic Press, 1968.

Skowron, J. F., "The continuous-cathode (emitting-sole) crossed-field amplifier," *Proc. IEEE*, Vol. 61, Mar. 1973.

CYCLOTRON RESONANCE TUBES (GYROTRONS)

Howard R. Jory

Introduction

Gyrotrons or cyclotron resonance masers are a class of microwave generators that make use of the cyclotron resonance condition to couple energy from an electron beam to a high-frequency electromagnetic (em) field. This type of coupling allows the beam-wave interaction region to be large compared to a wavelength. As a result, gyrotrons can produce orders of magnitude higher power at a given frequency compared to other microwave devices without exceeding power density limits.

The basic equation for cyclotron resonance coupling to an em field stationary in space is given by

$$\omega = \frac{neB}{\gamma m_0}$$

where ω = operating frequency
B = applied dc magnetic field
e/m_0 = charge to mass ratio for the electron
γ = relativistic mass factor
n = an integer

The strongest coupling occurs for the fundamental resonance where $n = 1$. Harmonic interactions where n has larger values are generally progressively weaker unless some special boundary geometry is used to shape the em fields. Second harmonic coupling can be very effective and harmonic interactions as high as $n = 12$ have been demonstrated.

The coupling equation illustrates one of the limitations of gyrotrons in that the required magnetic field is proportional to the frequency of operation. Operation at 30 GHz with fundamental resonance requires a magnetic field of about $1T$, which is near the limit of practical magnets based on room temperature copper technology. Higher-frequency gyrotrons generally employ superconducting magnets that require a supply of liquid nitrogen and liquid helium, but negligible electrical power.

In its simplest form, the cyclotron resonance interaction requires only electrons making circular orbits in a plane perpendicular to a dc magnetic field, and a time-varying electric field also in a direction perpendicular to the magnetic field. In a practical embodiment, the electrons will also have a component of axial velocity and the electric field will have both transverse variations in amplitude and possibly phase. The cyclotron resonance interaction is quite flexible. It can be realized with a single cavity (gyrotron), multiple cavities (gyroklystron), or traveling waves (gyro TWT). It can use simple cylindrical cavities using any TE (transverse electric) modes or quasi-optical cavities formed by mirror reflectors.

Gyrotron Oscillators

Most gyrotron oscillators have been built with the configuration shown in Fig. 7.2.29. The cavity where the interaction with the electron beam takes place is cylindrical with some tapers or steps in diameter to control the axial variation of electric field amplitude. The diameter can range in size from 1 to 20 wavelengths. The axial length is typically in the range of 5 to 10 wavelengths.

In the simplest case, the microwave output from the cavity propagates axially through the beam collector and out through a window at the top of the gyrotron. Since all of the structures are large compared to a wavelength, many waveguide modes could propagate. However, it is generally important to have all of the output power contained in a single waveguide mode. Therefore, care must be taken in designing the tapers in diameter such that mode conversion does not occur. Generally it is possible to achieve an output where 90 to 95 percent of the power is in a single mode.

The beam collector area must be large enough to avoid excessive beam impact density. A typical design value for long pulse or CW operation is 1 kW/cm². Hence a collector for a 1-MW beam should have a beam impact area of 1000 cm² or more. Collector magnet coils are often used to help distribute the beam in the collector.

Desirable properties for the output window are low microwave loss, high strength, and high thermal conductivity. Typical materials used are alumina, sapphire, berrylia, and boron nitride. Pulsed gyrotrons generally use single disc windows cooled only at the edges of the disc. CW gyrotrons normally use two discs with a modest axial space between the discs through which a dielectric cooling fluid flows. In this way one face of each window disc is cooled, and the window can transmit much higher CW power without overheating.

The type of electron gun that has been most successful for gyrotrons is the magnetron injection gun. The desired electron beam is one where electrons have helical motion. In the cavity region the electrons should have a transverse to axial velocity ratio of the order of 2. Then about 80 percent of the beam kinetic energy is transverse motion, which can couple in the cyclotron resonance interaction. The axial velocity in this case serves only to determine the transit time of the electrons in the cavity and to keep electron space charge fields to a reasonable value. The magnetron injection gun, combined with magnetic compression of the beam between the cathode and the cavity, effectively provides a good beam for interaction.

Figure 7.2.30 shows a simulation of the electron motion in the gun as well as through the gyrotron to the collector. The cathode (emitting portion) of the gun is a section of a cone. The cathode and its associated support structures operate below ground potential at the full beam voltage desired (typically 20 to 80 kV). The simulation shows the case where the conical anode of the gun is at ground potential along with the body and collector. This provides a large radial component of electric field at the cathode which results in a component of transverse velocity for the electrons. The simulation shows a projection of the electron motion on the R-Z plane. The true motion of the electron is helical. Figure 7.2.29 shows a version of the gun which has an intermediate electrode operated at a potential between cathode and ground. This electrode can be used to control the electric field at the cathode and, therefore, control the ratio of transverse to axial velocity in the final beam at the cavity.

The gyrotron will oscillate when the main magnetic field is adjusted so that the beam cyclotron resonance frequency and a resonant frequency of the cavity are nearly equal and when the electron transverse-to-axial velocity ratio is high enough. Since a cavity that is large compared to a wavelength will have many resonances spaced perhaps a few percent apart in frequency, it is often possible to step tune from one frequency to the next by changing the main magnetic field value. This has been demonstrated at MIT with a pulsed gyrotron covering a 2 to 1 frequency range with steps about 5 percent apart in frequency. High power CW gyrotrons, however, are generally limited to single frequency operation because of bandwidth limitations imposed by double disc windows and collector tapers. As the main magnetic field is varied a few percent about a normal operating value the oscillator frequency will change slightly and the output power will vary. The frequency variation is related to the Q of the cavity and typically has a value of 0.1 percent. The output power variation can be of the order of 10 to 1.

Output power can also be varied over a range of the order of 10 to 1 by changing the magnetic field at the cathode or by changing the gun anode voltage, if the design includes an intermediate anode. Beam voltage can be used to change power output over a range of about 2 to 1. Beam current can be used to change power output by the order of 10 to 1 also. In the magnetron injection gun beam current is controlled by changing the cathode heating power. The cathode is operated in the mode where emitted current is limited by temperature.

Peak operating efficiencies for gyrotron oscillators are in the range of 30 to 50 percent. When output power is varied by changing beam current the variation of efficiency will be minimized. When one of the other means discussed above is used, the efficiency will be reduced proportionally to the power output.

The high-power levels of gyrotrons require the use of oversize waveguides at the output. Various modes have been used such as TE_{01} or TE_{02} or whispering gallery modes such as $TE_{12,1}$ or $TE_{15,2}$ or $TE_{22,2}$. Waveguide diameters, large compared to a wavelength, are required to avoid breakdown as well as excessive loss. Typical sizes are 1 to 4 in. in diameter. Systems have been built, which transmit power for 50 to 100 m length with losses as low as 5 percent.

As gyrotron frequencies and power levels increase, it becomes particularly advantageous to separate the output waveguide from the beam collector. Figure 7.2.31 shows a technique to accomplish this, which has been used in Russia for about 10 years, and is now also being used in other countries. From the electron gun up through the output taper there are no essential changes. Above that point, the wall of the waveguide is cut in an appropriate way to cause the microwave power to radiate in the direction shown by the arrows. Perturbations in the wall of the waveguide just before the cut can be used to better control the radiation pattern. The radiated power is handled by a number of mirrors to form a Gaussion-like beam which passes through a window on the side of the gyrotron. The electron beam is confined magnetically to pass through the output coupler region and then is allowed to expand to hit a conveniently large collector structure. The output microwave beam can be transmitted further using quasi-optical techniques or by injection into an HE_{11} type of low loss waveguide.

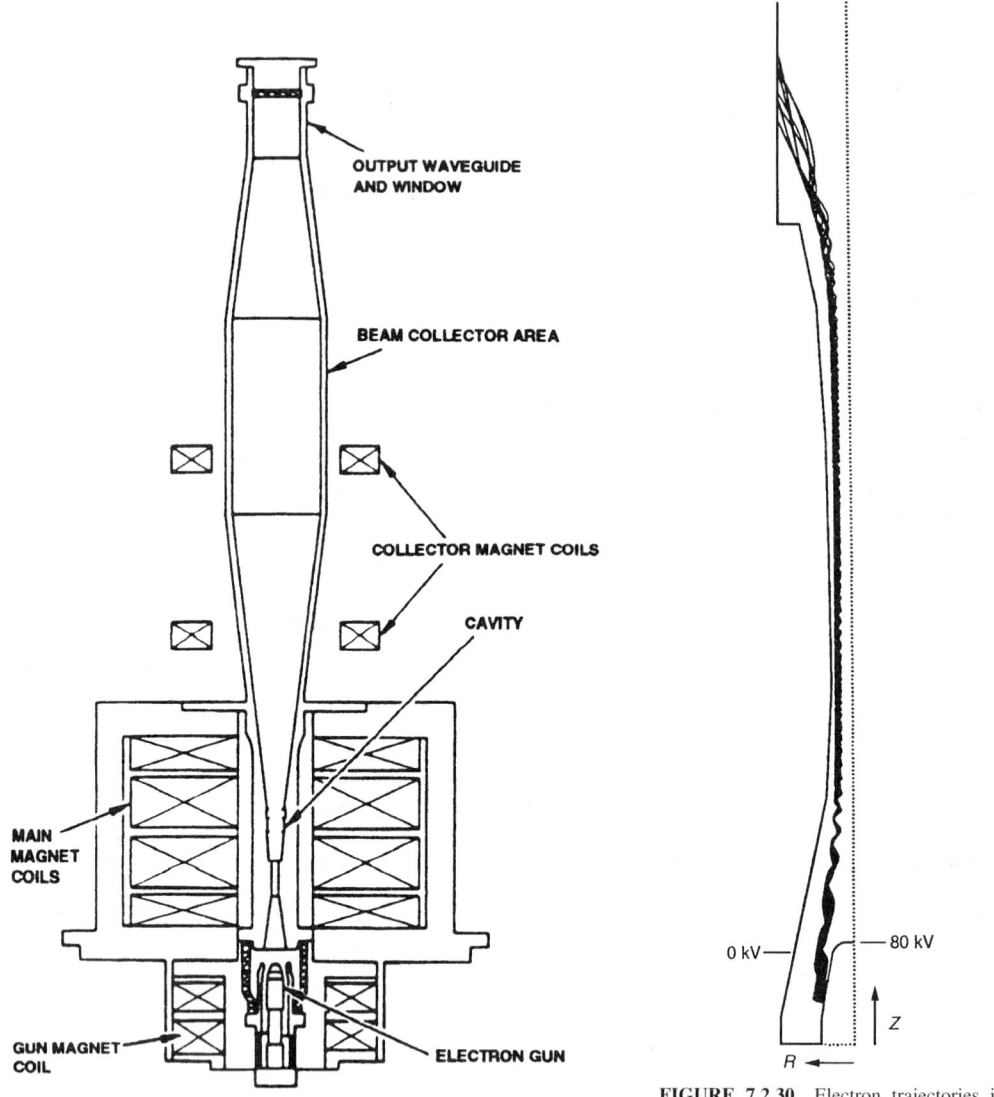

FIGURE 7.2.29 CW gyrotron with an axisymmetric RF output.

FIGURE 7.2.30 Electron trajectories in gyrotron.

Coaxial interaction cavities have produced good results recently as a means of achieving mode separation in very large cavities for higher power output. Coaxial modes such as the $TE_{28,16}$ have been used.

Gyrotrons for Fusion Applications

The gyrotrons used in connection with magnetic fusion have been mainly in the frequency range from 8 to 140 GHz. They have usually been operated under pulsed conditions with pulse lengths of 50 ms to 5 s and duty factors less than 1 percent. The measure of importance in this application is the output power in joules per pulse. Figure 7.2.32 shows the capability of various gyrotrons as a function of frequency. For each data point the first number gives the power output in MW and the second number the pulse length in seconds. Also given

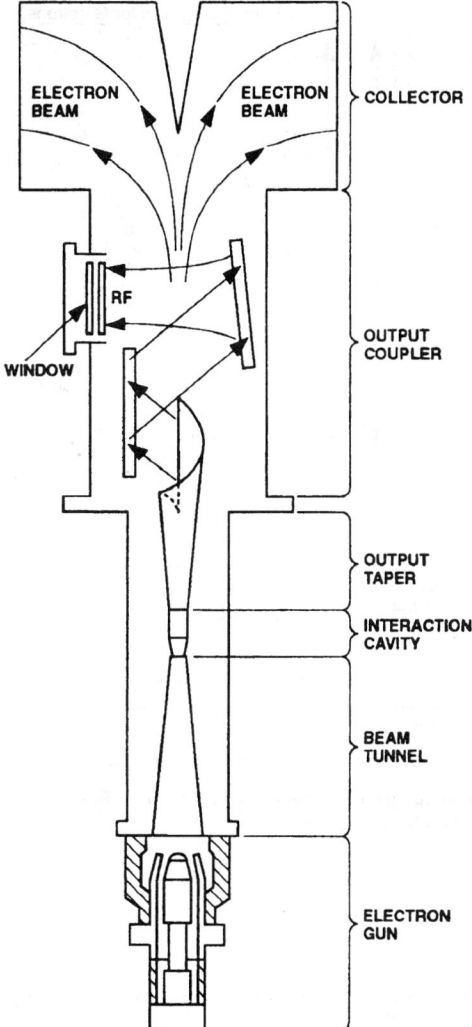

FIGURE 7.2.31   Gyrotron with coupler to separate RF output.

are the year of achievement and group involved. The dotted curves are for constant product of energy x frequency squared, which is a rough measure of technical difficulty.

Development programs are currently in progress in several countries to produce 1 MW at 110 GHz with pulse lengths of several seconds or CW. There is also development work for 1 MW at 170 GHz which corresponds to the current plan for the International Thermonuclear Experimental Reactor (ITER) project.

Other Gyrotrons

Gyrotrons have been built for operation at frequencies as high as 850 GHz with outputs at watt levels. Near 300 GHz, CW power of tens of watts has been produced and pulsed power of hundreds of kW. Some of these have been developed for plasma diagnostics or spectroscopy applications.

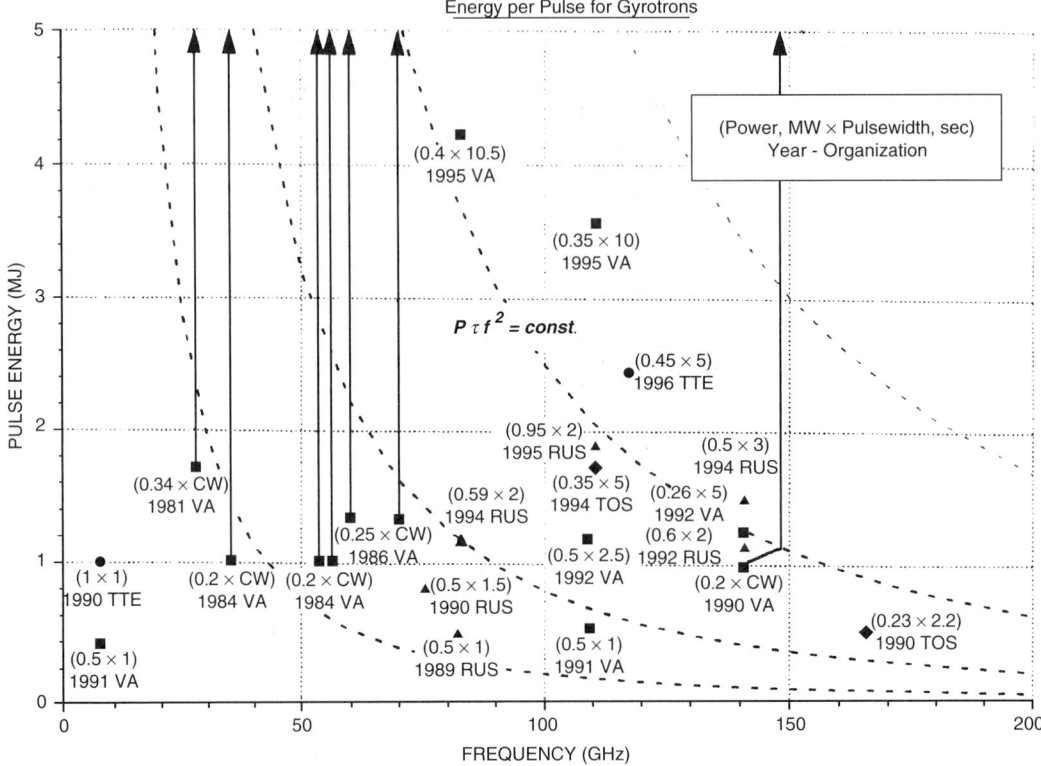

FIGURE 7.2.32   Energy per pulse for gyrotrons "VA = Varian/CPI, Rus = Russia, TTE = Thomson, TOS = Toshiba (data points for CW gyrotrons shown for 5-s pulse).

Gyrotrons using quasi-optical (confocal) resonators have been built near 100 GHz. Pulsed power outputs of hundreds of kW have been achieved but with efficiencies lower than the conventional cavity gyrotrons. Typical values are 10 to 15 percent.

Short-pulse, high-peak-power gyrotrons have been built based on cold-cathode flash x-ray technology. Outputs of hundreds of MW have been achieved at frequencies of 10 to 35 GHz. Lower power levels have been demonstrated up to 100 GHz. Typical pulse lengths are less than 100 ns. Beam voltages up to 1 MV and currents up to 10 kA have been employed. Typical efficiencies are 5 to 10 percent. The cold-cathode technology limits these devices to short pulse length and relatively short life compared to hot cathode devices. Repetitive pulsing has been demonstrated.

Gyroklystrons

The gyrotron can be made into an amplifier by using two or more cavities along the axis of the beam. The cavities are shorter to avoid self-oscillation. As in the conventional linear beam klystron, the RF input to the first cavity modulates the beam. But in this case it is an angular velocity modulation. As the beam drifts, the angular velocity modulation converts to bunching in angle. Intermediate cavities intensify the modulation and the output cavity extracts angular kinetic energy from the beam.

The most impressive results with gyroklystrons have been produced at the University of Maryland where 27 MW output at 9.85 GHz with μs pulse length was achieved. Beam voltage was 440 kV and efficiency 32 percent. This device was also modified to extract power with a second harmonic interaction in the output cavity. The results were 32 MW at 19.7 GHz with 29 percent efficiency.

A gyroklystron built by Varian demonstrated pulsed output of 75 kW at 28 GHz with 41 dB gain and 9 percent efficiency. Gyroklystrons have been built in Russia at XBand and other frequencies. For example, 60 kW pulsed output at 94 GHz with 34 percent efficiency and 40 dB gain is reported.

Major issues with gyroklystrons are stability and efficiency. The beam tunnels between cavities can be non-propagating for the normal cavity modes but will usually propagate for lower-order modes. Mode conversion and reflection back through the beam tunnel can easily result in spurious oscillation. This problem is generally controlled by careful cavity design and the inclusion of absorbing material in the beam tunnels.

Bandwidth of gyroklystron amplifiers is in the range of 0.1 to 1 percent because of the high cavity Q's required. Stagger tuning of cavities might allow small increases.

Gyro Traveling-Wave Tube

The gyro TWT is similar to the gyroklystron except that the resonant cavities are replaced by traveling-wave circuits. Stability in this case is even more difficult. In addition to spurious oscillations caused by reflections at the ends of each circuit, there are potential instabilities at the cutoff frequency of the traveling-wave circuit and with beam interactions with the backward circuit wave.

Nevertheless, reasonable results have been demonstrated for gyro TWTs. An experiment in Taiwan produced pulsed output of 25 kW at 35 GHz with 23 percent efficiency and gain of about 20 dB. Varian pulsed gyro TWTs have operated at 95 GHz with 30 kW output 16 dB gain, 8 percent efficiency and 2 percent bandwidth and at 5 GHz, 120 kW output, 16 dB gain, 26 percent efficiency, and 6 percent bandwidth.

BIBLIOGRAPHY

Benford, J., and J. Swegle, "High Power Microwaves," Artech House, 1992.

Felch, K., M. Blank, P. Borchard, T. S. Chu, J. Feinstein, H. R. Jong, T. A. Lorbeck, C. M. Loving, Y. M. Mizahara, J. M. Nelson, R. Schurmacher, and R. J. Temkin, "Long pulse and CW tests of a 110 GHz gyrotron with an internal, quasioptical converter," *IEEE Trans. Plasma Sci.*, Vol. 24, June 1996.

Felch, K., "Characteristics and applications of fast-wave gyrodevices," *Proc. IEEE*, Vol. 87, No. 5, May 1999.

Flyagin, V. A., and G. S. Nusinovich, "Gyrotron oscillators," *Proc. IEEE*, Vol. 76, No. 6, June 1988.

Granatstein, V. L., and W. Lawson, "Gyro-amplifiers as candidate RF drivers for TeV linear colliders," *IEEE Trans. Plasma Sci.*, Vol. 24, June 1996.

Granatstein, V. L., R. K. Parker, and C. M. Armstrong, "Scanning the technology: vacuum electronics at the dawn of the twenty-first century," *Proc. IEEE*, Vol. 87, No. 5, May 1999.

Graponov-Grekhov, A. V., and V. L. Granatstein, "Applications of High-Power Microwaves," Artech House, 1994.

IEEE Transactions on Plasma Science, Special Issue on High-Power Microwave Generation, Vol. 22, No. 5, October 1994 (also Vol. 20, No. 3, June 1992; Vol. 18, No. 3, June 1990; Vol. 16, No. 2, April 1988; and Vol. 13, No. 6, December 1985).

International Journal of Electronics, Special Issue on Gyrotrons, Vol. 72, Nos. 5a and b, May–June 1992 (also Vol. 65, No. 3, September 1988; also Vol. 64, No. 1, January 1988; Vol. 61, No. 6, December 1986; and earlier years).

Proceedings, 21st International Conference on Infrared and Millimeter Waves, SPIE, July 1996, Berlin, FRG (also earlier years).

Technical Digest, International Electron Devices Meeting (IEDM), December 1995, Washington, D.C. (also earlier years).

CHAPTER 7.3
MICROWAVE SEMICONDUCTOR DEVICES

George Bechtel, Joseph Feinstein

TYPES OF MICROWAVE SEMICONDUCTOR DEVICES

The number and variety of microwave semiconductor devices have greatly increased as new techniques, materials, and concepts have been applied. Silicon and compound semiconductor materials, such as gallium arsenide (GaAs) and indium phosphide (InP), are used for the fabrication of nearly all devices. The oldest structure is the tungsten-silicon *point-contact diode*, employing a metal-whisker contact, used for signal mixing and detection. The *Schottky barrier diode*, a rectifying metal-semiconductor junction, has supplanted the point contact because of its lower noise figure. These devices display a variable-resistance characteristic.

In contrast, the *variable-reactance (varactor) diode* makes use of the change in capacitance of a reverse-biased *pn* junction as a function of applied voltage. Physically, this capacitance change results from widening the depletion layer as the reverse-bias voltage is increased. By controlling the doping profile at the junction, the functional forms of this relation can be tailored to a specified application. Typical applications of varactor diodes are harmonic generation, parametric amplification, and electronic tuning.

pin diodes employ a wide intrinsic region which permits high-power-handling capability and offers an impedance at microwave frequencies controllable by a lower frequency (or dc) bias. They have proved useful for microwave switches, modulators, and protectors. In changing from reverse to forward bias the *pin* diode changes electrically from a small capacitance to a large conductance, approximating a short circuit.

For microwave power generation or amplification, a negative-resistance characteristic at microwave frequencies can be used. Beginning with the *tunnel diode* in the early 1960s and progressing to the higher-power *IMPATT diodes* and *Gunn diodes*, such negative-resistance devices experienced rapid development in the 1970s. The tunnel diode uses a heavily doped *pn* junction which is sufficiently abrupt for electrons to be able to tunnel through the potential barrier near zero applied voltage. Because this is a majority-carrier effect, the tunnel diode is very fast acting, permitting response in the millimeter-wave region. The very low power at which the tunnel diode saturates has limited its usefulness.

The *transferred-electron oscillator* (TEO), originally named for its discoverer J. B. Gunn, depends on a specific form of quantum-mechanical band structure for its negative resistance. This band structure is found in gallium arsenide, the semiconductor material generally associated with this class of device, and in a few other III-V compounds. *Gunn oscillators* have power output in the tens to hundreds of milliwatts at low dc operating voltage (9 to 28 V). They have a wide range of tunability and reasonably low AM and FM noise.

The *impact avalanche transit-time* (IMPATT) diode owes its negative resistance to the classical effects of phase shift introduced by the time lag between maximum field and maximum avalanche current multiplication and by the transit time through the device of this current. These effects can occur in all semiconductor *pn* junctions under sufficient reverse bias to initiate breakdown. While IMPATT diodes were originally developed as silicon devices (still the predominant type for millimeter waves), gallium arsenide IMPATTs are used because of their higher power (tens of watts). IMPATT diodes find applications in radar and millimeter wave transmitters.

Silicon bipolar transistors have penetrated the microwave region through the refinement of fabrication techniques based on the planar photolithography technology. By reducing the emitter width and base thickness to micrometer dimensions (and by paralleling stripes of structure to maintain power capability), transit times and charging times (resistance-capacitance product) can be kept low enough to provide useful devices with a maximum frequency of oscillation (f_{max}) of 22 GHz. Such transistors are also used in low-noise preamplifiers up to frequencies of about 4 GHz.

Heterojunction bipolar transistors (HBT) have been developed using GaAs and InP to take advantage of the superior electron transport in these materials. Heterojunction emitters using alloys of GaAs with ternary compound semiconductors such as aluminum gallium arsenide (AlGaAs) are used to improve the emitter injection efficiency for high current gain. Cutoff frequencies of 50 GHz have been achieved, about a factor of 2 better than silicon bipolar transistors.

Field-effect transistors (FETs) are microwave solid-state devices that use the higher mobility and saturated velocity of GaAs. The GaAs FET now dominates low-noise applications in the intermediate microwave region of the spectrum and has now taken over power-amplifier designs from the negative-resistance diodes discussed above. Noise figures less than 1 dB are obtained with low noise GaAs FETs at 10 GHz and below. Cutoff frequencies over 200 GHz have been achieved for new generation FETs, such as the *high electron mobility transistor* (HEMT). Power FETs have delivered power of 12 W at C-band for SATCOM amplifier applications.

Monolithic microwave integrated circuits (MMICs), using the monolithic integration of microwave transistors (both FET and bipolar) onto a single substrate, are now used in many high-frequency or high-speed applications. MMICs reduce the cost and size of RF and microwave components, in the same way as the use of silicon-integrated circuits have revolutionized computer and consumer electronics. GaAs and silicon MMICs provide circuit building blocks such as switches, amplifiers, oscillators, downconverters for such applications as cellular telephones, direct broadcast satellite TV receivers, and high-speed optical fiber communication systems.

Microwave Mixer and Detector Diodes

Mixing is defined as the conversion of a low-power-level signal from one frequency to another by combining it with a higher-power (local-oscillator) signal in a nonlinear device. In general, mixing produces a large number of sum and difference frequencies. Usually, the difference frequency between signal and local oscillator (the intermediate frequency) is of interest and is at a low-power level.

In *microwave detection,* solid-state devices are used as nonlinear elements to accomplish direct rectification of an applied RF signal. The sensitivity of a detector is usually much less than that of a superheterodyne receiver, but the detector circuit is simple and easy to adjust. The sensitivity has been improved considerably by the use of a wide-band low-noise microwave preamplifier preceding the detector.

Tunnel diodes, back diodes, point-contact diodes, and Schottky barrier diodes are majority-carrier devices that have nonlinear resistive characteristics and are useful for mixing and detecting. Noise figures lower than 6 dB are now possible at K_u band using Schottky barrier diodes. These improvements have resulted from better control of epitaxial material to achieve low series resistance and photolithographic techniques to achieve small-area Schottky diodes. Because of greater susecptibility to RF burnout, circuit complications, and fabrication difficulties, tunnel and back diodes have not found as wide acceptance as mixers and detectors at microwave frequencies. The point-contact (pressure contact between metal and semiconductor) and Schottky barrier diodes (formed by deposition of metal on semiconductor surface) are the primary mixer and detector microwave devices. However, back diodes are used in selected applications such as broadband low-level detectors and Doppler mixers.

Point-contact, Schottky barrier, and back diodes are used as mixers and detectors from UHF to millimeter frequencies. The construction and current-voltage characteristics of these diodes are shown in Fig. 7.3.1. The point is contact is fabricated by a metal whisker forming a rectifying junction in contact with the semiconductor. The Schottky is formed generally by an evaporated metal contact, and the back diode by an alloyed junction.

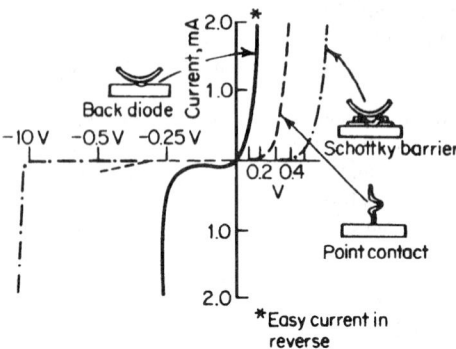

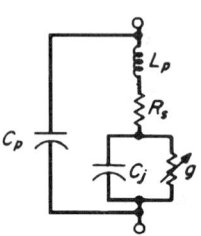

FIGURE 7.3.1   Current-voltage characteristics of back, point-contact, and Schottky barrier diodes.   (*Proc. IEEE, August 1971.*)

FIGURE 7.3.2   Equivalent circuit of microwave diode.

Point-Contact Diodes

The point-contact diode is the oldest structure. Until 1965 point-contact diodes were fabricated using moderately low resistivity material with the rectifying contact established by touching the semiconductor surface with a metal whisker (normally, tungsten for silicon and phosphorus bronze for germanium and GaAs).

Since that time the semiconductor epitaxial deposition has been applied to point-contact diodes (as well as to Schottky diodes) to maximize the frequency cutoff (minimize the $R_s C_j$ product). This is a significant consideration since conversion loss is directly proportional to the product of diode series resistance R_s and junction capacitance C_j, as shown in Fig. 7.3.2.

Such devices are generally used only in the millimeter-wavelength range because of inherent reproducibility problems and because they have poorer noise figures than Schottky diodes. Nevertheless they are more resistant to RF burnout than the Schottky devices.

Schottky Barrier Diodes

The Schottky barrier diode is a rectifying metal-semiconductor junction formed by plating, evaporating, or sputtering a variety of metals on *n*- or *p*-type semiconductor materials. Schottky diodes are fabricated from *n*-on-n^+ epitaxial material (*n* layer 0.5 to 1.0 μm thick), where the *n* layers are optimized in thickness and carrier concentration for minimum conversion loss and maximum RF burnout or power-handling capability. Because of their higher cutoff frequency, GaAs devices are preferred in applications above X-band frequencies. This results from the higher mobility of electrons in GaAs than in silicon. Although in practice this advantage is not as significant as predicted, a conversion-loss improvement of 0.5 dB at K_u band is readily obtainable with GaAs, compared with silicon.

Schottky diodes are fabricated by a planar technique. A SiO_2 layer (1 μm thick) is thermally grown or deposited on the semiconductor wafer, and windows are etched in the SiO_2 by photolithography techniques. Schottky junctions are formed by evaporation, sputtering, or plating techniques. Metal on the oxide is removed by a second photo step. Junction diameters as small as 5 μm are made by this technique. Schottky diodes are also fabricated with attached gold leads, using the beam lead process. After completion of the Schottky contact metallization step, thick gold contact leads are plated on the junction side of the wafer. The backside of the wafer is patterned and etched to form isolated diodes and to expose the gold beams on the frontside of the wafer. Figure 7.3.3*a* shows a silicon diode intended for operation at X band, and Fig. 7.3.3*b* shows a beam lead diode for millimeter waves.

Mixer-Diode Parameters

Figure 7.3.4 gives the noise figure for a variety of mixer diodes as a function of local oscillator power at 16 GHz. Below this frequency, silicon Schottky diodes exhibiting a minimum noise figure of 5.5 dB are generally used, while GaAs is preferred at higher frequencies.

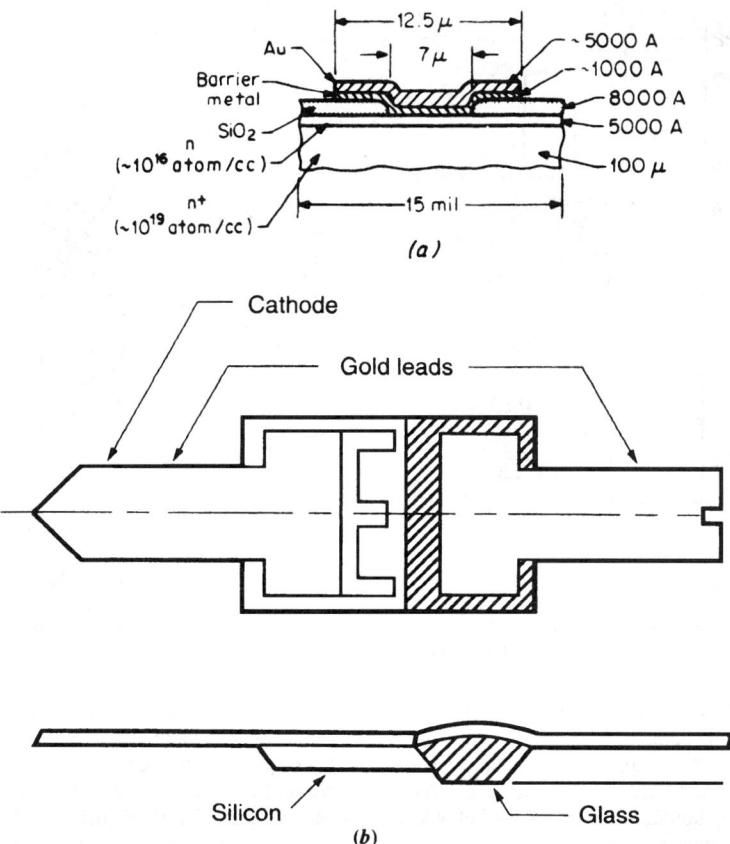

FIGURE 7.3.3 Microwave diodes: (*a*) silicon diode for operation at X band; (*b*) beam lead diode for stripline and microstrip circuits.

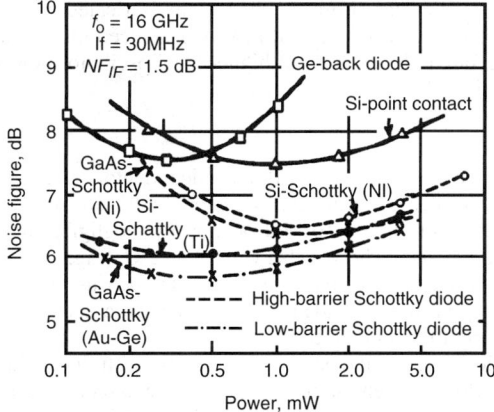

FIGURE 7.3.4 Noise figure vs. local oscillator power for point-contact diodes and *n*-type Schottky barrier diodes.

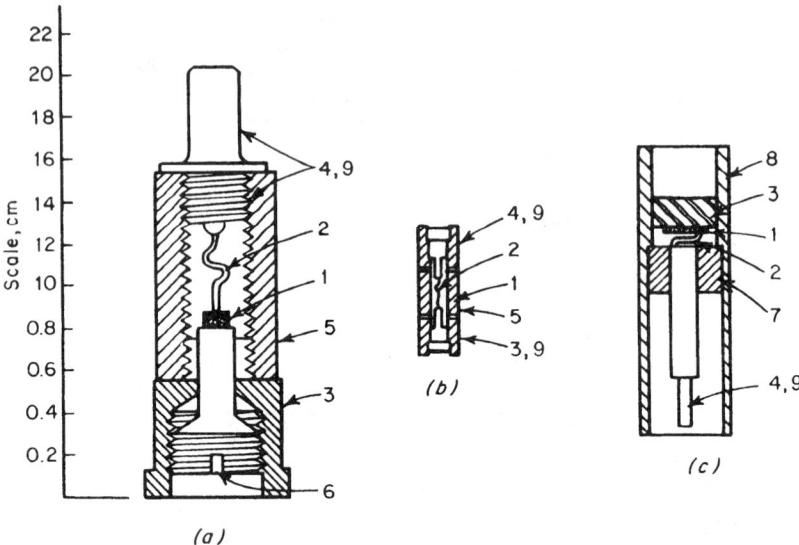

FIGURE 7.3.5 Mixer and detector diode packages: (1) semiconductor; (2) whisker; (3) wafer contact; (4) external whisker contact; (5) ceramic case; (6) adjustment screw; (7) insulating spacer; (8) outer conductor; (9) connection to i.f. amplifier or detector. (*After H. A. Watson, "Microwave Semiconductor Devices and Their Circuit Applications," McGraw-Hill, 1969*)

The *tangential signal sensitivity* (TSS) is the most widely used criterion for detector performance. It indicates the ability of the detector to detect a signal against a noise background and also includes the noise properties of the diode and video amplifier. This quantity varies with the square root of the amplifier bandwidth, since square-law response is obtained at low signal levels. As a rough basis of comparison, the TSS rating corresponds to a signal-to-noise of about 2.5. Biased silicon Schottky diodes have a typial TSS of −55 dBm at 10 GHz. For RF monitor applications, voltage sensitivity in microvolts per microwatt is important and a typical value of 7 mV/μW is achieved. Low barrier (or zero bias) detectors provide a higher sensitivity at higher video resistance.

Typical mixer- and detector-diode packages are shown in Fig. 7.3.5. They are designed to be compatible with a particular type of microwave circuitry, waveguide, coaxial, or strip line. The packages shown in Figs. 7.3.5*a* and *b* are generally used up to about 12 GHz. The coaxial package in Fig. 7.3.5*c* permits operation up to about 30 GHz. Stripline or microstrip mixer circuits generally use beam lead diodes, owing the planar lead configuration of the beam lead design. Multiple diodes are fabricated on the same chip, as monolithic pairs or quads for use in balanced or doubly balanced mixers. Balanced mixers provide noise and local oscillator cancellation at the IF port, but at the expense of higher local oscillator power.

Varactor Diodes

The active element of a varactor diode consists of a semiconductor wafer containing a *pn* junction of a well-defined geometry, usually formed by diffusion. Varactor diodes are normally operated under reverse bias, where the junction resistance (ordinarily 10 MΩ or more) is negligible in comparison with the microwave capacitive reactance of the junction. The variable capacitance C_j (Fig. 7.3.6) of the varactor diode as a function of applied voltage is used to tune microwave circuits such as voltage-controlled oscillators (VCOs). *Hyperabrupt junction* doping profiles provide a more rapid change in capacitance (and frequency tuning) as a function of voltage than abrupt junction diodes.

For forward bias the diode current increases exponentially with the applied voltage, and for reverse bias a small saturation current I_s flows. When the reverse bias is increased to the avalanche breakdown voltage V_B, the diode reverse current increases very rapidly, since it is limited only by the small diode resistance and any external resistance present in the circuit.

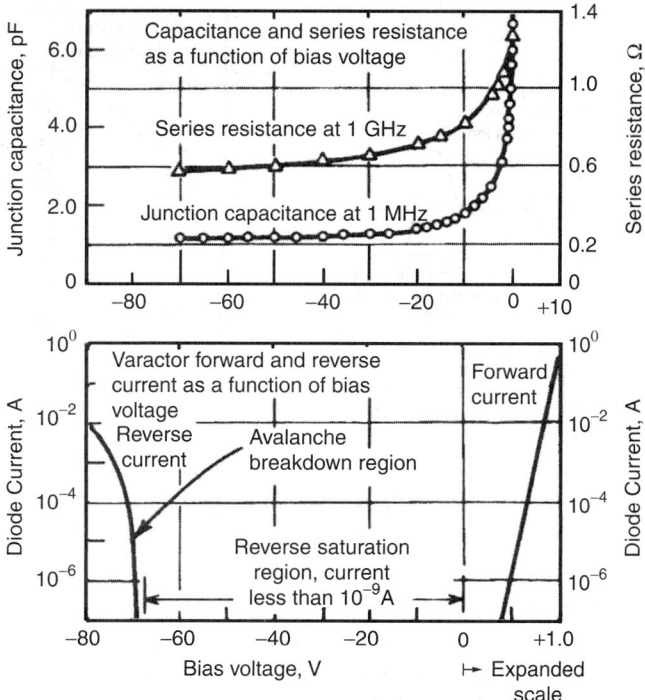

FIGURE 7.3.6 Varactor junction capacitance, series resistance, forward current, and reverse current as functions of bias voltage. (*After H. A. Watson, "Microwave Semiconductor Devices and Their Circuit Applications," McGraw-Hill, 1969*)

Figure 7.3.7 shows typical dc current-voltage characteristics, microwave series resistance, and 1-MHz capacitance-voltage characteristics.

Varactor Packages. Typical varactor diodes are mounted in hollow dielectric (usually alumina) cylinders with

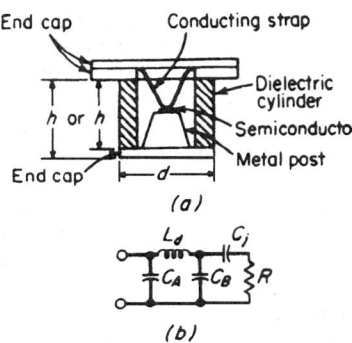

FIGURE 7.3.7 Low inductance diode construction and equivalent circuit. (*Microwave J., November 1970*).

Kovar or copper end caps, as shown in Fig. 7.3.7. A flexible connection is made to the diode mesa from one end cap by means of a thin gold strap. The equivalent circuit shown represents the coupling between the diode junction region and the package surface.

Charge-storage effects. In some applications the injected charge-storage capacitance is more important than the capacitance variation associated with the varying width of the depletion region. Charge-storage capacitance is produced by the injection of minority carriers during the forward-biased excursion of the varactor pump voltage and the withdrawal of this charge during the reverse-biased portion of the cycle. The resultant waveform is shown in Fig. 7.3.8. Efforts to maximize charge-storage effects in microwave diodes have led to a class of devices known as *snap-back*, or *step-recovery, diodes*. They feature steep doping profiles and narrow junctions, to give fast recovery of injected charge, typically in a transition period of a few tenths of a nanosecond, yielding a high-harmonic content. However, this design results in lower breakdown voltage, reducing the power capability of the

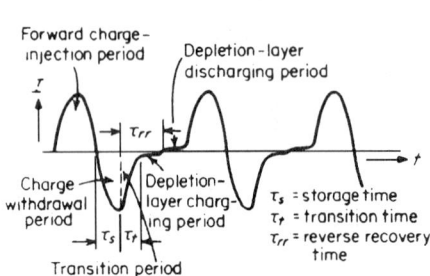

FIGURE 7.3.8 Current waveform of sinusoidally switched step-recovery diode. (*After H. A. Watson, "Microwave Semiconductor Devices and Their Circuit Applications," McGraw-Hill, 1969*)

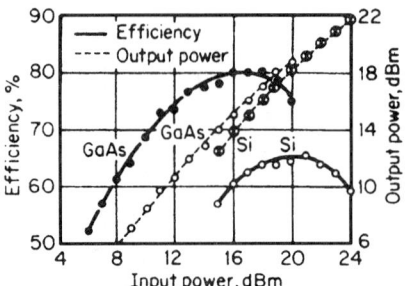

FIGURE 7.3.9 Conversion efficiency and output power of varactors used as frequency triplers, with bias adjusted to maximum output as each point. (*After H. A. Watson, "Microwave Semiconductor Devices and Their Circuit Applications," McGraw-Hill, 1969*)

diode. Frequency multipliers use these diodes when high-order multiplication (above eight) and circuit simplicity are desired, at the expense of power output.

Varactor Frequency Multipliers

Figure 7.3.9 gives the characteristics of diffused epitaxial GaAs and Si varactor diodes employed in a conventional tripler (4 to 12 GHz) frequency multiplier. Because of its higher cutoff frequency, GaAs gives higher efficiency but lower maximum output power than Si. Frequency triplers with output at 220 GHz have been fabricated for submillimeter wave sources.

pin Diodes

pin diodes consist of heavily doped p and n regions separated by a layer of high-resistivity intrinsic material. Typical construction of such a diode is shown in Fig. 7.3.10. Under zero and reverse bias this type of diode has a very high impedance, whereas at moderate forward current it has a very low impedance. This permits its use as a switch in microwave transmission lines. Generally, the diode is placed in shunt across a strip line, allowing unimpeded transmission when reverse-biased but short-circuiting the line to produce almost total reflection when forward-biased by as little as 1 V. In attenuator applications the diode behaves like a current-controlled resistance in parallel with the capacitance of the intrinsic region. For stripline or microstrip circuits, beam-leaded *pin* diodes similar to those of Fig. 7.3.3*b* are used, whereas the low-inductance package of Fig. 7.3.7 is used for high-power applications.

The wide intrinsic layer permits high microwave peak power to be controlled since the breakdown voltage is of the order of 1 kV. Very little power is dissipated by the diode itself because reflection switching is employed. *pin* diodes can be used as limiters, replacing TR tubes for peak powers smaller than 100 kW. At higher peak power, these diodes are useful, following the TR box to eliminate any spike leakage, although if fast response is required (less than 1 μs) a varactor diode is used.

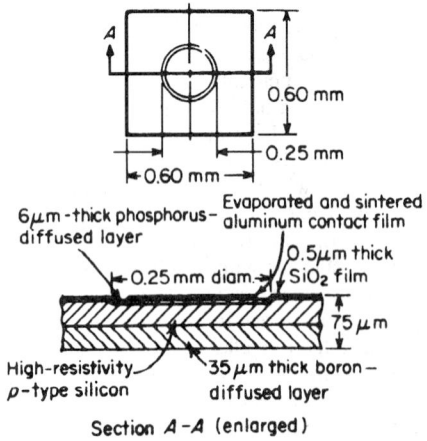

FIGURE 7.3.10 Typical planar *pin* microwave wafer.

Electrically controllable, rapid-acting microwave phase shifters are finding increasing use in phased-array systems. *pin* diodes are employed to switch lengths of transmission line, providing digital increments of phase in individual transmission paths, each capable of carrying many kilowatts of peak power.

Transferred-Electron (Gunn) Devices

The negative resistance that leads to oscillation for this class of device is a consequence of the band structure of certain semiconductors. An upper conduction band must exist in which carriers have lower mobility than in the initially occupied lower band. The transfer of carriers, generally electrons, from the lower to the upper conduction band takes place as a result of lattice collisions as the electric field strength across the material is increased.

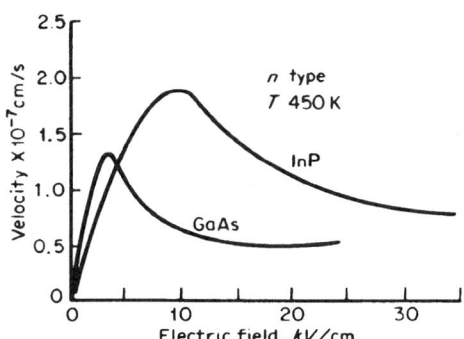

The transfer leads to a reduced current as the voltage increases and therefore represents a negative resistance on the *v-i* curve. The velocity-electric field characteristics of two materials that exhibit this behavior are shown in Fig. 7.3.11. By comparison, silicon has a monotonic characteristic which lies well below the curve for GaAs.

FIGURE 7.3.11 Velocity vs. electric field of indium phosphide and gallium arsenide.

Fabrication of Gunn Devices.   The fabrication of these devices requires stringent control over the material. Good electrical performance results from extremely pure and uniform material with a minimum of deep donor levels and traps and the use of very-low-loss contacts. Modern devices use an *n*-type epitaxial layer of GaAs or InP grown from the vapor phase on n^+ bulk material. Typical carrier concentrations range from 10^{14} to 10^{16} cm^{-3}, while device lengths range from a few to several hundred micrometers.

Noise.   One of the more important operating characteristics of the Gunn type of oscillator is its low-noise performance compared with other types of microwave sources at frequencies above 18 GHz. Phase noise for an InP Gunn diode oscillator is typically −87 dBc/Hz measured at 100 kHz offset from the oscillator frequency of 38 GHz.

Electronic Tuning.   Electronically tuned Gunn oscillators are available employing YIG spheres, varactor diodes, and *pin* diodes as the tuning element, in addition to mechanically tuned devices. Typical ranges covered are 8 to 12 and 12 to 18 GHz at the rather slow YIG magnetic-tuning speed of 100 Hz and of the order of 10 percent bandwidth at the much faster varactor rate of up to 10 MHz for oscillators up to 40 GHz.

Applications.   While Gunn diodes have been used in reflection amplifiers with a ferrite circulator to separate input from output, the advent of the GaAs FET, with its superior noise and power-amplification characteristics, has relegated the transferred-electron devices to oscillator (TEO) use.

The continuing development of indium phosphide (InP) Gunn devices has led to oscillators and reflection amplifiers at wavelengths in the 3- to 5-mm range, a region GaAs devices (TE or FET) cannot reach at present. The performance advantage is a consequence of the superior higher efficiency of InP devices. For example, InP Gunn diode oscillators have achieved a DC-to-RF conversion efficiency of nearly 3 percent at 140 GHz at an output power of 65 mW. Figure 7.3.12 displays the power and efficiency of cw InP Gunn diodes from 35 to 140 GHz.

Microwave Avalanche (IMPATT) Diodes

The power obtainable from IMPATT devices is greater than that available from the Gunn diodes described above but at the expense of higher noise and higher operating voltage. Two fundamental physical processes are pertinent to the operation of avalanche diodes: the drift velocity at which carriers travel under a reverse-biased electric field and the avalanche multiplication which occurs at sufficiently high fields.

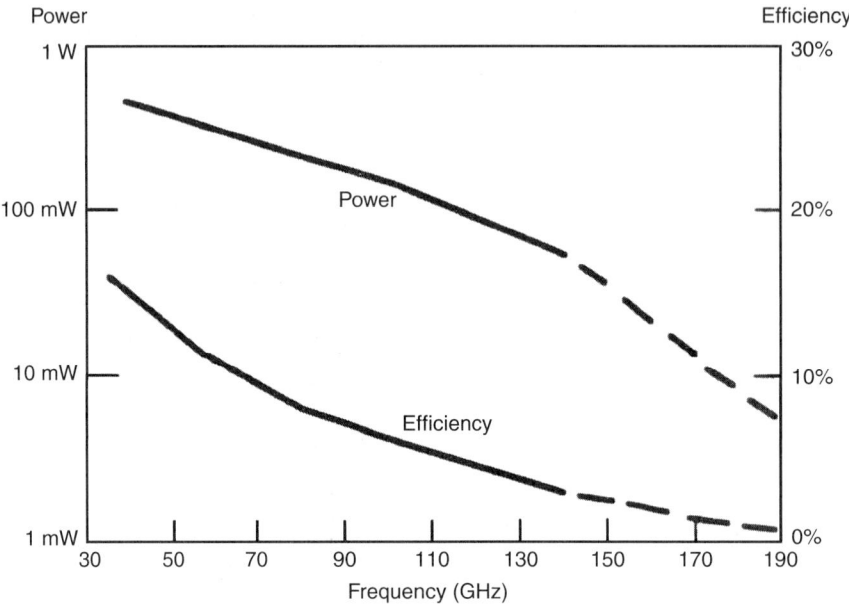

FIGURE 7.3.12   Power and efficiency of InP Gunn diodes.   (*Crowley, Litton Solid State*)

Theory of Operation.   An understanding of the dynamic operating characteristics of IMPATT diodes can be best obtained by considering the operation of the structure shown in Fig. 7.3.13. The IMPATT diode consists of two regions: a narrow avalanche region (*p* region), in which carrier multiplication by impact ionization occurs, and a drift region (*n* region), in which the carriers drift at saturated or field-independent velocities and where no impact ionization occurs.

The negative resistance or conductance of an IMPATT diode is attributed to phase shift between the current through the diode and the voltage across it. This phase shift consists of two components. There is a phase delay of the current caused by the avalanche multiplication process and by the finite transit time of the holes drifting through the drift region. If the diode is to operate as a stable oscillator, the negative conductance of the diode must decrease with increasing RF voltage. The RF voltage across the diode will grow until the admittance of the diode is balanced by the admittance of the microwave circuit.

At a sufficiently low frequency the phase delay of the transit time plus the phase lag of the avalanche process are not sufficient to have the fundamental component of the external current lag the RF voltage by more than 90°. Therefore, below a certain cutoff frequency, the conductance of the diode becomes positive.

At present the semiconductor materials used in commercial IMPATTs are silicon and gallium arsenide. The latter is operated in this case at much higher electric fields than corresponds to the region of negative differential mobility utilized in the Gunn effect. The basic structure of a typical *pn* junction silicon IMPATT is shown in Fig. 7.3.13. In operation the device is reverse-biased past the point of avalanche breakdown, so that a direct current of 50 to 500 mA flows through the diode.

Significant improvement in performance is obtained, however, by modifying the doping profile so that the

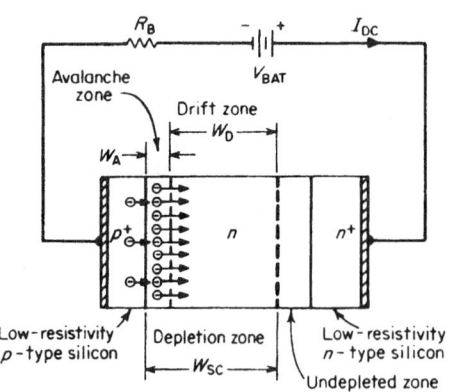

FIGURE 7.3.13   *pn* junction of IMPATT diode under reverse bias. (*After Cowley, WESCON, 1971*)

high-field avalanche region is narrowly confined and the electric field is optimized separately for the drift region. Such profiles, called high-low, low-high-low, and double-drift, yield efficiencies of 20 to 35 percent compared to 6 to 10 percent and 10 to 15 percent obtained from flat-profile Si and GaAs, respectively.

GaAs IMPATTs are preferred for frequencies up to 30 GHz because of their performance advantages, especially for pulsed radar transmitters at X band. Efficiencies of 20 percent are obtained at output powers of 15 W for low duty cycle RF waveforms. Multiple diodes are combined in an RF cavity to obtain 100 W. Si IMPATTs are used in the millimeter wave range and currently give the highest power available from a solid-state device in this frequency region.

Fabrication.   The IMPATT structure is fabricated by first growing a thin epitaxial layer of n-type silicon or GaAs on a heavily doped n-type (n^+) substrate and then adding a p layer by growing an epitaxial layer or by ion implantation. Finally, a thin platinum contact layer is formed by diffusion giving a p^+pnp^+ double-drift structure. For 94-GHz operation, the epitaxial-layer thickness is of the order of 0.5 μm, and the diameter of the diode is of the order of 50 μm.

Cooling.   A difficult technological problem in IMPATT diode packaging is efficient heat removal from the active portion of the device. These diodes operate at high dc power densities, typically 10 to 100 kW/cm², and since only a fraction (6 to 12 percent) is converted to RF power, the remainder must be removed as heat. Inverted mesa thermocompression bonding to a copper heat sink has been employed for this purpose. A better approach electroplates the heat sink at the wafer stage before individual diodes have been fabricated. Diamond has also been employed as a heat sink in experimental devices using inverted chips.

Microwave Bipolar Transistors

To achieve operation at microwave frequencies, individual transistor dimensions must be reduced to the micrometer range. To maintain current and power capability, various forms of internal paralleling on the chip are employed. These geometries fall into three general types, as shown in Fig. 7.3.14, interdigitated fingers forming emitter and base, overlay groupings of emitter and base stripes, and a mesh or matrix of emitter and base spots. All microwave transistors are now planar in form, and almost all are of the silicon *npn* type.

Construction and Fabrication.   Silicon microwave transistors are built up on an n-type epitaxial layer of the order of 1 Ω·cm resistivity (deposited on lower-resistivity silicon) to keep the collector depletion layer narrow. Diffusion of p-type dopant (typically boron) is used for the base. More recently ion implantation has replaced diffusion as the means of achieving base widths of less than 0.1 nm.

Following base formation, the emitter is then defined through photomasks, and the appropriate dopant (phosphorus or preferably arsenic) diffused in. Metal contacts are then evaporated or sputtered to interconnect the elements of the device.

Figure 7.3.15 illustrates the steps in the fabrication process. Reduction of strip width to increase frequency response has been made possible by improved lithography and etching, leading to state-of-the-art line definition of 1 nm.

Power Capability.   Most power microwave transistors include some form of integral emitter resistors to aid in equalizing the current over the distributed emitter structure.

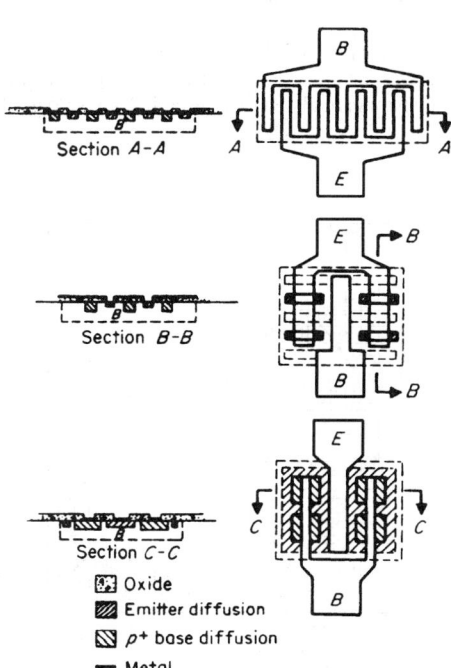

Oxide
Emitter diffusion
p^+ base diffusion
Metal

FIGURE   7.3.14 Typical geometrics of bipolar microwave transistors.

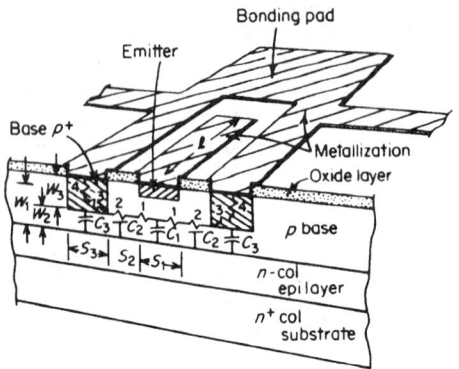

FIGURE 7.3.15   Cross section of planar transistor, showing (1)–(3) r'_b and (4) contact resistance R_C.   (*Proc. IEEE, August 1971*)

The *overlay transistor* has an integral diffused resistor as part of each emitter stripe, while *thin-film resistors* are deposited as part of the contacts on interdigitated devices. Guard rings are employed to raise the voltage breakdown limit of the collector to about half that of bulk silicon.

A figure of merit can be defined for the transistor in terms of the base resistance r'_b, the collector capacitance C, and the emitter-to-collector signal delay time τ_{ec}:

$$\text{(Power gain)}^{1/2}\text{(bandwidth)} \approx 1/4\pi(r'_b C\tau_{ec})^{1/2}$$

τ_{ec} is composed of the transit time of carriers across the base and collector depletion layer plus the charging time of the emitter-base junction capacitance and the collector capacitance. The maximum frequency of oscillation is obtained by setting the gain equal to unity in this expression.

Silicon bipolar transistors provide the highest pulsed output power of any solid-state microwave or RF device, up to 500 W at collector voltages of 50 VDC. In linear amplifier operation, bipolar transistors are used in personal communications network (PCN) and cellular base stations with output power capability of 30 W at 1.9 GHz. Figure 7.3.16 illustrates the output power capability of bipolar transistors from 900 MHz to 20 GHz. The projected maximum frequency of oscillation for silicon bipolar transistors is 22 GHz.

Noise.   The principal sources of noise in a microwave transistor are shot noise associated with the emitter and collector current and thermal noise in the base resistance. It is an increase in the latter and reduction in current gain which are responsible for the deterioration of noise figure with frequency. The current state of the art in the bipolar-transistor noise figure is about 2 dB at 4 GHz, increasing to 5 dB at 6 GHz.

Because of the high current gain at low currents, silicon bipolar transistors are used as amplifiers and oscillators in many portable applications up to 2.5 GHz. The transistors are also widely used in VCOs up to 20 GHz, because of their low $1/f$ noise.

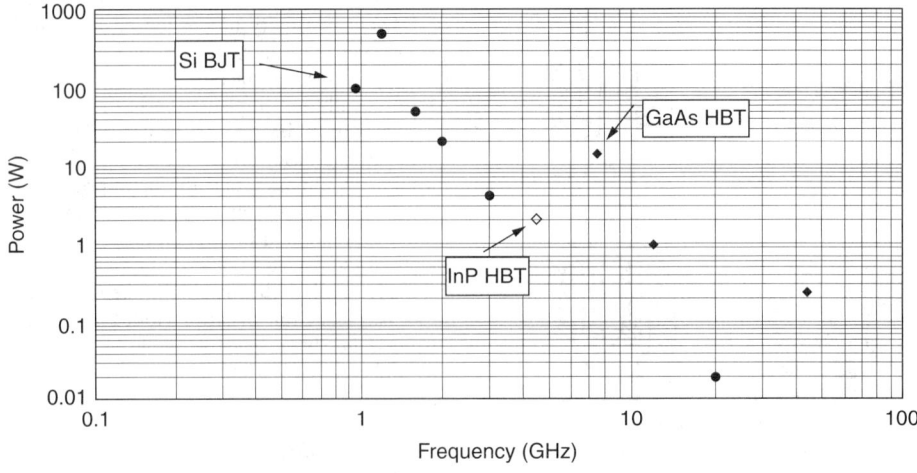

FIGURE 7.3.16   Performance of microwave power transistors as of 1979 and projected improvement.   (*Trans. IEEE, May 1979, Vol. MTT 27, No. 5*)

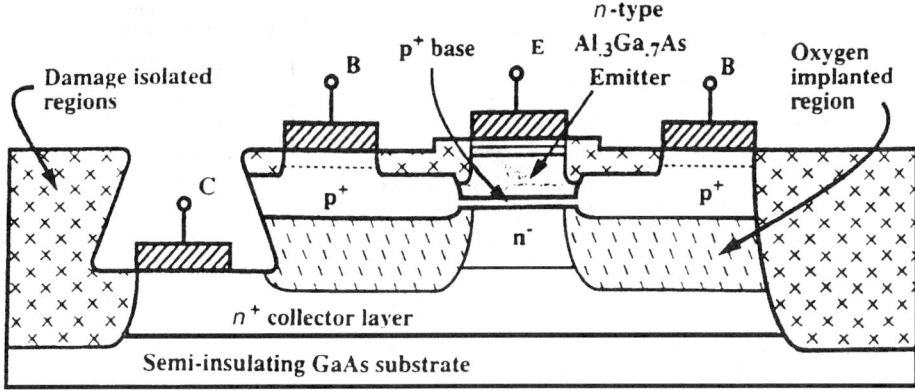

FIGURE 7.3.17   Cross section of GaAs HBT, showing AlGaAs emitter and low collector.

Heterojunction Bipolar Transistors

Heterojunction bipolar transistors (HBTs) fabricated on GaAs take advantage of the high electron mobility for electron transport across the base of the device. A cross section of an HBT is shown in Fig. 7.3.17. The emitter of highly doped, n-type AlGaAs forms a heterojunction with the highly doped GaAs p-type base (called the heterojunction due to different crystalline composition), resulting in favorable electron injection into the base. DC current gains are typically 20 to 200, with cutoff frequencies above 50 GHz. The devices are fabricated with all-epitaxial growth by means of molecular beam epitaxy (MBE) or metallo-organic chemical vapor deposition (MOCVD). HBT base and collector layers are similar in thickness to those for silicon bipolar transistors, but the emitter width can be wider (1 to 3 μm) because of the faster base transit time and lower capacitance per unit area. To achieve high gain, oxygen implantation is used to reduce the base-collector capacitance under the base contact.

The unity current gain figure-of-merit, f_T, and the maximum frequency of oscillation for GaAs HBTs range from 30 to 60 GHz, exceeding the same silicon bipolar transistor figures of merit by over two times. The power capability of the GaAs HBT from 7.5 to 44 GHz is compared to the silicon bipolar transistor in Fig. 7.3.16. Output power of over 10 W has been achieved at 7.5 GHz with 40 percent power-added efficiency.

HBT devices have also been fabricated on InP substrates, using an InAlAs emitter and InGaAs base, with f_{max} over 100 GHz. Silicon-based HBTs, using a silicon-germanium alloy base layer, have also been fabricated, with figures-of-merit over twice those of silicon homojunction devices. SiGe HBTs offer the promise of high performance with the lower cost of silicon processing, compared to GaAs and InP devices.

Field-Effect Transistors

The schematic in Fig. 7.3.18 illustrates the field effect transistor principle of operation. The flow of charge carriers from source to drain is controlled by the potential applied to the gate electrode. This flow takes place in a thin layer of n-type GaAs, called the channel. The gate is generally a Schottky barrier formed by deposition of an appropriate metal, leading to the acronym MESFET. The channel of the MESFET is fabricated by either epitaxy (MBE, MOCVD, VPE) or ion-implantation. The latter is more commonly used for general-purpose MESFETs, while epitaxy is used for low-noise devices.

The electron current from source to drain is determined by the depth of the depletion layer formed under the gate by the negative potential applied to it. The frequency response of this type of amplifier goes up as the gate length is reduced, while its power output increases linearly with gate width. Typical gate lengths range from $1/4$ to 1 μm, while gate widths vary from a fraction of a millimeter for low-noise FETs to tens of millimeters for power devices. About 0.5 W of microwave power per millimeter is a widely used figure of merit.

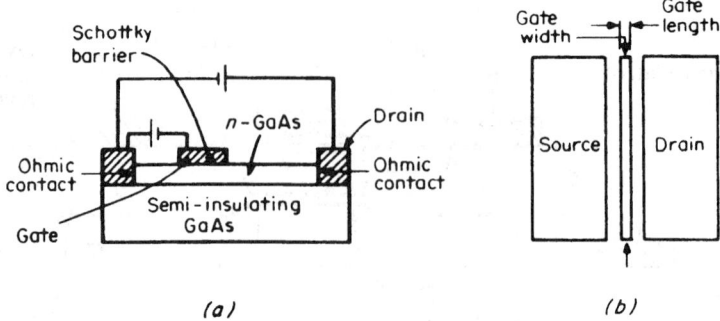

FIGURE 7.3.18   (*a*) Schematic and (*b*) dimensions of a single-gate Schottky barrier GaAs field-effect transistor.   (*Microwave J., November 1978*)

Noise of the MESFET.   The GaAs MESFET achieves its low-noise figure from the thermal noise characteristic of the majority carrier (electron) flow in the channel. The typical noise figure of a 0.3-μm gate length MESFET is shown in Fig. 7.3.19. The MESFET is 1.4 dB lower in noise figure at 4 GHz (0.6 vs. 2 dB) with over 14 dB gain, as compared to the silicon bipolar, and is widely used in low-noise amplifiers from 1 to 20 GHz. However, the heterojunction FET is now displacing the MESFET as a low-noise amplifier because of even lower noise figure.

The heterojunction FET has evolved from the earlier high electron mobility transistor (HEMT) to an improved (for lower noise and higher gain) epitaxial layer design called pseudomorphie HEMT (PHEMT). The noise figure of a 0.3-μm gate length PHEMT is also shown in Fig. 7.3.19; note that the noise figure is reduced by 0.7 dB over the MESFET noise figure at 12 GHz. Further improvements in gain have been achieved with the use of 0.1 μm gates and InP epitaxial structures, where f_{max} has exceeded 250 GHz.

Power of the MESFET.   A parallel interconnection of MESFET cells is used to increase the power handling capability in accordance with the power figure of merit of 0.5 W/mm of gate width. The number of gates is chosen on the basis of the cell gate length and width; typical cell dimensions are gate length of 0.5 μm by 125 μm width by 12 gates per cell. To achieve 25 W at 5 GHz, four chips of 12 cells each are connected to

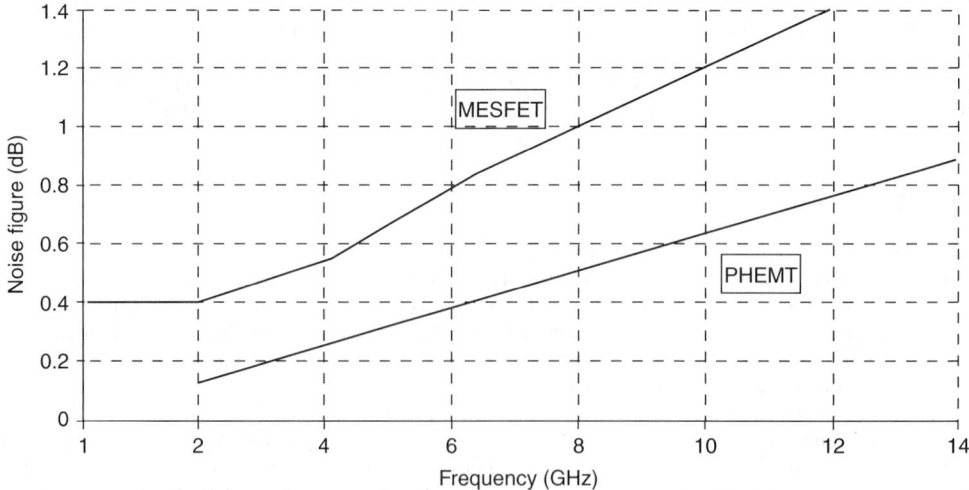

FIGURE 7.3.19   Noise figure of MESFET and PHEMT devices.

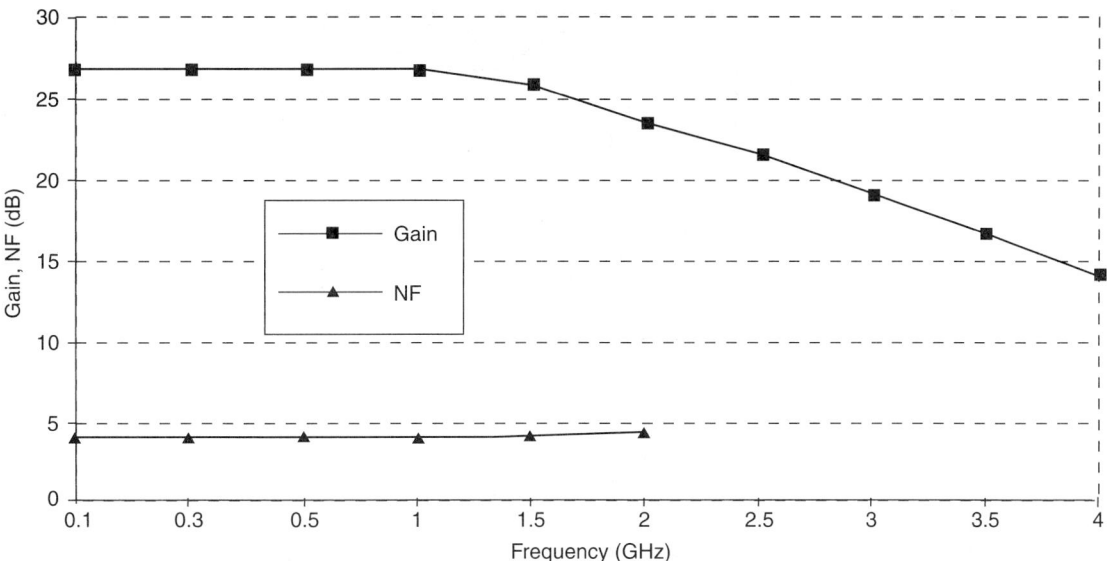

FIGURE 7.3.20   Gain and noise performance of silicon bipolar RFIC.

yield a total width of 72 mm. The chips are interconnected in a single package, using matching networks internal to the transistor package.

Silicon Bipolar RF and Microwave Integrated Circuits

Integrated circuits using high f_T silicon bipolar transistors are available with complexity ranging from simple single-stage amplifiers to functional blocks, such as vector modulators and downconverters. Resistors and capacitors are integrated onto the same semiconductor chip with a number of transistors and diodes to form microwave integrated circuits for applications from 100 MHz to 3 GHz. The RF gain and noise figure of a typical silicon RFIC, using feedback around a Darlington transistor to obtain a gain greater than 20 dB up to 3 GHz, are shown in Fig. 7.3.20.

More complex circuits have been developed that provide nearly the entire front end of a global positioning satellite receiver, comprising a dual downconversion design. Silicon bipolar integrated circuits have also been designed for new generation cellular telephone transmitters and receivers. Complementary MOS transistors have been combined with bipolar transistors on the same chip (using the *BiCMOS* process) to implement advanced functions for low-power portable telephones.

GaAs Monolithic Microwave Integrated Circuits

Monolithic microwave integrated circuits (MMICs) using GaAs MESFETs as the active elements and monolithically integrated with matching circuits have replaced the discrete circuit implementation in many RF and microwave designs. GaAs MMICs are available for applications from 500 MHz to 40 GHz, with developmental devices having gain to 100 GHz.

The fabrication of GaAs MMICs is facilitated by the use of a semi-insulating substrate as used for MESFETs, permitting the placement of microstrip matching elements and other passive components on the same semiconductor chip with minimum parasitic capacitance and loss. The functional diagram of an amplifier/switch MMIC for 2.4 to 2.5 GHz transmit/receive function is shown in Fig. 7.3.21. The receive path contains

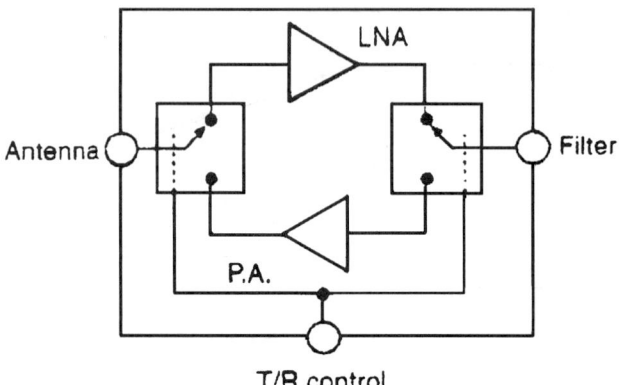

FIGURE 7.3.21 Functional diagram for 2.4 GHz GaAs transceiver. (TriQuint Semiconductor)

a high gain, low-noise amplifier, internally matched to 50 Ω at both ports. The transmit path contains an internally matched, class-A, medium power amplifier. The circuit also contains fully integrated T/R switches at input and output. The use of GaAs MESFETs as amplifiers provide both low-noise figure (3.5 dB) as well as good linearity for the transmitter in applications such as spread spectrum transceivers.

For millimeter wave amplifiers MESFETs can be replaced by PHEMTs, which have higher gain. A broadband power amplifier MMIC has been fabricated with a nominal gain of 13 dB and output power of +21 dBm from 20 to 50 GHz.

Other GaAs MMICs have been developed with greater complexity that provide an entire downconverter for satellite TV receivers at 12 GHz. Low-power radar transceiver MMICs at 94 GHz have been demonstrated in the laboratory that may provide collision warning sensors for automobiles by the year 2000.

BIBLIOGRAPHY

Bahl, I., and P. Bhartia, "Microwave Solid State Circuit Design," Wiley, 1988.

Chang, C.Y., and F. Kai, "GaAs High-Speed Devices: Physics, Technology and Circuit Applications," Wiley, 1994.

Chen, L., V. K. Varadan, V. V. Varadan, C. K. Ong, and C. P. Neo, "Microwave Electronics: Measurement and Materials Characterisation," Wiley Europe, 2004.

Freund, H. P., and G. R. Neil, "Free-electron lasers: vacuum electronic generators of coherent radiation," *Proc. IEEE*, Vol. 87, No. 5, 1999.

Hartnagel, H., R. Katilius, and A. Matulionis, "Microwave Noise in Semiconductor Devices," Wiley Europe, 2001.

Jalali, B., and S. J. Pearton (eds.), "InP HBTs: Growth, Processing and Applications," Artech House, 1995.

Shur, M., "GaAs Devices and Circuits," Plenum, 1987.

Sitch, J. E., "Microwave semiconductor devices," *Rep. Prog. Phys.*, March 1985.

Sze, S. M. (ed.), "High Speed Semiconductor Devices," Wiley, 1990.

Quantum electronics special issue on free-electron lasers, *IEEE J.* Vol. 23, Sep. 1987.

SECTION 8

TRANSDUCERS AND SENSORS

Measurements by direct comparison with a reference standard having the same characteristics as those of the quantity measured are called *direct measurements*, but most measurements yield results in a more indirect way. They are based on knowledge of the relationship between the quantity to be measured and the response of the measuring instrument or system influenced by it. Moving-coil meters are an example: under the influence of the applied voltage or current (or both) a mechanical torque is generated. The pointer attached to the moving coil acquires its final position indicative of the electrical quantity measured, the mechanical torque being balanced by a spring. Thus an electrical quantity is first translated into a torque, which in turn is translated into a position on a scale calibrated in units of the quantity measured.

In this example, the electrical quantities are translated into mechanical ones. In contrast, *transducers* translate primarily nonelectrical quantities into electric signals.

The term transducer has been applied to a variety of devices, including measuring instruments, acoustic-energy transmitters, signal converters, and phonograph cartidges. With the vast increase in the development and use of electronic measuring systems, however, instrumentation engineers found it necessary to devise a more limited definition of transducer as a device used for measurement purposes.

Transducers covered in the first four chapters of this section include those for mechanical, thermal, acoustic, optical, and electrical quantities.

A chapter on microsensors and microactuators based on microfabrication techniques emphasizes physical, chemical, and biomedical sensors. A microelectromechanical (MEMS) instrumentation system can be considered to consist of an input sensor, an output actuator, and a signal processing element.   H.N., D.C.

In This Section:

On the CD-ROM:

Norton, H. N., "Transducers for Nuclear Radiation," reproduced from the 4th edition of this handbook.

CHAPTER 8.1
TRANSDUCER CHARACTERISTICS AND SYSTEMS

Harry N. Norton

INTRODUCTION

American National Stand ard, ANSI MC6.1, Electrical Transducer Nomenclature and Terminology (see Ref. 1), defines a *transducer* as "a device which provides a usable output in response to a specified measurand." The *measurand* is "a physical quantity, property or condition which is measured." The *output* is "the electrical quantity, produced by a transducer, which is a function of the applied measurand." Only the last of these three definitions applies specifically to electrical transducers. It could apply equally well to transducers with pneumatic output if the word "electrical" in the definition were omitted. Only electrical transducers are covered in this handbook.

ANSI MC6.1 also applies to the construction of transducer nomenclature (see Table 8.1.1). When used in titles or for indexes, the sequence shown in the table should be used, e.g., "transducer, acceleration, potentiometric, ±5*g*." When the nomenclature is used in the text, the opposite of the sequence shown in the table should be used, e.g., "A 0- to 8-cm dc output reluctive displacement transducer was installed on the actuator."

Transducer Classes and Elements

Thermocouples are representatives of a transducer class in which transduction takes place in a single component. An electric signal is generated between the terminals of the two dissimilar wires forming the thermocouple when they are exposed to a temperature difference. This is the simplest class of transducer. It is also *self-generating,* the output signal being produced without an additional power source. Piezoelectric transducers are also of the self-generating type, electric charge or potential being generated when the crystal is exposed to stress.

The thermocouple (or *thermoelectric temperature transducer*) is one of the few transducer types in which the *sensing element* (the element on which the measurand acts directly) also acts as the *transduction element* (the element in which the output of a transducer originates). Most transducer types have separate sensing and transduction elements. For example, in a piezoelectric pressure transducer the pressure acts on a diaphragm (sensing element) which deflects with applied pressure and then stresses the piezoelectric crystal (transduction element) in which the output originates.

The self-generating piezoelectric transducer is an example of a transducer whose transduction element is "active." In the majority of transducer types the transduction element is "passive," i.e., it requires the application of *excitation* power from an external source. Some "passive" transduction elements can accept either ac or dc excitation (e.g., strain-gage bridges, potentiometric elements), whereas others can accept only ac excitation (e.g., reluctive, inductive, capacitive elements).

TABLE 8.1.1 Construction of Typical Transducer Nomenclature and Examples of Modifiers

Main noun	First modifier, measurand, examples	Second modifier, restricts measurand, examples	Third modifier, electrical transduction principle, examples	Fourth modifier[a] sensing element, special features or provisions, examples	Range, examples	Unit, examples
Transducer	Acceleration	Absolute	Capacitive	AC output	0 to 1000	A
	Air speed	Angular	Electromagnetic	Amplifying	±5	°C
	Attitude	Biaxial	Inductive	Bellows	−100 to +500	cm
	Attitude rate	Differential	Ionizing	Bondable	−430 to −415	cm/s
	Current	Gage	Photoconductive	Bonded		deg
	Displacement	Infrared	Photovoltaic	Bourdon tube		°F
	Flow rate	Intensity	Piezoelectric	Capsule[b]		ft/s
	Force	Linear	Potentiometric	DC output		g
	Heat flux	Mass	Reluctive	Diaphragm		Hz
	Humidity	Radiant	Resistive	Digital output		in./s
	Light	Relative	Strain gage	Discrete-increment		in.
	Liquid level	Surface	Thermoelectric	Dual-output		K
	Mach number	Total	Vibrating-element	Exposed-element		kg
	Nuclear radiation	Triaxial		Frequency output		lb/min
	Pressure	Volumetric		Gyro		m
	Speed[c]			Integrating		mmHg
	Sound pressure			Self-generating		N
	Strain			Semiconductor		% RH
	Temperature			Servo[d,e]		lb/in.²
	Torque			Switch		kPa
	Velocity			Toothed-rotor		mbar
				Turbine		rad/s
				Ultrasonic		
				Unbonded		
				Weldable		

[a] Nomenclature may include two of these terms.
[b] Preferred to "aneroid."
[c] Scalar quantity.
[d] Preferred to "force balance" or "null balance."
[e] When this modifier is used, the third modifier ("transduction principle") may be omitted.
[f] Vector quantity.

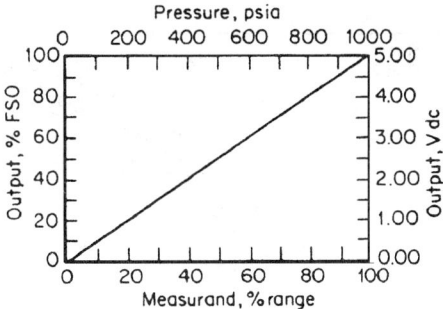

FIGURE 8.1.1   Output-measurand relationship of ideal linear-output transducer as exemplified for a dc-output transducer. (*From Ref. 2 by permission*)

Signal Conditioning

The widespread acceptance of industrial standards specific to a particular industry or application demands that the transducer output be in a specific form and within a specific range. Process industry standards, for instance, typically demand a transducer output signal either in the form of a direct current in the range from 4 to 20 mA or a dc voltage between 0 and 5 V. Transducers delivering control signals to microprocessors often require a digital output; this also applies to transducers used in digital data acquisition systems.

The process and the steps involved to provide an output signal in whatever specific form required are referred to as *signal conditioning.* The equipment may include ac or dc amplifiers, rectifiers, demodulators, circuits for square-root extraction, logarithmic amplifiers, and so forth, depending on the laws governing the relationship between the measurand and the desired output signal. For example, the radiation emanating from the surface of a body is an exponential function of its temperature. Thus to obtain a linear temperature scale at the output, logarithmic amplifiers are typically involved in the signal-conditioning process. In addition, excitation voltage or current regulation circuits are used in many instances to ensure accuracy and repeatability.

When a transducer system includes all the above-mentioned elements and circuits and is packaged in a single-housing, it is referred to as an *integrally conditioned transducer.* Recent examples include integrated-circuit semiconductor transducers for temperature-compensated pressure measurements contained in a single miniature housing.

Transfer Function

Every transducer design can be characterized by an ideal or theoretical output-measurand relationship (*transfer function*). This relationship is capable of being described exactly by a prescribed or known *theoretical curve* (see Fig. 8.1.1), stated in terms of an equation, a table of values, or a graphical representation. This applies primarily to the *static characteristics* of a transducer, i.e., the output-measurand relationship for a steady-state or very slowly varying measurand. It can also apply to the transducer's *dynamic characteristics,* i.e., the output-measurand characteristics for a relatively rapidly fluctuating measurand. However, this dynamic behavior is described by relationships other than the transducer's theoretical curve.

Transducer Errors

Because of a variety of factors, the behavior of a real transducer is nonideal. These factors include production variations, as well as the use of nonideal materials, production methods, ambient conditions during manufacture, and testing methods. It must also be recognized that many trade-offs enter into the design of a marketable transducer and that our knowledge (the *state of the art*) is limited with regard to producing an ideal transducer design and then compensating it perfectly for aging effects and a variety of environmental conditions the transducer may be subjected to during its operation.

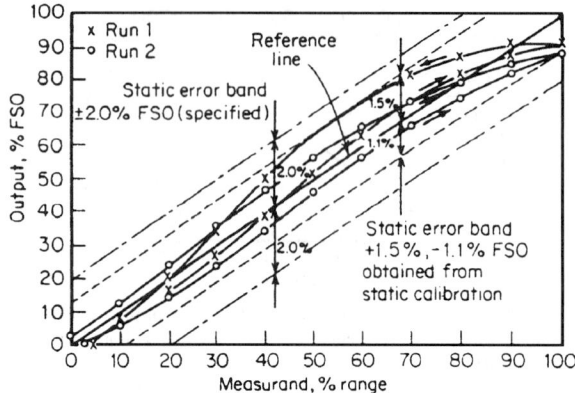

FIGURE 8.1.2 Static error band referred to terminal line (error scale 10:1).
(*From Ref. 2 by permission*)

Hence the measurand value indicated by the transducer may often differ from the true measurand value or the specified theoretical value. The algebraic difference between the indicated and the true value is the *error*.

Error Band

A convenient manner of determining or specifying transducer errors is to state them in terms of the band of maximum (or maximum allowable, for a specification) deviations from a specified reference line or curve. This band is defined as the *error band*. The *static error band* (see Fig. 8.1.2) is that error band obtained (or obtainable) by means of a *static calibration,* which is performed under "room conditions" (controlled room temperature, humidity, and atmospheric pressure) and in the absence of any vibration, shock, or acceleration (unless one of these is the measurand) by applying known values of measurand to the transducer and recording corresponding output readings. Other types of error band are applicable under somewhat different (and rigorously specified) conditions. Reference lines for error bands are not limited to the terminal line.

Dynamic Characteristics

When a step change in a measurand is applied to a transducer, the transducer output does not instantaneously indicate the new measurand level. Examples of such step changes are mechanical shock, a sudden pressure rise when a solenoid valve opens, or a temperature transducer is rapidly immersed in a very cold liquid. The lag between the time the measurand reaches its new level and the corresponding steady (final) transducer output reading is defined in various ways. The time required for the output change to reach 63 percent of its final value is the *time constant* of the transducer. The time required to reach a different specified percentage of this final value (say 90 or 98 percent) is the *response time*. The time in which the output changes from a small to a large specified percentage of the final value (usually from 10 to 90 percent) is the *rise time*. The output may rise beyond the final value before it stabilizes at that value. This *overshoot* depends on the *damping* characteristics of the transducer.

When the measurand fluctuates (sinusoidally) over a stated frequency range, the transducer output may not be able to indicate the correct amplitude of the measurand over these excursions. An example is the mechanical vibration (vibratory acceleration) of an engine housing. The output may be somewhat higher at certain measurand frequencies but usually drops off as the frequency increases until the output is essentially zero. The change with measurand frequency of the output-measurand amplitude ratio is the *frequency response* of the transducer, always stated for a specified frequency range. The above characteristics, as well as other transducer dynamic characteristics, are defined in the terminology section of ANSI Standard MC6.1-1975.

Environmental Characteristics

In most applications transducers are used only under the controlled room conditions of the facility where they are calibrated and where various static performance characteristics are determined. The external conditions to which a transducer is exposed not only while operating but also during shipping, storage, and handling can contribute additional errors, such as temperature error, acceleration error, or attitude error. Such environmental conditions (which can also include corrosive atmosphere, salt-water immersion, or nuclear radiation) may even cause a permanent deterioration or malfunction in the transducer.

Transduction Principles

The most essential determinant of any one transducer type is its transduction principle. How the electrical output is originated affects most other characteristics of the transducer. The most frequently used transduction principles are described below and illustrated in Fig. 8.1.3. It should be noted that photovoltaic, piezoelectric,

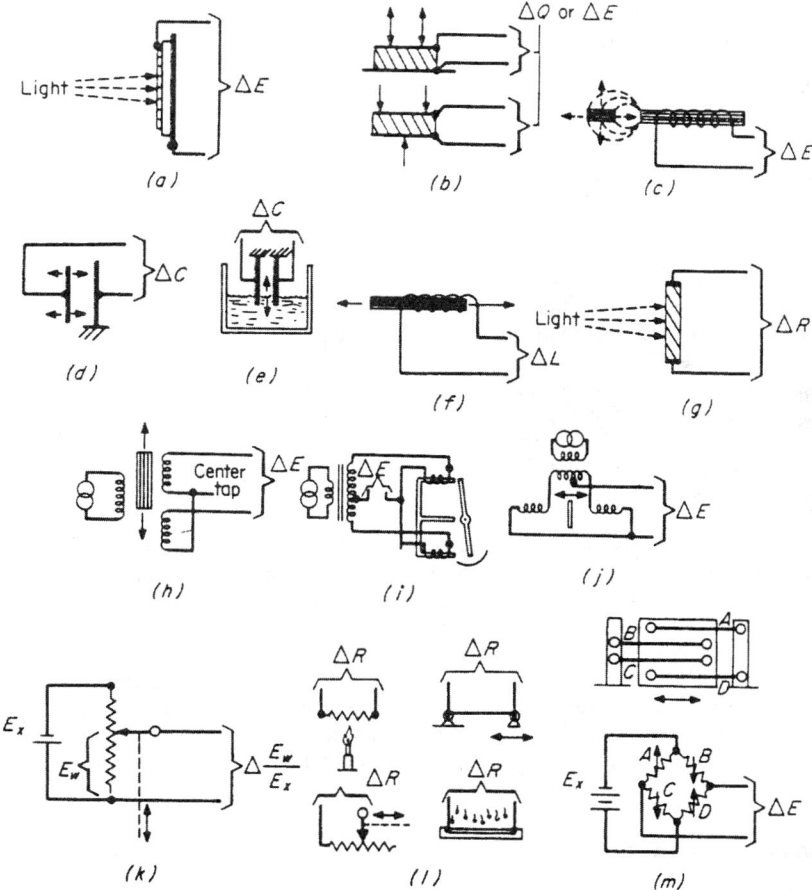

FIGURE 8.1.3   Transduction principles: (*a*) photovoltaic; (*b*) piezoelectric; (*c*) electromagnetic; (*d*), (*e*) capacitive; (*f*) inductive; (*g*) photoconductive; (*h*), (*i*), (*j*) reductive; (*k*) potentiometric; (*l*) resistive; (*m*) strain page.   (*From Ref. 2 by permission*)

and electromagnetic transduction are used in *self-generating* transducers, whereas all other transduction methods illustrated require some sort of external excitation power.

Photovoltaic Transduction.   The measurand is converted into a change in the voltage generated when a junction between certain dissimilar materials is illuminated. Used primarily in optical sensors, this principle has also been employed in transducers incorporating mechanical-displacement shutters to vary the intensity of a light beam between a built-in light source and the transduction element.

Piezoelectric Transduction.   The measurand is converted into a change in the voltage E or electrostatic charge Q generated by certain crystals when mechanically stressed by compression or tension forces or by bending forces. Either natural or synthetic crystals (usually ceramic mixtures) are used in such transduction elements.

Electromagnetic Transduction.   The measurand is converted into a voltage (electromotive force) induced in a conductor by a change in magnetic flux, usually because of a relative motion between a magnetic material and a coil having a ferrous core (electromagnet).

Capacitive Transduction.   The measurand is converted into a change of capacitance. This change occurs typically either by having a moving electrode move to or from a stationary electrode or by a change in the dielectric between two fixed electrodes.

Inductive Transduction.   The measurand is converted into a change of the self-inductance of a single coil.

Photoconductive Transduction.   The measurand is converted into a change in conductance (resistance change) of a semiconductive material because of a change in the illumination incident on the material. This transduction is implemented in a manner similar to that explained for the case of photovoltaic transduction, above.

Reluctive Transduction.   The measurand is converted into an ac voltage change by a change in the reluctance path between two or more coils while ac excitation is applied to the coil system. This transduction principle applies to a variety of circuits, including the differential transformer and the inductance bridge.

Potentiometric Transduction.   The measurand is converted into a change in the position of a movable contact on a resistance element. The displacement of the contact (wiper arm) causes a change in the ratio between the resistance from one end of the element to the wiper arm and the end-to-end resistance of the element. In its most common applications the resistance ratio is used in the form of a voltage ratio when excitation is applied across the resistance element.

Resistive Transduction.   The measurand is converted into a change of resistance. This change is typically effected in a conductor or semiconductor by heating or cooling, by the application of mechanical stresses, by sliding a wiper arm across a rheostat-connected resistive element, or by drying or wetting electrolytic salts.

Strain-Gage Transduction.   The measurand is converted into a resistance change, due to strain, usually in two or four arms of a Wheatstone bridge. This principle is a special version of resistive transduction. However, the output is always given by the bridge-output voltage change. In the typical configuration illustrated in Fig. 8.1.3 the upward arrows indicate increasing resistance, and the downward arrows decreasing resistance, in the respective bridge arms for sensing link motion toward the left.

Transducers in Process Control

The availability of an output signal responding to a measurand makes transducers key elements in the many areas of industrial process control. It permits direct comparison of a transducer output with an adjustable electric reference signal, calibrated in units of the measurand. Both the transducer output and the adjustable reference signal are fed into the input terminals of a differential amplifier. When any deviation of the transducer output from the adjustable reference signal (*set point*) occurs, an error signal is supplied to the controller, which in turn causes the control element to acquire such a position that the measurand again causes the transducer to deliver an output equal to the adjustable reference signal. This control process can be carried out in several alternative modes, e.g., on-off or proportional.

REFERENCES

1. "Electrical Transducer Nomenclature and Terminology," *ANSI MC 6.1/ISA-S37.1,* Instrument Society of America, 1982.
2. Norton, H. N., "Handbook of Transducers," Prentice Hall, Englewood Cliffs, 1989.

BIBLIOGRAPHY

Chesmond, C. J., "Control System Technology," Instrument Society of America, 1984.

"Dynamic Response Testing of Process Control Instrumentation." *ANSI MC 4.1/ISA-S26.* Instrument Society of America, 1975.

Engstrom, R. W., "Photomultiplier Handbook," Burle Industries, 1980.

Jones, B. E., "Instrumentation, Measurement and Feedback," McGraw-Hill (UK), 1977.

Kuo, B. C., "Automatic Control Systems." 3rd ed. Prentice Hall, 1975.

McCaw, L., "Industrial Measurements—A Laboratory Manual," Instrument Society of America, 1987.

Morrison, R., "Grounding and Shielding Techniques in Instrumentation," 2nd ed., Wiley, 1977.

Morrison, R., "Instrumentation Fundamentals and Applications," Instrument Society of America, 1984.

Murrill, P. W. "Fundamentals of Process Control Theory," Instrument Society of America, 1981.

Norton, H. N. (ed.), "Sensor and Transducer Selection Guide," Elsevier Science, 1990.

Norton, H. N., "Electronic Analysis Instruments," Prentice Hall, 1992.

"Process Instrumentation Terminology." *ANSI/ISA S51.1*, Instrument Society of America, 1979.

Strock, O. J., "Introduction to Telemetry," Instrument Society of America, 1986.

Sydenham, P. H., "Basic Electronics for Instrumentation," Instrument Society of America, 1982.

Travers, D., "Precision Signal Handling and Converter-Microprocessor Interface Techniques," Instrument Society of America, 1984.

CHAPTER 8.2
TRANSDUCERS FOR MECHANICAL QUANTITIES

Harry N. Norton

TRANSDUCERS FOR SOLID-MECHANICAL QUANTITIES

Transducers for solid-mechanical quantities sense and/or react to acceleration, velocity, vibration, shock, attitude, displacement, and position of a solid element. Transducers can also measure force, torque, weight or mass, and strain (deformation resulting from stress).

Acceleration, Vibration, and Shock

Acceleration, a vector quantity, is the time rate of change of velocity with respect to a reference system. When the term acceleration is used alone, it usually refers to *linear acceleration a*, which is then related to linear (translational) velocity v, and time t by $a = dv/dt$. *Angular acceleration* α is related to angular (rotational) velocity ω and time t by $\alpha = d\omega/dt$. Mechanical *vibration* is an oscillation wherein the quantity, varying in magnitude with time so that this variation is characterized by a number of reversals of direction, is mechanical in nature. This quantity can be stress, force, displacement, or acceleration; however, in measurement technology the term vibration is usually applied to *vibratory acceleration* and sometimes to *vibratory velocity*. Mechanical *shock* is a sudden nonperiodic or transient excitation of a mechanical system.

Acceleration Transducers (Accelerometers). Acceleration transducers (accelerometers) are used to measure acceleration as well as shock and vibration. Their sensing element is the *seismic mass,* restrained by a spring. The motion of the seismic mass in this acceleration-sensing arrangement is usually damped (see Fig. 8.2.1*a*). Acceleration applied to the transducer case causes motion of the mass relative to the case. When the acceleration stops, the mass is returned to its original position by the spring (see Fig. 8.2.1*b*). This displacement of the mass is then converted into an electrical output by various types of transduction elements in *steady-state acceleration transducers* whose frequency response extends down to essentially 0 Hz. In piezoelectric accelerometers the mass is restrained from motion by the crystal transduction element, which is thereby mechanically stressed when acceleration is applied to the transducer. Such *dynamic acceleration transducers* do not respond appreciably to acceleration fluctuating at a rate of less than 5 Hz. They are normally used for vibration and shock measurements.

Capacitive and photoelectric accelerometers have been produced at various times, and vibrating-element accelerometers (in which the mass, as it tends to move, applies tension to a wire or ribbon, thereby changing the frequency at which the wire can oscillate) have been used in some aerospace programs. However, the most commonly used steady-state acceleration transducers are the potentiometric, reluctive, strain-gage,

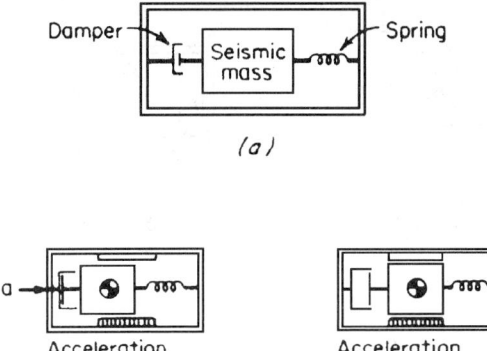

FIGURE 8.2.1 Basic operating principle of an acceleration transducer: (*a*) spring-mass system; (*b*) displacement of seismic mass.   (*From Ref. 1 by permission*)

and servo types. For vibration and shock measurement the piezoelectric accelerometers are most frequently used because of their inherently high frequency-response capability; some miniature semiconductor-strain-gage accelerometers are also used for these measurements since they can respond to fairly high acceleration frequencies.

Potentiometric Accelerometers.   Potentiometric accelerometers usually employ a mechanical linkage to amplify the motion of the seismic mass so as to produce the necessary extent of wiper-arm travel over the resistance element. The mass is supported by flexural springs or a cantilever spring in some models. In others it slides on a central coaxial shaft, restrained by calibrated coil springs. Magnetic, viscous, or gas damping is normally used in potentiometric accelerometers, primarily to reduce output noise due to wiper-arm whipping and transient wiper-contact resistance changes. Over-load stops keep the wiper arm from moving beyond the resistance-element ends in the presence of acceleration beyond the range of the accelerometer.

Reluctive Accelerometers.   Reluctive accelerometers require ac excitation power having a frequency greater than the upper limit of the transducer's frequency response. When moderately high frequency response is needed, the inductance-bridge version has been found most suitable. In a typical design the seismic mass is attached to a spring-restrained ferromagnetic armature plate, pivoted at its middle and placed above two coils so that the small seesaw motion of the plate, due to acceleration action on the mass, causes a decrease of inductance in one coil and an increase of inductance in the other. Since the coils are in opposite bridge arms, these inductance changes are additive and produce a bridge output voltage double that obtainable from having only one coil change its inductance. When a relatively low frequency response is needed, a differential-transformer, synchro or microsyn transduction circuit can be used to convert the seismic-mass displacement into the required electrical output.

Strain-Gage Accelerometers.   Strain-gage accelerometers are very popular and exist in several design versions. Some use unbonded metal wire stretched between the seismic mass and a stationary frame or between posts on a cross-shaped spring to whose center the seismic mass is attached and whose four tips are attached to a stationary frame. Other designs use bonded-metal wire, metal foil, or semiconductor gages bonded to one or two elastic members deflected by the displacement of the seismic mass. The recently developed micromachined accelerometers also employ strain-gage transductions.

Servo Accelerometers.   Servo accelerometers are closed-loop force-balance, torque-balance, or null-balance transducers. The displacement of the seismic mass is detected by a position-sensing element, usually reluctive

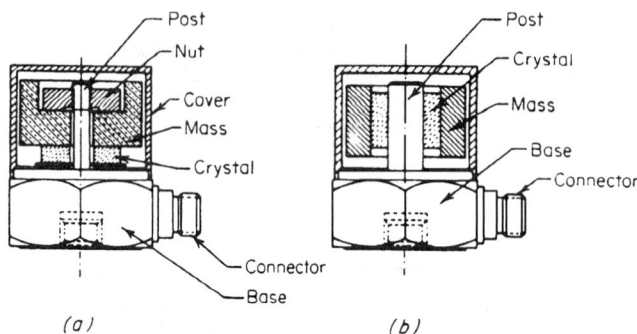

FIGURE 8.2.2 Piezoelectric acceleration transducers: (*a*) single-ended compression type; (*b*) shear type. (*Endevco, Dynamic Instrument Division*)

or capacitive, whose output is the error signal in the servo system. This signal is amplified and fed back to a torquer or restoring coil so that the restoring force is equal and opposite to the acceleration-induced force. The coil or torquer is attached to the seismic mass and returns the mass to its original position, when the feedback current is sufficient, so that the position error signal is reduced to zero. The current, which is proportional to acceleration, passes through a resistor. The *IR* drop across the resistor is the accelerometer output voltage, proportional to the acceleration.

Piezoelectric Accelerometers. Piezoelectric accelerometers exist in several design versions, two of which are illustrated in Fig. 8.2.2. Both contain a seismic mass that applies a force, due to acceleration, to a piezoelectric crystal. With acceleration acting perpendicular to the base, an output is generated by the crystal due to compression force in one design and to shear force in the other. Crystal materials include quartz and several ceramic mixtures such as titanates, niobates, and zirconates.

Ceramic crystals are used more frequently than natural crystals. They gain their piezoelectric characteristics by exposure to an orienting electric field during cooling after they are fired at a high temperature. If they are subsequently heated, as during transducer operation at elevated temperature, they can lose their piezoelectric qualities if that temperature is above the *Curie point*, which varies between about 100 and 600°C, depending on the materials used in the crystal. Piezoelectric accelerometers almost invariably require some signal-conditioning circuitry to provide a usable output since they have a relatively low output amplitude and a very high output impedance. In many designs, the necessary conditioning circuitry is included in the transducer. For other accelerometers, a separate charge or voltage amplifier is needed, connected to the transducer by a thin shielded coaxial cable of special low-noise construction to avoid noise pickup from within the cable itself.

Criteria for Selection. Criteria for selection of an acceleration transducer are primarily the required acceleration range and frequency response. They are mutually dependent; e.g., a typical ±2*g* potentiometric accelerometer design will have an upper frequency limit for flat response of about 12 Hz, whereas a ±20*g* accelerometer of the same design can have an upper frequency-response limit of about 40 Hz. As frequency-response requirements increase, the reluctive, servo, metal-strain-gage, semiconductor-strain-gage, and piezoelectric transducers successively become candidates for selection. The best accuracy characteristics are provided by servo accelerometers, which are also most suitable for low-range (±0.2*g* or lower) applications.

Attitude and Attitude-Rate Transducers

Attitude is the relative orientation of a vehicle or an object represented by its angles of inclination to three orthogonal reference axes. Attitude rate is the time rate of change of attitude.

The sensing methods employed by attitude transducers are best categorized by the kind of reference system to which the orientation to be measured is related. The *inertial* reference system is provided by a *gyroscope (gyro)* in which a rotating member will continue turning about a fixed axis as long as no forces are exerted on the member and the member is not accelerated. *Gravity* reference is used to establish a vertical reference axis. This principle is applied in *pendulum-type transducers*, in which a weight is attached to a wiper arm and a potentiometric or reluctive element is attached to the case, so that an output change is obtained when the object, to which the case is mounted, deflects from a vertical position.

A *magnetic reference axis* can be established by the poles of a magnetic field which remains fixed in position. This reference system is employed by certain navigational transducers related to the compass. *Flow-stream reference* refers to the direction of fluid flow past an object moving within that fluid, a reference system employed in *angle-of-attack* transducers mounted well forward of the nose of high-speed aircraft and rockets, so that the flow stream used is not altered in direction by the vehicle itself.

Optical reference systems are used by electrooptical transducers mounted (in a known attitude) so as to sense a remote light source or a light-dark interface whose position is known. This establishes a reference axis between the object on which the transducer is mounted and the target sensed by the transducer. *Optically referenced* transducers include such aerospace (primarily spacecraft) devices as the sun sensor, star tracker, and horizon sensor, as well as military target-locating equipment.

Gyros.   Gyros are the most widely used attitude and attitude-rate transducers. The operating principle of the gyro is illustrated in Fig. 8.2.3. A fast-revolving rotor turns about the *spin axis* of the gyro. This axis, which remains fixed in space as long as the rotor revolves, establishes the inertial reference axis. The rotor shaft ends are supported by a *gimbal* frame which is free to pivot about the gimbal axis. The pivot points are part of the gyro housing structure, which is attached to the object whose changes in attitude about the gimbal axis are to be measured. An angular-displacement transduction element (pick-off) is then used to provide an output proportional to attitude. A simple example of such an element is a wiper arm, attached to the gimbal frame at the pivot point, wiping over a ring-shaped potentiometric resistance element attached to the inside of the case. Potentiometric transduction as well as reluctive (especially synchro) and, occasionally, capacitive and photoconductive transduction are used in most gyros.

Gyro attitude transducers (free gyros) are often designed as two-degree-of-freedom gyros, i.e., those providing an output for each of two of a vehicle's three attitude planes (pitch, yaw, and roll, or x, y, and z axes). The design illustrated in Fig. 8.2.4 provides an inner gimbal for one axis and an outer gimbal for the other axis, with a separate pick-off for each axis. The caging mechanism (symbolized by the hand) is used to lock the inner gimbal to a reference position until the spin axis is to start serving as inertial reference axis. At this point the gyro is uncaged (after the rotor has come up to speed). AC or dc motors are commonly used to turn the rotor. Some gyros use a clock spring, wound before each use, or a pyrotechnic charge which, when activated, forces a stream of combustion gases into a small turbine.

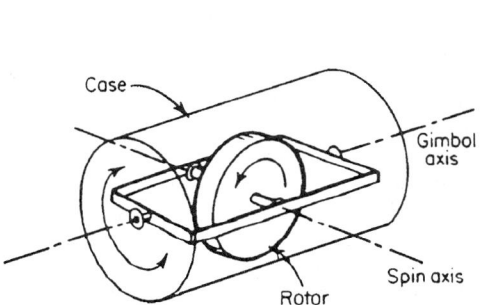

FIGURE 8.2.3 Basic single-degree-of-freedom gyro. (*From Ref. 1 by permission*)

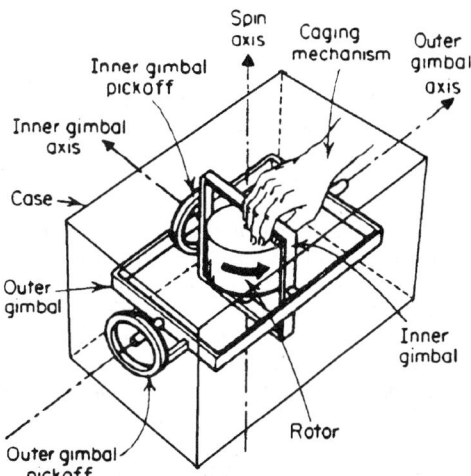

FIGURE 8.2.4 Two-degree-of-freedom gyro. (*Conrac Corp.*)

Rate gyros are attitude-rate transducers. They provide an output proportional to angular velocity (time rate of change of attitude). The operating principle of the rate gyro (see Fig. 8.2.5) is similar to that of the single-degree-of-freedom free gyro, except that the gimbal is elastically restrained and its motion is damped. The output is representative of gimbal deflection about the output axis in response to attitude-rate changes about the input axis. The deflection of the gimbal (precession) is caused by the torque T applied to it. The applied torque is the product of the instantaneous attitude rate about the input axis and the angular momentum of the gyro.

The more recently developed *ring laser gyro (RLG)* and *fiber-optic gyro* contain essentially no moving parts; the rotor is replaced by two laser beams traveling in opposite directions many times around a triangular or circular path; attitude rate changes then cause a frequency or phase difference at the optical path terminations.

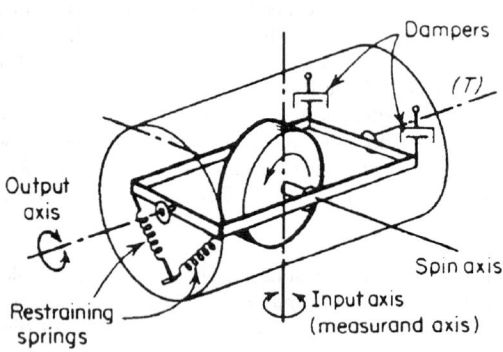

FIGURE 8.2.5   Basic-rate gyro.   *(From Ref. 1 by permission)*

When selecting a gyro, attention must be paid not only to the usual characteristics (weight, size, range, linearity, repeatability, threshold, and so on, and dynamic characteristics for rate gyros) but also to *drift,* the amount of precession, per hour, of the spin axis from its intended position because of internal unwanted torques or other instabilities, and the time period (after spin-motor runup or after uncaging) during which measurements must be obtained continuously.

Displacement and Position Transducers

Position is the spatial location of a body or point with respect to a reference point. *Displacement* is the vector representing a change in position of a body or point with respect to a reference point. Displacement transducers are used to measure linear and angular displacements, as well as to establish position from a displacement measurement.

The sensing element of most displacement transducers is the *sensing shaft* with its coupling device, which must be of a design suitable to make the motion of the sensing shaft truly representative of the motion of the measured point (driving point). A spring-loaded sensing shaft (without coupling device) is used for some applications. A number of *noncontacting* transducer designs are also in use. These require no coupling or sensing shaft. Various transduction principles are employed in displacement transducers.

Capacitive Displacement Transducers.   In these devices a linear or angular motion of the sensing shaft causes a change in capacitance either by relative motion between one or more moving (*rotor*) electrodes and one or more stationary (*stator*) electrodes or by moving a sleeve of insulting material, having a dielectric constant different from that of air, between two stationary electrodes.

Inductive Displacement Transducers.   Inductive displacement transducers can be of the coupled or the noncontacting types. Coupled designs contain a coil whose self-inductance is varied as a nonmagnetic sensing shaft moves a magnetically permeable core gradually into or out of the central hollow portion of the coil. Some designs incorporate an additional coil (balancing coil) having a fixed inductance value equal to the inductance of the transduction coil at a predetermined "zero" position of the sensing shaft. The two coils are connected as two arms of an inductance bridge. This two-coil principle is used in some noncontacting displacement transducers in which the transduction coil has a stationary core but changes its inductance with the distance between itself and a moving ferromagnetic object.

Photoconductive Displacement Transduction.   Photoconductive displacement transduction is employed in several nonconducting displacement measuring systems. Pulsed or continuous-wave lasers or light-emitting diodes are typically used as light sources. Noncontacting photoconductive sensors usually require an optical reflector mounted to the measured object. Various optical configurations are used to obtain an output from the photoconductive element as the intensity, the phase, or the position of the reflected light beam changes. The *laser interferometer* is included in this group of displacement-sensing devices.

Potentiometric Displacement Transducers.   Potentiometric displacement transducers are widely used because of their relative simplicity of construction and their ability to provide a high-level output. All these designs use a sensing shaft. The wiper arm is either attached directly to the shaft (but insulated from it) or mechanically connected to it through an amplification linkage. Straight potentiometric resistance elements are used in linear displacement transducers, circular or arc-shaped elements in angular displacement transducers. The elements are usually wire-wound, but conductive plastic, carbon film, metal film, or ceramic-metal mixtures (cermets) are also used. Some transducers have two or more wiper-element combinations moved by the same sensing shaft. A good sliding seal is needed at the point where the sensing shaft enters the transducer case, to protect the internally exposed resistance elements from atmospheric contaminants and moisture.

Reluctive displacement transducers are as commonly used as the potentiometric types. The reluctive transduction circuits employed in linear- and angular-displacement transducers are illustrated in Fig. 8.2.6. Only the linear-variable transformer (LVDT) and the inductance-bridge circuits are used for linear-displacement measurements. Many winding configurations exist for the LVDT transducers; one manufacturer offers 12 different "off-the-shelf" configurations, including several with two separate secondary windings. Alternating current excitation is required for all reluctive transducers. However, some designs are available with integral ac/dc output conversion and, in some cases, also integral dc/ac excitation conversion. Synchro-type transducers are often connected to a synchro-type receiver, which indicates the measured angle directly, e.g., on a dial.

A few strain-gage displacement transducers have been designed for the measurement of small linear and angular displacement. The gages are usually attached to the top and bottom surfaces of a cantilevered or end-supported beam which is deflected by the displacement.

Digital-Output Displacement Transducers (Encoders).   Digital-output displacement transducers (encoders) are frequently referred to as *linear encoders* and *angular* or *shaft-angle encoders*, respectively. These consist essentially of a strip (for linear displacements) or a disk (for angular displacements), coded so as to provide a digital readout for discrete (sometimes very small) displacement increments and a reading head. Two types of encoders are in common use:

1. Photoelectric encoders, in which the reading head consists of a light-source assembly on one side of the disk or strip and a corresponding light-sensor assembly facing it on the other side of the disk or strip; the coded pattern is partly translucent, partly opaque.

2. Magnetic encoders, with a magnetic reading head and a partly magnetized, partly nonmagnetized coded pattern.

Incremental encoders have a simple, alternately ON and OFF coded pattern. They provide an output in the form of number of *counts* between the start and end of the displacement. Hence the start position must be known if the end position is to be determined in absolute terms. *Absolute encoders* have a code pattern such that a unique digital word is formed for each discrete displacement increment. Various codes are used for this purpose, such as the binary, binary-coded-decimal (BCD), and the Gray code.

Among displacement-transducer *selection criteria*, the most critical are range, resolution, starting force, overtravel, and type and magnitude of full-scale output. Accuracy and dynamic characteristics, type of available excitation supply, and freedom from contamination by the ambient atmosphere or other fluids need to be considered as well, for all transducer applications.

Force, Torque, Mass and Weight Transducers

Force is the vector quantity necessary to cause a change in momentum. *Mass* is the inertial property of a body, a measure of the quantity of matter in the body and of its resistance to change in its motion. *Weight* is the gravitational force of attraction; where gravity exists, it is equal to mass times acceleration due to gravity. *Torque* is the moment of force, the product of force, and the perpendicular distance from the line of action of the force to the axis of rotation (lever arm).

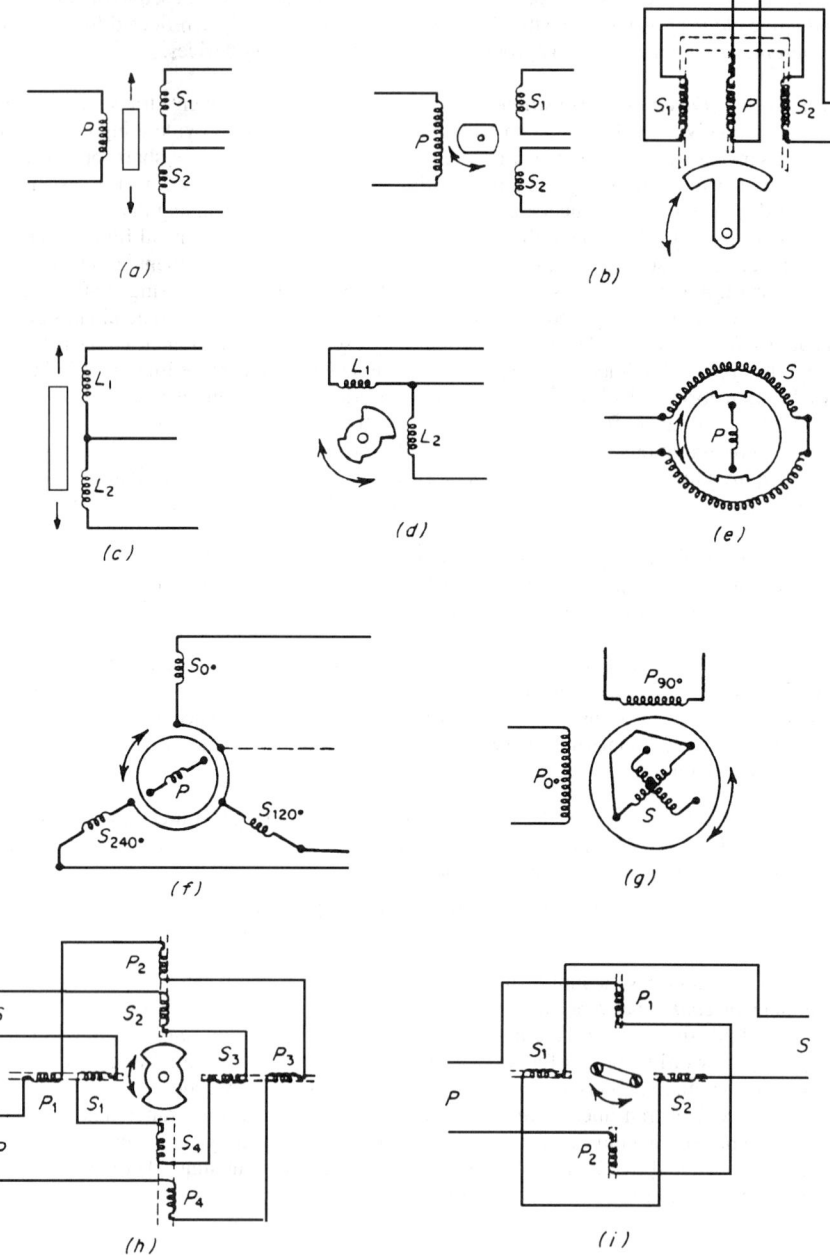

FIGURE 8.2.6 Transduction circuits of reluctive displacement transducers: (*a*) linear differential transformer; (*b*) angular differential transformer; (*c*) linear inductance bridge; (*d*) angular inductance bridge; (*e*) induction potentiometer; (*f*) synchro; (*g*) resolver; (*h*) microsyn; (*i*) shorted-turn signal generator. (*From Ref. 1 by permission*)

Force Transducers (Load Cells).   Force transducers (load cells) are used for force measurements (compression, tension, or both) as well as for weight determinations in any locality where gravity exists and the gravitational acceleration g is known. The standard g (on earth) is 9.80665 m/s^2. Mass can be determined from weight, which is expressed in force units. A mass of 1 kg, for example, "weighs" 2.205 lb (pounds force) on earth. Torque is measured by *torque transducers.*

The sensing elements of force and torque transducers usually convert the measured into a mechanical deformation of an elastic element. This deformation, in terms of either local strains or gross deflection, is then converted into a usable output by a suitable transduction element. Bending beams (cantilever, end-supported, or end-restrained), solid rings or frames (*proving rings*), and solid or hollow rectangular or cylindrical columns are the most commonly used force-sensing elements. Special solid or notched shafts are used as torque-sensing elements.

Piezoelectric Force Transducers.   Piezoelectric force transducers are used for dynamic compression-force measurements. A typical design has the shape of a thick washer. The annular piezoelectric crystal segments are sandwiched between two hollow cylindrical columns. Bidirectional force measurements can be obtained by preloading this *force washer.* An amplifier is used to boost the low-level output signals.

Reluctive Force Transducers.   Reluctive force transducers use proving-ring sensing elements in most design versions. The deflection of the proving ring is converted into an ac output by an inductance-bridge or differential-transformer transduction element. An entirely different design uses the permeability changes because of stresses in a laminated column to vary the voltage induced by a primary winding in a secondary winding.

Strain-Gage Force Transducers.   Strain-gage force transducers are the most widely used type. Bonded-metal foil and metal wire gages predominate, but unbonded wire gages and bonded semiconductor gages are used in some designs. Columns and proving rings are the usual sensing elements. The shear-web sensing element of the force transducer shown in Fig. 8.2.7 is related to the column, but is reported to offer greater transduction efficiency.

Torque Transducers.   Torque transducers are mostly of the reluctive, photoelectric, or strain-gage type. The last is more widely used than the first two. The metal-foil strain gages in the transducer shown in Fig. 8.2.8 are located on the sensing shaft, which is enclosed in a cylindrical *torque sensor* housing. The leads from the gages are carried through the shaft up to slip rings. Brushes ride on the slip rings to provide stationary external connections. The brush assembly can be lifted off the slip rings to increase brush life during periods when torque is not monitored. In most modern strain-gage torque transducers the slip rings and brushes are replaced by a rotary transformer or by RF coupling.

Other Torque Transducers.   Reluctive torque transducers use changes in shaft permeability, resulting from torque-induced stresses in the shaft, to change the voltage coupled from a primary winding to two secondary

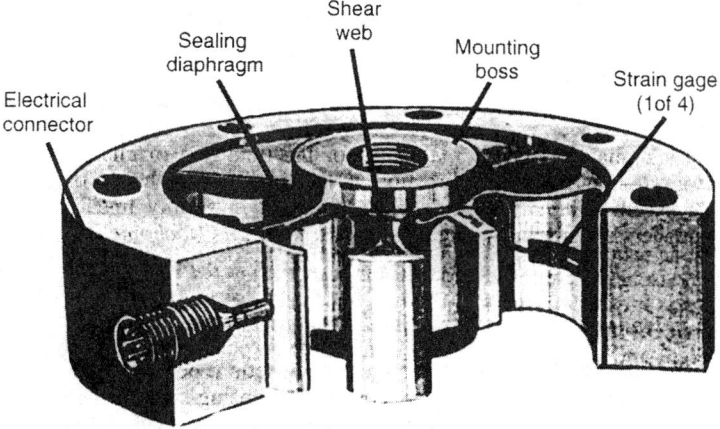

FIGURE 8.2.7   Strain-gage force transducer.   (*Interface, Inc.*)

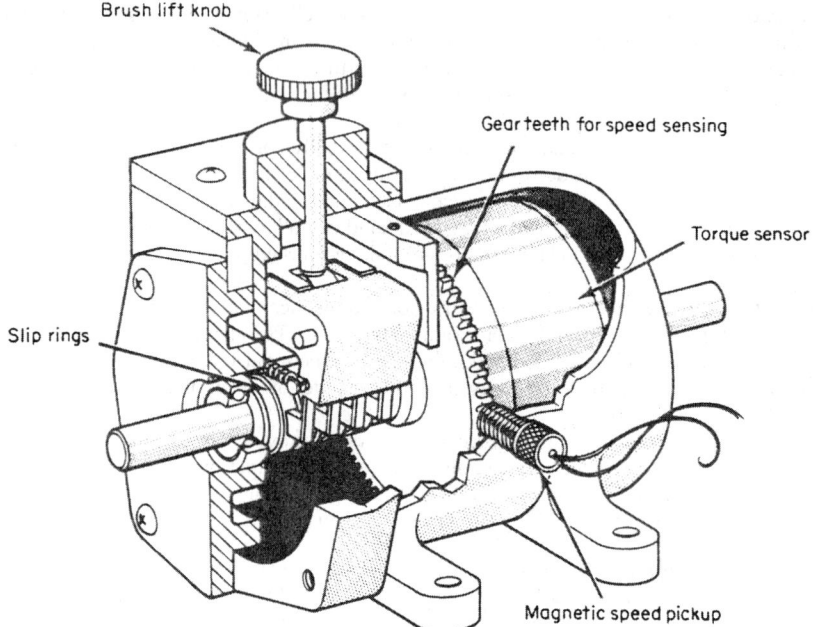

Brush lift knob

Gear teeth for speed sensing

Torque sensor

Slip rings

Magnetic speed pickup

FIGURE 8.2.8 Strain-gage torque transducer. (*Lebow Associates, Inc.*)

windings. *Photoelectric torque transducers* use two incremental-encoder disks, one on each end of the shaft, to change the illumination on a light sensor when one disk undergoes a small angular deflection, because of torque, relative to the other disk.

Selection criteria include the usual range, accuracy, excitation, and output characteristics, case configuration and dimensional constraints, overload rating, the thermal environment, and, for torque transducers, maximum shaft speed and proximity of any magnetic fields that may cause reading errors. A frequent application of force transducers is in automatic weighing systems.

Speed and Velocity Transducers

Speed (a scalar quantity) is the magnitude of the time rate of change in displacement. *Velocity* (a vector quantity— magnitude and direction) is the time rate of change of displacement with respect to a reference system. *Velocity transducers* are almost invariably linear-velocity transducers, whereas speed transducers are normally angular-speed transducers (*tachometers*).

Velocity Transducers. Velocity transducers are usually of the electromagnetic type, exemplified by a coil in which a permanent-magnet core moves freely. The core has a sensing-shaft extension, and the shaft is attached to the object whose (usually oscillatory) velocity is to be measured. The rate at which lines of magnetic flux from the core are cut by the coil turns determines the amount of electromotive force generated in the coil; hence the output is proportional to the velocity of the measured point. In some designs the coil moves within a fixed magnetic field instead.

Tachometers. Tachometers are also predominantly of the electromagnetic type. Such angular-speed transducers as the *dc tachometer generator,* the *ac induction tachometer,* and the *ac permanent magnet tachometers (ac magneto)* are electric generators. Their output amplitude increases with angular (rotational) speed. In the case of the ac magneto, the output frequency also increases with speed. The output of a *toothed-rotor tachometer* also varies in both amplitude and frequency, but the frequency variation is much greater than the amplitude variation and represents the angular speed much more accurately.

The speed-sensing gear teeth and sensing coil (pickup) incorporated in the torque transducer of Fig. 8.2.8 constitute a toothed-rotor tachometer. A pulse is generated in the electromagnetic sensing coil every time a ferromagnetic tooth passes by it. Since there are 60 teeth on the gear shown in the illustration, 60 pulses per revolution are provided by the sensing coil. By counting the pulses over a fixed time interval the angular speed can be determined with very close accuracy. *Photoelectric tachometers* provide the same degree of accuracy, typically by chopping a beam between the light source and a light sensor into equidistant pulses by an incremental-encoder disk attached to the sensing shaft. The pulse-frequency output can also be converted into a dc output voltage if the degraded accuracy, resulting from the conversion, can be tolerated.

Selection criteria include, besides range and accuracy characteristics, the mounting position and required frequency response for velocity transducers, and the type of available readout or signal-conditioning and telemetry equipment in the case of tachometers.

Strain Transducers

Strain is the deformation of a solid resulting from *stress,* the force acting on a unit area in a solid. Strain is measured as the ratio of dimensional change to the total value of the dimension in which the change occurs. Essentially all strain transducers are resistive and are referred to as *strain gages.* Their essential characteristic is their sensitivity (*gage factor*), the ratio of the unit change in resistance to the unit change in dimension (length).

Strain Gages.   Strain gages employ either a conductor or semiconductor, of small cross-sectional area, suitable for mounting to the measured surface so that it elongates or contracts with that surface and changes its resistance accordingly. Most types of metal gages are made of thin metal foil, die-cut or etched into the required pattern, or they can be deposited on an insulating substrate through a pattern mask by bombardment or evaporative methods. The metals used in strain gages are usually copper-nickel alloys; other alloys such as nickel-chromium, platinum-tungsten, and platinum-iridium are also used. *Semiconductor* gages are usually made from thin doped-silicon wafers or blocks.

Strain gages can be *bare* (surface-transferable, free-filament), bonded to an insulating carrier sheet on one side only, or completely *encapsulated* in a bondable (usually plastic) or weldable (metal) carrier, the latter insulated internally from the gage. Bare gages are normally supplied with a strippable insulating substrate (*carrier*). Since two or more gages are normally used to obtain a strain measurement, for temperature-compensation, linearity-compensation, and output-multiplication purposes, strain-gage *rosettes* are sometimes used, combining two, three, or four gages, mutually aligned as to their strain-sensing axes, on one carrier.

Selection criteria involve the desired type and size (always including gage length and width), type, and material of connecting leads and spacing between them on the gage itself, type of carrier or encapsulation, gage resistance, gage factor, transverse sensitivity tolerances, allowable overload (*strain limit*), and maximum excitation current for a given application. Semiconductor gages may have to be shielded from illumination, which can cause reading errors. Proper methods of attachment and of connection into a Wheatstone bridge circuit are very critical for strain gages.

TRANSDUCERS FOR FLUID-MECHANICAL QUANTITIES

Transducers for fluid-mechanical quantities can be used to measure density and flow of a homogeneous fluid. Transducers can also be used to measure humidity, moisture, and dew-point of a homogeneous gas. Liquid level sensors can determine several characteristics of a liquid or slurry within closed vessels.

Density Transducers

Density is the ratio of the mass of a homogeneous substance to a unit volume of that substance. Density transducers (*densitometers*) are used for the determination of the density of fluids (gases, liquids, and slurries).

They are, however, not related to densitometers used to measure optical density, as of a photographic image, or to equipment used to determine spectral density (e.g., power spectral density).

Three methods are primarily used for density sensing:

Sonic density sensing is achieved by an arrangement of piezoelectric sound (usually ultrasound) transmitters and receivers producing outputs proportional to the speed of sound in the fluid and to the acoustic impedance of the fluid. Since acoustic impedance varies with the product of speed of sound and density, a signal proportional to density can be derived from the transducer and signal conditioning system.

Radiation density sensing relies on the attenuation, owing to density, of the radiation passing from a radioisotope source, on one side of the fluid-carrying pipe or vessel, to a radiation detector on the opposite side.

Vibrating-element density sensing employs a simple mechanical structure, such as a cylinder or a plate, electromagnetically set into vibration at its resonant frequency. This frequency changes with density, and an output is produced, proportional to density, which is related directly to the square of the period of vibration.

Additional methods, used to infer density from other measurements, have also been employed in measurement systems.

Flow Transducers

Flow is the motion of a fluid. *Flow rate* is the time rate of motion expressed either as fluid volume per unit time (*volumetric flow rate*) or as fluid mass per unit time (*mass flow rate*). Transducers used for flow measurement (*flowmeters*) generally measure flow rate. Most flowmeters measure volumetric flow rate, which can be converted to mass flow rate by simultaneously measuring density and computing mass flow rate from the two measurements. Some flowmeters measure mass flow rate directly. Flow-sensing elements can be categorized as follows:

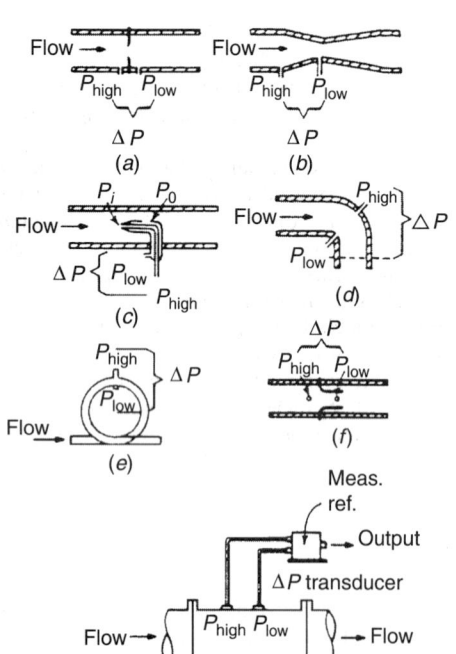

FIGURE 8.2.9   Flow measurement using differential pressure-sensing elements: (*a*) orifice plate; (*b*) venturi tube; (*c*) pitot tube; (*d*) centrifugal section (elbow); (*e*) centrifugal section (loop); (*f*) nozzle; (*g*) measurement of differential pressure due to flow rate. (*From Ref. 1 by permission*)

1. *Differential-pressure flow-sensing elements.* Sections of pipe provided with a restriction or curvature that produces a pressure differential ΔP proportional to flow rate across two points of the device (see Fig. 8.2.9). The output of a differential pressure transducer whose input ports are connected to these two points is representative of flow rate through the sensing element. Known relationships of ΔP versus flow rate exist for each type of element.

2. *Mechanical flow-sensing elements.* Freely moving elements, e.g., turbine or propeller, or mechanically restrained elements, e.g., a float in a vertical tapered tube, a spring-restrained plug, a hinged or cantilevered vane, whose displacement, deflection, or angular speed is proportional to flow rate.

3. *Flow sensing by fluid characteristics.* Certain transduction elements can be so designed and installed that they will interact with the moving fluid itself and produce an output relative to flow rate. The heated wire of a *hot-wire anemometer* transfers more of its heat to the fluid as the flow rate increases, thereby causing the resistance of the heated wire to decrease. When small amounts of radioisotope tracer material are added to the fluid, a radiation detector close to the moving fluid will respond with increasing output as the flow rate increases (*nucleonic flowmeter*).

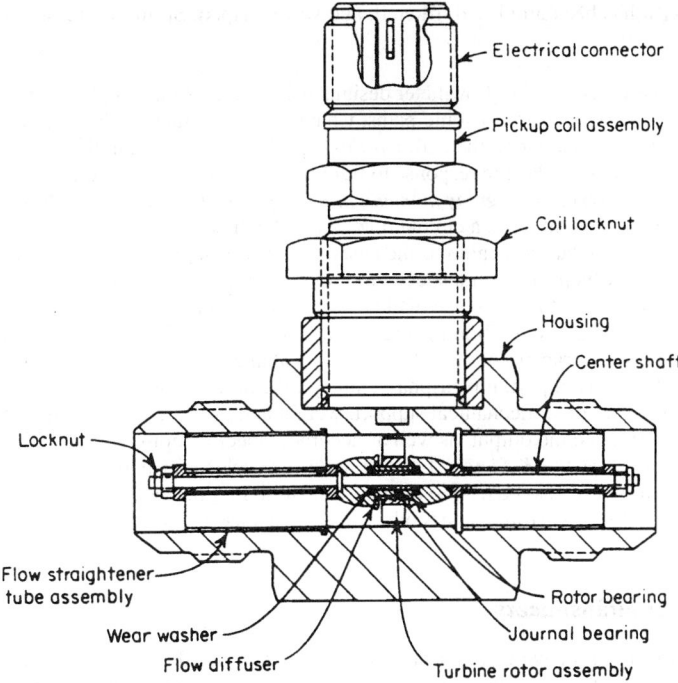

FIGURE 8.2.10 Turbine flowmeter. (*ITT Barton.*)

In the (*fluid-conductor*) *magnetic flowmeter* an increasing electromotive force is induced in an electrically conductive fluid, flowing through a transverse-magnetic field, as the flow rate increases. In the *thermal flowmeter* two thermocouple junctions are immersed in the moving fluid, one upstream, the other downstream, from an electric heater immersed in the same fluid, and the two junctions are connected as a differential thermocouple, the output of the latter increasing with mass flow rate. In a similar device, the *boundary-layer flowmeter,* only the portion of the fluid immediately adjacent to the inside wall of the pipe is heated and thermally sensed.

Turbine Flowmeters.   The turbine flowmeter (see Fig. 8.2.10) is among the most widely used flow-rate transducers. Its operating principle is similar to that of the toothed-rotor tachometer. The bladed rotor (turbine) rotates at an angular speed proportional to volumetric flow rate. Rotational friction is reduced as much as possible by special bearing design. As each magnetic rotor blade cuts the magnetic flux of the pickup coil's pole piece, a pulse is induced in the pickup coil (sensing coil). A frequency meter is used to display the frequency output of the flowmeter, or a frequency-to-dc converter can be used to provide a dc voltage increase with flow rate. The rotor blades can be so machined that the variable-frequency ac voltage across the sensing coil terminals is virtually sinusoidal. This permits use of an FM demodulator as a frequency-to-dc converter. The number of turbine blades, the pitch of the blades, and the internal geometry of the flowmeter determine the range of output frequencies for a given flowrate range.

Oscillating-Fluid Flowmeters.   In this device the fluid is the first forced into a swirling motion, then passes through a venturi-like cavity at a point of which the flow oscillates about the axis of the flowmeter. A fast-response temperature or force transducer at that point provides an output in terms of frequency of resistance changes. This frequency, proportional to flow rate and converted into voltage variations, can then be displayed on a counter after it has been amplified, filtered, and wave-shaped. Strain or pressure sensors are also used in such flowmeters.

Other Flowmeter Designs.   Other flowmeter designs include the *ultrasonic flowmeter,* typically using pairs of piezoelectric transducers to establish sonic paths. Changes in flow rate produce corresponding changes in the propagation velocity of sound along the path. *Strain-gage flowmeters* are cantilevered vanes or beam-supported drag bodies that deflect or displace in response to fluid flow. The strain in the deflecting beam is then transduced by strain gages. A few types of *angular-momentum mass flowmeters* have been developed in which the fluid either imparts angular momentum to a circular tube through which it flows or receives angular momentum by a rotating impeller. The angular momentum is then used to cause an angular displacement or a torque in a mechanical member, either of which can be transduced to provide an output proportional to mass flow.

Selection criteria involve, first, a choice of either a flowmeter alone or a complete flow-rate or flow-measuring system that can include signal conditioning and display equipment and, when required, a flow totalizer. Among the essential flowmeter characteristics are the (mass or volumetric) flow-rate range, the properties and type(s) of the measured fluid (gas, liquid, mixed-phase, slurry), the nominal and maximum pressure and temperature of the fluid, the configuration, mechanical support, weight and provisions for connection of the flowmeter, the required time constant, and the output, as well as accuracy, specifications. The sensitivity of a turbine flowmeter is usually expressed as the *K factor,* stated in hertz (or cycles) per gallon, per liter, per cubic foot, or per cubic meter. Attention must also be paid to the length of straight pipe upstream and downstream of the flowmeter and the necessity for flow straighteners other than those incorporated in the transducer itself.

Humidity and Moisture Transducers

Humidity is a measure of the water vapor present in a gas. It is usually measured as relative humidity or dew-point temperature, sometimes as absolute humidity. *Relative humidity,* which is temperature-dependent, is the ratio of the water-vapor pressure actually present to water-vapor pressure required for saturation at a given temperature; it is expressed in percent (percent RH). The *dew point* is the temperature at which the saturation water-vapor pressure is equal to the partial pressure of the water vapor in the atmosphere. Hence any cooling of the atmosphere, even a slight amount below the dew point, produces water condensation. The relative humidity at the dew point is 100 percent RH. *Moisture* is the amount of liquid adsorbed or absorbed by a solid; it is also the amount of water adsorbed, absorbed, or chemically bound in a nonaqueous liquid. Humidity and moisture measurements are made by one of three methods: hygrometry, psychrometry, and dew-point determination.

Hygrometers.   The hygrometer is a device that can measure humidity directly, with a single sensing element; it is usually calibrated in terms of relative humidity. Three types of hygrometric sensing elements are shown in Fig. 8.2.11. In the *resistive* humidity-transducer sensing element a change in ambient relative humidity produces a change in resistance of a conductive film between two electrodes. Carbon powder in a binder material has been used for such films, but hygroscopic salts, also in a binder material, are more common. Lithium chloride was used originally in such elements; more recently, sulfonated polystyrene and *electrolytic* elements (rather than resistive), such as phosphorous pentoxide, have come into use. The *mechanical* hygrometric element is the oldest type. It uses a material, such as human hair or animal membrane, which changes its dimensions with humidity. The resulting displacement on an attaching point on the material is then transduced into an output proportional to humidity. The *oscillating-crystal* hygrometric element consists of a quartz crystal with a hygroscopic coating, so that the total crystal mass changes as water is adsorbed on, or desorbed from, the coating. When the crystal is connected into an oscillator circuit, the oscillator output frequency will change with changes in humidity.

Several other types of hygrometric sensing elements have also been developed. In the *aluminium oxide element* an impedance (resistance and capacitive reactance) change occurs with changes in humidity. The *Brady array* also provides an ac output when excited with alternating current (at about 1 kHz). However, it differs from other devices in that it consists of an array of semiconducting crystal matrices that look electrically neutral to the water molecule. Vapor pressure then allows the molecules to drift in and out of the interstices, creating an exchange of energy within the structure. The *porous-glass-disk* hygrometric element has electrodes

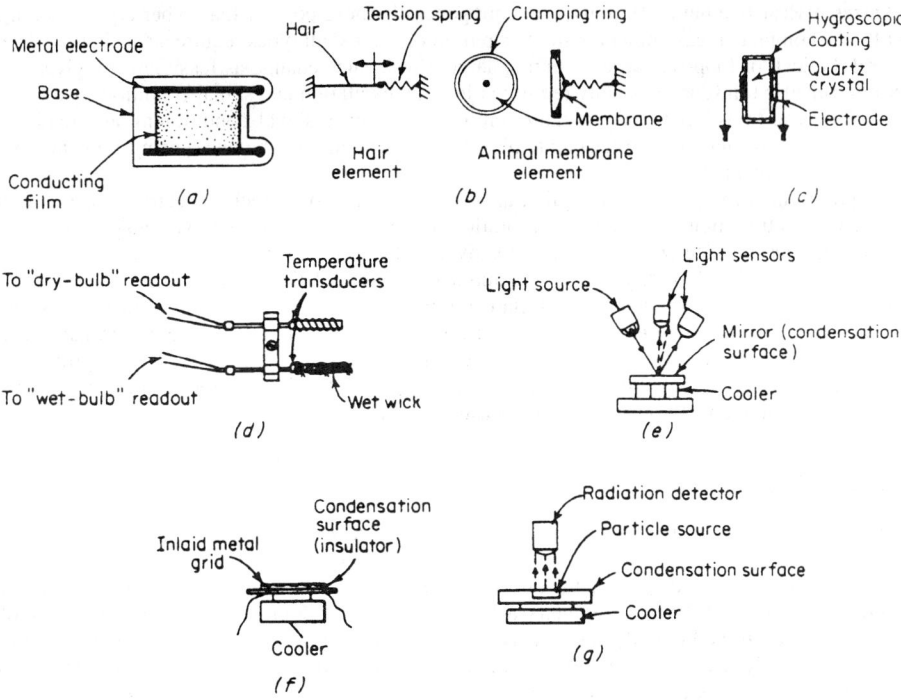

FIGURE 8.2.11   Sensing elements of humidity transducers: (*a*) resistive; (*b*) mechanical; (*c*) oscillating crystal; (*d*) psychometric; (*e*) photoelectric; (*f*) resistive; (*g*) nucleonic; (*e*) to (*g*) are dew-point sensors.   (*From Ref. 1 by permission*)

plated on the two surfaces of the disk. When water vapors permeate the pores in the glass, it is decomposed electrolytically when a voltage is applied across the electrodes. The current necessary to decompose the water is then a measure of relative humidity.

Psychrometers.   Psychrometers use two temperature-sensing elements (see Fig. 8.2.11*d*). One element, *dry bulb*, measures ambient temperature; the other, *wet bulb*, covered with a water-saturated wick or similar device, measures temperature reduction due to evaporative cooling. Relative humidity can be determined from the dry-bulb temperature reading, the differential temperature between dry-bulb and wet-bulb readings, and knowledge of the barometric pressure by referring to a *psychometric table* of numbers. Such tables are available from government agencies, e.g., weather service, as well as from manufacturers. The temperature-sensing elements are usually resistive (platinum- or nickel-wire windings or thermistors), sometimes thermoelectric.

Dew-Point Sensing Elements.   Dew-point sensing elements are dual elements. The condensation-detection element senses the first occurrence of dew on a surface whose temperature is being lowered. The temperature-sensing element measures the temperature of this surface so that the dew point (the temperature at which condensation first occurs as the temperature is lowered) can be determined by monitoring the output of both elements simultaneously. Typical condensation detectors (see Fig. 8.2.11) include a photoelectric device in which light sensors detect the difference in light, reflected from a mirror that serves as the condensation surface, when the dew point is reached; a resistive element in which a change in conductivity occurs in an inlaid metal grid at the condensation surface when condensation occurs; and a nucleonic device in which a drop in particle flux, emitted from a radiation source at the condensation surface, indicates the dew point.

Auxiliaries.   Resistive humidity transducers are generally more popular than other types when a transducer, rather than a complete measurement system, is required. Almost all types require ac excitation. The electrodes are spiral, helical, or loop-shaped to obtain as large a resistance change as feasible for a given element size. Other hygrometric transducers usually require at least an excitation and signal conditioning unit.

Psychrometric transducers are typically complemented by a signal conditioning and readout system. A small blower (*aspirator*) is often included to blow the ambient air over the two sensing elements so that a faster response can be obtained.

Dew-point humidity transducers require, as a minimum, a cooler (thermoelectric coolers are often used), its associated control circuit, and a power conditioning circuit, as well as the two sensing elements. However, several designs are miniaturized and require relatively little signal conditioning.

Selection criteria.   Humidity transducer applications should first be examined to see whether relative humidity or dew point is to be measured. Relative humidity can, of course, also be inferred from psychrometric and dew-point readings but not without a look-up table or calculations. Among performance characteristics the measurement range is the most important; measurement accuracy can usually be improved when only a partial range needs to be measured. Other important characteristics include the temperature and the chemical properties of the ambient atmosphere or the measured material.

Liquid-Level Sensing

A large variety of sensing approaches and transducer types have been developed for the determination of the level of liquids and quasi liquids, e.g., slurries and powdered or granular solids, in open or enclosed vessels (such as tanks and ducts). Not only is the knowledge of the *level* itself important, but other measurements can be inferred from level. If the tank geometry and dimensions are additionally known, the *volume* of the liquid can be determined. If, additionally, the density of the liquid is known, its *mass* can be calculated.

Level is generally sensed by one of two methods: obtaining a discrete indication when a predetermined level has been reached (*point sensing*) or obtaining an analog representation of the level as it changes (*continuous sensing*). Point sensing is also used when it is only desired to establish whether a liquid or a gas exists at a certain point, e.g., in a pipe. The different level-sensing methods can be classified into those lending themselves primarily to point sensing, to continuous sensing, or both. It should be understood, of course, that point-sensing systems are usually simpler and cheaper than continuous sensing systems and should be used when only a discrete indication has to be obtained. Even when two or more discrete levels must be established in one vessel, the use of two or more point sensors may be preferable to a continuous sensing system. On the other hand, electronic circuitry can be used to provide one or more discrete level indications from a continuous sensing system.

Point Level-Sensing Methods.   Point level-sensing methods are usually aimed at indicating the interface between a liquid and a gas, sometimes the interface between two different liquids. Three methods are illustrated in Fig. 8.2.12. *Heat-transfer sensing* is used by two types of sensors: the resistive sensor (wire-wound or thermistor) is heated to some degree by the current passing through it so that its resistance changes because of cooling when contacted by the liquid; the thermoelectric sensor detects the cooling, upon liquid contact, of a wire-wound heater it is in thermal contact with. *Optical sensing* relies either on the presence or absence of reflection of a light beam from the interface between a prism surface in contact with gas (reflection) or liquid (no reflection) or on the greater attenuation of a light beam when it passes through liquid on its way to a light sensor. In *damped-oscillation sensing* the mechanical vibration of an element, excited into such vibration electrically, is either stopped (in a magnetostrictive or piezoelectric element) or reduced in amplitude (e.g., in an oscillating-paddle element) in response to acoustic damping or viscous damping, respectively, when the measured fluid changes to a liquid.

Continuous Level Sensors.   Three classic continuous level-sensing methods are illustrated in Fig. 8.2.13. The level, volume, or mass of a liquid in a tank of known geometry can be determined by *weighing* the tank continuously, as by means of a load cell (force transducer), and subtracting the tare weight of the tank or compensating for the tare weight. *Pressure sensing* relies on the pressure (*head*) developed at the base of a liquid column. This pressure increases with the column height and hence with level above the point at which pressure is sensed. The differential pressure P_D, measured by the differential-pressure transducer, on the tank shown in the illustration, is equal to the difference in pressures between the bottom and top of the tank $(P_L - P_H)$. The *level-sensing float* mechanically actuates a transduction element, usually a potentiometer,

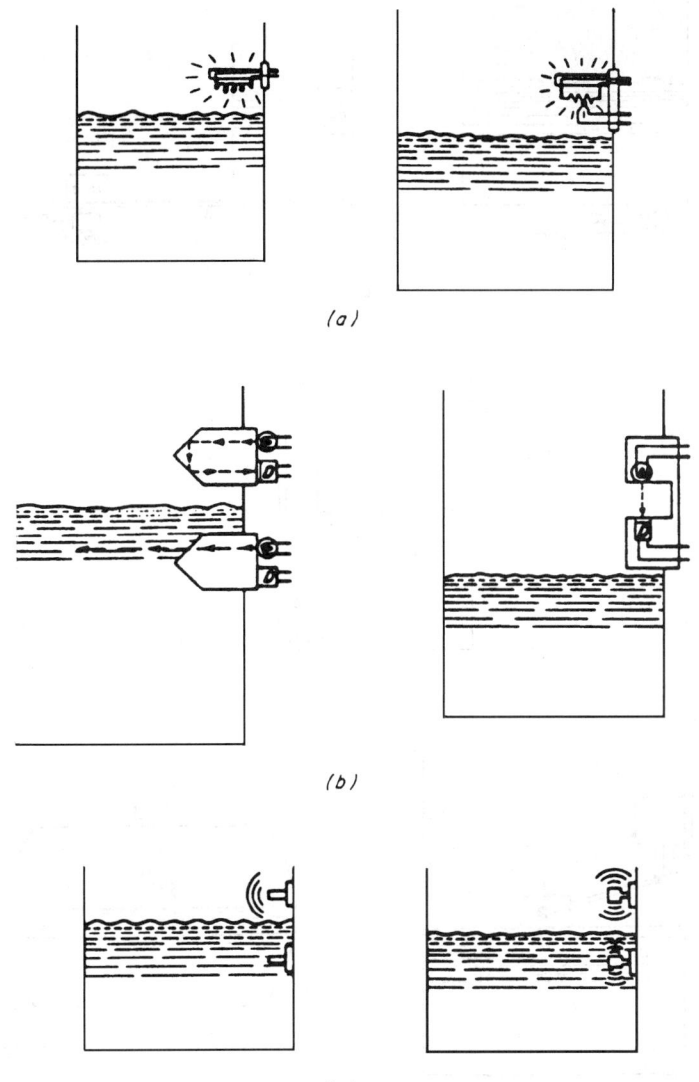

FIGURE 8.2.12   Point level-sensing methods: (*a*) by heat-transfer rate; (*b*) by optical means; (*c*) by oscillation damping.   (*From Ref. 1 by permission*)

sometimes a reluctive element or one or more magnetic reed switches. A radically different method (not illustrated) is *cavity-resonance sensing,* where electromagnetic oscillations are excited (from a coupling element at the tank top) within the gaseous cavity enclosed by the liquid surface and the upper tank walls, and the change in resonant frequency, as the liquid surface changes in location, becomes a measure of liquid level.

Several methods are equally useful for point and continuous level sensing (see Fig. 8.2.14). *Conductivity level sensing* is usable with even mildly conductive liquids. The resistance between two electrodes (the tank wall may serve as one of the two) changes continuously (or suddenly, in the case of the point-sensor version) as the liquid level rises or falls.

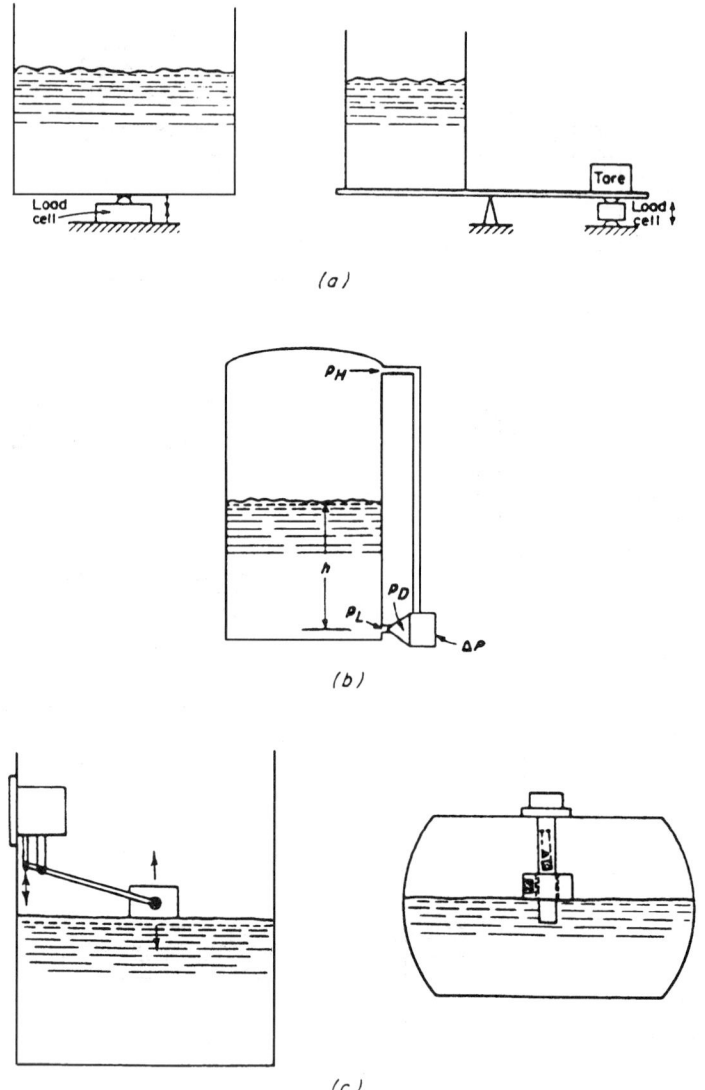

FIGURE 8.2.13   Continuous level-sensing methods: (*a*) by weighing; (*b*) by pressure sensing; (*c*) by float.   (*From Ref. 1 by permission*)

Dielectric-variation (capacitive) sensing is used primarily for nonconductive liquids, which then play the role of dielectric materials between two (sometimes four) concentric electrodes that are used (and electrically connected) as plates of a capacitor. The capacitance changes continuously (or suddenly, for the point sensor) as the vertical distance h of the level changes. If it is necessary to compensate for changes in the liquid's characteristics during measurement, a reference capacitor, always submerged, can be employed so that the ratio of the capacitance change equals the ratio of the measured level to the vertical dimension of the reference capacitor ($\Delta C / \Delta C_R = h/h_R$).

Sonic level sensing uses ultrasound either emitted from a sound projector and detected by a sound receiver or emitted and detected by a single sound transceiver operating alternately in the transmit and receive mode.

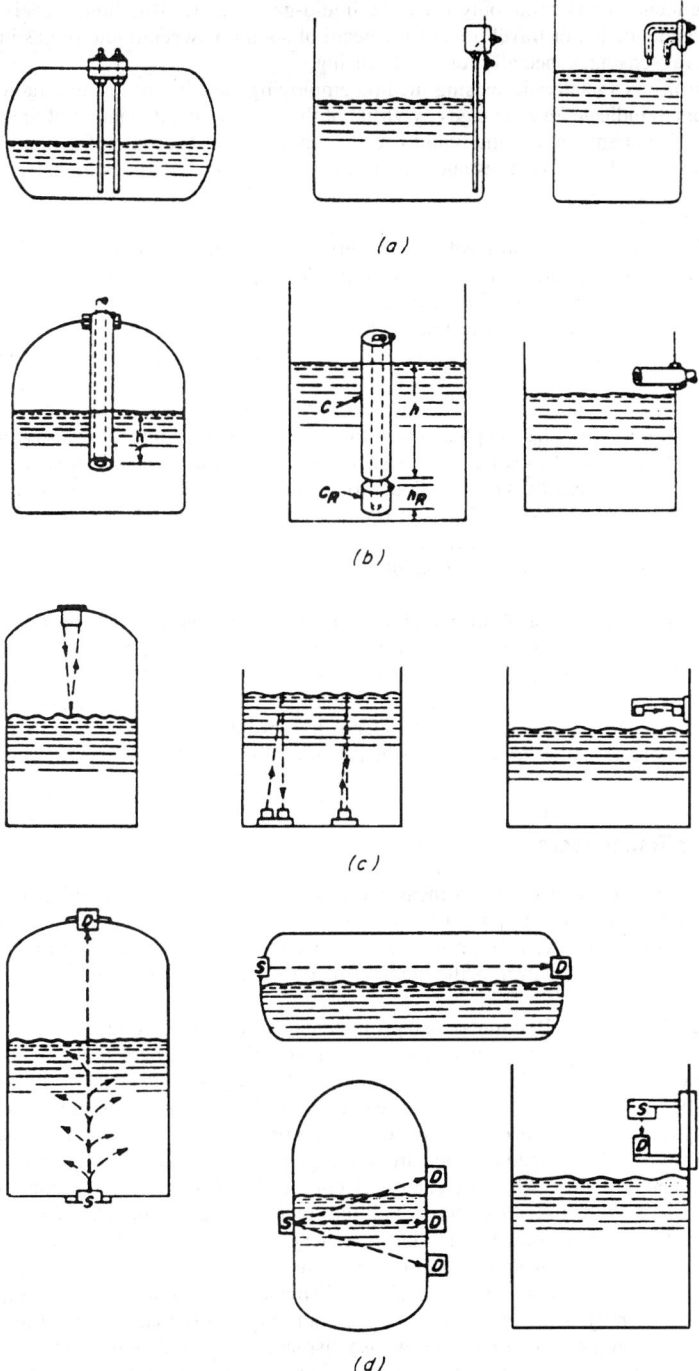

FIGURE 8.2.14 Continuous- and point-level sensing: (*a*) conductivity; (*b*) dielectric variation; (*c*) sonic sensing; (*d*) radiation sensing.   (*From Ref. 1 by permission*)

An echo-ranging technique is commonly used, the liquid-gas interface (the liquid level) acting as the target. The difference in attenuation or travel time of the beam of sound between liquid or gas in its path can also be used for sonic level sensing, especially for point sensing.

Radiation sensing is a nucleonic sensing method employing usually one or more radioisotope sources and radiation detectors to indicate level changes by virtue of the changes in attenuation of the radiation in response to level changes. The attenuation in the liquid is caused mainly by absorption. Such nucleonic methods have also been used to study density profiles and the location and extent of vortices in tanks and of gas bubbles in pipes.

Liquid-Level Transducers.   Liquid-level transducers, in their most common configuration, are probes, flange- or boss-mounted through the tank or duct wall. Some pipe-wall-mounted transducers are so designed that their sensing end is flush with the inside of the wall, to prevent obstructions to flow. Nucleonic transducer systems and some ultrasonic designs are attached to the outside of the wall.

The transduction principle of liquid-level transducers is given by the sensing technique employed. Dielectric-variation sensing demands capacitive transducers, using ac excitation having a frequency between 400 Hz and 200 kHz. Magnetostrictive and piezoelectric transducers, whose probe tip oscillates at a frequency in the vicinity of 40 kHz, find their application in the sonic, as well as the damped-oscillation, sensing techniques. Ionization-type, as well as solid-state, transducers are used in nucleonic systems. Photoelectric transducers are used in optical sensing systems. Potentiometric and reluctive transduction elements are found in float-actuated liquid-level transducers. Resistive transducers are used for heat-transfer sensing and, in a somewhat different form, for conductivity sensing. Thermoelectric elements are found in some heat-transfer sensors. Vibrating-element (notably vibrating-paddle) transducers find their use in damped-oscillation sensing systems.

Selection criteria involve, first of all, the choice of one or more point-level sensors or a continuous level sensor. After this choice has been made, together with an evaluation of end-to-end system requirements, the characteristics of the measured liquid are of primary importance. These include its conductivity, viscosity, temperature, chemical properties and, for installation in pipes or ducts, its flow rate and pressure. The transducer must also be designed and installed in such a manner as to prevent false level indications owing to slosh, spray, and splash or to adherence of liquid to the transducer with falling level.

Pressure and Vacuum Transducers

Pressure is force acting on a surface; it is measured as force per unit area, exerted at a given point. *Absolute pressure* is measured relative to zero pressure, *gage pressure* relative to ambient pressure, and *differential pressure* relative to a *reference pressure* or a range of reference pressures. A perfect *vacuum* is zero absolute pressure. Vacuum measurement, however, is the measurement of very low pressures.

Pressure-Sensing Elements.   Pressure-sensing elements are almost invariably mechanical in nature (see Fig. 8.2.15). They can be described generally as thin-walled elastic members which deflect when the pressure on one side of their wall is not balanced by a pressure on the opposite side. The former pressure is the measured pressure; the latter is either a vacuum or near vacuum (for absolute-pressure transducers), the ambient atmosphere (for gage-pressure transducers), or some other pressure (for differential-pressure transducers).

The *diaphragm* is a circular plate fastened around its periphery so that its center will deflect when pressure is applied to it. It can be flat or, when a greater deflection is required, contain a number of concentric corrugations that increase the effective area upon which the force (pressure) can act. Two corrugated diaphragms, welded, brazed, or soldered together around their periphery, form a *capsule* sensing element (aneroid). Two or more capsules can be fastened together so that the pressure acts on all. The displacement obtainable at the end of such a multiple-capsule element nearly equals the displacement of one capsule multiplied by the number of capsules in the assembly. The *bellows* sensing element is typically made from a thin-walled tube formed into deep convolutions and sealed at one end, whose displacement can then be made to act on a transduction element. In the *straight-tube* sensing element, again sealed at one end, applied pressure causes an expansion of the tube diameter. This expansion, though slight, can be converted into a usable output by a transduction element.

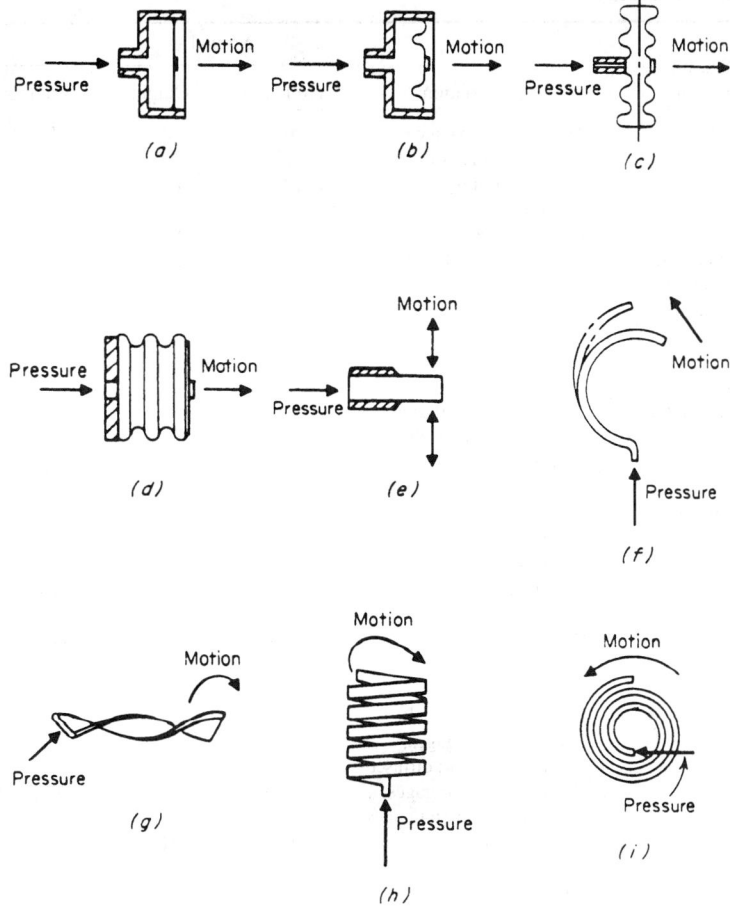

FIGURE 8.2.15   Pressure-sensing elements: (*a*) flat diaphragm; (*b*) corrugated diaphragm; (*c*) capsule; (*d*) bellows; (*e*) straight tube; (*f*) C-shaped Bourdon tube; (*g*) twisted Bourdon tube; (*h*) helical Bourdon tube; (*i*) spiral Bourdon tube.   (*From Ref. 1 by permission*)

The *Bourdon tube* is one of the most widely used sensing elements, particularly for pressure ranges higher than 2 MPa (about 300 lb/in.^{2}). The Bourdon tube, elliptical in cross section and sealed at its tip, tends to straighten from its curved, twisted, helical, or spiral shape, thus causing the tip to deflect sufficiently to act on a transduction element. The number of turns or twists in a Bourdon tube tends to multiply the tip travel.

Pressure Transducers.   Pressure transducers, using the sensing elements described above, provide their outputs by means of a large variety of transduction elements (see Table 8.2.1). Many designs are available with integrally packaged output- and excitation-conditioning circuitry. Certain designs, notably potentiometric, reluctive, and strain-gage transducers, are more prevalent than other types. Piezoelectric transducers are usable only for dynamic pressure measurements. Inductive transducers can be subject to severe temperature effects and are not used extensively.

A *potentiometric pressure transducer* features a dual-capsule sensing element that transfers its displacement to a lever-type wiper arm by means of a pushrod. The wiper then slides over the curved resistance

TABLE 8.2.1 Pressure Transducers

Transduction	Sensing elements	Type variations	Normal		Optional	
			Excitation	Output	Excitation	Output
Capacitive	Diaphragm	AC bridge unbalance	AC	AC	DC	DC
		Variable ionization	AC	DC		
		RF-tank-circuit detuning	AC	Freq.		
Inductive	Bellows	*LC* tank circuit	AC	Freq.		
	Diaphragm Bourdon tube	AC bridge unbalance	AC	AC		
Piezoelectric	Diaphragm	Natural crystal Synthetic crystal Ceramic	None	AC	DC	Amplified ac
Potentiometric	Capsule Bourdon tube	Wire-wound element Continuous-resolution element (metal film, cermet, plastic, carbon film)	DC	DC		
Reluctive	Diaphragm Bourdon tube	Inductance bridge Differential transformer	AC	AC	AC DC	DC DC
Strain gage	Diaphragm	Unbonded gages, metal wire	DC	DC	DC	Amplified dc
		Bonded gages, metal wire Micro-machined			DC	Digital
	Straight tube	Bonded gages, metal foil; bonded gages, semiconductor; diffused semiconductor gages; evaporated-metal gages				
Servo type	Capsule	Null balance	AC	AC		Encoder (digital)
	Bellows	Force balance	AC DC	DC DC		Synchro Potentiometric
Vibrating element	Diaphragm	Vibrating wire Vibrating diaphragm	AC	Freq.	DC	Freq.
	Straight tube	Vibrating cylinder				

element. Capsule elements are commonly used in such transducers for pressure ranges up to 2.5 MPa (about 360 lb/in.²).

Reluctive pressure transducers use either the inductance bridge circuit or, primarily when only the normal ac output is required, the differential-transformer circuit. When inductance bridge transducers use a diaphragm sensing element, the magnetic diaphragm itself, positioned between two coils, acts as the armature which increases the inductance of one coil while decreasing the inductance of the other coil. When inductance bridge transducers use a Bourdon tube sensing element, a flat armature plate, positioned over two coils, tilts more toward one coil than toward the other as the Bourdon tube tip rotates slightly with applied pressure. In differential-transformer transducers the sensing-element displacement is used to move a magnetic core within the transformer.

Most *strain-gage pressure transducers* use a diaphragm sensing element, although at least one good design uses a straight tube. Most designs have a four-active-arm strain-gage bridge, with the gages either on the diaphragm or on a beam actuated by the diaphragm. Included in this category are solid-state pressure transducers, micromachined from a silicon base, with integrally diffused gages.

When the sensing-element displacement is not sufficient for a given transduction element, a mechanical amplification linkage can be inserted between the two elements. Special design considerations apply to differential-pressure transducers when the measured fluid (at one of the two pressure ports) must not come in contact with the transduction element. One solution to this problem has been to fill the affected inside portion of the transducer with a *transfer fluid*, sealed off by a thin *membrane* to which the measured fluid can be applied safely. Gage-pressure transducers have the inside of their case (which usually acts as the *reference cavity*) vented to the outside through a small hole (*gage vent*), equipped with a fine-mesh screen, a porous plug, or another filter to prevent internal contamination.

Flush-diaphragm transducers are designed for high-frequency-response applications where use of tubing, or even the cavity formed by a mounting boss, may reduce response; these transducers are so designed that the diaphragm is flush (when installed) with the inside surface of the pipe wall (or other wall) through which they are mechanically fastened.

Specification characteristics of pressure transducers deserve particular attention since pressure is one of the two most common measurands (the other is temperature). Table 8.2.2 lists those characteristics which should be considered when preparing a specification for a pressure tarnsducer. Not all these characteristics need always be specified; some can be omitted when sufficient knowledge of the application permits.

Vacuum Transducers.   Vacuum transducers are an important subgroup of pressure transducers, though bearing little resemblance to them with regard to design and operation. The pressure constituting a practical dividing line between pressure and vacuum measurement is not well defined. Some pressure transducers are usable for very low pressure measurement. Generally, however, pressure measurements extending substantially below 133 Pa (= 1 torr) can be considered as vacuum measurements.

Vacuum transducers (see Table 8.2.3) exist in two major categories, given by their transduction principles.

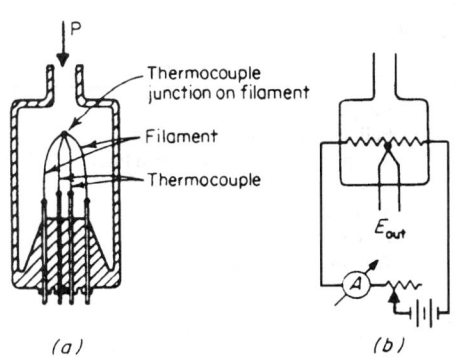

FIGURE 8.2.16   Thermoelectric thermoconductive vacuum transducer; (*a*) transducer; (*b*) typical circuit. (*From Ref. 1 by permission*)

Thermoconductive vacuum transducers measure pressure as a function of heat transfer by the measured gas. As the number of gas molecules within the transducer decreases, the quantity of heat transferred from a heated filament, through the gas, and to the case of the transducer, will decrease proportionally. The *Pirani gage,* as well as the *thermocouple gage* (which may use a thermopile instead of a single junction), both use this principle. A basic thermocouple gage is illustrated in Fig. 8.2.16.

Ionizing vacuum transducers measure pressure as a function of gas density by measuring ion current. Since different gases have different densities, the calibration of such a transducer will usually differ as well. The gas is usually ionized by electrons, except in one type using alpha particles for this purpose.

In thermionic vacuum transducers the electrons are emitted by a filamentary cathode, and positive ions are collected at the anode. Various modifications of the original triode type have helped to extend its lower range limit from 10^{-8} to 10^{-10} torr (*Bayard-Alpert gage,* by reducing internal x-ray effects) to 10^{-11} torr (*Nottingham gage,* by reducing electrostatic-charge effects) and to 10^{-12} torr (*Schuemann modification,* by virtually eliminating x-ray effects). The ion current, representative of pressure, is in the microampere region.

Several ionizing vacuum transducer types, whose electrons are emitted from either hot or ion-bombarded cold cathodes, use a magnetic field to increase the electron path length by forcing this path to be helical so that the probability of electron collisions with gas molecules is increased (*magnetron gages*). The hot-cathode versions include the *Lafferty gage.* The *Philips* (or *Penning*) *gage* and the *Redhead gage* are examples of the cold-cathode versions.

TABLE 8.2.2   Specification Characteristics for Pressure Transducers

Mechanical design characteristics	
Specified by user and manufacturer	Stated by manufacturer
Configuration and dimensions (shown on drawing)	Sensing element
Mountings (shown on drawing)	Transduction-element details
Type and location of pressure ports (shown on drawing)	Materials in contact with measured fluids
Type and location of electrical connections (shown on drawing)	Dead volume
Nature of pressure to be measured, including range	Type of damping (including type of damping
Measured fluids	oil if used)
Case sealing (explosion proof, burstproof, or waterproof enclosure)	
Isolation of transduction element	
Mounting and coupling torque	
Weight	
Identification	
Nameplate location (shown on drawing)	

Electrical design characteristics	
Excitation (nominal and limits)	
Power rating (optional)	
Input impedance (or element resistance)	
Output impedance	
Insulation resistance (or breakdown-voltage rating)	
Wiper noise (in potentiometric transducers)	
Output noise (in dc output transducers)	
Electrical connections and wiring diagram	
Integral provisions for simulated calibration (optional)	

Performance characteristics	
Individual specifications	Error-band specifications
Range	Range
End points	Full-scale output (nominal)
Full-scale output	End points (defining reference line)
Creep (optional)	Resolution (where applicable)
Resolution (where applicable)	Static error band
Linearity	Reference-pressure range*
Hysteresis	Warm-up period (optional)
Repeatability	Frequency response
Friction error	
Zero balance	
Zero shift	
Sensitivity shift	
Warm-up period (optional)	
Frequency response	
Operating temperature range	Operating temperature range
Temperature error or thermal zero shift and thermal sensitivity shift	Temperature error band
Temperature-gradient error	Temperature-gradient error
Ambient-pressure error	Ambient-pressure error band
Acceleration error	Acceleration error band
Vibration error	Vibration error band

(Continued)

TABLE 8.2.2 Specification Characteristics for Pressure Transducers (*Continued*)

Performance after exposure to:	Performance during and after exposure to:
Shock (triaxial)	High sound-pressure levels
Humidity	Sand and dust
Salt spray or salt atmosphere	Ozone
	Nuclear radiation
	High-intensity magnetic fields etc.

*Applies to differential-pressure transducers only.

Selection criteria for vacuum transducers (and any necessary ancillary equipment for them) are primarily the required measuring range; secondarily, size, weight, ruggedness, and complexity. Considerations for the selection of a pressure transducer are primarily range, type of excitation and output, accuracy and frequency response; secondarily, the properties of the measured fluid and environmental conditions.

Ranges of Pressure Transducers.   Some of the pressure transducers described in this section are available for ranges up to 100 MPa (about 15,000 lb/in.^{2}). Special sensing devices have been designed for pressures up to 7 GPa (about 1 million lb/in.^{2}). Pressure transducers are also used to measure altitude (a known nonlinear relationship exists between atmospheric pressure and altitude above sea level), water depth (pressure increases at the rate of approximately 1 KPa/m (0.44 lb/in.^{2} ft) when descending below the water surface), and air speed (by measuring the difference between impact pressure, obtained from a pitot tube, and static pressure, while in flight).

TABLE 8.2.3 Vacuum Transducers

Sensing element	Common name	Output	Nominal range, torr
	Thermoconductive transduction		
Heated filament	Pirani gage	Resistance change	10^{-3}–1
	Thermocouple gage	Thermoelectric emf	10^{-3}–1
	Ionizing transduction		
Thermionic, triode	Bayard-Albert gage	Direct current	10^{-10}–10^{-3}
Photomultiplier		Direct current	10^{-15}–10^{-3}
Hot cathode, magnetic field	Hot-cathode magnetron gage (Laffety gage)	Direct current	10^{-14}–10^{-5}
Cold cathode, magnetic field	Cold-cathode magnetron gage (Philips gage, Penning gage)	Direct current	10^{-7}–10^{-3}
With flash filament		Direct current	10^{-12}–10^{-3}
With auxiliary cathode	Redhead gage	Direct current	10^{-13}–10^{-4}
Radioactive, alpha particles	Alphatron	DC voltage (from amplifier)	10^{-5}–1000
Beta particles		DC voltage (from amplifier)	10^{-8}–1000

REFERENCE

1. Norton, H. N., "Handbook of Transducers," Prentice Hall, 1989.

BIBLIOGRAPHY

Cho, C. H., "Measurement and Control of Liquid Level," Instrument Society of America, 1982.

DeCarlo, J. P., "Fundamentals of Flow Measurement," Instrument Society of America, 1984.

Harris, C. M., and C. E. Crede (eds.), "Shock and Vibration Handbook," 2nd ed., McGraw-Hill, 1977.

McConnell, K. G., "Notes on Vibration Frequency Analysis," SEM Publications, 1978.

McConnell, K. G., "Dynamic Force, Motion and Pressure Measurements," SEM Publications, 1984.

Miller, R. W., "Flow Measurement Engineering Handbook," McGraw-Hill, 1983.

Roark, R. J., and Young, W. C., "Formulas for Stress and Strain," 5th ed., SEM Publications, 1975.

Szymkowiak, E., "Optimized Vibration Testing and Analysis," Institute of Environmental Sciences, 1983.

Window, A. L., and Holister, G. S., "Strain Gauge Technology," SEM Publications, 1982.

Moisture and Humidity—Measurement and Control in Science and Industry (Symposium papers), Instrument Society of America, 1985.

"Relative Humidity by Wet- and Dry-Bulb Psychrometer, Method of Test for," *ANSI Z110.3/ASTM E337*. American Society for Testing and Materials.

"Specification, Installation, and Calibration of Turbine Flowmeters," *ANSI/ISA RP31.1-1997*. Instrument Society of America, 1977.

Standards for Transducers, available from Instrument Society of America (All reaffirmed 1995).

a. "Guide for Specifications and Tests for Piezoelectric Acceleration Transducers for Aerospace Testing." *ISA RP37.2*.

b. "Specifications and Tests for Strain Gage Linear Acceleration Transducers," *ANSI/ISA S37.5-1975*.

c. "Specifications and Tests for Strain Gage Force Transducers," *ANSI/ISA S37.8-1977*.

d. "Specifications and Tests for Potentiometric Displacement Transducers," *ANSI/ISA S37.12-1977*.

e. "Specifications and Tests for Strain Gage Pressure Transducers," *ANSI/ISA S37.3-1975*.

f. "Specifications and Tests of Potentiometric Pressure Transducers," *ANSI/ISA S37.6-1976*.

g. "Specifications and Tests for Piezoelectric Pressure and Sound-Pressure Transducers," *ANSI MC6.4/ISA S37.10-1975*.

CHAPTER 8.3
TRANSDUCERS FOR THERMAL QUANTITIES

Harry N. Norton

INTRODUCTION

The *temperature* of a body or substance is (*a*) its potential of heat flow, (*b*) a measure of the mean kinetic energy of its molecules, and (*c*) its thermal state considered with reference to its power of communicating heat to other bodies or substances. *Heat* is energy in transfer, resulting from a difference in temperature between a system and its surroundings or between two systems, substances, or bodies. Heat energy is transferred by one or more of the following methods of *heat transfer:* (*a*) *conduction*, by diffusion through solid material or stagnant liquids or gases; (*b*) *convection*, by the movement of a liquid or gas between two points; and (*c*) *radiation*, by electromagnetic waves.

The sensing elements of temperature transducers typically act as transduction elements as well. The two most commonly used sensing-transduction elements are the *thermoelectric* element (*thermocouple*) and the *resistive* element (*resistance thermometer*). Among other sensing-transduction elements the only one that has found commercial acceptance is the *oscillating-crystal* element, essentially a quartz crystal (connected into an oscillator circuit) which has a substantial and highly linear temperature coefficient of frequency.

Thermocouples

A thermocouple is an electric circuit consisting of a pair of wires of different metals joined together at one end (*sensing junction*) and terminated at their other end in such a manner that the terminals (*reference junction*) are both at the same and known temperature (*reference temperature*). Connecting leads from the reference junction to some sort of load resistance (an indicating meter or the input impedance of other readout or signal-conditioning equipment) complete the thermocouple circuit. Both these connecting leads can be of copper or some other metals different from the metals joined at the sensing junction. Because of the *thermo-electric effect (Seebeck effect)*, a current is caused to flow through the circuit whenever the sensing junction and the reference junction are at different temperatures. In practice, the reference junction is either held at a known constant temperature (e.g., at 0°C) or is electrically compensated for variations from a preselected temperature.

The electromotive force (*thermoelectric emf*), which causes current flow through the circuit, is dependent in its magnitude on the sensing-junction wire materials, as well as on the temperature difference between the two junctions. Commonly used wire materials are Chromel (CR) and Alumel (AL) (both registered trade names of Hoskins Mfg. Co., Detroit, Michigan), Constantan (CN, an alloy of 53 percent copper and 45 percent nickel), copper (Cu), iron (Fe), platinum (Pt), an alloy of platinum and (either 10 or 13 percent) rhodium (Rh), tungsten (W), tungsten-rhenium (Re) alloys (5 or 26 percent rhenium content is typical), nickel (Ni), and ferrous nickel alloys.

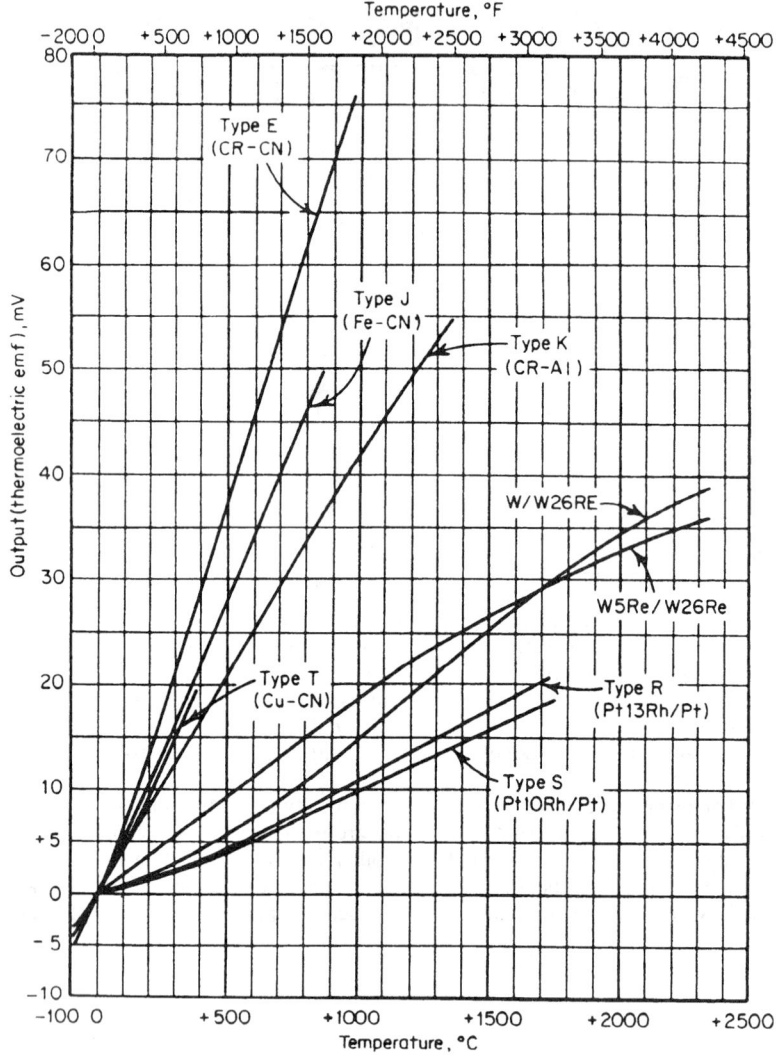

FIGURE 8.3.1 Thermocouple output vs. temperature characteristics (reference junction at 0°C).

The characteristics of certain combinations of wire materials, such as their thermoelectric emf versus temperature characteristics, their accuracy tolerances, and wire-insulation color codes, were standardized by ANSI Standard C96.1 (which is based on ISA Recommended Practice RP1) in such a manner that materials of different brand names can be used as long as the characteristics assigned to a specific type of thermocouple are maintained.

The names of the wire materials constituting, in their combination, a thermocouple sensing junction are now listed only as typical examples. Thus typical materials of a *type K* thermocouple are Chromel and Alumel. The ANSI Standard favors the use of type-letter designations in lieu of the names of the two metals used. Figure 8.3.1 shows the thermoelectric emf obtainable from various types of thermocouples when the reference temperature is held at 0°C.

Thermopiles (see Fig. 8.3.2) consists of several sensing junctions of the same material pairs, in close proximity to each other and connected in series so as to multiply the output obtainable from a single sensing junction.

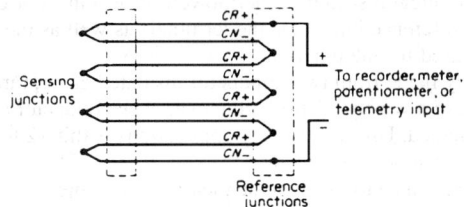

FIGURE 8.3.2 Thermopile schematic diagram: CR-CN combination shown as example.

The isothermal reference junctions are usually also in close proximity to each other to assure an equal temperature for each reference junction.

Resistive Temperature-Sensing Elements

Resistive temperature-sensing elements are either conductive or semiconductive. Conductive elements are usually wire-wound, sometimes made of metal foil or film. Elements wound of high-purity annealed platinum wire are best suited for most applications. Other metal-wire elements are wound of nickel or nickel alloy. Copper-wire elements are rarely used any more. Tungsten-wire elements have shown some promising characteristics but are generally considered too difficult to manufacture and too brittle to stay reliable.

A platinum-wire element has been used to define the International Practical Temperature Scale from − 183 to +630°C, and it is expected that this upper limit will be extended to the melting point of gold (+1063°C). The resistance versus temperature curve of such an element follows a well-defined theoretical relationship, making most points on the curve calculable within very close tolerances when only a few measured points have been established. Repeatabilities within about 0.01 K have been obtained at temperatures up to the freezing point of gold (1337.58 K). Semiconductive resistive temperature-sensing elements include thermistors, germanium and silicon crystals, carbon resistors, and gallium arsenide diodes. Thermistors have a nonlinear and negative temperature coefficient of resistance and an empirical resistance versus temperature relationship.

Temperature Transducers

Temperature transducers are classified into two general categories: surface-temperature transducers, which are cemented, welded, bolted, or clamped to a surface whose temperature is to be measured, and immersion probes, which are immersed into stagnant or moving fluids to measure their temperature. The fluid can be in a pipe, a duct, a tank, or other enclosed vessel, where the immersion probe is mounted through a pressure-sealed opening. It can also be freely moving, even at almost imperceptible rates of motion, e.g., an open body of water, an outdoor or indoor atmosphere.

Thermoelectric temperature transducers have the same sort of sensing junction, whether they are intended for surface temperature measurement or as immersion probes. The junctions between the two dissimilar-metal wire pairs are made by butt-welding the wire ends, by crossing them and welding them, by coiling one wire end around the other, or twisting the two ends about each other, then welding, brazing, or soldering the junction, or by welding both wire ends, in very close proximity to each other, to a metallic surface or to the metallic inside of an immersion-probe tip.

For surface measurements, the junctions are soldered, brazed, or welded to a surface (if it is metallic) or cemented to it (if it is not). If it is cemented, care must be taken to have the junction in solid thermal contact with the measured surface. Taping a junction to a surface is poor practice, since even a very small gap between junction and surface can introduce considerable errors. For immersion measurements, thermocouples are often produced with an integral sheath or inserted into a sealed immersion sheath (*thermowell*).

Junctions for thermoelectric immersion probes can be grounded (metallic contact from junction to sheath or thermowell) or isolated (ungrounded). In some cases, exposed junctions, at the tip of a probe, are immersed in the fluid without use of an integral sheath or thermowell. If terminals or connectors must be used between the sensing junction and the reference junctions, the terminals as well as the *extension wires* must be made of the same types of metals as used for the junction.

Thermocouples are usually made from two-conductor insulated cable, rarely from reels of individual bare-wire materials. The cables have a variety of insulation, over each conductor as well as over the conductor pair, and can be shielded or unshielded. Useful for many applications is thin (2 to 10 mm outside diameter) metal-sheathed, ceramic-insulated thermocouple cable.

Differential thermocouples can be used when the measurement objective is to measure the temperature difference between two points. In this case the sensing junction at the other measured point replaces the reference junction. The first wire of the first junction and the second wire of the second junction must still be brought to isothermal terminals; however, it is not necessary that the temperature of these terminals be known.

Resistive Temperature Transducers

Electrically conductive surface-temperature transducers are usually small and flat enough not to be influenced by convective heat transfer but only by conductive transfer from the measured surface. After installation they may be coated or covered to minimize any radiative heat transfer to them. The sensing element is usually a metal wire either wound around a thin insulating "card" or a coiled wire cemented to the base (see Fig. 8.3.3). Some metal-foil transducers (encapsulated or *free-grid*) are in the shape of a zigzag pattern. All designs are aimed at exposing the maximum sensing surface to the conductive heat transfer in an area of minimum size.

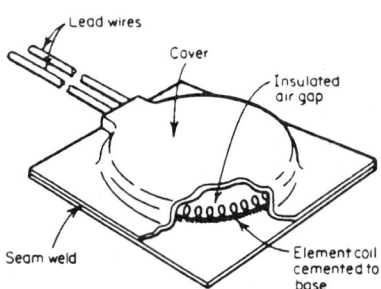

FIGURE 8.3.3   Platinum-wire resistive surface-temperature transducer.   (*Rosemount Inc.*)

Resistive metal-wire *immersion probes*, most commonly with a platinum-wire element but some-times with elements of nickel or nickel-alloy wire, are widely used for industrial and scientific fluid-temperature measurements. The probe-type transducer, illustrated in Fig. 8.3.4, has a ceramic encapsulated (coated) element in a perforated protective sheath so that it is usable for a variety of measured fluids over a wide temperature range. For applications in relatively stagnant fluids an unencapsulated (*exposed*) element is used to provide a shorter time constant. Some fluids require an element completely *enclosed* within a metallic well but with good thermal contact between well and element. The threaded mounting allows for compression sealing by means of a gasket or O ring between the housing and the mounting boss.

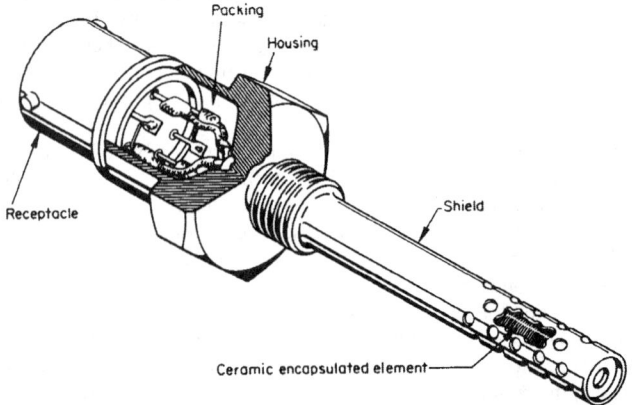

FIGURE 8.3.4   Platinum-wire resistive immersible-probe temperature transducer.   (*Rosemount Inc.*)

Thermistors

Thermistors are used for surface-temperature as well as fluid measurements. Because of their nonlinear (essentially negative exponential) resistance versus temperature characteristics, they are particularly useful when a large resistance change is needed for a narrow range of temperature. Where a short time constant is required, a glass-coated thermistor bead, as small as 0.3 mm in diameter, can be suspended on its 0.03-mm-diameter precious-metal-alloy leads. Where somewhat more ruggedness is required, a glass-encapsulated bead about 1.5 mm around the tip and 4 mm long can be used. Excitation power must be kept low to avoid errors due to self-heating. Thermistor-type temperature transducers are available in a large variety of configurations.

Semiconductor Thermometers

Germanium thermometers are made of germanium crystals with highly selected and controlled impurities (dopants). They are intended primarily for cryogenic temperature measurements (below −195°C). Carbon resistors have also been used for such applications, as have gallium-ărsenide junction diodes, which can be used to somewhat higher temperatures. Silicon-wafer transducers have been used for surface-temperature measurements in the range −50 to 275°C, where their resistance versus temperature characteristics are similar to those of some metal wires.

Quartz-Crystal Temperature Transducers

Quartz-crystal temperature transducers use oscillating-crystal sensing elements in such a manner that the change of oscillator frequency with temperature is nearly linear over a range from about −50 to 250°C. They are usually furnished with associated electronics and readout equipment. This tends to limit their usability for general telemetry application without, however, detracting from their advantages in laboratory applications.

The selection of a temperature transducer is more complex than the selection of most other types of transducers. The objective is to select a design whose sensing element will attain the temperature of the measured material within the time available to make the measurement. Among primary selection criteria are, then, the characteristics and properties of the measured solid or fluid, the measuring-range limits, the required response time (time constant), and the type of excitation and signal conditioning available or intended to be used.

Radiation Pyrometers

Radiation pyrometers are noncontacting temperature transducers which respond to radiative heat transfer from the measured surface or material. This radiation occurs primarily in the infrared portion of the electromagnetic spectrum (wavelengths between 0.75 and 1000 μm). Typical radiation pyrometers use an optical lens or mirror system (sensitive in the infrared region) which focuses the radiation on a thermoelectric or resistive (usually photoconductive) sensing element. The output of the sensing element can be correlated, by calibration, to the temperature of the measured surface. Radiation pyrometers are used primarily for high-temperature measurements (up to about 3500°C), but have also been found useful for noncontacting measurements in the medium temperature range (down to about −50°C). Some designs now use optical fibers in place of purely radiative coupling between source and instrument.

Heat-Flux Transducers

Two basic types of transducers have been developed to measure *heat flux*, heat transfer in terms of the total amount of thermal energy (heat flux is commonly expressed in W/cm^2 or Btu/ft^2·s). The *calorimeter* provides an output proportional to convective as well as radiant thermal energy (*total heat flux*). The *radiometer* responds to radiant thermal energy (*radiant heat flux*) only. Virtually all heat-flux transducers have thermoelectric sensing elements.

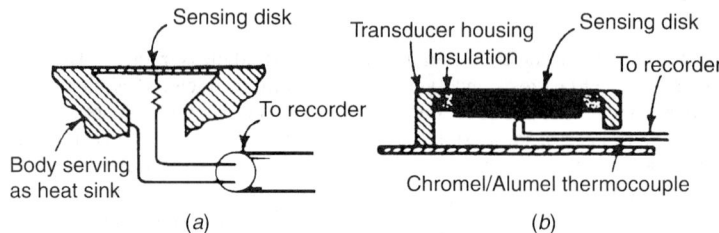

FIGURE 8.3.5   Calorimeters: (*a*) foil: (*b*) slug.

Calorimeters

The *foil calorimeter* (*membrance calorimeter*, Gardon gage) acts as a copper-Constantan differential thermocouple. When heat flux is received by the thin Constantan sensing disk (Fig. 8.3.5*a*), which is metallurgically bonded around its rim to a copper heat sink, the heat absorbed by the membrane is transferred radially to the heat sink. This causes a temperature difference between the center of the disk and its rim. A thin copper wire is attached to the bottom surface of the disk, at its exact center, thus forming one copper-to-Constantan sensing junction. The copper-to-Constantan contact around the rim of the disk forms the other junction. The output of the calorimeter is then proportional to the energy absorbed. When heat flux must be measured over long periods of time, the foil calorimeter can be provided with tubing and an internal flow path so that it can be water-cooled.

The *slug calorimeter (slope calorimeter)* uses a relatively thick thermal-mass sensing disk with an external high-emissivity (black) coating, which is thermally insulated from the transducer housing (Fig. 8.3.5*b*). A thin-wire thermocouple is attached to the bottom of the disk (slug), at its center. When heat flux is received by the slug, an output signal is produced by the thermocouple. The signal is proportional to the temperature rise of the slug.

Radiometers

A typical *radiometer* is essentially a foil calorimeter with a *window* (usually of quartz or synthetic sapphire) mounted over the sensing disk so that a disk can receive radiant heat flux but no convective heat flux. The cavity formed by window, transducer housing, and sensing disk is usually sealed, but provisions for gas purging of this cavity can be made to prevent window clouding when the radiometer is to be used in a contaminating atmosphere. Radiometers can also be water-cooled. The sensitivity of a radiometer can be increased by using a differential (multijunction) thermopile instead of the two-junction differential thermocouple.

BIBLIOGRAPHY

"American National Standard For Temperature Measurement Thermocouples," *ANSI MC96.1.* Instrument Society of America, 1982.

Benedict, R. P., "Fundamentals of Temperature, Pressure and Flow Measurements," 3d ed., Instrument Society of America, 1984.

Kerlin, T. W., and R. L. Shepard, "Industrial Temperature Measurement" (Student Text, Instructor's Guide, Slides). Instrument Society of America, 1982.

Temperature: Its Measurement and Control in Science and Industry (Symposium papers, 2 parts), American Institute of Physics, 1982.

CHAPTER 8.4
TRANSDUCERS FOR ACOUSTIC, OPTICAL, AND ELECTRICAL QUANTITIES

Harry N. Norton

TRANSDUCERS FOR ACOUSTIC QUANTITIES

Terminology

Sound is an oscillation in pressure, stress, particle displacement, or other physical characteristics, in an elastic or viscous medium. *Sound sensation* is the auditory sensation evoked by the oscillations associated with sound. *Sound pressure* is the total instantaneous pressure at a given point, in the presence of a sound wave, minus the static pressure at that point. *Sound pressure level* (SPL or L_p) is 20 times the logarithmic ratio of the mean-square sound pressure p to a mean-square reference pressure p_{ref}. It is normally expressed in decibels as SPL = 20 log ($p_{rms}/p_{ref.rms}$). The reference pressure is usually specified as 2×10^{-4} μbar, sometimes as 1 μbar (0.1 Pa). *Sound level* is a weighted sound-pressure-level reading obtained with a meter complying with a standard, e.g., ANSI Standard S1.4, Specification for General-Purpose Sound Level Meters.

Sound-Pressure Transducers

The sensing element of a sound-pressure transducer is almost invariably a diaphragm. The reference cavity behind the diaphragm is vented to the ambient atmosphere by means of a small hole in the transducer case so that static pressures on both sides of the diaphragm are equalized and only sound pressure is sensed.

A perforated cap over the diaphragm protects the diaphragm mechanically and, by its shape and geometry of perforations, provides some control over the transducer's directivity characteristics.

Sound-pressure transducers can be described, essentially, as special-purpose gage-pressure transducers. *Capacitive sound-pressure transducers* (usually called *condenser microphones*) use the sensing diaphragm as one electrode of a capacitor and a rigidly supported back plate, insulated from the rest of the structure but provided with a connecting lead or terminal, as the other electrode. A dc polarization voltage, applied across the two electrodes through a high-series resistance, maintains a constant charge on them. Capacitance changes due to diaphragm deflection cause changes in the voltage across the electrodes. The transducer output is first fed to an emitter follower so as to reduce the output impedance to a workable value. The output is the amplified. The emitter-follower (or cathode-follower) circuitry is sometimes built into the transducer case to keep the coupling path short. A shielded cable connects the transducer to the amplifier.

Piezoelectric and, to a limited extent, *inductive pressure transducers* have also been designed as sound-pressure transducers. Some piezoelectric designs have sealed cases, primarily to protect the internal components from atmospheric moisture and contaminants. The absence of a gage vent, however, necessitates correction of output readings when the transducer is used at low ambient pressures (e.g., high altitudes). Piezoelectric transducers do not require an excitation power supply but do require an amplifier.

The primary performance characteristics of sound-pressure transducers are range, output, frequency response, and directivity (directional response). Output is usually expressed as sensitivity or sensitivity level, sometimes as full-scale output for a stated range of sound pressures or sound-pressure levels.

Sound-Level Meters

Sound-level meters are complete, self-contained measuring systems, typically battery-operated and portable. A sound-level meter consists of a sound-pressure transducer (microphone), amplifier, standardized weighting networks, a calibrated attenuator, and an indicating meter. The sound-level range is always referred to a sound pressure of 10^{-4} μbar. The weightings denote different frequency-response characteristics of the measuring system. Referred to merely as A, B, or C, they are defined in a national standard as, for the United States, ANSI Standard S1.4.

Underwater Sound Detectors

Underwater sound detectors are used either for listening (*hydrophone*) or, in conjunction with an *underwater sound projector*, in sonar (*s*ound *n*avigation *a*nd *r*anging) systems. The transmitting and receiving function in a sonar system are frequently combined in a single device (sound transceiver). In the sonar field, underwater sound detectors, as well as projectors and transceivers, are commonly referred to as *transducers*.

TRANSDUCERS FOR OPTICAL QUANTITIES

Terminology

Light is a form of radiant energy, an electromagnetic radiation whose wavelength is between approximately 100 and 0.01 μm. By strict definition, only visible light (0.4 to 0.76 μm wavelength) can be considered as light, and infrared or ultraviolet light is then termed *radiation*. The light spectrum, in terms of wavelengths, frequency, photon energy, and blackbody temperature (all interrelated by physical laws), is illustrated in Fig. 8.4.1 with the visible-light spectrum (color spectrum) brought out in detail.

The transduction elements of light sensors (photocells, photosensors, *photodetectors*, light detectors) also act as sensing elements since they convert electromagnetic radiation into a usable electrical output. Four transduction principles are commonly used: photovoltaic, photoconductive, photoconductive junction, and photoemissive (see Fig. 8.4.2).

Photovoltaic Light Sensors

Photovoltaic light sensors are self-generating in that their output voltage is a function of the illumination of a junction between two dissimilar materials. These materials are semiconductive, either nonmetallic or metal compounds (Fig. 8.4.2a). Photons (particles of light) first pass through a thin conductive layer, and then impinge on the junction, causing an electron flow across the junction area so that the conductive layer becomes the negative terminal of the sensor. Various materials constitute the conductive and semiconductive portions of a photovoltaic light sensor.

The *silicon cell* (silicon photovoltaic cell, silicon solar cell) uses an arsenic-doped *n*-type silicon wafer. Boron is diffused into the upper (light-receiving) surface to create a thin *p*-type silicon layer. The *pn* junction between the layer and the wafer acts as a permanent electric field. Photons incident upon the junction cause a

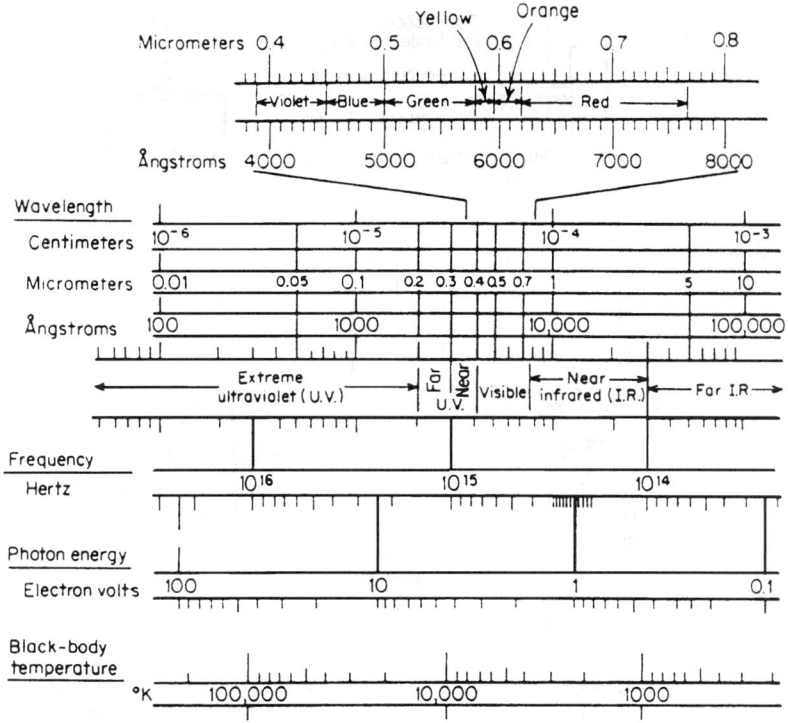

FIGURE 8.4.1 The light spectrum. *(From Ref. 1 by permission)*

flow of positive and negative charges. The *pn* junction, acting as an electric field, directs the positive charges into the *p*-type material (nickel plating around its edge forms the positive connecting terminal) while directing the negative charges into the *n*-type material (solder around the bottom edge of the silicon wafer forms the negative connecting terminal). The third connecting pin serves as a case connection.

Germanium photovoltaic light sensors are similar to silicon types but have a different spectral response (a peak near 1.55 μm, as compared with 0.8 to 0.9 μm for silicon). Special "blue-enhanced" silicon cells have spectral responses with their peak as low as 0.56 μm (the nominal peak of the human eye's response). *Indium-arsenide* (InAs) and *indium-antimonide* (InSb) photovoltaic sensors have spectral-response peaks near 3.5 and 6.8 μm, respectively, at room temperature. They are single-crystal *pn* junction semiconductors, used primarily for infrared-light sensing. In such applications they are often cooled artificially to increase their sensitivity.

Photoconductive Light Sensors

Photoconductive light sensors are widely used for conrol functions, such as automatic exposure control in cameras, in addition to their photometric (light-measurement) applications. The photoconductors are polycrystalline films or bulk single-crystal materials which, when contained between two conductive electrodes, act as light-sensitive resistors (Fig. 8.4.2*b*) whose resistance decreases as incident illumination increases. *Cadmium sulfide* (CdS) and *cadmium selenide* (CdSe) are popular because of their spectral-response peaks in the visible-light region (approximately 0.6 μm for CdS, 0.72 μm for CdSe) and because of their relatively high output without artificial cooling. Some photoconductive sensors use mixed CdS–CdSe crystals to obtain a response peak around 0.66 μm.

Lead sulfide (PbS) and *lead selenide* (PbSe) photoconductive cells are used for infrared-light sensing because their spectral-response peaks are close to 2.2 μm. The spectral-response curve of PbSe, however, is shallow enough to provide good sensitivity between about 1.8 and 3.6 μm, and its time constant is less than

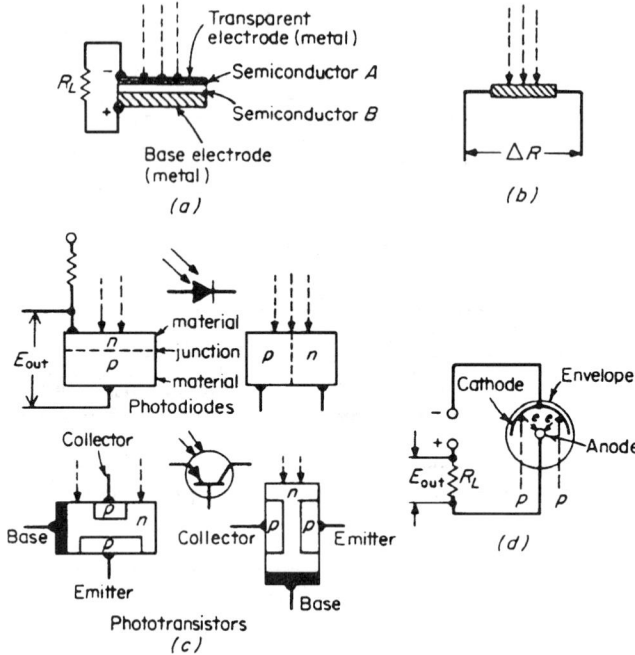

FIGURE 8.4.2 Basic methods of light transduction: (*a*) photovoltaic; (*b*) photoconductive; (*c*) photoconductive junction semiconductor; (*d*) photoemissive. (*From Ref. 1 by permission*)

one-tenth that of PbS. *Mercury-doped-germanium* (HgGe) photoconductive sensors have been used for far-infrared measurements while being cooled to cryogenic temperatures. *Lead-tin-telluride* (PbSnTE) sensors are always cooled to 77 K and have a spectral response in the 8- to 12-μm region. *Mercury-cadmium-telluride* sensors operate at 77 to 120 K and have a spectral response within the overall region of 2 to 14 μm adjustable by varying the proportions of the three materials.

Photoconductive Junction Sensors

In these devices the resistance across the junction in a semiconductor device changes as a function of incident light (Fig. 8.4.2*c*). Increasing incident illumination causes the junction photocurrent to increase. This category of photosensors includes *photodiodes* and *npn* as well as *pnp phototransistors*. They are made of silicon, with a spectral peak near 1.0 μm, except when germanium must be used to raise the spectral peak to about 1.6 μm. Time constants of photodiodes and phototransistors are less than 1 μs (compared with 10 μs for PbSe photoconductive sensors; 100 to 700 μs for PbS photoconductive sensors; 10 ms for CdSe photoconductive cells; and 100 ms for CdS cells but around 20 μs for silicon photovoltaic cells). Transistors also provide some amplification of the light-induced signal. Both types are usually sealed into a standard transistor can, sometimes furnished with a lens or window. A special silicon photodiode design has also been developed for ultraviolet measurements (0.06- to 0.25-μ region).

Photoemissive Sensors

The earliest *photoemissive light sensor* was the phototube, in which electrons are emitted by the cathode of a vacuum (or gas-filled) diode tube when photons impinge on the cathode surface (Fig. 8.4.2*d*). When a closely spaced anode is at a positive potential with respect to the cathode, the anode collects some of these electrons, and the resulting anode current can produce an output voltage as an *IR* drop across a suitable load resistor (R_L).

The photoemissive principle is now employed mostly by *photomultiplier* tubes, in which additional electrodes (*dynodes*) are placed between cathode (*photocathode*) and anode to amplify the electron current by means of secondary emission. An additional dynode is located behind the anode. A voltage divider network is used to apply a successively higher voltage to each of the dynodes as they approach the anode in proximity. Various spectral-response peaks can be obtained, depending on the photocathode material and material of that portion of the sealed envelope directly in front of the cathode (the *window*). Photomultiplier tubes are particularly useful in the visible-light and ultraviolet regions. They can provide very high sensitivities without artificial cooling.

Selection criteria for light sensors are, primarily, spectral-response characteristics and sensitivity and, secondarily, operating temperatures, relative complexity of associated circuitry, ruggedness, and cost. Where the spectral-response must be limited to only a portion of a sensor's basic response capabilities, a window can be placed in front of the sensing surface. Windows are optical filters which have spectral-response characteristics of their own, depending on their material. Typical window spectral responses are 0.2 to 1.4 μm (quartz crystal or fused silica), 0.4 to 1.2 μm (borosilicate glass), 0.15 to 1.6 μm (cultured sapphire), 0.11 to 1.8 μm (lithium fluoride), 0.12 to 11 μm (calcium fluoride), and 0.25 to 50 μm (cesium iodide).

Spectrometers and Colorimeters

Light sensors in conjunction with light sources are used in a number of measuring devices other than for photometry. In optical *spectrometers* the incident light is passed through a *monochromator,* which can be a filter or set of filters, or a grating or prism whose angular displacement relative to the incoming light beam can be closely correlated with the single wavelength (or narrow band of wavelengths) it sends on to the light sensor. The latter, in turn, is selected so that its spectral responses match the wavelegth (or wavelengths) of interest. The angular motion of the prism or grating can be mechanized so that a given spectrum is scanned at a known rate over a known time interval. The wavelength of an observed peak can then be determined from the time counted from the start of a scan. Spectrometers can be called *spectrophotometers* when their spectral range extends anywhere between ultraviolet to infrared. They are really a subclass of *spectroradiometers* whose spectra can include any electromagnetic radiation from gamma rays to microwaves. Spectroradiometers are widely used to determine chemical composition from the spectral distribution of electromagnetic energy.

Spectrophotometers exist in the following three basic categories: (1) In an *absorption spectrophotometer* the collimated beam from an artificial light source passes through a monochromator, then through the sample whose composition is to be determined, and finally to the light sensor (photodetector). (2) In an *emission spectrophotometer* the polychromatic light from a hot or heated sample (or body) is collimated and passed through the monochromator to the photodetector. (3) In a *fluorescence spectrophotometer* monochromatic wavelengths are selected from the polychromatic emissions of a light source (or other source of electromagnetic radiation) by the *excitation monochromator.* The collimated monochromatic beam is directed at the sample in which fluorescence will occur. The fluorescent light is then analyzed for spectral content by a second monochromator (*emission monochromator*) in conjunction with the photodetector. Spectroradiometers, including those operating in nonoptical wavelengths, are described in more detail in Ref. 2.

The *colorimeter* is a specialized type of spectrophotometer whose light-source-filter-photodetector combination simulates the tristimulus functions of human visual color perception (*tristimulus colorimeter*). Three filters and either one or three photodetectors are used to obtain three separate outputs which represent (in their simplest form) the amount of "red," "green," and "blue" reflected from a sample.

Interaction of Light with Substances

A light-source-photodetector combination is also used for material characterization in instruments other than colorimeters and spectrophotometers. The most common application of such instruments is in the detection of particulates in gases and liquids. The *turbidimeter, opacity sensor,* and *transmittance sensor* all employ the same operating principle: The collimated beam from a light source passes through the sample onto a photodetector. The latter sees maximum light intensity when the sample is perfectly clear. When the (gaseous or liquid) sample contains solid particles, the amount of light received by the photodetector is attenuated. The attenuation is due primarily to *absorption* or *scattering* or a combination of both. In a *nephelometer* the photodetector is mounted normal to the axis of the light beam entering the sample. Thus, the photodetector responds to light scattered from the sample at 90°, and its output increases with increasing turbidity in the sample. The detection threshold can be improved by adding a second photodetector at 180° to the first one. The transmitted light is either absorbed by a hood or sensed by an additional photodetector for reference purposes. The *fluorimeter* employs a source of ultraviolet light to cause *fluorescence* in a sample; i.e., the light is first absorbed and then is reemitted after a very short time (about 10^{-8} s). When the time of reemission is much longer, the phenomenon is known as *phosphorescence*. The incoming light is first passed through a filter for wavelength selection. The reemitted light is sensed (after filtering to exclude nonfluorescent products) by a photodetector mounted normal to the axis of the incident light beam. A *refractometer* measures the *index of refraction* of a sample, usually by having its photodetector sense the light reflected from the boundary between two media of different density (and, hence, a different index of refraction). As more light is lost because of refraction in the second medium, less light is reflected back to the photodetector. Its output is then typically compared with that of a reference photodetector that receives all the light from the light source. For more details on these, as well as other analysis instruments, see Ref. 2.

TRANSDUCERS FOR ELECTRICAL QUANTITIES

Current Sensors

Current measurements can be made by various devices acting as transducers. A *series resistance,* inserted into the conductor where current is to be measured, provides a usable voltage across it because of the *IR* drop caused by the resistance. When the series resistance is relatively low, it is referred to as a *shunt.*

If it is necessary to keep the measurement circuit electrically isolated from the measured circuit (which is always desirable), a differential amplifier can be used, with its input terminals closely coupled across the series resistance so that its output terminals can be isolated from the measured circuit, and so that a signal sufficient for telemetry can be provided without inserting too high a series resistance into the measured circuit. This method is usable for both ac and dc currents. Other isolating current transducers are the *saturable reactor,* an adjustable inductor in which the input-current versus output-voltage relationship is adjusted by controlled magnetomotive forces applied to the core, and the *Hall effect* current transducer, in which an output-voltage change is produced by measured-current-originated electromagnetic effects on a semiconductor placed in a magnetic field. The *current transformer* can be used to convert the measured current, with circuit isolation, into an output current or voltage.

Electrometers

Small currents (down to 10^{-15} A) can be measured by means of an *electrometer tube* or by special semiconductor devices such as an amplifier with varactor diodes, metal-oxide semiconductor field-effect transistors (MOSFET), or junction field-effect transistors. All these devices require output amplification to provide signals suitable for remote measurement. When current must be measured on the high-voltage secondary side of a transformer, the current in the low-voltage primary can often be measured instead, and a suitable calibration used to correlate the two currents.

Voltage Sensors

DC and ac voltages can be monitored by means of a voltage divider connected across the two terminals to be measured. A voltage divider consists of two resistors in series, with the output taken across only one of the two resistors when a signal lower than the actual voltage is required by the measurement system. AC voltages are most frequently sensed by a *voltage transformer* or by Hall effect devices.

Frequency Converters and Dividers

Frequency can be converted into a voltage signal by use of a tuned discriminator or of integrating circuitry. When a digital signal is required, the measured frequency can be passed through an electronic "gate" which is "opened" for a fixed period of time. The pulse count over the gated period is then indicative of the frequency. If the measuring system can accept a frequency, but one much lower than the measured frequency, a *frequency divider* circuit can be used to provide an output frequency which is a fixed fraction of the input frequency.

Power Sensors

Power measurements have often been derived from simultaneous but separate voltage and current measurements. *Hall multipliers* are now in common use for power measurements. In these devices voltage changes are transformed into excitation current changes; simultaneously, current changes are used to change the magnetic field applied to the Hall device. Power (especially at microwave frequencies) is also measured by using a portion of the power to raise the temperature of a resistive temperature transducer (e.g., a thermistor), then measuring the temperature change, which can be correlated to measured power by a suitable calibration. Such a sensor is often referred to as a *bolometer.*

REFERENCES

1. Norton, H. N., "Handbook of Transducers," Prentice Hall, 1989.
2. Norton, H. N., "Electronic Analysis Instruments," Prentice Hall, 1992.

BIBLIOGRAPHY

"American National Standard for Voltage or Current Reference Devices: Solid State Devices." *ANSI C100.6-3.* Instrument Society of America, 1984.

"Electrical Transducer Nomenclature and Terminology," *ANSI MC 6.1/ISA-S37.1.* Instrument Society of America, 1982.

"Instrumentation Grounding and Noise Minimization Handbook," *Technical Report No. AFRPL-TR-65-1,* reprinted by Instrument Society of America, Aerospace Div., 1973.

Keyes, R. J., "Optical and Infrared Detectors," 2nd ed., Springer-Verlag, 1980.

Knoll, G. F., "Radiation Detection and Measurement," Wiley, 1979.

Materials Technology for Infrared Detectors. SPIE—the International Society for Optical Engineering, 1986.

Standards for Sound Measurements, available from American National Standards Institute (ANSI), New York:

a. "Specification for Sound Level Meters," *ANSI S1.4.*

b. "Preferred Frequencies for Acoustical Measurements," *ANSI S1.6.*

c. "Preferred Reference Quantities for Acoustical Levels," *ANSI S1.8.*

d. "Methods for the Measurement of Sound Pressure Levels," *ANSI S1.13.*

e. "Procedures for Calibration of Underwater Electroacoustic Transducers," *ANSI S1.20.*

f. "Methods for Sound Power Determination," *ANSI S1.31 through S1.36.*

Wolfe, W. L., and Zissis, G. J. (eds.), "The Infrared Handbook," rev. ed., Environmental Research Institute of Michigan, Ann Arbor, 1985.

Yerges, L. F., "Sound, Noise and Vibration Control," Van Nostrand Reinhold, 1978.

CHAPTER 8.5
MICROSENSORS AND MICROACTUATORS

Wen H. Ko, C. C. Liu, M. Mehregany

INTRODUCTION

The term *microelectromechanical systems* (MEMS) was coined around 1987. Besides microfabricated sensors and actuators, the field includes micromechanical components and microassembled systems. The field of microtransducers traditionally has been application driven and technology limited, and has emerged as an interdisciplinary field that involves many areas of science and engineering. MEMS is expected to follow a similar trend.

Microelectromechanical devices and systems are inherently smaller, lighter, faster, and usually more precise than their macroscopic counterparts. However, development of micromechanical systems requires appropriate fabrication technologies that enable the definition of small geometries, precise dimensional control, design flexibility, interfacing with control electronics, repeatability, reliability and high yield, and low cost per device. IC fabrication technology meets all of the above criteria and has been the primary enabling technology for the development of micromechanical systems. Microfabrication provides a powerful tool for batch processing and miniaturization of mechanical systems into a dimensional domain not accessible by conventional (machining) techniques. It provides an opportunity for integration of mechanical systems with electronics to develop high-performance closed-loop functional MEMS. In its most general form, MEMS would consist of mechanical microstructures, microsensors, microactuators, and information processing electronics integrated in the same package. This chapter will emphasize microsensors and microactuators.

MICROMACHINING TECHNOLOGY

Micromachining is a key fabrication technology for solid-state sensors and actuators, as well as MEMS. There are two classifications of micromachining, bulk and surface, both of which are based on silicon IC technology. In surface micromachining, the shaping is done only on one or both surfaces of a substrate material. Bulk and surface micromachining include techniques for selective etching, deposition, bonding, and packaging. The deposition technology is the same as that used in IC fabrication; the packaging is a developing technology that differs with each situation. Therefore, the major discussion on micromachining usually concentrates on etching and bonding and other special processes.

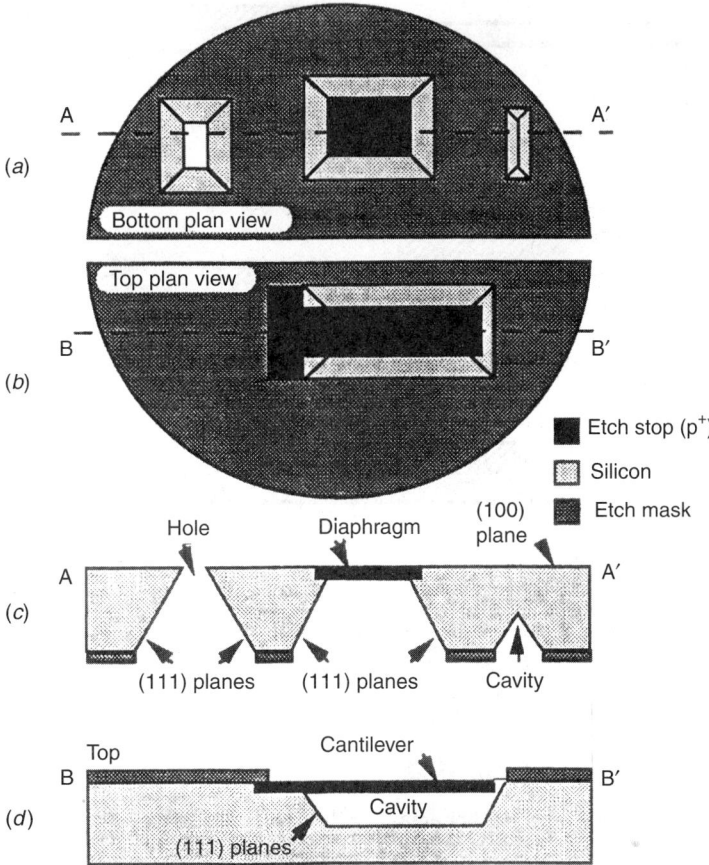

FIGURE 8.5.1 The basic concepts of bulk micromachining. (*a*) The bottom plan view of an anisotropic etched wafer showing the fabrication of cavities, diaphragms, and holes. (*b*) The top plan view of an anisotropically etched wafer showing the fabrication of a cantilever beam using etch stop layers. (*c*) The cross section, A-A′, showing the hole, diaphragm, and cavity of (*a*). (*d*) The cross section, B-B′, showing the cantilever beam of (*b*).

Bulk Micromachining

There are two techniques that make micromachining of silicon viable: (1) anisotropic etchants of silicon that preferentially etch single crystal silicon along selected crystal planes, and (2) etch masks and etch-stop techniques used in conjunction with silicon anisotropic etchants to selectively prevent regions of silicon from being etched. By appropriately combining etch masks and etch-stop patterns with anisotropic etchants, three-dimensional microstructures in a silicon substrate can be fabricated with high accuracy. Figure 8.5.1 shows two basic concepts of bulk micromachining.

Surface Micromachining

Surface micromachining relies on encasing the structural parts of the device in layers of a sacrificial material during the fabrication process. The sacrificial material (also called spacer material) is then dissolved away in

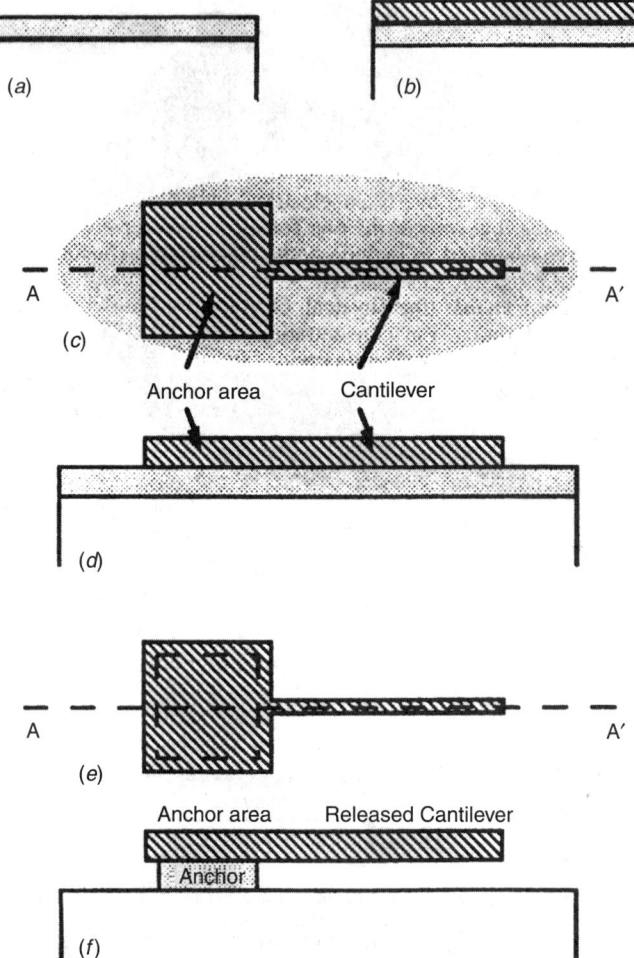

FIGURE 8.5.2 Schematics demonstrating surface micromachining in its simplest form: (*a*) cross-sectional view after the sacrificial layer is deposited; (*b*) cross-sectional view after the structural layer is deposited; (*c*) top plan view after patterning of the structural layer; (*d*) cross section AA′ of (*c*); (*e*) top plan view after release; and (*f*) cross section AA′ of (*e*).

a chemical etchant that does not attack the structural parts. The final stage of dissolving the sacrificial layer is called "release." In other words, there are two primary components in a surface micromachining process: (1) *structural* layers of which the final microstructures are made and (2) *sacrificial* layers that separate the structural layers and are dissolved in the final stage of device fabrication. The basic concepts and principles of surface micromachining are given below.

Figure 8.5.2 describes one of the simplest forms of surface micromachining in which only two film depositions (i.e., one sacrificial and one structural) followed by one patterning step (i.e., for the structural layer) are needed for device fabrication. The sacrificial layer is deposited first, followed by the deposition of the structural layer. The structural layer is then patterned, forming the cantilever and the anchor region in our example. At this point, the release step is performed, creating a cantilever suspended over the substrate at a height that is flush with the surface of the silicon.

Bonding

In micromachining, bonding techniques are used to assemble individually micromachined parts to form a complete structure. Usually entire wafers or individual dies are bonded together. Wafer bonding allows the fabrication of three-dimensional structures that are thicker than a single wafer. This is important for micropumps and microvalves where more than one cavity is needed in the thickness direction. Several processes have been developed for bonding silicon wafers. A review of bonding techniques is given by Ko, et al.[4] and Tong.[5] Fusion bonding was developed to bond silicon to silicon and to many other materials for MEMS (Kovacs 1998, Wolf 1986).

Dry Etch and Deep Reactive Ion Etch

Dry etch processes replace wet etches for precise pattern transfer and simple operation. There are many types of dry etch techniques developed for IC processes including gas discharge methods and ion beam methods. They are used in micromachining. The deep reactive ion etch can be used to fabricate structures with great depth and controlled wall profile. It is an important process for etching devices with high aspect ratio (ratio of height to width).

LIGA and High Aspect Ratio Processes

LIGA (a German acronym for Lithographie, Galvanoforming, Abformung) process uses lithography of thick photo resist and x-ray exposure to produce a desired structure, then uses metal electroplating to generate a complementary metal mold, and microinjection molding to fabricate multiple replicas of the original. This process is used to fabricate high aspect ratio structures that are difficult to make with conventional MEMS process (Ref. 7). A Synchrotron accelerator provides x-ray exposure. Other high-aspect ratio lithography employs ultraviolet and other techniques (Ref. 8).

Packaging

Hermetic seal, lapping, polishing, and assembly techniques for microsystems as well as protection package processes were developed to bring MEMS from laboratory to market place. However, the technology is developing and each technique is applicable to a small group of applications. There are no universal package methods available and this is a field of intense research.

Micromachining of Other Materials

Micromachining of compound semiconductors, quartz, glass, ceramic, and polymeric materials for optical communication, chemical, biological, and medical applications is possible. Large-scale manufacturing methods for microdevices including injection molding, precision plating, and microassembly are anticipated.

DEFINITIONS

As in the case of other instrumentation systems, MEMS can be deemed to consist of an input block (sensor), output block (actuator), and signal processing block (electronics). A *sensor* is a device that provides a usable electrical output signal in response to a specified measurand. An *actuator*, the reverse of a sensor, is a device that converts an electrical signal to an action, while a *transducer* can be considered as the device that transforms one form of signal or energy into another form. Therefore, the term transducer can be used to include both sensors and actuators. These definitions, though not universal, are used throughout this chapter. A number of criteria can be used to classify sensors, including *transduction principle* (the physical, chemical, or biological effects), *measurands* (e.g., pressure, acceleration, gas concentration, ion concentration), *fabrication technology and materials* (e.g., thin film, semiconductors, ceramics) *applications* (e.g., automotive, medical, aerospace), and *cost and performance*. In most of the literature, a combination of principle, measurand, and application is used to define sensors and actuators. This chapter will follow the popular approach of classifying sensors into three major groups: *physical, chemical,* and *biomedical*.

PHYSICAL MICROSENSORS

Integrated, Intelligent (Smart) Sensors

When a microsensor is integrated with signal processing circuits in a single package, it is referred to as an integrated sensor. A monolithic integrated sensor has the signal processing circuit fabricated on the same chip as the sensor, while a hybrid integrated sensor has the signal processing circuit on the same hybrid substrate as the sensor chip. The packaged sensor not only transduces the measurand into electrical signals, but also may have other signal processing and decision-making capabilities. Thus, the term *intelligent sensors* was coined.[9] In trade journals and popular news articles they are also called smart sensors. Integrated sensors with specific types of on-chip electronic circuits may have the following advantages:

Better signal-to-noise ratio.

Improved characteristics. Signal processing can provide: (a) on-chip feedback system or look-up tables to improve the output linearity, (b) compensation circuits to reduce cross-sensitivity to temperature, strain, or other known interfering effects, (c) on-chip accurate current and voltage sources and associated circuits to provide automatic and periodic self-checking and calibration, and (d) feedback and other circuits to improve or compensate for the frequency response of the sensor.

Signal conditioning and formatting. Signal conditioning can be performed by integrated circuits prior to the output, such as (a) A-to-D conversion, (b) impedance matching, (c) output formatting to a standard, and (d) signal averaging.

Improved signals representing the measurands. Decision-making and computing circuits as well as memory devices may be used with redundant sensors to exclude noisy or failed sensors from the array of devices, thus improving the signal beyond a single sensor's capability. Multiple sensors, each with poor selectivity to a family of measurands, can be used with pattern recognition circuits to obtain accurate signals for each of the measurands. Neural networks may be used to train the sensors to recognize the desired measurands among several interfering effects. Many of the signal processing techniques used so successfully in communications and defense applications may be incorporated with the sensors to improve the quality of the sensor output beyond the individual device's capability. A brief summary of commonly used physical sensors follows.

Thermal Sensors

The best-known thermal sensors are thermistors and thermocouples for measuring the temperature of the environment. Thermistors and resistive temperature detectors (RTDs) are based on the change of mobility and carrier density with temperature. These changes are represented by temperature coefficients that may be constants or nonlinear functions of temperature. The resistance of a thermistor is an exponential function of the temperature. Linearization networks may be used to make the output a linear function of the temperature over a specific range, with some sacrifice in sensitivity.

The thermocouple is based on the Seebeck effect, one of the three thermoelectric effects (Seebeck, Peltier, and Thompson effects). Two different materials (usually metals) are joined at one point to form a thermocouple. Various metal wires can be used as thermocouples for different temperature ranges and sensitivity. Common wire materials are: iron/constantan, chromel/constantan, and platinum/platinum +10 percent rhodium. Semiconductors, including silicon, may also be used with a metal to form a microthermocouple by micromachining techniques. Hundreds of these microthermocouples may be connected in series on a silicon chip to form a thermopile microsensor that is hundreds of times more sensitive than a single thermocouple. These microthermosensors are sensitive enough to measure the temperature rise in millidegree centigrade as a result of incident infrared radiation (IR), making them useful as IR imaging sensors and remote temperature sensors.[10]

If the current through a *p-n* junction is kept constant, the junction voltage is a linear function of temperature. This principle has been used for many commercial temperature sensors. When two bipolar transistors are operated at a constant ratio of emitter current densities, the difference in base emitter voltages is proportional to the absolute temperature. This principle is the basis of many precise commercial temperature sensor ICs.

The mass flow of fluids can be measured by the temperature difference between two sensors, one on the downstream and one on the upstream of a heating (or cooling) source. This is the principle of microanemometers and is also a good example of how the mechanical measurand (flow) is reduced to thermal parameters that are then transduced to electrical signals.

Electrical Sensors

Electrical signals can be picked up by *ideal* ohmic connections without special sensors. However, most connections and electrical probes are not *ideal* and, sometimes, the magnitude of the charge, currentm, and voltage may not be convenient to measure directly. For example, Hall effect sensors are used to measure the current and polarity in motor windings where the current value is very large or a contact is difficult to make. Similarly, electrical sensors may be used to measure very high voltages or very large charges.

The flow of electrical currents occurs through the motion of charge carriers that may be electrons (and holes), ions, and charged defects in materials. These different charge carriers can have very different charge-to-mass ratios and mobilities. When a current has to pass from a material with one type of charge carrier (e.g., electrons) through the interface to another material with a different type of charge carrier (e.g., ions), there are complex reactions occurring at the interface associated with the transfer of charges. The charge carriers in metals are electrons, and in semiconductors may be electrons or holes (the collective behavior of electrons). In electrolytes, the movement of ions carries the current. In dielectrics, it is much more complicated because electrons, ions, charged atom groups, and combinations of these can carry the current. The electrodes that are used to measure electrical parameters in electrolytes or dielectrics are electrical sensors. These electrodes are used in electrochemistry, material science, and biology. The equivalent circuit of such electrodes may include several resistance-capacitance (RC) networks to model the several layers of carriers near the interface. The characterization of the electrode may involve many techniques. A commonly used method is to plot the imaginary part of the impedance versus the real part. In all applications, it is important to know the RC equivalent circuit of the electrodes, the contact potential (work function difference) between the materials in contact, and the thermoelectric effects between the materials in the measurement loop. Using specialized computer programs, the equivalent circuit of the electrode may be found. Noble metal (platinum) wires coated with various materials may be used to measure the potential of the electrolyte, or the ionic concentration of the ions in the electrolyte. These chemical sensors are called *ion selective electrodes* and are used in many chemical and biomedical applications.

Microelectrodes were designed for the measurement of electrical potentials on the surfaces of tissues, cells, and other materials. Vibrating microelectrodes can be used to measure the potential or charge the surface without making direct contact to it.

Mechanical Sensors

There are many mechanical sensors described in the literature and commercially available.[3] Mechanical parameters may be converted to other energy domains and then sensed or measured directly. For direct sensing, the parameters are related to strain or displacement. Silicon, a brittle material, will fail (break) at a maximum strain of about 2 percent. However, below the elastic limit the strain is related to the stress by a nearly constant Young's modulus, which is less than but close to that of steel. The principles commonly used to sense strain are piezoelectricity, piezoresistivity, and capacitive or inductive impedance.

Piezoelectricity.   The piezoelectric effect relates the elastic strain S (or stress T) in one orientation to displacement charge density D (or electric field intensity E) in another orientation that may or may not be the same as the orientation of the strain, through the piezoelectric coefficients. The piezoelectric coefficients are, in general, a $(n \times m)$ matrix, where n is the number of orientations for strain—the mechanical parameter, and m is the number of orientations for D or E—the electrical parameters. The four parameters S, T, D, and E are interrelated by various coefficients or constants. For most applications, simplifying assumptions can be made so that only parallel and perpendicular orientation between the mechanical and electrical parameters are considered. For example, d_{ij} is the strain constant or charge constant of piezoelectricity relating D and T when E is constant, or S and E, when T is constant:

$$D_i = d_{ij}T_j + \epsilon E_i \tag{1}$$

and

$$S_j = YT_j + d_{ji}E_i \tag{2}$$

where ϵ is the dielectric constant or permittivity; and I/Y is the Young's modulus of the material. The coefficient $d_{33} = (\Delta D/\Delta T)_{E=k}$ or $d_{33} = (\Delta S/\Delta E)_{T=k}$, relates D and T or S and T when they are both oriented in the "$z - 3$" direction; thus, they are in parallel. d_{31} relates the electrical parameters D_3 and E_3 to the perpendicular mechanical parameters T_1 and S_1.

These coefficients can be used to estimate the characteristics of sensors and actuators. The piezoelectric effect of sensors is used to measure various forms of strain or stress. Examples are microphones for strains generated by acoustic pressure on a diaphragm; ultrasonic sensors for high-frequency strain waves arriving at or propagating through the sensors; and pressure sensors for AC pressures on a silicon diaphragm coated with piezoelectric materials. The piezoelectric effect can also be used to sense small displacements, bending, rotations, and so forth. These measurements require a high input impedance amplifier to measure the surface charges or voltages generated by the strain or stress. Piezoelectric thin films and micromachining technology are used to fabricate strain or stress microsensors of various types.

Piezoresistivity.   The piezoresistive effect in conductors and semiconductors is used for commercial pressure sensors and strain gauges. The strain on the crystal structure deforms the energy band structure and thus changes the mobility and carrier density that changes the resistivity or conductivity of the material. The strain in one orientation may affect the conductivity in another orientation. Therefore, similar to the piezoelectric effect, piezoresistivity can be described by an $n \times n$ matrix. For most applications, the piezoresistive material is a crystal and has symmetry in its structure. For cubic crystals, n is 6 (three for axial strain, three for sheer strain, three for voltage, and three for current). Using some simplifying assumptions, the matrix of piezoresistivity coefficients can be reduced to parallel and perpendicular coefficients just as for the piezoelectric coefficients. However, the coefficients are not only orientation dependent, but are also affected by doping and temperature. A practical piezoresistive pressure sensor can be built by fabricating four sensing resistors at the edges or at the center of a thin silicon diaphragm that acts as a mechanical amplifier to increase the stress and strain at the sensor site. Usually, the four sensing resistors are connected in a bridge configuration with push–pull signals to increase the sensitivity. The measurable pressure range for such a sensor can be from 10^{-3} to 10^6 Torr. Bulk and surface micromachined pressure sensors are currently at various stages of commercialization.

Capacitive/Inductive Impedance.   Capacitive or inductive impedances can also be used to measure displacements and strains. Capacitive pressure sensors present some advantages over piezoresistive devices.[11] Capacitive devices integrate the change of elementary capacitive areas, while piezoresistive devices take the difference of the resistance changes of the bridge arms. Therefore, capacitive sensors are less sensitive to the sideways forces and are more stable. Furthermore, capacitive changes can be much larger as a percentage of the reference value than the 5 percent maximum resistance change found in piezoresistive devices. However, capacitive sensors require a capacitance-to-voltage (C-to-V) converter on or near the chip to avoid the effects of stray capacitances.[12] Therefore, the device operation becomes complicated unless the required C-to-V converter can be fabricated on the chip or packaged in the same encapsulation. The measurement circuit also must be stable and have low noise.

Figure 8.5.3 shows a cross-sectional schematic drawing of pressure sensors based on three principles that can be used to measure acceleration, force, flow, and small displacements. Other sensors that can measure torque, rotation, touch, and so forth have also been reported in the literature.

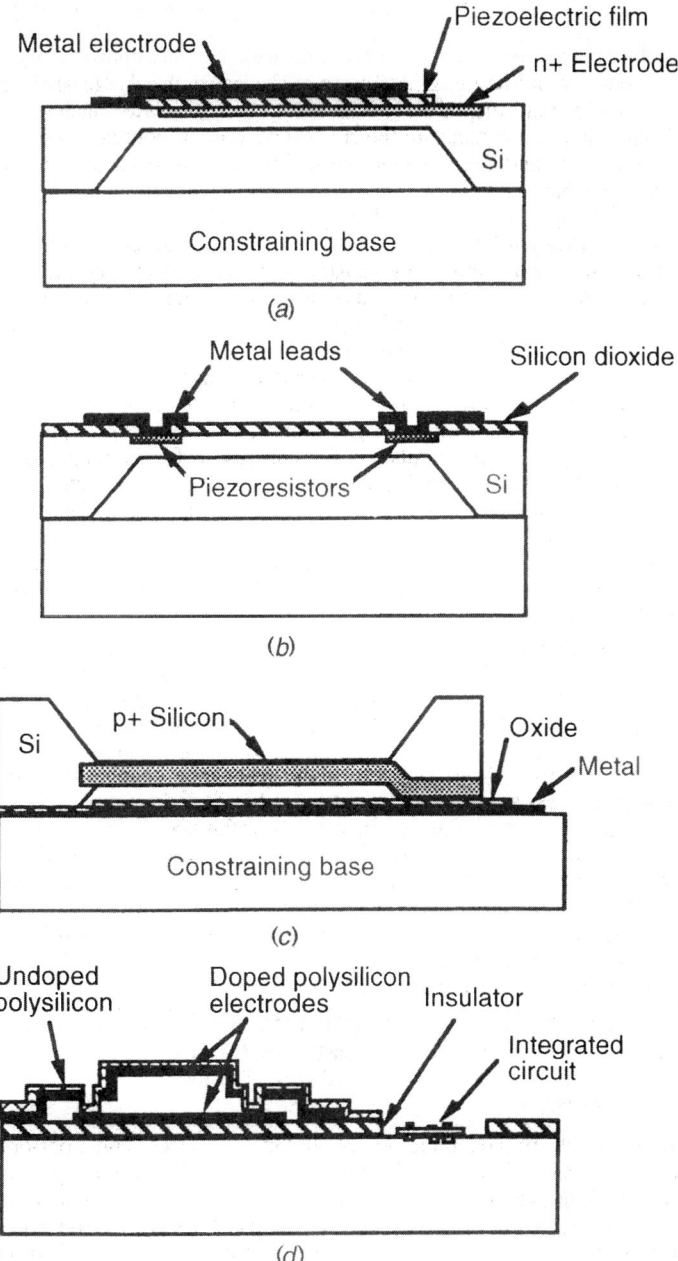

FIGURE 8.5.3   Schematic drawings of the different types of silicon pressure sensors: (*a*) piezoelectric; (*b*) piezoresistive; (*c*) capacitive; and (*d*) surface micromachined capacitive.

Optical Sensors

Optical sensors include photoconductors, photovoltaic devices, and fiber-optic sensors. The conductivity of photoconductors changes under optical radiation because of changes in the charge carrier population. Photovoltaic devices involve a *p-n* junction where radiation-generated carriers may cross the junction to form currents and a self-generated voltage. Photovoltaic devices, such as solar cells, can supply energy to external circuits. However, the current is proportional to the radiation intensity, but the voltage is not (it is very nonlinearly related to the intensity of the radiation). Many micro-optical sensors and imagers are commercial products now.

When strained, a fiber-optic cable changes the intensity or the phase delay of the output optical wave relative to a reference. Using an optical detector and an interference measuring technique, small strains can be measured with high sensitivity. Intensity sensors can detect changes in optical intensity, which is related to an applied measurand. Examples include: underwater acoustic sensors, fiber-microbend sensors, evanescent or coupled waveguide sensors, moving fiber-optic hydrophones, grating sensors, polarization sensors, and total internal reflection sensors.

Optical interference sensors have been developed for interferometer acoustic sensors, fiber-optic magnetic sensors with a magnetostrictive jacket, and fiber-optic gyroscopes. Specially doped or coated optical fibers have been shown to have great versatility for physical sensors of various types and configurations. They have been used for radiation dosimeters, current sensors, accelerometers, temperature sensors, as well as detectors for liquid level, displacement, strain, torque, and fluid flow. Many chemical and biomedical materials can be sensed with fiber-optic devices.[13]

Magnetic Sensors

Magnetic sensors may employ any of the following effects: (1) the magneto-optic effect based on the Faraday rotation of the polarization plane of linearly polarized light owing to the Lorentz force on bound electrons; (2) the magnetostrictive effect where the magnetic field causes strain on the material; (3) the galvanomagnetic effect that shows up as a Hall field and carrier deflection and magnetoresistance with different sample configurations. Sensing devices include: Hall effect devices—bulk Hall plate and magnetic field sensitive field effect transistors (MAGFETS), magnetotransistors, magnetodiodes, and current domain magnetometers. A comprehensive review paper by Baltes[14] is suggested for further reading.

Future Trends of Physical Sensors

Physical microsensors have been developed over the past 25 years and nearly all possible sensors have been explored. Future research and development will be focused on the following: (1) new materials, principles, and technology that push sensors beyond current limitations, such as high temperature devices, submicron devices, self-powered sensors, etc.; (2) sensor arrays and multiple sensors to improve the sensitivity, selectivity, and stability, as well as to sense the distribution of the measurands; (3) integrated sensors with built-in intelligence to recognize desired signals in noise; and (4) integrated sensors with actuators to provide self-checking and self-calibration functions, as a step toward integrated systems.

CHEMICAL AND BIOMEDICAL MICROSENSORS

General

The introduction of microelectrode assembly and the microfabrication of sensor elements are considered the most significant developments in chemical sensor technology in the past few decades. The potential advantages of microelectrodes for chemical sensor development include: (1) the small electrode size requires a small sample volume, even in biological sensing; (2) the very low level of Faradaic current results in the beneficial effect of very small ohmic potential drop, even in samples of very low conductivity; (3) the limiting current density increases as the electrode size decreases, thus improving the signal-to-noise ratio, since capacitive background currents are proportional to the surface area; (4) the quick response of microelectrodes allows monitoring of the low-frequency fluctuation of signals and rapid recording of steady-state polarization curves; and (5) the current output of ultramicroelectrodes with dimensions in the μm range is practically insensitive to conventional flow in solution.

Microelectronic fabrication techniques provide advantages for producing these microsize chemical sensors. These include reduced size, reduced sample volume, reduced sensor cost, and fast response. Furthermore, microfabrication processing produces identical, highly uniform, and geometrically well-defined sensor elements. This is particularly attractive for chemical sensors based on the amperometric mode of operation.

While the feasibility and advantages of microfabrication of chemical sensors have been well demonstrated, practical microfabricated chemical microsensors remain elusive. The recent advancements in micromachining technology add new dimensions and impetus to sensor development. These techniques permit the formation of three-dimensional structures and other desirable features and will have a significant and revolutionary impact in future chemical sensor development.[15,16]

Common microfabricated chemical sensors include electrochemical sensors (based on conductivity, potentiometric, and amperometric measurements), tin-oxide based sensors, and calorimetric devices. The operational principles of these sensors are described below.

Electromechanical Sensors

Electromechanical sensors have been used as a whole or as an integral part of many chemical and biosensors. The operational modes of electrochemical sensors are based on traditional laboratory electroanalytical principles. In recent years, the advancement of microelectronic fabrication techniques has provided impetus to the development of new types of electrochemical sensors for chemical and biosensor development. The introduction of solid electrolytes and conductive polymers further enhanced the applicability of electrochemical sensors, particularly in a gaseous environment. A parallel and significant development in sensor technology, though unrelated to conventional electroanalytical principles, is the development of semiconductive chemical sensors, such as the ion-sensitive field effect transistor (ISFET) depicted in Fig. 8.5.4.

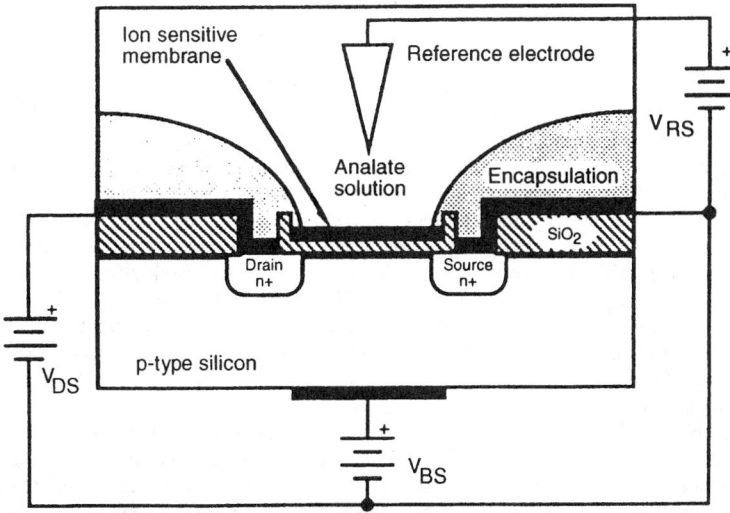

FIGURE 8.5.4   A schematic drawing of a typical ISFET.

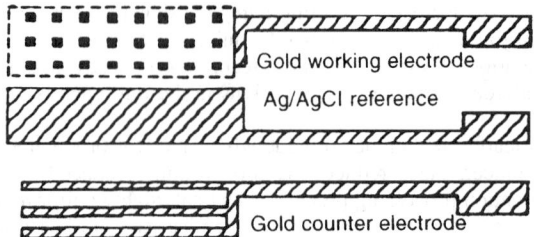

FIGURE 8.5.5 Typical planar, three-electrode electrochemical sensor. The working electrode is divided into many small cells which are in parallel to reduce the effects of flow on the sensor output.

Principles of Electromechanical Sensors.   Electromechanical sensors are essential electrochemical cells consisting of two or more electrodes in contact with a solid or a liquid electrolyte. A typical planar, three-electrode electrochemical sensor is shown in Fig. 8.5.5. Electrochemical sensors can be classified according to their mode of operation, e.g., conductivity/capacitance sensors, potentiometric sensors, voltammetric sensors. Amperometric sensors are considered a specific type of voltammetric sensor in this discussion. Measurements of capacitance or resistance, or potential or current of an electrochemical cell, are all feasible. When the cell is used as a sensor, the choice of a measurement parameter is based on the sensitivity, specificity, and accuracy provided with respect to the species to be sensed. Therefore, understanding the principles and limitations of each operational mode is essential in sensor design and development.

Conductivity and Capacitance Electromechanical Sensors.   An electromechanical conductivity sensor measures the change in the conductivity of the electrolyte in an electrochemical cell. It is different from the electrical (physical) sensor that measures the change in electrical resistance of the conductor in the presence of a given concentration of a solute. An electromechanical sensor may also involve a capacitive impedance resulting from the polarization of the electrodes and the Faradaic or charge transfer processes.

In a homogeneous electrolytic solution, the conductance of the electrolyte $G(\Omega^{-1})$ is inversely proportional to L (cm), the segment of the solution along the electrical field and directly proportional to $A(cm^2)$, the cross-sectional area perpendicular to the electric field.

$$G = \sigma \frac{A}{L} \qquad (3)$$

where $\sigma(\Omega^{-1}cm^{-1})$ is the specific conductivity of the electrolyte and is related quantitatively to the concentration and the magnitude of the charges of the ionic species.

The measurement of electrolytic conductance by an electromechanical conductivity sensor employs a Wheatstone bridge with the electromechanical cell (the sensor) forming one of the resistance arms of the bridge. However, the conductivity measurement is complicated by the polarization of the electrodes at the operating voltage. Charge transfer processes occur at the electrode surfaces and a double layer is formed adjacent to each of the electrodes when a potential is imposed on the cell. Both effects can be minimized by using a high-frequency, low-amplitude alternating current. The higher the frequency and the lower the amplitude of the imposed alternating current, the closer the measured value is to the true conductance of the electrolyte. A better technique would be to balance both the capacitance and the resistance of the cell by connecting a variable capacitor parallel to the resistance of the bridge arm adjacent to the cell. The basic measurement could be performed using an impedance measuring bridge in which the resistive and reactive components of the impedance of the electrochemical cell can be determined by balancing the bridge. Greater measurement precision with fewer demands on the operator can be achieved using special electronic instruments, such as lock-in amplifiers, to determine the magnitude of the cell impedance and its phase angle. One way of using the lock-in amplifier is to drive the conductivity cell with a constant current amplitude sinusoidal signal and measure the resulting potential across the cell and its phase angle with respect to the excitation. The impedance magnitude can then be found by taking the ratio between this measured voltage amplitude and the amplitude of the excitation current.

Potentiometric Sensors.   Potentiometric sensors use the effect of the concentration on the equilibrium of the redox reactions occurring at the electrode-electrolyte interface in an electrochemical cell. A potential may develop at this interface as a result of the redox reaction, $Ox + Ze^- = Red$, taking place at the electrode surface, where Ox denotes the oxidant, Red the reduced product, Z the number of electrons in the redox reaction and e^- an electron.

This reaction occurs at one of the electrodes (cathodic reaction in this case) and is called a half-cell reaction. At thermodynamical quasi-equilibrium conditions, the Nernst equation is applicable and can be expressed as

$$E = E^0 + \frac{RT}{ZF} \ln \left(\frac{a_{Ox}}{a_{Red}} \right) \tag{4}$$

where E and E^0 are the electrode potential and electrode potential at standard state, respectively, a_{Ox} and a_{Red} are the activities of Ox and Red, respectively, Z the number of electrons transferred, F the Faraday constant, R the gas constant, and T the operating temperature in Kelvin. In a potentiometric sensor, two half-cell reactions will take place simultaneously. However, only one of the two half-cell reactions should involve the sensing species of interest, and other half-cell reaction is preferably reversible and noninterfering.

Potentiometric sensors can be classified based on whether the electrode is inert or active. An inert electrode provides the surface for the electron transfer or provides a catalytic surface for the reaction and does not actively participate in the half-cell reaction, whereas an active electrode is either an ion donor or acceptor in the reaction. There are three types of active electrodes: the metal/metal ion, the metal/insoluble salt or oxide, and metal/metal chelate electrodes.

Each electrode type can be involved in a chemical cell or concentration cell. In a chemical cell, the two half-cell redox reactions are different. In a concentration cell, the two half-cell reactions are in redox fashion but with different reactant concentrations or activities.

Voltammetric and Amperometric Sensors.   Voltammetric measurement is characterized by the current and potential relationship of the electrochemical system, a technique commonly used by electrochemists. This method is often referred to as amperometric measurement. Amperometric sensors can be considered as a subclass of voltammetric sensors, since they are operated in one of the voltammetric modes. In this case, the current-solute concentration relationship is obtained at a fixed potential or overall cell voltage. On the other hand, voltammetric sensors can be operated in other modes such as linear voltammetric or cyclic voltammetric modes. Figure 8.5.6 shows the three different modes of operation and their respective current-potential responses.

Voltammetric sensors utilize the concentration effect on the current-potential characteristics of the electrochemical system for their sensing operation. The current-potential characteristics depend on the rate by which the components in the electrolyte are brought to the electrode surface (mass transfer) and the kinetics of the Faradaic or charge transfer reaction at the electrode surface. Three different modes of mass transfer can be identified in an electrolyte solution, namely, (a) ionic migration as a result of an electric potential gradient, (b) diffusion under a chemical potential difference or concentration gradient, and (c) convective or bulk transfer by virtue of natural convection (density difference) or forced convection.

Linear sweep voltammetry involves increasing the imposed potential linearly at a constant scanning rate from an initial value to a given upper potential limit. The current-potential curve usually shows a peak at a potential where the rate of reaction equals the mass transfer rate. Cyclic voltammetry is similar to the linear sweep voltammetry except that the electrode potential is returned to its initial value at the same scanning rate as shown in Fig. 8.5.6. The cathodic and anodic sweeps would normally generate two current peaks.

Semiconductive Gas-Sensing Microsensors

SnO_2-based Semiconductor Gas Sensors.   The surface conductance of semiconducting oxides, such as zinc oxide and tin dioxide, can be influenced by the composition of ambient gases. This phenomenon is the basis of many resistive semiconductor gas sensors.

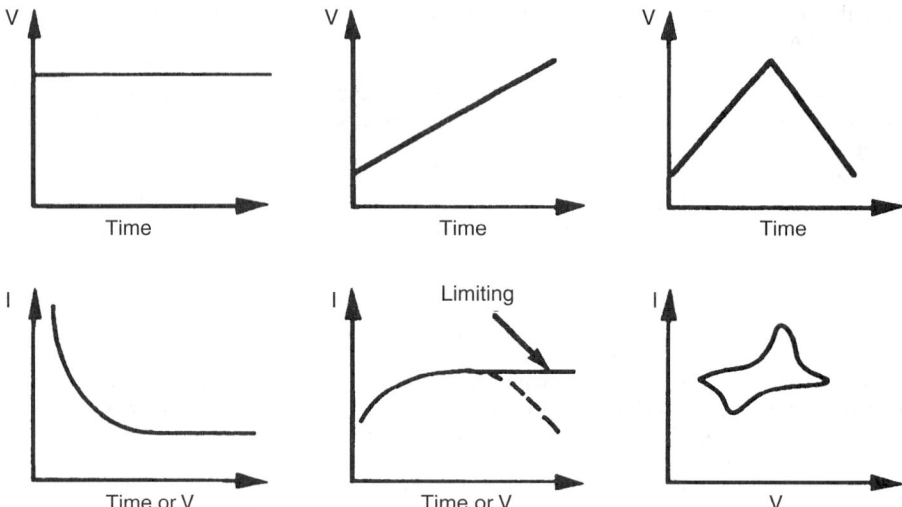

FIGURE 8.5.6   Current-potential relationship of various voltametric modes of operation.

Among the semiconducting oxides, SnO_2 is the most widely used for gas sensors. Such sensors have roles in the detection of flammable or reducing gases such as carbon monoxide, hydrogen and methane at low-level concentrations. SnO_2-based sensors are used to detect leaks of flammable and toxic gases, for combustion control and even for detecting odors. Millions of units of tin dioxide-based gas sensors are sold each year, mainly for automobile exhaust emission monitoring and for gas leak detection. Extensive literature on tin dioxide-based sensors can be accessed.[15,16,21]

As the active sensing part of the gas sensor, tin (IV) oxide (tin dioxide) is nonstoichiometric and is deficient in oxygen atoms. Charge neutrality is maintained by the presence of numbers of tin (II) ions (Sn^{2+}) instead of Sn^{4+} ions, and these Sn^{2+} ions act as electron donors during the process. Thus, the active material of tin dioxide is an *n*-type semiconductor. At elevated temperatures, adsorption of atmospheric oxygen takes place at the face of the tin-oxide film and oxygen accepts electrons that are donated by the *n*-type SnO_2 film and becomes ionic by forming O_2^-, O^-, or O^{2-} anions. If a reducing gas adsorption follows, the reducing gas may react with absorbed oxygen anions and become a positively charged species or the reducing gas may react with the absorbed oxygen atom and release bound electrons. As an example, the overall reaction of carbon monoxide on a tin dioxide film can be described by

$$2e^- + O_2 \leftrightarrow 2O^- \tag{5}$$

$$O^- + CO \leftrightarrow CO_2 + e^- \tag{6}$$

where e^- represents a conduction band electron. In the absence of reducing gas, electrons are removed from the conduction band via the reduction of molecular oxygen and an O^- species is formed such that in consequence the SnO_2 film becomes highly resistive. When carbon monoxide is introduced, it undergoes oxidation to carbon dioxide by surface oxygen anion (O^-) and electrons are reintroduced into the conductive band which leads to an increase in film conductivity. This example provides a general operation principle for tin dioxide-based gas sensors.

The main advantages of tin dioxide-based gas sensors include high sensitivity, low cost, fast response time and low power consumption. However, there are also significant drawbacks, such as long-term drift, relative low selectivity, and effects of humidity and temperature. Efforts have been made in recent years to improve the performance characteristics of tin dioxide-based gas sensors. For instance, noble metal doping of tin dioxide improves the catalysis process of the detecting gas, which greatly enhances the selectivity and decreases the operating temperature. Other modifications are described later.

Microfabrication of SnO$_2$-Based Gas Sensors.   Fabrication techniques for semiconductive gas sensors have been developed over many years. The microfabrication technologies previously described have added impetus to the advancement of semiconductive gas sensors. The use of microfabrication techniques in the production of tin dioxide type gas sensors provide a good illustration of the potential and future trends of this application.

The original Taguchi type tin dioxide gas sensor was adopted by Figaro Gas Sensor Inc. and by many Japanese groups. The tin dioxide layer is a porous thick film formed by tin dioxide powder sintering. The tin dioxide was prepared from stannic acid gel and followed by firing for 1 h at 500°C. A ratio of palladium and alumina was added to the oxide powder to get the resulting 2:1 ratio of Al_2O_3 to SnO_2. Distilled water was added to the above mixture to yield a paste that was then applied to the surface of an alumina tube with preprinted gold electrodes on the surface. A silica solution was used to cover the paste as a binder. Subsequent low-temperature baking resulted in the formation of a porous sintered thick film. A heater filament was placed inside the tube to provide the required operating temperature.

Thick-Film Technology.   Planar thick-film gas sensors have been developed in recent years. The fabrication procedures include sequential screen-printing of thick film gold electrodes, followed by printing a gas-sensitive layer onto an alumina substrate. The paste used for printing of the gas-sensing layer consists of tin dioxide powder along with an organic vehicle and a glass binder. Sometimes a catalytic additive is included. Normally, the thickness of the active film is around several micrometers to tens of micrometers. Schmitte and Wiegleb[17] proposed the composition of tin dioxide paste as 71 percent tin dioxide powder, 5 percent glass powder with solvents and adhesives, and a firing condition of 850°C for 1 hour. A thickness of 15 μm was used for this tin dioxide layer.

The advantages of screen printing technology in the fabrication of thick film gas sensors include the porosity of the sensitive layer, a simple sensor design and a variety of possible geometries for the sensor element (e.g., coplanar, interdigitated, or sandwich form). The heater, located on the back side, is helpful for saving power. The use of thick-film tin dioxide-based gas sensors for the detection of carbon monoxide, hydrogen sulfide, hydrogen, methane, and carbon tetrachloride have been reported.[18]

Thin-Film Technology.   This is another useful tool for fabricating tin dioxide-based gas sensors. Chemical vapor deposition (CVD) and reactive evaporation or sputtering can both be applied for this purpose.[19] A typical bulk micromachined, thin film tin oxide gas sensor is shown in Fig. 8.5.7.

Chemical vapor deposition technique involves two gaseous compounds that chemically react and form a compound on the substrate. For the tin dioxide gas sensor film, one of the compounds used is a tin-organic compound (e.g., dibutyl tin diacetate). It is dissolved in an organic solvent and then vaporized at a controlled temperature. The vapor is fed into the reacting chamber using nitrogen as the carrier gas. The tin-organic vapor will react with absorbed oxygen on the surface of the substrate (Al_2O_3 or SiO_2 (Si)) forming the tin dioxide film. A tin dioxide sensitive layer is formed after annealing at 500°C for 12 h or longer. The film thickness and the grain

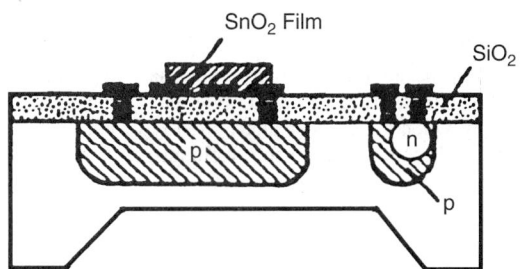

FIGURE 8.5.7   A micromachined tin oxide gas sensor. The tin oxide film is placed on a thin silicon region to reduce the power consumption when the film is heated to its operating temperature.

size of tin dioxide can be controlled by the fabrication conditions and are considered critical for the sensor response. In most cases, the thickness and grain size are in the order of several tens or hundreds of nanometers.

In the reactive evaporation method, the tin dioxide film is fabricated in an evaporator. Pure metallic tin (or tin alloy) is heated to a high temperature ($1200–1300°C$) and is deposited on a heated substrate on which the sensor heater and gold electrode have already been deposited. After deposition, the sensor unit is annealed in an oxygen-containing atmosphere for several hours at a designated temperature, allowing the tin metal to transform to the oxide. Generally, the resulting thickness of the oxide film is of the order of several hundred nanometers.

Response Characteristics of SnO$_2$-Based Gas Sensors.   *N*-type semiconductive tin dioxide-based gas sensors can be used to detect low-level concentrations of reducing gases such as hydrogen, carbon monoxide, methane, ethanol, hydrogen sulfide, and other inflammable gases, as well as some anesthetic agents like forane, halothane, and ethrane. This sensor is based on the conductance change through the tin dioxide grain boundaries on its surface in the absence of the detecting gas compared with its presence. The sensitivity S of the sensor is expressed as the ratio of the conductivity G under tested gas flow and ambient air, and is expressed as

$$S = G_g/G_a \tag{7}$$

where G_g is the conductivity of the sensor in the presence of tested gas flow and G_a is that in ambient air. The sensitivity of a tin dioxide sensor may be influenced by the layer composition, grain size, operating temperature, humidity, and various other factors.

Calorimetric Sensors

The calorimetric sensor is essentially a high-temperature resistance thermometer that measures the temperature change caused by the heat evolved during the catalytic oxidation of the combustible gases. The most commonly used calorimetric sensors are in the form of an encapsulated platinum coil, which serves both as a heater and as a resistance thermometer. This is normally referred to as a pellistor type calorimetric sensor, and was first developed in 1962. This type of calorimetric sensor can be used for the detection of combustible gas at low concentration. However, the results were not accurate because of the heat loss, a condition that may not be easy to improve with a conventionally sized sensor.

Microfabrication technology provides a means to fabricate miniature calorimetric sensors with improved characteristics. Recently, a thick-film calorimetric microsensor for combustible gas monitoring was fabricated on an alumina substrate.[20] Platinum-thick film was deposited using the screen-printing method. Two types of catalysts were employed, namely, platinum black and palladium. The noncatalytic compensating platinum film was left bare as printed. This microsensor can be used for the detection of hydrogen, carbon monoxide, methane, and ethane. Proper selection of the reaction temperature for catalytic oxidation may lead to possible identification of different gases. Catalytic metallic layer deposition onto silicon wafers or glass plates with the thin-film process has been reported by different groups.[21–24]

A thin-film thermopile sensor based on a bismuth-antimony junction array has been reported.[23] This thermopile consisted of an alternating junction of bismuth and antimony, with a gold contact between, on a kapton thick-film that was formed on a glass plate. Metal shadow masks were used in intimate contact with a kapton sheet to define the geometry for each of the metal layers. The thickness of the bismuth and antimony layer was around 1 μm or less.

An enzyme microsensor based on this thermopile has also been proposed.[24] Glucose oxidase was crosslinked above the active junctions of the thermopile with a thickness of 100 μm. It proved to have a rapid response and good stability when used either intermittently or continuously. Also, catalase and urease were used for immobilization, and both enzyme sensors yield attractive results.

Biosensors

Biosensors are a special class of chemical sensors for molecular detection that take advantage of the high selectivity and sensitivity of biologically sensitive materials. It is a device incorporating a biological sensing element with a traditional sensor, such as physical or chemical sensors. The biological sensing element selectively

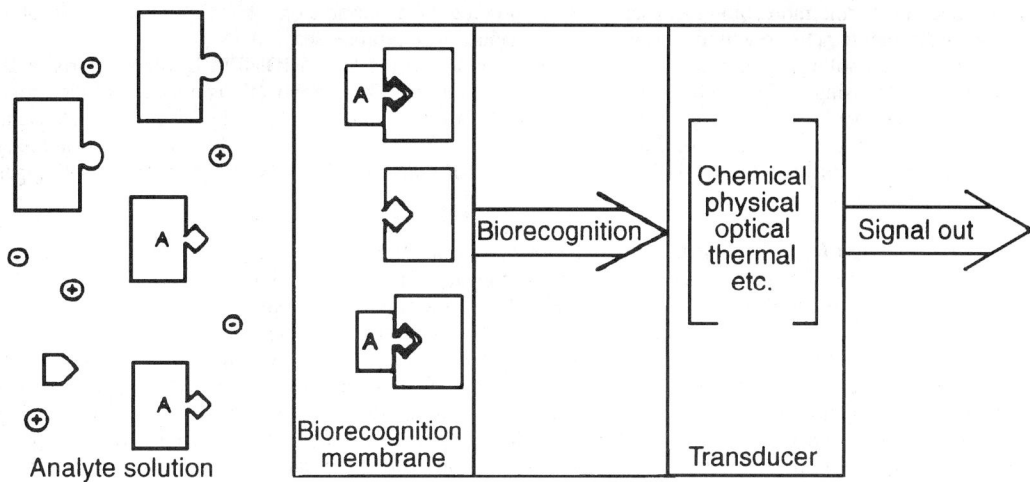

FIGURE 8.5.8 Schematic drawing of the general biosensor. It consists of a traditional transducer with a biological recognition membrane intimately in contact with the transducer. The biologically active material recognizes the analyte molecule, A, through a shape-specific recognition. In affinity-based biosensors, the binding of the analyte and bioactive molecule is the chemical signal that is detected by the transducer. In metabolic biosensors, the biologically active material converts the analyte, and any co-reactants, into product molecules. The transducer converts the result of that reaction into the output signal. The transduction can be through measurement of concentration changes in a product or coreactant concentration, or the heat liberated in the reaction.

recognizes a particular biological material (the measurand) through a reaction, specific adsorption or other physical or chemical process, and the sensor element converts the results of this recognition into a usable signal, usually electrical or optical. Figure 8.5.8 illustrates this definition.

Biosensors represent a specialized field in sensors. A selected reference list is provided.[23–25]

A biological sensor operates on similar principle of a chemical sensor. A biorecognition reaction with the analyte, through a transduction mechanism, results in a sensor output. This output can be electrical, optical, or as a change in color. The biorecognition reaction often involves biocatalysts such as enzymes. The sensory mechanism of a biological sensor is often based on a chemical sensor. Consequently, the microfabricated chemical sensors serve as the bases of microbiological sensors. For instance, the commercially available i-STAT portable clinical analyzer and the portable blood analysis system by Diametrics Medical Inc. are good examples of microchemical sensors serving as the core for biosensor applications.

Measurements of dissolved blood gases (oxygen and carbon dioxide) and blood electrolyte (pH, Na^+, K^+, Ca^{+2}, and others) can be accomplished using electrochemical sensors. For instance, the Clark oxygen sensor for measuring dissolved oxygen in biological fluids is based on an electrochemical-based amperometric sensor. In this electrochemical sensor, the reduction of oxygen results in a cathodic current, which can then be used to quantify the oxygen presented. This approach is applicable to measurements of dissolved oxygen in biological samples. Therefore, miniaturization of electrochemical sensors or chemical sensors can lead to the development of microsize biological sensors making portable biological analyzers a reality.

Biological sensors based on enzymatic reactions also frequently use a chemical sensor as the sensing mechanism. For example, a blood glucose sensor employs glucose oxidase to catalytically oxidize the glucose in the presence of oxygen. One of the resulting products is hydrogen peroxide, which can be detected electrochemically. The amount of hydrogen peroxide detected can then be used to quantify the blood glucose presented. Other enzymatic reactions have also been used for extensive biological sensor development, and the detection of a product or a co-reactant with a chemical or electrochemical sensor is then used to quantify the bio-analyte.

As anticipated, the application of MEMS technology to chemical sensors leads to the logical consideration of using MEMS technology in the development of biological sensors.[20,21] Many attempts in this direction have been made but with only limited success thus far. The fragility and relatively poor biocompatibility of silicon are concerns. Furthermore, biological sensors often prefer to be disposable or to have a limited number of uses. This is difficult to achieve with expensive MEMS manufacturing processes.

MEMS technology applied to biomedical fields has led to the term BIOMEMS. Major efforts in BIO-MEMS are focusing on DNA chips and drug delivery systems. These applications are beyond the scope of biological sensors for in vivo and in vitro applications. Development of biological sensors is a vast scientific endeavor. It is beyond the scope of this chapter to discuss it in detail. We wish only to point out the relationship between chemical and biological sensors, and the issues involved in MEMS technology to both types of microdevices.

Laboratory on a Chip.   One of the potential applications of MEMS technology is that of a complete chemical laboratory on a chip (LOC). The functional components of a simple analytical laboratory would be fabricated using silicon-based microfabrication technology. These functional components might include sample pretreatment, transportation, and separation of the sample, introduction of reagents, sensing and detection and data collection and analysis. The advantages of an LOC approach include the requirement of a small sample volume, minimum quantity of expensive reagents, and relatively portability. However, its success depends on many intricate microfabricated components such as microfluidic devices, micropumps, valves, channels, and sensors. *Fundamentals of Microfabrication* by M. Madou provides a summary of the commercial aspects of this technology. The transport properties of fluid in these microstructures can differ significantly from those in a larger-size environment. Fabrication costs are again an issue.

The term "μTAS," meaning "micro total analytical system," is used in the MEMS field. It is a part of the "system on a chip" when applied to the integrated biochemical analytical systems. The term biochip also is being used in popular science and news media in reference to the biomedical diagnostic systems fabricated on a chip.

Chemical and Biosensor Packaging

Packaging of chemical sensors is a critical issue that needs to be addressed. Differing from the packaging of electronics, that of chemical sensors must provide an area where the sensor is exposed to the environment. This means that on one hand the overall integrity of the sensors must be protected, yet on the other hand, the most critical part of the sensor must be exposed to the sensing environment, which can often be hostile. There is no simple or direct packaging method for chemical sensors, rather, it depends on the environment of the sensor applications, such as high temperature or the presence of corrosive materials.

MICROACTUATORS

Research on design, fabrication, and characterization of new microactuators is accelerating. A critical need is the development of actuation forms that are compatible with the materials and processing technologies of silicon microelectronics. Additionally, actuation is to be powered and controlled electrically, allowing full utilization of integration with on-chip electronics. Despite these stringent requirements, numerous physical phenomena have been demonstrated for microactuator applications.[26] However, the most commonly used microactuation methods have so far been either electromagnetic or thermal and are, therefore, the microactuation techniques described in this section as examples. Other microactuation techniques can be found in the journals and proceedings of the conferences in the field (see Bibliography).

Selected microactuators that exemplify the basic microactuator concepts and mechanical designs are described here. Two basic electrostatic microactuator designs, lateral resonant devices and micromotors, are discussed in more detail as examples.

Mechanical Design of Microactuators

The mechanical design of microactuators can be divided into two classes: (1) mechanisms and (2) deformable microstructures. Mechanism-type microactuators (e.g., the micromotor shown in Fig. 8.5.9) provide displacement

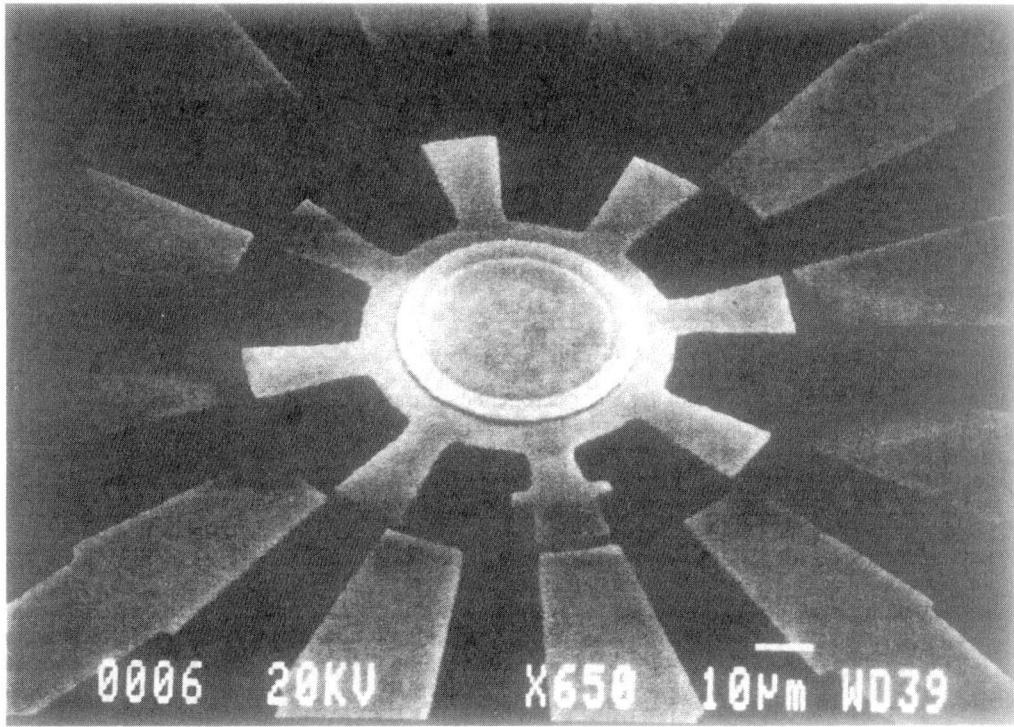

FIGURE 8.5.9   A photomicrograph of a salient-pole, variable-capacitance, side-drive micromotor. The rotor rotates in the plane of the motor/substrate.

and force through rigid-body motion while deformable microstructures (e.g., beams, diaphragms, and the lateral resonant microactuator of Fig. 8.5.10) provide displacement and force through mechanical deformation (or straining). The unrestrained, large motion capability of mechanism-type microactuators comes at the cost of friction, which is present in the bearings and joints of such devices. In contrast to the operation of macroscopic mechanical devices in which gravitational and inertial forces are often dominant, friction plays an important role in the operation of micromechanical devices. In micromechanical devices, gravitational forces (i.e., weight of the moving parts) are usually insignificant while inertial forces are often, but not always, negligible. Deformable microstructures eliminate friction since they only use joints and suspensions. Avoiding friction, however, comes at the cost of restrained (and often small) motion.

Electromagnetic and Thermal Microactuation

Most notable among electromagnetic microactuation methods have been electric (often referred to as electrostatic), piezoelectric, and magnetic actuation. Utilization of electrostatic actuation in microfabricated mechanical devices is readily possible since IC processes provide a wide selection of conductive and insulating materials. By using the conductors as electrodes and the insulators for electrical isolation of the electrodes, attractive electrostatic forces can be generated by applying a voltage across a pair of electrodes. To generate repulsive forces, *electrets* (dielectric materials that have a permanent charge) are required; however, electrets are not readily available in the IC industry and, as a result, have not as yet been used in microactuation.

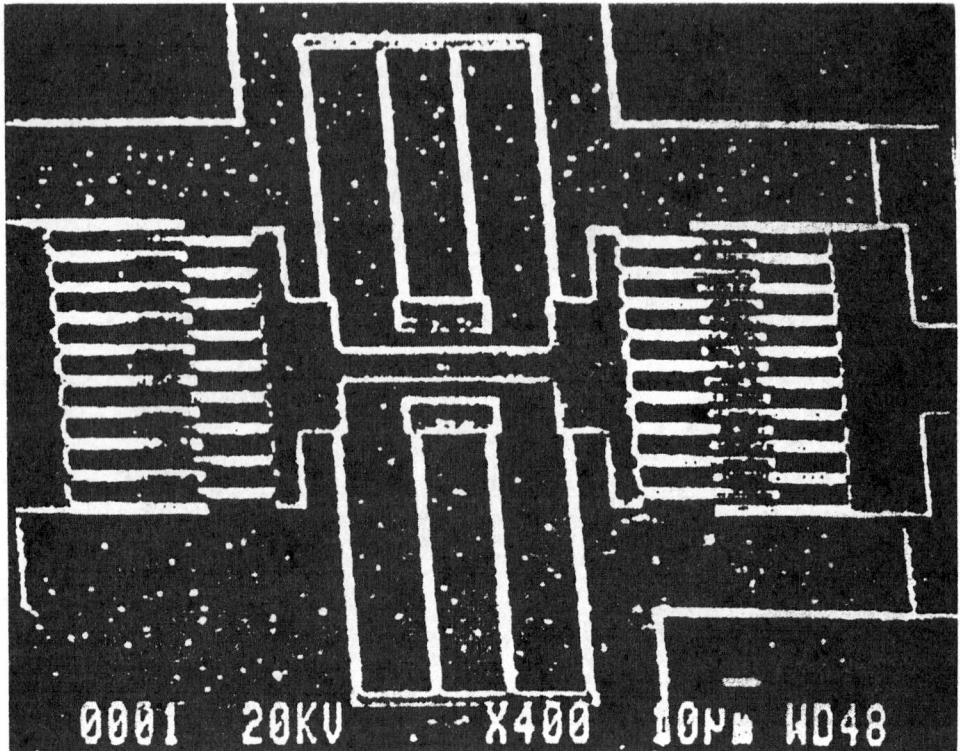

FIGURE 8.5.10 A photomicrograph of a lateral resonant device. The shuttle mass attached to the folded cantilever suspension moves in the plane of the device/substrate in the direction of the comb fingers.

Electrostatic forces can be used in conjunction with conventional deformable microstructures such as beams and diaphragms to generate force or provide motion. For this group of microactuators, the deformation is predominantly in the direction normal to substrate plane. Figure 8.5.11 is a schematic example of an electrostatic diaphragm microactuator. By applying a voltage across the diaphragm/electrode air gap (which is typically a few microns), the diaphragm can be deflected upward. The generated electric force results from the attractive force between the electrical charges of opposite sign on the electrically conductive silicon surfaces across the gap. As in the case of the lateral resonant microactuator in Fig. 8.5.11, electrostatic actuation can also be used in conjunction with deformable microstructures to provide motion in the plane of the substrate.

Electrostatic actuation can also be used in conjunction with a mechanism, for example, in the case of the micromotor in Fig. 8.5.9. In an electric micromotor, the attractive forces generated by electric charge distributions are used to convert electrical to mechanical energy. By proper commutation of these charge distributions (i.e., proper switching of excitation voltage) on a set of stationary electrodes, known as the stator, and a set of moving electrodes, known as the rotor, continuous motion of the rotor can be achieved.

Another form of microactuation has been based on the piezoelectric effect. Figure 8.5.12 is a schematic example of a possible cantilever microactuator design. In this design the piezoelectric thin film, which is sandwiched between two electrodes, is placed on top of a silicon cantilever beam. When a voltage is applied across the piezoelectric film, the film will expand or contract in the lateral direction, resulting in a downward or upward deflection, respectively, of the cantilever. The polarity of the applied voltage determines if the film expands or contracts. Piezoelectric films have been used to provide actuation in a variety of applications such

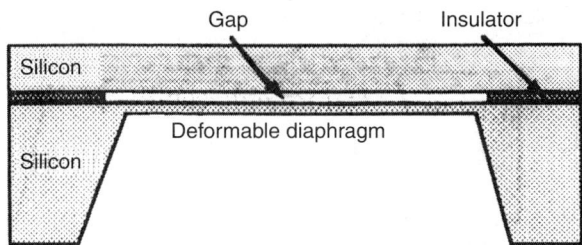

FIGURE 8.5.11   A schematic drawing of an electrostatic diaphragm microactuator.

as valves, pumps, positioning devices, and ultrasonic micromotors.[27,28] Typical piezoelectric thin films being used in microactuators are zinc oxide (ZnO), lead zirconate titanate (PZT), and polyvinylidene difluoride (PVDF). Of these materials, PZT has the largest piezoelectric coefficients.

Similar to electrostatic actuation, magnetic actuation can be used in conjunction with both deformable microstructures and mechanisms. Since most macroscopic electromagnetic motors use magnetic actuation, this form of actuation is most familiar. However, its utilization for microfabricated actuators has been limited because it requires magnetic materials and fabrications of windings. Nevertheless, a few studies[29–31] have addressed the development of magnetic microactuators. For example, magnetic micromotors made of nickel have been demonstrated by using the LIGA fabrication process.[32]

While electromagnetic microactuation has been demonstrated in conjunction with both classes of microactuators, thermal microactuation is primarily suitable for deformable microstructures. Most notable among thermal microactuation methods have been bimetallic, shape memory alloy (SMA), and thermopneumatic. Figure 8.5.13 shows a schematic example of a possible bimetallic or SMA cantilever microactuator. Bimetallic microactuators use the thermal coefficient of expansion mismatch between the material components of a sandwich layer to generate force or motion.[33] For example, applying a current through the heating resistor in the device of Fig. 8.5.13 will heat the composite metal/silicon cantilever beam. Since the metal expands more than the silicon, the beam will deflect downward. The downward deflection of the beam can be maintained by supplying power through the heating resistor. Since microfabrication techniques lend themselves more easily to the stacking of the materials on top of rather than beside each other, reported microactuators based on the bimetallic effect provide deformations that are predominantly in the direction normal to the plane of the substrate.

SMAs are metals that have shape-recovery characteristics. When these alloys are plastically deformed at one temperature (martensitic phase), they will completely recover their original shape on being raised to a higher temperature (austenitic phase).[34] In recovering their shape, the alloys can produce a displacement or force, or a combination of the two, as a function of temperature. The development of thin-film SMAs provides the potential for development of silicon-based MEMS that incorporate SMA microactuators. The specific SMA that has been studied for microactuator applications is Nitinol or TiNi, an alloy of titanium and nickel.[35,36] TiNi alloys undergo a martensitic transformation during cooling at a temperature M_s, and the transformation is completed at a lower temperature M_f. In the temperature interval between M_s and M_f, the alloy

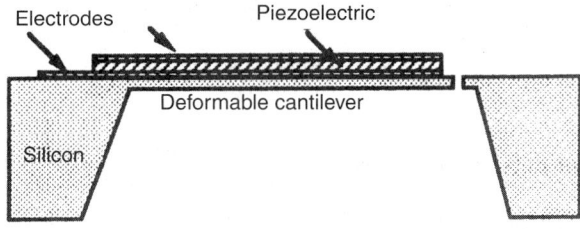

FIGURE 8.5.12   Schematic of a piezoelectric cantilever microactuator.

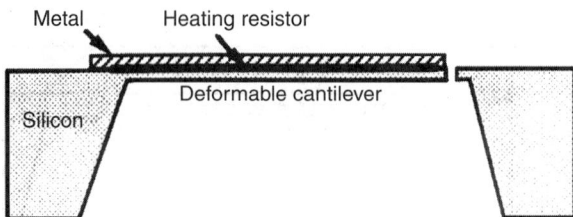

FIGURE 8.5.13 Schematic drawing of a bimetallic or shape-memory-alloy cantilever microactuator.

has a two-phase (austenite and martensite) microstructure; the transformation temperature is relatively narrow and is on the order of 15°C.[37,38] Therefore, one may speak of a transition temperature, M_t (average of M_s and M_f), below which the alloy shows the shape memory effect and above which the effect is absent. Associated with the martensitic transformation, TiNi shows a shape memory effect.

An SMA cantilever microactuator based on the construction shown in Fig. 8.5.13 would, for example, use TiNi for the metal component. Assume that the cantilever is fabricated such that it is flat (i.e., no deflection) with TiNi in its austenitic phase (well above M_t), and then cooled to below M_t to form the martensitic phase. Since the cantilever is also a bimetallic device, it will deflect upward on cooling (i.e., TiNi contracts more than the silicon). By applying current through the heating resistor, the cantilever will try to recover its flat shape once M_t is reached. Note that there would be some bimetallic contribution to the actuation until M_t is reached. However, the transition at M_t will be sudden and because of the shape memory effect.

In a thermopneumatic microactuator, the volume expansion of a fluid is used for actuation.[39] Figure 8.5.14 is a schematic of a possible implementation of a thermopneumatic diaphragm microactuator based on resistive heating of a fluid in a micromachined sealed cavity. As the fluid is heated, its volume expands, increasing the pressure inside the cavity. The increased pressure acts as a load on the thin diaphragm and results in its deflection. The fluid may be a liquid that turns to a gas or simply a gas that expands on heating.

Comparison of Actuation Methods

It is often of interest to compare the various microactuation methods for their relative advantages and disadvantages, but detailed comparisons are only realistic when performed in light of an application. Nevertheless, it is possible to outline some general points. An important point of interest for a microactuation method is the amount of force (or mechanical energy) that can be generated. One approach for general comparison is to estimate the amount of energy, W, available per unit volume. For example, the energy per unit volume, W, available from a TiNi SMA material can be calculated from the area under the operating curve on the stress-strain

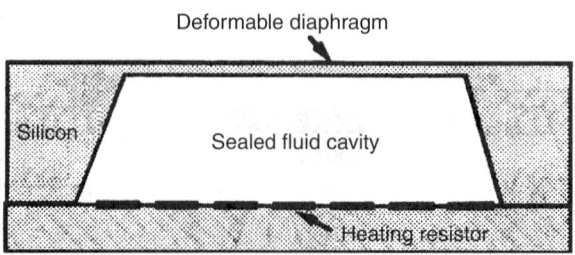

FIGURE 8.5.14 Schematic of a thermopneumatic diaphragm microactuator.

diagram. For an austenitic state with a yield strength of 420 MPa and an available shape memory strain of 8 percent, W is 4×10^7 J/m³. For bimetallic actuators, W would depend on the magnitude of temperature change but can be of comparable magnitude. For an electrostatic microactuator, $W = (1/2)\epsilon_0 E_{max}^2$ is nearly 4×10^5 J/m³, assuming an electric field breakdown limit (E_{max}) of 3×10^8 V/m for microscopic air gaps.[40] (ϵ_0 is the permittivity of free space.) For piezoelectric microactuators, W may be up to 10^5 J/m³, assuming PZT to be the piezoelectric material. For magnetic microactuators, $W = (1/2)(B_{max}^2/\mu_0)$ is about 10^6 J/m³, assuming a magnetic flux density saturation of 1.5T.[40] (μ_0 is the permeability of free space.)

Other factors must also be considered in comparing different microactuation methods. In general, thermal microactuators have a slow response time (e.g., on the order of tens of milliseconds) and high power consumption (e.g., on the order of tens of milliwatts). Comparatively, electromagnetic microactuators can be much faster (e.g., microsecond response time) and consume far less power, particularly electrostatic microactuators. The construction of thermal microactuators often requires the final freestanding part to be a laminate of layers with very different mechanical properties. This often complicates device design as evidenced by the fact that reported bimetallic actuators exhibit a preset deflection because of the residual stresses in the thin films. Piezoelectric microactuators require deposition of additional films since silicon is not piezoelectric. Furthermore, static operation of piezoelectric microactuators is limited by charge leakage. Electrostatic microactuators can be fabricated using conducting and insulating films that are common to microelectronics technology. Static excitation of electrostatic microactuators requires voltages across insulating gaps and nearly no loss. Magnetic microactuators require magnetic materials that are not common in IC technology and often require some type of manual assembly. Static excitation of magnetic microactuators requires current through windings and persistent conduction losses.

Microactuator Examples

While research and development of microactuators has been rapidly expanding in the last few years, the four classes of microactuators described below are representative of a large segment of the microactuator technology. The examples provide typical performance characteristics of a large cross section of microactuators.

Microvalves. Microactuator technology is being pursued for the development of microvalves for industrial (e.g., electronic pressure regulators) and biomedical applications (e.g., drug delivery and chemical analysis fluidic microsystems). The microvalve in Fig. 8.5.15 is an example of a bimetallic microactuator fabricated by bulk micromachining and bonding of two wafers.[41] Heating of the bimetallic diaphragm by passing a current

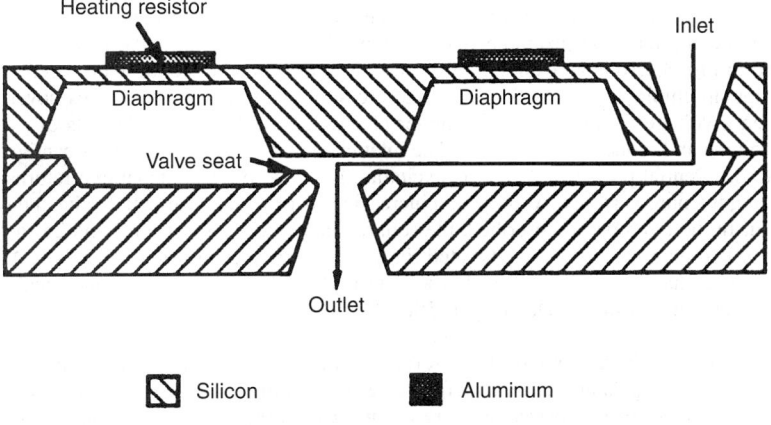

FIGURE 8.5.15 Schematic drawing of a bimetallic microvalve.[21]

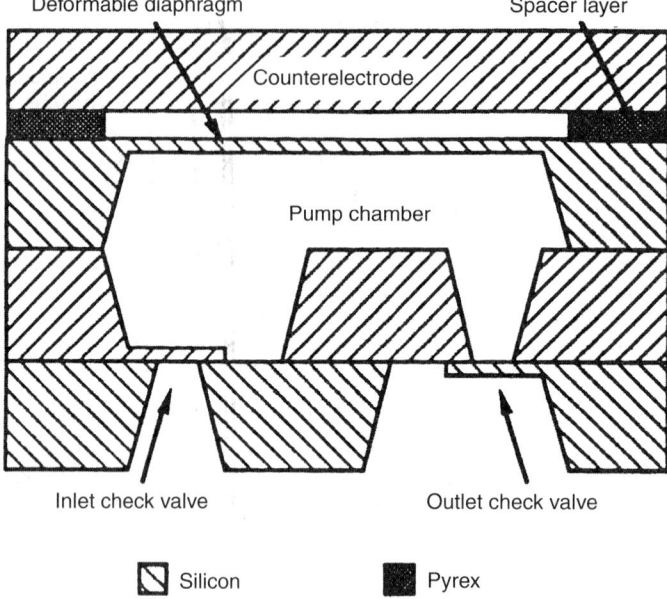

Deformable diaphragm   Spacer layer

Counterelectrode

Pump chamber

Inlet check valve   Outlet check valve

Silicon   Pyrex

FIGURE 8.5.16   Schematic drawing of an electrostatic micropump.[43]

through a heating resistor embedded between the metal and silicon sandwich causes the metal and silicon layers to expand. This expansion results in the downward deflection of the diaphragm because of the mismatch in the thermal expansion coefficients of the metal and silicon. If heated sufficiently, the diaphragm can deflect far enough to close off the valve. The diaphragm in this device is 2.5 mm in diameter and is 10 μm thick. A 5 μm-thick aluminum layer is used as the metal component. Fully proportional control of flows in the range of 0 to 300 cc/min has been demonstrated with this valve for input pressures from zero to 100 psi. On/off flow ratios have been greater than 1000. To close the valve at 20 psig input, 1.5 W of power is required.[41]

Electrostatic, shape memory metal (SMA)[42] and thermopneumatic[39] actuations are also being used in the development of microvalves similar in architecture and concept to that described above.

Micropumps.   Another application of microactuator technology is the development of micropumps. Such micropumps are of interest, for example, in developing chemical analysis and drug delivery microsystems. The micropump in Fig. 8.5.16 is an example of electric actuation in conjunction with a deformable diaphragm.[43] This micropump employs two check valves that are simply cantilever beam flaps covering micromachined holes. When a voltage is applied to the counter electrode, the diaphragm deflects up, increasing the pump chamber volume and reducing its pressure. The inlet check valve then opens as its cantilever flap bends up in response to differential pressure. When the excitation is turned off, the diaphragm returns to its normal position, reducing the pump chamber volume and increasing its pressure. The outlet valve then opens allowing the fluid to exit. In the micropump described, the square diaphragm is 4×4 mm^2 and 25 μm thick; the actuator gap is 4 μm. Pumping has been demonstrated for actuation frequencies of 1 to 100 Hz. At 25 Hz, a pumping rate of 70 μl/min has been demonstrated when the outlet and inlet pressures are equal. Typical forward to reverse flow rate ratio of the check valves is 5000:1.[43]

Lateral Resonant Devices.   Where the deformable microactuators above were fabricated by bulk micromachining and bonding, many current deformable microactuators rely on (polysilicon) surface micromachining. Lateral resonant microstructures are the newest among these;[44] they are actuated by electrostatic forces. Figure 8.5.17 shows a plan-view schematic of the lateral resonant microactuator of Fig. 8.5.10. While

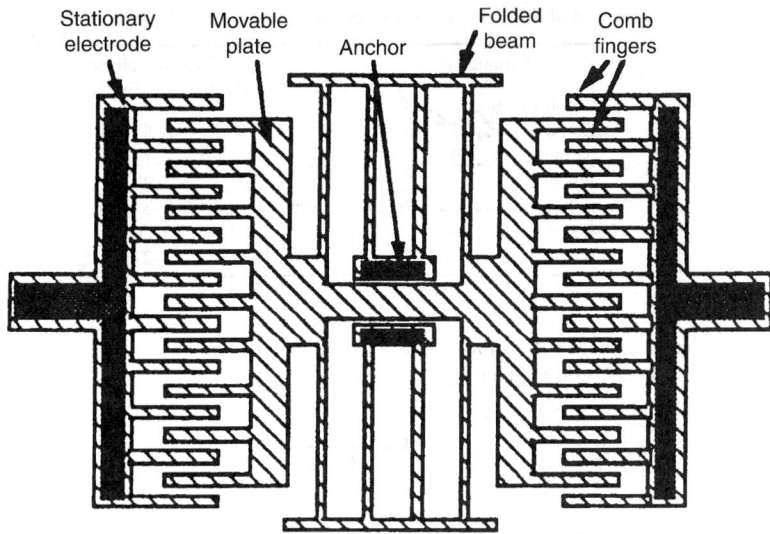

FIGURE 8.5.17   Plan view of a lateral resonant microactuator.

polysilicon surface micromachining has thus far been the primary technique for the fabrication of these devices, other fabrication techniques are emerging.[45] Lateral resonant structures typically consist of a movable plate that is suspended a couple of microns above the substrate by a flexible folded beam support. The folded beam support is anchored to the substrate at one end and attached to the movable plate at the other end. (In Fig. 8.5.17, the light-fill regions indicate parts that are suspended over the substrate and the dark-fill regions indicate the substrate anchor areas.) Electric forces through the use of comb-drive and/or side-drive electrodes are employed to provide actuation. The suspension and actuation are designed for lateral deformation of the suspension and, therefore, lateral movement of the movable plate. In resonant mode, these microactuators can provide lateral deformations near 10 μm; tens of kHz is typical of resonant frequencies for the designs reported.[44]

The electrostatic force generated by a single comb finger on the movable plate is

$$F = 2\epsilon_0 V^2 \left(\frac{t}{g} \right) \tag{8}$$

where F = force
ϵ_0 = permittivity of free space
V = applied voltage
t = beam thickness
g = comb finger gap

The factor of 2 accounts for the fact that on the stationary electrode there are always two comb fingers, one on each side of the comb finger on the movable plate. The above formula is based on simple parallel-plate capacitance calculations. For comparable values of t and g (which is often the case in actual devices based on polysilicon surface micromachining), the above formula provides a reasonable approximation. Its accuracy increases as the t to g ratio increases beyond 3 to 4 such that the contribution of the fringing fields becomes negligible. For t to g ratios below 1, the accuracy of the above formula may be quite poor.

This displacement x (along the comb finger direction) of the movable plate as a function of the applied force is

$$x = \frac{F}{K} \tag{9}$$

TABLE 8.5.1 Typical Lateral Resonant Device Parameters

Parameter	Typical values
Comb finger gap (μm)	1–4
Comb finger width (μm)	2–4
Number of fingers	9–11
Finger length (μm)	20–40
Polysilicon thickness (μm)	1.5–3
Beam length (μm)	80–200
Beam width (μm)	1–2

where x is the displacement of the shuttle, and K is the mechanical stiffness of the folded beam suspension. The stiffness of the suspension is given by

$$K = 2Et\left(\frac{w}{l}\right)^3 \tag{10}$$

where E = Young's modulus (near 160 GPa for polysilicon)
  υ = Poisson's ratio (assumed to be near 0.3 for polysilicon)
  l = beam length
  w = beam width

When the t to w ratio is greater than 1, wide beam effect can be accounted for by dividing the modulus by $(1 - \upsilon^2)$. The resonant frequency of the shuttle is given by

$$f = \frac{1}{2\pi}\sqrt{\frac{K}{M}} \tag{11}$$

where M is the mass of the shuttle.

Typical dimensions of lateral resonant devices are described in Table 8.5.1. Based on these typical dimensions, the generated forces per square volt, for a single comb finger, are usually in the range of 4 to 26 pN/V^2, while typical stiffness values for the suspension are in the range of 60 mN/m to 15 N/m.

To improve performance by increasing the generated actuation force[46] and displacement,[47] other polysilicon deformable microstructure mechanical designs that use electrostatic actuation have been tried. The reader is referred to the proceedings of recent conferences in the field for reported design variations.

Micromotors. In addition to deformable microstructures, MEMS require large-motion actuators (e.g., micromotors) to power linkages. Electrostatic actuation in conjunction with polysilicon surface micromachining has been used to develop variable-capacitance side-drive micromotors[48] similar to the one shown in Fig. 8.5.9. For variable-capacitance micromotors (which require conductive materials for the rotor and the stator, as well as insulators for electrical isolation), the material requirements are compatible with polysilicon surface micromachining. Heavily-phosphorous-doped polysilicon is used for the rotor and the stator, and silicon nitride is used for electrical isolation. Additionally, since the electric breakdown limit in air increases drastically for microscopic (order of 1 μm) air-gap sizes, high enough torques can be generated to start the motor. The electric field breakdown limit in air is 3×10^6 V/m for macroscopic dimensions and increases to above 10^8 V/m for microscopic, air-gap sizes.[44,49] For microscopic dimensions, the gap size approaches the mean free path in air. As a result, the field strength increases since avalanche breakdown, which is the field breakdown mechanism, requires a large number of collisions to ionize the air.

Figure 8.5.18 is a plan-view schematic of the micromotor in Fig. 8.5.9 and describes the micromotor operation. At the end of a phase excitation, the rotor is fully aligned with the excited stator phase (e.g., the configuration shown in Fig. 8.5.18 for alignment of rotor pole set 1 with stator phase B). To move the rotor clockwise,

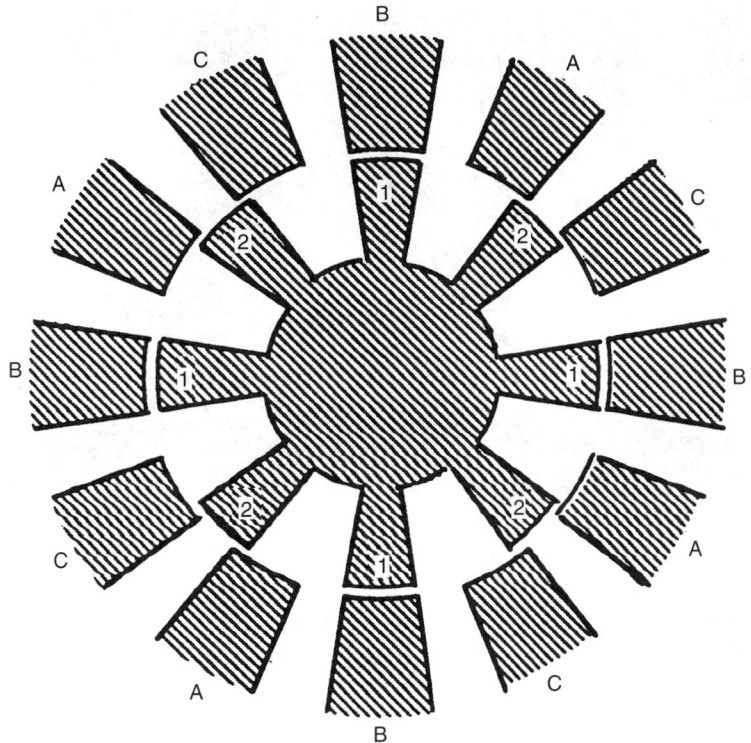

FIGURE 8.5.18 Plan-view schematic of the micromotor shown in Fig. 8.5.9.

stator phase C is excited next. To move the rotor counterclockwise, stator phase A would be excited next. Proper commutation of the excitation signal results in continuous rotation of the rotor. The operation of these micromotors relies on the storage of electrical energy in a variable rotor-stator capacitance. The change in this capacitance in the direction of motion is proportional to the output torque of the micromotor. In terms of design, one strives for optimum torque generation by enhancing the capacitance change in the direction of motion. Very rough estimates of the micromotor motive torque can be calculated by parallel-plate approximations. More accurate calculations of torque have been performed by finite element analysis.[50]

A detailed discussion of the evolution of polysilicon micromotors in terms of design and operational characteristics is documented elsewhere,[51–55] and a review of the micromotor technology is available in Ref. 49. In sum, the evolution in the design of variable-capacitance micromotors has been motivated by enhanced rotor stability, simpler fabrication processes, and increased torque. The former two are achieved at the cost of sacrificing the latter by using the side-drive architecture shown above. Some of the torque is regained by going from a salient-pole to a wobble (Fig. 8.5.19) micromotor design.[49,51,52,55] The structural design of a wobble micromotor is identical to that of the salient-pole (see Fig. 8.5.20) showing the micromotor cross-section), except that the rotor and stator designs are different. For the wobble micromotor, the rotor has no poles (or no saliency). The central feature of the wobble micromotor is that the rotor wobbles around the center bearing post as the excitation voltage is moved on the stator poles. Since there is a finite bearing clearance, the rotor's wobbling motion will also result in a rotation of the rotor.

Table 8.5.2 presents typical dimensions of electrostatic polysilicon micromotors. The reader is referred to Refs. 51, 53, and 56 for a detailed discussion of micromotor design and to Ref. 49 for a review of the micromotor technology.

Micromotor operation has been demonstrated in dielectric liquids (e.g., deionized water and silicone oil),[57] as well as in gaseous environments (e.g., nitrogen, argon, oxygen, and room air).[58] The performance and

FIGURE 8.5.19   A photomicrograph of a wobble micromotor having a 100-μm diameter.

lifetime of micromotors have been improved significantly over time. Excitation voltages as low as 30 V have been sufficient in operating some current micromotors.[57,58] Open-loop operational stepping speeds up to 15,000 rpm have been reported for operation in a gaseous environment.[51,57] The maximum attainable speeds for operation in gaseous environments have been limited by the power supply and not the micromotor. Micromotor lifetime has been extended to many millions of cycles over a period of several days for operation in room air or silicone oil.[55,57] For these micromotors, a clear failure point has not yet been identified even though some micromotors have been tested for a few hundred million start-stops. Dynamic friction torque is below 10 percent of the micromotor motive torque[54] and wear is not presently a limiting factor in micromotor operation.[55] The motive torque of the side-drive micromotors is typically on the order of tens of pico-Newton-meters (pN-m).

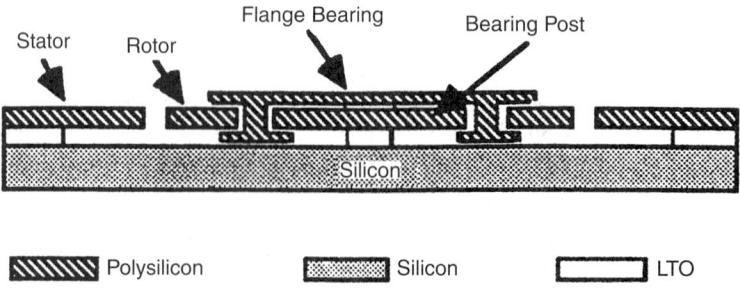

FIGURE 8.5.20   Schematic drawing of the cross section of salient-pole and wobble micromotors of Figs. 8.5.9 and 8.5.19.

TABLE 8.5.2 Typical Micromotor Dimensions

Parameter	Value
Rotor/stator gap (μm)	1.5–2.5
Rotor inner radius (μm)	10–20
Rotor outer radius (μm)	50–75
Bearing clearance (μm)	0.3–0.8
Rotor to substrate gap (μm)	2–2.5

REFERENCES

MEMS Technology

1. Muller, R. S., et al. (eds.) "Microsensors," *IEEE Press*, 1991.

2. Petersen K. E., "Silicon as a mechanical material," *Proc. IEEE*, 70, p. 420, 1982.

3. Mehregany, M., W. H. Ko, A. S. Dewa, and C. C. Liu, "Introduction to MEMs and Multiuser MEMs Processes," Case Western Reserve University, Short Course Handbook, <http://mems.cwru.edu/shortcourse/partITOC.html>.

4. Ko, W. H., J. T, Suminto, and G. J. Yeh, "Bonding Techniques for Microsensors," in Micromachining and Micropackaging of Transducers, C. D. Fung et al. (eds.), Elsevier, pp. 41–46, 1985.

5. Tong, Q. Y., and U. Gosele, "Semiconductor Wafer Bonding," Wiley, 1999.

6. Gui, Cheng-Qun, "Direct Wafer Bonding with Chemical Mechanical Polishing," Ph.D. Thesis, University of Twente, 1998.

7. Menz, W., W. Bacher, M. Harmening, and A. Michel, "The LIGA technique—a novel concept for microstructures and the combination with Si-techniques by injection molding," *MEMS* 90, p. 69, 1990.

8. Frozier, A. B., and M. G. Allen, "High aspect ration electroplated microstructures using a photosensitive polyimide Process," *MEMS* 92, p. 87, 1992.

Physical Sensors

9. Ko, W. H., and C. D. Fung, "VLSI and intelligent transducers," *Sensors and Actuators*, 2, pp. 239–250, 1982.

10. Lahiji, G. R., and K. D. Wise, "A monolithic thermopile detector fabricated using integrated-circuit technology," *Proc. Int. Electron. Dev. Meeting*, p. 676, 1980.

11. Ko, W. H., "Solid state capacitive pressure transducers," *Sensors and Actuators*, 10, pp. 303–320, 1986.

12. Ko, W. H., M. H. Bao, and Y. D. Hong, "A high sensitivity integrated circuit capacitive pressure transducer," *IEEE Trans. Elect. Dev.*, ED-29, pp. 48–56, 1982.

13. Seitz, W. R., "Chemical sensors based on fiber optics," *Anal. Chem.*, 56, 16A–34A, 1984.

14. Baltes, H. P., and R. S. Popovic, "Integrated semiconductor magnetic field sensors," *Proc. IEEE*, 74, pp. 1107–1132, 1986.

Chemical and Biomedical Sensors

15. Madou, M. J., and S. R. Morrison, "Chemical Sensing with Solid State Devices," Academic Press, 1989.

16. Microfabricated Systems and MEMS VI, Proceedings Vol. 2002–6, The Electrochemical Society, P. J. Hesketh et al. (ed.), 2002.

17. Schmitte, F. J., and G. Wiegleb, "Conductivity behavior of thick film Tin-dioxide gas sensors," *Sensors and Actuators*, 84, p. 473, 1991.

18. Torvela, H., A. Harkona-Mattila, and S. Leppavori, "Characterization of a combustion process using catalyzed Tin-oxide gas sensor to detect CO from emission gases," *Sensors and Actuators*. Bl, p. 83, 1990.

19. Gall, M., "The Si planar pellistor: a low-power pellistor sensor in Si thin film technology," *Sensors and Actuators*, B4, p. 533, 1991.

20. Accorsi, A., G. Delapierre, C. Vauchier, and D. Charlot, "A new microsensor for environmental measurements," *Sensors and Actuators*, B4, p. 539, 1991.

21. Hesketh, P. J., L. Smith, M. Gerber, T. Wulff, and J. Joseph, "Thermal stability of thin filmthermopile sensors," in Proc. 3rd Internat. Meeting on Chem. Sensors, p. 389, 1990.

22. Muehlbauer, M. J., E. J. Guilbeau, and B. C. Towe, "Model for a thermoelectric enzyme glucose sensor," *Anal. Chem.*, 61, p. 77, 1989.

23. Turner, A. F. P., I. Karube, and G. S. Wilson (eds.), "Biosensors: Fundamentals and Applications," Oxford Science Publications, 1987.

24. Blum, L. J., and P. R. Coulet (eds.), "Biosensor Principles and Applications," Marcel Dekker, 1991.

25. Turner, A. F. P. (ed.), "Advances in Biosensors: A Research Annual," JAI Press, 1991.

Microactuators

26. Fujita, H., and K. Gabriel, "New Opportunities for Microactuators," in Technical Digest, 6th International Conference Solid-State Sensors and Actuators, pp. 14–20, June 1991.

27. Feury, A. M., T. L. Poteat, and W. S. Trimmer, "A Micromachined Manipulator for Submicron Positioning of Optical Fibers," in Technical Digest, IEEE Solid-State Sensor Workshop, June 1986.

28. Moroney, R. M., R. M. White, and R. T. Howe, "Ultrasonic Micromotors: Physics and Applications," in Proceedings, IEEE Micro Electro Mechanical Systems Workshop, pp. 182–187, February 1990.

29. Guckel, H., T. R. Christenson, K. J. Skrobis, J. Klein, and M. Karnowsky, "Design and Testing of Planar Magnetic Micromotors Fabricated by Deep X-ray Lithography and Electroplating," in Technical Digest, 7th International Conference Solid-State Sensors and Actuators, pp. 60–64, June 1993.

30. Ahn, C., H. Y., J. Kim, and M. G. Allen, "A Planar Variable Reluctance Magnetic Micromotor with Fully Integrated Stator and Wrapped Coil," in Proceedings, IEEE Micro Electro Mechanical Systems Workshop, pp. 1–6, February 1993.

31. Wagner, B., M. Kreutzer, and W. Benecke, "Linear and Rotational Magnetic Micromotors Fabricated Using Silicon Technology," in Proceedings, IEEE Micro Electro Mechanical Systems Workshop, Travemunde, pp. 183–189, February 1992.

32. Guckel, H., K. J. Skrobis, T. R. Christenson, J. Klein, S. Han, B. Choi, E. G. Lovell, and T. W. Chapman, "Fabrication and testing of the planar magnetic micromotor," *J. Micromech. Microeng.* 1, pp. 135–138, September 1991.

33. Riethmuller, W., and W. Benecke, "Thermally excited silicon microactuators," *IEEE Trans. Electron Dev.*, 35, p. 758, 1988.

34. Schetky, L. M., "Shape-memory alloy," *Scientific American*, 241, p. 74, Nov. 1979.

35. Walker, J. A., K. J. Gabriel, and M. Mehregany, "Thin-film processing of TiNi shape memory alloy," *Sensors and Actuators*, A21–A23, p. 243, 1990.

36. Johnson, A. D., "Vacuum deposited TiNi shape memory film: Characterization and application in microdevices," *J. Micromech. Microeng.* 1, p. 34, 1991.

37. Buehler, W. J., J. V. Gilfrich, and R. C. Wiley, "Effect of low-temperature phase changes on the mechanical properties of alloys near composition TiNi," *J. Appl. Phys.*, 34, p. 1475, 1963.

38. Wang, F. E., and W. J. Buehler, "Additional unique property changes observed during TiNi transition," *Appl. Phys. Lett.*, 21, p. 105, 1972.

39. Zdeblick, M. J., and J. B. Angell, "A Microminiature Electro-Fluidic Valve," in Technical Digest, 4th International Conference on Solid-State Sensors and Actuators, p. 827, June 1987.

40. Bart, S. F., T. A. Lober, R. T. Howe, J. H. Lang, and M. F. Schlecht, "Design considerations for micromachined electric actuators," *Sensors and Actuators*, 14, p. 269–292, July 1988.

41. Jennan, H., "Electrically-Activated, Normally-Closed Diaphragm Valves," in Technical Digest, 6th International Conference on Solid-State Sensors and Actuators, pp. 1045–1048, June 1991.

42. Busch, J. D., and A. D. Johnson "Prototype Micro-Valve Actuator," in Proceedings, IEEE Micro Electro Mechanical Systems Workshop, p. 40, February 1990.

43. Zengerle, R., A. Richter, and H. Sandmaier, "A Micro Membrane Pump with Electrostatic Actuation," in Proceedings, IEEE Micro Electro Mechanical Systems Workshop, pp. 19–24. February, 1992.

44. Tang, W. C., T. H. Nguyen, and R. T. Howe, "Laterally driven polysilicon resonant microstructures," *Sensors and Actuators*, 20, pp. 25–32, 1989.

45. Gianchandani, Y. B., and K. Najafi, "A bulk silicon wafer dissolved process for microelectromechanical devices," *J. Microelectromech. Syst.*, 1, pp. 77–85, June 1992.

46. Takeshima, N., K. Gabriel, M. Ozaki, J. Takahashi, H. Horiguchi, and H. Fujita, "Electrostatic Parallelogram Actuators," in Technical Digest, 6th International Conference on Solid-State Sensors and Actuators, pp. 63–66, June 1991.

47. Brennen, R. A., M. G. Lim, A. P. Pisano, and A. T. Chou, "Large Displacement Linear Actuator," in Technical Digest, IEEE Solid-State Sensors and Actuators Workshop, pp. 135–139, June 1990.

48. Deng, K., M. Mehregany, and A. S. Dewa, "A Simple Fabrication Process for Side-Drive Micromotors," in Technical Digest, 7th International Conference Solid-State Sensors and Actuators, pp. 756–759, June 1993.

49. Mehregany, M., and Y. C. Tai, "Surface micromachined mechanisms and micromotors," *J. Micromech. Microeng.* l, pp. 73–85, June 1992.

50. Omar, M. P., M. Mehregany, and R. L. Mullen, "Modeling of electric and fluid fields in silicon microactuators," *Int. J. Appl. Electromagnet. Mater.*, 3, pp. 249–252, 1993.

51. Mehregany, M., "Microfabricated Silicon Electric Mechanisms," Ph.D. Thesis, Massachusetts Institute of Technology, June 1990.

52. Mehregany, M., S. F. Bart, L. S. Tavrow, J. H. Lang, S. D. Senturia, and M. F. Schlecht, "A study of three microfabricated variable-capacitance motors," *Sensors and Actuators*, A21–A23, pp. 73–179, February–April 1990.

53. Mehregany, M., S. F. Bart, L. S. Tavrow, J. H. Lang, and S. D. Senturia, "Principles in design and microfabrication of variable-capacitance side-drive motors," *J. Vac. Sci. Tech. A*, 8, pp. 3614–3624, July–August 1990.

54. Bart, S. F., M. Mehregany, L. S. Tavrow, J. H. Lang, and S. D. Senturia, "Electric micromotor dynamics," *IEEE Trans. Electron Dev.*, ED-39, pp. 566–575, March 1992.

55. Mehregany, M., S. D. Senturia, and J. H. Lang, "Measurement of wear in polysilicon micromotors," *IEEE Trans. Electron Dev.*, ED-39, pp. 1136–1143, May 1992.

56. Mehregany, M., S. D. Senturia, J. H. Lang, and P. Nagarkar, "Micromotor fabrication," *IEEE Trans. Electron Dev.* ED-39, pp. 2060–2069, September 1992.

57. Mehregay, M., and V. R. Dhuler "Operation of electrostatic micromotors in liquid environments," *J. Micromech. Microeng.*, 2, pp. 1–3, March 1992.

58. Dhuler, V. R., M. Mehregany, S. M. Phillips, and J. H. Lang, "A Comparative Study of Bearing Designs and Operational Environments for Harmonic Side-Drive Micromotors," Proceedings, IEEE Micro Electro Mechanical Systems Workshop. pp. 171–176, February 1992.

BIBLIOGRAPHY

A. Books

Gopel, W., J. Hesse, and J. N. Zemel (eds.), "Sensors—A Comprehensive Survey," VCH publisher, 1989.

Kovacs, G. T. A., "Micromachined Transducers—Sourcebook," McGraw-Hill, 1998.

Madou, M., "Fundamental of Microfabrication," 2nd ed., CRC Press, 2002.

Madou, M. J., and S. R. Morrison, "Chemical Sensing with Solid State Devices," Academic Press, 1989.

Maluf, N., "An Introduction to Microelectromechanical System Engineering," Artech House, 2000.

Norton, H. M., "Handbook of Transducers," Prentice Hall, 1989.

Senturia, S. D., "Microsystem Design," Kluwer Academic, 2001.

Sze, S. M. (ed.), "Semoconductor Sensors," Wiley, 1994.

Wolf, S., and R. N. Tauber, "Silicon Processing for the VLSI Era, Volume I—Process Technology," Lattice Press, 1986.

B. Journals

IEEE Transactions on Electron Devices. &. Special issues, ED-26 (1979), ED-29 (1982).

Journal of MEMS, IEEE and ASME (1992–present).

Journal of Micromechanics and Microengineering, Institute of Physics (1991–present).

Microsystems Technologies, Springer International, (1994–present).

Sensors and Actuators, Elsevier (1980–present) monthly, part A physical, part B chemical.

Sensors and Materials, MYU, (1989–present).

C. Conference Proceedings and Technical Digests

Asian Pacific Transducers and Micro/Nano Technologies Conference (2002).

Europe Sensors and Microsystems Conferences (1990, yearly).

IEEE/MEMS Workshop, IEEE (1987, 1989–2002, yearly).

International Meetings on Chemical Sensors (1983, 1986, 1990, 1992, 1994, 1996, 1998.).

Japan Sensor Symposia (annually 1981–2002, yearly).

Microfabricated Systems and MEMS VI, Proceedings Vol. 2002-4, The Electrochemical Society, P. J. Hesketh et al. (ed) (2002–2004, even years).

Sensors and Actuators Workshops (1986–2002, even years).

Sensor Expo in United States of America (1990–2002, yearly).

Transducer Conferences (1981–2001, odd years).

SECTION 9

RADIANT ENERGY SOURCES AND SENSORS

Radiant energy sources include both coherent and noncoherent sources. Noncoherent sources are typically traditional light sources for illuminating applications, but also include light-emitting diodes. Lasers, on the other hand, rely on coherent emission that produces a beam that is highly collimated and highly monochromatic.

Semiconductor laser diodes represent a class of lasers especially useful in optical computing applications, optical sensors, optical disc systems, and materials processing. They are small and highly efficient. When the laser diode is forward biased, incoherent light emission takes place until the drive current reaches a threshold value. Below threshold, the device is similar to a light-emitting diode (LED). Above threshold, lasing takes place and the light output increases rapidly with current. A completely new segment on semiconductor laser diodes begins on p. 9.31.

This section also includes a chapter on cathode-ray tubes and electroluminescent displays, with valuable complementary material on the accompanying CD-ROM covering electro-optics and nonlinear optics, phosphor screens, photoemissive electron tubes, and camera tubes.

Four chapters treat radiant energy sensors. Photoconductive and semiconductor junction detectors convert electromagnetic energy to electric energy. Charge-coupled-device and charge-injection-device image sensors are applicable to high- and low-light-level imaging, spatial character recognition, and facsimile reproduction. Infrared detectors include both thermal detectors and quantum detectors, and often require cryogenic cooling to reduce noise. Solar cells may be used in quantity for power generation or individually as light detectors in applications such as cameras.   D.C.

In This Section:

On the CD-ROM:

Sharp, E. J., "Electro-optics and Nonlinear Optics," reproduced from the 4th edition of this handbook.

Diakides, N. A., "Phosphor Screens," from the 4th edition of this handbook, provides a discussion of screens used to convert electron energy into radiant energy in image tubes, cathode-ray tubes, and storage CRTs.

Johnson, C. A., "Photoemissive Electron Tubes, Image Converters, and Intensifiers," from the 4th edition of this handbook.

Graft, R. D., and R. E. Franseen. "Television Camera Tubes," from the 4th edition of this handbook, covers theory of operation, construction, materials, and performance.

CHAPTER 9.1
NONCOHERENT RADIANT ENERGY SOURCES

**Clifton S. Fox, Mark Hodapp, David L. Evans,
David O'Brien, Jason Yorks**

INTRODUCTION

Glossary

Generation of light. Light is produced by the transition of electrons from states of higher energies to states of lower energies. The law of conservation of energy is satisfied in these transition processes by the emission of a photon, or quantum of light, whose energy corresponds to the difference in energy of the initial and final energy states of the electron.

Blackbody radiation. A blackbody is defined as a body which, if it existed, would absorb all and reflect none of the radiation incident on it. It is thus a perfect absorber and a perfect emitter. The blackbody curves for several values of T are plotted on a logarithmic scale in Fig. 9.1.1.

The total emissivity ϵ of a thermal radiator at a given temperature is the ratio of the total radiation output of that radiator to that of a blackbody of the same temperature.

The spectral emissivity $\epsilon(\lambda)$ of a thermal radiator is defined as the ratio of the output of the source at the wavelength λ to that of a blackbody at the same wavelength and operating temperature.

Graybody radiation. If the emissivity of a thermal radiator is a constant less than 1 for all wavelengths, the radiator is called a graybody.

Selective radiation. A thermal radiator whose spectral emissivity is not constant but is a function of wavelength is called a selective radiator.

The color temperature of a thermal radiator is the temperature of a blackbody chosen such that its output is the closest possible approximation to a perfect color match with the thermal radiator. Figure 9.1.2 shows the spectral distribution of a tungsten filament operating at a color temperature of 3000 K compared with a blackbody of the same temperature and a graybody whose emissivity is the same as tungsten in the visible spectrum.

The candela (cd) is the unit of luminous intensity. Luminous intensity is the amount of luminous flux per unit solid angle in a given direction. This is measured as the luminous flux on a target normal to the direction of incidence divided by the solid angle (measured in steradians, abbreviated sr) subtended by the target as viewed from the source.

The lumen (lm) is the unit of luminous flux. It is equal to the flux in a unit solid angle from a uniform point source of 1 cd intensity.

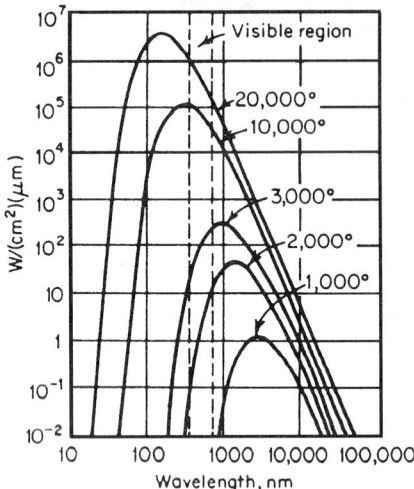

FIGURE 9.1.1 Blackbody distribution curves for several values of temperature in kelvins.

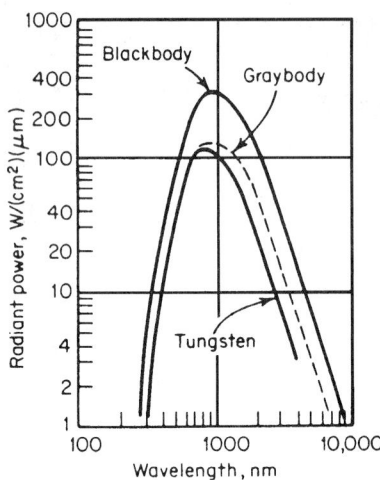

FIGURE 9.1.2 Spectral distribution of tungsten at a color temperature compared with blackbody and graybody of the same temperature.

The luminous efficacy of a light source is the measure of light-producing efficiency of the source. It is the ratio of the total luminous flux output to the total input power of the source. Luminous efficacy is measured in lumens per watt.

Radiative efficiency of a light source is the ratio (in percent) of total output power of the source measured in watts to the input power to the source.

Incandescence. Emission of radiation relating to the temperature of the source.

Luminescence. Emission of radiation relating to causes other than temperature of the source.

Fluorescence. Luminescence stimulated by radiation, not continuing more than about 10^{-8} s after the stimulating radiation is cut off.

The most commonly used light sources are the tungsten filament, electric discharge and electroluminescent lamps, and solid-state or light-emitting diodes. The first source is incandescent; the others are luminescent.

Tungsten-Filament Lamps

Filament. The higher the operating temperature of a solid filament, the higher the percentage of its radiation that falls in the visible portion of the electromagnetic spectrum. Tungsten, with its high melting point (3653 K), low vapor pressure, and other favorable characteristics, is the most frequently used filament material. In higher-power incandescent lamps (generally above 40 W) an inert gas instead of vacuum surrounds the filament to reduce the evaporation rate of the tungsten.

Lamp Types. Tungsten filament lamps are divided into the following categories: general-service lamps; high- and low-voltage lamps; series-burning lamps; projector and reflector lamps; showcase lamps; spotlight, floodlight, and projection lamps; halogen-cycle lamps; and infrared lamps.
   Tungsten halogen-cycle lamps have a quartz envelope and use a halogen fill, usually iodine, to keep the bulb clean by chemical reaction with sublimated tungsten. This reaction provides a high-lumen maintenance

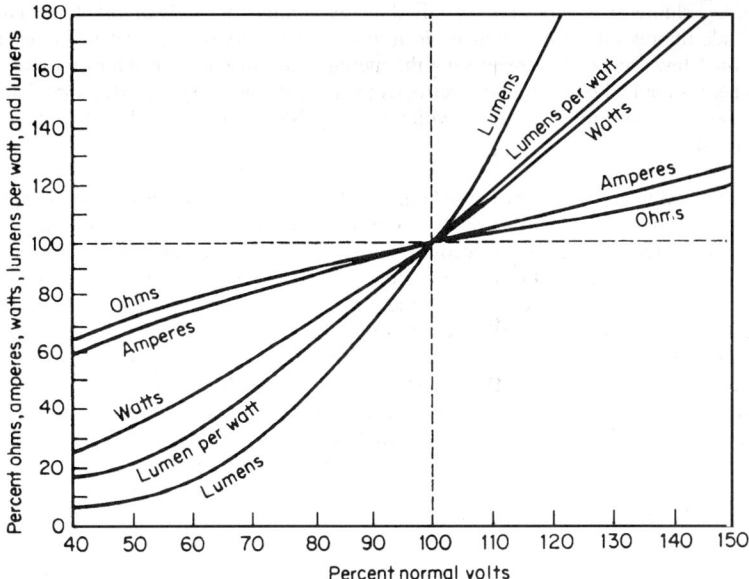

FIGURE 9.1.3 Interrelation of lamp parameters for large tungsten-filament lamps. (*General Electric Co.*)

throughout the life of the lamp by redepositing evaporated tungsten on the filament instead of on the bulb. *Infrared lamps* are tungsten-filament lamps that operate at low filament temperature.

Lamp Parameters.   The quantities voltage, current, resistance, temperature, watts, light output, efficacy, and life of a filament lamp are interrelated, and one cannot be changed without changing the others. Figure 9.1.3 shows how these quantities change typically as a function of the voltage for large gas-filled lamps.

Some useful exponential relations frequently applied to incandescent filament lamps, where capital letters indicate normal-rated values, are

$$\frac{\text{life}}{\text{LIFE}} = \left(\frac{\text{VOLTS}}{\text{volts}}\right)^{d} \qquad \frac{\text{lumens}}{\text{LUMENS}} = \left(\frac{\text{volts}}{\text{VOLTS}}\right)^{k}$$

$$\frac{\text{LM/W}}{\text{lm/w}} = \left(\frac{\text{VOLTS}}{\text{volts}}\right)^{g} \qquad \frac{\text{watts}}{\text{WATTS}} = \left(\frac{\text{volts}}{\text{VOLTS}}\right)^{n}$$

For approximate calculations the following average exponents may be used: $d = 13$, $g = 1.9$, $k = 3.4$, and $n = 1.6$.

Electric-Discharge Lamps

Fluorescent Lamps.   Fluorescent lamps are electric-discharge lamps in which light is produced through the excitation of phosphors by the ultraviolet energy from a mercury arc. The lamp usually consists of a phosphor-coated tubular bulb with electrodes sealed into each end and containing mercury vapor at low pressure along with an inert starting gas such as argon or an argon-neon mixture. Various colors, grades of white light, and even black light (near ultraviolet) are obtained by the choice of available phosphors. Cool white is the most widely used white-light fluorescent tube.

Lamp types.   Fluorescent lamps are classified as *hot-cathode* or *cold-cathode* type. There are three classes of hot-cathode fluorescent lamps: *preheat, instant-start*, and *rapid-start*. Preheat lamps allow preheating of the cathodes for a few seconds before striking the mercury arc. Instant-start lamps require no preheat because sufficient voltage is applied between the electrodes to strike the arc very quickly. Rapid-start lamps have continuously heated cathodes requiring a lower voltage than instant-start lamps. This feature also allows dimming and flashing of the lamp.

Mercury Lamps.   Mercury vapor discharge lamps of the wall-stabilized variety are used primarily for general lighting. Mercury lamps with additives such as the metal halide lamp and sodium vapor lamp and mercury compact arcs are treated in separate succeeding paragraphs. Most mercury vapor lamps have two bulbs. The inner bulb, called the *arc tube*, contains the arc and the electrodes between which the arc burns. The outer bulb protects the arc tube from drafts and stabilizes the operating temperature. Mercury lamps for general lighting are available in input power sizes from 50 to 3000 W.

In addition to mercury in the arc tube, an easily ionized inert gas such as argon is present to facilitate starting. The arc is generally ignited through use of a starting electrode and current-limiting starting resistor. An arc is first struck between the starting electrode and the adjacent main electrode. The heating and additional ionization resulting from this arc allow the large arc to form between the main electrodes.

Electrical and radiation characteristics.   The color-rendering properties of a clear mercury lamp are only fair, owing to the line structure in the blue end of the spectrum. A clear mercury lamp of 1000 W input power has a typical initial luminous efficacy of 56 lm/W.

Metal Halide-Mercury Vapor Lamps.   Metal halide-mercury vapor lamps are very nearly the same as regular mercury lamps except that additives such as the iodides of sodium, thallium, and indium are contained in the arc tube for the purpose of improving color rendition and efficiency. The typical initial luminous efficacy of a metal halide-mercury lamp of 1000 W input power is 90 lm/W.

High-Pressure Sodium Vapor Lamps.   High-pressure sodium vapor lamps are presently the most efficient source of artificial light. A typical 1000-W high-pressure sodium lamp has an initial luminous efficacy of 140 lm/W. The theoretical efficiency of white light in the visible region, assuming 100 percent conversion of power into a continuum output, is 220 lm/W. Typical life of high-pressure sodium lamps is 12,000 to 20,000 h.

High-pressure sodium lamps require a high-transmission ceramic envelope such as alumina to contain the alkali metal at high temperature and an alkali-resistant high-temperature metal seal. The corrosive effects of sodium at high operating temperature prohibit the use of quartz and other glasses as an arc-tube material. Xenon is used as the readily ionized starting gas. When the xenon starting gas is ionized, the arc is struck, producing heat, and the vapor pressure starts to rise.

Arc-Light Sources.   (a) *Short-arc or compact-arc lamps* are of the enclosed-bulb type. Most are made for dc operation, which results in long life and good arc stability. They have high operating pressure, a comparatively short arc gap, and very high luminance (photometric brightness). The arc length may vary from about $1/3$ to 17 mm. Since these lamps have the highest luminance of any continuous-operation light source and provide maintenance-free, clean operation, they have replaced the carbon arc in many applications.

Mercury short-arc lamps contain a low pressure of inert gas such as argon for starting. After the arc is struck, the lamp warms up and the mercury vapor reaches full operating pressure within a few minutes. Warmup time is reduced to approximately one-half of that of the mercury arc if xenon at 1 atm or more pressure is added. The spectra of mercury and mercury-xenon lamps are essentially the same. The luminous efficacy of these lamps is about 50 lm/W for a 1000-W lamp. These lamps are available in a power range from 30 to 5000 W.

Xenon short-arc lamps are filled to a cold pressure of several atmospheres with high-purity xenon gas. Operating pressure is roughly double the cold pressure. These lamps do not have as long a warmup time as the metal vapor types; 80 percent of the light output is obtained within 1 s of startup. Xenon has excellent color-rendering characteristics because of its continuous spectrum in the visible region. Luminous efficacy at 5000 W is approximately 45 lm/W.

(b) *Carbon arcs* are of three basic types: the low-intensity arc, the flame arc, and the high-intensity arc.

The *low-intensity arc* has as its source of light the incandescent tip of the positive carbon, which is maintained near the sublimation point of carbon (3700°C). The heat supplied to the positive carbon is from high-current-density electron bombardment originating from the negative carbon.

The *flame arc* is obtained by enlarging the core of the electrodes of a low-intensity arc and replacing the removed material with compounds of rare-earth elements such as cerium.

The *high-intensity arc* is obtained from the flame arc by increasing the core size and the current density so that the anode spot spreads over the entire tip of the carbon. A crater is formed, and this becomes the primary source of light.

Flashtubes.   Flashtubes are designed to produce high-luminance flashes of light of very short duration. They are used in optical pumping of lasers, stroboscopic work, photographic applications, and many other purposes requiring flashing lights.

Lamp construction.   The flashtube consists of a glass or quartz tube filled with gas and containing two or more electrodes. The fill gas preferred for most flashtube applications is xenon because of its high output of white light. Other gases, including argon, neon, krypton, and hydrogen, are frequently used.

Driving circuit.   Energy for flashing the tube is usually stored in a capacitor. This energy is determined by the equation $E = CV^2/2$, where E = energy (J), C = capacitance (μF), and V = voltage on capacitor (kV). The duration of the flash usually depends on the resistance of the discharge and the capacitance of the storage capacitor, the duration being approximately $3RC$, with R in ohms and C in farads. For short pulses of 1 μs or less duration, frequently the inductance of the tube or circuit is the dominant factor over the resistance in determining pulse length.

Many circuits have been used to flash lamps. One basic method is to hold off a voltage higher than the self-breakdown voltage of the lamp with an electronic switch, such as a thyratron, silicon-controlled rectifier, or spark gap, and trigger the switch when a flash is desired.

Electroluminescent Lamps

An electroluminescent lamp is a thin flat source in which light is produced through excitation of a phosphor by an alternating electric field. The lamp basically consists of a capacitor with a phosphor embedded in the dielectric material sandwiched between two conducting plates. The front plate is a transparent sheet of either plastic, glass, or ceramic with a thin transparent conductive film on it. The back conductive plate may be a metal sheet or film or a transparent material like the front plate.

LIGHT-EMITTING DIODES

Mark Hodapp, David L. Evans, David O'Brien, Jason Yorks

Pn Junction Injection Electroluminescence

The emission of light (photons) from a naturally occurring *pn* junction was first noted by Lassew in 1923. In 1962, studies of GaAs (gallium arsenide) revealed the feasibility of high-level electroluminescent emissions from *pn* junctions (see Holonyak, 1962). The same year, an intensive development program was begun at Hewlett-Packard and elsewhere to produce a useful manufacturable, visible-light-emitting *pn* junction device. The result was the light-emitting diode (LED). LEDs benefit from: (1) useful light output at low currents and voltages; (2) the accuracy with which the light-emitting area can be defined through the use of semiconductor photolithographic processes; (3) the high speed at which the devices can be switched; (4) a lifetime far in excess of an incandescent light source; and (5) the ability to withstand high degrees of shock and vibration.

Operating Mechanism

Light-emitting diode lamps operate much like a silicon or germanium diode. Although they both rectify current, neither silicon nor germanium emits light. When an LED is forward biased, photons are emitted in all directions; however, not all of these photons emerge from the lamps as useful light. The amount of emitted light is dependent upon the material, the geometry of the package, how many photons are absorbed, reflected, refracted, and finally emitted from the lamp.

LED Materials

There are numerous elements and elemental compounds that have bandgap energies capable of producing photon emissions ranging from ultraviolet to infrared. Very few of these materials, however, are practical for LED devices. Some of them are not useful because they cannot be easily doped to form a *pn* junction, some do not emit photons at a useful wavelength, and some have too low a conversion efficiency. Most commercially available LED devices are manufactured from Type 3 elements in the periodic table (A1, Ga, In) and Type 5 elements in the periodic table (N, P, As, Sb); however, a small number of LEDs use Type 4 elements (SiC) and Type 2 (Zn) and Type 6 (Se). Most commercial LED devices use the binary compounds GaAs, GaP, SiC, ternary compounds $GaAs_{1-x} P_x$, $Al_x Ga_{1-x} As$, or quaternary compounds $(Al_x G_{1-x})_{0.5} In_{0.5} P$.

Table 9.1.1 lists a few of the materials available for LEDs, their associated bandgaps, emission wavelengths, and transition types. In a direct transition, the electrons in the conduction band and the holes in the valence band both have zero momentum at the point where the energy bandgap between two bands is the smallest. Electrons can easily drop across the bandgap and emit a photon while conserving momentum.

In an indirect transition, the momenta are different at the narrowest point between the bands. To conserve momentum, a phonon (quantum of heat) is either generated or absorbed by the crystal structure. The addition of this extra component reduces the likelihood of recombination, resulting in a less efficient LED. The maximum possible energy of the emitted photon is determined by the bandgap energy of the solid in which the *pn* junction is formed.

Efficiency of LED Devices

The efficiency with which an electroluminescent material converts current flow into detectable photon emission determines whether usable devices can be manufactured from the material. The percentage of current flow, which results in recombination's yielding photons of the desired wavelength, is a measure of the internal conversion efficiency (n_{int}) of the diode, sometimes referred to as internal quantum efficiency (see Craford,

TABLE 9.1.1 LED Semiconductor Materials

Material	Band GaP Energy (eV)	Peak Wavelength (nm)	Transition Type
GaAs	1.43	910	Direct
$GaAs_{0.6}P_{0.4}$	1.91	650	Direct
$GaAs_{0.35}P_{0.65}$	2.09	635	Indirect
$GaAs_{0.20}P_{0.80}$	2.16	600	Indirect
$GaAs_{0.10}P_{0.90}$	2.21	583	Indirect
GaP:N	2.26	568	Indirect
GaP	2.26	555	Indirect
$Al_{0.35}Ga_{0.65}As$	1.93	645	Direct
$(Al_{0.08}Ga_{0.92})_{0.5}In_{0.5}P$	1.99	622	Direct
$(Al_{0.12}Ga_{0.88})_{0.5}In_{0.5}P$	2.02	615	Direct
$(Al_{0.27}Ga_{0.73})_{0.5}In_{0.5}P$	2.10	590	Direct
SiC	2.99	480	Indirect

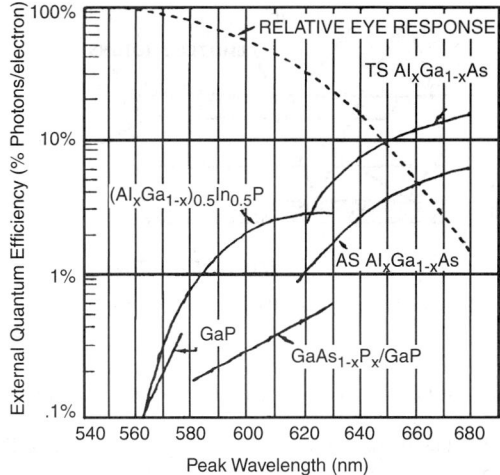

FIGURE 9.1.4 External quantum efficiency of LED materials.

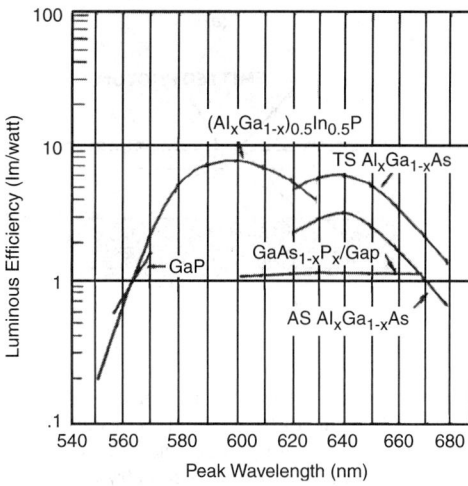

FIGURE 9.1.5 Luminous efficiency of LED materials.

1992). A material that has n_{int} of 100 percent may not be useful if the emitted photons cannot be efficiently coupled from the device.

Two major factors control the internal (n_{int}) to external coupling coefficient (n_{ext}). One factor is direct reabsorption of the emitted photon in the bulk material (basically a measure of the opacity of the material). The other factor is *internal reflection* at the crystal/air interface, causing the photon to be reflected back into the crystal and reabsorbed.

Figure 9.1.4 shows the external quantum efficiency (photons/electron) for a number of LED materials. $GaAs_{1-x}P_x$ grown on a GaP substrate has an external quantum efficiency of about 1 percent. $Al_{0.35}Ga_{0.65}As$ has an external quantum efficiency of 5 to 10 percent depending on whether the substrate is transparent or opaque. $(Al_xGa_{1-x})_{0.5}In_{0.5}P$ has an external quantum efficiency of about 6 percent in the red and orange spectrum.

It is desirable to optimize the total coupling efficiency between the emitting device's input signal and the detecting device's output. A standard emitter is matched to the human eye by picking the wavelength at which the product of the relative response of the eye and the relative efficiency of the diode is largest.

For diodes having shorter peak-emitting wavelengths, such as *n*-doped GaAsP emitters at 565 nm (green) and 585 nm (yellow), the relative sensitivity of the diode is substantially lower but the relative efficiency of the eye is higher. For visible applications, the efficiency of LEDs is usually specified in terms of lumens per watt or lumens per amp. The numerator, lumens, is a measurement of the total optical flux generated, scaled by the human eye response. The denominator, watts or amps, is the electrical power or current applied to the LED.

Figure 9.1.5 shows the luminous efficiency (lumens per watt) for a number of LED materials. $GaAs_{1-x}P_x$ grown on a GaP substrate has a luminous efficiency of about 1 lm/W. $Al_xGa_{1-x}As$ has a luminous efficiency that peaks at 8 lm/W in the red region. (This corresponds to the wavelength of $Al_{0.35}Ga_{0.65}As$. $(-Al_xGa_{1-x})_{0.5}In_{0.5}P$ has a luminous efficiency of about 8 lm/W in the red and orange regions. (For the sake of comparison, a red filtered incandescent bulb in an auto tail light has a luminous efficiency of about 6 lm/W.)

Real Versus Ideal LEDs

A *pn* junction in GaAsP material exhibits numerous discrepancies between the ideal and the real. Surface recombination, tunneling phenomena, space charge recombination, current crowding, and bulk recombination because of anomalous impurities tend to reduce the efficiency of an LED. The relative effects of these nonradiative phenomena tend to be dependent on the current density J (amps per square centimeter of junction area),

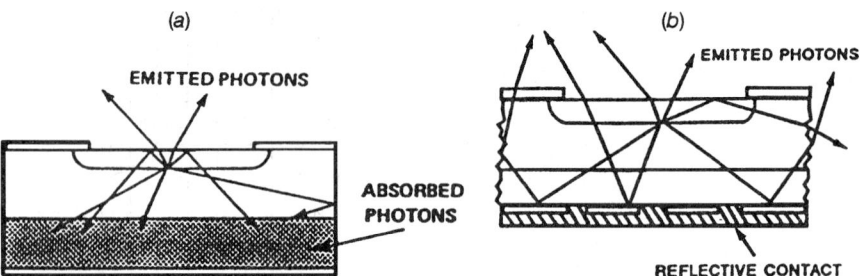

FIGURE 9.1.6 Cross sections of (*a*) GaAsP on opaque GaAs substrate; (*b*) GaAsP or GaP on transparent GaP substrate.

and the perimeter to area ratio of the diode. At low current ($< 1 \text{A/cm}^2$) nearly all the current in a diode may result from one of these nonideal mechanisms.

The net effect of nonideal mechanisms is that an LED has its peak operating efficiency at a current that is dependent on the area and geometry of the junction and the size of the electrical contact. A doubling of current at low current densities may lead to a light output increase of two to three times, whereas a doubling of current at high current densities may result in slightly less than a doubling of light output.

Transparent Versus Opaque Substrates

The photons generated at the junction of a *pn* electroluminescent diode are emitted in all directions. If the diode substrate is opaque, as is the case with GaAs, only those photons that are emitted upward within a critical angle defined by Snell's law will be emitted as useful light. All other photons emitted into or reflected into the crystal will be absorbed. This phenomenon is illustrated in Fig. 9.1.6*a*.

When compared to GaAs, GaP is nearly transparent. Diodes formed by an epitaxial layer grown on a GaP substrate will exhibit improved efficiency because of the emission of photons that would have been absorbed in the GaAs substrate. The resulting structure is shown in Fig. 9.1.6*b*.

Similar considerations exist for AlGaAs and AlInGaP LEDs. In the case of DH AlGaAs LEDs, the AlGaAs is grown on an opaque GaAs substrate. For the TS AlGaAs LEDs, the GaAs substrate is removed. The bottom layer of the $Al_{0.6} Ga_{0.4}$ As is transparent to the light emitted in the junction. This is the reason that TS AlGaAs LEDs have twice the luminous efficiency of DH AlGaAs LEDs. The present AlInGaP LEDs use an opaque GaAs substrate. If the substrate was removed, an additional factor of 2 in luminous efficiency could be achieved.

Temperature Effects on LED Parameters

Absolute temperature variation affects an LED's parameters. Those of greatest interest to the user are forward voltage, quantum efficiency, and emitted wavelength. These are discussed in some detail in Hewlett-Packard Application Note, "LED Theory" (see Evans, 1993).

Lamp Packaging

Figure 9.1.7 shows the cross section of a typical LED lamp. The light emitting element consists of a small chip of LED material about 0.010 in. on a side by 0.006 in., mounted in a reflective metal dish. The bottom of the LED is attached to the reflector using an electrically conducting epoxy. The top of the LED is bonded to the second pin of the package with a gold wire. The LED die and lead frame are cast in an epoxy housing. The epoxy package secures the pins of the lamp and contributes to the optical properties of the lamp. A wide range of optical properties can be created by using different lens shapes and positions of the reflector dish in the package, and by adding diffusant to the epoxy.

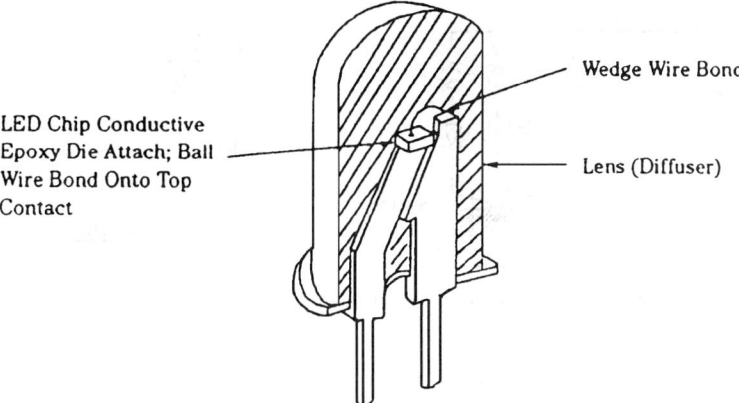

FIGURE 9.1.7   Cross section showing LED lamp construction.

The mechanical properties of the LED lamp are typically limited by the epoxy package rather than the LED die. The LED die, die attach material, and wire bond are very rugged and can withstand mechanical shock and vibration that would destroy a tungsten filament. However, the epoxy material typically has a higher temperature expansion coefficient than the rest of the lamp, so that temperature cycling at extremely high temperatures can put excessive stresses on the gold wire and wire bond. Today's LEDs are capable of withstanding hundreds of temperature cycles from –55/100°C; however, at temperatures above 100°C, the number of survivable temperature cycles is significantly reduced. Consequently most LED lamps recommend a maximum junction temperature of 110°C. The availability of higher temperature epoxies is expected to raise this limit to at least 125°C.

LED Drive Circuits

The LED lamp typically has an electrical characteristic shown in Fig. 9.1.8. Light is generated only when the *pn* junction is forward biased and current flows through the junction. This occurs at the turn-on voltage. At higher voltages, the current increases very quickly. The light output is a function of the current through the LED.

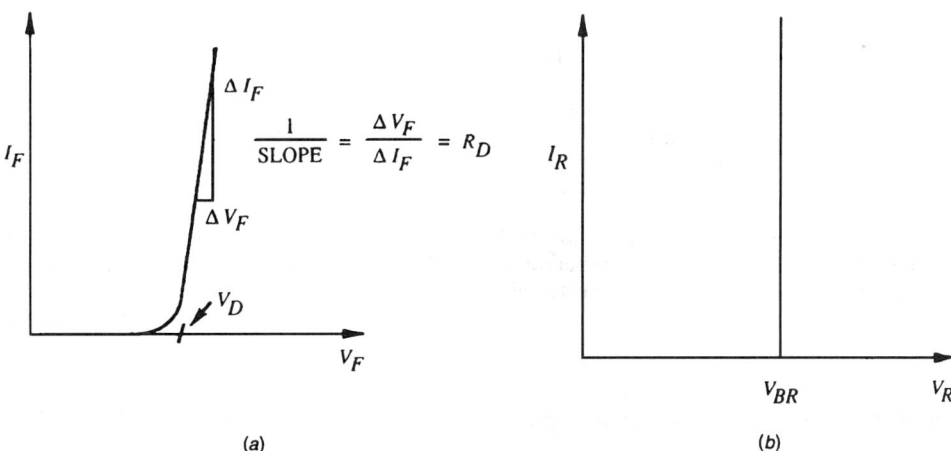

FIGURE 9.1.8   (*a*) Forward and (*b*) reverse characteristics of LED lamp.

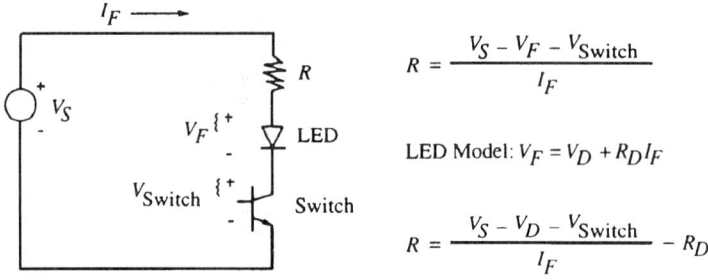

$$R = \frac{V_S - V_F - V_{\text{Switch}}}{I_F}$$

LED Model: $V_F = V_D + R_D I_F$

$$R = \frac{V_S - V_D - V_{\text{Switch}}}{I_F} - R_D$$

FIGURE 9.1.9   Typical resistive drive circuit for LED.

Under reverse bias conditions, negligible current flows through the LED until the breakdown voltage is exceeded. Higher negative voltages will force a current through the *pn* junction. This breakdown voltage depends on the doping of the *pn* junction and the type of LED materials technology used. Typically, the reverse breakdown voltage is greater than 10 V. If the current is not externally limited, the *pn* junction can be damaged by localized heating within the diode junction.

Most LED drive schemes use a resistor is series with the LED lamp (or string of LED lamps) as shown in Figure 9.1.9. The resistor sets the current to the LED lamp (or string). Most LEDs can be modeled in the forward direction as a resistor in series with a constant voltage, or by the equation described below:

$$V_F = V_D + I_F R_D$$

where V_F = voltage applied to the LED
  V_D = turn-on voltage of the LED
  R_D = reciprocal of the slope of the I-V characteristics
  I_F = current flowing through the LED

For most LEDs, V_D is in the range of 1.7 to 3.0 V, depending on the color of light generated. R_D typically varies from 1 to 20 Ω, depending on the LED material used.

Then, the current through the LED can be calculated as follows:

$$I_F = (V_S - V_{\text{SWITCH}} - nV_F)/R$$
$$= (V_S - V_{\text{SWITCH}} - nV_D)/(R + nR_D)$$

where I_F = current flowing through the LED
  V_S = power supply voltage
V_{SWITCH} = voltage across transistor switch
  V_F = voltage across the LED
  R = external resistance
  n = number of LED lamps in series

For most LEDs of a given materials technology, V_D is generally very well controlled. This means that the current through the circuit is determined primarily by R and R_D. If R is much larger than nR_D, then the current is controlled primarily by the value of the external resistance R.

LEDs can be driven by a number of other drive circuits including constant current drivers and a variety of pulsed drive circuits (see Hodapp, 1994).

Segmented Displays and Multi-indicator Applications

LED displays and individual lamps are commonly used as information or status indication devices. These products are typically categorized as intelligent or nonintelligent. An intelligent LED display is one that has an

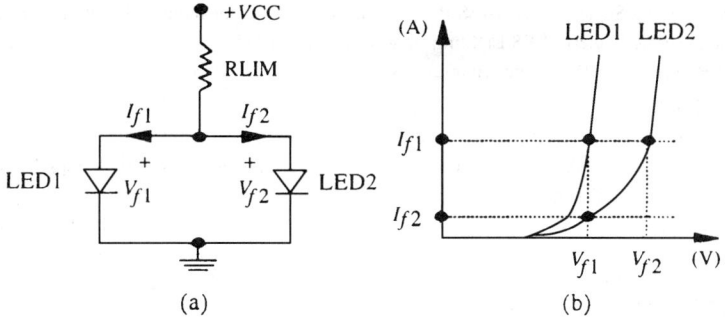

FIGURE 9.1.10 Application circuit: (a) two LEDs in parallel share one current limiting resistor; (b) forward voltage vs. forward current characteristics for the LEDs.

on-board IC integrated with the LEDs, where all the electro-optical biasing of the discrete LEDs has already been provided to the end user. The user varies the brightness of the LEDs by simply changing a combination of 1s and 0s in the control register of the IC. The active semiconductor chips have already been matched by the OEM and thus uniformity is provided. However, a nonintelligent display or individual lamp is made up of semiconductor chip(s), contact wires, and some packaging to provide mechanical stability, environmental protection, and optical lensing. These products are seven and sixteen segment displays, light bars, and discrete lamps. These nonintelligent devices require the designer to determine the electro-optical biasing configuration to achieve a desired brightness and uniformity.

There are important precautions in designing biasing circuits for nonintelligent applications. For example, in an application that requires the use of two or more LEDs of the same color, and luminous uniformity is required, it is recommended that the designer *not* place the LEDs in parallel with each other and in series with the same current limiting resistor (RLIM). Figure 9.1.10 depicts this situation. The mismatched LEDs can result in a discernible difference in brightness between the two. In this case LED 1 will appear to be brighter than LED 2 if the current ratio is greater than 2 to 1. This and other problems related to variations in the V_f versus I_f characteristics of a given LED are treated in the literature (see York, 1994).

LEDs and Light Guides

A light guide is a device designed to transport light from a light source to a point at some distance with minimal loss. Light is transmitted through a light guide by means of total internal reflection. Light guides are usually made of optical grade materials such as acrylic resin, polycarbonate, epoxies, and glass. A light guide can be used to transmit light from an LED lamp on a pc board to a front panel for use as status indication, can be used to collect and direct light to backlight an LCD display or legend, and can be used as the means to illuminate a grid pattern on a see-through window (see Evans, 1993).

BIBLIOGRAPHY

Noncoherent Sources

"Choosing Light Sources for General Lighting," IES Committee Report, IES, 1988.

"Colorimetry of Light Sources," IES, 1993.

Ditchburn, R. W., "Light," 2nd ed., Interscience, 1963.

Electrical and Photometric Measurements of General Service Incandescent Filament Lamps, IES, 1991.

Electrical and Photometric Measurements of Low Pressure Sodium (LPS) Lamps, IES, 1991.

Illuminating Engineering Society, "ANSI-IES Recommended Practices for Industrial Lighting," IES, 1991.

Illuminating Engineering Society, "IES Lighting Handbook," IES, 1987.

Meyer, C., and H. Nienhuis, "Discharge Lamps," IES, 1988.

Light-Emitting Diodes

Craford, M. G., "LEDs challenge the incandescents," *Circuits and Devices*, Sept. 1992.

Evans, D. L., Hewlett-Packard App. Brief I-003, "Light Guide Techniques Using LED Lamps," 1993.

Hodapp, M., Hewlett-Packard App. Note, "LED Theory," 1995.

Hodapp, M., "LEDs as Indicators, Illuminators, and Full Color Displays," Hewlett-Packard Co. App. Paper, 1994.

Holonyak, N. Jr., and S. F. Bevacqua, "Coherent (visible) light emission from $GaAs_{1-x}P_x$ junction," *Appl. Phys. Lett.* 1, 82, 1962.

York, J., Hewlett-Packard App. Brief D-007, "Solutions for Common LED Designer Errors in Segmented Display and Multi-Indicator Applications," 1994.

CHAPTER 9.2
LASERS

James E. Miller, Daniel S. Horowitz, Richard R. Shurtz II, Edward J. Sharp, Patrick R. Manzo, Gareth M. Janney, Bart H. Verbeek

LASER FUNDAMENTALS

James E. Miller, Daniel S. Horowitz

Laser Light Versus Nonlaser Light

The main characteristic of laser light is coherence, although laser light is usually more intense, more monochromatic, and more highly collimated than light from other sources.

Coherence is the property wherein corresponding points on the wavefront are in phase. A coherent beam can be visualized as an ideal wave whose spatial and time properties are clearly defined and predictable. Ordinary noncoherent light consists of random and discontinuous phases of varying amplitudes. The noncoherent beam has an average intensity and a predominant wavelength, but it is basically a superposition of different waves. The characteristic grainy appearance of laser light is a result of interference effects that result from coherence.

Intensity of laser light can be very high. For example, power densities of over 1000 MW/cm^2 can be obtained. A beam of such intensity can cut through and vaporize materials.

A laser beam is often highly *monochromatic* and highly *collimated*, both varying with the type of laser.

Stimulated Versus Spontaneous Emission

The laser operates on the principle of stimulated emission, an effect which is rarely observed except in connection with lasers.

Spontaneous Emission. This is the usual method whereby light is emitted from excited atoms or molecules. Assume that the laser material has energy levels, which can be occupied by electrons, that the lowest level or ground state is occupied, and that the next upper level is unoccupied. An excitation process can then raise an electron from the ground state to this upper state. The electron, after a variable time interval, returns to the ground state and emits a photon whose direction and phase of the associated wave are random and whose energy corresponds to the energy difference between the states. The upper-level lifetime may be comparatively short (less than 10 ps) or it may be long (greater than 1 μs), in which case the level is referred to as *metastable* and the light emission is *fluorescence*.

Stimulated Emission. When the electron is in the upper level, if a light wave of precisely the wavelength corresponding to the energy difference strikes the electron in the excited state, the light stimulates the electron to transfer down to the lower level and emit a photon. This photon is emitted precisely in the same direction, and its associated wave is in the same phase as that of the incident photon. Thus a travelling wave of the proper frequency is produced, passing through the excited material and growing in amplitude as it stimulates emission.

Pumping and Population Inversion

The process of exciting the laser material (raising electrons to excited states) is referred to as *pumping*. Pumping can be done optically using a lamp of some kind, by an electric discharge, a chemical reaction, or in the case of the semiconductor laser, by injecting electrons into an upper energy level by means of an electric current.

A *population inversion* is necessary to initiate and sustain laser action. Normally, the ground state is almost entirely occupied and the upper level or levels, assuming they are more than a few tenths of an electron volt above the ground state at room temperature, are essentially unoccupied. When the upper level has a greater electron population than the lower level, a population inversion is said to exist. This inverted population can support lasing, since a traveling wave of the proper frequency can stimulate downward transitions and the associated energy can be amplified.

Optical Resonators

The addition of a positive-feedback mechanism to a lasing medium permits it to serve as an oscillator.

The Fabry-Perot Resonator. The Fabry-Perot resonator that provides optical feedback consists of two parallel mirrors, the rear mirror fully reflecting and the front mirror partly reflecting and partly transmitting at the laser wavelength. The light reflected back from the front and rear mirrors serves as positive feedback to sustain oscillation, and the light transmitted through the front mirror serves as the laser output. Laser action is started by spontaneously emitted light with the proper direction to travel down the axis of the laser rod and be reflected on itself from the end mirrors. The two mirrors form an optical cavity which can be tuned by varying the spacing of the mirrors. The laser can operate only at wavelengths for which a standing-wave pattern can be set up in the cavity, i.e., for which the length of the cavity is an integral number of half wavelengths. Mirrors may be separate from the laser rod or deposited on its end faces.

Spectral Modes. Spectral modes, i.e., a multiplicity of radiation patterns, are permitted within the cavity. Longitudinal modes, determined by the spacing of the mirrors, occur at each wavelength for which a standing-wave pattern can be set up in the cavity. Transverse modes vary not only in wavelength but also in field strength in the plane perpendicular to the cavity axis. The longitudinal, or axial, mode structure determines the spectral characteristics of the laser, i.e., the coherence length and spectral bandwidth, while the transverse-mode structure determines the spatial characteristics, e.g., beam divergence and beam energy distribution.

If no attempt is made to control the mode of the laser, many longitudinal modes will be seen in the output. These modes lie within the natural spectral line width of the laser transition (Fig. 9.2.1). The transverse modes depend on the physical geometry of the cavity and are denoted similarly to waveguide modes at microwave frequencies. The lowest-order mode, characterized by a narrow, diffraction-limited beam spread, is the transverse electromagnetic (TEM_{00}) mode. Higher-order modes have multilobe intensity patterns and wider beams.

Mode control, or selection, is achieved by a variety of methods, e.g., varying the mirror curvatures, restricting the beam by apertures in the cavity, using resonant reflectors that reflect only a narrow band of wavelengths, and using Q-switching techniques.

Generation of Laser Pulses

The typical output of an optical laser consists of a series of spikes occurring during the major portion of the time the laser is pumped (Fig. 9.2.2). These spikes result because the inverted population is being alternately built up and depleted. *Q switching* (Q spoiling) is a means of obtaining all the energy in a single spike of very high peak power. As an example, an ordinary laser might generate 100 mJ over a time interval of 100 μs for a peak power (averaged over this time interval) of 1000 W. The same laser Q-switched might emit 80 mJ in a single 10-ns pulse for a peak power of 8 MW. The term Q switching is used by analogy to the Q of an electric circuit. Lowering the Q of the optical cavity means the laser cannot oscillate, and a large inverted population builds up.

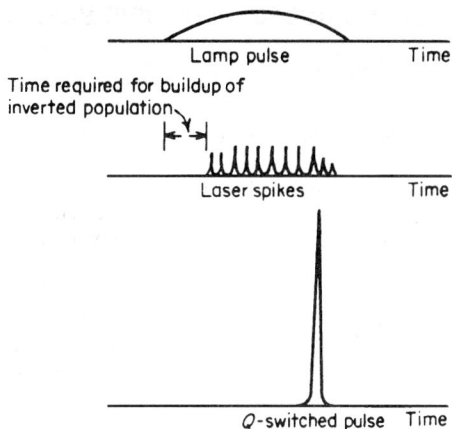

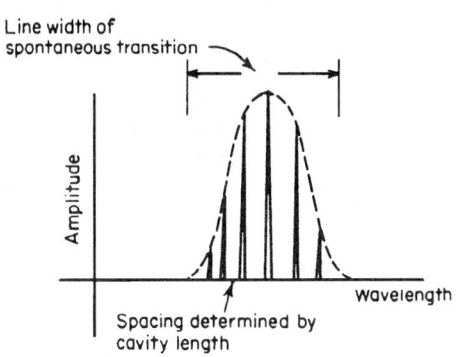

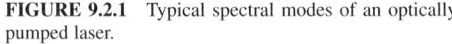

FIGURE 9.2.1 Typical spectral modes of an optically pumped laser.

FIGURE 9.2.2 Typical output of an optically pumped laser for Q-switched and non-Q-switched operation.

When the cavity Q is restored, a single "giant pulse" (see Fig. 9.2.2) is generated. This high-peak-power pulse is useful in optical ranging and communication and in producing nonlinear effects in materials.

Q *switches* use five techniques to inhibit laser action: the mechanical, or movable-optical-element, Q switch; the saturable organic-dye absorber; the electro-optic-effect crystal; the magneto-optic-effect crystal; and the acousto-optic-effect crystal.

All Q switches except the saturable absorber operate by deflecting radiation out of the optical cavity to prevent laser oscillation. The saturable absorber, called a *passive* Q switch because no external signal is required, consists of an organic compound usually in a liquid solution placed between the laser medium and one of the cavity mirrors. The dye is initially very highly absorbing at the laser wavelength, and the laser is isolated from one mirror. The absorbing transition in the dye is saturated as its ground state is depopulated, causing the dye to become transparent, or "bleach." The rate of bleaching is a strongly nonlinear function of incident radiation, and as the population inversion increases, the dye rapidly switches to a highly transparent state to allow lasing to occur.

Mode locking (phase locking).   This technique leads to even shorter pulses. A free-running laser output consists of a time average of many longitudinal and transverse modes with no fixed phase and amplitude relationship. If the laser is constrained to oscillate in only one transverse mode, there are still many longitudinal modes spaced in frequency at intervals of $c/2L$ (L = cavity length), which contribute randomly to the output. The time dependence of the output will be controlled by constraining these modes to maintain a fixed phase and amplitude relationship. A narrow pulse in the time domain requires a wide bandwidth in the frequency domain. The output of a mode-locked laser is a series of pulses, approximately $1/\Delta v$ wide, where Δv is the laser gain bandwidth. These pulses are spaced at intervals of the cavity round-trip transit time.

Frequency Selection

Methods of frequency (or wavelength) selection vary with the type of laser. The following techniques are typical. For solid-state lasers, the output frequencies are limited to a few sharp closely spaced lines. The wavelength of these lines depends on the active ion in the laser, and the specific wavelength is chosen by a highly selective dielectric mirror or a dispersive prism used as part of the optical cavity. Liquid lasers have broad emission bands, several tens of nanometers wide. The output wavelength can be continuously tuned within this band by an intracavity prism or diffraction grating.

Semiconductor lasers have narrow emission bands determined by the constituents. For a given semiconductor, the output wavelength can be varied by changing the temperature. Gas lasers emit discrete lines at wavelengths depending on the gas. Specific lines are chosen by a diffraction grating or prism. Some gas lasers are pumped by a second gas laser. The wavelength of this type of laser is controlled by the wavelength of the pump laser as well as the cavity elements described previously.

CLASSES OF LASERS AND THEIR BASIC APPLICATIONS

Richard R. Shurtz II

Primary Characteristics of the Laser Classes

Lasers can be divided into the four basic categories shown in Table 9.2.1: gas, solid-state optically pumped, liquid dye, and semiconductor. Combined, these lasers cover the spectral region from ultraviolet (0.1 μm) to far infrared (1000 μm), available wavelengths most densely populating the visible to infrared region.

The rapid development of new lasers over the past several years has created a dynamic and continually changing ensemble of devices available to the applications engineer. This set has stabilized as the most useful lasing systems have become known commodities and commercially available. In this section we quantify this established set of commercial devices and outline the primary applications base of each laser type. The reader should be able to identify the laser system best suited for a particular application and then refer to the appropriate paragraph for more specific information.

Gas lasers, shown in the top bar of Table 9.2.1, have the broadest spectral coverage.

The solid-state optically pumped laser category spans the visible to near infrared region. This is the only laser class that can be Q-switched or cavity-dumped. The electronic transitions are pumped either by flash lamps or by semiconductor diodes.

The family of *optically pumped organic-dye lasers* extends in wavelength from 0.4 to 1 μm. The major distinction of the category is its continuous tunability over the entire visible spectrum. Because of their

TABLE 9.2.1   Four Classes of Lasers

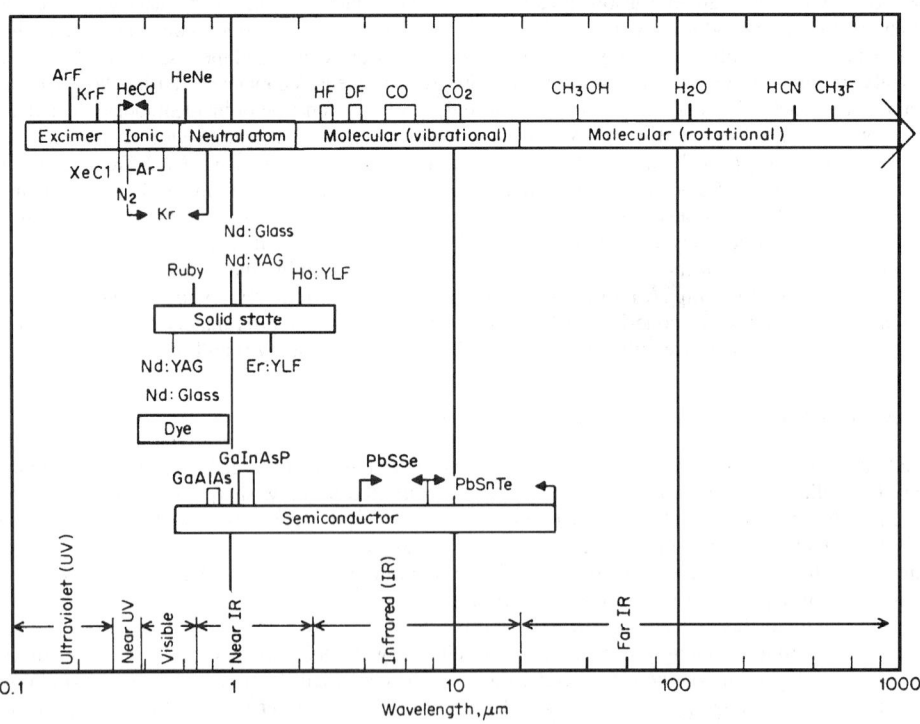

TABLE 9.2.2 Properties of Commercially Available Pulsed Lasers

Laser	Type	Wavelength, μm	Output, J TEM$_{00}$	Output, J Multimode	Pulse per second	Pulse length, μs	Beam diam, mm	Beam divergence, mrad
XeCl	Excimer	0.308	. . . . .	0.015–80	1–90	0.005–0.01	6×18–10×25	4–2×5
Nitrogen	Molecular gas	0.337–0.427	400–800 kw	0.002–0.01	1–100	0.005–0.01	6×20–10×25	1×7–5×10
Dye	Organic solution	0.39–1	. . . . .	10^2–10^6 W	1–10^3	10^5–5–0.8		
Ruby	Solid state	0.694	0.02–5	0.3–50	0.1–1	10^{-5}–10^4	1.5–20	0.4–8
GaAlAs	Semiconductor	0.8–0.9	. . . . .	0.1–3000 W	1–10^4	5×10^2–1	. . . . .	100–300
Nd:Yag	Solid state	1.06	5×10^{-3}–5	0.1–400	1–100	10^{-5}–10^4	1.5–75	0.3–12
Nd:Glass	Solid state	1.06	10^3–60	0.3–10^3	0.02–5	10^{-5}–10^4	2–120	0.1–10
Ho:YLF	Solid state	2.06	. . . . .	0.001–0.01	10–10^3	0.07–100	3	5
CO_2	Molecular gas	9–11	0.01–10^4	0.2–10^4	1–10^4	0.1–10^7	1–350	0.3–10
HCN	Molecular gas	337	0.001	. . . . .	1–100	30	10	40

broadband spectral output, these lasers can generate subpicosecond pulses when mode-locked. Their primary use is in spectroscopy and photochemistry.

Semiconductor lasers are pumped by the injection of excess electrons and holes into a thin semiconductor layer. Radiation is produced when the excess carriers recombine, producing photon energies equal to the bandgap energy. Lead salt laser diodes operate from 4 to 30 μm, continuously tunable either by varying the drive current or the temperature. They are used primarily for spectroscopy and must be cooled to nominally 77 K during operation. The shorter-wavelength semiconductor lasers are formed from III-V compounds: gallium arsenide (GaAs), gallium aluminum arsenide (GaAlAs), and gallium indium arsenide phosphide (GaInAsP). These devices, either pulsed or cw, operate at room temperature.

Single III-V compound laser diodes are used for fiber-optic communication, integrated optical processing, and rangefinding. Arrays of diodes are used for low-power infrared illuminators.

Characteristics of commercially available pulsed and cw lasers are summarized in Tables 9.2.2 and 9.2.3. The ranges shown are based on tables published in laser buyers' guides. A few key lasers have been selected to illustrate the essential properties of each laser category, as discussed in the previous paragraphs.

TABLE 9.2.3 Properties of Commercially Available Continuous Wave (cw) Lasers

Laser	Type	Wavelength, μm	Power, W TEM$_{00}$	Power, W Multimode	Beam diam, mm	Beam divergence, mrad
Argon (Ar)	Ionized gas	0.33–0.5145	4×10^{-3}–18	4×10^{-3}–22	0.8–2	0.5–1.5
Krypton (Kr)	Ionized gas	0.33–0.799	0.5–10	0.5–6	1–2	0.6–2
He-Cd	Ionized gas	0.325–0.442	10^{-3}–4×10^{-2}	1.5×10^{-3}–6×10^{-2}	0.82–1.5	0.5–1.6
Dye	Organic solution	0.39–1	. . . . .	0.1–1	0.5–0.7	1.4–2
He-Ne	Neutral gas	0.6328	10^{-3}–2	10^{-2}–1.3	0.45–30	0.71–2.1
GaAlAs	Semiconductor	0.8–0.89	. . . . .	10^{-3}–10^{-2}	. . . . .	5–610
Nd:YAG	Solid state	1.06	0.2–18	0.4–100	0.8–7	1.5–15
Ho:YLF	Solid state	2.06	. . . . .	5	3	25
CO_2	Molecular gas	9–11	1–10^4	4–10^4	1–70	1–10
HCN	Molecular gas	337	0.01–1	. . . . .	10	40

TABLE 9.2.4 Selected Laser Applications

Application	Technique	Mode of operation[*]	Desired laser property	Possible laser
Measurement				
Distance, short range	Interferometric	cw	Coherence	He-Ne
Long range	Time of flight	P	Short pulse length	Q-switch Nd:YAG, GaAS
Shape, thickness	Refraction measurement	cw	Collimation, transparency	He-Ne
Diameter	Interference	cw	Coherence	He-Ne
Overall dimension	Obscuration	cw	Collimation	He-Ne
Imperfections	Interference, obscuration	cw	Coherence, collimation	He-Ne or other cw for transparency
Alignment	Beam used as reference	cw	Collimation, visibility	He-Ne
Atmospheric monitoring				
Wind, turbulence	Doppler-shifted particulate backscatter	P[†]	Short pulse, narrow, bandwidth	CO_2, Nd:YAG
Particulate	Backscatter	P	Short pulse	CO_2, Nd_1:YAG
Gaseous pollution	Raman-shifted backscatter	P	Short pulse, narrow band width	CO_2, Nd_1:YAG
	Absorption	cw	Tunable, narrow bandwidth	PbSnTe diode, dye
	Fluorescence	cw	Tunable, narrow bandwidth	PbSnTe diode, dye
Material removal				
Drilling	Energy concentration	P	1–10 MW/cm^2	Ruby, Nd:YAG, Nd:glass
Cutting	Energy concentration	P, cw	1–10 MW/cm^2	CO_2, Nd:YAG
Film vaporization	Energy concentration	P, cw	100 kW/cm^2	CO_2, Nd:YAG, argon, xenon
Scribing wafers	Energy concentration	P, cw	100 kW/cm^2	CO_2 Nd:YAG
Illumination (infrared)	Pulse-gated for particulate penetration	P	Narrow pulse, efficient	GaAlAs diodearrays
Heating				
Welding	Energy concentration	P, cw	1–10 MW/cm^2	Nd:YAG, CO_2
Heat treating, annealing	Energy concentration	P, cw	Moderate to high power	Nd:YAG, CO_2, ruby
Fusion	Energy concentration	P	100 TW per 50-μm particle	Nd:glass, CO_2
Communications				
Fiber optic	Digital; analog	P, cw	Bandwidth; linear output	cw GaAlAs; cw GaInAsP
Space	Analog	cw	Stability collimation	CO_2
Optical information				
Processing	Integrated optics; three-dimensional optics	cw	Coherence	cw GaAlAs; coherent cw gas

TABLE 9.2.4   Selected Laser Applications (*Continued*)

Application	Technique	Mode of operation*	Desired laser property	Possible laser
Storage	Holography	cw	Coherence	He-Ne, argon
Chemical				
Laser-induced reactions	Selective breaking of bonds	P, cw	Tuning to molecular resonance	Variety of moderate-power lasers
Spectroscopy	Absorption; Raman emission	cw	Narrow bandwidth, tunable	Organic dye, PbSn Te diode
Isotope separation, e.g., uranium	Chemical isolation of excited species	cw	Laser tuned to excite one species	Organic dye, PbSn Te diode, others
Medical				
Skin treatment	Laser-induced reactions	P, cw	Tuned to resonant absorption	Ruby, tunable dyes, ultraviolet lasers
Surgery	Cutting, cauterization	P, cw	Collimation, power	Nd:YAG, CO_2, argon
Eye repair	Retina repair	P, cw	Collimation, power	Ruby, argon
Destructive weapons	Energy concentration	P, cw	High average power	CO_2, HF

*cw = continuous wave, P = pulsed.
†For range.

Laser Applications

Laser applications can be divided into several general categories: measurement of spatial parameters, material heating and/or removal, nondestructive probing of resonant phenomena, communications, optical processing, laser-induced chemical reactions, and weapons. Selected laser applications are shown in Table 9.2.4.

SOLID OPTICALLY PUMPED LASERS

Edward J. Sharp

Basic Principles

A solid optically pumped ionic laser consists of a solid material with optical gain, situated inside a resonator formed by two or more mirrors. Spontaneous emission in the material is amplified by stimulated emission. The resonator provides the feedback for the optical amplifier, resulting in a laser oscillator. The optical arrangement for a typical optically pumped laser is shown in Fig. 9.2.3.

These devices generally consist of a pump reflector enclosing the laser rod and optical pump in one of the number of efficient ways which is usually determined by the application or laser property to be exploited. The laser rod is pumped by flash lamp for pulsed operation and usually by arc lamps for cw (continuous-wave) operation, or in some cases an appropriate laser can be selected for the optical pump.

Two important conditions must be met: the wavelength region of the pump emission must overlap a significant portion of the absorption spectrum of the active ion or ions which are incorporated in the laser rod, and the optical pump must not damage the laser rod. Figure 9.2.4 illustrates the overlap of pump emissions and the absorption bands of a typical Nd^{3+} four-level laser scheme where pumping is accomplished by xenon-lamp pumping and LED laser pumping.

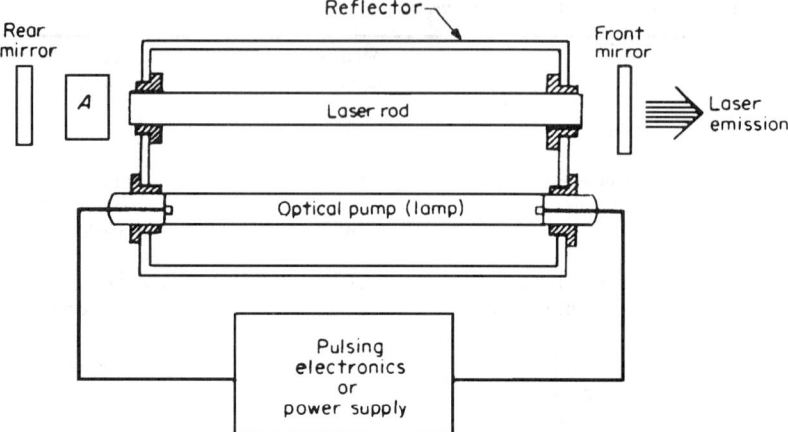

FIGURE 9.2.3 Typical optically pumped laser configuration. Component A can represent a Q switch, polarizer, or other optical element.

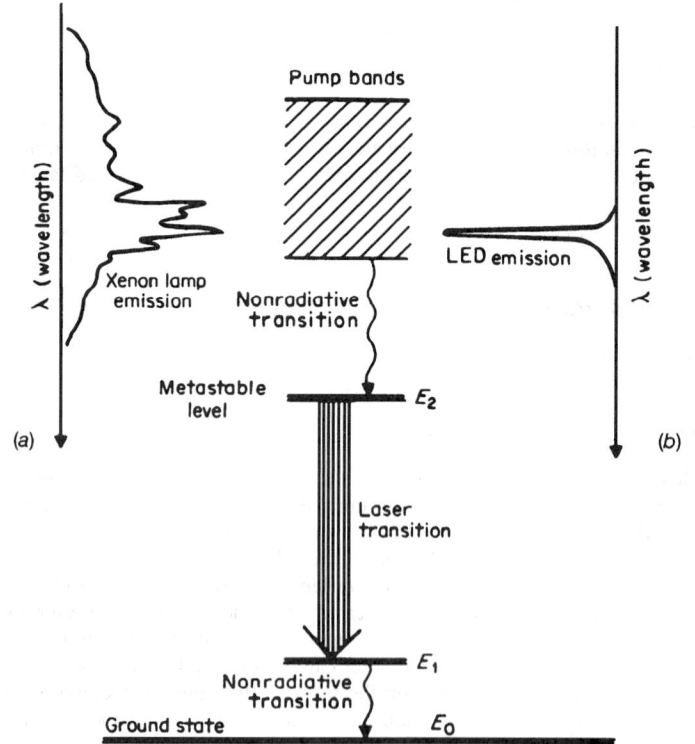

FIGURE 9.2.4 A typical four-level Nd^{3+} laser, in which the Nd^{3+} ions are optically excited by pumping the absorption bands of the ions with (*a*) a xenon lamp and (*b*) an LED laser.

Laser Components

Optical Pumps. The general technique of optical pumping has many uses. To achieve population inversion linear, low-pressure flash lamps are the most common pumping device. Typically they have tungsten electrodes and are filled to a few atmospheres pressure with xenon or krypton. For pulsed operation xenon is commonly used but the lines of krypton radiate at wavelengths more favorable than xenon for pumping a neodymium laser at low input levels. Thus it is a more efficient cw pump. At higher input levels, the blackbody spectrum becomes stronger than the emission lines, and this advantage is lost.

Pulsing Electronics. For pulsed operation the energy to be discharged into a flash lamp is stored in a capacitor and discharged through the lamp via a choke whose value is selected to give the desired pulse width with minimum ringing. The spectrum of the lamp discharge depends on the electrical parameters, pressure, and fill gases, so that it can be tailored to match the absorption bands of the active ion to a certain extent.

Coolants. Lasers that operate cw or at a high repetition rate need to be cooled, since most of the energy expended in the flash lamp is converted into heat and would quickly cause overheating and/or thermal cracking of the laser rod. Air and liquids have been successfully used as coolants, liquids being used when cooling requirements are more severe. Distilled or deionized water is often used, and mixtures of water and ethylene glycol are used for operation below 0°C. The coolant circuit includes a pump for circulation and a radiator, often with a fan, to dissipate the heat.

Pumping Reflectors. To obtain maximum use of the pump light, it must be efficiently collected and coupled into the rod. The reflectors used to do this are usually elliptical in cross section, with a rod at one focus and the lamp at the other. Sometimes two lamps are used, in a double elliptical cavity. Often a round cross section is used, resulting in an afocal system. In general, the smallest cross section is best, and the highest efficiency is obtained with a close-coupled arrangement, where the rod and lamp are almost in contact and the reflector encloses them closely. However, in this close-coupled configuration the thermal distortion of the rod owing to the asymmetrical pumping could be a serious problem.

Mirrors. Multilayer dielectric mirrors can be deposited on the laser-rod ends, which are finished flat and parallel. These mirrors are designed for specific reflectivities at the operating wavelengths and eliminate the problem of alignment. The rear-mirror reflectivity is normally 100 percent, but the front-mirror reflectivity is selected for the best performance, determined by laser-material properties, rod size, and operating power level. Where separate mirrors are used (see Fig. 9.2.3) the rod ends should be antireflection-coated or cut at Brewster's angle. Separate or external mirrors must be mounted in holders that permit fine adjustment about two axes to achieve the required parallelism.

Q-Switched Lasers

Q switching, a technique used for generating large output bursts of radiation from laser devices, is accomplished by effectively blocking the optical path to one of the mirrors for most of the time during which the rod is being pumped, causing the rod to store energy. The *Q* switch then quickly restores the optical path to the mirror and a giant pulse (see Fig. 9.2.2) is generated. The four main types of *Q* switches are the rotating mirror, acousto-optical, electro-optical, and saturable absorber.

Rotating Sector, Prism, or Mirror. The rotating prism or mirror requires careful mechanical design to provide rotation at the required speeds (30,000 r/min is common) and still maintain alignment. They are generally less reliable than other methods of *Q* switching, but the extinction ratio is infinite and the insertion loss is negligible, enabling this type of *Q* switch to attain high efficiencies.

Electro-Optic Q Switches. An electro-optical crystal, or liquid Kerr cell, can be placed in the optical resonator, usually between the rear mirror and the laser rod (position *A* in Fig. 9.2.3) along with an appropriate polarizer to effect the cavity losses. The polarization of the light passing through the electro-optic medium is changed by the application of an applied electric field and is rejected by the polarizer. When the voltage is removed, the cavity losses are reduced, permitting giant-pulse radiation.

Acousto-optic Q switch. An acousto-optic Bragg scattering switch deflects the beam, which is then used to provide feedback during the buildup time. In these devices a standing-wave pattern of ultrasonic waves is set up in a suitable cell positioned in the resonator. The refraction resulting from the passage of a plane-parallel light beam through this ultrasonic field is sufficient to provide shuttering action for infrared lasers, leading to giant-pulse radiation.

Saturable absorber Q switch. These switches are usually bleachable dyes that undergo decreased absorption at high light intensities. They are known as passive *Q* switches since they do not require any electrical or mechanical control. The material is opaque until the fluorescence of the laser rod bleaches it and the optical path is restored for a *Q*-switched pulse.

Ions and Hosts

The current list of ions which can be incorporated into a solid host and made to lase numbers about 20. The total number of different wavelengths available from these ions is approximately 150. The solid host is any crystal or glass that can accommodate trivalent or divalent rare-earth and iron-group ions. The iron-group ions in which laser oscillation has been achieved are divalent nickel and cobalt and trivalent chromium and vanadium. The rare-earth ions in which laser oscillation has been achieved are divalent samarium and dysprosium and trivalent europium, praseodymium, ytterbium, neodymium, erbium, thulium, and uranium.

Characteristics of the Hosts

The host material to which the active ion (dopant) is incorporated can be either crystalline or glass and should exhibit the following properties:

High thermal conductivity

Ease of fabrication

Hardness (to prevent degradation of optical finishes)

Resistance to solarization or radiation-induced color centers

Chemical inertness, i.e., not water soluble

High optical quality, which implies uniformity of refractive index and absence of voids, inclusions, or other scattering centers

Crystalline Hosts. Crystalline hosts offer as advantages in most cases their hardness, high thermal conductivity, narrow fluorescence line width, and, for some applications, optical anisotropy. Crystalline hosts usually have as disadvantages their poor optical quality, inhomogeneity of doping, and generally narrower absorption lines.

Glass Laser Hosts. Glass laser hosts are optically isotropic and easy to fabricate, possess excellent optical quality, and are hard enough to accept and retain optical finishes. In most cases glasses can be more heavily and more homogeneously doped than crystals, and, in general, glasses have broader absorption bands and exhibit longer fluorescent decay times. The primary disadvantages of glass are its broad fluorescence line widths (leading to higher thresholds), its significantly lower thermal conductivity (a factor of 10 leading to thermally induced birefringence and distortion when operated at high pulse-repetition rates or high average powers), and its susceptibility to solarization (darkening because of color centers that are formed in the glass as a result of the ultraviolet radiation from the flash lamps). These disadvantages limit the use of glass laser rods for cw and high-repetition-rate lasers.

Sensitized Lasers

Laser performance and efficiency can be enhanced through the technique of energy transfer. A second ion (sensitizer) is incorporated into the host in addition to the laser ion (activator) to accomplish this effect. The sensitizer may be a color center. Pump energy is absorbed by the sensitizer and is transferred to the activator, which then emits this energy at the laser wave-length.

Miniature Lasers

Several host materials that incorporate Nd as a constituent rather than a dopant, such as $NdAl_3(BO_3)_4$, neodymium aluminum borate (NAB); $LiNdP_4O_{12}$, lithium neodymium tetraphosphate (LNP); and NdP_5O_{14}, neodymium pentaphosphate (NdPP), have been studied recently, and since concentration quenching is not appreciable in these materials, they are excellent candidates for miniaturized lasers.

LIQUID LASERS

Patrick R. Manzo

Laser Media

Liquid Lasers. Liquid lasers use a liquid as the laser medium in place of a large single crystal or a gas (Fig. 9.2.5). Their properties are intermediate between those of gaseous and solid lasers. They are easy to prepare in large samples with excellent optical quality, and for certain types their energy output can be as high as 10^8 W peak and several tens of watts average. The wavelength coverage available for certain types of liquid lasers is considerably greater than that of both the solid and the gaseous lasers. There are two types of liquid lasers in common use.

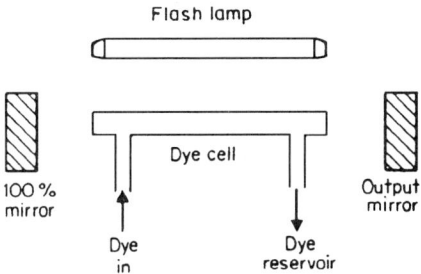

FIGURE 9.2.5 Typical dye-laser configuration.

The aprotic liquid laser consists of a rare-earth salt dissolved in an inorganic solvent that does not contain hydrogen. Energy from the excitation source is absorbed by the solvent and then transferred to the rare-earth ion, which lases. The absence of hydrogen in solution greatly increases the efficiency of this energy-transfer process because it lessens the possibility of this energy being transferred into vibrations of the molecule. The output wave-length is that of the rare-earth salt (see Table 9.2.1). In system gain, in output power levels, and in overall efficiencies, the aprotic liquid lasers are comparable with the solid-state lasers. They are capable of extremely high peak powers (10^8 W) and sustained high average powers (50 to 100 W). The principal aprotic liquid laser materials are Nd^{3+} in $SeOCl_3$ with $SnCl_4$ or in $POCl_3$ with $SnCl_4$.

The *dye laser* uses highly fluorescent organic molecules as the laser medium; unlike the aprotic liquid, these molecules do not contain the rare-earth salts. The radiative transition in the dye laser does not originate at a metastable energy level, as in most other lasers, but takes place instead between two singlet electronic states of the molecule. Consequently it is strongly allowed and extremely short-lived—approximately 10 ns compared with several hundreds of microseconds for the rare-earth ions. This requires quite different excitation techniques to achieve laser action. The first organic dye lasers required Q-switched lasers or flash lamps with extremely fast rise times in order to invert the population fast enough to achieve laser action. However, newer techniques alleviate the necessity for such fast excitation sources. Rhodamine 6G and several other dyes have been made to operate in a continuous mode with average power as high as 10 W.

The organic dyes exhibit excellent optical quality in solution, and they are extremely easy to prepare and handle. The absorption and fluorescent bands of the molecule are extremely broad (several tens of nanometers) because of the large number of vibrational and rotational energy levels associated with each electronic energy level. The laser output can be broadband (30 nm wide) or very narrow (0.001 nm).

The very short lifetime of the excited state means that sources must be of very high intensity to cause the dyes to lase. Therefore, the sources used to pump the dyes include high-intensity flash lamps and other lasers. Distortion in the dye because of thermal inhomogeneities requires that the dye be mixed rapidly and exchanged in order to minimize beam spreading and losses during cw and repetition-rate operation. The line width of typical dye materials is homogeneously broadened, and therefore all the stored energy in the upper laser level can be channeled into a spectrally narrow emission line, which can be tuned over the gain profile of the active medium.

Flashlamp-pumped dye lasers have demonstrated average powers as high as 100 W at high repetition rates. Single pulse energies to 10 J per pulse and peak powers to 10 MW have also been demonstrated. Present flashlamp-pumped systems are limited to approximately 1 percent efficiency while the efficiency of laser-pumped dyes is somewhat lower. Average power from a few milliwatts to over 10 W has been achieved in cw operation of laser dyes using pump laser configurations such as the one shown in Fig. 9.2.6a. Frequency-stabilized cw dyes have achieved a line width of less than 1 MHz and long-term stabilities of a few hundred hertz.

With an appropriate selection of dyes, this type of laser can be directly tuned from 340 to 1200 nm. This wavelength range has been extended into the ultraviolet and the infrared by frequency mixing of dye-laser radiation in nonlinear optical materials. Tunable ultraviolet radiation down to 196 nm has been obtained by second-harmonic and sum-frequency generation in nonlinear crystals. In the long-wavelength range the generation of tunable near and middle infrared, as the difference frequency of two laser oscillators, has been successful out to 12.7 μm. It is also possible to produce multiple wavelengths from the same or combination of dyes. Figure 9.2.6b illustrate one scheme for doing so.

It is possible to produce dye laser pulses with a duration of less than 1 ps. Durations as short as 0.3 ps have been reported. In addition, the broad emission bands of dyes make them the only sources capable of producing mode-locked picosecond pulses at high repetition rates which are wavelength tunable over several tens of nanometers.

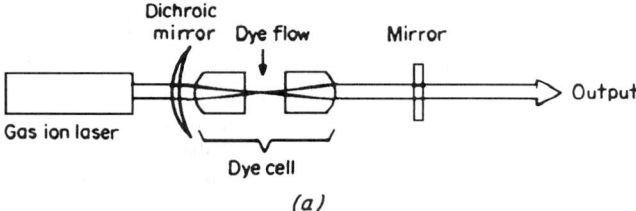

(a)

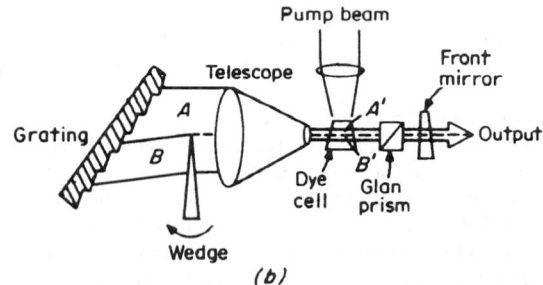

(b)

FIGURE 9.2.6 *(a)* Schematic of a continuous-wave dye laser; *(b)* double-wavelength dye laser. The lower portion of the beam is shifted toward the red by the wedge.

GAS LASERS

Gareth M. Janney

General Characteristics

Gas lasers can best be characterized by their variety. The laser medium may be a very pure, single-component gas or mixture of gases. If may be a permanent gas or a vaporized solid or liquid. The active species in a gas laser may be a neutral atom, an ionized atom, or a molecule (including excimers, which are "stable" molecules only in an excited state). The operating pressures range from a fraction of a torr to atmospheric pressure, and the operating temperature from −196 to 1600°C. Excitation methods include electric discharges (glow, arc, pulsed, rf, dc), chemical reactions, supersonic expansion of heated gases (gas dynamic), and optical pumping. The average output power of useful gas lasers range from a few microwatts to tens of kilowatts (10 orders of magnitude), and the peak power ranges from a fraction of a watt to 100 MW. The range of output wavelengths extends from 0.16 to 744 μm at discrete wavelengths.

Multiple Wavelengths

Most gas laser materials have a number of distinct laser transitions (i.e, different wavelengths). For example, the neon atom has more than 100, and the argon ion has more than 30. Lasers using these materials can operate with a multiwavelength output, or one transition at a time can be selected by a simple adjustment (rotating a prism or diffraction grating in the optical cavity). Table 9.2.5 (bottom) shows several commercially available lasers that provide more than one wavelength of operation.

Electrically Excited Electric-Discharge Gas Lasers

Most gas lasers are excited by electric discharges. Electrons that have been accelerated by an electric field transfer energy to the gas atoms and molecules by collisions. These collisions may excite the upper laser level directly. Indirect excitation is also possible by cascading from higher-energy levels of the same atom (or molecule) or resonant-energy transfer from one atom (or molecule) to another by collision.

Typical configuration (Fig 9.2.7).   The gas is contained in a glass tube having an electrode near either end. The ends are sealed by windows mounted at Brewster's angle to minimize reflections at the windows (for one plane of polarization). An optical cavity is formed by two mirrors (usually both are concave), at least one of which is partially transmitting. When an electric discharge is produced in the tube between the electrodes, the gas atoms or molecules are excited and laser action starts. Listed below are some typical parameters for several common types of electric-discharge gas lasers.

Laser species	Gas mixture	Pressure, torr	Current, density A/cm^2
Ne (neutral atom)	He	1.0	0.05–0.5
	Ne	0.1	
Ar (ion)	Ar	0.3	100–2000
CO_2	He	5–10	0.01–0.1
	N_2	1.5	
	CO_2	1.0	

The Transverse-Electric-Discharge Configuration (Fig. 9.2.8).   The transverse-electric-discharge configuration is used for some gas lasers, especially for high-average-power, fast-flowing gas lasers. It provides high electric fields at practical voltages while retaining long paths of excited gas.

TABLE 9.2.5 Representative Gas Lasers[*]

		Continuous-wave (cw)		
		Power, W		
λ, μm	Laser type	Typical	Max	Comment
0.3252	HeCd	10^{-3}–10^{-2}		Metal vapor, electric discharge
0.33–0.53	Ar	0.5–5	20	High-current electric discharge
0.33–1.09	Ar/Kr	0.5	6	High-current electric discharge
0.4416	HeCd	1–5×10^{-3}		Metal vapor electric discharge
0.46–0.65	HeSe	10^{-3}		Metal vapor electric discharge
0.6328	HeNe	10^{-3}	10^{-1}	Most common gas laser
1.15	HeNe	10^{-3}		
1.315	I_2		10^2	Experimental chemical laser
2.6–3.0	HF	5–50	10^4	Chemical laser
3.39	HeNe	10^{-3}	10^{-2}	
3.5	HeXe	5×10^{-3}		
3.6–4.0	DF	2–100	10^4	Chemical laser
5–6.5	CO	2–10	50	
9–11	CO_2	1–100	10^4	
27–374	H_2O	1–10×10^{-3}		Optically pumped by CO_2 or CO laser
34–388	NH_3	10^{-3}		
311, 337	HCN	10^{-2}	1.0	
37–1217	Methanol	1–10×10^{-4}		

		Pulsed lasers		
		Pulse energy, J		
λ, μm	Laser type	Typical	Max	Comment
0.157	F_2	10^{-2}		
0.193	ArF	10^{-1}		
0.222	KrCl	10^{-2}		
0.248	KrF	10^{-1}	10^2	e beam for high energy
0.308	XeU	10^{-1}		
0.337	N_2	10^{-3}–10^{-2}		
0.351	XeF	10^{-1}		
0.458–0.53	Ar	10^{-8}–10^{-6}		
0.5105, 0.5782	Cu	10^{-4}–10^{-3}		
1.5	Ba	10^{-3}		
2.8–3	HF	1	2×10^2	Chemical, electrically initiated
3.6–4.1	DF	10^{-1}–1		Chemical, electrically initiated
9–11	CO_2	10^{-1}–10^2	10^4	e beam for high energy
118.8	Methanol	10^{-3}		
496.1	Methyl fluoride	10^{-4}		

	Multiple-wavelength gas lasers			
Laser type	Wavelength, μm			Comment
Helium-neon	0.5939	0.6352	1.152	Rotate prism to select one wavelength at a time
	0.6046	0.6401	1.162	
	0.6118	0.7305	1.177	
	0.6294	1.080	1.199	
	0.6328	1.084	3.39	

TABLE 9.2.5   Representative Gas Lasers[*] (*Continued*)

Laser type	Multiple-wavelength gas lasers			Comment
	Wavelength, μm			
	Argon	Krypton	Xenon	
Noble-gas ions	0.3511	0.4619	0.4955	Rotate prism to select lines from one gas;
	0.3638	0.4762	0.5007	change gas in tube to change set of wave
	0.3795	0.5208	0.5160	lengths, mixtures of gases can be to extend
	0.4579	0.5682	0.5260	range; simultaneous oscillation on many
	0.4965	0.6471	0.5353	lines possible
	0.5017	0.6764	0.5395	
	0.5145		0.5959	
	0.5287			
Carbon dioxide	9.1–11.3			Several groups of closely spaced lines in this wavelength region; line separations within groups ~0.02 μm; rotate diffraction grating to select lines one at a time

[*]Wavelength ranges are sets of discrete wavelengths and simultaneous output at more than one wavelength may or may not be possible.

A pair of long electrodes (or linear arrays of electrodes) is located parallel to the optical axis, within the envelope that contains the gas. The discharge current flows transverse to the optical axis. Lasers employing this configuration include high peak power, pulsed N_2, Ne, and CO_2; high-average-power, fast-flowing CO_2 lasers; and excimer lasers (e.g., XeF, KrF).

E-Beam Lasers.   High-energy high-current electron beams are generated in an evacuated high-voltage electron gun. The electron beam passes through a thin metal foil into a chamber containing the gas mixture. As in electric-discharge lasers, some of the kinetic energy of the electrons is transferred directly or indirectly to the laser species (atom or molecule). Electron beams are sometimes used in conjunction with transverse electric discharges.

Chemical Lasers

Chemical lasers derive their energy from the free-energy change of a chemical reaction. The chemical reaction may be initiated by some other source of energy, such as light or electric discharges. Since the chemical reaction consumes the reactants, a flowing system is necessary for repetitive-pulsed or cw operation. Figure 9.2.9 shows an arrangement used to produce high-power, cw laser output from hydrogen fluoride (HF) and other

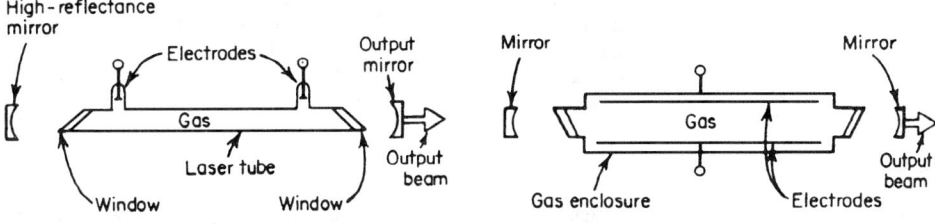

FIGURE 9.2.7   Electric-discharge gas laser.

FIGURE 9.2.8   Transverse-electric-discharge gas laser.

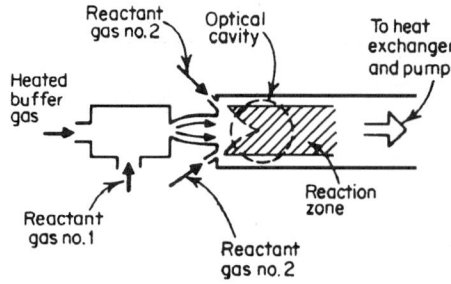

FIGURE 9.2.9   Fast-flowing chemical laser.

molecules. A series of chemical reactions is required to sustain the laser operation in practice, but in simplified form the reaction can be expressed as

$$F_2 + H_2 \rightarrow 2HF + \Delta E$$

where F_2 = reactant, 1
$\qquad H_2$ = reactant 2
$\qquad \Delta E$ = free-energy change, some of which is in the form of vibrationally excited HF molecules

Laser action occurs on vibrational transitions of the HF molecule in the wavelength region 2.6 to 2.9 μm. Chemical lasers of this general type are potential sources of high average power (multikilowatt).

Gas Dynamic Lasers

The expansion of a hot, high-pressure mixture of CO_2 and N_2 through a supersonic nozzle results in a lowering of the gas temperature in a time which is short compared with the vibrational relaxation time of the CO_2 molecule. A differential relaxation time between the upper and lower laser levels results in a population inversion for a short distance downstream from the supersonic nozzle, and laser operation in this region is possible. Average output power of 60 kW at 10.6 μm has been obtained from such a device.

Properties of the Gas-Laser Output Beam

Wavelength and Frequency.   It is customary in laser terminology to refer to a laser transition by its wavelength (in micrometers or nanometers) and to discuss the fine structure of the transition in terms of frequency. Wavelength λ and frequency v are related by $\lambda v = c$, where c is the velocity of light. The wavelength of a laser transition, as well as the wavelength interval or width of the transition over which optical gain exists, is a property of the laser atom or molecule. The transition width is also related to gas temperature and pressure. Typical widths of common gas lasers are:

Laser	λ, μm	Line width, MHz
HeNe	0.6328	1700
	1.15	920
	3.39	310
Ion laser	0.5	2500–3000
CO_2 low-pressure	10.6	60
High-pressure	10.6	800

Oscillation can occur at discrete frequencies (cavity modes) at frequency spacings of $\Delta v = c/2L$, where L is the separation of cavity mirrors. Depending on the nature of the laser medium and the geometry of the optical cavity, oscillation may occur at a number of different modes, within the line width of the transition. For example, for helium-neon at 0.6328 μm, with a cavity-mirror spacing of 100 cm, the mode spacing is 150 MHz, and 11 (axial) modes could oscillate. Single-axial-mode oscillation is generally achieved by reducing L so that only one axial mode lies within the line width of the laser transition. A stable optical cavity is required for both frequency and amplitude stability.

Available Gas Lasers

Of the many commercially available gas lasers (over 35 gas-laser species), four groups are most common: (1) low-power (few milliwatts) visible cw, including helium-neon (He-Ne) and helium-cadmium (He-Cd);

(2) medium-power (few watts), visible cw, including argon-ion (Ar$^+$) and krypton-ion (Kr$^+$); (3) medium- to high-power (10 to 10^4 W) infrared, cw, primarily carbon dioxide (CO_2) and ultraviolet wavelengths (rare gashalogen excimers, XeF, KrF, and so forth). These lasers along with a selected listing of other gas lasers are included in Table 9.2.5, which is arranged by wavelength, with cw and pulsed lasers listed separately. For many of the lasers, both the typical and the maximum available power or energy levels are given.

SEMICONDUCTOR LASER DIODES

Bart H. Verbeek

Introduction

Semiconductor laser diodes, or laser diodes in short, represent key components in the field of opto-electronics including optical communication, optical sensor technology, optical disc systems and material processing. They represent the smallest and most efficient of all lasers since it is based on the injection of carriers in a semiconductor *pn*-structure in combination with an optical cavity/resonator formed by the same semiconductor material. The *pn*-junction is operated in the forward direction and the injected carriers recombine at the *pn*-junction giving rise to the emission of photons. If the carrier density reaches a critical value, the light will be amplified and will yield in laser light emission. The injection current at which the laser start lasing is called the threshold current of the laser. The recombination of electrons and holes takes place in the active layer, sandwiched between a *p*-doped and *n*-doped cladding layer as shown in Fig. 9.2.10. These thin layers (= microns) are grown by epitaxial growth techniques on a semiconductor single crystal substrate with low defects. The dimensions of the laser chip are in the order of a few hundred micron (grain of salt size) but the active layer is much smaller (typically $1.5 \times 0.2 \times 250$ μm). Top and bottom of the laserchip are metallized to provide low ohmic contacts to the semiconductor material.

Since the laser diode is a small device and the aperture of the *optical beam* is in the order of the wavelength the light beam is divergent (Fraunhofer Diffraction). Semiconductor lasers require therefore a lens to couple the light from the chip into, e.g., an optical fiber or to generate a parallel beam. These lenses are usually provided in fiber-coupled packages together with a temperature stabilization unit (TEC). If the lens is not provided with the package (like TO-can type), external optics are required.

Laser diodes can be operated in the following modes:

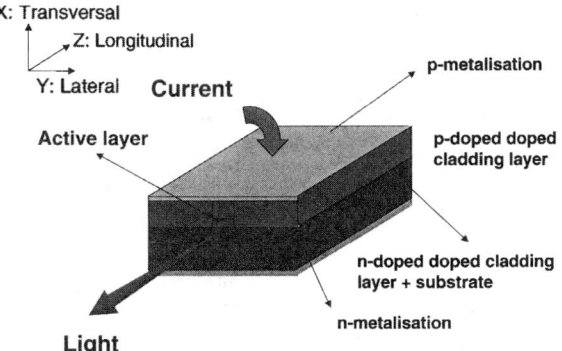

FIGURE 9.2.10   Schematic of semiconductor laser chip. The central area is the active layer surrounded by cladding layers. Top and bottom are metallized to provide electrical current injection. The laser light is a divergent beam because of diffraction emitted from the edge of the crystal. In the inset the definition of the transversal/lateral and longitudinal directions of the laser is given.

Continuous Wave (CW). A DC current flow will generate continuous emission of coherent light.

Amplitude Modulation. By modulation of the injection current, the optical output power can also be modulated. This is called direct modulation of the laser diode. Both analog as well as digital signals can be generated up to frequencies of 15 GHz.

Frequency Modulation. If a small ac injection current is applied to the dc current, the frequency of the emitted light is modulated (typically 100 MHz/mA), which can be employed in, e.g., FM transmission systems or suppressing Stimulated Brillouin Scattering in high-power optical transmission links.

Semiconductor lasers cover a wide spectrum in *emission wavelength* ranging from blue (450 nm) to be used in next-generation optical storage systems, via visible (red 633–650 nm), infrared (780–850 nm), IR (980 nm pump lasers), 1310, 1480, and 1520–1600 nm telecom wavelength) to FIR (>2000 nm for spectroscopy). Different semiconductor materials are used to achieve these wavelengths as will be discussed in the next sections.

Semiconductor Physics Background

Materials. In contrast to other lasers where the radiative transitions occur between atomic or molecular energy levels, semiconductor laser transitions occur between electronic bands, namely, the conduction band and valence bands of the semiconductor crystal. In particular, group III-V and II-VI compounds are *direct* semiconductors with efficient light-emitting characteristics (e.g., Si is an indirect semiconductor and does not radiate efficiently). Various III-V and II-VI materials are available; their ability to form solid solutions opened the possibility to vary the compositions in order to obtain the desired variations in bandgap and in refractive index, while maintaining lattice matching to the substrate. In Fig. 9.2.11 the solid dots represent the relationship between the bandgap E_g and the lattice constant of binary III-V compounds containing Al, Ga, or In and P, As, or Sb, and II-VI materials containing Cd or Zn and S, Se or Te, respectively. Since the photon energy E is approximately equal to the bandgap energy, the lasing wavelength λ can be obtained using $E_g = hc/\lambda$, where h is the Planck constant, and c is the speed of light in vacuum. If E_g is expressed in eV, the lasing wavelength in μm is given by $\lambda = 1.23889/E_g$.

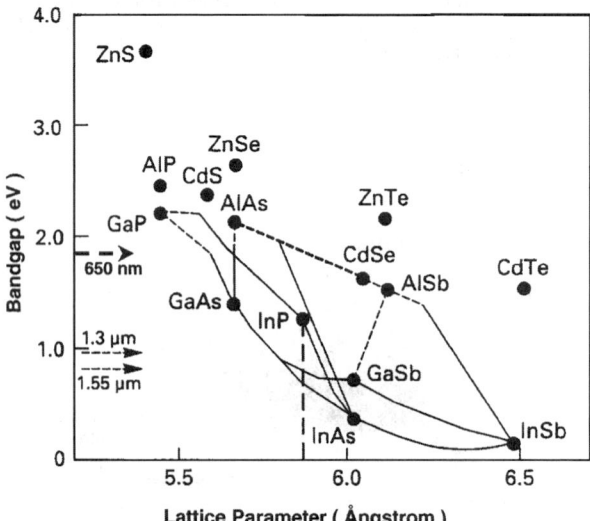

FIGURE 9.2.11 Graph showing the various semiconductor materials for making laser active and cladding layers. For each binary compound the lattice parameter and bandgap energy is given. Lines denote mixed compound solid solutions. As substrate material GaAs and InP are used for most lasers. Arrows indicate bandgap energy for three mostly used wavelength.

TABLE 9.2.6 Semiconductor Materials for Active/Cladding Layers and Their Wavelength Window/Applications

Substrate	Active/cladding layer	Wavelength range (μm)	Applications
GaAs	(Al)GaAs/AlGaAs	0.70–0.88	Optical disc
	(In)GaAs/AlGaAs	0.88–1.1	Pumping
	(In)GaAs/InGaP	0.88–1.1	High power
GaAs	(Al)GaInP/AlGaInP	0.60–0.67	Optical disc
	InGaAsP/AlGaInP	0.65–0.88	Laser printer
	InGaAsP/AlGaAs	0.65–0.88	Barcode/displays
InP	InGaAsP/InP	1.05–1.65	Fiber communication
	(Al)GaInAs/AlGaInAs		
Sapphire/SiC	InGaN/GaN	0.45	Optical disc
GaSb	GaInAsSb/AlGaAsSb	2–5	Spectroscopy laser radar

Figure 9.2.11 shows that the bandgaps and lattice parameters of $In_xGa_{1-x}As_yP_{1-y}$ (quaternary) compositions cover the area bound by connecting lines between the binaries InAs, InP, GaP, and GaAs, and can be grown lattice matched to InP and GaAs substrates for certain combinations of x and y. The interrelationships for the ternary $Al_xGa_{1-x}As$ follow the connecting line between the binary constituents GaAs and AlAs.

Lasers and other optoelectronic devices require a very low defect density in the crystal or else nonradiative processes will dominate over the radiative processes. Therefore, lattice matched crystal growth of the various compound layers on the substrate is essential and is achieved by modern epitaxial growth techniques such as MOVPE, MBE, and CBE. The semiconductor material can be doped with, eg., Zn (Si) to achieve p-type (n-type) doping. All modern semiconductor lasers consist of a double heterostructure in order to achieve high carrier densities at low currents by confining the recombining carriers in a small volume. The active layer is embedded between heterolayers having a larger bandgap—cladding layers. Wavelength ranges and applications for various substrates and active/cladding layers are given in Table 9.2.6. The pn-junction just occurs at the active layer, which has a lower bandgap. By injecting current in the diode, electrons from the n-side and holes from the p-side recombine in the active layer generating photons of the designed wavelength by the bandgap energy E_g (see Figs. 9.2.12a and 9.2.13). In order to reduce the total current, injection only takes place in a stripe of width w (typically 1–2 μm) (see also Fig. 9.2.10). The electrical diode characteristic of the laser is given by

$$I = I_0 \left[e^{(e(V - IR_s)/nkT)} - 1 \right] \tag{1}$$

where R_s is the series resistance of the diode and n is the ideality factor with $1 < n < 2$.

Now that carriers can recombine in the active layer, three processes will occur: (1) spontaneous emission, (2) stimulated emission, and (3) absorption of generated photons. At low injection current processes (1) and (3) dominate and the laser behaves as an LED. Only with sufficient carrier injection, the stimulated emission process will dominate and sufficient amplification of generated photons will occur. If the active layer is placed in an optical cavity preferred directional gain is obtained and laser action is possible. The optical material gain of the active layer is shown in Fig. 9.2.12b versus the wavelength, with increasing injection current as a parameter. With zero injection current, the material absorbs light for shorter wavelengths than λ_g, i.e., this material can be used as a photodetector. With increasing injection current gain is increasing and the maximum of the gain value shifts a little to lower wavelength because of bandfilling effects. For wavelengths beyond the bandgap energy (denoted as arrow) the material is transparent. The 3-dB bandwidth of the gain is typically 80 to 100 nm.

For efficient operation of the laser, the light that is generated in the active layer has to be guided within the laser structure. An optical waveguide is formed by the higher refractive index of the active layer with respect to the surrounding (cladding) layers. For stable laser operation it is required that this waveguide can only support the lowest-order mode (usually the TE_{00} mode) both in the transverse and lateral directions. Design parameters for obtaining a lowest-order mode are: thickness, width, and refractive index of the active layer and the cladding layers. In Fig. 9.2.13 the transversal index profile and the resulting optical mode are shown. Several numerical tools are available to solve the Maxwell equations for the dielectric waveguide and obtain the optical mode profile in the laser (Near Field), external beam (Far Field), and astigmatism. Usually the optical mode (near Gaussian shape) is larger than the active layer and the fraction of light that travels in the active layer is called the *confinement factor* Γ (see Fig. 9.2.13).

With the gain medium in the active layer and the waveguide description to guide the light in the laser, the next step is the longitudinal *optical cavity*. Since the refractive index of the semiconductor lasers is about 3.3 to 3.4, which is much higher than air ($n \approx 1$), the simplest cavity is formed by cleaving the crystal to form the endfacet-mirrors. The (power) reflection is calculated as $R = [(n-1)/(n+1)]^2$ and results in about 30 percent. This value can be modified by applying appropriate dielectric facet coatings like AR or HR coating depending on application. A semiconductor laser based on cleaved facets is referred to as Fabry Perot lasers. Other cavity structures will be discussed later. A FP-cavity allows a set of discrete longitudinal modi which are determined by the condition that in a round-trip an integer number of wavelengths fit within the cavity of length, $n_e L$ where n_e is the effective refractive index. Allowed wavelengths are $\lambda_m = 2n_e L/m$, where m is the mode index. For evaluating the spacing between adjacent modi, $\Delta\lambda$, one must take into account the dispersion of the refractive index and the effective group index $n_{e,g}$ is introduced. The mode distance is given by $\Delta\lambda = \lambda^2/2n_{e,g}L$, which corresponds to the inverse round-trip delay τ_L of the cavity. With an effective group index of 3.5 and a laser cavity of 300 to 400 μm a typical round-trip delay is about 10 ps.

When the semiconductor active medium with gain profile, as shown in Fig. 9.2.12b, is placed in the Fabry Perot cavity to build a laser, this laser will start lasing at a wavelength corresponding to the mode with the highest gain. The spectrum of the FP laser consists then of a set of modi around this highest gain mode to form the spectrum of the FP laser.

Semiconductor Laser Parameters

The threshold for lasing in a Fabry Perot laser can be obtained by considering the intensity of the laserfield in a round-trip within the cavity. If the semiconductor material gain is denoted by G, the mode gain is given by $g = \Gamma G$ where Γ is the confinement factor and internal loss α_i the reflection coefficients R_1 and R_2 and length L, the net round-trip gain g_r is given by

$$g_r = R_1 R_2 e^{[2(g-\alpha_i)L]} \tag{2}$$

The material gain can be expressed in a linear relation to the current density N by $G = A(N - N_{tr})$, where A is the differential gain ($\partial G/\partial N$) and N_{tr} is the transparency current density, i.e., the value for injecting carriers such that the absorption and gain are equal and the material is transparent. At the lasing threshold, the optical field is sustained after traveling on round-trip in the cavity, thus the net round-trip gain g_r is unity, resulting in the threshold condition given by

$$g_{th} = \alpha_i + \frac{1}{2L}\ln\frac{1}{R_1 R_2} = \alpha_i + \alpha_m \tag{3}$$

where α_m is called the mirror loss.

The current density is related to the carrier density by

$$J = \frac{ed_{act}N}{\tau_n} \tag{4}$$

where d_{act} is the thickness of the active layer, and τ_n the carrier lifetime. The threshold current density J_{th} is now given by

$$J_{th} = \frac{J_{tr}}{\eta_i} + \frac{d_{act}}{\Gamma\beta\eta_i}\left(\alpha_i + \frac{1}{2L}\ln\frac{1}{R_1 R_2}\right) \tag{5}$$

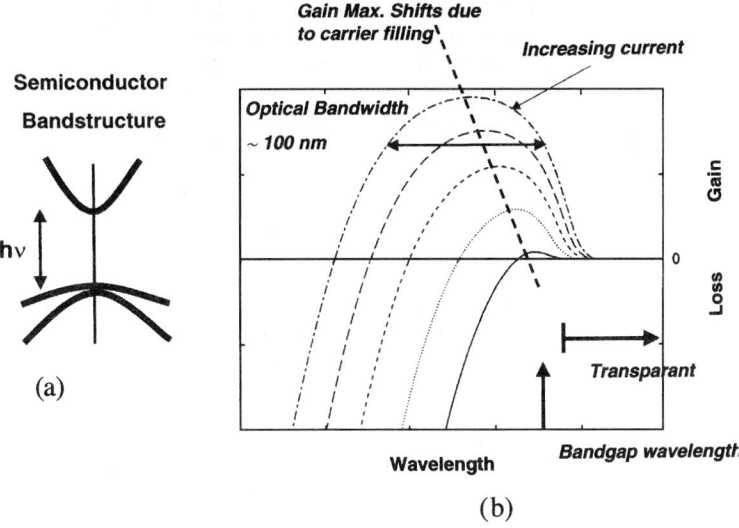

FIGURE 9.2.12 (a) Schematic representation of the conduction band and both valence bands of semiconductor laser crystals. The arrow denotes the bandgap energy determining the wavelength of the laser by $\lambda = 1.23889/E_g$. (b) Calculated gain spectrum of semiconductor materials as function of the wavelength. The vertical arrow indicates the bandgap wavelength, the horizontal arrow shows the transparent region of the material. The curves show the gain/absorption as a function of injection current density. Semiconductor lasers tend to operate near the gain maximum.

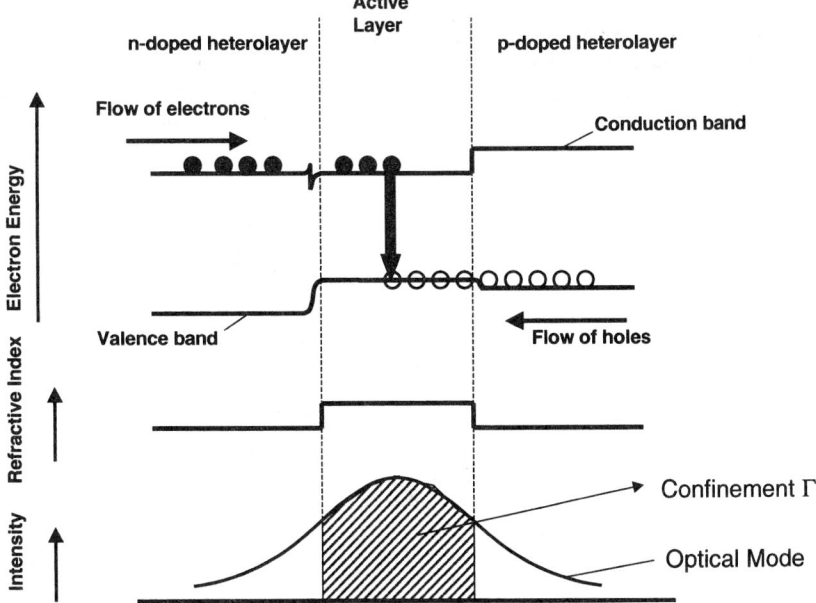

FIGURE 9.2.13 Top: Band diagram of a double heterostructure laser under forward current injection. Electrons (holes) flow from the n-(p-) doped cladding layer into the active layer with smaller bandgap giving radiative transitions emitting photons. Middle: Corresponding refractive index variation between cladding- and active region for waveguiding. Bottom: Optical intensity profile of the laser beam inside the laser cavity. The grey area is the part of the beam traveling in the active layer and defines the confinement factor Γ.

where J_{tr} is the current density to reach transparency, η_i is the internal efficiency with which electrons and holes recombine radiatively and β is the gain coefficient, which is related to the differential gain A by $\beta = A\tau_n/e$. Equation (5) shows that the threshold current consists of the transparency current density J_{tr} and a contribution to overcome the internal and mirror losses. By further increasing the injection current above threshold, the round-trip gain remains fixed, i.e., clamped to the value of g_{th} and the excess carriers are consumed to build up the laser oscillation intensity. The total optical power generated by stimulated emission in the cavity is given by

$$P = \eta_i \frac{hv}{e}(I - I_{th}) \tag{6}$$

where I and I_{th} are the driving and the threshold current, respectively, and hv the photon energy. Part of this power is dissipated inside the laser cavity, and the remaining is coupled out through the facets. These two powers are proportional to the effective internal and mirror loss. The total output power from both facets is thus given by

$$P = \eta_i hv \frac{(I - I_{th})}{e} \frac{\alpha_m}{\alpha_i - \alpha_m} \tag{7}$$

The external differential efficiency is the ratio of the variation in photon output rate that results from an increase in the injection current and is given by

$$\eta_d = \frac{d(P/hv)}{d[(I - I_{th})/e]} = \eta_i \left[1 + \frac{2\alpha_i L}{\ln \dfrac{1}{R_1 R_2}}\right]^{-1} \tag{8}$$

In Fig. 9.2.14 the typical optical output power per facet (mW) versus the injection current (mA) (so called *L/I* curve) of a semiconductor laser is shown for various temperature where a photodetector is placed in front of the output facet of the laser. Up to the threshold current only spontaneous emission is generated but at threshold current the output power increases dramatically by the stimulated photons.

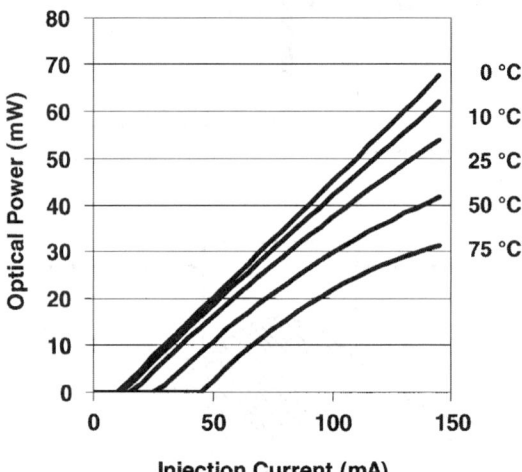

FIGURE 9.2.14 Light output power (mW) vs. injection current (mA)—L/I characteristic—of a semiconductor laser at various temperatures.

Several observations can be made from Fig. 9.2.14. First, the *L/I* curves are not exactly linear and deviate more at higher output power and high temperatures due to internal heating effects, increased internal losses and increased current (blocking) leakage. Second, the threshold current value increases with temperature owing to active layer temperature rise and additional loss mechanisms. The variation of the threshold current with temperature is expressed by an empirical relation:

$$I_{th}(T_2) = I_{th}(T_1)e^{(T_2-T_1)/T_0} \tag{9}$$

where T_0 is the characteristic temperature and T_1 and T_2 are expressed in Kelvin. T_0 values of 50 to 70 are found for InP-based lasers and >100 for GaAs-based lasers. It should be noted that T_0 is not constant over a wide temperature range.

The single-mode laser has a finite *linewidth* caused by spontaneous emission into the lasing mode adding phase noise. The well-known Shawlow-Townes expression for the linewidth of a semiconductor laser has been modified by Henry to include the dynamics of the carrier density on spontaneous emission events and the linewidth expression for a semiconductor laser is given by

$$\Delta v = \frac{hvv_g^2(\alpha_i + \alpha_m)\alpha_m n_{sp}}{8\pi P}(1+\alpha_H^2) \tag{10}$$

where hv is the photon energy, n_{sp} is the spontaneous emission coefficient and α_H is Henry's parameter. This equation shows that increasing power, reducing internal and mirror losses and low α_H values result in narrow (Lorentzian) linewidth values. The linewidth of the laser is also related to the coherence length L_c that can be measured using a Mach-Zehnder interferometer to obtain the autocorrelation function via $L_c = c/\pi\Delta v$. Example: laser with linewidth of 10 MHz has a coherence length of about 10 m.

Optical Waveguiding

The photons generated by stimulated emission in the active layer must be guided in the semiconductor laser structure in order to optimize the gain (high confinement Γ). Guiding is also needed to generate a proper beam from the laser that can be coupled into a fiber with high efficiency or can be handled with bulk optical lenses. Beam profiles that have a Gaussian shape are preferred and these can be obtained when a dielectrical waveguide is designed to support the lowest order mode only. Two-dimensional waveguiding in the laser structure is formed by designing both the transversal direction (*x*, perpendicular to the active layer) and the lateral directional (*y*, in plane of active layer).

Transversal Guiding. This is determined by the refractive index difference between the active layer and the cladding layers (see also Fig. 9.2.13) and the thickness of the active layer. The parameter that determines the cutoff condition for higher-order modes is called the normalized frequency or simply the *V* parameter.

$$V = \frac{\pi d}{\lambda}\sqrt{n_{act}^2 - n_{cladding}^2} \tag{11}$$

where *d* is the thickness of the active layer. In order to obtain single mode operation, the *V* parameter should obey *V* < 2.4, which is easily met with thin active layers and optimized cladding layers for carrier confinement.

Lateral Guiding. It is less straightforward mainly because of technological constraints. Three classes of lateral guided laser structures can be distinguished (Fig. 9.2.15).

1. *Gain guided structure.* A continuous active layer with no lateral index variation. Current injection in a stripe top contact results in local gain under the stripe and optical losses away from the stripe, hence the name gain guiding. The laser beam shows considerable astigmatism and a non-Gaussian lateral profile at high-power levels. It is used in very high-power lasers with a wide stripe where the beam properties are less

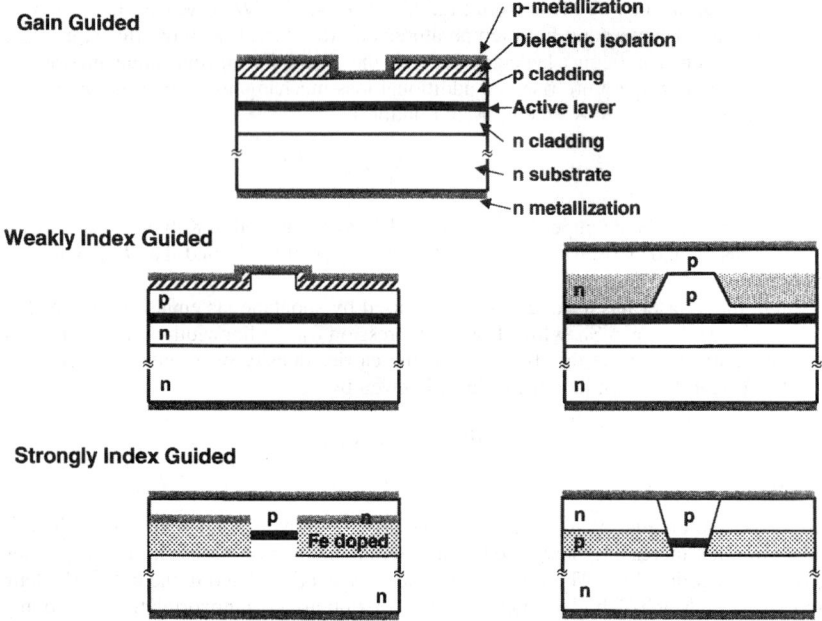

FIGURE 9.2.15 Lateral laser structures for lateral optical guiding. Top: Gain guided laser, Center: Weakly index guided structures (Ridge waveguide and buried Ridge structure). Bottom: Strongly index guided laser structures (SI-PBH, semi-insulating planar buried hetrostructure and *n-p-n* blocking structure).

important. Many lateral higher-order modes are present but this can be reduced by employing a longitudinal taper structure in the stripe.

2. *Weakly guided structure.* A continuous active layer but with a lateral index variation in the top/bottom cladding layer by processing, e.g., ridge structure or buried ridge structure. The lateral index guiding is small ([a]10-3) compared to the transversal confinement giving rise to elliptical beam profile.

3. *Strongly index guided structure.* The active layer is surrounded by cladding type material giving rise to both lateral and transversal strong index contrast and waveguiding with almost circular beam profiles. Strongly index guided laser structures are fabricated using several epitaxial growth and regrowth and processing steps. These structures have efficient current injection in the active layer of typically 1.5 to 2 mm width.

Longitudinal Laser Structures

Fabry-Perot Laser. The simplest longitudinal structure laser is a chip that is cleaved to form facets with reflectivity R to achieve optical feedback. These facets can be coated to modify the reflection value from low (AR) to high (HR). This structure allows for a discrete set of cavity modes and in combination with the gain profile, which is parabolic in shape, the laser operates in a series of modes around the gain maximum. The associated spectrum is called multilongitudinal mode and is depicted in Fig. 9.2.16. The number of modes (5 to 30) depends on the output power of the laser and decreases with increasing power. These modes operate simultaneously as is proven by the occurrence of frequency doubled components in the spectrum.

DFB Laser. In fiber optic long distance applications the multimode emission spectrum in FP lasers restricts the transmission distance because of the material dispersion of the fiber. To improve the link distance the need

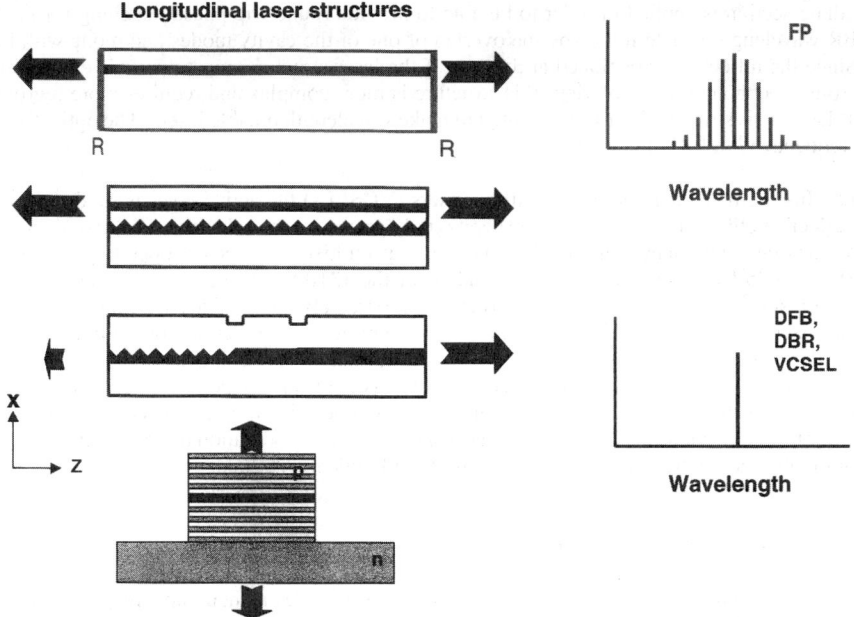

Longitudinal laser structures

FIGURE 9.2.16 Longitudinal laser structures and corresponding spectral behavior. From top to bottom: Fabry Perot laser, DFB (distributed Bragg reflector) laser, DBR (distributed Bragg reflector) laser, and VCSEL (vertical cavity surface emitting laser).

for a narrow spectral width laser has led to the development of the distributed feedback (DFB) laser emitting a narrow (few MHz) single longitudinal mode. The DFB laser is characterized by a grating near (above or below) the active layer in the longitudinal structure. The analysis of the coupled modes in the DFB laser is too lengthy to be discussed here and can be found in the reference books. The lasing wavelength of the DFB laser is to first approximation determined by the grating pitch Λ and the effective refractive index of the waveguide and is given by $\lambda = 2n_{eff}\ \Lambda$. Using $\lambda = 1550$ nm and $n_{eff} = 3.25$ this results in a submicron grating pitch of 0.2384 μm. This grating requires an additional transparent guiding layer in which the grating is made using holographic exposure or e-beam pattern generator. The accuracy of the laser wavelength needed in DWDM systems (e.g., 50 GHz spacing) requires a controlled grating pitch fabrication within 30 pm! Since DFB lasers have an internal feedback for laser operation, the facets need to be eliminated by applying AR coatings on both sides. The position of the laser wavelength with respect to the gain maximum is a design parameter and usually the negative detuning of 10 to 30 nm (i.e., laser wavelength is positioned on the short wavelength side of the gain peak) is applied for improved modulation properties. The spontaneous emission spectrum of DFB lasers (i.e., subthreshold) is characterized by a so-called stopband. The width of the stopband is linearly related to the coupling coefficient parameter κ of the grating to the optical mode. Standard DFB lasers have a nonuniform power distribution in the longitudinal cavity, which can be modified/improved by including phase shifts in the grating. Single $\lambda/4$-phase shift but also multiple $\lambda/8$ or $\lambda/16$ have been introduced to improve the uniform power density that is required for linear L/I characteristics in, e.g., analog transmission. The linewidth of DFB lasers is sub-MHz.

DBR Laser. In a distributed Bragg reflector (DBR) laser one of the facets is replaced by a waveguide structure with grating, which serves as a wavelength selective reflector. In contrast to the DFB laser, the active layer (= gain region) is now separated from the wavelength selection element (Bragg section). This Bragg section consists of a transparent guiding layer with an etched grating and cladding layers. This section is "butt-coupled" to the active section where the active layer provides the waveguiding. In many DBR lasers, an intermediate section is applied in order to be able to provide round-trip phase matching for the lasing mode. The DBR wavelength is determined by the overlap of one of the cavity modes and mode with highest reflectivity. Since the reflector is positioned at the rear of the laser structure, most power is emitted from the front facet. From a technology point of view, this structure is more complex and requires more regrowth steps than the DFB laser. This structure is also very suited to make wavelength tunable lasers. The optical spectrum shows single frequency with MHz linewidth.

VCSEL.   In a vertical cavity surface emitting lasers (VCSELs) the active layer is sandwiched between two highly reflective DBR mirrors (>90 percent) epitaxially grown on the substrate. The first VCSELs were made on GaAs substrate with many pairs of AlAs and GaAs as multistack reflectors operating at wavelength of 850 and 980 nm. Also InP-based VCSEL are available for the 1310 to 1550 nm telecom applications. The laser operates in a single longitudinal mode by virtue of an extremely small cavity length (about 1 μm) for which the mode spacing exceeds the gain bandwidth. Light is emitted in a direction normal to the active layer plane and since the output power has to pass the high reflector, VCSELs are characterized by low output power levels (few mWs). An important advantage of the VCSELs is the beam shape, which can easily be made narrow and symmetrical which allows direct fiber coupling without additional optics. This, along with the possibility of on-wafer testing, is viewed as considerable cost saving. Direct modulation has been achieved up to 2.5 Gb/s. Application areas are optical interconnect and short haul optical data links.

Dynamic Properties of Semiconductor Lasers

Rate Equations.   One of the virtues of semiconductor laser diodes is their capability to modulate the optical output power directly with the injection current. To analyze the modulation aspects it is necessary to consider the laser diode dynamics on the basis of the single mode rate equations for the photon density S in the optical mode and the carrier density N in the active medium. These equations can be written as

$$\frac{dS}{dt} = \Gamma v_g g(N,S)S - \frac{S}{\tau_p} + \Gamma \beta \frac{N}{\tau_n} \tag{12}$$

$$\frac{dN}{dt} = \frac{I(t)}{eV_{act}} - \frac{N}{\tau_n} - v_g g(N,S)S \tag{13}$$

Here v_g is the group velocity, Γ is the optical confinement factor, V_{act} is the volume of the active medium, e is the electronic charge, $I(t)$ is the (modulated) pump current, τ_p is the photon lifetime, τ_n is the carrier lifetime, and β is the fraction of the spontaneous emission that couples into the lasing mode. The first equation represents the photon density variation owing to the photon generation by stimulated and spontaneous emission and the photon loss at the mirrors and by absorption. The second equation describes the carrier density variation resulting from the carrier injection by the pump current $I(t)$ and the carrier annihilation by the stimulated and spontaneous recombination. The material gain $g(N,S)$ (per unit length) depends on both the carrier density and the photon density and is modeled as

$$g(N,S) = \frac{\partial g}{\partial N}(N - N_{tr})(1 - \epsilon S) \tag{14}$$

Here the gain is linearized at the transparency carrier density N_{tr} with a gain slope constant $\partial g/\partial N$ (differential gain) and the gain dependency on the photon density S is characterized by the gain suppression parameter ϵ.

The rate equations cannot be solved analytically; however, the small signal response of a laser diode can be derived by linearizing the rate equations and applying a small signal analysis. Then the intensity modulation transfer function normalized to the zero frequency response $M(0)$ is given by

$$\left| \frac{M(j\omega)}{M(0)} \right| = \frac{\omega_0^4}{(\omega^2 - \omega_0^2)^2 + \gamma^2 \omega^2} \tag{15}$$

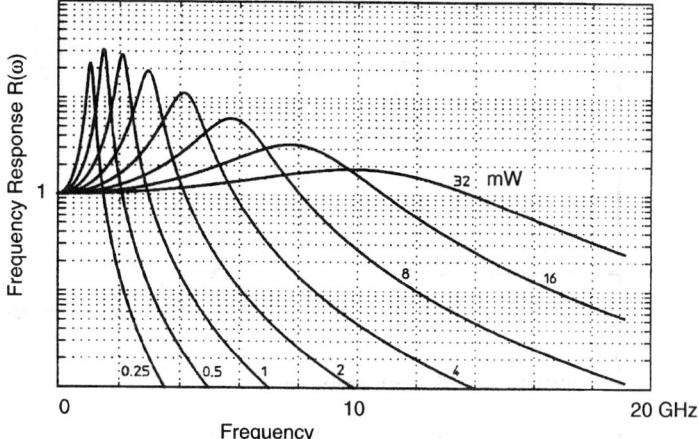

FIGURE 9.2.17 Calculated small signal laser diode response as function of the output power from 0.25 to 32 mW.

with the circular relaxation oscillation (RO) frequency given as

$$\omega_0^2 = \frac{v_g}{\tau_p} \frac{\partial g}{\partial N} S \tag{16}$$

and the damping factor γ

$$\gamma = \frac{\beta \Gamma N}{S \tau_n} + \frac{1}{\tau_n} + v_g \frac{\partial g}{\partial N} S + \frac{\epsilon S}{\tau_p} \tag{17}$$

As an example the calculated small signal response for a semiconductor laser diode at several output powers is given in Fig. 9.2.17. The intrinsic frequency response of a laser diode is completely described and shows a second-order low pass network nature. The transfer function is characterized by a clear resonance followed by a steep roll-off. The resonanace known as the relaxation oscillation results from the interplay between the photon field and the carriers. The resonance frequency ω_0, obtained from the small signal analysis, increases with the square root of the optical power. The damping of the resonance is determined by the spontaneous emission at low photon densities and by the gain suppression factor ϵ at high photon densities. In view of high frequency modulation an ultimate increase of the relaxation oscillation frequency is desired but will be severely limited at high photon densities by the damping introduced by the gain suppression ϵ.

In addition to the maximum intrinsic modulation frequency, parasitic capacitance in the laser structure and/or mounting are also determining the small signal response of laser diodes. To account for the chip and bonding parasitic capacitance, it should be expanded as

$$R(\omega) \approx \frac{\omega_0^4}{(\omega^2 - \omega_0^2)^2 + \gamma^2 \omega^2} \frac{1}{1 + \omega^2 (RC)^2} \tag{18}$$

In digital modulation schemes the large signal nonlinear modulation response should be taken into account and the small signal approach is not sufficient. A typical response on a rectangular current pulse is schematically given in Fig. 9.2.18. Important parameters here are the turn-on and the turn-off time. Moreover, frequency ringing will occur when the laser diode is switched from the "off" state to the "on" state (and back)

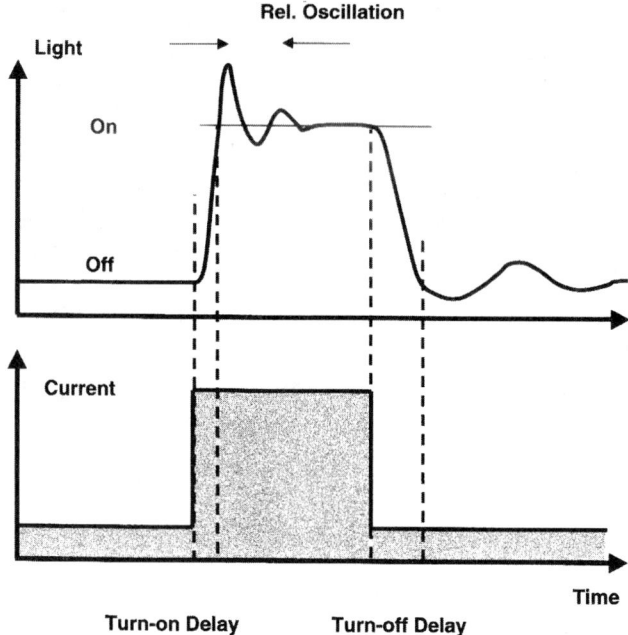

FIGURE 9.2.18   Large signal optical modulation response of a laser diode on rectangular current pulse.

with correspondingly different relaxation oscillation frequencies and damping factors. The turn-on time t_{on} for $I_{off} > I_{thr}$ can be calculated as

$$t_{on} = \frac{0.23}{f_0} \sqrt{\ln\left(\frac{S_{on}}{S_{off}}\right)} \qquad (19)$$

with $f_0 = \omega_0/2\pi$ is the relaxation oscillation frequency corresponding to the "on" state, S_{on} is the steady-state photon density in the "on" state, and S_{off} in the "off" state. Considering a high-speed laser diode with $f_0 = 10$ GHz and an extinction ratio of 6:1 yields a turn-on time t_{on} of 31 ps. The turn-off time t_{off} can be approximated as $(0.5–2)t_{on}$. As the relaxation oscillation frequency corresponding to the "off" state with $S_{off} << S_{on}$ is several factors lower than f_0, the turn-off time is larger than the turn-on time.

A powerful tool to predict the performance of laser diodes is given by simulation. By numerical integration of the rate equations the time domain response of the laser output power can be calculated when a time-varying input current $I(t)$ is inserted. The results of simulations are represented as eye patterns (Fig. 9.2.19) obtained with a 10 Gb/s2^7–1 PRBS data formats. The laser extinction ratio is approximately 6:1.

Optical Frequency Response of Laserdiodes.   When a modulated current is injected into a laserdiode, the varying carrier density will affect the refractive index of the active medium because of carrier-induced band-edge shifting. This implies that not only the photon density S will be modulated by the gain variations, but also the lasing resonance frequency by the varying refractive index. This optical frequency response (in intensity modulated direct detection systems regarded as highly undesirable) is denoted as *frequency chirp*. The so-called linewidth enhancement or Henry's factor α_H describes this coupling between the gain and the refractive index and is given by

$$\alpha_H = -\frac{4\pi}{\lambda} \frac{(\partial n_{eff}/\partial N)}{\Gamma(\partial g/\partial N)} \qquad (20)$$

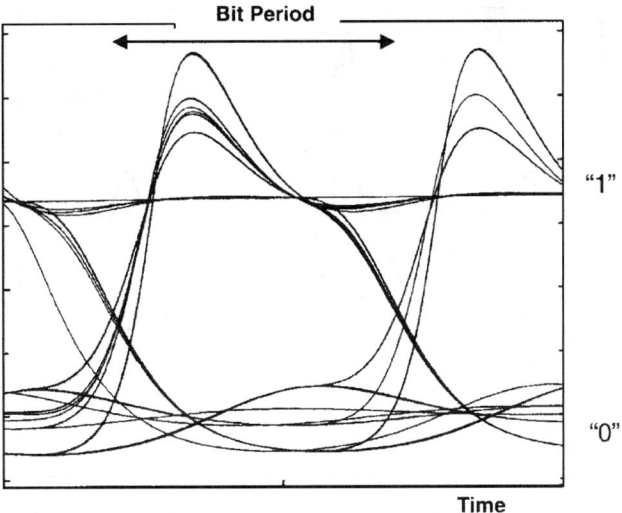

FIGURE 9.2.19   Simulated 10 Gb/s eye patterns (I_{off} = 18 mA, I_{on} = 54 mA) for SL-MQW active layer laser diode.

With this α_H parameter the expression for the frequency variation under injection current modulation is expressed as follows:

$$\Delta v(t) \cong \frac{\alpha_H}{4\pi} \left[\frac{1}{P(t)} \frac{dP(t)}{dt} + \frac{2\Gamma}{h v \eta_d V_{act}} \epsilon P(t) \right] \tag{21}$$

where η_d is the differential efficiency. In this equation, two chirp contributions are present. The first part (*transient chirp*) is associated with the transients and the ringing introduced by the relaxation oscillation when the laser is switched between the "on" and the "off" states. The second contribution is governed by nonlinear gain suppression coefficient ϵ and originates from the imperfect carrier density pinning because of the nonlinear gain. This *adiabatic chirp* depends on the laser structure and causes a frequency offset during modulation between the "on" state and the "off" state. The total chirp is directly proportional to the α_H-factor that highlights the significance of this parameter. Typical α_H-factors range from 2 to 6 depending on the detailed structure of the active layer and laser structure.

The laser frequency variations as function of the time can be calculated by numerical integration of the rate equations. Figure 9.2.20 shows the optical output power (dashed curve) and the corresponding frequency response (solid curve) for a single 2.5 Gb/s pulse (400 ps). The zero chirp value is related to the CW "on" state.

The optical frequency response of laser diodes poses severe problems in intensity modulated direct detection systems employing dispersive fibers. The addition of spectral components by chirp increases the total frequency bandwidth. In combination with fibers showing a substantial dispersion at the operating wavelength (–17 ps/nm/km) (@ 1550 nm), the frequencies that compose the signal travel with a different group velocity along the fiber. This results in a frequency dependent arrival time of the signal components and hence signal distortion at the receiver side. In digital systems information associated with one bit can occur in a neighbor bitslot giving rise to intersymbol interference (ISI).

Active Layer Structures

Quantum Well Lasers.   The development of advanced crystal growth techniques like metallo-organic chemical vapor phase epitaxy (MOVPE) and molecular beam epitaxy (MBE) allowed the growth of defect-free

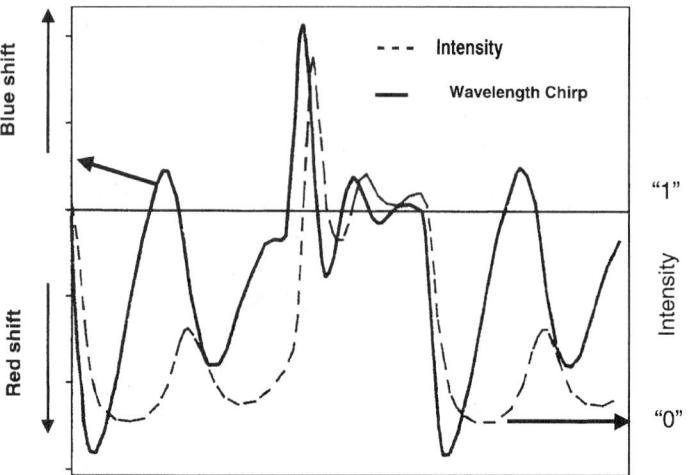

FIGURE 9.2.20   Simulation of the frequency response (solid curve) and the large signal response (dashed curve) of a bulk laser diode.

ultrathin layers with sharply defined heterointerfaces. The creation of these layers with a thickness close to the Broglie wavelength of electrons accelerated the development of single quantum well (SQW) and multiquantum well (MQW) laser diodes. For conventional bulk laser diodes the active layer thickness is about 150 nm or more. The active layer of quantum well lasers consists of one or more thin layers ($L_z < 30$ nm) which are separated by barrier layers with a larger bandgap material. For injected carriers confined to these wells, the energy for motion normal to the well is quantized into discrete energy levels. The motion of electrons/holes in the plane of the quantum well is not quantized and form continuous states. The heterostructure for bulk active layer, as shown in Fig. 9.2.13, is now modified as shown in Fig. 9.2.21. Here the conduction and valence band structure is shown for a MQW laser having four quantum wells. For every quantum number, one conduction band and two valence band (namely, the heavy and light holes) levels exist. These levels are degenerated. In the z direction the carriers are confined in the well as a particle in a box of width L_z and the calculation of the

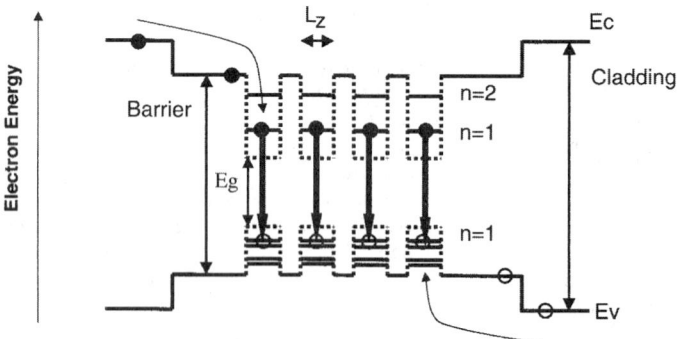

FIGURE 9.2.21   Schematic band energy diagram of a four quantum well active layer showing the conduction and valence band. The well width is L_z. The dashed lines indicate the energy corresponding to the material composition and the solid lines show the energy levels for ($n = 1, 0, \ldots$) because of the quantum effect ("particle in box"). Arrows show the radiative transitions between electrons (filled balls) and holes (open balls).

allowed energy states is a well-known quantum mechanical problem. The energy levels associated with this "particle-in-box" is given by

$$E_n = \frac{h^2 n^2}{8mL_z^2} \tag{22}$$

which is valid for both electrons and holes. Here n is the quantum number, L_z is the quantum well width (QW layer thickness), m is the effective mass for this level, and h is Planck's constant. The effective bandgap of quantum wells is given by $E_g = E_g^{\text{bulk}} + E^c_{n=1} + E^c_{n=1}$, and thus is a function of the well thickness L_z; i.e., by using the same composition in the QW, the emission wavelength can be adjusted by changing the well width. Furthermore, the gain spectrum and the gain current relations lead to improvements for laser operation such as a reduction of the threshold current and its temperature sensitivity, enhancement of the differential quantum efficiency, an increase in modulation bandwidth and a reduction of the spectral linewidth (lower α_H). These improvements have been observed mainly in AlGaAs-based lasers and to a lesser extent in InP-based lasers because of higher intervalence band absorption processes.

Strained Quantum Well Lasers.   To further improve the transition efficiency of MQW lasers, in particular for InP-based lasers, mechanical strain in the QWs can be introduced by growing lattice mismatched epitaxial layers on a thick substrate. These layers are subjected to a biaxial in-plane strain. As an example, the growth of $In_xGa_{1-x}As$ with $x = 0.53$ is lattice-matched to InP substrate. For $x > 0.53$, i.e., large In atom rich, the lattice is compressed in the x-y plane and is called compressive strain. For $x < 0.53$, i.e., smaller Ga atom rich, the strain is tensile. The lattice mismatch is bound by the critical thickness limitation depending on materials (elastic constants), degree of lattice mismatch and substrate orientation, and sets an upper limit to the applied strain before lattice defects such as misfit dislocations occur. Because of the deformation of the unit cell of the crystal, the bandstructure of the semiconductor material is modified. For InGaAs/InP materials the effect is most pronounced in the valence subbands, namely, the removal of the valence subband degeneracy and a larger split of the valence bands. This results in more efficient electron-hole transitions and reduction of the intervalence band absorption leading to much improved laser performance such as threshold current reduction, higher differential gain, lower α_H parameter, less temperature sensitivity, and TE or TM emission control.

Quantum Dot Lasers.   Quantum dot (QD) or quantum dash-lasers are characterized by an active layer consisting of small clusters of active material embedded in cladding material. These clusters are made of InAs using the self-assembly property in crystal growth using MBE or MOVPE. Self-assembled QDs form spontaneously on the GaAs surface during growth because of lattice mismatch between InAs and GaAs. The InAs islands are then capped by another GaAs layer. Since the GaAs conduction (valence) band minimum (maximum) has higher (lower) energy than that of InAs, both electrons and holes are trapped by the InAs island (QDs). The shape of these QDs depends on the growth conditions. Typically, the size is strongly asymmetric: the in-plane diameter of the QDs ranges from 20 to 40 nm, whereas the height ranges from 2 to 6 nm. Nevertheless, the electrons and holes are now trapped in three dimensions unlike the quantum well (one-dimensional) or the quantum wire (two-dimensional) giving rise to new physical properties when applied in an optical cavity. The transitions within the QD are now uncoupled from the surrounding QDs modifying the gain dynamics. Also, QD-lasers have been fabricated to obtain 1300-nm-emitting lasers using GaAs as substrate and allowing much higher T_0 values than for InP-based lasers.

Quantum Cascade Lasers.   The principle of quantum cascade laser is based on quantum confinement in quantum wells (with associated energy levels) and the tunneling property of electrons through energy barriers. In contrast to usual semiconductor lasers where the emission process is based on transitions between conduction band and valence band, in QCLs the transitions take place between the higher energy state of one QW and a lower energy state of the nearest neighbor QW as schematically depicted in Fig. 9.2.22. By applying many QWs in the active layer with carefully designed barriers, a single electron is able to generate more than one photon, thus allowing high power operation. The wide wavelength range in FIR (3 to 30 μm) can be obtained by properly designed QW and barrier layer thickness. The efficient photon generation allows high output power levels in the watt range. These QCLs are applied in environmental sensing and polluting monitoring, process control, automotive, and medical analysis.

Other Semiconductor Lasers

Analog Lasers.   Semiconductor lasers are used as the optical source in fiber optic cable television systems where a large number (up to 200) channels—having independent frequencies in the range between 25 and 301

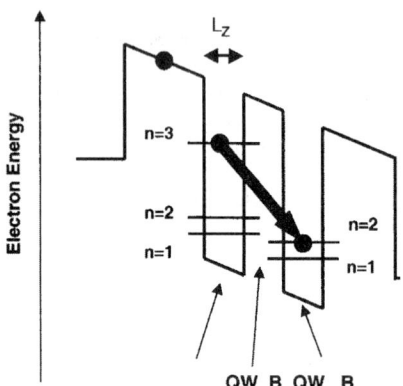

MHz or even to 860 MHz—are multiplexed on the optical carrier and broadcasted. This amplitude-modulated subcarrier lightwave CATV system requires extremely linear transfer characteristic of the laser, i.e., a linear *L-I* curve. If not, intermodulation products result from the modulation of the laser current with the frequency channels. If these intermodulation products have frequencies corresponding to one or more of the TV channel frequencies, distortion of the signals arises. A measure of the distortion is the so-called CSO (composite second order which is the sum of all mix frequencies by f_m and f_k at $|f_m \pm f_k|$) and CTB (composite triple beat; $|f_m \pm f_k \pm f_l|$). CATV lasers are specified by IM2 < −45 dBc and IM3 < −50 dBc. Also low noise in the optical power is important. The relative intensity noise (RIN) should be below −155 dB/Hz in a frequency interval between 100 kHz and 2.5 GHz. CATV lasers are characterized by high optical output power (>20 mW) allowing a higher split ratio in the broadcast mode.

FIGURE 9.2.22   Schematic conduction band energy diagram of a quantum cascade laser with three QWs of width L_z. In the left well the energy levels corresponding to $n = 1$, 2, and 3 is indicated by the horizontal lines. Tunneling of the electron takes place through the intermediate barrier B and an electron transition between $n = 3$ state to the $n = 2$ state of the nearest QW is shown by the arrow.

Tunable Lasers.   Semiconductor tunable lasers have the property that the emission wavelength can be tuned by thermal/mechanical/electronic means. This feature makes tunable lasers very attractive in DWDM applications in fiber optical systems, fiber sensor applications, environmental sensing, and instrumentation. A vast variety of technical solutions have been proposed and introduced in the market that offer wavelength tunability and maintaining the prime characteristics of nontunable lasers such as the DFB laser, i.e., high power, low noise, spectral purity, and small footprint. Also the tuning speed is an important differentiator and ranges between several ns for semiconductor solutions to ms to mechanical and thermal solutions. A good overview of all tunable laser concepts can be found in the textbook of Amann and Buus (1998) on this topic. Tunable semiconductor lasers can be classified into four groups:

Temperature tunable lasers (TTLs).   The wavelength of the diode laser increases with temperature because the material gain is temperature dependent with $d\lambda/dT = 0.5$ nm/K. For DFB lasers with an internal grating the wavelength tunability is much less because only the temperature dependence of the effective mode index has to be taken into account. The $d\lambda/dT$ for DFB lasers is 0.08 nm/K. Using the internal TEC in the laser package, the wavelength can be tuned thermally over typically 3.5 nm.

External cavity lasers (ECL).   This concept uses a FP laser chip of which one of the mirrors is AR-coated and the light from this facet is reflected back by an external grating that can rotate. The grating gives wavelength selective feedback and the achievable tuning range is usually determined by the gain bandwidth of the active layer of the FP laser. A tuning range of 100 nm can easily be obtained. The front mirror and the grating form the laser cavity, which is usually long (centimeters) so many "external cavity modes" are present. Laser linewidth is excellent (<100 kHz). This laser is mainly used in instrumentation applications. Recent developments in microelectromechanical systems (MEMS) have revealed miniature ECLs using a grating. An alternative to the grating is a set of etalons between the AR-coated facet and a retroreflector. If one of the etalons is made variable, then a differential etalon external cavity is obtained.

Multisection lasers.   Here the laser chip is split up into a number of electrically separated longitudinal segments where independent injection currents can be applied. The physical mechanism for tuning the wavelength is the free-carrier effect. The refractive index change Δn caused by injection of carriers is proportional to the carrier density, which in turn has a nonlinear relationship with the tuning current resulting from Auger recombination processes. The maximum achievable Δn is about −0.04, whereas the thermally induced index change is in the order of +0.01. Examples of multisection lasers based on tuning current injection only are 2-section DFB, 2- and 3-section DBR laser and tunable twin guide lasers. Maximum achievable tuning range is determined by $\Delta\lambda/\lambda \approx \Delta n/n$ and is typically 15 nm. For increased tuning ranges, so-called *widely tunable lasers*, use is made of the Vernier effect in sampled grating DBR lasers having a comb-like reflection spectrum from sampled gratings in two sections. The sampled grating consists of several sections of interrupted grating. The SG-DBR has four sections with the active and phase control section in between two sampled Bragg reflectors with slightly shifted periods. Tuning range of 50 nm can be achieved with this laser although with moderate output power because the beam has to propagate the SG reflector. Another structure is the super-structure grating DBR using a chirped grating giving more uniform reflection peaks. Tunable multisection lasers using codirectionally coupled waveguides provide an additional internal mode filter to discriminate between cavity modes. An example of this structure is the grating-assisted codirectional coupler with rear sampled grating reflector (GCSR) laser. This laser achieves over 100 nm tuning range. Multisection tunable lasers require sophisticated control loops to adjust the individual wavelengths and is usually delivered with the device.

Vertical cavity lasers.   If the top mirror of a VCSEL is made movable by applying MEMS technology, the single longitudinal mode that can exist in the small cavity can be tuned in wavelength. Up to 50 nm tuning range can be achieved with a simple control algoritm. However, the output power is low as a consequence of the VCSEL structure.

Pump Lasers.   Semiconductor lasers that can deliver high ouput power into the optical fiber for erbium-doped fiber amplifiers (EDFAs) and RAMAN amplifiers (for C- and L-band) are called pumplasers. Amplifier designs developed before 2001 incorporate the 14xx (1405 to 1520 nm), 1480 and 980 nm pump lasers with power levels of typically less than 200 mW. The 980-nm-based pumplasers are fabricated using InGaAs/AlGaAs/GaAs materials with a wide spot size to reduce catastrophic optical mirror damage because of facet erosion due to the high optical flux. Ridge waveguide (weakly index guided) laser structures are the common laser design. The other pump sources are based on InGaAs/InP material system where COD is no issue and an optimum fibre chip coupling can be achieved by beam design. Index-guided laser structures are used here. Pump lasers for the next generation of optical networks will typically have more than 300 mW of power. In order to increase the power level, hybrid integration of pump modules by, e.g., polarization multiplexing, allow for power levels up to 1 W.

Optical Storage Lasers.   Historically, the first optical disc lasers were made in GaAs/AlGaAs lasers operating at a wavelength of 820 nm. Since the spot size on the disc to read and/or write data is proportional to $\lambda/2NA$ where NA is the numerical aperture of the lens, a higher optical capacity can be obtained by developing semiconductor lasers operating at shorter wavelengths. This has led to weakly guided 780 nm GaAs/AlGaAs-lasers for the CD application and 633 nm lasers based on AlInGaP/InGaP/GaAs materials for DVD. To further increase the density, green lasers have been investigated based on II-VI materials using ZnSe/ZnSSe/GaAs materials but these lasers suffered from reliability problems because of crystal defects. Recent developments of InGaN/GaAlN/GaN on sapphire substrate-based lasers, emitting in blue 450 nm, have led to a capacity of 15 Gbyte of digital data on a single layer. Also based on InGaN materials, semiconductor lasers emitting 405 nm, i.e., blue-violet color have been made for optical data-storage systems such as DVD and future systems.

BIBLIOGRAPHY

Lasers

Cheo, P. K., "Far-infrared lasers for power cable manufacturing," *IEEE Circuits and Devices Magazine*, January 1986.

Dixon, J. A., "Laser Surgery and the Physical Sciences," 1986 Conf. Lasers and Optics.

Dye Lasers Conference Session, 1985 Conf. Lasers and Electro-Optics.

Gill, T. E., AVLIS Laser Data Acquisition and Control System, 1986 Southeastern Symp. System Theory.

Industrial Laser Review Buyers' Guide to Companies and Products Issue, Pennwell, July 1995.

Kujawski, A., and M. Lewenstein (eds.), "Quantum Optics," *Proc. Sixth Int. School of Coherent Optics*, Reidel, 1986.

Laser Focus World Buyers' Guide, Pennwell, 1995.

Laser Focus World Medical Laser Buyers' Guide, Pennwell, 1995.

Rhodes, C. K. (ed.), "Topics in Applied Physics," Vol. 30, Excimer Lasers, 2nd ed., Springer-Verlag, 1984.

Voyles, R. M., "Non-contrast Bent Pin Detection Using Laser Diffraction," 1986 Electronic Components Conference.

Young, M., "Optics and Lasers: Including Fibers and Integrated Topics," Springer, 1984.

Semiconductor Lasers

Agrawal, G. P., and N. K. Dutta, "Long Wavelength Semiconductor Lasers," Van Nostrand Reinhold Company, 1986.

Agrawal, G. P., "Fiber Optic Communication Systems," Wiley, 1997.

Amann, M. C., and J. Buus, "Tunable Laser Diodes," Artech House, 1998.

Botez, D., "Laser diodes are power packed," *IEEE Spectrum*, June 1985.

Chang, K. (ed.), "Handbook of Microwave and Optical Components," Vol. 3, Optical Components, Wiley Interscience, 1990.

Cheo, P. K. (ed.), "Handbook of Solid-State Lasers," Marcel Dekker, 1989.

Gillessen, K., and W. Schairer, "Light Emitting Diodes: An Introduction," Prentice Hall, 1987.

Hecht, J., "The Laser Guidebook," 2nd ed., McGraw-Hill, 1992.

"Laser Diode Operator's Manual and Technical Notes," SDL, 1994.

Peterman, K., "Laser Diode Modulation and Noise," Kluwer Academic Publishers, 1991.

Zeman, J., "Visible diode lasers show performance advantages," *Laser Focus World*, August 1989.

Zory, P. S., "Quantum Well Lasers," Academic Press, 1993.

CHAPTER 9.3
DISPLAY TECHNOLOGY

Munsey E. Crost, Irving Reingold, David A. Bosserman

CATHODE-RAY TUBES

Munsey E. Crost, Irving Reingold

Introduction

Generation of Radiation.   The cathode-ray tube (CRT) produces visible or ultraviolet radiation by bombardment of a thin layer of phosphor material by an energetic beam of electrons. The great preponderance of applications involves the use of a sharply focused electron beam directed time-sequentially toward relevant locations on the phosphor layer by means of externally controlled electrostatic or electromagnetic fields. In addition, the current in the electron beam can be controlled or modulated in response to an externally applied varying electrical signal.

General Design Principles.   The generalized modern CRT consists of an electron-beam-forming system, electron-beam deflecting system, phosphor screen, and evacuated envelope (see Figs. 9.3.1 and 9.3.2).

The electron beam is formed in the electron gun, where it is modulated and focused, and then travels through the deflection region, where it is directed toward a specific spot or sequence of spots on the phosphor screen. At the phosphor screen the electron beam gives up some of the energy of the electrons in producing light or other radiation, some in generating secondary electrons, and the remainder in producing heat.

The magnetic field imparts no kinetic energy to the electron, since it always acts in a direction perpendicular to the velocity.

Electrostatic Deflection.   In electrostatic deflection, metallic deflection plates are used in pairs within the neck of the CRT (see Fig. 9.3.2).

The simplest deflection plates are merely flat rectangular parallel plates facing each other, with the electron beam directed along the central plane between them. The deflection plates are located in the field-free space within the second-anode region, and the plates are essentially at second-anode voltage when no deflection signal is applied. Deflection of the electron beam is accomplished by establishing an electrostatic field between the plates.

The well-made modern electrostatic-deflection CRT does not exhibit excessive deflection defocusing until the beam deflection angle off axis exceeds the neighborhood of about 20°. Most electrostatic-deflection CRTs are used to display electric waveforms as a function of time.

To display the electric waveform it is necessary to generate a sweep representing passage of time and to superimpose on this an orthogonal deflection representing signal amplitude. This is most readily accomplished by the use of two pairs of deflection plates. The second pair of deflection plates must have an entrance window large enough to accept the maximum deflection of the beam produced by the first pair. This requires that although the plates may be close enough together at the entrance to afford high deflection sensitivity, they must

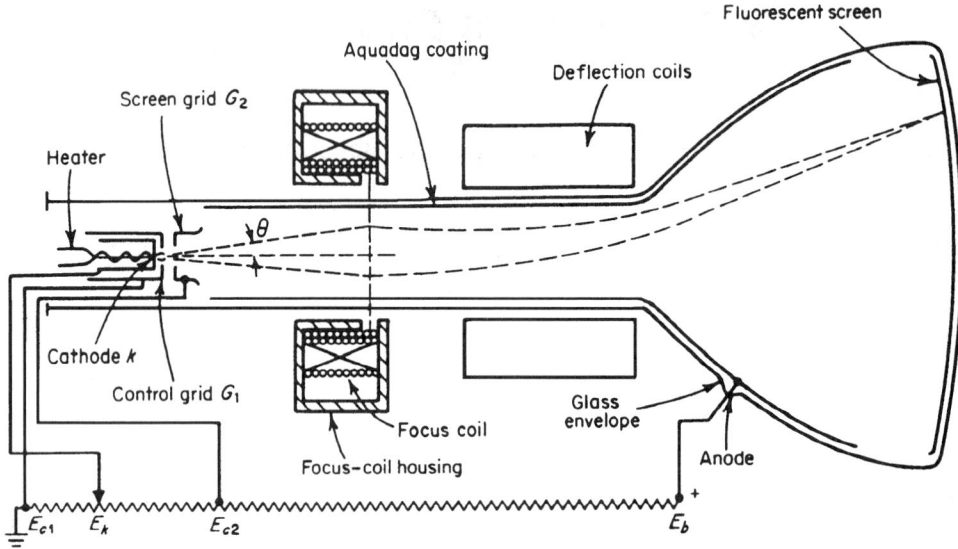

FIGURE 9.3.1   Generalized schematic of a cathode-ray tube with electromagnetic focus and deflection.

also have an appreciable width, which results in high capacitance. The plates must also diverge, to accommodate their own deflection of the beam.

To obtain an acceptable deflection sensitivity, the plates must be made long, and consequently the capacitance is increased.

Electrostatic deflection CRTs are particularly suited for the display of arbitrary waveforms, as opposed to electromagnetic-deflection CRTs, because the deflection plates generally have capacitances with reference to each other and to all other electrodes of the order of 10 pF or less.

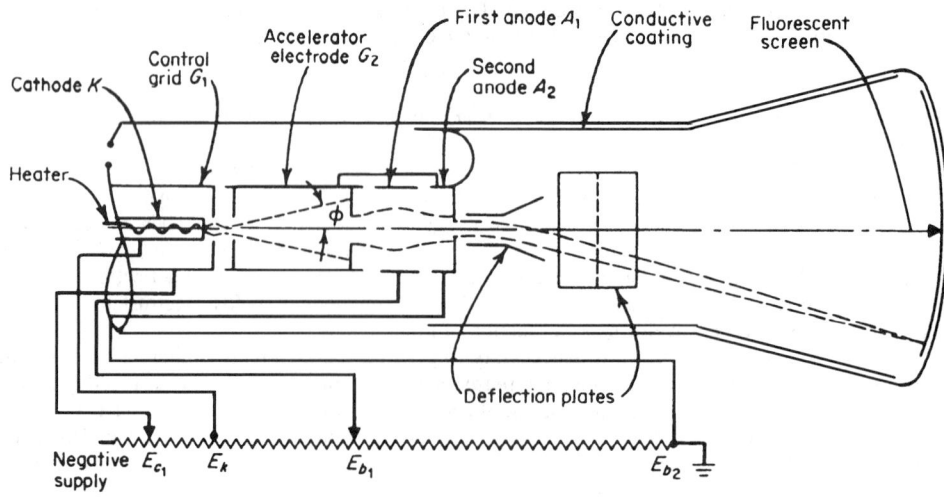

FIGURE 9.3.2   Generalized schematic of a cathode-ray tube with electrostatic focus and deflection.

Electromagnetic-Deflection Systems.   In contrast to electrostatic-deflection systems, the deflection components in electromagnetic-deflection systems are almost universally disposed outside the tube envelope, rather than inside the vacuum. Since the neck of the CRT beyond the electron gun is free of obstructions, a larger-diameter electron beam can be used in the magnetic-deflection CRTs than in the electrostatic-deflection CRTs, which permits a much greater beam current to the phosphor screen and consequently a much brighter picture than if electrostatic deflection were used. In fact, included deflection angles of 110° (55° off axis) are commonly used in television picture tubes without excessive spot defocusing. As is apparent, large deflection angles permit CRTs to be made with shorter bulb sections for any given screen size.

Cathode-Ray-Tube Faceplates.   The CRT envelope consists of the faceplate, bulb, funnel, neck, base press, base, faceplate safety panels, shielding, and potting. Not all CRTs will incorporate each of these components, of course.

The *faceplate* is the most critical component of the envelope, since the display on the phosphor must be viewed through it. Most faceplates are now pressed in molds from molten glass and are trimmed and annealed before further processing. Some specialized CRTs for photographic recording of flying-spot scanning use optical-quality glass faceplates sealed to the bulb section in such a way as to produce minimum distortion.

To minimize the return scattering of ambient light from the white phosphor, many CRT types, especially for television applications, use a neutral-gray-tinted faceplate. While the display information will be attenuated as it makes a single pass through this glass, ambient light will be attenuated both going in and coming out, thus squaring the attenuation ratio and increasing contrast.

Certain specialized CRTs have faceplates made wholly or partially of fiber optics, which may have extraordinary characteristics, such as high ultraviolet transmission. A fiber-optic region in the faceplate permits direct-contact exposure of photographic or other sensitive film without the necessity for external lenses or space for optical projection.

Categories of Cathode-Ray Tubes

Oscilloscope Tubes.   For oscilloscopic applications the general requirements on a CRT include a sharp, bright, rapidly deflectable, single-line trace with a minimum of deflection defocusing or astigmatism. The rapidity of deflection and the fact that arbitrary waveforms must be displayed dictate the use of ES deflection, at least for the vertical direction. For general use, both horizontal and vertical axes employ ES deflection. Since the included deflection angle must be small, usually less than 45°, to preserve good spot size and shape, these CRTs are relatively long compared with the face diameter.

The phosphors generally used for oscilloscope CRTs are P1 (green, medium persistence) or P2 (yellow-green, medium persistence, but with a much longer, low-level "tail" than P1).

Radar Display Tubes.   Except for the *A-scope* radar display, which is essentially the same as an oscilloscope display, most radar displays consist of a two-dimensional coordinate display with beam-intensity modulation. Since the coordinate scans are mathematically regular and at preselected rates, EM deflection is generally used, in as much as this permits greater deflection angles and consequently shorter tubes to be used for a given face diameter.

Especially in filtered radar displays, it is often necessary to include alphanumeric characters, symbols, and vectors in the display along with the radar information. Shaped-beam tubes, such as the Charactron, or a multiple-beam tube, in which one beam is devoted to the tracing of the characters of symbols and the other to the plan-position-indicator (PPI) display, are used for this purpose.

Long-persistence phosphors are generally used in CRTs for radar displays, since it is desirable to be able to see the radar situation in the entire area covered at any given time.

Television Picture Tubes

Monochrome tubes.   Since the standards for television transmission in the United States call for 30 frames of two interlaced fields each per second, producing the effect of 60 pictures per second, which is above the flicker fusion frequency for all light levels, there is no stringent limitation on phosphor persistence for monochrome

TV picture tubes so long as the persistence does not cause picture smearing. The white luminescence used for most applications is achieved by a mixture of phosphors rather than any single component. Several white-luminescing combinations, all designated P4, have been in common use, namely, the all-silicates, the silicate-sulfide mixture, and the all-sulfides.

Color tubes.   Many types of full-color CRTs have been developed for television use, but the shadow-mask tube is in most widespread use. This type of CRT uses a cluster of three electron guns in a wide neck, one gun for each of the colors red, green, and blue. All the guns are aimed at the same point at the center of the shadow mask, which is an iron-alloy grid with an array of perforations in triangular arrangement, generally spaced 0.025 in between centers for entertainment television. For high-resolution studio-monitor or computer-graphic readout monitor applications, color CRTs with shadow-mask aperture spacing as small as 0.012 in center-to-center are now readily available. This triangular arrangement of electron guns and shadow-mask apertures is known as the *delta-gun configuration.* Phosphor dots on the faceplate just beyond the shadow mask are arranged so that after passing through the perforations, the electron beam from each gun can strike only the dots emitting one color.

Because of the close proximity of the phosphor dots to each other and the strict dependence on angle of penetration of the electrons through the apertures to strike phosphor dots of the desired color, close *attention* must be paid to shielding the CRT from extraneous ambient magnetic fields and to degaussing of the shield and shadow mask, which is usually carried out automatically when the equipment is switched on or off. All three beams are deflected simultaneously by a single large-diameter deflection yoke, which is usually permanently bonded to the CRT envelope by the tube manufacturer. The three phosphors together are designated P22, individual phosphors of each color being denoted by the numbers P22R, P22G, and P22B. Most of the present color CRTs are made with rare-earth-element-activated phosphors, because of the superior colors and brightness compared with previously used phosphors.

Two other classes of multicolor CRTs are those with parallel-stripe phosphors and those with voltage-penetration phosphors. In the *parallel-stripe class* of CRTs, such as the *Trinitron,* sets of very fine stripes of red-, green-, and blue-emitting phosphors are deposited in continuous lines repetitively across the faceplate, generally in a vertical orientation. The Trinitron, unlike conventional color CRTs, has a single electron gun that emits three electron beams across a diameter perpendicular to the orientation of the phosphor stripes. This type of gun, also used by some United States CRT manufacturers, is called the *in-line gun.* Each beam is directed to the proper color stripe by means of the internal beam-aiming structure and a slitted *aperture grille.*

The *Lawrence tube,* or *Chromatron,* is another example of the parallel-stripe-phosphor class of color CRT. It employs a single electron beam, and color selection is accomplished solely by control voltages applied between the integrated combs of wires constituting the grille itself.

In the voltage-penetration type of phosphor screen, two or three unstructured layers of phosphors emitting different colors are deposited on each other, sometimes with a nonluminescing, transparent barrier layer between them for better color differentiation. A second important structure consists of individual phosphor grains built up in layers, called *onionskin phosphors.* The core phosphor is generally green-emitting. This is surrounded by a nonemitting layer which in turn is surrounded by an outer red-emitting layer. With both types of phosphor screen, a single electron beam is employed, and the resultant color of the screen is determined by preselected beam-accelerating voltages, which are changed to control the depth of beam penetration into the phosphor layers.

Intermediate colors are produced by the visual combination of the first color, resulting from the penetration of the first layer by the electron beam, with varying intensities of the second color, as more electrons penetrate into the second color-emitting layer. The range of colors producible is thus limited by the hues of the two phosphor layers, which must lie on the same side of the CIE chromaticity diagram and as close to the spectral locus as possible. A major problem associated with voltage-penetration color CRTs is the change in deflection sensitivity and focus of the electron beam as the screen voltage is changed to change the color displayed. This usually dictates operation at only a few preset screen voltages (and therefore colors) where the deflection amplifications and focus voltages are also preset to correspond. Another type of voltage-penetration CRT features a constant-potential mesh grid very close to the phosphor screen to separate the deflection and focusing space from the color-adjusting space. In this type of CRT a maximum residual deflection error of 1 to 2 percent can be automatically compensated for by means of a large weak electron lens formed between electrode bands deposited on the inside surface of the bulb.

Recording Tubes.   Cathode-ray tubes for recording or transcribing information on photographic or otherwise sensitized film are usually of the very-high-resolution (vhr) or ultrahigh resolution (uhr) types. The great majority of these types have nominal faceplate diameters of 4 or 5 in. The spot diameters of the vhr and uhr tubes range from approximately 0.0015 in. down to 0.00033 in.

The displayed information is transferred to the recording medium either by an external focusing optical system or by direct contact with a fiber-optic faceplate, requiring no focusing.

Computer-Terminal Display Tubes.   Cathode-ray tubes for computer display are very similar to tubes used in high-resolution video monitors, but since the display is principally alphanumeric and vector-graphical, the linearity of the beam-modulation characteristics is less important. Well-focused round spots with minimum spot growth or deflection aberrations from the center to the useful edges of the display area are required. High legibility is of primary importance, implying high contrast. White-emitting phosphors are not necessary, so that highly efficient, high-visual-response phosphors emitting in the yellow or green spectral regions are applicable. Most of these CRTs are made with rectangular faceplates.

ELECTROLUMINESCENT DISPLAYS

David A. Bosserman

Background

Destriau discovered intrinsic electroluminescence (EL) in ZnS among other materials. Sylvania's powdered SZnS EL panel dominated EL work in the 1950s and early to mid-1960s, when most of the projects were cancelled or deferred because of life and/or crosstalk-contrast problems. Attempts were made to correct crosstalk and contrast problems by placing addressable electronic components at each pixel site to serve as turn-on or threshold devices and to provide address-state memory, needed for bright displays where the addressing duty cycle is a factor. These attempts ran into manufacturing difficulties concerning the uniformity and reproducibility of the pixel circuit-elements.

Thin-Film Display

In 1972, Sigmatron, working on the thin-film EL (TFEL) [also called light-emitting film (LEF) and AC-TFEL], reported an *XY* matrix with inherent threshold and an optical light-absorbing layer in the otherwise transparent thin-film structure which enhanced the display contrast when viewed in bright ambient illumination. Shortly thereafter, Sharp Corporation in Japan achieved a breakthrough with a panel life of over 10,000 h coupled with a continuous ac excitation brightness of over 3500 cd/m^2 (over 1000 fL). Many laboratories now work on applying the TFEL structure to various types of matrix-addressed displays. Sharp has since reported on a memory phenomena intrinsic to the TFEL structure which can be used to freeze a frame or to increase the brightness of line-at-a-time addressed displays. Rockwell International has reported on a 500-pixel/in TFEL display.

The early successes of TFEL displays are attributable to the thin-film planar sandwich structure, consisting of electrode-dielectric-phosphor-dielectric-electrode, where at least one electrode is transparent. Since the thickness of this structure is less than 1 nm, high field strengths can be achieved with modest voltages across the structure. This high field strength produces hot electrons through tunneling from relatively deep traps. The hot electrons are fired across the ZnS film, exciting the color centers associated with the dopant, usually Mn. This process is relatively temperature-independent, which translates into a wide operational temperature range. The details of the tunneling and subsequent retrapping of activator electrons are being studied in various laboratories to determine the best structure for enhancing the various desirable operational parameters such as threshold, brightness, memory, and uniformity.

The low-cost production processes inherent in the planar TFEL structure make them eminently suitable for the commercial and consumer markets.

BIBLIOGRAPHY

Herold, E., "A history of color TV displays," *Proc. IEEE*, vol. 64, no. 9, September 1976.

Say, D. L., R. A. Hedler, L. L. Maninger, R. A. Momberger, and J. D. Robbins, "CRT display devices," *Standard Handbook of Video and Television Engineering*, 3rd ed. McGraw-Hill, 2000.

CHAPTER 9.4
PHOTOCONDUCTIVE AND SEMICONDUCTOR JUNCTION DETECTORS

William A. Gutierrez

INTRODUCTION

Photoconductors and junction devices constitute an important class of solid-state photodetectors that can operate somewhere within the 0.2- to 2-μm spectral region. These detectors convert electromagnetic energy directly into electric energy via the photoconductivity effect that occurs in semiconductors.

PHOTOCONDUCTORS

Operation

The simplest photoconductor detector is a bar of relatively low conductivity n- or p-type semiconductor (in bulk or thin-film form) with ohmic contacts at its ends (Fig. 9.4.1). The photoconductor varies its electrical resistance in accordance with the light wavelength and intensity it receives. Its operation depends on the photoconductivity that occurs in semiconducting materials. Electrons in bound states in the valence band (intrinsic) or in forbidden-gap levels (extrinsic) absorb the energy of the incident photons and are excited into the free states in the conduction band, where they remain for a characteristic lifetime. Electric conduction may take place either by the electrons in the conduction band or by the positive holes vacated in the valence band. The electrical resistance of the material thus decreases on illumination, and this resistance change can be translated into a change in the current that flows through the output circuit.

Performance

The performance of photoconductor detectors is measured not only in terms of D^* but also in terms of photoconductivity gain, response time, dark current, spectral response, and temperature coefficient. For a photoconductor in which the conductivity is dominated by one carrier (either holes or electrons) the gain is given by the ratio of free-carrier lifetime to the transit time of this carrier. It can also be expressed as

$$\text{Gain} = \tau \mu \, V/L^2$$

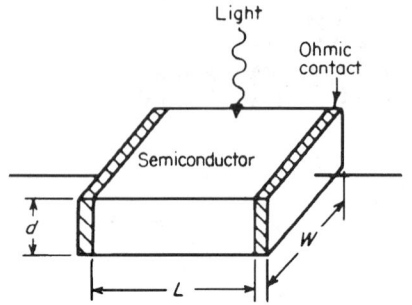

FIGURE 9.4.1   Diagram of a photoconductor detector.

where τ = free-carrier lifetime
μ = mobility
V = applied voltage
L = spacing between ohmic contacts, as shown in Fig. 9.4.1

The maximum D^* and spectral dependence of commercially available CdS and CdSe photodetectors are shown in Fig. 9.4.2. Table 9.4.1 lists typical gains, response times, and dark current.

Properties of Specific Photoconductors.   The long-wavelength threshold for photoconductivity is usually determined by the bandgap of the material according to the relationship

$$\lambda_c = 1.24/Eg$$

where λ_c = threshold wavelength (μm) and Eg = bandgap (eV).

Bandgap values of materials commonly used as photodetectors in the 0.1- to 2-μm region are given in Table 9.4.2. The normalized spectral response of some of them is shown in Fig. 9.4.3.

SEMICONDUCTOR JUNCTION DETECTORS

Semiconductor junction detectors (photodiodes and phototransistors) differ from photoconductive detectors in that their operation depends essentially on a reverse-biased diode whose leakage current is varied by electron-hole pairs generated near or at the depletion region by light absorption. Their response time is characteristically short.

Basic Classes of Photodiode Detectors

Photodiodes fall into two general categories, the *depletion-layer type* and the *avalanche type*. The distinguishing feature between them is the existence of a gain mechanism.

The Depletion-Layer Photodiode.   The depletion-layer photodiode family includes the *pn*-junction diode, the *pin* diode, the Schottky barrier (metal-semiconductor) diode, the point-contact diode, and the heterojunction diode.

1. *pn-junction diode*. Figure 9.4.4 is a diagram of a *pn*-junction diode. The junction is reverse-biased, and the diode is illuminated either at the *n* or *p* region, away from the depletion region (Fig. 9.4.4*a*), or right at the depletion region (Fig. 9.4.4*b*). Their built-in field enables them to be operated in the photovoltaic mode (i.e., no externally applied bias); however, the photoconductive mode, with a fairly large reverse bias, is usually the more common mode of operation.

2. *pin photodiode*. Figure 9.4.5 shows a cross-sectional diagram of a typical *pin* photodiode. The sensitivity range and frequency response of this type of diode depend principally on the thickness of the intrinsic layer (which defines the depletion layer). Light passes through the *p* region before it arrives at the depletion region, where it excites hole-electron pairs that are very quickly swept out by the large electric field present.

3. *Metal-semiconductor photodiode*. Figure 9.4.6 is a cross-sectional diagram of a metal-semiconductor (Schottky barrier) photodiode. In this case, light passes through a thin (~10 nm) metal film with a suitable antireflection coating to minimize large absorption and reflection losses. As with the *pn* and *pin* diodes, the photogenerated electron-hole pairs in the semiconductor give rise to an output-signal current.

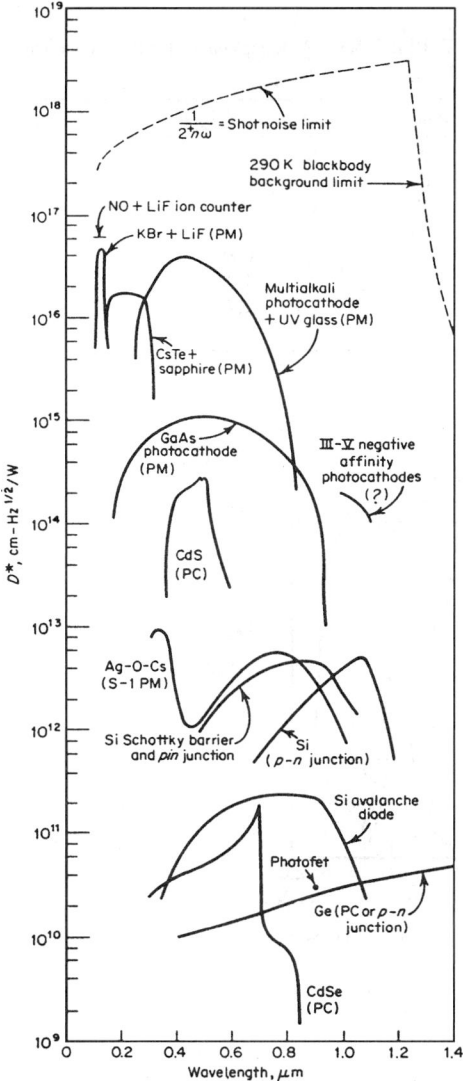

FIGURE 9.4.2 Detectivity vs. wavelength for various photoconductors. PC = photoconductive mode. PM = photomultiplier mode.

4. *Point-contact photodiode*. Figure 9.4.7 shows a diagram of a point-contact detector. Light is incident onto the Schottky barrier through an etched cavity in the semiconductor. This detector is extremely fast because of small dimensions and low capacitance.

5. *Heterojunction photodiode*. A depletion-layer photodiode can be constructed by forming a junction between two semiconductors of different bandgaps. Figure 9.4.8 shows a photodiode made up of n^- GaAs and p^- Ge. Light is absorbed almost completely in the low-bandgap material. Large dark currents could arise owing to spontaneous electron-hole generation in the depletion region from a large density of interface states. This could have deleterious effects on the signal-to-noise ratio at low light levels.

Avalanche Photodiodes. Depletion-layer photodiodes operated at higher-reverse-biased voltages give an increase in output signal. This is because of internal carrier multiplication via the avalanche effect. If the field in the depletion region can impart an energy equal to or greater than the bandgap energy to an electron, this electron can create another hole-electron pair by collision, and this pair can be accelerated to create an additional pair, and so on. This gives rise to carrier multiplication and to internal gain in the photodiode. The avalanche photodiode is therefore the counterpart of the photomultiplier tube, and its multiplication factor M is

$$M = K(1 - V/V_B)^{-1}$$

where K is a constant and V_B is the breakdown voltage.

Figure 9.4.9 shows two types of avalanche photodiodes with guard rings. The guard ring prevents a high-field breakdown region from reaching the surface.

Performance of Photodiodes. The spectral dependence of D^*, gain, speed of response, and dark current of the various photodiodes are shown in Table 9.4.1 and Fig. 9.4.2.

Phototransistors

A *pnp* or *npn* junction transistor can act as a photodetector, with the possibility of large internal gain. An *npn* structure, for example, is usually operated as a two-terminal device with the base floating and the collector positively biased.

Phototransistors are generally fabricated of Ge or Si in the same manner as conventional transistors, except that a lens or window is provided in the transistor to admit light at the base or base-collector junction. Response time of 10^{-8} and peak sensitivities of 30 A/lm are possible. Gains of several hundred have been attained with dark currents as low as nanoamperes.

TABLE 9.4.1 Parameters of Various Photoconductive Detectors

Photodetector	Gain	Response time, s	Dark current
Photoconductor	10^5	10^{-3}	1–10 mA
pn junction	1	10^{-11}	1–10 μA
Metal-semiconductor	1	10^{-11}	
Avalanche diode	10^4	10^{-10}	
Point contact	1	. . .	1–3 mA
Heterojunction photodiode	1	. . .	High
Phototransistor	10^2	10^{-8}	1 nA
Photofet	10^2	10^{-7}	1 μA

TABLE 9.4.2 Bandgap Values for Photoconductor Materials

Material	Threshold λ_c, μm	Bandgap, eV
CdS	0.52	2.4
CdSe	0.73	1.7
ZnS	0.33	3.7
ZnSe	0.48	2.6
GaAs	0.89	1.4
InP	1.03	1.2
Ge	1.77	0.7
Si	1.13	1.1

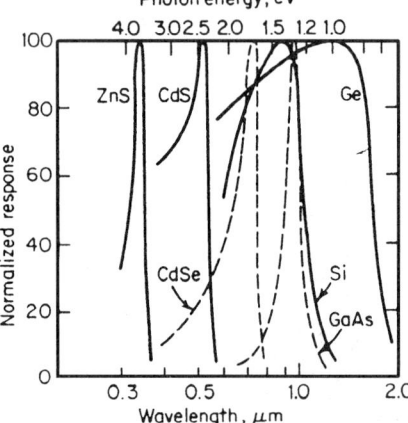

FIGURE 9.4.3 Normalized spectral response of some photoconductors.

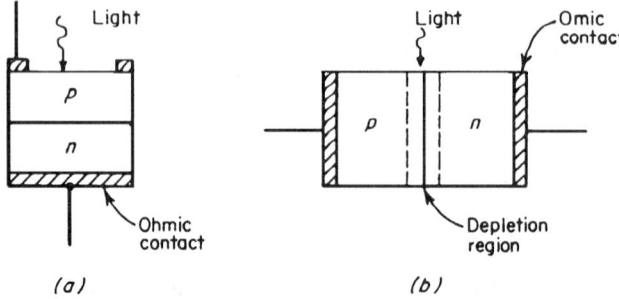

FIGURE 9.4.4 Diagram of a pn photodiode illuminated (a) away from the depletion region or (b) at the depletion region.

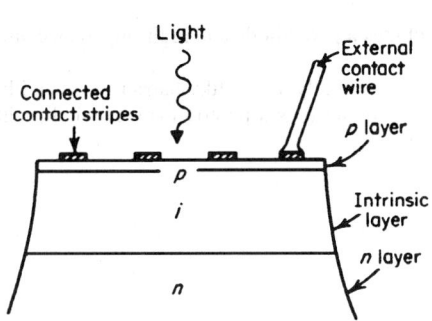

FIGURE 9.4.5   Cross section of a *pin* photodiode.

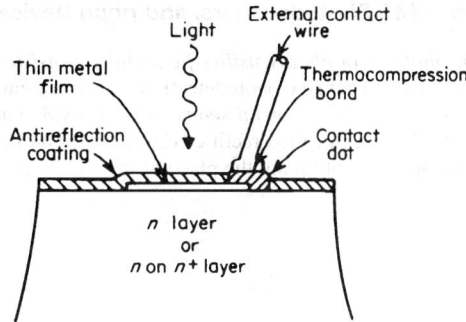

FIGURE 9.4.6   Cross section of a metal-semiconductor photodiode.

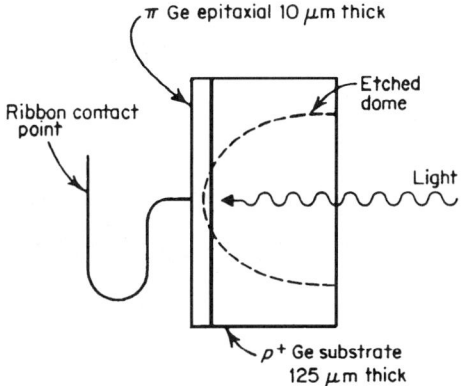

FIGURE 9.4.7   Diagram of a point-contact photodiode.

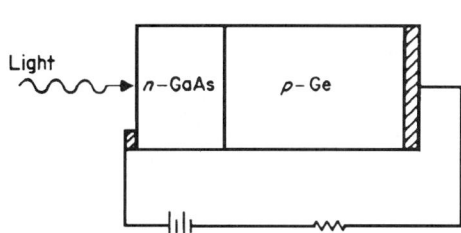

FIGURE 9.4.8   Diagram of a heterojunction photodiode with applied reverse bias.

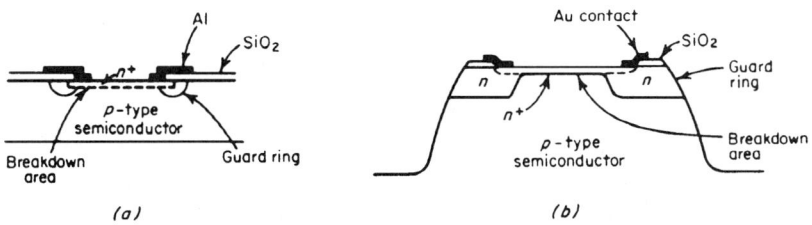

FIGURE 9.4.9   Cross section of avalanche photodiode with guard ring: (*a*) planar type; (*b*) mesa type.

Photofets, SMS Photodetectors, and *pnpn* Devices

The *photofet*, or *photosensitive field-effect transistor*, combines a photodiode and high-impedance amplifier in one device to achieve photodetection with large gain.

A *semiconductor-metal-semiconductor* (SMS) *device* is essentially a Schottky barrier device with gain.

A silicon controlled rectifier (SCR), or *pnpn device*, can be used as a photosensitive switch, with photo-generated current taking the place of the usual gate current.

CHAPTER 9.5
CHARGE TRANSFER DEVICE (CTD) IMAGERS

Larry V. Caldwell, Andrew J. Kennedy, John H. Pollard

MONOLITHIC SOLID-STATE SELF-SCANNED IMAGE SENSORS

Monolithic solid-state self-scanned image sensors of the charge-coupled-device (CCD) and charge-injection-device (CID) types are extremely versatile and can be applied to high- and low-light-level imaging, optical character recognition, and facsimile reproduction. Solid-state devices offer advantages over vidicons for imaging applications: they provide lag-free, burn-free imaging and operate at low power in a self-scanned mode; they are lightweight and have high sensitivity. Advanced metal-oxide-semiconductor (MOS) technology is employed to fabricate the closely spaced single- or multiple-capacitor imaging elements, called *pixels*. Linear or area configuration of the pixels, with the appropriate on-chip scanning circuit and low-noise preamplifier, constitute the focal-plane image sensor in a camera system.

CCD IMAGING

The principle of operation is based on the transfer of photogenerated minority-carrier charge packets to adjacent pixels. Sequentially clocked voltage pulses applied to adjacent pixels create potential wells, which transfer charge packets to adjacent positions across the surface. Each charge packet corresponds to a pixel-sized optical image element. Thus, the moving charge packets and transferred to a low-capacitance output diode that has typical values on the order of 0.2 pF to obtain the video signal. This is one of the major advantages of the CCD imagers and produces high signal-to-noise output signals. Two implementations of CCD imagers are possible, one using charge transport in a surface channel and the other in a buried or bulk channel.

In the surface-channel device (SCCD) the conduction channel is at the semiconductor-oxide interface and allows the signal charge packets to interact with interface states. The resultant interface-state trapping can be reduced by introducing a bias charge 5 to 20 percent of full well, referred to as "fat zero." The performance is limited by temporal noise resulting from interface state trapping and a spatial or fixed pattern noise caused by nonuniform fat-zero insertion.

For a buried-channel device (BCCD) the conduction channel is ion-implanted approximately 0.5 μm into the bulk with respect to the semiconductor-oxide interface. The charge packets interact with low density (of the order of 10^{11} cm^{-3}) bulk states, but the bulk trapping noise is insignificant even for signals as low as approximately 10 electrons per pixel per frame at –20°C. The charge transfer efficiency is also improved over the surface-channel device because of the larger fringing fields between the pixels present in the channel.

CID Imaging

The CID is an *XY* addressed matrix of MOS capacitor pairs in which each capacitor pair constitutes a pixel. Charge transfer occurs only between the capacitor pairs within a pixel, and the photogenerated charge packets are not transferred to a common output. The charge packets are stored by biasing at least one of the capacitors. When the capacitor pair of a pixel is simultaneously pulsed to zero, the potential well collapses and the stored minority carrier charge is injected into the substrate, where most of the carriers recombine. This injected charge provides the video signal. A reverse-biased epitaxial junction in the substrate collects the unrecombined minority carriers to prevent them from diffusing back into the same pixel or into neighboring pixels, which would degrade the resolution.

An alternative low signal-to-noise readout technique uses an output circuit that senses the relative magnitude of the stored charge in a given pixel by transferring the charge between the capacitor pairs. Even with this improvement, the effective output capacitance of a CID is still approximately an order of magnitude higher than that of a CCD. This is a definite disadvantage for low-light-level imaging, but the superior antiblooming control and the random-access features provided by the CID scheme are useful in many other applications.

CHARGE-TRANSFER-DEVICE ARCHITECTURE

Charge-coupled-device area imagers have been fabricated in two main configurations. One uses a frame-storage mode and the other in interline transfer mode. The former has nearly 100 percent optically sensitive imaging area with transparent electrodes or backside illumination but can suffer from optical transfer smearing effects. For certain applications not requiring TV readout rates, e.g., astronomical observations and space telescopes, full frame imagers can be used.

The frame-storage device is illustrated in Fig. 9.5.1. Charge is integrated in the imaging section and then rapidly moved into the frame-storage section for subsequent readout through the serial shift register. In this mode both frontside and backside illumination is possible. Interlaced readout is obtained by imaging under different electrode sets, giving alternate readout of frame A and frame B. The shaded regions are opaque.

The interline transfer device is shown in Fig. 9.5.2. The charge is integrated in the photosensitive areas and is shifted into opaque columns for subsequent transfer to the readout serial shift register. In this mode only frontside illumination is possible. Interlacing is obtained by integration under different electrodes in subsequent frames (*A* and *B* in the figure). Shaded regions are opaque.

In the charge-injection-device area imager (Fig. 9.5.3) a particular column is allowed to float, and the change in potential after a row transfer under that column provides the pixel signal readout. Area imaging is obtained by suitable scanning of the horizontal and vertical registers. The device has the advantage of random access and is also free of optical smearing effects. Transparent electrodes allow about 100 percent optically active areas with a sensitivity that depends on the readout scheme.

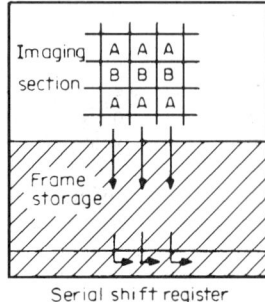

FIGURE 9.5.1 Frame-storage charge transfer device.

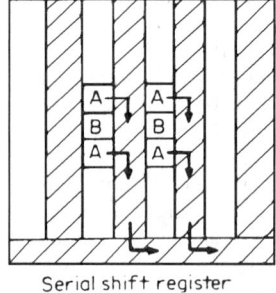

FIGURE 9.5.2 Interline charge transfer device.

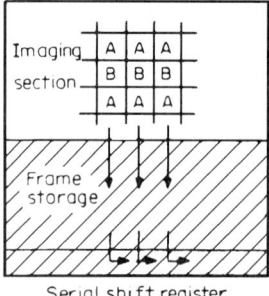

FIGURE 9.5.3 Charge injection area imager.

LOW-LIGHT-LEVEL IMAGING

Image-intensifier technology can be used with charge transfer devices for low-light-level applications, and two methods are available. In the EBS-CCD approach (electron-bombarded silicon-CCD), the phosphor screen, in a proximity-focused and/or inverter-type image intensifier, is replaced by a backside thinned and accumulated silicon CCD. The impinging 8- to 15-kV photoelectrons produce high gains in the approximately 10-μm-thick CCD substrate. The maximum theoretical gain is $V/3.5$, where V is the accelerating voltage of the incident electrons.

In this mode of operation the effective preamplifier noise of a cooled direct-view CCD is reduced by the approximately 1000 to 2500 electron gain in the CCD, and low-light level imaging has been demonstrated under overcast night-sky illuminance levels of 8 μlx. The second method uses image intensifiers, fiber optically coupled directly to either a frontside-illuminated CCD or CID area array. This hybrid approach allows the selection of conventional image intensifiers based on an optimum trade-off between gain, life expectancy, noise figure, and charge-transfer-device noise.

BIBLIOGRAPHY

Buss, D. D., and M. F. Tomposett (eds.), "Special issue on charge-transfer devices," *IEEE Trans. Electron Dev.*, February 1976.

Chamberlin, S. G., and M. Kuhn (eds.), "Joint special issue on optoelectronic devices and circuits," *IEEE Trans. Electron. Dev.*, February 1978.

Hynecek, J., "Virtual phase CCD technology," *Proc. IEDM*, December 1979.

Steckl, A. J., "Charge-coupled devices," Chap. 12 in W. L. Wolfe and G. J. Zissis (eds), "The Infrared Handbook," Government Printing Office, 1978.

CHAPTER 9.6
INFRARED DETECTORS AND ASSOCIATED CRYOGENICS

Wayne T. Grant, Randolph E. Longshore

DETECTORS

Infrared Detectors

Infrared detectors provide an electrical output, which is a useful measure of the incident infrared radiation. It is usually necessary to cool detectors to cryogenic temperatures to reduce the thermal noise inherent in an electrical transducer. Infrared detectors can be divided into two categories, *thermal detectors* and *quantum detectors*.

Thermal Detectors (Table 9.6.1)

Thermal detectors are of three types. The *bolometer* varies its electrical resistance as a result of temperature changes produced by absorbed infrared radiation.

The *thermocouple* is a junction of two dissimilar metals. When the junction is heated, it produces a voltage across the two open leads. A *thermopile* consists of several thermocouples combined in a single responsive element.

The *pyroelectric detector* produces an observable external electric field when heated by infrared radiation.

Quantum Detectors (Table 9.6.2)

Quantum detectors are of three types. A *photovoltaic detector* is a *pn* semiconductor junction. Absorbed infrared photons produce free charge carriers which, if near the junction, are separated by it, producing an external voltage (open circuit) or an electric current (closed circuit).

A *photoconductor* is a semiconductor in which absorbed infrared photons produce free charge carriers, resulting in a change in the electric conductivity.

A *photoemissive detector* is one in which incident photons impart sufficient energy to surface electrons to free them from the detector surface.

Detector Parameters

The parameters most often used in the description of infrared detectors are as follows.

TABLE 9.6.1 Thermal Detectors

Type	Operating temp, K	Detectivity D^*, cm·Hz$^{1/2}$/W	Wavelength region, μm	Response time, ms	Resistance
Pyroelectric	300	1×10^9	>1	1	10 TΩ
Thermistor	300	5×10^8	1–40	0.1	2 MΩ
Golay cell	300	1.5×10^9	1–2000	15	
Thermocouple	300	1×10^9	>1	10	10 Ω
Germanium bolometer	2.0	1×10^{12}	>10	10	100 kΩ
Tin bolometer	3.7	7×10^{10}	>10	10	100 Ω
Carbon bolometer	2.0	4×10^{10}	>10	1	100 kΩ

Responsivity R is the ratio of the rms signal voltage (or current) to the rms incident signal power, referred to an infinite load impedance or to the terminals of the detector, $R = V_{s,\text{rms}}/P_{s,\text{rms}}$. Spectral responsivity R_λ refers to monochromatic input signal, and blackbody responsivity R_{BB} refers to an input signal having a blackbody spectrum. The units of responsivity are volts per watt (or amperes per watt). Responsivity is a function of wavelength λ, signal frequency f, operating temperature T, and bias voltage V_B.

Impedance times area product ZA is a parameter used to characterize a photovoltaic detector. Here A is the area of the detector, and Z is the dynamic impedance taken from the I-V curve of the diode at some operating point; that is, $Z = \delta V/\delta I$. Usually the zero-bias impedance Z_0 is measured to determine the thermal noise and the proper matching to external electric circuits. The unit of ZA is Ω-cm^2.

TABLE 9.6.2 Quantum Detectors

Material	Type*	Operating temp, K	Peak wavelength, μm	Detectivity at peak wavelength, cm·Hz/W	Response time, μs	Resistance (size-dependent)
Si	PV[†]	300	0.9	5.6×10^{12}	10^{-2}	2 kΩ
Si	PV	300	0.9	6×10^{12}	0.2	5 kΩ
Ge	PV	300	1.6	4×10^{11}	1000	
PbS	PC	300	2.4	1×10^{11}	300	1 mΩ
PbS	PC	193	2.8	5×10^{11}	3000	5 MΩ
PbSe	PC	300	3.8	3×10^{10}	2	5 MΩ
PbSe	PC	193	4.8	2×10^{10}	30	50 MΩ
PbSe	PC	77	5.0	2×10^{10}	30	5 MΩ
InAs	PV	300	3.4	7×10^9	1	20 Ω
InAs	PV	195	3.2	1×10^{11}	1	5 kΩ
InAs	PV	77	3.0	7×10^{11}	1	100 kΩ
InSb	PV	77	5.0	1×10^{11}	0.1	1 MΩ
HgCdTe	PC	200	5	5×10^{10}	5	500 Ω
HgCdTe	PC	77	11.5	2×10^{10}	1	40 Ω
HgCdTe	PV	200	4.2	1×10^{11}	0.1	100 kΩ
HgCdTe	PV	77	10.6	2×10^{10}	0.1	1 kΩ
PbSnTe	PV	77	11.5	2×10^{10}	0.1	10 kΩ
Ge:Au	PC	77	5	7×10^9	0.1	300 kΩ
Ge:Hg	PC	28	11	2×10^{10}	0.1	100 kΩ
Ge:Cu	PC	5	25	3×10^{10}	0.1	300 kΩ
Si:Ga		30	15	4×10^{10}	0.1	100 kΩ
Si:In		50	5.6	3×10^{11}	0.1	300 kΩ

*PV = photovoltaic, PC = photoconductor.
[†]Avalanche.

Chopping frequency f_c is the rate at which the blackbody radiation source is mechanically interrupted to provide a strong periodic signal for separating ac components of the instantaneous power from the dc component.

Noise-equivalent power (NEP) is that value of incident rms signal power required to produce an rms signal-to-noise ratio of unity NEP = $V_{n,rms}/R$. Spectral noise-equivalent power (NEP)λ refers to a monochromatic input signal, and blackbody noise-equivalent power (NEP)$_{BB}$ refers to an input signal having a blackbody spectrum. The unit of NEP is the watt, NEP is a function of wavelength λ, detector area A, chopping frequency f_c, electric bandwidth Δf, temperature of blackbody T_{BB}, field of view Ω, and background temperature T_B.

Detectivity D is the reciprocal of the NEP. It can also be expressed as the rms signal-to-noise ratio per unit of rms power incident on the detector. Spectral detectivity D_λ and blackbody detectivity D_{BB} are the reciprocals of NEP and (NEP)$_{BB}$, respectively.

D star (D^*) is a normalization of the reciprocal of the noise-equivalent power to take into account the area and electric-bandwidth dependence, $D^* = \sqrt{A\Delta f}/\text{NEP} = \sqrt{A\Delta f}/D$. By using (NEP)$_{BB}$, D^*_{BB} can be determined, and from the definitions of (NEP)$_{BB}$ and R, $D^*_{BB} = \sqrt{A\Delta f}/V_{s,rms}/V_{n,rms} P_{BB,rms}$, where $P_{BB,rms}$ is the rms power incident on the detector from a blackbody at temperature T_{BB}. Since D^*_{BB} depends on T_{BB}, f_c, and Δf, it is sometimes written as $D^*(T_{BB}, f_c, \Delta f)$. Similarly, D^*_λ depends on λ, f_c, and Δf and is sometimes written as $D^*(\lambda, f_c, \Delta f)$. It can be determined from (NEP)$_\lambda$, $D^*_\lambda = \sqrt{A\Delta f} V_{s,rms}/V_{n,rms} P_{\lambda,rms}$ where $P_{\lambda,rms}$ is the rms power incident on the detector from a monochromatic source.

These figures of merit can be used to compare detectors of different types since the detector with the greater value of D^* is a better detector when the terms in the parentheses are identical. This does not hold for detectors limited by noise resulting from fluctuations in background photon flux. For such detectors, field of view and background temperature must be specified. The units of D^* are cm·Hz$^{1/2}$/W.

Quantum efficiency (QE or η) is the ratio of countable output events to the number of incident photons.

Time constant τ is a measure of the speed of response of a detector. It is usually defined as $\tau = (2\pi f_c)^{-1}$, where f_c is that signal frequency at which the responsivity has fallen to 0.707 times its maximum value.

Noise Mechanisms in Infrared Detectors

Noise mechanisms in infrared detectors are usually of five types.

Photon Noise (Background Noise). Random fluctuations in the arrival rate of background photons incident on the detector produce random fluctuations in its output signal. Photodetectors whose D^* is limited only by this type of noise are called *background-limited infrared photodetectors* (BLIP detectors). This value of D^* is the theoretical limit for a photon detector.

Figure 9.6.1 is a plot of relative improvement in D^* versus angular field of view for a BLIP detector. Cooled spectral filters also improve the D^*_λ of a BLIP or near-BLIP detector by attenuating radiation at wavelengths which are not of interest.

Johnson Noise (Nyquist or Thermal Noise).   The random motion of charge carriers in a resistive element at thermal equilibrium generates a random electric voltage across the element.

Generation-Recombination Noise (GR). Variations in the rate of generation and recombination of charge carriers in the detector create electric noise.

Shot Noise. Since the electric charge is discrete, there is a noise current flowing through the detector as a result of current pulses produced by individual charge carriers.

1/f Noise. The mechanism involved in this type of noise is not well understood. It is characterized by a $1/f^n$ noise-power spectrum, where n varies from 0.8 to 2.

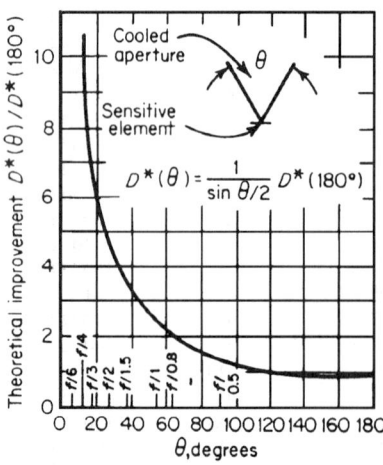

FIGURE 9.6.1   Relative increase in D^* for BLIP detectors obtained by using cold shielding.

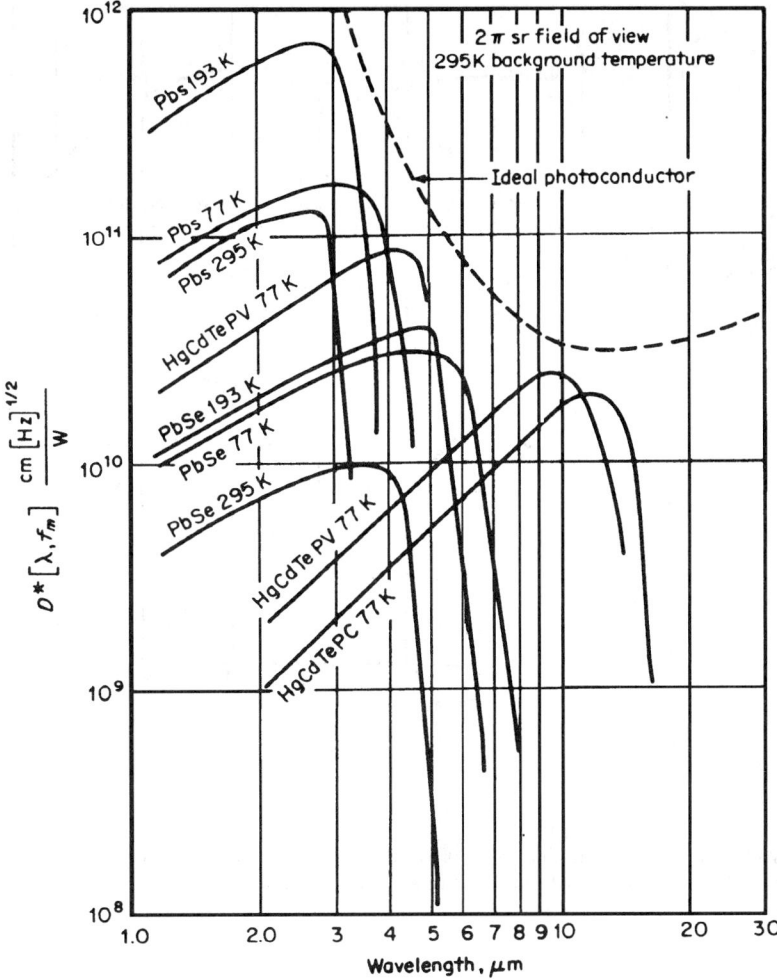

FIGURE 9.6.2   Spectral D^* of photodetectors.

Detector Data

Tables 9.6.1 and 9.6.2 list commercially available detectors. They include the operating temperature, cutoff wavelength, and peak D^*.

Figure 9.6.2 is a plot of D^*_λ versus wavelength for selected detectors. Since detector noise is a function of operating temperature, D^* will also vary with temperature. Figure 9.6.3 is a plot of peak D^* versus temperature for selected detectors. All values tabulated are data from above-average single-element detectors.

Charge Transfer Devices for Infrared Detection

In applications requiring many infrared detectors, e.g., thermal imaging, the detectors and signal processors have been integrated into a single structure. The signal processor provides rapid and efficient readout of the

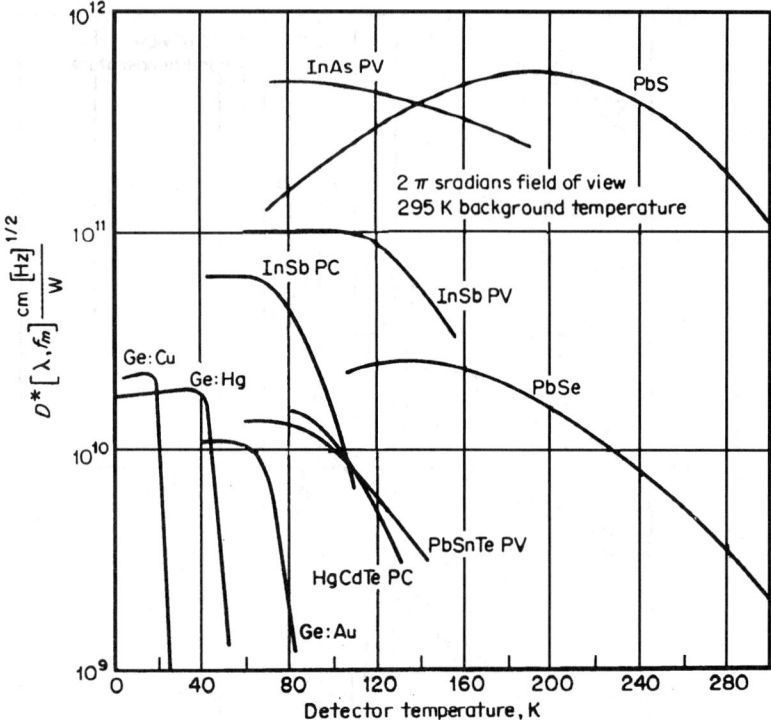

FIGURE 9.6.3 Peak D^* vs. temperature for selected detectors. PC = photoconductive mode, PV = photovoltaic mode.

many detected signals by using charge-coupled devices (CCDs) or charge-injection devices (CIDs). These structures have been fabricated with InSb, extrinsic silicon, and HgCdTe.

Detector Selection

Though no absolute guidelines for the choice of an infrared detector for a specific application are possible, certain criteria can be used to eliminate many of the available detectors. Among them are spectral region of interest, maximum signal frequency, required sensitivity, and available cooling.

COOLERS

Cryogenic Cooling

For BLIP performance, temperatures of 200 K or lower are required for intrinsic infrared detectors, and extrinsic detectors require temperatures of 80 K or less (Table 9.6.2).

The four basic types of cooling systems are:

1. *Open-cycle, expendable systems*, which operate by the transfer of stored cryogens
2. *Mechanical coolers*, which are closed-cycle, expansion engines providing cooling at low temperatures and rejecting heat at high temperatures

3. *Thermoelectric coolers*, which operate by using the Peltier cooling effect
4. *Radiation coolers*, which are passive and cool detectors by radiating heat to the low-temperature deep-space environment.

Cooling Specifications

The particular application and detector type determine the best-suited cooling system. Variables used for cooler specifications include cooling capacity, cooling temperature, cooldown time, temperature stability, reliability, duty cycle, environment, weight, configuration, noise (acoustical, electromagnetic, mechanical), and power (ac, dc, limits).

Open-Cycle Expendable Systems

Open-cycle systems include the use of high-pressure gas combined with a Joule-Thomson (J-T) expansion valve, cryogenic liquids, or solid cryogenics.

Joule-Thomson.   The J-T cooling process exploits the cooling effect obtained by the adiabatic expansion of a high-pressure (1000 to 6000 lb/in.^{2}) gas through an orifice. The cooled, expanded gas is used to cool incoming gas. This process of regenerative cooling continues until liquid forms at the orifice. Figure 9.6.4 shows a typical J-T cooler. Self-regulating throttle valves can be placed in the orifice of the J-T cooler to maintain constant temperature, reduce clogging, and improve gas economy. Some gases, e.g., helium, hydrogen, and neon, require precooling before the J-T cooling effect can occur.

Liquid-Cryogen Storage.   Cryogens can be stored as liquids in equilibrium with their vapors (subcritical) or as homogeneous fluids at high pressure and temperatures (supercritical). There are two basic types of storage, direct-contact and liquid-feed. In the direct-contact (or integral) system, the detector is built into a cryogenic dewar which stores the liquid cryogen. In the liquid-feed method, the cryogen is fed to the detector from a remote storage tank. The liquid is transferred by gravity or by gas pressure resulting from the natural pressure buildup in the storage tank.

Solid-Cryogen Storage.   Stored solid cryogen sublimes, causing an increase in pressure and temperature. The detector temperature and the dewar pressure are maintained at constant levels by venting the gas through specially designed ducts. Such coolers have operated successfully in space for a year or more.

Mechanical Coolers

Closed-cycle, expansion-engine refrigerators (described below) are used to cool infrared detectors. Cooling is produced by the expansion of a gas from a high pressure to a low pressure, with consequent reduction of working gas temperature. The figure of merit for coolers is the coefficient of performance (COP), the ratio of the produced cooling power to the power supplied. The Carnot cycle is used as a standard of comparison because for given temperature limits its COP is maximum. For a Carnot engine operating between the temperatures T_a and T_c, COP = $T_c/(T_a - T_c)$, where heat is absorbed at T_c and heat is rejected at a higher temperature T_a.

Stirling.   In the Stirling cycle, cooling is obtained by cyclic out-of-phase motion of a compression piston and a displacer-regenerator. The working gas is compressed while occupying the ambient space, at temperature T_a, by an upward motion of the compression piston, reducing the gas volume. Heat of compression is rejected to ambient. The COP for an ideal working gas is equal to that of the Carnot engine. This refrigeration cycle is well developed. Stirling cycle refrigerators have the best COP in practice (10^{-3} to 5×10^{-2}) and the best ratio of total weight per watt refrigeration compared with other refrigerators.

Vuilleumier (VM).   This is a heat-driven cycle which is exploited for its long life and low vibration, owing in part to inherently very low dynamic forces on moving parts. Coolers have been built that provide

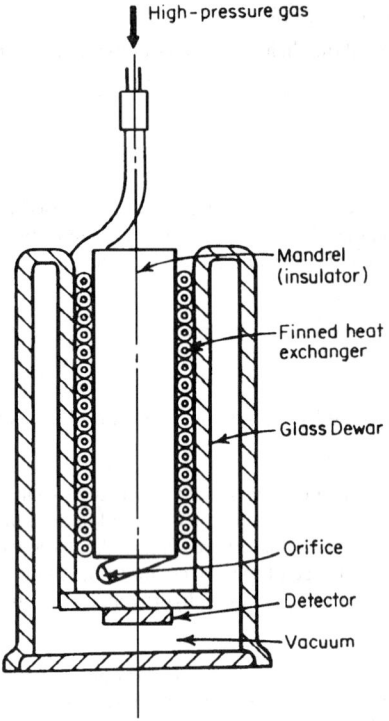

High-pressure gas

Mandrel (insulator)

Finned heat exchanger

Glass Dewar

Orifice

Detector

Vacuum

FIGURE 9.6.4 Single-stage, open-cycle Joule-Thomson cooling system.

refrigeration at 10 K. Cooling is obtained by cyclic out-of-phase motion of two displacer-regenerators. The working gas throughout the entire cooler is compressed by downward motion of the hot displacer-regenerator as it transfers part of the gas at the ambient end into the heated end. COP is equal to that for two Carnot heat engines in series: COP = $(T_c/T_h)(T_h - T_a)/(T_a - T_c)$. Typical COP values of 3×10^{-4} to 2×10^{-2} are obtained.

Gifford-McMahon (GM) and Solvay. By separating the expander from the compressor, a refrigeration system can be constructed that consists of a simple, lightweight cooling unit and a compressor which can be located remotely, connected to the expander with pressure lines. A piston displacer-regenerator is pneumatically moved up and down by timed valving of the high and low working-gas pressure. Cyclic charging and discharging of the expander working-gas pressures with time piston motion will pump heat from the cold to the ambient end. Heat pumped from the cold end using the GM cycle is rejected at the ambient end of the expander. Heat pumped from the cold end using the modified Solvay cycle is rejected along pressure lines and at the compressor. For these coolers COP values of 2×10^{-4} to 10^{-2} are measured.

J-T Closed Cycle. In this system a compressor is used to supply high-pressure gas to the J-T throttling valve. After expansion, the gas is recycled through the compressor. COP values of 3×10^{-3} to 10^{-2} are obtained for this system.

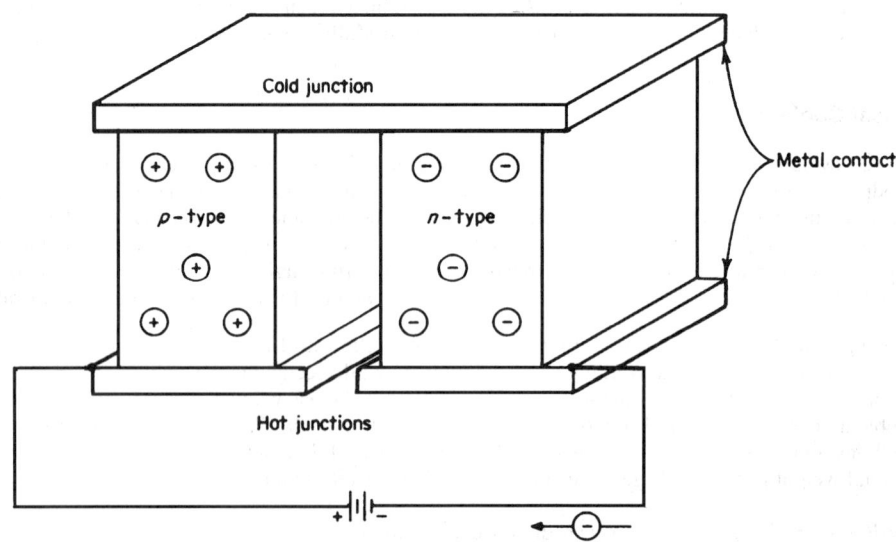

FIGURE 9.6.5 Single-stage thermoelectric refrigeration.

Thermoelectric (TE) Cooler

The basic operating principle of the thermoelectric cooler, the Peltier cooling effect, is the absorption or generation of heat as a current passes through a junction of two dissimilar materials (Fig. 9.6.5). Electrons passing across the junction absorb or give up an amount of energy equal to the transport energy and the energy difference between the dissimilar-materials conduction bands. Cryogenic temperatures are reached using heat rejected from one thermoelectric cooler stage to supply thermal input to the stage below. The maximum temperature difference attainable on a practical basis is about 150°C, which implies a minimum attainable temperature of approximately 150 K.

Radiation Coolers

Radiation coolers are used in spaceborne applications. These systems are passive and cool by radiating heat from the detector into the low-temperature (4-K) sink of deep space. The radiators consist of a suitably sized cold plate of high emissivity connected to the detectors. The high-vacuum in-space environment minimizes convective heating, but the radiator must be shielded from sunlight and (for near-earth orbits) from thermal emission and reflected sunlight from the earth and its atmosphere. Radiation coolers have been designed for cooling milliwatt-level loads at 85 K and 5-W loads at 135 K.

BIBLIOGRAPHY

Miller, J. L., "Principles of Infrared Technology," Van Nostrand Reinhold, 1994.

CHAPTER 9.7
SOLAR CELLS*

Kim W. Mitchell

INTRODUCTION

A solar cell is a semiconductor electric-junction device that absorbs the radiant energy of sunlight and converts it directly and efficiently into electric energy. Solar cells may be used individually as light detectors, e.g., in cameras, or connected in series and parallel to obtain the required values of current and voltage for electric-power generation.

Most solar cells are made from single-crystal silicon and have been to expensive for generating electricity, except for space satellites and remote areas where low-cost conventional power sources are unavailable. Recent research has emphasized lowering solar-cell cost by improving performance and by reducing materials and manufacturing costs. One approach is to use optical concentrators such as mirrors or fresnel lenses to focus the sunlight onto solar cells of smaller area. Other approaches replace the high-cost single-crystal silicon with thin films of amorphous or polycrystalline silicon, gallium arsenide, cadmium sulfide, or other compounds.

SOLAR RADIATION

The intensity and quality of sunlight is dramatically different outside the earth's atmosphere and on the surface of the earth, as shown in Fig. 9.7.1. The number of photons at each energy is reduced on entering the earth's atmosphere because of reflection, scattering, or absorption by water vapor and other gases. Thus, while the solar energy at normal incidence outside the earth's atmosphere is 1.36 kW/m^2, on the surface of the earth at noontime on a clear day the intensity is about 1 kW/m^2.

PRINCIPLES OF OPERATION

The conversion of sunlight into electric energy in a solar cell involves three major processes: (1) absorption of the sunlight in the semiconductor material; (2) generation and separation of free positive and negative charges which move to different regions of the solar cell, creating a voltage in it; and (3) transfer of these separated charges through electric terminals to the outside application in the form of electric current.

In the first step, the absorption of sunlight by a solar cell depends on the intensity and quality of the sunlight, the amount of light reflected from the front surface of the solar cell, the semiconductor bandgap energy,

---

*By permission of the copyright owners, this section is reprinted, with minor revisions, from the fifth edition of the "McGraw-Hill Encyclopedia of Science and Technology." Sybil P. Parker, Editor-in-Chief.

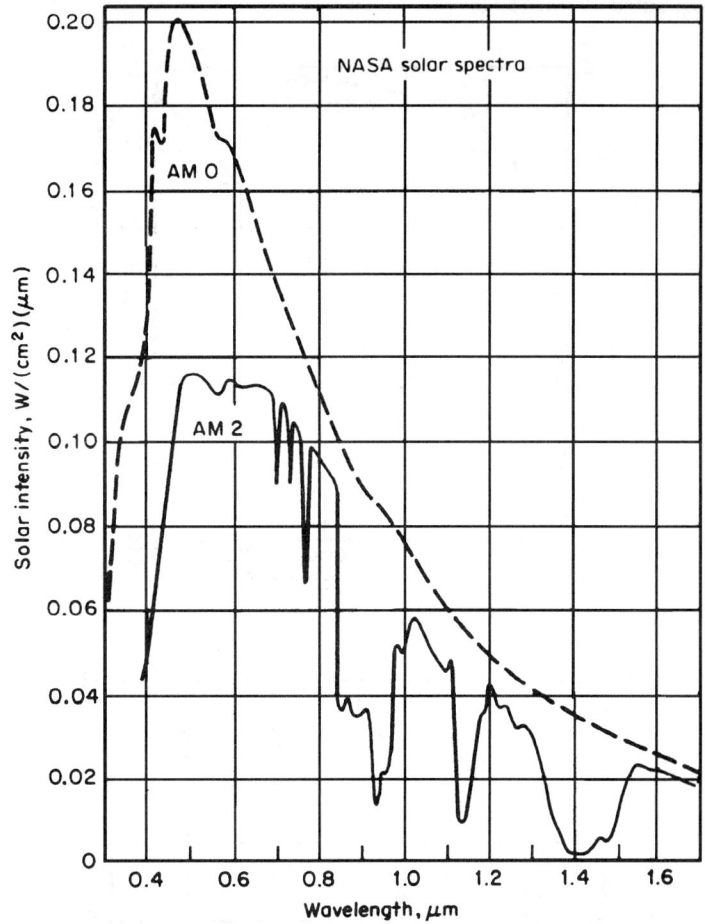

FIGURE 9.7.1 Variation of solar intensity with wavelength of photons for air mass O (AMO) outside the earth's atmosphere and for AM2, a typical spectrum on the surface of the earth.

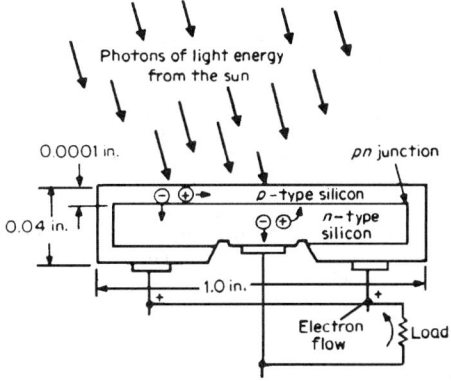

FIGURE 9.7.2 Cross-sectional view of a silicon *pn*-junction solar cell, illustrating the creation of electron pairs by photons of energy from the sun.

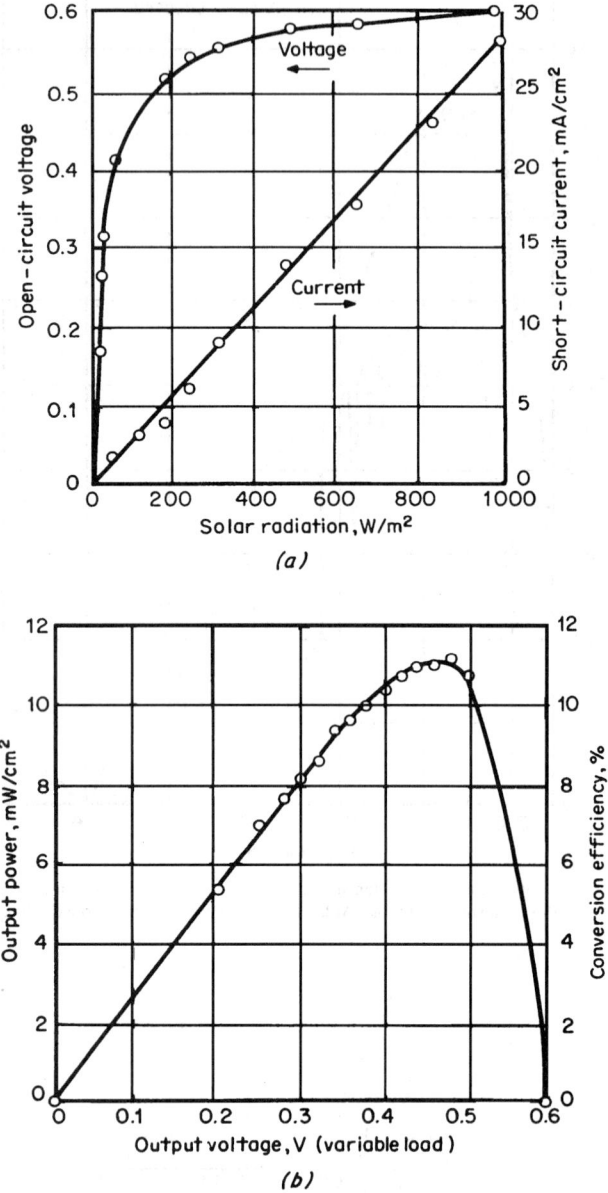

FIGURE 9.7.3 Electrical characteristics of silicon *pn*-junction solar cell at operating temperature of 17°C; (*a*) variation of open-circuit voltage and short-circuit current with light intensity; (*b*) variation in power output as load is varied from short to open circuit.

which is the minimum light (photon) energy the material absorbs, and the layer thickness. Some materials, e.g., silicon, require tens of micrometers thickness to absorb most of the sunlight, while others, e.g., gallium arsenide, cadmium telluride, and copper sulfide, require only a few micrometers.

When light is absorbed in the semiconductor, a negatively charged electron and positively charged hole are created. The heart of the solar cell is the electric junction which separates these electrons and holes from each other after they are created by the light. An electric junction can be formed by the contact of (1) a metal to a semiconductor (a Schottky barrier); (2) a liquid to a semiconductor (a photoelectrochemical cell); or (3) two semiconductor regions (a *pn* junction).

The fundamental principles of the electric junction can be illustrated with the silicon *pn* junction. Pure silicon, to which a trace amount of a column V element such as phosphorus has been added, is an *n*-type semiconductor where electric current is carried by free electrons. Each phosphorus atom contributes one free electron, leaving behind the phosphorus atom bound to the crystal structure with a unit positive charge. Similarly, pure silicon to which a trace amount of a column III element such as boron has been added is a *p*-type semiconductor, where the electric current is carried by free holes. Each boron atom contributes one hole, leaving behind the boron atom with a unit negative charge. The interface between the *p*- and *n*-type silicon is called the *pn* junction. The fixed charges at the interface owing to the bound boron and phosphorus atoms create a permanent-dipole charge layer with a high electric field. When photons of light energy from the sun produce electron-hole pairs near the junction, the built-in electric field forces the holes to the *p* side and the electrons to the *n* side, as illustrated in Fig. 9.7.2. This displacement of free charges results in a voltage difference between the two regions of the crystal, the *p* region being plus and the *n* region minus. When a load is connected at the terminals, an electron current flows in the direction shown by the arrow and useful electric power is available at the load.

SOLAR-CELL CHARACTERISTICS

The electrical characteristics of a typical silicon *pn*-junction solar cell are shown in Fig. 9.7.3. Figure 9.7.3*a* shows open-circuit voltage and short-circuit current as a function of light intensity from total darkness to full sunlight (1000 W/m^2). The short-circuit current is directly proportional to light intensity and amounts to 28 mA/cm^2 at full sunlight. The open-circuit voltage rises sharply under weak light and saturates at about 0.6 V for radiation between 200 and 1000 W/cm^2. The variation in power output from the solar cell irradiated by full sunlight as its load is varied from short circuit to open circuit is shown in Fig. 9.7.3*b*. The maximum power output is about 11 mW/cm^2 at an output voltage of 0.45 V.

Under these operating conditions, the overall conversion efficiency from solar to electric energy is 11 percent. The output power as well as the output current is proportional to the irradiated surface area, whereas the output voltage can be increased by connecting cells in series, as in a chemical storage battery. Experimental samples of silicon solar cells have been produced which operate at efficiencies up to 18 percent, but commercial cell efficiency is around 10 to 12 percent under normal operating conditions.

Using optical concentration to intensify the light incident on the solar cell, efficiencies above 20 percent have been achieved with silicon cells and above 25 percent with gallium arsenide cells. New concepts to split the solar spectrum and illuminate two optimized solar cells of different bandgaps have achieved efficiencies above 28 percent with expected efficiencies of 35 percent. Thin-film solar cells currently between 4 and 9 percent efficiency are expected in low-cost arrays to be above 10 percent.

BIBLIOGRAPHY

Solar Cells/Photovoltaics

Kreider, J. F., and F. Kreith (eds.), "Solar Energy Handbook," McGraw-Hill, 1981.

Maycock, P. D., and E. N. Stirewalt, "A Guide to the Photovoltaic Revolution," IES, 1985.

Partain, L. D. (ed.), "Solar Cells and Their Applications," Wiley, 1995.

P · A · R · T
3

CIRCUITS AND FUNCTIONS

FILTERS AND ATTENUATORS

To make communication systems, radar systems, and the like work, we need filters and attenuators. Filters are basic electronic building blocks that are passive and active, analog and digital. They basically allow us to condition electrical signals in order to accomplish the elements of most of our complex electrical systems in use today.

Active filters have led to the elimination of elements that prohibited miniaturization. Active filters easily fit into today's microcircuits and systems.

Digital filters represent a class of filters that do not have the limitations of analog filters. Although they require much care and attention in how they are designed and how they will be used, they most certainly have given us a significantly larger set of applications some of which could never be handled by analog filters. Because of the nature of digital filters, we need to fully understand the problems with phase.

In this section we look at the basic principles behind all of these filters. In Chap. 10.7, we look at attenuators that are used in matching impedances, critical in high-frequency systems.   C.A.

In This Section:

Section References and Bibliography:

1. Ruston, H., and J. Bordogna, "Electric Networks: Functions, Filters, Analysis," McGraw-Hill, 1966.

2. Darlington, S., "Synthesis of reactance 4-poles which produce prescribed insertion loss characteristics, including special applications to filter design," *J. Math. Phys.*, 1939, Vol. 18, pp. 257–353.

3. Saal, R., and E. Ulbrich, "On the design of filters by synthesis," *IRE Trans. Circuit Theory*, 1958, Vol. CT-5, pp. 284–327.

4. Papoulis, A., "A new class of filters," *Proc. IRE*, 1958, Vol. 46, pp. 649–653.

5. Thomson, W. E., "Delay networks having maximally flat frequency characteristics," *Proc. IEEE*, Pt. 3, November 1949, Vol. 96, pp. 487–490.

6. Zverev, A. I., "Handbook of Filter Synthesis," Wiley, 1967.

7. Weinberg, L., "Network Analysis and Synthesis," McGraw-Hill, 1962.

8. Humphreys, D. S., "The Analysis, Design, and Synthesis of Electric Filters," Prentice Hall, 1970.

9. Cauer, W., "Synthesis of Linear Communication Networks," McGraw-Hill, 1958.

10. Van Valkenburg, M. E., "Introduction to Modern Network Synthesis," Wiley, 1960.

11. Blinchikoff, H. J., and A. I. Zverev, "Filtering in the Time and Frequency Domains," Wiley, 1976.

12. Huelsman, L. P., "Active and Passive Analog Filter Design," McGraw-Hill, 1993.

13. Chen, W. K., "Passive and Active Filters: Theory and Implementations," Wiley, 1986.

14. Schaumann, R., Ghausi, M. S., and K. R. Laker, "Design of Analog Filters," Prentice Hall, 1990.

15. Tsividis, Y. P., and J. O. Voorman (ed.), "Integrated Continuous-Time Filters: Principles, Design, and Applications," IEEE Press, 1993.

16. Van Valkenburg, M. E., "Analog Filter Design," Holt, Rinehart, and Winston, 1982.

17. Lindquist, C. S., "Active Network Design," Steward and Sons, 1977.

18. Horowitz, P., and W. Hill, "The Art of Electronics," Cambridge University Press, 1980.

19. Geiger, R. L., Allen, P. E., and N. R. Strader, "VLSI Design Techniques for Analog and Digital Circuits," McGraw-Hill, 1988.

20. Mitra, S. K. (ed.), "Active Inductorless Filters," IEEE Press, 1971.

21. Moschytz, G. S., and P. Horn, "Active Filter Design Handbook," Wiley, 1981.

22. Orchard, H. J., "The roots of the maximally flat delay polynomials," *IEEE Trans. Circuit Theory*, 1965, Vol CT-12, pp. 452–454.

23. Van Valkenburg, M. E. (ed.), "Circuit Theory: Foundations and Classical Contributions," Dowden, Hutchinson, and Ross, 1974.

24. Tellegen, B. D. H., "The gyrator, a new electric network element," *Philips Res. Rep.* April 1948, Vol. 3, pp. 81–101.

25. Bruton, L. T., "Network transfer functions using the concept of frequency-dependent negative resistance," *IEEE Trans. Circuit Theory*, 1969, Vol. CT-16, pp. 406–408.

26. Friend, J. J., Harris, C. A., and D. Hilberman, "STAR: An active biquadratic filter section," *IEEE Trans. Circuits Syst.*, 1975, Vol. CAS-22, pp. 115–121.

27. Sheahan, D. F., and R. A. Johnson, "Modern Crystal and Mechanical Filters," IEEE-Wiley, 1979.

28. Girlin, F. E. J., and E. F. Good, "Active Filters 12–15: The leapfrog of active ladder synthesis, applications of active ladder synthesis, bandpass types, and simulated inductance," *Wireless World*. July–November, 1970.

29. Belevitch, V., "Summary of the history of circuit theory," *Proc. IEEE*, 1962, Vol. 50, pp. 848–855.

30. Chen, W. K., and T. Chaisrakeo, "Explicit formulas for the synthesis of optimum bandpass Butterworth and Chebyshev impedance matching networks," *IEEE Trans. Circuits and Systems* 1980, Vol. CAS-27, pp. 928–942.

31. Saal, R., "Handbook of Filter Design," AEG-Telefunken, 1979.

32. Stephenson, F. W., "RC Active Filter Design Handbook," Wiley, 1985.

33. Biey, M., and A. Premoli, "Cauer and MCPER Functions for Low-Q Filter Design," Georgi Publishing Company, 1980.

34. Biey, M., and A. Premoli, "Tables for Active Filter Design, Based on Cauer MCPER Functions," Artech House, 1985.

35. Christian, E., and E. Eisenmann, "Filter Design Tables and Graphs," Wiley, 1966. Also available from Transmission Networks, International, 1977.

36. Craig, J. W., "Design of Lossy Filters," MIT Press, 1970.

37. Genesio, R., A. Laurentini, V. Mauro, and A. R. Meo, "Butterworth and Chebyshev Digital Filters," Elsevier, 1973.

38. Hansell, G. E., "Filter Design and Evaluation," Van Nostrand Reinhold, 1969.

39. Johnson, D. E., J. R. Johnson, and H. P. Moore, "A Handbook of Active Filters," Prentice Hall, 1980.

40. Wetherhold, E., "Simplified passive LC filter design for the EMC engineer," *IEEE Int. Symp. Electromagnetic Compatibility, Conf. Proc. 1985*. IEEE, 1985.

41. Wetherhold, E., "Practical LC filter design," Parts 1–6, *Practical Wireless*, July–December 1984, and January, 1985.

42. Williams, A. B., "Electronic Filter Design Handbook," McGraw-Hill, 1981.

43. Temes, J. C., and J. W. LaPatra, "Introduction to Circuit Synthesis and Design," McGraw-Hill, 1977.

44. Daryanani, G., "Principles of Active Network Synthesis and Design," Wiley, 1976.

45. Mitra, S. K., "Analysis and Synthesis of Linear Active Networks," Wiley, 1969.

46. Oppenheim, A. V., and R. Schafer, "Digital Signal Processing," Prentice Hall, 1975.

47. Harris, J. F., "On the use of windows for harmonic analysis with discrete Fourier transforms," *Proc. IEEE*, January 1978.

48. McClellan, J. H., T. W. Parks, and L. R. Rabiner, "A computer program for designing optimum FIR linear phase filters," *IEEE Trans. Audio Electroacoustics*, December 1973.

49. Taylor, F. J., "Digital Filter Design Handbook," Marcel Dekker, 1983.

50. Taylor, F. J., and T. Stouraitis, "Digital Filter Design Software for the IBM PC," Marcel Dekker, 1987.

51. Zelniker, G., and F. Taylor, "Advanced Digital Signal Processing: Theory Applications," Marcel Dekker, 1994.

52. Williams, A. B., and F. J. Taylor, "Electronic Filter Design Handbooks," 3rd ed., McGraw-Hill, 1995.

CHAPTER 10.1
IDEAL FILTERS AND APPROXIMATIONS

Edwin C. Jones, Jr., Harry W. Hale

INTRODUCTION

Campbell and Wagner independently developed the concept of an electrical wave filter in 1915. The continuation of this work proceeded along two paths, *image-parameter filter design* and *insertion-loss filter design*. The latter technique is now dominant.

Insertion-loss filter design requires the designer to specify an appropriate network response as a part of a transfer function. The part might be magnitude, phase, or delay. Norton, Foster, Cauer, Bode, and Darlington developed procedures to determine the complete transfer function and to synthesize the network. Digital computers have made the techniques practical, and now the design of these filters is largely a matter of looking up element values in computer-generated tables and modifying them in a routine fashion. Virtually all techniques use a low-pass prototype and derive other filters from this prototype.

This chapter describes insertion loss techniques, develops common transfer functions, and gives structures and element values for passive networks. It also extends use of the transfer functions to several types of active networks.

Throughout, the notation is that of the Laplace transformation. The general frequency variable is $s = \sigma + j\omega$, where ω is the signal frequency in rad/s.

INSERTION-LOSS FILTER DESIGN

Figure 10.1.1 shows two general networks, often called two-ports. One has a current source, the other, a voltage source. The design procedure consists of these steps:

1. Determine filter specifications
2. Approximate the specifications with a network function part that will lead to a realizable circuit. Typical magnitude functions are

$$|G_{LS}(j\omega)| = \left|\frac{V_L}{V_S}(j\omega)\right| \tag{1}$$

Chapters 10.1 to 10.5 and Chap. 10.7 were contributed by Edwin C. Jones Jr. and Harry W. Hale. Portions were adapted from Fink and Beaty (eds.), "Standard Handbook for Electrical Engineers," 13th ed., McGraw-Hill, 1993. Chapter 10.6 was contributed by Arthur B. Williams and Fred J. Taylor, author's of the "Electronic Filter Design Handbook," 3rd ed., McGraw-Hill, 1995, from which portions were adapted.

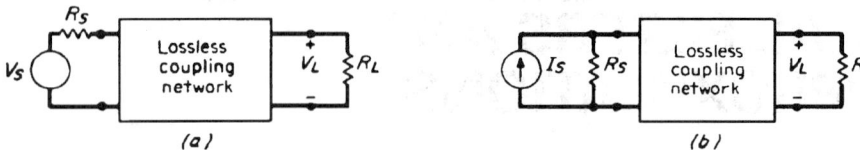

FIGURE 10.1.1   General form of network: (a) with voltage source; (b) with current source.

$$|Z_{LS}(j\omega)| = \left|\frac{V_L}{I_S}(j\omega)\right| \tag{2}$$

for networks of Fig. 10.1.1a and b, respectively. (For convenience, subsequent references to G_{LS} omit subscripts, and the discussion applies to Z_{LS} except where noted.) The phase function

$$\theta(\omega) = \arg G(j\omega) \tag{3}$$

is often the specification.

3. Synthesize the coupling network of Fig. 10.1.1a or b. Darlington[*] showed that this network may be lossless, and often that it is a ladder network, an attractive circuit structure.

CLASSIFICATION OF IDEAL FILTERS

It is convenient to have ideal filter characteristics for discussion purposes. Four are ideal magnitude characteristics, and one is for phase. They are:

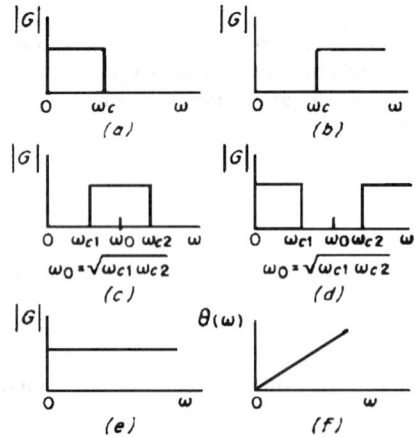

FIGURE 10.1.2   Ideal-filter characteristics: (a) low-pass; (b) high-pass; (c) bandpass; (d) band-elimination; (e) all-pass (magnitude); (f) all-pass (phase).

1. The *low-pass* filter transmits without attenuation or loss signal frequencies from zero to a cutoff frequency and stops all signal frequencies higher than the cutoff.

2. The *high-pass* filter stops all signal frequencies below its cutoff, and transmits without attenuation signal frequencies above the cutoff.

3. The *bandpass* filter passes all signal frequencies between the lower and upper cutoff frequencies, and stops all signal frequencies outside this range.

4. The *bandstop* (or reject) filter stops signal frequencies between its lower and upper cutoff frequencies, and transmits all signal frequencies outside this range.

5. The *all-pass* filter transmits all signal frequencies. It produces a predictable phase shift.

These ideal filter characteristics are shown in Fig. 10.1.2.

---

[*]In many cases, explicit bibliographic references are omitted. However, the bibliography entries are in about the same order as the topics that are discussed.

THE LOW-PASS PROTOTYPE

Synthesis techniques lead to low-pass prototypes, for which extensive tables exist. Modifications of these prototypes enable design of the three remaining filters. The technique is to make $R_L = 1$ and $\omega_C = 1$, to transform specifications to the low-pass domain, and to determine the prototype. Transformations of the low-pass prototype to the needed domain lead to the final network.

THE APPROXIMATION PROBLEM

The ideal low-pass filter is not realizable. A basic problem is to select a transfer function magnitude $|G(j\omega)|$ that *approximates* the ideal characteristic and results in a practical network. Figure 10.1.3 illustrates the concept, and defines the limits on the frequency response. The approximation for $|G(j\omega)|$ must lie within the shaded region.

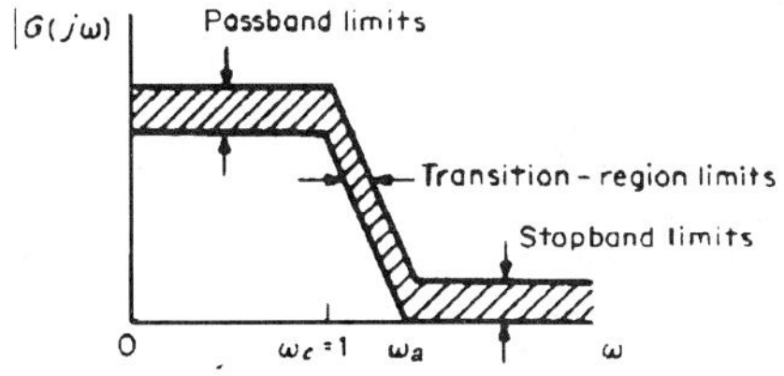

Often it is easier to work with the magnitude-squared function, which is of the form

$$|G(j\omega)|^2 = K^2 \frac{A(\omega^2)}{D(\omega^2)} \tag{4}$$

FIGURE 10.1.3   Limits on frequency response.

The numerator and denominator must be even, nonnegative polynomials, and other restrictions apply. Another form for Eq. (4) is

$$|G(j\omega)|^2 = K^2 \frac{N(\omega^2)}{N(\omega^2) + M(\omega^2)} = K^2 \frac{1}{1 + \dfrac{M(\omega^2)}{N(\omega^2)}} \tag{5}$$

A common technique is to let $N(\omega^2) = 1$. Equation (5) then becomes

$$|G(j\omega)|^2 = \frac{K^2}{1 + M(\omega^2)} \tag{6}$$

A requirement for the polynomial $M(\omega^2)$ is that

$$|M(\omega^2)| \begin{cases} \ll 1 & \omega \ll 1 \\ \gg 1 & \omega \gg 1 \end{cases}$$

which leads to all-pole approximations.

Low-pass filters based on Eq. (6) are ladder structures with inductors as series elements, and capacitors as shunt. The more general form of Eq. (5) leads to a filter that is a ladder with the series elements having parallel combinations of inductors and capacitors, while the shunt elements are series combinations of inductors and capacitors. The Inverse Chebyshev and elliptic approximations lead to such filters.

K of Eq. (4) is not arbitrary. The requirement is that, in terms of Fig. 10.1.1,

$$|G(0)| = \frac{R_L}{R_S + R_L} \tag{7}$$

for a voltage source, and

$$|Z(0)| = \frac{R_S R_L}{R_S + R_L} \tag{8}$$

for a current source. K must satisfy this constraint.

TRANSFER FUNCTION CONSTRUCTION

It is necessary to construct a complete transfer function from the approximation. The complete transfer function allows computation of time-domain responses. Several active realizations require them. The processes show construction of $G(s)$, in terms of Eq. (6). The technique for $Z(s)$ is identical. The Inverse Chebyshev shows an extension for the case of Eq. (5), where the nonconstant numerator imposes additional considerations.

From a suitable $G(\omega^2)$, use analytic continuation to obtain

$$|G(j\omega)|^2_{\omega^2 = -s^2} = G(s)G(-s) = \frac{K^2}{1 + M(-s^2)} \tag{9}$$

Assign the left-half-plane poles of $G(s)G(-s)$ to $G(s)$ and the right-half-plane poles to $G(-s)$, respectively, to ensure stability. The result is

$$G(s) = \frac{K}{\displaystyle\prod_{j=1}^{n}(s - s_j)} \tag{10}$$

where s_j are the left-half-plane poles.

The following sections describe six different approximations, five to the ideal low-pass filter and one to the ideal all-pass or time delay filter. Two approximations (the Inverse Chebyshev, and elliptic) are examples of Eq. (5), while the others all are examples of Eq. (6). Explicit equations give the poles for some approximations. Examples show use of these equations. Numerical solutions give the poles for other approximations. Tables show the roots, tabulated in terms of linear and quadratic factors.

THE BUTTERWORTH APPROXIMATION

The nth Butterworth approximation is

$$|G(j\omega)| = \frac{K}{(1 + \omega^{2n})^{1/2}} \tag{11}$$

Figure 10.1.4 shows the general form of the function. The first $2n - 1$ derivatives of $|G(j\omega)|$ are zero, and the magnitude of the function decreases monotonically with ω. Figure 10.1.4 also shows that increasing n improves the approximation in both the pass and stop bands, at the price of a more complex network. The appropriate n is the smallest value that will meet frequency domain specifications, as suggested in Fig. 10.1.5.

If the specifications require that

$$|G(j\omega)| \le A \qquad \omega_a \le \omega < \infty \tag{12}$$

then the value of n required is the smallest integer satisfying the inequality

$$A \ge \frac{K}{(1 + \omega_a^{2n})^{1/2}} \tag{13}$$

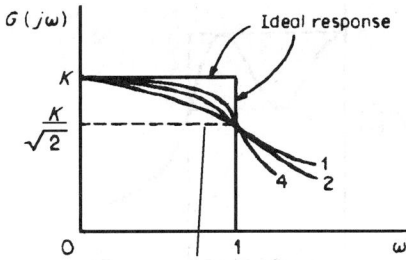

FIGURE 10.1.4 The Butterworth approximation.

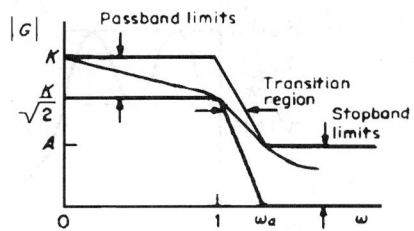

FIGURE 10.1.5 Determination of the minimum value of n for the Butterworth approximation.

In terms of attenuation, the frequency response is

$$\alpha = 20 \log \frac{|G(j\omega)|}{|G(0)|} = 10 \log(1 + \omega^{2n}) \quad \text{dB} \tag{14}$$

At $\omega = 1$,

$$\alpha = 10 \log 2 = 3.0103 \text{ dB} \tag{15}$$

This is a common interpretation of the cutoff frequency, frequently called the half-power or "3-dB" frequency. A more general specification of minimum attenuation $\alpha_{\min}$ (dB) for $\omega \geq \omega_a$ and a maximum attenuation $\alpha_{\max}$ for $\omega \leq \omega_b$ leads to a solution for n of

$$n = \log \frac{(10^{\alpha_{\min}/10} - 1)}{(10^{\alpha_{\max}/10} - 1)} \Bigg/ 2 \log\left(\frac{\omega_a}{\omega_b}\right) \tag{16}$$

with the next larger integer being chosen.

The stable poles lie on a unit circle in the left half s-plane, and are given by $s_k = \sigma_k + j\omega_k$, where

$$\sigma_k = -\sin\frac{2k-1}{2n}\pi \qquad \omega_k = \cos\frac{2k-1}{2n}\pi \qquad k = 1, 2, \ldots, n \tag{17}$$

Since complex poles occur in conjugate pairs, it is convenient to combine them into quadratic factors. When $n = 3$, Eq. (17) gives

$$G(s) = (s + 1)(s^2 + s + 1) \tag{18}$$

and for $n = 4$,

$$G(s) = (s^2 + 1.84776s + 1)(s^2 + 0.76537s + 1) \tag{19}$$

THE CHEBYSHEV APPROXIMATION

The nth order Chebyshev approximation is

$$|G(j\omega)| = \frac{K}{[1 + \epsilon^2 C_n^2(\omega)]^{1/2}} \tag{20}$$

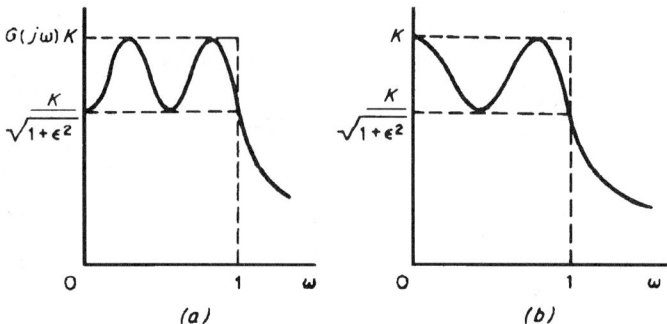

FIGURE 10.1.6 The Chebyshev approximation: (*a*) *n* even; (*b*) *n* odd.

where $C_n(\omega)$ is the *n*th-order Chebyshev polynomial and ϵ is a real constant less than 1. Specifically,

$$C_n(\omega) = \begin{cases} \cos(n\cos^{-1}(\omega)) & \text{for } 0 < \omega \le 1 \\ \cosh(n\cosh^{-1}(\omega)) & \text{for } 1 < \omega \end{cases} \qquad (21, 22)$$

The polynomials in the form of Eqs. (21) and (22) are convenient for calculations. To show a conventional polynomial appearance, note that $C_0(\omega) = 1$ and $C_1(\omega) = \omega$. Use the recursion formula, which is derived from a trigonometric identity,

$$C_{n+1}(\omega) = 2\omega C_n(\omega) - C_{n-1}(\omega) \qquad (23)$$

to develop higher-order polynomials. For example,

$$C_5(\omega) = 16\omega^5 - 20\omega^3 + 5\omega \qquad (24)$$

Figure 10.1.6 shows the frequency response for the Chebyshev approximation for *n* even and odd. In either case there are *n* half cycles (from maximum to minimum and the reverse) in the interval $0 \le \omega \le 1$. One effect of the relative minimum at $\omega = 0$ when *n* is even is that there are combinations of R_S, R_L, and ϵ that are not realizable. In particular, the even-order Chebyshev approximation is unrealizable for any value of ϵ when $R_S = R_L$. The modified Chebyshev approximation that leads to realizable networks in this case.

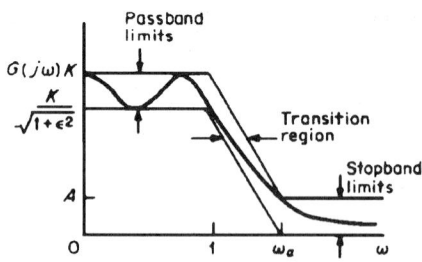

FIGURE 10.1.7 Identification of ϵ and *n*.

Two parameters, ϵ and *n*, define a particular Chebyshev approximation. Figure 10.1.7 defines the parameters for *n* odd (the results also apply to *n* even). The *ripple* factor ϵ controls the passband limits, while both ϵ and *n* control the stopband. The ratio of upper to lower passband limits is $(1 + \epsilon^2)^{1/2}$, and the value of this ratio is sufficient to determine ϵ. The logarithmic function

$$10 \log (1 + \epsilon^2) \quad \text{(dB)} \qquad (25)$$

describes this ratio. Thus, a Chebyshev approximation with $\epsilon = 0.7648$ has a 2.0 dB ripple. Equation (26) relates the ripple R, in dB, and ϵ, the ripple factor

$$\epsilon = \sqrt{10^{R/10} - 1} \qquad (26)$$

With ϵ known and a specification that

$$G(j\omega) \le A \quad \text{for } \omega_a \le \omega < \infty \qquad (27)$$

The attenuation requirement in the stopband leads to the inequality

$$A \geq \frac{K}{[1 + \epsilon^2 C_n^2(\omega_a)]^{1/2}} \tag{28}$$

From Eq. (28) and the known value of ϵ,

$$C_n(\omega_a) = \frac{(K^2 - A^2)^{1/2}}{A\epsilon} \tag{29}$$

This leads to the requirement on n, the order of the filter,

$$n = \frac{\cosh^{-1} C_n(\omega_a)}{\cosh^{-1} \omega_a} \tag{30}$$

with the next larger integer value being chosen.

The *half-power frequency* is a commonly used figure of merit for a filter. At this frequency, the power transmission to the load is 50 percent of the maximum, corresponding to a reduction of 3.0103 (or approximately 3) dB. Equation (31) gives the half-power frequency.

$$\omega_{hp} = \omega_{3dB} = \cosh\left(\frac{1}{n}\cosh^{-1}\frac{1}{\epsilon}\right) \tag{31}$$

To find the stable or left-half s-plane poles of $G(s)$, substitute $\omega \text{-} s/j$ and set the denominator of Eq. (20) to zero. From this,

$$C_n\left(\frac{s}{j}\right) = \cos\left(n\cos^{-1}\frac{s}{j}\right) = \pm\frac{j}{\epsilon} \tag{32}$$

This equation is a complex function, and the complex solution is

$$\sigma_k = -\sin u_k \sinh v \tag{33}$$

$$\omega_k = \cos u_k \cosh v \tag{34}$$

where

$$u_k = \frac{2k-1}{2n}\pi \tag{35}$$

$$k = 1, 2, 3, \ldots, n \tag{36}$$

$$v = \frac{1}{n}\sinh^{-1}\frac{1}{\epsilon} \tag{37}$$

Since complex roots occur in conjugate pairs, it is convenient to combine such pairs into quadratic factors. For example, with a ripple of 0.50 dB and third order,

$$G(s) = (s + 0.62646)(s^2 + 0.62646s + 1.14245) \tag{38}$$

while for a ripple of 1.0 dB and fourth order,

$$G(s) = (s^2 + 0.27907s + 0.98650)(s^2 + 0.67374s + 0.27940) \tag{39}$$

COMPARISON OF BUTTERWORTH AND CHEBYSHEV APPROXIMATIONS

A good comparison of the Butterworth and Chebyshev responses is possible when they have the same half-power frequency. In general, this requires a normalization of the Chebyshev approximation to a 3.0103 dB bandwidth before comparison. This occurs naturally when $\epsilon = 1$. Figure 10.1.8 shows the two-frequency responses for third order filters. Analysis shows that:

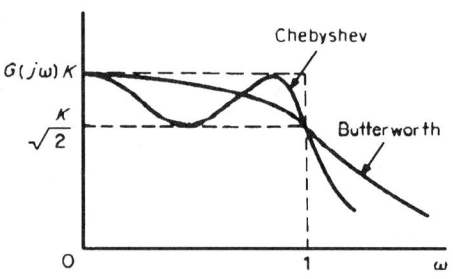

FIGURE 10.1.8 Butterworth and Chebyshev approximations for $n = 3$ and $\epsilon = 1$.

1. The Butterworth approximation is superior at and near $\omega = 0$. Of all polynomial approximations, it has the highest number of zero derivatives at the origin.

2. The Chebyshev approximation is superior at and near the cutoff frequency or passband edge.

3. The Chebyshev approximation is superior in the stopband.

4. The Chebyshev approximation sacrifices smoothness in the passband.

THE MODIFIED CHEBYSHEV APPROXIMATION

With passive networks, even-ordered Chebyshev networks with equal source and load resistances are unrealizable, because of the relative minimum at zero frequency in the frequency response. Saal has shown a modification of the Chebyshev polynomials that leads to even-ordered polynomials with a zero at $\omega = 0$, relative maxima and minima of +1 and −1 within the interval $0 < \omega \leq 1$, and that are monotonically increasing for $\omega > 1$. These polynomials are given in Table 10.1.1.

For a given n, the modified Chebyshev polynomials have the property that

$$C_n(\omega) > \bar{C}_n(\omega) > C_{n-1}(\omega) \qquad (40)$$

for $\omega > 1$. Thus a low-pass approximation based on these polynomials for even n will be better in the stopband than that using the Chebyshev polynomial of order $n - 1$, but not as good as that using the regular Chebyshev polynomial of order n. However, it will, in passive networks, permit equal source and load resistances.

Figure 10.1.9 compares $C_n(\omega)$ and $\bar{C}_n(\omega)$ by plotting the ratio as a function of ω.

Table 10.1.2 gives the poles of the transfer function $G(s)$, in terms of quadratic factors, for $n = 2, 4, 6, 8,$ and 10, and for ripples of 0.01, 0.10, 0.50, 1.00, and 3.00 dB. These poles assume that the end of the ripple band is the passband edge at $\omega = 1$. Table 10.1.11 gives the corresponding half-power frequencies.

TABLE 10.1.1 Modified Chebyshev Polynomials of Even Order

n	$C_n(\omega)$
2	ω^2
4	$5.82842713\omega^4 - 4.82842713\omega^2$
6	$25.9903811\omega^6 - 36.1865335\omega^4 + 11.1961524\omega^2$
8	$109.597711\omega^8 - 210.522708\omega^6$ $+ 122.034355\omega^4 - 20.109358\omega^2$
10	$452.344415\omega^{10} - 1102.492675\omega^8 + 926.297276\omega^6$ $- 306.717773\omega^1 + 31.568758\omega^2$

THE LEGENDRE–PAPOULIS APPROXIMATION

Papoulis derived an approximation using Legendre polynomials ($L_n(\omega^2)$) of the first kind. The nth order approximation is

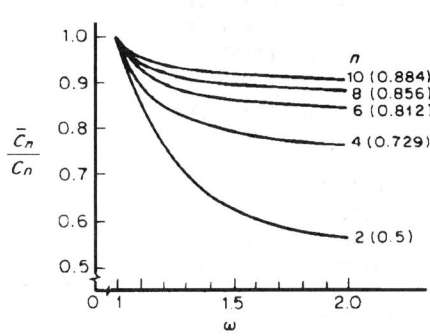

FIGURE 10.1.9 Ratio of modified Chebyshev polynomial to Chebyshev polynomial. Limiting value as ω approaches infinity is in parentheses for each value of n.

$$|G(j\omega)| = \frac{1}{\sqrt{1 + L_n(\omega^2)}} \qquad (41)$$

Table 10.1.3 shows the polynomials, and Table 10.1.4 shows the denominator roots (poles) for this function.

This approximation has three important properties:

1. $L_n(\omega^2)$ increases monotonically, and thus $|G(j\omega)|$ decreases monotonically.
2. $L_n(0) = 0$ and $L_n(1) = 1$, which means that $|G(0)| = 1$ and $|G(1)| = 0.50$.
3. Of all polynomials satisfying 1 and 2, $L_n(\omega^2)$ has the largest derivative at $\omega = 1$, and thus this filter has the steepest cutoff characteristic.

Figure 10.1.10 compares the Legendre–Papoulis approximation with the Butterworth, and Chebyshev (3 dB) approximations for $n = 3$.

The selection of n follows the same general procedure as for the Butterworth approximation. To satisfy the specification

$$A \geq |G(j\omega_a)| = 1/\sqrt{1 + L_n(\omega^2)} \qquad (42)$$

or

$$L_n(\omega_a^2) \geq (1/A^2) - 1 \qquad (43)$$

evaluate successively higher-ordered polynomials of Table 10.1.3 at ω_a.

THE BESSEL APPROXIMATION

The ideal all pass characteristics of Fig. 10.1.2.

$$|G(j\omega)| = K \qquad (44)$$

and

$$\theta(j\omega) = -T\omega \qquad (45)$$

implies that the output is a scaled (by K) replica of the input, delayed in time by T s. This notion leads to the definition of group delay, which is the negative derivative of the phase function

$$-\frac{\partial}{\partial \omega} \theta(j\omega) = \tau_d(\omega) \qquad (46)$$

In the ideal case, the group delay is a constant T.

TABLE 10.1.2 Quadratic Factors Giving Poles of Modified Chebyshev Transfer Functions

n	Ripple, dB	$[G(s)]^{-1}$
2	0.01	$s^2 + 6.4541054s + 20.8277385$
	0.10	$s^2 + 3.6200009s + 6.5522033$
	0.50	$s^2 + 2.3928122s + 2.8627752$
	1.00	$s^2 + 1.9825371s + 1.9652267$
	3.00	$s^2 + 1.4158936s + 1.0023773$
4	0.01	$s^2 + 2.2680440s + 1.6170715$
		$s^2 + 0.9235542s + 2.2098435$
	0.10	$s^2 + 1.5452285s + 0.7991383$
		$s^2 + 0.6059475s + 1.4067406$
	0.50	$s^2 + 1.1191759s + 0.4523364$
		$s^2 + 0.4086395s + 1.0858612$
	1.00	$s^2 + 0.9486848s + 0.3398608$
		$s^2 + 0.3272401s + 0.9921108$
	3.00	$s^2 + 0.6845838s + 0.1932927$
		$s^2 + 0.2013994s + 0.8897428$
6	0.01	$s^2 + 1.4118889s + 0.5893450$
		$s^2 + 1.0058571s + 0.9699759$
		$s^2 + 0.3640166s + 1.4018414$
	0.10	$s^2 + 0.9995998s + 0.3173248$
		$s^2 + 0.6819299s + 0.6966158$
		$s^2 + 0.2448454s + 1.1404529$
	0.50	$s^2 + 0.7362729s + 0.1875012$
		$s^2 + 0.4662515s + 0.5727968$
		$s^2 + 0.1663135s + 1.0255810$
	1.00	$s^2 + 0.6267633s + 0.1428406$
		$s^2 + 0.3748110s + 0.5343433$
		$s^2 + 0.1333343s + 0.9906678$
	3.00	$s^2 + 0.4534072s + 0.0825380$
		$s^2 + 0.2315851s + 0.4909174$
		$s^2 + 0.0820877s + 0.9518235$
8	0.01	$s^2 + 1.0335548s + 0.3097540$
		$s^2 + 0.8491714s + 0.5489161$
		$s^2 + 0.5591534s + 0.9283901$
		$s^2 + 0.1954150s + 1.2038909$
	0.10	$s^2 + 0.7417790s + 0.1718214$
		$s^2 + 0.5828511s + 0.4083754$
		$s^2 + 0.3806989s + 0.7943820$
		$s^2 + 0.1328521s + 1.0725554$
	0.50	$s^2 + 0.5496051s + 0.1030188$
		$s^2 + 0.4007060s + 0.3416991$
		$s^2 + 0.2600622s + 0.7328336$
		$s^2 + 0.0906706s + 1.0125586$
	1.00	$s^2 + 0.4685593s + 0.0788470$
		$s^2 + 0.3226466s + 0.3205446$
		$s^2 + 0.2088595s + 0.7137508$
		$s^2 + 0.0727952s + 0.9940103$
	3.00	$s^2 + 0.3392712s + 0.0458028$
		$s^2 + 0.1997071s + 0.2963792$
		$s^2 + 0.1288387s + 0.6922959$
		$s^2 + 0.0448874s + 0.9731910$
10	0.01	$s^2 + 0.8175750s + 0.1921666$
		$s^2 + 0.7133475s + 0.3528340$

TABLE 10.1.2   Quadratic Factors Giving Poles of Modified Chebyshev Transfer Functions (*Continued*)

n	Ripple, dB	$[G(s)]^{-1}$
		$s^2 + 0.5571617s + 0.6426718$
		$s^2 + 0.3555630s + 0.9397252$
		$s^2 + 0.1222554s + 1.1244354$
	0.10	$s^2 + 0.5905036s + 0.1080675$
		$s^2 + 0.4925060s + 0.2665100$
		$s^2 + 0.3816272s + 0.5602769$
		$s^2 + 0.2431732s + 0.8592520$
		$s^2 + 0.0835800s + 1.0446869$
	0.50	$s^2 + 0.4387194s + 0.0652281$
		$s^2 + 0.3394908s + 0.2246543$
		$s^2 + 0.2614400s + 0.5216326$
		$s^2 + 0.1664374s + 0.8217378$
		$s^2 + 0.0571933s + 1.0075592$
	1.00	$s^2 + 0.3742846s + 0.0500294$
		$s^2 + 0.2735736s + 0.2112433$
		$s^2 + 0.2101536s + 0.5095395$
		$s^2 + 0.1337446s + 0.8100361$
		$s^2 + 0.0459557s + 0.9959854$
	3.00	$s^2 + 0.2711265s + 0.0291328$
		$s^2 + 0.1694797s + 0.1958431$
		$s^2 + 0.1297678s + 0.4958770$
		$s^2 + 0.0825542s + 0.7968406$
		$s^2 + 0.0283639s + 0.9829386$

A series of Bessel polynomials provides a transfer function that yields a maximally flat approximation to the ideal delay. The form of the function that gives a group delay of 1 s at $\omega = 0$ is

$$G(s) = \frac{Kb_0}{s^n + b_{n-1}s^{n-1} + \cdots + bs + b_0} \tag{47}$$

A recursion formula relates these polynomials

$$B_n = (2n - 1)B_{n-1} + s^2 B_{n-2} \tag{48}$$

and Table 10.1.5 gives the coefficients of the first eight Bessel polynomials. Table 10.1.6 gives the pole locations for $G(s)$ in terms of quadratic factors.

TABLE 10.1.3   The Polynomials $L_n(\omega^2)$

n	
2	ω^4
3	$3\omega^6 - 3\omega^4 + \omega^2$
4	$6\omega^8 - 8\omega^6 + 3\omega^4$
5	$20\omega^{10} - 40\omega^8 + 28\omega^6 - 8\omega^4 + \omega^2$
6	$50\omega^{12} - 120\omega^{10} + 105\omega^8 - 40\omega^6 + 6\omega^4$
7	$175\omega^{14} - 525\omega^{12} + 615\omega^{10} - 355\omega^8 + 105\omega^6 - 15\omega^4 + \omega^2$
8	$490\omega^{16} - 1680\omega^{14} + 2310\omega^{12} - 1624\omega^{10} + 615\omega^8 - 120\omega^6 + 10\omega^4$
9	$1764\omega^{18} - 7056\omega^{16} + 11704\omega^{14} - 10416\omega^{12} + 5376\omega^{10} - 1624\omega^8 + 276\omega^6 - 24\omega^4 + \omega^2$
10	$5292\omega^{20} - 23520\omega^{18} + 44100\omega^{16} - 45360\omega^{14} + 27860\omega^{12} - 10416\omega^{10} + 2310\omega^8 - 280\omega^6 + 15\omega^4$

TABLE 10.1.4   Linear and Quadratic Factors for Poles of Legendre-Papoulis Approximation

n	$[G(s)]^{-1}$	n	$[G(s)]^{-1}$
2	$s^2 + 1.4142136s + 1.0000000$	8	$s^2 + 0.1378844s + 0.9808397$
3	$s + 0.6203318$		$s^2 + 0.3885518s + 0.7179832$
	$s^2 + 0.6903712s + 0.9307119$		$s^2 + 0.6005680s + 0.3828971$
4	$s^2 + 0.4633774s + 0.9476701$		$s^2 + 0.7343526s + 0.1675357$
	$s^2 + 1.0994868s + 0.4307915$	9	$s + 0.3256878$
5	$s + 0.4680899$		$s^2 + 0.1101944s + 0.9844435$
	$s^2 + 0.3071734s + 0.9608963$		$s^2 + 0.3145676s + 0.7666498$
	$s^2 + 0.7762796s + 0.4971406$		$s^2 + 0.4971058s + 0.4635058$
6	$s^2 + 0.2303854s + 0.9696012$		$s^2 + 0.6187708s + 0.2089807$
	$s^2 + 0.6179218s + 0.5828947$		
	$s^2 + 0.8778030s + 0.2502256$	10	$s^2 + 0.0918020s + 0.9869313$
7	$s + 0.3821033$		$s^2 + 0.2650376s + 0.8012497$
	$s^2 + 0.1724170s + 0.9764158$		$s^2 + 0.4283460s + 0.5282527$
	$s^2 + 0.4748794s + 0.6621299$		$s^2 + 0.5548108s + 0.2702425$
	$s^2 + 0.6984636s + 0.3060005$		$s^2 + 0.6344130s + 0.1217699$

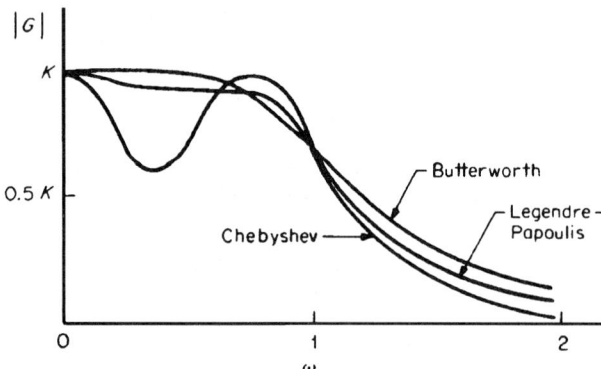

FIGURE 10.1.10   Comparison of Butterworth, Legendre-Papoulis, and Chebyshev ($\epsilon = 1$) approximations for $n = 3$.

TABLE 10.1.5   Coefficients of the Bessel Polynomials

n	b_0	b_1	b_2	b_3	b_4	b_5	b_6	b_7
1	1							
2	3	3						
3	15	15	6					
4	105	105	45	10				
5	945	945	420	105	15			
6	10,395	10,395	4,725	1,260	210	21		
7	135,135	135,135	62,370	17,325	3,150	378	28	
8	2,027,025	2,027,025	945,945	270,270	51,975	6,930	630	36

TABLE 10.1.6 Linear and Quadratic Factors for Poles of Bessel Approximation

n	$[G(s)]^{-1}$	n	$[G(s)]^{-1}$
2	$s^2 + 3s + 3$	8	$s^2 + 11.175772s + 31.977224$
3	$s + 2.322185$		$s^2 + 10.409682s + 33.934741$
	$s^2 + 3.677814s + 6.459432$		$s^2 + 8.736578s + 38.569256$
4	$s^2 + 5.792422s + 9.140133$		$s^2 + 5.677968s + 48.432015$
	$s^2 + 4.207578s + 11.487799$		
5	$s + 3.646739$	9	$s + 6.297019$
	$s^2 + 6.703912s + 14.272476$		$s^2 + 12.258736s + 40.589268$
	$s^2 + 4.649348s + 18.156314$		$s^2 + 11.208844s + 43.646648$
6	$s^2 + 8.496718s + 18.801128$		$s^2 + 9.276880s + 49.788507$
	$s^2 + 7.471416s + 20.852819$		$s^2 + 5.958522s + 62.041443$
	$s^2 + 5.031864s + 26.514025$	10	$s^2 + 13.844090s + 48.667550$
7	$s + 4.971787$		$s^2 + 13.230582s + 50.582362$
	$s^2 + 9.516582s + 25.666449$		$s^2 + 11.935056s + 54.839151$
	$s^2 + 8.140278s + 28.936544$		$s^2 + 9.772440s + 62.625584$
	$s^2 + 5.371354s + 36.596784$		$s^2 + 6.217832s + 77.442692$

Figure 10.1.11 shows the general forms of the delay and magnitude functions for this approximation. The magnitude behavior is that of a low-pass function, and Table 10.1.11 gives the resulting half-power frequencies.

The parameter n is selected to satisfy either a minimum group delay requirement or a minimum magnitude (or the equivalent maximum attenuation) requirement at some frequency ω_a, that is,

$$\tau_d(\omega_a) \geq T_a \tag{49}$$

or

$$|G(j\omega_a)| \geq A \tag{50}$$

with the more stringent condition being chosen. Figure 10.1.12 shows the attenuation versus ω and the group delay versus ω for $n = 2$ to 8, and can be used to select n.

THE STEP RESPONSE

The primary design consideration for a filter is its frequency response, but often the step response is of interest. Computation of the step response requires computation of the inverse Laplace transform of the transfer function

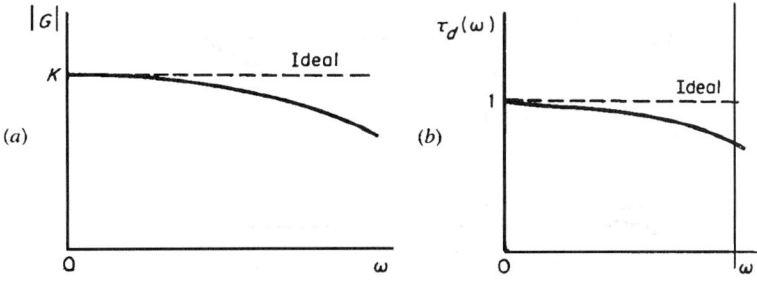

FIGURE 10.1.11 (*a*) Magnitude and (*b*) group delay for maximally flat group-delay approximation.

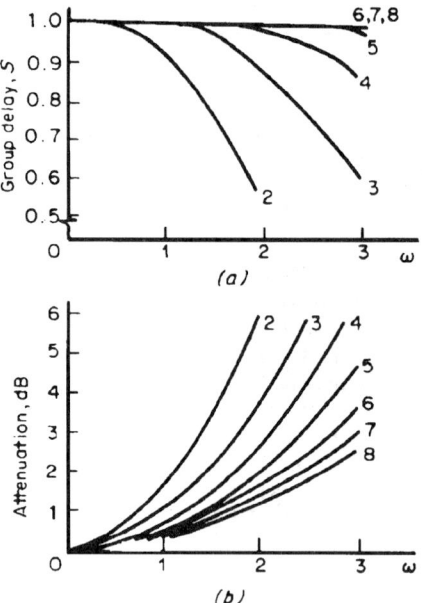

FIGURE 10.1.12   Group delay and attenuation for the Bessel approximation.

multiplied by $1/s$, readily done with modern computers and suitable software. Three figures of merit commonly characterize the step response. Figure 10.1.13 defines these, which are *overshoot, rise time*, and *delay time*. While the general form of Fig. 10.1.13 is common, the odd order, high ripple Chebyshev and modified Chebyshev responses have unusual features, as suggested in Figs. 10.1.14 and 10.1.15. Often several relative maxima occur before the peak value. Tables 10.1.7, 10.1.8, and 10.1.9 give the percent overshoot, rise time, and delay time, respectively, for the all pole approximations described.

THE PROTOTYPE FILTER NETWORKS

Passive networks that realize all pole functions are lossless ladder networks terminated in resistance at both ends. Figure 10.1.16 shows four networks classified by the type of transfer function, Z_{LS} or G_{LS}, being realized

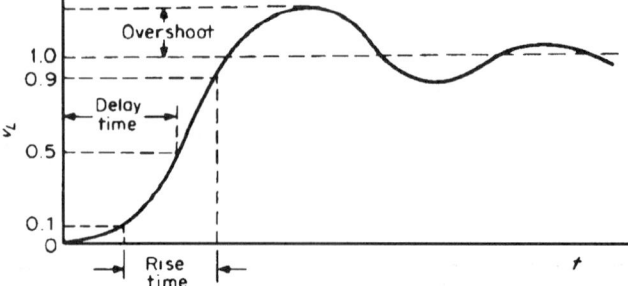

FIGURE 10.1.13   General form of the step response and definitions of figures of merit.

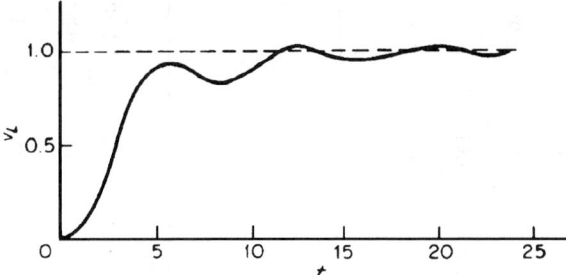

FIGURE 10.1.14   Step response of third-order Chebyshev approximation with a 3-dB ripple.

and whether n is odd or even. The networks of Figs. 10.1.16c and d are duals of those in Figs. 10.1.16a and b, respectively. An application of source transformations to the networks of Fig. 10.1.16 yields the four networks of Fig. 10.1.17.

For a given type of source and value of n there are two networks to choose from. When the source is a current source and n is odd, Figs. 10.1.16b and 10.1.17d apply. In the first case, the first and last lossless elements are capacitors; in the second case, they are inductors.

For a given type of approximation and order of filter, numerical element values apply to four different networks. For example, a Butterworth approximation with $n = 2$ and R_S or $G_S = 1/2$ according to the specific networking being used, yields the four networks of Fig. 10.1.18. Here, the numerical values for the lossless elements are the same when they are taken in order from the source end. This fact makes it possible to construct a table of element values, which uses as parameters for entry (1) the type of approximation, including the ripple width in dB, Eq. (25), for Chebyshev approximations; (2) the value of n; and (3) the value of R_S or G_S, according to the particular network for Figs. 10.1.16 and 10.1.17 that is being used. Table 10.1.10 is such a table.

THE TABLE OF ELEMENT VALUES

Table 10.1.10 gives element values, numbered in order from the source end as in Figs. 10.1.16 and 10.1.17, for the five all-pole approximations, $n = 2$ to 10, and for R_S or G_S equal to 0 and 1. Five different ripple widths are tabulated for the Chebyshev approximations.

The synthesis process for determining the element values in Table 10.1.10 often involves a choice of location for the zeros of the reflection coefficient. In this table when choices were necessary, the networks have left-half s-plane zeros.

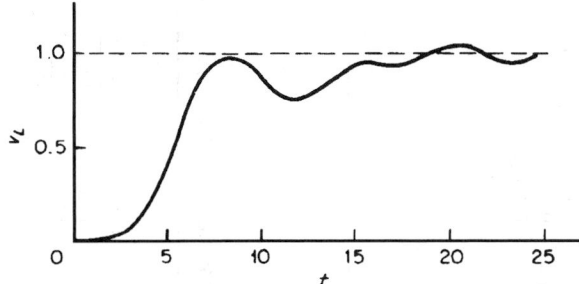

FIGURE 10.1.15   Step response of fifth-order Chebyshev approximation with a 3-dB ripple.

TABLE 10.1.7 Percent Overshoot

Order	Bessel, %	Butterworth, %	Legendre-Papoulis, %	Chebyshev, %				Modified Chebyshev, %			
				3.0 dB	1.0 dB	0.5 dB	0.1 dB	3.0 dB	1.0 dB	0.5 dB	0.1 dB
2	0.4	4.3	4.3	27.2	14.6	10.7	6.7	4.3	4.3	4.3	4.3
3	0.7	8.1	7.5	2.7	6.4	8.9	10.2				
4	0.8	10.8	11.2	35.7	21.9	18.1	14.5	8.0	10.5	12.1	12.9
5	0.8	12.8	13.3	1.7	10.2	13.2	15.2				
6	0.6	14.3	15.2	38.7	24.9	21.2	18.0	8.5	12.8	15.1	16.5
7	0.5	15.4	16.4	2.0	12.1	15.3	17.7				
8	0.3	16.3	17.6	40.4	26.5	23.0	19.8	8.7	14.0	16.6	18.5
9	0.2	17.1	18.3	3.1	13.3	16.7	19.1				
10	0.1	17.8	19.1	41.5	27.6	24.1	21.0	9.5	14.8	17.6	19.7

TABLE 10.1.8 Rise Time for Polynomial Filters

Order	Bessel	Butterworth	Legendre-Papoulis	Chebyshev				Modified Chebyshev			
				3.0 dB	1.0 dB	0.5 dB	0.1 dB	3.0 dB	1.0 dB	0.5 dB	0.1 dB
2	1.58	2.2	2.2	1.7	1.6	1.5	1.1	2.2	1.5	1.3	0.8
3	1.25	2.3	2.5	3.2	2.4	2.2	1.7				
4	1.04	2.4	2.7	2.5	2.5	2.4	2.2	3.2	2.6	2.4	2.1
5	0.91	2.6	2.9	3.7	3.0	2.8	2.5				
6	0.82	2.7	3.0	2.9	3.0	2.9	2.8	3.7	3.2	3.0	2.7
7	0.75	2.8	3.2	4.0	3.4	3.3	3.0				
8	0.69	2.9	3.4	3.2	3.3	3.3	3.2	4.1	3.6	3.4	3.2
9	0.64	3.0	3.5	4.3	3.8	3.6	3.4				
10	0.60	3.1	3.6	3.5	3.6	3.6	3.5	4.4	3.9	3.7	3.5

TABLE 10.1.9 Delay Times for Polynomial Filters

Order	Bessel	Butterworth	Legendre-Papoulis	Chebyshev				Modified Chebyshev			
				3.0 dB	1.0 dB	0.5 dB	0.1 dB	3.0 dB	1.0 dB	0.5 dB	0.1 dB
2	0.90	1.4	1.4	1.5	1.2	1.1	0.8	1.4	1.0	0.8	0.6
3	0.96	2.1	2.4	3.0	2.4	2.2	1.7				
4	0.98	2.8	3.3	3.6	3.3	3.2	2.7	3.9	3.3	3.1	2.6
5	0.99	3.5	4.2	5.2	4.6	4.3	3.8				
6	0.99	4.2	5.2	5.6	5.4	5.2	4.8	6.1	5.5	5.2	4.7
7	1.00	4.8	6.2	7.3	6.7	6.4	5.8				
8	1.00	5.5	7.1	7.7	7.5	7.3	6.8	8.2	7.6	7.3	6.8
9	1.00	6.2	8.1	9.4	8.8	8.5	7.9				
10	1.00	6.8	9.0	9.7	9.6	9.4	8.9	10.3	9.7	9.4	6.9

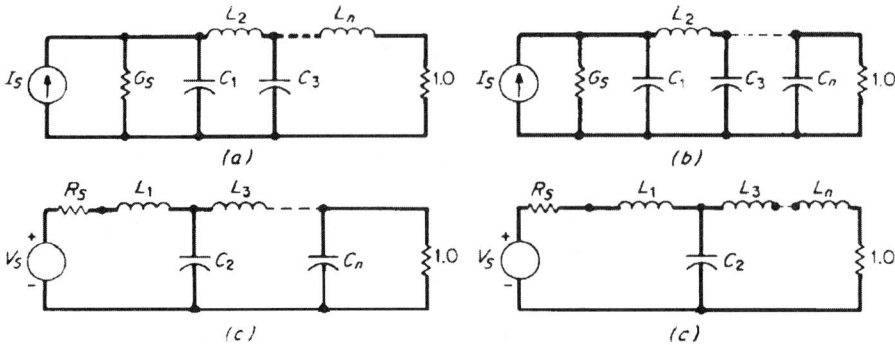

FIGURE 10.1.16 Four filter networks classified according to type of transfer function and n even or odd: (a) Z_{LS}, n even; (b) Z_{LS}, n odd; (c) G_{LS}, n even; (d) G_{LS}, n odd.

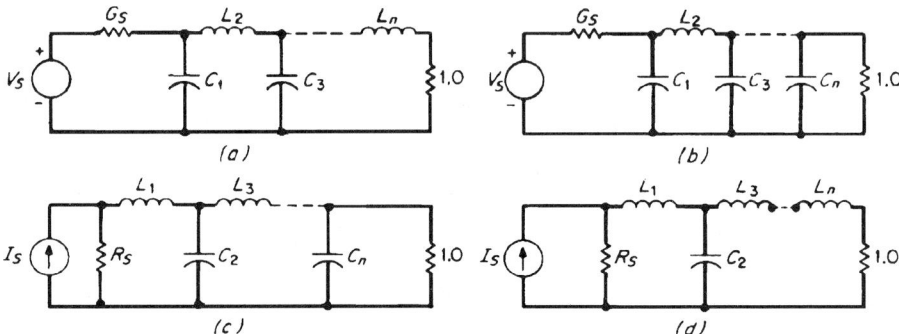

FIGURE 10.1.17 Filter networks derived from those in Fig. 10.1.16 by source transformations: (a) G_{LS}, n even; (b) G_{LS}, n odd; (c) Z_{LS}, n even; (d) Z_{LS}, n odd.

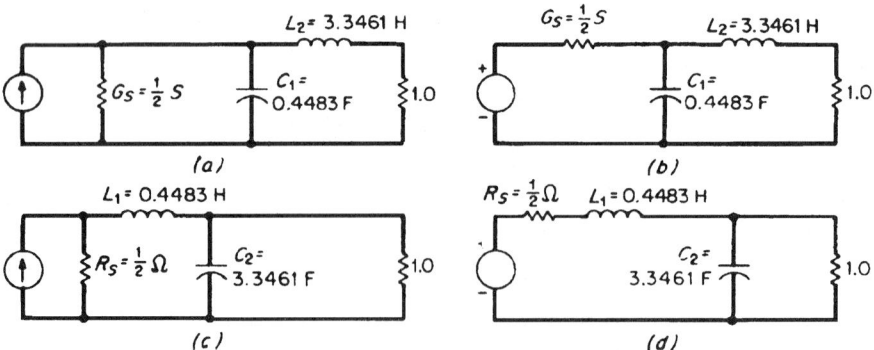

FIGURE 10.1.18 Four networks resulting from a Butterworth approximation with $n = 2$ and R_S or G_S (as appropriate to the type of network) equal to $1/2$.

TABLE 10.1.10 Low-Pass Filter Circuit Element Values*

(Table 10.1.11 gives half-power frequencies for Bessel and Chebyshev circuits, the half-power frequency is 1.000 rad/s for Butterworth and Legendre-Papoulis; empty spaces indicate unrealizable networks.)

Filter type	Element number R_s or G_s	Second-order network $n = 2$		Third-order network $n = 3$			Fourth-order network $n = 4$				Fifth-order network $n = 5$				
		1	2	1	2	3	1	2	3	4	1	2	3	4	5
Bessel	0	1.0000	0.3333	0.8333	0.4800	0.1667	0.7101	0.4627	0.2899	0.1000	0.6231	0.4215	0.3103	0.1948	0.0667
	1	1.5774	0.4227	1.2550	0.5528	0.1922	1.0598	0.5116	0.3181	0.1104	0.9303	0.4577	0.3312	0.2089	0.0718
Butterworth	0	1.4142	0.7071	1.5000	1.3333	0.5000	1.5307	1.5772	1.0824	0.3827	1.5451	1.6944	1.3820	0.8944	0.3090
	1	1.4142	1.4142	1.0000	2.0000	1.0000	0.7654	1.8478	1.8478	0.7654	0.6180	1.6180	2.0000	1.6180	0.6180
Legendre-Papoulis	0	1.4142	0.7071	1.5909	1.4270	0.7629	1.6120	1.6616	1.4292	0.6399	1.6372	1.7509	1.7358	1.3945	0.6445
	1	1.4142	1.4142	2.1801	1.3538	1.1737	1.5645	1.9584	1.4769	1.0826	1.9990	1.5395	2.0673	1.4780	0.9512
Chebyshev 0.01 dB	0	0.4274	0.2244	0.7997	0.7634	0.3146	1.0421	1.1547	0.8945	0.3564	1.1978	1.3902	1.2739	0.9576	0.3782
	1			0.6292	0.9703	0.6292					0.7563	1.3049	1.5773	1.3049	0.7563
Chebyshev 0.10 dB	0	0.7159	0.4215	1.0895	1.0864	0.5158	1.2453	1.4576	1.1994	0.5544	1.3759	1.5924	1.5562	1.2490	0.5734
	1			1.0316	1.1474	1.0316					1.1468	1.3712	1.9750	1.3712	1.1468
Chebyshev 0.50 dB	0	0.9403	0.7014	1.3465	1.3001	0.7981	1.3138	1.7279	1.3916	0.8352	1.5388	1.6426	1.8142	1.4291	0.8529
	1			1.5963	1.0967	1.5963					1.7058	1.2296	2.5408	1.2296	1.7058
Chebyshev 1.0 dB	0	0.9957	0.9110	1.5088	1.3332	1.0118	1.2817	1.9093	1.4126	1.0495	1.6652	1.5908	1.9938	1.4441	1.0674
	1			2.0236	0.9941	2.0236					2.1349	1.0911	3.0009	1.0911	2.1349
Chebyshev 3.0 dB	0	0.9109	1.5506	2.0302	1.1739	1.6744	1.0578	2.5272	1.2292	1.7195	2.1489	1.3016	2.6224	1.2502	1.7406
	1			3.3487	0.7117	3.3487					3.4813	0.7619	4.5375	0.7619	3.4813
Modified Chebyshev 0.01 dB	1	0.3099	0.3099				0.6266	1.1938	1.1938	0.6266					
0.10 dB	1	0.5525	0.5525				0.9297	1.4346	1.4346	0.9297					
0.50 dB	1	0.8358	0.8358				1.3091	1.5415	1.5415	1.3091					
1.0 dB	1	1.0088	1.0088				1.5675	1.5537	1.5537	1.5675					
3.0 dB	1	1.4125	1.4125				2.2574	1.5107	1.5107	2.2574					

(Continued)

TABLE 10.1.10 Low-Pass Filter Circuit Element Values* (*Continued*)

Filter type	Element number R_s or G_s	Sixth-order network $n = 6$						Seventh-order network $n = 7$						
		1	2	3	4	5	6	1	2	3	4	5	6	7
Bessel	0	0.5595	0.3821	0.3005	0.2246	0.1400	0.0476	0.5111	0.3487	0.2827	0.2288	0.1704	0.1055	0.0357
	1	0.8376	0.4116	0.3158	0.2364	0.1480	0.0505	0.7677	0.3744	0.2944	0.2378	0.1778	0.1104	0.0375
Butterworth	0	1.5529	1.7593	1.5529	1.2016	0.7579	0.2588	1.5576	1.7988	1.6588	1.3972	1.0550	0.6560	0.2225
	1	0.5176	1.4142	1.9319	1.9319	1.4142	0.5176	0.4450	1.2470	1.8019	2.0000	1.8019	1.2470	0.4450
Legendre-Papoulis	0	1.6348	1.8088	1.8223	1.6795	1.3486	0.5793	1.6391	1.8312	1.8911	1.7988	1.6845	1.3290	0.5787
	1	1.5763	1.9040	1.7442	1.9857	1.4852	0.9160	1.8640	1.5895	2.1506	1.7270	1.9394	1.4770	0.8394
Chebyshev 0.01 dB	0	1.2931	1.5400	1.4922	1.3319	0.9925	0.3907	1.3615	1.6303	1.6290	1.5412	1.3650	1.0137	0.3985
	1							0.7969	1.3924	1.7481	1.6331	1.7481	1.3924	0.7969
Chebyshev 0.10 dB	0	1.4035	1.7236	1.6749	1.5999	1.2752	0.5841	1.4745	1.7395	1.7987	1.7107	1.6236	1.2908	0.5906
	1							1.1812	1.4228	2.0967	1.5734	2.0967	1.4228	1.1812
Chebyshev 0.50 dB	0	1.4042	1.9018	1.7101	1.8494	1.4483	0.8627	1.5983	1.7252	1.9713	1.7369	1.8677	1.4595	0.8686
	1							1.7373	1.2582	2.6383	1.3443	2.6383	1.2582	1.7373
Chebyshev 1.0 dB	0	1.3457	2.0491	1.6507	2.0270	1.4601	1.0773	1.7120	1.6488	2.1194	1.6735	2.0438	1.4692	1.0833
	1							2.1666	1.1115	3.0936	1.1735	3.0936	1.1115	2.1666
Chebyshev 3.0 dB	0	1.0876	2.6309	1.3455	2.6578	1.2606	1.7522	2.1828	1.3281	2.7143	1.3613	2.6752	1.2665	1.7593
	1							3.5185	0.7722	4.6390	0.8038	4.6390	0.7722	3.5185
Modified 0.01 dB	1	0.7190	1.3728	1.6006	1.6006	1.3728	0.7190							
Chebyshev 0.10 dB	1	1.0382	1.5163	1.7892	1.7892	1.5163	1.0382							
0.50 dB	1	1.4611	1.5085	1.9333	1.9333	1.5085	1.4611							
1.0 dB	1	1.7623	1.4528	2.0089	2.0089	1.4528	1.7623							
3.0 dB	1	2.6073	1.2711	2.1729	2.1729	1.2711	2.6073							

TABLE 10.1.10 Low-Pass Filter Circuit Element Values* (*Continued*)

Filter type	R_s or G_s Element number	Eight-order network $n = 8$								Ninth-order network $n = 9$								
		1	2	3	4	5	6	7	8	1	2	3	4	5	6	7	8	9
Bessel	0	0.4732	0.3212	0.2639	0.2227	0.1806	0.1338	0.0823	0.0278	0.4424	0.2986	0.2465	0.2129	0.1811	0.1463	0.1077	0.0660	0.0222
	1	0.7125	0.3446	0.2735	0.2297	0.1867	0.1387	0.0855	0.0289	0.6678	0.3203	0.2547	0.2184	0.1859	0.1506	0.1112	0.0682	0.0230
Butterworth	0	1.5607	1.8246	1.7287	1.5283	1.2588	0.9371	0.5776	0.1951	1.5628	1.8424	1.7772	1.6202	1.4037	1.1408	0.8414	0.5155	0.1736
	1	0.3902	1.1111	1.6629	1.9616	1.9616	1.6629	1.1111	0.3902	0.3473	1.0000	1.5321	1.8794	2.0000	1.8794	1.5321	1.0000	0.3473
Legendre-Papoulis	0	1.6345	1.8542	1.9102	1.8673	1.8019	1.6437	1.2869	0.5372	1.6341	1.8625	1.9349	1.8961	1.8815	1.7860	1.6397	1.2740	0.5358
	1	1.5564	1.8501	1.8411	2.0515	1.7672	1.9115	1.4688	0.8205	1.7645	1.6134	2.1585	1.7816	2.0662	1.7755	1.8674	1.4555	0.7695
Chebyshev 0.01 dB	0	1.4036	1.6971	1.7097	1.6708	1.5697	1.3858	1.0275	0.4036	1.4392	1.7371	1.7701	1.7457	1.6949	1.5878	1.3998	1.0370	0.4072
	1									0.8145	1.4271	1.8044	1.7125	1.9058	1.7125	1.8044	1.4271	0.8145
Chebyshev 0.10 dB	0	1.4660	1.8163	1.8070	1.8302	1.7302	1.6380	1.3008	0.5949	1.5182	1.7991	1.8814	1.8343	1.8473	1.7423	1.6476	1.3076	0.5978
	1									1.1957	1.4426	2.1346	1.6167	2.2054	1.6167	2.1346	1.4426	1.1957
Chebyshev 0.50 dB	0	1.4379	1.9571	1.7838	1.9980	1.7508	1.8786	1.4666	0.8725	1.6238	1.7571	2.0203	1.8055	2.0116	1.7591	1.8856	1.4714	0.8752
	1									1.7504	1.2690	2.6678	1.3673	2.7239	1.3673	2.6678	1.2690	1.7504
Chebyshev 1.0 dB	0	1.3691	2.0922	1.7021	2.1453	1.6850	2.0537	1.4751	1.0872	1.7317	1.6707	2.1574	1.7213	2.1582	1.6918	2.0601	1.4790	1.0899
	1									2.1797	1.1192	3.1214	1.1897	3.1746	1.1897	3.1214	1.1192	2.1797
Chebyshev 3.0 dB	0	1.0982	2.6618	1.3687	2.7436	1.3690	2.6852	1.2701	1.7638	2.1970	1.3380	2.7413	1.3827	2.7576	1.3733	2.6915	1.2726	1.7670
	1									3.5339	0.7760	4.6691	0.8118	4.7270	0.8118	4.6691	0.7760	3.5339
Modified Chebyshev 0.01 dB	1	0.7584	1.4274	1.7122	1.7503	1.7503	1.7122	1.4274	0.7584									
0.10 dB	1	1.0880	1.5265	1.9029	1.8301	1.8301	1.9029	1.5265	1.0880									
0.50 dB	1	1.5372	1.4679	2.1033	1.8436	1.8436	2.1033	1.4679	1.5372									
1.0 dB	1	1.8642	1.3833	2.2357	1.8319	1.8319	2.2357	1.3833	1.8642									
3.0 dB	1	2.8062	1.1473	2.5784	1.7813	1.7813	2.5784	1.1473	2.8062									

(*Continued*)

TABLE 10.1.10 Low-Pass Filter Circuit Element Values* (*Continued*)

Tenth-order network $n = 10$

Filter type	R_s or G_s	Element number 1	2	3	4	5	6	7	8	9	10
Bessel	0	0.4170	0.2797	0.2311	0.2021	0.1770	0.1504	0.1209	0.0886	0.0541	0.0182
	1	0.6305	0.3002	0.2384	0.2066	0.1808	0.1539	0.1240	0.0911	0.0556	0.0187
Butterworth	0	1.5643	1.8552	1.8121	1.6869	1.5100	1.2921	1.0406	0.7626	0.4654	0.1564
	1	0.3129	0.9080	1.4142	1.7820	1.9754	1.9754	1.7820	1.4142	0.9080	0.3129
Legendre-Papoulis	0	1.6298	1.8741	1.9405	1.9223	1.9102	1.8629	1.7785	1.6082	1.2386	0.5065
	1	1.5286	1.8122	1.8953	2.0409	1.8453	2.0327	1.7839	1.8537	1.4454	0.7575
Chebyshev 0.01 dB	0	1.4598	1.7731	1.8054	1.8020	1.7664	1.7106	1.6003	1.4096	1.0439	0.4098
	1										
Chebyshev 0.10 dB	0	1.4964	1.8585	1.8600	1.9068	1.8489	1.8579	1.7503	1.6542	1.3124	0.6000
	1										
Chebyshev 0.50 dB	0	1.4539	1.9816	1.8119	2.0432	1.8165	2.0197	1.7645	1.8905	1.4748	0.8771
	1										
Chebyshev 1.0 dB	0	1.3801	2.1111	1.7215	2.1803	1.7307	2.1658	1.6962	2.0645	1.4817	1.0918
	1										
Chebyshev 3.0 dB	0	1.1032	2.6753	1.3774	2.7682	1.3893	2.7655	1.3761	2.6958	1.2744	1.7692
	1										
Modified 0.01 dB	1	0.7795	1.4503	1.7608	1.7903	1.8495	1.8495	1.7903	1.7608	1.4503	0.7795
Chebyshev 0.10 dB	1	1.1165	1.5249	1.9621	1.8183	1.9346	1.9346	1.8183	1.9621	1.5249	1.1165
0.50 dB	1	1.5832	1.4380	2.2072	1.7663	2.0029	2.0029	1.7663	2.2072	1.4380	1.5832
1.0 dB	1	1.9273	1.3374	2.3826	1.7118	2.0410	2.0410	1.7118	2.3826	1.3374	1.9273
3.0 dB	1	2.9356	1.0731	2.8661	1.5644	2.1269	2.1269	1.5644	2.8661	1.0731	2.9356

*Values for the Legendre-Papoulis approximation by permission from A. Zverev, "Handbook of Filter Synthesis," John Wiley & Sons, 1967.

TABLE 10.1.11   Half-Power Frequencies for Various Chebyshev and Bessel Filters[*]

	Order of filter								
Type of filter	$n=2$	$n=3$	$n=4$	$n=5$	$n=6$	$n=7$	$n=8$	$n=9$	$n=10$
Modified Chebyshev 0.01 dB	4.563742		1.532784		1.212468		1.114760		1.072036
Modified Chebyshev 0.10 dB	2.559727		1.246004		1.099300		1.053929		1.033948
Modified Chebyshev 0.50 dB	1.691974		1.108290		1.043913		1.023911		1.015073
Modified Chebyshev 1.00 dB	1.401865		1.061830		1.025105		1.013681		1.008628
Modified Chebyshev 3.00 dB	1.001188		1.000174		1.000071		1.000039		1.000024
Bessel half-power frequency	1.3617	1.7557	2.1140	2.4274	2.7034	2.9517	3.1797	3.3917	3.5910
Delay at half-power frequency	0.8090	0.9349	0.9819	0.9960	0.9993	0.9999	0.9999	1.0000	1.0000

[*]The Chebyshev filter prototypes have the ripple specified at 1 rad/s. The Bessel filter has a delay of 1 s at very low frequency. The second figure given is the delay at the half-power frequency.

Table 10.1.10 assumes:

1. $R_L = 1$.
2. $\omega_{hp} = 1$ for the Butterworth and Legendre-Papoulis approximations.
3. $\omega = 1$ is the end of the ripple band for the Chebyshev approximations.
4. The group delay at $\omega = 0$ is 1 s for the Bessel approximation.

The half-power frequencies for the (regular) Chebyshev approximation are found from Eq. (31), and, for the modified Chebyshev and Bessel approximations, are tabulated in Table 10.1.11.

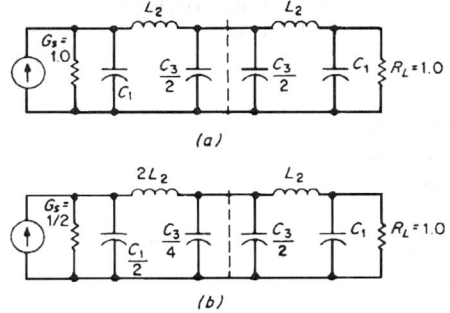

FIGURE 10.1.19   The scaling of symmetrical networks to change the value of G_S.

The two values of R_S or G_S, 0 and 1, used in Table 10.1.10, represent the two most commonly encountered situations. Zverev and Weinberg give tables for additional values of R_S or G_S.

When n is odd, $G_S = R_S = 1$, and the approximation is either Butterworth or (regular) Chebyshev, the networks are symmetrical; that is, $C_1 = C_n$, $L_2 = L_{n-1}$, etc. In these cases, there is a technique to realize other values of R_S or G. Consider Fig. 10.1.16b with $n = 5$, redrawn as Fig. 10.1.19a to emphasize the symmetry. Scale the half of the network containing G_S, and recombine the two center elements. Making $G_S = \frac{1}{2}$ leads to the network of Fig. 10.1.19b, where the two parts of the center capacitor are shown separately before recombination. This procedure leads to right-half s-plane zeros of the reflection coefficient.

EXPLICIT BUTTERWORTH AND CHEBYSHEV FORMULAS

Closed-form solutions for doubly terminated Butterworth and Chebyshev filters were derived by Takahasi and restated by Humpherys.[8] They assume that the zeros of the reflection coefficient lie in the left half plane, and so give different networks than the symmetrical network procedure. This gives the designer a choice in cases where both techniques may be applicable.

The Butterworth formulas make use of the poles of the transfer functions and require also a factor λ that relates the load and source resistances. The source resistance is assumed to be 1.0, and the formulas are

$$s_i = 2\sin(\pi i/2\pi) \tag{51}$$

$$c_i = 2\cos(\pi i/2\pi) \tag{52}$$

$$\lambda = -\left(\frac{R_L - 1}{R_L + 1}\right)^{1/n} \tag{53}$$

when the first reactive element is a shunt capacitor, and

$$\lambda = -\left(\frac{G_L - 1}{G_L + 1}\right)^{1/n} \tag{54}$$

when the first reactive element is a series inductor. Recursive equations give the element values

$$C_1 = \frac{s_1}{1-\lambda} \tag{55}$$

$$C_n = \frac{s_1}{(1+\lambda)R_L} \qquad n \text{ odd} \tag{56}$$

$$L_n = \frac{s_1 R_L}{1+\lambda} \qquad n \text{ even} \tag{57}$$

$$C_{2m-1}L_{2m} = \frac{s_{4m-1}s_{4m+1}}{1-\lambda c_{4m-2}+\lambda^2} \tag{58}$$

$$L_{2m-1}C_{2m+1} = \frac{s_{4m-1}s_{4m+1}}{1-\lambda c_{4m}+\lambda^2}$$

$$\text{where } m = \begin{cases} 1,2,\ldots,(n-1)/2 & n \text{ odd} \\ 1,2,\ldots,n/2 & n \text{ even} \end{cases}$$

When the first element is a series inductor, the roles of L and C are interchanged. An example follows. Develop the prototype network for third-order Butterworth, $R_L = 3$, $R_s = 1$. It follows that

$$\lambda = -0.7937 \tag{59}$$

$$s_1 = 1.0000 = s_5 \tag{60}$$

$$s_2 = 1.7321 = s_4 \tag{61}$$

$$s_3 = 2.0000 \tag{62}$$

$$c_1 = 1.7321 \tag{63}$$

$$c_2 = 1.0000 \tag{64}$$

$$c_3 = 0.0000 \tag{65}$$

$$c_4 = 1.0000 \tag{66}$$

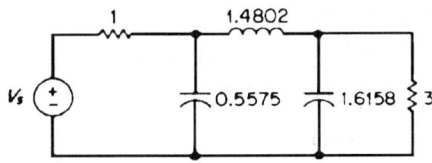

FIGURE 10.1.20 Prototype Butterworth network, $n = 3$, $R_S = 1$, $R_L = 3$.

$$C_1 = \frac{1.0000}{1 - (-0.7937)} = 0.5575 \tag{67}$$

$$L_2 = \frac{s_1 s_3}{1 - (-0.7937)(1.0000) + (0.7937)^2} \frac{1}{C_1} = 1.4802 \tag{68}$$

$$C_3 = \frac{s_3 s_5}{1 - (-0.7937)(-1.0000) + (0.7937)^2} \frac{1}{L_2} = 1.6158 \tag{69}$$

The prototype network thus becomes that of Fig. 10.1.20.

The results are similar if the equation for C_n is used first and the network is developed from the load end; this serves as a check on the work.

The Chebyshev equations are similar. Define the following terms:

$$\epsilon = \sqrt{10^{r/10} - 1} \quad \epsilon = \text{ripple factor}$$
$$r = \text{ripple, dB} \tag{70}$$

$$A = \begin{cases} 4R_L R_S/(R_L + R_S)^2 & n \text{ odd} \\ 4(1 + \epsilon^2)R_L R_S/(R_L + R_S)^2 \le 1 & n \text{ even} \end{cases} \tag{71a, 71b}$$

$$s_1 = 2\sin(\pi i/2n) \tag{72}$$

$$c_1 = 2\cos(\pi i/2n) \tag{73}$$

$$k = \left(\frac{1}{\epsilon} + \sqrt{\frac{1}{\epsilon^2} + 1} \right)^{1/n} \tag{74}$$

$$h = -\left(\sqrt{\frac{1-A}{\epsilon^2}} + \sqrt{\frac{1-A}{\epsilon^2} + 1} \right)^{1/n} \tag{75}$$

$$k' = k - 1/k \tag{76}$$

$$h' = h - 1/h \tag{77}$$

Recursive equations give the element values

$$C_1 = \frac{2s_1/R_s}{k' - h'} \tag{78}$$

$$C_n = \frac{2s_1/R_L}{k' + h'} \quad n \text{ odd} \tag{79}$$

$$L_n = \frac{2s_1 R_L}{k' - h'} \quad n \text{ even} \tag{80}$$

$$C_{2m-1} L_{2m} = \frac{4s_{4m-3}s_{4m-1}}{k'^2 - c_{2i}k'h' + h'^2 + s_{2i}^2} \tag{81}$$

$$L_{2m}C_{2m+1} = \frac{4S_{4m-1}S_{4m+1}}{k'^2 - c_{2i}k'h' + h'^2 + s_{2i}^2} \tag{82}$$

$$\text{where } m = \begin{cases} 1,2,\ldots,(n-1)/2 & n \text{ odd} \\ 1,2,\ldots,n/2 & n \text{ even} \end{cases}$$

When the first element is a series inductor, the roles of L and C are interchanged, while G_s and G_L are substituted for R_S and R_L. As an example, develop the prototype when $R_S = 3$, $R_L = 1$, 0.5 dB ripple, and a fourth-order network is needed. It follows that

$$\epsilon = \sqrt{10^{0.05} - 1} = 0.3493 \tag{83}$$

$$A = 0.75[1 + (0.3493)^2] = 0.8415 \tag{84}$$

$$s_1 = 0.7654 = s_7 \tag{85}$$

$$s_2 = 1.4142 = s_6 \tag{86}$$

$$s_3 = 1.8478 = s_5 \tag{87}$$

$$s_4 = 2.0000 \tag{88}$$

$$c_0 = 2.0000 \tag{89}$$

$$c_2 = 1.4142 \tag{90}$$

$$c_4 = 0.0000 \tag{91}$$

$$c_6 = -1.4142 \tag{92}$$

$$k = 1.5582 \tag{93}$$

$$h = -1.2766 \tag{94}$$

$$k' = 0.9164 \tag{95}$$

$$h' = -0.4933 \tag{96}$$

$$C_1 = \frac{2(0.7654)}{3[0.9164 - (-0.4933)]} = 0.3620 \tag{97}$$

$$L_2 = \frac{4s_1 s_3}{C_1(k'^2 - c_2 k'h' + h'^2 + s_2^2)} = 4.1985 \tag{98}$$

$$C_3 = \frac{4s_3 s_5}{L_2(k'^2 - c_4 k'h' + h'^2 + s_4^2)} = 0.6399 \tag{99}$$

$$L_4 = \frac{4s_5 s_7}{C_3(h'^2 - c_6 k'h' + h'^2 + s_6^2)} = 3.6172 \tag{100}$$

The prototype network is shown in Fig. 10.1.21.

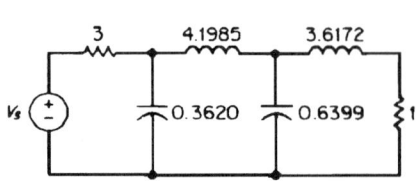

FIGURE 10.1.21 Prototype Chebyshev network, $R_S = 3$, $R_L = 1$, $n = 4$, 0.5 dB ripple.

THE INVERSE CHEBYSHEV APPROXIMATION

The Inverse Chebyshev approximation uses the Chebyshev polynomials in an approximation that decreases monotonically in the passband, and has ripples in the stopband. Specifically,

$$|G_{IC}(\omega)| = \sqrt{\frac{\epsilon^2 C_n^2(1/\omega)}{1+\epsilon^2 C_n^2(1/\omega)}} \tag{101}$$

where Eqs. (21) and (22) define C_n. Figure 10.1.22 shows this approximation and defines appropriate design criteria. K_s is the minimum attenuation in the stopband; K_p is the maximum attenuation in the passband, and it occurs at $\omega \geq \omega_p$. The edge of the stopband is at $\omega_a = 1$, which requires scaling of the specifications differently from the previous approximations.

When K_s is expressed in dB,

$$\epsilon = \sqrt{\frac{1}{10^{0.1k_s}-1}} \tag{102}$$

Equation (103) gives the order n_{IC} required to meet the specifications, with the next larger integer being chosen.

$$n_{IC} = \frac{\cosh^{-1}\sqrt{\dfrac{(10^{0.1k_s}-1)}{(10^{0.1k_p}-1)}}}{\cosh^{-1}(1/\omega_p)} \tag{103}$$

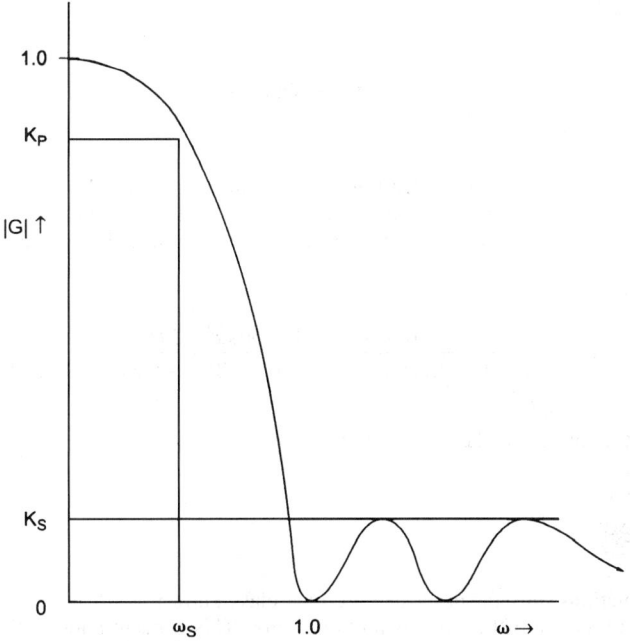

FIGURE 10.1.22 The Inverse Chebyshev low-pass approximation. A fifth-order function is illustrated.

In general, the order of inverse Chebyshev filter required to meet a set of specifications is identical with the order required for a conventional Chebyshev filter. It has improved step response and phase response characteristics when compared with the (regular) Chebyshev approximation, and a more complex network realization.

The stopband ripples and the zero magnitudes at finite frequencies in the stopband mean that the Inverse Chebyshev filter is not an *all-pole* approximation as the preceding have been, but rather it has roots in the numerator as well as in the denominator. Explicit formulas for these roots exist.

The numerator roots, or zeros, are

$$z_k = \alpha_k + j\beta_k \tag{104}$$

where

$$\alpha_k = 0 \quad \beta_k = \frac{1}{\cos u_k} \quad k = 1, 2, 3, \dots, n \tag{105}$$

and

$$u_k = \frac{2k-1}{2n}\pi \tag{106}$$

The denominator factors, or poles, are

$$p_k = \frac{1}{\sigma_k + j\omega_k} \quad k = 1, 2, 3, \dots n \tag{107}$$

where

$$\sigma_k = -\sin u_k \sinh v \quad \omega_k = \cos u_k \cosh v \tag{108}$$

and

$$v = \frac{1}{n}\sinh^{-1}\frac{1}{\epsilon} \tag{109}$$

and Eq. (106) gives u_k.

For example, assume that, using scaled frequency parameters, a designer needs a filter with a maximum passband attenuation K_p of 2.0 dB at frequencies below $\omega_p = 0.58$ rad/s, and a minimum stopband attenuation K_s of 40 dB above 1.0 rad/s. Equation (103) shows that a fifth-order filter is adequate. Application of Eqs. (104) through (109) shows that

$$G(s) = \frac{0.0500(s^2 + 1.1056)(s^2 + 2.8944)}{(s + 0.7878)(s^2 + 1.0496s + 0.5110)(s^2 + 0.3118s + 0.3975)} \tag{110}$$

where the numerator constant makes $G(0) = 1.0000$.

THE ELLIPTIC FILTER

The five approximations thus far described are of a class known as *all-pole approximations* because the magnitude-squared function of Eq. (6) has no finite zeros. This concentration of the zeros at infinity usually gives more attenuation than required at the higher frequencies and less in the vicinity of the cutoff frequency. If the more general form of Eq. (5) is used for the magnitude-squared function, some of the zeros can be placed

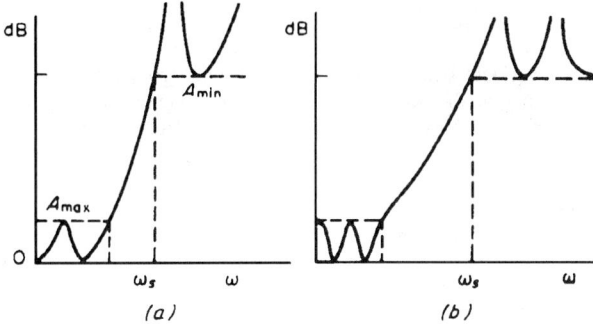

FIGURE 10.1.23   The elliptic approximation for (a) $n = 3$ and (b) $n = 4$.

close to the stopband edge frequency, with an improvement in the cutoff rate. Specifically, if

$$|G(j\omega)|^2 = K^2 \frac{1}{1 + \epsilon^2 R_n^2(\omega)} \tag{111}$$

where

$$R_n(\omega) = \begin{cases} r\dfrac{\omega\prod\limits_{i=1}^{(n-1)/2}(\omega^2 - \omega_{pi}^2)}{\prod\limits_{i=1}^{(n-1)/2}(\omega^2 - \omega_{si}^2)} & n \text{ odd} \tag{112} \\[4em] r\dfrac{\prod\limits_{i=1}^{n/2}(\omega^2 - \omega_{pi}^2)}{\prod\limits_{i=1}^{n/2}(\omega^2 - \omega_{si}^2)} & n \text{ even} \tag{113} \end{cases}$$

r being a multiplicative constant, the approximation can be made equiripple in both the passband and the stopband. Such an approximation is called an *elliptic approximation* because elliptic functions are used in its determination. Figure 10.1.23 shows such an approximation in a plot of attenuation versus frequency for $n = 3$ and $n = 4$, illustrative of the general characteristics for even and odd n, respectively.

Since the even-order elliptic approximation does not have infinite attenuation at infinite frequency, it cannot be realized by an LC ladder. A modified even-order elliptic approximation can be obtained by a frequency transformation. This transformation sacrifices some stopband attenuation to shift a pole of attenuation to ∞. A further modification of the even-order approximation can be used to make the attenuation zero to zero frequency. This permits an LC ladder realization to be equally terminated.

The elliptic approximation is characterized by four parameters.

1. The passband ripple A_{max} or, equivalently, the reflection coefficient ρ, related to A_{max} by $A_{max} = -10 \log (1 - \rho^2)$

2. The order n

3. The minimum stopband attenuation A_{min}

4. The stopband edge frequency ω_s or, equivalently, the modular angle θ[6]

Any three of these four parameters can be independently specified.

As for the all-pole approximations, tabulations of pole-zero locations and tables of element values for normalized low-pass filters based on the elliptic approximation are possible. Since three parameters are required to specify in elliptic approximation, these tabulations are voluminous.

DELAY EQUALIZATION

The ideal low-pass filter characteristic would have a constant group delay as well as a constant amplitude throughout the passband so that the various frequency components of a signal would arrive at the output unattenuated and with the proper phase relationship. Deviations from constant amplitude and group delay produce amplitude and phase distortion, respectively. While phase distortion is not a problem in many applications, it is significant when pulse transmission is involved.

With the exception of the Bessel approximation, the various approximations have focused on the amplitude characteristic. While the Bessel approximation has a good group-delay characteristic, it has a significantly poorer amplitude characteristic. One approach to the problem of obtaining filters with good amplitude *and* group-delay characteristics is through the use of delay equalizers. This consists of using one of the approximations with desirable amplitude characteristics in conjunction with an all-pass function that does not affect the amplitude characteristic but modifies the group delay beneficially.

The first-order all-pass function is

$$G(s) = \frac{s - \delta}{s + \delta} \tag{114}$$

This function has an emplitude of 1 for all frequencies and a group-delay function

$$\tau_d(\omega) = \frac{2}{\delta} \frac{1}{1 + (\omega/\delta)^2} \tag{115}$$

The specific properties of the group-delay function are controlled by the selection of δ.

The second-order all-pass function is

$$G(s) = \frac{(s - \delta)^2 + \beta^2}{(s + \delta)^2 + \beta^2} \tag{116}$$

with

$$|G(j\omega)| = 1 \tag{117}$$

and

$$\tau_d(\omega) = \frac{4\delta(\omega^2 + \delta^2 + \beta^2)}{(\delta^2 + \beta^2 - \omega^2)^2 + 4\delta^2\omega^2} \tag{118}$$

The specific properties of this group-delay function are controlled by the selection of δ and β.

Except for some simple situations, the problem of delay equalization is best approached by using a computer to optimize the group delay in some sense. Blinchikoff and Zverev[11] give an excellent discussion of a least-squares optimization as well as the basic principles of delay equalization.

TABLES

Many authors have published extensive tables of prototype element values for filters. Table 10.1.12 lists some of the readily available tables, with comments.

TABLE 10.1.12 Books Containing Element Values for Filters

First-named author	Reference number in references and bibliography	Length, pages	Comment
Biey	33	624	Cauer (elliptic and multicritical-pole equal-ripple rational (MCPER) functions
Biey	34	561	Active Cauer and MCPER functions
Christian	35	310	Nomographs, poles, and zeros for Butterworth, Chebyshev, inverted Chebyshev, and Cauer filters
Craig	36	197	Nomographs and tables for lossy Butterworth and Chebyshev filters
Genesio	37	598	Constants for digital representations of Butterworth and Chebyshev filters
Hansell	38	203	Passive filter parameters, attenuation and phase data
Johnson	39	244	60 pages of tables. Design formulas
Moschytz	21	316	Calculator and computer programs
Saal	31	662	Element value tables. Design formulas
Weinberg	7	692	70 pages of element value tables. Design formulas
Wetherhold		9	Tables using standard capacitor sizes for 50-Ω,
	40	22	5- and 7-element Chebyshev, and for 5-element elliptic filters
Williams	41	540	116 pages of tables. Design techniques and
	42		formulas for active networks
Zverev	6	576	About 150 pages of tables for Butterworth, Chebyshev, Bessel, linear phase, Gaussian, Legendre, Cauer (elliptic) filters. Design formulas. Crystal filters

CHAPTER 10.2
FILTER DESIGN

Edwin C. Jones, Jr., Harry W. Hale

PRACTICAL FILTER DESIGN

Table 10.1.10 gives element value for normalized networks. Three general steps lead to practical filters.

1. Statement of the filter specifications
2. Translation of the specifications to equivalent statements for the normalized prototype and their use to determine the parameters necessary to enter Table 10.1.10
3. Application of frequency transformations and impedance-level scaling to the normalized prototype to determine the filter network

This process is described in the following paragraphs for the various types of filters, accompanied by examples.

THE LOW-PASS FILTER

The low-pass filter is related to the normalized prototype by frequency and impedance-level scaling. The frequency in the prototype is related to the frequency in the low-pass filter by

$$\omega = \omega'/\omega'_c \tag{1}$$

where the primed quantities refer to the practical low-pass filter, with ω'_c being its cutoff frequency. The application of the frequency scaling implied by Eq. (1) and the impedance-level scaling required to change the load resistance from 1.0 to R_L results in the elements of the low-pass filter becoming those indicated in Fig. 10.2.1.

> **Example.** A low-pass filter is to be designed to have (a) a maximally flat amplitude-versus-frequency characteristic at $f = 0$, (b) a cutoff (half-power) frequency of 2 kHz, (c) a load resistance of 200 Ω, (d) a voltage source input with zero resistance, and (e) an attenuation of not less than 20 dB at frequencies greater than 3.2 kHz. A Butterworth approximation is indicated. The frequency in the prototype equivalent to 3.2 kHz is
>
> $$\omega = \frac{(3.2 \times 10^3)(2\pi)}{(2 \times 10^3)(2\pi)} = 1.6 \tag{2}$$

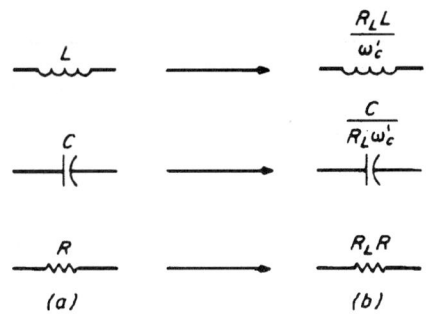

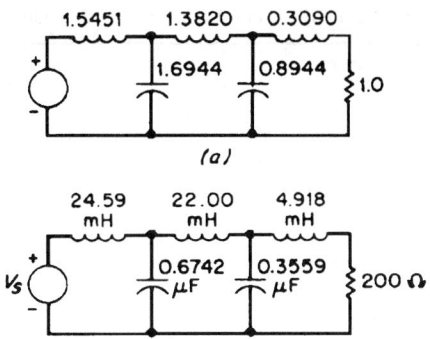

(a)

(b)

FIGURE 10.2.1 Relations between elements of (a) a normalized prototype and those of (b) a practical low-pass filter.

FIGURE 10.2.2 (a) A normalized prototype and (b) resulting low-pass filter.

and the equivalent design specifications for the prototype require that the attenuation be no less than 20 dB for $\omega \geq 1.6$. This, using Eq. (16), results in

$$n \geq \frac{20}{20 \log 1.6} = 4.899 \tag{3}$$

and the next larger integer value of 5 is used. The appropriate prototype appears in Fig. 10.2.2a with element values taken from Table 10.1.10 for the Butterworth approximation with $n = 5$ and $R_s = 0$. The application of frequency and magnitude scaling by means of the relations in Fig. 10.2.1 results in the low-pass filter of Fig. 10.2.2b.

A TIME-DELAY NETWORK

The frequencies in the time-delay network and its normalized prototype are related by

$$\omega = \tau'_a \omega' \tag{4}$$

where the primed quantities again refer to the practical network, τ'_d being the group delay evaluated at $\omega' = 0$. The element values of the practical network are related to those of the normalized prototype by frequency and impedance-level scaling. These relations are shown in Fig. 10.2.3.

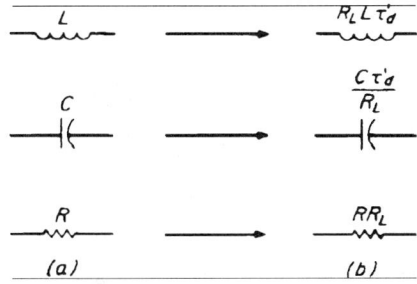

(a)   (b)

FIGURE 10.2.3 Relations between elements for (a) a normalized prototype and those of (b) a time-delay network.

Example. A time-delay network is to be designed to have (a) a time delay of 1 ms, the error being no greater than 5 percent at 500 Hz; (b) a voltage input with a resistance of 500 Ω; and (c) a load resistance of 500 Ω. The prototype frequency equivalent to 500 Hz is

$$\omega = 10^{-3}(2\pi \times 500) = 3.142 \tag{5}$$

and the equivalent specifications for the prototype require a time delay of 1 s, the error being no greater than 5 percent at $\omega = 3.142$. Figure 10.1.12 is used to determine that $n = 5$ is required to satisfy this requirement. The prototype can be of the form of either Fig. 10.1.16d or Fig. 10.1.17b. The latter is chosen with the element values taken from Table 10.2.1 for

TABLE 10.2.1   Element Choices for GIC Biquadratic Filters

Filter type	Y_0	Y_1	Y_2	Y_3	Y_4	Y_5	Y_A	Y_B
Low-pass	0	$G_1 + sC_1$	G_2	sC_3	G_4	G_5	0	G_B
High-pass	G_0	G_1	sC_2	G_3	G_4	G_5	0	sC_B
Bandpass	sC_0	G_1	sC_2	G_3	G_4	G_5	0	G_B
Band-stop, version 1	0	sC_1	G_2	G_3	G_4	G_5	sC_A	G_B
Version 2	G_0	G_1	G_2	G_3	sC_4	0	G_A	sC_B
All-pass	G_0	G_1	sC_2	G_3	G_4	0	$G_A = G_4$	sC_B

the Bessel approximation with $n = 5$ and $G_s = 1$. The prototype is shown in Fig. 10.2.4a and the application of the relation is Fig. 10.2.3 results in the time-delay network of Fig. 10.2.4b.

THE HIGH-PASS FILTER

The frequency in the high-pass filter is related to that in the normalized prototype by

$$\omega = \omega_c' / \omega' \tag{6}$$

where the primed quantities refer to the high-pass filter, ω_c' being the cutoff frequency. The relations between the elements of the normalized prototype and those of the high-pass filter are given in Fig. 10.2.5.

> **Example.**   A high-pass filter is to be designed to have (a) a cutoff frequency of 7.5 kHz, (b) an equiripple characteristic in the passband with a 0.5-dB ripple ($\epsilon = 0.3493$), (c) an attenuation of at least 35 dB at frequencies below 3 kHz, (d) equal load and source resistance of 2000 Ω, and (e) a voltage source input. The prototype frequency equivalent to 3 kHz is
>
> $$\omega = \frac{2\pi(7.5 \times 10^3)}{2\pi(3 \times 10^3)} = 2.5 \tag{7}$$

Thus the prototype is based on a Chebyshev approximation with a 0.5-dB ripple, equal load and source terminations, voltage source input, and an attenuation of at least 35 dB for frequencies greater than $\omega = 2.5$. The last requirement, used with Eq. (29), results in $C_n(2.5) \approx 161.0$. This result, used in Eq. (30), yields $n = 3.69$, and $n = 4$ is used. Since the regular Chebyshev approximation is not available, the modified Chebyshev approximation is used. The networks of Figs. 10.1.16c and 10.1.17a are possible. The former is chosen, and the

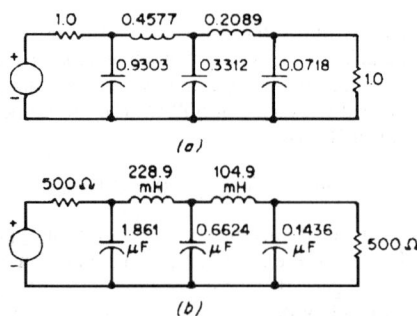

FIGURE 10.2.4   (a) A normalized prototype and (b) resulting time-delay network.

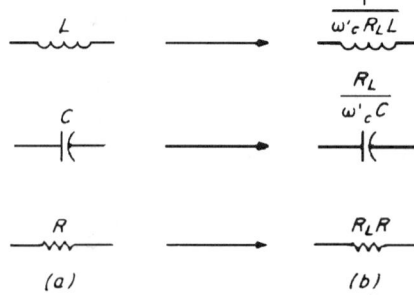

FIGURE 10.2.5   Relations between elements of (a) a normalized prototype and those of (b) a high-pass filter.

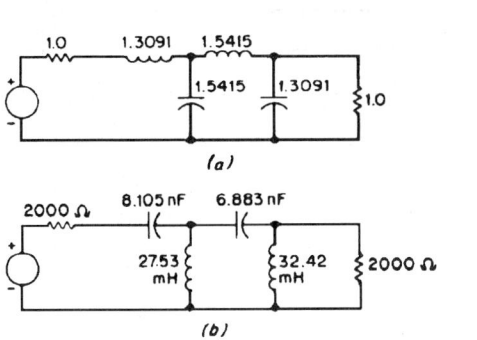

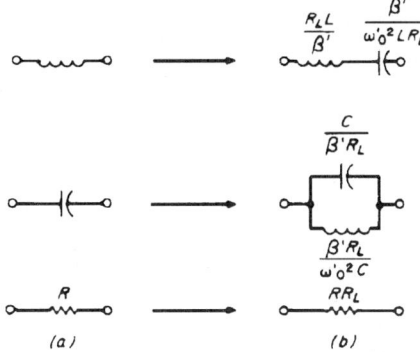

FIGURE 10.2.6 (*a*) A normalized prototype and (*b*) resulting high-pass filter.

FIGURE 10.2.7 Relations between elements of (*a*) a normalized prototype and those of (*b*) a bandpass filter.

resulting prototype is shown in Fig. 10.2.6*a*. The relations of Fig. 10.2.5 are then used to determine the element values of the high-pass filter of Fig. 10.2.6*b*.

THE BANDPASS FILTER

The frequency in the bandpass filter is related to that in the normalized prototype by

$$\omega = \frac{\omega'_0}{\beta'}\left(\frac{\omega'}{\omega'_0} - \frac{\omega'_0}{\omega'}\right) \tag{8}$$

where the primed quantities refer to the bandpass filter, ω'_0 being the center frequency and β' the bandwidth, $\omega'_{c2} - \omega'_{c1}$, as defined in Fig. 10.1.2*c*. The elements of the normalized prototype and those of the bandpass filter are related as shown in Fig. 10.2.7.

Example. A bandpass filter is to be designed to have (*a*) a center frequency of 4.0 kHz, (*b*) a bandwidth of 900 Hz, (*c*) an equiripple characteristic in the passband with a 1-dB ripple ($\epsilon = 0.5088$), (*d*) an attenuation of at least 18 dB at 4.9 kHz, (*e*) equal load and source resistances of 200 Ω, and (*f*) a current source input. The prototype frequency equivalent to 4.9 kHz is, from Eq. (8), 1.816 and the prototype is based on a Chebyshev approximation with a 1-dB ripple, equal load and source terminations, current source input, and an attenuation of at least 18 dB at $\omega = 1.816$. The last requirement, used with Eq. (29), results in $C_n(1.8163) \approx 19.65$. This result, used in Eq. (30), yields $n = 2.86$, and $n = 3$ is used. Two networks, those of Figs. 10.1.16*a* and 10.1.17*d*, are possible. The former is chosen, and the resulting prototype is shown in Fig. 10.2.8*a*. The relations of Fig. 10.2.7 are then used to determine the element values of the bandpass filter of Fig. 10.2.8*b*.

The results of the preceding example illustrate a problem in the transformation from prototype to bandpass filter. For even the moderate ratio of center frequency to bandwidth of the example, the ratio of series-arm to shunt-arm inductances in the bandpass filter is large. This creates some practical problems arising from the stray capacitances associated with large inductances. The reader is referred to Humpherys[8] for an excellent

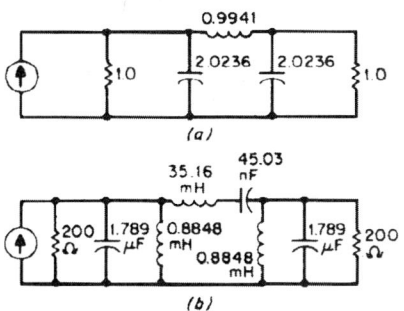

FIGURE 10.2.8 (*a*) A normalized prototype and (*b*) resulting bandpass filter.

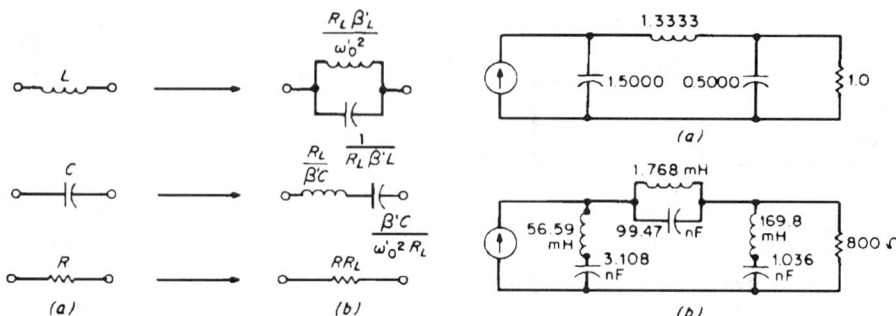

FIGURE 10.2.9 Relations between elements of (*a*) a normalized prototype and those of (*b*) a band-elimination filter.

FIGURE 10.2.10 (*a*) A normalized prototype and (*b*) resulting band-elimination filter.

discussion of this problem and possible ways of overcoming the difficulty. A similarly large ratio of shunt-arm to series-arm capacitances is also present, although the practical problems are not so severe.

THE BAND-ELIMINATION FILTER

The frequency in the band-elimination filter is related to that in the normalized prototype by

$$\omega = \frac{1}{(\omega_0'/\beta')[(\omega_0'/\omega') - \omega'/\omega_0']} \tag{9}$$

where the primed quantities refer to the band-elimination filter, ω_0' being the center frequency and β' the bandwidth, $\omega_{c2}' - \omega_{c1}'$, as defined in Fig. 10.1.2*d*. The elements of the normalized prototype and those of the band-elimination filter are related as shown in Fig. 10.2.9.

> **Example.** A band-elimination filter is to be designed to have (*a*) a center frequency of 12 kHz, (*b*) a bandwidth (half-power) of 1.5 kHz, (*c*) a maximally flat characteristic at $\omega' = 0$, (*d*) an attenuation of at least 16 dB at 11.6 kHz, (*e*) a current source input, and (*f*) $R_I = 800\ \Omega$ and $G_S = 0$. Since the prototype frequency equivalent to 11.6 kHz is, from Eq. (9), 1.8432, the prototype is based on a Butterworth approximation with an attenuation of at least 16 dB at this frequency. This requirement, used with Eq. (16), results in $n = 2.99$, and $n = 3$ is used. Since $G_S = 0$, the only network available is that of Fig. 10.2.17*b*, and the resulting prototype is shown in Fig. 10.2.10*a*. The relations of Fig. 10.2.9 are then used to determine the element values of the band-elimination filter of Fig. 10.2.10*b*.
>
> This example illustrates the large ratios of shunt-arm to series-arm inductances and of series-arm to shunt-arm capacitances that result with even a moderate ratio of center frequency to bandwidth. The reader is again referred to Humpherys[8] for a discussion of this difficulty and steps that can be taken to overcome it.

CHAPTER 10.3
ACTIVE FILTERS

Edwin C. Jones, Jr., Harry W. Hale

INTRODUCTION

Active filters are electronic filter circuits that incorporate one or more electronic devices in their realizations. They include operational amplifiers, operational transconductances amplifiers, and, in some cases, special transistors. Many are available in integrated circuit form. Active filters eliminate inductors and usually provide a voltage or current gain. Three basic techniques are synthetic inductance networks, infinite gain networks, and controlled source (including Sallen and Key) realizations.

The factors computed from Eqs. (17) (Butterworth), (33), and (44) (Chebyshev), or tabulated in Tables 10.1.2 (modified Chebyshev), 10.1.4 (Legendre-Papoulis), or 10.1.6 (Bessel) are all low-pass functions, and are all denominator polynomials. In the functions, the numerators are constants. To transform these functions to high pass, use the change of variable

$$s = 1/p \tag{1}$$

followed by frequency scaling.

To transform to bandpass functions, use the transformation

$$s = (p^2 + \omega_0^2)/Bp \tag{2}$$

on a factor-by-factor basis to avoid later factoring of high-ordered polynomials. Use the quadratic equation on Eq. (2) to transform bandpass variables to equivalent low-pass variables. An nth order, low-pass function is of order $2n$ when this transform is used, and the numerator includes a factor p^n and a constant. In this expression, B = bandwidth, $\omega_{c2} - \omega_{c1}$, and ω_0 = center frequency (Fig. 10.1.2, rad/s).

To transform to a band reject function, use the reciprocal of Eq. (2).

FIRST-ORDER STAGES

Most active network realizations require the cascading of second-order networks, each of which realizes one of the second-order denominator polynomials. To realize a first-order factor (real poles), use either of the networks of Fig. 10.3.1 as the initial or final stage in the cascade. In Fig. 10.3.1,

$$\frac{V_2}{V_1}(s) = -K_1\left(\frac{s+a}{s+b}\right) \tag{3}$$

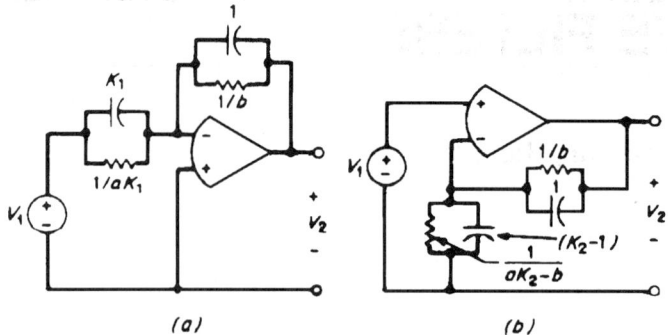

FIGURE 10.3.1 (*a*) Inverting and (*b*) noninverting amplifiers to realize real poles.

and in Fig. 10.3.1,

$$\frac{V_2}{V_1}(s) = K_2\left(\frac{s+a}{s+b}\right)$$

(4)

In Fig. 10.3.1, $K_2 \geq 1$, and $aK_2 \geq b$.

SYNTHETIC INDUCTANCE FILTERS—GYRATORS

It is possible to build filters with active elements that replace inductors. These realizations use the published tables for passive low-pass networks and the networks that are transformed from them. They also have the low-sensitivity properties of passive networks. One technique is based on the gyrator,[24] a two-port with a z-matrix representation

$$\begin{bmatrix} V_1 \\ V_2 \end{bmatrix} = \begin{bmatrix} 0 & -a \\ a & 0 \end{bmatrix}\begin{bmatrix} I_1 \\ I_2 \end{bmatrix}$$

(5)

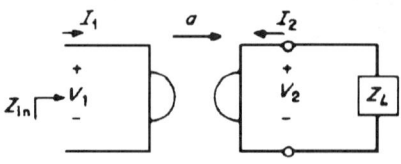

FIGURE 10.3.2 Gyrator terminated with Z_L.

where a is called the *gyration resistance*. A gyrator terminated in an impedance Z_L is shown in Fig. 10.3.2, and it is possible to show that

$$V_1/I_1 = Z_{IN} = a^2/Z_L$$

(6)

when $Z_L = 1/sC$,

$$Z_{IN} = a^2 sC$$

(7)

which is equivalent to an inductance of a^2C. The problem then becomes one of choosing a and C.

Gyrators are available in integrated-circuit form from several electronics manufacturers; the user simply adds the appropriate capacitor. Gyrators can also be built with operational amplifiers. One useful circuit is the Riordan gyrator, shown in Fig. 10.3.3 with Z_L as one of the five impedances it requires.

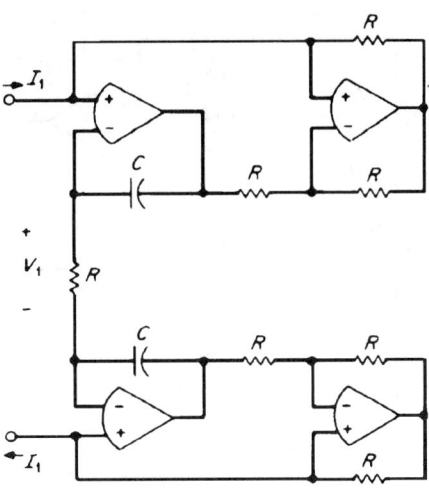

FIGURE 10.3.4   Riordan-type back-to-back gyrator for floating inductors.

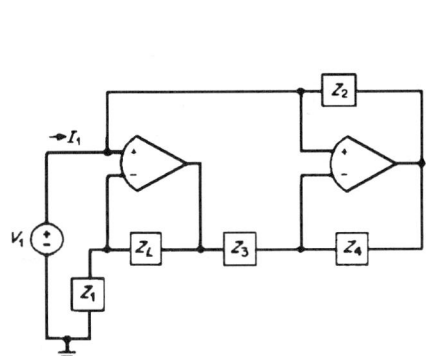

FIGURE 10.3.3   Riordan gyrator.

It can be shown that

$$\frac{V_1}{I_1} = Z_{IN} = \frac{Z_1 Z_2 Z_3}{Z_4} \frac{1}{Z_L} \tag{8}$$

When all Z's are resistors R, and $Z_L = 1/sC$,

$$Z_{IN} = R^2 sC \tag{9}$$

This circuit must be used to replace a grounded inductor, e.g., in a high-pass network. If an ungrounded inductor is needed, the modification shown in Fig. 10.3.4 can be used, for which

$$Z_{IN} = R^2 sC \tag{10}$$

Other circuits, for which one side of the added capacitor C may be grounded, can also be used, though they may be more difficult to align.

SYNTHETIC INDUCTANCE FILTERS—FREQUENCY-DEPENDENT NEGATIVE RESISTORS

Bruton[25] introduced a new type of circuit, called the *frequency-dependent negative resistor* (FDNR); the symbol is shown in Fig. 10.3.5. To use this idea, consider a new type of impedance scaling, in which passive elements become scaled by A/s; this does not affect the transfer function:

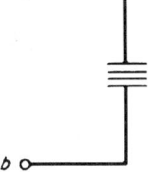

FIGURE 10.3.5   FDNR representation.

Passive impedance		Scaled impedance	
sL	$\rightarrow$	AL	(11a)
R	$\rightarrow$	AR/s	(11b)
$1/sC$	$\rightarrow$	A/Cs^2	(11c)

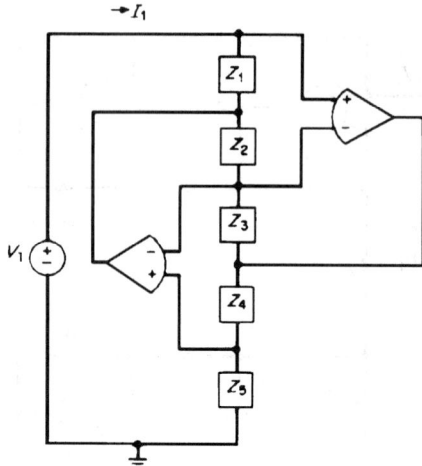

FIGURE 10.3.6 Generalized impedance converter.

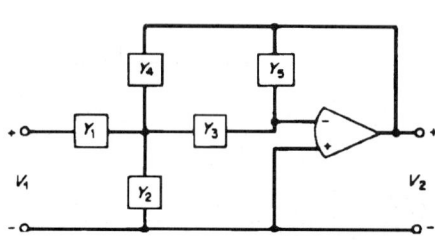

FIGURE 10.3.7 A multiple-feedback, infinite-gain active realization for low-pass, high-pass, and bandpass filters. Table 10.3.1 indicates the choice of the five active elements for each case.

When $s = j\omega$ or $s^2 = -\omega^2$, the last element becomes

$$Z(j\omega) = -A/D\omega^2 \tag{12}$$

a real, negative, frequency-dependent resistor. Thus, inductors are replaced by resistors, resistors by capacitors, and capacitors by FDNRs.

Bruton also gives a circuit for an FDNR, as a special case of the generalized impedance converter (GIC). It is similar to the gyrator and is shown in Fig. 10.3.6. For this circuit,

$$V_1/I_1 = Z_{IN} = Z_1 Z_3 Z_5 / Z_2 Z_4 \tag{13}$$

when

$$Z_1 = Z_3 = 1/sC \tag{14}$$

$$Z_2 = Z_4 = Z_5 = R \tag{15}$$

$$Z_{IN} = 1/s^2 R C^2 \tag{16}$$

which is an FDNR. GICs are available from integrated-electronic-circuit manufacturers, and the user adds resistors and capacitors to make FDNRs. They can also be used to make gyrators. FDNRs are grounded when replacing capacitors in low-pass networks; floating FDNRs can be achieved by back-to-back GICs, as with gyrators.

INFINITE-GAIN, MULTIPLE-FEEDBACK REALIZATION

The circuit of Fig. 10.3.7 shows an operational amplifier with five passive elements, which are either resistors or capacitors. The general voltage transfer function is

$$\frac{V_2}{V_1} = \frac{-Y_1 Y_3}{Y_5(Y_1 + Y_2 + Y_3 + Y_4) + Y_3 Y_4} \tag{17}$$

TABLE 10.3.1 Element Choices for Active Filter Circuit of Fig. 10.3.7.

Filter desired	Y_1	Y_2	Y_3	Y_4	Y_5
Low-pass	Resistor	Capacitor	Resistor	Resistor	Capacitor
High-pass	Capacitor	Resistor	Capacitor	Capacitor	Resistor
Bandpass	Resistor	Resistor	Capacitor	Capacitor	Resistor

Table 10.3.1 describes how the five passive elements can be chosen to implement a low-pass, high-pass, or bandpass network.

STATE-VARIABLE REALIZATION

This network is a special but important type of infinite-gain realization. It has the advantage that low-pass, high-pass, and bandpass configuration can be realized simultaneously, and it is also easy to adjust and to produce in quantity. The network is shown in Fig. 10.3.8, and the three possible transfer functions are

Low-pass:
$$\frac{V_{lp}}{V_1} = \frac{\dfrac{R_4(R_5 + R_6)}{R_1 R_2 R_5 C_1 C_2 (R_3 + R_4)}}{s^2 + s\left[\dfrac{R_3(R_5 + R_6)}{R_1 C_1 (R_3 + R_4)}\right] + \dfrac{R_6}{R_1 R_2 R_5 C_1 C_2}} \tag{18}$$

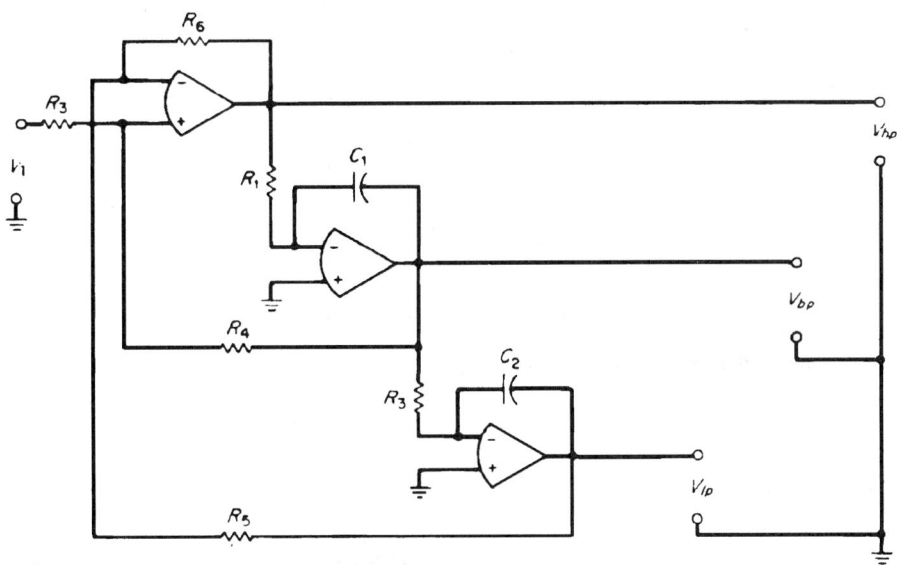

FIGURE 10.3.8 State-variable realization of second-order active filters.

High-pass:

$$\frac{V_{hp}}{V_1} = \frac{\dfrac{s^2 R_4 (R_5 + R_6)}{R_5 (R_3 + R_4)}}{s^2 + s\left[\dfrac{R_3 (R_5 + R_6)}{R_1 C_1 (R_3 + R_4)}\right] + \dfrac{R_6}{R_1 R_2 R_5 C_1 C_2}} \tag{19}$$

Bandpass:

$$\frac{V_{bp}}{V_1} = \frac{\dfrac{-s R_4 (R_5 + R_6)}{R_1 R_5 C_1 (R_3 + R_4)}}{s^2 + s\left[\dfrac{R_3 (R_5 + R_6)}{R_1 C_1 (R_3 + R_4)}\right] + \dfrac{R_6}{R_1 R_2 R_5 C_1 C_2}} \tag{20}$$

The network has a low output impedance at each terminal, so that RC sections can be added for odd-ordered networks. These sections can be cascaded so that networks of order 4 and higher can be built. They have the property that performance variations with parameter changes are comparable with those of strictly passive networks, but the disadvantage of requiring three operational amplifiers.

As with the other networks, design is a matter of choosing the appropriate quadratic factors, matching coefficients with other constraints that arise from technological and economic considerations, and finally, scaling impedance and frequency.

DELYIANNIS BANDPASS CIRCUIT

Daryanani[44] describes a useful bandpass circuit shown in Fig. 10.3.9. For this circuit

$$\frac{V_2}{V_1}(s) = \frac{-s/[R_1 C_2 (1 - 1/k)]}{s^2 + (\omega_p/Q_p)s + \omega_p^2} \tag{21}$$

where

$$k = 1 + \frac{R_B}{R_A} \tag{22}$$

$$\omega_p^2 = \frac{1}{R_1 R_2 C_1 C_2} \tag{23}$$

$$\frac{\omega_p}{Q_p} = \frac{1}{R_2 C_1} + \frac{1}{R_2 C_2} - \frac{1}{k-1} \frac{1}{R_1 C_2} \tag{24}$$

The circuit can be used with $R_A = 0$; in this case $k \to \infty$, and Eq. (21) is readily simplified. Here it is necessary to limit Q_p to about 5.

When Q_p greater than 5 is desired, R_A is included and the difference term in the denominator permits a greater Q_p.

FRIEND BIQUADRATIC

Friend[26] has described a generalization of the Delyiannis circuit of the preceding section that can be used for high-pass, band-reject, and all-pass networks with proper choice of components. It is shown in Fig. 10.3.10. For this circuit

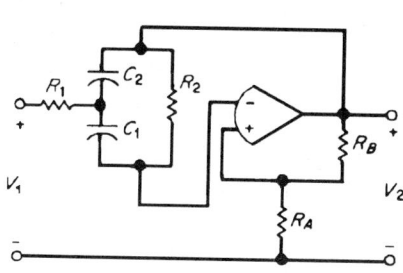

FIGURE 10.3.9 Delyiannis bandpass circuit.

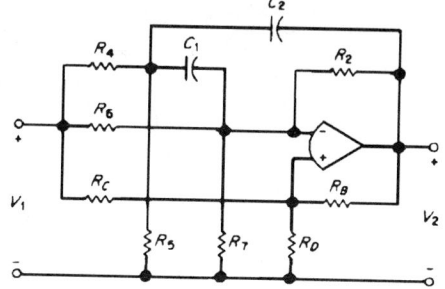

FIGURE 10.3.10 Friend biquadratic circuit.

$$\frac{V_2}{V_1}(s) = \frac{K_2 s^2 + as + b}{s^2 + (\omega_p / Q_p)s + \omega_p^2} \tag{25}$$

where

$$a = \frac{K_2}{C_2}\left(\frac{1}{R_1} + \frac{1}{R_2} + \frac{1}{R_3}\right) + \frac{K_2}{C_1}\left(\frac{1}{R_2} + \frac{1}{R_3}\right) - \frac{K_1}{R_1 C_2}\left(1 + \frac{R_A}{R_B}\right)$$

$$- \frac{K_3}{R_3}\left(\frac{1}{C_1} + \frac{1}{C_2}\right)\left(1 + \frac{R_A}{R_B}\right)$$

$$b = \frac{1}{C_1 C_2}\left[\frac{K_2}{R_1}\left(\frac{1}{R_2} + \frac{1}{R_3}\right) - \frac{K_3}{R_1 R_3}\left(1 + \frac{R_A}{R_B}\right)\right] \tag{26}$$

$$\frac{\omega_p}{Q_p} = \frac{C + C_2}{C_1 C_2}\left(\frac{1}{R_2} - \frac{R_A}{R_B R_3}\right) - \frac{R_A}{R_B R_1 C_2}$$

$$\omega_p^2 = \frac{1}{R_1 C_1 C_2}\left(\frac{1}{R_2} - \frac{R_A}{R_B R_3}\right)$$

and

$$K_1 = \frac{R_5}{R_1 + R_5} \qquad K_2 = \frac{R_D}{R_C + R_D} \qquad K_3 = \frac{R_7}{R_6 + R_7} \tag{27}$$

$$R_4 = \frac{R_C R_D}{R_C + R_D} \qquad R_1 = \frac{R_4 R_5}{R_4 + R_5} \qquad R_3 = \frac{R_6 R_7}{R_6 + R_7}$$

With this circuit, normal practice is to make $C_1 = C_2$. If a high-pass network is needed, $a = b = 0$; this leads to constraints on K_1, K_3, and the resistors. For band-reject or notch filters, $a = 0$, other constraints follow from Eqs. (26), but it is usually possible to adjust them so that all element values are positive.

BIQUADRATICS WITH GENERALIZED IMPEDANCE CONVERTERS

Temes describes a general method for designing any biquadratic transfer function using two operational amplifiers and eight impedances. The GIC, introduced earlier, can be used. A complete circuit is shown in Fig. 10.3.11. For this circuit,

$$\frac{V_2}{V_1}(s) = \frac{Y_A(Y_1 Y_3 - Y_0 Y_3) + Y_B Y_2(Y_4 + Y_5)}{Y_1 Y_3(Y_A + Y_5) + Y_2 Y_4(Y_B + Y_0)} \tag{28}$$

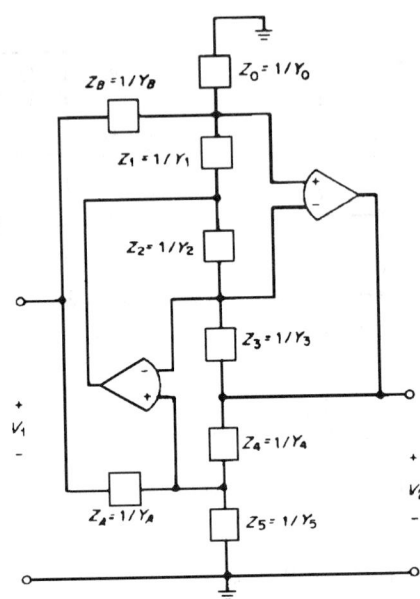

FIGURE 10.3.11   GIC biquadratic filter.

Table 10.2.1 shows to choose the elements for various types of transfer function. In version 1, the notch frequency is greater than the resonant frequency of the circuit, while in version 2, the notch frequency is less than the resonant frequency. In addition, version 2 requires $G_1G_3 > G_0G_2$. These circuits have the property of low sensitivity at the expense of two amplifiers.

SALLEN AND KEY NETWORKS

The circuits of Figs. 10.3.12, 10.3.13, and 10.3.14 are low-pass, high-pass, and bandpass circuits, respectively, having a positive gain K. Design of any of these circuits requires choice of suitable linear and quadratic denominator factors, transformation and frequency scaling, and coefficient matching. Since there are more elements

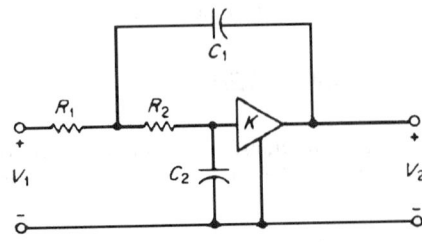

FIGURE 10.3.12   A low-pass active filter network with gain K greater than 0.

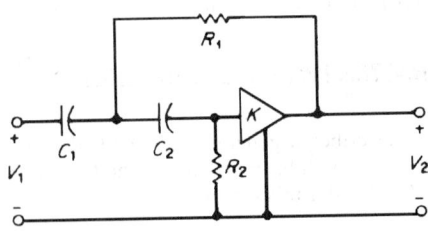

FIGURE 10.3.13   A high-pass active filter network with gain K greater than 0.

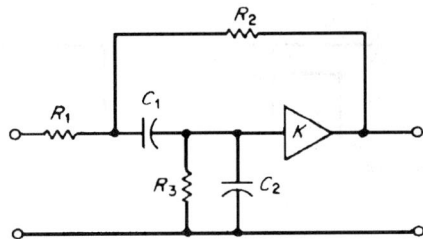

FIGURE 10.3.14   A bandpass active filter network with gain K greater than 0.

to be specified than there are constraints, two elements may be chosen arbitrarily. Often, $K = 1$ or $K = 2$ leads to a good network. For Fig. 10.3.12,

$$\frac{V_2}{V_1}(s) = \frac{K/(R_1R_2C_1C_2)}{s^2 + \left[(1-K)\dfrac{1}{R_2C_2} + \dfrac{1}{R_1C_1} + \dfrac{1}{R_2C_1}\right]s + \dfrac{1}{R_1R_2C_1C_2}} \tag{29}$$

For Fig. 10.3.13,

$$\frac{V_2}{V_1}(s) = \frac{Ks^2}{s_2 + \left[(1-K)\dfrac{1}{R_1C_1} + \dfrac{1}{R_2C_2} + \dfrac{1}{R_2C_1}\right]s + \dfrac{1}{R_1R_2C_1C_2}} \tag{30}$$

For Fig. 10.3.14,

$$\frac{V_2}{V_1}(s) = \frac{\dfrac{Ks}{R_1C_2}}{s^2 + \left[\dfrac{(1-K)}{R_2C_2} + \dfrac{1}{R_3C_2} + \dfrac{1}{R_1C_1} + \dfrac{1}{R_2C_1} + \dfrac{1}{R_1C_2}\right]s + \dfrac{1}{R_3C_1C_2}\left(\dfrac{1}{R_1} + \dfrac{1}{R_2}\right)} \tag{31}$$

CHAIN NETWORK

Figure 10.3.15 shows a chain network that realizes low-pass functions and is easily designed. For this circuit,

$$\frac{V_2}{V_1}(s) = \frac{\omega_1\omega_2\omega_3\cdots\omega_n}{s^n + \omega_1 s^{n-1} + \omega_1\omega_2 s^{n-2} + \cdots + \omega_1\omega_2\omega_3\cdots\omega_n} \tag{32}$$

where

$$\omega_i = 1/R_iC_i \tag{33}$$

As an example, consider a third-order Bessel filter, for which

$$\frac{V_2}{V_1}(s) = \frac{15}{s^3 + 6s^2 + 15s + 15} \tag{34}$$

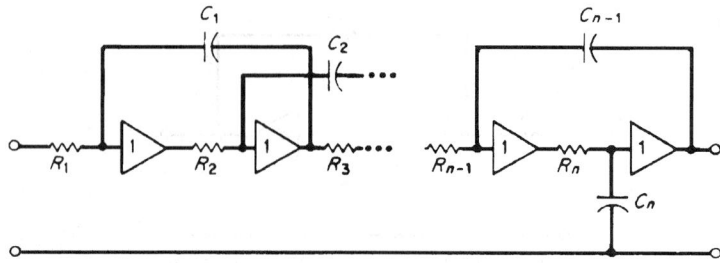

FIGURE 10.3.15   An RC-unity-gain amplifier realization of an active low-pass filter.

Choose $1/R_1C_1 = 6$, $6(1/R_2C_2) = 15$, and $15(1/R_3C_3) = 15$. If all C's are set to 1.0, then $R_1 = 1/6$, $R_2 = 2/5$, and $R_3 = 1$. Use frequency and impedance scaling as required.

LEAPFROG FILTERS

Also called active ladders or multiple feedback filters, these circuits use the tabulated element values from Table 10.1.10 to develop a set of active networks that have the sensitivity characteristics of passive ladders. The process may be extended from low-pass to bandpass networks using the transformation from prototype to bandpass filter disc under "Bandpass Filter" and techniques that will be discussed in this paragraph. The term "leapfrog" was suggested by Girling and Good,[28] the inventors, because of the topology.

Figure 10.3.16*a* shows a conventional fourth-order low-pass prototype network, and Fig. 10.3.16*b* shows a block diagram of a simulation with the same equations, which follow, using Laplace notation. In writing these equations and preparing the block diagram, current terms have been multiplied by an arbitrary constant R so that all variables appear to have the dimensions of voltage. This simplifies the block diagram and later examples.

$$I_1 = (V_1 - V_3)[R/(R_1 + sL_1)]$$
$$V_3 = (RI_1 - RI_3)(1/sC_2R)$$

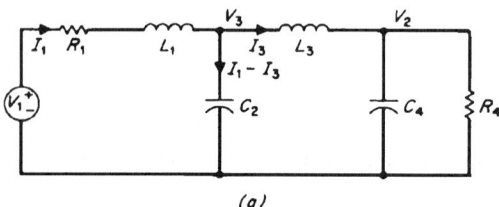

(a)

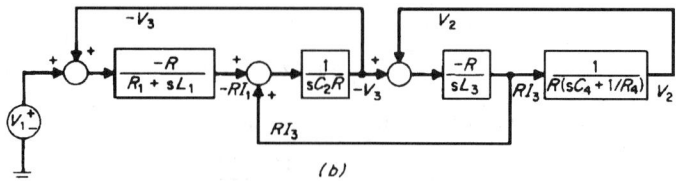

(b)

FIGURE 10.3.16   (*a*) Low-pass prototype ladder filter; (*b*) block-diagram simulation.

$$RI_3 = (V_3 - V_2)(R/sL_3)$$ (35)

$$V_2 = RI_3 \frac{1}{R(1/R_4 + sC_4)}$$

Though shown for a specific case, the technique is general and may be extended for any order of ladder network. In the simulation it should be noted that the algebraic signs of the blocks alternate. This variation is important in the realization.

In Fig. 10.3.16b, the currents are simulated by voltages, and this suggests the use of operational amplifiers as realization elements. The blocks in simulation require integrations, which the readily achieved with operational amplifiers, resistors, and capacitors. Figure 10.3.17 shows suitable combinations that will realize integrators, both inverting and noninverting, and also lossy integrators, which have a pole in the left half plane, rather than at the origin. Bruton[25] shows that, for integration, the circuit of Fig. 10.3.17b has superior performance compared with Fig. 10.3.17a when the imperfections of nonideal operational amplifiers are considered.

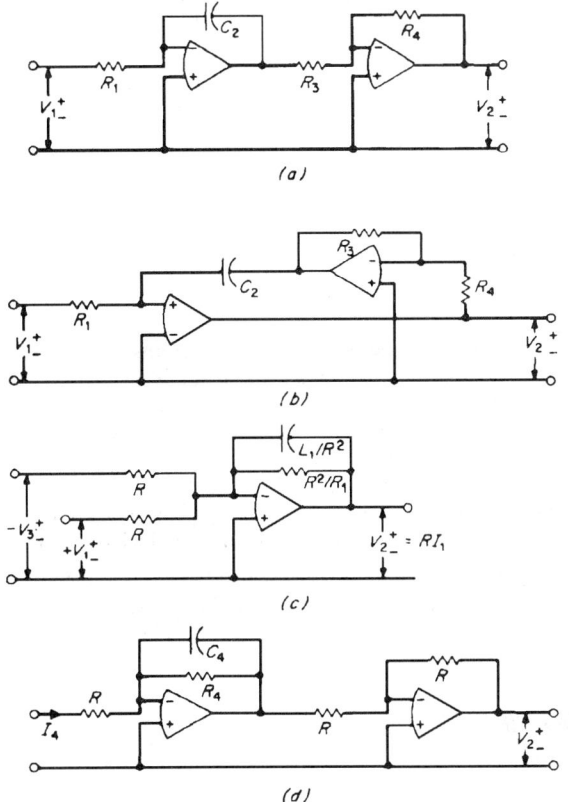

FIGURE 10.3.17   Building blocks for leapfrog filters: (a) noninverting integrator, for which $V_2/V_1 = R_4/sC_2R_1R_3$; (b) noninverting integrator, for which $V_2/V_1 = R_4/sC_2R_1R_3$; (c) lossy, summing integrator to realize $(V_1 - V_3)R/(R_1 + sL_1) = -RI_1$; (d) lossy, noninverting integrator to realize $V_2/RI_4 = 1/R(1/R_4 + sC_4)$.

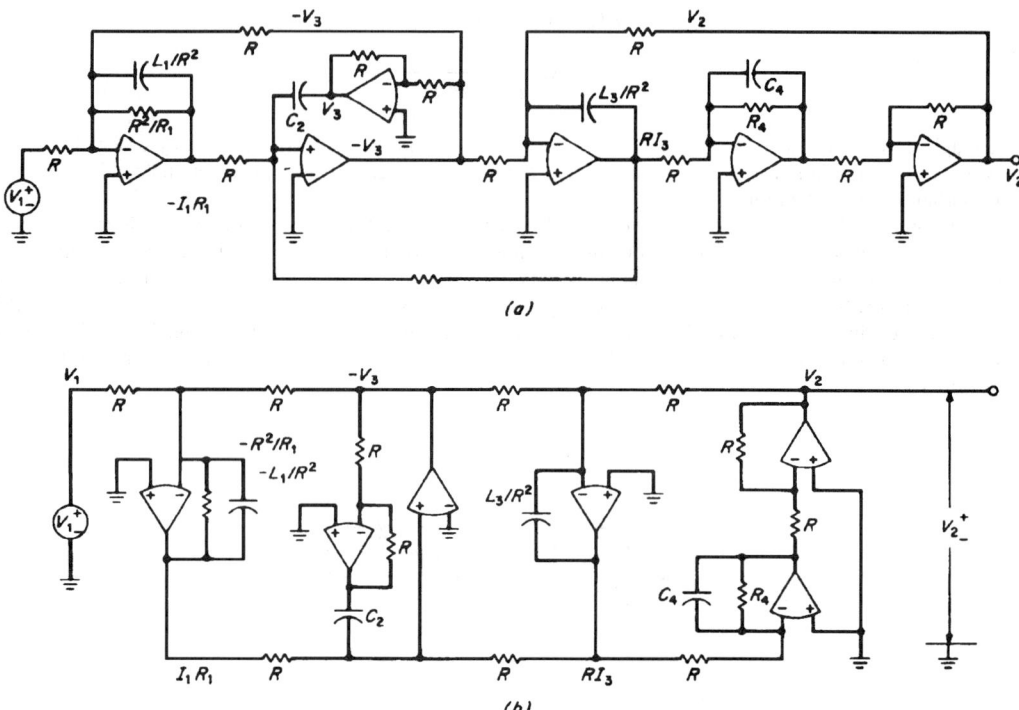

FIGURE 10.3.18 Two arrangements of a leapfrog low-pass circuit: (*a*) block-diagram arrangement; (*b*) ladder arrangement.

In Fig. 10.3.18, the combination of these blocks into a circuit is shown. In the preparation of this drawing, the integrator of Fig. 10.3.16*b* has been used, and the drawing is given in two forms. Figure 10.3.18*a* follows from the simulation, while Fig. 10.3.18*b* is a rearrangement that emphasizes the ladder structure.

The design of a low-pass leapfrog ladder may be summarized in these steps.

1. Select a normalized low-pass filter from Table 10.1.10
2. Identify the integrations represented by inductors, capacitors, and series resistor-inductor or parallel resistor-capacitor combinations. For each, determine an appropriate block diagram.
3. Connect together, using inverters, summers, and gain adjustment as needed.
4. Apply techniques of frequency and magnitude scaling to achieve a practical circuit.

BANDPASS LEAPFROG CIRCUITS

The technique of the previous section may be extended to bandpass circuits. The basic idea follows from the low-pass to bandpass transformation introduced in "Bandpass Filters" and from the recognition that it is possible to build second-order resonators from operational amplifiers, capacitors, and resistors. When the transformation is applied to an inductor, the circuit of Fig. 10.3.19*a* results. The resistor is added to allow for losses or source/load resistors that may be present. Similarly, the transformation applied to a

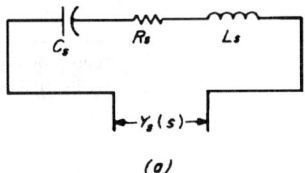

(a)

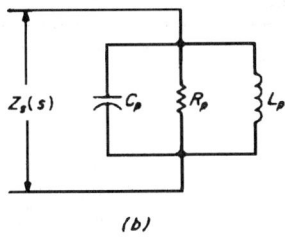

(b)

FIGURE 10.3.19 Passive resonant circuits: (*a*) series resonator; (*b*) parallel resonator.

capacitor yields Fig. 10.3.19*b*. It is to be noted that the forms of the equations are identical, and since the leapfrog technique makes use of simulation, the realizations will be similar. The necessary equations are given by Eq. (36).

Figure 10.3.20 shows an active simulation of a resonant circuit. This circuit is similar to that of Fig. 10.3.8, though the first two stages are interchanged. The new circuit has the advantage that both inverted and noninverted resonant outputs are available. Further, the input allows for summing operations, which may be needed in the leapfrog realization. Appropriate equations are given by Eq. (37).

$$Y_s(s) = \frac{(1/L_s)s}{s^2 + (R_s/L_s)s + 1/L_sC_s}$$

$$Z_p(s) = \frac{(1/C_p)s}{s^2 + (1/R_pC_p)s + 1/L_pC_p} \tag{36}$$

$$V_{o2}(s) = -V_{o1}(s) = \frac{(1/R_3C_1)s}{s^2 + (1/R_1C_1)s + 1/R_2R_4C_1C_2}(V_{11} + V_{12})(s) \tag{37}$$

The implementation of this circuit is substantially the same as that of Fig. 10.3.18, with the resonators being used as the blocks of the simulation. Since both inverted and noninverted signals are available, the one needed is chosen. Table 10.3.2 gives the parameters of the active resonators in terms of the transformed series or parallel resonant circuits. Frequency and magnitude scaling may be done either before or after the elements of the active resonator are determined, but it generally is more convenient to do it afterward.

An example is given to show this technique. It begins with a third-order Butterworth normalized low-pass filter, Fig. 10.3.21*a*, which has been taken from Table 10.1.10. The circuit is transformed to a bandpass network

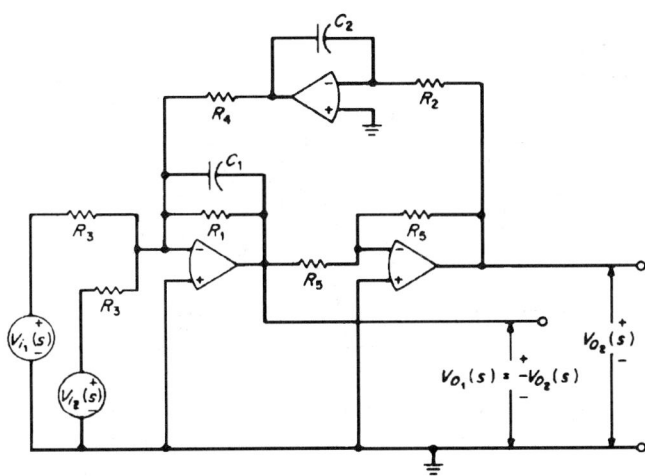

FIGURE 10.3.20 Active resonator.

TABLE 10.3.2   Resonator Design Relationships[*]

Circuit parameters from Fig. 10.3.20	Series circuit prototype values from Fig. 10.3.19a	Parallel circuit prototype values from Fig. 10.3.19b
$R_2 = R_4 = R$	$(1/C)\sqrt{L_s C_s}$	$(1/C)\sqrt{L_p C_p}$
R_1	$R\sqrt{L_s/C_s}$	$R\sqrt{C_p/L_p}$
R_3	$R\sqrt{L_s/C_s}$	$R\sqrt{C_p/L_p}$
R_5		Choose any convenient value

[*]In this table it is presumed that $C_1 = C_2 = C$ and that this is chosen to be some convenient value. It is further presumed that $R_2 = R_4$.

for which the center frequency ω_0 is 1.0 rad/s and the bandwidth β' is 0.45 rad/s, corresponding to upper and lower half-power frequencies of 1.25 and 0.80 rad/s. This result is shown as Fig. 10.3.21b.

For the first and third stages of the circuit, application of the equations from Table 10.3.2 shows that $R_1 = R_3 = {}^{20}/_9\ \Omega$, and that all other components have unit value. For the second stage, R_1 is infinite, as indicated by the table, and $R_3 = {}^9/_{40}\ \Omega$. Other components have unit value. The complete circuit is shown as Fig. 10.3.21c. This circuit has been left in normalized form. Impedance and frequency denormalization techniques must be used to achieve reasonable values.

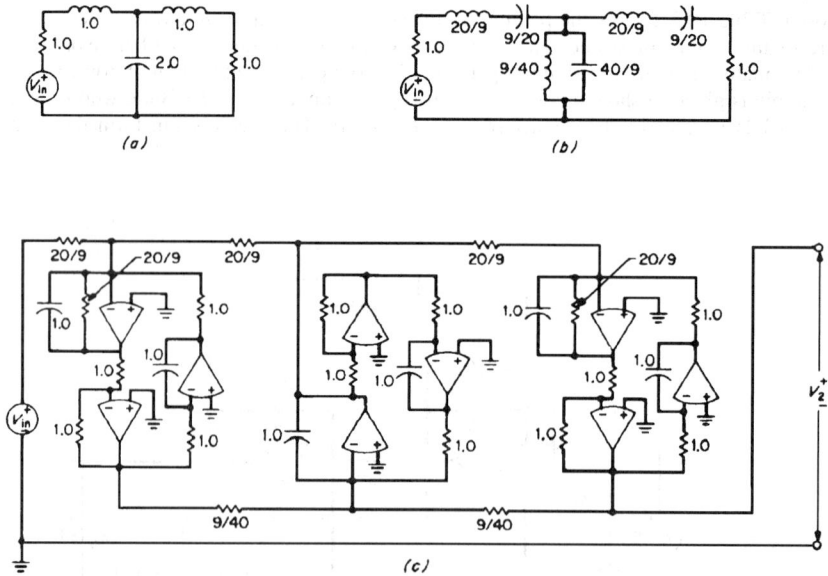

FIGURE 10.3.21   Leapfrog active resonator realization: (a) low-pass prototype; (b) bandpass transformation, $\omega_0' = 1.0\ \beta' = 0.45$; (c) complete circuit.

CHAPTER 10.4
SWITCHED CAPACITOR FILTERS

Edwin C. Jones, Jr., Harry W. Hale

Switched capacitor filters, also known as analog sampled data filters, result from a new technology that builds on the passive network theory of Darlington and implements the circuits in active-network integrated-circuit forms. Essentially, the switched capacitor replaces the resistor in operational-amplifier circuits, including the resonator. Early work was reported by Allstot, Broderson, Fried, Gray, Hosticka, Jacobs, Kuntz, and others. Huelsman[12] and Van Valkenburg[16] give additional information.

Consider the circuit of Fig. 10.4.1a and the two-phase clock signal of Fig. 10.4.1b. The circuit has two MOS switches and a capacitor C. The clock cycles the MOS switches between their low- and high-resistance states. In the analysis that follows, it is assumed that the clock speed is sufficiently high that a simplified analysis is valid. It is also assumed that the Nyquist sampling theorem is satisfied. It is possible to use discrete circuit analysis and z transforms if some of the approximations are not met.

Let the clock be such that switch A is closed and B is open. This may be modeled by Fig. 10.4.1c. The source will charge the capacitor to V_1. When the clock cycles so that B is closed and A is open, the capacitor will discharge toward V_2, transferring a charge

$$q_C = C(V_1 - V_2)$$

This will require a time $T_C = 1/f_C$, yielding an average current

$$i_{av} = C(V_1 - V_2)/T_C$$

corresponding to a resistor $R_{eq} = (V_1 - V_2)/i_{av}$, or

$$R_{eq} = T_C/C = 1/(Cf_C) \tag{1}$$

Figure 10.4.2 shows a conventional integrator, a damped integrator, and several sum and difference integrators, along with realizations and transfer functions of circuits implemented with switched capacitors. It is noted that the transfer functions are functions of the ratios of two capacitors, and this fact makes them useful. It is possible in integrated-circuit design to realize the ratio of two capacitors with high precision, leading to accurate designs of filters. With similar techniques it is possible to realize many of the second-order circuits given in earlier sections.

Possibly the most important application is in the realization of leapfrog filters. As discussed in the previous section, leapfrog filters use integrators to realize the resonators that are the basic building blocks. In this technology, the resistors of the resonators are replaced by switched capacitors. In essence, the technique is to realize the circuit with resonators, as was done in Fig. 10.3.21, and then to replace the resistors with switched capacitors. Though slightly less complex, the technique for low-pass filters is similar.

Consider the low-pass prototype filter of Fig. 10.4.3a. A simulation is shown in Fig. 10.4.3b. This simulation has equations that are identical with those of the prototype. While the simulation is similar to that of Fig. 10.3.16; two important differences may be noted. The first is that the termination resistors are separate,

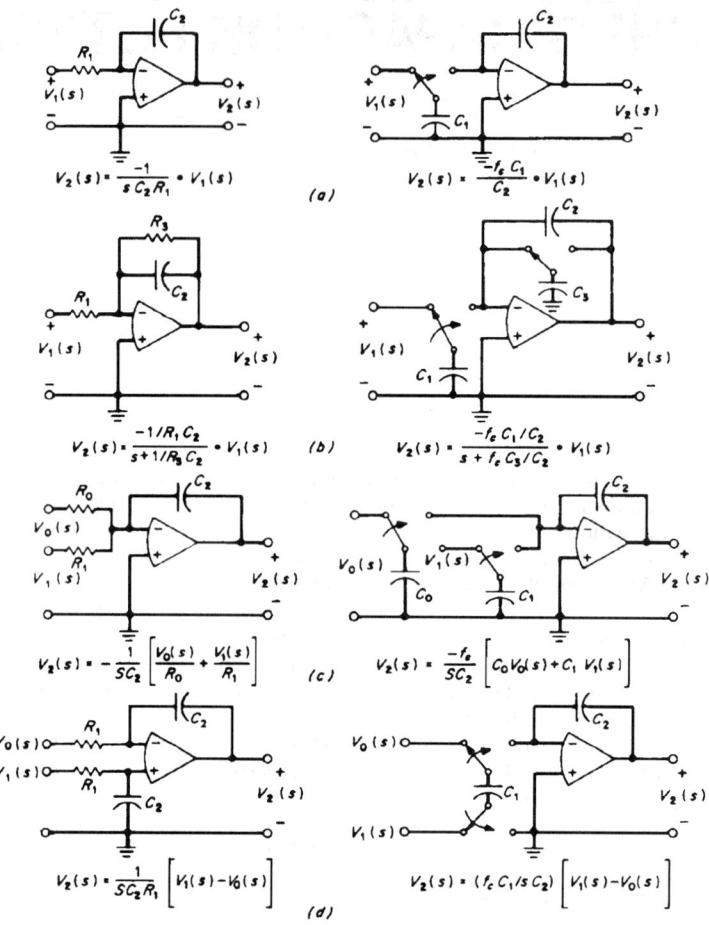

FIGURE 10.4.1 Integrators with switched capacitor realizations: (*a*) conventional integrator; (*b*) damped or lossy integrator; (*c*) summing integrator; (*d*) difference integrator.

rather than being incorporated with the input and output elements. The second is that all the elements have positive signs, and the amplifiers used are different types. This is more convenient. Figure 10.4.3*c* shows a switched capacitor equivalent for the low-pass filter. The equations for the simulation and for the development of the switched capacitor version follow.

$$RI_3 = (V_1 - V_4)(R/R_1)$$
$$V_4 = I_5(1/sC_3) = (I_3 - I_6)(R)(1/sC_3R)$$
$$RI_6 = (V_4 - V_6)(R/sL_4)$$
$$V_2 = R(I_6 - I_8)(1/sC_5R)$$
$$RI_8 = V_2(R/R_2)$$

$$(2)$$

As was done previously, the current equations have been nominally multiplied by R so that all terms appear to be voltages. In practice, R may be set to 1.0 as scaling will take care of it.

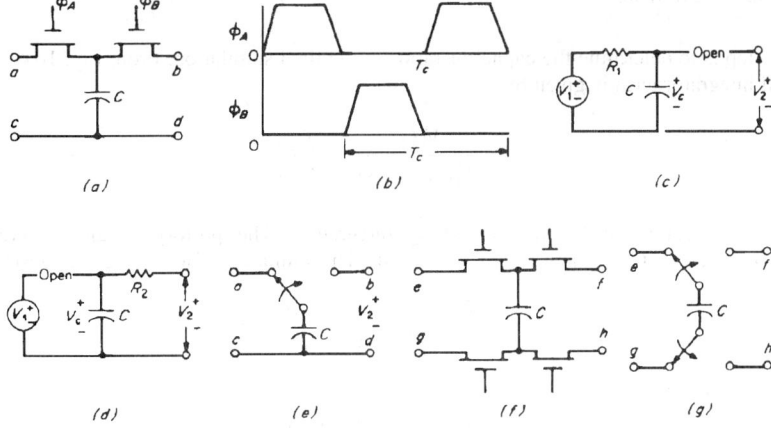

FIGURE 10.4.2 Development of equivalent circuit for a switched capacitor: (*a*) double MOS switch; (*b*) two-phase clock; (*c*) switch A closed; (*d*) switch B closed; (*e*) representation; (*f*) double-pole double-throw switch; and (*g*) representation.

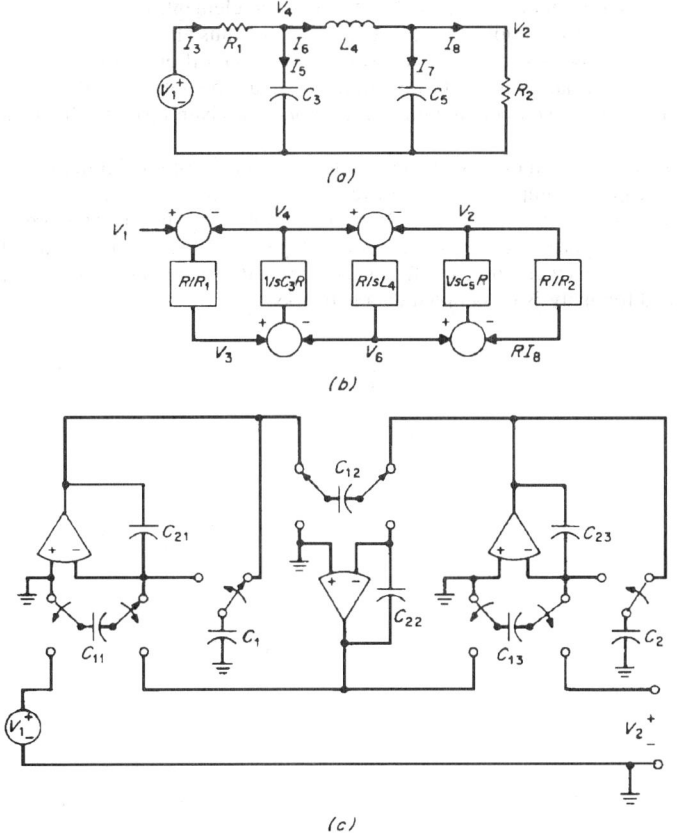

FIGURE 10.4.3 Low-pass switched capacitor filter development; (*a*) low-pass prototype and definition of equation symbols; C_3, L_4, and C_5 would be obtained from Table 10.1.10; (*b*) simulation of low-pass prototype; R may be set to 1.0; (*c*) switched capacitor equivalent network.

The next step is to determine the capacitor ratios in the final simulation. From Fig. 10.4.2d it may be seen that a typical integrator term is given by

$$\frac{V_2(s)}{V_1(s) - V_0(s)} = \frac{f_c C_1}{s C_2}$$

Similar results are obtained for the remaining integrators. The prototype values were obtained from Table 10.1.10, including C_3, L_4, and C_5 for this circuit. The similarity of terms then suggests that

$$C_3 = (1/f_C)(C_{23}/C_{13}) \tag{3}$$
$$C_5 = (1/f_C)(C_{25}/C_{15})$$

Extension to the inductors shows that

$$L_4 = (1/f_C)(C_{24}/C_{14}) \tag{4}$$

As used here, C_3, L_4, and C_5 are prototype values, but they may be magnitude- and frequency-scaled as desired to achieve realistic values. In Eqs. (3) and (4), the ratios C_2/C_1 are computed after the clock speed is known, and the second subscripts on both numerator and denominator denote which prototype the ratio has been derived from. In general they will differ for each element. In design, it is likely that one of these would be fixed and have a common value for all integrators, thus allowing the other capacitor to vary in each case. Figure 10.4.3c shows the circuit that results. It uses the difference-type integrator of Fig. 10.4.1d. It also shows a method for handling the terminations. It should be noted that the clock phases are adjusted so that alternate integrators have open and closed resistors at a given instant. This is necessary to avoid delay problems.

The extension of this technique to bandpass filters is a matter of combining the principles of leapfrog filters and switched capacitor implementation of resistors. From the specifications, a transformation to equivalent low-pass requirements is made, and the low-pass prototype is chosen. This prototype is transformed to a bandpass network, an appropriate simulation is developed, and the network is developed using integrators. Finally, scaling in frequency and magnitude is needed. It is desirable to test these simulations using computer programs designed for analysis of sampled data networks.

CHAPTER 10.5
CRYSTAL, MECHANICAL, AND ACOUSTIC COUPLED-RESONATOR FILTERS

Edwin C. Jones, Jr., Harry W. Hale

In applications such as single-sideband communications it is often necessary to have a bandpass filter with a bandwidth that is a fractional percentage of the center frequency and in which one or both transition regions are very short. Meeting such requirements usually requires a filter in which the resonators are not electrical. Two types of resonator are quartz crystals and mechanical elements, such as disks or rods. Transducers from the electric signal to the mechanical device, output transducers, and resonator-coupling elements are needed.

Crystal filters include resonators made from piezoelectric quartz crystals. The transducers are plates of a conductor deposited on the appropriate surfaces of the crystal, and coupling from one crystal to the next is electrical. The center frequency depends on the size of the crystal, its manner of cutting, and the choice of frequency determining modes of oscillation. It can vary from about 1.0 kHz to 100 MHz. If extreme care is taken, equivalent quality factors (Q's) can be greater than 100,000. These filters can also be very stable with regard to temperature and age.

Mechanical filters use rods or disks as resonating elements, which are coupled together mechanically, usually with wires welded to the resonators. The transducers are magnetostrictive. The frequency range varies from as low as 100 Hz to above 500 kHz. Quality factors above 20,000 are possible and, with proper choice of alloys, temperature coefficients of as low as 1.0 ppm/°C are possible.

Acoustic filters use a combination of crystal and mechanical filter principles. The resonators are monolithic quartz crystals; the transducers are similar to those of crystal filters, but the coupling is mechanical (referred to as acoustic coupling). These filters have many of the properties of crystal filters, but the design techniques have much in common with those of mechanical filters.

Coupled-resonator filters are usually described in terms of an electric equivalent circuit. The direct or mobility analogy (mass to capacitance, friction to conductance, and springs to inductance) is more useful, because the "across" variables of velocity and voltage are analogous, as are the "through" variables of force and current. Equivalent capacitances or inductances and center frequencies are among the common parameters specified for filter elements.

The following paragraphs discuss, in general terms, the design procedure used for coupled-resonator filters, the equivalent circuits used, and some network transformations that enable the designer to implement the design procedure. References 6, 8, and 27 give much additional information, and in particular, Ref. 27 contains an extensive discussion and bibliography. Manufacturer's catalogs are a good source of current data.

COUPLED-RESONATOR DESIGN PROCEDURE

The insertion-loss low-pass prototype filters can be used to design coupled-resonator bandpass filters. Five steps can be identified in the process, though in some cases the dividing lines become indistinct.

1. Transform the bandpass specifications to a low-pass prototype, using Eq. (8). This will take the center frequency to $\omega = 0$ and, usually, the band edge to $\omega = 1$.

2. Choose the appropriate low-pass response, e.g., Chebyshev, elliptic, or Butterworth, that meets the transformed specifications. Zeros of transmission are fixed at this time. From this characteristic function determine the transfer function that is needed. The tables presented earlier may be useful.

3. Determine the short-circuit y or open-circuit z parameters from the transfer function.

4. If possible, look up or synthesize the appropriate ladder or lattice network needed. At this point, it is still a low-pass prototype. The technique chosen may depend on the expected form of the final network.

5. Use Fig. 10.2.7 to transform the network into a bandpass network and then use network theorems to adjust the network to a configuration and a set of element values that is practical, i.e., one that matches the resonators.

This process is not one in which success is assured. It may require a variety of attempts before a suitable design is achieved. Equivalent circuits and network theorems are summarized in the following paragraphs.

EQUIVALENT CIRCUITS

The most common equivalent circuit for a piezoelectric crystal shows a series-resonant RLC circuit in parallel with a second capacitor, as shown in Fig. 10.5.1. The parallel capacitor C_p is composed of the mounting hardware and electric plates on the crystal. In practice, the ratio C_p/C cannot be reduced below about 125, but it may be increased if needed. When a filter contains more than one crystal, the coupling is electrical, usually with capacitors.

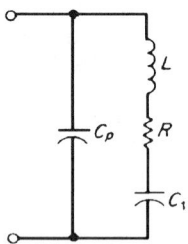

Mechanical filters have an equivalent circuit, as indicated in Fig. 10.5.2. The resonant circuits L_0, C_R represent the transducer magnetostrictive coils and their tuning capacitances. (In cases of small R_L, it may be more accurate to place C_R in series with L_0.) The resonant circuit L_1, C_1, R_1 and L_n, C_n, R_n include the motional parameters of the transducers. Elements L_2, C_2, R_2, . . . , L_{n-1}, C_{n-1}, R_{n-1} represent the motional parameters of the resonant elements, and L_{12}, . . . , $L_{n-1,n}$ represent the compliances of the coupling wires.

FIGURE 10.5.1 Equivalent circuit for piezoelectric crystal. Because coupling is electrical, a one-port representation is sufficient.

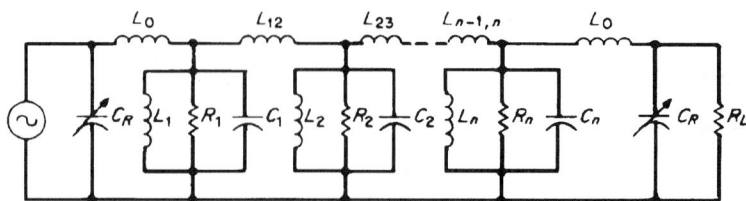

FIGURE 10.5.2 Equivalent circuit for a mechanical filter. A two-port representation allows an electric equivalent circuit for the entire filter.

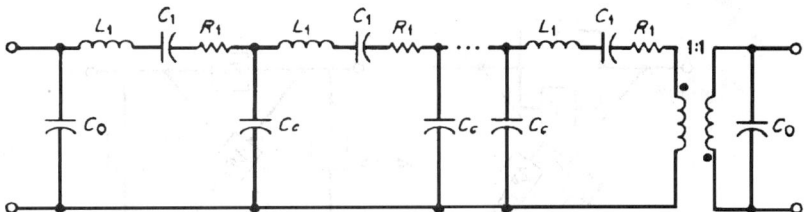

FIGURE 10.5.3 Equivalent circuit for a monolithic crystal or acoustic filter. The one-to-one ideal transformer models the 180° phase shift observed in these filters.

The acoustic filter is represented, after substantial development, by the circuit shown in Fig. 10.5.3. The development has made the circuit easy to use, but the association between the electrical elements and the filter elements is less apparent than in the previous circuits. The ideal transformer at the output accounts for the 180° phase shift observed in these filters. In some analyses, it may be omitted.

NETWORK TRANSFORMATIONS

In the process of changing a bandpass circuit to meet the configuration of the equivalent circuit of a coupled resonator a variety of equivalent networks may be useful. At one step negative elements may appear. These can be absorbed later in series or parallel with positive elements so that the overall result is positive.

The impedance inverters of Fig. 10.5.4 can be used to invert an impedance, as indicated. Over a very narrow frequency range they can often be approximated with three capacitors, two of which are negative. An inverter can be used to convert an inductance into a capacitance provided the negative elements can then be absorbed. Other similar reactive configurations can also be used.

Lattice networks (Fig. 10.5.5) are often used in crystal filters. If the condition prevailing in Fig. 10.5.6 exists, the equivalent can be used in either direction to affect a change. In particular, the ladder can be transformed into a lattice, which then has the crystal equivalent circuit.

Two Norton transformations and networks derived from them are shown in Figs. 10.5.7 and 10.5.8. They lead to negative elements, and it is expected that they will later be absorbed into positive elements. Humpherys[8] gives another derived Norton transformation that can be used to reduce inductance values. It changes the impedance level on one side of the network. When this is applied to a symmetrical network, the new impedance levels will eventually become directly connected, so that no transformer is needed.

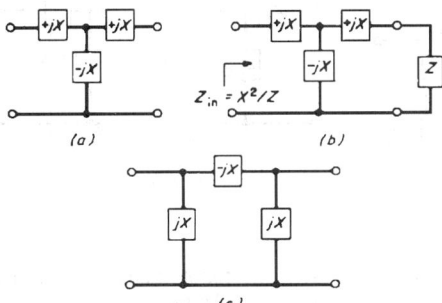

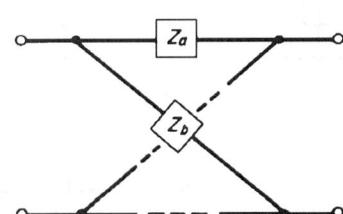

FIGURE 10.5.4 Reactive impedance inverters: (a) *T* inverter; (b) *T* inverter with load Z; $Z_{in} = X^2/Z$; (c) π inverter.

FIGURE 10.5.5 Symmetrical lattice. The dotted diagonal line indicates a second Z_b; the dotted horizontal line, a second Z_a.

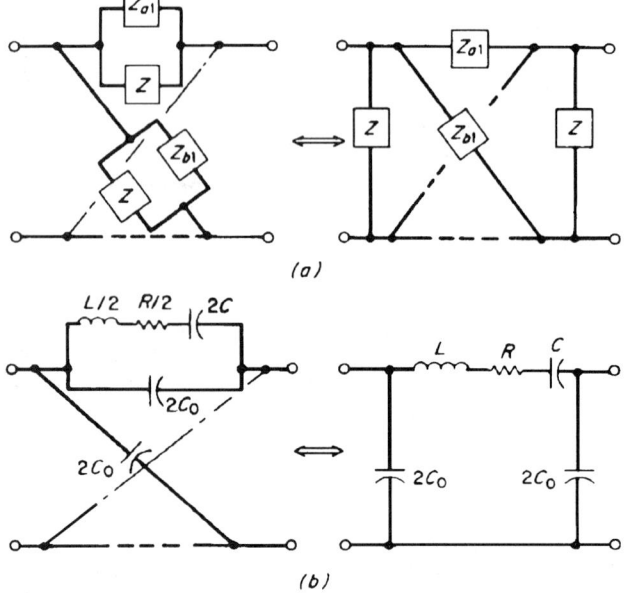

FIGURE 10.5.6   Lattice and ladder: (*a*) general lattice and equivalent circuit; (*b*) application to crystal filters.

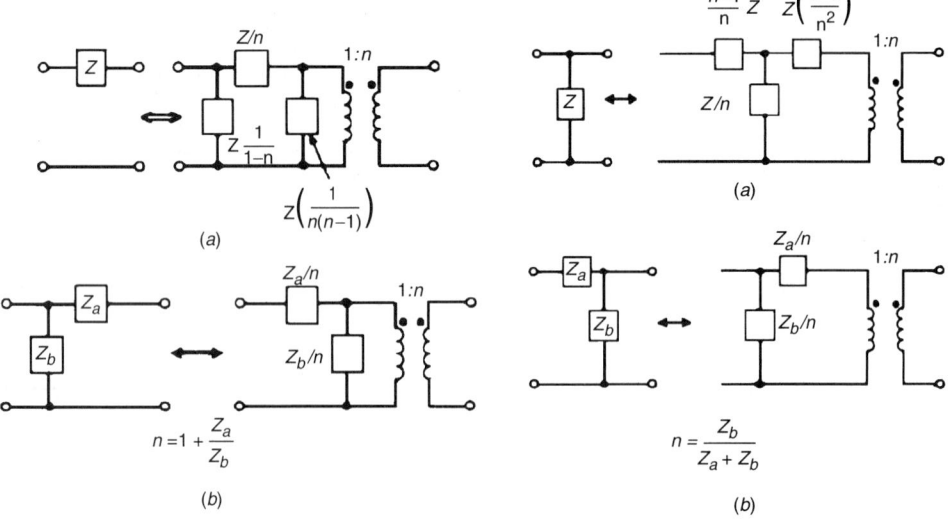

FIGURE 10.5.7   Norton's first transformation and a derived network.

FIGURE 10.5.8   Norton's second transformation and a derived network.

CHAPTER 10.6
DIGITAL FILTERS*

Arthur B. Williams, Fred J. Taylor

RELATION TO ANALOG FILTERS

Digital filters provide many of the same frequency selective services (low, high, bandpass) expected of analog filters. In some cases, digital filters are defined in terms of equivalent analog filters. Other digital filters are designed using rules unique to this technology. The hardware required to fabricate a digital filter are basic digital devices such as memory and arithmetic logic units (ALUs). Many of these hardware building blocks provide both high performance and low cost. Coefficients and data are stored as digital computer words and, as a result, provide a precise and noise free (in an analog sense) signal processing medium. Compared to analog filters, digital filters generally enjoy the following advantages:

1. They can be fabricated in high-performance general-purpose digital hardware or with application-specific integrated circuits (ASIC).
2. The stability of certain classes of digital filters can be guaranteed.
3. There are no input or output impedance matching problems.
4. Coefficients can be easily programmed and altered.
5. Digital filters can operate over a wide range of frequencies.
6. Digital filters can operate over a wide dynamic range and with high precision.
7. Some digital filters provide excellent phase linearity.
8. Digital filters do not require periodic alignment and do not drift or degrade because of aging.

DATA REPRESENTATION

In an analog filter, all signals and coefficients are considered to be real or complex numbers. As such, they are defined over an infinite range with infinite precision. In the analog case, filter coefficients are implemented with lumped R, L, C, and amplifier components of assumed absolute precision. Along similar lines, the designs of digital filters typically begin with the manipulation of idealized equations. However, the implementation of a digital filter is accomplished using digital computational elements of finite precision (measured in bits). Therefore, the analysis of a digital filter is not complete until the effects of finite precision arithmetic has been

---

*This section is based on the author's *Electronic Filter Design Handbook*, 3rd ed., McGraw-Hill, 1995.

determined. As a result, even though there has been a significant sharing of techniques in the area of filter synthesis between analog and digital filters, the analyses of these two classes of filters have developed separate tools and techniques.

Data, in a digital system, are represented as a set of binary-valued digits. The process by which a real signal or number is converted into a digital word is called analog-to-digital conversion (ADC). The most common formats used to represent data are called *fixed* and *floating* point (FXP and FLP). Within the fixed-point family of codes, the most popular are binary-coded decimal sign magnitude (SM) and diminished radix (DR) codes. Any integer X such that $|X| < 2^{n-1}$ has a unique sign-magnitude representation, given by

$$X = X_{n-1}: (2^{n-2} X_{n-2} + \cdots + 2X_1 + X_0) \tag{1}$$

where X_i is the ith bit and X_0 is referred to as the least significant bit (LSB). Similarly X_{n-2} is called the most significant bit (MSB) and X_{n-1} is the sign bit. The LSB often corresponds to a physically measurable electrical unit. For example, if a signed 12-bit ADC is used to digitize a signal whose range is ±15 V, the LSB represents a quantization step size of $Q = 30$ V (range)/2^{12} – bits = 7.32 mV/bit.

Fractional numbers are also possible simply by scaling X by a power of 2. The value of $X' = X/2^m$ has the same binary representation of X except that the m LSBs are considered to be fractional bits.

SIGNAL REPRESENTATION

An analog filter manipulates real signals of assumed infinite precision. In a discrete system, analog signals of assumed infinite precision are periodically sampled at a rate of f samples per second. The same period is therefore given by $t_s = 1/f_s$ second(s). A string of contiguous samples is called a *time series*. If the samples are further processed by an ADC, a digital time series results. A digital filter can be used to manipulate this time series using digital technology. The hardware required to implement such a filter is the product of the microelectronics revolution.

SPECTRAL REPRESENTATION

Besides representing signals in the continuous or discrete time domain, signals can also be modeled in the frequency domain. This condition is called *spectral representation*. The principal tools used to describe a signal in the frequency domain are: (1) Fourier transforms, (2) Fourier series, and (3) discrete Fourier transforms (DFT).

A Fourier transform will map an arbitrary transformable signal into a continuous frequency spectrum consisting of all frequency components from $-\infty$ to $+\infty$. The Fourier transform is defined by an indefinite integral equation whose limits range from $-\infty$ to $+\infty$. The Fourier series will map a continuous but periodic signal of period T [i.e., $x(t) = x(t + kT)$ for all integer values of k] into a discrete but infinite spectrum consisting of frequency harmonics located at multiples of the fundamental frequency $1/T$. The Fourier series is defined by a definite integral equation whose limits are $[0, T]$. The discrete Fourier transform differs from the first two transforms in that it does not accept data continuously but rather from a time series of finite length. Also, unlike the first two transforms, which produce spectra ranging out to $\pm\infty$ Hz, the DFT spectrum consists of a finite number of harmonics.

The DFT is an important and useful tool in the study of digital filters. The DFT can be used to both analyze and design digital filters. One of its principal applications is the analysis of a filter's impulse response. An impulse response database can be directly generated by presenting a one-sample unit pulse to a digital filter that is initially at a zero state (i.e., zero initial conditions). The output is the filter's impulse response, which is observed for N contiguous samples. The N-sample database is then presented to an N-point DFT, transformed, and analyzed. The spectrum produced by the DFT should be a reasonable facsimile of the frequency response of the digital filter under test.

FILTER REPRESENTATION

A transfer function is defined by the ratio of output and input transforms. For digital filters, it is given by $H(z) = Y(z)/U(z)$ where $U(z)$ is the z transform of the input signal $u(n)$ and $Y(z)$ is for the output signal $y(n)$. The frequency response of a filter $H(z)$ can be computed using a DFT of the filter's impulse response.

Another transform tool that also is extensively used to study digital filters is the *bilinear z transform*. While the standard z transform can be related to the simple sample and hold circuit, the bilinear z transform is analogous to a first-order hold. The bilinear z transform is related to the familiar Laplace transform through

$$s = \frac{2(z-1)}{t_s(z+1)} \qquad z = \frac{(2/t_s)+s}{(2/t_s)-s} \tag{2}$$

Once an analog filter $H(s)$ is defined, it can be converted into a discrete filter $H(z)$ by using the variable substitution rule.

FINITE IMPULSE-RESPONSE (FIR) FILTERS

Linear constant coefficient filters can be categorized into two broad classes known as finite impulse-response (FIR) or infinite impulse-response (IIR) filters. An FIR filter can be expressed in terms of a simple discrete equation:

$$y(n) = c_0 x(n) + c_1 x(n-1) + \cdots + c_{N-1} x(n-N+1) \tag{3}$$

where the coefficients $\{C_i\}$ are called *filter tap weights*. In terms of a transfer function, Eq. (3) can be restated as

$$H(z) = \sum_{i=0}^{n-1} C_i z^{-1} \tag{4}$$

As an example, a typical $N = 111$th-order FIR is shown in Fig. 10.6.1. The FIR exhibits several interesting features:

1. The filter's impulse response exists for only $N = 111$ (finite) contiguous samples.
2. The filter's transform function consists of zeros only (i.e., no poles). As a result, an FIR is sometimes referred to as an all-zero, or transversal, filter.
3. The filter has a very simple design consisting of a set of word-wide shift registers, tap-weight multipliers, and adders (accumulators).
4. If the input is bounded by united (i.e., $|x(i)| \leq 1$ for all i), the maximum value of the output $y(i)$ is $\Sigma|C_i|$. If all the tap weights C_i are bounded, the filter's output is likewise bounded and, as a result, stability is guaranteed.
5. The phase, when plotted with respect to frequency (plot shown over the principal angles $\pm \pi/2$), is linear with constant slope.

LINEAR PHASE BEHAVIOR

The FIR is basically a shift-register network. Since digital shift registers are precise and easily controlled, the FIR can offer the designer several phase domain attributes that are difficult to achieve with analog filters. The most important of these are: (1) Potential for linear phase versus frequency behavior and (2) potential for constant

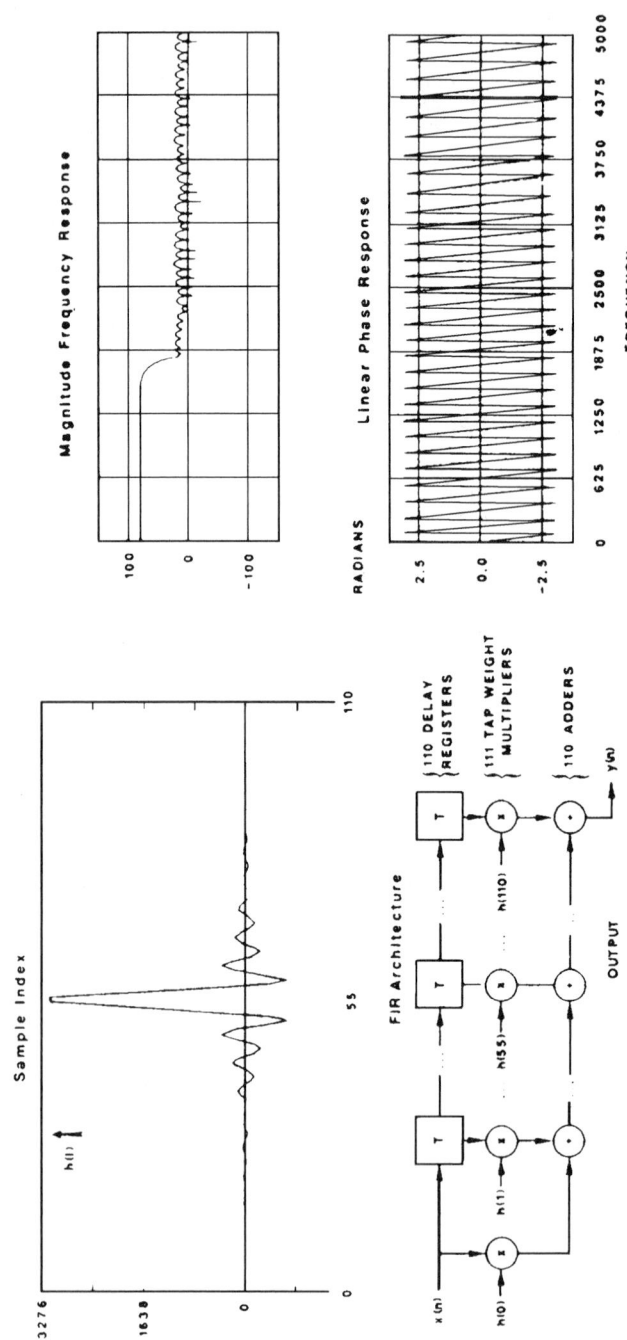

FIGURE 10.6.1 Typical FIR architecture, impulse response, and frequency response.

group-delay behavior. These properties are fundamentally important in the fields of digital communications systems, phase synchronization systems (e.g., phase-locked loops), speech processing, image processing, spectral analysis (e.g., Fourier analysis), and other areas where nonlinear phase distortion cannot be tolerated.

FIR DESIGN METHODS

The design of an FIR entails specifying the filter's impulse response, tap weights $\{C_i\}$. As a result, the design of an FIR can be as simple as prespecifying the desired impulse response. Other acceptable analytical techniques used to synthesize a desired impulse response are the inverse Fourier transform of a given frequency domain filter specification or the use of polynomial approximation techniques. These methods are summarized below.

A simple procedure for designing an FIR is to specify an acceptable frequency domain model, invert the filter's spectral representation using the inverse Fourier transform, and use the resulting time series to represent the filter's impulse response. In general, the inverse Fourier transform of a *desired* spectral waveshape would produce an infinitely long time domain record. However, from a hardware cost or throughput standpoint, it is unreasonable to consider the implementing of an infinitely or extremely long FIR. Therefore, a realizable FIR would be defined in terms of a truncated Fourier series. For example, the Fourier transform of the "nearly ideal" $N = 101$ order low-pass filter has a sin $(x)/x$ type impulse-response envelope. For a large value of N, the difference between the response of an infinitely long impulse response and its N-sample approximation is small; however, when N is small, large approximation errors can occur.

Optimal Modeling Techniques

Weighted Chebyshev polynomials have been successfully used to design FIRs. In this application, Chebyshev polynomials are combined so that their combined sum minimizes the maximum difference between an ideal and the realized frequency response (i.e., mini-max principle). Because of the nature of these polynomials, they produce a "rippled" magnitude frequency-response envelope of equal minima and maxima in the pass- and stopbands. As a result, this class of filters is often called an *equiripple* filter. Much is known about the synthesis process, which can be traced back to McClellan et al.[48] Based on these techniques, a number of software-based CAD tools have been developed to support FIR design.

WINDOWS

Digital filters usually are expected to operate over long, constantly changing data records. An FIR, while being capable of offering this service, can only work with a limited number of samples at a time. A similar situation presents itself in the context of a discrete Fourier transform. The quality of the produced spectrum is a function of the number of transformed samples. Ideally, an infinitely long impulse response would be defined by an ideal filter. A *uniform window* of length T will pass N contiguous samples of data. The windowing effect may be modeled as a multiplicative switch that multiplies the presented signal by zero (open) for all time exclusive of the interval $[0, T]$. Over $[0, T]$, the signal is multiplied by unity (closed). In a sampled system, the interval $[0, T]$ is replaced by N samples taken at a sample rate f_s where $T = N/f_s$. When the observation interval (i.e., N) becomes small, the quality of the spectral estimate begins to deteriorate. This consequence is called the *finite aperture effect*.

Windowing is a technique that tends to improve the quality of a spectrum obtained from a limited number of samples. Some of the more popular windows found in contemporary use are the rectangular or uniform window, the Hamming window, the Hann window, the Blackman window, and the Kaiser window.

Windows can be directly applied to FIRs. To window an N-point FIR, simply multiply the tap weight coefficients C_i with the corresponding window weights w_i. Note that all of the standard window functions have even symmetry about the midsample. As a result, the application of such a window will not disturb the linear phase behavior of the original FIR.

MULTIRATE SIGNAL PROCESSING

Digital signal processing systems accept an input time series and produce an output time series. In between, a signal can be modified in terms of its time and/or frequency domain attributes. One of the important functions that a digital signal processing system can serve is that of sample rate conversion. As the name implies, a sample rate converter changes a system's sample rate from a value of f_{in} samples per second to a rate of f_{out} samples per second. Such devices are also called multirate systems since they are defined in terms of two or more sample rates. If $f_{in} > f_{out}$ then the system is said to perform decimation and is said to be decimated by an integer M if

$$M = \frac{f_{out}}{f_{in}} \tag{5}$$

In this case, the decimated time series $x_d[n] = x[Mn]$, or every Mth sample of the original time series is retained. Furthermore, the effective sample rate is reduced from f_{in} to $f_{dec} = f_{in}/M$ samples per second.

Applications of decimation include audio and image signal processing involving two or more subsystems having dissimilar sample rates. Other applications occur when a high data rate ADC is placed at the front end of a system and the output is to be processed parameters that are sampled at a very low rate by a general-purpose digital computer. At other times, multirate systems are used simply to reduce the Nyquist rate to facilitate computational intensive algorithms, such as a digital Fourier analyzer, to be performed at a slower arithmetic rate. Another class of applications involves processing signals, sampled at a high data rate, through a limited bandwidth channel.

QUADRATURE MIRROR FILTERS (QMF)

We have stated that multirate systems are often used to reduce the sample rate to a value that can be passed through a band-limited communication channel. Supposedly, the signal can be reconstructed on the receiver side. The amount of allowable decimation has been established by the Nyquist sampling theorem. When the bandwidth of the signal establishes a Nyquist frequency that exceeds the bandwidth of a communication channel, the signal must be decomposed into subbands that can be individually transmitted across band-limited channels. This technique uses a bank of band-limited filters to break the signal down into a collection of subbands that fit within the available channel bandwidths.

Quadrature mirror filters (QMF) are often used in the subband application described in Fig. 10.6.2. The basic architecture shown in that figure defines a QMF system and establishes two input-output paths that have a bandwidth requirement that is half the input or output requirements. Using this technique, the channels can be subdivided over and over, reducing the bandwidth by a factor of 2 each time. The top path consists of low-pass filters and the bottom path is formed by high-pass filters.

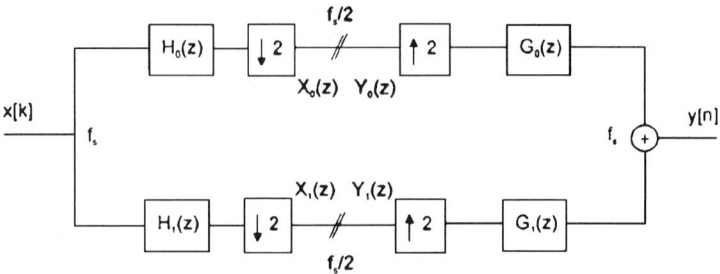

FIGURE 10.6.2   Quadrature mirror filter (QMF).

Designing QMF is, unfortunately, not a trivial process. No meaningful flat response linear phase QMF filter exists. Most QMF designs represent some compromise.

INFINITE IMPULSE-RESPONSE FILTER

The FIR filter exhibits superb linear phase behavior; however, in order to achieve a high-quality (steep-skirt) magnitude frequency response, a high-order FIR is required. Compared to the FIR, the IIR filter

1. Generally satisfies a given magnitude frequency-response design objective with a lower-order filter.
2. Does not generally exhibit linear phase or constant group-delay behavior.

If the principal objective of the digital filter design is to satisfy the prespecified magnitude frequency response, an IIR is usually the design of choice. Since the order of an IIR is usually significantly less than that of an FIR, the IIR would require fewer coefficients. This translates into a reduced multiplication budget and an attendant saving in hardware and cost. Since multiplication is time consuming, a reduced multiplication budget also translates into potentially higher sample rates.

From a practical viewpoint, a realizable filter must produce bounded outputs if stimulated by bounded inputs. The magnitude is bounded on an IIR's impulse response, namely,

$$\sum_{n=0}^{\infty} |h(n)| < M \tag{6}$$

If M is finite (bounded), the filter is stable, and if it is infinite (unbounded), the filter is unstable. This condition can also be more conveniently related to the pole locations of the filter under study. It is well known that a causal discrete system with a rational transfer function $H(z)$ is stable (i.e., bounded inputs produce bounded outputs) if and only if its poles are interior to the unit circle in the z domain. This is often referred to as the circle criterion and it can be tested using general-purpose computer root-finding methods. Other algebraic tests—Schur-Cohen, Routh-Hurwitz, and Nyquist—may also be used. The stability condition is implicit to the FIR as long as all N coefficients are finite. Here the finite sum of real bounded coefficients will always be bounded.

DESIGN OBJECTIVES

The design of an IIR begins with a magnitude frequency-response specification of the target filter. The filter's magnitude frequency response is specified as it would be for an analog filter design. In particular, assume that a filter with a magnitude-squared frequency response given by $|H(\omega)|^2$, having passband, transition band, and stopband behavior as suggested by Fig. 10.6.3, is to be synthesized. The frequency response of the desired filter is specified in terms of a cutoff critical frequency ω_p a stopband critical frequency ω_s and stop- and passband delimiters ϵ and A. The critical frequencies ω_p and ω_a represent the end of the passband and the start of the stopband, respectively, for the low-pass example. In decibels, the gains at these critical frequencies are given by (passband ripple constraint) and $-A_a = -10 \log (A^2)$ (stopband attenuation). For the case where $\epsilon = 1$, the common 3-dB passband filter is realized.

FIR AND IIR FILTERS COMPARED

The principal attributes of FIR are its simplicity, phase linearity, and ability to decimate a signal. The strength of the IIR is its ability to achieve high-quality (steep-skirt) filtering with a limited number order design. Those positive characteristics of the FIR are absent in the IIR and vice versa.

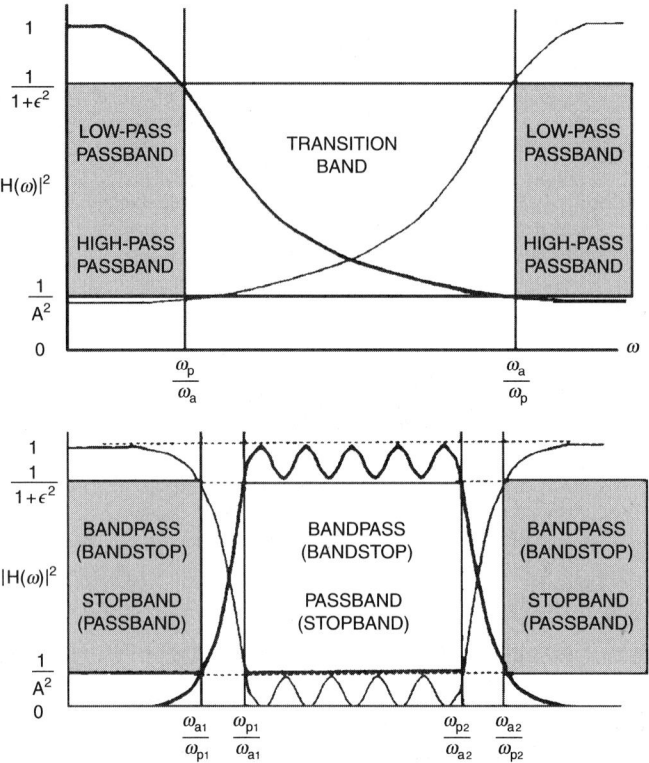

FIGURE 10.6.3   Typical design objective for low-pass, high-pass, and bandstop IIR filters.

The estimated order of an FIR, required to achieve an IIR design specification, was empirically determined by Rabiner. It was found that the approximate order n of an FIR required to meet a design objective

$$(1-\delta_1)^2 = \frac{1}{1+\varepsilon^2} \quad \delta_1 \quad \text{(passband-ripple)}$$

$$\delta_2^2 = \frac{1}{A^2} \quad \delta_2 \text{ (stopband bound)} \tag{7}$$

$$\Delta f \text{ (transition frequency range/} f_s)$$

is given by

$$n \sim \frac{-10\log((1-\delta_1)\delta_2)-15}{14\Delta f}+1 \tag{8}$$

STANDARD FILTER FORMS

The standard filter forms found in common use are: (1) Direct II, (2) Standard, (3) Cascade, and (4) Parallel. These basic filter cases are graphically interpreted in Ref. 52. The direct II and standard architectures are somewhat similar in their structure. Both strategies possess information feedback paths ranging from one delay to n delays. The transfer function denominator is an nth-order polynomial. The cascade and parallel models are constructed using a system of low-order subsections or subfilters. In the cascade design, the low-order subsections are serially interconnected. In the parallel filter, these sections are simply connected in parallel. The low-order subfilters, in both cases, are the result of factoring the nth-order transfer function polynomial into lower-order polynomials.

The design and analysis of all four classes of filters can be performed using manually manipulated equations or a digital computer. The most efficient method of formulating the filter design problem, whether using tables, calculators, or a computer, is called the *state-variable* technique. A state variable is a parameter that represents the information stored in a system. The set of state variables is called a *state vector*. For an analog system, information is stored on capacitors or in inductors. In earlier chapters, state variables were used to specify and facilitate the manipulation of the R, L, and C components of an analog filter. In these cases, capacitive voltage and inductive current were valid state variables. Since resistors have no memory, they would not be the source of a state variable.

In digital filters, the memory element, which stores the state information, is simply a delay (shift) register. The realization of digital filters is described in Ref. 52.

FIXED-POINT DESIGN

An IIR, once designed and architected, often needs to be implemented in hardware. The choices are fixed- or floating-point. Of the two, fixed-point solutions generally provide the highest real-time bandwidth at the lowest cost. Unfortunately, fixed-point designs also introduce errors that are not found in more expensive floating-point IIR designs. The fixed-point error sources are either low-order inaccuracies, because of finite precision arithmetic and data (coefficient) roundoff effects or potentially large errors because of run-time dynamic range overflow (saturation).

Additional precision can be gained by increasing the number of fractional bits assigned to the data and coefficients fields with an attendant decrease in dynamic range and an increased potential for runtime overflow. On the other hand, the overflow saturation problem can be reduced by enlarging the dynamic range of the system by increasing the integer bit field with an accompanying loss of precision. The problem facing the fixed-point filter design, therefore, is achieving a balance between the competing desire to maximize precision and to simultaneously eliminate (or reduce) run-time overflow errors. This is called the *binary-point assignment problem*.

CHAPTER 10.7
ATTENUATORS

Arthur B. Williams, Fred, J. Taylor

ATTENUATOR NETWORK DESIGN

Attenuators are passive circuits that introduce a fixed power loss between a source and load while matching impedances. The power loss is independent of the direction of power flow. Figure 10.7.1 shows T, Π, and bridged-T networks. The first two are unbalanced and unsymmetrical, unless $Z_L = Z_S$. In this case, $Z_1 = Z_2$, and the network is symmetrical. To build an unbalanced network, divide Z_1 and Z_2 by 2, and put half of each element in each series arm. The bridged-T shown is only for symmetrical networks.

These design equations are valid for resistive and complex impedances.

$$Z_S = \text{source impedance}$$
$$Z_L = \text{load impedance}$$
$$A = \text{ratio of available power to desired load power (dB)} = 10^{B/10} \qquad (1)$$
$$B = 10 \log A$$
$$\theta = 1/2 \ln A = 1/2 \ln 10^{B/10}$$

As an example, design an attenuator to match a 75-Ω source to a 300-Ω load and to introduce a 14.0-dB loss. Use a T section. In terms of Eq. (1),

$$Z_s = 75\ \Omega \qquad Z_L = 300\ \Omega \qquad B = 14.0\ \text{dB} \qquad A = 25.12$$

$$\theta = 1.612 \qquad Z_3 = 62.34\ \Omega \qquad Z_1 = 18.88\ \Omega \qquad Z_2 = 262.54\ \Omega$$

Figure 10.7.2 shows the network.

Circuit name	Circuit diagram	Equations
T-Network	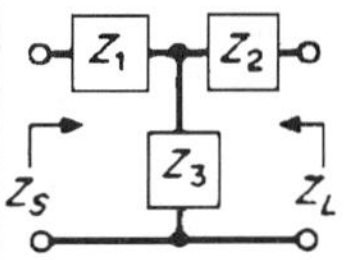 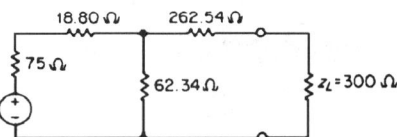	$Z_3 = \sqrt{Z_s Z_L}/\sinh\theta = \dfrac{2\sqrt{Z_s Z_L \cdot A}}{A-1}$ $Z_1 = Z_s\coth\theta - Z_3 = Z_s\left(\dfrac{A+1}{A-1}\right) - Z_3$ $Z_2 = Z_L\coth\theta - Z_3 = Z_L\left(\dfrac{A+1}{A-1}\right) - Z_3$
π-Network	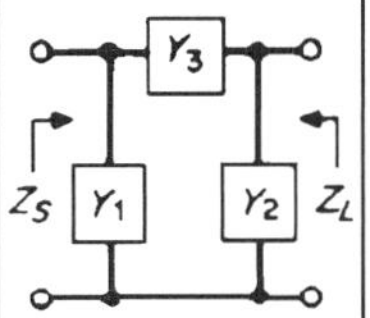	$Y_3 = \dfrac{1}{\sqrt{Z_s Z_L}\,\sinh\theta}$ $= \dfrac{2}{A-1}\sqrt{\dfrac{A}{Z_s Z_L}}$ $Y_1 = \dfrac{\tanh\theta}{Z_s} - Y_3 = \dfrac{1}{Z_s}\left(\dfrac{A+1}{A-1}\right) - Y_3$ $Y_2 = \dfrac{\tanh\theta}{Z_L} - Y_3 = \dfrac{1}{Z_L}\left(\dfrac{A+1}{A-1}\right) - Y_3$
Bridged-T	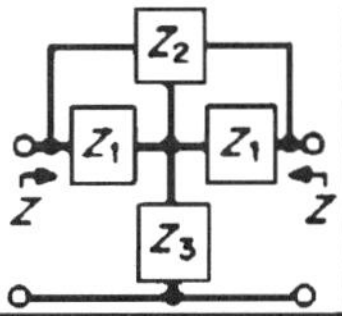	$Z_1 = Z$ $Z_2 = Z(A-1)$ $Z_3 = \dfrac{Z}{A-1}$

FIGURE 10.7.1 Attenuator networks and equations.

FIGURE 10.7.2 A 14.0-dB attenuator between a 75-Ω source and a 300-Ω load.

SECTION 11

AMPLIFIERS AND OSCILLATORS

Amplifiers serve a number of purposes from allowing us to hear beautiful music to accurately positioning elements of complicated systems using control technologies. Oscillators are found in a number of applications from the watch on your wrist to the transmitter and receiver in your cell phone. We look at audio-frequency amplifiers and oscillators and radio-frequency amplifiers and oscillators.

The most versatile amplifier has to be the operational amplifier (op amp). The key to its success is that it is perhaps the most ideal device in analog electronics. Because of this it is found in a number of amplifier designs.

High-power amplifiers are necessary where significant amounts of power need to be used to accomplish activities such as radio and television broadcasts. Just imagine what a rock concert might sound like without power amplifiers. Microwave amplifiers and oscillators represent a special part of the high-power amplifier and oscillator field. C.A.

In This Section:

Section Bibliography:

Classic General References

Bode, H. W., "Network Analysis and Feedback Amplifier Design," Van Nostrand, 1959. (Reprinted 1975 by R. E. Krieger).

Ghausi, M. S., and D. O. Pederson, "A new approach to feedback amplifiers," *IRE Trans. Circuit Theory*, Vol. CT-4, September 1957.

Ginzton, E. L., W. R. Hewlett, J. H. Jasberg, and J. D. Noe, "Distributed amplification," *Proc. IRE*, Vol. 20, August 1948.

Glasford, G. M., "Fundamentals of Television Engineering," McGraw-Hill, 1955.

Hines, M. E., "High-frequency negative-resistance circuit principles for Esaki diodes," *Bell Syst. Tech. J.*, Vol. 39, May 1960.

Hutson, A. R., J. H. McFee, and D. L. White, "Ultrasonic amplification in CdS," *Phys. Rev. Lett.*, September 15, 1961.

Kim, C. S., and A. Brandli, "High frequency high power operation of tunnel diodes," *IRE Trans. Circuit Theory*, December 1962.

Millman, J., "Vacuum Tube and Semiconductor Electronics," McGraw-Hill, 1958.

Read, W. T., "A proposal high-frequency negative resistance diode," *Bell Syst. Tech. J.*, Vol. 37, 1958.

Reich, H. J., "Functional Circuits and Oscillators," Van Nostrand, 1961.

Seely, S., "Electron Tube Circuits," McGraw-Hill, 1950.

Shea, R. F. (ed.), "Amplifier Handbook," McGraw-Hill, 1968.

Singer, J. R., "Masers," Wiley, 1959.

Storm, H. F., "Magnetic Amplifiers," Wiley, 1955.

Truxal, J. C., "Automatic Feedback Control System Synthesis," McGraw-Hill, 1955.

Specific-Topic and Contemporary References

Bahl, I. (ed.), "Microwave Solid State Circuit Design," Wiley, 1988.

Wilson, F. A., "An Introduction to Microwaves," Babani, 1992.

Blackwell, L. A., and K. L. Kotzebue, "Semiconductor-Diode Parametric Amplifiers," Prentice Hall, 1961.

Blotekjaer, K., and C. F. Quate, "The coupled modes of acoustic waves and drifting carriers in piezoelectric crystals," *Proc. IEEE*, Vol. 52, No. 4, pp. 360–377, April 1965.

Cate, T., "Modern techniques of analog multiplication," *Electron. Eng.*, pp. 75–79, April 1970.

Chang, K. K. N., "Parametric and Tunnel Diodes," Prentice Hall, 1964.

Coldren, L. A., and G. S. Kino, "Monolithic acoustic surface-wave amplifier," *Appl. Phys. Lett.*, Vol. 18, No. 8, p. 317, 1971.

Cunningham, D. R., and J. A. Stiller, "Basic Circuit Analysis," Houghton Mifflin, 1991.

Curtis, F. W., "High-Frequency Induction Heating," McGraw-Hill, 1964.

Datta, S., "Surface Acoustic Wave Devices," Prentice Hall, Vol. 72, 1986.

Duenas, J. A., and A. Serrano, "Directional coupler design graphs for parallel coupled lines and interdigitated 3 dB couplers," *RF Design*, pp. 62–64, February 1986.

Evaluating, Selecting, and Using Multiplier Circuit Modules for Signal Manipulation and Function Generation, Analog Devices, 1970.

Garmand, P. A., "Complete small size 2 to 30 GHz hybrid distributed amplifier using a novel design technique," *IEEE MITT-S Digest*, pp. 343–346, 1986.

Hagt, W. H., Jr., and J. E. Kemmerly, "Engineering Circuit Analysis," 5th ed., McGraw-Hill, 1993.

Helms, H. L., "Contemporary Electronics Circuit Deskbook," McGraw-Hill, 1986.

Helszajn, J., "Microwave Planar Passive Circuits and Filters," Wiley, 1994.

Ingebritsen, K. A., "Linear and nonlinear attenuation of acoustic surface waves in a piezoelectric coated with a semiconducting film," *J. Appl. Phys.*, Vol. 41, p. 454, 1970.

Inglis, A. F., "Video Engineering," McGraw-Hill, 1993.

Kino, G. S., and T. M. Reeder, "A normal mode theory for the Rayleigh wave amplifier," *IEEE Trans. Electron Devices*, Vol. ED-18, p. 909, 1971.

Kotelyanski, I. M., A. I. Kribunov, A. V. Edved, R. A. Mishkinis, and V. V. Panteleev, "Fabrication of LiNbO3-InSb layer structures and their use in amplification of surface acoustic waves," *Sov. Phys. Semicon*, Vol. 12, No. 7, pp. 751–754, July 1978.

Kouril, F., "Non-linear and Parametric Circuits: Principles, Theory, and Applications," Chichester/Wiley, 1988.

Ladbrooke, P. H., "MMIC Design: GaAs FETs and HEMTs," Artech House, 1989.

Lakin, K. M., and H. J. Shaw, "Surface wave delay line amplifiers, *IEEE Trans. Microwave Theory Techniques*, MTT-17, p. 912, 1969.

Lange, L., "Interdigitated stripline quadrature hybrid," *IEEE Trans. MTT*, December 1969.

Liff, A. A., "Color and Black and White Television," Regents/Prentice Hall, 1993.

Lin, Y., "Ion Beam Sputtered InSb Thin Films and Their Application to Surface Acoustic Wave Amplifiers," Ph.D. Dissertation, Polytechnic University of 1995.

May, J. E., Jr., "Electronic signal amplification in the UHF range with the ultrasonic traveling wave amplifier," *Proc. IEEE*, Vol. 53, No. 10, October 1965.

McFee, J. H., "Transmission and amplification of acoustic waves," in: *Physical Acoustics*, Vol. 4A, Academic Press, 1964.

Middlebrook, R. D., "Differential Amplifiers," Wiley, 1963.

Mizuta, H., "The Physics and Applications of Resonant Tunnelling Diodes," Cambridge University Press, 1995.

Nelson, J. C. C., "Operational Amplifier Circuits: Analysis and Design," Butterworth-Heinemann, 1995.

Optical Pumping and Masers, *Appl. Opt.*, Vol. 1, No. 1, January 1962.

Pauley, R. G., P. G. Asher, J. M. Schellenberg, and H. Yamasaki, "A 2 to 40 GHz monolithic distributed amplifier," *GaAs IC Symp.*, pp. 15–17, November 1985.

Penfield, P. Jr., and R. P. Rafuse, "Varactor Applications," The MIT Press, 1962.

Petruzzela, F. D., "Industrial Electronics," Glencoe/McGraw-Hill, 1996.

"Power Op-amp Handbook," Apex Microtechnology, 85741, 1987.

Pucel, R. A., "Monolithic Microwave Integrated Circuits," IEEE Press, 1985.

Robertson, I. D., "MMIC Design," IEE, London, 1995.

Rutkowski, G. B., "Operational Amplifiers: Integrated and Hybrid Circuits," Wiley, 1993.

Simpson, C. D., "Industrial Electronics," Prentice Hall, 1996.

Southgate, P. D., and H. N. Spector, "Effect of carrier trapping on the Weinreich relation in acoustic amplification," *J. Appl. Phys.*, vol. 36, pp. 3728–3730, December 1965.

Tehon, S. W., "Acoustic wave amplifiers," Chap. 30, *Amplifier Handbook*, McGraw-Hill, 1968.

Tobey, G. E., L. P. Huelsman, and J. G. Graeme, "Operational Amplifiers," McGraw-Hill, 1971.

Traister, R. J., "Operational Amplifier Circuit Manual," Academic Press, 1989.

Vizmuller, P., "RF Design Guide: Systems, Circuits and Equations," Artech House, 1995.

Wang, W. C., "Strong electroacoustic effect in CdS," *Phys. Rev. Lett.*, Vol. 9, No. 11, pp. 443–445, December 1, 1962.

Wang, W. C., and Y. Lin, "Acousto-electric Attenuation Determined by Transmission Line Technique," International Workshop on Ultrasonic Application, September 1–3, 1996.

Wanuga, S., "CW acoustic amplifier, *Proc. IEEE* (Corres.), Vol. 53, No. 5, p. 555, May 1965.

White, D. L., Amplification of ultrasonic waves in piezoelectric semiconductors," *J. Appl. Phys.*, Vol. 33, No. 8, pp. 2547–2554, August 1962.

White, R. M., "Surface elastic-wave propagation and amplification," *IEEE Trans. Electron Devices*, ED-14, 181 (1967).

Wilkinson, E. J., "An N-way hybrid power divider," *IRE MTT*, January 1960.

Wilson, T. G., "Series connected magnetic amplifier with inductive loading," *Trans. AIEE*, Vol. 71, 1952.

CHAPTER 11.1
AMPLIFIER AND OSCILLATOR PRINCIPLES OF OPERATION

G. Burton Harrold

AMPLIFIERS: PRINCIPLES OF OPERATION

Gain

In most amplifier applications the prime concern is gain. A generalized amplifier is shown in Fig. 11.1.1. The most widely applied definitions of gain using the quantities defined there are:

$$\text{Voltage gain } A_v = e_{22}/e_{11} \qquad \text{Current gain } A_i = i_2/i_1$$

$$\text{Available power from source } P_{avs} = \frac{|e_s|^2}{4 \text{ Re } Z_s} \qquad \text{where Re} = \text{real part of complex impedance}$$

$$\text{Output load power } P_L = \frac{|e_{22}|^2}{\text{Re } Z_L} \qquad \text{Input power } P_I = \frac{|e_{11}|^2}{\text{Re } Z_{in}}$$

$$\text{Available power at output } P_{avo} = \frac{|e_{22}|^2}{4 \text{ Re } Z_{out}} \qquad \text{Transducer gain } G_T = P_L/P_{avs}$$

$$\text{Available power gain } G_A = P_{avo}/P_{avs} \qquad \text{Power gain } G = P_L/P_I$$

$$\text{Insertion power gain } G_I = \frac{\text{power into load with network inserted}}{\text{power into load with source connected to load}}$$

Bandwidth and Gain-Bandwidth Product

Bandwidth is a measure of the range of frequencies within which an amplifier will respond. The frequency range (passband) is usually measured between the half-power (3-dB) points on the output-response-versus-frequency curve, for constant input. In some cases it is defined at the quarter-power points (6 dB). See Fig. 11.1.2.

The gain-bandwidth product of a device is a commonly used figure of merit. It is defined for a bandpass amplifier as

$$F_a = A_r B$$

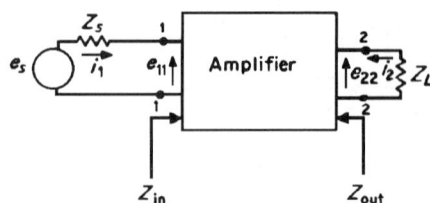

FIGURE 11.1.1   Input and output quantities of generalized amplifier.

where F_a = figure of merit (rad/s)
A_r = reference gain, either the maximum gain or the gain at the frequency where the gain is purely real or purely imaginary
B = 3-dB bandwidth (rad/s)

For low-pass amplifiers

$$F_a = A_r W_H$$

where F_a = figure of merit (rad/s)
A_r = reference gain
W_H = upper cutoff frequency (rad/s)

In the case of vacuum tubes and certain other active devices this definition is reduced to

$$F_a = g_m / C_T$$

where F_a = figure of merit (rad/s)
g_m = transconductance of active device
C_T = total output capacitance, plus input capacitance of subsequent stage

Noise

The major types of noise are illustrated in Fig. 11.1.3. Important relations and definitions in noise computations are:

Noise factor
$$F = \frac{S_i / N_i}{S_o / N_o}$$

where S_i = signal power available at input
S_o = signal power available at output
N_i = noise power available at input at $T = 290$ K
N_o = noise power available at output

Available noise power
$$P_{n,av} = \frac{e_n^2}{4R} = KTB \quad \text{for thermal noise}$$

where the quantities are as defined in Fig. 11.1.3.

Excess noise factor
$$F - 1 = N_e / N_i$$

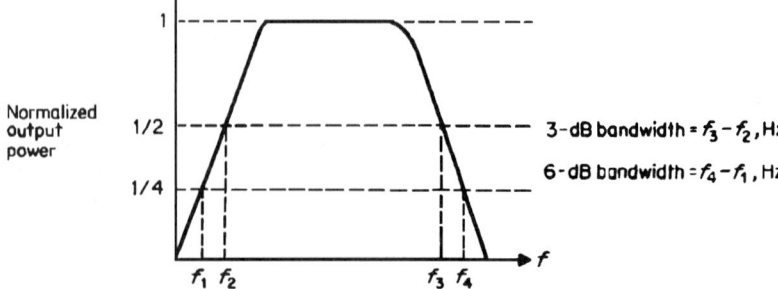

FIGURE 11.1.2   Amplifier response and bandwidth.

Thermal noise

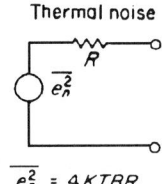

$\overline{e_n^2}$ = 4KTBR

$\overline{e_n^2}$ = mean-square open-circuit noise
voltage from a resistor R

$K = 1.38 \times 10^{-23}\,\text{J/K}$

T = temperature, K

B = bandwith, Hz

R = resistance, Ω

Shot noise

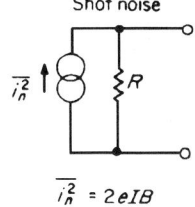

$\overline{i_n^2}$ = 2eIB

$\overline{i_n^2}$ = mean-square short-circuit noise
current

$e = 1.6 \times 10^{-19}\,C$

I = dc current amps through R

R = resistance, Ω

B = bandwith, Hz

1/f noise
(flicker noise)

$\overline{i_{nf}^2}$ = $k\,\dfrac{I^\alpha}{f^n}\,\Delta f$

$\overline{i_{nf}^2}$ = mean-square short-circuit flicker
noise current

R = resistance, Ω

I = dc current

f = frequency, Hz

Δf = frequency interval

k, α, n = empirical constants depending on
device and mode of operation

FIGURE 11.1.3   Noise-equivalent circuits.

where $F - 1$ = excess noise factor
N_e = total equivalent device noise referred to input
N_i = thermal noise of source at standard temperature

Noise temperature $T = P_{n,\text{av}}/KB$

where $P_{n,\text{av}}$ is the average noise power available.
   At a single input-output frequency in a two-port,

Effective input noise temperature $T_e = 290(F - 1)$

Noise Factor of Transmission Lines and Attenuators.   The noise factor of two ports composed entirely of resistive elements at room temperature (290 K) and an impedance matched loss of $L = 1/G_A$ is $F = L$.

Cascaded noise factor $F_T = F_1 + (F_2 - 1)/G_A$

where F_T = overall noise factor
$\quad\quad F_1$ = noise factor of first stage
$\quad\quad F_2$ = noise factor of second stage
$\quad\quad G_A$ = available gain of first stage

System Noise Temperature. Space probes and satellite communication systems using low-noise amplifiers and antennas directed toward outer space make use of system noise temperatures. When we define T_A = antenna temperature, L = waveguide numeric loss (greater than 1), T_{E1} = amplifier noise temperature, G_A = amplifier available gain, F = postamplifier noise factor, and B = postamplifier bandwidth, this temperature can be calculated as

$$T_{\text{sys}} = T_A + |L - 1|\,290° + LT_{E1} + \frac{(F-1)(290L)}{G_{A1}}$$

The quantity of interest is the output signal-to-noise ratio where S_A is available signal power at the antenna (assuming the antenna is matched to free space)

$$\text{S/N} = S_A / KT_{\text{sys}}B \quad K = 1.38 \times 10^{-23}$$

Generalized Noise Factor. A general representation of noise performances can be expressed in terms of Fig. 11.1.4. This is the representation of a noisy two-port in terms of external voltage and current noise sources with a correlation admittance. In this case the noise factor becomes

$$F = 1 + \frac{G_u}{G_s} - \frac{R_N}{G_s}[(G_s + G_\gamma)^2 + (B_s + B_\gamma)^2]$$

where F = noise factor
$\quad\quad G_s$ = real part of Y_s
$\quad\quad B_s$ = imaginary part of Y_s
$\quad\quad G_u$ = conductance owing to the uncorrelated part of the noise current
$\quad\quad Y_\gamma$ = correlation admittance between cross product of current and voltage noise sources
$\quad\quad G_\gamma$ = real part of Y_γ
$\quad\quad B_\gamma$ = imaginary part of Y_γ
$\quad\quad R_N$ = equivalent noise resistance of the noise voltage

The optimum source admittance is $Y_{\text{opt}} = G_{\text{opt}} + jB_{\text{opt}}$

$$G_{\text{opt}} = \left(\frac{G_u + R_N G_\gamma^2}{R_N}\right)^{1/2}$$

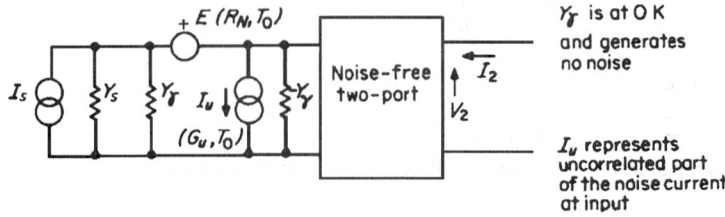

FIGURE 11.1.4 Noise representation using correlation admittance.

where $B_{opt} = -B_{gg}$ and the value of the optimum noise factor F_{opt} is

$$F_{opt} = 1 + 2R_N(G_\gamma + G_0)$$

The noise factor for an arbitrary source impedance is

$$F = F_{opt} + \frac{R_N}{G_s}[(G_s - G_0)^2 + (B_s - B_0)^2]$$

The values of the parameters of Fig. 11.1.4 can be determined by measurement of (1) noise figure versus B_s with G_s constant and (2) noise figure versus G_s with B_s at its optimum value.

Dynamic Characteristic, Load Lines, and Class of Operation

Most active devices have two considerations involved in their operation. The first is the dc bias condition that establishes the operating point (the *quiescent point*). The choice of operating point is determined by such considerations as signal level, uniformity of the device, and temperature of operation.

The second consideration is the ac operating performance, related to the slope of the dc characteristic and to the parasitic reactances of the device. These ac variations give rise to the *small-signal parameters*. The ac parameters may also influence the choice of dc bias point when basic constraints, such as gain and noise performance, are considered.

For frequencies of operation where these parasites are not significant, the use of a load line is valuable. The class of amplifier operation is dependent on its quiescent point, its load line, and input signal level. The types of operation are shown in Fig. 11.1.5.

Distortion

Distortion takes many forms, most of them undesirable. The basic causes of distortion are nonlinearity in amplitude response and nonuniformity of phase response. The most commonly encountered types of distortion are as follows:

Harmonic distortion is a result of nonlinearity in the amplitude transfer characteristics. The typical output contains not only the fundamental frequency but integer multiples of it.

Crossover distortion is a result of the nonlinear characteristics of a device when changing operating modes (e.g., in a push-pull amplifier). It occurs when one device is cut off and the second turned on if the crossover is not smooth between the two modes.

Intermodulation distortion is a spurious output resulting from the mixing of two or more signals of different frequencies. The spurious output occurs at the sum or difference of integer multiples of the original frequencies.

Cross-modulation distortion occurs when two signals pass through an amplifier and the modulation of one is transferred to the other.

Phase distortion results from the deviation from a constant slope of the output-phase–versus–frequency response of an amplifier. This deviation gives rise to echo responses in the output that precede and follow the main response, and a distortion of the output signal when an input signal having a large number of frequency components is applied.

Feedback Amplifiers

Feedback amplifiers fall into two categories: those having positive feedback (usually oscillators) and those having negative feedback. The positive-feedback case is discussed under oscillators. The following discussion is concerned with negative-feedback amplifiers.

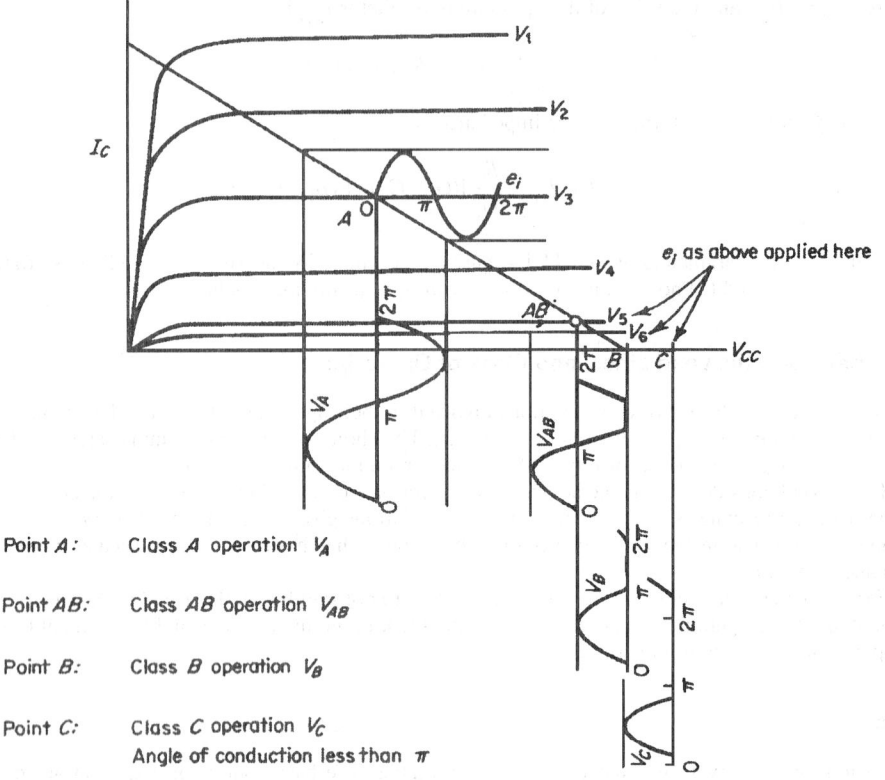

FIGURE 11.1.5 Classes of amplifier operation. Class S operation is a switching mode in which a square-wave output is produced by a sine-wave input.

Negative Feedback

A simple representation of a feedback network is shown in Fig. 11.1.6. The closed-loop gain is given by

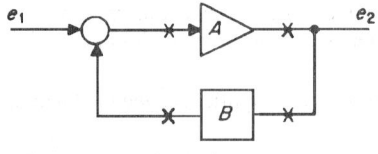

FIGURE 11.1.6 Amplifier with feedback loop.

$$e_2/e_1 = A/(1 - BA)$$

where A is the forward gain with feedback removed and B is the fraction of output returned to input.

For negative feedback, A provides a 180° phase shift in midband, so that

$$1 - AB > 1 \qquad \text{in this frequency range}$$

The quantity $1 - AB$ is called the *feedback factor*, and if the circuit is cut at any X point in Fig. 11.1.6, the open-loop gain is AB.

It can be shown that for large loop gain AB the closed-loop transfer function reduces to

$$e_2/e_1 \approx 1/B$$

The gain then becomes essentially independent of variations in A. In particular, if B is passive, the closed-loop gain is controlled only by passive components. Feedback has no beneficial effect in reducing unwanted signals at the input of the amplifier, e.g., input noise, but does reduce unwanted signals generated in the amplifier chain (e.g., output distortion).

The return ratio can be found if the circuit is opened at any point X (Fig. 11.1.6) and a unit signal P is injected at that X point. The return signal P' is equal to the return ratio, since the input P is unity. In this case the return ratio T is the same at any point X and is

$$T = -AB$$

The minus sign is chosen because the typical amplifier has an odd number of phase reversals and T is then a positive quantity. The return difference is by definition

$$F = 1 + T$$

It has been shown by Bode that

$$F = \Delta / \Delta^0$$

where Δ is the network determinant with XX point connected and Δ^0 is the network determinant of amplifier when gain of active device is set to zero.

Stability

The stability of the network can be analyzed by several techniques. Of prime interest are the Nyquist, Bode, Routh, and root-locus techniques of analyzing stability.

Nyquist Method.   The basic technique of Nyquist involves plotting T on a polar plot as shown in Fig. 11.1.7 for all values $s = j\omega$ for ω between minus and plus infinity. Stability is then determined by the following method:

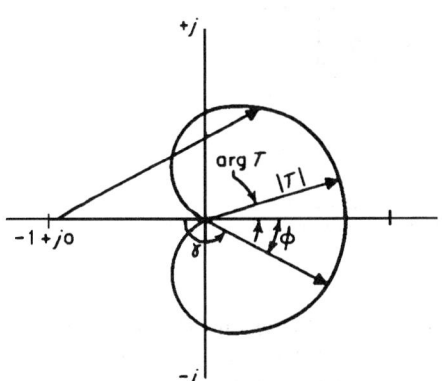

1. Draw a vector from the $-1 + j0$ point to the plotted curve and observe the rotation of this vector as ω varies from $-\infty$ to $+\infty$. Let R be the net number of counterclockwise revolutions of this vector.

2. Determine the number of roots of the denominator of $T = -AB$ which have positive real parts. Call this number P.

3. The system is stable if and only if $P = R$. Note that in many systems A and B are stable by themselves, so that P becomes zero and the net counterclockwise revolution N becomes zero for stability.

Bode's Technique.   A technique that has historically found wide use in determining stability and performance, especially in control systems, is the Bode diagram. The assumptions used here for this method are that $T = -AB$, where A and B are stable when the system is open-circuited and consists of minimum-phase networks. It is also necessary to define a phase margin γ such that $\gamma = 180 + \phi$, where ϕ is the phase angle of T and is positive when measured counterclockwise from zero, and γ, the phase mar-

FIGURE 11.1.7 Nyquist diagram for determining stability.

gin, is positive when measured counterclockwise from the 180° line (Fig. 11.1.7). The stability criterion under these conditions reads: Systems having a positive phase margin when their return ratio equal to 20 log |T| goes through 0 dB (i.e., where |T| crosses the unit circle in the Nyquist plot) are stable; if a negative γ exists at 0 dB, the system is unstable.

Bode's theorems show that the phase angle of a system is related to the attenuation or gain characteristic as a function of frequency. Bode's technique relies heavily on straight-line approximation.

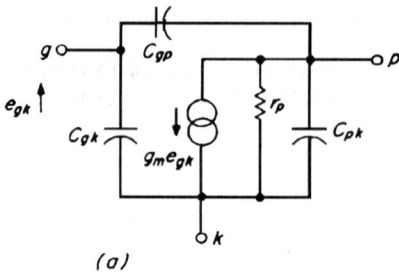

(a)

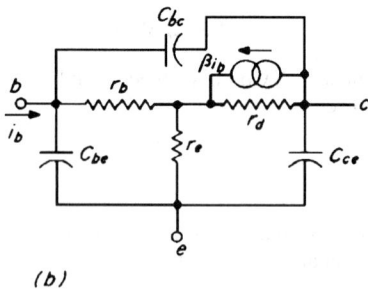

(b)

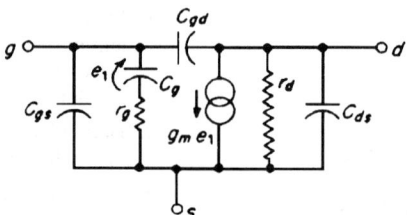

FIGURE 11.1.8 Equivalent circuits of active devices: (a) vacuum tube; (b) bipolar transistor; (c) field-effect transistor (FET).

Routh's Criterion for Stability. Routh's method has also been used to test the characteristic equations or return difference $F = 1 + T = 0$, to determine whether it has any roots that are real and positive or complex with positive real parts that will give rise to growing exponential responses and hence instability.

Root-Locus Method. The root-locus method of analysis is a means of finding the variations of the poles of a closed-loop response as some network parameter is varied. The most convenient and commonly used parameter is that of the gain K. The basic equation then used is

$$F = 1 + KT(s) = 1 - K\frac{(S - S_2)(S - S_4)\cdots}{(S - S_1)(S - S_3)\cdots} = 0$$

This is a useful technique in feedback and control systems, but it has not found wide application in amplifier design. A detailed exposition of the technique is found in Truxal.

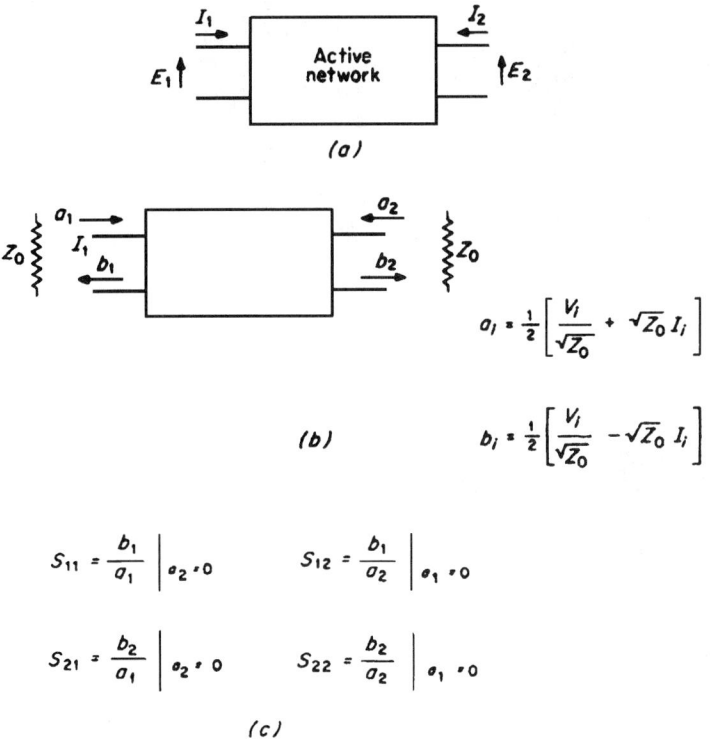

$$a_i = \frac{1}{2}\left[\frac{V_i}{\sqrt{Z_0}} + \sqrt{Z_0}\,I_i\right]$$

$$b_i = \frac{1}{2}\left[\frac{V_i}{\sqrt{Z_0}} - \sqrt{Z_0}\,I_i\right]$$

$$S_{11} = \frac{b_1}{a_1}\bigg|_{a_2=0} \qquad S_{12} = \frac{b_1}{a_2}\bigg|_{a_1=0}$$

$$S_{21} = \frac{b_2}{a_1}\bigg|_{a_2=0} \qquad S_{22} = \frac{b_2}{a_2}\bigg|_{a_1=0}$$

(c)

FIGURE 11.1.9 Definitions of active-network parameters: (*a*) general network; (*b*) ratios a_i and b_i of incident and reflected waves (square root of power); (*c*) *s* parameters.

Active Devices Used in Amplifiers

There are numerous ways of representing active devices and their properties. Several common equivalent circuits are shown in Fig. 11.1.8. Active devices are best analyzed in terms of the *immittance* or *hybrid matrices*. Figures 11.1.9 and 11.1.10 show the definition of the commonly used matrices, and their interconnections are shown in Fig. 11.1.11. The requirements at the bottom of Fig. 11.1.11 must be met before the interconnection of two matrices is allowed.

The matrix that is becoming increasingly important at higher frequencies is the *S* matrix. Here the network is embedded in a transmission-line structure, and the incident and reflected powers are measured and reflected coefficients and transmission coefficients are defined.

Cascaded and Distributed Amplifiers

Most amplifiers are cascaded (i.e., connected to a second amplifier). The two techniques commonly used are shown in Fig. 11.1.12. In the cascade structure the overall response is the product of the individual responses: in the distributed structure the response is one-half the sum of the individual responses, since each stage's output is propagated in both directions. In cascaded amplifiers the frequency response and gain are determined by the active device as well as the interstage networks. In simple audio amplifiers these interstage networks may become simple *RC* combinations, while in rf amplifiers they may become critically coupled double-tuned circuits. Interstage coupling networks are discussed in subsequent sections.

$$\begin{bmatrix} E_1 \\ E_2 \end{bmatrix} = \begin{bmatrix} z_{11} & z_{12} \\ z_{21} & z_{22} \end{bmatrix} \times \begin{bmatrix} I_1 \\ I_2 \end{bmatrix}$$

$$z_{11} = \left(\frac{E_1}{I_1}\right)_{I_2=0} \qquad z_{12} = \left(\frac{E_1}{I_2}\right)_{I_1=0}$$

$$z_{21} = \left(\frac{E_2}{I_1}\right)_{I_2=0} \qquad z_{22} = \left(\frac{E_2}{I_2}\right)_{I_1=0}$$

$$\begin{bmatrix} I_1 \\ I_2 \end{bmatrix} = \begin{bmatrix} y_{11} & y_{12} \\ y_{21} & y_{22} \end{bmatrix} \times \begin{bmatrix} E_1 \\ E_2 \end{bmatrix}$$

$$y_{11} = \left(\frac{I_1}{E_1}\right)_{E_2=0} \qquad y_{12} = \left(\frac{I_1}{E_2}\right)_{E_1=0}$$

$$y_{21} = \left(\frac{I_2}{E_1}\right)_{E_2=0} \qquad y_{22} = \left(\frac{I_2}{E_2}\right)_{E_1=0}$$

$$\begin{bmatrix} E_1 \\ I_2 \end{bmatrix} = \begin{bmatrix} h_{11} & h_{12} \\ h_{21} & h_{22} \end{bmatrix} \times \begin{bmatrix} I_1 \\ E_2 \end{bmatrix}$$

$$h_{11} = \frac{1}{y_{11}} \qquad h_{12} = \left(\frac{E_1}{E_2}\right)_{I_1=0}$$

$$h_{21} = \left(\frac{I_2}{I_1}\right)_{E_2=0} \qquad h_{22} = \frac{1}{z_{22}}$$

$$\begin{bmatrix} I_1 \\ E_2 \end{bmatrix} = \begin{bmatrix} l_{11} & l_{12} \\ l_{21} & l_{22} \end{bmatrix} \times \begin{bmatrix} E_1 \\ I_2 \end{bmatrix}$$

$$l_{11} = \frac{1}{z_{11}} \qquad l_{12} = \left(\frac{I_1}{I_2}\right)_{E_1=0}$$

$$l_{21} = \left(\frac{E_2}{E_1}\right)_{I_2=0} \qquad l_{22} = \frac{1}{y_{22}}$$

$$\begin{bmatrix} E_1 \\ I_1 \end{bmatrix} = \begin{bmatrix} a_{11} & a_{12} \\ a_{21} & a_{22} \end{bmatrix} \times \begin{bmatrix} E_2 \\ -I_2 \end{bmatrix}$$

$$a_{11} = \frac{1}{l_{21}} \qquad a_{12} = -\frac{1}{y_{21}}$$

$$a_{21} = \frac{1}{z_{21}} \qquad a_{22} = -\frac{1}{h_{21}}$$

$$\begin{bmatrix} E_2 \\ I_2 \end{bmatrix} = \begin{bmatrix} b_{11} & b_{12} \\ b_{21} & b_{22} \end{bmatrix} \times \begin{bmatrix} E_1 \\ -I_1 \end{bmatrix}$$

$$b_{11} = \frac{1}{h_{12}} \qquad b_{12} = -\frac{1}{y_{12}}$$

$$b_{21} = \frac{1}{z_{12}} \qquad b_{22} = -\frac{1}{l_{12}}$$

$$\begin{bmatrix} b_1 \\ b_2 \end{bmatrix} = \begin{bmatrix} S_{11} & S_{12} \\ S_{21} & S_{22} \end{bmatrix} \times \begin{bmatrix} a_1 \\ a_2 \end{bmatrix}$$

$$S_{11} = \left(\frac{b_1}{a_1}\right)_{a_2=0} \qquad S_{12} = \left(\frac{b_1}{a_2}\right)_{a_1=0}$$

$$S_{21} = \left(\frac{b_2}{a_1}\right)_{a_2=0} \qquad S_{22} = \left(\frac{b_2}{a_2}\right)_{a_1=0}$$

FIGURE 11.1.10   Network matrix terms.

In distributed structures (Fig. 11.1.12b), actual transmission lines are used for the input to the amplifier, while the output is taken at one end of the upper transmission line. The propagation time along the input line must be the same as that along the output line, or distortion will result. This type of amplifier, noted for its wide frequency response, is discussed later.

OSCILLATORS: PRINCIPLES OF OPERATION

Introduction

An oscillator can be considered as a circuit that converts a dc input to a time-varying output. This discussion deals with oscillators whose output is sinusoidal, as opposed to the relaxation oscillator whose output exhibits abrupt transitions (see Section 14). Oscillators often have a circuit element that can be varied to produce different frequencies.

An oscillator's frequency is sensitive to the stability of the frequency-determining elements as well as the variation in the active-device parameters (e.g., effects of temperature, bias point, and aging). In many instances

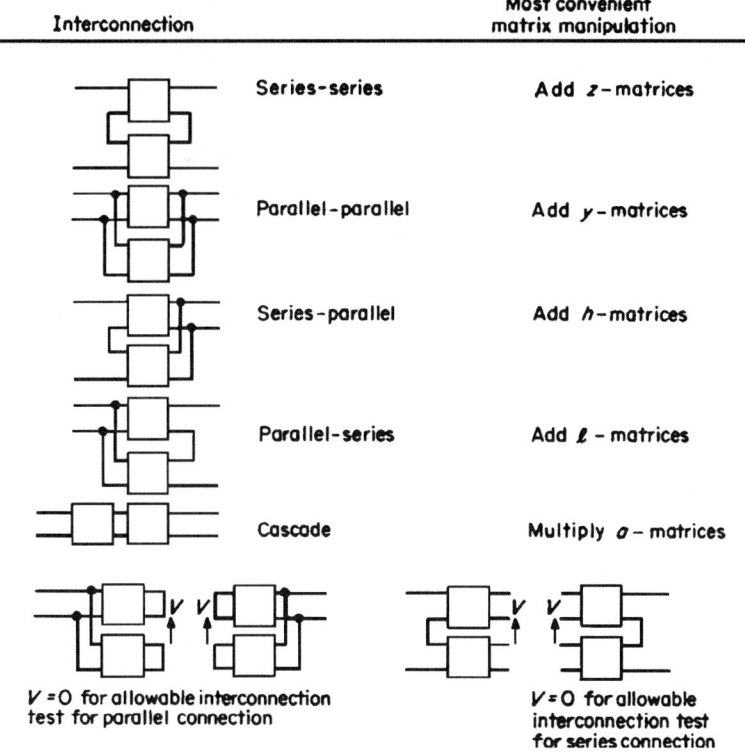

FIGURE 11.1.11   Matrix equivalents of network interconnections.

the oscillator is followed by a second stage serving as a buffer, so that there is isolation between the oscillator and its load. The amplitude of the oscillation can be controlled by automatic gain control (AGC) circuits, but the nonlinearity of the active element usually determines the amplitude. Variations in bias, temperature, and component aging have a direct effect on amplitude stability.

Requirements for Oscillation

Oscillators can be considered from two viewpoints: as using positive feedback around an amplifier or as a one-port network in which the real component of the input immittance is negative. An oscillator must have frequency-determining elements (generally passive components), an amplitude-limiting mechanism, and sufficient closed-loop gain to make up for the losses in the circuit. It is possible to predict the operating frequency and conditions needed to produce oscillation from a Nyquist or Bode analysis. The prediction of output amplitude requires the use of nonlinear analysis.

Oscillator Circuits

Typical oscillator circuits applicable up to ultra high frequencies (UHF) are shown in Fig. 11.1.13. These are discussed in detail in the following subsections. Also of interest are crystal oscillators. In this case the crystal is used as the passive frequency-determining element. The frequency range of crystal oscillators extends from a few hundred hertz to over 200 MHz by use of overtone crystals. The analysis of crystal oscillators is best done using the equivalent circuit of the crystal.

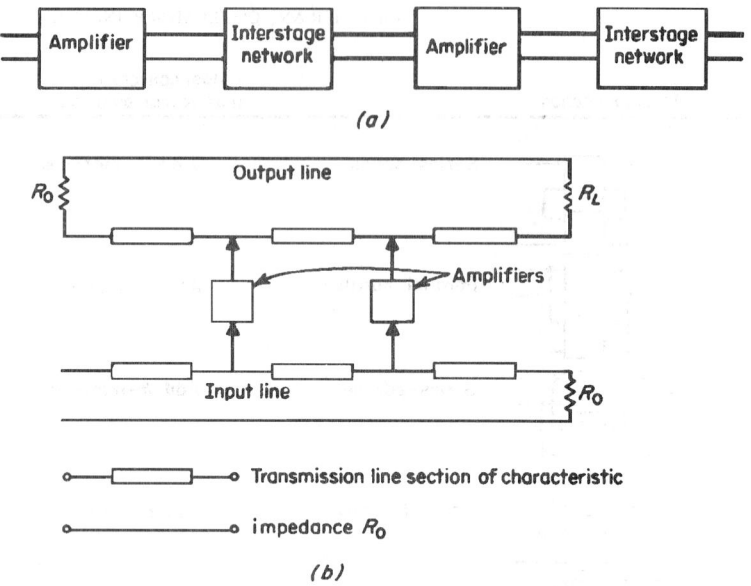

FIGURE 11.1.12 Multiamplifier structures: (*a*) cascade; (*b*) distributed.

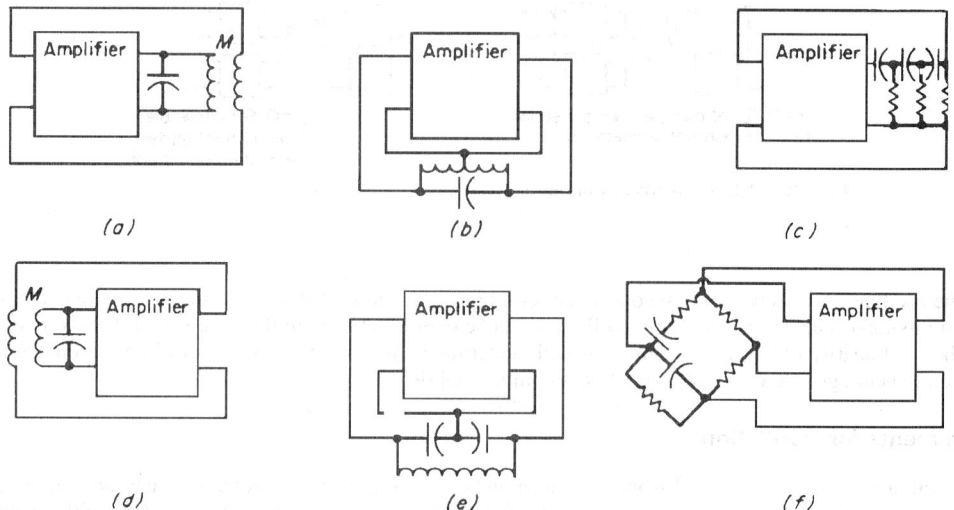

FIGURE 11.1.13 Types of oscillators: (*a*) tuned-output; (*b*) Hartley; (*c*) phase-shift; (*d*) tuned-input; (*e*) Colpitts; (*f*) Wien bridge.

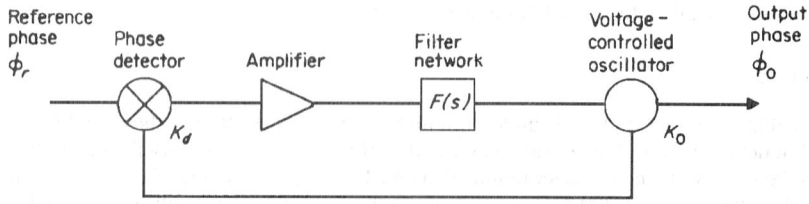

FIGURE 11.1.14 Phase-locked-loop oscillator.

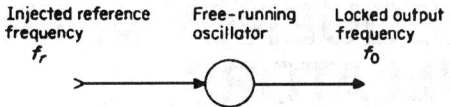

FIGURE 11.1.15   Injection-locked oscillator.

Synchronization

Synchronization of oscillators is accomplished by using phase-locked loops or by direct low-level injection of a reference frequency into the main oscillator. The diagram of a phase-locked loop is shown in Fig. 11.1.14 and that of an injection-locked oscillator in Fig. 11.1.15.

Harmonic Content

The harmonic content of the oscillator output is related to the amount of oscillator output power at frequencies other than the fundamental. From the viewpoint of a negative-conductance (resistance) oscillator, better results are obtained if the curve of the negative conductance (or resistance) versus amplitude of oscillation is smooth and without an inflection point over the operating range. Harmonic content is also reduced if the oscillator's operating point Q is chosen so that the range of negative conductance is symmetrical about Q on the negative conductance-versus-amplitude curve. This can be done by adjusting the oscillator's bias point within the requirement of $|G_C| = |G_D|$ for sustained oscillation (see Fig. 11.1.16).

Stability

The stability of the oscillator's output amplitude and frequency from a negative-conductance viewpoint depends on the variation of its negative conductance with operating point and the amount of fixed positive conductance in the oscillator's associated circuit. In particular, if the change of bias results in vertical translation of the conductance-(resistance)-versus-amplitude curve, the oscillator's stability is related to the change of slope at the point where the circuit's fixed conductance intersects this curve (point Q in Fig. 11.1.16). If the $|G_D|$ curve is of the shape of $|G_D|_2$, the oscillation can stop when a large enough change in bias point occurs for $|G_D|$ to be less than $|G_C|$ for all amplitudes of oscillation. Stabilization of the amplitude of oscillation may occur in the form of modifying G_C, G_D, or both to compensate for bias changes.

Particular types of oscillators and their parameters are discussed later in this section.

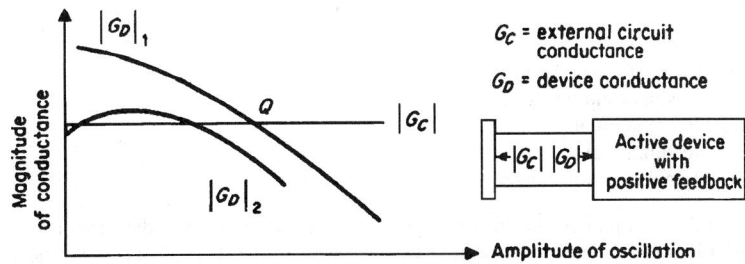

FIGURE 11.1.16   Device conductance vs. amplitude of oscillation.

CHAPTER 11.2
AUDIO-FREQUENCY AMPLIFIERS AND OSCILLATORS

Samuel M. Korzekwa, Robert J. McFadyen

AUDIO-FREQUENCY AMPLIFIERS

Samuel M. Korzekwa

Preamplifiers

General Considerations. The function of a preamplifier is to amplify a low-level signal to a higher level before further processing or transmission to another location. The required amplification is achieved by increased signal voltage and/or impedance reduction. The amount of power amplification required varies with the particular application. A general guideline is to provide sufficient preamplification to ensure that further signal handling adds minimal (or acceptable) signal-to-noise degradation.

Signal-to-Noise Considerations. The design of a preamplifier must consider all potential signal degradation from sources of noise, whether generated externally or within the preamplifier itself.

Examples of externally generated noise are hum and pickup, which may be introduced by the input-signal lines or the power-supply lines. Shielding of the input-signal lines often proves to be an acceptable solution. The preamplifier should be located close to the transmitting source, and the preamplifier power gain must be sufficient to override interference that remains after these steps are taken.

A second major source of noise is that internally generated in the amplifier itself. The noise figure specified in decibels for a preamplifier, which serves as a figure of merit, is defined as the ratio of the available input-to-output signal-to-noise power ratios:

$$F = \frac{S_i / N_i}{S_o / N_o}$$

where F = noise figure of preamplifier
S_i = available signal input power
N_i = available noise input power
S_o = available signal output power
N_o = available noise output power

Design precautions to realize the lowest possible noise figure include the proper selection of the active device, optimum input and output impedance, correct voltage and current biasing conditions, and pertinent design parameters of devices.

Low-Level Amplifiers

The low-level designation applies to amplifiers operated below maximum permissible power-dissipation, current, and voltage limits. Thus many low-level amplifiers are purposely designed to realize specific attributes other than delivering the maximum attainable power to the load, such as gain stability, bandwidth, optimum noise figure, and low cost.

In an amplifier designed to be operated with a 24-V power supply and a specified load termination, for example, the operating conditions may be such that the active devices are just within their allowable limits. If operated at these maximum limits, this is not a low-level amplifier; however, if this amplifier also fulfills its performance requirements at a reduced power-supply voltage of 6 V, with resulting much lower internal dissipation levels, it becomes a low-level amplifier.

Medium-Level and Power Amplifiers

The medium-power designation for an amplifier implies that some active devices are operated near their maximum dissipation limits, and precautions must be taken to protect these devices. If power-handling capability is taken as the criterion, the 5- to 100-W power range is a current demarcation line. As higher-power-handling devices come into use, this range will tend to shift to higher power levels.

The amount of power that can safely be handled by an amplifier is usually dictated by the dissipation limits of the active devices in the output stages, the efficiency of the circuit, and the means used to extract heat to maintain devices within their maximum permissible temperature limits. The classes of operation (A, B, AB, C) are discussed relative to Fig. 11.1.5. When single active devices do not suffice, multiple series or parallel configurations can be used to achieve higher voltage or power operation.

Multistage Amplifiers

An amplifier may take the form of a single stage or a complex single stage, or it may employ an interconnection of several steps. Various biasing, coupling, feedback, and other design alternatives influence the topology of the amplifier. For a multistage amplifier, the individual stages may be essentially identical or radically different. Feedback techniques may be used at the individual stage level, at the amplifier functional level, or both, to realize bias stabilization, gain stabilization, output-impedance reduction, and so forth.

Typical Electron-Tube Amplifier

Figure 11.2.1 shows a typical electron-tube amplifier stage. For clarity the signal-source and load sections are shown partitioned. For a multistage amplifier the source represents the equivalent signal generator of the preceding stage. Similarly, the load indicated includes the loading effect of the subsequent stage, if any.

The voltage gain from the grid of the tube to the output can be calculated to be

$$A_{v1} = -\frac{\mu R_l}{r_p + R_l}$$

Similarly, the voltage gain from the source to the tube grid is

$$A_{v2} = \frac{R_1}{(R_1 + R_g) + 1/j\omega C}$$

Combining the above equations gives the composite amplifier voltage gain

$$A_v = \frac{\mu R_1 R_l}{(r_p + R_l)[(R_1 + R_g) + 1/j\omega C]}$$

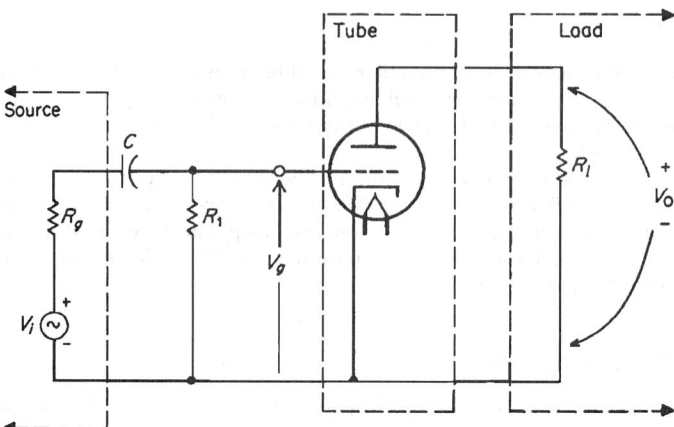

FIGURE 11.2.1   Typical triode electron-tube amplifier stage (biasing not shown).

This example illustrates the fundamentals of an electron-tube amplifier stage. Many excellent references treat this subject in detail.

Typical Transistor Amplifier

The analysis techniques used for electron-tube amplifier stages generally apply to transistorized amplifier stages. The principal difference is that different active-device models are used.

The typical transistor stage shown in Fig. 11.2.2 illustrates a possible form of biasing and coupling. The source section is partitioned and includes the preceding-stage equivalent generator, and the load includes subsequent stage-loading effects. Figure 11.2.3 shows the generalized h-equivalent circuit representation for transistors. Table 11.2.1 lists the h-parameter transformations for the common-base, common-emitter, and common-collector configurations.

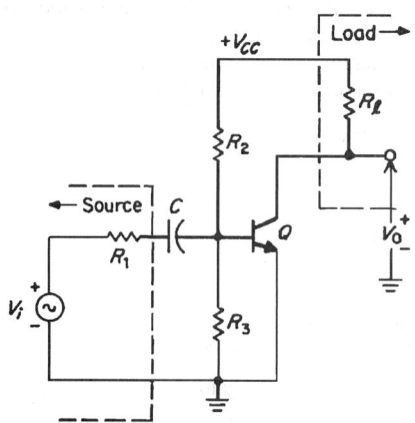

FIGURE 11.2.2   Typical bipolar transistor-amplifier stage.

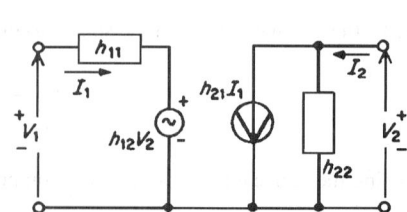

FIGURE 11.2.3   Equivalent circuit of transistor, based on h parameters.

TABLE 11.2.1   *h* Parameters of the Three Transistor Circuit Configurations

	Common-base	Common-emitter	Common-collector
h_{11}	h_{ib}	$h_{ib}(h_{fe} + 1)$	$h_{ib}(h_{fe} + 1)$
h_{12}	h_{rb}	$h_{ib}h_{ob}(h_{fe} + 1) - h_{rb}$	1
h_{21}	h_{fb}	h_{fe}	$-(h_{fe} + 1)$
h_{22}	h_{ob}	$h_{ob}(h_{fe} + 1)$	$h_{ob}(h_{fe} + 1)$

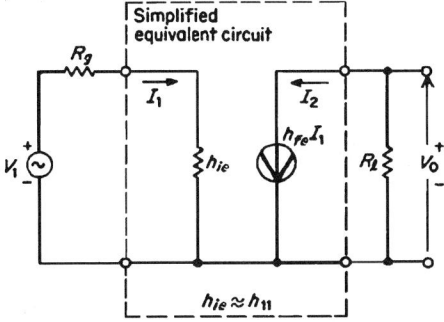

FIGURE 11.2.4   Simplified equivalent circuit of transistor amplifier stage.

While these parameters are complex and frequency-dependent, it is often feasible to use simplifications. Most transistors have their parameters specified by their manufacturers, but it may be necessary to determine additional parameters by test.

Figure 11.2.4 illustrates a simplified model of the transistor amplifier stage of Fig. 11.2.2. The common-emitter *h* parameters are used to represent the equivalent transistor. The voltage gain for this stage is

$$A_v = \frac{V_o}{V_i} = -\frac{h_{fe}R_l}{R_g + h_{ie}}$$

The complexity of analysis depends on the accuracy needed. Currently, most of the more complex analysis is performed with the aid of computers. Several transistor-amplifier-analysis references treat this subject in detail.

Typical Multistage Transistor Amplifier

Figure 11.2.5 is an example of a capacitively coupled three-stage transistor amplifier. It has a broad frequency response, illustrating the fact that an audio amplifier can be useful in other applications. The component values are

$$R_1 = 16,000 \ \Omega \quad R_2 = 6200 \ \Omega \quad R_3 = 1600 \ \Omega \quad R_4 = 1000 \ \Omega$$
$$R_L = 560 \ \Omega \quad Q_1, Q_2, Q_3 = 2N1565 \quad C_1 = 10 \ \mu F \quad C_2 = 100 \ \mu F$$

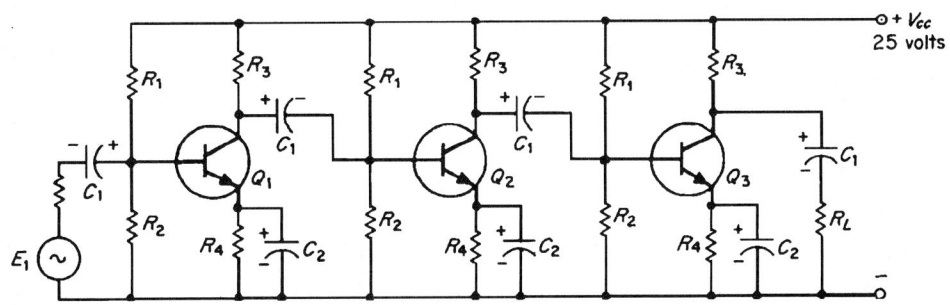

FIGURE 11.2.5   Typical three-stage transistor amplifier.

This amplifier is designed to operate over a range of −55 to +125°C, with an output voltage swing of 2 V peak to peak and frequency response down 3 dB at approximately 200 Hz and 2 MHz. The overall gain at 1000 Hz is nominally 88 dB at 25°C.

Biasing Methods

The biasing scheme used in an amplifier determines the ultimate performance that can be realized. Conversely, an amplifier with poorly implemented biasing may suffer in performance, and be susceptible to catastrophic circuit failure owing to high stresses within the active devices. In view of the variation of parameters within the active devices, it is important that the amplifier function properly even when the initial and/or end-of-life parameters of the devices vary.

Electron-Tube Biasing

Biasing is intended to maintain the quiescent currents and voltages of the electron tube at the prescribed levels. The tube-plate characteristics represent the biasing relations between the tube parameters.

The principal bias parameters (steady-state plate and grid voltages) can be readily identified by the construction of a load line on the plate characteristic. The operating point Q is located at the intersection of the selected plate characteristic with the load line.

Transistor Biasing

Although the methods of biasing a transistor-amplifier stage are in many respects similar to those of an electron-tube amplifier, there are many different types of transistors, each characterized by different curves. Bipolar transistors are generally characterized by their collector and emitter families, while field-effect transistors have different characterizations. The *npn* transistor requires a positive base bias voltage and current (with respect to its emitter) for proper operation; the converse is true for a *pnp* transistor.

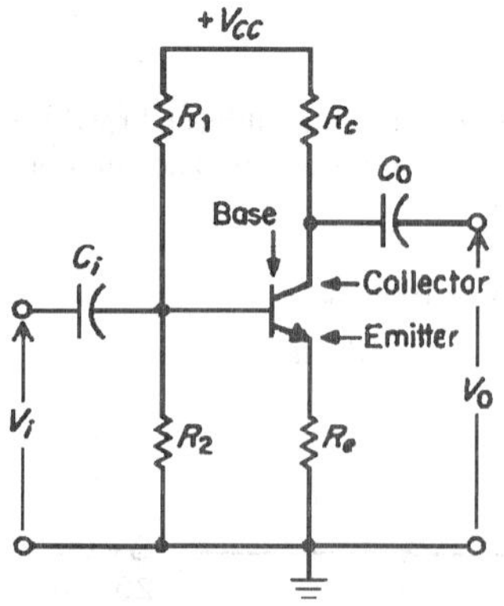

FIGURE 11.2.6 Capacitively coupled *npn* transistor-amplifier stage.

Figure 11.2.6 illustrates a common biasing technique. A single power supply is used, and the transistor is self-biased with the unbypassed emitter resistor R_e. Although a graphical solution of the value of R_e could be found by referring to the collector-emitter curves, an iterative solution, described below, is also commonly used.

Because the performance of the transistors depends on the collector current and collector-to-emitter voltage, they are often selected as starting conditions for biasing design. The unbypassed emitter resistor R_e and collector resistor R_c, the primary voltage-gain-determining components, are determined next, taking into account other considerations such as the anticipated maximum signal level and available power supply V_{cc}. The last step is to determine the R_1 and R_2 values.

Coupling Methods

Transformer coupling and capacitance coupling are commonly used in transistor and electron-tube audio amplifiers. Direct coupling is also used in transistor stages and particularly in integrated transistor amplifiers. Capacitance coupling, referred to as *RC* coupling, is the most common method of coupling stages of an audio amplifier. The discrete-component transistorized amplifier stage shown in Fig. 11.2.6 serves as an example of *RC* coupling, where C_i and C_o are the input and output coupling capacitors, respectively.

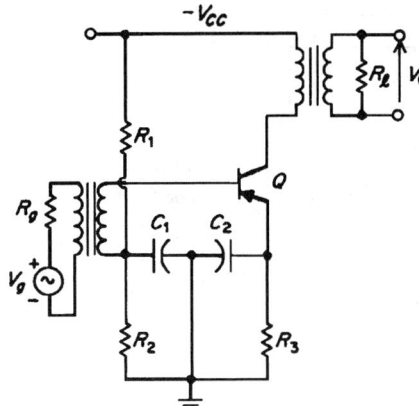

FIGURE 11.2.7 Transformer-coupled *pnp* transistor-amplifier stage.

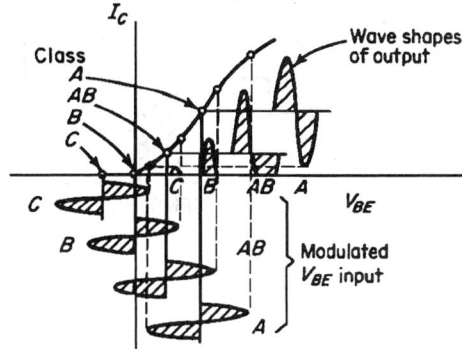

FIGURE 11.2.8 Classes of amplifier operation, based on transistor characteristics.

Transformer coupling is commonly used to match the input and output impedances of electron-tube amplifier stages. Since the input impedance of an electron tube is very high at audio frequencies, the design of an electron-tube stage depends primarily on the transformer parameters. The much lower input impedances of transistors demand that many other factors be taken into account, and the design becomes more complex. The output-stage transformer coupling to a specific load is often the optimum method of realizing the best power match. Figure 11.2.7 illustrates a typical transformer-coupled transistor audio-amplifier stage.

The direct-coupling approach is now also used for discrete-component transistorized amplifiers, and particularly in integrated amplifier versions. The level-shifting requirement is realized by selection from the host of available components, such as *npn* and *pnp* transistors and zener diodes. Since it is difficult to realize large-size capacitors via integrated-circuit techniques, special methods have been developed to direct-couple integrated amplifiers.

Classes A, B, AB, and C Operation

The output or power stage of an amplifier is usually classified as operating class A, B, AB, or C, depending on the conduction characteristics of the active devices (see Fig. 11.1.5). These definitions can also apply to any intermediate amplifier stage. Figure 11.2.8 illustrates relations between the class of operation and conduction using transistor parameters. This figure would be essentially the same for an electron-tube amplifier with the tube plate current and grid voltage as the equivalent device parameters.

Subscripts may be used to denote additional conduction characteristics of the device. For example, the electron-tube grid conduction can also be further classified as A_1, to show that no grid current flows, or A_2, to show that grid-current conduction exists during some portion of the cycle.

Push-Pull Amplifiers

In a single-ended amplifier the active devices conduct continuously. The single-ended configuration is generally used in low-power applications, operated in class A. For example, preamplifiers and low-level amplifiers are generally operated single-ended, unless the output power levels necessitate the more efficient power handling of the push-pull circuit.

In a push-pull configuration there are at least two active devices that alternately amplify the negative and positive cycles of the input waveform. The output connection to the load is most often transformer-coupled. An example of a transformer input and output in a push-pull amplifier is illustrated in Fig. 11.2.9. Direct-coupled push-pull amplifiers and capacitively coupled push-pull amplifiers are also feasible, as illustrated in Fig. 11.2.10.

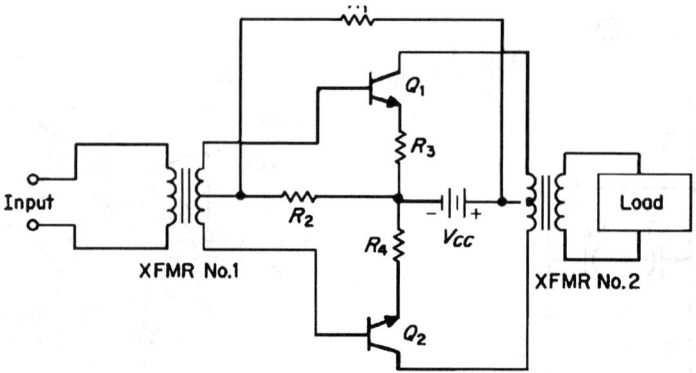

FIGURE 11.2.9 Transformer-coupled push-pull transistor stage.

The active devices in push-pull are usually operated either in class B or AB because of the high power-conversion efficiency. Feedback techniques can be used to stabilize gain, stabilize biasing or operating points, minimize distortion, and the like.

Output Amplifiers

The function of an audio output amplifier is to interface with the preceding amplifier stages and to provide the necessary drive to the load. Thus the output-amplifier designation does not uniquely identify a particular amplifier class. When several different types of amplifiers are cascaded between the signal source and its load, e.g., a high-power speaker, the last-stage amplifier is designated as the output amplifier. Because of the high power requirements, this amplifier is usually a push-pull type operating either in class B or AB.

Stereo Amplifiers

A stereo amplifier provides two separate audio channels properly phased with respect to each other. The objective of this two-channel technique is to enhance the audio reproduction process, making it more realistic

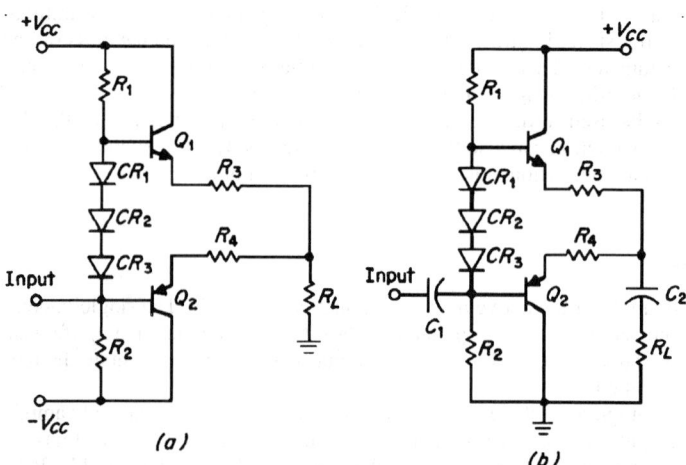

FIGURE 11.2.10 (*a*) Direct- and (*b*) capacitively coupled push-pull stages.

and lifelike. It is also feasible to extend the system to contain more than two channels of information. A stereo amplifier is a complete system that contains its power supply and other commonly required control functions.

Each channel has its own preamplifier, medium-level stages, and output power stage, with different gain and frequency responses for each mode of operation, e.g., for tape, phonograph, CD, and so forth. The input signal is selected from the phonograph input connection, tape input, or a turner output. Special-purpose trims and controls are also used to optimize performance on each mode. The bandwidth of the amplifier extends to 20 kHz or higher.

AUDIO OSCILLATORS

Robert J. McFadyen

General Considerations

In the strict sense, an audio oscillator is limited to frequencies from about 15 to 20,000 Hz, but a much wider frequency range is included in most oscillators used in audio measurements since knowledge of amplifier characteristics in the region above audibility is often required.

For the production of sinusoidal waves, audio oscillators consist of an amplifier having a nonlinear power gain characteristic, with a path for regenerative feedback. Single- and multistage transistor amplifiers with *LC* or *RC* feedback networks are most often used. The term *harmonic oscillator* is used for these types. *Relaxation oscillators*, which may be designed to oscillate in the audio range, exhibit sharp transitions in the output voltages and currents. Relaxation oscillators are treated in Section 14.

The instantaneous excursions of the operating point in a harmonic oscillator is restricted to the range where the circuit exhibits an impedance with a negative real part. The amplifier supplies the power, which is dissipated in the feedback path and the load. The regenerative feedback would cause the amplitude of oscillation to grow without bound were it not for the fact that the dynamic range of the amplifier is limited by circuit nonlinearities. Thus, in most sine-wave audio oscillators; the operating frequency is determined by passive-feedback elements, whereas the amplitude is controlled by the active-circuit design.

Analytical expressions predicting the frequency and required starting conditions for oscillation can be derived using Bode's amplifier feedback theory, and the stability theorem of Nyquist. Since this analytical approach is based on a linear-circuit model, the results are approximate but usually suitable for design of sinusoidal oscillators. No prediction on waveform amplitude results, since this is determined by nonlinear-circuit characteristics. Estimates of the waveform amplitude can be made from the bias and limiting levels of the active circuits. Separate limiters and AGC techniques are also useful for controlling the amplitude to a prescribed level. Graphical and nonlinear analysis methods can also be used for obtaining a prediction of the amplitude of oscillation.

A general formulation suitable for a linear analysis of almost all audio oscillators can be derived from the feedback diagram in Fig. 11.2.11. Note that the amplifier internal feedback generator has been neglected: that is, y_{12A} is assumed to be zero. This assumption of unilateral amplification is almost always valid in the audio range even for single-stage transistor amplifiers.

The stability requirements for the circuit are derived from the closed-loop-gain expression

$$A_c = A/(1 - A\beta) \tag{1}$$

where the gain A is treated as a negative quantity for an inverting amplifier. Infinite closed-loop gain occurs when AB is equal to unity, and this defines the oscillatory condition. In terms of the equivalent circuit parameters used in Fig. 11.2.1,

$$1 - A\beta = 1 - y_{21A} \frac{y_{12\beta}}{(y_{11A} + y_{11\beta})(y_{22A} + y_{22\beta}) - y_{12\beta}y_{21\beta}} \tag{2}$$

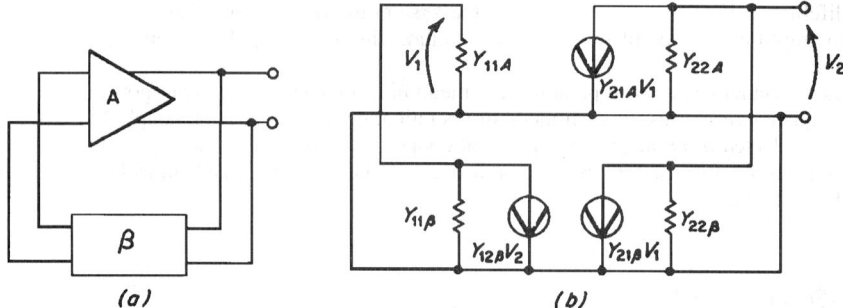

FIGURE 11.2.11   Oscillator representations: (*a*) generalized feedback circuit; (*b*) equivalent *y*-parameter circuit.

In the audio range, y_{21A} remains real, but the fractional portion of the function is complex because β is frequency-sensitive. Therefore, the open-loop gain $A\beta$ can be expressed in the general form

$$A\beta = y_{21A} \frac{A_r + jA_i}{B_r + jB_i} \tag{3}$$

It follows from Nyquist's stability theorem that this feedback system will be unstable if, first, the phase shift of $A\beta$ is zero and, second, the magnitude is equal to or greater than unity. Applying this criterion to Eq. (3) yields the following two conditions for oscillation:

$$A_i B_r - A_r B_i = 0 \tag{4}$$

$$y_{21}^2 \geq \frac{B_r^2 + B_i^2}{A_r^2 + A_i^2} \tag{5}$$

Equation (4) results from the phase condition and determines the frequency of oscillation. The inequality in Eq. (5) is the consequence of the magnitude constraint and defines the necessary condition for sustained oscillation. Equation (5) is evaluated at the oscillation frequency determined from Eq. (4).

A large number of single-stage oscillators have been developed in both vacuum-tube and transistor versions. The transistor circuits followed by direct analogy from the earlier vacuum-tube circuits. In the following examples, transistor versions are illustrated, but the *y*-parameter equations apply to other devices as well.

LC Oscillators

The *Hartley oscillator* circuit is one of the oldest forms: the transistor version is shown in Fig. 11.2.12. With the collector and base at opposite ends of the tuned circuit, the 180° phase relation is secured, and feedback occurs through mutual inductance between the two parts of the coil. The frequency and condition for oscillation are expressed in terms of the transistor *y* parameters and feedback inductance *L*, inductor coupling coefficient *k*, inductance ratio *n*, and tuning capacitance *C*. The frequency of oscillation is

$$\omega^2 = \frac{1}{LC(1 + 2k\sqrt{n} + n) + nL^2(1 - k^2)(y_{11A} y_{22A})}$$

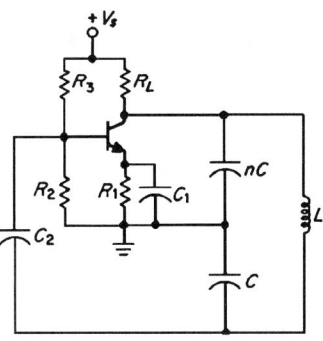

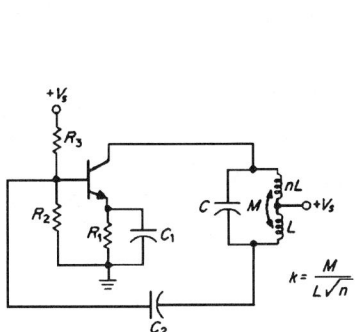

FIGURE 11.2.12 Hartley oscillator circuit.

FIGURE 11.2.13 Colpitts oscillator circuit.

The condition for oscillation is

$$y_{21A} \geq \frac{y_{11A} + ny_{22A} + n\omega^2 LC(1-k^2)(y_{11A}y_{22A})}{k\sqrt{n} + n\omega^2 LC(1-k^2)}$$

The admittance parameters of the bias network R_1, R_2, and R_3, as well as the reactance of bypass capacitor C and coupling capacitor C_2, have been neglected. These admittances could be included in the amplifier y parameters in cases where their effect is not negligible.

If

$$\frac{C}{L} \gg \frac{n(1-k^2)(y_{11A}y_{22A})}{1+2k\sqrt{n}+n} \tag{6}$$

the frequency of oscillation will be essentially independent of transistor parameters.

The transistor version of the *Colpitts oscillator* is shown in Fig. 11.2.13. Capacitors C and nC in combination with inductance L determine the resonant frequency of the circuit. A fraction of the current flowing in the tank circuit is regeneratively fed back to the base through the coupling capacitor C_2. Bias resistors R_1, R_2, R_3, and R_L, as well as capacitors C_1 and C_2, are chosen so as not to affect the frequency or conditions for oscillation. The frequency of oscillation is

$$\omega^2 = \frac{1}{LC}\left(1+\frac{1}{n}\right) + \frac{1}{nC^2}(y_{11A}y_{22A})$$

The condition for oscillation is

$$y_{21A} \geq \omega^2 LC(nY_{11A} + Y_{22A}) - (y_{11A} + y_{22A})$$

Alternatively, the bias element admittances may be included in the amplifier y parameters.

In the Colpitts circuit, if the ratio of C/L is chosen so that

$$\frac{C}{L} \gg \frac{y_{11A}y_{22A}}{1+n} \tag{7}$$

the frequency of oscillation is essentially determined by the tuned-circuit parameters.

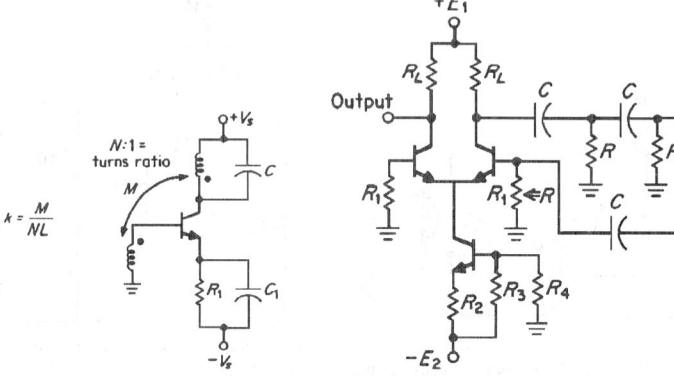

FIGURE 11.2.14 Tuned-collector oscillator.

FIGURE 11.2.15 *RC* oscillator with high-pass feedback network.

Another oscillator configuration useful in the audio-frequency range is the tuned-collector circuit shown in Fig. 11.2.14. Here regenerative feedback is furnished via the transformer turns ratio N from the collector to base. The frequency of oscillation is

$$\omega^2 = \frac{1}{LC + N^2 L^2 y_{11A} y_{22A}(1-k^2)}$$

The condition for oscillation is

$$y_{22A} \geq \frac{1}{Nk}(N^2 Y_{11A} + Y_{22A}) - \frac{\omega^2 NLCY_{11A}}{k}(1-k^2)$$

If the ratio of C/L is such that

$$\frac{C}{L} \gg N^2 y_{11A} y_{22A}(1-k^2) \tag{8}$$

the frequency of oscillation is specified by $\omega^2 = 1/LC$. This circuit can be tuned over a wide range by varying the capacitor C and is compatible with simple biasing techniques.

RC Oscillators

Audio sinusoidal oscillators can be designed using an *RC* ladder network (of three or more sections) as a feedback path in an amplifier. This scheme originally appeared in vacuum-tube circuits, but the principles have been directly extended to transistor design. *RC* phase-shift oscillators can be distinguished from tuned oscillators in that the feedback network has a relatively broad frequency-response characteristic.

Typically, the phase-shift network has three *RC* sections of either a high- or a low-pass nature. Oscillation occurs at the frequency where the total phase shift is 180° when used with an inverting amplifier. Figures 11.2.15 and 11.2.16 show examples of high-pass and low-pass feedback-connection schemes. The amplifier is a differential pair with a transistor current source, a configuration which is common in integrated-circuit amplifiers. The output is obtained at the opposite collector from the feedback connection, since this minimizes external loading on the phase-shift network. The conditions for, and the frequency of, oscillation are derived, assuming that the input resistance of the amplifier, which loads the phase-shift network, has been adjusted to equal the resistance R. The load resistor R_L is considered to be part of the amplifier output resistance, and it is included in y_{22A}. The frequency of oscillation for the high-pass case is

$$\omega^2 = \frac{y_{22A}}{2C^2R(2+3Ry_{22A})}$$

The condition for oscillation for the high-pass case is

$$y_{21A} \geq \frac{1}{R}\left(\frac{1+5R/R_L}{\omega^2R^2C^2} - \frac{R}{R_L} - 3\right)$$

The frequency of oscillation for the low-pass case is

$$\omega = \frac{1}{RC}\sqrt{6+4\frac{R}{R_L}}$$

The condition for oscillation for the low-pass case is

$$y_{21A} \geq \frac{1}{R}\left(23\frac{R}{R_L} + 29 + 4\frac{R^2}{R_L^2}\right)$$

Null-Network Oscillators

In almost all respects null-network oscillators are superior to the RC phase-shift circuits described in the previous paragraph. While many null-network configurations are useful (including the bridged-T and twin-T), the Wien bridge design predominates.

The general form for the Wien bridge oscillator is shown in Fig. 11.2.17. In the figure, an ideal differential voltage amplifier is assumed, i.e., one with infinite input impedance and zero output impedance.

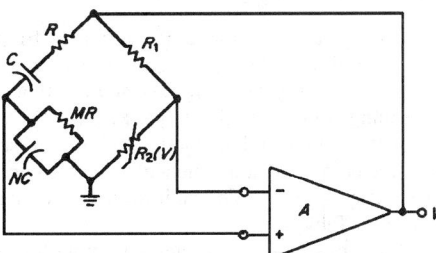

Frequency of oscillation ($M = N = 1$):

$$\omega_0 = \frac{1}{RC}$$

Condition for oscillation:

$$A \geq 8 = \frac{3(R_1 + R_2)}{R_1 - 2R_2}$$

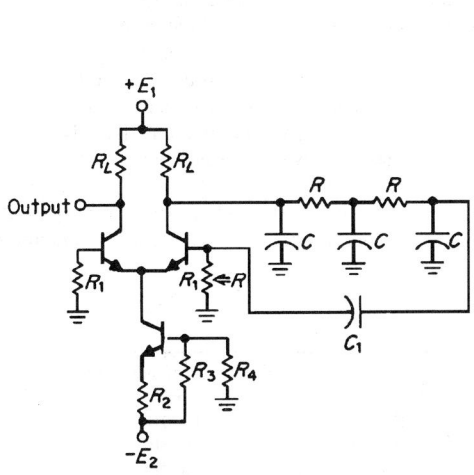

FIGURE 11.2.16 *RC* oscillator with low-pass feedback network.

FIGURE 11.2.17 Wien bridge oscillator circuit.

An integrated-circuit operational amplifier that has a differential input stage is a practical approximation to this type of amplifier and is often used in bridge-oscillator designs.

The Wien bridge is used as the feedback network, with positive feedback provided through the RC branches for regeneration and negative feedback through the resistor divider. Usually the resistor-divider network includes an amplitude-sensitive device in one or both arms which provides automatic correction for variation of the amplifier gain. Circuit elements such as a tungsten lamp, thermistor, and field-effect transistor used as the voltage-sensitive resistance element maintain a constant output level with a high degree of stability. Amplitude variations of less than ±1 percent over the band from 10 to 100,000 Hz are realizable. In addition, since the amplifier is never driven into the nonlinear region, harmonic distortion in the output waveform is minimized. For the connection shown in Fig. 11.2.17, an increase in V will cause a decrease in R_2, restoring V to the original level.

The lamp or thermistor have thermal time constants that set at a lower frequency limit on this method of amplitude control. When the period is comparable with the thermal time constant, the change in resistance over an individual cycle distorts the output waveform. There is an additional degree of freedom with the field-effect transistor, since the control voltage must be derived by a separate detector from the amplifier output. The time constant of the detector, and hence the resistor, are set by a capacitor, which can be chosen commensurate with the lowest oscillation frequency desired.

At ω_0 the positive feedback predominates, but at harmonics of ω_0 the net negative feedback reduces the distortion components. Typically, the output waveform exhibits less than 1 percent total harmonic distortion. Distortion components well below 0.1 percent in the mid-audio-frequency range are also achieved.

Unlike LC oscillators, in which the frequency is inversely proportional to the square root of L and C, in the Wien bridge ω_0 varies as $1/RC$. Thus, a tuning range in excess of 10:1 is easily achieved. Continuous tuning within one decade is usually accomplished by varying both capacitors in the reactive feedback branch. Decade changes are normally accomplished by switching both resistors in the resistive arm. Component tracking problems are eased when the resistors and capacitors are chosen to be equal.

Almost any three-terminal null network can be used for the reactive branch in the bridge; the resistor divider network adjusts the degree of imbalance in the manner described. Many of these networks lack the simplicity of the Wien bridge since they may require the tracking of three components for frequency tuning. For this reason networks such as the bridged-T and twin-T are usually restricted to fixed-tuned applications.

Low-Frequency Crystal Oscillators

Quartz-crystal resonators are used where frequency stability is a primary concern. The frequency variations with both time and temperature are several orders of magnitude lower than obtainable in LC or RC oscillator circuits. The very high stiffness and elasticity of piezoelectric quartz make it possible to produce resonators extending from approximately 1 kHz to 200 MHz. The performance characteristics of crystal depend on both the particular cut and the mode of vibration (see Section 5). For convenience, each "cut-mode" combination is considered as a separate piezoelectric element, and the more commonly used elements have been designated with letter symbols. The audio-frequency range (above 1 kHz) is covered by elements J, H, N, and XY, as shown in Table 11.2.2.

The temperature coefficients vary with frequency, i.e., with the crystal dimensions, and except for the H element, a parabolic frequency variation with temperature is observed. The H element is characterized by a

TABLE 11.2.2   Low-Frequency Crystal Elements

Symbol	Cut	Mode of vibration	Frequency range, kHz
J	Duplex 5°X	Length-thickness flexure	0.9–10
H	5°X	Length-width flexure	10–50
N	NT	Length-width flexure	4–200
XY	XY	XY flexure	8–40

negative temperature coefficient on the order of –10 ppm/°C. The other elements have lower temperature coefficients, which at some temperatures are zero because of the parabolic nature of the frequency-deviation curve. The point where the zero temperature coefficient occurs is adjustable and varies with frequency. At temperatures below this point the coefficient is positive, and at higher temperatures it is negative. On the slope of the curves the temperature coefficients for the N and XY elements are on the order of 2 ppm/°C, whereas the J element is about double at 4 ppm/°C.

Although the various elements differ in both cut and mode of vibration, the electric equivalent circuit remains invariant. The schematic representation and the lumped constant equivalent circuit are shown in Fig. 11.2.18. As is characteristic of most mechanical resonators, the motional inductance L resulting from the mechanical mass in motion is large relative to that obtainable from coils. The extreme stiffness of quartz makes for very small values of the motional capacitance C, and the very high order of elasticity allows the motional resistance R to be relatively low. The shunt capacitance C_0 is the electrostatic capacitance existing between crystal electrodes with the quartz plate as the dielectric and is present whether or not the crystal is in mechanical motion. Some typical values for these equivalent-circuit parameters are shown in Table 11.2.3.

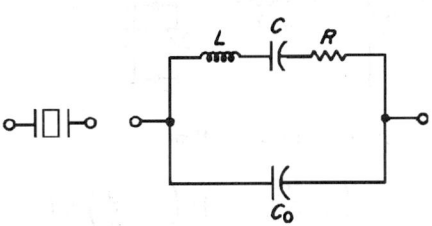

FIGURE 11.2.18   Symbol and equivalent circuit of a quartz crystal.

The H element can have a high Q value when mounted in a vacuum enclosure; however, it then has the poorest temperature coefficient. The N element exhibits an excellent temperature characteristic, but the piezoelectric activity is rather low, so that special care is required when it is used in oscillator circuits. The J and XY elements operate well in low-frequency oscillator designs, the latter having lower temperature drift. For the same frequency the XY crystal is about 40 percent longer than the J element. Where extreme frequency stability is required, the crystals are usually controlled to a constant temperature.

The reactance curve of a quartz resonator is shown in Fig. 11.2.19. The zero occurs at the frequency f_s, which corresponds to series resonance of the mechanical L and C equivalences. The antiresonant frequency f_p is dependent on the interelectrode capacitance C_0. Between f_s and f_p the crystal is inductive and this frequency range is normally referred to as the *crystal bandwidth*

$$\text{BW} = f_s/(2C_0/C) \tag{9}$$

In oscillator circuits the crystal can be used as either a series or a parallel resonator. At series resonance the crystal impedance is purely resistive, but in the parallel mode the crystal is operated between f_s and f_p and is therefore inductive. For oscillator applications the circuit capacitance shunting the crystal must also be included when specifying the crystal, since it is part of the resonant circuit. If a capacitor C_L, that is, a negative reactance, is placed in series with a crystal, the combination will series-resonate at the frequency f_R of zero reactance for the combination.

$$f_R = f_s\left[1+\frac{1}{(2C_0/C)(1+C_L/C_0)}\right] \tag{10}$$

TABLE 11.2.3   Typical Crystal Parameter Values

Element	Frequency, kHz	L, H	C, pF	R, kΩ	C_0, pF	Q, approx
J	10	8,000	0.03	50	6	20,000
H	10	2,500	0.1	10	75	20,000
N	10	8,000	0.03	75	30	10,000
XY	10	12,000	0.02	30	20	30,000

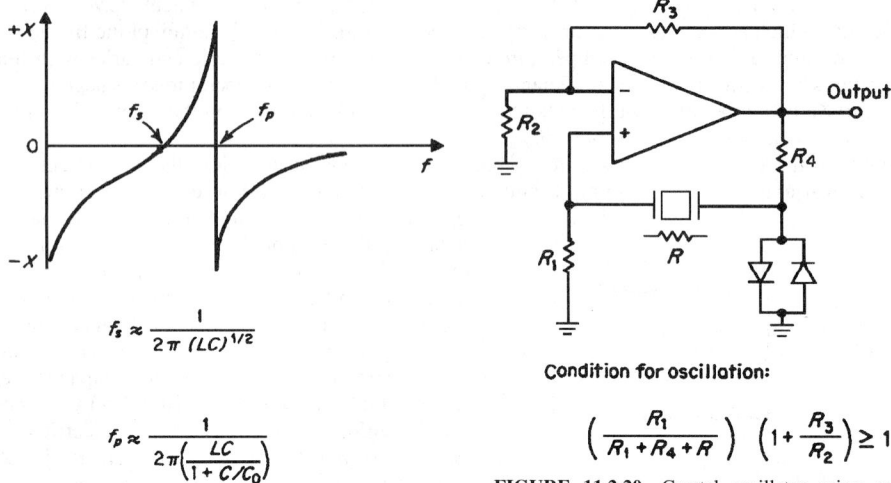

$$f_s \approx \frac{1}{2\pi (LC)^{1/2}}$$

$$f_p \approx \frac{1}{2\pi\left(\dfrac{LC}{1+C/C_0}\right)}$$

FIGURE 11.2.19   Quartz-crystal reactance curve.

Condition for oscillation:

$$\left(\frac{R_1}{R_1+R_4+R}\right)\left(1+\frac{R_3}{R_2}\right) \geq 1$$

FIGURE 11.2.20   Crystal oscillator using an integrated-circuit operational amplifier.

The operating frequency can vary in value due to changes in the load capacitance, and this variation is prescribed by

$$\Delta f_R = \frac{f_s}{2C_0/C}\frac{\Delta C_L/C_0}{(1+C_L/C_0)^2} \tag{11}$$

This effect can be used to "pull" the crystal for initial alignment, or if the external capacitor is a voltage-controllable device, a VCO with a range of about ±0.01 percent can be constructed. Phase changes in the amplifier will also give rise to frequency shifts since the total phase around the loop must remain at 0° to maintain oscillation.

Although single-stage transistor designs are possible, more flexibility is available in the circuit of Fig. 11.2.20, which uses an integrated-circuit operational amplifier for the gain element. The crystal is operated in the series mode, and the amplifier gain is precisely controlled by the negative-feedback divider R_2 and R_3. The output will be sinusoidal if

$$\frac{V_D R_1}{R_1+R}\left(1+\frac{R_3}{R_2}\right) < V_{\text{lim}} \tag{12}$$

where V_D is the limiting diode forward voltage drop and V_{lim} is the limiting level of amplifier output.

Low-cost electronic wristwatches use quartz crystals for establishing a high degree of timekeeping accuracy. A high-quality mechanical watch may have a yearly accuracy on the order of 20 min, whereas many quartz watches are guaranteed to vary less than 1 min/year.

Generally the XY crystal is used, but other types are continually being developed to improve accuracy, reduce size, and lower manufacturing cost. The active gain elements for the oscillator are part of the integrated circuit that contains the electronics for the watch functions. The flexure or tuning-fork frequency is set generally to 32,768 Hz, which is 2^{15} Hz. This frequency reference is divided down on the integrated circuit to provide seconds, minutes, hours, day of the week, date, month, and so forth.

A logic gate or inverter is often used as the gain element in the oscillator circuit. A typical configuration is shown in Fig. 11.2.21. The resistor R_1 is used to bias the logic inverter for class A amplifier operation. The resistor R_2 helps reduce both voltage sensitivity of the network and crystal power dissipation. The combination of R_2 and C_2 provides added phase shift for good oscillator startup. The series combination of capacitors

C_1 and C_2 provides the parallel load for the crystal. C_1 can be made tunable for precise setting of the crystal oscillation frequency. The inverter provides the necessary gain and 180° phase shift. The π network consisting of the capacitors and the crystal provides the additional 180° phase shift needed to satisfy the conditions for oscillation.

Frequency Stability

Many factors contribute to the ability of an oscillator to hold a constant output frequency over a period of time and range from short-term effects, caused by random noise, to longer-term variations, caused by circuit parameter dependence on temperature, bias voltage, and the like. In addition to the temperature and aging effects of the frequency-determining elements, nonlinearities, impedance loading, and amplifier phase variations also contribute to instability.

Harmonics generated by circuit nonlinearities are passed through the feedback network, with various phase shifts, to the input of the amplifier. Intermodulation of the harmonic frequencies produces a fundamental frequency component that differs in phase from the amplifier output. Since the condition $A\beta = 1$ must be satisfied, the frequency of oscillation will shift so that the network phase shift cancels the phase perturbation caused by the nonlinearity. Therefore, the frequency of oscillation is influenced by an unpredictable amplifier characteristic, namely, the saturation nonlinearity. This effect is negligible in the Wien bridge oscillator, where automatic level control keeps harmonic distortion to a minimum.

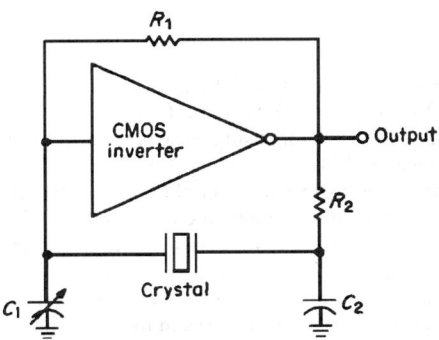

FIGURE 11.2.21 Crystal oscillator using a logic gate for the gain element.

The relationships shown in Fig. 11.2.17 were derived assuming that the amplifier does not load the bridge circuit on either the input or output sides. In the practical sense this is never true, and changes in the input and output impedances will load the bridge and cause frequency variations to occur.

Another source of frequency instability is small phase changes in the amplifier. The effect is minimized by using a network with a large stability factor, defined by

$$S = \frac{d\phi}{d\omega/\omega_0}\bigg|_{\omega=\omega_0} \tag{13}$$

For the Wien bridge oscillator, which has amplitude-sensitive resistive feedback, the RC impedances can be optimized to provide a maximum stability factor value. As shown in Fig. 11.2.17, this amounts to choosing proper values for M and N. The maximum stability-factor value is $A/4$, and it occurs for $N = 1/2$ and $M = 2$. Most often the bridge is used with equal resistor and capacitor values; that is, $M = N = 1$, in which case the stability factor is $2A/9$. This represents only a slight degradation from the optimum.

Synchronization

It is often desirable to lock the oscillator frequency to an input reference. Usually this is done by injecting sufficient energy at the reference frequency into the oscillator circuit. When the oscillator is tuned sufficiently close to the reference, natural oscillations cease and the synchronization signal is amplified to the output. Thus the circuit appears to oscillate at the injected signal frequency. The injected reference is amplitude-stabilized by the AGC or limiting circuit in the same manner as the natural oscillation. The frequency range over which locking can occur is a linear function of the amplitude of the injected signal. Thus, as the synchronization frequency is moved away from the natural oscillator frequency, the amplitude threshold to maintain lock increases. The phase error between the input reference and the oscillator output will also deviate as the input frequency varies from the natural frequency.

Methods for injecting the lock signal vary and depend on the type of oscillator under consideration. For example, *LC* oscillators may have signals coupled directly to the tank circuit, whereas the lock signal for the Wien network is usually coupled into the center of the resistive side of the bridge, i.e., the junction of R_1 and R_2 in Fig. 11.2.17.

If the natural frequency of oscillation can be voltage controlled, synchronization can be accomplished with a phase-locked loop. Replacing both R's with field-effect transistors, or alternatively shunting both C's with var-icaps, provides an effective means for voltage controlling the frequency of the Wien bridge oscillator. Although more complicated in structure, the phase-locked loop is more versatile and has many diverse applications.

Piezoelectric Annunciators

Another important class of audio oscillators uses piezoelectric elements for both frequency control and audible-sound generation. Because of their low cost and high efficiency these devices are finding increasing use in smoke detectors, burglar alarms, and other warming devices. Annunciators using these elements typically pro-duce a sound level in excess of 85 dB measured at a distance of 10 ft.

Usually the element consists of a thin brass disk to which a piezoelectric material has been attached. When an electric signal is applied across its surfaces, the piezoceramic disk attempts to change diameter. The brass disk to which it is bonded acts as a spring restraining force on one surface of the ceramic. The brass plate serves as one electrode for applying the electric signal to the ceramic. On the other surface a fired-on silver paste is used as an electrode. The restraining action of the brass disk causes the assembly to change from a flat to a convex shape. When the polarity of the electric signal reverses, the assembly flexes in the other direction to a concave shape. When the device is properly mounted in a suitable horn structure, this motion is used to pro-duce high-level sound waves. One useful method is to clamp the disk at nodal points, i.e., at a distance from the center of the disk where mechanical motion is at a vibrational null.

The piezoelectric assembly will produce sound levels more efficiently when excited near the series-resonant frequency. The simple equivalent circuit used for the quartz crystal (Fig. 11.2.18) also applies to the piezoce-ramic assembly for frequencies near resonance. Generally the piezoelectric element is used as the frequency-determining element in an audio oscillator. The advantage of this method is that the excitation frequency is inherently near the optimum value, since it is self-excited. A typical piezoceramic 1-in diameter mounted on a $1^3/_4$-in brass disk would have the following equivalent values: $C_0 = 0.02$ μF, $C = 0.0015$ μF, $L = 2$ H, $R = 500$ Ω, $Q = 75$, $f_s = 2.9$ kHz, and $f_p = 3.0$ kHz.

A basic oscillator, capable of producing high-level sound, is shown in Fig. 11.2.22. The inductor L_1 provides a dc path to the transistor and broadly tunes the parallel input capacitance of the piezoelectric element. C_1 is an optional capacitor which adds to the input shunt capacitance for optimizing the drive impedance to the element. Resistor R_1 provides base-current bias to the transistor so that oscillation can start.

The element has a third small electrode etched in the silver pattern. It is used to derive a feedback signal which, when resis-tively loaded by R_1, provides an in-phase signal to the base for sustaining circuit oscillation. The circuit operates like a blocking oscillator in that the transistor is switched on and off and the col-lector voltage can fly above B-plus because of the inductor L_1.

The collector load consisting of L_1 and C_1 can be replaced with a resistor, in which case the audio output will be less.

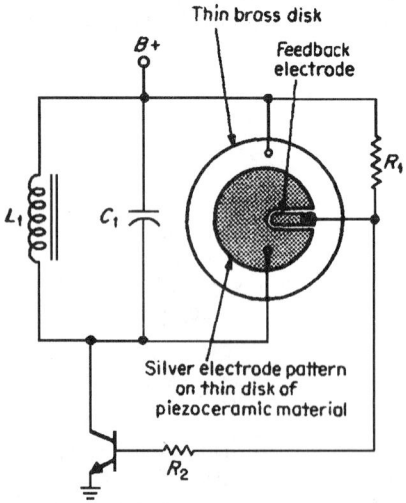

FIGURE 11.2.22   Basic audio annunciator oscilla-tor circuit using a thin-disk piezoelectric transducer.

CHAPTER 11.3
RADIO-FREQUENCY AMPLIFIERS AND OSCILLATORS

G. Burton Harrold, John W. Lunden, Jennifer E. Doyle, Chang S. Kim, Conrad E. Nelson, Gunter K. Wessel, Stephen W. Tehon, Y. J. Lin, Wen-Chung Wang, Harold W. Lord

RADIO-FREQUENCY AMPLIFIERS

G. Burton Harrold

Small-Signal RF Amplifiers

The prime considerations in the design of first-stage rf amplifiers are gain and noise figure. As a rule, the gain of the first rf stage should be greater than 10 dB, so that subsequent stages contribute little to the overall amplifier noise figure. The trade-off between amplifier cost and noise figure is an important design consideration. For example, if the environment in which the rf amplifier operates is noisy, it is uneconomic to demand the ultimate in noise performance. Conversely, where a direct trade-off exists in transmitter power versus amplifier noise performance, as it does in many space applications, money spent to obtain the best possible noise figure is fully justified.

Another consideration in many systems is the input-output impedance match of the rf amplifier. For example, TV cable distribution systems require an amplifier whose input and output match produce little or no line reflections. The performance of many rf amplifiers is also specified in handling large signals, to minimize cross- and intermodulation products in the output. The wide acceptance of transistors has placed an additional constraint on first-stage rf amplifiers, since many rf transistors having low noise, high gain, and high frequency response are susceptible to burnout and must be protected to prevent destruction in the presence of high-level input signals.

Another common requirement is that first rf stages be gain-controlled by automatic gain control (AGC) voltage. The amount of gain control and the linearity of control are system parameters. Many rf amplifiers have the additional requirement that they be tuned over a range of frequencies. In most receivers, regardless of configuration, local-oscillator leakage back to the input is strictly controlled by government regulation. Finally, the rf amplifier must be stable under all conditions of operation.

Device Evaluation for RF Amplifiers

An important consideration in an rf amplifier is the choice of active device. This information on device parameters can often be found in published data sheets. If parameter data are not available or not a suitable operating point, the following characterization techniques can be used.

Network Analyzers. The development of the modern network analyzer has eliminated much of the work in device and circuit evaluation. These systems automate sweep frequency measurements of the complex device or

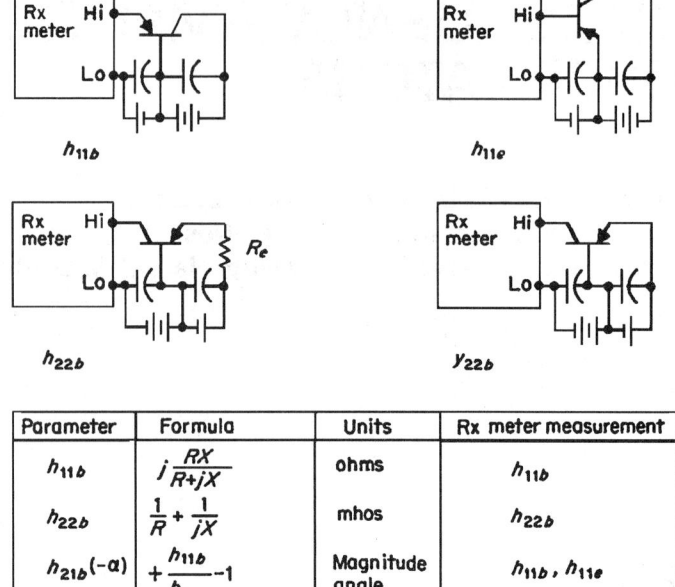

Parameter	Formula	Units	Rx meter measurement
h_{11b}	$j\dfrac{RX}{R+jX}$	ohms	h_{11b}
h_{22b}	$\dfrac{1}{R}+\dfrac{1}{jX}$	mhos	h_{22b}
$h_{21b}(-\alpha)$	$+\dfrac{h_{11b}}{h_{11e}}-1$	Magnitude angle	h_{11b}, h_{11e}
h_{12b}	$(y_{22b}-h_{22b})\dfrac{h_{11b}}{-h_{21b}}$	Magnitude angle	h_{11b}, h_{11e}, y_{22b}, h_{22b}

Assumes: Determinate of $|h| \ll h_{21}$ and $h_{12} \ll 1$

R and X are Rx meter's reading of parallel resistance and reactance

FIGURE 11.3.1 Use of the *Rx* meter in device characterization.

circuit parameters and avoid the tedious calculations that were previously required. The range of measurement frequencies extends from a few hertz to 60 GHz.

Network analyzers perform the modeling function by measuring the transfer and impedance function of the device by means of sine-wave excitation. These transfer voltages/currents and the reflected voltages/currents are then separated, and the proper ratios are formed to define the device parameters. There results are then displayed graphically and/or in a digital form for designer use. Newer systems allow these data to be transferred directly to computerized design programs, thus automating the total design process. The principle of actual operation is similar to that described below under *Vector Voltmeter*.

Rx Meter.[*] This measurement technique is usually employed at frequencies below 200 MHz for active devices that have high input and output impedance. The technique is summarized in Fig. 11.3.1 with assumptions tacit in these measurements. The biasing techniques are shown. In particular, the measurement of h_{22b} requires a very large resistor R_e to be inserted in the emitter, and this may cause difficulty in achieving the proper biasing. Care should be taken to prevent burnout of the bridge when a large dc bias is applied. The bridge's drive to the active device may be reduced for more accurate measurement by varying the B-plus voltage applied to the internal oscillator.

Vector Voltmeter.[*] This characterization technique measures the *S* parameters; see Fig. 11.1.9. The measurement consists in inserting the device in a transmission line, usually 50 Ω characteristic impedance, and measuring the incident and reflected voltages at the two ports of the device.

---

[*]Trademark of the Hewlett Packard Co.

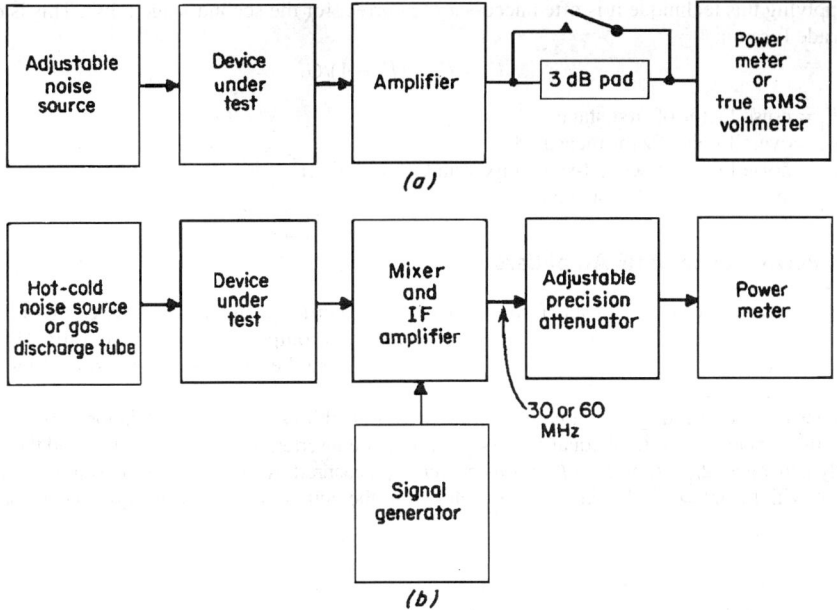

FIGURE 11.3.2   Noise-measurement techniques: (*a*) at low frequencies; (*b*) at high frequencies.

Several other techniques include the use of the H-P 8743 reflectometer, the general radio bridge GR 1607, the Rhode-Schwartz diagraph, and the H-P type 8510 microwave network analyzer to measure device parameters automatically from 45 MHz to 100 GHz with display and printout features.

Noise in RF Amplifiers

A common technique employing a noise source to measure the noise performance of an rf amplifier is shown in Fig. 11.3.2. Initially the external noise source (a temperature-limited diode) is turned off, the 3-dB pad short-circuited, and the reading on the output power meter recorded. The 3-dB pad is then inserted, the noise source is turned on, and its output increased until a reading equal to the previous one is obtained. The noise figure can then be read directly from the noise source, or calculated from 1 plus the added noise per unit bandwidth divided by the standard noise power available KT_0, where $T_0 = 290$ K and $K = $ Boltzmann's constant = 1.38×10^{-23} J/K.

At higher frequencies, the use of a temperature-limited diode is not practical, and a gas-discharge tube or a hot-cold noise source is employed. The *Y*-factor technique of measurement is used. The output from the device to be measured is put into a mixer, and the noise output converted to a 30- or 60-MHz center-frequency (i.f.) output. A precision attenuator is then inserted between this i.f. output and the power-measuring device. The attenuator is adjusted to give the same power reading for two different conditions of noise power output represented by effective temperatures T_1 and T_2. The *Y* factor is the difference in decibels between the two precision attenuator values needed to maintain the same power-meter reading. The noise factor is

$$F = \frac{(T_2/290) - T_1 Y/290}{Y - 1} + 1$$

where T_1 = effective temperature at reference condition 1
T_2 = effective temperature of reference condition 2
Y = decibel reading defined in the text, converted to a numerical ratio

In applying this technique it is often necessary to correct for the second-stage noise. This is done by use of the cascade formula

$$F_1 = F_T - (F_2 - 1)/G_1$$

where F_1 = noise factor of first stage
F_T = overall noise factor measured
F_2 = noise factor of second-stage mixer and i.f. amplifier
G_1 = available gain of first stage

Large-Signal Performance of RF Amplifiers

The large-signal performance of an rf amplifier can be specified in many ways. A common technique is to specify the input where the departure from a straight-line input-output characteristic is 1 dB. This point is commonly called the *1-dB compression point*. The greater the input before this compression point is reached, the better the large-signal performance.

Another method of rating an rf amplifier is in terms of its third-order intermodulation performance. Here two different frequencies, f_1 and f_2, of equal powers, p_1 and p_2, are inserted into the rf amplifier, and the third frequency, internally generated, $2f_1 - f_2$ or $2f_2 - f_1$, has its power p_{12} measured. All three frequencies must be in the amplifier passband. With the intermodulation power p_{12} referred to the output, the following equation can be written:

$$P_{12} = 2P_1 + P_2 + K_{12}$$

where P_{12} = intermodulation output power at $2f_1 - f_2$ or $2f_2 - f_1$
P_1 = output power at input frequency f_1
P_2 = output power at input frequency f_2, all in decibels referred to (0 dBm)
K_{12} = constant associated with the particular device

The value of K_{12} in the above formula can be used to rate the performance of various device choices. Higher orders of intermodulation products can also be used.

A third measure of large-signal performance commonly used is that of cross-modulation. In this instance, a carrier at f_D with no modulation is inserted into the amplifier. A receiver is then placed at the output and tuned to this unmodulated carrier. A second carrier at f_1 with amplitude-modulation index M_1 is then added. The power of P_1 of f_1 is increased, and its modulation is partially transferred to f_D. The equation becomes

$$10 \log (M_K/M_1) = P_1 + K$$

where M_K = cross-modulation index of originally unmodulated signal at f_D
M_1 = modulation index of signal F_1
P_1 = output power of signal at f_1, all in decibels referred to 1 mW (0 dBm)
K = cross-modulation constant

Maximum Input Power

In addition to the large-signal performance, the maximum power of voltage input into an rf amplifier is specified, with a requirement that device burnout must not occur at this input. There are two ways of specifying this input: by a stated pulse of energy or by a requirement to withstand a continuously applied large signal. It is also common to specify the time required to unblock the amplifier after removal of the large input. With the increased use of field effect transistors (FETs especially) having good noise performance, these overload characteristics have become a severe problem. In many cases, conventional or zener diodes, in a back-to-back configuration shunting the input, are used to reduce the amount of power the input of the active devices must dissipate.

RF Amplifiers in Receivers

RF amplifiers intended for the first stages of receivers have additional restrictions placed on them. In most cases, such amplifiers are tunable across a band of frequencies with one or more tuned circuits. The tuned circuits must track across the frequency band, and in the case of the superheterodyne, tracking of the local oscillator is necessary so that a constant frequency difference (i.f.) is maintained. The receiver's rf section can be tracked with the local oscillator by the two- or the three-point method, i.e., with zero error in the tracking at either two or three points.

A second consideration peculiar to rf amplifiers used for receivers is the AGC. This requirement is often stated by specifying a low-level rf input to the receiver and noting the power out. The rf signal input is then increased with the AGC applied until the output power has increased a predetermined amount. This becomes a measure of the AGC effectiveness. The AGC performance can also be measured by plotting a curve of rf input versus AGC voltage needed to maintain constant output, compared with the desired performance.

A third consideration in superheterodynes is the leakage of the local oscillator in the receiver to the outside. This spurious radiation is specified by the Federal Communications Commission (FCC) in the United States.

Design Using Immittance and Hybrid Parameters

The general gain and input-output impedance of an amplifier can be formulated, in terms of the Z or Y parameters, to be

$$Y_{in} = y_{11} - \frac{y_{12}y_{21}}{y_{22} + y_L} \qquad Y_{out} = y_{22} - \frac{y_{12}y_{21}}{y_{11} + y_s}$$

where y_L = load admittance
$\quad y_s$ = source admittance
$\quad Y_{in}$ = input admittance
$\quad Y_{out}$ = output admittance
$\quad G_T$ = transducer gain

and the transducer gain is

$$G_T = \frac{4 \operatorname{Re} y_s \operatorname{Re} y_L |y_{21}|^2}{|(y_{11} + y_s)(y_{22} + y_L) - y_{12}y_{21}|^2}$$

for the y parameters, and interchange of z or y is allowed.

The stability of the circuit can be determined by either Linvill's C or Stern's k factor as defined below. Using the y parameters, $y_{ij} = g_{ik} + jB_{ik}$, these are

Linvill:

$$C = \frac{|y_{12}y_{21}|}{2g_{11}g_{22} - \operatorname{Re} y_{12}y_{21}}$$

where $C < 1$ for stability does not include effects of load and source admittance.

Stern:

$$k = \frac{2(g_{11} + g_s)(g_{22} + g_L)}{|y_{12}y_{21}| + \operatorname{Re} y_{12}y_{21}}$$

where $k > 1$ for stability
$\quad g_L$ = load conductance
$\quad g_s$ = source conductance

The preceding C factor defines only unconditional stability; i.e., no combination of load and source impedance will give instability. There is an invariant quantity K defined as

$$K = \frac{2 \operatorname{Re} \gamma_{11} \operatorname{Re} \gamma_{22} - \operatorname{Re} \gamma_{12}\gamma_{21}}{|\gamma_{21}\gamma_{12}|} \qquad \begin{array}{l} \operatorname{Re} \gamma_{11} > 0 \\ \operatorname{Re} \gamma_{22} > 0 \end{array}$$

where γ represents either the y, z, g, or h parameters, and $K > 1$ denotes stability.
This quantity K has then been used to define maximum available power gain G_{max} (only if $K > 1$)

$$G_{max} = |\gamma_{21}/\gamma_{12}|(K - \sqrt{K^2 - 1})$$

To obtain this gain, the source and load immittance are found to be $(K > 1)$

$$\gamma_s = \frac{\gamma_{12}\gamma_{21} + |\gamma_{12}\gamma_{21}|(K + \sqrt{K^2 - 1})}{2\,\text{Re}\,\gamma_{22}} - \gamma_{11} \qquad \gamma_s = \text{source immittance}$$

$$\gamma_L = \frac{\gamma_{12}\gamma_{21} + |\gamma_{12}\gamma_{21}|(K + \sqrt{K^2 - 1})}{2\,\text{Re}\,\gamma_{11}} - \gamma_{22} \qquad \gamma_L = \text{load immittance}$$

The procedure is to calculate the K factor, and if $K > 1$, calculate G_{max}, γ_s, and γ_L. If $K < 1$, the circuit can be modified either by use of feedback or by adding immittances to the input-output.

Design Using S Parameters

The advent of automatic test equipment and the extension of vacuum tubes and transistors to be gigahertz frequency range have led to design procedures using the S parameters. Following the previous discussion, the input and output reflection coefficient can be defined as

$$p_{in} = S_{11} + p_L\frac{S_{12}S_{21}}{1 - p_L S_{22}} \qquad p_L = \frac{Z_L - Z_0}{Z_L + Z_0}$$

$$p_{out} = S_{22} + p\frac{S_{12}S_{21}}{1 - pS_{11}} \qquad p_s = \frac{Z_s - Z_0}{Z_s + Z_0}$$

where Z_0 = characteristic impedance
p_{in} = input reflection coefficient
p_{out} = output reflection coefficient

The transducer gain can be written

$$G_{transducer} = \frac{|S_{21}|^2\,(1 - |p_s|^2)(1 - |p_L|^2)}{|(1 - S_{11}p_s)(1 - S_{22}p_L) - S_{21}S_{12}p_s p_L|^2}.$$

The unconditional stability of the amplifier can be defined by requiring the input (output) impedance to have a positive real part for any load (source) impedance having a positive real part.
This requirement gives the following criterion:

$$|S_{11}|^2 + |S_{12}S_{21}| < 1 \qquad |S_{22}|^2 + |S_{12}/S_{11}| < 1$$

and

$$\eta = \frac{1 - |\Delta|^2 - |S_{11}|^2 - |S_{22}|^2}{2|S_{12}S_{21}|} > 1 \qquad \Delta_s = S_{11}S_{22} - S_{12}S_{21}$$

Similarly, the maximum transducer gain, for $\eta > 1$, becomes

$$G_{\text{max transducer}} = |S_{21}/S_{12}|(\eta \pm \sqrt{\eta^2 - 1})$$

(positive sign when $|S_{22}|^2 - |S_{11}|^2 - 1 + |\Delta_s|^2 > 0$) for conditions listed above.

The source and load to provide conjugate match to the amplifier when $\eta > 1$ are the solutions of the following equations, which give $|p_s|$, and $|p_L|$ less than 1

$$p_{ms} = C_1^* \frac{B_1 \pm \sqrt{B_1^2 - 4|C_1|^2}}{2|C_1|^2} \qquad p_{mL} = C_2^* \frac{B_2 \pm \sqrt{B_2^2 - 4|C_2|^2}}{2|C_2|^2}$$

where

$$B_1 = 1 + |S_{11}|^2 - |S_{22}|^2 - |\Delta_s|^2 \qquad B_2 = 1 + |S_{22}|^2 - |S_{11}|^2 - |\Delta_s|^2$$
$$C_1 = S_{11} - \Delta_s S_{22}^* \qquad C_2 = S_{22} - \Delta_s S_{11}^*$$

the star ($*$) denoting conjugate.

If $|\eta| > 1$ but η is negative or $|\eta| < 1$, it is not possible to match simultaneously the two-port with real source and load admittance. Both graphical techniques and computer programs are available to aid in the design of rf amplifiers.

Intermediate-Frequency Amplifiers

Intermediate-frequency amplifiers consist of a cascade of a number of stages whose frequency response is determined either by a filter or by tuned interstages. The design of the individual active stages follows the techniques discussed earlier, but the interstages become important for frequency shaping. There are various forms of interstage networks; several important cases are discussed below.

Synchronous-Tuned Interstages. The simplest forms of tuned interstages are synchronously tuned circuits. The two common types are the single- and double-tuned interstage.

The governing equations are:

1. Single-tuned interstage (Fig. 11.3.3*a*):

$$A(j\omega) = -A_r \frac{1}{1 + jQ_L(\omega/\omega_0 - \omega_0/\omega)}$$

where Q_L = loaded Q of the tuned circuit greater than 10
ω_0 = resonance frequency of the tuned circuit = $1\sqrt{LC}$
ω = frequency variable
A_r = midband gain equal to g_m times the midband impedance level

For an n-stage amplifier with n interstages,

$$A_T = A^n(j\omega) = A_r^n \left[1 + \left(\frac{\omega^2 - \omega_0^2}{B\omega}\right)^2\right]^{-n/2}$$

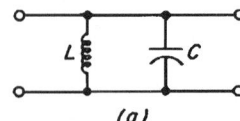

(a)

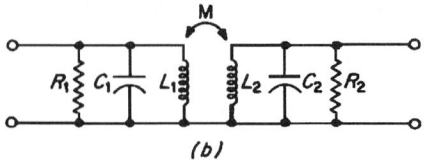

(b)

FIGURE 11.3.3 Interstage coupling circuits: (a) single-tuned; (b) double-tuned.

where $B = \omega_0/Q_L$ = single-stage bandwidth
n = number of stages
ω_0 = center frequencies
Q_L = loaded Q
$B_n = B\sqrt{2^{1/n}-1}$ is the overall bandwidth reduction owing to n cascades

2. Double-tuned interstage (Fig. 11.3.3b):

$$A(j\omega) = \frac{g_m k}{C_1 C_2 (1-k^2)\sqrt{L_1 L_2}} \frac{j\omega}{\omega^4 - ja_1\omega^3 - a_2\omega^2 + ja_3\omega + a_4}$$

(for a single double-tuned stage), where

$$a_1 = \omega_r\left(\frac{1}{Q_1}+\frac{1}{Q_2}\right) \qquad a_2 = \frac{\omega_r^2}{Q_1 Q_2} + \frac{1}{1-k^2}\left(\omega_1^2 + \omega_2^2\right)$$

$$a_3 = \frac{\omega_r}{1+k^2}\left(\frac{\omega_2^2}{Q_1}+\frac{\omega_1^2}{Q_2}\right) \qquad a_4 = \frac{\omega_1^2\omega_2^2}{1-k^2}$$

The circuit parameters are

R_1 = total resistance primary side
C_1 = total capacitance primary side
L_1 = total inductance primary side
R_2 = total resistance secondary side
C_2 = total capacitance secondary side
L_2 = total inductance secondary side
M = mutual inductance = $k\sqrt{L_1 L_2}$
k = coefficient of coupling

ω_r = resonant frequency of amplifier
$\omega_1 = 1/\sqrt{L_1 C_1}$
$\omega_1 = 1/\sqrt{L_2 C_2}$
Q_1 = primary Q at $\omega_r = \omega_r C_1 R_1$
Q_2 = secondary Q at $\omega_r = \omega_r C_2 R_2$
g_m = transconductance of active device at midband frequency

Simplification. If $\omega_1 = \omega_2 = \omega_0$, that is, primary and secondary tuned to the same frequency, then

$$\omega_r = \omega_0/\sqrt{1-k^2}$$

is the resonant frequency of the amplifier and

$$A(j\omega_r) = \frac{+jkg_m\sqrt{R_1 R_2}}{\sqrt{Q_1 Q_2}\,(k^2+1/Q_1 Q_2)}$$

is the gain at this resonant frequency.
For maximum gain,

$$k_c = \frac{1}{\sqrt{Q_1 Q_2}} = \text{critical coupling}$$

and for maximum flatness,

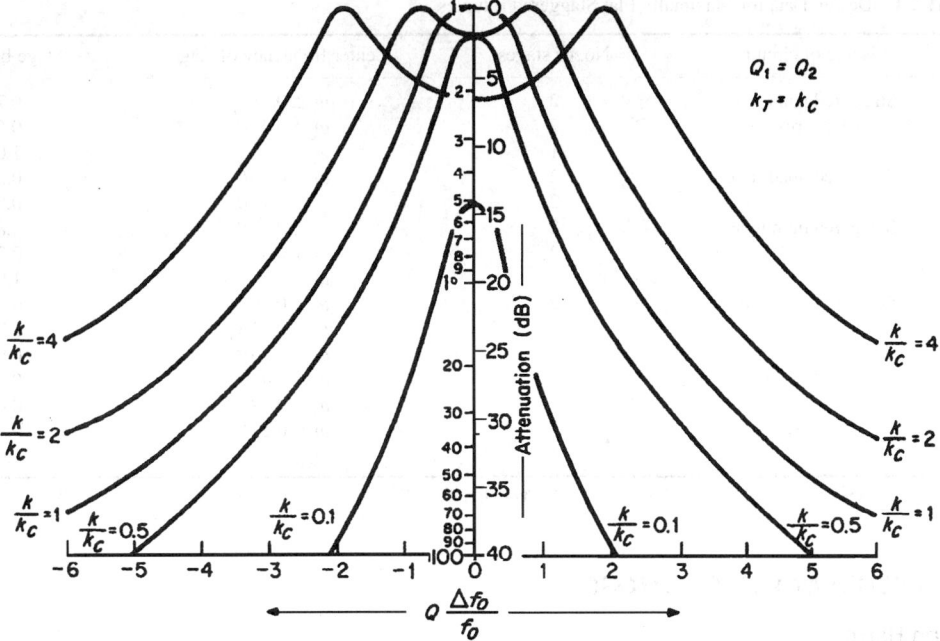

FIGURE 11.3.4   Selective curves for two identical circuits in a double-tuned interstage circuit, at various values of k/k_c.

$$k_T = \sqrt{\frac{1}{2}\left(\frac{1}{Q_1^2} + \frac{1}{Q_2^2}\right)} = \text{transitional coupling}$$

If k is increased beyond k_T, a double-humped response is obtained.

Overall bandwidth of an n-stage amplifier having equal Q circuits with transitional coupled interstages whose bandwidth is B is

$$B_n = B(2^{1/n} - 1)^{1/4}$$

The governing equations for the double-tuned-interstage case are shown above. The response for various degrees of coupling related to $k_T = k_C$ in the equal-coil-Q case is shown in Fig. 11.3.4.

Maximally Flat Staggered Interstage Coupling

This type of coupling consists of n single-tuned interstages that are cascaded and adjusted so that the overall gain function is maximally flat. The overall cascade bandwidth is B_n, the center frequency of the cascade is ω_c, and each stage is a single-tuned circuit whose bandwidth B and center frequency are determined from Table 11.3.1. The gain of each stage at cascade center frequency is $A(j\omega_c) = -g_m/C_T[B + j(\omega_c^2 - \omega_0^2/\omega_c)]$, where C_T = sum of output capacitance and input capacitance to next stage and wiring capacitance of cascade, B = stage bandwidth, ω_0 = center frequency of stage, and ω_c = center frequency of cascade.

TABLE 11.3.1   Design Data for Maximally Flat Staggered n-tuples

n	Name of circuit	No. of stages	Center frequency of stage	Stage bandwidth
2	Staggered pair	2	$\omega_c \pm 0.35B_n$	$0.71B_n$
3	Staggered triple	2	$\omega_c \pm 0.43B_n$	$0.50B_n$
		1	ω_c	$1.00B_n$
4	Staggered quadruple	2	$\omega_c \pm 0.46B_n$	$0.38B_n$
		2	$\omega_c \pm 0.19B_n$	$0.92B_n$
5	Staggered quintuple	2	$\omega_c \pm 0.29B_n$	$0.81B_n$
		2	$\omega_c \pm 0.48B_n$	$0.26B_n$
		1	ω_c	$1.00B_n$
6	Staggered sextuple	2	$\omega_c \pm 0.48B_n$	$0.26B_n$
		2	$\omega_c \pm 0.35B_n$	$0.71B_n$
		2	$\omega_c \pm 0.13B_n$	$0.97B_n$
7	Staggered septuple	2	$\omega_c \pm 0.49B_n$	$0.22B_n$
		2	$\omega_c \pm 0.39B_n$	$0.62B_n$
		2	$\omega_c \pm 0.22B_n$	$0.90B_n$
		1	ω_c	$1.00B_n$

For $Q_L > 20$

RADIO-FREQUENCY OSCILLATORS

G. Burton Harrold

General Considerations

Oscillators at rf frequencies are usually of the class A sine-wave-output type.

RF oscillators (in common with audio oscillators) may be considered either as one-port networks that exhibit a negative real component at the input or as two-port-type networks consisting of an amplifier and a frequency-sensitive passive network that couples back to the input port of the amplifier. It can be shown that the latter type of feedback oscillator also has a negative resistance at one port. This negative resistance is of a dynamic nature and is best defined as the ratio between the fundamental components of voltage and current.

The sensitivity of the oscillator's frequency is directly dependent on the effective Q of the frequency-determining element and the sensitivity of the amplifier to variations in temperature, voltage variation, and aging. For example, the effective Q of the frequency-determining element is important because the percentage change in frequency required to produce the compensating phase shift in a feedback oscillator is inversely proportional to the circuit Q, thus the larger the effective Q the greater the frequency stability. The load on an oscillator is also critical to the frequency stability since it affects the effective Q and in many cases the oscillator is followed by a buffer stage for isolation.

It is also desirable to provide some means of stabilizing the oscillator's operating point, either by a regulated supply, dc feedback for bias stabilization, or oscillator self-biasing schemes such as grid-leak bias. This stabilizes not only the frequency but also the output amplitude, by tending to compensate any drift in the active device's parameters. It is also necessary to eliminate the harmonics in the output since they give rise to cross-modulation products producing currents at the fundamental frequency that are not necessarily in phase with the dominant oscillating mode. The use of high-Q circuits and the control of the nonlinearity helps in controlling harmonic output.

Negative-Resistance Oscillators

The analysis of the negative-impedance oscillator is shown in Fig. 11.3.5. The frequency of oscillation at buildup is not completely determined by the LC circuit but has a component that is dependent upon the circuit resistance. At steady state, the frequency of oscillation is a function of $1 + R/R_{iv}$ or $1 + R_{ic}/R$, depending on the particular circuit where the ratios R/R_{ic}, R_{iv}/R are usually chosen to be small. While R is a fixed function of the

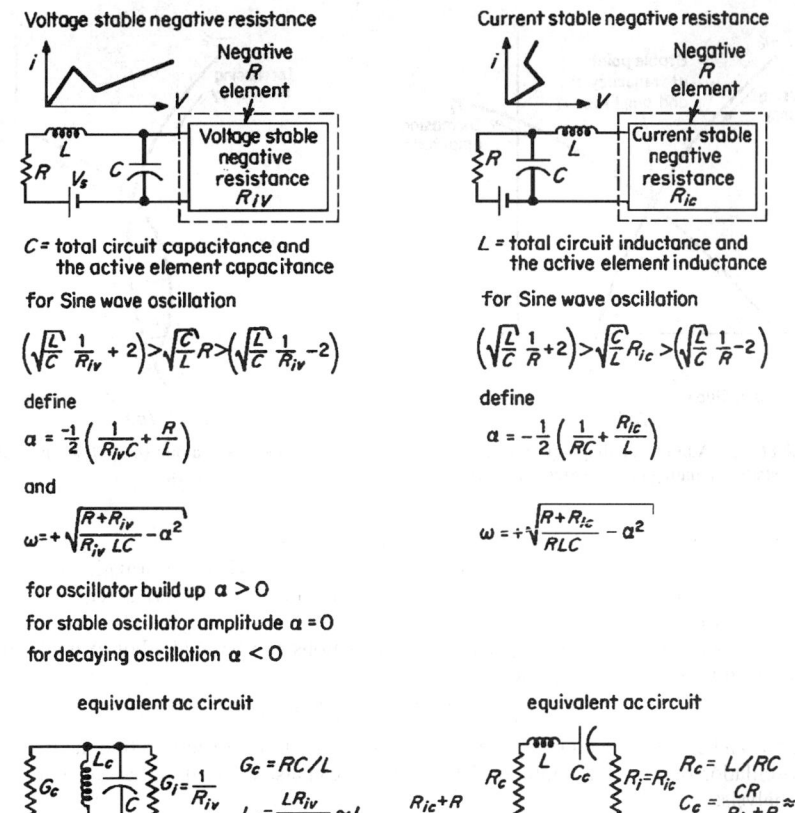

FIGURE 11.3.5   General analysis of negative-resistance oscillators.

loading, R_{ic} or R_{iv}/R must change with amplitude during oscillator buildup, so that the condition of $\alpha = 0$ can be reached. Thus R_{iv}, R_{ic} cannot be constant but are dynamic impedances defined as the ratio of the fundamental voltage across the elements to the fundamental current into the element.

The type of dc load for biasing and the resonant circuit required for the proper operation of a negative-resistance oscillator depend on the type of active element. R must be less than $|R_{iv}|$ or R must be greater than $|R_{ic}|$ in order for oscillation to build up and be sustained.

The detailed analysis of the steady-state oscillator amplitude and frequency can be undertaken by graphical techniques. The magnitude of G_i or R_i is expressed in terms of its voltage dependence. Care must be taken with this representation, since the shape of the G_i or R_i curve depends on the initial bias point.

The analysis of negative-resistance oscillators can now be performed by means of admittance diagrams. The assumption for oscillation to be sustaining is that the negative-resistance element, having admittance y_i, must equal $-y_c$, the external circuit admittance. This can be summarized by $G_i = -G_c$ and $B_i = -B_c$. A typical set of admittance curves is shown in Fig. 11.3.6. In this construction, it is assumed that $B_i = -B_c$, even during the oscillator buildup. Also shown is the fact that G_i at zero amplitude must be larger than G_c so that the oscillator can be started, that is, $\alpha > 0$, and that it may be possible to have two or more stable modes of oscillation.

Feedback Oscillators

Several techniques exist for the analysis of feedback oscillators. In the generalized treatment, the active element is represented by its y parameters whose element values are at the frequency of interest, having magnitudes

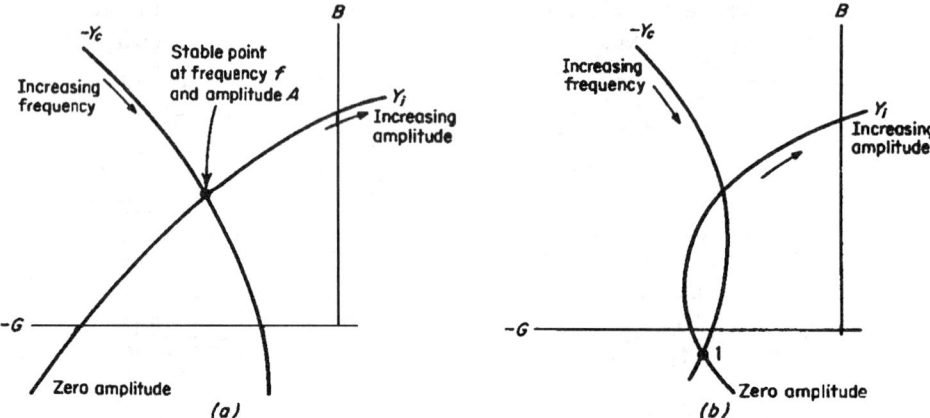

FIGURE 11.3.6   Admittance diagram of voltage-stable negative-resistance oscillators: (*a*) self-starting case, $\alpha > 0$; (*b*) circuit starts oscillating only if externally excited beyond point 1.

defined by the ratio of the fundamental current divided by fundamental voltage. The general block diagram and equations are shown in Fig. 11.3.7. Solution of the equations given yields information on the oscillator's performance. In particular, equating the real and imaginary parts of the characteristic equation gives information on amplitude and frequency of oscillation.

In many instances, many simplifications to these equations can be made. For example, if y_{11} and y_{12} are made small (as in vacuum-tube amplifiers), then

$$y_{21} = -(1/z_{21})(y_{22}z_{11} + 1) = -1/Z$$

This equation can be solved by equating the real and imaginary terms to zero to find the frequency and the criterion for oscillation of constant amplitude. This equation can also be used to draw an admittance diagram for oscillator analysis.

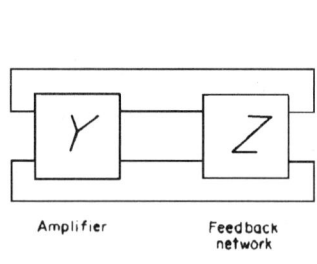

Characteristic equation:

$$y_{21}\, z_{21} + y_{11}\, z_{22} + y_{22}\, z_{11} + y_{12}\, z_{12} + \Delta_y\, \Delta_z + 1 = 0$$

$$\Delta_y = y_{11}\, y_{22} - y_{12}\, y_{21} \qquad \Delta_z = z_{11}\, z_{22} - z_{12}\, z_{21}$$

If $y_{21} \gg y_{12}$ $\qquad y_{12} \approx 0$ $\qquad$ and $[z]$ passive $z_{12} = z_{21}$

Then:

$$y_{21}\, z_{12} + y_{11}\, z_{22} + y_{22}\, z_{11} + \Delta_y\, \Delta_z + 1 = 0$$

FIGURE 11.3.7   General analysis of feedback oscillators.

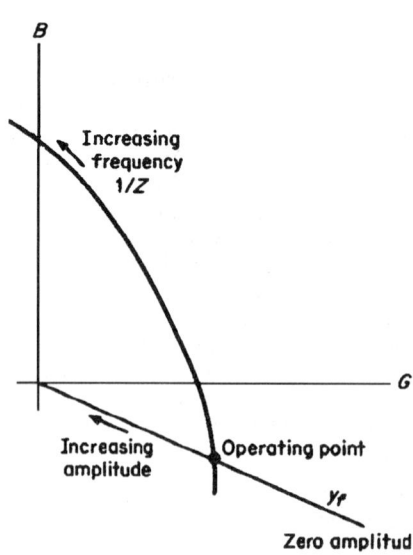

FIGURE 11.3.8   Admittance diagram of feedback oscillator.

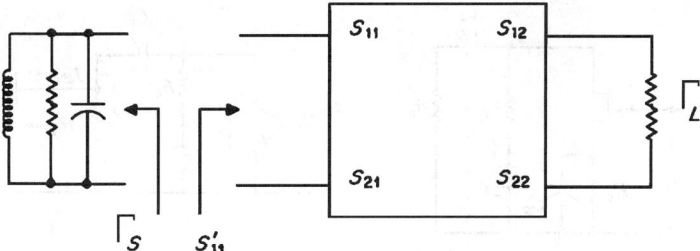

FIGURE 11.3.9   *S*-parameter and analysis of oscillators.

These admittance diagrams are similar to those discussed under negative-resistance oscillators. The technique is illustrated in Fig. 11.3.8.

At higher frequencies, the *S* parameters can also be used to design oscillators (Fig. 11.3.9). The basis for the oscillator is that the magnitude of the input reflection coefficient must be greater than unity, causing the circuit to be potentially unstable (in other words, it has a negative real part for the input impedance). The input reflection coefficient with a Γ_L output termination is

$$S'_{11} = S_{11} + \frac{S_{12}S_{21}\Gamma_L}{1 - S_{22}\Gamma_L}$$

Either by additional feedback or adjustment of Γ_L it is possible to make $|S'_{11}| > 1$. Next, establishing a load Γ_s such that it reflects all the energy incident on it will cause the circuit to oscillate. This criterion is stated as

$$\Gamma_L S'_{11} = 1$$

at the frequency of oscillation.

This technique can be applied graphically, using a Smith chart as before. Here the reciprocal of S'_{11} is plotted as a function of frequency since $S'_{11} > 1$. Now choose either a parallel or a series-tuned circuit and plot its Γ_s. If f_1 is the frequency common to $1/S'_{11}$ and Γ_s, and satisfies the above criterion, the circuit will oscillate at this point.

BROADBAND AMPLIFIERS

John W. Lunden, Jennifer E. Doyle

Introduction

In broadband amplifiers signals are amplified so as to preserve over a wide band of frequencies such characteristics as signal amplitude, gain response, phase shift, delay, distortion, and efficiency. The width of the band depends on the active device used, the frequency range, and power level in the current state of the art. As a general rule, above 100 MHz, a 20 percent or greater bandwidth is considered broadband, whereas an octave or more is typical below 100 MHz. As the state-of-the-art advances, it is becoming more common to achieve octave-bandwidth or wider amplifiers well into the microwave region using bipolar and FET active devices.

Hybrid-integrated-circuit techniques and new monolithic techniques eliminate many undesired package and bonding parasitics which can limit broadband amplifier performance. Additionally, distributed amplifiers, and other approaches which use multiple devices, have become more economical with increasing levels of integration.

It has become uncommon to use tube devices for new amplifier designs. Solid-state devices have replaced tubes in most amplifier applications because of superior long-term reliability and lower noise figures. In the following discussion both field-effect and bipolar transistor notations appear for generality.

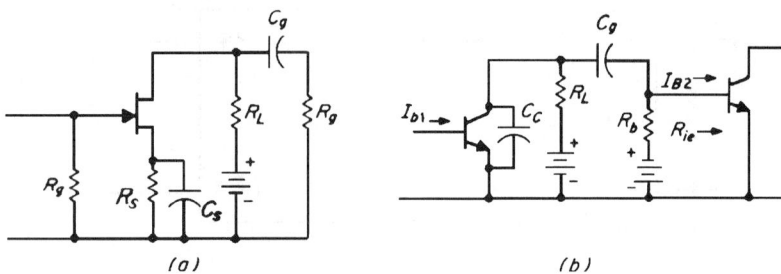

FIGURE 11.3.10   *RC*-coupled stages: (*a*) field-effect transistor form; (*b*) bipolar junction transistor form.

Low-, Mid-, and High-Frequency Performance

Consider the basic common-source and common-emitter-broadband *RC* coupled configurations shown in Fig. 11.3.10. Simplified low-frequency small-signal equivalent circuits are shown in Fig. 11.3.11.

The voltage gain of the FET amplifier stage under the condition that all reactances are negligibly small is the midband value (at frequency *f*)

$$(A_{\text{mid}})_{\text{FET}} = \frac{-g_m}{1/r_{ds} + 1/R_L + 1/R_g} \approx -g_m R_L$$

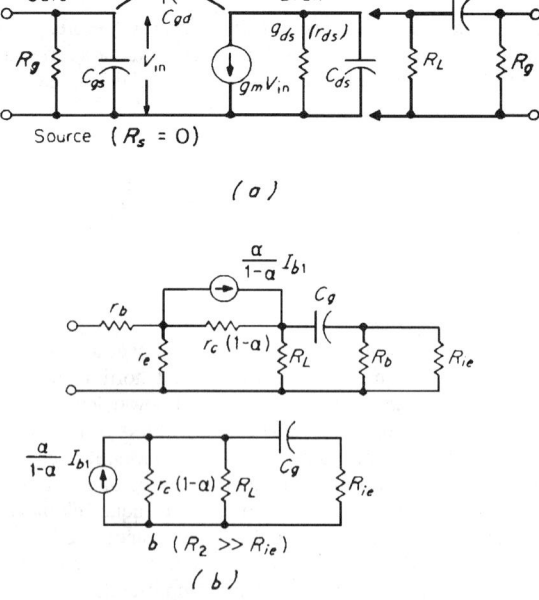

FIGURE 11.3.11   Equivalent circuits of the stages shown in Fig. 11.3.10: (*a*) FET form; (*b*) bipolar form.

If the low-frequency effects are included, this becomes

$$(A_{low})_{FET} = \frac{-g_m R_L}{1+1/j\omega R_g C_g} \frac{1+1/j\omega R_S C_S}{1+(1+g_m R_S)/j\omega C_S R_S} \left(= \frac{-g_m R_L}{1+1/j\omega R_g C_g} \text{ for } R_S = 0 \right)$$

The low-frequency cutoff is because principally of two basic time constants, $R_g C_g$ and $R_S C_S$. For C_S values large enough for the time constant to be much longer than that associated with C_g, a low-frequency cutoff or half-power point can be determined as

$$(f_1)_{FET} = \frac{1}{2\pi C_g [R_g + r_{ds} R_L/(r_{ds} + R_L)]}$$

If the coupling capacitor is very large, the low-frequency cutoff is a result of C_S. The slope of the actual rolloff is a function of the relative effect of these two time constants. Therefore, the design of coupling and bypass circuits to achieve very-low-frequency response requires very large values of capacitance.

Similarly, for a bipolar transistor stage, the midband current gain can be determined as

$$(A_{mid})_{BJT} = \frac{-\alpha r_c R_L}{[R_L + r_c(1-\alpha)]\left[R_{ie} + \dfrac{R_L r_c(1-\alpha)}{R_L + r_c(1-\alpha)}\right]} \approx \frac{-\alpha}{1-\alpha} \frac{R_L}{R_L + R_{ie}}$$

where

$$R_{ie} = r_b + \frac{r_e}{1-\alpha}$$

When low-frequency effects are included, this becomes

$$(A_{low})_{BJT} \approx \frac{-\alpha}{1-\alpha} \frac{R_L}{R_L + R_{ie} - j/\omega C_g} \quad \text{for } R_L \ll r_c(1-\alpha)$$

and

$$(f_1)_{BJT} = \frac{1}{2\pi C_g} \frac{1}{R_{ie} + \dfrac{R_L r_c(1-\alpha)}{R_L + r_c(1-\alpha)}} \approx \frac{1}{2\pi C_g} \frac{1}{R_{ie} + R_L}$$

If the ratio of low- to midfrequency voltage or current gain is taken, its reactive term goes to unity at $f = f_1$, that is, the cutoff frequency.

$$\frac{A_{low}}{A_{mid}} = \frac{1}{1 - j(f_1/f)} \qquad \phi_{low} = \tan^{-1}\frac{f_1}{f}$$

These quantities are plotted in Fig. 11.3.12 for a single time-constant rolloff.

Caution should be exercised in assuming that reactances between input and output terminals are negligible. Although this is generally the case, gain multiplicative effects can result in input or output reactance values greater than the values assumed above, e.g., by the Miller effect:

$$C_{in} = C_{gs} + C_{gd}(1 + g_m R'_L)$$

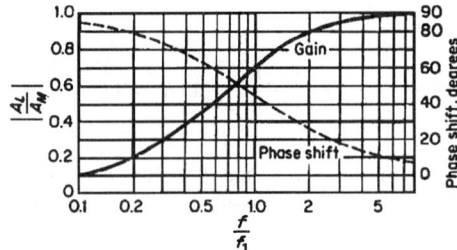

FIGURE 11.3.12 Gain and phase-shift curves at low frequencies.

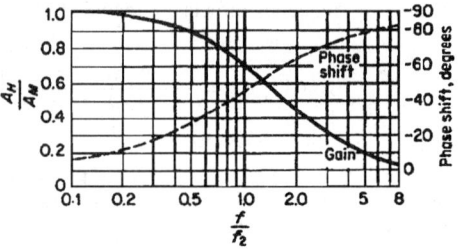

FIGURE 11.3.13 Gain and phase-shift curves at high frequencies.

Typically, the midfrequency gain equation can be used for frequencies above that at which $X_c = R_g/10$ below that at which $X_{cg} = 10R_gR_L/(R_g + R_L)$ (for the FET circuit).

If the frequency is increased further, a point is reached where the shunt reactances are no longer high with respect to the circuit resistances. At this point the coupling and bypass capacitors can be neglected. The high-frequency gain can be determined as

$$(A_{high})_{FET} = \frac{-g_m}{1/r_{ds} + 1/R_L + j\omega C_L}$$

where C_L is the effective total interstage shunt capacitance.

$$(A_{high})_{BJT} \approx \frac{-\alpha}{1-\alpha} \frac{1}{1 + R_{ie}\left(\dfrac{1}{R_L} + \dfrac{j\omega C_c}{1-\alpha}\right)} \quad \text{for } R_L \ll r_c(1-\alpha)$$

The ratio of high- to midfrequency gains can be taken and upper cutoff frequencies determined

$$\left(\frac{A_{high}}{A_{mid}}\right)_{FET} = \frac{1}{1 + j\omega C_L \dfrac{1}{(1/r_{ds}) + (1/R_L) + (1/R_g)}}$$

$$(f_2)_{FET} = \frac{1}{2\pi C_L}\left(\frac{1}{r_{ds}} + \frac{1}{R_L} + \frac{1}{R_g}\right)$$

$$\left(\frac{A_{high}}{A_{mid}}\right)_{BJT} = \frac{1}{1 + \dfrac{j\omega C_c r_c R_L R_{ie}}{R_{ie}[R_L + r_c(1-\alpha)] + R_L r_c(1-\alpha)}}$$

$$(f_2)_{BJT} \approx \frac{1-\alpha}{2\pi C_c}\left(\frac{1}{R_L} + \frac{1}{R_{ie}}\right) \quad \text{and} \quad \phi_{high} = -\tan^{-1}(f/f_2)$$

Dimensionless curves for these gain ratios and phase responses are plotted in Fig. 11.3.13.

Compensation Techniques

To extend the cutoff frequencies f_1 and f_2 to lower or higher values, respectively, compensation techniques can be used.

Figure 11.3.14 illustrates two techniques for low-frequency compensation. If the condition $R_gC_g = C_XR_XR_L/(R_X + R_L)$ is fulfilled (in circuit a or b), the gain relative to the midband gain is

$$\frac{A_{\text{low}}}{A_{\text{mid}}} = \frac{1}{1 - j(1/\omega R_g C_g)[R_L/(R_L + R_X)]} \quad \text{and} \quad f_1 = \frac{1}{2\pi R_g C_g}\frac{R_L}{R_L + R_X}$$

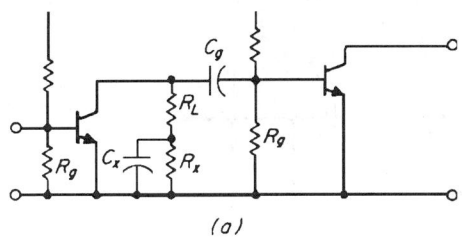

(a)

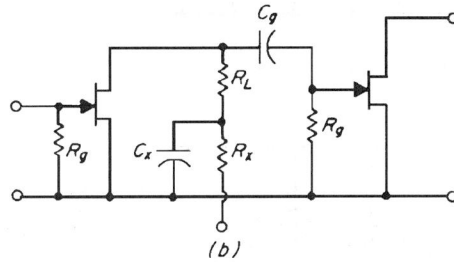

(b)

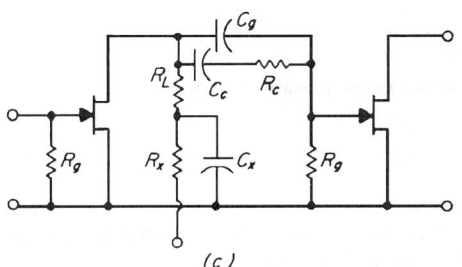

(c)

FIGURE 11.3.14 Low-frequency compensation networks: (a) bipolar transistor version; (b), (c) FET versions.

Hence, improved low-frequency response is obtained with increased values of R_X. This value is related to R_L and restricted by active-device operating considerations. Also, R_L is dependent on the desired high-frequency response. It can be shown that equality of time constants $R_LC_X = R_gC_g$ will produce zero phase shift in the coupling circuit (for $R_X > 1/\omega C_X$). The circuit shown in Fig. 11.3.14c is more critical. It is used with element ratios set to $R_L/R_X = R_g/R_c$ and $C_X/C_g = R_c/R_X$.

Various compensation circuits are also available for high-frequency-response extension. Two of the most common, the series- and shunt-compensation cases, are shown in Fig. 11.3.15. The high-frequency-gain expressions of these configurations can be written

$$\left|\frac{A_{\text{high}}}{A_{\text{mid}}}\right| = \sqrt{\frac{1 + a_1(f/f_2)^2 + a_2(f/f_2)^4 + \cdots}{1 + b_1(f/f_2)^2 + b_2(f/f_2)^4 + b_3(f/f_2)^6 + \cdots}}$$

The coefficients of the terms decrease rapidly for the higher-order terms, so that if $a_1 = b_1$, $a_2 = b_2$, etc., to as high an order of the f/f_2 ratio as possible, a maximally flat response curve is obtained.

For the phase response, $d\phi/d\omega$ can also be expressed as a ratio of two polynomials in f/f_2, and a similar procedure can be followed. A flat time-delay curve results. Unfortunately, the sets of conditions for flat gain and linear phase are different, and compromise values must be used.

Shunt Compensation. The high-frequency gain and time delay for the shunt-compensated stage are

$$\left|\frac{A_{\text{high}}}{A_{\text{mid}}}\right| = \sqrt{\frac{1 + \alpha^2(f/f_2)^2}{1 + (1 - 2\alpha)(f/f_2)^2 + \alpha^2(f/f_2)^4}} \quad \phi = -\tan^{-1}\frac{f}{f_2}\left[1 - \alpha + \left(\frac{f}{f_2}\right)^2\alpha^2\right]$$

where $a = L/C_g R_L^2$ and $R_g \gg R_L$.

A case when R_g cannot be assumed to be high, such as the input of a following bipolar transistor stage, is considerably more complex, depending on the transistor equivalent circuit used. This is particularly true when operating near the transistor f_T and/or above the VHF band.

Series Compensation. In the series-compensated circuit, the ratio of C_s to C_g is an additional parameter. If this can be optimized, the circuit performance is better than in the shunt-compensated case. Typically, however, control of this parameter is not available due to physical and active-device constraints.

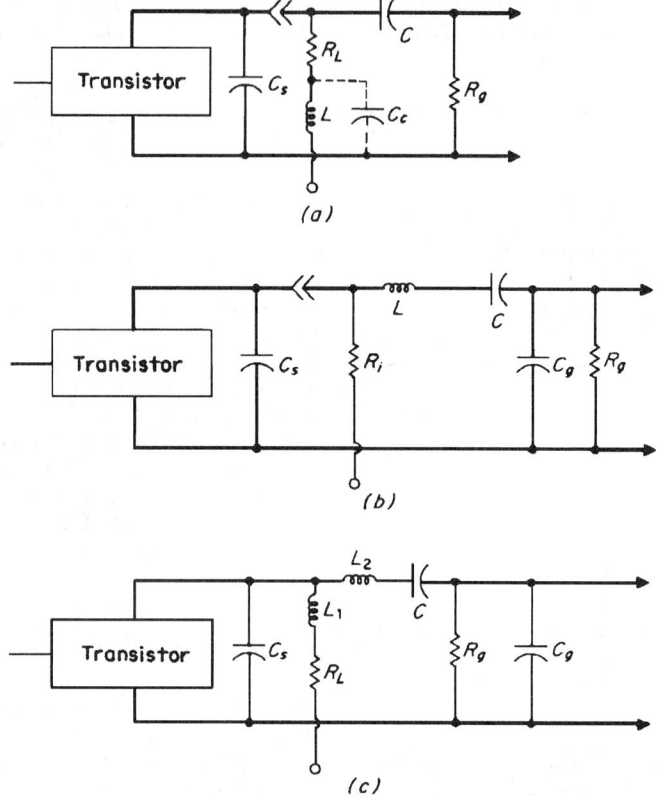

FIGURE 11.3.15 High-frequency compensation schemes: (*a*) shunt; (*b*) series; (*c*) shunt-series.

These two basic techniques can be combined to improve the response at the expense of complexity. The shunt-series-compensation case and the so-called "modified" case are examples. The latter involves a capacitance added in shunt with the inductance L or placing L between C_s and R_L.

For the modified-shunt case, the added capacitance C_c permits an additional degree of freedom, and associated parameter, $k_1 = C_c/C_s$.

Other circuit variations exist for specific broadband compensation requirements. Phase compensation, for example, may be necessary as a result of cascading a number of minimum-phase circuits designed for flat frequency response. Circuits such as the lattice and bridged-T can be used to alter the system response by reducing the overshoot without severely increasing the overall rise time.

Cascaded Broadband Stages

When an amplifier is made up of n cascaded RC stages, not necessarily identical, the overall gain A_n can be written

$$\left|\frac{A_n}{A_{\text{mid}}}\right| = \left[\frac{1}{1+(f/f_a)^2}\right]^{1/2}\left[\frac{1}{1+(f/f_b)^2}\right]^{1/2}\cdots\left|\frac{1}{1+(f/f_n)^2}\right|^{1/2}$$

where $f_a, f_b, \ldots, f_n$ are the f_1 or f_2 values for the respective stages, depending on whether the overall low- or high-frequency gain ratio is being determined. The phase angle is the sum of the individual phase angles. If the stages are identical, $f_a = f_b = f_x$ for all, and

$$\left| \frac{A_n}{A_{\text{mid}}} \right| = \left| \frac{1}{1 + (f/f_x)^2} \right|^{n/2}$$

Stagger Peaking.   In stagger tuning a number of individual bandpass amplifier stages are cascaded with frequencies skewed according to some predetermined criteria. The most straightforward is with the center frequencies adjusted so that the f_2 of one stage concludes with the f_1 of the succeeding stage, and so forth. The overall gain bandwidth then becomes

$$(\text{GBW})_n = \sum_{n=1}^{N} (\text{GBW})_n$$

A significant simplifying criterion of this technique is stage isolation. Isolation, in transistor stages particularly, is not generally high, except at low frequencies. Hence the overall design equations and subsequent overall alignment can be significantly complicated because of the interactions. Complex device models and computer-aided design greatly facilitate the implementation of this type of compensation. The simple shunt-compensated stage has found extensive use in stagger-tuned pulse-amplifier applications.

Transient Response

Time-domain analysis is particularly useful for broadband applications. Extensive theoretical studies have been made of the separate effects of nonlinearities of amplitude and phase response. These effects can be investigated starting with a normalized low-pass response function.

$$A(j\omega)/A(0) = \exp(a^m w^m - jb^n w^n)$$

where a and m are constants describing amplitude-frequency response and b and n are constants describing phase-frequency response. Figure 11.3.16 illustrates the time response to an impulse and a unit-step forcing function for various values of m, with $n = 0$. Rapid change of amplitude with frequency (large m) results in

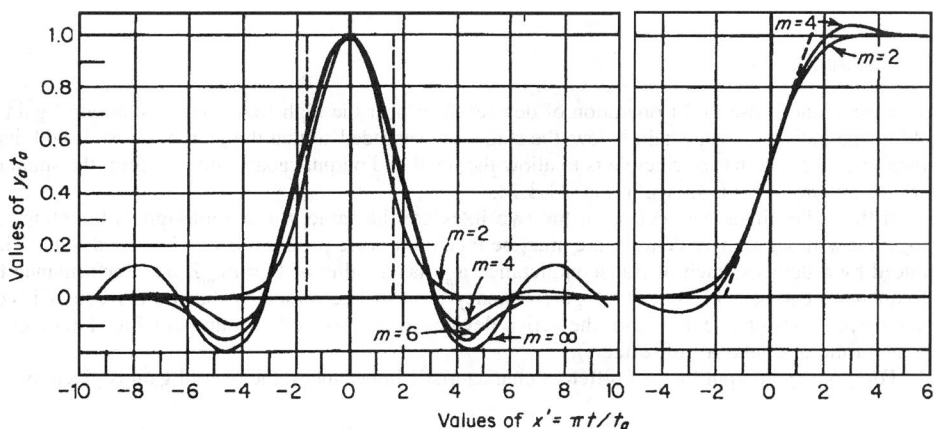

FIGURE 11.3.16   Transient responses to unit impulse (*left*) and unit step (*right*) for various values of m.

overshoot. Nonzero, but linear, phase-frequency characteristics ($n = 1$) result in a delay of these responses, without introducing distortion. Further increase in n results in increased ringing and asymmetry of the time function.

An empirical relationship between rise time (10 to 90 percent) and bandwidth (3 dB) can be expressed as

$$t_r \cdot \mathrm{BW} = K$$

where K varies from about 0.35 for circuits with little or no overshoot to 0.45 for circuits with about 5 percent overshoot. K is 0.51 for the ideal rectangular low-pass response with 9 percent overshoot; for the Gaussian amplitude response with no overshoot, $K = 0.41$.

The effect on rise time of cascading a number of networks n depends on the individual network pole-zero configurations. Some general rules follow.

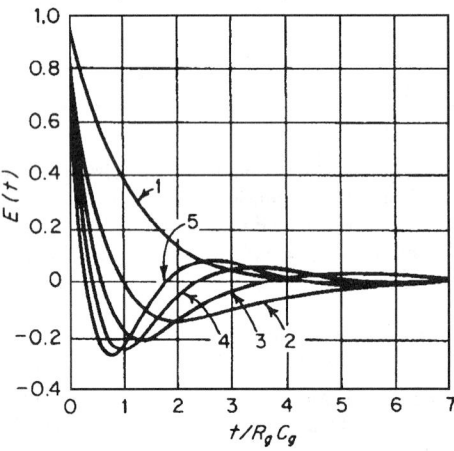

1. For individual circuits having little or no overshoot, the overall rise time is

$$t_{rt} = (t_{r1}^2 + t_{r2}^2 + t_{r3}^2 + \cdots)^{1/2}$$

2. If $t_{r1} = t_{r2} = t_{rn}$,

$$t_{rt} = 1.1\sqrt{n}\,t_{r1}$$

3. For individual stage overshoots of 5 to 10 percent, total overshoot increases as $\sqrt{n}$.

4. For circuits with low overshoot (~1 percent), the total overshoot is essentially that of one stage.

The effect of insufficient low-frequency response of an amplifier is sag of the time response. A small amount of sag (<10 percent) can be described by the formula

$$E_{\mathrm{sag}}/E_{\mathrm{tot}} = T/2.75RC$$

FIGURE 11.3.17   Response to a unit step of n capacitively coupled stages of the same time constant.

where T is one-half period of a square-wave voltage.

Figure 11.3.17 illustrates the response to a step function of n capacitor-coupled stages with the same time constant. The effect of arithmetic addition of individual stage initial slopes determining the net total initial transition slope can be seen.

Distributed Amplifiers

This technique is useful for operation of devices at or near the high-frequency limitation of gain bandwidth. While the individual stage gain is low, the stages are cascaded so that the gain response is additive instead of multiplicative. The basic principle is to allow the input and output capacitances to form the shunt elements of two delay lines. This is shown in Fig. 11.3.18.

If the delay times per section in the two lines are the same, the output signals traveling forward add together without relative delay. Care must be taken to ensure proper terminating conditions. The gain produced by n devices, each of transconductance g_m, has a value of $G = ng_m Z_{02}/2$. Performance to very low frequencies can be achieved. The high-frequency limit is determined by the cutoff frequencies of the input and output lines or effects within the active devices themselves other than parasitic shunt capacities (e.g., transit time or alpha-fall-off effects).

For input and output lines of different characteristic impedances, the overall gain is given by

$$G = \frac{2Z_{01}}{Z_{01} + Z_{02}} \frac{ng_m Z_{02}}{2}$$

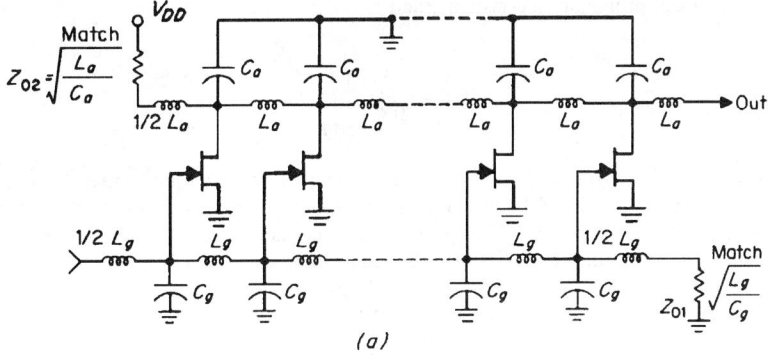

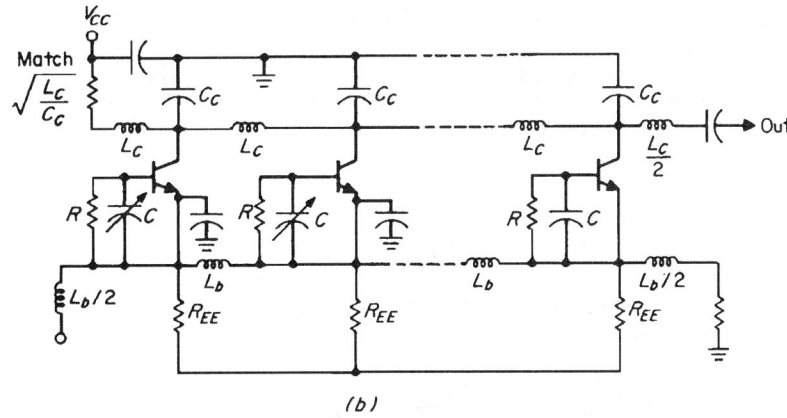

FIGURE 11.3.18   Distributed amplifier circuits: (a) FET version; (b) bipolar transistor version.

The characteristic impedances and cutoff frequencies are given as

$$Z_{01} = \sqrt{L_g/C_g} \quad Z_{02} = \sqrt{L_a/C_a}$$
$$f_{c1} = \pi/\sqrt{L_gC_g} \quad f_{c2} = \pi/\sqrt{L_aC_a}$$

There does not seem to be an optimum choice for Z_{02}; a device with a high figure of merit is simply chosen, and the rest follows. There is, however, an optimum way in which N devices can be grouped, namely, in m identical cascaded stages, each with n devices, so that N is a minimum for a given overall gain A. N is a minimum when $m = \ln A$. Consequently, the optimum gain per stage $G = e = 2.72$.

Various techniques are utilized to determine the characteristics of the lumped transmission lines. Constant-K and m-derived filter sections are most common, augmented by several compensation circuit variations. The latter include paired gate-drain (base-collector), bridged-T, and resistive loading connections. The constant-K lumped line has several limitations, including the fact that the termination is not a constant resistance and that impedance increases with frequency and time delay also changes with frequency.

m-derived terminating half sections and time-delay equalizing sections result in a frequency-amplitude response that is quite flat at very high frequencies.

The effects of input loading and/or line loss modify the gain expression to

$$G = \frac{2(GB)}{f_c} \frac{e^{-a}(1 - e^{-n\infty})}{1 - e^{-\infty}}$$

where $\propto$ = the real part of propagation constant γ and

$$GB = \frac{g_m}{2\pi C_{\text{eff}}}\bigg|_{\text{FET}}$$

The design of distributed amplifiers using bipolar transistors is more difficult because of the low input imped-ance. In addition, the intrinsic gain of the transistor (β or α) is a decreasing function of frequency.

The simplest approach in overcoming this problem is to connect a parallel RC in series with the base, as shown in Fig. 11.3.18b. By setting $RC = \beta_0/2\pi f_T$, the gain is made essentially independent of frequency up to about $f_T/2$, with an overall voltage gain of

$$A_v = (n\beta_0 Z_0/2R)^m$$

where m is the number of stages and n is the number of transistors per stage.

The increasing impedance of the constant-K line helps keep the frequency response flat, compensating for the increased loading of the bipolar transistor.

Another bipolar transistor approach involves the division of the frequency range into three regions to account for the losses. Each region is associated with a linear-Q variation with frequency of the transistor input-output circuits, approximated by shunt RC elements.

$$Q_{\text{in}} = w(C_{\text{in}} + C)/(G_{\text{in}} + G_k)$$

The lossless voltage gain is

$$A_L = f_T Z_0/\{2f[1 - (f/f_c)^2]^{1/2}\}$$

In region I, the input impedance is high and the transistor supplies high voltage gain. In regions II and III, the voltage gains are given by

$$A_{\text{II}} = A_L n\frac{1 - nK_{\text{II}}Z_0(f/f_c)^2}{4r_b'[1 - (f/f_c)^2]^{1/2}} \qquad A_{\text{III}} = A_L n\frac{1 - nK_{\text{III}}Z_0}{4r_b'[1 - (f/f_c)^2]^{1/2}}$$

where $K_{\text{II}} = (f_c/f_Q)^2$
$\quad\;\; K_{\text{III}} = 1$
$\quad\;\; f_Q$ = frequency of min Q

Broadband Matching Circuits

Broadband impedance transformation interstage-coupling matching can be achieved with balun transformers, quarter-wave transmission-line sections, lumped reactances in configurations other than discussed above, and short lengths of transmission lines.

Balun Transformers.   In conventional coupling transformers, the interwinding capacitance resonates with the leakage inductance, producing a loss peak. This limits the high-frequency response. A solution is to use transmission-line transformers in which the turns are arranged physically to include the interwinding capaci-tance as a component of the characteristic impedance of a transmission line. With this technique, bandwidths of hundreds of megahertz can be achieved. Good coupling can be realized without resonances, leading to the use of these transformers in power dividers, couplers, hybrids, and so forth.

Typically, the lines take the form of twisted wire pairs, although coaxial lines can also be used. In some configurations, the length of the line determines the upper cutoff frequency. The low-frequency limit is deter-mined by the primary inductance. The larger the core permeability the fewer the turns required for a given low-frequency response. Ferrite toroids have been found to be satisfactory with widely varying core-material characteristics. The decreasing permeability with increasing frequency is offset by the increasing reactance of the wire itself, causing a wide-band, flat-frequency response.

Quarter-Wave Transformers.   The quarter-wavelength line transformer is another well-known element. It is simply a transmission line one-quarter wavelength long, with a characteristic impedance

$$Z_{\text{line}} = \sqrt{Z_{\text{in}} Z_{\text{out}}}$$

where Z_{in} and Z_{out} are the terminating impedances. The insertion loss of this line section is

$$10 \log \left[1 + \frac{(r-1)^2}{4r} \cos^2 \theta \right] \quad \text{(dB)}$$

where $r = Z_{\text{in}}/Z_{\text{out}}$ and $\theta = 2\pi L/\lambda = 90°$ at f_0.

Figure 11.3.19 shows the bandwidth performance of the quarter-wave line for several matching ratios. Such lines may be cascaded to achieve broader-band matching by reducing the matching ratio required of each individual section.

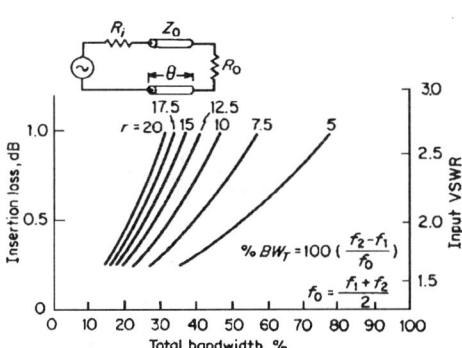

FIGURE 11.3.19   Bandwidth of a quarter-wave matching transformer.

Lumped and Pseudo-Lumped Transformations.   The familiar lumped-element ladder network, depending on the element values and realization, approximates a short-step transmission-line transformer or a tapered transmission line.

Feedback and Gain Compensation

Feedback.   The bandwidth of amplifiers can be increased by the application of inverse feedback. A multiplicity of feedback combinations and analysis techniques are available and extensively treated in the literature. In addition, the related concept of feedforward has been investigated and successfully implemented. Figure 11.3.20 shows four feedback arrangements and formulas describing their performance. A major consideration, particularly for rf applications, is the control of impedance and loop-delay characteristics to ensure stability.

Gain Compensation.   The power gain of a bipolar transistor amplifier typically falls 6 dB/octave with increasing frequency above the f_β value. The gain can be leveled (bandwidth-widened) by exact matching of the source impedance to the device at the upper frequency only, causing increasing mismatch with decreasing frequency and the associated gain loss. The overall flat gain is the value at the high best-match frequency. Sufficient resistive loss is usually required in this interstage to prevent instabilities in either driver or driven stage because of the mismatch unloading.

Power-Combining Amplifiers

Many circuit techniques have been developed to obtain relatively high output powers with given modest-power devices. Two approaches are the direct-paralleling and hybrid splitting-combining techniques. The direct-paralleling approach is limited by device-to-device variations and the difficulties in providing balanced conditions because of the physical wavelength restrictions. A technique commonly used at UHF and microwave frequencies to obtain multioctave response incorporates matched stages driven from hybrid couplers in a balanced configuration. The coupler offers a constant-source impedance (equal to the driving-port

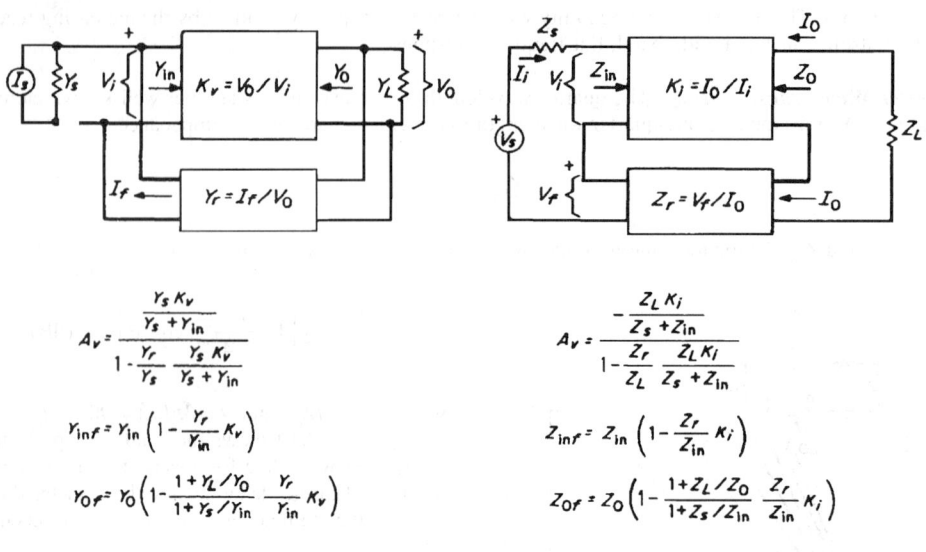

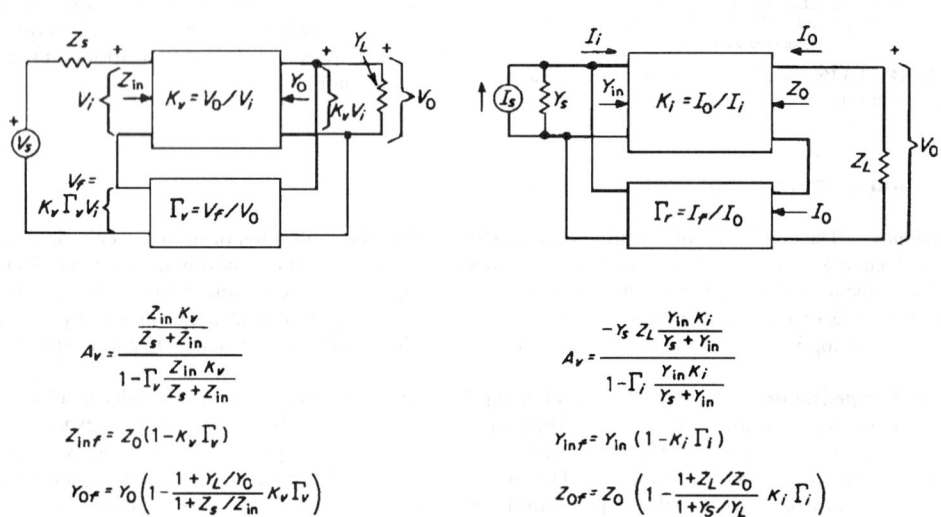

FIGURE 11.3.20 Four methods of applying feedback, showing influence on gain and input and output immittances (*From Hakim "Junction Transistor Circuit Analysis," by permission of John Wiley & Sons, New York*).

impedance) to the two stages connected to its output ports. Power reflected because of the identical mismatches is coupled to the difference port of the hybrid and dissipated in the idler load. The hybrid splitting-combining approach has proved quite effective in implementing high-output-level requirements (e.g., kilowatts at 1.4 GHz with bipolar transistors).

The hybrid splitting-combining approach enhances circuit operation. In particular, quadrature hybrids affect a voltage standing wave ratio (VSWR)-canceling phenomenon that results in extremely well-matched

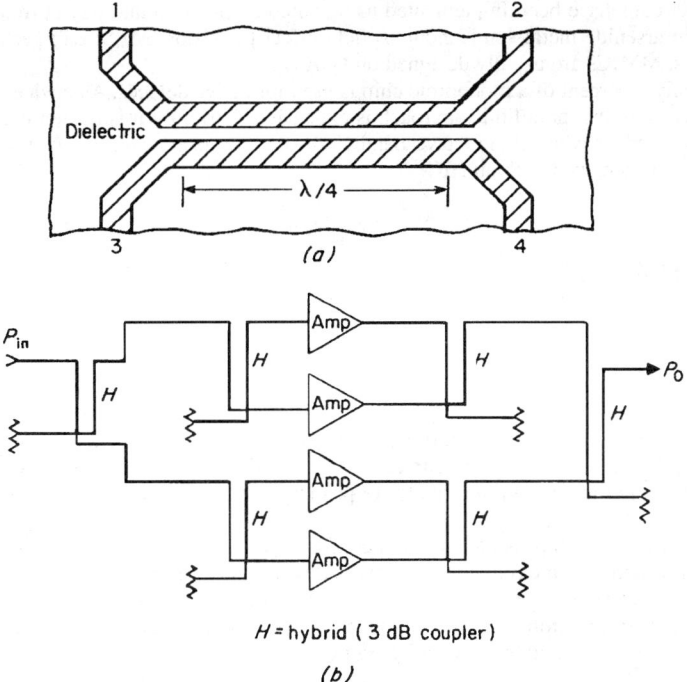

H = hybrid (3 dB coupler)

(b)

FIGURE 11.3.21 Wide-band combining techniques: (*a*) wide-band quarter-wave coupler; (*b*) balanced combining amplifier configuration.

power-amplifier inputs and outputs that can be broadbanded on proper selection of particular hybrid trees. Also, the excellent isolation between devices makes reliable power-amplifier service possible.

Several hybrid-directional-coupler configurations are possible, including the split-T, branch-line, magic-T, backward-wave, and lumped. Important factors in the choice of hybrid for this application are coupling bandwidth, isolation, fabrication ease, and form. The equiamplitude, quadrature-phase, reverse-coupled transverse electromagnetic wave (TEM) $\lambda/4$ coupler is a particularly attractive implementation owing to its bandwidth and amenability to strip-transmission-line circuits. Figure 11.3.21 illustrates this coupler type. Other commonly used broadband microstrip hybrids include the Wilkinson and the Lange couplers.

Integrated Amplifiers

As broadband amplifier designs are extended into microwave frequencies, package parasitics begin to limit high-frequency performance. A hybrid integrated circuit (HIC) has transmission paths and passive components formed together on a low-loss substrate by etching and deposition techniques. Active devices, and some discrete passive components, are then connected to the substrate transmission paths by beam-leading, soldering, or wire bonding. The active devices may be chips, packaged devices, or chips mounted on carriers. HICs show improved high-frequency performance because of minimal lead lengths, and transmission lines which provide impedance matching.

The monolithic microwave integrated circuit (MMIC) includes all active devices, as well as passive components and interconnections, in the deposition process, which may require several steps. The elimination of wire bonds and package parasitics can result in multioctave bandwidths at microwave and even millimeter wave frequencies.

Monolithic circuits have been implemented using silicon bipolar technology at frequencies below 4 GHz. Because gallium arsenide metal-semiconductor field-effect transistor (MESFET) performance is superior at high frequencies, MMICs are usually designed on GaAs.

Generally, only 5 percent of a monolithic chip is used for active devices. As device yields are low and processing costs are high, the monolithic approach is expensive. Currently it is economical only when very high quantities are needed or when multioctave bandwidth or millimeter wave frequency operation is required, beyond what can be achieved with an HIC.

TUNNEL-DIODE AMPLIFIERS

Chang S. Kim

Introduction

Tunnel-diode (TD) amplifiers are one-port negative-conductive devices. Hence the problems associated with them are quite different from those encountered in conventional amplifier design. Circuit stabilization and isolation between input and output terminals are primary concerns in using this very wide-frequency-range device.

Although there are several possible amplifier configurations, the following discussion is limited to the most practical design, which uses a circulator for signal isolation. The advantage of the circulator-coupled form of tunnel-diode amplifier resides in the fact that it is thereby possible to convert a bilateral one-port amplifier into an ideal unilateral two-port amplifier. Tunnel diodes can provide amplification at microwave frequencies with a relatively simple structure and at a low noise figure.

Tunnel Diodes

Three kinds of tunnel diodes are available, namely, those using Ge, GaAs, and GaSb. *V-I* characteristics and corresponding small-signal conductance-voltage relationships are shown in Figs. 11.3.22 and 11.3.23, respectively. A typical small-signal tunnel-diode equivalent circuit is shown in Fig. 11.3.24. Here g_j, C_j, r_s, L, and C_p are the small-signal conductance, junction capacitance, series resistance, series inductance, and shunt capacitance, respectively. Noise generators e_s and i_j are included for subsequent discussion.

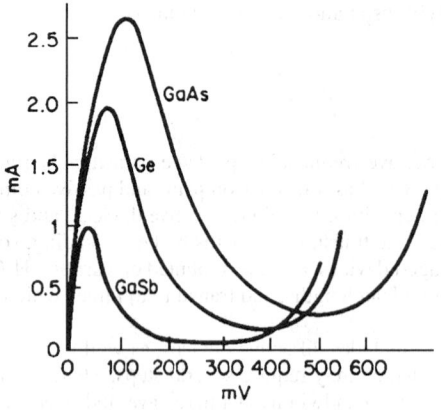

FIGURE 11.3.22 Characteristics of tunnel diodes.

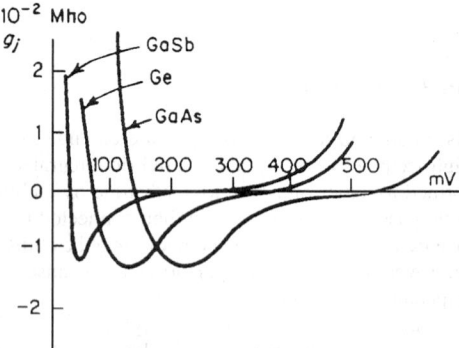

FIGURE 11.3.23 Voltage-vs.-g_j characteristics of tunnel diodes of Fig. 11.3.22.

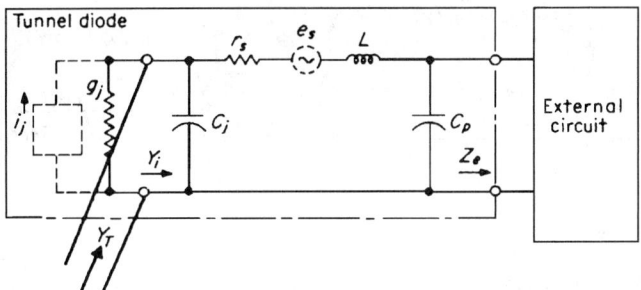

FIGURE 11.3.24   Small-signal equivalent circuit of tunnel-diode amplifier.

Stability

The stability criteria are derived from the immittance expression across the diode terminals and are quite complicated.

Using the short-circuit stable condition, the stability criteria can be simplified considerably. With reference to Fig. 11.3.24, with the external circuit connected, the total admittance Y_T across g_j can be expressed as

$$Y_T(s) = Y_i(s) - g_j = p(s)/q(s)$$

where $Y_i(s)$, the admittance facing g_j, is a positive real function. Since Y_i is connected in parallel with a short-circuit stable device with negative conductance g_i, Y_T is short-circuit-stable. This implies that, since $q(s)$ is always a Hurwitz polynomial, the stability condition is that $p(s)$ must also be Hurwitz polynomial.

A simple graphical interpretation of this stability condition is as follows. The plot[*] of $Y_T(\omega)$ can be obtained from the plot of $Y_i(\omega)$ by shifting the imaginary axis by $|g_j|$ along the real axis, as shown in Fig. 11.3.25. Since $q(s)$ has no roots in the right half plane, any encirclement of the origin of $Y_T(\omega)$ must come from the right-half-plane roots of $p(s)$ only. Therefore, the circuit will be stable if and only if $Y_T(\omega)$ does not encircle the origin (Fig. 11.3.25a). If g_j becomes large so that the origin is encircled by $Y_T(\omega)$ (Fig. 11.3.25b), the circuit is unstable.

Tunnel-Diode-Amplifier Design

A simplified block diagram of a circulator-coupled TD amplifier is shown in Fig. 11.3.26. The three basic circuit parts are the tunnel diode, the stabilizing circuit, and the tuning circuit, which includes a four-port circulator. The following conditions are imposed on the amplifier design:

1. In the band
    a. $G_i = \mathrm{Re}\, Y_i$ is slightly larger than $|g_j|$.
    b. G_i is contributed by the tuning circuit only.
    c. $B_i = \mathrm{Im}\, Y_i = 0$ at the center frequency f_0 and small in the band.
2. Outside the band
    a. $G_i = \mathrm{Re}\, Y_i$ is larger than $|g_j|$
    b. If $G_i \le |g_j|$, B_j should not be zero.

To satisfy these conditions, a stabilizing circuit, shown in Fig. 11.3.26, is required. This circuit is designed so that the following relationships are satisfied:

$$Y_1(f_0) = Y_1(3f_0) = Y_2(f_0) = 0 \qquad Y_1(2f_0) = Y_1(4f_0) = Y_2(3f_0) = 1/R \qquad Y_s = Y_1 + Y_2$$

where f_0 is the center frequency.

---

[*]Here, the plot of $YT_T(\omega)$ represents the case of $Z_e(\omega)$ short-circuited; however, similar plots can be obtained for more general cases.

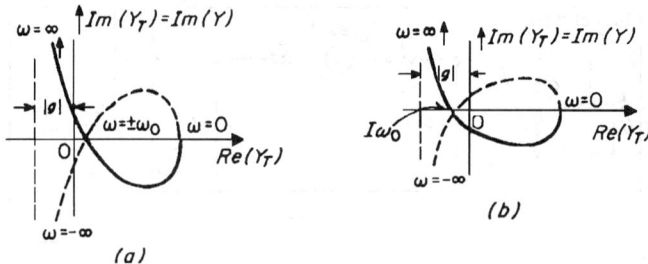

FIGURE 11.3.25   Representative plots of real and imaginary parts of Y_T: (a) stable condition; (b) unstable condition.

The equivalent circuit of Fig. 11.3.24 can be transformed into the parallel equivalent circuit of Fig. 11.3.27. The following identities relate the parameters of Figs. 11.3.24 and 11.3.27:

$$g_{jp} = 1/|Z_s|^2 \, g_j X \qquad g_{sp} = r_s / |Z_s|^2$$

$$B = \omega C_p - \frac{\omega L - (1 - 1/X)1/\omega C_j}{|Z_s|^2} \qquad \text{where } X = 1 + \omega^2 c_j^2 / g_j^2$$

$$|Z_s|^2 = \gamma_s - 1/|g_j| X + [\omega L - (1 - 1/X)1/\omega C_j]^2$$

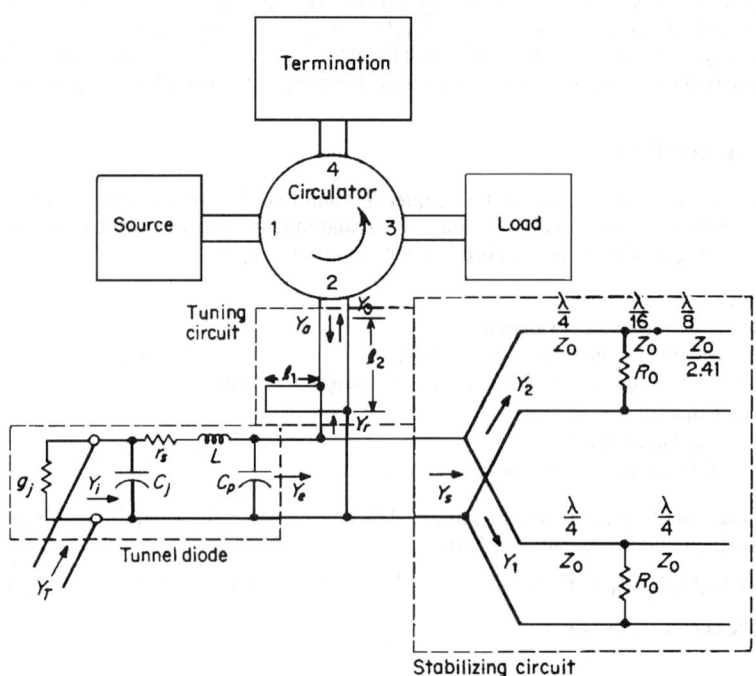

FIGURE 11.3.26   Tunnel-diode amplifier using circulator.

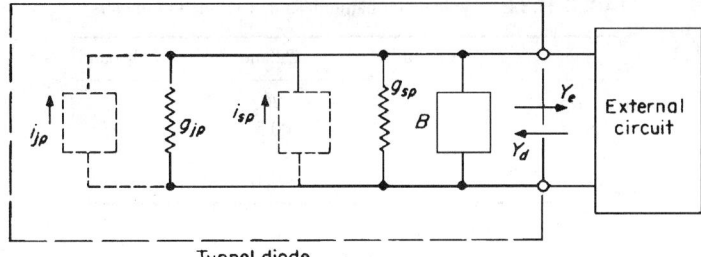

Tunnel diode

FIGURE 11.3.27 Parallel version of equivalent circuit of Fig. 11.3.24.

The gain can be expressed by

$$|\Gamma|^2 = \left|\frac{Y_e - Y_d}{Y_e + Y_d}\right|^2 \qquad |\Gamma|^2_{f=f_0} = \left|\frac{G_e - g_{sp} + |g_{jp}|}{G_e + g_{sp} - |g_{jp}|}\right|^2$$

where $B_e - B = 0$ at $f = f_0$, in which B_e is the reactive part of Y_e.

In this case, the tuning circuit is a more general matching circuit using combinations of parallel and series transmission lines. It should be noted that $|\Gamma|$ becomes 1, or $g_{sp} = |g_{jp}|$, as the operating frequency increases to f_r, the resistive cutoff frequency, which is defined by

$$f_r = \frac{|g_j|}{2\pi C}\sqrt{1 - \frac{1}{r_s|g_j|}}$$

It is desirable to have a device with f_r several times (at least three) larger than f_0. Furthermore, it is desirable to make the self-resonance frequency f_x,

$$f_x = \frac{1}{2\pi}\sqrt{\frac{1}{LC_j} - \left(\frac{g_j}{C_j}\right)^2}$$

as high as possible (higher than f_c) to improve the stability margin.

The gain expression can be modified to include the input amplifier admittance Y_a of Fig. 11.3.26 as follows:

$$|\Gamma|^2 = \left|\frac{Y_0 - Y_a}{Y_0 + Y_a}\right|^2$$

Similarly, the bandwidth can be determined from the expression for Γ^2 and f_x, above.

Typical germanium tunnel-diode parameters pertinent to S-, C-, and X-band amplifiers with approximate gains to 10 dB are shown in Table 11.3.2.

Noise Figure in TD Amplifiers

A noise equivalent circuit can be completed by inserting a current generator, $\overline{i_j^2}$ and a voltage generator $\overline{e_s^2}$, as shown in Fig. 11.3.24 (dotted lines). The mean-square values are determined from

$$\overline{i_j^2} = 2eI_{eq}\Delta f \qquad \overline{e_s^2} = 2KTr_s\Delta f$$

where I_{eq} = equivalent shot-noise current = I_{dc} = dc current in negative-conductance region.

TABLE 11.3.2 Typical Germanium Tunnel-Diode Parameters

	S band	C band	X band
$|r_j| = 1/g_j$, Ω	70	70	70
C_j, pF	<1	<0.5	<0.2
L, H	<0.1	<0.1	<0.2
r_s, Ω	<$0.1r_j$	<$0.1r_j$	<$0.1r_j$

The noise equivalent circuit of Fig. 11.3.24 can be transformed into a parallel-equivalent circuit of Fig. 11.3.27 having current generators i_{jp}^2 and i_{sp}^2 (dotted lines); i_{jp}^2 and i_{sp}^2 can be derived from the two equivalent circuits to be

$$\overline{i_{jp}^2} = 4KT(G_{eq}/|g_j|)\,|g_{jp}|\,\Delta f \quad \overline{i_{sp}^2} = 4KTg_{sp}\Delta f$$

where $G_{eq} = eI_{eq}/2KT = 20I_{eq}$ at room temperature.

Tunnel diodes provide good noise performance with a relatively simple amplifier structure requiring only a dc source for bias. It is therefore a useful low-noise, small-signal amplifier for microwave applications.

PARAMETRIC AMPLIFIERS

Conrad E. Nelson

Introduction

The term *parametric amplifier* (paramp) refers to an amplifier (with or without frequency conversion) using a nonlinear or time-varying reactance. Development of low-loss variable-capacitance (varactor) diodes resulted in the development of varactor-diode parametric amplifiers with low noise figure in the microwave frequency region. Types of paramps include one-port, two-port (traveling-wave), degenerate (pump frequency twice the signal frequency), nondegenerate, multiple pumps, and multiple idlers. The most widely used amplifier is the nondegenerate one-port paramp with a circulator, because it achieves very good noise figures without undue circuit complexity.

One-Port Paramp with Circulator

The one-port paramp with circulator is illustrated in Fig. 11.3.28, and the simplified circuit diagram is shown in Fig. 11.3.29. The input signal and the amplified output signal (at the same frequency) are separated by the circulator. The cw pump source is coupled to a back-biased varactor to drive the nonlinear junction capacitance at the pump frequency. The signal and pump currents mix in the nonlinear varactor to produce voltages at many frequencies. The additional (idler) filter allows only the difference or idler current to flow at the idler frequency; i.e.,

$$f_i = f_p - f_s$$

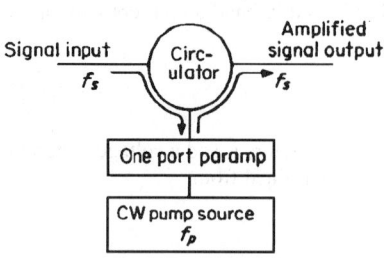

FIGURE 11.3.28 One-port parametric amplifier and circulator.

The idler current remixes with the pump current to produce the signal frequency again. The phasing of this signal due to nonlinear reactance mixing is such that the original incident signal is reenforced (i.e., amplified) and reflected back to the circulator. The one-port paramp at band center is essentially a negative-resistance device at the signal frequency.

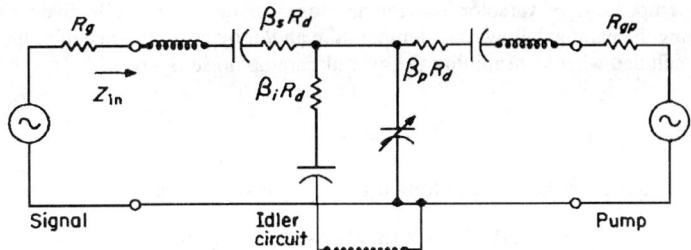

FIGURE 11.3.29   Circuit of one-port paramp using single-tuned resonators.

Power Gain and Impedance Effects

The one-port paramp power gain at the signal frequency is

$$G = \left| \frac{Z_{in} - R_g}{Z_{in} + R_g} \right|^2$$

where R_g is the signal-circuit equivalent generator resistance and Z_{in} is the input impedance at the signal frequency.

For a single-tuned signal circuit and a single-tuned idler circuit the input impedance at the signal frequency is

$$Z_{in} = \beta_s R_d + jX_s - \frac{\sigma \beta_s \beta_i R_d^2}{\beta_i R_d - jX_i} \quad \text{where } \sigma = \frac{m_1^2 f_c^2}{\beta_s \beta_i f_s f_i}$$

The diode loss resistance R_d has been modified at the signal and idler frequencies to include circuit losses (that is, $\beta \geq 1$). The signal and idler reactances are X_s and X_i, respectively, and must include the varactor junction capacitance. (In the simplified circuit the diode package capacitance has been neglected.) For a fully pumped varactor the cutoff frequency is

$$f_c = 1/2\pi R_d C_R$$

where C_R is the junction capacitance at reverse breakdown. The modulation ratio m_1 for an abrupt-junction diode is 0.25.

Bandwidth and Noise

Factors that determine the overall paramp bandwidth include the varactor characteristics (cutoff frequency, junction and package capacitance, and lead inductance), choice of idler (pump) frequency, the nature of the signal and idler resonant circuits and the choice of band-center gain. Multiple-tuned signal and idler circuits are often used to increase the overall paramp bandwidth.

At band center the effective noise temperature of the one-port paramp (due to the diode and circuit loss resistances at the signal and idler frequencies) is

$$T_e = T_d \frac{G-1}{G} \left(\frac{f_p}{f_i} \frac{\sigma}{\sigma - 1} - 1 \right)$$

where T_d is the temperature of varactor junction and loss resistances. The effective noise temperature can be reduced by cooling the paramp below room temperature and/or by proper choice of pump frequency. Circulator losses must be included when determining the overall paramp noise figure.

Pump Power

The pump power required at the diode to fully drive a single varactor is

$$P_p = k_l \beta_p R_d [\omega_p C_R (\phi - V_R)]^2 \quad \text{(watts)}$$

where the pump circuit loss resistance and diode resistance $= \beta_p R_d$, and $k_1 = 0.5$ for an abrupt-junction varactor.

Gain and Phase Stability

The stability of the cw pump source (amplitude and frequency) is a significant factor in the overall paramp gain and phase stability. For small pump-power changes an approximate expression for the paramp power-gain variation is

$$\frac{\Delta G}{G} \approx \frac{G-1}{\sqrt{G}} \frac{\Delta P_p}{P_p}$$

Environmental temperature changes often require a temperature-regulated enclosure for the paramp and pump source.

At low levels of nonlinear distortion, the amplifier third-order relative power intermodulation product at band center is a function of the one-port amplifier gain and the incident signal power level; i.e.,

$$\Delta(\text{IMP}) \propto \frac{(G-1)^4 P_s^2}{G} \approx G^3 P_s^2 \approx G P_{\text{out}}^2$$

Thus low-band-center amplifier gains reduce this nonlinear distortion.

MASER AMPLIFIERS

Gunter K. Wessel

Introduction

A *maser* is a microwave active device whose name is an acronym derived from microwave amplification by stimulated emission of radiation. A *laser* is an amplifier or oscillator also based on stimulated emission but operating in the optical part of the spectrum. The expressions microwave, millimeter, submillimeter, infrared, optical, and ultraviolet maser are also in common use. Here the word maser is used for the whole frequency range of devices, the *m* standing for molecular.

A description of the laser principle, some of its properties and means of achieving laser operations, is given in Sec. 9.

In these devices, no tubes or transistors are employed in the radiation process, but the properties of atoms or molecules—as gases, liquids, or solids—serve for the amplification of the signals. Oscillators require the addition of a feedback mechanism. The interaction of the electromagnetic radiation with the maser material occurs in a suitable cavity or resonance structure. This structure often serves the additional purpose of generating a desired phase relationship between different spatial parts of the signal. The application of external energy, required for the amplification or oscillation process, is referred to as *pumping*. The pumping power consists, in many cases, of external electromagnetic radiation of a different frequency (usually higher) from that for the signal.

Microwave masers are used as low-noise preamplifiers and as time and frequency standards. The stimulated emission properties of atoms and molecules at optical frequencies, with their relatively high noise content, make the laser more useful for high-power light amplification and oscillation.

Maser Principles

In the processes of emission and absorption, atoms or molecules interact with electromagnetic radiation. It is assumed that the atoms possess either sharp internal-energy states or broader energy bands. A change of internal energy from one state to another is accompanied by absorption or emission of electromagnetic radiation, depending on the direction of the process. The difference in energy from the original to the final state is proportional to the frequency of the radiation, the proportionality constant being Planck's constant h.

The processes of the spontaneous emission and absorption were defined by Einstein.

Spontaneous emission:

$$-\frac{dN_2}{dt} = A_{21}N_2 \tag{1}$$

Absorption:

$$-\frac{dN_1}{dt} = B_{12}\mu(v_{12},T)N_1 \tag{2}$$

where N_1 and N_2 are population densities of atoms or molecules of material having energy states E_1 and E_2 and an energy separation $E_2 - E_1 = hv_{12}$ between them, and A_{21} and B_{12} are the Einstein coefficients.

At high temperature, an inconsistency in the equilibrium system arises unless a second emission process, stimulated emission, takes place. The rate of stimulated emission is defined very similarly to that of absorption (the stimulated emission is sometimes called *negative absorption*):

$$-\frac{dN_2}{dt} = B_{21}\mu(v_{12},T)N_2 \tag{3}$$

In temperature equilibrium, the rates of emission must be equal to the rate of absorption

$$A_{21}N_2 + B_{21}\mu N_2 = B_{12}\mu N_1 \tag{4}$$

Assuming that the ratio of the population densities is equal to the Boltzmann factor,[*]

$$N_1/N_2 = \exp(hv_{12}/kT) \tag{5}$$

Equation (4) yields, for the radiation density,

$$\mu(v_{12},T) = \frac{A_{21}/B_{21}}{B_{12}/B_{21}\exp(hv_{12}/kT) - 1} \tag{6}$$

Equation (6) is identical with Planck's radiation law if we set

$$B_{12} = B_{21} \tag{7a}$$

and

$$A_{21}/B_{21} = 8\pi hv_{12}^3/c^3 \tag{7b}$$

where c is the velocity of light.

Equation (7a) shows the close relationship between stimulated emission and absorption. The rates of population decrease, for these two processes [Eqs. (2) and (3)] depend only on their respective population densities.

---

[*]It is assumed that the statistical weights are equal to 1.

Neglecting the spontaneous emission, the net absorption or the net stimulated emission of an incoming radiation depends, therefore, only on the difference in population density: for absorption, if $(N_1 - N_2) > 0$; for amplification, if $(N_1 - N_2) < 0$. In particular, if a system is in quasiequilibrium such that the upper energy state E_2 is more populated than the lower one, E_1, it is capable of amplifying electromagnetic radiation.

To create and maintain the quasiequilibrium requires application of external energy since the natural equilibrium has the opposite population excess ($N_1 > N_2$). Different masers differ widely in the methods of how to accomplish the reverse in population density.

Properties of Masers

The properties of masers can be understood by analyzing Eqs. (1) to (7b), as follows.

Signal-to-Noise Ratio.   From Eqs. (1) and (3) it follows that the Einstein coefficients A and B have different dimensions but the ratio $B_{21}\mu/A_{21}$ is dimensionless. Here μ is proportional to the strength of the incoming signal and $B_{21}\mu$ is proportional to the amplified signal strength. A_{21} is proportional to the noise contribution because of spontaneous emission. After rewriting Eq. (7b),

$$B_{21}\mu/A_{21} = c^3\mu/8\pi h\nu_{12}^3 \qquad (8)$$

the ratio is found to be proportional to the signal-to-noise ratio of the amplifier. It thus becomes clear that the noise contribution (thermal or Johnson noise) is very small at microwave frequencies, whereas it becomes very large at optical frequencies (an increase of 15 orders of magnitude for an increase of 5 orders for the frequency). Microwave masers are therefore very useful as low-noise microwave preamplifiers.

Infrared, optical, and ultraviolet masers (lasers), on the other hand, are commonly used for power amplification and as powerful light sources (oscillators). The high content of spontaneous-emission noise does not make them easily applicable for low-noise amplification or sensitive detection of light, except in a few special cases.

Linearity and Line Width

The proportionality between the incoming signal strength, the radiation density μ, and the amplified signal strength means that a maser is a linear amplifier as long as it is not driven into saturation. The latter occurs when the excess population becomes $N_2 - N_1 \approx N_2$.

Line Width.   In the absence of a strong interaction of the individual atoms of the material, either with the environment or with each other, the line width is determined by the average length of time of interaction of an atom with the radiation field. However, if regeneration because of feedback is taking place, the line width may be much narrower because of the following effect. It is assumed that the incoming radiation has a frequency distribution, as shown in Fig. 11.3.30. The amplified signal is proportional to the rate of stimulated emission dN_2/dt. According to Eq. (3), the latter is proportional to $\mu(\nu_{12})$. Thus the center portions of the line will be more amplified than the wings, leading to increasingly narrower line widths. In an oscillator the final width is ultimately limited by statistical fluctuations because of the quantum nature of the radiation.

The narrow line width coupled with the low-noise property of a microwave maser makes the latter useful as a time and frequency standard device. To reduce interaction, gases are used as microwave maser materials.

In low-noise microwave amplifiers one is usually interested in a relatively large bandwidth (typically 50 MHz or more). For such amplifier solid-state materials are used, with strong spin-spin interactions of atoms (for example, Cr^{3+} in ruby).

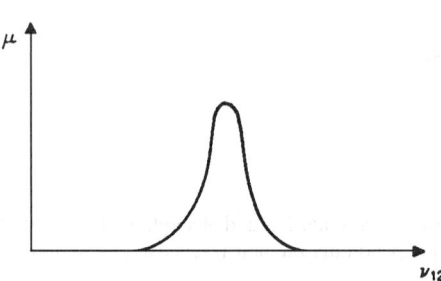

FIGURE 11.3.30   Frequency distribution of radiation incoming to maser.

Tuning

Solid-state microwave masers can be tuned over a wide range of frequencies by adjustment of an external magnetic field, while the relative tuning range by a magnetic field is smaller by a factor of 10^{-4} to 10^{-5} in optical masers. Generally, in lasers, one has to depend on fixed frequencies wherever a suitable spectral line occurs in the material; however, modern liquid-dye lasers can be tuned over a relatively wide range in the visible spectrum.

Power Output

The power output owing to stimulated emission depends on the number of excess atoms $(N_2 - N_1)$ available for the transition. Assuming p to be the probability of a transition per second, the power output is given by

$$P = (N_2 - N_1)h\nu_{12}p \tag{9}$$

Equation (9) shows that the power output of a microwave gas maser must be very low (for example, 100 to 1 pW) because the population density and the frequency are small. On the other hand, solid-state lasers and some gas lasers with very high efficiencies may have cw power outputs of 1000 W and more.

Coherence

The similarity between absorption and stimulated emission [as expressed in Eqs. (2), (3), and (7a)] means that the stimulated emission is a coherent process. In practice, this means that a beam of radiation will be amplified only in the direction of propagation of the incoming beam and when all atoms participate coherently in the amplification. This is very different from the noise-producing spontaneous emission, where the radiation is emitted isotropically and no phase relationship exists between the radiation coming from different atoms.

The coherence of the stimulated emission enables one to produce coherent light and radiation patterns similar to those which are well known and applied in the radio and microwave parts of the spectrum. In particular, it is possible by a suitable geometry of the device to produce a plane-parallel light beam whose divergence angle is limited only by diffraction. The divergence angle is given by

$$\alpha \approx \lambda/d \tag{10}$$

Time and Frequency Standards

The ammonia and hydrogen atomic-beam masers are used as time and frequency standards with operating frequencies of about 24 GHz and about 1.4 GHz, respectively. The reverse of population difference is achieved by focusing atoms in the excited state into a suitable microwave cavity, whereas the atoms in the lower state are defocused by a special focuser and do not reach the interior of the cavity. The microwave field in the cavity causes stimulated emission and amplification. If the cavity losses can be overcome, oscillations will set in. Time and frequency accuracy of 1 part in 10^{12} or higher can be obtained with modern maser standard devices.

Three- and Four-Level Devices

In many of the masers of all frequency regions, population reverse is obtained by a *three-level-maser* method first described by Bloembergen. Radiation from an external power source at the pumping frequency ν_p (Fig. 11.3.31) is applied to the material. E_3 may be a band to make the pumping more economical. The material is contained in a suitable cavity, a slow-wave structure, or other resonance structure which is resonant at the signal frequency ν_s. Under favorable conditions of the pumping power, the frequency ratio ν_s/ν_p, and the involved relaxation times, an excess population in the excited state E_2 over the ground state E_1 can be obtained. The interaction with the electromagnetic field of frequency ν_s causes stimulated emission. Thus amplification (and if required, oscillation) is produced.

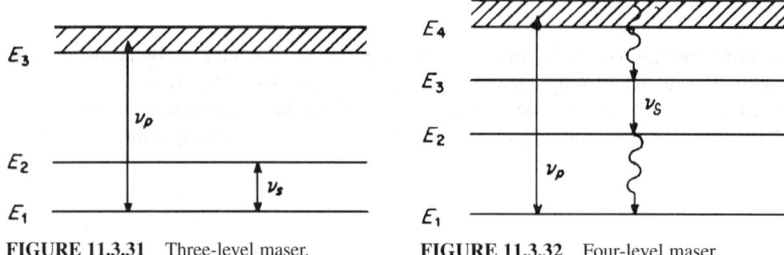

FIGURE 11.3.31 Three-level maser.　　　**FIGURE 11.3.32** Four-level maser.

The four-level maser (Fig. 11.3.32) percent has the additional advantage that the population of the ground state does not have to be reduced to less than 50 percent of its original equilibrium value and the level E_4 can remain a relatively wide band. The transfer of energy from E_4 to E_1, and E_2 to E_1, is by spontaneous decay or some other means like spin-lattice relaxation. Most solid-state microwave and optical masers are of the three- or four-level variety.

Semiconductor Lasers

Stimulated light emission can be obtained by carrier injection in certain semiconductor materials. Upon application of an external dc potential, recombination light emission can take place at the interface of an *n*- and *p*-type semiconductor (for example, Zn and Te in GaAs). The semiconductor laser requires only a very simple power source (a few volts of direct current). However, the laser material has to be cooled to liquid-nitrogen temperature and the power output is relatively low.

Applications

The particular properties of devices based on stimulated emission have resulted in numerous applications which often surpass the capabilities of standard devices. In some instances, e.g., holography, new fields have opened up through the emergence of lasers and masers.

Solid-state microwave preamplifiers for communication, radar, and radio astronomy have by far the best signal-to-noise ratio of all microwave amplifiers. The hydrogen-beam maser is the most accurate of all frequency and time standards.

Lasers have revolutionized the field of optical instrumentation and spectroscopy. They have made possible the new field of nonlinear optics. The laser is or may be used in many applications where large energy densities are required, as in microwelding, medical surgery, or even in cracking rocks in tunnel building. Lasers can be used as communication media where extremely large bandwidths are required. Optical radar, with its high precision owing to the short wavelength, uses lasers as powerful, well-collimated light sources. Their range extends as far as the moon.

ACOUSTIC AMPLIFIERS

Stephen W. Tehon, Y. J. Lin, Wen-Chung. Wang

Acoustoelectric Interaction

The acoustic amplifier stems from the announcement by Hutson, McFee, and White, in 1961, that they had observed a sizable influence on acoustic waves in single crystals of CdS caused by a bias current of change carriers. CdS is both a semiconductor and piezoelectric crystal, and the interaction was found to involve an

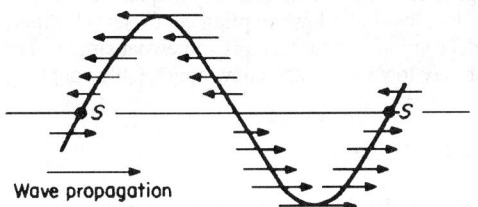

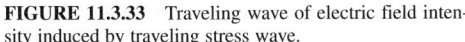

Wave propagation

FIGURE 11.3.33 Traveling wave of electric field intensity induced by traveling stress wave.

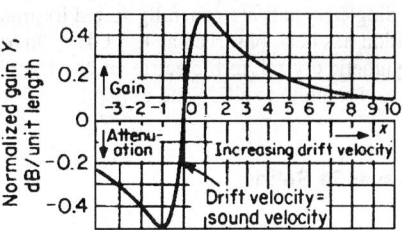

FIGURE 11.3.34 Normalized gain as a function of bias field.

energy transfer via traveling electric fields generated by acoustic waves, producing, in turn, bunching of the drift carriers. Quantitative analyses showing that either loss or gain in wave propagation could be selected by controlling the drift field were published by White.

Figure 11.3.33 illustrates the nature of the interaction. As an acoustic wave propagates in the crystal its stresses induce a similar pattern of electric field through piezoelectric coupling. Since the coupling is linear, compressive stresses in half the wave induce a forward-directed field, and tensile stresses in the remainder of the wave induce a backward-directed field. Drifting charge carriers tend to bunch at points of zero field to which they are forced by surrounding fields. When a charge carrier loses velocity in bunching, it gives its excess kinetic energy to the acoustic wave; when it gains velocity, it extracts energy from the wave. Therefore the drift field is effective for determining attenuation or amplification in the range near which carriers move at the speed of sound. Figure 11.3.34 illustrates the gain characteristic as a function of drift velocity. At the zero-crossover point, the carrier velocity is just equal to the velocity of propagation for the acoustic wave; the gain ranges from maximum attenuation up to maximum amplification over a relatively small change in bias. Beyond this range of drift velocity, bunching becomes decreasingly effective, and the interaction has less effect on the acoustic wave.

Piezoelectric Materials

Piezoelectricity is the linear, reversible coupling between mechanical and electric energy because of displacement of charges bound in molecular structure. Pressure applied to a piezoelectric material produces a change in observed surface density of charge, and conversely, charge applied over the surfaces produces internal stress and strain. If S is the strain, T the stress, E the electric field intensity, and D the dielectric displacement, the piezoelectric effect at a point in a medium is described by the pair of linear equations

$$S = sT + dE \qquad D = dT + \epsilon E$$

where constant s = elastic compliance
ϵ = permittivity
d = piezoelectric constant

The ratio $d^2/s\epsilon$, calculated from these constants for a particular piezoelectric material, is defined as K^2, where K^2 is the coefficient of electromechanical coupling. As a consequence of conservation of energy, it can be shown that K is a number less than unity and that a fraction K^2 of applied energy (mechanical or electric) is stored in the other form (electric or mechanical). Since D and E are vectors and S and T are tensors, the piezoelectric equations are tensor equations and are equivalent to nine algebraic equations, describing three vector and six tensor components.

Materials that are appreciably piezoelectric are either crystals with anisotropic properties or ceramics with ferroelectric properties which can be given permanent charge polarization through dielectric hysteresis. Ferroelectric ceramics, principally barium titanate and lead zirconate titanate, are characterized by relative dielectric constants ranging from several hundred to several thousand, by coupling coefficients as high as 0.7, and by polycrystalline grain structure which will propagate acoustic waves with moderate attenuation at frequencies extending up to the low-megahertz range.

Single crystals are generally suited for much higher acoustic frequencies, and in quartz acoustic-wave propagation has been observed at 125 GHz. Quartz has low loss but a low coupling coefficient. Lithium niobate is a synthetic crystal, ferroelectric and highly piezoelectric; lithium tantalate is somewhat similar. The semiconductors cadmium sulfide and zinc oxide are moderately low in loss and show appreciable coupling.

Stress Waves in Solids

Acoustic waves propagate with low loss and high velocity (5000 m/s) in solids. Solids also have shear strength, whereas sound waves in gases and fluids are simple pressure waves, manifested as traveling disturbances measurable by pressure and longitudinal motion.

Sound waves in solids involve either longitudinal or transverse particle motion. The transverse waves may be propagation of simple shear strains or may involve bending in flexural waves. Since different modes of waves travel with different velocities, and since both reflections and mode changes can occur at material discontinuities, the general pattern of wave propagation in a bounded solid medium is quite complicated.

Bulk waves are longitudinal or transverse waves traveling through solids essentially without boundaries; e.g., the wavefronts extend over many wavelengths in all directions. A solid body supporting bulk waves undergoes motion and stress throughout its volume. A *surface wave* follows a smooth boundary plane, with elliptical particle motion which is greatest at the surface and drops off so rapidly with depth that almost all the energy is carried in a one-wavelength layer at the surface.

Ideally, the wave medium for surface-wave propagation is regarded as infinitely deep; practically if it is many wavelengths deep, its properties are equivalent. A surface wave following a surface free from forces is a Rayleigh wave; if an upper material with different elastic properties bounds the surface, the motion may be a Stonely wave.

Bulk-Wave Devices

Most acoustic amplifiers utilizing bulk waves have the components shown in Fig. 11.3.35. The input and output transducers are piezoelectric crystals or deposited thin layers of piezoelectric material, used for energy conversion at the terminals. The amplifier crystal is generally CdS, which is not only piezoelectric but also an *n*-type semiconductor. Electrodes are attached at the input and output surfaces of the CdS crystal, for bias current. Since the mobility of negative-charge carriers in CdS is only about 250 cm²/V·s, large bias fields are required to provide a drift velocity equal to acoustic velocity: in CdS, 4500 m/s for longitudinal waves and 1800 m/s for shear waves. The buffer rods shown in Fig. 11.3.35 are therefore added for electrical isolation of the bias supply. Furthermore, the transducers and amplifying crystal are cut in the desired orientation, to couple to either longitudinal or shear waves.

In the analysis of bulk-wave amplification the crystal properties are characterized by f_D = diffusion frequency = $\omega_D/2\pi$; f_C = dielectric relaxation, or conductivity frequency = $\omega_C/2\pi$; $\gamma = (1 - f)$ times the drift velocity divided by the acoustic phase velocity, where f is the fraction of space charge removed from the conduction band by trapping; k = coefficient of electromechanical coupling; and ω = radian frequency of maximum gain = $\sqrt{\omega_c \omega_d}$. Figure 11.3.36 shows the computed gain curves, using these constants, for four values of drift velocity. The frequency of maximum gain can be selected by control of crystal conductivity, which is

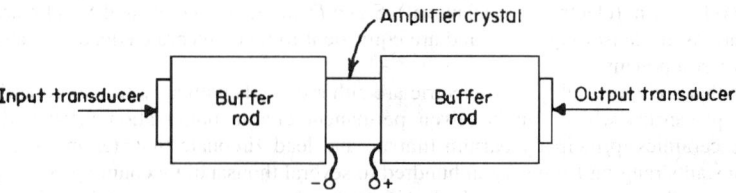

FIGURE 11.3.35 Typical structure for acoustic amplification measurements.

easily accomplished in CdS by application of light. The amount of gain, and to some extent the bandwidth, are controlled by the bias field.

Figure 11.3.37 shows characteristics specifically calculated for CdS with shear acoustic-wave amplification. Large bias voltages (800 to 1300 V/cm) are required, and extremely high gains are possible. Generally, heat dissipation owing to the power supplied by bias is so high (up to 13 W/cm^3 at $\gamma = 0.76$) that only pulsed operation is feasible.

Surface-Wave Devices

The disadvantages in bulk-wave amplifiers, evidenced as excessive bias requirements, could be alleviated if materials with high mobility were available. However, the amplifier crystal must show low acoustic loss, high piezoelectric coupling, and high mobility; a suitable material combining these properties has not been found. Figure 11.3.38 shows an acoustic amplifier structure which operates with acoustic surface waves and provides charge carriers in a thin film of silicon placed adjacent to the insulating piezoelectric crystal. Coupling for the interaction takes place in the electric field across the very small gap between piezoelectric and semiconductor surfaces. Transducers are formed by metallic fingers, interlaced to provide signal field for piezoelectric coupling between adjacent fingers. This interdigital-array technique is flexible, providing means for complicated transducer patterns useful in signal processing.

The composite structure makes it possible to combine separate materials, one chosen for strong piezoelectric coupling and the other chosen for high mobility, to cooperatively produce interactive amplification, or attenuation, of acoustic waves propagating in the piezoelectric component. The variety of material combinations and structural modifications possible has given rise to extensive analysis and experimental development in many countries. Lithium niobate crystals have the strongest piezoelectric coupling

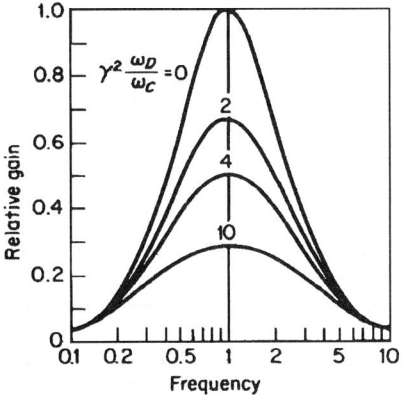

FIGURE 11.3.36 Normalized plots of gain vs. frequency. Note symmetry of frequency response and effects of bias and critical frequency values.

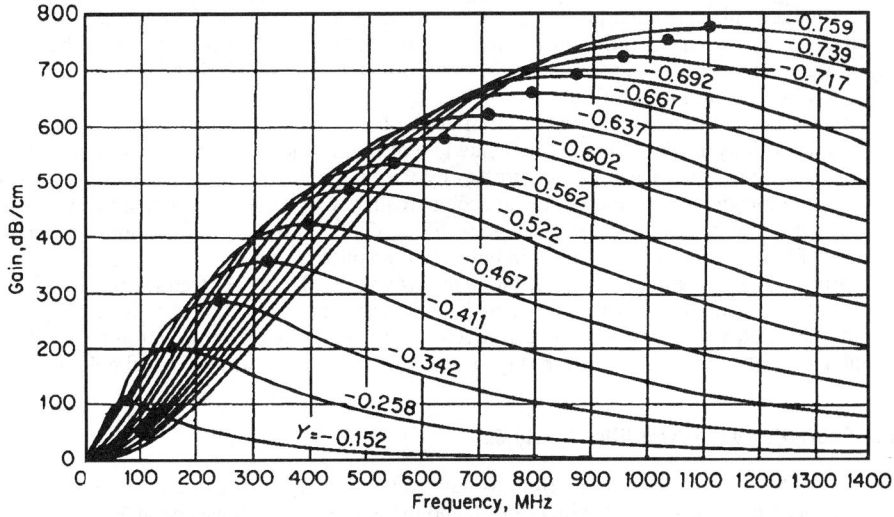

FIGURE 11.3.37 Gain vs. frequency for CdS shear-wave amplification at $f_D = 796$ MHz.

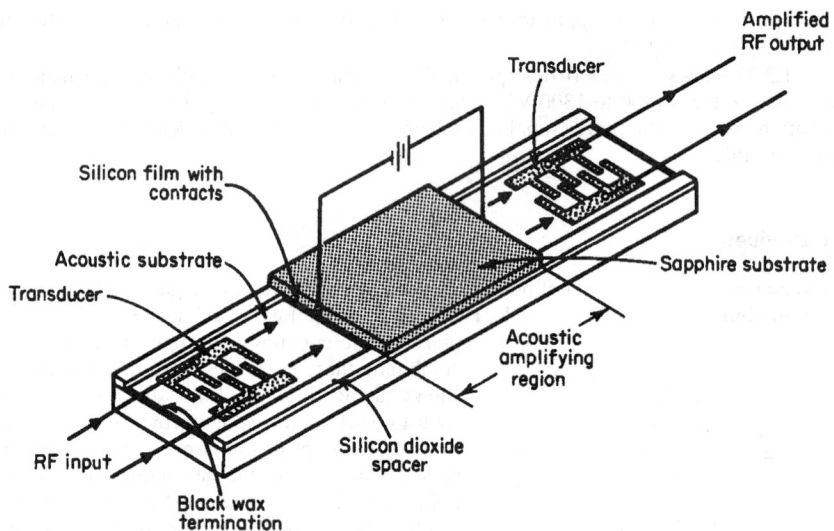

FIGURE 11.3.38 Structure of a surface-wave acoustic amplifier. (See Lakin and Shaw)

of materials acoustically suitable for high-frequency operation, and indium antimonide semiconductor films have the highest drift mobility. Since high-frequency performance requires extremely accurate control of narrow air gaps, amplifier structures employing surface waves that propagate along the interface between a piezoelectric crystal and a deposited semiconductor film have been developed.

Surface-acoustic-wave (SAW) devices, with interdigital transducers designed for wide-band or filter frequency-response characteristics, have values of signal insertion loss owing primarily to acoustic attenuation and to the conversion loss of the input and output transducers. One role of acoustic amplification can be to offset that basic insertion loss. This is illustrated in Fig. 11.3.39, which shows the gain characteristic for a Rayleigh surface-wave amplifier employing lithium niobate, separated by thin dielectric rails from a silicon-on-sapphire semiconductor film. Continuous operation with cw signals is facilitated by the surface-wave geometry, specifically by high-mobility semiconductor films backed by dielectric substrate heat sinks. Amplified signal power, carried by bunched charge carriers, flows in the semiconductor films, so the level of signal saturation is a function of film thickness, compounded by reduction of drift mobility at high temperature. A saturation level of 15 dBm, with cw signals at 340 MHz, has been observed using a 550-Å indium antimonide film 25 μm wide.

Typically, gain in a surface-wave amplifier can be as much as 100 dB in a crystal less than 1 cm long, operating with as much as 50 percent bandwidth at frequencies as high as several hundred megahertz. The amount and direction of gain are controlled by the bias field, in some cases in conjunction with incident optical illumination. The bandwidth and center frequency of the transducers are determined by the geometry of the interdigital arrays. Semiconductor films of silicon or indium antimonide operate at much lower values of bias field than are required in bulk-wave amplifiers and can be deposited on a dielectric substrate in narrow strips to permit parallel excitation at low voltage. The upper frequency limits are set by resolution of the photolithographic processes used to form the transducer fingers and by the ability to provide air gaps, in the Rayleigh wave configurations, of much less than a wavelength. Electron-beam lithography of interdigital arrays has been developed for high resolution, extending transducer technology to more than 3 GHz.

Surface-Acoustic-Wave-Amplifier Analysis

When the SAW amplifier evolved from its bulk wave counterpart, the amplification medium was piezoelectric semiconductor CdS [see White (1967)], and the gain expression was obtained by solving the field equations. Ingebrigtsen (1970 and Lakin and Shaw (1969) were the first ones to investigate the gain of the

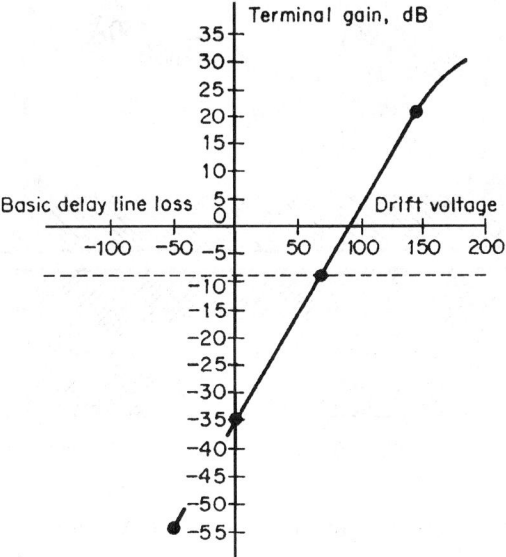

FIGURE 11.3.39 Terminal gain of an air-gap Rayleigh wave amplifier. The semiconductor, Si, was segmented to reduce the required drift voltage. [*After Lakin and Shaw (1969)*]

separate-medium SAW amplifiers by using perturbation techniques. Finally, a general form, similar to the expressions derived by White (1962) for bulk wave amplifiers, is obtained by Kino and Reeder (1971) using normal mode theory. Hunsinger and Datta (1986) introduced the concept of capacitance perturbation into the transmission line model of SAW propagation. This technique has extensively simplified the analytical process. We have further utilized the transmission line model to derive a SAW amplifier gain expression, in which the effects of the passivation and the encapsulation layers are included [see Lin (1995)]. Because of the space limitation, we will present the simplified one-dimensional analysis; however, the result of the two-dimensional analysis for the SAW amplifier will be given.

If we define the surface potential ϕ of a SAW as the voltage V and select the current I to satisfy the power P of the SAW such that $P = VI^*$, then the familiar relationships for a transmission line can be used to describe SAW propagation:

$$V_a = \frac{1}{\sqrt{LC}} \qquad C = \frac{1}{Z_a V_a} \tag{11}$$

where L and C are the distributed inductance and capacitance, both defined by the Mason model, and $V_a = \omega/k$ is the acoustic wave velocity. Z_a, the characteristic impedance of the propagating SAW, is defined as

$$Z_a = \frac{\phi_a \phi_a^*}{2P} \tag{12}$$

where ϕ_a is the electric potential generated by the SAW and P is the total power of that wave. If a layer of charge is placed above the surface of the SAW propagation path, separated by h (see Figs. 11.3.40 and

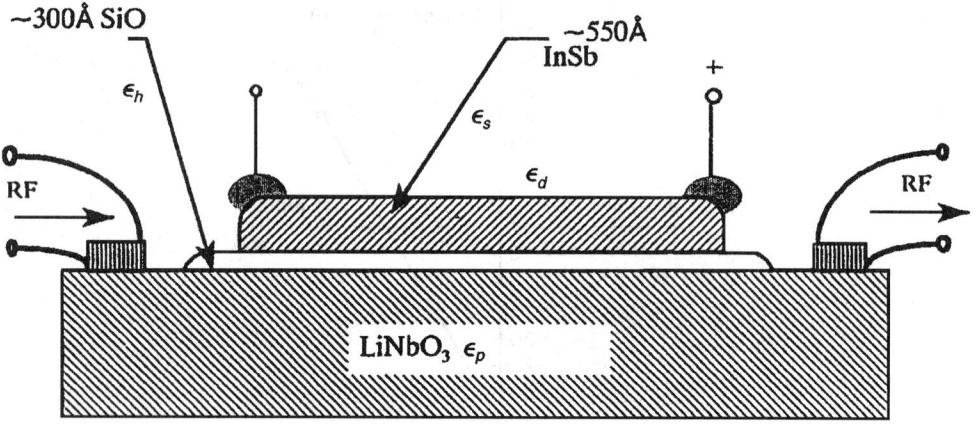

FIGURE 11.3.40 Monolithic acoustic surface-wave amplifier. [*After Coldren and Kino (1971)*]

11.3.41), it is equivalent to have an additional capacitance ΔC on the transmission line, which can be defined as

$$\Delta C = -\frac{\rho_s W}{\phi_{ah}}$$

(13)

where ρ_s = induced surface charge
W = beam width
ϕ_{ah} = acoustic potential value at $y = h$, where the charge is located

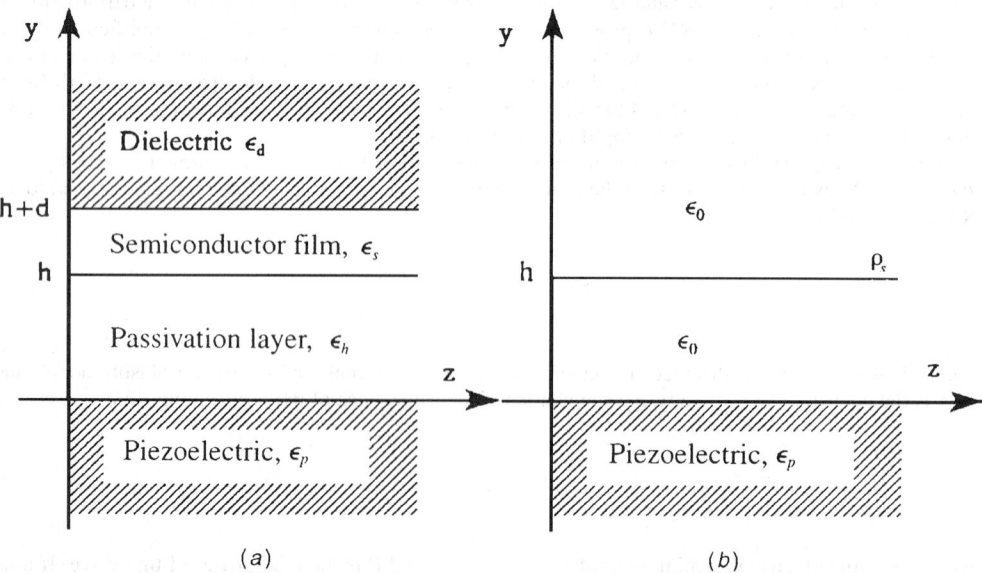

FIGURE 11.3.41 Analytical model: (*a*) two-dimensional semiconductor layer, and (*b*) simplified one-dimensional.

Because of ΔC, V_a will change accordingly (assume that the SAW propagates as $e^{j(\omega t - kz)}$):

$$\frac{\Delta V_a}{V_a} = -\frac{\Delta k}{k} = -\frac{1}{2}\frac{\Delta C}{C} \tag{14}$$

The imaginary part of Δk ($\Delta k = \beta + j\alpha$) is the gain of the SAW amplifier, which can be found by combining Eqs. (11), (12), and (14) is

$$\alpha = k\,\mathrm{Im}\left(\frac{\Delta k}{k}\right) = \frac{k}{2}\,\mathrm{Im}\left(\frac{\Delta C}{C}\right) = \frac{k}{2}\,\mathrm{Im}(\Delta C Z_a V_a) \tag{15}$$

In the following, ΔC and Z_a are to be derived. For a very thin semiconductor layer, a good approximation is to consider the charge layer infinitesimally thin. Thus, one-dimensional analysis, such as the model in Fig. 11.3.41b, can be employed.

In order to express ρ_s in terms of ϕ_{ah}, we start from the current equation and the continuity equation for the charge layer of the semiconductor:

$$\frac{\partial J_z}{\partial z} + \frac{\partial \rho}{\partial t} = 0 \qquad J_z = \sigma E_z - D_n \frac{\partial \rho}{\partial z} \tag{16}$$

where J_z = current density
 $E_z = -\partial(\phi_a + \phi_c)/\partial z = jk(\phi_a + \phi_c)$ the electrical field
 σ = conductivity of the semiconductor
 D_n = diffusion coefficient of the electron carriers
 ρ = bulk charge density
 ϕ_c = potential induced by the charge layer. Then we have (at $y = h$)

$$\rho = \frac{-\phi_{ah}}{\dfrac{\phi_{ch}}{\rho} + \dfrac{D_n}{\sigma} + j\dfrac{V_a}{\sigma k}} \tag{17}$$

For a very thin semiconductor layer, ρ_s can be approximated with ρd (d is the thickness of the layer). The above equation leads to

$$\rho_s = \frac{-\phi_{ah}}{\dfrac{\phi_{ch}}{\rho_s} + \dfrac{1}{\sigma d}\left(D_n + j\dfrac{V_a}{k}\right)} \tag{18}$$

First, we need to find ϕ_{ch}/ρ_s. To simplify the analysis, we assume that the thin semiconductor layer is right on the piezoelectric substrate, i.e., $h \to 0$, and there is no encapsulation layer, $\epsilon_d \to \epsilon_0$. Then

$$\phi_{ch} = \begin{cases} Ae^{-ky}, & y \geq 0 \\ De^{ky}, & y < 0 \end{cases} \tag{19}$$

where the decay factor is k because the potential must satisfy the Laplace equation. With the following boundary conditions:

$$\phi_{ch}(y = 0^+) = \phi_{ch}(y = 0^-)$$

$$-\epsilon_0 \frac{\partial \phi_{ch}}{\partial y}(y = 0^+) + \epsilon_p \frac{\partial \phi_{ch}}{\partial y}(y = 0^-) = \rho_s \tag{20}$$

we obtain

$$\frac{\phi_{ch}}{\rho_s} = \frac{1}{k(\epsilon_0 + \epsilon_p)} \tag{21}$$

Therefore,

$$\Delta C = -\frac{\rho_s W}{\phi_{ah}} = \frac{k(\epsilon_0 + \epsilon_p)W}{1 + \frac{(\epsilon_0 + \epsilon_p)}{\sigma d}(kD_n + jV_a)} \tag{22}$$

Now we proceed to find Z_a in terms of the electromechanical coupling coefficient K^2. By definition,

$$\frac{K^2}{2} = \frac{\Delta k}{k}\bigg|_{\substack{\sigma=\infty \\ h=0}} = \frac{1}{2}\frac{\Delta C}{C}\bigg|_{\substack{\sigma=\infty \\ h=0}} = \frac{1}{2}Z_a V_a \Delta C\bigg|_{\substack{\sigma=\infty \\ h=0}}$$

$$= \frac{1}{2}k(\epsilon_0 + \epsilon_p)WZ_a V_a = \frac{1}{2}(\epsilon_0 + \epsilon_p)W\omega Z_a \tag{23}$$

Therefore,

$$Z_a = \frac{K^2}{(\epsilon_0 + \epsilon_p)W\omega} \tag{24}$$

Substituting Z_a and ΔC into Eq. (15)

$$\alpha = -\frac{K^2}{2}\frac{\frac{k\sigma d V_a}{\epsilon_0 + \epsilon_p}}{\left(\frac{k\sigma d}{\epsilon_0 + \epsilon_p} + kD_n\right)^2 + V_a^2} \tag{25}$$

Taking the applied electrical field E_0 into account, the current density in Eq. (16) and the charge density in Eq. (17) become

$$J_z = \sigma E_z + \rho\mu E_0 - D_n\frac{\partial\rho}{\partial z} \qquad \rho = \frac{-\phi_{ah}}{\frac{\phi_{ch}}{\rho} + \frac{D_n}{\sigma} + j\frac{V_a - \mu E_0}{\sigma k}} \tag{26}$$

where μE_0 gives the carrier drift velocity. It is clear that the gain is analogous to Eq. (25) except that the term V_a should be replaced with $V_a - \mu E_0$,

$$\alpha_{SAW} = -\frac{K^2}{2}\frac{\frac{\sigma(dk)}{\epsilon_0 + \epsilon_p}(V_a - \mu E_0)}{(V_a - \mu E_0)^2 + \left[\frac{\sigma(dk)}{(\epsilon_0 + \epsilon_p)k} + kD_n\right]^2} \tag{27}$$

The gain for bulk acoustic wave amplifier can be expressed as

$$\alpha_{BAW} = -\frac{K^2}{2}\frac{\dfrac{\sigma}{\epsilon_p}(V_a - \mu E_0)}{(V_a - \mu E_0)^2 + \left(\dfrac{\sigma}{\epsilon_p k} + k D_n\right)^2} \tag{28}$$

Obviously, σ in the bulk wave amplifier has been replaced by $\sigma(kd)$ in the SAW amplifier because of the thin-film form semiconductor material.

In order to explain the approximate experimental linear relationship between gain and drift field illustrated by Fig. 11.3.39, the gain expression [Eq. (27)] can be further simplified. For InSb/LiNbO$_3$ SAW amplifier, σd is typically about 45×10^{-6} S/□, and $\epsilon_p = 50\epsilon_0 \approx 4.5 \times 10^{-12}$ F/cm, so that $\sigma d(\epsilon_0 + \epsilon_p) \gg k D_n$ and $V_0 - \mu E_0$. Thus, by dropping the negligible terms, we have

$$\alpha \approx \frac{K^2}{2}(\mu E_0 - V_a)\frac{\epsilon_p k}{\sigma d} \tag{29}$$

This is essentially the electronic gain equation given by Kotelyanski et al. (1978).

In the above derivation, we have assumed that $kd \ll 1$ and d is much smaller than the semiconductor's Debye length, which can be written as $(D_n \epsilon_s/\sigma)^{1/2} = 1/\zeta$. The assumptions imply that the potential and charge distribution inside the semiconductor is homogeneous along the y axis. To remove those restrictions and cover more general cases, the variation in y direction must be considered, i.e., analytical model Fig. 11.3.41a must be used.

A thicker semiconductor layer affects the SAW amplifier gain in two ways. First, the decay of acoustic potential in the semiconductor from the $y = h$ surface will become significant. The induced charge and the associated potential will decay accordingly, although at different rates. Therefore a field owing to the charge gradient in the y-direction will appear. The boundary condition now must be matched by the total potential from each side.

The second factor is the charge per unit area, which is needed in order to find ΔC. ρ_s is no longer ρd. It has to be calculated by the integration $\int \rho dy$.

The modification [see Lin (1995)] leads to

$$\alpha = -\frac{K^2}{2}\frac{\dfrac{\sigma \tanh(kd)(V_a - \mu E_0)}{\epsilon_h + \epsilon_p}\dfrac{e^{-2kh}}{F^2}}{\left[\dfrac{\sigma \tanh(kd)}{k(\epsilon_h + \epsilon_p)} + \zeta D_n \dfrac{\tanh(kd)}{\tanh(\zeta d)}\right]^2 + (V_a - \mu E_0)^2} \tag{30}$$

where

$$F = \frac{1}{2}\left[\left(\frac{\epsilon_d}{\epsilon_h} - 1\right)\frac{\epsilon_h - \epsilon_p}{\epsilon_h + \epsilon_p}e^{-2kh} + \left(\frac{\epsilon_d}{\epsilon_h} + 1\right)\right] \tag{31}$$

If $\epsilon_d = \epsilon_h$, $F = 1$.

Equation (55) is derived in Wang and Lin (1996). Equation (30) reduces to Eq. (27), when the semiconductor thickness d is extremely thin such that both $kd \ll 1$ and $\zeta d \ll 1$.

MAGNETIC AMPLIFIERS

Harold W. Lord

Static Magnetic Amplifiers

Static magnetic amplifiers can be divided into two classes, identified by the terms saturable reactor and self-saturating magnetic amplifier. A *saturable reactor* is an adjustable inductor in which the current-versus-voltage relationship is adjusted by control magnetomotive forces applied to the core. A *magnetic amplifier* is a device using saturable reactors either alone or in combination with other circuit elements to secure amplification or control. A *simple magnetic amplifier* is a magnetic amplifier consisting only of saturable reactors. The abbreviation SR is used in this section to denote a *saturable reactor* and/or simple magnetic amplifier.

A self-saturating magnetic amplifier is a magnetic amplifier in which half-wave rectifying circuit elements are connected in series with the output windings of saturable reactors. It has been shown that saturable reactors can be considered to have negative feedback. Half-wave rectifiers in series with the load windings will block this intrinsic feedback. A self-saturating magnetic amplifier is therefore a parallel-connected saturable reactor with blocked intrinsic feedback. This latter term avoids the term self-saturation, which, although extensively used, does not have a sound physical basis. The abbreviation MA is used here to denote this type of high-performance magnetic amplifier.

Saturable-Reactor (SR) Amplifiers

The SR can be considered to have a very high impedance throughout one part (the *exciting interval*) of the half cycle of alternating supply voltage and to abruptly change to a low impedance throughout the remainder of the half cycle (the *saturation interval*). The phase angle at which the impedance changes is controlled by a direct current. This is the type of operation obtained when the core material of the SR has a highly rectangular hysteresis loop. Two types of operation, representing limiting cases, are discussed here, namely, *free even-harmonic currents* and *suppressed even-harmonic currents*. Intermediate cases are very complex, but use of one or the other of the two extremes is sufficiently accurate for most practical applications.

The present treatment is limited to resistive loads, the most usual type for SR applications. [See Storm (1955), for inductive dc loads and Wilson (1952), for inductive ac loads]. The basic principles of operation of SRs and MAs are given in more detail in Shea (1968).

Series-Connected SR Amplifiers

Basically, an SR circuit consists of the equivalent of two identical single-phase transformers. Figure 11.3.42 shows two transformers, SR_A and SR_B, interconnected to form a rudimentary series-connected SR circuit.

The two series-connected SR windings in series with the load are called *gate windings*, and the other two series-connected windings are called *control windings*. Note, from the dots that indicate relative polarity, that the gate windings are connected in series additive and the control windings are connected in series subtractive. By reason of these connections, the fundamental power frequency and all *odd* harmonics thereof will not appear across the total of the two control windings but any *even* harmonic induced in one control winding will be additive with respect to a corresponding even harmonic induced in the other control winding.

For this connection, SR_A and SR_B are normally so designed that each gate winding will accommodate one-half the alternating voltage of the supply without producing a peak flux density in the core that exceeds the knee of the magnetization curve, assuming no direct current is flowing in the control winding. Under these conditions, if SR_A is identical with SR_B, one-half of the supply voltage appears across each gate winding, and the net voltage induced in the control circuit is zero. The two SRs operate as transformers over the entire portion of each half cycle.

When a direct current is supplied to the control circuit, each SR will have a saturation interval during a part of each cycle, SR_A during half cycles of one polarity and SR_B during half cycles of the opposite polarity. The ratio of the saturation interval to the exciting interval can be controlled by varying the direct current in the control winding.

When an SR core has a saturation interval during part of half cycles of one polarity, and there is a load or other impedance in series with the gate winding, even-harmonic voltages are induced in all other windings on that core. In the series connection, one SR gate winding can be the

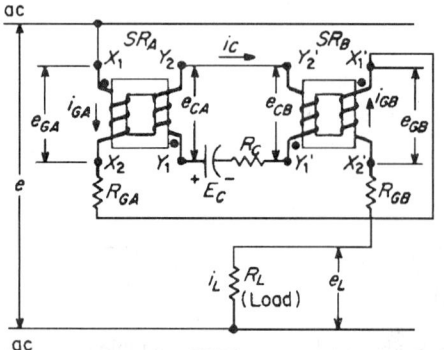

FIGURE 11.3.42 Series-connected SR amplifier.

series impedance for the SR, and large even-harmonic voltages will appear across the individual gate windings and control windings when direct current flows in the control windings. The amount of even-harmonic current which flows in the control circuit of Fig. 11.3.42 will depend on the impedance of the control circuit to the harmonic voltages. If the control-circuit impedance between terminals Y_1 and Y_1' is low with respect to the induced harmonic voltages of the control windings (usually referred to as a *relatively low* control-circuit impedance), the harmonic currents can flow freely in this circuit, and the SR circuit is identified by the term *free even-harmonic currents*. If the control-circuit impedance is high with respect to the induced harmonic voltages (a *relatively high* control-circuit impedance), the harmonic-current flow is suppressed and the SR circuit is identified by the term *suppressed even-harmonic currents*.

If R_C in Fig. 11.3.42 is relatively low and the source impedance of the dc control current is low, the circuit is of the free even-harmonic-currents type and the two SRs are tightly coupled together by the control circuit. If one SR is in its saturation interval, it will reflect a low impedance to the gate winding of the other SR, even though it is then operating in its exciting interval. Thus, during the saturation interval of any core, transformer action causes both gate windings to have low impedances and current can flow from the ac source to the load, with correspondingly high harmonic circulating currents in the control circuit. When both cores are in their exciting intervals, only the low core-exciting current can flow and the current circulating in the control circuit is substantially zero.

If the exciting current is so low as to be negligible, and if N_G are the turns in each gate winding, N_C are the turns in each control winding, I_G is the rectified average of the load current, and I_C is the average value of the control current, then applying the law of equal ampere-turns for transformers provides the following expression for the circuit in Fig. 11.3.42: $I_C N_C = I_G N_G$. This is the law of equal ampere-turns for the series-connected SR with resistive load. It applies to operation in the so-called *proportional region*, the upper limit of this region being that point in the control characteristic where the load current is limited solely by the load circuit resistance. Figure 11.3.43 shows the gate-current circuit and control-circuit current at one operating point in the proportional region. An increase in control current causes a decrease in the angle α, thereby increasing I_L, the rectified average of the current to the load.

Parallel-Connected SR Amplifiers

Figure 11.3.44 shows the circuit diagram for the parallel-connected SR. Each gate winding is connected directly between the ac supply and the load resistance, thus providing two parallel paths through the SR. There is therefore a low-impedance path for the free flow of even-harmonic currents, even though the impedance of the control circuit happens to be relatively high. As a result, the parallel-connected SR is always of the *free flow of even-harmonic-currents* type, the cores operate in the same manner as described in the previous paragraph, and the waveshapes of currents to the *load* are as shown in Fig. 11.3.43. However, the gate winding currents are different, as shown in Fig. 11.3.45.

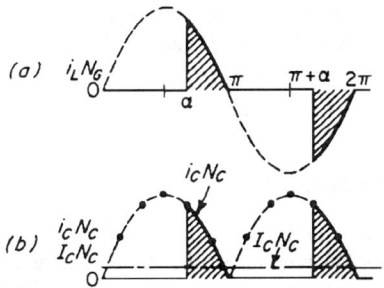

FIGURE 11.3.43 Currents in circuit of Fig. 11.3.42, for free even-harmonic-current conditions: (*a*) gate current; (*b*) control-circuit current.

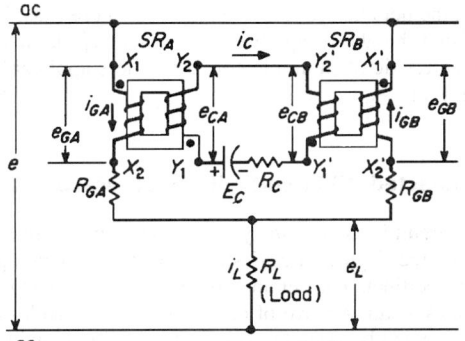

FIGURE 11.3.44 Parallel-connected SR amplifier.

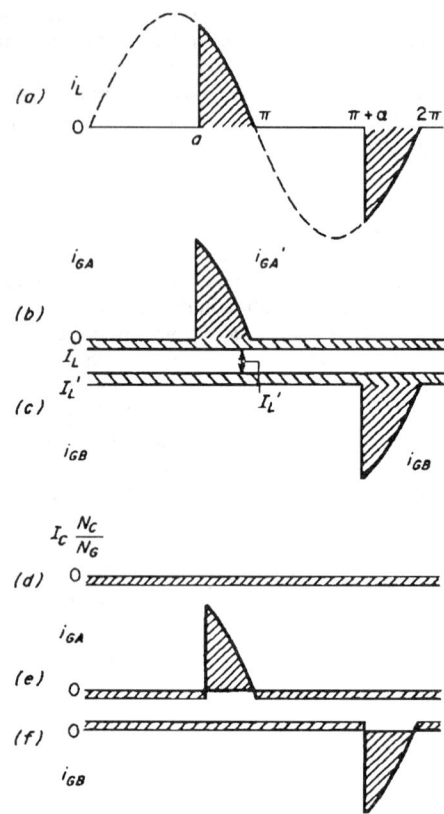

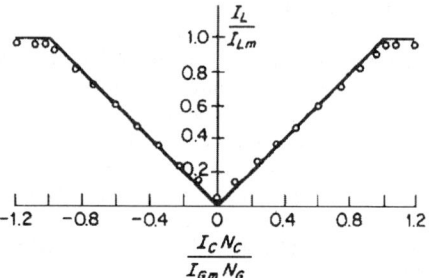

FIGURE 11.3.46 Control characteristic of SR amplifier applying both to free and suppressed even-harmonic conditions.

FIGURE 11.3.45 Currents in the circuit of Fig. 11.3.44 for free even-harmonic-current conditions: (*a*) load current; (*b*) components of load current in SR$_A$ gate winding; (*c*) same in SR$_B$; (*d*) control-circuit current; (*e*) gate-current in SR$_A$; (*f*) same in SR$_B$.

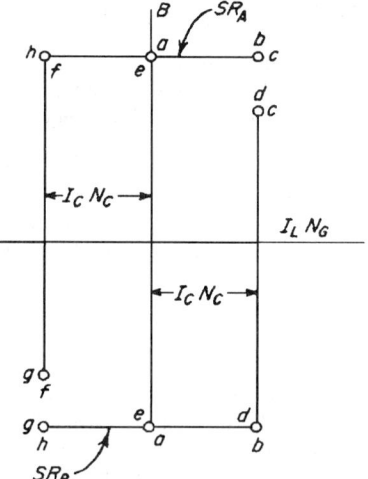

FIGURE 11.3.47 Idealized operating hysteresis loops for suppressed even-harmonic-current condition.

It is obvious from Fig. 11.3.45 that $I_L' = -I_L/2$. Since, except for the low-excitation requirements for the core, the net ampere-turns acting on the core must be zero during the exciting interval, then $I_L' N_G + I_C N_C = 0$. Using the above equation to eliminate I_L', the result is $I_C N_C = I_L N_G/2$. Figure 11.3.46 shows the control characteristic of saturable core amplifiers applicable both to free and suppressed even-harmonic modes of operation.

Series-Connected SR Circuit with Suppressed Even Harmonics

This circuit is usually analyzed by assuming operation into a short circuit and the control *current* from a current source. Figure 11.3.47 shows idealized operating minor hysteresis loops for this circuit, and Fig. 11.3.48 shows pertinent current and voltage waveshapes lettered to correspond to Fig. 11.3.47. Note that this current supplies a square wave of current to the load so long as it is operating in the proportional region. Energy is interchanged between the power-supply circuit and the control circuit so as to accomplish this type of operation. If a large inductor is used to maintain substantially ripple-free current in the control circuit, the energy interchanged between the supply voltage and control circuit is alternately stored in, and given up by, the control-circuit inductor.

Gain and Speed of Response of SRs

It is obvious from the generalized control characteristic that the current gain is directly proportional to the ratio of the control winding turns to the gate winding turns. Thus $I_L/I_C = N_C/N_G$. If R_C is the control-circuit resistance and R_L is the load resistance, the power gain GP is $N_C^2 R_L^2/N_G^2 R_C^2$.

The response time of saturable reactors is the combination of the time constants of the control circuit and of the gate-winding circuit. Expressed in terms of the number of cycles of the supply frequency, the time constant of the control winding of a series-connected SR is $\tau_C = R_0 N_C^2/4R_C N_C^2$ cycles. If the SR is operating under-excited, a transportation lag may cause an additional delay in response.

High-Performance Magnetic Amplifier (MA)

If a rectifier is placed in series with each gate winding of a parallel-connected SR and the rectifiers are poled to provide an ac output, it becomes a type of high-gain MA circuit called the *doubler circuit*. When this is done, the law of equal ampere-turns no longer applies, and the transfer characteristic is mainly determined by the magnetic properties of the SR cores. The design of an MA therefore requires more core-materials data than are required for SRs in simple magnetic amplifier circuits. The text of this section assumes that the designer has the required magnetic-core-materials data on hand, since such data are readily available from the manufacturers of cores for use in high-performance MAs.

Descriptions of the operation of MAs require several terms which are defined here for reference.

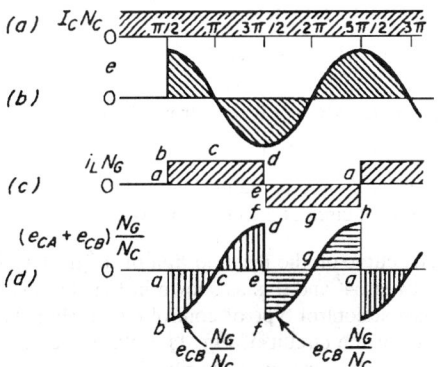

FIGURE 11.3.48   Series-connected suppressed even-harmonic condition: (*a*) control-circuit current; (*b*) applied input voltage; (*c*) gate winding and load current; (*d*) voltage across series-connected control coils.

Firing.   In a magnetic amplifier, the transition from the unsaturated to the saturated state of the saturable reactor during the conducting or gating alternation. Firing is also used as an adjective modifying phase or time to designate when the firing occurs.

Gate angle (firing angle).   The angle at which the gate impedance changes from a high to a low value.

Gating.   The function or operation of a saturable reactor or magnetic amplifier that causes it, during the first portion of the conducting alternation of the ac supply voltage, to block substantially all the supply voltage from the load and during a later portion allows substantially all the supply voltage to appear across the load. The "gate" is said to be virtually closed before firing and substantially open after firing.

Reset, degree of.   The reset flux level expressed as a percentage or fraction of the reset flux level required to just prevent firing of the reactor in the subsequent gating alternation under given conditions.

Reset flux level.   The difference in saturable-reactor core flux level between the saturation level and the level attained at the end of the resetting alternation.

Resetting (presetting).   The action of changing saturable-reactor core flux level to a controlled ultimate reset level, which determines the gating action of the reactor during the subsequent gating alternation. The terms resetting and presetting are synonymous in common usage.

Resetting half cycle.   The half cycle of the MA ac supply voltage at which resetting of the saturable reactor may take place.

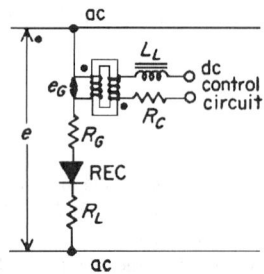
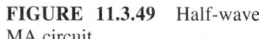

FIGURE 11.3.49 Half-wave MA circuit.

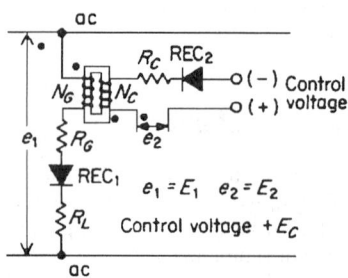

FIGURE 11.3.50 Voltage-controlled MA circuit.

Half-Wave MA Circuits

Figure 11.3.49 shows a half-wave MA circuit with control from a source of controllable direct *current*. Figure 11.3.50 shows a half-wave MA circuit with control from a controllable source of direct *voltage*. The following sequence of operation is the same for both circuits and is described in connection with Fig. 11.3.51.

1. During the gating half cycle (diode REC or REC_1 conducting) the core flux increases toward a saturation level (3′) from some reset flux level (0) and the current to the load R_L is very low.

2. When saturation of the SR occurs at 2′, firing occurs and current flows to the load for the rest of the gating half cycle, leaving the core flux at 4′.

3. During the resetting half cycle (diode REC or REC_1 blocking) the SR core is reset from 4′ through 5′ to a value of reset flux level corresponding to B′ and 0.

The waveshape of the current to the load is the same for both circuits, being a portion of one polarity of sine wave (phase-controlled half-wave). The two types of control circuits differ as follows:

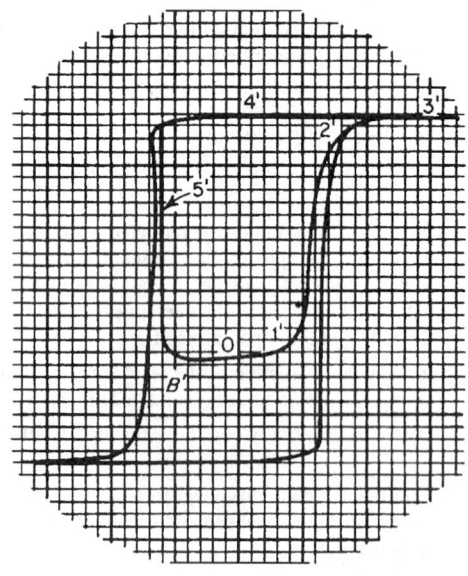

FIGURE 11.3.51 Major and operating minor dynamic hysteresis loops for circuit of Fig. 11.3.49.

1. The curve on the reset portion of the hysteresis loop between 4′ and 0 is as shown in Fig. 11.3.51 for the current control type of control circuit (Fig. 11.3.49), but for the circuit of Fig. 11.3.50, the resetting portion of the curve in the region 5′ is not a vertical line and may coincide with the outer major hysteresis loop for a substantial portion of the resetting period.

2. The output of the circuit of Fig. 11.3.49 is a maximum for zero dc control current, but for Fig. 11.3.50 the output is a minimum for zero dc control voltage.

The operation of a half-wave MA can be summarized as follows: (1) During the gating half cycle, the core acts to withhold current from the load for a portion of each half cycle, the length of the withholding period being determined by the degree of reset of core flux provided during the immediately preceding resetting half cycle. (2) During the resetting half cycle, the degree of reset can be controlled by varying the amount of direct current of proper polarity in the control winding, or by varying the amount of average voltage of proper polarity applied to the control winding. (3) The amount of current and power required to reset the core is primarily a function of the excitation requirements of the core and bears no direct relationship to the power delivered to the load.

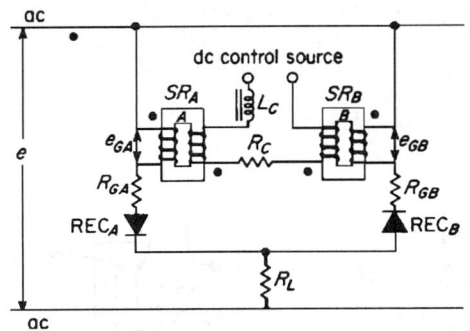

FIGURE 11.3.52 Full-wave ac-output MA circuit with high-impedance control circuit.

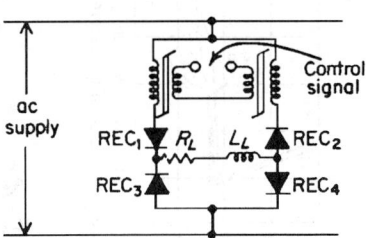

FIGURE 11.3.53 Full-wave bridge MA circuit with inductive load.

Full-Wave MA Circuits

Two half-wave MA circuits can be combined to provide either full-wave ac output (Fig. 11.3.52) or full-wave dc output (Fig. 11.3.53). During the gating half cycle, all gate windings act in the same manner as described for the half-wave MA circuit. Because of an interaction of the common load resistor voltage with the availability of inverse voltage across a rectifier during its resetting half cycle, each half-wave part of Fig. 11.3.52 cannot be assumed to operate independently of its mate. The differences in control characteristics are shown by Fig. 11.3.54, where curve A applies to the circuit of Fig. 11.3.49 and curve B applies to the circuit of Fig. 11.3.52.

Equations for MA Design

For SRs in simple MA circuits, since the ampere-turn balance rules apply, their design is so similar to transformer design that ordinary transformer design procedures are applicable. For SRs in MA circuits, the gate windings must carry rms load currents that are related to the ac and dc load currents in the same manner that rectifier transformer secondary rms currents are related to the load current. Gate windings must be able to withstand a full half cycle of ac supply voltage, without saturation, for a total flux swing from $-B\mu_m$ to $+B\mu_m$, where $B\mu_m$ is the flux density at which the permeability of the core material is a maximum. In the cgs system of units:

Maximum ac supply voltage, rms: $E_e = 4.44 N_G B\mu_m A f \times 10^{-8}$, where N_G is gate winding turns, A is core area, and f is supply frequency.

Maximum load current for firing angle $\alpha = 0°$ (rectified average): $I_{LM} = E_e/1.11(R_G + R_L)$, where R_G is total gate winding resistance, including diode drops, and R_L is the effective load resistance.

For current-controlled MA circuits the following equations apply:

Minimum load current ($\alpha = 180°$); $I_{LX} = 2I_X = 2H_{csf}l/0.4\pi N_G$, where I_X = exciting current (A) of one SR core. H_{csf} = sine flux coercive force of the core material (Oe) and l = mean length of magnetic circuit of one core.

Control current for upper end of control current: $I_C = H_C l/0.4\pi N_C$, where H_C = dc coercive force of the core material.

Control current for minimum load-current point: $I_C = H_{csf}l/0.4\pi N_C$. Ampere-turn gain $G_{AT} = (I_{LM}/2I_X)\pi$.

Time constant $\tau_C = 0.9\pi E_e R N_C^2/4 f I_X R_C N_G^2$ seconds, where R_C is the control circuit resistance and assuming $\alpha = 90°$. This equation does not take into account a transportation or any underexcited effects which increase the delay time.

Power gain $G_p = G_{AT}^2 (N_C/N_G)^2 (R_L/R_C)$, where k_f = form factor (ratio of rms to average values). If only average values are used, k_f is unity. At $\alpha = 90°$, dynamic power gain $G_D = G_p/\tau_C = \pi f I_{LM} R_L/I_X (R_G + R_L)$ per second.

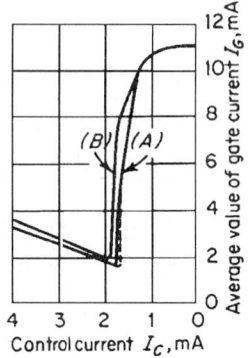

FIGURE 11.3.54 MA circuit control characteristics: (A) half-wave circuit; (B) full-wave ac-output circuit.

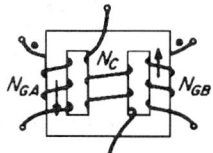

FIGURE 11.3.55 Three-legged SR core-and-coil configuration.

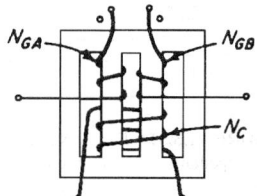

FIGURE 11.3.56 Four-legged SR core-and-coil configuration.

For voltage-controlled MA circuits, the following equations apply to Fig. 11.3.50:

AC bias voltage $E_2 > E_1 N_C/N_G$ and of indicated relative polarities. A value of $E_2 = 1.2E_1 N_C/N_G$ is usually adequate. Since the dc control voltage E_C acts to inhibit the resetting effect of E_2, a firing angle $\alpha = 0°$ occurs when $E_C \geq \sqrt{2E_2}$.

A single-stage amplifier has a half-cycle transport lag and no other time delay unless the resetting is inhibited by common-load interactions with other half-wave-amplifier elements.

Maximum average load current $I_{LM} = 0.9E_1/(R_G + R_L)$.

Minimum average load current $I_{LX} = I_X = H_{csf} l/0.4\pi N_G$.

The calculation of the power gain of voltage-controlled MA circuits largely depends on the control circuit. The dc control voltage of Fig. 11.3.50 actually must *absorb* power from the ac bias circuit during much of its control range. Assuming that $E_2 = 1.2E_1 N_C/N_G$ and a resistor R'_C is connected across the control source, which has a drop of $0.2E_2$ at cutoff, then $R'_C = 0.24E_1 N_C^2/N_{GX}^2$. The maximum dc control voltage is then $1.2\sqrt{2}\ E_1 N_C/N_G$.

When the dc control voltage, $E_C > 2E_2$, rectifier REC$_2$ is blocking all the time, and so the only load on the control source is R'_C. Therefore the maximum power from the control source $P_C = E_C^2 / R_C^1 = 5\sqrt{2E_1 I_X}$.

The maximum power output is $P_0 = I_{LM}^2 R_L (0.9E_1)^2/(R_G + R_L)^2$.

The power gain is then $G_p = P_0/P_C$ if the gain is assumed to be linear over most of the control characteristic. This can also be written $G_p = I_{LM}^2 G_p - I_{LM}^2 R_L/5\sqrt{2E_1 I_X}$.

Core Configurations

Figures 11.3.55 and 11.3.56 show two coil-and-core configurations commonly used for SRs. The best core-and-coil geometry for the SRs and MAs is the toroid-shaped core with the gate winding uniformly wound over the full 360° of the core. Full-wave operation requires two such cores. After winding a gate winding on each core of a matched pair, the two cores and coils can be stacked together coaxially and the required control coils wound over the stack. The gate coils are so connected into the load circuit that no fundamental or odd-harmonic voltage is induced in the control windings.

CHAPTER 11.4
OPERATIONAL AMPLIFIERS

Samuel M. Korzekwa, Joseph P. Hesler, Clayton R. Roberts

DIRECT-COUPLED AMPLIFIERS

Samuel M. Korzekwa

A direct-coupled amplifier has frequency response that starts at zero frequency (dc) and extends to some specified upper limit. To obtain the zero-frequency capability, such amplifiers are normally direct-coupled throughout; i.e., they do not use capacitive or transformer coupling (except for auxiliary higher-frequency compensation or signal transmission).

The primary sources of error of a direct-coupled amplifier are initial offset, drift, and gain variations, errors usually dependent on temperature, aging, and so forth. The gain-variation problem can be minimized by feedback gain-stabilization techniques, but offset and drift errors are usually not handled so effectively by feedback techniques. A bias shift or drift error cannot be distinguished from a signal response because their output responses are identical. Various techniques are available to minimize this drift problem, and several different methods may be used in an amplifier.

One method of minimizing drift is to use a balanced topology, as in differential amplifiers, whereby the drift errors tend to cancel out. A more complicated but effective method is the modulated-carrier-amplifier approach: the signal is first converted into a carrier signal, amplified using an ac-coupled amplifier, and then demodulated to a baseband signal. The function performed by the modulated-carrier amplifier is identical with that of a direct-coupled amplifier, but the signal processing technique differs.

An example of a direct-coupled transistor amplifier is illustrated in Fig. 11.4.1. The primary signal path, shown in heavy lines, directly couples the input signal to the amplifier output. This configuration also provides gain and bias stabilization via the direct-coupled R_3 feedback path. Usually the low-frequency amplifier voltage gain is primarily determined via the R_1 and R_3 components, while the function of the C_1 capacitor and R_6 resistor is to provide high-frequency compensation or stabilization. Sources of error for this topology include the initial V_{BE} offset of the Q_1 input transistor and the subsequent drift caused by the temperature dependence of this offset.

Differential Amplifiers

A differential amplifier is a dual-input amplifier that amplifies the difference between its two signal inputs. This amplifier may have an output that is single-ended (one output) or it may have a differential output.

The differential amplifier eliminates or greatly minimizes many common sources of error. The drift problem encountered in direct-coupled amplifiers can be handled more effectively by the differential approach. A second major advantage of a differential amplifier is its ability to reject common-mode signals, i.e., unwanted signals present at both of the amplifier inputs or other common points. Common-mode performance is usually a critical requirement of instrumentation amplifiers.

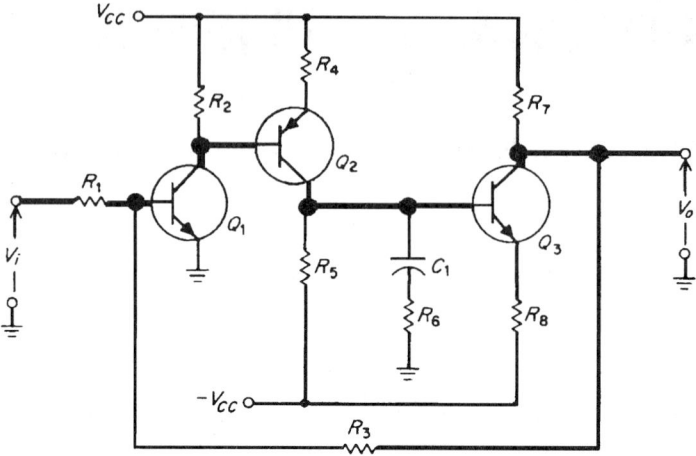

FIGURE 11.4.1 Direct-coupled transistor amplifier.

The basic circuit commonly used in differential amplifiers is shown in Fig. 11.4.2. Such differential pairs can be constructed using separate devices or in integrated-circuit form. The integrated package yields additional advantages since the parameter differences between the units of the integral differential pair are usually much less than if separate devices are used. Thus the units of such integral pairs tend to track differentially more closely, even though their individual parameters may vary in absolute value. Also, many of the passive components in the integrated amplifiers track better. Figure 11.4.3 shows a typical differential transistor amplifier. For further detailed information, including analysis procedures, design techniques, and application data, refer to the literature.

Chopper (Modulated-Carrier) Amplifiers

The chopper amplifier performs the function of a direct-coupled amplifier by using a carrier frequency and an ac-coupled amplifier. The modulated-carrier approach is used specifically to minimize drift and offset types of errors. Although this approach is often more complex to implement, the performance improvement makes this technique desirable for applications demanding low-drift performance.

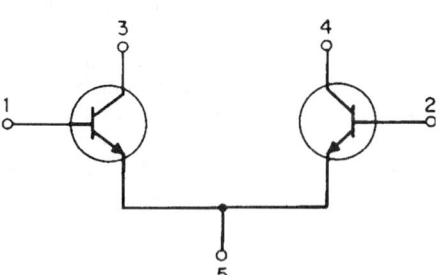

FIGURE 11.4.2 Differential amplifier pair.

Figure 11.4.4 shows the block diagram of a modulated-carrier amplifier. While the input and output signals are dc-coupled, the interstage coupling between the modulator, amplifier, and demodulator may be either a capacitor or a transformer.

Since modulator and demodulator are usually operated synchronously, the carrier delay must be maintained at acceptable levels. Low-pass filters are generally included in the modulator and demodulator. The choice of carrier frequency depends on the application. Typical examples of low-frequency carriers are 60 and 400 Hz. Carrier frequencies above 10 kHz are also feasible. Since the chopper and demodulator generate noise at their carrier frequency and its harmonics, the carrier frequency should normally be chosen above the baseband frequency range of interest. Early chopper modulators and demodulators were mechanical devices, but modern types employ solid-state electronics.

A variation of the modulated-carrier technique uses the chopper-stabilized approach illustrated in Fig. 11.4.5, which includes an additional parallel ac-coupled high-frequency signal path. The two signal paths

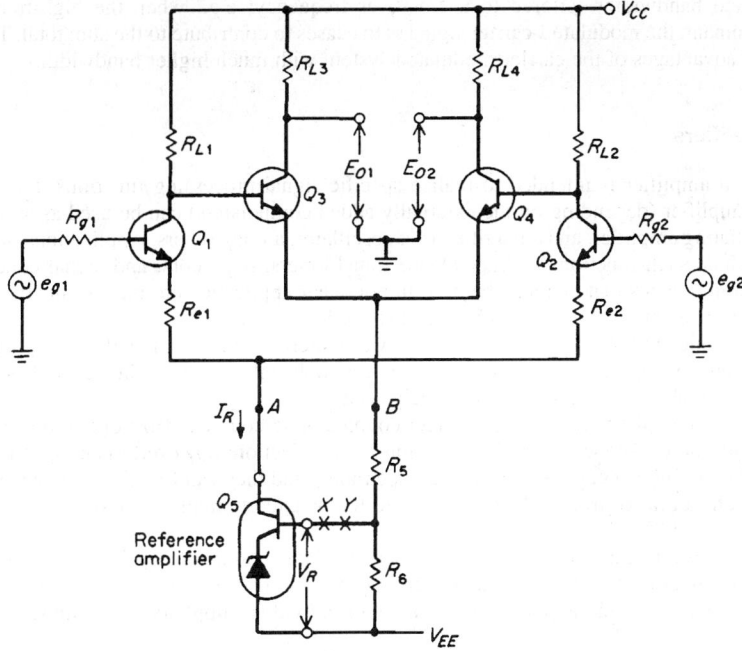

FIGURE 11.4.3 Two-stage differential amplifier with common-mode feedback.

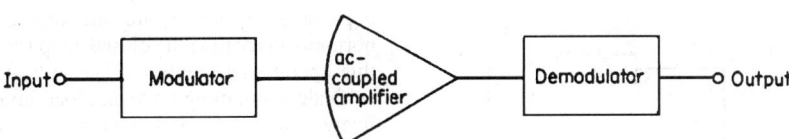

FIGURE 11.4.4 Modulated-carrier (chopper) amplifier.

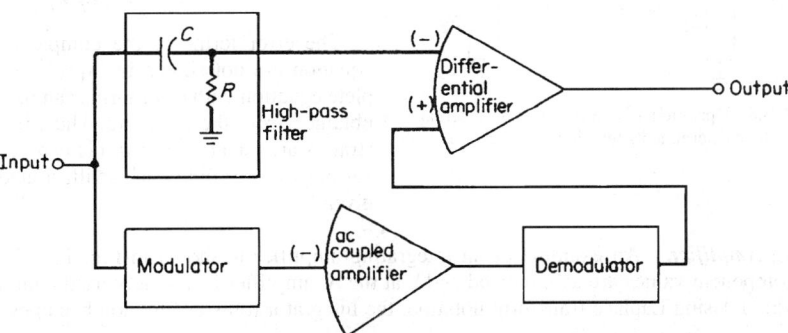

FIGURE 11.4.5 Chopper-stabilized amplifier.

have gains and bandwidths tailored to a crossover frequency; e.g., when the high-frequency signal path becomes dominant, the modulated-carrier signal path ceases to contribute to the sum total. This variation offers the low-drift advantages of the carrier-modulated system with much higher bandwidth.

Operational Amplifiers

An operational amplifier is intended to realize specific signal processing functions. For example, the same operational amplifier (depending on the externally added components) can be used as an integrating amplifier, a differentiating amplifier, an active filter, or an oscillator, among others. Applications of operational amplifiers also include such functions as impendance transformers, regulators, and signal conditioning. They are versatile building blocks that can also be used in nonlinear applications to realize functions such as logarithmic amplifiers, comparators, ideal rectifiers, and so forth.

The early application of operational amplifiers was largely in the area of analog computations. The requirements placed on the amplifiers were severe, and the cost was high. Currently, however, these high-performance functions are available at low cost and are widely used.

An operational amplifier can be either direct-coupled or ac-coupled. Most operational amplifiers have differential inputs and consequently realize common-mode rejection; however, many operational amplifiers are single-ended. Their power capability covers a wide range, and they can function as a power driver. The open-loop bandwidth can range from below 1 kHz up to the megahertz range. Voltage gain can be designed from unity to above 100,000.

To design or use the many excellent models now available commercially, it is necessary to understand the attributes and limitations of these versatile building blocks. The reader is referred to the literature and, of course, to the data sheets and applications literature of individual suppliers of op amps.

Application of Operational Amplifiers

Figure 11.4.6 shows the block diagram for an operational amplifier with associated external elements. A_1 represents the transfer function of the amplifier, either current gain or voltage gain, which is generally frequency-dependent. Z_i and Z_f are the primary elements that normally determine the closed-loop transfer function for this operational amplifier. The indicated Z_{in} and Z_l are equivalent summing node and load impedances, respectively, which are usually factored into the error terms of the composite transfer function. All the elements are general impedances that may be real or complex. Thus the simplified transfer function of this operational amplifier can be written

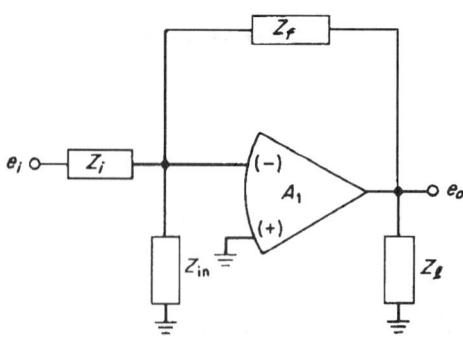

FIGURE 11.4.6   Operational amplifier A_1 with external impedances which determine its functional application.

$$e_o/e_i \approx -Z_f/Z_i \tag{1}$$

The error terms of the complete transfer function equation are not shown in Eq. (1); however, the complete equation with error terms can be readily derived or obtained from the literature. Thus, if the required constraints are adhered to, Eq. (1) can be used to generate various transfer functions, as illustrated in the examples given below.

Integrating Amplifier.   An example of an integrating amplifier is shown in Fig. 11.4.7. Assume that the external-component values are as indicated and that the A_1 amplifier is a widely used commercial integrated circuit op amp. Using Laplace transform notation, the integrator transfer function becomes

$$(e_o/e_i)S = (-)1/RCS \tag{2}$$

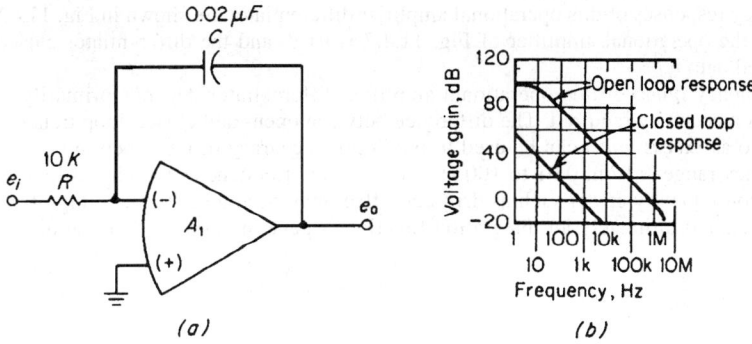

FIGURE 11.4.7 Operational amplifier used for integration: (*a*) basic schematic; (*b*) frequency response.

Using the component values specified in the figure, the integrator transfer function can then be written

$$(e_o/e_i)S = (-)5000/S \tag{3}$$

An alternative frequency-domain representation of the above integrator transfer function can also be used.

$$(e_o/e_i)f = (-)800/jf \tag{4}$$

The closed- and open-loop frequency responses of this operational integrator amplifier are illustrated in Fig. 11.4.7*b*.

The frequency range of application depends primarily on the accuracy desired. In this example, 1 Hz to 10 kHz frequency is considered a realistic range of operation. For very low frequencies the error tends to increase because of inadequate excess gain within the operational amplifier, whereas for high frequencies the closed-loop gain becomes low and drift and offset errors may then become important.

Differentiating Amplifier.   An example of a differentiating amplifier is shown in Fig. 11.4.8. The following simplified closed-loop equations are applicable:

$$(e_o/e_i)S \approx (-)RCS \tag{5}$$

$$(e_o/e_i)S \approx (-)0.165S \tag{6}$$

$$(e_o/e_i)f \approx (-)jf \tag{7}$$

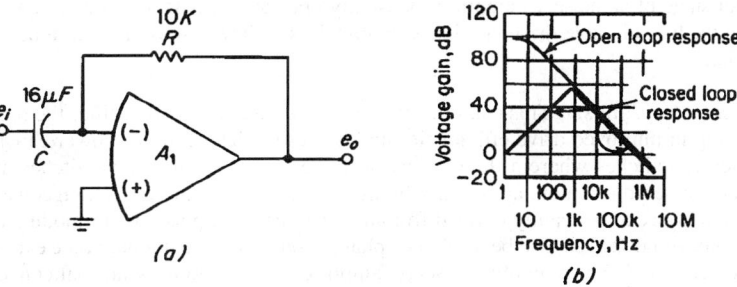

FIGURE 11.4.8 Operational amplifier used for differentiation: (*a*) basic schematic; (*b*) frequency response.

The frequency responses of this operational amplifier differentiator are shown in Fig. 11.4.8*b*. The open-loop response of the operational amplifier of Fig. 11.4.7 is used, and the differentiator closed-loop response is superimposed onto it.

The frequency range of this operational amplifier differentiator depends primarily on the closed-loop differentiation accuracy required. The difference between open- and closed-loop transfer functions (usually referred to as *excess gain*) can be used to predict the accuracy of the function being generated. A realistic frequency range is from 0.01 to 100 Hz. At very low frequencies the closed-loop gain becomes very small, and consequently errors such as drift and offset may become critical. At high frequencies the accuracy degrades, and ultimately an integration function is generated rather than the differentiation function intended.

Servo Amplifiers

The function of a servo amplifier, one of the principal components in a control feedback system, is to amplify the input (usually low-level) error signals and to provide rated drive power to the load (the servo actuator). Servo amplifiers can be dc- or ac-operated, and can be linear or nonlinear.

Direct-Coupled Servo Amplifiers. A direct-coupled servo amplifier can operate on dc error signals, i.e., zero frequency signals. The bandwidth of a dc servo amplifier is often quite large, and thus it has the capability of dynamic operation. In practice it is often operated in the dynamic mode, and the transient-response characterization is often used.

A servo amplifier is intended to drive or control its load to some prescribed reference level. It is the drift and offset errors associated with this control function that are the main concern in this dc mode of operation. Thus, if the actuator is positioned at its exact reference level, the amplifier should provide zero drive power. However, because of initial offsets or subsequent temperature or aging effects, the amplifier may still provide unwanted drive to the load. The usual solution is to minimize these problems by circuit design, e.g., tracking or balanced configurations, or by use of an intermediate-carrier modulated by the dc error input signals, subsequently demodulated for use as the direct-coupled drive to the actuator.

AC Servo Amplifiers. An ac servo amplifier operates at a selected fixed frequency and is consequently a carrier-system amplifier. The most commonly used carrier frequencies are 60 and 400 Hz, but ac servo systems can be designed to operate at almost any frequency compatible with the servo actuator used.

The principal advantage of ac servo amplifiers is that the previously discussed drift and offset problems present in dc servo amplifiers are virtually eliminated. Another advantage is that the prime power from the actuator now can be supplied directly from the line, for example, 60-Hz 115-V source.

For applications wherein the servo actuator or load is a servomotor, the load is often tuned to the carrier frequency via the addition of a capacitor in parallel with the motor. Note that the servomotor impedance is a function of the motor speed; however, the losses (usually resistive) change, whereas the parallel inductance remains virtually constant. Thus tuning the load causes it to become real (resistive), and consequently the efficiency of the amplifier output stage improves significantly.

The output stage of an ac servo amplifier is usually operated class B or AB and in a push-pull configuration to obtain the best possible drive efficiency. Figure 11.4.9 shows an example of a push-pull tuned-load ac servo amplifier.

Nonlinear Servo Amplifiers. This class of servo amplifier uses the load to filter the highly nonlinear drive signals resulting in improved drive efficiencies and higher realizable driver-power capabilities. The drivers essentially act as switches wherein the dissipation losses are low when the switches are either on or off. Since the loads or servo actuators are usually highly reactive, they can be advantageously used in this manner. The drive power can be readily derived from a dc source via a pulse-width-modulated scheme. In addition, an ac source of power can also be used via a phase-modulation technique. Some examples of high-power switching devices used for ac nonlinear servo amplifiers are thyratrons and silicon controlled rectifiers (SCRs).

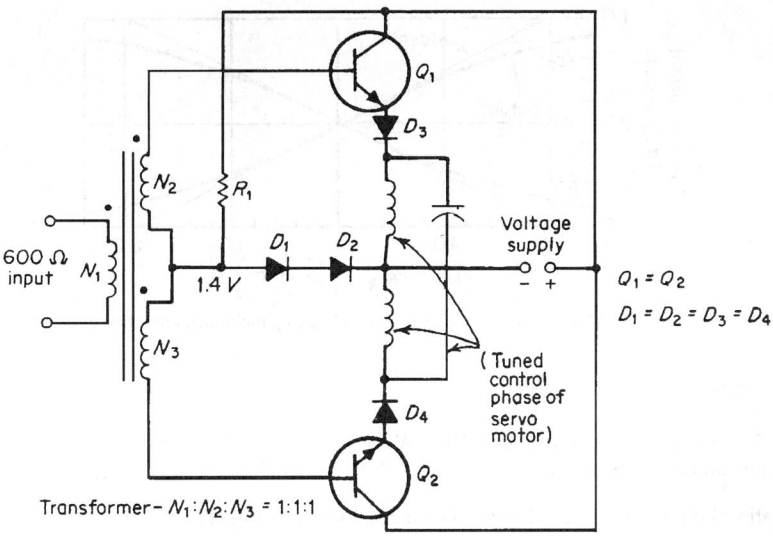

FIGURE 11.4.9   Servo amplifier used as a motor drive.

OPERATIONAL AMPLIFIERS FOR ANALOG ARITHMETIC

Joseph P. Hesler

Analog Multiplier Circuits

Circuits used for analog multiplication fall into three categories: transconductance multipliers, averaging-type multipliers, and exponential multipliers.

Transconductance Multipliers

The most prominent type of transconductance multiplier uses the property of the bipolar transistor; i.e., its collector current and transconductance are linearly related. The balanced transistor differential amplifier used in these circuits offers good accuracy, wide bandwidth, and low cost.

In the differential amplifier as shown in Fig. 11.4.10, the collector currents I_{c1} and I_{c2} are functions of the differential input voltage $V_{b1} - V_{b2}$. It is assumed that the current transfer ratios of the two transistors (Q_1 and Q_2), the junction ambient temperatures, and the base-emitter junction saturation currents are the same, as is probable in integrated-circuit fabrication.

The diode equation governs the relationship between the transistor emitter currents I_{Ei} and the base-to-emitter voltage V_{BE1}

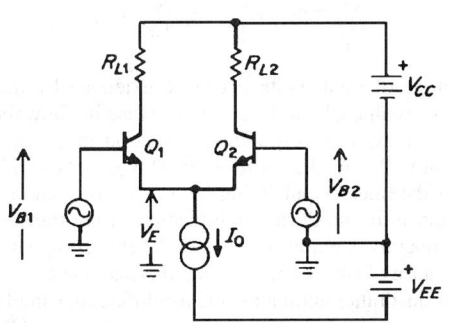

FIGURE 11.4.10   Basic differential amplifier for analog multiplication.

$$I_E = I_S \exp \left(\frac{V_{BE}q}{kT} - 1 \right)$$

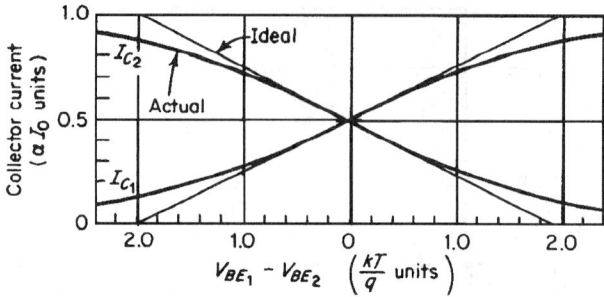

FIGURE 11.4.11   Transfer curves on which analog multiplication is based.

where I_S = junction saturation current
$\quad q$ = electron charge = 1.6×10^{19} C
$\quad k$ = Boltzmann's constant = 1.38×10^{-23} W/s·K
$\quad T$ = junction temperature (K)

The relationships of collector currents, shown graphically in Fig. 11.4.11 are

$$I_{c1} = \frac{\alpha_1 I_0}{1 + \exp\left[(V_{BE_2} - V_{BE_1})q/kT\right]} \qquad I_{c2} = \frac{\alpha_2 I_0}{1 + \exp\left[(V_{BE_1} - V_{BE_2})q/kT\right]}$$

For zero differential input, $V_{BE_1} = V_{BE_2}$ and $\alpha_1 = \alpha_2$, and so

$$I_{c_1} = I_{c_2} = \alpha I_0/2$$

At zero input, $\Delta V_{BE} = 0$, the maximum single-ended transconductance occurs at

$$gm_{\text{max}} = \alpha I_0/(4kt/q)$$

For a differential output connection the maximum transconductance is twice this value. The linear range of transconductance extends over a differential input signal range of about 50 mV peak to peak at room temperature.

Linear multiplication occurs when one input varies the I_0 term and the other input is used to vary ΔV_{BE}.

$$\Delta I_c = gm \, \Delta V_{BE} = \frac{\alpha I_0}{2kT/q} \Delta V_{BE}$$

The linear range of input voltage can be extended by inserting resistance in series with each emitter. This increase in allowable signal is accompanied by a corresponding reduction in transconductance. For I_0 equal to 2 mA, the addition of 50 X in series with each emitter increases the linear input swing by a factor of 3 and reduces the transconductance to one-third. Optimization of the transconductance multiplier operation requires a trade-off between linearity and error sources because of offset voltages and thermal noise.

A typical four-quadrant multiplier using the differential amplifier as building blocks is shown in Fig. 11.4.12. This circuit is generally used with differential inputs to obtain maximum linearity and common-mode signal rejection.

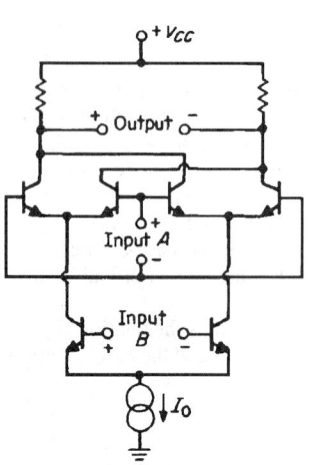

FIGURE 11.4.12   Four-quadrant transconductance multiplier.

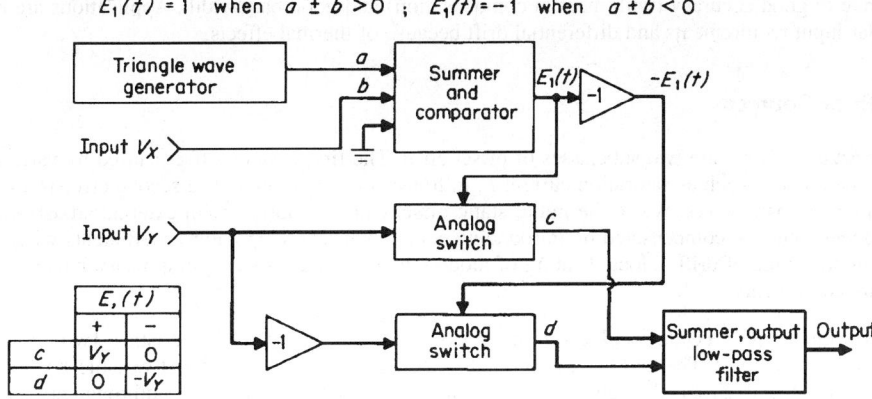

$E_1(t) = 1$ when $a \pm b > 0$; $E_1(t) = -1$ when $a \pm b < 0$

FIGURE 11.4.13 Block diagram of an averaging-type multiplier.

Averaging-Type Multipliers

Two types of averaging multipliers have found extensive use. The first type, the pulse height and width multiplier, uses one of the input variables to modulate the height or amplitude of a pulse train, while the pulse width is modulated with the other input variable. The pulse area, averaged over a suitable period, is proportional to the product of the input variables. High pulse rates permit short averaging intervals and fast response. These multipliers offer very good accuracy and stability but are more expensive than the transconductance multipliers. A block diagram of a typical pulse height and width averaging multiplier is shown in Fig. 11.4.13.

A second version of the averaging multiplier is the triangle averaging multiplier. In this type a high-frequency triangular waveform is generated and combined with the input variables to form an averaged output proportional to the product of the input signals. These multipliers are less accurate than the pulse height and width multipliers and are being displaced by transconductance multipliers.

A third type of averaging multiplier is the time-base multiplier. This approach uses a comparator to sense the time interval required for the integral of a reference input voltage to equal the amplitude of a sample of one input variable. During the same interval, 0 to T, the other input variable is integrated to produce an output V. This type of circuit can be built with a few components to provide moderate accuracy, 1 to 5 percent, at low cost.

Exponential Multipliers

The first electronic multipliers were of the exponential type. In these circuits, resistor-diode networks are designed to provide a current or voltage output approximating the square of the input. These multipliers are also called *quarter-square* multipliers based on the identity

$$XY = \frac{1}{4}[(X + Y)^2 - (X - Y)^2]$$

The piecewise-linear approximations to the squared response result in "lumpy" error characteristics. Also, the amount of circuitry required to compute the quarter-square algorithm is expensive. Although these multipliers are capable of good accuracy and bandwidth, they are becoming obsolete.

A second type of exponential multiplier uses logarithmic amplifiers, a summer, and an antilog amplifier to implement the relationship

$$XY = \text{antilog}_a (\log_a X + \log_a Y)$$

Semiconductor diodes are available with excellent logging characteristics over many decades of bias current. These diodes are used with operational amplifiers to realize the necessary functions. The circuits can provide moderate to good accuracy (with thermal compensation) and good bandwidth. Applications are restricted by unipolar input requirements and differential drift because of thermal effects.

Multiplier Error Sources

Offset Error. There are two subclasses of offset error. The first is static offset caused by variances in component parameters such as saturation current I_s, in transistors and diodes. The second error is a result of drift in component parameters. While the initial static offset can be trimmed with external adjustments, the drift components must be compensated by introducing complementary temperature coefficients within the circuit. A common source of drift is local heating of diode junctions and resistive components having nonzero temperature coefficients.

Feedthrough Error. Feedthrough errors result from nonideal transfer characteristics. Two feedthrough error contributions can exist. The first, E_{FY}, is defined as the output owing to the input variable Y when the X input is zero. The second, E_{FX}, is the complementary function owing to an X input. Feedthrough erros may result in dc, fundamental, and harmonic components of the contributing input signal.

Gain and nonlinearity errors. Gain errors produce output deviations from the expected scale factor of the multipliers. Nonlinearity of the transfer functions of the multiplier can produce additional error contributions, as previously discussed. Gain is most apt to vary as a function of the combined input-signal values because the internal components are operated over a range of bias conditions.

In certain multiplier applications, required transient responses may overtax the bandwidth capabilities of the circuits. In these instances additional error terms may appear as a result of limited slew rate of the circuits and differential phase response between the input channels.

Squaring Circuits

Squaring circuits are readily implemented by introducing the variable to both inputs of a multiplier circuit. Alternatively, the resistor-diode squaring circuits used in exponential multipliers may be applied directly.

Dividing and Square-Root Circuits

Division and square-root functions can be implemented with basic multiplier circuits by altering the interconnections. A typical dividing-circuit connection is shown in Fig. 11.4.14. The multiplier output is fed back

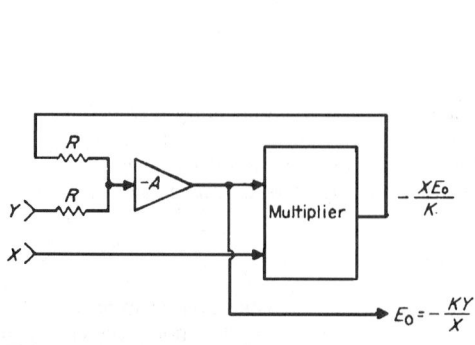

FIGURE 11.4.14 Dividing circuit.

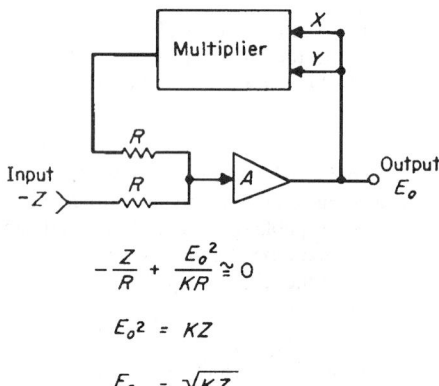

FIGURE 11.4.15 Square-root circuit.

through a summing amplifier to one of the multiplier inputs. The summing amplifier maintains an equivalence between the numerator Y and the multiplier output XE_0/K.

$$Y/R = -XE_0/KR \quad \text{or} \quad E_0 = -KY/X$$

Square-root circuits can be implemented in a similar method with another feedback connection involving a multiplier. In this instance, the output is used as each of the two multiplier-input variables, and the multiplier output is summed with the variable whose square root is desired. The square-root connection is shown in Fig. 11.4.15. The nonnegative input limitations should be stated for division and square-root connections.

LOW-NOISE OPERATIONAL AMPLIFIERS

Clayton R. Roberts

Low-Noise Op-Amp Design

The use of the standard operational-amplifier module or chip has grown into many areas of design because of its low cost, ease of design, and low number of supporting components. Some of the well-known basic amplifier uses are shown in Fig. 11.4.16. Low-noise amplifier developments have moved into the op-amp category for the same reasons. Applications in this area include microphone pre-amplifiers, threshold detectors, instrumentation amplifiers, and sensor amplifiers. Good low-noise op-amps are available, and understanding some of the basic design principles is helpful when using them.

There are many types of noise derived from a number of sources, but the types concerning amplifiers include $1/f$ or pink noise and thermal and other current- and voltage-associated effects within the amplifier. Pink noise is of concern at the very low end of the spectrum and involves dc amplifiers and some types of sensors. Generally the amplifier noise density can be expressed as follows:

$$E_n = G\sqrt{e_n^2 + (i_n R_{eq})^2 + e_t^2}$$

where e_n = op-amp noise voltage
i_n = op-amp current noise
R_{eq} = equivalent input source resistance (including feedback network)
e_t = thermal noise of source network (resistance)

and since

$$e_t = \sqrt{4KTR_{eq}}$$

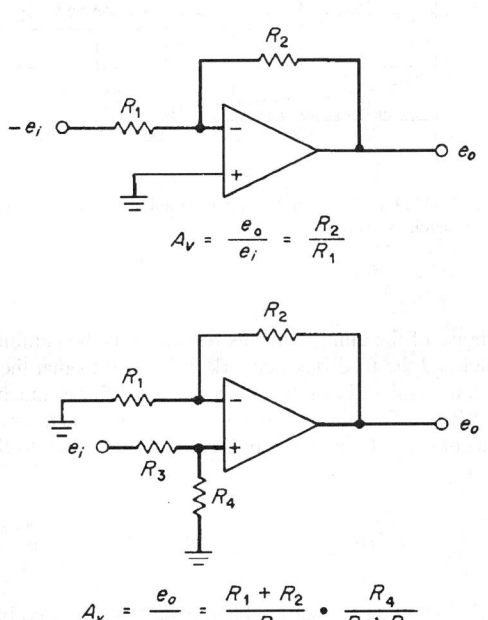

FIGURE 11.4.16 Basic operational-amplifier configurations.

where $K = 1.38 \times 10^{-23}$, $T = 300$ K, then

$$E_n = G\sqrt{e_n^2 + (i_n R_{eq})^2 + 4KTR_{eq}}$$

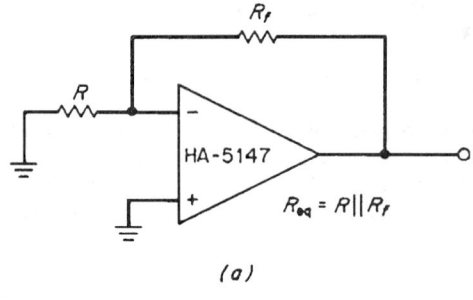

(a)

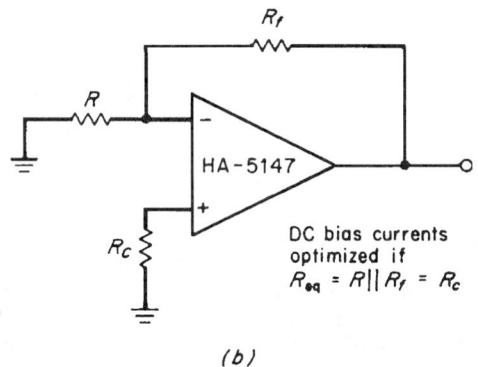

(b)

FIGURE 11.4.17 Simple op-amp configuration with and without bias.

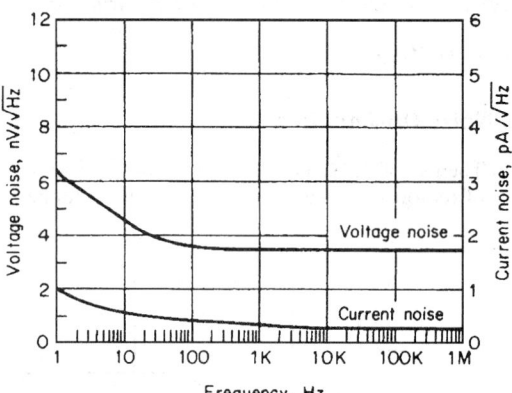

FIGURE 11.4.18 Voltage and current noise characteristics of a commercially available op amp.

The R_{eq} can become a predominant influence in the noise of the amplifier. This resistance is the combination of the resistive component of the input to the amplifier and the feedback network. It is obvious that the input to the amplifier should be kept as low as possible without loading down the device the amplifier is attached to. Figure 11.4.17 shows the R_{eq} of the simple amplifier. When offset bias is added to the op amp, the thermal and current noise components must be added to the noise equation. Usually the bias resistance is equal to the R_{eq} making the noise density

$$E_n = G\sqrt{e_n^2 + 2(i_n R_{eq})^2 + 8KTR_{eq}}$$

The amplifier design should have bandwidth restrictions on it so that the band extent does not go beyond that which is necessary. When the amplifier design is the first stage of a larger amplifier and it becomes the noise determining circuit, the bandpass restrictions do not have to be repeated further on.

There is one other consideration regarding the R_{eq} and the source resistances. If truly low noise is desired, noncarbon resistors are needed in this part of the circuit. Metal film resistors are best. Wire-would resistors are also a possibility, but not if high frequencies are involved. Carbon resistors have a larger "particle component" of noise beside the normal thermal noise.

Generally, the current noise i_n and voltage noise e_n remain relatively constant above 1 kHz and rise somewhat below 100 Hz, as shown in Fig. 11.4.18. If high values of R_{eq} or source resistances are necessary, op-amp selection is possible to reduce noise in parts of the spectrum. This is apparently a result of the biasing arrangement of different op amps with their input transistors ($e_n = f/\sqrt{i_c}$, $i_n = f\sqrt{i_c}$). With R_{eq} less than 1 kΩ, the thermal noise is expected to be low and the total noise is dominated by the voltage noise e_n.

POWER OPERATIONAL AMPLIFIERS

Clayton R. Roberts

Power Op Amps

The advent of the *power op amp* has provided us with another versatile design approach when high-current high-voltage drive requirements are needed. It is useful in motor drive circuits, deflection sweep circuits, many control functions, and high-quality audio amplifiers. Distortion levels below 0.05 percent can be achieved. In addition to the high-voltage high-current ratings the power op amps usually provide better matched components from internal control of the devices and laser trimming of the stages. Standard design approaches for op amps are used except for some considerations mentioned below. The definition of where the power op amp takes over from the standard op amp appears to be when the total supply voltage must be above 44 V and the delivered current is above 0.1 A.

Power Dissipation.   In addition to the maximum voltage rating concerns, the power op-amp design must include operating temperature and dissipation values. Mounting of the device as well as the operating region of the parameters is a major design criterion. Figure 11.4.19 shows the effect of junction temperature of the power devices on failure rates as was derived from the base failure rate tables of MIL-HDBK-217C. Some power op-amp manufacturers design the cases so that they can be directly mounted to grounded heat sinks so that there is minimum thermal resistance to the sink. The required thermal resistance ϕ from a heat sink is calculated by the following:

$$\phi = (T_j - T_a)/P_{\text{dis}} - \phi_{\text{JC}}$$

where T_j = maximum junction temperature
T_a = maximum ambient temperature
P_{dis} = internal power dissipation
ϕ_{JC} = thermal resistance of the amplifier

Another heat-sink-design approach (Fig. 11.4.20) offers the approximate area required of a convection cooled surface.

The complexity of the power-dissipation calculation is very much dependent on the type of load that is being driven. Taking dc or instantaneous values of current and voltage, it may be expressed as

$$P_{\text{dis}} = (V_s - V_o)I_o + V_{ss}(I_q)$$

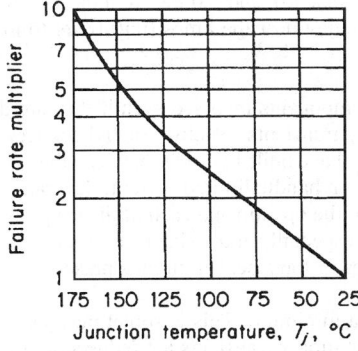

FIGURE 11.4.19   Failure rate (MTTF) vs. junction temperature.

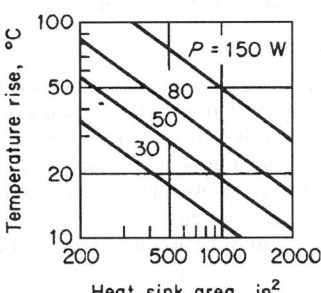

FIGURE 11.4.20   Heat-sink area estimated for temperature rise and several values of power dissipation.

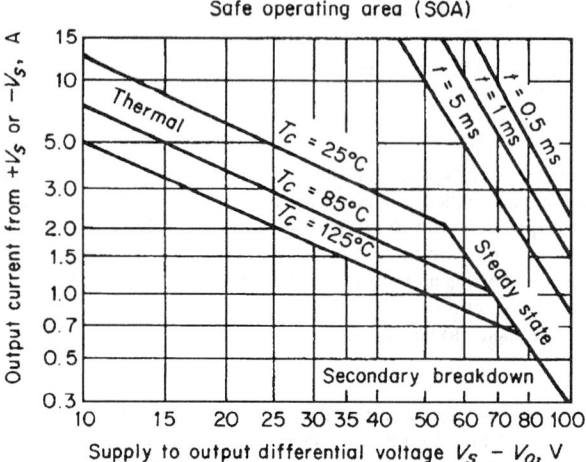

FIGURE 11.4.21 Safe operating curves for a power operational amplifier.

where V_s = one rail of the supply voltage
V_o = output voltage
I_o = output current
V_{ss} = rail-to-rail supply voltage
I_q = quiescent current

For sinusoidal waveforms the peak value of P_{dis} may be expressed as

$$P_{dis} = (V_s/2Z_l)[1 - \cos(60° - \theta)]$$

where θ is the absolute value of the phase angle of Z_l and Z is the magnitude of the load impedance. This function is approximate.

Safe-operating-area (SOA) curves are supplied by some manufacturers to aid in limiting the power dissipation to within specifications. Figure 11.4.21 shows a curve for a commercial power op amp. The secondary breakdown region shown results from excess current densities in the base region of the output devices when high collector current occurs simultaneously with high collector-emitter voltage. These curves include all contributors to the dissipation problem. The three curves at the right-hand end of Fig. 11.4.21 are the limits of the short-pulse power applications for the SOA curve. Pulse width values of 0.5, 1.0, and 5.0 ms are indicated. Additional foldover or limiting current protection can be obtained when operating near the limits (rail) (refer to manufacturers product data).

Output Loads. Output loads can have reactance components involved even if they are designed to be resistive. With the higher currents present, and especially transients, reactive, or flyback (reversed polarity) voltages can exceed specified limits. Most power op amps have built-in diodes between the two rails and the output to help suppress such transients. By necessity these can handle limited current. Several manufacturers recommend external diodes be placed between the output of the op amp and each of the supplies. The back emf must be considered in the design (SOA curves). This is especially true with loads such as motors. Start-up and reverse conditions must be accounted for. Some types of loudspeaker circuits present a similar situation.

Circuit Layout. Lead inductance must be kept to a minimum. This is sometimes taken care of by use of a microstrip-type low-impedance line approach. Power-supply transients are reduced to a minimum by bypassing right at the op amp as well as at the supply terminals. With high currents involved, bypass capacitors must have low reactance leads and several times their normal values. With low-distortion amplifiers, deteriorization

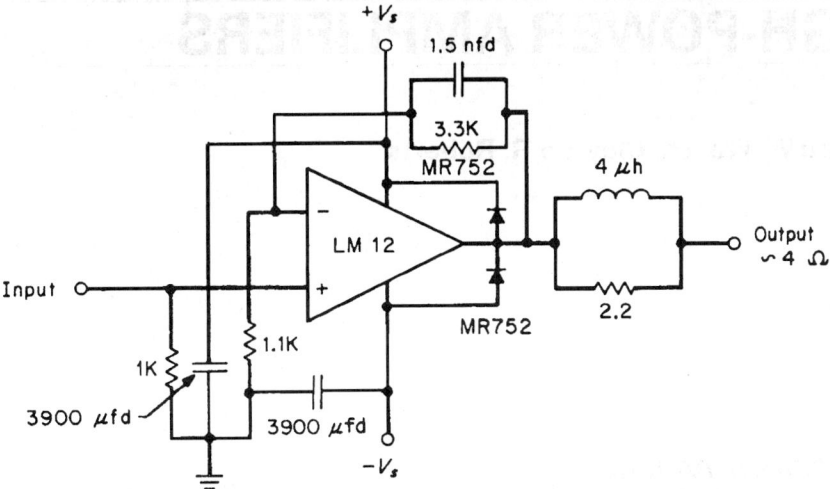

FIGURE 11.4.22   National Semiconductor LM12 power amplifier.

at the high-frequency end of the spectrum (higher distortion) can be the result of insufficient effective bypass-ing from the supply sources, but this is best taken care of by large short-lead bypasses at the device location. Ground loops and ground return interaction effects are also emphasized with the high-current devices. Input-output ground return coupling is to be avoided.

An audio power amplifier shown in Fig. 11.4.22 has some of the features indicated in a low-cost high-performance design. The LM12 (National Semiconductor) is a 150-W op amp capable of delivering that power to a 4-Ω load. The design includes Motorola MR752 diodes in its output to the rails. These diodes are capable of handling up to 400 A of surge current, 28 A average forward current, and a repetitive reverse voltage of 200 V. These specifications provide good protection against transients and back emf. The small capacitor in the feedback network aids in reducing the distortion at the high end of the spectrum and pro-vides stability. The distortion is claimed to be about 0.01 percent. Good supply bypassing is obtained right at the device terminals. A central grounding point is used to avoid ground coupling of output to input. Thermal limiting, output current limiting, and over voltage shutdown is designed into the device. Note the parts count associated with circuit.

CHAPTER 11.5
HIGH-POWER AMPLIFIERS

Richard W. French, Clayton R. Roberts

THERMAL CONSIDERATIONS

A problem common to all high-power amplifier and oscillator equipment is removal of the excess thermal energy produced in the active devices and other circuit components so that operating temperatures consistent with reliable performance can be maintained. Available cooling methods are radiation, natural convection, forced convection, liquid, evaporative, and conduction. Radiation and evaporative cooling depend for suitable operation on a rather high temperature for the device being cooled and are thus generally restricted to use with vacuum tubes. The remaining methods are suitable for use in both solid-state and vacuum-tube systems.

HIGH-POWER BROADCAST-SERVICE AMPLIFIERS

Transmitters for amplitude modulation broadcast service may employ high-level (plate) modulation or low-level (grid or screen) modulation of the output stage or modulation of an intermediate stage, followed by linear amplification in the following stage(s). Generally, the last approach is used in very-high-power transmitters, because of the difficulty and expense in designing and building a modulation transformer to handle the high audio modulating power required. High-level modulation has been used in AM transmitters up to at least 250 kW carrier power with a modulator output power requirement of 125 kW. By a unique design in which the positive terminal of the modulator plate supply is grounded, and autotransformer is used as the modulation transformer, with a significant reduction in size and cost.

CLASS B LINEAR RF AMPLIFIERS

The conventional means for achieving linear amplification of an AM signal is the class B rf amplifier circuit, often referred to simply as a *linear amplifier*. The plate efficiency, plate dissipation, and output power are highly dependent on the drive level. It is convenient, therefore, to define a *drive ratio* or normalized drive level k.

$$k = E_{pm}/E_{bb}$$

where E_{bb} is the dc plate supply voltage and E_{pm} is the peak ac plate signal voltage.

Class B amplifiers using transistors can approach the theoretical maximum collector efficiency of $\pi/4$, or 78.54 percent, as a consequence of the very low collector saturation voltage of transistors, which gives a correspondingly high value for K.

EFFICIENCY POWER AMPLIFIERS

The plate efficiency at carrier level in a practical linear amplifier circuit is about 33 percent, while that of a high-level (plate) modulated stage is 65 to 80 percent. When the efficiency of the modulator is taken into account, the net efficiency of the high-level modulated stage is still higher than that of the linear amplifier, since the output power level of the modulator is at most one-third of the total output power of the modulated stage.

Because of the difficulty and expense involved in the design and manufacture of very large modulation transformers and chokes, several linear amplifier circuits having much greater efficiency than the class B amplifier while amplifying an AM waveform have been developed. Low-level modulation can be employed, and the need for the large, expensive modulator components is circumvented.

Such high-efficiency amplifier circuits include the Chireix outphasing modulated amplifier, the Dome high-efficiency modulated amplifier, the Doherty high-efficiency amplifier, and the Terman-Woodyard high-efficiency modulated amplifier.

INDUCTION HEATING CIRCUITS

Induction heating is achieved by placing a coil carrying alternating current adjacent to a metal workpiece so that the magnetic flux produced by the current in the coil induces a voltage in the workpiece, which produces the necessary current flow.

Power sources for induction heating, in addition to direct use of commercial power, include spark-gap converters, motor-generator sets, vacuum-tube oscillators, and inverters. Motor-generator sets generally provide outputs from 1 kW to more than 1 MW and from 1 to 10 kHz. Spark-gap converters are generally used for the frequency range from 20 to 400 kHz and provide output power levels up to 20 kW. Vacuum-tube oscillators operate at frequencies from 3 kHz to several MHz and provide output levels from less than 1 kW to hundreds of kilowatts. Inverters using mercury-arc tubes have been used up to about 3 kHz. Solid-state inverters operate at frequencies up to about 10 kHz and at power levels of several megawatts. These solid-state inverters generally employ thyristors (silicon controlled rectifiers) and are replacing motor-generator sets and mercury-arc inverters.

DIELECTRIC HEATING

Whereas induction heating is used to heat materials which are electrical conductors, dielectric heating is used to heat nonconductors or dielectric materials. The basic arrangement for a dielectric heating system is that of a capacitor in which the material to be heated forms the dielectric or insulator. The heat generated in the material is proportional to the loss factor, which is the product of the dielectric constant and the power factor. Because the power factor of most dielectric materials is quite low at low frequencies, the range of frequencies employed for dielectric heating is higher than for induction heating, extending from a few megahertz to a few gigahertz.

The power generated in a material is given by

$$P = 141AV^2f\frac{K\cos\phi}{t} \times 10^{-6} \quad (\text{W})$$

where V = voltage across material (V)
f = frequency of power source (MHz)
A = area of material (in^2)
K = dielectric constant
t = material thickness (in)
$\cos\phi$ = power factor of dielectric material

The voltage that can be applied to a particular material is limited by the insulation properties of the material at the required process temperature. The frequency that can be used is limited by voltage standing waves on the electrodes, which will be appreciable when the electrode dimensions are comparable with one-eighth wavelength (10 percent voltage variation).

TRANSISTORS IN HIGH-POWER AMPLIFIERS

Many high-power amplifier circuits can be implemented by using solid-state power devices. Bipolar transistors have been successfully applied in induction heaters, sonar transmitters, and broadcast transmitters. The sonar transmitters have employed class B, class C, and the so-called class D power amplifier circuits. In the class D circuit the active device is used as an on-off switch, and output power variations are achieved by pulse-width modulation or supply-voltage control. A tuned load or one incorporating a low-pass filter is used with the class D amplifier.

Power MOSFETs are advantageous compared with bipolar transistors in a number of applications. These include HF and VHF transmitters, high-speed power switching, and certain audio or low-frequency amplifiers.

The power MOSFET is a majority carrier device (contrast with a minority carrier device for the bipolar transistor). The input drive circuit requires current to supply the parasitic capacitances only. There is no minority carrier storage time or secondary breakdown. Switching time is usually much less than that of the bipolar transistor of the same die. The power MOSFET device uses a gate voltage-controlled circuit and requires very little current to maintain it in its OFF, ON, or an intermediate state. The ON resistance (r_{DS}) can vary from less than 0.5 Ω to several ohms depending on the device selected. There is a trade-off between high-breakdown drain-source voltage and low ON-resistance. For a given die size the ON-resistance varies as follows:

$$r_{DS} = K(V_{DS})^2$$

where K is a constant related to the die and V_{DS} is the maximum breakdown voltage between drain and source. A typical drain current (I_D)-drain-source voltage (V_{DS}) transfer characteristic curve of a TMOS power MOSFET is shown in Fig. 11.5.1. The ON-resistance r_{DS} is, of course, a function of the conduction state of the device and does have a threshold value depending on the gate voltage which controls the conduction (see Fig. 11.5.2). Also, if the design is dependent on the r_{DS} value, one must keep in mind a potential shortcoming which is its variation with temperature. Figure 11.5.3 shows this function with a Motorola MTM8N15. Temperature coefficients are provided as a function of I_D with a number of device data sheets. Switching speeds of this device are not influenced by temperature.

Considerations for Broadcast Transmitters. There are some advantages to using the power FET in this type of application through the VHF spectrum. The power FET offers stability in the drive circuit as well as its own transfer characteristics. With the bipolar transistor there is a dependent relationship between the collector current i_C and the current gain β. If the drive level is decreased, causing the i_C to decrease, the β will increase and can cause instability in an amplifier. The input impedance can also change as the drive level is changed. Both of these factors can become a stability design problem. The input and output impedances of the

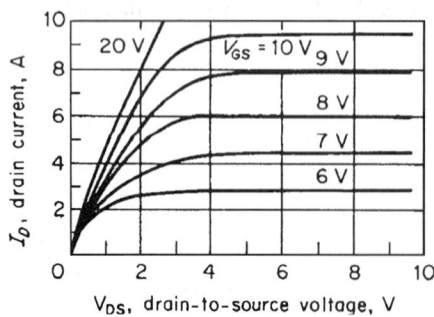

FIGURE 11.5.1 Typical on-region transfer characteristics for the MTP 7N05.06. *Copyright by Motorola, Inc. Used by permission.*

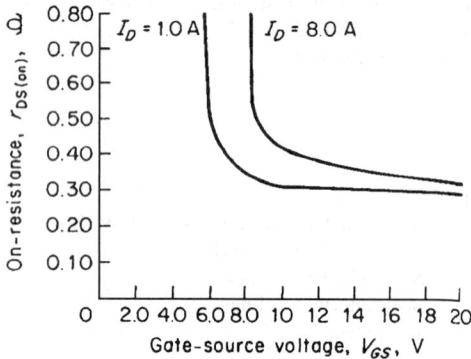

FIGURE 11.5.2 Variation of r_{DS} with V_{GS} and I_D for MTP8N15. *Copyrigth by Motorola, Inc. Used by permission.*

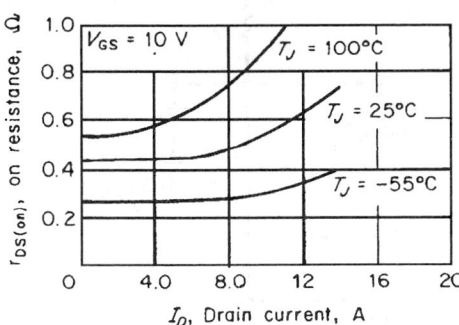

FIGURE 11.5.3 Variation of r_{DS} with drain current and temperature for MTM8N15. *Copyrigth by Motorola, Inc. Used by permission.*

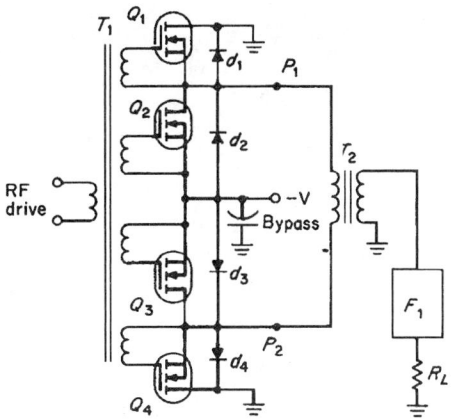

FIGURE 11.5.4 Power amplifier for Nautel broadcast transmitter.

device are reasonably independent of drive. The input impedance is somewhat higher in magnitude and more capacitive than bipolar devices, allowing simpler matching circuits to be used.

There are a number of approaches used in transmitter design. Which one to use depends on the number of modulation and carrier uses. Usually, a slightly positive (adjustable) bias is required on the gate unless strictly class C is needed. The amplifier design can be made to withstand high VSWR conditions in the output without damage. This is accomplished by selection of the device so that maximum rated drain-source voltage ($V_{(BR)DSS}$) is not exceeded. Likewise, the gate-to-source voltage (V_{GS}) cannot be exceeded. This is sometimes protected using a zener diode across the gate to source. Remember that using such devices tends to generate undesirable harmonics when the drive signal is clipped and a low-pass filter or equivalent is required in the output.

While power FET amplifiers can be operated in class A, B, AB, and C, it is of strong interest to consider class D for AM broadcast service. Power FETs can have a wide linear operating range and swinging them hard to an ON condition allows the efficiency to be quite high. Nautical Electronic Laboratories Limited of Nova Scotia has developed a very efficient AM transmitter using power FETs as switches (class D operation). Figure 11.5.4 shows the basic configuration of the final rf amplifier. A low-power rf drive signal at the carrier frequency is applied via $T2$ to bias FET pairs $Q1/Q3$ and commutated between input terminals of $T2$ as a square wave of reversing polarity at the carrier frequency. The resulting high-power signal at $T2$ is filtered by low-pass filter $F1$ to remove unwanted harmonics. The final rf amplifier is thus simply a device that converts from dc to rf. Efficiencies of greater than 90 percent are achievable.

The pulse width modulator also uses FETs in a switched mode to vary the negative supply to the rf power amplifier in sympathy with an audio modulating signal to produce high level modulation with a similar high efficiency. Nautical Electronic Laboratories has developed a 50-kW solid-state transmitter of this type with distortion typically less than 1.0 percent.

MOSFET AUDIO AMPLIFIERS AND SWITCHING APPLICATIONS

The power MOSFET offers a very linear transfer characteristic. This can be used effectively with the design of audio power amplifiers with proper gate biasing.

The gain characteristic is identified with the transconductance of the MOSFET. The voltage gain is as follows:

$$A_V = g_{fs}R_L$$

where g_{fs} is the transconductance and R_L is the equivalent load. The transconductance is defined as the ratio of the change in drain current relating to a change in gate voltage. This rating on a device is typically taken at half

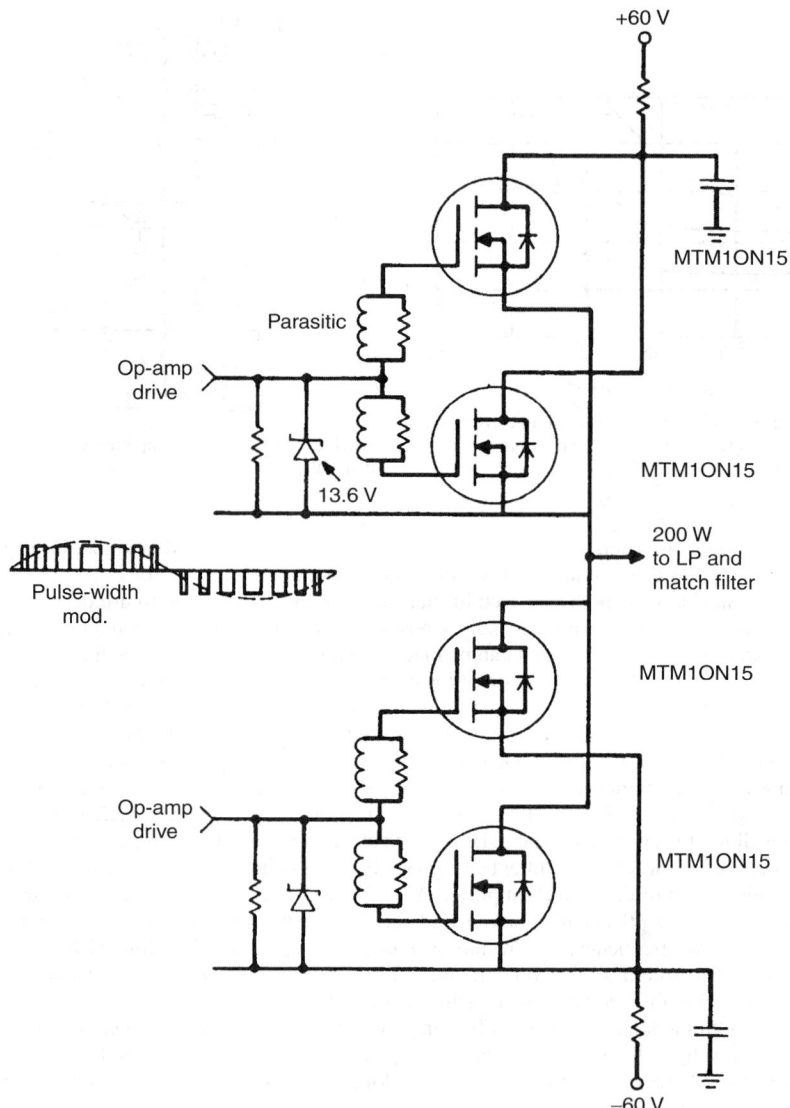

FIGURE 11.5.5   Class D audio amplifier using power MOSFET.

the rated continuous drain current at a V_{gs} of 15 V. The linear increase in g_{fs} with V_{gs} occurs in the square law region but levels off to a constant in the velocity-saturated region.

Considering the very high switching efficiency the power MOSFET has, it can be effectively used as a "class D" audio amplifier or one that is pulse-width modulated at the audio rate. A high-frequency carrier is used which is on the order of 6 to 10 times the highest audio-modulated frequency. The waveform is demodulated at the output for driving large speaker systems. Very high audio power amplifiers can be designed with high fidelity, small size, and high efficiency. Power FETs can be placed in parallel with few problems of current sharing. Figure 11.5.5 shows a simple version of this approach. With the high carrier frequency, coupling capacitors and supply decoupling become much less severe.

CHAPTER 11.6
MICROWAVE AMPLIFIERS AND OSCILLATORS

John W. Lunden, Jennifer E. Doyle

MICROWAVE SOLID-STATE DEVICES

Since the 1950s, solid-state devices have replaced tubes in most amplifier and oscillator applications. Solid-state devices are preferred because of better long-term reliability, lower cost, and improved noise characteristics. In the 1960s and 1970s, avalanche and Gunn diodes were used extensively in the generation of microwave power.

More recently, junction and field-effect transistors have been increasingly used in microwave amplifiers and oscillators. Gallium arsenide (GaAs) FETs have replaced other solid-state devices in many microwave applications; however, impact ionization avalanche transit-time (IMPATT) and Gunn diodes are still commonly used at frequencies above 20 GHz.

IMPATT DIODE CIRCUITS

The generation of microwave power in a reverse-biased *pn* junction was originally suggested by Read (1958). He proposed that the finite delay between applied rf voltage and the current generated by avalanche breakdown, with the subsequent drift of the generated carriers through the depletion layer of the junction, would lead to negative resistance at microwave frequencies. The diode is biased in the avalanche-breakdown region. As the rf voltage rises above the dc breakdown voltage during the positive half cycle, excess charge builds up in the avalanche region, reaching a peak when the rf voltage is zero. Hence this charge waveform lags the rf voltage by 90°. Subsequently, the direction of the field in the diode causes the multiplied carriers to drift across the depletion region. This, in turn, induces a positive current in the external circuit while the diode rf voltage is going through its negative half cycle. This is equivalent to negative resistance, which is maximum when the transit angle is approximately 0.74π.

A simplified equivalent circuit of the IMPATT diode circuit is shown in Fig. 11.6.1. The resistance R_D includes both the parasitic positive resistance because of contacts, bulk material, and so forth, and the dynamic negative resistance. The net magnitude is typically in the range −0.5 to −4.0 Ω and varies with current. The capacitance C_D is the voltage-sensitive depletion-layer capacitance and can be approximated sufficiently accurately with the value at breakdown. The diode resistance variation results in a stable operating point for any positive load resistance equal to or less than the diode's peak negative value.

Oscillations will build up and be maintained at the frequency for which the net inductive reactance of the package parasitics and the external load equals the capacitive reactance of C_D. The values for L_p and C_p vary with package or mounting style. Typical values range from 0.3 to 0.6 nH and 0.2 to 0.4 pF, respectively. It is important to minimize these parasitics since they limit the operating frequency and band-width.

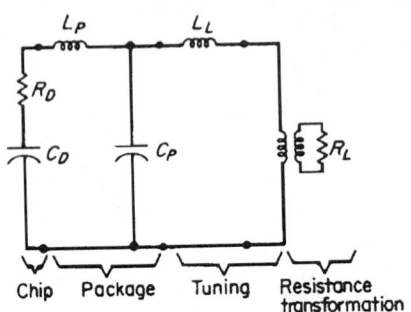

FIGURE 11.6.1   Approximate rf equivalent circuit of IMPATT oscillator.

IMPATT diodes can be used in several mounting configurations, including coaxial, waveguide, strip-line, or microstrip. It is important to ensure that a good heat flow path is provided (owing to the typically low efficiency) and that low-electrical-resistance contacts be made to both anode and cathode.

In avalanche breakdown, the diode tends to look like a voltage source. Hence a current source is desirable for dc bias. Several circuits are possible. The *RC* bias circuit (Fig. 11.6.2*a*) is the simplest but is inefficient, and the transistor current regulator (Fig. 11.6.2*b*) may be more desirable. In either case, the loading of the diode with shunt capacitance or a resonance path to ground (at some frequency) must be avoided to prevent instabilities (noise and/or spurious frequencies).

Two broadly tunable diode loading circuits are the multiple-slug cavity and the variable-package-inductance types. Coaxial implementations of these two techniques are shown in the cross sections of Fig. 11.6.3. In Fig. 11.6.3*a* slugs $\lambda/8$ and/or $\lambda/4$ long at the desired center frequency of operation with characteristic impedance of between 10 and 20 Ω are adjusted in position along the centerlines to provide a load reactance equal to the negative of the diode reactance and a load resistance equal to or less than the magnitude of the diode net negative resistance. Circuit *b* tunes the diode by recessing it into the holder, effectively decreasing the net series inductance, which is resonant with the diode capacitance. Single- or multisection transformers can also be included for resistive matching to the load R_L. Similar waveguide-circuit implementations can be used with typically narrower bandwidths, better frequency stability, and lower FM noise.

The IMPATT diode can function as an amplifier if the load resistance presented to it is larger in magnitude than the negative resistance of the diode. Typically, a circulator is used in conjunction with the tuned diode circuit to separate input and output signals. At the center frequency, the power gain of the amplifier is given by

$$G_0 = \left(\frac{R_D - R_L}{R_D + R_L} \right)^2$$

where R_D and R_L are as shown in Fig. 11.6.1.

In general, amplifier operation for a given diode requires a smaller shunt tuning capacitor or a large transformer characteristic impedance. An estimate of the diode rf current and voltage can be obtained from

$$V_D \approx \frac{\sqrt{2P_o R_L}}{\omega C_j(V_b)} \left(1 + \frac{1}{\sqrt{G_0}} \right) \qquad I_D \approx \omega C_j(V_b) V_D$$

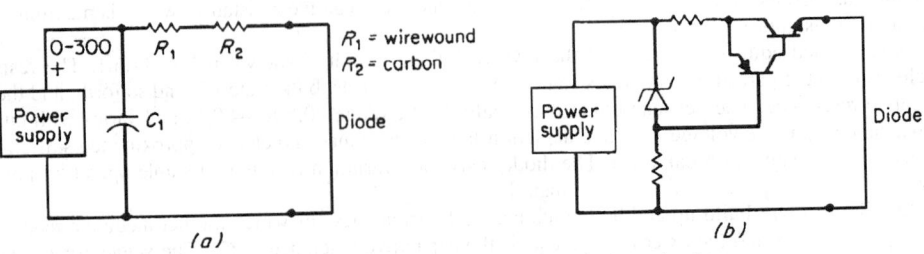

FIGURE 11.6.2   Bias circuits for IMPATT diodes: (*a*) *RC* type; (*b*) transistor-regulated type.

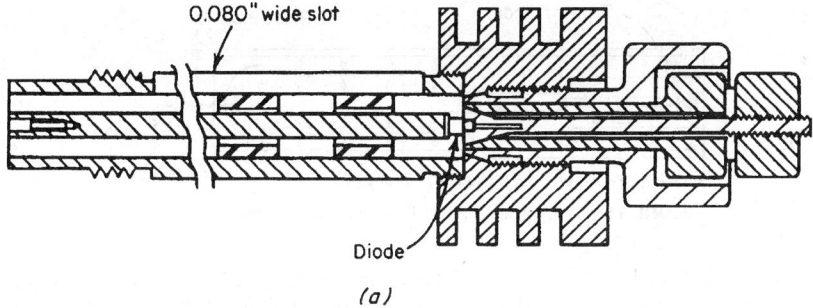

(a)

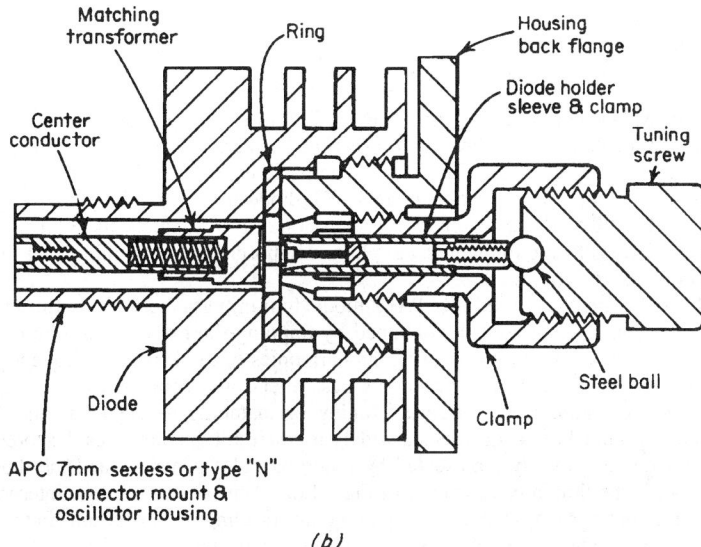

(b)

FIGURE 11.6.3   Mechanical tuning of IMPATT circuit: (*a*) multiple-slug type; (*b*) variable-inductance (diode-recess) type.   (*Hewlett-Packard Company.*)

where V_b = diode breakdown voltage
C_j = junction capacitance
P_o' = power output
G_0 = gain

Higher output powers have been achieved by multiple-diode series-parallel configurations and/or hybrid combining techniques.

In general, properly designed IMPATT oscillator circuits can have noise performance comparable with reflex klystron or Gunn oscillators. The noise performance of both Si and GaAs IMPATT diode amplifiers and oscillators has been extensively treated in the literature.

IMPATT diodes are usable at higher frequencies than any other solid-state device currently available. Silicon IMPATT diodes can be used at frequencies as high as 300 GHz. Some of the disadvantages of IMPATTS are a low gain-bandwidth product, difficult adjustments necessary for long-term performance reliability, bias circuit instabilities, and inherent nonlinearity. The cw power output achievable with a single

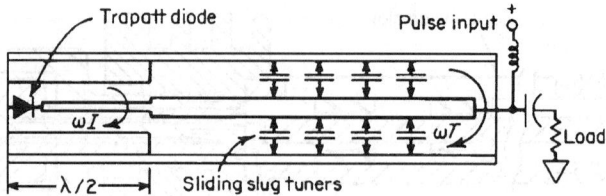

FIGURE 11.6.4   Typical TRAPATT diode circuit configuration.

IMPATT diode is 3 W at 11 GHz, 2 W at 15 GHz, or 1 W at 20 GHz. Diodes can be paralleled to obtain higher power, for example, 10 W at 41 GHz, with 30 dB gain and 250 MHz bandwidth, by combining 12 diodes in one cavity.

TRAPATT DIODE CIRCUITS

The trapped plasma avalanche triggered transit (TRAPATT) mode of operation is characterized by high efficiency, operation at frequencies well below the transit-time frequency, and a significant change in the dc operating point when the diode switches into the mode. The basic understanding of this high-efficiency mode of operation has proceeded from consideration of IMPATT diode behavior with large signals. Two saturation mechanisms in the IMPATT diode, space-charge suppression of the avalanche and carrier trapping, reduce the power generated at the transit-time frequency, but play an important role in establishing the "trapped-plasma" states for high-efficiency operation.

To manifest the high efficiency of the TRAPATT mode, four important circuit conditions must be met: large IMPATT-generated voltage swings must be obtained by trapping the IMPATT oscillation in a high-Q-cavity circuit; selective reflection and/or rejection of all subharmonics of the IMPATT frequency except the desired subharmonics must be realized (typically with a low-pass filter); sufficient capacitance must be provided near the diode to sustain the high-current state; and tuning or matching to the load must be provided at the TRAPATT frequency. Figure 11.6.4 illustrates a widely used circuit that achieves the foregoing conditions.

The TRAPATT diode is typically represented by a current pulse generator and the diode's depletion-layer capacitance, as shown in the simplified schematic of Fig. 11.6.5. This permits a simple interpretation of operation. Initially, a large voltage pulse reaches the diode, triggering the traveling avalanche zone in the diode. This initiates a high-current low-voltage state. The drop in voltage propagates down the line l, is inverted (owing to the –1 reflection coefficient of the low-pass filter), and travels back toward the diode. The process then repeats. Consequently, the frequency of operation is inversely proportional to the length of line l. The period of oscillation is slightly modified by the finite time required for the diode voltage to drop to zero. The inductance L is a result of the bond lead and is helpful in driving the diode voltage high enough to initiate the TRAPATT plasma state. The capacitance C is provided to supply the current required by the diode to the extent that the transmission line is insufficient. The total capacitance which can be discharged during the short interval of high conduction current in the diode is

$$C_T = C + \tau_P/Z_0$$

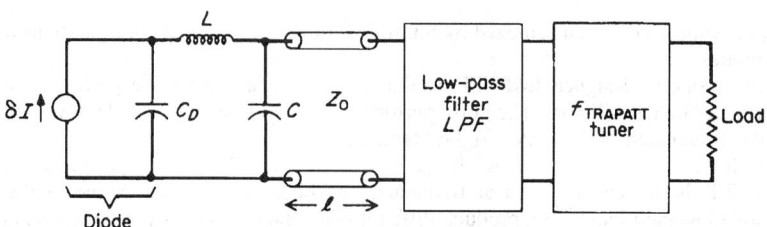

FIGURE 11.6.5   Simplified schematic diagram of TRAPATT circuit.

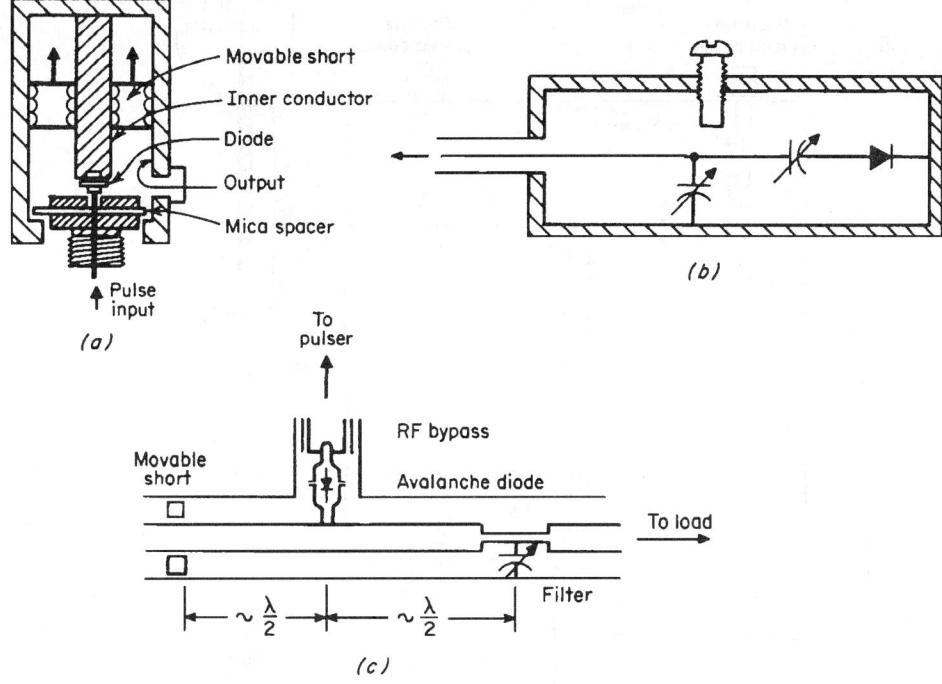

FIGURE 11.6.6   TRAPATT diode circuit arrangements: (*a*) tuned coaxial cavity; (*b*) lumped circuit; (*c*) coaxial circuit.

where τ_p is the time high-current or trapped plasma state exists. τ_p must be at least at large as the transit time of carriers through the diode, thus putting an upper limit on C. This is contrary to the high-current requirements to initially drive the avalanche zone through the diode (10 to 20 kA/cm^2) in large-area devices.

Since the TRAPATT frequency is generally an integral submultiple of the diode transit-time frequency, the line length l ($l = \lambda/2$ at $f_{TRAPATT}$) presents the low-pass-filter short circuit to the diode as a series resonance at the IMPATT frequency. This net series resonance, however, includes the diode series-reactive elements. Further, this circuit should have a high Q at the transit-time frequency to reduce the buildup time to TRAPATT initiation.

Several TRAPATT circuit configurations are shown in Fig. 11.6.6. The coaxial cavity circuit in Fig. 11.6.6*a* places the diode in the reentrant gap of the half-wave cavity resonator. The output coupling loop passes only the fundamental to a triple stub tuner to match to the load. Proper dc biasing and bypassing are included. This circuit is good into the lower *L* band. The lumped circuit (Fig. 11.6.6*b*) is compact and very useful for VHF and UHF. The series capacitor is resonated with the inductance of a copper bar and the self-inductance of the diode. The trimmer controls the resonance of the third and fifth harmonics. The circuit of Fig. 11.6.6*c* is a variation on the circuit of Fig. 11.6.4. The use of additional filter sections and/or lumped elements provides better harmonic filtering and results in higher efficiencies. Circuits analogous to those of Fig. 11.6.4, can be implemented in waveguide for the higher frequencies. In all these circuits, the presence of the third and fifth harmonics has been found essential for stable and high-efficiency performance. Higher power levels can be achieved by operating multiple diodes in series and/or parallel configurations.

Another useful technique for extending both power and frequency is the antiparallel diode configuration. The circuit consists of two diodes, placed with opposite polarity approximately one-half fundamental wavelength apart in a transmission line. Operation is similar to a free-running multivibrator. Output may be extracted with a transmission line connected to the midpoint of the diodes, followed by the usual low-pass filter. The position of this filter should be adjusted so that the roundtrip delays from midpoint to diodes and the filter are equal. A microstrip circuit realization for antiparallel operation is given in Fig. 11.6.7.

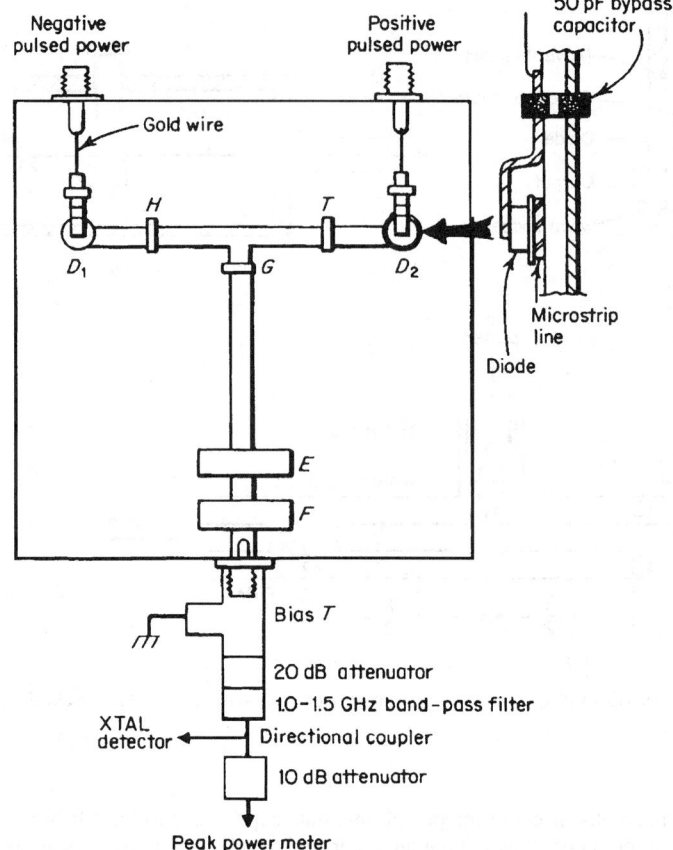

FIGURE 11.6.7 Microstrip version of antiparallel TRAPATT circuit.

Operation of the anomalous avalanche diode for microwave amplification has also been established. A 10-dB dynamic range is typical, with a low-level threshold gain decreasing with increasing power level to a saturated condition. The pulsed bias can be replaced by a simple dc source and storage capacitor. Unlike the *locked-oscillator* mode of operation, only a small residual output is present without the input signal. The locked oscillator will typically display only a 3:1 power change between locked and unlocked cases.

TRAPATT operation has yielded output power levels of 10 to 500 W with efficiencies of 20 to 75 percent in the frequency range from 0.5 to 10 GHz.

BARITT AND DOVETT DIODES

The barrier-injected transit-time (BARITT) diode is the newest active microwave diode. It is similar to the IMPATT diode; however, carriers are injected into the drift zone by a forward-biased junction rather than being extracted from an avalanche region. Holes emitted by the *pn* junction cross the intrinsic region at their saturation velocity and are collected at the *p* contact. The diode behaves as a negative resistance with a phase shift of π rad. BARITT diodes operate at a lower bias voltage and are much lower in noise figure than an IMPATT; however, BARITT diode operating bandwidths are narrow, and output power capability is very low. For C band operation the following results have been attained: 15 dB noise figure and 50 mW output power at 1.8 percent efficiency.

The double velocity transit-time (DOVETT) diode is a new type of BARITT diode. The carrier velocity within a DOVETT diode increases between the injection point and the collection contact. The DOVETT diode has a larger negative resistance than the BARITT diode and therefore operates at higher current density.

TRANSFERRED ELECTRON EFFECT DEVICE (TED) CIRCUITS

This class of circuit, using both the Gunn and limited-space-charge accumulation (LSA) devices depends on the internal negative resistance owing to carrier motion in the semiconductor at high electric fields. When the material is biased above the critical threshold field, a negative dielectric relaxation time is exhibited, which results in amplification of any carrier concentration fluctuations, causing a deviation from space-charge neutrality. The resultant *domain* drifts toward the anode and is extinguished, and a new domain is formed at the cathode. The current through the device consists of a series of narrow spikes with a period equal to the transit time of the domain.

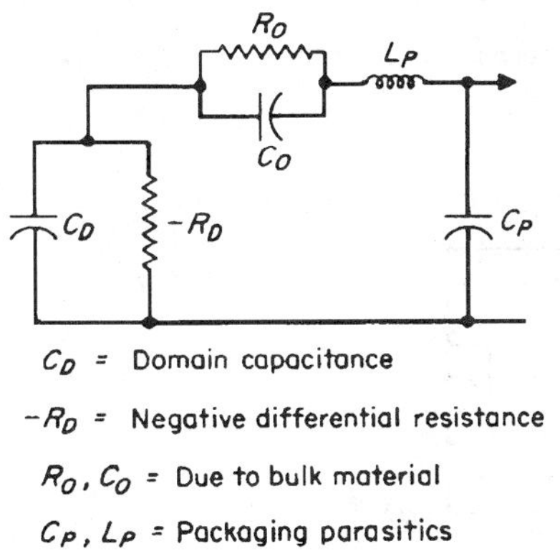

When an rf voltage is superimposed on the bias, in a given period of time, the terminal voltage can be below both the threshold voltage V_{th} and the domain-sustaining voltage V_s. The domain is quenched at any place in the device when the latter occurs, and the nucleation of a new domain is delayed until the voltage again exceeds V_{th}. Therefore the frequency of oscillation is determined by the resonant circuit, including the impedance of the device. Experimental results and computer modeling have shown that the device can be tuned over greater than an octave bandwidth by the external circuit cavity. Although other modes of operation are possible, depending on the characteristics of the external circuit, the LSA mode appears to be the most important.

C_D = Domain capacitance

$-R_D$ = Negative differential resistance

R_O, C_O = Due to bulk material

C_P, L_P = Packaging parasitics

FIGURE 11.6.8 Approximate equivalent circuit of a TED device and its package.

An approximate equivalent circuit is given in Fig. 11.6.8 with values dependent on frequency, bias, and power level. The capacitance includes the diode static capacitance, in addition to that because of traveling high-field domains.

One of the simplest tuned circuits is the coaxial-line cavity, as shown in Fig. 11.6.9a. The diode is mounted concentric with the line, at one end to facilitate heat sinking. The frequency of oscillation is determined

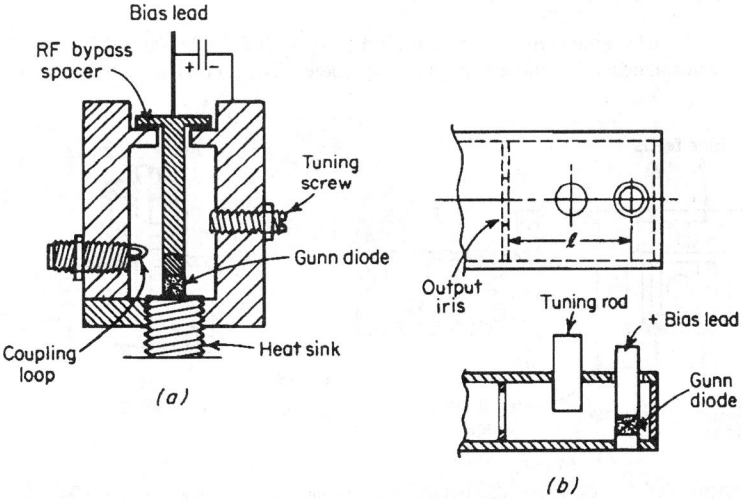

FIGURE 11.6.9 Gunn diode cavity circuits: (*a*) coaxial form; (*b*) waveguide-cavity form.

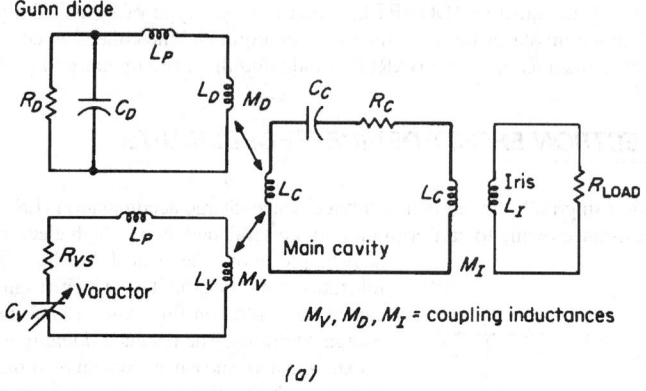

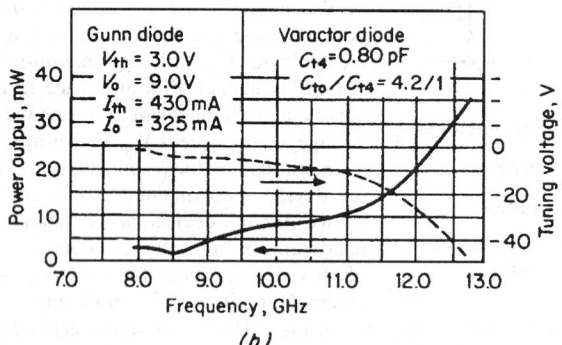

FIGURE 11.6.10 Varactor tuning of Gunn oscillator: (*a*) simplified equivalent circuit; (*b*) tuning characteristic.

primarily by the length of the cavity. The position of the output coupling loop (or plate) determines the load impedance.

A rectangular waveguide cavity configuration (Fig. 11.6.9*b*) is more widely used because of its higher Q and better performance at X band and higher frequencies. The diode post acts as a large inductive susceptance,

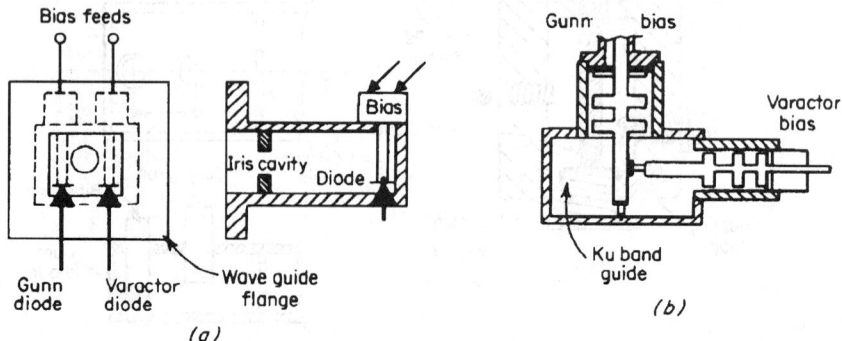

FIGURE 11.6.11 Varactor tuning techniques for Gunn diodes: (*a*) front and side views of double-port waveguide-type; (*b*) type used at K_u band.

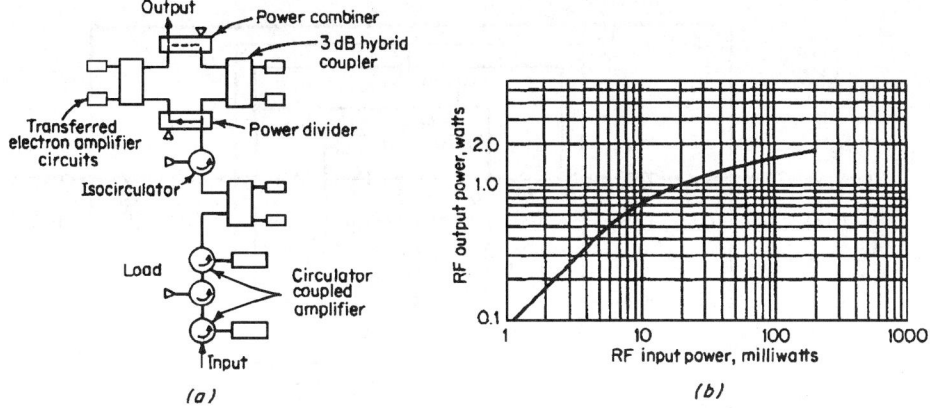

FIGURE 11.6.12   Four-stage ganged TED amplifier chain for 5.7 GHz: (*a*) block diagram; (*b*) power-transfer curve.

which, with the inductive iris, produces the resonant frequency for which the length *l* is $\lambda/2$. The tuning rod lowers the frequency as its insertion length increases.

In addition to mechanical tuning, both YIG and varactor tuning techniques are applicable. YIG tuning has the potential for the widest tuning range but is limited in tuning speed, hysteresis, and physical bulk. Figure 11.6.10 illustrates a typical varactor-tuned Gunn oscillator equivalent circuit and characteristics. The varactor diode Q is highest at the maximum reverse voltage and decreases with decreasing voltage because of an increase in R_{vs}. Figure 11.6.11 illustrates two varactor-tuned implementation techniques.

The noise characteristic of a GaAs TED is comparable with that of a klystron and is 10 to 20 dB better than an IMPATT diode. Various noise-source models and measuring equipments are discussed in the literature. Several methods can be employed to reduce the AM and FM noise of a Gunn oscillator, including increasing the loaded Q of the cavity circuit: biasing at or near the frequency and/or power turnover points (i.e., bias at which $df/dv = 0$ and $dP_o/dv = 0$); minimizing power-supply ripple; and diode selection.

The TED can also be operated as an amplifier, typically using circulator or hybrid techniques similar to the IMPATT circuits. Parameters of importance are saturation characteristics, bandwidth, gain and phase tracking, linearity, efficiency, and dynamic range. The block diagram of a four-stage chain is shown in Fig. 11.6.12*a*, with operating characteristics in Fig. 11.6.12*b*. Hybrid coupling was found to provide greater linear power output. Gains in excess of 20 dB can be realized. Efficiency and bandwidth are typically less than 10 percent and greater than 35 percent, respectively.

The use of Gunn and LSA diodes extends to frequencies in excess of 90 Hz, with higher peak power capability for the LSA diode than for the IMPATT or for any other Gunn-effect device. TEDs can be operated with a pulsed bias, allowing extremely high power density without damage to the device. Over 1000 W at 10 GHz has been achieved in the LSA mode with short low-duty cycle pulses. Reactive termination and unloading of the circuit at the harmonic frequencies can significantly improve the efficiency. A problem associated with pulsed oscillators is the significant frequency change during each pulse caused by rapid temperature rise. Starting-time jitter can be alleviated by *priming*, i.e., injecting a weak cw signal into the circuit.

TRANSISTOR AMPLIFIER AND OSCILLATOR MICROWAVE CIRCUITS

Silicon bipolar and GaAs FETs are available with cutoff frequencies extending well into X and K_a bands, respectively. Equivalent circuits for these transistor types are given in Fig. 11.6.13. The intrinsic chip element values, with variations considered at low frequencies, are further modified by high-frequency effects.

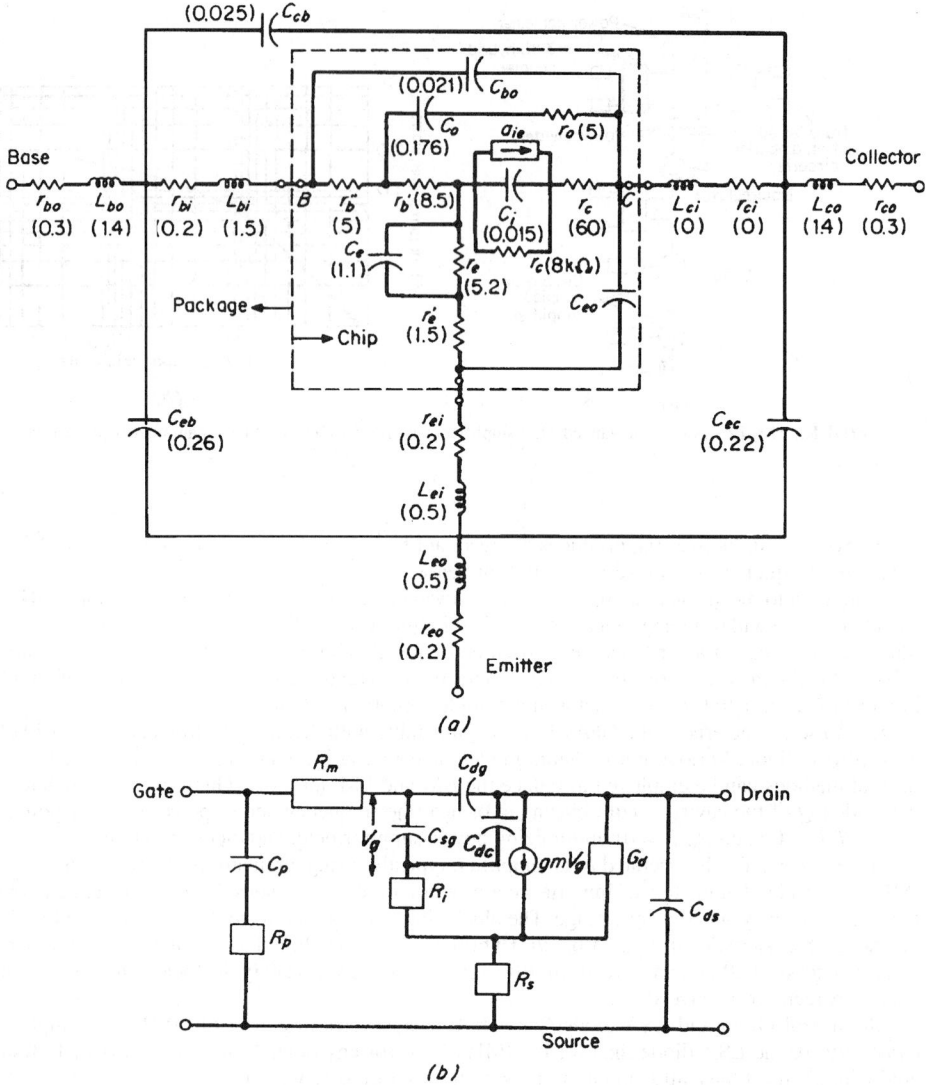

FIGURE 11.6.13 Equivalent circuits of microwave solid-state devices: (*a*) bipolar transistor; (*b*) field-effect transistor (chip only).

A small-signal figure of merit has been defined in terms of the contributing time constants, for a bipolar transistor,

$$K = (\text{power gain})^{1/2}(\text{bandwidth}) = 1/4\pi(r_b'C_c\tau_{ec})^{1/2}$$

where $\tau_{ec} = \tau_e + \tau_b + \tau_l + \tau_c$, τ_e = emitter barrier charging time, τ_b = base transit time, τ_l = collector transit time, and τ_c = collector depletion-layer charging time. The maximum frequency of oscillation is defined as the frequency for which the power gain is unity

$$f_{\max} \approx 1/4\pi(r_b'C_c\tau_{ec})^{1/2}$$

The application of transistors at microwavelengths demands that considerable attention be given to packaging, fixturing, and impedance characterization. Historically, characterization has taken the form of f_T, $r_b' C_c$ specification and/or h-y-parameter techniques. A more desirable method for high-frequency characterization is by scattering parameters. Scattering parameters describe the relationship between the incident and reflected power waves in any N-port network. As such, this technique offers substantial advantages, including remote measurement, broadband (no turning), stability (no short-circuited or open terminations), accuracy, and ease of measurement.

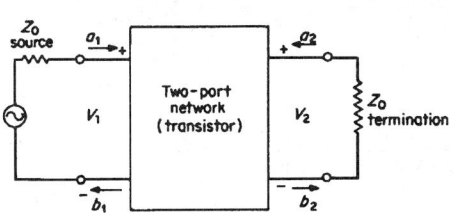

FIGURE 11.6.14 Basic two-port configuration for a microwave transistor.

From Fig. 11.6.14, the scattering equations describing the two-port network can be written.

$$b_1 = S_{11}a_1 + S_{12}a_2 \quad b_2 = S_{21}a_1 + S_{22}a_2$$

Solving for the S parameters yields

$$S_{11} = \frac{b_1}{a_1}\bigg|_{a_2=0} = \text{input reflection coefficient with } Z_L = Z_0$$

$$S_{12} = \frac{b_1}{a_2}\bigg|_{a_1=0} = \text{reverse transmission gain with } Z_L = Z_0$$

$$S_{21} = \frac{b_2}{a_1}\bigg|_{a_2=0} = \text{forward transmission gain with } Z_L = Z_0$$

$$S_{22} = \frac{b_2}{a_2}\bigg|_{a_1=0} = \text{output reflection coefficient with } Z_L = Z_0$$

Other linear two-port parameters can be calculated from S parameters (for example, y parameters for feedback analysis). Either manual or complex automatic measurement techniques can be used. In either case, the transistor chip package is embedded in a system with a given reference impedance Z_0 and well-defined reference wave planes.

A number of useful relationships can be calculated from the S parameters and used in amplifier-oscillator design.

Reflection coefficient-impedance relationship:

$$S_{11} = \frac{Z - Z_0}{Z + Z_0}$$

where Z_0 is the reference impedance.

Input reflection coefficient with arbitrary Z_L:

$$S_{11}' = S_{11} + \frac{S_{12}S_{21}\Gamma_L}{1 - S_{22}\Gamma_L}$$

Output reflection coefficient with arbitrary Z_S:

$$S_{22}' = S_{22} + \frac{S_{12}S_{21}\Gamma_S}{1 - S_{11}\Gamma_S}$$

Stability factor:

$$K = \frac{1 + |D|^2 - |S_{11}|^2 - |S_{22}|^2}{2(S_{12}S_{21})}$$

Transducer power gain:

$$G_T = \frac{|S_{21}|2(1 - |\Gamma_S|^2)(1 - |\Gamma_L|^2)}{|(1 - S_{11}\Gamma_S)(1 - S_{22}\Gamma_L) - S_{12}S_{21}\Gamma_L\Gamma_S|^2}$$

Maximum available power gain:

$$G_{max} = \left| \frac{S_{21}}{S_{12}} (k \pm \sqrt{k^2 - 1}) \right| \qquad \text{for } k > 1$$

Source and load reflection coefficients for simultaneous match:

$$\Gamma_{ms} = M^* \left[\frac{B_1 \pm \sqrt{B_1^2 - 4|M|^2}}{2|M|^2} \right] \qquad \Gamma_{mL} = N^* \left[\frac{B_2 \pm \sqrt{B_2^2 - 4|N|^2}}{2|N|^2} \right]$$

where Γ_S, Γ_L = source and load reflection coefficients
$M = S_{11} - DS_{22}^*$
$N = S_{22} - DS_{11}^*$
$D = S_{11}S_{22} - S_{12}S_{21}$
$B_1 = 1 + |S_{11}|^2 - |S_{22}|^2 - |D|^2$
$B_2 = 1 + |S_{22}|^2 - |S_{11}|^2 - |D|^2$

The maximum power gain is obtained only if the transistor is terminated with the Γ_{ms} and Γ_{mL} resultant impedances. A lossless transforming network is placed between the source and load to realize this transformation.

Generally, the embedding circuits used take the form of simple ladder networks: series-shunt combinations of L's and C's or their transmission-line equivalents. These elements can be determined by moving on the Smith chart from the value of the terminating impedance to the center of the chart along constant resistance-conductance, reactance-susceptance contours for series-shunt elements, respectively. The circuits are typically implemented in strip-line or lumped form, as shown in Fig. 11.6.15.

Optimum design at more than one frequency requires plotting of gain circles at each frequency, with subsequent terminating impedance iteration to obtain the best compromise across the band. Several computer-aided optimization programs are available to simplify this routine. A plot of gain circles and impedance loci for a typical S-band design is shown in Fig. 11.6.16, which corresponds to the collector circuit of Fig. 11.6.15a.

NOISE PERFORMANCE OF MICROWAVE BIPOLAR TRANSISTOR CIRCUITS

The noise performance of a well-designed amplifier depends almost entirely on the noise figure of the first transistor. The noise factor of a bipolar transistor amplifier is related to its equivalent circuit parameters and external circuit by

$$NF = 1 + \frac{r_{bb}'}{R_g} + \frac{r_e}{2R_g} + \frac{(r_{bb}' + r_e + R_g)^2}{2r_e R_g h_{feo}} \left[1 + \left(\frac{f}{f_\alpha} \right)^2 (1 + h_{feo}) \right]$$

As a result, r'_{bb} should be as low as possible, and the alpha cutoff frequency should be high. The source resistance providing minimum noise figure is

$$R_g\Big|_{F\min} = (r'_{bb} + r_e)^2 + \frac{(r'_{bb} + 0.5r_e)(2h_{feo}r_e)}{1 + (f/f_\alpha)^2(1 + h_{feo})}$$

This value is typically close to the value providing maximum power gain for the common-emitter configuration. Care must be taken in the matching-circuit implementation to minimize any losses in the signal path, e.g., by using high-Q elements and isolated bias resistances. Bipolar junction transistors are available with less than 1 dB noise figure for frequencies below 1 GHz. At 4 GHz the best bipolar transistor noise figure is approximately 3 dB.

HIGH-POWER MICROWAVE TRANSISTOR AMPLIFIERS (USING BIPOLAR TRANSISTORS)

The difficulties arising in power-amplifier operation are because of the nonlinear variation of device parameters as a function of time and to bias conditions. Saturation, junction capacitance-voltage dependence, h_{fe} current (and voltage) level dependence, and charge-storage effects are the prime contributors to this situation. Class A operation is normally not used, implying collector-current conduction angles less than 360°, resulting in further complication of the time-averaging effects. With these qualifications, the basic equivalent circuit given in Fig. 11.6.13a applies. Figure 11.6.17a shows a greatly simplified equivalent circuit useful for first-order design. Of particular note is the low input resistance, high output capacitance, and a nonnegligible feedback element causing bandwidth, gain, and stability limitations.

Several methods have been used to determine large-signal-device characteristics with varying degree of success. One method involves the measurement of the embedding circuitry at the plane of the transistor terminals, with the transistor removed and the source and load properly terminated. The circuit is tuned for maximum power gain before the transistor is removed to make the measurement. An average or effective device impedance is then the complex conjugate of the measured circuit impedance at that frequency. The maximum power obtainable from a particular transistor is determined by thermal considerations (cw operation), avalanche-breakdown voltages, current-gain falloff at high current levels, and second-breakdown effects.

Bandwidth.   The input impedance has been found to be the primary bandwidth-limiting element. R_S varies inversely with the area of the transistor. Hence, for a given package L_S, the Q increases and bandwidth decreases with higher-power transistors

$$Q = \omega_0 L_S/R_S \qquad \text{BW}|_{3\text{ dB}} \approx f_0/Q$$

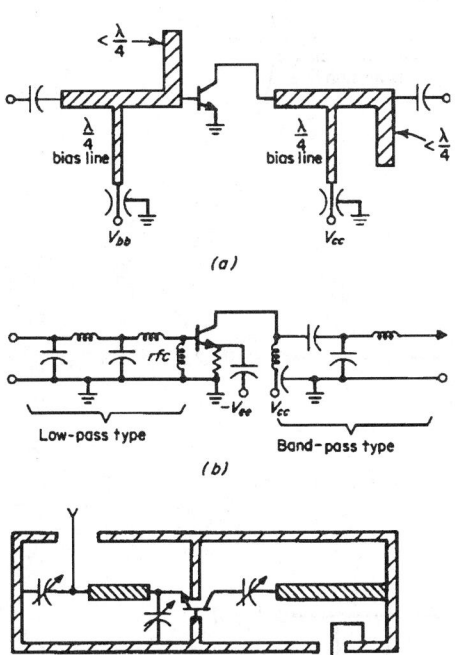

FIGURE 11.6.15   Circuit realizations for microwave transistor circuits: (*a*) strip-line; (*b*) lumped design; (*c*) coaxial design.

Impedance matching.   The problem of matching complex impedances over a wide band of frequencies has been treated by a number of authors. Essentially, high-order networks can be used to achieve nearly rectangular

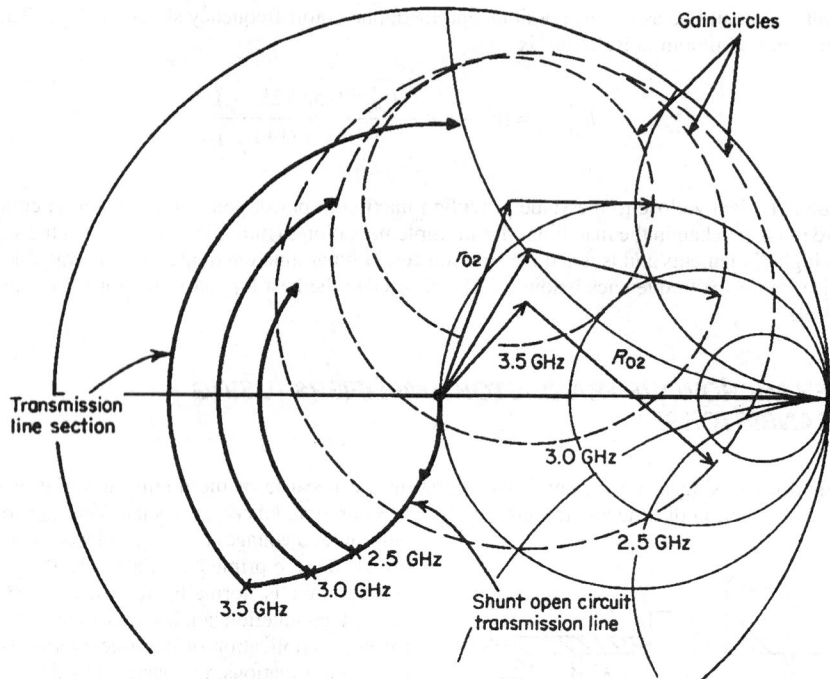

FIGURE 11.6.16 Use of Smith chart in microwave transistor circuit design.

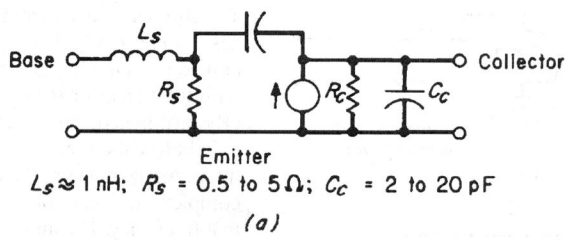

$L_S \approx 1 \, \text{nH}; \quad R_S = 0.5 \, \text{to} \, 5 \, \Omega; \quad C_C = 2 \, \text{to} \, 20 \, \text{pF}$

(a)

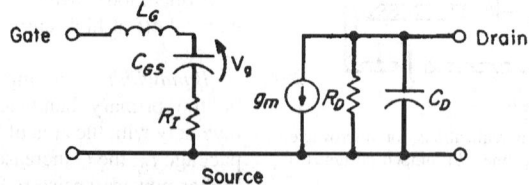

$C_{GS} = 0.2 \, \text{to} \, 0.5 \, \text{pF}; \quad R_I = 10 \, \text{to} \, 70 \, \Omega; \quad R_D = 200 \, \text{to} \, 500 \, \Omega$

(b)

FIGURE 11.6.17 Simplified equivalent circuits of microwave power transistors: *(a)* bipolar transistor; *(b)* FET.

bandpass characteristics. However, without mismatching, the 3-dB bandwidth determined above cannot be extended. In fact, the greater the ratio of generator resistance R_g to transistor input resistance R_S the greater will be the reduction of the intrinsic bandwidth, for a given ripple and number of circuit elements. Hence the external circuit design must consider both the transistor-input-circuit Q value and the impedance level relative to the driving source for a given bandwidth.

Either lumped- or transmission-line-element networks of relative simplicity are typically used to achieve the necessary input-output matching. Although the bandpass type yields somewhat better performance, the low-pass configuration is more convenient to realize physically. Quarter-wave line sections are particularly useful for broadband impedance transformations and bias feed-bypassing functions. Eighth-wave transformers are useful to match the small complex impedances directly without tuning-out mechanisms. The input impedance to a $\lambda/8$ section is real if it is terminated in an impedance with magnitude equal to the Z_0 of the line.

Load resistance.   The desired load resistance can be determined to first order by

$$R_L \approx (V_{CC} - V_{CE,\text{sat}})^2/2P_o$$

where P_o is the desired fundamental power output and V_{CC} is the collector supply voltage. This expression is altered by several factors, including circuit Q, harmonic frequencies, leakage, current conduction angle, and so forth. Assuming only the presence of the fundamental frequency, V_{CC} is limited to $1/2 BV_{CBO}$ by resonant effects. Recently, it has been shown that BV_{CBO} can be exceeded for short pulses without causing avalanche.

Power gain.   The power gain depends on the dynamic f_T or large-signal current gain-bandwidth capability, the dynamic input impedance, and the collector load impedance. A simple expression for power gain is

$$PG = \frac{(f_T/f)^2 R_L}{4R_e(Z_{\text{in}})}$$

The current gain at f_T is of particular importance. The effect of parasitic common-terminal inductance is to reduce this gain in the common-emitter and to cause regeneration in the common-base connection. The latter configuration is more commonly used at frequencies near or above the f_T value, whereas the former generally results in a more stable circuit below f_T. This situation is highly dependent on the parasitic-element situation with respect to the specific frequency. Various forms of instabilities, such as hysteresis (jump modes), parametric, low-frequency, and thermal, can occur because of the parameter values changing with time-varying high-signal levels. Usually, most of these difficulties can be eliminated or minimized by careful design of bias circuits (including fewer elements), ground returns, parasitics, and out-of-band terminating impedances.

The *collector efficiency* of a transistor amplifier is the ratio of rf power output to dc power input. High efficiency implies low circuit losses, high ratio of output resistance to load resistance, high f_T, and low collector saturation voltage. In addition, experiments and calculations show that for high efficiency, the impedance presented to the collector by the output network should be inductive for the favored generation of second harmonic. If the phase is correct, the amplitude of the fundamental is raised beyond the limit otherwise set by the difference between the supply voltage and $V_{C,\text{sat}}$. Figure 11.6.18 illustrates this effect.

A high value of f_T relative to the operating frequency improves efficiency by causing the operating point to spend less time (per cycle) in the high-dissipation active region between cutoff and saturation. The integrity of the transistor die bond has been found to have a substantial effect on efficiency because of the effects of low intrinsic bulk collector resistance and thermal gradient.

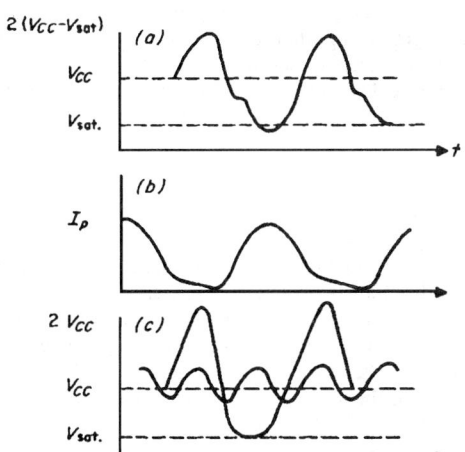

FIGURE 11.6.18 Typical voltage-current waveforms in power applications: (*a*) collector voltage; (*b*) collector current; (*c*) second-harmonic enhancement.

GaAs FIELD-EFFECT TRANSISTOR CIRCUITS

The state of the art of GaAs field-effect transistors for both small-signal and power applications is rapidly improving, thanks to the intrinsically higher carrier mobility in GaAs, coupled with significantly improved fabrication techniques, e.g., in the area of photolithography. Many devices fabricated today use feature sizes as small as 0.25 μm. GaAs FETs are particularly attractive since the parameter values are less variable as a function of operating conditions and their impedances are generally higher than in comparable bipolar devices. As a result, GaAs FET circuits are easier to design and typically achieve lower noise, wider bandwidth, and higher linearity. The common-source configuration has been found to work well for both amplifiers and oscillators.

NOISE PERFORMANCE OF MICROWAVE FET CIRCUITS

The noise figure of an FET can be represented as a minimum noise figure plus a noise factor which depends on the signal source impedance presented to the FET. For an optimum signal source impedance of $R_{SS} + jX_{SS} = R_{OP} + jX_{OP}$, the noise figure is minimized.

$$\text{NF} = F_{min} + \frac{K_2}{g_m^2} \frac{(R_{SS} - R_{OP})^2 + (X_{SS} - X_{OP})^2}{R_{SS}(R_{OP}^2 + X_{OP}^2)}$$

$$F_{min} = 1 + K_f(f/f_T)\sqrt{g_m(R_G + R_S)} \qquad K_2 \approx 0.03 \qquad K_f \approx 2.5$$

K_2 and K_f are fitting factors, with K_2 depending on the channel carrier concentrations and dimensions, and K_f depending on the quality of the channel material. The optimum source impedance is

$$R_{SS} = R_{OP} \approx 2.2[1/4g_m + R_G + R_S] \qquad X_{SS} = X_{OP} \approx 160/fC_{GS}$$

It is evident that a low-noise figure requires a high cutoff frequency f_T, and the cutoff frequency is inversely proportional to the gate length. Therefore, a short gate length is important to low-noise operation. The noise figure can also be improved by minimizing the parasitic resistances R_G and R_S.

GaAs FETs are commercially available with noise figures less than 0.5 dB at 4 GHz, less than 1 dB at 8 GHz, and less than 2 dB at 16 GHz.

HIGH ELECTRON MOBILITY TRANSISTORS

The performance of low-noise GaAs FETs is approaching a lower limit because of the finest gate geometry currently attainable (0.25 μm). The new high electron mobility transistor (HEMT) on GaAs offers superior performance because of its carrier transport mechanism, which resembles transport within undoped GaAs, which has little impurity scattering. The structure of an HEMT is similar to a GaAs MESFET, using a heterojunction between GaAs and AlGaAs. The carrier velocity and mobility are twice that of carriers in a MESFET, which results in a higher cutoff frequency and lower noise figure for the same gate geometry. The following results have been achieved with HEMTs: 2 dB noise figure up to 20 GHz without cooling, and 5 dB gain with 7.8 dB noise figure at 71 GHz.

HIGH-POWER MICROWAVE FET AMPLIFIERS

High-power field-effect transistors, like bipolar transistors, exhibit nonlinear device parameter variation with bias conditions. Figure 11.6.17b shows a greatly simplified equivalent circuit of a power FET. The parameters C_{GS}, R_I, g_m, R_D, and C_D are bias voltage dependent. The dominant nonlinear effects on power FET operation are the variation of the channel capacitance C_{GS} with gate and drain bias voltages, the variation of the transconductance g_m with gate voltage, and the variation in drain resistance R_D with drain voltage.

The high-power performance of an FET depends mainly on the gate width, the common lead inductance and other parasitics, the transistor thermal impedance, and the maximum drain voltage capability of the device. The total gate width can be increased by combining a number of individual cells in parallel on one chip. Cells are connected by wire bonding, "flip-chip" mounting techniques, or via-hole connections. A single FET cell may consist of many gate fingers connected in parallel by crossover structures, or wrap-around-plating. Using these techniques total gate widths of 30 mm or more have been attained.

The bias point of the FET has an important effect on the output power capability of the amplifier circuit. The optimum load resistance is determined from the device I-V characteristic. V_{GD} is the gate-drain avalanche breakdown voltage when the FET gate is biased at pinch-off (V_P). The limiting drain-source voltage is $V_T = V_{GD} - V_P$. Beyond this voltage, excess current will flow which is unaffected by the input signal at the gate. Therefore, V_L is the level of V_{DS} which corresponds to the maximum rf output power. If V_K is the "knee" drain voltage above which the drain current is approximately flat (independent of drain voltage), then the maximum drain voltage swing can be expressed as $V_L - V_K$ or $V_{GD} - V_P - V_K$. If I_F is the maximum drain current available with a forward-biased gate, then the optimum load resistance for maximum output power is $(V_{GD} - V_P - V_K)/I_F$ and the maximum rf power available to the load is $I_F(V_{GD} - V_P - V_K)/8$.

With a single high-power FET the following cw output power levels can be achieved at 1 dB gain compression: 12 W at 2 GHz, 8 W at 12 GHz, and 1 W at 24 GHz. Efficiencies are typically 25 to 35 percent for class A cw operation. GaAs FETs are commonly used in moderate power applications at microwave frequencies up to 26 GHz. Devices are frequently combined in balanced amplifier circuits to achieve higher output powers. Internally matched chip carrier structures are often used to reduce the effects of package parasitics, and to provide low-loss matching close to the chip. One type of input circuit for an internally matched FET comprises a small SiO_2 chip capacitor connected to the FET gate with a bond wire, forming a π-matching section with the FET input capacitance. Input and output circuits and typically LC ladder networks implemented in both discrete and semidistributed circuit forms.

MONOLITHIC MICROWAVE INTEGRATED CIRCUITS

Monolithic circuits, or MMICs, consist of active devices and associated circuitry fabricated together on the same substrate. GaAs MMICs are used to achieve very broad bandwidths and high-frequency operation. Because of the high area of GaAs required and low production yields, MMICs are very expensive. They are best suited to low- and medium-power applications above 20 GHz, where high production volume is expected. The advantages of monolithic circuitry are multioctave bandwidths, millimeter-frequency operation, small size and weight. With MMICs on GaAs the following results have been attained: 4 dB gain, 9 dB noise figure, and +12 dBm output power from 2 to 40 GHz with a distributed amplifier.

TRANSISTOR OSCILLATORS

Transistor oscillators can be designed by choosing the source (or load) terminating impedances such that $S'_{11}\Gamma_S \geq 1$ ($S'_{22}\Gamma_L \geq 1$). The design is facilitated by plotting stability circles on the Smith chart.

A number of circuit configurations are appropriate, including the standard Colpitts, Hartley, Clapp, coupled-hybrid, and so forth. The major difference, however, is the proper inclusion of device parasitic reactances into the intended configuration. The frequency-determining element(s) may be in the input, output, or feedback circuit but should have a high Q for good frequency stability. Care must be taken to decouple or resistively load the circuit at frequencies outside the band where oscillatory conditions are satisfied. This is particularly true for lower frequencies where the current gain is much better.

For bipolar transistors, the common-base and common-collector connections are generally more unstable than common-emitter and most commonly used, depending on the power requirements and frequency of oscillation.

For a GaAs FET oscillator, the device may be in common-source, gate, or drain configuration. The common grounded source arrangement provides the lowest thermal resistance which makes this configuration the

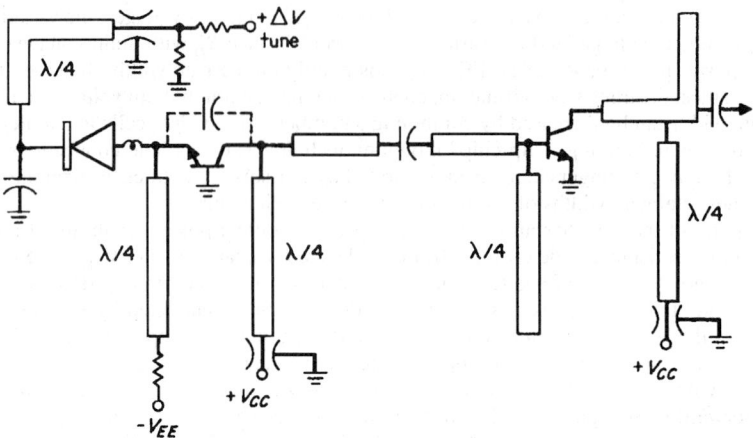

FIGURE 11.6.19   Varactor-tuned microstrip oscillator-buffer circuit.

most likely choice for a power FET oscillator. In the grounded source case a parallel feedback element is used to couple the gate and drain terminals and must be small enough to minimize parasitics.

Transistor oscillators have the advantage of the third terminal which may be used for compensation, modulation, or stabilization. A high-Q cavity resonator is often used in the feedback path for stabilization. The resonator may also be placed between the transistor and the load where it acts as a bandpass filter, or the resonator may be used as a band-reject filter coupled to one oscillator port with the load coupled to the other oscillator port. A high dielectric constant ceramic is usually used for the cavity material to result in a resonator of very small physical size. Barium nanotitanate, which provides a relative dielectric constant of 39, and a Q of up to 8000, is often used as a resonator. Oscillators that use this type of stabilization are called dielectric resonator oscillators, or DROs.

The frequency of an oscillator which is stabilized by a dielectric resonator can be mechanically tuned by varying the gap between the resonator and a metal tuning plate.

Several methods are available to vary the frequency of an oscillator dynamically. Bias variation is effective only over a narrow band and results in substantial power variation. The YIG sphere provides very high Q but is somewhat bulky and of limited tuning speed. The varactor diode provides very low power, octave-tuning bandwidth, and fast tuning in a form compatible with hybrid integration. Figure 11.6.19 shows the schematic of an S-band varactor-tuned wide-band oscillator implemented in hybrid thin-film form. The frequency-determining elements are connected in the high-Q common-base input circuit. This also allows maximum isolation from load mismatch effects (pushing) and permits the balance of the circuitry to be low-Q (broadband).

Both high- and low-power oscillators are large-signal; i.e., the output power is limited primarily by beta falloff, with increasing current at a given frequency and collector voltage.

Bipolar transistors are generally used for oscillators up to X band. Output powers of +16 dBm at 8 GHz and +7 dBm at 13 GHz have been achieved with bipolar transistor oscillators by using a finline mount. The near carrier noise for the silicon bipolar transistor was 10 to 20 dB better than the noise characteristic of an FET oscillator operating at the same frequency.

The maximum frequency of oscillation of a GaAs FET is about 80 GHz for a 0.25 μm gate length. The following results have been attained at 35 GHz: +1 dBm output power and noise-to-carrier ratios of –78 dBc/Hz at 10 kHz and –87 dBc/Hz at 25 kHz.

TRAVELING-WAVE-TUBE CIRCUITS

The traveling-wave tube is a unique structure capable of providing high amplification of rf signals varying in frequency over several octaves without the need for any tuning or voltage adjustment (Sec. 7). Figure 11.6.20

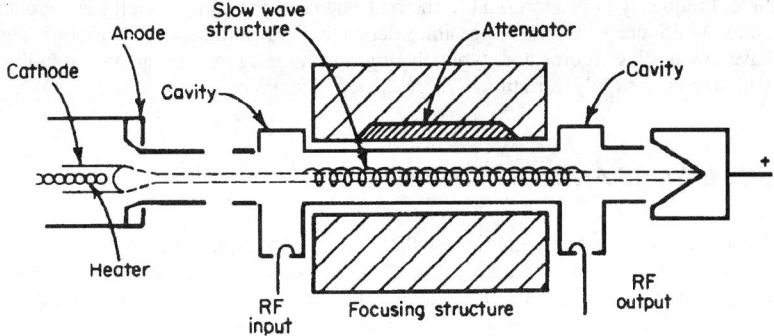

FIGURE 11.6.20   Traveling-wave amplifier circuit.

shows the principal components of an amplifier using this tube. Electrons emitted by the electron-gun assembly are sharply focused, drawn through the length of the slow-wave rf structure, and eventually dissipated in the collector. Synchronism between the rf electromagnetic wave and the beam electrons results in a cumulative interaction which transfers energy from the dc beam to the rf wave. For details on this type of amplifier see Sec. 7.

Figure 11.6.21 shows the gain characteristics of a typical broadband TWT amplifier. As more energy is extracted from the electron beam, it slows down. This loss of synchronism results in lower gains at higher power levels. One advantage of this overdrive characteristic is the protection of following stages against strong signals.

Traveling-wave tube amplifiers provide high gain for signals up to about 30 GHz. Power TWT technology is still advancing. TWTs are available with as high as 30 W cw output power at 12 GHz and 42 percent efficiency. Long-term reliability and noise figure remain poor in comparison to most solid-state devices.

The *backward-wave oscillator* is essentially a TWT device making use of the interaction of the electron stream with an electromagnetic wave whose phase and group velocities are 180° apart. At a sufficient beam current, oscillations are produced as discussed above, without a reverse-wave attenuator. These devices are

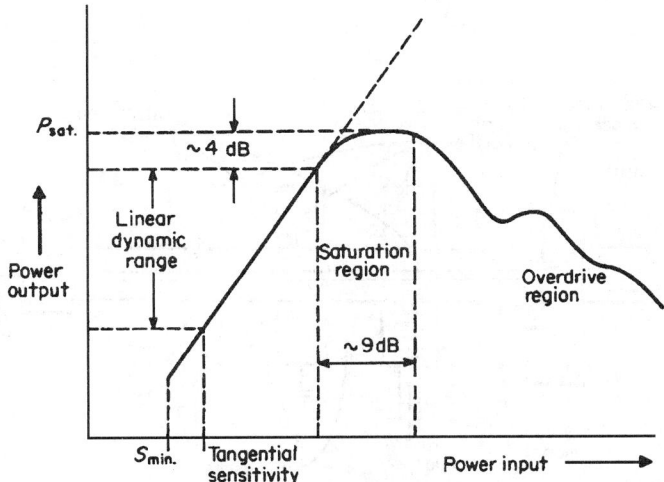

FIGURE 11.6.21   Typical gain characteristic of a broadband TWT amplifier.

voltage-tunable. Frequency is proportional to the half power of the cathode-helix voltage and the dimensions of the structure. Multioctave tuning is possible, depending on output-power-variation requirements. These oscillators have low pulling figures and, typically, high pushing figures. Frequency stability is excellent, usually dependent on power-supply variations. These devices are typically low power (<1 W).

KLYSTRON OSCILLATORS AND AMPLIFIERS

The chief advantages of the klystron amplifier oscillator are that it is capable of large stable output power (10 MW) with good efficiency (40 percent) and high gain (70 dB). Basically, the mechanism involves modulation of the velocity of electrons in a beam by an input rf signal. This is converted into a density-modulated (bunching) beam from which a resonant cavity extracts the rf energy and transforms it to a useful load.

Klystron amplifiers can be conveniently divided into three categories: (1) two-resonator single-stage high- and low-noise voltage amplifiers, (2) two-resonator single stage (*optimum bunching*) power amplifiers, and (3) multiresonator cascade-stage voltage and power amplifiers. The power gain of a two-cavity voltage amplifier can be given by

$$G = \frac{M_1^2 N_2^2}{240\beta} \left(\frac{\pi a}{\lambda}\right)^2 \frac{G_0}{(G_{BR})^2}$$

where M_1, M_2 = beam coupling factors
  a = beam radius
  G_0 = beam conductance
  G_{BR} = cavity shunt conductance contributions because of beam loading and ohmic losses
  β = (electron velocity)/(velocity of light)

A simplified schematic representation of a klystron amplifier is shown in Fig. 11.6.22. Multicavity tubes are typically used for high pulse power and cw applications. The intermediate cavities serve to remodulate the beam, causing additional bunching and higher gain-power output. Optimum power output is obtained with the second cavity slightly detuned. Further, loading of this cavity serves to increase the bandwidth (at the expense of gain). These tubes typically use magnetically focused high-perveance beams.

The broadbanding of a multicavity klystron is accomplished in a manner analogous to that of multistage i.f. amplifiers. A common technique is stagger tuning, which is modified somewhat, because of nonadjacent cavity interactions.

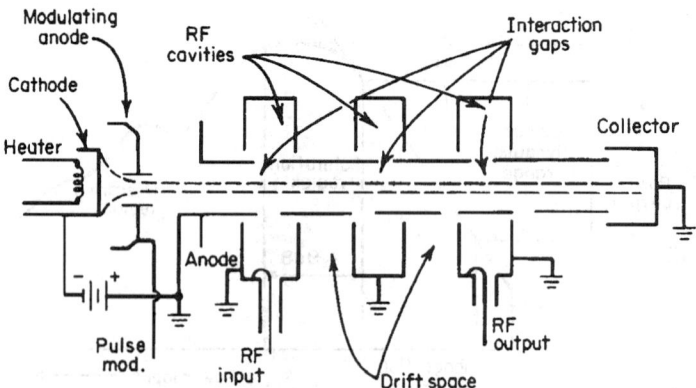

FIGURE 11.6.22 Klystron amplifier structure.

The two-cavity amplifier can be made to oscillate by providing a feedback loop from output to input with proper phase relationship. Klystrons can also be used for frequency multiplication, using the high harmonic content of the bunched-beam current waveforms.

Reflex Klystron

A simple klystron oscillator results if the electron-beam direction is reversed by a negative electrode, termed the *reflector*. A schematic diagram of such a structure is given in Fig. 11.6.23. Performance data for a reflex klystron are usually given in the form of a reflector-characteristic chart. This chart displays power output and frequency deviation as a function of reflector voltage. Two distinct classes of reflex klystrons are low-power tubes for oscillator, pump, and test applications and higher-power tubes (10 W) for frequency-modulator applications. Operating voltage varies from 300 to 2000 V with bandwidths up to ~200 MHz.

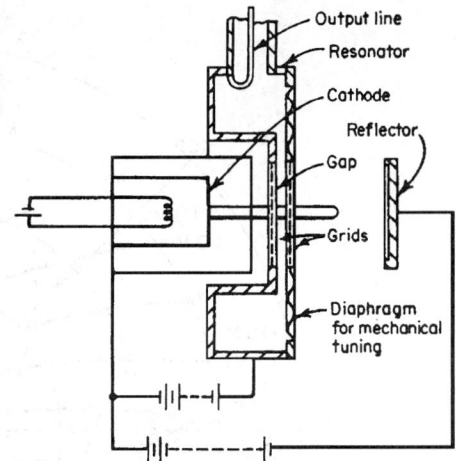

FIGURE 11.6.23   Reflex klystron with coaxial-line loop output.

CROSSED-FIELD-TUBE CIRCUITS

Practically all crossed-field tubes have integrally attached distributed constant circuits. Operation within a critical range of beam current is necessary to maintain the proper bandpass characteristics.

Magnetron

The original microwave tube was a magnetron diode switch with oscillations owing to the cyclotron resonance frequency. Several oscillator circuits were used until the standard cavity-resonator magnetron was introduced in 1940.

The relations between power, frequency, and voltage versus the load admittance are shown in the Rieke diagram (Fig. 11.6.24). Such charts illustrate the compromises necessary to obtain desired operating conditions. In general, good efficiency results from increasing anode current and magnetic field strength.

Various load effects can be considered with the Rieke diagram. The pulling figure [measure of frequency change for a defined load mismatch – (SWR) = 1.5 at all phase angles] and long-line effect are examples.

The area of highest power on the Rieke diagram is called the *sink* and represents the tightest load coupling to the tube. The highest efficiency results in this region; however, a poor spectrum or instability typically results. The buildup of oscillations in the antisink region is closer to ideal; however, this lightly loaded condition may also result in instability. Stability is a measure of the percentage of missing pulses, usually defined at 30 percent energy loss.

Tuning

Methods used to tune a magnetron are classified as mechanical, electronic, and voltage tuning. In the mechanical methods, the frequency of oscillation is changed by the motion of an element in the resonant circuit. Two types are shown in Fig. 11.6.25. An electron beam injected into the cavities will change the effective dielectric constant and hence the frequency. By using the frequency pushing effect it has been possible to tune the magnetron over a frequency range of 4 to 1 (typically, 0.1 to 2 MHz/V) using voltage tuning. The frequency change is usually linear, but the power output is not constant over the tuning range. Magnetrons with power output greater than 5 MW and efficiencies 50 percent or more are available.

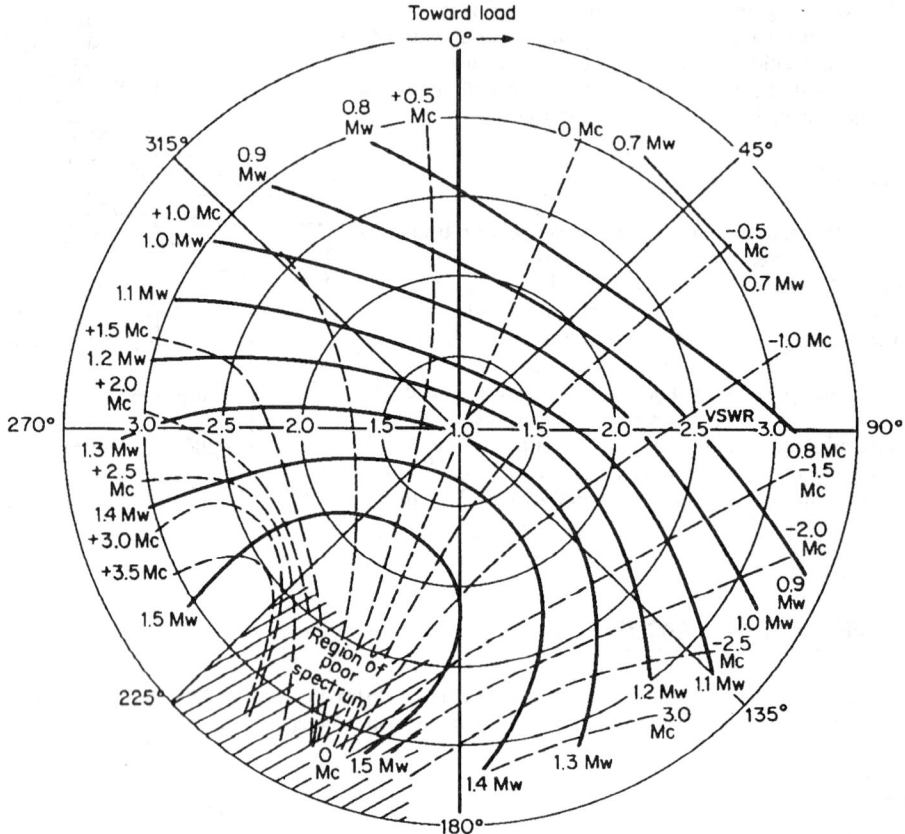

FIGURE 11.6.24 Rieke diagram for an L-band magnetron at 1250 MHz. Solid curves are contours of constant power; dashed curves of constant frequency. (*Raytheon Company.*)

Amplitron

The Amplitron is essentially a magnetron with two external couplings, enabling amplifier operation. It is characterized by high power, broad bandwidth, very high efficiency, and low gain. The output is independent of rf input but dependent on dc input. It acts as a low-loss passive transmission line in the absence of high voltage. A typical plot of power for an Amplitron is shown in Fig. 11.6.26. Conversion efficiency of an Amplitron can be as high as 85 percent. The gain can be increased by inserting mismatches into the input and output transmission lines.

GYROTRON CIRCUITS

Beyond 30 GHz, the power available from classical microwave tubes declines sharply. The gyrotron offers the possibility of high power at millimeter wave frequencies.

The gyrotron is a cyclotron resonance maser. The structure of the gyrotron includes a cathode, a collector, and a circular waveguide of a gradually varying diameter. Electrons are emitted at the cathode with small variations in speed. The electron current density is very high. The electrons are accelerated by an electric field and guided by a static magnetic field. The nonuniform induction field causes the rotational

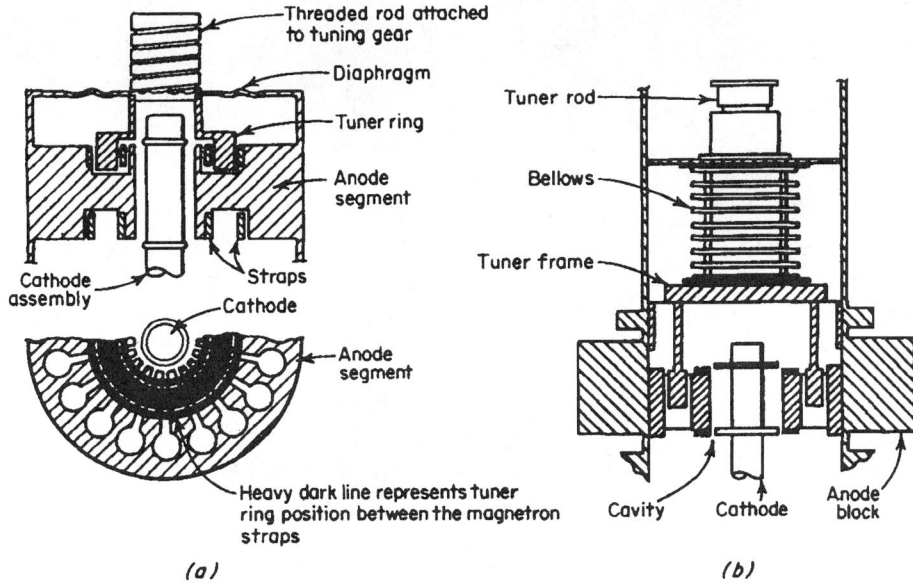

FIGURE 11.6.25 Mechanical magnetron tuning mechanisms: (*a*) capacitance type; (*b*) inductance type.

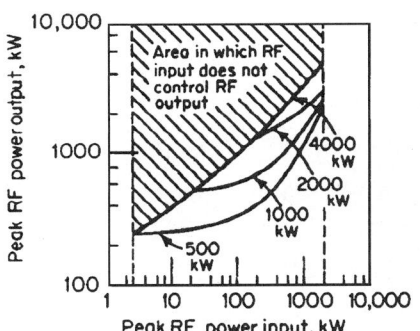

FIGURE 11.6.26 Gain characteristic of model QK520 L-band amplifier.

speed of the electrons to increase, which in turn causes the linear velocity of the electrons to decrease, while they follow a helical trajectory. The microwave field exists within the circular waveguide, where the cutoff of the mode is only slightly above the microwave signal frequency. The electrons give up energy because of interaction with the field of the rotating mode. This causes bunching of the electrons, similar to the bunching within a klystron. Finally there is a decompression zone where the induction decreases and the electrons are collected.

The smooth shape of the gyrotron circular waveguide produces less loss than other microwave tubes. The power available with a gyrotron is 100 or more times greater than that available from classical microwave tubes at the same frequency. The output power levels which can be attained with a gyrotron in cw operation are as follows: 12 kW at 107 GHz and 31 percent efficiency, or 1.5 kW at 330 GHz and 6 percent efficiency.

SECTION 12

MODULATORS, DEMODULATORS, AND CONVERTERS

Understanding modulation is important to understanding AM and FM broadcasting and receiving. The advantages of both are presented and methods of improving both are discussed. The key features of modulation and demodulation are discussed in this section.

Pulse modulation and demodulation is widely used in communications and control. Various techniques are discussed and compared.

Although optical radiation is just another form of electromagnetic radiation, the transmitters and receivers are considerably different from those used at radio frequencies. The use of these devices has increased rapidly necessitating a better understanding of the devices themselves.

We conclude the section with a discussion of frequency converters and detectors. Characteristics that lead to effective design are presented and discussed.   C.A.

In This Section:

Section Bibliography:

Bach Andersen, J. (ed.), "Modern Radio Science 1990," International Union of Radio Science, Oxford University Press, 1990.

Bedrosian, E., "Amplitude and Phase Demodulation of Filtered AM/PM Signals," Rand Corporation, 1986.

Carden, F., "Telemetry Systems Design," Artech House, 1995.

Dixon, R. C., "Spread Spectrum Systems," Wiley-Interscience, 1976.

Dixon, R. C. (ed.), "Spread Spectrum Techniques," IEEE Press, 1976.

Engberg, J., and T. Larse, "Noise Theory of Linear and Nonlinear Circuits," Wiley, 1995.

Freeman, M. H., "Optics," 10th ed., Butterworth-Heinemann, 1995.

Gibson, J. D., and J. L. Melsa, "Introduction to Nonparametric Detection with Applications," IEEE Press, 1996.

Graf, R. F., "The Modern Converter and Filter Circuit Encyclopedia," TAB Books, 1993.

Green, D. C., "Radio Systems Technology," Longman Scientific & Technical, 1990.

Isermann, R., K. Lachman, and D. Matko, "Adaptive Control Systems," Prentice Hall, 1992.

Kahn, D., "Cryptology and the Origins of Spread Spectrum," *IEEE Spectrum*, Vol. 21, No. 9, pp. 70–80, September 1984.

Kaminow, I. P., and E. H. Turner, "Electo-Optic Light Modulators," *Proc IEEE*, Vol. 54, p. 1374, October 1966.

Mass, S. A., "Microwave Mixers," 2nd ed., Artech House, 1993.

Patterson, E. W., et al., "Frequency-Hopped Waveform Synthesis by Using S. A. W. Chirp Filters," *Electro. Lett.*, Vol. 13, No. 21, pp. 633–635, October 13, 1977.

Peterson, R. L., R. E. Ziemer, and D. E. Borth, "Introduction to Spread-Spectrum Communications," Prentice Hall, 1995.

Price, R., "Further Notes and Anecdotes on Spread-spectrum Origins," *IEEE Trans. Comm.*, Vol. Com 31, No. 1, pp. 85–97, January 1983.

Price, R., and P. E. Green, Jr., "A Communication Technique for Multipath Channels," *Proc. IRE*, Vol. 29, pp. 555–570, March 1953.

Purser, M., "Introduction to Error-Correcting Codes," Artech House, 1995.

Schoenbeck, R. J., "Electronic Communications: Modulation and Transmission," 2nd ed., Merrill, Maxwell Macmillan, Maxwell Macmillan International, 1992.

Scholtz, R. A., "The Origins of Spread-spectrum Communications," *IEEE Trans. Comm.*, Vol. Com 30, No. 5, pp. 822–854, May 1982.

Scholtz, R. A., "Notes on Spread-spectrum History," *IEEE Trans. Comm.*, Vol. Com 31, No. 1, pp. 82–84, January 1983.

Schwartz, M., "Information Transmission, Modulation, and Noise," 4th ed., McGraw-Hill, 1990.

Simon, M. K., et al., "Spread Spectrum Communications Handbook," Rev. ed., McGraw-Hill, 1994.

Titsworth, R. C., "Correlation Properties of Cyclic Sequence," *Calif. Inst. Technol. Jet Propulsion Lab. Tech. Rep.* 32–338, July 1963.

Van Trees, H. L., "Detection, Estimation, and Modulation Theory: Radar-Sonar Signal Processing and Gaussian Signals in Noise," Kreiger Pub. Co., 1992.

Walker, B. H., "Optical Engineering Fundamentals," McGraw-Hill, 1994.

Webb, W., "Modern Quadrature Amplitude Modulation: Principles and Applications for Fixed and Wireless Communications," Pentech Press, London; IEEE Press, 1994.

Zlemer, R. E., and W. H. Tranter, "Principles of Communications: Systems, Modulation, and Noise," 4th ed., Houghton Mifflin, 1995.

CHAPTER 12.1
AMPLITUDE MODULATION/ DEMODULATION

Joseph P. Hesler

INTRODUCTION

Frequency translation is fundamental to radio communications in that it produces signal energy, in proportion to variations of an information source, at frequencies that have desirable transmission characteristics, such as antenna size, freedom of interference from similar information sources, line-of-sight to long-range propagation, and freedom of interference from particular noise sources. Frequency translation permits the efficient utilization of open and closed propagation media by many simultaneous users and/or signals.

One of the most used forms of frequency translation is linear modulation, the most common of which is amplitude modulation. In general, amplitude modulation consists in varying the magnitude of a carrier signal in direct correspondence to the instantaneous fluctuations of a modulating signal source, as illustrated in Fig. 12.1.1.

Variations of the basic amplitude-modulation process have been developed to achieve more efficient spectrum utilization and to reduce transmitter power requirements. These include suppressed-carrier systems such as vestigial-sideband, single-sideband, and double-sideband modulation systems. The companion form of frequency translation used in amplitude-modulation systems in *detection*. This is the process whereby the originally translated information is recovered as a baseband signal. Linear amplitude modulation has been the most widely used form of frequency translation in general communications for three reasons: relative ease of implementation, efficient utilization of bandwidth, and availability of devices to implement a simple detection procedure.

It can be argued that amplitude modulation includes such modulation methods as pulse-code keying, pulse-amplitude modulation, frequency-shift keying (the sequential keying of multiple carrier signals), and variations of them, such as pulse-position modulation and pulse-width modulation. This subsection, however, is restricted to the amplitude modulation by signal sources whose outputs are continuous time functions. The special types of modulation mentioned above are discussed in Chap. 12.3.

The general expression for the output of a linear amplitude modulator with a sinusoidal modulation input is

$$E = E_0(1 + m \sin \omega_m t) \sin (\omega_c t - \phi) \qquad (1)$$

where E_0 = peak amplitude of carrier signal
ω_m = modulating signal frequency (rad/s)
ω_c = carrier frequency (rad/s)
m = modulation index
ϕ = arbitrary carrier phase angle (rad)
t = time (s)

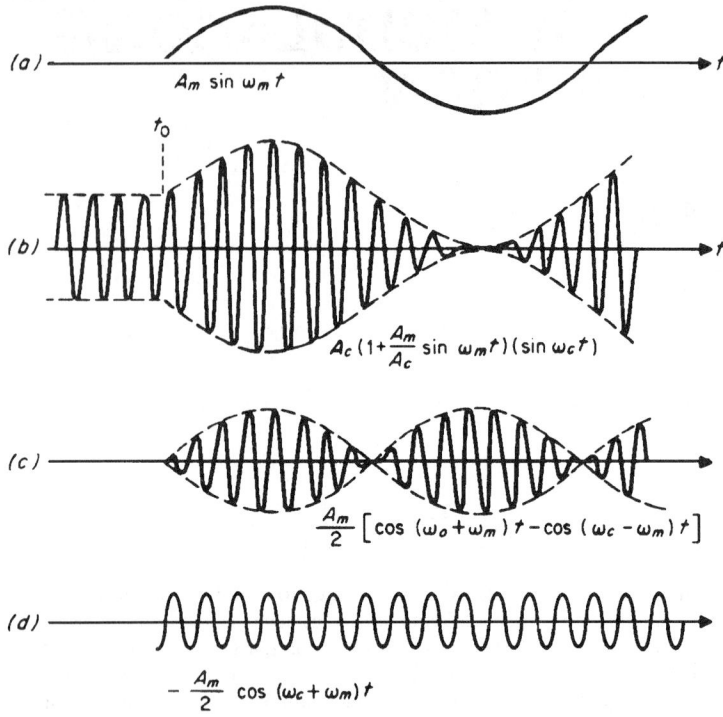

FIGURE 12.1.1   Amplitude modulation: (*a*) modulating signal; (*b*) double-sideband amplitude-modulated signal; (*c*) double-sideband suppressed-carrier amplitude-modulated signal; (*d*) single-sideband suppressed-carrier amplitude-modulated signal.

Expansion of Eq. (1) provides

$$E = E_0 \sin(\omega_c t + \phi) + \frac{mE_0}{2}\cos[(\omega_c - \omega_m)t + \phi] - \frac{mE_0}{2}\cos[(\omega_c + \omega_m)t + \phi] \tag{2}$$

Note that the carrier signal is reproduced exactly as if it carried no modulation. The carrier in itself does not carry any information. The second and third terms in Eq. (2) represented sideband signals produced in the modulation process. These signals are displaced from the carrier signal in the frequency spectrum, on each side of the carrier, by a frequency difference equal to the modulating-signal frequency. The magnitudes of the sideband signals are equal and are proportional to the modulating index m.

Both positive and negative amplitude modulation can be produced in an unsymmetrical manner. For each case the amplitude-modulation index m is defined as

$$m = \begin{cases} \dfrac{E_{\max} - E_0}{E_0} & \text{positive modulation} \\[2ex] \dfrac{E_0 - E_{\min}}{E_0} & \text{negative modulation} \end{cases} \tag{3}$$

where $E_{\max}$ is the peak amplitude of modulated carrier and E_0 is the peak amplitude of unmodulated carrier.

The maximum negative-modulation index of unity results from the reduction of the instantaneous amplitude modulation that can be produced corresponds to a modulation index of unity.

For a complex modulating signal $G(t)$ the modulated carrier spectrum is

$$F(\omega) = \mathfrak{F}\{E_0[1 + mG(t)] \sin (\omega_c t + \phi)\}$$

$$= \frac{E_0}{2\pi} \int_{-\infty}^{\infty} [1 + mG(t)] \sin (\omega_c t + \phi)e^{-j\omega t} dt$$

(4)

where $F(\omega)$ is the Fourier transform of the time function $\mathcal{E}[f(t)]$.

TYPES OF AMPLITUDE MODULATION

Generation of an amplitude-modulated waveform requires the multiplication of signals, $f_1(t)f_2(t)$. Both signals need not be time-variant, e.g., a carbon microphone modulates a dc potential with voice signals. The main applications, however, concern the modulation of *carrier* signals to exploit desirable transmission characteristics, that is, $f_1(t)[A \sin (\omega_c t + \phi)]$. Two classes of circuits are used as modulators, *square-law* devices and *linear* modulators.

Square-Law Modulation

Any device having a nonlinear transfer function can be expressed in a power series form, e.g., the current in a diode, $i(e) = a_0 + a_1 e + a_2 e^2 + a_3 e^3 + \cdots$.

When the diode characteristics and bias conditions are chosen so as to enhance the coefficient a_2 with respect to the other coefficients, the device is considered to be a square-law nonlinear element.

Under these conditions, when two signal inputs $f_1(t)$ and $f_2(t)$ are summed and used as the driving function e, a significant portion of the output will exist as $a_2 e^2$

$$e = f_1(t) + f_2(t)$$

(5)

$$e^2 = 2f_1(t)f_2(t) + [f_1(t)]^2 + [f_2(t)]^2$$

(6)

Suitable nonlinear characteristics are exhibited by various types of rectifiers, triodes, and transistors.

Linear Modulation

Linear modulators are devices with transfer functions that are linearly related to a control parameter. Examples include the outputs of class C rf amplifiers as a function of the B-plus supply voltage, the transconductance gain of transistor differential amplifiers versus emitter current-source magnitude, and Hall effect devices whose transconductance is proportional to the applied magnetic field.

METHODS OF AMPLITUDE MODULATION

Square-Law Amplitude Modulators

A square-law modulator requires three features: a method of summing the two input signals $f_1(t)$ and $f_2(t)$, a device with a nonlinear transfer function, and a tuned circuit and coupling network for extracting the desired modulation products. Voltage summing or current summing are used depending on the transfer characteristic of interest. The nonlinear device is biased in a region that enhances the second-order coefficient of the power series that represents the nonlinear transfer function. The most common devices used for this type of modulation are semiconductor diodes and vacuum-tube triodes. An example of a typical circuit is shown in Fig. 12.1.2.

The efficiency of this type of amplitude modulation is generally low, and all the output energy is supplied by the driving functions.

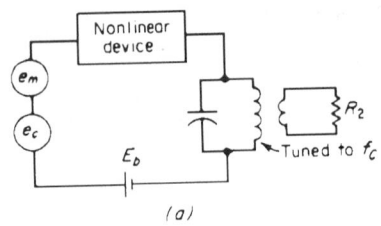

(a)

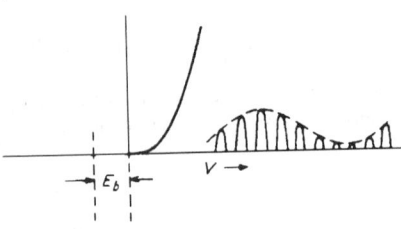

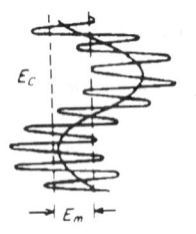

(b)

FIGURE 12.1.2 Square-law modulator.

Consider two input signals:

$$e_m = E_m \cos \omega_m t \quad \text{modulating signal} \qquad (7)$$

$$e_c = E_c \cos \omega_c t \quad \text{carrier signal} \qquad (8)$$

The input applied to the nonlinear device is

$$e_s = E_m \cos \omega_m t + E_c \cos \omega_c t \qquad (9)$$

If the transfer function is represented by the two terms of interest from a Taylor series,

$$e_0 = a_1 e_s + a_2 e_s^2 \qquad (10)$$

The output components resulting are

$$e_0 = (a_2/2)(E_m^2 - e_c^2) \qquad \text{dc rectified component}$$

$$-a_1 E_m \cos \omega_m t \qquad \text{modulating signal}$$

$$-a_1 E_c \cos \omega_c t \qquad \text{carrier}$$

$$-(a_2/2)E_m^2 \cos^2 2\omega_m t \qquad \text{second harmonic of modulation} \quad (11)$$

$$-(a_2/2)E_c^2 \cos^2 2\omega_c t \qquad \text{second harmonic of carrier}$$

$$-a_2 E_c E_m \cos (\omega_c - \omega_m)t \quad \text{lower sideband}$$

$$-a_2 E_c E_m \cos (\omega_c + \omega_m)t \quad \text{upper sideband}$$

There would be other terms, also, from the higher-order coefficients of the Taylor series. The degree of modulation is expressed as

$$\text{Modulation index} = 2(a_2/a_1)E_m \qquad (12)$$

The desired outputs for double-sideband amplitude modulation are

$$e_0' = a_1 E_c \cos \omega_c t + a_2 E_c E_m \cos (\omega_c \pm \omega_m)t \qquad (13)$$

The square-law devices are reciprocal in that the modulating frequency will appear as an output if a modulated signal is applied as the input. Thus the square-law device can also be used as a demodulator or detector.

LOW- AND MEDIUM-POWER LINEAR MODULATORS

Many applications exist in mobile equipment for amplitude modulators with output powers from milliwatts to tens of watts. Transistor circuits are used almost exclusively for these circuits. Carrier frequencies above 1 GHz can be used in the lower-power transistor circuits. The most common methods of amplitude modulation used are class C collector-modulated stages with the rf applied in the common-emitter or common-base configuration, as shown in Fig. 12.1.3. The common-emitter configuration provides the maximum power gain and excellent efficiency. Common-base stages are used to increase the upward modulation capabilities, where the maximum modulation indices are important. Increased linearity can be achieved at the expense of efficiency by biasing the amplifier class B so that a nominal collector current flows under no-modulation conditions.

For class C operation the transistor transfer characteristics of the modulated amplifier are determined from the large-signal input and output parallel equivalent-impedance data. These are determined experimentally or provided on device specification sheets. The transistors are operated in a very nonlinear manner as class C amplifiers; therefore the experimental data should be representative of the expected operating point, because the small-signal transistor parameters are not adequate.

POWER RELATIONSHIPS

The instantaneous rms output voltage E varies about the unmodulated carrier rms level E_0. The maximum rms output is $E_0(1 + m)$, where m is the modulation index that can vary from zero to unity for symmetrical modulation. The minimum rms output is $E_0(1 - m)$. The unmodulated power into the load R is

$$P_o = E_0^2/R \qquad (14)$$

The peak power into the load is

$$P_{\max} = [E_0(1 + m)]^2/R \qquad (15)$$

The minimum is

$$P_{\min} = [E_0(1 - m)]^2/R \qquad (16)$$

The average power into the load for sinusoidal modulation is

$$P_{av} = P_o(1 + m^2/2) = (E_0^2/R)(1 + m^2/2) \qquad (17)$$

The unmodulated output power E_0^2/R is supplied by the class C amplifier, and the sideband energy $(E_0^2/R)(m^2/2)$ is supplied by the modulator. The class C amplifier can be biased very close to the peak modulated-output envelope swing, $V_{CC} \approx \sqrt{2}mE_0$. For 100 percent modulation, $V_{CC} \approx \sqrt{2}E_0$. The dissipation in the output voltage stage is the difference between the total input power, consisting of the dc collector bias and the input rf drive, and the output rf power. The input rf drive for an amplifier with power gain A is

$$P_{\text{drive}} = E_0^2/RA \qquad (18)$$

The input dc bias, unmodulated, is slightly greater than

$$P_{\text{dc}} = \sqrt{2}E_0\overline{I}_c \qquad (19)$$

where $\overline{I}_c$ is the average dc collector current.

The output transistor dissipation unmodulated is

$$P_{TR_O} \approx \sqrt{2}E_0\overline{I}_c + P_{\text{drive}} - E_0^2/R \qquad (20)$$

$$P_{TR_O} \approx \sqrt{2}E_0\overline{I}_c - \frac{E_0^2}{R}\frac{A-1}{A} \qquad (21)$$

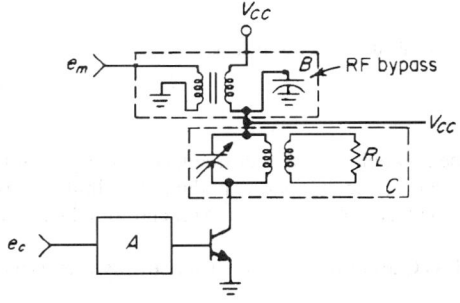

A = input impedance matching network

B = modulator circuit for producing V_{CC} which follows the modulating signal, e_m, and provides a low source impedance at f_c

C = Frequency selective network, $f_c \pm f_m$, and output impedance matching network

FIGURE 12.1.3 Collector-modulated transistor (class C) rf amplifier.

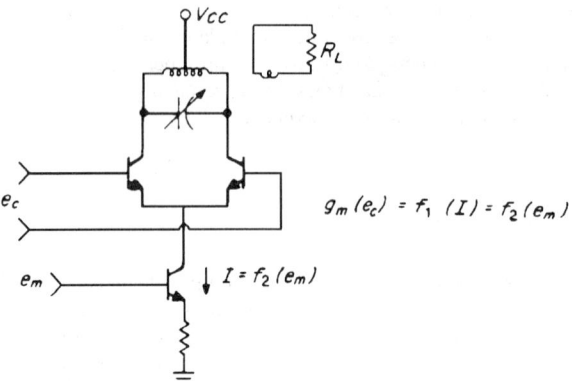

FIGURE 12.1.4 Differential amplifier amplitude modulator (without dc bias details).

In higher-power systems the output transistors can be paralleled. At higher frequencies the gain of the output transistors may be such that the output power is limited by the dissipation in the driver. In these instances the driver may also be collector-modulated to achieve adequate upward modulation and to reduce the power dissipation in the driver.

The input and output impedances of the transistor class C amplifiers are characteristically low. Typical parallel equivalent input impedances are

$$2 < R_{in} < 50 \ \Omega \qquad 30 < C_{in} < 5000 \ \text{pF}$$

The collector load impedance-resistance component is determined from the bias voltage and output power

$$R'_L = (V_{CC})^2 / 2 P_o \qquad (22)$$

where V_{CC} is the dc collector bias voltage and the output voltage swing $2V_{CC}$ peak to peak and P_o is the unmodulated output power.

The collector parallel output capacitance depends on the device geometry and may range from a few picofarads for very-high-frequency lower-power devices to a few hundred picofarads for large-geometry high-power devices.

Interstage and output networks are used to obtain conjugate matches for maximum power gain.

Other types of linear transistor modulators can be used where efficiency is less critical or where very-wide-band operation precludes effective output filtering for the elimination of harmonics. A differential amplifier circuit, as shown in Fig. 12.1.4, will have a transconductance gain that is very nearly proportional to the emitter current-source magnitude. Modulation of the current source will produce a nearly ideal multiplication of the rf input signal and the modulation signal for a wide range of low current levels. The differential amplifier must be biased class A with minimum dc offset at the base inputs. A single transistor multiplier can also be used, as shown in Fig. 12.1.5 for very-low-level outputs. An emitter bypass capacitor for the rf signal is used in place of the additional transistor for the rf return.

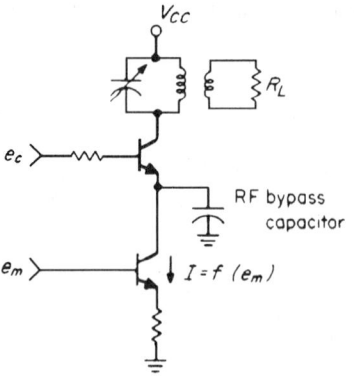

FIGURE 12.1.5 Two-transistor transconductance modulator.

HIGH-POWER LINEAR MODULATORS

High-power linear modulators, 50 W and up, are generally constructed using class C plate-modulation vacuum-circuits. Some of the intermediate power and frequency applications use paralleled transistor configurations with class C collector modulation.

Triodes, tetrodes, and pentodes are used in the vacuum-tube circuits. Multigrid tubes require screen-grid modulation in conjunction with the control-grid modulation to achieve space-charge modulation and to minimize screen current and screen dissipation. The two methods of screen modulation commonly used are: (1) self-bias of the screen grid with a bypassed dropping resistor or inductor from the screen to the plate supply or the screen-grid supply and (2) screen modulation via a separate winding on the modulation transformer.

GRID MODULATION

The amplitude of a class C rf amplifier output can also be modulated by changing the grid bias with the modulating signal. The modulating signal is added to the rf input signal. The effect is to change the magnitude of the plate-current pulses, and hence the fundamental component of the plate current.

The disadvantage associated with grid modulation is that the fixed-plate supply voltage must be twice the peak rf voltage without modulation. This causes high plate dissipation and lowers the plate efficiency to the range of 35 to 45 percent when unmodulated.

The carrier power obtained from a plate-modulated class C amplifier is about three times that available from a grid-modulated circuit using the same tube.

CATHODE MODULATION OF CLASS C RF AMPLIFIERS

Cathode modulation can be used with a class C rf amplifier. The modulation transformer output varies the grid-cathode as well as the plate-cathode voltages. The ratio of grid and plate modulation can be selected by varying the tap: thus the circuit provides a means of producing varying combinations of grid and plate modulation. Some grid-leak bias is normally used to improve linearity.

MODIFIED AM METHODS

The information transmitted by an amplitude-modulated carrier is contained wholly in the modulation sidebands. The transmission of the carrier energy simplifies the receiver-detector implementation but adds no information. In addition, each sideband contains the same information, and only one is required to transmit the intelligence. Elimination of the carrier and/or one sideband can affect a substantial transmitter power saving. For 100 percent amplitude modulation the carrier power is two-thirds of the transmitter power and each sideband one-sixth. Elimination of the carrier will result in *double-sideband suppressed-carrier modulation*. Elimination of one sideband while retaining the carrier or a substantial portion of the carrier results in *vestigial-sideband transmission*. Elimination of the carrier and one sideband is called *single-sideband suppressed-carrier modulation*.

The easiest of these modulation schemes to implement is the vestigial-sideband transmission, both from a transmitter and a receiver viewpoint. The unwanted sideband is generally filtered out at low levels in the transmitter chain and is known as *transmitter attenuation* (TA). One sideband can also be eliminated in the receiver by selective filtering and is called *receiver attenuation* (RA). The latter scheme is not practical from a spectrum-utilization sense; hence it is normally used only as in television broadcast to complete unwanted sideband rejection performed primarily at the transmitter.

A vector notation can be used to illustrate the phenomena of various types of amplitude modulation. The vector is a complex function. The sinusoidal function of time that is of interest is the real part or the vector projection on the real axis of the complex plane. Thus the real part of $Ae^{j\omega t}$ is $A \cos \omega t$, as shown in Fig. 12.1.6a.

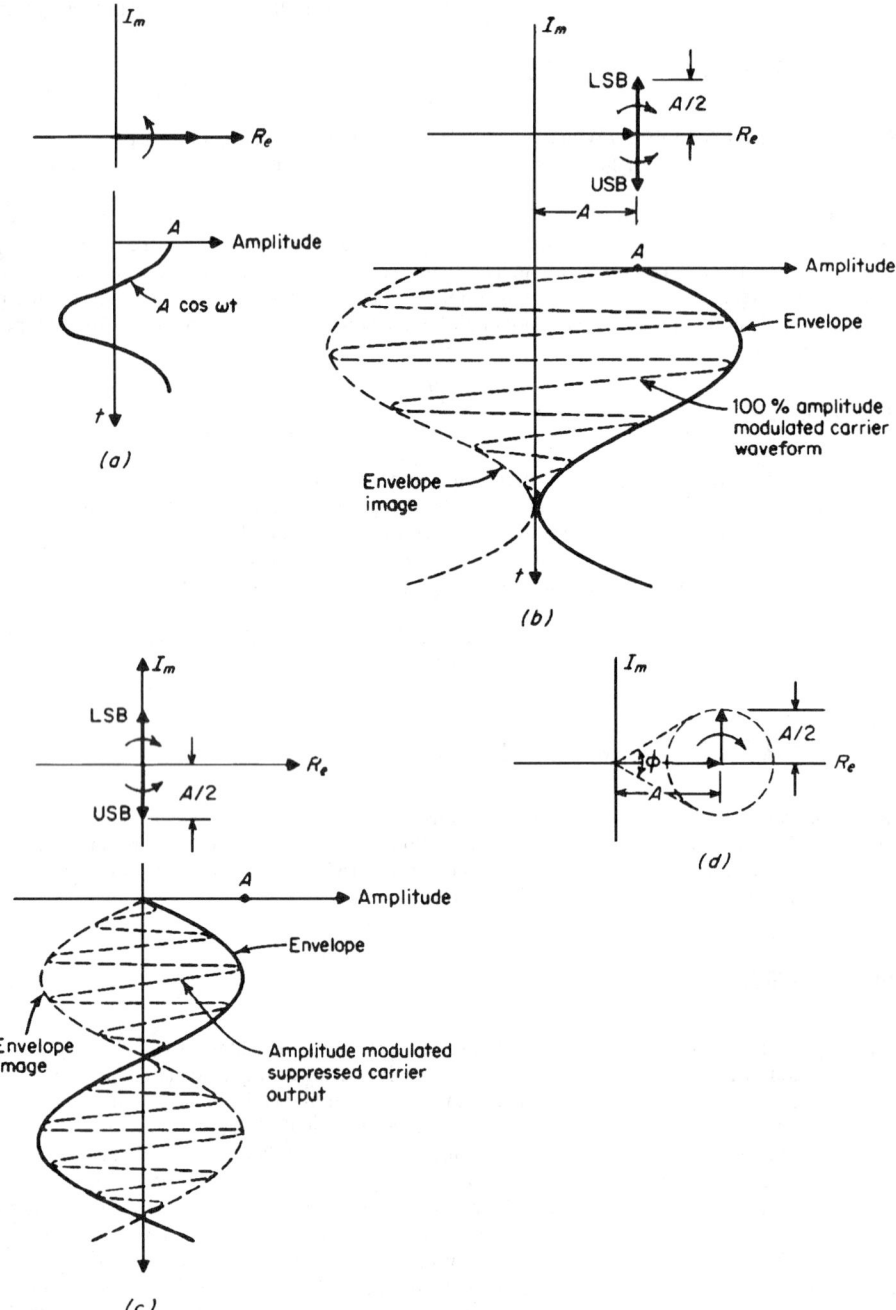

FIGURE 12.1.6 Vector representation of amplitude modulation, (*a*) Rotating carrier vector and its real projection. (*b*) Vector representation with sideband vectors resulting from 100 percent amplitude modulation. Envelope function is locus of projection on the real axis with a nonrotating carrier vector. (*c*) Double-sideband suppressed-carrier vector representation. Envelope function is locus of projection on the real axis with coordinate rotation as in (*b*). (*d*) Vector representation of a single-sideband amplitude-modulated waveform showing peak-to-peak carrier phase modulation ϕ resulting from the absence of the other sideband signal.

The projection on the real axis of the vector $Ae^{j\omega t}$ can be considered as the carrier signal in subsequent amplitude-modulation (AM) discussions. Since the unmodulated carrier signal in amplitude-modulation processes is a fixed peak amplitude and fixed frequency function of time, a modified vector representation can be used to describe the envelope of the modulated waveform. The modified vector diagram maintains the carrier vector as a fixed, nonrotating vector. By this means subsequent illustrations are referenced to a complex plane rotating at the carrier angular rate, and the projection on the real axis corresponds to the *modulation envelope variations with time*.

For an amplitude-modulation system using a sinusoidal modulating function at 100 percent modulation, $m = 1$, the addition of two vectors to the basic vector diagram is required. The two additional vectors represent the sideband signals produced in the modulation process. The two sideband signals are displaced on either side of the carrier signal in the modulated signal spectrum by a frequency equal to the modulating frequency. Therefore, with respect to the modified reference system, the lower sideband vector will rotate clockwise at the modulation-signal angular rate, and the upper sideband will rotate counterclockwise at the same rate.

The relative phases of the two sideband vectors are such that no change in carrier phase angle is introduced when they are both present, i.e., the imaginary parts of the sideband contributions cancel. The initial angles are set by the phase of the modulating function. As the sideband vectors rotate in the modified complex plane, the vector sum of the carrier plus sidebands describes sinusoidal projection on the real axis at the modulating frequency rate.

Note that the unmodulated carrier projection is constant and equal to the peak carrier magnitude. The 100 percent modulated carrier projection is nonnegative. If the original complex plane were used, the entire vector system would rotate at the carrier angular rate, and the projection on the real axis would be the actual time function produced by the modulator. This function would have negative portions and would fill the envelope function and its image with amplitude-modified sinusoids at the carrier frequency, as shown in Fig. 12.1.6b with the dashed lines within the envelope.

VESTIGIAL-SIDEBAND SYSTEMS

Vestigial-sideband transmission introduces angle modulation on the carrier because the symmetry of the contrarotating sideband vectors is lost, as illustrated in Fig. 12.1.6d. A standard envelope detector can still be used to recover the modulating signal, however. The primary objective in the application of vestigial-sideband transmission is to conserve spectrum in the transmission medium.

SUPPRESSED-CARRIER SYSTEMS

Suppressed-carrier systems for AM transmission and reception require modifications to the receiver. The double-sideband suppressed-carrier signal cannot be envelope-detected without the reinsertion of a carrier signal. The frequency and the phase of the reinserted carrier are critical. This type of transmission is used with special phase-locked receivers or with the transmission of low-level carrier to permit the reconstitution of the carrier frequency and phase at the receiver. A double-sideband suppressed-carrier signal can be generated with a balanced modulator, as shown in Fig. 12.1.7a.

The nonlinear devices can be diodes or modulated class C rf amplifiers. The balanced modulator simplifies the tuned-output-circuit design because all even harmonics of the modulating process tend to be canceled along with the carrier.

Single-sideband suppressed-carrier modulation simplifies the receiver design in some applications. The phase of the reinserted carrier at the receiver is not critical as in double-sideband suppressed-carrier systems. Also, a frequency error in the reconstituted carrier will result only in a corresponding shift in the demodulated signal frequencies, which may be tolerable. If accurate modulation-frequency preservation is required, a low-level carrier can be transmitted to aid in the reconstruction of the carrier signal at the receiver. The reinserted carrier amplitude with respect to the received sideband signal is generally made large to minimize angle modulation of the carrier at the detector.

Two methods are usually employed to generate suppressed-carrier single-sideband transmissions. The most direct method is to filter out the undesired sideband and carrier at low levels in the transmitter chain.

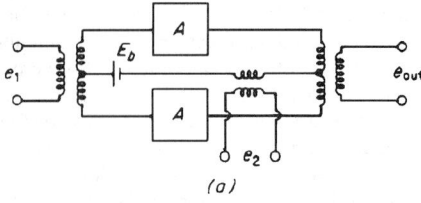

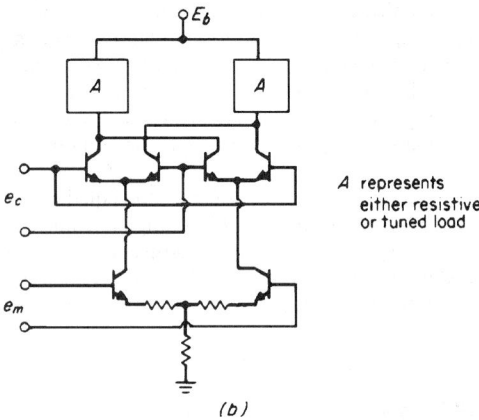

FIGURE 12.1.7 Double-sideband suppressed-carrier modulators: (*a*) general form of a balanced modulator for suppressed-carrier output using nonlinear elements *A* and bias voltage E_m, (*b*) four-quadrant multiplier.

This filtering problem is difficult when the modulation sidebands of interest are close to the carrier and the carrier frequency is high.

An alternative method of suppressed-carrier single-sideband modulation is by unwanted sideband cancellation. Two balanced modulators are used, with their outputs combined in push-pull.

MODULATED OSCILLATORS

A direct modulated class C oscillator can be used as an AM transmitter. The linearity of such circuits is generally as good as or better than the plate-modulated class C amplifiers. The main disadvantage is the tendency for carrier-frequency pulling that results from the changes in the oscillator operating point as the modulation signal varies.

A very useful circuit for the generation of low-level double-sideband suppressed-carrier signals is the four-quadrant multiplier. The same circuit will also operate as a double-sideband modulator with carrier and as low-level demodulator in phase-locked receiver systems. Multipliers are available in integrated-circuit form, and they can be used at frequencies from dc to beyond 100 MHz (Fig. 12.1.7*b*).

DETECTORS

The most common AM detector or demodulator is a diode rectifier. The ideal diode detector passes current in only one direction and will essentially follow the envelope of an amplitude-modulated waveform when used in a circuit as shown in Fig. 12.1.8.

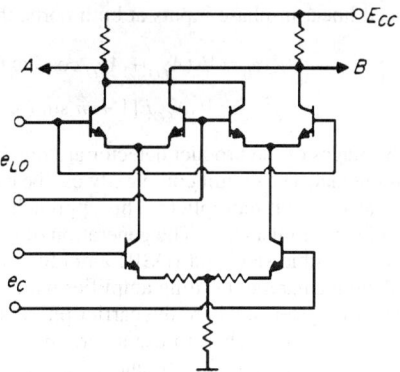

FIGURE **12.1.9** Four-quadrant multiplier used as a product detector. The local-oscillator input e_{LO} must be phase-coherent with the carrier of the modulated input signal e_C. Complementary outputs are obtained at A and B.

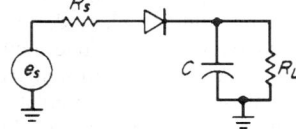

FIGURE **12.1.8** Diode envelope detector.

The charging time constant R_sC must be short, so that the capacitor voltage follows the input signal E_{rf} when the diode is forward-biased or conducting. Conversely, the discharge time constant R_LC must be long enough to retain most of the rectified voltage between cycles of the carrier signal but not so large that the capacitor voltage will not discharge at the maximum rate of change of the input signal envelope. The envelope detector is essentially insensitive to residual angle modulation of the carrier, and it is therefore usable in single-sideband receivers.

Practical diode rectifiers have nonlinear resistance characteristics in the conduction bias region and are therefore operated at fairly high input signal levels, on the order of 2 to 10 V peak for semiconductor diodes.

PRODUCT DETECTORS

Another type of amplitude demodulator is the product detector, or multiplier circuit (Fig. 12.1.9). This circuit has distinct advantages and disadvantages. The advantages include the ability to detect lower-level signals with a linear response; the ability to differentiate phase reversals in the modulated waveform, resulting from balanced amplitude modulation with suppressed carrier; and the ability to produce, in some designs, error signals for automatic-frequency-control systems in receivers.

The analytical expression for the output of a product detector is

$$e(t) = E \underbrace{[1 + m \underbrace{\sin(\omega_m t + \phi_m)}] \cos(\omega_c t + \phi_c)}_{\text{amplitude-modulated signal}} \underbrace{\cos(\omega_c t + \phi_d)}_{\text{local-oscillator signal}}$$

(23)

Expansion of Eq. (23) shows that the product of the incoming carrier signal, $\cos(\omega_c t + \phi_c)$, and the local-oscillator signal, $\cos(\omega_c t + \phi_d)$, produces a dc term except when these two inputs are in quadrature phase. The output dc term is proportional to the cosine of the relative phase angle of the carrier and local-oscillator signals. A four-quadrant multiplier circuit capable of performing this type of detection is shown in Fig. 12.1.9.

The details of the dc bias network are not included in the elementary schematic of the multiplier circuit. The input signals, rf and local-oscillator, are applied to either input port. Balanced-differential or single-ended inputs can be used, although the balanced inputs give the added performance of increased linearity and common-mode signal rejection. Two outputs of opposite polarity are available at the collectors of the upper-rank transistors. Design options are available to increase the efficiency and linearity of the circuit. Normally, an overdrive is applied to the local-oscillator port to make that section of the multiplier operate in a switching mode. The effect is to multiply the rf signal with a square wave at the carrier frequency instead of a sine wave. This type of operation also produces additional outputs at the higher harmonics of the carrier frequency.

For balanced linear in-phase inputs at both ports, the outputs consist of

$$e_0 = \tfrac{1}{2}(A_{LO} + A_{LO} \cos 2\omega_c t)E[1 + m \sin (\omega_m t + \phi_m)]$$
$$= \tfrac{1}{2}A_{LO}E[1 + m \sin (\omega_m t + \phi_m)] + \tfrac{1}{2}A_{LO}E \cos 2\omega_c t \tag{24}$$

The disadvantages of the product detector are relative circuit complexity and the need for a phase-coherent local-oscillator signal. The circuit complexity can be circumvented for input signal frequencies up to 100 MHz by using integrated-circuit multipliers. This approach also minimizes the possibility of serious dc offset problems because of device mismatch. The generation of the coherent local-oscillator signal can be achieved in two ways. For simple double-sideband (DSB) amplitude-modulated signals, the carrier can be stripped from the input rf signal, with a parallel limiting amplifier with narrow bandwidth. This approach cannot be used in suppressed-carrier AM systems, where the carrier phase reversals are introduced in the modulation process.

The more general approach is to use an additional multiplier circuit as a phase detector. When the rf and local-oscillator signals are equal in frequency and in phase quadrature, the multiplier output has no dc component. Any relative phase shift from quadrature will produce an odd-function-error signal at dc. This signal can be used with a voltage-controlled oscillator to correct the phase of the local oscillator or input rf signal. The system described is a phase-locked loop. The bandwidth of this loop can be controlled independently from the rf bandwidth for optimum acquisition and noise-suppression characteristics. This type of system is capable of producing a stable and noise-free local-oscillator signal that tracks any variations in the frequency and phase of the input rf signal.

With additional modifications the phase-locked receiver system is capable of reinserting the desired carrier in suppressed-carrier double- and vestigial-sideband systems.

CHAPTER 12.2
FREQUENCY AND PHASE MODULATION

Nobel R. Powell

ANGLE MODULATION

The representation of angle modulation is conveniently made in terms of the notion of the analytic function. For real continuous functions of time $x(t)$, consider

$$m(t) = x(t) + jHx(t)$$

where $x(t)$ = real continuous function
$\qquad j = (-1)^{1/2}$
$\qquad Hx$ = Hilbert transformation of x

$$Hx(t) = \pi^{-1} \int_{-z}^{z} x(\tau)(t - \tau)^{-1} \, d\tau$$

The angle of $m(t)$ is said to be the angle θ, defined by

$$\theta = \tan^{-1} (Hx/x)$$

Angle modulation may be considered as the change in θ with time or as that portion of the total change in θ which can be associated with the phenomenon of interest. If the relationship between the changes in θ and the effect of interest is direct, the modulation is called *phase modulation* (PM) and the devices producing this relationship *phase modulators*. If the relationship between the changes in the derivative $d\theta/dt$ and the effect of interest is direct, the modulation is called *frequency modulation* (FM), and the devices producing this relationship are called *frequency modulators*.

The derivative $d\theta/dt = \dot\theta$ is related to the components of the analytical function $m(t)$ by

$$\dot\theta = \frac{\begin{vmatrix} x & Hx \\ \dot x & H\dot x \end{vmatrix}}{x^2 + (Hx)^2} \tag{1}$$

for functions $x(t)$ for which the differential and Hilbert operators commute.

Thus in angle modulators, whether implemented functionally in the general form (as with a general-purpose computer) suggested by the representation of θ and $\dot\theta$, or as some special combination of electronic networks,

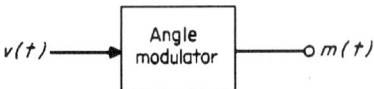

FIGURE 12.2.1   Linear angle modulator.

a direct or proportional relationship is established within the device between either of these two functions and an effect, call it the input $v(t)$, of interest. Diagrammatically, a linear angle modulator can be considered to be a device that transforms $v(t)$, as shown in Fig. 12.2.1.

Demodulators simply perform the inverse of this operation, providing an output function proportional to $\theta(t)$ from a function proportional to the angle of $m(t)$ as an input.

ANGLE-MODULATION SPECTRA

The spectral distribution of power for angle-modulated waveforms varies widely with the nature of the input function $v(t)$.

Random Modulation

For a random process having sample functions

$$x(t) = b \, \sin \, (2\pi f t + \phi) \quad b\text{-const}$$

where f and ϕ are independent random variables, ϕ being uniformly distributed over $-\pi \le \phi \le \pi$ and f having a symmetric probability density $p(f)$, this stationary random process has a spectral density function

$$S_{xx}(f) = (b^2/2)p(f) \tag{2}$$

Note that if $p(f)$ is not a discrete distribution, the process is not in general periodic and $S_{xx}(f)$ must be considered continuous.

Periodic Modulation

For a random process having sample functions

$$x(t) = b \, \cos \, [\omega_c t - \phi(t) + \theta] \quad b, \omega_c\text{-const}$$

θ uniformly distributed over $-\pi \le \theta \le \pi$. $\phi(t)$ is a stationary process independent of θ; that is,

$$\phi(t) = d \, \cos \, (\omega_m t + \theta') \quad d, \omega_m\text{-const}$$

For θ' uniformly distributed over $-\pi \le \theta' \le \pi$, the spectral density function is

$$S_{xx}(f) = (b^2/4)\Bigg(J_0^2(d)[\delta(f - f_c) + \delta(f + f_c)]$$

$$+n\sum_{n=1}^{\infty} J_n^2(d)\{\delta[f - (f_c \pm nf_m)] + \delta[f + (f_c \pm nf_m)]\}\Bigg) \tag{3}$$

where $J_n(d)$ = nth-order Bessel function of first kind evaluated at d
  δ = Kronecker delta function
  $S_{xx}(f)$ = Fourier transformation of autocorrelation function of x

Deterministic Modulation

For a function of a completely specified type, e.g.,

$$x(t) = b \sin(\omega_c t + d \sin \omega_m t) \tag{4}$$

frequently it is possible to reexpress $x(t)$ in terms of $J_n(d)$ as

$$x(t) = \sum_{n=-\infty}^{\infty} J_n^2(d) \sin(\omega_c t + n\omega_m t) \tag{5}$$

Examination of Eqs. (1) and (3) and tables of Bessel functions permits the construction of Fig. 12.2.2 showing the implied increase in bandwidth versus modulation index for FM. Modulation index for sinusoidal modulation can be defined as

$$d = |\Delta f|/f_m$$

where $(\Delta f) = df_m$ = amount of instantaneous frequency change $\dot{\theta}$ to be associated with the input $v(t)$.

Such a set of curves can be readily constructed for most deterministic modulation waveforms, since $J_n(d)$ decreases monotonically and rapidly with n for $n > d > 1$. Using the criteria for n indicated for curves A, B, and C, the frequency range (or bandwidth) centered about ω_c, which contains all such spectral components, is indicated. These are the components that are found to be below the value of nf_m for which $J_n(d)$ is monotonically decreasing and equal to the value for each case. Thus the bandwidth (BW) required at the frequency ω_c is

$$BW = 2df_m(1 + I) \tag{6}$$

Such criteria should be used with care, since the relationship which these measures bear to distortion of the input $v(t)$ when carried at frequency ω_c through linear-tuned circuits as angle modulation is rather indirect.

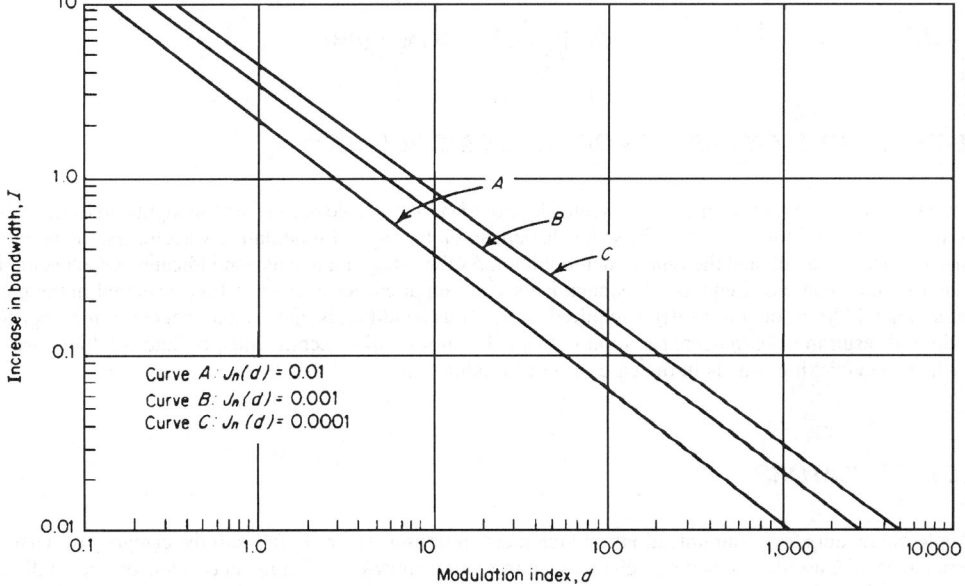

FIGURE 12.2.2 Bandwidth increase vs. modulation index.

ANGLE-MODULATION SIGNAL-TO-NOISE IMPROVEMENT

One of the principal reasons for using frequency modulation in communications and telemetry systems is that it provides a convenient and power-efficient method of trading power for bandwidth while providing high-quality transmission of the input. This is expressed in phase modulation by the relationship

$$\left.\frac{S}{N}\right|_{fm} = ad^2 \left.\frac{C}{N}\right|_{fm} \tag{7}$$

where $(S/N)|_{fm}$ = demodulated output signal-to-noise ratio measured in a bandwidth f_m
  $(C/N)|_{fm}$ = demodulator input signal-to-noise ratio measured in a bandwidth f_m
  d = modulation index
  a = a constant of proportionality

The constant of proportionality a is unity for sinusoidal phase modulation of constant-amplitude sinusoidal carrier. The constant varies between 0.5 and 3.0 with class of modulation waveforms, type of network compensation, and noise spectrum; however, Eq. (7) can be conservatively applied to FM single-channel voice systems for $a = {}^3/_2$ and preemphasis and deemphasis networks that preshape the spectrum of the modulation to match the sloped noise spectrum and to restore the original spectrum after demodulation.

The trade-off between the signal-to-noise improvement and the required bandwidth corresponding to curve A in Fig. 12.2.2 is shown for FM in Fig. 12.2.3. These curves have been prepared for a conventional demodulator operating at an input carrier-to-noise ratio 1 dB above the threshold (13 dB) measured in the input noise bandwidth to the demodulator, with a constant of proportionality a equal to 0.5. Output signal-to-noise ratio referred to the output information bandwidth $(S/N)_{fm}$ is

$$(S/N)|_{fm} = 10 + G \text{ (dB)} \tag{8}$$

where G is obtained from the figure along with r, the ratio of premodulator bandwidth to output information bandwidth. The carrier-to-noise ratio in a noise bandwidth equal to the output information bandwidth is

$$(C/N)|_{fm} = 13 + 10 \log r \text{ (dB)} \tag{9}$$

NOISE-THRESHOLD PROPERTIES OF ANGLE MODULATION

The signal-to-noise improvement represented by Eq. (7) is achievable only when the input carrier-to-noise ratio is above certain minimum levels. These levels depend on the type of modulating waveforms, the type of noise interference prevalent, and the type of demodulator. As the foregoing discussion indicates, whenever a demodulator without phase or frequency feedback is used, an input carrier-to-noise ratio (measured in the premodulator bandwidth) of roughly 12 dB is required. Unless this condition is met, a small decrease in carrier-to-noise ratio will result in a sharp decrease in output signal-to-noise ratio, accompanied by undesirable noise effects, such as load clicking sounds in the case of voice modulation.

ANGLE MODULATORS

Angle modulators for communications and telemetry purposes generally fall into the category of what may be termed "hard" oscillators having relatively high-Q frequency-determining networks; or they fall into the category of "soft" oscillators having supply and bias sources as the frequency-determining networks. Examples of each are shown in Fig. 12.2.4.

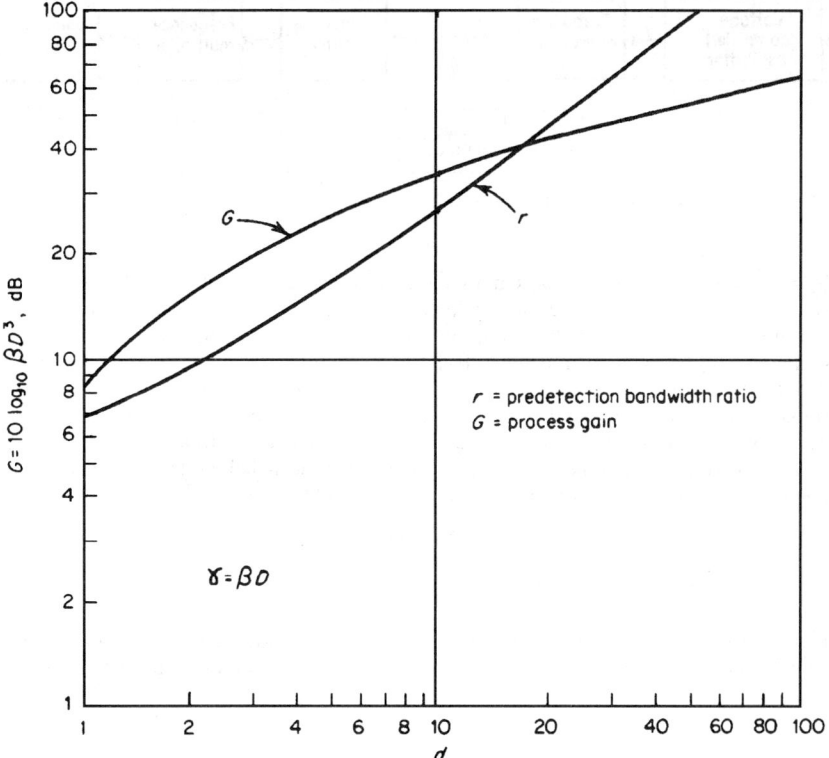

FIGURE 12.2.3   Process gain and rf bandwidth vs. deviation ratio.

Control of the hard oscillator is executed by symmetrical incremental variation of the reactive components. For the case of a Hartley oscillator, Z_1 and Z_2 are inductors and Z_3 is a capacitor, allowing the use of varicaps paralleling, Z_3 as the voltage-controllable reactance. In such a case

$$f_o = (2\pi)^{-1}[C_3(L_1 + L_2)]^{-1/2}$$

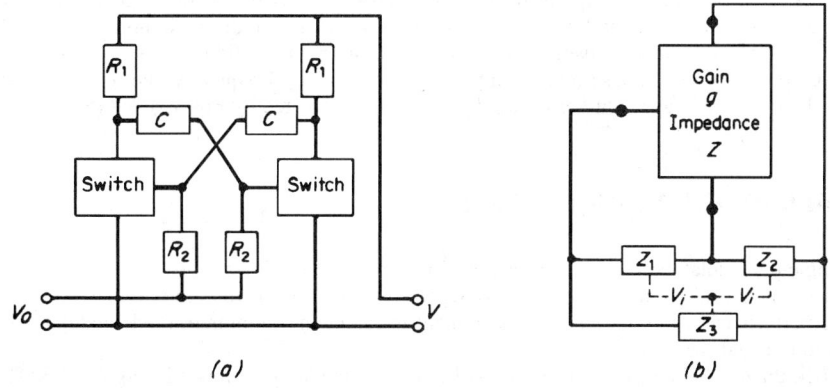

FIGURE 12.2.4   Voltage-controlled oscillators: (*a*) soft oscillator; (*b*) hard oscillator.

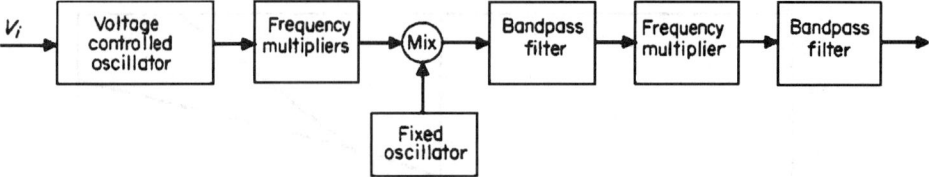

FIGURE 12.2.5   Frequency-modulator configuration.

where C_3 = total capacitance of varicaps and fixed capacitor of Z_3. Note that the bandwidth of this modulator is determined by the frequency-determining impedances of the network, i.e., the overall Q and center frequency. In view of the need for certain minimum bandwidth requirements from Fig. 12.2.2 and the need for good oscillator stability, the total frequency deviation required is sometimes obtained by following the oscillator with a series of frequency multipliers and frequency translators, as shown in Fig. 12.2.5. This configuration permits the attainment of good oscillator stability, constant proportionality between output frequency change and input voltage change, and the necessary modulator bandwidth to achieve wide-band FM.

Control for the soft oscillator is introduced as a change in the switching level of the active-device switches. The frequency of a transistor version of the oscillator is, roughly,

$$f = [2R_1 C \ln (1 + V/V_i)]^{-1} \tag{10}$$

Since this type of oscillator is a relaxation oscillator, the rate at which the frequency of oscillation can be changed is limited only by the rate at which the switching points can be altered by voltage control. Modulators of this kind can be designed with bandwidths greater than the frequency of oscillation. The disadvantage of such networks is the relatively poor frequency stability compared with the high-Q hard oscillators.

DISCRIMINATORS

Basic angle demodulators can be designed using balanced tuned networks with suitably connected nonlinearities (such as diode switches); or angle demodulators can be designed using voltage-controlled oscillators (VCO), multipliers, and appropriate feedback networks.

The former are simpler to implement than the latter but require much higher input signal-to-noise ratios for operation above the noise threshold at which the angle modulation produces signal process gain. Two popular versions of the discriminator type of frequency demodulator are shown in Fig. 12.2.6. The diode discriminator is designed so that one diode conducts more with increases, the other with decreases, in frequency. For greatest linearity, the mutual coupling between the tuned circuits is generally greater than unity. The other basic type of conventional demodulator simply implements a version of the definition [Eq. (1)] of FM for the case of a constant-amplitude sinusoidal waveform. The mixing operation can be replaced by a simple phase shifter that provides a 90° phase relationship between x and Hx. Frequently, the reference oscillator will be phase-locked to the average value of the $J_0(d)$ component indicated by Eqs. (3) and (5).

FM FEEDBACK (FMFB) DEMODULATORS

These angle demodulators have the advantage of lower distortion, lower noise threshold, and little or no drift in center frequency of operation and cab be designed to be less sensitive to interference. No limiter is required for good performance, and there is no requirement to maintain a minimum predemodulator input carrier-to-noise ratio to avoid noise threshold.

A block diagram of a basic synchronous-filtering demodulator is shown in Fig. 12.2.7. The synchronous filter is indicated by the dashed lines that enclose a phase-locked loop designed to follow the instantaneous

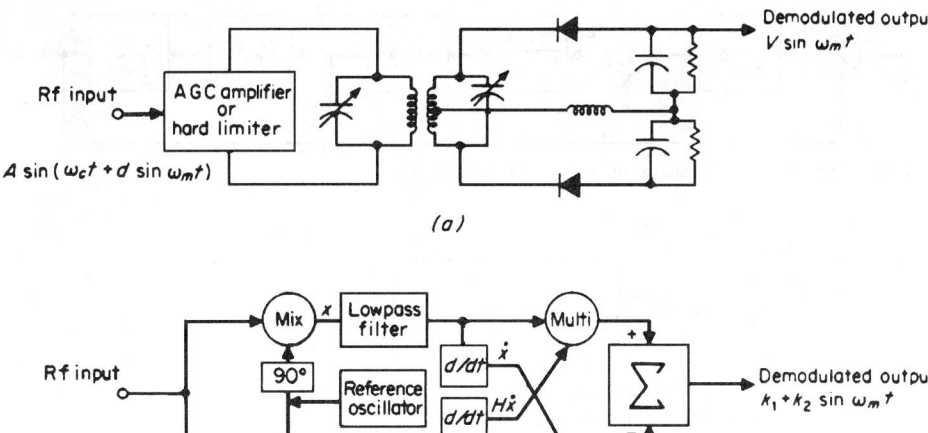

FIGURE 12.2.6 Conventional demodulators: (*a*) diode discriminator; (*b*) phase-shift discriminator.

excursions in the phase of the angle-modulated signal ϕ_2. The mixer, i.f. amplifier, discriminator, filter 2, and the VCO-2 form a frequency feedback loop for the demodulator.

Aside from the synchronous filter, the basic configuration can be considered that of a conventional FMFB demodulator which compresses the wide-band FM input signal so that it can be passed through a relatively narrow bandwidth fixed-tuned filter and to a discriminator for detection. It should be noted that the configuration shown here can also be considered simply as a phase-locked loop (PLL) with frequency feedback around it.

The significant advantages of each of these techniques (FMFB and PLL) can be combined. It is important to consider the design from both the synchronous and frequency-feedback viewpoints. Note in this regard Figs. 12.2.8 and 12.2.9 in which $F_1(S)$ and $F_2(S)$ have been assigned. Observe that Fig. 12.2.10 is an

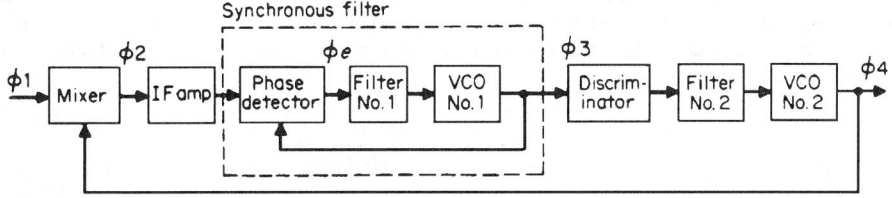

FIGURE 12.2.7 Block diagram of basic feedback demodulator.

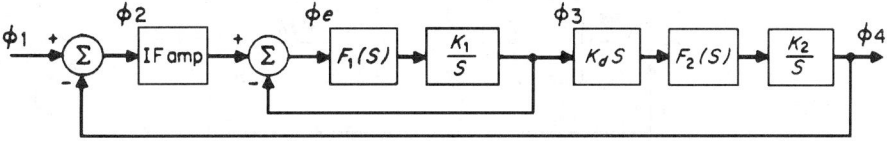

FIGURE 12.2.8 Equivalent network of basic demodulator shown in Fig. 12.2.7.

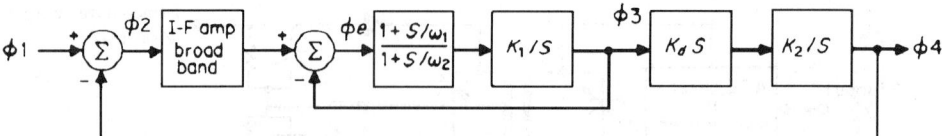

FIGURE 12.2.9   Direct linear equivalent form of basic demodulator.

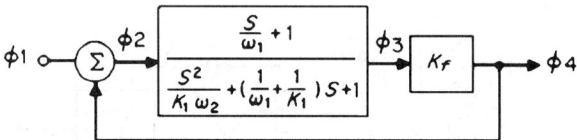

FIGURE 12.2.10   FMFB demodulator.

FMFB equivalent linear form obtained by substituting the closed-loop transfer function for the synchronous filter; however, by retaining the inner loop and combining phase comparators, we obtain the synchronous phase-locked-loop form, as shown in Fig. 12.2.11. To the extent that linear analysis and quasi-linear substitutions can be made in these block diagrams, the remarks that follow are thus relevant to both PLL and FMFB forms implemented with synchronous filters or with broadband amplifiers in cascade with single-tuned filters.

PHASE-LOCKED-LOOP DEMODULATORS

The introduction of the phase-locked loop between the i.f. amplifier and discriminator may be viewed simply as a means by which the i.f. signal can be tracked, limited, and filtered regardless of Doppler shift or oscillator drift. The synchronous filter is employed in conjunction with the relatively wide-bandwidth i.f. amplifier shown on the block diagram to perform this critical filtering function, as well as to take advantage of the phase coherence between the FM signal sidebands.

The operation of the demodulator can be understood from an examination of the equivalent network shown in Fig. 12.2.8, in which the transfer functions relate to phase as the input and output variables. If the bandwidth of the i.f. amplifier is broad by comparison with both the synchronous-filter bandwidth and significant modulation sidebands, it can be ignored in the equivalent linear representation of the demodulator. Since the synchronous-filter function is

$$\frac{\phi_3(s)}{\phi_2(s)} = \frac{(K_1/s)F_1(s)}{1+(K_1/s)F_1(s)} \tag{11}$$

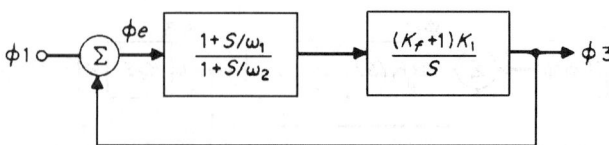

FIGURE 12.2.11   PLL demodulator.

the signal ϕ_2 is related to the input by

$$\frac{\phi_2(s)}{\phi_1(s)} = \frac{1}{1 + K_1 K_2 K_d F_1(s) F_2(s) / \{s[1 + (K_1/s) F_1(s)]\}} \tag{12}$$

For frequency components of $\phi_1(s)$ lying well within the bandwidths of $F_2(s)$ and the synchronous filter, this transfer function reduces to the familiar form of a type-zero feedback network: i.e.,

$$\frac{\phi_2(s)}{\phi_1(s)} = \frac{1}{1 + K_2 K_d} = \frac{s\phi_2(s)}{s\phi_1(s)} \tag{13}$$

Since this is also the transfer function with respect to frequency variations, the compression of frequency excursions is evident. If

$$\phi_1(t) = (\Delta\omega/\omega_m)\sin \omega_m t = D \sin \omega_m t \tag{14}$$

where $\phi(t)$ represents the instantaneous variation of the phase of the input signal relative to some reference carrier phase, and if the synchronous filter follows this instantaneous variation, the effective phase excursion to the discriminator is reduced to

$$\phi_3(t) = \frac{D}{1 + K_f} \sin \omega_m t \qquad K_t = K_2 K_d \tag{15}$$

if ω_m is well within the passband of $F_2(s)$ and the phase-locked loop. This reduction in deviation ratio by the use of frequency feedback gain K_f suggests that an optimum gain and synchronous-filter bandwidth combination should be sought for given modulation and noise characteristics, just as the proper i.f. amplifier and K_f must be chosen in a conventional FMFB demodulator.

Figure 12.2.10 shows the closed-loop transfer function of the synchronous filter along with the frequency-feedback loop. It has been assumed that the i.f. amplifier bandwidth is very broad compared with the bandwidth occupied by the significant portions of the signal spectrum as it appears at i.f. frequencies. This amounts to assuming that the dispersive effect produced by the i.f. amplifier is negligible. If the phase-locked-loop gain can be considered large compared with the filter zero ($K_1 \gg \omega_1$ is usually satisfied in practice), the following relationships between the synchronous-filter and the complete feedback-demodulator parameters can be written

$$\zeta_f^2 = (K_f + 1)\zeta_\phi^2 \qquad \omega_{nf}^2 = (K_f + 1)\omega_{n\phi}^2 \qquad B_{nf} = B_{n\phi}\left(1 + K_f \frac{1}{1 + 1/4\zeta_\phi^2}\right) \tag{16}$$

where ζ_f = damping of demodulator
$\quad \omega_{nf}$ = natural frequency of demodulator
$\quad \zeta_\phi$ = damping of synchronous filter
$\quad \omega_{n\phi}$ = natural frequency of synchronous filter
$\quad B_{nf}$ = noise bandwidth of demodulator
$\quad B_{n\phi}$ = noise bandwidth of synchronous filter

and in terms of the actual network parameters

$$\zeta_\phi^2 = \frac{K_1 \omega_2}{4\omega_1^2} \qquad \omega_{n\phi}^2 = K_1 \omega_2 \qquad B_{n\phi} = \frac{1}{2}\left(\frac{K_1 \omega_2}{\omega_1} + \omega_1\right) \quad (\text{Hz}) \tag{17}$$

Note that all the demodulator response variables can be controlled by simple *RC* adjustments.

FM FEEDBACK-DEMODULATOR DESIGN FORMULAS

Acquisition. The rate at which the VCO in a phase-locked demodulator can be swept through the frequency and phase at which pull-in and phase lock will occur is indicated by the representative curves of Fig. 12.2.12.

Pull-In

The frequency difference $\Delta\omega$ (between an unmodulated carrier and the VCO) within which a phase-locked loop will pull into phase synchronism is, roughly,

$$|\Delta\omega|_p = (2pK\omega_n)^{1/2} \quad \text{for } K/\omega_n \gg 1$$

where K = total open-loop gain (rad/s)
ω_n = loop natural frequency (rad/s)
ρ = dimensionless constant ≈ 1

The time required for pull-in is given by

$$T = 4(\Delta f)^2/B_n^3 \quad \text{(seconds)}$$

for loop damping of 0.5 and $|\Delta f| < 0.8|\Delta f|_p$, where Δf = difference frequency (Hz) and B_n = closed-loop noise bandwidth (Hz).

Stability

The condition of sustained beat-note stability without the loop capacity to be swept into lock, which is exhibited by high-gain narrow-bandwidth phase-locked demodulators, is a condition of loop oscillation arising from the presence of the phase detector nonlinearity and extraneous memory, such as that in the tuned circuits in

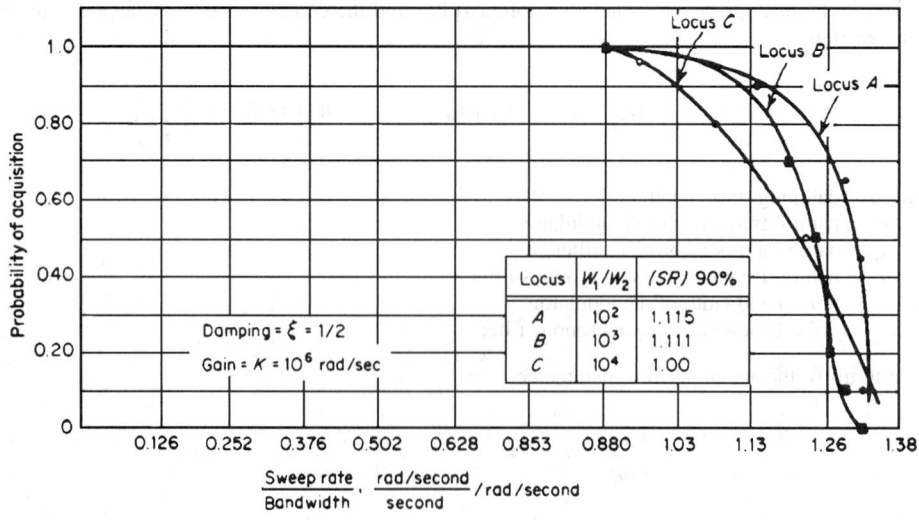

FIGURE 12.2.12 Probability of acquisition vs. sweep-rate per unit bandwidth.

phase detectors and in voltage-controlled oscillators. This condition can be predicted from the approximate condition

$$\frac{K^2 G |\omega_0|}{2|\omega_0|^2}[\cos\phi(\omega_0)] + 1 = 0 \tag{18}$$

where K = total open-loop gain
  G = complete transfer function between phase detector and VCO
  ϕ = angle of G

This condition can be used to predict to nonlinear network oscillations preventing lockup.

Parameter Variations

The rate at which second-order demodulators damped for minimum-noise bandwidth can have the loop natural frequency changed at constant damping is given by

$$\frac{d\omega_n/dt}{\omega_n^2} \leq 0.1 \tag{19}$$

for peak phase errors less than 5°, where ω_n is the natural frequency of the loop.

Minimum-Threshold Parameters

The parametric relationships for minimum-noise threshold for the demodulator in Fig. 12.2.9 are shown in Fig. 12.2.13.

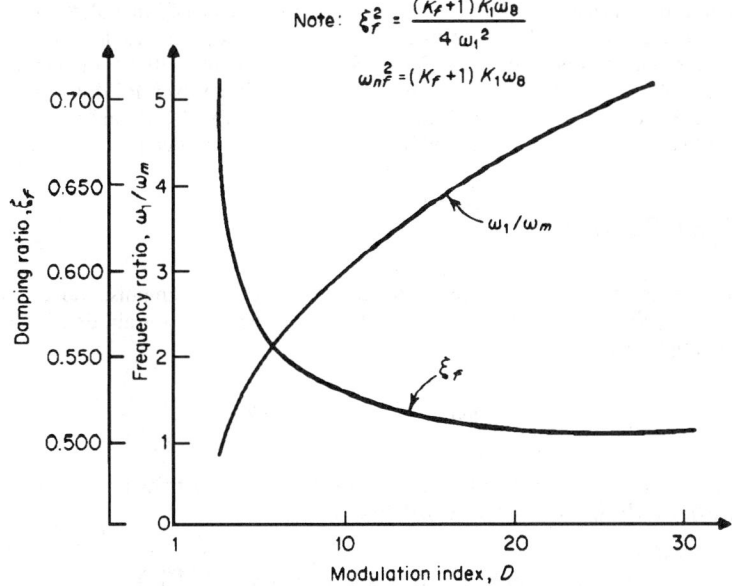

FIGURE 12.2.13   Parametric relationships for minimum threshold (voice).

CHAPTER 12.3
PULSE MODULATION/ DEMODULATION

George F. Pfeifer

PULSE MODULATION

Pulse modulation is, in general, the encoding of information by means of varying one or more pulse parameters. It finds application in both the communication and the control fields. The control applications are usually confined to the use of pulse-time modulation (PTM) and pulse-frequency modulation (PFM), where on-off control power can be used to minimize device dissipation. All pulse modulation schemes require sampling analog signals, and some, such as pulse-code modulation (PCM) and delta modulation, require the additional quantization of the analog signals.

In communications, the chief application of pulse modulation is found where it is desired to time-multiplex by interleaving a number of single-channel, low-duty-cycle pulse trains. The pulse trains may, in turn, be used for compound modulation by amplitude or angle modulation of a continuous carrier. In usual applications, subcarriers are pulsed, time-division-multiplexed, and then used to frequency-modulate a carrier.

Since noise is present in all systems, a prime consideration in modulation selection is the choice of a waveform based on its signal-to-noise efficiency. For instance, PTM is more efficient than pulse amplitude modulation (PAM), which offers no improvement over continuous AM; however, PTM is less efficient than PCM or delta modulation. A chief advantage of pulsed systems such as PTM, PCM, and delta is improved signal-to-noise ratio in exchange for increased bandwidth, in the same manner as continuous FM improves over AM.

SAMPLING AND SMOOTHING

An ideal impulse sampler can be considered as the multiplication of an impulse train, period T seconds, with the continuous signal $f(t)$. This is shown in Fig. 12.3.1a for the impulse train defined as $p_T(t) = \mathrm{rep}_T[\delta(t)]$, where $\delta(t)$ is an impulse at $t = 0$ and

$$\mathrm{rep}_T[u(t)] = \sum_{n=-\infty}^{\infty} u(t - nT) \tag{1}$$

The output spectrum function is the convolution of $F(f)$, the Fourier transform of the input signal, and the transform of $p_T(t)$, which is $(1/T)\,\mathrm{comb}_{1/T}$ (1), defined as

$$\mathrm{comb}_{1/T}[U(f)] = \sum_{n=-\infty}^{\infty} U\left(\frac{n}{T}\right) \delta\left(f - \frac{n}{T}\right) \tag{2}$$

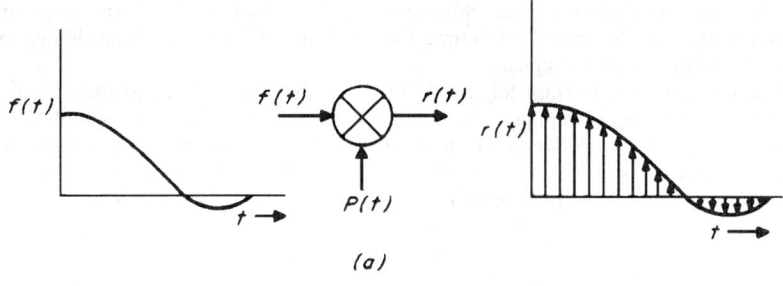

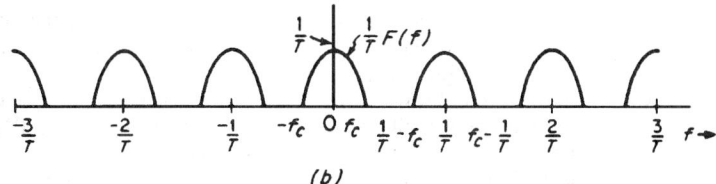

FIGURE 12.3.1   Pulse modulation: (*a*) output spectrum; (*b*) sampling configuration.

Thus the transform $R(f) = F(f) * (1/T) \, \text{comb}_{1/T}$ (1) with spectrum is as shown in Fig. 12.3.1*b*. The result of ideal impulse sampling has been to repeat the original signal spectrum, assumed to be band-limited, each $1/T$ Hz and multiply each by a $1/T$ scale factor. Since all the signal information is present in each lobe of Fig. 12.3.1*b*, it is only necessary to recover a single lobe through filtering in order to recover the signal function reduced by a scale factor.

Consider an ideal low-pass rectangular filter of bandwidth of f_f Hz defined as $T(jf) = A(f) \exp [-j\theta(jf)]$, where

$$A(f) = \begin{cases} 1 & |f| < f_f \\ 0 & |f| > f_f \end{cases} \quad \text{and} \quad \theta(jf) = 2\pi\alpha f \quad \text{for all } f$$

The cutoff frequency f_f is adjusted to select the output spectral lobe about zero $f_c < f_f < 1/T - f_c$ and will fall in the guard band between lobes. That portion of filter output $R(f)$ selected is

$$R_0(f) = (1/T)F(f) \exp (-j2\pi\alpha f) \tag{3}$$

that will inverse transform as

$$r_0(t) = (1/T)f(t - \alpha) \tag{4}$$

which is identical with the signal function, with the amplitude reduced by a scale factor and function shifted by α seconds. If $\alpha = 0$, signifying no delay, the filter is termed a "cardinal data hold"; otherwise, it is an "ideal low-pass filter." Unfortunately, these filters cannot be realized in practice, since they are required to respond before they are excited.

Examination of Fig. 12.3.1*b* gives rise to the sampling theorem accredited to Shannon and/or Nyquist, which states that when a continuous time function with band-limited spectrum $-f_c < f < f_c$ is sampled at twice the highest frequency, $f_s = 2f_c$, the original time function can be recovered. This corresponds to the point where the sampling frequency $f_s = 1/T$ is decreased so that the spectral lobes of Fig. 12.3.1*b* are just touching. To decrease f_s beyond the value of $2f_c$ would cause spectral overlap and make recovery with an ideal filter impossible. A more general form of the sampling theorem states that any $2f$ independent samples per second will completely describe a band-limited signal, thus removing the restriction of uniform sampling, as long as independent samples are used. In general, for a time-limited signal of T seconds, band-limited to f_c Hz, only $2f_cT$ samples are needed to specify the signal completely.

In practice, the signal is not completely band-limited, so that it is common to allow for a greater separation of spectral lobes, called the *guard band*. This guard band is generated simply by sampling at greater than $2f_c$, as in the case for Fig. 12.3.1*b*. Although the actual tolerable overlap depends on the signal spectral slope, setting the sampling rate at about $3f_c = f_s$ is usually adequate to recover the signal.

In practice, narrow but finite-width pulse trains are used in place of the idealized impulse sampling train.

PULSE-AMPLITUDE MODULATION

Pulse-amplitude modulation is essentially a sampled-data type of encoding where the information is encoded into the amplitude of a train of finite-width pulses. The pulse train can be looked upon as the carrier in much the same way as the sine wave is for continuous-amplitude modulation. There is no improvement in signal-to-noise when using PAM, and furthermore, PAM is not considered wideband in the sense of FM or PTM. Thus PAM would correspond to continuous AM, while PTM corresponds to FM. Generally, PAM is used chiefly for time-multiplex systems employing a number of channels sampled, consistent with the sampling theorem.

There are a number of ways of encoding information as the amplitude of a pulse train. They include both bipolar and unipolar pulse trains for both instantaneous or square-topped sampling and for exact or top sampling. In top sampling, the magnitude of the individual pulses follows the modulating signal during the pulse duration, while for square-topped sampling, the individual pulses assume a constant value, depending on the particular exact sampling point that occurs somewhere during the pulse time. These various waveforms are shown in Fig. 12.3.2.

The top-modulation bipolar sampling case is shown in Fig. 12.3.2*c*; it is simply sampling with a finite-pulse-width train. Carrying out the convolution yields

$$R_{\text{STB}}(f) = \frac{\tau}{T} \sum_{n=-\infty}^{\infty} \left(\text{sinc}\, \frac{\tau n}{T} \right) F\left(f - \frac{n}{T} \right) \tag{5}$$

The spectrum for top-modulation bipolar sampling, using a square-topped rectangular spectrum for the original signal spectrum, is shown in Fig. 12.3.3*a*. The signal spectrum repeats with a $(\sin x)/x$ scale factor determined by the sampling pulse width, with each repetition a replica of $F(f)$.

Unipolar sampling can be implemented by adding a constant bias A to $f(t)$, the signal, to produce $f(t) + A$, where A is large enough to keep the sum positive; that is, $A > |f(t)|$. Sampling the new sum signal by multiplication with the pulse train results in the unipolar top-modulated waveform of Fig. 12.3.2*e*. The spectrum is

$$R_{\text{STU}}(f) = \frac{\tau}{T} \sum_{n=-\infty}^{\infty} \left(\text{sinc}\, \frac{\tau n}{T} \right) \left[F\left(f - \frac{n}{T} \right) + A\delta\left(f - \frac{n}{T} \right) \right] \tag{6}$$

The delta-function part of the summation reduces to the spectrum function of the pulse train $S(f)$

$$R_{\text{STU}}(f) = AS(f) + \frac{\tau}{T} \sum_{n=-\infty}^{\infty} \left(\text{sinc}\, \frac{\tau n}{T} \right) F\left(f - \frac{n}{T} \right) \tag{7}$$

The resulting spectrum of top-modulation unipolar sampling is the same as with bipolar sampling plus the impulse spectrum of the sampling pulse train, as shown in Fig. 12.3.3*b*. For square-topped-modulation bipolar sampling, the time-domain result is

$$r_{\text{SSB}}(t) = \text{rect}\,(t/\tau) * \text{comb}_T\, f(t) \tag{8}$$

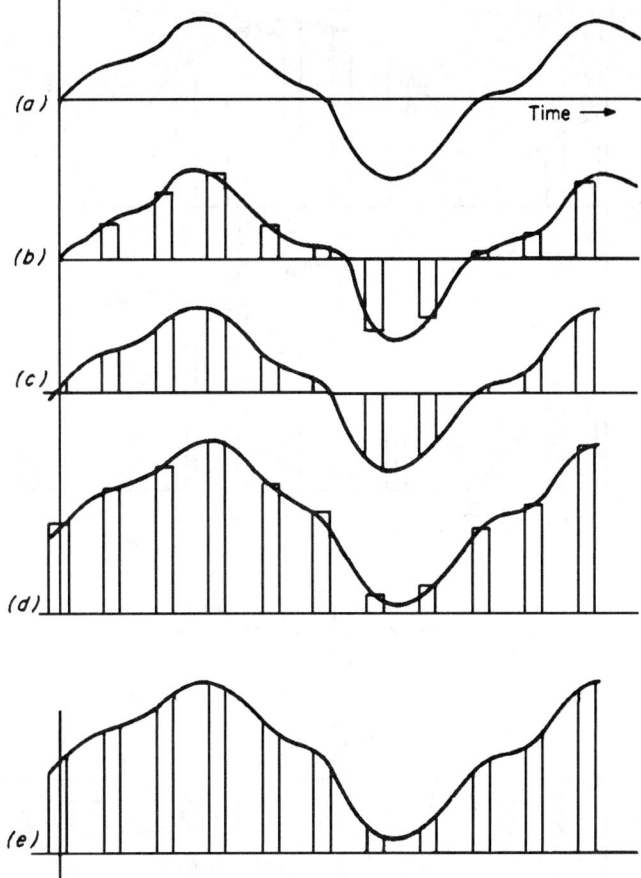

FIGURE 12.3.2 PAM waveforms: (*a*) modulation; (*b*) square-top sampling, bipole pulse train; (*c*) top sampling, bipole pulse train; (*d*) square-top sampling, unipolar pulse train; (*e*) top sampling, unipolar pulse train.

with spectrum function

$$R_{\text{SSB}}(f) = \frac{\tau}{T}(\text{sinc}\, f\tau) \sum_{n=-\infty}^{\infty} F\left(f - \frac{n}{T}\right) \tag{9}$$

In this case, the signal spectrum is distorted by the sinc $f\tau$ envelope, as shown in Fig. 12.3.2*c*. This frequency distortion is referred to as *aperture effect* and may be corrected by use of an equalizer sinc $f\tau$ form, following the low-pass reconstruction filter.

As in the previous case of unipolar sampling, the resulting spectrum for square-topped modulation will contain the pulse-train spectrum, as shown in Fig. 12.3.2*d*. The expression is

$$R_{\text{SSU}}(f) = AS(f) + \frac{\tau}{T}(\text{sinc}\, f\tau) \sum_{n=-\infty}^{\infty} F\left(f - \frac{n}{T}\right) \tag{10}$$

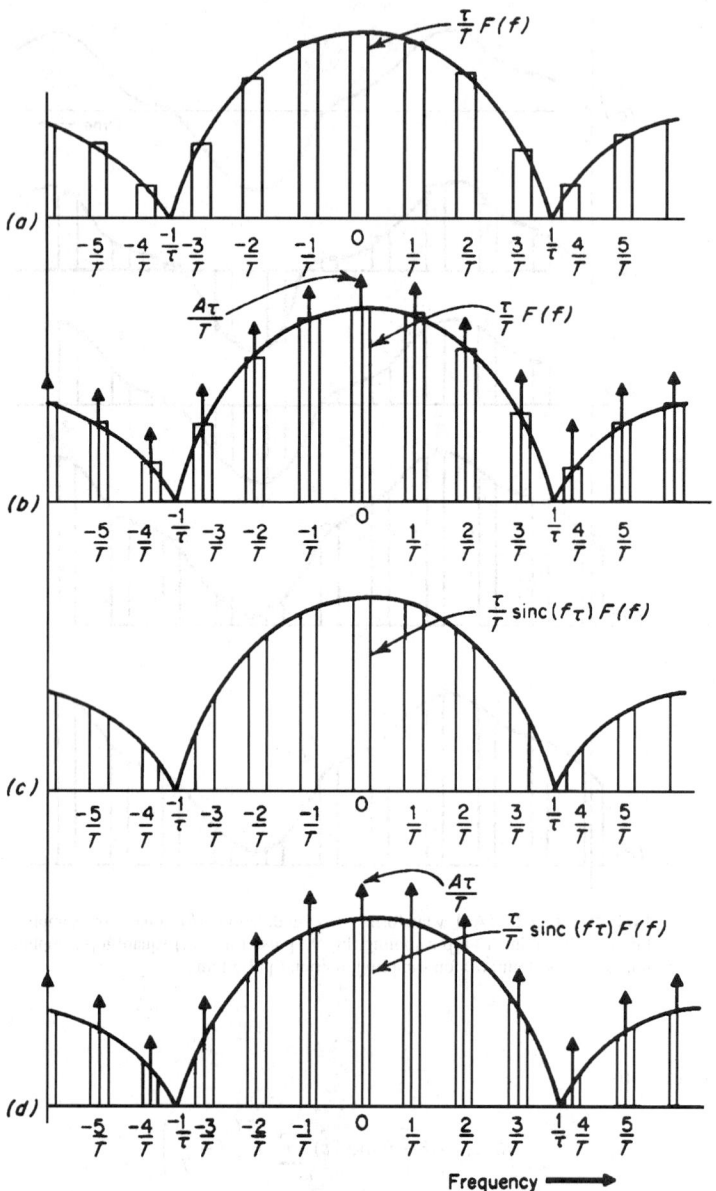

FIGURE 12.3.3   PAM spectra: (*a*) top modulation, bipolar sampling; (*b*) top modulation, unipolar sampling; (*c*) square-top modulation, bipolar sampling; (*d*) square-top modulation, unipolar sampling.

The signal information is generally recovered, in PAM systems, by use of a low-pass filter that acts on the reduced signal energy around zero frequency, as shown in Fig. 12.3.3.

PULSE-TIME, PULSE-POSITION, AND PULSE-WIDTH MODULATION

In PTM the information is encoded into the time parameter instead of, for instance, the amplitude, as in PAM. There are two basic types of PTM: pulse-position modulation (PPM) and pulse-width modulation (PWM), also known as pulse-duration (PDM) or pulse-length (PLM) modulation. The PTM allows the power-driver circuitry to operate at saturation level, thus conserving power loss. Operating driver circuitry full on, full off, is especially important for heavy-duty high-load control applications, as well as for communication applications.

In PPM the information is encoded into the time position of a narrow pulse, generally with respect to a reference pulse. The basic pulse width and amplitude are kept constant, while only the pulse position is changed, as shown in Fig. 12.3.4. There are three cases of PWM which are the modulation of the leading edge, trailing edge, or both edges, as displayed in Fig. 12.3.5. In this case the information is encoded into the width of the pulse, with the pulse amplitude and period held constant. The derivative relationship existing between PPM and PWM can be illustrated by consideration of trailing-edge PWM modulation. The pulses of PPM can be derived from the edges of trailing-edge PWM (Fig. 12.3.5b) by differentiation of the PWM signal and a sign change of the trailing-edge pulse. Pulse-position modulation is essentially the same as PWM, with the information-carrying variable edge replaced by a pulse. Thus, when that part of the signal power of PWM that carries no information is deleted, the result is PPM.

Generally, in PTM systems a guard interval is necessary because of the pulse rise times and system responses. Thus 100 percent of the interpulse period cannot be used without considerable channel cross-talk because of pulse overlap. It is necessary to trade off crosstalk versus channel utilization at the system design level.

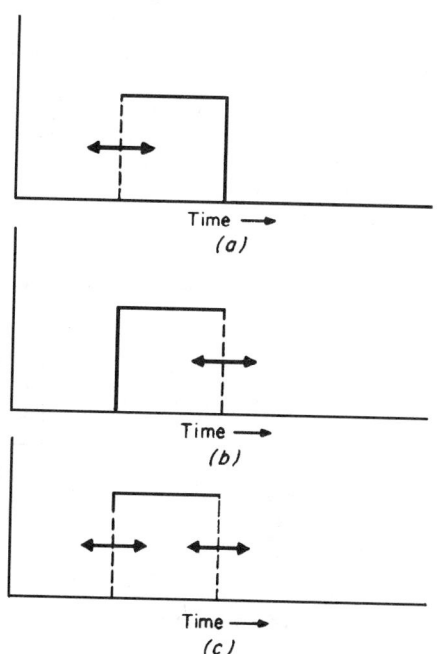

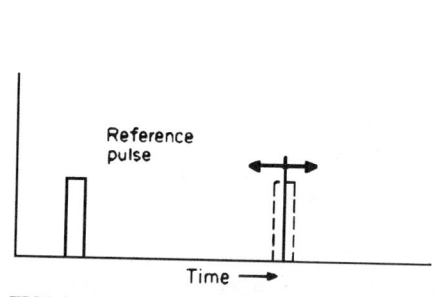

FIGURE 12.3.4 PPM time waveform.

FIGURE 12.3.5 PWM time waveforms: (a) leading-edge modulation; (b) trailing-edge modulation; (c) both-edge modulation.

Another consideration is that the information sampling rate cannot exceed the pulse repetition frequency and would be less for a single channel of a multiplexed system where channels are interwoven in time.

Generation of PTM

There are two basic methods of pulse-time modulation: (1) based on uniform sampling in which the pulse-time parameter is directly proportional to the modulating signal at uniformly sampled points and (2) in which there is some distortion of the pulse-time parameter because of the modulation process. Both methods of modulation are illustrated in Fig. 12.3.6 for PWM. Basically, PPM can be derived from trailing-edge PWM, as shown in Fig. 12.3.6c by use of an edge detector or differentiator and a standard narrow-pulse generator.

In the uniform sampling case for PWM of Fig. 12.3.6a, the modulating signal is sampled uniformly in time and the special PAM derived by a sample-and-hold circuit as shown in Fig. 12.3.7a. This PAM signal provides a pedestal for each of the three types of sawtooth waveforms producing leading, trailing, or double-edge PWM, as shown in Fig. 12.3.7c, e, and g, respectively. The uniform sampled PPM is shown in Fig. 12.3.7h, as derived from the trailing-edge modulation of g.

Nonuniformly sampled modulation, termed *natural sampling* by some authors, is shown in Fig. 12.3.8, and results from the method of Fig. 12.3.8b, where the sawtooth is added directly to the modulating signal. In this case the modulating waveform influences the time when the samples are actually taken. This distortion is small when the modulating-amplitude change is small during the interpulse period T. The distortion is caused by the modulating signal distorting the sawtooth wave-form when they are added, as indicated in Fig. 12.3.6b. The information in the PPM waveform is similarly distorted because it is derived from the PWM waveform, as shown in Fig. 12.3.7h.

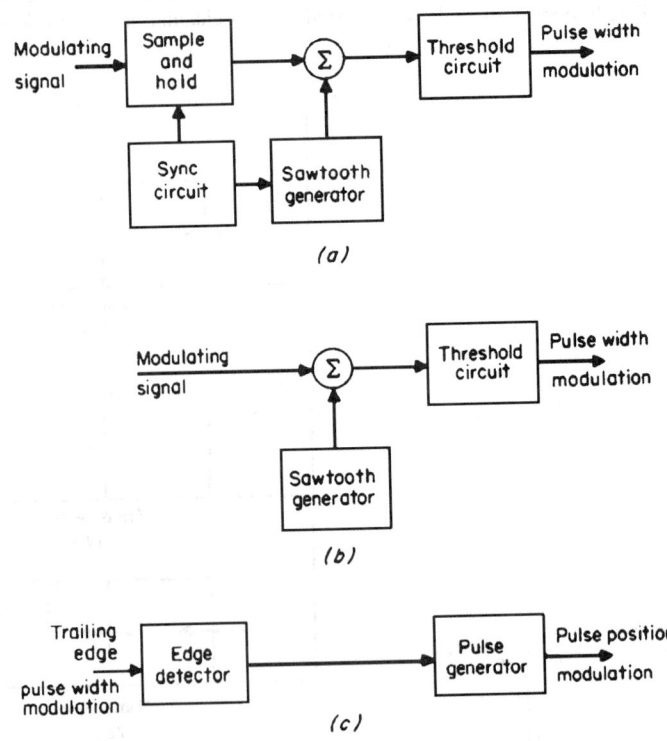

FIGURE 12.3.6   PTM generation: (a) pulse-width-modulation generation, uniform sampling; (b) pulse-width-modulation generation, nonuniform sampling; (c) pulse-position-modulation generation.

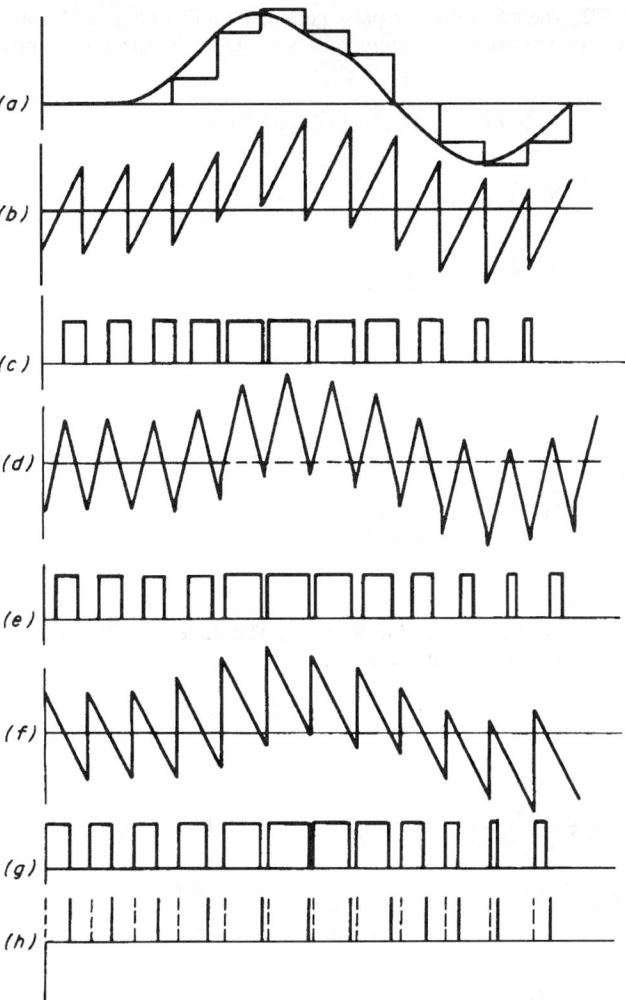

FIGURE 12.3.7 Pulse-time modulation, uniform sampling: (*a*) modulating signal and sample-and-hold waveform; (*b*) sawtooth added to sample-and-hold waveform; (*c*) leading-edge modulation; (*d*) sawtooth added to sample-and-hold waveform; (*e*) double-edge modulation; (*f*) sawtooth added to sample-and-hold waveform; (*g*) trailing-edge modulation; (*h*) pulse-position modulation (reference pulse dotted).

PULSE-TIME MODULATION SPECTRA

The spectra are smeared in general, for most modulating signals, and are difficult to derive; however, it is possible to get some idea of what happens to the spectra with modulation by considering a sinusoidal modulation of form

$$A \cos 2\pi f_s t \tag{11}$$

The amplitude $A < T/2$, where T is the interpulse period, assuming no guard band.

For PPM with uniform sampling and unity pulse amplitude, the spectrum is given by

$$
\begin{aligned}
x(t) = \frac{\tau}{T} + \frac{2\tau}{T} \sum_{m=1}^{\infty} (\text{sinc } mf_0) J_0(2\pi Amf_0) \cos 2\pi \, mf_0 t \\
+ \frac{2\tau}{T} \sum_{n=1}^{\infty} \text{sinc } (nf_s) J_n(2\pi Anf_s) \cos\left(2\pi \, nf_s t - \frac{n\pi}{2} \right) \\
+ \frac{2\tau}{T} \sum_{m=1}^{\infty} \sum_{n=1}^{\infty} \left\{ \text{sinc } (mf_0 + nf_s) J_n[2\pi A(mf_0 + nf_s)] \cos\left[2\pi(mf_0 + nf_s)t - \frac{n\pi}{2} \right] \right. \\
\left. + \text{sinc } (nf_s - mf_0) J_n[2\pi A(nf_s - mf_0)] \cos\left[2\pi(nf_s - mf_0)t - \frac{n\pi}{2} \right] \right\}
\end{aligned}
\tag{12}
$$

where τ = pulse width
T = pulse period
f_s = modulation frequency
J_n = Bessel function of first kind, nth order
$f_0 = 1/T$

As is apparent, all the harmonics of the pulse-repetition frequency and the modulation frequency are present, as well as all possible sums and differences. The dc level is τ/T, with the harmonics carrying the modulation. The pulse shape effects the line amplitudes as a sinc function, reducing the spectra for higher frequencies.

The spectrum for PWM is similar to that of PPM, and for uniformly sampled trailing-edge sinusoidal modulation is given by

$$
\begin{aligned}
x(t) = \frac{1}{2} + \frac{1}{\pi T} \sum_{m=1}^{\infty} \frac{1}{mf_0} \cos\left[2\pi \, mf_0 t + \frac{\pi}{2}(2m-1) \right] \\
+ \frac{1}{\pi T} \sum_{m=1}^{\infty} \frac{1}{mf_0} J_0(2\pi Amf_0) \cos\left(2\pi \, mf_0 t - \frac{\pi}{2} \right) \\
+ \frac{1}{\pi T} \sum_{n=1}^{\infty} \frac{1}{nf_s} J_n(2\pi Anf_s) \cos\left[2\pi \, nf_s t - (n+1)\frac{\pi}{2} \right] \\
+ \frac{1}{\pi T} \sum_{\substack{m=1 \\ n=1}}^{\infty} \left\{ \frac{1}{mf_0 + nf_s} J_n[2\pi A(mf_0 + nf_s)] \cos\left[2\pi(mf_0 + nf_s)t - (n+1)\frac{\pi}{2} \right] \right. \\
\left. + \frac{1}{nf_s - mf_0} J_n[2\pi A(nf_s - mf_0)] \cos\left[2\pi(nf_s - mf_0)t - (n+1)\frac{\pi}{2} \right] \right\}
\end{aligned}
\tag{13}
$$

The same comments apply for PWM as for PPM.

A more compact form is given for PPM and PWM, respectively, as

$$
x(t) = \frac{1}{T} \sum_{\substack{m = \infty \\ n = \infty}} (-j)^n J_n[2\pi A(mf_0 + nf_s)] P(mf_0 + nf_s) \exp [j2\pi(mf_0 + nf_s)t]
\tag{14}
$$

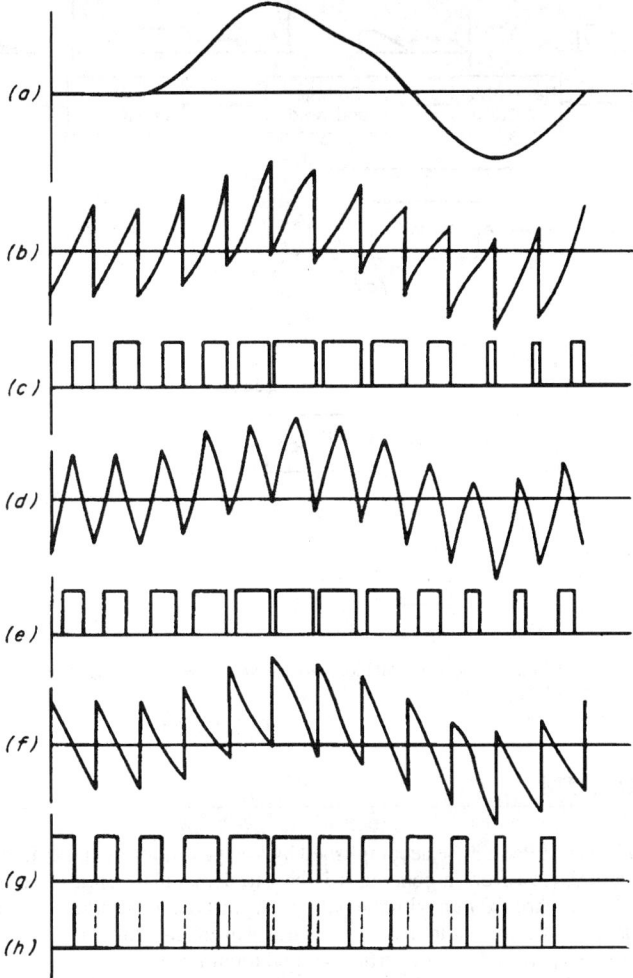

FIGURE 12.3.8 Pulse-time modulation, nonuniform sampling; (*a*) modulating signal; (*b*) sawtooth added to modulation; (*c*) leading-edge modulation; (*d*) sawtooth added to modulation; (*e*) double-edge modulation; (*f*) sawtooth added to modulation; (*g*) trailing-edge modulation; (*h*) pulse-position modulation.

where $P(f)$ is the Fourier transform of the pulse shape $p(t)$, and

$$x(t) = \frac{1}{2} + \frac{1}{T} \sum_{\substack{m=-\infty \\ m \neq 0}}^{\infty} j^{2m-1} \frac{e^{j2\pi \, mf_0 t}}{2\pi \, mf_0} - \frac{1}{T} \sum_{\substack{m=-\infty \\ n=-\infty \\ |m| + |n| \neq 0}}^{\infty}$$

$$(-j)^{n+1} \frac{J_n[2\pi A(mf_0 + nf_s)]}{2\pi(mf_0 + nf_s)} \exp\left[j2\pi(mf_0 + nf_s)t\right] \tag{15}$$

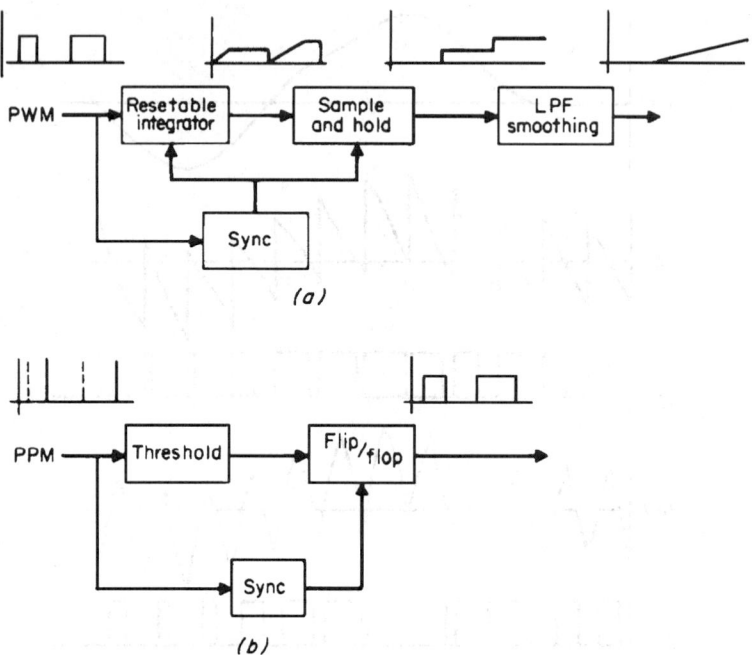

FIGURE 12.3.9 Pulse-time demodulation: (*a*) PWM demodulation; (*b*) PPM to PWM for demodulation.

DEMODULATION OF PTM

Demodulation of PWM or PPM can be accomplished by low-pass filtering if the modulation is small compared with the impulse period. However, in general, it is best to demodulate on a pulse-to-pulse basis that usually requires some form of synchronization with the pulses. The distortion introduced by nonuniform sampling cannot be eliminated and will be present in the demodulated waveform. However, if the modulation is small compared with the interpulse period T, the distortion will be minimized.

To demodulate PWM each pulse can be integrated and the maximum value sampled and held and low-pass-filtered, as shown in Fig. 12.3.9*a*. To sample and reset the integrator, it is necessary to derive sync from the PWM waveform, in this case trailing-edge-modulated.

Generally, PPM is demodulated by conversion to PWM and then demodulated as PWM. Although in some demodulation schemes the actual PWM waveform may not exist as such, the general demodulation scheme is the same. PPM can be converted to PWM by the configuration of Fig. 12.3.9*b*. The PPM signal is applied to an amplitude threshold, usually termed a *slicer*, that rejects noise except near the pulses. The pulses are applied to a flip-flop synchronized to one particular state by the reference pulse, and it generates the PWM as its output.

PULSE FREQUENCY MODULATION

In PFM the information is contained in the frequency of the pulse train, which is composed of narrow pulses. The highest frequency possible ideally occurs when there is no more interpulse spacing left for finite-width pulses. This frequency, given by $1/\tau$, where τ is the pulse width, will not be achieved in practice, owing to the pulse rise time. The lowest frequency is determined by the modulator, usually a voltage-controlled oscillator

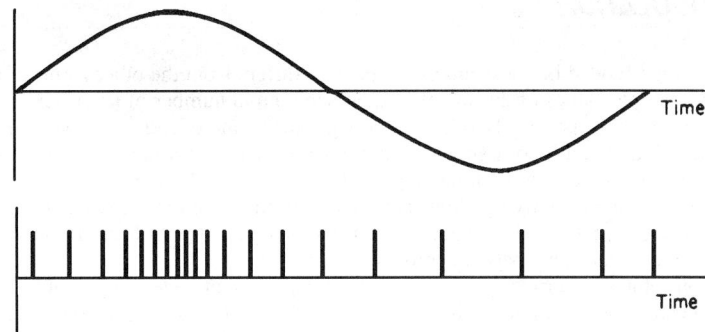

FIGURE 12.3.10 PFM modulation.

(VCO), in which in practice a 100:1 ratio of high to low frequency is easily achievable. Examination of Fig. 12.3.10 indicates why PFM is used mostly for control purposes rather than communications. The wide variation and uncertainty of pulse position do not lend themselves to time multiplexing, which requires the interweaving of channels in time. Since one of the chief motivations of pulse modulation in communication systems is to be able to time-multiplex a number of channels, PFM is not used. On the other hand, PFM is a good choice for on-off control applications, especially where fine control is required. A classic example of PFM control is for the attitude control of near-earth satellites that have on-off gas thrusters where a very close approximation to a linear system response is achievable.

Generation of PFM

Basically, PFM is generated by modulation of a VCO as shown in Fig. 12.3.11a. A constant reference voltage is added to the modulation so that the frequency can swing above and below the reference-determined value. For control applications it is usually required that the frequency follow the magnitude of the modulation, its sign determining which actuators are to be turned on, as shown in Fig. 12.3.11b.

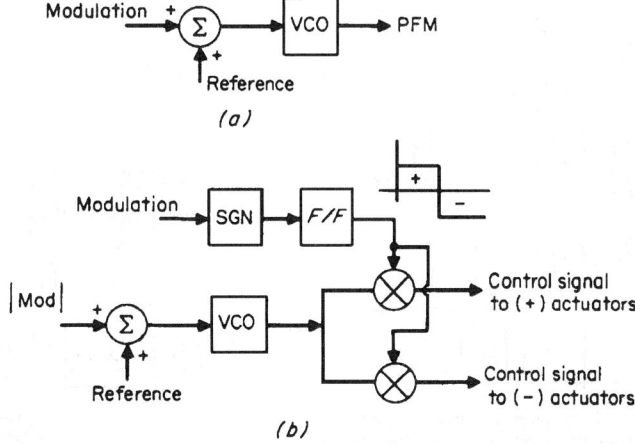

FIGURE 12.3.11 Generation of PFM: (a) PFM modulation; (b) PFM for control.

PULSE-CODE MODULATION

In PCM the signal is encoded into a steam of digits. This differs from the other forms of pulse modulation by requiring that the sample values of the signal be quantized into a number of levels and subsequently coded as a series of pulses for transmission. By selecting enough levels, the quantized signal can be made to approximate closely the original continuous signal at the expense of transmitting more bits per sample. The PCM scheme lends itself readily to time multiplexing of channels and will allow widely different types of signals; however, synchronization is strictly required. This synchronization of the system can be on a single-sample or code-group basis. The synchronizing signal is most likely inserted with a group of samples from different channels, on a frame or subframe basis to conserve space.

The motivation behind modern PCM is that improved implementation techniques of solid-state circuitry allow extremely fast quantization of samples and translation to complex codes with reasonable equipment constraints. PCM is an attractive way to trade bandwidth for signal-to-noise and has the additional advantage of transmission through regenerative repeaters with a signal-to-noise ratio that is substantially independent of the number of repeaters. The only requirement is that the noise, interference, and other disturbances be less than one-half a quantum step at each repeater. Also, systems can be designed that have error-detecting and error-correcting features.

PCM CODING AND DECODING

Coding is the generation of a PCM waveform from an input signal, and decoding is the reverse process. There are many ways to code and many code groups to use: hence standardization is necessary when more than one user is considered. Each sample value of the signal waveform is quantized and represented to sufficient accuracy by an appropriate code character. Each code character is composed of a specified number of code elements. The code elements can be chosen as two-level, or binary; three-level, or ternary; or n-ary. However, general practice is to use binary, since it is not affected as much by interference introduced by the required increased bandwidth. An example of binary coding is shown in Fig. 12.3.12 for 3-bit or eight levels of quantization. Each code group is composed of three pulses, with the pulse trains shown for on-off pulses in Fig. 12.3.12b and bipolar pulses in Fig. 12.3.12c.

A generic diagram of a complete system is shown in Fig. 12.3.13. The recovered signal is a delayed copy of the input signal degraded by noise because of sources such as sampling, quantization, and interference. For this type of system to be efficient, both sending and receiving terminals must be synchronized. The synchronism is

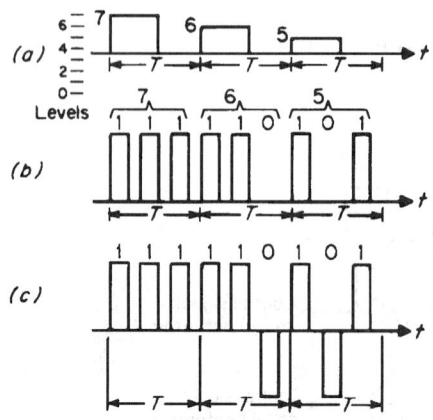

FIGURE 12.3.12 Binary pulse coding: (a) quantized samples; (b) on-off coded pulses; (c) bipolar coded pulses.

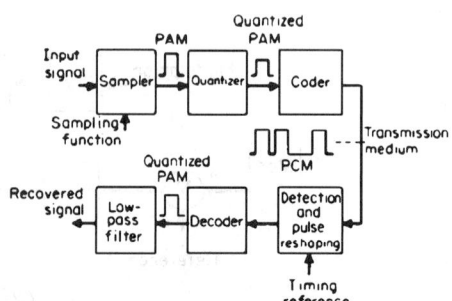

FIGURE 12.3.13 Basic operations of a PCM system.

required to be monitored continuously and be capable of establishing initial synchronism when the system is out of frame. The synchronization is usually accomplished by use of special sync pulses that establish frame, subframe, or word sync.

There are three basic ways to code, namely, feedback and subtraction, pulse counting, and parallel comparison. In *feedback subtraction* the sample value is compared with the most significant code-element value and that value subtracted from the sample value if the element value is less. This process of comparison and subtraction is repeated for each code-element value down to the least significant bit. At each subtraction the appropriate code element or bit is selected to complete the coding. In *pulse counting* a gate is established by using the PWM pulse corresponding to a sample value. Clock pulses are gated using the PWM gate and are connected in a counter. The output of a decoding network attached to the counter is read out as the PCM. *Parallel comparison* is the fastest method since the sampled value is applied to a number of different threshold values. The thresholders are read out as the PCM.

SYSTEM CONSIDERATIONS FOR PCM

Quantization introduces an irremovable error into the system, referred to as *quantization noise*. This kind of noise is characterized by the fact that its magnitude is always less than one-half a quantum step, and it can be treated as uniformly distributed additive noise with zero mean value and rms value equal to $1/\sqrt{12}$ times the total height of a quantum step. When the ratio of signal power to quantization noise power at the quantizer output is used as a measure of fidelity the improvement with quantizer levels is as shown in Fig. 12.3.14 for different kinds of signals.

In general, using an n-ary code with m pulses allows transmission of n^m values. For the binary code this reduces the 2^m values which approximate the signal to 1 part in $2^m - 1$ levels. Encoding into pulse and minus pulses, assuming either pulse is equally likely, results in an average power of $A^2/4$, which is half the on-off power of $A^2/2$, where the total pulse amplitude, peak to peak, is A. The channel capacity for a system sampled at the Nyquist rate of $2f_m$ and quantized into s levels is

$$C = 2f_m \log_2 s \quad \text{(bits/s)} \qquad (16)$$

or for m pulses of n values each

$$C = mf_m \log_2 n^2 \quad \text{(bits/s)} \qquad (17)$$

Since the encoding process squeezes one sample into m pulses, the pulse widths are effectively reduced by $1/m$; thus the transmission bandwidth is increased by a factor of m, or $B = mf_m$.

The maximum possible ideal rate of transmission of binary bits is

$$C = B \log_2 (1 + S/N) \quad \text{(bits/s)} \qquad (18)$$

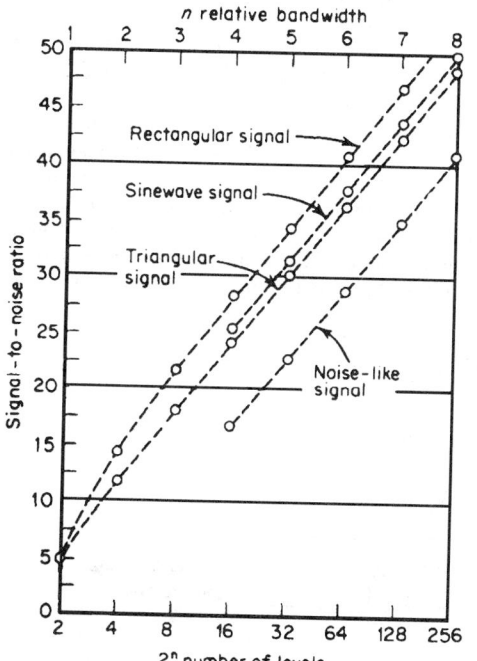

n relative bandwidth

FIGURE 12.3.14   PCM signal-to-noise improvement with number of quantization levels.

according to Shannon. For a system sampled at the Nyquist rate, quantized to $K\sigma$ per level and using the plus and minus pulses, the channel capacity is

$$C = B \log_2(1 + 12S/K^2N) \quad \text{(bits/s)} \quad N = \sigma^2 \qquad (19)$$

where S is the average power over large time interval and σ is the rms noise voltage at decoder input.

DELTA MODULATION

Delta modulation (DM) is basically a one-digit PCM system where the analog waveform has been encoded in a differential form. In contrast to the use of n digits in PCM, simple DM uses only one digit to indicate the changes in the sample values. This is equivalent to sending an approximation to the signal derivative. At the receiver the pulses are integrated to obtain the original signal. Although DM can be simply implemented in circuitry, it requires a sampling rate much higher than the Nyquist rate of $2f_m$ and a wider bandwidth than a comparable PCM system. Most of the other characteristics of PCM apply to DM.

Delta modulation differs from differential PCM in which the difference in successive signal samples is transmitted. In DM only 1 bit is used to express and transmit the difference. Thus DM transmits the sign of successive slopes.

Coding and Decoding DM

There are a number of coding and decoding variations in DM, such as single-integration, double-integration, mixed-integration, delta-sigma, and high-information DM (HIDM). In addition, companding the signal which is compressing the signal at transmission and expanding it at reception is also used to extend the limited dynamic range. The simple single-integration DM of the coding-decoding scheme is shown in Fig. 12.3.15. In the encoder the modulator produces positive pulses when the sign of the difference signal $\epsilon(t)$ is positive and negative pulses otherwise; and the output pulse train is integrated and compared with the input signal to provide an error signal $\epsilon(t)$, thus closing the encode feedback loop. At the receiver the pulse train is integrated and filtered to produce a delayed approximation to the signal, as shown in Fig. 12.3.16. The actual circuit implementation with operational amplifiers and logic circuits is very simple.

By changing to a double-integration network in the decoder, a smoother replica of the signal is provided. This decoder has the disadvantage, however, of not recognizing changes in the slope of the signal. This gave rise to a scheme to encode differences in slope instead of amplitude, leading to coders with double integration; however, systems of this type are marginally stable and can oscillate under certain conditions. Waveforms of a double-integrating delta coder are shown in Fig. 12.3.17. Single and double integration can be combined to give improved performance while avoiding the stability problem. These mixed systems are often referred to in the literature as *delta modulators* with double integration. A comparison of waveforms is shown in Fig. 12.3.18.

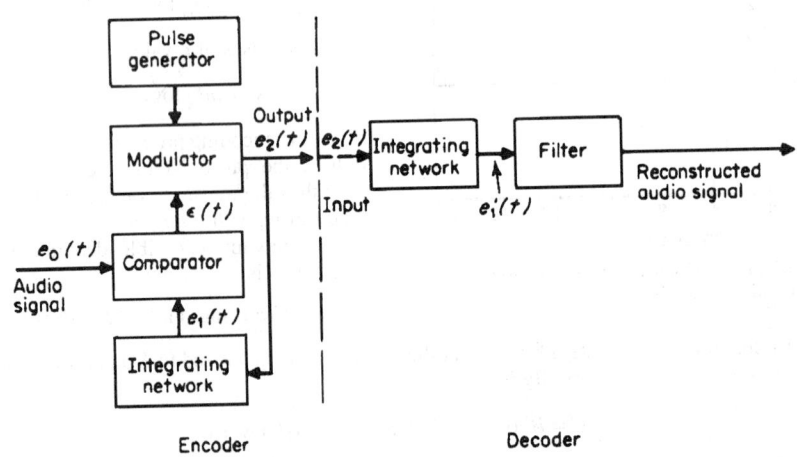

FIGURE 12.3.15   Basic coding-decoding diagram for DM.

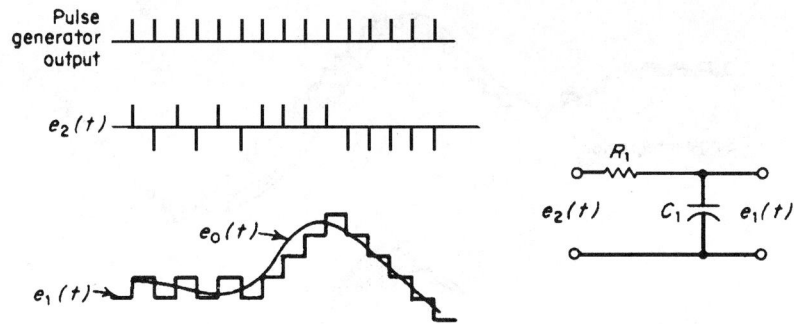

FIGURE 12.3.16   Delta-modulation waveforms using single integration.

System Considerations for DM

The synthesized waveform can change only one level each clock pulse; thus DM overloads when the slope of the signal is large. The maximum signal power will depend on the type of signal, since the greatest slope that can be reproduced is the integration of one level in one pulse period. For a sine wave of frequency f, the maximum-amplitude signal is

$$A_{max} = f_s \sigma / 2\pi \tag{20}$$

where f_s is the sampling frequency and σ is one quantum step.

It has been observed that a DM system will transmit a speech signal without overloading if the amplitude of the signal does not exceed the maximum permissible amplitude of an 800-Hz sine wave. The DM

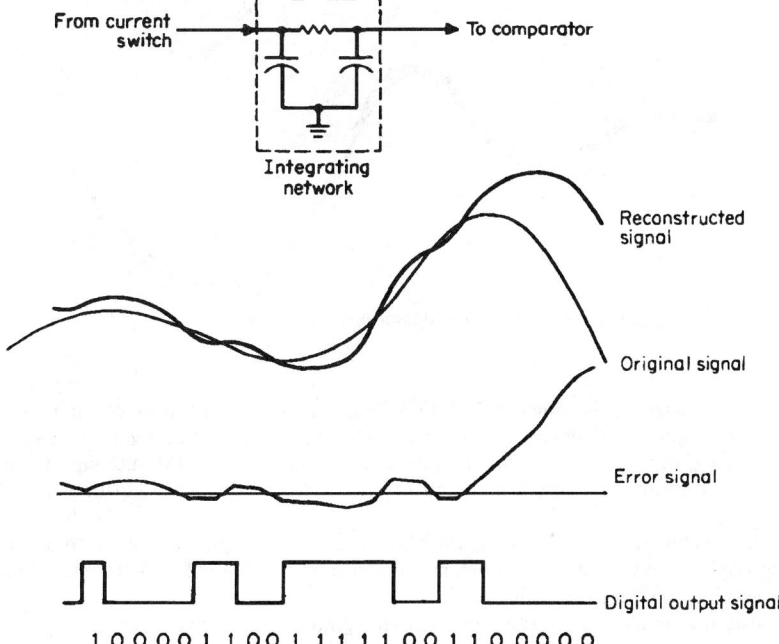

FIGURE 12.3.17   Waveforms for delta coder with double integration.

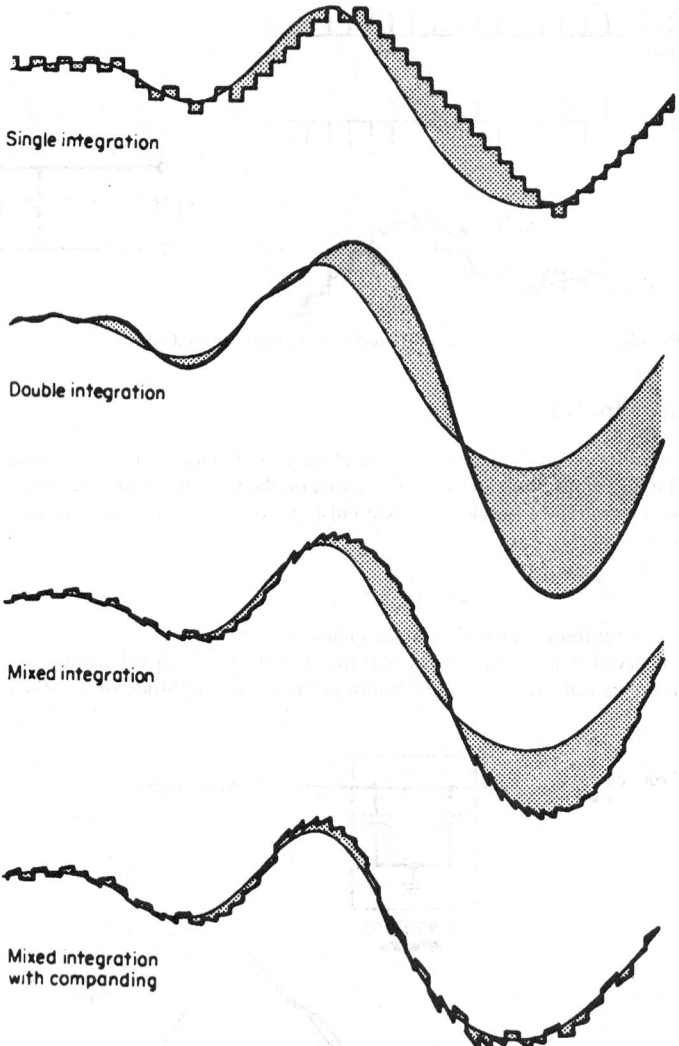

FIGURE 12.3.18   Waveforms for various integrating systems.

coder overload characteristic is shown in Fig. 12.3.19 along with the spectrum of a human voice. Notice that they decrease in frequency together, indicating that DM can be used effectively with speech transmission. Generally speaking, transmission of speech is the chief application of DM, although various modifications and improvements are being studied to extend DM to higher frequencies and transmission of the lost dc component.

Among these techniques is delta-sigma modulation, where the signal is integrated and compared with an integrated approximation to form the error signal similar to $\epsilon(t)$ of Fig. 12.3.15. The decoding is accomplished with a low-pass filter and requires no integration.

The signal-to-quantization noise ratio for single-integration DM is given by

$$S/N = 0.2 f_s^{3/2}/f f_0^{1/2}$$

(21)

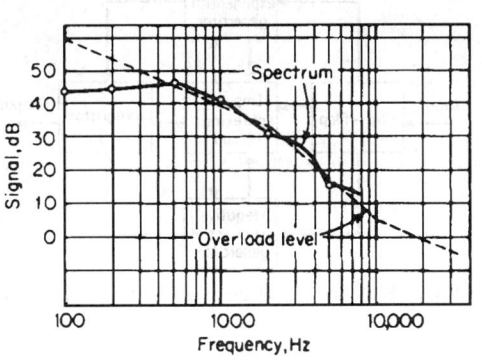

FIGURE 12.3.19   Spectrum of the human voice compared with delta-coder overload level.

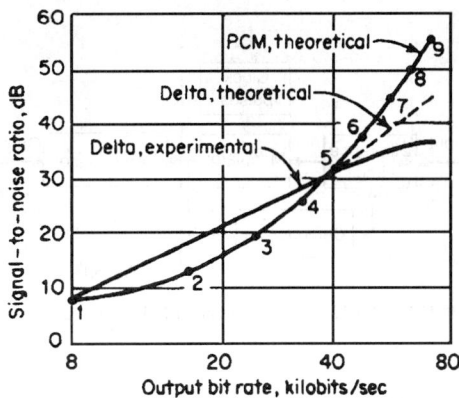

FIGURE 12.3.20   Signal-to-noise ratio for DM and PCM.

where f_s = sampling frequency
f = signal frequency
f_0 = signal bandwidth

For double or mixed DM

$$S/N = 0.026 f_s^{5/2} / f\, f_0^{3/2}$$

(22)

A comparison of signal-to-noise ratio (SNR) for DM and PCM is shown in Fig. 12.3.20, along with an experimental DM system for voice application. Note that DM at 40 kbits/s sampling rate is equal in performance with a 5-bit PCM system.

Extended-Range DM

A system termed *high-information* DM (HIDM, developed by M. R. Winkler in 1963) falls in the category of companded systems and encodes more information in the binary sequence than normal DM. Basically, the method doubles the size of the quantization step when two identical, consecutive binary values appear and takes one-half of the step after each transition of the binary train. The HIDM system is capable of reproducing the signal with smaller quantization and overload errors. This technique also increases the dynamic range. The response of HIDM compared with that of DM is shown in Fig. 12.3.21.

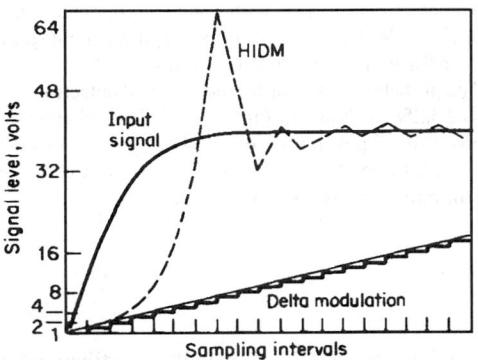

FIGURE 12.3.21   Step response for a high-information delta modulation.

Implementation of HIDM is similar to that of DM, as shown in Figs. 12.3.22 and 12.3.23, with the difference only in the demodulator. The flip-flop of Fig. 12.3.23 changes state on the polarity of the input pulses. While the impulse generator initializes the experimental generators each pulse time, the flip-flop selects either the positive or negative one. The integrator adds and smooths the exponential waveforms to form the output signal. The scheme has a dynamic range with slope limiting of 11.1 levels per pulse period, which is much greater than DM and is equivalent to a 7-bit linear-quantized PDM system.

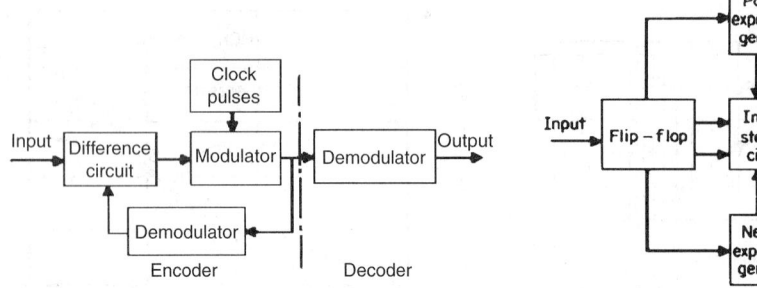

FIGURE 12.3.22 Block diagram of HIDM system. **FIGURE 12.3.23** Block diagram of HIDM demodulator.

DIGITAL MODULATION

Digital modulation is concerned with the transmission of a binary pulse train over some medium. The output of, say, a PCM coder would be used to modulate a carrier for transmission. In PCM systems, for instance, the high-quality reproduction of the analog signal is a function only of the probability of correct reception of the pulse sequences. Thus the measure of digital modulation is the probability of error resulting from the digital modulation. The three basic types of digital modulation, amplitude-shift keying (ASK), frequency-shift keying (FSK), and phase-shift (PSK), are treated below.

AMPLITUDE-SHIFT KEYING

In ASK the carrier amplitude is turned on or off, generating the waveform of Fig. 12.3.24 for rectangular pulses. Pulse shaping, such as raised cosine, is sometimes used to conserve bandwidth. The elements of a binary ASK receiver are shown in Fig. 12.3.25. The detection can be either coherent or noncoherent; however, if the added complexity of coherent methods is to be applied, a higher performance can be achieved by using one of the other methods of digital modulation.

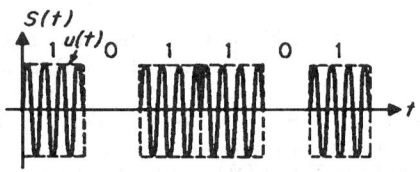

FIGURE 12.3.24 ASK modulation.

The error rate of ASK with noncoherent detection is given in Fig. 12.3.26. Note that the curves approach constant values of error for high signal-to-noise ratios.

The probability of error for the coherent detection scheme of Fig. 12.3.25c is shown in Fig. 12.3.27. The coherent-detection operation is equivalent to bandpass filtering of the received signal plus noise, followed by synchronous detection, as shown. At the optimum threshold shown in Fig. 12.3.27, the probability of error of marks and spaces is the same. The curves also tend toward a constant false-alarm rate, as in the noncoherent case.

FREQUENCY-SHIFT KEYING

In FSK the frequency is shifted rapidly between one of two frequencies. Generally, two filters are used in favor of a conventional FM detector to discriminate between the marks and spaces, as illustrated in Fig. 12.3.28. As with ASK, either noncoherent or coherent detection can be used, although in practice coherent detection is not often used. This is because it is just as easy to use PSK with coherent detection and achieve superior performance.

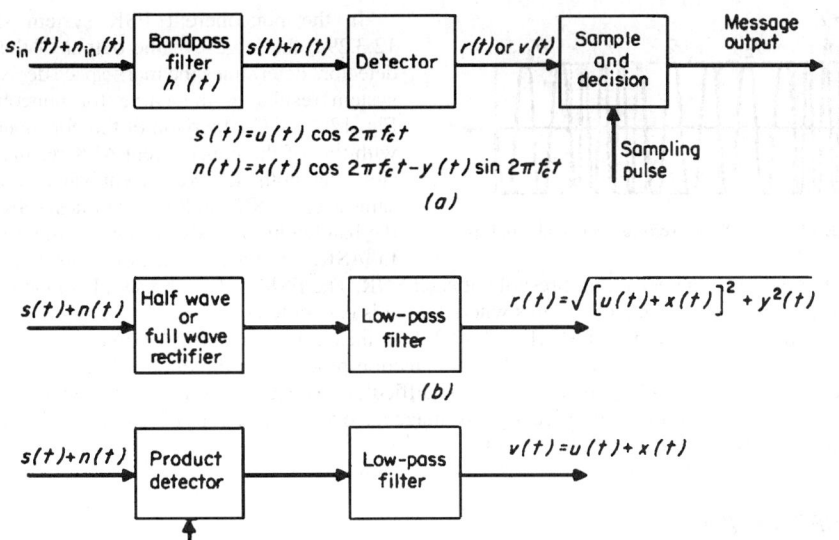

FIGURE 12.3.25   Elements of a binary digital receiver: (*a*) elements of a simple receiver; (*b*) noncoherent (envelope) detector; (*c*) coherent (synchronous) detector.

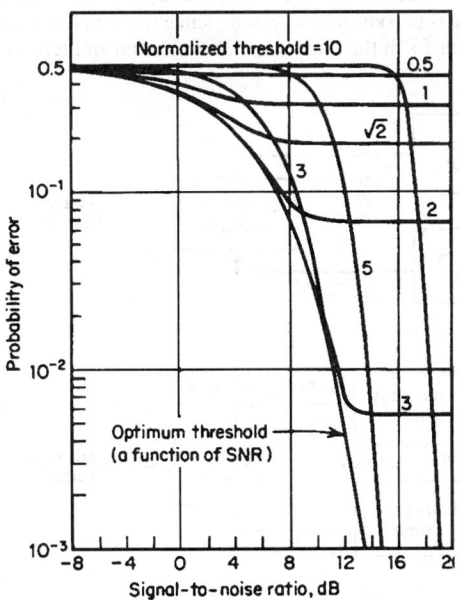

FIGURE 12.3.26   Error rate for on-off keying, noncoherent detection.

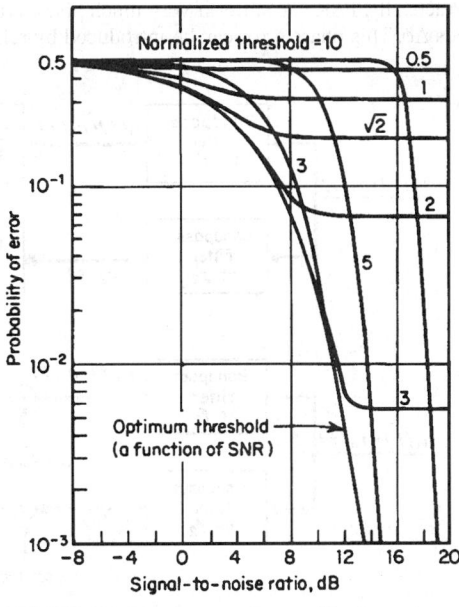

FIGURE 12.3.27   Error-rate for on-off keying, coherent detection.

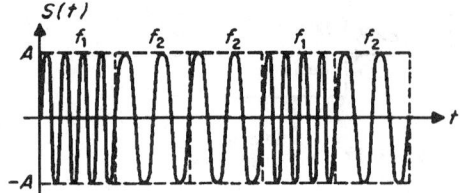

FIGURE 12.3.28 FSK waveform, rectangular pulses.

In the noncoherent FSK system shown in Fig. 12.3.29*a*, the largest of the output of the two envelope detectors determines the mark-space decision. Using this system results in the curve for noncoherent FSK in Fig. 12.3.30. Comparison of the noncoherent FSK error with that of the noncoherent ASK results in the conclusion that both achieve an equivalent error rate at the same average SNR at low error rates. FSK requires twice the bandwidth of ASK because of the use of two tones. In ASK, in order to achieve this performance, it is required to optimize the detection threshold at each SNR. The FSK system threshold is independent of SNR, and thus is preferred in practical systems where fading is encountered.

By synchronous detection of FSK (Fig. 12.3.29*b*) is meant the availability of an exact replica of each possible transmission at the receiver. The coherent-detection process has the effect of rejecting a portion of the bandpass noise. Coherent FSK involves the same difficulties as phase-shift keying but achieves poorer performance. Also, coherent FSK is significantly advantageous over noncoherent FSK only at high error rates. The probability of error is shown in Fig. 12.3.30.

PHASE-SHIFT KEYING

Phase-shift keying is optimum in the minimum-error-rate sense from a decision-theory point of view. The PSK of a constant-amplitude carrier is shown in Fig. 12.3.31, where the two states are represented by a phase difference of π rad. Thus PSK has the form of a sequence of plus and minus rectangular pulses of a continuous sinusoidal carrier. It can be generated by double-sideband suppressed-carrier modulation by a bipolar rectangular waveform or by direct phase modulation. It is also possible to phase-modulate more complex signals than a sinusoid.

There is no performance difference in binary PSK between the coherent detector and the normal phase detector, both of which are shown in Fig. 12.3.32. Reference to Fig. 12.3.32 shows that there is a 3-dB design advantage for ideal coherent PSK over ideal coherent FSK, with about the same equipment requirements. Practically, PSK can suffer if very much phase error $\Delta\phi$ is present in the system, since the signal is reduced by $\cos \Delta\phi$. This phase error can be introduced by relative drifts in the master oscillators at transmitter or receiver

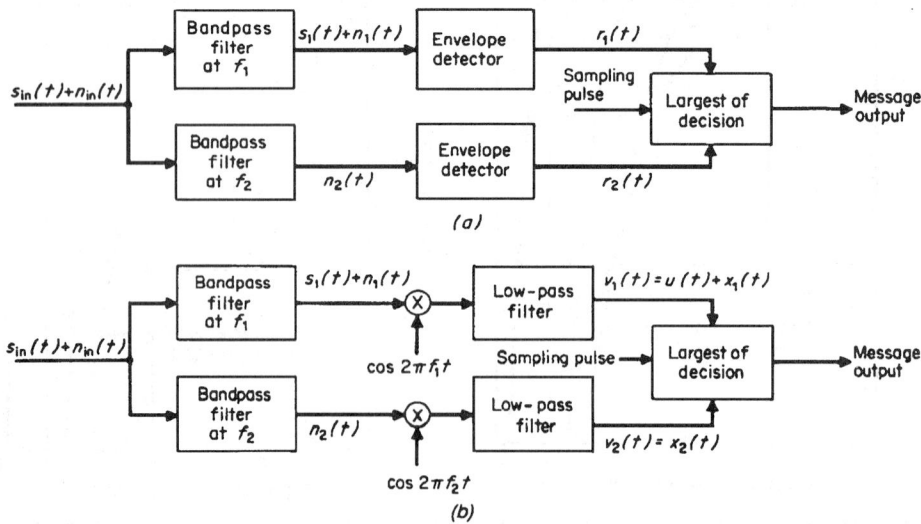

FIGURE 12.3.29 Dual-filter detection of binary FSK signals: (*a*) noncoherent detection tone f_1 signaled; (*b*) coherent detection tone f_1 signaled.

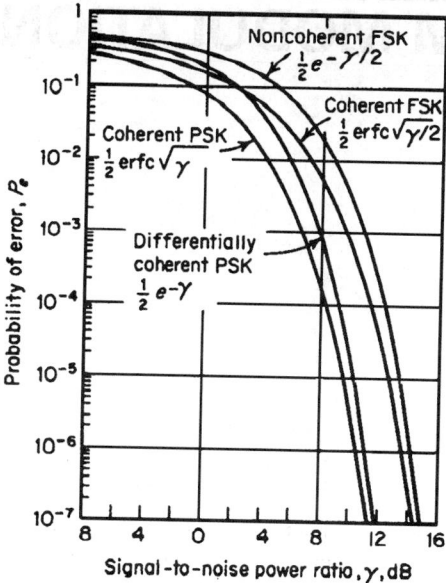

FIGURE 12.3.30 Error rates for several binary systems.

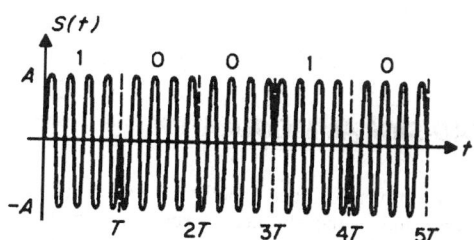

FIGURE 12.3.31 PSK signal, rectangular pulses.

or be a result of phase drift or fluctuation in the propagation path. In most cases this phase error can be compensated at the expense of requiring long-term smoothing.

An alternative to PSK is differential phase-shift keying (DPSK), where it is required that there be enough stability in the oscillators and transmission path to allow negligible phase change from one information pulse to the next. Information is encoded differentially in terms of phase change between two successive pulses. For instance, if the phase remains the same from one pulse to the next (0° phase shift), a mark would be indicated; however, a phase shift of π from the previous pulse to the next would indicate a space. A coherent detector is still required where one input is the current pulse with the other input the previous pulse.

The probability of error is shown in Fig. 12.3.30. At all error rates DPSK requires 3 dB less SNR than noncoherent FSK for the same error rate. Also, at high SNR, DPSK performs almost as well as ideal coherent PSK at the same keying rate and power level.

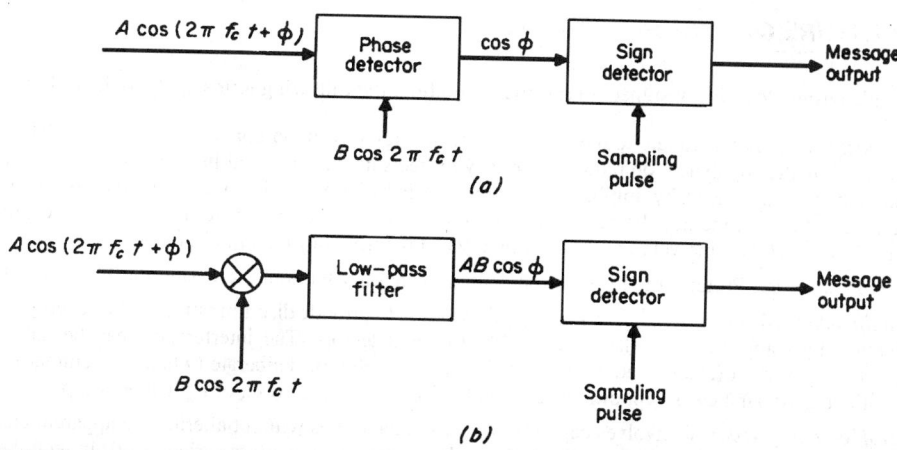

FIGURE 12.3.32 Two detection schemes for ideal coherent PSK: (*a*) phase detection; (*b*) coherent detection.

CHAPTER 12.4
SPREAD-SPECTRUM MODULATION

Myron D. Egtvedt

SPREAD-SIGNAL MODULATION

In a receiver designed exactly for a specified set of possible transmitted waveforms (in the presence of white noise and in the absence of such propagation defects as multipath and dispersion), the performance of a matched filter or cross-correlation detector depends only on the ratio of signal energy to noise power density E/n_o, where E is the received energy in one information symbol and $n_o/2$ is the rf noise density at the receiver input. Since signal bandwidth has no effect on performance in white noise, it is interesting to examine the effect of spreading the signal bandwidth in situations involving jamming, message and traffic-density security, and transmission security. Other applications include random-multiple-access communication channels, multipath propagation analysis, and ranging.

The information-symbol waveform can be characterized by its time-bandwidth (TW) product. Consider a binary system with the information symbol defined as a *bit* (of time duration T), while the fundamental component of the binary waveform is called a *chip*. For this *direct-sequence* system, the ratio (chips per bit) is equal to the TW product. An additional requirement on the symbol waveforms is that their cross-correlation with each other and the noise or extraneous signals be minimal.

Spread-spectrum systems occupy a signal bandwidth much larger (>10) than the information bandwidth, while the conventional systems have a TW of well under 10. FM with a high modulation index might slightly exceed 10 but is not optimally detectable and has a processing gain only above a predetection signal-to-noise threshold.

NOMENCLATURE OF SECURE SYSTEMS

While terminology is not subject to rigorous definition, the following terms apply to the following material:

Security and privacy. Relate to the protection of the signal from an unauthorized receiver. They are differentiated by the sophistication required. Privacy protects against a casual listener with little or no analytical equipment, while security implies an interceptor familiar with the principles and using an analytical approach to learn the *key*. Protection requirements must be defined in terms of the interceptor's applied capability *and* the time value of the message. Various forms of *protection* include:

Crypto security. Protects the information content, generally without increasing the TW product.

Antijamming (AJ) security. Spreads the signal spectrum to provide discrimination against energy-limited interference by using cross-correlation or matched-filter detectors. The interference may be natural (impulse noise), inadvertent (as in amateur radio or aircraft channels), or deliberate (where the jammer may transmit continuous or burst cw, swept cw, narrow-band noise, wide-band noise, or replica or deception waveforms).

Traffic-density security. Involves capability to switch data rates without altering the apparent characteristics of the spread-spectrum waveform. The TW product (processing gain) is varied inversely with the data rates.

Transmission security. Involves spreading the bandwidth so that, beyond some range from the transmitter, the transmitted signal is buried in the natural background noise. The process gain (TW) controls the reduction in detectable range vis-à-vis a "clear" signal.

Use in Radar

It is usual to view radar applications as a variation on communication; that is, the return waveforms are known except with respect to noise, Doppler shift, and delay. Spectrum spreading is applicable to both cw and pulse radars. The major differentiation is in the choice of cross-correlation or matched-filter detector. The TW product is the key performance parameter, but the covariance function properties must frequently be determined to resolve Doppler shifts as well as range delays.

CLASSIFICATION OF SPREAD-SPECTRUM SIGNALS

Spread-spectrum signals can be classified on the basis of their spectral occupancy versus time characteristics, as sketched in Fig. 12.4.1. Direct-sequence (DS) and pseudo-noise (PN) waveforms provide continuous full

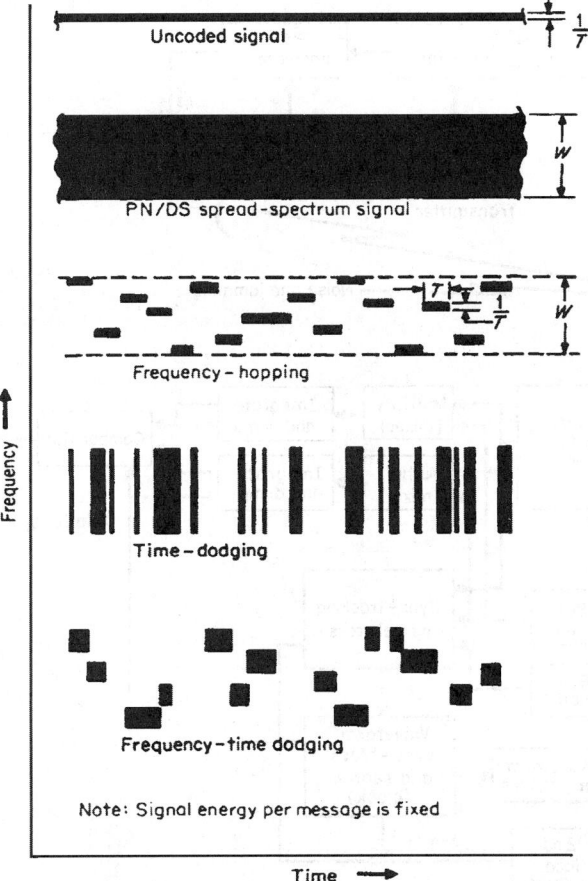

FIGURE 12.4.1   Spectral occupancy vs. time characteristics of spread-spectrum signals.

coverage, while frequency-hopping (FH), time-dodging, and frequency-time dodging (F-TD), fill the frequency-time plane only in a long-term averaging sense.

DS waveforms are pseudo-random digital streams generated by digital techniques and transmitted without significant spectral filtering. If heavy filtering is used, the signal amplitude statistics become quite noiselike, and this is called a *PN waveform*. In either case correlation detection is generally used because the waveform is dimensionally too large to implement a practical matched filter, and the sequence generator is relatively simple and capable of changing codes.

In FH schemes the spectrum is divided into subchannels spaced orthogonally at $1/T$ separations. One or more (e.g., two for FSK) are selected by pseudo-random techniques for each data bit. In time-dodging schemes the signal burst time is controlled by pulse repetition methods, while F-TD combines both selections. In each case a jammer must either jam the total spectrum continuously or accept a much lower effectiveness (approaching 1/TW). Frequency-hopped signals can be generated using SAW chirp devices.

CORRELATION-DETECTION SYSTEMS

The basic components of a typical DS type of link are shown in Fig. 12.4.2. The data are used to select the appropriate waveform, which is shifted to the desired rf spectrum by suppressed-carrier frequency-conversion

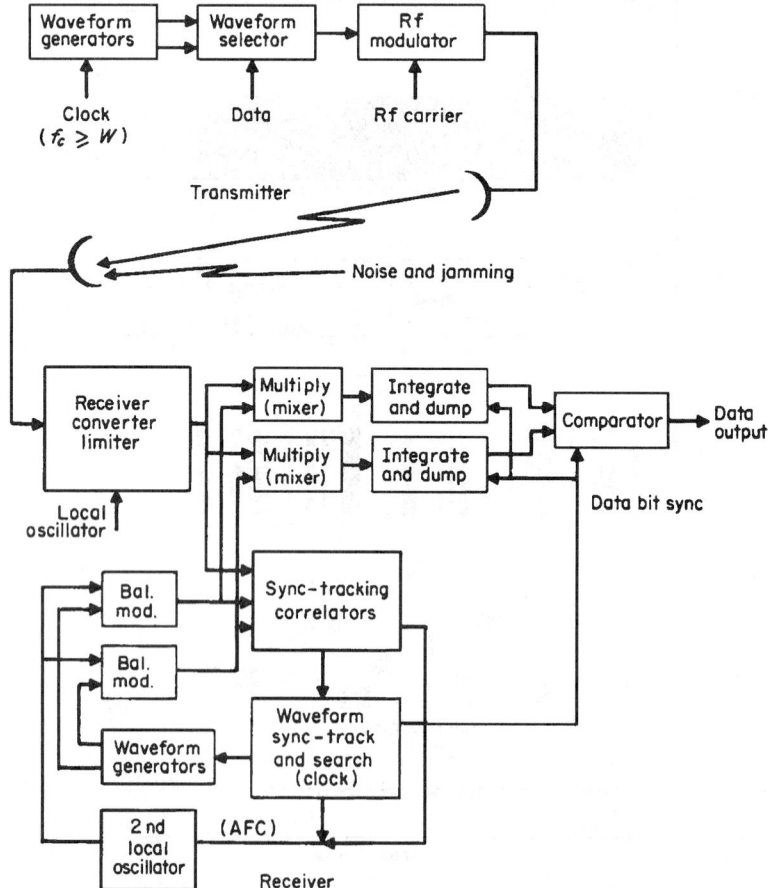

FIGURE 12.4.2 Direct-sequence link for spread-spectrum system.

techniques, and transmitted. At the receiver identical locally generated waveforms multiply with the incoming signal. The stored reference signals are often modulated onto a local oscillator, and the incoming rf may be converted to an intermediate frequency, usually with rf or i.f. limiters.

The mixing detectors are followed by linear integrate-and-dump filters, with a "greatest of" decision at the end of each period. The integrator is either a low-pass or bandpass quenchable narrowband filter. Digital techniques are increasingly being used.

Synchronization is a major design and operational problem. Given a priori knowledge of the transmitted sequences, the receiver must bring its stored reference timing to within $\pm 1/(2W)$ of the width of the received signal and hold it at that value. In a system having a 19-stage *pn* generator, a 1-MHz *pn* clock, and a 1-kHz data rate, the width of the correlation function is $\pm \frac{1}{2}$ μs, repeating $\frac{1}{2}$ s separations, corresponding to 524.287 clock periods. In the worst case, it would be necessary to inspect each sequence position for 1 ms; that is, 524 s would be required to acquire sync. If oscillator tolerances and/or Doppler lead to frequency uncertainties equal to or greater than the 1-kHz data rate, then parallel receivers or multiple searches are required.

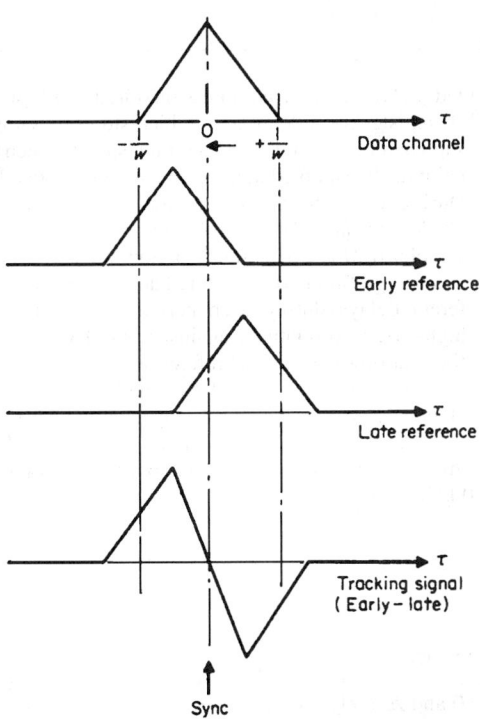

FIGURE 12.4.3 Sync tracking by early-late correlators.

Ways to reduce the sync acquisition time include using jointly available timing references to start the *pn* generators, using shorter sequences for acquisition only; "clear" sync triggers; and paralleling detectors. Titsworth (see bibliography) discusses composite sequences which allow acquiring each component sequentially, searching $N_1 + N_2 + N_3$ delays, while the composite sequence has length $N_1 N_2 N_3$. These methods have advantages for space-vehicle ranging applications but have reduced security to jamming.

Sync tracking is usually performed by measuring the correlation at early and late times, $\pm \tau$, where $\tau \leq 1/W$, as shown in Fig. 12.4.3. Subtracting the two provides a useful time discrimination function, which controls the *pn* clock. The displaced values can be obtained by two tracking-loop correlators or by time-sharing a single unit. "Dithering" the reference signal to the signal correlator can also be used, but with performance compromises.

The tracking function can also be obtained by using the time derivative of one of the inputs

$$\frac{d}{d\tau}\, \varphi_{XY}(\tau) = \frac{dX(t)}{dt} \cdot Y(t+\tau) \tag{1}$$

A third approach has been to add by mod 2 methods the clock to the transmitted *pn* waveform. The spectral envelope is altered, but very accurate peak tracking can be accomplished by phase locking to the recovered clock.

LIMITERS IN SPREAD-SPECTRUM RECEIVERS

Limiters are frequently used in spread-spectrum receivers to avoid overload saturation effects, such as circuit recovery time, and the incidental phase modulation. In the usual low-input signal-noise range, the limiter tends to normalize the output noise level, which simplifies the decision circuit design. In repeater applications (e.g., satellite), a limiter is desirable to allow the transmitter to be fully modulated regardless of the input-signal strength. When automatic gain control (AGC) is used, the receiver is highly vulnerable to pulse jamming, while the limiter causes a slight reduction of the instantaneous signal-to-jamming ratio and a proportional reduction of transmitter power allocated to the desired signal.

DELTIC-AIDED SEARCH

The sync search can be accelerated by use of deltic-aided (delay-line time compression) circuits if logic speeds permit. The basic deltic consists of a recirculating shift register (or a delay line) which stores M samples, as shown in Fig. 12.4.4. The incoming spread-spectrum signal must be sampled at a rate above W (W = bandwidth). During each intersample period the shift register is clocked through $M + 1$ shifts before accepting the next sample. If $M \geq 2W$, a signal period at least equal to the data integration period is stored and is read out at M different delays during each period T, permitting many high-speed correlations against a similarly accelerated (but not time-advancing) reference.

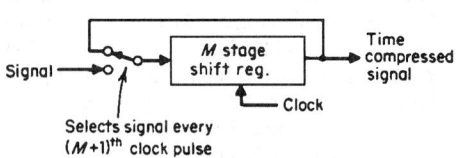

FIGURE 12.4.4 Delay-line time compression (deltic) configuration.

For a serial-deltic and shift-register delay line the clock rate is at least $4TW^2$. Using a deltic with K parallel interleaved delay lines, the internal delay lines are clocked at $4TW^2/K^2$ and the demultiplexed output has a bit rate of $4TW^2/K$, providing only M/K discrete delays. This technique is device-limited to moderate signal bandwidths, primarily in the acoustic range up to about 10 kHz.

WAVEFORMS

The desired properties of a spread-spectrum signal include:

An autocorrelation function, which is unity at $\tau = 0$ and zero elsewhere

A zero cross-correlation coefficient with noise and other signals

A large library of orthogonal waveforms

Maximal-Length Linear Sequences

A widely used class of waveforms is the maximal-length sequence (MLS) generated by a tapped refed shift register, as shown in Fig. 12.4.5a and as one-tap unit in Fig. 12.4.5b. The mod 2 half adder ($\oplus$) and EXCLUSIVE-OR logic gate are identical for 1-bit binary signals.

If analog levels +1 and −1 are substituted, respectively, for 0 and 1 logic levels, the circuit is observed to function as a 1-bit multiplier.

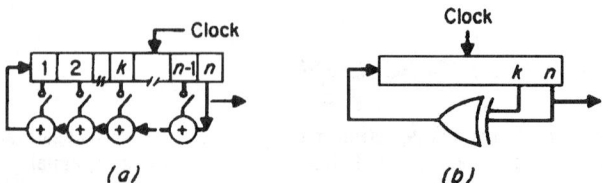

FIGURE 12.4.5 Maximal-length-sequence (MLS) system.

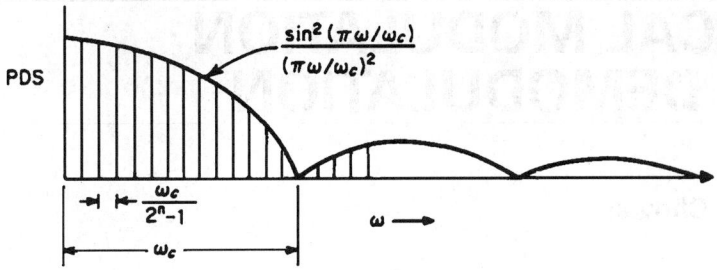

FIGURE 12.4.6   Spectrum of MLS system.

Pertinent properties of the MLS are as follows. Its length, for an n-stage shift register, is $2^n - 1$ bits. During $2^n - 1$ successive clock pulses, all n-bit binary numbers (except all zeros) will have been present. The autocorrelation function is unity at $\tau = 0$, and at each $2^n - 1$ clock pulses displacement, and $1/(2n - 1)$ at all other displacements. This assumes that the sequences repeat cyclically, i.e., the last bit is closed onto the first. The autocorrelation function of a single (noncyclic) MLS shows significant time side lobes. Titsworth (see bibliography) has analyzed the self-noise of incomplete integration over p chips, obtaining for MLSs,

$$\sigma^2(t) = (p - t)\,(p^2 - 1)/p^3 t \tag{2}$$

which approaches $1/t$ for the usual case of $p \gg t$. Since $t \approx TW$, the self-noise component is usually negligible.

Another self-noise component is frequently present owing to amplitude and dispersion differences, caused by filtering, propagation effects, and circuit nonlinearities. In addition to intentional clipping, the correlation multiplier is frequently a balanced modulator, which is linear only to the smaller signal, unless deliberately operated in a bilinear range. The power spectrum is shown in Fig. 12.4.6. The envelope has a $(\sin^2 X)/X^2$ shape ($X = \pi\omega/\omega_{\text{clock}}$), while the individual lines are separated by $\omega_{\text{clock}}/(2^n - 1)$.

An upper bound on the number of MLS for an n-stage shift register is given in terms of the Euler ϕ function:

$$N_u = \phi(2^n - 1)/n \le 2^{(n - \log 2n)} \tag{3}$$

where $\phi(k)$ is the number of positive integers less than k, including 1, which are relatively prime to k.

CHAPTER 12.5
OPTICAL MODULATION AND DEMODULATION

Joseph L. Chovan

MODULATION OF BEAMS OF RADIATION

This discussion of optical modulators is restricted to devices that operate on a directed beam of optical energy to control its intensity, phase, or frequency, according to some time-varying modulating signal. Devices that deflect a light beam or spatially modulate a light beam, such as light-valve projectors, are treated in Chap. 21.

Phase or frequency modulation requires a coherent light source, such as a laser. Optical heterodyning is then used to shift the received signal to lower frequencies, where conventional FM demodulation techniques can be applied.

Intensity modulation can be used on incoherent as well as coherent light sources. However, the properties of some types of intensity modulators are wavelength-dependent. Such modulators are restricted to monochromatic operation but not limited to the extremely narrow laser line widths required for frequency modulation.

Optical modulation depends on either perturbing the optical properties of some material with a modulating signal or mechanical motion to interact with the light beam. Modulation bandwidths of mechanical modulators are limited by the inertia of the moving masses. Optical-index modulators generally have a greater modulation bandwidth but typically require critical and expensive optical materials.

Optical-index modulation can be achieved with electric or magnetic fields or by mechanical stress. Typical modulator configurations are presented below, as in heterodyning, which is often useful in demodulation. Optical modulation can also be achieved using semiconductor junctions.

OPTICAL-INDEX MODULATION: ELECTRIC FIELD MODULATION

Pockels and Kerr Effects

In some materials, an electric field vector **E** can produce a displacement vector **D** whose direction and magnitude depend on the orientation of the material. Such a material can be completely characterized in terms of three independent dielectric constants associated with three mutually perpendicular natural directions of the material. If all three dielectric constants are equal, the material is *isotropic*. If two are equal and one is not, the material is *uniaxial*. If all three are unequal, the material is *biaxial*.

The optical properties of such a material can be described in terms of the *ellipsoid of wave normals* (Fig. 12.5.1). This is an ellipsoid whose semiaxes are the square roots of the associated dielectric constants. The behavior of any plane monochromatic wave through the medium can be determined from the ellipse formed by the intersection of the ellipsoid with a plane through the center of the ellipsoid and perpendicular to the direction of wave travel. The instantaneous electric field vector **E** associated with the optical wave has components along the two axes of this ellipse. Each of these components travels with a phase velocity that is inversely proportional to the length of the associated ellipse axis.

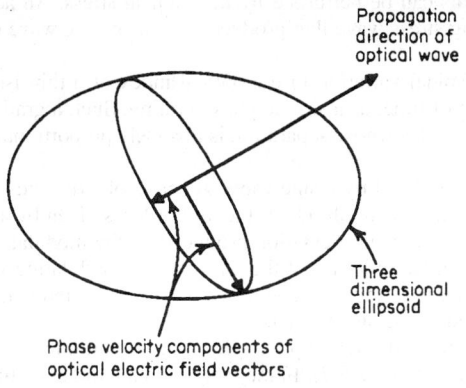

FIGURE 12.5.1   Ellipsoid of wave normals.

Thus there is a differential phase shift between the two orthogonal components of the electric field vector after it has traveled some distance through such a birefringent medium. The two orthogonal components of the vector vary sinusoidally with time but have a phase difference between them, which results in a vector whose magnitude and direction vary to trace out an ellipse once during each optical cycle. Thus linear polarization is converted into elliptical polarization in a birefringent medium.

In some materials it is possible to induce a perturbation in one or more of the ellipsoid axes by applying an external electric field. This is the electrooptical effect. The electrooptical effect is most commonly used in optical modulators presently available. More detailed configurations using these effects are discussed later. Kaminow and Turner (1966), present design considerations for various configurations and tabulates material properties.

Stark Effect

Materials absorb and emit optical energy at frequencies which depend on molecular or atomic resonances characteristic of the material. In some materials an externally applied electric field can perturb these natural resonances. This is known as the Stark effect.

Kaminow and Turner (1966) discusses a modulator for the CO_2 laser on the 3- to 22-μm region. The laser output is passed through an absorption cell whose natural absorption frequency is varied by the modulating signal, using the Stark effect. Since the laser frequency remains fixed, the amount of absorption depends on how closely the absorption cell is tuned to the laser frequency—intensity modulation results.

MAGNETIC FIELD MODULATION

Faraday Effect

Two equal-length vectors circularly rotating at equal rates in opposite directions in space combine to give a nonrotating resultant whose direction in space depends on the relative phase between the counterrotating components. Thus any linearly polarized light wave can be considered to consist of equal right and left circularly polarized waves.

In a material which exhibits the Faraday effect, an externally applied magnetic field causes a difference in the phase velocities of right and left circularly polarized waves traveling along the direction of the applied magnetic field. This results in a rotation of the electric field vector of the optical wave as it travels through the material. The amount of the rotation is controlled by the strength of a modulating current producing the magnetic field.

Zeeman Effect

In some materials the natural resonance frequencies at which the material emits or absorbs optical energy can be perturbed by an externally applied magnetic field. This is known as the Zeeman effect.

Intensity modulation can be achieved using an absorption cell modulated by a magnetizing current in much the same manner as the Stark effect absorption cell is used. The Zeeman effect has also been used to tune the frequency at which the active material in a laser emits.

MECHANICAL-STRESS MODULATION

In some materials the ellipsoid of optical-wave normals can be perturbed by mechanical stress. An acoustic wave traveling through such a medium is a propagating stress wave that produces a propagating wave of perturbation in the optical index.

When a sinusoidal acoustic wave produces a sinusoidal variation in the optical index of a thin isotropic medium, the medium can be considered, at any instant of time, as a simple phase grating. Such a grating diffracts a collimated beam of coherent light into discrete angles whose separation is inversely proportional to the spatial period of the grating.

This situation is analogous to an rf carrier phase-modulated by a sine wave. A series of sidebands results which correspond to the various orders of diffracted light. The amplitude of the mth order is given by an mth-order Bessel function whose argument depends on the peak phase deviation produced by the modulating signal. The phases of the sidebands are the appropriate integral multiples of the phase of the modulating signal.

The mth order of diffracted light has its optical frequency shifted by m times the acoustic frequency. The frequency is increased for positive orders and decreased for negative orders.

Similarly, a thick acoustic grating refracts light mainly at discrete input angles. This condition is known as *Bragg reflection* and is the basis for the *Bragg modulator* (Fig. 12.5.2). In the Bragg modulator, essentially all the incident light can be refracted into the desired order, and the optical frequency is shifted by the appropriate integral multiple of the acoustic frequency.

Figure 12.5.2 shows the geometry of a typical Bragg modulator. The input angles for which Bragg modulation occurs are given by

$$\sin \theta = m\lambda/2\Lambda$$

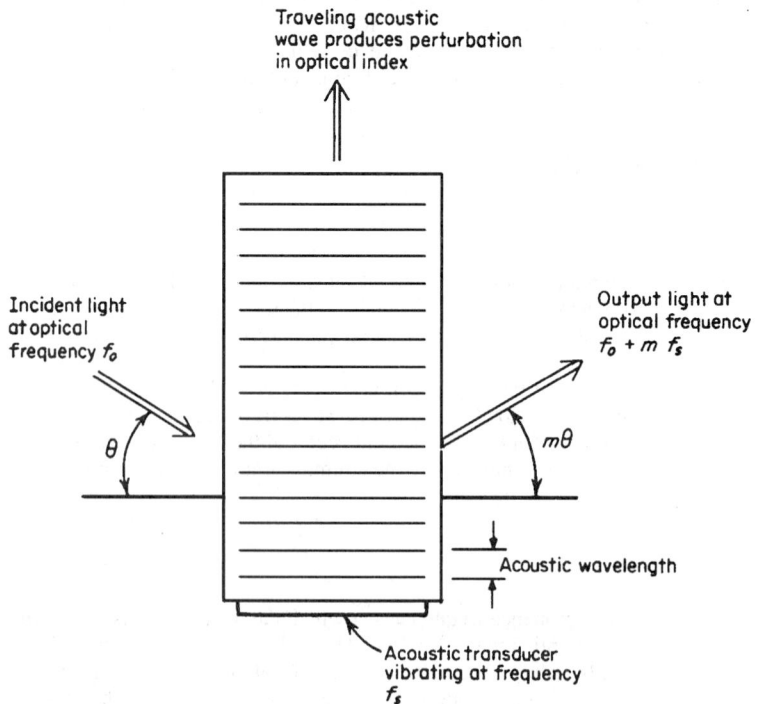

FIGURE 12.5.2   The Bragg modulator.

where θ = angle between propagation of input optical beam and planar acoustic wavefronts
λ = optical wavelength in medium
Λ = acoustic wavelength in medium
$m = \pm 1, \pm 2, \pm 3, \ldots$
$m\theta$ = angle between propagation direction of output optical beam and planar acoustic wavefronts

The ratio of optical to acoustic wavelength is typically quite small, and m is a low integer, so that the angle θ is very small. Critical alignment is thus required between the acoustic wavefronts and the input light beam.

If the modulation bandwidth of the acoustic signal is broad, the acoustic wavelength varies, so that there is a corresponding variation in the angle θ for which Bragg reflection occurs. To overcome this problem, a phased array of acoustic transducers is often used to steer the angle of the acoustic wave as a function of frequency in the desired manner.

A limitation on bandwidth is the acoustic transit time across the optical beam. Since the phase grating in the optical beam at any instant of time must be essentially constant frequency if all the light is to be diffracted at the same angle, the bandwidth is limited so that only small changes can occur in this time interval.

MODULATOR CONFIGURATIONS: INTENSITY MODULATION

Polarization Changes

Linearly polarized light can be passed through a medium exhibiting an electrooptical effect and the output beam passed through another polarizer. The modulating electric field controls the eccentricity and orientation of the elliptical polarization and hence the magnitude of the component in the direction of the output polarizer. Typically, the input linear polarization is oriented to have equal components along the fast and slow axes of the birefringent medium, and the output polarizer is orthogonal to the input polarizer. The modulating field causes a phase differential varying from 0 to π rad. This causes the polarization to change from linear (at 0) to circular (at $\pi/2$) to linear normal to the input polarization (at π). Thus the intensity passing through the output polarizer varies from 0 to 100 percent as the phase differential varies from 0 to π rad.

Figure 12.5.3 shows this typical configuration. The following equations relate the optical intensity transmission of this configuration to the modulation.

$$I_o/I_i = {}^1\!/_2(1 - \cos \phi)$$

where I_o = output optical intensity
I_i = input optical intensity
ϕ = differential phase shift between fast and slow axes.

In the Pockels effect the differential phase shift is linearly related to applied voltage; in the Kerr effect it is related to the voltage squared.

$$\phi = \begin{cases} \pi v/V & \text{Pockels effect} \\ \pi(v/V)^2 & \text{Kerr effect} \end{cases}$$

where v is the modulation voltage and V is the voltage to produce π rad differential phase shift.

Figure 12.5.4 shows the intensity transmission given by the above expression. The most linear part of the modulation curve is at $\phi = \pi/2$. Often a quarter-wave plate is added in series with the electrooptical material to provide this fixed bias at $\pi/2$. A fixed-bias voltage on the electrooptical material can also be used.

This arrangement is probably the most commonly used broadband intensity modulator. Early modulators of this type used a uniaxial Pockels cell with the electric field in the direction of optical propagation. In this arrangement, the induced phase differential is directly proportional to the opticalpath length, but the electric field is inversely proportional to this path length (at fixed voltage). Thus the phase differential is independent of the path length and depends only on applied voltage. Typical materials require several kilovolts for a differential phase shift of π in the visible-light region.

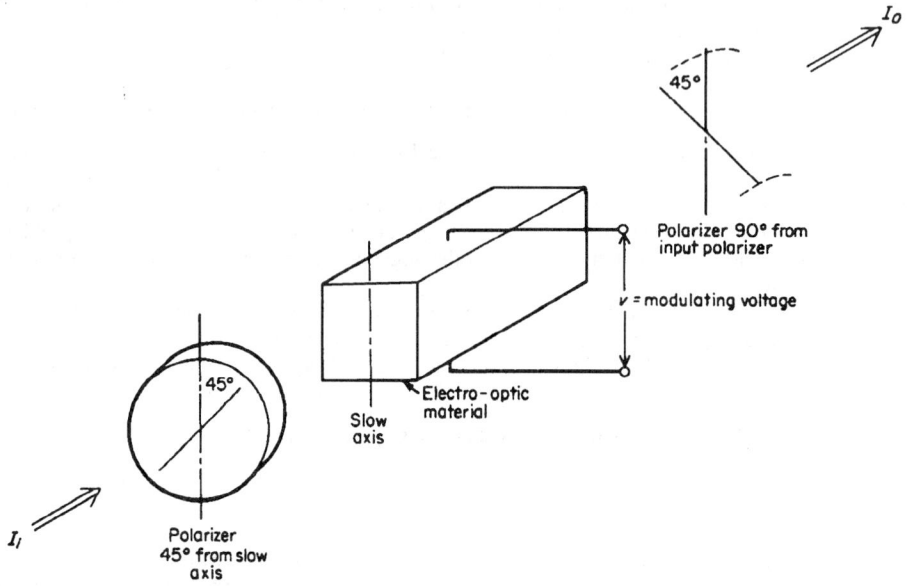

FIGURE 12.5.3 Electrooptical intensity modulator.

Since the Pockels cell is essentially a capacitor, the energy stored in it is $^1/_2 CV^2$ where C is the capacitance and V is the voltage. This capacitor must be discharged and charged during each modulation cycle. Discharge is typically done through a load resistor, where this energy is dissipated. The high voltages involved mean that the dissipated power at high modulation rates is appreciable.

The high-voltage problem can be overcome by passing light through the medium in a direction normal to the applied electric field. This permits a short distance between the electrodes (so that a high-E field is obtained from a low voltage) and a long optical path in the orthogonal direction (so that the cumulative phase differential is experienced).

Unfortunately, materials available are typically uniaxial, having a high eccentricity in the absence of electric fields. When oriented in a direction that permits the modulating electric field to be orthogonal to the propagation direction, the material has an inherent phase differential which is orders of magnitude greater than that induced by the modulating field. Furthermore, minor temperature variations cause perturbations in this phase differential which are large compared with those caused by modulation.

This difficulty is overcome by cascading two crystals which are carefully oriented so that temperature effects in one are compensated for by temperature effects in the other. The modulation electrodes are then connected so that their effects add. Commercially available electrooptical modulators are of this type.

The Kerr effect is often used in a similar arrangement. Kerr cells containing nitrobenzene are commonly used as high-speed optical shutters.

Polarization rotation produced by the Faraday effect is also used in intensity modulation by passing through an output polarizer in a manner similar to that discussed above. The Faraday effect is more commonly used at wavelengths where materials exhibiting the electrooptical effect are not readily available.

Controlled Absorption

As noted above, the frequency at which a material absorbs energy because of molecular or atomic resonances can be tuned over some small range in materials exhibiting the Stark or Zeeman effect. Laser spectral widths are typically narrow compared with such an absorption line width. Thus the absorption of the narrow laser line can be modulated by tuning the absorption frequency over a range near the laser frequency. Although such modulators have been used, they are not as common as the electrooptical modulators discussed above.

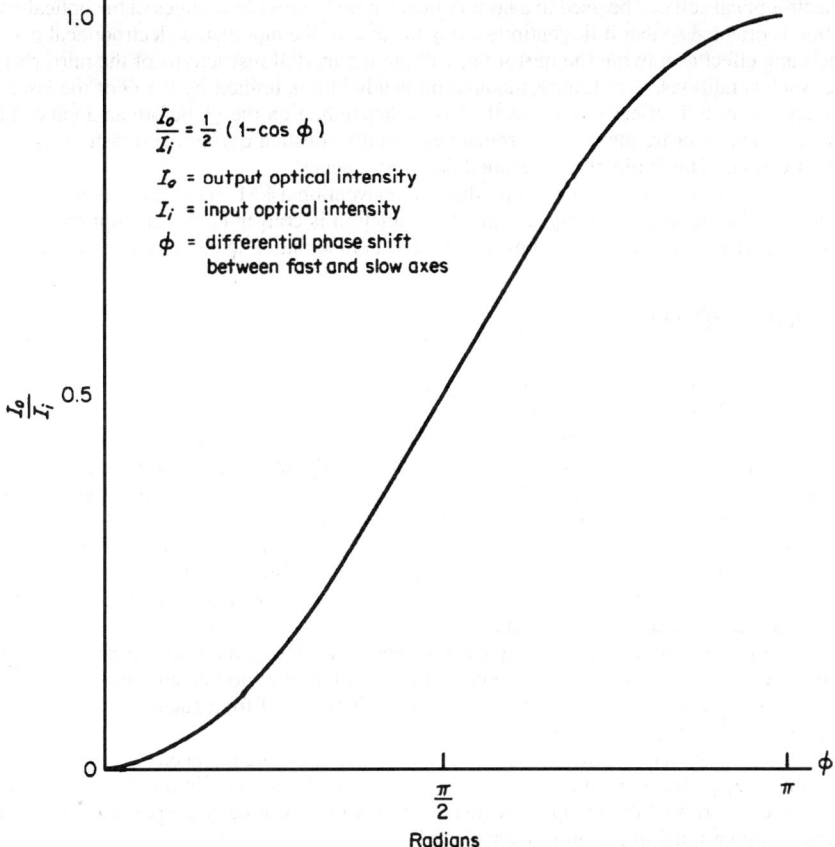

The plotted curve with annotations:

$$\frac{I_o}{I_i} = \frac{1}{2}(1 - \cos\phi)$$

I_o = output optical intensity

I_i = input optical intensity

ϕ = differential phase shift between fast and slow axes

Vertical axis: $\frac{I_o}{I_i}$ marked at 0.5 and 1.0

Horizontal axis: ϕ marked at $\frac{\pi}{2}$ and π

Radians

FIGURE 12.5.4 Transmission of electrooptical intensity modulator.

PHASE AND FREQUENCY MODULATION OF BEAMS

Laser-Cavity Modulation

The distance between mirrors in a laser cavity must be an integral number of wavelengths. If this distance is changed by a slight amount, the laser frequency changes to maintain an integral number. The following equation relates the change in cavity length to the change in frequency:

$$\Delta f = \frac{C}{L}\frac{\Delta L}{\lambda}$$

where Δf = change in optical frequency

ΔL = change in laser-cavity length

L = laser-cavity length

λ = optical wavelength of laser output

C = velocity of light in laser cavity.

In a cavity 1 m long, a change in mirror position of one optical wavelength produces about 300 MHz frequency shift. Thus a laser can be frequency-modulated by moving one of its mirrors with an acoustic transducer, but the mass of the transducer and mirror limit the modulation bandwidths that can be achieved.

An electrooptical cell can be used in a laser optical cavity to provide changes in the optical-path length. The polarization is oriented so that it lies entirely along the axis of the modulated electrooptical material. This produces the same effect as moving the mirror but without the inertial restrictions of the mirror's mass.

Under such conditions, the ultimate modulation bandwidth is limited by the Q of the laser cavity. A light beam undergoes several reflections across the cavity, depending on the Q, before an appreciable portion of it is coupled out. The laser frequency must remain essentially constant during the transit time required for these multiple reflections. This limits the upper modulation frequency.

Modulation of the laser-cavity length produces a conventional FM signal with modulating signal directly proportional to change in laser-cavity length. Demodulation is conveniently accomplished by optical heterodyning to lower rf frequencies where conventional FM demodulation techniques can be used.

EXTERNAL MODULATION

The Bragg modulator (Fig. 12.5.2) is commonly used to modulate the optical frequency. As such it produces a single-sideband suppressed-carrier type of modulation.

Demodulation can be achieved by optical heterodyning to lower rf frequencies, where conventional techniques can be employed for this type of modulation. It is also possible to reinsert the carrier at the transmitter for a frequency reference. This is done by using optical-beam splitters to combine a portion of the unmodulated laser beam with the Bragg modulator output.

Conventional double-sideband amplitude modulation has also been achieved by simultaneously modulating two laser beams (derived from the same source) with a common Bragg modulator to obtain signals shifted up and down. Optical-beam splitters are used to combine both signals with an unmodulated carrier. Conventional power detection demodulates such a signal.

Optical phase modulation is commonly accomplished by passing the laser output beam through an electrooptical material, with the polarization vector oriented along the modulated ellipsoid axis of the material. Demodulation is conveniently achieved by optical heterodyning to rf frequencies, FM demodulation, and integrating to recover the phase modulation in the usual manner.

For low modulation bandwidths, the electrooptical material can be replaced by a mechanically driven mirror. The light reflected from the mirror is phase modulated by the changes in the mirror position. This effect is often described in terms of the Doppler frequency shift, which is directly proportional to the mirror velocity and inversely proportional to the optical wavelength.

TRAVELING-WAVE MODULATION

In the electrooptical and magnetooptical modulators described thus far, it is assumed that the modulating signal is essentially constant during the optical transit time through the material. This sets a basic limit on the highest modulating frequency that can be used in a lumped modulator.

This problem is overcome in a traveling-wave modulator. The optical wave and the modulation signal propagate with equal phase velocities through the modulating medium, allowing the modulating fields to act on the optical wave over a long path, regardless of how rapidly the modulating fields are changing. The degree to which the two phase velocities can be matched determines the maximum interaction length possible.

OPTICAL HETERODYNING

Two collimated optical beams, derived from the same laser source and illuminating a common surface, produce straight-line interference fringes. The distance between fringes is inversely proportional to the angle between the beams. Shifting the phase of one of the beams results in a translation of the interference pattern, such that a 2π-rad phase shift translates the pattern by a complete cycle. An optical detector having a sensing area small compared with the fringe spacing has a sinusoidal output as the sinusoidal intensity of the interference pattern translates across the detector.

A frequency difference between the two optical beams produces a phase difference between the beams that changes at a constant rate with time. This causes the fringe pattern to translate across the detector at a constant rate, producing an output at the difference frequency. This technique is known as *optical heterodyning* in which one of the beams is the signal beam, the other the local oscillator.

The effect of the optical alignment between the beams is evident. As the angle between the two collimated beams is reduced, the spacing between the interference fringes increases, until the spacing becomes large compared with the overall beam size. This permits a large detector which uses all the light in the beam. If converging or diverging beams are used instead of collimated beams, the situation is similar, except that the interference fringes are curved instead of straight. Making the image of the local-oscillator point coincide with the image of the signal-beam point causes the desired infinite fringe spacing.

Optical heterodyning provides a convenient solution to several possible problems in optical demodulation. In systems where a technique other than simple amplitude modulation has been used (e.g., single-sideband, frequency, or phase modulation), optical heterodyning permits shifting to frequencies where established demodulation techniques are readily available.

In systems where background radiation, such as from the sun, is a problem, heterodyning permits shifting to lower frequencies, so that filtering to the modulation bandwidth removes most of the broadband background radiation. The required phase front alignment also eliminates background radiation from spatial positions other than that of the signal source.

Many systems are limited by thermal noise in the detector and/or front-end amplifier. Cooled detectors and elaborate amplifiers are often used to reduce this noise to the point that photon noise in the signal itself dominates. This limit also can be achieved in an optical heterodyne system with noncooled detector and normal amplifiers by increasing the local-oscillator power to the point where photon noise in the local oscillator is the dominant noise source. Under these conditions, the signal-to-noise power ratio is given by the following equation:

$$S/N = \eta \lambda P / 2hBC$$

where S/N = signal-power-noise-power ratio
η = quantum efficiency of photo detector
λ = optical wavelength
h = Planck's constant
C = velocity of light
B = bandwidth over which S/N is evaluated
P = optical signal power received by detector

CHAPTER 12.6
FREQUENCY CONVERTERS AND DETECTORS

Glenn B. Gawler

GENERAL CONSIDERATIONS OF FREQUENCY CONVERTERS

A frequency converter usually consists of an oscillator (called a *local oscillator* or LO) and a device used as a mixer. The mixing device is either nonlinear or its transfer parameter can be made to vary in synchronism with the local oscillator. A signal voltage with information in a frequency band centered at frequency f_s enters the frequency converter, and the information is reproduced in the intermediate-frequency (i.f.) voltage leaving the converter. If the local-oscillator frequency is designated f_{LO}, the i.f. voltage information is centered about a frequency $f_{if} = f_{LO} \pm f_s$. The situation is shown pictorially in Fig. 12.6.1. Characteristics of interest for design in systems using frequency converters are gain, noise figure, image rejection, spurious responses, intermodulation and cross-modulation capability, desensitization, area local-oscillator to rf, and to i.f. isolation. These characteristics will be discussed at length in the descriptions of different types of frequency-converter mixers and their uses in various systems. First, explanations are in order for the above terms.

Frequency-Converter Gain. The available power gain of a frequency converter is the ratio of power available from the i.f. port to the power available at the signal port. Similar definitions apply for transducer gain and power gain.

Noise Figure of Frequency Converter. The noise factor is the ratio of noise power available at the i.f. port to the noise power available at the i.f. port because of the source alone at the signal port.

Image Rejection. For difference mixing $f_{if} = f_{LO} - f_s$ and the image is $2f_{LO} - f_s$. For sum mixing $f_{if} = f_{LO} + f_s$, and the image is $2f_{LO} + f_s$. An undesired signal at the difference mixing frequency $2f_{LO} - f_s$ results in energy at the i.f. port. This condition is called *image response* and attenuation of the image response is image rejection, measured in decibels.

Spurious Responses. Spurious external signals reach the mixer and result in generation of undesired frequencies that may fall into the intermediate-frequency band. The condition for an interference in the i.f. band is

$$mf'_s \pm nf_1 = \pm f_{if}$$

where m and n are integers and f'_s represents spurious frequencies at the signal port of the mixer.

> **Example.** There is a strong local station in the broadcast band at 810 kHz and a weak distant station at 580 kHz. A receiver is tuned to the distant station, and a whistle, or beat, at 5 kHz is heard on the receiver (refer to Fig. 12.6.2).

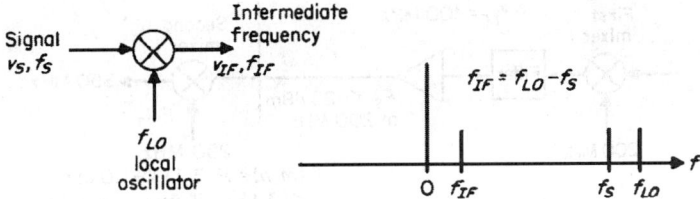

FIGURE 12.6.1 Frequency-converter terminals and spectrum.

An analysis shows that the second harmonic of the local oscillator interacts with the second harmonic of the 810-kHz signal to produce a mixer output at 450 kHz in the i.f. band of the receiver:

$$580 + 455 = 1035 \text{ kHz} = \text{LO frequency}$$
$$2 \times 1035 - 2 \times 810 = 450 \text{ kHz} = \text{i.f. interference frequency}$$

The interference at 450 kHz then mixes with the 455-kHz desired signal in the second detector to produce the 5-kHz whistle. Notice that if the receiver is slightly detuned upward by 5 kHz, the whistle will zero-beat. Further upward detuning will create a whistle of increasing frequency.

INTERMODULATION

Intermodulation is particularly troublesome because a pair of strong signals that pass through a receiver preselector can cause interference in the i.f. passband, even though the strong signals themselves do not enter the passband.

Consider two undesired signals at 97 MHz passing through a superheterodyne receiver tuned to 100 MHz. Suppose, further, that the i.f. is so selective that a perfect mixer allows no response to the signals (see Fig. 12.6.3). Third-order intermodulation in a physically realizable mixer will result in interfering signals at the i.f. frequency and 9 MHz away (corresponding to 100 and 91 MHz rf frequencies, respectively). Fifth-order intermodulation will produce interferences 3 and 12 MHz from the intermediate frequency (103 and 88 MHz rf frequencies).

There is a formula for variation of intermodulation products that is quite useful. Figure 12.6.4 shows typical variations of desired output and intermodulation with input power level. Desired output increases 1 dB for each 1-dB increase of input level, whereas third-order intermodulation increases 3 dB for each 1-dB increase of input level. At some point the mixer saturates and the above behavior no longer obtains. Since the interference of the intermodulation product is primarily of interest near the system sensitivity limit (usually somewhere below –20 dBm), the 1 dB per 1 dB and 3 dB per 1 dB patterns hold. The formula can be written

$$P_{21} = 2P_N + P_F - 2P_{/21}$$

where P_{21} = level of intermodulation product (dBm)
 P_N = power level of interfering signal nearest P_{21}
 P_F = power interfering signal farthest from P_{21}

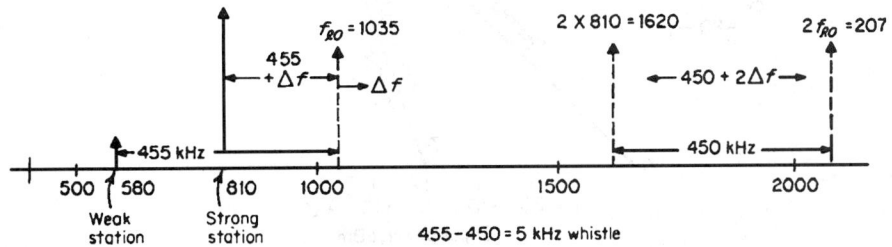

FIGURE 12.6.2 Spurious response in AM receiver.

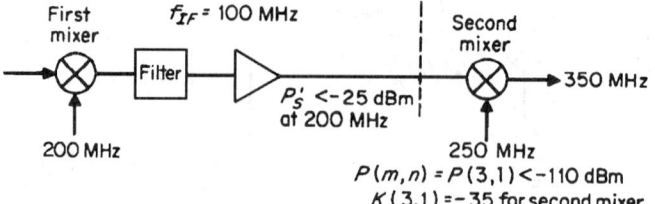

FIGURE 12.6.3   Spurious-response analysis.

$P_{/21}$ is the third-order intercept power. For proper orientation, $f_N = 97$ MHz, $f_F = 94$ MHz, $f_{21} = 100$ MHz in Fig. 12.6.5. The intercept power is a function of frequency. It can be used for comparisons between mixer designs and for determining allowable preselector gain in a receiving system.

FREQUENCY-CONVERTER ISOLATION

There are two paths in a mixer where isolation is important. The so-called *balanced mixers* give some isolation of the local-oscillator energy at the rf port. This keeps the superheterodyne receiver from radiating excessively. The doubly balanced mixers also give rf-to-i.f. isolation. This keeps interference in the receiver rf environment from penetrating the mixer directly at the i.f. frequency. Less important, but still significant, is the LO-to-i.f. isolation. This keeps LO energy from overloading the i.f. amplifier. Also, in multiple-conversion receivers low LO-to-i.f. leakage minimizes spurious responses in subsequent frequency converters.

Desensitization

A strong signal in the rf bandwidth, not directly converted to i.f., drives the operating point of the mixer into a nonlinear region. The mixer gain is then either decreased or increased. In radar, the characteristic of concern is pulse desensitization. In television receivers the characteristic is called cross-modulation. Here the strong

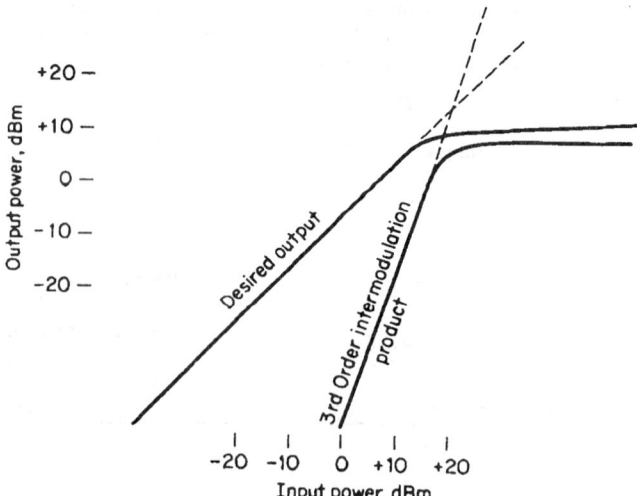

FIGURE 12.6.4   Third-order intermodulation intercept power.

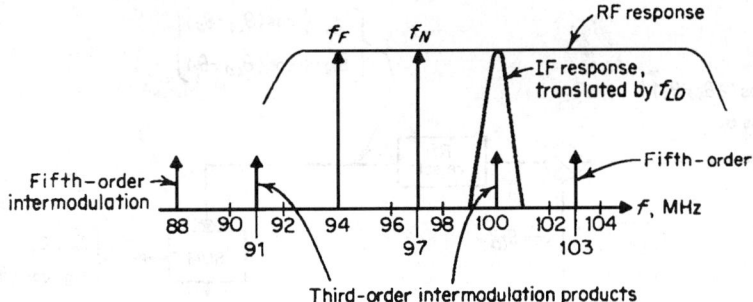

FIGURE 12.6.5 Intermodulation in a superheterodyne receiver.

undesired adjacent TV station modulates the mixer gain, especially during synchronization intervals, where the signal is strongest. The result appears in the desired signal as a contrast modulation of picture with the pattern of the undesired sync periods, corresponding to mixer gain *pumping* by the strong adjacent channel.

SCHOTTKY DIODE MIXERS

The Schottky barrier diode is an improvement over the point-contact diode. The Schottky diode has two features that make it very valuable in high-frequency mixers: (1) it has low series resistance and virtually no charge storage, which results in low conversion loss; (2) it has noise-temperature ratio very close to unity. The noise factor of a mixed-i.f. amplifier cascade is

$$F = L_M(t_D + F_{if} - 1)$$

where L_M = mixer loss
t_D = diode noise-temperature ratio
F_{if} = i.f. noise factor

Since t_D is near unity and L_M is in the range of 2.4 to 6 dB, overall noise factor is quite good, with F_{if} near 1.5 dB in well-designed systems.

The complete conversion matrix involves LO harmonic sums and differences, as well as signal, i.f., and image frequencies. They restrict their treatment of crystal rectifiers to the third-order matrix

$$\begin{bmatrix} I_1 \\ I_2 \\ I_3^* \end{bmatrix} = [Y] \begin{bmatrix} V_1 \\ V_2 \\ V_3^* \end{bmatrix} \quad Y = \begin{bmatrix} y_{11} & y_{12} & y_{13} \\ y_{21} & y_{22} & y_{23} \\ y_{31} & y_{32} & y_{33} \end{bmatrix}$$

where 1 denotes signal port; 2, i.f. port; and 3, image port.

With point-contact diodes, the series resistance is so large that not much improvement is realized by terminating the image frequency, and terminating the other frequencies involved is less significant.

With the advent of Schottky barrier diodes, which have much smaller series resistances, proper termination of pertinent frequencies, other than signal and i.f. frequencies, results in a minimizing of conversion loss. This, in turn, leads to a minimizing of noise figure.

Several different configurations are used with Schottky mixers. Figure 12.6.6 shows an image-rejection mixer, which is used for low i.f. frequency systems where rf filtering of the image is impractical.

There is a general rule of thumb for obtaining good intermodulation, cross-modulation, and desensitizable performance in mixers. It has been found experimentally that pumping a mixer harder extends its range of

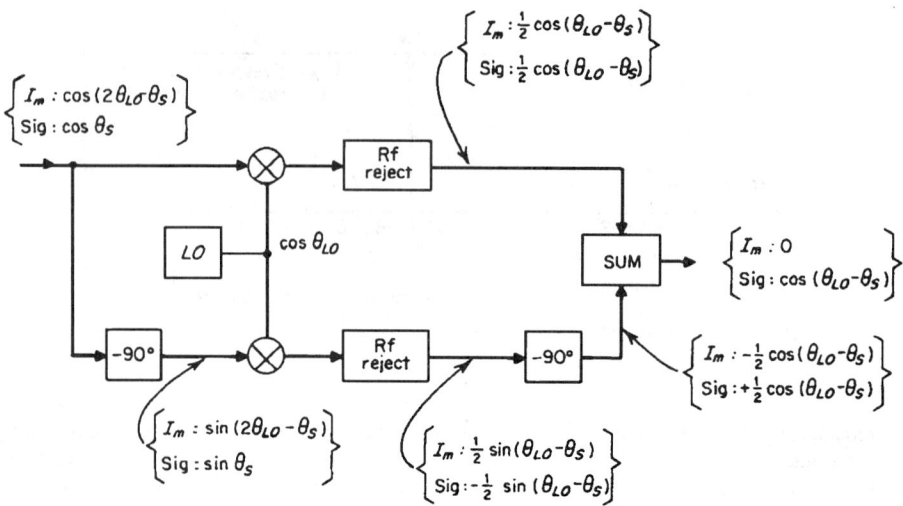

FIGURE 12.6.6   Mixer designed for image rejection.

linear operation. The point-contact diode had a rapidly increasing noise figure with high LO power level and could easily burn out with too much power. The Schottky diode, however, degrades in noise figure relatively slowly with increasing LO power, and it can tolerate quite large amounts of power without burnout. There is a limit to this process of increasing LO power: the Schottky diode series resistance begins to appear nonlinear. This leads to another rule of thumb: pump the diode between two linear regions, and spend as little time as possible in the transition region. Application of these two rules leads to the doubly balanced Schottky mixer. The reason for this is that one pair of diodes conducts hard and holds the other pair off. Hence large LO power is required, and one diode pair is conducting well into its linear region while the other diode pair is held in its linear nonconducting region.

DOUBLY BALANCED MIXERS

The diode doubly balanced mixer, or ring modulator, is shown in Fig. 12.6.7. The doubly balanced mixer is used up to and beyond 1 GHz in this configuration. The noise-figure optimization process previously discussed applies to this type of mixer. It exhibits good LO-to-rf and LO-to-i.f. isolation, as shown in Fig. 12.6.8. Typical published data on mixers quote 30 dB rf-to-i.f. isolation below 50 MHz and 20 dB isolation from 50 to 500 MHz.

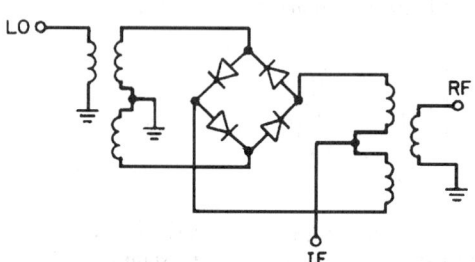

FIGURE 12.6.7   Doubly balanced mixer.

Another feature of balanced mixers is their LO noise-suppression capability. Modern mixers using Schottky diodes in a ring modulator provide somewhat better LO noise suppression. The ring modulator provides suppression only to AM LO noise, not FM noise.

PARAMETRIC CONVERTERS

Parametric converters make use of time-varying energy-storage elements. Their operation is in many ways similar to that of parametric amplifiers (see Section 11). The difference is that output and input frequencies are the same in parametric amplifiers, while the frequencies differ in parametric converters. The device most widely

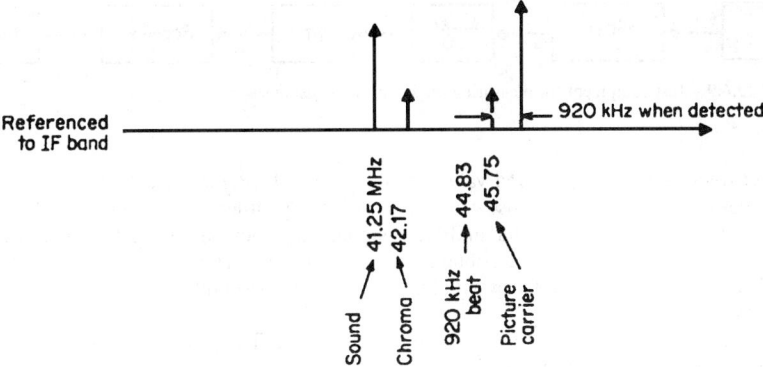

FIGURE 12.6.8   Generation of 920-kHz beat in TV tuners.

used for microwave parametric converters today is the varactor diode, which has a voltage-dependent junction capacitance. The time variation of varactor capacitance is provided by a local oscillator, usually called the *pump*.

Attainable gain of a parametric converter is limited by the ratio of output to input frequencies. Therefore up conversion is generally used to achieve some gain. Because lower-sideband up conversion results in negative resistance, the upper sideband is generally used. This results in simpler circuit elements to achieve stability.

There is a distinct advantage to up conversion; image rejection is easily achievable by a simple low-pass filter.

TRANSISTOR MIXERS

One of the original concerns in transistor mixers was their noise performance. The base spreading resistance r_b is very important in noise performance. The reason is that mixing occurs across the base-emitter junction; then the i.f. signal is amplified by transistor action; however, r_b is a lossy part of the termination at the i.f. signal, image, and all other frequencies present in the mixing process. Hence r_b dissipates some energy at each of the frequencies present, and all these contributions add to appear as a loss in the signal-to-i.f. conversion. This loss, in turn, degrades noise figure.

Manufacturers do not promote transistors used as mixers, probably because of their intermodulation and spurious-response performance. Estimates of intermodulation intercept power go as high as +12 dBm, while one measurement gave +5 dBm at 200 MHz; however, a cascade transistor mixer is used in a commercial VHF television tuner.

MEASUREMENT OF SPURIOUS RESPONSES

Figure 12.6.9 shows an arrangement for measuring mixer spurious responses. The filter following the signal generator implies that generator harmonics are down, say 40 dB. This ensures that frequency-multiplying action is owing only to the mixer under test. The attenuator following the mixer can be used to be sure that a spurious response of the receiver is not being measured. That is, a 6-dB change in attenuator setting should be accompanied by a 6-dB change on the indicator.

Generally the most convenient way of performing the spurious-response test is first to obtain an indication of the indicator. Then tune the signal generator to the desired frequency and record the level required to obtain the original response. This should be repeated at one or two more levels of the undesired signal to ensure that the spur follows the appropriate laws. For example, if the response is fourth-order (four times the signal frequency $\pm n$

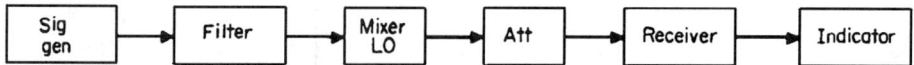

FIGURE 12.6.9   Test equipment for measuring mixer spurious responses.

times the LO frequency), the measured value should change 4 dB for a 1-dB change in undesired frequency level. The order of the spurious response can be determined by either of two methods. The first method is simply by knowing with some accuracy the undesired signal frequency and the LO frequency and then determining the harmonic numbers required to obtain the i.f. frequency. The other technique entails observing the incremental changes of the i.f. frequency with known changes in the undesired signal frequency and the LO signal frequency.

This completes the measurement for one spurious response. The procedure should be repeated for each of the spurious responses to be measured.

The intermodulation test setup is shown in Fig. 12.6.10. In general, a diplexer is preferable to a directional coupler for keeping generator 1 signal out of generator 2. This is necessary so that the measurement is not limited by the test setup. A good idea would be to establish that no third-order intermodulation occurs becuase of the setup alone. To do this, initially remove the mixer-LO circuit. Then tune generator 1 off from center frequency to about 10 or 20 dB down on the skirt of the receiver preselector. Tune generator 2 twice this amount from the receiver center frequency. Set generator levels equal and at some initial value, say –30 dBm. Then vary one generator frequency slightly and look for a response peak on the indicator. If none is noticed, increase the generator level to –20 dBm and repeat the procedure. Usually, except for very good receivers, the third-order intermodulation response is found. Vary the attenuator by 6 dB, and look for a 6-dB variation in the indicator reading. If the latter is not 6 dB but 18 dB, intermodulation is occurring in the receiver. If the indicator variation is between 6 and 18 dB, intermodulation is occurring in the circuitry preceding the attenuator and in the receiver. To obtain trustworthy measurements with a mixer in the test position, the indicator should read at least 20 dB greater than without the mixer, while the generator levels should be lower by mixer gain +10 dB than they were without the mixer. This ensures that the test setup is contributing an insignificant amount to the intermodulation measurement.

With the mixer in test position and the above conditions satisfied, obtain a reading on the indicator and let the power referred to the mixer input be denoted by P (dBm). Turn down both generator levels, and retune generator 1 to center frequency. Adjust generator 1 level to obtain the previous indicator reading. This essentially calibrates the measurement setup. Denote generator 1 level referred to the mixer input by P_{21} dBm. Then the intermodulation intercept power is given by

$$P_{/21} \text{ dBm} = (3P - P_{21})/2$$

The subscripts on intercept power $P_{/21}$ refer to second order for the near frequency and first order for the far frequency (see Fig. 12.6.4).

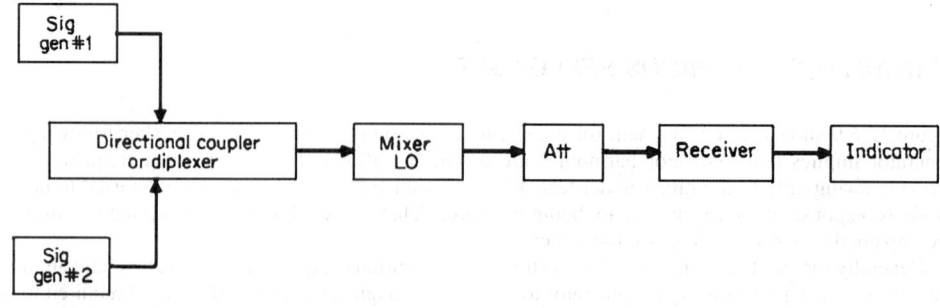

FIGURE 12.6.10   Test equivalent for measurement of mixer intermodulation.

The procedure should be repeated for one or two lower values of P. The corresponding values of $P_{/21}$ should asymptomatically approach a constant value. The constant value of $P_{/21}$ so obtained is then a valid number for predicting behavior of the mixer near its sensitivity limit.

DETECTORS (FREQUENCY DECONVERTERS)

Detectors have become more complex and versatile since the advent of integrated circuits. Up to the mid-1950s most radio receivers used the standard single-diode envelope detector for AM and a Foster-Seeley discriminator or ratio detector for FM. Today, integrated circuits are available with i.f. amplifier, detector, and audio-amplifier functions in a single package.

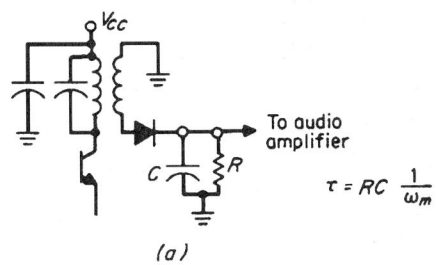

(a)

$$\tau = RC \quad \frac{1}{\omega_m}$$

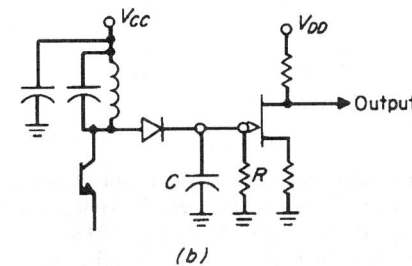

(b)

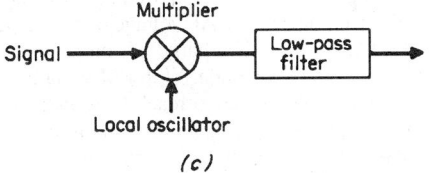

(c)

FIGURE 12.6.11 AM detectors: (*a*) AM envelope detector; (*b*) peak detector; (*c*) product detector.

Figure 12.6.11 shows three conventional AM detectors. In Fig. 12.6.11*a* an envelope detector is shown. In order for the detected output to follow the modulation envelope faithfully, the *RC* time constant must be chosen so that $RC < 1/\omega_m$, where ω_m is the maximum angular modulation frequency in the envelope. Figure 12.6.11*b* shows a peak detector. Here the *RC* time constant is chosen large, so that *C* stays charged to the peak voltage. Usually, the time constant depends on the application. In a television field-strength meter, the charge on *C* should not decay significantly between horizontal sync pulses separated by 62.5 μs. Hence a time constant of 1 to 6 ms should suffice. On the other hand, an AGC detector for single-sideband use should have a time constant of 1 s or longer.

Figure 12.6.11*c* shows a product (synchronous) detector. This type of detector has been used since the advent of single-sideband transmission. The product detector multiplies the signal with the LO, or beat frequency oscillator (BFO), to produce outputs at sum and difference frequencies. Then the low-pass filter passes only the difference frequency. The result is a clean demodulation with a minimum of distortion for single-sideband signals.

The two classical FM detectors widely used up to the present are the Foster-Seeley discriminator and the ratio detector. Figure 12.6.12 shows the Foster-Seeley discriminator and its phasor diagrams. The circuit consists of a double-tuned transformer, with primary and secondary voltages series-connected. The diode connected to point *A* detects the peak value of $V_1 + V_2/2$, and the diode at *B* detects the peak value of $V_1 - V_2/2$. The audio output is then the difference between the detected voltages. When the incoming frequency is in the center of the passband, V_2 is in quadrature with V_1, the detected voltages are equal, and audio output is zero. Below the center frequency the detected voltage from *B* decreases, while that from *A* increases, and the audio output is positive. By similar reasoning, an incoming frequency above band center produces a negative audio output. Optimum linearity requires that $KQ = 2$, where K is the transformer coupling and Q is the primary and secondary quality factor.

Figure 12.6.12*c* shows a ratio detector, which has an advantage over the Foster-Seeley discriminator in being relatively insensitive to AM. The ratio detector uses a tertiary winding (winding 3) instead of the primary voltage, and one diode is reversed; however, the phasor diagrams also apply to the ratio detector. The AM

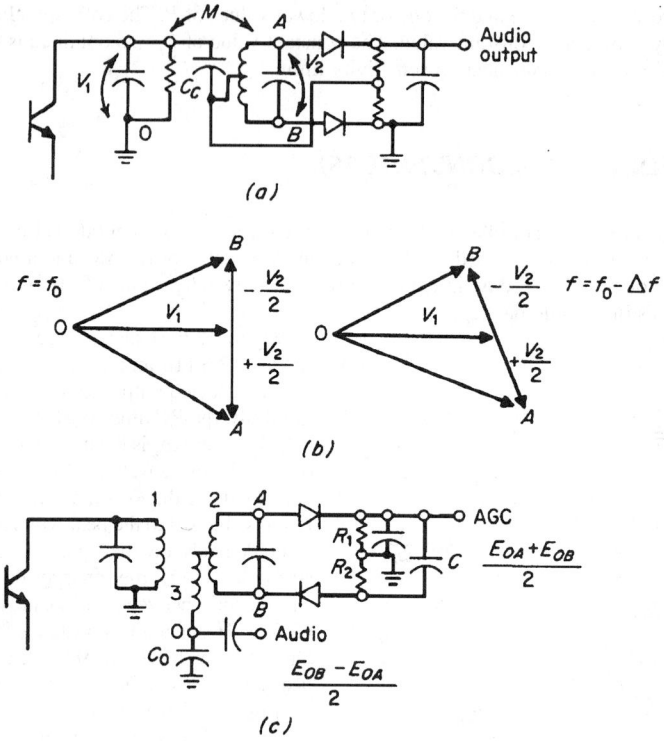

FIGURE 12.6.12   FM detectors: (*a*) Foster-Seeley FM discriminator; (*b*) phasor diagrams; (*c*) ratio detector.

rejection feature results from choosing the $(R_1 + R_2)C$ time constant large compared with the lowest frequency to be faithfully reproduced. The voltages E_{OA} and E_{OB} represent the detected values of rf voltages across OA and OB, respectively. With the large time constant above, voltage on C changes slowly with AM and the conduction angles of the diodes vary, loading the tuned circuit so as to keep the rf amplitudes relatively constant.

Capacitor C_0 is chosen to be an rf short circuit but small enough to follow the required audio variations. In the AM rejection process, AF voltage on C_0 does not follow the AM because the charge put on by one diode is removed by the other diode. With FM variations on the rf, voltage on C_0 changes to reach the condition, again, that charge put on C_0 by one diode is removed by the other diode. The ratio detector is generally used with little or no previous limiting of the rf, while the Foster-Seeley discriminator must be preceded by limiters to provide AM rejection.

With the recent trend toward integrated circuits, there has been increased interest in using phase-locked loops and product detectors. These techniques have been selected because they do not require inductors, which are not readily available in integrated form. Figure 12.6.13 shows a phase-locked loop (PLL) as an

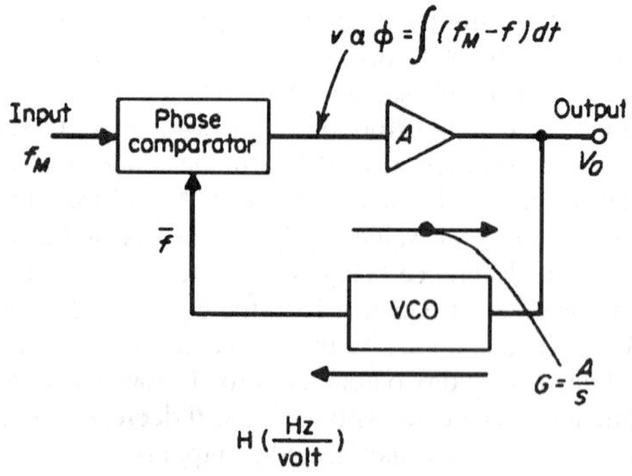

$$v \alpha \phi = \int (f_M - f)\,dt$$

$$H\left(\frac{\text{Hz}}{\text{volt}}\right)$$

$$\frac{V_0}{f_M} = \frac{G}{1 + GH} = \frac{1}{H}\,\frac{1}{1 + \frac{1}{GH}}$$

FIGURE 12.6.13   FM detector using phase-locked loop.

FM detector. The phase comparator merely provides a dc voltage proportional to the difference in phase between signals represented by f_M and f. Initially, f and f_M are unequal, but because of high loop gain, $GH \gg 1$, f and f_M quickly become locked and stay locked. Then as f_M varies, f follows exactly. But because of the high loop gain, response is essentially $1/H$, which is the voltage-controlled oscillator (VCO) characteristic. Hence the PLL serves as an FM detector.

AM product detectors also make use of the PLL to provide a carrier locked to the incoming signal carrier. The output of the VCO is used to drive the product detector. Probably one of the most stringent uses of the product detector is in an FM stereo decoder. The *left minus right* (L − R) subcarrier is located at 38 kHz with sidebands from 23 to 53 kHz. There may also be an SCA signal centered about 67 kHz which is used to provide a music service for restaurants and commercial offices. The L − R product detector is driven by a 38-kHz VCO, the output of which also goes to a 2-to-1 counter. The counter output is compared with the 19-kHz pilot signal in a phase comparator, and the phase-comparator output then controls the VCO. Because of the relatively small pilot signal and the presence of L + R, L − R, and SCA information, the requirement for phase locking is stringent.

SECTION 13

POWER ELECTRONICS

Power electronics deals with the application of electronic devices and associated components to the conversion, control, and conditioning of electric power. The primary characteristics of electric power, which are subject to control, include its basic form (ac or dc), its effective voltage or current (including the limiting cases of initiation and interruption of conduction), and its frequency and power factor (if ac). The control of electric power is a means for achieving control or regulation of one or more nonelectrical parameters, e.g., the speed of a motor, the temperature of an oven, the rate of an electrochemical process, or the intensity of lighting.

Aside from the obvious difference in function, power-electronics technology differs markedly from the technology of low-level electronics for information processing in that much greater emphasis is required on achieving high-power efficiency. Few low-level circuits exceed a power efficiency of 15 percent, but few power circuits can tolerate a power efficiency less than 85 percent. High efficiency is vital, first, because of the economic and environmental value of wasted power and, second, because of the cost of dissipating the heat it generates. This high efficiency cannot be achieved by simply scaling up low-level circuits; a different approach must be adopted.

This different approach is attained by using electronic devices as switches, e.g., approximating ideal closed (no voltage drop) or open (no current flow) switches. This differs from low-level digital switching circuits in that digital systems are primarily designed to deliver two distinct small voltage levels while conducting small currents (ideally zero). Power electronic circuits, though, must have the capability of delivering large currents and be able to withstand large voltages. Power can be controlled and modified by controlling the timing of repetitive switch action. Because of wear and limited switching speed, mechanical switches are ordinarily not suitable, but electronic switches have made this approach feasible into the multigigawatt power region while maintaining high-power efficiencies over wide ranges of control. However, the inherent nonlinearity of the switching action leads to the generation of transients and spurious frequencies that must be considered in the design process.

Reliability of the power electronics circuits is just as important as efficiency. Modern power converter and control circuits must be extremely robust, with MTBF (mean time between failure) for typical systems in the order of 1,000,000 h of operation.

Power electronic circuits are often divided into categories depending on their intended function. Converter circuits that change ac into dc are called rectifiers, circuits that change the dc operating voltage or current are called dc-to-dc converters, circuits that convert dc into ac power are called inverters, and those that change the amplitude and frequency of the ac voltage and/or current without using an intermediate dc stage are ac-to-ac converters (also called cycloconverters).

Rectifiers are used in many power electronics applications because of the widespread availability of ac power sources, and rectification is often a first step in the power conditioning scheme. Rectifiers are used in very low voltage systems (e.g., 3 V logic circuits) as well as very high voltage applications of commercial utilities. The control and circuit topology can vary according to the application requirements.

Dc–dc converters have many implementations that depend on the intended application, and can make use of different types of input power sources. Often, ac power is rectified and filtered to supply the requisite input dc levels. An inverter section is then used to transform the dc power to high frequency ac voltage or current, which a transformer then steps up or down. The new ac from the transformer secondary is then rectified and filtered to provide the desired output dc level. Other dc–dc converters step voltage up or down without the intervening transformer.

Inverters convert dc into ac power. Many applications require the production of three-phase power waveforms for speed control of large motors used in industry. The reconstruction of single-frequency, near-sinusoidal voltage, or current waveforms requires precisely controlled switching circuits. The exact mode and timing of the switching action in the associated power electronic devices can be complex, especially when regenerative schemes are employed to recover energy from the mechanical system and convert it back to electrical energy for more efficient operation. Inverter circuit design and control has been the subject of much research and development over the past several decades.

Ac–ac power control, without changing frequency, is accomplished by simple converters that allow conduction to begin at a time past the zero-crossing of the voltage or current waveform (referred to as phase control), or more complex converters that create completely new amplitudes and frequencies for the output ac power.

Note: Original contributions to this section were made by W. Newell. Portions of the material on diodes were contributed by P. F. Pittman, J. C. Engel, and J. W. Motto.   D.C.

In This Section:

CHAPTER 13.1
POWER ELECTRONIC DEVICES

Jerry L. Hudgins

POWER ELECTRONIC DEVICE FAMILIES

Power electronic devices have historically been separated into three broad categories: diodes, transistors, and thyristors. Modern devices can still be classified in this way, though there is increasing overlap in device design and function. Also, new materials as well as novel designs have increased the suitability and broadened the applications of semiconductor switches in energy conversion circuits and systems. Diodes are two-terminal devices that perform functions such as rectification and protection of other components. Diodes are not controllable in the sense that they will conduct current when a positive forward voltage is applied between the anode and cathode. Transistors are three-terminal devices that include the traditional power bipolar (two types of charge carriers), power MOSFETs (metal-oxide-semiconductor field-effect transistor), and hybrid devices that have some aspect of a control-FET element integrated with a bipolar structure, such as an IGBT (insulated-gate bipolar transistor). Thyristors are also three-terminal devices that have a four-layer structure (several p-n junctions) for the main power handling section of the device. All transistors and thyristor types are controllable in switching from a forward blocking state (very little current flows) into a forward conduction state (large forward current flows). All transistors and most thyristors (except SCRs) are also controllable in switching from forward conduction back to a forward blocking state.

Typically, thyristors are used at the highest energy levels in power conditioning circuits because they are designed to handle the largest currents and voltages of any device technology (systems approximately with voltages above 3 kV or currents above 100 A). Many medium-power circuits (systems operating at less than 3 kV or 100 A) and particularly low-power circuits (systems operating below 100 V or several amperes) generally make use of transistors as the main switching elements because of the relative ease in controlling them. IGBTs are also replacing thyristors (e.g., GTEs) in industrial motor drives and traction applications as the IGBT voltage blocking capability improves. Diodes are used throughout all levels of power conditioning circuits and systems.

COMMON DEVICE CHARACTERISTICS

A high-resistivity region of silicon is present in all power semiconductor devices. It is this region that must support the large applied forward voltages that occur when the switch is in its off state (nonconducting). The higher the forward blocking voltage rating of the device, the thicker this region must be. Increasing the thickness of this high-resistivity region results in slower turn-on and turn-off (i.e., longer switching times and/or lower frequency of operation). For example, a device rated for a forward blocking voltage of 5 kV will by its physical construction switch much more slowly than one rated for 100 V. In addition, the thicker high-resistivity region of the 5 kV device will cause a larger forward voltage drop during conduction than the 100 V device carrying the same current. There are other effects associated with the relative thickness and layout of the various regions that make up modern power devices, but the major trade-off between forward blocking voltage rating and switching times and between forward blocking voltage and forward voltage drop during conduction should be kept in mind. Another physical aspect of the semiconductor material is that the maximum breakdown voltage achievable using a semiconductor is proportional to the energy difference between the conduction and valence bands (bandgap). Hence, a material with a larger bandgap energy than that in silicon (Si) can in principle achieve the same blocking voltage rating with a thinner high-resistivity region. This is one of the reasons that new semiconductor devices are being designed and are recently becoming available in materials such as silicon carbide (SiC).

The time rate of rise of device current (*di/dt*) during turn-on and the time rate of rise of device voltage (*dv/dt*) during turn-off are important parameters to control for ensuring proper and reliable operation. Many power electronic devices have maximum limits for *di/dt* and *dv/dt* that must not be exceeded. Devices capable of conducting large currents in the on-state are necessarily made with large surface areas through which the current flows. During turn-on, localized regions of a device begin to conduct current. If the local current density becomes too large, then heating will damage the device. Sufficient time must be allowed for the entire area to begin conducting before the localized currents become too high and the device's *di/dt* rating is exceeded. The circuit designer sometimes adds series inductance to limit *di/dt* below the recommended maximum value.

During turn-off, current is decreasing while voltage across the device is increasing. If the forward voltage becomes too high while sufficient current is still flowing, then the device will drop back into its conduction mode instead of completing its turn-off cycle. Also, during turn-off, the power dissipation can become excessive if the current and voltage are simultaneously too large. Both of these turn-off problems can damage the device as well as other portions of the circuit. Another problem that occurs is associated primarily with thyristors. Thyristors can self-trigger into a forward conduction mode from a forward blocking mode if their *dv/dt* rating is exceeded (because of excessive displacement current through parasitic capacitances). Protection circuits, known as snubbers, are used with power semiconductor devices to control *dv/dt*.

The snubber circuit specifically protects devices from a large *di/dt* during turn-on and a large *dv/dt* during turn off. A general snubber topology is shown in Fig. 13.1.1. The turn-on snubber is made by inductance L_1 (often L_1 is stray inductance only). This protects the device from a large *di/dt* during the turn-on process. The auxiliary circuit made by R_1 and D_1 allows the discharging of L_1 when the device is turned off. The turn-off snubber is made by resistor R_2 and capacitance C_2. This circuit protects the power electronic device from large *dv/dt* during the turn-off process. The auxiliary circuit made by D_2 and R_2 allows the discharging of C_2 when the device is turned on. The circuit of capacitance C_2 and inductance L_1 also limits the value of *dv/dt* across the device during forward blocking. In addition, L_1 protects the device from reverse overcurrents.

All power electronic devices must be derated (e.g., power dissipation levels, current conduction, voltage blocking, and switching frequency must be reduced), when operating above room temperature

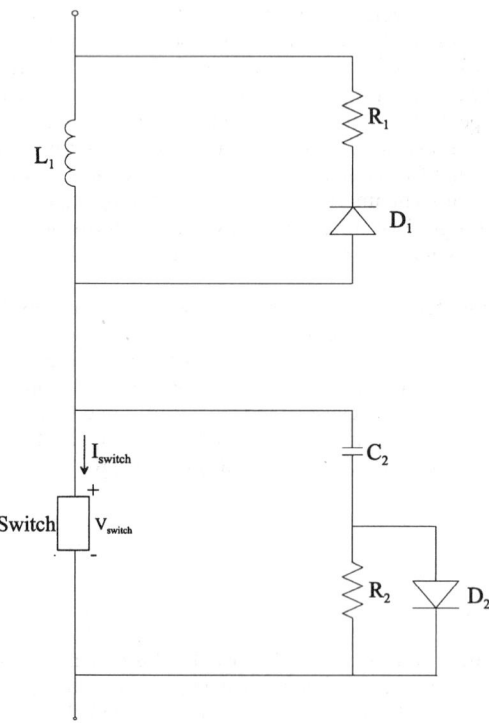

FIGURE 13.1.1   Turn-on (top elements) and turn-off (bottom elements) snubber circuits for typical power electronic devices.

(defined as about 25°C). Bipolar-type devices have thermal runway problems, in that if allowed to conduct unlimited current, these devices will heat up internally causing more current to flow, thus generating more heat, and so forth until destruction. Devices that exhibit this behavior are *pin* diodes, bipolar transistors, and thyristors. MOSFETs must also be derated for current conduction and power dissipation when heated, but they do not suffer from thermal runaway as other device types do. IGBTs fall in between the behavior of MOSFETs and bipolar transistors. At low current levels they behave similar to bipolar transistors, whereas operating at high currents causes them to behave more like MOSFETs.

There are many subtleties of power device fabrication and design that are made to improve the switching times, forward voltage drop during conduction, dv/dt, di/dt, and other ratings. Many of these improvements cause the loss of the ability of the device to hold-off large applied reverse voltages. In other devices the inherent structure itself precludes reverse blocking capability. In general, only some versions of two types of thyristors have equal (symmetric) forward and reverse voltage hold-off capabilities: GTOs (gate turn-off thyristor) and SCRs (silicon controlled rectifier).

A simple diagram of the internal structure of the major power semiconductor devices, the corresponding circuit symbols, some simple equivalent circuits, and a summary of the principal characteristics of each device are shown in Fig. 13.1.2. A comparison between types of devices illustrating the useable switching frequency range and switched power capability is shown in Fig. 13.1.3. Switched power capability is defined here as the maximum forward hold-off voltage obtainable multiplied by the maximum continuous conduction current. Further information on power electronic devices can be obtained from manufacturer's databooks and applications notes, textbooks (Baliga, 1987, 1996; Ghandi, 1977; Sze, 1981), and in many technical journal publications (including Azuma and Kurata, 1988; Hower, 1988; Hudgins, 1993).

DIODES

Diode Types

Schottky and *pin* diodes are used extensively in power electronic circuits. Schottky diodes are formed by placing a metal layer directly on a lightly doped (usually *n*-type) semiconductor. The naturally occurring potential barrier at the metal-semiconductor interface gives rise to the rectifying properties of the device. A *pin* diode is a *pn*-junction device with a lightly doped (near intrinsic) region placed between the typical diode *p*- and *n*-type regions. The lightly doped region is necessary to support large applied reverse voltages. The diode characteristic is such that current easily flows in one direction while it is blocked in the other. Power Schottky diodes are limited to about 200 V reverse blocking capability because the forward voltage drop becomes excessive in the high-resistivity region of the semiconductor and the lowering of the interface potential barrier (and associated increase in reverse leakage current) owing to the applied reverse voltage also increases (Sze, 1981). However, new Schottky structures made from SiC material are commercially available that have much higher voltage blocking capability. It is likely that these SiC diodes will be available with multi-kV ratings soon.

Reverse blocking of up to 10 kV is obtainable with a *pin* structure in Si. These types of diodes can easily handle surge currents of tens of thousands of amperes and rms currents of several thousand amperes. The *pin* diode has the advantages of much higher voltage and current capabilities than the Schottky diode, though the new SiC Schottky diodes are moving into higher power ratings all the time. Also, *pin* diodes are inherently slower in switching speed than Schottky devices, and for low reverse-blocking values, they have a larger forward voltage drop than Schottky diodes. For devices rated for 50 V reverse blocking, a Schottky diode has a forward drop of about 0.6 V as compared to a *pin* diode's forward drop of about 0.9 V at the same current density. The fast switching of the Schottky structure can be used to advantage in high-frequency power converter circuits. The Schottky's low forward drop can be used to advantage on the output rectifiers of low-voltage converter circuits, also. There have been several structures proposed that merge the features of the *pin* (for high reverse-blocking capability) and Schottky (for fast switching and low forward drop) diodes (Hower, 1988; Baliga, 1987). This concept is beginning to be implemented into the newer SiC diodes.

Circuit Symbol	Structure	Characteristics

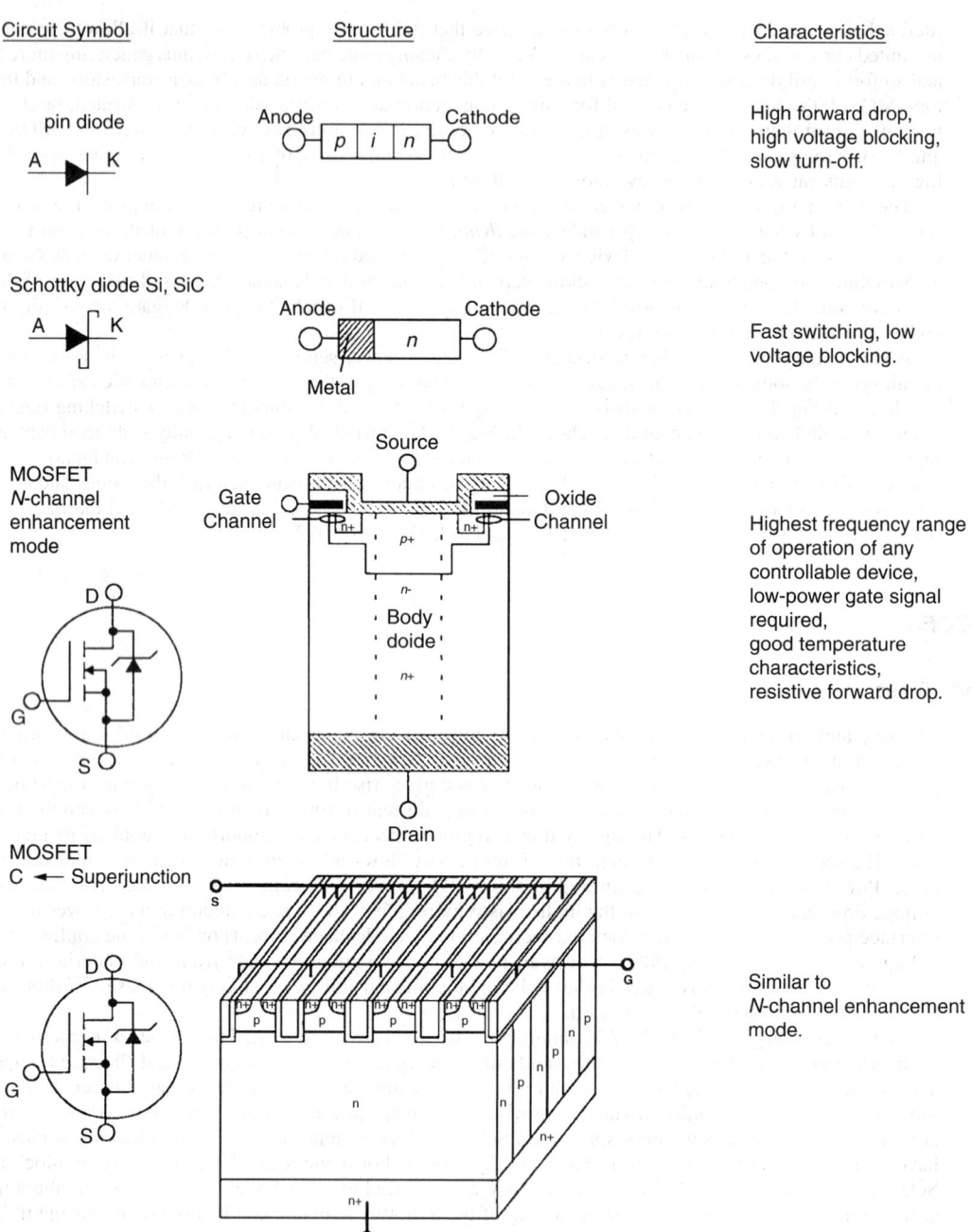

pin diode

Anode p i n Cathode

High forward drop, high voltage blocking, slow turn-off.

Schottky diode Si, SiC

Anode — Metal — n — Cathode

Fast switching, low voltage blocking.

MOSFET
N-channel enhancement mode

Source — Gate — Channel — Oxide — Channel — n+ — p+ — n- — Body doide — n+ — Drain

Highest frequency range of operation of any controllable device, low-power gate signal required, good temperature characteristics, resistive forward drop.

MOSFET
C ← Superjunction

Similar to *N*-channel enhancement mode.

FIGURE 13.1.2 Commonly used power electronic devices showing their circuit symbols, simplified internal structures, equivalent circuits (where appropriate), and some major characteristics.

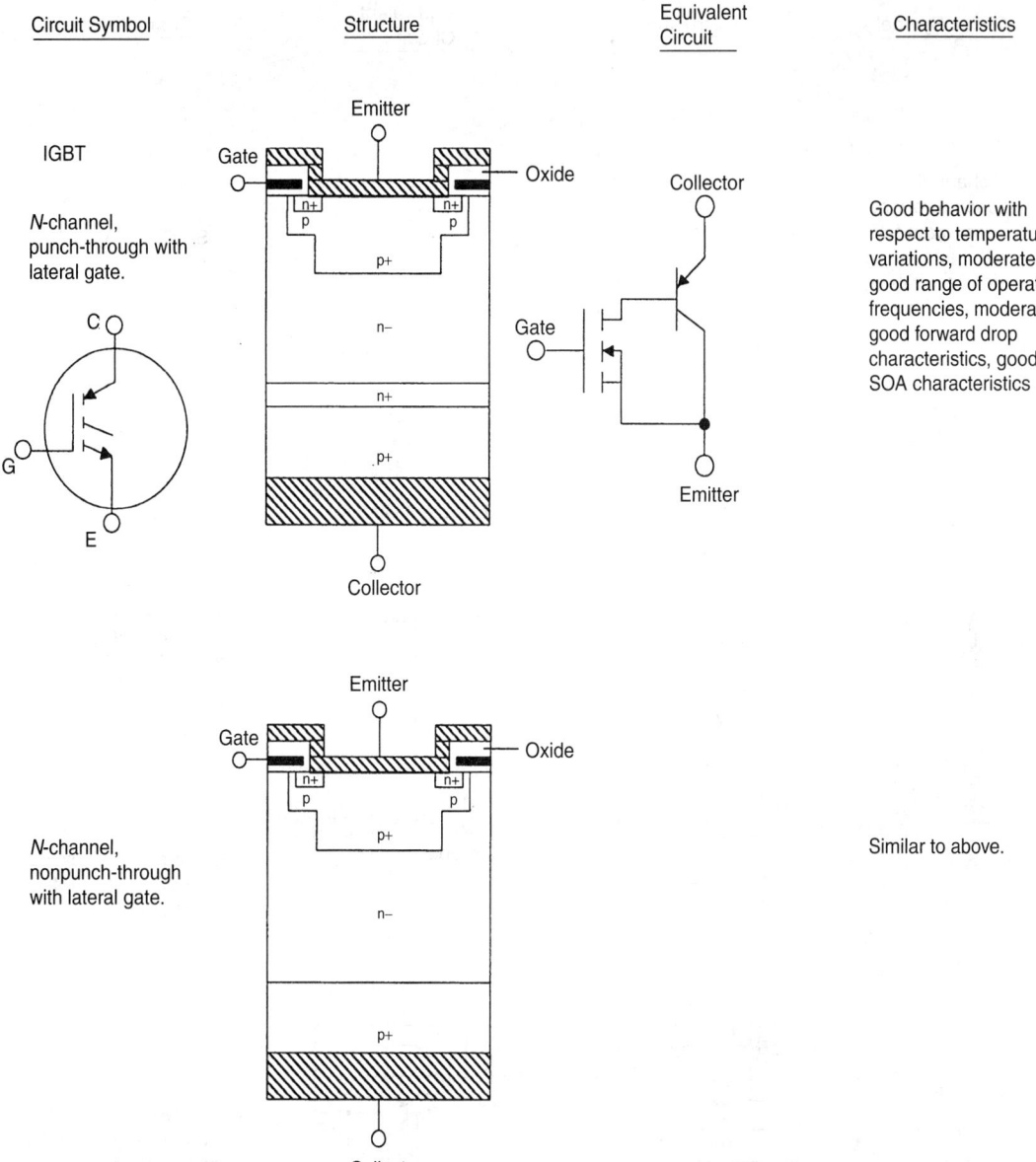

Circuit Symbol

Structure

Equivalent Circuit

Characteristics

IGBT

N-channel, punch-through with lateral gate.

Good behavior with respect to temperature variations, moderately good range of operating frequencies, moderately good forward drop characteristics, good SOA characteristics

N-channel, nonpunch-through with lateral gate.

Similar to above.

FIGURE 13.1.2   (*Continued*)

Circuit Symbol	Structure	Equivalent Circuit	Characteristics
N-channel, punch-through, trench-gate IGBT with transparent emitter			Similar to above

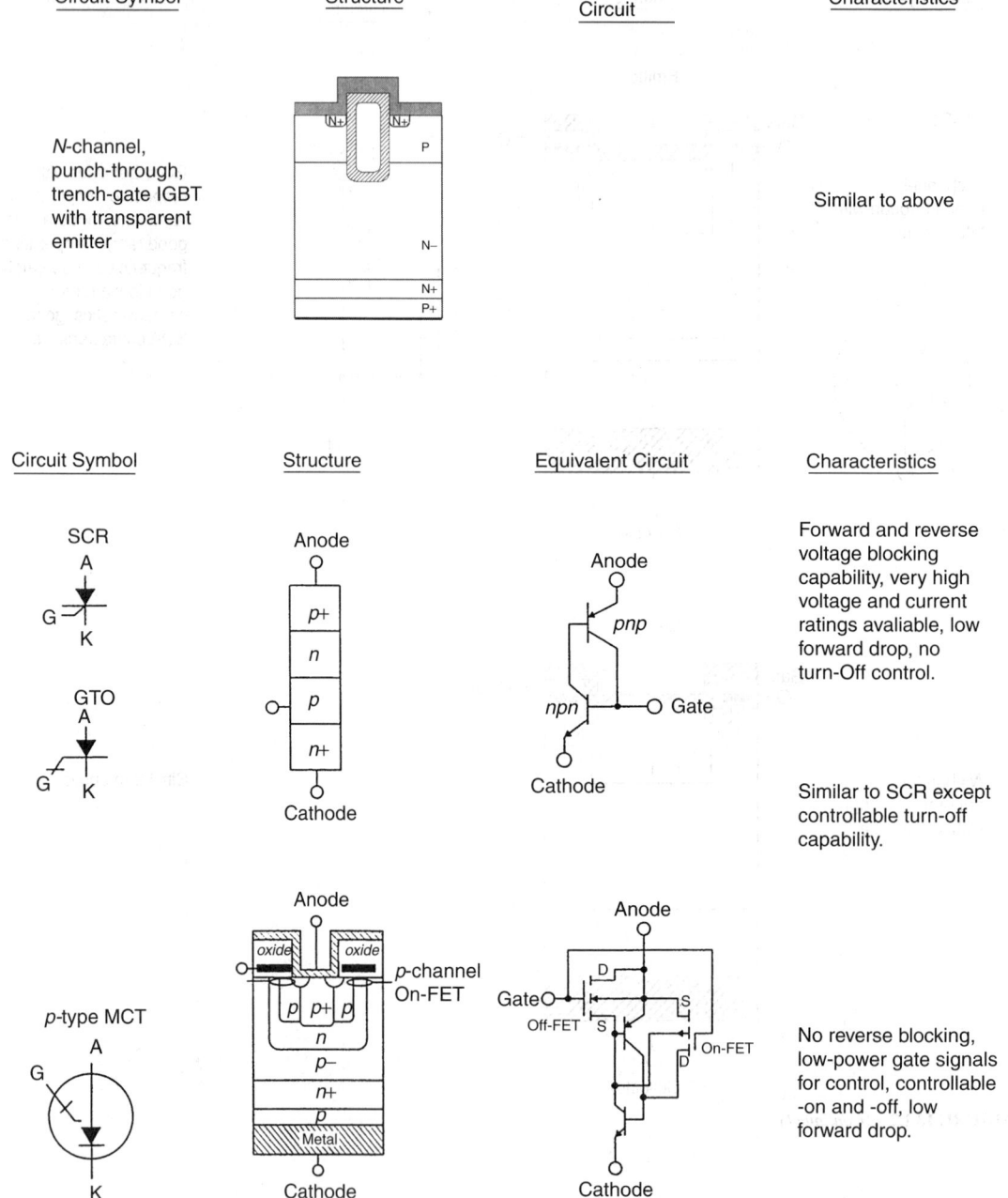

FIGURE 13.1.2 (*Continued*)

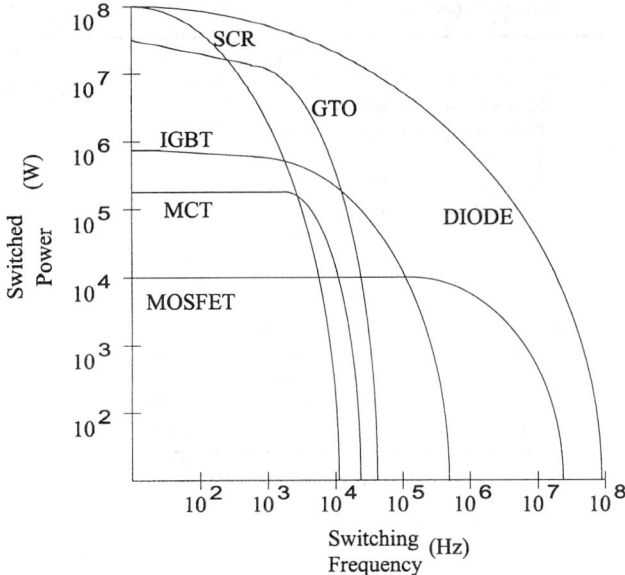

FIGURE 13.1.3   Comparison between major power electronic devices of their maximum switched power capability in terms of associated switching frequency. The switched power refers to the maximum forward blocking voltage multiplied by the maximum forward conduction current that each device is capable of handling.

Diode Ratings*

Silicon diode ratings include voltage, current, and junction temperature. A list of some of the more important parameters is shown in Table 13.1.1. The device current rating I_F is primarily determined by the area of the silicon die, power dissipation, and the method of heat sinking, while the spread of voltage ratings V_{RRM} is determined by silicon resistivity and die thickness.

Reverse voltage ratings are designated as repetitive V_{RRM} and nonrepetitive V_{RSM}. The repetitive value pertains to steady-state operating conditions, while the nonrepetitive peak value applies to occasional transient or fault conditions. Care must be exercised when applying a device to ensure that the voltage rating is never exceeded, even momentarily. When the blocking capability of a conventional diode is exceeded, leakage currents flow through the localized areas at the edge of the crystal. The resulting localized heating can cause rapid device failure.

Although even low-energy reverse overvoltage transients are likely to be destructive, the silicon diode is remarkably rugged with respect to forward current transients. This property is demonstrated by the I_{FSM} rating that permits one-half-cycle peak surge current of over ten times the I_F rating. For shorter current pulses, less than 4 ms, the surge current is specified by an I^2t rating similar to that of a fuse.

Proper circuit design must ensure that the maximum average junction temperature will never exceed its design limit of typically 150°C. Good design practice for high reliability, however, limits the maximum junction temperature to a lower value. The average junction-temperature rise above ambient is calculated by multiplying the average power dissipation, given approximately by the product of V_F and I_F, by the thermal resistance $R_{\theta JC}$. Transient junction temperatures can be computed from the transient thermal-impedance curve.

Device ratings are normally specified at a given case temperature and operating frequency. The proper use of a device at other operating conditions requires an appreciation of certain basic device characteristics.

---

*Major portions of this subsection were originally contributed by P. F. Pittman, J. C. Engel, and J. W. Motto.

TABLE 13.1.1   Symbols for Some Diode Ratings and Characteristics

Maximum Ratings

V_{RRM}	Peak repetitive reverse voltage
V_{RSM}	Peak nonrepetitive reverse voltage
$I_{F(RMS)}$	RMS forward current
$I_{F(AV)}$	Average forward current
I_{FSM}	Surge forward current
I^2t	Nonrepetitive pulse overcurrent capability
T_j	Junction temperature

Characteristics

V_F	Forward voltage drop (at specified temperature and forward current)
I_R	Maximum reverse current (at specified temperature and reverse voltage)
t_{RR}	Reverse recovery time (under specified switching of forward and reverse currents)
$R_{\theta jC}$	Junction-to-case thermal resistance

This is especially true in applications where the operating conditions of a number of devices are interdependent, as in series and parallel operation. For example, the forward voltage drop of a silicon diode has a negative temperature coefficient of 2 mV/°C for currents below the rated value. This variation in forward drop must be considered when devices are to be operated in parallel.

The reverse blocking voltage of a diode, at a specified reverse current, effectively decreases with an increase in temperature. The tendency to decrease comes from the fact that the reverse leakage current of a junction increases with temperature, thereby decreasing the voltage attained at a given measuring-current level. If the leakage current is very low, the maximum reverse voltage will be determined by avalanche breakdown (which has a coefficient of approximately 0.1 percent per °C in silicon). Thus, the voltage required to cause avalanche actually increases as the temperature rises. It should be noted that the reverse blocking voltage of a conventional diode is usually determined by imperfections at the edge of the die, and thus an ideal avalanche breakdown is usually not observed.

The reverse recovery time of a diode causes its performance to degrade with increasing frequency. Because of this effect, the rectification efficiency of a conventional diode used in a power circuit at high frequency is poor. In order to serve this application, a family of fast-recovery diodes has been developed. The stored charge of these devices is low, with the result that the amplitude and duration of the sweep-out current are greatly reduced compared with those of a conventional diode. However, improved turnoff characteristics of the fast-recovery diodes are obtained at some sacrifice in blocking voltage and forward drop compared with a conventional diode.

TRANSISTORS

Power MOSFETs

MOSFETs and IGBTs have an insulating oxide layer separating the gate contact and the silicon substrate. This insulating layer provides a large effective input resistance so that the control power necessary to switch these devices is considerably lower than that for a comparable bipolar transistor. The oxide layer also makes MOSFETs and IGBTs subject to damage from electrostatic charge build-up at the gate so that care must be exercised in their handling. Because of the internal structure of the power MOSFET, a *pn* junction (referred to as the "body diode") is present that conducts when a reverse voltage is applied across the drain and source. Power MOSFETs do not suffer from second breakdown as bipolar transistors do and generally switch much faster, particularly during turn-off. Power MOSFETs have a large, voltage-dependent, effective input capacitance (combination of the gate-to-source and gate-to-drain capacitances) that can interact with stray circuit inductance in the gate-drive circuit to create oscillations. An external, small-valued resistor is usually placed in series with the gate lead to damp the oscillatory behavior. Even with a fairly large input capacitance, power MOSFETs can be made to turn on and off faster than any other type of power electronic device.

Power MOSFETs are enhancement-type devices; a nonzero gate-to-source voltage must be applied to form a conducting channel between the drain and source to allow external current to flow. N-channel MOSFETs require a positive applied voltage at the gate with respect to the source for turn-on, while p-channel MOSFETs require a negative gate-source voltage. The gate electrode must be externally shorted to the source electrode for the device to support the maximum drain-source voltage V_{DS}, and keep the device in its forward-blocking mode. Drain current will flow if the gate-source voltage V_{GS} is above some minimum value (threshold voltage $V_{GS(TH)}$) necessary to form the conducting channel between the drain and source. In the saturated-mode of operation (i.e., drain current I_D, primarily dependent only on the gate-source voltage V_{GS}) the most important device characteristic is the forward transconductance g_{fs} usually specified by the manufacturer with a graph showing the value as a function of I_D.

The linear-mode of operation is preferred for switching applications. Here, V_{GS} is typically in the range of 10 to 20 V. In this mode, I_D is approximately proportional to the applied V_{DS} for a given value of V_{GS}. The proportionality constant defines the on-resistance $r_{DS(ON)}$. The on-resistance is the total resistance between the source and drain electrodes in the on-state and it determines the maximum I_D rating (based on power dissipation restrictions). As temperature increases, the ability of charge to move through the conduction channel from source to drain decreases. The effect appears as an increase in $r_{DS(ON)}$. The increase in $r_{DS(ON)}$ as a function of temperature goes approximately as $T^{2.3}$. Because of the positive temperature exponent, power MOSFETs can be operated in parallel, for increased current capacity, with relative ease. In addition, the safe operating area (SOA) of MOSFETs is relatively large and the devices can be operated reliably near the SOA limits.

Power MOSFETs can be obtained with a forward voltage hold-off capability BV_{DSS} of around 1.2 kV (n-channel) and current handling capacity of up to 100 A at lower BV_{DSS} values. P-channel devices typically have less spread in ratings and are generally not available in extremes of current handling or hold-off voltage values like n-channel devices. MOSFETs can be obtained as discretely packaged parts or with several die configured together to form various half-bridge or full H-bridge topologies in a module. Advanced MOSFETs have integrated features that provide capabilities such as current limiting, voltage clamping, and current sensing for more intelligent system design. Trench- or buried-gate technology has contributed to the reduction of the $R_{on} \times$ Area product in power MOSFETs (and IGBTs) by a factor of 3 or more compared to surface gate devices (see Fig. 13.1.2). The trench-gate technology has been further adapted into the newest structure called the Superjunction MOSFET (see Fig. 13.1.2). The horizontal distribution of alternating p- and n-regions modifies the electric field distribution in the forward blocking mode such that the n-regions can be designed with a smaller vertical dimension, for the same blocking capability, as the trench-gate structure. Hence, the shorter current path causes the forward drop to be greatly reduced during conduction. At 100 A/cm² the SJ-MOSFET has been shown to have a forward drop of 0.6 V as compared to 0.9 V for the traditional MOSFET (Fujihira, 1998). Table 13.1.2 lists some of the more important power MOSFET ratings and characteristics.

IGBTs. Insulated-gate bipolar transistors are designated as n-type or p-type. The n-type of device dominates the marketplace because of its ease of use (it is controlled by a positive gate-emitter voltage). The n-type device can be thought of as an n-channel enhancement-mode MOSFET controlling the base current of a pnp bipolar transistor, as shown in Fig. 13.1.2. The naming convention is somewhat confusing because the external leads are labeled with the idea of an IGBT being a direct replacement for an npn transistor with a gate lead replacing the base lead (i.e., the emitter of the equivalent pnp, in Fig. 13.1.2, is the collector of the IGBT, and so forth).

Applying a positive gate voltage above the threshold value, $V_{GE(TH)}$, turns the IGBT on. For switching applications, V_{GE} is typically in the range of 10 to 20 V. The IGBT has a saturated mode of operation (similar to a MOSFET), where the collector current is relatively independent of collector-to-emitter voltage V_{CE}. The base-collector junction, of the equivalent pnp, can never become forward biased because of drain current flow through the equivalent MOSFET. Therefore, the IGBT always has a forward drop, during conduction, of at least one pn junction (typically around 1 V). This is why the forward voltage drop $V_{CE(ON)}$ of the IGBT is greater than a comparable bipolar transistor, but less than a pure MOSFET structure at rated current flow. The switching times of the IGBT are shorter than comparable bipolar transistors (resulting in higher frequency of operation) and are not as susceptible to failure modes as are bipolars. The turn-off of an IGBT is characterized by two distinct portions of its current waveform. The first portion is characterized by a steep drop associated with the interruption of base-current to the equivalent pnp transistor (i.e., the internal MOSFET turns off). The second

TABLE 13.1.2   Symbols for Some MOSFET Ratings and Characteristics

Maximum Ratings	
V_{DS}	Drain-source voltage
I_D	Continuous drain current
I_{DM}	Pulsed drain current
T_j	Junction temperature
P_D	Maximum power dissipation

Characteristics	
BV_{DSS}	Drain-source breakdown voltage
$V_{GS(TH)}$	Gate threshold voltage
I_{DSS}	Zero gate-voltage drain current
$I_{D(on)}$	On-state drain current
$r_{DS(ON)}$	Static drain-source on-state resistance
g_{fs}	Common-source forward transconductance
C_{ISS}	Input capacitance
C_{OSS}	Output capacitance
C_{RSS}	Reverse transfer capacitance
$t_{d(on)}$	Turn-on delay time
t_r	Rise time
$t_{d(off)}$	Turn-off delay time
t_f	Fall time
Q_g	Total gate charge (gate-source + gate-drain)
Q_{gs}	Gate-source charge
Q_{gd}	Gate-drain ("Miller") charge
L_D	Internal drain inductance
L_S	Internal source inductance
$R_{\theta JC}$	Junction-to-case thermal resistance

Body Diode Ratings	
I_S	Continuous source current
I_{SM}	Pulse source current
V_{SD}	Diode forward voltage drop
t_{rr}	Reverse recovery time
Q_{RR}	Reverse recovered charge
t_{ON}	Forward turn-on time

portion is known as the current-tail and can be very long in time. This is associated with final turn-off of the bipolar transistor structure. Much of the IGBT design efforts are aimed at modifying this current-tail to control switching time and/or power dissipation during turn-off.

If a large collector current is allowed to flow, the self-heating can cause the internal parasitic thyristor structure to latch into conductance (the gate thus loses the ability to turn the device off). This behavior is known as the short-circuit, shoot-through, or latch-up current limit. The maximum current that can flow (limited only by the device impedance), before latch-up occurs, must usually be limited to less than 10 μs duration. The behavior as a function of temperature is complicated for IGBTs. At low collector current values, the forward drop dependency as a function of temperature is similar to bipolar transistors. At high collector current values, the forward drop dependency on temperature is closer to that of a MOSFET. The exact design and fabrication steps used in the production of the device plays a strong role in the exact behavior because of temperature changes. Further details are available from Baliga (1987) and Hefner (1992).

IGBTs can now be obtained with hold-off voltage ratings of up to 6.3 kV and pulsed forward current capability of over 200 A. These devices can be obtained as discrete components or with several parallel die (to form one switch) and then several sets of switches configured into bridge or half-bridge topologies in modules. They are also available with an integrated current sensing feature for active monitoring of device performance. There are two types of IGBT designs—punch through (PT) and non-punch through (NPT). NPT structures have no n^+ buffer layer next to the p^+ emitter (see Fig.13.1.2). This means that the applied forward blocking voltage can extend the associated depletion region all the way across the n^- base causing breakdown at the p-emitter/n-base junction if the applied voltage is high enough. In a PT structure (shown in Fig. 13.1.2) the depletion region is pinned to the n^+ buffer layer, thus allowing a thinner n^- base (high-resistivity region) to be used in the device design.

Previous generation IGBTs have a punch-through structure designed around a p^+ Si substrate with two epitaxial regions (n^- base region and n^+ buffer layer). Carrier lifetime reduction techniques are often used in the drift region to modify the turn-off characteristics. Recently, trench-gate devices have been designed with local lifetime control in the buffer layer (Motto, 1998). High-voltage devices (>1.2 kV) have been created using a non-punch-through structure beginning with the n^- base region as the substrate upon which a shallow (transparent) p^+ emitter is formed (Cotorogea, 2000). Cross-sections of typical unit cells for planar-gate IGBTs are shown in Fig. 13.1.2.

Third-generation IGBTs make use of improved cell density and shallow diffusion technologies that create fast switching devices with lower forward drops than have been achieved with previous devices. These lateral channel structures have nearly reached their limit for improvements. New trench-gate technologies offer the promise of greatly improved operation (Santi, 2001). Trench technologies can create an almost ideal IGBT structure because it connects in series the MOSFET and a p-n diode. There is no parasitic JFET as is created by the diffused p-wells in a lateral channel device (see Fig. 13.1.2). A simplified cross-section of the trench-gate IGBT is shown in Fig. 13.1.2. The forward drop in a trench-gate device is reduced significantly from the value in a third-generation lateral-gate IGBT. For example, in devices rated for 100 A and 1200 V, the forward drop, V_{CE}, is 1.8 V in a trench-gate IGBT as compared to 2.7 V in a lateral-gate (third generation) IGBT at the same current density, gate voltage, and temperature (Motto, 1998.) Local lifetime control is obtained in the n^+ base layer by using proton irradiation. This helps decrease the effective resistance in the n^- base by increasing the on-state carrier concentration. The surface structure of the gate is such that the MOS-channel width is increased (causing a decrease in channel resistance). The trend is for devices to be of the PT type as processing technology is improved. Table 13.1.3 lists some of the more important IGBT ratings and characteristics.

Bipolar Transistors. Power bipolar transistors and bipolar Darlingtons are seldom used in modern converter systems because of the amount of power required by the control signal and the limited SOA of the traditional

TABLE 13.1.3 Symbols for Some IGBT Ratings and Characteristics

Maximum Ratings	
V_{CES}	Collector-emitter voltage
V_{CGR}	Collector-gate voltage
V_{GE}	Gate-emitter voltage
I_C	Continuous collector current
T_j	Junction temperature
P_D	Maximum power dissipation

Characteristic	
BV_{CES}	Collector-emitter breakdown voltage
$V_{GE(TH)}$	Gate threshold voltage
I_{CES}	Zero gate-voltage collector current (at specified T_j and V_{CE} value)
$V_{CE(ON)}$	Collector-emitter on-voltage (at specified T_j, I_C, and V_{GE} values)
$Q_{G(ON)}$	On-state gate charge (at specified I_C and V_{CE} values)
$t_{D(ON)}$	Turn-on delay time (for specified test)
t_{RI}	Rise time (for specified test)
$t_{D(OFF)}$	Turn-off delay time (for specified test)
t_{FI}	Fall time (or specified test)
W_{OFF}	Turn-off energy loss per cycle
$R_{\theta JC}$	Junction-to-case thermal resistance

power bipolar transistor. Because of their declining use, no further discussion will be given. Further details should be obtained from manufacturers' databooks.

THYRISTORS

There are four major types of thyristors: (*i*) silicon-controlled rectifier (SCR), (*ii*) gate turn-off (GTO) thyristor, (*iii*) MOS-controlled thyristor (MCT) and related forms, and (*iv*) static induction thyristor (SITh). MCTs are so-named because many parallel enhancement-mode, MOSFET structures of one charge type are integrated into the thyristor for turn-on and many more MOSFETs of the other charge type are integrated into the thyristor for turn-off. A static induction thyristor (SITh), or field-controlled thyristor (FCTh), has essentially the same construction as a power diode with a gate structure that can pinch-off anode current flow. The advantage of using MCTs, derivative forms of the MCT, or SIThs is that they are essentially voltage-controlled devices, (e.g., little control current is required for turn-on or turn-off) and therefore require simplified control circuits attached to the gate electrode (Hudgins, 1993). Less important types of thyristors include the Triac (a pair of low-power, anti-parallel SCRs integrated together to form a bi-directional current switch) and the programmable unijunction transistor (PUT).

A thyristor used in some ac power circuits (50 or 60 Hz in commercial utilities or 400 Hz in aircraft) to control ac power flow can be made to optimize internal power loss at the expense of switching speed. These thyristors are called phase-control devices, because they are generally turned from a forward-blocking into a forward-conducting state at some specified phase angle of the applied sinusoidal anode-cathode voltage waveform. A second class of thyristors is used in association with dc sources or in converting ac power at one amplitude and frequency into ac power at another amplitude and frequency, and must generally switch on and off relatively quickly. The associated thyristors used are often referred to as inverter thyristors.

SCRs and GTOs.   The voltage hold-off ratings for SCRs and GTOs is above 6 kV and continuing development will push this higher. The pulsed current rating for these devices is easily tens of kiloamperes. A gate signal of 0.1 to 100 A peak is typical for triggering an SCR or GTO from forward blocking into forward conduction. These thyristors are being produced in silicon with diameters greater than 100 mm.

The large wafer area places a limit on the rate of rise of anode current, and hence a *di/dt* limit (rating) is specified. The depletion capacitances around the *pn* junctions, in particular the center junction, limit the rate of rise in forward voltage that can be applied even after all the stored charge, introduced during conduction, is removed. The associated displacement current under application of forward voltage during the thyristor blocking state sets a *dv/dt* limit. Some effort in improving the voltage hold-off capability and over-voltage protection of conventional SCRs is underway by incorporating a lateral high-resistivity region to help dissipate the energy during breakover. Most effort, though, is being placed in the further development of high-performance GTO thyristors because of their controllability and to a lesser extent in optically triggered structures that feature gate circuit isolation.

Optically gated thyristors have traditionally been used in power utility applications where series stacks of devices are necessary to achieve the high voltages required. Isolation between gate-drive circuits for circuits such as static VAR compensators and high voltage dc to ac inverters have driven the development of this class of devices. One of the most recent devices can block 6 kV forward and reverse, conduct 2.5 kA average current, and maintain a *di/dt* capability of 300 A/μs and a *dv/dt* capability of 3000 V/μs, with a required trigger power of 10 mW.

High-voltage GTO thyristors with symmetric blocking capability require thick *n*-base regions to support the high electric field. The addition of an n^+ buffer-layer next to the p^+-anode allows high voltage blocking capability and yet produces a low forward voltage drop during conduction because of the thinner n^--base required. Many other design modifications have been introduced by manufacturers so that GTOs with a forward blocking capability of around 8 kV and anode conduction of 1 kA have been produced. Also, a reverse conducting GTO has been fabricated that can block 6 kV in the forward direction, interrupt a peak current of 3 kA, and has a turn-off gain of about 5.

A modified GTO structure, called a gate-commutated thyristor (GCT), has been designed and manufactured that commutates all of the cathode current away from the cathode region and diverts it out the gate contact. The GCT is similar to a GTO in structure except that it has a low-loss *n*-buffer region between the *n*-base and *p*-emitter. The GCT device package is designed to result in very low parasitic inductance and is integrated with a specially designed gate-drive circuit (IGCT). The specially designed gate drive and ring-gate package circuit allows the GCT to be operated without a snubber circuit and switch with higher anode *di/dt*, than a similar GTO. At blocking voltages of 4.5 kV and higher, the IGCT seems to provide better performance than a conventional GTO. The speed at which the cathode current is diverted to the gate (di_{GQ}/dt) is directly related to the peak snubberless turn-off capability of the GCT. The gate-drive circuit can sink current for turn-off at di_{GQ}/dt values in excess of 7000 A/µs. This hard gate drive results in a low-charge storage time of about 1 µs. The low storage time and the fail-short mode makes the IGCT attractive for high-voltage series applications.

The bi-directional control thyristor (BCT) is an integrated assembly of two anti-parallel thyristors on one Si wafer. The intended application for this switch is VAR compensators, static switches, soft starters, and motor drives. These devices are rated up to 6.5 kV blocking. Cross-talk between the two halves has been minimized. The small gate-cathode periphery necessarily restricts the BCT to low-frequency applications because of its *di/dt* limit.

The continuing improvement in GTO performance has caused a decline in the use of SCRs, except at the very highest power levels. In addition, the improvement in IGBT design further reduces the attractiveness of SCRs. These developments make the future use of SCRs seemingly diminish.

MCTs.   There is a *p*-channel and an *n*-channel MOSFET integrated into the MCT, one FET-structure for turn-on and one for turn-off. The MCT itself comes in two varieties: *p*-type (gate voltage applied with respect to the anode) and an *n*-type (gate voltage applied with respect to the cathode). Just as in a GTO, the MCT has a maximum controllable cathode current value. The inherent optimization for good switching and forward conduction characteristics make the MCT unable to block reverse applied voltages.

MCTs are presently limited to operation at medium power levels. The seeming variability in fabrication of the turn-off FET structure continues to limit the performance of MCTs, particularly current interruption capability, though these devices can handle two to five times the conduction current density of IGBTs. All MCT device designs center on the problem of current interruption capability. Turn-on is relatively simple, by comparison, with it and conduction properties approaching the one-dimensional thyristor limit. Other types of integrated MOS-thyristor structures can be operated at high power levels, but these devices are not commonly available or are produced for specific applications. Typical MCT ratings are for 1 kV forward blocking and a peak controllable current of 75 A. A recent version of the traditional MCT design is a diffusion-doped (instead of the usual epitaxial growth) device. They are rated for 3 kV forward blocking, have a forward drop of 2.5 V at 100 A, and are capable of interrupting around 300 A with a recovery time of 5 µs.

An MCT that uses trench-gate technology, called a depletion-mode thyristor (DMT), has been designed. A similar device is the base resistance controlled thyristor (BRT). Here, a *p*-channel MOSFET is integrated into the *n*-drift region of the MCT. These devices operate in an "IGBT" mode until the current is large enough to cause the thyristor structure to latch.

Another new MCT-type structure is called an emitter switched thyristor (EST), and uses an integrated lateral MOSFET to connect a floating thyristor *n*-emitter region to an n^+ thyristor cathode region. All thyristor current flows through the lateral MOSFET so it can control the thyristor current. Integrating an IGBT into a thyristor structure has been proposed. One device, called an IGBT triggered thyristor (ITT), is similar in structure and operation to the EST. The best designed EST, however, is the dual gate emitter switched thyristor (DG-EST). The device has two gate electrodes. One gate controls an integrated IGBT section, while the other gate controls a thyristor section. The DG-EST is intended to be switched in IGBT mode, to exploit the controllability and snubberless capabilities of an IGBT. During forward conduction, the thyristor section takes over and thus the DG-EST takes advantage of a low forward drop and the latching nature of a thyristor.

Static Induction Thyristors.   A static induction thyristor (SITh) or field controlled thyristor (FCTh) is essentially a *pin* diode with a gate structure that can pinch-off anode current flow. High-power SIThs have a subsurface gate (buried-gate) structure to allow larger cathode areas to be used, and hence larger current densities can be conducted. Other SITh configurations have surface gate structures.

Planar gate devices have been fabricated with blocking capabilities of up to 1.2 kV and conduction currents of 200 A, while step-gate (trench-gate) structures have been produced that are able to block up to 4 kV and conduct 400 A. Similar devices with a "Verigrid" structure have been demonstrated that can block 2 kV and conduct 200 A, with claims of up to 3.5 kV blocking and 200 A conduction. Buried gate devices that block 2.5 kV and conduct 300 A have also been fabricated.

An integrated light-triggered and light-quenched SITh has been produced that can block 1.2 kV and conduct up to 20 A (at a forward drop of 2.5 V). This device is an integration of a normally off buried-gate static induction photothyristor and a normally off *p*-channel Darlington surface-gate static induction phototransistor. The optical trigger and quenching power required is less than 5 and 0.2 mW, respectively.

Thyristor Behavior. The thyristor is a three-terminal semiconductor device comprising four layers of silicon so as to form three separate *pn* junctions. In contrast to the linear relation that exists between load and control currents in a transistor, the thyristor is bistable. The four-layer structure of the thyristor is shown in Fig. 13.1.2. The anode and cathode terminals are connected in series with the load to which power is to be controlled. The thyristor is turned on by application of a low-power control signal between the third terminal, or gate, and the cathode (between gate and anode for *p*-type MCT).

The reverse characteristic is determined by the outer two junctions, which are reverse-biased in this operating mode. With zero gate current, the forward characteristic in the off- or blocking-state is determined by the center junction, which is reverse biased. However, if the applied voltage exceeds the forward blocking voltage, the thyristor switches to its on- or conducting state. The effect of gate current is to lower the blocking voltage at which switching takes place.

This behavior can be explained in terms of the two-transistor analog shown in Fig. 13.1.2. The two transistors are regeneratively coupled so that if the sum of their current gains (α's) exceeds unity, each drives the other into saturation. In the forward blocking-state, the leakage current is small, both α's are small, and their sum is less than unity. Gate current increases the current in both transistors, increasing their α's. When the sum of the two α's equals 1, the thyristor switches to its on-state (latches).

The form of the gate-to-cathode *VI* characteristic of SCRs and GTOs is similar to that of a diode. With positive gate bias, the gate-cathode junction is forward-biased and permits the flow of a large current in the presence of a low voltage drop. When negative gate voltage is applied to an SCR, the gate-cathode junction is reverse-biased and prevents the flow of current until the avalanche breakdown voltage is reached. In a GTO, a negative gate voltage is applied to provide a low impedance for anode current to flow out of the device instead of out of the cathode. In this way the cathode region turns off, thus pulling the equivalent *npn* transistor out of conduction. This causes the entire thyristor to return to its blocking state. The problem with the GTO is that the gate-drive circuitry is typically required to sink from 5 to 10 percent of the anode current to achieve turn-off. The MCT achieves turn-off by internally diverting current through an integrated MOSFET. Switching the equivalent MOSFET only requires a voltage signal to be applied at the gate electrode.

A summary is provided in Table 13.1.4 of some of the ratings which must be considered when choosing a thyristor for a given application. Both forward and reverse repetitive and nonrepetitive voltage ratings must be considered, and a properly rated device must be chosen so that the maximum voltage ratings are never exceeded. In most cases, either forward or reverse voltage transients in excess of the nonrepetitive maximum ratings result in destruction of the device.

The maximum rms or average current ratings given are usually those which cause the junction to reach its maximum rated temperature. Because the maximum current will depend on the current waveform and on thermal conditions external to the device, the rating is usually shown as a function of case temperature and conduction angle. The peak single half-cycle surge-current rating must be considered, and in applications where the thyristor must be protected from damage by overloads, a fuse with an I^2t rating smaller than the maximum rated value for the device must be used. Maximum ratings for both forward and reverse gate voltage, current, and power also must not be exceeded.

The maximum rated operating junction temperature T_J must not be exceeded, since device performance, in particular voltage-blocking capability, will be degraded. Junction temperature cannot be measured directly but must be calculated from a knowledge of steady-state thermal resistance $R_{\theta JC}$ and the average power dissipation. For transients or surges, the transient thermal impedance ($Z_{\theta JC}$) curve must be used. The maximum average power dissipation P_T is related to the maximum rated operating junction temperature and the case temperature by the steady-state thermal resistance. In general, both the maximum dissipation and its derating with increasing case temperature are provided.

The number of thyristor characteristics specified varies widely from one manufacturer to another. Some characteristics are given only as typical values of minima or maxima, while many characteristics are displayed graphically. Table 13.1.4 summarizes some of the characteristics provided. Thyristor types shown in parentheses

TABLE 13.1.4 Symbols for Some Thyristor Ratings and Characteristics

Maximum Ratings

V_{RRM}	Peak repetitive reverse voltage
V_{RSM} (SCR & GTO)	Peak nonrepetitive reverse voltage
V_{DRM}	Peak repetitive forward off-state voltage
V_{DSM} (SCR & GTO)	Peak nonrepetitive forward off-state voltage
$I_{T(RMS)}$	RMS forward current
$I_{T(AV)}$ (I_K for MCT)	Average forward current
I_{TSM} (I_{KSM} for MCT)	Surge forward current
I_{TGO} (I_{KC} for MCT)	Peak controllable current
I^2t (SCR & GTO)	Nonrepetitive pulse overcurrent capability
P_T (MCT)	Maximum power dissipation
di/dt	Critical rate of rise of on-state current
dv/dt	Critical rate of rise of off-state voltage
P_{GM}(P_{FGM} for GTO)	Peak gate forward power dissipation
P_{RGM} (GTO)	Peak gate reverse power dissipation
V_{FGM}	Peak forward gate voltage
V_{RGM}	Peak reverse gate voltage
I_{FGM} (SCR & GTO)	Peak forward gate current
I_{RGM} (GTO)	Peak reverse gate current
T_j	Junction temperature

Characteristics

V_{TM}	On-state voltage drop (at specified temperature and forward current)
I_{DRM}	Maximum forward off-state current (at specified temperature and forward voltage)
I_{RRM}	Maximum reverse blocking current (at specified temperature and reverse voltage)
C_{ISS} (MCT)	Input capacitance (at specified temperature and gate and anode voltages)
V_{GT} (SCR & GTO)	Gate trigger voltage (at specified temperature and forward applied voltage)
V_{GD} (SCR & GTO)	Gate nontrigger voltage (at specified temperature and forward applied voltage)
I_{GT} (SCR & GTO)	Gate trigger current (at specified temperature and forward applied voltage)
t_{gt} (GTO)	Turn-on time (under specified switching conditions)
t_q (SCR & GTO)	Turn-off time (under specified switching conditions)
$t_{D(ON)}$ (MCT)	Turn-on delay time (for specified test)
t_{Rl} (MCT)	Rise time (for specified test)
$t_{D(OFF)}$ (MCT)	Turn-off delay time (for specified test)
t_{Fl} (MCT)	Fall time (for specified test)
W_{OFF} (MCT)	Turn-off energy loss per cycle
$R_{\theta JC}$	Junction-to-case thermal resistance

indicate a characteristic unique to that device or devices. Gate conditions of both voltage and current to ensure either nontriggered or triggered device operation are included.

The turn-on and turn-off transients of the thyristor are characterized by switching times like the turn-off time listed in Table 13.1.4. The turn-on transient can be divided into three intervals—gate-delay interval, turn-on of initial area, and spreading interval. The gate-delay interval is simply the time between application of a turn-on pulse at the gate and the time the initial area turns on. This delay decreases with increasing gate drive current and is of the order of a few microseconds. The second interval, the time required for turn-on of the initial area, is quite short, typically less than 1 μs. In general, the initial area turned on is a small percentage of the total useful device area. After the initial area turns on, conduction spreads (spreading interval) throughout the device in tens of microseconds.

It is during this spreading interval that the *di/dt* limit must not be exceeded. Typical *di/dt* values range from 100 to 1000 A/μs. Special inverter-type SCRs and GTOs are made that have increased switching speed (at the expense of higher forward voltage drop during conduction) with *di/dt* values in the range of 2000 A/μs. The rate of application of forward voltage is also restricted by the *dv/dt* characteristic. Typical values range from 100 to 1000 V/μs.

Thyristors are available in a wide variety of packages, from small plastic ones for low-power (i.e., TO-247), to stud-mount packages for medium-power, to press-pack (also called flat-pack) for the highest power devices. The press-packs must be mounted under pressure to obtain proper electrical and thermal contact between the device and the external metal electrodes. Special force-calibrated clamps are made for this purpose.

OTHER POWER SEMICONDUCTOR DEVICES

Semiconductor materials such as silicon carbide (SiC), gallium arsenide (GaAs), and gallium nitride (GaN) are being used to develop *pn*-junction and Schottky diodes, power MOSFET structures, some thyristors, and other switches. SiC diodes are commercially available now. No other commercial power devices made from these materials yet exist, but will likely be available in the future. Further information about advanced power semiconductor materials and device structures can be found in (Baliga, 1996) and (Hudgins, 1993, 1995, 2003).

BIBLIOGRAPHY

Azuma, M. and M. Kurata, "GTO thyristors," *IEEE Proc.*, pp. 419–427, April, 1988.

Baliga, B. J., "Modern Power Devices," Wiley, 1987.

Baliga, B. J., "Power Semiconductor Devices," PWS, 1996.

Busatto, G., G. F. Vitale, G. Ferla, A. Galluzzo, and M. Melito, "Comparative analysis of power bipolar devices," *IEEE PESC Rec.*, pp. 147–153, 1990.

Cotorogea, M., A. Claudio, and J. Aguayo, "Analysis by measurements and circuit simulations of the PT- and NPT-IGBT under different short-circuit conditions," *IEEE APEC Rec.*, pp. 1115–1121, 2000.

Fujihira, T., and Y. Miyasaka, "Simulated superior performances of semiconductor superjunction devices," *IEEE Proc. ISPSD*, pp. 423–426, 1998.

Ghandhi, S. K., "Semiconductor Power Devices," Wiley, 1977.

Hefner, A. R., "A dynamic electro-thermal model for the IGBT," *IEEE Industry Appl. Soc. Ann. Mtg. Rec.*, pp. 1094–1104, 1992.

Hower, P. L., "Power semiconductor devices: An overview," *IEEE Proc.*, Vol. 76, pp. 335–342, April, 1988.

Hudgins, J. L., "A review of modern power semiconductor devices," *Microelect. J.*, Vol. 24, pp. 41–54, 1993.

Hudgins, J. L., "Streamer model for ionization growth in a photoconductive power switch," *IEEE Trans. PEL*, Vol. 10, pp. 615–620, September, 1995.

Hudgins, J. L., G. S. Simin, E. Santi, and M. A. Khan, "A new assessment of wide bandgap semiconductors for power devices," *IEEE Trans. PEL*, Vol. 18, pp. 907–914, May 2003.

Motto, E. R., J. F. Donlon, H. Takahashi, M. Tabata, and H. Iwamoto, "Characteristics of a 1200 V PT IGBT with trench gate and local lifetime control," *IEEE IAS Annual Mtg. Rec.*, pp. 811–816, 1998.

Nishizawa, J., T. Terasaki, and J. Shibata, "Field effect transistor versus analog transistor: static induction transistor," *IEEE Tran. ED*, Vol. ED-22, pp. 185–197, 1975.

Santi, E., A. Caiafa, X. Kang, J. L. Hudgins, P. R. Palmer, D. Goodwine, and A. Monti, "Temperature effects on trench-gate IGBTs," *IEEE IAS Annual Mtg. Rec.*, pp. 1931–1937, 2001.

Sze, S. M., "Physics of Semiconductor Devices," 2nd ed., Wiley, 1981.

Venkataramanan, G., A. Mertens, H. Skudelny, and H. Grunig, "Switching characteristics of field controlled thyristors," *Proc. EPE—MADEP '91*, pp. 220–225, 1991.

CHAPTER 13.2
NATURALLY COMMUTATED CONVERTERS

Arthur W. Kelley

INTRODUCTION

The applications for this family of naturally commutated converters embrace a very wide range, including dc power supplies for electronic equipment, battery chargers, dc power supplies delivering many thousands of amperes for electrochemical and other industrial processes, high-performance reversing drives for dc machines rated at thousands of horsepower, and high-voltage dc transmission at the gigawatt power level.

The basic feature common to this class of converters is that one set of terminals is connected to an ac voltage source. The ac source causes natural commutation of the converter power electronic devices. In these converters, a second set of terminals operates with dc voltage and current. This class of converters is divided in function depending on the direction of power flow. In *ac-to-dc rectification*, the ac source, typically the utility line voltage, supplies power to the converter, which in turn supplies power to a dc load. In *dc-to-ac inversion*, a dc source, typically a battery or dc generator, provides power to the converter, which in turn transfers the power to the ac source, again, usually the utility line voltage. Because natural commutation synchronizes the power semiconductor device turn on and turn off to the ac source, this converter is also known as a *synchronous inverter* or a *line-commutated inverter*. This process is different from supplying power to an ac load, which usually requires forced commutation.

The power electronic devices in these converters are typically either *silicon controlled rectifiers* (SCRs) *diodes*. To simplify the discussion that follows, the SCRs and diodes are assumed to (1) conduct forward current with zero forward voltage drop, (2) block reverse voltage with zero leakage current, and (3) switch instantaneously between conduction and blocking. Furthermore, stray resistive loss is ignored and balanced three-phase ac sources are assumed.

Converter Topologies

Basic Topologies. The number of different converter topologies is very large (Schaeffer, 1965; Pelly, 1971; Dewan, 1975; Rashid, 1993). Using SCRs as the power electronic devices, Table 13.2.1 illustrates four basic topologies from which many others are derived. These ac-to-dc converters are rectifiers that provide a dc voltage V_O to a load. The rectifier often uses an output filter inductor L_O and capacitor C_O, but one or the other or both are often omitted. Rectifiers are usually connected to the ac source through a transformer. Note that the transformer is often utility equipment, located separately from the rectifier. The transformer adds a series leakage inductance L_S, which is often detrimental to rectifier operation.

Rectifier topologies are classified by whether the rectifier operates from a single- or three-phase source and whether the rectifier uses a bridge connection or transformer midpoint connection. The *single-phase bridge rectifier* shown in Table 13.2.1a requires four SCRs and a two-winding transformer. The *single-phase midpoint*

TABLE 13.2.1   Basic Converter Topologies

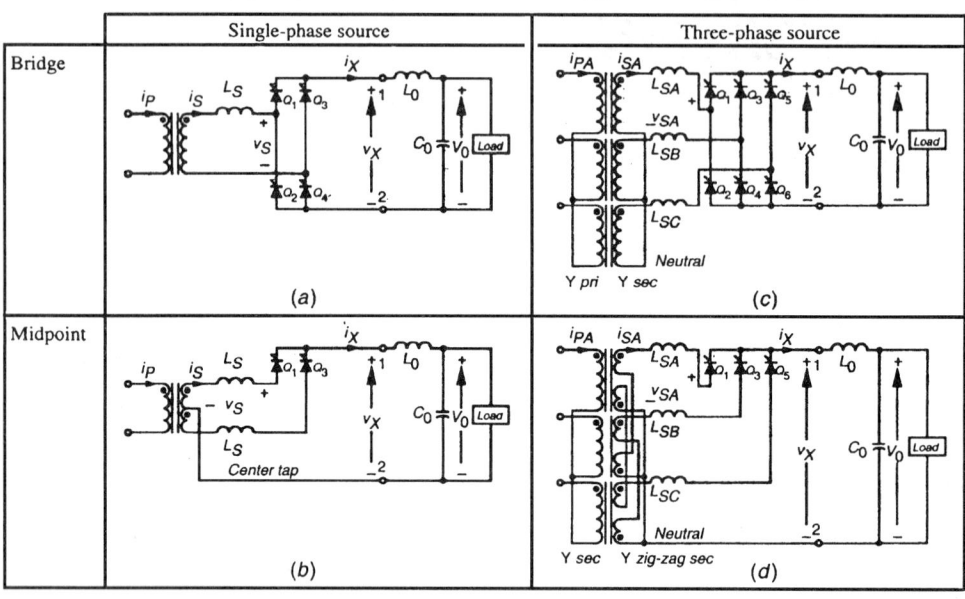

rectifier shown in Table 13.2.1*b* requires only two SCRs but requires a transformer with a center-tapped secondary to provide the midpoint connection. The *three-phase bridge rectifier* shown in Table 13.2.1*c* requires six SCRs and three two-winding transformers. The *three-phase midpoint rectifier* shown in Table 13.2.1*d* requires three SCRs and three transformers using a Y-connected "zig-zag" secondary. The Y-connected secondary provides the necessary midpoint connection and the zig-zag winding prevents unidirectional secondary winding currents from causing magnetic saturation of the transformers.

   The bridge rectifier is better suited to using the simple connection provided by the typical utility transformer. For the same power delivered to the load, the bridge rectifier often requires a smaller transformer. Therefore, in the absence of other constraints, the bridge rectifier is often preferred over the midpoint rectifier.

Pulse Number.   Converters are also classified by their pulse number q, an integer that is the number of current pulses appearing in the rectifier output current waveform i_X per cycle of the ac source voltage. Higher pulse number rectifiers generally have higher performance but usually with a penalty of increased complexity. Of the rectifiers shown in Table 13.2.1, both single-phase rectifiers are two-pulse converters ($q = 2$) with one current pulse in i_X for each half-cycle of the ac source voltage. The three-phase midpoint rectifier is a three-pulse converter ($q = 3$) with one current pulse in i_X for each cycle of each phase of the three-phase ac source voltage. The three-phase bridge rectifier is a six-pulse converter ($q = 6$) with one current pulse in i_X for each half cycle of each phase of the three-phase ac source voltage.

BASIC CONVERTER OPERATION

Given a certain operating point, rectifier operation and performance are dramatically influenced by the values of source inductance L_S, output filter inductance L_O, and output filter capacitance C_O.

Operation with Negligible Ac Source Inductance.   Figures 13.2.1 and 13.2.2 show example time waveforms for the single- and three-phase bridge rectifiers of Table 13.2.1*a* and 13.2.1*c*, respectively. In these

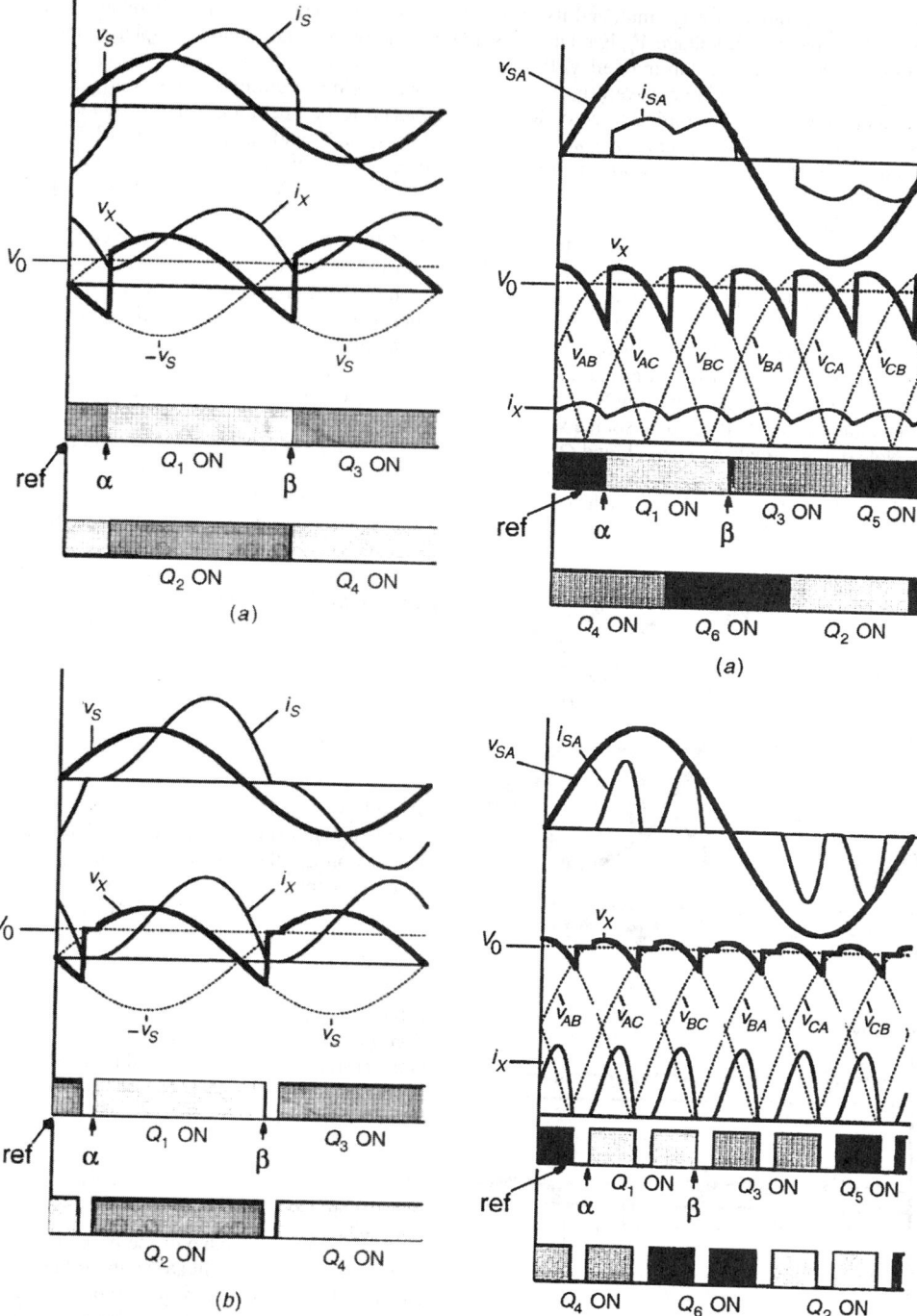

FIGURE 13.2.1 Time waveforms for single-phase bridge rectifier with $\alpha = 40°$: (a) CCM and (b) DCM.

FIGURE 13.2.2 Time waveforms for three-phase bridge rectifier with $\alpha = 20°$: (a) CCM and (b) DCM.

examples L_S is comparatively small and its influence is neglected. The value of C_O is relatively large so that the ripple in the output voltage V_O is relatively small. Operation of single- and three-phase phase-controlled rectifiers is described in detail in (Kelley, 1990).

Figure 13.2.1a shows time waveforms for the single-phase bridge rectifier when the current i_X in L_O flows continuously without ever falling to zero. The rectifier is said to be operating in the *continuous conduction mode* (CCM). The CCM occurs for relatively large L_O, heavy loads, and small α. Figure 13.2.1b shows time waveforms for the single-phase bridge rectifier when i_X drops to zero twice each cycle and the rectifier is said to be operating in the *discontinuous conduction mode* (DCM). The DCM occurs for relatively small L_O, light loads, and large α.

Figure 13.2.1 also shows the conduction intervals for SCRs Q_1 to Q_4 and the rectifier voltage, v_X. A controller, not shown in Fig. 13.2.1, generates gating pulses for the SCRs. The controller gates each SCR at a *firing angle α* (alpha) with respect to a *reference* that is the point in time at which the SCR is first forward biased. The SCR ceases conduction at the *extinction angle β* (beta). The reference, α, and β for Q_1 are illustrated in Fig. 13.2.1. The SCR *conduction angle γ* (gamma) is the difference between β and α. In DCM the SCR ceases conduction because i_X falls naturally to zero, while in the CCM the SCR ceases conduction even though i_X is not zero because the opposing SCR is gated and begins conducting i_X. Therefore in CCM, γ is limited to a maximum of one-half of an ac source voltage cycle, while in DCM γ depends on L_O, load, and α.

Note that v_X equals v_S when Q_1 and Q_4 are conducting and that v_X equals $-v_S$ when Q_2 and Q_3 are conducting. The output filter L_O and C_O reduces the ripple in v_X and delivers a relatively ripple-free voltage V_O to the load. The firing angle α determines the composition of v_X and ultimately the value of V_O. Increasing α reduces V_O and is the mechanism by which the controller regulates V_O against changes in ac source voltage and load.

This method of output voltage regulation is referred to as *phase control*, and a rectifier using it is said to be a *phase-controlled rectifier*.

Since in CCM the conduction angle is always one half of a source voltage cycle, the dc output voltage is easily found from v_X as

$$V_O = \frac{2}{\pi} \sqrt{2}\, V_S \cos \alpha \qquad (1)$$

where V_S is the rms value of the transformer secondary voltage v_S. Unfortunately, the conduction angle in DCM depends on L_O, the load, and α, and V_O cannot be calculated except by numerical methods.

For the three-phase bridge rectifier, Figures 13.2.2a and 13.2.2b show time waveforms for CCM and DCM, respectively. Operation is similar to the single-phase rectifier except that v_X equals each of the six line-to-line voltages—v_{AB}, v_{AC}, v_{BC}, v_{BA}, v_{CA}, and v_{CB}—in succession. In CCM, the SCR conduction angle γ is one-third of an ac source voltage cycle, and in DCM γ depends on L_O, load, and α. In CCM the dc output voltage V_O is found from v_X as

$$V_O = \frac{3}{\pi} \sqrt{3} \sqrt{2}\, V_S \cos \alpha \qquad (2)$$

where V_S is the rms value of the transformer secondary line-to-neutral voltage. In DCM, the value of V_O must be calculated by numerical means. To produce a ripple-free output voltage V_O, the time waveform of v_X for the three-phase rectifier naturally requires less filtering than the time waveform of v_X for the single-phase rectifier.

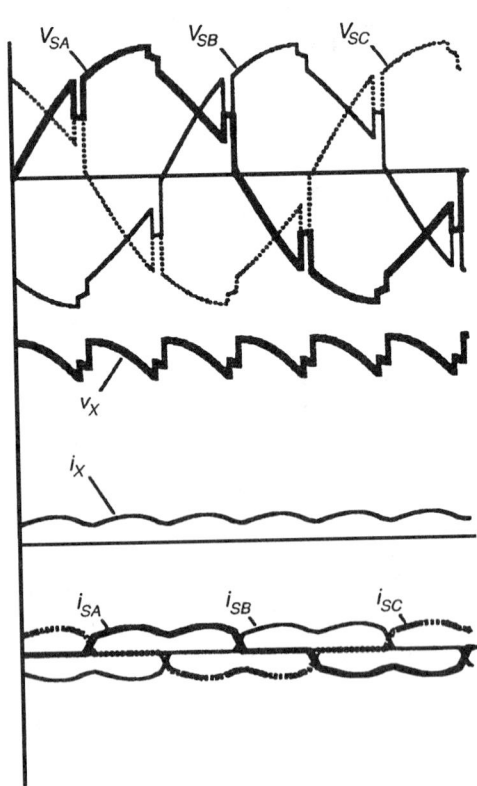

FIGURE 13.2.3 Time waveforms for three-phase bridge rectifier with appreciable L_S.

Therefore, if a three-phase ac source is available, a three-phase rectifier is always preferred over a single-phase rectifier.

Operation with Appreciable Ac Source Inductance.   The preceding discussion assumes that the value of L_S is small and does not influence circuit operation. In practice the effect of L_S must often be considered. The three-phase rectifier CCM time waveforms of Fig. 13.2.2 are repeated in Fig. 13.2.3 but for an appreciable L_S. Since i_X is always nonzero in CCM, the principal effect of L_S is to prevent instantaneous transfer of i_X from one transformer secondary winding to the next transformer secondary winding as the SCRs are gated in succession. This process is called *commutation* and the interval during which it occurs is called *commutation overlap*.

For example, at some point in time Q_1 is conducting i_{SA} equal to i_X and Q_3 is gated by the controller. Current i_{SA} through Q_1 falls while i_{SB} through Q_3 rises. During this interval both Q_1 and Q_3 conduct simultaneously and the sum of i_{SA} and i_{SB} is equal to i_X. As a result transformer secondary v_{SA} is directly connected to transformer secondary v_{SB} effectively creating a line-to-line short circuit. This connection persists until i_{SA} falls to zero and Q_1 ceases conduction. The duration of the connection is the *commutation angle* μ (mu). During this interval v_{SA} experiences a positive-going voltage "notch" while v_{SB} experiences a negative-going voltage notch. The enclosed area of the positive-going notch equals the enclosed area of the negative-going notch and represents the flux linkage or "volt seconds" necessary to produce a change in current through L_S equal to i_X. If L_O is sufficiently large so that i_X is relatively constant with value I_X during the time that both SCRs conduct, then the notch area is used to find

$$\cos\alpha - \cos(\mu + \alpha) = \sqrt{\frac{2}{3}}\,(2\pi f L_S I_X / V_S) \tag{3}$$

which can be solved numerically for μ. Note that the commutation angle is always zero in DCM since i_X is zero when each SCR is gated to begin conduction.

CONVERTER POWER FACTOR

Source Current Harmonic Composition.   The time waveforms of the prior section show that the rectifier is a nonlinear load that draws a highly distorted nonsinusoidal waveform i_S. Fourier series is used to decompose i_S into a fundamental-frequency component with rms value $I_{S(1)}$ and phase angle $\phi_{S(1)}$ with respect to v_S, and into harmonic-frequency components with rms value $I_{S(h)}$ where h is an integer representing the harmonic number. In general, the $I_{S(h)}$ are zero for even h. Furthermore, depending on converter pulse number q, certain $I_{S(h)}$ are also zero for some odd h. Apart from $h = 1$ for which $I_{S(1)}$ is always nonzero, the $I_{S(h)}$ are nonzero for

$$h = kq \pm 1 \ (k \text{ integer} \geq 1) \tag{4}$$

Therefore harmonic currents for certain harmonic numbers are eliminated for higher pulse numbers. For example, the single-phase bridge rectifier with $q = 2$ produces nonzero $I_{S(h)}$ for $h = 1, 3, 5, 7, 9, \ldots$, while the three-phase bridge rectifier with $q = 6$ produces nonzero $I_{S(h)}$ for $h = 1, 5, 7, 11, 13, \ldots$. If rectifier operation is unbalanced, then harmonics are produced for all h. An unbalanced condition can result from asymmetrical gating of the SCRs or from voltage or impedance unbalance of the ac source. The effect is particularly pronounced for three-phase rectifiers with a comparatively small L_O and a comparatively large C_O since these rectifiers act like "peak detectors" and C_O charges to the point where V_O approaches the peak value of the line-to-line voltage. One phase needs to be only several percent below the other two phases for it to conduct a greatly reduced current and shift most of the current to the other two phases. An unbalanced condition is always evident from the waveform of i_X because the heights of the pulses are not all the same.

Power Factor.   The rms value I_S of i_S is found from

$$I_S = \sqrt{I_{S(1)}^2 + \sum_{h>1} I_{S(h)}^2} \tag{5}$$

The ac source is rated for apparent power S, which is the product of V_S and I_S (in volt-amperes, VA). However, the source delivers real input power P_I (in watts, W), which the rectifier converts to dc and supplies to the load. The *total power factor PF* is the ratio of the real input power and the apparent power supplied by the ac source

$$PF = \frac{P_I}{S} = \frac{P_I}{V_S I_S} \qquad (6)$$

and measures the fraction of the available apparent power actually delivered to the rectifier. The source voltage v_S is an undistorted sine wave only if L_S is negligible. In this case power is delivered only at the fundamental frequency so that

$$P_I = V_S I_{S(1)} \cos \phi_{S(1)} \qquad (7)$$

Note that the harmonics $I_{S(h)}$ draw apparent power from the source by increasing I_S but do not deliver real power to the rectifier. Using this assumption, the expression for power factor reduces to

$$PF = \cos \phi_{S(1)} \frac{I_{S(1)}}{I_S} \qquad (8)$$

The *displacement power factor* $\cos\phi_{S(1)}$ is the traditional power factor used in electric power systems for sinusoidal operation and is unity when the fundamental of i_S is in phase with v_S. The *purity factor* $I_{S(1)}/I_S$ is unity when i_S is a pure sine wave and the rms values of $I_{S(h)}$ are zero so that I_S equals $I_{S(1)}$. The distortion of i_S is often and equivalently represented by the *total harmonic distortion for current THD_i*

$$THD_i = 100\frac{\sqrt{\sum_{h>1} I_{S(h)^2}}}{I_{S(1)}} = 100\sqrt{\frac{1}{(I_{S(1)}/I_S)^2} - 1} \quad \text{(expressed in percent)} \qquad (9)$$

The purity factor $I_{S(1)}/I_S$ is also called the distortion power factor, which is easily confused with the total harmonic distortion THD_i.

The theoretical maximum power factor for the single-phase bridge rectifier is 0.90, which occurs for $\alpha = 0°$ and usually requires an uneconomically large value of L_O. The actual power factor often ranges from 0.5 to 0.75. The theoretical maximum power factor for the three-phase bridge rectifier is 0.96 which also occurs for $\alpha = 0°$. Because the three-phase bridge rectifier requires less filtering, it is often possible to approach this theoretical maximum power factor with an economical value of L_O. However, for cost reasons, L_O is often omitted in both the single- and three-phase rectifiers which dramatically reduces the power factor and leaves it to depend on the value of L_S.

Source Voltage Distortion and Power Quality.   The time waveforms of Fig. 13.2.3 show that with appreciable L_S the rectifier distorts the voltage source v_S supplying the rectifier. Fourier series is also used to represent v_S as a fundamental voltage of rms value $V_{S(1)}$ and harmonic voltages of rms value $V_{S(h)}$. The distortion of v_S is often represented by the total harmonic distortion for voltage THD_v

$$THD_v = 100\frac{\sqrt{\sum_{h>1} V_{S(h)^2}}}{V_{S(1)}} \quad \text{(expressed in percent)} \qquad (10)$$

Note that the definition of power factor (Eq. (7)) is valid for appreciable L_S and distorted v_S but (Eq. (8)) is strictly valid only when L_S is negligible and v_S is undistorted.

Voltage distortion can cause problems for other loads sharing the rectifier ac voltage source. Computer-based loads, which have become very common, appear to be particularly sensitive. Issues of this kind have been receiving increased attention and fall under the general heading of *power quality*. Increasingly strict power factor and harmonic current limits are being placed on ac-to-dc converters (IEEE-519, 1992; IEC-1000, 1995). In particular, limits on the total harmonic distortion of the current THD_i, the rms values $I_{S(h)}$ of the harmonics, and the rms values of the harmonics relative to the fundamental $I_{S(h)}/I_{S(1)}$ are often specified. These limits present new challenges to the designers of ac-to-dc converters.

ADDITIONAL CONVERTER TOPOLOGIES

This section summarizes the large number of converters that are based on the rectifiers of Table 13.2.1. These converters are shown in Table 13.2.2.

Uncontrolled Diode Rectifier. Replacing SCRs with diodes produces an uncontrolled rectifier as shown in Table 13.2.2a. In contrast to the SCRs, which are gated by a controller, the diodes begin conduction when initially forward biased by the circuit so that an uncontrolled diode rectifier behaves like a phase-controlled rectifier operated with $\alpha = 0°$. Details of uncontrolled diode rectifier operation are described in Kelley (1992).

Half-Controlled Bridge Rectifier. In the half-controlled bridge rectifier the even-numbered SCRs (Q_2 and Q_4 for the single-phase rectifier, and Q_2, Q_4, and Q_6 for the three-phase rectifier) are replaced with diodes as shown in Table 13.2.2b. The remaining odd-numbered SCRs (Q_1 and Q_3 for the single-phase rectifier, and Q_1, Q_3, and Q_5 for the three-phase rectifier) are phase controlled to regulate the dc output voltage V_O. This substitution is advantageous because diodes are cheaper than SCRs and the cathodes of the remaining SCRs are connected to a common point that simplifies SCR gating.

Note that the diodes begin conduction when first forward biased while the SCRs begin conduction only after being gated while under forward bias. As a result, during a certain portion of each cycle, i_X *freewheels* through the series connection of a diode and SCR, thereby reducing i_S to zero. For example, in the single-phase bridge rectifier i_X freewheels through Q_1 and D_2 for one part of the cycle and through Q_3 and D_4 for another part of the cycle. In the three-phase rectifier i_X freewheels through Q_1 and D_2, Q_3 and D_4, and Q_5 and D_6 during different parts of the cycle. This freewheeling action prevents v_X from changing polarity and improves rectifier power factor as α increases and V_O decreases.

Freewheeling Diode. The same effect is achieved if a *freewheeling diode* D_X is connected across terminals 1 and 2 of the rectifier as shown in Table 13.2.2c. The freewheeling diode is used with both the bridge and midpoint rectifier connections.

Dc Motor Drive. Any of the phase-controlled rectifiers described above can be used as a dc motor drive by connecting the motor armature across terminals 1 and 2 as shown in Table 13.2.2d. Phase control of SCR firing angle α controls motor speed.

Battery Charger. Phase-controlled rectifiers are widely used as battery chargers as shown in Table 13.2.2e. Phase control of SCR firing angle α regulates battery charging current.

Line Commutated Inverter. A line-commutated inverter transfers power from the dc terminals 1 and 2 of the converter to the ac source. As shown in Table 13.2.2f, the dc terminals are connected to a dc source of power such as a dc generator or a battery. The polarity of each SCR is reversed and the rectifier is operated with $\alpha > 90°$. This circuit is called a line-commutated inverter or a synchronous inverter. Note that the half-controlled bridge and the freewheeling diode cannot be used with a line-commutated inverter because they prevent a change in the polarity of v_X.

Operation with $\alpha > 90°$ causes the majority of the positive half cycle of i_S to coincide with the negative half cycle of v_S. Similarly, the negative half cycle of i_S coincides with the positive half cycle of v_S. It is this mechanism that, on average, causes power flow from the dc source into the ac source. In principal α could approach 180°, but in practice α must be limited to 160° or less to permit sufficient time for the SCRs to stop conducting and regain forward voltage blocking capability before forward voltage is reapplied to them. This requirement is particularly important when L_S is appreciable.

TABLE 13.2.2   Additional Converter Topologies

	Single-Phase Source	Three-Phase Source
(a) Uncontrolled diode rectifier	Replace all SCRs ⊻ with diodes ⊻	
(b) Half-controlled bridge rectifier	Replace SCRs Q_2, Q_4 ⊻ with diodes D_2, D_4 ⊻	Replace SCRs Q_2, Q_4, Q_6 ⊻ with diodes D_2, D_4, D_6 ⊻
(c) Freewheeling diode	Retain L_O, C_O, and load, adding diode D_X	
(d) Dc motor drive	Replace L_O, C_O, and load with dc motor	
(e) Battery charger	Replace L_O, C_O, and load with L_O and battery	
(f) Line commutated inverter	Reverse polarity of all SCRs ⊻ changing to ⊻ Replace L_O, C_O, and load with dc generator or battery and L_O	

TABLE 13.2.2   Additional Converter Topologies (*Continued*)

	Single-Phase Source	Three-Phase Source
(*g*) Alternate three-phase transformer connections	Not applicable	Replace Y primary and/or secondary with Δ primary and/or secondary
(*h*) Bidirectional converter	Replace all SCRs ⎬ with two SCRs in antiparallel Replace L_O, C_O, and load with dc generator or battery and L_O	
(*i*) Active power factor correction	Replace all SCRs ⎬ with diodes ⎬ Replace L_O, C_O, and load with high-frequency filter, dc-to-dc converter, C_O and load	

(*Continued*)

TABLE 13.2.2   Additional Converter Topologies (*Continued*)

	Single-Phase Source	Three-Phase Source
(*j*) Series-connected 12-pulse rectifier	Not-applicable	
(*k*) Parallel-connected 12-pulse rectifier	Not-applicable	

Alternate Three-Phase Transformer Connections.   Both primary and secondary transformer windings may be either Y-connected or Δ-connected, as shown in Table 13.2.2*g* except for the midpoint connection, which requires a Y-connected secondary winding. If the connection is Y-Y or Δ-Δ the waveform of secondary current i_S is scaled by the transformer turn ratio to become the waveform of primary current i_P. Therefore, the secondary current fundamental $I_{S(1)}$ and harmonics $I_{S(h)}$, when scaled by the turn ratio, become the primary fundamental $I_{P(1)}$ and harmonics $I_{P(h)}$.

   Similarly, if the transformer connection is Y-Δ or Δ-Y, the secondary current fundamental $I_{S(1)}$ and harmonics $I_{S(h)}$, when scaled by the turn ratio, become the primary fundamental $I_{P(1)}$ and harmonics $I_{P(h)}$. However, the Y-Δ and Δ-Y transformer connections introduce different phase shifts for each harmonic so that the primary current waveform i_P differs in shape from the secondary current wave-form i_S. Rectifier power factor remains unchanged; however, this phase shift is used to produce harmonic cancellation in rectifiers with high pulse numbers as described subsequently.

Bidirectional Converter.   Many applications require bidirectional power flow from a single converter. Table 13.2.2*h* illustrates one example in which a phase-controlled rectifier is, effectively speaking, connected in parallel with a line commutated inverter by replacing each SCR with a *pair* of SCRs connected in antiparallel. The load is replaced either by a battery or by a dc motor. In the bidirectional converter, one polarity of SCRs is used to transfer energy from the ac source to the battery or motor while the opposite polarity of SCRs is used to reverse the power flow and transfer energy from the battery or motor to the ac source.

   For example, using the battery, the converter operates as a battery charger to store energy when demand on the utility is low and at a subsequent time the converter operates as a line commutated inverter to supply energy when demand on the utility is high. Using a dc motor, the converter operates as a dc motor drive to supply power to a rotating load. Depending on the direction of motor rotation and on which polarity of SCRs is used, the bidirectional converter can both brake the motor efficiently by returning the energy stored in the rotating momentum back to the ac source and subsequently reverse the motors direction of rotation.

Active Power Factor Corrector. In many instances the basic converter topologies of Table 13.2.1 and the additional converter topologies of Table 13.2.2a to 13.2.2g cannot meet increasingly strict power factor and harmonic current limits without the addition of expensive passive filters operating at line frequency. The active power factor corrector, illustrated in Table 13.2.2i, is one solution to this problem (Rippel, 1979; Kocher, 1982; Latos, 1982). The output filter inductor L_O is replaced by a high-frequency filter and a dc-to-dc converter. The dc-to-dc converter uses high-frequency switching and a fast control loop to actively control the waveshape of i_X, and therefore control the waveshape of i_S, for near unity displacement power factor and near unity purity factor resulting in near unity power factor ac-to-dc conversion (Huliehel, 1992). A high-frequency filter is required to prevent dc-to-dc converter switching noise from reaching the ac source.

A slower control loop regulates V_O against changes in source voltage and load. Because the dc-to-dc converter regulates V_O over a wide range of source voltage, the active power factor corrector can be designed for a *universal input* that allows the corrector to operate from nearly any ac voltage source. The active power factor corrector is used most commonly for lower powers.

Higher Pulse Numbers. When strict power factor and harmonic limits are imposed at higher power levels, and the active factor corrector cannot be used, the performance of the basic rectifier is improved by increasing the pulse number q and elimination of current harmonics $I_{S(h)}$ for certain harmonic numbers as shown by Eq. (4). Table 13.2.2j and 13.2.2k illustrate two examples based on the three-phase six-pulse bridge rectifier ($q = 6$) of Table 13.2.1c. The six-pulse rectifiers are shown connected in series in Table 13.2.2j and in parallel in Table 13.2.2k. The parallel connection in Table 13.2.2k requires an *interphase reactor* to prevent commutation of the SCRs in one rectifier from interfering with commutation of the SCRs in the other rectifier. The interphase reactor also helps the two rectifiers share the load equally.

Both approaches use a Y-Y transformer connection to supply one six-pulse rectifier and a Δ-Y transformer connection to supply the second six-pulse rectifier. The primary-to-secondary voltage phase shift of the Δ-Y transformer means the two rectifiers operate out of phase with each other producing 12-pulse operation ($q = 12$). As described previously, the Δ-Y transformer also produces a secondary-to-primary phase shift of the current harmonics. As a result, at the point of connection to the ac source, harmonics from the Δ-Y connected six-pulse rectifier cancel the harmonics from the Y-Y connected six-pulse rectifier for certain harmonic numbers. For example, the harmonics cancel for $h = 5$ and 7, but not for $h = 11$ and 13. Thus the total harmonic distortion for current THD_i and the total power factor for the 12-pulse converter is greatly improved in comparison to either six-pulse converter alone. The 12-pulse output voltage ripple filtering requirement is also greatly reduced compared to a single six-pulse rectifier. This principle can be extended to even higher pulse numbers by using additional six-pulse rectifiers and transformer phase-shift connections.

High Voltage Dc Transmission. High Voltage dc (HVDC) Transmission is a method for transmitting power over long distances while avoiding certain problems associated with long distance ac transmission. This requirement often arises when a large hydroelectric power generator is located a great distance from a large load such as a major city. The hydroelectric generator's relatively low ac voltage is stepped up by a transformer, and a phase-controlled rectifier converts it to a dc voltage of a megavolt or more. After transmission over a long distance, a line commutated inverter and transformer convert the dc back to ac and supply the power to the load. Alternately, the rectifier and inverter are co-located and used as a tie between adjacent utilities. The arrangement can be used to actively control power flow between utilities and to change frequency between adjacent utilities operating at 50 and 60 Hz.

With power in the gigawatt range, this is perhaps the highest power application of a power electronic converter. To ensure a stable system, the control algorithms of the rectifier and inverter must be carefully coordinated. Note that since the highest voltage rating of an individual SCR is less than 10 kV, many SCRs are connected in series to form a *valve* capable of blocking the large dc voltage. Both the rectifier and inverter use a high pulse number to minimize filtering at the point of connection to the ac source.

REFERENCES

Dewan, S. B., and A. Straughen, "Power Semiconductor Circuits," Wiley, 1975.

Huliehel, F. A., F. C. Lee, and B. H. Cho, "Small-signal modeling of the single-phase boost high power factor converter with constant frequency control," *Record of the 1992 IEEE Power Electronics Specialists Conference (PESC '92)*, pp. 475–482, June 1992.

IEC-1000 Electromagnetic Compatibility (EMC), Part 3: Limits, Section 2: Limits for Harmonic Current Emissions (formerly IEC-555-2), 1995.

IEEE Standard 519, "IEEE Recommended Practices and Requirements for Harmonic Control in Electrical Power Systems," 1992.

Kelley, A. W., and W. F. Yadusky, "Phase-controlled rectifier line-current harmonics and power factor as a function of firing angles and output filter inductance," *Proc. IEEE Applied Power Electronics Conf.,* pp. 588–597, March 1990.

Kelley, A. W., and W. F. Yadusky, "Rectifier design for minimum line-current harmonics and maximum power factor," *IEEE Trans. Power Electronics*, Vol. 7, No, 2, pp. 332–341, April 1992.

Kocher, M. J., and R. L. Steigerwald, "An ac-to-dc converter with high-quality input waveforms," *Record of the 1982 IEEE Power Electronics Specialists Conference (PESC '82)*, pp. 63–75, June 1982.

Latos, T. S., and D. J. Bosak, "A high-efficiency 3-kW switchmode battery charger," *Record of the 1982 IEEE Power Electronics Specialists Conference (PESC '82)*, pp. 341–349, June 1982.

Pelley, B. R., "Thyristor Phase-Controlled Converters and Cycloconverters," Wiley, 1971.

Rashid, M. "Power Electronics: Circuits, Devices, and Applications," 2nd ed., Prentice Hall, 1993.

Rippel, W. E., "Optimizing boost chopper charger design," *Proceedings of the Sixth National Solid-State Power Conversion Conference (POWERCON6)*, pp. D1-1–D1-20, 1979.

Schaeffer, J., "Rectifier Circuits: Theory and Design," Wiley, 1965.

CHAPTER 13.3
DC-DC CONVERTERS

Philip T. Krein

INTRODUCTION

Power conversion among different dc voltage and current levels is important in applications ranging from spacecraft and automobiles to personal computers and consumer products. Power electronics technology can be used to create a *dc transformer* function for power processing. Today, most dc power supplies rectify the incoming ac line, then use a dc-dc converter to provide a transformer function and produce the desired output voltages. Dc-dc designs often use input voltages near 170 V (the peak value of rectified 120-V ac) or 300–400 V (the peak values of 230, 240 V ac, and many three-phase sources). For direct dc inputs, 48-V sources and 28-V sources reflect practice in the telecommunications and aerospace industries. *Universal input power* supplies commonly handle rectified power from sources ranging between 85 and 270 V ac.

There are a number of detailed treatments of dc-dc converters in the literature. The book by Severns and Bloom (1985) explores a wide range of topologies. Mitchell (1988) offers detailed analysis and extensive treatment of control issues. Chryssis (1989) compares topologies from a practical standpoint, and addresses many aspects associated with actual implementation. Middlebrook has published several exhaustive treatments of various topologies; one example (Middlebrook, 1989) addresses some key attributes of analysis and control. The discussion here follows the treatment in Krein (1998). More recently, Erickson and Maksimovich (2001) have detailed many operation and control aspects of modern dc-dc converters.

DIRECT DC-DC CONVERTERS

The most general dc-dc conversion process is based on a switch matrix that interconnects two dc ports. The two dc ports need to have complementary characteristics, since Kirchhoff's laws prohibit direct interconnection of unlike voltages or currents. A generic example, called a *direct converter*, is shown in Fig. 13.3.1a. A more complete version is shown in Fig. 13.3.1b, in which an inductor provides the characteristics of a current source. In the figure, only four switch combinations can be used without shorting the voltage source or opening the current source:

Combination	Result
Close 1,1 and 2,2	$V_d = V_{IN}$
Close 2,1 and 2,2	$V_d = 0$
Close 1,1 and 1,2	$V_d = 0$
Close 1,2 and 2,1	$V_d = -V_{IN}$

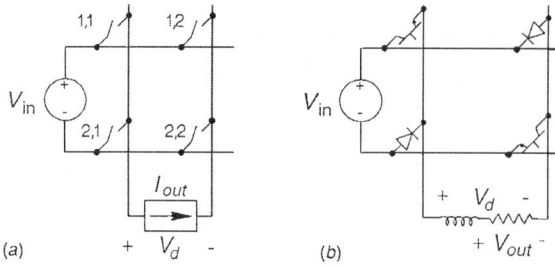

FIGURE 13.3.1   Dc voltage to dc current direct converter: (*a*) general arrangement; (*b*) circuit realization.

Switch action selects among $-V_{in}$, 0, and $+V_{in}$ to provide a desired average output. The energy flow is controlled by operating the switches periodically, then adjusting the *duty ratio* to manipulate the average behavior. Duty-ratio control, or *pulse width modulation*, is the primary control method for most dc-dc power electronics. The switching frequency can be chosen somewhat arbitrarily with this method. Typical rates range from 50 kHz to 500 kHz for converters operating up to 200 W, and from 20 kHz to 100 kHz for converters operating up to 2 kW. In dc-dc circuits that use soft switching or *resonant switching* techniques, the switching frequency is usually adjusted to match internal circuit resonances.

Switching functions $q(t)$ can be defined for each switch in the converter. A switching function has the value 1 when the associated switch is on, and 0 when it is off. The converter voltage v_d in Figure 13.3.1*b* can be written in terms of the switching functions in the compact form

$$v_d(t) = q_{1,1}q_{2,2}V_{in} - q_{1,2}q_{2,1}V_{in} \tag{1}$$

For power flow in one direction, two switches suffice. The usual practice is to establish a common ground between the input and output, equivalent to permanently turning on switch 2,2 and turning off switch 1,2 in Fig. 13.3.1. This simplified circuit is shown in Fig. 13.3.2. The transistor is generic: a BJT, MOSFET, IGBT, or other fully controlled device can be used. The voltage $v_d(t)$ in this circuit becomes $v_d(t) = q_1(t)V_{in}$. Of interest is the dc or average value of the output, indicated by the angle brackets as $\langle v_{out}(t) \rangle$. The inductor cannot sustain an average voltage in the periodic steady state, so the resistor voltage average value $\langle v_{out}(t) \rangle = \langle v_d(t) \rangle$. The voltage $v_d(t)$ is a pulse train with period T, amplitude V_{in}, and an average value related to the duty ratio of the switching function. Therefore,

$$\langle v_d(t) \rangle = \langle v_{out}(t) \rangle = \frac{1}{T} \int_0^T q_1(t)V_{in}dt = D_1 V_{in} \tag{2}$$

where T is the switching period and D_1 is the duty ratio of the transistor. (Typically, an average value in a dc-dc converter is equivalent to substituting a switch duty ratio D for a switching function q.) If the L-R pair serves

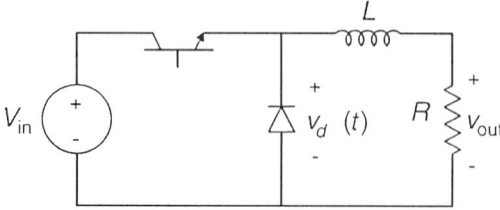

FIGURE 13.3.2   Common-ground direct converter (*buck converter*).

TABLE 13.3.1 Buck Converter Characteristics

Characteristic	Value
Input–output relationships	$V_{out} = D_1 V_{in}$, $I_{in} = D_1 I_{out}$
Device ratings	Must handle V_{in} when off, and I_{out} when on.
Open-loop load regulation	Load has no direct effect on output.
Open-loop line regulation	No line regulation: line changes are reflected directly at the output. Feedback control is required.
Ripple	Governed by choice of inductor. Typically 50 mV $_{peak–to–peak}$ or 1 percent (whichever is higher) can be achieved.

as an effective low-pass filter, the output voltage will be a dc value $V_{out} = \langle v_d(t) \rangle$. The circuit has the basic characteristics of a transformer, with the restriction that $V_{out} \le V_{in}$. The name *buck converter* is used to reflect this behavior.

The buck converter, sometimes called a *buck regulator*, or a *step-down converter*, is the basis for many more sophisticated dc-dc converters. Some of its characteristics are summarized in Table 13.3.1. Load regulation is perfect in principle if the inductor maintains current flow. Line regulation requires closed-loop control. Although these relationships are based on ideal, lossless switches, the analysis process applies to more detailed circuits. The following example illustrates the approach.

Example: Relationships in a dc-dc converter.

The circuit of Fig. 13.3.3 shows a dc-dc buck converter with switch on-state voltage drops taken into account. What is the input–output voltage relationship? How much power is lost in the converter?

To analyze the effect of switch voltages, KVL and KCL relations can be written in terms of switching functions. When averages are computed, variables such as inductor voltages and capacitor currents are eliminated since these elements cannot sustain dc voltage and current, respectively. Circuit laws require

$$v_d(t) = q_1(t)(V_{in} - V_{s1}) - q_2(t)V_{s2}, \qquad V_{out} = v_d(t) - v_L \tag{3}$$
$$i_{in}(t) = q_1(t)I_L, \qquad I_{out}(t) = I_L - i_C(t)$$

In this circuit, the inductor will force the diode to turn on whenever the transistor is off. This can be represented with the expression $q_1(t) + q_2(t) = 1$. When the average behavior is computed, the duty ratios must follow the relationship $D_1 + D_2 = 1$. The average value of $v_d(t)$ must match V_{out}, and the average input will be the duty ratio of switch 1 multiplied by the inductor current. These relationships reduce to

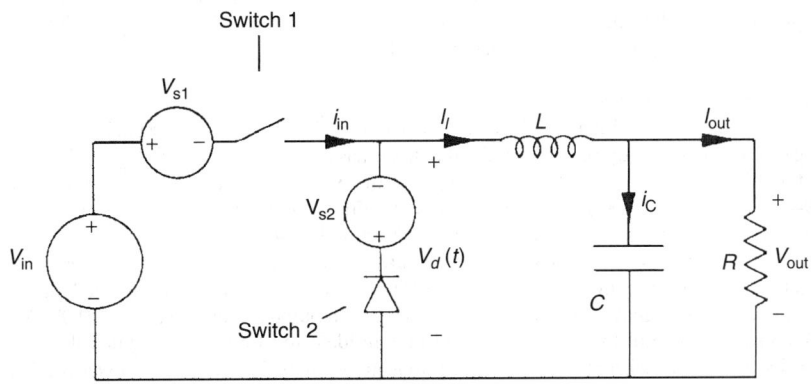

FIGURE 13.3.3 Buck converter with switch forward drop models.

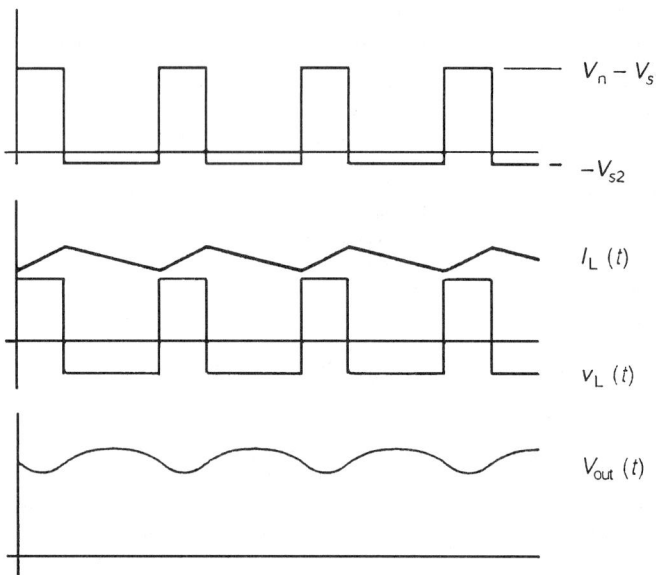

FIGURE 13.3.4   Buck converter waveforms.

$$V_{out} = D_1(V_{in} - V_{s1} + V_{s2}) - V_{s2}$$
$$\langle i_{in}(t) \rangle = D_1 I_{out}$$
$$P_{in} = \langle i_{in}(t) V_{in} \rangle = D_1 V_{in} I_{out} \tag{4}$$
$$P_{out} = V_{out} I_{out} = D_1(V_{in} - V_{s1} + V_{s2}) I_{out} - V_{s2} I_{out}$$

When the switch voltage drops V_{s1} and V_{s2} have similar values, the loss fraction is approximately the ratio of the diode drop to the output voltage.

The primary design considerations are to choose an inductor and capacitor to meet requirements on output voltage ripple. The design process is simple if a *small ripple assumption* is used: since V_{out} is nearly constant, the voltage across the inductor will be a pulsed waveform at the switching frequency. The current $i_L(t)$ will exhibit triangular ripple. Some waveform samples appear in Fig. 13.3.4. The current swings over its full peak-to-peak ripple during either the transistor on time or the diode on time. The example below takes advantage of the triangular variation to compute the expected ripple.

Example: Buck converter analysis.
A buck converter circuit with an R-L load and a switching frequency of 200 kHz is shown in Fig. 13.3.5. The transistor exhibits on-state drop of 0.5 V, while the diode has a 1 V forward drop. Determine the output ripple for 15 V input and 5 V output. From Eq. (5), the duty ratio of switch 1 should be 6/15.5 = 0.387. Switch 1 should be on for 1.94 μs, then off for 3.06 μs. At 5 V output, the inductor voltage is 9.5 V with switch 1 on and −6 V with switch 1 off. The inductor current has di/dt = (9.5 V)/(200 μH) = 47.5 kA/s when #1 is on and di/dt = −30 kA/s when #2 is on. During the on time of switch 1, the current increases (47500 A/s)·(1.93 μs) = 0.092 A. Here, the time constant L/R = 200 μs, which is 65 times the switch 2 on time. It would be expected that the current change is small and linear. The current change of 0.092 A produces an output voltage change of 0.092 V for this 1 Ω load. Figure 13.3.6 shows some of the important waveforms in idealized form. The output voltage is nearly constant, with V_{out} = 5 ± 0.046 V, consistent with the assumptions in the analysis. The output power is 25 W. The input average current is $D_1 I_{out}$ = 0.387(5 A) = 1.94 A. The input power therefore is 29.0 W, and the efficiency is 86 percent. This neglects any energy consumed in the commutation process.

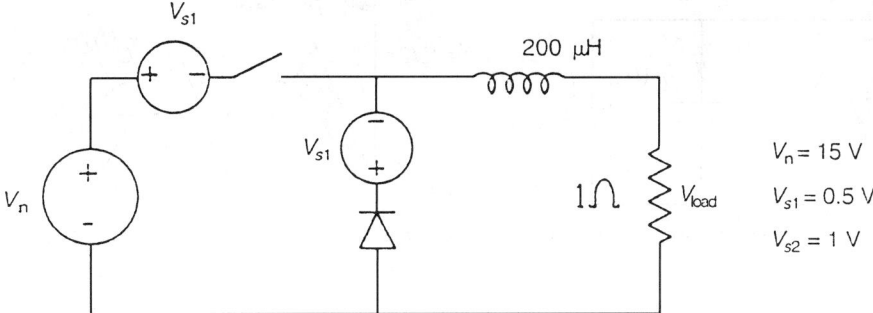

FIGURE 13.3.5   Buck converter example.

More generally, the buck converter imposes $V_{in} - V_{out}$ on the inductor while the transistor is on. The current derivative di/dt over a given time interval is the linear change $\Delta i_L/\Delta t$. For the on-time interval $\Delta t = D_1 T$, the peak-to-peak ripple Δi_L is

$$\Delta i_L = \frac{(V_{in} - V_{out})D_1 T}{L} = \frac{V_{in}(1 - D_1)D_1 T}{L} \tag{5}$$

If only inductive filtering is used, the output voltage ripple is the load resistance times the current ripple.

If an output capacitor is added across the load resistor, its effect can be found by treating the inductor as an equivalent triangular current source, then solving for the output voltage. Assuming that the capacitor handles the full ripple current, it is straightforward to integrate the triangular current to compute the ripple voltage. The process is illustrated in Fig. 13.3.7. Voltage $v_C(t)$ will increase whenever $i_C(t) > 0$. The voltage increase is given by

$$\Delta v_C = \frac{1}{C}\int_0^{T/2} i_C(t)\,dt \tag{6}$$

The integral is the triangular area $\frac{1}{2}(T/2)(\Delta i_L/2)$, so

$$\Delta v_C = \frac{T\Delta i_L}{8C} \tag{7}$$

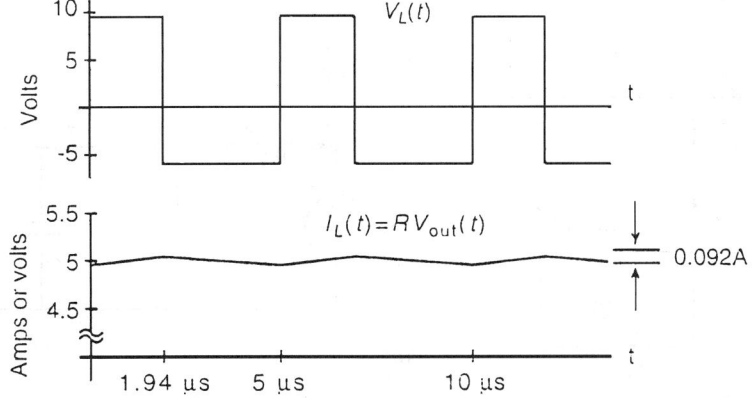

FIGURE 13.3.6   Buck converter inductor voltage and output current.

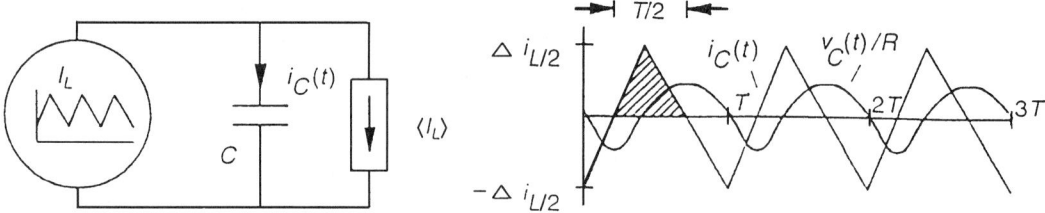

FIGURE 13.3.7   Output ripple effect given the addition of capacitive filter.

This expression is accurate if the capacitor is large enough to provide significant voltage ripple reduction.

An alternative direct dc-dc converter has a current source input and voltage source output. The relationships for this *boost converter* are dual to those of the buck circuit. The input current and output voltage act as fixed source values, while the input voltage and output current are determined by switch matrix action. For the common-ground version in Fig. 13.3.8*b*, the transistor and diode must operate in complement, so that $q_1 + q_2 = 1$ and $D_1 + D_2 = 1$. For ideal switches,

$$q_1 + q_2 = 1$$
$$v_t(t) = q_2 V_{out} = (1 - q_1)V_{out}$$
$$i_{out} = q_2 I_{in} = (1 - q_1)I_{in} \tag{8}$$
$$\langle v_t \rangle = D_2 V_{out} = (1 - D_1)V_{out}$$
$$\langle i_{out} \rangle = (1 - D_1)I_{in}$$

With these energy storage devices, notice that $V_{in} = \langle v_t \rangle$ and $I_{out} = \langle i_{out} \rangle$. The relationships can be written

$$V_{out} = \frac{1}{1 - D_1} V_{in} \qquad \text{and} \qquad I_{in} = \frac{1}{1 - D_1} I_{out} \tag{9}$$

The output voltage will be higher than the input. The boost converter uses an inductor at the input to create current-source behavior and a capacitor at the output to provide voltage source characteristics. The capacitor

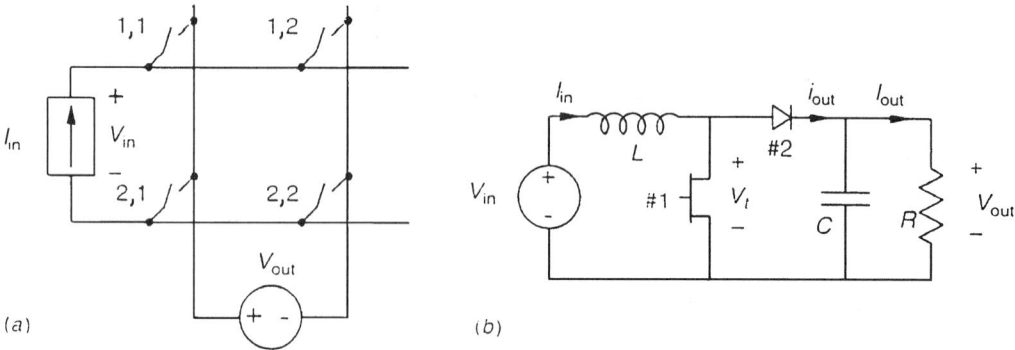

(a)

(b)

FIGURE 13.3.8   Boost dc-dc converter: (*a*) general arrangement; (*b*) common-ground version.

TABLE 13.3.2 Relationships for the Boost Converter

Characteristic	Value
Input–output relationships	$V_{out} = V_{in}/(1 - D_1)$, $I_{out} = I_{in}(1 - D_1)$
Open-loop load regulation	Load does not alter output
Open-loop line regulation	Unregulated without control
Device ratings	Must handle V_{out} when off or I_{in} when on
Input–output relationships with switch drops	$V_{out} = (V_{in} - V_{s2} - D_1V_{s1} + D_1V_{s2})/(1 - D_1)$, $I_{out} = I_{in}(1 - D_1)$
Inductor ripple relationship	$\Delta i_L = V_{in}D_1T/L$ (if Δi_L is small compared to the input current)
Capacitor ripple relationship	$\Delta v_C = I_{out}D_1T/C$

is exposed to a square-wave current signal, and produces a triangular ripple voltage in response. Table 13.3.2 provides a summary of relationships, based on ideal switches.

INDIRECT DC-DC CONVERTERS

Cascade arrangements of buck and boost converters are used to avoid limitations on the magnitude of V_{out}. A buck-boost cascade is developed in Fig. 13.3.9. This is an example of an *indirect converter* because at no point in time does power flow directly from the input to the output. Some of the switches in the cascade are redundant, and can be removed. In fact, only two switches are needed in the final result, shown in Fig. 13.3.10. The current source, called a *transfer current source*, has been replaced by an inductor.

The voltage across the inductor, v_t, is V_{in} when switch 1 is on, and $-V_{out}$ when switch 2 is on. The transfer current source value is I_s. The inductor cannot sustain dc voltage drop, so $\langle v_t \rangle = 0$. The voltage relationships are

$$q_1 + q_2 = 1$$
$$v_t = q_1V_{in} - q_2V_{out} \tag{10}$$
$$\langle v_t \rangle = 0 = D_1V_{in} - D_2V_{out}$$

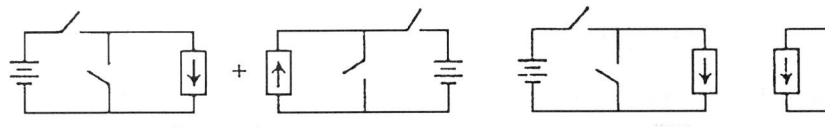

(a) Buck and boost.   (b) Boost upside down

(c) Cascade them.   (d) Remove obvious redundancy.

FIGURE 13.3.9 Cascaded buck and boost converters. (From Krein (1998), copyright © 1998 Oxford University Press, Inc., U.S.; used by permission.)

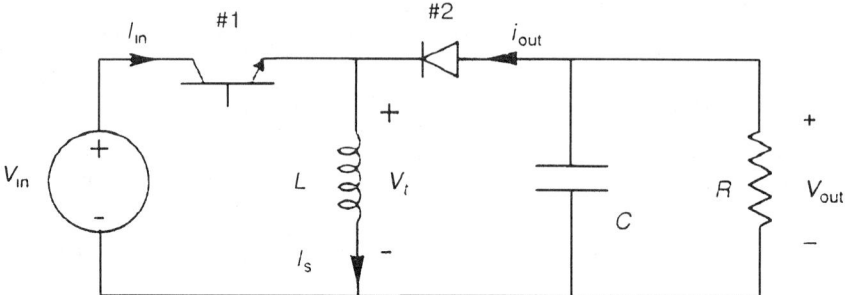

FIGURE 13.3.10   Buck-boost converter.

The last part of Eq. (10) requires $D_1 V_{\text{in}} = D_2 V_{\text{out}}$ in steady state. The switches must act in complement, so $D_1 + D_2 = 1$, and

$$V_{\text{out}} = \frac{D_1}{1 - D_1} V_{\text{in}} \tag{11}$$

A summary of results is given in Table 13.3.3. The cascade process produces a negative voltage with respect to the input. This polarity reversal property is fundamental to the buck-boost converter.

A boost-buck cascade also allows full output range with a polarity reversal. As in the buck-boost case, many of the switches are redundant in the basic cascade, and only two switches are required. The final circuit, with energy storage elements in place, as shown in Fig. 13.3.11. The center capacitor serves as a *transfer voltage source*. The transfer source must exhibit $\langle i_t \rangle = 0$, since a capacitor cannot sustain dc current. Some of the major relationships are summarized in Table 13.3.4. In the literature, this arrangement is called a Ćuk converter, after the developer who patented it in the mid-1970s (Middlebrook and Ćuk, 1977). The transfer capacitor must be able to sustain a current equal to the sum of the input and output currents. The RMS capacitor current causes losses in the capacitor's internal equivalent series resistance (ESR), so low ESR components are usually required.

Figure 13.3.12 shows the *single-ended primary inductor converter* or SEPIC circuit (Massey and Snyder, 1977). This is a boost-buck-boost cascade. As in the preceding cases, the cascade arrangement can be simplified to require only two switches. The transfer sources C_t and L_t carry zero average power to be consistent with a capacitor and an inductor as the actual devices. The relationships are

$$v_{\text{in}} = q_2(V_{\text{out}} + V_{t1}), \quad \langle v_{\text{in}} \rangle = D_2(V_{\text{out}} + V_{t1})$$
$$i_{\text{out}} = q_2(I_{\text{in}} + I_{t2}), \quad \langle i_{\text{out}} \rangle = D_2(I_{\text{in}} + I_{t2})$$
$$i_{t1} = -q_1 I_{t2} + q_2 I_{\text{in}}, \quad \langle i_{t1} \rangle = 0 = -D_1 I_{t2} + D_2 I_{\text{in}} \tag{12}$$

TABLE 13.3.3   Buck-Boost Converter Relationships

Characteristic	Value
Input–output voltage relationship	$|V_{\text{out}}| = V_{\text{in}} D_1 / 1 - D_1)$, $I_{\text{in}} = |I_{\text{out}}| D_1/(1 - D_1)$, output is negative with respect to input
Device ratings	Must handle $|V_{\text{in}}| + |V_{\text{out}}|$ when off, $|I_{\text{in}}| + |I_{\text{out}}|$ when on
Inductor current	$I_L = |I_{\text{in}}| - |I_{\text{out}}|$
Regulation	Perfect load regulation. No open-loop line regulation
Inductor ripple current	$\Delta i_L = V_{\text{in}} D_1 T / L$
Capacitor ripple voltage	$\Delta v_C = |I_{\text{out}}| D_1 T / C$
Output range	Ideally, any negative voltage can be produced

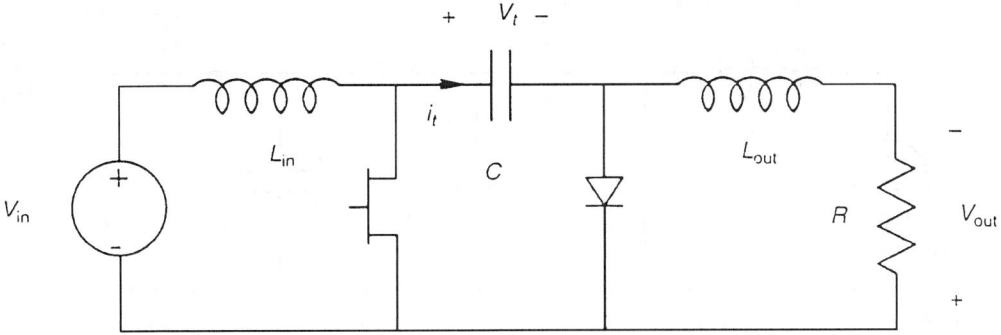

FIGURE 13.3.11 Boost-buck converter.

$$v_{t2} = -q_1 V_{t1} + q_2 V_{out}, \quad \langle v_{t2} \rangle = 0 = -D_1 V_{t1} + D_2 V_{out}$$
$$q_1 + q_2 = 1, \quad D_1 + D_2 = 1$$

Some algebra will bring out the transfer source values and input-output ratios:

$$I_{t2} = \frac{D_2}{D_1} I_{in} = \frac{1-D_1}{D_1} I_{in}$$

$$V_{t1} = \frac{D_2}{D_1} V_{out} = \frac{1-D_1}{D_1} V_{out} \tag{13}$$

$$\langle v_{in} \rangle = D_2 \left(V_{out} + \frac{1-D_1}{D_1} V_{out} \right) = \frac{1-D_1}{D_1} V_{out}$$

This is the same input–output ratio as the buck-boost converter, except that there is no polarity reversal.

These and related indirect converters provide opportunities for the use of more sophisticated magnetics such as *coupled inductors*. In the Ćuk converter, for example, the input and output filter inductors are often coupled on a single core to cancel out part of the ripple current (Middlebrook and Ćuk, 1981). In a buck-boost converter, the transfer source inductor can be split by providing a second winding. One winding can be used to inject energy into the inductor, while the other can be used to remove it. The two windings provide isolation. This arrangement is known as a *flyback converter* because diode turn-on occurs when the inductor output coil voltage "flies back" as the input switch turns off. An example is shown in Fig. 13.3.13.

The flyback converter is one of the most common low-power dc-dc converters. It is functionally equivalent to the buck-boost converter. This is easy to see if the turns ratio between the windings is unity. The possibility of a nonunity turns ratio is a helpful extra feature of the flyback converter. Extreme step-downs, such as the 170-V to 5-V converter, often used in a dc power supply, can be supported with reasonable duty ratios by selecting an appropriate turns ratio. In general, flyback converters are designed to keep the nominal duty ratio close to 50 percent. This tends to minimize the energy storage requirements, and keeps the sensitivity to variation as

TABLE 13.3.4 Relationships for Boost-Buck Converter

Characteristic	Value
Input–output relationships	$|V_{out}| = D_1 V_{in}/(1 - D_1)$, $I_{in} = D_1 |I_{out}|(1 - D_1)$
Device ratings	Must handle $|V_{in}| + |V_{out}|$ when off. Must handle $|I_{in}| + |I_{out}|$ when on
Regulation	Perfect load regulation, no line regulation

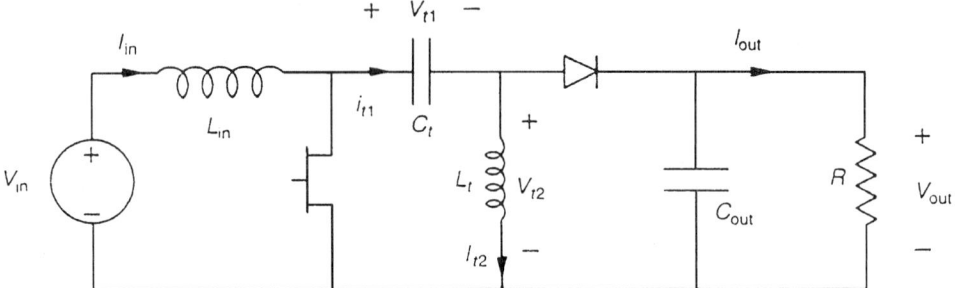

FIGURE 13.3.12 The SEPIC converter. Two transfer sources permit any input-to-output ratio without polarity reversal.

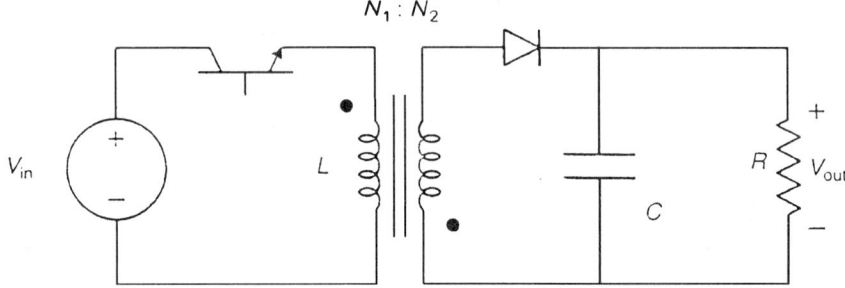

FIGURE 13.3.13 Flyback converter.

low as possible. An additional advantage appears when several different dc supplies are needed: if it is possible to use two separate coils on the magnetic core of the inductor, it should be just as reasonable to use three, four, or even more coils. Each can have its own turns ratio with mutual isolation. This is the basic for many types of multi-output dc power supplies.

One challenge with a flyback converter, as shown in Fig. 13.3.14, is the primary leakage inductance. During transistor turn-off, the leakage inductance energy must be removed. A capacitor or other snubber circuit is used to avoid damage to the active switch during turn-off.

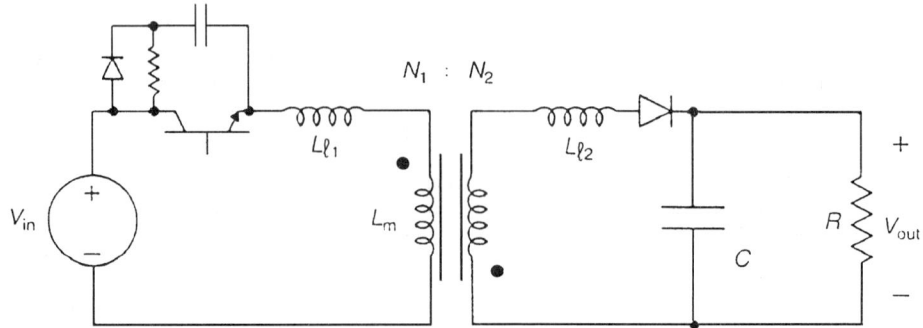

FIGURE 13.3.14 Leakage inductance issue in flyback converter.

FORWARD CONVERTERS

Coupled inductors in indirect converters, unlike transformers, must store energy and carry a net dc current. Basic buck and boost circuits lack a transfer source, so a coupled inductor will not give them isolation properties. Instead, a buck or boost converter can be augmented with a transformer, inserted at a location with only ac waveforms. Circuits based on this technique are called *forward converters*. A transformer can be added either by providing a *catch winding* tertiary or other circuitry for flux resetting, or by using an *ac link* arrangement. With either alternative, the objective is to avoid saturation because of dc current.

Figure 13.3.15 shows the catch-winding alternative in a buck converter. The tertiary allows the core flux to be reset while the transistor is off. Operation is as follows: the transistor carries the primary current i_1 and also the magnetizing current i_m when it is on. Voltage V_{in} is imposed on the primary, and the flux increases. When the transistor turns off, the magnetizing inductance will maintain the current flow in coil 1, such that $i_1 = -i_m$. The diode D_3 permits current $i_3 = i_m(N_1/N_3)$ to flow. The tertiary voltage v_3 flies back to $-V_{in}$, resetting the flux. If $N_1 = N_3$, the duty ratio of switch 1 must not exceed 50 percent so that there will be enough time to bring the flux down sufficiently. If it is desired to reach a higher duty ratio, the ratio N_1/N_3 must be at least $D_1/(1 - D_1)$.

With a catch winding, the primary carries a voltage $v_1 = -N_1/N_3$ after the transistor turns off. The transistor must be able to block $V_{in}(1 + N_1/N_3)$ to support this voltage. For power supplies, this can lead to extreme ratings. For example, an off-line supply designed for 350 V_{dc} input with $N_1/N_3 = 1.5$ to support duty ratios up to 60 percent requires a transistor rating of about 1000 V. This extreme voltage rating is an important drawback of the catch-winding approach.

The secondary voltage v_2 in this converter is positive whenever the transistor is on. The diode D_1 will exhibit the same switching function as the transistor, and the voltage across D_2 will be just like the diode voltage of a buck converter except for the turns ratio. The output and its average value will be

$$v_{out} = q_1 V_{in} \frac{N_2}{N_1}, \quad \langle v_{out} \rangle = D_1 V_{in} \frac{N_2}{N_1} \tag{14}$$

so this forward converter is termed a *buck-derived* circuit.

The ac link configuration comprises an inverter-rectifier cascade, such as the buck-derived half-bridge converter in Fig. 13.3.16. With adjustment of duty ratio, a waveform such as that shown in Fig. 13.3.17*a* is typically used as the inverter output. The signal has no dc component, and therefore a transformer can be used. Once full-wave rectification is performed, the result will be the square wave of Fig. 13.3.17*b*. The average output is

$$\langle v_{out} \rangle = 2aDV_{in} \tag{15}$$

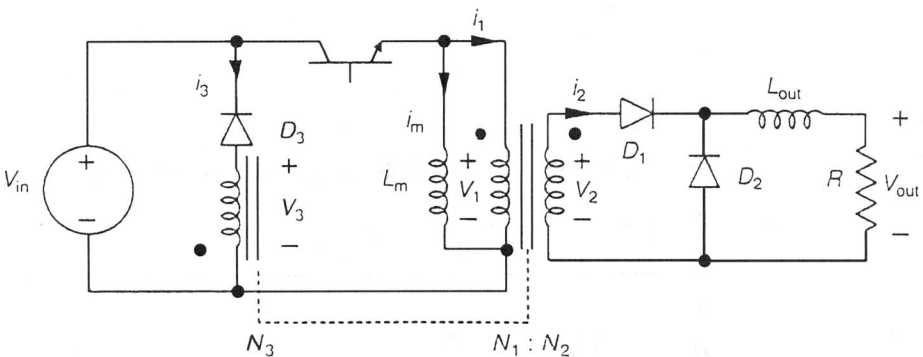

FIGURE 13.3.15 Catch-winding forward converter. (From Krein (1998), copyright © 1998 Oxford University Press. Inc., U.S.; used by permission.)

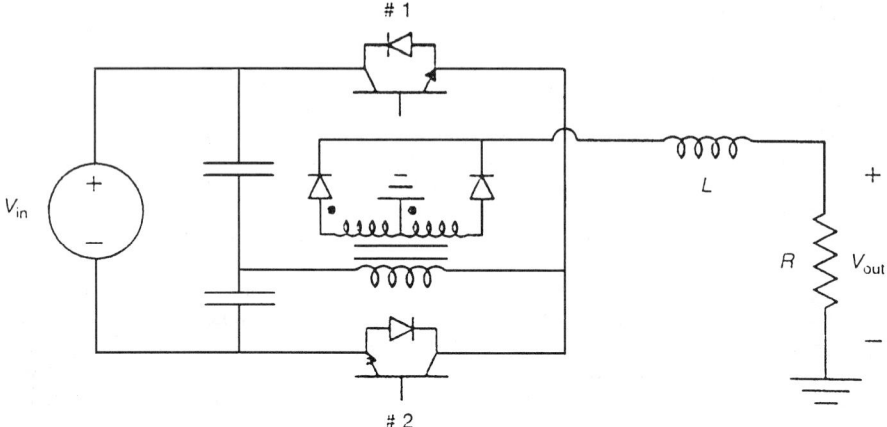

FIGURE 13.3.16   Half-bridge forward converter.

reflecting the fact that there are two output pulses during each switching period. No switch on the inverter side will be on more than 50 percent of each cycle, and the transistors block only V_{in} when off.

Four other forward converter topologies are shown in Fig. 13.3.18. The full bridge circuit in particular has been used successfully for power levels up to a few kilowatts. The others avoid the complexity of four active switches, although possibly with a penalty. For example, the *push-pull* converter in the figure has the important advantage that both switch gate drives share a common reference node with V_{in}. Its drawback is that the transistor must block $2V_{in}$ when off, because of an autotransformer effect of the center-tapped primary. The topologies are compared in Table 13.3.5.

The full-bridge converter perhaps offers the most straightforward operation. The switches always provide a path for magnetizing and leakage inductance currents, and circuit behavior is affected little by these extra inductances. In other circuits, these inductances are a significant complicating factor. For example, magnetizing inductance can turn on the primary-side diodes in a half-bridge converter, altering the operation of duty

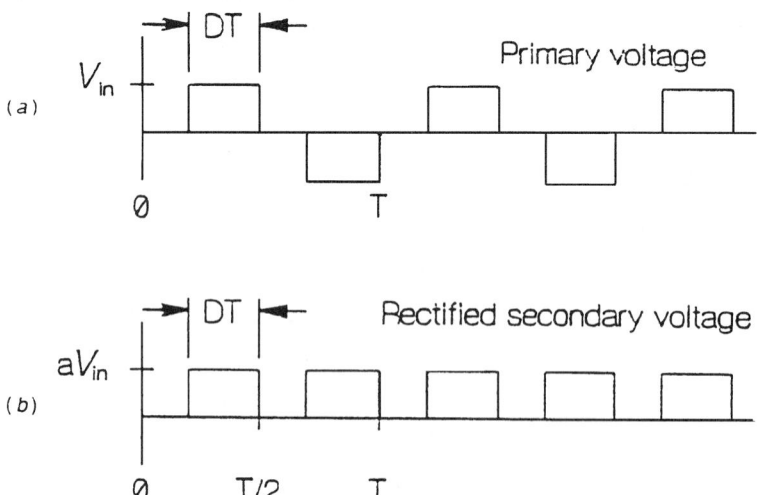

FIGURE 13.3.17   Typical waveforms in inverter-rectifier cascade.   (From Krein (1998), copyright © 1998 Oxford University Press, Inc., New York, U.S.; used by permission.)

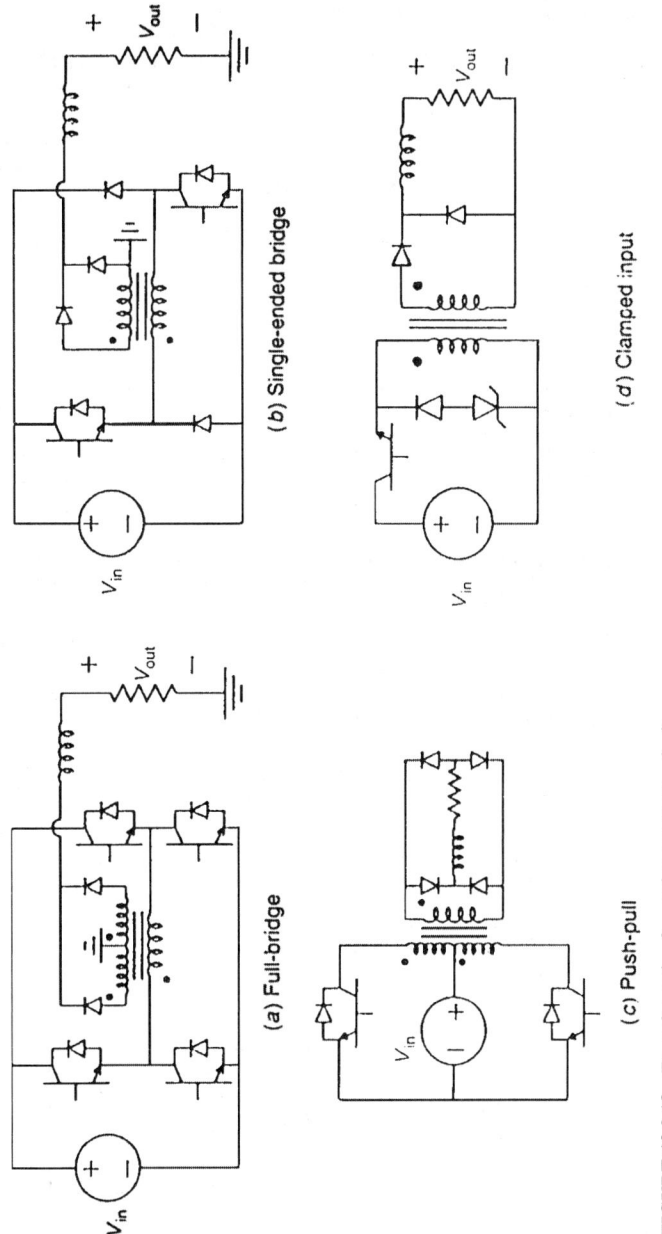

FIGURE 13.3.18 Four alternative forward converter topologies.

(a) Full-bridge

(b) Single-ended bridge

(c) Push-pull

(d) Clamped input

TABLE 13.3.5 Characteristics of Common Buck-Derived Forward Converters

Forward Converter Topology	Transistor Off-State Voltage	Flux Behavior	Comments
Full-bridge	V_{in}	Full variation from $-\phi_{max}$ to $+\phi_{max}$	Preferred for high power levels by many designers
Half-bridge	V_{in}	Full variation	Capacitive divider avoids any dc offset in flux. Preferred at moderate power levels by many designers
Single-ended	V_{in}	Variation only between 0 and $+\phi_{max}$	Less effective use of core, but only two transistors
Push-pull	$2V_{in}$	Full variation	Common-ground gate drives. Timing errors can bias the flux and drive the core into saturation
Clamp	$V_{in} + V_z$	Variation only between 0 and $+\phi_{max}$	Similar to catch winding circuit, except that energy in magnetizing inductance is lost

ratio control. Leakage inductance is a problem in the push-pull circuit, particularly since both transistors must be off to establish the portion of time when no energy is transferred from input to output. Snubbers are necessary parts of this converter. These issues are discussed at length in at least one text (Kassakian, Schlecht, and Verghese, 1991).

The boost converter also supports forward converter designs. A boost-derived push-pull forward converter is shown in Fig. 13.3.19. Like the boost converter, this circuit has an output higher than the input but now with a turns ratio. The operation differs from a buck-derived converter in an important way: both transistors must turn on to establish the time interval when no energy flows from input to output. In effect, each transistor has a *minimum* duty ratio of 50 percent, and control is provided by allowing the switching functions to overlap. The output duty ratio associated with each diode becomes $1 - D$, where D is the duty ratio of one of the transistors over a full switching period. The output voltage is

$$V_{out} = \frac{N_2}{N_1} \frac{V_{in}}{2(1-D)} \tag{16}$$

The other forward converter arrangements also have boost-derived counterparts.

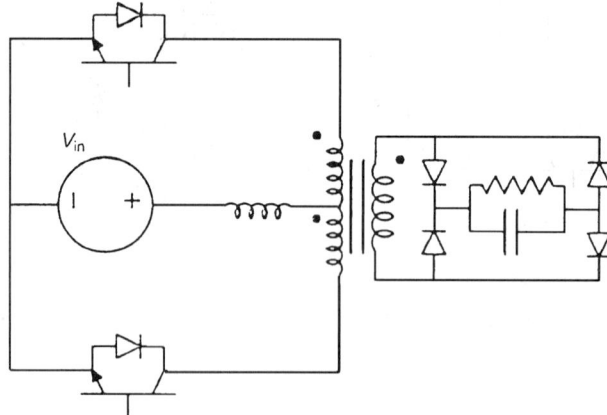

FIGURE 13.3.19 Boost-derived push-pull converter.

In each of the converters discussed so far, it has been assumed that energy storage components have been sufficiently large to be treated as approximate current or voltage sources. This is not always the case. If values of inductance or capacitance are chosen below a certain value, the current or voltage, respectively, will reach zero when energy is extracted. This creates *discontinuous mode* behavior in a dc-dc converter. The values of L and C sufficient to ensure that this does not occur are termed *critical inductance* and *critical capacitance*, respectively. The usual effect of subcritical inductance is that all switches on the converter turn off together for a time. Subcritical capacitance in a boost-buck converter creates times when all switches are on together.

In discontinuous mode, converter load regulation degrades, and closed-loop output control becomes essential. However, discontinuous mode can be helpful in certain situations. It implies fast response times since there is no extra time required for energy buildup. It provides an additional degree of freedom—the extra configuration when all switches are off or on—for control purposes.

Discontinuous mode behavior can be analyzed through the techniques of this subsection, with the additional constraint that all energy in the storage element is removed during each switching period. The literature (Mitchell, 1988; Mohan, Undeland, and Robbins, 1995) provides a detailed analysis, including computation of critical inductance and capacitance.

RESONANT DC-DC CONVERSION TECHNIQUES

The dc-dc converters examined thus far operate their switches in a *square-wave* or *hard-switched* mode. Switch action is strictly a function of time. Since switch action requires finite time, hard-switched operation produces significant switching loss. Resonant techniques for *soft switching* attempt to maintain voltages or currents at low values during switch commutation, thereby reducing losses. *Zero-current switching* (ZCS) or *zero-voltage switching* (ZVS) can be performed in dc converters by establishing resonant combinations. Resonant approaches for soft switching are discussed extensively in Kazimierczuk and Czarkowski (1995).

The SCR supports natural zero-current switching, since turn-off corresponds to a current zero crossing. A basic arrangement, given in Fig. 13.3.20, is often used as the basis for soft switching in transistor-based dc-dc converters as well as for inverters. Starting from rest, the top SCR is triggered. This applies a step dc voltage to the RLC set. If the quality factor of the RLC set is more than $\frac{1}{2}$, the current is underdamped, and will oscillate. When the current swings back to zero, the SCR will turn off with low loss. After that point, the lower SCR can be triggered for the negative half-cycle.

The SCR inverter represents a *series resonant switch* configuration and provides ZCS action. However, SCRs are not appropriate for high-frequency dc-dc conversion because of their long switching times and control limitations. The circuit of Fig. 13.3.21 shows an interesting arrangement for dc-dc conversion, similar to the SCR circuit, but based on a fast MOSFET. In this case, an inductor and a capacitor have been added to a standard buck converter to alter the transistor action. Circuit behavior will depend on the relative values.

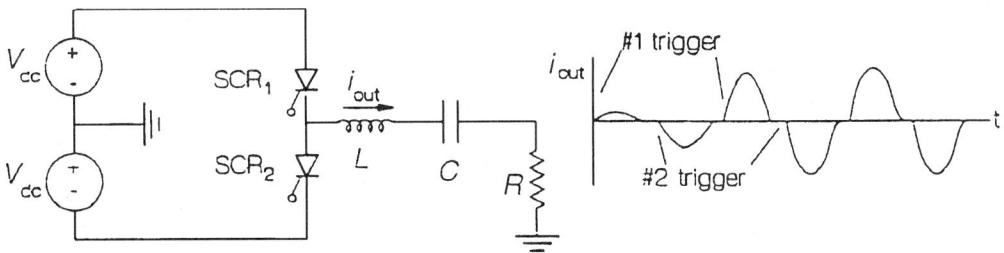

FIGURE 13.3.20   SCR soft-switching inverter.   (From Krein (1998), copyright © 1998 Oxford University Press, Inc., U.S.; used by permission.)

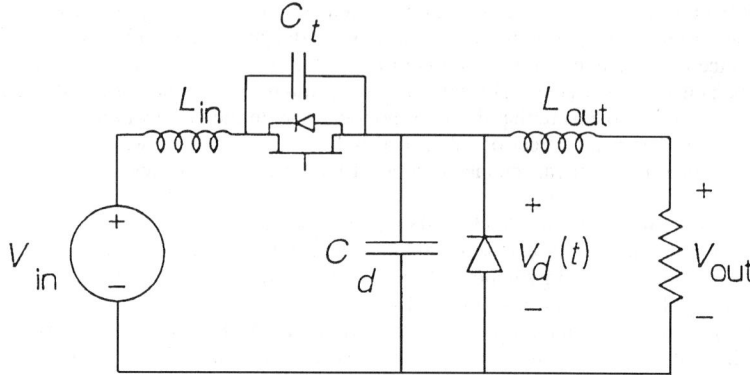

FIGURE 13.3.21   A soft-switching arrangement for a buck converter.

If the capacitor C_t is large, its behavior during the transistor's off interval will introduce opportunities for resonant switching. In this case, the basic circuit action is as follows:

- When the transistor turns off, the input voltage excites the pair L_{in} and C_t. The input inductor current begins to oscillate with the capacitor voltage. The capacitor can be used to keep the transistor voltage low as it turns off.
- The capacitor voltage swings well above V_{in}, and the main output diode turns on.

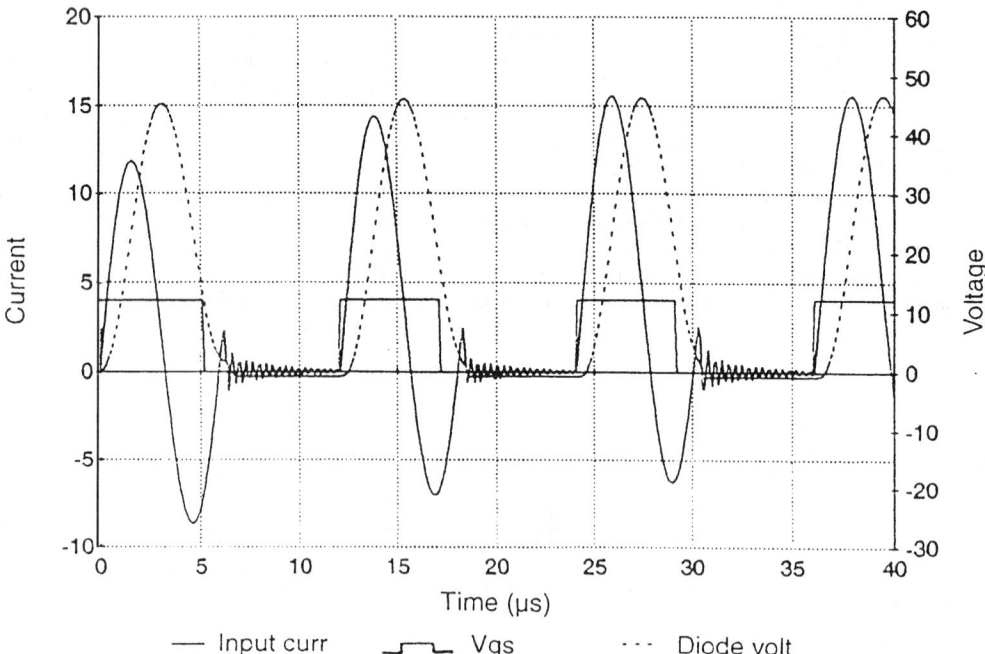

FIGURE 13.3.22   Input current and diode voltage in a resonant buck converter.   (From Krein (1998), copyright © 1998 Oxford University Press. Inc., U.S.; used by permission.)

- The capacitor voltage swings back down. There might be an opportunity for zero-voltage turn-on of the transistor when the capacitor voltage swings back to zero.

This represents ZVS action.

When the transistor is on and the main diode is off, the input inductor forms a resonant pair with C_d. This pair provides an opportunity for zero-current switching at transistor turn-off, very much like the zero-current switch action in the SCR inverter. The action is as follows in this case:

- When the transistor is turned on, L_{in} limits the rate of rise of current. The diode remains on initially, and the current builds up linearly because V_{in} appears across the inductor.
- When the current arises to the level I_{out}, the diode shuts off and the transistor carries the full current. The pair L_{in} and C_d form a resonant LC pair, and the current oscillates.
- The current rises above I_{out} because of resonant action, but then swings back down toward the origin.
- When the current swings negative, the transistor's reverse body diode begins to conduct, and the gate signal can be shut off. When the current tries to swing positive again, the switch will turn off.
- The transistor on-time is determined by the resonant frequency and the average output current.

Figure 13.3.22 shows the input current and main diode voltage for a choice of parameters that gives ZCS action in the circuit of Fig. 13.3.21.

Resonant action changes the basic control characteristics substantially. The gate control in both ZVS and ZCS circuits must be properly synchronized to match the desired resonance characteristics. This means that pulse-width modulation is not a useful control option. Instead, resonant dc converters are adjusted by changing the switching frequency—in effect setting the portion of time during which resonant action is permitted. In a ZCS circuit, for example, the average output voltage can be reduced by dropping the gate pulse frequency.

In general, ZCS or ZVS action is very beneficial for loss reduction. Lower switching losses permit higher switching frequencies, which in turn allow smaller energy storage elements to be used for converter design. Without resonant action, it is difficult to operate a dc-dc converter above perhaps 1 MHz. Resonant designs have been tested to frequencies as high as 10 MHz (Tabisz, Gradzki, and Lee, 1989). Designs even up to 100 MHz have been considered for aerospace applications. In principle, resonance provides size reduction of more than an order of magnitude compared to the best nonresonant designs. However, there is one important drawback: The oscillatory behavior substantially increases the on-state currents and off-state voltages that a switch must handle. Under some circumstances, the switching loss improvements of resonance are offset by the extra on-state losses caused by current overshoot. There are magnetic techniques to help mitigate this issue (Erickson, Hernandez, and Witulski, 1989), but they add complexity to the overall conversion system. The switching loss trade-offs tend to favor resonant designs at relatively low voltage and current levels. More sophisticated resonant design approaches, based on Class E methods (Kasimierczuk and Czarkowski, 1995), can further reduce losses.

Example: Input-output relationships in a ZCS dc-dc converter.
Let us explore ZCS switching in a dc-dc converter and analyze the results. The approach in this example follows an analysis in Kassakian, Schlecht, and Verghese (1991). The circuit in Fig. 13.3.21 is the focus, with C_t selected to be small and L_{out} selected to be large. Parameters are $V_{in} = 24$ V, $C_t = 200$ pF, $L_{in} = 2$ μH, $C_d = 0.5$ μF, $L_{out} = 50$ μH, and a 10-Ω load in parallel with an 8 μF filter capacitor. The FET is supplied with a 5-μs pulse with a period of 12 μs.

In periodic steady-state operation, the inductor L_{out} will carry a substantial current. The output time constant is long enough to ensure that the current will not change very much. As a result, the output inductor can be modelled as a current source, with value I_{out}. The diode provides a current path while the transistor is off. Consider the moment at which the transistor turns on. Since the current in L_{in} cannot change instantly, the diode remains on for a time while the input current rises. We have

$$i_{in}(t) = \frac{V_{in}}{L_{in}} t$$

$$(17)$$

until the current i_{in} reaches the value I_{out}. At the moment $t_{on} = I_{out}L_{in}/V_{in}$, the diode current reaches zero and the diode turns off. The input circuit becomes an undamped resonant tank determined by L_{in} and C_d. For C_d, circuit laws require

$$V_{in} - v_{Cd}(t) - L_{in}C_d\ddot{v}_{Cd}(t) = 0, \qquad v_{Cd}(t_{on}) = 0, \qquad \dot{v}_{Cd}(t_{on}) = 0 \tag{18}$$

This has the solution

$$v_{Cd}(t) = V_{in}\{1 - \cos[\omega_r(t - t_{on})]\}, \qquad \text{where } \omega_r = \frac{1}{\sqrt{L_{in}C_d}} \tag{19}$$

For the input current, the corresponding solution is

$$i_{in}(t) = I_{out} + \frac{V_{in}\sin[\omega_r(t - t_{on})]}{Z_c}, \qquad \text{where } Z_c = \sqrt{\frac{L_{in}}{C_d}} \tag{20}$$

With the selected parameters, $Z_c = 2\ \Omega$ and $\omega_r = 10^6$ rad/s, corresponding to about 160 kHz. The inductor current will cross zero again a bit more than one half-period after t_{on}. In this example, I_{out} might be on the order of 1 A, so t_{on} corresponds to only about 83 ns. The half-period of the resonant ring signal will be about 3.2 μs. Therefore, a 5-μs gate pulse should ensure that the transistor remains on until the zero crossing point.

When the current crosses zero, the FET turns off, but its reverse body diode turns on and maintains negative flow for approximately another half-cycle, at approximately $t = 6.4\ \mu$s. Since the gate signal is removed between the zero crossings, the FET and its diode will both turn off at the second zero crossing. Figure 13.3.22 is a SPICE simulation for these circuit parameters. The ZCS action should be clear: since the gate pulse is removed while the FET's reverse body diode is active, the complete FET will shut off at a rising current zero crossing. The shut-off point t_{off} is determined by

$$0 = I_{out} + \frac{V_{in}\sin[\omega_r(t_{off} - t_{on})]}{Z_c}, \qquad \omega_r(t_{off} - t_{on}) = \sin^{-1}\left(\frac{-I_{out}Z}{V_{in}}\right) \tag{21}$$

In solving this expression, it is crucial to be careful about the quadrant for $\sin^{-1}(x)$. The rising zero crossing is sought.

Once the FET is off, capacitor C_d carries the full current I_{out}. The voltage $v_{Cd}(t)$ will fall quickly with slope $-I_{out}/C_d$ until the diode becomes forward biased and turns on. The voltage $v_{Cd}(t)$ is of special interest, since the output is $V_{out} = \langle v_{Cd}(t)\rangle$. The average is

$$\langle v_{Cd}\rangle = \frac{1}{T}\left(\int_{t_{on}}^{t_{off}} V_{in}(1 - \cos[\omega_r(t - t_{on})])\,dt + \int_{t_{off}}^{t_{(\text{diode on})}} v_{Cd}(t_{off}) - \frac{I_{out}}{C_d}t\,dt\right) \tag{22}$$

The second integral is a triangular area $\frac{1}{2}v_{Cd}(t_{off})^2(C_d/I_{out})$. For $I_{out} \approx 1$ A, the time $t_{off} - t_{on}$ can be found from Eq. (21) to be 6.20 μs. The value $v_{Cd}(t_{off})$ is therefore 83 mV. The average value is $V_{out} = 12.6$ V. This corresponds to $I_{out} = 1.26$ A. In this example, the average value comes out very close to

$$V_{out} = \frac{4.8 \times 10^{-6}\pi}{T} \tag{23}$$

for $T > 6.4\ \mu$s. The solution comes out quite evenly because the current zero-crossing times nearly match those of the resonant sine wave.

In the ZCS circuit, the on time of the transistor is determined by resonant action, provided the gate pulse turns off during a time window when reverse current is flowing through the device's diode. The gate pulses need to have fixed duration to tune the circuit, but the pulse period can be altered to adjust the output voltage.

BIBLIOGRAPHY

Baliga, B. J., "Modern Power Devices," Wiley, 1987.

Baliga, B. J., "Power Semiconductor Devices," PWS, 1966.

Chryssis, G. C., "High-Frequency Switching Power Supplies," McGraw-Hill, 1989.

Erickson, R. W., A. F. Hernandez, and A. F. Witulski, "A nonlinear resonant switch," *IEEE Trans. Power Electronics*, Vol. 4, No. 2, pp. 242–252, 1989.

Erikson, R. W., and D. Maksimovich, *"Fundamentals of Power Electronics,"* 2nd ed., Kluwer Academic Publishers, 2001.

Hower, P. L., "Power semiconductor devices: An overview," *IEEE Proc.,* Vol. 82, pp. 1194–1214, 1994.

Hudgins, J. L., "A review of modern power semiconductor devices," *Microelect. J.,* Vol. 24, pp. 41–54, 1993.

Kassakian, J. G., M. F. Schlecht, and G. C. Verghese, "Principles of Power Electronics," Addison-Wesley, 1991.

Kazimierczuk, M. K., and D. Czarkowski, "Resonant Power Converters." Wiley, 1995.

Krein, P. T., "Elements of Power Electronics," Oxford University Press, 1998. Portions used by permission.

Massey, R. P., and E. C. Snyder, "High voltage single-ended dc-dc converter," *Record, IEEE Power Electronics Specialists Conf.,* pp. 156–159, 1977.

Middlebrook, R. D., "Modeling current-programmed buck and boost regulators," *IEEE Trans. Power Electronics*, Vol. 4, No. 1, pp. 36–52, 1989.

Middlebrook, R. D., and S. Ćuk, "A new optimum topology switching dc-to-dc converter,*" Record, IEEE Power Electronics Specialists Conf.,* pp. 160–179, 1977.

Mitchell, D. M., "Dc-Dc Switching Regulator Analysis," McGraw-Hill, 1988.

Mohan, N., T. M. Undeland, and W. P. Robbins, "Power Electronics: Converters, Applications and Design," 2nd ed., Wiley, 1995.

Severns, R. P., and E. J. Bloom, "Modern dc-to-dc Switchmode Power Converter Circuits," Van Nostrand Reinhold, 1985.

Tabisz, W. A., P. M. Gradzki, and F. C. Y. Lee, "Zero-voltage-switched quasi-resonant buck and flyback converters—experimental results at 10 MHz," *IEEE Trans. Power Electronics*, Vol. 4, pp. 194–204, 1989.

Vithayathil, J., "Power Electronics: Principles and Applications," McGraw-Hill, 1995.

CHAPTER 13.4
INVERTERS

David A. Torrey

INTRODUCTION

Inverters are used to convert dc into ac. This is accomplished through alternating application of the source to the load, achieved through proper use of controllable switches. This section reviews the basic principles of inverter circuits and their control. Four major applications of inverter circuits are also reviewed.

Both voltage- and current-source inverters are used in practice. The trend, however, is to use voltage-source inverters for the vast majority of applications. Current-source inverters are still used at extremely high power levels, though voltage-source inverters are gradually filling even these applications. Because of the dominance of voltage-source inverters, this section focuses exclusively on this type of inverter.

There are many issues involved in the design of an inverter. The more prominent issues involve the interactions among the power circuit, the source, the load, and the control. Other subtle issues involve the control of parasitics and the protection of controllable switches through the use of snubber and clamp circuits, and the juxtaposition of controller speed with the desire for increased switching frequency, while maintaining high efficiency.

The technical literature contains abundant information on inverters. A set of technical papers is found in Bose (1992). In addition to technical papers, most power electronics textbooks have a section on inverters (Mohan, Undeland, and Robbins, 1995; Kassakian, Schlecht, and Verghese, 1991; Krein, 1998).

AN INVERTER PHASE-LEG

An inverter phase-leg is shown in Fig. 13.4.1. It comprises two fully controllable switches and two diodes in antiparallel to the controllable switches. This phase-leg is placed in parallel with a voltage source. The center of the phase-leg is taken to the load. The basic circuit shown in Fig. 13.4.1 is usually augmented with a snubber circuit or clamp to shape the switching locus of the controllable switches. Insulated gate bipolar transistors (IGBTs) are shown as the controllable switches in Fig. 13.4.1. While the IGBT finds significant application in inverters, any fully controllable device, such as an FET or a GTO, may be used in its place; see Chapter 13.1 for a description of the fully controllable switches that can be used in an inverter.

Basic Principles.　In the phase-leg of Fig. 13.4.1, there are two restrictions on the use of the controllable switches. First, at most one controllable switch may be conducting at any time. The dc supply is shorted if both switches are conducting. In practice, one switch is turned off before the other is turned on. This blanking time, also known as dead time, compensates for the tendency of power devices to turn on faster than they turn off. Second, at least one controllable switch (or an associated diode) must be on at all times if the load current is to be nonzero.

If the upper switch is conducting, the load is connected to the positive side of V_{dc}. If the lower switch is conducting, the load is connected to the negative side of V_{dc}. It follows that the voltage applied to the load will, on

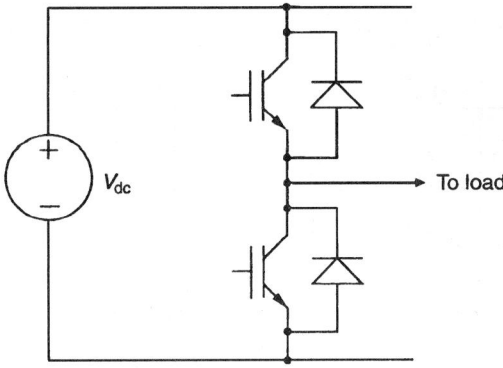

FIGURE 13.4.1   An inverter phase-leg.

average, fall somewhere between 0 and V_{dc}. When the phase-leg of Fig. 13.4.1 is used with one or more additional phase legs, the load voltage can be made to alternate. The details of how it alternates is the responsibility of the controller.

The peak voltage seen by each switch is the total dc voltage across the phase-leg. This is determined by recognizing that one of the two switches is always conducting. The peak current that must be supported by each switch is the peak current of the load. Under balanced control of the two switches, each switch must support the same peak current; each diode must support the same peak current as the switches.

In an effort to improve the efficiency and spectral performance of inverters, the use of inverters with a resonant dc link has been reported in the technical literature (Divan, 1989; Murai and Lipo, 1988). Through periodic resonance of the dc bus voltage to zero, or the dc bus current to zero, the inverter switches can change states in synchronism with these zero crossings in order to reduce the switching losses in the power devices. Figure 13.4.2 shows a schematic for the basic resonant dc link converter (Divan, 1989). The resonance of L_r and C_r forces the bus voltage (the voltage applied to the controllable switches) to swing between zero and $2V_{dc}$. The inverter switches change states when the voltage across C_r is zero. It is often necessary to hold the bus voltage at zero for a brief time to ensure that sufficient energy has been put into L_r to force resonance back to zero voltage across C_r. The bus can be clamped at zero voltage by turning on both switches in an inverter phase-leg. One drawback of the resonant dc link is the increased voltage or current imposed on the power devices. Auxiliary clamp circuits have been implemented to minimize this drawback (Divan and Skibinski, 1989; He and Mohan, 1991; Simonelli and Torrey, 1994). The control of a resonant link inverter is complicated by the simultaneous need to manage energy flow to the load while managing energy in the resonant link.

Snubber Circuits and Clamps.   Snubber circuits are used to control the voltage across and the current through a controllable switch as that device is turning on or off. A complete snubber will typically limit the rate of rise in current as the device is turning on and limit the rate of rise in voltage as the device is turning off. Additional circuit components are used to accomplish this shaping of the switching locus. Figure 13.4.3 shows three snubber circuits that are used with inverter legs, one of which has auxiliary switches (McMurray, 1987; McMurray, 1989;

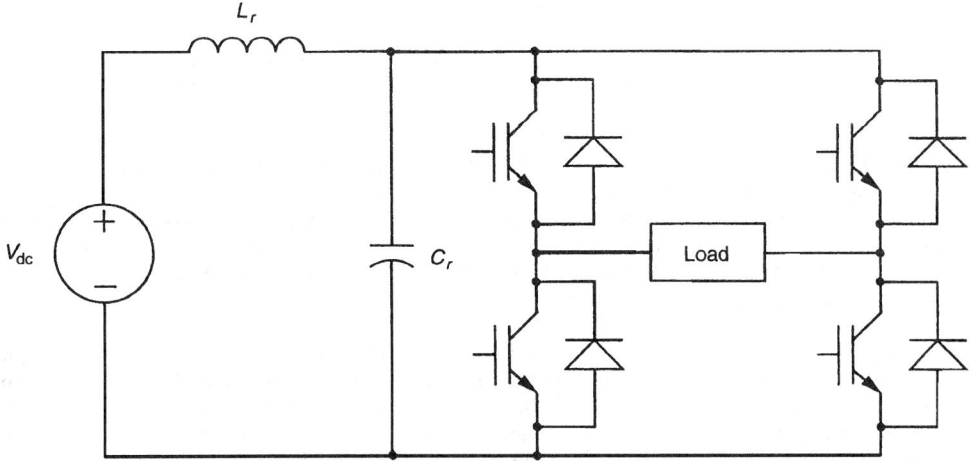

FIGURE 13.4.2   A basic resonant dc link inverter system.

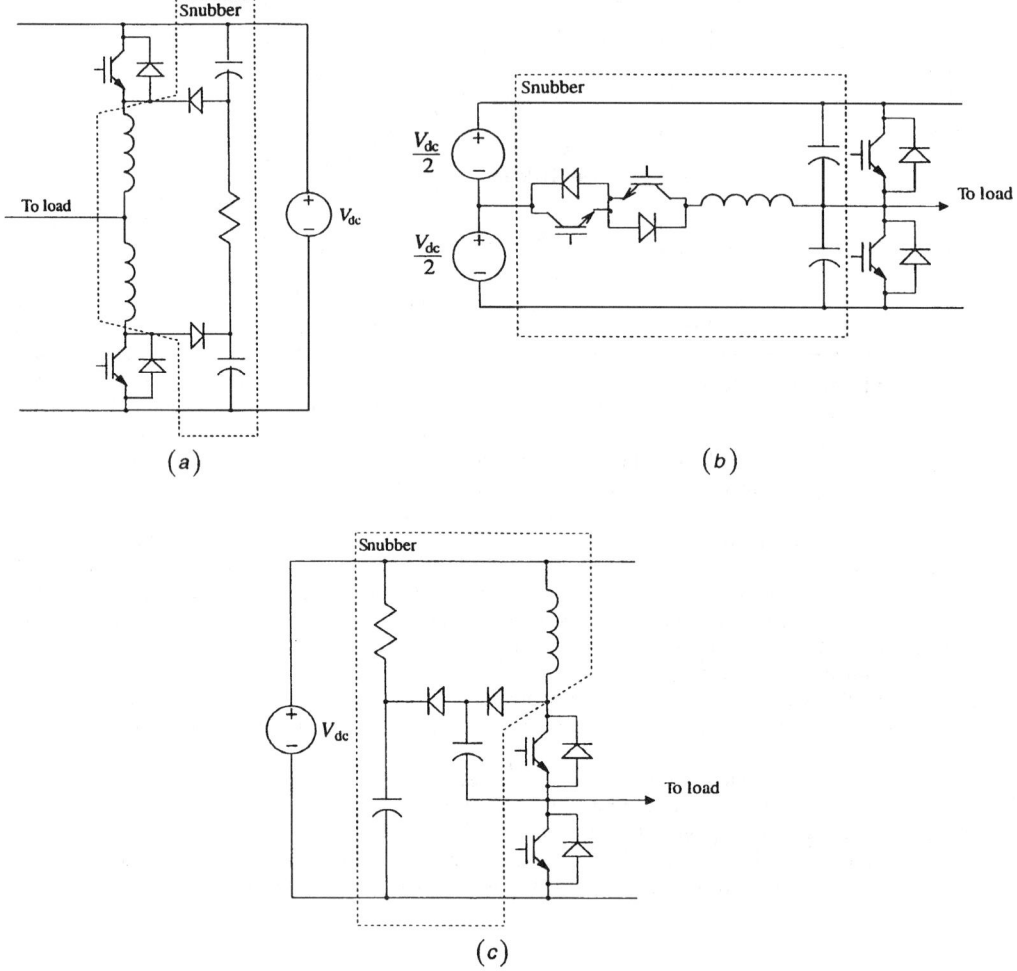

FIGURE 13.4.3 A number of snubber circuits for an inverter phase-leg: (*a*) a McMurray snubber (McMurray, 1987); (*b*) a resonant snubber with auxilliary switches (McMurray, 1989); and (*c*) the Undeland snubber (Undeland, 1976).

Undeland, 1976). Snubber circuits become increasingly important as the power rating of the inverter increases, where the additional cost of small auxiliary devices is justified by improved efficiency and spectral performance.

Clamps differ from snubbers in that a clamp is used only to limit a switch variable, usually the voltage, to some maximum value. The clamp does not dictate how quickly this maximum value is attained. Figure 13.4.4 shows three clamp circuits that are commonly used with inverter legs. The clamp circuits generally become more complex with increasing power level.

Interfacing to Controllable Switches. The interface to the controllable switches within the inverter phase-leg of Fig. 13.4.2 requires careful attention. Power semiconductor devices are generally controlled by manipulating the control terminal, usually relative to one of the power terminals. For example, Figure 13.4.2 shows insulated gate bipolar transistors (IGBTs) as the controllable switches. These devices are turned on and off through the voltage level applied to the gate relative to the emitter. Because the emitter of the upper IGBT moves around, the control of the upper IGBT must accommodate this movement. The emitter voltage of the upper IGBT moves from the positive side of the dc bus when the IGBT is conducting to the negative side of the dc bus when

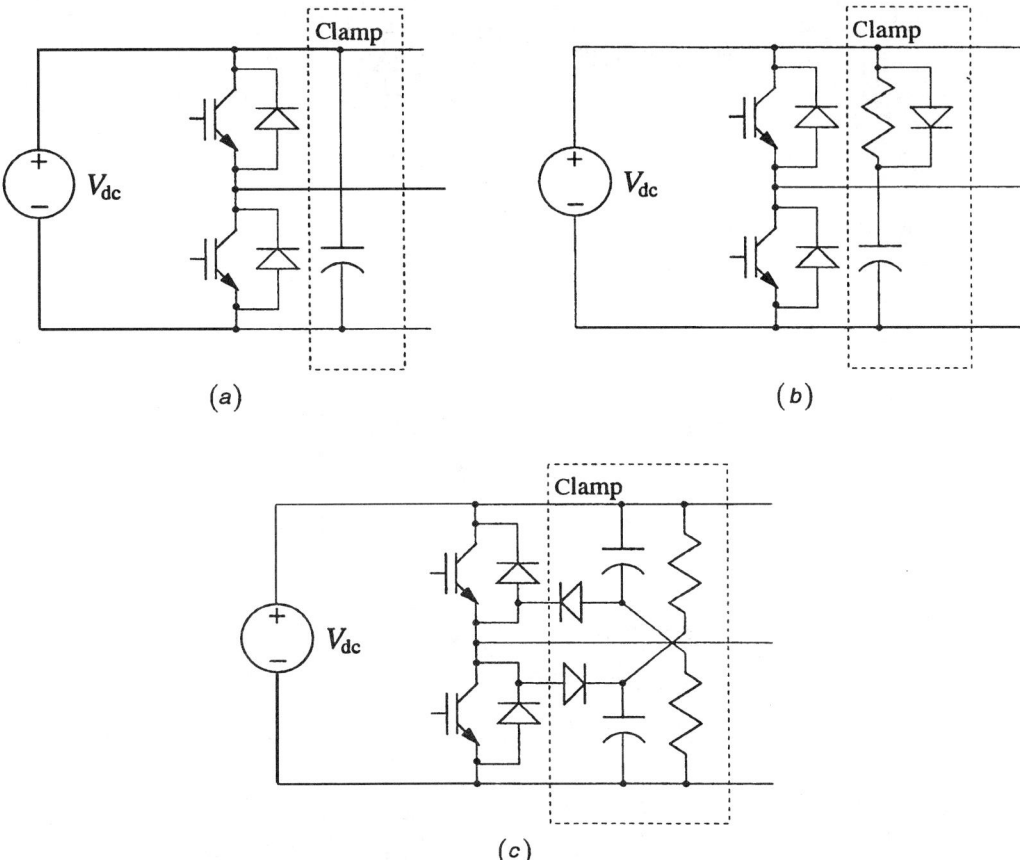

FIGURE 13.4.4 Clamp circuits for an inverter phase-leg: (*a*) low power (~ 50 A); (*b*) medium power (~ 200 A); and (*c*) high power (~300 A).

the lower IGBT is conducting. The circuit responsible for turning the power semiconductor on and off as commanded by a controller is usually known as a gate-drive circuit. Some gate-drive circuits incorporate protection mechanisms for over current or over temperature.

There are a number of approaches that are used to control the semiconductor devices within the inverter. The choice among these approaches is often dictated by the power levels involved, the switching frequency supported by the switches, and the preference of the designer, among others. It is possible to purchase high-voltage integrated circuits (HVICs) that perform the level shifting necessary to take a logic signal referenced to the negative side of the dc bus and control the upper controllable switch in the phase-leg. This approach is generally limited to applications where the controllable switches do not require a negative bias to hold them in the blocking state. High-frequency applications may use transformer-coupled gate drives that are insensitive to the common-mode voltage between the primary and secondary. This approach runs into problems at lower frequencies because the size of the transformer core begins to get large. High-power applications may use optocouplers to optically couple the control information to the gate drive, where the gate drive is supported by an isolated power supply. Often the power supply is the dominant factor in the overall cost of the gate drive.

Figure 13.4.5*a* shows how an HVIC is interfaced to a phase-leg. The capacitors are used as local power supplies for the upper and lower gate drives. In this implementation, the upper capacitor is charged through the diode while the lower switch is conducting. Figure 13.4.5*b* shows a transformer-coupled gate drive for a high-frequency application (Internatonal Rectifier). It is important to design a mechanism for resetting the core in

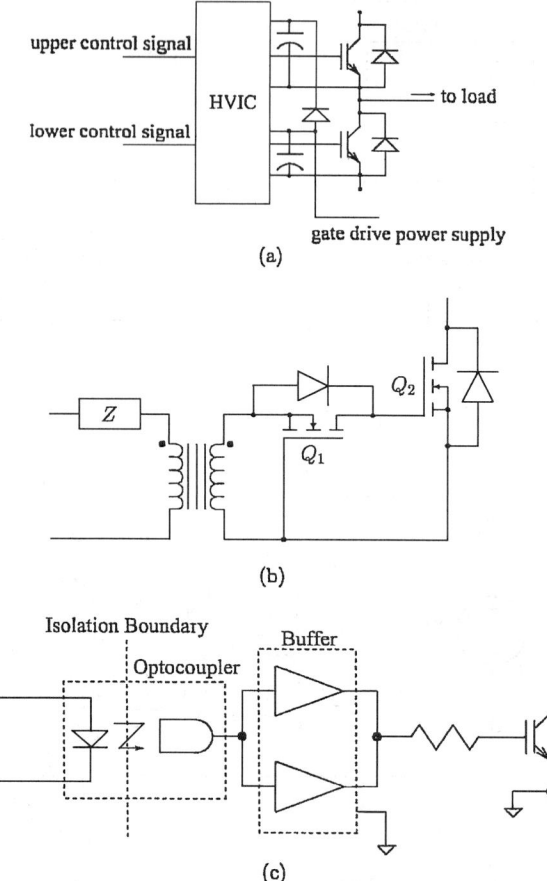

FIGURE 13.4.5   Three common approaches to interfacing to controllable switches: (a) the use of an HVIC; (b) the use of transformer coupling; and (c) the use of an optocoupler.

a transformer-coupled gate drive. In Fig. 13.4.5*b*, the core is reset by driving the transformer primary with a bipolar voltage that provides sufficient volt-seconds to drive the transformer into saturation. This need for core reset may place unacceptable limitations on the duty ratio of the switches for some applications. Figure 13.4.5*c* shows the use of an optocoupler to provide isolation of the control signal going to the gate drive. The isolated power supply required to support the use of the optocoupler is not shown.

SINGLE-PHASE INVERTERS

There are two ways to form a single-phase inverter. The first way is shown in Fig. 13.4.6, where the phase leg of Fig. 13.4.1 is used to control the voltage applied to one side of the load. The other side of the load is connected to the common node of two voltage sources. The half-bridge inverter of Fig. 13.4.6 applies positive voltage to the load when the upper switch is conducting and negative voltage to the load when the lower switch is conducting. It is not possible for the half-bridge inverter to apply zero voltage to the load.

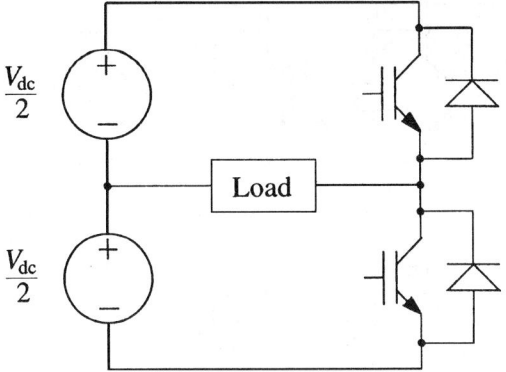

FIGURE 13.4.6   A single-phase inverter using one phase-leg
and two dc voltage sources.

A single-phase inverter is also formed by placing the load between two inverter phase-legs, as shown in
Fig. 13.4.7. This circuit is often referred to as a full- or H-bridge inverter. Through appropriate control of the
inverter switches, positive, negative, and zero voltage can be applied to the load. The zero voltage state is
achieved by having both upper switches, or both lower switches, conducting at the same time. Note that this
state requires one switch and one diode to be supporting the load current.

THREE-PHASE INVERTERS

A three-phase inverter is used to support both Δ- and Y-connected three-phase loads. The three-phase inverter
topology can be derived by using three single-phase full-bridge inverters, with each inverter supporting one
phase of the load. Upon careful examination of the resulting connection of the 12 controllable switches with
the load, it is seen that there are six redundant switches because phase-legs are connected in parallel.
Elimination of the six redundant switches yields the topology shown in Fig. 13.4.8.

There are six switches in the three-phase inverter topology of Fig. 13.4.8. Considering all combinations of
the switch states, there are seven possible voltages that may be applied to the load; the cases of all three upper
switches being closed and all three lower switches being closed are functionally indistinguishable. The
controller is responsible for applying the appropriate voltage to the load according to the method used for synthesizing the output voltage.

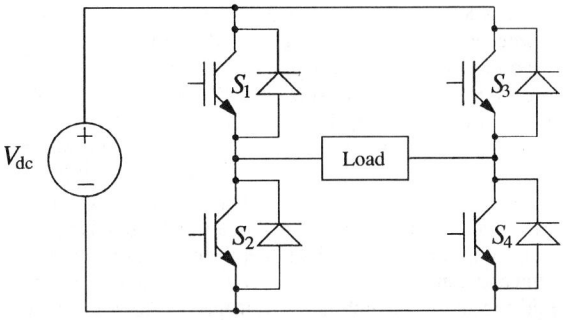

FIGURE 13.4.7   A single-phase inverter using two phase-legs and one
dc voltage source.

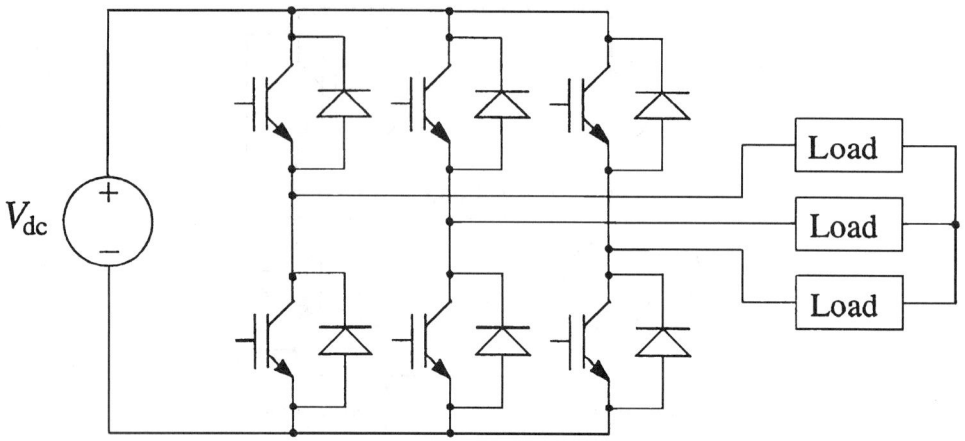

FIGURE 13.4.8 The three-phase inverter topology.

MULTILEVEL INVERTERS

The three-phase inverter of Fig. 13.4.8 applies one of two voltages to each output load terminal. The output voltage is V_{dc} when the upper switch is conducting, and it is zero when the lower switch is conducting. Accordingly, this inverter could be called a two-level inverter. Multilevel inverters are based on extending this concept and are becoming the inverter of choice for higher-voltage and higher-power applications. The discussion here focuses on a three-level inverter; extensions to five or more levels can be found in the technical literature (Lai and Peng, 1996; Peng, 2001).

Among their advantages, multilevel inverters allow the synthesis of voltage waveforms that have a lower harmonic content than a two-level inverter for the same switching frequency. This is because each output terminal is switched among at least three voltages, not just two. In addition, the input dc bus voltage can be higher because multiple devices are connected in series to support the full bus voltage.

Figure 13.4.9 shows one phase-leg of a three-level inverter. In the three-level inverter the dc bus is partitioned into two equal levels of $V_{dc}/2$. The four controllable switches with antiparallel diodes, S_1 through S_4, are connected in series to form a phase-leg. In addition, two steering diodes, D_5 and D_6, are used to support current flow to and from the midpoint of the dc bus. Additional phase-legs would be connected in parallel across the full dc bus.

Operation of four switches is used to connect the load output terminal to either V_{dc}, $V_{dc}/2$, or zero. Switches S_1 and S_2 are used to connect the load to V_{dc}, switches S_2 and S_3 are used to connect the load to $V_{dc}/2$, and switches S_3 and S_4 are used to connect the load to zero.

While switches S_1 and S_2 are conducting, diode D_6 ensures that the voltage across switch S_4 does not exceed $V_{dc}/2$. Similarly, when switches S_3 and S_4 are conducting, diode D_5 ensures that the voltage across switch S_1 does not exceed $V_{dc}/2$. While switches S_2 and S_3 are conducting, the voltage across both S_1 and S_4 is clamped at $V_{dc}/2$; diode D_5 and switch S_2 support positive load current, while diode D_6 and switch S_3 support negative load current.

A three-phase, three-level inverter provides substantially increased flexibility for voltage synthesis over the conventional three-phase inverter of Fig. 13.4.8. The three-phase inverter of Fig. 13.4.8 offers eight switch combinations that support seven different voltage combinations among the three output terminals. A three-phase, three-level inverter offers 27 switch combinations, supporting 19 different voltage combinations among the three output terminals. Further, the increased redundancy of certain voltages provides additional degrees of freedom in designing the voltage synthesis algorithm. These additional degrees of freedom could, for example, be used to minimize the common mode voltage between the three outputs.

Issues within the design and control of the multilevel inverter include the dynamic balancing of the voltages within each level of the dc bus and the switching patterns needed to best synthesize the desired output voltage. One would expect that symmetric operation of the phase-leg should be sufficient for maintaining balanced voltages across each level of the dc bus. The variation in capacitor values, however, will cause the midpoint of the

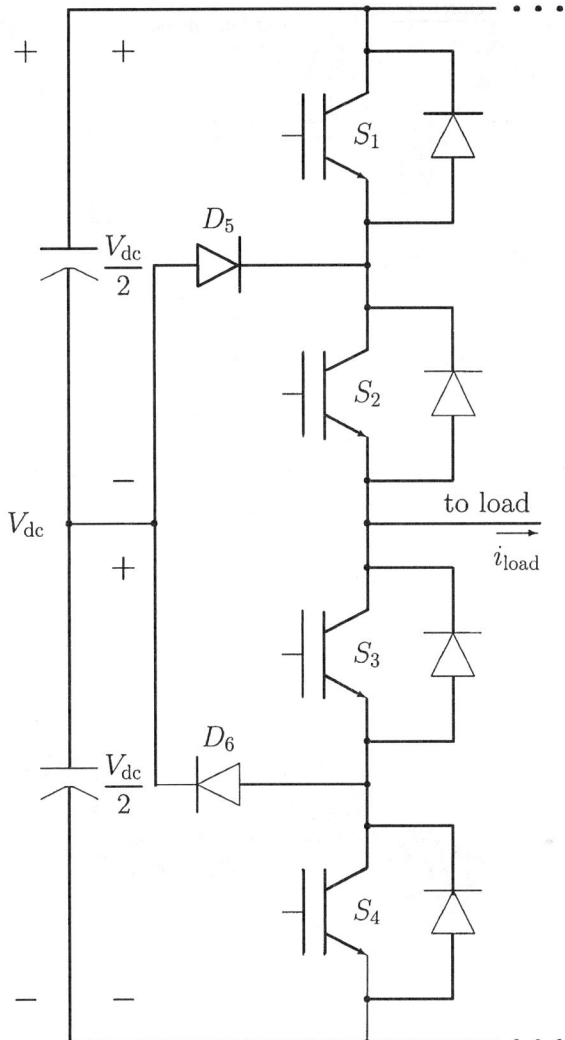

FIGURE 13.4.9   A phase-leg for a three-level inverter.

dc bus to move to a voltage other than $V_{dc}/2$ for symmetric load currents. This shift in voltage will have repercussions on the synthesis of the output voltage.

VOLTAGE WAVEFORM SYNTHESIS TECHNIQUES

There are three principal ways to synthesize the output voltage waveform in an inverter: harmonic elimination, harmonic cancellation, and pulse-width modulation. The synthesis technique that is applied is generally driven by consideration of the required output quality, the inverter power rating (which is closely tied to the speed of the controllable switches), the computational power of the available controller, and the acceptable cost of the inverter. This subsection reviews some of the common techniques used to synthesize the inverter output voltage.

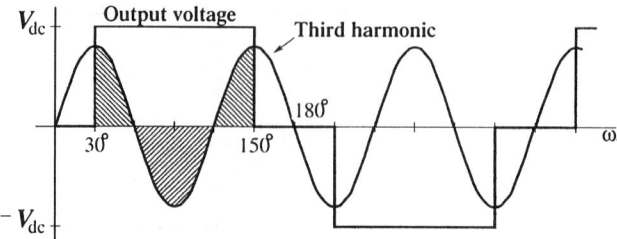

FIGURE 13.4.10 Elimination of the third harmonic.

Harmonic Elimination. Harmonic elimination implies that the output waveform shape is controlled to be free of specific harmonics through the selection of switch transitions (Patel and Hoft, 1973; Patel and Hoft, 1974). That is, the switches are controlled so that one or more harmonics are never generated. This is often accomplished by notching the output waveform. Examples of harmonic elimination are shown in Figs. 13.4.10 and 13.4.11, which respectively show the elimination of only the third harmonic and simultaneous elimination of the third and fifth harmonics from the output of a single-phase inverter.

As suggested by Figs. 13.4.10 and 13.4.11, additional switch transitions must be inserted in the output waveform for each harmonic that is to be eliminated. As the number of notches gets very large, the output voltage waveform begins to resemble something which could be produced by pulse-width modulation. Harmonic

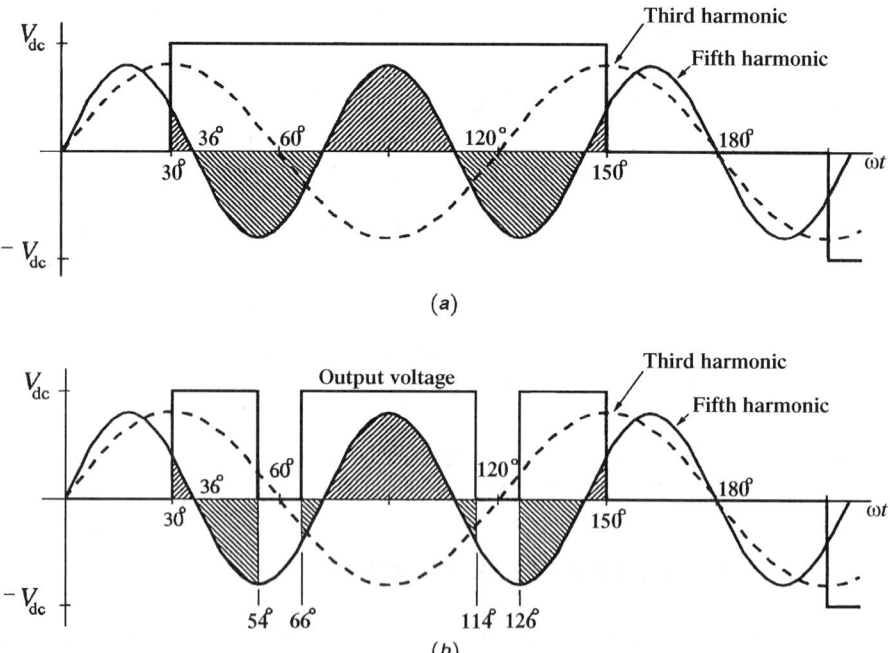

FIGURE 13.4.11 Simultaneous elimination of the third and fifth harmonics: (*a*) shows the fifth harmonic superimposed on the waveform of Fig. 13.4.10: (*b*) shows the switch transitions introduced to eliminate the fifth harmonic without reintroducing the third harmonic.

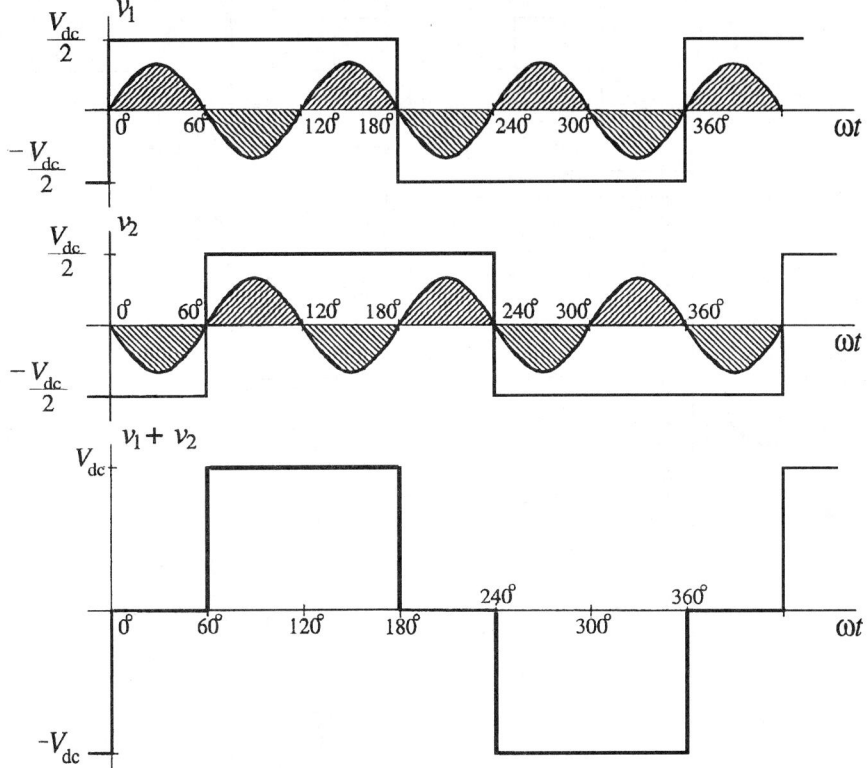

FIGURE 13.4.12   The superposition of two waveforms to cancel the third harmonic.

elimination is sometimes referred to as programmed PWM because the switching angles of the output voltage are programmed according to the intended harmonic content (Enjeti, Ziogas, and Lindsay, 1985).

Harmonic Cancellation.   Harmonic cancellation uses the superposition of two or more waveforms to cancel undesired harmonics (Kassakian, Schlecht, and Verghese, 1991). Figure 13.4.12 shows how two waveforms that contain the third harmonic may be phase-shifted and superimposed in order to create a waveform that is free of the third harmonic. The circuit of Fig. 13.4.13 can be used to synthesize the waveform of Fig. 13.4.12.

By combining harmonic cancellation and harmonic elimination, it is possible to create relatively high-quality voltage waveforms. This quality comes at the expense of a more complicated circuit. This additional complexity may be warranted depending on the power level and the specified quality.

Pulse-Width Modulation.   Pulse-width modulation (PWM) is a method of voltage synthesis through which high-frequency voltage pulses are applied to the inverter load (Holtz, 1994). The width of the pulses are made to vary at the desired frequency of the output voltage. Successful application of PWM generally involves a wide frequency separation between the carrier frequency and the modulation frequency. This frequency separation moves the distortion in the output voltage to high frequencies, thereby simplifying the required filtering. This subsection reviews two of the more common techniques used for synthesizing voltage waveforms using modulation.

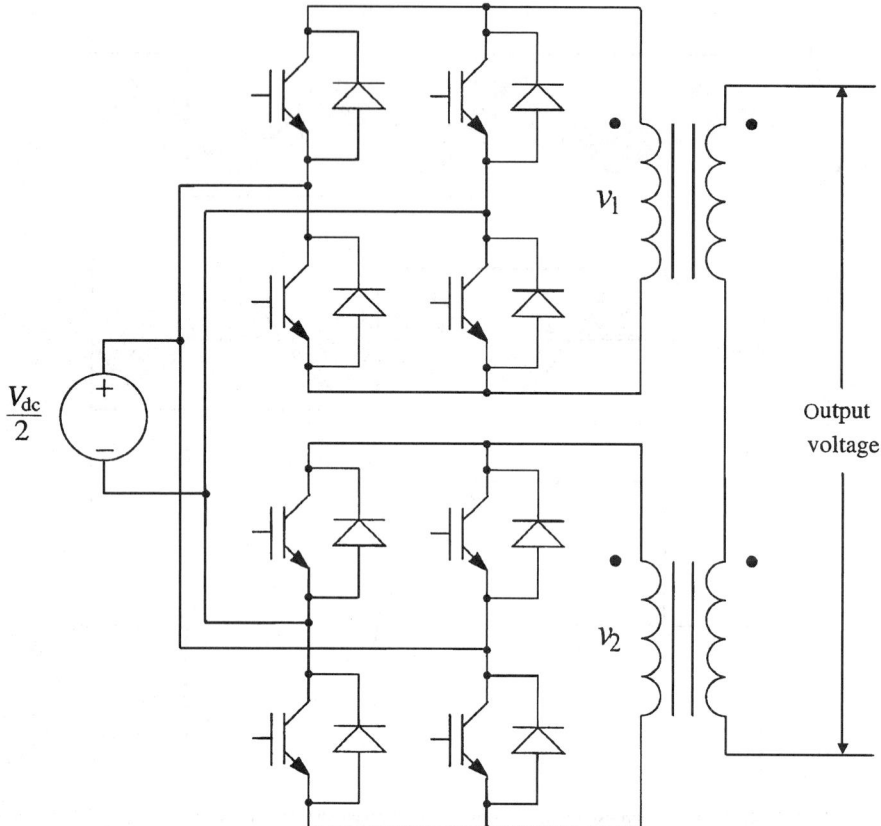

FIGURE 13.4.13   A circuit capable of synthesizing the waveforms of Fig. 13.4.12.

Sinusoidal PWM.   Sinusoidal PWM implies that the pulse widths of the output voltage are distributed sinusoidally. The pulse widths are generally determined by comparing a sinusoidal reference waveform with a triangular waveform. The sinusoidal waveform sets the modulation (output) frequency and the triangular waveform sets the switching frequency. Sinusoidal PWM is routinely applied to single- and three-phase inverters.

In a single-phase inverter, the implementation of sinusoidal PWM depends on whether or not both phase-legs are operated at high frequency. Referring to Fig. 13.4.7, we see that it is not necessary for both phase-legs to operate at high frequency. We could, for example, operate switches S_1 and S_2 at high frequency to control the shape of the voltage, while switches S_3 and S_4 are operated at the frequency of the reference sinusoid to dictate the polarity of the output voltage. One advantage of this approach is that the inverter is more efficient because only two of the switches are operated at high frequency. Figure 13.4.14 shows how the sinusoidal pulse widths are created by a comparator and a unipolar triangular carrier (Kassakian, Schlecht, and Verghese, 1991). An alternative approach is to use switches S_2 and S_4 to control the polarity of the output voltage. While switch S_4 is conducting, switches S_1 and S_2 are used to control the shape of the voltage. Similarly, switches S_3 and S_4 are used to control the shape of the voltage, while switch S_2 is conducting. This approach tends to equalize the stress on the two phase-legs, and can simplify the control logic necessary to implement the PWM.

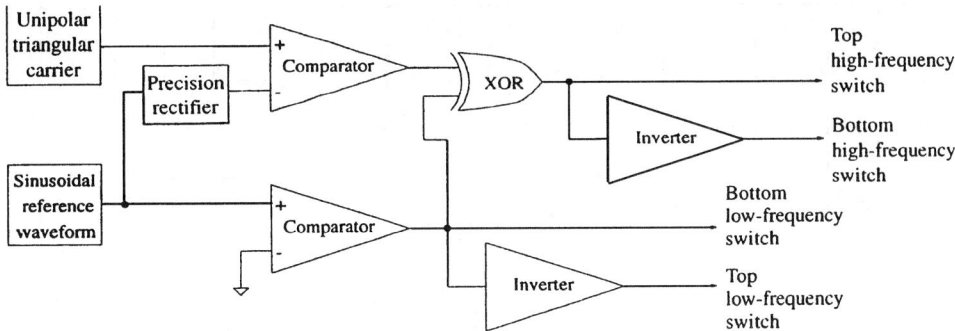

FIGURE 13.4.14 The generation of sinusoidally distributed pulse widths using a unipolar triangular carrier.

A second way of implementing sinusoidal PWM in a single-phase inverter is to operate both phase-legs at high frequency (Vithayathil, 1995). This method of control gives the same output voltage waveform as the high-frequency/low-frequency approach. The basic difference in control structures between the two is that the first method uses a unipolar triangular carrier, while the second way uses a bipolar triangular carrier. Figure 13.4.15 shows how the bipolar triangular carrier is used to create the sinusoidally distributed pulse widths.

Space-Vector PWM.   Space-vector modulation is a technique that is becoming the standard method for controlling the output voltage of three-phase inverters. The technique bears great similarity to the field-oriented control techniques which are applied to ac electric machines (Van der Broeck, Skedulny, and Stanke, 1988; Holtz, Lammert, and Lotzkat, 1986; Trzynadlowski and Legowski, 1994).

A balanced set of three-phase quantities can be transformed into direct and quadrature components through the transformation

$$\begin{bmatrix} x_d \\ x_q \end{bmatrix} = \sqrt{\frac{2}{3}} \cdot \begin{bmatrix} 1 & -1/2 & -1/2 \\ 0 & \sqrt{3}/2 & -\sqrt{3}/2 \end{bmatrix} \begin{bmatrix} x_a \\ x_b \\ x_c \end{bmatrix} \tag{1}$$

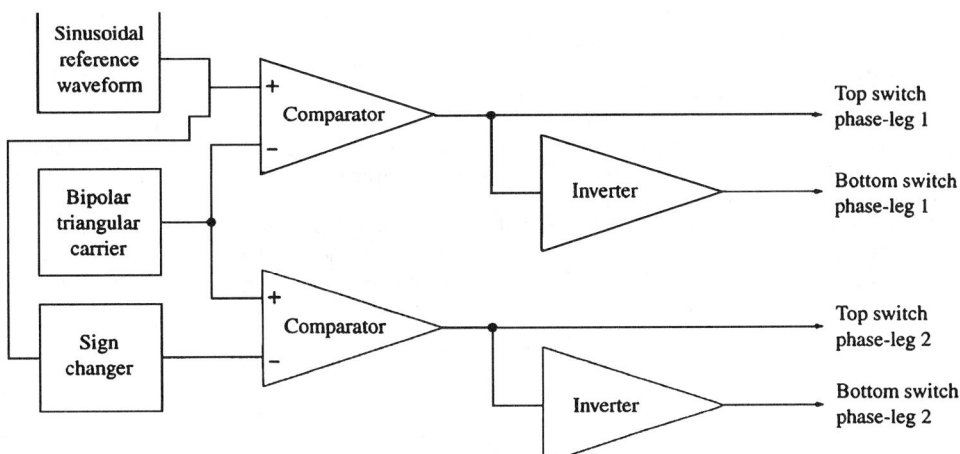

FIGURE 13.4.15 The generation of sinusoidally distributed pulse widths using a bipolar triangular carrier.

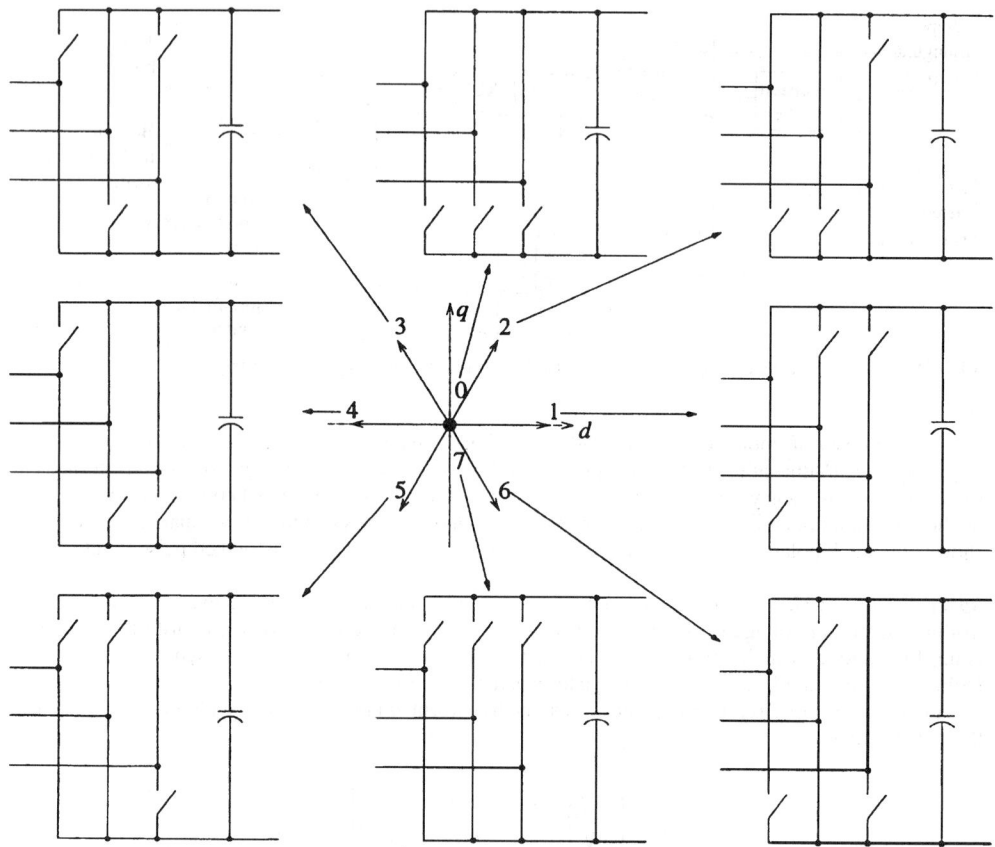

FIGURE 13.4.16 The eight space vectors that can be produced by the three-phase inverter of Fig. 13.4.8.

where x is a voltage or current. A similar transformation exists for taking direct and quadrature components back to phase quantities, though space-vector modulation does not need to use the inverse transformation.

Applying Eq. (1) to the three-phase inverter of Fig. 13.4.8, we see that each distinct switch state of the inverter corresponds with a different space vector. The zero vector can be produced by the electrically equivalent topologies of all upper switches conducting and all lower switches conducting. It is important to note that the six nonzero space vectors created by the inverter states are of the same magnitude ($\sqrt{2/3}V_{dc}$) and are symmetrically displaced. Figure 13.4.16 shows the connection between the three-phase inverter of Fig. 13.4.8 and the generation of the eight space vectors.

Any desired output voltage, up to the magnitude of $V_{dc}/\sqrt{2}$ may be synthesized by taking the three adjacent space vectors in proper proportion. Figure 13.4.17 shows how the desired voltage V^* is synthesized from the space vectors V_1, V_2, and V_0. Over one sampling interval, the duty ratios of V_1, V_2, and V_0 are, respectively,

$$d_1 = \frac{2/\sqrt{3}|V^*|\sin(60° - \gamma)}{\sqrt{2/3}V_{dc}} \tag{2}$$

$$d_2 = \frac{2/\sqrt{3}|V^*|\sin\gamma}{\sqrt{2/3}V_{dc}} \tag{3}$$

$$d_0 = 1 - d_1 - d_2 \tag{4}$$

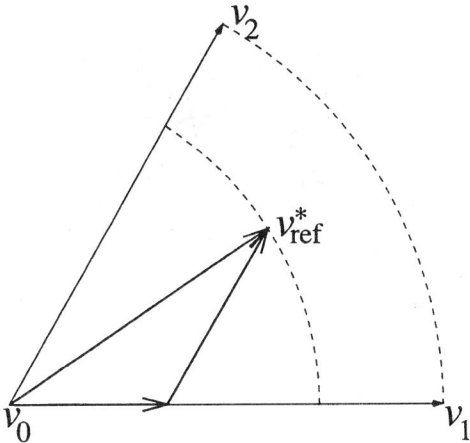

FIGURE 13.4.17 The synthesis of voltage V^* using space vector modulation.

The order used in implementing the space vectors is driven by the desire to minimize the number of switching operations. Careful examination of Figs. 13.4.16 and 13.4.17 reveals that within any of the six segments delimited by space vectors, the move from one vector to the next requires changing the state of only one switch. In practice, the switching in adjacent sampling intervals would apply the sequence $\cdots |V_1|V_2|V_0|V_0|V_2|V_1| \cdots$. Different approaches use different criteria for selecting the best implementation of V_0 (Trzynadlowski and Legowski, 1994).

Extension of the space-vector concept is possible with multilevel inverters (Holmes and McGrath, 2001; Tolbert, Peng, and Habetler, 2000). Multilevel inverters, however, offer a substantially greater number of space vectors from which to choose. With three-level inverters, for example, there are now 19 different space vectors. Additional levels would increase the number of space vectors still further. The number of redundant space vectors also increases in multilevel inverters, thereby offering additional degrees of freedom within the voltage synthesis algorithm. Figure 13.4.18 shows the space vectors that can be created by a three-phase three-level inverter based on the phase-leg of Fig. 13.4.12. The numbers adjacent to each space vector represent the switch configuration of phases a, b, and c, respectively. Space vectors with more than one set of numbers can be achieved with any of the switch combinations indicated. Referring to Fig. 13.4.12, a 0 indicates that switches S_3 and S_4 are connecting the output terminal to the negative side of the dc bus. Similarly, a 1 indicates switches S_2 and S_3 are connecting the output terminal to the midpoint of the dc bus. Finally, a 2 indicates that switches S_1 and S_2 are connecting the load terminal to the positive side of the dc bus.

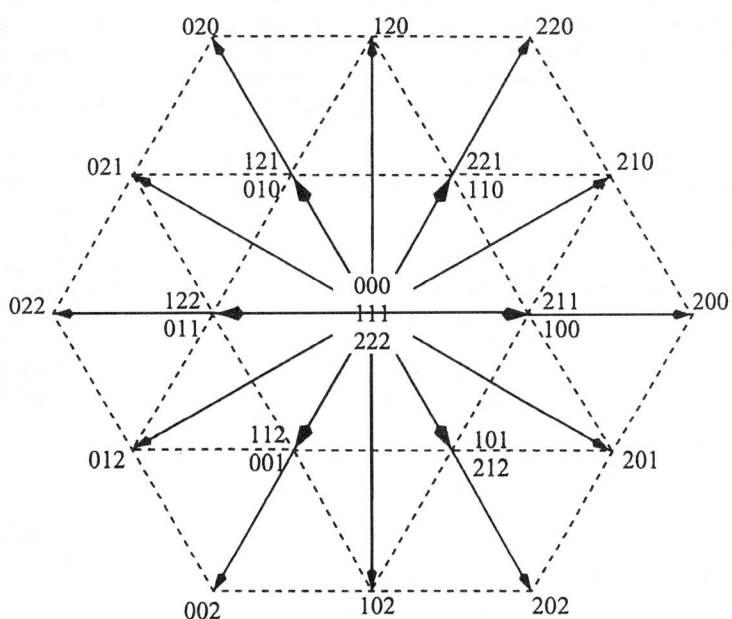

FIGURE 13.4.18 The achievable space vectors associated with a three-phase three-level inverter.

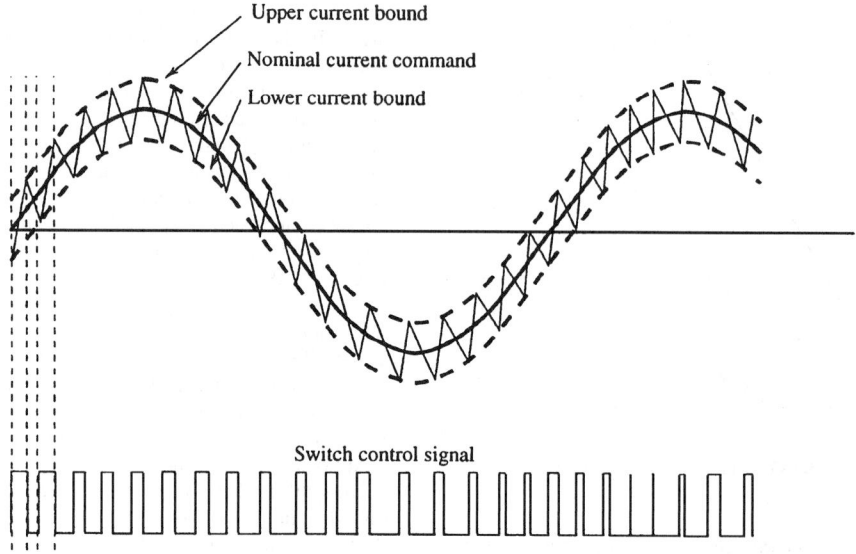

FIGURE 13.4.19   The principles of hysteretic current control.

CURRENT WAVEFORM SYNTHESIS TECHNIQUES

While voltage-source inverters always output a voltage waveform, there are many applications where the details of the voltage waveform are driven by the creation of a current with a specific shape. In this context, the controller is determining the state of each inverter switch based on how well the inverter output currents are tracking commanded output currents. While the PWM techniques of the previous subsection can often be applied to transform a voltage-source inverter into a controlled current source (Brod and Novotny, 1985; Habetler, 1993), there are some additional techniques that are useful in this type of operation. The control techniques described in this subsection are amenable to synthesizing current waveforms with inverters.

Hysteresis and Sliding-Mode Control.   Hysteresis and sliding-mode control are very similar in nature. In both of these control approaches, a reference current waveform is established and the switching of the inverter is tied to the relative location of the actual current and the reference waveform (Brod and Novotny, 1985; Bose, 1990; Slotine and Li, 1991; Torrey and Al-Zamel, 1995). Under hysteretic control, a hysteresis band is introduced around the reference waveform in order to limit the switching frequency. Figure. 13.4.19 illustrates the principles of hysteretic control. Sliding-mode control can be implemented in a manner which is indistinguishable from hysteretic control, or it can be implemented as shown in Fig. 13.4.20 where there is a known upper limit on the switching frequency.

A common problem with hysteretic control is that the switching frequency is not fixed and may vary widely over one cycle of the output. This can complicate the design of filters and may raise reliability concerns relative

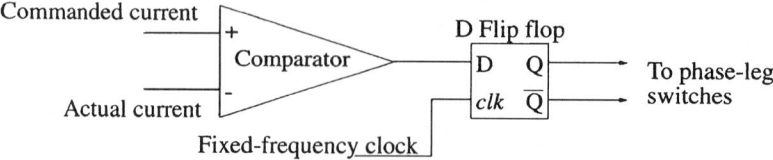

FIGURE 13.4.20   One method of implementing sliding-mode control.

to the safe operation of the switches. Fixing the switching frequency is possible (Kazerani, Ziogas, and Joos, 1991), at the expense of a hysteresis band that changes throughout each cycle of the output.

Predictive Current Regulation.   Predictive current regulation is similar to hysteresis and sliding-mode control in the establishment of a reference current and acceptable error bounds (Holtz, 1994; Wu, Dewan, and Slemon, 1991). The predictive controller uses a model of the system in conjunction with measurements of the system state in order to predict how long the next switching state is to be maintained so that the actual current remains within the established error bounds. In contrast to hysteresis and sliding-mode control, the predictive controller is always looking ahead one sampling interval into the future.

INVERTER APPLICATIONS

This subsection reviews four important applications of inverters: uninterruptible power supplies, motor drives, active power filters, and utility interfaces for distributed generation. Uninterruptible power supplies have become an extremely large market in support of the expanding use of personal computers and other critical loads. These systems are able to support computer operation in the face of unreliable utility power, thereby preventing the loss of data. Motor drives allow for adjustable speed operation of electric motors, thereby providing a better match between the motor output and the power required by the load. The proper application of adjustable speed drives can result in significant energy savings. The increasing application of active power filters and active power line conditioners is a reflection of increasing harmonic distortion on power systems, and the regulatory responses to this distortion. Distributed generation sources (fuel cells, solar photovoltaics, wind turbines, and microturbines) often function as sources of dc power, thereby requiring an inverter to deliver this power to the ac utility grid.

Uninterruptible Power Supplies.   An uninterruptible power supply (UPS) uses a battery to provide energy to a critical load in the event of a power system disturbance. There are two basic topologies used in UPS systems, as shown in the block diagrams of Figs. 13.4.21 and 13.4.22 (Mohan, Undeland, and Robbins, 1995). Both single- and three-phase UPS systems are available.

In Fig. 13.4.21, the critical load is always supplied through an inverter. This inverter is fed from either the utility or the battery bank, depending on the availability of the utility. The battery is continually charged when the utility is present. This type of UPS provides effective protection of the critical load by isolating it from utility under- and over-voltages.

In Fig. 13.4.22, the functions of battery charging and the inverter are combined. While the utility is present, the inverter is run as a controlled rectifier to support battery charging. When the utility fails, the load is supplied from the battery-fed inverter. Because an inverter does not have the ability to deliver a larger voltage than that of the dc source, it may be necessary to include a bidirectional dc/dc converter if a low voltage battery is used in the UPS.

Motor Drives.   Motor drives can be found in a very wide range of power levels, from fractional horsepower up to hundreds of horsepower (Bose, 1986; Murphy and Turnbull, 1988). These applications range from very

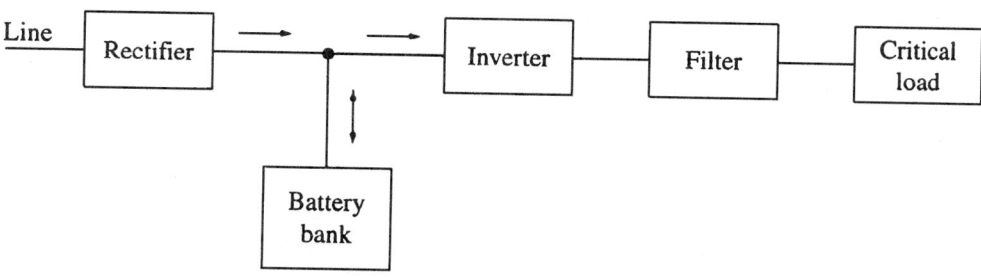

FIGURE 13.4.21   A block diagram for one configuration of a UPS system.

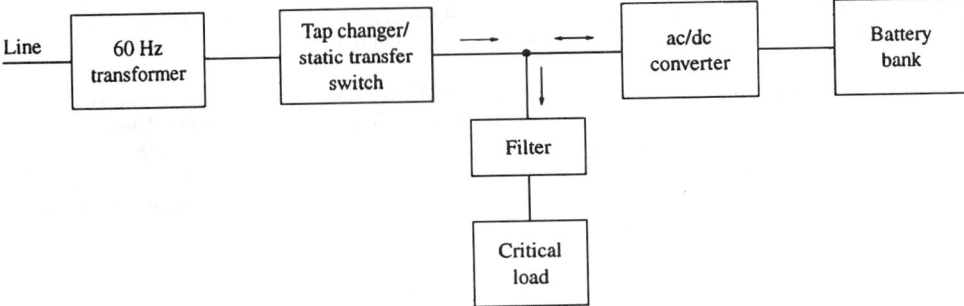

FIGURE 13.4.22 A block diagram for a second UPS system configuration.

precise motion control to adjustable speed operation of pumps and compressors for saving energy. In a motor drive, the inverter is used to provide an adjustable frequency ac voltage to the motor, thereby enabling the motor to operate over a wide range of speeds without derating the torque production of the motor. In order to prevent the motor from being pushed into magnetic saturation, the amplitude of the synthesized voltage is usually tied to the output frequency. In the simplest adjustable speed drives, the ratio of peak output voltage to output frequency is maintained nominally constant, with a small boost at low frequencies to compensate to the resistance of the motor windings. More sophisticated adjustable speed drives implement sensorless flux vector control (Bose, 1997).

Active Power Filters. Active power filters are able to simultaneously provide compensation for the reactive power drawn by other linear loads while compensating for the harmonic currents being drawn by still other nonlinear loads (Gyugyi and Strycula, 1976; Akagi, Kanazawa, and Nabae, 1984; Akagi, Nabae, and Atoh, 1986; Torrey and Al-Zamel, 1995). It is possible to compensate for multiple loads at one point. The basic idea of an active power filter is shown in Fig. 13.4.23. The active power filter is formed by putting an inverter in parallel with the loads for which compensation is needed. The inverter switches are then controlled to force the line current drawn from the utility to be of the desired quality. The inverter is controlled to draw currents that precisely compensate for the undesired components in the currents drawn by the loads. The undesired components may be either reactive or harmonic in nature. In Fig. 13.4.23, i_{utility} is forced to be of the desired quality and phase through the superposition of i_{filter} and $i_{\text{nonlinear loads}}$. With control over i_{filter}, the utility current can be forced to track its intended shape and amplitude.

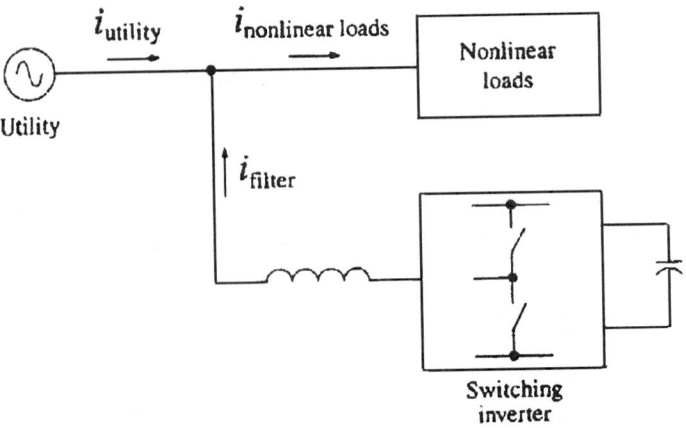

FIGURE 13.4.23 A one-line diagram of an active filter system.

Distributed Generation.   There is an ever-increasing interest in the integration of distributed generation sources into the electric utility system. Distributed generation sources include fuel cells, solar photovoltaics, wind energy, hydroelectric, and microturbines, among others. Their application is sometimes driven by the utility in an effort to use a local resource such as hydroelectric energy, to increase generating capacity without having to increase the capacity of their transmission lines, to add generation capacity incrementally without the large capital investment required of a more traditional generating station, or to increase the reliability of the supply for critical customers. Distributed generation sources are sometimes used by electricity customers to reduce their electricity costs or to increase the reliability of their electricity supply.

Some distributed generation sources naturally provide energy through dc, thereby requiring an inverter to deliver the energy to the utility grid. Fuel cells and solar photovoltaics fall into this category. Other distributed generation sources provide energy through ac with variable frequency and amplitude. Wind turbines, microturbines, and some hydroelectric systems fall into this category. This ac energy is usually delivered to the utility by first rectifying the variable ac into dc and then using an inverter to provide fixed frequency ac with a fixed amplitude to the utility grid. In some cases the rectification process is facilitated by an inverter structure where the flow of energy is from the ac side to the dc side. In this way phase currents can be controlled far more precisely than would be possible with an uncontrolled rectifier.

A significant issue with the deployment of distributed generation sources is the prevention of a situation known as islanding. An inverter that is unable to detect the presence or absence of the utility may continue to feed a section of the utility system even after the utility has taken actions to deenergize that section. This creates a serious safety issue for any utility workers who may be working on the utility system. For this reason, any inverter that is designed to interact with the ac utility must include anti-islanding controls that actively and continuously verify the presence of the larger ac utility system. Techniques for accomplishing this are described in Stevens et al. (2000) for photovoltaic systems, but the techniques described are applicable to other energy sources.

Inverters for distributed generation systems are designed to be either utility interactive or utility independent. The difference between them is found in the control. Utility interactive inverters are controlled to behave as a current source, delivering power to the utility with near-unity power factor. Utility independent inverters behave as voltage sources, where the phase difference between the output voltage and the load current is dictated by the load on the inverter.

REFERENCES

1. Akagi, H., Y. Kanazawa, and A. Nabae, "Instantaneous reactive power compensators comprising switching devices without energy storage components," *IEEE Trans. Ind. Appl.*, Vol. IA-20, pp. 625–630, 1984.

2. Akagi, H., A. Nabae, and S. Atoh, "Control strategy of active power filters using multiple voltage-source PWM converters," *IEEE Trans. Ind. Appl.*, Vol. IA-22, pp. 460–465, 1986.

3. Bose, B. K., "An adaptive hysteresis-band current control technique of a voltage-fed PWM inverter for machine drive system," *IEEE Trans. Ind. Electron.*, Vol. 37, pp. 402–408, 1990.

4. Bose, B. K. ed., "Modern Power Electronics: Evolution, Technology, and Applications," IEEE Press, 1992.

5. Bose, B. K. ed., "Power Electronics and Variable Frequency Drives," IEEE Press, 1997.

6. Brod, D. M., and D. W. Novotny, "Current control of VSI-PWM inverters," *IEEE Trans. Ind. Appl.*, Vol. IA-21, pp. 562–570, 1985.

7. Divan, D. M., "The resonant dc link converter—a new concept in static power conversion," *IEEE Trans. Ind. Appl.*, Vol. 25, pp. 317–325, 1989.

8. Divan, D. M., and G. Skibinski, "Zero-switching-loss inverters for high-power applications," *IEEE Trans. Ind. Appl.*, Vol. 25, pp. 634–643, 1989.

9. Enjeti, P. N., P. D. Ziogas, and J. L. Lindsay, "Programmed PWM techniques to eliminate harmonics: A critical evaluation," *IEEE Trans. Ind. Appl.*, Vol. 26, pp. 302–316, 1985.

10. Gyugyi, L., and E. C. Strycula, "Active ac power filter," *IEEE/IAS Annual Meeting Conference Record*, pp. 529–535, 1976.

11. He, J., and N. Mohan, "Parallel resonant dc link circuit—a novel zero switching loss topology with minimum voltage stresses," *IEEE Trans. Power Electron.*, Vol. 6, pp. 687–694, 1991.

12. Habetler, T. G., "A space vector-based rectifier regulator for ac/dc/ac converters," *IEEE Trans. Power Electron.*, Vol. 8, pp. 30–36, 1993.

13. Holmes, D. G., and B. P. McGrath, "Opportunities for harmonic cancellation with carrier-based PWM for a two-level and multilevel cascaded inverters," *IEEE Trans. Ind. Appl.*, Vol. 37, pp. 574–582, 2001.

14. Holtz, J., P. Lammert, and W. Lotzkat, "High-speed drive system with ultrasonic MOSFET PWM inverter and single-chip microprocessor control," *IEEE/IAS Annual Meeting Conference Record*, pp. 12–17, 1986.

15. Holtz, J., "Pulsewidth modulation for electronics power conversion," *IEEE Proc.*, Vol. 82, pp. 1194–1214, 1994.

16. International Rectifier, "Transformer-isolated gate driver provides very large duty cycle ratios," Application Note AN-950. Available through URL http://www.irf.com/technical-info/appnotes.htm.

17. Kazerani, M., P. D. Ziogas, and G. Joos, "A novel active current waveshaping technique for solid-state input power factor conditioners," *IEEE Trans. Ind. Electron.*, Vol. 38, pp. 72–78, 1991.

18. Kassakian, J. G., M. F. Schlecht, and G. C. Verghese, "Principles of Power Electronics," Addison-Wesley, 1991.

19. Krein, P. T., "Elements of Power Electronics," Oxford University Press, 1998.

20. Lai, J.-S., and F. Z. Peng, "Multilevel converters—a new breed of power converters," *IEEE Trans. Ind. Appl.*, Vol. 32, pp. 509–517, 1996.

21. McMurray, W., "Efficient snubbers for voltage-source GTO inverters," *IEEE Trans. Power Electron.*, Vol. PE-2, pp. 264–272, 1987.

22. McMurray, W. "Resonant snubbers with auxiliary switches," *IEEE/IAS Annual Meeting Conference Record*, pp. 829–834, 1989.

23. Mohan, N., T. M. Undeland, and W. P. Robbins, "Power Electronics: Converters, Applications and Design," 2nd ed., John Wiley & Sons, 1995.

24. Murai, Y., and T. A. Lipo, "High frequency series resonant dc link power conversion," *IEEE/IAS Annual Meeting Conference Record*, pp. 772–779, 1998.

25. Patel, H. S., and R. G. Hoft, "Generalized techniques of harmonic elimination and voltage control in thyristor inverters: Part I–Harmonic elimination techniques," *IEEE Trans. Ind. Appl.*, Vol. IA-9, pp. 310–317, 1973.

26. Patel, H. S., and R. G. Hoft, "Generalized techniques of harmonic elimination and voltage control in thyristor inverters: Part II–Voltage control techniques," *IEEE Trans. Ind. Appl.*, Vol. IA-10, pp. 666–673, 1974.

27. Peng, F. Z., "A generalized multilevel inverter topology with self voltage balancing," *IEEE Trans. Ind. Appl.*, Vol. 37, pp. 611–618, 2001.

28. Simonelli, J. M., and D. A. Torrey, "An alternative bus clamp for resonant dc link converters," *IEEE Trans. Power Electron.*, Vol. 9, pp. 56–63, 1994.

29. Slotine, J. J., and W. Li, "Applied Nonlinear Control," Prentice Hall, 1991.

30. Stevens, J., R. Bonn, J. Ginn, S. Gonzalez, and G. Kern, "Development and testing of an approach to anti-islanding in utility-interconnected photovoltaic systems," Report SAND 2000-1939, Sandia National Laboratories, August 2000.

31. Tolbert, L. M., F. Z. Peng, and T. G. Habetler, "Multilevel PWM methods at low modulation indices," *IEEE Trans. Power Electron.*, Vol. 15, pp. 719–725, 2000.

32. Torrey, D. A., and A. M. Al-Zamel, "Single-phase active power filters for multiple nonlinear loads," *IEEE Trans. Power Electron.*, Vol. 10, pp. 263–272, 1995.

33. Trzynadlowski, A. M., and S. Legowski, "Minimum-loss vector PWM strategy for three-phase inverters," *IEEE Trans. Power Electron.*, Vol. 9, pp. 26–34, 1994.

34. Undeland, T. M., "Switching stress reduction in power transistor converters," *IEEE/IAS Annual Meeting Conference Record*, pp. 383–392, 1976.

35. Van der Broeck, H. W., H. C. Skudelny, and G. Stanke, "Analysis and realization of a pulse width modulator based on space vector theory," *IEEE Trans. Ind. Appl.*, Vol. IA-24, pp. 142–150, 1988.

36. Vithayathil, J., "Power Electronics: Principles and Applications," McGraw-Hill, 1995.

37. Wu, R., S. Dewan, and G. Slemon, "Analysis of a PWM ac to dc voltage source converter under predicted current control with fixed switching frequency," *IEEE Trans. Ind. Appl.*, Vol. 27, pp. 756–764, 1991.

CHAPTER 13.5
AC REGULATORS

Peter Wood

CIRCUITS FOR CONTROLLING POWER FLOW IN AC LOADS

Switch configurations such as those in Fig. 13.5.1 can be used to control ac waveforms. The control may be merely *transistory*, as in soft starting an induction motor or limiting the inrush current to a transformer, or *perpetual*, as in the control of resistive heating elements, incandescent lamps, and the reactors of a static reactive volt-ampere (VAR) generator.

The basic single-phase ac regulator is depicted in Fig. 13.5.2, using a triac as the ac switch (but any of the ac switch combinations shown in Fig. 13.5.1 is applicable). The various three-phase arrangements possible are shown in Fig. 13.5.3. The first of these, the wye-connected regulator with a neutral connection (Fig. 13.5.3a), exhibits behavior identical to that of the single-phase regulator, since it is merely a threefold replica of the single-phase version.

The delta-connected regulator arrangement of Fig. 13.5.3b is also essentially similar in behavior to the single-phase regulator insofar as load voltages and currents are concerned. Because of the delta connection, however, any symmetrical zero sequence components of the load currents will not flow in the supply lines but will only circulate in the delta-connected loads.

The three-phase three-wire wye-switched regulator of Fig. 13.5.3c behaves differently because two switches must be closed for current to flow in any load. Shown delta-loaded, it may also have the loads wye-connected without a neutral return. In this connection, each ac switch may consist of the antiparallel combination of a thyristor and a diode. The normal wye-delta transformations apply to load voltages and currents.

The "British delta" circuit of Fig. 13.5.3d behaves in the same way as a wye-switched regulator in which thyristors with inverse parallel connected diodes are used as the switches and is unique in that only unidirectional current capability is required of its switches.

When the loads are essentially resistive, two methods of control are currently employed. The technique known as *integral-cycle control* operates the regulator by keeping the switch(es) closed for some number m of complete cycles of the supply and then keeping the switch(es) open for some number n of cycles. The power delivered to the load(s) is then simply $m/(m + n)$ times the power delivered if the switch(es) are kept permanently closed, for the single-phase, wye with neutral, and delta-connected regulators.

For the wye-switched (without neutral) and British delta regulators, the power delivered is slightly greater than $m/(m + n)$ times the power at full switch conduction, and dc and unbalanced ac components develop in the supply unless special control techniques are used. These phenomena arise because of the transient conditions, inevitably attending the first cycle of operation of these circuits.

An undesirable consequence of integral-cycle control is that the load voltages and currents, and hence the supply currents, contain sideband components having frequencies $f_s[1 \pm pm/(m + n)]$, where f_s is the supply frequency and p is any integer, 1 to infinity. Many of these frequencies are obviously lower than the supply frequency and may create problems for the supply system and other connected loads. The existence of this type of unwanted components in the voltage and current spectra makes integral-cycle control unsuitable for inductive loads (including loads fed by transformers). Since none of the sidebands are zero sequence, the line currents of an integral-cycle-controlled delta regulator are identical to the properly transposed load currents.

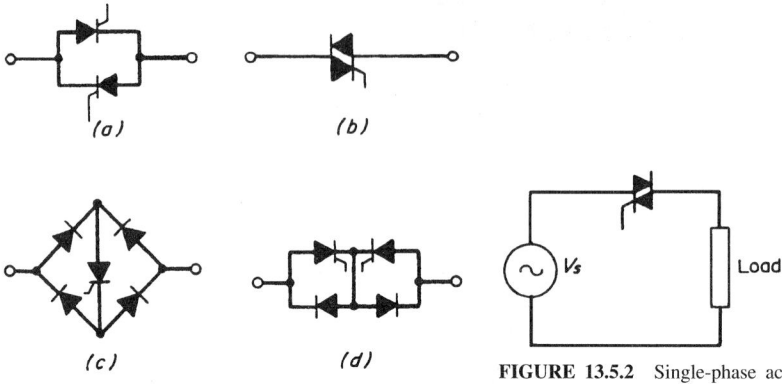

FIGURE 13.5.1 Single-phase ac switches.

FIGURE 13.5.2 Single-phase ac regulator.

Integral-cycle control results in unity displacement factor (the cosine of the angle between the fundamental component of supply current and the supply voltage). The power factor of the burden they impose on the supply with pure resistive loads is $[m/(m + n)]^{0.5}$. This is true because any regulator that forces the load current to flow in the supply while reducing the rms voltage applied to a resistive load has a power factor equal to the rms load voltage divided by the rms supply voltage.

The other method of control commonly used is termed *phase-delay control*. It is implemented by delaying the closing of the switch(es) by an angle α (called the firing angle) from each zero crossing of the supply voltage(s) and allowing the switch(es) to open again on each succeeding current zero. The load voltages and currents in this case contain only harmonics of the supply frequency as unwanted components, and except for the regulator shown in Fig. 13.5.3c, using thyristor-inverse diode switches, only odd-order harmonics are present. Thus, this control technique can be used with inductive loads.

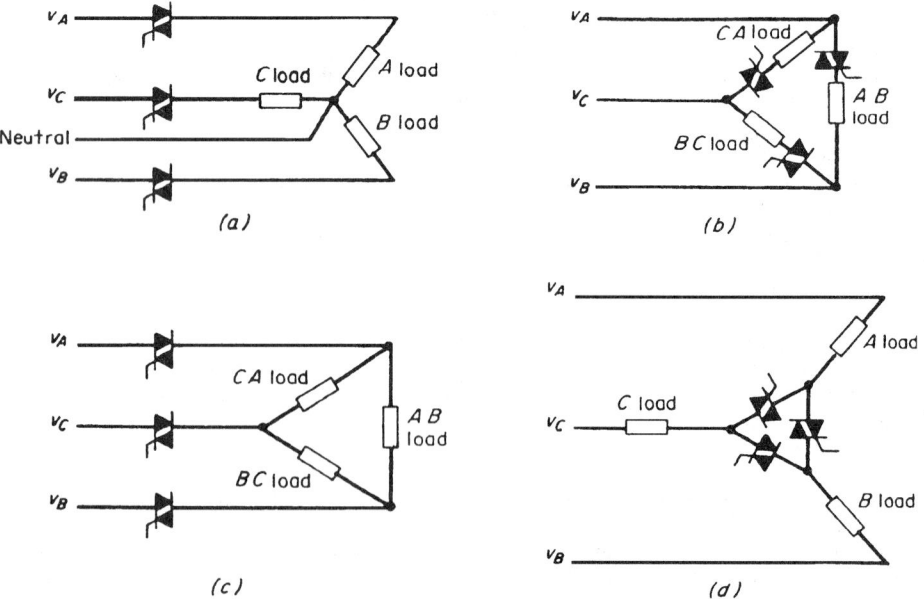

FIGURE 13.5.3 Three-phase ac regulators.

The general expressions for the load voltages and currents produced by the single-phase regulator are very cumbersome but simplify considerably for the two cases of greatest practical importance, pure resistive and pure inductive loads. For a pure resistive load with a supply voltage $V \cos \omega_s t$, the fundamental component of load voltage is given by

$$V_{DIR} = \left(1 - \frac{\alpha}{\pi} + \frac{\sin 2\alpha}{2\pi}\right) V \cos \omega_s t + \frac{\sin^2 \alpha}{\pi} V \sin \omega_s t \tag{1}$$

and the total rms load voltage by

$$V_{RMSR} = \frac{V}{\sqrt{2}} \left(1 - \frac{\alpha}{\pi} + \frac{\sin 2\alpha}{2\pi}\right)^{1/2} \tag{2}$$

where α is the firing angle measured from the supply-voltage zero crossings. For pure inductive load it is convenient to define the firing angle $\alpha' = \alpha - \pi/2$, so that at full output $\alpha' = 0$. The fundamental voltage component is then given by

$$V_{DIL} = \left(1 - \frac{2\alpha'}{\pi} - \frac{\sin 2\alpha'}{\pi}\right) V \cos \omega_s t \tag{3}$$

and the total rms voltage by

$$V_{RMSL} = \frac{V}{\sqrt{2}} \left(1 - \frac{2\alpha'}{\pi} - \frac{\sin 2\alpha'}{\pi}\right)^{1/2} \tag{4}$$

The same relationships apply to the three-phase circuits, which are in effect triplicates of the single-phase circuit (Fig. 13.5.3a and 13.5.3b); more complex relationships exist for the remaining three-phase circuits.

The use of phase-delay control results in decreasing lagging displacement factor with increasing firing angle. Maximum displacement factor is obtained at full output, equaling the power factor of the given load. At a reduced power setting the power factor is less than the displacement factor; the ratio of the two equals the ratio of the fundamental line currents versus the total rms line currents. This ratio is less than unity because of the presence of harmonic currents.

The load voltages and currents and, more importantly, the line currents of phase-delay-controlled regulators have lower total rms distortion than those of integral-cycle-controlled regulators. Among the circuits shown, the delta regulator of Fig. 13.5.3b is most beneficial; since the triple n harmonics (those of orders which are integer multiples of 3) in its load currents are zero sequence, they do not flow in the supply lines and the circuit has both a better power factor and lower total line-current distortion than integral-cycle regulators or the phase-delay-controlled wye regulators with neutral.

For the wye regulator without neutral, the range of α is 0 to $7\pi/6$ rad, provided fully bilateral switches are used; for the British delta regulator and the wye regulator without neutral using thyristor-inverse diode switches, the range is 0 to $5\pi/6$ rad. When phase-delay regulators are used with inductive loads, the range of α used for control is reduced because current-zero crossings lag voltage-zero crossings and thus abrogate part of the delay obtained with resistive loads. The regulators most commonly used with inductive loads are the single-phase, wye with neutral and the delta, for which the range of α becomes ϕ to π, where ϕ is the load-phase angle.

STATIC VAR GENERATORS

The delta regulator with purely inductive loading finds extensive use in the *static VAR generator* (SVG) (Gyugyi et al., 1978, 1980). A basic SVG consists of three delta-connected inductors with phase-controlled switches ($\pi/2 \leq \alpha \leq \pi$) and three fixed capacitive branches which may be delta-or wye-connected. The capacitive branches draw a fixed current from the supply, leading the voltage by $\pi/2$ rad. The fundamental current in the inductors is lagging the voltage by $\pi/2$ rad. Its amplitude can be varied, by phase controlling the switches, from the full inductor current to zero.

Hence the net reactive volt-ampere burden on the supply can be continuously controlled from the full capacitive VAR, when $\alpha = \pi$ and the inductor currents are zero, to the difference between the capacitive- and inductive-branch VARs when $\alpha = \pi/2$ and full inductor currents flow. This difference will be zero if inductive-branch VARs are made equal to capacitive-branch VARs and become an inductive burden if inductive VARs at full conduction exceed the capacitive VARs. Since the firing angle α can be varied on a half-cycle-to-half-cycle basis, extremely rapid changes in VAR supply (capacitive burden) or demand (inductive burden) can be accomplished.

SVGs can be used to supply shunt-reactive compensation on ac transmission and distribution systems, helping system stability and voltage regulation. They can also be used to provide damping of the subsynchronous resonances, which often prove troublesome during transient disturbances on series capacitor-compensated transmission systems, and to reduce the voltage fluctuations (flicker) produced by arc-furnace loads. In the latter application, their ability to accomplish dynamic load balancing is especially valuable.

An SVG which can provide control of reactive power supply or demand can obviously compensate for an unbalanced reactive load. It can also act as a Steinmetz balancer, providing the reactive power exchange between phases necessary to transform an unbalanced resistive load into a perfectly balanced and totally active (real) power load on the supply system.

This action can be explained as follows. Suppose a single-phase resistive load is connected between lines A and B of a three-phase system. Then the current it draws will be in phase with the AB voltage, and thus the A-line current created will lead the A-phase (line-to-neutral) voltage by $\pi/6$ rad and the B-line current will lag the B-phase voltage by $\pi/6$ rad.

If equal-impedance purely reactive loads are now connected to the BC and CA line pairs, capacitance on BC and inductive on CB, they create currents with the following phase relationships to the phase voltages:

In the A line, lagging by $2\pi/3$ rad

In the B line, leading by $2\pi/3$ rad

In the C line, one leading by $\pi/3$ rad and the other lagging by $\pi/3$ rad

The result in the C line is clearly an in-phase, wholly real current. If the impedances are of appropriate magnitude, their lagging and leading quadrature contributions in the A and B lines, respectively, can be made to cancel the lagging and leading quadrature currents created therein by the single-phase resistive load. The impedance required is $\sqrt{3}$ times the resistance. Obviously an SVG capable of providing either leading or lagging line-to-line loading on any of the line pairs can be used to balance a single-phase resistive load on any one line pair; by extension, it can be used to balance any unbalanced load. It can respond rapidly to changes in the degree of imbalance existing and thus dynamically balance the load despite the fluctuating imbalance typically created by an arc furnace.

In addition to a varying reactive fundamental current, an SVG operating other than at full or zero conduction in its reactive branches generates harmonic currents. Thus at least part of the capacitive branch is usually realized in the form of tuned harmonic filters to limit harmonic injection to the ac supply system. Maximum harmonic amplitudes relative to maximum fundamental are:

Harmonic order	3d	5th	7th	9th	11th	13th
Maximum amplitude percent	13.8	5.05	2.59	1.57	1.05	0.752

with diminishing amplitudes of the higher-order components.

When the SVG is in balanced operation, the triple n harmonics (3d and 9th in the table above) do not flow in the supply, being zero sequence. When operation is unbalanced in order to balance an unbalanced real load, positive and negative sequence components of the triple n harmonics develop and of course do flow in the supply unless filtering is provided for them.

REFERENCES

Gyugyi, L., R. A. Otto, and T. H. Putnam, "Principles and applications of static, thyristor controlled shunt compensators," *IEEE Trans. Power Apparatus and Systems*, Vol. PAS-97, No. 5, 1978.

Gyugyi, L., and E. C. Strycula, "Active ac power filter," *IEEE Ind. Appl. Soc. Annual Meeting Rec.*, pp. 529–535, 1976.

Gyugyi, L., and E. R. Taylor, "Characteristics of static, thyristor-controlled shunt compensators for power transmission system applications," *IEEE Trans. Power Eng. Soc.*, Vol. F 80, p. 236, 1980.

SECTION 14

PULSED CIRCUITS AND WAVEFORM GENERATION

Pulsed circuits and waveform generation are very important to testing and identification in a whole range of electrical and electronic circuits and systems. There are essentially two types of such networks, those that are considered passive and the rest that can be lumped into active wave shaping (which includes those done digitally).

Passive circuits are lumped into linear and nonlinear. Linear are most commonly, single pole RC and RL networks. Nonlinear networks are usually designed around diodes with or without capacitors and inductors.

A common element used in waveform generation is the switch. Mechanical switches are cleaner giving better electrical characteristics; however, they do have serious limitations. Electronic switches can be compensated so that they can come close to the mechanical switches without the serious limitations such as contact bounce. In addition, electronic switches can be made smaller and are able to work at much higher frequencies.

Active networks are either analog or digital. Analog networks have been in use for a long period of time and still have many practical uses. Digital networks have advantages especially in the area of noise, speed, and accuracy and have successfully replaced most of the analog networks in most applications.   C.A.

In This Section:

On the CD-ROM:

Dynamic Behavior of Bipolar Switches

Section References:

1. Ebers, J. J., and J. L. Moll, "Large-signal behavior of junction transistors," *Proc. IRE*, December 1954, Vol. 42, pp. 1761–1772.

2. Moll, J. L., "Large-signal transient response of junction transistors," *Proc. IRE*, December 1954, Vol. 42, pp. 1773–1784.

3. Glaser, L. A., and D. W. Dobberpuhl, "The Design and Analysis of VLSI Circuits," Addison-Wesley, 1985.

4. Horowitz, P., and W. Hill, "The Art of Electronics," Cambridge University Press, 1990.

5. Eccles, W. H., and F. W. Jordan, "A trigger relay utilizing three electrode thermionic vacuum tubes," *Radio Rev.*, 1919, Vol. 1, No. 3, pp. 143–146.

6. Schmitt, O. H. A., "Thermionic trigger," *J. Sci. Instrum.*, 1938, Vol. 15, p. 24.

7. Tietze, U., and C. Schenk, "Advanced Electronic Circuits," Springer, 1978.

8. Masakazu, S., "CMOS Digital Circuit Technology, ATT," Prentice Hall, 1988.

9. Stein, K. U., and H. Friedrich, "A 1-m/1² single-transistor cell in n-silicon gate technology," *IEEE J. Solid-State Circuits*, 1973, No. 8, pp. 319–323.

10. Jespers, P. G. A., "Integrated Converters, D. to A. and A. to D. Architecture, Analysis and Simulation," Oxford University Press, 2001.

11. Van den Plassche, R. J., "Dynamic element matching for high-accuracy monolithic D/A converters," *IEEE J. Solid-State Circuits*, December 1976, Vol. SC-11, No. 6, pp. 795–800; Van den Plassche, R. J., and D. Goedhart, "A monolithic 14 bit D/A converter," *IEEE J. Solid-State Circuits*, June 1979, Vol. SC-14, No. 3, pp. 552–556.

12. Schoeff, J. A., "An inherently monotonic 14 bit DAC," *IEEE J. Solid-State Circuits*, December 1979, Vol. SC-14, pp. 904–911.

13. Tuthill, M. A, "16 Bit monolithic CMOS D/A converter," *ESSCIRC*, *Digest of Papers*, 1980, pp. 352–353.

14. Caves, J., C. H. Chen, S. D. Rosenbaum, L. P. Sellars, and J. B. Terry, "A PCM voice codec with on-chip filters," *IEEE J. Solid-State Circuits*, February 1979, Vol. SC-14, pp. 65–73.

15. McCreavy, J. L., and P. R. Gray, "All-MOS charge redistribution analog-to-digital conversion techniques—Part I," *IEEE J. Solid-State Circuits*, December 1975, Vol. SC-10, No. 6, pp. 371–379.

16. Chao, K. C.-H., S. Nadeem, W. Lee, and C. Sodini, "A higher order topology for interpolative moderators for oversampling A/D converters," *IEEE Trans. Circuits Syst.*, March 1990, Vol. 37, No. 3, pp. 309–318.

17. Peterson, J. G., "A monolithic, fully parallel, 8 bit A/D converter," *ISSCC Digest*, 1979, pp. 128–129.

18. Song, B. S., S. H. Lee, and M. F. Tompsett, "A 10 bit 15 MHz CMOS recycling two-step A-D convertor," *IEEE JSSC*, December 1990, Vol. 25, No. 6., pp. 1328–1338.

19. Wegmann, G., E. A. Vittoz, and F. Rahali, "Charge injection in MOS switches," *IEEE JSSC*, December 1987, Vol. SC-22, No. 6. pp. 1091–1097.

20. Norsworthy, S. R., R. Schreier, and G. C. Temes, "Delta–Sigma Data Converters, Theory, Design and Simulation," *IEEE Press*, 1997.

Section Bibliography:

Wegmann, G., E. A. Vittoz, and F. Rahali, "Charge injection in MOS switches." *IEEE JSSC,* Vol. SC-22, No. 6. December 1987.

CHAPTER 14.1
PASSIVE WAVEFORM SHAPING

Paul G. A. Jespers

LINEAR PASSIVE NETWORKS

Waveform generation is customarily performed in active nonlinear circuits. Since passive networks, linear as well as nonlinear, enter into the design of pulse-forming circuits, this survey starts with the study of the transient behavior of passive circuits.

Among linear passive networks, the single-pole *RC* and *RL* networks are the most widely used. Their transient behavior in fact has a broad field of applications since the responses of many complex higher-order networks are dominated by a single pole; i.e., their response to a step function is very similar to that of a first-order system.

Transient Analysis of the *RC* Integrator

The step-function response of the *RC* circuit shown in Fig. 14.1.1*a*, after closing of the switch *S*, is given by

$$V(t) = E[1 - \exp(-t/T)] \tag{1}$$

where T = time constant = RC. The inverse of T is called the cutoff pulsation ω_0 of the circuit.

The Taylor-series expansion of Eq. (1) yields

$$V(t) = E\frac{t}{T}\left(1 - \frac{t}{2!T} + \frac{t^2}{3!T^2} - \cdots\right) \tag{2}$$

When the values of t are small compared with T, a first-order approximation of Eq. (2) is

$$V(t) \approx Et/T \tag{3}$$

In other words, the *RC* circuit of Fig. 14.1.1 behaves like an imperfect integrator. The relative error ϵ with respect to the true integral response is given by

$$\epsilon = -\frac{t}{2!T} + \frac{t^2}{3!T^2} - \frac{t^2}{4!T^3} + \cdots$$

The theoretical step-function response of Eq. (1) and the ideal-integrator output of Eq. (3) are represented in Fig. 14.1.1*b*.

Small values of t with respect to T correspond in the frequency domain (Fig. 14.1.1*c*) to frequency components situated above ω_0, that is, the transient signal whose spectrum lies to the right of ω_0 in the figure. In that case, the difference is small between the response curve of the *RC* filter and that of an ideal integrator

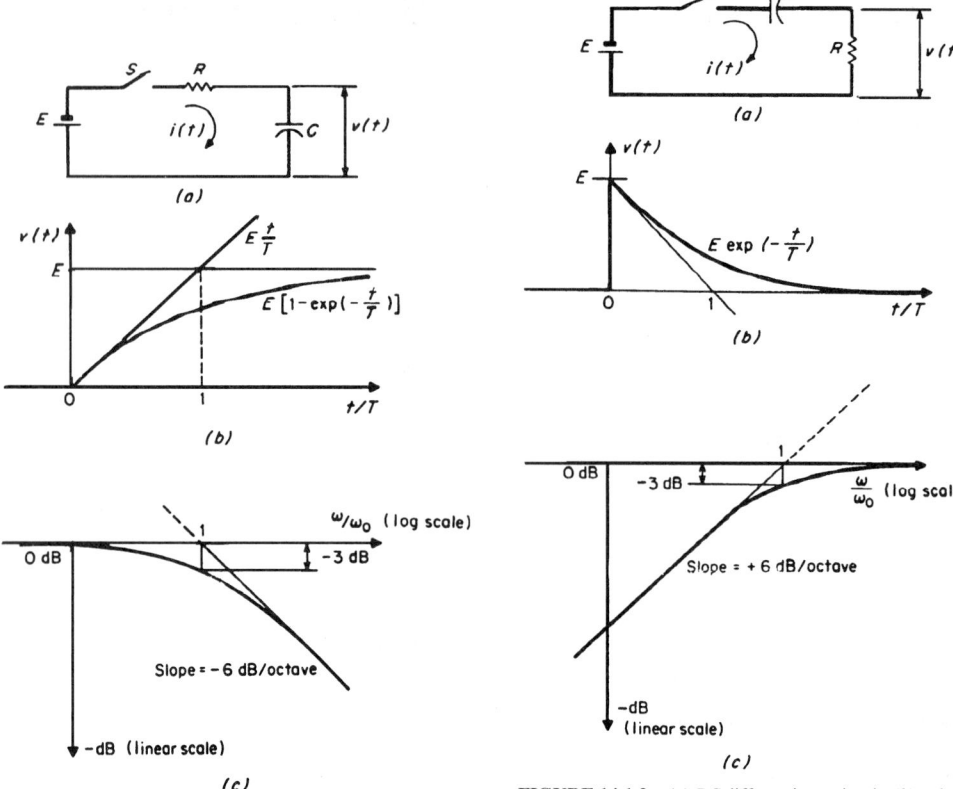

FIGURE 14.1.1   (*a*) *RC* integrator circuit; (*b*) voltage vs. time across capacitor; (*c*) attenuation vs. angular frequency.

FIGURE 14.1.2   (*a*) *RC* differentiator circuit; (*b*) voltage across resistor vs. time; (*c*) attenuation vs. angular frequency.

(represented by the –6 dB/octave line in the figure). The circuit shown in Fig. 14.1.1*a* thus approximates an integrator, provided either of the following conditions is satisfied: (1) the time under consideration is much smaller than T or (2) the spectrum of the signal lies almost entirely above ω_0.

Transient Analysis of the *RC* Differentiator

When the resistor and the capacitor of the integrator are interchanged, the circuit (Fig. 14.1.2*a*) is able to differentiate signals. The step-function response (Fig. 14.1.2*b*) of the *RC* differentiator is given by

$$v(t) = E \exp(-t/T) \tag{4}$$

The time constant T is equal to the product RC, and its inverse ω_0 represents the cutoff of the frequency response of the circuit. As the values of t become large compared with T, the step-function response becomes more like a sharp spike; i.e., it increasingly resembles the delta function.

The response differs from the ideal delta function, however, because both its amplitude and its duration are always finite quantities. The area under the exponential pulse, equal to ET, is the important quantity in applications where such a signal is generated to simulate a delta function, as in the measurement of the impulse response of a system. These considerations may be transported in the frequency domain (Fig. 14.1.2*a*).

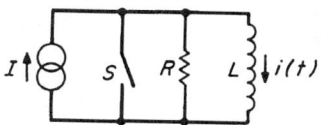

FIGURE 14.1.3 *RL* current-integrator circuit, the dual of the circuit in Fig. 14.1.1*a*.

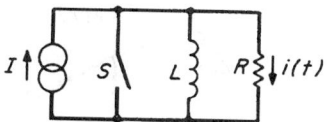

FIGURE 14.1.4 *RL* current-differentiator circuit, the dual of the circuit in Fig. 14.1.2*a*.

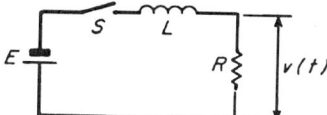

FIGURE 14.1.5 *RL* voltage integrator.

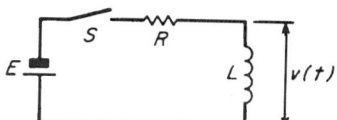

FIGURE 14.1.6 *RL* voltage differentiator.

Transient Analysis of *RL* Networks

Circuits involving a resistor and an inductor are also often used in pulse formation. Since integration and differentiation are related to the functional properties of first-order systems rather than to the topology of actual circuits, *RL* networks may perform the same function as *RC* networks. The duals of the circuits represented in Figs. 14.1.1 and 14.1.2, respectively, are shown in Figs. 14.1.3 and 14.1.4 and exhibit identical functional properties. In the first case, the current in the inductor increases exponentially from zero to *I* with a time constant equal to *L/R*, while in the second case it drops exponentially from the initial value *I* to zero, with the same time constant. Similar behavior can be obtained regarding voltage instead of current by changing the circuit from Fig. 14.1.3 to that of Fig. 14.1.5 and from Fig. 14.1.4 to Fig. 14.1.6, respectively. This duality applies also to the *RC* case.

Compensated Attenuator

The compensated attenuator is a widely used network, e.g., as an attenuator probe used in conjunction with oscilloscopes. The compensated attenuator (Fig. 14.1.7) is designed to perform the following functions:

1. To provide remote sensing with a very high input impedance, thus producing a minimum perturbation to the circuit under test.

2. To deliver a signal to the receiving end (usually the input of a wide-band oscilloscope) which is an accurate replica of the signal at the input of the attenuator probe. These conditions can be met only by introducing substantial attenuation to the signal being measured, but this is a minor drawback since adequate gain to compensate the loss is usually available.

Diagrams of two types of oscilloscope attenuator probes are given in Fig. 14.1.9, similar to the circuit of Fig. 14.1.7. In both cases, the coaxial-cable parallels the input capacitance of the receiver end; C_p represents the sum of both capacitances.

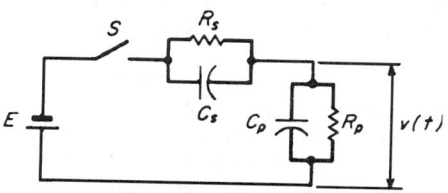

FIGURE 14.1.7 Compensated attenuator circuit.

The shunt resistor R_p has a high value, usually 1 MΩ, while the series resistor R_s is typically 9 MΩ. The dc attenuation ratio of the attenuator probe therefore is 1:10, while the input impedance of the probe is 10 times that of the receiver.

At high frequencies the parallel and series capacitors C_p and C_s play the same role as the resistive attenuator. Ideally these capacitors should be kept as low as possible to achieve a high input impedance even at high frequencies. Since it is impossible to reduce C_p below the capacitance of the coaxial cable, there is no alternative other than

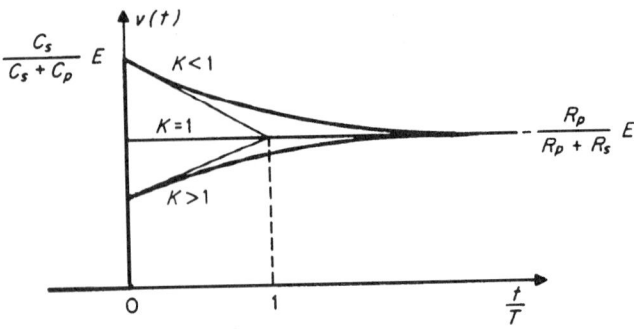

FIGURE 14.1.8   Voltage vs. time responses of attenuator, showing correctly compensated condition at $K = 1$.

to insert the appropriate value of C_s to achieve a constant attenuation ratio over the required frequency band. In consequence, as the frequency increases, the nature of the attenuator changes from resistive to capacitive. However, the attenuation ratio remains unaffected, and no signal distortion is produced. The condition that ensures constant attenuation ratio is given by

$$R_p C_p = R_s C_s \tag{5}$$

The step-function response of the compensated attenuator, which is illustrated in Fig. 14.1.8, clearly shows how distortion occurs when the above condition is not met. The output voltage $V(t)$ of the attenuator is given by

$$V(t) = \frac{C_s}{C_s + C_p} \left\{ 1 - (1-K) \left[1 - \exp\left(-\frac{t}{T}\right) \right] \right\} E \tag{6}$$

where K represents the ratio of the resistive attenuation factor to that of the capacitive attenuation factor

$$K = \frac{R_p}{R_p + R_s} \Big/ \frac{C_s}{C_p + C_s}$$

and

$$T = (R_p \parallel R_s)(C_s + C_p) \tag{7}$$

The $\parallel$ sign stands for the parallel combination of two elements, e.g., in the present case $R_p \parallel R_s = R_p R_s / (R_p + R_s)$. Only when K is equal to 1, in other words when Eq. (5) is satisfied, will no distortion occur, as shown in Fig. 14.1.8.

In all other cases there is a difference between the initial amplitude of the step-function response (which is controlled by the attenuation ratio of the capacitive divider) and the steady-state response (which depends on the resistive divider only).

A simple adjustment to compensate the attenuator consists of trimming one capacitor, either C_p and C_s, to obtain the proper step-function response. Adjustments of this kind are provided in attenuators like those shown in Fig. 14.1.9.

Compensated attenuators may be placed in cascade to achieve variable levels of attenuation. The conditions imposed on each cell are like those enumerated above, but an additional requirement is introduced, namely, the requirement for constant input impedance. This introduces a different structure compared with the compensated attenuator, as shown in Fig. 14.1.10. The resistances R_p and R_s must be chosen so that the impedance is kept constant and equal to R. The capacitor C_s is adjusted to compensate the attenuator, while C_p provides the required additional capacitance to make the input susceptance equal to that of the load.

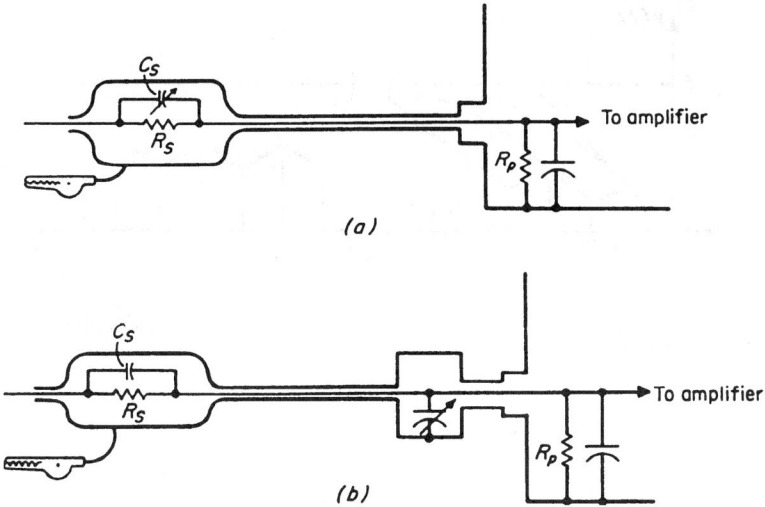

FIGURE 14.1.9 Coaxial-cable type of attenuator circuit: (*a*) series adjustment; (*b*) shunt adjustment.

Periodic Input Signals

Repetitive transients are typical input signals to the majority of pulsed circuits. In linear networks there is no difficulty in predicting the response of circuits to a succession of periodic step functions, alternatively positive and negative, since the principle of superposition holds. We restrict our attention here to two simple cases, the square-wave response of an *RC* integrator and an *RC* differentiator.

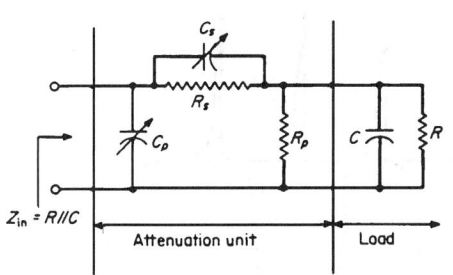

FIGURE 14.1.10 Compensated attenuator suitable for use in cascaded circuits.

Figure 14.1.11 represents, at the left, the buildup of the response of the *RC* integrator, assuming that the period τ of the input square wave is smaller than the time constant of the circuit *T*. On the right in the figure the steady-state response is shown. The triangular waveshape represents a fair approximation to the integral of the input square wave. The triangular wave is superimposed on a dc pedestal of amplitude *E*/2. Higher repetition rates of the input reduce the amplitude of the triangular wave without affecting the dc pedestal. When the frequency of the input square wave is high enough, the dc component is the only remaining signal; i.e., the *RC* integrator then acts like an ideal low-pass filter.

A similar presentation of the behavior of the *RC* differentiator is shown in Fig. 14.1.12*a* and *b*. The steady-state output in this case is symmetrical with respect to the zero axis because no dc component can flow through the series capacitor. When, as shown in Fig.14.1.12*b*, no overlapping of the pulses occurs, the steady-state solution is obtained from the first step.

Pulse Generators

The step function and the delta function (Dirac function) are widely used to determine the dynamic behavior of physical systems. Theoretically the delta function is, a pulse of infinite amplitude and infinitesimal duration but having a finite area (product of amplitude and time). In practice the question of the equivalent physical impulse arises. The answer involves the system under consideration as well as the impulse itself.

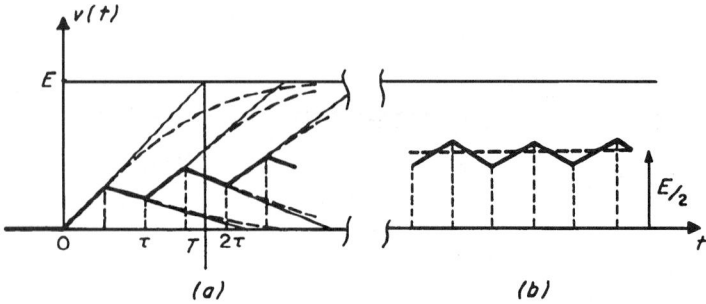

FIGURE 14.1.11   *RC* integrator with square-wave input of period smaller than *RC*: (*a*) initial buildup; (*b*) steady state.

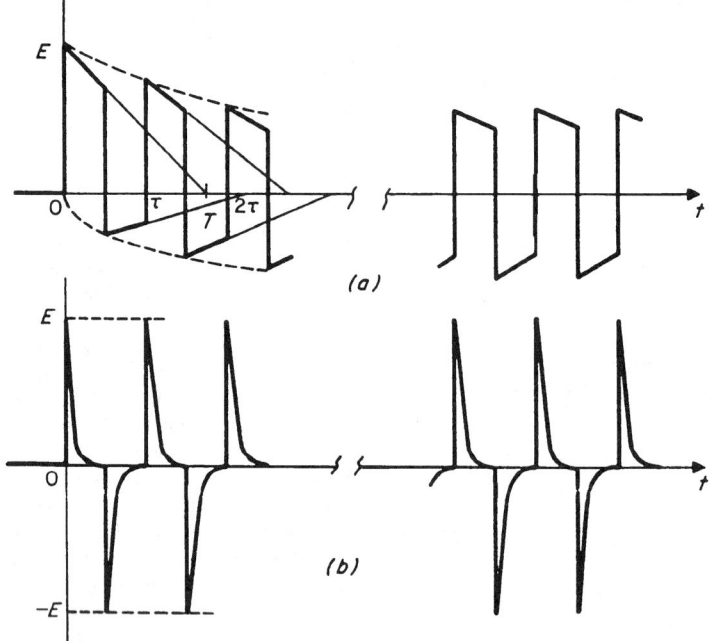

FIGURE 14.1.12   *RC* differentiator with square-wave input: (*a*) period of input signal smaller than *RC*; (*b*) input period longer than *RC*.

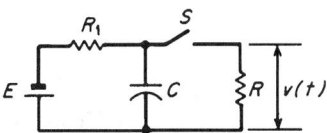

FIGURE 14.1.13   *RC* pulse-generator circuit with large series resistance R_1.

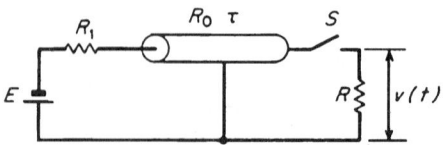

FIGURE 14.1.14   Coaxial-cable version of *RC* pulse generator.

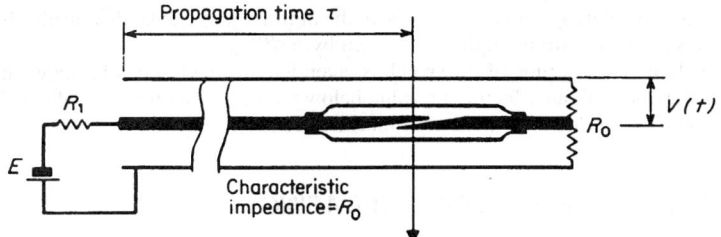

FIGURE 14.1.15   Use of mercury-wetted switch contacts in coaxial pulse generator.

The spectrum of the delta function has a constant amplitude over the whole frequency spectrum from zero to infinity. Other signals of finite area (amplitude × time) have different spectral distributions. On a logarithmic scale of frequency, the spectrum of any finite-area transient signal tends to be constant between zero and a cutoff frequency that depends on the shape of the signal. The shorter the duration of the signal, the wider the constant-amplitude portion of the spectrum.

If such a signal is used in a system whose useful frequency band is located below the cutoff frequency of the signal spectrum, the system response is indistinguishable from its delta impulse response. Any transient signal with a finite area, whatever its shape, can thus be considered as a delta function relative to the given system, provided that the flat portion of its spectrum embraces the whole system's useful frequency range. A measure of the effectiveness of a pulse to serve as a delta function is given by the approximation of useful spectrum bandwidth $B = 1/\tau$, where τ represents the midheight duration of the pulse.

Very short pulses are used in various applications in order to measure their delta-function response. In the field of radio interference, for instance, the basic response curve of the CISPR receiver[*] is defined in terms of its response to regularly repeated pulses. In this case, the amplitude of the uniform portion of the pulse spectrum must be calibrated, i.e., the area under the pulse must be a known constant which is a function of a limited number of circuit parameters.

The step-function response of an RC differentiator provides such a convenient signal. Its area is given by the amplitude of the input step multiplied by the time constant RC of the circuit. Moreover, since the signal is exponential in shape, its -3-dB spectrum bandwidth is equal to $1/RC$. In the circuit of Fig. 14.1.13, R_1 is much larger than R; when the switch S is open, the capacitor charges to the voltage E of the dc source. When the switch is closed, the capacitor discharges through R, producing an exponential signal of known amplitude and duration (known area).

A circuit based on the same principle is shown in Fig. 14.1.14. Here the coaxial line plays the role of energy storage source. If the line is lossless, its characteristic impedance is given by R_0, the propagation delay is equal to τ, and the Laplace transform of the voltage drop across R is

$$V(p) = (1/p)E[1 + (R_0/R) \coth p\tau]^{-1} \tag{8}$$

When the line is matched to the load, Eq. (8) reduces to

$$V(p) = (1/2p)E(1 - e^{-p2\tau}) \tag{9}$$

which indicates that $V(t)$ is a square wave of amplitude $E/2$ and duration 2τ. The area of the pulse is equal to the product of E and the time constant τ. Both quantities can be kept reasonably constant. The bandwidth is larger than that of an exponential pulse of the same area (Fig. 14.1.13) by the factor π.

Very wide bandwidth pulse generators based on this principles use a coaxial mercury-wetted switch built into the line (Fig. 14.1.15) to achieve low standing-wave ratios. A bandwidth of several GHz can be obtained in this manner.

In coaxial circuits, any impedance mismatch causes reflections to occur at both ends of the line, replacing the desired square-wave signal by a succession of steps of decreasing amplitude. The cutoff frequency of the

---

[*]International Electrotechnical Commission (IEC), "Specification, de l'apparelloge de mesure CISPR pour les frequences comprises entre 25 et 300 MHz," 1961.

spectrum is lowered thereby, and its shape above the uniform part can be drastically changed. Below cutoff frequency, however, the spectrum amplitude is given by $E\tau R/R_0$.

When the finite closing time of the switch is taken into account, it can be shown that only the width of the spectrum is reduced without affecting its value below the cutoff frequency. Stable calibrated pulse generators can also be built using electronic instead of mechanical switches.

NONLINEAR-PASSIVE-NETWORK WAVESHAPING

Nonlinear passive networks offer wider possibilities for waveshaping than linear networks, especially when energy-storage elements such as capacitors or inductors are used with nonlinear devices. Since the analysis of the behavior of such circuits is difficult, we first consider purely resistive nonlinear circuits.

Diode Networks without Storage Elements

Diodes provide a simple means for clamping a voltage to a constant value. Both forward conduction and avalanche (zener) breakdown are well suited for this purpose. Avalanche breakdown usually offers sharper nonlinearity than forward biasing, but consumes more power.

Clamping action can be obtained in many different ways. The distinction between series and parallel clamping is shown in Fig. 14.1.16. Clamping occurs in the first case when the diode conducts; in the second when it is blocked.

Since the diode is not an ideal device, it is useful to introduce an equivalent network that takes into account some of its imperfections. The complexity of the equivalent network is a trade-off between accuracy and ease of manipulation.

The physical diode is characterized by

$$I = I_s[\exp{(V/V_T)} - 1] \tag{10}$$

where I_S is the leakage current and $V_T = kT/q$, typically 26 mV at room temperature.

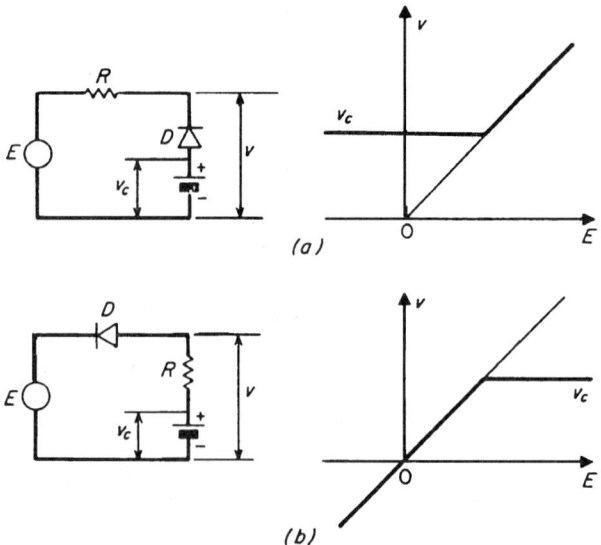

FIGURE 14.1.16   Diode clamping circuit and voltage vs. time responses:
(a) shunt diode; (b) series diode.

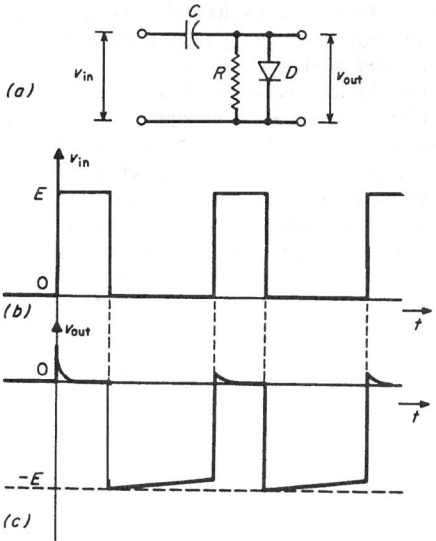

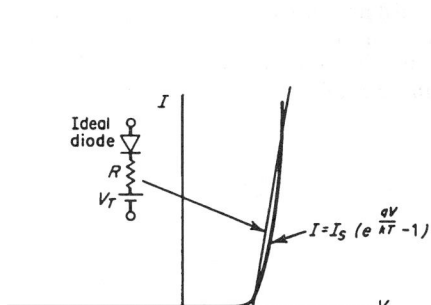

FIGURE 14.1.17 Actual and approximate current-voltage characteristics of ideal and real diodes.

FIGURE 14.1.18 (*a*) DC restorer circuit; (*b*) input signal; (*c*) output signal.

The leakage current is usually quite small, typically 100 pA or less. Therefore, V must be at least several hundred millivolts, typically 600 mV or more, to attain values of forward current I in the range of milliamperes. A first approximation of the forward-biased real diode consists therefore of a series combination of the ideal diode and a small emf (Fig. 14.1.17). Moreover, to take into account the finite slope of the forward characteristic, a better approximation is obtained by inserting a small resistance in series.

Diode Networks with Storage Elements

There is no simple theory to represent the behavior of nonlinear circuits with storage elements, such as capacitors or inductances. Acceptable solutions can be found, however, by breaking the analysis of the circuit under investigation into a series of linear problems. A typical example is the dc restorer circuit hereafter.

The circuit shown in Fig. 14.1.18 resembles the RC differentiator but exhibits properties that differ substantially from those examined previously. The diode D is assumed to be ideal first to simplify the analysis of the circuit, which is carried out in two steps, i.e., with the diode forward- and reverse-biased. In the first step, the output of the circuit is short-circuited; in the second, the diode has no effect, and the circuit is identical to the linear RC differentiator.

When a series of alternatively positive and negative steps is applied at the input, after the first positive step is applied, no output voltage is obtained. The first positive step causes a large transient current to flow through the diode and charges the capacitor. Since D is assumed to be an ideal short circuit, the current will be close to a delta function as long as the internal impedance of the generator connected at the input is zero.

In practice, the finite series resistance of the diode must be added to the generator internal impedance, but this does not affect the load time constant significantly, since it is assumed to be much smaller than the time between the first positive step and the following negative step. This allows the circuit to attain the steady-state conditions between steps. When the input voltage suddenly returns to zero, the output voltage undergoes a large negative swing whose magnitude is equal to that of the input step. The diode is then blocked, and the capacitor discharges slowly through the resistor. If the time constant is assumed to be much larger than the period of the input wave, the output voltage swings back to zero when the second positive voltage step is applied and only a small current flows through the forward-biased diode to restore the charge lost when the diode was under the reverse-bias condition.

If the finite resistance of the diode is taken into consideration, a series of short positive exponential pulses must be added to the output signal, as shown in the lower part of Fig. 14.1.18. The first pulse, which corresponds to the initial full charge on the capacitor, is substantially higher than the next pulse, but this is of little importance in the operation of the circuit.

An interesting feature of the dc restorer circuit lies in the fact that although no dc component can flow from input to output, the output signal has a well-defined nonzero dc level, although determined only by the amplitude of the negative steps (assuming, of course, that the lost charge between two steps is negligible). This circuit is used extensively in video systems to prevent the average brightness level of the image from being affected by its varying video content. In this case, the reference steps are the line-synchronizing pulses.

CHAPTER 14.2
SWITCHES

Paul G. A. Jespers

THE IDEAL SWITCH

An ideal switch is a two-pole device that satisfies the following conditions:

Closed-switch condition. The voltage drop across the switch is zero whatever the current flowing through the switch may be.

Open-switch condition. The current through the switch is zero whatever the voltage across the switch may be.

Mechanical switches are usually electrically ideal, but they suffer from other drawbacks; e.g., their switching rate is low, and they exhibit jitter. Moreover, bouncing of the contacts may be experienced after closing, unless mercury-wetted contacts are used. Electronic switches do not exhibit these effects, but they are less ideal in their electrical characteristics.

BIPOLAR-TRANSISTOR SWITCHES

The bipolar transistor approximates an open switch between emitter and collector when its base terminal is open or when both junctions are reverse-biased or even only slightly forward-biased. Inversely, under saturated conditions, the transistor resembles a closed switch with a small voltage drop in series, typically 50 to 200 mV. This drop may be considered negligible in many applications.

Static Characteristics

A more rigorous approach to the transistor static characteristics is based on the Ebers and Moll transport equations

$$\begin{bmatrix} I_E \\ I_C \end{bmatrix} = I_S \begin{bmatrix} -\dfrac{1}{\beta_F}-1 & 1 \\ 1 & -1-\dfrac{1}{\beta_R} \end{bmatrix} \begin{bmatrix} \exp\,(V_E/V_T)-1 \\ \exp\,(V_C/V_T)-1 \end{bmatrix} \tag{1}$$

where V_E, V_C = voltage drops across emitter and collector junctions, respectively (positive voltages stand for forward bias, negative for reverse bias); $V_T = kT/q$ (typically 26 mV at room temperature); I_S = saturation current; and β_F, β_R represent forward (I_C/I_B) and reverse (I_E/I_B) current gains, respectively, with $V_E > 0$, $V_C < 0$ in the first case and $V_E < 0$, $V_C > 0$ in the second.

The saturation current I_S governs the leakage current flowing through the transistor under blocked conditions. It is always exceedingly small, and since it usually amounts to 10^{-14} or 10^{-15} A, it is difficult to measure. A standard procedure is to draw the plot representing the collector current in log scale versus the emitter forward bias V_E in linear scale. To find the saturation current, one must extrapolate the part of the curve, which can be assimilated to a straight line with a slope of 60 mV/decade to the intercept with the vertical axis for which $V_E = 0$. The saturation current can also be obtained with emitter and collector terminals permutated.

The current gains β_F and β_R can be evaluated by means of the same experimental setup. An additional ammeter is required to measure I_B.

It is common practice to rewrite Eq. (1) so that the emitter and collector currents are expressed as functions of

$$I_F = I_S[\exp (V_E/V_T) - 1] \tag{2}$$

and

$$I_R = I_S[\exp (V_C/V_T) - 1] \tag{3}$$

With these definitions Eq. (1) can be expressed as

$$I_C = I_F - I_R - I_R/\beta_R \tag{4}$$

$$I_E = -I_F - (I_F/\beta_F) + I_R \tag{5}$$

Hence, the Ebers and Moll transport model is found. This is illustrated by the equivalent circuit of Fig. 14.2.1. The leakage currents of the two diodes D_1 and D_2 are given respectively by I_S/β_F and I_S/β_R. With this model, it is possible to compute the currents flowing through the transistor under any circumstance.

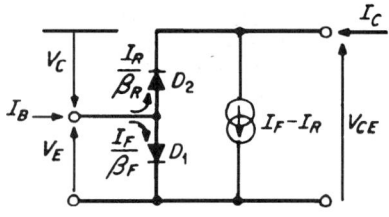

FIGURE 14.2.1   Equivalent circuit of the Ebers and Moll transport model of the bipolar transistor.

For instance, if the collector junction is reverse-biased and a small positive bias of, for example, +100 mV is established across the emitter-junction, the reverse current I_R is almost equal to $-I_S$ and I_F is equal to $I_S \exp (100/26)$ or $46.8\,I_S$. Hence, from Eqs. (4) and (5), the collector current is found to be equal to $48\,I_S$, and the emitter current is approximately the same with opposite sign. With the assumption that I_S is equal to 10 pA, both I_C and I_E are essentially negligible. A fortiori, I_B as derived from Eqs. (4) and (5) is also small:

$$I_B = (I_F/\beta_F) + (I_R/\beta_R) \tag{6}$$

To drive current through the transistor, the voltage across one of the two junctions must reach at least 0.5 V, according to:

$$V_E = V_T \ln (I_F/I_S) \tag{7}$$

or

$$V_C = V_T \ln (I_R/I_S) \tag{8}$$

derived from Eqs. (2) and (3).

The transistor operates in the saturation region when it approximates a closed switch. The voltage drop between the emitter and collector terminals is then given by

$$V_{CE,\text{sat}} = V_T \ln \frac{n + (n + \beta_F)/\beta_R}{n - 1} \tag{9}$$

where n represents the ratio $\beta_F I_B/I_C$, assumed larger than 1. For most transistors, this voltage drop lies between 50 and 200 mV. The inevitable resistance in series with the collector increases this voltage by a few tens of millivolts.

An interesting situation arises when I_C is almost equal to zero; e.g., when the bipolar transistor is used to set the potential across a capacitor. In this case, Eq. (9) becomes

$$V_{CE,\text{sat}} = V_T \ln (1 + 1/\beta_R) \tag{10}$$

Similarly, with interchanged emitter and collector terminals, the voltage drop is given by

$$V_{EC,\text{sat}} = V_T \ln (1 + 1/\beta_F) \tag{11}$$

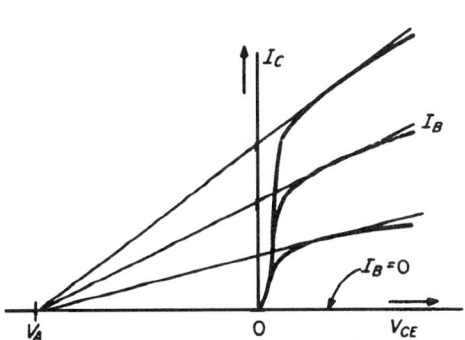

FIGURE 14.2.2 Typical common-emitter characteristics of the bipolar transistor. V_A is called the Early voltage.

Since β_F is normally much larger than β_R, $V_{EC,\text{sat}}$ may be smaller than 1 mV provided β_F is at least equal to 25. Consequently, inverted bipolar transistors are switches with a very small series voltage drop, provided that the current flowing through the transistor is kept small.

The two situations examined so far (open or closed switch) correspond in Fig. 14.2.2, respectively, to $I_B = 0$ and to the part of the curves closely parallel to the collector-current axis. The fact that all the curves have nearly the same vertical shape means that the series resistance of the saturated transistor is quite small. Since the characteristics do not coincide with the vertical coordinate axis, a small series emf must be considered, however, as previously stated.

A third region exists where the transistor plays the role of a current switch instead of a voltage switch. It concerns the switching from blocked conditions to any point within the active region or vice versa. Conceptually, the transistor may be compared to a controlled current source, which is switched on or off. However, because of the Early effect, the current is a function of the collector to emitter voltage. The Ebers and Moll model is inappropriate to describe this effect. A better expression of I_C is

$$I_C = I_S \exp (V_E/V_T)(1 + V_{CE}/V_A) \tag{12}$$

where V_A is called the *Early voltage*. Equation (12) is illustrated in Fig. 14.2.2. The finite output conductance of the transistor is given by I_C/V_A.

Dynamic Characteristics

The dynamic behavior of bipolar transistors suffers from a drawback called "charge storage in the base," which takes place every time transistors are driven in or out of saturation. The phenomenon is related to the majority carriers supplementing the minority carrier in the base to guarantee neutrality. The question is how to remove these extra carriers when the transistor is supposed to desaturate. Zeroing the base current is not a satisfactory solution for the majority carriers can only recombine with minority carriers. This requires a lot of time for lifetimes in the base are generally large in order to maximize the current gain. A better technique is to reverse the polarity of the base current. The larger this current, the more rapidly the majority carriers disappear and the faster the transistor gets out of saturation. Current continues to flow, however, until one of the two junctions gets reverse biased (usually the collector junction). Only then the transistor enters the active region and the collector current may start to decrease. When all majority carriers are swept away, both junctions are reverse-biased and the impedance of the base contact unfolds from a low impedance to an open circuit. A quantitative analysis of the desaturation mechanism can be found in the CD-ROM together with an example.

Charge storage is generally associated with the base of bipolar transistors, although the same phenomenon takes place near the depleted region in the emitter neutral region also. The reason why this is not considered is related to the doping profile in the emitter comparatively to the base region. The emitter doping is much larger

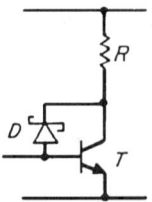

FIGURE 14.2.3   A Schottky diode D prevents T from going into saturation.

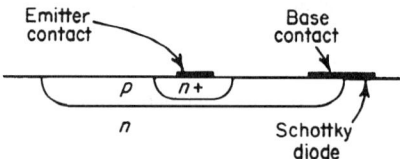

FIGURE 14.2.4   Planar *npn* transistor and Schottky diode in integrated-circuit form.

than the base doping because it increases the emitter's efficiency, which controls the current gain (the emitter efficiency is the ratio of emitter-base current over base-emitter current).

Junction diodes suffer from the same drawback. The diffusion length in the less-doped region being much longer than the base width of a bipolar transistor, charge storage problems should be worse. This is not the case, however, as diodes are generally bipolar transistors with their base and collector terminals shorted, making their charge storage effects similar to those of bipolar transistors.

In order to reduce the delay between the control signal applied to the base and the moment the collector current begins to decay, several techniques have been developed. Two of these are reviewed hereafter.

The first takes advantage of Schottky diodes, which exploit field emission and, therefore, ignore charge storage phenomena. Shottky diodes exhibit a slightly smaller voltage drop under forward bias than junction diodes (of the order of 0.4 V instead of 0.7 V), which is currently exploited to prevent bipolar transistors from getting saturated. The idea is illustrated in Fig. 14.2.3, which represents a bipolar transistor whose collector junction is paralleled by a Schottky diode. When the transistor is nearing saturation, the Schottky diode starts conducting before the collector junction does. This prevents the transistor from entering saturation. The base current in excess to what is needed to sustain the actual collector current flows directly to ground through the series combination of the forward-biased Schottky diode and emitter junction. Figure 14.2.4 shows how the combination of a Shottky diode and a bipolar transistor can be implemented. The Schottky diode consists of the metal contact that overlaps the lightly doped collector, whereas in the base region the metal to the P-type semiconductor resumes to an Ohmic contact. Such combination is currently used in Schottky logic, a family of fast bipolar logic circuits.

The second way to avoid charge storage is to devise circuits that never operate in saturation. The switch shown in Fig. 14.2.5, which involves two transistors controlled by a pair of square waves with opposite polarities is a good example. When Q1 is on, Q2 is off or vice versa. Current is switched either left or right. Although the circuit looks like a differential pair, it operates in a quite different manner: the conducting transistor is in the common base configuration, while the other is blocked. Since the output current is taken from the collector of the common base transistor, the output impedance is very large. This means that the switch is a current-mode

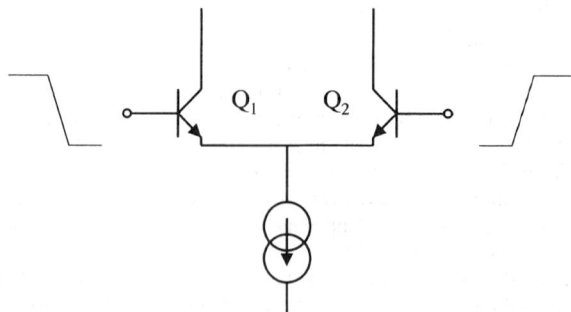

FIGURE 14.2.5   A bipolar current switch.

switch instead of a voltage-mode switch. Very short switching times are feasible this way for none of the two transistors ever saturates. Emitter-coupled logic (ECL) takes advantage of this circuit configuration.

MOS SWITCHES

Insulated gate field-effects (IGFETs, also called MOS transistors) and junction field-effect transistors (JFETs) can be put to use in order to mimic switches. They resemble an ideal closed switch in series with a linear resistor when "on" and an open switch when "off." The leakage current, however, is larger than with bipolar transistors.

Static Characteristics

When a field-effect transistor is turned on, its characteristics differ substantially from those of a bipolar switch. Since no residual emf in series with the switch is experienced, the transistor is comparable to a resistor whose conductance G is given by

$$G = \mu C_{\text{ox}} \frac{W}{L} (V_G - V_{T0} - \lambda V) \tag{13}$$

where μ is the mobility in the inversion layer, C_{ox} the gate oxide capacitance per unit area, V_G the gate-to-substrate voltage, V_{T0} the gate threshold voltage under zero bias, V the source or drain voltage, and λ a dimensionless factor that does take into account the so-called substrate effect (the value of λ lies somewhere between 1.3 and 1.5). The dependance on source and drain voltages of the conductance G represents a problem that impairs severely the performances of MOS switches. Consider, for instance, a MOS switch in series with a grounded capacitor to implement a simple sample-and-hold circuit. Since the MOS transistor is similar to a resistor when its gate voltage is high, the circuit may be assimilated to an RC network with a time constant that varies like the reciprocal of the difference in the right part of Eq. (13). Hence, when the input voltage V equals $(V_G - V_{T0})/\lambda$ the conductance becomes equal to zero and the switch resumes to an open circuit. In practice, V must be well below this limit to sample the input within a small enough time window. Single MOS switches are not favored therefore. For instance, in logic circuits, where the logic 1s and 0s are set by the power supply and ground, respectively, the logic high signal may be substantially corrupted and the speed reduced. CMOS switches are preferred threfore to single MOS switches. A typical CMOS transmission switch is shown in Fig. 14.2.6. It consists of the parallel combination of an

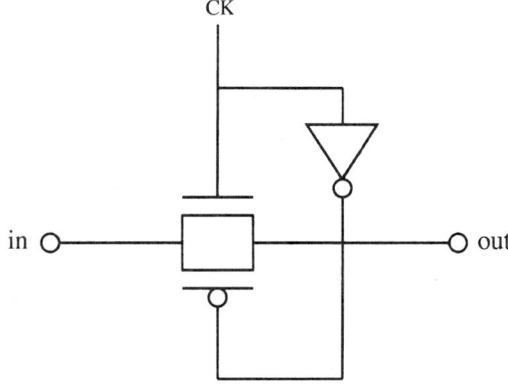

FIGURE 14.2.6 The complementary MOS switch.

N-MOS and a P-MOS transistor controlled by complementary logic signals. The idea is simply to counterbalance the decreasing conductance of the N-MOS transistor when the input goes from low to high by the increasing conductance of the P-MOS transistor. Thanks to the parallel combination, the series resistance is kept large and almost unchanged for any input voltage. The same holds true as long as the time is constant.

Dynamic Characteristics

MOS transistors ignore charge storage phenomena for they are unipolar devices. Their transient behavior is controlled by the parasitic capacitances associated with their gate, source, and drain. Source and drain are reverse-biased junctions, which exhibit parasitic capacitances with respect to the substrate. The gate capacitance relates to the inversion layer and to the regions overlaping source and drain. These capacitances control the dynamic behavior of the switch in conjunction with the parasitics associated to the elements connected to the MOS transistor terminals.

What happens with the inversion layer charge when the transistor is switched off is considered hereafter. Since charge cannot simply vanish, it must go somewhere, either to the source or drain or to both. This introduces generally a short spike in memoryless circuits that does not affect the performances significantly except at high frequency. In circuits that exhibit memory, like in the MOS sampling network discussed earlier, the impact is more serious. The part of the inversion layer charge left on the capacitive terminal is "integrated," which leads to a DC offset.

The charge partition problem in memory circuits is illustrated by the simple circuit shown in Fig 14.2.7, which consists of a MOS switch between two capacitors C_1 and C_2. We start from the situation where the voltages V_1 and V_2 across the capacitors are equal and no current is flowing through the transistor, supposed to be "on." As soon as the gate voltage starts to decrease, the charge in the inversion layer tends to divide equally between MOS terminals for these are at the same potential. If the capacitors are not identical, a voltage difference starts to build up as soon as charge is being transferred. This causes current to flow in the MOS transistor,

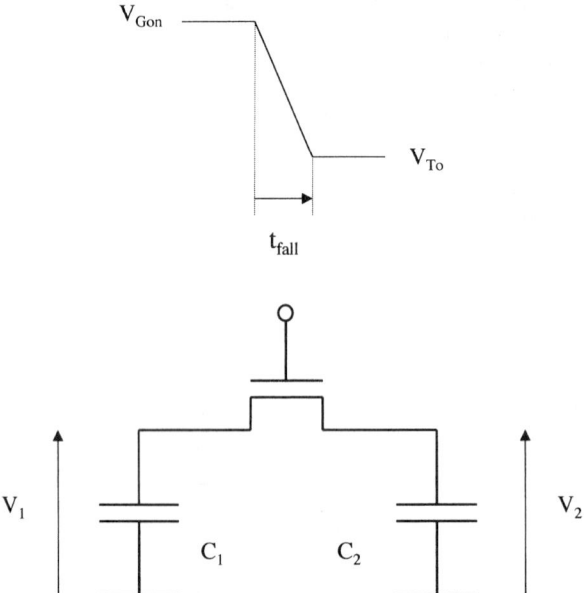

FIGURE 14.2.7   The inversion layer charge divides between C_1 and C_2 after cutoff.

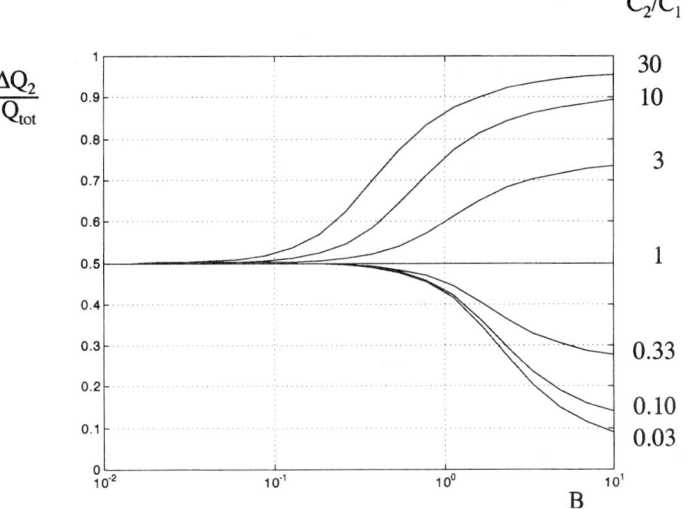

FIGURE 14.2.8   Fraction of the inversion layer charge left in C_2 after cutoff vs. the parameter B defined under Eq. (14).

which tend to reduce the voltage difference between the capacitors. This re-equilibration mechanism holds on as long as the gate voltage exceeds the effective threshold voltage although it is getting weaker and weaker as the transistor is nearing cut-off. When finally the MOS transistor is cut-off, a nonzero voltage difference is left over, which may be assimilated to an offset. The size of this offset is a function of several factors including the gate slewing-rate. It is obvious that an abrupt gate voltage step, which does not leave time for re-equilibration, will split the inversion layer charge equally between the two capacitors, whereas slow cut-off will tend to keep the voltages across the capacitors more alike.

The problem is addressed in Ref. 19. The fraction of the total inversion layer charge that is stored in capacitor C_2 versus the parameter B defined below is illustrated in Fig. 14.2.8:

$$B = (V_{\text{Gon}} - V_T) \bullet \sqrt{\frac{\beta}{aC_2}} \tag{14}$$

V_{Gon} is the gate voltage prior to switching, V_T the effective threshold voltage of the MOS transistor equal to $(V_{T0} + \lambda V_{\text{in}})$, β the well-known factor $\mu C_{\text{ox}} W/L$ and a the gate voltage slewing rate defined as $(V_{\text{Gon}} - V_{T0})$ divided by the fall time. Notice that fast switching yields small values of B, whereas long switching times lead to large values of B. When B is small the inversion layer charge divides equally. Voltage equalization tends to prevail when B is large, as can be found from the large differences experienced once the ratio C_2/C_1 departs from one. Let us consider, for instance, a MOS transistor with a β equal to 10^{-4} A/V^2, a gate capacitance C_G of 0.1 pF, V_{Gon} and V_{T0}, respectively, equal to 5 and 0.7 V, a large capacitor C_1 to mimic a voltage generator and a load capacitance C_2 equal to 1 pF. For fall times between 1 ps and 1 ns, the factor B varies from 0.021 until 0.626. The offset voltage is large and varies little from 215 to only 200 mV since re-equilibration cannot take place in such a short time. A 10 ns fall time reduces the final offset to 125 mV, and 100 ns, 1 μs, and 10 μs fall times yield, respectively, 41, 13, and 4 mV offset. In any case, these are still large offsets in terms of analog signals. To get smaller offsets the switching times must be very long. Hence, switching noise cannot be avoided as such.

A second important problem is nonlinear distortion. Less time is needed to block the MOS transistor when the input voltage is close to V_{Gon} for the effective threshold voltage becomes quite large. The amount of charge stored in C_2 varies with the magnitude of the input signal and the offset is thus a nonlinear replica of the input.

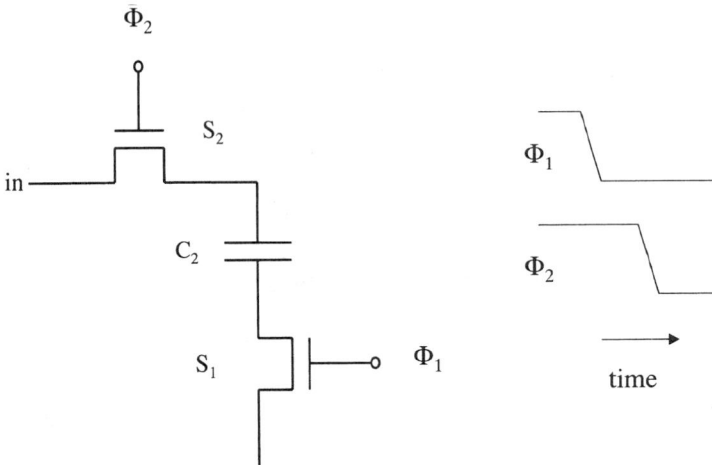

FIGURE 14.2.9   Switching noise nonlinearity can be lessened by means of the transistor S_2 in series with the storage capacitor.

This makes nonlinear distortion figures less than –70 dB hard to achieve unless a technique such as the one described hereafter is put to use.

In the circuit illustrated in Fig 14.2.9, the lower end of capacitor C_2 is tied to the ground by means of a second switch S_2. During the acquisition time, both S_2 and S_1 are conducting. Sampling occurs when the switch S_2 opens, shortly before S_1 opens. Suppose switching times of both transistors are short enough to avoid charge re-equilibration. When S_2 opens, the charge in the inversion layer divides equally between C_2 and ground. When S_1 opens, since C_2 is already open-ended, the signal dependent charge of S_1 has no other way out than to flow back to the generator. Thus, C_2 stores only charge from S_2, which is constant since the lower end of C_2 is tied always to ground. In fact, one exchanges a signal-dependent offset against a constant offset, which does not impair linearity. This offset moreover can be compensated easily by taking advantage of differential architectures that turn a constant offset into a common mode signal, which is ignored further on.

TRANSISTOR SWITCHES OTHER THAN LOGIC GATES

Transistor switches are extensively used in applications other than logic gates, covering a wide variety of both digital and analog applications. A typical illustration is the circuit converting the frequency of a signal into a proportional current, the so-called *diode pump*. This circuit (Fig. 14.2.10) consists of a capacitor C, two diodes D_1 and D_2, and a switch formed by a transistor T_1 and a resistor R. The transistor is assumed to be driven periodically by a square-wave source, alternatively on and off. When T_1 is blocked, the capacitor C charges through the diode D_1, while D_2 has no effect. As soon as the voltage across C has reached its steady-state value E_{cc}, T_1 may be turned on abruptly. The voltage with respect to ground at point A becomes negative, and D_1 is blocked, while D_2 is forward-biased. The capacitor thus discharges itself in the load (in Fig. 14.2.10 an ammeter, but it could be any other circuit element that does not exhibit storage), allowing V_A to reach 0 V before T_1 is again turned

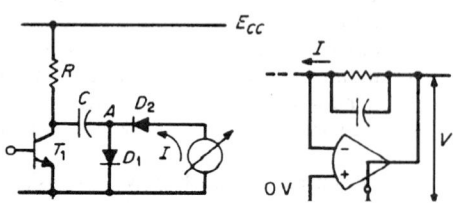

FIGURE 14.2.10   Diode-pump circuit.

off. The charge fed to the load thus amounts to CE_{cc} coulombs. If we suppose that this process is repeated periodically, the average current in the load is given by

$$I = fCE_{cc} \tag{15}$$

where f represents the switching repetition rate.

The diode-pump circuit provides a pulsed current whose average value is proportional to the frequency of the square-wave generator controlling the switching transistor T_1. The proportionality would of course be lost if the load exhibited storage, e.g., if the load were a parallel combination of a resistor and a capacitor in order to obtain the average current. Using an operational amplifier, as shown in the right side of Fig. 14.2.10, circumvents the problem.

The requirements on the switching transistor in this application are different and in many respects more stringent than for logic gates. The transistor in a logic circuit provides a way of defining two well-distinguished states, logic 1 and 0. Nothing further is required whether these states approach an actual short circuit or an open circuit. In the diode-pump circuit, however, the actual switching characteristics are important, since the residual voltage drop across the saturated transistor of Fig. 14.2.10 influences the charge transfer from C to the load, thereby also introducing unwanted temperature sensitivity. The main difference lies in the fact that while T_1 is operated as a logical element, the purpose of the circuit actually is to deliver an analog signal.

There are many other examples where the characteristics of switching transistors influence the accuracy of given circuits or instruments. An even more critical problem pertains to amplitude gating, since this class of applications requires switches which correctly transfer analog signals without introducing too much distortion. Furthermore, positive and negative signals must be transmitted equally well, and noise introduced by the gating signals must be minimized.

Analog gating.   A typical high-frequency gating network for analog signals is shown in Fig. 14.2.11. Gating is performed by means of the diode bridge in the center. All remaining circuitry controls the on-off switching.

In order to transmit the analog signal, the current sources Qbar and Q must, respectively, be on and off. The current $2I$ from transistor T_1 is split into two equal components, one that flows through T_3, the other that flows vertically through T_4 and the bridge. The second forward biases all the diodes. Current, moreover, is injected horizontally from the signal source, left, to the output terminal, right. Those in- and out-currents representing the analog signal are equal since the sum of all currents injected in the bridge must necessarily be zero and the vertical current components though the bridge are balanced by the network. Voltage drops across the diodes are supposed to compensate each other.

When the path between source and load must be interrupted, the current sources Q and Qbar take opposite states. No current then flows through the bridge and the extra-currents supplied by T_2 and T_3 are diverted, respectively, through T_6 and T_5. The two vertical nodes of the bridge are now connected to low impedance nodes so that the equivalent high-frequency network between in- and output terminals consist actually of two branches, each with two small parasitic capacitances representing series reverse-biased diodes short circuited in their middle to ground. This ensures an excellent separation between in- and output terminals making this type of gating network well suited for the sampling of high-frequency signals, like those used in sampling oscilloscopes.

Field-effect transistors also are extensively used to perform analog gating. A typical application is switched-capacitor filters. Figure 14.2.12a illustrates an elementary switched-capacitor network. In this circuit, the capacitor C is connected alternatively between the two terminals so that a charge $C(V_1 - V_2)$ is transferred at each cycle. Hence, if the repetition rate is f, the switched-capacitor network allows an average direct-current $C(V_1 - V_2)f$ to flow from one terminal to the other. It is thus equivalent to a resistor whose value is $1/Cf$.

If another capacitor C_o is connected at the output port, an elementary sampled-data RC circuit is built with a time constant equal to C_o/Cf. This time constant depends only on the clock frequency f and on the ratio of two capacitors. Hence, relatively large time constants can be achieved with good accuracy using very small capacitors and MOS transistor switches. In practice, MOS capacitors of a few picofarads match better than 0.1 percent. In addition, a slight modification (see Fig. 14.2.12b) of the circuit avoids the stray capacitance, which would otherwise affect the accuracy adversely. Hence, fully integrated switched-capacitor RC active filters can be designed to tight specifications, e.g., for telephone applications.

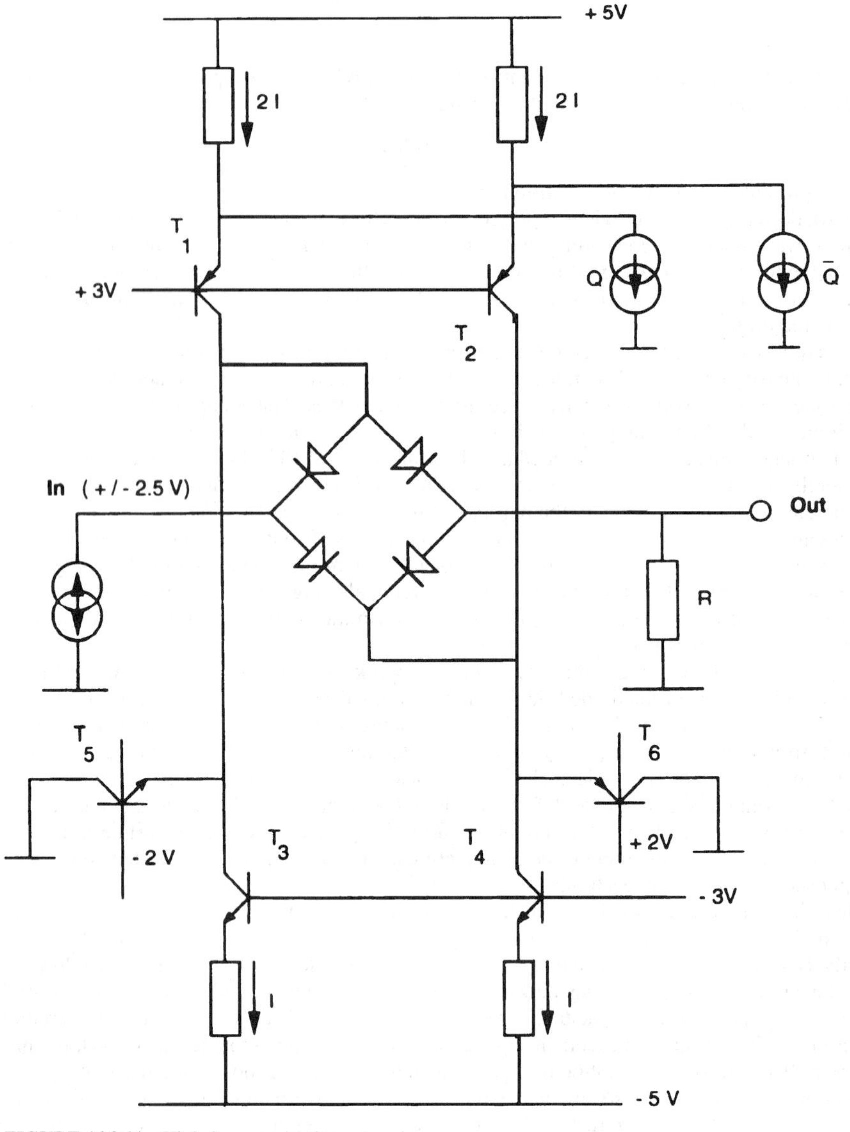

+ 5V
2 I
2 I
T 1
+ 3V
Q
Q̄
T 2
In (+ / - 2.5 V)
Out
R
T 5
T 6
- 2 V
+ 2V
T 3
T 4
- 3V
I
I
- 5 V

FIGURE 14.2.11 High-frequency gating network.

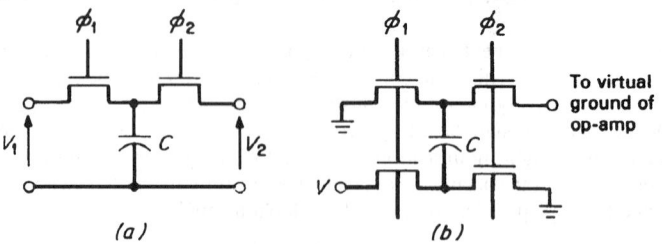

ϕ1
ϕ2
V1
C
V2
(a)
ϕ1
ϕ2
To virtual
ground of
op-amp
C
V
(b)

FIGURE 14.2.12 (*a*) Switched capacitor resistor; (*b*) the circuit is not affected by stray capacitances.

CHAPTER 14.3
ACTIVE WAVEFORM SHAPING

Paul G. A. Jespers

ACTIVE CIRCUITS

Linear active networks used currently for waveshaping take advantage of negative or positive feedback circuits to improve the performance. Of the linear negative feedback active wave-shaping circuits, the operational amplifier-integrator is widely used.

RC *OPERATIONAL AMPLIFIER-INTEGRATOR*

In Fig. 14.3.1 it is assumed that the operational amplifier has infinite input impedance, zero output impedance, and a high negative gain A. The overall transfer function is

$$\frac{A}{1 + p(1 - A)T} \quad \text{where } T = RC \tag{1}$$

This function represents a first-order system with gain A and a cutoff frequency, which is approximately $|A|$ times lower than the inverse of the time constant T of the RC circuit. In Fig. 14.3.1b the frequency response of the active circuit is compared with that of the passive RC integrator. The widening of the spectrum useful for integration is clearly visible. For instance, an integrator using an operational amplifier with a gain of 10^4 and an RC network having a 0.1 s time constant has a cutoff frequency as low as 1.6 MHz.

In the time domain, the Taylor expansion of the amplifier-integrator response to the step function is

$$V(t) = E\frac{t}{T}\left[1 - \frac{t}{2!|A|T} + \frac{t^2}{3!(|A|T)^2} - \cdots\right] \tag{2}$$

This shows that almost any desired degree of linearity of $V(t)$ can be achieved by providing sufficient gain.

SWEEP GENERATORS[4]

Sweep generators (also called time-base circuits) produce linear voltage or current ramps versus time. They are widely used in applications such as oscilloscopes, digital voltmeters, and television. In almost all circuits the linearity of the ramp results from charging or discharging a capacitor through a constant-current source. The difference between circuits used in practice rests in the manner of realizing the constant-current source. Sweep generators may also be looked upon as integrators with a constant-amplitude input signal. The latter point of view shows that RC operational amplifier-integrators provide the basic structure for sweep generation.

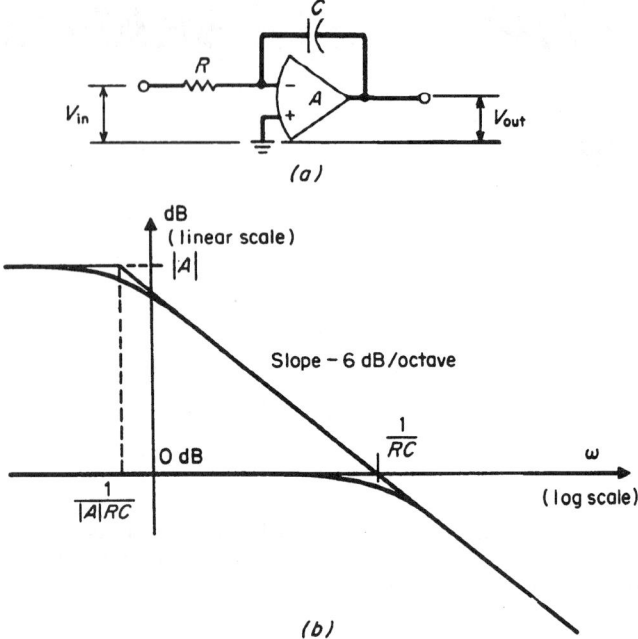

FIGURE 14.3.1   (*a*) Operational amplifier-integrator; (*b*) gain vs. angular frequency.

Circuits delivering a linear voltage sweep fall into two categories, the Miller time base and bootstrap time base. A simple Miller circuit (Fig. 14.3.2) comprises a capacitor C in a feedback loop around the amplifier formed by T_1. Transistor T_2 acts like a switch. When it is on, all the current flowing through the base resistor R_B is driven to ground, keeping T_1 blocked, since the voltage drop across T_2 is lower than the normal base-to-emitter voltage of T_1. The output signal V_{CE} of T_1 is thereby clamped at the level of the power-supply voltage E_{cc}, and the voltage drop across the capacitor C is approximately the same. When T_2 is turned off, it drives T_1 into the active region and

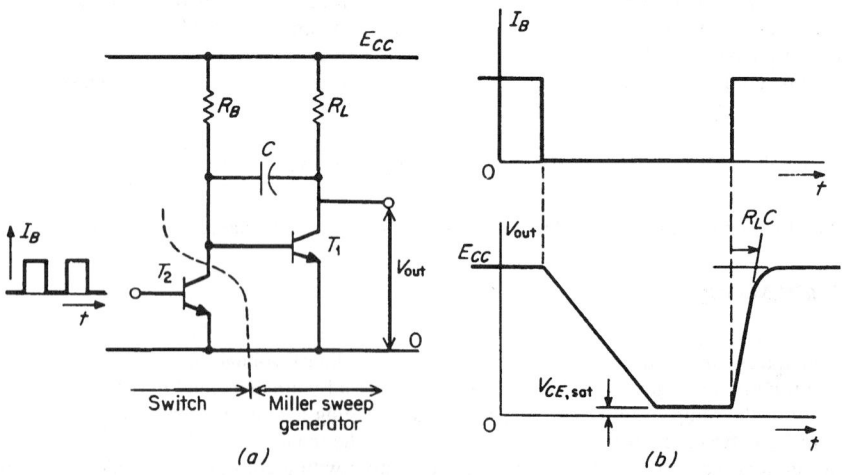

FIGURE 14.3.2   Miller sweep generator: (*a*) circuit; (*b*) input and output vs. time.

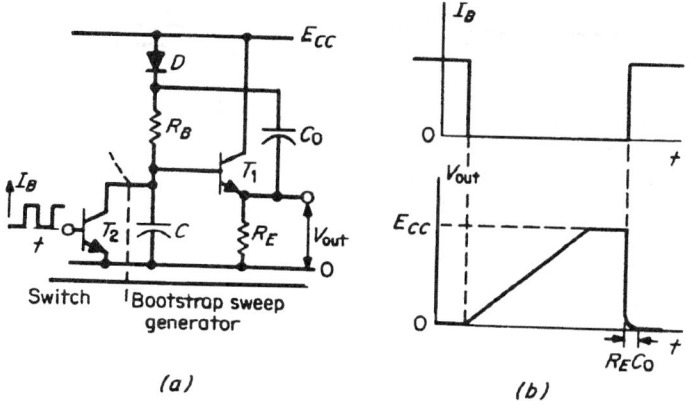

FIGURE 14.3.3 Bootstrap sweep generator: (*a*) circuit; (*b*) input and output vs. time.

causes collector current to flow through R_L. The resulting voltage drop across R_L is coupled capacitively to the base of T_1, tending to minimize the base current; i.e., the negative-feedback loop is closed. The collector-to-emitter voltage V_{CE} of T_1 subsequently undergoes a linear voltage sweep downward, as illustrated in Fig. 14.3.2*b*.

The circuit behaves in the same manner as the RC operational amplifier above. Almost all the current flowing through R_B is derived through the feedback capacitor, and only a very small part is used for controlling the base of T_1. The feedback loop opens when T_1 enters into saturation, and the voltage gain of the amplifier becomes small.

When T_2 is subsequently turned on again, blocking T_1 and recharging C through R_L and the saturated switch, the output voltage V_{CE} rises again according to an exponential with time constant $R_L C$.

Figure 14.3.3 shows a typical bootstrap time-base circuit. It differs from the Miller circuit in that the capacitor C is not a part of the feedback loop. Instead the amplifier is replaced by an emitter-follower delivering an output signal V_{out} which reproduces the voltage drop across the capacitor. C is charged through resistor R_B from a floating voltage source formed by the capacitor C_0 (C_0 is large compared with C).

First, we consider that the switch T_2 is on. Current then flows through the series combination formed by the diode D, the resistor R_B, and the saturated transistor T_2. The emitter follower T_1 is blocked since T_2 is saturated. Moreover, the capacitor C_0 can charge through the path formed by the diode D and the emitter resistor R_E, and the voltage drop across its terminals is equal to E_{CC}. When T_2 is cut off, the current through R_B flows into the capacitor C, causing the voltage drop across its terminals to rise gradually, driving T_1 into the active region. Because T_1 is a unity-gain amplifier, V_{out} is a replica of the voltage drop across C.

Since C_0 acts as a floating dc voltage source, diode D is reverse-biased immediately. The current flowing through R_B is supplied exclusively by C_0. Since $C_0 \gg C$, the voltage across R_B remains practically constant and equal to the voltage drop across C_0 minus the base-to-emitter voltage of T_1.

Considering that the base current of T_1 represents only a small fraction of the total current flowing through R_B, it is evident that the charging of capacitor C occurs under constant-current and that therefore a linear voltage ramp is obtained as long as the output voltage of T_1 is not clamped to the level of the power-supply voltage E_{CC}.

The corresponding output waveforms are shown in Fig. 14.3.3*b*. After T_2 is switched on again, C discharges rapidly, causing V_{out} to drop, while the diode D again is forward-biased and the small charge lost by C_0 is restored. In practice, C_0 should be at least 100 times larger than C to ensure a quasi-constant voltage source.

More detailed analysis of the Miller and bootstrap sweep generators reveals that they are in fact equivalent. We redraw the Miller circuit as shown at the left of Fig. 14.3.4. Remembering that the operation of the sweep generator is independent of which output terminal is grounded, we ground the collector of T_1 and redraw the corresponding circuit. As shown at the right in the figure, this is a bootstrap circuit, so that the two circuits are equivalent.

Any sweep generator can be regarded as a simple loop (Fig. 14.3.5) comprising a capacitor C delivering a voltage ramp, a loading resistor R_B, and the series combination of two sources: a constant voltage source E_{cc} and a variable source whose emf E reproduces the voltage drop V across the capacitor. The voltage drop across

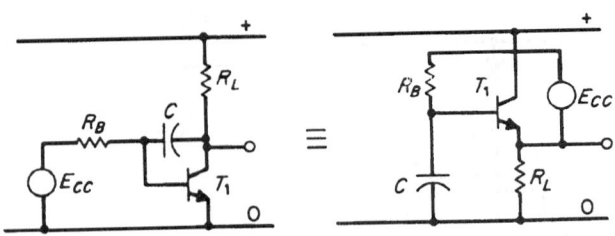

FIGURE 14.3.4 Equivalency of the Miller and bootstrap sweep generators.

R_B consequently remains constant and equal to E_{cc} making the loop current also constant. The voltage ramp consequently is given by

$$E = V = (E_{cc}/R_B C)t \tag{3}$$

Grounding terminal 1 yields the Miller network, while grounding terminal 2 leads to the bootstrap circuit.

Since linearity is one of the essential features of sweep generators, we consider the equivalent networks represented in Fig. 14.3.6. Starting with the Miller circuit, we determine the impedance in parallel with C

$$|A|(R_B \| h_{11}) \tag{4}$$

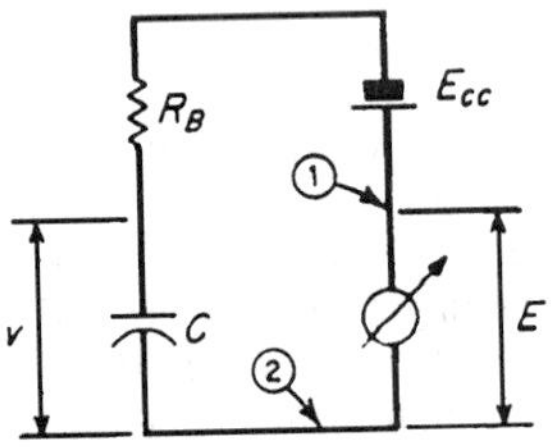

where $|A|$ is the absolute value of the voltage gain of the amplifier

$$|A| = (h_{21}/h_{11})R_L$$

Next, considering the bootstrap circuit, we calculate the input impedance of the unity-gain amplifier to determine the loading impedance acting on C. This impedance is

$$R_L h_{21} R_B/(R_B + h_{11}) \tag{5}$$

which turns out to be the same as that given in Eq. (4); i.e., the two circuits are equivalent. To determine the degree of linearity it is sufficient to consider the common equivalent circuit of Fig. 14.3.7 and to calculate the Taylor expansion of the voltage V

FIGURE 14.3.5 Basic loop of sweep-generator circuits.

$$V = \frac{E_{CC}}{R_B C}t\left[1 - \frac{t}{2!|A|(R_B\|h_{11})C} + \frac{t^2}{3![|A|(R_B\|h_{11})C]^2} - \cdots\right] \tag{6}$$

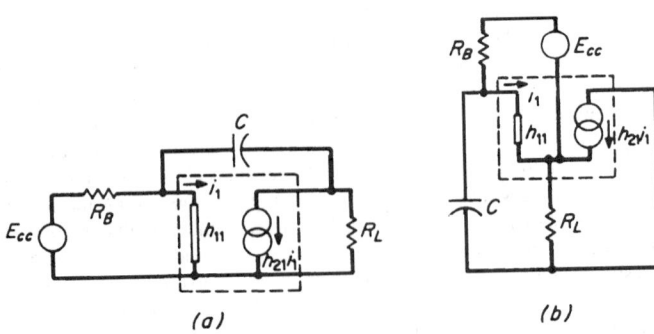

FIGURE 14.3.6 Equivalent forms of (*a*) Miller and (*b*) bootstrap sweep circuits.

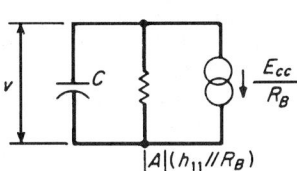

FIGURE 14.3.7   Common equivalent circuit of sweep generators.

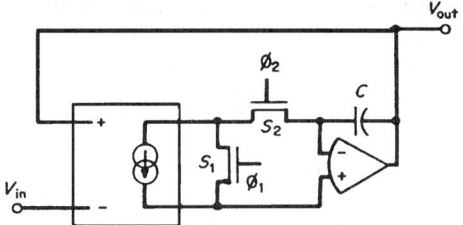

FIGURE 14.3.8   Typical sample-and-hold circuit.

The higher the voltage gain $|A|$, the better the linearity. Thus, an integrated operational amplifier in place of T_1 leads to excellent performance in both the Miller and the bootstrap circuit. Voltage gains as high as 10,000 are easily obtained for this purpose.

SAMPLE-AND-HOLD CIRCUITS[4]

Sample-and-hold circuits are widely used to store analog voltages accurately over time ranging from microseconds to minutes. They are basically switched-capacitor networks, but since the analog voltage across the storage capacitor in the hold mode must be sensed under low impedance, a buffer amplifier is needed. Op-amps with FET input are commonly used for this purpose to minimize the hold-mode droop. The schematic of a widely used integrated circuit is shown in Fig. 14.3.8.

Storage and readout are achieved by the FET input op-amp in the hold mode. During the acquisition time, transistor S_2 is conducting, while S_1 is blocked. Current is supplied by the voltage-dependent current source to minimize the voltage difference between input and output terminals. As soon as S_1 and S_2 change their states, V_{out} ceases to follow V_{in} and remains unchanged.

The main requirements for sample-and-hold circuits are low hold-mode droop, short settling time in the acquisition mode, low offset voltage, and small hold-mode feedthrough. The hold-mode droop is dependent on the leakage current of the op-amp inverting node. Short settling times require high-slew-rate op-amps and large current-handling capabilities for both the current source and the op-amp. The offset voltage is determined by the differential amplifier which controls the current source. Finally, feedthrough is a result of imperfect isolation between the current source and the op-amp. For this reason, a double switch is preferred to a single series switch.

Another important feedthrough problem is related to the unavoidable gate-to-source or drain-overlap capacitance of the MOS switch S_2. When the gate-control signal is switched off, some small charge is always transferred capacitively to the storage capacitor and a small voltage step is superimposed on the output terminal when the circuit enters the hold state. Minimization of this effect can be achieved by increasing the ratio of the storage capacitance to the switch-overlap capacitance. Since the latter cannot be made equal to zero, the storage capacitance must be chosen sufficiently large, but this inevitably lengthens the settling time. One means of alleviating the problem is to compensate the switching charge because of the control signal by injection of an equal and opposite charge on the inverting input node of the op-amp. This can be achieved by means of a dummy transistor controlled by the inverted signal.

NONLINEAR NEGATIVE FEEDBACK WAVEFORM SHAPING

The use of nonlinear devices with negative feedback accentuates the character of waveshaping networks. In many circumstances, this leads to an idealization of the nonlinear character of the devices considered. A good example of this is given by the ideal rectifier circuit.

The negative-feedback loop in the circuit shown in Fig. 14.3.9 is formed by two parallel branches. Each comprises a diode connected in such manner that if V_1 is positive, the current injected by resistor R_1 flows through D_1, and if V_1 is negative, through D_2. A resistor R_2 is placed in series with D_1, and the output voltage V_2 is taken at the node between R_2 and D_1. Hence, V_2 is given by $-(R_2/R_1)V_1$ when V_1 is positive, independently of the forward voltage drop across D_1.

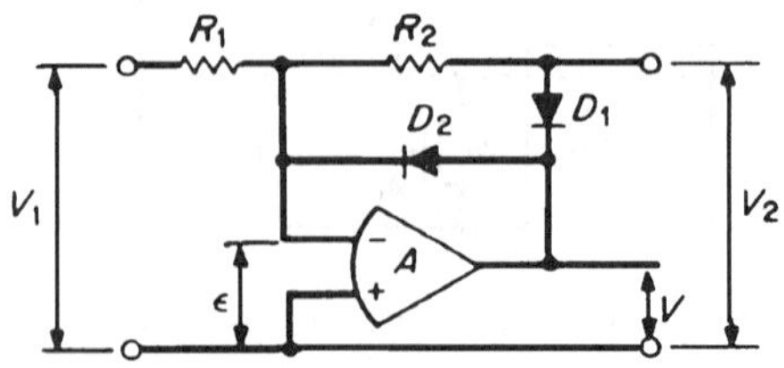

When D_1 is forward-biased, the voltage V at the output of the op-amp adjusts itself to force the current flowing through D_1 and R_L to be exactly the same as through R_1. This means that V may be much larger than V_2, especially when V_2 (and thus also V_1) is of the order of millivolts. In fact, V exhibits approximately the same shape as V_2 plus an additional pedestal of approximately 0.6 to 0.7 V. Typical waveforms obtained with a sinusoidal voltage of a few tens of millivolts are shown in Fig. 14.3.10.

FIGURE 14.3.9   The precision rectifier using negative feedback is almost an ideal rectifier.

The quasi-ideal rectification characteristic of this circuit is readily understood by considering the Norton equivalent network seen from R_2 and D_1 in series. In consists of a current source delivering the current V_1/R_1 in parallel with an almost infinite resistor $|A| R$, where A represents the voltage gain of the op-amp. Hence, the current flowing through the branch formed by R_2 and D_1 is delivered by a quasi-ideal current source, and the voltage drop across R_2 is unaffected by the series diode D_1. As for D_2, it is required to prevent the feedback loop from being opened when V_1 is negative. If this could happen, the artificial ground at the input of the op-amp would be lost and V_2 would not be zero.

Other negative-feedback configurations leading to very high output impedances are equally powerful in achieving ideal rectification characteristics. For instance, the unity-gain amplifier used in instrumentation has wide linear ac measurement capabilities (Fig. 14.3.11).

POSITIVE FEEDBACK WAVEFORM SHAPING

Positive feedback is used extensively in bistable, monostable, and astable (free-running) circuits. Astable networks include free-running relaxation circuits whether self-excited or synchronized by external trigger pulses. Monostable and bistable circuits also exist, with one and two distinct stable states, respectively.

The degree to which positive feedback is used in harmonic oscillators differs substantially from that of astable, monostable, or bistable circuits. In an oscillator the total loop gain must be kept close to 1. It needs to compensate only for small losses in the resonating tank circuit. In pulsed circuits, positive feedback permits

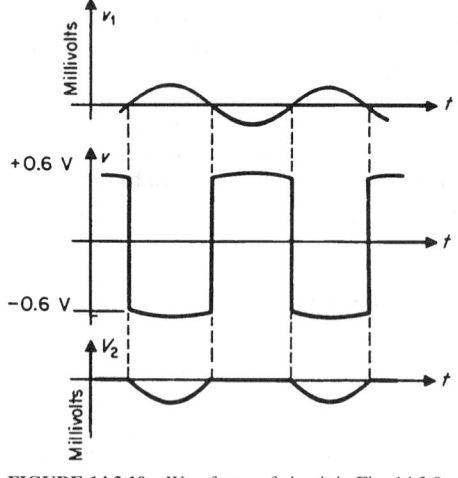

FIGURE 14.3.10   Waveforms of circuit in Fig. 14.3.9.

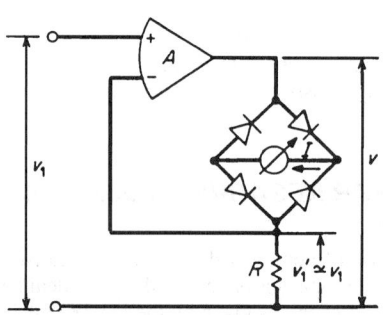

FIGURE 14.3.11   Feedback rectification circuit used in precision measurements.

fast switching from one state to another, e.g., from cutoff to saturation and vice versa. Before and after these occur, the circuit is passive. Switching occurs in extremely short times, typically a few ns. After switching, the circuit evolves more slowly, approaching steady-state conditions.

It is common practice to call the switching time the *regeneration time* and the time needed to reach final steady-state conditions the *resolution time*. The resolution time may range from tens of nanoseconds to several seconds or more, depending on the circuit.

An important feature of triggered regenerative circuits is that their switching times are essentially independent of the steepness of the trigger-signal waveshape. Once instability is reached the transition occurs at a rate fixed by the total loop gain and the reactive parasitics of the circuit itself but independent of the rate of change of the trigger signal itself. Regenerative circuits, therefore, provide means of restoring short rise times.

Positive-feedback pulse circuits are necessarily nonlinear. The most conventional way to study their behavior is to take advantage of piecewise-linear analysis techniques.

Bistable Circuits[5] (Collector Coupled)

Two cascaded common-emitter transistor stages implement an amplifier with a high positive gain. Connecting the output to the input (Fig. 14.3.12) produces an unstable network known as the *Eccles-Jordan bistable circuit* or *flip-flop*. Under steady-state conditions one transistor is saturated and the other is blocked.

Suppose the circuit of Fig. 14.3.12 has the value $R_L = 1\ k\Omega$, $R = 2.2\ k\Omega$, and $E_{cc} = 5\ V$. Suppose T_1 is at cutoff, and consider the equivalent network connected to the base of T_2. It can be viewed as an emf of 5 V and series resistances of 3.2 kΩ. The base current of T_2 is given by

$$I_{B2} = (5 - 0.7)/3.2 = 1.34\ mA \tag{7}$$

T_2 being saturated, the collector current is equal to E_{CC}/R_L or 5 mA. A current gain of 4 would be sufficient to ensure saturation of T_2. Hence the collector-to-emitter voltage across T_2 will be very small ($V_{CE,sat}$), and consequently T_1 will be blocked, as stated initially. The reverse situation, with T_1 saturated and T_2 cut off, is governed by identical considerations for reasons of symmetry. Two distinct stable states thus are possible.

When one of the transistors is suddenly switched from one state to the opposite, the other transistor automatically undergoes an opposite transition. At a given time both transistors conduct simultaneously, which increases the loop gain from zero to a high positive value. This corresponds to the regenerative phase, during which the circuit becomes active.

It is difficult to compute the regeneration time since the operating points of both transistors move through the entire active region, causing large variations of the small-signal parameters. Although determination of the regeneration time on the basis of a linear model is unrealistic and leads only to a rough approximation, we briefly examine this problem since it illustrates how the regeneration phase of unstable networks may be analyzed.

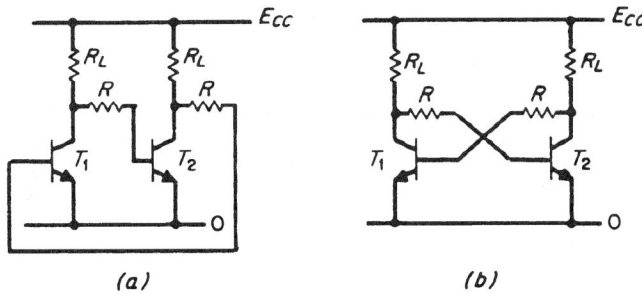

FIGURE 14.3.12 The Eccles-Jordan bistable circuit (flip-flop): (*a*) in the form of two cascaded amplifiers with output connected to input; (*b*) as customarily drawn, showing symmetry of connections.

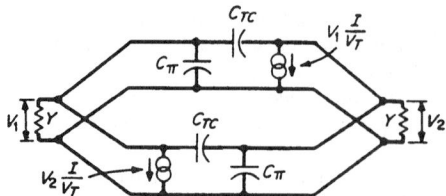

FIGURE 14.3.13 Flip-flop circuit showing capacitances that determine time constants.

First, we introduce two capacitors in parallel with the two resistors R. These capacitors provide a direct connection from collector to base under transient conditions and hence increase the high-frequency loop again. The circuit can now be described by the network of Fig. 14.3.13, which consists of a parallel combination of two reversed transistors without extrinsic base resistances (for calculation convenience) and with two load admittances Y which combine the load and resistive input of each transistor. Starting from the admittance matrix of one of the transistors with its load, we equate the determinant of the parallel combination

$$\begin{vmatrix} p(C_\pi + C_{TC}) & -pC_{TC} \\ \dfrac{I}{V_T} - pC_{TC} & pC_{TC} + Y \end{vmatrix} \tag{8}$$

to zero to find the natural frequencies of the circuit. This leads to

$$pC_\pi + (I/V_T) + Y = 0 \tag{9}$$

and

$$p(C_\pi + 4C_{TC}) + Y - I/V_T = 0 \tag{10}$$

where C_π stands for the parallel combination of C_{TE} and the diffusion capacitance $\tau_F I/V_T$. Only Eq. (10) has a zero with a real positive pole, producing an increasing exponential function with time constant approximately equal to

$$\tau = (C_\pi + 4C_{TC}) V_T/I \tag{11}$$

Since the diffusion capacitance overrules the transition capacitances at high current, Eq. (11) reduces finally to τ_F. This yields extremely short switching times. For instance, a transistor with a maximum transition frequency f_T of 300 MHz and a τ_F equal to 0.53 ns, exhibits a regeneration time (defined as the time elapsing between 10 and 90 percent of the total voltage excursion from cutoff to saturation or vice versa) equal to $2.2\tau_F$, or 1.2 ns.

A more accurate but much more elaborate analysis, taking into account the influence of the extrinsic base resistance and nonlinear transition capacitances in the region of small-collector current, requires a computer simulation based on the dynamic large-signal model of the bipolar transistor. Nevertheless, Eq. (11) clearly pinpoints the factors controlling the regeneration time; the transconductance and unavoidable parasitic capacitances. This is verified in many other positive-feedback switching circuits.

We consider next which factors control the resolution time still with the same numerical data. We suppose T_1 initially nonconducting and T_2 saturated. The sudden turnoff of T_2 is simulated by opening the short-circuit switch S_2 in Fig. 14.3.14a. Immediately, V_{CE2} starts increasing toward E_{cc}. The base voltage V_{BE1} of T_1 consequently rises with a time constant fixed only by the total parasitic capacitance C_T at the collector of T_2 and base of T_1 times the resistor R_L. Hence this time constant is

$$\tau_1 = R_L C_T \tag{12}$$

This time is normally extremely short; e.g., a parasitic capacitance of 1 pF yields a time constant of 1 ns. The charge accumulated across C evidently cannot change, for C is much larger than C_T. So V_{BE1} and V_{CE2} increase at the same rate. This situation is illustrated in Fig. 14.3.14b. When V_{BE1} reaches approximately 0.7 V, T_1 starts conducting and a new situation arises, illustrated in Fig. 14.3.14 by the passage from (b) to (c). This is when regeneration actually takes place, forcing T_1 to go into saturation very rapidly. With the regeneration period neglected, case (c) is characterized by the time constant

$$\tau_2 = (R_L/R)C \tag{13}$$

For instance, if C is equal to 10 pF, τ_2 yields 7 ns. Although this time constant is much longer than τ_1, it is still not the longest, for we have not yet considered the evolution of V_{BE2}.

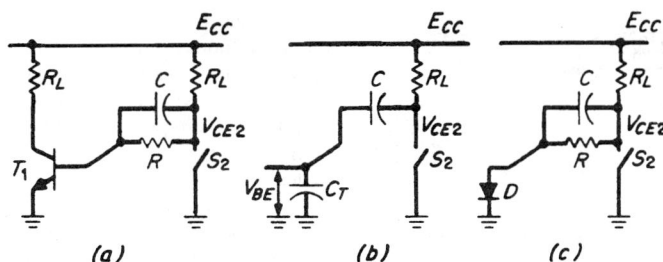

FIGURE 14.3.14   Piecewise analysis of flip-flop switching behavior. Transistor T_2 is assumed to be on before (a). The opening of S_2 simulates cutoff. The new steady-state conditions are reached in (c).

It is considered in Fig. 14.3.15, where the saturated transistor T_1 is replaced by the closing of S_1. The problem is the same as for the compensated attenuator. Since overcompensation is achieved, V_{BE2} undergoes a large negative-voltage swing almost equal to E_{cc} before climbing toward its steady-state value 0 V. The time constant of this third phase is given by

$$\tau_3 = RC \tag{14}$$

In the present case, τ_3 equals 22 ns.

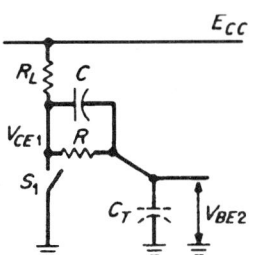

FIGURE 14.3.15   The longest time constant is experienced when T_1 is turned on. This is simulated by the closure of switch S_1.

The voltage variations versus time of the flip-flop circuit thus far analyzed are reviewed in Fig. 14.3.16 with the assumption that the regeneration time is negligible. Clearly C plays a double role. The first is favorable since it ensures fast regeneration and efficiently removes excess charges from the base of the saturated transistor, but the second is unfavorable since it increases the resolution time and sets an upper limit to the maximum repetition rate at which the flip-flop can be switched. The proper choice of C as well as of R_L and R must take this fact into consideration. Small values of the resistances make high repetition rates possible at the price of increased dc power consumption.

INTEGRATED-CIRCUIT FLIP-FLOPS

The Eccles-Jordan circuit (Fig. 14.3.12) is the basic structure of integrated bistable circuits. The capacitor C is not present.

The integrated flip-flop can be viewed as two cross-coupled single-input NOR or NAND circuits. In fact, integrated flip-flops vary only in the way external signals act upon them for control purposes. A typical example is given in Fig. 14.3.17 with the corresponding logic-symbol representation. The triggering inputs are called *set S* and *reset R* terminals. Transistors T_3 and T_4 are used for triggering.

The truth for NOR and NAND bistable circuits are

		NOR bistable			NAND bistable		
R	S	Q_1	Q_2	Line	Q_1	Q_2	Line
0	0	Q	Q	1	1	1	5
0	1	1	0	2	1	0	6
1	0	0	1	3	0	1	7
1	1	0	0	4	Q	Q	8

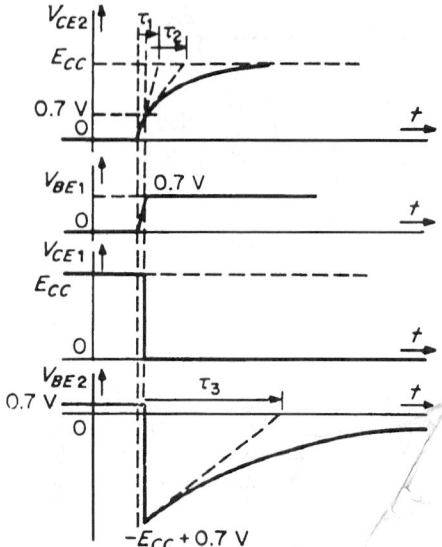

FIGURE 14.3.16   Voltage variations vs. time of flip-flop circuit.

Lines 1 and 8 correspond to situations where the S and R inputs are both inactive, leaving the bistable circuit in one of its two possible states indicated in the tables above by the letters Q and $\bar{Q}$ (Q may be either 1 or 0). If a specified output state is required, a pair of adequate complementary dc trigger signals is applied to the S and R inputs simultaneously.

For instance, if the output pair is to be characterized by $Q_1 = 1$ and $Q_2 = 0$, the necessary input combination, for NOR and NAND bistable circuits, is $S = 1$ and $R = 0$. Changing S back from 1 to 0 does not change anything in the output state in the NOR bistable. The same is true if S is made equal to 1 in the NAND bistable. In both cases, the flip-flop exhibits infinite memory of the stored state. The name *sequential circuit* is given to this class of networks as opposed to previous circuits, which are called combinational circuits.

Lines 4 and 5 must be avoided, for the passage from line 4 to line 1 or from line 5 to line 8 leads to uncertainty regarding the final state of the bistable circuit. In fact, the final transition is entirely out of the control of the input, since in both cases it results solely from small imbalances between transistor parasitics that allow faster switching of one or another inverter.

SYNCHRONOUS BISTABLE CIRCUITS[4]

Sequential networks may be either synchronous or asynchronous. The asynchronous class describes circuits in which the application of an input control signal triggers the bistable circuit immediately. This is true of the circuits thus far considered. In the synchronous class, changes of state occur only at selected times, after a clock signal has occurred.

Synchronous circuits are less sensitive to hazard conditions. Asynchronous circuits may be severely troubled by this effect, which results from differential propagation delays. These delays, although individually very small (typically of the order of a few nanoseconds), are responsible for introducing skew between signals that travel through different logic layers. Unwanted signal combinations may therefore appear for short periods and be interpreted erroneously.

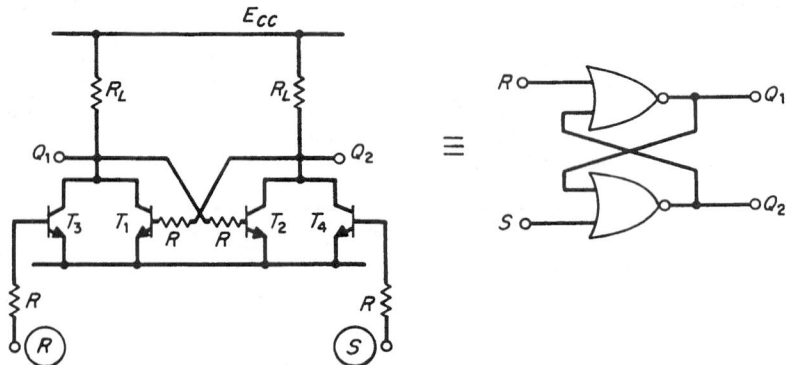

FIGURE 14.3.17   DC-coupled version of flip-flop, customarily used in integrated-circuit versions of this circuit.

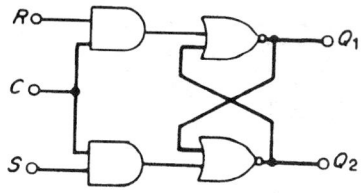

FIGURE 14.3.18 Synchronous flip-flop.

Synchronous circuits do not suffer from this limitation because they conform to the control signals only when the clock pulse is present, usually after the transient spurious combinations are over. A simple synchronous circuit is shown in Fig. 14.3.18. The inhibition action provided by the absence of the clock signal is provided by a pair of input AND circuits. Otherwise nothing is changed with respect to the bistable network.

A difficulty occurs in cascading bistable circuits, to achieve time-division. Instead of each circuit controlling its closest neighbor, when a clock signal is applied, the set and reset signals of the first bistable jump from one circuit to the next, traveling throughout the entire chain in a time which may be shorter than the duration of the clock transition. To prevent this, a time delay must be introduced between the gating NAND circuits and the actual bistable network, so that changes of state can occur only after the clock signal has disappeared.

One approach is to take advantage of storage effects in bipolar transistors, but the so-called *master-slave* association, shown in Fig. 14.3.19, is preferred. In this circuit, intermediate storage is realized by an auxiliary clocked bistable network controlled by the complement of the clock signal. The additional circuit complexity is appreciable, but the approach is practical in integrated-circuit technology.

The master-slave bistable truth table can be found from that of the synchronous circuit in Fig. 14.3.18, which in turn can be deduced from the truth table given in the previous section. One problem remains, however, the forbidden 1,1 input pair, which is responsible for ambiguous states each time the clock goes to zero. To solve this problem, the JK bistable was introduced (see Fig. 14.3.20). The main difference is the introduction of a double feedback loop. Hence the S and R inputs become, respectively, JQ and KQ. As long as the J and K inputs are not simultaneously equal to 1, nothing in fact is changed with respect to the behavior of the SR synchronous circuit.

When J and K are both high, the cross-coupled output signals fed back to the input gates cause the flip-flop to toggle under control of the clock signal. The truth table then becomes

J	K	Q_{n+1}	Line
0	0	Q_n	1
1	0	1	2
0	1	0	3
1	1	Q_n	4

Q_{n+1} stands for Q at the clock time $n + 1$ and Q_n for Q at clock time n. Lines 1 to 3 match the corresponding lines of the NOR bistable truth table. Line 4 indicates that state transitions occur each time the clock signal goes from high to low. The corresponding logic equation of the JK flip-flop, therefore, is

$$Q_{n+1} = J\bar{Q}_n + \bar{K}Q_n \tag{15}$$

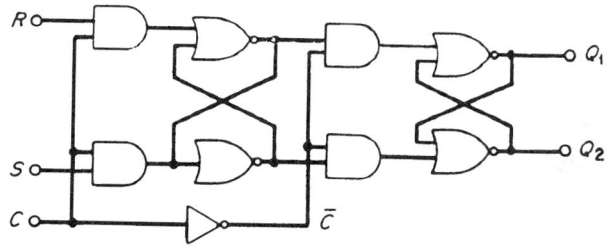

FIGURE 14.3.19 Master-slave synchronous flip-flop.

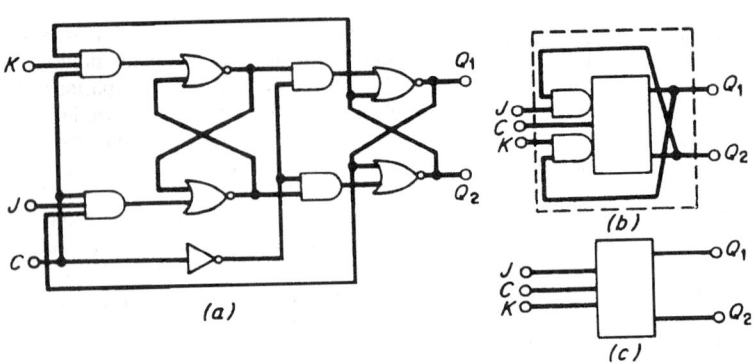

FIGURE 14.3.20 *JK* Flip-flop.

When only one control signal is used, J, for instance, and K is obtained by negation of the J signal, a new type of bistable is found, which is called the D flip-flop. The name given to the J input is D. Since K is equal to J, Eq. (15) reduces to

$$Q_{n+1} = D \qquad (16)$$

Hence, in this circuit the output is set by the input D after a clock cycle has elapsed. Notice that the flip-flop is insensitive to changes of D occurring while the clock is high.

D flip-flops without the master-slave configuration also exist, but their output state follows the D signal if changes occur while the clock is high. These bistables can be used to latch data. Several D flip-flops controlled by the same clock form a register for data storage. The clock signal then is called an enable signal.

Bistable Circuits, Emitter-Coupled Bistables (Schmitt Circuits)[6]

In the basic Schmitt circuit represented in Fig. 14.3.21 bistable operation is obtained by a positive-feedback loop formed by the common-base and common-collector transistor pair (respectively T_1 and T_2). The Schmitt circuit can be considered as a differential amplifier with a positive-feedback loop, which is a series-parallel association.

Emitter-coupled bistables are fundamentally different from Eccles-Jordan circuits, since no transistor saturates in either of the two stable states. Storage effects therefore need not be considered.

The two permanent states are examined in Fig. 14.3.22. In each state, (*a*) as well as (*b*), one transistor operates in the common-collector configuration while the other is blocked. In Fig. 14.3.22*a*, the collector voltage V_{C1} of T_1 and base voltage V_{B2} of T_2 are given by

$$V_{C1} = E_{cc} \frac{R_1 + R_2}{R_1 + R_2 + R_c}$$

$$V_{B2} = E_{cc} \frac{R_1}{R_1 + R_2 + R_c} = V_h \qquad (17)$$

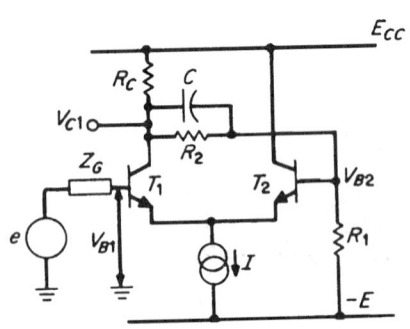

FIGURE 14.3.21 Emitter-coupled Schmitt circuit, showing positive-feedback loop.

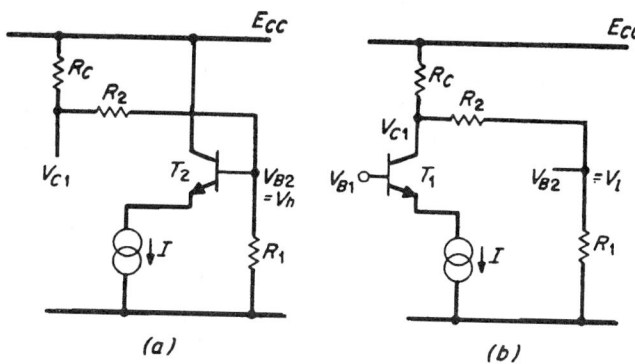

FIGURE 14.3.22   Execution of transfer in Schmitt circuit: (*a*) with T_1 blocked; (*b*) with T_1 conducting.

When the other stable state (*b*) is considered,

$$V_{C1} = (E_{cc} - R_c I)\frac{R_1 + R_2}{R_1 + R_2 + R_c}$$

$$V_{B2} = (E_{cc} - R_c I)\frac{R_1}{R_1 + R_2 + R_c} = V_l \qquad (18)$$

The situation depicted in Fig. 14.3.22 remains unchanged as long as the input voltage V_{B1} applied to T_1 is kept below the actual value V_h. In the other state (*b*), T_2 will be off as long as V_{B1} is larger than V_l. A range of input voltages between V_h and V_l thus exists where either of the two states is possible.

To alleviate the ambiguity, let us consider an input voltage below the smallest of the two possible values of V_{B2} so that the transistor T_1 necessarily is blocked. This corresponds to the situation of Fig. 14.3.22*a*. Now let the input voltage be gradually increased. Nothing will happen until V_{B1} approaches V_h. When the difference between the two base voltages is reduced to 100 mV or less, T_1 will start conducting and the voltage drop across R_c will lower V_{B2}. The emitter current of T_2 will consequently be reduced, and more current will be fed back to T_1. Hence, an unstable situation is created, which ends when T_1 takes over all the current delivered by the current source and T_2 is blocked.

Now the situation depicted in Fig. 14.3.22*b* is reached. The base voltage of T_2 becomes V_l, and the input voltage may either continue to increase or decrease without anything else happening as long as V_{B1} has not reached V_l. When V_{B1} approaches V_l, another unstable situation is created causing the switching from (*b*) to (*a*). Hence, the input-output characteristic of the Schmitt trigger is as shown in Fig. 14.3.23 with a hysteresis loop.

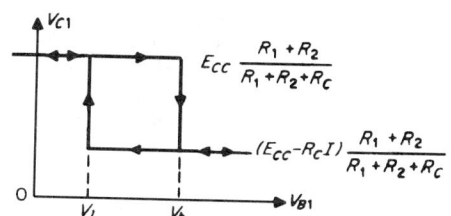

FIGURE 14.3.23   Input-output characteristic of Schmitt circuit, showing rectangular hysteresis.

Schmitt triggers are suitable for detecting the moment when an analog signal crosses a given *DC* level. They are widely used in oscilloscopes to achieve time-base synchronization. This is illustrated in Fig. 14.3.24, which shows a periodic signal triggering a Schmitt circuit and the corresponding output waves. It is possible to modify the switching levels by changing the operating points of the transistors electrically, e.g., by modifying the current delivered by the current source.

In many applications, the width of the hysteresis does not play a significant role. The width can be decreased, however, by increasing the attenuation of the resistive divider formed by R_1 and R_2, but one should not go below 1 V because sensitivity to variations in circuit components or supply voltage may occur.

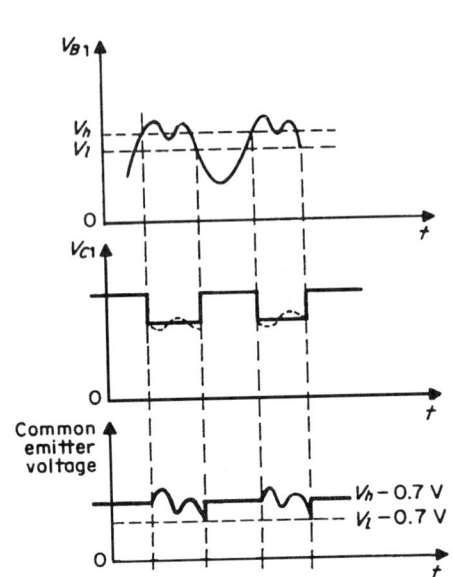

FIGURE 14.3.24 Trigger input and output voltage of Schmitt circuit; solid line delivered by a current source; broken line delivered by a resistor.

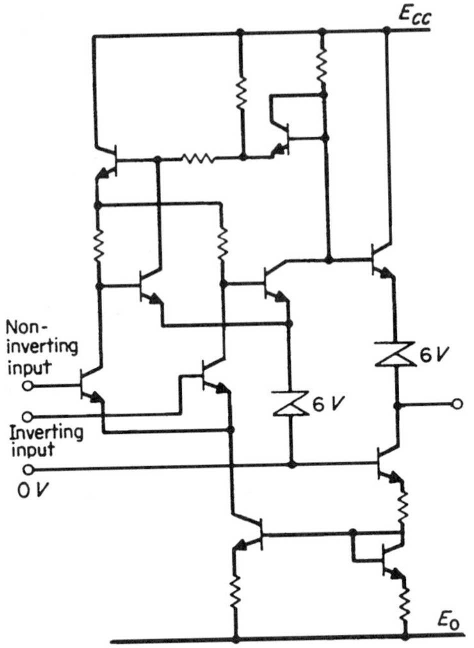

FIGURE 14.3.25 Bipolar integrated version of a comparator.

Furthermore, the increased attenuation in the feedback loop must be compensated for by a corresponding increase in the differential amplifier gain. Otherwise the loop gain may fall below 1, preventing the Schmitt circuit from functioning. A hysteresis of a few millivolts is therefore difficult to achieve.

A much better solution is to use *comparators* instead of Schmitt triggers when hysteresis must be avoided. A typical integrated comparator is shown in Fig. 14.3.25. It is a medium-gain amplifier (10^3 to 10^4) with a very fast response (a few nanoseconds) and an excellent slew rate. Comparators are not designed to be used as linear amplifiers like op-amps. Their large gain-bandwidth product makes them inappropriate for feedback configurations. They inevitably oscillate in any type of closed loop. In the open-loop configuration they behave like clipping circuits with an exceedingly small input range, which is equal to the output-voltage swing, usually 5 V, divided by the gain. The main difference compared with Schmitt triggers is the fact that comparators do not exhibit hysteresis. This makes a significant difference when considering very slowly varying input signals.

In the circuit of Fig. 14.3.21 the common-emitter current source can be replaced by a resistor. This solution introduces some common-mode sensitivity. The output signal does not look square, as shown in Fig 14.3.24 by the dashed lines. If unwanted, this effect can be avoided by taking the output signal at the collector of T_2 through an additional resistor, since the current flowing through T_2 is constant. An additional advantage of the latter circuit is that the output load does not interfere with the feedback loop.

INTEGRATED-CIRCUIT SCHMITT TRIGGERS[7]

Basically a Schmitt trigger can always be implemented by means of an integrated differential amplifier and an external positive-feedback loop. If the amplifier is an op-amp, poor switching characteristics are obtained unless an amplifier with a very high slewing rate is chosen. If a comparator is considered instead of an op-amp, switching will be fast but generally the output signal will exhibit spurious oscillations during the transition period. The oscillatory character of the output signal is due to the trade-off between speed and stability which is typical of comparators compared with op-amps. Any attempt to create a dominant pole, in fact, inevitably would ruin their speed performance.

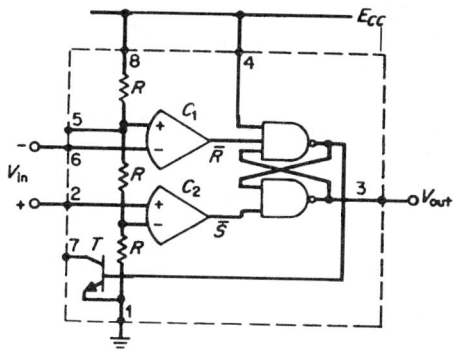

FIGURE 14.3.26   Precision integrated Schmitt trigger.

The integrated-circuit counterpart of the Schmitt trigger is shown in Fig. 14.3.26. It consists of two comparators connected to a resistive divider formed by three equal resistors R. Input terminals 2 and 6 are normally tied together. The output signals of the two comparators control a flip-flop. When the input voltage is below $E_{cc}/3$, the flip-flop is set. Similarly when the input voltage exceeds $2E_{cc}/3$, the circuit is reset.

The actual state of the flip-flop in the range between $E_{cc}/3$ and $2E_{cc}/3$ will depend on how the input voltage enters the critical zone. For instance, if the input voltage starts below $E_{cc}/3$ and is increased so that it changes the state of comparator C_2 but not that of comparator C_1, both S and R are equal to 1 and the flip-flop remains set. The state changes only if the input voltage exceeds the limit $2E_{cc}/3$. Similarly, if the input voltage is lowered, setting the flip-flop will occur only when $E_{cc}/3$ is reached. Hence the circuit of Fig. 14.3.26 behaves like a Schmitt trigger with a hysteresis width $E_{cc}/3$ depending only on the resistive divider $3R$ and the offset voltages of the two comparators C_1 and C_2. This circuit can therefore be considered as a precision Schmitt trigger and is widely used as such.

Monostable and Astable Circuits (Discrete Components)[4]

Figures 14.3.27 and 14.3.28 show monostable and astable collector-coupled pairs, respectively. The fundamental difference between these circuits and bistable networks lies in the way DC biasing is achieved. In Fig. 14.3.27a, T_2 is normally conducting except when a negative trigger pulse drives this transistor into the cutoff region. T_1 necessarily undergoes the inverse transition, suddenly producing a large negative voltage step at the base of T_2 shown in Fig. 14.3.27b. V_{BE2}, however, cannot remain negative since its base is connected to the positive-voltage supply through the resistor R_1. The base voltage rises toward E_{cc}, with a time constant R_1C. As soon as the emitter junction of T_2 becomes forward-biased, the monostable circuit changes its state again and the circuit remains in that state until another trigger signal is applied.

The time T between the application of a trigger pulse and the instant T_2 saturates again is given approximately by

$$T = \tau \ln 2 = 0.693\tau \tag{19}$$

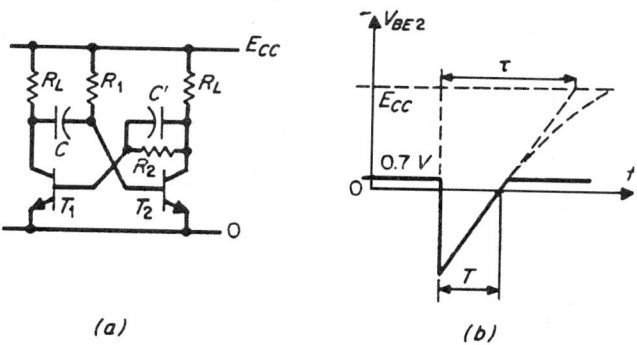

(a)   **(b)**

FIGURE 14.3.27   Monostable collector-coupled pair: (*a*) circuit; (*b*) output vs. time characteristics.

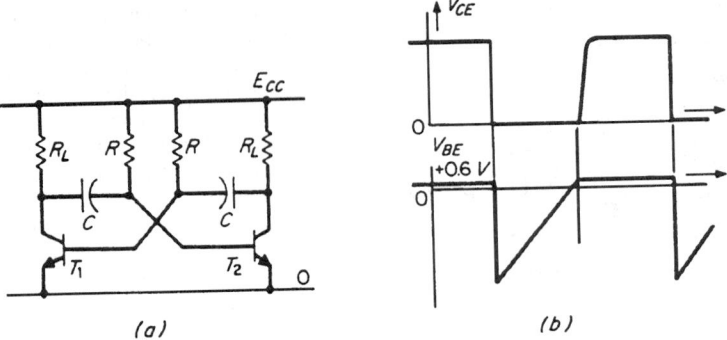

FIGURE 14.3.28   Astable (free-running) flip-flop: (*a*) circuit; (*b*) output vs. time character-istics.

where $\tau = R_1 C$. The supply voltage E_{CC} is supposed to be large compared with the forward-voltage drop of the emitter junction of T_2 for this expression to apply.

The astable collector-coupled pair, or free-running multivibrator (Fig. 14.3.28), operates according to the same scheme except that steady-state conditions are never reached. The base-bias networks of both transistors are connected to the positive power supply. The period of the multivibrator thus equals $2T$ if the circuit is symmetrical, and the repetition rate F_r is given by

$$F_r = \frac{1}{2\tau \ln 2} \approx \frac{0.7}{RC} \tag{20}$$

INTEGRATED MONOSTABLE AND ASTABLE CIRCUITS[7]

The discrete component circuits discussed above are interesting only because of their inherent simplicity and exemplative value. Improved means to integrate monostable and astable are shown below.

The circuit shown in Fig. 14.3.29 is derived from the Schmitt trigger. It is widely used in order to implement high-frequency (100 MHz) relaxation oscillators. The capacitor C provides a short circuit between the emitters of the two transistors, closing the positive-feedback loop during the regeneration time. As long as one or the other of the two transistors is cut off, C offers a current sink to the current source connected to the emitter of the blocked transistor. The capacitor thus is periodically charged and discharged by the two current sources, and the voltage across its terminal exhibits a triangular waveform. The collector current of T_1 is either zero or $I_1 + I_2$, so that the resulting voltage step across R_c is $(R_B \parallel R_C)(I_1 + I_2)$. Since the base of T_2 is directly connected to the collector of T_1, the same voltage step controls T_2 and determines the width of the input hysteresis, i.e., the maximum amplitude of the voltage sweep across C. The period of oscillation is computed from

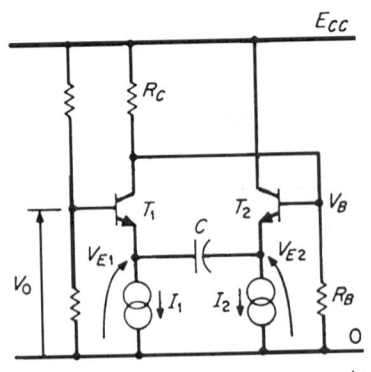

FIGURE 14.3.29   Discrete-component emitter-coupled astable circuit.

$$T = C\left(\frac{1}{I_1} + \frac{1}{I_2}\right)(R_B \parallel R_C)(I_1 + I_2) \tag{21}$$

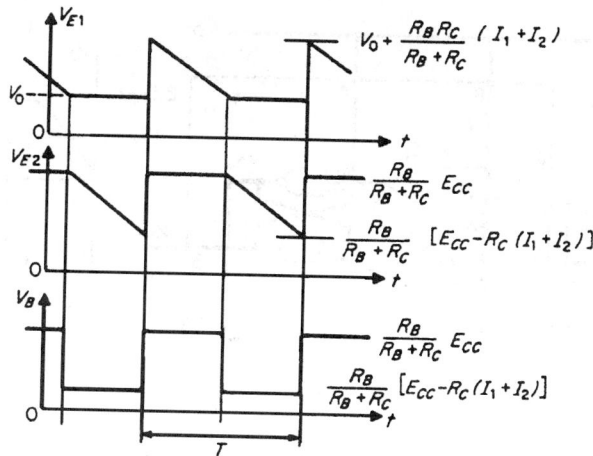

FIGURE 14.3.30   Waveforms of circuit in Fig. 14.3.29.

When, as is usual, both current sources deliver equal currents, the expression for T reduces to

$$T = 4(R_B \parallel R_C)C \qquad (22)$$

T does not depend, in this case, on the amplitude of the current because changes in current in fact modify the amplitude and the slope of the voltage sweep across C in the same manner. A review of the waveforms obtained at various points of the circuit is given in Fig. 14.3.30.

Integrated monostable and astable circuits can be derived from the precision Schmitt trigger circuit shown in Fig. 14.3.26. The monostable configuration is illustrated in Fig. 14.3.31.

Under steady-state conditions, the flip-flop is set and the transistor T saturated. The input voltage V_{in} is kept somewhere between the two triggering levels $E_{cc}/3$ and $2E_{cc}/3$. To initiate the monostable condition, it is sufficient that V_{in} drops below $E_{cc}/3$ even for a very short time, in order to reset the flip-flop and prevent T from conducting. The current flowing through R_1 then charges C until the voltage V_C reaches the triggering level

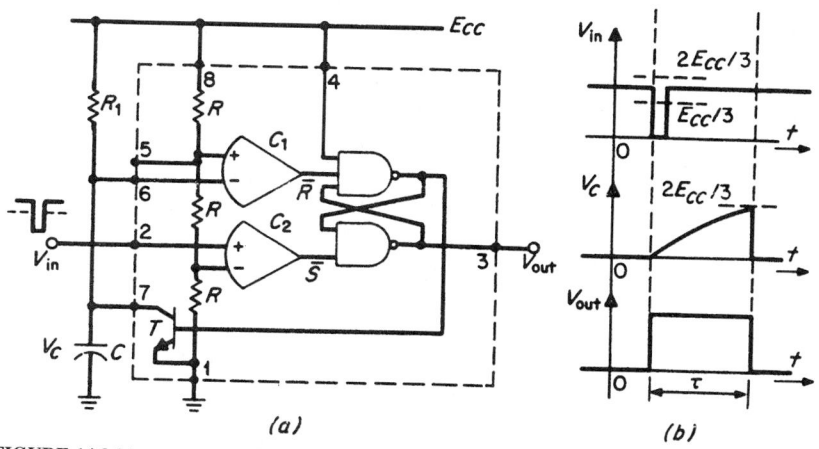

FIGURE 14.3.31   Monostable precision Schmitt trigger: (*a*) circuit; (*b*) waveforms.

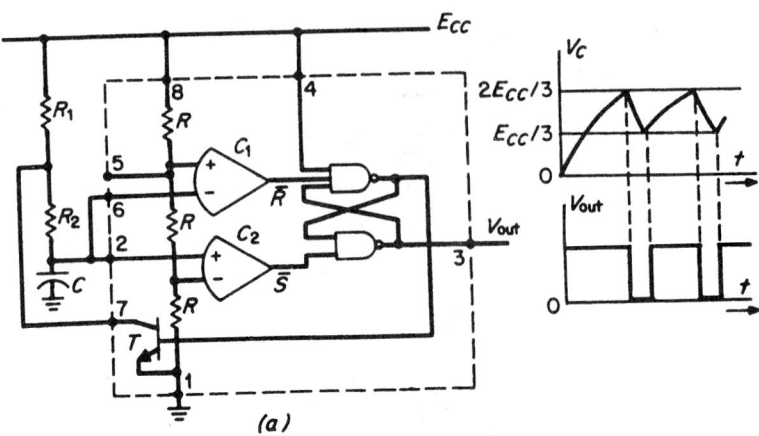

FIGURE 14.3.32 Astable precision Schmitt trigger: (a) circuit; (b) waveforms.

$2E_{cc}/3$. Immediately thereafter, the circuit switches to the opposite state and transistor T discharges C. The monostable circuit remains in that state until a new triggering pulse V_{in} is fed to the comparator C_2. The waveforms V_{in}, V_C, and V_{out} are shown in Fig. 14.3.31b. This circuit is also called a *timer* because it provides constant-duration pulses, triggered by a short input pulse. A slight modification of the external control circuitry may change this monostable into a retriggerable timer.

The astable version of the precision Schmitt trigger is shown in Fig. 14.3.32. Its operation is easily understood from the preceding discussion. The capacitor C is repetitively discharged through R_2 in series with the saturated transistor T and recharged through $R_1 + R_2$. The voltage V_C therefore consists of two distinct exponentials clamped between the triggering levels $E_{cc}/3$ and $2E_{cc}/3$. The frequency is equal to $1.44/(R_1 + 2R_2)C$. Because of the precision of the triggering levels (10^{-3} to 10^{-4}) short-term frequency stability can be achieved.

CHAPTER 14.4
DIGITAL AND ANALOG SYSTEMS

Paul G. A. Jespers

INTEGRATED SYSTEMS

With the trend toward ever higher integration levels, an increasing number of *ICs* combine some of the circuits seen before in order to build large systems, digital as well as analog, or mixed, such as wave generators and A/D and D/A converters. Some of these are reviewed below.

COUNTERS[4,7]

To count any number N of events, at least k flip-flops are required, such that

$$2^k \geq N \tag{1}$$

Ripple Counters

JK flip-flops with J and K inputs equal to 1 are *divide-by-2* circuits. Hence, a cascade of k flip-flops with each output Q driving the clock of the next circuit forms a divide-by-2^k chain, or a binary counter (see Fig. 14.4.1). The main drawback of this circuit is its increasing propagation delay with k. When all the flip-flops switch, the clock signal must ripple through the entire counter. Hence, enough time must be allowed to obtain the correct count. Furthermore, the delays between the various stages of the counter may produce glitches, e.g., when parallel decoding is achieved.

Synchronous Binary Counters

Minimization of delay and glitches can be achieved by designing synchronous instead of asynchronous counters. In a synchronous counter all the clock inputs of the *JK* flip-flops are driven in parallel by a single clock signal. The control of the counter is achieved by driving the J input by means of the AND combination of all the preceding Q outputs, as shown in Fig. 14.4.2. In this manner, all state changes occur on the same trailing edge of the clock signal. The only remaining requirement is to allow enough time between clock pulses for the propagation through a single flip-flop and an AND gate. The drawback of course is increased complexity with the order k.

Divide-by-*N* Synchronous Counters

When the number N of counts cannot be expressed in binary form, auxiliary decoding circuitry is required. This is true also for counters that provide truncated and irregular count sequences. Their synthesis is based

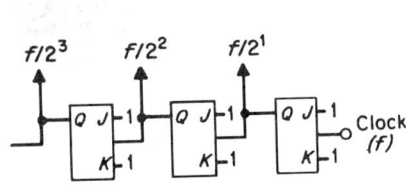

FIGURE 14.4.1 Ripple counter formed by cascading flip-flops.

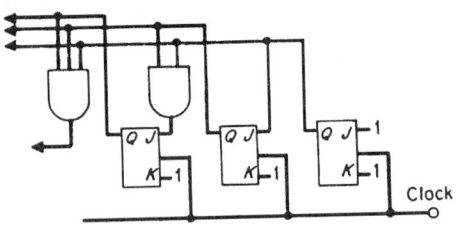

FIGURE 14.4.2 Synchronous counter.

on the so-called transition tables. The basic transition table of the JK flip-flop is derived easily from its truth table.

Q_n	Q_{n+1}	J	K	Line
0	0	0	×	1
0	1	1	×	2
1	0	×	1	3
1	1	×	0	4

Line 1, for instance, means that in order to maintain Q equal to 0 after a clock signal has occurred, the J input must be made equal to 0 whatever K may be (X stands for "don't care"). This can be easily verified with the truth table (lines 1 and 3). Hence, the synthesis of a synchronous counter consists simply of determining the J and K inputs of all flip-flops that are needed to obtain a given sequence of states.

Once the J and K truth tables have been obtained, classical minimization procedures can be used to synthesize the counter. For instance, consider the design of a divide-by-5 synchronous counter for which a minimum of three flip-flops is required. First, the present and next states of the flip-flops are listed. Then the required J and K inputs are found by means of the JK transition table:

	Present state			Next state			JK inputs					
	Q_3	Q_2	Q_1	Q_3	Q_2	Q_1	J_3	K_3	J_2	K_2	J_1	K_1
0	0	0	0	0	0	1	0	×	0	×	1	×
1	0	0	1	0	1	0	0	×	1	×	×	1
2	0	1	0	0	1	1	0	×	×	0	1	×
3	0	1	1	1	0	0	1	×	×	1	×	1
4	1	0	0	0	0	0	×	1	0	×	0	×

Using Karnaugh minimization techniques, one finds

$$J_3 = Q_1 Q_2 \quad K_3 = 1$$
$$J_2 = Q_1 \quad K_2 = Q_1$$
$$J_1 = Q_3 \quad K_1 = 1$$

The corresponding counter is shown in Fig. 14.4.3. With this procedure it is quite simple to synthesize a decimal counter with four flip-flops. The same method also applies to the synthesis with D flip-flops.

Up-Down Counters

Upward and downward counters differ from the preceding ones only by the fact that an additional bit is provided to select the proper J and K controls for up or down count. The same synthesis methods are applicable.

Presettable Counters

Since a counter is basically a chain of flip-flops, parallel loading by any number within the count sequence is readily possible. This can be achieved by means of the set terminals. Hence, the actual count sequence can be initiated from any arbitrary number.

SHIFT REGISTERS[3]

Shift registers are chains of flip-flops connected so that the state of each can be transferred to its next left or right neighbor under control of the clock signal. Shift registers can be built with *JK* as well as *D* flip-flops. An example of a typical bidirectional shift register is shown in Fig. 14.4.4. Shift registers, such as counters, may be loaded in parallel or serial mode. This is also true for reading out.

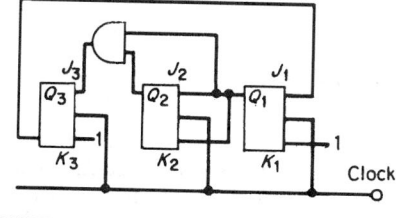

FIGURE 14.4.3   Synchronous divide-by-5 circuit.

In MOS technology, dynamic shift registers can be implemented in a simple manner. Memorization of states occurs electrostatically rather than by means of flip-flops. The information is stored in the form of charge on the gate of an MOS transistor. The main advantage of MOS dynamic shift registers is area saving resulting from the replacement of flip-flops by single MOS transistors. A typical dynamic 2-phase shift register (Fig. 14.4.5) consists of two cascaded MOS inverters connected by means of series switches. These switches, T_3 and T_6, are divided into two classes: *odd* switches controlled by the clock ϕ_1, and *even* switches controlled by ϕ_2.

The control signals determine two nonoverlapping phases. When ϕ_1 turns on the odd switches, the output signal of the first inverter controls the next inverter but T_6 prevents the information from going further. When ϕ_2 turns on, data are shifted one half cycle further. The signal jumps from one inverter to the next until it reaches the last stage.

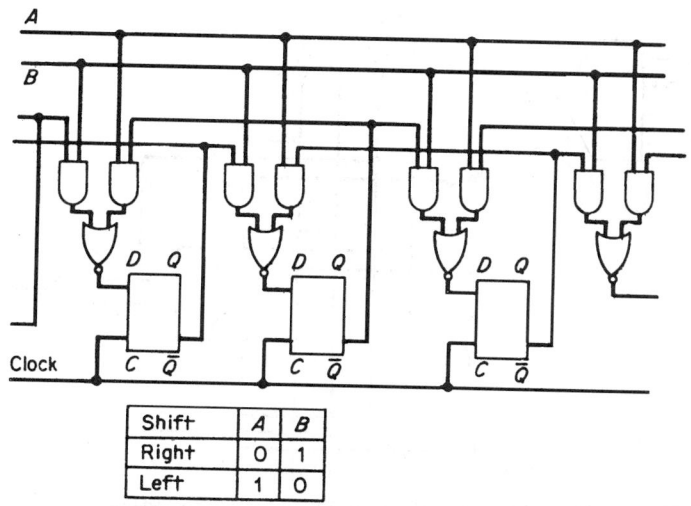

Shift	A	B
Right	0	1
Left	1	0

FIGURE 14.4.4   Bidirectional shift register.

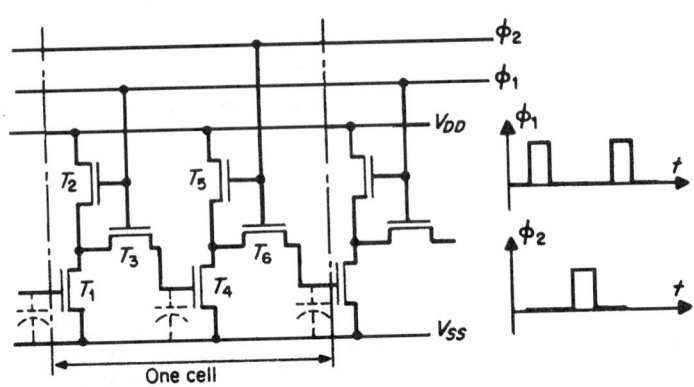

FIGURE 14.4.5   A 2-phase MOS dynamic shift register.

MULTIPLEXERS, DEMULTIPLEXERS, DECODERS, ROMS, AND PLAS[3-8]

A multiplexer is a combinatorial circuit that selects binary data from multiple input lines and directs them to a single output line. The selected input line is chosen by means of an address word. A representation of a 4-input multiplexer (MUX) is shown in Fig. 14.4.6 as well as a possible implementations. MOS technology lends itself to the implementation of multiplexers based on pass transistors. In the example illustrated by Fig. 14.4.7 only a single input is connected to the output through two conducting series transistors according to the $S_1 \, S_0$ code address.

Multiplexers may be used to implement canonical logical equations. For instance, the 4-input MUX considered above corresponds to the equation

$$y = x_0 \overline{S}_0 \overline{S}_1 + x_1 S_0 \overline{S}_1 + x_2 \overline{S}_0 S_1 + x_3 S_0 S_1 \tag{3}$$

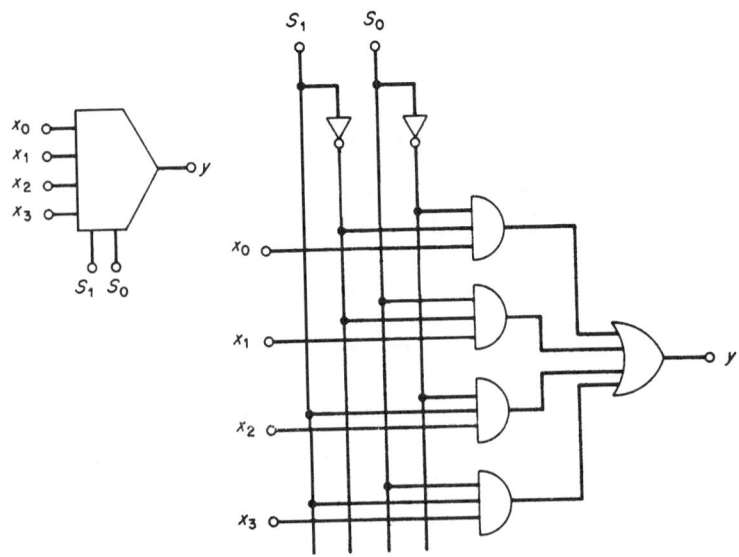

FIGURE 14.4.6   A 4-input multiplexer (MUX) with a 2-bit $(S_1 S_0)$ address.

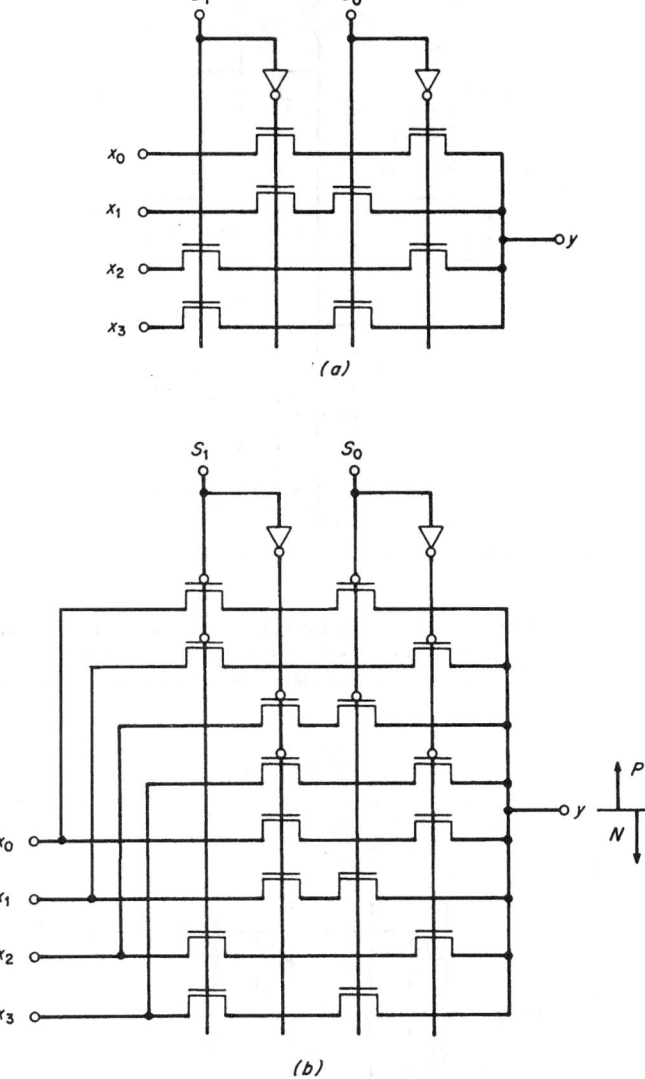

FIGURE 14.4.7 (*a*) NMOS and (*b*) CMOS implementations of multiplexers.

Hence, logical functions of the variables S_0 and S_1 can be synthesized by means of multiplexers. For instance, an EXOR circuit corresponds to $x_0 = x_3 = 0$ and $x_1 = x_2 = 1$.

Demultiplexers (DEMUX) perform the inverse passage from a single input line to several output lines under the control of an address word. They implement logic functions that are less general than those performed by multiplexers because only miniterms are involved. The symbolic representation of a 4-input DEMUX is shown in Fig. 14.4.8 with a possible implementation. The MUX represented in Fig. 14.4.7 may seem attractive for the purpose to achieve demultiplexation but one should not forget that when a given channel is disconnected, the data across the corresponding output remain unchanged. In a DEMUX, unselected channels should take a well-defined state, whether 0 or 1.

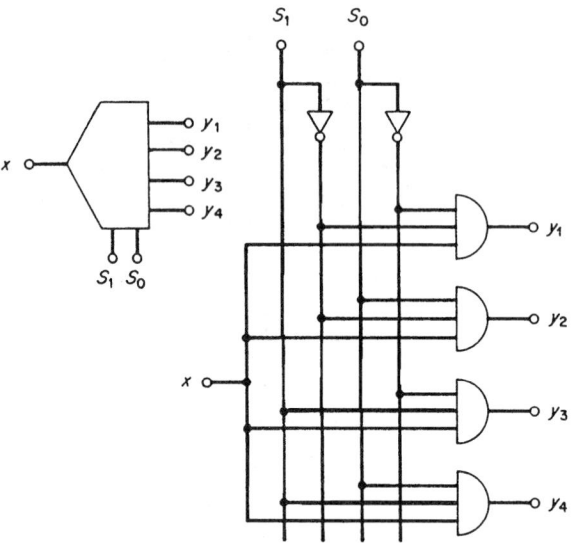

FIGURE 14.4.8   A 4-output demultiplexer (DEMUX).

A decoder is a DEMUX with a constant input. Decoders are currently used to select data stored in memories. Figure 14.4.9 shows a simple NMOS decoder. In this circuit, all unselected outputs are grounded while the selected row is at the logical 1. Any row may be viewed as a circuit implementing a miniterm. For instance, the first row corresponds to

$$y_0 = \overline{S_1 + S_0} = \overline{S}_1 \overline{S}_0 \tag{4}$$

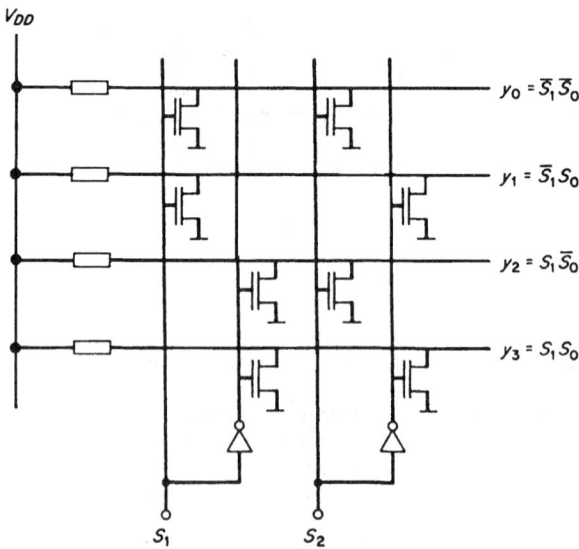

FIGURE 14.4.9   An NMOS decoder circuit.

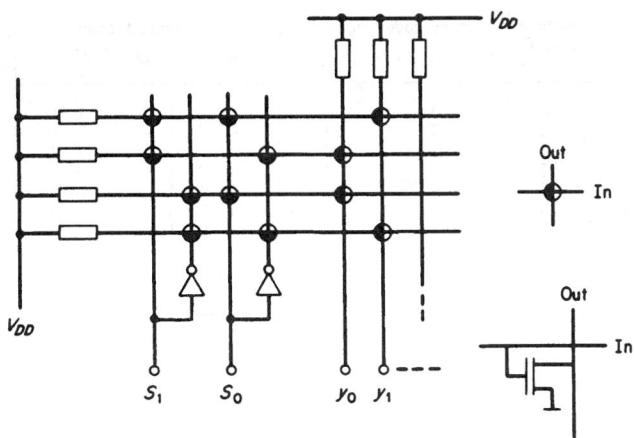

FIGURE 14.4.10 A ROM memory.

When a decoder drives a NOR circuit, a canonical equation is obtained again. Figure 14.4.10 shows a decoder driving several NOR circuits which are implemented along vertical columns. This circuit is called a read-only memory (ROM). One can easily recognize an EXOR function (y_1) and its complement (y_0) in the example. The actual silicon implementation strictly follows the pattern illustrated by Fig. 14.4.10. Notice that the decoder block and ROM column block both have the same structure after turning them by 90°.

When a ROM is used to implement combinatorial logic, the number of outputs usually is restricted to only those miniterms which are necessary to achieve the desired purpose. The ROM is then called a programmable logic array (PLA)

ROMs as well as PLAs are extensively used in integrated circuits because they provide the means to implement logic in a very regular manner. They contribute substantially to area minimization and lend themselves to automatic layout, reducing design time (silicon compilers).

A large number of functions can be implemented by means of the circuits described above: circuits converting the format of digital data, circuits coding pure binary into binary-coded decimal (BCD) or decimal data, and so forth. The conversion from a 4-bit binary-coded number into a 7-segment display by means of a PLA is illustrated by Fig. 14.4.11.

All the required OR functions are obtained by means of the right-plane decoder, while the left one determines the AND functions (they are called, respectively, OR-AND planes). In applications where multiple-digit displays are required, rather than repeating the circuit of Fig. 14.4.11 as many times as there are digits, a combination of MUX and DEMUX circuit and a single PLA can be used. An example is shown in Fig. 14.4.12.

The 4-input codes representing 4 digits are multiplexed in a quadruple MUX in order to drive a PLA 7-segment generator. Data appear sequentially on the seven common output lines connected to the four displays. Selection of a given display is achieved by a decoder driven by the same address as the MUX circuit. If the cyclic switching is done at a speed of 60 Hz or more, the human eye cannot detect the flicker.

MEMORIES[3]

Memories provide storage for large quantities of binary data. Individual bits are stored in minimum-size memory cells that can be accessed within two- or three-dimensional arrays. Read-only memories (ROMs) provide access only to permanently stored data. Programmable ROMs (PROMs) allow data modifications only during special write cycles, which occur much less frequently than readout cycles. Random access memories (RAMs) provide equal opportunities for read and write operations. RAMs store data in bistable circuits (flip-flops, SRAMs), or in the form of charge across a capacitor (DRAM).

Decimal display	Binary coded input				7 segment code						
	x_3	x_2	x_1	x_0	A	B	C	D	E	F	G
0	0	0	0	0	1	1	1	1	1	1	0
1	0	0	0	1	0	1	1	0	0	0	0
2	0	0	1	0	1	1	0	1	1	0	1
3	0	0	1	1	1	1	1	1	0	0	1
4	0	1	0	0	0	1	1	0	0	1	1
5	0	1	0	1	1	0	1	1	0	1	1
6	0	1	1	0	1	0	1	1	1	1	1
7	0	1	1	1	1	1	1	0	0	0	0
8	1	0	0	0	1	1	1	1	1	1	1
9	1	0	0	1	1	1	1	1	0	1	1
10–15	—				x						

Seven segment truth table

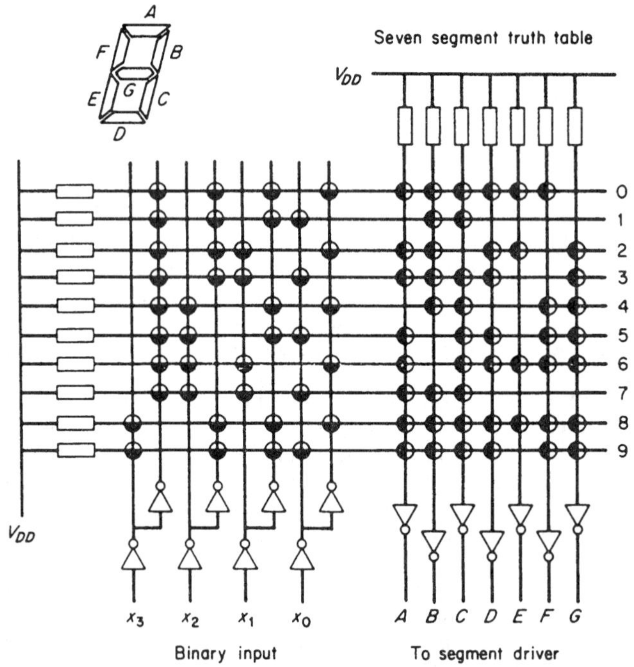

FIGURE 14.4.11   A 7-segment PLA driver.

Static Memories

Static memory cells are arrays of integrated flip-flops. The state of each flip-flop represents a stored bit. Read-out is achieved by selecting a row (word-line: WL) and a column (but or bit-lines: BL) crossing each other over the chosen cell. Writing occurs in the same manner. A typical six-transistor MOS memory cell is shown in Fig. 14.4.13. The word-line controls the two transistors that connect the flip-flop to the read-write bus (BL and BLbar). In order to read out nondestructively the data stored in the cell, the read-write bus must be precharged. The inverter that is in the low-state discharges the corresponding bit-line. Writing data either confirms or changes the state of the flip-flop. When a change must happen, the high bit-line must overrule the corresponding low-state inverter, while the low bit-line simply discharges the output node of the high inverter. The second event takes less time than the first for the big differences between bus and inverter node capacitances. Bus capacitances are at least 10 times larger than cell node capacitances. The rapid discharge of the high inverter output node blocks the other half of the flip-flop before it has a chance to discharge the high bus. In order to prevent a read cycle from

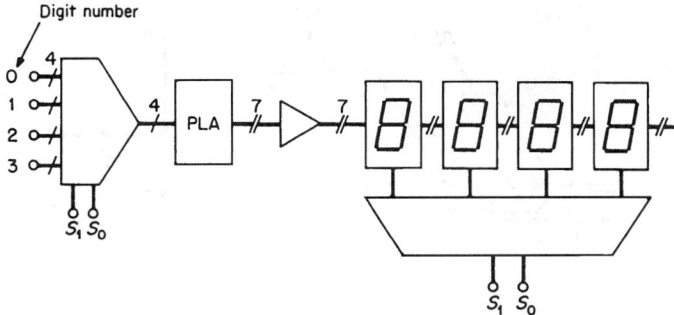

FIGURE 14.4.12 A 4-digit, 7-segment multiplexed display.

becoming a write-cycle, it is important to equalize the voltages of the bit-lines prior to any attempt to read out. This is achieved by means of a single equalizing transistor on top of the read-write bus. This transistor shorts the two bit-lines for just long enough time to neutralize any bit-line voltage difference whatever the mean voltage may be. The pull-up circuitry precharging the read-write bus accommodates conflicting requirements. It must load the bit-lines as fast as possible but not counteract the discharge of the bit-line, which is tied to the low data during read out. Usually, the pull-up circuitry is clocked and driven by the same clock that controls the equalizing transistor.

Cell size and power consumption are the two key items that determine the performances and size of present memory chips which may count as much as several million transistors. Cell sizes have shrunken continuously until they are a few tens of microns square. Static power is minimized by using CMOS instead of resistively loaded inverters. Most of the power needed is to provide fast load and discharge cycles of the line capacitances. Therefore a distinction is generally made between standby and active conditions. The load-discharge processes imply short but very large currents. The design of memories (static as well as dynamic) has always been at the forefront of the most advanced technologies.

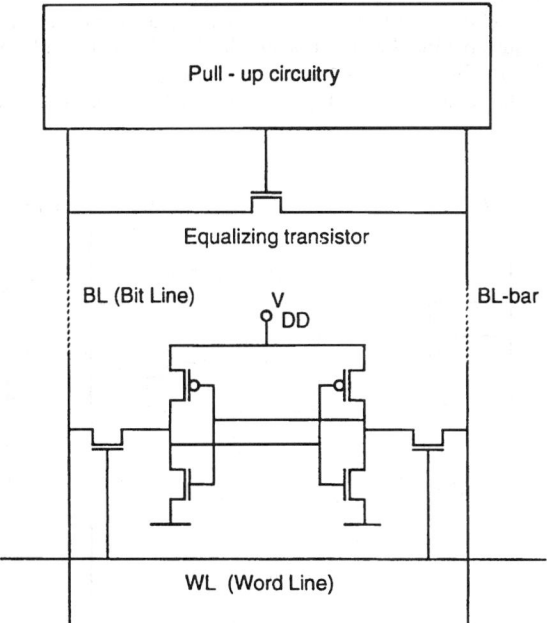

FIGURE 14.4.13 The basic six-transistor circuit of a static MOS memory cell.

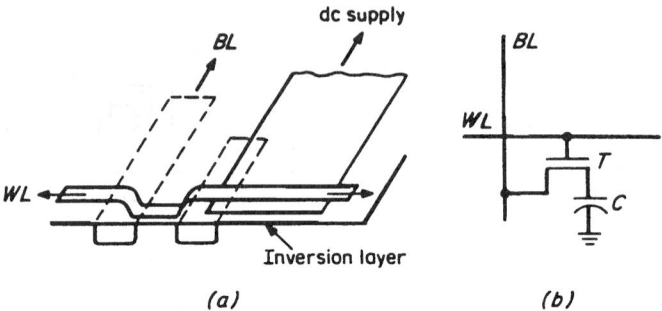

FIGURE 14.4.14   The one-transistor MOS memory cell; (*a*) IC implementation; (*b*) equivalent circuit.

Dynamic Memories[8,9]

The trend toward ever larger memory chips has led to the introduction of single transistor memory cells. Here the word-line controls the gate of the MOS switch that connects the single bit line to a storage capacitor (Fig. 14.1.14). Charge is pumped from (or fed back to) the bit-line by properly choosing the voltage of the bit-line. The actual data consist of charge (or absence of charge) stored in the capacitor in the inversion layer below a field plate. Typical storage capacitances are 100 to 50 fF. The useful charge is quite small, around 0.1 or 0.2 pC. In order to read data, this charge is transferred to the bit-line capacitance, which is usually one order of magnitude larger. The actual voltage change of the bit-line is thus very small, typically 100 to 200 mV. In order to restore the correct binary data, amplification is required. The amplifier must fulfill a series of requirements: It must fit in the small space between cells, be sensitive and respond very rapidly, but also distinguish data from inevitable switching noise injected by the overlap capacitance of the transistor in series with the storage capacitor. To solve this problem, dynamic memories generally use two bit-line output lines instead of a single one. Read-out of a cell always occurs in the same time as the selection of a single identical dummy cell. This compensates switching noises as far as tracking of parasitic capacitances of both switching transistors is achieved. The key idea behind the detection amplifier is the metastable state of a flip-flop. The circuit is illustrated in Fig. 14.4.15. Before read-out, the common source terminal is left open. Hence, assuming the bit-line voltages are the same, the flip-flop behaves like two cross-coupled diode-connected transistors.

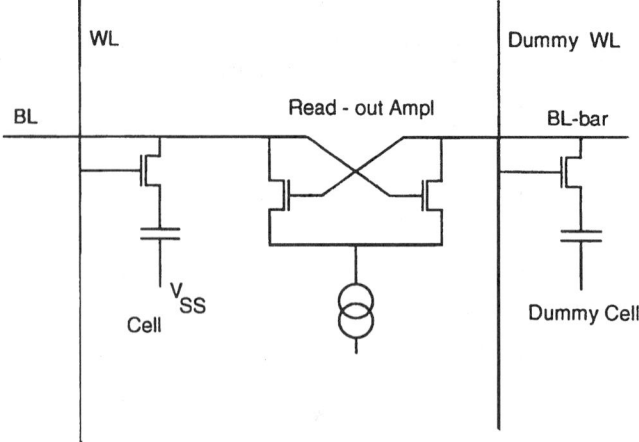

FIGURE 14.4.15   The readout amplifier of dynamic one-transistor-per-cell MOS memories consists of a flip-flop in the metastable state in order to enhance sensitivity and speed and to minimize silicon area.

The common source voltage adjusts itself to the pinch-off voltage of the transistors. After read-out has occurred, a small voltage imbalance is experienced between bit-lines. The sign of this imbalance is determined by the actual data previously stored in the cell. If then the common source node voltage is somewhat reduced, one of the two transistors becomes slightly conducting while the other remains blocked. The conducting transistor slightly discharges the bit-line with the lowest voltage, increasing the voltage imbalance. The larger this difference, the faster the common source node can be lowered without risk to switch-on of the second transistor. After a short time, the imbalance is large enough to connect the common source to ground. The final state of the flip-flop reproduces the content of the selected cell. Notice that data read during the read-out cycle are now available for rewriting in the cell. Permitting the transistor in series with the storage capacitor to conduct longer performs automatic rewriting. Because storage is likely to be corrupted by leakage current, this same procedure is repeated regularly, even when data are not demanded, in order to keep the memory active. Dynamic memories are the densest circuits designed. Memory sizes currently attain 16 Mbits, and 256 Mbits memories are being developed.

Commercially Available Memory Chips

Memory arrays usually consist of several square arrays packed on a single chip. In order to access data, the user provides an address. The larger the memory size, the longer the address. In order to reduce pin count, address words are divided in fields controlling less pins in a sequential manner. Many memories are accessed by a row address strobe (RAS) followed by a column address strobe (CAS), each one requiring only half of the total number of address bits. Memories represent a very substantial part of the semiconductor market. Besides RAMs (random access memories), a large share of nonvolatile memories is available comprising ROMs (read-only memories), programmable read-only memories (PROMs), and electrically alterable read-only memories (EAROMs), which are considered below.

RAM.   Random access memories (RAMs) are either static or dynamic memory arrays like those described above. They are used to read and write binary data. Storage is warranted as long as the power supply remains on.

ROM.   In read-only memories, the information is frozen. Information can only be read out. Access is the same as in RAMs. Memory elements are single transistors. Whether a cell is conducting or not has been determined during the fabrication process. Some MOS transistors have a thin gate oxide layer; others have a thick oxide layer, or some are connected to the access lines, while others are not.

PROM.   The requirements for committing to a fixed particular memory content, inherent in the structure of ROMs, is a serious disadvantage in many applications. PROMs (programmable ROMs) allow the manufacturer or the user to program ROMs. Various principles exist:

* With mask programmable ROMs, a single mask is all the manufacturer needs to implement the client code.
* With fuse link programmable ROMs, the actual content is written in the memory by blowing a number of microfuses. This allows the customer to program the ROM, but the final stage is irreversible as in mask programmable ROMs.

EPROM.   In electrically programmable ROMs the data are stored as shifts of the threshold voltages of the memory transistors. Floating-gate transistors are currently used for this purpose. Loading the gate occurs by hot electrons tunneling through the oxide from an avalanching junction toward the isolated gate. The charges trapped in the gate can only be removed by ultraviolet light, which provides the energy required to cross the oxide barrier.

EEPROM.   Electrically erasable PROMs use metal-nitride-oxide-silicon transistors in which the charges are trapped at the nitride oxide interface. The memory principle is based on Fowler-Nordheim tunneling to move charges from the substrate in the oxide or vice versa. The electric field is produced by a control gate. EAROMs (electrically alterable ROMs) use the same principle to charge or discharge a floating gate with some additional features providing better yields.

DIGITAL-TO-ANALOG CONVERTERS (D/A OR DAC)[10]

Converting data from the digital to the analog domain may be achieved in a variety of ways. One of the most obvious is to add electrical quantities such as voltages, currents or charges following a binary weighted scale. Circuits illustrating this principle follow. They are of two kinds: first, converters aiming at integral linearity and, second, converters exchanging integral linearity for differential linearity.

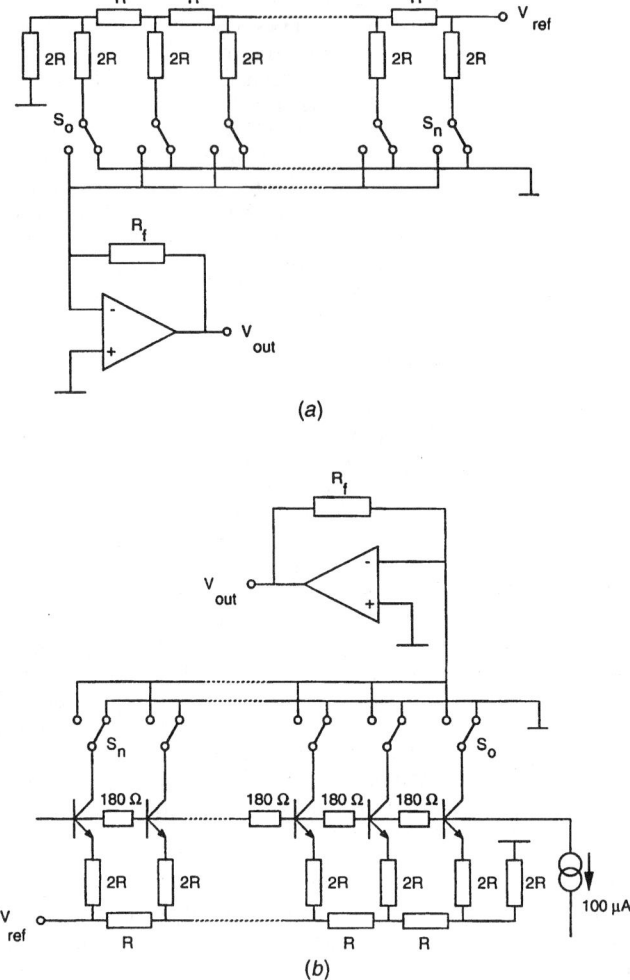

FIGURE 14.4.16 Integral linearity converters. Switching is implemented by (*a*) MOS transistors and (*b*) bipolar circuits.

Integral Linearity Converters

A binary scale of voltages and currents is easily obtained by means of the well-known *R*-2*R* network. In the circuit shown in Fig. 14.4.16*a*, binary weighted currents flowing in the vertical resistances are either dumped to ground or injected into the summing node of an operational amplifier. The positions of the switches reproduce the code of the digital word to be converted. Practical switches are implemented by means of MOS transistors. Their widths double going from left to right to keep the voltage drops constant between drains and sources. The resulting constant offset voltage can be compensated easily. Another approach, better suited for bipolar circuits, is illustrated in Fig. 14.4.16*b*. Here the currents are injected into the emitters of transistors whose outputs feed bipolar current switches performing the digital code conversion. To keep all emitters at the same potential, small resistors are placed between bases of the transistors. They introduce voltage drops of 18 mV (UT.ln2) to compensate for the small base to emitter voltages changes resulting from the systematic halving of collector current going from left to right.

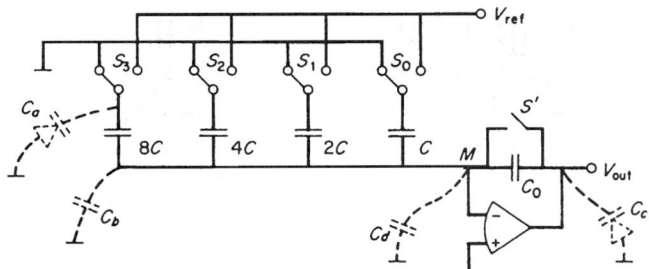

FIGURE 14.4.17 Capacitive D/A converter with charge integration op-amp.

The accuracy of *R-2R* circuits is based on the matching of the resistances. It does not exceed 10 bits in practice, and may reach 12 bits when laser trimming is used.

Switched capacitors are used also to perform D/A conversion. Capacitors are easy to integrate and offer superior temperature characteristics compared to thin-film resistors. However, they suffer from several drawbacks and require special care in order to provide good matching. Integrated capacitors are sandwiches of metal-oxide-silicon layers (MOS technology) or polysilicon-oxide-polysilicon (double poly technology). The latter exhibit a highly linear behavior and extremely small temperature coefficient (typically 10 to 20 ppm/°C). In order to improve geometrical tracking, any capacitor must be a combination of "unit" capacitors that represent the minimum-sized element available (typically 100 fF). MOS capacitors have relatively large stray capacitances to the substrate through their inversion layer, about 10 to 30 percent of their nominal value, depending on the oxide thickness. Doubly poly capacitors offer better figures, but their technology is more elaborate. Whatever choice is made, the circuits must be designed to be inherently insensitive to stray capacitance. This is the case for the circuit shown in Fig. 14.4.17.

In this circuit, a change of the position of a switch results in a charge redistribution among a weighted capacitor and the feedback capacitor C_0. The transferred charge is stored on the feedback capacitor so that the output voltage stays constant, provided leakage currents are small (typically 10^{-15} A). The stray capacitors are illustrated in Fig. 14.4.17. They have no influence on the converter accuracy. Indeed, C_a and C_c are loaded and discharged according to position changes of switch S_3, but the currents through those capacitors do not flow through the summing junction. On the other hand, C_b as well as C_d is always discharged and therefore do not influence the accuracy.

Another version of a capacitive D/A converter is shown in Fig. 14.4.18. In this network, changing the position of switches S_3 and S' produces a charge redistribution between the capacitor at the extreme left and the parallel combination of all other capacitors that represent a total of 8 C. Hence the upper node undergoes a voltage swing equal to $V_{ref}/2$. A binary scale of voltages consequently may be produced according to the various switches. Switch S' fixes the initial potential of the upper node and allows the discharge of all capacitors prior to conversion. There is no specific requirement that the upper node be connected to ground or to any fixed potential V_0, provided V_{out} is evaluated against V_0.

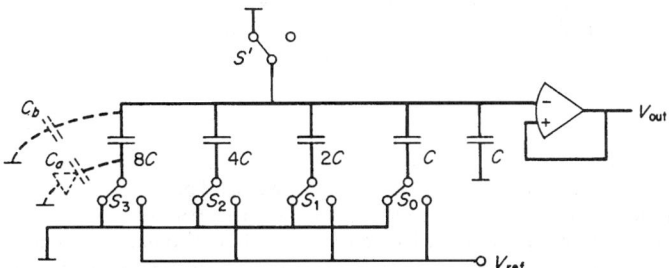

FIGURE 14.4.18 Capacitive D/A converter with unity-gain buffer.

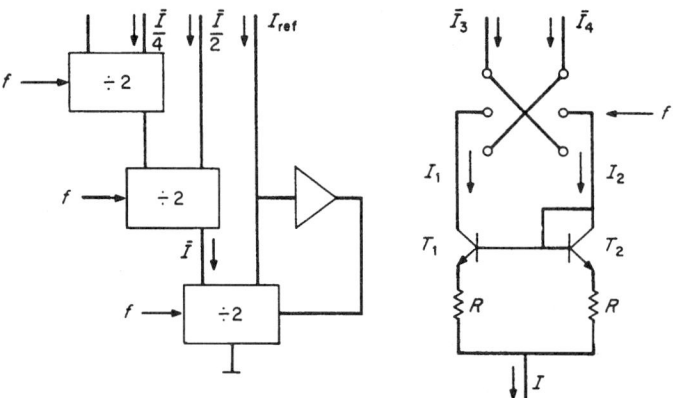

FIGURE 14.4.19   Dynamic element matching converter principle.

Notice that the stray capacitance C_b degrades the converter accuracy, while C_a has no effect at all. Careful layout therefore is required to minimize C_b, unless this stray capacitance is included in the evaluation of the capacitor C at the extreme right. For this reason, only small numbers of bits should be considered. The unavoidable exponential growth of capacitances with the number of bits is relieved to some extent because the capacitance area increases as the square of dimensions.

Another widely used technique to perform D/A conversion is paralleling of many identical transistors controlled by a single base-to-emitter or gate-to-source voltage source. The output terminals are tied together in order to implement banks of binary weighted current sources. This requires $2^{(N+1)}$ transistors to make an N-bit converter, which is very efficient for transistors are the smallest on-chip devices available. Usually all the transistors are placed in an array and controlled by a row and column decoder, as in memories. One of the interesting features of this type of converter is its inherent monotonicity. Furthermore, if access to the individual transistors follows a pseudorandom pattern, many imperfections resulting from processing, such as an oxide gradient in the MOS process, have counterbalancing effects. Hence, an accuracy of 10 bits may be achieved with transistors whose standard deviation is much worse.

A converter[11] with excellent absolute accuracy is illustrated in Fig. 14.4.19. It is based on high-accuracy divide-by-2 blocks like those shown in the right part of the illustration. Each block consists of a Widlar circuit with equal resistances in the emitters of transistors T_1 and T_2 in order to split the current I approximately into two equal currents: I_1 and I_2 (1 percent tolerance). A double-throw switch is placed in series with the Widlar circuit in order to alternate I_1 and I_2 at the rate of the clock f. Provided the half clock periods t_1 and t_2 can be made equal within 0.1 percent, a condition that can easily be met, one obtains two currents whose averages I_3 and I_4 represent $I/2$ with an accuracy approaching 10^{-5}. This technique, known as the dynamic element matching method, has been successfully integrated in bipolar technology offering an accuracy of 14 bits without expensive element trimming. A band-gap current generator provides the reference current I_{ref}, which is compared to the right-side output current of the first divide-by-2 block. A high current-gain amplifier closes the feedback loop controlling the base terminals of T_1 and T_2 of the first block.

Segment Converters

In some applications, accuracy does not imply absolute linearity but rather differential linearity, which is very critical because errors between two successive steps larger than a half LSB cannot be tolerated. The difficulty occurs when changes of MSBs occur. To overcome the problem, segment converters were designed. The idea is to divide the full conversion scale into segments. A D/A converter with a maximum of 10 bits is used within segments. Passing from one segment to another is achieved in a smooth manner as in the circuit illustrated by Fig. 14.4.20.[12] In this circuit, the 10-bit segment converter is powered by one of the four bottom current sources. A 2-bit decoder,

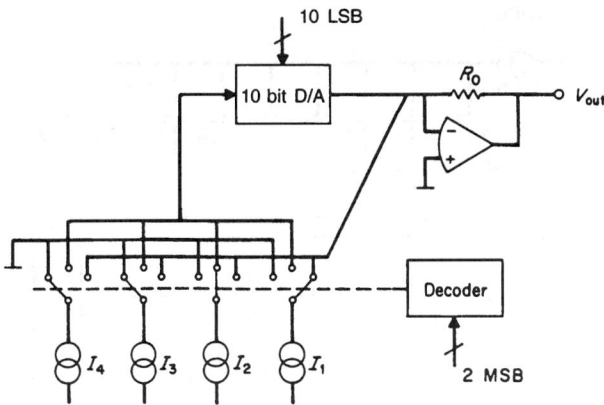

FIGURE 14.4.20   Segment D/A converter.

which is controlled by the two MSBs, makes the appropriate connections. When the two MSBs are 00, current source I_1 is chosen and the three others are short-circuited to ground. When the MSB pattern changes to 01, the switches take the positions illustrated in the figure. Then I_2 supplies the 10-bit converter while I_1 is injected directly into the summing node of the op amp. Hence, no change occurs in the main current component. The differential linearity is entirely determined by the 10-bit converter. The price paid for this improvement is a degradation of absolute linearity. Small discrepancies among the four current sources introduce slope or gain changes between segments. The absolute linearity specifications thus do not reach the differential linearity specs of the converter.

Voltage segment converters can also be integrated. Instead of parallel current sources, series voltage sources are required. This is achieved by means of a chain of equal resistors dividing the reference voltage into segments. The D/A converter is connected along the chain by means of two unity-gain buffers in order not to affect input voltages. Passing from one segment to the next implies that the buffers are interchanged in order to avoid the degradation of the differential linearity, which could result from different offsets in the buffers. An integrated 16-bit D/A converter has been built along these lines.[13]

When a switched-capacitor D/A converter is embedded in a voltage segment converter, no buffer amplifiers are needed because no dc current drain is required. This property simplifies greatly the design of moderate accuracy converters and has been extensively used in the design of integrated codecs.

Codecs.   Codecs are nonlinear converters used in telephony. To cover the wide dynamic range of speech satisfactorily with only 8 bits, steps of variable size must be considered: small steps for weak signals, and large steps when the upper limit of the dynamic range is approached. The four LSBs of a codec-coded word define a step size within a segment whose number is given by the three following bits. The polarity of the sample is given by the MSB. Going from one segment to the next implies doubling the step size in order to generate the nonlinear conversion law (μlaw). In the codec circuit of Fig. 14.4.21,[14] bottom plates are connected either in the middle of the resistive divider considered as an ac ground (position 1), or to V_{low} or V_{high} (position 2 plus switch S_1 or S_2) representing, respectively, the negative and positive reference voltage sources, or to a third position (position 3) connecting a single capacitor to an appropriate node of the resistive voltage divider. Hence, the resistive divider determines the step size (four LSBs), the capacitive divider defines the segment (three following bits), and switches S_1 or S_2 control the sign (MSB).

ANALOG-TO-DIGITAL (A/D) CONVERTERS (ADC)[10]

Some analog-to-digital converters are in fact D/A converters embedded in a negative feedback loop with appropriate logic (see Fig. 14.4.22). The analog voltage to be converted is sensed against the analog output of a D/A converter, which in turn is controlled by a logic block so as to minimize the difference sensed by the

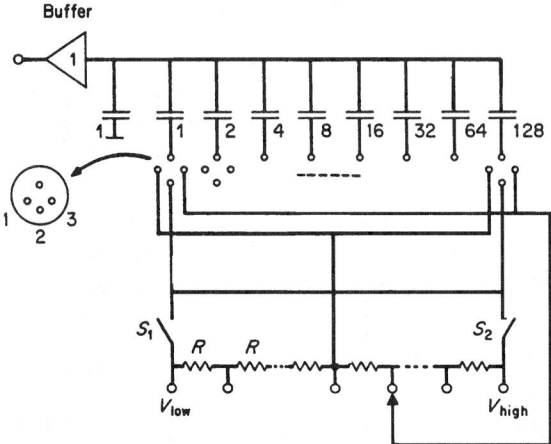

FIGURE 14.4.21   Segment type codec.

comparator. Various strategies have been proposed for the logic block, depending on the type of D/A converter used. Most of the converters described previously have been implemented as ADCs in this manner. In particular, ladder and capacitive D/A converters are used in the so-called successive approximation D/A converter. In those devices, the logic block first sets the MSB; then the comparator determines whether the analog input voltage is larger or smaller than the half-reference voltage. Depending on the result, the next bit is introduced, leaving the MSB unchanged if the analog signal is larger and resetting the MSB in the opposite state. The algorithm is repeated until the LSB is reached.

Successive approximation ADCs are moderately fast, for conversion time is strictly proportional to the number of bits and is determined by the settling time of the op-amp. The algorithm implies that the analog input voltage remains unaltered during the conversion time; otherwise it may fail. To avoid this, a sample-and-hold (SH) circuit must be placed in front of the converter.

A capacitive A/D converter is shown in Fig. 14.4.23.[15] The conversion is accomplished by a sequence of three operations. During the first, the unknown analog input voltage is sampled. The corresponding positions of the switches are as illustrated in Fig. 14.4.23a. The total stored charge is proportional to the input voltage V_{in}. In the second step, switch S_A is opened and the positions of all S_i switches are changed (Fig. 14.4.23b). The bottom plates of all capacitors are grounded, and consequently the voltage at the input node of the comparator

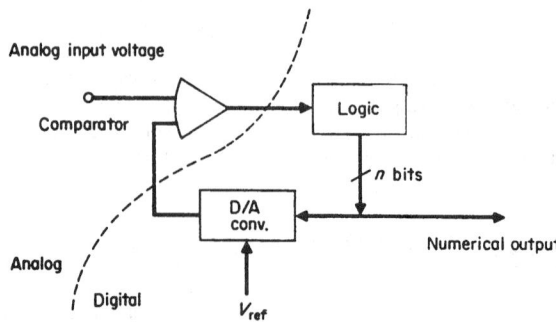

FIGURE 14.4.22   A D/A converter within a feedback loop forms an A/D converter.

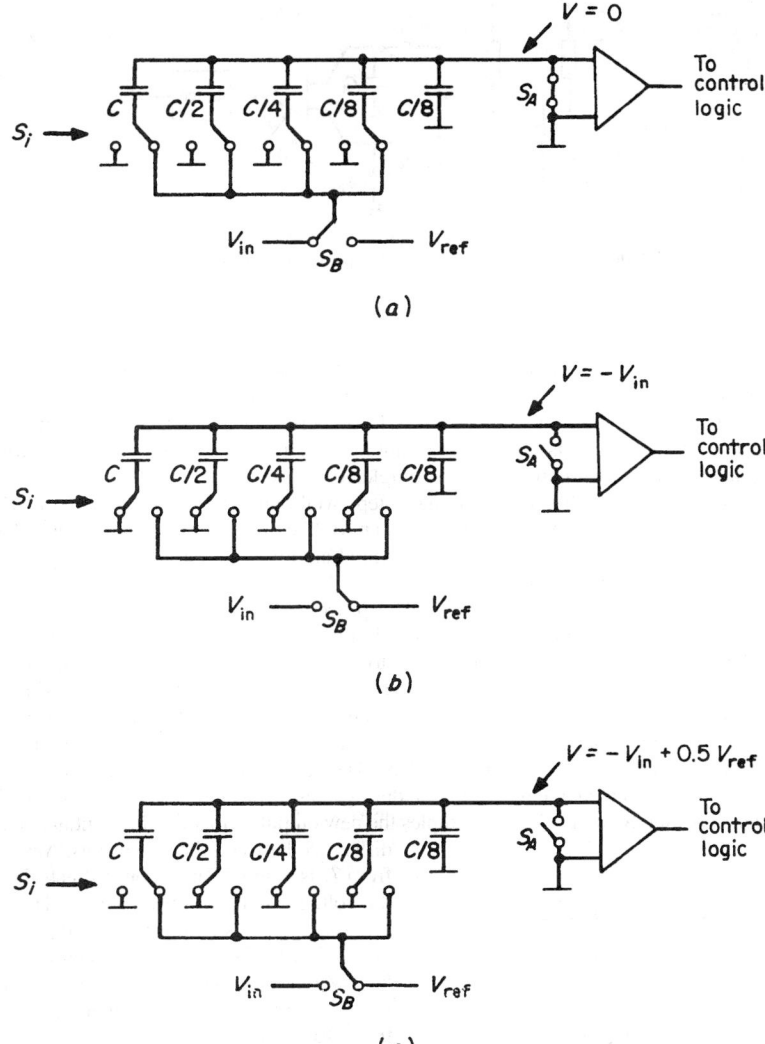

FIGURE 14.4.23   An integrated switched-capacitor A/D converter: (*a*) sample mode; (*b*) hold mode; and (*c*) redistribution mode.

equals $-V_{in}$. The third step is initiated when raising the bottom plate of the MSB capacitor from ground to the reference voltage V_{ref} (Fig. 14.4.23*c*). This is done by changing again the position of the MSB switch and connecting S_b to V_{ref} instead of V_{in}. The voltage at the input node of the comparator is thus increased by $V_{ref}/2$, so that the comparator's output is a logic 1 or 0, according to the sign of $(V_{ref}/2 - V_{in})$.

The circuit operates similarly to any successive approximations converter. That is, V is compared to $(V_{ref}/2 + V_{ref}/4)$ when the result of the previous operation is negative, or the MSB switch returns to its initial position and only the comparison with $V_{ref}/4$ is considered. After having carried out the same test until the LSB is reached, the conversion is achieved and the digital output may be determined from the position of the switches. The voltage of the common capacitive node is approximately equal to, or smaller than, the smallest incremental step.

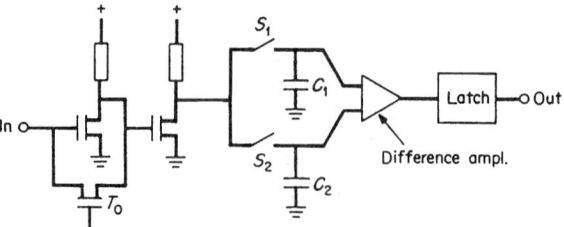

FIGURE 14.4.24   A typical MOS comparator for an A/D converter.

Since this is almost negligible, the stray capacitance of the upper common node is practically discharged. It thus has no effect on the overall accuracy of the converter. Hence the problem in the circuit of Fig. 14.4.18 does not occur. The name "charge redistribution A/D converter" was given to this circuit because the charge stored during the first sequence is redistributed among only those capacitors whose bottom plates are connected to V_{ref} after the conversion algorithm is completed.

The design of comparators able to discriminate steps well below the offset voltage of MOS amplifiers is a problem deserving careful attention. An illustration of an MOS comparator is shown in Fig. 14.4.24.[15] The amplifier comprises three parts: a double inverter, a differential amplifier, and a latch. The input-output nodes of the first inverter may be short-circuited by means of transistor T_0 to bias the gates and drains of the identical two first inverters at approximately half the supply voltage. This procedure occurs during the "sampling" phase of the converter and corresponds to the grounding of switch S_A in Fig. 14.4.23. As stated above in the comments concerning the circuit of Fig. 14.4.17, there is no necessity to impose zero voltage on this node during sampling; any voltage is suitable as long as the impedance of the voltage source is kept small. This is what occurs when T_0 is on, since the input impedance of the converter is then minimum because of the negative-feedback loop across the first stage. Once the double inverter is preset, T_0 opens and the output voltage of the second inverter is stored across the capacitor C_1, momentarily closing switch S_1. Then the floating input node of the comparator senses the voltage change resulting from charge redistribution during a test cycle and reflects the amplified signal at the output of the second inverter. Switch S_2 in turn samples the new output so that the differential amplifier sees only the difference between the two events. Any feed through noise from T_0 is ignored since it only affects the common mode of the voltages sampled on C_1 and C_2. Hence, the overall offset voltage of this comparator is equal to the offset voltage of the differential amplifier divided by the gain of the double inverter input stage. Signals in the millivolt range can be sensed correctly regardless of the poor behavior of MOS transistors' offset voltages.

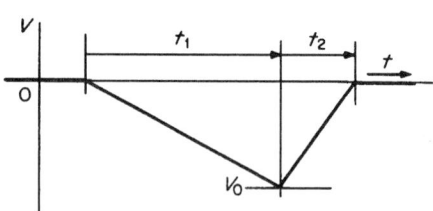

FIGURE 14.4.25   The dual-slope A/D converter principle.

Another integrated converter widely accepted in the field of electrical measurements is the dual-ramp A/D converter.[13] The principle of this device is first to integrate the unknown voltage V_x during a fixed duration t_1 (see Fig. 14.4.25). Then the input signal V_x is replaced by a reference voltage V_{ref}. Since the polarities of both signals are opposite, the integrator provides an output ramp with opposite slope. It takes a time t_2 for the integrator output voltage to return from V_0 to zero. Hence

$$V_0 = -\frac{1}{RC} \int_0^{t1} V_x \, dt \qquad (5)$$

and

$$V_0 = (V_{ref}/RC)t_2 \qquad (6)$$

Consequently, if V_x is a constant

$$V_x = -V_{ref} t_2/t_1 \tag{7}$$

the time t_1 is determined by a counter. At the moment the counter overflows, switching of the input signal from V_x to V_{ref} occurs. The counter is automatically reset, and a new counting sequence initiated. When the integrator output voltage has returned to zero, the comparator stops the counting sequence. Thus the actual count is a direct measure of t_2.

The dual-ramp converter has a number of interesting features. Its accuracy is not influenced by the value of the integration time constant RC; neither is it sensitive to the long-term stability of the clock-frequency generator. The comparator offset can easily be compensated by autozeroing techniques. The only signal that actually controls the accuracy of the A/D converter is the reference voltage V_{ref}. Excellent thermal stability of V_{ref} can be achieved by means of band-gap reference sources.

Another interesting feature of the dual-ramp A/D converter is the fact that since V_x is integrated during a constant time t_1, any spurious periodic signal with zero mean whose period is a submultiple of t_1 is automatically canceled. Hence, by making t_1 equal to an entire number of periods of the power supply one obtains excellent hum rejection.

DELTA-SIGMA CONVERTERS[10,20]

Because the development of integrated systems is driven mainly by digital applications, mixed analog-digital circuits, especially converters, should be implementable equally well without loss of accuracy in digital technologies, even the most advanced ones. Short channels, however, do not just improve bandwidth, they also negatively affect other features, such as dynamic range and $1/f$ noise, because both supply voltage and gate area of the MOS transistors get even smaller. Therefore, a trade-off of speed and digital complexity for resolution in signal amplitude is needed.

Delta-Sigma converters illustrate this trend. Their object is to abate the quantization noise to enhance the signal-to-noise ratio (SNR) of the output data (quantization noise is the difference between the continuous analog data and their discrete digital counterpart). When the number of steps of a converter increases the quantization noise decreases. Similarly, if we enhance the SNR the resolution increases. Delta-Sigma converters take advantage of *oversampling* and *noise shaping* techniques to improve the SNR.

Oversampling refers to the Nyquist criterion, which requires a sampling frequency twice the baseband to keep the integrity of sampled signals. Oversampling does not improve the signal. Neither does it increase the quantization noise power, nor is it a way to spread the quantization noise power density over a large spectrum, thereby lessening its magnitude to keep the noise power unchanged. If we restrict the bandwidth of the oversampled signal to the signal baseband in compliance with the Nyquist criterion, the amount of noise power left in the baseband is divided automatically by the ratio of the actual sampling frequency over the Nyquist frequency. We improve thus the SNR and consequently increase the resolution.

Noise shaping is the other essential feature of Delta-Sigma converters. It refers to the filtering step oversampled data must undergo. The purpose is to further decrease the amount of noise left in the baseband by shifting a large part of the remaining quantization noise outside the baseband. During the step, the SNR is substantially improved.

Both A/D and D/A converters lend themselves to Delta-Sigma architectures. The same principles prevail, but their implementation differs. We consider both separately.

A/D Delta-Sigma Converters[16,26]

In A/D converters, oversampling is done simply by increasing the sampling frequency of the analog input signal. Noise shaping is achieved by means of a nonlinear filter like the one illustrated in the upper part of Fig. 14.4.26. As stated already above, the goal is to shift most quantization noise out of the baseband. The noise shaper consists of a feedback loop similar to the linear circuit shown in the lower part. In the upper figure, the quantization noise is produced by the A/D converter. This converter is followed by a D/A converter, which is required because the difference between the continuous analog input signal and the discrete digital output signal delivered by the

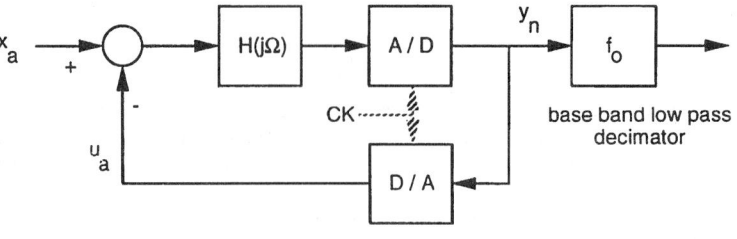

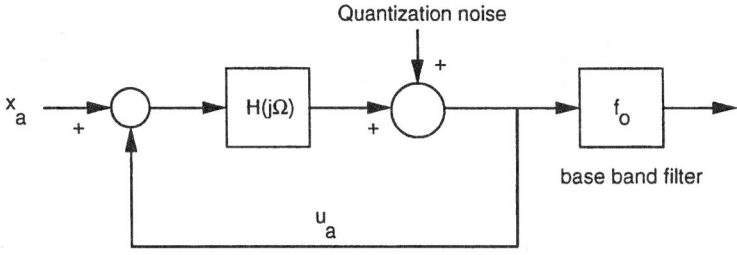

FIGURE 14.4.26   Principle of sigma-delta converters.

A/D converter must be sensed in the analog domain. This difference is amplified and low-passed by means of the analog loop filter H. The idea is truly to minimize the signal fed to the amplifier as in the continuous circuit shown below. Consequently, the output signal becomes a replica of the input. This requires some caution, however, because matching continuous input data and discrete output data is not feasible. The situation is very different from the one illustrated in the lower part of Fig. 14.4.26 where the quantization noise is simulated by means of an independent analog continuous noise source. Clearly, if the loop gain is high, the signal outputted by the amplifier should be the sum of the input signal minus the noise in order to make u_a look like x_a. In the upper circuit, however, the amplifier senses steps whose magnitude is determined by the A/D and D/A quantizer. The signal delivered by the amplifier is an averaged image over time of the difference between the input and the discrete signal fed back. The latter tracks the input the best it can—thanks to the feedback loop. A look at the signals displayed in Fig. 14.4.27 confirms this statement. The figure represents output data of a third-order noise shaper with a 3-bit quantizer. The loop filter consists of three integrators in cascade. Their outputs are respectively illustrated in the three upper plots. The signal delivered by the third integrator is applied to the quantizer whose output is shown below. It is obvious that notwithstanding the poor resolution of the A/D converter, the noise shaper tracks the input sine wave pretty well by correcting its output data continuously.

Once the bandwidth of the signal outputted by the noise shaper has been restricted to the baseband, we get a signal whose SNR may be very high. To illustrate this, consider Fig. 14.4.28, which shows a plot of the signal-to-noise improvement that can be obtained with the linear circuit shown in Fig. 14.4.26. Although the actual noise shaper differs from its analog counterpart because the quantization noise is correlated with the input and the quantizer is a nonlinear device, results are comparable. Large SNR figures, 60 or 80 dB and even better, are readily achievable. It is obvious that large OSRs are not the only way to get high SNRs; these can also be obtained by increasing the order of the loop filters (assuming of course stability is achieved, which is not always easy in a nonlinear system). Another interesting feature that stems from the above figure is that A/D and D/A converters with few bits resolution only do not impair the resolution of the Delta-Sigma converter given the large SNRs noise shapers can achieve. Even a single-bit quantizer, notwithstanding its large amount of quantization noise, suffices.

In practice the above converters are parallel devices in order not to slow down needlessly the conversion rate. The A/D converter is a flash converter and the D/A converter a unit-elements converter. The accuracy of the A/D converter is not critical as it is a part of the forward branch and it does not control the overall accuracy like

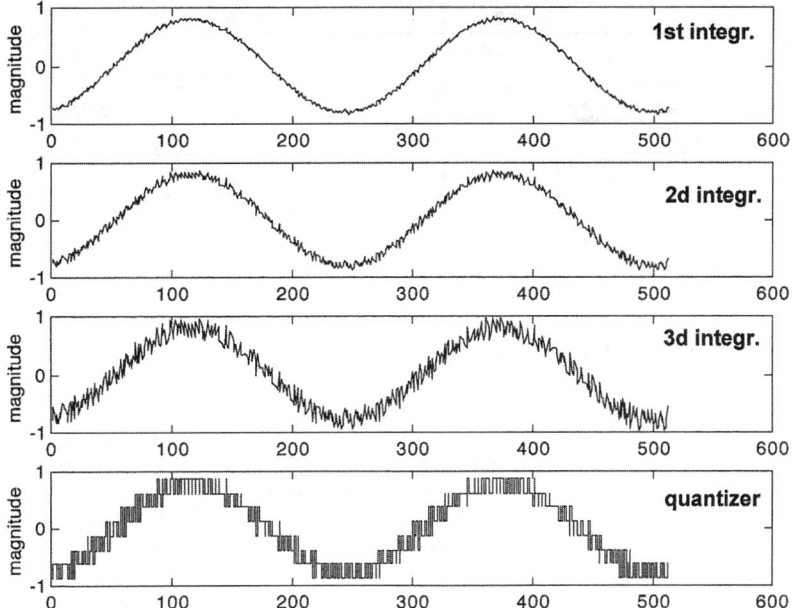

FIGURE 14.4.27   Waveforms observed in a third-order noise shaper making use of a 3-bit quantizer.

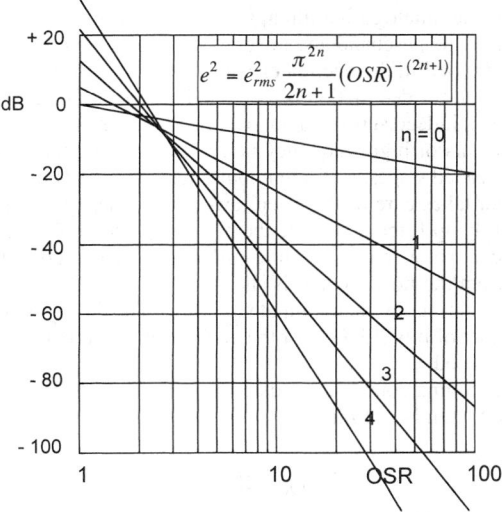

FIGURE 14.4.28   Plot of the quantization noise attenuation vs. the oversampling rate (OSR) and the loop filter order. The case $n = 0$ corresponds to oversampling without noise shaping.

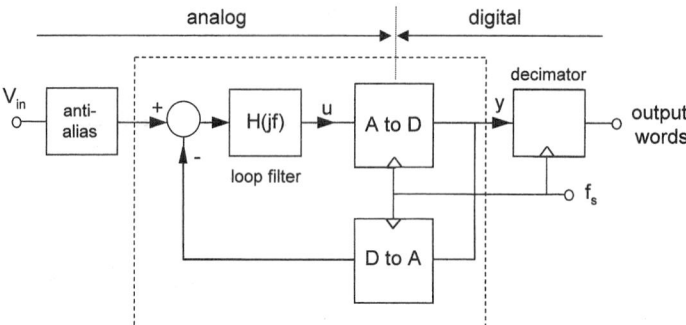

FIGURE 14.4.29   The generic Delta-Sigma A/D converter.

in any feedback loops. The D/A accuracy is more critical because it is located in the feedback branch and impairs the overall accuracy since the feedback loop minimizes only the difference between the oversampled input signal and the signal delivered by the D/A converter. This puts, of course, a heavy toll on the D/A converter unless we use a single-bit architecture. In single-bit quantizers, the A/D converter resumes to a comparator and the D/A converter to a device whose output can take only two states. No possibility exists that steps of unequal heights are found as in multi-bit D/As. This explains why single-bit quantizers are preferred generally to multi-bit. Multi-bit quantizers, however, are not evaded; they offer better alternatives for large bandwidth converters where the sampling frequency can be bound by the technology and the OSRs may take large values. To improve the performances of the D/A converter, randomization of their unit elements is generally recommended. This spreads the impairments of the unit elements mismatch over a large bandwidth. Without randomization, the same impairments produce harmonic distortion, which is a lot more annoying than white noise in terms of SNRs.

A generic A/D Delta-Sigma converter is shown in Fig. 14.4.29. Besides the noise shaper, two additional items are visible—an anti-alias filter at the input and a decimator at the output. The anti-alias filter is not specific to Delta-Sigma converters, but its implementation is much less demanding than in Nyquist converters. The reason therefore stems from the fact that the sampling frequency is much larger than the signal baseband so that a third- or second-order analog filter already is enough. At the output of the noise shaper, the decimator restricts the bandwidth to the baseband and lowers the sampling rate from oversampled to the Nyquist frequency. Since the signal inputted in the decimator is taken after the A/D converter, the decimator is in fact a digital filter. This is important for decimation is a demanding operation that cannot be done easily in the analog domain. The decimator is indeed supposed not only to restrict the bandwidth of the converted data but also to get rid of the large amounts of high-frequency noise lying outside the baseband. Therefore the decimator consists of two or three cascaded stages, an finite impulse response (FIR) filter and one or two accumulate-and-dump filters. The FIR filter takes care of the steep low-pass frequency characteristic that fixes the baseband, while the accumulate-and-dump filters attenuate the high-frequency noise. An illustration of the input-output signals of a fourth-order decimator fed by the third-order noise shaper considered in Fig. 14.4.27 is shown in Fig. 14.4.30. The signal from the noise shaper is shown in Fig. 14.4.27. The output of the decimator is illustrated below together with continuous analog input sine wave. The down-sampling rate in the example is equal to 8 and the effective number of bits (ENOB) of the decimated signal reaches 9.6 bits. In practice, resolutions of 16, even 24 bits, are obtained.

The resolution of Delta-Sigma converters is determined by the ENOB, which is derived from the relation linking quantization noise to number of bits:

$$\text{ENOB} = \frac{\text{SNR}_{dB} - 1.8}{6} \tag{8}$$

The SNR is measured by comparing the power of a full-scale pure sine wave to the quantization noise power taking advantage of the fast Fourier transform (FFT) algorithm. The measured noise consists of two

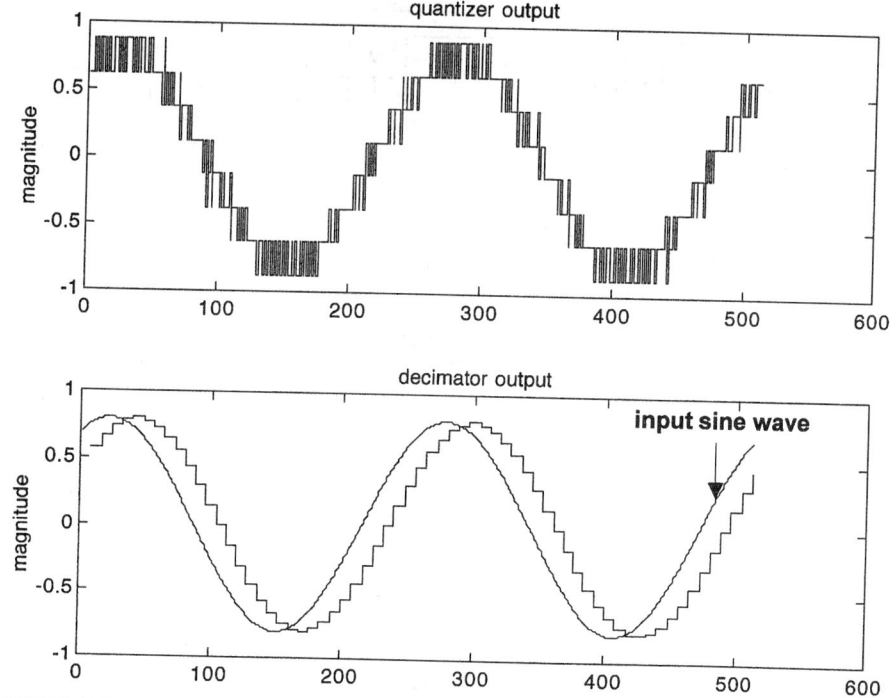

FIGURE 14.4.30 (Above) The quantizer output signal of Fig. 14.4.28. (Below) The same after decimation (order 4) compared to the input signal.

components—the actual quantization noise, which is a measure of the resolution, and the noise caused by the converter impairments. The first does not represent a defect; the second is unwanted and inevitable. The magnitude of this extra noise generally increases very rapidly when the input signal gets large because more and more nonlinear distortion tends to enter the picture. In order to avoid the impact of distortion on the ENOB evaluation, the SNR is evaluated generally as follows. The magnitude of the input signal is varied from a very low level, for instance, 60 dB below full scale, until full scale and the measured SNRs are plotted versus the magnitude of the input signal. As long as distortion does not prevail, the SNR varies like the power of the input signal, but beyond this it departs from ideal. The figure that must be considered in the above equation is the SNR obtained after extrapolating the linear portion of the SNR plot until full scale.

D/A Delta-Sigma Converters[10,20]

The principles underlying A/D converters can be transposed in D/A Delta-Sigma converters (Fig. 14.4.31). Oversampling and noise shaping are applied concurrently but in a different way since the input data are digital words and the output is an analog signal. An interpolation step is needed first in order to generate additional samples interleaved with the input data. The output of the interpolator is then noise shaped before being applied to a low-resolution D/A converter whose analog output is filtered by means of a low-order analog low-pass filter. The quantization noise is evaluated and fed back to the loop filter before closing the feedback loop of the noise shaper. One of the differences with respect to A/D Delta-Sigma converters is the manner quantization noise is measured. All that is needed is to split the words delivered by the noise shaper in two fields—an MSB field and an LSB field. The MSB field, which consists of a few bits, or even a single-bit, controls the D/A converter, while the LSB field closes the feedback loop after the digital loop filter.

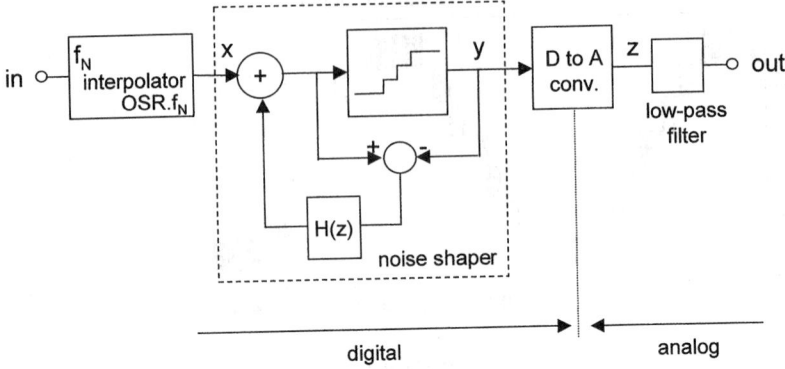

FIGURE 14.4.31 Block diagram of a Delta-Sigma D/A converter.

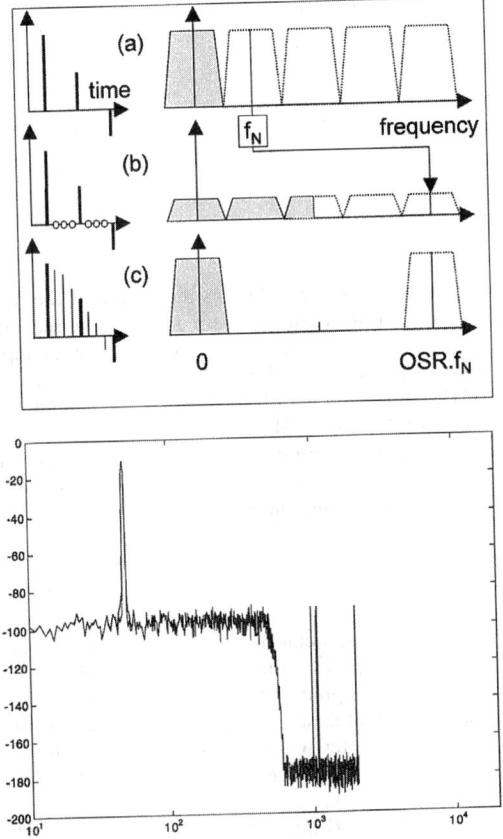

FIGURE 14.4.32 Interpolation (above) the principle (below) spectrum of a fourfold interpolated sine wave.

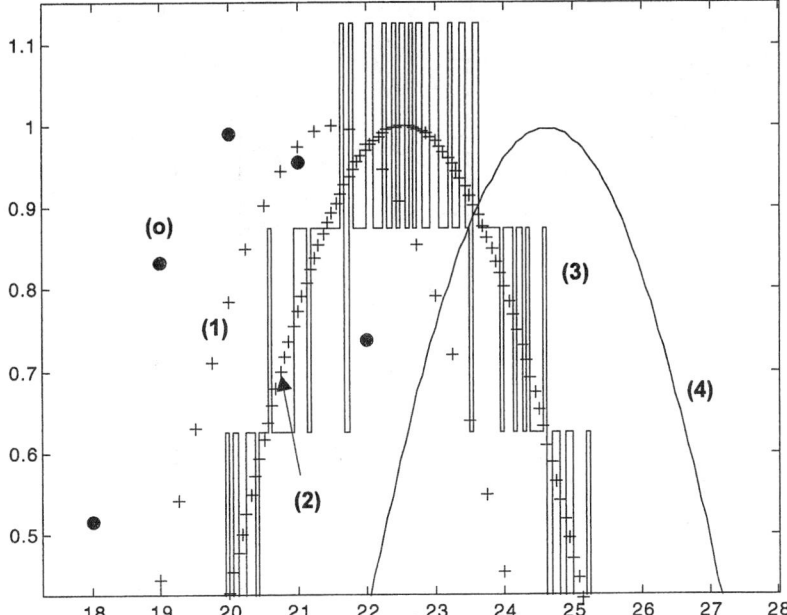

FIGURE 14.4.33 Timing of signals in a D/A Delta-Sigma converter: (0) the 13-bit digital input sine wave, (1) after a first fourfold interpolation, (2) after the second interpolation, (3) the noise shaped 3-bit signal, and (4) the analog output signal.

Interpolators are the counterpart of decimators. Instead of down-sampling they add data between samples as shown in the upper part of Fig. 14.4.32. In the figure, three zeros are placed between every sample. This multiplies already the sampling frequency by 4 but does not suffice because the spectrum of the resulting over-sampled signal is like the one shown in Fig. 14.4.32b. One must erase part of the spectrum and multiply the signal by 4 in order to get the correct spectrum shown under Fig. 14.4.32c. This is done by means of a digital filter with a sharp cutoff frequency at the edge of the baseband. In practice this filter consists of several cascaded filters like in the decimator. An FIR filter takes care of the sharp cutoff frequency, whereas one or two low-pass filters perform the rest. The FFT of a sine wave after a fourfold interpolator with a single FIR filter is shown in the lower part of Fig. 14.4.32. The fundamental ray that is reproduced three times lies approximately 80 dB below the quantization noise floor and does no harm.

A second more elaborate example is shown in Fig. 14.4.33. It illustrates the changes signals undergo versus time. The input is a 13-bit sine wave represented by large black dots. After a first interpolation, which multiplies the sampling frequency by 4, we get the data marked (1). A second interpolation by 4 yields data (2). The signal is then noise shaped. The three MSBs control the D/A converter whose analog output (3) consists of large steps that approximate the signal from the second interpolator. When the signal outputted by the D/A converter is filtered by means of a third-order low-pass, the continuous sine wave (4) is obtained, which is the actual output signal of the converter. With an SNR of nearly 80 dB, the 13-bit accuracy of the digital input signal is still met.

VIDEO A/D CONVERTERS[10]

Video applications require extremely short conversion times. Few of the previous converters meet the required speed. An obvious solution is full parallel conversion with as many comparators and reference levels as there are quantization steps. Fortunately, most video applications do not require accuracies higher than 8 bits, so 256 identical channels are sufficient. Only integrated circuits offer an economical solution in this respect.

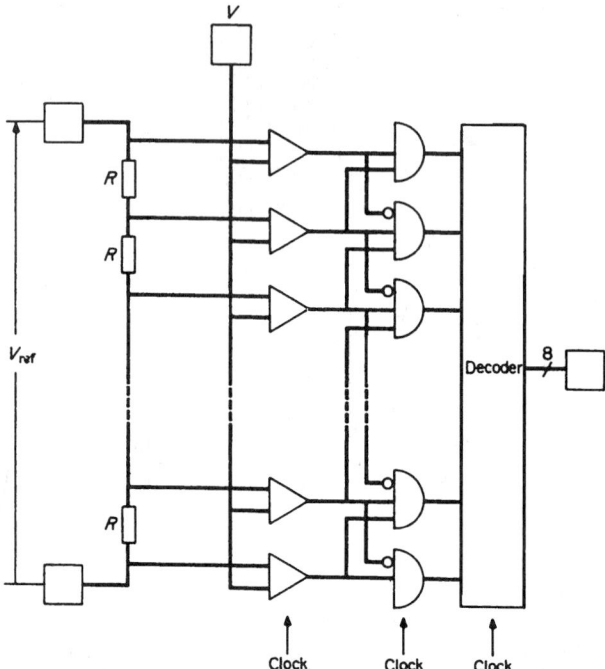

FIGURE 14.4.34   A flash converter.

A typical parallel video converter, called a flash converter, is shown in Fig. 14.4.34.[17] The architecture of the converter is simple. A string of 255 identical thin-film aluminum or polysilicon resistors is used in order to produce 2^8 reference levels. The input voltage V produces at the outputs of the comparators a vector which may be divided into two fields of 1s and 0s. An additional layer of logic transforms this vector into a new vector with 0s everywhere except at the boundary between the two fields. The new vector is further decoded to produce an 8-bit coded output. No sample-and-hold (SH) circuit is required since all comparators are synchronously driven by the same clock, but extremely fast comparators are required. An 8-bit resolution and a 5-MHz bandwidth imply binary decisions in a time as short as 250 ps. This is achieved by means of ECL logic drivers and storage flip-flops. Signal propagation across the chip also requires careful layout. The resistive divider impedance may not exceed 50 to 100 Ω. Another particular feature is the inevitable high-input capacitance, which results from the paralleling of many comparators.

The huge area and power consumption inherent to flash devices has stimulated the design of less greedy converters. Although these cannot compete with flash converters as far as speed is concerned, they offer conversion times short enough to comply with the requirements of a number of "fast" applications. Their principle comes to fragment the conversion process into several cycles, eventually two, during which strings of bits are evaluated by means of a subconverter. The process starts with the bits representing the MSB field. The other sets are evaluated one after another until the correct output word can be reconfigured by concatenating the individual segment codes. Each conversion step requires a single clock. This means, of course, that all subconverters must be fast devices, like flash converters. A 9-bit converter, for instance, operating by means of 3-bit segments requires three cycles for full conversion. Although only three clock cycles are required, the real conversion time is always much longer than three times what is needed for the 9-bit flash converter. Segmentation supposes indeed that some kind of analog memory be used in order to store intermediate results. Op-amps are thus required, which is a drawback with respect to flash converters, for the latter don't need op-amps, and consequently ignore dominant poles.

The most stringent difference between flash converters and segmented A/D converters is the number of comparators. In a flash converter, the number of comparators increases exponentially with the number of bits.

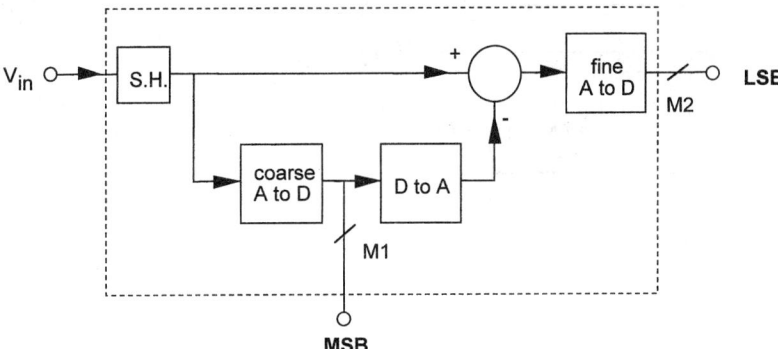

FIGURE 14.4.35   A subranging converter.

In a segmented converter, less comparators are required. Not only is the area much smaller, but the power consumption is drastically reduced.

Figure 14.4.35 shows the implementation of a two-cycle segmented A/D converter called a *subranging converter*. The coarse A/D converter evaluates first the M1 MSBs. For the remaining M2 LSBs, an analog replica of the MSBs is subtracted from the analog input signal. This is done by means of the M1-bit wide D/A converter whose output is subtracted from the input signal. The difference, which is nothing but coarse conversion quantization noise, is fed to the fine A/D flash converter, which evaluates the LSBs. The output code word is obtained by concatenation of the M1 and M2 segment codes. The total number of comparators in this type of converter is drastically reduced. For instance, in a 10-bit converter taking advantage of two 5-bit segments, each flash converter requires 31 comparators, that is, a total of 62 comparators, which is very little compared to the 1023 comparators required by a true 10-bit flash converter. Naturally the coarse D/A converter, which is a unit-elements parallel device, and the sample-and-hold, which holds the input while A/D and D/A conversions take place, must be brought into the picture also. In any case, segmented converters save a lot of area and power.

The number of comparators can be decreased further if we consider the *recycling converter*[24] shown in Fig. 14.4.36. In this circuit the difference between the input and its coarse quantized approximation is fed back to the input, stored in the sample-and-hold device, and recycled through the A/D converter. Eventually a second cycle may be envisaged. Each cycle, a new set of lower rank bits is generated. The fine flash converter is not needed, but an interstage amplifier is required. Its purpose is to amplify the quantization noise so that it spans exactly over the full dynamic range of the A/D and D/A converters. If this were not the case, we would be forced to adapt continuously the resolution of the converters to cope with the decreasing magnitude of the difference signal.

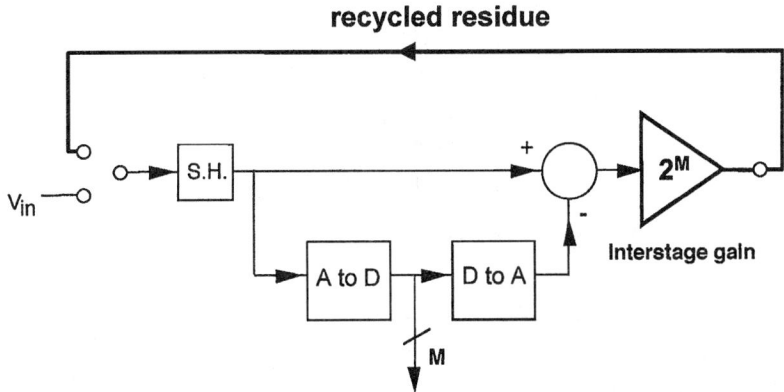

FIGURE 14.4.36   A recycling converter.

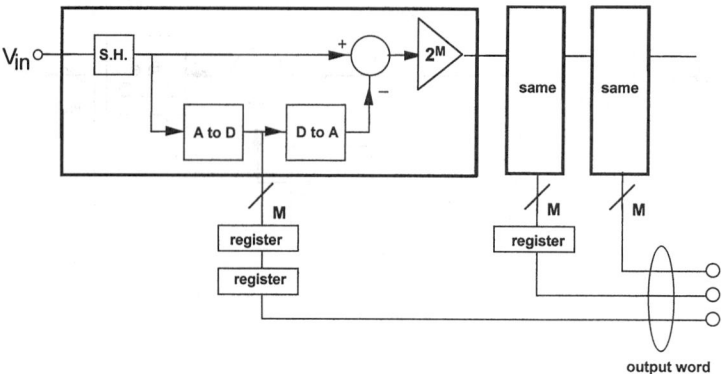

FIGURE 14.4.37   A pipelined converter.

Of course, it is useless to try to repeat this procedure over and over again because the errors generated in every new cycle pile up and corrupt the difference signal. It is very important to determine which sources of errors are important. The main ones are the interstage gain error, the D/A and the A/D nonlinearities. The A/D errors can easily be corrected by extending the dynamic range of both converters. Errors from the flash converter are corrected automatically during recycling. The other errors are more difficult to correct but one should not overlook the fact that they affect only bits generated after the first cycle.

The conversion time of recycling converters varies, of course, like the number of cycles. Pipelined converters offer a good means to keep the conversion time equal to a single cycle time, however, at the expense of area. Such a converter is shown in Fig. 14.4.37. The idea is simply to exchange time for space. In other words, we cascade blocks identical to the circuit of Fig. 14.4.36, but each circuit feeds its neighbor instead of recycling its own data. Every bloc thus deals with data that belong to different time samples. In order to reconfigure the correct output words, one must reshuffle the code segments in order to recover time consistency. This is done by means of registers.

Recycling and pipelined converters that operate with segments only 1-bit wide are currently designated *algorithmic converters*. They are not fast devices since they require as many cycles as number of bits to

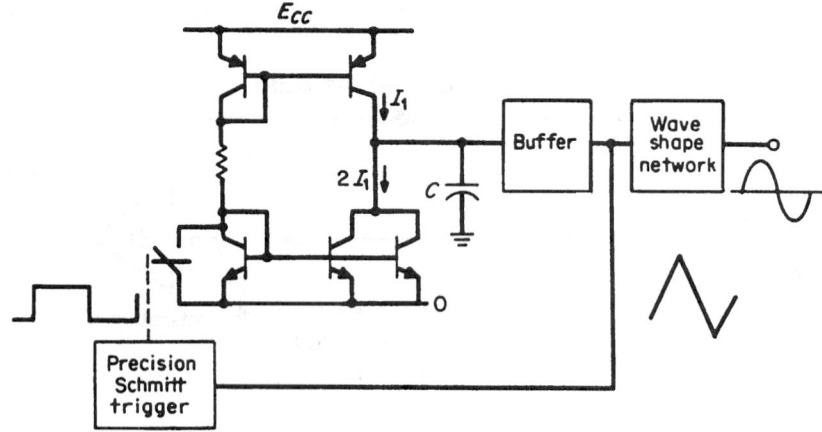

FIGURE 14.4.38   A typical integrated wave generator that can deliver a square wave, a triangular wave, and a sine wave.

output code words. Algorithmic converters are similar to successive approximation converters although they operate differently.

FUNCTION GENERATORS

Integrated function generators consist generally of a free-running relaxation oscillator controlling a nonlinear shaping circuit. A typical block diagram of a function generator is shown in Fig. 14.4.38. The relaxation oscillator is a combination of a time-base generator and a Schmitt trigger. The time base in the present case is obtained by successive loading and discharging of a capacitor using two current sources. One is a constant current source I_1 and the other a controlled current source delivering a current step equal to $-2I_1$ or zero. Hence, the voltage across the capacitor C is a sawtooth.

The switching of the controlled current source is monitored by the logical output signal of the Schmitt trigger. This last circuit is in fact a precision Schmitt trigger. The oscillating voltage across C is obtained in the same manner. The output sawtooth signal is buffered and drives a network consisting of resistors and diodes which changes the sawtooth into a more or less sinusoidal voltage.

The advantage of function generators over RC or op-amp oscillators is their excellent amplitude stability versus frequency. Also frequency modulation can easily be achieved by changing the current delivered by the two current sources. This type of function generator can be frequency-swept over a wide dynamic range without spurious amplitude transients since no selective network is involved.

SECTION 15

MEASUREMENT SYSTEMS

Measurement circuits are critical in the analysis, design, and maintenance of electronic systems. For those who work in electronics, these circuits and systems become the eyes into the world of the electron that cannot be directly seen. The main objective of such systems is to not influence what is being measured or observed. This is accomplished by a range of types of measurement circuits. All of these are considered in this section.

A key element of systems that measure is how the measurement is actually made. Without an understanding of how the measurement is made, one cannot understand the limitations. It is possible to make a measurement and be off by several orders of magnitude. We look at the process of making measurements and what to look for so that one can have a level of confidence in the measurement.

Substitution and analog measurements have been an important mainstay of this field. Unlike measurements that involve some digital systems, the accuracy and precision of the measurements depend totally on the precision of elements used in the measurement systems. We look at a variety of measurement techniques using substitution and look at analog devices like ohmmeters. Digital instruments have many advantages especially when used in data acquisition systems.

An important component of measurement systems is the transducer. The transducer converts a physical quantity into an electrical signal that can be measured by a measurement system. Knowing the characteristics, especially its limitations, helps in understanding the precision with which measurements can be made. It is in the area of transducers and sensors that we have seen some of the most dramatic advances.

Bridge circuits gave us the first opportunity to actually make measurements without "loading" the circuit being measured. The accuracy of the measurements merely depended on the precision of the elements used in the bridge.

We need AC impedance measurements to develop the small signal characteristics of a circuit or system and to evaluate the stability of the circuit or system. These kinds of measurements are important in a variety of applications from the space station to how well your television works.   C.A.

In This Section:

Section Bibliography:

Andrew, W. G., "Applied Instrumentation in the Process Industries," Gulf Pub. Co., 1993.

Bell, D. A., "Electronic Instrumentation and Measurements," Prentice Hall, 1994.

Carr, J. J., "Elements of Electronic Instrumentation and Measurement," 3rd ed., Prentice Hall, 1996.

Considine, D. M., and S. D. Ross (eds.), "Handbook of Applied Instrumentation," Krieger, 1982.

Coombs, C. F., Jr., "Electronic Instrument Handbook," 2nd ed., McGraw-Hill, 1995.

Decker, T., and R. Temple, "Choosing a phase noise measurement technique," H-P RF and Microwave Measurement Symposium, 1989.

Erickson, C., Switches in Automated Test Systems, Chap. 41 in Coombs's "Electronic Instrument Handbook," 2nd ed., McGraw-Hill, 1995.

"Direct Current Comparator Potentiometer Manual, Model 9930," Guideline Instruments, April 1975.

Harris, F. K., "Electrical Measurements," Wiley, 1952.

IEEE Standard Digital Interface for Programmable Instrumentation, ANSI/IEEE Std. 488.1, 1987.

Keithley, J. R., J. R. Yeager, and R. J. Erdman, "Low Level Measurements," 3rd ed., Keithley Instruments, June 1984.

Manassewitsch, V., "Frequency Synthesizers: Theory and Design," Wiley, 1980.

McGillivary, J. M., Computer-Controlled Instrument Systems, Chap. 43 in Coombs's "Electronic Instrument Handbook," 2nd ed., McGraw-Hill, 1995.

Mueller, J. E., Microprocessors in Electronic Instruments, Chap. 10 in Coombs's "Electronic Instrument Handbook," 2nd ed., McGraw-Hill, 1995.

Nachtigal, C. L., "Instrumentation and Control: Fundamentals and Applications," Wiley, 1990.

"Operation and Service Manual for Model 4191A RF Impedance Analyzer," Hewlett-Packard, January 1982.

"Operation and Service Manual for Model 4342A Q Meter," Hewlett-Packard, March 1983.

Reissland, M. V., "Electrical Measurement: Fundamentals, Concepts, Applications," Wiley, 1989.

Santoni, A., "IEEE-488 Instruments," *EDN*, pp. 77–94, October 21, 1981.

Schoenwetter, H. K., "A high-speed low-noise 18-bit digital to analog converter," *IEEE Trans. Instrum. Meas.*, Vol. IM-27, No. 4, pp. 413–417, December, 1978.

Souders, R. M., "A bridge circuit for the dynamic characteristics of sample/hold amplifiers," *IEEE Trans. Instrum. Meas.*, Vol. IM-27, No. 4, December, 1978.

Walston, J. A., and J. R. Miller (eds.),"Transistor Circuit Design," McGraw-Hill, 1963.

Witte, R. A., "Electronic Test Instruments: Theory and Practice," Prentice Hall, 1993.

Workman, D. R., "Calibration status: a key element of measurement systems management," 1993 National Conference of Standards Laboratories Symposium.

CHAPTER 15.1
PRINCIPLES OF MEASUREMENT CIRCUITS

Francis T. Thompson*

DEFINITIONS AND PRINCIPLES OF MEASUREMENT

Precision is a measure of the spread of repeated determinations of a particular quantity. Precision depends on the resolution of the measurement means and variations in the measured value caused by instabilities in the measurement system. A measurement system may provide precise readings, all of which are inaccurate because of an error in calibration or a defect in the system.

Accuracy is a statement of the limits that bound the departure of a measured value from the true value. Accuracy includes the imprecision of the measurement along with all the accumulated errors in the measurement chain extending from the basic reference standards to the measurement in question.

Errors may be classified into two categories, systematic and random. *Systematic errors* are those which consistently recur when a number of measurements are taken. Systematic errors may be caused by deterioration of the measurement system (weakened magnetic field, change in a reference resistance value), alteration of the measured value by the addition or extraction of energy from the element being measured, response-time effects, and attenuation or distortion of the measurement signal. *Random errors* are accidental, tend to follow the laws of chance, and do not exhibit a consistent magnitude or sign. Noise and environmental factors normally produce random errors but may also contribute to systematic errors.

The arithmetic average of a number of observations should be used to minimize the effect of random errors. The arithmetic average or mean X of a set of n readings $X_1, X_2, \ldots, X_n$ is

$$X = \sum X_i / n$$

The dispersion of these reading about the mean is generally described in terms of the standard deviation σ, which can be estimated for n observations by

$$s = \sqrt{\frac{\sum (X_i - X)^2}{n-1}}$$

where s approaches σ as n becomes large.

*The author is indebted to I. A. Whyte, L. C. Vercellotti, T. H. Putnam, T. M. Heinrich, T. I. Pattantyus, and R. A. Mathias for suggestions and constructive criticism for Chap. 15.1.

A *confidence interval* can be determined within which a specified fraction of all observed values may be expected to lie. The *confidence level* is the probability of a randomly selected reading falling within this interval. Detailed information on measurement errors is given in Coombs (Chap. 15.5).

Standardization and calibration involve the comparison of a physical measurement with a reference standard. Calibration normally refers to the determination of the accuracy and linearity of a measuring system at a number of points, while standardization involves the adjustment of a parameter of the measurement system so that the reading at one specific value is in correspondence with a reference standard. The numerical value of any reference standard should be capable of being traced through a chain of measurements to a National Reference Standard maintained by the National Institute of Standards and Technology (formerly the National Bureau of Standards).

The range of a measurement system refers to the values of the input variable over which the system is designed to provide satisfactory measurements. The range of an instrument used for a measurement should be chosen so that the reading is large enough to provide the desired precision. An instrument having a linear scale, which can be read within 1 percent at full scale, can be read only within 2 percent at half scale.

The resolution of a measuring system is defined as the smallest increment of the measured quantity which can be distinguished. The resolution of an indicating instrument depends on the deflection per unit input. Instruments having a square-law scale provide twice the resolution of full scale as linear-scale instruments. Amplification and zero suppression can be used to expand the deflection in the region of interest and thereby increase the resolution. The resolution is ultimately limited by the magnitude of the signal that can be discriminated from the noise background.

Noise may be defined as any signal that does not convey useful information. Noise is introduced in measurement systems by mechanical coupling, electrostatic fields, and magnetic fields. The coupling of external noise can be reduced by vibration isolation, electrostatic shielding, and electromagnetic shielding. Electrical noise is often present at the power-line frequency and its harmonics, as well as at radio frequencies.

In systems containing amplification, the noise introduced in low-level stages is most detrimental because the noise components within the amplifier passband will be amplified along with the signal. The noise in the output determines the lower limit of the signal that can be observed.

Even if external noise is minimized by shielding, filtering, and isolation, noise will be introduced by random disturbances within the system caused by such mechanisms as the Brownian motion in mechanical systems, Johnson noise in electrical resistance, and the Barkhausen effect in magnetic elements. Johnson noise is generated by electron thermal agitation in the resistance of a circuit. The equivalent rms noise voltage developed across a resistor R at an absolute temperature T is equal to $\sqrt{4kTR\,\Delta f}$, where k is Boltzmann's constant $(1.38 \times 10^{-23}\ J/K)$ and Δf is the bandwidth in hertz over which the noise is observed.

The bandwidth Δf of a system is the difference between the upper and lower frequencies passed by the system (Chap. 15.2). The bandwidth determines the ability of the system to follow variations in the quantity being measured. The lower frequency is zero for dc systems, and their response time is approximately equal to $1/(3\Delta f)$. Although a wider bandwidth improves the response time, it makes the system more susceptible to interference from noise.

Environmental factors that influence the accuracy of a measurement system include temperature, humidity, magnetic and electrostatic influences, mechanical stability, shock, vibration, and position. Temperature changes can alter the value of resistance and capacitance, produce thermally generated emfs, cause variations in the dimensions of mechanical members, and alter the properties of matter. Humidity affects resistance values and the dimensions of some organic materials. DC magnetic and electrostatic fields can produce an offset in instruments which are sensitive to these fields, while ac fields can introduce noise. The lack of mechanical stability can alter instrument reference values and produce spurious responses. Mechanical energy imparted to the system in the form of shock or vibration can cause measurement errors and, if severe enough, can result in permanent damage. The position of an instrument can affect the measurements because of the influence of magnetic, electrostatic, or gravitational fields.

TRANSDUCERS, INSTRUMENTS, AND INDICATORS

Transducers   are used to respond to the state of a quantity to be measured and to convert this state into a convenient electrical or mechanical quantity. Transducers can be classified according to the variable to be measured. Variable classifications include mechanical, thermal, physical, chemical, nuclear-radiation, electromagnetic-radiation, electrical, and magnetic, as detailed in Sec. 8.

Instruments   can be classified according to whether their output means is analog or digital. Analog instruments include the d'Arsonval (moving-coil) galvanometer, dynamometer instrument, moving-iron instrument, electrostatic voltmeter, galvanometer oscillograph, cathode-ray oscilloscope, and potentiometric recorders. Digital-indicator instruments provide a numerical readout of the quantity being measured and have the advantage of allowing unskilled people to make rapid and accurate readings.

Indicators   are used to communicate output information from the measurement system to the observer.

MEASUREMENT CIRCUITS

Substitution circuits   are used in the comparison of the value of an unknown electrical quantity with a reference voltage, current, resistance, inductance, or capacitance. Various potentiometer circuits are used for voltage substitution, and divider circuits are used for voltage, current, and impedance comparison. A number of these circuits and the reference components used in them are described in Chap. 15.2.

Analog circuits   are used to embody mathematical relationships, which permit the value of an unknown electrical quantity to be determined by measuring related electrical quantities. Analog-measurement techniques are discussed in Chap. 15.2, and a number of special-purpose measurement circuits are described in Chap. 15.4.

Digital instruments   combine analog circuits with digital processing to provide a convenient means of making rapid and accurate measurements. Digital instruments are described in Chaps. 15.2 and 15.4. Digital processing using the computational power of microprocessors is discussed in Chap. 15.3.

Bridge circuits   provide a convenient and accurate method of determining the value of an unknown impedance in terms of other impedances of known value. The circuits of a number of impedance bridges and amplifiers and detectors used for bridge measurements are described in Chap. 15.4.

Transducer amplifying and stabilizing circuits   are used in conjunction with measurement transducers to provide an electric signal of adequate amplitude, which is suitable for use in measurement and control systems. These circuits, which often have severe linearity, drift, and gain-stability requirements, are described in Chap. 15.3.

CHAPTER 15.2
SUBSTITUTION AND ANALOG MEASUREMENTS

Francis T. Thompson

VOLTAGE SUBSTITUTION

The constant-current potentiometer, which is used for the precise measurement of unknown voltages below 1.5 V, is shown schematically in Fig. 15.2.1. For a constant current, the output voltage V_o is proportional to the resistance included between the sliding contacts. In this circuit all the current-carrying connections can be soldered, thereby minimizing contact-resistance errors. When the sliding contacts are adjusted to produce a null, V_o is equal to the unknown emf and no current flows in the sliding contacts. At null, no current is drawn from the unknown emf, and therefore the measured voltage is independent of the internal resistance of the source.

The circuit of a multirange commercial potentiometer is shown in Fig. 15.2.2. The instrument is standardized with the range switch in the highest range position as shown and switch S connected to the standard cell. The calibrated standard-cell dial is adjusted to correspond to the known voltage of the standard cell, and the standardizing resistance is adjusted to obtain a null on the galvanometer. This procedure establishes a constant current of 20 mA through the potentiometer. The unknown emf is connected to the emf terminals, and switch S is thrown to the emf position. The unknown emf can be read to at least five significant figures by adjusting the tap slider and the 11-turn 5.5-Ω potentiometer for a null on the galvanometer. The range switch reduces the potentiometer current to 2 or 0.2 mA for the 0.1 and the 0.01 ranges, respectively, thereby permitting lower voltages to be measured accurately. Since the range switch does not alter the battery current (22 mA), the instrument remains standardized on the lower ranges. When making measurements, the current should be checked using the standard cell to ensure that the current has not drifted from the standardized value. The Leeds and Northrup Model 7556-B six-dial potentiometer operates in a similar manner and provides an accuracy of $\pm(0.001$ percent of reading $+ 0.1 \mu V)$.

The constant-resistance potentiometer of Fig. 15.2.3 uses a variable current through a fixed resistance to generate a voltage for obtaining a null with the unknown emf. The constant-resistance potentiometer is used primarily for measurements in the millivolt and microvolt range.

The microvolt potentiometer, or low-range potentiometer, is designed to minimize the effect of contact resistance and thermal emfs. Thermal shielding is used to minimize temperature differences. The galvanometer is connected to the circuit through a special Wenner thermo-free reversing key of copper and gold construction to eliminate thermal effects in the galvanometer circuit.

A typical microvolt potentiometer circuit consisting of two constant-current decades and a constant-resistance element is shown in Fig. 15.2.4. The constant-current decades use Diesselhorst rings, in which the constant current entering and leaving the ring divides two paths. The *IR* drop across the resistance in the isothermal shield increases in 10 equal increments as the dial switch is rotated. The switch contacts are in the constant-current

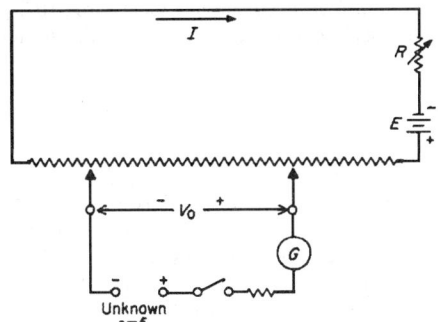

FIGURE 15.2.1   Constant-current potentiometer.

supply circuit, and therefore the effects of their *IR* drops and thermal emfs are minimized. A 100-division milliameter associated with the constant-resistance element provides nearly three additional decades of resolution. Readings to 10 nV are possible with this type of potentiometer.

The direct-current comparator potentiometer used for precise measurement of unknown voltages below 2.1 V is shown in Fig. 15.2.5. Feedback from the zero-flux detector winding is used to adjust the current supply for ampere-turn balance between the primary and secondary windings as in the dc-comparator ratio bridge of Chap. 15.4.

Standardization on the 1X range is obtained in a two-step procedure. First, the external standard cell and galvanometer are connected in series across resistor EH by the selector switch, and the constant current source is adjusted for zero galvanometer deflection. This transfers the standard cell voltage across resistor EH. Second, the seven measuring dials are set to the known standard voltage, and the selector switch is used to connect the galvanometer across the opposing voltages AD and EH. Trimmer turns n_s are used to obtain zero galvanometer deflection. This results in the standard cell voltage being generated across resistor AD with the dials set to the known standard cell value, and therefore calibration of the 1X range.

The unknown emf is measured on the 1X range by using the selector switch to connect the unknown emf and the voltage generated across resistor AD in series opposition across the galvanometer. The seven

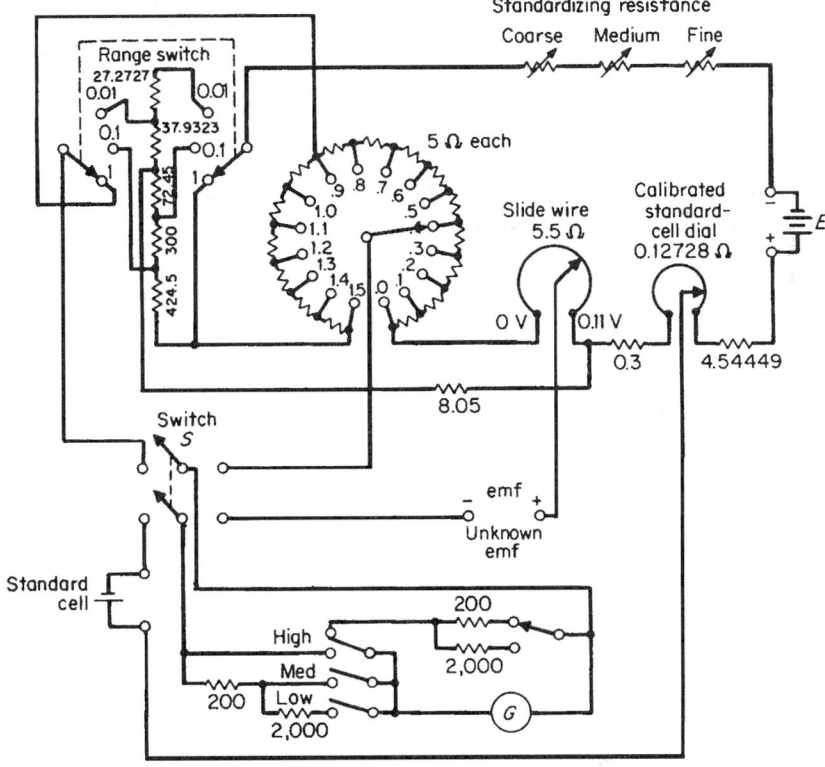

FIGURE 15.2.2   *K2 potentiometer.*   (*Leeds and Northrup*)

measuring dials are adjusted to obtain zero deflection. Specified measurement accuracies are $\pm(0.5$ ppm of reading $+ 0.05 \ \mu V)$ on the 1X range (2.111111 V full scale), $\pm(1$ ppm of reading $+ 0.01 \ \mu V)$ on the 0.1X range, and $\pm(2$ ppm of reading $+ 0.005 \ \mu V)$ on the 0.01X range.

DIVIDER CIRCUITS

The volt box   (Fig. 15.2.6) is used to extend the voltage range of a potentiometer. The unknown voltage is connected between 0 and an appropriate terminal, for example, ×100. The potentiometer is connected between the 0 and P output terminals. When the potentiometer is balanced, it draws no current, and therefore the current drawn from the source flows through the resistor between terminals 0 and P. The unknown voltage is equal to the potentiometer reading multiplied by the selected tap multiplier. Unlike the potentiometer, the volt box does load the voltage source. Typical resistances range from about 200 to 1000 Ω/V. The higher resistance values minimize self-heating and do not load the source as heavily. Errors due to leakage currents, which could flow through the insulators supporting the resistors, are minimized by using a guard circuit (see Chap. 15.4).

Decade voltage dividers   provide a wide range of precisely defined, and very accurate voltage ratios. The Kelvin-Varley vernier decade circuit is shown in Fig. 15.2.7. The slide arms in the first three decades are arranged so that they always span two contacts. The shunting effect of the second gang resistance across the slide arms of the first decade is equal to $2R$, thereby giving a net resistance of R between

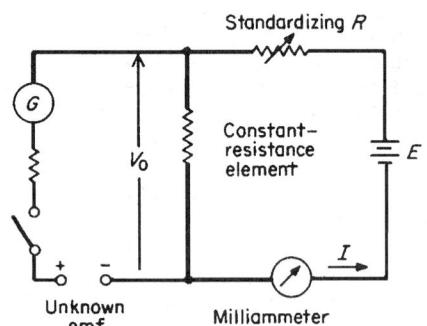

FIGURE 15.2.3   Constant-resistance potentiometer.

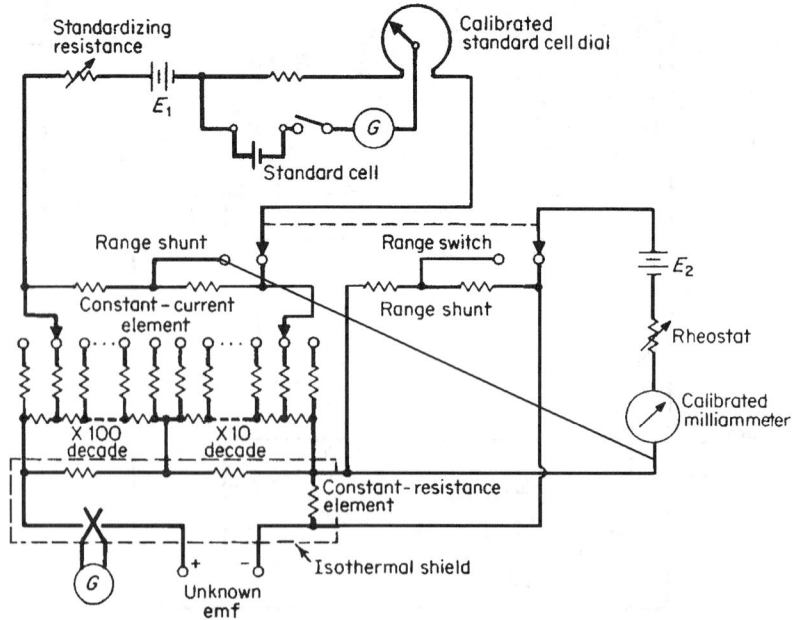

FIGURE 15.2.4   Microvolt potentiometer.

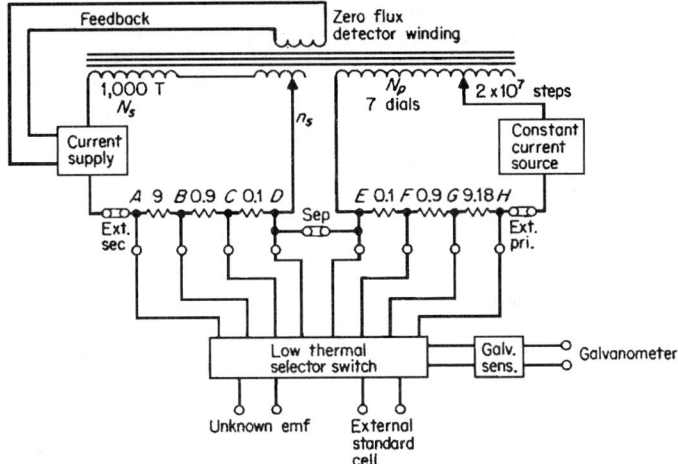

FIGURE 15.2.5 Direct-current comparator potentiometer.

the slide-arm contacts. With no current drawn from the output, the resistance loading on the input is equal to $10R$ and is independent of the slide-arm settings. In each of the first three decades, 11 resistors are used, while only 10 resistors are used in the final decade, which has a single sliding contact. Potentiometers with six decades have been constructed using the Kelvin-Varley circuit.

DECADE BOXES

Decade resistor boxes contain an assembly of resistances and switches, as shown in Fig. 15.2.8. The power rating of each resistance step is approximately constant; therefore, each decade has a different maximum current rating, which should not be exceeded. Boxes having four to seven decades are available with accuracies of 0.02 percent. Two typical seven-decade boxes provide resistance values from 0 to 1,111,111 Ω in 0.1-Ω steps and values from 0 to 11,111,110 Ω in 1-Ω steps. The accuracy at higher frequencies is affected by skin effect, series inductance, and shunt capacitance. The equivalent circuit of a resistance decade is shown in Fig. 15.2.9, where ΔL is the undesired incremental inductance added with each resistance step ΔR. Silver contacts are used to obtain a zero resistance R_o, as low as 1 mΩ/decade at dc. Zero inductance values L_o as low as 0.1 μH/decade are obtainable. The shunt capacitance for the configuration of Fig. 15.2.8 is a function of the highest decade in use, i.e., not set at zero. The shunt capacitance with the low terminal connected to the shield is typically 10 to 15 pF for the highest decade in use plus an equal value for each higher decade not in use.

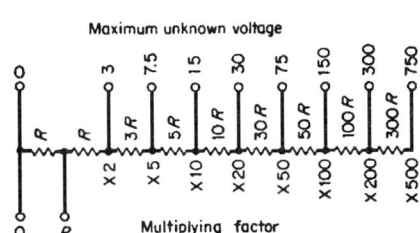

FIGURE 15.2.6 Volt-box circuit.

Some applications, e.g., the determination of small inductances at audio frequency and the determination of resistance at radio frequency by the substitution method, require that the equivalent series inductance of the resistance box remain constant, independent of the resistance setting. In the inductively compensated decade resistance box small copper-wound coils each having an inductance equal to the inductance of an individual resistance unit are selected by the decade switch so as to maintain a constant total inductance.

Decade capacitor units generally consist of four capacitances which are selectively connected in parallel by a four-gang 11-position switch (Fig. 15.2.10). The individual capacitors and their associated switch are shielded to

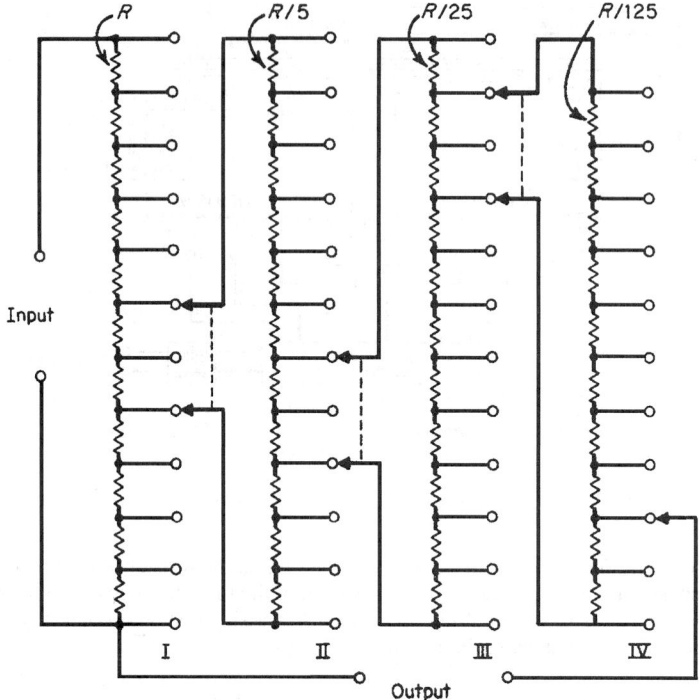

FIGURE 15.2.7   Decade voltage divider.

ensure that the selected steps add properly. *Decade capacitor boxes* are available with six-decade resolution, which provides a range of 0 to 1.11111 μF in increments of 1 pF and with an accuracy of 0.05 percent. Air capacitors are used in the 1- and 10-pF decades, and silver-mica capacitors in the higher ranges. Polystyrene capacitors are used in some less-precise decade capacitors.

Decade inductance units   can be constructed using four series-connected inductances of relative units 1, 2, 3, 4 or 1, 2, 2, 5. A four-gang 11-position switch is used to short-circuit the undesired inductances. Care must be

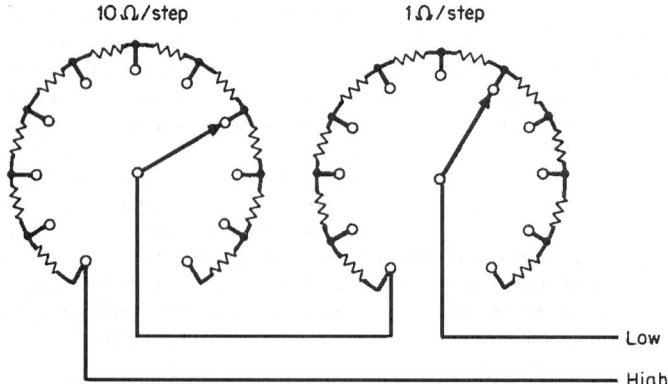

FIGURE 15.2.8   Decade resistance box.

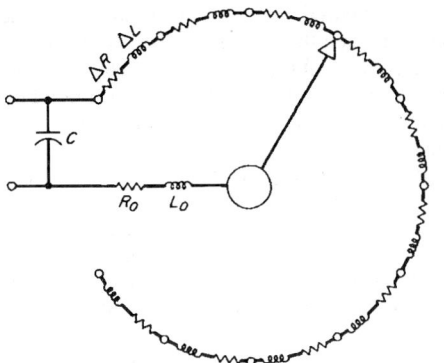
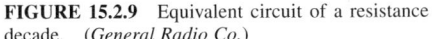

FIGURE 15.2.9   Equivalent circuit of a resistance
decade.   (*General Radio Co.*)

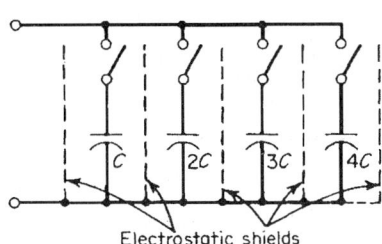

Electrostatic shields

FIGURE 15.2.10   Capacitor decade.

taken to avoid mutual coupling between the inductances. *Decade inductance boxes* are available with individual decades ranging from 1 mH to 10 H total inductance. A commercial single-decade unit consists of an
assembly of four inductors wound on molybdenum-Permalloy dust cores and a switch which enables consecutive values to be selected. Typical units have an accuracy of 1 percent at zero frequency. The effective series
inductance of a typical decade unit increases with frequency. The inductance is also a function of the ac current and any dc bias current. The Q of the coils varies with frequency.

ANALOG MEASUREMENTS

Ohmmeter circuits   provide a convenient means of obtaining an approximate measurement of resistance.
   The basic series-type ohmmeter circuit of Fig. 15.2.11a consists of an emf source, series resistor R_1 and
d'Arsonval milliammeter. Resistor R_2 is used to compensate for changes in battery emf and is adjusted to provide
full-scale meter deflection (0-Ω indication) with terminals X_1 and X_2 short-circuited. No deflection (infinite
resistance indication) is obtained with X_1 and X_2 open-circuited. When an unknown resistor R_x is connected

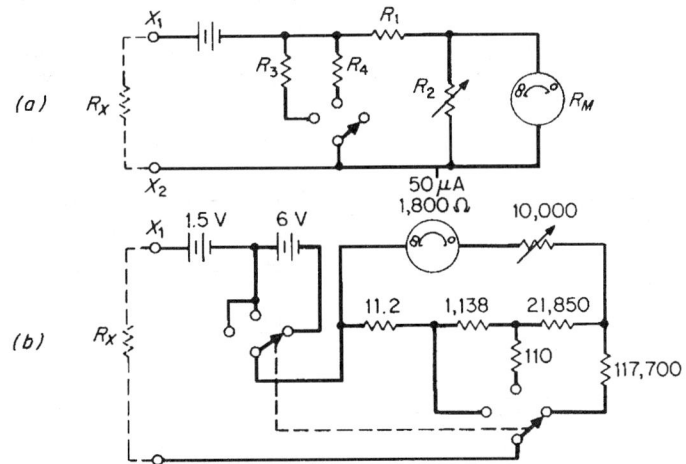

FIGURE 15.2.11   Series-type ohmmeters: (*a*) basic circuit; (*b*) commercial circuit
(*Simpson Electric Co.*)

across the terminals, the meter deflection varies inversely with the unknown resistance. With the range switch in the position shown, half-scale deflection is obtained when the external resistance is equal to $R_1 + R_2R_M/(R_2 + R_M)$. A multirange meter can be obtained using current-shunting resistors R_3 and R_4. A typical commercial ohmmeter circuit (Fig. 15.2.11b) having midscale readings of 12 Ω, 1200 Ω, and 120 kΩ uses an Ayrton shunt for range selection and a higher battery voltage for the highest resistance range.

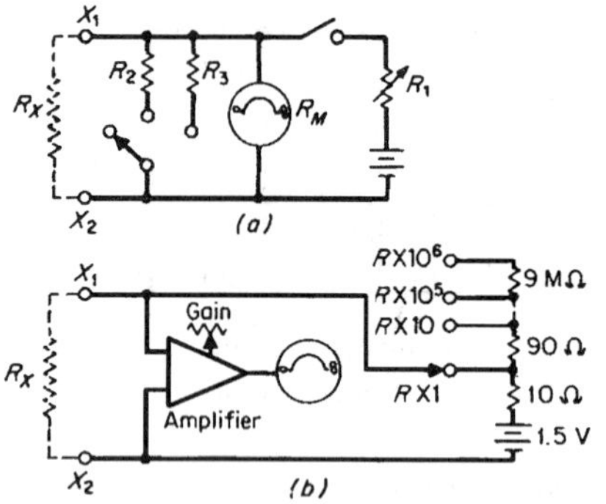

In the shunt-type ohmmeter the unknown resistor R_x is connected across the d'Arsonval milliammeter, as shown in Fig. 15.2.12a. The variable resistance R_1 is adjusted for full-scale deflection (infinite-resistance indication) with terminals X_1 and X_2 open-circuited. The ohm scale, with 0 Ω corresponding to zero deflection, is the reverse of the series-type ohmmeter scale. The resistance range can be lowered by switching a shunt resistor across the meter. With the range switch selecting shunt resistor R_2, half-scale deflection occurs when R_x is equal to the parallel combination of R_1, R_M, and R_2. The shunt-type ohmmeter is therefore most suited to low-resistance measurements.

The use of a high-input impedance amplifier between the circuit and the d'Arsonval meter permits the shunt-type ohmmeter to be used for high- as well as low-resistance measurements. A commercial ohmmeter (Fig. 15.2.12b) uses a field-effect-amplifier input stage which draws negligible current. The amplifier gain is adjusted to provide full-scale deflection with terminals X_1 and X_2 open-circuited. Half-scale deflection occurs when R_x is equal to the total selected tap resistance.

FIGURE 15.2.12   Shunt-type ohmmeter.   (*Triplett Electrical Instrument Co.*)

Voltage-drop   (or fall-of-potential) methods for determining resistance involve measuring the current flowing through the resistor with an ammeter, measuring the voltage drop across the resistor with a voltmeter, and calculating the resistance using Ohm's law. The circuit of Fig. 15.2.13a should be used for low-resistance measurements since the current drawn by the voltmeter V/R_v will be small with respect to the total current I. The circuit of Fig. 15.2.13b should be used for high-resistance measurements since the resistance of the ammeter R_A will be small with respect to the unknown resistance R_x. An accuracy of 1 percent or better can be obtained using 0.5 percent accurate instruments if the voltage source and instrument ranges are selected to provide readings near full scale.

Resonance methods   can be used to measure the inductance, capacitance, and Q factor of components at radio frequencies. In Fig. 15.2.14, resistors R_1 and R_2 couple the oscillator voltage e to a series-connected known capacitance and an unknown inductance represented by effective inductance L' and effective series resistance r'. Resistor R_2 is chosen to be small with respect to resistance r', thereby minimizing the effect of source resistance of the injected voltage.

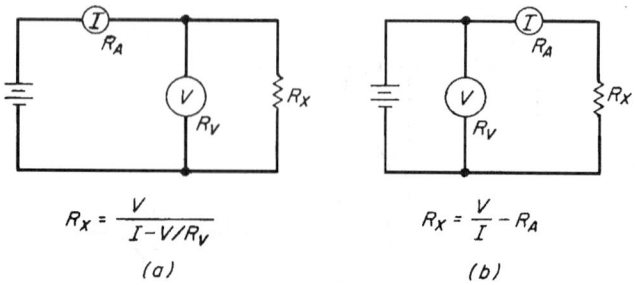

$$R_x = \frac{V}{I - V/R_V}$$

$$(a)$$

$$R_x = \frac{V}{I} - R_A$$

$$(b)$$

FIGURE 15.2.13   Fall-of-potential method: (*a*) for low resistances; (*b*) for high resistances.

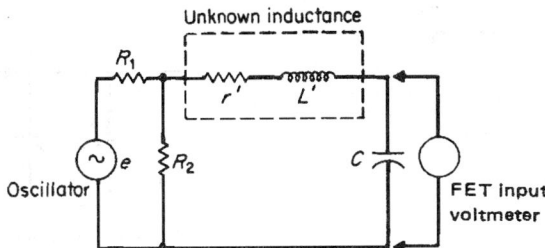

FIGURE 15.2.14   Inductance measurement.

A circuit containing reactive components is in resonance when the supply current is in phase with the applied voltage. The series circuit of Fig. 15.2.14 is in resonance when the inductive reactance $X_{L'}$ is equal to the capacitive X_C, which occurs when

$$\omega^2 = \omega_0^2 = 1/L'C$$

where $X_{L'} = \omega L'$, $X_C = 1/\omega C$, $\omega = 2\pi f$, $\omega_0 = 2\pi f_0$, f_0 = resonant frequency (Hz), L' = effective inductance (H), and C = capacitance (F).

If L', r', C, and e are constant and the oscillator frequency ω is adjusted until the voltage read across C by the FET input voltmeter is maximum, the frequency ω will be slightly less than the resonant frequency ω_0:

$$\omega^2 = \frac{1}{L'C} - \frac{(r' + R_s)^2}{2L'^2} = \omega_0^2 \left(1 - \frac{1}{2Q^{*2}}\right)$$

where $R_s = R_1 R_2/(R_1 + R_2)$ and $Q^* = \omega_0 L'/(r' + R_s)$. If $Q^* \geq 10$, ω and ω_0 will differ by less than 0.3 percent. The ratio m of the voltage across C to the voltage across R_2 can be measured while operating at ω. If $R_s \ll r'$, the value of the effective Q of the unknown inductance is related to m by

$$m = \frac{2Q'^2}{\sqrt{4Q'^2 - 1}} \quad Q' = \sqrt{\frac{m}{2(m - \sqrt{m_2 - 1})}}$$

where $Q' = \omega_0 L'/r'$ and $m = V_C/V_{R2}$ measured at ω. The values of m and Q' are nearly equal for high values of Q'; the difference is less than 0.4 percent for $Q' > 10$. If R_s is not small with respect to r', its value affects the determination of Q' only indirectly through its effect on ω. If $R_s = r$ and $Q' \geq 10$, the determination of Q' by the above equation is in error by less than 1 percent.

If ω, L', r', and e are constant and the capacitance C is adjusted until the voltage across it is maximum, the capacitance value C will be slightly less than the capacitance value C_R needed for resonance at the frequency ω:

$$C = \frac{1}{\omega^2 L'}\left(\frac{1}{1 + 1/Q^{*2}}\right) = C_R \frac{1}{1 + 1/Q^{*2}}$$

where $\omega = \omega_0$, $Q^* = \omega_0 L'/(r' + R_s)$, and $R_s = R_1 R_2/(R_1 + R_2)$. For $Q^* \geq 10$, C differs from C_R by less than 1 percent. If $R_s \ll r'$, the value of the effective Q of the unknown inductance is related to m by

$$m = \sqrt{Q'^2 + 1} \quad Q' = \sqrt{m^2 - 1}$$

where $Q' = \omega L'/r'$, $m = V_C/V_{R2}$, $\omega = \omega_0 = 2\pi f_0$, and f_0 is the resonant frequency in hertz for L and C_R.

The circuit of Fig. 15.3.14 can be used to find the value of an unknown capacitance if a stable inductance of known value is available. Similar circuits are used in Q meters (see Chap. 15.4). Discussions of series resonance, parallel resonance, and Q factor are given in Refs. 3, 12, and 13.

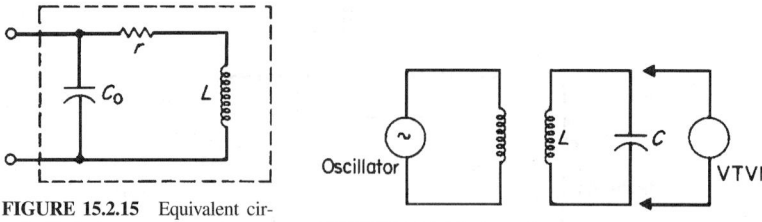

FIGURE 15.2.15 Equivalent circuit of inductance.

FIGURE 15.2.16 Loosely coupled tuned circuit.

Lead lengths should be kept short when making measurements at high frequencies, and the capacitance of the voltmeter must be added to capacitor C if it is not negligible.

The self-capacitance C_0 of an inductance is generally not negligible and can cause discrepancies in measurements. An inductance can be represented by the equivalent circuit of Fig. 15.2.15, where C_0 is the self-capacitance, L is the true inductance, and r is the true series resistance. The self-capacitance causes the effective inductance L' and the effective series resistance r' of the equivalent unknown inductance as measured in the circuit of Fig. 15.2.14 to differ somewhat from L and r, that is

$$r' = r\left(\frac{C+C_0}{C}\right)^2 \quad \text{and} \quad L' = L\frac{C+C_0}{C}$$

The value of C_0 can be measured with the aid of a Q meter (Chap. 15.4) having a calibrated variable capacitance.

The bandwidth of a tuned circuit can be determined by measuring the frequencies f_1 and f_2 at which the capacitor voltage in Fig. 15.2.16 is 0.707 times the maximum capacitor voltage. The oscillator, which maintains a constant voltage, is loosely coupled to the tuned circuit. The bandwidth Δf is equal to the difference $f_2 - f_1$ and is a function of the resonant frequency f_0 and the Q of the tuned circuit: $\Delta f = f_0/Q$.

The circuit Q of a tuned circuit for sinusoidal excitation is defined as 2π(maximum stored energy)/(energy dissipated per cycle). Circuit Q can be measured directly using the circuit of Fig.15.2.16. In this circuit the self-capacitance C_0 appears directly in parallel with the tuning capacitor C. The bandwidth is determined as explained above, and the Q is equal to $f_0/(f_2 - f_1)$. If the circuit of Fig. 15.2.14 is used, the self-capacitance no longer appears in parallel with the tuning capacitor C and the effective Q is measured. The real Q can be calculated if the self-capacitance C_0 is known; $Q = Q'(C + C_0)/C$, where Q' is the effective Q.

DIGITAL INSTRUMENTS

Digital multimeters provide a convenient means of making rapid and accurate measurements of voltage, current, and resistance. These instruments rely on the accuracy of A/D converters (see Chap. 15.3) and built-in precision voltage references. Many contain microprocessor-based intelligence and can be incorporated in automated-measurement and data-collection systems by means of the IEEE-488 bus (see Chap. 15.3). A typical block diagram is shown in Fig. 15.2.17. The performance is largely dependent on the A/D converter. Low-cost handheld units use a single-chip $3\frac{1}{2}$-digit A/D converter, while high-accuracy $5\frac{1}{2}$- and $6\frac{1}{2}$-digit meters use microprocessor-controlled discrete A/D converters. Typical performance specifications are given in Table 15.2.1. Unlike analog meters that have accuracies specified in terms of full scale, accuracy figures for digital meters are given in percent of the reading plus the number of counts of the least significant digit. Therefore, a 5.00 V reading on a $3\frac{1}{2}$-digit meter with a specified accuracy of $\pm(0.1$ percent $+1$ digit) would be accurate to ±0.3 percent or ±0.015 V.

The input resistance R_{in} of the meter can introduce an error in the measurement of an unknown voltage V_x by loading the unknown as shown in Fig. 15.2.18. The voltage V_m indicated by the meter is equal to $V_x R_{in}/(R_{in} + R_x)$.

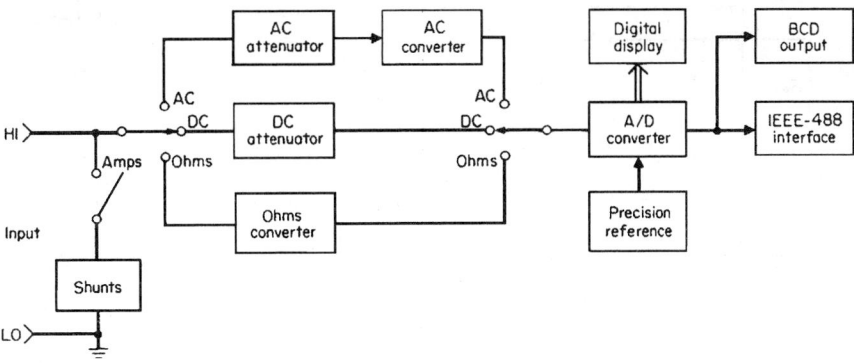

FIGURE 15.2.17 Digital multimeter. (*Keithley Instruments*)

TABLE 15.2.1 Typical Digital Multimeter Performance

	A/D digits		
	3½	4½	6½
Counts	1999	19,999	1,999,999
DC voltage			
Max. range	1000 V	1000 V	1000 V
Min. range	200 mV	200 mV	200 mV
Sensitivity	100 μV	10 μV	100 nV
Basic accuracy	± (0.1% + 1)	± (0.04% + 1)	±(0.0007% + 10)
Input resistance	10 MΩ	10 MΩ	≤ 20 V: 1 GΩ
			>20 V: 10 MΩ
CMRR (1 kΩ unbalance)	>100 dB	>120 dB	>120 dB
AC voltage			
Max. range	750 V	750 V	750 V
Min. range	200 mV	200 mV	2 V
Sensitivity	100 μV	10 μV	1 μV
Basic accuracy (<1 kHz)	±(1% + 2)	±(0.5% + 10)	±(0.25% + 1000)
Input resistance/cap	10 MΩ/100 pF	10 MΩ/100 pF	2 MΩ/50 pF
CMRR (1 kΩ unbalance)	>60 dB	>60 dB	>60 dB
Ohms			
Max. range	200 Ω	200 Ω	200 Ω
Min. range	20 mΩ	200 mΩ	200 mΩ
Sensitivity	100 mΩ	10 mΩ	100 $\mu\Omega$
Basic accuracy	±(0.2% + 1)	±(0.05% + 2)	±(0.01% + 2)
DC current			
Max. range	2A	2A	2 A
Min. range	2 mA	200 μA	200 μA
Sensitivity	1 μA	10 nA	1 nA
Basic accuracy	±(0.75% + 1)	±(0.2% + 2)	±(0.09% + 10)
AC current			
Max. range	2 A	2 A	2 A
Min. range	2 mA	200 μA	200 μA
Sensitivity	1 μA	10 nA	1 nA
Basic accuracy	±(1.5% + 2)	±(1% + 20)	±(0.6% + 300)

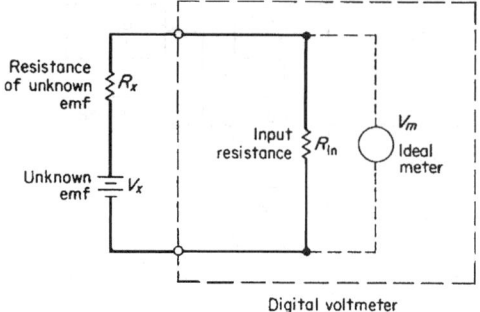

FIGURE 15.2.18   Effect of loading.

For example, if a meter having a 10-MΩ input resistance is used to measure an unknown emf having an equivalent resistance of 100 KΩ, the meter will read 99 percent of the correct value.

Resistance measurements rely on the accuracy of an internal current reference, shown as I_r in Fig.15.2.19. For an ideal meter with infinite input resistance R_{in}, the unknown resistance R_x is equal to V_m/I_r and the meter is calibrated to read the resistance value directly. For finite values of R_{in}, the indicated resistance value of R_x is equal to the parallel resistance of R_x and R_{in}, which can introduce a significant error in the measurement of high-value unknown resistances. The resistance of the leads connecting the unknown resistance to the meter of Fig. 15.2.19 can introduce an error in the measurement of low-value resistances. This problem is overcome in high-precision meters by using separate terminals and leads to connect the precision current reference to the unknown resistance. In this case the voltage measurement leads carry only the current drawn by the input resistance R_{in}.

Digital electrometers   are highly refined dc multimeters which provide exceptionally high input resistance and sensitivity. The block diagram of a typical microcomputer-controlled digital electrometer is shown in Fig. 15.2.20. Performance depends on the high-input-resistance, low-offset-current JFET preamp and operational amplifier as well as the A/D converter. A state-of-the-art 4$^1/_2$-digit autoranging meter[*] provides voltage ranges from 200 mV to 200 V, current ranges from 2 pA (2×10^{-12} A) to 20 mA, resistance ranges from 2 kΩ to 200 GΩ (200×10^9 Ω), and Coulomb ranges from 200 pC to 200 nC. Specified accuracies for the most-sensitive ranges are $\pm(0.05$ percent + 4 counts) for voltage, $\pm(1.6$ percent + 66 counts) for current, $\pm(0.20$ percent + 4 counts) for resistance, and $\pm(0.4$ percent + 4 counts) for charge. The input impedance is greater than 200×10^{12} Ω in parallel with 20 pF, and the input bias current is less than 5×10^{-15} A at 23°C. A bipolar 100-V, 0.2 percent-accuracy voltage source programmable in 50-mV steps and a 1-nA to 100-μA decade current source are built into the meter. Two techniques may be used to measure resistance or generate $I - V$ curves: The decade current source can be used to force a known current through the unknown impedance with the voltage measured by the high-input-impedance voltmeter, or the voltage source can be applied across the unknown with the resulting current measured by the current meter. The latter method is preferred for characterizing voltage-dependent materials.

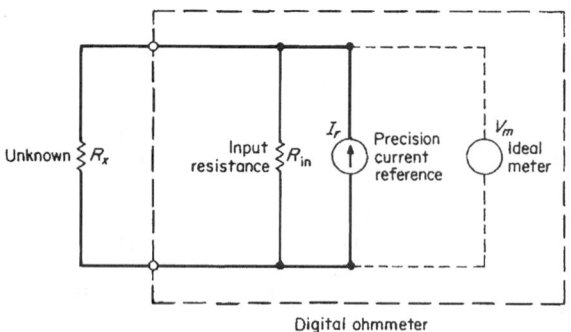

FIGURE 15.2.19   Effect of loading.

[*]Keithley Model 617 Electrometer.

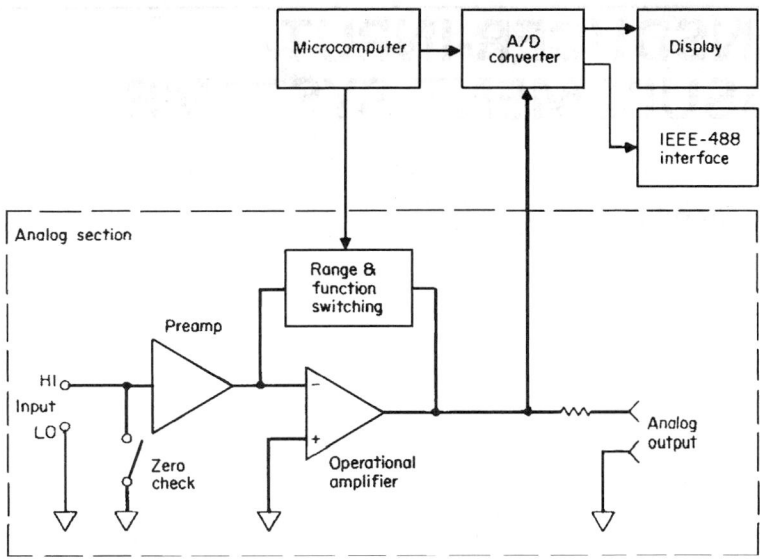

FIGURE 15.2.20   Typical digital electrometer.

A triaxial input cable, Fig. 15.2.21, is used with the electrometer to minimize input lead leakage and cable capacitance problems. The outer shield connected to the LO input can be grounded. The inner shield is driven by electrometer unity gain low-impedance output which maintains the guard voltage essentially equal to the high-impedance input signal HI. Insulator leakage through r becomes negligible since there is essentially no voltage difference between the center conductor and the guard.

Digital nanovoltmeters   are digital voltmeters that are optimized for nanovolt measurements. A state-of-the-art meter uses a JFET input amplifier to obtain an input resistance of 10^9 Ω and a 5-nF input capacitance. Voltage ranges of 2 mV to 1000 V are available. Specified accuracy for a 24-h period is $\pm$ (0.006 percent + 5 counts) when properly zeroed. A special connector having a low thermoelectric voltage is used for the 200-mV and lower ranges.

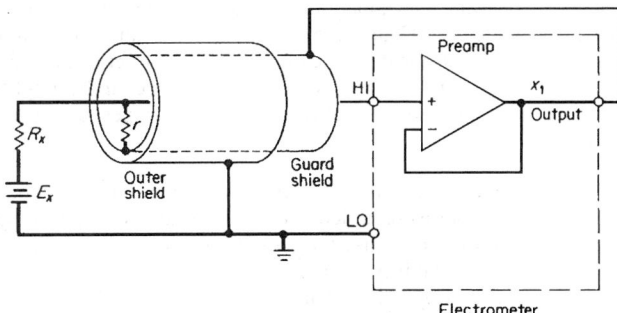

FIGURE 15.2.21   Guard shield minimizes input leakage.

CHAPTER 15.3
TRANSDUCER-INPUT MEASUREMENT SYSTEMS

Francis T. Thompson

Transducers are used to convert the quantity to be measured into an electric signal. Transducer types and their input and output quantities are discussed in Sec. 8.

TRANSDUCER SIGNAL CIRCUITS

Amplifiers are often required to increase the voltage and power levels of the transducer output and to prevent excessive loading of the transducer by the measurement system. The design of the amplifier is a function of the performance specifications, which include required amplification in terms of voltage gain or power gain, frequency response, distortion permissible at a given maximum signal level, dynamic range, residual noise permissible at the output, gain stability, permissible drift (for dc amplifiers), operating-temperature range, available supply voltage, permissible power consumption and dissipation, reliability, size, weight, and cost.

Capacitive-coupled amplifiers (ac amplifiers) are used when it is not necessary to preserve the dc component of the signal. AC amplifiers are used with transducers that produce a modulated carrier signal. Low-level amplifiers increase the signal from millivolts to several volts. The two-stage class A capacitor-coupled transistor amplifier of Fig. 15.3.1 has a power gain of 64 dB and a voltage gain of approximately 1000. Design information, an explanation of biasing, and equations for calculating the input impedance and various gain values are given in Walston and Miller (see Chap. 15.5). An excellent ac amplifier can be obtained by connecting a coupling capacitor in series with resistor R_1 of the operational amplifier of Fig. 15.3.4. The capacitor should be selected so that $C > 1/2\pi f R_1$, where f is the lowest signal frequency to be amplified. Class B transformer-coupled amplifiers, which are often used for higher power-output stages, are also discussed in Walston and Miller (see Chap. 15.5).

Direct-coupled amplifiers are used when the dc component of the signal must be preserved. These designs are more difficult than those of capacitive-coupled amplifiers because changes in transistor leakage currents, gain, and base-emitter voltage drops can cause the output voltage to change for a fixed input voltage, i.e., cause a dc-stability problem. The dc stability of an amplifier is determined primarily by the input stage since the equivalent input drift introduced by subsequent stages is equal to their drift divided by the preceding gain. Balanced input stages, such as the differential amplifier of Fig. 15.3.2, are widely used because drift components tend to cancel. By selecting a pair of transistors, Q_1 and Q_2, which are matched for current gain within 10 percent and base-to-emitter voltage within 3 mV, the temperature drift referred to the input can be held to within 10 μV/°C. Transistor Q_3 acts as a constant-current source and thereby increases the ability of the amplifier to reject common-mode input voltages. For applications where the generator resistance r_g is

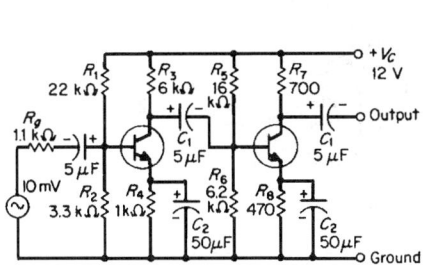

FIGURE 15.3.1 Two-stage cascaded common-emitter capacitive audio amplifier.

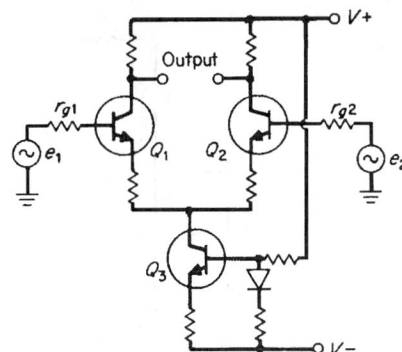

FIGURE 15.3.2 Differential amplifier.

greater than 50 kΩ, current offset becomes dominant, and lower overall drift can be obtained by using field-effect transistors in place of the bipolar transistors Q_1 and Q_2. Voltage drifts as low as 0.4 μV/°C can be obtained using integrated-circuit operational amplifiers (see Table 15.3.1).

Operational amplifiers are widely used for amplifying low-level ac and dc signals. They usually consist of a balanced input stage, a number of direct-coupled intermediate stages, and a low-impedance output stage. They provide high open-loop gain, which permits the use of a large amount of gain-stabilizing negative feedback. High-gain bipolar and JFET transistors are often used in the balanced input stage. The bipolar input provides lower voltage offset and offset voltage drift while the JFET input provides higher input impedance, lower bias current, and lower offset current as can be seen in Table 15.3.1 which compares the typical specifications of three high-performance operational amplifiers. The chopper-stabilized CMOS operational amplifier provides the low offset and bias currents of FET transistors while using internal chopper stabilization to achieve excellent offset voltage characteristics. With input voltages, e_1, e_2, and e_{cm} equal to zero, the output of the amplifier of Fig. 15.3.3 will have an offset voltage E_{os} defined by

$$E_{os} = V_{os} \frac{R_1 + R_2}{R_1} + I_{b1}R_2 - I_{b2} \frac{R_3 R_4 (R_1 + R_2)}{R_1 (R_3 + R_4)}$$

TABLE 15.3.1 Operational Amplifier Comparison

Typical parameter (except where noted)	Temp.	Bipolar input (Harris 5130)	JFET input (Harris 5170)	Chopper-stabilized CMOS (Siliconix 7652)
Input characteristics				
Offset voltage	25°C	10 μV	100 μV	0.7 μV
Maximum offset voltage	25°C	25 μV	300 μV	5 μV
Avg. offset voltage drift	Full	0.4 μV/°C	2 μV/°C	0.01 μV/°C
Bias current	25°C	1 nA	0.02 nA	0.015 nA
Bias current avg. drift	Full	20 pA/°C	3 pA/°C	
Maximum offset current	25°C	2 nA	0.03 nA	0.06 nA
Offset current avg. drift	Full	20 pA/°C	0.3 pA/°C	
Differential input resistance	25°C	30×10^6 Ω	6×10^{10} Ω	10^{12} Ω
Transfer characteristics				
Voltage gain	25°C	140 dB	116 dB	150 dB
Common-mode rejection ratio	25°C	120 dB	100 dB	130 dB

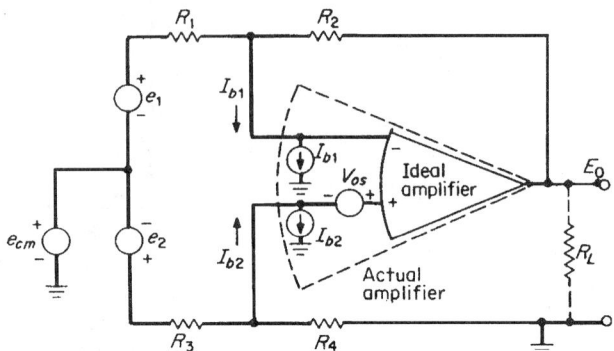

FIGURE 15.3.3   Operational amplifier.

where I_{b1} and I_{b2} are bias currents that flow into the amplifier when the output is zero and V_{os} is the input off-set voltage that must be applied across the input terminals to achieve zero output. The input bias current specified for an operational amplifier is the average of I_{b1} and I_{b2}. Since the bias currents are approximately equal, it is desirable to choose the parallel combination of R_3 and R_4 equal to the parallel combination of R_1 and R_2. For this case, $E_{os} = V_{os}(R_1 + R_2)/R_1 + I_{os}R_2$, where the offset current $I_{os} = I_{b1} - I_{b2}$.

In the ideal case, where V_{os} and I_{os} are zero, the output voltage E_0 as a function of signal voltage e_1 and e_2 and common-mode voltage e_{cm} is

$$E_0 = -e_1\frac{R_2}{R_1} + e_2\frac{R_4(R_1 + R_2)}{R_1(R_3 + R_4)} + e_{cm}\frac{R_4(R_1 + R_2) - R_2(R_3 + R_4)}{R_1(R_3 + R_4)}$$

Maximum common-mode rejection can be obtained by choosing $R_4/R_3 = R_2/R_1$, which reduces the above equation to $E_0 = R_2(e_2 - e_1)/R_1$. The common-mode signal is not entirely rejected in an actual amplifier but will be reduced relative to a differential signal by the common-mode rejection ratio of the amplifier. Minimum drift and maximum common-mode rejection, which are important when terminating the wires from a remote transducer, can be obtained by selecting $R_3 = R_1$ and $R_4 = R_2$.

Where common-mode voltages are not a problem, the simple inverting amplifier (Fig. 15.3.4) is obtained by replacing e_{cm} and e_2 with short circuits and combining R_3 and R_4 into one resistor, which is equal to the parallel equivalent of R_1 and R_2. The input impedance of this circuit is equal to R_1.

Similarly, the simple noninverting amplifier (Fig. 15.3.5) is obtained by replacing e_{cm} and e_1 with short circuits. The voltage follower is a special case of the noninverting amplifier where $R_1 = \infty$ and $R_2 = 0$. The input impedance of the circuit of Fig. 15.3.5 is equal to the parallel combination of the common-mode input impedance of the amplifier and impedance $Z_{id}[1 + (AR_1)/(R_1 + R_2)]$, where Z_{id} is the differential-mode amplifier input impedance and A is the amplifier gain. Where very high input impedances are required, as in electrometer circuits, operational amplifiers having field-effect input transistors are used to provide input resistances up to $10^{12}\ \Omega$.

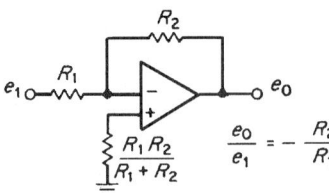

FIGURE 15.3.4   Inverting amplifier.

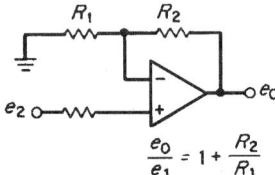

FIGURE 15.3.5   Noninverting amplifier.

An ac-coupled amplifier can be obtained by connecting a coupling capacitor in series with the input resistor of Fig. 15.3.4. The capacitor value should be selected so that the capacitive reactance at the lowest frequency of interest is lower than the amplifier input impedance R_1.

Operational amplifiers are useful for realizing filter networks and integrators. Other applications include absolute-value circuits, logarithmic converters, nonlinear amplification, voltage-level detection, function generation, and analog multiplication and division. Care should be taken not to exceed the maximum supply voltage and maximum common-mode voltage ratings and also to be sure that the load resistance R_L is not smaller than that permitted by the rated output.

The charge amplifier is used to amplify the ac signals from variable-capacitance transducers and transducers having a capacitive impedance such as piezoelectric transducers. In the simplified circuit of Fig. 15.3.6a, the current through C_s is equal to the current through C_1, and therefore

$$C_s \frac{\partial e_s}{\partial t} + e_s \frac{\partial C_s}{\partial t} = -C_1 \frac{de_o}{dt}$$

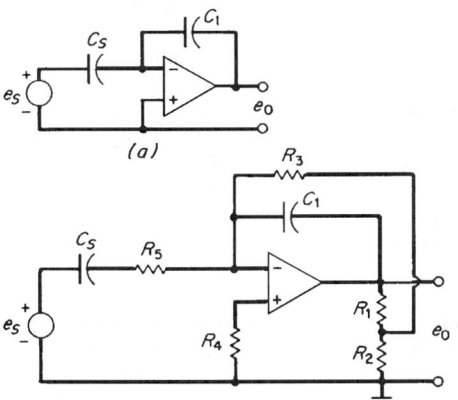

FIGURE 15.3.6   Charge amplifier.

For the piezoelectric transducer, C_s is assumed constant, and the gain $\delta e_o/\delta e_s = -C_s/C_1$. For the variable-capacitance transducer, e_s is constant, and the gain $de_o/dC_s = -e_s/C_1$. A practical circuit requires a resistance across C_1 to limit output drift. The value of this resistance must be greater than the impedance of C_1 at the lowest frequency of interest. A typical operational amplifier having field-effect input transistors has a specified maximum input current of 1 nA, which will result in an output offset of only 0.1 V if a 100-MΩ resistance is used across C_1. It is preferable to provide a high effective resistance by using a network of resistors, each of which has a value of 1 MΩ or less.

The effective feedback resistance R' in the practical circuit of Fig. 15.3.6b is given by $R' = R_3(R_1 + R_2)/R_2$, assuming that $R_3 > 10R_1R_2/(R_1 + R_2)$. Output drift is further reduced by selecting $R_4 = R_3 + R_1R_2/(R_1 + R_2)$. Resistor R_5 is used to provide an upper frequency rolloff at $f = 1/2\pi R_5 C_s$, which improves the signal-to-noise ratio.

Amplifier-gain stability is enhanced by the use of feedback since the gain of the amplifier with feedback is relatively insensitive to changes in the open-loop amplifier gain G provided that the loop gain GH is high. For example, if the open-loop gain G changes by 10 percent from 100,000 to 90,000 and the feedback divider gain H remains constant at 0.01, the closed-loop gain $G/(1 + GH) \approx 99.9$ changes only 0.011 percent.

If a high closed-loop gain is required, simply decreasing the value of H will reduce the value of GH and thereby reduce the accuracy. The desired accuracy can be maintained by cascading two or more amplifiers, thereby reducing the closed-loop gain required from each amplifier. Each amplifier has its own individual feedback, and no feedback is applied around the cascaded amplifiers. In this case, it is unwise to cascade more stages than needed to achieve a reasonable value of GH in each individual amplifier, since excessive loop gain will make the individual stages more prone to oscillation and the overall system will exhibit increased sensitivity to noise transients.

Chopper amplifiers are used for amplifying dc signals in applications requiring very low drift. The dc input signal is converted by a chopper to a switched ac signal having an amplitude proportional to the input signal and a phase of 0 or 180° with respect to the chopper reference frequency, depending on the polarity of the input signal. This ac signal is amplified by an ac amplifier, which eliminates the drift problem, and then is converted back into a proportional dc output voltage by a phase-sensitive demodulator.

The frequency response of a chopper amplifier is theoretically limited to one-half the carrier frequency. In practice, however, the frequency response is much lower than the theoretical limit. High-frequency components in the input signal exceeding the theoretical limit are removed to avoid unwanted beat signals with the chopper frequency.

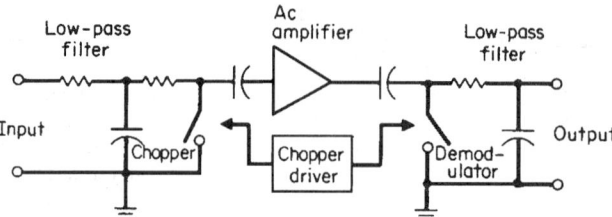

FIGURE 15.3.7   Chopper amplifier.

The chopper amplifier of Fig. 15.3.7 consists of a low-pass filter to attenuate high frequencies, an input chopper, an ac amplifier, a phase-sensitive demodulator, and a low-pass output filter to attenuate the chopper ripple component in the output signal. The frequency response of this amplifier is limited to a small fraction of the chopper frequency.

The frequency-response limitation of the chopper amplifier can be overcome by using the chopper amplifier for the dc and low-frequency signals and a separate ac amplifier for the higher-frequency signals, as shown in Fig. 15.3.8. Simple shunt field-effect-transistor choppers Q_1 and Q_2 are used for modulation and detection, respectively. Capacitor C_T is used to minimize spikes at the input to the ac amplifier.

A CMOS auto-zeroed operational amplifier, which contains a built-in chopper amplifier, has typical specifications of 0.7 μV input offset voltage, 0.01 μV/°C average offset voltage drift, and 15 pA input bias current (see Table 15.3.1).

Modulator-demodulator systems   avoid the drift problems of dc amplifiers by using a modulated carrier which can be amplified by ac amplifiers (Fig. 15.3.9). Inputs and outputs may be either electrical or mechanical.

The varactor modulator (Fig. 15.3.10) takes advantage of the variation of diode-junction capacitance with voltage to modulate a sinusoidal carrier. The carrier and signal voltages applied to the diodes are small, and the diodes never reach a low-resistance condition. Input bias currents of the order of 0.01 pA are possible. For zero signal input, the capacitance values of the diodes are equal, and the carrier signals coupled by the diodes cancel. A dc-input signal will increase the capacitance of one diode while decreasing the capacitance of the other and thereby produce a carrier imbalance signal which is coupled to the ac amplifier by capacitor C_2. A phase-sensitive demodulator, such as field-effect transistor Q_2 of Fig. 15.3.8, may be used to recover the dc signal.

The magnetic amplifier and second-harmonic modulator can also be used to convert dc-input signals to modulation on a carrier, which is amplified and later demodulated. Mechanical-input modulators include ac-driven potentiometers, linear variable differential transformers, and synchros. The amplified ac carrier can be converted directly into a mechanical output by a two-phase induction servomotor.

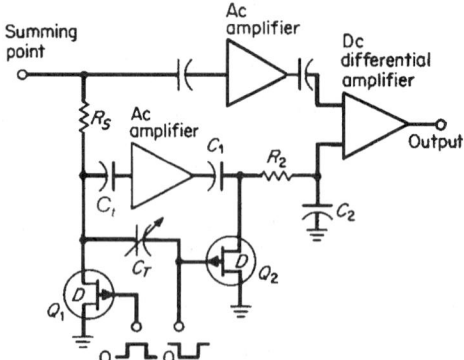

FIGURE 15.3.8   Chopper-stabilized dc amplifier.

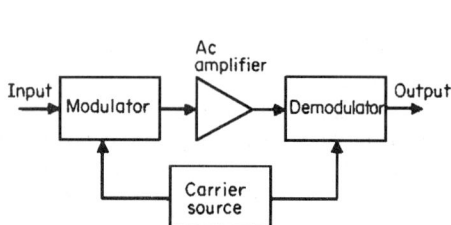

FIGURE 15.3.9   Modulator-demodulator system.

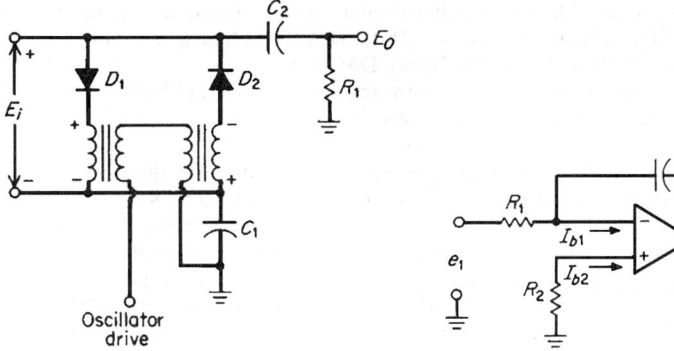

FIGURE 15.3.10 Basic varactor modulator.   **FIGURE 15.3.11** Analog integrator.

Integrators   are often required in systems where the transducer signal is a derivative of the desired output, e.g., when an accelerometer is used to measure the velocity of a vibrating object. The output of the analog integrator of Fig. 15.3.11 consists of an integrated signal term plus error terms caused by the offset voltage V_{os} and the bias currents I_{b1} and I_{b2}

$$e_0 = \frac{1}{R_1 C} \int e_1 \, dt + \frac{1}{R_1 C} \int (V_{os} + I_{b1} R_1 - I_{b2} R_2) \, dt$$

These error terms will cause the integrator to saturate unless the integrator is reset periodically or a feedback path exists, which tends to drive the output toward a given level within the linear range. In the accelerometer integrator, accurate integration may not be required below a given frequency, and the desired stabilizing feedback path can be introduced by incorporating a large effective resistance across the capacitor using the technique shown in Fig. 15.3.6*b*.

Digital processing   of analog quantities is frequently used because of the availability of high-performance analog-to-digital (A/D) converters, digital-to-analog (D/A) converters, microprocessors, and other special digital processors. The analog input signal (Fig. 15.3.12) is converted into a sequence of digital values by the A/D converter. The digital values are processed by the microprocessor (see "The microprocessor") or other digital processor, which can be programmed to provide a wide variety of functions including linear or nonlinear gain characteristics, digital filtering, integration, differentiation, modulation, linearization of signals from nonlinear transducers, computation, and self-calibration. The Texas Instruments TMS 320 series of digital processors are specially designed for these applications.

The digital output may be used directly or converted back into an analog signal by the D/A converter. Commercial A/D converter modules are available with 15-bit resolution and 17-μs conversion times using the successive-approximation technique (Analog Devices AD 376, Burr Brown ADC 76). Monolithic integrated-circuit A/D converters are available with 12-bit resolution and 6-μs conversion times (Harris HI-774A). An 18-bit D/A converter with a full-scale current output of 100 mA and a compliance voltage range of ±12 V has been

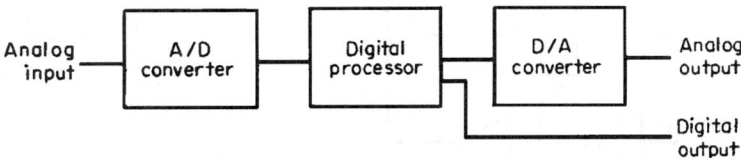

FIGURE 15.3.12 Digital processing of analog signals.

built at the National Bureau of Standards (Schoenwetter). It exhibits a settling a time to $1/2$ least significant bit (2 ppm) of less than 20 μs. Commercial 16- and 18-bit D/A converters are available with 2-mA full-scale outputs (Analog Devices DAC1136/1138, Burr Brown DAC 70).

The application of A/D converters, D/A converters, and sample-and-hold (S/H) circuits requires careful consideration of their static and dynamic characteristics.

The microprocessor   is revolutionizing instrumentation and control. The digital processing power that previously cost many thousands of dollars has been made available on a silicon integrated-circuit chip for only a few dollars. A microcomputer system (Fig. 15.3.13) consists of a microprocessor central processing unit (CPU), memory, and input-output ports. A typical microprocessor-based system consists of the 8051 microprocessor, which provides 4 kbytes of read-only memory (ROM), 128 bytes of random-access memory (RAM), 2 timers and 32 input-output lines, and a number of additional RAM and ROM memory chips as needed for a particular application. The microprocessor can provide a number of features at little incremental cost by means of software modification. Typical features include automatic ranging, self-calibration,

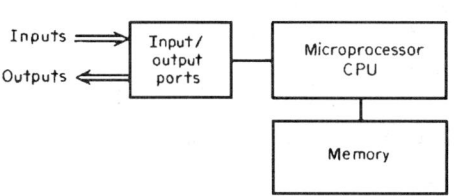

FIGURE 15.3.13   Microcomputer system.

self-diagnosis, data conversion, data processing, linearization, regulation, process monitoring, and control. Instruments using microprocessors are described in Chaps. 15.2 and 15.4.

Instrument systems   can be automated by connecting the instruments to a computer using the IEEE-448 general purpose interface bus (Santoni). The bus can connect as many as 14 individually addressable instruments to a host computer port. Full control capability requires the selection of instruments that permit bus control of front panel settings as well as bus data transfers.

Voltage comparators   are useful in a number of applications, including signal comparison with a reference, positive and negative signal-peak detection, zero-crossing detection, A/D successive-approximation converter systems, crystal oscillators, voltage-controlled oscillators, magnetic-transducer voltage detection, pulse generation, and square-wave generation. Comparators with input offset voltages of 2 mV and input offset currents of 5 nA are commercially available.

Analog switches   using field-effect-transistor (FET) technology are used in general-purpose high-level switching ($\pm$10 V), multiplexing, A/D conversion, chopper applications, set-point adjustment, and bridge circuits. Logic and schematic diagrams of a typical bilateral switch are shown in Fig. 15.3.14. The switch provides the required isolation between the data signal and the control signal. Typical switches provide zero offset voltage, on resistance of 35 Ω, and leakage current of 0.04 nA. Multiple-channel switches are commercially available in a single package.

Output Indicators.   A variety of analog and digital output indicators can be used to display and record the output from the signal-processing circuitry.

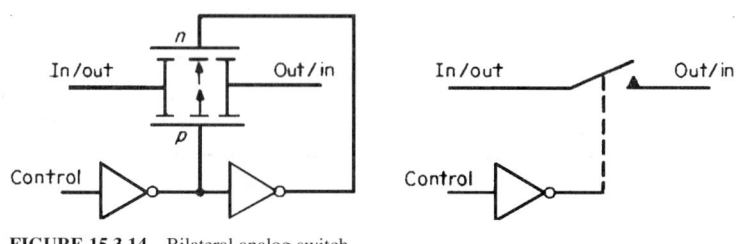

FIGURE 15.3.14   Bilateral analog switch.

CHAPTER 15.4
BRIDGE CIRCUITS, DETECTORS, AND AMPLIFIERS

Francis T. Thompson

PRINCIPLES OF BRIDGE MEASUREMENTS

Bridge circuits   are used to determine the value of an unknown impedance in terms of other impedances of known value. Highly accurate measurements are possible because a null condition is used to compare ratios of impedances.

The most common bridge arrangement (Fig. 15.4.1) contains four branch impedances, a voltage source, and a null detector. Galvanometers, alone or with chopper amplifiers, are used as null detectors for dc bridges; while telephone receivers, vibration galvanometers, and tuned amplifiers with suitable detectors and indicators are used for null detection in ac bridges. The voltage across an infinite-impedance detector is

$$V_d = \frac{(Z_1 Z_3 - Z_2 Z_x)E}{(Z_1 + Z_2)(Z_3 + Z_x)}$$

If the detector has a finite impedance Z_5, the current in the detector is

$$I_d = \frac{(Z_1 Z_3 - Z_2 Z_s)E}{Z_5(Z_1 + Z_2)(Z_3 + Z_x) + Z_1 Z_2(Z_3 + Z_x) + Z_3 Z_x(Z_1 + Z_2)}$$

where E is the potential applied across the bridge terminals.

A null or balance condition   exists when there is no potential across the detector. This condition is satisfied, independent of the detector impedance, when $Z_1 Z_3 = Z_2 Z_x$. Therefore, at balance, the value of the unknown impedance Z_x can be determined in terms of the known impedances Z_1, Z_2, and Z_3:

$$Z_x = Z_1 Z_3 / Z_2$$

Since the impedances are complex quantities, balance requires that both magnitude and phase angle conditions be met: $|Z_x| = |Z_1| \cdot |Z_3| / |Z_2|$ and $\theta_x = \theta_1 + \theta_3 - \theta_2$. Two of the known impedances are usually fixed impedances, while the third impedance is adjusted in resistance and reactance until balance is attained.

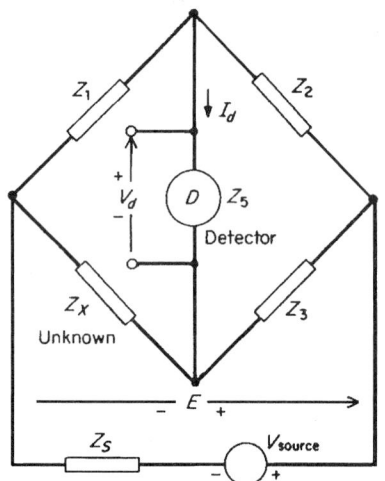

FIGURE 15.4.1   Basic impedance bridge.

The sensitivity of the bridge can be expressed in terms of the incremental detector current ΔI_d for a given small per-unit deviation δ of the adjustable impedance from the balance value. If Z_1 is adjusted, $\delta = \Delta Z_1/Z_1$ and

$$\Delta I_d = \frac{Z_3 Z_x E \delta}{(Z_3 + Z_x)^2 [Z_5 + Z_1 Z_2/(Z_1 + Z_2) + Z_3 Z_x/(Z_3 + Z_x)]}$$

when Z_5 is the detector impedance.

If a high-input-impedance amplifier is used for the detector and impedance Z_5 can be considered infinite, the sensitivity can be expressed in terms of the incremental input voltage to the detector ΔV_d for a small deviation from balance

$$\Delta V_d = Z_3 Z_x E \delta/(Z_3 + Z_x)^2 = Z_1 Z_2 E \delta/(Z_1 + Z_2)^2$$

where $\delta = \Delta Z_1/Z_1$ and ΔZ_1 is the deviation of impedance Z_1 from its balance value Z_1. Maximum sensitivity occurs when the magnitudes of Z_3 and Z_x are equal (which for balance implies that the magnitudes of Z_1 and Z_2 are equal). Under this condition, $\Delta V_d = E\delta/4$ when the phase angles θ_3 and θ_x are equal; $\Delta V_d = E\delta/2$ when the phase angles θ_3 and θ_x are in quadrature; and ΔV_d is infinite when $\theta_3 = -\theta_x$, as is the case with lossless reactive components of opposite sign. In practice, the value of the adjustable impedance must be sufficiently large to ensure that the resolution provided by the finest adjusting step permits the desired precision to be obtained. This value may not be compatible with the highest sensitivity, but adequate sensitivity can be obtained for an order-of-magnitude difference between Z_3 and Z_x or Z_1 and Z_2, especially if a tuned-amplifier detector is used.

Interchanging the source and detector can be shown to be equivalent to interchanging impedances Z_1 and Z_3. This interchange does not change the equation for balance but does change the sensitivity of the bridge. For a fixed applied voltage E higher sensitivity is obtained with the detector connected from the junction of the two high-impedance arms to the junction of the two low-impedance arms.

The source voltage must be carefully selected to ensure that the allowable power dissipation and voltage ratings of the known and unknown impedances of the bridge are not exceeded. If the bridge impedances are low with respect to the source impedance Z_s, the bridge-terminal voltage E will be lowered. This can adversely affect the sensitivity, which is proportional to E. The source for an ac bridge should provide a pure sinusoidal voltage since the harmonic voltages will usually not be nulled when balance is achieved at the fundamental frequency. A tuned detector is helpful in achieving an accurate balance.

Balance Convergence. The process of balancing an ac bridge consists of making successive adjustments of two parameters until a null is obtained at the detector. It is desirable that these parameters do not interact and that convergence be rapid.

The equation for balance can be written in terms of resistances and reactances as

$$R_x + jX_x = (R_1 + jX_1)(R_3 + jX_3)/(R_2 + jX_2)$$

Balance can be achieved by adjusting any or all of the six known parameters, but only two of them need be adjusted to achieve the required equality of both magnitude and phase (or real and imaginary components). In a ratio-type bridge, one of the arms adjacent to the unknown, either Z_1 and Z_3, is adjusted. Assuming that Z_1 is adjusted, then to make the resistance adjustment independent of the change in the corresponding reactance, the ratio $(R_3 + jX_3)/(R_2 + jX_2)$ must be either real or imaginary but not complex. If this ratio is equal to the real number k, then for balance $R_x = kR_1$ and $X_x = kX_1$. In a product-type bridge, the arm opposite the unknown Z_2

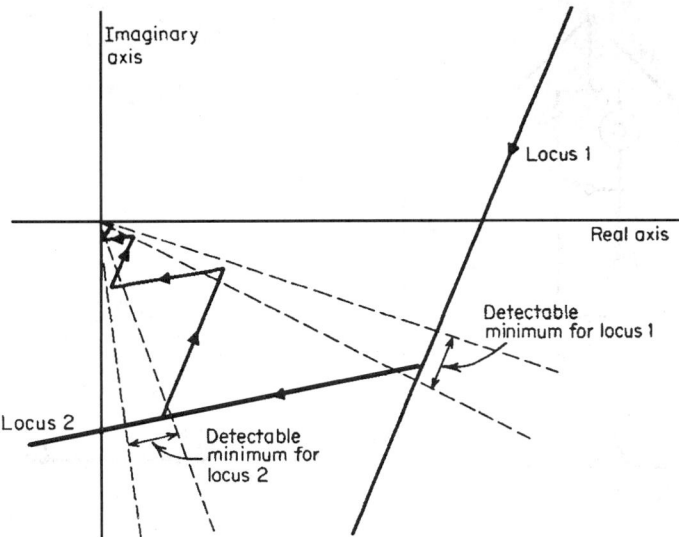

FIGURE 15.4.2 Linearized convergence locus.

is adjusted for balance, and the product of Z_1Z_3 must be either real or imaginary to make the resistance adjustment independent of the reactance adjustment.

Near balance, the denominator of the equation giving the detector voltage (or current) changes little with the varied parameter, while the numerator changes considerably. The usual convergence loci, which consist of circular segments, can be simplified to obtain linear convergence loci by assuming that the detector voltage near balance is proportional to the numerator, $Z_1Z_3 - Z_2Z_x$. Values of this quantity can be plotted on the complex plane. When only a single adjustable parameter is varied, a straight-line locus will be produced as shown in Fig. 15.4.2. Varying the other adjustable parameter will produce a different straight-line locus. The rate of convergence to the origin (balance condition) will be most rapid if these two loci are perpendicular, slow if they intersect at a small angle, and zero if they are parallel. The cases of independent resistance and reactance adjustments described above correspond to perpendicular loci.

RESISTANCE BRIDGES

The Wheatstone bridge is used for the precise measurement of two-terminal resistances. The lower limit for accurate measurement is about 1 Ω, because contact resistance is likely to be several milliohms. For simple galvanometer detectors, the upper limit is about 1 MΩ, which can be extended to 10^{12} by using a high-impedance high-sensitivity detector and a guard terminal to substantially eliminate the effects of stray leakage resistance to ground.

The Wheatstone bridge (Fig. 15.4.3) although historically older, may be considered as a resistance version of the impedance bridge of Fig. 15.4.1, and therefore the sensitivity equations are applicable. At balance

$$R_x = R_1R_3/R_2$$

Known fixed resistors, having values of 1, 10, 100, or 1000 Ω, are generally used for two arms of the bridge, for example, R_2 and R_3. These arms provide a ratio R_3/R_2, which can be selected from 10^{-3} to 10^3. Resistor R_1, typically adjustable to 10,000 Ω in 1- or 0.1-Ω steps, is adjusted to achieve balance. The ratio R_3/R_2 should be chosen so that R_1 can be read to its full precision. The magnitudes of R_2 and R_3 should be chosen to maximize the sensitivity while taking care not to draw excessive current.

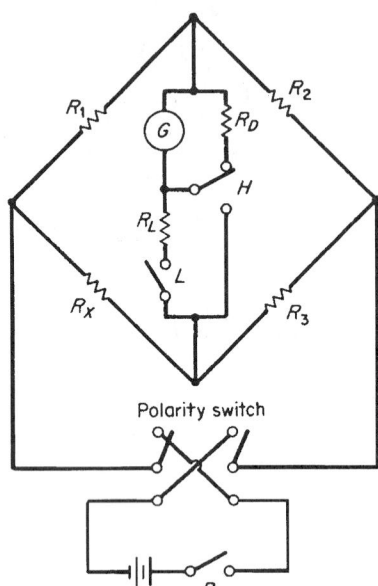

FIGURE 15.4.3   Wheatstone bridge.

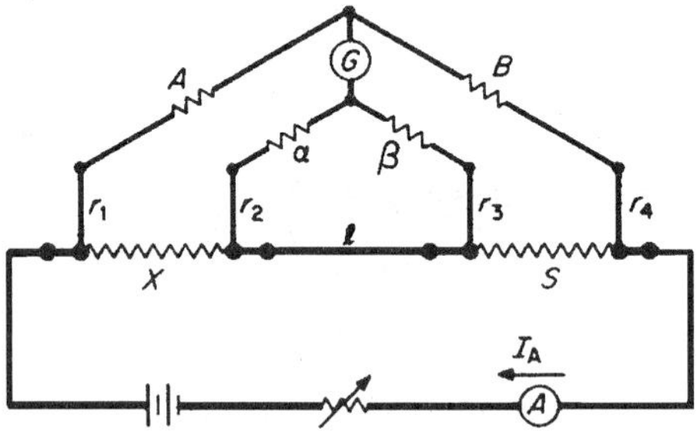

FIGURE 15.4.4   Kelvin double bridge. $A + B$ is typically 1000 Ω and $\alpha + \beta$ is typically 1000 Ω.

An alternate arrangement using R_1 and R_2 for the ratio resistors and adjusting R_3 for balance will generally provide a different sensitivity.

The battery key B should be depressed first to allow any reactive transients to decay before the galvanometer key is depressed. The low-galvanometer-sensitivity key L should be used until the bridge is close to balance. The high-sensitivity key H is then used to achieve the final balance. Resistance R_D provides critical damping between the galvanometer measurements. The battery connections to the bridge may be reversed and two separate resistance determinations made to eliminate any thermoelectric errors.

The Kelvin double bridge   (Fig. 15.4.4) is used for the precise measurement of low-value four-terminal resistors in the range 1 $\mu\Omega$ to 10 Ω. The resistance to be measured X and a standard resistance S are connected by means of their current terminals in a series loop containing a battery, an ammeter, an adjustable resistor, and a low-resistance link l. Ratio-arm resistances A and B and α and β are connected to the potential terminals of resistors X and S as shown. The equation for balance is

$$X = S\frac{A}{B} + \frac{\beta l}{\alpha + \beta + l}\left(\frac{A}{B} - \frac{\alpha}{\beta}\right)$$

If the ratio α/β is made equal to the ratio A/B, the equation reduces to $X = S(A/B)$.

The equality of the ratios should be verified after the bridge is balanced by removing the link. If $\alpha/\beta = A/B$, the bridge will remain balanced. Lead resistances r_1, r_2, r_3, and r_4 between the bridge and the potential terminals of the resistors may contribute to ratio imbalance unless they have the same ratio as the arms to which they are connected. Ratio imbalance caused by lead resistance can be compensated by shunting α or β with a high resistance until balance is obtained with the link removed.

In some bridges a fixed standard resistor S having a value of the same order of magnitude as resistor X is used. Fixed resistors of 10, 100, or 1000 Ω are used for two arms, for example, B and β, with B and β having equal values. Bridge balance is obtained by adjusting tap switches to select equal resistances for the other two arms, for example, A and α, from values adjustable up to 1000 Ω in 0.1-Ω steps. In other bridges, only decimal ratio resistors are provided for A, B, α, and β, and balance is obtained by means of an adjustable standard having nine steps of 0.001 Ω each and a Manganin slide bar of 0.0011 Ω.

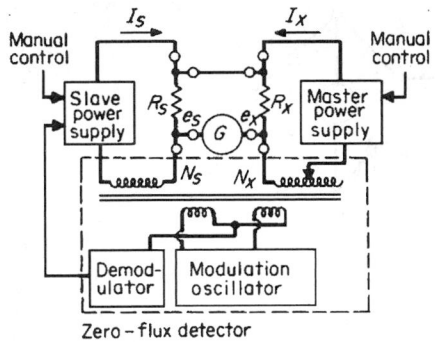

FIGURE 15.4.5 Comparator ratio bridge.

The battery connection should be reversed and two separate resistance determinations made to eliminate thermoelectric errors.

The dc-comparator ratio bridge   (Fig. 15.4.5) is used for very precise measurement of four-terminal resistors. Its accuracy and stability depend mainly on the turns ratio of a precision transformer. The master current supply is set at a convenient fixed value I_x. The zero-flux detector maintains an ampere-turn balance, $I_x N_x = I_s N_s$, by automatically adjusting the current I_s from the slave supply as N_x is manually adjusted. A null reading on the galvanometer is obtained when $I_s R_s = I_x R_x$. Since the current ratio is precisely related to the turns ratio, the unknown resistance $R_x = N_x R_s/N_s$. Fractional turn resolution for N_x can be obtained by diverting a fraction of the current I_x as obtained from a decade current divider through an additional winding on the transformer. Turns ratios have been achieved with an accuracy of better than 1 part in 10^7. The zero-flux detector operates by superimposing a modulating mmf on the core using modulation and detector windings in a second-harmonic modulator configuration. The limit sensitivity of the bridge is set by noise and is about 3 μA turns.

Murray and Varley bridge   circuits are used for locating faults in wire lines and cables. The faulted line is connected to a good line at one end by means of a jumper to form a loop. The resistance r of the loop is measured using a Wheatstone bridge. The loop is then connected as shown in Fig. 15.4.6 to form a bridge in which one arm contains the resistance R_x between the test set and the fault and the adjacent arm contains the remainder of the loop resistance. The galvanometer detector is connected across the open terminals of the loop, while the voltage supply is connected between the fault and the junction of fixed resistor R_2 and variable resistor R_3. When balance is attained

$$R_x = rR_3/(R_2 + R_3)$$

where r is the resistance of the loop. Resistance R_x is proportional to the distance to the fault. In the Varley loop of Fig. 15.4.7, variable resistor R_1 is adjusted to achieve balance and

$$R_x = \frac{rR_3 - R_1 R_2}{R_2 + R_3}$$

where r is the resistance of the loop.

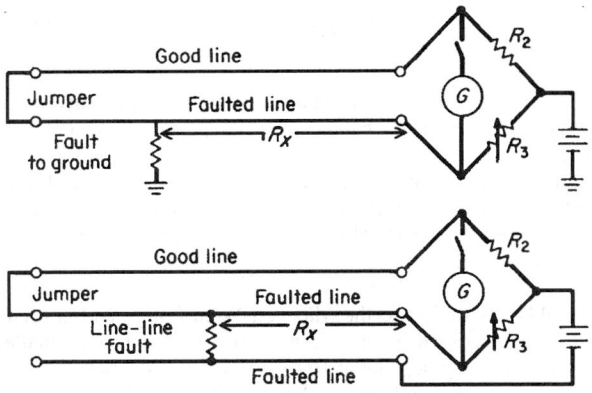

FIGURE 15.4.6 Murray loop-bridge circuits.

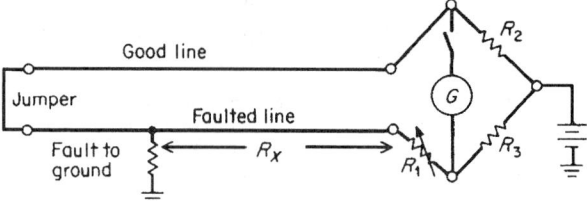

FIGURE 15.4.7   Varley loop circuit.

INDUCTANCE BRIDGES

General.   Many bridge types are possible since the impedance of each arm may be a combination of resistances, inductances, and capacitances. A number of popular inductance bridges are shown in Fig. 15.4.8. In the balance equations L and M are given in henrys, C in farads, and R in ohms; ω is 2π times the frequency in hertz. The Q of an inductance is equal to $\omega L/R$, where R is the series resistance of the inductance.

The symmetrical inductance bridge   (Fig. 15.4.8a) is useful for comparing the impedance of an unknown inductance with that of a known inductance. An adjustable resistance is connected in series with the inductance having the higher Q, and the inductance and resistance values of this resistance are added to those of the associated inductance to obtain the impedance of that arm. If this series resistance is adjusted along with the known inductance to obtain balance, the resistance and reactance balances are independent and balance convergence is rapid. If only a fixed inductance is available, the series resistance is adjusted along with the ratio R_3/R_2 until balance is obtained. These adjustments are interacting, and the rate of convergence will be proportional to the Q of the unknown inductance. Care must be taken to avoid inductive coupling between the known and unknown inductances since it will cause a measurement error.

The Maxwell-Wien bridge   (Fig. 15.4.8b) is widely used for accurate inductance measurements. It has the advantage of using a capacitance standard which is more accurate and easier to shield and produces practically no external field. R_2 and C_2 are usually adjusted since they provide a noninteracting resistance and inductance balance. If C_2 is fixed and R_2 and R_1 or R_3 are adjusted, the balance adjustments interact and balancing may be tedious.

Anderson's bridge   (Fig. 15.4.8c) is useful for measuring a wide range of inductances with reasonable values of fixed capacitance. The bridge is usually balanced by adjusting r and a resistance in series with the unknown inductance. Preferred values for good sensitivity are $R_1 = R_2 = R_3/2 = R_x/2$ and $L/C = 2R_x^2$. This bridge is also used to measure the residuals of resistors using a substitution method to eliminate the effects of residuals in the bridge elements.

Owen's bridge   (Fig. 15.4.8d) is used to measure a wide range of inductance values in terms of resistance and capacitance. The inductance and resistance balances are independent if R_3 and C_3 are adjusted. The bridge can also be balanced by adjusting R_1 and R_3.

   This bridge is useful for finding the incremental inductance of iron-cored inductors to alternating current superimposed on a direct current. The direct current can be introduced by connecting a dc-voltage source with a large series inductance across the detector branch. Low-impedance blocking capacitors are placed in series with the detector and the ac source.

Hay's bridge   (Fig. 15.4.8e) is similar to the Maxwell-Wien bridge and is used for measuring inductances having large values of Q. The series $R_2 C_2$ arrangement permits the use of smaller resistance values than the parallel arrangement. The frequency-dependent $1/Q_x^2$ term in the inductance equation is inconvenient since the dials cannot be calibrated to indicate inductance directly unless the term is neglected, which causes a 1 percent error for $Q_x = 10$.

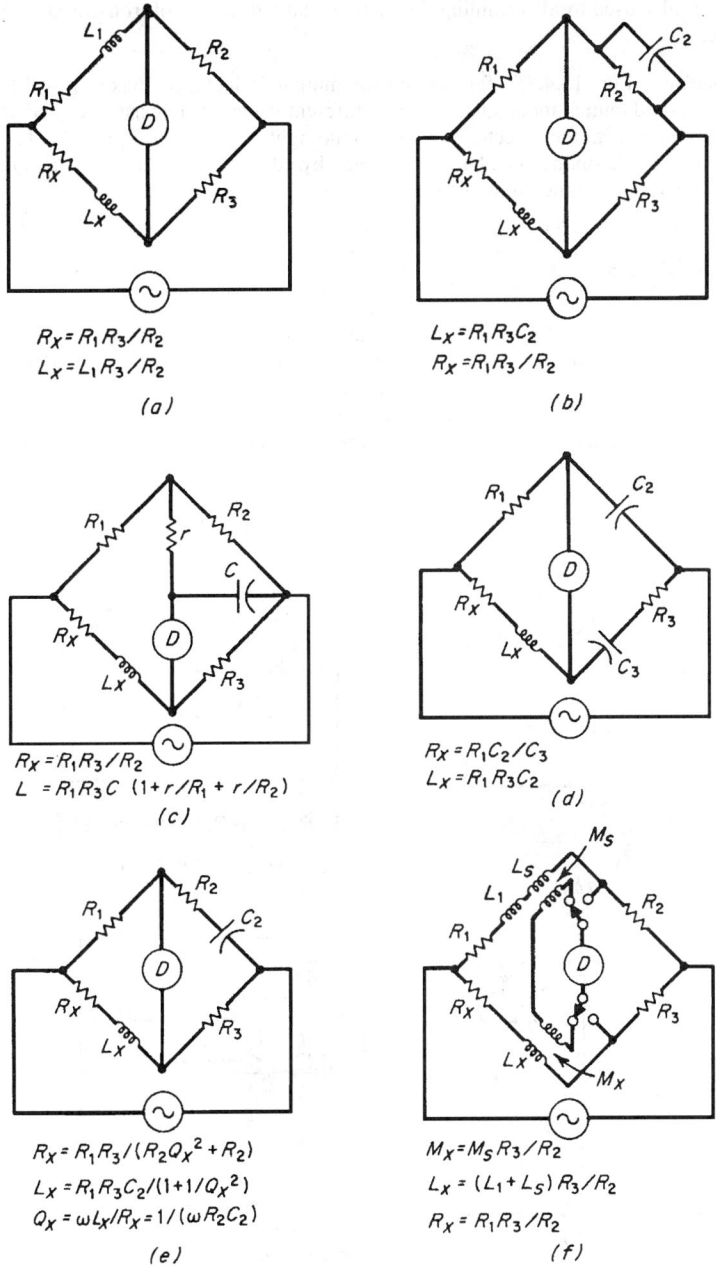

$$R_x = R_1 R_3 / R_2$$
$$L_x = L_1 R_3 / R_2$$

(a)

$$L_x = R_1 R_3 C_2$$
$$R_x = R_1 R_3 / R_2$$

(b)

$$R_x = R_1 R_3 / R_2$$
$$L = R_1 R_3 C \ (1 + r/R_1 + r/R_2)$$

(c)

$$R_x = R_1 C_2 / C_3$$
$$L_x = R_1 R_3 C_2$$

(d)

$$R_x = R_1 R_3 / (R_2 Q_x^2 + R_2)$$
$$L_x = R_1 R_3 C_2 / (1 + 1/Q_x^2)$$
$$Q_x = \omega L_x / R_x = 1 / (\omega R_2 C_2)$$

(e)

$$M_x = M_S R_3 / R_2$$
$$L_x = (L_1 + L_S) R_3 / R_2$$
$$R_x = R_1 R_3 / R_2$$

(f)

FIGURE 15.4.8 Inductance bridges: (*a*) symmetrical inductance bridge; (*b*) Maxwell-Wien bridge; (*c*) Anderson's bridge; (*d*) Owen's bridge; (*e*) Hay's bridge; (*f*) Campbell's bridge.

This bridge is also used for determining the incremental inductance of iron-cored reactors, as discussed for Owen's bridge.

Campbell's bridge (Fig. 15.4.8*f*) for measuring mutual inductance makes possible the comparison of unknown and standard mutual inductances having different values. The resistances and self-inductances of the primaries are balanced with the detector switches to the right by adjusting L_1 and R_1. The switches are thrown to the left, and the mutual-inductance balance is made by adjusting M_s. Care must be taken to avoid coupling between the standard and unknown inductances.

CAPACITANCE BRIDGES

Capacitance bridges are used to make precise measurements of capacitance and the associated loss resistance in terms of known capacitance and resistance values. Several different bridge circuits are shown in Fig. 15.4.9. In the balance equations R is given in ohms and C in farads, and ω is 2π times the frequency in hertz. The loss angle δ of a capacitor may be expressed either in terms of its series loss resistance r_s, which gives $\tan \delta = \omega C r_s$, or in terms of the parallel loss resistance r_p, in which case, $\tan \delta = 1/\omega C r_p$.

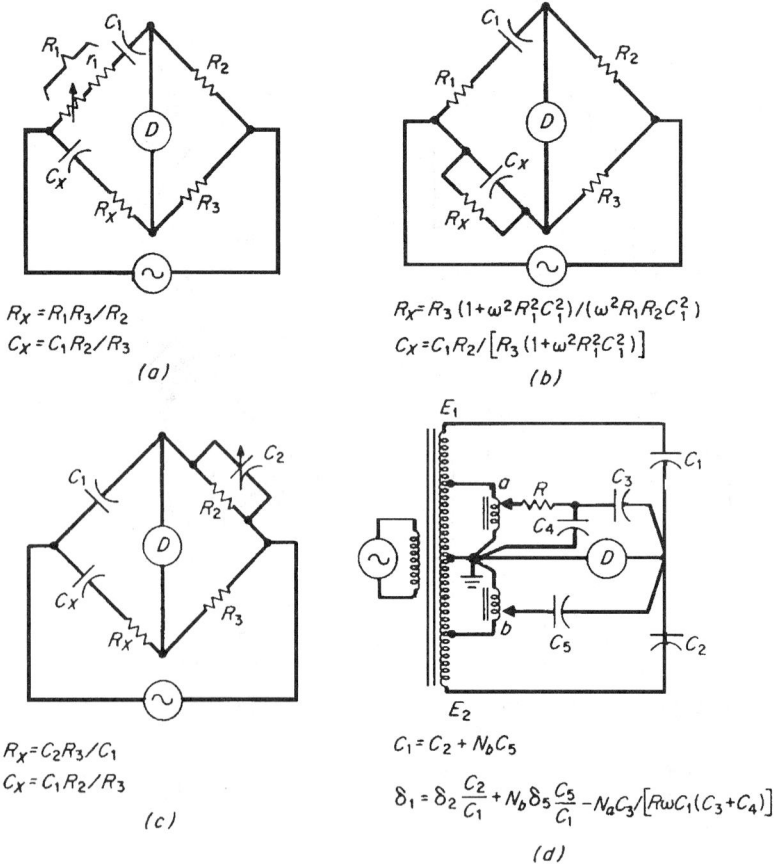

$$R_x = R_1 R_3 / R_2$$
$$C_x = C_1 R_2 / R_3$$
$$(a)$$

$$R_x = R_3 (1+\omega^2 R_1^2 C_1^2)/(\omega^2 R_1 R_2 C_1^2)$$
$$C_x = C_1 R_2 / [R_3 (1+\omega^2 R_1^2 C_1^2)]$$
$$(b)$$

$$R_x = C_2 R_3 / C_1$$
$$C_x = C_1 R_2 / R_3$$
$$(c)$$

$$C_1 = C_2 + N_b C_5$$
$$\delta_1 = \delta_2 \frac{C_2}{C_1} + N_b \delta_5 \frac{C_5}{C_1} - N_a C_3 / [R\omega C_1 (C_3 + C_4)]$$
$$(d)$$

FIGURE 15.4.9 Capacitance bridges: (*a*) series-resistance-capacitance bridge; (*b*) Wien bridge; (*c*) Schering's bridge; (*d*) transformer bridge.

The Series* RC *bridge   (Fig. 15.4.9a) is a resistance-ratio bridge used to compare a known capacitance with an unknown capacitance. The adjustable series resistance is added to the arm containing the capacitor having the smaller loss angle δ.

The Wien bridge   (Fig. 15.4.9b) is useful for determining the equivalent capacitance C_x and parallel loss resistance R_x of an imperfect capacitor, e.g., a sample of insulation or a length of cable.

An important application of the Wien bridge network is its use as the frequency-determining network in RC oscillators.

Schering's bridge   (Fig. 15.4.9c) is widely used for measuring capacitance and dissipation factors. The unknown capacitance is directly proportional to known capacitance C_1. The dissipation factor $\omega C_x R_x$ can be measured with good accuracy using this bridge. The bridge is also used for measuring the loss angles of high-voltage power cables and insulators. In this application, the bridge is grounded at the R_2/R_3 node, thereby keeping the adjustable elements R_2, R_3, and C_2 at ground potential.

The transformer bridge   is used for the precise comparison of capacitors, especially for three-terminal shielded capacitors. A three-winding toroidal transformer having low leakage reactance is used to provide a stable ratio, known to better than 1 part in 10^7. In Fig. 15.4.9d capacitors C_1 and C_2 are being compared, and a balance scheme using inductive-voltage dividers a and b is shown. It is assumed that $C_1 > C_2$ and loss angle $\delta_2 > \delta_1$. In-phase current to balance any inequality in magnitude between C_1 and C_2 is injected through C_5 while quadrature current is supplied by means of resistor R and current divider $C_3/(C_3 + C_4)$. The current divider permits the value of R to be kept below 1 MΩ. Fine adjustments are provided by dividers a and b. N_a is the fraction of the voltage E_1 that is applied to R, while N_b is the fraction of the voltage E_2 applied to C_5, δ_1 is the loss angle of capacitor C_1 and $\tan \delta_1 = \omega C_1 r_1$, where r_1 is the series loss resistance associated with C_1. The impedance of C_3 and C_4 in parallel must be small compared with the resistance of R.

The substitution-bridge   method is particularly valuable for determining the value of capacitance at radtio frequency. The *shunt-substitution method* is shown for the series RC bridge in Fig. 15.4.10. Calibrated adjustable standards R_s and C_s are connected as shown, and the bridge is balanced in the usual manner with the unknown capacitance disconnected. The unknown is then connected in parallel with C_s, and C_s and R_s are readjusted to obtain balance. The unknown capacitance C_x and its equivalent series resistance R_x are determined by the rebalancing changes ΔC_s and ΔR_s in C_s and R_s, respectively: $C_x = \Delta C_s$ and $R_x = \Delta R_s (C_{s1}/C_x)^2$, where C_{s1} is the value of C_s in the initial balance.

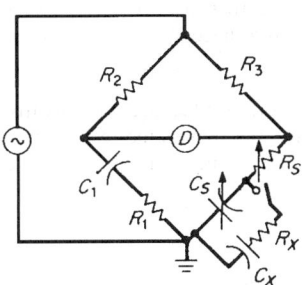

FIGURE 15.4.10   Substitution measurement.

In *series substitution* the bridge arm is first balanced with the standard elements alone, the standard elements having an impedance of Z_{s1}, and then the unknown is inserted in series with the standard elements. The standard elements are readjusted to an impedance Z_{s2} to restore balance. The unknown impedance Z_x is equal to the change in the standard impedance, that is, $Z_x = Z_{s1} - Z_{s2}$.

Measurement accuracy depends on the accuracy with which the changes in the standard values are known. The effects of residuals, stray capacitance, stray coupling, and inaccuracies in the impedances of the other three bridges arms are minimal, since these effects are the same with and without the unknown impedance. The proper handling of the leads used to connect the unknown impedance can be important.

FACTORS AFFECTING ACCURACY

Stray Capacitance and Residuals.   The bridge circuits of Figs. 15.4.8 and 15.4.9 are idealized since stray capacitances which are inevitably present and the residual inductances associated with resistances and connecting leads have been neglected. These spurious circuit elements can disturb the balance conditions and

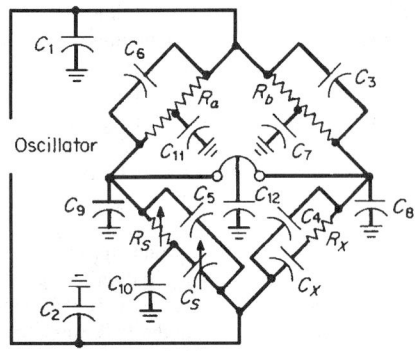

FIGURE 15.4.11 Stray capacitances in unshielded and ungrounded bridge.

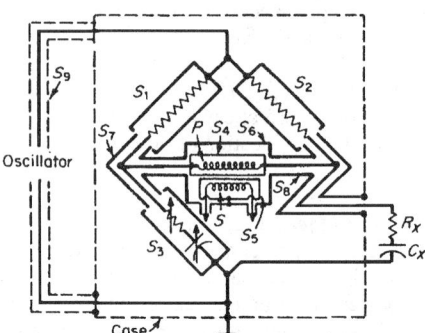

FIGURE 15.4.12 Bridge with shields and ground.

result in serious measurement errors. Detailed discussions of the residuals associated with the various bridges are given in Souders.

Shielding and grounding can be used to control errors caused by stray capacitance. Stray capacitances in an ungrounded, unshielded series *RC* bridge are shown schematically by C_1 through C_{12} in Fig. 15.4.11. The elements of the bridge may be enclosed in the grounded metal shield, as shown schematically in Fig. 15.4.12. Shielding and grounding eliminate some capacitances and make the others definite localized capacitances which act in a known way as illustrated in Fig. 15.4.13. The capacitances associated with terminal *D* shunt the oscillator and have no adverse effect. The possible adverse effects of the capacitance associated with the output diagonal *EF* are overcome by using a shielded output transformer. If the shields are adjusted so that $C_{22}/C_{21} = R_a/R_b$, the ratio of the bridge is independent of frequency. Capacitance C_{24} can be taken into account in the calibration of C_s, and capacitance C_{23} can be measured and its shunting effect across the unknown impedance can be calculated. Shielding, which is used at audio frequencies, becomes more necessary as the frequency and impedance levels are increased.

Guard circuits (Fig. 15.4.14) are often used at critical circuit points to prevent leakage currents from causing measurement errors. In an unguarded circuit surface leakage current may bypass the resistor *R* and flow through the detector *G*, thereby giving an erroneous reading. If a guard ring surrounds the positive terminal

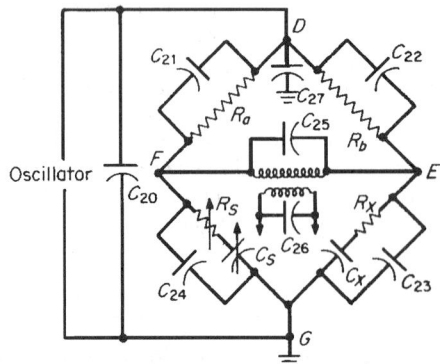

FIGURE 15.4.13 Schematic circuit of shielded and grounded bridge.

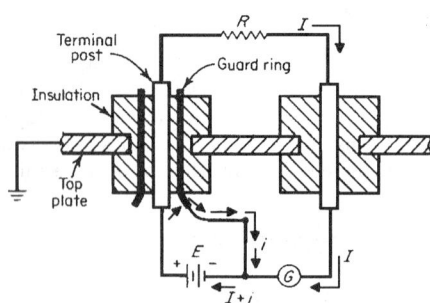

FIGURE 15.4.14 Leakage current in guarded circuit. (*Leeds and Northrup*)

post (as in the circuit of Fig. 15.4.14), the surface leakage current flows through the guard ring and a noncritical return path to the voltage source. A true reading is obtained since only the resistor current flows through the detector.

Coaxial leads and twisted-wire pairs   may be used in connecting impedances to a bridge arm in order to minimize spurious-signal pickup from electrostatic and electromagnetic fields. It is important to keep lead lengths short, especially at high frequencies.

BRIDGE DETECTORS AND AMPLIFIERS

Galvanometers   are used for null decision in dc bridges. The permanent-magnet moving-coil d'Arsonval galvanometer is widely used. The suspension provides a restoring torque so that the coil seeks a zero position for zero current. A mirror is used in the sensitive suspension-type galvanometer to reflect light from a fixed source to a scale. This type of galvanometer is capable of sensitivities on the order of 0.001 μA per millimeter scale division but is delicate and subject to mechanical disturbances. Galvanometers for portable instruments generally have indicating pointers and use taut suspensions which are less sensitive but more rugged and less subject to disturbances. Sensitivities are typically in the range of 0.5 μA per millimeter scale division. Galvanometers exhibit a natural mechanical frequency which depends on the suspension stiffness and the moment of inertia. Overshoot and oscillatory behavior can be avoided without an excessive increase in response time if an external resistance of the proper value to produce critical damping is connected across the galvanometer terminals.

Null-detector amplifiers   incorporating choppers or modulators (see Chap. 15.3) are used to amplify the null output signal from dc bridges to provide higher sensitivity and permit the use of rugged, less-sensitive microammeter indicators. Null-detector systems such as the L&N 9838 are available with sensitivities of 10 nV per division for a 300-Ω input impedance. The Guideline 9460A nanovolt amplifier uses a light-beam-coupled amplifier that can provide 7.5-mm deflection per nV when used with a sensitive galvanometer. The input signal polarity to this amplifier may be reversed without introducing errors because of hysteresis or input offset currents. This reversal-capability is useful to balance out parasitic or thermal emfs in the measured circuit.

Frequency-selective amplifiers   are extensively used to increase the sensitivity of ac bridges. An ac amplifier with a twin-T network in the feedback loop provides full amplification at the selected frequency but falls off rapidly as the frequency is changed. Rectifiers or phase-sensitive detectors are used to convert the amplified ac signal into a direct current to drive a dc microammeter indicator. The General Radio 1232-A tuned amplifier and null detector, which is tunable from 20 Hz to 20 kHz with fixed-tuned frequencies of 50 and 100 kHz, provides a sensitivity better than 0.1 μV. Cathode-ray-tube displays using Lissajous patterns are also used to indicate the deviation from null conditions. Amplifier and detector circuits are described in Chap. 15.3.

MISCELLANEOUS MEASUREMENT CIRCUITS

Multifrequency LCR meters   incorporate microprocessor control of ranging and decimal-point positioning, which permits automated measurement of inductance, capacitance, and resistance in less than 1 s. Typical of the new generation of microprocessor instrumentations is the General Radio 1689 Digibridge, which automatically measures a wide range of L, C, R, D, and Q values from 12 Hz to 100 kHz with a basic accuracy of 0.02 percent. This instrument compares sequential measurements rather than obtaining a null condition and therefore is not actually a bridge.

Similar performance is provided by the Hewlett-Packard microprocessor-based 4274A (100 Hz to 100 kHz) and 4275A (10 kHz to 10 MHz) multifrequency *LCR* meters, which measure the impedance of the device under test at a selected frequency and compute the value of L, C, R, D, and Q as well as the impedance, reactance, conductance, susceptance, and phase angle with a basic accuracy of 0.1 percent.

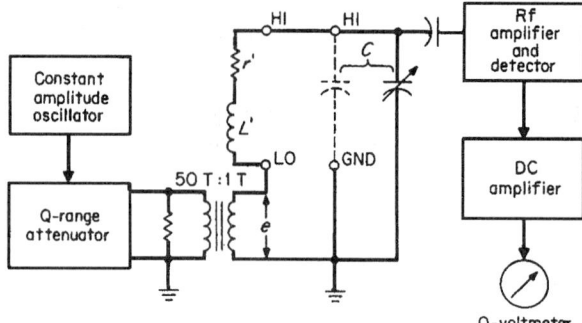

FIGURE 15.4.15 Q meter.   (*Hewlett-Packard*)

The Q meter   is used to measure the quality factor Q of coils and the dissipation factor of capacitors; the dissipation factor is the reciprocal of Q. The Q meter provides a convenient method of measuring the effective values of inductors and capacitors at the frequency of interest over a range of 22 kHz to 70 MHz.

The simplified circuit of a Q meter is shown in Fig. 15.4.15, where an unknown impedance of effective inductance L' and effective resistance r' is being measured. A sinusoidal voltage e is injected by the transformer secondary in series with the circuit containing the unknown impedance and the tuning capacitor C. The transformer secondary has an output impedance of approximately 1 mΩ. Unknown capacitors can be measured by connecting them between the HI and GND terminals while using a known standard inductor for L'.

Either the oscillator frequency or the tuning-capacitor value is adjusted to bring the circuit to approximate resonance, as indicated by a maximum voltage across capacitor C. At resonance $X_{L'} = X_c$ where $X_{L'} = 2\pi f L'$, $X_c = 1/2\pi fC$, L' is the effective inductance in henrys, C is the capacitance in farads, and f is the frequency in hertz. The current at resonance is $I = e/R$, where R is the sum of the resistances of the unknown and the internal circuit. The voltage across the capacitor C is $V_C = IX_C = eX_C/R$, and the indicated circuit Q is equal to V_C/e. In practice, the injected voltage e is known for each Q-range attenuator setting and the meter is calibrated to indicate the value of Q. Corrections for residual resistances and reactances in the internal circuit become increasingly important at higher frequencies (see Chap. 15.2). For low values of Q, neglecting the difference between the resonance and the approximate resonance achieved by maximizing the capacitor voltage may result in an unacceptable error. Exact equations are given in Chap. 15.2.

The rf impedance analyzer   is a microprocessor-based instrument designed to measure impedance parameter values of devices and materials in the rf and UHF regions. The basic measurement circuit consists of a signal source, an rf directional bridge, and a vector voltage ratio detector as shown in Fig. 15.4.16. The measurement source produces a selectable 1-MHz to 1-GHz sinusoidal signal using frequency synthesizer techniques. The unknown impedance Z_x is connected to a test port of an rf directional bridge having resistor values Z equal to the 50-Ω characteristic impedance of the measuring circuit. The test-channel and reference-channel signal frequencies are converted to 100 kHz by the sampling i.f. converters in order to improve the accuracy of vector ratio detection. The vector ratio of the test channel and reference channel i.f. signals is detected for both the real and imaginary component vectors. The vector ratio, which is equal to e_2/e_1, is proportional to the reflection coefficient Γ, where $\Gamma = (Z - Z_x)/(Z + Z_x)$. The microprocessor computes values of L, C, D, Q, R, X, θ, G, and B from the real and imaginary components Γ_x and Γ_y of the reflection coefficient Γ. The basic accuracy of the magnitude of Γ is better than 1 percent, while that of other parameters is typically better than 2 percent.

The twin-T measuring circuit   of Fig. 15.4.17 is used for admittance measurements at radio frequencies. This circuit operates on a null principle similar to a bridge circuit, but it has an advantage in that one side of the oscillator and detector are common and therefore can be grounded. The substitution method is used with this circuit, and therefore the effect of stray capacitances is minimized. The circuit is first balanced to a null condition

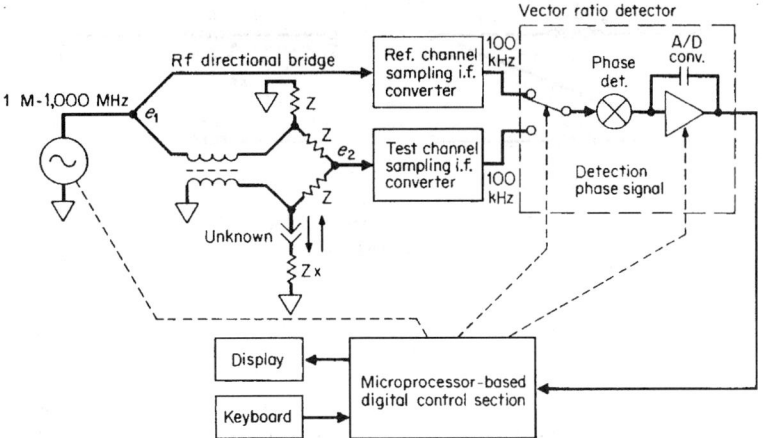

FIGURE 15.4.16   Rf impedance analyzer.   *(Hewlett-Packard)*

with the unknown admittance $G_x + jB_x$ unconnected.

$$G_L = \omega^2 RC_1C_2(1 + C_o/C_3)$$
$$L = 1/[\omega^2(C_b + C_1 + C_2 + C_1C_2/C_3)]$$

The unknown admittance is connected to terminals a and b, and a null condition is obtained by readjusting the variable capacitors to values C'_a and C'_b. The conductance G_x and the susceptance B_x of the unknown are proportional to the changes in the capacitance settings:

$$G_x = \omega^2 RC_1C_2(C'_a - C_a)/C_3 \qquad B_x = \omega(C_b - C'_b)$$

Measurement of Coefficient of Coupling.   Two coils are inductively coupled when their relative positions are such that lines of flux from each coil link with turns of the other coil. The mutual inductance M in henrys can be measured in terms of the voltage e induced in one coil by a rate of change of current di/dt in the other coil; $M = -e_1/(di_2/dt) = -e_2/(di_1/dt)$. The maximum coupling between two coils of self-inductance L_1 and L_2 exists when all the flux from each of the coils links all the turns of the other coil; this condition produces the maximum value of mutual inductance, $M_{max} = \sqrt{L_1L_2}$. The *coefficient of coupling* k is defined as the ratio of the actual mutual inductance to its maximum value; $k = M/\sqrt{L_1L_2}$.

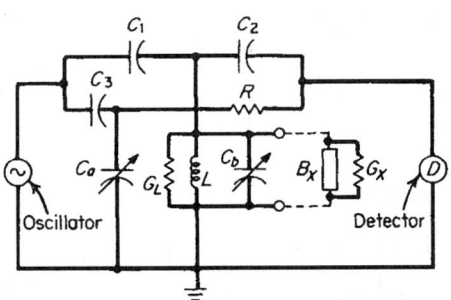

FIGURE 15.4.17   Twin-T measuring circuit.   *(General Radio Co.)*

The value of mutual inductance can be measured using Campbell's mutual-inductance bridge. Alternately, the mutual inductance can be measured using a self-inductance bridge. When the coils are connected in series with the mutual-inductance emf aiding the self-inductance emf (Fig. 15.4.18a), the total inductance $L_a = L_1 + L_2 + 2M$ is measured. With the coils connected with

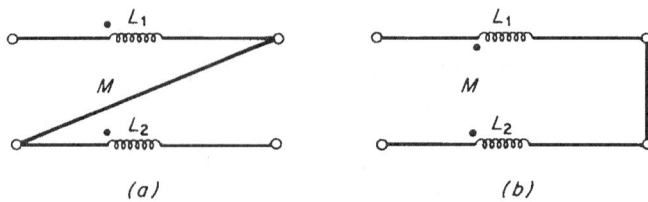

FIGURE 15.4.18 Mutual inductance connected for self-inductance measurement: (*a*) aiding configuration; (*b*) opposing configuration.

the mutual-inductance emf opposing the self-inductance emf (Fig. 15.4.18*b*), inductance $L_b = L_1 + L_2 - 2M$ is measured. The mutual inductance is $M = (L_a - L_b)/4$.

Permeameters   are used to test magnetic materials. By simulating the conditions of an infinite solenoid, the magnetizing force H can be computed from the ampere-turns per unit length. When H is reversed, the change in flux linkages in a test coil induces an emf whose time integral can be measured by a ballistic galvanometer. The *Burrows permeameter* (Fig. 15.4.19) uses two magnetic specimen bars, S_1 and S_2, usually 1 cm in diameter and

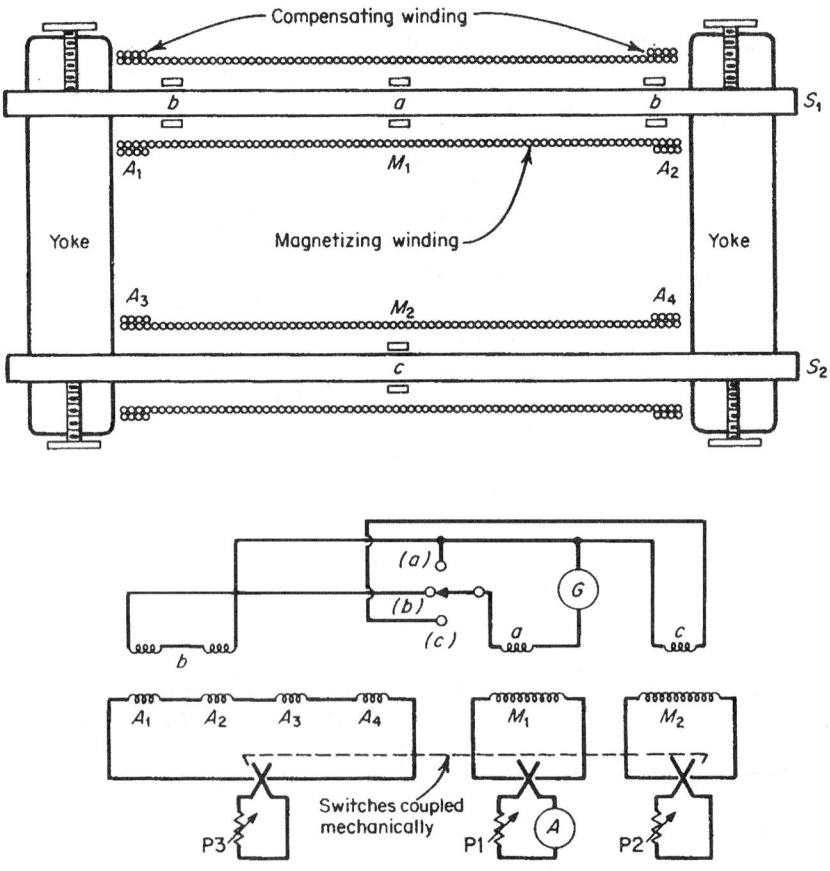

FIGURE 15.4.19 Burrows permeameter.

30 cm long, joined by soft-iron yokes. High precision is obtainable for magnetizing forces up to 300 Oe. The currents in magnetizing windings M_1 and M_2 and in compensating windings A_1, A_2, A_3, and A_4 are adjusted independently to obtain uniform induction over the entire magnetic circuit. Windings A_1, A_2, A_3, and A_4 compensate for the reluctance of the joints. The reversing switches are mechanically coupled and operate simultaneously. Test coils a and c each have n turns, while each half of the test coil b has $n/2$ turns. Coils a and b are connected in opposing polarity to the galvanometer when the switch is in position b, while coils a and c are opposed across the galvanometer for switch position c.

Potentiometer P1 is adjusted to obtain the desired magnetizing force, and potentiometers P2 and P3 are adjusted so that no galvanometer deflection is obtained on magnetizing current reversal with the switches in either position b or c. This establishes uniform flux density at each coil. The switch is now set at position a and the galvanometer deflection d is noted when the magnetizing current is reversed.

The values of B in gauss and H in oersteds can be calculated from

$$H = \frac{0.4\pi NI}{l} \qquad B = 10^8 \frac{dkR}{2an} - \frac{A-a}{a}H$$

where N = turns of coil M_1
    I = current in coil M_1 (A)
    l = length of coil M_1 (cm)
    d = galvanometer deflection
    k = galvanometer constant
    R = total resistance of test coil a circuit
    a = area of specimen (cm^2)
    A = area of test coil (cm^2)
    n = turns in test coil a

The term $(A - a)H/a$ is a small correction term for the flux in the space between the surface of the specimen and the test coil.

Other permeameters such as the Fahy permeameter, which requires only a single specimen, the Sandford-Winter permeameter, which uses a single specimen of rectangular cross section, and Ewing's isthmus permeameter, which is useful for magnetizing forces as high as 24,000 G, are discussed in Harris.

The frequency standard of the National Institute of Standards and Technology (formerly the National Bureau of Standards) is based on atomic resonance of the cesium atom and is accurate to 1 part in 10^{13}. The second is defined as the duration of 9,192,631,770 periods of the radiation corresponding to the transition between the two hyper-fine levels of the ground state of the atom of cesium 133. Reference frequency signals are transmitted by the NIST radio stations WWV and WWH at 2.5, 5, 10, and 15 MHz. Pulses are transmitted to mark the seconds of each minute. In alternate minutes during most of each hour, 500- or 600-Hz audio tones are broadcast. A binary-coded-decimal time code is transmitted continuously on a 100-Hz subcarrier. The carrier and modulation frequencies are accurate to better than 1 part in 10^{11}.

These frequencies are offset by a known and stable amount relative to the atomic-resonance frequency standard to provide "Coordinated Universal Time" (UTC), which is coordinated through international agreements by the International Time Bureau. UTC is maintained within ±0.9 s of the UT1 time scale used for astronomical measurements by adding leap seconds about once per year to UTC, depending on the behavior of the earth's rotation.

Quartz-crystal oscillators are used as secondary standards for frequency and time-interval measurement. They are periodically calibrated using the standard radio transmissions.

Frequency measurements can be made by comparing the unknown frequency with a known frequency, by counting cycles over a known time interval, by balancing a frequency-sensitive bridge, or by using a calibrated resonant circuit.

Frequency-comparison methods include using Lissajous patterns on an oscilloscope and heterodyne measurement methods. In Fig. 15.4.20, the frequency to be measured is compared with a harmonic of the 100-kHz reference oscillator. The difference frequency lying between 0 and 50 kHz is selected by the low-pass filter and

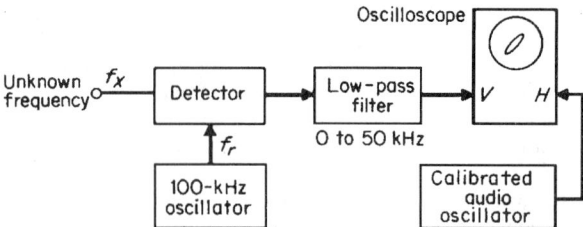

FIGURE 15.4.20 Heterodyne frequency-comparison method.

compared with the output of a calibrated audio oscillator using Lissajous patterns. Alternately, the difference frequency and the audio oscillator frequency may be applied to another detector capable of providing a zero output frequency.

Digital frequency meters provide a convenient and accurate means for measuring frequency. The unknown frequency is counted for a known time interval, usually 1 or 10 s, and displayed in digital form. The time interval is derived by counting pulses from a quartz-crystal oscillator reference. Frequencies as high as 50 MHz can be measured by using scalers (frequency dividers). Frequencies as high as 110 GHz are measured using heterodyne frequency-conversion techniques. At low frequencies, for example 60 Hz, better resolution is obtained by measuring the period $T = 1/f$. A counter with a built-in computer is available, which measures the period at low frequencies and automatically calculates and displays the frequency.

A frequency-sensitive bridge can be used to measure frequency to an accuracy of about 0.5 percent if the impedance elements are known. The Wien bridge of Fig. 15.4.21 is commonly used, R_3 and R_4 being identical slide-wire resistors mounted on a common shaft. The equations for balance are $f = 1/2\pi \sqrt{R_3 R_4 C_3 C_4}$ and $R_1/R_2 = R_4/R_3 + C_3/C_4$.

In practice, the values are selected so that $R_3 = R_4$, $C_3 = C_4$, and $R_1 = 2R_2$. Slide wire r, which has a total resistance of $R_1/100$, is used to correct any slight tracking errors in R_3 and R_4. Under these conditions $f = 1/2\pi R_4 C_4$. A filter is needed to reject harmonics if a null indicator is used since the bridge is not balanced at harmonic frequencies.

Time intervals can be measured accurately and conveniently by gating a reference frequency derived from a quartz-crystal oscillator standard to a counter during the time interval to be measured. Reference frequencies of 10, 1, and 0.1 MHz, derived from a 10-MHz oscillator, are commonly used.

Analog frequency circuits that produce an analog output proportional to frequency are used in control systems and to drive frequency-indicating meters. In Fig. 15.4.22, a fixed amount of charge proportional to $C_1(E - 2d)$, where d is the diode-voltage drop, is withdrawn through diode D_1 during each cycle of the input. The current

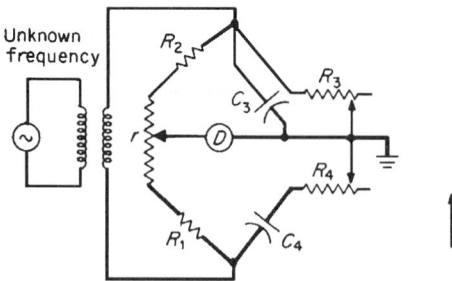

FIGURE 15.4.21 Wien frequency bridge.

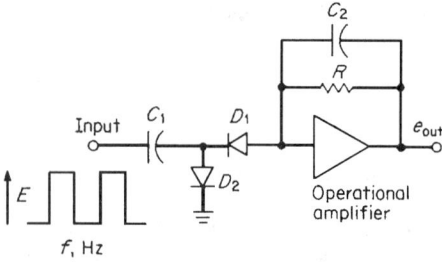

FIGURE 15.4.22 Frequency-to-voltage converter.

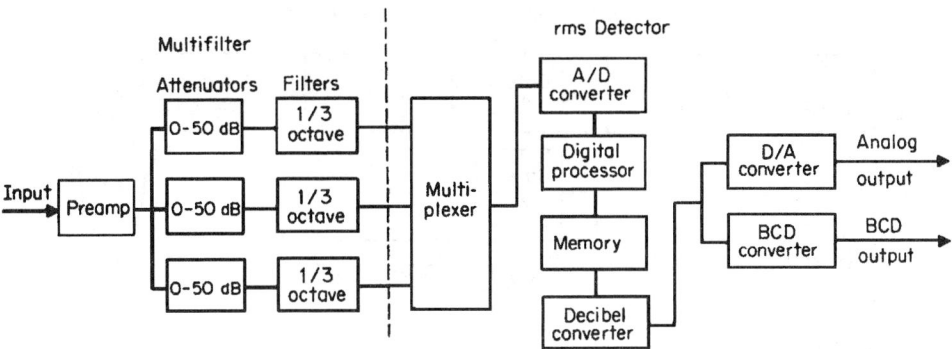

FIGURE 15.4.23   Real-time analyzer using 30 attenuators and filters.   (*General Radio Co.*)

through diode D_1, which is proportional to frequency, is balanced by the current through resistor R, which is proportional to e_{out}. Therefore, $e_{out} = fRC_1(E - 2d)$. Temperature compensation is achieved by adjusting the voltage E with temperature so that the quantity $E - 2d$ is constant.

Frequency analyzers   are used for measuring the frequency components and analyzing the spectra of acoustic noise, mechanical vibrations, and complex electric signals. They permit harmonic and intermodulation distortion components to be separated and measured. A simple analyzer consists of a narrow-bandwidth filter, which can be adjusted in frequency or swept over the frequency range of interest. The output amplitude in decibels is generally plotted as a function of frequency using a logarithmic frequency scale. Desirable characteristics include wide dynamic range, low distortion, and high stop-band attenuation. Analog filters that operate at the frequency of interest exhibit a constant bandwidth, for example, 10 Hz. The signal must be averaged over a period inversely proportional to the filter bandwidth if the reading is to be within given confidence limits of the long-time average value.

Real-time frequency analyzers   are available which perform $1/3$-octave spectrum analysis on a continuous real-time basis. The analyzer of Fig. 15.4.23 uses 30 separate filters each having a bandwidth of $1/3$ octave to achieve the required speed of response. The multiplexer sequentially samples the filter output of each channel at a high rate. These samples are converted into a binary number by the A/D converter. The true rms values for each channel are computed from these numbers during an integration period adjustable from $1/8$ to 32 s and stored in the memory. The rms value for each channel is computed from 1,024 samples for integration periods of 1 to 32 s.

Real-time analyzers are also available for analyzing narrow-bandwidth frequency components in real time. The required rapid response time is obtained by sampling the input waveform at 3 times the highest frequency of interest using an A/D converter and storing the values of a large number of samples in a digital memory. The frequency components can be calculated in real time by a microprocessor using fast Fourier transforms.

Time-compression systems   can be used to preprocess the input signal so that analog filters can be used to analyze narrow-bandwidth frequency components in real time. The time-compression system of Fig. 15.4.24 uses a recirculating digital memory and a D/A converter to provide an output signal having the same waveform as the input with a repetition rate which is k times faster. This multiplies the output-frequency spectrum by a factor of k and reduces the time required to analyze the signal by the same factor. The system operates as follows. A new sample is entered into the circulating memory through gate A during one of each k shifting periods. Information from the output of the memory recirculates through gate B during the remaining $k - 1$ periods. Since information experiences k shifts between the addition of new samples in a memory of length $k - 1$, each new sample p is entered directly behind the previous sample $p - 1$, and therefore the correct order is preserved. $(k - 1)/n$ seconds is required to fill an empty memory, and thereafter the oldest sample is discarded when a new sample is entered.

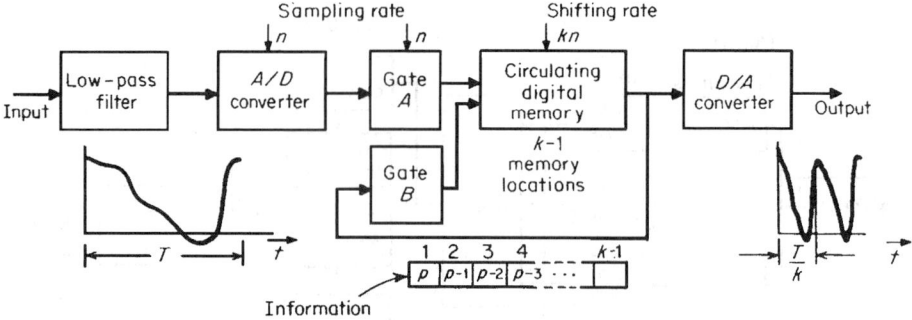

FIGURE 15.4.24 Time-compression system.

Frequency synthesizers/function generators provide sine wave, square wave, triangle, ramp, or pulse voltage outputs, which are selectable over a wide frequency range and yet have the frequency stability and accuracy of a crystal oscillator reference. They are useful for providing accurate reference frequencies and for making measurements on filter networks, tuned circuits, and communications equipment. High-precision units feature up to 11-decade digital frequency selection and programmable linear or logarithmic frequency sweep. A variety of units are available, which cover frequencies from a fraction of a hertz to tens of gigahertz.

Many synthesizers use the indirect synthesis method shown in Fig. 15.4.25. The desired output frequency is obtained from a voltage-controlled oscillator, which is part of a phase-locked loop. The selectable frequency divider is set to provide the desired ratio between the output frequency and the crystal reference frequency. Fractional frequency division can be obtained by selecting division ratio R alternately equal to N and $N + 1$ for appropriate time intervals.

Time-domain reflectometry is used to identify and locate cable faults. The cable-testing equipment is connected to a line in the cable and sends an electrical pulse that is reflected back to the equipment by a fault in the cable. The original and reflected signals are displayed on an oscilloscope. The type of fault is identified by the shape of the reflected pulse, and the distance is determined by the interval between the original and reflected pulses. Accuracies of 2 percent are typical.

A low-frequency voltmeter using a microprocessor has been developed that is capable of measuring the true rms voltage of approximately sinusoidal inputs at voltages from 2 mV to 10 V and frequencies from 0.1 to 120 Hz. A combination of computer algorithms is used to implement the voltage- and harmonic-analysis functions. Harmonic distortion is calculated using a fast Fourier transform algorithm. The total autoranging, settling, and measurement time is only two signal periods for frequencies below 10 Hz.

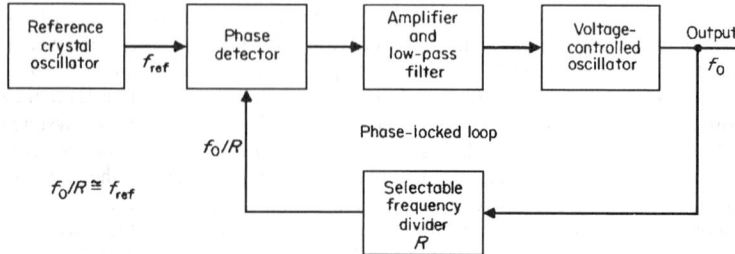

FIGURE 15.4.25 Indirect frequency synthesis.

CHAPTER 15.5
AC IMPEDANCE MEASUREMENT

Ramon C. Lebron

The generation of Impedance Gain and Phase versus frequency plots of an electrical circuit (passive or active) is extremely important to characterize small signal behavior and evaluate stability. In space dc power distribution systems, such as the International Space Station power system, each of the power system components and dc-to-dc converters has to meet strict input and output impedance requirements over a specific frequency range to ensure end-to-end system stability.

Figures 15.5.1 and 15.5.2 show the technique used to measure ac impedance. This test can be performed with the device under test (DUT) operating at rated voltage and rated load. The network analyzer output provides a sinusoidal signal with a frequency that will vary over the desired range. The signal is amplified and fed into the primary of an audio isolation transformer. Figure 15.5.1 shows the method of ac voltage injection by connecting the transformer secondary in series with the power source. The voltage and current amplifiers are ac coupled and the network analyzer is set to generate the magnitude and phase plot of Channel 2 (ac voltage) divided by Channel 1 (ac). Therefore the impedance magnitude plot is $|Z| = |Vac|/|Iac|$ and the impedance phase plot is $\theta_Z = \theta_{vac} - \theta_{iac}$ for the required frequency values.

Figure 15.5.2 shows the method of ac injection where a capacitor and a resistor are connected in series with the secondary of the transformer. The transformer and RC series combination is connected in parallel with the terminals of the DUT to inject a small ac into the DUT. The network analyzer performs the same computations for $|Z|$ and θ_z at the desired frequency range. The voltage injection method is used for high-impedance measurements such as the input impedance of a dc-to-dc converter, and the current injection method is better suited for low-impedance measurements such as the output impedance of a dc-to-dc converter.

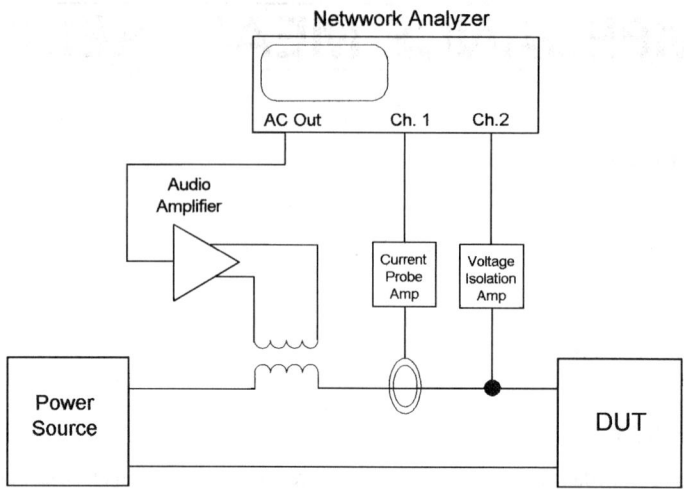

FIGURE 15.5.1 Impedance measurement with voltage injection method.

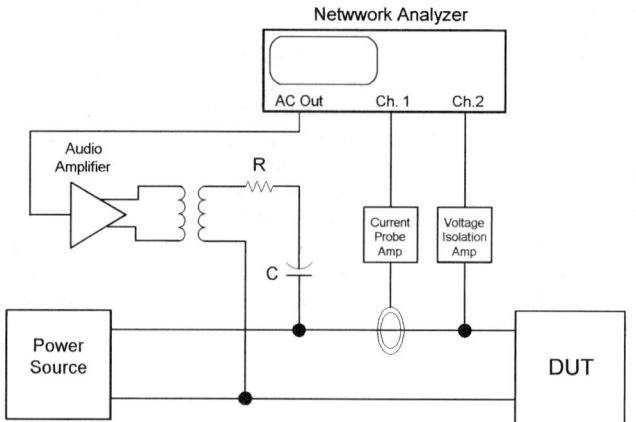

FIGURE 15.5.2 Impedance measurement with current injection method.

SECTION 16

ANTENNAS AND WAVE PROPAGATION

An antenna can be thought of as the control unit between a source and some medium that will propagate an electromagnetic wave. In the case of a wire antenna, it can be seen as a natural extension of the two-wire transmission line, and, in a similar way, a horn antenna can be considered a natural extension of the waveguide that feeds the horn.

The radiation properties of antennas are covered in Chap. 16.1. They include gain and directivity, beam efficiency, and radiation impedance. Depending on the antenna application, certain parameters may be more important. For example, in the case of receiving antennas that measure noise signals from an extended source, beam efficiency is an important measure of its performance.

Chapter 16.2 examines the various types of antennas, including simple wire antennas, waveguide antennas useful in aircraft and spacecraft applications, and low-profile microstrip antennas.

Finally, Chap. 16.3 treats the propagation of electromagnetic waves through or along the surface of the earth, through the atmosphere, and by reflection or scattering from the ionosphere or troposphere. More details of propagation over the earth through the nonionized atmosphere and propagation via the ionosphere are covered on the accompanying CD-ROM.   D.C.

In This Section:

On the CD-ROM:

Kirby, R. C., and K. A. Hughes, *Propagation over the Earth Through the Nonionized Atmosphere*, reproduced from the 4th edition of this handbook.

Kirby, R. C., and K. A. Hughes, *Propagation via the Ionosphere*, reproduced from the 4th edition of this handbook.

CHAPTER 16.1
PROPERTIES OF ANTENNAS AND ARRAYS

William F. Croswell

ANTENNA PRINCIPLES

The radiation properties of antennas can be obtained from source currents or fields distributed along a line or about an area or volume, depending on the antenna type. The magnetic field H can be determined from the vector potential as

$$\mathbf{H} = \frac{1}{\mu} \nabla \times \mathbf{A} \tag{1}$$

To determine the form of $\mathbf{A}$ first consider an infinitesimal dipole of length L and current I aligned with the z axis and placed at the center of the coordinate system given in Fig. 16.1.1.

$$A = z[\mu IL \exp(-jkr)]/4\pi r \tag{2}$$

where $k = 2\pi/\lambda$, and r is the radial distance away from origin in Fig. 16.1.1. From Eqs. (1) and (2) and Maxwell's equations, the fields of a short current element are

$$H_\phi = \frac{jkIL \sin\theta}{4\pi r} \left(1 + \frac{1}{jkr}\right) \exp(-jkr)$$

$$E_\theta = \frac{jkLI\eta \sin\theta}{4\pi r} \left(1 + \frac{1}{jkr} - \frac{1}{k^2 r^2}\right) \exp(-jkr) \tag{3}$$

$$E_r = \frac{IL\eta}{2\pi r} \cos\theta \left(1 + \frac{1}{jkr}\right) \exp(-jkr)$$

when $\eta = \sqrt{\mu/\epsilon}$ and $\epsilon = $ permitivity of source medium.

By superposition, these results can be generalized to the vector of an arbitrary oriented volume-current density $\mathbf{J}$ given by

$$A(x, y, z) = \frac{\mu}{4\pi} \int_{r'} \mathbf{J}(x', y', z') \frac{\exp(-jkR)}{R} dx' \, dy' \, dz' \tag{4}$$

For a surface current, the volume-current integral Eq. (4) reduces to a surface integral of $\mathbf{J}_s[\exp(-jkR)]/R$, and for a line current reduces to a line integral of $\mathbf{I} [\exp(-jkR)]/R$. The fields of all physical antennas can be

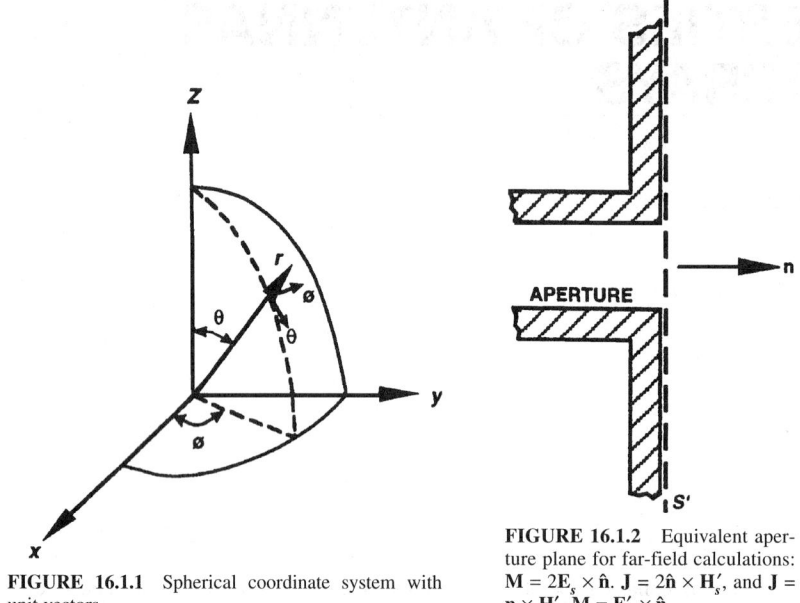

FIGURE 16.1.1 Spherical coordinate system with unit vectors.

FIGURE 16.1.2 Equivalent aperture plane for far-field calculations: $\mathbf{M} = 2\mathbf{E}_s \times \hat{\mathbf{n}}$. $\mathbf{J} = 2\hat{\mathbf{n}} \times \mathbf{H}'_s$, and $\mathbf{J} = \mathbf{n} \times \mathbf{H}'_s$, $\mathbf{M} = \mathbf{E}'_s \times \hat{\mathbf{n}}$.

obtained from the knowledge of **J** alone. However, in the synthesis of antenna fields the concept of a magnetic volume current **M** is useful, even though the magnetic current is physically unrealizable. In a homogeneous medium the electric field can be determined by

$$\mathbf{E} = \frac{1}{\epsilon}\nabla \times \mathbf{F} \qquad F = \frac{\epsilon}{4\pi}\int_{r'} M(x', y', z')\frac{\exp\,(-jkR)}{R}dx'\,dy'\,dz' \tag{5}$$

Examples of antennas that have a dual property are the thin dipole in free space and the thin slot in an infinite ground plane. The fields of an electric source **J** can be determined using Eqs. (1) and (5) and Maxwell's equations. From the far-field conditions and the relationships between the unit vectors in the rectangular and spherical coordinate systems, the far fields of an electric source **J** are

$$\begin{aligned}\eta H^J_\theta = \frac{-jnk\exp(-jkr)}{4\pi r}&\int_v (J_{x'}\cos\theta\ \cos\phi \\ &+ J_{y'}\cos\theta\ \sin\phi - J_{z'}\sin\theta)\exp[jk(x'\sin\theta\cos\phi \\ &+ y'\sin\theta\ \sin\phi + z'\cos\theta)]\ dx'\,dy'\,dz'\end{aligned} \tag{6}$$

$$\begin{aligned}-\eta H^J_\theta = E^J_\phi = \frac{j\eta k\exp(-jkr)}{4\eta r}&\int_{v'} (J_{x'}\sin\phi \\ &- J'_y\cos\phi)\exp[\,jk(x'\sin\theta\cos\phi \\ &+ y'\sin\theta\ \sin\phi\ + z'\cos\theta]\ dx'\,dy'\,dz'\end{aligned} \tag{7}$$

In a similar manner the radiated far fields from a magnetic current **M** are

$$\begin{aligned}\eta H^M_\phi = E^M_\theta = \frac{-jk\,\exp(-jkr)}{4\pi r}&\int_{v'} (M_{y'}\cos\phi \\ &- M_{x'}\sin\phi)\ \exp[\,jk(x'\sin\theta\cos\phi \\ &+ y'\sin\theta\ \sin\phi\ + z'\cos\theta]\ dx'\,dy'\,dz'\end{aligned} \tag{8}$$

$$\eta H_\theta^M = E_\phi^M = \frac{-jk \ \exp(-jkr)}{4\pi r} \int_{v'} (M_{x'} \cos\phi \ \cos\theta$$
$$+ M_{y'} \sin\phi \ \cos\theta - M_{z'} \sin\theta) \ \exp[\, jk(x' \sin\theta \ \cos\phi$$
$$+ y' \sin\theta \ \sin\phi + z \cos\theta] \ dx' \, dy' \, dz' \tag{9}$$

Currents and Fields in an Aperture

For aperture antennas such as horns, slots, waveguides, and reflector antennas, it is sometimes more convenient or analytically simpler to calculate patterns by integrating the currents or fields over a fictitious plane parallel to the physical aperture than to integrate the source currents. Obviously, the fictitious plane can be chosen to be arbitrarily close to the aperture plane. If the integration is chosen to be an infinitesimal distance away from the aperture plane, the fields to the right of s' in Fig. 16.1.2 can be found using either of the equivalent currents

$$\mathbf{M}_{s'} = 2\mathbf{E}_{s'} \times \hat{\mathbf{n}} \tag{10a}$$

$$\mathbf{J}_{s'} = 2\hat{\mathbf{n}} \times \mathbf{H}_{s'} \tag{10b}$$

$$\mathbf{J}_{s'} = \hat{\mathbf{n}} \times \mathbf{H}_{s'} \quad \text{and} \quad \mathbf{M}_{s'} = -\hat{\mathbf{n}} \times \mathbf{E}_{s'} \tag{10c}$$

The combined electric and magnetic current given in Eq. (10c) is the general Huygens' source and is generally useful for aperture problems where the electric and magnetic fields are small outside the aperture: In limited cases, the waveguide without a ground plane, a small horn, and a large tapered aperture can be approximated this way.

Far Fields of Particular Antennas

From the field equations stated previously or coordinate transformations of these equations, the far-field pattern of antennas can be determined when the near-field or source currents are known. Approximate forms of these fields or currents can often be estimated, giving good pattern predictions for practical purposes.

Electric Line Source

Consider an electric line source (current filament) of length L centered on the z' axis of Fig. 16.1.1 with a time harmonic-current $I(z')e^{j\omega t}$. The fields of this antenna are, from Eq. (6),

$$E_\theta = \frac{jnk \ \sin\theta \ \exp(-jkr)}{4\pi r} \int_{-L/2}^{L/2} I(z') \ \exp[-jkz' \ \cos\theta)] dz' \quad E_\phi = 0$$

For the short dipole where $kL \ll 1$ and $I(z') = I_0$

$$E_\theta = [\, jnkLI_0 \ \exp(-jkr) \ \sin\theta]/4\pi r$$

which agrees with Eq. (3).

Electric Current Loop

The far fields of an electric current loop of radius a, centered in the xy plane of Fig. 16.1.1, which has a current flowing on it can be obtained by returning to the vector potential $\mathbf{A}$ and deriving the expressions similar to Eqs. (6) and (7) using a potential $A_\phi = A_r$. The resulting field is

$$E\theta = \frac{-j\eta\,\exp(-jkr)\cos\theta}{4\pi r}\int_0^{2\pi} I(\phi')\sin(\phi-\phi')\exp[\,jka\,\cos(\phi-\phi^1)\sin\theta]d\phi'$$

$$E_\phi = \frac{-jk\,\exp(-jkr)\cos\theta}{4\pi r}\int_0^{2\pi} I(\phi')\cos(\phi-\phi')\exp[\,jka\,\cos(\phi-\phi')\sin\theta]d\phi'$$

The fields for the constant current loop. $I(z') = I_0$ with a radius $a \ll \lambda$, are

$$E\phi = (k^2a^2\eta/r)I_0\,\exp(-jkr)\sin\theta$$

Note that $E_\theta = 0$. The field of the small loop is similar to the field produced by a constant magnetic current source of length L, where $L \ll \lambda$.

Elementary Huygens' Source

Assume that constant electric and magnetic current sources $J_{x'} = J_0$ and $M_{y'} = M_0$ of equal length $L \ll \lambda$ are simultaneously placed at the origin of Fig. 16.1.1. If the currents are adjusted such that $\eta J_0 L = M_0 L$, the far fields of this source are

$$E_\theta = \frac{-jk\,\exp(-jkr)}{4\pi r}\cos\phi(1+\cos\theta)J_0 L$$

$$E\phi = \frac{jk\,\exp(-jkr)}{4\pi r}\sin\phi(1+\cos\theta)J_0 L$$

The unique feature of this fictitious source compared with the electric or magnetic current element alone is the obliquity factor $(1 + \cos\theta)$, which tends to cancel the far-field radiation pattern in the region $\pi/2 \le \theta \le \pi$ due to its cardioid shape. Aperture antennas have field distributions which can be constructed from Huygens' source elements having the patterns described above.

Aperture in Infinite Ground Plane

With the equivalent current $\mathbf{M}_{s'} = -2\mathbf{z} \times \mathbf{E}_{s'}$, the far field of a waveguide aperture opening onto an infinite ground plane can be obtained by integrating the aperture field since the tangential electric field is zero on the ground plane. From the magnetic current and Eqs. (6) to (9), the far-field patterns of the dominant-mode circular and rectangular waveguides can be derived.

Simple Arrays

Consider a linear array of radiating elements which, for simplicity, are assumed to be equally spaced at a distance d apart, as illustrated in Fig. 16.1.3. The field at a large distance away from the mth element can be written

$$E_m = f_m(\theta, \phi)\frac{\exp(-jkr)}{r}\exp(jk_m d\cos\theta) \tag{11}$$

By superposition the field of an array of N elements is given by

$$E_N = \frac{-\exp(-jkr)}{r}\sum_{m=1}^{N} f_m(\theta, \phi)\exp[-j(m-1)kd\cos\theta] \tag{12}$$

If the element at the origin is chosen as a reference. If each element has an identical pattern, $f_m(\theta, \phi) = a_m E(\theta, \phi)$ and

$$E_n(\theta, \phi) = E(\theta, \phi), f(\varphi)$$

$$f(\varphi) = \sum_{m=1}^{N} a_m \exp j\varphi \tag{13}$$

where a_m is a complex number representing the excitation current (voltage in the case of slots) for the mth element and $\varphi = -(m - 1)kd \cos \theta$. The function $f(\varphi)$ is commonly called the *array factor* or *array polynomial*, and the factorization process given in Eq. (13) is called *pattern multiplication*.

Uniform Linear Array

Suppose the array of point sources in Fig. 16.1.3 is fed uniformly in amplitude and has a phase shift δ between adjacent elements. Here $a_m = e^{-jm\delta}$ and $\varphi = kd \cos \theta - \delta$, and consequently $|f(\theta, \phi)|^2$ is as follows:
General:

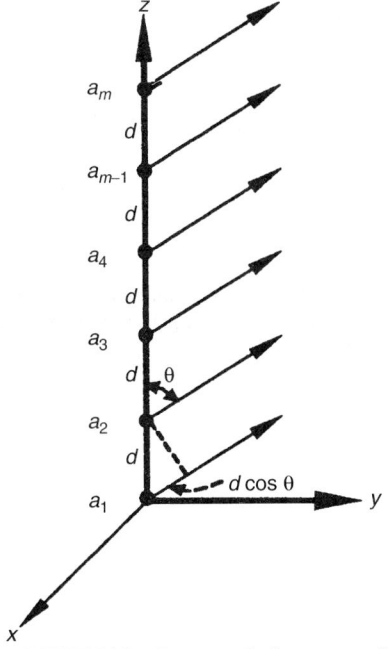

FIGURE 16.1.3 Geometry of a linear array of equally spaced elements.

$$|f(\theta, \phi)|^2 = \left| \frac{\sin^2[(N/2)(kd \cos\theta - \delta)]}{N^2 \sin^2[(kd/2)\cos\theta - \delta]} \right| \tag{14}$$

Broadside $\delta = 0$:

$$|f(\theta, \phi)|^2 = \left| \frac{\sin^2[(N/2)(kd \cos\theta)]}{N^2 \sin^2[(kd/2)\cos\theta]} \right| \tag{15}$$

End fire $\delta = kd$:

$$|f(\theta, \phi)|^2 = \left| \frac{\sin^2[(Nkd/2)(\cos\theta - 1)]}{N^2 \sin^2[kd/2)(\cos\theta - 1)]} \right| \tag{16}$$

Owing to the ϕ rotational symmetry in Eq. (14), the pattern of the line source has rotational symmetry about the z axis. As the cone angle of the pattern decreases with scan angle from broadside, the pattern directivity remains constant at the value N regardless of scan angle. By choosing the phase shift between elements so that $\delta = kd + 2.94/N$ the sharpest pattern in the end-fire direction ($\theta = 0$) is obtained. In this case

$$|f(\theta, \phi)|^2 = \left| \frac{\sin^2[(Nkd/2)(k \cos\theta - k')]}{N^2 \sin^2[(d/2)(k \cos\theta - k')]} \right| \tag{17}$$

where $k' = k + 2.94/Nd$. This phase condition, which is determined graphically, is called the *Hansen-Woodward condition* for superdirectivity. This extra directivity may be several decibels in practical antennas.

Circular Arrays

Consider an array of N equally spaced elements about a circle of radius a in the xy plane of Fig. 16.1.1. If the azimuthal location of the mth element is $\phi_m = 2\pi n/N$, then for an element excitation of the form $a_m = A_{mn} e^{j\alpha m}$ the array factor is given by

$$f(\theta, \phi) = \sum_{m=1}^{N} A_m \exp\{j[\alpha_m + ka\, \sin\theta\, \cos(\phi - \phi_m)]\} \tag{18}$$

For arrays having a large number of elements, $|A_m| = $ const, the array-factor pattern in the xy plane can be approximated by

$$|f(\theta,\ \phi = \pi/2)| \approx |\mathbf{J}_0[2ka\, \sin 1/2(\phi - \phi_0)]|$$

which is a directional beam with a maximum of $\phi = \phi_0$. If the pattern of each antenna in the circular array is of the form $F(\phi) = \Sigma_{m=0}^{\infty} A_m \cos^m \phi$, the pattern of a uniformly excited circular array of N such elements is approximately given by

$$\Phi(\theta, \phi\ ) \approx N \sum_{m=0}^{M} A_m (-i)^m \frac{d^m}{dz^m} [J_0(z) + 2j^N J_m(z)\cos N\phi] \tag{19}$$

where $M < N$ and $Z = ka\, \sin\theta$. A design curve for determining the number of sources N required in a circular array of circumference Z to produce an omnidirectional pattern with 0.5 dB ripple or less is given in Ref. 1.

Planar Arrays

Now consider a planar array of equally spaced elements in the yz plane of Fig. 16.1.3, where the elements in the y direction are a distance d from the elements on the axis. With the origin as a phase reference, the array factor is given by

$$f(\theta, \phi) = \sum_{n=1}^{N} \sum_{m=1}^{M} a_{mn} \exp[jk(nd\, \cos\theta + md\, \sin\theta\, \sin\phi)] \tag{20}$$

where the excitation coefficient $a_{mn} = A_{mn} \exp[jk(m\delta_y + n\delta_z)]$. If $A_{mn} = 1$ and $\delta_y = \delta_z = 0$, the array has a pattern maximum at broadside and has the array factor

$$|f(\theta, \phi)|^2 = \left| \frac{\sin^2(Nkd/2)\cos\theta}{N^2 \sin[(kd/2)\cos\theta]} \right| \left| \frac{\sin^2(Mkd/2)\cos\theta\, \sin\phi}{M^2 \sin^2[(kd/2)\cos\theta\, \sin\phi]} \right| \tag{21}$$

The pattern in the two principal planes of the uniformly excited array is identical in form to the linear array in the same plane. As the array is scanned from broadside, the pattern broadens and becomes asymmetrical.

Gain and Directivity

Antennas that are several wavelengths or larger in dimension have far-field patterns for which most of the radiated energy is restricted to narrow angular regions. Several useful measures of how the pattern is concentrated are gain, directivity, effective area, and beam efficiency. The beam efficiency is important for low-noise communication systems, microwave radiometry, and radio-astronomy antennas. The definitions of directional directivity $D(\theta, \phi)$, directivity D, gain G, and directional $G(\theta, \phi)$ are:

$$G(\theta, \phi) = N_A(1 - |\Gamma|^2)[D(\theta, \phi)] = N_A(1 - |\Gamma|^2) \left[4\pi \frac{\text{power radiated in direction } (\theta, \phi)}{\text{total power radiated by antenna}} \right]$$

$$G = N_A(1 - |\Gamma|^2)[D = N_A(1 - |\Gamma|^2) \left(4\pi \frac{\text{maximum power radiated by antenna}}{\text{total power radiated by antenna}} \right)$$

The term N_A is the antenna efficiency related to I^2R losses, and Γ is the reflection coefficient as seen at the antenna input terminals. In equation form

$$D(\theta, \phi) = \frac{4\pi |E(\theta, \phi)|^2}{\int_0^{2\pi} \int_0^{\pi} |E(\theta, \phi)|^2 \sin\theta \, d\theta \, d\phi} \tag{22}$$

where $E(\theta, \phi)$ = electric field of the antenna, and

$$D = \frac{4\pi |E(\theta, \phi)|^2 \max}{\int_0^{2\pi} \int_0^{\pi} |E(\theta, \phi)|^2 \sin\theta \, d\theta \, d\phi} \tag{23}$$

An example of the directivity of two-dimensional arrays is given in Fig. 16.1.4. The element pattern is of the form $2J_1(c_\theta)/C_\theta$, which approximates the elements such as slots and dipoles. Other calculations for different element patterns are given in Eqs. (2) and (3).

Beam Efficiency

In addition to the gain of an antenna, the beam efficiency is a very useful parameter for judging the quality of receiving antennas intended for measurement of noise signals from an extended source. The beam efficiency is a measure of the ability of an antenna to discriminate between the received signal in the main beam and unwanted signals received through the side lobes in other directions. Assuming that the antenna aperture is in the xy plane in Fig. 16.1.1, the beam efficiency is defined as

$$BE = \frac{\text{power radiated in a cone angle } \theta_1}{\text{total power radiated by the antenna}} = \frac{\int_0^{2\pi} \int_0^{\theta_1} |E(\theta, \phi)|^2 \sin\theta \, d\theta \, d\phi}{\int_0^{2\pi} \int_0^{\pi} |E(\theta, \phi)|^2 \sin\theta \, d\theta \, d\phi} \tag{24}$$

Beam efficiency varies significantly dependent on the shape of the aperture and the aperture distribution. Calculations for circular apertures with a symmetrical distribution $f(p) = [1 - (p/a)^2]N$ are given in Fig. 16.1.5. This distribution approximates the aperture field of parabolic antennas. Calculations for square apertures with field distribution $f(x) = \cos^n(x/a)$ are given in Fig. 16.1.5. Further beam efficiencies for circular apertures with different field distributions are given in Eqs. (4) and (5).

Antenna Temperature

An antenna located on the earth and pointing at an angle (θ, ϕ) to the sky will receive noise from all directions. The amplitude of this noise as seen at the antenna terminals will depend on the noise source (warm earth, cosmic noise,

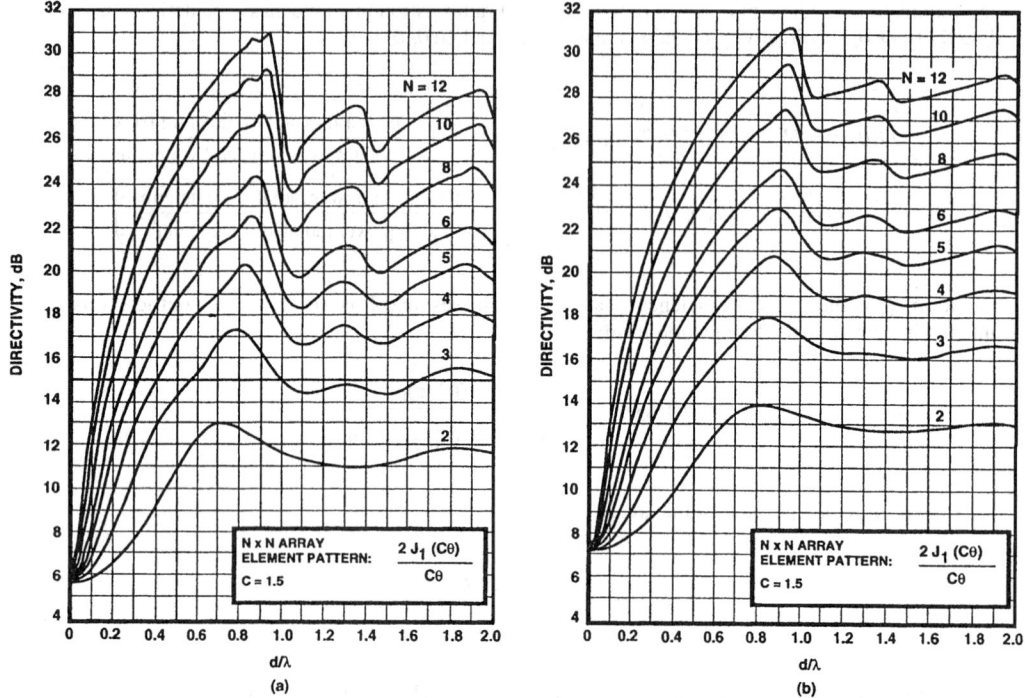

FIGURE 16.1.4   Directivities for uniformly excited $M \times N$ arrays with the element pattern of $2J_1 (C_\theta)/C_\theta$ (after Ref. 2): (a) $c = 1.5$, (b) $c = 2$.

water vapor, radio stars, and so forth), the antenna orientation, and the operating frequency and polarization. The equivalent received noise power in a receive matched to the antenna terminal impedance by a lossless transmission line is

$$P_n = kTB \tag{25}$$

where k = Boltzmann's constant
T = temperature (K)
B = bandwidth (Hz)

Noise power for a fixed bandwidth may be thought of an equivalent temperature. Consequently, if it is assumed that the various noise sources that make up the antenna noise environment have an equivalent temperature $T(\theta, \phi)$ the apparent antenna temperature is given by

$$T_A = \frac{\int_0^{2\pi} \int_0^{\pi} G(\theta, \phi)\, T(\theta, \phi) \sin\theta\, d\theta\, d\phi}{\int_0^{2\pi} \int_0^{\pi} G(\theta, \phi) \sin\theta\, d\theta\, d\phi} \tag{26}$$

The most important natural emitter of noise at microwave frequencies is the ground at 377 K compared with the sky temperature at a few degrees. Therefore, antennas that have low side and back lobes will have low apparent antenna temperatures T_A.

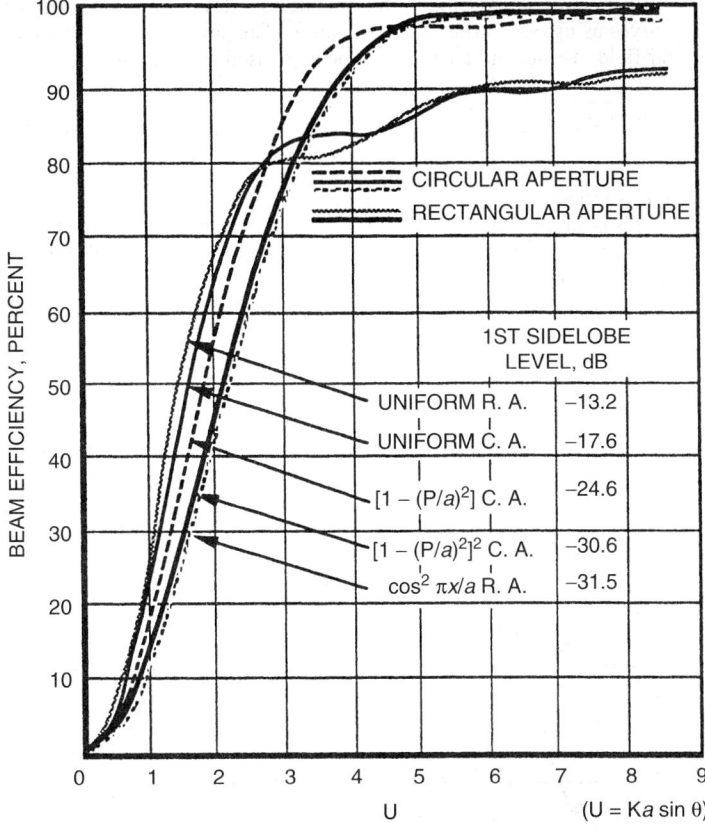

FIGURE 16.1.5 Beam efficiencies of rectangular apertures (RA) and circular apertures (CA) with various aperture distributions.

If the antenna has losses and is not matched to the receiver for maximum power transfer, the antenna temperature will be higher than predicted by Eq. (26). The contribution due to the particular mismatch and losses of every component must be analyzed for each particular receiving system in detail. An outline of an excellent method of analysis is available.[6] For a receiving system with no mismatch loss, the apparent temperature T_a is given by

$$T_a = (1 - L)T_A + LT_0 \qquad (27)$$

where T_A is the antenna temperature given by Eq. (27) and T_0 is the physical temperature of lossy device.

Friis Transmission Formula

Assume that there is a source antenna and an antenna under test located at a distance r apart such that

$$r \geq 2(d_t)^2/\lambda \qquad (28)$$

where d_t is the maximum aperture dimension of the antenna under test. The distance specified by Eq. (28) is the so-called *far-field distance*. The far-field distance is commonly specified as the distance where the phase front of a spherical wave over a planar aperture will not exceed $\pi/8$ rad. For special purposes, such as the measurement of deep nulls or extremely precise side-lobe levels, the far-field distance may have to be extended further; curves using other criteria are available.[7] The power received at the terminals of one antenna located in the far field of a second antenna can be expressed as a fraction of the transmitted power as

$$P_R = P_T \{\lambda/4\pi r\}^2 N_{AT} N_{AR} D_T(\theta_T, \phi_T) D_R(\theta_R, \phi_R)(1-|\Gamma_r|^2)(1-|\Gamma_r|^2)|\rho_R \cdot \rho_T|^2 \tag{29}$$

where N_{AT}, N_{AR} are loss efficiencies of antennas and $D_T(\theta_T, \phi_T)$, $D_R(\theta_R, \phi_R)$ are directivities of antennas in the direction one antenna is pointing toward the other. $|\Gamma_T|^2$ and $|\Gamma_R|^2$ are the reflected power due to mismatch of the antenna terminals and $|\rho_R \cdot \rho_T|^2$ is the polarization loss. The term in brackets in Eq. (29) is the so-called free-space loss, which is a result of spherical spreading of the energy radiated by an antenna. In the far field all antennas appear as a spherical wave emanating from a point source located at the phase center of the antenna, where the phase center may be a function of the observation angle.

Polarization

Consider a plane wave propagating in the z direction, which has an arbitrary plane polarization with an axial ratio r_A is defined as the ratio of the major to minor axis of the ellipse referenced to a coordinate system (Fig. 16.1.6). The field expression for this arbitrary plane-polarized wave is

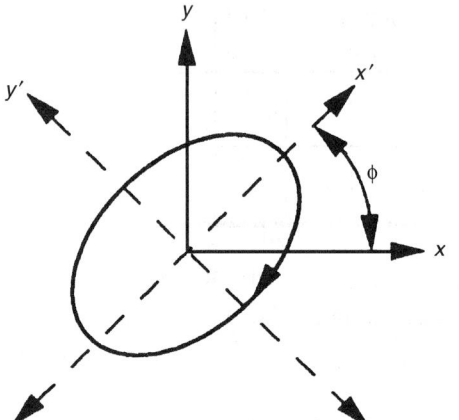

FIGURE 16.1.6 Polarization ellipse.

$$E = C[x(r_a \cos\phi + j \sin\phi) + y(r_a \sin\phi - j\cos\phi)] \exp(-jkz)$$

When the z-phase dependence is neglected except for the sign, the normalized fields of two different plane-polarized waves with the wave number 2 propagating in the negative z direction (two antennas pointing at one another) are given by

$$E_1 = E_{1p1} = E_1 \frac{x(r_1 \cos\phi_1 + j\sin\phi_1) + y(r_1 \sin\phi_1 - j\cos\phi_1)}{\sqrt{r_1^2 + 1}} \tag{30}$$

$$E_2 = E_{2p2} = E_2 \frac{x(r_2 \cos\phi_2 + j\sin\phi_2) + y(r_2 \sin\phi_2 + j\cos\phi_2)}{\sqrt{r_2^2 + 1}} \tag{31}$$

A chart of polarization loss between two elliptically polarized waves with arbitrary axial ratio is given in Fig. 16.1.7. Table 16.1.1 enables the calculation of polarization loss between plane-polarized waves for three different configurations, and for the general case.

Radiation Impedance

The complex Poynting vector $P = E \times H^*$ can be integrated over a closed surface about an antenna to give

$$\iint_{s'} \mathbf{E} \times \mathbf{H}^* \cdot ds = 4j\omega(\mathrm{W}_m - \mathrm{W}_f) + \mathbf{P}_s = \mathbf{VI}^* \tag{32}$$

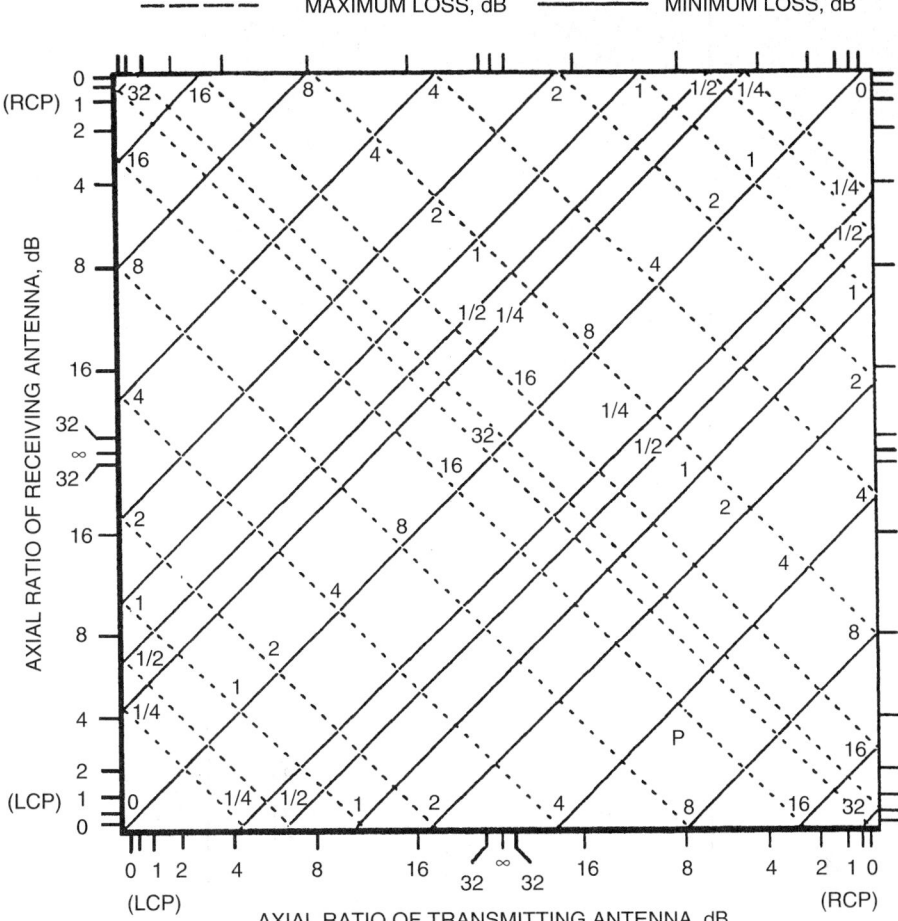

FIGURE 16.1.7 Polarization loss between two elliptically polarized wares.

Where $W_m - W_f$ = time-average net reactive power stored within the volume enclosed by S' and P_s = net real power flow through S'. As a result, the real power can be related to a radiation resistance and the reactive power to a radiation reactance by equating this total complex power to an equivalent voltage V and current I at a defined set of terminals. This radiation impedance, which is a lumped-circuit-element description of an antenna, is defined at a specific set of terminals or terminal plane which may be the aperture of a waveguide antenna. The simplest term to determine in Eq. (32) is the radiated power P_s and the corresponding radiation resistance Rr. This is true because the surface S' may be chosen in the far field of the antenna, where fields with only $1/r$ dependence are important. The radiation reactance is determined by choosing S' close to the antenna and integrating fields where $1/r^2$ and $1/r^3$ terms are important. A convenient surface to choose for these calculations is a sphere of radius r surrounding the antenna.

TABLE 16.1.1 Polarization Loss Between Plane-Polarized Waves

Type	ρ_1	ρ_2	$|\rho_1 \cdot \rho_2|^2$
Linear to linear $r_1 = r_2 \to \infty$	$x\cos\phi_1 + y\sin\phi_1$	$x\cos\phi_2 + y\sin\phi_2$	$\cos^2(\phi_1 - \phi_2)$
Linear to circular ($r_1 = 1,\ \phi = 0$ to π)	$\dfrac{x \pm jy}{\sqrt{2}}$	$x\cos\phi_2 + y\sin\phi_2$	$\dfrac{1}{2}$
Circular to circular	$\dfrac{x+jy}{\sqrt{2}}$	$\dfrac{x+jy}{\sqrt{2}}$	1
$r_1 = r_2 = 1$ $\phi_1 = 0$ or η $\phi_1 = 0$ or η	$\dfrac{x+jy}{\sqrt{2}}$	$\dfrac{x+jy}{\sqrt{2}}$	0
General case	$\dfrac{x(r_1\cos\phi_1 + j\sin\phi_1)}{\sqrt{r_1^2+1}}$ $+ \dfrac{y(r_1\sin\phi_1 + j\cos\phi_1)}{\sqrt{r_1^2+1}}$	$\dfrac{x(r_2\cos\phi_2 - j\sin\phi_2)}{\sqrt{r_1^2+1}}$ $+ \dfrac{y(r_2\cos\phi_2 + j\cos\phi_2)}{\sqrt{r_1^2+1}}$	$\dfrac{(1+r_1^2+r_2^2+r_1^2r_2^2+4r_1r_2)+(1+r_1^2r_2^2-r_1^2-r_2^2)\cos^2(\phi_1-\phi_2)}{2(r_1^2+1)(r_2^2+1)}$

Radiation Resistance of Short Current Filament

The far fields of a current filament from Eq. (3) are

$$E_\theta = \frac{jkI_0\eta L \sin\theta \exp(-jkr)}{4\pi r} \qquad H_\phi = \frac{jkI_0 L \sin\theta \exp(-jkr)}{4\pi r} \tag{33}$$

When these fields are used, P_s from Eq. (33) is given by

$$P_s = \int_0^{2\pi}\int_0^\pi E_\theta H_\phi^* \sin\theta\, r^2\, d\theta\, d\phi = \frac{\eta\pi I_0^2}{3}\left(\frac{L}{\lambda}\right) = I_0^2 R_r \tag{34}$$

Array Impedance

The pattern expressions given earlier for arrays have neglected coupling between elements in the array. Mutual coupling will not only affect the input impedance of each element input terminal as a function of scan angle but will also change the pattern of each element in a manner dependent on the element location in the array. The determination of the mutual and self-impedance of an element in an array and the effect of these changes on array patterns is a specialized problem. In general, however, arrays of antennas radiating into a linear medium will have to satisfy the same equations as any linear system with n pairs of terminals, or

$$V_1 = Z_{11}I_1 + Z_{12}I_2 + \cdots + Z_{in}I_n \quad V_n = Z_{n1}I_1 + Z_{n2}I_2 + \cdots + Z_{nn}I_n \tag{35}$$

Input Impedance as a Function of Scan Angle

As a first approximation a large array can be treated as large continuous aperture. Wheeler[9] has shown that the impedance of a large aperture approximation in an array may be thought of as an impedance sheet whose normalized impedance and reflection coefficient vary as

$$\eta_{aperture} = (1 - \sin^2\theta\, \cos^2\phi)/\cos\theta \tag{36}$$

which results in

$$|\Gamma| = \tan^2(\theta/2) \tag{37}$$

for $\phi = 0°$ or $\pi/2$. This simple impedance-sheet concept acts as an upper bound on the input-impedance variation of the central element of a large array as a function of scan angle. Perhaps this model inspired the erroneous description of the scan-angle reflection peak as related to a surface wave for uncoated arrays. An indication of the accuracy of this simple bound is given in Fig. 16.1.8 from dipole calculations by Allen.[10] It should be noted that the reflection peak is the result of a null in the element pattern of each element in the infinite array at the scan angle corresponding to the reflection peak. For waveguide arrays the addition of fences in between waveguides will extend the scan-angle range to wider angles by filling in the reflection null in the element patterns. The impedance variation as a function of scan angle, for infinite arrays, can be obtained using waveguide simulators.[11,12]

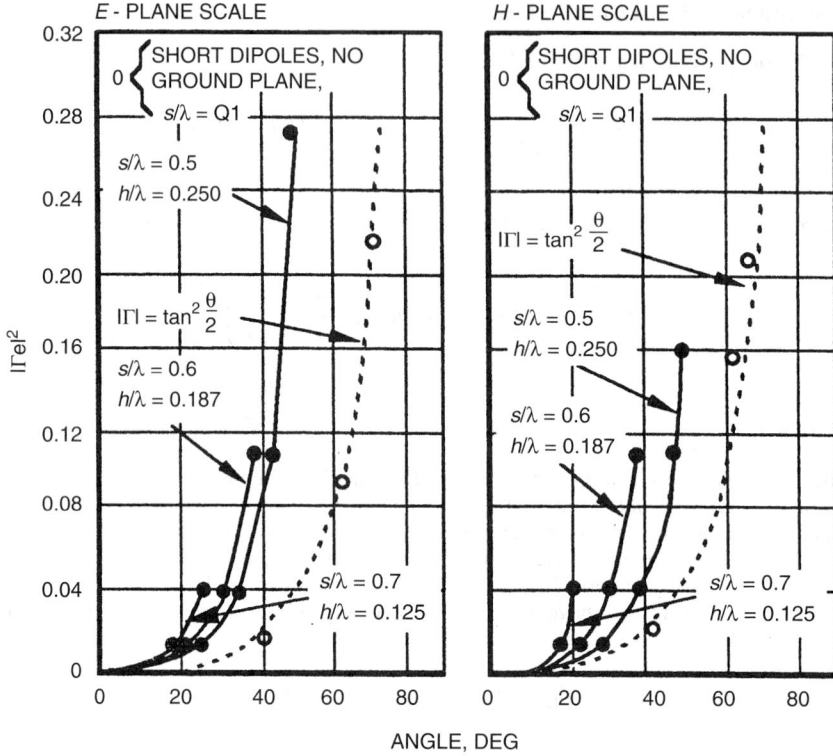

FIGURE 16.1.8 Reflection coefficient as a function of scan angle.[10]

REFERENCES

1. Croswell, W. F., and C. R. Cockrell, "An omnidirectional microwave antenna for use on spacecraft," *IEEE Trans. Antennas Propag.*, July 1969, Vol. AP-17, pp. 459–466.

2. Wong, J. L., and H. E. King, "Directivity of a uniformly excited M × N array of directive elements," *IEEE Trans. Antennas Propag.*, May 1975, Vol. AP-23, pp. 401–405.

3. H. E. King, et al., "Directivity of a Uniformly Excited N × N Array of Directive Elements," The Aerospace Corp., Aerospace TR-0074 46241-1, November 21, 1973.

4. Milligan, T. A., "Modern Antenna Design," McGraw-Hill, 1985, p. 177.

5. Sciambi, A. F., "The effect of the aperture illumination on the circular aperture antenna pattern characteristics," *Microwave J.*, August, 1965, pp. 79–84.

6. Otoshi, T. Y., "The effects of mismatched components on microwave noise-temperature calibrations," *IEEE Trans. Microwave Theory Tech.*, September 1968, Vol. MTT-16, No. 9, pp. 675–687.

7. Hollis, J. S., et al., "Microwave Antenna Measurements," *Scientific Atlanta*, 1970.

8. Offutt, W. B., and L. K. DeSize, "Methods of Polarization Synthesis, Chapter 23, Antenna Engineering Handbook," McGraw-Hill, 1993, pp. 23–29.

9. Wheeler, H. A., "Simple relations derived from a phased-array made of an infinite current sheet," *IEEE Trans. Antennas Propag.*, July 1965, Vol. AP-13, pp. 506–514.

10. Allen, J. L., "On array element impedance variation with spacing," *IEEE Trans. Antennas Propag.*, May 1964, Vol. AP-12, p. 371.

11. Hannan, P. W., and M. A. Balfour, "Simulation of a phased array antenna in a waveguide," *IEEE Trans. Antennas Propag.*, May 1965, Vol. AP-13, pp. 342–353.

12. Balfour, M. A., "Phased array simulators in waveguide for a triangular array of elements," *IEEE Trans. Antennas Propag.*, May 1965, Vol. AP-13, pp. 475–476.

CHAPTER 16.2
TYPES OF ANTENNAS

William F. Croswell

WIRE ANTENNAS

Analysis of Wire Antennas

The development of wire antennas has been extensive, since such antennas are simple to construct. The classical analysis of wire antennas such as dipoles, loops, and loaded-wire antennas has been developed by Hallen and R. W. P. King and his students; a good summary of theoretical and experimental results, including specific impedance curves and design data, is given in Ref. 1. The unfortunate drawback of this analysis method is that each new wire-antenna configuration presents another analytical problem which must be solved before design computations can be made. A systematic method of solving wire-antenna problems using computerized matrix methods has been developed by extending the analysis of scattering by wire objects by Richmond[2–5] and Harrington.[6–9] These matrix analysis methods have been applied to wire antennas to determine the input impedance, current distribution, and radiation patterns by subdividing any particular wire antenna into segments and determining the mutual coupling between any one segment and all other segments. The method therefore can treat any arbitrary wire configuration, including loading and arrays, the limitation being the storage capacity of available digital computers and the patience of the programmer. In the last 10 years, these numerical methods have been extended to include wire antenna wear obstacles such as towers and mounted on aircraft and missiles.

Wire Antennas over Ground Planes

The wire antenna mounted over a ground plane forms an image in the ground plane such that its pattern is that of the real antenna and the image antenna and the impedance is one-half of the impedance of the antenna and its image when fed as a physical antenna in free space. For example, the quarter-wave monopole mounted on an infinite ground plane has an impedance equal to one-half the free-space impedance of the half-wave dipole. The advantage of the ground-plane-mounted wire antenna is that the coaxial feed can be used without disrupting the driving-point impedance. In practice an antenna mounted on a 2- to 3-wavelength ground plane has about the same impedance as the same antenna mounted on an infinite ground plane. The finite-ground-plane edges produce pattern ripples whose depth and angular extent depend on the ground-plane size.

V Dipole

The V dipole is constructed by bending a wire dipole antenna into a V. The impedance of the half-wave V-dipole antenna has been calculated as a function of the V angle, as given in Fig. 16.2.1. Note that the V dipole is equivalent to the bent-wire monopole antenna over a ground plane. The effect of bending the dipole is to tune

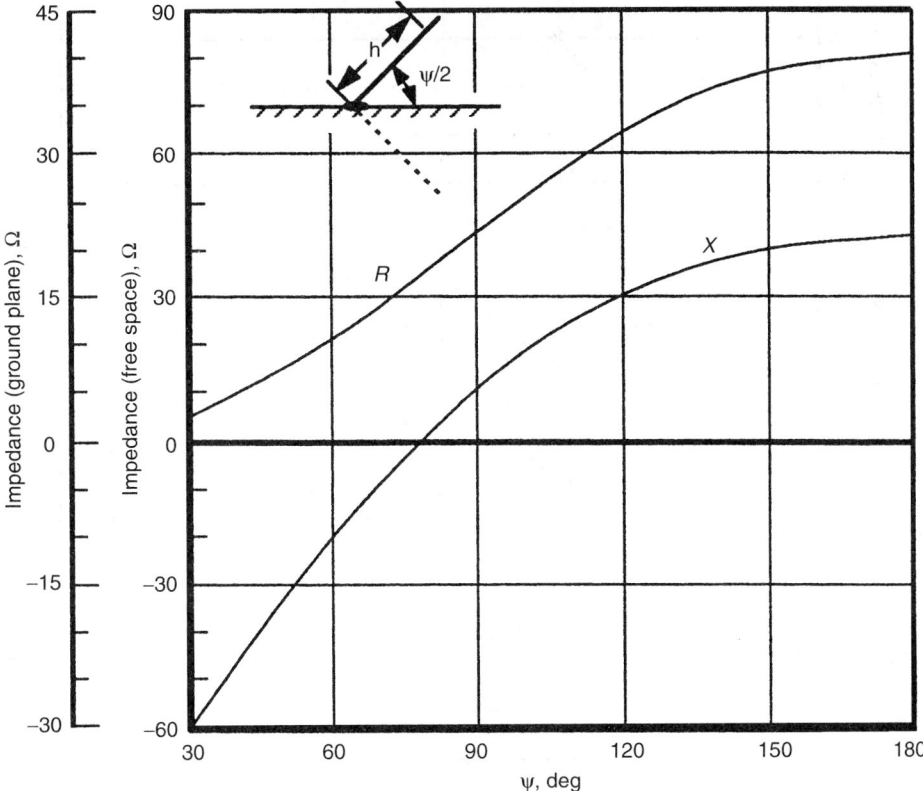

FIGURE 16.2.1 Impedance of a V-dipole antenna as a function of V angle, $h/a = 1000$ n – arm length = $\lambda/4$, a = wire radius.   (*Courtesy of J. E. Jones*)

it, giving another practical tuning method in addition to adjusting its length. The patterns of the V dipole (vertical polarization) are nearly identical to those of the straight dipole for φ angles as small as 120°. With decreasing tilt angle this antenna excites a horizontal polarized field component which tends to fill in the pattern, making the antenna a popular communication antenna for aircraft.

Bent Dipole

Another form of the dipole antenna that has practical application, particularly for ground-plane or airplane applications, is the bent-wire dipole formed by bending the wire 90° some distance out from the feed point. The impedance of the bent wire is given in Fig. 16.2.2 for the free-space and ground-plane case. Note that this antenna can also be tuned by adjusting the lengths perpendicular and parallel to the driving point. The radiation pattern in the plane of this antenna is nearly omnidirectional for values of $H_1 \leq 0.10$, after which the pattern approaches that of the vertical half-wave dipole. Other forms of this antenna can be constructed, including loading to reduce the effective length.

Loop Antenna

Another useful classical antenna is the loop antenna. As stated earlier, many investigators have erroneously designated this antenna as a magnetic dipole when indeed it is just another form of the wire antenna.

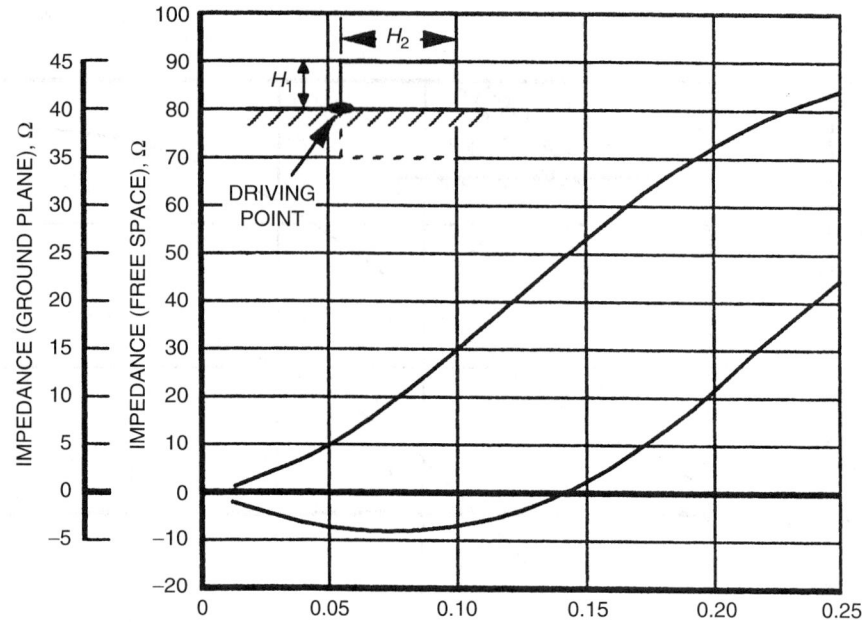

FIGURE 16.2.2 Impedance of half-wave bent-dipole antenna, $H_1 + H_2 = \lambda/4$. (*Courtesy of J. E. Jones*)

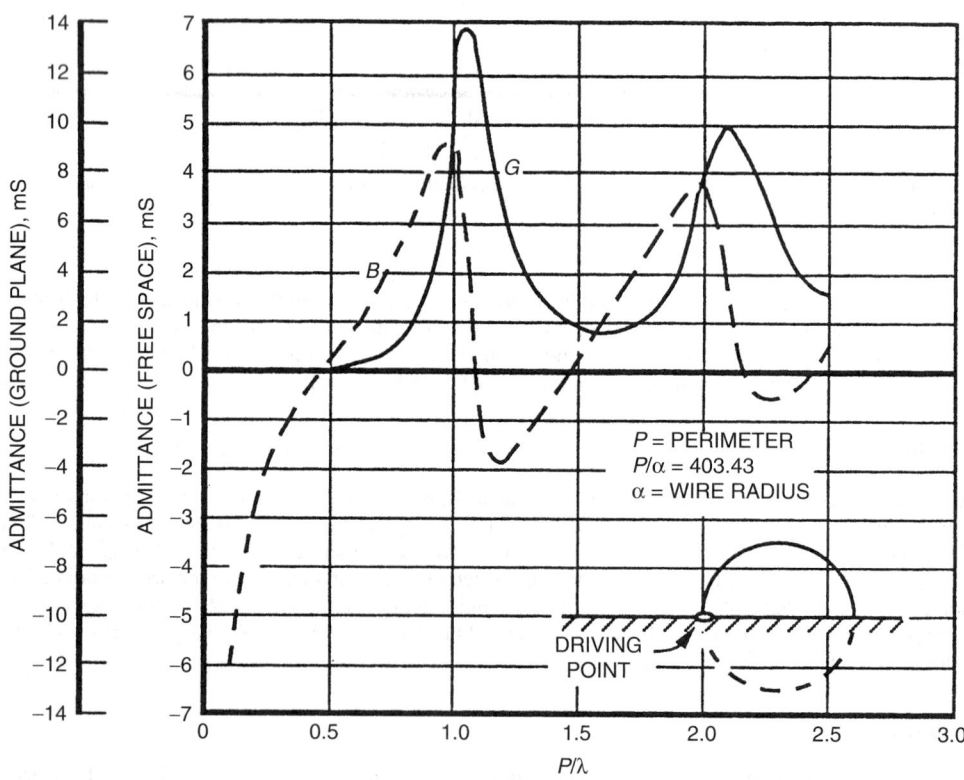

FIGURE 16.2.3 Admittance of the loop antenna. (*Courtesy of J. E. Jones*)

The admittance of the loop antenna can be computed using the matrix method by approximating the loop with a polygon having the same electrical length. The admittance of a 12-sided polygon, which is identical to the admittance of a loop of the same length, has been computed and is given in Fig. 16.2.3. These results have been verified in the ground-plane case experimentally.[10] Indeed the square loop or any other multisided loop with the same electrical-perimeter length has approximately the same admittance as the circular loop. Another method of improving the impedance of a loop is to add turns or load the loop with discrete lumped capacitances. The patterns and impedance of loop antennas mounted on aircraft structures have been studied.

Wire Antennas near Ground Planes

Although the impedance of wire antennas mounted on ground planes several wavelengths in dimension for practical purposes is similar to the impedance of the same antenna mounted on an infinite ground plane, the patterns of wire antennas on finite ground planes strongly depend on the ground-plane size. In recent years, the geometrical theory of diffraction (GTD) has been successfully applied to such problems.[11-16] These published results and the method of analysis are of great interest to antenna engineers since the geometry is the practical one of interest.

Loop above a Finite Ground Plane

The loop above a finite ground plane is an antenna commonly used as an array element in VHF omnirange stations located near all airports as an aircraft landing aid. Since the pattern of the small loop is symmetrical in azimuth, the elevation pattern of a loop over a finite circular ground plane can be computed using a pair of closely spaced line sources fed out of phase and located over a finite-width conducting strip.[14] The geometry of the line-source pair a distance d above a ground plane $2x_0$ in width is shown in Fig. 16.2.4.

The pattern of this antenna has a null at $\phi = 90°$ and a maximum value at some angle above the ground plane, which depends on the spacing d and the ground-plane size $2x_0$. The value of the field along the ground-plane edge ($\phi = 0$) is also of interest to the antenna designer. The variation of these field parameters is given in Fig. 16.2.5.

Horizontal Dipole over a Finite Ground Plane

The GTD method can be applied to the horizontal dipole over a finite ground plane,[15] the geometry of which is shown in Fig. 16.2.6. Also shown in Fig. 16.2.6 is the geometry of the same dipole placed over a cylinder. The purpose of this antenna design is to minimize the ripple or field variation in the pattern above the ground plane and simultaneously achieve a low back-lobe level. These field parameters are plotted as a function of ground-plane size and dipole spacing in Fig. 16.2.7. Also plotted in Fig. 16.2.7 are similar design curves for the dipole spaced the same parametric distances above a perfectly conducting cylinder having a diameter equal to the finite-ground-plane width. These calculations were made by programming available formulas. The cylinder curvature allows one to obtain a better back-lobe level for a given pattern variation or ripple in the forward region. Experimentally, it has been determined that the rear part of the metal cylinder can be substantially removed with little effect.

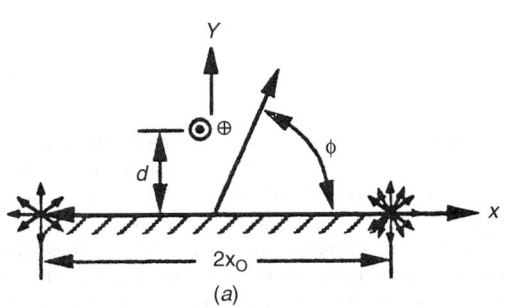

FIGURE 16.2.4 Linear array of line sources and its diffraction mechanism, a two-element array above a ground plane. (*Adapted from Balanis*[14]).

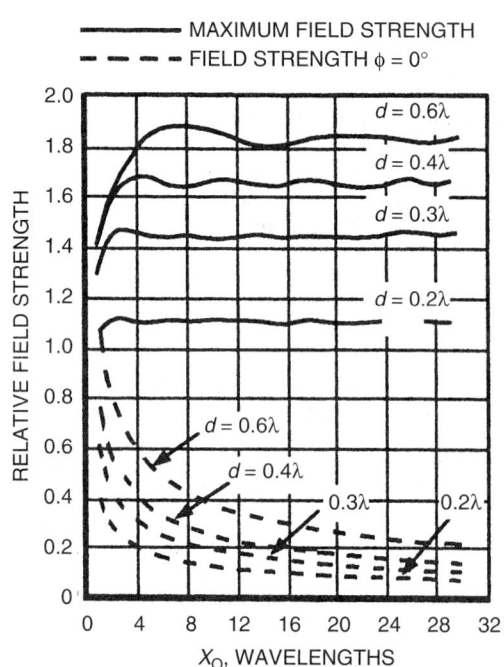

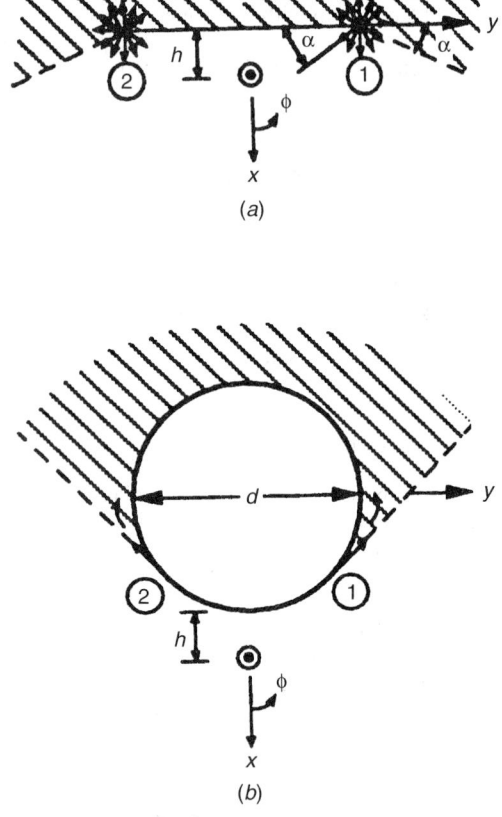

FIGURE 16.2.5   Variations of the maximum field strength and the field strength at the angle of the ground-plane edge as a function of line-source spacing and ground-plane size. (*Adapted from Balanis*[14])

FIGURE 16.2.6   Radiation mechanism of dipole near finite ground plane and circular conducting cylinder. (*a*) ground plane; (*b*) circular cylinder.   (*From Balanis and Cockrell*[15])

WAVEGUIDE ANTENNAS

General Considerations

The waveguide antenna, which consists of a dominant-mode-fed waveguide opening onto a conducting ground plane, is very useful for many applications such as a feed for reflector antennas or a flush-mounted antenna for aircraft or spacecraft. For flush-mounting purposes it is sometimes desirable or necessary to cover the ground plane with dielectric layers to protect the aperture from the external environment or in some instances to put dielectric plugs in the feed-waveguide section. The impedance properties of waveguide antennas have been studied extensively both theoretically and experimentally, particularly for the rectangular waveguide, the circular waveguide, and the coaxial waveguide or so-called annular slot. For unloaded apertures, the assumptions of the single-mode trial field in the impedance variational solution has proved adequate for practical purposes. The impedance of these antennas is relatively independent of ground-plane size so long as the ground plane is 2λ in dimension or greater. However, the radiation pattern of the waveguide antenna mounted on finite ground

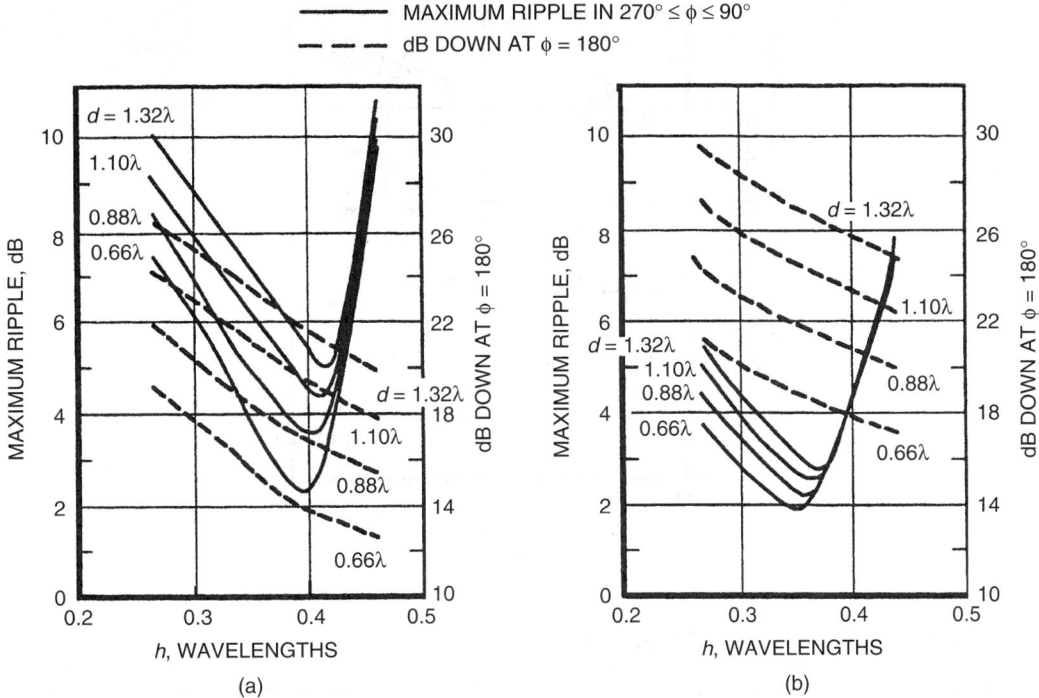

FIGURE 16.2.7 Variations of maximum ripple in $270° \leq \phi \leq 90°$ region and radiation $\phi = 180°$ as functions of dipole position h near (*a*) the ground plane and (*b*) circular conducting cylinder. (*From Balanis and Cockrell*[15])

planes is very dependent on the ground-plane size. The effects of diffraction by ground-plane edges can be treated by the GTD in a manner similar to that used for wire antennas above a finite ground plane.

Aperture Admittance of Rectangular Waveguides

Calculations have been made assuming the dominant mode as a trial function. These calculations are compared with measured results in Fig. 16.2.8, where the waveguide flange was used as a ground plane. (Note that the X-band flange was 1.62 by 1.62 in. and the S-band flange was 6.42 by 6.42 in.) Other measurements were made with up to 10λ ground planes with similar results, as given in Fig. 16.2.8. Calculations of the aperture admittance of square waveguides are given in Fig. 16.2.9. Note that the square waveguide aperture is more nearly matched to the characteristics admittance of the waveguide than the rectangular one. This is also true for the circular aperture.

Aperture Admittance of Circular Waveguides

Calculations for the circular waveguide radiating both into free space and into dielectric slabs have been performed and compared with measurements.[18–19] Calculations for a 1.5 in.-diameter waveguide operating at C band are given in Fig. 16.2.10. Note that the circular waveguide is nearly matched when radiating into free space. Like that of the rectangular waveguide, the aperture admittance of the 2λ ground-plane antenna closely approximates the infinite-ground-plane model.

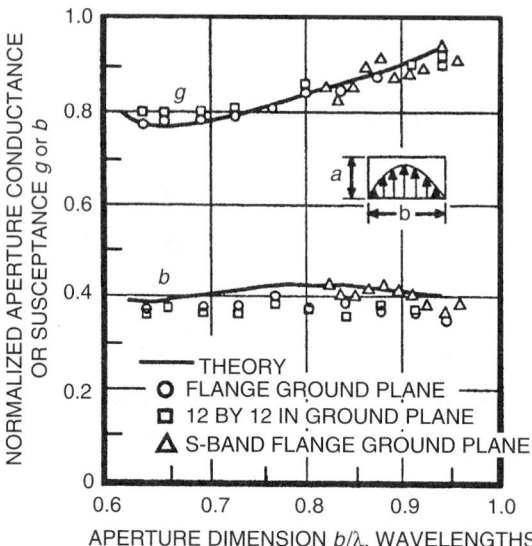

FIGURE 16.2.8 Aperture admittance of a square waveguide.

Patterns of Waveguides on Finite Ground Planes

The design of waveguide antennas on finite ground planes can also be treated by the geometrical theory of diffraction in a manner similar to that for the wire antennas above a finite groundplane. Edge or diffraction effects will primarily occur only in the E-plane (yz plane) of the circular or rectangular waveguide.

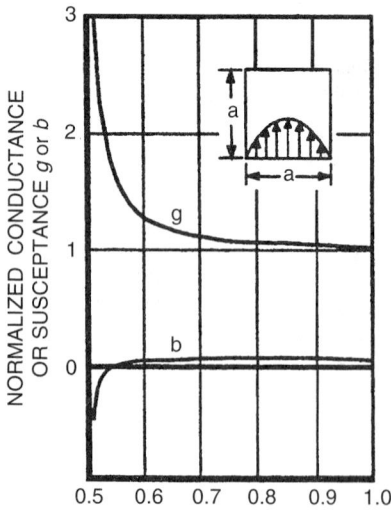

FIGURE 16.2.9 Measured and computed aperture admittance of a circular waveguide. (*From Bailey and Swift*[19])

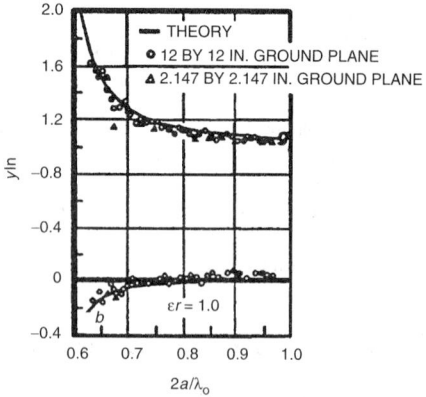

FIGURE 16.2.10 *E*-plane radiation patterns of a TE01-mode-excited rectangular aperture opening onto a finite ground plane vs. aperture size in wavelengths. Ground-plane size = 4λ.

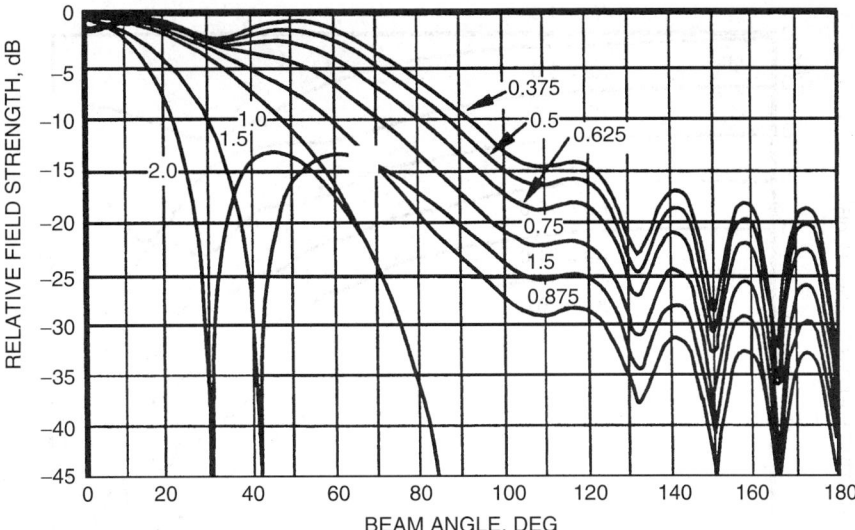

FIGURE 16.2.11   *E*-plane radiation patterns of a TE11-mode-excited circular aperture opening onto a finite ground plane vs. aperture size in wavelengths. Ground-plane size = 4λ.

A summary of calculations for waveguides with different aperture sizes is given in Fig. 16.2.11. The radiation patterns differ for different ground-plane sizes. It should be noted that no diffractions occur in the *E* plane for a 1λ-wide rectangular aperture or a 1.22λ-diameter circular aperture. Indeed, for these aperture dimensions the *E*- and *H*-plane patterns are nearly identical.

The impedance properties of thin-slot antennas are relatively independent of ground-plane size.

Patterns of Narrow Slots in Cylinders

The radiation pattern of a thin circumferential waveguide fed slot on a cylinder is about the same as a similar slot on an infinite ground plane if the cylinder $C = ka$ is greater than about 8.0 to 9.0. The pattern of the axial slot on a cylinder is quite sensitive to the mounting cylinder size, as shown in Fig. 16.2.12.

HORN ANTENNAS

The horn antenna may be thought of as a natural extension of the dominant-mode waveguide feeding the horn in a manner similar to the wire antenna, which is a natural extension to the two-wire transmission line. The most common type of horns are the *E*-plane sectoral, *H*-plane sectoral, and pyramidal horn, formed by expanding the walls of the TE_{01}-mode-fed rectangular waveguide or the conical horn formed by expanding the wall of the TE_{11}-mode-fed circular waveguide. Early work concerned the determination of the forward radiation patterns, directivity, and approximate impedance of sectoral, pyramidal, and conical horns, including comprehensive experimental studies.[21–30] An excellent summary of this work is given by Compton and Collin.[31] In later work, the input impedance, wide-angle side lobes, and back lobes of certain horn antennas have been determined to a high precision using the GTD.[32–36] The remainder of this chapter on horn antenna will follow the work by Milligan[89] and his chapter on horn antennas.

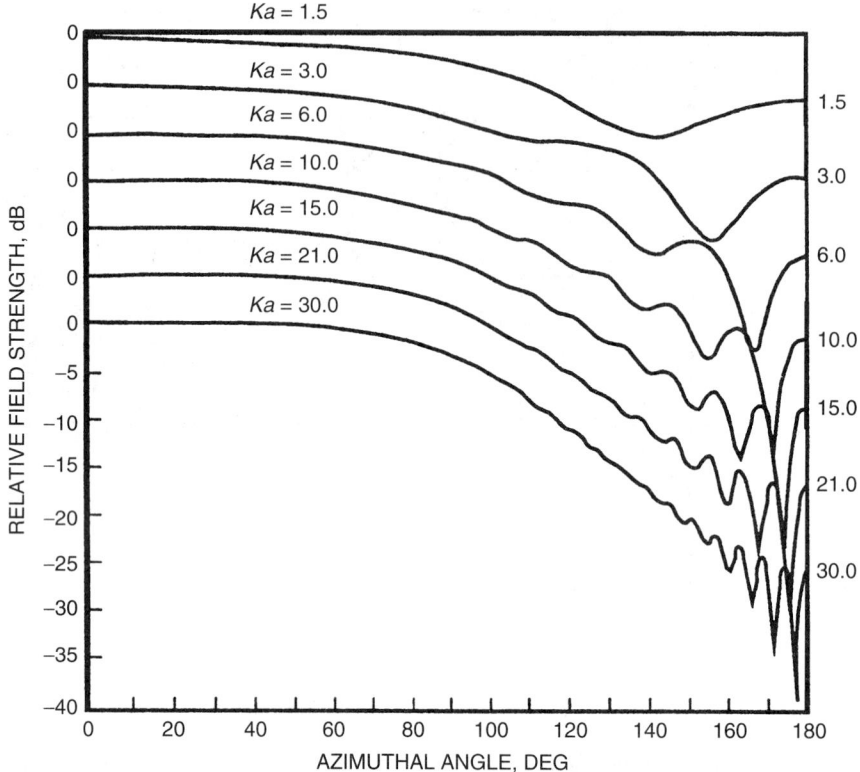

FIGURE 16.2.12   Radiation patterns of an axial infinitesimal slot on a cylinder vs. cylinder circumference *Ka* in wavelengths. Note that the vertical scale for each pattern is displaced 5 dB for clarity.

The general geometry of the waveguide fed horn is given in Fig. 16.2.13. The general phase error in the aperture is given by

$$S = \frac{\Delta}{\lambda} = \frac{W^2}{8\lambda R} \tag{1}$$

Sectoral and Pyramid Horns

The general geometry for the pyramidal horn is given in Fig. 16.2.14. For the *E* plane and *H* plane of the sectoral horn, or the *E* and *H* planes of the pyramidal horn, the quadratic phase error can be expressed as

$$S_E = \frac{a^2}{8\lambda R_E}, \quad S_H = \frac{b^2}{8\lambda R_H} \tag{2a,b}$$

The quadratic phase error change the TE_{10} mode excited horn aperture radiation pattern that is dependent on the amount of phase error S_E and/or S_H. Universal patterns for the sectoral and pyramidal horn are given in Figs. 16.2.15 and 16.2.16.

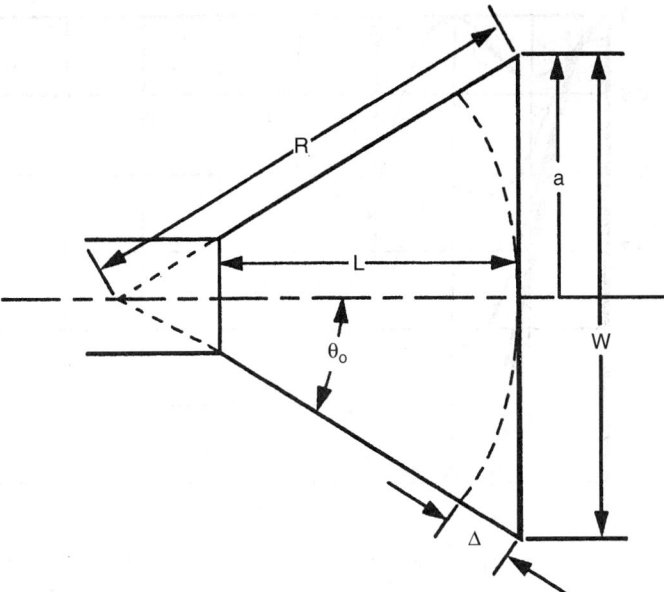

FIGURE 16.2.13 General geometry of a waveguide horn. (*Adapted from Milligan,*[89] *p. 180*)

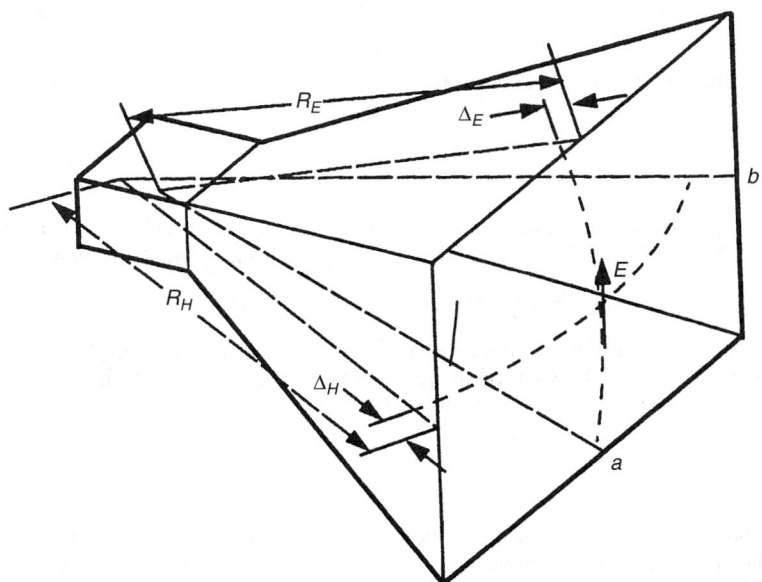

FIGURE 16.2.14 The geometry of a pyramidal horn. (*Adapted from Milligan,*[89] *p. 181*)

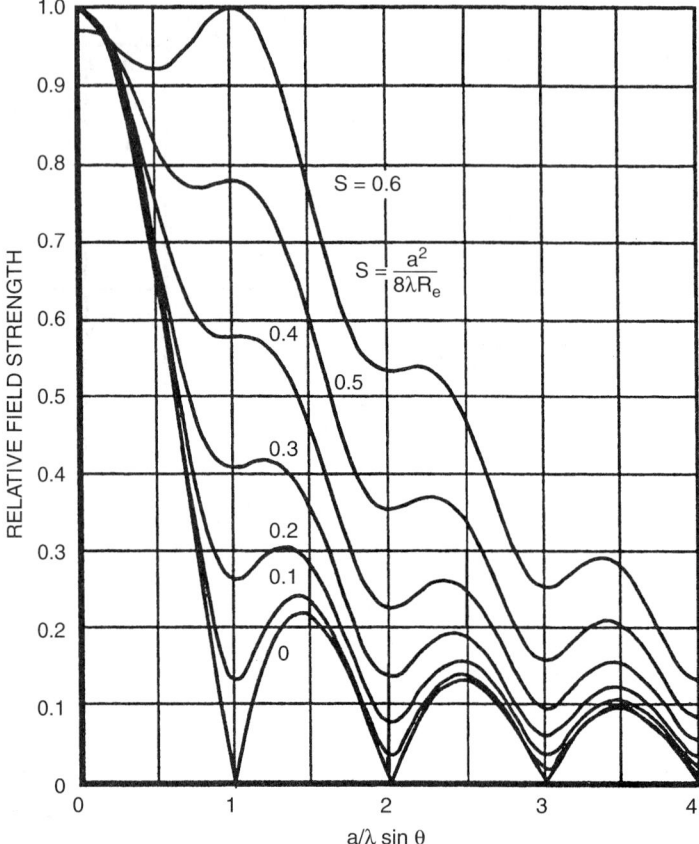

FIGURE 16.2.15   Universal pattern of a pyramidal horn, *E* plane. (*Adapted from Milligan,*[89] *p. 182*)

The gain of pyramidal and sectoral horns has been determined to accuracies of about ±0.4 dB even for frequencies as high as 38 GHz.[37] The most common gain standard is the pyramidal horn, which has a gain equal to

$$\text{Gain}_{dB} = 10(1.008) + \log a_\lambda b_\lambda) - (L_E - L_H) \tag{3}$$

where a_λ, b_λ are aperture dimensions in wavelengths and L_e, L_h are the loss due to phase error in *E* and *H* planes of horn as given in Fig. 16.2.17. Gain curves for other horns and an excellent summary of horn-design information are given by Jakes[38] and Compton and Collin.[31]

Conical Horns

The conical horn formed as an extension of the circular waveguide excited in the TE_{11} mode has been thoroughly studied by King.[29] The radiation patterns of his antenna can be obtained by integrating the dominant TE_{11}-mode field, with quadratic phase error (Fig. 16.2.17). As in rectangular-waveguide-fed horns, the fields outside the aperture are neglected, and therefore, the wide-angle side lobes and back lobes will be computed incorrectly with such a procedure. The universal radiation patterns of conical horns excited by a circular

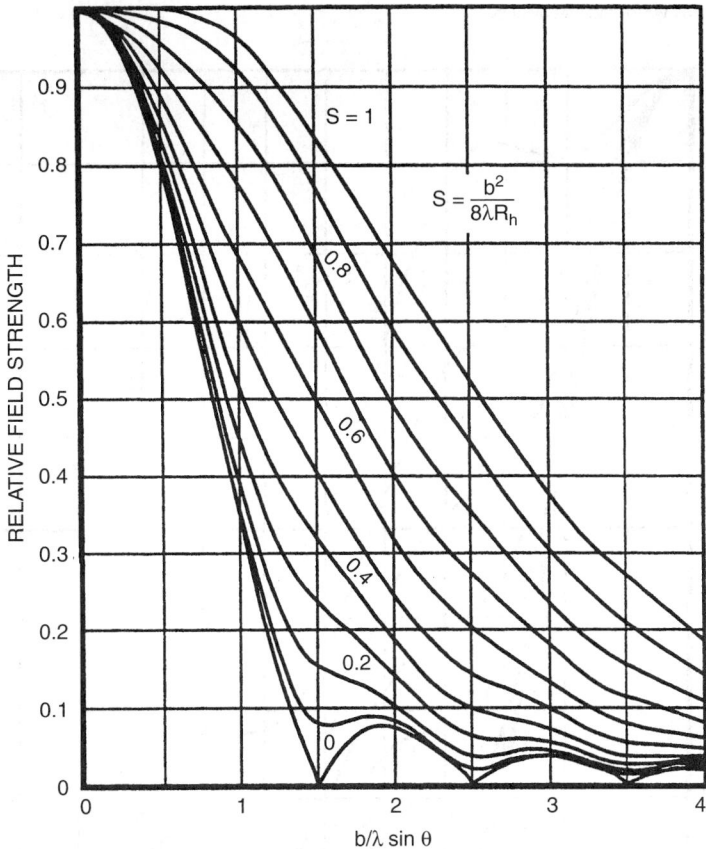

FIGURE 16.2.16 Universal pattern of a pyramidal horn, *H*-plane. (*Adapted from Milligan,*[89] *p. 183*)

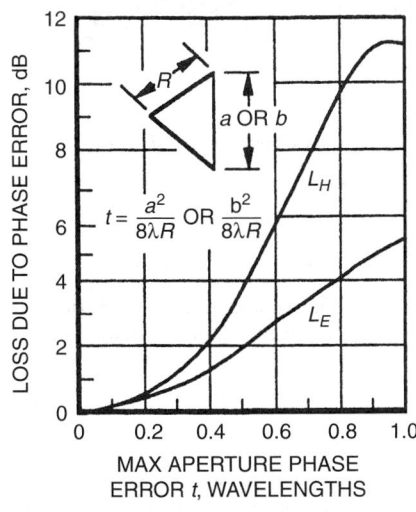

FIGURE 16.2.17a Loss correction for phase error in sectoral and pyramidal horns. (*From Jakes*[38])

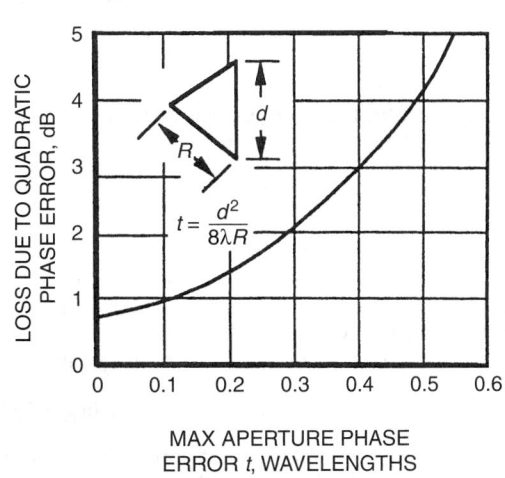

FIGURE 16.2.17b Loss correction for phase error in conical horns. (*From Jakes*[38])

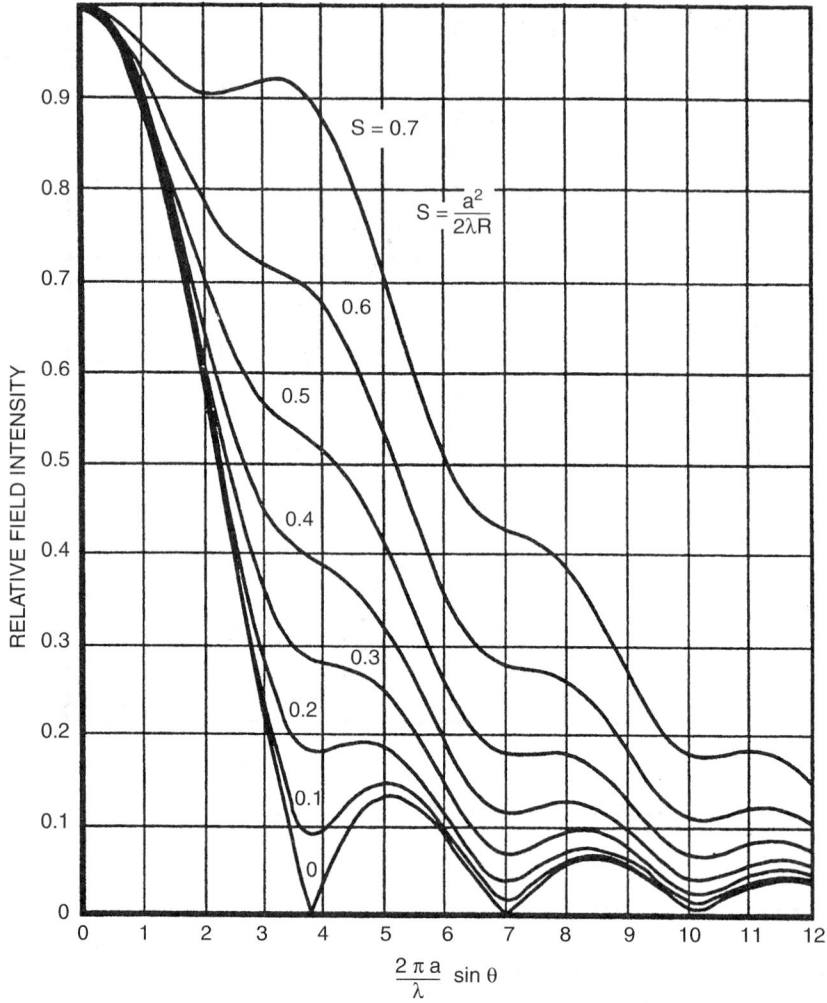

FIGURE 16.2.18   *E*-plane universal pattern of a TE_{11} mode excited conical horn.   (*After Milligan,*[89] *p. 192*)

waveguide excited in the TE_{11} dominant mode are given in Figs. 16.2.18 and 16.2.19 for various values of quadratic phase error. Care must be used in exciting this horn since any feed asymmetry may excite the TM_{01} mode in the feed waveguide.

The gain of the conical horn has been determined to be

$$\text{Gain}_{dB} = 20 \log C\lambda - L \tag{4}$$

where C is the circumference of horn aperture and L is the gain loss due to phase error given by curve in Fig. 16.2.17. It should be carefully noted that the gain given by Eqs. (3) and (4) neglects the losses in the conducting walls of the entire horn and the VSWR.

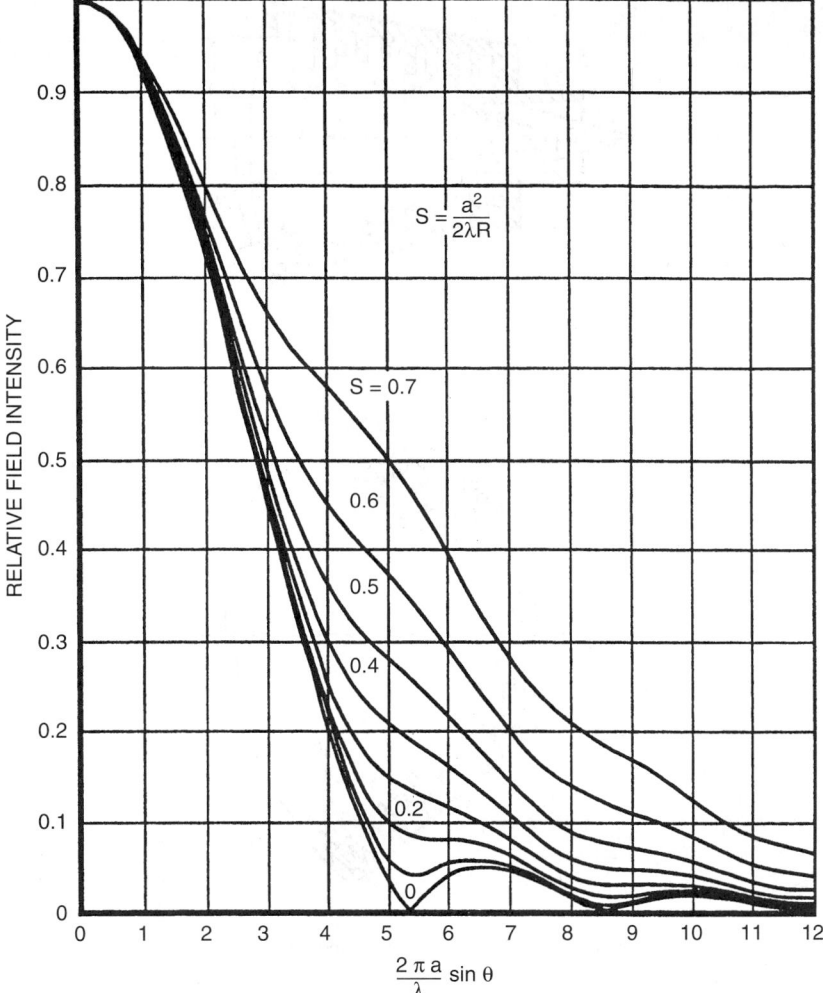

FIGURE 16.2.19   *H*-plane universal pattern of a TE$_{11}$ mode excited conical horn.   (*After Milligan,*[89] *p. 193*)

Corrugated Horns

The conical horn can have undesired pattern ripples, particularly in the *E* plane. Such patterns are undesirable for horn feeds used in high-efficiency parabolic antennas. The most commonly used horn feed is the corrugated conical horn described[39] in Fig. 16.2.20. The teeth in these horns generally are between $\lambda/4$ and $\lambda/2$ deep, although the best operating band is about 1.5 to 1. These horns convert the dominant waveguide feed TE$_{11}$ mode into the hybrid HE$_{11}$ which tend to make the *E*-plane pattern similar to the H-plane pattern. Therefore, the narrow angle corrugated horn (10 dB beamwidth, less than 74°) has a very symmetrical pattern in the ϕ plane as given in Fig. 16.2.21. The so-called scalar horn is of sufficiently wide angle that the phase center is at the horn throat. This horn is particularly useful for wideband applications.[40]

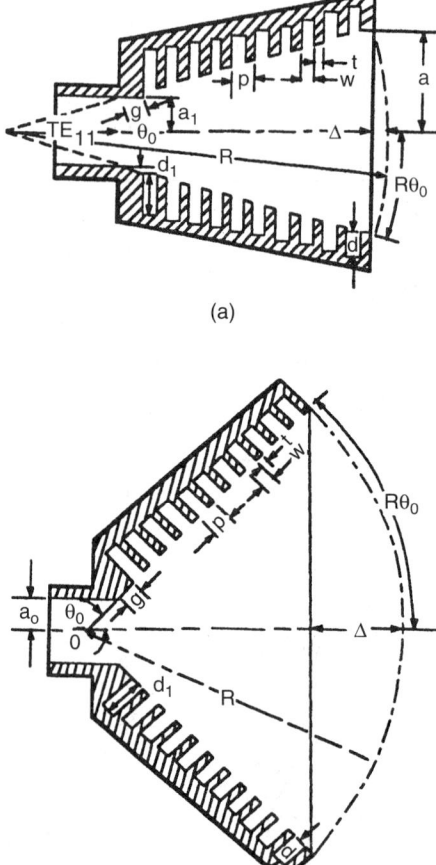

(a)

(b)

FIGURE 16.2.20   (*a*) Corrugated horn and (*b*) Scalar horn.   (*After Milligan*[89] *and Thomas*[39])

REFLECTOR ANTENNAS

The theory commonly used for the direct-fed parabolic antenna is that of Silver,[41] using physical optics. As outlined by Silver, this theory is adequate to predict the gain, radiation pattern, and the level of the first few side lobes of the secondary pattern, neglecting blockage and scattering by feed struts. The effect of feed-strut blockage can be estimated using geometrical optics;[42] however, an analysis of the diffraction of struts using a more rigorous formulation is necessary to improve the quantitative understanding of the problem. The radiation-pattern characteristics of the offset-fed parabola have been determined approximately using an extension of Silver's formulation to include this geometry.[43–44]

The spherical reflector is a good design for a scanning reflector antenna—thanks to its geometrical symmetry; however, a point source at the focal region will not produce a set of parallel rays from the secondary reflector. To correct this spherical-aberration error a line-source feed is employed.[45–46] By phasing this line source, the beam can be scanned to other positions.

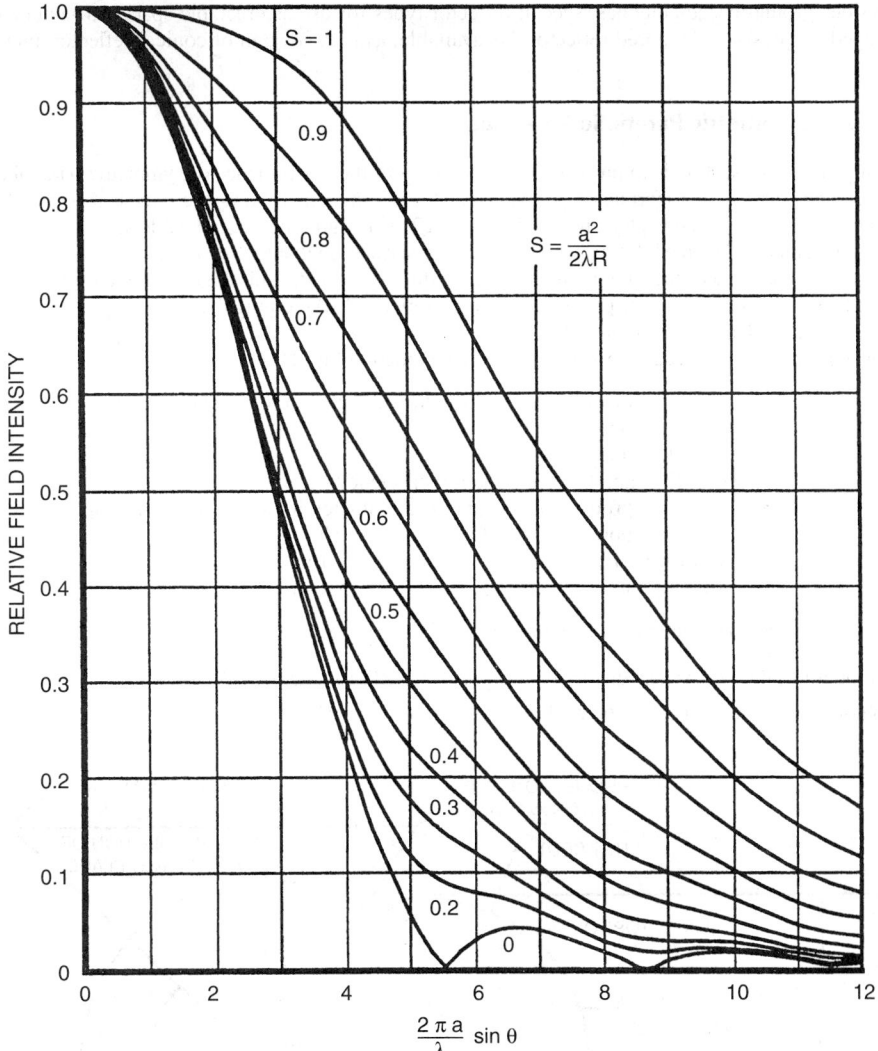

FIGURE 16.2.21   Universal pattern of a conical corrugated horn, HE_{11} mode.   (*After Milligan*[89])

Another way of feeding the parabolic reflector antenna is to use a subreflector in the focal region of the parabola and illuminate the subreflector from the parabolic surface. The principal advantages of the Cassegrain system or other methods of folded optics are the increase in effective f/d ratio and the simple mechanical location of the feed so that cooled receivers used with such antennas can be serviced in a more practical manner. The chief disadvantage is the aperture blockage of the subreflector, which restricts the application of this principle to large aperture. The Cassegrain antenna has been analyzed extensively, and computer programs are available. This design effort resulted in the design and construction of the 210-ft dish antenna[47–56] used in the worldwide space-receiving network. In order to achieve all-weather operating capability a precision 120 ft Cassegrain dish under a 150-ft radome has been constructed at Millstone Hill, Massachusetts. The radome produces about 1- to 2.8-dB loss in the microwave to millimeter-wavelength region.[58]

Besides ground-based antennas, special reflector types for use as erectable spacecraft antennas have been developed. Analysis of the gored reflector[59] is available, as is the design of conical-reflector antenna.[60]

Performance of Symmetric Parabolic Antennas

The purpose of this section is to present a brief summary of the performance of symmetric parabolic antennas in terms of geometric and performance parameters. The information follows the excellent work of Knop[61] and Milligan.[89] Consider the geometry given in Fig. 16.2.22. This figure depicts a feed located at the focal point of a symmetric parabola. The feed illuminates the reflecting surface, so that the secondary radiation pattern is obtained by integrating the surface current until the edge of the dish is reached. (This edge field or edge taper is a design parameter that affects the secondary pattern and gain performance.) Assuming that the feed pattern is symmetrical in both the E and H planes, a set of universal design curves can be generated as given in Fig. 16.2.23. These data are presented as a function of edge taper in dB. The data plotted in Fig. 16.2.24 is defined in the following manner:

η_B = beam efficiency, which is the percent of the horn energy captured by the parabolic dish antenna

η_A = total aperture efficiency

$\eta A_I = \eta A/\eta B$ = aperture illumination efficiency

$2\theta_3 D/\lambda$, degrees = product of the aperture diameter in wavelengths, multiplied by the full half power beamwidth in degrees $(2\theta_3)$

FSL, dB down = first sidelobe level in dB down from the main lobe peak

$U_{FSL} = \pi(D/\lambda)\sin\theta_{FSL}$ = universal angle at which the first sidelobe occurs.

Other performance parameters are given in Fig. 16.2.23. For example:

(a) amount of phase error loss due to axial defocus $\pm DZ/\lambda$ for various f/D ratios

(b) phase error loss for phase center differential for different f/D ratios

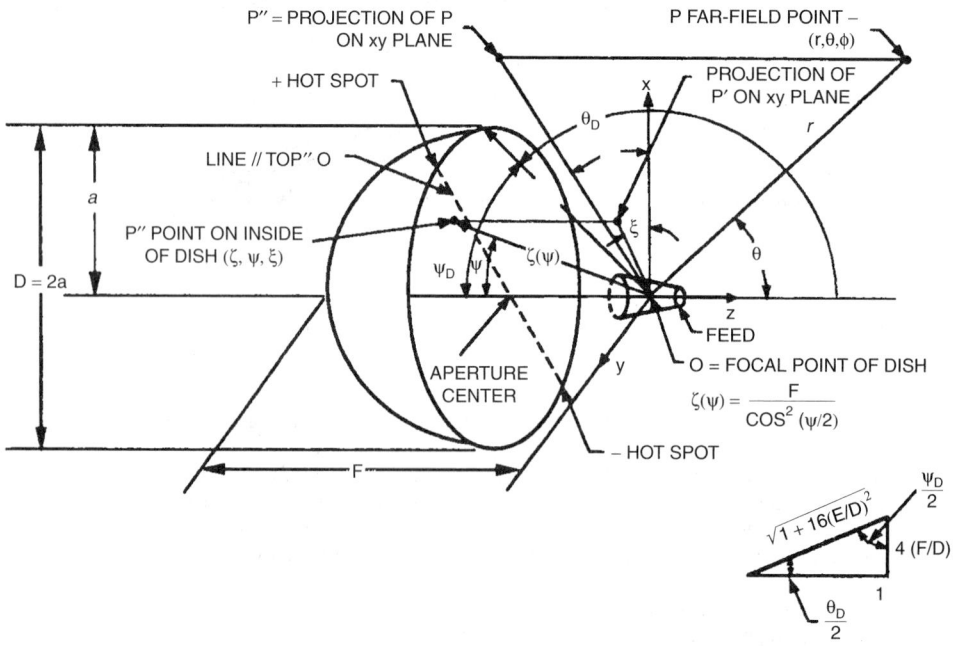

FIGURE 16.2.22   Geometry of a symmetric parabolic dish antenna.   (*After Knop*[61])

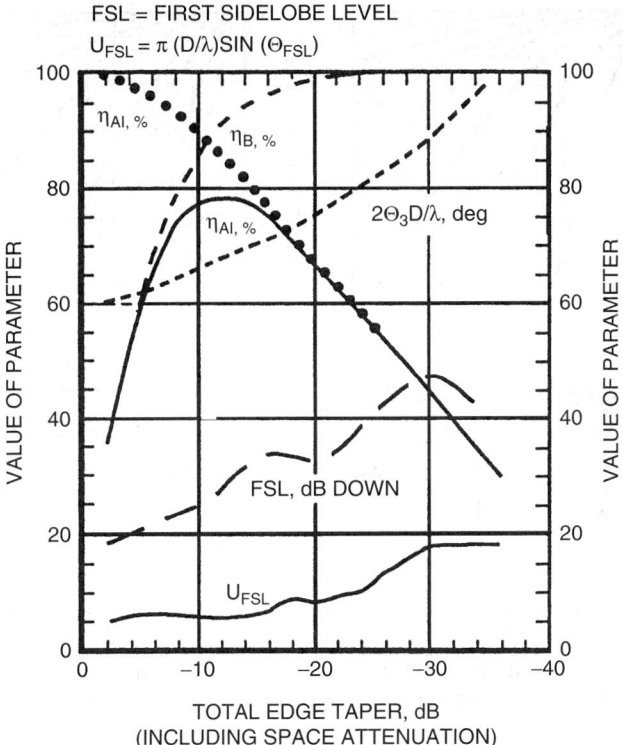

FIGURE 16.2.23   Universal performance curves for a symmetric parabolic dish antenna fed with a feed having symmetrical radiation pattern.   (*After Knop*[61])

(**c**) scan loss versus beamwidths of scan for different *f/D* ratios

(**d**) beam deviation factor as a function of *f/D* ratio.

The most important performance parameter, the gain loss as a function of the surface roughness, is given in Fig. 16.2.25.

LOG-PERIODIC ANTENNAS

The Frequency Independent Antenna

The frequency independent antenna is specified only by angles. It was suggested by Rumsey[62] in 1954. The simplest form of such antennas is the equiangular spiral,[63] although early models with the frequency-independent idea included the tapered helix.[64–65] All antenna shapes that are completely specified by angles must extend to infinity; thus any physically realizable frequency-independent antenna has bandwidth limitations because of end effects. A simple modification of the frequency-independent antenna is the logarithmically periodic antenna,[66] whose properties vary periodically with the logarithm of the frequency. This modification tends to minimize the end effect, although the impedance will vary as a function of frequency; such variations are sometimes small. From these early designs a number of log-periodic antennas have been developed, including conical log spirals,[67] the log-periodic V,[68] the log-periodic dipole,[69–70] and the log-periodic Yagi-Uda array.[71]

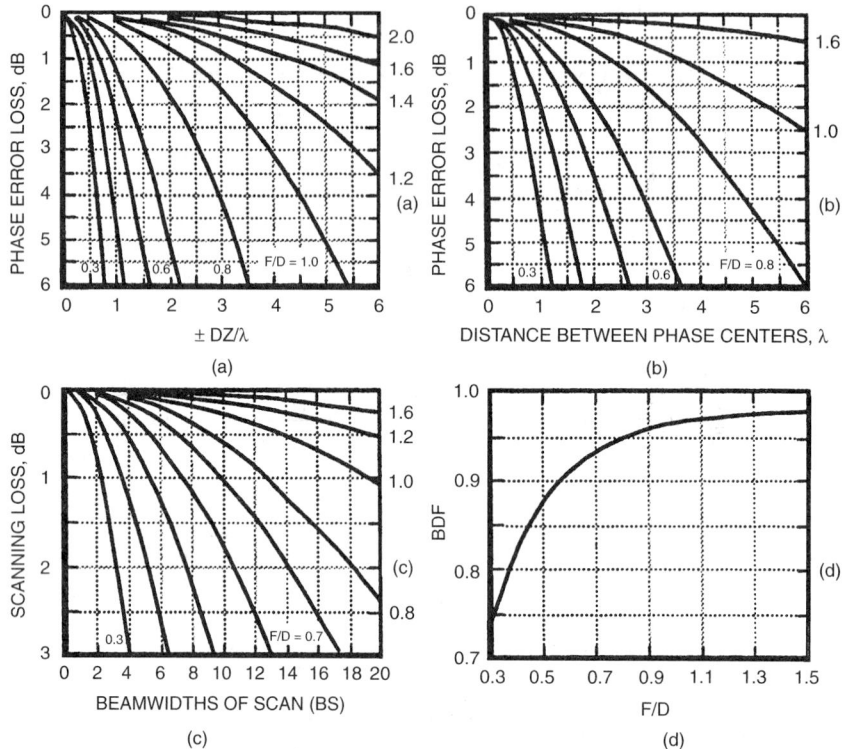

FIGURE 16.2.24 Performance parameters for a symmetrical parabola dish antenna: (*a*) Gain loss due to phase error caused by axial defocusing; (*b*) Gain loss due to the difference between *E*-plane and *H*-plane phase errors; (*c*) Gain loss due to feed scan; and (*d*) the beam deviation factor vs. *f/D* ratio.

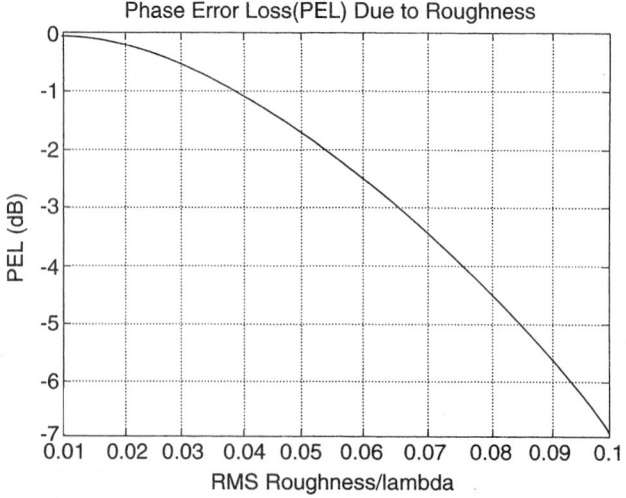

FIGURE 16.2.25 Gain loss due to random RMS error (PEL = −685.8. (RMS Roughness)^2, db).

DIRECTION OF BEAM

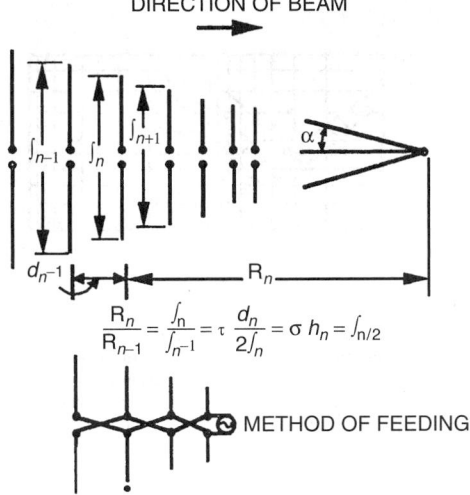

FIGURE 16.2.26 The log-periodic dipole antenna with definition of parameters. *(From Carrel[70])*

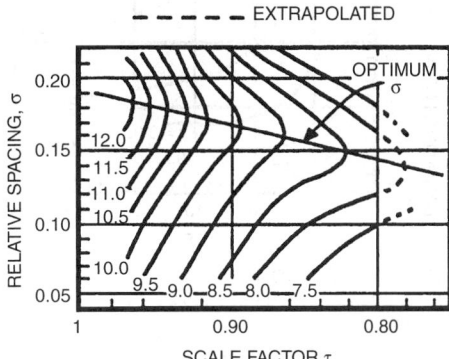

FIGURE 16.2.27 Constant-directivity contours in decibels versus τ and σ. Optimum indicates maximum directivity for a given value of τ. $Z_0 = 100\ \Omega$, $h/a = 100$, $Z_T = 100\ \Omega$.

Log-Periodic Dipole Design

One of the most popular antennas of this type is the log-periodic dipole antenna, which has the geometry depicted in Fig. 16.2.26. This antenna can be fed either by using alternating connections to a balanced line, as indicated in Fig. 16.2.26, or by a coaxial line running through one of the feeders from front to back. A simple procedure determined by Carre[70] which can be used for designing this antenna is outlined here. The number of elements is primarily determined by τ, and the antenna size is determined by boom length, which depends primarily on σ. The procedure is as follows:

1. An estimate of τ and σ based on the desired gain can be obtained from Fig. 16.2.27.
2. The bandwidth of the structure B_s is given by $B_s = B\,B_{ar}$ where B is the operating bandwidth and B_{ar} is determined in Fig. 16.2.28 using the parameter $\tan \alpha = (1 - \Gamma)/4\sigma$.
3. The length of the first elements is always made $\lambda_{max}/2$, so that the boom length L between the largest and smallest elements can be found from $L/\lambda_{max} = 1/4(1 - 1/B_s) \cot \alpha$.
4. The number of elements required is given by $N = 1 + [(\log B_s)/\log(1/\tau)]$.

By several iterations of this design procedure a minimum boom length can be obtained. The relative feeder impedance of the design can be found using available data.[70]

SURFACE-WAVE ANTENNAS

General Description

A wide class of *surface-wave antennas* has been devised, e.g., the Yagi, backfire, helix, cigar, and polyrod antenna. The surface-wave nomenclature is related to the idea that these antennas, if infinite in length, will support a wave that travels along the structure at a velocity slower than the velocity of light in free space. Data for the phase velocity along such antenna structures are available.[72–74] If the parameters of the antenna structure are chosen so

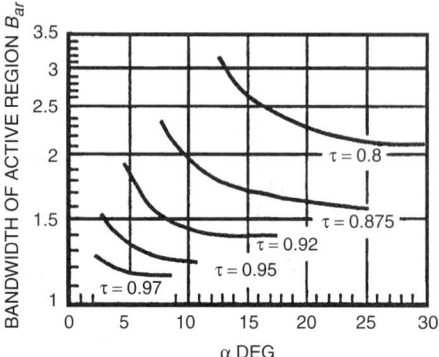

FIGURE 16.2.28   Bandwidth of active region B_{ar} vs. for several values of, for $Z_0 = 100\ \Omega$, $h/a = 125$.   (*From Carrel[70]*)

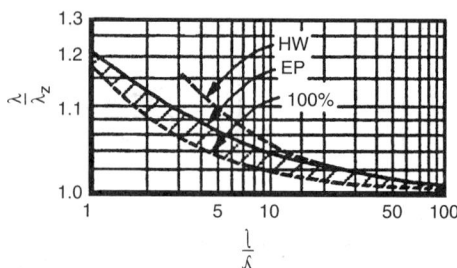

FIGURE 16.2.29   Relative phase velocity $c/v = \lambda/\lambda_z$ for maximum-gain surface-wave antennas as a function of relative antenna length $1/\lambda$. HW = Hansen-Woodward condition; EP = Ehrenspeck-Poehler experimental values; 100 percent = idealized perfect excitation.   (*From Zuckel[75]*)

that the resultant phase velocity causes the Hansen-Woodward condition to be met on the finite length of the antenna, an increased or supergain condition occurs. The relative phase velocity $c/v = \lambda/\lambda_z$ to maximize the gain as a function of antenna length is given in Fig. 16.2.29. A typical phase-velocity variation as a function of specific antenna parameters is given in Fig. 16.2.30 for the Yagi. Choosing particular antenna parameters so that the optimum phase-velocity conditions are met will result in a good first-cut design. Improved designs require extensive parametric experimental studies where the antenna elements are varied about the initial dimensions. An estimate of how much gain can be expected from surface-wave antennas is given in Fig. 16.2.31. An excellent summary of surface-wave antenna design and literature has been compiled by Zucker.[75,76]

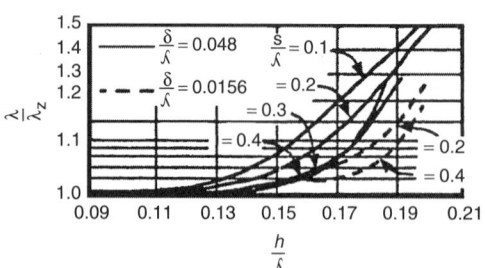

FIGURE 16.2.30   Relative phase velocity on a Yagi antenna (data from Ehrenspeck and Poehler and Frost); δ = diameter of wire element, s = spacing between elements, and h = half length of element.   (*From Zucker[76]*)

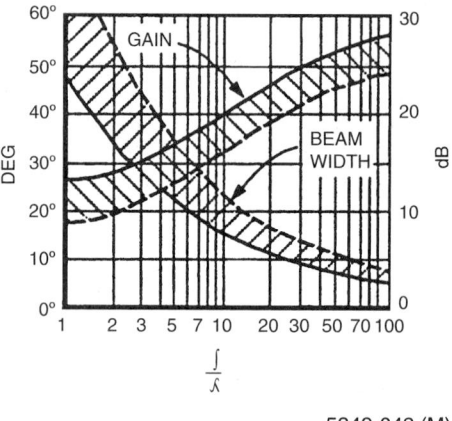

5349-042 (M)

FIGURE 16.2.31   Gain and beam width of surface-wave antenna as a function of a relative antenna length $1/\lambda$. For gain (in decibels above an isotropic source) use right-hand coordinate; for beam width left-hand coordinate. Solid lines are optimum values; dashed lines are for low-side-lobe and broadband design.   (*From Zucker[76]*)

The surface-wave antenna is a misnomer since surface waves do not exist on finite antennas. Consequently, aside from the use of the general concepts mentioned above, most surface-wave antennas are designed experimentally because of the importance of the feed radiation and end effects.

Use of Feed Shields

Nearly all surface-wave antennas suffer from side-lobe and beam asymmetry if not fed from special launchers. For example, the helix mounted on a flat ground plane has pattern asymmetry and poor axial ratio (about 3 dB) on axis.[77] Short helices in particular have been found to exhibit this property. A way to improve this helix performance is to use a conical feed shield and to taper the beginning and end turns as in the helicone antenna.[78] The feed shield that improves the performance of the cigar antenna is a conical horn[79] or a square cavity or bucker.[80]

MICROSTRIP ANTENNAS

General Considerations

In many applications where low-profile antennas are required and bandwidths less than a few percent are acceptable microstrip antennas may have the desired characteristics. Microstrip antennas are constructed on a thin dielectric sheet over a ground plane using printed-circuit-board and photo-etching techniques. The most common board is dual-copper-coated Teflon-fiberglass as it allows the microstrip antenna to be curved to conform to the shape of the mounting surface. The antenna itself may be square, rectangular, round, elliptical, and the like; the two most common elements (rectangular and round) are illustrated in Fig. 16.2.32. Circular-polarized radiation can be obtained by exciting the square or round element at two feed points 90° apart and in phase quadrature. Circular polarization can also be obtained over a limited frequency range by making the element slightly rectangular ($W/L \approx 1.03$) or slightly elliptical (eccentricity ≈ 0.2) and using a single feed point on the 45° diagonal.

Resonant Frequency

The frequency response for the basic elements (rectangular or round) is similar to a resonant tuned circuit or cavity. When viewed as a thin cavity, the rectangular element should be resonant when the length is equal to a half wavelength, and the round element should be resonant when the radius is equal to 0.293 wavelength; however, owing to fringing fields at the edges of the element, the actual resonant size is smaller by a few percent, that is, $L_e = 0.5\lambda_e$ and $\alpha_e = 0.23\lambda_e$ where the effective length L_e and effective radius α_e are given by

$$L_e = L + 0.824h[(\epsilon_e + 0.3)/(\epsilon_e - 0.258)][(\alpha L + 0.262h)/(\alpha L + 0.813h)] \tag{5}$$

$$\alpha_e = a[1 + 2h(\pi\alpha\epsilon_r)^{-1} \ln(9.246a/h)]^{1/2} \tag{6}$$

where $\alpha = W/L$, and

$$\epsilon_e = 0.5 (\epsilon_r + 1) + 0.5(\epsilon_r - 1)(1 + 12h/\alpha L)^{-1/2} \tag{7}$$

The resonant size (for minimum VSWR) decreases for an increase in the thickness h, as shown in Fig. 16.2.33 for the square ($W = L$) and in Fig. 16.2.34 for the round element. The resonant length of the rectangular element also decreases with an increase in the width.[81] A TE-excited strip provides a lower bound on the resonant length of rectangular elements with a large width-to-length ratio.

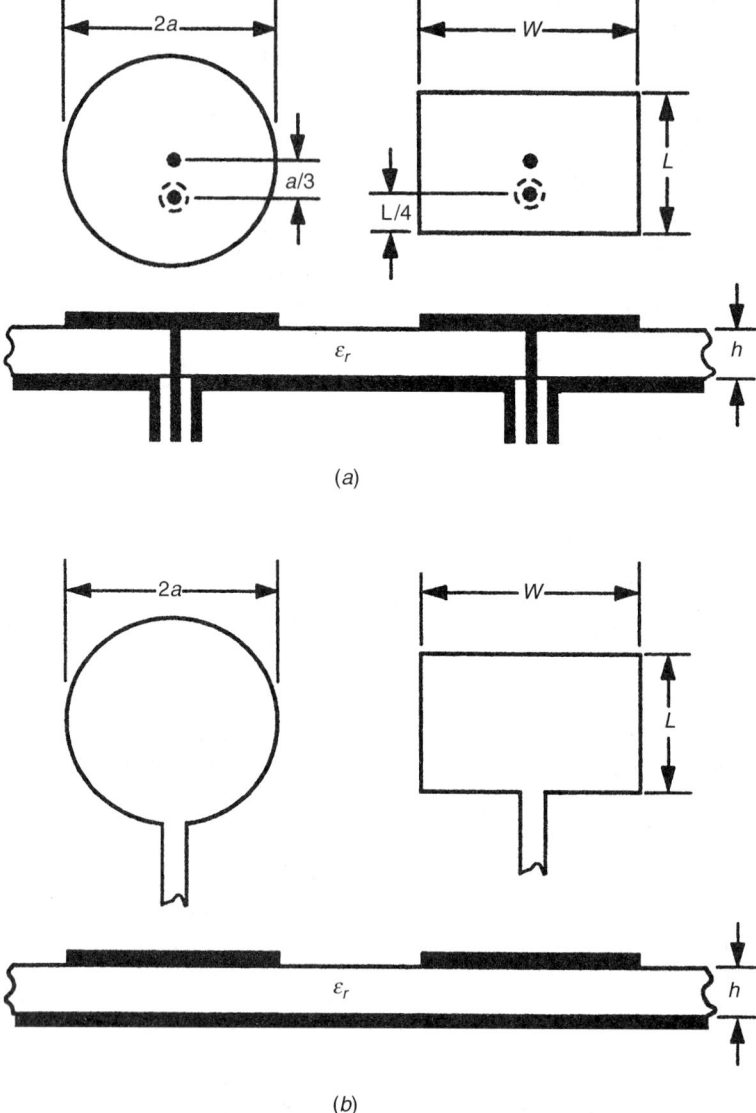

FIGURE 16.2.32 (*a*) Coaxial-fed microstrip antennas; (*b*) line-fed microstrip antennas.

Bandwidth

The bandwidth of microstrip antennas is determined primarily by the thickness of the dielectric, increasing with an increase in thickness, as shown in Fig. 16.2.35 for square and round elements on Teflon-fiberglass with coaxial feed probes. The data in Fig. 16.2.35 can be used to select a board thickness to meet the bandwidth requirement, and the resonant size of the element is then determined for this thickness by using the effective length or radius described earlier.

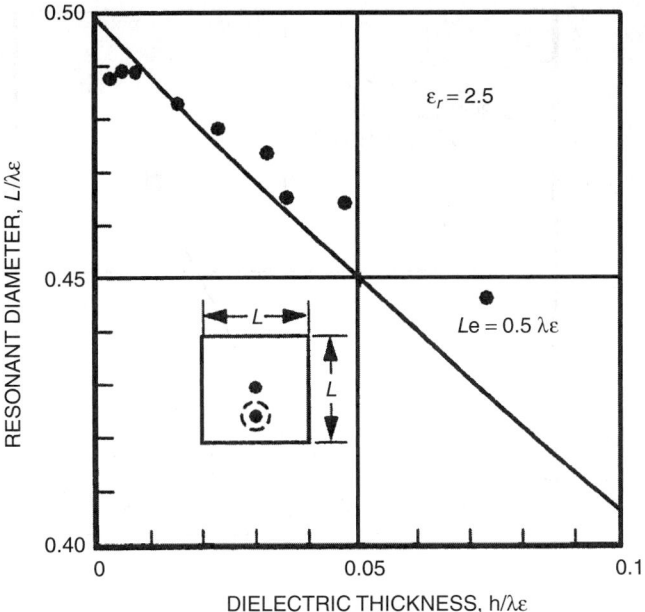

FIGURE 16.2.33   Resonant size of square microstrip antenna.

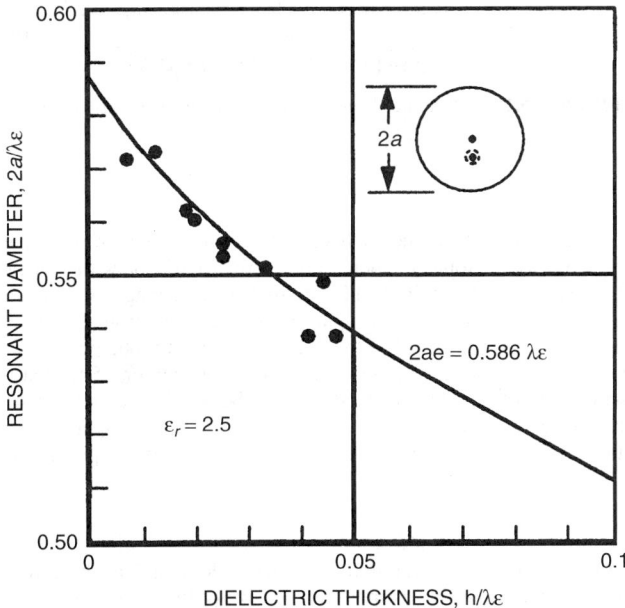

FIGURE 16.2.34   Resonant diameter of round microstrip antenna.

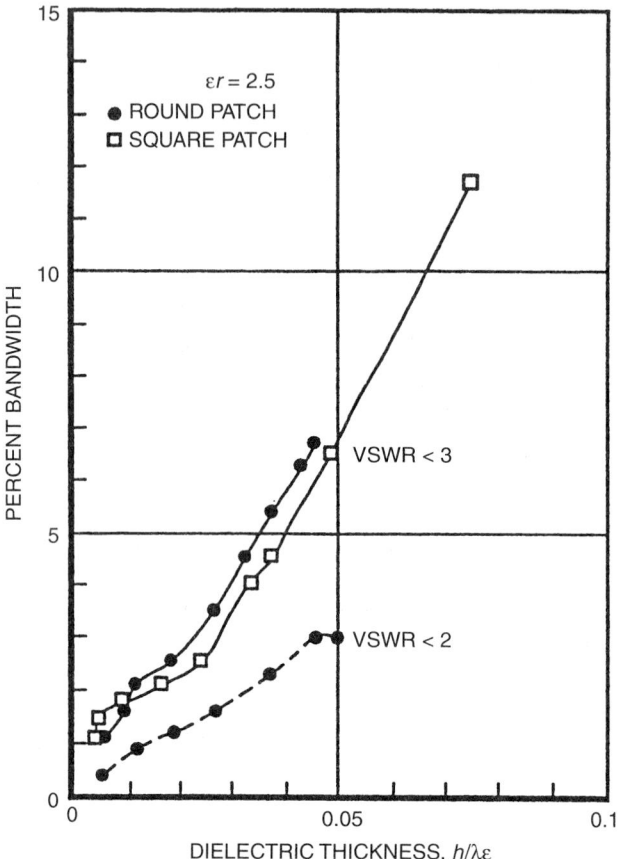

FIGURE 16.2.35 Measured bandwidth for coaxial-fed microstrip antenna.

Impedance

The resonant input impedance for a feed line connected at the edge of a microstrip antenna is about 120 Ω for the rectangular and about 300 Ω for the round element. A quarter-wave section with a characteristic impedance of 77.5 or 122.5 Ω can be inserted in the microstrip feed line to match the rectangular or round element to 50 Ω. In coaxial probe-fed microstrip antennas, the resonant input impedance varies from zero to the center of the element to about 120 Ω (rectangular) or 240 Ω (round) to the edge; therefore, a probe position can always be found, which matches the antenna to the 50 Ω coaxial feed. The exact probe position for a 50 Ω match depends on the thickness of the dielectric; however, it has been found that a coaxial probe positioned $L/4$ from the edge of a rectangular element or $a/3$ from the center of a round element will usually yield an input impedance quite close to 50 Ω. Approximate analytical models are also available[81–86] for calculating the impedance of microstrip antennas.

Radiation Patterns

The radiation patterns of round or rectangular microstrip elements can be found from a complementary-waveguide-fed aperture of the same size as the effective size of the microstrip antenna[87] or from slots located around the perimeter of the element.[88]

REFERENCES

1. King, R. W. P. Cylindrical Antennas and Arrays, in R. E. Collin and F. J. Zucker (eds.), "Antenna Theory," Pt. I, pp. 352–420, McGraw-Hill, 1969.

2. Richmond, J. H. Scattering by an Arbitrary Array of Parallel Wires, *Ohio State Univ. Antenna Lab., Rep.* 1522–1528, Contract N123 (1953)-31663A, April 1964.

3. Richmond, J. H. Scattering by an Arbitrary Array of Wires, *IEEE Trans. Microwave Theory Tech.*, July 1965, Vol. MTT-13, pp. 408–412.

4. Richmond, J. H. A Wire-Grid Model for Scattering by Conducting Bodies, *IEEE Trans. Antennas Propag.*, November 1966, Vol., AP-14, pp. 782–786.

5. Richmond, J. H. Scattering by Imperfectly Conducting Wires, *IEEE Trans. Antennas Propag.*, November 1967, Vol. AP-15, pp. 802–806.

6. Harrington, R. F. Theory of Loaded Scatterers, *Proc. IEEE*, April 1964, Vol. 111, pp. 617–628.

7. Harrington, R. F., and J. Mautz Matrix Methods for Solving Field Problems, *Syracuse Univ. Rep.* RADC TR-66-351, Vol. II, August 1966.

8. Harrington, R. F. Matrix Methods for Field Problems, *Proc. IEEE*, February 1967, Vol. 55, pp. 136–149.

9. Harrington, R. F. "Field Computation by Moment Methods," Macmillan, 1968.

10. Richards, G. A. Reaction Formulation and Numerical Results for Multiturn Loop Antennas and Arrays, Ph.D. dissertation, Ohio State University, 1970.

11. Sengupta, D. L., and V. H. Weston Investigation of the Parasitic Loop Counterpiece Antennas, *IEEE Trans. Antennas Propag.*, March 1969, Vol. AP-17, pp. 180–191.

12. Sengupta, D. L., and J. E. Ferris On the Radiation Patterns of Parasitic Loop Counterpoise Antennas, *IEEE Trans. Antennas Propag.*, January 1970, Vol. AP-18, pp. 34–41.

13. Balanis, C. A. Radiation Characteristics of Current Elements near a Finite-Length Cylinder, *IEEE Trans. Antennas Propag.*, May 1970, Vol. AP-18, pp. 352–359.

14. Balanis, C. A. Analysis of an Array of Line Sources above a Finite Groundplate, *IEEE Trans. Antennas Propag.*, March 1971, Vol. AP-19, pp. 181–185.

15. Balanis, C. A., and C. R. Cockrell Analysis and Design of Antennas for Air Traffic Collision Avoidance Systems, *IEEE Trans. Aerosp. Electron. Syst.*, September 1971, Vol. AES-7, pp. 960–967.

16. Balanis, C. A. Radiation from Conical Surfaces Used for High-Speed Aircraft, *Radio Sci.*, February 1972, Vol. 7, pp. 339–343.

17. Croswell, W. F., et al. The Admittance of a Rectangular Waveguide Radiating into a Dielectric Slab, *IEEE Trans. Antennas Propag.*, September 1967, Vol. AP-15, pp. 627–633.

18. Bailey, M. C., et al. Electromagnetic Properties of a Circular Aperture in a Dielectric Covered or Uncovered Groundplane, *NASA Langley Tech. Note* D-7452, October 1968.

19. Bailey, M. C., and C. T. Swift Input Admittance of a Circular Waveguide Aperture Covered by a Dielectric Slab, *IEEE Trans. Antennas Propag.*, July 1968, Vol. AP-16, pp. 386–391.

20. Balanis, C. A. Pattern Distortion Due to Edge Diffractions, *IEEE Trans. Antennas Propag.*, July 1970, Vol. AP-18, pp. 561–563.

21. Southworth, G. C., and A. P. King Metal Horns as Directive Receivers of Ultrashort Waves, *Proc. IRE*, 1939, Vol. 27, pp. 95–102.

22. Barrow, W. L., and L. J. Chu Theory of the Electromagnetic Horn. *Proc. IRE*, January 1939, Vol. 27, pp. 51–64.

23. Chu, L. J., and W. L. Barrow Electromagnetic Horn Design, *Trans. AIEE*, July 1939, Vol. 58, pp. 333–338.

24. Barrow, W. L., and F. D. Lewis The Sectoral Electromagnetic Horn, *Proc. IRE*, January 1939, Vol. 27, pp. 41–50.

25. Chu, L. J. Calculation of the Radiation Properties of Hllow Pipes and Horns, *J. App. Phys.*, 1940, Vol. 11, pp. 603–610.

26. Schelkunoff, S. A. "Electromagnetic Waves," Van Nostrand, 1943.

27. Rhodes, D. R. An Experimental Investigation of the Radiation Properties of Electromagnetic Horn Antennas, *Proc. IRE*, September 1948, Vol. 36, pp. 1101–1105.

28. Schelkunoff, S. A., and H. T. Friis "Antennas: Theory and Practice," Wiley, 1952.

29. King, A. P. The Radiation Characteristics of Conical Horn Antennas, *Proc. IRE*, March 1952, Vol. 38, pp. 249–251.

30. Schorr, M. G., and J. J. Beck Electromagnetic Field of the Conical Horn, *J. App. Phys.*, August 1950, Vol. 21, pp. 795–801.

31. Compton, R. T., Jr., and R. E. Collin Open Waveguides and Small Horns, in R. E. Collin and F. J. Zucker "Antenna Theory," Pt. I, pp. 621–655, McGraw-Hill, 1969.

32. Russo, P. M., R. C. Rudduck, and L. Peters A Method for Computing *E*-Plane Patterns of Horn Antennas, *IEEE Trans. Antennas Propag.*, March 1965, Vol. AP-13, pp. 219–224.

33. Yu, J. S., R. C. Rudduck, and L. Peters Comprehensive Analysis for *E*-Plane of Horn Antennas by Edge Diffraction Theory, *IEEE Trans. Antennas Propag.*, March 1966, Vol. AP-14, pp. 138–149.

34. Yu, J. S., and R. C. Rudduck *H*-Plane Pattern of a Pyramidal Horn, *IEEE Trans. Antennas Propag.*, September 1969, Vol. AP-17, pp. 651–652.

35. Thomas, D. T. A Half Blinder for Reducing Certain Side-Lobes in Large Horn, Reflector Antennas, *IEEE Trans. Antennas Propag.*, November 1971, Vol. AP-19, pp. 774–776.

36. Jull, E. V. Reflection from the Aperture of a Long *E*-Plane Sectoral Horn, *IEEE Trans. Antennas Propag.*, January 1972, Vol. AP-20, pp. 62–68.

37. Wrixom, G. T., and W. J. Welch Gain Measurements of Standard Electromagnetic Horns in the K and K_a Bands, *IEEE Trans. Antennas Propag.*, March 1972, Vol. AP-20, pp. 136–142.

38. Jakes, W. C. Horn Antennas, in H. Jasik (ed.), "Antenna Engineering Handbook," pp. 10-1 to 10–18, McGraw-Hill, 1961.

39. Mac A. Thomas, B. Design of Corrugated Conical Horns, *IEEE Trans. Antennas Propag.*, Vol. AP-26, No. 2, March 1978, pp. 698–703.

40. Simmons, A. J. and A. F. Kay "The Scalarfeed—A High-Performance Feed for Large Paraboloid Reflectors," Design and Construction of Large Steerable Aerials, *IEEE Conf. Pub. 21*, 1966, pp. 762–773.

41. Silver, S. "Microwave Antenna Theory and Design," M.I.T. Radiation Laboratory Series, Vol. 12, McGraw-Hill, 1949.

42. Gray, C. Larry Estimating the Effect of Feed Support Member Blocking on Antenna Gain and Sidelobe Level, *Microwave J.*, March 1964, pp. 88–91.

43. Ruze, J. Lateral-Feed Displacement in a Paraboloid, *IEEE Trans. Antennas Propag.*, September 1965, Vol. AP-16, pp. 660–665.

44. Pagones, M. J. Gain Factor of an Offset-Fed Paraboloidal Reflector, *IEEE Trans. Antennas Propag.*, September 1965, Vol. AP-16, pp. 536–541.

45. Love, A. W. Spherical Reflecting Antennas with Corrected Line Sources, *IEEE Trans. Antennas Propag.*, September 1962, Vol. AP-13, pp. 529–537.

46. Schell, A. C. The Diffraction Theory of Large-Aperture Spherical Reflector Antennas, *IEEE Trans. Antennas Propag.*, July 1963, Vol. AP-14, pp. 428–532.

47. Potter, P. D. The Aperture Efficiency of Large Paraboloidal Antennas as a Function of Their Feed System Radiation Characteristics, *Jet Prop. Lab. Tech. Rep.* 32–149, September 25, 1961.

48. Rusch, W. V. T. Phase Error and Associated Cross-Polarization Effects in Cassegrainian-Fed Microwave Antennas, *Jet Prop. Lab. Tech. Rep.* 32-612, May 30, 1962.

49. Potter, P. D. The Application of the Cassegrainian Principle to Ground Antennas for Space Communications, *Jet Prop. Lab. Tech. Rep.* 32-295, June 1962.

50. Potter, P. D. A Simple Beamshaping Device for Cassegrainian Antennas, *Jet Prop. Lab. Tech. Rep.* 32-214, Jan. 31, 1962.

51. Potter, P. D. A Computer Program for Machine Design of Cassegrain Feed Systems, *Jet Prop. Lab. Tech. Rep.* 32-1202, December 15, 1967.

52. Rusch, W. V. T. Edge Diffraction from Truncated Paraboloids and Hyperboloids, *Jet Prop. Lab. Tech. Rep.* 32-113, June 1, 1967.

53. Ludwig, A., and W. T. T. Rusch Digital Computer Analysis and Design of a Subreflector of Complex Shape, *Jet Prop. Lab. Tech. Rep.* 32-1190, November 15, 1967.

54. Williams, W. F. High Efficiency Antenna Reflector, *Microwave J.*, July 1965, pp. 79–82.

55. Potter, P. D. Application of Spherical Wave Theory to Cassegrain-Fed Paraboloids, *IEEE Trans. Antennas Propag.*, November 1967, Vol. AP-15, pp. 727–736.

56. Space Program Summary, *Jet Prop. Lab. Tech. Rep.* 37-50, January 1, 1968-March 31, 1968.

57. The Deep Space Network, *Jet Prop. Lab. Space Programs Summ.* 37–52, Vol. II, July 31, 1968, pp. 78–105.

58. Meeks, M. L., and J. Ruze Evaluation of the Haystack Antenna and Radome, *IEEE Trans. Antennas Propag.*, November 1971, Vol. AP-19, pp. 723–728.

59. Ingerson, P. G., and W. C. Wong The Analysis of Deployable Umbrella Parabolic Reflectors, *IEEE Trans. Antennas Propag.*, July 1972, Vol. AP-20, pp. 409–415.

60. Ludwig, A. C. Conical-Reflector Antennas, *IEEE Trans. Antennas Propag.*, November 1972, Vol. AP-20, pp. 146–152.

61. Knop, C. M., "Microwave Relay Antenna, Chapter 31, Antenna Engineering Handbook," McGraw-Hill, 1993.

62. Rumsey, V. H. The Equiangular Spiral, *IRE Nat. Conv. Rec.*, 1957, Pt. I, pp. 114–118.

63. Dyson, J. D. The Equiangular Spiral, *IRE Trans. Antennas Propag.*, April 1959, Vol. AP-7, pp. 181–187.

64. Springer, P. S. End-Loaded and Expanding Helices as Broadband Circularly Polarized Radiators, *Proc. Natl. Electron. Conf.*, 1949, Vol. 5, pp. 161–171.

65. Chatterjee, J. S. Radiation Characteristics of a Conical Helix of Low Pitch Angles, *J. Appl. Phys.*, March 1955, Vol. 26, pp. 331–335.

66. Duhamel, R. H., and R. E. Isbell Broadband Logarithmically Periodic Antenna Structures, *IRE Natl. Conv. Rec.*, 1957, Pt. 1, pp. 119–128.

67. Dyson, J. D. The Unidirectional Equiangular Spiral Antenna, *IRE Trans. Antennas Propag.*, October 1959, Vol. AP-7, pp. 329–334.

68. Mayes, P. E., and R. L. Carrel Log-Periodic Resonant V-Arrays, *IRE West. Conv.*, 1961.

69. Isbell, D. E. Log-Periodic Dipole Arrays, *IRE Trans. Antennas Propag.*, May 1960, Vol. AP-8, pp. 260–267.

70. Carrel, R. L. The Design of Logarithmically Periodic Dipole Antennas, *IRE Natl. Conv. Rec.*, 1961, Pt. 1, pp. 61–75.

71. Barbano, N. Log-Periodic Yagi-Uda Array, *IEEE Trans. Antennas Propag.* March 1966, Vol. AP-14, pp. 235–238.

72. Mailloux, R. J. The Long Yagi-Uda Array, *IEEE Trans. Antennas Propag.* March 1966, Vol. AP-14, p. 128.

73. Collin, R. E. "Field Theory of Guided Waves," McGraw-Hil, 1960.

74. Brunstein, S. A., and R. F. Thomas Characteristics of a Cigar Antenna, *Jet Prop. Lab. Q. Rev.*, July 1972, Vol. 1, No. 2, pp. 87–95.

75. Zucker, F. J. Surface Wave Antennas, in R. E. Collin and F. J. Zucker (eds.), "Antenna Theory," Pt. II, Chap. 21, pp. 298-348, McGraw-Hill, 1969.

76. Zucker, F. J. Surface and Leaky-Wave Antennas, in H. Jasik (ed.), "Antenna Engineering Handbook," Chap. 16, pp. 16–1 to 16–57, McGraw-Hill, 1961.

77. Kraus, J. D. "Antennas," McGraw-Hill, 1950.

78. Angelakos, D. J., and Kajfez Darko Modifications on the Axial Mode Helical Antenna, *Proc. IEEE*, April 1967, Vol. 55, No. 4, pp. 558–559.

79. Carver, K. R., and B. M. Potts Some Characteristics of the Helicone Antenna, *1970 IEEE G-AP Symp. Dig.*, pp. 142–150.

80. Croswell, W. F., and M. C. Gilreath Erectable Yagi Disk Antennas for Space Vehicle Applications, *NASA Langley Tech. Note* D-1401, October 1962.

81. Lo, Y. T., D. Solomon, and W. F. Richards Theory and Experiment on Microstrip Antennas, IEEE *Trans. Antennas Propag.*, March 1979, Vol. AP-27, pp. 137–145.

82. Bailey, M. C. Resonant Frequency of Microstrip Antennas Calculated from TE-Excitation of an Infinite Strip Embedded in a Grounded Dielectric Slab, *NASA Langley Tech. Mem.* 80190, November 1979.

83. Derneryd, A. G. Linearly Polarized Micorstrip Antennas, *IEEE Trans. Antennas Propag.*, November 1976, Vol. AP-24, pp. 846–851.

84. Agrawal, P. K., and M. C. Bailey An Analysis Technique for Micorstrip Antennas, *IEEE Trans. Antennas Propag.*, November 1977, Vol. AP-25, pp. 756–759.

85. Carver, K. R. A Modal Expansion Theory for the Microstrip Antenna, *IEEE AP-S Symp. Dig.*, 1979, pp. 101–104.

86. Richards, W. F., and Y. T. Lo Am Improved Theory for Microstrip Antennas and Applications, *IEEE AP-S Symp. Dig.*, 1979, pp. 113–116.

87. Bailey, M. C., and F. G. Parks Design of Microstrip Disk Antenna Arrays, *NASA Langley Tech. Mem.* 78631. February 1978.

88. Hammer, P., D. Van Bauchaute, D. Verschraeven, and S. Van de Capelle A Model for Calculating the Radiation Field of Microstrip Antennas, *IEEE Trans. Antennas Propag.*, March 1979, Vol. AP-27, pp. 167–270.

89. Milligan, T. A., Modern Antenna Design, McGraw-Hill, 1985.

CHAPTER 16.3
FUNDAMENTALS OF WAVE PROPAGATION

Richard C. Kirby, Kevin A. Hughes

INTRODUCTION: MECHANISMS, MEDIA, AND FREQUENCY BANDS

This chapter deals with radio waves propagated through or along the surface of the earth, through the atmosphere, and by reflection or scattering from the ionosphere or troposphere. The particular propagation mechanism around which a given application is designed depends on the distance to be spanned, the type of information to be transmitted or the service to be provided, and the reliability required. Other propagation mechanisms can affect the performance of the system or lead to interference with and from other systems, depending on frequency and distance.

Over a line-of-sight path within the nonionized atmosphere, transmission is much as through free space, though atmospheric refraction causes bending, reflection, scattering, and possibly fading. At frequencies above about 10 GHz, there may be attenuation because of rainfall and absorption by air and water vapor. The conductivity and permittivity (dielectric constant) of the earth are markedly different from those of the atmosphere. A wave mainly diffracted along the surface of the ground encounters increasing loss with increasing frequency. Very low frequency waves are propagated with little attenuated over thousands of kilometers. At high frequencies, losses along the ground become so great that the usefulness of the ground wave is limited to short distances. At medium and high frequencies, ionospheric reflections permit radio communication to great distances. At frequencies much above 30 MHz, ionospheric reflections are not dependable, and most communications depend on line-of-sight propagation or tropospheric scattering beyond the horizon.

Because of the dependence of propagation characteristics on frequency, much of the discussion of these sections will be in terms of frequency bands, and abbreviations such as VLF for very low frequencies and VHF for very high frequencies will be used.

The International Telecommunication Union (ITU) has defined nine frequency bands designated by integer band numbers; for example, 1 MHz is the approximate midband of band 6, and so on, as listed in Table 16.3.1.

Certain frequency bands, however, are sometimes unofficially designated by letter rather than by the abbreviations given in Table 16.3.1. Since there is no standard correspondence between the letters and the frequency bands concerned, it is advisable that their use be clarified by reference to the approximate limits of the band or to a frequency within the band. For information, letter designations used essentially in the areas of radar and space communications, are given in Table 16.3.2.

Propagation characteristics are discussed according to somewhat different bands, such as 10 to 150 kHz and 150 to 1500 kHz, corresponding to bands of relatively homogeneous propagation characteristics.

References 1 to 6 are basic texts on electromagnetic wave propagation. References 7 to 22 are comprehensive texts on tropospheric and ionospheric radiowave propagation which, in some cases, adopt a systems approach in describing the particular propagation mechanisms of concern. Recent results of theoretical and experimental radio science worldwide are outlined in Ref. 23.

TABLE 16.3.1 Frequency Bands Defined by ITU

ITU band number[*]	Frequency range (lower limit exclusive, upper limit inclusive)	Corresponding metric subdivision	Abbreviation
4	3–30 kHz	Myriametric waves	VLF
5	30–300 kHz	Kilometric waves	LF
6	300–3000 kHz	Hectometric waves	MF
7	3–30 MHz	Decametric waves	HF
8	30–300 MHz	Metric waves	VHF
9	300–3000 MHz	Decimetric waves	UHF
10	3–30 GHz	Centimetric waves	SHF
11	30–300 GHz	Millimetric waves	EHF
12	300–3000 GHz (3 THz)	Decimillimetric waves	

[*]Band number N extends from 0.3×10^N to 3×10^N Hz.

The emphasis here is mainly descriptive, key formulas indicating the behavior of parameters and references to significant publications providing material for engineering calculations. Maxwell's uniform plane-wave equations are cited only to show the role of the electrical constants and the vector relationships of electric and magnetic field and power flux.

Wave Propagation in Homogeneous Media

Electromagnetic radiation is composed of two mutually dependent vector fields, electric and magnetic. The electric field is characterized by the vectors **E**, electric field strength in volts per meter, and **D**, dielectric displacement in coulombs per square meter. The magnetic field is characterized by **H**, the magnetic field strength in ampere-turns per meter (or amperes per meter) and **B**, flux density, in webers per square meter. The vector current density **J** is in amperes per square meter.

The relationship between the members of the various pairs of field vectors is characterized by the constitutive parameters, or *electrical constants*, of the medium:

$$\epsilon = \text{permittivity (dielectric constant), F/m}$$
$$\sigma = \text{conductivity, S/m} \quad \mu = \text{permeability, H/m}$$

(1)

In some cases these constants are functions of the coordinate. Locally, however, they are always considered to be constant. Nearly always the time factor in these sections is $\exp(+i\omega t)$, where ω is the angular frequency $2\pi f$ (f in hertz) and t the time. The *electric field*, then, is the real part of $E \exp i\omega t$.

TABLE 16.3.2 Unofficial Letter Designations for Certain Bands (Recomm. ITU-R V.431)

Letter symbols	Radar (GHz)		Space radiocommunications	
	Spectrum regions	Examples	Nominal designations	Examples (GHz)
L	1–2	1.215–1.4	1.5 GHz band	1.525–1.710
		2.3–2.5	2.5 GHz band	
S	2–4	2.7–3.4		
C	4–8	5.25–5.85	4/6 GHz band	3.7–4.2
				5.925–6.425
X	8–12	8.5–10.5	—	
		13.4–14.0	11/14 GHz band	10.7–13.25
Ku	12–18	15.3–17.3	12/14 GHz band	14.0–14.5
K	18–27	24.05–24.25	20 GHz band	
Ka	27–40	33.4–36.0	30 GHz band	20–30 (approx.)

To explain very basic notation,* a short outline of plane electromagnetic waves in a homogeneous medium is given.

Ohm's law in the complex form is

$$\mathbf{J} = (\sigma + i\epsilon\omega)\mathbf{E} \tag{2}$$

where $\mathbf{J}$ is the current density vector and $\mathbf{E}$ is the electric field vector.

The analogous relation for magnetic quantities is

$$\mathbf{B} = \mu\mathbf{H} \tag{3}$$

In source-free media the above vector quantities are related by

$$\text{curl } \mathbf{E} = -i\mu\omega\mathbf{H} \tag{4}$$

and

$$\text{curl } \mathbf{H} = (\sigma + i\epsilon\omega\mathbf{E} \tag{5}$$

These are Maxwell's equations.

For a homogenous medium

$$\text{curl curl } \mathbf{E} = \text{grad div } \mathbf{E} - \text{div grad } \mathbf{E} = -i\mu\omega(\sigma + i\epsilon\omega)\mathbf{E} \tag{6}$$

Since div $\mathbf{E} = 0$,

$$(\nabla^2 - \gamma^2)\mathbf{E} = 0 \tag{7}$$

where ∇^2 = div grad = Laplacian operator (which operates on the rectangular components of $\mathbf{E}$) and $\gamma^2 = i\mu\omega$ $(\sigma + i\epsilon\omega)$. The quantity γ is called the *propagation constant*.

As a simple illustration of the role of the electrical constants of the medium, the field of a wave is assumed to vary only in the z direction in space (time factor exp $i\omega t$ understood), and the electric field is assumed to have only an x component E_x. For this case Eq. (7) reduces to

$$\left(\frac{d^2}{dz^2} - \gamma^2\right)E_x = 0 \tag{8}$$

and the solutions are exp $(+\gamma z)$ and exp $(-\gamma z)$ or, in general,

$$E_x = A \exp \gamma z + B \exp (-\gamma z) \tag{9}$$

where A and B are constants. The magnetic field then has only a y component given by

$$H_y = -\frac{1}{i\mu\omega}\frac{\partial E_x}{\partial z} = -\eta^{-1}[A\exp(+\gamma z) - B\exp(-\gamma z)] \tag{10}$$

where

$$\eta = \left(\frac{i\mu\omega}{\sigma + i\epsilon\omega}\right)^{1/2} \tag{11}$$

η is defined as the *characteristic impedance* of the medium for plane-wave propagation. Remembering that the time factor is exp $i\omega t$, we see that the term B exp $(-\gamma z)$ is a wave traveling in the positive z direction with diminishing amplitude and the term A exp γz is a wave traveling in the negative z direction with diminishing amplitude.*[†] The electric and magnetic fields are both transverse to the direction of propagation and orthogonal to each other.

---

*From Wait[3] by permission.
[†]The geometry of subsequent paragraphs uses a different convention for x, y, and z directions, shown in figures.

Such radiation is termed *plane-polarized*. It is by convention designated as *horizontal* or *vertical* according to the orientation of the plane containing the **E** vector.

The quantity η is equal to the complex ratio of the electric and magnetic field components in the x and y directions, respectively, for plane waves in an unbounded homogeneous medium, i.e.,

$$H_y = (i\omega\epsilon/r)E_x = E_x/\eta, \qquad H_x = (-r/i\omega\mu)E_y = -E_y/\eta \tag{12}$$

For a perfect dielectric ($\sigma = 0$)

$$\eta = \sqrt{\mu/\epsilon} \quad \text{and} \quad r = ik \tag{13}$$

where $k = \omega\sqrt{\mu\epsilon} = 2\pi/\lambda$ and λ is the wavelength.

The velocity of this wave is

$$v = 1/\sqrt{\mu\epsilon} \tag{14}$$

v is called the *phase velocity* of the wave. It represents the velocity of propagation of phase and does not necessarily coincide with the velocity with which the energy of a wave of signal is propagated, known as the group velocity. In fact, v may exceed free-space wave velocity without violating relativity in any way.

The wavelength is defined as the distance the wave propagates in one period.

$$\lambda = \frac{2\pi}{\omega\sqrt{\mu\epsilon}} = \frac{v(\text{m/s})}{f(\text{Hz})} \ \text{m} \tag{15}$$

For free space

$$\epsilon = \epsilon_0 = 8.854 \times 10^{-12}\,\text{F/m}, \qquad \mu = \mu_0 = 4\pi \times 10^{-7}\,\text{H/m}, \qquad \sigma = 0 \tag{16}$$

Then

$$\gamma = ik = i\omega\sqrt{\mu_0\epsilon_0} = i2\pi/\lambda$$

The *velocity of the wave for free space* is

$$v_0 = 1/\sqrt{\mu_0\epsilon_0} \approx 3 \times 10^8 \ \text{m/s} \tag{17}$$

The *characteristic impedance of free space* is

$$\eta_0 = \sqrt{\mu_0/\epsilon_0} \approx 120\pi \ \Omega \approx 4\pi v_0 \times 10^{-7}\,\Omega \tag{18}$$

Energy flow in the electromagnetic field is described by the Poynting vector

$$\mathbf{P} = \mathbf{E} \times \mathbf{H}^* \tag{19}$$

where the complex representation of the time-periodic quantities is implied and the asterisk denotes the complex conjugate. The *real part* of the Poynting vector represents the average power flow over a cycle of the time variation per unit area in the direction of transmission.

$$\mathbf{P}_{\text{av}} = \tfrac{1}{2}\,\text{Re}\,(\mathbf{E} \times \mathbf{H}^*) = E^2/2\eta_0 \ \ \text{W/m}^2 \tag{20}$$

$\mathbf{P}_{\text{av}}$ is called the *power flux density* or *field intensity*. Note that **E** and **H** are peak values.

The permeability and permittivity of any medium relative to free space are called the *relative permeability* and *relative permittivity*. These are usually the values given in tables of physical constants; they are dimensionless and designated by μ_r and ϵ_r, respectively.

Polarization of the Wave.   *Polarization* is a term characterizing the orientation of the field vector in its travel. In radio, polarization usually refers to the electric vector. In the simplest case E_z and H_z (field components in the direction of propagation) are zero, and **E** and **H** lie in a plane transverse to the direction of propagation and orthogonal to each other. Such a plane wave is *elliptically polarized* when the electric vector **E** describes an ellipse in the plane perpendicular to the direction of propagation over one cycle of the wave.

When the amplitudes of the rectangular components are equal and their phases differ by some odd integral multiple of $\pi/2$, the polarization ellipse becomes a circle and the wave is *circularly polarized*. It is customary to describe as *right-handed* circularly polarized a clockwise rotation of **E** when viewed in the direction of propagation; counterclockwise rotation is *left-handed* polarization.[*]

An important case for many radio problems is that in which the polarization is a straight line. The wave is then *linearly polarized*. In *horizontal polarization* the electric vector lies in a plane parallel to the earth's surface.

To obtain maximum transfer of power between two antennas the polarization should match. If the transmitting antenna is horizontally polarized, the receiving antenna must likewise be horizontally polarized. If the transmitting antenna is elliptically polarized with a given degree of ellipticity and a specified direction of rotation, the receiving antenna should have the proper direction of rotation and degree of ellipticity in order to maximize the path antenna gain.

It should be noted that in the process of propagation, except through free space, the polarization may be altered. This can be caused by reflections from surfaces and, for frequencies at SHF and above, by hydrometers in the neutral atmosphere such as rain. Passage through the ionosphere in the presence of magnetic field is likely to impart elliptical polarization to a plane-polarized incident wave and rotation of the major axis. For MF or HF ionospheric propagation, the downcoming wave may be randomly polarized.

Reflection

Most problems in wave propagation involve reflection from a boundary between media of different refractive properties, often between air and the ground or between air and the ionosphere. In general, such a boundary may involve dissipative media (finite conductivity), curvature, finite dimensions, roughness, and stratification.

The *complex index of refraction* for a conducting medium is

$$n^2 = [i\omega\mu(\sigma + i\omega\epsilon)]/[i\omega\mu_0(i\omega\epsilon_0)] \qquad (21)$$

When $\sigma = 0$,

$$n = \sqrt{\mu_r\epsilon_r} \quad \text{where } \mu_r = \mu/\mu_0, \epsilon_r = \epsilon/\epsilon_0$$

For many applications $\mu = 1$ and $\sigma = 0$, and the index of refraction is simply the square root of ϵ_r.

Figure 16.3.1 illustrates Snell's law for refraction of plane waves at an infinite plane interface. The angle ϕ between the direction of propagation and the normal to the boundary is called the *angle of incidence*. The angle ψ between the direction of propagation and the boundary, called the *grazing angle*

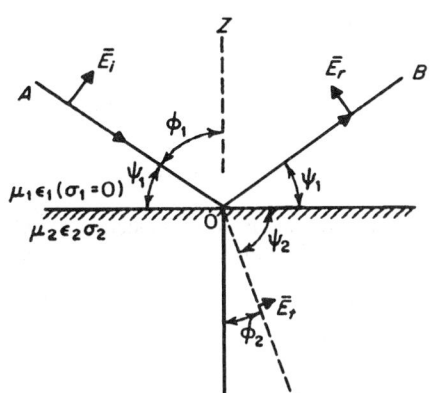

FIGURE 16.3.1   Geometry of reflection and transmission.

or *Elevation angle*, is often more convenient. If the medium containing the incident wave is lossy, the angle of incidence is complex and can be defined in various ways.[3] At the boundary, the tangential components of **E** and **H** must be continuous; the phase of the reflected wave is in step with the phase of the incident wave to satisfy this requirement.

Snell's law of refraction for the direction of the transmitted wave toward C is

$$n_1 \sin \phi_1 = n_2 \sin \phi_2 \quad \text{or} \quad n_1 \cos \psi_1 = n_2 \cos \psi_2 \tag{22}$$

The *penetration depth* δ, or the depth at which the transmitted wave E_t has attenuated to $1/e$ of its incident value (for a conducting medium where $\sigma \gg \omega\epsilon$), is

$$\delta = \frac{1}{\sqrt{\omega\mu\sigma/2}} \quad \text{m} \tag{23}$$

Ground Reflection, Reflection Coefficients, Fresnel Zones

A wave incident on a plane surface can be resolved into two components, one polarized normal and the other parallel to the plane of incidence. The reflection coefficients for the two components differ, and consequently the polarization of the reflected wave depends on the angle of incidence. Consider an air-earth boundary, taking the media to be nonmagnetic; for the case where the **H** vector is parallel to the ground surface the complex reflection coefficient[1] is

$$\mathbf{R}_v = \frac{(\epsilon_r - i60\sigma\lambda)\sin\psi - (\epsilon_r - \cos^2\psi - i60\sigma\lambda)^{1/2}}{(\epsilon_r - i60\sigma\lambda)\sin\psi + (\epsilon_r - \cos^2\psi - i60\sigma\lambda)^{1/2}} \tag{24}$$

where σ = conductivity (S/m)
 ψ = grazing angle (Fig. 16.3.1)
 ϵ_r = relative permittivity of earth to air or free space
 $\epsilon_r - i60\sigma\lambda$ is referred to as the *complex permittivity*.

If E is parallel to the ground surface and H is in the plane of incidence,

$$\mathbf{R}_h = \frac{\sin\psi - (\epsilon_r - \cos^2\psi - i60\sigma\lambda)^{1/2}}{\sin\psi + (\epsilon_r - \cos^2\psi - i60\sigma\lambda)^{1/2}} \tag{25}$$

These are reflection coefficients for vertical and horizontal polarization, respectively. Curves of values for a range of ϵ and σ are given in Ref. 1.

An important property for vertical polarization is that there exists an angle of incidence for which the reflection coefficient approaches zero (for purely dielectric media it equals zero). This is the *Brewster angle*, also called the *polarizing angle*, given by

$$\phi_0 = \tan^{-1}\sqrt{\epsilon_1/\epsilon_2} \tag{26}$$

It is equal to the angle of incidence for which the reflected and refracted (transmitted) rays are at right angles. If the incidence occurs at the Brewster angle, the reflected wave is polarized entirely in the direction normal to the plane of incidence.

Wave tilt is a property frequently used to determine the electrical constants of the earth. For waves traveling at nearly grazing incidence along the surface of the earth, wave tilt may be interpreted geometrically as the angle between the normal to the wavefront and the tangent to the earth's surface. Wave tilt is defined to be the ratio of the horizontal to the vertical component of the electric field in the air just above the ground:

$$W = \mathbf{E}_h/\mathbf{E}_v \tag{27}$$

Wave tilt is related to the electrical constants of a homogeneous earth by

$$W = \frac{\sqrt{\mu_1/\epsilon_{1c}}}{\sqrt{\mu/\epsilon_0}} \sqrt{1 - \frac{\mu_0\epsilon_0}{\mu_1\epsilon_{1c}}} \tag{28}$$

The subscript 1 refers to earth constants and 0 refers to free space; ϵ_{1c} = complex permittivity = $\epsilon_1 - i\sigma_1/\omega$.

This procedure assumes that $\mu_0 = \mu_1$, generally a valid assumption; if it is not, μ_1 must be determined by some other procedure.

An important consideration in many propagation problems is the interference pattern generated by vector addition of the fields corresponding to the direct ray from an antenna to a point within line of sight plus the ground-reflected ray. In fact, the ground acts as a partial reflector and as a partial absorber, and the resultant field strength at a receiving point is given by

$$E_R = E_1[1 + Re^{i\Delta} + (1 - R)Ae^{i\Delta} + \cdots] \tag{29}$$

$$\underbrace{\qquad}_{a} \quad \underbrace{\qquad}_{b} \quad \underbrace{\qquad}_{c}$$

Term a corresponds to the direct wave while term b represents the reflected wave with R as the reflection coefficient and Δ the phase difference (in radians) corresponding to the path difference between the direct and reflected rays.

Term c corresponds to the surface wave, with the attenuation factor A a function of frequency, polarization, and electrical constants of the ground. The surface wave is particularly important at medium and low frequencies.

Using the notation in Fig. 16.3.2a the phase difference Δ may be written as

$$\Delta = 4\pi h_1 h_2/\lambda d \quad \text{rad} \tag{30}$$

for a wavelength λ and for grazing angles less than about 0.5 rad, Eq. (29) may be written:

$$E_R = E_1 \frac{4\pi h_1 h_2}{\lambda d} \tag{31}$$

omitting any contribution from the surface wave.

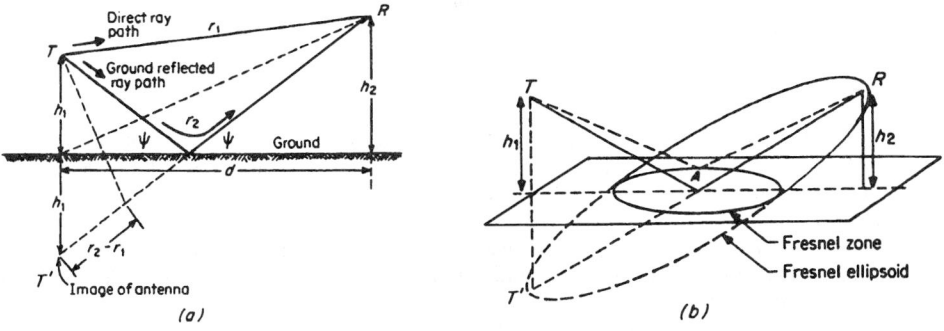

FIGURE 16.3.2 Geometry of ground reflection, image antennas, and Fresnel zones for plane earth: (a) ray paths; (b) Fresnel zone and ellipsoid.

The received field strength at a given distance will therefore display maxima and minima around the free-space value (corresponding to the direct wave alone) as the height of either antenna is altered. The angle at which maxima and minima occur are given by

$$\sin\psi = n\lambda/4h_1 \quad \text{maxima for } n \text{ odd, minima for } n \text{ even} \tag{32}$$

For the first maximum to occur at a specified elevation ψ_1,

$$h_1 = \lambda/(4\ \sin\psi_1) \tag{33}$$

In Fig. 16.3.2*a* the ray reflections are shown as though they occurred at a point. Actually, the surface of the earth is illuminated over a wide region corresponding to the radiation patterns of the two antennas and, in accordance with Huygens' principle, reradiates elementary wavelets in all directions. In any particular direction, as toward R, these elementary wavelets arrive with strength and phase such that the waves from an elliptical zone in the neighborhood of the ray reflection add nearly in phase. From successive ring areas, similarly bounded by larger ellipses, the waves alternately cancel and add.

These zones of physical reflection are called *Fresnel zones*, since they are closely related to the Fresnel zones of diffraction theory.[5,10] Most of the energy can be thought of as being reflected from the first Fresnel zone; it is defined, with the aid of Fig. 16.3.2*b*, for reflection paths between points such as T and R, as the area from which all the reradiated elementary wavelets arrive, according to geometric optics, within half wavelength of the phase of the direct ray. Thus the length of the geometric ray path at the edge of the nth Fresnel zone is n half wavelengths greater than the geometric ray path.

From Fig. 16.3.2*b*, the boundary of the first Fresnel zone is defined as the locus of points A such that $TA + AR$ differs by half a wavelength from TR. This locus is an ellipsoid of revolution with foci at T' and R. The minor axis of the first Fresnel zone has the same dimension as the diameter of the first Fresnel ellipsoid, which can be easily shown to be $\sqrt{\lambda d}$, where d is the length of the propagation path. The length of the major axis will depend on the antenna heights h_1 and h_2 and is given by

$$L = \frac{d\sqrt{1 + 4h_1 h_2/\lambda d}}{1 + (h_1 + h_2)^2/\lambda d} \tag{34}$$

For a well-developed ground reflection, the ground should be flat over an area that includes at least the first Fresnel zone. The degree of flatness depends on the wavelength and angle of incidence; assuming that phase-path changes less than $\lambda/16$ are unimportant, Rayleigh's criterion limits height deviations in terrain from a smooth surface to a magnitude less than $\Delta h = \lambda/(16 \sin\psi)$ over the area of the first Fresnel zone for waves incident at angle ψ. Methods allowing for surface roughness, finite conductivity, and divergence owing to the spherical shape of the earth are outlined on the accompanying CD-ROM under "Propagation Over the Earth Through the Nonionized Atmosphere."

The general topic of reflection from the surface of the earth and a description of the influence-reflected signals have on the performance of telecommunication systems is given in Refs. 7 and 8.

Diffraction and Scattering

The spherical shape of the earth and irregularities of terrain give rise to *diffraction* as an important part of the ground wave and as a mechanism for propagation beyond the optical horizon. Diffraction is included in the methods for calculation given on the accompanying CD-ROM under "Propagation Over the Earth Through the Nonionized Atmosphere."

Scattering takes place from the rough surface of the earth and from small-scale irregularities in the index of refraction of the atmosphere or the ionosphere. It is analogous to scattering of light, although the radio problem is complicated by the wide range of relationships of radio wavelength to size of irregularity.

Scattering is discussed on the accompanying CD-ROM (Tropospheric forward scattering under "Propagation Over the Earth Through the Nonionized Atmosphere," and ionospheric scattering under "Propagation Via the Ionosphere").

Reciprocity

Reciprocity in wave propagation means that the source and receiver can be interchanged, with the transmission loss and phase unaffected by direction of propagation. In most radio-wave-propagation problems, with the notable exception of those involving an ionized medium with magnetic field, such reciprocity obtains. The refraction index of the ionosphere depends on magnetic field effects; the direction of propagation of the wave affects attenuation, phase, and bending, and the medium is called *anisotropic*. Thus, especially at very low and medium frequencies propagated via the ionosphere, reciprocity does not obtain. Reciprocity does not in any case imply the same signal-to-noise ratio in both directions. The noise environment may be very different at the transmitting and receiving locations.

Transmission Loss; Free-Space Attenuation; Field Strength; Power Flux Density

The concept of transmission loss for radio links consisting of a transmitting antenna, a receiving antenna, and the intervening propagation medium is addressed and defined in *Recomm. ITU-R P.341.*[69]

In general terms, the transmission loss on a radio link between a transmitter and a receiver is defined by the ratio between the power supplied by the transmitter and the power available at the receiver input. The transmission loss depends on several factors such as the losses in the antennas or in the transmission feed lines, the attenuation in the propagation medium, the losses because of faulty adjustment of the impedance or polarization, and so forth. As a consequence there are several definitions employed to characterize transmission loss and its components.

The *system loss* (L_s) on a radio link is the ratio, usually expressed in decibels, of the radio-frequency power input to the terminals of the transmitting antenna p_t, and the resultant radio-frequency signal power available at the terminals of the receiving antenna p_a. The available power is the maximum real power that a source can deliver to a load, i.e., the power that would be delivered to the load if the impedances were conjugately matched. The system loss may be expressed by[*]

$$L_s = 10 \log(p_t/p_a) = P_t - P_a \quad \text{dB} \tag{35}$$

and as defined here excludes losses in feeder lines but includes all losses in radio-frequency circuits associated with the antennas, such as ground losses, dielectric losses, antenna loading coil losses, and terminating resistor losses.

The *transmission loss* (L) on a radio line is the ratio, usually expressed in decibels, between the power radiated by the transmitting antenna and the power that would be available at the receiving antenna output if there were no loss in the radio-frequency circuits, assuming that the antenna radiation diagrams are retained. The transmission loss may be expressed by

$$L = L_s - L_{tc} - L_{rc} \quad \text{dB} \tag{36}$$

where L_{tc} and L_{rc} are the losses, expressed in decibels, in the transmitting and receiving antenna circuits, respectively, excluding the dissipation associated with antenna radiation, i.e., L_{tc} and L_{rc} may be expressed by $10 \log (r'/r)$, where r' is the resistive component of the antenna circuit and r is the radiation resistance.

The *basic transmission loss* (L_b) on a radio link is the transmission loss that would occur if the antennas were replaced by isotropic antennas with the same polarization as the real antennas, the propagation path being retained, but the effects of obstacles close to the antennas being disregarded. An isotropic antenna is one that radiates (or receives radio energy equally in (or from) all directions. In determining the basic transmission loss, any local features, such as the ground or nearby structures, which affect the power gain and directivity of the antenna, but which do not affect the overall propagation path, are assumed to be removed. The effect of the local ground is included in computing the antenna gain, but not in L_b. For instance, in the case of ionospheric propagation using an antenna near the ground which has a strong influence on the effective gain for the sky-wave path,

---

[*]Capital letters are used to denote quantities in decibels.

the ground is removed for the calculation of L_b so as to maintain the gain in the desired direction. In the case of a tropospheric propagation path involving diffraction over a distance obstacle, that obstacle is not removed in estimating L_b.

The *free-space basic transmission loss* (L_{bf}) is the transmission loss that would occur if the antennas were replaced by isotropic antennas located in a perfectly dielectric, homogeneous, isotropic, and unlimited environments, the distance between the antennas being retained. At a distance d very much greater than the wavelength λ, the power flux density (field intensity), expressed in watts per square meter, is simply $p_t'/4\pi d^2$ since the power p_t' is radiated uniformly in all directions. The effective absorbing area of the isotropic receiving antenna is $\lambda^2/4\pi$, and the available power at the terminals of the loss-free isotropic receiving antenna is given by

$$p_a' = \frac{\lambda^2}{4\pi} \frac{p_t'}{4\pi d^2} \tag{37}$$

Consequently, the free-space *basic transmission loss* can be expressed by

$$L_b = 20 \log \frac{4\pi d}{\lambda} \quad \text{dB} \tag{38}$$

d and λ are expressed in the same units.

A practical form of this equation is

$$L_b = 32.45 + 20 \log f_{\text{MHz}} + 20 \log d_{\text{km}} \quad \text{dB} \tag{39}$$

In many cases, it is important to know the *loss relative to free space* (L_m) on a radio link which is the difference between the basic transmission loss and the free-space basic transmission loss, expressed in decibels, and may be expressed by

$$L_m = L_b - L_{bf} \quad \text{dB} \tag{40}$$

The loss relative to free space may be divided into losses of different types, such as

- Absorption loss (ionospheric, atmospheric gases or precipitation)
- Diffraction loss as for ground waves
- Effective reflection or scattering loss, as for ionospheric propagation and including the results of any focusing or defocusing because of curvature of a reflection layer
- Polarization coupling loss, which can arise from any polarization mismatch between the antennas for the particular ray path considered
- Aperture to medium coupling loss or antenna gain degradation, which may be a result of the presence of substantial scatter phenomena on the path
- Effect of wave interference between the direct ray and rays reflected from the ground, other obstacles, or atmospheric layers

For broadcasting and mobile services, where characteristics and locations of receiving installations vary, it is convenient to calculate the electric *field strength* at some distance form the transmitter. The rms field strength e (V/m) of a plane wave of wavelength λ (m) is related to the power p_a' (W) available from an ideal loss-free isotropic receiving antenna by

$$e = (480 \, \pi^2 p_a'/\lambda^2)^{1/2} \tag{41}$$

It then follows from Eq. (37) that the field strength produced by a transmitter having an *equivalent isotopically radiated power* (EIRP) of p_t (W), at the distance d (m) sufficiently large for the wavefront to be considered plane, is given by

$$e = (30 p_t'/d^2)^{1/2} \tag{42}$$

or in more practical units,

$$e(\text{mV/m}) = 173[p_t'\ (\text{kW})]^{1/2}/d\ (\text{km}) \tag{43}$$

In decibels, the equivalent expression is

$$E\,[\text{dB}(\mu\text{V/m})] = 105 - 20\log d\ (\text{km}) + P_t' \tag{44}$$

where p_t is the EIRP in dB(kW).

In the above expressions, the EIRP includes the gain of the transmitting antenna with respect to an isotropic antenna. Often it is more convenient to express antenna gain with respect to a standard antenna other than an isotope, and an appropriate conversion must be made in order to determine the EIRP. For example, if the gain is given with respect to a half-wave dipole, the gain must be multiplied by 1.64 or, in decibels, be increased by 2.1 dB to obtain the EIRP. Alternatively, the following formula may be used

$$E \approx 7(p_t'')^{1/2}/d \tag{45}$$

where E is the field strength in V/m, d is the distance in m, and p_t'' is the effective radiated power (ERP), which includes the transmitter gain with respect to a half-wave dipole.

Similarly, the antenna gain may be expressed with respect to a short vertical monopole above a perfectly conducting ground plane. In this case, the gain must be multiplied by 3 or, in decibels, be increased by 4.8 dB to obtain the EIRP.

For microwave and satellite services, power flux density is a convenient term. The *power flux density* is given by the characteristic relations of a plane wave

$$p = e^2/\eta\quad \text{W/m}^2 \tag{46}$$

where η is the characteristic impedance of the medium in which the measurement is made ($\eta_0 = 120\pi$ in free space). In free space

$$p = e^2/120\pi \quad \text{and} \quad p = 4\pi p_a/\lambda^2 \tag{47}$$

where p_a is available power received by an isotropic antenna in the field.

The relations between field strength, power flux density, and available power in the receiving antenna are outlined below.

The absorbing area of a receiving antenna with gain g_r relative to an isotropic antenna can be written

$$a_e = \lambda^2 g_r r_f/4\pi r \tag{48}$$

where λ = wavelength in medium
   r = radiation resistance of antenna
   r_f = radiation resistance of antenna in free space

Combining the above two equations, we find the following formula for the available power p_a' from a lossless receiving antenna:

$$p_a' = e^2\lambda^2 g_r r_f/4\pi\eta r = v^2/4r \tag{49}$$

The v in this equation denotes the open-circuit voltage induced in the receiving antenna. The field strength is related to the open-circuit voltage by

$$v = e\sqrt{\lambda^2 g_r r_f/\pi\eta_0} = el \tag{50}$$

Field-strength meters usually are calibrated in terms of the effective length l of the antenna.

The relation between the available power p_a from the receiving antenna (neglecting losses) and the field strength e can be expressed in decibels as

$$E = 10 \log[4\pi\eta_0 p_a \times 10^{12})/\lambda^2 g_r] = P_a + 20 \log f - G_r + 107.22 \text{ dB } (\mu\text{V/m}) \tag{51}$$

where f is in megahertz.

For field strength E in decibels referred to 1 μV/m, referred to 1 kW radiated from a half-wave dipole over perfectly conducting earth, propagation loss is

$$L_p = 139.4 - G_r + 20 \log f - E \quad \text{dB} \tag{52}$$

If the reference radiation is a short electric dipole, the constant becomes 136.0.

Power Fading and Time-Variant Multipath Fading

Random variations appear in the signal received via various transmission media, especially at frequencies above about 100 kHz when propagation is by the troposphere or ionosphere. Such variation is usually of two types: one is *attenuation*, or *power fading*, which may be quite slow (minute to minute, hour to hour, and so forth) and is associated with comparatively large-scale changes in the medium, such as absorption; the other is *variable-multipath* or *phase-interference fading*.

Power fading is usually allowed for in the power margin designed into the system. Phase-interference fading, on the other hand, affects not only the amplitude but also the variable phase-vs.-frequency characteristic of the channel, limiting coherence bandwidth and introducing extraneous fluctuation in received-signal parameters. Alleviation of the effects of variable multipath is possible by diversity techniques, signal design, and signal receiving, processing, and detection operations.

The amplitude probability distribution of the fading envelope is usually determined from samples of duration much shorter than the shortest fade duration; observation intervals over which statistical averages are taken are about 1000 times the reciprocal of the nominal fading rate. The fit of experimental distributions of envelope fading to the Rayleigh distribution is often excellent for ionospheric and tropospheric scatter propagation, and similarly to a Nakagami-Rice[8] distribution for situations where a specular component is mixed with scattered components. (See also Recomm. ITU-R PN. P.1057).

It is often necessary, however, to consider the combination of long-term (power) fading, represented by a log-normal distribution, with short-term variations, represented by a Rayleigh distribution. The distribution of instantaneous values over a long period can be obtained from the Rayleigh law, whose mean is itself a random variable having a log-normal distribution.

A description of the fundamental properties of the most common probability distribution used in the statistical study of radio-wave propagation is given in Recomm. ITU-R PN. P.1057.

Most theoretical treatments of communication performance in the presence of variable multipath fading resulting from several signal components have been carried out for channels characterized by Rayleigh envelope distribution. The probability density function is given by

$$p(V) = (2V/v^2)\exp[-(V/v)^2] \tag{53}$$

where V is the fluctuating envelope and v^2 is the mean square value of V over distribution. For the Rayleigh fading channel, the probability that the received signal envelope will fall at or below some specified value of V is given by the cumulative distribution

$$p(V \leq V') = \int_0^{V'} \frac{2V}{v^2}\exp\left[-\left(\frac{V}{v^2}\right)^2\right]dV = 1 - \exp\left[-\left(\frac{V'}{v}\right)^2\right] \tag{54}$$

The Rayleigh probability distribution function Eq. (54) is often used in the form

$$p(V \leq V') = 1 - \exp[-0.693(V'/V_M)^2] \tag{55}$$

where V_M is the median value, about 1.6 dB below the rms value. *Fading rate* is important to certain systems. One measure of fading rate is the number of times per second (hertz) the carrier envelope crosses its median with a positive or negative slope. Another measure is the width of the received carrier-envelope spectral density.

It is also important to obtain information on fade durations; the effect of a large number of very short fades is quite different from that of a few long-duration fades, the former being serious for digital links, the latter for analog links. The rate of change of fade is important in the design of switching systems for diversity reception.

For some time, design concern (besides its emphasis on the amplitude variation) has centered on the dispersion and multipath characteristics of the medium, in terms of linear time-variant amplitude and phase- and frequency-distortion parameters, often referred to as *multiplicative noise*, which cannot be overcome by power increase.

A more comprehensive characterization is in terms of the *system function* or *impulse function*. This approach relates the response and excitation of a channel at its input and output terminals.[12,24,25]

The expressions for output are formulated in terms of operations on the input time function $x(t)$ or the spectral function (Fourier transform) $X(\omega)$, to produce the output function $y(t)$ or $Y(\omega)$. Each path is characterized by a system function $h(t, \tau)$ that operates on the replica of $x(t)$ traversing it; t is the time at which the observation is made, and τ is the delay or transit time for the path. The spread of delays between the input and output is determined by the system function $h(t, \tau)$, which can be called the *delay-spread system function*. For any particular elemental path x in a distribution of paths that covers some range of delays, the output for a range of delay $\Delta\tau$ centered on τ is given by $h(t, \tau)x(t - \tau) \Delta\tau$, and the total output of the channel is the sum of all such weighted and delayed contributing paths, namely.

$$y(t) = \int_{-\infty}^{\infty} x(t - \tau)h(t, \tau) \, d\tau \qquad (56)$$

For a Fourier transformable input $x(t)$, the output can be expressed in terms of the input spectral function $X(i\omega)$:

$$x(t) = \int_{-\infty}^{\infty} X(i\omega)\exp(i\omega t)d(\omega/2\pi) \qquad (57)$$

Here we characterize the channel by stating that it modifies the contribution to the structure of $x(t)$ from spectral components in the range $\Delta\omega$ centered at ω by multiplying it by the transfer function $H(i\omega, t)$. The total channel response to $x(t)$ is then

$$x(t) = \int_{-\infty}^{\infty} H(i\omega, t)X(i\omega)\exp(i\omega t)d(\omega/2\pi) \qquad (58)$$

which is the limit of the sum of the channel responses to the components of $x(t)$ from various infinitesimally wide spectral elements. The time-variant, *frequency-dependent transfer function $H(i\omega, t)$* can be shown to be the *Fourier transform over the delay-spread variable τ* of the *delay-spread function $h(t, \tau)$*.

The randomness of the channel with time is reflected in the treatment of the system functions $h(t, \tau)$ and $H(i\omega, t)$ as sample functions of processes that are random over the space of the time variable t. The autocorrelation function of the channel response process is given by an inverse Fourier transform operation on the product of the spectral density function of the input process and the autocorrelation function of the time-variant, frequency-dependent transfer function $H(i\omega, t)$ of the channel. Further transformations produce a combined time-shift and frequency-shift correlation function and a so-called *scattering function $S(\tau_0, f_0)$*, which has the physical significance of a function that determines the weighting of the signal power as a function of the time delay τ_0 and Doppler shift f_0 incurred in transmission.

On the basis of the above functions, a set of transmission parameters for random time-variant linear filters is defined.[25]

1. The *multipath spread* or *delay spread* is determined by the relative delays of the component paths, or the "duration" of $h(t, \tau)$ over the delay variable τ.

2. The *coherence bandwidth*, the bandwidth over which correlation of amplitude fading (or coherence of phase for some applications) remains to a desired degree is usually defined in terms of specified degradation of error rate, distortion, or other parameter.

3. *Diversity bandwidth* is the frequency separation between two sinusoidal inputs which results in a specified decorrelation of the fluctuating responses, usually taken to be a correlation coefficient of $1/e$.

4. The *fading rate, fading bandwidth,*[*] *frequency smear,* or *Doppler spread* is a measure of the bandwidth of the received signal when the input to the channel is a stable single-frequency signal.

5. The *diversity time* (or decorrelation time) is a measure of the time separation between input signals to yield correlation of less than $1/e$ between the envelopes of the responses.

These parameters are not all independent; the coherence bandwidth and delay spread are inversely proportional to each other, as are fading bandwidth and decorrelation time.

Most channel characteristics can be measured. The delay-spread response $h(t, \tau)$ can be measured directly by transmitting very short, widely spaced pulses; each received replica will correspond to one path, which can be resolved to examine the relative amplitudes and delays. The amplitude characteristic of the frequency-dependent transfer function $|H(\omega, t)|$ can be measured for short intervals by transmitting a constant-amplitude test signal with repetitive linear sweep covering the desired frequency range. The envelope of the received signal will give a very close approximation to $|H(\omega, t)|$.

Diversity techniques for counteracting short-term fading are used extensively for HF ionospheric communication, forward scatter systems, microwave line-of-sight systems, and Earth-space systems, where high reliability is required. The most common mechanism is to use *spaced antennas*, taking advantage of the fact that fading at one antenna tends to be independent of the signal fluctuation received on another antenna; provision is made to switch between signals or to combine two or more of them. Depending on the propagation mechanism, other kinds of useful diversity include *frequency, angle of arrival, polarization*, and *time*. Line-of-sight links frequently employ vertical space diversity where vertical separation of the antennas is found to be more effective than horizontal spacing. A simple, generally effective design procedure for this so-called height diversity gives the required vertical spacing between the centers of the antennas as

$$\Delta h = 0.3\sqrt{\lambda d} \tag{59}$$

where the path length d, wavelength λ, and the spacing Δh are all in the same units. A more thorough treatment is found in Recomm. ITU-R P.530.[69]

At frequencies affected by rain, i.e., above about 10 GHz, route diversity is used in which the spacing between the terminals is sufficiently large so that rain affecting one route is unlikely to affect the other.

The performance of a diversity system may be quantified by two parameters, *diversity gain* or *diversity advantage*. The diversity gain of a system is the ratio of the power output at a given time percentage to the corresponding power that would be obtained from a single channel. The diversity advantage is the ratio of the time percentage for which a given fade level is exceeded for a single channel to that for a diversity system. Both parameters may be obtained from cumulative distributions of fade statistics.

Recommendations ITU-R P.530 and 618[69] give expressions for diversity improvement under different propagation conditions for line-of-sight and Earth-space links, respectively, with further discussion given elsewhere.[7,8,14]

The outline of fading and diversity improvement given here has assumed *flat* (nonfrequency-selective) fading and Gaussian (white) noise. References 11, 14, and 25 discuss frequency-selective fading for the various types of system. Non-Gaussian noise effects are considered in the next paragraph.

Noise: Signal-to-Noise Ratio

Several types of radio noise must be considered in any design, though, in general, one type will be the dominant factor. In broad categories, the noise can be divided into two types: noise internal to the receiving system and noise external to the receiving antenna.

The *noise of the receiving system* is often the controlling noise in systems operating above about 300 MHz. This type of noise is a result of antenna losses, transmission-line losses, and the circuit noise of the receiver itself and has the characteristics of thermal (i.e. white) noise and thus is Gaussian in nature.[10]

---

[*]*Fading bandwidth* is also often used in the same sense as diversity bandwidth, or coherence bandwidth, above.

The second of the broad categories, *external radio noise*, can be subdivided further into natural and artificial sources. Natural sources of radio noise are: (1) atmospheric, (2) galactic, (3) solar noise from antennas pointing at the sun, (4) precipitation (blowing snow or dust), (5) corona, and (6) noise reradiating from any absorbing medium through which the wanted radio signal passes. Very low noise systems used in space communications can be limited by such absorption by clouds, water vapor, and oxygen. Such *sky noise* has the Gaussian characteristic of receiver noise. Examples of artificial noise sources are: (1) power lines or generating equipment, (2) automotive ignition systems, (3) fluorescent lights, (4) switching transients, and (5) electrical equipment in general. Unlike internal noise, external noise is generally non-Gaussian, being impulsive in nature.

Noise power is generally the most significant single parameter in relating the interference value of the noise to system performance. This parameter, however, is seldom sufficient in the case of impulsive noise, and a more detailed statistical description of the received-noise waveform is generally required.

A useful parameter for expressing the noise power external to the antenna is the effective antenna noise factor f_a which is defined by

$$f_a = p_n/Kt_0b = t_a/t_0 \tag{60}$$

where p_n = noise power available from an equivalent loss-free antenna (W)
K = Boltzmann's constant, 1.38×10^{-23} J/K
t_0 = reference temperature, taken as 288 K
b = effective receiver noise bandwidth, Hz
t_a = effective antenna temperature in the presence of external noise

The noise factor f_a is commonly given as the corresponding noise figure F_a, which is given by $F_a = 10 \log f_a$(dB).

Figure 16.3.3 shows the median value of the *available noise-power*, F_a (dB above Kt_0), from various sources. While the solar noise, galactic noise, and sky noise are Gaussian, the atmospheric and artificial noises are very impulsive.

Based on measurements from a worldwide network of stations, Recomm. ITU-R P.372[69] gives detailed estimates of F_a for atmospheric noise as a function of geographic location, season, frequency, and time. Additional data are also provided on noise variability and character which include the standard deviation of F_a, upper and lower deciles of F_a with standard deviations, and estimates of V_d, the ratio of the rms envelope voltage to the average noise envelope voltage. The values of F_a are for a lossless short vertical antenna over a perfectly conducting ground plane; they are related to the vertical rms field strength by

$$E_n = F_a - 95.5 + 20 \log f_{\text{MHz}} + 10 \log b \tag{61}$$

where E_n is the rms noise field strength [dB(μV/m)] in bandwidth b (Hz), and F_a is the noise figure for the center frequency, f_{MHz}.

Estimates of the noise-power spectral density from artificial noise expected in business, residential, rural, and quiet rural areas have been developed from measurements.[27–29] These expected values are the means of a number of location medians, with variation from location to location in each type of area and temporal variation indicated. Generally the noise below 20 MHz is associated with power lines. At 20 MHz and above, automotive electrical systems, especially ignition systems, are the dominant sources in all but rural locations. Median values of artificial noise power expressed in terms of F_a are given in Recomm. ITU-R P.372[69] for different categories of environment, of which curve 3 and 4 in Fig. 16.3.3 are examples.

Impulsive atmospheric or *artificial noise* disturbs communications in a way quite different from Gaussian noise. Figure 16.3.4 shows the amplitude probability distribution of Gaussian noise and a sample of atmospheric noise. The parameter V_d is the ratio in decibels of the rms voltage to the average voltage and is commonly used as an *impulsive index*. The two distributions are plotted relative to their rms level; i.e., both noises shown have the same energy or power. The probability distribution of the noise envelope determines the performance of most basic digital receivers, as indicated by the two error-rate curves for a binary coherent phase-shift-keying system.

Digital receivers are frequently designed for optimum performance in white Gaussian noise. Their performance in impulsive artificial or atmospheric noise can be summarized as follows, with comparisons made on the basis of equal noise power:

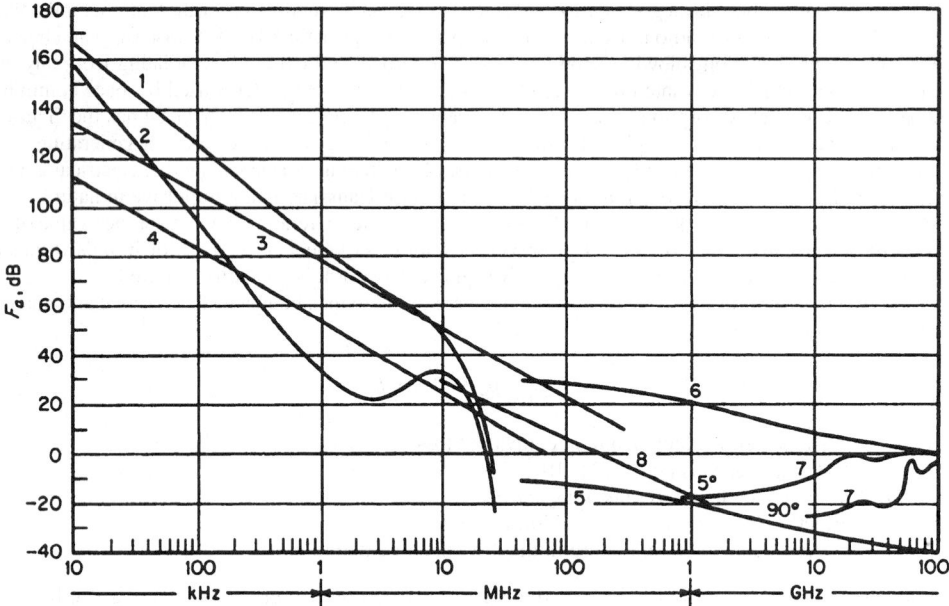

FIGURE 16.3.3 Median radio noise-power spectral density from various sources. Curve 1 = atmospheric noise, summer, 2000 to 2400 h. Washington, D.C., omnidirectional antenna near ground; curve 2 = atmospheric noise, winter 0800 to 1200 h. Washington, D.C., omnidirectional antenna near ground; curve 3 = artificial noise, business area, omnidirectional antenna near ground; curve 4 = artificial noise, quiet rural areas, omnidirectional antenna near ground; curve 5 = quiet sun, isotropic (0 dB gain) antenna; curve 6 = disturbed sun, isotropic (0 dB gain) antenna; curve 7 = sky noise, narrow-beam antenna (degrees from vertical); curve 8 = galactic noise, omnidirectional antenna near ground.

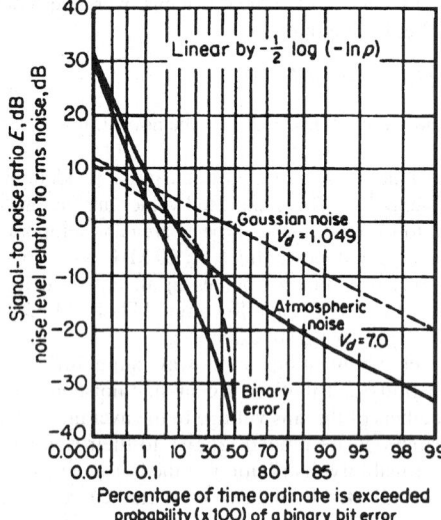

FIGURE 16.3.4 Comparison of noise distribution and error probabilities for Gaussian and non-Gaussian noise (same noise power, coherent phase-shift keying).

1. For constant signal, at high signal-to-noise (S/N) ratio, impulsive noise causes more errors than Gaussian noise; at lower S/N ratio, Gaussian noise causes more errors.

2. For Rayleigh-fading signals, Gaussian noise causes more errors at all S/N ratios; flat-fading cases do arise, for which impulsive noise will cause more errors than Gaussian nose; for diversity reception, impulsive noise is more harmful.

3. While pairing of errors in differentially coherent phase-shift keying (DCPSK) becomes more unlikely as the S/N ratio increases, pairing of errors increases as the noise becomes more impulsive.

4. For systems with time-bandwidth products in the order of unity, the standard matched filter-receiver is also optimum for impulsive noise.

5. Noise-suppression schemes, such as wide-band limiting and smear-desmear, are not particularly effective at high S/N ratios.

6. Receivers especially designed to reject a particular type of impulsive noise perform substantially better then receivers using the "standard" noise-suppression techniques.

TABLE 16.3.3 ITU-R Recommendations

	Title	Year
Recommendation ITU-R		
P.313	Exchange of information for short-term forecasts and transmission of ionospheric disturbance warnings	1995
F.339	Bandwidths, signal-to-noise ratios and fading allowances in complete systems	1990
P.368	Ground-wave propagation curves for frequencies between 10 kHz and 30 MHz	1992
P.369	Reference atmosphere for refraction	1994
P.370	VHF and UHF propagation curves for the frequency range from 30 to 1000 MHz. *Broadcasting services*	1995
P.371	Choice of indices for long-term ionospheric predictions	1995
P.372	Radio noise	1994
V.431	Nomenclature of the frequency and wavelength bands used in telecommunications	1994
P.434	ITU-R reference ionospheric characteristics and methods of basic MUF, operational MUF and ray-path prediction	1995
P.435	Sky-wave field-strength prediction method for the broadcasting service in the frequency range 150 to 1600 kHz	1992
P.452	Prediction procedure for the evaluation of microwave interference between stations on the surface of the earth at frequencies above about 0.7 GHz	1995
P.453	The radio refractive index: its formula and refractivity data	1995
P.526	Propagation by diffraction	1995
P.527	Electrical characteristics of the surface of the earth	1992
P.529	Prediction methods for the terrestrial land mobile service in the VHF and UHF bands	1995
P.530	Propagation data and prediction methods required for the design of terrestrial line-of-sight systems	1995
P.531	Ionospheric effects influencing radio systems involving spacecraft	1994
P.532	Ionospheric effects and operational considerations associated with artificial modification of the ionosphere and the radio-wave channel	1992
P.533	HF propagation prediction method	1995
P.534	Method for calculating sporadic-E field strength	1990
P.617	Propagation prediction techniques for data required for the design of trans-horizon radio-relay systems	1992
P.618	Propagation data and prediction methods required for the design of Earth-space telecommunications systems	1995
P.676	Attenuation by atmospheric gases	1995
P.684	Prediction of field strength at frequencies below about 500 kHz	1994
P.832	World atlas of ground conductivities	1992
P.834	Effects of tropospheric refraction on radio-wave propagation	1994
P.837	Characteristics of precipitation for propagation modeling	1994
P.838	Specific attenuation model for rain for use in prediction methods	1992
P.840	Attenuation due to clouds and fog	1994
P.842	Computation of reliability of HF radio systems	1994
P.843	Communication by meteor-burst propagation	1992
P.844	Ionospheric factors affecting frequency sharing in the VHF (30–300 MHz) band	1994
P.1057	Probability distributions relevant to radio-wave propagation modeling	1994

Other texts
CCIR Report 340-4 (Geneva, 1983) CCIR Atlas of ionospheric characteristics (deleted)
CCIR: Handbook of Curves for Radio Wave Propagation over the Surface of the Earth (Geneva, 1991)

The performance of analog voice systems in impulsive noise can be summarized as follows:

1. For a given articulation index, a much lower S/N ratio is required for impulsive noise than for Gaussian noise.
2. Various forms of limiting in AM systems (pre-i.f. limiting, i.f. limiting, postdetection limiting) are quite effective in further reducing the required S/N ratio when the noise is impulsive.

 As mentioned above, noise power is also generated from atmospheric gases, clouds, and rain. Recomm. ITU-R P.372[69] discusses radio emission at frequencies above about 50 MHz from natural sources, not only in the atmosphere but also from extraterrestrial sources and from the surface of the earth. This topic is discussed on the accompanying CD-ROM under "Propagation Over the Earth Through the Nonionized Atmosphere."

Minimum external noise levels to be expected at terrestrial receiving sites owing to natural and artificial noise sources are also specified in Recomm. ITU-R P.372.[69] With such data, the appropriate minimum receiver noise figure can be determined so that a terrestrial receiving system can be designed to be almost limited by external noise. Finally, Recomm. ITU-R F.339[69] gives ratios of required signal energy to noise power spectral density for various systems operating in the presence of atmospheric noise.

Ionospheric Modification by High-Power Radio Transmissions

High-power radio waves can modify the ionospheric plasma by classical ohmic heating, which changes the density and distribution of the ionization, and by generating parametric instabilities which in turn leads to field-aligned ionospheric irregularities (see Recomm. ITU-R P.532, Table 16.3.3).[69] At HF, ionospheric modification experiments generally use purpose-built transmitters, operating close to the F-region critical frequency, to modify the upper ionosphere (150–400 km). The ionosphere can also be appreciably modified by oblique high-power transmission at frequencies considerably in excess of the critical frequency. However, transmitters operating over the range from VLF to UHF can give rise to modifications in all regions of the ionosphere, with the resulting modified region having an effect on other radio signals passing through it. In particular, the scattering properties of ionospheric irregularities, artificially produced in the F-region, have been used to establish communications up to UHF between two points on the earth's surface. Such experiments demonstrate the interference potential to existing services from such mechanisms.

REFERENCES

1. Jordan, E. C., and K. G. Balmain "Electromagnetic Waves and Radiating Systems," 2nd ed., Prentice Hall, 1968.
2. Bremmer, H. "Terrestrial Radio Waves: Theory of Propagation," Elsevier, 1949.
3. Wait, J. R. "Electromagnetic Waves in Stratified Media," Pergamon, 1962 (2nd. ed. 1970).
4. Fock, V. A. "Electromagnetic Diffraction and Propagation Problems," Pergamon, 1965.
5. Rohan, P. "Introduction to Electromagnetic Wave Propagation," Artech House, 1991.
6. Beckmann, P., and A. Spizzichino "The Scattering of Electromagnetic Waves from Rough Surfaces," Artech House, 1987.
7. Hall, M. P. M. "Effects of the Troposphere on Radio Communication," *IEEE Electromagnetic Waves*, Series 8, 1980.
8. Boithias, L. "Radio Wave Propagation," North Oxford Academic, 1987.
9. Bean, B. R., and E. J. Dutton "Radio Meteorology," *Natl. Bur. Stand. Monogr. 92.* 1966; also Dover, 1968.
10. Shibuya, S. "A Basic Atlas of Radio-Wave Propagation," Wiley, 1987.
11. Giger, A. J., "Low-Angle Microwave Propagation: Physics and Modelling," Artech House, 1991.
12. Parsons, J. D. "The Mobile Radio Propagation Channel," Pentech Press, 1992.
13. Roda, G. "Troposcatter Radio Links," Artech House, 1988.
14. Allnutt, J. E. "Satellite-to-Ground Radiowave Propagation," Peter Peregrins, 1989.
15. Davies, K. "Ionospheric Radio," *IEEE Electromagnetic Waves* Series 31, Peter Peregrins, 1990.

16. Budden, K. G. "The Propagation of Radio Waves; the Theory of Radio Waves of Low Power in the Ionosphere and Magnetosphere," Cambridge University Press, 1985.

17. Ratcliffe, J. A. "Physics of the Upper Atmosphere," Academic, 1960.

18. Ratcliffe, J. A. "The Magneto-Ionic Theory," Cambridge University Press, 1962.

19. Rishbeth, H., and O. K. Garriott "Introduction to Ionospheric Physics," Academic, 1969.

20. Smith, E. K., and S. Matsushita "Ionospheric Sporadic E," Macmillan, 1962.

21. Maslin, N. M. "HF Communications: A Systems Approach," Pitman, 1987.

22. Schanker, J. Z. "Meteor Burst Communications," Artech House, 1990.

23. URSI, Review of Radio Science 1990–1992, R. W. Stone and H. Matsumoto (eds.), *Int. Union Radio Sci.*, Brussels, 1993.

24. Baghdady, E. J. "Lectures on Communications System Theory," McGraw-Hill, 1961.

25. Baghdady, E. J. Models for Signal Distorting Media, in R. E. Kalman and N. DeClaris (eds.), "Aspects of Network and System Theory," pp. 337–381, Holt, 1971.

26. Spaulding, A. D., and J. S. Washburn Atmospheric Radio Noise; Worldwide Levels and Other Characteristics, *NTIA Report 85–173*, 1985, ITS/NTIA, Dept. of Commerce.

27. Spaulding, A. D., and R. T. Disney Man-Made Noise, Pt. I., *OT Report 44-38*, U.S. Govt. Printing Office, 1974.

28. Spaulding, A. D., R. T. Disney, and A. G. Hubbard Man-Made Noise, Pt. II, *OT Report 75–63*, U.S. Govt. Printing Office, 1975.

29. Skomal. E. N. "Man-Made Radio Noise," Van Nostrand Reinhold, 1978.

30. Dougherty, H. T. A. Survey of Microwave Fading Mechanisms, Remedies and Applications, *Environ. Sci. Serv. Admin. Tech. Rep.* ERL-69-WPL 4, 1968.

31. Bean, B. R., B. A. Cahoon, C. A. Samson, and G. D. Thayer A World Atlas of Atmospheric Radio Refractivity, *Environ. Sci. Serv. Admin. Monogr.* 1, 1966.

32. Rotheram, S. Ground-Wave Propagation, Part I: Theory for Short Distances; Part II: Theory for Medium and Long Distances and Reference Propagation Curves, *Proc. IEE*, 1981, Vol. 128 (Part F), 5, pp. 275–284, 285–195.

33. Millington, G. Ground-Wave Propagation over an Inhomogeneous Smooth Earth, *J. IEE (Lond.)* January 1949, Pt. III, Vol. 96, p. 53.

34. Stokke, K. N. Some Graphical Considerations on Millington's Method for Calculating Field Strength over Inhomogeneous Earth, *Telecomm. J.*, 1975, Vol. 42, No. III, pp. 157–163.

35. Epstein, J., and D. W. Peterson An Experimental Study of Wave Propagation at 850 Mc, *Proc. IRE*, 1953, 41, pp. 595–611.

36. Deygout, J. Multiple Knife-Edge Diffraction of Microwaves, *IEEE Trans. Ant. Prop.*, 1966, Vol. AP-14, 4, pp. 480–489.

37. Ott, R. H. An Alternative Integral Equation for Propagation over Irregular Terrain, *Radio Sci.*, Vol. 5, May 1970, pp. 767–771; see also part 2, April 1971, pp. 429–435.

38. Hufford, G. A. An Integral Equation Approach to the Problem of Wave Propagation over an Irregular Terrain, *Quart. J. Appl. Math.* 1952, Vol. 914, pp. 391–404.

39. Blomquist, A., and L. Ladell Prediction and Calculation of Transmission Loss in Different Types of Terrain, in A.N. Ince (ed.) "Electromagnetic Wave Propagation Involving Irregular Surfaces and Inhomogeneous Media," *AGARD Conf. Proc.* No. 144, 1975, paper 32.

40. FCC Rules and Regulations, Radio Broadcast Services, Secs. 73.333 and 73.699, March 1980.

41. COST Project 210 [1991] "Influence of the atmosphere on interference between radio communications systems at frequencies above 1 GHz." Final Report of the Management Committee for COST Project 210, Report EUR 13407, ISBN 92-826-2400-5, CEC-COST.

42. Samson, C. A. Refractivity Gradients in the Northern Hemisphere, *OT Report 75–59*, U.S. Govt. Printing Office, 1975.

43. Samson, C. A. Refractivity and Rainfall Data for Radio Systems Engineering, *OT Report 76-105*, U.S. Govt. Printing Office, 1976.

44. Liebe, H. J., K. C. Allen, G. R. Hand, R. H. Espeland, and E. J. Violet Millimeterwave Propagation in Moist Air; Model verses Path Data, *NTIA Report 85-171*, 1985, ITS/NTIA, Dept. of Commerce.

45. Ito, S. A. Method for Estimating Atmospheric Attenuation on Earth-space Paths in Fair and Rainy Weather, *Trans. IEICE*, 1987, Vol. J. 70-B, No. 11, pp. 1407–1414.

46. Webber, R. V., and K. S. McCormick. Low Angle Measurements of the ATS-6 Beacons at 4 and 30 GHz, *Proc. URSI* (*Comm. F*), *Int. Symp. Effects of Lower Atmosphere on Radio Propagation at Frequencies above* 1 *GHz*, 1980.

47. Bryant, D. L. Low Elevation Angle 11 GHz Beacon Measurements at Goonhilly Earth Station, BT Tech. J., 1992, Vol. 10, 4, pp. 68–75.

48. Vilar, E., J. Haddon, P. Lo, and T. J. Moulsley Measurement and Modelling of Amplitude and Phase Scintillations in an Earth-Space Path, *J. IERE*, 1985, Vol. 55, No. 3, pp. 87–96.

49. Crane, R. K. Low Elevation Angle Measurement Limitations Imposed by the Troposphere; An Analysis of Scintillation Observations Made at Haystack and Millstone, *MIT Lincoln Tech.* 518, 1976.

50. Olsen, R. L., D. V. Rogers, and D. B. Hodge The AR^b Relation in the Calculation of Rain Attenuation, *IEEE Trans. Antennas Propag.* 1978, Vol. AP-26, No. 2, pp. 318–329.

51. Fedi, F. Attenuation due to Rain on a Terrestrial Path, *Alta Frequenza*, 1979, Vol. 66, pp. 167–184.

52. Maggiori, D. Computed Transmission through Rain in the 1–400 GHz Frequency Range for Spherical and Elliptical Drops and any Polarization, FVR Rept. IC379, *Alta Frequenza*, 1981, L, 5, pp. 262–273.

53. Laws, V. O., and P. A. Parsons The Relation of Raindrop Size to Intensity, *Trans. Am. Geophys. Union*, 1943, Vol. 24, pp. 165–166.

54. Howell. R. G., J. W. Harris, and M. Mehler Satellite Crosspolar Measurements at BT Laboratories, *BT Tech. J.*, 1992, Vol. 10, 4, pp. 52–67.

55. Olsen, R. L. Cross-Polarization During Clear Air Conditions on Terrestrial Links: A Review, *Radio Sci.*, 1981, Vol. 16, No. 5, pp. 631–647.

56. Van Zandt, T. E., and R. W. Knecht The Structure and Physics of the Upper Atmosphere, in A. Rosen and D. P. LeGallery (eds.), "Space Physics," Wiley, 1964.

57. Whitehead, J. D. Report on the production and prediction of sporadic E, *Rev. Geophys. Space Phys.* 8, 65, 1970.

58. Whitehead, J. D. Recent Work on Mid-Latitude and Equatorial Sporadic-E, *JATP*, 1989, Vol. 51, 5, pp. 401–424.

59. Piggott, W. R. and K. Rawer "URSI Handbook of Ionogram Interpretation and Reduction," 2nd ed., World Data Center A for Solar Terrestrial Physics, *Report UAG-23*, NOAA, 1972.

60. Barghausen, A. F. Medium Frequency Sky Wave Propagation in Middle and Low Latitudes, *IEEE Trans. Broadcast.*, June 1966, Vol. 12, pp. 1–14.

61. Wang, J. C. H. Prediction of Medium Frequency Skywave Field Strength in North America, *IEEE Trans. Broadcast.*, 1977, Vol. BC-23, pp. 43–49.

62. Wang, J. C. H. Sky-Wave Propagation Study in Preparation for the 1605–1705 kHz Broadcasting Conference, *IEEE Trans. Broadcast.*, 1985, Vol. BC-31, pp. 10–17.

63. Jones, W. B., and R. M. Gallet Ionospheric Mapping by Numerical Methods, *ITU Telecommun. J.*, December 1960, Vol. 27, No. 12, pp. 280–282.

64. Jones, W. B. and R. M. Gallet Methods of Applying Numerical Maps of Ionospheric Characteristics, *J. Res. Natl. Bur. Stand. (Radio Propag.)*, November-December 1962, Vol. 66D, No. 6, pp. 649–662.

65. Leftin, M., W. M. Roberts, and R. K. Rosich "Ionospheric Predictions," 4 Volts., U.S. Dept. Commer., Off. Telecommunic. OR/TRER 13, 1971.

66. Leftin, M., S. M. Ostrow, and C. Preston Numerical Maps of $f_d E_s$ for Solar Cycle Minimum and Maximum, *Environ. Sci. Serv. Admin. Tech. Rep.* ERL 73-ITS 63, 1968.

67. Kirby, R. C. Review of VHF Forward Scatter, *AGARD Proc. 37 Scatter Propag. Radio Waves*, 1968.

68. Edwards, K. J., L. Kersley, and L. F. Shrubsole Sporadic-E Propagation at Frequencies around 70 MHz, *Rad. Elec. Eng.*, 1984, Vol. 54, 5, pp. 231–237.

69. ITU-R See Table 16.3.3 for a list of Recommendations and other texts of the ITU Radiocommunication Sector.

ON THE CD-ROM:

Kirby, R. C., and K. A. Hughes, *Propagation over the Earth Through the Nonionized Atmosphere*, reproduced from the 4th edition of this handbook.

Kirby, R. C., and K. A. Hughes, *Propagation via the Ionosphere*, reproduced from the 4th edition of this handbook.

P · A · R · T
4

SYSTEMS AND APPLICATIONS

 On the CD-ROM

Electronics in Medicine and Biology

SECTION 17

TELECOMMUNICATIONS

Telecommunication systems and applications deal with the actual technology by which devices communicate with each other. This includes transmission systems, switching systems, local area networks (LANs), wide area networks (WANs), broadband systems, cellular and mobile communication systems, and wireless digital communications. It is interesting to note that in areas of the world where the infrastructure for wire-based communication systems does not exist, it seems that development of such systems will be skipped and they will go directly to wireless.

Future trends dictate that we must communicate more information much faster, much more reliably, and send more transmissions over the same system. We will continue to see wireless technology becoming the dominant communication system. A very good example is the increase in ways of wireless connection to the Internet such as cell phones and PDAs. Clearly, security will also be a dominant force in determining how we communicate. Such security will need to be achieved without degrading the communication process and without making communicating more difficult.

Chapter 17.5, Wireless Networks, deals with technologies that combine Internet access with cell phone communication, with picture and email transmission, and walkie-talkie communication. These technologies must allow us to communicate and use information seamlessly, across a wide variety of devices and platforms.

In Chap. 17.6, we look at data networks and the Internet. LANs, metropolitan area networks (MANs), and WANs must be able to cope with the explosion in volume of information communicated at extremely high speeds over hard wire, fiber, and wireless systems.

This explosion in wireless technology and all the resulting capabilities it enables requires more and more sophisticated micro-miniaturized circuits and systems, thus becoming a major driver for MEMS technology. This is fully covered in Chap. 17.8.   C.A.

In This Section:

Section Bibliography:

Agosta, J., and T. Russel "CDPD: Cellular Digital Packet Data Standards and Technology," McGraw-Hill, 1997.

Alber, J., and J. N. Pelton (eds.) "The INTELSAT Global Satellite System," American Institute of Aeronautics and Astronautics, 1984.

Aveyard, R. L. (ed.) Special Issue on Digital Cross-Connect Systems, *IEEE J. Selected Areas Commun.*, January 1987, Vol. SAC-5.

Barth, P. A. Compatibility Considerations When Interfacing Network Services and Network Functions, *GLOBECOM 1987.*

Basch, E. E., and R. G. Brown Introduction to Coherent Optical Fiber Transmission, *IEEE Commun. Magazine*, May 1985, Vol. 23.

Bates, B. "Wireless Networked Communications," McGraw-Hill, 1993.

Bates, C. P., D. H. Skillman, and M. A. Skinner A New Microwave Radio System Expands Digital Transmission Capacity, *Bell Labor. Record*, January 1986, Vol. 64, pp. 26–31.

Bates, R. J. "Wireless Networked Communications," McGraw-Hill, 1993.

Bender, W. J., G. Campbell, Jr., E. T. Harkless, and T. F. McMaster Telstar 3 Spacecraft Design Summary, *AIAA 9th Commun. Satellite Syst. Conf.*, March 7–11, 1982, pp. 653–663.

Benvenuto, N., G. Bertocci, W. R. Daumer, and D. K. Sparell The 32 kb/s ADPCM Coding Standard, *AT&T Tech. J.*, December 1986, Vol. 65.

Black, U. "Network Management Standards (OSI, SNMP and CMOL Protocols)," McGraw-Hill, 1994.

Black, U. "Network Management Standards (SNMP, CMIP, TMN, MIBs, and Object Libraries)," 2nd ed., McGraw-Hill, 1994.

Black, U. "The X Series Recommendations (Protocols for Data Communication)," McGraw-Hill, 1991.

Broadband Wireless Techniques, *IEEE J. Selected Areas Commun.*, October 1999, Vol. 17, No. 10.

Carroll, R. L., and P. S. Miller Loop Transients at the Customer Station, *Bell Sys. Tech. J.*, Vol. 59, November 1980.

Conti, M. E. Gregori, and L. Lenzini "Metropolitan Area Networks," Springer-Verlag, 1997.

Cotterhill, D. A. User Requirements and Provisions of Digital Circuit Multiplication Systems for TAT-8, *Br. Telecom. Engin.*, July 1986, Vol. 5, pp. 158–164.

Davis, P. T., and C. R. McGriffin "Wireless Local Area Networks," McGraw-Hill, 1995.

Davis, P. T., and C. R. McGuffin "Wireless Local Area Networks," McGraw-Hill, 1995, pp. 41–117.

Durand, A. Deploying IPv6, *IEEE Internet Comput.*, January/February 2001, pp. 79–81.

Early, S. H., A. Kuzma, and E. Dorsey The VideoPhone 2500—Video Telephony on the Public Switched Telephone Network, *AT&T Tech. J.*, January/February 1993, pp. 22–32.

Ebel, H., and E. Martin State of the Art of Telephone Transmitters and Receivers: A Century After Bell's Invention, *Telefon Report*, 1976, Vol. 12, No. 1, pp. 30–39.

Error Performance of an International Digital Connection Forming Part of an Integrated Services Digital Network, Recommendation G.821, Red Book, *CCITT*, 1985.

Faruque, S. "Cellular Mobile Systems Engineering," Artech House, 1996.

Feher, K. "Wireless Digital Communications," Prentice Hall, 1995.

Fish, P. J. "Electronic Noise and Low Noise Design," McGraw-Hill, 1994.

Fortier, P. J. "Handbook of LAN Technology," 2nd ed., McGraw-Hill, 1992.

Freeny, S. L. TDM/FDM Translation as an Application of Digital Signal Processing, *IEEE Commun. Magazine*, January 1980, Vol. 18, pp. 5–15.

Halsall, F. "Data Communications, Computer Networks and Open Systems," 4th ed., Addison-Wesley, 1995.

Heldman, R. K. "Global Telecommunications (Layered Networks' Layered Services)," McGraw-Hill, 1992.

Heldman, R. K. "Information Telecommunications (Networks, Products, and Services)," McGraw-Hill, 1993.

Henry, P. S. Introduction to Lightwave Transmission, *IEEE Commun. Magazine*, May 1985, Vol. 23.

Hinton, H. S., and D. A. B. Miller Free-Space Photonics in Switching, *AT&T Tech. J.*, January/February 1992, pp. 84–92.

IEEE Commun. Magazine, Sept. 1999, special issue on "Wireless Mobile and ATM Technologies for Third-Generation Wireless Communications."

IEEE Communication Magazine, Nov. 1997, special issue on "Mobile and Wireless ATM."

IEEE J. Selected Areas Commun. Vol. 15, No. 1, January 1997, special issue on "Wireless ATM."

IEEE Personal Communications, August 1996, special issue on "Wireless ATM."

IEEE Personal Communications, special issue on "Cellular-Based Personal Communications Services," Vol. 4. No. 3, June 1997.

IEEE Standard Method for Measuring Transmission Performance of Telephone Sets, *ANSI/IEEE Std* 269–1983, "Internetworking Technology Overview," Cisco Systems, Inc., San Jose, CA., 1993.

Jain, B. N. "Open Systems Interconnection (Architecture and Protocols)," McGraw-Hill, 1993.

Johannes, V. I. Improving on Bit Error Rate, *IEEE Commun. Magazine*, December 1984, Vol. 22, pp. 18–20.

Kellerman, W. "Analysis and Design of Multirate Systems for Cancellation of Acoustic Echoes," *ICASSP*, 1988.

Kessler, G. C., and D. A. Train, "Metropolitan Area Networks," McGraw-Hill, 1992.

Klessig, R. W., and K. Tesink, "SMDS: Wide-Area Data Networking with Switched Multi-megabit Data Service," Prentice Hall, 1995.

Knightson, K. G. "OSI Protocol Conformance Testing (ISO 9646 in Practice)," McGraw-Hill, 1993.

Kumar, B. "Broadband Communications (Guide to ATM, Frame Relay, SMDS, SONET, and B-ISDN)," McGraw-Hill, 1994.

Kumar, B. "Broadband Communications," McGraw-Hill, 1995, pp. 111–140.

Li, V. O. K. and X. Qiu Personal Communication Systems (PCS), *Proc. IEEE*, Vol. 83, No. 9, Sept. 1995, pp. 1210–1243.

Li, V. O. K., et al. A Survey of Research and Standards in High-Speed Networks, *Int. J. Digital Analog Commun. Syst.*, Vol. 4, No. 4, 1991, pp. 269–309.

Lin, Y. B. Paging Systems: Network Architectures and Interfaces, *IEEE Network*, July/August 1997, pp. 56–61.

Lipoff, S. J. Personal Communications Networks Bridging the Gap between Cellular and Cordless Phones," *Proc. IEEE*, Vol. 82, No. 4, April 1994, pp. 564–571.

Littlewood, M. Metropolitan Area Networks and Broadband ISDN: a Perspective, *Telecom. J. Australia*, Vol. 39, No. 2, 1989, pp. 37–44.

Macaris, R. C. V. "Cellular Radio, Principles and Design," McGraw-Hill, 1993.

Majeti, V. C., and R. A. Prasad Advanced Intelligent Network Directions, *IEEE Globecom '93—Conference Record*, 1993 pp. 1938–1943.

McDysan, D. E., and D. L. Spohn "ATM—Theory and Application," McGrw-Hill, 1995.

McNulty, J. J. A 150-km Repeaterless Undersea Lightwave System Operating at 1.55 μm, *IEEE J. Lightwave Tech.*, December 1984, Vol. LT-2, pp. 787–791.

Miller, M. A. "Analyzing Broadband Networks: Frame Relay, SMDS, and ATM," M&T, 1994.

Minoli, D. "1st, 2nd, and Next Generation LANs," McGraw-Hill, 1993.

Mollenauer, J. F. Standards Metropolitan Area Networks, *IEEE Commun. Magazine*, Vol. 26, No. 4, April 1988, pp. 15–19.

Muller, N. J. "Mobile Telecommunications Factbook," McGraw-Hill, 1998.

Murato, Y., S. Kobayashi, M. Oohara, R. Tanihati, J. Sasaguri, H. Nemoto, K. Abe, K. Nagatomi, and K. Nagashima New Digital Key Telephone Systems DK-16/DK-32/DK-64, NEC Res. Dev., April 1986, No. 81, pp. 78–85.

Naugle, M. G. "Network Protocol Handbook," McGraw-Hill, 1993.

Nemzow, M. "Implementing Wireless Networks," McGraw-Hill, 1995.

Okumura, Y., E. Ohmori, T. Kawano, and K. Fukuda Field Strength and Its Variability in VHF and UHF Land Mobile Service, *Rev. Elec. Commun. Lab.*, 1968, Vol. 16, p. 825.

Padgett, J. E., et al. Overview of Wireless Personal Communications, *IEEE Commun. Magazine*, Jan. 1995, pp. 28–41.

Pandya, R. Emerging Mobile and Personal Communication Systems, *IEEE Commun. Magazine*, June 1995, pp. 44–52.

Park, K. "Personal and Wireless Communications: Digital Technology and Standards," Kluwer Academic Publishers, 1996.

Perlman, R. "Interconnections," Addison-Wesley, 1992 (bridges and routers).

Personick, S. D. "Optical Fiber Transmission Systems," Plenum Press, 1983.

Radicati, S. "Electronic Mail (Introduction to the X.400 Message Handling Standards)," McGraw-Hill, 1992.

Rappaport, T. S. "Wireless Communications: Principles and Practice," Prentice Hall, 1996.

Ray, R. F. (ed.) "Engineering and Operations in the Bell System," 2d ed., AT&T Bell Laboratories, Murray Hill, 1983.

Raychaudhuri, D. Wireless ATM Networks: Technology Status and Future Directions, *IEEE Proc.*, Vol. 87, No. 10, October 1999, pp. 1790–1806.

Sadiku, M. N. O. "Optical and Wireless Communications: Next Generation Networks," CRC Press, 2002.

Sadiku, M. N. O. "Elements of Electromagnetics," 3rd ed., Oxford University Press, 2001.

Sadiku, M. N. O. "Metropolitan Area Networks," CRC Press, 1995.

Sadiku, M. N. O., and A. S. Arvind, An Annotated Bibliography of Distributed Queue Dual Bus (DQDB), *Computer Commun. Rev.*, Vol. 24, No. 1, January 1994, pp. 21–36.

Samuelsson, M. Intelligent Network Products Bring New Services Faster, *AT&T Tech.*, 1991 Vol. 6, No. 2, pp. 2–7.

Silverio, V. J., J. M. Bennett, S. J. Tillman, and J. Girardi The MERLIN Communications System, *IEEE J. Selected Areas Commun.*, July 1985, Vol. SAC-3, No. 4, pp. 584–594.

Sondhi, M. M., and A. J. Presti A Self-Adaptive Echo Canceler, *Bell Syst. Tech. J.*, 1966, Vol. 45, pp. 1850–1854.

Special Issue on Computer Communications, *IEEE Trans. Commun.*, January 1977, Vol. COM-25, No. 1.

Special Issue on Computer Network Architectures and Protocols, *IEEE Trans. Commun.*, April 1980, Vol. COM-28.

Special Issue on Wireless Networks for Mobile and Personal Communications, *Proc. IEEE*, September 1994, Vol. 82, No. 9.

Spohn, D. L. "Data Network Design (Packet Switching, Frame Relay, 802.6/DQDB (SMDS), ATM (B-ISDN), SONET)," McGraw-Hill, 1993.

Spohn, D. L. "Data Network Design," McGraw-Hill, 1993.

Sreetharan, M., and R. Kumar "Cellular Digital Packet Data," Artech House, 1996.

Stallings, W. "Data and Computer Communications," 4th ed., Macmillan, 1994.

Stallings, W. "Local and Metropolitan Area Networks," 5th ed., Prentice Hall, 1997.

Stallings, W. "Network Standards: a Guide to OSI, ISDN, LAN, and MAN Standards," Addison-Wesley, 1993, pp. 398–433.

Stallings, W. "Optical and Wireless Communications: Next Generation Networks," CRC Press, 2002.

Stallings, W. "SNMP, SNMPv2, and CMIP: The Practical Guide to Network Management," Addision-Wesley, 1993.

Stallings, W. IPv6: The New Internet Protocol, *IEEE Commun. Magazine*, July 1996, pp. 96–108.

Stallings, W., and R. Van Slyke "Business Data Communications," Macmillan College Publishing Company, 1994.

Stevens, W. "TCP/IP Illustrated, Volume 1," Addison-Wesley, 1994.

Taylor, D. P., and P. R. Hartman Telecommunications by Microwave Digital Radio, *IEEE Commun. Magazine*, August 1986, Vol. 24, pp. 11–16.

Tuttlebee, W. H. W. Cordless Telephones and Cellular Radio: Synergies of DECT and GSM, Electron. Commun. Eng. J., October. 1996, pp. 213–223.

Walker, M. CCS7 Offers New Paths to Revenue Generating Services, *AT&T Tech.*, 1991 Vol. 6, No. 2, pp. 8–15.

Winch, R. G. "Telecommunications Transmission Systems (Microwave, Fiber Optic, Mobile Cellular Radio, Data, and Digital Multiplexing)," McGraw-Hill, 1993.

Wireless Communication Series, *IEEE J. Selected Areas Commun.*, Vol. 17, No. 11, November 1999.

Wireless Communication Series, *IEEE J. Selected Areas Commun.*, Vol. 17, No. 3, March 1999.

Wireless Communication Series, *IEEE J. Selected Areas Commun.*, Vol. 17, No. 7, July 1999.

Wireless Communication Series, *IEEE J. Selected Areas Commun.*, Vol. 18, No. 3, March 2000.

Wireless Communication Series, *IEEE J. Selected Areas Commun.*, Vol. 18, No. 7, July 2000.

Wireless Communication Series, *IEEE J. Selected Areas Commun.*, Vol. 18, No. 7, July 2001.

Wireless Communication Series, *IEEE J. Selected Areas Commun.*, Vol. 19, No. 2, February 2001.

Wireless Communication Series, *IEEE J. Selected Areas Commun.*, Vol. 19, No. 6, June 2001.

Wireless Communication Series, *IEEE J. Selected Areas Commun.* Vol. 18, No. 11, November 2000.

Wyndrum, R. W., Jr., and Y. Mochida The Impact of New Technologies on Subscriber Loop Networks, *IEEE Commun. Magazine*, March 1987, Vol. 25, pp. 24–27.

Zimmerman, H. OSI Reference Model: The ISO Model of Architecture for Open Systems Interconnection, *IEEE Trans. Commun.*, April 1980, Vol. COM-28, pp. 425–432.

Recommended Periodicals:

Business Communications Review (monthly) BCR Enterprises.

Data Communications ($17 \times$ yr.) McGraw-Hill.

IEEE Communications Magazine (monthly) IEEE.

IEEE Network (*Magazine*) IEEE, (bimonthly).

Network World (weekly) Network World Publishing Co.

CHAPTER 17.1
TRANSMISSION SYSTEMS

A. B. Brown, Jr., Virgil I. Johannes

INTRODUCTION

A. B. Brown, Jr.

Service Networks

Communication entails the transfer of information from one point to another; the information may be in any of several forms, including voice, data, video, and facsimile. The facilities used for any particular form of service are known as a *service network*. A service network is made up of *terminals*, through which the information enters and leaves the network; *transmission facilities*, which provide the transfer of information from place to place, and *switching*, which connects the appropriate transmission facilities to cause the information to be delivered to the desired place. Some service networks do not include terminals, and some do not include switching. Terminals are of many types, including telephones, teletypewriters, facsimile terminals, and computer ports.

Interfaces

The boundary of the service network, where the user interacts with it (as by means of some piece of terminal equipment) or connects equipment to it is the *interface*. If the interface is digital, the use of digital transmission and switching obviates the need for digital-to-analog (D/A) and analog-to-digital (A/D) conversion of the signals. Contemporary networks with digital interfaces are used for digitally encoded video or voice as well as for high-speed data. Some older networks are used only for data.

Codes and Protocols

Communication of digital data is usually done in the form of characters, each of which is represented as a sequence of bits. Less commonly, the signals may be a bit stream generated and interpreted by the terminal equipment. The number of bits per character and the correspondence between the bit sequence and the character represented is called a *code*. Several codes have been used at various times for telegraphic and computer communications, starting with the Morse dot-dash codes used by telegraphers. The most common standard code at the present time is International Alphabet no. 5 (known in the United States as the American Standard Code for Information Interchange, ASCII). Other codes have been in use for older types of transmission equipment (teleprinters), and some codes have been introduced by individual manufacturers.

In coded transmission, two methods of maintaining synchronism between the transmitting and receiving points are commonly used. In *start-stop transmission*, the interval between characters is represented by a steady 1 signal; the transmission of a single 0 bit signals the receiving terminal that a character is starting. The information bits follow the start bit and are followed by the stop pulse. It is the same as the signal between characters and has a minimum length that is part of the specification of the terminal: 1.0, 1.42, 1.5, or 2.0 b are commonly used. In the synchronous method, the bits are sent at a uniform rate; if there is no character ready to be sent when it is time to send one, a synchronous idle pattern is used to maintain the timing. The synchronous method is used at higher speeds.

Protocols are standard procedures for the operations of communication; their purpose is to coordinate the equipment and processes at interfaces and at the ends of the communication channel. The International Telegraph and Telephone Consultative Committee (CCITT) recommendations in the X series, which apply to public data networks, include X.1, speeds; X.2, user options; X.20, interface for start-stop transmission services; X.21, interface for synchronous operation; X.22, multiplex interface for synchronous terminals; X.24, definitions of interface circuits; and X.121, international numbering plan for public data networks. Additional recommendations apply to packet networks.

Network Planning Objectives

The objectives that form the basis of network planning are derived from the service the network is to provide. There may, however, be different objectives for a network corresponding to different intended uses by the customers. For example, most of the objectives of the public telephone network are derived from the needs of voice communication; however, some objectives, e.g., the limits on delay distortion and impulse noise, concern its use for data communication. In turn, network objectives are the bases for the design objectives of the facilities which will compose the network. In practice networks evolve while they are in service, usually with more stringent objectives to meet the rising expectations of the users and with improved performance made possible by improved technology and designs. Thus, additions to the network may be designed according to objectives different from those of the earlier parts.

Packet Networks

Among the data networks provided up to the mid-1960s, both the circuit-switched and message-switched types were in use. In the first, a channel was assigned full time for the duration of a call. In the second, a message or section of a serial message was transmitted to the next switch if a path (loop or trunk) was available; if not, it was stored until a path became available. The use of trunks between the message switches was often very efficient. In many circuit-switched applications, however, data were transmitted only a fraction of the time the circuit was in use. In order to make more efficient use of facilities and to make it possible to charge users only according to the amount of data transmitted, packet networks were developed. In a packet network, a message from one terminal to another is divided into *packets* of some definite maximum length, often 128 octets (corresponding to bytes in a computer) of data; the packets are sent from the origination point to the destination individually. Each packet contains a header, which provides the network with necessary information to handle the packet. The packets transmitted by one terminal to another are interleaved on the facilities between the packets transmitted by other users to their addresses, so that the idle time of one source can be used by another. At the destination switching center, the message is reassembled and formatted before delivery to the called station. In general, a network has an internal protocol to control the movement of data within the network. The internal speed of the network is in general higher than any of the terminal speeds, so that there is a change of speed from source to network and from network to destination.

The same physical interface circuit can be used for communication with more than one other terminal or computer at the same time by the use of *logical channels*. At any given time, each logical channel is used for communication with some particular addressee; each packet includes the identification of its logical channel, and the packets for the various logical channels are interleaved on the physical-interface circuit.

Three methods of handling messages are in common use. *Datagrams* are one-way messages sent from an originator to a destination; the packets are delivered independently, not necessarily in the order they were sent. Delivery and nondelivery notifications may be provided. In *virtual calls*, packets may be exchanged between two users of the network; at the destination they are delivered to the addressee in the same order as they were originated. *Permanent virtual circuits* also provide for exchange of packets between two users of the network; each assigns a logical channel, by arrangement with the provider of the network, for exchange of packets with the other. No setup or clearing of the channel is then needed. Some packet networks support terminals that are not capable of formulating the data in packets, by means of a packet assembler and disassembler included in the network.

Other packet networks have been set up since ARPANET (the earliest major packet network in the United States) to provide packet communication service to commercial users, in the United States, Canada, France, Spain, Japan, and elsewhere. An early commercial network in the United States provides service at speeds up to 56 kb/s. The service is either common-user or with closed user groups; the latter service prevents communication into or out of a given group of stations, the equivalent of a private network. The network is arranged in a hierarchy of switching nodes, the higher nodes being redundantly connected by 56-kb/s trunks. The transit delay for a packet is less than 200 ms. The network operates in the virtual call mode.

Protocols for packet networks are the subject of several Recommendations of the CCITT. Recommendations X.3, X.28, and X.29 cover start-stop mode terminal handling; X.25 covers the interface for packet-mode data terminal equipment, X.75 covers the interface for packet-network interconnection, and X.244 covers exchange of protocol identification during virtual call establishment.

In early packet networks, routing of each packet in a message was independent; each packet carried the address of its destination, as well as a number to permit arranging the packets in the proper order at the destination. Later networks use a *virtual circuit*, which is set up at the beginning of a call and which contains the routing information for all the packets of that call. The packets after the first contain the designation of the virtual circuit. In some networks, the choice of route is based on measurements, received from all other nodes, of the delay to every other node in the network. It is sufficient for a node to measure and transmit to the other nodes the delay to nodes to which it is directly connected.

Integrated Service Digital Networks (ISDNs)

An ISDN uses a single digital transmission network to provide a wide variety of services, such as voice, text, facsimile, videotex, and video, both switched and nonswitched, and both circuit- and packet-made. See Chap. 17.4.

TRANSMISSION SYSTEM OPERATIONS

Virgil I. Johannes

Transmission System Principles

Transmission Systems. A telecommunications transmission link can be either a *loop*, which connects a subscriber with a switch, or a *trunk*, which connects two switches. A loop may consist of a *feeder* portion, connecting the switch at the central office with a terminal near the subscriber, and a *distribution* portion connecting the terminal to the subscriber. Transmission can be at voice frequency over wire pairs, or a number of voice-frequency channels can be multiplexed together using frequency-division techniques (analog carrier) or time-division techniques (digital carrier). The multiplexed signal can then be transmitted over guided wave media, such as wire and optical fibers, or through free space, as in radio systems.

Digital carrier provides higher quality transmission than analog, as the signal can be completely regenerated when necessary without the addition of noise and distortion that is characteristic of analog carrier. Digital carrier over fiber-optic cables is also more economical than analog carrier on cable. As a result, new carrier systems being installed throughout the world are almost exclusively digital, and in the United States at least, most analog carrier systems have been removed from service, leaving only vestiges of a once ubiquitous analog carrier network.

Transmission at voice frequency is commonly used in loops, and for some short trunks between analog switches, such as the 1A ESS, as the cost of converting the signal to and from digital form can be avoided. Loops to customers with digital PBX (Private Branch Exchanges) are often digital carrier, and there is increasing use of digital carrier on loops longer than 3 to 5 km. In this latter technique, signals are converted to and from digital form in equipment near the customer, and the final connection to the customer is at voice frequency over wire pair. Direct digital connection to the residential and small business customer is not widely available, but there are a few installations, and many proposals, both technical and regulatory, for bridging this "last mile." These include use of wire pairs from the central office to provide a 144 kb/s digital link, in connection with the ISDN, "fiber to the home" schemes with much higher capacity, and "fiber to the curb" schemes in which the feeder portion of the loop is fiber, while the distribution part is copper.

Voice-Band Transmission and Analog Carrier. Voice-band (200 to 3500 Hz) transmission over a pair of wires in a cable is widely used for loops, and for trunks between nearby switches. For lengths of a few miles or more, the cable is often *loaded* to improve voice-band performance.

In a loop, wire gauge and loading are normally selected to keep the loop resistance below 1300 Ω, which permits proper operation of the supervisory and transmission circuits. Loops up to about 3 mi can be 26-gauge pairs. Loops up to 5 mi can be 24-gauge, and longer loops are often of mixed gauges with the thinner wire closer to the exchange. An alternative approach (*long-route design* or *unigauge*) is to use 26-gauge only (up to 2800 Ω loops) and compensate for the additional resistance with audio gain, equalization, and increased signaling power at the office.

Voice-band trunks were once widely used among local and tandem offices but with digital switches, digital carrier connection is generally preferred. Voice-band trunks are still used for shorter connections among analog switches. Such trunks can be either two-wire, in which both directions of transmission are carried on a single wire pair, or four-wire, preferred for longer trunks, in which the two directions of transmission are carried on separate pairs. Use of amplifiers (repeaters) along the transmission path to compensate for the loss of longer cables is now rare, any gain required being included in a terminal near the switch, which also incorporates any other functions required, such as two-wire to four-wire conversion, and equalization of the frequency response. Voice-band trunks are typically designed to have a loss of about 6 dB (after any amplification) between local offices, and less for tandem office connections.

Typical analog carrier transmission involves single-sideband suppressed-carrier modulation of a voice-band signal, thus shifting the signal to a higher frequency. Such shifted signals, each typically occupying a 4-kHz bandwidth, are then combined into groups of 12 or 24 voice channels, and then further modulated and combined into 300 or 600 channel mastergroups. Short haul systems that carry groups, and long-haul systems carrying many mastergroups on a single coaxial, microwave radio, or satellite facility, while no longer being installed, are still in service in some areas, mostly outside North America. The SG submarine cable system, for example, carries 4200 two-way voice circuits in a single coaxial cable with dielectric diameter of 1.7 in. and has been in service across the Atlantic since 1976.

Digital Transmission Formats and Hierarchies

The first telecommunications digital transmission system, (T1) put into service in the United States in 1962, was designed to carry 24 voice channels over distances up to 50 miles. Each voice channel was sampled 8000 times per second, and each sample represented by an 8-bit byte, thus occupying 64 kb/s. The line signal was assembled by taking 1 byte from each channel in sequence (for a total of 192 bits) and then adding a single bit, the framing bit, to make a 193-bit frame, with a line rate of 1.544 Mb/s. At the receiving end, the framing bit was used to identify the other bytes so that they could be associated with the appropriate voice channel. In this initial design, an alarm was given, and the system was taken out of service, if more than 1 percent of the received framing bits failed to match the transmitted 1-0 pattern for about 300 ms. T1 systems were point-to-point, independent of each other, and the exact line rate of each system was determined by its own crystal clock, with an accuracy of about ±50 ppm. This system became widespread within a few years in the United States, Canada, and Japan. Subsequently, a similar system, with 8 kHz sampling, and 8 bits per sample, but carrying 30 channels at a line rate of 2.048 MHz, and differing in coding technique as well, was developed extensively in Europe. These are referred to as *primary rate* systems.

As the use of the primary rate systems increased, higher bit-rate systems were developed for longer distances and more economical transmission. The line signals for these higher rate systems are assembled from the primary rate signals in one or more steps using bit-by-bit time division multiplexing, and pulse stuffing. Two separate hierarchies of rates, each based on one of the two primary rates, along with rules for multiplexing and demultiplexing have been standardized. These are shown in Table 17.1.1, along with a variant used only in Japan. In the United States, transmission systems at rates other than the hierarchical rates are more the rule than the exception, and the

TABLE 17.1.1 Hierarchical Rates in the Plesiochronous Digital Hierarchy—kb/s

North America	Japan	Europe
64	64	64
1544	1544	2048
6312	6312	8448
44736	32064	34368
	97728	139264

6.312 Mb/s rate is little used. Higher rate signals for transmission are assembled from whatever number of DS-1 or DS-3 signals may be appropriate for the application at hand, using the pulse-stuffing technique. Thus the existing U.S. network has transmission systems carrying one, two, and four primary rate signals, in which the standardized primary rate signals appear only at the ends, as well as systems operating at various multiples of 44.736 Mb/s in which again the standardized signals appear only at the ends. The termination of these non-hierarchical rate systems in standardized signals allows for ready interconnection and rearrangement. All these rates and variations in both hierarchies are sometimes referred to, in the aggregate, as the *plesiochronous digital hierarchy* (PDH), in reference to the derivation of the various rates from independent timing sources that are not necessarily synchronized.

While systems above the primary rate generally include parity check bits for monitoring of the performance, the primary rate systems were initially monitored by examining the signal for violations of the line code and these violations (but not the errors!) disappear when the signal is multiplexed to a higher rate. To provide some means of determining the end-to-end performance of a DS-1 signal that may be multiplexed and demultiplexed many times between source and destination, newer DS-1 sources use the 193rd bit not only for framing, but also for error detection using a cyclical redundancy check code. A similar provision has been added to the 2.048 Mb/s signal. Even with these additions, the provisions for monitoring, maintenance, and network control provided in the PDH are not well suited to the worldwide digital network, and the existence of two separate hierarchies makes interconnection cumbersome.

In 1988, a new set of rates and multiplexing rules for higher bit-rate systems was standardized by the CCITT.* This new hierarchy includes a single set of rates intended for worldwide use, to transport signals from either of the two older hierarchies. There is also extensive provision of additional bits for monitoring and network operation. Recognizing the present availability of synchronized timing sources, the new hierarchy is nominally synchronous (hence the name, *synchronous digital hierarchy*, or SDH) although there is provision for small frequency variations, to account for the possibility that synchronism among parts of the network may be lost from time to time. In the United States, this hierarchy is part of a more extensive set of standards, referred to as the Synchronous Optical Network (SONET). SONET is intended also to encourage use of line systems and a standardized set of optical interfaces at the established rates of the hierarchy. The lowest rate envisioned is 51.84 Mb/s, appropriate for carrying a DS-3 signal.

Media for Transmission

Paired Cable.   Paired telephone cables have been described earlier. Trunk and loop cables are similar, except that to reduce loss trunk cables often use thicker wire and therefore include fewer pairs (900–pair cable with 22 gage pairs is typical). The performance of carrier systems on paired cable is generally limited by crosstalk. Near-end crosstalk is more significant than far-end crosstalk. Some trunk cables are available with an internal shield (screen) to separate the directions of transmission and thus reduce near-end crosstalk, and in some applications a separate cable is used in each direction (two-cable operation).

Coaxial Cable, Waveguide.   Coaxial cable for telecommunication trunks has been made with as many as 22 coaxial tubes, stranded and sheathed as for paired cable, and often including some wire pairs for maintenance. Tube diameters have ranged from 2.9 mm (in Europe), through 0.375 in. (once widespread in the United States), to 1.7 in. (for single tube submarine cable). Such cables were widely used for medium- and long-haul transmission before the advent of optical fiber, and many remain in service outside the United States. Circular waveguide systems operating at mm wavelengths had been used in some experimental installations before being obsoleted by optical fiber. (See also Chap. 17.3.)

Optical-Fiber Cables.   Optical fibers for telecommunications are covered in Chap. 17.3 and they are typically about 125 μm thick and are made of very pure silica with dopants added to control the index of refraction. They consist of a doped silica glass core with a glass cladding having a 0.3 to 1 percent lower index of refraction, which

---

*International Consultative Committee on Telephone and Telegraph, recently renamed the Standards Sector of the ITU (International Telecommunications Union). The CCITT formerly issued its standards in the form of "Recommendations" every four years in volumes identified by their color. Current practice is to issue or revise individual Recommendations from time to time.

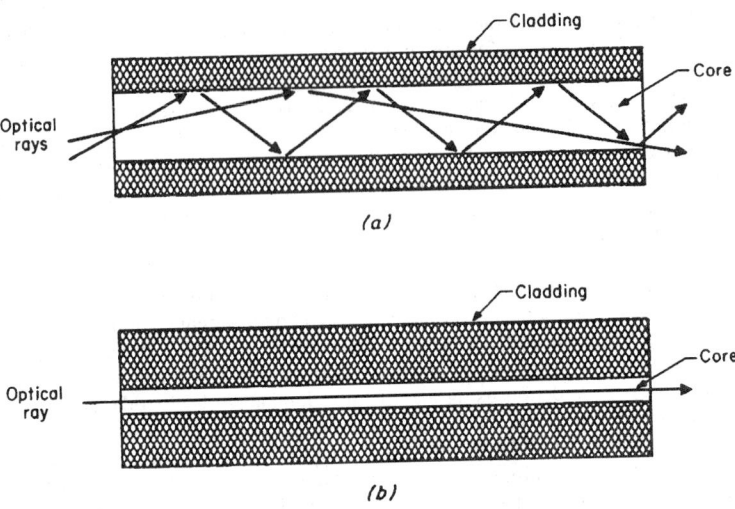

FIGURE 17.1.1 Optical fibers: (*a*) Multimode fiber; (*b*) single-mode fiber. (*From Paul S. Henry, "Introduction to Lightwave Transmission," IEEE Commun. Mag., Vol. 23, No. 5, p. 13. © 1985 IEEE. Used by permission*)

serves to confine the light within the fiber. If the core is thick enough (50 μm), light at the wavelengths used (0.8 to 1.55 μm) will propagate in several modes, as shown in the multimode fiber of Fig. 17.1.1*a*. A thinner core (0.85 μm) can restrict propagation to a single mode as shown in Fig. 17.1.1*b*. Single-mode fibers are more difficult to splice and couple, but usually have lower loss because of lower dopant concentration in the core.

Radio and Free Space. Microwave frequency bands allocated to telecommunications within the United States are shown in Table 17.1.2. (Except for mobile applications there is little present use of other frequency bands.) At these frequencies propagation is line-of-sight, so is limited by the curvature of the earth. At typical relay tower spacings of 15 to 30 mi, the free space loss is about 140 dB in the 4- and 6-GHz bands. With antenna gains of about 40 dB, the normal loss from antenna input to output is about 60 dB. Loss well above the normal can occur because of propagation effects, as discussed in Sec. 16. At frequencies above 10 GHz, the additional attenuation resulting from heavy rainfall is often the limiting factor. Path lengths 10 to 15 mi in the 11-GHz band, and 1 to 3 mi in the 18-GHz band are typical. The deployment of microwave systems is often limited by the need to avoid interference with existing systems using the same frequencies. These same microwave bands are used for satellite links.

Rainfall and other propagation problems have mitigated against the use of optical, or near optical, wavelengths in free-space transmission.

TABLE 17.1.2 Selected United States Common-Carrier Microwave Frequency Allocations

Band, GHz	Allotted frequencies, MHz	Bandwidth, MHz
2	2,110–2,130	20
	2,160–2,180	20
4	3,700–4,200	500
6	5,925–6,425	500
11	10,700–11,700	1,000
18	17,700–19,700	2,000
30	27,500–29,500	2,000

Voice Encoding, Circuit Multiplication, Echo Control

Voice Encoding.   To prepare a voice signal for digital transmission or switching, it is band-limited to about 3500 Hz and sampled at 8 kHz; each sample is encoded into 8-b PCM, producing a 64-kb/s signal. The quantization of the signal to one of $2^8 = 256$ levels that is inherent in the coding process produces noise, termed *quantizing noise* or *quantizing distortion*, in the decoded signal. (This is the major impairment suffered by a voice signal transmitted digitally.)

To improve the signal-to-distortion ratio at small-signal levels, a logarithmic encoding law is employed in order to use a greater portion of the available levels for weak signals. In this way, a certain *percentage* change in the instantaneous signal corresponds closely to a change of one encoding level, and the signal-to-distortion ratio is approximately constant over a wide range of talker volume. This process can be thought of as compressing the signal in amplitude according to one of the curves of Fig. 17.1.2 and then performing a linear A/D conversion. In current practice, however, a piecewise linear approximation to the logarithmic curve is incorporated in the coder proper and, of course, in the decoder as well. The two approximations in current use (A-law in Europe, μ-law in North America and Japan) are shown in Table 17.1.3. Both meet the CCITT requirement of 33-dB signal-to-quantizing-distortion ratio for sine waves from 0 to –30 dBm0, with overload point at

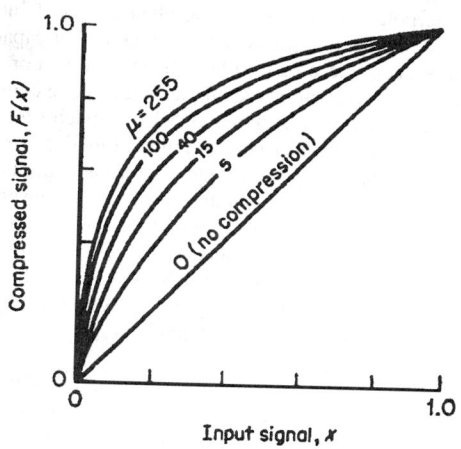

FIGURE 17.1.2   Logarithmic compression characteristics. (*Copyright 1982, Bell Telephone Laboratories, Inc. Reprinted by permission*)

about 3 dBm0. Idle channel noise for the μ-law coder is specified as less than 23 dBrnc0. (In this, as in other sine-wave specifications of coders, exact submultiples of the 8-kHz sampling frequency are excluded.) In the μ-law code, the all-zero code word is not used, slightly increasing the quantizing noise.

Some coding schemes for voice which use less than 64 kb/s found use in special applications even prior to the CCITT standardizing a 32-kb/s ADPCM (adaptive differential pulse code modulation) method. In differential PCM, the *difference* between sampled values of the voice signal is coded rather than the values themselves. The adaptive feature, an improvement on the logarithmic coding law, changes the code in accord with a complex algorithm dependent on past signal amplitudes. A major problem in finding an acceptable 32-kb/s coding method was providing acceptable performance for voice-band data signals from data modems. This is even more of a problem with 16-kb/s coding, and along with decreasing cost of high-capacity transmission systems, and increasing nonvoice traffic on the network, has limited the application of these lower bit-rate coding schemes.

TABLE 17.1.3   CCITT Recommended Coding Laws for Voice

Name	Basic law $0 \le |x| \le 1$	Parameter value	Segments in linear approximation	Primary multiplex rate, Mb/s
μ law	$y = \dfrac{\ln(1+\mu\,|x|)}{\ln(1+\mu)}$	$\mu = 255^*$	15	1.544
A law	$\dfrac{1+\ln A\,|x|}{1+\ln A} \quad \dfrac{1}{A} \le x \le 1$	$A = 87.6$	13	2.048
	$\dfrac{A\,|x|}{1+\ln A} \quad 0 \le x \le \dfrac{1}{A}$			

$^*\mu = 100$ was used in original T1 system.

Circuit Multiplication, Time-Assignment Speech Interpolation.   In a typical telephone conversation, only one party is talking at a time, so each direction of transmission is idle half the time. When pauses are also taken into account, the actual utilization is about 40 percent. Equipment taking advantage of this phenomenon to increase the number of voice channels carried by a transmission link is widely used on long distance international transmission links on submarine cable and satellite. It was first introduced on analog submarine cable systems, and termed TASI (time assignment speech interpolation), as the analog channels available were assigned to various talkers as they became active. The TASI systems achieved a 2:1 increase in capacity in practice. Similar systems, sometimes called DSI (digital speech interpolation) have been deployed on digital satellite links as well, with similar results. The combination of interpolation techniques with a code conversion from the 64 kb/s PCM signal to 32 kb/s ADPCM allows an increase of as much as 5:1 in capacity for digital voice applications. (The signal is restored to 64 kb/s at the receiving end.) This combined function is referred to as *circuit multiplication*, and is in general used on satellite and intercontinental submarine cable digital links.

Circuit multiplication introduces some slight degradations to voice signals, introduced by the conversion to and from ADPCM, the finite time to recognize the beginning of speech (processing clip), and the occasional need for more channels than are available. (The effect of the latter can be mitigated by reducing the number of bits for all conversations, rather than cutting off some conversations for short intervals.) Circuit multiplication equipment must detect nonvoice signals, and pass them without alteration, as they would not be restored correctly. This detection may be based on the 2000–2200 Hz tone emitted by voice-band data modems.

Echo Control.   If the round trip delay is less than 45 ms, acceptable echo performance can be achieved by designing all trunks to have a loss that increases the return loss of the echo. Voice-band and other analog trunks were traditionally assigned losses in the range of 0.5 to 8.9 dB, according to an overall plan called the Via Net Loss plan. With digital transmission and switching, which are inherently lossless, 3 dB is typically inserted at both ends of a connection, and no attempt is made to add loss to intermediate trunks. Above a 45-ms delay, the loss required for acceptable echo performance would be excessive, and echo cancelers are used. An echo canceler incorporates an internal delay line with taps that are adjusted to produce a signal duplicating the echo, which is then subtracted from the received signal, canceling the effect of the echo. The adjustment is done rapidly and continuously. One echo canceler is necessary for each voice channel, but economical integrated circuit realizations are available.

The Digital Transmission Network

Network Structure.   The structure of the existing digital transmission network, based on the plesiochronous digital hierarchy, is illustrated in Fig. 17.1.3, which shows a cross-section of equipment arrangements and signal flows applicable to any of the hierarchies shown in Table 17.1.1.* Primary rate signals may be coded voice signals multiplexed together, with the encoding and multiplexing done in a unit called a *channel bank* or *Primary PCM Multiplex* as described above for the original T1 system. Digital switches typically include the coding and multiplexing functions as well, and primary rate signals can originate from other sources, such as computer data and coded video. Primary rate signals destined for a remote location may either be transmitted directly, typically over a wire pair, or further combined into higher-rate signal by the multiplexes for more economical transmission, as illustrated in Fig. 17.1.3. The *Line Terminating Equipment* of the figure include such functions as power feeding for a repeated line, error monitoring, switching from a failed line to an operating one (protection switching), and conversion of the signal to the line code. (These functions may actually be packaged with the multiplex.) In the United States, the 6.312 Mb/s line rate and crossconnect capability have not found much use, so direct multiplexing of DS-1 signals into DS-3 signals in a single piece of equipment without access to the DS-2 signal is the norm.

Figure 17.1.3 is representative of equipment found in telephone company offices. The offices are connected by transmission line systems using cable or radio, and often include regularly spaced repeaters, which serve to amplify, retime, and regenerate the received digital signal. The line systems do not necessarily operate at a hierarchical rate. They do, however, have interfaces at hierarchical levels (in the United States at DS-1 and DS-3), and at standardized signal levels and format, and thus are readily interconnectable with other line systems,

---

*Interconnection between the two hierarchies is based on the common 64 kb/s rate. Conversation between A and μ-law coding and remultiplexing into appropriate format is performed at a connection point in the μ-law country, and often associated with circuit modulation equipment. Some United States–Europe links use a "hybrid hierarchy" in which the multiplexing sequence is 64 kb/s (A-law), 2.048 Mb/s, 6.312 Mb/s, 44.736 Mb/s, and 139.264 Mb/s.

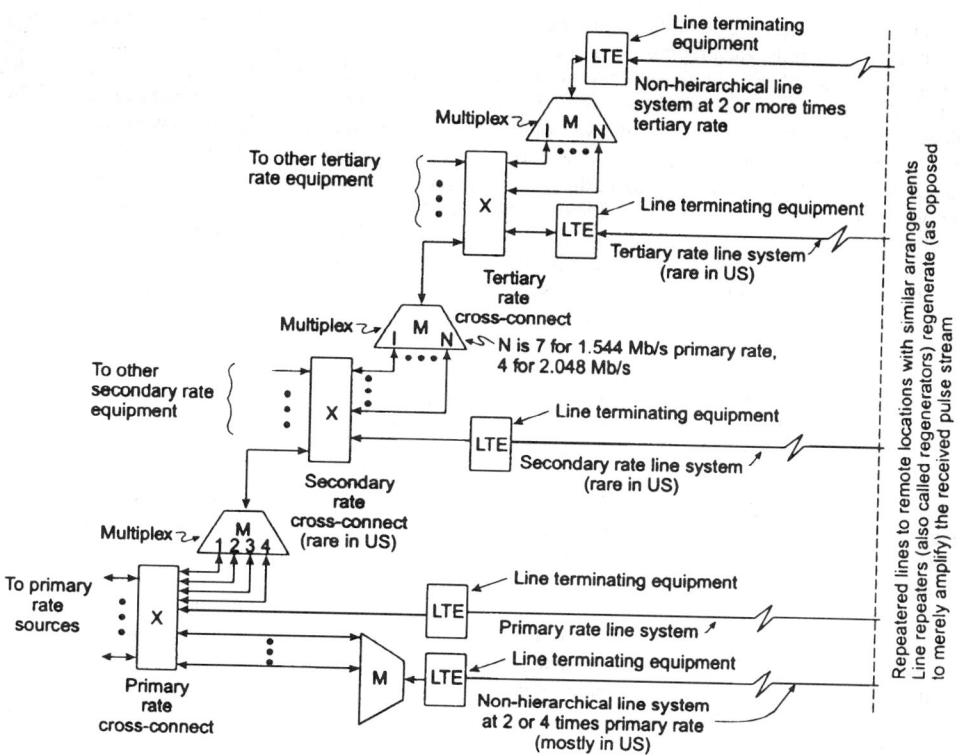

FIGURE 17.1.3 Functions and equipment in the plesiochronous digital hierarchy.

signal sources, and so forth. The line systems proper are proprietary, and terminal equipment at both ends, as well as intermediate repeaters, are generally from the same manufacturer, or at least a cooperating group of manufacturers. The repeaters may be powered locally, if they are located in a telephone company office, for example, or via power fed over the cable. Paired cable systems are typically powered over the pairs that carry the signals, and conductors for the purpose may be provided in optical fiber cables.

Crossconnects.   To facilitate interconnection and rearrangements, all signals of a given hierarchial rate in a telephone company office are brought together at the *crossconnects* of Fig. 17.1.3. These crossconnects may be simple wiring panels at which the various sources, multiplexes, and line terminating equipments are interconnected and accessed for testing by manually placed wiring. In the United States, such a crossconnect is referred to, for example, as a DSX-1 for a crossconnect at the primary level, and so on. A more elaborate alternative to simple wiring panels is the digital crossconnect system (DCS, or DACS for digital access and crossconnect system). These provide electronic switching of the signals and include multiplexes so that *components* of the signals may be interchanged. Thus a primary level DCS can reroute entire 1.544 Mb/s signals, and interchange 64 kb/s channels freely among such signals. (This latter function is similar to that performed by digital switches, but the DCS is too slow to be used on a call-by-call basis, and does not incorporate the signaling capability of a digital telephone switch.) This requires synchronism of the various signals, which is achieved by synchronizing all connected switches and channel banks from a master clock.

Equipment incorporating the line terminating functions, multiplex, and electronically switched crossconnect in a single unit is also available, offering space, cost, and operational savings.

Time-Division Multiplexing: Framing.   If two or more synchronous digital signals are to be multiplexed, they can be adjusted in phase and pulse width and combined directly, with some method for identification of the signals for later demultiplexing. A common method of identification is to group a number of input bits into a *frame* with an additional framing bit at the beginning of the frame. For example, in the 24-channel PCM channel bank, twenty-four 8-b words, each representing one sample, constitute 192 b of a 193-b frame, and the 193d bit is the framing bit. If no other features were desired, the framing bit could be a simple pattern such as 101010, which would allow frame to be recovered at the receiving channel bank by finding a position in the received pulse stream which alternates between 0 and 1 at intervals of 193 b. To provide additional functions while maintaining the T1 line rate, an *extended superframe format* involving a framing pattern that identifies groups of twenty-four 193-b frames is used. In this format, only six of the 24 slots allocated to framing bits have a fixed pattern used for framing, while six slots are used for cyclic redundancy check bits (e.g., an error-detecting code covering the entire superframe of 24 frames) and 12 slots are used for 4-kb/s data link for maintenance communication.

This use of a single framing bit at the beginning of a frame is common in the United States at all rates. Another method, common in Europe, is to use a multibit framing alignment word at the beginning of a much longer frame. In Synchronous Digital Hierarchy, framing bytes 8 b are included in section overhead, and distributed throughout the frame.

Synchronization, Pulse Stuffing. Multiplexing to the primary level (1.544 or 2.048 Mb/s) is synchronous, and entire 8-b bytes are multiplexed. In channel banks and switches the locally encoded voice signals are synchronous and appear 8 b at a time, so that this is a natural mode of operation. Where 64-kb/s signals from more than one source are to be intermingled, e.g., in a digital switch, they are synchronized to a central reference which is distributed via designated digital facilities. In a digital channel bank connected to a digital switch, e.g. via a digital line, the digital switch drives the line toward the channel bank at a rate synchronous with the references and the channel bank is loop-timed; i.e., it derives its transmit frequency from the signal it receives from the switch, which in turn is timed from the network reference.

Above the primary level, signals normally arrive at the multiplex a bit at a time, and it is not easy to assure that all signals that arrive at a multiplex will be synchronous. Present practice, therefore, is to stuff pulses into each input digital stream to synchronize them all to a common, higher rate. The synchronous streams are then multiplexed bit by bit. This process, called *pulse stuffing* or *positive justification*, is illustrated in Fig. 17.1.4. The location of the stuffed pulses is signaled to the receiving end (to allow for removal) on still other bits added to the frames for the purpose.

The pulse stuffing-destuffing process introduces some jitter (undesired phase modulation) in the demultiplexed pulse stream, which is usually not troublesome as it is of low amplitude and frequency.

Superframes. The pulse-stuffing scheme described above requires a low-data-rate digital channel between multiplexer and demultiplexer to signal the presence or absence of stuffed time slots. Low-data-rate channels may also be required to provide communication between terminals for maintenance functions such as switching to a spare or for parity bits for detection of transmission errors. These low-data-rate channels are provided by adding bits that are located by establishing a superframe structure encompassing many information-bit frames. The extended superframe mentioned above is an example.

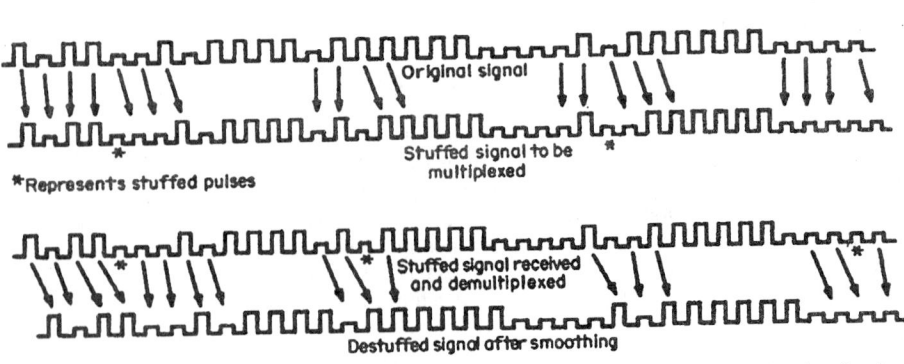

FIGURE 17.1.4 Pulse-stuffing synchronization. (*Copyright 1982, Bell Telephone Laboratories, Inc. Reprinted by permission*)

Regenerative Repeaters. A block diagram of a regenerative repeater is shown in Fig 17.1.5. The input signal is amplified and equalized, a timing wave is extracted in the clock-extraction circuit, and then the signal is regenerated; i.e., in each time slot a decision is made whether a 1 or 0 is present and a pulse accordingly applied (or not) to the output. The stylized waveshapes of Fig. 17.1.5 illustrate this action, and Fig. 17.1.6*a* shows an idealized "eye pattern" at the regenerator input, for a ternary repeater, i.e., one that can produce either a positive or negative pulse as well as a zero. An eye pattern is the superposition of waveshapes resulting from all possible pulse sequences $(0 + 0, - + -, \cdots)$.

The frequency response of the equalized channel is deliberately rolled off, often with a cosine shape, to produce a pulse that may be as wide as two time slots at the base (as in Fig. 17.1.6*a*) rather than a close replica of the transmitted pulse. This reduces the noise and crosstalk on the equalized pulse. The adaptive equalizer acts to bring the peak of the received pulse to a standard amplitude and thus provides the proper equalization for any cable length within its range. (Several fixed equalizers may be required to span the complete range of cable lengths.) The adaptive equalizer also compensates for variations in cable temperature, and is usually analog, involving only a few singularities. Misequalization and other circuit imperfections can result in closing the eye (Fig. 17.1.6*b*), increasing the probability of error because of thermal or other noise.

Jitter in a repeatered line results from pattern variations in the signal transmitted. As a result of imperfect equalization and other idiosyncrasies of the repeater, these pattern variations appear as phase variations applied to the timing extraction filter (see below), and any components of the phase variation at a frequency within the bandpass of this filter appear as jitter on the timing signal and hence on the repeater output. The rms jitter of a line is approximately proportional to the square root of the number of repeaters. (For T1, for example, with a repeater tank Q of 80, the rms jitter for 10 repeaters is about 3°.) Jitter in the amounts usually encountered has little effect on 64-kb/s encoded voice signals. Jitter (unlike errors) can be reduced or eliminated completely at the endpoints by writing the receiving information into a buffer memory and reading it out under control of a stable clock. Thus the major concern in practice is to assure an adequate size of buffer memory where dejitterization intentionally or inadvertently takes place. (An example of inadvertent dejitterization would be the connection of a repeatered line to a terminal with an input repeater of higher Q.)

Clock extraction for a ternary repeater involves full-wave rectification of the signal, filtering the result in an LC tank or equivalent, in order to obtain a sine wave at the symbol rate, and then shaping the sine wave to obtain a sampling pulse or edge. The full-wave rectification produces a strong component at the symbol rate. The required Q of the tank circuit, or equivalent, depends on the line code selected. In any case, a higher Q in the filter reduces the jitter of the timing wave, which results from pattern variations in the digital stream. On the other hand, a higher Q increases the static offset of the sampling pulse because of temperature variations of the frequency of the tank. Arrangements used have included an LC tank with Q of 80 (T1), a monolithic crystal filter, a surface-acoustic-wave (SAW) filter, and phase-locked loop with crystal-controlled VCO (voltage-controlled oscillator).

The regenerator proper is a clocked flip-flop or similar arrangement with carefully controlled threshold. The output pulse is often clocked to about 60 percent of the time slot in ternary systems, but nonreturn to zero (NRZ) is usual in binary systems, such as fiber-optic systems.

Most regenerative repeaters have some provision for fault location, i.e., determining from an office which of the many repeaters in a failed line is at fault. This may be accomplished by feeding a small part of the output back to the originating station through a filter that passes only a particular audio frequency assigned to that manhole. To locate a faulty repeater, a digital pattern containing the particular audio frequency associated with a certain manhole is applied to the line. If the audio frequency then appears on the fault-locating pair, the repeater in that manhole and all repeaters upstream of that line are known to be operating. The procedure is then repeated with other audio frequencies until the defective repeater is located. All repeaters at a given manhole share the same audio filter, and all manholes share the same voice-frequency fault-locate pair. Recent designs for high-capacity fiber-optic and radio systems incorporate error detectors (based on parity bits or line code redundancy) in each repeater, with a method for interrogation from the terminals.

The probability that an ideal digital repeater with gaussian noise added to its input will make an error is the probability that the noise will exceed half the eye height, which for m-level transmission is

$$P_E = \frac{m-1}{m} \operatorname{erfc} \left| \frac{1}{(m-1)\sqrt{2}} \frac{\text{peak signal}}{\text{rms value of noise}} \right|$$

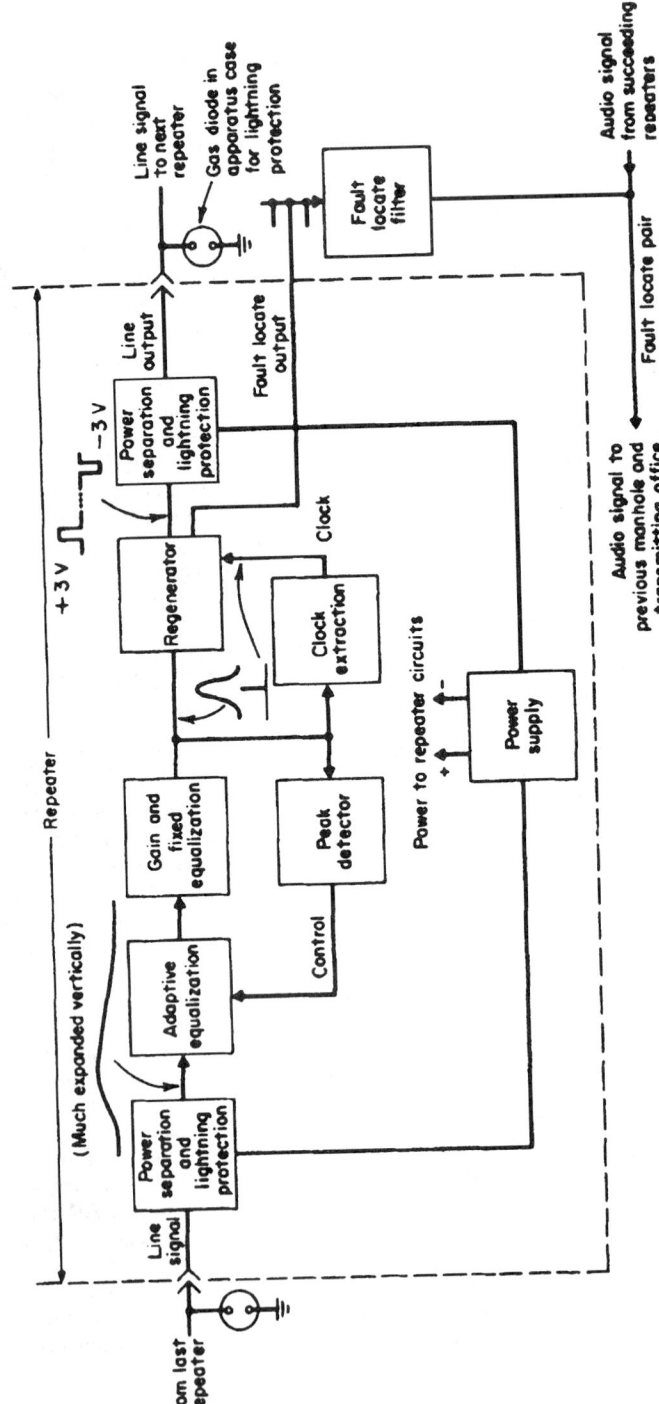

FIGURE 17.1.5 Representative regenerative repeater for cable system with associated external components.

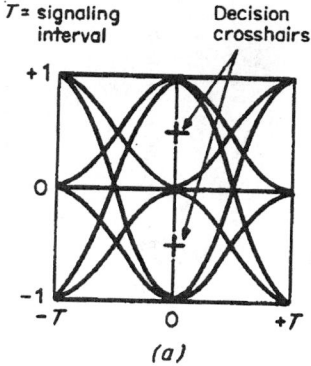

T = signaling interval

Decision crosshairs

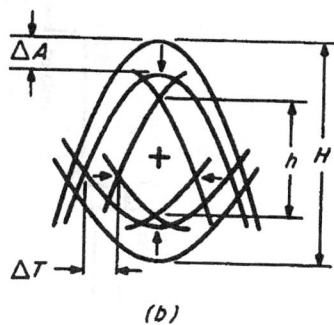

(a)

(b)

FIGURE 17.1.6 Eye diagrams for a ternary repeater: (*a*) ideal eye; (*b*) closing the eye to account for practical degradations (upper eye only). (*Copyright 1982, Bell Telephone Laboratories, Inc. Reprinted by permission*)

This is plotted in Fig. 17.1.7. It can be seen that in the range of usual interest, 10^{-6} or better, variation of a few decibels of signal-to-noise ratio produces a variation of many orders of magnitude in error probability. This is generally true in practice as well, even though crosstalk or some other interference, rather than gaussian thermal noise, may be the major cause of errors.

The features described above are typical of regenerative repeaters for use on cable systems. Repeaters on metallic cable systems normally are powered from a direct current carried on the same conductors as the signal, and they also require a power-separation filter, as shown in Fig. 17.1.5, while radio and fiber-optic repeaters do not. Fiber-optic regenerators are similar to regenerators for metallic cable, except that a light source and optical detector are used at the output and input. The light source may be a solid-state laser or light-emitting diode (LED). Early systems were restricted to wavelengths of 0.82 to 0.85 μm for which lasers and LEDs were available. (LEDs are less expensive but provide less light power to the fiber.) Once lasers at 1.3 μm (where the fiber has lower loss) with satisfactory lifetimes became available, they were widely incorporated in new systems, and some systems have appeared with 1.55 μm lasers, the wavelengths at which

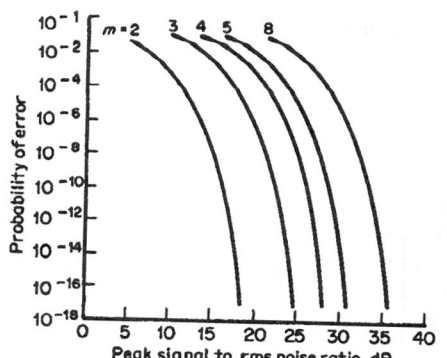

FIGURE 17.1.7 Probability of error vs. peak signal to rms gaussian noise for random *m*-level polar transmission. (*Copyright 1982, Bell Telephone Laboratories, Inc. Reprinted by permission*)

present fibers have minimum loss (but not minimum dispersion). Optical detectors are typically *pin* diodes or avalanche diodes. For both lasers and detectors, compound semiconductor alloys, such as InGaAsP are widely used. Coherent detection, in which the incoming signal is mixed or compared with a local light source of about the same wavelength, has not yet proved practical in optical systems (although it is the norm in digital radio systems) because of the lack of light sources with sufficient spectral purity and stability. Regenerative repeaters for radio use require modulators and demodulators as well as the functions described above. Figure 17.1.8 shows the block diagram of a regenerator for a 135-Mb/s (actually 3×44.736-Mb/s) system using 64 QAM (64 quadrature amplitude modulation). The symbol rate is 22.76×10 Mb/s. This regenerator operates on a 70-MHz i.f. signal, and is used on a 30-MHz channel with radio systems in the 6-GHz band, and on a 40-MHz channel with radio systems in the 11-GHz band. In 64 QAM, the transmitted signal has 64 possible states, that is, combinations of amplitude and phase as shown in Fig. 17.1.9. Other modulation methods, such as QPSK (quaternary phase shift keying) and 8PSK, are also in use. In QPSK the transmitted amplitude is constant, but

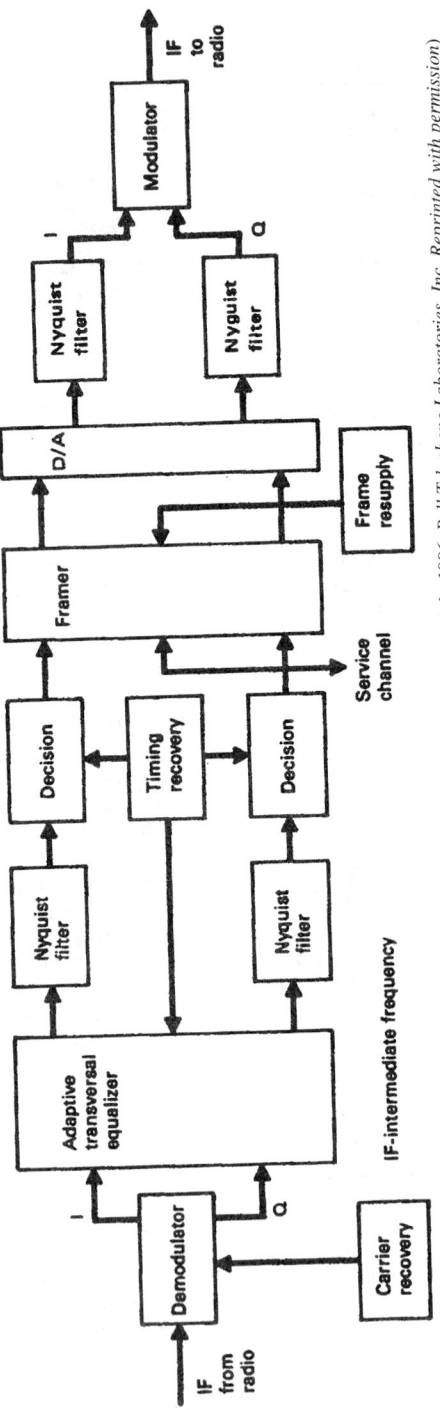

FIGURE 17.1.8 Digital regenerator for 135-Mb/s digital radio system using 64 QAM. (*Copyright 1986, Bell Telephone Laboratories, Inc. Reprinted with permission*)

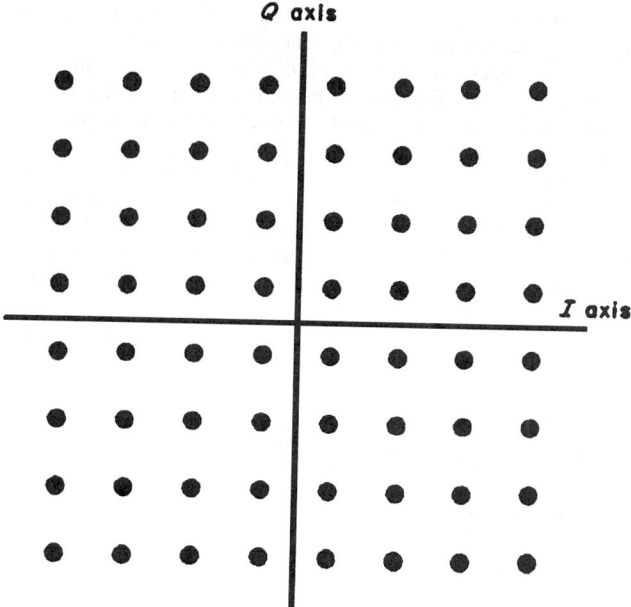

FIGURE 17.1.9 Phase plane representation of possible signals in 64 QAM. (The signals are sinusoids with amplitude given by distance from origin and phase given by angle with *I* axis.) (*From AT&T Bell Laboratories RECORD, January 1986, p. 31. Copyright 1986, Bell Telephone Laboratories, Inc. Reprinted with permission*)

the phase can take on one of four values. In regenerators for radio use, the equalizer is compensating for distortion introduced by propagation effects such as multipath, and it is the development of improved equalizers that has allowed the use of high-capacity modulation such as 64 QAM.

Scramblers and Line Codes for Digital Systems. Normally it is not possible to transmit an arbitrary digital stream directly as a binary signal because of the possibility that some patterns might cause the line system to misfunction. For example, too long a string of zeros could cause loss of timing signal in the repeaters, repetitive patterns could produce strong discrete frequencies in the output of radio systems that might violate emission limits, and a variation of the average density of 1s would cause a wander of the baseline of the signal after passing through a transformer or power-separation filter. To counter these effects, the signal is usually scrambled or coded (sometimes both).

A digital signal is scrambled by adding to it, in mod 2,[*] a long predetermined pseudo-random pattern of 1s and 0s. It can be unscrambled at the receiving end by adding (mod 2) the same pattern, in the same phase. The pseudo-random pattern is obtained from a shift register clocked at the signal rate with some taps added and fed back to the input. Appropriately selected taps produce a maximal-length pattern $2^N - 1$ long, where N is the number of stages in the register. The phase of the sequence is normally synchronized to the framing pattern, allowing ready descrambling at the receiving end.

Scrambling may reduce the probability (but cannot *guarantee* the absence) of any unfortunate pattern such as a long string of zeros. Therefore, if scrambling is to serve this purpose, high-Q timing recovery circuits such as phase-locked loops are required, and this is typical of fiber optic and radio repeaters. Another use of scrambling is encryption of a digital signal to maintain its confidentiality. This can be accomplished by scrambling

---

[*]In mod 2 (modulo 2) addition $0 + 0 = 0$, $1 + 0 = 1$, $1 + 1 = 0$.

the signal with a very long sequence known only to the parties involved. This practice was formerly largely confined to government communications, but heightened commercial interest in privacy has led to a standard method of encrypting, the data encryption standard (DES) promulgated by the National Bureau of Standards, and reasonably priced silicon integrated circuits to implement it. Scrambling for encryption is, of course, best applied at the source and destination of the signal to be encrypted, while scrambling to control signal statistics is applied within a particular transmission system.

Coding, as opposed to scrambling, adds enough redundancy to the signal to guarantee some desired properties, regardless of the pattern to be transmitted. T1 uses a ternary (three-level) code called bipolar (or AMI, for alternate mark inversion) with all zero limitation, used on T1. In this code alternate 1s are transmitted as + and – pulses, assuring dc balance and avoiding baseline wander. Further, an average density of one pulse in eight slots with a maximum of 15 zeros between 1s is required.[*] This maintains timing signal sufficient for the inexpensive, low-Q (80) timing circuits in the repeaters.

Another scheme to guarantee timing density with a ternary code is to replace strings of zeros with a pattern with two successive pulses of the same polarity, allowing its identification and removal at the receiving end. The arrangement, called BNZS (for bipolar with N zero substitution) is also in considerable use. In Europe, it is called HDB or CHDB, for high-density bipolar or compatible high-density bipolar. BNZS codes all carry one information bit per ternary symbol, and codes with $N = 3$, 4, and 6 are in use (see Tables 17.1.4 and 17.1.5). Greater information capacity can be obtained using block codes such as 4B3T (each block consists of 4 b on three ternary symbols). Generally, as the information capacity is increased, the reduction in redundancy results in poorer timing and baseline-wander performance. There are may variants, but all have spectra that are zero at both zero frequency and the symbol rate, as do the bipolar and BNZS codes.

TABLE 17.1.4 B3ZS Choice of Sequences to Substitute for Three Zeros

Last pulse transmitted	Last substitute sequence used	
	00 + or + 0 +	00 – or – 0 –
+	– 0 –	00 +
–	00 –	+ 0 +

Another type of code is epitomized by the *duobinary* code, which is a 1-b-per-symbol ternary code in which adjacent pulses of opposite polarity cannot occur. Thus the redundancy is used to reduce the transmitted energy at high frequencies rather than to control baseline wander and assure timing. Scrambling, high-Q tanks, and circuit design to minimize baseline wander are normally required. The duobinary spectrum is maximum at zero frequency and zero at half the symbol rate. The reduction of high-frequency energy makes it attractive for sharply band-limited media such as radio channels or, on wire pairs, to minimize crosstalk. Variants with larger numbers of transmitted levels have been dubbed *partial-response codes* and can also limit the transmitted spectrum in a similar fashion.

Various ternary codes are illustrated by example in Table 17.1.5. The redundancy added in any of these codes is sufficient to allow reasonably accurate estimates of errors in transmission by detection of violations

TABLE 17.1.5 Coding Examples

Code name	Coding rule	Example
Binary		10111001100000001111
Bipolar, AMI	Invert alternate 1s	+0 – + –00 + –0000000 + – + –
B3ZS, CHDB2	Invert alternate 1s, but for three 0s substituting according to Table 17.1.4	+0 – + –00 + –00 – + 0 + 0 – + – +
4B3T	Code Table	+00 + –00 + –0 – + – – –
Duobinary	From binary input sequence a first form sequence[*] C $C_n = a'_n \oplus C_{n-1}$ then output = $C_n + C_{n-1} - 1$	01111011101010100000 –0 + + + 00 + + 0000000 – – – –

[*] $\oplus$ = addition modulo 2, and a'_n is the logical inverse of a_n, that is, substitute 0 for 1 and vice versa. The first step avoids a long string of decoding errors resulting from a single transmission error.

[*]This is readily obtained in voice-band coding by simply not using the all-zero word, and the standard μ-law coder has this feature.

TABLE 17.1.6   CCITT Error Performance Objectives for a 27,500-km Connection at 64 kb/s

Performance classification	Objective
Errored seconds	Fewer than 8% of one-second intervals to have any errors (equivalent to 92% error-free-seconds)
Severely errored seconds	Fewer than 0.2% of one-second intervals to have a bit error ratio worse than 10^{-3}

of the constraints of the particular code. In uncoded systems parity bits may be added to accomplish this purpose. Forward-acting error correcting codes, in which redundancy is added to allow detection and correction of errors at the receiving end, have not been widely used on cable systems. They are, however, widely used on radio and satellite systems, where the capacity is limited by the transmitter power, rather than the bandwidth.

Performance Issues.   Digital transmission systems consist of terminals connected by cable or radio links that include regenerative repeaters. Each repeater detects, regenerates, and retimes incoming pulses, so the only impairments introduced are errors and jitter (unwanted phase modulation). Error performance is sometimes characterized in terms of average error rate (or ratio) which is the number of errors divided by the number of bits received, with the implicit assumption that errors are more or less randomly distributed. This has not proved to be a satisfactory measure, as most systems have enough design margin against thermal noise and crosstalk (the normal causes of randomly distributed errors) that the errors on actual systems in service occur in infrequent bursts, as a result of noise impulses or other disturbances. Consequently, the CCITT states error performance objectives for 64 kb/s in terms of error free seconds and severely errored seconds as shown in Table 17.1.6.* Objectives for performance at higher rates are shown in Table 17.1.7. This table is stated in terms of blocks of bits, intended to correspond to block sizes used for error monitoring.

Jitter can be removed by reading the signal into a buffer memory, and reading it out at a constant rate. It is only necessary to organize the network so that the jitter does not exceed the capacity of the buffer memories, and the CCITT has recommended jitter limits to this end.

Maintenance Techniques in Digital Systems.   Faults in the basic T1 system are detected by the inability of a channel bank to find framing pulses in its 1.544-Mb/s input signal. When such a condition persists long

TABLE 17.1.7   CCITT Error Performance Objectives for a 27,500-km Path at or Above the Primary Rate

Rate Mb/s	1.5 to 5	>5 to 15	>15 to 55	>55 to 160	>160 to 3500
Bits/block	2000 to 8000	2000 to 8000	4000 to 20,000	6000 to 20,000	15,000 to 30,000
Errored seconds	4%	5%	7.5%	16%	not specified
Severely errored seconds	0.2%	0.2%	0.2%	0.2%	0.2%
Background errored blocks	0.03%	0.02%	0.02%	0.02%	0.01%

An errored block is a block in which 1 or more bits are in error.
An errored second is a 1-s period with one or more errored blocks.
A severely errored second is a 1-s period which contains 30 percent or more errored blocks, or four contiguous blocks each of which has more than 1 percent of its bits in error, or a period of loss of signal.
A background errored block is an errored block not occurring as part of a severely errored second.

*An additional requirement based on a one-minute interval will likely be deleted as redundant in practice.

enough (perhaps 2 s), the channel bank causes the trunks it serves to be declared "busy," lights a red alarm light, and sounds an office alarm. It also transmits a special code that causes the other channel bank to take the trunks out of service, light a yellow alarm light, and sound an alarm. Maintenance personnel then clear the trouble, perhaps by patching in a spare T1 line, and proceed with fault location and repair. Channel banks can be checked by looping, i.e., connecting digital output and input. Repeatered line-fault location techniques were discussed earlier.

In higher-speed systems, automatic line and multiplex protection switching is often provided. A typical line-protection switch monitors and removes violations of the redundancy rules of the line signal on the working line at the receiving (tail) end. When violations in excess of the threshold are detected, a spare line is bridged on at the transmitting (head) end and if a violation-free signal is received on this line, the tail-end switch to spare is completed. If the spare line also has violations, there is probably an upstream failure and no switch is performed. A multiplex-protection switch is typically based on a time-shared monitor that evaluates each of several multiplexers and demultiplexers in turn by pulse-by-pulse comparison of the actual output with correct output based on current input. Multiplex monitors also usually check the incoming high- and low-speed signals using the line-code redundancy.

As the digital network has grown, there has been an increasing use of maintenance centers, to which all alarms in a geographic region are remoted, and which are responsible for dispatching and supervising maintenance personnel, directing restoration of failed facilities over other facilities or routes, and rearrangements for other purposes as well. There is also increasing provision, in network design, of alternative routes, sometimes by routing lines to form a ring, so that there are two physically separate paths between any two offices.

Synchronous Digital Hierarchy and SONET.   The existing plesiochronous digital network has grown piecemeal over decades, with the parameters of new systems reflecting the technology and needs at the time of their development. In the late 1980s, a worldwide effort brought forth a new hierarchy for higher rate systems to provide capabilities not possible in the existing network. In this new hierarchy, multiplexing is by interleaving of 8-b bytes, as in the primary rate multiplexes, as opposed to the bit interleaving used elsewhere in the existing network. Further, similar formats are used for multiplexing at all levels, and it is intended that new transmission systems will be at hierarchical levels. Another important feature of the new hierarchy is an overall plan for monitoring and controlling a complex network, and the inclusion of enough *overhead* in the formats to support it. In spite of the name, the new hierarchy allows nonsynchronous signals based on the existing hierarchy to enter, and multiplexing throughout includes enough justification capability to accommodate the small frequency deviations characteristics of reference clocks.

The new hierarchy starts at 51.84 Mb/s, and all higher rates are an integral multiple of this lowest rate. Multiples up to 255 have been envisioned, and structures for several rates have been standardized within the United States. The rates of most interest are shown in Table 17.1.8.

A single frame of the STS-1 signal consists of 810 bytes, as shown in Fig. 17.1.10. The bytes appear on the transmission line read from left to right, starting with the first row. The transported signal occupies 774 bytes of the frame, with the remainder of the frame dedicated to overhead. The transported signal plus the path overhead, the *payload*, are intended to be transported across the network without alteration as the signal is multiplexed to, and recovered from, higher levels. In order to accommodate frequency and phase variations in the network, the 783 bytes of payload can start anywhere within the 783 byte locations allocated to the payload, and continue into the next frame. The starting point is signaled by a *payload pointer* included in the line overhead, so the proper alignment can be recovered at the receiving end. This pointer, as well

TABLE 17.1.8   Major Rates in the Synchronous Digital Hierarchy

Line Rate Mb/s	Designation US (SONET)	Designation CCITT	Comment
51.84	STS-1		Used to carry one 44.736 Mb/s (DS-3) signal
155.52	STS-3	STM-1	Used to carry one 139.254 Mb/s signal
622.08	STS-12	STM-4	Used for fiber optic systems
2488.32	STS-48	STM-16	Used for fiber optic systems

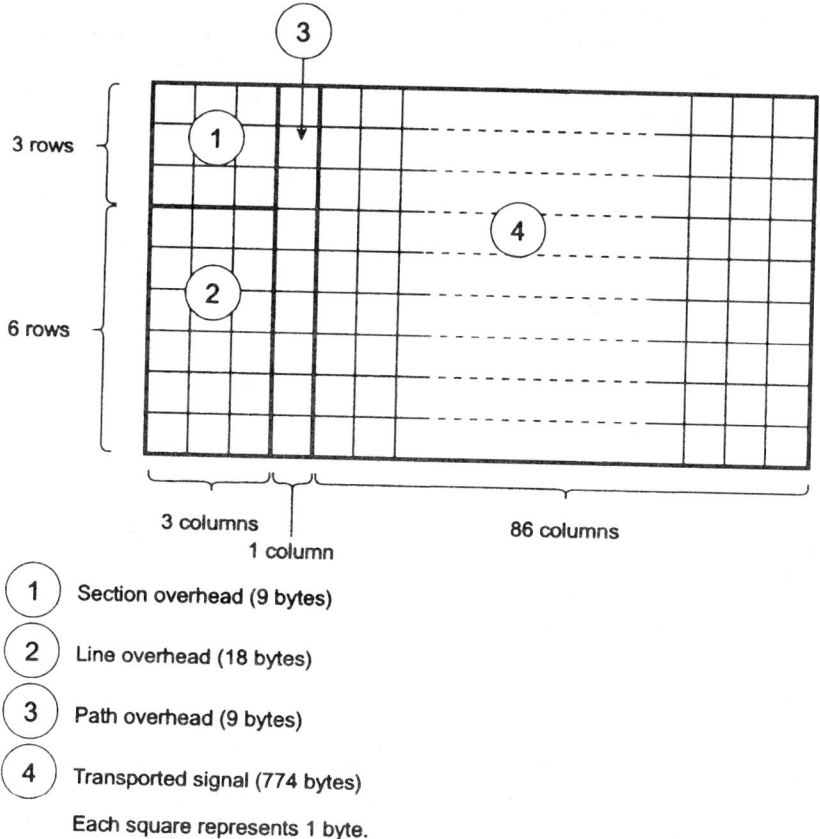

3 rows

6 rows

3 columns

1 column

86 columns

1) Section overhead (9 bytes)

2) Line overhead (18 bytes)

3) Path overhead (9 bytes)

4) Transported signal (774 bytes)

Each square represents 1 byte.

FIGURE 17.1.10   An STS-1 frame: (1) Section overhead (9 bytes); (2) line overhead (18 bytes); (3) path overhead (9 bytes); (4) transported signal (774 bytes). Each square represents 1 byte.

as the remainder of the line and section overhead bytes, are provided for the use of multiplex and line equipment, and will normally be changed several times as the frame passes through the network.

The *path* overhead is placed on the signal at the *path terminating equipment*, where the transported signal is assembled and embedded in the frame. It is not intentionally modified as the STS-1 frame passes through subsequent multiplexes and line systems, but can be read and used at intermediate points. This overhead is, from the point of view of the hierarchy, *end-to-end* information. It contains signals identifying the structure of the frame to aid in retrieving the embedded signal, status of maintenance indications, such as loss of signal for the opposite direction of transmission, a parity check on the previous frame, as well as provision for a message channel for use by the path terminating equipment.

Line overhead information may be inserted or modified when the STS-1 signal is multiplexed to a higher rate, or transferred between higher rate signals. Normally the capability to switch the higher rate signal to a standby facility in case of failure will be provided at such points, so the line overhead includes signaling for coordinating the operation of this protection switching, as well as functionality similar to the path overhead, but for use over the shorter "line."

The *section* overhead includes the framing alignment pattern by which the frame is located, and functionality similar to that of the path overhead, but for use and modification within individual sections, which end at regenerators or multiplexes.

TABLE 17.1.9   Representative North American Digital Systems for Paired Cable

Line rate, Mb/s	64-kb/s voice channel capacity	Widely used designation	Line formats	Usual medium	Typical repeater spacing, mi	Typical section loss, dB
1.544	24	T1 T10S (T1-outstate)	Bipolar with 1-in-8 1s density, 15 0s maximum	Wire pairs in single cable (but two directions in different units)	1 (on 22 gauge)	32
3.152	48	T1C, T1D, T148	Bipolar, 4B3T; duobinary; modified duobinary	As for T1, but also with shielded (screened) units	1 (on 22 gauge)	48
6.443	96	T1G	Quaternary	As for T1C		48

Frames for higher rate signals are generally similar. An STM-N frame has nine rows, and $N \times 270$ columns of which $N \times 3$ are for section overhead. The rather complex structure is based on *virtual containers, tributary units*, and *administrative units*, which are combinations of user signal with the overheads defined above appropriate to the administration of various types of paths.

Line Systems for Transmission

Systems on Wire Cable.   Systems providing trunks on wire pair are generally designed to operate on the same cable types used for voice trunks, and to share such cables with voice trunks. Large numbers of such systems that have characteristics indicated in Table 17.1.9 are in service in North America and Japan, although the fiber systems are increasingly being used in new installations. Wire pair systems at 2.048 Mb/s are common in Europe. All the above are four-wire systems, using one pair for each direction of transmission. Two-wire systems providing 144 kb/s in both directions on a single pair have been specified and developed for use as ISDN loops, but little deployed as yet. These two-wire systems use echo cancelers, or time compression multiplexing in which the pair is used alternately in each direction (at about twice the average bit rate) with buffering at the ends.

Systems on Fiber Optic Cable.   Fiber-optic systems, operating digitally, and using one fiber for each direction of transmission have developed extremely rapidly since their introduction in 1977, with steadily increasing capacity and, correspondingly, decreased per-channel cost. Systems for trunks and loops have been installed at many of the hierarchical rates (Table 17.1.1), but systems at even higher rates are most prevalent. The characteristics of such systems are summarized in Table 17.1.10, and some specific systems are shown on in Table 17.1.11. A branching unit, including 296 Mb/s regenerative repeaters, used in TAT-8 (Trans ATlantic cable 8) is shown in Fig. 17.1.11. Similar branching units in TAT-9 operate at 591 Mb/s, and include some multiplexing functions as well.

All terrestrial and submarine systems have customarily used intermediate regenerators when the system length requires gain between the terminals. Systems using optical amplifiers instead of regenerators have recently appeared, and the characteristics of one of these is also included in Table 17.1.10. In such systems, erbium doped optical amplifiers are used at intermediate points to overcome loss of the dispersion-shifted fiber with regeneration only at the ends. Figure 7.1.12 shows an amplifier designed for the system shown in the table. Typical output power of such amplifiers is +1 to +3 dBm.

Even these systems do not come close to exploiting the theoretical capacity of the fibers, and further developments are to be expected. Wavelength-division multiplexing, in which two or more transmitter-receiver pairs operate over a single fiber but at different wavelengths, is one way of tapping this capacity, and has seen limited use. The use of solitrons is being explored, and the record as of early 1993 for simulated long-distance transmission in the laboratory, 20 Gb/s over 13,000 km, used this technology.

TABLE 17.1.10 Parameters of Fiber-Optic Systems

Wavelength, nm	Fiber type	Bit rate, Mb/s	Maximum regenerator spacing, km[*]
850	Graded index	2–140	15–20
1300	Graded index	2–140	45–60
1300	Single mode	140–1700	25–60
1550	Single mode	1.5–2500	50–150

[*]Lower spacings generally correspond to higher bit rates.

TABLE 17.1.11 Some Fiber-Optic Systems

Name	Primary application	Bit rate per fiber	Wavelength, nm	Repeater spacing, km	Comment
TAT-10	Long undersea routes	591.2 Mb/s	1550	110	
FT-2000	Short and long terrestrial routes	2488.32 Mb/s	1310 or 1550	60 at 1310 nm 84 at 1550 nm	A variety of terminals is available
TAT-12	Long undersea routes	5 Gb/s	1550	33–45	Uses optical amplifiers

FIGURE 17.1.11 TAT-8 (transatlantic telephone cable no. 8) branching repeater with cable-laying ship in background. (*From AT&T. Used with permission*)

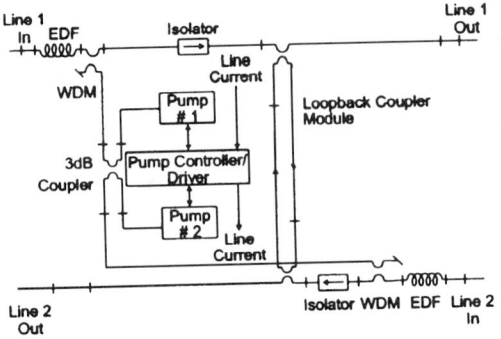

FIGURE 17.1.12 Amplifier pair for TAT-12 (AT&T section).

Terrestrial Radio. Frequencies allocated to telecommunications in the United States are shown in Table 17.1.2. Typical systems for analog signals modulate a carrier using low index FM with a signal consisting of one or more multiplexed mastergroups, and occupy a bandwidth of two or more times 4 kHz for each voice channel. Systems with very linear amplifiers have also used single-sideband AM, with a resulting bandwidth of closet to 4 kHz per voice channel.

While analog microwave radio once carried the bulk of long-haul telecommunications in the United States and many other countries, it has been mostly displaced in long-haul applications by optical fiber, particularly in the United States, and in short-haul applications by digital radio, owing to the need to interconnect with digital switches and other digital transmission systems.

A block diagram of a digital radio regenerator is shown in Fig. 17.1.8. The regenerator operates at i.f., and a complete station involves frequency conversion to and from rf, as well as receiving and transmitting antennas. An end station uses the transmitting and receiving portions of the regenerator separately to create and demodulate the transmitted and received signals.

Communications Satellite Systems. While the first experimental communications satellites were in low earth orbit, commercial satellites have almost uniformly been in a geostationary orbit, 22,300 miles above the equator. (The exceptions were in polar orbits, for better visibility from the northern polar regions.) In such an orbit, the satellite appears stationary from the earth, so a permanent communication link can be established using a single satellite, and earth stations with very directive stationary (or almost stationary) antennas. The disadvantages of this orbit are the high loss and delay resulting from the long distance the signal must travel. Table 17.1.12 lists representative applications of communication satellites, including current proposals for new low earth orbit systems.

Communications satellites receive signals from an earth station, and include *transponders*, which amplify and translate the signal in frequency and retransmit it to the receiving earth station, thus making effective use of the line-of-sight microwave bands without requiring erection of relay towers. The transponders are powered from solar cells, with batteries for periods of eclipse. *Spin-stabilized* satellites are roughly cylindrical and spin at about 60 r/min, except for a "despun" portion, including the antennas, that is pointed at the earth. *Three-axis-stabilized* satellites have internal high-speed rotating wheels for stability, and solar cells on appendages which unfold after they are in orbit. Adjustments in the position and orientation of a satellite in orbit are accomplished under control of the telemetry tracking and control (TTC) station on the earth, and the exhaustion of fuel for this purpose is the normal cause of end of life of the satellite, typically 10 to 12 years. (At end of life, the TTC moves the satellites to an unused portion of the orbit, where it remains, an archeological resource for future generations.) High reliability is necessary, and on-board spares for the electronics, switchable from the TTC, at a ratio of 50 to 100 percent, are typically provided. Table 17.1.13 gives the characteristics of two current satellites.

Most civilian communication satellites have used the common-carrier bands of 5925 to 6425 MHz in the uplinks, and the 3700 to 4200-MHz band in the downlinks. Now, direct broadcast satellites (DBS) use the 11- and 14-GHz bands for down- and uplinks, respectively. Since these bands are not so widely used in terrestrial

TABLE 17.1.12   Representative Applications of Communication Satellites

Application	Type of Service	Technical Characteristics	Status
Intercontinental telephone trunking	Point-to-point, 2-way	Earth station antennas to 30 m. FDMA, TDMA, and FM with a single channel per transponder used	The Intelsat system. Widely used where fiber optic cables are not available, also for handling peak loads on cables, and during repair of failed cables
Intercontinental TV transmission	Point-to-point, 1-way	Analog TV signals using FM with a single channel per transponder	Carried along with voice on the Intelsat satellites. Primary way of providing this service
National telephone trunks	Point-to-point, 2-way	Wide variety of antenna sizes, access methods have been used	No longer used in the United States, primarily because voice users don't like delay. Still used in countries with difficult terrain, long distances between population centers, or sparse networks
Distribution of TV signals to local broadcast stations or CATV distribution centers	Point-to-multipoint, 1-way	Smaller receiving antennas. Analog TV signals using FM with a single channel per transponder	Major provider of this service. Economics generally favorable compared to cable and microwave radio
Business and educational TV distribution, typically directly to viewing site	Point-to-multipoint, 1-way	Originally analog TV using FM, but increasingly digital, using coders, which, by removing redundancy encode the signal into 6 Mb/s or less, allowing multiple channels per transponder	Major provider of this comparatively new service
Data links, international and domestic	Point-to-point, 2-way	Low rate data channels can be multiplexed to a high rate to fill a transponder, or FDMA or TDMA can be used	Has seen considerable use, as with proper protocols, delay is not a problem in most applications. Fiber optic cables are eroding market
Maritime Mobile telephone	Fixed-point to mobile, 2-way	Operates at 1.5 GHz, with geosynchronous satellite	Via the INMARSAT system, the major modality for ship-to-shore telephony
Paging, short message	Fixed-point to mobile, 1-way	Would operate at 150 MHz with a total bandwidth of 1 MHz, using low-earth-orbit satellite	Proposal. Intent is to provide paging and limited message capability to personal receivers
Terrestrial mobile	Fixed-point to mobile, 2-way	Would operate at about 1.5 GHz, with a total bandwidth of about 20 MHz. Would use from 12 to 30 satellites in low earth orbit, using circular polarization and low directivity antennas on the mobile stations	Proposal. Intent is to provide mobile service roughly comparable to cellular, but available without the necessity for local terrestrial construction and network access

TABLE 17.1.13 Representative Communications Satellites

Satellite	Intelsat VI	Telstar 4
Type	Spin stabilized	Three-axis stabilized
Mass	4000 kg	4212 lb
Size	11.6 m high, 3.6 m diameter	Body 7.25 × 8.33 × 13.4 ft, extended length 80.4 ft
First launch	1989	1993
Launch vehicle	Ariane 4, Titan III	Atlas Centaur II AS
4/6 GHz Transponders and bandwidths	12 @ 36 MHz, 2 @ 41 MHz, 26 @ 72 MHz	24 @ 6 MHz
11/14 GHz Transponders and bandwidths	1 @ 36 MHz, 6 @ 72 MHz, 2 @ 77 MHz, 2 @ 150 MHz	8 @ 54 MHz, 16 @ 27 MHz (Can be used in pairs as 54 MHz also)
Primary power, W	2204	3744
Main applications	International telephony, TV, data	US domestic TV, data
Transponder output power, W	1.3–16 @ 4 GHz, 10 @ 11 GHz	12 @ 4 GHz, 60 @ 14 GHz (Can be doubled on some transponders)
Other features	SSTDMA (6×6 switch at microwave frequencies)	Can switch to a smaller receive spot beam to reduce and locate interference. Compensates for uplink rain attenuation by increasing transponder gain

microwave relay, interference problems are less although rain attenuation is much higher. All frequencies are subject to sun-transit outage when the satellite is directly between the sun and the receiving earth station, so that the receiving antenna is pointing directly at the noisy sun. This occurs for periods of up to $1/2$ h/d for several days around the equinoxes. The effect can be avoided by switching to another distant satellite during the sun-transit period. The propagation delay between earth stations is about 0.25 s in each direction for geostationary

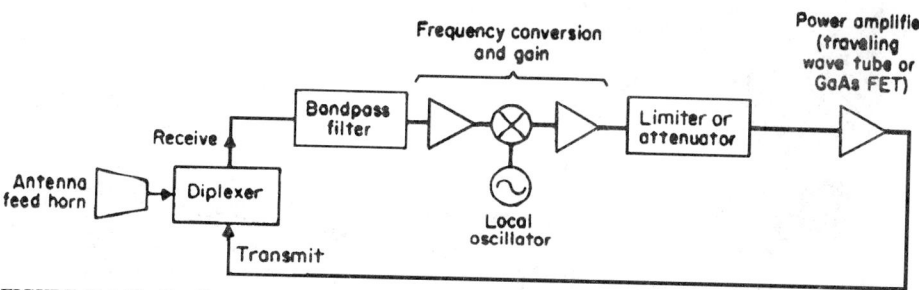

FIGURE 17.1.13   Satellite transponder.

satellites. This delay is of no consequence for one-way television transmission, but is disturbing to telephone users.

Some satellites use each frequency band twice, once in each polarization. Earth antenna directivity permits reuse of the same frequencies by different satellites as long as satellites are not too close in orbit (2° is the limit for domestic U.S. satellites in the 4- to 6-GHz band). Further frequency reuse is possible in a single satellite by using more directive satellite antennas which direct separate beams to different earth areas. Although Intelsat has made use of spot beams, most present satellite antennas have beam widths covering upward of 1000 mi on earth. A simplified transponder block diagram is given in Fig. 17.1.13.

Transponder utilization and multiple access.   A transponder can be used for a single signal (single carrier operation) which may be either frequency- or time-division-multiplexed. Such signals have included a single TV signal, two or three 600-channel analog master groups multiplexed together and used to frequency modulate a carrier, thirteen 600-channel master groups using compounders and single-sideband amplitude modulation, and a digital signal with rates up to 14.0 Mb/s used to modulate a carrier using quaternary phase-shift keying (QPSK), or coded octal phase-shift keying. Single-carrier operation can be either point-to-point (as for normal telecommunication) or broadcast (as for distributing TV programs). Transponders can also be used in either of two multiple-access modes in which the same transponder carries (simultaneously) signals from several different earth stations.

In frequency-division multiplex access (FDMA) the frequency band of each transponder is subdivided and portions assigned to different earth stations. Each station can then transmit continuously in its assigned frequency band without interfering with the other signals. All earth stations receive all signals but demodulate only signals directed to that station. In the limit of subdivision, one voice channel can be placed on a single carrier (single channel per carrier or SCPC). As the high-power amplifiers (HPA) in the earth station and the satellite are highly nonlinear, power levels must be reduced considerably ("backed off") below the saturation level to reduce intermodulation distortion between the several carriers. It is also possible to use *demand assignment* in which a given frequency slot can be reassigned among several earth stations as traffic demands change.

In TDMA (time-division multiple access) each earth station uses the entire bandwidth of a transponder for a portion of the time, as illustrated in Fig. 17.1.14. This arrangement implies digital transmission (such as QPSK) with buffer memories at the earth stations to form the bursts. A synchronization arrangement that controls the time of transmission of each station is also required. As at any given time only a single carrier is involved, less backoff is required than with FDMA, allowing an improved signal-to-noise ratio. Demand assignment can be realized by reassigning burst times among the stations in the network. Satellite-switched TDMA (SSTDMA) in which a switch in the satellite routes bursts among spot beams covering different terrestrial areas is a feature of Intelsat VI.

Transmission considerations.   The free-space loss between a geostationary satellite and the earth is about 200 dB. To overcome this large loss, earth stations for telecommunications trunks have traditionally used large parabolic antennas (10 to 30 m in diameter), high output power (up to several kilowatts), and low-noise receiving amplifiers (cryogenically cooled in some cases). Transponder output power is limited to the power available from the solar cells, and therefore, downlink thermal noise often accounts for most of the system noise with intermodulation in the transponder power amplifier a significant limiting factor. Consequently, the capacity of

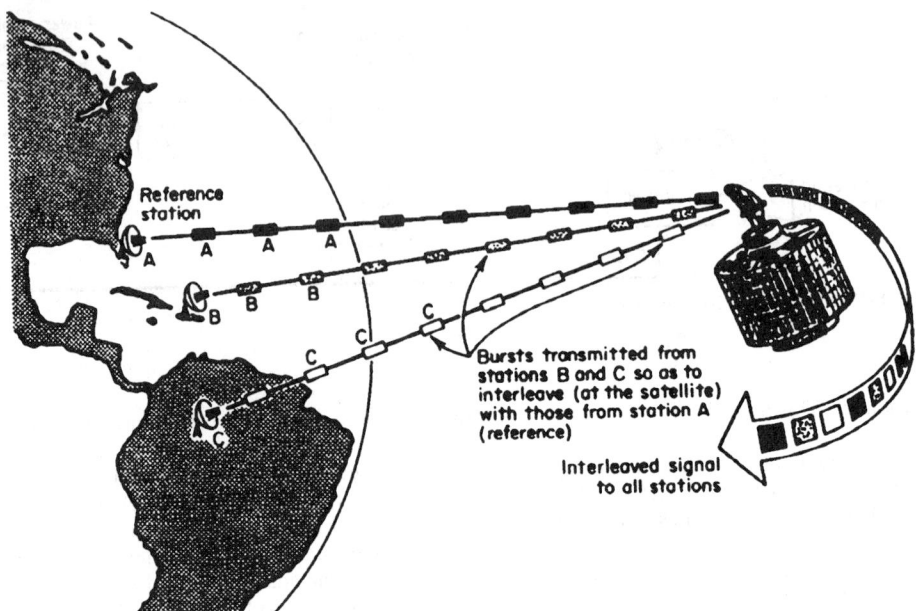

FIGURE 17.1.14 Satellite time-division multiplex access (TDMA). From Digital Communications Corporation; used by permission.

a satellite channel is often limited by the received signal-to-noise ratio (power-limited) rather than by the bandwidth of the channel.

For applications other than high-capacity trunking, the cost of large antennas at the earth stations is often prohibitive, so lower capacity is accepted, and received power may be increased by dedicating more power to the transponder or by use of spot beams, as the economics of the application dictate. Smaller antennas are less directive, possibly causing interference to adjacent satellites, unless the station is receive-only. Therefore VSATs (very small aperture terminals), which may have antennas as small as 1 m, typically operate in the higher frequency bands where directivity of smaller antennas may be adequate. Some applications, including the proposals included in Table 17.1.12, use much lower frequencies, and accept, or even exploit, the lesser directivity.

CHAPTER 17.2
SWITCHING SYSTEMS

Amos E. Joel, Jr.

A telecommunication service that includes directing a message from any input to one or more selected outputs requires a switching system. The terminals are connected to the switching system by *loops*, which together with the terminals are known as *lines*. The switching systems at nodes of a network are connected to each other by channels called *trunks*. This section deals primarily with systems that provide *circuit switching*, i.e., provision of a channel that is assigned for the duration of a call. Other forms of switching are noted later. Switching systems find application throughout a communication network. They range from small and simple manual key telephone systems or PBXs to the largest automatic local and toll switching systems.

SWITCHING FUNCTIONS

Introduction

A switching system performs certain basic functions plus others that depend on the type of services being rendered. Generally switching systems are designed to act on each message or call, although there are some switches that perform less often, e.g., to switch spare or alternate facilities. Each function is described briefly here and in greater detail in specific paragraphs devoted to each function.

A basic function of a circuit telecommunication switching system is connected by the *switching fabric*,[*] the transfer of communication from a source to a selected destination. Vital to this basic function are the additional functions of *signaling* and *control* (call processing) (Fig. 17.2.1). Other functions are required to *operate, administer*, and *maintain* the system.

Signaling

Automatic switching is remote-controlled switching. Transfer of control information from the user to the switching office and between offices requires electrical technology and a format. This is known as signaling, and it is usually a special form of data communication. Voice recognition is also used.

---

[*]The term *switching fabric* will be used in these paragraphs to identify the implementation of the connection function within a switching system. The term *communications* or *switched network* will refer to the collection of switching systems and transmission systems that constitute a communications system.

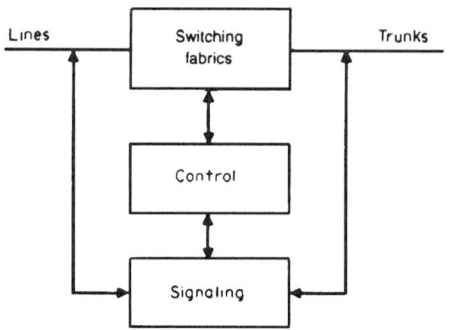

FIGURE 17.2.1   Basic switching functions in circuit switching.

Originally, signaling was developed to accommodate the type of switching technology used for local switching. Most of these systems used dc electric signals. Later, with the advent of longer distances, signaling using single- and multiple-frequency tones in the voice band was developed. Most recently, signaling between offices using digital signals has been introduced over dedicated networks, distinct from the talking channels.

As dialing between more distant countries became feasible, specific international signaling standards were set. These standards were necessarily different from national signaling standards since it was necessary to provide for differences in calling devices (dials) and call handling such as requests for language assistance or restrictions in routing.

Control

Control of switching systems and their application is called *system control*, the overall technique by which a system receives and interprets signals to take the required actions and to direct the switching fabric to carry them out.

In the past, the control of switching systems was accomplished by logic circuits. Virtually all systems now employ stored-program control (SPC). By changing and adding to a program one can modify the behavior of a switching system faster and more efficiently than with wired logic control.

Switching Fabrics

The switching fabric provides the function of connecting channels within a circuit-switching system. Store-and-forward or packet-switching systems do not need complex switching fabrics but do require connecting networks such as a bus structure.

Switching systems have generally derived their names or titles from the type of switching technology used in the switching fabric, e.g., step-by-step, panel, and crossbar. These devices constitute the principal connective elements of switching fabrics. Two-state devices that change in electrical impedance are known as *crosspoints*. Typically electromechanical crosspoints are metallic and go from almost infinite to zero impedance; electronic crosspoints change impedance by several orders of magnitude. The off-to-on impedance ratio must be great enough to keep intelligible signals from passing into other paths in the network (crosstalk). A plurality of crosspoints accessible to or from a common path or *link* is known as a *switch* or, for a rectangular array, a *switch matrix*. A crosspoint may contain more than one gate or contact. The number depends on the information switched and the technology.

Generally a number of stages of switches are used to provide a network in order to conserve the total number of crosspoints required. For connecting 100 inputs to 100 outputs, a single switch matrix requires $100 \times 100 = 10,000$ crosspoints. A two-stage fabric requires only 2000 crosspoints when formed with twenty 10×10 matrices. In a two-stage fabric, an output of each first-stage switch is connected to an input of a second-stage switch via a link. There is a connectable path for each and every input to each and every output. Since each input has access to every output, the network is characterized as having *full* access. However, two paths may not simultaneously exist between two inputs on the same first-stage switch and two outputs of a single-output-stage switch. (There is only one link between any first- and second-stage switch.) A second call cannot be placed, and this network is said to be a *blocking network*. By making the switches larger and adding links to provide parallel paths, the chance of incurring a blocking condition is reduced or eliminated. A three-stage Clos *nonblocking fabric* can be designed requiring only 5700 crosspoints. Even fewer crosspoints are needed if existing internal paths can be *rearranged* to accommodate a new connection that would otherwise encounter blocking. The design of most practical switching fabrics includes a modest degree of blocking in order to provide an economical design.

Large central-office switching networks may have more than 100,000 lines and trunks to be interconnected and provide tens of thousands of simultaneous connections. Such networks typically require six to eight stages of switches and are built to carry loads, which result in less than 2 percent of call attempts in the peak traffic period being blocked.

Network Control

While the switching system as a whole requires a control, the control required for a switching fabric may be separated in part or in its entirety from the system control function. The most general network control accepts the address of the input(s) and output(s) for which an interconnection is required and performs all the logic and decision functions associated with the process of establishing (and later releasing) connections. The control for some networks may be common to many switches or individual to each switch.

Self-routing is also used in fabrics where terminal addresses are transmitted through and acted on by the switches.

Some form of memory is involved with all networks. It may be intimately associated with the crosspoint device employed, e.g., to hold it operated, or it may be separated in a bulk memory. The memory keeps a record of the device in use and of the associated switch path. (In some electronic switching systems it may also designate a path reserved for future use.)

Operation, Administration, and Maintenance (OAM)

When switching systems are to be used by the public, a high-quality continuous service day in and day out over every 24-h period is required.

A system providing such reliable service requires additional functions and features. Examples are continuity of service in the presence of device or component failure and capability for growth while the system is in service.

Separate maintenance and administrative functions are introduced into systems to monitor, test, and record and to provide human control of the service-affecting conditions of the system. These functions together with a human input/output (I/O) interface constitute the basic maintenance functions needed to detect, locate, and repair system and component faults.

In addition to specific maintenance functions, *redundancy* in the switching system is usually necessary to provide the desired quality of service. Complete duplication of an active system with a standby system will protect against one or more failures in one system but presents severe recovery problems in the event of a simultaneous failure of both systems. Judicious subdivision of the system into parts that can be reconfigured (e.g., either of a pair of central processors may work with either of a pair of program memories) can greatly increase the ability of the system to continue operation in the presence of multiple faults.

Where there are many switching entities in a telecommunications network and as systems have become more reliable and training more expensive, the centralization of maintenance has become a more efficient technique. It ensures better and more continuous use of training and can also provide access to more extensive automated data bases that benefit from more numerous experiences.

For public operation, a basic subset of administration and operation features has become accepted as required features. These include the collecting of traffic data, service-evaluation data, and data for call billing.

SWITCHING FABRICS

Three different aspects will be considered in the design of switching fabrics: (1) the types of switching fabrics, (2) the technology of the devices, and (3) the topology of their interconnection.

Types of Switching Fabrics

The three types of switching fabrics are known by the manner in which the message passes through the network.

In *space-division* fabrics analog or digital signals representing messages pass through a succession of operated crosspoints that are assigned to the call for all or most of its duration. In *virtual circuit-switching* systems previously assigned crosspoints are reoperated and released during successive message segments.

In *time-division* fabrics analog or digital signals representing periodically sampled message segments from a plurality of time multiplexed inputs are switched to the same number of outputs. Using equal length segments assigned in time to *time slots* identifies them for address purposes in the system control.

There are two kinds of time-division switching elements, referred to as *space* switches and *time* switches (or time-slot interchanges, TSI). The space switch (also known as a time-multiplexed switch, TMS), shown in

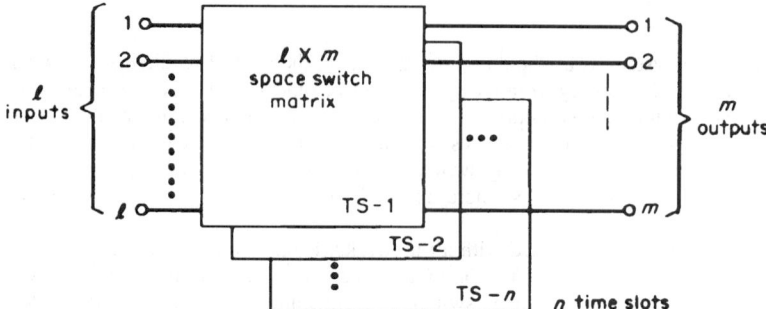

FIGURE 17.2.2 Time-multiplex switch (TMS): space switch.

Fig. 17.2.2, operates like the normal space-switch matrix but with each new time slot the electronic gates are reconfigured to provide a new set of input and output connections. The two-dimensional space switch now has an added third dimension of time.

The time-slot interchange uses a buffer memory, into which one frame of input information is stored. Under direction of the contents of the control memory, transfer logic recorders the sequence of information stored in the buffer, as shown in Fig. 17.2.3. To ensure the timely progression of signals through the TSI, two memories are used, one being loaded while the other is being read. The TSI necessarily creates delays in handling the information stream. Also with storage of message (voice) samples in T-stages, delay of at least one frame (e.g., 125 μs) is introduced into transmission by each switch through which the message passes.

Channels arriving at the switch in time-multiplexed form can be further multiplexed (and demultiplexed) into frames of greater (or lesser) capacity, i.e., at a higher rate and with more time slots. This function is generally used before using TSI so that channels from different input multiplexes can be interchanged.

Time-division switch fabrics are designated by the sequence of time and space stages through which the samples pass, e.g., TSST. The most popular general form of fabric is TST. The choice of others, e.g., STS, is dependent on the size of the fabric and growth patterns.

Analog samples can be switched in both directions through bilateral gates. An efficient and accurate transfer of the pulse is effected by a technique known as *resonant transfer*. For most analog and all digital time-division networks, however, the two directions of signals to be switched are separated. Therefore two reciprocal connections or what is known as four-wire connections (the equivalent of two wires in each direction) must be established in the network. When connections transmit in only one direction, amplification and other forms of signal processing can more readily be switched into the network.

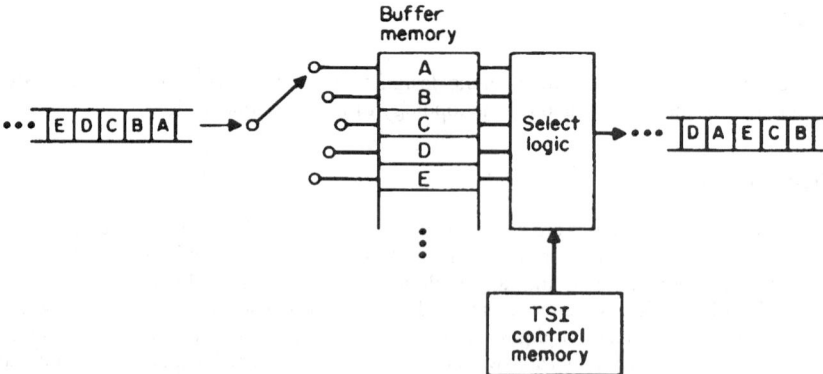

FIGURE 17.2.3 Time-slot interchange (TSI): time switch.

If multiplexing is performed in such a way that samples from an incoming circuit can be assigned arbitrarily to any of a number of time slots on an outgoing circuit, time-slot interchange and multiplexing are effectively achieved in a single operation.

With the application of digital facilities throughout telecommunications and in particular with the digitalization of speech, *digital time-division fabrics* are currently the most popular form of switching found in public networks. Digital voice communication exists throughout the public network. The *ISDN* (see Chap. 17.4) becomes a reality with digital access line interfaces in the local central offices, completing the end-to-end digital capability. As a result, switched 64,000 b/s clear digital channels are now available not only for voice but also for data.

The number of time slots provided for in a time-division fabric depends on the speed employed. Typically in a voice-band fabric there may be from 32 to 1024 time slots. The coded information in digitized samples may be sent serially (typically 8 b per sample for voice signals), in parallel, or combinations. Extra bits are sometimes added as they pass through the switch for checking parity, for other signals, or to allow for timing adjustments. For digital transmission, the crosspoints used in S stages of a switching fabric need not be linear.

Figure 17.2.4 shows the block diagram of the switching fabric for a no. 4 ESS, a large digital time-division switching system presently being deployed mainly in North America. Incoming digital T carrier streams (five T1 lines with 24 channels each) are further multiplexed to frames of 120 (DS120). The information is buffered in registers to permit synchronization of all inputs. The TSIs on the right side of the figure reverse the order of selecting and buffering; selected input sequences driven by a control memory (not shown) and sequentially gated out of the buffer attain the desired interchange in time. Note that the fabric shown is unilateral (left to right); the complete fabric includes a second unilateral fabric to carry the right-to-left portion of the conversation. This fabric has a maximum of 107,000 input channels, which can accommodate over 47,000 simultaneous conversations with essentially no blocking.

When digital time division fabrics are designed to work with digital carrier (T carrier in the United States) systems either in the loop as pair gain systems, line concentrators, or as interoffice trunks, carrier multiplexed bit streams can be synchronized and applied directly to the switch fabrics requiring no demultiplexing. This represents a cost advantage synergy between switching and transmission.

Frequency Division.   Since frequency-multiplex carrier has been used successfully for transmission, its use for switching has been proposed. Connections are established by assigning the same carrier frequency to the two terminals to be connected. Generally to achieve this requires a tunable modulator and a tunable demodulator to be associated with each terminal, and therefore frequency-division switching has had little practical application. *Wave Division* switching is a version of frequency division used in optical or photonic transmission and switching. Other forms of *photonic switching* use true space division to switch optical paths en masse in *free space* (see Hinton and Miller, 1992).

Switching Fabric Technology

Broadly speaking, basically three types of technology have been used to implement switching networks. (1) From the distant past comes the *manually operated switch*, where wires, generally with plug ends, can be moved within the reach of the operator. (2) *Electromechanical* switches can be remotely controlled. They may be *electromagnetically operated* or *power-driven*. Another classification is by the contact movement distance, *gross* motion and *fine* motion. Gross-motion switches inherently have limitations in their operating speeds and tend to provide noisy transmission paths. Consequently, they have seen little recent development. (3) The *electronic* switch is prevalent in modern design.

Electronic Crosspoints.   Gross- and fine-motion switches can be used only in space-division systems. Electronic crosspoints achieve much higher operating speeds. Although they can be used in space-, time-, and frequency-division systems, they have the disadvantage of not having as high an open-to-closed impedance ratio as metallic contacts. Steps must therefore be taken to ensure that excessive transmission loss or crosstalk is not introduced into connections.

The crosspoint devices are either externally triggered or are self-latching diodes of the four-layer *pnpn* type. The external trigger may be an electric or optical pulse. The devices have a negative resistance characteristic

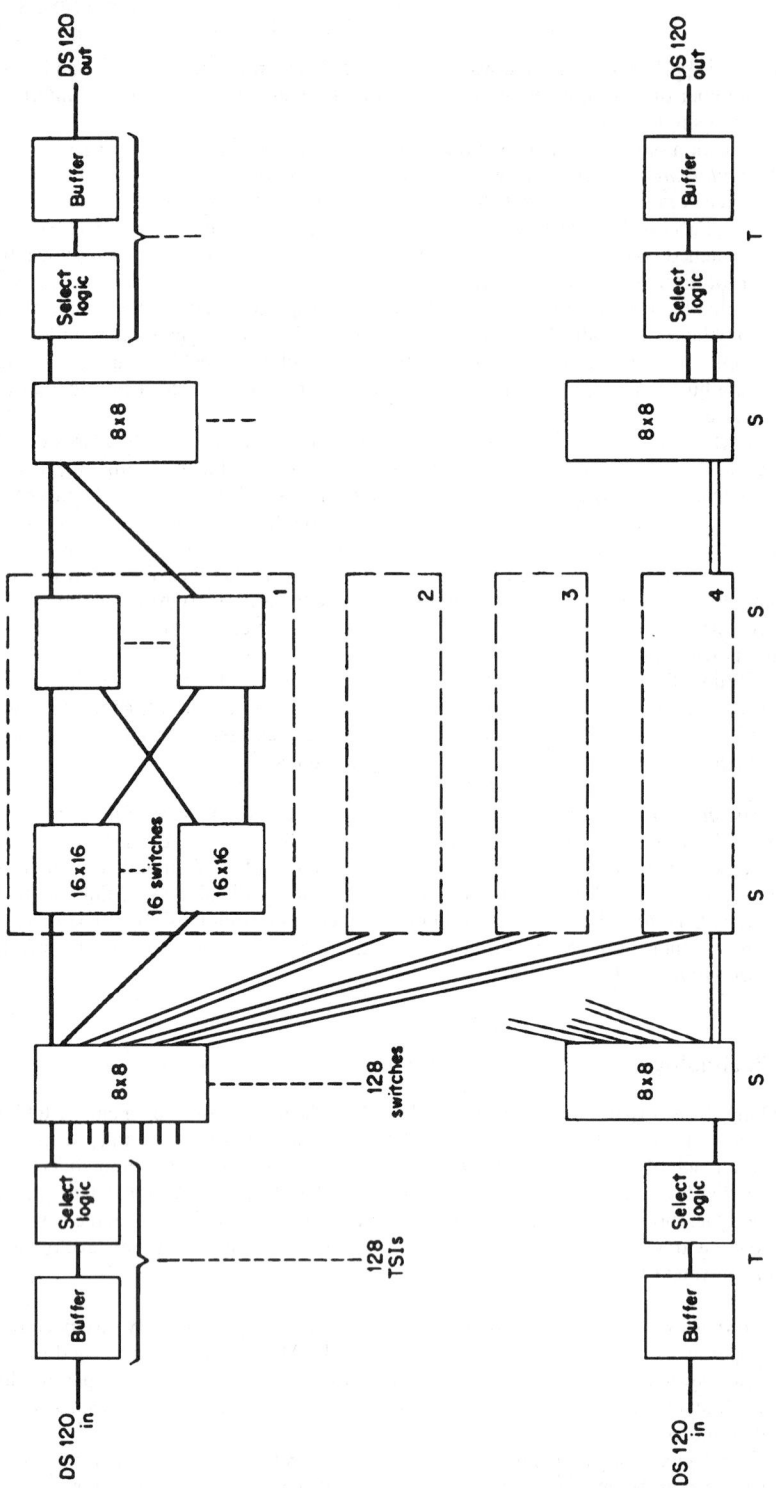

FIGURE 17.2.4 Block diagram of no. 4 ESS digital time-division fabric.

and are operated in a linear region if they are to pass analog voice or wideband signals. For fixed-amplitude pulse transmission, as in PCM, the devices need not be operated over a linear region.

Electronic crosspoints are generally designed to pass low-level signals at high speed. Recently a new class of high-energy integrated-circuit crosspoints has been developed, which can pass signals used in telephone circuit switching such as ringing and coin control.

Switching Fabric Topology

Of all the switching functions, the topology and traffic aspects of fabrics have been most amenable to normal analytical treatment, although many less precise engineering and technology considerations are also involved.

The simplest fabric is one provided by a single-stage rectangular switch (or, equivalently, a TSI) so that any idle input can reach any idle output. If some of the contacts are omitted, *grading* has been introduced and not every input can reach every output. With the advent of electronic crosspoints and time division, grading has become less important and will not be pursued further here. When inputs to a rectangular switch exceed the outputs, *concentration* is achieved; the converse is *expansion*.

A switching fabric is usually arranged in stages. Input lines connect to a concentration stage, several stages of distribution follow, and a last expansion stage connects to trunks or other lines. Within the design of a switching system, provision is usually made for installation of switches in only the quantity required by the traffic and number of inputs and outputs of each particular application. To achieve this, the size of each stage and sometimes the number of stages is made adjustable. Consideration of control, wiring expense, transition methods (for rearranging the system during growth without stopping service), and technology leads to the configurations selected for each system.

In order to achieve acceptable blocking in networks that are smaller than their maximum designed size, more parallel paths are provided from one stage to the next. In this case, because the distribution need is also reduced, the connections between stages are rewired so that those switch inputs and outputs which are required for distribution in a large network are used for additional parallel paths instead.

It is convenient to divide the fabric into groups of stages or subfabrics according to the direction of the connection. (Calls are considered as flowing from an originating circuit, associated with the request for a connection, to a terminating circuit.) Local interoffice telephone trunks, for example, are usually designed to carry traffic in only one direction. The trunk circuit appearances at a tandem office are then either originating (incoming) or terminating (outgoing). Figure 17.2.5a illustrates such an arrangement where the whole network is *unidirectional*.

Telephone lines are usually *bidirectional:* they can originate or terminate calls. For control or other design purposes, however, they can be served by unidirectional stages, as shown in Fig. 17.2.5b. Concentration and expansion are normally used with line switching to increase the internal network occupancy above that of lines. In smaller systems a bidirectional network can serve all terminal needs: lines, trunks, service circuits, and so forth (Fig. 17.2.5c). When interconnection between trunks, as in a combined local-tandem office, is required, line stages can be kept bidirectional while trunk stages are unidirectional (Fig. 17.2.5d). When the majority of trunks are bidirectional, as may occur in a toll office, a bidirectional switching fabric is used (Fig. 17.2.5e). Many other configurations are possible.

SYSTEM CONTROLS

Stored Program Control

As discussed earlier, most modern systems use some form of general-purpose stored-program control (SPC). Full SPC implies a flexibility of features, within the capability of existing hardware, by changes in the program.

SPC system controls generally include two memory sets, one for the program and other semipermanent memory requirements and one for information that changes on a real-time basis, such as progress of telephone calls, or the busy-idle status of lines, trunks, or paths in the switching network. These latter writable memories are *call stores* or *scratch-pad memories*. The two memories may be in the same storage medium, in which case there is a need for a nonvolatile backup store such as disc or tape. Sometimes the less frequently used programs are also retrieved from this type of bulk storage when needed.

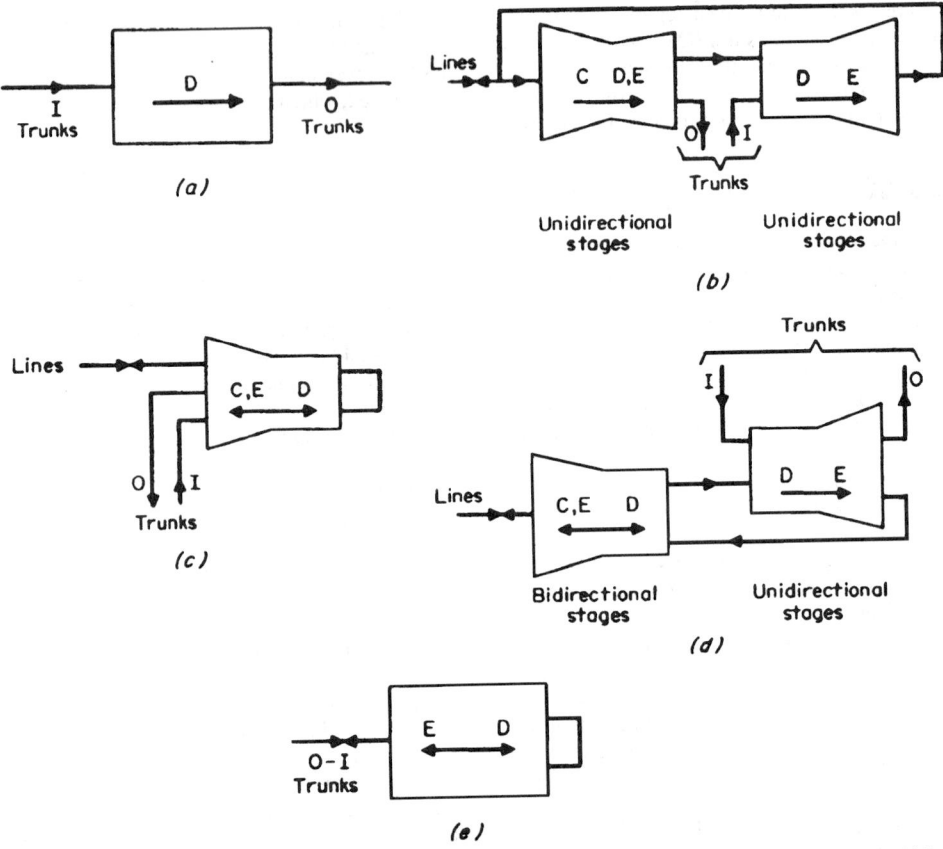

FIGURE 17.2.5 Switching fabrics: (*a*) unidirectional network (tandem); (*b*) unidirectional network (local); (*c*) bidirectional network (local); (*d*) combined network (local-toll); (*e*) bidirectional network; line stages: C = concentration, E = expansion, D = distribution; trunk stages: O = outgoing, I = incoming. Arrows indicate direction of progress of setup of call.

Nonprogram semipermanent memory is required for such data as *parameters* and *translations*. A switching system is generally designed to cover a range of applications; memory, fabric, and other equipment modules are provided in quantities needed for each particular switching office. Parameters define for the program the actual number of these modules in a particular installation. The translation data base provides relations between signal addresses and physical addresses as well as other service and feature class identification information.

Central Control

Single Active. The simplest system control in concept is the *common* or *centralized control*. Before the advent of electronics, the control of a large telephone switching system generally required up to 10 or more central controls. The application of electronics has made it possible for a system to be fully serviced by a single control. This has the advantage of greatly simplifying the access circuits between the control and the remainder of a switching system. It also presents a single point in the system for introducing additional service capabilities. It has the disadvantage that a complete control must be provided regardless of system size, and complete redundancy is often required so that the system can continue to operate in the presence of a single trouble or while changes are being made in the control.

Redundancy is usually provided by a duplicate central control that is idle and available as a standby to replace the active unit (the unit actually in control) if it has a hardware failure. The duplicate may carry out each program step in synchronism with the active unit; matching circuits observe both units and almost instantaneously detect the occurrence of a fault in either unit. Otherwise central-control faults can be detected by additional redundant self-checking logic built into each central control unit or by software checks. In these latter modes of operation the central controls may be designed to operate independently and share the workload. *Load sharing* allows the two (or more) central controls to handle more calls per unit time than would be possible with a single unit. However, in the event of a failure of one unit, the remaining unit(s) must carry on with reduced system capacity.

Load-sharing represents *independent multiprocessing* where two or more processors may have full capability of handling calls and do not depend on each other. At least part of the call store memory (containing, for example, the busy-idle indications of lines) must be accessible, either directly or through another processor, to more than one processor in order to avoid conflicting actions among processors.

A small office would require less than the maximum number of processors, so that the control cost is lower for that office; as the office grows, more processors can be added. Increasing the number of processors results in decreasing added capacity per processor. Conflicts on processor access to memory and other equipment modules, with accompanying delays, accelerate with the number of processors. Independent multiprocessing or load sharing rapidly reaches its practical limit.

Functional Multiprocessing. Another way to allocate central-control workload is to assign different functions to different processors. Each carries out its task; together they are responsible for total capability of the switching system. This functional or *dependent* multiprocessing arrangement can also evolve from a single central control. A small office can start with the entire program in one processor. When one or more functional processing units are added, the software is modified and apportioned on a functional basis. As in load sharing, the mutually dependent processors must communicate with each other directly or through common memory stores.

In handling calls, each processor may process a portion and hand the next step to a succeeding processor, as in a factory assembly line. This *sequential multiprocessing* has been used in wired-logic switching systems. Virtually all SPC-dependent multiprocessing arrangements are *hierarchical*. A master processor assigns the more routine tasks to subsidiary processors and maintains control of system.

The one or more subsidiary processors may be *centralized* or *distributed*. If the subsidiary processors are centralized, they have full access to network and other peripheral equipment. Distributed controls are dedicated to segments of the switching network and associated signaling circuits. As network and associated signaling equipment modules are added, the control capability is correspondingly enlarged. Most newer switching systems use distributed controls.

TYPES OF SWITCHING SYSTEMS

In the preceding paragraphs the various switching functions were described. A variety of switching systems can be assembled using these functions. The choice of system type depends on the environment and the quantity of the services the system is required to provide. Combining the various types of systems within one embodiment is also possible.

Circuit Switching

Circuit switching is generally used where visual, data, or voice messages must be delivered with imperceptible delay (<0.050 s) and are relatively long. For these applications connections are established through a switch that will pass the required bandwidth. The Nyquist criterion is used in time-division networks by choosing the sample rate to be at least twice the desired maximum bandwidth.

Circuit switching usually (but not always) implies message dialogue, i.e., reciprocal communication in both directions.

Circuit switching has the advantage of being able to switch a very broad spectrum of signal rates. For example, analog or digital video signals can be switched using either metallic or nonmetallic crosspoints.

Systems Other than Circuit Switching

Services that provide one-way transmission and can accept deferred delivery of data messages have been available in both public and private versions for many years; the switching involved is sometimes called *message switching*. Messages are stored in the switching system until a transmission channel is available, providing efficient use of channels, or until it is convenient to deliver the message to the recipient's station. Further, messages can be provided from the source to the communications system in bulk, or over a single high-volume *port*, which permits efficiencies in the nature and operation of such a source. Semipermanent record storage and message numbering can be provided by the switching system. Message switching lends itself particularly well to *multiple-address messages.*

The implementation of message switching service has evolved to systems that are computer-based, employing a large electronic local telephone switching system augmented with large disk stores, special input-output equipment, and special programs. The airlines industry has long been a user of computer-based electronic message switching for applications ranging from operations traffic to agent traffic.

Message switching, like circuit switching, provides for full duplex operation. A one-way service is known as *file transfer.*

Long data messages or bursts of real-time generated data are divided into packets and transmitted over networks especially arranged to take advantage of these message bursts. The packets are usually of uniform maximum length and prefixed with an address header. Switching takes place to select available network facilities over which to transmit the packets. Packets are reassembled at the receiving terminal or switching node. Retransmission of missing or defective packets are implemented at the switch.

This *packet switching* of data is used in a wide variety of private and public networks; local, campus, metropolitan, wide-area, and so forth, most with their own standard digital protocols. Analog data generally uses voice circuit switching. Packet switching may be enhanced by the use of virtual circuit switching.

Fast packet switching uses locally generated address headers that interpret incoming packet addresses so that they may pass through unique high-speed self-routing switching fabrics. As a result, packets from different inputs destined for the same output route are switched to the same fabric output(s).

By assigning packets to time slots in digital transmission systems and eliminating the checks and retransmission required by packet protocols a technique known as *frame relay* is popular. A switched version is becoming available.

To take advantage of higher speed, broadband digital transmission, such as SONET digitized visual, data, or voice services may be reduced to a fixed size (53 bytes) packet standard called a "cell." This method provides for the delivery of several or multimedia services intermixed over the same transmission facilities. Like packets, cells include headers that are interpreted by the switch in several ways depending on the type of service represented by the information contained in each cell. Networks functioning with these cells are said to employ the *Asynchronous Transfer Mode (ATM)* See McDysan and Spohn (1995).

ATM or *Cell Switching* enables ATM to become a ubiquitous switched services network or broadband ISDN for both private and public applications. Like fast packet switches, switch fabrics for ATM cells are designed to forward cells to the route indicated by the address. Cells are self-routed through space stage fabrics or by a time stage using a shared memory. Paths and channels through the network are identified to enable switches to act as both a combined virtual circuit and packet switch depending on the adapted service represented by particular cells. To reduce the probability of blocking in the switch, cell buffers are needed.

ATM switching is also useful in joining together backbone and other networks that use different protocols internally, such as LANs.

SIGNALING

Basic Purposes

Signaling has three basic purposes: *supervising* the call or message, *addressing* the call or message, and conveying *supplementary* information relative to the call. Supervision consists of indicating a call origination, answer or end, and station alerting. The application of these signals may be between the stations and the switch, in which case it is known as *station signaling*, or between switching offices, when it is known as *interoffice signaling*. There are two basic approaches to signaling, *per channel* and *common channel*. Per channel signaling may be in or out of the bandwidth used to carry the messages. In the case of digital time-division multiplex transmission, signals may occur in each channel time slot (*in slot*) or in a separate signaling channel (*out of slot*). Common-channel signaling, on the other hand, is carried over one or more channels dedicated to signaling which usually serve many trunks. The electrical characteristics of signals are divided into three categories, *dc*, *ac*, and *digital signals*.

Standards and Compatibility

Only by following standards for the transmission and use of signaling can the various elements of a telecommunication network function together. Signaling standards for public networks operating between nations are generally established by the CCITT, now known as the "International Telecommunications Union-Telecommunications Standardization Sector (ITU-TSS)." The deliberations leading to the setting of these standards involve telecommunication administrations and recognized private operating agencies with the assistance of scientific and industrial organizations. Over the years, as new technology and requirements have appeared, new and revised signaling standards have followed.

National systems are more dependent on the type of signaling required by the local switching systems, and conditions vary widely throughout the world. In the United States the "Alliance for Telecommunications Industry Solutions (ATIS)" is responsible for devising national standards.

Station Signaling

There are few varieties of station signaling in public networks. One is the universal dc loop supervision and dial pulsing. Other signals sent over the loop are those for ringing, coin control, party identification, toll denial (signals indicating limitation of access to the DDD network of a particular station), metering, and so forth. Worldwide they vary in voltage, frequency, and how ground is used as a conductor. An important attribute is the distance or electrical range over which each signal functions satisfactorily.

A standard ac station-address signaling using two-out-of-eight frequencies (twice one-out-of-four) is known as *dual-tone multifrequency* (DTMF) (or "TOUCH TONE" as trademarked by AT&T). For digital access ISDN national standards have been issued by ATIS. *Voice recognition* is also being introduced as a means for addressing telephone calls. For purposes of interconnection to the United States network an Electronic Industries Association standard has been issued on station signaling as applied to PBXs.

Distributed Switching

The range of dc signaling has been particularly important in station signaling since it has determined the location and the number of wire centers required to serve a given area. Various signaling arrangements were developed to extend the dc loop and ringing range of central switching systems. The lower cost of electronics for small transmission and switching systems has made it economical to extend the range further and to decentralize switching.

Interoffice Signaling

Many forms of dc interoffice signaling have been developed and used. They were designed to accommodate specific electromechanical switching systems.

Signaling systems designed for use between offices, particularly over long distances and between countries, must be designed to be transmitted over carrier systems either analog or digital. For analog carrier trunks international standards have been set involving codes of two-out-of-six frequencies (multifrequency code, MFC) for address signaling and one or two frequencies for supervisory signaling. In digital carrier systems a means of carrying encoded supervisory signals in the bit stream is provided to be used with the MFC address signaling.

The per channel signaling systems have a limited number of signals that can be transmitted and usually lengthen the time of use of the transmission paths. To overcome these limitations and provide other advantages, common channel signaling (CCS) systems have been introduced. Common-channel signaling systems consist of a full-time data link between two signaling points and the necessary terminal equipment. By appropriately encoding the data stream, a variety of signals needed for both address and supervisory as well as service enhancements can be transmitted at higher speed between SPC switches. If the CCS link carries the signals for the group of trunks between two offices only, it is said to be operating in the *associated* mode. To signal between two offices with too few trunks to justify an associated link, it is possible to signal over two or more signaling links in tandem via intermediate signaling points. These act as packet switches, routing messages between offices depending on label information contained in each packet. The links carry the signals for more than one trunk group, are not associated with any one, and are said to be operating in the *nonassociated* mode.

For common-channel signaling, two international standards were adopted, one optimized for analog transmission (CCITT signaling system 6) and one optimized for use with digital data links (CCITT signaling system 7). With digital transmission becoming dominant, signaling system 7 is gradually being introduced into all networks.

Signaling Fabrics

With common channel signaling becoming the basic signaling method, its signaling links form a separate *signaling network*. The switching systems appear as users on this network.

Each office eventually will rely on this network for all its interoffice or internodal signaling needs. *Signal transfer points* (STPs) serve offices in a geographical *signal region*. An associated signaling link between two offices at any point in the signaling network, it can be added where justified by traffic and reliability considerations. To ensure service reliability, each office has access to two or more STPs in its region (see Fig. 17.2.6). Alternate signal messages from served offices are sent over A links to a different STP in the same region. Signaling networks interconnect not only signaling regions of a particular carrier but also between carriers. The

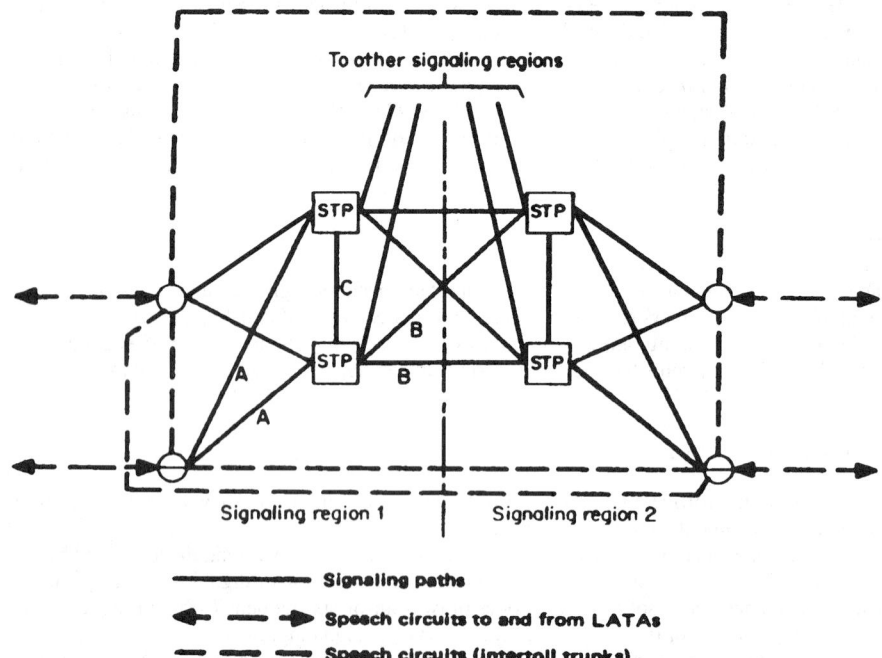

FIGURE 17.2.6 Signaling-network concept. (*Adapted from R. C. Nance and B. Kaskey, ISS Conf. Rec., 1976, p. 413. Copyright 1976 Inst. of Elec. and Comm. Engrs. of Japan. Used by permission*)

message switches that serve as STPs employ redundancy of their own and achieve the same quality of service continuity as the switching offices.

Separate data links can also be used for signaling from customer premises, e.g., for passing the calling PBX station identification to the central office in what is known as automatic identified outward dialing (AIOD).

Tones and Announcements

Switching systems need to inform the users of progress being made in serving a particular call. Tones and verbal announcements are used for this purpose. To enable tones to be automatically detected by call originators and to prevent interference with DTMF signals, a precise (frequency) tone plan has been adopted in the United States.

The tones range from *dial tone*, which prompts the caller to start dialing, to *busy tone*, which indicates that a called line is unavailable. Similarly, where more than symbolic information is needed, verbal announcements are given to the caller.

Generally tones and announcements reach the caller through a regular switching network connection. For tones, one or more network terminals are used. If the same source supplies all terminations, low impedance is employed to prevent crosstalk between calls simultaneously reaching the same tone source.

Announcements may be composed from a recorded vocabulary; e.g., intercept systems announce number changes by concatenating a limited vocabulary of phrases and digits.

SERVICES AND OAM FEATURES

A *service* is what the user perceives as being delivered by a switching system. OAM *features* of central-office telephone switching systems are those uses of system functions which are needed to operate, administer, and maintain the switching, customer terminal, and transmission equipment to provide services.

For example, to provide telephone service (see below) a directory number or address is assigned to a line. Because directory numbers are not permanently associated with switching-network terminations, an administrative feature is required to store this association as a translation in system-control memory.

Telephone Service

The switching functions required to establish a telephone call were listed earlier. Other services also use these functions, plus such additional functions as may be needed.

Telephone service includes the ability to place calls in a telephone network. Users reach other users' telephones by dialing a sequence of digits and can similarly be reached. To respond and alert users, switching systems in a network provide the appropriate signals: dial tone, ringing, busy tone, and so on.

In many networks, operators provide additional services: assisting in placing toll calls (person-to-person), calculating charges and collecting coins in certain calls originating from public coin telephones, providing telephone numbers for users given a name and address (directory assistance), informing users whenever they have dialed a number that has been changed or is not in service (intercept), and so on. These operator services may be fully or partially automated with SPC systems connected to the network. Depending on the operating entity and the service provided, users may be billed for the use of such operator services.

Additional services are provided (again depending on the country and/or the operating entity) for additional one-time or periodic charges. These include DTMF dialing, key telephone or PBX services, or private user networks embedded within the public switched communication network, special telephone sets, and special calling services.

A user wishing to purchase or subscribe to services such as the above makes appropriate arrangements with the business office of the telephone administration or operating company. Some of the other services may be activated, used, and deactivated by dialing directly into an SPC switching system. As the connectivity of SPC systems pervades a communication network, more and more services become available to meet user needs and the trend is toward further automation of the direct control by the user in obtaining such services.

Call waiting is an example of a calling service available in the North American network where SPC switching systems are in service. When a user who has call-waiting service is engaged in a telephone conversation, a distinctive tone will be received if a new call arrives. The user can be connected to the new call by "flashing," or briefly operating the switch hook. The system responds by placing the existing connection in a hold state and reconnects the subscribing user to the new call. By subsequent flashes of the switch hook the subscribing user can alternate connections between the two calls as needed.

New Services

With the introduction of SPC and associated bulk memory, many new services have been devised for switching offices. Usually they can be implemented with little or no additional hardware except for memory to store additional programs. Switching systems generally include a dynamic set of service capabilities.

Because of their location in networks switching systems provide for control implementation of most services. They serve party and mobile telephones, private branch exchanges, and many different rate categories (flat rate or measured service, coin, and so on.). Some central office systems include capabilities known as "CENTREX" service for serving intramural business needs directly, avoiding the need for key and PBX systems.

Just as the use of SPC in central offices has increased the number and flexibility of the services offered, the use of common channel signaling has created new opportunities to provide new services and features in public and private networks. A popular service made possible by the deployment of the common channel signaling networks is *caller identification*. Except where the caller blocks its application on specific calls or on all calls, the calling telephone number appears on a device at the called telephone after the first ring. Eventually it is expected that this service will be extended nationwide. (See Walker, 1991.) By adding centralized data bases, known as network control points (NCP) or service control points (SCP), to some of the STPs of a signaling network, it is possible to offer new network-based services. Some of those currently in service include translation of called addresses when there are special charge treatments, credit-card calling, and virtual private networks using public network facilities.

By adding *triggers* to the call processing software in each local central office, calls requiring special treatment are directed to an SCP that is able to continue call processing and addressing that is tailored to the needs of specific subscribers or organizations within an area or nationwide. This concept is known as the *Advanced Intelligent Network (AIN)*. *See* Majeti and Prasad, 1993. Services changes are implemented centrally through a *service management system (SMS)* using *service creation environment (SCE)* software.

Customer premises equipment has been at the forefront of offering new services. General business users need both voice and data services. Most of the PBXs designed since 1980 have included provision for switching data. With the growth of such data services have come new switching system architectures that use the same switching networks for connecting digitized voice and packetized data terminations. Many can also serve local area networks (LANs) through separate gateways or with LAN rings built into PBXs.

OAM Features

Network Management.   Automatic and manual routing control is provided in large modern networks to route calls around portions of the network that are temporarily congested or where disasters or other types of problems have reduced or eliminated the ability to reach or receive traffic from portions of the network. This is known as *network management*. Network management also includes turning back calls to reduce congestion to a particular destination and the temporary augmentation of facilities to serve overloads. Further, it includes threshold measurements and alerting when offered calls do not appear to be reaching their intended specific destinations. The application of self-healing loops of digital optical facilities for trunks is improving network management and reliability.

Traffic Measurement.   To observe how the switching system is operating, an administration needs indications of what loads are being carried by the system components. Typical are counts of calls, measurements of call delays, counts of call dispositions, duration of all circuits busy, and circuit use. The last measurement is best reported in terms of call hours per hour, or *erlangs*. It is often made by periodic counting of busy circuits and reporting the average or by summing the actual hours of circuit use for all circuits in a group. Output of data may be printed out directly at a switching system or may be transmitted to a centralized support system.

Billing. Broad allocation of costs to various users of a network can be determined by traffic measurements. In public service systems, more detailed information is needed. Two basic methods are used, *bulk billing* and *detailed billing.* In bulk billing, charge units are allocated to call setup, duration of call, and distance called. In Europe, the pulse metering system is used, where the local office generates pulses on a per line basis. In North America detailed billing is used for toll calls. Call details are recorded, either centrally or locally, and charges are later computed in centralized data processing centers where bills are prepared. This process, called *automatic message accounting* (AMA), has the advantage of more flexibility and full reporting of charges to the customer at the expense of more data processing than is required of bulk billing.

For coin telephones, with pulse metering systems, the charge pulses can be used to control coin collection directly. With detailed billing systems, operators or centralized charge calculation and control capabilities are required to quote charges and handle the collection of coins.

Calls in public data networks are usually billed on the basis of duration of call or number of packets plus a network access charge. The rates may change with the time of day and priority required.

Maintenance

The place of the switching system in a communication network puts unusually severe requirements on its maintenance. Loss of service for any but very short periods is unacceptable. Detection, recovery, diagnosis, and repair of trouble must be carried out while the system continues to process calls. Central processor design therefore incorporates a considerable amount of attention to internal trouble detection and the ability to reassign faulty system elements automatically so that processing can continue without loss of calls. Tests diagnosing the nature of the trouble may also be automatic or subject to request by maintenance personnel.

Alarms and diagnosis results must be generated; most indications appear as lighted lamps or typewritten characters. A standard *user-machine language* is being adopted by the ITV-TSS.

Contents of translation and other semipermanently stored data bases change daily in a large public system. Provision is made to change this information locally or remotely. Care is needed to ensure that changes are free from error and that the data bases will not be lost. In some systems changes are made in two steps. The information is first stored in a temporary *recent change* location and later relocated to a regular data base address.

Another class of feature deals with the cutover of a new system, recovery of a failed system, and change of hardware or software in a working system.

In addition to the tests internal to the switching system, separate test sets, consoles, display boards, and test access switchboards are used to varying degrees and constitute maintenance features. The switching system is also used for connecting test circuits to remote switching and transmission systems.

In the design of systems, a critical factor is the objective in-service time. For public switching systems for two-way voice service, an objective of 2 h in 40 years downtime has been used.

Centralized Operation Support. In the past, operational features have been largely contained within the design of the switching system. There is a trend toward providing these features, as well as maintenance and administrative facilities, at a centralized point where they can service a number of switching systems. The centralized systems are known as *operations support systems* (OSS). Operations includes maintenance and administration as well as operation. The design of switching system hardware and/or software includes features needed to provide and interact with OSSs.

In some centralized support systems, programs and other information for infrequently used real-time call processing can be accessed at a centralized call processor or OSS designed with the required reliability objective.

Applications

This section gives the reader an understanding of typical switching systems. To determine whether a given switching system meets a specific application need one must first know the basic dimensions and capacity requirements.

Dimensions

Limiting the size of a system are the number of physical terminations for lines and trunks, the call-carrying capacity of the control, the traffic capability of the network(s), and the address range of the memories.

The *terminations* are those lines, trunks, service, and similar circuits, which are principal service inputs and outputs of the system. The service circuits do not extend out of the system but are used to provide a function such as ringing, or call signal receiving or transmitting.

Small data and PBX switching systems may serve tens or hundreds of terminations while large central-office voice-switching systems can have a termination capacity of over 150,000. Most switching systems are designed so that they can grow over a range of terminations, typically an order of magnitude or more. Ideally a switching system should be able to grow over a greater range, and new technology continues to expand this range.

System capacity depends on many factors, including not only the properties of the switching system but also customer traffic characteristics and customer expectation of grade of service. Grade-of-service criteria are usually defined for the average busy season busy hour (ABS) as well as for the 10 highest busy hours and for the highest busy hour. Examples of ABS criteria are 1.5 percent of calls delayed over 3 s in receiving dial tone and 2 percent of incoming calls blocked in the switching network. Capacity must be determined for each central office installation, as it depends on the amount of equipment provided.

Call Capacity. The limit to the maximum office size is frequently designed to be the capacity of the central processor(s). Processor capacity is usually stated in terms of busy hour originating plus incoming calls, although any particular office capacity calculation must consider the proportions of all types of calls. In determining engineered capacity, consideration must be given to the maximum capacity of the central processor. With the growth of services and features designed into the programs of SPCs, the more real time required to process all calls, and the lower the call capacity. Usually this corresponds to almost 100 percent use of the time available for call processing. An allowance must then be made for safety (typically 5 percent) on a high-day engineering basis, and beyond that for the ratio of high day to average business day. (A common ABS use is 70 percent of available processor time.) Then all delay criteria must be checked; if any are not met, the capacity must be lowered further.

Systems have been built with distributed microprocessors serving as many as 100,000 busy-hour call attempts (BHCA), and with a central SPC using multiprocessors and high-speed integrated-circuit technology of over 1,200,000 BHCA. Generally BHCAs are at least twice the number of successfully processed calls, owing to the large number of calls where a receiver–off-hook signal is detected without the completion or initiation of dialing.

Data-message or packet-switching capacity is measured in terms of maximum rate of *throughput*. The capacity of the control includes overhead as well as message handling. The control capacity is generally considered to be the major factor in throughput. Processing capacity can be limited by congestion delays within a system delivering or receiving calls for processing.

Switching-Fabric Capacity. The switching-fabric capacity is expressed in the number of erlangs that can be carried at an objective blocking. Fabrics for smaller offices are partially equipped; each arrangement will have its own capacity. If the average call duration is long, the erlang capacity may limit call capacity below that of the processor:

$$\text{Network calls/h} = \frac{\text{erlang capacity}}{\text{call holding time (h)}}$$

Because processors are not partially equipped and fabrics can be, designers choose to design fabrics with a smaller chance of being the limiting dimension of system capacity.

Memory Capacity. The total memory requirements depend on many components. An important factor is whether there is only one storage subsystem or separate storage subsystems are used for different storage needs. Typically, separate subsystems might be used for program and call data storage. In a message-switching system separate subsystems might be used for call processing and message storage.

One limit on memory size is the amount of memory that can be directly addressed. By having larger program words (more bits per word), more memory can be accessed, but this means that larger, more expensive program memories are required.

As in most SPC systems, software techniques can be used to extend the address range at the expense of real time. One address can refer to a table containing another range of addresses.

Physical memory modules are used to form a storage subsystem. Only as many memory modules are used as are needed to provide for the memory requirements for a particular installation.

Host/Remote Systems

The dimensional limits represent the individual maxima in each case (call attempts per hour, erlangs, and terminations). Depending on the specific environment, any one of these may limit further growth of the system, and the remaining limits would be unattainable. These limits represent approximate values, which in themselves depend on assumptions of the system environment, e.g., the ratio of intraoffice to interoffice calls in a local central office. The limits change with time as additional features and services are added to a system or when improvements in hardware or software are introduced.

Two basic switching entities are used at points distant from the host central wire center. One is the *remote line concentrator* (RLC) and the other the *remote switching unit* (RSU). Both the RLC and RSU are used to concentrate traffic at a point closer to the lines they serve. The RLC provides only for remoting this network function. Generally, in the central office an equivalent expansion function is provided, and consequently each RLC line is given a central office appearance. Some RLC systems eliminate the need for this expansion by connecting RLC trunks to links within the switching network. Generally the line range is not extended with the use of RLCs. Generally digital subscriber line carrier (SLC) systems are used to connect the host with the RLCs and RSUs. This provides a range as much as 100 mi. Recently fiber-optic links with digital transmission have also been employed by extending internal system links as in the No. 5 ESS.

Should all the trunks between the RLC and the central office be busy or the facilities carrying the trunks be severed, calls to and from the RLC lines cannot be served. Since the lines served are in a confined area, they are subject to higher traffic variation. Concentrators are generally small, serving no more than a few hundred lines.

With the advent of microprocessors, not only the network but also the control can be remoted. This means that more intelligence can be designed into the remote switching. Also, the remote switch may be able to complete intra-RSU calls without using trunks to the host office. The RSUs have been developed with SPC and limited call-processing capability so that they can provide basic service if the link to the host is lost. This is known as *stand-alone* capability.

Trends in Switching Systems

Switching systems have made the transition from those relays and other forms of electromechanical switches to all-electronic fabrics and controls. More than half of the switches in service are digital time-division systems. Whether time or space division, the major ingredient to stimulate this transition has been stored-program control. SPC has provided the flexibility and power to add to and change switching system capabilities easily. As administrations make the transition to national and international networks of SPC systems, capabilities are being extended further. The advent of common-channel signaling has provided a separate fast communication network between the processors of SPC switching systems, further increasing the capability of telecommunication networks. New services concepts that are possible within an SPC office may be extended to the entire intelligent network.

ISDN will be in full blossom by the end of this century. Other digital loop technologies will further extend the end-to-end digital capabilities of the public network. The inherent high-speed capability of optical fiber for digital transmission will be used for *Broadband ISDN (B-ISDN)*. Initially ATM as part of a B-ISDN for data, file transfers and images will further extend private backbone networks to the public B-ISDN.

Switching provides access to the public network for wireless telephones and terminals. For cellular radio it also provides not only the means for access but follows mobiles as they move from one area to another.

As mobiles move from one system to another, Intelligent Network switches and databases will provide seamless connections. *Personal Communication System (PCS)* will provide wireless switched access for users who do not move significantly during a call.

As more telecommunications services vendors enter the switched public network many new switching solutions will be needed to deal with such things as *number portability*. The telephone number addresses used by switches in the worldwide public network are the immutable factor in extending telecommunications services. Competition generally places new requirements on public networks that are manifested and implemented in the switches.

CHAPTER 17.3
LOCAL ACCESS NETWORKS

George T. Hawley

INTRODUCTION

The invention of the telephone in 1876 began an era of communications by voice over great distances. In order for telephone communications to be effected there needed to be low cost connections between telephones that could transmit voice signals with minimum attenuation and interference. Copper wires had the best balance of cost, tensile strength, durability, and transmission performance for the purpose. For economy it was thought that one wire between each pair of telephones with ground return would be adequate. This was not the case because of high direct and induced interference in the wires so connected. The use of a balanced pair of wires improved performance although it increased the cost of a connection by over 50 percent.

It was not practical to connect every pair of telephones with dedicated wires. Central switching offices were created where dedicated lines from telephones in the vicinity of the office could be terminated and flexibly connected to lines associated with other telephones as required to place a call. By the 1890s initial installations of automatic switching equipment began in local central offices to make connections between local *loops*, the name given to the pair of wires between the telephone and the central office and between loops and *trunks*, the name given to wire pairs between central offices. Over the years intercentral office or, simply, interoffice trunks of many varieties have been developed to support toll, tandem, equal access toll, and other applications.

The first telephone loops and trunks were pairs of uninsulated wires strung along aerial pole lines between terminations. As the number of telephone lines grew in the 1880s, these *open wires* proved to be impractical in urban applications. Cables of pairs of insulated wires were invented in the 1880s to improve the physical efficiency of aerial telephone lines and to make possible the placement of underground lines. After much trial and error, paired cables, consisting of wires insulated by paper wrappings, were perfected and standardized by 1888.

By the early 1900s the modern wireline telephone networks had taken shape and have not changed much since (Fig. 17.3.1). The term local access in the context of a wireline network means the local loop facilities between the central office and the end user's premises. Until 1976, telephones and other end-user line-terminating equipment were part of the local access network and were provided by the locally franchised telephone company. In 1976 the FCC separated line terminating equipment such as telephones, private branch exchanges, data modems, and the like into an unregulated category of equipment, called customer premises equipment or CPE under Part 68 rules. Although the various types of CPE play an integral role in the transmission and signaling performance of telephone networks, they are not considered part of the local access network. Part 68 incorporates rules governing the protection of the network and so-called third parties (parties not participating in a telephone connection) from nonconforming CPE or combinations of CPE. Part 68 contains no network or end-to-end compatibility information.

The Exchange Carrier Standards Association (ECSA) codifies network interface standards between CPE and local access networks in the form of American National Standards Institute standards adopted from standards developed under the ECSA T1 Committee. Customer premises equipment standards are established by the Telecommunications Industry Association (TIA) and are generally based on pre-1976 telephone set and other CPE properties.

TELEPHONE LOCAL ACCESS-LOOP DESIGN AND TECHNOLOGIES

Design Based on Wire Pairs

Design of wireline local access facilities or loops must economically accommodate two fundamental characteristics of the society being served: (1) the tendency for the population of end users to be clustered around population centers with decreasing density as a function of increasing distance from the center and (2) the presumption that service will be provided within a day or so of a request. The first characteristic suggests the location of a central office near the center of population with decreasing numbers of lines in the local loop plant as distance from the central office increases. The second characteristic suggests a continuous inventory of unused lines available everywhere in the community to meet unscheduled demands for service.

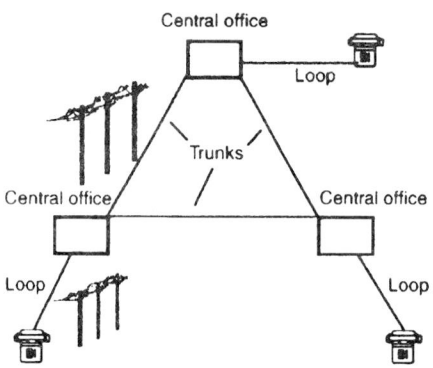

FIGURE 17.3.1   Local telephone network, basic layout.

The characteristics of human hearing and voice have a further influence on the design of the local loop plant. Copper wire pairs have dc resistance and attenuate alternating current signals. The frequency range of human hearing is generally considered to be from 100 Hz to about 20,000 Hz and is sensitive to sounds over many orders of magnitude of sound pressure intensity. The human voice contains 80 percent of its energy at frequencies below 1000 Hz and over 80 percent of articulation in frequencies below about 3000 to 4000 Hz. Faithful transmission of the human voice using energy in the frequency range between 300 and 3000 Hz provides efficient, intelligible, and recognizable communication over telephone networks.

Copper wires of larger diameter provide lower attenuation per unit length but are more expensive. In the early days of telephony, before the advent of electronic signal amplification, load coils were used to alter the transmission properties of wire pairs to reduce the attenuation of signals in the voice frequency band. The most common arrangement in the local loop is to place 88 mH coils in series with wire pairs every 6000 ft beginning 3000 ft from the central office. This technique, referred to as H88 loading, allows finer gauge wires to be used for economy at distances greater than about 18,000 ft from central offices, about 25 percent of local loops.

Similarly, there was no need to design telephone transmitters and receivers to operate effectively outside of the range of about 300 Hz to 3000 Hz. The technology employed in telephone sets for the first 100 years for the conversion of acoustic speech to electrical signals and vice versa remained essentially unchanged. The carbon microphone transmitters and electromagnetic diaphragm receivers used had limited efficiency and dictated, along with average acuity and comfortable speaking volume, the maximum voice frequency signal attenuation that could be tolerated on a connection experiencing acceptable background noise. The last major improvement in telephone set efficiency occurred in 1951 with the introduction by AT&T of the model 500 set, which properties are the basis for the current TIA standards for telephone set performance. Although new technologies exist that might further improve telephone set efficiency, 500 set equivalence is all that is required to match the characteristics of network connections and human speech and hearing comfort levels.

The design of local loops was influenced by voice frequency transmission across telephone connections as well as by early telephone properties. A typical telephone connection in a metropolitan area might involve two loops and one trunk connection. To minimize the total cost of network construction it was necessary to balance the performance of such connections against the cost of the networks needed to create the connections. Larger wire gauges improved transmission performance but increased cost. More central offices closer to customers reduced loop lengths and costs but increased the number of trunk routes required between more numerous central offices. Early designs balancing these factors resulted in loops with an average length under 2 mi and paired cables with wire gauges of 19, 22, and 24 AWG. Wire of 26 AWG was introduced in the 1920s. To achieve satisfactory voice transmission local loop designs were adopted over time that limited the maximum insertion loss at 1000 Hz to less than 8 dB with terminations simulating a telephone set at one end of the loop and a switching system line termination at the other end. Insertion loss is

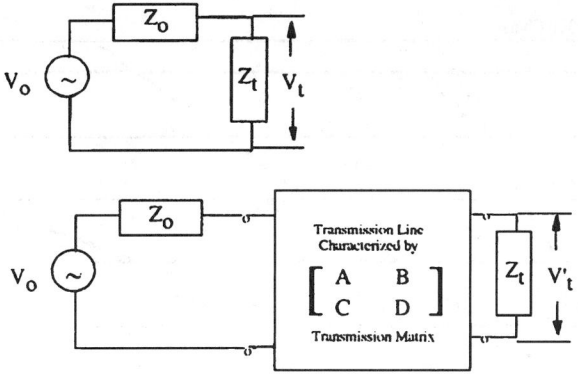

Insertion Loss $= 20 \log (V_t/V'_t)$

$= 20 \log (|(AZ_t + B + (CZ_t + D)Z_o)/(Z_o + Z_t)|)$

FIGURE 17.3.2   Definition of insertion loss of a transmission line.

defined in Fig. 17.3.2, which refers to A, B, C, D parameters to characterize transmission lines. These parameters are defined by the characteristic impedance and propagation constant of a transmission line and are discussed later in this section.

Loop design was also influenced by the operating characteristics of electromagnetic relays employed in central office switching system line circuits. The signaling protocol between telephone sets and switches is based on the flow or not of direct current from batteries in the central office through the off-hook telephone set or other CPE. Batteries with a nominal voltage of 48 V dc had a minimum value of 45 V dc in the 1950s. The relays commonly used in the early 1950s operated reliably on currents as low as 23 mA. This dictated a loop design that presented a maximum dc resistance, including the resistance of the line circuit and the telephone set to 45/.023 = 1956 Ω. In the 1950s maximum telephone set resistance was less than 220 Ω and line relay winding resistance was less than 440 Ω with high probability, leaving about 1300 Ω for the wire pair between the central office and the customer's premises. In 1954 AT&T adopted loop design rules referred to as Resistance Design rules using 1300 Ω as the maximum loop resistance, assuming the use of 500-type telephone sets or equivalent.

Resistance Design rules used a combination of H88 loading, maximum loop resistance, and 500-type telephone set properties to achieve the goal of bounding the performance of the loop at satisfactory loss level while achieving economy in loop design. The Resistance Design rules are summarized as follows:

1. Maximum loop resistance less than 1300 Ω at operating temperatures.

2. Maximum nonloaded loop length 18,000 ft with bridged tap < 6000 ft.

3. All loops greater than 18,000 ft use H88 loading rules.

Bridged tap is a pair of wires that branches off another pair at a junction point. At voice frequencies bridged tap acts as a capacitive load and increases attenuation.

Resistance Design rules promote voice frequency transmission performance that is consistent with residential telephone service quality requirements (less than about 8 dB insertion loss at 1000 Hz). These rules also ensure satisfactory operation of central office switch dc and voice frequency signaling circuits to the low voltage discharge level of central office battery supplies in the event of an extended power outage and the failure of emergency generators to operate.

Figure 17.3.3 illustrates application of Resistance Design principles applied to a cable route of 18,000 ft. A line is drawn parallel to the 24-gauge line so that it intersects the 1300-Ω ordinate at the 18,000 ft abscissa value. The intercept with the 26-gauge line shows that a combination of 11,700 ft of 26-gauge cable and 6300 ft of 24-gauge cable will result in an 18,000-ft cable route that just meets 1300-Ω loop resistance.

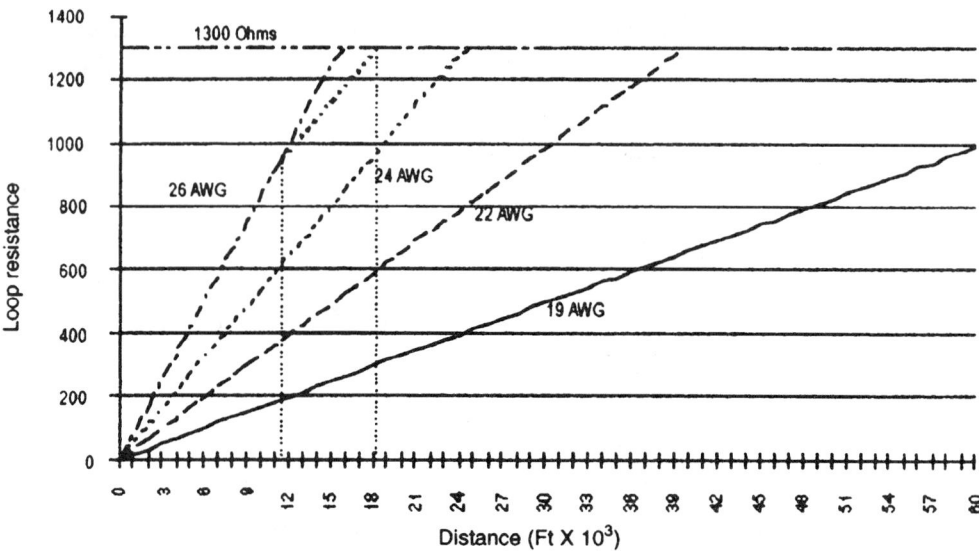

FIGURE 17.3.3   Example resistance design application.

The Resistance Design limit was increased to 1500 Ω in the early 1980s based on the improved dc current sensitivity of electronic switching system line circuits and the finding that the 500-type telephone set transmitter was more efficient at a loop current of 20 mA by about 2 dB than had previously been thought. Figure 17.3.4 shows the statistics of telephone loop length over the period from 1963 to 1983. It can be seen that average loop length increased slightly over the period. This resulted from the growth of suburbia with reduced housing density and the consolidation of smaller central offices into fewer, larger central offices.

Structure Based on Cables of Wire Pairs

In planning new paired cables for telephone local loops, engineers anticipate potential future demand. An economical design will provide just enough copper pairs at the right locations to satisfy demand for some time into the future so that the goal of near instant satisfaction of requests for service can be met. The new cables must be added before existing cables are full. Yet, the design must not place excessive capacity that will never be required and thereby become stranded investment. Elaborate computer aids have been developed over the years, principally by AT&T, using economic theory based on the time value of money with constraints of technology choices, tax policies, and regulatory rules to assist engineers in developing plans for building or adding to local loop cable plant. Discussion of the use of these aids and the underlying economic theory is beyond the scope of this section. However, there are two considerations that limit the choices of solutions using paired cables: the performance of twisted pair copper cable technology and the administrative structure of the areas to be served.

Copper Cable Technology

Cables consisting of copper wire pairs began to be manufactured at the end of the 1880s. The wires in each pair were twisted around one another to reduce the electromagnetic coupling, called crosstalk, with nearby pairs. Bundles of 50 or 25 pairs, called units or binder groups, are assembled into cables of up to 2700 pairs. The number of pairs possible to be bundled into a single cable is constrained by the maximum allowed cable diameter (less than 3 in. due to the diameter of underground ducts through which cables may be pulled), wire

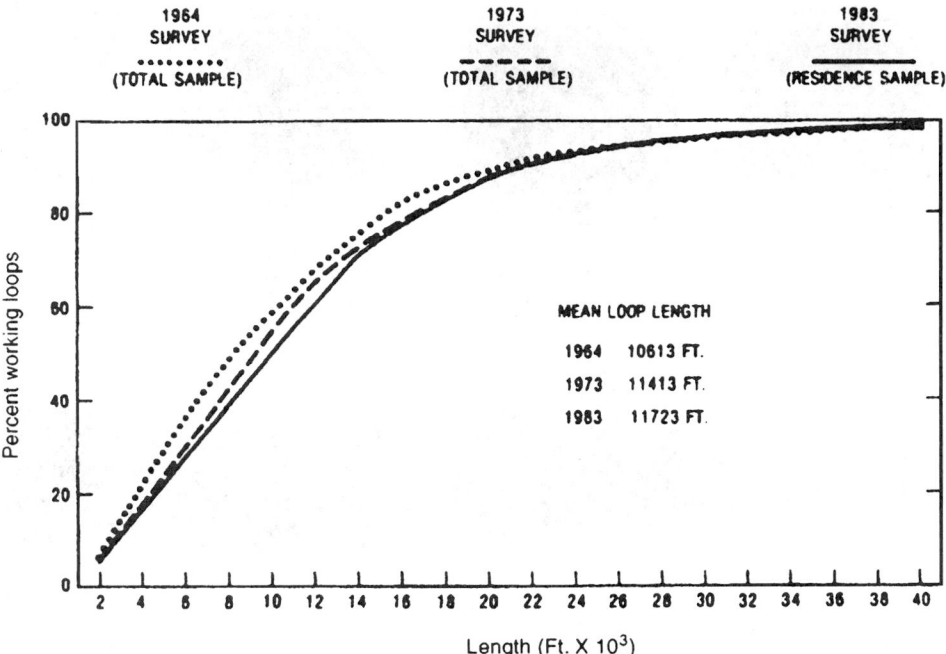

FIGURE 17.3.4   Loop length statistics resulting from economic design principles.

gauge, wire insulation thickness, and sheath thickness. The wires are insulated from each other using a variety or materials. Paper insulation proved to be most practical in the early days of telephony. Paper produced lower attenuation than materials with a higher dielectric constant as long as the paper was kept dry. Wires were later coated with wood pulp, called pulp insulation, to allow the design of compact cable structures. Paper insulation was used in a majority of cables into the 1930s. Plastic insulation materials such as polyethylene became predominant in the 1960s. The density of pairs in polyethylene insulated cables was improved in the 1970s by introducing air bubbles into the plastic, reducing its effective dielectric constant so that the insulation coating on the wires could be of reduced diameter while maintaining the standard capacitance between the wires in the pair. This form of insulation is called "expanded" insulation.

The bundle of twisted pairs form the core of the cable, which is surrounded by a sheath. Early sheaths consisted mainly of an extruded lead jacket that kept water out of the core, provided some shielding against external electromagnetic interference and diverted lightning discharges to earth. Modern cable sheaths typically have an extruded outer polyethylene or polyvinyl chloride jacket covering a metal shield made of corrugated aluminum or steel wrapped around an inner jacket of polyethylene extruded over the core of the cable. Some typical lead sheath cable cross-sections of the past are shown in Fig. 17.3.5.

Early cable designs in the 1880s suffered from lack of standards and from increased attenuation compared to open wire pairs. By the late 1880s, standards were established that ensured predictable performance in dry cables. The two principal attributes for paired cable standardization have been wire diameter (gauge), which dictates the resistance of the wire, and capacitance per unit length. Resistance and capacitance largely control the voice frequency transmission performance of the pairs and resistance controls the dc signaling range of operation. The attenuation limitations of twisted pairs were first addressed by improvements in telephone set transmitter and receiver efficiencies. This was augmented by the use of inductive loading which trades lower voice frequency attenuation for greatly increased attenuation at frequencies above the voice range. Finally, electronic circuits came to be used to compensate for wire pair attenuation. In the local loop the use of electronics became commonplace in the 1970s.

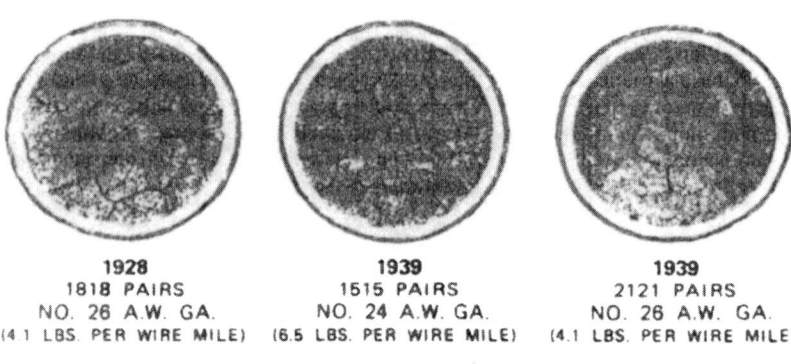

1928	1939	1939
1818 PAIRS	1515 PAIRS	2121 PAIRS
NO. 26 A.W. GA.	NO. 24 A.W. GA.	NO. 26 A.W. GA.
(4.1 LBS. PER WIRE MILE)	(6.5 LBS. PER WIRE MILE)	(4.1 LBS. PER WIRE MILE)

FIGURE 17.3.5   Representative larger size twisted pair cables.

Table 17.3.1 gives example measured values of the resistance, capacitance, and variability of those quantities per 1000-ft length of typical paired copper cables.

The nominal value of capacitance used in most engineering studies is 15.7 nF/1000 ft or 83 nF/mi for local loop applications. The nominal value of inductance of twisted pairs is about 1 mH/mi or 189 μH/1000 ft.

Twisted pairs in a local access application experience three fundamental transmission impairments: propagational attenuation and distortion, crosstalk, and induced noise (other than crosstalk). Induced noise from external sources of interference is mitigated by the cable shield and its ground connections. Crosstalk is reduced by staggered twist lengths from pair to pair and matched impedances between each wire in a pair and the cable shield (balance). Signal propagation is a function of the primary constants of a cable (resistance R,

TABLE 17.3.1   Electrical Characteristics of Paired Copper Cables

Gauge	Insulation	No. Pairs	R (Ω/1000 ft)	Std. Dev. (R)	C (nF/1000 ft)	Std. Dev. (C)
26 AWG	pulp	300	83.6	0.65	15.38	1.57
26 AWG	PIC	100	86.0	0.69	15.12	1.23
24 AWG	PIC	600	53.4	0.56	15.38	0.66
24 AWG	DEPIC	50	51.9	0.36	15.54	0.61
22 AWG	PIC-wp	100	33.9	0.14	15.31	0.79
22 AWG	PIC	50	34.7	0.14	15.52	0.68
19 AWG	PIC-wp	50	16.7	0.14	15.63	0.70
19 AWG	DEPIC-wp	25	16.3	0.15	15.79	0.62

PIC = polyethylene insulated conductor
DEPIC = dual, expanded polyethylene insulated conductor
wp = waterproof (filled cable)

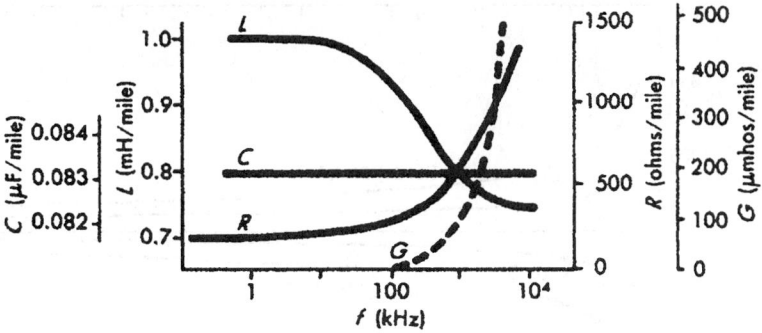

FIGURE 17.3.6   Representative primary constants for 22-gauge PIC copper pair.

capacitance C, inductance L, and conductance G, per unit length). Propagation of signals is characterized by the characteristic impedance Z_c and propagation constant Γ with the following relationship to the primary constants:

$$\Gamma = \sqrt{(G + i\omega C)(R + i\omega L)} \qquad \text{and} \qquad Z_c = \sqrt{\frac{(R + iwL)}{(G + iwC)}}$$

R, L, and G and functions of frequency. At frequencies below 0.1 MHz, G is negligible. R and L are approximately constant in the voice frequency range. At higher frequencies R increases as the square root of frequency and L decreases as the square root of frequency. Figure 17.3.6 illustrates the behavior of these parameters with increasing frequency for a typical 22 AWG copper twisted pair over a wide range of frequencies. Figure 17.3.7

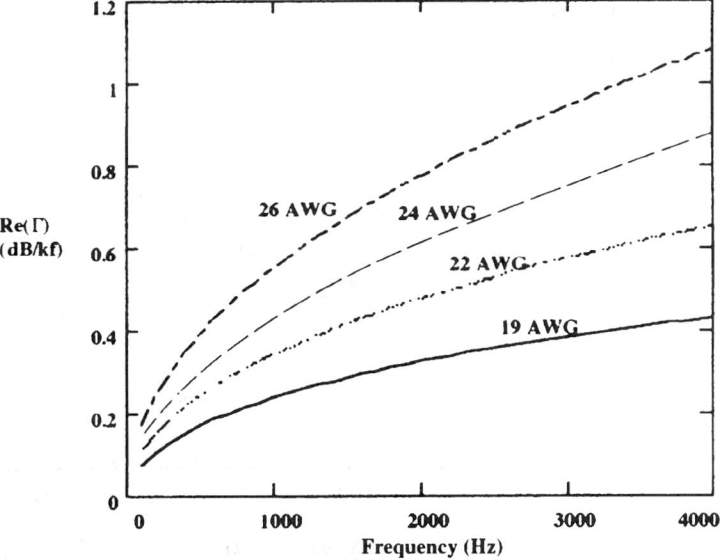

FIGURE 17.3.7   Attenuation of 19, 22, 24, and 26 AWG PIC copper pairs.

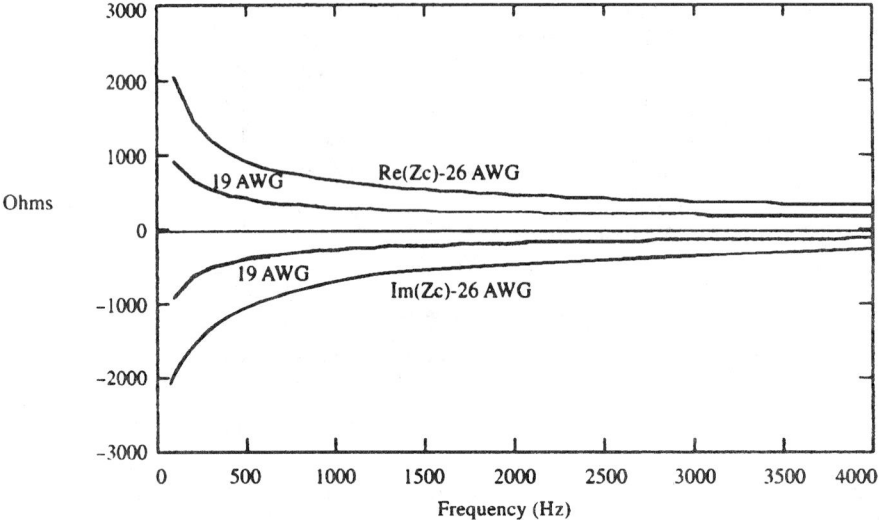

FIGURE 17.3.8 Characteristic impedance of 19 and 26 AWG pairs at voice frequencies.

shows the attenuation (real part of Γ) of 19-, 22-, 24-, and 26-gauge pairs in the voice frequency range. Z_c, the characteristic impedance of a twisted pair, is illustrated for 19- and 26-gauge pairs in Fig. 17.3.8. At frequencies above the voice band Z_c approaches a resistive value of about 100 ohms.

The effect of H88 loading is illustrated in Fig. 17.3.9 for the case of an 18,000 ft long, 26 AWG loop. The figure shows the insertion loss (see Fig. 17.3.2 for a definition) of a loop with and without load coils at 3000 ft,

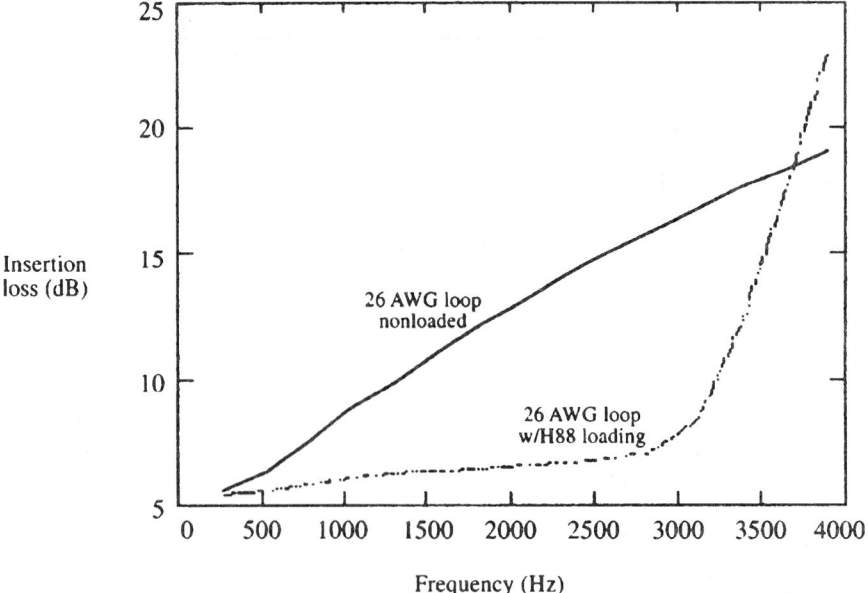

FIGURE 17.3.9 Insertion loss of 18,000 ft of 26 AWG loop, nonloaded and with H-88 loading.

9000 ft, and 15,000 ft from the central office. Terminations at each end of the loop were assumed to be 900-Ω resistors. The transmission matrix for a transmission line of length, 1, is given by

$$A = D = \cosh \Gamma 1, \quad B = Z_c \sinh \Gamma 1, \quad \text{and} \quad C = \sinh \Gamma 1/Z_c$$

The transmission matrix for a load coil is given by

$$A = D = 1, C = 0, \quad \text{and} \quad B = 5.4 + i\omega 0.088$$

As shown in Fig. 17.3.9, the load coils cause the insertion loss to be lower and less variable to about 3000 Hz and to increase much more rapidly with frequency above 3000 Hz. In practice the terminations will not be resistive. During the course of a telephone call the impedance at the central office will be that designed into the line circuit of the switching system, typically a 900-Ω resistor in series with a 2.16 μF capacitor. At the user end of the loop the impedance will be that of an off-hook telephone, a quantity that is only loosely characterized and that typically varies with the dc current supplied by the switching system or other line circuit through the loop. Figure 17.3.10 illustrates a model for the impedance of an off-hook 500-type telephone set at two representative values of dc loop current. A resistive value of 600 Ω is often used as an approximation.

Crosstalk is the coupling of electromagnetic energy from one pair to another. The two most important forms of crosstalk are near end crosstalk (NEXT) and far end crosstalk (FEXT). NEXT is the coupled signal sensed at the same end of a circuit as the source of the unwanted signal. FEXT is the coupled signal sensed at the end of the circuit distant from the source. These types of crosstalk are illustrated in Fig. 17.3.11. NEXT is most important as an impairment in circuits that transmit signals in the same frequency range in both directions, e.g., T1 carrier lines. FEXT is more important where signals in the two directions through adjacent pairs use separate frequency bands, e.g., analog carrier systems. NEXT increases approximately as the 3/2 power of frequency and relatively insensitive to loop length. FEXT increases approximately as the square of frequency and increases with the length of the loop.

The attenuation of crosstalk interference into a pair from another pair in a cable is called crosstalk loss. Crosstalk loss exhibits a random distribution between any pair in an n-pair cable and the other $n-1$ pairs.

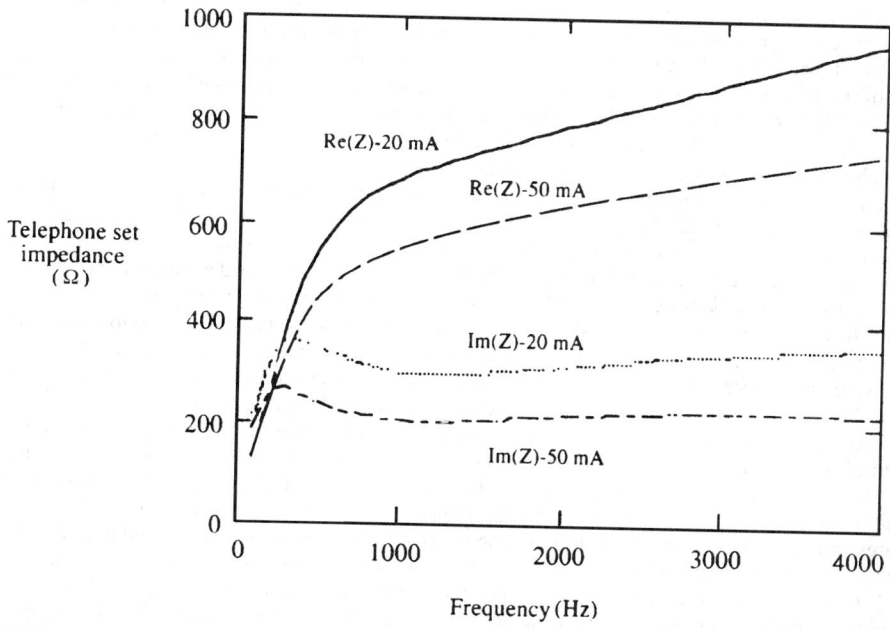

FIGURE 17.3.10   Model for 500-type Telephone Set impedance at 20- and 50-mA loop current.

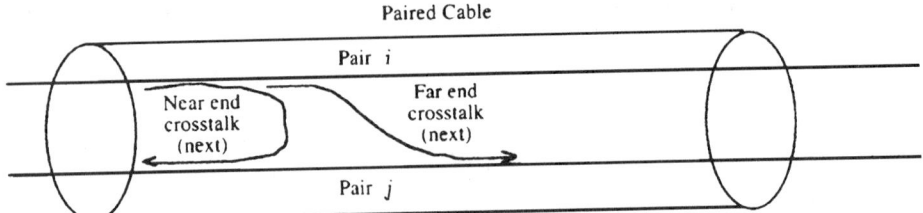

FIGURE 17.3.11   Near-end and far-end crosstalk paths.

Adjacent pairs generally experience the lower crosstalk losses. Separation of directions of transmission to pairs in nonadjacent binder groups is a way to reduce crosstalk levels.

Near-end crosstalk is more important than FEXT in voice frequency circuits. If NEXT is excessive in such a circuit, the unwanted crosstalk signal may have sufficient power to be understood by a listener. This is known as *intelligible* crosstalk. This form of impairment is generally unacceptable in the local loop because the crosstalk loss between two loops is fixed by the physical proximity of the two pairs dedicated to the two loops and not a function of call connections as it is in interoffice trunks with random call-to-call crosstalk associations. The principal way to repair intelligible crosstalk in the loop is to move one of the affected parties to a different pair in a cable.

Local Access Cable Organization

It is not practical to arrange loops terminating in a central office so that 100 percent of them would pass each home and business being served. Estimates must be made of the number of lines expected to be needed in each neighborhood or serving area to satisfy the expected demand for some time into the future. It has been especially important in the United States to have sufficient excess pairs available to any neighborhood to allow new service to be initiated within a day or two.

For many years telephone company engineers used a method called *multiple plant* to provide copper pairs to neighborhoods in a way that would be both flexible and economical. The use of multiple plant provides a *feeder* cable of a given number of pairs part way from a central office to a neighborhood. At a strategic point the cable is then spliced to two or more *distribution* cables that pass the homes and businesses to be served. Some of the pairs in multiple distribution cables are connected to the same feeder pairs, creating *bridged tap*. The voice frequency impairments due to the bridged tap are bounded by limiting the sum of the lengths of all bridged taps on a loop to less than 6000 ft. These impairments are traded for the flexibility needed to supply future circuits where the location of the potential demand is uncertain.

The long-term use of multiple plant caused some problems. Foremost among these was the need to open splices to rearrange pairs to provide more feeder pair capacity than originally planned to a distribution cable. Over long periods of time, the rearrangement activity caused damage to splices and corrupted cable records through accumulated errors. In the late 1960s, the serving area concept was developed to provide a more orderly and accessible method for connecting feeder cable pairs to distribution cable pairs.

The distribution cables are designed to have a capacity of up to two pairs dedicated to each living unit anticipated to exist in a neighborhood over a period of 20 or more years into the future. The number of pairs per ultimate living unit in distribution cables varies with telephone company local practice. Feeder cables to areas served by distribution cables are designed to have sufficient capacity for 5 to 7 years of anticipated demand and, therefore, have fewer pairs than the total numbers of pairs in the distribution cables being served. The feeder cables are terminated in an aboveground housing called serving area interface or, alternatively, feeder-distribution interface. The serving area interface also terminates all of the distribution cables that pass through the serving area and provides a cross-connection field to allow any feeder pair to be connected to any distribution pair using short pairs of wires called jumpers. Typically distribution areas have 600 or fewer living units and have distribution cables less than a mile in length. Distribution cables customarily are designed with 1.2 to two pairs per ultimate living unit. It is not uncommon for a serving area interface to terminate 600 feeder pairs and 1200 distribution pairs. Historically, about half of the capital investment in the local loop has been devoted to the feeder plant and half to the distribution plant. Figure 17.3.12 illustrates the multiple plant and serving area plant concepts.

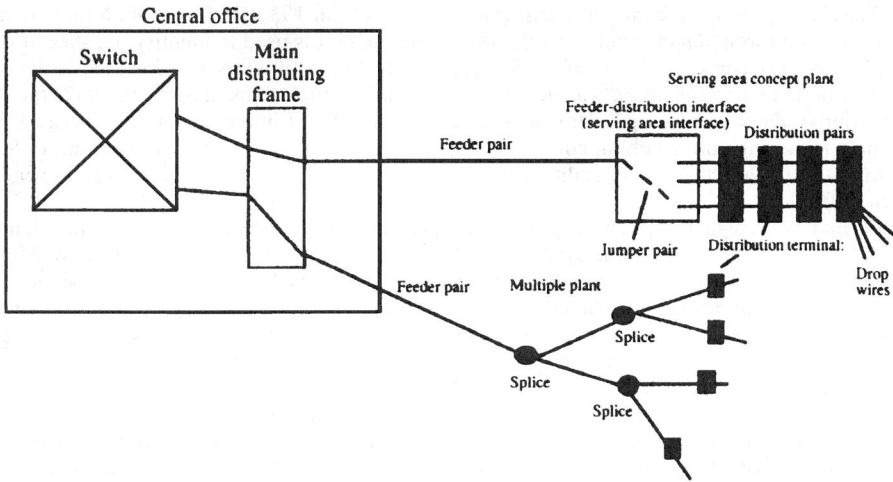

FIGURE 17.3.12   Multiple plant and serving area concept designs.

The Use of Electronics and Optics in the Local Loop

In the early 1920s electronic amplifiers began to be used in long-distance telephony. In the 1930s analog carrier systems were developed using frequency division multiplex techniques to carry multiple voice channels on a single-wire pair. These technologies were far too expensive to be used in the local loop, which continued to consist exclusively of wire pairs with those pairs longer than 18,000 using H88 loading to improve transmission.

After World War II, there was a surge of demand for telephone service that could not be met rapidly with added cables in the loop. This occasioned the examination of various electronic systems to allow a single-wire pair in the local loop to carry more than one conversation at a time. Early analog loop carrier systems, based on frequency division multiplex techniques that had been employed in telephone trunk transmission since the 1930s, were designed in the 1950s. Circuits based on the use of transistors, invented in 1947, were needed but the technology was immature. Systems cost too much and were not reliable enough to deploy widely.

By the late 1960s solid-state device technology was sufficiently economical and robust for application in the local loop. It became practical to build analog frequency division multiplex carrier systems for rural loop applications where the addition of a few circuits could obviate the construction of new cable plant. These systems typically carried six voice channels on a single pair of wires using frequency division duplex to achieve bidirectional operation with minimum NEXT. During that period it was found that investment in new cable construction for long rural loops could be reduced by employing voice frequency range extension electronics and finer gauge wires.

The principal of range extension is to aid the local switch in detecting telephone switch-hook transitions on loops of resistance higher than 1300 Ω and to amplify voice and tone signals to compensate for the added attenuation of the high-resistance loops. The amount of gain that may be inserted at the central office end of a loop is limited to 6 dB at 1 kHz to maintain the probability of encountering intelligible crosstalk at an acceptable level.

The use of a voice frequency amplifier with 6 dB of insertion gain allows loaded loops of up to 2800 Ω resistance to be used. Higher loop resistances are possible but, if transmission objectives of 8 dB insertion loss are to be maintained and crosstalk is to be acceptable, amplification is required at the customer premises end of the loop. Loop designs of up to 3600 Ω have been achieved with approximately 3 dB gain in the telephone handset. Deregulation of telephone sets by the FCC in 1976 made this option impractical, limiting the maximum copper loop resistance to 2800 Ω. A 2800-Ω loop resistance corresponds to about 82,000 ft of 22-gauge wire, a distance that reaches all but a small fraction of 1 percent of all loops. The 2800-Ω design virtually eliminated the need for 19-gauge cable in the loop, a 50 percent savings in copper.

T1 carrier line installation began in 1962. The T1 carrier line transmits digital signals at a rate of 1.544 Mb/s, capable of carrying simultaneously 24 pulse code modulated (PCM) voice channels at 64,000 b/s each. The T1 carrier signal contains a sequence of frames with 193 data bits each, 8 bits are assigned to each of the 24 information channels, and the remaining data bit is used to identify the start of a frame and for other, auxiliary purposes. A T1 carrier line requires two pairs of wires, one for each direction of transmission. Initial T1 line operation was devoted to interoffice trunk transmission. In 1972, the first digital loop carrier systems, employing T1 transmission, were introduced into commercial service. The first systems included a stage of switching concentration at each end of the T1 line, allowing up to 80 customer lines to share the 24-channel T1 facility with acceptable probability of blocking calls, depending on the offered traffic load.

In the mid-1970s digital loop carrier systems were introduced that used adaptive delta modulation techniques, taking advantage of knowledge of the dynamics of the human voice to reduce the data rate needed for satisfactory voice transmission to about 37,000 b/s. These systems enabled a T1 line to carry up to 40 simultaneous voice conversations with no blocking. Delta modulation was not compatible with PCM trunk transmission and gave way to the use of PCM in the next generation of loop carrier systems, the standard practice to the present time.

Digital loop carrier systems based on T1 lines are called "pair-gain" systems because they allow the simultaneous transmission of 24 voice signals, that would ordinarily require 24 pairs, on two pairs of wire thereby gaining the equivalent of 22 pairs. Early digital loop carrier systems were relatively expensive compared to wire pairs, depending on the length of the wire pair. Their use in the 1970s was largely confined to long rural loop cable routes. The value of digital loop carrier systems was the deferral of the construction of a new cable in an area of low demand for new lines. The savings due to the deferral of a major cable construction project for a few years more than offset the cost of purchasing and installing the carrier systems.

In 1979 the next generation of digital loop carrier systems was introduced using PCM at about the time of the introduction of digital local switching systems that also employed PCM transmission. These developments allowed the T1 lines from a digital loop carrier remote terminal to terminate directly on the switch without the need to convert the loop signals back to analog voice format at the switch interface. The efficiency of integrating digital loop carrier systems into the line side of digital switching systems added to the lower cost of manufacturing digital systems in the 1980s. These cost savings encouraged the routine use of digital loop carrier systems in conjunction with digital switches on much shorter than customary loop feeder cable routes in urban and suburban applications. By the mid-1980s over half of *new* loops were on digital loop carrier systems even though fewer than 10 percent of *all* loops were on digital loop carrier systems.

The density of loops increases exponentially as distance from the central office decreases. The cross-section of lines served by a digital loop carrier system remote terminal increases proportionately as they are installed on shorter loops. Typical digital loop carrier system remote terminals in the 1980s served from 96 to 544 lines each depending on the manufacturer. Typical high-density applications require more than 20 such terminals at a remote terminal location. It is not uncommon for digital loop carrier systems to be used on loops shorter than 2 mi.

Rural applications of digital loop carrier remote terminals require a cabinet of less than 30 ft^3 to serve 96 lines. Urban applications required underground vaults of over 1000 ft^3. The wide variety of applications of digital loop carrier remote terminals has required a large array of alternative enclosures to be deployed to meet local needs. Urban applications of digital loop carrier systems have also required a large variety of "special service" line interfaces to be supported in addition to the basic residential telephone line and coin line interface circuits.

Special services include private branch exchange analog and digital trunks, private line analog and digital data lines, foreign exchange two-wire and four-wire lines and trunks, and other specialized business services that have come to be offered over the years. These services each have specific transmission and signaling requirements to be met including the more stringent allocation of impairments to the local loop than is required for residential telephone service. Where special services are supported by copper pairs alone, signaling and transmission equipment has been installed in local central offices to compensate for the properties of wire loops in cables designed for residential service. This compensation has been designed into individual digital loop carrier plug-in "channel units" to enable telephone companies to provide special services through digital loop carrier systems.

The first trials of optical fiber in local loop transmission were conducted in the late 1970s. By the early 1980s the first commercial installations of digital loop carrier in association with optical fibers took place. The first digital loop carrier installation used multimode fibers with 50 or 62.5 μm core diameters. Multimode fiber was displaced by single-mode fiber by 1984, taking advantage of the superior cost and performance benefits of single-mode fibers. The type of single-mode fiber used typically has a core diameter of about 9 μm and is designed for minimum chromatic dispersion at an infrared wavelength of about 1300 nm.

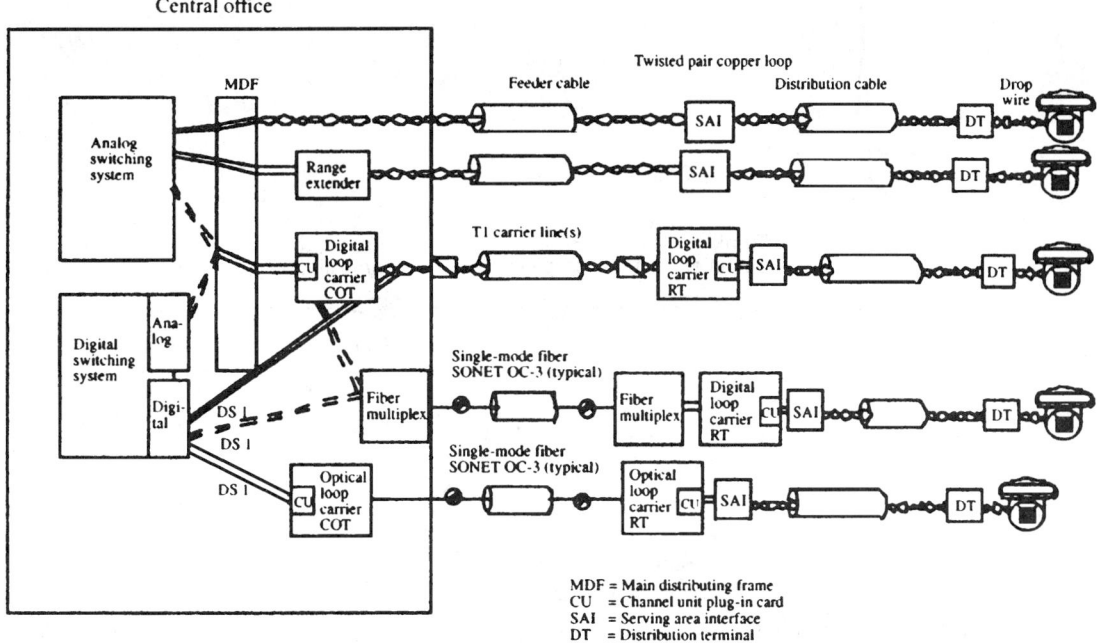

FIGURE 17.3.13   Overview of range extension, digital carrier, and optical carrier systems.

Single-mode fiber transmission was at first at 45 Mb/s and higher while typical digital loop carrier systems were based on T1 lines at 1.544 Mb/s. The use of single-mode fiber required the addition of multiplexers that combined many T1 signals from multiple digital loop carrier systems into a higher rate digital signal suitable for transmission through fibers.

In 1988 the T1 Committee of the Exchange Carrier Standards Association published the first Synchronous Optical Network (SONET) rate and format standards, creating a standard, synchronous digital signal hierarchy with standard payload mappings suitable for transmission through single-mode optical fibers. It then became practical to design a digital loop carrier system that could connect directly to single-mode fibers, eliminating the need for separate multiplexers. The current generation of digital loop carrier systems is of this type. The most popular system can serve up to 2016 voice lines over four fibers (two operating fibers and two backup fibers) using the SONET OC-3 signal format at 155.52 Mb/s.

Figure 17.3.13 summarizes the generations of loop electronic systems that have been deployed from the late 1960s to the present time.

CABLE TELEVISION LOCAL ACCESS-LOOP DESIGN AND TECHNOLOGIES

Introduction

Beginning in the late 1940s community antenna television (CATV) companies were created to address the problem of weak signal strength from distant very high frequency (VHF) television broadcast stations. The 12 television channels of the day were assigned 6 MHz bands in the frequency space between 54 MHz and 216 MHz, skipping frequencies previously allocated for FM commercial radio and other applications. The over-the-air modulation plan is amplitude modulation with vestigial sideband (AM-VSB) transmission. Early CATV systems consisted of a head end and a distribution system. The head-end usually had a tower up to

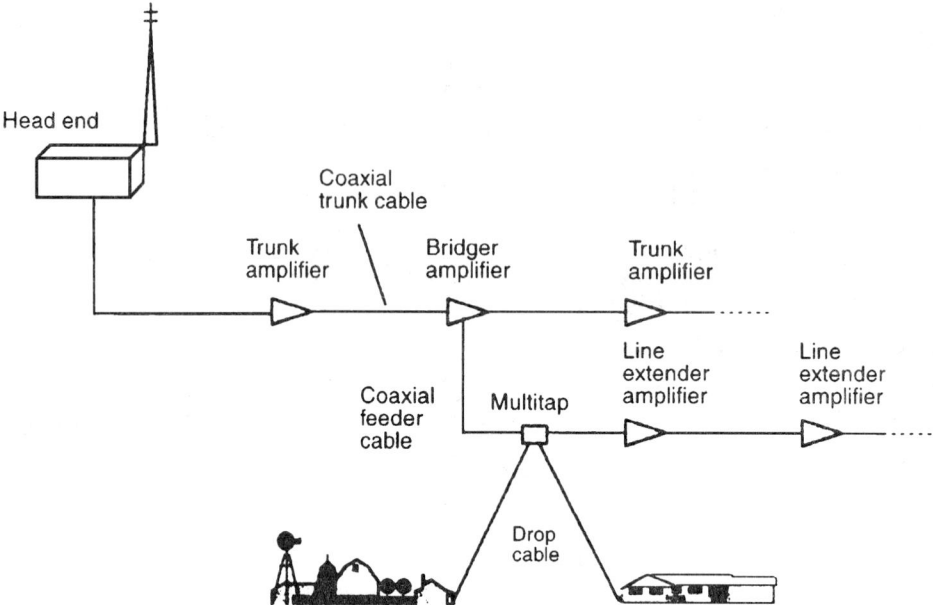

FIGURE 17.3.14 CATV Local Access Network, all coaxial cable.

several hundreds of feet high at the outskirts of a community equipped with antennas aimed at distant transmitters. The distribution system carried a composite signal through trunk coaxial cables to the neighborhoods of homes to be served. Trunk amplifiers were placed at intervals along the cable route to compensate for signal attenuation and changes in attenuation versus frequency due to temperature fluctuations. The trunk cables were connected to feeder cables through bridging amplifiers. Customers were connected to feeder coaxial cables that passed their homes via in-line multitaps. The multitaps were connected to customers homes via coaxial drop cables that terminated at a grounding block at the side of the home for connection to inside coaxial cables terminated in television sets. As the CATV transmission system carried unaltered off-air signals, television set tuners could directly access the broadcast programs. By the mid-1960s the CATV local access distribution network architecture was well established as illustrated in Fig. 17.3.14. Many CATV systems carry FM radio signals as well as television signals, since the 88- to 108-MHz FM radio band is in the middle of the VHF television band.

In the 1970s, satellite transmissions of television signals to CATV head-ends became economical. C-Band satellites were the more common type using spectrum at 5.9 to 6.4 GHz for the uplink and at 3.7 to 4.2 GHz for the downlinks with capacity for 24 analog video channels. Each channel is served by a "transponder" in the satellite. Each channel occupies 36 MHz of bandwidth using frequency modulation. The addition of satellite antennas and receivers added considerably to the capital costs of headend electronics and suggested connecting more than one local distribution network to a single master headend. The links between local headends and distant "master" headends became known as "super-trunks." Single-mode fiber optics came into use in the super trunk and trunk plant in 1984 and 1989, respectively. CATV systems now pass more than 90 percent of the homes in the United States and serve more than 60 percent.

There are three standard channel frequency assignment plans in use in CATV systems in the United States, as specified in FCC Bulletin 21006: (1) The standard (STD) plan, (2) the interval related carriers (IRC) plan, and (3) the harmonic related carriers (HRC) plan. The IRC and HRC plans lock the carrier frequency for each channel to a precision reference oscillator to reduce some of the second- and third-order harmonic distortion noise in the multiplexed CATV signal. Most systems use the STD plan. Newly installed commercial distribution systems now carry up to 110 channels in a band of frequencies up to 750 MHz.

Table 17.3.2 gives the frequency assignments up to channel 99 for the three plans.

TABLE 17.3.2 CATV Channel Frequency Plans (MHz)

Channel ID	STD plan	HRC plan	IRC plan
2	55.25	54.0027	55.25
3	61.25	60.0030	61.25
4	67.25	66.0033	67.25
5	77.25	78.0039	79.25
6	83.25	84.0042	85.25
7	175.25	174.0087	175.25
8	181.25	180.0090	181.25
9	187.25	186.0093	187.25
10	193.25	192.0096	193.25
11	199.25	198.0099	199.25
12	205.25	204.0102	205.25
13	211.25	210.0105	211.25
14(A)	121.2625	120.0060	121.2625
15(B)	127.2625	126.0063	127.2625
16(C)	133.2625	132.0066	133.2625
17(D)	139.25	138.00669	139.25
18(E)	145.25	144.0072	145.25
19(F)	151.25	150.0075	151.25
20(G)	157.25	156.0078	157.25
21(H)	163.25	162.0081	163.25
22(I)	169.25	168.0084	169.25
23(J)	217.25	216.0108	217.25
24(K)	223.25	222.0111	223.25
25(L)	229.2625	228.0114	229.2625
26(M)	235.2625	234.0017	235.2625
27(N)	241.2625	240.0120	241.2625
28(O)	247.2625	246.0123	247.2625
29(P)	253.2625	228.0114	253.2625
30(Q)	259.2625	258.0129	259.2625
31(R)	265.2625	264.0132	265.2625
32(S)	271.2625	270.0135	271.2625
33(T)	277.2625	276.0138	277.2625
34(U)	283.2625	282.0141	283.2625
35(V)	289.2625	288.0144	289.2625
36(W)	295.2625	294.0147	295.2625
37(AA)	301.2625	300.0150	301.2625
38(BB)	307.2625	306.0153	307.2625
39(CC)	313.2625	312.0156	313.2625
40(DD)	319.2625	318.0159	319.2625
41(EE)	325.2625	324.0162	325.2625
42(FF)	331.2750	330.0165	331.2750
43(GG)	337.2625	336.0168	337.2625
44(HH)	343.2625	342.0171	343.2625
45(II)	349.2625	348.0174	349.2625
46(JJ)	355.2625	354.0177	355.2625
47(KK)	361.2625	360.0180	361.2625
48(LL)	367.2625	366.0183	367.2625
49(MM)	373.2625	372.0186	373.2625
50(NN)	379.2625	378.0189	379.2625
51(OO)	385.2625	384.0192	385.2625
52(PP)	391.2625	390.0195	391.2625
53(QQ)	397.2625	396.0198	397.2625

(*continued*)

TABLE 17.3.2 CATV Channel Frequency Plans (MHz) (*Continued*)

Channel ID	STD plan	HRC plan	IRC plan
54(RR)	403.25	402.0201	403.25
55(SS)	409.25	408.0204	409.25
56(TT)	415.25	414.0207	415.25
57(UU)	421.25	420.0210	421.25
58(VV)	427.25	426.0213	427.25
59(WW)	433.25	432.0216	433.25
60(AAA)	439.25	438.0219	439.25
61(BBB)	445.25	444.0222	445.25
62(CCC)	451.25	450.0225	451.25
63(DDD)	457.25	456.0228	457.25
64(EEE)	463.25	463.0231	463.25
65(FFF)	469.25	468.0234	469.25
66(GGG)	475.25	474.0237	475.25
67(HHH)	481.25	480.0240	481.25
68(III)	487.25	486.0243	487.25
69(JJJ)	493.25	492.0246	493.25
70(KKK)	499.25	498.0249	499.25
71(LLL)	505.25	504.0252	505.25
72(MMM)	511.25	510.0255	511.25
73(NNN)	517.25	516.0258	517.25
74(OOO)	523.25	522.0261	523.25
75(PPP)	529.25	528.0264	529.25
76(QQQ)	535.25	534.0267	535.25
77(RRR)	541.25	540.0270	541.25
78(SSS)	547.25	546.0273	547.25
79(TTT)	553.25	552.0276	553.25
80(UUU)	559.25	558.0279	559.25
81(VVV)	565.25	564.0282	565.25
82(WWW)	571.25	570.0285	571.25
83	577.25	576.0288	577.25
84	583.25	582.0291	583.25
85	589.25	588.0294	589.25
86	595.25	594.0297	595.25
87	601.25	600.0300	601.25
90(FM1)	89.25	NA	NA
91(FM2)	95.25	NA	NA
92(FM3)	101.25	NA	NA
93(FM4)	107.25	NA	NA
94(FM5)	113.2750	NA	NA
95	NA	72.0036	73.25
96(A-5)	91.25	90.0045	91.25
97(A-4)	97.25	96.0048	97.25
98(A-3)	103.25	102.0051	103.25
99(A-2)	109.2750	108.0054	109.2750
1(A-1)	115.2750	114.0057	115.2750

In 1972, the FCC required among other regulations that all CATV systems serving more than 500 subscribers be "two-way capable," with a band of frequencies from 5 to 35 MHz being assigned to the reverse direction in a "subsplit" system. For many years this requirement was honored by installed trunk, bridging, and line amplifiers that included plug-in connectors for the attachment of diplex filters and reverse channel amplifiers to allow a system to be upgraded should the occasion arise. For the most part, this band has gone unused.

CATV System Transmission Considerations

CATV system transmission suffers from the attenuation of signals through coaxial cables, the accumulation of thermal noise through a succession of amplifiers, the accumulation of second-order and third-order harmonic noise caused by nonlinearities in active and passive components, and the ingress of external interference.

Coaxial cable propagation constant and characteristic impedance may be written in the same form as for twisted pair transmission lines:

$$\Gamma = \sqrt{(G + i\omega C)(R + i\omega L)} \qquad \text{and} \qquad Z_c = \sqrt{\frac{(R + i\omega L)}{(G + i\omega C)}}$$

where $C = 2\pi\epsilon/(\ln b/a)$
b = inside diameter of outer conductor
a = outside diameter of the inner conductor

$G \sim 0$. R and L vary depending on the size and material of the inner and outer conductors.

The attenuation of a coaxial cable decreases with increasing diameter and increases with the square root of frequency. Typical trunk coaxial cables are 3/4-in. or 1-in. diameter with attenuation generally in the range of 1.0 to 1.5 dB/100 ft at 550 MHz. Feeder cables are similar in design to trunk cables but smaller in diameter, typically 1/2- to 5/8-in. diameter with attenuation in the range of 1.3 to 2.0 dB/100 ft at 550 MHz.

Trunk design typically allows about 20 dB loss between amplifiers creating spacings of 1400 to 2000 ft between trunk amplifiers at the high end of the frequency band supported by the system. Trunk amplifiers often include automatic gain and gain-slope adjustment capabilities to compensate for component aging and temperature variations. The carrier-to-noise (C/N) ratio is an industry metric for signal quality. C/N should be in excess of 46 dB. This limits the maximum number of amplifiers in tandem to less than about 20, thereby limiting the length of a distribution system. However, some applications may require a greater number of amplifiers because of the distance to be covered from the headend to the most distant customers, sacrificing noise performance.

The feeder cable design must compensate for the loss of multitaps as well as the feeder cable. Line extender amplifiers are usually limited to about three in cascade and are set for 25 to 30 dB of gain. The line extender amplifiers contribute significantly to harmonic distortion in the signal because they operate at high output levels. Composite second (CS) order and composite triple beat (CTB) are metrics governing signal distortion. Each is maintained at less than –53 dBc (dB with respect to carrier power) in a well-designed system.

Subscriber drop cables are typically braided outer conductor coaxial cables with attenuation as high as 15 dB. Almost half of the total length of coaxial cable in a distribution system is composed of drop cables. The most common drop cables are RG-59 and RG-6 equivalent types.

The drop cable is connected to the feeder cable through a passive component called a multitap. Multitaps contain directional coupler circuitry with two, four, or eight drop cable connections. The attenuation toward the subscriber is fixed and is selected by the engineer/designer of the plant to match the signal level at that point along a feeder cable so that the subscriber signal level will be as close to the desired 0 dBmv level as possible. The reverse direction from the customer experiences greater attenuation to limit the interference that is transmitted back toward the headend from the subscriber's equipment.

Customer wiring may be customer-installed or CATV company-installed RG-59 coaxial cable with passive splitters to allow service to branch to different rooms in a home. In-home cables terminate either directly on television sets or VCRs or on set-top converters supplied by the CATV company. CATV services are categorized as basic, premium, and pay-per-view. Premium and pay-per-view signals are encrypted by the CATV service provider and require a set-top converter to decipher the signal. In the case of pay-per-view services, the set-top converter is addressable by the CATV provider and is instructed to decipher signals on request of individual subscribers. Basic services may be viewed without the aid of the set-top converter by television sets and through VCRs with tuner circuits designed to correctly interpret the standard, IRC, and HRC channel formats. Appliances that can do this are called cable-ready.

CATV distribution systems are powered by distributed in-line 60-V, 60-Hz supplies connected to public utility power distribution networks. CATV power supplies feed ac current over the coaxial cables in both directions from the location of the supply. As much as 10 A may be passed through the cables and associated amplifiers. Set-top converters are powered by the subscriber.

The total costs, including labor and materials of a coaxial cable distribution system, consist of about 38 percent for coaxial cable, 24 percent for electronics, 20 percent for physical hardware, and 18 percent for passive components.

The Use of Fiber Optics in CATV Distribution Systems

The use of single-mode optical fibers in CATV trunk plant has advanced very rapidly with advances in the design of laser transmitters and optical receivers capable of the linearity required to satisfactorily carry the CATV amplitude-modulated, frequency-multiplexed extremely wide bandwidth signals. The use of optical fiber trunk plant benefit the CATV operator in three ways: (1) It reduces the number of amplifiers in cascade between the headend and the subscribers, improving system performance; (2) it improves the reliability of the distribution system by eliminating many active elements; and (3) it moves the bandwidth constraint on the trunk plant to the ends of the fibers, requiring simpler engineering to upgrade the plant to greater bandwidths. Laser transmitter and optical receivers are commercially available to carry more than 100 channels in a 700-MHz bandwidth from 50 to 750 MHz on a single fiber.

Early CATV applications of optical fibers used Fabry-Perot lasers that lacked the linearity to carry amplitude modulated signals. Modern CATV trunk systems mainly use distributed feedback (DFB) lasers that exhibit superior linearity of operation. The lasers are sensitive to reflected signals from splices and connectors. Typical transmitters include an optical isolator between the laser and the trunk fiber to block reflections coming back toward the laser. DFB laser transmitters are relatively expensive but costs have been reduced to allow an optical transmitter to be dedicated to clusters of as few as 2000 homes. Optical couplers may be used to allow a laser transmitter to be shared by two to four trunk fibers. Figure 17.3.15 illustrates the use of optical fiber in a CATV distribution system trunk application. This arrangement of fiber trunks and coaxial feeder cables is commonly referred to as Hybrid Fiber-Coaxial or HFC plant.

The remote optical node or, simply, remote node that terminates the fiber at the end of the trunk plant typically contains a receiver using a *pin* photo diode detector. The remote node may contain amplifiers capable of

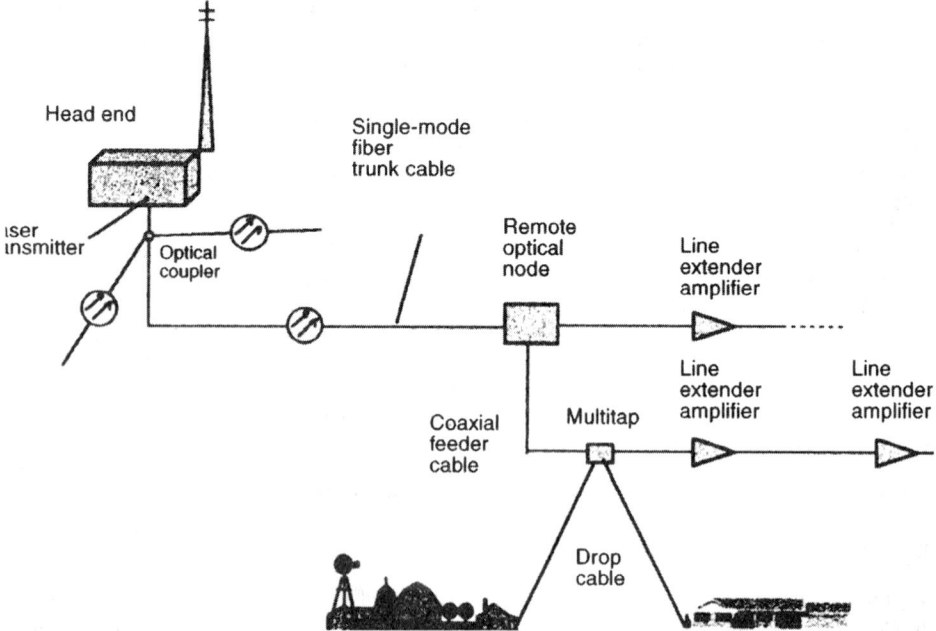

FIGURE 17.3.15   Cable television local access with optical fiber trunk.

serving up to four feeder cables. The attenuation of the optical signal between the laser transmitter and a remote node is commonly in the 10- to 12-dB range. If a coupler is used, over half of the allocated optical loss budget will be consumed by the coupler.

EMERGING TRENDS IN TELEPHONE AND CATV LOCAL ACCESS

New Technologies for Wire Pair Transmission

In the early 1990s, developments in integrated circuit technology have enabled digitally encoded signal formats to be defined that make more efficient use of the available frequency spectrum for wire pair transmission. As a group, these techniques are referred to as *digital subscriber line* (DSL) technologies. The first of the standard DSLs to employ full duplex transmission over a single-wire pair in the United States is the integrated services digital network (ISDN) basic rate access transmission line that enables the bidirectional transmission of 160 kb/s over nonloaded pairs over 18,000 ft in length. The ISDN basic rate access line uses a multilevel line code referred to as 2B1Q for two binary, one quaternary. The 2B1Q line code places 4 bits of information in each pulse by using four pulse amplitudes, two positive and two negative. Complex echo cancellation and signal equalization techniques are used to compensate for bridged tap and impedance discontinuity impairments.

More recently, the 2B1Q encoding has been extended to allow bidirectional signals with data rates of 784 kb/s to be transmitted over twisted pairs meeting Resistance Design criteria to a distance of 12,000 ft. This technique is referred to as high bit rate digital subscriber line or HDSL. By using HDSL, telephone companies can provide services requiring T1 data rates over two pairs that are twice as long as the spacing between regenerators on T1 carrier lines and require no conditioning, i.e., removal of bridged tap, or complex design procedures.

Another technology that has been developed recently is the asymmetric digital subscriber line (ADSL) using digital modulation and frequency division multiplexing to transport high data rate digital signals over unconditioned twisted pairs. Early modulation techniques used include discrete multitone (DMT) and carrierless amplitude and phase modulation (CAP). Transmission is asymmetric over a single twisted pair to support the possible delivery of compressed, digital video signals to residential consumers. Experimental systems offer up to 6 Mb/s transmission toward the customer with up to 640 kb/s in the reverse direction over loops up to 12,000 ft in length. There is promise that ADSL techniques may allow the transmission of signals up to 52 Mb/s data rate over much shorter, but useful distances.

Digital Video Services over Fiber and Coaxial Cables

Telephone companies have been experimenting with "fiber-to-the-curb" systems since 1988. These systems are based on the technology of digital loop carrier systems but extend optical fibers into the telephone distribution plant. There have been limited commercial deployments of systems that followed requirements in Bellcore Technical Reference (IR) TSY-000909. To date the cost of purchasing and installing FTTC systems has been too high to allow economical mass deployment for switched and private line telephone services in this way.

TR-909 compliant FTTC systems consist of a host digital terminal (HDT), a fiber distribution system, and optical network units (ONU). The HDT provides the connections to the local telephone switch, private lines that do not terminate on the switch, and network management systems. The HDT may be located in the central office or at a remote terminal site distant from the central office. ONUs provide the service interface circuits for connection to the end users and test and maintenance circuits to allow remote testing to be done on the drop wire connection to the home from a centralized maintenance system.

The distribution connection from the ONU may be a simple point-to-point time division multiplex arrangement or, alternatively, a point-to-multipoint arrangement having more than one ONU sharing the optical interface at the HDT. Point-to-multipoint connections may use a variety of multiple access protocols. Most common to date are time division multiple access (TDMA) systems that broadcast the same time division multiplex signal to all ONUs but receive signals from each ONU in turn, allowing multiple ONUs to share the single HDT optical receiver and transmitter. Figure 17.3.16 illustrates one form of FTTC connections with tandem HDTs that are located in the outside plant as well as in the central office. The HDT may also serve the function of an optical loop carrier terminal in this example, remote or central office, since all

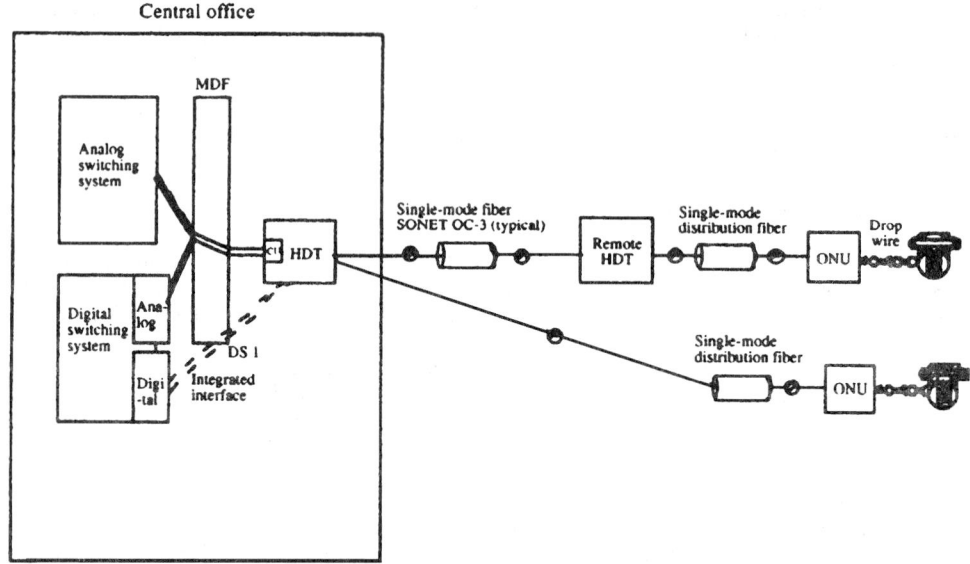

FIGURE 17.3.16   Telephone fiber-to-the-curb example.

of the service, switch, maintenance, and network management interfaces are the same. This may be accomplished if the optical loop carrier system has an architecture that enables the provision of optical distribution interfaces to ONUs as well as twisted pair distribution interfaces to end users. Otherwise, the remote HDT must be connected to a digital switch via a pair of back-to-back fiber multiplexers and to an analog switch via a pair of multiplexers and a digital loop carrier COT.

The telephone local access challenges have increased since the FCC Video Dial Tone order, issued in June 1992. After lengthy deliberations, the FCC reaffirmed the order in November 1994 and has since begun approving telephone company Section 214 construction applications. Video digital tone is regarded as a common carrier video services delivery service that telephone companies may provide. Heretofore, telephone companies have been restrained from providing cable television services and have not been allowed to own cable television companies in their franchise territories.

After considerable study, telephone companies have arrived at different conclusions regarding the best combination of technologies to use for video dial tone services. If these services are to include multicast analog video channels such as the local broadcast stations, then hybrid fiber-coaxial plant is considered to be the most economical approach. Some companies have decided to replace their telephone loop plant with HFC plant for video dial tone applications and have decided to include telephony services to be transported over the same plant.

This has stimulated a number of equipment suppliers to begin the development of systems that drive telephone services over HFC plant, illustrated in Fig. 17.3.17. Although not absolutely required, most systems in development use a second fiber from the HDT in the central office for the return or upstream direction of transmission. The application of telephony to HFC plant significantly increases the reliability requirements of the plant. In particular power supplies must be equipped with 8 h of battery reserve operation in the event of power failures in a neighborhood in order to provide telephone service of comparable reliability to existing wireline services.

Adding telephony services requires that coaxial termination units that derive the telephony service interfaces, much like those in optical network units and digital loop carrier systems, must be installed at or near subscriber homes. The coaxial termination units communicate toward the HDT using frequencies under the downstream 50 MHz lower bound for subsplit systems. Emerging systems are carrying the upstream telephony signals in the range of approximately 15 to 40 MHz using various modulation and multiplexing designs to maximize the efficiency of use of the limited spectrum. Some systems incorporate frequency shifting circuitry to avoid, as needed, the significant levels of ingress noise found in the lower frequencies of the upstream band. Some systems also assign digital channels on demand for added efficiency rather than dedicating upstream

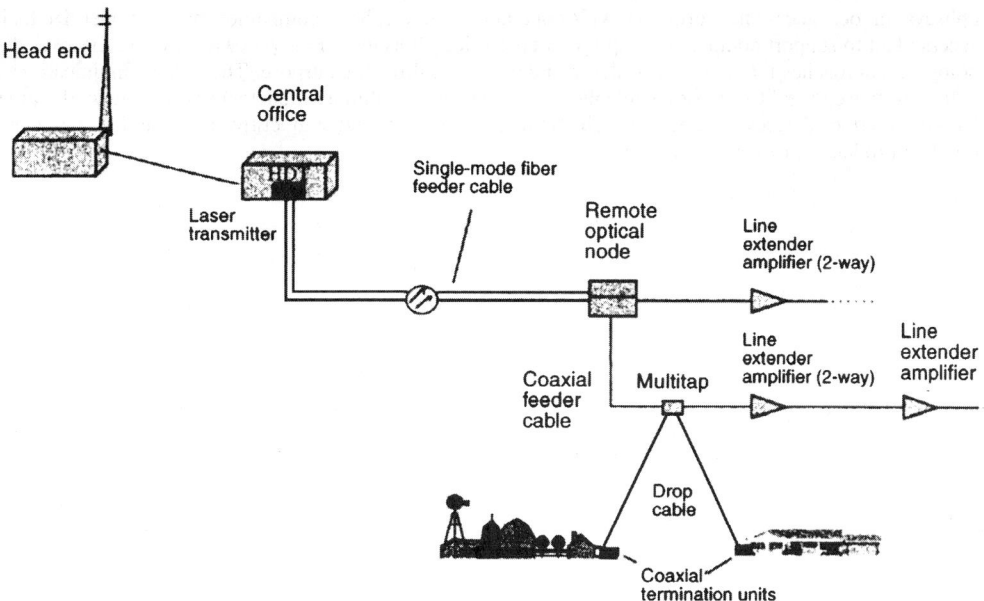

FIGURE 17.3.17   Hybrid fiber-coaxial plant with video and telephony capabilities.

channels full time to customer lines. CATV system operators, interested in providing alternate local telephone access, are also likely to employ such systems to derive telephony services over HFC plant.

Great strides have been made in the development of algorithms for reducing the amount of data required to carry television signals in digital form. Standards for these techniques have been drafted by the International Standards Organization Motion Picture Experts Group (MPEG). The emerging standards are referred to as MPEG1 and MPEG2. These algorithms take advantage of the frame-to-frame redundancy in motion pictures and television signals to remove more than 90 percent to the data contained in digitally encoded versions of the material. The MPEG1 standard was developed primarily for CD-ROM applications that allow fixed data transfer rates of up to about 1.5 Mb/s but has been extended to provide higher quality signals at higher data rates for satellite and potential CATV distribution. MPEG2 incorporates a packet transport protocol that allows the multiplexing of multiple images in a single data stream and supports variable data rate encodings. The MPEG standards are asymmetric, favoring simple decoding and complex encoding processes in support of uni-directional transmission.

MPEG encoding of video, coupled with efficient digital signal modulation techniques such as quadrature amplitude modulation (QAM) allows several video signals to be carried in the 6 MHz channel space of a single analog video signal. The number of compressed, digital signals that can be carried in a channel can be 10 or more and is a function of the compression ratio achieved for a given level of quality and the modulation method chosen. With MPEG and digital modulation a modern CATV distribution system can carry many hundreds of programs simultaneously.

Advances have also been achieved in the area of Asynchronous Transfer Mode (ATM) standards that allow the flexible multiplexing of voice, video, and data signals together despite great disparities in the data rates of the individual signals. ATM standards support the mapping of signals into fixed length packets, 53 bytes in length, called cells, that can be transported and switched readily through high-speed optical transmission lines. The SONET standards include a payload mapping to support the transport of ATM signals through the SONET digital signal hierarchy. It is believed by many that video dial tone networks will rely on ATM-based switches to allow end users to call up digitally compressed video program material on demand from archives of material.

Both hybrid fiber-coaxial and fiber-to-the-curb access networks can support the transmission of digitally compressed video signals, multicast and switched on demand. The latter part of the 1990s will witness a great upheaval in local access networks as CATV companies and telephone companies race to modernize their local access plant to support compressed, digital video service offerings. In the process they will discover the technological approaches that are most suitable and economical for the purpose. The Telecommunications Act of 1996 has caused the FCC to order telephone companies to unbundle their local access networks to support facility-based local access competition. This will add regulatory and competitive uncertainties, perhaps retarding the introduction of new technology.

CHAPTER 17.4
BROADBAND SYSTEMS

David H. Su, David E. Cypher, John J. Knab, William W. Wu

Broadband systems refer in general to Broadband Integrated Services Digital Network (B-ISDN), which provides user access at a speed over the T1 rate (1.5 Mb/s). Integrated Services Digital Networks are defined to be networks that provide end-to-end digital connectivity to support a wide range of services, including voice and nonvoice services. Nonvoice services include image and video, as well as high-speed data transmission. B-ISDN allows the traditional interactive computer applications to run at higher speeds and enables the development of innovative distributed applications using computing and storage resources distributed over the network. It also enables new applications that were not possible because of bandwidth limitations, such as distribution or broadcasting of digital video information.

The concept of ISDN began to emerge in the 1970s, and the set of international standards were issued in the early 1980s by the ITU-T (formerly CCITT). The initial ISDN consists of multiples of fixed 64 kb/s channels, up to a total of 1.5/2.0 Mb/s. Subsequently, the idea of integrated services was extended to broadband networks, such as Synchronous Optical Network (SONET) at a speed of 155 Mb/s or higher, and was named Broadband ISDN (B-ISDN) to contrast it with the original lower bandwidth ISDN. In this section the term ISDN (without the B prefix) refers to the narrowband ISDN.

In addition to the speed difference, ISDN uses synchronous time division multiplexing for 64 kb/s channels, while B-ISDN uses an asynchronous fast packet switching technique called Asynchronous Transfer Mode (ATM) in which data channels are identified by a virtual path/virtual channel identifier in each packet. The size of the packets is fixed and small; the packets are called ATM cells.

This section first introduces the basic concepts of narrowband ISDN, then covers the physical interfaces and services of broadband ISDN.

THE NARROWBAND ISDN

From a telephone user's perspective, there have been very few changes in the way a telephone set works ever since it was invested. The few major evolutions include powering from the central offices, automatic switching, and tone dialing. The underlying telecommunications network, however, has gone through major changes such as digitalization of transmission facilities and development of intelligent network architecture. ISDN provides a revolutionary change of the capability and functionality of telephone sets. The definition given above shows several unique characteristics of ISDN: (1) It allows end-to-end digital connections by providing digital access on the local subscribers loop from the customer's premises to the central office; (2) it provides an integrated service by carrying voice and nonvoice data, including text, image, and motion video, over a single physical interface. In addition, it also provides a signaling channel separate from the user's data channels, permitting access to advanced features of an intelligent network.

Digital Access to Customer Premises

Over the years, the public telephone network has evolved from an analog transmission network into a digital network. Most of the central offices use digital switching, and trunk lines employ digital transmission facilities. The link from the central office to the customers' premises, called the Local Subscribers Loop, however, remains analog. ISDN standards define two types of digital interfaces: the *Basic Rate Interface* (BRI) and the *Primary Rate Interface* (PRI). Each of them provides a signaling channel called the D channel, and a number of user data channels called B channels. The BRI D-channel can also carry limited user data.

The basic rate interface consists of two 64 kb/s B-channels and a 16-kb/s D-channel (2B+D). The B-channel uses 64 kb/s rate since that is the basic rate at which all analog voice signals are digitalized. The physical interface operates at a 192 kb/s rate, using time division multiplexing for the B and D channels and other overhead bits. The primary rate interface has two formats. One is based on the T1 (U.S.) digital hierarchy and the other is based on the E1 (European) digital hierarchy. The T1-based PRI consists of 23 B-channels and one D-channel (23B+D) while the E1 rate consists of 30B+D. The bandwidth of each B- or D-channel is 64 kb/s. The B-channels could be combined into multirate circuit mode ($N \times 64$) channels at increments of 64 kb/s. Some special rates are defined as H channels, for example, the H_0 channel at 384 kb/s, and the H_{10} channel at 1472 kb/s.

Out-of-Band Signaling

In the Plain Old Telephone System (POTS), the signaling between a user and the network for control of a connection is by means of on-off hook current levels and dialing pulses/tones in line with user data. In ISDN, a data link connection (based on Q.921) is established between a terminal and the central office switch on the D channel. Call control signals (based on Q.931) could be exchanged at any time, independent of the use of the data stream on the B channels. Call control information includes the called and calling numbers, bandwidth requirements and B-channel assignment, service request (voice or data, circuit or packet), end-to-end protocol selection, and so forth. Additional services could be invoked or provided, such as call redirect from the terminal, call waiting notification, conference call, or additional end-to-end-user data for call setup.

ISDN Services

Very often multiple phone lines are needed in an office, one for voice and one or more for data (using modems). ISDN enables multiple connections and services over a single physical interface. Not only can it provide high-speed connections at 64 kb/s, which are not possible with modems, it also allows multiple connections to be synchronized or coordinated. Instead of dedicating a phone line for a specific service, ISDN allows a greater freedom in the sharing of phone facilities. The bearer services, or the basic services to which users can subscribe to run their applications include circuit mode and packet mode services, as described below.

Circuit-Mode Data Service.   This service provides an unrestricted channel so that the bit stream is passed unchanged end to end. The actual application is controlled by user equipment at each end and could thus, in principle, transport voice or data with necessary protocols such as X.25 or Frame Relay.

Circuit-Mode Speech and Voiceband Services.   The Speech and Voiceband services carry voice signals as currently provided by POTS. The bit stream is assumed to be digitalized analog signals using appropriate encoding standards. Conversion may take place when passing through networks employing different encoding standards. The Voiceband service is for 3.1 kHz audio, which is used for modems.

Packet-Mode Data Service.   This service allows access to the X.25 packet network service offered by the network to which the terminal is directly connected. From the users' perspective, the packet switching network is fully integrated with the circuit switching network. The service is available only on the BRI B and D channels.

BROADBAND ISDN AND ASYNCHRONOUS TRANSFER MODE NETWORKS

The concept of broadband ISDN was a logical extension of narrowband ISDN in the evolution of telecommunications networks. The standards development started in the late 1980s, and ITU-T reached its first agreement on general aspects of B-ISDN in 1988. While B-ISDN was originally developed as a telecommunications technology, it has been adapted for data communications as well. In particular, it has been used as a new Local Area Network (LAN) technology. The ATM Forum, an industry consortium, has speeded up the development and made major contributions in this area.

Since the major feature of B-ISDN is the use of ATM, the terms B-ISDN and ATM are used synonymously.

Asynchronous Transfer Mode Cell Switching

The use of ATM technology is a major paradigm shift from synchronous networks where user data are carried in fixed time slots in the transmission facilities. ATM uses an asynchronous time division multiplexing technique to carry information in fixed size ATM cells. An ATM cell consists of two fields: a payload field for user data and a header field for identification of the user channel. Cells are assigned to user channels on demand, therefore the bandwidth occupied by a given user may vary from zero to the capacity of the physical link.

The size of an ATM cell is 53 octets, the first five for the header and remaining 48 for the payload. The header, among other things, contains an 8-bit virtual path identifier (VPI), a 16-bit Virtual Channel Identifier (VCI), and an 8-bit Header Error Check (HEC). The switch simply relays the cells based on the VPI/VCI number. The HEC is needed to prevent misrouting. There is no error control for the cell payload. Error recovery must be performed by the end systems. All that is required of the network is the delivery of cells in their original sequence.

There are two types of connections: *Virtual Path Connection* (VPC) and *Virtual Channel Connection* (VCC). VPC is analogous to a big pipe which allows end points to do their own distribution and assignment of individual virtual channels. The network will route cells based on the VPI. The VCI is for individual connections in which the network will route cells based on the combined value of VPI/VCI.

Physical Layer

Since B-ISDN was developed for public telecommunications network, it is expected that the standard-based SONET and synchronous digital hierarchy interfaces will be the most important physical media for B-ISDN/ATM networks. For the OC-3 interface the SONET STS-3c synchronous payload envelope is continually filled with ATM cells, byte aligned. This gives an effective rate of about 150 Mb/s over the 155.52 Mb/s interface. The mapping for the OC-12 interface has also been defined.

Currently not all network transmission facilities are SONET based; the existing digital hierarchy, the DS3, DS1 (for the United States) and E3, E1 (for Europe) are also to be used for ATM interfaces. There are two ways to map ATM cells into the DS3 payload: the DS3 Physical Layer Convergence Protocol (PLCP) format and the direct mapping format. The PLCP frame consists of 12 rows of ATM cells, each preceded by 4 octets of overhead. At the end the frame is stuffed with a variable number of nibbles (4 bits) to fill up the 125 μs frame. The direct mapping format continuously fills the DS3 payload with ATM cells with nibble alignment.

For use in the LAN environment, other less expensive media have also been accepted. These include 155.52 Mb/s over Category 5 Unshielded or Shielded Twisted Pair (UTP and STP), 51.84 Mb/s over Category 3 UTP, and 25.6 and 25.2 Mb/s over various twisted pairs. Additional fiber formats could also be used; this includes 100 Mb/s over multimode fiber using the FDDI 4 bit/5 bit (4B/5B) code, and 155 Mb/s over multimode fiber using 8B/10B code.

Another important transmission medium for B-ISDN is the use of communications satellites. The subject is covered later.

B-ISDN Services

B-ISDN provides integrated services for voice, text, image, and video just like the narrowband ISDN, but at a wider range of bandwidth. Since B-ISDN uses packet mode for information transfer, it needs to simulate the services provided in the traditional circuit mode transfer. Therefore, cell transfer delay and delay variation become important service parameters. It is a major challenge for the service provider to take full advantage of statistical multiplexing while maintaining the desired level of quality of services.

Constant Bit Rate Service (CBR).   This service is for applications that generate a stream of bits at a constant rate and have a real-time requirement, such as voice and video applications. These applications require tightly constrained cell transfer delay and delay variations.

Variable Bit Rate Service (VBR).   This service supports applications that generate bursty traffic. Some applications may have real-time requirements, such as compressed voice and video applications. Other applications may be non-real-time but still require a bound on cell transfer delay.

Unspecified and Available Bit Rate Service (UBR and ABR).   This service supports the traditional data communication applications, such as e-mail, file transfer, and interactive traffic. The source is expected to be bursty. The network provides the best effort delivery service, and may discard cells when congestion occurs. The ABR service requires a flow control mechanism that provides feedback to the source from the destination as well as from network nodes.

ATM Adaptation, Signaling, and Interworking Functions

ATM Adaptation.   Because of the diversity of services that must be supported in a B-ISDN/ATM network, some type of interface must be provided between the ATM layer protocol and the applications on the end systems. The ATM adaptation layer (AAL) protocol was developed for this purpose as well as for the end-to-end error control, which is not provided by the ATM layer cell switching protocol. The major functions of AAL at the sender side include: (1) packing of user messages into AAL message(s) with insertion (when necessary) of sequence numbers, error protection (parity, CRC or FEC), clocking signal, and padding; (2) segmentation of messages into ATM cells and delivery for transmission of cells at the specified rate. The receiver side performs the reverse functions, including error checking of cells received, recovery of source clock and handling of cell transfer delay (for CBR traffic), reassembly and delivery of messages to the application, and notification of transmission errors. There are several types of AALs defined, each designed to meet the requirement of a particular service, for example, type 1 for CBR service and type 5 for ABR/UBR and some VBR services.

ATM Signaling.   Since ATM is a connection-oriented protocol, a B-ISDN/ATM network must provide connection setup. The procedure uses the signaling protocol Q.2931, which is an extension of the one used for narrowband ISDN. During the call setup, the calling and called parties and the network must check or negotiate the relevant service parameters such as cell rate (peak, sustainable, minimum), cell transfer delay (mean, maximum, variation), and cell loss ratio, depending on the service requested.

Interworking with other Networks.   The higher bandwidth of B-ISDN/ATM networks enable better use of applications involving multimedia and digital video. In addition to addressing new applications, we must still make these networks interwork seamlessly with existing POTS, ISDN, Frame Relay, LANs, and the Internet. One important issue is support of connectionless IP Internet in a connection-oriented ATM network. The LAN emulation service developed at the ATM forum and the adaption of Internet router functions to the ATM network are examples of solutions for interworking.

Satellite Links for B-ISDN/ATM Networks

As was mentioned earlier, the standard-based SONET and Synchronous Digital Hierarchy will be widely used for ATM. However, for regions not served with fiber, satellite communications can be used to provide the physical layer for B-ISDN/ATM networks. It is shown here that data rates of 155.52 Mb/s are possible using satellites and modems available today. It is shown that higher data rates could be supported but only in restricted cases.

TABLE 17.4.1 Modulation and Coding Performance for Various Modulation Types Used in Satellite Communications.*

Modulation (Coding Rate)	Eb/No dB	Rb dB* Mb/s	Rb Mb/s	BW MHz
QPSK($^1/_2$)	9.0	84.6	288	374
QPSK($^1/_2$,RS)	5.2	88.4	692	990
QPSK($^3/_4$)	11.0	82.6	182	158
QPSK($^3/_4$, RS)	6.5	87.1	512	488
8PSK($^5/_6$)	12.0	81.6	146	76
8PSK($^2/_3$, RS)	7.5	86.1	407	291
16 QAM($^7/_8$, RS)	10.9	82.7	186	76

*For C band zone, Example: 8PSK(2/3,RS) is 8 phase shift keying with rate 2/3 inner code with Reed-Solomon outer code.

Link Budget. Link budgets, which are fundamental to satellite communications, were performed for both C and Ku band with INTELSAT satellite systems to determine the maximum practical data rate that could be supported with the Standard A antenna (15 m) for C band and the Standard C antenna (11 m) for Ku band. These link budgets yielded a system carrier to noise density C/No of about 93.6 dB*Hz for the C band zone beam with a saturated effective isotropic radiated power (e.i.r.p.) of 34 dBw and a system C/No of about 2 dB greater for the Ku spot beam with a saturated e.i.r.p. of 46 dBw. These links are operated several dB backoff to accommodate phase and amplitude modulation.

Supportable Data Rates. With these values of C/No, the maximum data rate supportable can be calculated. Table 17.4.1 gives typical values for a bit error rate (BER) = 10^{**} (–11) when using the C band zone beam. The bit energy to noise density Eb/No values were taken from the EFData Corporation Modem manufacturer's literature and are based on measured values, but at lower data rates than are considered below. (It is reasoned that the modem performance is independent of data rate if all other parameters are equal.) Table 17.4.1 also gives the approximate required bandwidth to support the modulation and coding used. These modems use either a convolutional code with the Viterbi decoder or the concatenated code scheme with an outer code of the Reed-Solomon (of rate approximately 9/10) and an inner code of the convolutional type just mentioned. The bandwidth (BW) required is given by the formula:

$$BW = Rb^*(1/R)^*(1/f)^*1.3 \text{ Hz}$$

where *Rb* = data rate
    R = inner code rate
    f = modulation factor and is 2 for quadrature phase shifting keying (QPSK), 3 for 8 phase shift keying (8PSK), and 4 for 16 quadrature amplitude modulation (16 QAM)

if the Reed-Solomon (RS) outer code is used when BW is increased by 10 percent.

There are three commonly used data rates considered here: 44.736 Mb/s (DS-3), 155 Mb/s (OC-3), and 622 Mb/s (OC-12). These data rates are used in the ATM architecture. Using the modulations given in Table 17.4.1 above, Table 17.4.2 yields the required bandwidth for the three data rates mentioned.

From Table 17.4.2, it seems that the DS-3 can be supported on 36 MHz transponders only for 8PSK and 16 QAM modulation. A transponder with a bandwidth of about 43 MHz can support DS-3 using all combinations except QPSK($^1/_2$) and QPSK(1/2,RS). Consulting Table 17.4.1 shows that a DS-3 link can be supported by any entry in the table, as far as the required Eb/No is concerned. If a 72 MHz bandwidth transponder is considered, then Table 17.4.2 shows that a DS-3 link can be supported by all modulation/coding combinations.

The OC-3 (155.5 Mb/s) rate can be supported on about an 81-MHz transponder using the 8PSK(5/6) as shown in Table 17.4.2. Table 17.4.1 shows that the maximum data rate supportable from the Eb/No viewpoint is about 146 Mb/s. If one has a C band zone 72-MHz transponder, available on INTELSAT, this combination of modulation, coding, and transponder bandwidth will give performance close to the goal of the target BER and is recommended for OC-3 transmission. Results using the Ku band zone with the Standard C terminal will yield improved performance because of the 2-dB link improvement.

TABLE 17.4.2 Bandwidth Required for Certain Data Rates and Modulation

Modulation (coding rate)	44.736 Mb/s DS-3 (BW, MHz)	155.5 Mb/s OC-3 (BW, MHz)	622 Mb/s OC-12 (BW, MHz)
QPSK(1/2)	58	202	809
QPSK(1.2, RS)	64	222	890
SPSK(3/4)	39	135	539
QPSK(3/4, RS)	43	149	593
8PSK(5/6)	23	81	324
8PSK(2/3, RS)	32	111	445
16QAM(7/8, RS)	18	64	150

Tables 17.4.1 and 17.4.2 show that there is no combination of modulation/coding and available transponders that will support the OC-12 data rate. However, Table 17.4.1 shows that the QPSK(1/2,RS) will support OC-12 from the required Eb/No viewpoint but requires a bandwidth of 890 MHz. This clearly presents a problem since the fixed satellite service bands are only 500 MHz wide. The only other parameter in this discussion is the size of the earth station required to support OC-12 in a transponder of wider bandwidth. For example, a transponder of 500 MHz would support 8PSK(2/3,RS) but requires an Eb/No of about 7.5 dB (see Table 17.4.1). This would require a C/No of about 98.8 dB*Hz, equivalent to an antenna at C band of about 26 m in diameter. This is clearly a very large antenna by today's standards.

The other main parameter on the satellite besides the transponder bandwidth is the downlink e.i.r.p. This parameter could be increased allowing for one of the higher order modulations to be used. A several dB increase in the downlink e.i.r.p. would allow for 16QAM(7/8,RS) to be used to support the OC-12 data rate but the bandwidth would still be larger than 150 MHz.

It has been shown that the maximum data rate that can be supported is about 155 Mb/s using what is available now and in the near future. The present designs of satellites do not permit the support of OC-12 (622-Mb/s) rate. If this rate is to be supported by commercial satellite links, then the satellites of the future must increase both their transponder power and bandwidth.

CHAPTER 17.5
WIRELESS NETWORKS

Matthew. N. O. Sadiku

The recent past has seen fast-moving developments in the field of telecommunications. Wireless communications in particular continue to experience growth at an unprecedented rate. This is partly caused by our growing dependence on computer, cellular phone, pagers, fax machines, e-mail, and the Internet. The demand for real-time information exchange is being made on an increasingly mobile workforce. Many jobs require workers to be mobile, e.g., inventory clerks, healthcare workers, policemen, and emergency care specialists. It is estimated that roughly 48 million U.S. workers cannot do without one form of tetherless communication. Their work tools must be mobile.

Wireless refers to any communication that uses electromagnetic or acoustic waves as a medium, rather than a wire connection. It could be the technology that allows seamless and instant access to information across any number of devices and platforms. Wireless technologies are targeted to meeting certain needs, which include:

- Bypass physical barriers (roads, railroads, buildings, rivers, and so forth)
- Remote data entry
- Mobile objects (car, airplanes, ships, and so forth)
- Connectivity to hard-to-wire places
- Worldwide connectivity for voice, video, and data communications
- Satellite communications
- Inexpensive setup

Wireless communication works on the same set of fundamental principles, whether it is computer nodes on a local area network (LAN) or the common cordless telephone. In this chapter, those common fundamentals are presented first. We will then discuss cordless phones, paging, cellular networks, personal communications systems (PCSs), and wireless data networks such as cellular digital packet data (CDPD), wireless LAN, and wireless asynchronous transfer mode (ATM).

PROPAGATION CHARACTERISTICS

The major factors that affect the design and performance of wireless networks are the characteristics of radio or electromagnetic wave propagation over the geographical area. The concept of propagation refers to the various ways by which an electromagnetic (EM) wave travels from the transmitting antenna to the receiving antenna. Propagation of EM wave may be regarded as a means of transferring energy or information from one point (a transmitter) to another (a receiver). Here we describe the free space propagation model and the empirical path loss formula.

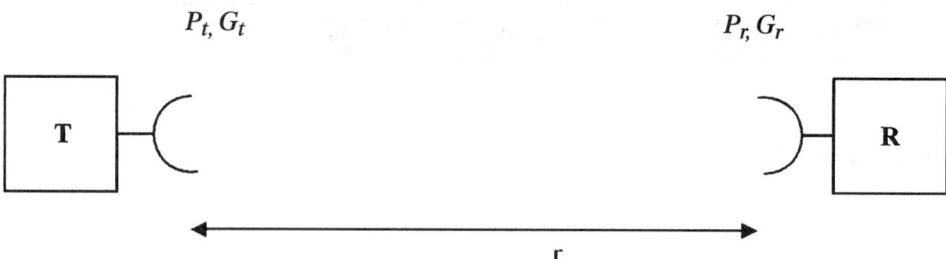

FIGURE 17.5.1   Basic wireless systems.

Free Space Propagation Model

Wireless links typically experience free space propagation. The free space propagation model is used in predicting received signal strength when the transmitter and receiver have a clear line-of-sight path between them. If receiving antenna is separated from the transmitting antenna in free space by distance r, as shown in Fig. 17.5.1, the power received P_r by the receiving antenna is given by Friis equation

$$P_r = G_r G_t \left(\frac{\lambda}{4\pi r} \right)^2 P_t \tag{1}$$

where P_t = transmitted power
$\quad\quad G_r$ = receiving antenna gain
$\quad\quad G_t$ = transmitting antenna gain
$\quad\quad \lambda$ = wavelength (=c/f) of the transmitted signal

Friis equation relates the power received by one antenna to the power transmitted by the other, provided that the two antenna are separated by $r > 2d^2/\lambda$, where d is the largest dimension of either antenna. Thus, Friis equation applies only when the two antennas are in the far-field of each other. It also shows that the received power falls off as the square or the separation distance r. The power decay as $1/r^2$ in wireless systems as exhibited in Eq. (1) is better than the exponential decay in power in a wired link. In actual practice, the value of the received power given in Eq. (1) should be taken as the maximum possible because some factors can serve to reduce the received power in a real wireless system.

From Eq. (1), we notice that the received power depends on the product $P_t G_t$. The product is defined as the *effective isotropic radiated power* (EIRP), i.e.

$$\text{EIRP} = P_t G_t \tag{2}$$

The EIRP represents the maximum radiated power available from a transmitter in the direction of maximum antenna gain relative to an isotropic antenna.

Empirical Path Loss Formula

In addition to the theoretical model presented in the preceding section, there are empirical models for finding path loss. Of the several models in the literature, Okumura et al.'s model is the most popular choice for analyzing mobile-radio propagation because of its simplicity and accuracy. The model is based on extensive measurements in and around Tokyo between 200 MHz and 2 GHz, compiled into charts, which can be applied to VHF and UHF mobile-radio propagation. The medium path loss is given by

$$L_p = \begin{cases} A + B\log_{10}(r), & \text{for urban area} \\ A + B\log_{10}(r) - C, & \text{for suburban area} \\ A + B\log_{10}(r) - D, & \text{for open area} \end{cases} \tag{3}$$

base station

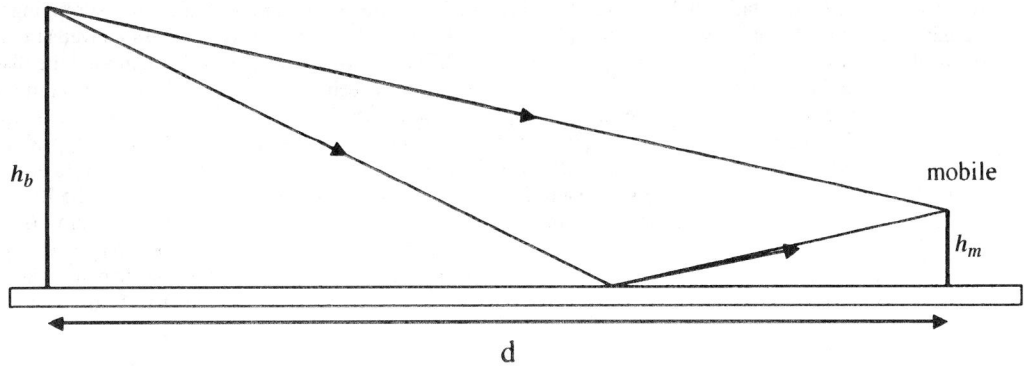

FIGURE 17.5.2   Radio propagation over a flat surface.

where r (in kilometers) is the distance between the base and mobile stations, as illustrated in Fig. 17.5.2. The values of A, B, C, and D are given in terms of the carrier frequency f, the base station antenna height h_b (in meters), and the mobile station antenna height h_m (in meters) as

$$A = 69.55 + 26.16 \log_{10}(f) - 13.82 \log_{10}(h_b) - a(h_m) \tag{4a}$$

$$B = 44.9 - 6.55 \log_{10}(h_b) \tag{4b}$$

$$C = 5.4 + 2\left[\log_{10}\left(\frac{f}{28}\right)\right]^2 \tag{4c}$$

$$D = 40.94 - 19.33 \log_{10}(f) + 4.78[\log_{10}(f)]^2 \tag{4d}$$

where

$$a(h_m) = \begin{cases} 0.8 - 1.56 \log_{10}(f) + [1.1 \log_{10}(f) - 0.7]h_m, & \text{for medium/small city} \\ 8.28[\log_{10}(1.54 h_m)]^2 - 1.1, & \text{for } f \geq 200\,\text{MHz} \\ 3.2[\log_{10}(11.75 h_m)]^2 - 4.97, & \text{for } f < 400\,\text{MHz} \\ & \text{for large city} \end{cases} \tag{5}$$

The following conditions must be satisfied before Eq. (3) is used: $150 < f < 1500$ MHz; $1 < r < 80$ km, $30 < h_b < 400$ m; $1 < h_m < 10$ m. Okumura's model has been found to be fairly good in urban and suburban areas, but not as good in rural areas.

CORDLESS TELEPHONY

Cordless telephones first became widespread in the mid-1980s as products became available at an affordable price. The earliest cordless telephone used narrow band technology and used separate channels for frequency channel for transmission to/from the base station. They had limited range, poor sound quality, and poor security—people could easily intercept signals from another cordless phone because of the limited number of channels. The Federal Communications Commission (FCC) granted the frequency range of 47 to 49 MHz for cordless phones in 1986 and the frequency range of 900 MHz in 1990. This improved their interference problem, reduced the power needed to run them, and allowed cordless phones to be clearer, broadcast a longer distance, and choose from more channels. However, cordless phones were still quite expensive.

The use of digital technology transformed the cordless phone. Digital technology represents the voice as a series of 0s and 1s, just as a CD stores music. Digital cordless phones in the 900-MHz frequency range were introduced in 1994. Digital signals allowed the phones to be more secure and decreased eavesdropping. With the introduction of digital spread spectrum (DSS) in 1995, eavesdropping on the cordless conversations was practically made impossible. The opening up of the 2.4-GHz range by the FCC in 1988 increased the distance over which a cordless phone can operate and further increased security. With the cordless phone components getting smaller, more and more features and functions can be placed in phones without making them any bigger. Such functions may include voice mail, call screening, and placing outside calls. With many appealing features, there continues to be a strong market interest in cordless telephones for residential and private office use.

As shown in Fig. 17.5.3, the cordless telephone has gone through an evolution. This started with the 46/49 MHz telephones. Although earlier cordless telephones existed, the 46/49 MHz cordless telephones were the first to be produced in substantial quantities. The second generation used 900-MHz frequency range resulting in longer range. The third generation introduced the spread spectrum telephones in the 900 MHz band. The fourth generation changed from 900-MHz to 2.4-GHz band, which is accepted worldwide. The fifth generation of cordless telephones is emerging now and employs time division multiple access (TDMA).

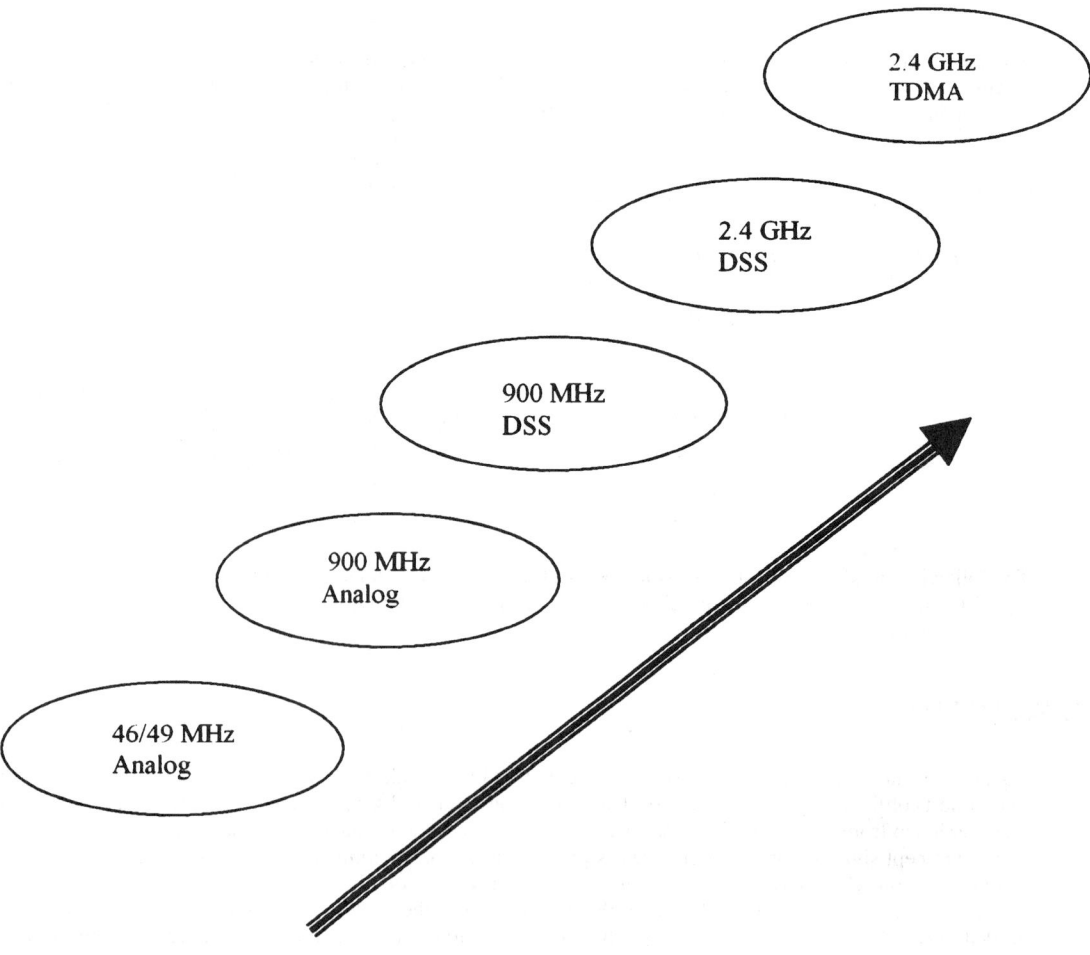

FIGURE 17.5.3 Evolution of cordless phone.

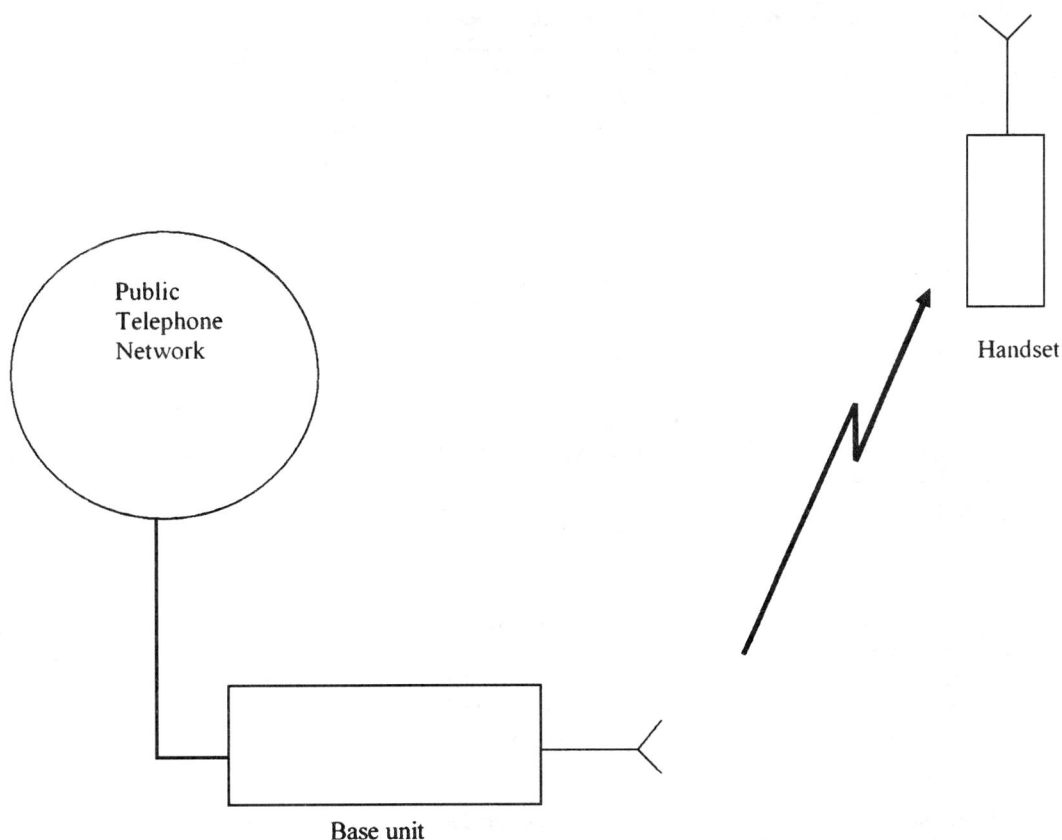

FIGURE 17.5.4   Cordless telephone system configuration.

Basic Features

A cordless phone basically combines the features of telephone and radio transmitter/receiver. As shown in Fig. 17.5.4, it consists of two major units: base and handset. The base unit interfaces with the public telephone network through the phone jack. It receives the incoming call through the phone line, converts it to an FM radio signal, and then broadcasts that signal. The handset receives the radio signal from the base, converts it to an electrical signal, and sends that signal to the speaker, where it is converted into the sound wave. When someone talks, the handset broadcasts the voice through a second FM radio signal back to the base. The base receives the voice signal, converts it to an electrical signal, and sends that signal through the phone line to the other party. The base and handset operate on a frequency pair (duplex frequency) that allows one to talk and listen simultaneously.

Types of Cordless Telephone

Over the years, several types of cordless have been developed. These include:

- *CT1*: This first generation cordless telephone was introduced in 1983. It provides a maximum range of about 200 m between handset and base station. It is an analog phone that is primarily designed for domestic use. It employs analog radio and uses eight RF channel and frequency division multiple access (FDMA) scheme.

TABLE 17.5.1 CT1 Cordless Telephone Duplex Frequencies

Channel number	Base unit transmission frequency (kHz)	Handset transmission frequency (MHz)
1	1642.00	47.45625
2	1662.00	47.46875
3	1682.00	47.48125
4	1702.00	47.49375
5	1722.00	47.50625
6	1742.00	47.51875
7	1762.00	47.53125 or 47.44375
8	1782.00	47.54375

Operation has to be on not more than one of the pair of frequencies shown in Table 17.5.1 at any one time. As the number of users grew, so did the co-channel interference levels, while the quality of the service (customer satisfaction) deteriorated.

- *CT2*: This second generation cordless telephone uses digitized speech and digital transmission, thereby offering a clearer voice signal than analog CT1. Another advantage is that CT2 does not suffer from the inherent interference problems associated with CT1.

- *DECT*: DECT stands for digital enhanced cordless telecommunications. The DECT specification was developed by European Telecommunications Standards Institute (ETSI) and operates throughout Europe in the frequency band 1880 to 1900 MHz. DECT provides cordless telephones with the greater range, up to several hundred meters, allows encryption, provides for greater number of handsets, and even allows data communication. It uses high-frequency signals (1.88 to 1.9 GHz) and also employs time division multiple access (TDMA), which allows several conversations to share the same frequency. Although CT1, CT2, and DECT are European standards, the US PCS standards have followed these models too. DECT is being adopted increasingly worldwide.

- *PHS*: The personal hand-phone system (PHS) was introduced in Japan in 1995 for private use as well as for PCS. Unlike conventional cellular telephone systems, the PHS system employs ISDN technology. With PHS, a subscriber can have two separate telephone numbers: one for the home and the other for outside the home. The PHS system uses TDMA format because of the flexibility for call control and economy—characteristics common to the cellular system. To allow for two-way communication, forward and reverse channels are located on the same frequency by employing time-division duplex (TDD). It employs carrier spaced 300 kHz apart over a 23-MHz band from 1895 to 1918 MHz. Each carrier supports four channels—one control channel broadcast on a carrier, while three speech channels broadcast on other carrier waves. PHS is attracting attention around the world, particularly in Asian nations.

- *ISM*: The 900-MHz digital spread spectrum (DSS) cordless telephone operates in the 902 to 928 MHz industrial-scientific-medical (ISM) band. The spread spectrum systems have the additional advantage of enhanced security. The channel spacing is 1.2 MHz and there are 21 nonoverlapping channels in the band. The system is operated using TDD at a frame rate of 250 Hz. It provides clear sound, superb range, and security. It has a greater output power than other cordless phones. This increased power dramatically boosts range. The 2.4-GHz DSS cordless telephone is an upgrade of this.

Cordless telephones are categorized by the radio frequency used and whether transmission between the handset and base unit is in the form of analog or digital signals. Generally speaking, the clarity of a cordless telephone improves with the use of higher frequencies and digital technology. Regulatory authorities in each country also specify and allocate the frequencies that may be used by cordless telephones in their respective countries and all telephones intended for use in their countries must receive their approvals. All cordless telephones are approved in the respective markets in which they are sold. Table 17.5.2 shows the common cordless telephone standards and their respective frequency range. In common with all areas of communications, the trend is away from analog systems to digital systems.

TABLE 17.5.2   Comparison of Cordless Telephone Standards

	Analog cordless telephone			Digital cordless telephone		
Standard	CT1	JCT	900 MHz	DECT	PHS	ISM
Region	Europe	Japan	worldwide	Europe	Japan	USA
Frequency (MHz)	914/960	254/380	900	1880–1990	1895–1918	2400–2485
Range (km)	Up to 7	0.3	0.25	0.4	0.2	0.5

JCT—Japanese cordless telephone; PHS—Personal Hand-phone System.

PAGING

Paging started as early as 1921 when the concept of one-way information broadcasting was introduced. The 1930s saw the widespread use of radio paging by government agencies, police departments, and the armed forces in the United States. Paging systems have undergone dramatic development. Radio transmission technology has advanced and so are the computer hardware and firmware (computer program) used in radio-paging systems.

One-Way Pagers

A paging system is a one-way wireless messaging system that allows continuous accessibility to someone away from the wired communications network. In its most basic form, the person on-the-move carries a palm-sized device (the pager) that has an identification number. The calling party inputs this number, usually through the public telephone network, to the paging system which then signals the pager to alert the called party. Early paging systems were nonselective and operator assisted. Not only did it waste airtime, the system was inconvenient, labor-intensive, and offered no privacy. With automatic paging, a telephone number is assigned to each pager and the paging terminal can automatically signal for voice input from the calling party.

The basic paging system consists of the following components:

- *Input Source*: A caller enters a page from a phone or through an operator. Once it is entered, the page is sent through the public switched telephone network (PSTN) to the paging terminal for encoding and transmission through the paging system.

- *Encoder*: The encoder typically accepts the incoming page, checks the validity of the pager number, looks up the database for the subscriber's pager address, and converts it into the appropriate paging signaling protocol. The encoded paging signal is then sent to the transmitters (base stations).

- *Base Station*: This transmits page codes on an assigned radio frequency. Most base stations are designed specifically for paging but some of those designed for two-way voice can also be used.

- *Page Receivers*: These are the pagers, which are basically FM receivers turned to the same RF frequency as the paging base station in the system. A decoder in each pager recognizes the unique code assigned to the pager and rejects all other codes for selective alerting. The most basic function of the pager is alerting. On receiving its own paging code, the receiver sets off an alert that can be audible (tone), visual (flashing indicator), or silent (vibrating). Messaging functions can also include voice and/or display (numeric/alphanumeric) messaging.

Today's paging systems offer much more than the basic system described above. A paging system subscriber can be alerted anytime and at almost any place as coverage can be easily extended, even across national borders. Paging systems are increasingly migrating from tone and numeric paging, to alphanumeric paging. Alphanumeric pagers display alphabetic or numeric messages entered by the calling party. The introduction of alphanumeric pagers also enables important information/data (e.g., business, financial news) to be constantly updated and monitored. Pagers that can display different ideographic languages, e.g., Chinese and Japanese, are now available in the market. The specific language supported is determined by the computer program installed in the pager.

Two-Way Pagers

The conventional paging systems are designed for one-way communication—from network toward pagers. Such systems provide the users one or more of the following services: beep, voice messaging, numeric messaging, and alphanumeric messages. With the recent development in paging systems, it is possible to supply a reverse link and thus allow two-way paging services. Two-way paging offers some significant capabilities with distinct advantages.

Two-way paging is essentially alphanumeric paging that lets the pager send messages, either to respond to received messages or to originate its own messages. The pagers come in various shapes and sizes. Some look almost like today's alphanumeric pagers, while some add small keyboards and larger displays. Two-way paging networks employ the 900-MHz band, a small fraction of the spectrum originally meant for PCS.

Networks are built around protocols. Two-way messaging network is based on Reflex, which is basically an extension of Motorola's Flex protocol for one-way paging. Reflex is a new generation of paging protocols. But reflex is a proprietary protocol. In view of the fact that two-way paging is the dominant application for wide area wireless networks and that a set of open, nonproprietary protocols is required to enable the convergence of the two-way paging and the Internet, efficient mail submission and delivery (EMSD) has been designed. EMSD is an open, efficient, Internet messaging protocol that is highly optimized for short messages. Devices that provide two-way paging capabilities should use this protocol (e.g., dedicated pagers, cell phones, palm PCs, handheld PCs, laptops, and desktops).

The majority of pagers are used for the purpose of contacting someone on the move. The most popular type of pager used for this application is the alphanumeric pager that displays the telephone number to call back after alerting the paging subscriber. Although they are not common or cheap, the trend toward alphanumeric paging is inevitable with improved speed and better pagers. There will be more varied applications of paging such as the sending of e-mail, voice mail, faxes, or other useful information to a pager, which will also take on more attractive and innovative forms. Future pagers may compete more aggressively with other two-way technologies, such as cellular and PCS. Although paging does not provide real-time interactive communications between the caller and the called party, it has some advantages over other forms of PCS, such as cellular telephone. These include less bandwidth requirement, larger coverage area, lower cost, and lighter weight. Owing to these advantages, paging service is bound to be in a strong competitive PCS services in years to come.

With more than 51 million paging subscribers worldwide, major paging markets in the world, especially in Asia, continue to expand rapidly. With the use of pagers getting more and more integrated into our daily lives, we will be seeing a host of new and exciting applications. There will emerge satellite pagers, which will send and receive messages through satellite systems such as Iridium, ICO, and Globalstar. With the help of such pagers, it is possible to supply paging services in global scale.

CELLULAR NETWORKS

The conventional approach to mobile radio involved setting up a high-power transmitter on top of the highest point in the coverage area. The mobile telephone must have a line-of-sight to the base station for proper coverage. Line-of-sight transmission is limited to as much as 40 to 50 miles on the horizon. Also, if a mobile travels too far from its base station, the quality of the communications link becomes unacceptable. These and other limitations of conventional mobile telephone systems are overcome by cellular technology.

Areas of coverage are divided into small hexagonal radio coverage units known as *cells*. A cell is the basic geographic unit of a cellular system. A cellular communications system employs a large number of low-power wireless transmitters to create the cells. These cells overlap at the outer boundaries, as shown in Fig. 17.5.5. Cells are base stations transmitting over small geographic areas that are represented as hexagons. Each cell size varies depending on the landscape and tele-density. Those stick towers one sees on hilltops with triangular structures at the top are cellular telephone sites. Each site typically covers an area of 15 miles across, depending on the local terrain. The cell sites are spaced over the area to provide a slightly overlapping blanket of coverage. Like the early mobile systems, the base station communicates with mobiles via a channel. The channel is made of two frequencies—one frequency (the forward link) for transmitting information to the base station and the other frequency (the reverse link) to receive from the base station.

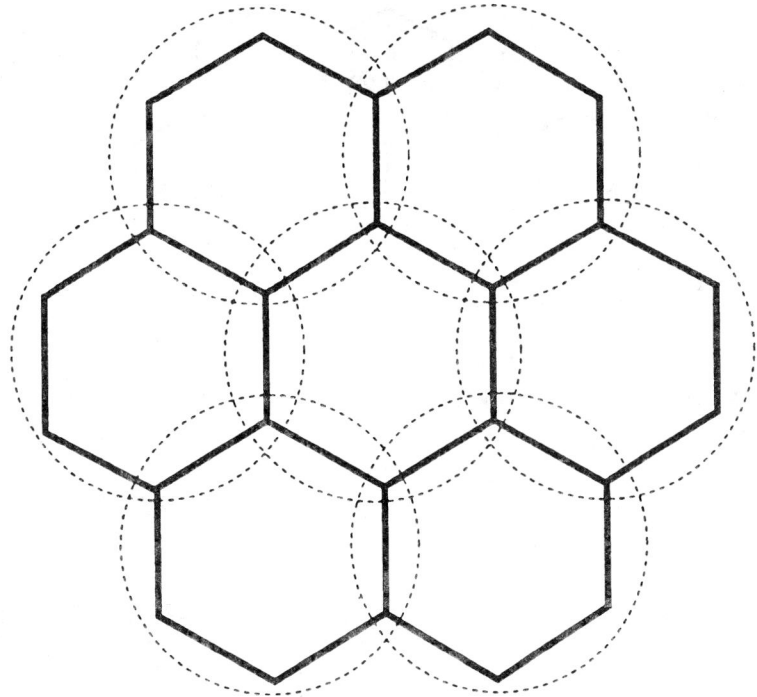

FIGURE 17.5.5   A typical wireless seven-cell pattern; cells overlap to provide greater coverage.

Fundamental Features

Besides the idea of cells, the essential principles of cellular systems include cell splitting, frequency reuse, handover, capacity, spectral efficiency, mobility, and roaming.

- *Cell Splitting*: As a service area becomes full of users, the single area is split into smaller ones. This way, urban regions with heavy traffic can be split into as many areas as necessary to provide acceptable service, while large cell can be used to cover remote rural regions. Cell splitting increases the capacity of the system.

- *Frequency Reuse*: This is the core concept that defines the cellular system. The cellular-telephone industry is faced with a dilemma: services are growing rapidly and users are demanding more sophisticated call-handling features, but the amount of the EM spectrum allocation for cellular service is fixed. This dilemma is overcome by the ability to reuse the same frequency (channel) many times. Several frequency-reuse patterns are in use in the cellular industry, each with its advantages and disadvantages. A typical example is shown in Fig. 17.5.6, where all the available channels are divided into 21 frequency groups numbered 1 to 21. Each cell is assigned three frequency groups. For example, the same frequencies are reused in cell designated 1 and adjacent locations do not reuse the same frequencies. A cluster is a group of cells; frequency reuse does not apply to clusters.

- *Handoff*: This is another fundamental feature of the cellular technology. When a call is in progress and the switch from one cell to another becomes necessary, a handoff takes place. Handoff is important because as a mobile user travels from one cell to another during a call, as adjacent cells do not use the same radio channels, a call must be either dropped or transferred from one channel to another. Dropping the call is not acceptable. Handoff was created to solve the problem. A number of algorithms are used to generate and process a handoff request and eventual handoff order. Handing off from cell to cell is the process of transferring the mobile unit that has a call on a voice channel to another voice channel, all done without interfering with the call. The need for handoff is determined by the quality of the signal, whether it is weak or strong. A handoff threshold is predefined. When the received signal level is weak and reaches the threshold, the system provides

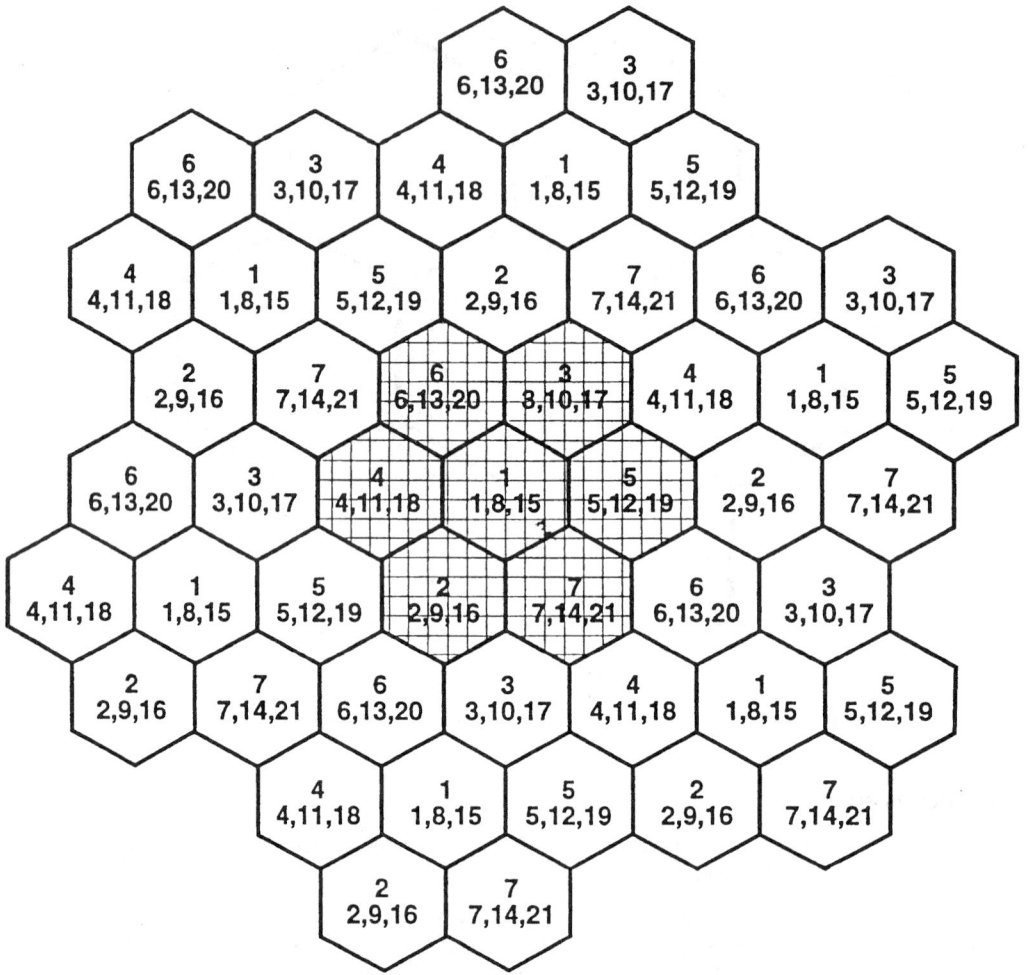

FIGURE 17.5.6 Frequency reuse in a seven-cell pattern cellular system.

a stronger channel from an adjacent cell. This handoff process continues as the mobile moves from one cell to another as long as the mobile is in the coverage area.

- *Mobility and Roaming*: Mobility implies that a mobile user while in motion will be able to maintain the same call without service interruption. This is made possible by the built-in-handoff mechanism that assigns a new frequency when the mobile moves to another cell. Because of several cellular operators within the same region using different equipment and a subscriber is only registered with one operator, some form of agreement is necessary to provide services to subscribers. Roaming is the process whereby a mobile moves out of its own territory and establishes a call from another territory. If we consider a cell (an area) with a perimeter L where ρ mobile units per unit area are located, the average number of users M crossing the cell boundaries per unit time is

$$M = \frac{\rho V L}{\pi} \tag{6}$$

where V is the average velocity of the mobile units.

- *Capacity*: This is the number of subscribers that can use the cellular system. For an FDMA system, the capacity is determined by the loading (no. of calls and the average time per call) and system layout (size of cells and amount of frequency reuse utilized). Capacity expansion is required because cellular systems must serve more subscribers. It takes place through frequency reuse, cell splitting, planning, and redesigning of the system.

- *Spectral Efficiency*: This a performance measure of the efficient use of the frequency spectrum. It is the most desirable feature of a mobile communication system. It produces a measure of how efficiently space, frequency, and time are used. Expressed in channels/MHz/km^2, channel efficiency is given by

$$\eta = \frac{\text{total no. of channels available in the system}}{\text{bandwidth} \times \text{total coverage area}}$$

$$\eta = \frac{\dfrac{B_w}{B_c} \times \dfrac{N_c}{N}}{B_w \times N_c \times A_c} = \frac{1}{B_c \times N \times A_c} \tag{7}$$

where B_w = bandwidth of the system in MHz
B_c = channel spacing in MHz
N_c = number of cells in a cluster
N = frequency reuse factor of the system
A_c = area covered by a cell in km^2

Cellular System

A typical cellular network is shown in Fig. 17.5.7. It consists of the following three major hardware components [3]:

- *Cell Site (Base Stations)*: The cell site acts as the user-to-MTSO interface, as shown in Fig. 17.5.7. It consists of a transmitter and two receivers per channel, an antenna, a controller, and data links to the cellular office. Up to 12 channels can operate within a cell depending on the coverage area.

- *Mobile Telephone Switching Office (MTSO)*: This is the physical provider of connections between the base stations and the local exchange carrier. MTSO is also known as mobile switching center (MSC) or digital multiplex switch-mobile telephone exchange (DMS-MTX) depending on the manufacturer. It manages and controls cell site equipment and connections. It supports multiple-access technologies such as AMPS, TDMA, CDMA, and CDPD. As a mobile moves from one cell to another, it must continually send messages to the MTSO to verify its location.

- *Cellular (Mobile) Handset*: This provides the interface between the user and the cellular system. It is essentially a transceiver with an antenna and is capable of tuning to all channels (666 frequencies) within a service area. It also has a handset and a number assignment module (NAM), which is a unique address given to each cellular phone.

Cellular Standards

Because of the rapid development of cellular technology, different standards have resulted. These include:

- *Advanced Mobile Phone System (AMPS)*: This is the standard introduced in 1979. Although it was developed and used in North America, it has also been used in over 72 countries. It operates in the 800-MHz frequency band. It is based on FDMA. The mobile transmit channels are in the 825- to 845-MHz range, while the mobile receive channels are in the 870- to 890-MHz range. There is also the digital AMPS, which is also known as TDMA (or IS-54). FDMA systems allow for a single mobile telephone to call on a radio channel;

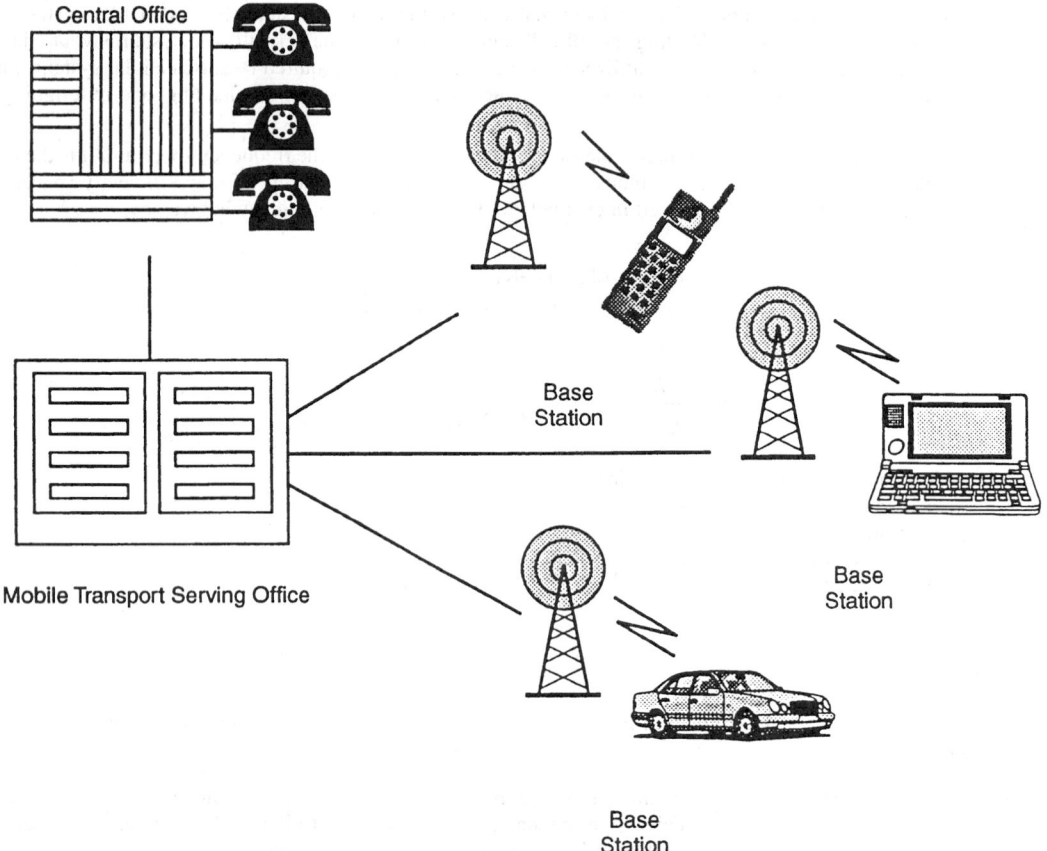

FIGURE 17.5.7   A typical cellular network.

each voice channel can communicate with only one mobile telephone at a time. TDMA systems allow several mobile telephones to communicate all the same time on a single radio carrier frequency. This is achieved by dividing their signal into time slots.

- *IS-54 and IS-95*: The IS-54 is a North American standard developed by the Electronic Industries Association (EIA) and the Telecommunications Industry Association (TIA) to meet the growing demand for cellular capacity in high-density areas. It is based on TDMA and it retains the 30-kHz channel spacing of AMPS to facilitate evolution from analog to digital systems. The IS-95 standard was also adopted by EIA/TIA. It is based on CDMA, a spread-spectrum technique that allows many users to access the same band by assigning a unique orthogonal code to each user.

- *Global System for Mobile Communications (GSM)*: This is a digital cellular standard developed in Europe and designed to operate in the 900-MHz band. It is a globally accepted standard for digital cellular communication. It uses a 200-kHz channel divided into eight time slots with frequency division multiplexing (FDM). The technology allows international roaming and provides integrated cellular systems across different national borders. GSM is the most successful digital cellular system in the world. It is estimated that many countries outside Europe will join the GSM partnership.

- *Personal Digital Cellular (PDC)*: This is a digital cellular standard developed in Japan. It was designed to operate in 800-MHz and 1.5-GHz bands.

• *Future Public Land Mobile Telecommunication Systems (FPLMTS)*: This is a new standard being developed in ITU to form the basis for third-generation wireless systems. It will consolidate today's increasingly diverse and incompatible mobile environments into a seamless infrastructure that will offer a diverse portfolio of telecommunication services to an exponentially growing number of mobile users on a global scale. It is a digital system based on 1.8- to 2.2-GHz band. It is being tested to gain valuable user and operator experience. In many European countries, the use of GSM has allowed cross-country roaming. However, global roaming has not been realized because there are too many of these incompatible standards.

PERSONAL COMMUNICATION SYSTEMS

The GSM digital network has pervaded Europe and Asia. A comparable technology known as PCS is beginning to make inroads in the United States. According to FCC, "PCS is the system by which every user can exchange information with anyone, at anytime, in any place, through any type of device, using a single personal telecommunication number (PTN)." PCS is an advanced phone service that combines the freedom and convenience of wireless communications with the reliability of the legacy telephone service. Both GSM and PCS promise clear transmissions, digital capabilities, and sophisticated encryption algorithms to prevent eavesdropping.

PCS is a new concept that will expand the horizon of wireless communications beyond the limitations of current cellular systems to provide users with the means to communicate with anyone, anywhere, anytime. It is called PCS by the FCC or personal communications networks (PCN) by the rest of the world. Its goal is to provide integrated communications (such as voice, data, and video) between nomadic subscribers irrespective of time, location, and mobility patterns. It promises near-universal access to mobile telephony, messaging, paging, and data transfer.

PCS/PCN networks and the existing cellular networks should be regarded as complimentary rather than competitive. One may view PCS as an extension of the cellular to the 1900-MHz band, using identical standards. Major factors that separate cellular networks from PCS networks are speech quality, complexity, flexibility of radio-link architecture, economics of serving high-user-density or low-user-density areas, and power consumption of the handsets. Table 17.5.3 summarizes the differences between the two technologies and services.

PCS offers a number of advantages over traditional cellular communications:

• A truly personal service, combining lightweight phones with advanced features such as paging and voice mail that can be tailored to each individual customer.

• Less background noise and fewer dropped calls

• An affordable fully integrated voice and text messaging that works just about anywhere, anytime

• A more secure all-digital network that minimizes chances of eavesdropping or number cloning

• An advanced radio network that uses smaller cell sites

• A state-of-the-art billing and operational support system

TABLE 17.5.3 Comparison of Cellular and PCS Technologies

Cellular	PCS
Fewer sites required to provide coverage	More sites required to provide coverage (e.g., a 20:1 ratio)
More expensive equipment	Less expensive cells
Higher costs for airtime	Airtime costs dropping rapidly
High antenna and more space needed for site	Smaller space for the microcell
Higher power output	Lower power output

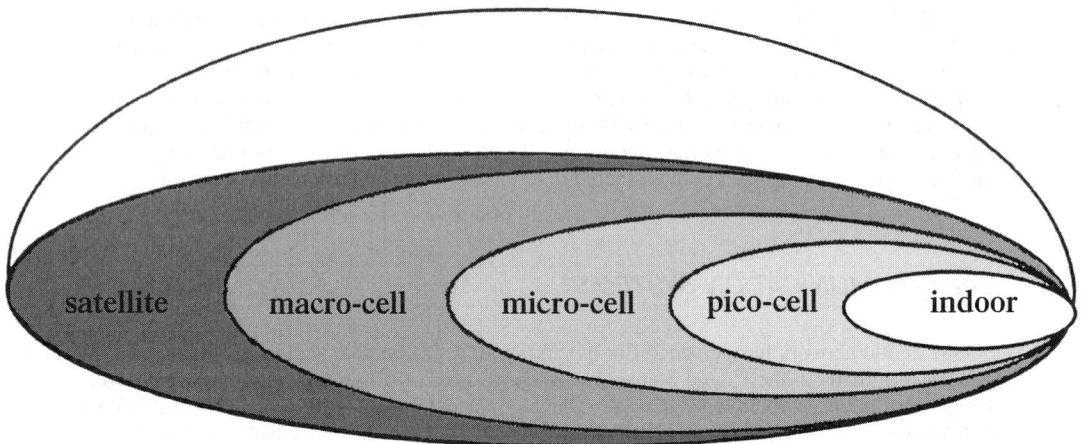

FIGURE 17.5.8   Various cell sizes.

Basic Features

PCS refers to digital wireless communications and services operating at broadband (1900 MHz) or narrowband (900 MHz) frequencies. Thus there are three categories of PCS: broadband, narrowband, and unlicensed. Broadband PCS addresses both cellular and cordless handset services, while narrowband PCS focuses on enhanced paging functions. Unlicensed service is allocated from 1890 to 1930 MHz and is designed to allow unlicensed short-distance operation.

The salient features that enable PCS to provide communications with anyone, anywhere, anytime include:

- *Roaming Capability*: The roaming service should be greatly expanded to provide universal accessibility. PCS will have the capability to support global roaming.
- *Diverse Environment*: Users must be able to use the PCS in all types of environments, e.g., urban, rural, commercial, residential, mountains, and recreational area.
- *Various Cell Size*: With PCS, there will be a mix of broad types of cell sizes: the picocell for low power indoor applications, the microcell for lower power outdoor pedestrian application, macrocell for high power vehicular applications, and supermacrocell with satellites, as shown in Fig. 17.5.8. For example, a picocell of a PCS will be in the 10 to 30 m range; a microcell may have a radius of 50 to 150 m; and a macrocell may have a radius of 1 km.
- *Portable Handset*: PCS provides a low-power radio, switched access connection to the PSTN. The user should be able to carry a single, small, universal handset outside without having to recharge its batter.
- *Single PTN*: The user can be reached through a single PTN regardless of the location and the type of service used.

The FCC frequency allocation for PCS usage is significant. FCC allocated 120 MHz for licensed operation and another 20 MHz for unlicensed operation, amounting to a total of 140 MHz for PCS, which is three times the spectrum currently allocated for cellular network. The FCC's frequency allocation for PCS is shown in Tables 17.5.4 and 17.5.5 for licensed and unlicensed operators. To use the PCS licensed frequency band, a company must obtain a license from FCC. To use the unlicensed (or unregulated) PCS spectrum, a company must use equipment that will conform with the FCC unlicensed requirements that include low power transmission to prevent interference with other users in the same frequency band.

PCS Architecture

A PCS network is a wireless network that provides communication services to PCS subscribers. The service area of the PCS network is populated with base stations, which are connected to a fixed wireline network

TABLE 17.5.4 The PCS Frequency Bands for Licensed Operation

Block	Spectrum low side (MHz)	Spectrum high side (MHz)	Bandwidth (MHz)
A	1850–1865	1930–1945	30
D	1865–1870	1945–1950	10
B	1870–1885	1950–1965	30
E	1885–1890	1965–1970	10
F	1890–1895	1970–1975	10
C	1895–1910	1975–1990	30
Total			120

through mobile switch centers (MSCs). Like a cellular network, the radio coverage of a base station is called a cell. The base station locates a subscriber or mobile unit and delivers calls to and from the mobile unit by means of paging within the cell it serves.

PCS architecture resembles that of a cellular network with some differences. The structure of the local portion of a PCS network is shown in Fig. 17.5.9. It basically consists of five major components:

• Terminals installed in the mobile unit or carried by pedestrians

• Cellular base stations to relay signals

• Wireless switching offices that handle switching and routing calls

• Connections to PSTN central office

• Database of customers and other network-related information

Since the goal of PCS is to provide anytime-anywhere communication, the end device must be portable and both real-time interactive communication (e.g., voice) and data services must be available. PCS should be able to integrator or accommodate the current PSTN, ISDN, the paging system, the cordless system, the wireless PBX, the terrestrial mobile system, and the satellite system. The range of applications associated with PCS is depicted in Fig. 17.5.10.

PCS Standards

The Joint Technical Committee (JTC) has been responsible for developing standards for PCS in the United States. The JTC committee worked cooperatively with the TIA committee working on the TR-46 reference model and ATIS committee working on the T1P1 reference model. Unlike GSM, PCS is unfortunately not a single standard but a mosaic consisting of several incompatible versions coexisting rather uneasily with one another. One major obstacle to PCS adoption in the United States has been the industry's failure to sufficiently convince customers on the advantages of PCS over AMPS, which already offers a single standard. This places the onus on manufacturers to inundate phones with features that attract market attention without compromising the benefits inherent in cellular phones. However, digital cellular technology enjoys distinct advantages. Perhaps the most significant advantage involves security because one cannot adequately encrypt AMPS signals.

TABLE 17.5.5 The PCS Frequency Bands for Unlicensed Operation

Block	Spectrum (MHz)	Bandwidth (MHz)
Isochronous	1910–1920	10
Asynchronous	1920–1930	10
Total		20

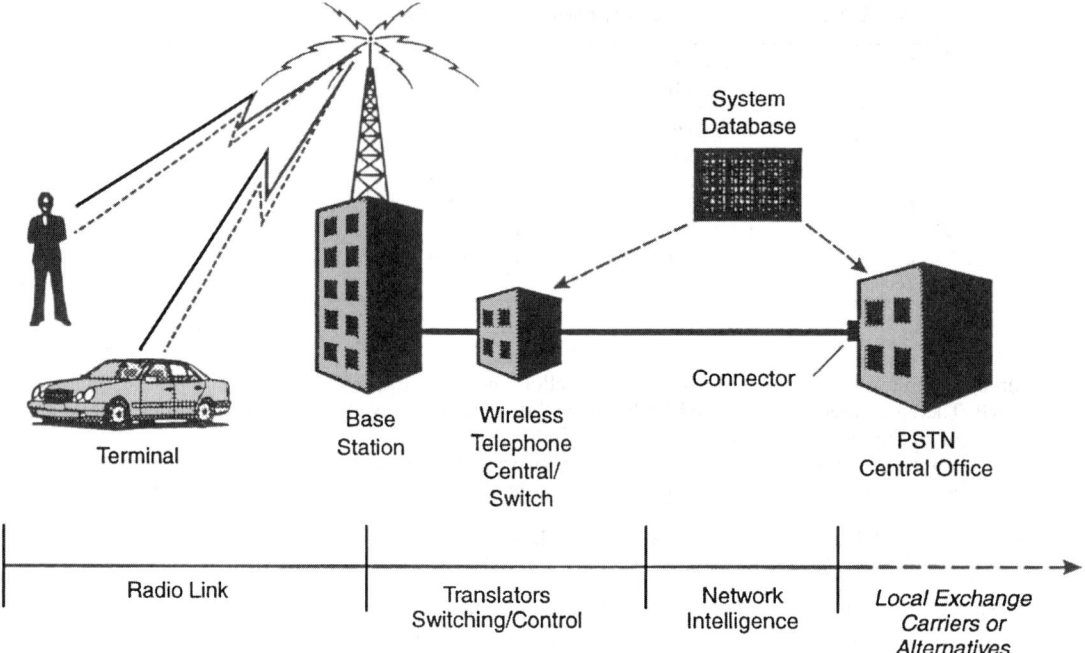

FIGURE 17.5.9 Structure of PCS network.

Satellites are instrumental in achieving global coverage and providing PCS services. Mobile satellite communications for commercial users is evolving rapidly toward PCS systems to provide basic telephone, fax, and data services virtually anywhere on the globe. Satellite orbits are being moved closed to the earth, improving communication speed and enabling PCS services. Global satellite systems are being built for personal communications. In the United States, the FCC licensed five such systems: Iridium, Globalstar,

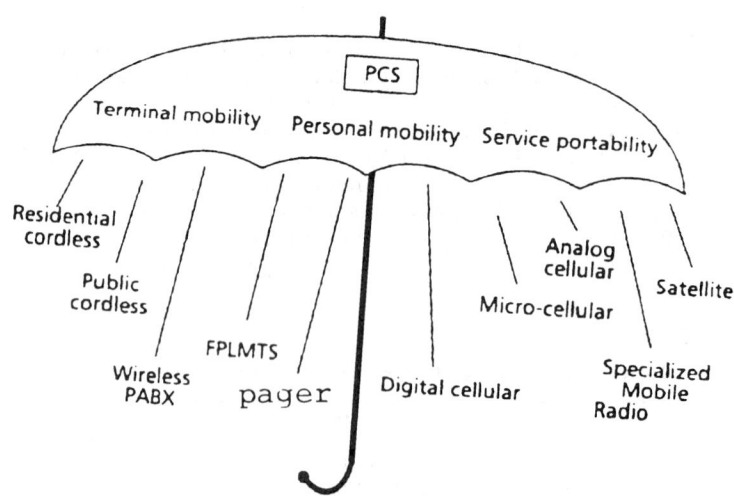

FIGURE 17.5.10 Range of applications associated with PCS.

Odyssey, Ellipso, and Aries. In Europe, ICO-Global is building ICO. Japan, Australia, Mexico, and India are making similar effort.

Future growth and success of PCS services cannot be taken for granted. Like any new technology, the success of PCS system will depend on a number of factors. These include initial system overall cost, quality, and convenience of the service provided, and cost to subscribers.

WIRELESS DATA NETWORKS

Wireless data networks are designed for low speed data communications. The proliferation of portable computers coupled with the increasing usage of the Internet and the mobile user's need for communication is the major driving force behind these networks. Examples of such networks include CDPD, wireless LAN, and wireless ATM.

Cellular Digital Packet Data

Cellular digital packet data is the latest in wireless data communication. CDPD systems offer one of the most advanced means of wireless data transmission technology. CDPD is a cellular standard aimed at providing Internet protocol (IP) data service over the existing cellular voice networks and circuit switched telephone networks. The technology solves the problem of business individuals on the move who must communicate data between their work base and remote locations.

The idea of CDPD was formed in 1992 by a development consortium with key industry leaders including IBM, six of the seven regional Bell operating companies, and McCaw Cellular. The goal was to create a uniform standard for sending data over existing cellular telephone channel. The Wireless Data Forum (www.wirelessdata.org), formerly known as CDPD Forum, has emerged as a trade association for wireless data service providers and currently has over 90 members. CDPD has been defined by the CDPD Forum CDPD Specification R1.1 and operates over AMPS.

By building CDPD as an overlay to the existing cellular infrastructure and using the same frequencies as cellular voice, carriers are able to minimize the capital expenditures. It costs approximately $1 million to implement a new cellular cell site and only about $50,000 to build the CDPD overlay to an existing site.

CDPD is designed to exploit the capabilities of the advanced cellular mobile services (AMPS) infrastructure throughout North America. One weakness of cellular telephone channels is that there are moments when the channels are idle (roughly 30 percent of the air time is unused). CDPD exploits this by detecting and using the otherwise wasted moments by sending packets during the idle time. As a result, data are transmitted without affecting voice system capability. CDPD transmits digital packet data at 19.2 kbps using idle times between cellular voice calls on the cellular telephone network.

CDPD has the following features:

- It is an advanced form of radio communication operating in the 800- and 900-MHz bands.
- It shares the use of the AMPS radio equipment on the cell site.
- It supports multiple, connectionless sessions.
- It uses the Internet protocol (IP) and the open systems interconnection (OSI) connectionless network protocol (CLNP).
- It is fairly painless for users to adopt. To gain access to CDPD infrastructure, one only requires a special CDPD modem.
- It supports both the TCP/IP protocols as well as the international set of equivalent standards.
- It was designed with security in mind unlike other wireless services. It provides for encryption of the user's data as well as conceals the user's identity over the air link.

CDPD provides the following services:

- Data rate of 19.2 kbps.

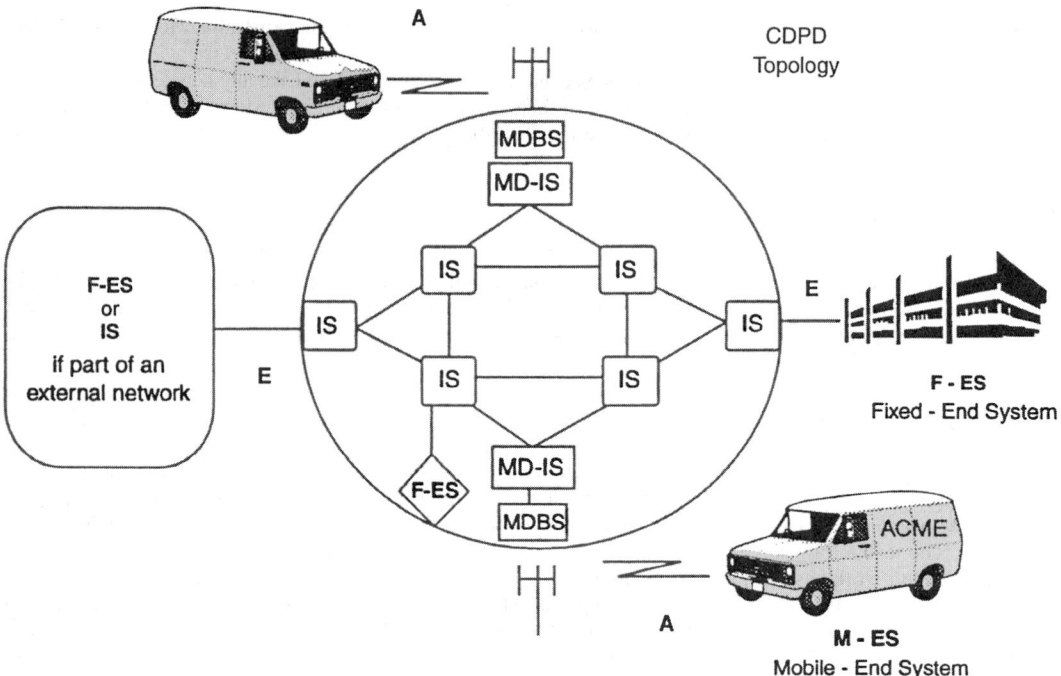

FIGURE 17.5.11   Major components of a CDPD network.

- Connectionless as the basic service; a user may build a connection-oriented service on top of that if desired.
- All three modes of point-to-point, multicast, and broadcast are available.
- Security that involves authentication of users and data encryption.

CDPD is a packet switched data transfer technology that employs radio frequency (RF) and spectrum in existing analog mobile phone system such as AMPS. The CDPD overlay network is made of some major components that operate together to provision the overall service. The key components that define CDPD infrastructure are illustrated in Fig. 17.5.11. They are as follows:

- *Mobile End System (MES)*: This is the subscriber's device for gaining access to the wireless communication services offered by a CDPD service. It is any mobile computing device, which is an equipment with a CDPD modem. Examples of an MES are laptop computers, palmtop computers, and personal digital assistants (PDAs), or any portable computing devices.
- *Fixed End System (FES)*: This is a stationary computing device (e.g., a host computer, a UNIX workstation, and so forth) connected to landline networks. The FES is the final destination of the message sent from an MES.
- *Intermediate System (IS)*: This is made up of routers that are CDPD compatible. It is responsible for routing data packets into and out of the CDPD service provider network. It may also perform gateway and protocol conversion functions to aid network interconnection.
- *Mobile Data Base Station (MDBS)*: CDPD uses a packet switched system that splits data into small packets and sends them across the voice channel. This involves detecting idle time on the voice channel and sending the packets on the appropriate unoccupied voice frequencies. This detection of unoccupied frequencies and sending of packets is done by the MDBS. Thus, the MDBS is responsible for relaying data between the mobile units and the telephone network. In other words, it relays data packets from the MES to the mobile data intermediate system (MDIS) and vice versa.

- *Mobile Data Intermediate System*: MDBSs that service a particular cell can be grouped together and connected to the backbone router, also known as the MDIS. The MDIS units form the backbone of the CDPD network. All mobility management functions are taken care of by MDIS. In other words, the MDIS is responsible for keeping track of the MES's location and routing data packets to and from the CDPD network and the MES appropriately.

Very little new equipment is needed for CDPD service since existing cellular networks are used. Only the MDBSs are to be added to each cell. One can purchase CDPD cellular communication systems for Windows or MS-DOS computers. The hardware can be a handheld AMPS telephone or a small modem which can be attached to a notebook computer. One would need to put up the antenna on the modem.

In order to effectively integrate voice and data traffic on the same cellular network without degrading the service provided for the voice customer, the CDPD network employs a technique known as *channel hopping*. When a mobile unit wants to transmit, it checks for an available cellular channel. Once a channel is found, the data link is established and the mobile unit can use the assigned channel to transmit as long as the channel is not needed for voice communication. Because voice is king, data packets are sent after giving priority to voice traffic. Therefore, if a cellular voice customer needs the channel, it will take priority over the data transmission. In that case, the mobile unit is advised by the MDBS to "hop" to another available channel. If there are no other available channels, then extra frequencies purposely set aside for CDPD can be used. This is a rare situation because each cell typically has 57 channels and each channel has an average idle time of 25 to 30 percent. The process of establishing and releasing channel links is called channel hopping and it is completely transparent to the mobile data unit. It ensures that the data transmission does not interfere with the voice transmission. It usually occurs within the call setup phase of the voice call. The major disadvantage of channel hopping is the potential interference to the cellular system.

CDPD has been referred to as an "open" technology because it is based on the OSI reference model, as shown in Fig. 17.5.12. The CDPD network comprises many layers: layer 1 is the physical layer; layer 2 is the data link layer; and layer 3 is the network layer; and so forth. For example, the physical layer corresponds to a functional entity that accepts a sequence of bits from the medium access control (MAC) layer and transforms them into a modulated waveform for transmission onto a physical 30 kHz RF channel. The network can use either the ISO

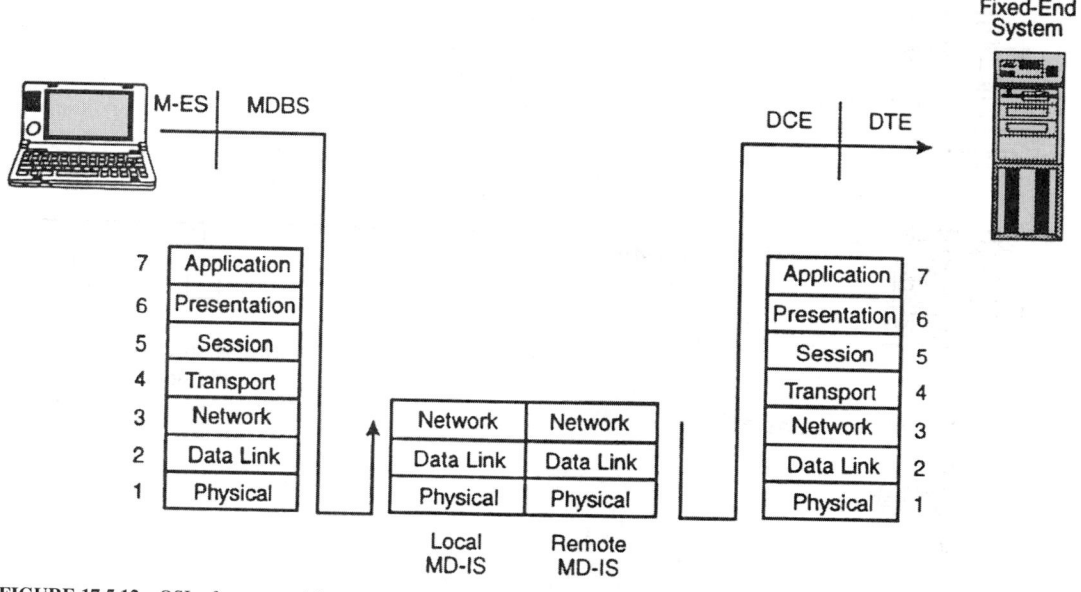

FIGURE 17.5.12   OSI reference model.

connectionless network protocol (CLNP) or the transmission control protocol/Internet protocol (TCP/IP). For now, CDPD can coexist with PCS and CDMA-based infrastructure.

Wireless LAN

Wireless local area network (WLAN) is a new form of communication system. It is basically a local area network, confined to a geographically small area such as a single building, office, store, or campus, that provides high data connectivity to mobile stations. Using electromagnetic airwaves (radio frequency or infrared), WLANs transmit and receive data over the air. A WLAN suggests less expensive, fast, and simple network installation and reconfiguration.

WLAN does not compete with wired LAN. Rather, WLANs are used to extend wired LANs for convenience and mobility. Wireless links essentially fill in for wired links using electromagnetic radiation at radio or light frequencies between transceivers. A typical WLAN consists of an access point and the WLAN adapter installed on the portable notebook. The access point is a transmitter/receiver (transceiver) device; it is essentially the wireless equivalent of a regular LAN hub. An access point is typically connected with the wired backbone network at a fixed location through a standard Ethernet cable and communicates with wireless devices by means of an antenna. WLANs operate within the prescribed 900-MHz, 2.4-GHz, and 5.8-GHz frequency bands. Most LANs use 2.4-GHz frequency bands because it is most widely accepted.

A wireless link can provide services in several ways. One way is to act as a stand-alone WLAN for a group of wireless nodes. This can be achieved using topologies similar to wired LAN, namely, a star topology can be formed with central hub controlling the wireless nodes, a ring topology with each wireless node receiving or passing information sent to it or a bus topology with each wireless capable of hearing everything said by all the other nodes. A typical WLAN configuration is shown in Fig. 17.5.13.

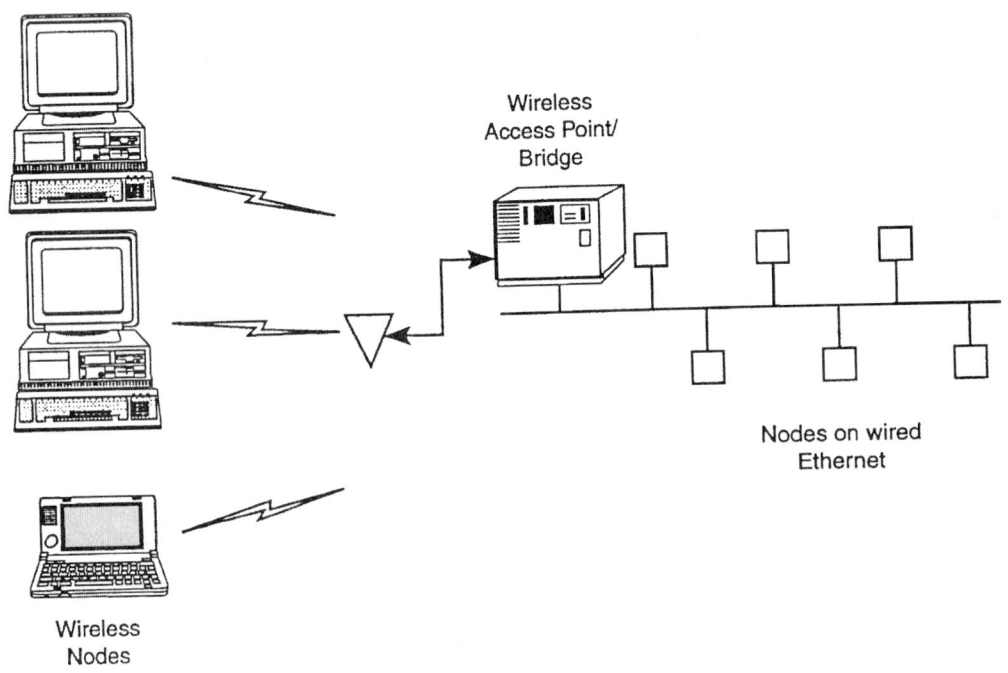

FIGURE 17.5.13   Connection of a wired LAN to wireless nodes.

When designing WLANs, manufacturers have to choose from two main technologies that are used for wireless communications today: radio frequency (RF) and infra red (IR). Each technology has its own merits and demerits. RF is used for applications where communications are over long distances and are not line-of-sight. In order to operate in the license free portion of the frequency spectrum known as the ISM band (industrial, scientific, and medical), the RF system must use a modulation technique called *spread spectrum* (SS). The second technology used in WLAN is infra red, where the communication is carried by light in the invisible part of the spectrum. It is primarily used for very short distance communications (less than 1 m), where there is a line-of-sight connection. Since IR light does not penetrate solid materials (it is even attenuated greatly by window glass), it is not really useful in comparison to RF in WLAN system. However, IR is used in applications where the power is extremely limited such as a pager.

Wireless ATM

Asynchronous transfer mode technology is the result of efforts to devise a transmission and networking technology to provide high-speed broadband integrated services: a single infrastructure for data, voice, and video. Until recently, the integration of wireless access and mobility with ATM has received little attention.

The concept of wireless ATM (WATM) was first proposed in 1992. It is now regarded as the potential framework for next generation wireless broadband communications that will support integrated quality-of-service (QoS) multimedia services. WATM technology is currently migrating from research stage to standardization and early commercialization.

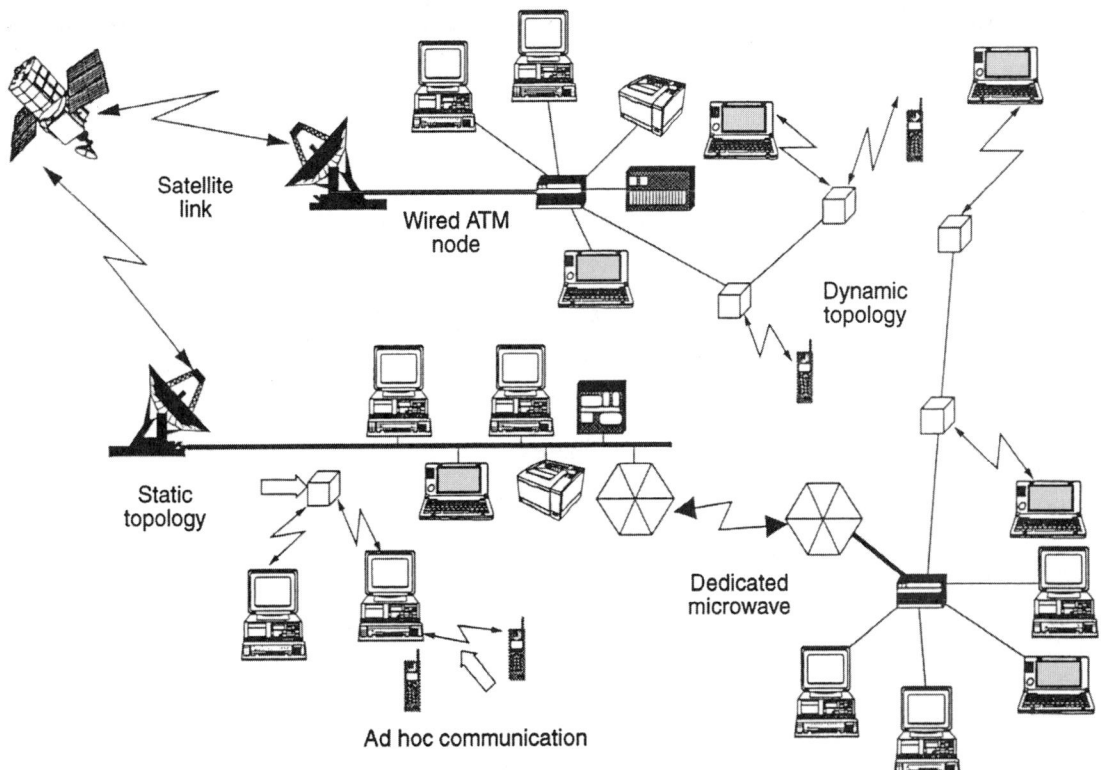

FIGURE 17.5.14   A typical wireless ATM network.

Wireless ATM network is basically the wireless extension of fixed ATM network. The 53-byte ATM cell is too big for wireless ATM network. Therefore, WATM networks may use 16 or 24 bytes payload. Thus, in a wireless ATM network, information is transmitted in the form of a large number of small transmission cells called *picocells*. Each picocell is served by a base station, while all the base stations in the network are connected via the wired ATM network. The ATM header is compressed or expanded to standard ATM cell at the base station. Base stations are simple cell relays that translate the header formats from the wireless ATM network to the wired ATM network. ATM cells are transmitted via radio frames between a central station (B-CS) and user radio modules (B-RM) as shown in Fig. 17.5.14. All base stations operate on the same frequency so that there is no hard boundary between picocells.

Reducing the size of the picocells helps in mitigating some of the major problems related to wireless LANs. The main difficulties encountered are the delay spread because of multipath effects and the lack of a line-of-sight path that results in high attenuation. Also, small cells have some drawbacks compared to large cells.

From Fig. 17.5.14, we notice that a wireless ATM typically consists of three major components: (1) ATM switches with standard UNI/NNI capabilities, (2) ATM base stations, and (3) wireless ATM terminal with a radio network interface card (NIC). There are two new hardware components: ATM base station and WATM NIC. The new software components are the mobile ATM protocol extension and WATM UNI driver.

In conventional mobile networks, transmission cells are "colored" using frequency-division multiplexing or code-division multiplexing to present interference between cells. Coloring is considered a waste of bandwidth because in order for it to be successful there must be areas between reuse of the color in which it is idle. These inactive areas are wasted rather than be used for transmission.

Wireless ATM architecture is based on integration of radio access and mobility features. The idea is to fully integrate new wireless physical layer (PHY), medium access control (MAC), and data link control (DLC), wireless control and mobility signaling functions into the ATM protocol stack.

Wireless ATM is not as matured as wireless LAN. No standards have been defined by either ITU-T or ATM Forum. However, the ATM Forum's WATM Working Group (started in June 1996) is developing specifications that will facilitate deployment of WATM.

CHAPTER 17.6
DATA NETWORKS AND INTERNET

Matthew N. O. Sadiku

The coming of the information age has brought about unprecedented growth in telecommunications-based services, driven primarily by the Internet, the information superhighway. Within a short period of time, the volume of data traffic transported across communications networks has grown rapidly and now exceeds the volume of voice traffic. While voice networks, such as the ubiquitous telephone network, have been in use for over a century, computer data networks are a recent phenomenon.

A computer communications network is an interconnection of different devices to enable them to communicate among themselves. Computer networks are generally classified into three groups on the basis of their geographical scope: local area networks (LANs), metropolitan area networks (MANs), and wide area networks (WANs). These networks differ in geographic scope, type of organization using them, types of services provided, and transmission techniques. LANs and WANs are well-established communication networks. MANs are relatively new. On the one hand, LAN is used in connecting equipments owned by the same organization over relatively short distances. Its performance degrades as the area of coverage becomes large. Thus LANs have limitations of geography, speed, traffic capacity, and the number of stations they are able to connect. On the other hand, WAN provides long-haul communication services to various points within a large geographical area, e.g., a nation or continent. With some of the characteristics of LANs and some reflecting WANs, the MAN embraces the best features of both.

We begin this chapter by looking at the open systems interconnection (OSI) reference model, which is commonly used to describe the functions involved in data communication networks. We then examine different LANs, MANs, and WANs including the Internet.

OSI REFERENCE MODEL

There are at least two reasons for needing a standard protocol architecture such as the OSI reference model. First, the uphill task of understanding, designing, and constructing a computer network is made more manageable by dividing it into structured smaller subtasks. Second, the proliferation of computer systems has created heterogeneous networks—different vendors, different models from the same vendor, different data formats, different network management protocols, different operating systems, and so on. A way to resolve this heterogeneity is for vendors to abide by the same set of rules. Attempts to formulate these rules have preoccupied standards bodies such as International Standards Organization (ISO), Consultative Committee for International Telephone and Telegraph (CCITT) [now known as International Telecommunication Union (ITU)], Institute of Electrical and Electronics Engineers (IEEE), American National Standards Institute (ANSI), British Standards Institution (BSI), and European Computer Manufacturers Association (ECMA). Here we consider the more universal standard protocol architecture developed by ISO.

The ISO divides the task of networking computers into seven layers so that manufacturers can develop their own applications and implementations within the guidelines of each layer. In 1978, the ISO set up a committee

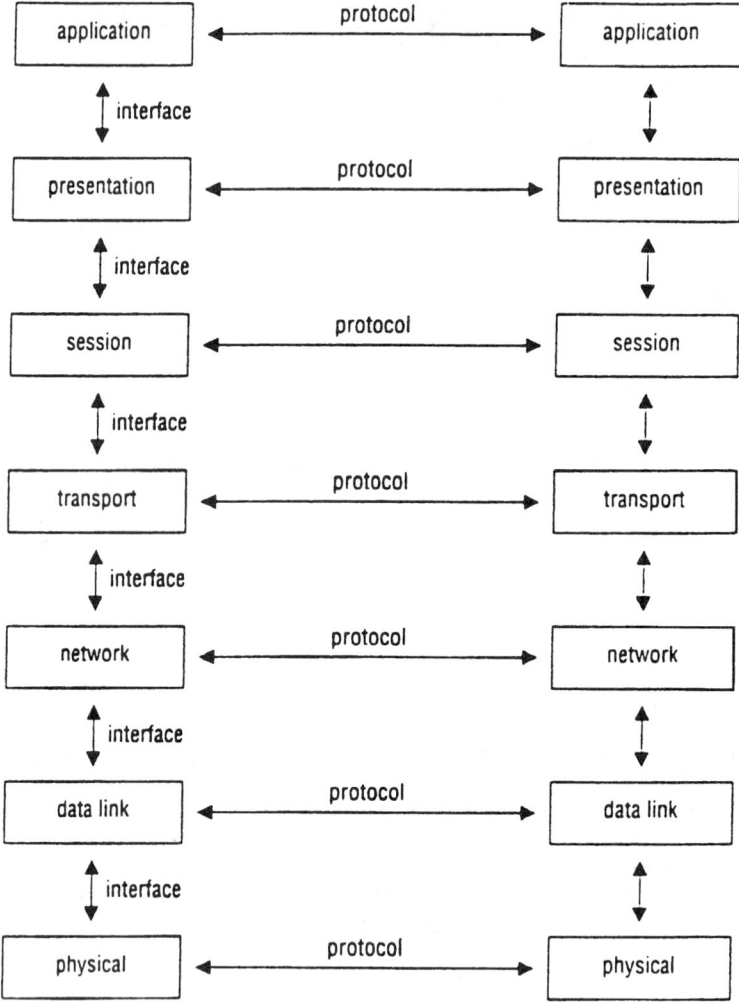

FIGURE 17.6.1   OSI reference model.

to develop a seven-layer model of network architecture (initially for WANs), known as the OSI. The model serves as a means of comparing different layers of communication networks. Also, the open model is standard-based rather than proprietary-based; one system can communicate with another system using interfaces and protocols that both systems understand. Network users and vendors have "open systems" in which any standard computer device would be able to interoperate with others.

The seven layers of the OSI model are shown in Fig. 17.6.1 and briefly explained as follows. We begin with the application layer (layer 7) and work our way down.

- *Application Layer*: This layer (layer 7) allows transferring information between application processes. It is implemented with host software. It is composed of specific application programs and its content varies with individual users. By application, we mean a set of information-processing desired by the user. Typical applications (or user programs) include login, password check, wordprocessing, spreadsheet, graphics program, document transfer, electronic mailing system, virtual terminal emulation, remote database access, network management, bank balance, stock prices, credit check, inventory check, and airline reservation. Examples of application layer protocols are Telnet (remote terminal protocol), file transfer protocol (FTP), simple mail transfer protocol (SMTP), remote login service (rlogin), and remote copy protocol (rcp).

- *Presentation Layer*: This layer (layer 6) presents information in a way that is meaningful to the network user. It performs functions such as translation of character sets, interpretation of graphic commands, data compression/decompression, data reformatting, and data encryption/decryption. Popular character sets include American Standard Code for Information Interchange (ASCII), Extended Binary Coded Decimal Interchange Code (EBCDIC), and Alphabet 5.

- *Session Layer*: A session is a connection between users. The session layer (layer 5) establishes the appropriate connection between users and manages dialog between them i.e., controlling starting, stopping, and synchronization of the dialog. It decides the type of communication such as two-way simultaneous (full duplex), two-way alternate (half-duplex), one-way, or broadcast. It is also responsible for checking for user authenticity and providing billing. For example, login and logout are the responsibilities of this layer. IBM's network basic input/output system (NetBIOS), sequenced packet exchange (NetWare's SPX), manufacturing automation protocol (MAP), and technical and office protocol (TOP) operate at this layer.

- *Transport Layer*: This layer (layer 4) uses the lower layers to establish reliable end-to-end transport connections for the higher layers. Its other function is to provide the necessary functions and protocols to satisfy a quality of service (QoS) (expressed in terms of time delay, throughput, priority, cost, and security) required by the session layer. It creates several logical connections over the same network by multiplexing end-to-end user addresses onto the network. It fragments messages from the session layer into smaller units (packets or frames) and reassembles the packets into messages at the receiving end. It also controls the end-to-end flow of packets, performs error control and sequence checking, acknowledges successful transmission of packets, and requests retransmission of corrupted packets. For example, the transmission control protocol (TCP) of TCP/IP and Internet transport protocol (ITP) of Xerox operate at this level.

- *Network Layer*: This layer (layer 3) handles routing procedure and flow control. It establishes routes (virtual circuits) for packets to travel and routes the packets from their source to destination and controls congestion. (Routing is of greater importance on MANs and WANs than on LANs.) It carries addressing information that identifies the source and ultimate destination. It also counts transmitted bits for billing information. It ensures that packets arrive at their destination in a reasonable amount of time. Examples of protocols designed for layer 3 are X.25 packet switching protocol and X.75 gateway protocol, both by CCITT. Also, the Internet protocol (IP) of TCP/IP and NetWare's Internetwork Packet Exchange (IPX) operate at this layer.

- *Data Link Layer*: This layer (layer 2) specifies how a device gains access to the medium specified in the physical layer. It converts the bit pipe provided by the physical layer into a packet link, which is a facility for transmitting packets. It deals with procedures and services related to the node-to-node data transfer. A major difference between the data link layer and the transport layer is that the domain for the data link layer is between adjacent nodes, whereas that of the transport layer is end-to-end. In addition, the data link layer ensures error-free delivery of data; hence it is concerned with error-detection, error correction, and retransmission. The error control is usually implemented by performing checksums on all bits of a packet after a cyclic redundancy check (CRC) process. This way, any transmission errors can be detected. The layer is implemented in hardware and is highly dependent of the physical medium. Typical examples of data link protocols are binary synchronous communications (BSC), synchronous data link control (SDLC), and high-level data link control (HDLC). For LANs and MANs, the data link layer is decomposed into the media-access control (MAC) and the logical link control (LLC) sublayers.

- *Physical Layer*: This layer (layer 1) consists of a set of rules that specifies the electrical and physical connection between devices. It is implemented in hardware. It is responsible for converting raw bits into electrical signal and physically transmitting them over a physical medium such as coaxial cable or an optical fiber between adjacent nodes. It provides standards for electrical, mechanical, and procedural characteristics required to transmit the bit stream properly. It handles frequency specifications, encoding the data, defining voltage or current levels, defining cable requirements, defining the connector size, shapes, and pin number, and so on. RS-232, RS-449, X.21, X.25, V.24, IEEE 802.3, IEEE 802.4, and IEEE 802.5 are examples of physical-layer standards.

TABLE 17.6.1   Summary of the Functions of OSI Layers

Layer	Name	Function
7	Application layer	Transfers information between application processes
6	Presentation layer	Syntax conversion, data compression, and encryption
5	Session layer	Establishes connection and manages a dialog
4	Transport layer	Provides end-to-end transfer of data
3	Network layer	End-to-end routing and flow control
2	Data link layer	Medium access, framing, and error control
1	Physical layer	Electrical/mechanical interface

A summary of the functions of the seven layers is presented in Table 17.6.1. The seven layers are often subdivided into two. The first consists of the lower three layers (physical, data link, and network layers) and is known as the communications subnetwork. The upper three layers (session, presentation, and application layers) are termed the host process. The upper layers are usually implemented by networking software on the node. The transport layer is the middle layer, separating the data-communication functions of the lower three layers and the data-processing functions of the upper layers. It is sometimes grouped with the upper layers as part of the host process or grouped with the lower layers as part of data transport.

LOCAL AREA NETWORKS

A LAN is a computer network that spans a geographically small area. It consists of two or more computers that are connected together to share expensive resources such as printers, exchange files, or allow electronic communications. Most LANs are confined to a single building or campus. They connect workstations, personal computers, printers, and other computer peripherals. Users connected to the LAN can use it to communicate with each other. LANs are capable of transmitting data at very fast rates, much faster than data can be transmitted over a telephone line; but the distances are limited. Also, since all the devices are located within a single establishment, LANs are usually owned and maintained by an organization. A key motivation for using LANs is to increase the productivity and efficiency of workers.

LANs differ from MANs and WANs by geographic coverage, data transmission and error rates, topology and data routing techniques, ownership, and sometimes by the type of traffic. Unique characteristics that differentiate LANs include:

- LANs generally operate within a few kilometers, spanning only a small geographical area.
- LANs usually have very high bit rates, ranging from 1 Mbps to 10 Gbps.
- LANs have a very low error rate, say $1:10^8$.
- A LAN is often owned and maintained by a single private company, institution, or organization using the facility.

There are different kinds of LANs. The following features differentiate one LAN from another:

- *Topology*: The geometric arrangement of devices on the LAN. As shown in Fig. 17.6.2, this can be bus, ring, star, or tree.
- *Protocols*: These are procedures or rules that govern the transfer of information between devices connected to a LAN. Protocols are to computer networks what languages are to humans.

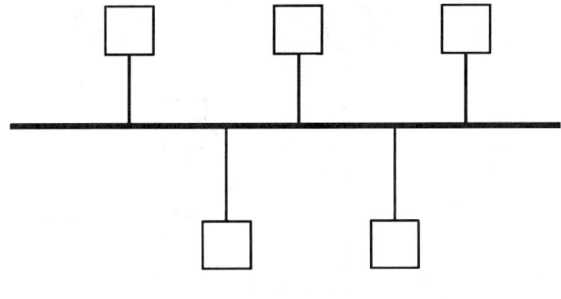

(a) Bus

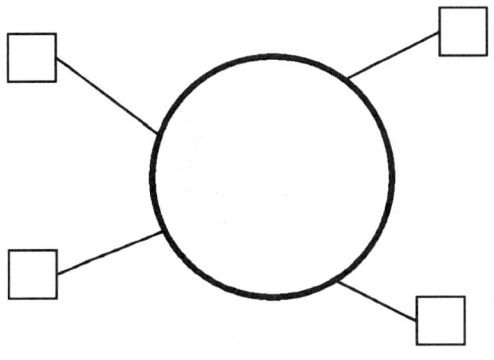

(b) Ring

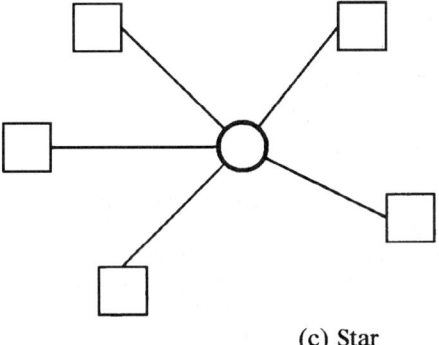

(c) Star

FIGURE 17.6.2   Typical LAN topologies.

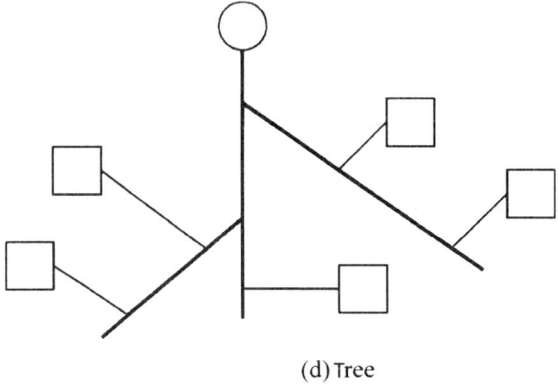

(d) Tree

FIGURE 17.6.2 (*Continued*)

- *Media*: The transmission medium connecting the devices can be twisted-pair wire, coaxial cables, or fiber optic cables. Wireless LAN use radio waves as media. Of all these media, optic fiber is the fastest but the most expensive. Common LANs include Ethernet, token ring, token bus, and star LAN. For bus or tree LANs, the most common transmission medium is coaxial cable. The two common transmission methods used on coaxial cable are baseband and broadband. A baseband LAN is characterized by the use of digital technology; binary data are inserted into the cable as a sequence of pulses using Manchester or Differential encoding scheme. A broadband LAN employs analog signaling and a modem. The frequency spectrum of the cable can be divided into channels using frequency division multiplexing (FDM). One of the most well-known applications of broadband transmission is the community antenna television (CATV). However, baseband LANs are more prevalent.

The Institute of Electrical and Electronics Engineers (IEEE) has established the following eight committees to provide standards for LANs:

- IEEE 802.1—standard for LAN/MAN bridging and management
- IEEE 802.2—standard for logical link control protocol
- IEEE 802.3—standard for CSMA/CD protocol
- IEEE 802.4—standard for token bus MAC protocol
- IEEE 802.5—standard for token ring MAC protocol
- IEEE 802.7—standard for broadband LAN
- IEEE 802.10—standard for LAN/MAN security
- IEEE 802.11—standard for wireless LAN

Token ring is a network architecture that uses token passing technology and ring-type network structure. Although token ring is standardized in IEEE 802.5 standard, its use has quite much faded to few organizations. Ethernet (IEEE 802.3) is the most popular and the least expensive high-speed LAN.

Ethernet is a LAN architecture developed by Xerox Corp. in cooperation with DEC and Intel in 1976. The IEEE 802.3 standard refined the Ethernet and made it globally accepted. Ethernet has since become the most popular and most widely deployed LAN in the world.

Conventional Ethernet uses a bus or star topology and supports data transfer rates of 10 Mbps. It uses a protocol known carrier sense multiple access with collision detection (CSMA/CD) as an access method to handle simultaneous demands. Each station or node attached to the Ethernet must sense the medium before transmitting data to see if any other station is already sending something. If the medium appears to be idle, then the station can begin to send data. If two stations sense the medium idle and transmit at the same time, collision may take place. When such a collision occurs, the two stations stop transmitting, wait, and try again later after a randomly chosen delay period. The delay period is determined using Binary Exponential Backoff.

Ethernet is one of the most widely implemented LAN standards. A newer version of Ethernet, called *Fast Ethernet* (or 100Base-T) supports data transfer rates of 100 Mbps. Gigabit Ethernet (or 1000Base-T) delivers at 1 Gbps speed. Upcoming 10 Gbps version of Ethernet is expected to be ready by 2002.

Security is an important issue with LANs since they are designed to provide access to many users. Network security is a measure designed to protect LAN users against attacks that originate from the network and other networks such as Internet connected to it. When individuals send private communications through a LAN, they desire secure communications. Currently, there are no systems in wide use that will keep data secure as they transit a public network. Several methods are being used to prevent attacks. One approach is to encrypt data as they leave one machine and decrypt it at the destination. Encryption is the fundamental tool for ensuring security in data networks. Another approach is to regulate which packets can go between two sites. For example, firewalls are placed between an organization's LAN and the Internet. A firewall is simply a group of components that collectively form a barrier between two networks.

METROPOLITAN AREA NETWORKS

Metropolitan area networks are basically an outgrowth of LANs. A variety of users and applications drive the requirements for MANs. The requirements include cost, scalability, security, reliability, compatibility with existing and future networks, and management issues. To meet these requirements, several proposals have been made for MAN protocols and architectures. Of these proposed MANs, fiber distributed data interface (FDDI) and distributed queue dual bus (DQDB) have emerged as standards that compete for use as backbones.

FDDI

In the mid 1970s, it was recognized that the existing copper technology would be unsuitable for future communication networks. Optical fibers offer some benefits over copper in that they are essentially immune to electromagnetic interference (EMI), have low weight, do not radiate, and reduce electrical safety concerns.

FDDI was proposed by the American National Standard Institute (ANSI) as a dual token ring that supports data rates of 100 Mbps and uses optical fiber media. An optical fiber is a thin, flexible glass or plastic structure (or waveguide) through which light is transmitted.

The FDDI specification recommends an optical fiber with a core diameter of 62.5 μm and a cladding diameter of 125 μm. There are two types of optical-fiber mode: single mode and multimode. A mode is a discrete optical wave or signal that propagates down the fiber. In a single mode fiber, only the fundamental mode can propagate. In multimode fiber, a large number of modes are coupled into the cable, making it suitable for the less costly light-emitting diode (LED) light source. The advantages of fiber optics over electrical media and the inherent advantages of a ring design contribute to the widespread acceptance of FDDI as a standard.

FDDI is a collection of standards formed by ANSI X3T9.5 task group over a period of 10 years. The standards produced by the task group cover physical hardware, physical and data link protocol layers, and a conformance testing standard. The original standard, known as *FDDI-I*, provides the basic data-only operation. An extended standard, *FDDI-II*, supports hybrid data and real-time applications.

FDDI is a follow-on to IEEE 802.5 (token ring) in that FDDI is based on token ring mechanics. Although the FDDI MAC protocol is similar (but not identical) to token ring, there are some differences. Unlike in token ring, FDDI performs all networking monitoring and control algorithms in a distributed way among active stations and does not need an active monitor. (Hence the term "distributed" in FDDI.) Whenever any device is down, other devices reorganize and continue to function, including token initialization, fault recovery, clock synchronization, and topology control.

The key highlights of FDDI are summarized as follows:

- ANSI standard through the X3T9.5 committee
- Dual counter-rotating ring topology for fault tolerance
- Data rate of 100 Mbps
- Total ring loop of size 100 km
- Maximum of 500 directly attached stations or devices
- 2 km maximum distance between stations
- Variable packet size (4500 bytes, maximum)
- 4B/5B data encoding scheme to ensure data integrity
- Shared medium using a timed-token protocol
- Variety of physical media, including fiber and twisted pair
- 62.5/125 μm multimode fiber-optic-based network
- Low bit error rate of 10^{-9} (one in one billion)
- Compatibility with IEEE 802 LANs by use of IEEE 802.2 LLC
- Distributed clocking to support large number of stations
- Support for both synchronous and asynchronous services

FDDI has two types of nodes: stations and concentrators. The stations transmit information to other stations on the ring and receive from them. Concentrators are nodes that provide additional ports for attachments of stations to the network. A concentrator receives data from the ring and forwards it to each of the connected ports sequentially at 100 Mbps. While a station may have one or more MAC, a concentrator may or may not have a MAC. As shown in Fig. 17.6.3, each FDDI station is connected to two rings, primary and secondary simultaneously. Stations have active taps on the ring and operate as repeaters. This allows the FDDI network to be so large without signal degradation. The network uses its primary ring for data transmission, while the secondary ring can be used either to ensure fault tolerance or for data. When a station or link fails, the primary and secondary rings form a single one-way ring, isolating the fault while maintaining a logical path among users, as shown in Fig. 17.6.4. Thus, FDDI's dual-ring topology and connection management functions establish a fault-tolerance mechanism.

FDDI was developed to conform with the OSI reference model. FDDI divides the physical layer of the OSI reference model into two sublayers: physical layer dependent (PMD) and physical layer (PHY), while the data link layer is split into two sublayers: media access control (MAC) and IEEE 802.2 LLC. A comparison of the FDDI architectural model to the lower two layers of the OSI model along with the summary of the functions of the FDDI standards is illustrated in Fig. 17.6.5. The FDDI MAC uses a timed-token rotation (TTR) protocol for controlling access to the medium. With the protocol, the MAC in each station measures the time that has elapsed since the station last received a token. Each station on the FDDI ring uses three timers to regulate its operation. The station management (SMT) controls the other three layers (PMD, PHY, and MAC) and ensures proper operation of the station. It handles such functions as initial FDDI ring initialization, station insertion and removal, ring's stability, activation, connection management, address administration, scheduling policies, collection of statistics, bandwidth allocation, performance and reliability monitoring, bit error monitoring, fault detection and isolation, and ring reconfiguration.

Though the original FDDI, described above, provides a bounded delay for synchronous services, the delay can vary. FDDI was initially envisioned as a data-only LAN. The full integration of isochronous and bursty data traffic is obtained with the enhanced version of the protocol, known as FDDI-II. FDDI-II is described by the hybrid ring control (HRC) standard that specifies an upward-compatible extension of FDDI. FDDI-II adds one document, HRC, to the existing four documents that specify FDDI standard. FDDI-II builds on original FDDI capabilities and supports integrated voice, video, and data capabilities but maintains the same transmission rate of 100 Mbps. FDDI-II therefore expands the range of applications of FDDI. FDDI-II supports both packet switched (synchronous and asynchronous) and circuit switched (isochronous) traffic. It can connect high-performance workstations, processors, and mass storage systems with bridges, routers, and gateways to other LANs, MANs, and WANs.

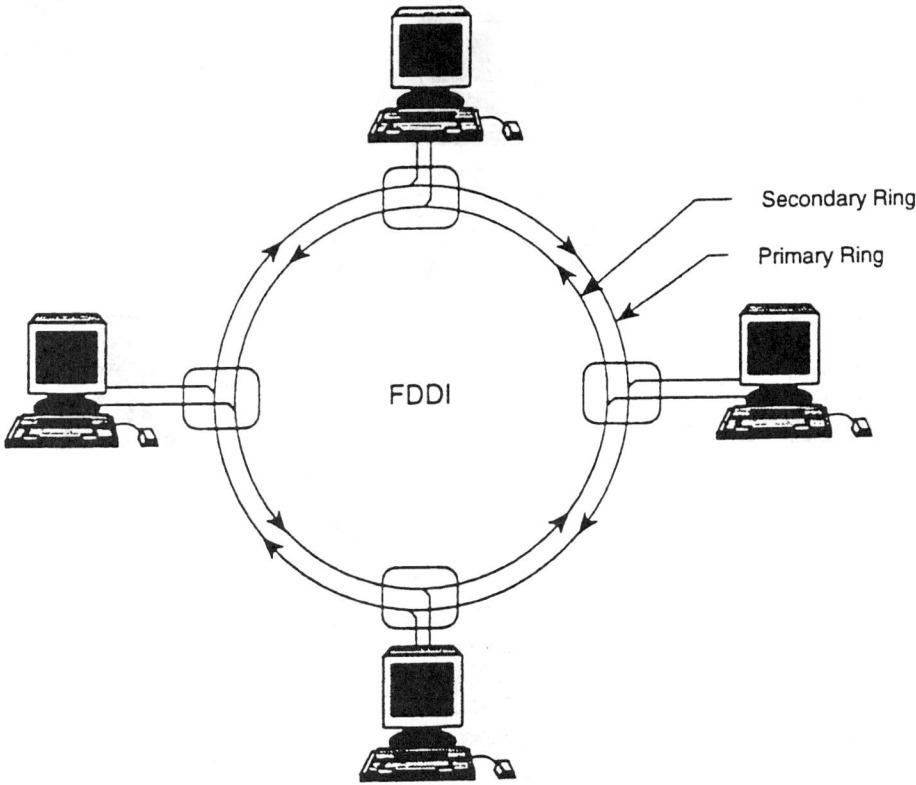

FIGURE 17.6.3   FDDI rings.

DQDB and SMDS

The IEEE 802 committee perceived the need for high-speed services over wide areas and formed the IEEE 802.6 MAN committee in 1982. The committee reached a consensus to use the DQDB as the standard MAC protocol. A by-product of DQDB is the switched multimegabit data service (SMDS).

The DQDB standard is both a protocol and a subnetwork. It is a subnetwork in that it is a component in a collection of networks to provide a service. The term "distributed-queue dual-bus" refers to the use of a dual-bus topology and a MAC technique based on the maintenance of distributed queues. In other words, each station connected to the subnetwork maintains queues of outstanding requests that determine access to the MAN medium. The DQDB subnetwork provides all stations on the dual bus with the knowledge of the frames queued at all other stations, thereby eliminating the possibility of collision and improving data throughput.

The DQDB subnetwork has many features, some of which make it attractive for high-speed data services. Such features include:

- *Shared media*: It extends the capabilities of shared media systems over large geographical areas.
- *Dual bus*: Its use of two separate buses carrying data simultaneously and independently makes it distinct from IEEE 802 LANs.
- *High speed*: It operates at a variety of data rates, ranging from 34 Mbps to 155 Mbps.

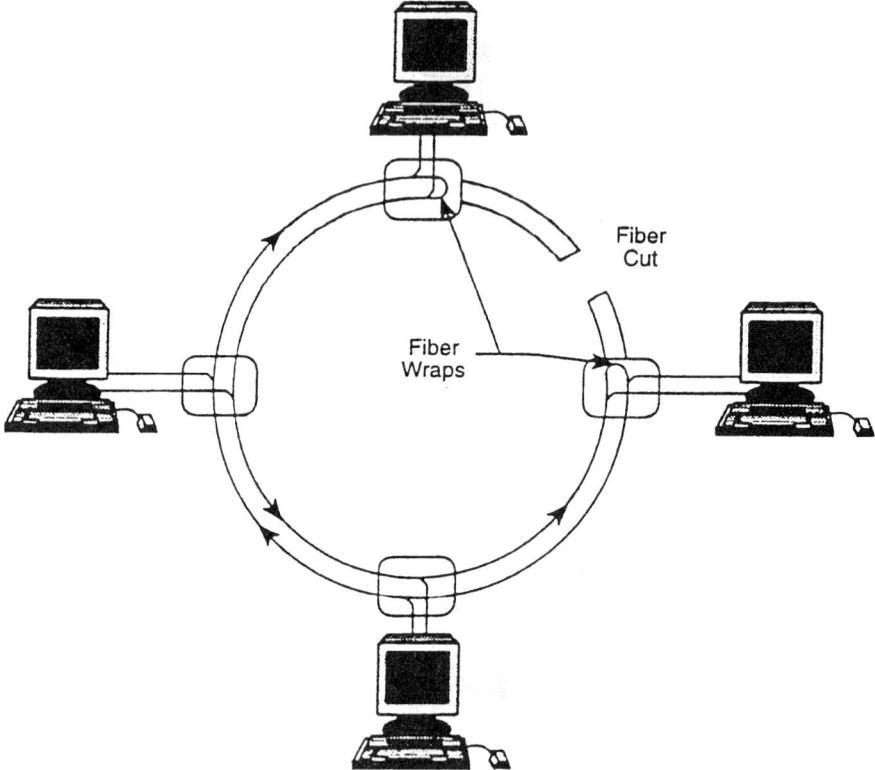

FIGURE 17.6.4 FDDI isolates fault without impairing the network.

- *Compatibility with legacy LANs*: It is compatible with IEEE 802.X LAN standards. A DQDB station should recognize the 16-bit and 48-bit addresses used by IEEE 802.X LAN standards. DQDB is designed to support data traffic under connectionless IEEE 802.2 LLC.
- *Fault tolerance*: It is tolerant to transmission faults when the system is configured in a loop.
- *Congestion control*: It is based on a distributed queuing algorithm as a way of resolving congestion.
- *Segmentation*: Its use of ATM technique allows long variable length packets to be segmented into short fixed-length segments. This provides efficient and effective support for small and large packets and for isochronous data.
- *Flexibility*: It uses a variety of media including coaxial cable and fiber optics. It can simultaneously support both circuit switched and packet switched services.
- *Compatibility*: It is compatible with current IEEE 802 LANs and future networks such as BISDN.

The DQDB network span of about 50 km, transmission rate of about 150 Mbps, and slot size of 53 bytes allow many slots to be in transit between the nodes. DQDB supports different types of traffic, which may be classified into two categories, isochronous and nonisochronous (asynchronous).

The DQDB dual-bus topology is shown in Fig. 17.6.6. As both buses are operational at all times, the capacity of the subnetwork is twice the capacity of each bus. In this network, nodes are connected to two unidirectional buses, which operate independently and propagate in opposite directions as shown in Fig. 17.6.6. Every node is able to send information on one bus and receive on the other bus. The head station (frame generator) generates a frame every 125 μs to suit digitized voice requirement. The frames are continuously generated on

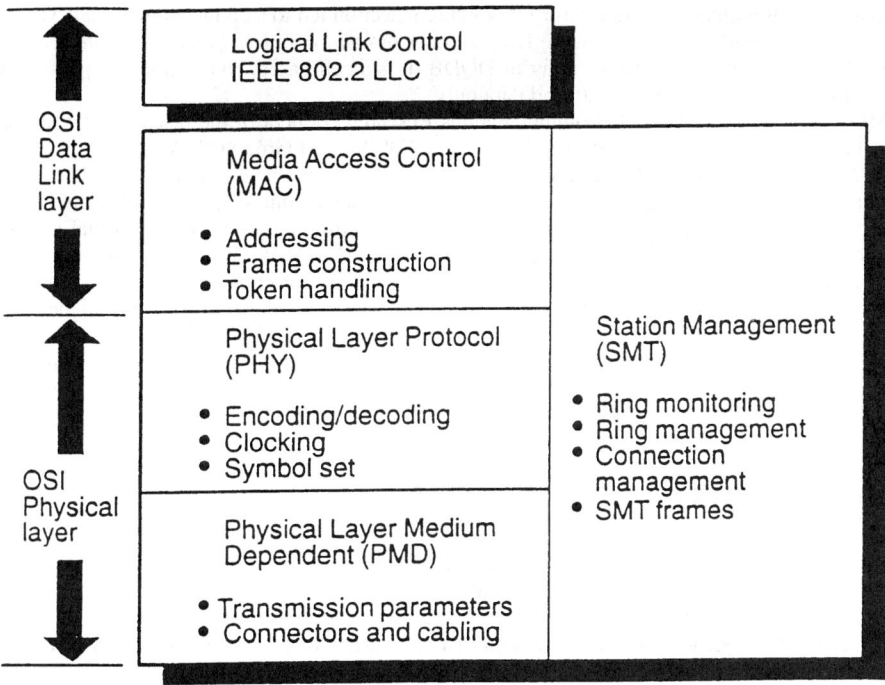

FIGURE 17.6.5 Summary of the functions of the FDDI standards.

each bus so that there is never any period of silence on the bus. The frame is subdivided into equal-sized slots. The empty slots generated can be written into by other nodes. The end station (slave frame generator) terminates the forward bus, removes all incoming slots, and generates the same slot pattern at the same transmission rate on the opposite bus. The slots are 53 octets long, the same as ATM cells, to make DQDB MANs compatible to BISDN.

SMDS represents the first broadband service to make use of DQDB MAN standard and technologies. The need for a high-speed, connectionless data service that provides both high transmission speed, low delay, and

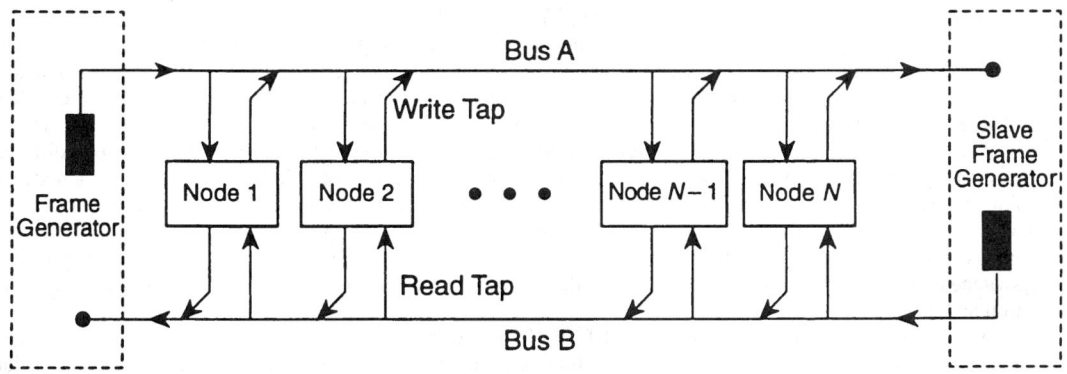

FIGURE 17.6.6 Open bus topology of DQDB network.

a simple, efficient protocol adaptation for LAN interconnection led to a connectionless data service known as *switched multisegment data service* in the United States or *connectionless broadband data service* (CBDS) in Europe. SMDS is the first service offering of DQDB. It is a cell-based, connectionless, packet-switched network that focuses on transmitting data and data only.

SMDS was developed by Bell Communications Research (Bellcore), the research arm of the seven Bell regional holding companies and popularized by the SMDS Interest Group (SIG).

SMDS is not a technology but a service. Although a DQDB can be configured as either a loop bus or an open bus, SMDS uses the open bus topology. SMDS is a connectionless, public, cell-switched data service. The service is connectionless because there is no need for setting up a physical or virtual path between two sites. SMDS offers services characteristically equivalent to LAN MAC. It operates much the same way as a LAN, but over a greater geographical area with a larger number of users.

Compared with other competing high-speed technologies such as FDDI, SMDS has no theoretical distance limitation as FDDI. FDDI's use of tokens limits the perimeter of the FDDI ring to about 60 mi. The data rate of FDDI (100 Mbps) does not match any of the standardized public transmission rate, whereas SMDS is based on standard public network speeds. FDDI will probably be used for high-speed LANs and complement SMDS rather than compete with it.

WIDE AREA NETWORKS

A WAN is an interconnected network of LANs and MANs. A WAN connects remote LANs and ties remote computers together over long distances. Computers connected to a WAN are often connected through public networks, such as the telephone system. They can also be connected through leased lines or satellites. WANs are, by default, heterogeneous networks that consist of a variety of computers, operating systems, topologies, and protocols. The largest WANs in existence is the Internet.

Because of the long distance involved, WANs are usually developed and maintained by a nation's public telecommunication companies (such as AT&T in the United States), which offer various communication services to the people. Today's WANs are designed in the most cost-effective way using optical fiber. Fiber-based WANs are capable of transporting voice, video, and data with no known restriction to bandwidth. Such WANs will remain cutting edge for years to come. There is also the possibilitiy of connecting networks using wireless technologies.

Circuit and Packet Switching

For a WAN, communication is achieved by transmitting data from the source node to the destination node through a network of intermediate switching nodes. Thus, unlike a LAN, a WAN is a switched network. There are many types of switched networks, but the most common methods of communication are circuit switching and packet switching. Circuit switching is a much older technology than packet switching. Circuit switching systems are ideal for communications that require data to be transmitted in real time. Packet-switching networks are more efficient if some amount of delay is acceptable.

Circuit switching is a communication method in which a dedicated path (channel or circuit) is established for the duration of a transmission. This is a type of point-to-point network connection. A switched circuit is maintained while the sender and recipient are communicating, as opposed to a dedicated circuit, which is held open regardless of whether data are being sent or not. The most common circuit-switching network is the telephone system.

Packet switching is a technique whereby the network routes individual packets of data between different destinations based on addressing within each packet. A packet is a segment of information sent over a network. Any message exceeding a network-defined maximum length (a set size) is broken up into shorter units, known as packets. Packet-switching is the process by which a carrier breaks up messages (or data) into these segments, bundles, or packets by the source data terminal equipment (DTE) before they are sent. Each packet is switched and transmitted individually through the network and can even follow different routes to its destination and may arrive out of order.

Most modern WAN protocols, such as TCP/IP, X.25, and frame relay, are based on packet-switching technologies. Besides data networks such as the Internet, wireless services such as cellular digital packet data (CDPD) employ packet switching.

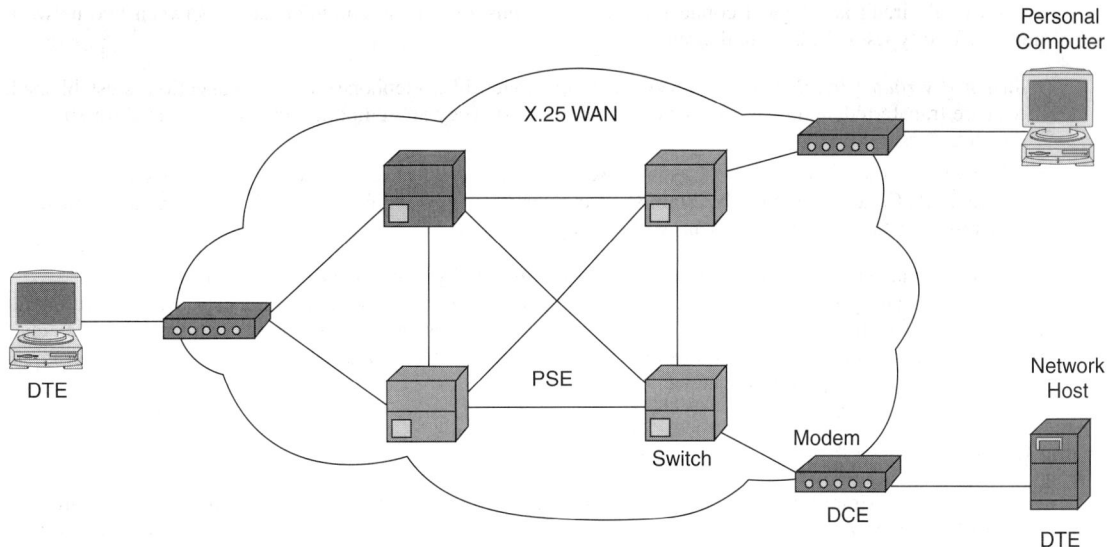

FIGURE 17.6.7   DTEs, DCEs, and PSEs make up an X.25 network.

X.25

For roughly 20 years, X.25 was the dominant player in the WAN packet-switching technology until frame relay, SMDS, and ATM appeared. X.25 has been around since the mid-1970s and so is pretty well debugged and stable. It was originally approved in 1976 and subsequently revised in 1977, 1980, 1984, 1988, 1992, and 1996. It is currently one of the most widely used interfaces for data communication networks. There are literally no data errors on modern X.25 networks.

X.25 is a communications packet-switching protocol designed for the exchange of data over a WAN. It is regarded as a standard, a network, or an interface protocol. It is a popular standard for packet-switching networks approved in 1976 by the International Telecommunication Union—Telecommunication Standardization Sector (ITU-T) for WAN communications. It defines how connections between user devices and network devices are established and maintained. X.25 uses a connection-oriented service that ensures that packets are transmitted in order. Through statistical multiplexing, X.25 enables multiple users to share bandwidth, as it becomes available, therefore ensuring flexible use of network resources among all users. X.25 is also an interface protocol in that it spells the required interface protocols that enable a DTE to communicate with data circuit-terminating equipment (DCE), which provides access to the network. The DTE-DCE link provides full-duplex multiplexing allowing a virtual circuit to transmit in either direction.

X.25 network devices fall into three general categories: DTE, DCE, and packet-switching exchange (PSE). DTE devices are user end systems that communicate across the X.25 network. They are usually terminals, personal computers, or network hosts, and are located on the premises of individual subscribers. DCE devices are carrier's equipment, such as modems and packet switches, that provide the interface between DTE devices and a PSE and are generally located in the carrier's facilities. PSEs are switches that compose the bulk of the carrier's network. They transfer data from one DTE device to another. Figure 17.6.7 illustrates the relationships between the three types of X.25 network devices.

The packet assembler/disassembler (PAD) is a device commonly found in X.25 networks. PADs are used when a DTE device is too simple to implement the full X.25 functionality. The PAD is located between a DTE device and a DCE device, and it performs three primary functions: buffering, packet assembly, and packet disassembly. The PAD buffers data sent to or from the DTE device. It also assembles outgoing data into packets and forwards them to the DCE device; this includes adding an X.25 header. Finally, the PAD disassembles incoming packets before forwarding the data to the DTE; this includes removing the X.25 header.

A virtual circuit is a logical connection created to ensure reliable communication between two network devices. Two types of X.25 virtual circuits exist:

- *Switched Virtual Circuits (SVCs)*:   SVCs are very much like telephone lines; a connection is established, data are transferred, and then the connection is released. They are temporary connections used for sporadic data transfers.

- *Permanent Virtual Circuits (PVCs)*:   A PVC is similar to a leased line in that the connection is always present. PVCs are permanently established connections used for frequent and consistent data transfers. Therefore, data may always be sent, without any call setup.

Maximum packet sizes vary from 64 to 4096 bytes, with 128 bytes being a default on most networks.

X.25 users are typically large organizations with widely dispersed and communications-intensive operations in sectors such as finance, insurance, transportation, utilities, and retail. For example, X.25 is often chosen for zero-error tolerance applications by banks involved in large-scale transfers of funds, or by government uses that manage electrical power networks.

Frame Relay

Frame relay is a simplified form of packet switching (similar in principle to X.25) in which synchronous frames of data are routed to different destinations depending on header information. It is basically an interface used for WAN. It is used to reduce the cost of connecting remote sites in any application that would typically use expensive leased circuits.

Frame relay is an *interface*, a method of multiplexing traffic to be submitted to a WAN. Carriers build frame relay networks using switches. The physical layout of a sample frame relay network is depicted in Fig. 17.6.8. The CSU/DSU is the channel service unit/data service unit. This unit provides a "translation" between the telephone company's equipment and the router. The router actually delivers information to the CSU/DSU over a serial connection, much like the computer uses a modem, only at a much higher speed.

All major carrier networks implement PVCs. These circuits are established via contract with the carrier and typically are built on a flat-rate basis. Although SVCs have standards support and are provided by the major frame relay backbone switch vendors, they have not been widely implemented in customer equipment or carrier networks.

Two major frame relay devices are frame relay access devices (FRADs) and routers. Stand-alone FRADs typically connect small remote sites to a limited number of locations. FRAD is also known as *frame relay assembler/disassembler*. Frame relay routers offer more sophisticated protocol handling than most FRADs. They may be packaged specifically for frame relay use, or they may be general-purpose routers with frame relay software.

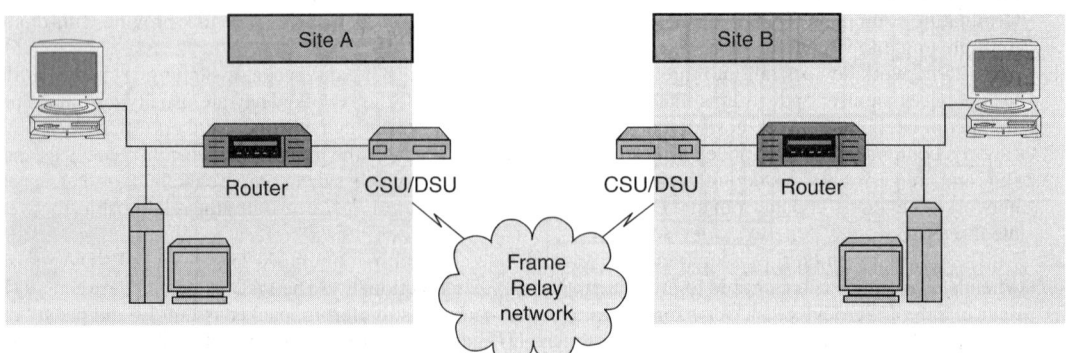

FIGURE 17.6.8   Physical layout of a typical frame relay network.

Frame relay is the fastest growing WAN technology in the United States. In North America it is fast taking on the role that X.25 has had in Europe. It is used by large corporations, government agencies, small businesses, and even Internet service providers (ISPs). The demand for frame relay services is exploding, and for two very good reasons—speed and economics. Frame relay is consistently less expensive than equivalent-leased services and provides the bandwidth needed for other services such as LAN routing, voice, and fax.

INTERNET

The Internet is a global network of computer networks (or WAN) that exchange information via telephone, cable television, wireless networks, and satellite communication technologies. It is being used by an increasing number of people worldwide. As a result, the Internet has been growing exponentially with the number of machines connected to the network and the amount of network traffic roughly doubling each year. The Internet today is fundamentally changing our social, political, and economic structures, and in many ways obviating geographic boundaries.

Internet Protocol Suite

The Internet is a combination of networks, including the Arpanet, NSFnet, regional networks such as NY sernet, local networks at a number of universities and research institutions, and a number of military networks. Each network on the Internet contains anywhere from two to thousands of addressable devices or nodes (computers) connected by communication channels. All computers do not speak the same language, but if they are going to be networked they must share a common set of rules known as *protocols.* That is where the two most critical protocols, transmission control protocol/Internet-working protocol (TCP/IP), come in. Perhaps the most accurate name for the set of protocols is the *Internet protocol suite.* (TCP and IP are two of the protocols in this suite.) TCP/IP is an agreed-upon standard for computer communication over Internet. The protocols are implemented in software that runs on each node.

The TCP/IP is a layered set of protocols developed to allow computers to share resources across a network. Figure 17.6.9 shows the Internet protocol architecture. The figure is by no means exhaustive, but shows the major protocols and application components common to most commercial TCP/IP software packages and their relationship.

As a layered set of protocols, Internet applications generally use four layers:

- *Application Layer:* This is where application programs that use the Internet reside. It is the layer with which end users normally interact. Some application-level protocols in most TCP/IP implementations include FTP, TELNET, and SMTP. For example, FTP (file transfer protocol) allows a user to transfer files to and from computers that are connected to the Internet.

- *Transport Layer:* It controls the movement of data between nodes. TCP is a connection-based service that provides services need by many applications. User datagram protocol (UDP) provides connectionless services.

- *Internet Layer:* It handles addressing and routing of the data. It is also responsible for breaking up large messages and reassembling them at the destination. IP provides the basic service of getting datagrams to their destination. Address resolution protocol (ARP) figures out the unique address of devices on the network from their IP addresses.

- *Network Layer:* It supervises addressing, routing, and congestion control. Protocols at this layer are needed to manage a specific physical medium, such as Ethernet or a point-to-point line.

TCP/IP is built on connectionless technology. IP provides a *connectionless, unreliable, best-effort* packet delivery service. Information is transferred as a sequence of datagrams. Those datagrams are treated by the network as completely separate.

TCP sends datagrams to IP with the Internet address of the computer at the other end. The job of IP is simply to find a route for the datagram and get it to the other end. In order to allow gateways or other intermediate

Application Layer	TELNET, FTP, Finger, Http, Gopher, SMTP, and so forth	DNS, RIP, SNMP, and so forth	
Transport Layer	TCP	UDP	
Internet Layer	IP		ARP
Network Layer	Ethernet, Token ring, X.25, FDDI, ISDN, SMDS, DWDM, Frame Relay, ATM, SONET/SDH, Wireless, xDSL, and so forth		

FIGURE 17.6.9 Abbreviated Internet protocol suite.

systems to forward the datagram, it adds its own header, as shown in Fig. 17.6.10. The main things in this header are the source and destination Internet address (32-bit addresses, such as 128.6.4.194), the protocol number, and another checksum. The source Internet address is simply the address of your machine. The destination Internet address is the address of the other machine. The protocol number tells IP at the other end to send the datagram to TCP. Although most IP traffic uses TCP, there are other protocols that can use IP, so one has to tell IP which protocol to send the datagram to. Finally, the checksum allows IP at the other end to verify that

Bit 0   **31**

Version (4)	IHL (4)	Service Type (8)	Total Length (16)	
Identification (16)			Flags (3)	Fragment Offset (13)
Time to Live (8)		Protocol (8)	Header Checksum (16)	
Source Address (32)				
Destination Address (32)				
Options (Variable)			Padding (Variable)	

FIGURE 17.6.10 IP header format (20 bytes).

the header was not damaged in transit. IP needs to be able to verify that the header did not get damaged in transit, or it could send a message to the wrong place. After IP has tacked on its header, the message looks like what is in Fig. 17.6.10.

Addresses and Addressing Scheme

For IP to work, every computer must have its own number to identify itself. This number is called the IP address. You can think of an IP address as similar to your telephone number of postal address. All IP addresses on a particular LAN must start with the same numbers. In addition, every host and router on the Internet has an address that uniquely identifies it and also denotes the network on which it resides. No two machines can have the same IP address. To avoid addressing conflicts, the network numbers have been assigned by the InterNIC (formerly known simply as NIC).

Blocks of IP addresses are assigned to individuals or organizations according to one of three categories—Class A, Class B, or Class C. The *network* part of the address is common for all machines on a local network. It is similar to a postal zip code that is used by a post office to route letters to a general area. The rest of the address on the letter (i.e., the street and house number) are relevant only within that area. It is only used by the local post office to deliver the letter to its final destination. The *host* part of the IP address performs this same function. There are five types of IP addresses:

- *Class A format*: 126 networks with 16 million hosts each; an IP address in this class starts with a number between 0 and 127
- *Class B format*: 16,382 networks with up to 64K hosts each; an IP address in this class starts with a number between 128 and 191
- *Class C format*: 2 million networks with 254 hosts each; an IP address in this class starts with a number between 192 and 223
- *Class D format*: Used for multicasting, in which a datagram is directed to multiple hosts
- *Class E format*: Reserved for future use

The IP address formats for the three classes are shown in Fig. 17.6.11.

IPv6

Most of today's Internet uses Internet Protocol Version 4 (IPv4), which is now nearly 25 years old. Because of the phenomenal growth of the Internet, the rapid increase in palmtop computers, and the profusion of smart cellular phones and PDAs, the demand for IP addresses has outnumbered the limited supply provided by IPv4. In response to this shortcoming of IPv4, the Internet Engineering Task Force (IETF) approved IPv6 in 1997.

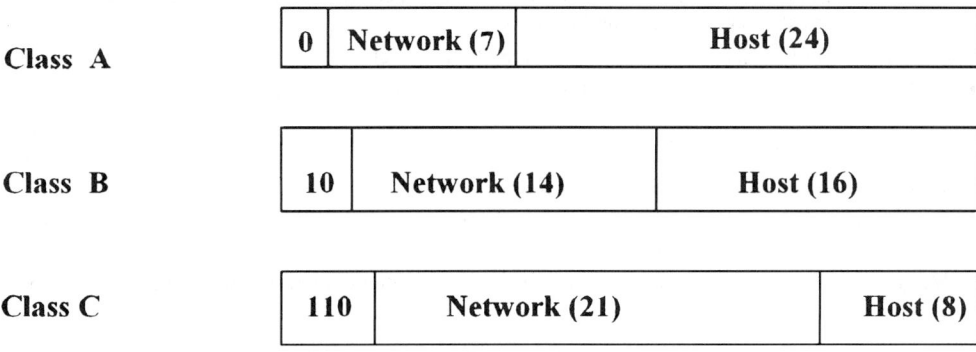

FIGURE 17.6.11   IP Address formats.

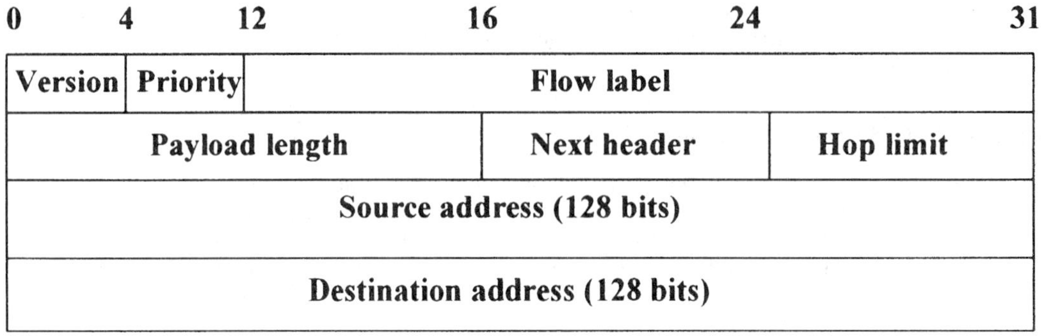

0	4	12	16	24	31

Version	Priority	Flow label		
Payload length		Next header		Hop limit
Source address (128 bits)				
Destination address (128 bits)				

FIGURE 17.6.12   IPv6 header format.

IPv4 will be replaced by Internet Protocol Version 6 (IPv6), which is sometimes called the Next Generation Internet Protocol (or IPng). IPv6 adds many improvements and fixes a number of problems in IPv4, such as the limited number of available IPv4 addresses.

With only a 32-bit address field, IPv4 can assign only 2^{32} different addresses, i.e., 4.29 billion IP addresses, which are inadequate in view of rapid proliferation of networks and the two-level structure of the IP addresses (network number and host number). To solve the problem of severe IP address shortage, IPv6 uses 128-bit addresses instead of the 32-bit addresses of IPv4. That means IPv6 can have as many as 2^{128} IP address, which is roughly 3.4×10^{38} or about 340 billion billion billion billion unique addresses.

The IPv6 packet consists of the IPv6 header, routing header, fragment header, the authentication header, TCP header, and application data. The IPv6 packet header is of fixed length, whereas the IPv4 header is of variable length. The IPv6 header consists of 40 bytes as shown in Fig. 17.6.12. It consists of the following fields:

- *Version (4 bits)*: This is the IP version number, which is 6.
- *Priority (4 bits)*: This field enables a source to identify the priority of each packet relative to other packets from the same source.
- *Flow Label (24 bits)*: The source assigns the flow label to all packets that are part of the same flow. A flow may be a single TCP connection or a multiple of TCP connections.
- *Payload Length (16 bits)*: This field specifies the length of the remaining part of the packet following the header.
- *Next Header (8 bits)*: This identifies the type of header immediately following the header.
- *Hop Limit (8 bits)*: This is to set some desired maximum value at the source and the field denotes the remaining number of hops allowed for the packet. It is decremented by 1 at each node the packet passes and the packet is discarded when the hop limit becomes zero.
- *Source Address (128)*: The address of the source of the packet.
- *Destination Address (128 bits)*: The address of the recipient of the packet.

There are three types of IPv6 addresses:

1. Unicast is used to identify a single interface.
2. Anycast identifies a set of interfaces. A source may use an anycast address to contact any node from a group of nodes.
3. Multicast identifies a set of interfaces. A packet with a multicast address is delivered to all members of the group.

IPv6 is expected to gradually replace IPv4, with the two coexisting for a number of years during a transition period. IPv6 may be most widely deployed in mobile phones, PDAs, and other wireless terminals in the future.

BISDN AND ATM

ISDN is a high-speed communication network, which allows voice, data, text, graphics, music, video, and other source material to be transmitted simultaneously across the world using end-to-end digital connectivity. ISDN stands for Integrated Services Digital Network. "Digital network" means that the user is given access to a telecom network ensuring high-quality transmission via digital circuits, while "integrated services" refers to the simultaneous transmission of voice and data services over the same wires. This way, computers can connect directly to the telephone network without first converting their signals to an analog form using modems. This integration brings with it a host of new capabilities combining voice, data, fax, and sophisticated switching. And because ISDN uses the existing local telephone wiring, it is equally available to home and business customers. ISDN was intended to eventually replace the traditional plain old telephone service (POTS) phone lines with a digital network that would carry voice, data, and video.

ISDN service is available today in most major metropolitan areas and probably will be completely deployed throughout the United States very soon. Many ISPs now sell ISDN access. However, the idea of using existing copper wiring to provide this network decreased ISDN capabilities in reality. When the digital video systems started to develop in the 1980s, it was soon noticed that the maximum bandwidth (2.048 Mbps) of the ISDN is not enough. That is why broadband ISDN (BISDN) was born.

BISDN is a digital network operating at data rates in excess of 2.048 Mbps—the maximum rate of standard ISDN. BISDN is a second generation of ISDN. BISDN is not only an improved ISDN but also a complete redesign of the "old" ISDN, now called narrowband ISDN. It consists of ITU-T communication standards designed to handle high-bandwidth applications such as video. The key characteristic of broadband ISDN is that it provides transmission channels capable of supporting rates greater than the primary ISDN rate. Broadband services are aimed at both business applications and residential subscribers.

BISDN's foundation is cell switching, and the international standard supporting it is *Asynchronous Transfer Mode* (ATM). Because BISDN is a blueprint for ubiquitous worldwide connectivity, standards are of the utmost importance. Major strides have been made in this area by the International Telecommunications Union-Telecommunications (ITU-T) during the past decade. More recently, the ATM Forum has advanced that agenda.

ATM is a fast packet-oriented transfer mode based on asynchronous time-division multiplexing. The words *transfer mode* say that this technology is a specific way of transmitting and switching through the network. The term *asynchronous* refers to the fact that the packets are transmitted using asynchronous techniques (e.g., on demand), and the two end points need not have synchronized clocks. ATM will support both circuit switched and packet switched services. ATM can handle any kind of information, i.e., voice, data, image, text, and video in an integrated manner.

An ATM network is made up of an ATM switch and ATM end points. An ATM switch is responsible for cell transit through an ATM network. An ATM end point (or end system) contains an ATM network interface adapter. Examples of ATM end points are workstations, routers, digital service units (DSUs), LAN switches, and video coder-decoders (CODECs). An ATM network consists of a set of ATM switches interconnected by point-to-point ATM links or interfaces. ATM switches support two primary types of interfaces: user-network interface (UNI) and network-network interface (NNI). The UNI connects ATM end systems (such as hosts and routers) to an ATM switch. The NNI connects two ATM switches.

In ATM the information to be transmitted is divide into short 53 byte packets or cells. There are reasons for such a short cell length. First, ATM must deliver real-time service at low bit rates. Thus the size allows ATM to carry multiple forms of traffic. Both time-sensitive traffic (voice) and time-insensitive traffic (data) can be carried with the best possible balance between efficiency and minimal packetization delay. Second, using short, fixed-length cells allows for time-efficient and cost-effective hardware such as switches and multiplexers.

Each ATM cell consists of 48 bytes for information field and 5 bytes for header. The header is used to identify cells belonging to the same virtual channel and thus used in appropriate routing. The ATM cell structure is shown in Fig. 17.6.13. The cell header comes in two forms: the UNI header and the NNI header. The UNI is described as the point where the user enters the network. The NNI is the interface between networks. The typical header therefore looks like that shown in Fig. 17.6.14 for the UNI. The header is slightly different for NNI, as shown in Fig. 17.6.15.

ATM is connection-oriented and connections are identified by the virtual channel identifier (VCI). A virtual channel (VC) represents a given path between the user and the destination. A virtual path (VP) is created by multiple virtual channels heading to the same destination. The relationship between virtual channels and

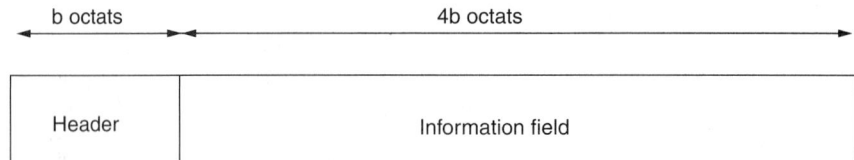

FIGURE 17.6.13   ATM cell structure.

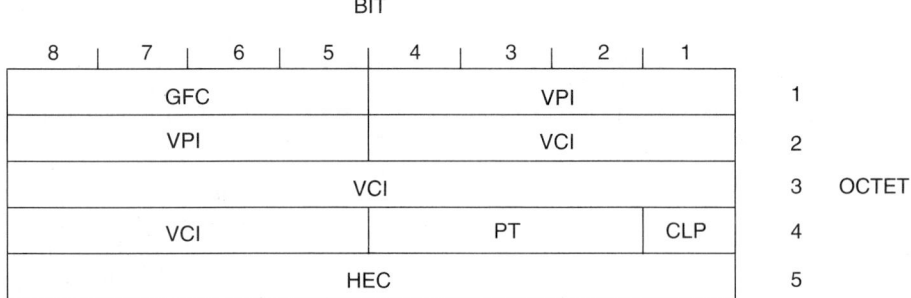

VPI	virtual path identifier	PT	payload type
VCI	virtual channel indentifier	CLP	cell loss priority
HEC	header error control	GFC	ganaric flow control

FIGURE 17.6.14   ATM cell header for UNI.

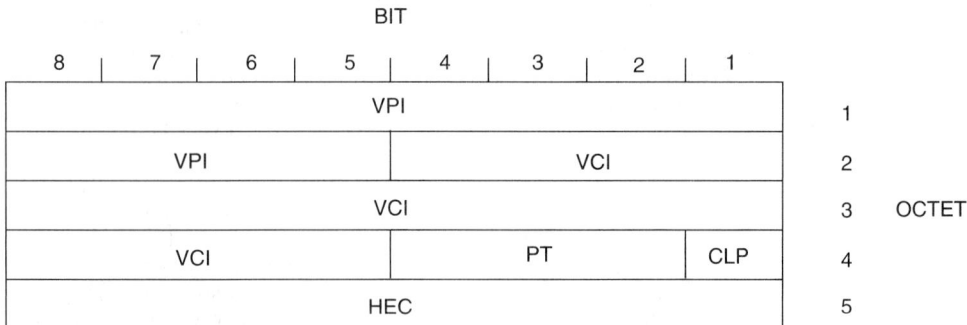

VPI	virtual path identifier	PT	payload type
VCI	virtual channel indentifier	CLP	cell loss priority
HEC	header error control		

FIGURE 17.6.15   ATM cell header for NNI.

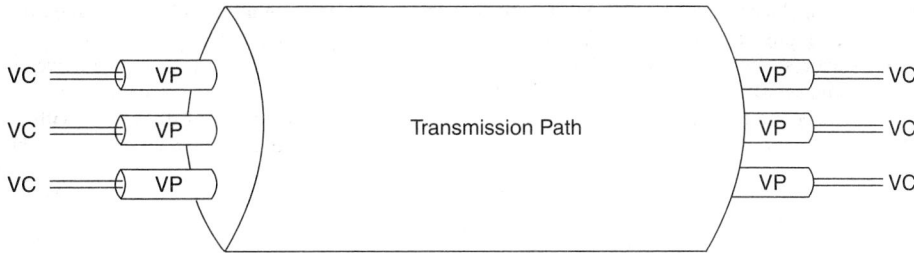

FIGURE 17.6.16 Relationship between virtual channel, virtual path, and transmission path.

virtual paths is illustrated in Fig. 17.6.16. A virtual channel is established at connection time and torn down at termination time. The establishment of the connections includes the allocation of a virtual channel identifier and/or virtual path identifier (VPI) and also includes the allocation of the required resources on the user access and inside the network. These resources, expressed in terms of throughput and quality of service (QoS), can be negotiated between user and network either before the call set up or during the call. Having both virtual paths and channels make it easy for the switch to handle many connections with the same origin and destination.

ATM can be used in existing twisted pair, fiber-optic, coaxial, and hybrid fiber/coax (HFC), SONET/SDH, T1, E1, T3, E3, E4, and so on, for LAN and WAN communications. ATM is also compatible with wireless and satellite communications.

Figure 17.6.17 depicts the architecture for the BISDN protocol. It is evident that the BISDN protocol uses a three-plane approach. The user plane (U-plane) is responsible for user information transfer including flow control and error control. The U-plane contains all of the ATM layers. The control plane (C-plane) manages the call-control and connection-control functions. The C-plane shares the physical and ATM layers with the U-plane, and contains ATM adaptation layer (AAL) functions dealing with signaling. The management plane (M-plane) includes plane management and layer management. This plane provides the management functions and the capability to transfer information between the C- and U-planes. The layer management performs layer-specific management functions, while the plane management deals with the complete system.

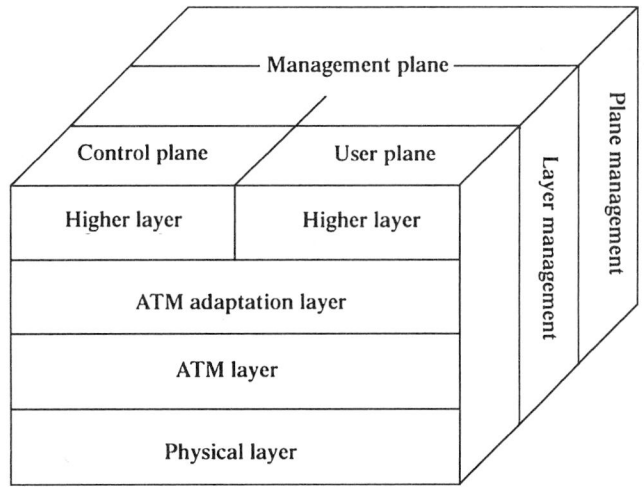

FIGURE 17.6.17 BISDN protocol reference model.

Figure 17.6.17 also shows how ATM fits into BISDN. The ATM system is divided into three functional layers, namely, the physical layer, the ATM layer, and the ATM adaptation layer.

BISDN access can be based on a single optical fiber per customer site. A variety of interactive and distribution broadband services is contemplated for BISDN: high-speed data transmission, broadband video telephony, corporate videoconferencing, video surveillance, high-speed file transfer, TV distribution (with existing TV and/or high-definition television), video on demand, LAN interconnection, hi-fi audio distribution, and so forth.

CHAPTER 17.7
TERMINAL EQUIPMENT

C. A. Tenorio, E. W. Underhill, J. C. Baumhauer, Jr., L. A. Marcus, D. R. Means, P. J. Yankura, Herbert M. Zydney, R. M. Sachs, W. J. Lawless

TELEPHONES

C. A. Tenorio, E. W. Underhill, J. C. Baumhauer, Jr., L. A. Marcus, D. R. Means, P. J. Yankura

Telephone equipment ranges from the familiar desk or wall telephone set to the versatile communications system terminal of the information age. Telecommunications has merged with compute technologies in telephones to make available the entire spectrum of voice, data, video, and graphics. Terminal equipment now allows the exchange of this information over the telephone network.

The Telephone Set

The basic functions of the telephone set include signaling, alerting, central office supervision, and transmission of voice communications. In a typical call sequence, when the caller (the near-end party) picks up the handset, the telephone draws loop current (the telephone line is known as a *loop*) from the central office battery, which signals the central office (CO) that it wants service. The loop current also provides power for telephone functions. The caller then dials, sending address signals to the central office by either pulse or tone dialing. The CO collects the address signal in registers and sets up a transmission path with the CO for the number being called. The called CO sends an alerting signal to the called telephone, causing it to ring. When the called or far-end party picks up their handset, loop current is drawn signaling the CO to trip (interrupt) ringing and complete the talking circuit.

The functional elements of traditional telephones (Fig. 17.7.1) include a carbon transmitter to convert acoustic energy to an electrical voice signal, an electromagnetic receiver to convert the electrical voice signal back into acoustic energy, a switch hook to turn the telephone on and off, rotary dial contacts, which make and break loop current, a loop-equalizer circuit to compensate for loop resistance, a balance circuit, a hybrid transformer for coupling the transmitter and receiver to the telephone line, and an electromechanical ringer. The two-wire telephone line connections are known as Tip and Ring. The loop equalizer, balance circuit, and hybrid transformer are collectively known as the speech network. Such traditional speech networks are called *passive* networks. Electronic speech networks using solid-state components are called *active* networks.

The ringer is shown bridged across the telephone line. The capacitor C_1 blocks the flow of loop current through the ringer. Resistor R_1 and varistor V_1 constitute the loop-equalizer circuit. On long loops with low loop current, varistor V_1 maintains a high resistance and takes little current away from the rest of the speech network. On short loops, higher levels of loop circuit result in a lower resistance of V_1, thereby reducing the transmit and receive levels.

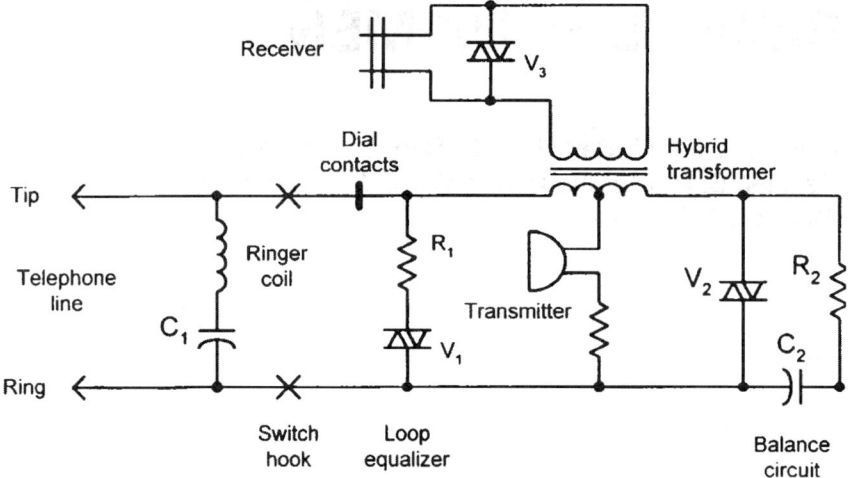

FIGURE 17.7.1   Traditional passive network telephone set.

The combination of a three-winding hybrid transformer and impedance balancing circuitry provides the means of coupling the transmitter and receiver to the loop independently. This is called an *antisidetone* network. Sidetone is that portion of the transmitted signal that is heard in the receiver while talking. Sidetone is subjectively desirable because it provides the live quality of face-to-face conversation. The antisidetone network is designed to provide a sidetone signal at about the same level as received speech. If the sidetone level is too high, the talker tends to speak softly to keep the sidetone level pleasant, which results in signal strength too low for good transmission. If the sidetone level is too low, the talker perceives the telephone as dead or inoperative.

Incoming voice signals from the telephone loop are transformer coupled to the receiver. The induction voltages are such that most of the incoming signal power is delivered to the receiver with little power to the balance network. Outgoing voice signals generated by the transmitter induce voltages in two of the transformer windings that cancel each other, so that most of the signal power is divided between the balance circuit resistor R_2 and loop impedances with little to the receiver. The choice of impedance and turns ratios provides a compromise in sidetone balance and impedance matching to the telephone line. Capacitor C_2 prevents dc power from being dissipated in R_2. Varistor V_2 helps match the balance circuit impedance to the loop impedance.

The main advantage of an active over a passive network is its smaller physical volume, lower cost, and greater versatility. An active network also provides power gain, thus allowing the use of microphones such as electrets. In the active network the gain of the transmit and receiver amplifiers can be automatically adjusted, depending on the loop current, to provide loop equalization.

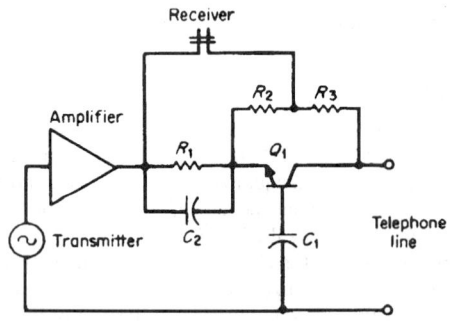

FIGURE 17.7.2   Typical active-network circuit.

A basic active network is shown in Fig. 17.7.2. The base of Q_1 is returned to common (at voice frequencies) by capacitor C_1. The emitter of Q_1 is virtual ground, since its low base impedance is divided by the transistor's beta. A received signal appearing on the telephone line is routed to the receiver through the voltage divider consisting of R_2 and R_3; R_2 is connected to a virtual ground. The other end of the receiver is returned to common through the low output impedance of the transmit amplifier.

The transmit signal is first amplified by the amplifier and further amplified by transistor Q_1. The voltage gain of this common-base state is determined by the input impedance of the telephone line and the impedance of R_1 in parallel with C_2. The antisidetone balance is achieved by adjusting the voltage divider (R_2 and R_3) to compensate for the gain of the common-base stage, leaving about the same potential at both

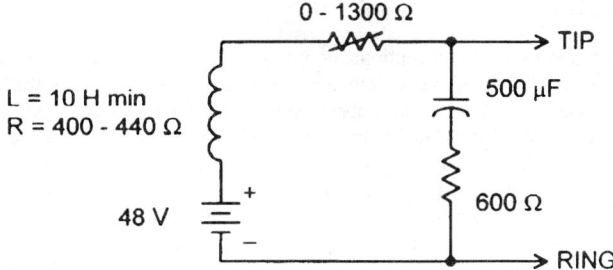

FIGURE 17.7.3 Loop simulator.

ends of the receiver. Capacitor C_2 is added to minimize any phase shift through this stage caused by the capacitance of the telephone line.

Transmission

Terminal equipment is part of a transmission and signaling circuit set up by a network provider for voice frequency transmission in the range of 300 to 3300 Hz. Four important characteristics a telephone must have in order to work properly in this circuit are dc resistance (for dc powering and loop supervision), ac impedance at 1000 Hz, signal power level at both the receiver and transmitter, and audio frequency response.

A simple loop simulator is shown in Fig. 17.7.3. A ring generator (86 V, 20 Hz, 400 Ω), present only during ringing, is not shown. DC power is provided by a nominal 48-V battery. Loop current is limited by the resistance of relay coils or current limiting circuits in the central office, and by the resistance of the loop itself. Maximum loop resistance is 1300 Ω, which is 15,000 ft of 26 AWG cable. With a 300-Ω telephone, loop current is about 20 to 80 mA. At 1000 Hz the transmission cable has a characteristic impedance of 600 Ω, which the telephone should match for maximum energy transfer and minimum echo.

Typical signal levels are shown in Fig. 17.7.4. Desirable sound power at the receiver was determined by subjective testing of people. The transmitter, while converting acoustic power into electrical energy, must also amplify the energy by about 20 dB to compensate for the 20-dB loss in converting the electrical energy back into acoustic power. The loop resistance provides an attenuation of 2.8 dB/m for 26 AWG cable. Today, virtually all central offices' trunks are digital, so there is no transmission loss between them.

Telephone receivers and transmitters are designed to achieve desired frequency characteristics. For example, telephone handset microphones have a rising response characteristic near 3 kHz to compensate for capacitive shunting loss in the loop and simulate the effects of acoustic diffraction about the human head that is present in face-to-face conversation.

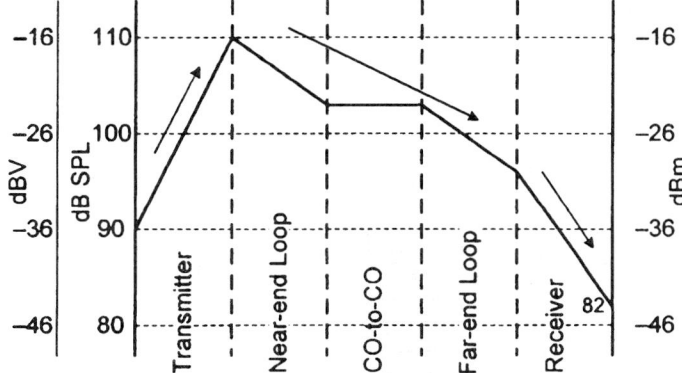

FIGURE 17.7.4 Transmission circuit signal levels.

Multifunction Electronic Telephones

Most new telephones are electronic. Conventional components, such as the bell ringer, the hybrid transformer, and the mechanical dial (rotary or push button) are replaced by active electronic devices that are incorporated into large-scale integrated (LSI) circuits. A typical dial-pulse electronic telephone set contains at tone ringer, an active network, an electronic dial keypad, a dial-pulsing transistor, and a low-power microphone. Several advantages result. The overall reliability of the telephone increases, automated manufacturing assembly is possible, telephone-set weight and size are reduced, and finally, overall transmission performance of the telephone is improved.

A diagram of a typical microcomputer-controlled multifunction electronic telephone is given in Fig. 17.7.5. Telephone features include last-number redial, repertory dialing, dial-tone detectors, speakerphone, integrated answer/record, hold, conference (for two-line telephones), and display of the dialed digits. Repertory dialing permits the user to store several telephone numbers in an auxiliary memory for automatic dialing. The dial circuit can produce pulse dialing or dual-tone multifrequency (DTMF) tones. The architecture often includes both general-purpose and custom LSI circuits, such as DTMF generator chips, clock (timer) chips, and display-driver chips, or it may contain one very large-scale integrated (VLSI) circuit. The microcomputer controls the operation of the various LSI circuits. The microcomputer receives information from the ringing detector, the dial keypad, function buttons, the electronic line switch, and the active network, and controls such items as the tone generator (ringer), the integrated answer/record system, the dial circuits, the display, and the speakerphone.

Electronic logic performs a variety of common switching functions, such as switching out the transmitter and lowering the gain to the receiver during dialing, functions performed by mechanical switches in traditional telephone sets. The line switch may also be electronic. The switch hook, rather than closing line circuit contacts when the handset is lifted, turns on a solid-state line switch. The user can also turn the telephone on and off electronically without having to lift the handset. Speakerphone, answer/record, and hold operations are common uses for an electronic line switch.

The media for message storage is either tape or solid-state electronic memory. Audio storage on tape uses standard tape deck recording and playback techniques. Storage in solid-state memory requires conversion to a digital format with the use of a CODEC. This digital data is stored and retrieved under control of a microprocessor. To further conserve relatively expensive memory, a DSP can be used to massage the data. Dead time is removed, and various compression algorithms are used to conserve memory. Here a trade-off is made between the amount of memory needed and the quality of the speech desired.

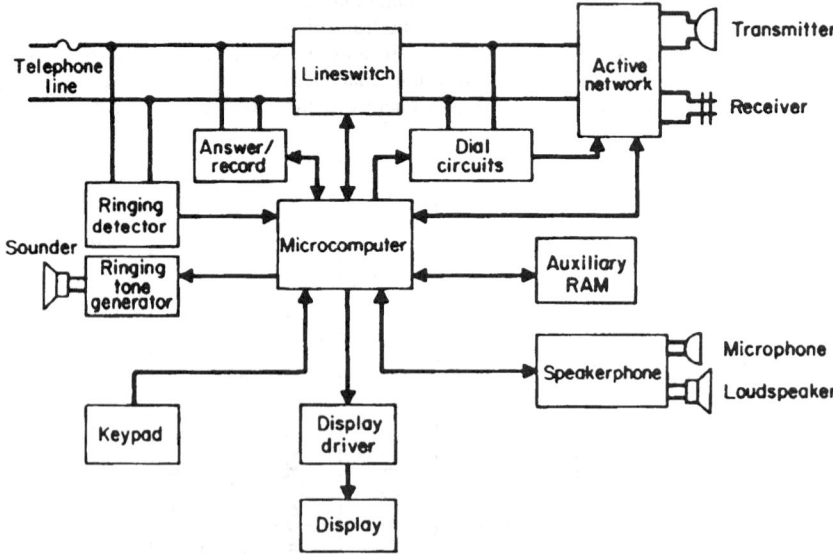

FIGURE 17.7.5   Multifunction electronic telephone.

Speakerphones must pick up much lower speech signals than a handset transmitter and must generate more acoustic energy than a handset receiver can, so it often requires more power than can be drawn over the telephone line. To prevent feedback from the loudspeaker to the microphone, the loudspeaker is muted when speech (or noise) is detected. Because the microphone must be very sensitive to pick up ordinary conversation it also picks up room reverberation, which causes speech to sound as if the speaker is in a tin can. New highly directional microphones can minimize unwanted echo and produce more natural speech.

Cordless Telephones

In a cordless telephone, the usual telephone functions are performed over a radio link, thereby eliminating the handset cord and providing the user with added mobility. A cordless telephone block diagram (Fig. 17.7.6) shows a portable handset unit, used for talking, listening, dialing, and ringing, and a fixed base unit, used for interfacing between the telephone line and radio link. More sophisticated applications include units with duplex intercoms and base units with integrated speakerphones or telephone answer/record devices.

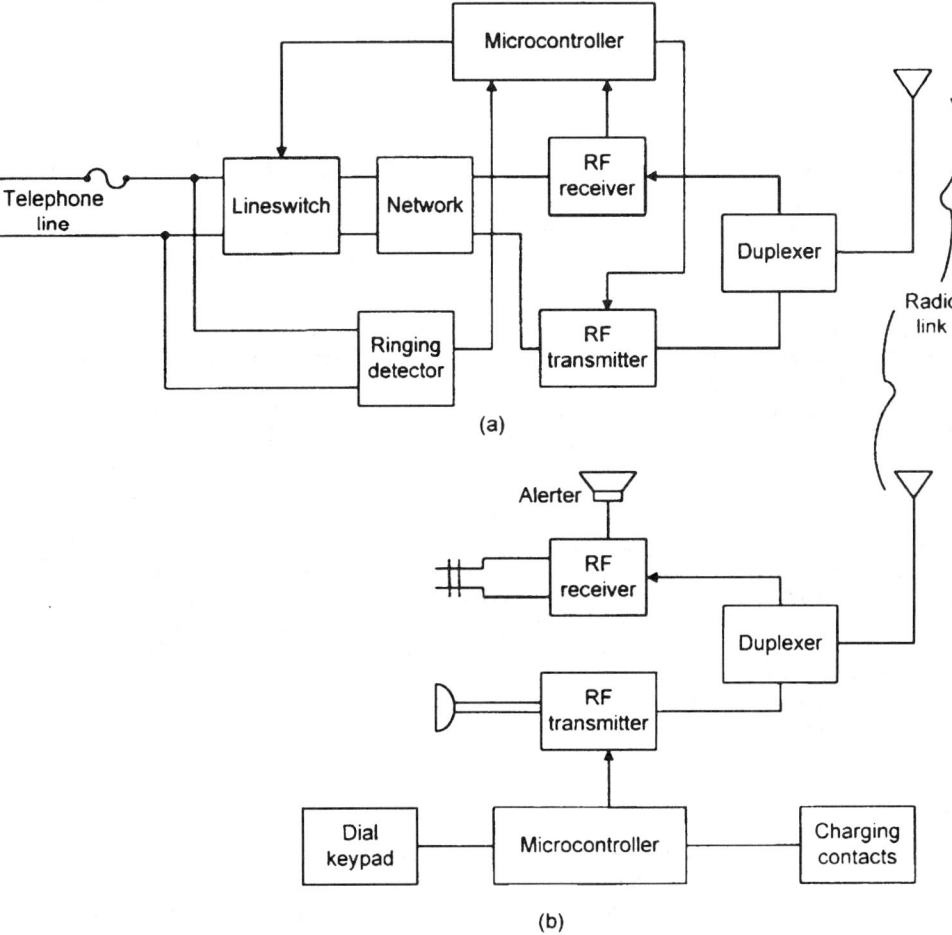

FIGURE 17.7.6 Cordless telephone: (*a*) base unit and (*b*) handset unit.

The normal listening and talking signals as well as ringing and DTMF signaling are transmitted over a radio-frequency link as a frequency-modulated carrier. This carrier is centered in one of ten 20-kHz-wide channels with the base-to-handset channels in the 46.6- to 47-MHz band and the handset-to-base channels in the 49.6- to 50-MHz band. Before 1983 the base-to-handset link used the 1.6- to 1.7-MHz band and there were only five FCC allocated channels. As cordless telephones became more popular, interference between people using the same channel became common. To minimize the probability of hearing conversations from nearby cordless telephones on the same channel, the FCC limits the maximum radiated field strength to 10,000 (V/m at 3 m, which allows satisfactory cordless telephone performance up to 1000 ft from the base station under ideal conditions. To minimize interference, some telephones scan the available channels and choose any vacant channel found.

Typical cordless telephones employ full duplex signaling between handset and base using *frequency shift keying (FSK)* modulation of the carrier. By embedding a digital code in all exchanges of information between handset and base, it is possible to virtually eliminate false operations and ringing. A security code is also embedded to prevent access of the base unit by a neighbor's handset.

Increasing user traffic in the 46/49 MHz frequency bands has caused renewed congestion. Therefore, the FCC has allowed use of one of the commercial bands with frequencies ranging from 902 to 928 MHz, using either analog or digital modulation schemes. Units using digital communication provide a higher degree of security because they do not allow simple FM demodulation by scanners or FM receivers, thereby preventing eavesdropping. This band also has less congestion and allows better RF propagation throughout the usable area, decreasing noise and interference.

Video Telephones

New video telephones provide motion video in color over the public switched telephone network. Advancements made in audio and video compression algorithms have made this possible. See Early et al. (1993). Previous video systems required a special transmission line, making them suitable only for business users. The video telephone first establishes a talking connection with another video telephone, then video mode is entered by pressing a "video" button on each telephone. In "video" mode compressed audio and video signals are transmitted between the two telephones.

The full-duplex DSP-based modem uses a four-dimensional, 16-state trellis code and supports data rates of 19.2 and 16.8 kb/s. The 19.2 kb/s mode transmits 6 bits per baud at 3200 baud, and the 16.8 kb/s mode transmits 6 bits per baud at 2800 baud. The modem will automatically drop back to the slower speed if the error rate indicates that the connection will not support the higher speed. The speech compression is achieved using a code-excited linear prediction (CELP) algorithm that has been enhanced by incorporating fractional-pitch prediction and constrained stochastic excitation. The speech encoder operates at 6.8 kb/s.

The video signal is preprocessed to produce three separate video frames. One frame contains the luminance information with a resolution of 128 pixels by 112 lines. The other two frames contain chrominance information with resolutions of 32 pixels by 28 lines. The frames are each segmented into 16 by 16 blocks for processing. A motion estimator compares the block being processed with blocks from the previous frame, and generates an error signal based on the best match. This signal is converted to the frequency domain using the discrete cosine transform. The transformed signal is quantized, encoded, and sent to the output buffer for transmission. The transmitted frame rate is 10 frames per second using an overlay scan instead of a TV raster scan. If there is a large amount of motion, the frame rate is reduced so that the output buffer does not overflow. A button on the telephone can be used to adjust the frame rate transmitted from the distance set. However, a higher frame rate decreases the resolution of the received image. A self-view mode is also provided to allow the near-end party to view the image that is being transmitted to the far-end party.

Bell and Tone Ringers

In traditional telephones an electromechanical bell ringer is used to alert the customer to an incoming call. The typical ringer has two bells of different pitch that produce a distinctive sound when struck by a clapper driven by a moving-armature motor. The ringer coil inductance in series with a capacitor resonates at 20 Hz to provide a low-impedance path for a 20-Hz ringing signal. The high inductance of the ringer coil prevents loading of the speech network or DTMF generator when the handset is off-hook.

Other ringer connections are used when customers are on party lines and must be rung selectively. Selective ringing schemes include the connection of the ringer between the tip or ring conductors and ground, or ringers tuned to different ringing frequencies (16 to 68 Hz).

Electronic tone ringers are used in most new telephones. A resistor-capacitor circuit is bridged to the telephone line to provide the proper input impedance (defined by FCC rules) for a ring-detect chip. Tone ringers can have equivalent effectiveness and acceptability to the customer when compared to bell ringers if acoustic spectral content and loudness are adequate. Typically, the tone ringer consists of a detector circuit, which distinguishes between valid ringing signals and transients on the telephone line, and a tone generator and amplifier circuit that drives an efficient electroacoustic transducer. The transducer may be a small ordinary loudspeaker, a modified telephone receiver, or a special piezoelectric-driven sounder (see Fig. 17.7.10).

Tone and Pulse Dialers

Dial-pulse signaling interrupts the telephone line current with a series of breaks. The number of breaks in a string represents the number being dialed; one break is a 1 and 10 breaks is a 0. These breaks occur at a nominal rate of 10 pulses per second, with 600 ms between pulse trains. The ratio of the time the line current is broken to the total cycle time (percent break) is nominally 61 percent. Dial pulse signaling can be used with all central offices.

The mechanical rotary dial in the traditional telephone uses a single cam to open and close the dial contacts. The dial is driven by a spring motor that is wound up by the user as each digit is dialed. The return motion is controlled by a speed governor mechanism to maintain the proper dial-pulsing rate.

DTMF signaling consists of sending simultaneously two audio frequencies of at least 50-ms duration representing a single number, separated by at least 45-ms intervals between numbers. On the standard 4-by-3 dial format, each column and each row is associated with a different frequency, as shown in Fig. 17.7.7. This method of signaling permits faster dialing for the user and more efficient use of the switching systems. Since the frequencies are in the audio band, they can be transmitted throughout the telephone network.

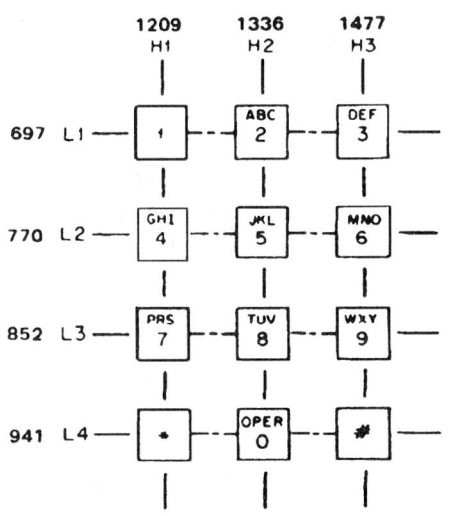

FIGURE 17.7.7 Basic arrangement of pushbuttons for dual-frequency dialing.

Pushbutton dials originally used for DTMF signaling were laid out in a rectangular format to accommodate the cranks and levers necessary to operate the mechanical switches. With modern electronic push-button dials any layout can be used, but the 4-by-3 format is still popular for its dialing speed. Pushbutton dials can perform the dial-pulse function electronically. These electronic "rotary dials" interrupt the line current with transistors or relays. Since the user can enter a number into the dial faster than the number can be pulsed out, a first-in, first-out memory is used to store the number as it is dialed. The dial-pulse timing is generated using an internal clock.

Several methods have been used to generate DTMF signals. Early methods used an inductor-capacitor oscillator with two independently tuned transformers. Different values of inductances are switched into the circuit to obtain different frequencies. Another method is a resistor-capacitor oscillator. Here a twin-tee notch filter in the feedback loop of a high-gain amplifier gives the desired frequency. Two amplifier-filter units are used, one for each frequency group. A modern method to generate DTMF signals uses CMOS integrated circuits to use digit synthesis techniques (Fig. 17.7.8). The keypad information is combined with a master clock to generate the desired frequency. This information is fed to a D/A converter, whose output is a stair-step waveform. The waveform is filtered and fed to a driver circuit, which provides the desired sine-wave frequency signals to the telephone line.

The latest method used to generate DTMF signals is in software. If a digital speech processor is available in the product, a subroutine can be written to generate the appropriate DTMF waveform. The output is a digital word that is periodically fed to a CODEC for conversion to an analog signal. This is fed to a buffer amplifier that drives the telephone line.

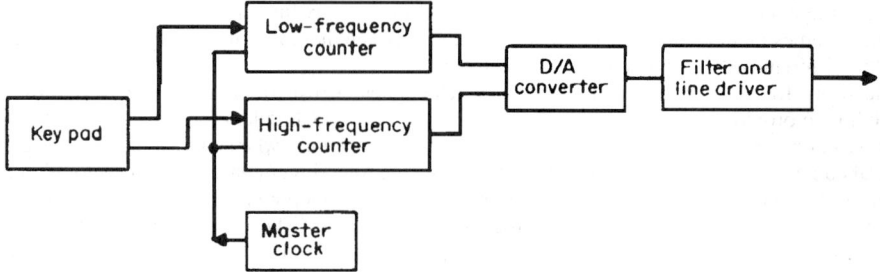

FIGURE 17.7.8   Digital-synthesis circuit.

To combine the versatility of both pulse and DTMF signaling, many telephones have the ability to switch between the two systems. This permits the user to use pulse dialing for making the telephone call and then switch to DTMF for end-to-end signaling for services such as bank-by-phone.

Microphones

The granular carbon microphone, often called a transmitter in telephony, dates back over 100 years to the birth of telephony. Sound striking the diaphragm imparts a pressure fluctuation on the carbon aggregate (Fig. 17.7.9a). Since granule contact force and dc resistance R_0 are inversely related, a modulation of the telephone loop current I_0 results. Carbon transmitters offer 20 to 25 dB of inherent signal-power gain and have a nonlinear input/output relationship that advantageously favors speech over low level background noise, but their low resistance consumes loop power. Electronic telephones require a microphone that consumes much less power.

The electret microphone is a small capacitive microphone that is widely used today. It has low sensitivity to mechanical vibrations, a low power requirement, and high reliability. An electret has no inherent gain, so requires a preamplifier. An effective bias voltage V_0 depends on a polymer diaphragm's trapped electret charge (Fig. 17.7.9b). The piezoelectric ceramic unimorph element (Fig. 17.7.10) is used as a microphone in cordless and cellular handsets. Its piezoelectric activity owes to an electrically polarized, synthetic (as opposed to natural crystal) ferroelectric ceramic. The flexural neutral axis of the structural composite is not at the ceramic's midplane; thus, vibration results in variation of the ceramic's diameter that induces a voltage across its thickness as defined through the piezoelectric constant d_{31}.

Receivers

The receiver converts the electrical voice signal back into acoustic sound pressure. Either one of the following designs is often used. The electromagnetic (moving-iron) receiver uses voice coil currents to modulate the dc flux,

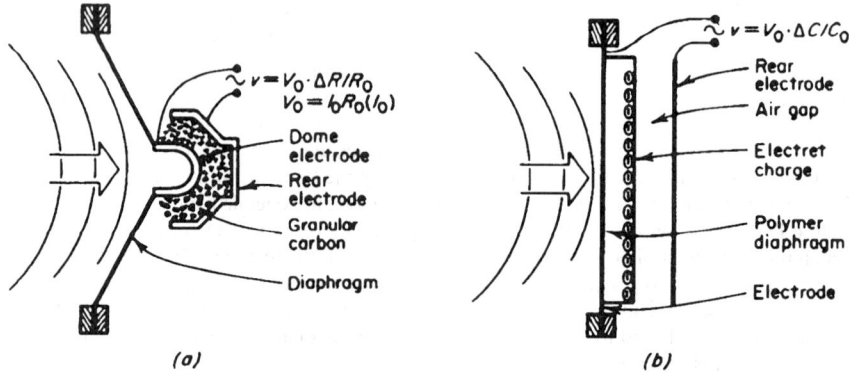

                    (a)   *(b)*

FIGURE 17.7.9   Microphones: (*a*) Carbon and (*b*) electret.

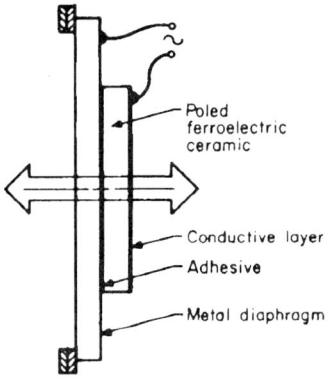

FIGURE 17.7.10   Piezoelectric ceramic unimorph.

which produces variable force on the iron armature (Fig. 17.7.11*a*). The electrodynamic (moving-coil) receiver uses coil current perpendicular to the dc magnetic field to generate an axial force on the movable coil (Fig. 17.7.11*b*). The constant (dc) reluctance of the coil air gap results in less distortion than in the electromagnetic receiver. The moving-coil unit has a lower, more nearly real impedance compared with the moving-iron design. Both of these permanent magnet devices can be used as a microphone, since their input/output is reversible.

Certain hearing aids can inductively couple to the leakage flux from receivers that generate an external magnetic filed. Some receiver types (e.g., piezoelectric) must have an induction coil added to the design to provide an adequate magnetic field. Most telephones are required to provide a magnetic field (see Public Law 100-394, 1988) in order to provide the hearing impaired with access to the telephone network.

Handsets

The handset holds the microphone and receiver. It may also contain a line switch, dial keypad, and other circuits to make it a complete one-piece telephone. The handset positions the transmitter in proper location with respect to the mouth when the receiver is held on the ear. Standard dimensions for the relative mouth and ear locations have been defined (Fig. 17.7.12) for a head that is statistically modal for the population.

The handset should provide an acoustic seal to the ear, provide proper acoustic coupling for its transmitter and receiver, and be heavy enough to operate the telephone switch hook when placed on it. Handsets for hearing impaired users may also contain an amplifier and a volume control.

Protection

The user must be protected against contact with ringing voltages, lightning surges, and test signals applied to the telephone line. Telephone service personnel are trained to work on live telephone lines, but users are not. Lightning surges that are induced onto the telephone might be 1000 V peak but are of limited energy (Fig. 17.7.13). (See Carroll and Miller, 1980). Test signals applied from the central office can be up to 202 V dc.

Telephone cables, often strung on the same poles as power cables, are subject to power crosses if the power cable breaks (as in a storm) and falls on the telephone cables. The telephone company installs

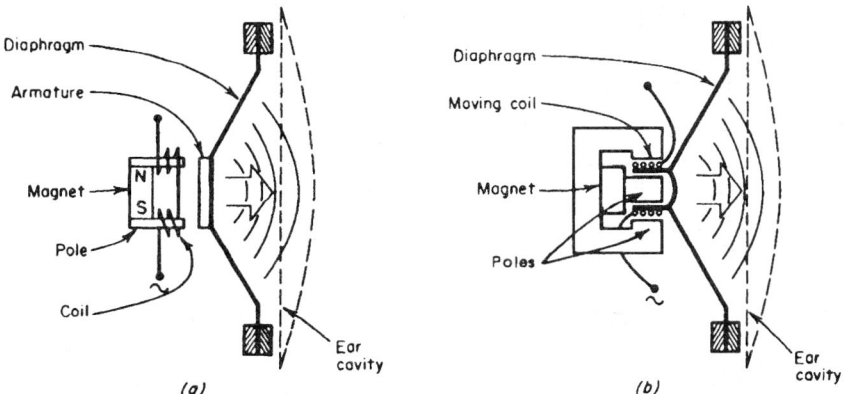

FIGURE 17.7.11   Receivers: (*a*) central armature magnetic (moving armature) and (*b*) dynamic (moving coil).

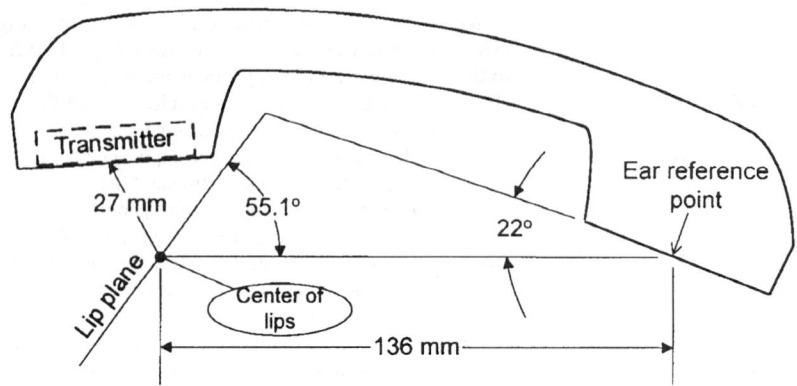

FIGURE 17.7.12 Handset modal dimensions.

primary protectors at building entrances to shunt voltages from power crosses and direct lightning strikes to ground.

Extraneous signals can cause very high acoustic output from the receiver. Either a varistor, V_3 in Fig. 17.7.1, placed directly across the receiver terminals, or the telephone's circuit design is used to limit maximum acoustic output to 125 dBA.

Finally, the telephone network itself needs protection from excessive signal power, fault voltages, and other disturbances caused by the terminal equipment. Requirements for telephone network protection are contained in FCC Rules and Regulations, Part 68.

TELEPHONES WITH ADDED FUNCTIONS

Herbert M. Zydney, R. M. Sachs

Key Telephone Sets

Key telephones are designed for users who need access to more than one central-office (CO) line or PBX extension. In almost all cases, this is accomplished by the addition or illuminated keys (hence the term "key"

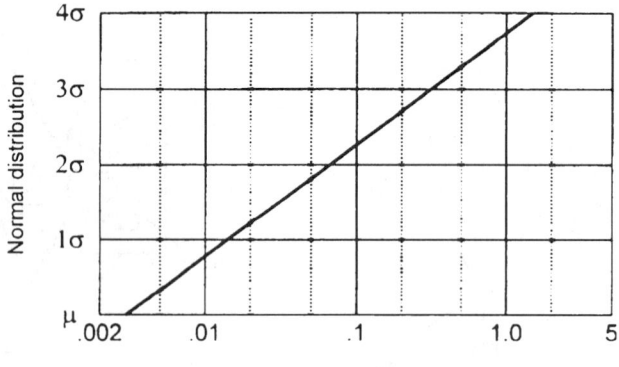

FIGURE 17.7.13 Induced lightning energy distribution.

telephone) to a telephone instrument. These are arranged to correspond to the line or extension and are generally illuminated to identify their status, either directly or by adjacent LEDs or LCDs. The keys can be operated independently to select the desired line. To permit switching between two calls, a hold key is provided so that the active call can be maintained in a "hold" state before operating the key associated with a different call. To allow the user to be alerted, a common audible ringer is provided, which sounds when a new call occurs on any line. Where the telephone is used as a part of a self-contained system, a key is also included so that internal, or "intercom," calls can be made between individual key telephone sets. As technology has evolved, more features have been included in key telephones that improve the efficiency of telephone system operation. Because these newer systems are software controlled, a number of the features are not fully resident in the telephone itself, but depend on a distributed architecture to fully implement them.

Examples include:

(a) *Memory Dialing*: Prestored telephone numbers can be dialed at either the touch of one button or by an abbreviated code from the dial pad.
(b) *Speaker/Microphone Services*: A loudspeaker powered at the station can support hands-free dialing or, when connected to the telephone channel, permit hands-free intercom services or speaker phone operations.
(c) *Display-Based Features*: The most advanced key telephone sets offer a display of either numeric or alphanumeric characters, sometimes augmented by graphic symbols. In conjunction with system software support, these permit users to identify who is calling them, determine the status of other telephones in the same system, and retrieve messages such as the intercom extensions of unanswered calls.

The technology used to implement key telephones presently in use is quite varied. Four categories are worth singling out:

Electromechanical Key Telephone Sets.   These early-style telephones generally use electromechanical keys with a relatively large number of electrical contacts for most functions. Individual CO lines and control leads are brought to each telephone, and the actual switching of lines occurs within the telephone set. Additional wires provide control of illumination and ringing. They rely on external line-holding relay circuits and power for their operation, although standard telephone operation is possible on a stand-alone basis. A common hold button activates the external relay holding bridge for all telephones and is released when any other telephone activates its line key. Many key stations are arranged to offer visual and audible signals for status identification using varying rates of interruption for different states. For ease of installation, dedicated wires are usually grouped into 25-pair (50-conductor) cables with a standard connector. Some of the larger telephones, with dozens of line keys, can have four or more such connectors.

Key Telephone Sets with Internal Control.   By adding electronic circuitry within the key telephone, many of the functions accomplished by the external relay circuitry can occur internally. Every CO line is brought to each telephone. Internally, ringing signals can be sensed and holding bridges can be applied. Complementary electronics in other telephones can sense the holding bridges and provide the necessary visual signals for multiple line states and access. Because of limitations associated with the power and sensitivity of CO lines, these key sets generally are limited to two, or at the most, four lines. External power, often rectified in the set, is required for the larger systems; smaller systems can operate from the power provided over CO loops.

Electronic Key Telephone Sets.   Electronic sets (see Silverio et al., 1985) are different from the first two classes because the CO lines terminate in a centralized switch (often called a key service unit or control unit) rather than in the telephone itself. One, or sometimes two, voice paths are terminated in the telephone itself. A separate digital data link between the telephone and the control unit is used to exchange information, including what keys have been pressed and what visual and audible status information is to be presented to the user. There is little standardization in the functional definition of the conductor pairs. The voice pair generally operates at 600 Ω or 900 Ω, although the actual transmission levels may differ from standard loop levels. The digital data link may be provided on either one or two pairs and can operate from less than 1 to over 200 kb/s. Power is derived in a number of ways: the simplest is to dedicate one pair to power; alternatively, the power may be impressed on a center tap of paired transformers carrying either voice or data and removed and filtered in the telephones; lastly, the digital signals can be shaped so that they have no dc component, and the power may then be sent along with the data. Voltage of +24, +36, and –48 V are commonly used. The basis of operations within the telephone is to scan input stimuli such as keys in real time, format messages, and then send them serially to the key service unit. Messages from the key service unit are demultiplexed and are used either to drive audible ringers or to flash lights. Advanced systems embed a cyclic redundancy check with each message to minimize false operations. Somewhat more complex message structures are involved where displays are to be updated. This approach minimizes the requirement for changes within the telephone to customize it to the user's needs. For example, software in the key service unit is responsible for determining the meaning of most keys on the telephone sets. If it is changed, only the set labels are varied. The circuitry for operating the set is implemented in a number of ways. Small 4- and 8-bit microprocessors are common, although the smallest sets use custom VLSI for their operation. Wiring is reduced to as few as two pairs, although up to four pairs are also used where a second voice path is required. Depending on the speed of transmission, twisted wire pairs without bridge taps are usually required, which makes these sets of limited use in residential locations unless the locations are rewired.

Digital Key Telephones.   The most technologically advanced key telephones send and receive voice signals in digitally encoded format. See Murato et al. (1986). The standard network encoding of 64 kb/s in mu-law format is most common. The telephone contains a compatible codec that converts from analog to digital formats. As with the electronic key telephones, many designs have nonstandard interfaces. The most basic digital key telephones combine the digital data stream and the encoded voice stream on two pairs. More advanced telephones add an additional channel for binary data at speeds from 9.2 to 64 kb/s. As an option, or at times built-in, this channel appears at a standard EIA RS 232 interface, which can directly interface with compatible terminals or desktop computers. The key service unit can switch this data channel independently of the voice channel. The conductor formats vary. Some systems use two pairs, each carrying the combined voice and other signals in different directions at speeds up to about 200 kb/s. Other systems use just a single pair at speeds approximating 500 kb/s. These operate by sending information to the telephone preceded by a flag bit; the telephone synchronizes to this data stream and, at its conclusion, replies with a return message. This format is defined as time-compression multiplex in transmission terminology; informally, it is referred to as "ping-pong" because the signals go back and forth constantly. In recent years, international standards bodies have defined a new network standard, ISDN (integrated services digital network). The protocols formulated for this service network are broad enough to embrace the needs of digital key telephones. The implementation costs and flexibility of this new standard have become low enough that these protocols are now appearing in digital key telephone sets.

PC-Emulation of Key Telephone Sets

With the increasing number of desktops that have both telephones and PCs, the two are being integrated in a number of forms. In the simplest arrangement, logical connections are made between the PC and the telephone. Graphics software recreates the telephone on the face of the screen, allowing keyboard commands or mouse-and-click operations to replace buttons. Graphic displays replace the illumination of traditional telephones. A more advanced configuration builds circuitry into the PC, eliminating the requirement for a physical instrument entirely, if the PC has audio capability. User benefits include simpler operation for complex features, built-in help screens, and better integration with data bases. For example, if the PC is able to receive the incoming telephone number of the calling party, the PC screen can automatically display information about the incoming caller.

Speakerphone

A speakerphone is basically the transmission portion of a telephone set in which the handset has been replaced by a microphone and loudspeaker, typically located within a few feet of the user. This arrangement provides the user freedom to move about with hands free during a telephone conversation, as well as some reduction in fatigue on long calls. It also facilitates small-group participation on a telephone call and can be of benefit in special cases involving hearing loss and physical handicaps.

The lengthened transmitting and receiving acoustical paths in the speakerphone arrangement, compared with those of a conventional telephone, introduce loss, which can be on the order of 20 dB or more in each path. This requires that gain be added to both the transmit and receive channels over that provided in a conventional telephone. The amount of gain that can be added in each channel to compensate for the loss in the acoustical paths is limited by two problems. A "singing" problem may occur if an acoustic signal picked up by the microphone is

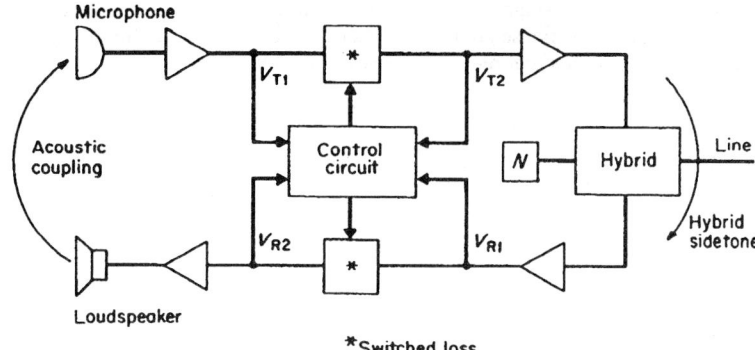

FIGURE 17.7.14   Block diagram of a voice-switched speakerphone.   (*Copyright American Telephone and Telegraph Co. Used by permission*)

fed back to the loudspeaker via the sidetone path and returns to the microphone through acoustic coupling in the room. Singing can occur when too much gain is added in this loop. Even before reaching this condition, however, the return of room echoes to the distant talker can become highly objectionable. Echoes occur when coupling from loudspeaker to microphone causes the incoming speech to be returned to the distant party with delay.

A solution to these problems can be found in *voice switching*, in which only one direction of transmission is fully active at a time. With voice switching, a switched-loss or switched-gain element is provided in both the transmit and receive channels, which operate in a complementary fashion. In this manner, full gain is realized in the chosen direction of transmission, while margin is provided against singing and distant talker echo. Voice switching, however, results in one-way-at-a-time communication. Also, there can be a problem of clipping of a portion of the speech, since control of the voice-switching operation is derived from the speech energy itself. A functional diagram of the essential elements for a voice-switched speakerphone is shown in Fig. 17.7.14, in which a measure of the speech energy is provided to the control circuit from four distinct locations in the transmission paths. Signals V_{TI} and V_{RI} that are a measure of the speech energy in the transmit and receive paths, respectively, are compared to determine the direction of transmission to be enabled. Signal V_{TZ} is used by the control circuit to guard against switching falsely into receive because of transmit energy arriving in the receive path through the hybrid sidetone circuit. Similarly, the control circuit uses V_{R2} to guard against switching falsely into transmit because of receive energy arriving in the transmit path from acoustic coupling. Many speakerphone designs do not use all four control signals directly, but equivalent functions are generally provided by other means.

Another solution to echo control and reduction that enables full duplex performance of speaker-phones is acoustic echo cancellation (AEC). With this technique, the signal driving the speakerphone's loudspeaker is compared to the signal generated by the microphone. The speakerphone builds an acoustic model to remove the acoustic echo from the microphone signal before it is transmitted over the telephone network. The use of AEC in speakerphones has been enabled by the availability of lower cost, high function digital signal processors (DSPs), and advances in adaptive signal processing needed to handle dynamic changes of the acoustic environment as people move about and room conditions change. Once the AEC is adapted, the need for switched loss is diminished and the speakerphone can attain a full-open, or full duplex, condition. This allows for very natural and fluid voice communications.

An AEC is typically placed between the control circuits of a speakerphone and the signals to and from its transducers. This effectively eliminates echoes before they pass the control circuit. Typically there is a signaling connection between the AEC and the control circuit to communicate the status of the AEC. This allows the control circuit to switch less loss when the AEC has adapted, and to prevent the AEC from adapting during periods of double-talking. Hybrid echo cancellers (see Sondhi and Presti, 1966) is typically placed between the hybrid and the control circuit of the speakerphone.

While voice switching and acoustic echo cancellation is effective in eliminating the problems of singing and distant-talker echo, it does not relieve the higher transmitted levels of room ambient noise and reverberant (barrel effect) speech caused by increased gain in the transmit channel. Both of these problems can be reduced by one or more of the following methods: (1) install the speakerphone in a quiet, nonreverberant room; (2) reduce the transmit gain and the actual talker-to-microphone distance; (3) reduce the transmit gain and the effective talker-to-microphone distance with the use of a directional microphone; and (4) roll off the low-frequency end of the transmit response slightly to reduce the pickup of low-frequency noise and reverberant room modes, at the expense of a slight reduction in voice naturalness.

DATA TERMINALS

W. J. Lawless

Data Terminal Equipment (DTE)

Data terminals are used in a variety of data communication systems. Terminals can be classified into two categories: general-purpose terminals and special-purpose terminals. General-purpose terminals contain many elements found in a modern day personal computer (PC) and thus it is very common to use a PC to emulate a data terminal. In some applications, where cost is particularly important, a teletypewriter data terminal containing only the basic terminal elements is used as a general-purpose data terminal.

Special-purpose data terminals have been designed for a number of applications. In some cases the functionality is limited as in the case of a send-only or a receive-only terminal. Likewise, special purpose terminals are used in applications requiring special functionality such as in handheld applications and point-of-sale applications.

In either case the data terminal is used to communicate messages. Some common applications include inquiry-response, data collection, record update, remote batch, and message switching.

Data terminals are typically made up of combinations of the following modules: keyboard, CRT display, printer, storage device, and controller. These modules can be organized into the categories of operator *input-output*, terminal *control*, and *storage*.

Keyboards are available in a number of formats. In some cases, two formats are provided on the same keyboard, side by side or integrated, e.g., typewriter with a numeric pad. Future designs call for multifunction keyboards whose designations or functions can be easily changed by the user, because in many cases differently trained operators will be using the same terminal to enter different types of data for different applications. There is also a trend toward special-application keyboards, such as the cash register keyboard used by a fast food chain in which the name of each item on the menu appears on a proximity-switch plastic overlay.

CRT displays allow the operator to enter, view, and edit information; editing information with a CRT display is typically much faster than if a mechanical printer were used. The most popular CRT displays exhibit 24 lines of up to 80 characters each. Other size variations include 12 lines of up to 80 characters and a full page display (approximately 66 lines of up to 80 characters). Other features found on CRT displays include blinking, half-intensity, upper- and lowercase character sets, foreign-character sets, variable character sizes, graphic (line-drawing) character sets, and multicolor displays.

Printers for data terminals generally are classified into three types—dot-matrix, inkjet, and laser printers. Dot-matrix printers are lowest in cost but are also lowest in print speed and print quality. Laser printers are highest in cost but provide the highest print speed and quality. Inkjet devices are intermediate in cost, speed, and quality.

Storage devices for data terminals include electronic (RAM, ROM), magnetic (floppy disk, hard disk), and optical (CD-ROM). Optical storage is emerging as the most popular medium for storage and retrieval of large amounts of information, particularly in applications such as online encyclopedias, image retrieval, and large databases.

Controllers interconnect the channel interface and the various terminal components and make them interact to perform the terminals' specific functions. They also perform other functions, such as recognizing and acting on protocol commands, e.g., polling and selecting sequences in selective calling systems; code translation; and error detection and correction. Controller designs typically use microprocessors to increase their functions, including programmability, which greatly increases terminal versatility.

Standardized codes, protocols, and interfaces provide a uniform framework for the transmission and reception of data by data terminals. Codes and some character-oriented protocols and interfaces are discussed in Chap. 1.

IBM's binary synchronous communication (bisynch) protocol has been implemented widely. It accommodates half-duplex transmission and is character-oriented. Bit-oriented protocols that provide both half- and full-duplex transmission are also in widespread use. The American National Standards Institute's (ANSI) Advanced Data Communications Control Procedures (ADCCP), the International Organization for Standardization's High Level Data Link Control (HDLC), and Synchronous Data Link Control (SDLC) are the current standards for bit-oriented protocols. Although there are differences, they are generally compatible, and the trend is for other bit-oriented protocols to establish compatibility with them.

Display Terminals (Text and Graphics)

Terminals for the entry and electronic display of textual and or graphic information generally are used for communicating between people and computers and, increasingly, between people. Typical application areas include computer program preparation, data entry, inquiry-response, inventory control, word processing, financial transactions, computer-aided design, and electronic mail. A terminal is divided functionally into control, display, user input, and communication-line-interface portions (see Fig. 17.7.15). The first three portions are detailed in the following sections. Interface to a communication line is typically through either an analog or digital modem.

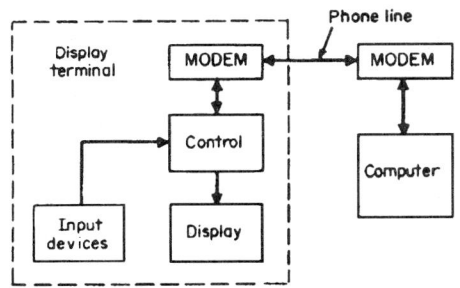

FIGURE 17.7.15 Display-terminal organization.

Terminals have differing amounts of control or decision-making logic built into them. A popular classification scheme is based on terminal control capabilities. A nonintelligent ("dumb") terminal is a basic I/O and control device. Although it may have some information buffering, it must rely on the computer to which it is connected for most processing of any external information and editing of output displays. A "smart" terminal has both information entry and editing capabilities, and although it too is generally connected to a computer, it can perform information processing locally. The terminal usually contains a microcomputer that is programmed by the terminal manufacturer to meet the general and special needs of a user. An "intelligent" terminal has a microcomputer that can be programmed by a user to meet user-specific needs (as mentioned earlier, it is very common to use a PC as a smart or intelligent terminal). Terminals can be connected to a supporting computer either directly or through a cluster controller that supervises and services the data-communication needs of a number of terminals.

The most common terminal display device is a cathode-ray tube (CRT) similar to that used in a home TV receiver. In a home TV receiver, the refreshing is done by the broadcast station, which sends 30 complete images per second. In a terminal, the refreshing must be provided by the control logic from information stored electronically in the terminal. This storage may be in a separate electronic memory or memory that is part of the microcomputer. In either case the microcomputer can address, enter, or change the information.

Flat panel displays are an alternative to the CRT. Since these devices require much less space and power, they are ideal for laptop and notebook terminals and computers. The most common technology is liquid-crystal displays (LCD). For LCD, liquid-crystal molecules in most displays align themselves parallel to an electric field and lie flat when no field is present. In another type of display, the crystals tilt in different directions in response to a field. Depending on the device construction and the type of liquid crystals used, some of those orientations allow light to pass; others block it out. The result is either a dark image on a light background or the reverse.

Most display terminals have a typewriterlike keyboard and a few extra buttons as input devices for the entry of textual and control information. Text is entered at a position on the display indicated by a special symbol called a *cursor*. The control moves this cursor along much as a typewriter head moves along as text is entered via the keyboard. Different manufacturers use different cursors (underlines, blinking squares, video-inverted characters).

To allow editing of existing images or flexible entry of additions, auxiliary control devices are provided to change the cursor position. The simplest is a set of five buttons, four of which move the cursor up, down, right, or left one character position from the current position. The fifth indicates that the cursor is to be "homed" to the upper left-hand corner of the image, where textual entry usually starts. Another popular position-controlling device is a joystick, a lever mounted on gimbals. Movement of the top of the lever is measured by potentiometers attached to the gimbals. The control senses the potentiometer outputs and moves the cursor correspondingly.

Another form of control for the cursor is the mouse. Movement of the mouse on the desktop provides a direct relationship with the movement of the cursor on the screen. A precision mouse will typically use either a magnetic or optical tablet that maps the position of the mouse on the tablet to a particular spot on the screen. A less-accurate form of the mouse uses friction between the desk surface and a ball contact on the bottom of the mouse to indicate the *relative* motion of the cursor. This form of mouse typically uses two optically sensed shafts mounted perpendicular to each other to detect movement of the ball contact. A mouse can have one to three buttons on top.

Additional input methods include track balls, pressure-sensitive pads, and even eye-position trackers. These find utility in specific areas such as artwork generation, computer image processing, industrial applications, training, and games.

Terminals intended for display of complex graphic information, e.g., for computer-aided design, generally have either tablet-stylus devices, light pens, or mice for input of new information or indication of existing information. One popular tablet-stylus has a surface area under which magneto-strictive waves are alternately propagated horizontally and vertically. A stylus with a coil pickup in the tip senses the passing of a wave under the stylus position. Electronic circuitry measures the time between launching of a wave and its sensing by the stylus and computes the position of the stylus from that time and the known velocity of the wave.

A light pen senses when light is within its field of view. In a raster-scan display, the time from the start of a displayed image until the light-pen signal is received indicates where the pen is in the raster-scan pattern and, with suitable scaling factors, gives the position of the pen over the image. In a directed-beam display, pen-position locating is more complicated. The centering of a special tracking pattern under the pen is sensed, and through a feedback arrangement controlled by the terminal computer the pattern is moved to keep it centered. The position of the pattern is then the same as the pen position.

Multimedia terminals are now becoming more commonplace. These terminals combine text, image, and voice capabilities. In the office environment, tighter coupling between data, e-mail, fax, and voice will take place. Similarly, with more of the workforce working at home, either full-time or part-time, multimedia services in the home will be required.

Data Transmission on Analog Circuits

The devices used for DTE generate digital signals. These signals are not compatible with the voice circuits of the public telephone network, partly because of frequency range but more importantly because these circuits are likely to produce a small frequency offset in the transmitted signal. The offset causes drastic changes in the received waveform. This effect, while quite tolerable in voice communications, destroys the integrity of individual pulses of a baseband signal. Compatible transmission is obtained by modulating a carrier frequency within the channel passband by the baseband signal in a modem (modulator-demodulator).

For a band-limited system, Nyquist showed that the maximum rate for sending noninterfering pulses is two pulses (usually called symbols) per second per hertz of bandwidth. The bit rate depends on how these pulses are encoded. For example, a two-level system transmits 1 b with each pulse. A four-level system transmits 2 b per pulse, an eight-level system transmits 3 b, and so forth. Unfortunately, we cannot go to an arbitrarily large number of levels because, assuming that the total power is limited, the levels will become so closely spaced that the random disturbances in the transmission medium will make one level indistinguishable from the next. Shannon's fundamental result states that there is a maximum rate, called the *channel capacity,* up to which one can send information reliably over a given channel. This capacity is determined by the random disturbances on the channel. If these random disturbances can be characterized as white Gaussian noise, the channel capacity C is given by

$$C = W \log_2 (1 + S/N)$$

where W is the bandwidth of the channel and S and N are the average signal and noise powers, respectively.

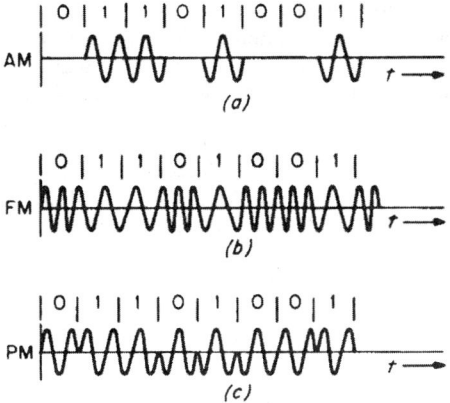

FIGURE 17.7.16 Binary (*a*) amplitude-, (*b*) frequency-, and (*c*) phase-modulated carrier waves.

With today's more elaborate commercial modulation and coding techniques it is possible to transmit data at approximately 70 percent of the capacity of the channel. With the most complex modulation and coding techniques, it is possible to get very close to the capacity of the channel, but this is achieved at the expense of considerable processing complexity and delay.

Data transmission employs all three modulation methods (amplitude, frequency, and phase) plus combinations of them. A description of these methods is given in Sec. 12.

An on-off amplitude-modulated (AM) signal is shown in Fig. 17.7.16*a*. This modulation scheme is little used for voice-band modems because of its inferior performance compared to other modulation schemes.

An example of a binary frequency-modulated (FM) carrier wave, sometimes called frequency-shift keying (FSK), is shown in Fig. 17.7.16*b*. While FM or FSK requires somewhat greater bandwidth than AM for the same symbol rate, it gives much better performance in the presence of impulse noise and gain change. It is used extensively in low- and medium-speed (voice-band) telegraph and data systems.

A binary phase-modulated (PM) carrier wave is shown in Fig. 17.7.16*c*, where a phase change of 180° is depicted. However, this modulation method is usually employed in either four-phase or eight-phase systems. In a four-phase-change system, the binary bits are formed in pairs, called *dibits*. The dibits determine the phase change from one signal element to the next. The four phases are spaced 90° apart. In effect, this method employs a four-state signal, and such a system is inherently capable of greater transmission speeds for the same bandwidth, as is obvious from the Nyquist-Shannon criteria stated above. With improvement of voice channels, the four-phase (quaternary-phase-modulated) scheme is being employed increasingly for medium-speed (voice-band) data transmission systems to give higher transmission speeds than FM for the same bandwidth. The system is useful only in synchronous transmission.

In present-day modems FSK is the preferred modulation technique for bit rates below 1800 b/s. At 2400 b/s, the commonly used modulation technique is PSK using four phases, and at 4800 b/s it is PSK using eight phases. The latter requires the use of an adaptive equalizer, an automatically adjustable filter that minimizes the distortion of the transmitted pulses resulting from the imperfect amplitude and phase characteristics of the channel. At 9600 b/s, the preferred modulation technique is quadrature amplitude modulation, a combination of amplitude and phase modulation.

Above 9600 b/s (e.g., 19,200 and 28,800 b/s) the preferred modulation technique is quadrature amplitude modulation combined with a coding technique called trellis-coded modulation (TCM). TCM provides much improved performance at the expense of more complex implementation. CCITT Recommendation V.34 specifies this type of modulation for 19.2 kb/s.

Data Transmission on Digital Circuits

The techniques outlined in the previous section apply for transmission over the analog telephone network, sometimes referred to as public switched telephone network (PSTN). While local access to PSTN is over analog facilities (typically, twisted copper wires), the interexchange and long-distance telephone network consists principally of digital facilities. Analog signals are converted via pulse-code-modulation into digital signals for transmission over these digital facilities. If digital access to the long-distance digital facilities is provided, then end-to-end digital transmission is possible. Data signals then need not go through the digital to analog conversion outlined in the previous section, but can remain digital end-to-end (from source to destination).

An early example of an end-to-end digital system was AT&T's Digital Data System, which was deployed during the mid-1970s. Here, digital access lines together with Central Office digital multiplexers enabled transmission at rates of 2.4, 4.8, 9.6, and 56 kb/s.

Switched network digital systems are provided today via ISDN, which provide circuits at 64 and 128 kb/s. Higher bandwidth circuits and services are now being offered by various entities. Modern systems use packet-switching techniques rather than circuit-switching to enable very efficient use of the digital transmission facilities. These packet-switched systems typically use frame relay and asynchronous transfer mode (ATM) protocols.

In addition to the advances in the long-distance marketplace, changes are also taking place in the local access area. Regional Bell operating companies as well as alternate access providers are beginning to deploy local distribution systems using optical fiber and coax cable combinations. While the initial thrust for these systems is to provide multichannel video to the home, the broadband facilities also provide an excellent vehicle for broadband multimedia data services to the home.

Local Area Networks

As with wide area networks, a local network is a communication network that interconnects a variety of devices and provides a means for information exchange among those devices. See Stallings and Van Slyke (1994). There are several key distinctions between local networks and wide area networks:

1. The scope of the local network is small, typically a single building or a cluster of buildings. This difference in geographic scope leads to different technical solutions.

2. It is usually the case that the local network is owned by the same organization that owns the attached devices.

3. The internal data rates of local networks are much greater than those of wide area networks.

The key elements of a local area network are the following:

- Topology: bus or ring
- Transmission medium: twisted pair, coaxial cable, or optical fiber

TABLE 17.7.1 LAN Technology Elements

Element	Options	Restrictions	Comments
Topology	Bus	Not with optical fiber	No active elements
	Ring	Not CSMA/CS or broadband	Supports fiber, high availability with star wiring
Transmission medium	Unshielded twisted pair	—	Inexpensive; prewired; noise vulnerability
	Shielded twisted pair	—	Relatively inexpensive
	Baseband coaxial cable	—	—
	Broadband coaxial cable	Not with ring	High capacity; multiple channels; rugged
	Optical fiber	Not with bus	Very high capacity; security
Layout	Linear	—	Minimal cable
	Star	Best limited to twisted pair	Ease of wiring; availability
Medium access control	CSMA/CD	Bus, not good for broadband or optical fiber	Simple
	Token passing	Bus or ring, best for broadband	High throughout, deterministic

Source: Stallings, W. and Van Slyke, R. "Business Data Communications," Macmillan College Publishing Company, 1994.

- Layout: linear or star
- Medium access control: CSMA/CD or token passing

Together, these elements determine not only the cost and capability of the LAN but also the type of data that may be transmitted, the speed and efficiency of communications, and even the kinds of applications that can be supported. Table 17.7.1 provides an overview of these elements.

Data Communication Equipment (DCE) Trends

Most modern businesses of significant size depend heavily on their data-communications networks. Large businesses usually have a staff of specialists whose job it is to manage the network. To assist in this network-management function, DCE manufacturers have provided various testing capabilities in their products. Sophisticated DCEs now are capable of automatically monitoring their own "health" and reporting it to the centralized location where the network management staff resides. These new capabilities are frequently implemented through the use of microprocessors. Some DCEs can also establish whether a trouble is in the modem itself or the interconnecting channel. If it is the channel that is in trouble, equipment is available that automatically sets up dialed connections to be used as backup for the original channel. Another capability is to send in a trouble report automatically when a malfunction is detected.

CHAPTER 17.8
MEMS FOR COMMUNICATION SYSTEMS

D. J. Young

MEMS FOR WIRELESS COMMUNICATIONS

Introduction

The increasing demand for wireless communication applications, such as cellular telephony, cordless phone, and wireless date networks, motivates a growing interest in building miniaturized wireless transceivers with multistandard capabilities. Such transceivers will greatly enhance the convenience and accessibility of various wireless services independent of geographic location. Miniaturizing current single-standard transceivers, through a high-level of integration, is a oritical step toward building transceivers that are compatible with multiple standards. Highly integrated transceivers will also result in reduced package complexity, power consumption, and cost. At present, most radio transceivers rely on a large number of discrete frequency-selection components, such as radio-frequency (RF) and intermediate-frequency (IF) band-pass filters, RF voltage-controlled oscillators (VCOs), quartz crystal oscillators, and solid-state switches, to perform the necessary analog signal processing. These off-chip devices severely hinder transceiver miniaturization. MEMS technology, however, offers a potential solution to integrate these discrete components onto silicon substrates with microelectronics, achieving a size reduction of a few orders of magnitude. It is therefore expected to become an enabling technology to ultimately miniaturize radio transceivers for future wireless communications.

MEMS Variable Capacitors

Integrated high-performance variable capacitors are critical for low noise VCOs, antenna tuning, tunable matching networks, and so on. Capacitors with high quality factor (Q), large tuning range, and linear characteristics are crucial for achieving system performance requirements. On-chip silicon *pn* junction and MOS-based variable capacitors suffer from low quality factors, limited tuning range, and poor linearity, and are thus inadequate for building high-performance transceivers. MEMS technology has demonstrated monolithic variable capacitors achieving stringent performance requirements. These devices typically reply on an electrostatic actuation method to vary the air gap between a set of parallel plates or vary the capacitance area between a set of conductors or mechanically displace a dielectric layer in an air-gap capacitor. Improved tuning ranges have been achieved with various device configurations. Capacitors fabricated through using metal and metalized silicon materials have demonstrated superior quality factors compared to solid-state semiconductor counterparts. Besides the above advantages, micromachined variable

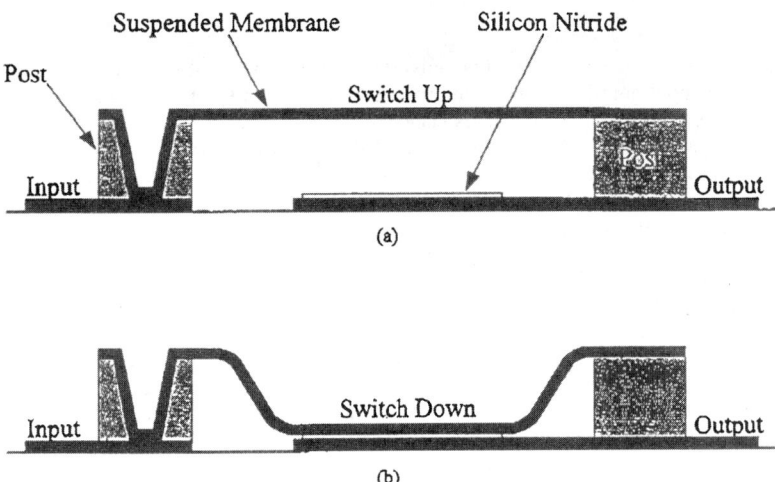

FIGURE 17.18.1   Micromachined RF switch: (*a*) switch up; (*b*) switch down.

capacitors suffer from a reduced speed, potentially a large tuning voltage, and mechanical thermal vibration commonly referred to as Brownian motion, which deserves great attention when used to implement low phase noise VCOs.

MEMS Switches

The microelectromechanical switch is another potentially attractive miniaturized component offered by micromachining technologies. These switches offer superior electrical performance in terms of insertion loss, isolation, linearity, and so on and are intended to replace off-chip solid-state counterparts switching between the receiver and transmitter signal paths. They are also critical for building phase shifters, tunable antennas, and filters. The MEMS switches can be characterized into two categories: capacitive and metal-to-metal contact types. Figure 17.8.1 presents the cross-sectional schematic of an RF MEMS switch. The device consists of a conductive membrane, typically made of aluminum or gold alloy suspended above a coplanar electrode by a few micrometers air gap. For RF or microwave applications, actual metal-to-metal contact is not necessary; rather, a step change in the plate-to-plate capacitance realizes the switching function. A thin silicon nitride layer with a thickness on the order of 1000 Å is typically deposited above the bottom electrode. When the switch is in on-state, the membrane is high, resulting in a small plate-to-plate capacitance; hence, a minimum high-frequency signal coupling (high isolation) between the two electrodes. The switch in the off-state with a large enough applied dc voltage, however, provides a large capacitance owing to the thin dielectric layer, thus causing a strong signal coupling (low insertion loss). The capacitive switch consumes near-zero power dissipation, attractive for low power portable applications. Superior linearity performance has also been demonstrated because of the electromechanical behavior of the device. Metal-to-metal contact switches are important for interfacing large bandwidth signals including dc. This type of device typically consists of a cantilever beam or clamped-clamped bridge with a metallic contact pad positioned at the beam tip or underneath bridge center. Through an electrostatic actuation, a contact can be formed between the suspended contact pad and an electrode on the substrate underneath. High performance on a par with the capacitive counterparts has been demonstrated. Microelectromechanical switches, either capacitive or metal contact versions, exhibit certain drawbacks including low switching speed, high actuation voltage, sticking phenomena due to dielectric charging, metal-to-metal contact welding, and so on, thus limiting device lift time and power handling capability. Device packaging with inert atmosphere (nitrogen, argon, and so on) and low humidity is also required.

MEMS Resonators

Microelectromechanical resonators based on polycrystalline silicon comb-drive fingers, suspended beams, and center-pivoted dick configurations have been proposed for performing analog signal processing. These micro-resonators can be excited into mechanical resonance through an electrostatic drive. The mechanical motion will cause a device capacitance change resulting in an output electrical current when a proper dc bias voltage is applied. This output current exhibits the same frequency as the mechanical resonance, thus achieving an electrical filtering function through the electromechanical coupling. The resonators can obtain high-quality factors close to 10,000 in vacuum with operating frequencies above 150 MHz reported in literatures and size reduction by a few orders of magnitude compared to discrete counterparts. These devices with demonstrated performance are attractive for potentially implementing low-loss IF band-pass filters for wireless transceivers design. Future research effort is needed to increase the device operating frequency up to gigahertz (GHz) range. As with other MEMS devices, the micromachined resonators also have certain drawbacks. For example, vacuum packaging is required to achieve a high-quality factor for building low loss filters. The devices may also suffer from a limited dynamic range and power-handling capability. The mechanical resonant frequency is strongly dependent on the structure dimensions and material characteristics. Thus, a reliable tuning method is needed to overcome the process variation effect and inherent temperature sensitivity.

Micromachined Inductors

Integrated inductors with high-quality factors are the key components for implementing low noise oscillators, low loss matching networks, and so forth. Conventional on-chip spiral inductors suffer from limited quality factors around 5 at 1 GHz, an order of magnitude lower than the required values from discrete counterparts. The poor performance is mainly caused by substrate loss and metal resistive loss at high frequencies. Micromachining technology provides an attractive solution to minimize these loss contributions; hence enhancing the device quality factors. Q factors around 30 have been achieved at 1 GHz matching the discrete component performance. Three-dimensional coil inductors have been fabricated on silicon substrates by micromachining techniques. Levitated spiral inductors have also been demonstrated. All these devices exhibit three common characteristics: (1) minimizing device capacitive coupling to the substrate, (2) reducing winding resistive loss through employing highly conductive materials, and (3) nonmovable structures upon fabrication completion.

MEMS FOR OPTICAL COMMUNICATIONS

Introduction

High-speed communication infrastructures are desirable for transferring and processing real-time information such as voice and video. Optical fiber communication technology has been identified as the critical backbone to support such systems. High-performance optical data switching network, which routes various optical signals from their sources to destinations, is one of the key building blocks for system implementation. At present, optical signal switching is performed by using hybrid optical-electronic-optical (O-E-O) switches. These devices convert incoming lights from input fibers to electrical signals first and then route them to the proper output ports after signal analyses. At the output ports, the electrical signals are converted back to streams of photons or optical signals for further transmission over the fibers to their next destinations. The O-E-O switches are expensive to build, integrate, and maintain. Furthermore, they consume a substantial amount of power and introduce additional latency. It is therefore highly desirable to develop all-optical switching network in which optical signals can be routed without intermediate conversion into electrical form, thus minimizing power dissipation and system delay. While a number of approaches are being considered for building all-optical switches, MEMS technology is attractive for providing arrays of tiny movable mirrors, which can redirect incoming beams from input fibers to corresponding output fibers. These micromirrors can be batch fabricated using silicon micromachining technologies, achieving a low-cost integrated solution. A significant reduction in power dissipation is also expected.

MEMS Mirrors

Various micromachined mirrors have been developed over the years. They can be typically divided into two categories: (1) out-of-plane mirrors and (2) in-plane mirrors. The out-of-plane mirrors are usually fabricated by polycrystalline silicon surface micromachining techniques. After sacrificial release, the mirror structures can be folded out of the substrate and position secured by silicon hinges. The mirror surface can be moved by an electrostatic vibromotor, comb-drive fingers, and other electrostatic means. These mirrors can achieve one degree of freedom and thus are attractive for routing optical signals in a two-dimensional switching matrix and also for raster-scanning display applications. The in-plane mirrors are typically fabricated using a thick single crystal silicon layer on the order of a few tens micrometers from a SOI wafer by deep RIE and micromachining techniques. The thick structural layer minimizes mirror warping, critical for high-performance optical communication applications. Self-assembly technique relying on deposited film stress has also been employed to realize lifted-up micromirror structures. The mirror position can be modulated by a vertical actuation of comb-drive fingers and electrostatic pads or a lateral push-pull force. Micromirrors with two degrees of freedom have also been demonstrated by similar techniques. These mirrors with an analog actuation and control scheme are capable of directing optical beams to any desired position, and are thus useful for implementing large three-dimensional optical switching arrays to establish connections between any set of fibers in the network.

DIGITAL COMPUTER SYSTEMS

**Murray J. Haims, Stephen C. Choolfaian, Daniel Rosich,
Richard E. Matick, William C. McGee,
Benton D. Moldow, Robert A. Myers,
George C. Stierhoff, Claude E. Walston**

No other invention in the twentieth century has impacted virtually every aspect of our life like the digital computer. The hardware along with software have forever changed how we work with everything from the automobile to our checking accounts. It goes to the deepest parts of our oceans and planet to well beyond our solar system. In spite of how pervasive and sophisticated they are, they remain fairly simple and straight forward devices (although anyone who works with them regularly could easily argue with us on this point). An understanding of the material in this section should allow us to work with many hardware and software challenges.

Computers are organized into basic activities. This is referred to as the architecture of the computer as well as the software. The first part to look at is data processing. Data are merely the representation of something in the real world such as money by their binary equivalent. After we look at the rest of the architecture of the computer we will look at how these can be varied in the design process to produce different types of computer activities.

Next we look at how software can control the interactions of the hardware to produce the desired results. Keep in mind that even though there are a number of different software packages that we use, they all do essentially the same things and all that is different is really just the syntax of the specific application.

Future trends in this field will essentially focus on two areas. The first will be in the area of how we interact with computer and input and output data. We already have the ability to use handwritten interaction and voice interaction. It is in the area of verbal interaction with the computer where we will most likely see the end of the keyboard as we know it. Current voice interactive software can achieve accuracies of 97 percent with training. This compares favorably with keyboard input and is decidedly faster for everyone but the most skilled typist.

The other area will be in the basic architecture of both the hardware and software. Increasingly we are developing portions of the computer into parallel processors that will enable a major shift in the way software will be developed. Eventually we will have computers that come from a manufacturer with a defined hardware and software architecture. Once the computer is turned on it will begin reconfiguring its hardware and software to adapt to the needs of the user.   C.A.

In This Section:

Section Bibliography:

Aiken, A. H., and G. M. Hopper "The automatic sequence controlled calculator," *Elec. Eng.,* 1948, Vol. 65, p. 384.

Babbage, C. "Passages from the Life of a Philosopher," Longmans, 1864.

Babbage, H. P. "Babbage's Calculating Engines," Spon, 1889.

Bjorner, D., E. F. Codd, K. L. Deckert, and I. L. Traiger "The GAMMA-O *n*-ary relational data base interface specification of objects and operations," *Research Report RJ1200*, IBM Research Division, 1973.

Black, V. D. "Data Communications and Distributed Networks," 2nd ed., Prentice Hall, 1987.

Boole, G. "The Mathematical Analysis of Logic," 1847, reprinted Blackwell, 1951.

Brainerd, J. G., and T. K. Sharpless "The ENIAC," *Elec. Eng.,* February 1948, pp. 163–172.

Brooks, F. P., Jr. "The Mythical Man-Month," Addison-Wesley, 1975.

Codd, E. F. "A data base sublanguage founded on the relational calculus," *Proc. ACM SIGFIDET Workshop on Data Description, Access, and Control*, Association for Computing Machinery, 1971.

Cypser, R. J. "Communications Architecture for Distributed Systems," Addison-Wesley, 1978.

Deitel, H. M. "Operating Systems," 2nd ed., Addison-Wesley, 1990.

Dijkstra, E. W. *Commun. Ass. Comput. Mach.*, 1968, Vol. II, No. 3, p. 341.

Dijkstra, E. W. "A Discipline of Programming," Prentice Hall, 1976.

Enslow, P. H., Jr. "Multiprocessor organization—a survey," *Comput. Surv.,* March 1977, Vol. 9, No. 1, pp. 103–129.

Feldman, J. M., and C. T. Retter "Computer Architecture, A Designer's Text Based on a Generic RISC," McGraw-Hill, 1994.

Flynn, M. J. "Some computer organizations and their effectiveness," *IEEE Trans. Comput.*, September 1972, Vol. C-21, No. 9, pp. 948–960.

Gear, C. W. "Computer Organization and Programming," 2nd ed., McGraw-Hill, 1978.

Gilmore, C. M. "Microprocessor Principles and Applications," McGraw-Hill, 1989.

Godman, J. E. "Applied Data Communication," Wiley, 1995.

Hamming, W. R. "Error detecting and error correcting codes," *Bell Syst. Tech. J.*, 1947, Vol. 29.

Hancock, L., and M. Krieger "The C Primer," McGraw-Hill, 1986.

Hellerman, H. "Digital Computer System Principles," 2nd ed., McGraw-Hill, 1973.

Horowitz, E., and S. Sani "Fundamentals of Data Structures," Computer Science Press, 1976.

Lam, S. L. "Principles of Communication and Networking Protocols," Computer Society Press, 1984.

Mano, M. M. "Computer System Architecture," 2nd ed., Prentice Hall, 1982.

Mano, M. M. "Digital Logic and Computer Design," Prentice Hall, 1979.

Morris, D. C. "Relational Systems Development," McGraw-Hill, 1987.

O'Connor, P. J. "Digital and Microprocessor Technology," 2nd ed., Prentice Hall, 1989.

Smith, J. T. "Getting the Most from TURBO PASCAL," McGraw-Hill, 1988.

Stallings, W., and R. Van Slyke "Business Data Communications," 2nd ed., Macmillan, 1994.

Van de Goor, A. J. "Computer Design and Architecture," Addison-Wesley, 1985.

Wegner, P. "Programming Languages, Information Structures and Machine Organization," McGraw-Hill, 1968.

Whitten, J. L., L. D. Bentley, and V. M. Barlow "Systems Analysis and Design Methods," 3rd ed., Irwin, 1994.

CHAPTER 18.1
COMPUTER ORGANIZATION

PRINCIPLES OF DATA PROCESSING

Memory, Processing, and Control Units

The basic subsystems in a computer are the input and output sections, the store, arithmetic logic unit (processing unit), and the control section. Each unit is described in detail in this chapter. Generally, a computer operates as follows: An external device such as a disc file delivers a program and data to specific locations in the computer store. Control is then transferred to the stored program, which manipulates the data and the program itself to generate the output. These output data are delivered to a device, such as a CD-ROM/RW, DVD, printer, or display, where the information is used in accordance with the purpose of the digital manipulation.

Historical Background

There has been a line of development of mechanical calculator components, beginning with Babbage in the early 1800s and leading to a variety of mechanical desk and larger mechanical calculators. Another line of development has used relays as computing circuit elements. Today's computers have benefited from these lines of development, but especially they are based on electronic components, the vacuum tube, and the transistor. The transistor, first described by Shockley, Bardeen, and Brattain in 1974, began a line of development that is today characterized by the miniaturization and low-power operation of very large-scale integration (VLSI).

VLSI permits the interconnection of large numbers of computing elements by means of microscopic layered structures on a semiconductor substrate (usually silicon) or *chip* sometimes as small as $1/4$ in. square. Since the entire arithmetic and logic circuit of a computer can be built on a single chip (microprocessor), computers incorporating VLSI are often called *minicomputers* or *microcomputers*.

Binary Numbers

Most transistors display random variations of their operating parameters over relatively wide limits. Similarly, passive circuit elements experience a considerable degree of variation, and noise and power-supply variations, and so forth, limit the accuracy with which quantities can be represented. As a result, the preferred method is to use each circuit in the manner of an on-off switch, and representation of quantities in a computer is thus almost always on a *binary* basis. Figure 18.1.1 shows the binary numbers equivalent to the decimal numbers between 1 and 10. Figure 18.1.2 shows the addition of binary 6 to binary 3 to obtain binary 9.

The process of addition can be dissected into digital, logical, or boolean operations upon the binary digits, or bits (b). For example, a first step in the procedure for addition is to form the so-called EXCLUSIVE-OR addition between bits in each column. This function of two binary numbers is expressed in Fig. 18.1.3a in tabular form. This table is called a *truth table*. In Fig. 18.1.3b is the table used to generate the *carries* of a binary

Decimal	Binary
0	0
1	1
2	10
3	11
4	100
5	101
6	110
7	111
8	1000
9	1001
10	1010

FIGURE 18.1.1   Decimal and binary numbers between 0 and 10.

Decimal	Binary
6	110
+3	+ 11
9	1001

FIGURE 18.1.2   Addition of 6 and 3 in decimal and binary.

	0	1
0	01	
1	01	10

Binary addition table excluding the carry
(*a*)

	0	1
0	01	
1	00	01

Binary addition carry generation
(*b*)

FIGURE 18.1.3   Addition tables for decimal and binary numbers. The binary addition table (*a*) is called the EXCLUSIVE-OR or module-2 truth table; the carry table (*b*) performs the AND or *intersection* operation.

bit from one column to another. This latter function of two binary numbers is variously called the AND function, *intersection*, or *product*. The entries at each intersection in each table are the result of the combination of two binary numbers in the respective row and column. Figure 18.1.3 also shows the decimal addition tables. They illustrate the relative simplicity of the binary number system.

The names truth table and logical function arise from the fact that such manipulations were first developed in the *sentential calculus*, a subsection of the calculus of logic, dealing with the truth or falsity of combinations of true or false sentences.

Binary Encoding.   Information in a digital processing machine is not restricted to numerical information since a different specific numeric code can be assigned to each letter of the alphabet. For example, A in the *EBCDIC code* (see next paragraph) is given by the binary sequence 11000010. When alphanumeric information is specified, such a code sequence represents the symbol A, but in the numeric context the same entry is the binary number equal to decimal 194.

Computer Codes

Alphanumeric information is stored in a computer via coded binary bits. Some of the more useful codes are:

ASCII (American Standard Code for Information Interchange), a seven-level alphanumeric code comprising 32 control characters, an uppercase and lowercase alphabet, numerals, and 34 special characters (Fig. 18.1.4). This code is used in personal computers and non-IBM machines. "A" in ASCII is given by the 7-bit binary codes 1000001.

BCDIC (binary-coded decimal interchange code), a six-level alphanumeric code that provides alphabetic (caps), numeric, and 28 special characters.

Binary code, a representation of numbers in which the only two digits used are 0 and 1, and each position's value is twice that of its right-hand neighbor (with the rightmost place having a value of 1).

Binary-coded decimal (BCD) code, in which the first ten 4-bit (hexadecimal) codes are used for the decimal digits, and each nibble represents one decimal place. Codes A through F are never used.

EBCDIC (expanded BCD interchange code), an eight-level alphanumeric code comprising control codes, an uppercase and lowercase alphabet, numerals, and special characters (Fig. 18.1.5). This code is used in IBM mainframe computers.

Gray code, a binary code that does not follow the positional notation of true binary code. Only one bit changes from any Gray number to the next.

Hexadecimal byte code, a two-digit hexadecimal number for each byte, with values ranging from 00 to FF.

Hexadecimal code, a base-16 number code that uses the letters A, B, C, D, E, and F as one-digit numbers for 10 through 15.

Hollerith code, a 12-level binary code used on punchcards to represent alphanumeric characters. Holes on the punchcard are ones, unpunched positions are zeros.

"back" 5 bits	0 0 0 0 0	0 0 0 0 1	0 0 0 1 0	0 0 0 1 1	0 0 1 0 0	0 0 1 0 1	0 0 1 1 0	0 0 1 1 1	0 1 0 0 0	0 1 0 0 1	0 1 0 1 0	0 1 0 1 1	0 1 1 0 0	0 1 1 0 1	0 1 1 1 0	0 1 1 1 1	1 0 0 0 0	1 0 0 0 1	1 0 0 1 0	1 0 0 1 1	1 0 1 0 0	1 0 1 0 1	1 0 1 1 0	1 0 1 1 1	1 1 0 0 0	1 1 0 0 1	1 1 0 1 0	1 1 0 1 1	1 1 1 0 0	1 1 1 0 1	1 1 1 1 0	1 1 1 1 1
"front" 2 bits																																
00	Control code for typewriter control which don't print anything																															
01	b	!	"	#	$	%	&	'	(	)	*	+	,	-	.	/	0	1	2	3	4	5	6	7	8	9	:	;	<	=	>	?
10	@	A	B	C	D	E	F	G	H	I	J	K	L	M	N	O	P	Q	R	S	T	U	V	W	X	Y	Z	[	/	]	^	←
11	'	a	b	c	d	e	f	g	h	i	j	k	l	m	n	o	p	q	r	s	t	u	v	w	x	y	z	¦	|	¦	~	DEL

FIGURE 18.1.4   American Standard Code for Information Interchange (ASCII).

Most computers are designed to work internally in binary fashion but at each juncture of input or output to translate the codes, either by programming or by hardware, so as to accept and offer decimal numeric information. Such systems are complicated, and failures in the coding and decoding system can prevent interaction with the program.

Communication needs lead to a compromise between the human requirement for a decimal system and the machine requirement for binary. An example is the use of the base-16 (hexadecimal) system, which is relatively amendable to human recognition and manipulation. The binary code is broken into 4-bit groups. Each 4-bit group is represented by a decimal numeral or letter, as indicated in Fig. 18.1.6. In this case, 0011 is three, and 1010 is ten or A.

Error-Correction Codes

Though the circuits in modern digital systems have reached degrees of reliability undreamed of in the relatively recent past, errors can still arise. Hence it is desirable to detect and, if possible, correct such errors. It is

Some characters of the expanded B.C.D. interchange code (EBCDIC)

No #	Numeric 1	Numeric 2	Numeric 3	Numeric 4	Numeric 5	Numeric 6	Numeric 7	Numeric 8	Numeric 9	Zone bits	Binary code
No zone ∅	1	2	3	4	5	6	7	8	9	111	$D \cdot C \cdot B \cdot A$
12 zone &	A	B	C	D	E	F	G	H	I	1100	$D \cdot C \cdot \bar{B} \cdot \bar{A}$
11 zone –	J	K	L	M	N	O	P	Q	R	1101	$D \cdot C \cdot \bar{B} \cdot A$
0 zones 0	/	S	T	U	V	W	X	Y	Z	1110	$D \cdot C \cdot B \cdot \bar{A}$
12 • 0 zones	a	b	c	d	e	f	g	h	i	1000	$D \cdot \bar{C} \cdot \bar{B} \cdot \bar{A}$
12 • 11 zones	j	k	l	m	n	o	p	q	r	1001	$D \cdot \bar{C} \cdot \bar{B} \cdot A$
11 • 0 zone	?	s	t	u	v	w	x	y	z	1010	$D \cdot \bar{C} \cdot B \cdot \bar{A}$
Numeric bits	0001	0010	0011	0100	0101	0110	0111	1000	1001		
Binary code	$\bar{8} \cdot \bar{4} \cdot 2 \cdot 1$	$\bar{8} \cdot \bar{4} \cdot 2 \cdot \bar{1}$	$\bar{8} \cdot 4 \cdot \bar{2} \cdot \bar{1}$	$\bar{8} \cdot 4 \cdot \bar{2} \cdot 1$	$\bar{8} \cdot 4 \cdot 2 \cdot \bar{1}$	$\bar{8} \cdot 4 \cdot 2 \cdot 1$	$\bar{8} \cdot \bar{4} \cdot \bar{2} \cdot \bar{1}$	$8 \cdot \bar{4} \cdot \bar{2} \cdot \bar{1}$	$8 \cdot \bar{4} \cdot \bar{2} \cdot 1$		

FIGURE 18.1.5   Hollerith/EBCDIC code chart.

Names of 4 bit groups

Digit name	4-bit binary	Digit symbol	Digit name	4-bit binary	Digit symbol
Zero	0000	0	Eight	1000	8
One	0001	1	Nine	1001	9
Two	0010	2	Ten	1010	A
Three	0011	3	Eleven	1011	B
Four	0100	4	Twelve	1100	C
Five	0101	5	Thirteen	1101	D
Six	0110	6	Fourteen	1110	E
Seven	0111	7	Fifteen	1111	F

FIGURE 18.1.6 Hexadecimal code.

possible by appropriate selection of binary codes to detect errors. For example, if a 6-b code is used, a seventh bit can be added to maintain the number of 1 bits in the group of 7 as an odd number. When any group of 7 with an even number of 1s is found by appropriate circuits in the machine, an error is detected. Such a procedure is known as *parity checking*. Although these error-control coding schemes were originally developed for noisy transmission channels, they are also applicable to storage devices in data-processing systems.

Boolean Functions

Figure 18.1.7 illustrates truth tables for functions of one, two, and three binary variables. Each x entry in each table can be either 0 or 1. Hence for one variable x four functions $f(x)$ can be formed; for two variables x_1 and x_2, 16 functions $f(x_1, x_2)$ exist; for three variables, 256 functions, and so on. In general, if $f(x_1, \ldots, x_n)$ is a function of n binary variables, 2^{2n} such functions exist.

For functions of one variable, the most important is the *inverse function*, defined in Fig. 18.1.8. This is called NOT A or $\bar{A}$, where A is the binary variable. Also illustrated in Fig 18.1.8 are the two most important functions of two binary variables, the AND (*product* or *intersection*) and the OR (*sum* or *union*). If A and B are the two variables, the AND function is usually represented as AB and the OR as $A + B$.

Figure 18.1.9 shows how the products AB, $\bar{A}B$, $A\bar{B}$ and AB are summed to yield any function of two binary variables. Each of these products has only one 1 in the four positions in their respective truth tables, so that appropriate sums can generate any function of two binary variables. This concept can be expanded to functions of more than two variables, i.e., any function of n binary variables can be expanded into a sum of products of the variables and their negatives. This is the general theorem of boolean algebra. Such a sum is called the *standard sum* or the *disjunctive normal form*.

The fact that any binary function can be so realized implies that mechanical or electrical simulations of the AND, OR, and NOT functions of binary variables can be used to represent any binary function whatever.

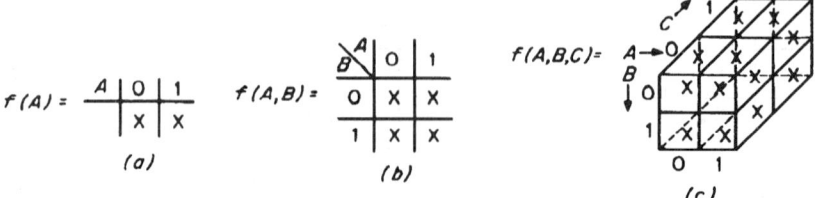

FIGURE 18.1.7 Binary functions of (*a*) one, (*b*) two, and (*c*) three binary variables.

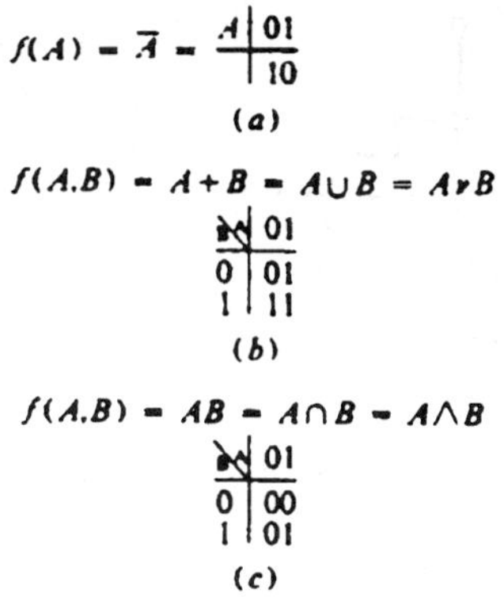

$$f(A) = \overline{A} = \begin{array}{c|c} A & 01 \\ \hline & 10 \end{array}$$

(a)

$$f(A,B) = A + B = A \cup B = A \vee B$$

$$\begin{array}{c|c} & 01 \\ \hline 0 & 01 \\ 1 & 11 \end{array}$$

(b)

$$f(A,B) = AB = A \cap B = A \wedge B$$

$$\begin{array}{c|c} & 01 \\ \hline 0 & 00 \\ 1 & 01 \end{array}$$

(c)

FIGURE 18.1.8 Significant functions of one and two binary variables: (*a*) the negation (NOT) function of one binary variable; (*b*) the OR function of two binary variables; (*c*) the AND function of two binary variables.

$$AB = \begin{array}{c|c} & 01 \\ \hline 0 & 00 \\ 1 & 01 \end{array} \qquad A\overline{B} = \begin{array}{c|c} & 01 \\ \hline 0 & 00 \\ 1 & 10 \end{array}$$

$$\overline{A}B = \begin{array}{c|c} & 01 \\ \hline 0 & 01 \\ 1 & 01 \end{array} \qquad \overline{A}\,\overline{B} = \begin{array}{c|c} & 01 \\ \hline 0 & 10 \\ 1 & 00 \end{array}$$

$$A\overline{B} + \overline{A}B = A \oplus B = \begin{array}{c|c} & 01 \\ \hline 0 & 01 \\ 1 & 10 \end{array}$$

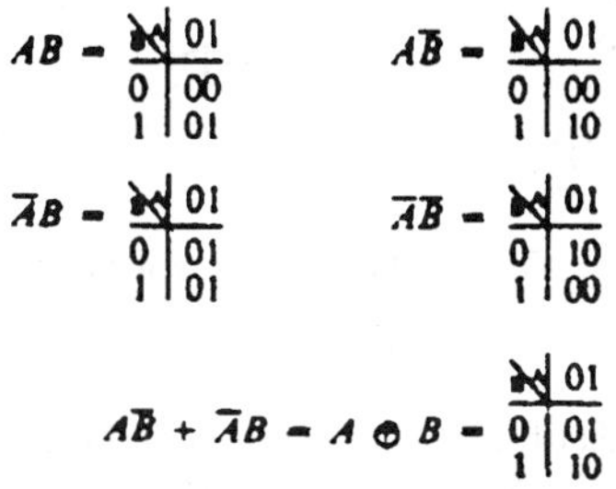

FIGURE 18.1.9 The four products of two binary variables (top). The realization of the EXCLUSIVE-OR function is shown below.

Electronic Realization of Logical Functions

Logical functions can be realized using electronic circuits. Figure 18.1.10 illustrates the realization of the OR function and Fig. 18.1.11 illustrates the realization of the AND circuit using diodes. Each input lead is associated with a boolean variable, and the upper level of voltage represents the logical 1 for that variable; in the OR circuit, any input gives rise to an output. Thus for a three-variable input, the output is $A + B + C$. With the AND function no output is realized unless all inputs are positive; the output function generated is ABC.

The inverse function (NOT) of a boolean variable cannot be readily realized with diodes. The circuit shown in Fig 18.1.12 uses the inverting property of a grounded-emitter transistor amplifier to perform the inverse function. Also shown in Fig. 18.1.12 is an example of how the OR function and the NOT function are combined to form the NOT-OR (NOR) function. In this case, since the transistor circuit provides both voltage and current gain, the signal-amplitude loss associated with transmission through the diode can be compensated, so that successive levels of logic circuits can be interconnected to form complex switching nets.

Figure 18.1.13 illustrates the realization of the EXCLUSIVE-OR function. Note that the variables are represented by the wiring of interconnected circuit blocks, while the function is realized by the circuit blocks themselves.

Levels of Operation in Data Processing

A detailed sequence of operations is generally required in a data-processing system to realize even simple operations. For example, in carrying out addition, a machine typically performs the following sequence of operations:

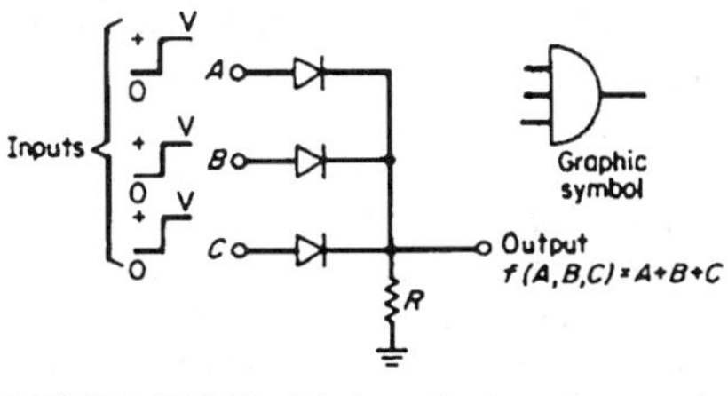

FIGURE 18.1.10 Diode realization of an OR circuit. A positive input on any line produces an output.

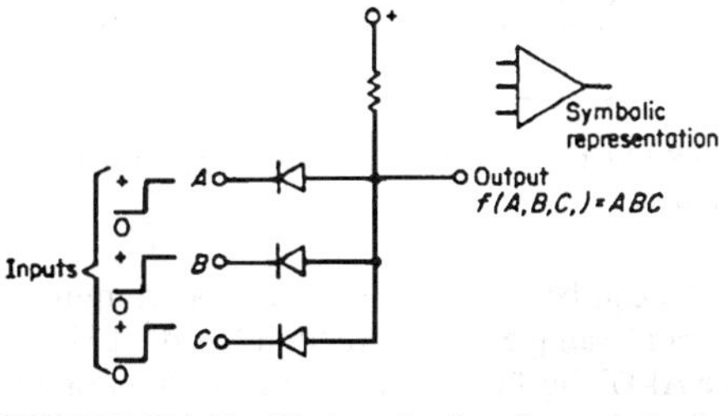

FIGURE 18.1.11 Diode realization of an AND circuit. All inputs must be positive to produce an output.

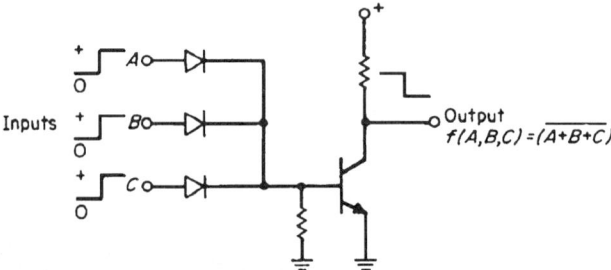

FIGURE 18.1.12   Use of a transistor circuit for inverting a function. The circuit shown forms the NOT-OR (NOR) of the inputs.

1. Fetch a number from a specific location in storage.
   a. Decode the address of the program instruction to activate suitable memory lines. Such decoding is accomplished by activating appropriate AND and OR gates to apply voltage to the lines in storage specified by the instruction address.
   b. Sequence storage to withdraw the information and place it in a storage output register.
   c. Transmit information from the storage output register into the appropriate ALU.

2. Withdraw a number from storage and add it to the number in the ALU. These operations break down into:
   a. Decode the instruction address, activate storage lines, and transmit the information to the ALU input for addition.
   b. Form the EXCLUSIVE-OR of the second number with the number in the ALU to form the sum less the carry. Form the AND of the two numbers to develop the first-level carry.
   c. Form the second-level EXCLUSIVE-OR sum.
   d. AND the first-level carry with the first-level EXCLUSIVE-OR sum to form the second-level carry.
   e. Generate the third-level EXCLUSIVE-OR by forming the EXCLUSIVE-OR of the second-level carry with the second-level EXCLUSIVE-OR sum, AND the second-level carries with the second-level EXCLUSIVE-OR for the third-level carry and so forth until no more carries are generated.

3. Store the result of the addition into a specified location in storage.

This sequence illustrates two basic types of operation in a data-processing machine. Operations denoted above by numbers are of specific interest to the programmer, since they are concerned with the data stored and the operations performed thereupon. The second level, denoted above by letters, are operations at the logical-circuit level within the machine. These operations depend on the particular configurations of circuits and other hardware in the machine at hand.

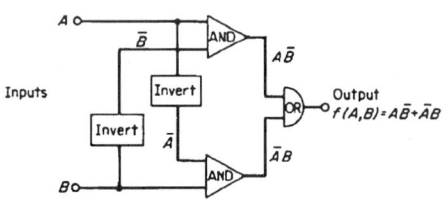

FIGURE 18.1.13   Circuit realization of the EXCLUSIVE-OR function.

If only the higher-level (numbered) instructions are used, some flexibility in machine operation is lost. For example, only an add operation is possible at the higher level. At the lower-level (lettered) operations the AND or EXCLUSIVE-OR of the data words can be formed and placed in storage.

The organization of current digital computers follows the lines of these two divisions (numbered and lettered, above). The *macroinstruction set* associated with each machine can be manipulated by the programmer. These instructions are usually implemented in a numerical code. For example, the instruction "load ALU" might be 01 in binary. "Add ALU" might be given by 10 and "store ALU" by 11. Similarly, each instruction has an associated storage address to provide source data. The microinstruction set comprises a series of suboperation that is combined in various sequences to realize a given macroinstruction.

Two methods of realizing the sequence of suboperations specified by the operations portion of the instruction have been used in machine design. In one such method a direct decoding of the information from the instruction occurs when it is placed in an instruction address register. Specific clock sequences turn on the successively required lines that have been wired in place to realize the action sought.

An alternative for actuating a subprogram is to store a number of information bits, called *microinstructions*, that are successively directed to the appropriate control circuits to activate selectively and sequentially individual wires to gate sequential actions for the realization of the requisite instruction.

The first method of computer design is called *hard-wired*, and the second is the *microprogrammed*. The microprogram essentially specifies a sequence of operations at the individual circuit level to specify the operations performed by macroinstructions. Microprogramming is preferred in modern computer designs.

Types of Computer Systems

There is a wide variety of computer-system arrangements, depending on the type of application. One type of installation is that associated with *batch processing*. A computer in a central job location receives programs from many different sources and runs the programs sequentially at high speed. An overall supervisory program, called an *operating system*, controls the sequence of programs, rejecting any that are improperly coded and completing those which are correct.

Another type of system, the *time-shared system*, provides access to the computer from a number of remote input-output stations. The computer scans each remote station at high speed and accepts or delivers information to that location as required by the operator or by its internal program. Thus a small terminal can gain access to a large high-speed system.

Still another type of installation, the *microcomputer*, involves an individual small computer that, though limited in power, is dedicated to the service of a single operator. Such applications vary from those associated with a small business, with limited computational requirements, to an individual engaged in scientific operations.

Other computers are used for dedicated control of complex industrial processes. These are individual, once-programmed units that perform a *real-time* operation in *systems control*, with sensing elements that provide the inputs. Highly complex interrelated systems have been developed in which individual computers communicate with and control each other in an overall major *systems network*. Among the first of such systems was the SAGE network, developed in the 1950s for defense against missile or aircraft attack.

Computers that are interconnected to share workload or problems are said to form a *multiprocessing system*. A computer system arranged so that more than program can be executed simultaneously is said to be *multiprogrammed*.

Interactive systems allow users to communicate directly with the computer and have the computer respond. The development of these systems parallels that of the keyboard and of the video display. The systems are used commercially (e. g., airline reservations) and scientifically (users input data at their terminals or telephones and get a response to their input). The *terminal*, a widely used input/output device, has a keyboard and a visual display. A terminal may be *dumb*, which means that it has no computing power, or *smart*, which indicates computing capabilities, such as those provided by a personal computer. *Client–server* systems are interactive systems where the data are at a remote computer called a server.

Internal Organization of Digital Computers

The internal organization of a data-processing system is called the *system architecture*. Such matters as the minimum addressable field in memory, the interrelations between data and instruction word size, the instruction format and length or lengths, parallel or serial (by bit or set of bits) ALU organization, decimal or binary internal organization, and so forth, are typical questions for the system architect. The answers depend heavily on the application for which the computer is intended.

Two broad classes of computer systems are *general-purpose* and *special-purpose* types. Most systems are in the general-purpose class. They are used for business and scientific purposes. General-purpose computers of varying computer power and memory size can be grouped, sharing a common architecture. These are said to constitute a *computer family*.

A computer scientifically designed for, and dedicated to the control of, say, a complex refinery process is an example of a special-purpose system.

A number of design methods have been adopted to increase the speed and functional range for a small increase in cost. For example, in the instruction sequence, the next cell in storage is likely to be the location of the next instruction. Since an instruction can usually be executed in a time that is short compared with storage access, the store is divided into subsections. Instructions are called from each subsection independently at high speed and put into a queue for execution. This type of operation is called *look-ahead*. If the instructions are not sequential, the queue is destroyed and a new queue put in its place.

Since instructions and data tend to be clustered together in storage, it is advantageous to provide a small, high-speed store (local store) to work with a larger, slower-speed, lower-cost unit. If the programs in the local store need information from the larger store, a least-used piece of the local store reverts to the larger store and a batch of data surrounding the information sought is automatically brought into the high-speed unit. This arrangement is called a *hierarchical memory* and the high-speed store is often called a *cache*.

NUMBER SYSTEMS, ARITHMETIC, AND CODES

Representation of Numbers

A set of codes and names for numbers that meets the requirements of expendability and convenience of operation can be obtained using the following power series:

$$N = A_n X^n + A_{n-1} X^{n-1} + \cdots + A_1 X + A_0 + A_{-1} X^{-1} + \cdots + A_{-m} X^{-m} \tag{1}$$

Here the number is represented by the sum of the powers of an integer X, each having a coefficient A_i. A_i may be an integer equal to or greater than zero and less than X. In the decimal system, X equals 10 and the coefficients A_i range from 0 to 9.

Note that Eq. (1) can be used to represent X^{m+n+1} numbers ranging between 0 and $X^{n+1} - X^m$ with an accuracy limited by X_m. Thus m and n must be of reasonable size to be useful in most applications. A useful property of the power series is the fact that its multiplication by X^k can be viewed as a shift of the coefficients of any given term by the number of positions specified by the value of k. These results are independent of the choice of X in the series representation.

There is little reason to write the value of the number in the form shown in Eq. (1) since complete information on the value can be readily deduced from the coefficient A_i. Thus, a number can be represented merely by a sequence of the values of the coefficients. To determine the value of the implied exponents on X, it is customary to mark the position of the X_0 term by a period immediately to the right of its coefficient. The power series for a number represented in the decimal system ($X = 10$) and its normal decimal notation are

$$3 \times 10^3 + 0 \times 10^2 + 2 \times 10^1 + 4 \times 10^0 + 6 \times 10^{-1} + 2 \times 10^{-2} = 3{,}024.62 \tag{2}$$

The value of X is called the *radix* or *base* of the number system. Where ambiguity might arise, a subscript to indicate the radix is attached to the low-order digit, as in $1000_2 = 8_{10} = 10_8$ (1000 binary equals 8 decimal equals 10 octal). The power series for a number in base 2 and its representation in binary notation is

$$1 \times 2^4 + 1 \times 2^3 + 0 \times 2^2 + 1 \times 2^1 + 1 \times 2^0 + 0 \times 2^{-1} + 1 \times 2^{-2} + 1 \times 2^{-3} = 11011.011 \tag{3}$$

Number-System Conversions

Since computer systems, in general, use number systems other than base 10, conversion from one system to another must be carried out frequently. Equation (4) shows the integer N represented by a power series in base 10 and base 2

$$\sum_{i=0}^{n} A_i 10^i = \sum_{j=0}^{m} B_j 2^j \tag{4}$$

The problem is to find the correlation between the coefficients A_i and B_j. In the binary series, if N is divisible by 2, then B_0 must be 0. Similarly, if N is divisible by 4, B_1 must be 0, and so forth. Thus if the decimal coefficients A_i are given, successive divisions of the decimal number by 2 will yield the binary number, the binary digits depending on the value of the remainder of each successive division. This process is shown in Fig. 18.1.14.

The conversion of a binary integer to a decimal integer is

$$100011011 = 1 \times 2^8 + 0 \times 2^7 + 0 \times 2^6 + 0 \times 2^5$$
$$+ 1 \times 2^4 + 1 \times 2^3 + 0 \times 2^2 + 1 \times 2 + 1 \times 2^0$$
$$= 283$$

In the case of conversion of an integer in binary to an integer in decimal, the powers of 2 are written in decimal notation and a decimal sum is formed from the contribution of each term of the binary representation. For conversion from a binary fraction to a decimal fraction, a similar procedure is used since the value of terms as multiplied by the A_i can be added together in decimal form to form the decimal equivalent.

The conversion of a decimal fraction to a binary fraction is defined by

$$0.5764_{10} = A_{-1}2^{-1} + A_{-2}2^{-2} + \cdots + A_n 2^{-n} \tag{5}$$

To determine the values of the A_i, first multiply both sides of Eq. (27.5) by 2 to give

$$1.1528_{10} = A_{-1} + A_{-2}2^{-1} + \cdots + A_{-n}2^{n-1} \tag{6}$$

Since the position of the decimal point (more accurately called the *radix point*) is invariant, and since in a binary series each successive term is at most half of the maximum value of the preceding term, the leading 1 in the decimal number in Eq. (6) indicates that A_{-1} must have been 1. A second multiplication by 2 can similarly determine the coefficient A_{-2}. This process of conversion of a base-10 fraction to a base-2 fraction is illustrated in Fig. 18.1.15.

Conversion from binary integers to octal (base 8) and the reverse can be handled simply since the octal base is a power of 2. Binary to octal conversion consists of grouping the terms of a binary number in threes and replacing the value of each group with its octal representation. The process works on either side of a decimal point. The octal-to-binary conversion is handled by converting each octal digit, in order, to binary and retaining the ordering of the resulting groups of three bits.

Since there are not enough symbols in decimal notation to represent the 16 symbols required for the hexadecimal system, it is customary in the data processing field to use the first six letters of the alphabet to complete the set.

Conversions from decimal to octal or hexadecimal can proceed indirectly by first converting decimal to binary and then binary to octal or hexadecimal. Similarly, a reverse path from octal or hexadecimal to binary to decimal can be used. Direct conversions, however, between hexadecimal and octal and decimal exist and are widely used. In going from hexadecimal or octal to decimal, each term in the implied power series is expressed directly in decimal, and the result is summed. In converting from a decimal integer to either hexadecimal or octal, the decimal is divided by either 16 or 8, respectively, and the remainder becomes the next higher-order digit in the converted number. Examples of four common number representations are shown in Table 18.1.1.

Binary-Arithmetic Operations

Figure 18.1.16 shows an example of the addition of two binary numbers, 1001 and 1011 (9 and 11 in decimal). The rules for manipulation are similar to those in decimal arithmetic except that only the two symbols 1 and 0 are used and the addition and carry tables are greatly simplified.

Figure 18.1.17 shows an example of binary multiplication with a multiplication table. This process is also simple compared with that used in the decimal system. The rule for multiplication in binary is as follows: if a particular digit in the multiplier is 1, place the multiplicand in the product register; if 0, do nothing; in either case shift the product register to the right by one position; repeat the operations for the next digit of the multiplier.

Figure 18.1.18 shows an example of binary subtraction and the subtraction and borrow tables. The subtraction table is the same as the addition table, a feature unique to the binary system. The borrow operation is handled in

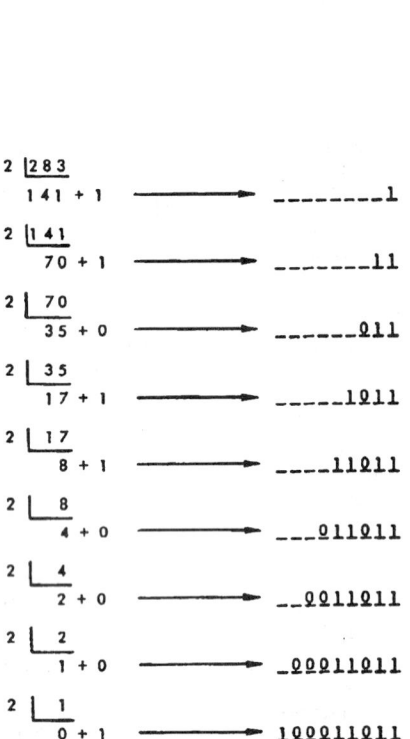

FIGURE 18.1.14 Conversion from a decimal to binary by repeated division of the decimal integer. At each division the remainder becomes the next higher-order binary digit.

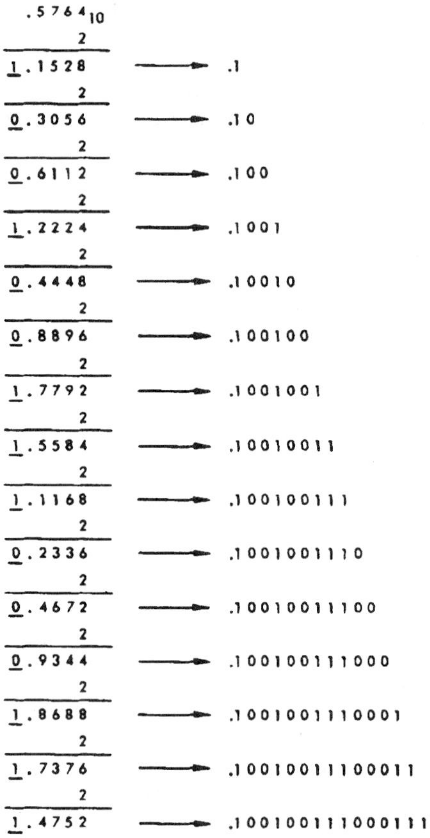

FIGURE 18.1.15 Conversion of a decimal fraction into a binary fraction. At each stage the number to the right of the decimal is multiplied by 2. The resulting number to the left of the decimal point is entered as the next available lower-order position of the binary fraction to the right of the binary radix point.

TABLE 18.1.1 Comparison of Decimal, Binary, Octal, and Hexadecimal Numbers

Decimal	Binary	Octal	Hexadecimal	Decimal	Binary	Octal	Hexadecimal
0	0	0	0	8	1000	10	8
1	1	1	1	9	1001	11	9
2	10	2	2	10	1010	12	A
3	11	3	3	11	1011	13	B
4	100	4	4	12	1100	14	C
5	101	5	5	13	1101	15	D
6	110	6	6	14	1110	16	E
7	111	7	7	15	1111	17	F

```
Binary              Decimal
1011                  11
× 1001               ×  9
1011                  99
101100
1100011
```

```
        | 0   1
      ──┼──────
      0 | 0   0
      1 | 0   1
```

FIGURE 18.1.17 Binary multiplication. The binary multiplication table is the AND function of two binary variables. The process of multiplication consists of merely replicating and adding the multiplicand, as shown, if a 1 is found in the multiplier. If 0 is found, a single 0 is entered and the next position to the left in the multiplier is taken up.

```
Binary              Decimal
1011        -          11
+ 1001      -           9
10100       -          20
```

FIGURE 18.1.16 Binary addition and corresponding decimal addition.

a fashion analogous to that in decimal. If a 1 is found in the preceding column of the subtrahend, it is borrowed, leaving a 0. If a 0 is found, an attempt is made to borrow from the next higher-order position, and so forth.

An example of binary division is

$$
\begin{array}{r}
101 \\
101\,\overline{)11110} \\
101 \\
\hline
101 \\
101 \\
\hline
0
\end{array}
\qquad
\begin{array}{r}
6 \\
5\,\overline{)30}
\end{array}
$$

The procedure is as follows:

1. Compare the divisor with the leftmost bits of the dividend.
2. If the divisor is greater, enter a 0 in the quotient and shift the dividend and quotient to the left.
3. Try subtraction again.
4. When the subtraction yields a positive result, i.e., the divisor is less than the bits in the dividend, enter a 1 in the quotient and shift the dividend and the quotient left one position.
5. Return to step 1 and repeat.

Binary division, like binary multiplication, is considerably simpler than the decimal operation.

Subtraction by Complement Addition

If subtraction were performed by the usual method of borrowing from the next higher order, a separate subtraction circuit would be required. Subtraction can be performed, however, by the method of *adding complements*

```
Binary      Decimal     A − B, less borrow      A − B, borrow
100110         38           | 0   1               | 0   1
1001            9         ──┼──────             ──┼──────
11101          29         0 | 0   1             0 | 0   0
                          1 | 1   0             1 | 1   0
```

FIGURE 18.1.18 Binary subtraction and corresponding decimal subtraction. The subtraction table is the same as the addition table. The borrow operation is handled analogously to decimal subtraction.

(or adding 1's complements, as the method is also called). By this method, the *subtrahend*, i.e., the number that is to be subtracted, is *inverted*, changing the 0s to 1s and the 1s to 0s. Then the inverted subtrahend is added to the *minuend*, i.e., the number that is to be subtracted from, and an additional 1 is added to find the difference. As an example, consider the subtraction 1101 − 1001. The subtrahend (1001) is first inverted to form the complement (0110). The difference is formed by adding the minuend and the complement of the subtrahend (plus 0001) as follows: 1101 − 1001 = 1101 + 0110 (complement) + 0001 = (1)0011 + 0001 = 0100. Note that in subtraction by complement addition, a leading 1 (in parentheses) in the result (difference) must be suppressed, and that 1 must be added to obtain the result. The result of this operation can be verified by observing that the decimal equivalent of this operation is 13 − 9 = 3 + 1 = 4.

Floating-Point Numbers

In a computer having a fixed number of bits that define a word, the bits represent the maximum size of a numerical value. For example, if 40-bit positions are provided for a word, the maximum decimal number that can be represented is in the order of $1.009 \times 10.^{12}$ Though this number is large, it does not suffice for many applications, especially in science, where a greater range of magnitudes may be routinely encountered. To extend the range of values that can be handled, numbers are represented in floating-point notation. In floating point the most significant digits of the number are written with an assumed radix point immediately to the left of the highest-order digit. This number is called the *fraction*. The intended position of the radix point is identified by a second number, called the *characteristic*, which is appended to the fraction: The characteristic denotes the number of positions that the assumed radix point must be shifted to achieve the intended number. For example, the number 146.754 in floating point might be 146754.03 where 146754 would be equivalent to 0.146754 and the 0.03 would denote a shift of the decimal point three places to the right. In binary notation the number 11011.011 (27.375 in decimal) might be represented in floating point as 11011011.101 with the fraction again to the left of the decimal and the characteristic to the right.

With floating-point addition and subtraction, a shift register is required to align the radix points of the numbers. To perform multiplication or division, the fraction fields are appropriately multiplied or divided and the exponents summed or subtracted, respectively. As with fixed-point addition or subtraction, provision is usually made to detect an overflow condition in the characteristic fields. In some systems provision is made to note when an addition or subtraction occurs with such widely differing characteristics that justification destroys one of the two numbers (by shifting it out the end of a shift register).

Numeric and Alphanumeric Codes

The numeric codes used to represent numerical values previously discussed include the hexadecimal, octal, binary, and decimal codes. In many applications the need arises for the coding of nonnumeric as well as numeric information, and such coding must use the binary scheme. A code embracing numbers, alphabetic characters, and special symbols is known as an *alphanumeric code*.

A widely used code with its roots in the past is the telegraph code (the Baudot code). Other alphanumeric codes have been devised for special purposes. One of the most significant of these, because of its present use and its contribution to the design of other codes, is the Hollerith code, developed in the 1890s. Hollerith's equipment contributed to the development of electromechanical accounting machines that provided the foundation for electronic computers.

Another code of importance in the United States is the American Standard Code for Information Interchange (ASCII) (see Fig. 18.1.19). This code, developed by a committee of the American National Standards Institute (ANSI), has the advantage over most other codes of being *contiguous*, in the sense that the binary combination used to represent alphanumeric information is sequential. Hence alphabetic sorting can be easily accomplished by arithmetic manipulation of the code values.

Codes used for data transmission generally have both data characters and *control characters*. The latter perform control functions on the machine receiving information. In more sophisticated codes, such as

	000	001	010	011	100	101	110	111
0000	NULL	① DC_0	♭	0	@	P	↑	↑
0001	SOM	DC_1	!	1	A	Q		
0010	EOA	DC_2	"	2	B	R		
0011	EOM	DC_3	'	3	C	S		
0100	EOT	DC_4 (STOP)	$	4	D	T		
0101	WRU	ERR	%	5	E	U		
0110	RU	SYNC	&	6	F	V		
0111	BELL	LEM	'	7	G	W	Unassigned	
1000	FE_0	S_0	(	8	H	X		
1001	HT SK	S_1	)	9	I	Y		
1010	LF	S_2	*	:	J	Z		
1011	V TAB	S_3	+	;	K	[		
1100	FF	S_4	,	<	L	\		ACK
1101	CR	S_5	-	=	M	]		②
1110	SO	S_0	.	>	N	↑		ESC
1111	SI	S_7	/	?	O	←	↓	DEL

Identification of control symbols and some graphics

NULL	Null/idle	V_{TAB}	Vertical tabulation
SOM	Start of message	FF	Form feed
EOA	End of address	CR	Carriage return
EOM	End of message	SO	Shift out
EOT	End of transmission	SI	Shift in
WRU	"Who are you?"	DC_0	Device control 1 Reserved for data link escape
RU	"Are you...?"		
BELL	Audible signal	DC_1–DC_3	Device control
FE_0	Format effector	DC_4 (STOP)	Device control (stop)
HT	Horizontal tabulation	ERR	Error
SK	Skip (punched card)	SYNC	Synchronous idle
LF	Line feed	LEM	Logical end of media

S_0–S_7	Separator (information)
♭	Word separator (space, normally nonprinting)
<	Less than
>	Greater than
↑	Up arrow (exponentiation)
←	Left arrow (implies/ replaced by)
\	Reverse slant
ACK	Acknowledge
②	Unassigned control
ESC	Escape
DEL	Delete/idle

FIGURE 18.1.19 The ASCII code has a contiguous alphabet, so that numeric ordering permits alphabetic sorting.

ASCII, these control functions are greatly extended and hence are applicable to machines of different design.

Other Numeric Codes. Not all numeric information is represented by binary numbers. Other codes are also used for numeric information in special applications. Figure 18.1.20 shows a widely used code called the *reflected* or *Gray* code. It has the property that only 1 b is changed between any two successive values, irrespective of number size. This code is used in digital-to-analog systems since there is no need for propagation of carry integers in sequential counting as in a binary code.

Word parity	Binary code
1	0 0 0 0 0 0
0	0 0 0 0 0 1
0	0 0 0 0 1 0
1	0 0 0 0 1 1
0	0 0 0 1 0 0
1	0 0 0 1 0 1
1	0 0 0 1 1 0
0	0 0 0 1 1 1
0	0 0 1 0 0 0
1	0 0 1 0 0 1
1	0 0 1 0 1 0
0	0 0 1 0 1 1
1	0 0 1 1 0 0
0	0 0 1 1 0 1
0	0 0 1 1 1 0
1	0 0 1 1 1 1
0	0 1 0 0 0 0
1	0 1 0 0 0 1
1	0 1 0 0 1 0
0	0 1 0 0 1 1
1	0 1 0 1 0 0
0	0 1 0 1 0 1
0	0 1 0 1 1 0
1	0 1 0 1 1 1
1	0 1 1 0 0 0
0	0 1 1 0 0 1
0	0 1 1 0 1 0
1	0 1 1 0 1 1
0	0 1 1 1 0 0
1	0 1 1 1 0 1
1	0 1 1 1 1 0
0	0 1 1 1 1 1
0	1 0 0 0 0 0
1	1 0 0 0 0 1
1	1 0 0 0 1 0
0	1 0 0 0 1 1
1	1 0 0 1 0 0
0	1 0 0 1 0 1
0	1 0 0 1 1 0
1	1 0 0 1 1 1
1	1 0 1 0 0 0

List parity: 0 1 0 1 1 1

Decimal	Gray code
0	0000
1	0001
2	0011
3	0010
4	0110
5	0111
6	0101
7	0100
8	1100
9	1101
10	1111
11	1110
12	1010
13	1011
14	1001
15	1000

FIGURE 18.1.20 The Gray code, used in analog-to-digital encoding systems. There is only a 1-b change between any two successive integers.

FIGURE 18.1.21 Two-dimensional parity checking, in which a single error can be corrected and a triple error detected.

Error Detection and Correction Codes

The integrity of data in a computer system is of paramount importance because the serial nature of computation tends to propagate errors. Internal data transmission between computer-system units takes place repeatedly and at high speed. Data may also be sent over wires to remote terminals, printers, and other such equipment. Because imperfections of transmission channels inevitably produce some erroneous data, means must be provided to detect and correct errors whenever they occur.

A basic procedure for error detection and correction is to design a code in which each word contains more bits than are needed to represent all the symbols used in a data set. If a bit sequence is found that is not among those assigned to the data symbols, an error is known to have occurred.

One such commonly used error-detection code is called the *parity check*. Suppose that 8 bits are used to represent data and that an additional bit is reserved as a check bit. A simple electronic circuit can determine whether an even or odd number of 1 bits is included in the eight bit positions. If an even number exists, a 1 can be inserted in the check position. If an odd number of 1s exist, the check position contains a 0. As a result all code words must contain an odd number of 1 bits. If a 9-b sequence is found to contain an even number of 1s, an error can be presumed.

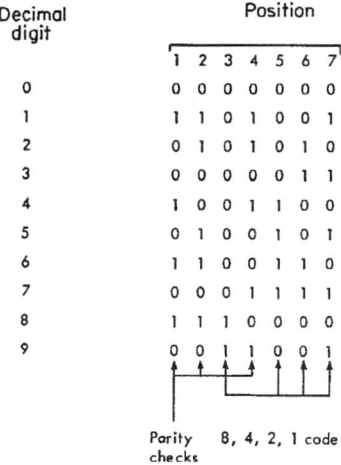

Decimal digit	Position						
	1	2	3	4	5	6	7
0	0	0	0	0	0	0	0
1	1	1	0	1	0	0	1
2	0	1	0	1	0	1	0
3	0	0	0	0	0	1	1
4	1	0	0	1	1	0	0
5	0	1	0	0	1	0	1
6	1	1	0	0	1	1	0
7	0	0	0	1	1	1	1
8	1	1	1	0	0	0	0
9	0	0	1	1	0	0	1

Parity checks   8, 4, 2, 1 code

FIGURE 18.1.22 A Hamming code. The parity bit in column 1 checks parity in columns 1, 3, 5, and 7; the bit in column 2 checks 2, 3, 6, and 7; and the bit in column 4 checks 4, 5, 6, and 7. The overlapping structure of the code permits the correction of a single error or the detection of single or double errors in any code word.

There are limitations in the use of the simple parity check as a mechanism for error detection, since in many transmission channels and storage systems there is a tendency for a failure to produce simultaneous errors in two adjacent positions. Such an error would not be detected since the parity of the code word would remain unchanged.

To increase the power of the parity check, a list of code words in a two-dimensional array can be used, as shown in Fig. 18.1.21. The code words in the horizontal dimension have a parity bit added, and the list in the vertical dimension also has an added parity bit, in each column. If one bit is in error, errors appear in both the row and the column. If simultaneous errors occur in two adjacent positions of a code word, no parity error will show up in that row, but the column checks will detect two errors. This code can detect any 3-b errors.

It is possible to design codes that can detect directly whether errors have occurred in two bit positions in a single code word. Figure 18.1.22 shows such a code, an example of a *Hamming* code. The code positions in columns 1, 2, and 4 are used to check the parity of the respective bit combinations. Two code words of a Hamming code must differ at three or more bit positions, and therefore any 2-b error patterns can be detected. The pattern formed by the particular parity bits that show errors indicates which bit is in error in the case of a single bit failure.

In general, if two code words must differ at D or more bit positions, the code can detect up to $D - 1$ bit errors. For $D = 2t + 1$, the code can detect $2t$ bit errors or correct t bit errors.

COMPUTER ORGANIZATION AND ARCHITECTURE

Introduction

Over the past 40 years great progress has been made as component technology has moved from vacuum tubes to solid-state devices to large-scale integration (LSI). This has been achieved as a result of increased understanding of semiconductor materials along with improvements in the fabrication processes. The result has been significant enhancement in the performance of the logic and memory components used in computer construction along with significant reductions in cost and size. Figure 18.1.23, for example, indicates the reduction in volume of main memory that has occurred in the last 40 years. The volume required to store 1 million characters has been reduced by a factor of 3 million in that period. Smilarly, during that same period, the system cost to execute a specific mix of instructions has also decreased by a factor of 5000. Integrated circuit manufacturers are able to incorporate millions of transistors in the area of a square inch. Large-scale integration (LSI) and VLSI have led to small-size, lower cost, large-memory, ultrafast computers ranging from the familiar PC to the high-performance, high-priced supercomputers. There is no reason to expect that this progress will not continue into the future as new technology improvements continue to occur.

These advances in component technology have also had a major impact on computer organization and its realization. Functions and features that were too expensive to be included in earlier designs are now feasible, and the trade-offs between software and hardware need to be reevaluated as hardware costs continue to decrease. New approaches to computer organization must also be considered as technology continues to improve. The advances in component technology also have had a major impact on such aspects of computer realization as packaging, coding, and power.

Basic Computer Organization

The basic organization of a digital computer is shown in the block diagram of Fig. 18.1.24. This structure was proposed in 1946 by von Neumann. It is a tribute to his genius that this design, which was intended for use in

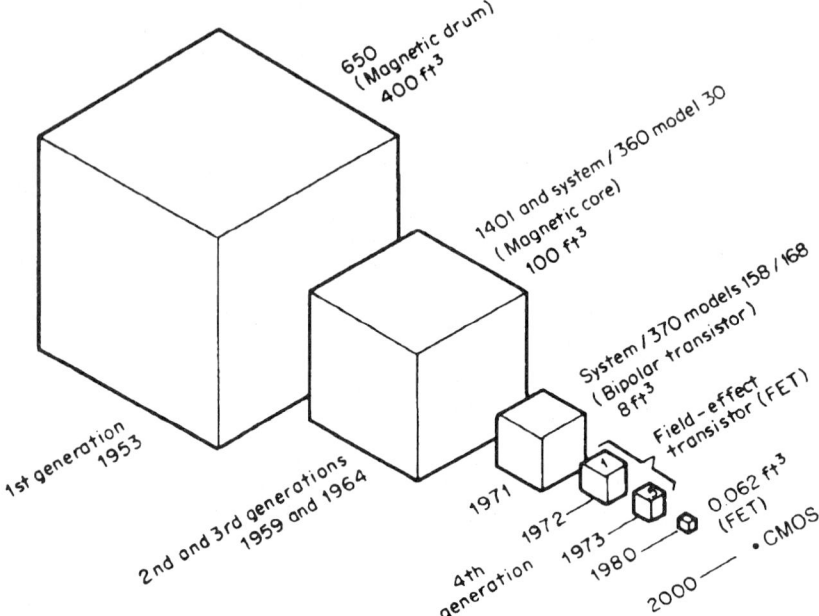

FIGURE 18.1.23 The reduction by a factor of 6400 in memory size from the first- to the fourth-generation families of IBM computers.

solving differential equations, has also been applicable in solving other types of problems in such diverse areas as business data processing and real-time control. Von Neumann recognized the value of maintaining both data and computer instructions in storage and in being able to modify instructions as well as data. He recognized the importance of branch or jump instructions to alter the sequence of control of computer execution. His contributions were so significant that the vast majority of computers in use today are based on his design and are called von Neumann computers.

The four basic elements of the digital computer are its *main storage, control unit, arithmetic-logic unit* (ALU), and *input/output* (I/O). These elements are interconnected as shown in Fig. 18.1.24. The ALU, or *processor*, when combined with the control unit, is referred to as the *central processing unit* (CPU).

Main storage provides the computer with directly addressable fast-access storage of data. The storage unit stores programs as well as input, output, and intermediate data. Both data and programs must be loaded into main storage from input devices before they can be processed.

The control unit is the controlling center of the computer. It supervises the flow of information between the various units. It contains the sequencing and processing controls for instruction decoding and execution and for handling interrupts. It controls the timing of the computer and provides other system-related functions.

The ALU carries out the processing tasks specified by the instruction set. It performs various arithmetic operations as well as logical operations and other data processing tasks.

Input/output devices, which permit the computer to interact with users and the external world, include such equipment as card readers and punches, magnetic-tape units, disc storage units, display devices, keyboard terminals, printers, teleprocessing devices, and sensor-based devices.

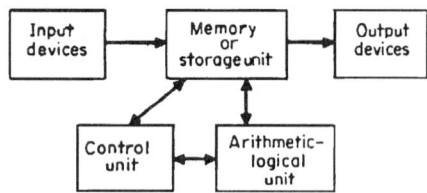

FIGURE 18.1.24 Block diagram of a digital computer illustrating the main elements of the von Neumann architecture.

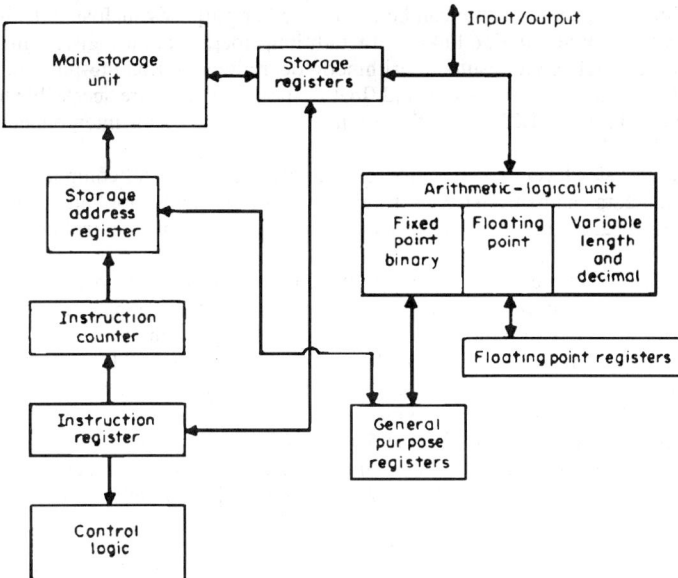

FIGURE 18.1.25   Basic structure of a digital computer showing a typical register organization and interconnections.

Detailed Computer Organization

The block diagram of Fig. 18.1.25 provides an overview of the basic structure of the digital computer. Computer systems are complex, and block diagrams cannot describe the computer in sufficient detail for most purposes. One is therefore forced to go to lower levels of description. There are at least five levels[66a] that can describe the implementation of a computer system:

1. Processor-memory-switch (block-diagram) level
2. Programming level (including the operating system)
3. Register-transfer level
4. Switching-circuits level
5. Circuit or realization level

Each of these levels is an abstraction of the levels beneath it. A number of computer-hardware description languages have been developed to represent the components used in each level along with their modes of combination and their behavior.

A *register* is a device capable of receiving information, holding it, and transferring it as directed by control circuits. The actual realization of registers can take a number of forms depending on the technology used. Registers store data temporarily during a program's execution. Some registers are accessible to the user through instructions.

Registers are found in every element of the computer system. They are an integral part of main storage, being used as storage registers to contain the information being transferred from memory (read) or into memory (write) as well as storage address registers (SAR) to hold the address of the location in storage involved in the information transfer. In the control unit, the instruction (or program) counter contains the storage address of the instruction to be executed while the instruction register holds the instruction being decoded and executed.

In the ALU internal registers are used to hold the operands and partial results while arithmetic and logical operations are being performed. Other ALU registers, called *general-purpose registers*, are used to accumulate the results of arithmetic operations but can be used for other purposes such as indexing, addressing, counting looping operations, or subroutine linkage. In addition, floating-point registers may be provided to hold the operands and accumulate the results of arithmetic operations on floating-point numbers. Of all the registers mentioned, only the general-purpose and floating-point registers are accessible to program control and to the programmer. Figure 18.1.25 shows the primary registers and their interconnections in the basic digital computer.

At the register transfer level of abstraction one can describe a digital computer as a collection of registers between which data can be transferred. Logical operations can be applied to the data during the transfers. The sequencing and timing of the transfers are scheduled and controlled by logic circuits in the control unit.

The data transferred between registers within the computer consist of groups of binary digits. The number of bits in the group is determined by the computer architecture and in particular by the organization of its main storage. Main storage is structured into segments, called bytes, and each storage word is uniquely identified by a number, called its *address*, assigned to it. A byte consists of 8 binary bits and is the standard for describing memory elements. Many computers read groups of bytes (2, 4, 6, 8) from memory in one access. A group of bytes is referred to as a word, which can vary in length from one computer to another. A memory access is a sequence to read data from memory or store it into memory. The 1s and 0s can be interpreted in various ways. These bit patterns can be interpreted as: (1) a pure binary word, (2) a signed binary number, (3) a floating-point number, (4) a binary-coded decimal number, (5) data characters, or (6) an instruction word.

In a signed binary number the high-order (leftmost) bit indicates the sign of the number. If the bit is 0, the number is positive; a 1 indicates it is negative. Thus

$$0bbbbbbb \text{ represents a positive 7-bit number}$$

$$1bbbbbbb \text{ represents a negative 7-bit number}$$

A negative number is carried in 2's-complement (inverted) form. For example,

$$11111110_2 = 2^{10}$$

A binary-coded decimal code uses 4 b to represent a decimal digit. It uses the binary digit combinations for 0 to 9; combinations greater than 9 are not allowed. Thus

$$0101_2 = 5_{10} \quad 1010_2 = \text{illegal}$$

The sign of the decimal number can be indicated in several ways. One technique uses the low-order bits to indicate the sign; for example,

$$bbbbbbbbbbbb1100 \text{ represents a positive 3-digit number}$$

$$bbbbbbbbbbbb1011 \text{ represents a negative 3-digit number}$$

Thus, $0001010100111011_2 = -153_{10}$.

For external communication, as well as text processing and other nonnumeric functions, the digital computer must be able to handle character sets. The byte has been accepted as the data unit for representing character codes. The two most common codes, described earlier, are the American Standard Code for Information Interchange (ASCII) and the Extended Binary-Coded Decimal Interchange Code (EBCDIC). The 16-b word 1100011111001011 coded in EBCDIC represents the two-letter word "go." (See Figs. 18.1.4, 18.1.5, and 18.1.9).

The *instruction word* is composed of two major parts, an *operation part* and an *operand part*. The length of the instruction word is determined by the computer architecture. Some computers have a single format for their instruction words and thus a single length, whereas other computers have several formats and several different lengths. The operation part consists of the operation code that describes the particular operation to be performed by the computer as a result of executing that instruction. The operand part usually contains the addresses of the two operands involved in the operation. For example, the RR (register-to-register) instruction format in the System/370 is

Op Code	Reg 1	Reg 2
bbbbbbbb	bbbb	bbbb
0 7	8 11	12 15

The instruction 1AB4 (in hexadecimal), for example, instructs the computer to add the contents of register 11 to the contents of register 4 and to put the resulting sum in register 11, replacing its original contents. Most computers use the two-address instruction with three sections: The first section consists of the op-code and the second and third sections each contain an address of an operand. Different computers use these addresses differently. In some cases, the operand section is used as a modifier or to extend the instruction length. A single-operation computer has two sections—an op code and an operand, usually a memory address, with the accumulator the implied source or destination.

Two facts should be noted. First, the discussion thus far may have implied that digital computers can deal only with fixed-length words. That is true for some computers, but other families of computers can also deal with variable-length words. For these, the operand part of the instruction contains the address of the first digit or character of each variable-length word plus a measure of its length, i.e., the number of characters it contains. The second fact is that it is impossible to distinguish between the various data representations when they are stored. For example, there is nothing to indicate whether a word of memory contains a binary number or a binary-coded decimal (BCD) number. Programmers must make the distinction in the programs they develop and not attempt meaningless operations such as adding a binary number to a decimal number. The only way the computer distinguishes an instruction word from other data words is by the time when, as discussed in the next section, it is read from storage into the control unit.

Instruction Execution

The digital computer operates in a cyclic fashion. Each cycle is called a *machine cycle* and consists of two main subcycles, the *instruction* (I) *cycle* (sometimes called the *fetch cycle*) and the *execution* (E) *cycle*. During the machine cycle, the following basic steps occur in sequence (see Fig. 18.1.25):

1. The cycle begins with the I cycle:
   a. The contents of the instruction counter are transferred to the storage address register (SAR). (The instruction counter holds the address of the next instruction to be executed.)
   b. The specified word is transferred from storage to the instruction register. (The control unit assumes that this storage word is an instruction.)
   c. The contents of the instruction register are decoded by logical circuits in the control unit. This identifies the type of operation to be performed and the locations of the operands to be used in the operation.

2. At this point, the E cycle begins:
   a. The specified computer operation is performed using the designated operands, and the result is transferred to the location indicated by the instruction.
   b. The instruction counter is advanced to the address of the next instruction in the sequence. (If a branch, or change in execution control sequence, is to occur, the contents of the instruction counter are replaced by an address as directed by the instruction currently being executed.)

3. At this point, the I cycle is repeated.

To indicate in more specific terms what happens in the CPU during instruction execution it is necessary to go to the switching level of description. The following paragraphs describe of the operations of the ALU and the control section in more detail.

Arithmetic Logic Unit

The ALU performs arithmetic and logical operations between two operands, such as OR, AND, EXCLUSIVE-OR, ADD, MULTIPLY, SUBTRACT, or DIVIDE. The unit may also perform operations such as INVERT on only one operand, and it tests for minus or zero and forms a complement.

Adders and multipliers are at the heart of the ALU. In Fig. 18.1.26 one bit position of an ALU is shown as part of a wider data path. One latch (part of a register A) feeds an AND circuit that is conditioned by a CONTROL A. The output feeds INPUT A of the adder circuits. One latch of register B is also ANDed with CONROL B and feeds the other input into the adder. A true-complement circuit is shown on the B line. This latter circuit has to do with subtraction and can be assumed to be a direct connection when adding. Each adder stage is a combinatorial circuit that accepts a carry from the stage representing the next lower digit in the binary number (assumed to be on the right). The collection of outputs from all adder stages is the sum. This sum is ANDed into register D by CONTROL D.

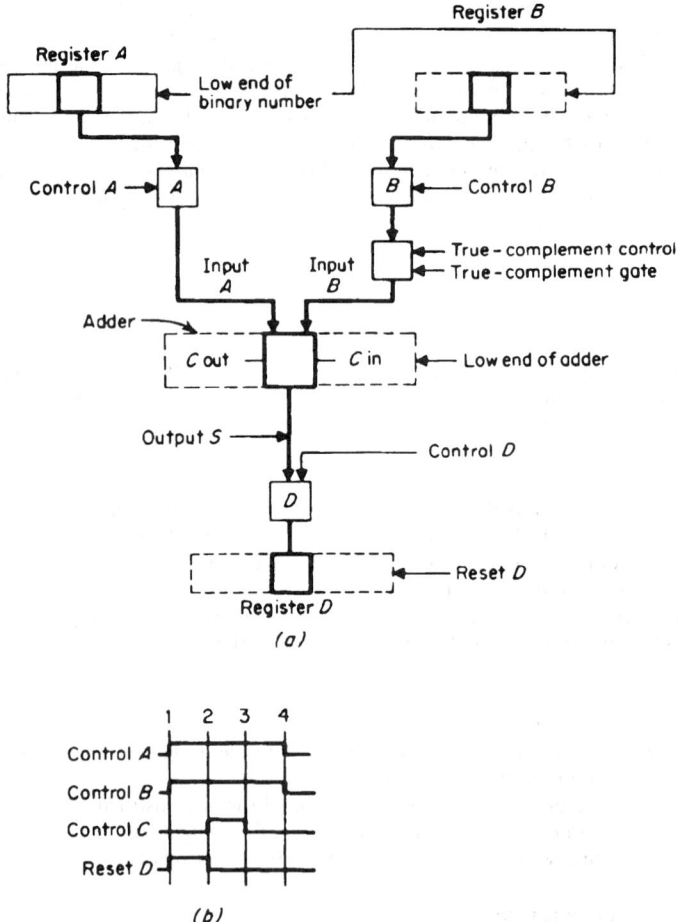

FIGURE 18.1.26   Basic addition logic. The heart of the ALU is the adder circuit shown in functional form in (a). The control section applies the appropriate time sequence of pulses (b) on the control to perform addition. Heavy lines indicate a repeat of circuitry in each bit position to form a machine adder.

All bit positions of each of the registers are gated by a single control line. If the gate is closed (control equal to 0), all outputs are 0s. If the gate is open (control equal to 1), the bit pattern appearing in the register is transmitted through the series of AND circuits to the input of the adders. Thus, a gate is a two-way AND circuit for each bit position. The diagram of Fig. 18.1.26 illustrates all positions of an n-position adder since all positions are identical. In such a case heavy lines, as shown, indicate that this one line represents a line in each bit position.

Binary Addition

At the outset of an addition, it is assumed that registers A and B (Fig. 18.1.26a) contain the addends. An addition is performed by pulsing the control lines with signals originating in the control section of the CPU. Time is assumed to be metered into fixed intervals by an oscillator (clock) in the control section. These time slots are numbered for easier identification (Fig. 18.1.26b). At time 1 the inputs are gated to the adder, and the adders begin to compute the sum. At the same time, register D is *reset* (all latches set to 0). At time 2 the outputs of the adders have reached steady state, and control line D is raised, permitting those bit positions for which the sum was 1 to set the corresponding latches in register D. Between times 2 and 3, the result is latched up in register D, and at time 3, control D is lowered. Only after the result is locked into D and cannot change, may control A and B be lowered. If they were lowered earlier, the change might propagate through the adder to produce an incorrect answer.

The length of the pulses depends on the circuits used. The times from 2 to 3 and 3 to 4 are usually equal to a few logic delays (time to propagate through an AND, OR, or INVERTER). The time from 1 to 2 depends on the length of the adder and is proportional to the number of positions in a parallel adder because of potential carry-propagation times. This delay can be reduced by *carry look-ahead* (some-times called *carry bypass*, or *carry anticipation*).

Binary Subtraction

Subtraction can be accomplished using the operation of addition, by forming the complement of a number. Negative numbers are represented throughout the system in complement form. To subtract two numbers such as B from A, a set of logic elements may be put into the line shown in Fig. 18.1.26a an input B. Using 2's complement, the sign of a number is changed by complementing each bit and adding 1 to the result. The inversion of the bit is performed by the logic element interposed on the input B line in Fig. 18.1.26a known as a *true-complement* (T/C) *gate*. This unit gates the unmodified bit if the control is 0 and inverts each bit if the control is 1. The boolean equation for the output of the T/C gate is

$$\text{Output} = \overline{\text{T/C}} \cdot B + \text{T/C} \cdot \overline{B}$$

The T/C gate is a series of EXCLUSIVE-ORS with one leg, common to all bit positions, connected to the T/C control line. The other leg of each EXCLUSIVE-OR is connected to one bit of the circuit containing the number to be complemented.

The T/C gate produces the 1's complement; a 1 must be added in the low-bit position to produce the true complement. The low stage of an adder may be designed to have an input for a carry-in, designed to accommodate the 1 bit automatically produced from the high-order position of the T/C gate. Such a logical interconnection accomplishes the required 1 input for a true-complement system when a positive number B is subtracted from a positive number and is called an *end-around carry*. Consistency of operation is obtained by entering the appropriate high-order T/C gate into the low-order carry position.

Decimal Addition

In some systems the internal organization of a computer is such that decimal representations are used in arithmetic operations. In BCD, a conventional binary adder can be used to add decimal digits with a small amount of additional hardware. Adding two 4-b binary numbers produces a 4- or 5-b binary result. When two BCD

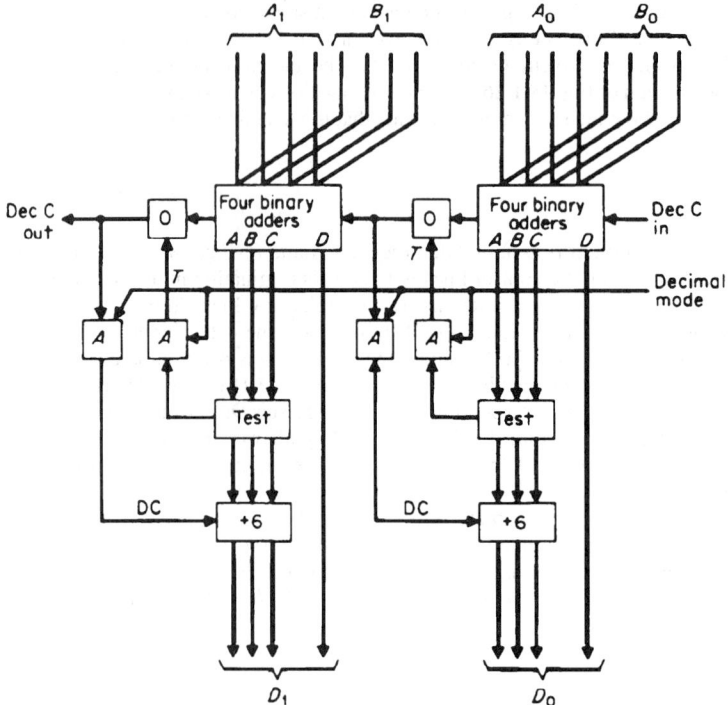

FIGURE 18.1.27 Binary adder with decimal-add feature. If the input of the binary add is 1010 or greater, the addition of a binary 6 produces the correct result by modulo-10 addition.

numbers are added, the result is correct if it lies in the range 0 to 9. If the result is greater than 9, that is the resulting bit pattern is 1010 to 10010, the answer must be adjusted by adding 6 to that group of 4 b, this number being the difference between the desired base (10) and the actual base ($2^4 = 16$). The binary carry from a block of 4 b must also be adjusted to generate the appropriate decimal carry.

The circuits to accomplish decimal addition are shown in Fig. 18.1.27. A test circuit generates an output if the binary sum is 1010 or greater. This output causes a 6 to be added into the sum and also is ORed with the original binary carry to produce a decimal carry. The added circuits needed to perform decimal additions with a binary adder represent one-half to two-thirds of the circuits of the original binary adder.

Decimal Subtraction

In most computers that provide for decimal operation, decimal numbers are stored with a sign and magnitude. To perform subtraction the true complement is formed by subtracting each decimal digit in the number from 9 and adding back 1. Once the complement is formed, addition produces the desired difference.

In a machine that provides for decimal arithmetic a *decimal true-complement* switch may be incorporated in each group of four BCD bits to form the complement. As with binary, provision must be made for the addition of an appropriate low-order digit and for the occurrence of overflows.

Shifting

All computers have shift instructions in their instruction repertoire. Shifting is required, for example, for multiply and divide operations.

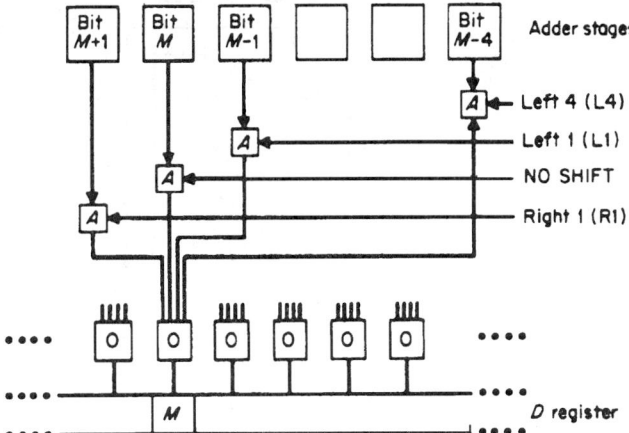

FIGURE 18.1.28   Shift gates. In many systems shifting is accomplished in conjunction with the output of the adder, i.e., position shifts can be accomplished with only one circuit delay.

The minimum is a shift of one position left or right, but most computers have circuits permitting a shift of one or more bits at a time, that is, 1, 4, or 8.

In shifting, a problem arises with the bit(s) shifted out at the end of the register and the new (open) bit position(s). Those shifted out at one end can be inserted in the other end (referred to as *end-around shift* or *circulate*), or they can be discarded or placed in a special register. The newly created vacancies can be filled with all 0s, all 1s, or from another special register.

In a typical computer, shifting is not performed in a shift register but with a set of gates that follow the adders. The outputs of the shift gates are connected to the output register, with an offset by providing a separate gate for each distance and direction of shift. One bit position of the output register can therefore be fed from several adder outputs, but only one gate is opened at a time, as in Fig. 18.1.28. The pattern shown is repeated for every bit position.

Multiplication

Figure 18.1.29 shows a possible data-flow system for multiplication, i.e., the data flow of the adder shown in Fig. 18.1.26 with the addition of register C to hold the multiplier and the shift registers as shown in Fig. 18.1.28. An extra register E holds any extra bits that might be generated in the process of multiplication. Register E in the particular system shown also receives the contents of C after transmission through C's shift register.

Computers also have an extension of the shift instruction called *rotate*. A shift usually occurs as a result, but the leading bit of the shift right or left is rotated to the opposite end of the register. This feature can be used for bit detection without losing data.

The process of binary multiplication involves decoding successive bits in the multiplier, starting with its lowest-order position. If the bit is a 1, the multiplier is added into an accumulating sum; if 0, no addition takes place. In either case the sum is moved one position to the right and the next higher-order bit in the multiplier considered.

In Fig. 18.1.29 the multiplier is stored in register C, the multiplicand in A, and all others are reset to zero. If the low-order position of C is a 1, the contents of B and C are added and shifted one position to the right into register D. After the addition, register C is shifted one position to the right and stored in register E. If the low-order bit in C had been a zero, only register B would have been gated into the adders; i.e., the addition of the contents of A would have been suppressed, but subsequent operations would have remained the same.

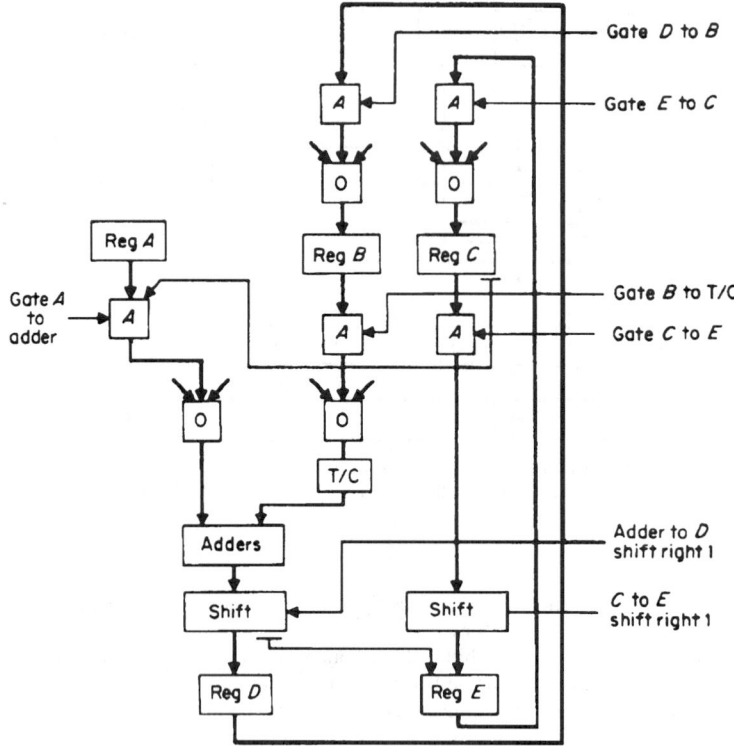

FIGURE 18.1.29   Data flow for multiplication.

Each add and shift operation subsequent to an add may generate low-order bits that may be shifted into register E, since as the contents of C are shifted to the right, unused positions become successively available. After the add and shift cycles, registers D and E are transferred into B and C, respectively, and the process is repeated until all positions of the multiplier are used. The content of D and E is the product.

Division

To provide for the division of two numbers the functions of the registers in Fig. 18.1.29 must be rearranged as shown in Fig. 18.1.30. A gating loop is provided from the shift register to register A. Initially the divisor is placed in register B and the dividend in A. The T/C gate is used to subtract B from A, and if the result is zero or positive, a 1 is placed in the low-order position of register E. If the result is negative, a 0 is placed in E and the input from B ungated to reset to the contents of A. The shift register then shifts the output of the adder one position to the right and gates it back to A. E is gated through C and shifted on to the right. The whole process is repeated until the dividend is exhausted. E contains the quotient and A the remainder.

Floating-Point Operations

In some applications, it is convenient to represent numerical values in floating-point form. Such numbers consist of a *fraction* that represents the number's most significant digits, in a portion of the field of a word in the machine, and a *characteristic* in the remaining portion. The characteristic denotes the decimal-point position

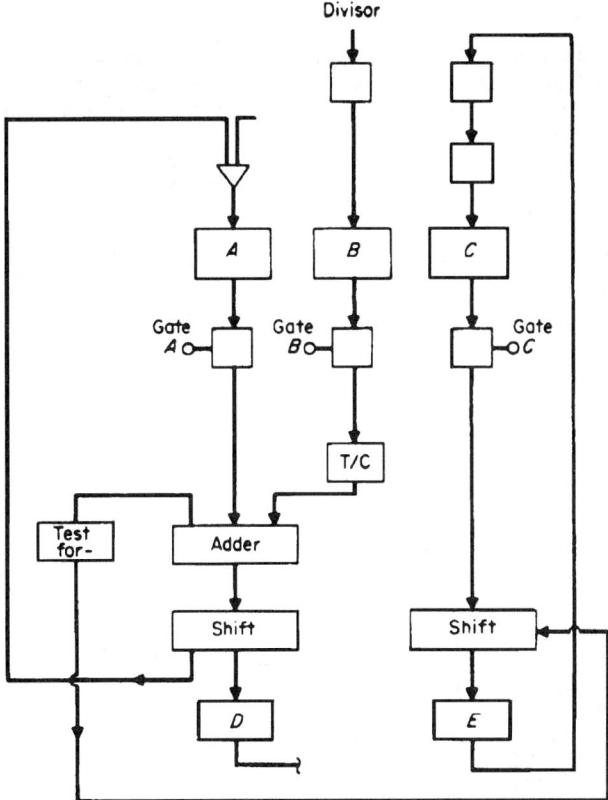

FIGURE 18.1.30   Data flow for division. *B* holds the divisor and *A* the dividend. A trial subtraction is made of the high-order bits of *B* from *A*, and if the results are 0 or positive, a 1 is entered into the lower-order bit of *E* and the result of the subtraction shifted left and reentered into *A* (with gate *A* closed). If the result had been negative the *B* gate would have closed to negate the subtraction, and a 0 would have been entered into *E*. The output of the adder would then have been shifted left one position and reentered in *A*.

relative to the assumed decimal point in the characteristic field. In floating-point addition and subtraction, justification of the fractions according to the contents of the characteristic field must take place before the operation is performed; i.e., the decimal points must be lined up.

In an ALU such as in Fig. 18.1.29, the operation proceeds in the following way. Two numbers *A* and *B* are placed in registers *A* and *B*, respectively. The control selection then gates only the characteristic fields into the adder, in a subtract mode of operation. The absolute difference is stored in an appropriate position of the control section. Controlled by the sign of the subtraction operation, the fraction of the least number is placed through the adder into the shift register. The control section then shifts the fraction the required number of positions, i.e., according to the stored difference of characteristic fields, and places the result back in the appropriate register. Addition or subtraction can then proceed according to the generating machine instruction.

This procedure is costly of machine time. Instead of using the ALU adder for the characteristic-field difference, provision can be made for subtraction of the characteristics in the control section. In such a case, only the characteristics *A* and *B* need be entered into registers *A* and *B* and the lesser number can be placed in *B* so that shifting can be accomplished by the circuits normally used in multiplication.

Control Section

The control section is the part of a CPU that initiates, executes, and monitors the system's operations. It contains clocking circuits and timing rings for opening and closing gates in the ALU. It fetches instructions from main storage, controls execution of these instructions, and maintains the address for the next instruction to be executed. The ALU also initiates I/O operations.

Basic Timing Circuits

Basic to any control unit is a continuously running oscillator. The speed of this oscillator depends on the type of computer (parallel or serial), the speed of the logic circuits, the type of gating used (the type of registers), and the number of logic levels between registers. The oscillator pulses are usually grouped to form the basic operating cycle of the computer, referred to as *machine cycles*. In this example four pulses are combined into such a group.

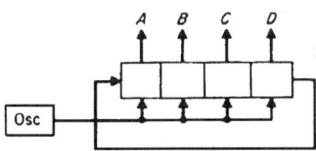

In Fig. 18.1.31 an oscillator is shown that drives a four-stage ring. At any one time, only one stage is on. Suppose an addition is to be performed between registers A and B and the result is to be placed back in B. The addition circuitry described in Fig. 18.1.29 uses three registers, two for the addends (registers A and B), and one for temporary storage (register D). The operation to be performed is to add the content of register A to the content of register B and store the result in register B. Register D is required for temporary storage since if B were connected back on itself, an unstable situation would exist; i.e., its output (modified by A) would feed its input.

The operation of the four-stage clock ring in controlling the addition and transfer is shown in Fig. 18.1.32. Action is initiated by the coincidence of an ADD signal and clock pulse A, which starts the *add ring*. This latter ring has two stages and advances to the next stage upon occurrence of the next A clock pulse. The timing chart in Fig. 18.1.32 describes one sequence of actions required and the gates needed for the addition. The circuit diagram shows a realization of these gates. Each register is reset before it is reused, and all pulses are derived from the four basic clock pulses shown in Fig. 18.1.31. The add ring initiates the add by opening the gates between registers A and B and the adder. The add latch is then reset, D is reset, the ring stage transferred, and so forth. An ADD-FINISHED signal is furnished, and may be used elsewhere in the system.

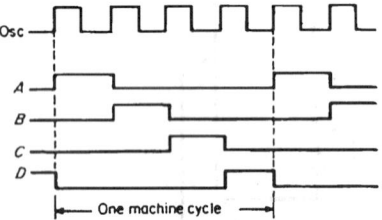

FIGURE 18.1.31 Oscillator and ring circuit. Many control functions can be performed by using an oscillator in conjunction with a timing ring that sequentially sends signals on separate lines, in synchronism with the oscillator.

In the timing diagram, it is assumed that the time required for transmission through the adder is about one machine cycle and that one clock cycle is sufficient for the signal to propagate through the necessary gating and set the information in the target register. These times must include the delay in the circuits and any signal-propagation delay.

Control of Instruction Execution

The following approach for the design of a control section is straightforward but extravagant of hardware. For each instruction in the computer a timing diagram is developed similar to the one shown for the add operation in the previous section. These timings are implemented in rings that vary in length, according to the complexity of the instruction. The concept is simple and has been widely used, but it is costly. To reduce cost, rings are used repeatedly within one instruction and/or by several instructions. Subtraction, for example, might use the ADD ring, except for an extra latch that might be set at ring time 1, clock time A (denoted $1.A$) and reset at $2.C$. This new latch feeds the T/C gates. Another latch that might be added to denote decimal arithmetic would also be set at $1.A$ time and reset at $2.C$ time. The addition of two latches and a few additional ANDS and ORS permits the elimination

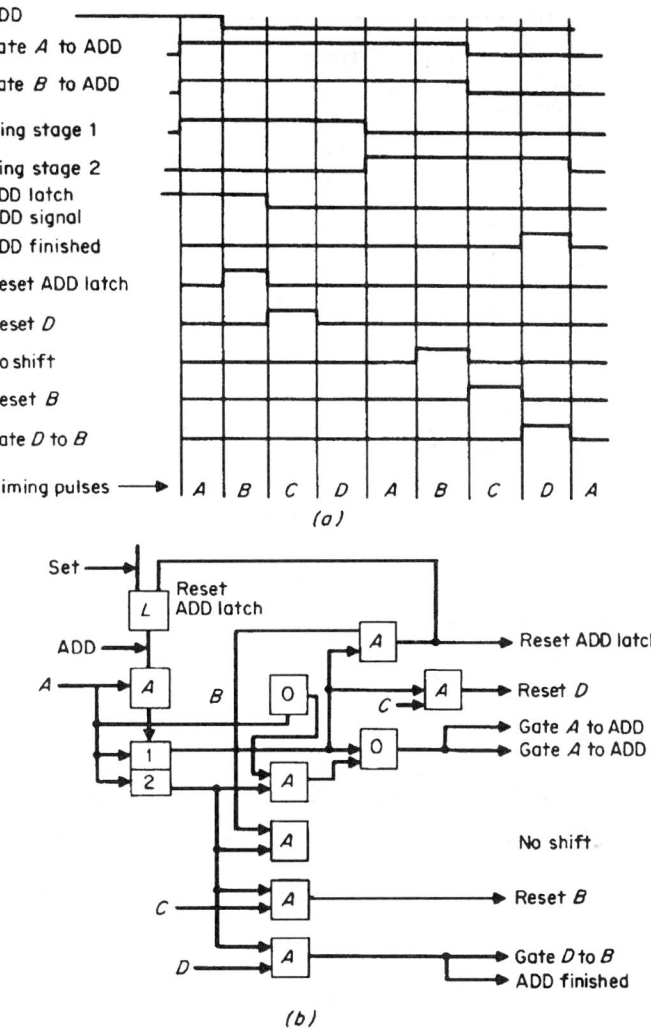

FIGURE 18.1.32   Control circuit (*b*) and timing chart (*a*) for addition. The timing ring shown in Fig. 18.1.31 to control the adder circuit shown in Fig. 18.1.29 with the help of additional switching circuits.

of three two-stage rings and associated logic. Further reductions in the number of required circuits can be achieved by considering the iterative nature of some instructions, such as multiplication, division, or multiple shifting.

Controls Using Counters (Multiplication)

To exemplify this approach multiplication is considered. Multiplication can be implemented as many add-shift cycles, one per digit in the multiplier (say N). The control for such an instruction can be implemented using one $2N + 1$ position ring, the first position initializing and the next $2N$ positions for N add-shift cycles (two positions are needed per add cycle). Such an approach requires not only an unnecessarily long ring but is relatively inflexible for other than N-b multipliers.

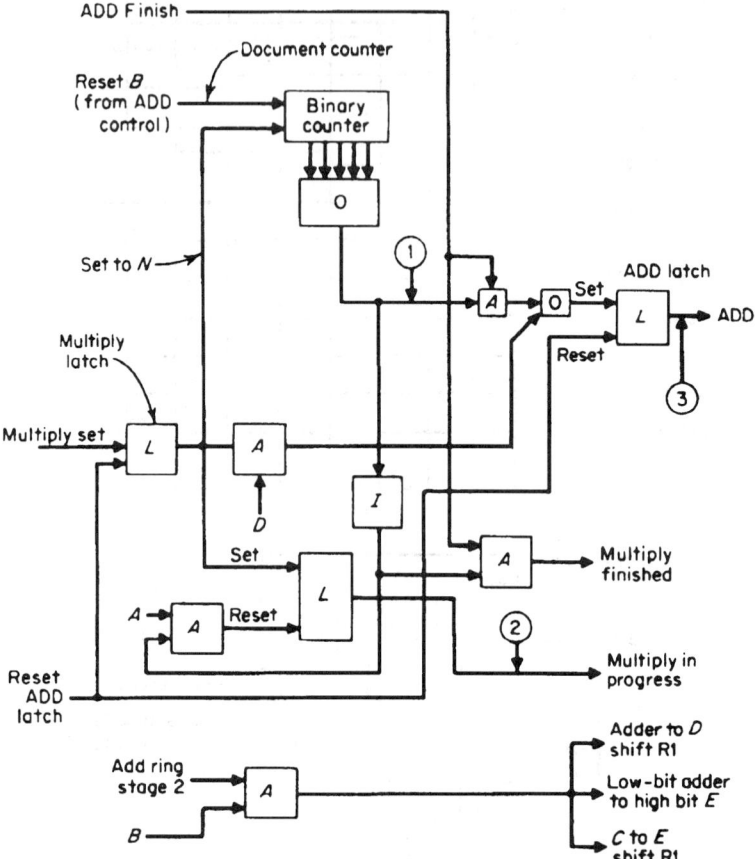

FIGURE 18.1.33 Control circuit for multiplication. Repetitive operations such as add and shift are combined in cyclical fashion. The timing diagram of Fig. 18.1.32*a* is assumed.

The alternate approach, requiring considerably less hardware, uses the basic operation in multiplication, an add and a shift. Therefore a multiply can be implemented by using the controls for the add-shift instruction, plus a binary counter, plus some assorted logic gates to control the gates unique to the multiply, as in Fig. 18.1.33. In the figure some of the less important controls needed for multiply have not been included to simplify the presentation. For example, the NO SHIFT signal for add must be conditioned with a signal that multiply is not in progress, and during multiplication an ADD-FINISHED signal should be ignored (nor start an I cycle), the reset *B* also resets *C*, the reset *D* also resets *E*, gate *D* to *B* also gates *E* to *C*, gating of *A* to the adder is conditioned on the last bit of *C*, and so forth.

In Fig. 18.1.33 the action is started by raising the MPY line that sets a *multiply* latch. This in turn sets the binary counter to the desired cycle count. Also set is the ADD-LATCH, and the addition cycles start. When the counter goes to 0, the ADD-LATCH is no longer set and the MULTIPLY-FINISHED signal is raised.

Microprogramming

The control of the E cycle, in the preceding descriptions, is performed by a set of sequential circuits designed into the machine. Once an execution is initiated, the clocking circuits complete the operation, through wired paths in a specific manner. An alternative method of design for a control unit is *microprogramming*. The concept is not sharply defined but has as its objective implementation of the control section at low cost and with high flexibility.

In many cases it is desirable to design compatible computers, i.e., units with the same instruction set, with widely varying performance and cost. To provide a slower version at a lower cost, the width of the data path is reduced to lower circuit counts in the ALU. On the other hand, to operate with the same instruction set, the reduced data-path width usually implies more iterative operations in the control section, at added cost.

Considerable investment is required for programming development, and normally such systems run only on computers with identical instruction sets. If appropriate flexibility is provided, one computer can mimic the instruction set of another. Microprogramming provides for such operations. The process by which one system mimics another is called emulation.

Control lines are activated not by logic gates in conjunction with counters but by words in a storage system that are translated directly into electric signals. The words selected are under the control of the instruction decoder, but the sequence of words may be controlled by the words themselves by provisions of a field that does not directly control gates but specifies the location of the next word. The name given to these control words is *microinstruction* as opposed to machine instruction or instruction.

When two computers are designed for the same instruction set but with different data-path widths, the microinstruction sets of the two computers are radically different. For the small computer, the program for a given machine instruction is considerably longer than that for the large computer. The same instructions can be used in both computers. The difference in control-system cost between the two is not large. Although the microprogram is longer with the smaller computer, the difference is in the number of storage places provided, not in the control-section hardware.

The design of the sequence of microprogramming words is conceptually little different from other programming. The microprogram implementation, however, requires a thorough knowledge of a machine at the gate level. A *microinstruction counter* is used to remember the location of the next microinstruction, or a field can be provided to allow branching or specification of a next instruction.

A microprogram generally does not reside in main store but in a special unit called a *control store*. This may not be addressable by the user, or it may be a separate storage unit. In many cases, the microprogram is stored in a read-only store (ROS); i.e., it is written at the factory. ROS units are faster and may be cheaper per bit of storage and do not require reloading when power is applied to the computer. Alternatively the microprogram may be stored in medium that can be written into, called a *writable control store* (WCS). By reloading the WCS, entirely different macroinstruction can be implemented, using the same microinstruction set in a different microprogram. By such means emulation is achieved at minimal expense.

CPU Microprogramming

Figure 18.1.34 shows an ALU and Fig. 18.1.35 a control section with microprogram organization. The microinstructions embodied in these two units are shown in Table 18.1.2.

To simplify the program several provisions have been made:

1. Each microinstruction contains the control-store address of the next microinstruction. If omitted, the address in one microinstruction is the current address of the one written just below it. It may *not* be next in numeric sequence.

2. An asterisk at the beginning of a line in a program indicates a comment. This means that the entire line contains information about the program and does not translate into a microinstruction.

3. To simplify the drawing, a gate in a path is indicated by placing an X in the line and omitting from the drawing any control lines controlling these gates. It is also assumed that where two lines join, OR circuits are implied.

4. Rather than listing a numeric value for each field, a shorthand description for the desired action is invented. All actions not so described are assumed to be zero. For example, A to ADD implies that register A is gated to the adder, or T/C means raise T/C gate. These charges do not in any way modify the concept of microprogramming but make the result more readable.

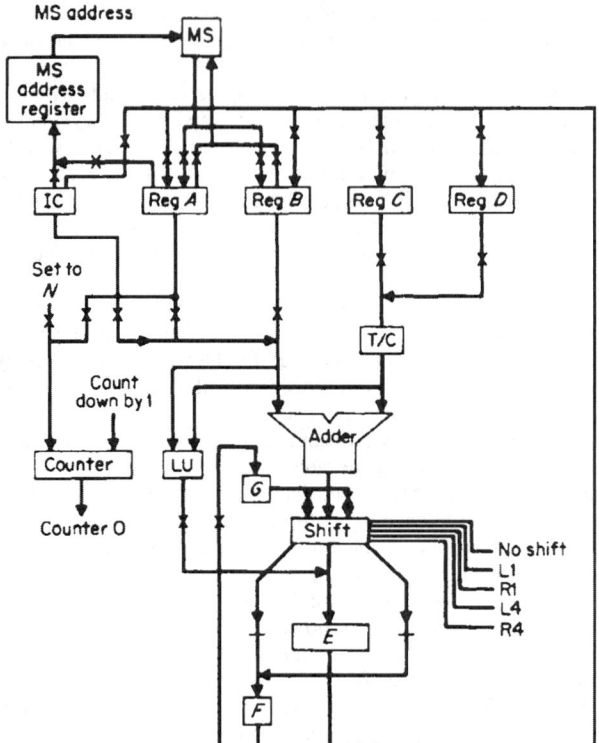

FIGURE 18.1.34 Microprogrammed ALU.

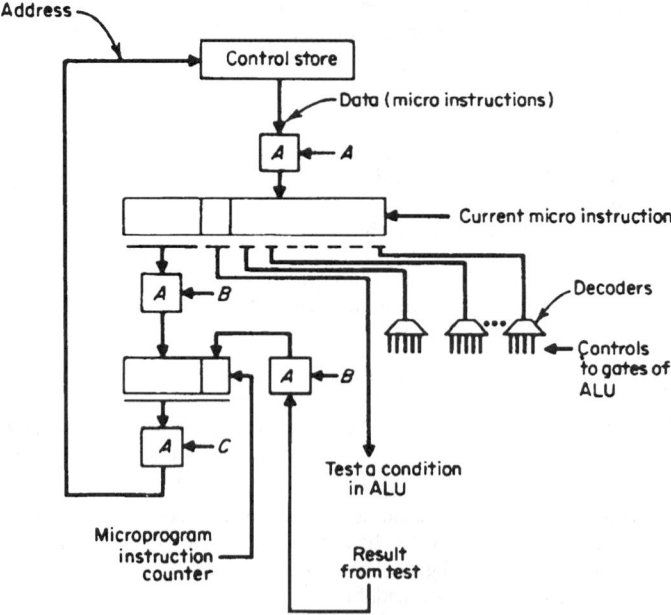

FIGURE 18.1.35 Microprogrammed control unit.

TABLE 18.1.2 Microinstructions Embodied in an ALU and a Control Section

Current address	Microinstruction	Comment
	•ADD	
51	B to ADD, C to ADD, NO-SHIFT	Add two operands
8	E to B	
9	A(2) to MS-ADR-REG, B to MS, GO TO 1	Store result branch to 1, next micro-instruction executed to be taken from control-store location 1
	•SUBTRACT	
52	B to ADD, C to ADD, NO-SHIFT T/C, GO TO 8	Go to 9, where result is stored
	•BRANCH (unconditional)	
53	A(3) to 1C, GO TO 1	This is the macroinstruction branch
	•MULTIPLY	
54	Set an N into counter, C to ADD, NO-SHIFT	Initialize
17	E to D, set C to 0s	
18	If last bit D = 1, then (B to ADD), C to ADD shift-R 1, 0 to input of high end of shifter, output of low end of shifter to F	Perform one add shift if last bit D = 1, only shift if last D = 0
19	E to C, F to G, COUNT down by 1	Increment counter
20	D to ADD, SHIFT-R1, G to input of high end of shifter	Shift D
21	E to D, if counter is not 0 then GO TO 17	Close loop
22	C to ADD	Store result in two MS locations
23	E to B, force 1 into C	
24	A(2) to ADD, C to ADD NO-SHIFT; A(2) to MS-ADR-REG B to MS	Store first half of result, increment result address
25	E to A(2)	
26	D to ADD, NO-SHIFT	
27	E to B	
28	A(2) to MS-ADR-REG, B to MS GOT TO 1	Store second half of result Branch to 1-fetch
	•SHIFT LEFT	
55	A(2) to COUNTER, C to ADD, NO-SHIFT	Operand 1 is number of bits operand 2 is to be shifted
10	If COUNTER = 0 then GO TO 9 E to B	Test if shift count was 0
11	B to ADD, SHIFT L1, COUNT down by 1	Shift loop
12	IF COUNT not 0, then GO TO 11 E to B	
13	GO TO 9	Completed shift

NOTE 1: (Location 7) This special test places the op-code bit pattern into the low part of the address for the next instruction, causing a branch to the appropriate microroutine for each op code. Branches are as follows:

Op code	Instruction	Address
1	ADD	51
2	SUBTRACT	52
3	BRANCH	53
4	MULTIPLY	54
5	SHIFT LEFT	55

NOTE 2: (Location 1) It is assumed that START button sets microinstruction counter to 1.

Instruction FETCH

In the preceding paragraphs the execution of instructions (E cycles) is discussed. In these cases, operations are initiated by setting an appropriate latch for the function to be performed. The signals that set the latch are in turn generated by circuits that interpret the information of the operation-code part of an instruction cycle (I cycle).

Whenever an instruction has completed execution (or when the computer operator presses the start button on the console), a *start-cycle* signal is generated at the next *A* time of the master-clock timing ring. This signal starts the I-cycle ring. The first action of this ring is to fetch the instruction to be executed from the main store by gating the instruction counter (IC) (sometimes called *program counter*) to the address lines of main storage and initiating a main-store cycle by pulsing a line called *start MS*.

These operations are illustrated in Fig. 18.1.36. At ring time 2, the instruction arrives back and is placed in the instruction register (IR). The instruction typically contains three main fields: the *operation code* (op code),

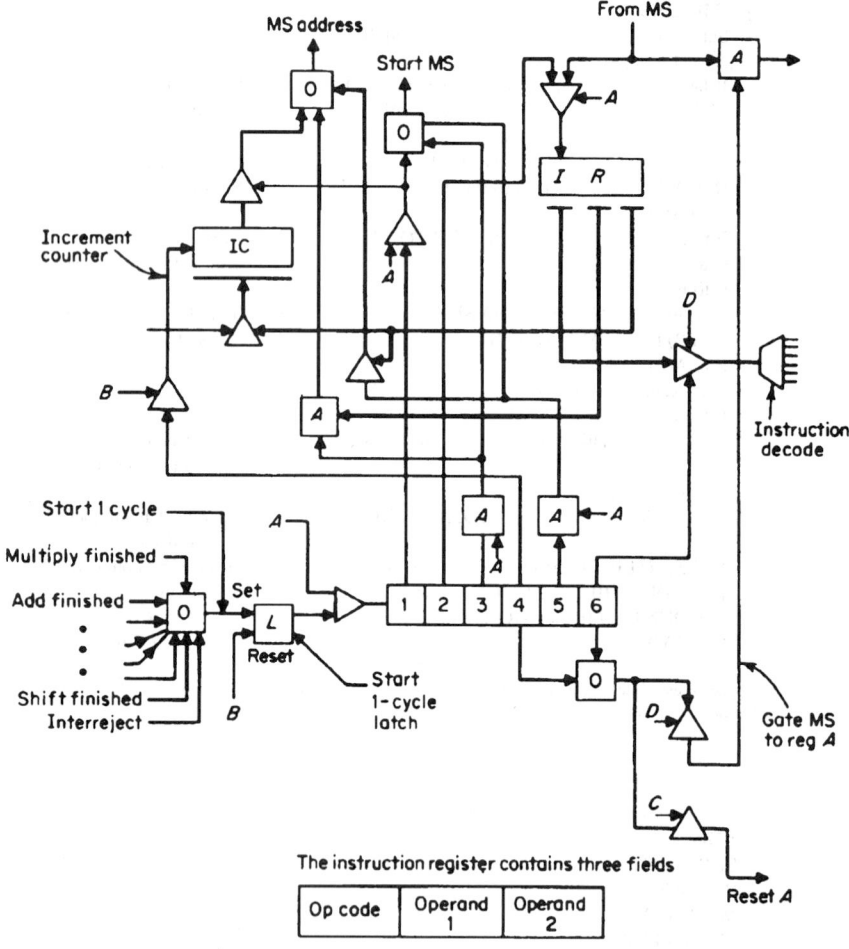

FIGURE 18.1.36 Implementation of equipment for an I cycle in a two-address machine. The instruction is first brought from main store and deposited in the instruction register. The operation proceeds by successively gating the information associated with two address fields into register *A*. The first is moved out of *A*, while the second is being sought from main store.

that determines what instructions is to be executed (ADD, SUBTRACT, MULTIPLY, DIVIDE, BRANCH), and the two addresses of the two operands participating in the operation. For certain classes of instruction, the operands must be delivered to appropriate locations before the E cycle begins.

During ring time 3 and 4, the first operand is fetched from main store and stored in A. During ring time 5, this operand must be transferred from A to an alternate location, depending on the nature of the instruction being executed. During ring time 5 and 6, the second operand is fetched and stored in A. Ring time 6 is also used to gate the op code to the *instruction decoder*, which is a combinatorial logic circuit accepting the *operation code*, or *instruction code*, from the P b in the op code. The decoder has 2^P input wires, one for each unique input combination. Thus for each bit combination entering the decoder, one output line is activated. These signals represent the start of an E cycle with each wire initiating the execution of one instruction by setting some latch, e.g., an add or multiply latch, or by initiating an appropriate microprogram.

Some op-code bit combinations may not be valid instruction codes. The outputs of the decoder are ORed together and fed into a circuit that ultimately interrupts the normal processing of the computer and indicates the invalid condition. At the beginning of the I cycle, the content of the instruction counter (IC) points to the instruction to be executed next. An *increment-counter* signal is generated during the I cycle in order to increment the counter, so that the address stored in IC points to the next sequential instruction in the program.

Instruction- and Execution-Cycle Control

In normal program operation, I and E cycles alternate. The I cycle brings forth an instruction from storage, sets up the ALU for execution, resets the instruction counter for the next I cycle, and initiates an E cycle. A number of conditions serve to interrupt this orderly flow, as follows:

1. When no more work is to be done, the end of a program is reached and the computer goes to a WAIT state. Such a state is reached, for example, by a specific instruction that terminates operations at the end of the I cycle in question, by a signal from the instruction counter when a predetermined limit is reached.

2. A STOP button may prevent the next setting of the *Start I cycle* hatch. A START button resets this latch.

3. When starting up after a shutdown, e.g., in the morning, activity is usually initiated by depressing an INITIAL PROGRAM LOAD button on the operator's console (the name varies from system to system, e.g., IPL, LOAD, START). The button usually performs three functions: a reset of all rings, latches, and registers to some predefined initial condition; a read-in operation from an I/O device into main store (usually the first few locations) of a short program; and an initiation of an I cycle. Program execution generally starts at a fixed location so that the IC initially is set to this value. In most computers, including PCs, this value is called IMPL (initial microprogrammed program load), which initializes the machine and performs self-testing on memory, I/O ports, and internal status conditions.

4. In multiprogramming, i.e., concurrent operations on more than one program, only one program at a time is in operation in the CPU, but transfers occur from one to another as required by I/O accesses, and so forth. The program to be transferred is handled by an *interrupt*. Under interrupt, an address is forced into the instruction counter so that on completion of the E cycle of the current program, a new instruction is referenced that starts the interruption. This instruction initiates program steps that store the data of the old program, e.g., in special registers or in special main-store locations. The contents of the IC, part of the IR, the contents of any registers, and the reason for the interruption are stored. The collection of these fields together is called a *program status word* (PSW). In most computers, this information, plus general registers must be saved when switching machine states. It can be referenced to reinitiate action at later time on the program interrupted.

Branch and Jump Instructions

Two kinds of instruction permit change in program sequence: *conditional* and *unconditional*. The purpose of such instructions is to permit the system to make some decision, so as to alter the flow of program, and to continue execution at some point not in the original sequence. In nonconditional branches the original program instruction provides for a branch or jump whenever the particular instruction occurs.

Conditional branches take the extra step of determining if some condition to be tested has been satisfied. Either the op code or the operand 1 field normally defines the test and/or the condition to be tested for. If the specified test has been satisfied, the branch is executed as described above. Otherwise no action is taken, and the next normally sequenced instruction is executed.

Advanced Architectural Features

The basic structure of the digital computer and its operation have been described in the previous paragraphs. This structure has proved to be flexible and adaptable to the solution of many different applications. There is, however, a continuing need to increase the performance of the computer and to make it easier to program so that even more applications can be handled. This has been achieved through a number of different approaches. One has been to develop sophisticated operating systems and to couple them closely to the hardware design of the computer. The net effect is that a programmer viewing the computer does not distinguish between the hardware and the operating system but perceives them as an integrated whole.

A second solution has been the development of architectural features that permits overlapped processing operations within the computer. This is in contrast to earlier computers that were strictly sequential and resulted in reduced performance and low throughput because vulnerable computer resources could remain idle for relatively long periods of time. The newer architectural features include such concepts as data channels, storage-organization enhancements, and pipelining described briefly in the following paragraphs. Another totally different approach to improved computer performance has been the development of computers with non-von Neumann architecture, described in the next section.

An increase in computer performance can be achieved by overlapping input/output (I/O) operations and processor operations. *Channels* have been introduced in large systems to permit concurrent reading, writing, and computing. The channel is in effect a special processor that acts as a data and control buffer between the computer and its peripheral devices. Figure 18.1.37 shows the organization of the computer when channels are introduced. Each channel can accommodate one or more I/O device control units. The channel is designed with a standard interface that is designed to permit a standard set of control status signals and data signals and sequences to be used to control I/O devices. This permits a number of different I/O devices to be attached to each channel by using I/O device control units that also meet the standard interface requirements. Each device control unit is usually designed to function with only one I/O device type, but one control unit may control several devices.

The channel functions independently of the CPU. It has its own program that controls its operations. The CPU controls channel activity by executing I/O instructions. These cause it to send control information to the channel and to initiate its operation. The channel then functions independently by being given a channel program to execute. This program contains the commands to be executed by the channel as well as the addresses of storage locations to be used in the transfer of data between main storage and the I/O devices. The channel in turn issues orders to the device control unit, which in turn controls the selected I/O device. When the I/O operation is completed, the channel interrupts the CPU by sending it a signal indicating that the channel is

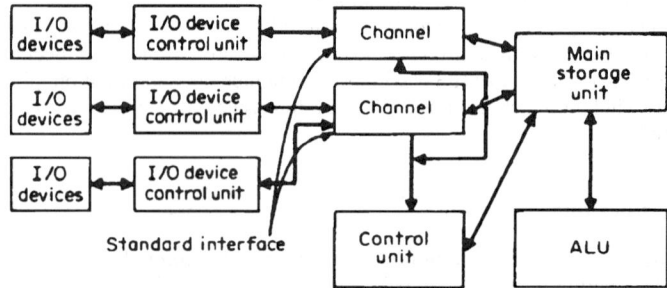

FIGURE 18.1.37 Computer organization with channels, separate logical processors that permit simultaneous input, output, and processing.

again free to perform further I/O operations. Several channels can be attached to the CPU and can operate concurrently. In PCs, I/O operation is controlled by adapter cards that fit into the motherboard.

Cache Storage

Cache storage was introduced to achieve a significant increase in the performance of the CPU at only a modest increase in cost. It is, in effect, a very high-speed storage unit that is added to the computer but is designed to operate in a unique way with the main storage unit. It is transparent to the program at the instruction level and can thus be added to a computer design without changing the instruction set or requiring modification to existing programs. Cache storage was first introduced commercially on the IBM System/360 model 85 in 1968.

Virtual Storage

Properly using and managing the memory resources available in the computer has been a continuing problem. The programmer never seems to have enough high-speed main storage and has been forced to use fairly elaborate procedures such as overlays to make programs fit into main storage and run efficiently. *Virtual-storage* systems were introduced to permit the programmer to think of memory as one uniform *single-level storage* unit but to provide a *dynamic address-translation unit* that automatically moves program blocks on pages between auxiliary storage and high-speed storage on demand.

Pipelining

A further improvement in computer performance was achieved through the use of *pipelining*. This technique consists of decomposing repetitive processes within the computer into subprocesses that can be executed concurrently and in an overlapped fashion.

Instruction execution in the control unit of the CPU lends itself to pipelining. As discussed earlier, instruction execution consists of several steps that can be executed relatively independently of each other. There is instruction fetch, decoding fetching the operands, and then execution of the instruction. Separate units can be designed to perform each one of these steps. As each unit finishes its activity on an instruction, it passes it on to the next succeeding unit and begins to work on the next instruction in succession. Even though each instruction takes as long to execute overall as it does in a conventional design, the net effect of pipelining is to increase the overall performance of the computer. For example, under optimal conditions, once the pipeline is full, when one instruction finishes, the next instruction is only one unit behind it in the pipeline. In this four-unit example, the net effect would be to increase the speed of instruction execution by a factor of 4.

This approach can be carried to the point where 20 or 30 instructions are at various stages of execution at one time. This type of processing is called *pipeline processing*. No difficulties arise during uninterrupted processing. When an interrupt does occur, however, it is difficult to determine which instruction has caused the interrupt since the interrupt may arise in a subsystem sometime after the IC has initiated action. In the meantime, the IC may have started a number of subtasks elsewhere by stepping through subsequent cycles. Operands within subunits may not be saved in an arbitrary intermediate state, since information is in the process of being generated for return to the main program. Because of the requirement that no further I cycles be started, interrupt is signaled when the pipeline is empty. At the time the interrupt is signaled, the IC does not point at the instruction causing the interrupt but somewhat past it. This type of interrupt is called *imprecise*.

Advanced Organizations

In addition to the von Neumann computer and its enhancements, a number of other computer organizations have been developed to provide alternative approaches to satisfying the needs for improved computational performance, increased throughput, and improved system reliability and availability. In addition, certain unique architectures have been proposed to solve specific problems or classes of problems.

TABLE 18.1.3 Classification Scheme for Computer Architectures

Acronym	Meaning	Instruction streams	Data streams	Examples
SISD	Single instruction stream, single data stream	1	1	IBM 370, DEC VAX, Macintosh
SIMD	Single instruction stream, multiple data stream	1	>1	ILLIAC IV, Connection Machine, NASA's MPP
MISD	Multiple instruction stream, single data stream	>1	1	Not used
MIMD	Multiple instruction stream, multiple data stream	>1	>1	Cray X/MP, Cedar, Butterfly

Computer organizations can be categorized according to the number of procedure (instruction) streams and the number of data streams processed: (1) a single-instruction-stream-single-data-stream (SISD) organization, which is the conventional computer; (2) a multiple-instruction-stream-multiple-data-stream (MIMD) organization, which includes multiprocessor or multicomputer systems; (3) a single-instruction-stream-multiple-data-stream (SIMD) organization; this uses a single control unit that executes a single instruction at a time, but the operation is applied across a number of processing units each of which acts in a synchronous, concurrent fashion on its own data set (parallel and associative processors fall into this category); and (4) a multiple-instruction-stream-single-data-stream (MISD) organization. (Pipeline processors fall into this category.) Table 18.1.3 summarizes the categories.

One way to achieve an improvement in performance and to improve reliability at the same time is to use multiprocessors. The American National Standard Vocabulary for Information Processing defines a multiprocessor as "a computer employing two or more processing units under integrated control." Enslow[61a] amplifies this definition by pointing out that a multiprocessor contains two or more processors of approximately comparable capabilities. Furthermore, all processors share access to common storage, to I/O channels, control units, and devices. Finally, the entire system is controlled by one operating system that provides interaction between processors and their programs.

A number of different multiprocessor system organizations have been developed: (1) time-shared, or common-bus, systems that use a communication path to connect all functional units, (2) crossbar-switch systems that use a crossbar switching matrix to interconnect various system elements, (3) multiport storage systems in which the switching and control logic is concentrated at the interface to the memory units, and (4) networking of computers interconnected via high-speed buses.

A number of large problems require high throughput rates on structured data, e.g., weather forecasting, nuclear-reactor calculations, pattern recognition, and ballistic-missile defense. Problems like these require high computation rates and may not be solved cost-effectively using a general-purpose (SISD) computer. Parallel processors, which are SIMD organization types, were designed to address problems of this nature. A parallel processor consists of a series of process elements (cells) each having data memories and operand registers. The cells are interconnected. A central control unit accesses a program, interprets each program step, and broadcasts the same instructions to all the processing elements simultaneously.

Distributed Processing

One of the newest concepts in computer organizations is that of *distributed processing*. The term distributed processing has been loosely applied to any computer system that has any degree of decentralization. The consensus seems to be that a distributed processing system consists of a number of processing elements (not necessarily identical), which are interconnected but which operate with a distributed, i.e., decentralized, control of all resources. With the advent of the less expensive micro- and miniprocessors, distributed processing is receiving much attention since it offers the potential for organizing these processors so that they can handle problems that would otherwise require more expensive supercomputers. Through resource sharing and decentralized control, distributed processing also provides for reliability and extensibility since processors can be removed or added to the system without disrupting system operations. The types of distributed processing can be described by configurations in terms of the processing elements, paths, and switching elements: (1) loop, (2) complete interconnection, (3) central memory, (4) global bus, (5) star, (6) loop with central switch, (7) bus with central switch, (8) regular network, (9) irregular network, and (10) bus window. Distributed systems have been designed that fall into each of these categories. Hybrid forms use combinations of two or more of these architectural types.

Distributed and parallel processing systems can be thought of as being loosely coupled (each has its own CPU and memory) or tightly coupled (each has its own CPU and shares the same memory). Master–slave and client–server systems are part of the distributed and cooperative processing. In master–slave, one computer controls another, whereas in client–server, the requesting computer is the client and the requested resource is the server.

Stack Computers

In the CPUs discussed thus far, the instructions store all results, so that the next time an operand is used it has to be fetched. For example, a program to add A, B, and C and put result into E appears as

$$MOVE\ A\ to\ E \qquad \text{1 fetch, 1 store}$$
$$ADD\ E\ to\ B\ store\ in\ E \qquad \text{2 fetch, 1 store}$$
$$ADD\ E\ to\ C\ store\ in\ E \qquad \text{2 fetch, 1 store}$$

In languages such as PL/I or FORTRAN this program might be written as the single statement

$$E = A + B + C$$

This equation describes a sequence of actions, as in the case of the program, but the specific sequence is not described. Since addition is commutative, a correct result is achieved by $E = [(A + B) + C]$, $E = [A + (B + C)]$, or $E = [(A + C) + B]$, each step occurring in any order. The computer, however, uses a specific program in achieving a result so that the method of writing the equation must generate a specific sequence of actions.

A method of writing an equation that specifies the order of operation is called *Polish notation*. For the above example of addition, one possible Polish string would be

$$AB + C + E =$$

In this string, the system would find A and B and, as determined by the plus sign *following* the two operands, add them. The result is then combined with C under addition called for by the second plus sign. The $E =$ symbols indicates that the result is to be stored in E. The plus sign appears *after* the A and B, and the specific string shown is called *postfixed*. An equivalent convention could place the operator first and would be called *prefixed*.

Any complex expression can be translated into a Polish string. For example in PL/I language, the statement

$$M = (A + B)*(C + D*E) - F;$$

means evaluate the right-hand side of the equation and store the result in the main-store location corresponding to variable M (asterisks indicate multiplication). The Polish string translation for this statement is

$$AB + DE*C + *F - M =$$

In translation from the types of expression permitted by higher-level languages, a machine can be programmed to analyze successively an arithmetic expression of the types shown above. In so doing, first, the outermost expressed or implied parentheses are aggregated and successively broken down until not more quantities remain. The first such quantities analyzed are generally the last computed, so that in the development of a Polish string from an algebraic expression, a first-in, last-out situation prevails.

Stacks

Evaluation of a Polish string in a machine is best performed using a *stack* (push-down list). A stack has the property that it holds numbers in order. A PUSH command places a value on the stack; i.e., it stores a number and an operation at the top of the stack and, in the process, lowers all previous items by one position. Numbers are retrieved from the stack by issuing a POP command. The number returned by the stack on a POP command is the most recently PUSHED one. The following example illustrates the behavior of a stack. The value in parentheses is the value *placed* on the stack for PUSH and returned by the stack for POP (assume the stack is initially empty):

PUSH (*A*)	stack contains	*A*
PUSH (*B*)	stack contains	*B A*
POP (*B*)	stack contains	*A*
PUSH (*C*)	stack contains	*C A*
POP (*C*)	stack contains	*A*
POP (*A*)	stack contains	nothing

Such a stack lends itself very well to the evaluation of Polish strings. The rules for evaluation are:

1. Scan the string from left to right.

2. If a variable (or constant) is encountered, fetch it from main store and place its value on the stack.

3. If an operator is encountered, POP the operands and PUSH the result.

4. Stop at the end of the string. If executed correctly, the stack is in the same state at the end of execution as it was at the start.

The advantage of using a stack is that intermediate results never need storing and therefore no intermediate variables are needed. In sequences of instructions where there are no branches, the operations can be stored in a stack. A program becomes a series of such stacks put together between branches.

This approach is called *stack processing*. In stack processing, a program consists of many Polish strings that are executed one after another. In some cases the entire program may be considered to be one long string.

Stacks are implemented by using a series of parallel shift registers, one per bit of the character code. The input is placed into the leftmost set of register positions. PUSH moves an entry to the right, and POP moves it to the left. The length of the shift registers is finite and fixed. The stack, however, usually must appear to the user as though it were infinitely deep. The stack is thus implemented so that the most active locations are in the shift register, and if the shift register overflows, the number shifted out at the right on a PUSH is placed in main storage. There the order is maintained by hardware, microprogramming, or a normal system program.

Trends in Computer Organization

Several trends in computer architecture and organization may have a significant impact on future computer systems. The first of these are the *data-flow computers*, which are data-driven rather than control-driven. In the data-flow computer, an instruction is ready for execution when its operands have arrived; there is no concept of control flow, and there is no program counter. A data-flow program can feature concurrent processing since many instructions can be ready for execution at the same time. In another area *capability systems* are receiving increased attention because their inherent protection facilities make them ideal for implementing secure operating systems. A capability is a protected token (key) authorizing the use of the object named in the token. One approach to implementing capabilities is through a *tagged* architecture, where tag bits are added to each word in storage and to each register. This tag specifies whether the contents represent a capability or not.

Special-Purpose Processors

Certain classes of problems require unique processing capabilities not found in general-purpose computers. Special-purpose processors have been designed to solve these problems. In some cases they have been designed as stand-alone processors, but often they are designed to be attached to a general-purpose computer that acts as the host. One such class of special-purpose processors is the *associative processor.* It uses an associative store. Unlike the storage units described earlier, which require explicit addresses, an associative store retrieves data from memory locations based upon their content. The associative store does its searching in parallel over its entire storage in approximately the same time as required to access a single word in a conventional storage unit.

In digital signal processing (DSP), many repetitive mathematical operations must be performed, e.g., fast Fourier transforms, and require a large number of multiplications and summations. DSP algorithms are being used extensively in graphics processing and modem transmission. The *array processor* has been designed for these types of operations. It has a high-speed arithmetic processor and its own control unit and can operate on a number of operands in parallel. It attaches to a host CPU, from which it receives its initiation commands and data and to which it returns the finished results of its computation.

The *hybrid processor* uses a host digital CPU to which is attached an analog computer. These systems operate in a digital-analog mode and provide the advantages of both methods of computation. Hybrid processors are being replaced by very high-speed digital processors.

HARDWARE DESIGN

Design and Packaging of Computer Systems

Even a small computer may have as many as 3 million integrated circuits on a single chip, whereas a large system may have up to 100 million by the year 2000.

Nanosecond circuit speeds are common in high-performance computer systems. Since light travels approximately $1/3$ m in 1 ns, system configurations must be kept small to take advantage of the speed potential of available circuits. Thus the emphasis is on the use of microcircuit fabrication and packaging techniques. The layout of the system must minimize the length and complexity of these interconnections and must be realized, without error, from detailed manufacturing instructions. To permit these requirements to be met a *basic* circuit package must be available. The upper limit to the size of such a basic unit, e.g., an integrated circuit, is set by the number of crystal defects per unit area of silicon. If these defects are distributed at random, the selection of too large a chip size results in some inoperative circuits on a majority of chips. There is thus an economic balance between the number of circuits that can be fabricated in an integrated circuit and the yield of the manufacturing process.

Another limit on the size of the basic package is set by the number of interconnections between the integrated circuits. Reduced VLSI-chip power requirements and shrinking device dimensions have increased the number of circuits on a chip.

CHAPTER 18.2
COMPUTER STORAGE

BASIC CONCEPTS

The main memory attached to a processor represents the most crucial resource of the computing system. In most instances, the storage system determines the bulk of any general-purpose computer architecture. Once the word size and instruction set have been specified, the rest of computer architecture and design deals mainly with optimization of the attached memory and storage hierarchy. Early computers were designed by first choosing the main memory hardware. This not only specified much of the remaining architecture, e.g., serial or parallel machine, but also dictated the *processor cycle time*, which was chosen equal to the *memory cycle time*. As computer technology evolved, the logic circuits and hence processor cycle time[*] improved dramatically. This improved speed demanded more main-memory capacity to keep the processor busy, but the need for increasingly larger capacity at higher speed placed a difficult and usually impossible demand on memory technology. In the early 1960s, a gap appeared between the main-memory and processor cycle times, and the gap grew with time. Fundamentally, it is desirable that main-memory cycle time be approximately equal to the processor cycle time, so this gap could not continually widen without serious consequences. In the late 1960s, this gap became intolerable and was bridged by the "cache" concept, introduced by IBM in the System/36 Model 85.

The cache concept proved to be so useful and important that by the late 1970s it became quite common in small, medium, and large machine architectures. The *cache* is a relatively small high-speed random-access memory that is paged out of main memory and holds the most recently and frequently used instructions and data. The same fundamental concepts that provide the basis for cache design apply equally well to "virtual memory" systems; only the methods of implementation are different.

In terms of implementation methods, there are five types of storage systems used in computers.

1. *Random-access memory* is one for which any location (word, bit, byte, record) of relatively small size has a unique, physically wired-in addressing mechanism and is retrieved in one memory-cycle time interval. The time to retrieve from any given location is made to be the same for all locations.

2. *Direct-access storage* is a system for which any location (word, record, and so on) is not physically wired in and addressing is accomplished by a combination of direct access to reach a general vicinity plus sequential searching, counting, or waiting to find the final location. *Access time* depends on the physical location of the record at any given time; thus access time can vary considerably, from record to record and to a given record when accessed at a different time. Since addressing is not wired in, the storage medium must contain a certain amount of information to assist in the location of the desired data. This is referred to as *stored addressing information.*

3. *Sequential-access storage* designates a system for which the stored words or records do not have a unique address and are stored and retrieved entirely sequentially. Stored addressing information in the form of

---

[*]Processor cycle time is roughly 10 logic-gate delays with appropriate circuit and package loads.

simple interrecord gaps is used to separate records and assist in retrieval. Access time varies with the record being accessed, as with direct access, but sequentially accessing may require a search of every record in the storage medium before the correct one is located.

4. *Associative (content-addressable) memory* is a random-access type of memory that in addition to having a conventional wired-in addressing mechanism also has wired-in logic that makes possible a comparison of desired bit locations for a specified match for all words simultaneously during one memory-cycle time. Thus, the specific address of a desired word need not be known since a portion of its contents can be used to access the word. All words that match the specified bit locations are flagged and can then be addressed on subsequent memory cycles.

5. *Read-only memory (ROM)* is a memory that has permanently stored information programmed during the manufacturing process and can only be read and never destroyed. There are several variations of ROM. *Postable* or *programmable* ROM (PROM) is one for which the stored information need not be written in during the manufacturing process but can be written at any time, even while the system is in use, i.e., it can be posted at any time. However, once written, the medium cannot be erased and rewritten. Another variation is a *fast-read, slow-write* memory for which writing is an order of magnitude slower than reading. In one such case, the writing is done much as in random-access memory but very slowly to permit use of low-cost devices. Several types of ROM chips are available. The erasable programmable ROM (EPROM) allows the use of ultraviolet light (UV) to erase its contents. This UV-EPROM can take up to 20 min to erase its contents; a ROM burner may then be used to implant new data. An electrically erasable programmable ROM (EEPROM) allows the use of electrical energy to clear its memory. The flash memory EPROM is a high-speed erasable EPROM.

The various computer-storage types are related to each other through access time. An approximate rule of thumb for access time comparisons is as follows (where T = access time):

$$T_c = 10^{-1}T_m = 10^{-6}T_d = 10^{-9}T_t$$

where T_c = cache time
$\quad T_m$ = main time
$\quad T_d$ = disc time
$\quad T_t$ = tape time

STORAGE-SYSTEM PARAMETERS

In any storage system the most important parameters are the capacity of a given module, the access time to any piece of stored information, the data rate at which the stored information can be read out (once found), the cycle time (how frequently the system can be accessed for new information), and the cost to implement all these functions.

Capacity is simply the maximum number of bits (b), bytes (B), or words that can be assembled in one basic self-contained operating module.

Access time can vary depending on the type of storage. For random-access memory the *access time* is the time from the instant a request appears in an address register until the desired information appears in an output register, where it can subsequently be further processed. For nonrandom-access storage, the access time is the time from the instant an instruction is decoded asking for information until the desired information is found but not read. Thus, access time is a different quantity for random and nonrandom-access storage. In fact, it is the access time that distinguishes the two, as is evident by the definitions above. Access time is made constant on random-access memory whereas on nonrandom storage access time may vary substantially, depending on the location of information being sought and the current position of the storage system relative to that information.

Data rate is the rate (usually bits per second, bytes per second, or words per second) at which data can be read out of a storage device. *Data transfer time* for reading or writing equals the product of the data rate and the quantity of the information being transferred. Data rate is usually associated with nonrandom-access storage where large pieces of information are stored and read serially. Since an entire word is then read out of random-access memory in parallel, data rate has no significance for such memories.

Cycle time is the rate at which a memory can be accessed, i.e., the number of accesses per unit time, and is applicable primarily to random-access storage.

FUNDAMENTAL SYSTEM REQUIREMENTS FOR STORAGE AND RETRIEVAL

In order to be able to store and subsequently find and retrieve information, a storage system must have the following four basic requirements:

Medium for storing energy

Energy source for writing the information, i.e., write transducers on word and bit lines

Energy sources and sensors to read, i.e., read and sense transducers

Information addressing capability, i.e., address-selection mechanism for reading and writing

The fourth requirement implicitly includes some coincidence mechanisms within the system to bring the necessary energy to the proper position on the medium for writing and a coincidence mechanism for associating the sensed information with the proper location during reading. In random-access memory, it is provided by the coincidence of electric pulses within the storage cell, whereas in nonrandom-access storage, it is commonly provided by the coincidence of an electric signal with mechanical position. In many cases, the write energy source serves as the read energy source as well, thus leaving only sense transducers for the third requirement. Nevertheless, a read energy source is still a basic requirement.

The differences between storage systems lie only in how these four requirements are implemented and, more specifically, in the number of transducers required to achieve these necessary functions. Here a *transducer* denotes any device (such as magnetic head, laser, transistor circuits) that generates the necessary energies for reading and writing, senses restored energy and generates a sense signal, or provides the decoding for address selection.

Since random-access memory is so crucial to processor architecture, we discuss this type first and continue in the order defined above.

The organization of random-access memory systems is intimately dependent on the number of functional terminals inherent in the memory cells. The cell and overall array organization are so interwoven that a discussion of one is difficult without the other. We discuss first the fundamental building block, the memory cell, and subsequently build the array structure from this.

RANDOM-ACCESS MEMORY CELLS

To be useful in random-access memory, the cell must have at least two independent functional terminals consisting of a word line and a common bit/sense line, as shown in Fig 18.2.1a. For writing into the cell, the coincidence of a pulse on the word-select line with the desired data on the bit-select line places the cell into one of two stable binary states. For reading, only a pulse on the word-select line is applied, and a sense signal indicating the binary state of the cell is obtained on the sense line. A more versatile but more complex cell is one having three functional terminals, as in Fig. 18.2.1b. A coincidence of pulses on both the x and y lines is necessary to select the cell for both reading and writing. The appearance of data on the bit/sense line puts the cell in the proper binary state. If no data are applied, a sense signal indicating the binary state of the cell is obtained. The use of coincidence for reading as well as writing allows this cell to be used in complex array organizations.

Magnetic ferrite cores, which played a key role in the early history of computers, were first introduced as three-terminal cells. In fact, the first ferrite core cells used four wires through each core (separate bit and sense lines) to achieve a three-functional terminal cell as in Fig. 18.2.1b. Refinements gradually allowed the bit and sense line to be physically combined to give a three-wire cell and eventually even a two-wire two-terminal cell.

The introduction of *integrated-circuit cells* in the early 1970s for random-access memory led to the disuse of ferrite cores. There are two general types of integrated-circuit cells, static and dynamic. *Static cells* remain in the state they are set until they are reset or the power is removed and are read nondestructively; i.e., the information is not lost when read. A *dynamic cell* gradually loses its stored information and must be refreshed periodically. In addition, the cell state is destroyed when read, and the data must be rewritten after each read cycle. However, dynamic cells can be made much smaller than static cells, which gives a greater density of bits per semiconductor chip. The resulting lower cost more than compensates for the additional complexity.

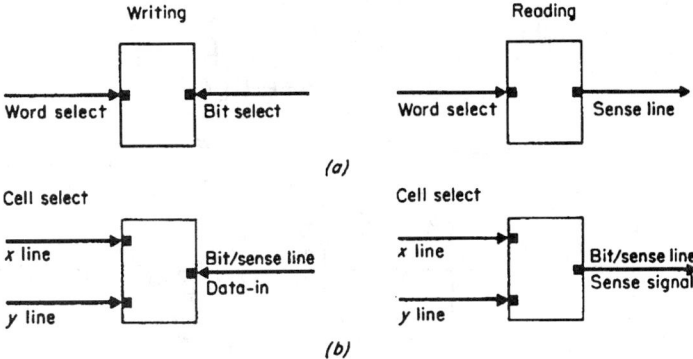

FIGURE 18.2.1   Reading and writing operations for random-access memory cells having two and three functional terminals: (*a*) cell with two functional terminals, and (*b*) cell with three functional terminals.

STATIC CELLS

All static cells use the basic principles of the Eccles-Jordan flip-flop circuit to store binary information in a cross-coupled transistor circuit like that shown in Fig. 18.2.2. Either *junction (bipolar)* or *field-effect transistors (FET)* can be used in the same configuration; only the voltages, currents, and load resistors will be different. To store a 0, transistor T_0 is turned on and T_1 turned off. A stored 1 is just the opposite condition, T_0 off and T_1 on. To achieve these states, it is necessary to control the node voltages at A and B. For instance, if node A voltage V_A is made sufficiently low, the base of T_1 will be low enough to turn T_1 off, causing node B to rise in voltage. This turns the base and hence the collector of T_0 on and holds node A at a low voltage so that a 0 is stored in the flip-flop. If the voltage at node B is made sufficiently low, the opposite state occurs, giving a stored 1. For reading, it is only necessary to sample the voltages at nodes A or B to see which is high or low. This gives a nondestructive read cell since the state is not changed, only sampled. Hence, nodes A and B are the access ports to the cell for both reading and writing.

The basic difference between all *static* cells is in how nodes A and B are accessed. Note that although the flip-flop of Fig. 18.2.2 has two access nodes A and B, it is functionally only a one-terminal cell since nodes

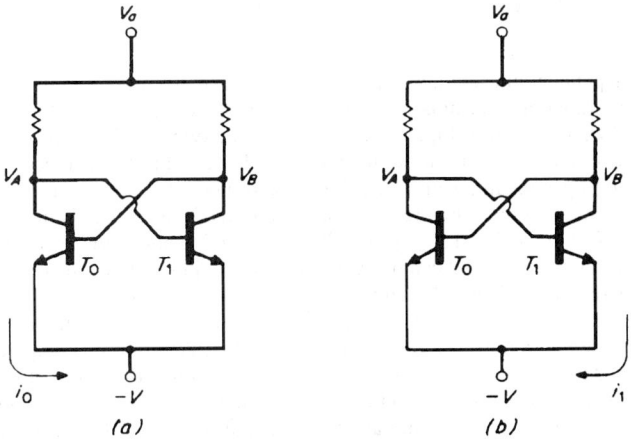

FIGURE 18.2.2   Transistor flip-flop circuits: (*a*) stored 0, T_0 conducts, $V_A = 0$, $V_B = 1$ V; (*b*) stored 1, T_1 conducts, $V_A = 1$ V, $V_B = 0$.

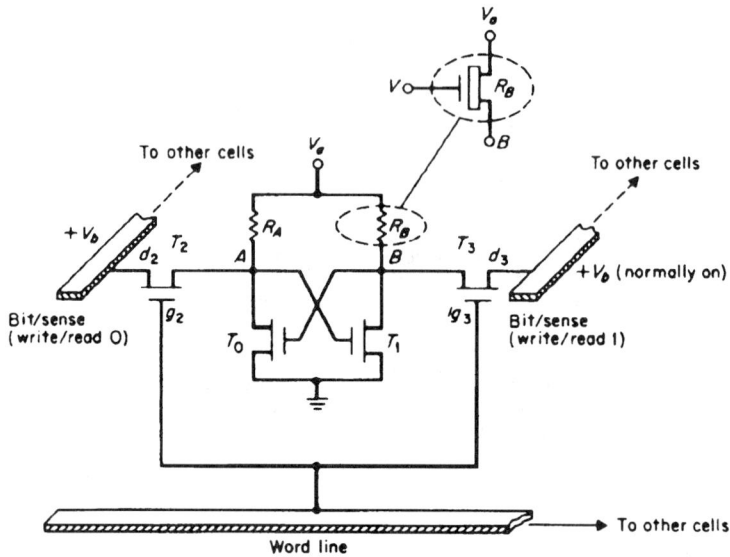

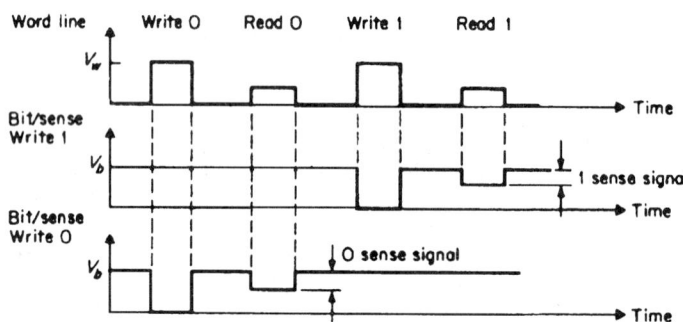

FIGURE 18.2.3   MOSFET two-terminal storage cell.

A and *B* are not independent but operate in a cooperative mode. To make a cell suitable for a random-access array, at least another functional terminal must be added. This can be achieved by the addition of another FET to each node, *A* and *B*, as shown in Fig. 18.2.3. Although the two transistors provide a total of four physical connections to the cell, that is, T_2 provides one gate g_2 and one drain d_2 for node *A* and T_3 provides g_3 and d_3 for node *B*, the circuit operates in a symmetrical, balanced mode. Since these four terminals are not independent, only two functional terminals are present. The operation of the cell can be understood by tracing through the pulsing sequence of Fig. 18.2.3. The peripheral circuits used to obtain these pulse sequences are not shown. An equivalent two-terminal static cell can be achieved by using junction transistors with some charges, but the operating principles remain basically the same.

A two-terminal static cell that was very popular in early integrated-circuit memories used multiemitter transistors. The cell required high power, however, a condition that greatly limits chip density.

Higher density and lower power can be obtained with the Schottky diode flip-flop cell of Fig. 18.2.4, a very popular cell for high-speed, low-power junction transistor memories. Resistors R_c are part of the internal collector contact resistance of the devices and are much smaller than R_L. The pulses on the word and bit/sense lines are used to forward- or back-bias the diodes D_0 and D_1, thus controlling or sensing the voltages and nodes *A* and *B*, as can be seen by tracing through the pulse sequence shown.

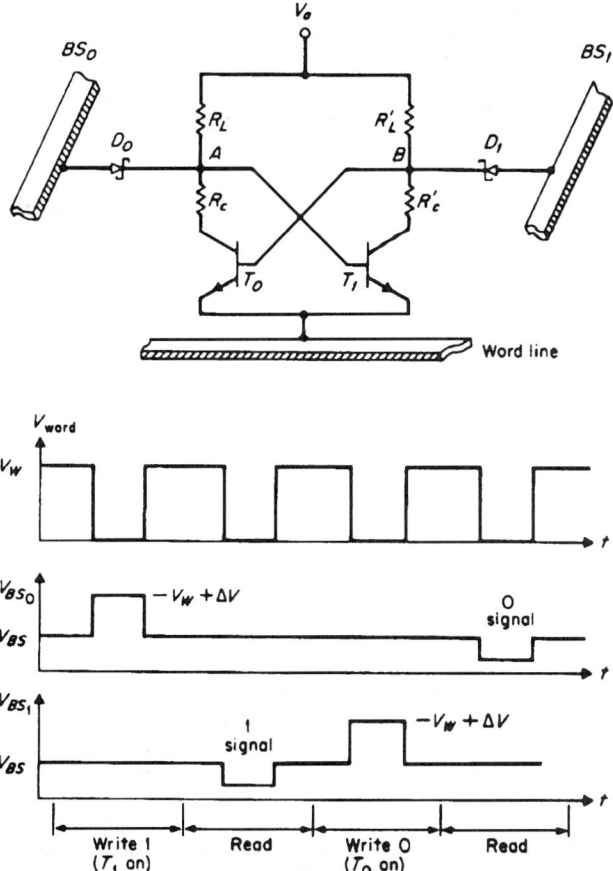

FIGURE 18.2.4 Schottky diode storage cell.

DYNAMIC CELLS

Since the static cells described above operate in a balanced, differential mode, it would seem reasonable that a nondifferential mode would be able to achieve at least a twofold reduction in component count per cell. This is indeed the case, and by sacrificing other properties of static cells, such as their inherent nondestructive read capability, a very significant reduction in cell size is possible. The most widely used such cell is the one-FET dynamic cell shown in a nonintegrated form in Fig. 18.2.5. The essential principle is simply the storing of charge on capacitor C_s for a 1 and no charge for a 0. A single capacitor by itself is sufficient to accomplish this, but an array of such devices requires a means of selecting the desired capacitors for reading and writing and of isolating nonselected capacitors from the array lines. If isolation is not provided, the charge stored on half-selected capacitors may inadvertently be removed, making the scheme inoperable. The isolation and selection means is provided by the simple one-FET device in series with the capacitor, as shown. For operation in an array, the terminal c can be either at ground or $+V_c$, depending on the technology and details of the cell. Assume that c is at $+V_c$, as indicated. To write either state, the word line is pulsed high to turn the FET on. If the bit/sense line is at ground, V_c will charge C_s to a stored 1 state. However, if the bit/sense line is at $+V_c$, any charge on C_s will be removed or no charge is stored if none was there originally, giving a stored 0. For reading, the word line is pulsed high to turn the FET on with the sense line at a normally high voltage. If there was

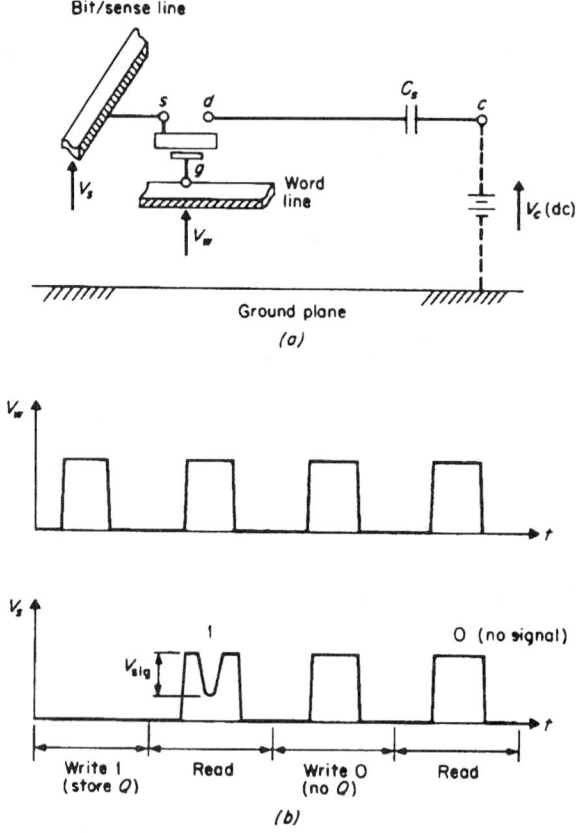

FIGURE 18.2.5   One-FET-device dynamic storage cell: (*a*) general equivalent circuit; (*b*) pulsing sequence.

charge on the capacitor, it will discharge through the bit/sense line to give a signal as shown. If there was no stored charge, no signal would be obtained.

Note that the reading is destructive since a stored 1 is discharged to a 0 and requires regeneration after each read operation in an array. Note also that the FET must carry current in both directions; i.e., the current charging C_s during writing is in the opposite direction of the current during reading. This cell has one further disadvantage: in an integrated structure, the charge on C_s will unavoidably leak off in a time typically measured in milliseconds. Hence the cell requires periodic refreshing, which necessitates additional peripheral circuits and complications. This feature gives rise to the term *dynamic cell*. Despite these disadvantages, this technique allows a very substantial improvement in cell density and cost, at very adequate cycle times and has become very popular.

RANDOM-ACCESS MEMORY ORGANIZATION

A memory is organized into bytes or byte groups known as words. If N is a word of fixed length, each word is assigned an address, or location, in memory. Each word has the same number of bits, called the word length. Each access usually retrieves the entire word. The addresses in memory run consecutively, from address 0 to the largest address.

In any memory system, if E units are to be selected, an address length of N bits is required which satisfies

$$2^N = E$$

In most cases of interest E is equal to W, the number of logical words in the system.

DIGITAL MAGNETIC RECORDING

Magnetic recording is attractive for data processing since it is inexpensive, easily transportable, unaffected by normal environments, and can be reused many times with no processing or developing steps.

The essential parts of a simplified but complete magnetic recording system are shown in Fig. 18.2.6. They consist of a *controller* to perform all the logic functions as well as write-current generation and signal detection; serial-to-parallel conversion registers; a read-write head with an air gap to provide a magnetic field for writing and sensing the flux during reading; and finally the medium. The wired-in cells, array, and transducers of random-access memory have been replaced by one read-write transducer, which is shared by all stored bits, and a shared controller. Coincident selection is still required for reading and writing, and this is obtained by the coincidence of electric signals in the read-write head with physical positioning of the medium under the head.

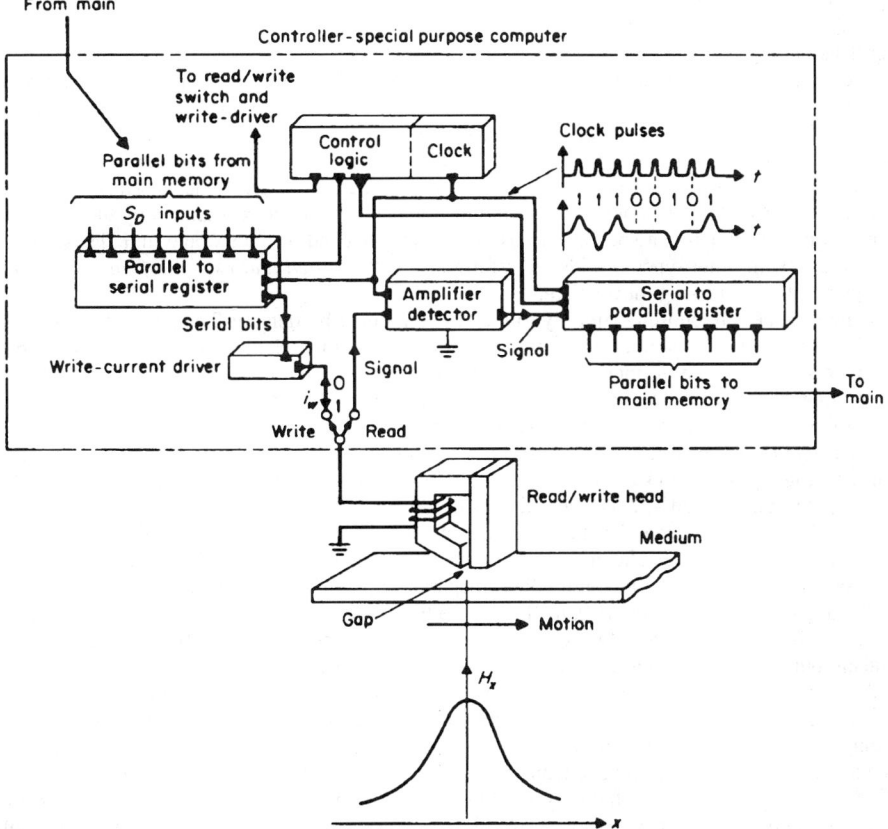

FIGURE 18.2.6 Schematic of simplified complete magnetic recording system.

The recording medium is very similar in principle to ferrite material used in cores. The common material for digital magnetic recording is ferric oxide, Fe_2O_3. It has remained essentially unchanged for many years except for reductions in the particle size, smoother surfaces, and thinner, more uniform coating, all necessary for high density. This material remained the sole medium for discs and tapes until the late 1970s when NiCo (nickel cobalt) was introduced. These new materials have a higher coercive force and can be deposited thinner and smoother, allowing higher recording densities. Operation of these media requires a reasonably rectangular magnetic hysteresis loop with two stable residual states $+B_r$ and $-B_r$ for storing binary information. The media must be capable of being switched an infinite number of times by a magnetic field produced by the write head, which exceeds the coercive force. Stored information is sensed by moving the magnetized bits at constant velocity under the read head to provide time-changing flux and hence an induced sense signal.

The essence of magnetic recording consists of being able to write very small binary bits, to place these bits as close together as possible, to obtain an unambiguous read-back voltage from these bits, and to convert this continuously varying voltage into discrete binary signals. The writing is done by the trailing edge of the write field.

The minimum size of one stored bit is determined by the minimum transition length required within the medium to change from $+B_r$ to $-B_r$ without self-demagnetizing. The smaller the transition length, the larger the self-demagnetizing field. The minimum spacing at which adjacent bits can now be placed with respect to a given bit is governed mainly by the distortion of the sense signal when adjacent bits are too close, referred to as *bit crowding*. This results from the overlapping of the fringe field from adjacent bits when they are too close and this total, overlapped magnetic field is picked up in the read head as a different induced signal from that produced by a single transition. Conversion of the analog read-back signal to digital form requires accurate clocking; this means that clocking information must be built into the coded information, particularly at higher densities.

Neglecting clocking and analog-to-digital conversion problems for the moment, the signals obtained during a read cycle are just a continuous series of 1s and 0s. A precise means of identifying the exact beginning and end of the desired string of data is necessary, and furthermore, some means for identifying specific parts within the data string is often desirable. Since the only way to recognize particular pieces of stored information is through the sequence of pulse patterns, special sequences of patterns such as gaps, address markers, and numerous other coded patterns are inserted into the data. These can be recognized by the logic hardware built into the controller, which is a special-purpose computer attached to the storage unit. These special recorded patterns along with other types of aids are referred to as the *stored addressing information* and constitute at least a part of the addressing mechanism.

Coding schemes are chosen primarily to increase the linear bit density. The particular coding scheme used determines the frequency content of the write currents and read-back signals. Different codes place different requirements on the mode of operation and frequency response of various parts of the system such as clocking technique, timing accuracy, head time constant, medium response, and others. Each of these can influence the recording density in different ways, but in the overall design the trade-offs are made in the direction of higher density. Thus special coding schemes are not fundamentally necessary but only meet practical needs. For instance, it is possible to store bits by magnetizing the medium over a given region where say $+M_r$ (magnetization) is a stored 1 and $-M_r$ is a stored 0, as in Fig. 18.2.7a. The transition region in between 1s and 0s is assumed to have $M = 0$ except for the small regions of north and south poles on the edges as shown. As the medium is moved past the read head, a signal proportional to dM/dt or dM/dx is induced, so that the north poles induce, say, a positive signal and south poles a negative signal as shown (polarity arbitrary).

This code is known as *return to zero* (RZ) since the magnetization returns to zero after each bit. Each bit has one north- and one south-pole region, so that two pulses per bit result. Not only are two pulses per bit redundant, but considerable space is wasted on the medium for regions separating stored bits. It is possible to push these bits closer together, as in Fig. 18.2.7b, so that the magnetization does not return to 0 when two successive bits are identical, as shown for the first two 1s; hence the name *nonreturn to zero* (NRZ). The result is then only one transition region and one signal pulse per bit. By adjusting the clocking pulses to coincide with the signal peaks, we have a coding scheme in which 0s are always negative signals and 1s are always positive (Fig. 18.2.7b). The difficulty is that only a change from 1 to 0 or 0 to 1 produces a pulse; a string of only 1s or 0s produces no signals. This requires considerable logic and accurate clocking in the controller to avoid accumulated errors as well as to separate 1s from 0s.

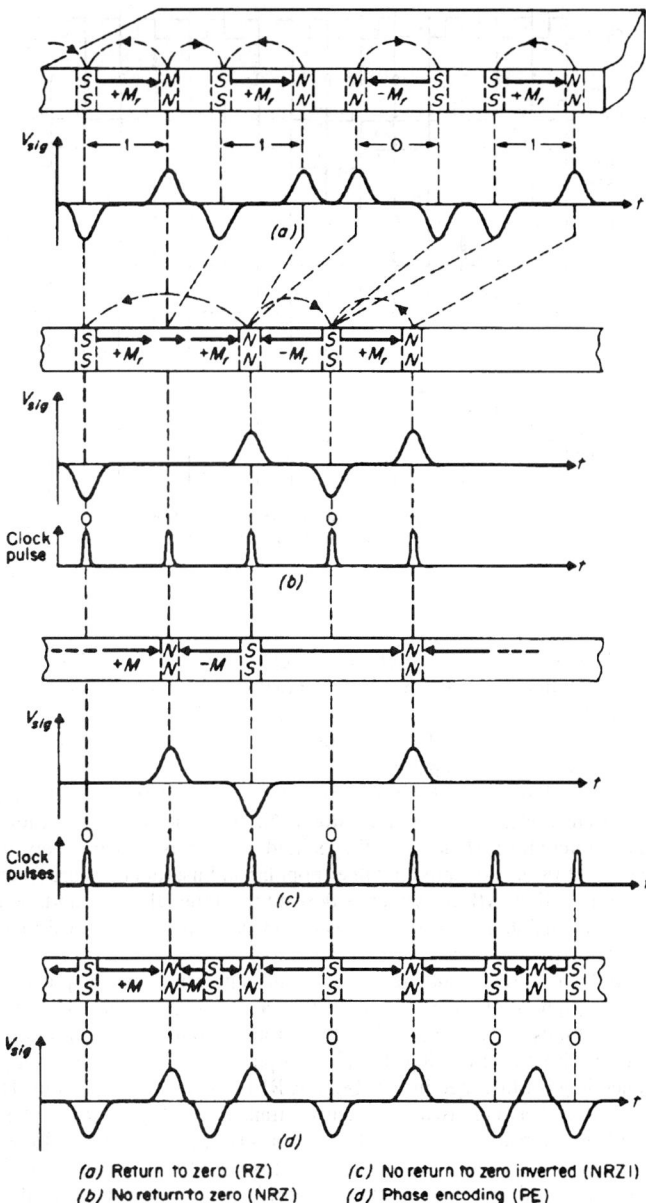

FIGURE 18.2.7 Cross-sectional view of magnetic recording medium showing stored bits, signal patterns, and clocking for various codes.

One popular coding scheme is a slightly revised version of the above, namely, *nonreturn to zero inverted* (NRZI), in which all 1s are recorded as a transition (signal pulse) and all 0s as no transition, as shown in Fig. 18.2.7c. There is no ambiguity between 1s and 0s, but again a string of 0s produces no pulses. A *double-clock* scheme, with two clocks, each triggered by the peaks of alternate signals, is used to set the clock timing period to the following strobe point. NRZI is a common coding scheme for magnetic tapes used at medium density.

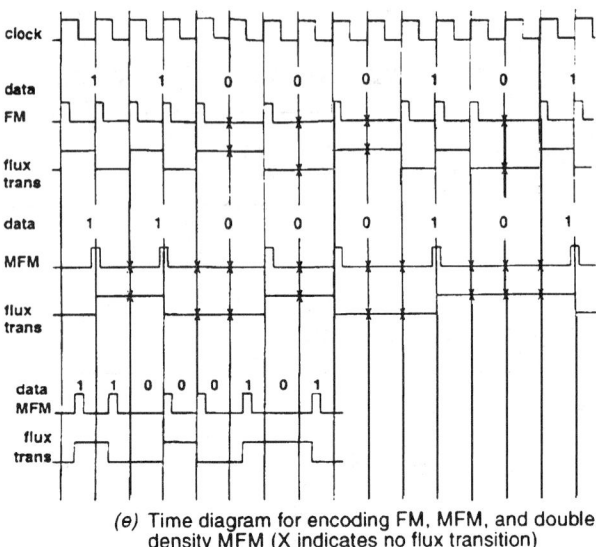

(e) Time diagram for encoding FM, MFM, and double density MFM (X indicates no flux transition)

FIGURE 18.2.7 *(Continued)*

For greater than density 1600 b/in. the clocking and sensing become critical, and *phase encoding* (PE), shown in Fig. 18.2.7*d*, is often used. Since 1s give a positive signal and 0s give a negative signal, a signal is available for every bit. Phase encoding requires additional transitions within the medium, e.g., between successive 1s or successive 0s, as shown. Density, however, is usually limited by sensing, clocking, and other problems rather than by the medium capability.

For magnetic-disc recording, a double-frequency NRZI code is often used. This is obtained from NRZI by adding an additional transition just before each stored bit. The additional transition generates an additional pulse to serve as a clocking pulse. Hence a well-specified window is provided between bits to avoid clocking problems when a string of 0s is encountered. Other popular and useful codes are frequency modulation (FM), modified frequency-modulation (MFM), which is derived from the DF code, and the run-length-limited code. Since the latter does not maintain a distinction between data and clock transitions, a special algorithm is required to retrieve the data.

FM encodes 1 and 0 in different frequencies, but entails a minimum of one pulse per digit for both. A pulse always appears at the beginning of a clock cycle. If the data are 0s, there are no further pulses; if the data are 1s, there is an additional pulse. MFM eliminates the automatic pulse for each digit, and so is more efficient. MFM retains the pulse for 1; for encoding 0, there is a pulse at the beginning of a period unless it was preceded by a 0. The encoding is illustrated in Fig. 18.2.7*e*. RLL is even more efficient. The run length is the number of no-flux transitions between two consecutive transitions. Figure 18.2.7*e* shows the encoding for 11000101. FM has 12 flux transitions, while MFM requires only six transitions. In personal computers, MFM is used to achieve *double density* recording, whereas FM is used for single density.

Writing the transition (north or south) in the medium is done by the trailing edge of the fringe field produced by the write head. Since writing is rather straightforward and is not a limiting factor, we dwell here on the more difficult problems of reading and clocking.

Read-back signals can best be understood in terms of the reciprocity theorem of mutual inductance. This theorem states that for any two coils in a linear medium, the mutual inductance from coil 1 to 2 is the same as that from coil 2 to 1. When applied to magnetic recording, the net result is that the signal, as a function of time observed across the read-head winding induced by a step-function magnetization transition, has a shape that is identical to the curve H_x versus x for the same position of the medium below the head gap (Fig. 18.2.6). It is necessary only to replace x by vt, where v = velocity, for the translation of the x scale on H_x to the time scale on V_{sig} versus t. The H_x-versus-x curve with a multiplication factor is often referred to as the *sensitivity function*.

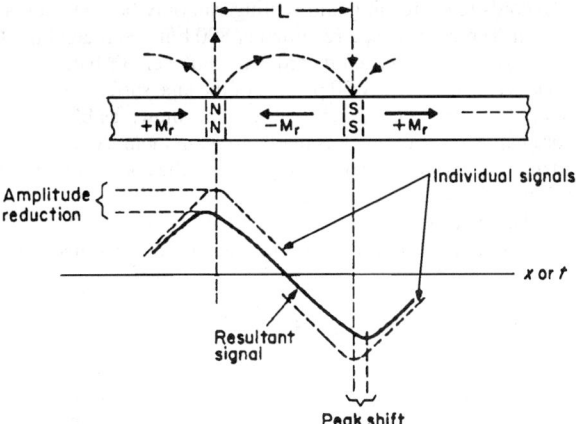

FIGURE 18.2.8 Bit crowding on read-back showing amplitude reduction and peak shift.

The writing of a transition is done by only one small portion of the H_x-versus-x curve, whereas the sense signal is determined by the entire shape of H_x versus x; that is, the signal is spread out.

This fact gives rise to *bit crowding,* which makes the read-back process more detrimental in limiting density than the writing process. To understand bit crowding, suppose there are two step-function transitions of north and south poles separated by some distance L, as in Fig. 18.2.8. When these transitions are far apart, their individual sense signals shown by the dashed lines appear at the read winding. However, as L becomes small, the signals begin to overlap and, in fact, subtract from each other,[*] giving both a reduction in peak amplitude and a time shift in the peak position as shown. This represents, to a large extent, the actual situation in practice. The transitions can be written closer together than they can be read.

Clocking or strobing of the serial data as they come from the head to convert them into digital characters is another fundamental problem. If perfect clock circuits with no drift and hence no accumulated error could be made, the clocking problem would disappear. But all circuits have tolerances, and as the bit density increases, the time between bits becomes comparable to the drift in clock cycle times. Since the drift can be different during reading and writing, serious detection errors can result. For high density, it is necessary to have some clocking information contained within the stored patterns as in the PE and double-frequency NRZI codes discussed previously.

MAGNETIC TAPE

The most common use of magnetic tape systems is providing backup for hard disk drives.

Since there are either seven or nine tracks written across the width of the tape, either seven or nine read-write heads are required to store one complete character or byte at a time. The bit spacing along a track is approximately the reciprocal of the linear density, or 0.00125 for an 800 b/in. system. The actual transition lengths are generally about half this bit-cell spacing. In many systems there are separate read and write gaps in tandem to check the reliability of the recording by reading immediately after writing. The tape is mechanically moved back and forth in contact with the heads, all under the direction of a controller. Tape transports in digital systems must be able to start and stop quickly, and achieve high tape speed rapidly. The streaming tape drive used in PCs does not start and stop quickly; it is used to backup disk systems.

The stored addressing information in tapes is relatively simple, consisting of specially coded bits and tape marks in addition to *interrecord gaps* (IRG). The latter are black space on the tape to provide space to accelerate

---

[*]Linear superposition is possible since the air gap makes the read head linear.

and decelerate between records since reading and writing can only be done at constant velocity. The common gap sizes are 0.6 and 0.75 in. Typically, a tape recorded at 800 b/in. with eight tracks plus parity storing records of 1K B each and a gap of 0.6 in between each record can hold over 10^8 b of data. Even though this represents a large capacity, the gap spaces consume nearly 50 percent of the tape surface, a rather extravagant amount. In order to increase efficiency, records are often combined into groups known as blocks. Since the system can stop and start only at an interrecord gap, the entire block is read into main memory for further processing during one read operation. The highest-density tapes can hold nearly eight times the above amount. Half-inch reel tapes are beginning to be replaced by shorter reels mounted in self-contained cartridges. One tape drive can hold multiple cartridges, which can be mechanically selected and mounted. Typical cartridge tapes are 165 in. long and record 18 tracks on $1/2$-in. tape at about 25K flux transitions per inch (972 flux transitions/mm).

DIRECT-ACCESS STORAGE SYSTEMS—DISCS

The major type of direct-access storage system is a disc. The system's recording head usually consists of one gap, which is used for both reading and writing. The head is "flown" on an air cushion above the disc surface at separations in the neighborhood of 5 to 100 μin., depending on the system. A well-controlled separation is vital to reliable recording.

For discs, the medium consists of very thin coatings of about 10 μin. or less of the same magnetic material used on tapes but applied to polished aluminum discs. Several discs are usually mounted on one shaft, all rotated in unison. Each surface is serviced by one head. The arms and heads are moved mechanically along a radial line, and each fixed position sweeps out a track on each surface, the entire group of heads sweeping out a cylinder. A typical bit cell is a rectangle 0.005 in. wide by 0.0005 in. long for a 2000 b/in. linear density. The transition length is about half this size.

The fundamental difference between various disc systems centers on the stored addressing information and addressing mechanisms built into the system. Some manufacturers provide a rather complex track format permitting the files to be organized in many different ways, using keys, identifying numbers, stored data, or other techniques for finding a particular word. Thus, the user can "program" the tracks and records to retrieve a particular word "on the fly." This provides a very versatile system but only with additional cost, since considerable function must be built into the controller. Other systems use a very simple track format consisting mainly of gaps and sector marks that do not permit the user to include programmable information about the data. This scheme is more suitable for well-organized data, such as scientific data; it still can be used in other applications with more user involvement.

Floppy discs are used in wide application as inexpensive high-density, medium-speed peripherals. Floppy discs often consist of the same flexible medium (hence the name floppy) as magnetic tape but cut in the form of a disc. Such discs straighten out when spun. The read-write mechanism is usually identical to that above except that the head is in contact with the medium, as in tape. This causes wear that is more significant than in ordinary discs, required frequent replacement of the disc and occasional replacement of the head, particularly when heavily used.

Floppy discs use tracks in concentric circles with movable heads and track-following servo. The systems record on both sides. The major deviations from flying-head discs are much smaller track density and data rate. The linear bit densities are quite comparable to those of tape. Typical parameters are 77 to 150 tracks per disc surface, 1600 to 6800 b/in., rotation from 90 to 3600 r/min.

Optical discs provide more storage than magnetic discs and are commonly in use. One optical disc can contain the same information as 20 or more floppy discs.

VIRTUAL-MEMORY SYSTEMS

Virtual memory is a term usually applied to the concept of paying a gain memory out of a disc. This concept makes the normal-sized main memory appear to the user as large as the virtual-address space* while still appearing to run at essentially the speed of the actual memory. Virtual memories are particularly useful

---

*Represented by a register holding the virtual address.

in multiprogrammed systems. On the average, virtual memory provides for better management of the memory resource with little wasted space because of fragmentation, which can otherwise be quite severe.

To understand *fragmentation,* suppose that a number of programs to be processed require small, medium, and large amounts of memory on a multiprogrammed machine. Suppose further that a small and a large program are both resident in main memory and that the small one has been completed. The operating system attempts to swap into memory a medium-sized program, to be processed, but it cannot fit in the space freed by the small program. If none of the other waiting programs is small enough to fit, this memory space is wasted until the large program has been completed. It can be seen that with many different-sized programs, fitting several of them into available memory space becomes difficult and leads to much unusable memory at any one time, i.e., fragmentation. Virtual memory avoids this by breaking all programs into *pages* of equal size, and dynamically paging them into main memory on demand. The identical concepts used in a virtual memory are applicable to a cache speed paged out of main memory; only the details of implementation are different, as shown later. We discuss first the basic concepts as applied to a main memory paged out of a disc and indicate, whenever possible, the differences applicable to a cache.

All virtual memories start with a *virtual address* that is larger than the address of the available main memory. Such being the case, the desired information may not be resident in the memory. Hence it is necessary to find out *if*, in fact, it is present. If the information is resident, it is necessary to determine *where* it is residing because the physical address cannot bear a one-to-one correspondence to the virtual address. In fact, in the most general case, there is no relationship between the two, so that an address translation scheme is necessary to find *where* the information does reside.

If the requested page is not resident in memory, a page *fault* results. This requires a separate process to find the page on the disc, remove some page from memory, and bring the new page into this open spot, called a *page frame*. Which page to remove from main storage is determined by a page-replacement algorithm that replaces some page "not recently used." The address-translation and page-replacement functions are shown conceptually in Fig. 18.2.9.

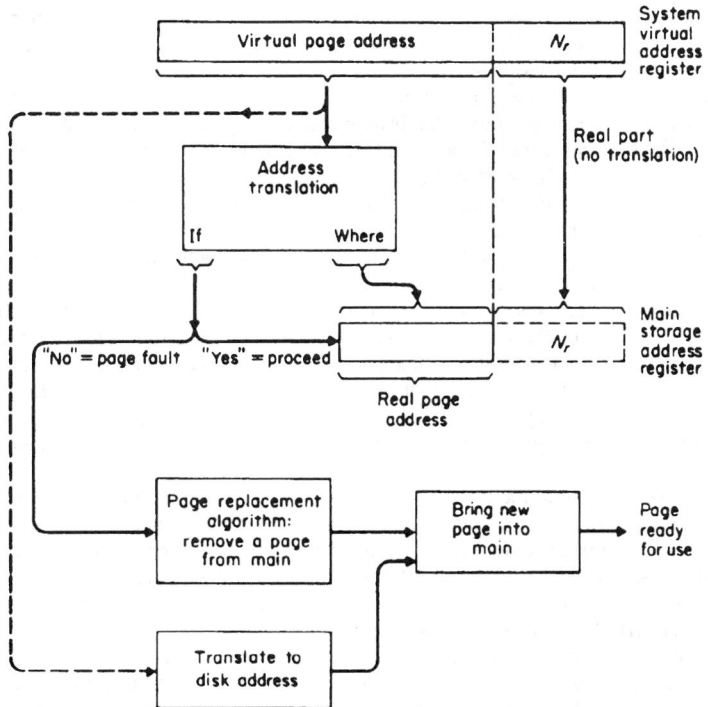

FIGURE 18.2.9   Block diagram of address-translation and page-replacement process.

Thus there are at last three fundamental requirements: (1) a mapping function to specify how pages from the disc are to be mapped into physical locations in memory, (2) an address translation function to determine *if* and *where* a virtual page is located in main memory, and (3) a replacement algorithm to determine which page in memory is to be removed when the *if* translation is a "no," i.e., when a page fault occurs. These are the three fundamental requirements needed to implement a virtual memory system, either a virtual main memory as here described, or a cache.

MAPPING FUNCTION AND ADDRESS TRANSLATION

The mapping function is a logical construct, whereas the address-translation function is the physical implementation of the mapping function. Mapping functions cover a range from *direct mapping* to fully *associative mapping*, with a continuum of *set-associative* mapping functions in between. A very simple way to understand maps is to consider the example of building one's own personal telephone directory "paged' out of larger telephone books. Assume that the personal directory is of fixed size, say 4(26) = 104 names, addresses, and associated telephone numbers. A direct mapping from the large books to the personal director is an alphabetical listing of the 104 names. Given any name, we can go directly to the directory entry, if present. Such an address translation could be hard-wired if desired. There are two difficulties with direct mapping: (1) it is very difficult to change, and (2) suppose we allow one entry for Jones. If later we wish to include another Jones, there is no room. If both Joneses are needed, there is a conflict unless we restructure the entire directory. Because of such conflicts, direct maps are seldom used.

At the other end of the spectrum is a *fully associative directory* in which 104 names in any combinations are placed in any positions of the directory. This directory is very easily changed because a name not frequently used can simply be removed and a new name entered in its place without regard to the logical (alphabetical) structure of the two names. For instance, if the directory is full and we wish to make a new entry Zeyer, we first find a name not used much recently. If Abas is in position 50 of the directory, remove Abas and replace the entry with Zeyer. The major difficulty, obviously, is in searching the directory. If we wish to know the number for Smith, we must associatively search the entire directory (worst case) to find the desired information. This is very time-consuming and impractical in most cases.

Imagine the usual telephone directory that is associatively organized. There are several ways to resolve the fundamental conflict between ease of search and ease of change. The fully associative directory can be augmented with a separate, directly organized and accessed table that contains a list of all names. However, the only other piece of data is a number indicating the entry this name now occupies in the associative directory. If a directory entry is changed, the new entry number must be placed in this table. If we wish to access a given name, a direct access to the table gives the entry number . A subsequent direct access to this entry number gives the desired address and telephone number.

The penalty is the two accesses plus the storage and maintenance of the translation table. Nevertheless, this is exactly the scheme used in all virtual main memories paged out of disc, drum, or tape. This table is typically broken into a hierarchy of two tables called *segment table* and *page table* to facilitate the sharing of segments among users and to allow the tables to also be pages as units of, say, 4K B. These tables are built, manipulated, and maintained by supervisory programs and generally are invisible to the user. Although these tables can consume large amounts of main memory and of system overhead, the saving greatly exceeds the loss.

An example of such a virtual memory with a fully associative mapping using a two-level table-translation scheme is illustrated in Fig. 18.2.10. Each user has a separate segment and page table (the two are, in principle, one table) stored in main memory along with the user's data and programs. When an access is required to a virtual address, say $N_v N_r$, as in Fig. 18.2.10, a sequence of several accesses is required. The user ID register bits μ give a direct address to the origin of that user's segment table in main memory. The higher-order segment-index bits (SI) of the virtual address (typically 4 to 8 b) specify the index (depth) into this table for the required entry. This segment table entry contains a flag specifying *if* this entry is valid or not, and a *where* specifying the origin of that user's page table in memory, as shown. The lower-order page index bits (PI) of the virtual address specify the index into the page table as shown. The page-table entry so accessed contains an *if* bit to indicate whether the entry is valid (*if* page is present in main memory) and a *where* address that

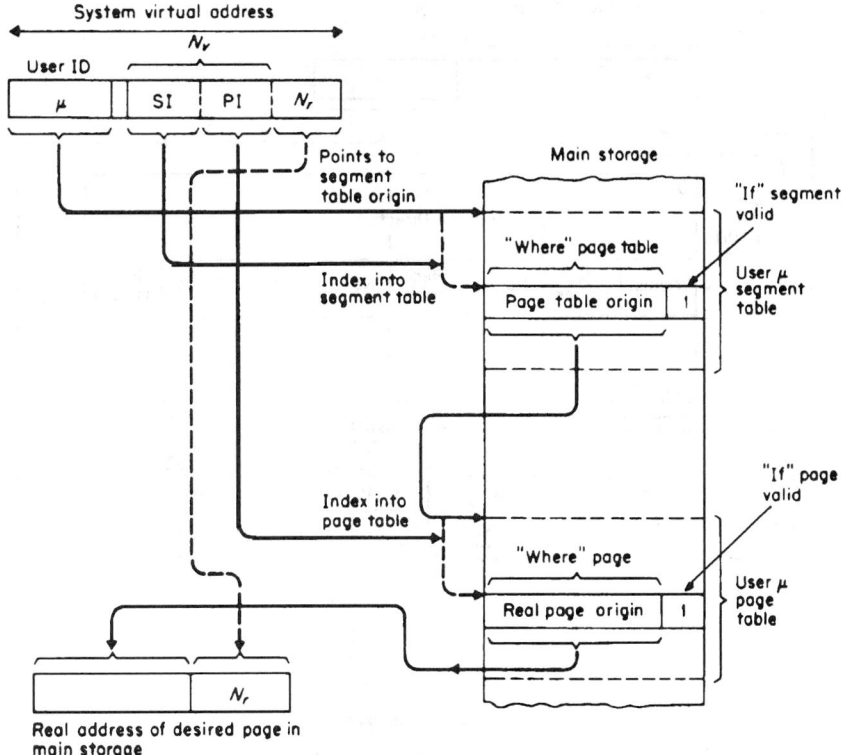

FIGURE 18.2.10   Virtual-storage-address translation using a two-level table (segment and page) for each user.

gives the real main-memory address of the desired page. The lower-order N_r bits of the address, typically 11 or 12 b for 2K- or 4K-B pages, are real, representing the word or byte of the page and hence do not require translation.

In cache memories, speed is so important that a fully associative mapping with a table-translation scheme is impractical. Instead a set-associative mapping and address translation is implemented directly in hardware. Although it is not generally realized, set-associative mapping and translation is commonly used in everyday life. It is a combination of a partially direct mapping and selection, with a partially associative mapping and translation. Examples include the 4-way set associative organization in Fig. 18.2.11 and the most common type of personal telephone directory as shown in Fig. 18.2.12, where a small index knob is moved to the first letter of the name being sought. Suppose there is one known position on the directory for each letter of the alphabet and we organize it with a set associativity of four. This means that for each letter there are exactly four entries or names possible, and these four can be in any order, as shown by the insert of Fig. 18.2.12. To find the telephone number for any given name such as Becker, we first do a direct selection to the letter B followed by an associative search over the four entries. Thus it is apparent that a set-associative access is a combination of the two limiting cases, namely, part direct and part associative. Many different combinations are possible with various advantages and disadvantages.

This set-associative directory (Fig. 18.2.12) with four names or entries per set is exactly the same fundamental type used in many cache-memory systems. The directory is implemented in a random-access memory array as shown in Fig. 18.2.11, where there are 128(4) entries total, requiring nine total virtual page address bits. Each set of 4 is part of one physical word (horizontal) of the random-access array, so there are 128 such

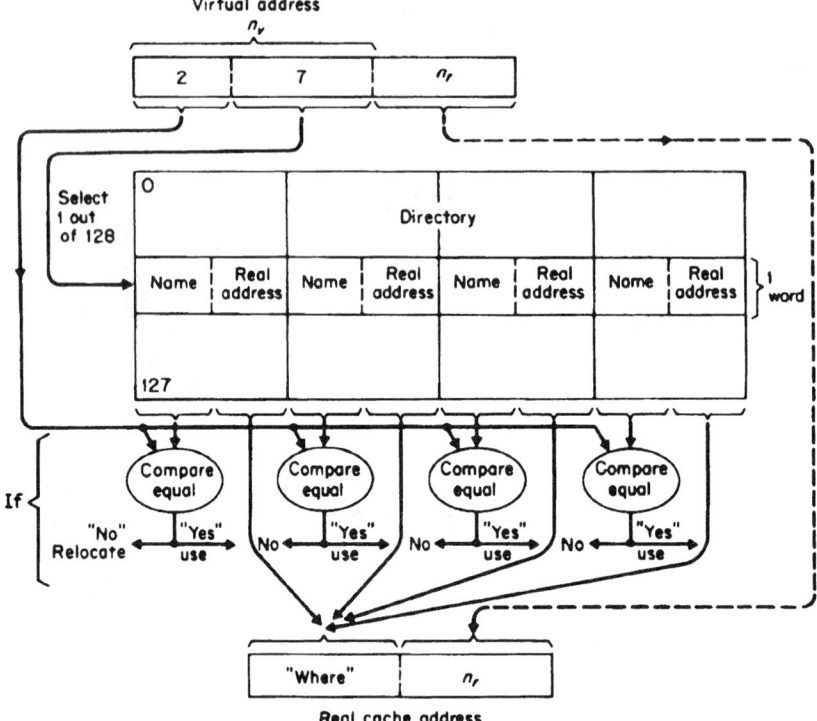

FIGURE 18.2.11 Cache directory using 4-way set associative organization.

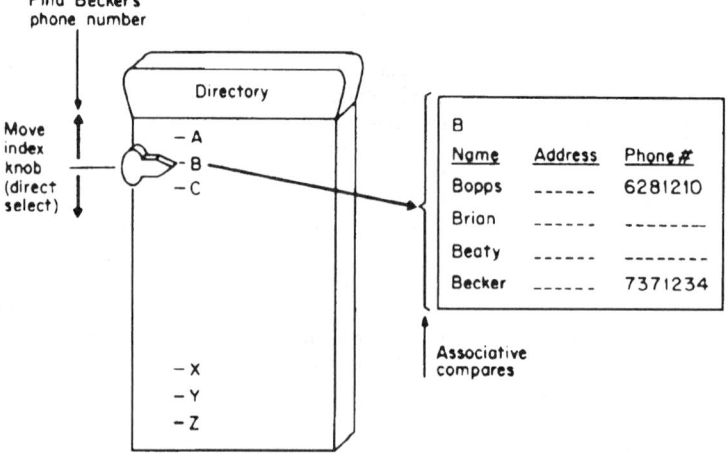

FIGURE 18.2.12 Fundamentals of a general set-associative directory showing direct and associative parts of the addressing process.

words, requiring seven address bits. The total virtual address $n_v = 9$ must be used in the address translation to determine *if* and *where* the cache page resides. As before, the lower-order bits n_r, which represent the byte within the page, need not be translated. Seven virtual bits are used to select directly one of the 128 sets as shown. This is analogous to moving the index knob in Fig. 18.2.12 and reads out all four names of the set. The NAME part is 2 b long, representing one of the four of the set. All four are compared simultaneously with the 2 b of the virtual address. *If* one of these gives a "yes" on compare equal, then the correct "real" address of the page in the cache, which resides in the directory with the correct NAME, is gated to the "real" cache-address register.

CHAPTER 18.3
INPUT/OUTPUT

INPUT-OUTPUT EQUIPMENT

Input-output (I/O) equipment includes cathode-ray tubes and other display devices; printers; keyboards; character, handwriting, voice, and other recognition devices; optical scanners and facsimile equipment; speech synthesizers; process control sensors and effectors; various pointing devices such as "mice," wands, joysticks, and touch screens; card readers and punches; paper tapes; magnetic storage devices such as tape cassette units, floppy and hard disk drivers, drums, and various other types of tape drives; and several different types of optical storage devices, mainly disks.

Such equipment presents a wide range of characteristics that must be taken into account at their interface with the central processing unit (CPU) and its associated main memory. A magnetic tape unit usually operates serially by byte, writing or reading information of varying record length. Different types of printers deal with information on a character, line, or page basis, both in a coded (character) and noncoded mode. Most cathode-ray tube display devices build up the image in a bit-by-bit raster, but others handle data in the form of vectors or even directly coded as characters. Different types of I/O gear operate over widely different speed ranges, ranging from a few bytes per second to millions of bytes per second. Many I/O devices do little more than pass bits to a processor, while others have substantial internal processing power of their own.

In the many systems making use of telecommunications, an ever-growing array of I/O equipment can be attached to a variety of telecommunications links, ranging from twisted-pair copper wire to long-haul satellite and optical fiber links. Virtually any I/O device can be attached both locally and remotely to other elements of a system.

Data are entered by human operators, read from storage devices, or collected automatically from sensors or instruments.

Keyboards, wands, and character recognition equipment are typical *input devices*, while printers and displays are the most common *output devices*. Storage devices that can function in a read-write (R/W) mode naturally can fill both functions.

I/O CONFIGURATIONS

Input-output devices can be attached to other elements of the system in a wide variety of ways. When the attachment is *local*, the attachment may be an I/O bus, or a channel. Figures 18.3.1 and 18.3.2 show a typical bus attachment configuration of small systems such as personal computers. A program gains access to the I/O gear through the logic circuitry of the ALU. For example, to transfer information to I/O gear, a program might extract requested information from storage, edit and format it, add the address of the I/O device to the message, and deliver it to the bus, from which the proper I/O device can read it and deliver it to the requestor. In this simple

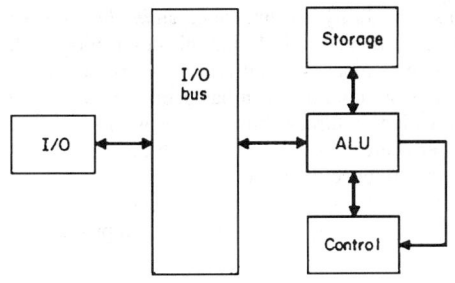

FIGURE 18.3.1 Machine organization with an I/O bus.

configuration, time must be spent by the CPU waiting for delivery of the information to the I/O gear and, if so programmed, for reception of an acknowledgement of the reception. If several types of I/O gear are used, the program or—more commonly—the operating system must take into account variations in formatting control, and speed of each type. A common method of improving the performance of the *I/O subsystem* is I/O buffering. A buffer may be part of the CPU or a segment of main memory, which accumulates information at machine speeds so that subsequent information transfers to the I/O gear can take place in an orderly fashion without holding up the CPU. In other cases, the I/O gear itself will be buffered, thus enabling, for example, synchronous input devices to operate at maximum throughput without having to repeatedly interrupt the CPU. A buffered system is desirable in all but the simplest configurations and is all but essential in any case in which the speed of the I/O device is not matched to the data rate of the receiving communications link or I/O bus.

I/O MEMORY–CHANNEL METHODS

The arrangement shown in Fig 18.3.2 involves information transfer from the I/O gear into the ALU and thence to main store. With a modularized main store, I/O data can be directly entered in, and extracted from, main storage. Direct access to storage implies control interrelationships at the interface between the I/O and the CPU to ensure coordination of accesses between the two units. The basic configuration used for direct access of I/O into storage is shown in Fig. 18.3.2. The connecting unit between the main store and the I/O gear is called a *channel*. A channel is not merely a data path but a logical device that incorporates control circuits to fulfill the relatively complex functions of timing, editing, data preparation, I/O control, and the like.

TERMINAL SYSTEMS

Terminal systems use key-entry devices with a display and/or printer for direct access to a computer on a time-shared basis. The programming system in the CPU is arranged to scan a number of such terminals through a set of ports. Each character, line, or block of data from each terminal is entered into a storage area in the computer until an appropriate action signal is received. Then by means of an internal program the computer generates such responses as error signals, requests for data, or program outputs. Terminal systems extend the power of a large computer to many users and give each user the perception of using an independent system. Terminals remotely connected to host CPUs have the same overall description, except that communication is mediated by means of various network architectures and protocols (q.v.).

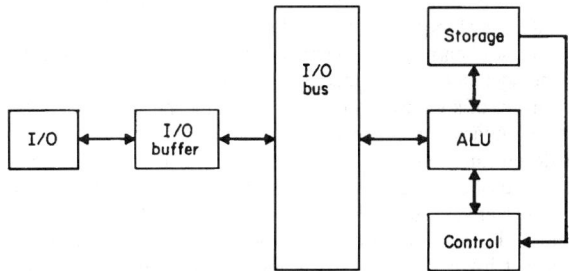

FIGURE 18.3.2 Machine organization with an I/O bus and I/O buffer.

A terminal that contains a high level of logic operational capability is often called an *intelligent terminal.* A personal computer with a communications adapter or modem is increasingly filling this description. If there are many terminals and a centralized computer or computer complex in a system, it is sometimes more economical to use simpler terminals and use the logic in the computer. More commonly, it has been the practice to attach a number of *dumb* terminals to an intelligent control unit, which, in turn, communicates with a host processor. Economic factors such as communication line cost and hardware cost, as well as such intangibles as system reliability and data security, are among those factors which influence the choice.

The proliferation of communicating personal computers has formed the basis of local and wide area networks (LANs and WANs). These networks provide both connectively between sites and local processing power.

PROCESS-CONTROL ENTRY DEVICES

In process control, e.g., chemical plants and petroleum refineries, inputs to the computer generally come from sensors (see Sec. 8) that measure such physical quantities as temperature, pressure, rate of flow, or density. These units operate with a suitable analog-to-digital converter that forms direct inputs to the computer. The CPU may in turn communicate with such devices as valves, heaters, refrigerators, pump, and so forth, on a time-shared basis, to feed the requisite control information back to the process-control system.

MAGNETIC-INK CHARACTER-RECOGNITION EQUIPMENT

A number of systems have been developed to read documents by machine. *Magnetic-ink character recognition* is one. Magnetic characters are deposited on paper or other carrier materials in patterns designed to be recognized by machines and by operators. A change in reluctance, associated with the presence of magnetic ink, is sensed by a magnetic read head. The form of the characters is selected so that each yields a characteristic signature.

OPTICAL SCANNING

Information to be subjected to data processing comes from a variety of sources, and often it is not feasible to use magnetic-ink characters. To avoid the need for retranscription of such information by key-entry methods, devices to read documents optically have been developed.

Character recognition by optical means occurs in a number of sequential steps. First, the characters to be recognized must be located and initial starting points on the appropriate characters found. Unless the characters have been intentionally constrained to be placed in well-defined locations, it is then necessary to segment the characters, that is, to determine where one character leaves off and the next one begins. Third, the characters must be scanned to generate a sequence of bits that represents the character. The resulting bit pattern must then be compared with prestored reference patterns in order to identify the pattern in question.

The scanning can be performed in a number of ways. In one technique the pattern to be recognized is transported past a linear photodetector array and the output is sampled at periodic intervals to determine the bit pattern. A second way is to scan a laser beam across the page using a rotating mirror or holographic mirror device, with the light scattered from the document being detected by an assembly of photodetectors. Other approaches that use various combinations of arrays of sources and detectors with mechanical light deflection devices are also used.

As an alternative to comparing the scanned bit pattern with stored references, a *correlation function* between the input and stored functions can be computed. Optical spatial filtering is occasionally used, too.

Another common method of data entry is through the encoding of characters into a digital format, permitting scan by a wand or, alternatively, by moving the pattern past a "stationary" scanning laser pattern. Typical of this encoding scheme is the bar code, known as the *universal product code* (UPC), which has been standardized in the supermarket industry by industry agreement to the code selection by the Uniform Grocery Product Code Council, Inc.

BATCH-PROCESSING ENTRY

One computer can be used to enter information into another. For example, it is often advantageous for a small computer system to communicate with a larger one. The smaller system may receive a high-level language problem, edit and format it, and then transmit it to a larger system for translation. The larger system, in turn, may generate machine language that is executed at the remote location when it is transmitted back.

Remote systems may receive data, operate on them, and generate local output, with portions transmitted to a second unit for filing or incorporation into summary journals. In other cases, two CPUs may operate generally on data in an independent fashion but may be so interconnected that in the event of failure one system can assume the function of the other. In other cases, systems may be interconnected to share the work load. Computer systems that operate to share CPU functions are called *multiprocessing* systems.

PRINTERS

Much of the output of computers takes the form of printed documents, and a number of *printers* have been developed to produce them. The two basic types of printer are *impact printers*, which use mechanical motion of type slugs or dot-forming hammers driven against a carbon film or ink-loaded fabric ribbon to mark paper, and *nonimpact* printers, which use various physical and chemical phenomena to produce the characters.

Printers are additionally characterized by the sequence in which characters are placed on the page. *Serial* printers print one character at a time, by means of a moving print head. After a line of characters has been printed, the paper is indexed to the next print line. *Line* or *parallel* printers are designed such that they print all the characters on a print line at (nearly) the same time, after which the paper is indexed. Finally, *page* printers are limited by the technology to marking an entire piece of paper before the output is accessible to a user. Most such printers actually generate a page one dot at a time using some type of raster scanning means.

Printers may either be *fully formed character* printers, or *matrix* (or *all-points-addressable*) printers. Matrix printers build up characters and graphics images dot by dot; some matrix-printing mechanisms allow the dots to be placed anywhere on a page, while others are constrained to printing on a fixed grid. Dot size (resolution) ranges from 0.012 in. (0.3 mm) in many wire matrix printers to 0.002 in. (0.05 mm) in some nonimpact printers. Dot spacing (addressability) ranges from as few as 60 dots per inch (24 dots per centimeter) to more than 800 dots per inch.

The speed of serial printers typically ranges from 10 to over 400 characters per second. Line printer speeds range from 200 to about 4000 lines per minute. Page printers print from 4 to several hundred pages per minute.

IMPACT PRINTING TECHNOLOGIES

Figure 18.3.3 is a sketch of a band printer, the most common high speed impact line printer. Band printers are used where price and throughput (price/performance) are the dominant criteria. In this printer, the characters are engraved on a steel band, which is driven continuously past a bank of hammers—usually, 132—at speeds of the order of 500 cm/s. A ribbon is suspended between the band and the paper, behind which hammer bank is located. This is known as *back printing*. When a desired character is located at a specific print position, that hammer is actuated without stopping the band. Several characters may be printed simultaneously, although activating too many hammers at one time must be avoided, since that would slow or stop the band and overload the hammer power supply. Timing marks (and an associated sensor) provide precise positional information for the band.

Figure 18.3.4 is a schematic sketch of a print head for a wire matrix printer. These printers are most commonly used with personal computers. The print head contains nine hammers, each of which is a metal wire. The wires are activated in accordance with information received from a character generator so that a desired character can be printed in an array of dots. The wires strike a ribbon, and the print head is positioned on a carrier. Printing is done while the carrier is in continuous motion, and it is necessary to know the position to an

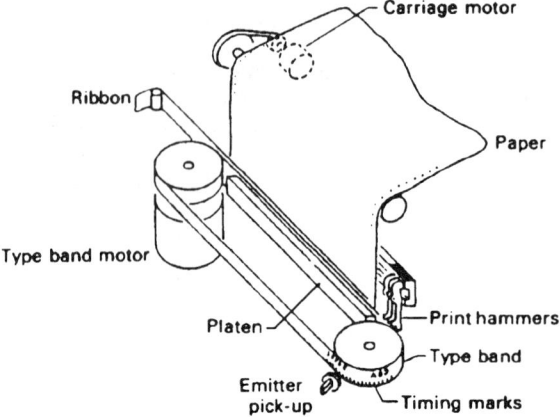

FIGURE 18.3.3 An engraved band print mechanism.

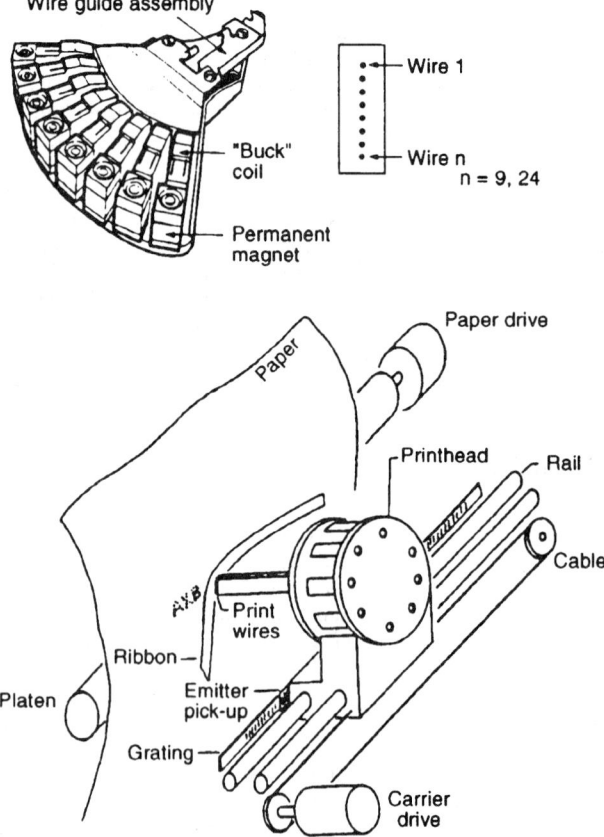

FIGURE 18.3.4 Schematic drawing of wire-matrix print head and print mechanism.

accuracy of a fraction of a dot spacing, that is, to much better than 0.1 mm. To do this, an optical grating is placed on the printer frame, and the motion of the carrier is detected by the combination of a simple light source and photodetector. Other sensing and control schemes are also used. Quite attractive printing can be generated with a wire matrix printer if the dots are more closely spaced. This can be done in the serial printer shown here by printing the vertical rows of dots in an overlapping fashion, and then indexing the page vertically only one-half or one-third the dot spacing in the basic matrix, so that the dots also overlap in the other direction. Many variations of this print mechanism are available. Among them are the use of 18 or 24 wires, and the use of multicolor ribbons to enable color printing.

NONIMPACT PRINTERS

Nonimpact printers use various mechanisms to imprint characters on a page. There are two major classes of such devices, those using *special paper* and those using *ordinary paper.* In one type of special-paper device, the paper is coated with chemicals that form a dark or colored dye when heated. Characters are formed by contact with selective heated wires or resistors that form a matrix character. Another type of special paper, known as *electro-erosion paper*, is coated with a thin aluminum metal film that can be evaporated when subjected to an intense electric spark, obtained from a suitably actuated array of wires.

Another method of nonimpact printing on special paper, called *electrography*, or, more often, *electrostatic printing*, uses a dielectric coated paper that is directly charged by ion generation. In this case an electrostatic latent image is generated by an array of suitably addressed styli. The charged paper can be toned and fixed as in *electrophotography*, also known as *xerography.* In the xerographic process, the image is developed by exposing the latent image to a toner, which may be very small solid powder particles or a suspension of solid particles or a dye in a neutral liquid. The toner particles, suitably charged, adhere either to the charged or the neutral portions of the image. This pattern is then fixed, by heat, pressure, or evaporation of the liquid carrier.

Most plain-paper nonimpact page printers combine laser and electrophotographic techniques to produce high-speed, high-quality computer printout. The printing process uses a light-sensitive photoconductive material (such as selenium, or various organic compounds) wrapped around a rotating drum. The photoconductor is electrically charged and then exposed to light images of alphanumeric or graphic (image) information). These images selectively discharge the photoconductor where there is light and leave it charged where there is no light. A powdered black toner material is then distributed over the photoconductor where it adheres to the unexposed areas and does not adhere to the exposed areas, thus forming a dry-powder image. This image is then electrostatically transferred to the paper where it is fixed by fusing it with heat or pressure. In some printers the toner is suspended in a liquid carrier, and the image is fixed by evaporation of the carrier liquid. A different choice of toner materials allows development of the unexposed rather than the exposed areas. The various elements of this process can all be quite complex and many variations are used. (See Fig. 18.3.5.)

IMAGE FORMATION

Electrophotography, transfer electrophotography, and thermal methods require that optical images of characters be generated. In some types of image generation, the images of the characters to be printed are stored. Other devices use a linear sweep arrangement with an on-off switch. The character image is generated from a digital storage device that supplies bits to piece together images. In the latter system any type of material can be printed, not just the character sets stored; the unit is said to have a *noncoded-information* (NCI) capability or *all-points-addressable* (APA) capability.

Patterns are generated either by scanning a single source of energy, usually a laser in printers and an electron beam in display devices, and selectivity modulating the output, or else by controlling an array of transducers. The array can be the actuators in a wire matrix printer, light-emitting diodes, ink-jet nozzles (see below), an array of ion sources, thermal elements, or magnetic or electrical styli.

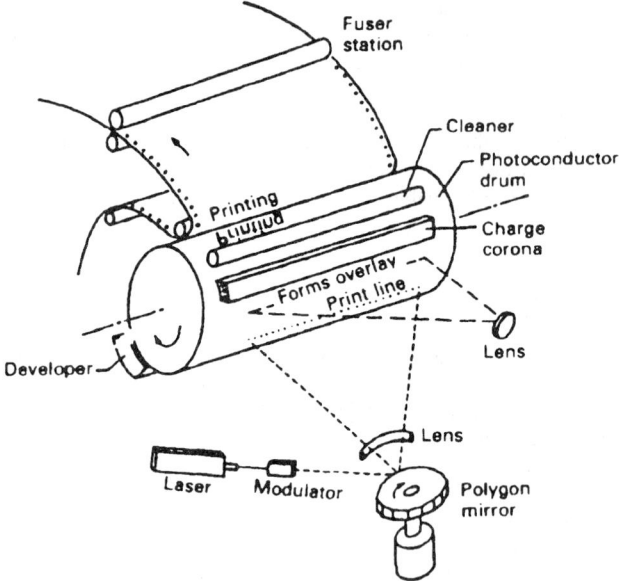

FIGURE 18.3.5   Xerographic printer mechanism.

The output terminals for digital image processing are high-resolution monochrome or color monitors for real-time displays. For off-line processing, a high-resolution printer is required. The inputs to digital image-processing requirements are generally devices such as the vidicon, flying-spot scanner, or color facsimile scanner.

INK JETS

Another method of direct character formation, usually used on untreated paper, employs ink droplets.

An example of this technique is continuous-droplet ink-jet printers, which have been developed for medium-to high-quality serial character printing and is also usable for very high speed line printing. When a stream of ink emerges from a nozzle vibrated at a suitable rate, droplets tend to form in a uniform, serial manner. Figure 18.3.6 shows droplets emerging from a nozzle and being electrostatically charged by induction as they

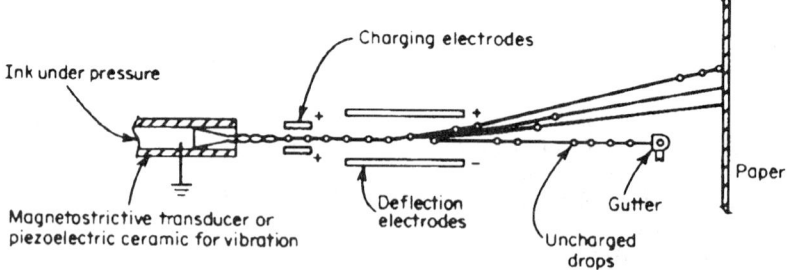

FIGURE 18.3.6   Ink-jet printing. Ink under pressure is emitted from a vibrating nozzle, producing droplets that are charged by a signal applied to the charging electrodes. After charging, each drop is deflected by a fixed field, the amount of deflection depending on the charge previously induced by the charging electrodes.

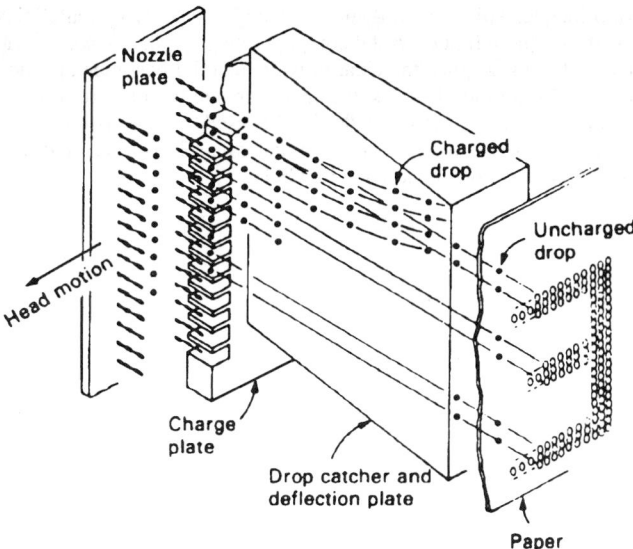

FIGURE 18.3.7   Nozzle-per-spot binary ink-jet printer.

break off from the ink stream. In subsequent flight through an electrostatic field, the droplets are displaced according to the charge they received from the charging electrode. Droplets generated at high rates (of the order of 100,000 droplets per second) are guided in one dimension and deposited upon untreated paper. The second dimension is furnished by moving the nozzle relative to the paper. Since the stream of droplets is continuous, it is necessary to dispose of most of the ink before it reaches the paper. This is usually done by selectively charging only those drops that should reach the paper; uncharged drops are intercepted by a gutter and recirculated.

An array of nozzles can be used in combination with paper displacement to deposit characters. The simplest form of such a printer has a separate nozzle for each dot position extending across the page (Fig. 18.3.7). Uncharged droplets proceed directly to the paper, while unwanted drops are charged such that they are deflected into a gutter for recirculation. Moving the paper past the array of nozzles provides full two-dimensional coverage of the paper. Such a binary nozzle-per-spot printer can print tens of thousands of lines per minute. Ink-jet systems have also been developed using very fine matrices for character generation, producing high document quality.

VISUAL-DISPLAY DEVICES

Visual-display devices associated with a computer system range from console lights that indicate the internal state of the system to *cathode-ray-tube* displays that can be used for interactive problem solving. In the cathode-ray tube, a raster scan, in conjunction with suitable bit storage, generates the output image.

For certain graphics applications, it is preferable to write the vectors making up the image directly, rather than by means of raster deflection. Such a vector mode provides higher-quality output, but can be overcome by refresh limitations when there is a large number of vectors to be displayed. Such displays are usually used with a keyboard for information entry, so that the operator and computer can operate in an interactive mode. Graphic input devices such as mice, joysticks, touch panels, and the like are also often used.

A relatively low cost, low power, and high-performance display is the *liquid-crystal display* (LCD). Liquid crystals differ from most other displays in that they depend on external light sources for visibility; i.e., they employ a *light-valve* principle. This type of display is composed of two parallel glass plates with conductive lines on their inner surfaces and a liquid-crystal compound (e.g., of the *nematic* variety) sandwiched between them. In the dynamic scattering display, the clear organic material becomes opaque and reflective when subjected to an electric field. As in other displays, the characters are built up from segments or dots.

Color LCDs are available. In one version, a dot triad is used, with thin-film color filters deposited directly on the glass. Again, the complexity and cost are such that where low power and the thin-form factors are not critical, the technology has been uncompetitive with CRTs.

CHAPTER 18.4
SOFTWARE

NATURE OF THE PROBLEM

Even though hardware costs have been declining dramatically over a 30-year period, the overall cost of developing and implementing new data processing systems and applications has not decreased. Because developing software is a predominantly labor-intensive effort, overall costs have been increasing. Furthermore, the problems being solved by software are becoming more and more complex. This creates a real challenge to achieve intellectual and management control over the software development process.

The successful development of software requires discipline and rigor coupled with appropriate management control arising from adequate visibility into the development process itself. This has led to the rise of *software engineering*, defined as the application of scientific knowledge to the design and construction of computer programs and the associated documentation and to the widespread use of standardized commercially available software packages. In addition, a set of software tools has been developed to assist in system analysis of designs. The tools, often called computer assisted systems engineering, or CASE tools, mechanize the graphic and textual descriptions of processes, test interrelationships, and maintain cross-referenced data dictionaries.

THE SOFTWARE LIFE-CYCLE PROCESS

In the earlier history of software the primary focus was on its development, but it has become evident that many programs are not one-shot consumables but are tools intended to be used repetitively over an extended time. As a result, it is obvious that the entire software life cycle must be considered. The software life cycle is that period of time over which the software is defined, developed, and used. Figure 18.4.1 shows the traditional model of the software life-cycle process and its five major phases. It begins with the *definition phase*, which is the key to everything that follows. During the definition phase, the system requirements to be satisfied by the system are developed and the system specifications, both hardware and software, are developed. These specifications describe *what* the software product must accomplish. At the same time, test requirements should also be developed as a requisite for systems acceptance testing.

The *design phase* is concerned with the design of a software structure that can meet the requirements. The design describes *how* the software product is to function. During the *development phase*, the software product is itself produced, implemented in a programming language, tested to a limited degree, and integrated. During the *test phase*, the product is extensively tested to show that it does in fact satisfy the user's requirements. The *operational phase* includes the shipment and installation of the data-processing system in the user's facility. The system is then employed by the user, who usually embarks on a maintenance effort, modifying the system to improve its performance and to satisfy new requirements. This effort continues for the remainder of the life of the system.

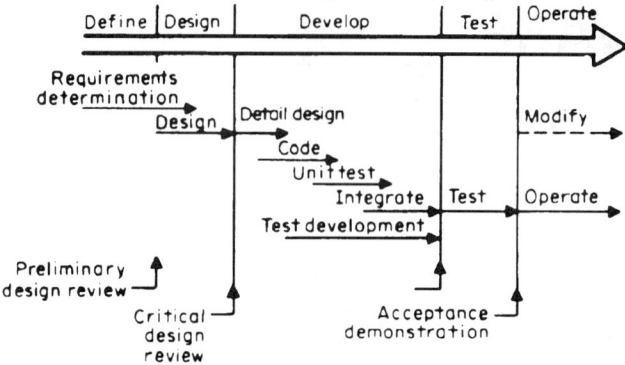

FIGURE 18.4.1   Traditional model of the software life-cycle process showing its five major phases.

PROGRAMMING

When a stored-program digital computer operates, its storage contains two types of information: the data being processed and program instructions controlling its operations. Both types of information are stored in binary form. The control unit accesses storage to acquire instructions; the ALU makes reference to storage to gain access to data and modify it. The set of instructions describing the various operations the computer is designed to execute is referred to as a *machine language*, and the act of constructing programs using the appropriate sequences of these computer instructions is called *machine-language programming*. It is possible but expensive to write them directly in machine languages, and maintenance and modification is virtually impossible. *Programming languages* have been created to make the code more accessible to its writers.

A programming language consists of two major parts: the language itself and a translator. The language is described by a set of *symbols* (the *alphabet*) and a *grammar* that tells how to assemble the symbols into correct strings. The *translator* is a machine-language program whose main function is to translate a program written in the *programming language* (the *source code*) into machine language (*object code*) that can be executed in the computer. Before describing some of the major programming languages currently in use, we consider two important programming concepts, *alternation and iteration*, and also see by examples some of the difficulties associated with machine-language programming.

ALTERNATION AND ITERATION

These techniques are illustrated here using a computer whose storage consists of 10,000 words each containing 4 bytes numbered 1 to 4. The instruction format is

1	2	3	4
op code	0	Address	

Op code	Name	Description
01	LOAD	Loads value from addressed word into data register
02	COMP	Compares value of addressed word with data-register value
03	ADD	Adds value of addressed word to data register
08	STORE	Copies value of data register into addressed storage word
20	BRLO	Branches if data-register value from last previously executed COMP was less than comparand

The computer used in this simplified example contains a separate nonaddressable data register that contains one word of data. Further, each instruction is accessed at an address 1 more than that of the previously executed instruction unless that instruction was BRLO instruction with a low COMP condition, in which case the address part of the BRLO instruction is the address at which the next instruction is to be accessed.

Consider the following program instructions (beginning at address 0100) to select the lower value of two items (in words 0950 and 0951) and place the selected value in a specific place (word 0800):

Address	Instruction	Effect
0100	01000950	Place first-item value in data register
0101	02000951	Compare second-item value with data-register value
0102	20000104	Branch to next instruction at address 0104 if data-register value was lower
0103	01000951	Place second item value in data register
0104	08000800	Store lower value in result (word 0800)

FLOWCHARTS

One way to depict the logical structure of a program graphically is by the use of *flowcharts*. Flowcharts are limited in what they can convey about a computer program, and with the advent of modern programming design languages they are becoming less widely used. However, they are used here to portray these simple programs graphically. The program of the preceding example is depicted by the flowchart shown in Fig. 18.4.2.

The flowchart contains boxes representing processes (rectangular boxes) and decisions (alternations—diamond-shaped boxes). The arrows connecting the boxes represent the paths and sequences of instruction execution. An alternation represents an instruction (or a sequence of instructions) with more than one possible successor depending on the result of some processing test (this is commonly a conditional branch). In the example, instruction 0103 is or is not executed depending on the values of the two items.

If the example is extended to require finding the least of four item values, the flowchart is that shown in Fig. 18.4.3. If the example is further extended to find the largest value of 1000 items (in locations 0336 through 0790 inclusive in hexadecimal), the flowchart and the corresponding program become very large if analogous extensions of the flowcharts are used.

The alternative is to use the technique known as the *program loop*. A program loop for this latter example is:

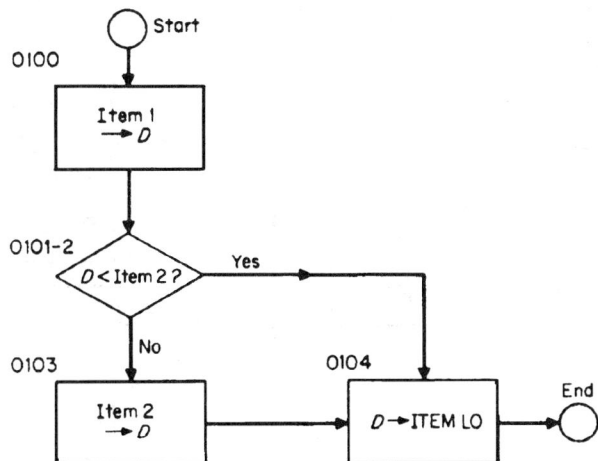

FIGURE 18.4.2 Flowchart of a simple program. The boxes represent processes; the diamonds represent decisions.

Address	Instruction	Effect
0100	01000950	Move first item as initial value of result (ITEMHI)
0101	08000800	
0102	01000900	Initialize loop to begin with item 2
0103	08000104	
0104	(00000000)	Loop, Nth item to data register
0105	02000800	Compare with prior ITEMHI value
0106	20000108	Branch to 108 if Nth item value low
0107	08000800	Store Nth item value as ITEMHI
0108	01000104	Increment value, of N by 1
0109	03000901	
010A	08000104	
010B	02000902	Compare against N = 1001
010C	2000104	Branch for looping if N < 1001
010D	end	
0900	01000951	Load item 2; initial instruction
0901	00000001	Address increment of 1
0902	010003E9	Limit test; load 1001st item

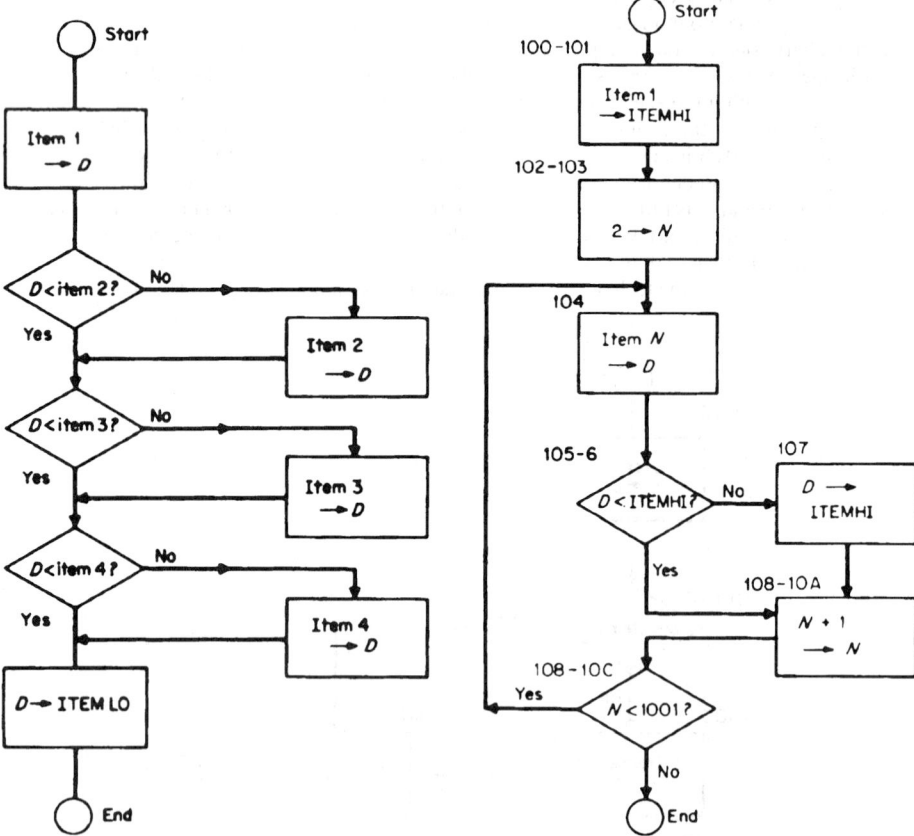

FIGURE 18.4.3 Flowchart of a repetitive task.

FIGURE 18.4.4 Flowchart showing a program loop.

The corresponding flowchart appears in Fig. 18.4.4. The loop proper (instructions 0104 to 010C) is executed 999 times. The instruction at 0104 accesses the *N*th item and is indexed each time the program flows through the loop so that on successive executions successive words of the item table are obtained. After each loop execution, a test is made to determine whether processing is complete or a branch should be made back to the beginning of the loop to repeat the loop program.

The loop proper is preceded by several instructions that *initialize* the loop, presetting ITEMHI and the instruction 0104 value for the first time through. A loop customarily has a *process* part, and an *induction* part to make changes for the next loop iteration, and an *exit test* or termination to determine whether an additional iteration is required.

ASSEMBLY LANGUAGES

The previous example illustrates the difficulty of preparing and understanding even simple machine-language programs. One help would be the ability to use a symbolic (or mnemonic) representation of the operations and addresses used in the program. The actual translation of these symbols to specific computer operations and addresses is a more or less routine clerical procedure. Since computers are well suited to performing such routine operations, it was quite natural that the first automatic programming aids, *assembly languages* and their associated assembly programs, were developed to take advantage of that fact. Assembly languages permit the critical addressing interrelations in a program to be described regardless of the storage arrangement, and they can produce therefrom a set of machine instructions suitable for the specific storage layout of the computer in use. An assembly-language program for the 1000-value program of Fig. 18.4.4 is shown in Fig. 18.4.5.

The program format illustrated is typical. Each line has four parts: location, operation, operand(s), and comments. The location part permits the programmer to specify a symbolic name to be associated with the address

LOC	OP	Operand	Comment
	ORG	100	Start at ADDR 100
START	LOAD	ITEM 1	Move first item value to ITEMHI
	STORE	ITEM HI	
	LOAD	CONST 1	Set LOOP to start with second item
	STORE	LOOP ST	
LOOP ST	CONST		* Load *N*TH item
	COMP	ITEM HI	Compare with previous HI
	BRLO	LOOP INC	Skip if lower
	STORE	ITEM HI	Store on equal or high
LOOP INC	LOAD	LOOP ST	Increment *N* by 1
	ADD	ONE	
	STORE	LOOP ST	Modify storage
	COMP	CONST 2	Exit test
	BRLO	LOOP ST	Repeat if *N* <1001
			(End of example)
	. . .		
	ORG	700	
CONST 1	LOAD	ITEM 1 + 1	Initial LOOP ST instruction
ONE	CONST	+ 1	Incrementation constant
CONST 2	LOAD	ITEM 1 + 1000	LOOP end test constant
	. . .		
	ORG	800	
ITEM HI	RESRV	1	Word where high value left
	. . .		
	ORG	950	
ITEM 1	RESRV	1	First item value
	RESRV	999	Second through thousandth items

FIGURE 18.4.5 An assembly program. The program statements are in a one-to-one correspondence with machine instructions. Hence the procedure is fully supplied by the programmer according to the particular macroinstruction set of the system. The assembly language alleviates housekeeping routines, such as specific assignments, and makes user-oriented symbols possible instead of numeric or binary code.

of the instruction (or datum) defined on that line. The operation part contains a mnemonic designation of the instruction operation code. Alternatively, that line may be designated to be a datum constant, a reservation of data space, or a designation of an assembly *pseudo operation* (a specification to control the assembly process itself). Pseudo operations in the example are ORG for origin and END to designate the end of the program.

The operand field(s) give the additional information needed to specify the machine instruction, e.g., the name of a constant, the size of the data reservation, or a name associated with a pseudo operation. The comment part serves for documentation only; it does not affect the assembly-program operation.

After a program is written in assembly language, it is processed by an assembler. The assembly program reads the symbolic assembly-language input and produces (1) a machine instruction program with constants, usually in a form convenient for subsequent program loading, and (2) an assembly listing that shows in typed or printed form each line of the symbolic assembly-language input, together with any associated machine instructions or constants produced therefrom.

The assembly pseudo operation ORG specifies that the instructions and/or constant entries for succeeding lines are to be prepared for loading at successive addresses, beginning at the specified load origin (value of operand field or ORG entry). Thus the 13 symbolic instructions following the initial ORG line in Fig. 18.4.5 are prepared for loading at addresses 0100 through 010C inclusive, with the following symbolic associations established:

Location symbol	(Local) address
START	100
LOOP ST	104
LOOP INC	108

Four instructions of this group of 13 contain the symbol LOOP ST in the operand field, and the corresponding machine instructions will contain 0104 in their address parts.

The operation of a typical assembly program therefore consists of (1) collecting all location symbols and determining their values (addresses), called *building the symbol table*, and (2) building the machine instructions and/or constants by substituting op codes for the OP mnemonics and location symbol values for their positions in the operand field. The symbol table must be formed first since, as the first instruction in the example shows, a machine instruction may refer to a location symbol that appears in the location field near the program end. Thus most assembly programs process the program twice; the *first pass* builds the symbol table, and the *second pass* builds the machine-language program. Note in the example the use of the operation RESRV to reserve space (skipping in the load-address sequence) for variable data.

Assembly language is specific to a particular computer instruction repertoire. Hence, the basic unit of assembly language describes a single machine instruction (so-called one-for-one assembly process).

Most assembly languages have a *macroinstruction* facility. This permits the programmer to define *macros* that can generate desired sequences of assembly-language statements to perform specific functions. These macro definitions can be placed in macro libraries, where they are available to all programmers in the facility.

The term *procedure* (also *subroutine* and *subprogram*) is used to refer to a group of instructions that perform some particular function used repeatedly in essentially the same *context*. The quantities that vary between contexts may be regarded as parameters (or arguments) of the procedure. The method of adaptation of the procedure determines whether it is an *open* or *closed* procedure.

An open subroutine is adapted to its parameter values during code preparation (assembly or compilation) in advance of execution, and a separate copy of the subroutine code is made for each different execution context. A closed subroutine is written to adapt itself during execution to its parameter values; hence, a single copy suffices for several execution contexts in the same program. The open subroutine executes faster since tailoring to its parameter values occurs before execution begins. The closed subroutine not only saves storage space, since one copy serves multiple uses, but is more flexible, in that parameter values derived from the execution itself can be used.

A closed subroutine must be written to determine its parameter values in a standard way (including the return point after completion). The conventions for finding the values and/or addresses of values are called the subroutine linkage *conventions*. Quite commonly, a single address is placed in a particular register, and this address in turn points to a consecutively addressed list of addresses and/or values to be used. Subroutines commonly use (or *call*) other closed subroutines, so that there are usually a number of levels of subroutine control available at any point during execution. That is, one routine is currently executing, and others are waiting at various points in partially executed condition.

HIGH-LEVEL PROGRAMMING LANGUAGES

On general-purpose digital computers, *high-level programming languages* have largely superseded assembly languages as the predominant method of describing application programs. Such programming languages are said to be *high-level* and *machine-independent.* High-level means that each program function is such that several or many machine instructions must be executed to perform that function. Machine-independent means that the functions are intended to be applied to a wide range of machine-instruction repertoires and to produce for each a specific machine representation of data.

The high-level language translator is known as a *compiler*, i.e., a program that converts an input program written in a particular high-level language (*source program*) to the machine language of a particular machine type (*object program*) each time the source code is executed.

HIGH-LEVEL PROCEDURAL LANGUAGES

Most of the high-level programming languages are said to be *procedural.* The programmer writing in a high-level procedural language thinks in terms of the precise sequence of operations, and the program description is in terms of sequentially executed *procedural statements.* Most high-level procedural languages have statements for *documentation, procedural execution, data declaration*, and various compiler and execution *control specifications.*

The program in Fig. 18.4.6, written in the FORTRAN high-level language, describes the program function given in Fig. 18.4.5 in assembly language. The first six lines are for documentation only, as indicated by C in the first column. The DIMENSION statement defines ITEM to consist of 1000 values. The assignment statement ITEMHI = ITEM (1) is read as "set the value of ITEMHI to the value of the first ITEM." The next statement is a loop-control statement meaning: "do the following statements through the statement labeled 1 for the variable *N* assuming every value from 2 through 1000." The statement labeled 1 causes a test to be made to "see if the *N*th ITEM is greater than .GT. the value of ITEMHI, and if so, set the ITEMHI value equal to the value of the *N*th item.

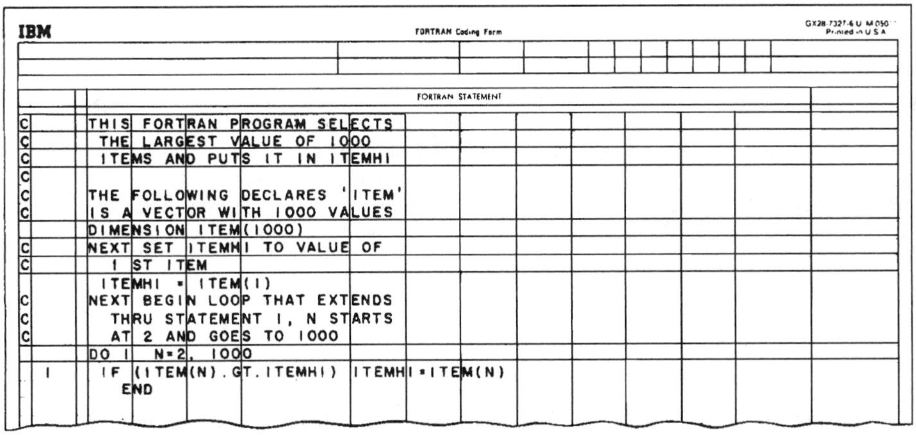


```
C      THIS FORTRAN PROGRAM SELECTS
C       THE LARGEST VALUE OF 1000
C       ITEMS AND PUTS IT IN ITEMHI
C
C      THE FOLLOWING DECLARES 'ITEM'
C      IS A VECTOR WITH 1000 VALUES
       DIMENSION ITEM(1000)
C      NEXT SET ITEMHI TO VALUE OF
C       1 ST ITEM
       ITEMHI = ITEM(1)
C      NEXT BEGIN LOOP THAT EXTENDS
C       THRU STATEMENT 1, N STARTS
C       AT 2 AND GOES TO 1000
       DO 1  N=2, 1000
    1  IF (ITEM(N).GT.ITEMHI) ITEMHI=ITEM(N)
        END
```

FIGURE 18.4.6 An example of a FORTRAN program, corresponding to the flowchart of Fig. 18.4.4 and assembly program of Fig. 18.4.5.

FORTRAN

The high-level programming languages most commonly used in engineering and scientific computation are C++, FORTRAN, ALGOL, BASIC, APL, PL/I, and PASCAL FORTRAN, the first to appear was developed during 1954 to 1957 by a group headed by Backus of IBM. Based on algebraic notation, it allows two types of numbers: integers (positive and negative) and floating point. Variables are given character names of up to six positions. All variables beginning with the letters I, J, K, L, M, or N are integers; otherwise they are floating point. Integer constants are written in normal fashion, 1, 0, −4, and so on. Floating-point constants must contain a decimal point, 3.1, −0.1, 2.0, 0.0, and so on. For example, 6.02×10^{24} is written 6.02E24. This standard notation was adopted to accommodate the limited capability of computer input-output equipment.

READ and WRITE statements permit values of variables to be read into or written from the ALU, from or to input, output, or intermediate storage devices. The latter may operate merely by transcribing values or may be accompanied by conversions or editing specified in a separate FORMAT statement. Some idea of the range of operations provided in FORTRAN is shown by the following value-assignment statement:

$$\text{ROOT} = (-(B/2.0) + \text{SQRT} ((B/2.0) ** 2 - A*C))/A$$

This is the formula for the root of a quadratic equation with coefficients A, B, and C. The asterisk indicates multiplication, / stands for division, and ** exponentiation.

The notation: *name (expression)* and *name (Expression, expression)*, and so forth, is used in FORTRAN with two distinct meanings depending on whether or not the specific name appears in a DIMENSION statement. If so, the *expression*(s) are subscript values; otherwise the *name* is considered to be a function name, and the expressions are the values of the arguments of the function. SQRT ((B/2.0) **2 − A*C) in the preceding assignment statement requires the expression (B/2.0)**2 − A*C to be evaluated, and then the function (square root here) of that value is determined. Square root and various other common trigonometric and logarithmic functions and their respective inverses are standardized in FORTRAN, typically as closed subroutines.

The same notations may be employed for a function defined by a FORTRAN programmer in the FORTRAN language. This operation is performed by writing a separate FORTRAN program headed by the statement

FUNCTION name (arg 1, arg 2, etc.)

where arg represents the name that stands for the actual argument value at each evaluation of that function. Similarly, any action or set of actions described by a closed FORTRAN subroutine is called for by "CALL subroutines (args)" together with a defining FORTRAN subroutine headed by "SUBROUTINE subroutine name (args)."

BASIC

BASIC is high-level programming language based on algebraic notation that was developed for solving problems at a terminal; it is particularly suitable for short programs and instructional purposes. The user normally remains at the terminal after entering his program in BASIC, while it compiles, executes, and types the output, a process that typically requires only a few seconds. Widely used in PCs, BASIC is usually bundled with PC operating systems. It is available in both interpretive and compiled versions, and may include an extensive set of programming capabilities. Visual BASIC is currently being used on display-oriented operating systems such as Windows.

APL

A programming language (APL) is high-level language that is often used because it is easy to learn and has an excellent interactive programming system supporting it. Its primitive objects are arrays (lists, tables, and so forth). It has a simple syntax, and its semantic rules are few. The usefulness of the primitive functions is further enhanced by operations that modify their behavior in a systematic manner. The sequence control is simple because one statement type embraces all types of branches and the termination of the execution of any function always returns control to the point of use. External communication is established by variables shared between APL and other systems.

PASCAL

An early high-level programming languages is PASCAL, developed by Niklaus Wirth. It has had widespread acceptance and use since its introduction in the early 1970s. The language was developed for two specific purposes: (1) to make available a language to teach programming as a systematic discipline and (2) to develop a language that supports reliable and efficient implementations. PASCAL provides a rich set of both control statements and data structuring facilities. Six control statements are provided: BEGIN-END, IF-THEN-ELSE, WHILE-DO, REPEAT-UNTIL, FOR-DO, and CASE-END. Similar control statements can be found in virtually all high-level languages.

In addition to the standard scalar data types, PASCAL provides the ability to extend the language via user-defined scalar data types. In the area of higher-level structured data types, PASCAL extends the array facility of ALGOL 60 to include the record, set, file, and pointer data types. In addition to these, PASCAL contains a number of other features that make it useful for programming and teaching purposes. In spite of this, PASCAL is a systematic language and modest in size, attributes that account for its popularity.

ADA PROGRAMMING LANGUAGE

(Ada is a registered trademark of the Department of Defense.) Ada is named after Lord Byron's daughter. This language was developed by the U.S. Department of Defense to be a single successor to a number of high-level languages in use by the armed forces of the United States. It was finalized in 1980.

The Ada language was designed to be a strongly typed language, with features from modern programming language theory and software engineering practices. It is a block-structured language providing mechanisms for data abstraction and modularization. It supports concurrent processing and provides user control over scheduling and interrupt handling.

C PROGRAMMING LANGUAGE

Research is continuing in the development of new languages that support the concepts growing out of modern software technology development. One such language is the C programming language (a registered trademark of AT&T). C is a general-purpose programming language designed to feature modern control flow and data structures and a rich set of operators, yet provide an economy of expression. Although it was not specialized for any one area of application, it has been found especially useful for implementing operating systems and is being more widely used in communications and other areas. C++ is a later version of this language. An extension of this language has been called JAVA, developed by Sun Microsystems.

OBJECT-ORIENTED PROGRAMMING LANGUAGES

A second area of programming language development has been the creation of object-oriented languages. These are used for message-object programming, which incorporates the concepts of objects that communicate by messages. An object includes data, a set of procedures (methods) that operate on the data, and a mechanism for translating messages. These languages should contribute to improved reusability of software—a highly sought after goal to increase programming productivity and reduce software costs. These languages are based on the concept that "objects," once defined, are henceforth available for reuse, without reprogramming. Programs can then be viewed as mechanisms that employ the appropriate object at the appropriate time to accomplish the task at hand.

COBOL AND RPG

High-level programming languages used for business data-processing applications emphasize description and handling of files for business record keeping. Two widely used programming languages for business applications are COBOL (*common business-oriented language*) and RPG (*report program generation*). Compilers for these languages, with generalized sorting programs, form the fundamental automatic programming aids of many computer installations primarily used for business data processing. COBOL and RPG have comparable file, record, and field-within-record descriptive capabilities, but the specification of processing and sequence control device from basically different concepts.

OPERATING SYSTEMS

There are many reasons for developing and using an *operating system* for a digital computer. One of the main reasons is to optimize the scheduling and use of computer resources, so as to increase the number of jobs that can be run in a given period. Creation of a multiprogramming environment means that the resources and facilities of the computing system can be shared by a number of different programs, each written as if it were the only program in the system.

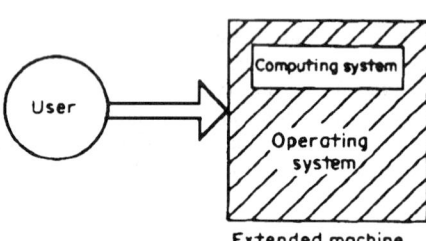

Another major objective for an operating system is to provide the full capability of the computing system to the user while minimizing the complexity and depth of knowledge of the computer system required. This is accomplished by establishing standard techniques for handling system functions like *program calling* and *data management* and providing a convenient and effective interface to the user. In effect the user is able to deal with the operating system as an entity rather than having to deal with each of the computer's features. As indicated in Fig. 18.4.7 each user will be thought of conceptually as a unit consisting of both the hardware and the programs and procedures that make up the operating system.

FIGURE 18.4.7   The user's view of the operating system as an extension of the computing system yet an integral part of it.

GENERAL ORGANIZATION OF AN OPERATING SYSTEM

There are many ways to structure operating systems, but for the purpose of this discussion the organization shown in Fig. 18.4.8 is typical. The operating system is composed of two major sets of programs, control (or supervision) programs and processing programs. *Control programs* supervise the execution of the support programs (including the user application programs), control the location, storage, and retrieval of data, handle interrupts, and schedule jobs and resources needed in processing. *Processing programs* consist of language translators, service programs, and user-written application programs, all of which are used by the programmer in support of program development.

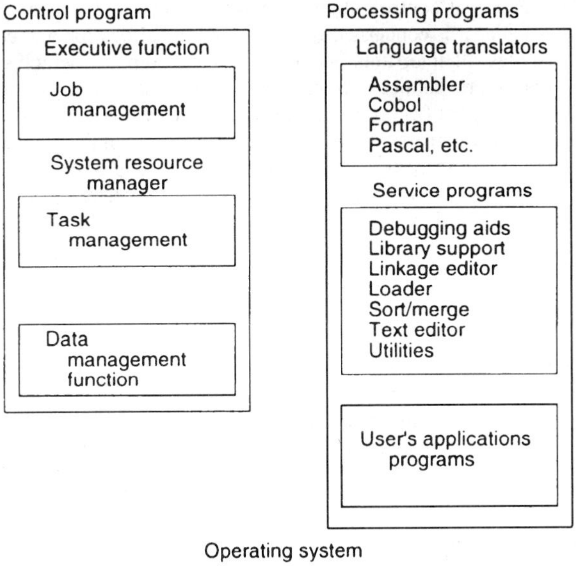

FIGURE 18.4.8 A typical operating system and its constituent parts.

The work to be processed by the computer can be viewed as a stack of jobs to be run under the management of the control program. A *job* is a unit of computational work that is independent of all other jobs concurrently in the system. A single job may consist of one or a number of *steps*.

TYPES OF OPERATING SYSTEMS

There are many basic types of operating systems including multiprogramming, time-sharing, real-time, and multiprocessing.

The multiprocessing system must schedule and control the execution of jobs that are distributed across two or more coupled processors. These processors may share a common storage, in which case they are said to be *tightly* (or *directly*) *coupled*, or they may have their own private storage and communicate via other means such as sending messages over networks, in which case they are said to be *loosely coupled*.

Operating systems were generally developed for a specific CPU architecture or for a family of CPUs. For example, MS-DOS and WINDOWS apply to xx86-based systems. However, one operating system, the UNIX system (a registered trademark of AT&T), has been *transported* to a number of different manufactures' systems and is in very wide use today. UNIX was developed as a unified, interactive, multiuser system. It consists of a kernel that schedules tasks and manages data, a shell that executes user commands—one at a time or in a series called a pipe—and a series of utility programs.

TASK-MANAGEMENT FUNCTION

This function, sometimes called the *supervisor*, controls the operation of the system as it executes units of work known as *tasks* or *processes*. (The performance of a task is requested by a job step.) The distinction between a task and a program should be noted. A *program* is a static entity, a sequence of instructions, while a *task* is a dynamic entity, the work to be done in execution of the program. Task management initiates and controls the execution of tasks. If necessary it controls their synchronization. In allocates system resources to tasks and monitors their use. In particular it is concerned with the dynamic allocation and control of main storage space.

Task management handles all interrupts occurring in the computer, which can arise from five different sources: (1) supervisor call interrupts occur when a task needs a service from the task-management function, such as initiating an I/O operation; (2) program interrupts occur when unusual conditions are encountered in the execution of a task; (3) I/O interrupts indicate that an I/O operation is complete or some unusual condition has occurred; (4) machine-check interrupts are initiated by the detection of hardware errors; and (5) external interrupts are initiated by the timer, by the operator's console, or other external devices.

DATA MANAGEMENT

This function provides the necessary I/O control system services needed by the operating system and the user application programs. It frees the programmer from the tedious and error-prone details of I/O programming and permits standardization of these services. It constructs and maintains various file organization structures, including the construction and use of index tables. It allocates space on disc (auxiliary) storage. It maintains a directory showing the locations of data sets (files) within the system. It also provides protection for data sets against unauthorized entry.

OPERATING SYSTEM SECURITY

One of the major concerns in the design of operating systems is to make certain that they are reliable and that they provide for the protection and the integrity of the data and programs stored within the system. Work is under way to develop secure operating systems. These systems use the concept of a security kernel—a minimal set of operating system programs that are formally specified and designed so that they can be proved to implement the desired security policy correctly. This assures the corrections of all access-controlling operations throughout the system.

SOFTWARE-DEVELOPMENT SUPPORT

There have been great strides in software engineering technology. Out of the research and development efforts in universities, industry, and government have emerged a number of significant ideas and concepts that can have significant and long-lasting influence on the way that software is developed and managed. Those concepts are just now starting to find their way into the software-development process but should become more widely used in the future. They are briefly reviewed below.

REQUIREMENTS AND SPECIFICATIONS

This has been one of the problem areas through the years. Analysis and design errors are, by far, the most costly and crucial types of errors, and a number of attempts are being made to develop methods for recording and analyzing software requirements and developing specifications. Most requirements and specifications documents are still recorded in English narrative form, which introduces problems of inconsistency, ambiguity, and incompleteness. These problems are addressed with CASE tools and structured programming.

SOFTWARE DESIGN

The work of Dijkstra, Hoare, and Mills (1968, 1976) had a major influence on software-design methodology by introducing a number of concepts that led to the development of *structured programming*. Structured programming is a methodology based on mathematical rigor. It uses the concept of top-down design and

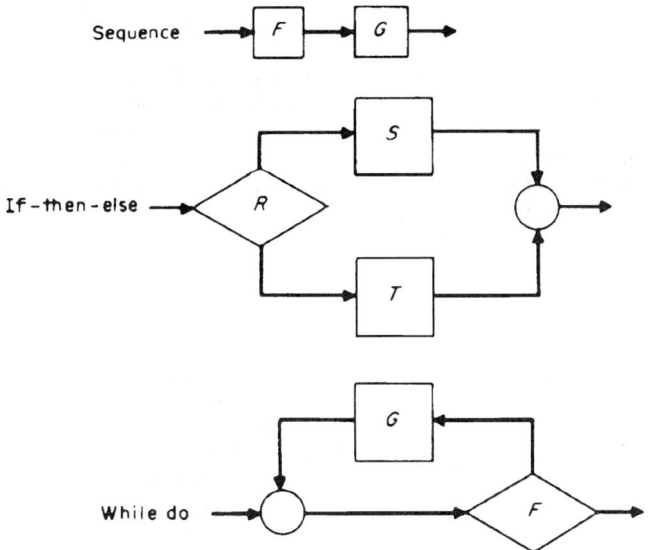

FIGURE 18.4.9 Basic set of structured control primitives.

implementation by describing the program at a high level of abstraction and then expanding (refining) this abstraction into more detailed representations through a series of steps until sufficient detail is present for implementation of the design in a programming language to be possible. This process is called *stepwise refinement*. The design is represented by a small, finite set of primitives, such as shown in Fig. 18.4.9. These three primitives are adequate for design purposes but for convenience several others have been introduced; namely, the *indexed alternation* or *case*, the *do-until*, and the *indexed sequence* or *for-do* structure.

It also recognized that the organization and representation of software systems are clearer if certain data and the operations permitted on those data are organized into data abstractions. The internal details of the organization of the data are hidden from the user.

The result of applying this methodology is the organization of a sequential software process into a hierarchical structure through the use of stepwise refinement. The software system structure is then defined at three levels—the *system*, the *module*, and the *procedure*. The system (or *job*) describes the highest level of program execution. The system is decomposed into modules. The module is composed of one or more procedures and data that persist between successive invocations of the module. The procedure is the lowest level of system decomposition, the executable unit of the stored program.

Another important aspect of the design process is its documentation. There is a critical need to record the design as it is being developed, from its highest level all the way to its lowest level of detail, before its implementation in a programming language, using language that can be used not only to communicate software designs between specialists in software development but also between specialists and nonspecialists in rigorous logical terms.

Important elements of the software development process are reviews, walk-throughs, and inspections, which can be applied to the products of the software-development process to ensure that they are complete, accurate, and consistent. They are applied to such areas as the design (design inspections), the software source code (code inspections), documentation, test designs, and test-results analysis. The goals of the software review and inspection process are to ensure that standards are met, to check on the quality of the product, and to detect and correct errors at the earliest possible point in the software life cycle. Another important value of the review process is that it permits progress against development-plan milestones to be measured more objectively and rework for error correction to be monitored more closely.

TESTING

An important activity in the software-development cycle that is often ignored until too late in the process is testing. It is also important to note what testing can and cannot do for the software product. Quality cannot be tested into the software; it must be designed into it. Test planning begins with the requirements analysis at the beginning of the project. Requirements should be testable; i.e., they should be stated in a form that permits the final product to be tested to assure that it satisfies the requirements. Test planning and test designs should be developed in parallel with the design of the software.

EXPERT SYSTEMS

One important area of research in computer science has been that of *artificial intelligence.* The most successful application of artificial intelligence techniques has been in the development of *expert systems*, or *knowledge-based systems*, as they are often called. These are human-machine interactive systems with specialized problem-solving expertise that are used to solve complex problems in such specific areas as medicine, chemistry, mathematics, and engineering. This expertise consists of knowledge about the particular problem domain that the expert system is designed to support (e.g., diagnosis and therapy for infectious diseases) and planning and problem-solving rules for processes used to identify and solve the particular problem at hand. The two main elements of an expert system are its knowledge base, which contains the domain knowledge for the problem area being addressed, and the inference engine, which contains the general problem-solving knowledge. A key task in constructing an expert system is knowledge acquisition: the extraction and formulation of knowledge (facts, concepts, rules) from existing sources, with special attention paid to the experience of experts in the particular problem domain being addressed.

CHAPTER 18.5
DATABASE TECHNOLOGY

DATABASE OVERVIEW

Around 1964 a new term appeared in the computer literature to denote a new concept. The term was "database," and it was coined by workers in military information systems to denote collections of data shared by end users of time-sharing computer systems. The commercial data-processing world at the time was in the throes of "integrated data processing," and quickly appropriated "database" to denote the data collection that results from consolidating the data requirements of individual applications. Since that time, the term and the concept have become firmly entrenched in the computer world. Today, computer applications in which users access a database are called *database applications*. The *database management system*, or DBMS, has evolved to facilitate the development of database applications. The development of DBMS, in turn, has given rise to new languages, algorithms, and software techniques, which together make up what might be called a *database technology*. An overview of a typical DBMS is shown in Fig. 18.5.1.

Traditional data-processing applications used *master files* to maintain continuity between program runs. Master files "belonged to" applications, and the master files within an enterprise were often designed and maintained independently of one another. As a result, common data items often appeared in different master files, and the values of such items often did not agree. There was thus a requirement to consolidate the various master files into a single database, which could be centrally maintained and shared among various applications. Data consolidation was also required for the development of certain types of "management information" applications that were not feasible with fragmented master files. There was a requirement to raise the level of languages used to specify application procedures, and also to provide software for automatically transforming high-level specifications into equivalent low-level specifications. In the database context, this property of languages has come to be known as *data independence*. The consolidation of master files into databases had the undesirable side effect of increasing the potential for data loss and unauthorized data use. The requirement for data consolidation thus carried with it a requirement for tools and techniques to control the use of databases and to protect against their loss.

A DBMS is characterized by its *data-structure class*, that is, the class of data structures that it makes available to users for the formulation of applications. Most DBMSs distinguish between structure *instances* and structure *types*, the latter being abstractions of sets of structure instances. A DBMS also provides an implementation of its data-structure class, which is conceptually a mapping of the structures of the class into the structures of a lower-level class. The structures of the former class are often referred to as *logical* structures, whereas those of the latter are called *physical* structures. The data-structure classes are early systems, were derived from punched-card technology, and thus tended to be quite simple. A typical class was composed of *files* of *records* of a single type, with the record type being defined by an ordered set of fixed-length fields. Because of their regularity, such files are now referred to as *flat files*.

Database technology has produced a variety of improved data-structuring methods, many of which have been embodied in DBMS. Although many specific data-structure classes have been produced (essentially one class per system), these classes have tended to cluster into a small number of families, the most important of which are the *hierarchic*, the *network*, the *relational*, and the *semantic* families. These families have evolved more or less in the

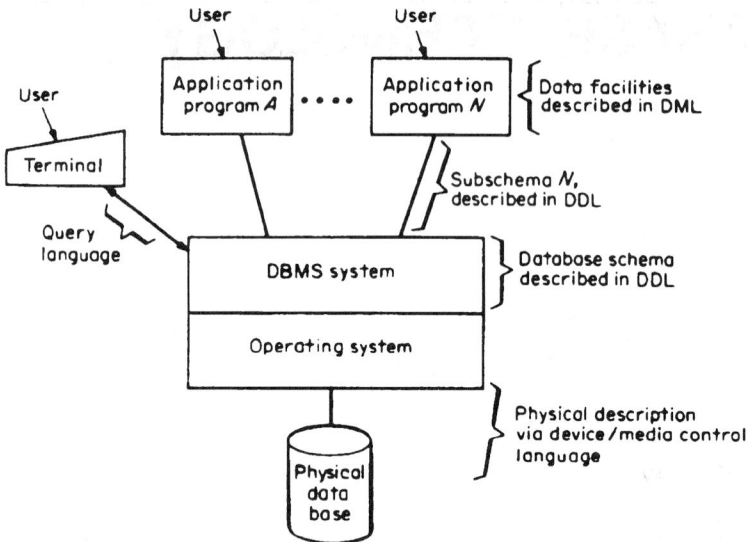

FIGURE 18.5.1   Overview of a typical database management system (DBMS).

order indicated, and all are represented in the data-structure classes of present-day DBMS. The use of large databases in distributed computing systems is sometimes called *Data Repository* or *Data Warehouse*.

HIERARCHIC DATA STRUCTURES

The *hierarchic data-structuring methods* that began to appear in the early 1960s provided some relief for the entity association problem. These methods were developed primarily to accommodate the variability that frequently occurs in the records of a file. For example, in the popular two-level hierarchic method, a record was divided into a *header* segment and a variable number of *trailer* segments of one or more types. The header segment represented attributes common to all entities of a set, while the trailer segments were used for the variably occurring attributes. The method was also capable of representing one-many associations between two sets of entities, by representing one set as header segments and the other as trailers and thus provided a primitive tool for data consolidation.

This two-level approach was expanded to *n*-level structures. These structures have also been implemented extensively on direct-access storage devices (DASD), which afford numerous additional representation possibilities. IMS was one of the first commercial systems to offer hierarchic data structuring and is often cited to illustrate the hierarchic structuring concept. The IMS equivalent of a file is the *physical database*, which consists of a set of hierarchical structured records of a single type. A record type is composed according to the following rules. The record type has a single type of *root* segment. The root segment type may have any number of *child* segment types. Each child of the root may also have any number of child segment types, and so on.

NETWORK DATA STRUCTURES

The first network structuring method to be developed for commercial data processing had its origins in the bill-of-materials application, which requires the representation of many-many associations between a set of parts and itself; e.g., a given part may simultaneously act as an assembly of other parts and as a component of other parts.

This concept was used as the basis of a database language developed by the database task group (DBTG) of CODASYL in the late 1960s and early 1970s. This language introduced some new terminology and generalized some features, and has been implemented in a number of DBMSs.

RELATIONAL DATA STRUCTURES

In the mid-1960s, a number of investigators began to grow dissatisfied with the hardware orientation of then extent data-structuring methods and, in particular, with the manner in which pointers and similar devices for implementing entity associations were being exposed to the users. These investigators sought a way of raising the perceived level of data structures and at the same time bringing them closer to the way in which people look at information. Their efforts resulted in an *entity-set*-structuring method, wherein information is represented in a set of tables, with each table corresponding to a set of entities of a single type. The rows of a table correspond to the entities in the set, and the columns correspond to the attributes that characterize the entity set type.

Tables can be used to represent associations among entities. In this case, each row corresponds to an association, and the columns correspond to entity identifiers, i.e., entity attributes that can be used to uniquely identify entities. Additional columns may be used to record attributes of the association itself (as opposed to attributes of the associated entities). The key new concepts in the entity set methods were the simplicity of the structures it provided and the use of entity identifiers (rather than pointers or hardware-dictated structures) for representing entity associations. These concepts represented a major step forward in meeting the general goal of *data independence.*

Codd (1971) noted that an entity set could be viewed as a mathematical relation on a set of domains, where each domain corresponds to a different property of the entity set. Associations among entities could be similarly represented, with the domains in this case corresponding to entity identifiers. Codd defined a (data) *relation* to be a time-varying subset of the cartesian product of the members of the set of domain *n*-tuples (or simply *tuples*). Codd proposed that relations be built exclusively on domains of elementary values—integers, character strings, and so forth. He called such relations *normalized relations* and the process of converting relations to normalized form, *normalization.* Virtually all work done with relations has been with normalized relations.

Codd postulated levels of normalization called *normal forms.* An unconstrained normalized relation is in *first normal form* (1NF). A relation in 1NF in which all nonkey domains are functionally dependent on (i.e., have their values determined by) the entire key are in *second normal form* (2NF), which solves the problem of parasitic entity representation. A relation in 2NF in which all nonkey domains are dependent *only* on the key is the *third normal form* (3NF), which solves the problem of masquerading entities. As part of the development of the relational method, Codd postulated a *relational algebra*, i.e., a set of operations on relations that is closed in the sense of a traditional algebra, and thereby provided an important formal vehicle for carrying out a variety of research in data structures and systems. In addition to the conventional set operations, the relational algebra provides such operations as *restriction*, to delete selected tuples of a relation; *projection*, to delete selected domains of a relation; and *join*, to join two relations into one. Codd also proposed a *relational calculus,* whose distinguishing feature is the method used to designate sets of tuples. The method is patterned after the predicate calculus and makes use of free and bound variables and the universal and existential quantifiers.

Codd characterized his methodology as a *data model*, and thereby provided a concise term for an important but previously unarticulated database concept, namely, the *combination* of a class of data structures and the operation allowed on the structures of the class. (A similar concept, the *abstract data type* or *data abstraction*, has evolved elsewhere in software technology.) The term "model" has been applied retroactively to early data-structuring methods, so that, for example, we now speak of hierarchic models and network models, as well as the relational model. The term is now generally used to denote an abstract data-structure class, although there is a growing realization that it should embrace operations as well as structures. The relational model has been implemented in a number of DBMSs.

SEMANTIC DATA STRUCTURES

During the evolution of the hierarchic, network, and relational methods, it gradually became apparent that building a database was in fact equivalent to building a model of an enterprise and that databases could be developed more or less independently of applications simply by studying the enterprise. The notion of a *conceptual schema* for the application-independent modeling of an enterprise and various *external schemata* derivable from the conceptual schema for expressing data requirements of specific applications is the outgrowth of this view.

Application-independent modeling has produced a spate of *semantic* data models and debate over which of these is best for modeling "reality." One of the most successful semantic models is the *entity-relationship model*, which provides data constructs at two levels: the *conceptual* level, whose constructs include entities, relationships (*n*-ary associations among entities), value sets, and attributes; and the *representation* level, in which conceptual constructs are mapped into tables.

DATA DEFINITION AND DATA-DEFINITION LANGUAGES

The history of computer applications has been marked by a steady increase in the level of the language used to implement applications. In database technology, this trend is manifested in the development of high-level data-definition languages and data-manipulation languages. A *data-definition language* (DDL) provides the DBMS user with a way to declare the attributes of structure types within the database, and thus enable the system to perform implicitly many operations (e.g., name resolution, data-type checking) that would otherwise have to be invoked explicitly. A DDL typically provides for the definition of both logical and physical data attributes as well as for the definition of different *views* of the (logical) data. The latter are useful in limiting or tailoring the way in which specific programs or end users look at the database. A *data-manipulation language* (DML) provides the user with a way to express operations on the data structure instances of a database, using names previously established through data definitions. Data-manipulation facilities are of two general types: host-language and self-contained.

A *host-language* facility permits the manipulation of databases through programs written in conventional procedural languages such as COBOL or PL/I. It provides statements that the user may embed in a program at the points were database operations are to be performed. When such a statement is encountered control is transferred to the database system, which performs the operation and returns the results (data and return codes) to the program in prearranged main-storage locations.

A *self-contained* facility permits the manipulation of the database through a high-level, nonprocedural language, which is independent of any procedural language, i.e., whose language is self-contained. An important type of self-contained facility is the *query facility*, which enables "casual" users to access a database without the mediation of a professional programmer. Other types of self-contained facility are available for performing generalizable operations on data, such as sorting report generation, and data translation.

REPORT PROGRAM GENERATORS

The use of fixed files for reporting is found in the *report program generator*, a software package intended primarily for the production of reports from formatted files. Attributes of the source files and the desired reports are described by the user in a simple declarative language, and this description is then processed by a compiler to generate a program that, when run, produces the desired reports. A key concept of the report program generator is the use of a fixed structure for the generated program, consisting of input, calculation, and output phases. Such a structure limits the transformations that can be carried out with a single generated program, but it has nevertheless proved to be remarkably versatile. (Report program generators are routinely used for file maintenance as well as for report generation.) The fixed structure of the generated program imposes a discipline on the user, which enables the user to produce a running program much more quickly than could otherwise be done with conventional languages. Report program generators were especially popular in smaller installations where conventional programming talent is scarce, and in some installations it was the only "programming language" used.

The report program generators and the formatted file systems were the precursors of the contemporary DBMS query facility. A query processor is in effect a generalized routine that is particularized to a specific application (i.e., the user's query) by the parameters (data names, Boolean predicates, and so forth) appearing in the query. Query facilities are more advanced than most early generalized routines in that they provide online (as opposed to batch) access to databases (as opposed to individual files). The basic concept is unchanged, however, and the lessons learned in implementing the generalized routines, and especially in reconciling ease of use with acceptable performance, have been directly applicable to query-language processors.

PROGRAM ISOLATION

Most DBMSs permit a database to be accessed concurrently by a number of users. If this access is not controlled, the consistency of the data can be compromised (e.g., lost updates) or the logic of programs can be affected (e.g., nonrepeatable read operations). With program isolation, records are locked for a program upon updating any item within the record and unlocked when the program reaches a *synchpoint*, i.e., a point at which the changes made by the program are committed to the database. Deadlocks can occur and are resolved by selecting one of the deadlocked programs and restarting it at its most recent synchpoint.

AUTHORIZATION

Consolidated data often constitute sensitive information, which the user may not want divulged to other than authorized people, for reasons of national security, competitive advantage, or personal privacy. DBMSs, therefore, provide mechanisms for limiting data access to properly authorized persons. A system administrator can grant specific capabilities with respect to specific data objects to specific users. Grantable capabilities with respect to relations include the capability to read from the relation, to insert tuples, to delete tuples, to update specific fields, and to delete the relation. The holder of a capability may also be given authority to grant that capability to others, so that authorization tasks may be delegated to different individuals within an organization.

CHAPTER 18.6
ADVANCED COMPUTER TECHNOLOGY

BACKGROUND

Early computer systems were tightly integrated with locally attached I/O facilities, such as card readers and punches, tape readers and punches, tape drives, and discs. Data had to be moved to and from the site of the programming system physically. For certain systems, the turnaround time in such processes was considered excessive and it was not long before steps were taken to interconnect *remotely located* I/O devices to a computer site via telephone and telegraph facilities.

Telecommunications today means any form of communicating that exists to support the exchange of information between people and application programs in one of the three following combinations: person to person, person to process, or process to process. Thus, today, telecommunications encompasses telephony, facsimile transmission, television, and data communications. The impetus for this change has been continual advancement in digital technology, a common technological thread running through all the previously mentioned disciplines.

TERMINALS

Terminals are the mechanisms employed by individuals to interface to communications mechanisms and to represent them in their information exchange with other individuals or processes. Since the early advent of telecommunications, many terminal device types have come in the market. The earliest and most popular devices were the *keyboard printers* made by Teletype Corporation and other manufacturers. These electro-mechanical machines operated at speeds from 5 to 11 characters per second.

Another popular type of terminal device is the *cathode-ray-tube* (CRT) *visual display* and keyboard. This *dumb terminal* operates at speeds from 1200 to 9600 b/s across common-carrier telephone facilities or at megabit rates directly attached to the local channel of a processor. Displays generally are alphanumeric or graphic, the preponderance of terminals being alphanumeric. Many such displays can be attached to a single *control unit*, or *node*, which can greatly reduce the overall costs of the electronic devices required to drive the multiple devices. The control unit can also reduce communications cost by concentrating the multiple-terminal data traffic from the locally connected drives into a single data stream and sharing a single channel between the controller and a *host processor.* Some controllers support the attachment of hard-copy printers and have sufficient logic to permit printout of information of a display screen without communicating with a host computer. Displays are generally used in conversational or interactive modes of operation.

In recent years PCs have had a profound impact on the quantity of this type of display. Although the personal computer really is an *intelligent workstation*, most PCs can employ communications software subsystems to emulate nonintelligent visual display and keyboard devices. In this way, they can communicate with host software.

Many specialized *transaction-type terminals* are designed for specific industry applications and form another category of terminals. Special point-of-sale terminals for the retail industry, bank teller and cash dispensers, and airline ticketing terminals are examples of transaction-type terminals.

A category of terminals that incorporates many of the features of previous categories, the *intelligent workstation*, is really a remote mini- or microcomputer system that permits the distribution of application program functions to remote sites. Such *intelligent terminals* are capable of supporting a diversity of functions from interactive processing to remote batched job entry. Some of these terminals contain such a high level of processing capability that they are, in fact, true host processors. With the expanded scope of telecommunications, television and telephones are beginning to play an important role as terminals. Not that they were not terminals before, but rather, they are now recognized as an important part of the terminal types to be found in the world of information. Terminals now support voice, data, and video communications within one single integrated package.

HOSTS

When people communicate with machines via their terminals, they are communicating with application programs. These programs offer services to the user, such as data processing (computation, word processing, editing, and so forth) and data storage or retrieval. In such applications as message switching and conferencing, hosts often act as intermediates between two terminals. Thus they may contain application software that stores a sender's message until the designated receiver actively connects to the message application or to some program that has been made aware of the person's relationship to the host service.

Host applications also may have a need to cooperate with other host applications in *distributed processors.* Distributed processing and distributed database applications typically involve host-to-host communications. Host-to-host communications typically require much higher-speed communication channels than terminal-to-host traffic. Within recent years, host channel speeds have increased to tens of megabits per second, commensurate with the increased millions-of-instructions-per-second (MIPS) rate of processors.

Personal computers can also assume the role of a host. The difference between being a terminal or a host is defined by the telecommunications software supporting the PC. Acceptance of this statement should be easy when one recognizes that PCs today come with the processing power and storage capacity attributable to what was once viewed as a sizable host processor.

COMMUNICATIONS SYSTEMS

For communications to occur between end users, it is essential that two aspects be present: (1) There must be interconnectivity between the communicating parties to permit the transference of signals and bits that represent the information transfer, and (2) there must be commonality of representation and interpretation of the bits that represent the information.

In the early years of telecommunications, the world of *connectivity* was divided and classified as *local* and *remote.* Local implied limited distances from a host, usually with relatively high-speed connections (channel speeds). Terminals were connected to controllers that concentrated the data traffic. Remote connections depended on the services available from common carriers. These worlds were separate and distinct.

Evolving standards obscure the earlier distinction between local and remote and introduce the new concept of *networking.* In the telecommunications world, networking is the interconnection and interoperability of systems for the purpose of providing communications service. The types of communications services to be found in modern networks vary with the needs of subscribers and the service characteristics of the media available. The end users impose requirements in the form of service classifications. Examples of requirements are capacity, integrity, delay characterizations, acceptable cost, security, and connectivity. Each medium imposes constraints for which the systems may have to compensate. Thus a medium may have a limited capacity, it may be susceptible to error, it may introduce a propagation delay, it may also lack security, and it may limit connectivity.

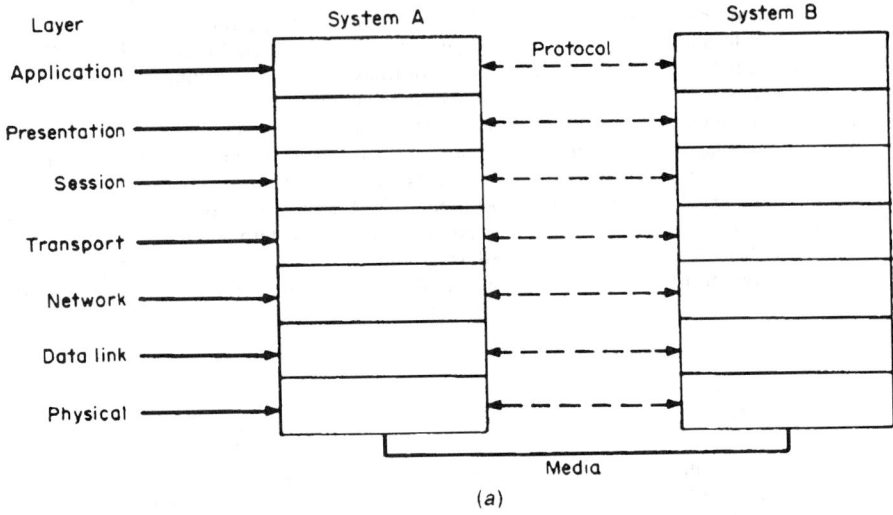

FIGURE 18.6.1a   ISO-OSI model.

Most early communications networks were designated to meet specific subscriber needs, as was the case with the early telephone system, airline reservation systems, and the early warning system. Computer and transmission technology have changed this concept. The introduction of digital technology introduces a common denominator to the field of information communications that eases the sharing of resources used in the creation of communications networks. However, the existence of millions of terminals and tens of thousands of processors designed to operate within the construct of their own private communications world, and the need for innovation to accommodate new and heretofore unanticipated subscriber needs, presents a formidable challenge to the information communications industry.

In an attempt to bring about order, the International Standards Organization (ISO) has developed standards recommendations for an *open system*. An open system is one that conforms to the standards defined by ISO and thus facilitates connectivity and interoperability between systems manufactured by different vendors. Figure 18.6.1 depicts the ISO open system interconnect (OSI) model.

OSI REFERENCE MODEL

The OSI reference model describes a communication system in the seven hierarchic layers shown in Fig. 18.6.1. Each layer provides services to the layer above and invokes services from the layer below. Thus the *end users* of a communications system interconnect to the *application layer*, which provides the user interface and interprets user service requests. This layer is often thought of as a *distributed operating system*, because it supports the interconnectivity and communicability between end users that are distributed. The model is a general one and even permits two end users to use the application layer interface of a common system to communicate, instead of a shared local operating system. By hiding the difference between locally connected and remotely connected end users, the interconnected and interrelated application layer entities assume the role of a *global operating system*, as shown in Fig. 18.6.2.

In a single system, the operating system contains *supervisory control logic* for the resources—both logical and physical—that are allocated to provide services to the end user. In a distributed system, such as a communications system, the global supervisory service for all the layers resides in the application layer, hence, the reason to view the layer as a global operating system.

Note that the model applies equally as well to telephony. An end user, in requesting a connection service from the communications system, communicates with a supervisory controller. This controller is in the local

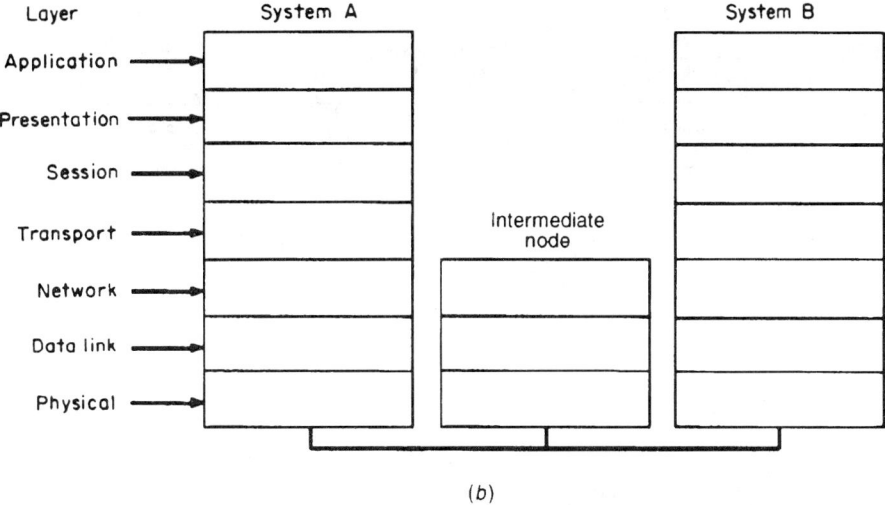

FIGURE 18.6.1b   ISO-OSI model with an intermediate node.

exchange of a telephone operating company. The cooperating efforts of the distributed supervisory control components of the telephone systems are an analog of the distributed application layers of a data network, as shown in Fig. 18.6.1b.

A user views the application layer as a global server. Requests for service are passed from the user to the communications system through an interface that typically has been defined by the serving system. Even though systems differ in their programs and hardware architectures, there is a growing recognition of the necessity for the introduction of standardization at this critical interface. Consider the difficulty in making a telephone call when traveling, if every telephone company had chosen a different interface mechanism. This would require a knowledge of all the different methods, as is the case for persons who travel to foreign countries.

The services available from the communications systems can be quite varied. In telephony, a basic transport and connection service is prevalent. However, as greater intelligence was placed within the systems of

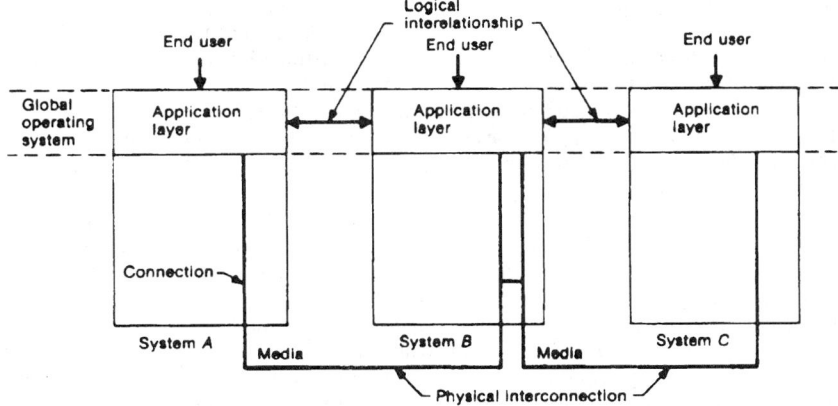

FIGURE 18.6.2   Application layers as a global operating system.

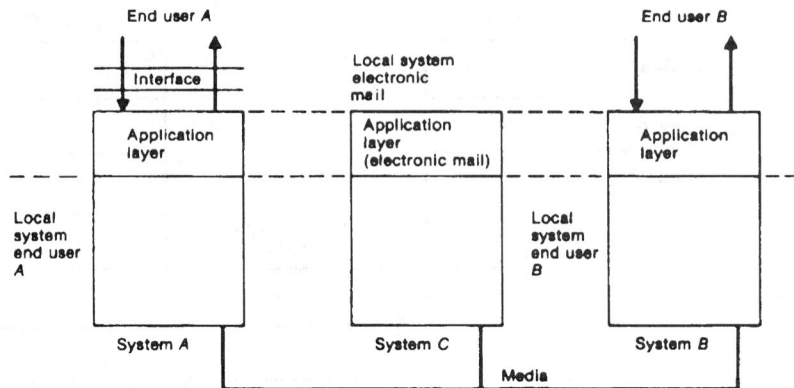

FIGURE 18.6.3 Example use of the application layer.

the telephone networks, the operating companies have increased the types of services they provide. Systems are operational in the United States, which integrate voice and data information through a common interface mechanism and across a shared medium. That medium is the existing twisted-pair wire that once carried only analog voice signals, or optical lines, which carry a much greater volume of data. Carrier networks offer *electronic mail*, *packet switching*, *telemetry services*, *videotex*, and many other services. The limit on what can or will be provided will be determined by either government regulations or market demand.

The layers that lie below the application layer exist to support communications between end users or the communications services of an application layer that exist in a distributed system, as is shown in Fig. 18.6.3. The application layer of system *A* can contain knowledge of the electronic mail server in system *C* and, on behest of end user *A* (EUA), invoke that service. The application layer of system *A* can also support a connect service that enables it to create a communications path between system *A* and its equivalent partner (*functional pair*) in system *B*, for the purpose of supporting communications between EUA and EUB.

Each of the layers of the OSI model in Fig. 18.6.1 contributes some value to the communications service between communicating partners, be they end users or communication systems distributed entities.

The *application layer* is a user of the *presentation service layer* and is concerned with the differences that exist in the various processors and operating systems in which each of the distributed communications systems is implemented. The presentation service layer provides the service to overcome differences in coded representation, format, and the presentations of information. To use an analogy, if one system's machine talked and understood Greek and another system's machine talked and understood Latin, the presentation services layer would perform the necessary service to permit the comprehension of information exchanged between the two systems.

The *presentation service layer* is a user of the session layer which manages the dialogue between two communicating partners. The session layer assures that the information exchange conforms to the rules necessary to satisfy the end user needs. For example, if the exchange is to be by the *two-way alternate* mode, the session layer monitors and enforces this mode of exchange. It regulates the rate of flow to end users and assures that information is delivered in the same form and sequence in which it had been transmitted, if that had been a requirement. It is the higher layer's port into the transmission subsystem that encompasses the lower four layers of the OSI reference model in Fig. 18.6.1*a*.

The *session layer* is the user of the *transport layer*, which creates a logical pipe between the session layer of its system and that of any other system. The transport layer is responsible for selecting the appropriate lower-layer network to meet the service requirement of the session-layer entities, and where necessary, to enhance lower-level network services by providing system end-to-end integrity, information resequencing, and system-to-system flow control to assure no underrun or overrun of the receiving system's resources.

The transport layer uses the *network layer* to create a logical path between two systems. Systems may be attached to each other in many different ways. The simplest way is through a single point-to-point connection. The network layer in such an example is very simple. However, witness the topology depicted in Fig. 18.6.4. To go from system *A* to system *B* involves passing through several different networks, each of which exhibits

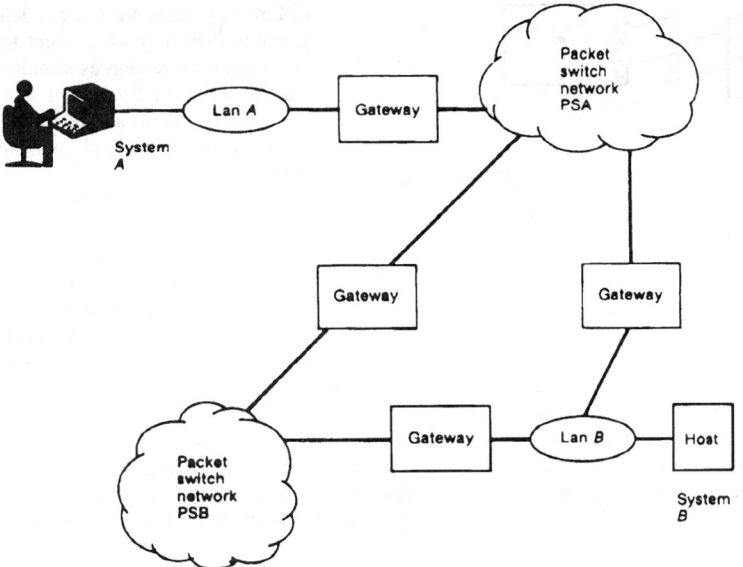

FIGURE 18.6.4   Creating a logical path between two systems.

different characteristics of transmission service and each of which may differ radically in its software and hardware design, topology, and *protocol* (rules of operation).

Such a network interconnection is rather typical. In fact, many interconnections can be far more complex. Consider the effect of the inclusion of an alternate packet-switch network PSB between local area network (LAN) *A* and PSA. The logic to decide which network or sequence of networks to employ must be dealt with within the network layer. This function is referred to as *routing*. However, to hide the presence of network complexity from the transport layer and thus provide the appearance of a single network, a unifying sublayer, called the *internet layer*, is inserted between the transport layer and subnetworks, as shown in Fig. 18.6.5.

Because networks differ with regard to the size of data units they can handle, the network layer must deal with the breaking of information frames into the size required by individual subnetworks within the path. The network layer must ultimately be capable of reassembling the information frames at the target system before passing them to that system's transport layer. The network layer must also address the problem of congestion caused by network sources being shared by large numbers of users attached to different systems. Because the end users of these

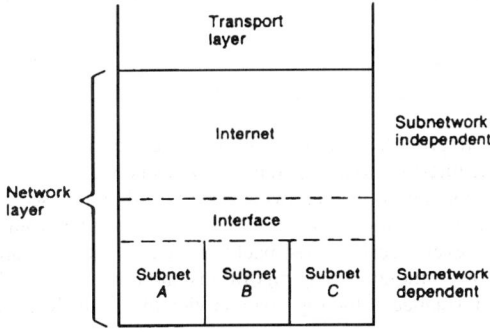

FIGURE 18.6.5   Network layer structure.

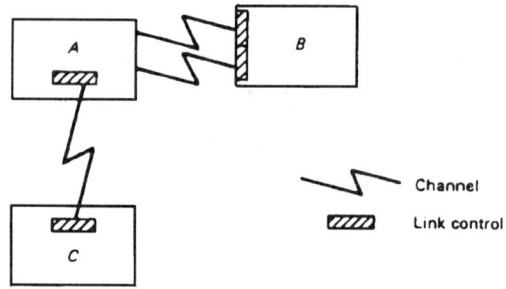

FIGURE 18.6.6   A new link control for each new channel.

different systems are independent of each other, it is possible they may all attempt to access and use the same network resources simultaneously. Congestion can result in deadlock, and such an occurrence can bring a network to a halt. Thus network layers must offer means of deadlock avoidance or detection and correction.

The network layer is a user of the *link-control layer*, which is responsible for building a point-to-point connection between two system nodes that share a common communications circuit. Link control is aware only of its neighboring node(s) on the shared channel. Each new circuit requires a new link control, as illustrated in Fig. 18.6.6.

It is possible to hide the existence of multiple circuits between two nodes by the inclusion of a multilink sublayer above the individual link-control layers. Link control performs such functions as error detection and correction, framing, flow-control sequencing, and channel-access control.

Link control is the user of the *physical layer*. The physical layer is responsible for transforming the information frame into a form suitable for transmission onto the medium. Thus its major function is signaling, i.e., putting information onto the medium, removing it, and retransforming it into the code structure understandable by the link control.

REAL SYSTEMS

It is important to understand that OSI describes a layered model that allocates functions necessary in the process of communications to each of the layers. However, the actual implementation of any system to attain conformance with the model is left to individual system vendors. Thus, whether there are layers and where a function is actually executed within a single system are matters beyond the scope of ISO. The test of conformance is the successful interoperability between systems. However, one can use the model to place in proper context existing communications protocols, associated entities, services, and products. For example, modems are entities of the physical layer. A data service unit is another such entity, and LAN adapter is still another.

For the network layer, the list is still growing. There are many different subnetworks, each with its own routing schemes to effect the switching of information through the nodes of its unique communications subnetwork. Thus, when one looks at some of the network service providers, such as Telenet or Tymnet, different logical and physical entities performing the same services are observed. Agreement exists among vendors and carriers that if systems are to interconnect, there is a need for a unifying network address structure to avoid ambiguity in identifying systems. There must also be a way of shielding the transport layer entities from various differences within the possible networks that could be involved in achieving end-to-end communications. Such shielding is implemented in the protocols of the Internet and associated software, such as TCP/IP.

PACKET SWITCH

A packet switch is designed to provide the three lower-layer services of the OSI model to its subscribers (physical, link, and network). The interface to access these services is X.25. (See Fig. 18.6.1*b*). When one observes the implementation of various vendor packet switches, it is difficult to see the layers. Most vendors tend to integrate their layers and thus do not clearly define the interfaces that would demonstrate the layer-to-layer independence required. However, because most packet-switched networks are in themselves closed systems, and each generally has its own unique routing algorithms and congestion and flow-control mechanisms, OSI is concerned only with conformance at the top-layer service interface (X.25) and at the gateway between different packet networks (X.75).

SECTION 19

CONTROL SYSTEMS

Control is used to modify the behavior of a system so it behaves in a specific desirable way over time. For example, we may want the speed of a car on the highway to remain as close as possible to 60 miles per hour in spite of possible hills or adverse wind; or we may want an aircraft to follow a desired altitude, heading, and velocity profile independent of wind gusts; or we may want the temperature and pressure in a reactor vessel in a chemical process plant to be maintained at desired levels. All these are being accomplished today by control methods and the above are examples of what automatic control systems are designed to do, without human intervention. Control is used whenever quantities such as speed, altitude, temperature, or voltage must be made to behave in some desirable way over time.

This section provides an introduction to control system design methods. D.C.

In This Section:

 On the CD-ROM:

"A Brief Review of the Laplace Transform Useful in Control Systems," by the authors of this section, examines its usefulness in control systems analysis and design.

CHAPTER 19.1
CONTROL SYSTEM DESIGN

Panos Antsaklis, Zhiqiang Gao

INTRODUCTION

To gain some insight into how an automatic control system operates we shall briefly examine the speed control mechanism in a car.

It is perhaps instructive to consider first how a typical driver may control the car speed over uneven terrain. The driver, by carefully observing the speedometer, and appropriately increasing or decreasing the fuel flow to the engine, using the gas pedal, can maintain the speed quite accurately. Higher accuracy can perhaps be achieved by looking ahead to anticipate road inclines. An automatic speed control system, also called *cruise control*, works by using the difference, or error, between the actual and desired speeds and knowledge of the car's response to fuel increases and decreases to calculate via some algorithm an appropriate gas pedal position, so to drive the speed error to zero. This decision process is called a *control law* and it is implemented in the *controller*. The system configuration is shown in Fig. 19.1.1. The car dynamics of interest are captured in the *plant*. Information about the actual speed is fed back to the controller by *sensors*, and the control decisions are implemented via a device, the *actuator*, that changes the position of the gas pedal. The knowledge of the car's response to fuel increases and decreases is most often captured in a mathematical model.

Certainly in an automobile today there are many more automatic control systems such as the antilock brake system (ABS), emission control, and tracking control. The use of feedback control preceded control theory, outlined in the following sections, by over 2000 years. The first feedback device on record is the famous Water Clock of Ktesibios in Alexandria, Egypt, from the third century BC.

Proportional-Integral-Derivative Control

The proportional-integral-derivative (PID) controller, defined by

$$u = k_P e + k_I \int e + k_d e \qquad (1)$$

is a particularly useful control approach that was invented over 80 years ago. Here k_P, k_I, and k_d are controller parameters to be selected, often by trial and error or by the use of a lookup table in industry practice. The goal, as in the cruise control example, is to drive the error to zero in a desirable manner. All three terms in Eq. (1) have explicit physical meanings in that e is the current error, $\int e$ is the accumulated error, and $\dot{e}$ represents the trend. This, together with the basic understanding of the causal relationship between the control signal (u) and the output (y), forms the basis for engineers to "tune," or adjust, the controller parameters to meet the design specifications. This intuitive design, as it turns out, is sufficient for many control applications.

To this day, PID control is still the predominant method in industry and is found in over 95 percent of industrial applications. Its success can be attributed to the simplicity, efficiency, and effectiveness of this method.

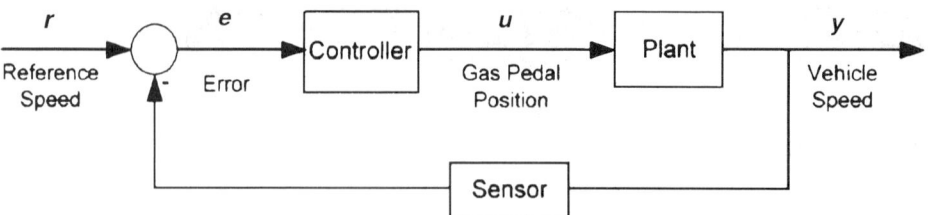

FIGURE 19.1.1   Feedback control configuration with cruise control as an example.

The Role of Control Theory

To design a controller that makes a system behave in a desirable manner, we need a way to predict the behavior of the quantities of interest over time, specifically how they change in response to different inputs. Mathematical models are most often used to predict future behavior, and control system design methodologies are based on such models. Understanding control theory requires engineers to be well versed in basic mathematical concepts and skills, such as solving differential equations and using Laplace transform. The role of control theory is to help us gain insight on how and why feedback control systems work and how to *systematically* deal with various design and analysis issues. Specifically, the following issues are of both practical importance and theoretical interest:

1. Stability and stability margins of closed-loop systems.

2. How fast and how smoothly the error between the output and the set point is driven to zero.

3. How well the control system handles unexpected external disturbances, sensor noises, and internal dynamic changes.

In the following, modeling and analysis are first introduced, followed by an overview of the classical design methods for single-input single-output plants, design evaluation methods, and implementation issues. Alternative design methods are then briefly presented. Finally, for the sake of simplicity and brevity, the discussion is restricted to linear, time invariant systems. Results maybe found in the literature for the cases of linear, time-varying systems, and also for nonlinear systems, systems with delays, systems described by partial differential equations, and so on; these results, however, tend to be more restricted and case dependent.

MATHEMATICAL DESCRIPTIONS

Mathematical models of physical processes are the foundations of control theory. The existing analysis and synthesis tools are all based on certain types of mathematical descriptions of the systems to be controlled, also called plants. Most require that the plants are linear, causal, and time invariant. Three different mathematical models for such plants, namely, linear ordinary differential equations, state variable or state space descriptions, and transfer functions are introduced below.

Linear Differential Equations

In control system design the most common mathematical models of the behavior of interest are, in the time domain, linear ordinary differential equations with constant coefficients, and in the frequency or transform domain, transfer functions obtained from time domain descriptions via Laplace transforms.

Mathematical models of dynamic processes are often derived using physical laws such as Newton's and Kirchhoff's. As an example consider first a simple mechanical system, a spring/mass/damper. It consists of a weight m on a spring with spring constant k, its motion damped by friction with coefficient f (Fig. 19.1.2).

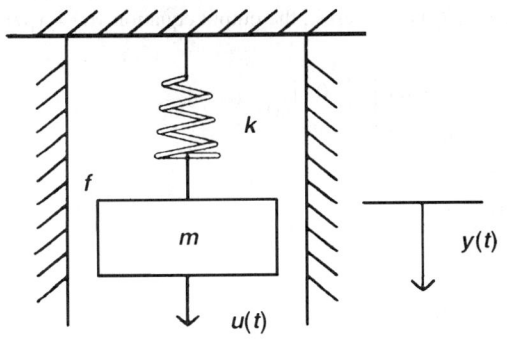

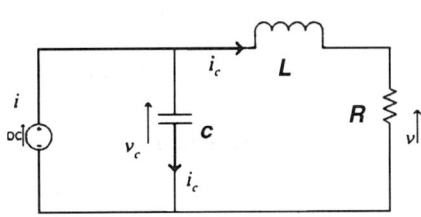

FIGURE 19.1.2   Spring, mass, and damper system.   **FIGURE 19.1.3**   RLC circuit.

If $y(t)$ is the displacement from the resting position and $u(t)$ is the force applied, it can be shown using Newton's law that the motion is described by the following linear, ordinary differential equation with constant coefficients:

$$\ddot{y}(t) + \frac{f}{m}\dot{y}(t) + \frac{k}{m}y(t) = \frac{1}{m}u(t)$$

where $\dot{y}(t) \overset{\Delta}{=} dy(t)/dt$ with initial conditions

$$y(t)\big|_{t=0} = y(0) = y_0 \quad \text{and} \quad \frac{dy(t)}{dt}\bigg|_{t=0} = \frac{dy(0)}{dt} = \dot{y}(0) = y_1$$

Note that in the next subsection the trajectory $y(t)$ is determined, in terms of the system parameters, the initial conditions, and the applied input force $u(t)$, using a methodology based on Laplace transform. The Laplace transform is briefly reviewed in the CD-ROM.

For a second example consider an electric RLC circuit with $i(t)$ the input current of a current source, and $v(t)$ the output voltage across a load resistance R (Fig. 19.1.3).

Using Kirchhoff's laws one may derive:

$$\ddot{v}(t) + \frac{R}{L}\dot{v}(t) + \frac{1}{LC}v(t) = \frac{R}{LC}i(t)$$

which describes the dependence of the output voltage $v(t)$ to the input current $i(t)$. Given $i(t)$ for $t \geq 0$, the initial values $v(0)$ and $\dot{v}(0)$ must also be given to uniquely define $v(t)$ for $t \geq 0$.

It is important to note the similarity between the two differential equations that describe the behavior of a mechanical and an electrical system, respectively. Although the interpretation of the variables is completely different, their relations are described by the same linear, second-order differential equation with constant coefficients. This fact is well understood and leads to the study of mechanical, thermal, fluid systems via convenient electric circuits.

State Variable Descriptions

Instead of working with many different types of higher-order differential equations that describe the behavior of the system, it is possible to work with an equivalent set of standardized first-order vector differential equations that can be derived in a systematic way. To illustrate, consider the spring/mass/damper example. Let $x_1(t) = y(t)$, $x_2(t) = \dot{y}(t)$ be new variables, called *state variables*. Then the system is equivalently described by the equations

$$\dot{x}_1(t) = x_2(t) \quad \text{and} \quad \dot{x}_2(t) = \frac{-f}{m}x_2(t) - \frac{k}{m}x_1(t) + \frac{1}{m}u(t)$$

with initial conditions $x_1(0) = y_0$ and $x_2(0) = y_1$. Since $y(t)$ is of interest, the output equation $y(t) = x_1(t)$ is also added. These can be written as

$$\begin{bmatrix} \dot{x}_1(t) \\ \dot{x}_2(t) \end{bmatrix} = \begin{bmatrix} 0 & 1 \\ -k/m & -f/m \end{bmatrix} \begin{bmatrix} x_1(t) \\ x_2(t) \end{bmatrix} + \begin{bmatrix} 0 \\ 1/m \end{bmatrix} u(t)$$

$$y(t) = \begin{bmatrix} 1 & 0 \end{bmatrix} \begin{bmatrix} x_1(t) \\ x_2(t) \end{bmatrix}$$

which are of the general form

$$\dot{x}(t) = Ax(t) + Bu(t), \qquad y(t) = Cx(t) + Du(t)$$

Here $x(t)$ is a 2×1 vector (a column vector) with elements the two state variables $x_1(t)$ and $x_2(t)$. It is called the *state vector*. The variable $u(t)$ is the *input* and $y(t)$ is the *output* of the system. The first equation is a vector differential equation called the *state equation*. The second equation is an algebraic equation called the *output equation*. In the above example $D = 0$; D is called the direct link, as it directly connects the input to the output, as opposed to connecting through $x(t)$ and the dynamics of the system. The above description is the *state variable* or *state space description* of the system. The advantage is that system descriptions can be written in a standard form (the state space form) for which many mathematical results exist. We shall present a number of them in this section.

A state variable description of a system can sometimes be derived directly, and not through a higher-order differential equation. To illustrate, consider the circuit example presented above, using Kirchhoff's current law

$$i_c = C \frac{dv_c}{dt} = i - i_L$$

and from the voltage law

$$L \frac{di_L}{dt} = -Ri_L + v_c$$

If the state variables are selected to be $x_1 = v_c$, $x_2 = i_L$, then the equations may be written as

$$\begin{bmatrix} \dot{x}_1 \\ \dot{x}_2 \end{bmatrix} = \begin{bmatrix} 0 & -1/C \\ 1/L & -R/L \end{bmatrix} \begin{bmatrix} x_1 \\ x_2 \end{bmatrix} + \begin{bmatrix} 1/C \\ 0 \end{bmatrix} i$$

$$v = \begin{bmatrix} 0 & R \end{bmatrix} \begin{bmatrix} x_1 \\ x_2 \end{bmatrix}$$

where $v = Ri_L = Rx_2$ is the output of interest. Note that the choice of state variables is not unique. In fact, if we start from the second-order differential equation and set $\bar{x}_1 = v$ and $\bar{x}_2 = \dot{v}$, we derive an equivalent state variable description, namely,

$$\begin{bmatrix} \dot{\bar{x}}_1 \\ \dot{\bar{x}}_2 \end{bmatrix} = \begin{bmatrix} 0 & 1 \\ -1/LC & -R/L \end{bmatrix} \begin{bmatrix} \bar{x}_1 \\ \bar{x}_2 \end{bmatrix} + \begin{bmatrix} 0 \\ R/LC \end{bmatrix} i$$

$$v = \begin{bmatrix} 1 & 0 \end{bmatrix} \begin{bmatrix} \bar{x}_1 \\ \bar{x}_2 \end{bmatrix}$$

Equivalent state variable descriptions are obtained by a change in the basis (coordinate system) of the vector state space. Any two equivalent representations

$$x = Ax + Bu, \qquad y = Cx + Du \quad \text{and} \quad \bar{x} = \bar{A}\bar{x} + \bar{B}u, \qquad y = \bar{C}\bar{x} + \bar{D}u$$

are related by $\bar{A} = PAP^{-1}, \bar{B} = PB, \bar{C} = CP^{-1}, \bar{D} = D$, and $\bar{x} = Px$ where P is a square and nonsingular matrix. Note that state variables can represent physical quantities that may be measured, for instance, $x_1 = v_c$ voltage, $x_2 = i_L$ current in the above example; or they can be mathematical quantities, which may not have direct physical interpretation.

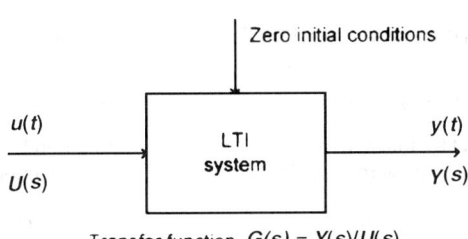

Transfer function $G(s) = Y(s)/U(s)$

FIGURE 19.1.4   The transfer function model.

Linearization. The linear models studied here are very useful not only because they describe linear dynamical processes, but also because they can be approximations of nonlinear dynamical processes in the neighborhood of an operating point. The idea in linear approximations of nonlinear dynamics is analogous to using Taylor series approximations of functions to extract a linear approximation. A simple example is that of a simple pendulum $x_1 = x_2, x_2 = -k \sin x_1$, where for small excursions from the equilibrium at zero, $\sin x_1$ is approximately equal to x_1 and the equations become linear, namely, $x_1 = x_2, x_2 = -kx_1$.

Transfer Functions

The *transfer function* of a linear, time-invariant system is the ratio of the Laplace transform of the output $Y(s)$ to the Laplace transform of the corresponding input $U(s)$ with all initial conditions assumed to be zero (Fig. 19.1.4).

From Differential Equations to Transfer Functions.   Let the equation

$$\frac{d^2 y(t)}{dt^2} + a_1 \frac{dy(t)}{dt} + a_0 y(t) = b_0 u(t)$$

with some initial conditions

$$y(t)\big|_{t=0} = y(0) \quad \text{and} \quad \frac{dy(t)}{dt}\bigg|_{t=0} = \frac{dy(0)}{dt} = y(0)$$

describe a process of interest, for example, a spring/mass/damper system; see previous subsection. Taking the Laplace transform of both sides we obtain

$$[s^2 Y(s) - sy(0) - y(0)] + a_1[sY(s) - y(0)] + a_0 Y(s) = b_0 U(s)$$

where $Y(s) = L\{y(t)\}$ and $U(s) = L\{u(t)\}$. Combining terms and solving with respect to $Y(s)$ we obtain:

$$Y(s) = \frac{b_0}{s^2 + a_1 s + a_0} U(s) + \frac{(s + a_1)y(0) + y(0)}{s^2 + a_1 s + a_0}$$

Assuming the initial conditions are zero,

$$Y(s)/U(s) = G(s) = \frac{b_0}{s^2 + a_1 s + a_0}$$

where $G(s)$ *is the transfer function of the system* defined above.

We are concerned with transfer functions $G(s)$ that are rational functions, that is, ratios of polynomials in s, $G(s) = n(s)/d(s)$. We are interested in *proper* $G(s)$ where $\lim_{s \to \infty} G(s) < \infty$. Proper $G(s)$ have degree $n(s) \le$ degree $d(s)$.

In most cases degree $n(s) <$ degree $d(s)$, in which case $G(s)$ is called *strictly proper*. Consider the transfer function

$$G(s) = \frac{b_m s^m + b_{m-1} s^{m-1} + \quad + b_1 s + b_0}{s^n + a_{n-1} s^{n-1} + \quad + a_1 s + a_0} \quad \text{with } m \le n$$

Note that the system described by this $G(s)$ ($Y(s) = G(s)U(s)$) is described in the time domain by the following differential equation:

$$y^{(n)}(t) + a_{n-1} y^{(n-1)}(t) + \cdots + a_1 y^{(1)}(t) + a_0 y(t) = b_m u^{(m)}(t) + \cdots + b_1 u^{(1)}(t) + b_0 u(t)$$

where $y^{(n)}(t)$ denotes the nth derivative of $y(t)$ with respect to time t. Taking the Laplace transform of both sides of this differential equation, assuming that all initial conditions are zero, one obtains the above transfer function $G(s)$.

From State Space Descriptions to Transfer Functions.   Consider $\dot{x}(t) = Ax(t) + Bu(t)$, $y(t) = Cx(t) + Du(t)$ with $x(0)$ initial conditions; $x(t)$ is in general an n-tuple, that is, a (column) vector with n elements. Taking the Laplace transform of both sides of the state equation:

$$sX(s) - x(0) = AX(s) + BU(s) \quad \text{or} \quad (sI_n - A)X(s) = BU(s) + x(0)$$

where I_n is the $n \times n$ *identity* matrix; it has 1 on all diagonal elements and 0 everywhere else, e.g.,

$$I_2 = \begin{bmatrix} 1 & 0 \\ 0 & 1 \end{bmatrix}$$

Then

$$X(s) = (sI_n - A)^{-1} BU(s) + (sI_n - A)^{-1} x(0)$$

Taking now the Laplace transform on both sides of the output equation we obtain $Y(s) = CX(s) + DU(s)$. Substituting we obtain,

$$Y(s) = [C(sI_n - A)^{-1} B + D]U(s) + C(sI - A)^{-1} x(0)$$

The response $y(t)$ is the inverse Laplace of $Y(s)$. Note that the second term on the right-hand side of the expression depends on $x(0)$ and it is zero when the initial conditions are zero, i.e., when $x(0) = 0$. The first term describes the dependence of $Y(s)$ on $U(s)$ and it is not difficult to see that the transfer function $G(s)$ of the systems is

$$G(s) = C(sI_n - A)^{-1} B + D$$

Example   Consider the spring/mass/damper example discussed previously with state variable description $\dot{x} = Ax + Bu, y = Cx$. If $m = 1, f = 3, k = 2$, then

$$A = \begin{bmatrix} 0 & 1 \\ -2 & -3 \end{bmatrix}, \quad B = \begin{bmatrix} 0 \\ 1 \end{bmatrix}, \quad C = [1 \quad 0]$$

and its transfer function $G(s)$ ($Y(s) = G(s)U(s)$) is

$$G(s) = C(sI_2 - A)^{-1} B = [1 \quad 0]\left[\begin{bmatrix} s & 0 \\ 0 & s \end{bmatrix} - \begin{bmatrix} 0 & 1 \\ -2 & -3 \end{bmatrix}\right]^{-1}\begin{bmatrix} 0 \\ 1 \end{bmatrix}$$

$$= [1 \quad 0]\begin{bmatrix} s & -1 \\ 2 & s+3 \end{bmatrix}^{-1}\begin{bmatrix} 0 \\ 1 \end{bmatrix} = [1 \quad 0]\frac{1}{s^2 + 3s + 2}\begin{bmatrix} s+3 & 1 \\ -2 & s \end{bmatrix}\begin{bmatrix} 0 \\ 1 \end{bmatrix}$$

$$= \frac{1}{s^2 + 3s + 2}$$

as before.

Using the state space description and properties of the Laplace transform an explicit expression for $y(t)$ in terms of $u(t)$ and $x(0)$ may be derived. To illustrate, consider the *scalar case* $\dot{z} = az + bu$ with $z(0)$ initial condition. Using the Laplace transform:

$$Z(s) = \frac{1}{s-a}z(0) + \frac{b}{s-a}U(s)$$

from which

$$z(t) = L^{-1}\{Z(s)\} = e^{at}z(0) + \int_0^t e^{a(t-\tau)}bu(\tau)d\tau$$

Note that the second term is a convolution integral. Similarly in the *vector case*, given

$$x(t) = Ax(t) + Bu(t), \qquad y(t) = Cx(t) + B(t)u(t)$$

it can be shown that

$$x(t) = e^{At}x(0) + \int_0^t e^{A(t-\tau)}Bu(\tau)d\tau$$

and

$$y(t) = Ce^{At}x(0) + \int_0^t Ce^{A(t-\tau)}Bu(\tau)d\tau + Du(t)$$

Notice that $e^{At} = L^{-1}\{(sI - A)^{-1}\}$. The *matrix exponential* e^{At} is defined by the (convergent) series

$$e^{At} = I + e^{At} + \frac{A^2 t^2}{2!} + \quad + \frac{A^k t^k}{k!} + \quad = I + \sum_{k=1}^{\infty} \frac{t^k}{k!}A^k$$

Poles and Zeros. The n roots of the denominator polynomial $d(s)$ of $G(s)$ are the *poles of $G(s)$*. The m roots of the numerator polynomial $n(s)$ of $G(s)$ are (finite) *zeros of $G(s)$*.

Example (Fig. 19.1.5)

$$G(s) = \frac{s+2}{s^2 + 2s + 2} = \frac{s+2}{(s+1)^2 + 1} = \frac{s+2}{(s+1-j)(s+1+j)}$$

$G(s)$ has one (finite) zero at -2 and two complex conjugate poles at $-1 \pm j$. In general, a transfer function with m zeros and n poles can be written as

$$G(s) = k\frac{(s-z_1) \quad (s-z_m)}{(s-p_1) \quad (s-p_n)}$$

where k is the *gain*.

Frequency Response

The *frequency response of a system* is given by its transfer function $G(s)$ evaluated at $s = j\omega$, that is, $G(j\omega)$. The frequency response is a very useful means characterizing a system, since typically it can be determined experimentally, and since control system specifications are frequently expressed in terms of the frequency response. When the poles of $G(s)$ have negative real parts, the system turns out to be bounded-input/bounded-output

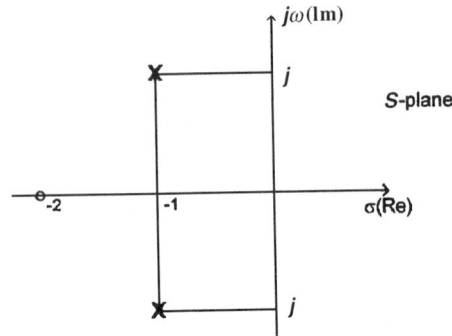

FIGURE 19.1.5   Complex conjugate poles of $G(s)$.

(BIBO) stable. Under these conditions the frequency response $G(j\omega)$ has a clear physical meaning, and this fact can be used to determine $G(j\omega)$ experimentally. In particular, it can be shown that if the input $u(t) = k \sin(\omega_o t)$ is applied to a system with a stable transfer function $G(s)$ ($Y(s) = G(s)U(s)$), then the output $y(t)$ at steady state (after all transients have died out) is given by

$$y_{ss}(t) = k \, |G(\omega_o)| \sin[\omega_o t + \theta(\omega_q)]$$

where $|G(\omega_o)|$ denotes the magnitude of $G(j\omega_o)$ and $\theta(\omega_o) = \arg G(j\omega_o)$ is the argument or phase of the complex quantity $G(j\omega_o)$. Applying sinusoidal inputs with different frequencies ω_o and measuring the magnitude and phase of the output at steady state, it is possible to determine the full frequency response of the system $G(j\omega_o) = |G(\omega_o)| \, e^{j\theta(\omega_o)}$.

ANALYSIS OF DYNAMICAL BEHAVIOR

System Response, Modes, and Stability

It was shown above how the response of a system to an input and under some given initial conditions can be calculated from its differential equation description using Laplace transforms. Specifically, $y(t) = L^{-1}\{Y(s)\}$ where

$$Y(s) = \frac{n(s)}{d(s)} U(s) + \frac{m(s)}{d(s)}$$

with $n(s)/d(s) = G(s)$, the system transfer function; the numerator $m(s)$ of the second term depends on the initial conditions and it is zero when all initial conditions are zero, i.e., when the system is initially at rest.

In view now of the partial fraction expansion rules, $Y(s)$ can be written as follows:

$$Y(s) = \frac{c_1}{s - p_1} + \cdots + \frac{c_{i1}}{s - p_i} + \frac{c_{i2}}{(s - p_i)^2} + \cdots + \frac{b_1 s + b_0}{s^2 + a_1 s + a_0} + \cdots + I(s)$$

This expression shows real poles of $G(s)$, namely, $p_1, p_2, \ldots$, and it allows for multiple poles p_i; it also shows complex conjugate poles $a \pm jb$ written as second-order terms. $I(s)$ denotes the terms due to the input $U(s)$; they are fractions with the poles of $U(s)$. Note that if $G(s)$ and $U(s)$ have common poles they are combined to form multiple-pole terms.

Taking now the inverse Laplace transform of $Y(s)$:

$$y(t) = L^{-1}\{Y(s)\} = c_1 e^{p_1 t} + \cdots + c_{i1} e^{p_i t} + (\cdot) t e^{p_i t} + \cdots + e^{at}[(\cdot)\sin bt + (\cdot)\cos bt] + \cdots + i(t)$$

where $i(t)$ depends on the input. Note that the terms of the form $c t^k e^{p_i t}$ are the *modes of the system*. The system behavior is the aggregate of the behaviors of the modes. Each mode depends primarily on the location of the pole p_i; the location of the zeros affects the size of its coefficient c.

If the input $u(t)$ is a *bounded signal*, i.e., $|u(t)| < \infty$ for all t, then all the poles of $I(s)$ have real parts that are negative or zero, and this implies that $I(t)$ is also bounded for all t. In that case, the response $y(t)$ of the system will also be bounded for any bounded $u(t)$ if and only if all the poles of $G(s)$ have strictly negative real parts. Note that poles of $G(s)$ with real parts equal to zero are not allowed, since if $U(s)$ also has poles at the same locations, $y(t)$ will be unbounded. Take, for example, $G(s) = 1/s$ and consider the bounded step input $U(s) = 1/s$; the response $y(t) = t$, which is not bounded.

Having all the poles of $G(s)$ located in the open left half of the s-plane is very desirable and it corresponds to the system being stable. In fact, *a system is bounded-input, bounded-output (BIBO) stable if and only if all poles of its transfer function have negative real parts*. If at least one of the poles has positive real parts, then the system in *unstable*. If a pole has zero real parts, then the term *marginally stable* is sometimes used.

Note that in a BIBO stable system if there is no forcing input, but only initial conditions are allowed to excite the system, then $y(t)$ will go to zero as t goes to infinity. This is a very desirable property for a system to have, because nonzero initial conditions always exist in most real systems. For example, disturbances such as interference may add charge to a capacitor in an electric circuit, or a sudden brief gust of wind may change the heading of an aircraft. In a stable system the effect of the disturbances will diminish and the system will return to its previous desirable operating condition. For these reasons a control system should first and foremost be guaranteed to be stable, that is, it should always have poles with negative real parts. There are many design methods to stabilize a system or if it is initially stable to preserve its stability, and several are discussed later in this section.

Response of First- and Second-Order Systems

Consider a system described by a first-order differential equation, namely, $\dot{y}(t) + a_0 y(t) = a_0 u(t)$ and let $y(0) = 0$. In view of the previous subsection, the transfer function of the system is

$$G(s) = \frac{a_0}{s + a_0}$$

and the response to a *unit step input* $q(t)$ ($q(t) = 1$ for $t \geq 0$, $q(t) = 0$ for $t < 0$) may be found as follows:

$$y(t) = L^{-1}\{Y(s)\} = L^{-1}\{G(s)U(s)\} = L^{-1}\left\{\frac{a_0}{s + a_0}\frac{1}{s}\right\}$$

$$= L^{-1}\left\{\frac{1}{s} + \frac{-1}{s + a_0}\right\} = [1 - e^{-a_0 t}]q(t)$$

Note that the pole of the system is $p = -a_0$ (in Fig. 19.1.6 we have assumed that $a_0 > 0$). As that pole moves to the left on the real axis, i.e., as a_0 becomes larger, the system becomes faster. This can be seen from the fact that the steady state value of the system response

$$y_{ss} = \lim_{t \to \infty} y(t) = 1$$

is approached by the trajectory of $y(t)$ faster, as a_0 becomes larger. To see this, note that the value $1 - e^{-1}$ is attained at time $\tau = 1/a_0$, which is smaller as a_0 becomes larger. τ is *the time constant* of this first-order system; see below for further discussion of the time constant of a system.

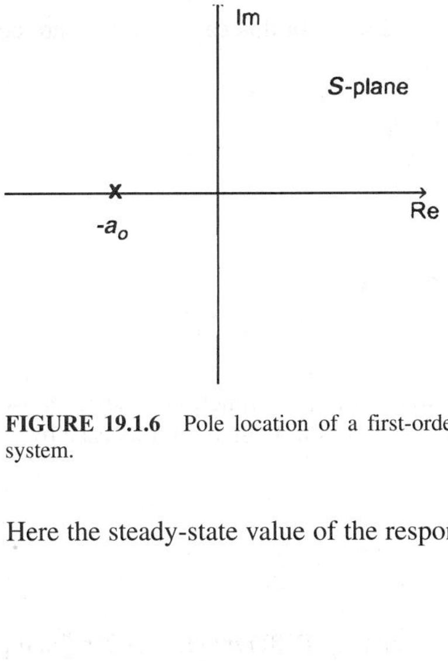

We now derive the response of a second-order system to a unit step input (Fig. 19.1.7). Consider a system described by $\ddot{y}(t) + a_1\dot{y}(t) + a_0 y(t) = a_0 u(t)$, which gives rise to the transfer function:

FIGURE 19.1.6 Pole location of a first-order system.

$$G(s) = \frac{a_0}{s^2 + a_1 s + a_0}$$

Here the steady-state value of the response to a unit step is

$$y_{ss} = \lim_{s \to 0} sG(s)\frac{1}{s} = 1$$

Note that this normalization or scaling to 1 is in fact the reason for selecting the constant numerator to be a_0. $G(s)$ above does not have any finite zeros—only poles—as we want to study first the effect of the poles on the system behavior. We shall discuss the effect of adding a zero or an extra pole later.

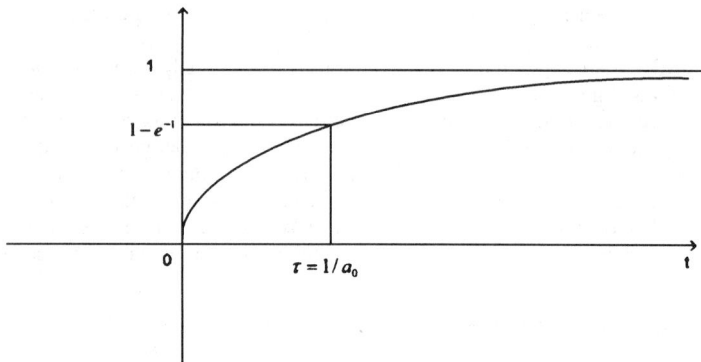

FIGURE 19.1.7   Step response of a first-order plant.

It is customary, and useful as we will see, to write the above transfer function as

$$G(s) = \frac{\omega_n^2}{s^2 + 2\zeta\omega_n s + \omega_n^2}$$

where ζ is the *damping ratio* of the system and ω_n is the *(undamped) natural frequency* of the system, i.e., the frequency of oscillations when the damping is zero.

The poles of the system are

$$p_{1,2} = -\zeta\omega_n \pm \omega_n\sqrt{\zeta^2 - 1}$$

When $\zeta > 1$ the poles are real and distinct and the unit step response approaches its steady-state value of 1 without overshoot. In this case the system is *overdamped*. The system is called critically damped when $\zeta = 1$ in which case the poles are real, repeated, and located at $-\zeta\omega_n$.

The more interesting case is when the system is *underdamped* ($\zeta < 1$). In this case the poles are complex conjugate and are given by

$$p_{1,2} = -\zeta\omega_n \pm j\omega_n\sqrt{1 - \zeta^2} = \sigma + j\omega_d$$

The response to a unit step input in this case is

$$y(t) = \left[1 - \frac{e^{-\zeta\omega_n t}}{\sqrt{1-\zeta^2}}\sin(\omega_d t + \theta)\right]q(t)$$

where $\theta = \cos^{-1}\zeta = \tan^{-1}(\sqrt{1-\zeta^2}/\zeta), \omega_d = \omega_n\sqrt{1-\zeta^2}$, and $q(t)$ is the step function. The response to an *impulse input* ($u(t) = \delta(t)$) also called the *impulse response $h(t)$* of the system is given in this case by

$$h(t) = \left[\omega_n \frac{e^{-\zeta\omega_n t}}{\sqrt{1-\zeta^2}}\sin\omega_n\left(\sqrt{1-\zeta^2}t\right)\right]q(t)$$

The second-order system is parameterized by the two parameters ζ and ω_n. Different choices for ζ and ω_n lead to different pole locations and to different behavior of (the modes of) the system. Fig. 19.1.8 shows the relation between the parameters and the pole location.

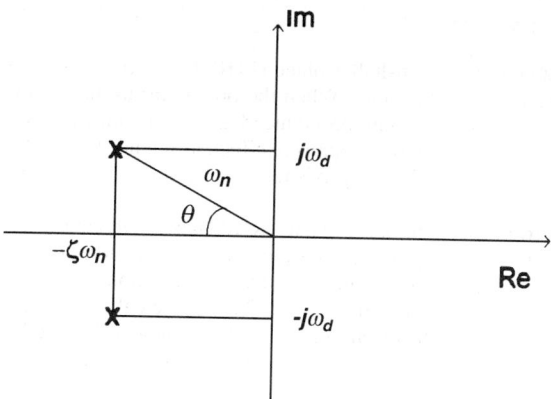

FIGURE 19.1.8   Relation between pole location and parameters.

Time Constant of a Mode and of a System.   The time constant of a mode ce^{pt} of a system is the time value that makes $|pt| = 1$, i.e., $\tau = 1/|p|$. For example, in the above first-order system we have seen that $\tau = 1/a_0 = RC$. A pair of complex conjugate poles $p_{1,2} = \sigma \pm j\omega$ give rise to the term of the form $Ce^{\sigma t}\sin(\omega t + \theta)$. In this case, $\tau = 1/|\sigma|$, i.e., τ is again the inverse of the distance of the pole from the imaginary axis. The time constant of a system is the time constant of its dominant modes.

Transient Response Performance Specifications for a Second-Order Underdamped System

For the system

$$G(s) = \frac{\omega_n^2}{s^2 + 2\zeta\omega_n s + \omega_n^2}$$

and a unit step input, explicit formulas for important measures of performance of its transient response can be derived. Note that the steady state is

$$y_{ss} = \lim_{s \to 0} sG(s)\frac{1}{s} = 1$$

The *rise time* t_r shows how long it takes for the system's output to rise from 0 to 66 percent of its final value (equal to 1 here) and it can be shown to be $t_r = (\pi - \theta)/\omega_n$, where $\theta = \cos^{-1}\zeta$ and $\omega_d = \omega_n\sqrt{1 - \zeta^2}$. The *settling time* t_s is the time required for the output to settle within some percentage, typically 2 or 5 percent, of its final value. $t_s \cong 4/\zeta\omega_n$ is the 2 percent settling time ($t_s \cong 3/\zeta\omega_n$ is the 5 percent settling time). Before the underdamped system settles, it will overshoot its final value. The *peak time* t_p measures the time it takes for the output to reach its first (and highest) peak value. M_p measures the actual *overshoot* that occurs in percentage terms of the final value. M_p occurs at time t_p, which is the time of the first and largest overshoot.

$$t_p = \frac{\pi}{\omega_d}, \quad M_p = 100e^{-\zeta\pi/\sqrt{1-\zeta^2}}\%$$

It is important to notice that the overshoot depends only on ζ. Typically, tolerable overshoot values are between 2.5 and 25 percent, which correspond to damping ratio ζ between 0.8 and 0.4.

Effect of Additional Poles and Zeros

The addition of an extra pole in the left-half s-plane (LHP) tends to slow the system down—the rise time of the system, for example, will become larger. When the pole is far to the left of the imaginary axis, its effect tends to be small. The effect becomes more pronounced as the pole moves toward the imaginary axis.

The addition of a zero in the LHP has the opposite effect, as it tends to speed the system up. Again the effect of a zero far away to the left of the imaginary axis tends to be small. It becomes more pronounced as the zero moves closer to the imaginary axis.

The addition of a zero in the right-half s-plane (RHP) has a delaying effect much more severe than the addition of a LHP pole. In fact a RHP zero causes the response (say, to a step input) to start toward the wrong direction. It will move down first and become negative, for example, before it becomes positive again and starts toward its steady-state value. Systems with RHP zeros are called *nonminimum phase systems* (for reasons that will become clearer after the discussion of the frequency design methods) and are typically difficult to control. Systems with only LHP poles (stable) and LHP zeros are called *minimum phase systems*.

CLASSICAL CONTROL DESIGN METHODS

In this section, we focus on the problem of controlling a single-input and single-output (SISO) LTI plant. It is understood from the above sections that such a plant can be represented by a transfer function $G_p(s)$. The closed-loop system is shown in Fig. 19.1.9.

The goal of feedback control is to make the output of the plant, y, follow the reference input r as closely as possible. Classical design methods are those used to determine the controller transfer function $G_c(s)$ so that the closed-loop system, represented by the transfer function:

$$G_{CL}(s) = \frac{G_c(s)G_p(s)}{1 + G_c(s)G_p(s)}$$

has desired characteristics.

Design Specifications and Constraints

The design specifications are typically described in terms of step response, i.e., r is the set point described as a step-like function. These specifications are given in terms of transient response and steady-state error, assuming the feedback control system is stable. The transient response is characterized by the rise time, i.e., the time it takes for the output to reach 66 percent of its final value, the settling time, i.e., the time it takes for the output to settle within 2 percent of its final value, and the percent overshoot, which is how much the output exceeds the set-point r percentagewise during the period that y converges to r. The steady-state error refers to the difference, if any, between y and r as y reaches its steady-state value.

There are many constraints a control designer has to deal with in practice, as shown in Fig. 19.1.10. They can be described as follows:

1. *Actuator Saturation*: The input u to the plant is physically limited to a certain range, beyond which it "saturates," i.e., becomes a constant.

2. *Disturbance Rejection and Sensor Noise Reduction*: There are always disturbances and sensor noises in the plant to be dealt with.

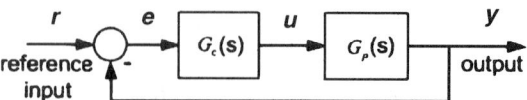

FIGURE 19.1.9 Feedback control configuration.

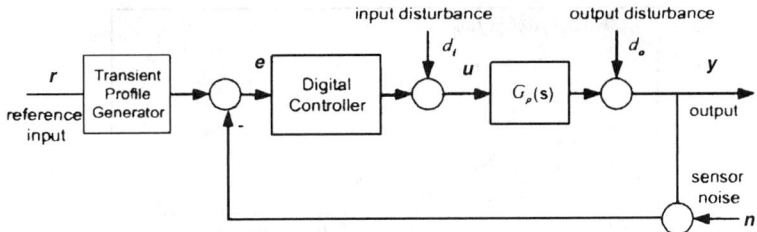

FIGURE 19.1.10 Closed-loop simulator setup.

3. *Dynamic Changes in the Plant*: Physical systems are almost never truly linear nor time invariant.

4. *Transient Profile*: In practice, it is often not enough to just move *y* from one operating point to another. How it gets there is sometimes just as important. Transient profile is a mechanism to define the desired trajectory of *y* in transition, which is of great practical concern. The smoothness of *y* and its derivatives, the energy consumed, the maximum value, and the rate of change required of the control action are all influenced by the choice of transient profile.

5. *Digital Control*: Most controllers are implemented today in digital forms, which makes the sampling rate and quantization errors limiting factors in the controller performance.

Control Design Strategy Overview

The control strategies are summarized here in ascending order of complexity and, hopefully, performance.

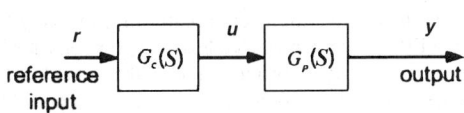

FIGURE 19.1.11 Open-loop control configuration.

1. *Open-Loop Control*: If the plant transfer function is known and there is very little disturbance, a simple open loop controller, where $G_c(s)$ is an approximate inverse of $G_p(s)$, as shown in Fig. 19.1.11, would satisfy most design requirements. Such control strategy has been used as an economic means in controlling stepper motors, for example.

2. *Feedback Control with a Constant Gain:* With significant disturbance and dynamic variations in the plant, feedback control, as shown in Fig. 19.1.9, is the only choice; see also the Appendix. Its simplest form is $G_c(s) = k$, or $u = ke$, where *k* is a constant. Such proportional controller is very appealing because of its simplicity. The common problems with this controller are significant steady-state error and overshoot.

3. *Proportional-Integral-Derivative Controller*: To correct the above problems with the constant gain controller, two additional terms are added:

$$u = k_p e + k_i \int e + k_d e \quad \text{or} \quad G_c(s) = k_p + k_i / s + k_d s$$

This is the well-known PID controller, which is used by most engineers in industry today. The design can be quite intuitive: the proportional term usually plays the key role, with the integral term added to reduce/eliminate the steady-state error and the derivative term the overshoot. The primary drawbacks of PID are that the integrator introduces phase lag that could lead to stability problems and the differentiator makes the controller sensitive to noise.

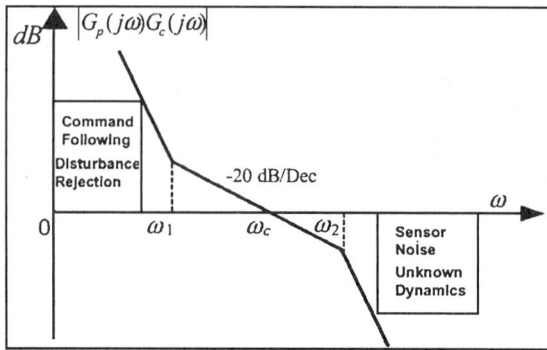

FIGURE 19.1.12   Loop-shaping.

4. *Root Locus Method*: A significant portion of most current control textbooks is devoted to the question of how to place the poles of the closed-loop system in Fig.19.1.9 at desired locations, assuming we know where they are. Root locus is a graphical technique to manipulate the closed-loop poles given the open-loop transfer function. This technique is most effective if disturbance rejection, plant dynamic variations, and sensor noise are not to be considered. This is because these properties cannot be easily linked to closed-loop pole locations.

5. *Loop-Shaping Method*: Loop-shaping [5] refers to the manipulation of the *loop gain* frequency response, $L(j\omega) = G_p(j\omega)G_c(j\omega)$, as a control design tool. It is the only existing design method that can bring most of design specifications and constraints, as discussed above, under one umbrella and systematically find a solution. This makes it a very useful tool in understanding, diagnosing, and solving practical control problems. The loop-shaping process consists of two steps:

a. Convert all design specifications to loop gain constraints, as shown in Fig.19.1.12.
b. Find a controller $G_c(s)$ to meet the specifications.

Loop-shaping as a concept and a design tool helped the practicing engineers greatly in improving the PID loop performance and stability margins. For example, a PID implemented as a lead-lag compensator is commonly seen in industry today. This is where the classical control theory provides the mathematical and design insights on why and how feedback control works. It has also laid the foundation for modern control theory.

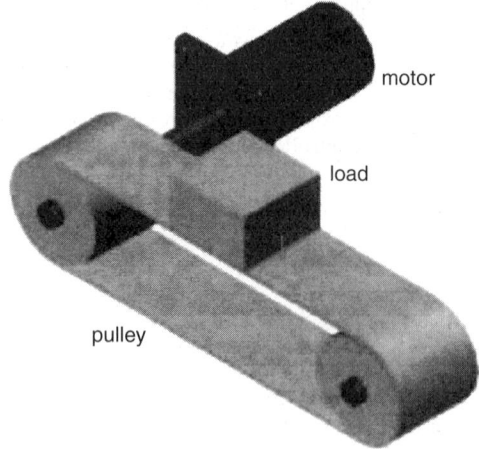

FIGURE 19.1.13   A Digital servo control design example.

Example   Consider a motion control system as shown in Fig. 19.1.13. It consists of a digital controller, a dc motor drive (motor and power amplifier), and a load of 235 lb that is to be moved linearly by 12 in. in 0.3 s with an accuracy of 1 percent or better. A belt and pulley mechanism is used to convert the motor rotation a linear motion. Here a servo motor is used to drive the load to perform a linear motion. The motor is coupled with the load through a pulley.

The design process involves:

1. Selection of components including motor, power amplifier, the belt-and-pulley, the feedback devices (position sensor and/or speed sensor)

2. Modeling of the plant

3. Control design and simulation

4. Implementation and tuning

The first step results in a system with the following parameters:

1. *Electrical:*
   - Winding resistance and inductance: $R_a = 0.4$ mho $L_a = 8$ mH (the transfer function of armature voltage to current is $(1/R_a)/[(L_a/R_a)s + 1]$
   - back emf constant: $K_E = 1.49$ V/(rad/s)
   - power amplifier gain: $K_{pa} = 80$
   - current feedback gain: $K_{cf} = 0.075$ V/A

2. *Mechanical:*
   - Torque constant: $K_t = 13.2$ in-lb/A
   - Motor inertia $J_m = .05$ lb-in.s^2
   - Pulley radius $R_p = 1.25$ in.
   - Load weight: $W = 235$ lb (including the assembly)
   - Total inertia $J_t = J_m + J_l = 0.05 + (W/g)R_p^2 = 1.0$ lb-in.s^2

With the maximum armature current set at 100 A, the maximum torque $= K_t I_{a,\max} = 13.2 \times 100 = 1320$ in.-lb; the maximum angular acceleration $= 1320/J_t = 1320$ rad/s^2, and the maximum linear acceleration $= 1320 \times R_p = 1650$ in./s$^2 = 4.27$ g's (1650/386). As it turned out, they are sufficient for this application.

The second step produces a simulation model (Fig. 19.1.14).

A simplified transfer function of the plant, from the control input, v_c (in volts), to the linear position output, x_{out} (in inches), is

$$G_p(s) = \frac{206}{s(s+3)}$$

An open loop controller is not suitable here because it cannot handle the torque disturbances and the inertia change in the load. Now consider the feedback control scheme in Fig. 19.1.9 with a constant controller, $u = ke$. The root locus plot in Fig. 19.1.15 indicates that, even at a high gain, the real part of the closed-loop poles does not exceed -1.5, which corresponds to a settling time of about 2.7 s. This is far slower than desired.

In order to make the system respond faster, the closed-loop poles must be moved further away from the $j\omega$ axis. In particular, a settling time of 0.3 s or less corresponds to the closed-loop poles with real parts smaller than -13.3. This is achieved by using a PD controller of the form

$$G_c(s) = K(s + 3); \quad K \geq 13.3/206$$

will result in a settling time of less than 0.3 s.

The above PD design is a simple solution in servo design that is commonly used. There are several issues, however, that cannot be completely resolved in this framework:

1. Low-frequency torque disturbance induces steady-state error that affects the accuracy.

2. The presence of a resonant mode within or close to the bandwidth of the servo loop may create undesirable vibrations.

3. Sensor noise may cause the control signal to be very noisy.

4. The change in the dynamics of the plant, for example, the inertia of the load, may require frequency tweaking of the controller parameters.

5. The step-like set point change results in an initial surge in the control signal and could shorten the life span of the motor and other mechanical parts.

These are problems that most control textbooks do not adequately address, but they are of significant importance in practice. The first three problems can be tackled using the loop-shaping design technique introduced above. The tuning problem is an industrywide design issue and the focus of various research and development efforts. The last problem is addressed by employing a smooth transient as the set point, instead of a step-like set point. This is known as the "motion profile" in industry.

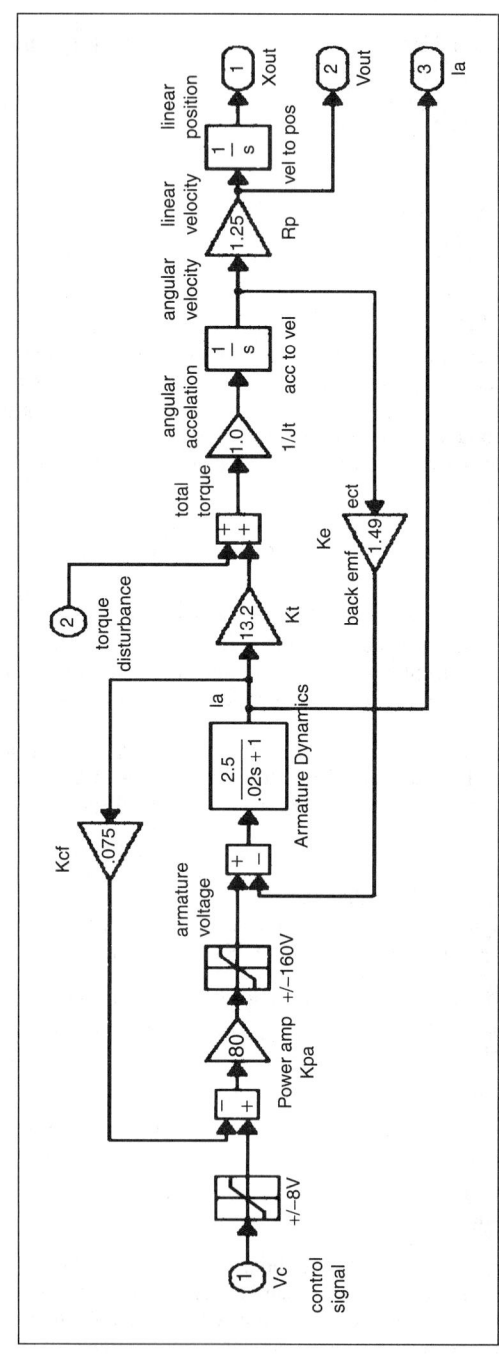

FIGURE 19.1.14   Simulation model of the motion control system.

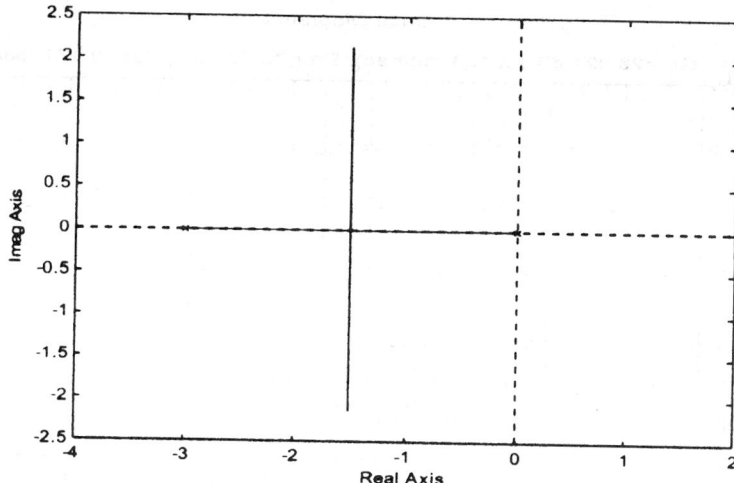

FIGURE 19.1.15   Root locus plot of the servo design problem.

Evaluation of Control Systems

Analysis of control system provides crucial insights to control practitioners on why and how feedback control works. Although the use of PID precedes the birth of classical control theory of the 1950s by at least two decades, it is the latter that established the control engineering discipline. The core of classical control theory are the frequency-response-based analysis techniques, namely, Bode and Nyquist plots, stability margins, and so forth.

In particular, by examining the loop gain frequency response of the system in Fig. 19.1.9, that is, $L(j\omega) = G_c(j\omega)G_p(j\omega)$, and the sensitivity function $1/[1 + L(j\omega)]$, one can determine the following:

1. How fast the control system responds to the command or disturbance input (i.e., the bandwidth).
2. Whether the closed-loop system is stable (Nyquist Stability Theorem); if it is stable, how much dynamic variation it takes to make the system unstable (in terms of the gain and phase change in the plant). It leads to the definition of gain and phase margins. More broadly, it defines how robust the control system is.
3. How sensitive the performance (or closed-loop transfer function) is to the changes in the parameters of the plant transfer function (described by the sensitivity function).
4. The frequency range and the amount of attenuation for the input and output disturbances shown in Fig. 19.1.10 (again described by the sensitivity function).

Evidently, these characteristics obtained via frequency-response analysis are invaluable to control engineers. The efforts to improve these characteristics led to the lead-lag compensator design and, eventually, loop-shaping technique described above.

Example: The PD controller in Fig. 19.1.10 is known to be sensitive to sensor noise. A practical cure to this problem is to add a low pass filter to the controller to attenuate high-frequency noise, that is,

$$G_c(s) = \frac{13.3(s+3)}{206\left(\dfrac{s}{133}+1\right)^2}$$

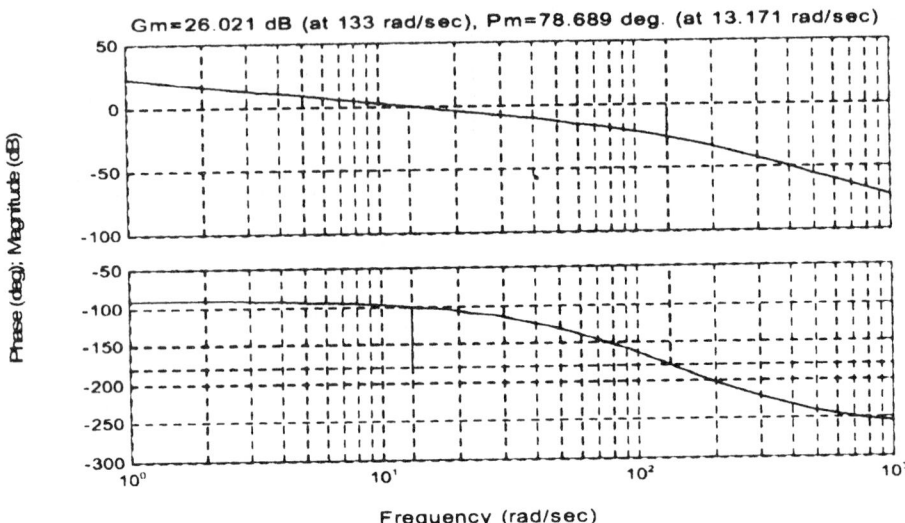

Bode Diagrams

Gm=26.021 dB (at 133 rad/sec), Pm=78.689 deg. (at 13.171 rad/sec)

Frequency (rad/sec)

FIGURE 19.1.16 Bode plot evaluation of the control design.

The loop gain transfer function is now

$$L(s) = G_p(s)G_c(s) = \frac{13.3}{s\left(\dfrac{s}{133} + 1\right)^2}$$

The bandwidth of the low pass filter is chosen to be one decade higher than the loop gain bandwidth to maintain proper gain and phase margins. The Bode plot of the new loop gain, as shown in Fig. 19.1.16, indicates that (a) the feedback system has a bandwidth 13.2 rad/s, which corresponds to a 0.3 s settling time as specified and (b) this design has adequate stability margins (gain margin is 26 dB and phase margin is 79°).

Digital Implementation

Once the controller is designed and simulated successfully, the next step is to digitize it so that it can be programmed into the processor in the digital control hardware. To do this:

1. Determine the sampling period T_s and the number of bits used in analog-to-digital converter (ADC) and digital-to-analog converter (DAC).
2. Convert the continuous time transfer function $G_c(s)$ to its corresponding form in discrete time transfer function $G_{cd}(z)$ using, for example, Tustin's method, $s = (1/T)(z-1)/(z+1)$.
3. From $G_{cd}(z)$, derive the difference equation, $u(k) = g(u(k-1), u(k-2), \ldots y(k), y(k-1), \ldots)$, where g is a linear algebraic function.

After the conversion, the sampled data system, with the plant running in continuous time and the controller in discrete time, should be verified in simulation first before the actual implementation. The quantization error and sensor noise should also be included to make it realistic.

The minimum sampling frequency required for a given control system design has not been established analytically. The rule of thumb given in control textbooks is that $f_s = 1/T_s$ should be chosen approximately 30 to 60 times the bandwidth of the closed-loop system. Lower-sampling frequency is possible after careful tuning but the aliasing, or signal distortion, will occur when the data to be sampled have significant energy above the

Nyquist frequency. For this reason, an antialiasing filter is often placed in front of the ADC to filter out the high-frequency contents in the signal.

Typical ADC and DAC chips have 8, 12, and 16 bits of resolution. It is the length of the binary number used to approximate an analog one. The selection of the resolution depends on the noise level in the sensor signal and the accuracy specification. For example, the sensor noise level, say 0.1 percent, must be below the accuracy specification, say 0.5 percent. Allowing one bit for the sign, an 8-bit ADC with a resolution of $1/2^7$, or 0.8 percent, is not good enough; similarly, a 16-bit ADC with a resolution of 0.003 percent is unnecessary because several bits are "lost" in the sensor noise. Therefore, a 12-bit ADC, which has a resolution of 0.04 percent, is appropriate for this case. This is an example of "error budget," as it is known among designers, where components are selected economically so that the sources of inaccuracies are distributed evenly.

Converting $G_c(s)$ to $G_{cd}(z)$ is a matter of numerical integration. There have been many methods suggested; some are too simple and inaccurate (such as the Euler's forward and backward methods), others are too complex. Tustin's method suggested above, also known as trapezoidal method or bilinear transformation, is a good compromise. Once the discrete transfer function $G_{cd}(z)$ is obtained, finding the corresponding difference equation that can be easily programmed in C is straightforward. For example, given a controller with input $e(k)$ and output $u(k)$, and the transfer function

$$G_{cd}(z) = \frac{z+2}{z+1} = \frac{1+2z^{-1}}{1+z^{-1}}$$

the corresponding input-output relationship is

$$u(k) = \frac{1+2q^{-1}}{1+q^{-1}} e(k)$$

or equivalently, $(1 + q^{-1})u(k) = (1 + 2q^{-1})e(k)$. $[q^{-1}u(k) = u(k-1)$, i.e., q^{-1} is the unit time delay operater.] That is, $u(k) = -u(k-1) + e(k) + 2e(k-1)$.

Finally, the presence of the sensor noise usually requires that an antialiasing filter be used in front of the ADC to avoid distortion of the signal in ADC. The phase lag from such a filter must not occur at the crossover frequency (bandwidth) or it will reduce the stability margin or even destabilize the system. This puts yet another constraint on the controller design.

ALTERNATIVE DESIGN METHODS

Nonlinear PID

Using a nonlinear PID (NPID) is an alternative to PID for better performance. It maintains the simplicity and intuition of PID, but empowers it with nonlinear gains. An example of NPID is shown in Fig. 19.1.17. The need for the integral control is reduced, by making the proportional gain larger, when the error is small. The limited

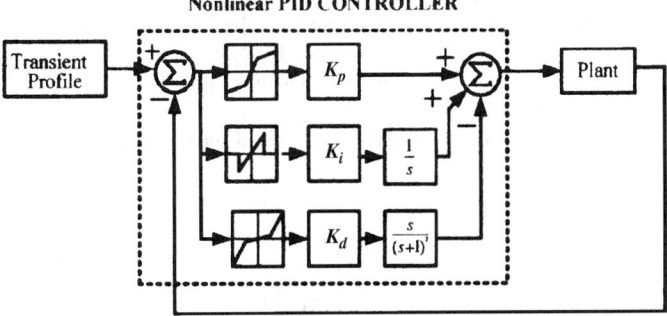

FIGURE 19.1.17   Nonlinear PID for a power converter control problem.

authority integral control has its gain zeroed outside a small interval around the origin to reduce the phase lag. Finally the differential gain is reduced for small errors to reduce sensitivities to sensor noise. See Ref. 8.

State Feedback and Observer-Based Design

If the state space model of the plant

$$\dot{x} = Ax + Bu$$
$$y = Cx + Du$$

is available, the pole-placement design can be achieved via state feedback

$$u = r + Kx$$

where K is the gain vector to be determined so that the eigenvalues of the closed-loops system

$$\dot{x} = (A + BK)x + Br$$
$$y = Cx + Du$$

are at the desired locations, assuming they are known. Usually the state vector is not available through measurements and the state observer is of the form

$$\dot{\hat{x}} = A\hat{x} + Bu + L(y - \hat{y})$$
$$\hat{y} = C\hat{x} + Du$$

where $\hat{x}$ is the estimate of x and L is the observer gain vector to be determined.

The state feedback design approach has the same drawbacks as those of Root Locus approach, but the use of the state observer does provide a means to extract the information about the plant that is otherwise unavailable in the previous control schemes, which are based on the input-output descriptions of the plant. This proves to be valuable in many applications. In addition, the state space methodologies are also applicable to systems with many inputs and outputs.

Controllability and Observability.   Controllability and observability are useful system properties and are defined as follows. Consider an nth order system described by

$$x = Ax + Bu, \qquad y = Cx + Du$$

where A is an $n \times n$ matrix. The system is *controllable* if it is possible to transfer the state to any other state in finite time. This property is important as it measures, for example, the ability of a satellite system to reorient itself to face another part of the earth's surface using the available thrusters; or to shift the temperature in an industrial oven to a specified temperature. Two equivalent tests for controllability are:

The system (or the pair (A, B)) is *controllable* if and only if the controllability matrix $C = [B, AB, \ldots, A^{n-1}B]$ has full (row) rank n. Equivalently if and only if $[s_i I - A, B]$ has full (row) rank n for all eigenvalues s_i of A.

The system is *observable* if by observing the output and the input over a finite period of time it is possible to deduce the value of the state vector of the system. If, for example, a circuit is observable it may be possible to determine all the voltages across the capacitors and all currents through the inductances by observing the input and output voltages.

The system (or the pair (A, C)) is *observable* if and only if the observability matrix

$$\theta = \begin{bmatrix} C \\ CA \\ \vdots \\ CA^{n-1} \end{bmatrix}$$

has full (column) rank n. Equivalently if and only if

$$\begin{bmatrix} s_i I - A \\ C \end{bmatrix}$$

has full (column) rank n for all eigenvalues s_i of A.

Consider now the transfer function

$$G(s) = C(sI - A)^{-1}B + D$$

Note that, by definition, in a transfer function all possible cancellations between numerator and denominator polynomials are assumed to have already taken place. In general, therefore, the poles of $G(s)$ are some (or all) of the eigenvalues of A. It can be shown that when the system is both controllable and observable no cancellations take place and so in this case the poles of $G(s)$ are exactly the eigenvalues of A.

Eigenvalue Assignment Design. Consider the equations: $\dot{x} = Ax + Bu$, $y = Cx + Du$, and $u = r + Kx$. When the system is controllable, K can be selected to assign the closed-loop eigenvalues to any desired locations (real or complex conjugate) and thus significantly modify the behavior of the open-loop system. Many algorithms exist to determine such K. In the case of a single input, there is a convenient formula called Ackermann's formula

$$K = -[0, \ldots, 0, 1]\, C^{-1}\, \alpha_d(A)$$

where $C = [B, \ldots, A^{n-1}B]$ is the $n \times n$ controllability matrix and the roots of $\alpha_d(s)$ are the desired closed-loop eigenvalues.

Example Let

$$A = \begin{bmatrix} 1/2 & 1 \\ 1 & 2 \end{bmatrix}, \qquad B = \begin{bmatrix} 1 \\ 1 \end{bmatrix}$$

and the desired eigenvalues be $-1 \pm j$.

Here

$$C = [B, AB] = \begin{bmatrix} 1 & 3/2 \\ 1 & 3 \end{bmatrix}$$

Note that A has eigenvalues at 0 and 5/2. We wish to determine K so that the eigenvalues of $A + BK$ are at $-1 \pm j$, which are the roots of $\alpha_d(s) = s^2 + 2s + 2$.

Here

$$C = [B, AB] = \begin{bmatrix} 1 & 3/2 \\ 1 & 3 \end{bmatrix}$$

and

$$\alpha_d(A) = A^2 + 2A + 2I = \left(\begin{bmatrix} 1/2 & 1 \\ 1 & 2 \end{bmatrix}^2 + 2\begin{bmatrix} 1/2 & 1 \\ 1 & 2 \end{bmatrix} + 2\begin{bmatrix} 1 & 0 \\ 0 & 1 \end{bmatrix} \right) = \begin{bmatrix} 17/4 & 9/2 \\ 9/2 & 11 \end{bmatrix}$$

Then

$$K = -[0 \ \ 1]C^{-1}\alpha_d(A) = [-1/6 \ \ -13/3]$$

Here

$$A + BK = \begin{bmatrix} 1/3 & -10/3 \\ 5/6 & -7/3 \end{bmatrix}$$

which has the desired eigenvalues.

Linear Quadratic Regulator (LQR) Problem.   Consider

$$x = Ax + Bu, \qquad y = Cx$$

We wish to determine $u(t)$, $t \geq 0$, which minimizes the quadratic cost

$$J(u) = \int_0^\infty \left[x^T(t)(M^T Q M)x + u^T(t)Ru(t) \right] dt$$

for any initial state $x(0)$. The weighting matrices Q and R are real, symmetric ($Q = Q^T$, $R = R^T$), Q and R are positive definite ($R > 0$, $Q > 0$) and $M^T Q M$ is positive semidefinite ($M^T Q M \geq 0$). Since $R > 0$, the term $u^T Ru$ is always positive for any $u \neq 0$, by definition. Minimizing its integral forces $u(t)$ to remain small. $M^T Q M \geq 0$ implies that $x^T M^T Q M x$ is positive, but it can also be zero for some $x \neq 0$; this allows some of the states to be treated as "do not care states." Minimizing the integral of $x^T M^T Q M x$ forces the states to become smaller as time progresses. It is convenient to take Q (and R in the multi-input case) to be diagonal with positive entries on the diagonal. The above performance index is designed so that the minimizing control input drives the states to the zero state, or as close as possible, without using excessive control action, in fact minimizing the control energy. When $(A, B, Q^{1/2}M)$ is controllable and observable, the solution $u^*(t)$ of this optimal control problem is a state feedback control law, namely,

$$u^*(t) = K^* x(t) = -R^{-1} B^T P_c^* x(t)$$

where P_c^* is the unique symmetric positive definite solution of the *algebraic Riccati equation:*

$$A^T P_c + P_c A - P_c BR^{-1}B^T P_c + M^T Q M = 0$$

Example.   Consider

$$\dot{x} = \begin{bmatrix} 0 & 1 \\ 0 & 0 \end{bmatrix} x + \begin{bmatrix} 0 \\ 1 \end{bmatrix} u, \; y = [1 \quad 0]x$$

And let

$$J = \int_0^\infty \left(y^2(t) + 4u^2(t) \right) dt$$

Here

$$M = C, \qquad Q = 1, \qquad M^T Q M = C^T C = \begin{bmatrix} 1 \\ 0 \end{bmatrix}[1 \quad 0] = \begin{bmatrix} 1 & 0 \\ 0 & 0 \end{bmatrix}, \qquad R = 4$$

Solving the Riccati equation we obtain

$$P_c^* = \begin{bmatrix} 2 & 2 \\ 2 & 2\sqrt{2} \end{bmatrix}$$

and

$$u^*(t) = K^* x(t) = -\frac{1}{4}[0 \quad 1]\begin{bmatrix} 2 & 2 \\ 2 & 2\sqrt{2} \end{bmatrix}x(t) = -\frac{1}{2}\left[1, \sqrt{2}\right]x(t)$$

Linear State Observers. Since the states of a system contain a great deal of useful information, knowledge of the state vector is desirable. Frequently, however, it may be either impossible or impractical to obtain measurements of all states. Therefore, it is important to be able to estimate the states from available measurements, namely, of inputs and outputs.

Let the system be

$$x = Ax + Bu, \qquad y = Cx + Du$$

An asymptotic state estimator of the full state, also called *Luenberger observer*, is given by

$$\dot{\hat{x}} = A\hat{x} + Bu + L(y - \hat{y})$$

where L is selected so that all eigenvalues of $A - LC$ are in the LHP (have negative real parts). Note that a L that arbitrarily assigns the eigenvalues of $A - LC$ exists if and only if the system is observable. The observer may be written as

$$\hat{x} = (A - LC)\hat{x} + [B - LD, K]\begin{bmatrix} u \\ y \end{bmatrix}$$

which clearly shows the role of u and y; they are the inputs to the observer. If the error $e(t) = x(t) - \hat{x}(t)$ then $e(t) = e^{[(A - LC)t]}e(0)$, which shows that $e(t) \to 0$ or $\hat{x}(t) \to x(t)$ as $t \to \infty$.

To determine appropriate L, note that $(A - LC)^T = A^T + C^T(-L) = \overline{A} + \overline{B}\,\overline{K}$, which is the problem addressed above in the state feedback case. One could also use the following observable version of Ackermann's formula, namely,

$$L = \alpha_d(A)\, O^{-1}[0, \dots 0, 1]^T$$

where

$$O = \begin{bmatrix} C \\ CA \\ \\ CA^{n-1} \end{bmatrix}$$

The gain L in the above estimator may be determined so that it is optimal in an appropriate sense. In the following, some of the key equations of such an optimal estimator (*Linear Quadratic Gaussian* (LQG)), also known as the *Kalman-Bucy filter*, are briefly outlined.

Consider

$$x = Ax + Bu + \Gamma w, \qquad y = Cx + v$$

where w and v represent process and measurement noise terms. Both w and v are assumed to be white, zero-mean Gaussian stochastic processes, i.e., they are uncorrelated in time and have expected values $E[w] = 0$ and $E[v] = 0$. Let $E[ww^T] = W, E[vv^T] = V$ denote the covariances where W, V are real, symmetric and positive definite matrices. Assume also that the noise processes w and v are independent, i.e., $E[wv^T] = 0$. Also assume that the initial state $x(0)$ is a Gaussian random variable of known mean, $E[x(0)] = x_0$, and known covariance $E[(x(0) - x_0)(x(0) - x_0)^T] = P_{e0}$. Assume also that $x(0)$ is independent of w and v.

Consider now the estimator

$$\dot{\hat{x}} = (A - LC)\hat{x} + Bu + Ly$$

and let $(A, \Gamma\, W^{1/2}, C)$ be controllable and observable. Then the error covariance $E[(x - \hat{x})(x - \hat{x})^T]$ is minimized when the filter gain $L^* = P_e^* C^T V^{-1}$, where P_e^* denotes the symmetric, positive definite solution of the (dual to control) *algebraic Riccati equation*

$$P_e A^T + AP_e - P_e C^T V^{-1} CP_e + \Gamma W \Gamma^T = 0$$

The above Riccati is the *dual* to the Riccati equation for optimal control and can be obtained from the optimal control equation by making use of the substitutions:

$$A \rightarrow A^T, B \rightarrow C^T, M \rightarrow \Gamma^T, R \rightarrow V, Q \rightarrow W$$

In the state feedback control law $u = Kx + r$, when state measurements are not available, it is common to use the estimate of state $\hat{x}$ from a Luenberger observer. That is, given

$$x = Ax + Bu, \qquad y = Cx + Du$$

the control law is $u = K\hat{x} + r$, where $\hat{x}$ is the state estimate from the observer

$$\hat{x} = (A - LC)\hat{x} + [B - KD, K] \begin{bmatrix} u \\ y \end{bmatrix}$$

The closed-loop system is then of order $2n$ since the plant and the observer are each of order n. It can be shown that in this case, of linear output feedback control design, the design of the control law and of the gain K (using, for example, LQR) can be carried out independent of the design of the estimator and the filter gain L (using, for example, LQG). This is known as the *separation property*.

It is remarkable to notice that the overall transfer function of the compensated system that includes the state feedback and the observer is

$$T(s) = (C + DK)[sI - (A + BK)]^{-1} B + D$$

which is exactly the transfer function one would obtain if the state x were measured directly and the state observer were not present. This is of course assuming zero initial conditions (to obtain the transfer function); if nonzero initial conditions are present, then there is some deterioration of performance owing to observer dynamics, and the fact that at least initially the state estimate typically contains significant error.

ADVANCED ANALYSIS AND DESIGN TECHNIQUES

The foregoing section described some fundamental analysis and design methods in classical control theory, the development of which was primarily driven by engineering practice and needs. Over the last few decades, vast efforts in control research have led to the creation of modern mathematical control theory, or advanced control, or control science. This development started with optimal control theory in the 50s and 60s to study the optimality of control design; a brief glimpse of optimal control was given above. In optimal control theory, a cost function is to be minimized, and analytical or computational methods are used to derive optimal controllers. Examples include minimum fuel problem, time-optimal control (Bang-Bang) problem, LQ, H_2, and H_∞, each corresponding to a different cost function. Other major branches in modern control theory include multi-input multi-output (MIMO) control systems methodologies, which attempt to extend well-known SISO design methods and concepts to MIMO problems; adaptive control, designed to extend the operating range of a controller by adjusting automatically the controller parameters based on the estimated dynamic changes in the plants; analysis and design of nonlinear control systems, and so forth.

A key problem is the robustness of the control system. The analysis and design methods in control theory are all based on the mathematical model of the plants, which is an approximate description of physical processes. Whether a control system can tolerate the uncertainties in the dynamics of the plants, or how much uncertainty it takes to make a system unstable, is studied in robust control theory, where H_2, H_∞, and other analysis and design methods were originated. Even with recent progress, open problems remain when dealing with real world applications. Some recent approaches such as in Ref. 8 attempt to address some of these difficulties in a realistic way.

APPENDIX: OPEN- AND CLOSED-LOOP STABILIZATION

It is impossible to stabilize an unstable system using open-loop control, owing to system uncertainties. In general, closed-loop or feedback control is necessary to control a system—stabilize if unstable and improve performance—because of uncertainties that are always present. Feedback provides current information about the system and so the controller does not have to rely solely on incomplete system information contained in a nominal plant model. These uncertainties are system parameter uncertainties and also uncertainties induced on the system by its environment, including uncertainties in the initial condition of the system, and uncertainties because of disturbances and noise.

Consider the plant with transfer function

$$G(s) = \frac{1}{s-(1+\varepsilon)}$$

where the pole location at 1 is inaccurately known.

Plant

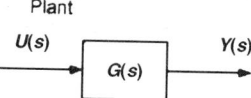

The corresponding description in the time domain using differential equations is $\dot{y}(t) - (1+\varepsilon)y(t) = u(t)$. Solving, using Laplace transform, we obtain $sY(s) - y(0) - (1+\varepsilon)Y(s) = U(s)$ from which

$$Y(s) = \frac{y(0)}{s-(1+\varepsilon)} + \frac{1}{s-(1+\varepsilon)}U(s)$$

Consider now the controller with transfer function

$$G_c(s) = \frac{s-1}{s+2}$$

Controller

The corresponding description in the time domain using differential equations is $\dot{u}(t) + 2u(t) = \dot{r}(t) - r(t)$. Solving, using Laplace transform, we obtain $sU(s) - u(0) + 2U(s) = sR(s) - R(s)$ from which

$$U(s) = \frac{u(0)}{s+2} + \frac{s-1}{s-2}R(s)$$

Connect now the plant and the controller in series (open-loop control)

Connecting in Series - Open loop

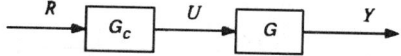

The overall transfer function is

$$T = GG_c = \frac{s-1}{[s-(1+\varepsilon)](s+2)}$$

Including the initial conditions

$$Y(s) = \frac{(s+2)y(0)+u(0)}{[s-(1+\varepsilon)](s+2)} + \frac{s-1}{[s-(1+\varepsilon)](s+2)}R(s)$$

It is now clear that open-loop control cannot be used to stabilize the plant:

1. First, because of the uncertainties in the plant parameters. Note that the plant pole is not exactly at +1 but at $1 + \varepsilon$ and so the controller zero cannot cancel the plant pole exactly.
2. Second, even if we had knowledge of the exact pole location, that is, $\varepsilon = 0$, and

$$Y(s) = \frac{(s+2)y(0)+r(0)}{(s-1)(s+2)} + \frac{1}{s+2}R(s)$$

still we cannot stabilize the system because of the uncertainty in the initial conditions. We cannot, for example, select $r(0)$ so as to cancel the unstable pole at +1 because $y(0)$ may not be known exactly.

We shall now stabilize the above plant using a simple feedback controller.

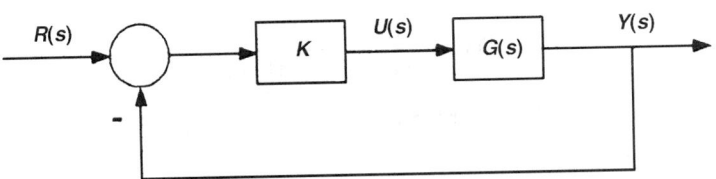

Consider a unity feedback control system with the controller being just a gain k to be determined. The closed-loop transfer function is

$$T(s) = \frac{kG(s)}{1+kG(s)} = \frac{k}{s-(1+\varepsilon)+k}$$

Working in the time domain,

$$\dot{y} - (1+\varepsilon)y = u = k(r-y)$$

from which

$$\dot{y} + [k-(1+\varepsilon)]y = kr$$

Using Laplace transform we obtain

$$sY(s) - y(0) + [k-(1+\varepsilon)]Y(s) = kR(s)$$

and

$$Y(s) = \frac{y(0)}{s+k-(1+\varepsilon)} + \frac{k}{s+k-(1+\varepsilon)}R(s)$$

It is now clear that if the controller gain is selected so that $k > 1 + \varepsilon$, then the system will be stable. Is fact, we could have worked with the nominal system to derive $k > 1$ for stability. For stability robustness, we take k somewhat larger than 1 to have some safety margin and satisfy $k > 1 + \varepsilon$ for some unknown small ε.

REFERENCES

1. Dorf, R. C., and R. H. Bishop, "Modern Control Systems," 9th ed., Prentice Hall, 2001.
2. Franklin, G. F., J. D. Powell, and A. Emami-Naeimi, "Feedback Control of Dynamic Systems," 3rd ed., Addison-Wesley, 1994.
3. Kuo, B. C., "Automatic Control Systems," 7th ed., Prentice Hall, 1995.
4. Ogata, K., "Modern Control Engineering," 3rd ed., Prentice Hall, 1997.
5. Rohrs, C. E., J. L. Melsa, and D. G. Schultz, "Linear Control Systems," McGraw-Hill, 1993.
6. Antsaklis, P. J., and A. N. Michel, "Linear Systems," McGraw-Hill, 1997.
7. Goodwin, G. C., S. F. Graebe, and M. E. Salgado, "Control System Design," Prentice Hall, 2001.
8. Gao, Z., Y. Huang, and J. Han, "An Alternative Paradigm for Control System Design," Presented at the 40th IEEE Conference on Decision and Control, Dec 4–7, 2001.

ON THE CD-ROM:

"A Brief Review of the Laplace Transform Useful in Control Systems," by the authors of this section, examines its usefulness in control systems analysis and design.

SECTION 20

AUDIO SYSTEMS

Although much of this section contains fundamental information on basic audio technology, extensive information on new evolving digital audio formats and recording and reproduction systems has been added since the last edition.

As noted herein, DVD-Audio, for example, is based on the same DVD technology as DVD-Video discs and DVD-ROM computer discs. It has a theoretical sampling rate of 192 kHz with 24-bit processing and can store 4.7 gigabytes on a disc with a choice of two- or six-channel audio tracks or a mix of both. Super Audio CD (SACD) has the same storage capacity. It uses direct stream digital (DSD) with 2.8 MHz sampling in three possible disc types. The first two contain only DSD data (4.7 gigabytes of data on a single-layer disc and slightly less than 9 gigabytes on the dual layer disc). The third version, the SACD hybrid, combines a single 4.7 gigabyte layer with a conventional CD that can be played back on conventional CD players.

MPEG audio coding variations continue to evolve. For example:

- MPEG-1 is a low-bit-rate audio format.
- MPEG-2 extends MPEG-1 toward the audio needs of digital video broadcasting.
- MPEG-2 Advanced Audio Coding (AAC) is an enhanced multichannel coding system.
- MP3 is the popular name for MPEG-1 Layer III.
- MPEG-4 adds object-based representation, content-based interactivity, and scalability.
- MPEG-7 defines a universal standardized mechanism for exchanging descriptive data.
- MPEG-21 defines a multimedia framework to enable transparent and augmented use of multimedia services across a wide range of networks and devices used by different communities.   R.J.

In This Section:

On the CD-ROM:

The following are reproduced from the 4th edition of this handbook:

"Ambient Noise and Its Control," by Daniel W. Martin;

"Acoustical Environment Control," by Daniel W. Martin;

"Mechanical Disc Reproduction Systems," by Daniel W. Martin;

"Magnetic-Tape Analog Recording and Reproduction," by Daniel W. Martin.

CHAPTER 20.1

SOUND UNITS AND FORMATS

Daniel W. Martin, Ronald M. Aarts

STANDARD UNITS FOR SOUND SPECIFICATION[1,2]

Sound Pressure

Airborne sound waves are a physical disturbance pattern in the air, an elastic medium, traveling through the air at a speed that depends somewhat on air temperature (but not on static air pressure). The instantaneous magnitude of the wave at a specific point in space and time can be expressed in various ways, e.g., displacement, particle velocity, and pressure. However, the most widely used and measured property of sound waves is *sound pressure*, the fluctuation above and below atmospheric pressure, which results from the wave.

An *atmosphere* (atm) of pressure is typically about 10^5 pascals (Pa) in the International System of units. Sound pressure is usually a very small part of atmospheric pressure. For example, the minimum audible sound pressure (threshold of hearing) at 2000 Hz is 20 μPa, or $2(10)^{-10}$ atm.

Sound-Pressure Level

Sound pressures important to electronics engineering range from the weakest noises that can interfere with sound recording to the strongest sounds a loudspeaker diaphragm should be expected to radiate. This range is approximately 10^6. Consequently, for convenience, sound pressures are commonly plotted on a logarithmic scale called *sound-pressure level* expressed in *decibels* (dB).

The decibel, a unit widely used for other purposes in electronics engineering, originated in audio engineering (in telephony), and is named for Alexander Graham Bell. Because it is logarithmic, it requires a reference value for comparison just as it does in other branches of electronics engineering. The reference pressure for sounds in air, corresponding to 0 dB, has been defined as a sound pressure of 20 μPa (previously 0.0002 dyn/cm^2). This is the reference sound pressure p_0 used throughout this section of the handbook. Thus the sound-pressure level L_p in decibels corresponding to a sound pressure p is defined by

$$L_p = 20 \log (p/p_0) \text{ dB} \tag{1}$$

The reference pressure p_0 approximates the weakest audible sound pressure at 2000 Hz. Consequently most decibel values for sound levels are positive in sign. Figure 20.1.1 relates sound-pressure level in decibels to sound pressure in micropascals.

Sound power and sound intensity (power flow per unit area of wavefront) are generally proportional to the square of the sound pressure. Doubling the sound pressure quadruples the intensity in the sound field, requiring four times the power from the sound source.

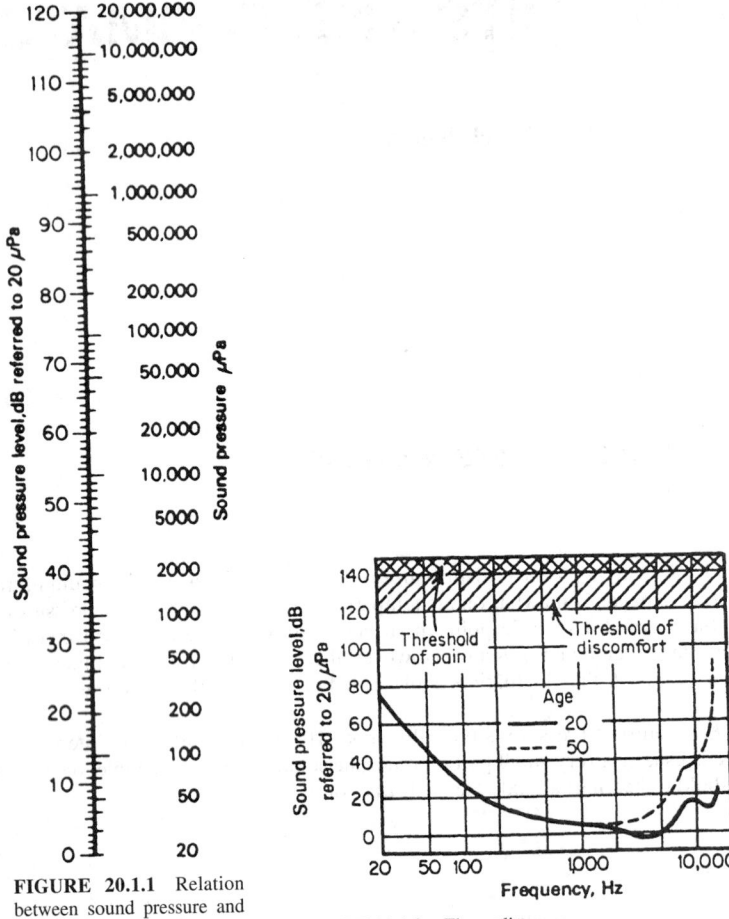

FIGURE 20.1.1   Relation between sound pressure and sound-pressure level.[1]

FIGURE 20.1.2   The auditory area.

Audible Frequency Range

The international abbreviation Hz (hertz) is now used (instead of the former cps) for audible frequencies as well as the rest of the frequency domain. The limits of audible frequency are only approximate because tactile sensations below 20 Hz overlap aural sensations above this lower limit. Moreover, only young listeners can hear pure sounds near or above 20 kHz, the nominal upper limit.

Frequencies beyond both limits, however, have significance to audio-electronics engineers. For example, near-infrasonic (below 20 Hz) sounds are needed for classical organ music but can be noise in turntable rumble. Near-ultrasonic (above 20 kHz) intermodulation in audio circuits can produce undesirable difference-frequency components, which are audible.

The *audible sound-pressure level range* can be combined with the audible frequency range to describe an *auditory area*, shown in Fig. 20.1.2. The lowest curve shows the weakest audible sound-pressure level for listening with both ears to a pure tone while facing the sound source in a free field. The minimum level depends greatly on the frequency of the sound. It also varies somewhat among listeners. The levels that quickly produce discomfort or pain for listeners are only approximate, as indicated by the shaded and cross-hatched areas of Fig. 20.1.2. Extended exposure can produce temporary (or permanent) loss of auditory area at sound-pressure levels as low as 90 dB.

Wavelength effects are of great importance in the design of sound systems and rooms because wavelength varies over a 3-decade range, much wider than is typical elsewhere in electronics engineering. Audible sound waves vary in length from 1 cm to 15 m. The dimensions of the sound sources and receivers used in electroacoustics also vary greatly, e.g., from 1 cm to 3 m.

Sound waves follow the principles of geometrical optics and acoustics when the wavelength is very small relative to object size and pass completely around obstacles much smaller than a wavelength. This wide range of physical effects complicates the typical practical problem of sound production or reproduction.

Loudness Level

The simple, direct method for determining experimentally the loudness *level* of a sound is to match its observed loudness with that of a 1000-Hz sinewave reference tone of calibrated, variable sound-pressure level. (Usually this is a group judgment, or an average of individual judgments, in order to overcome individual observer differences.)

When the two loudnesses are matched, the *loudness level* of the sound, expressed in *phons*, is defined as numerically equal to the sound-pressure level of the reference tone in decibels. For example, a series of observers, each listening alternately to a machine noise and to a 1000-Hz reference tone, judge them (on the average) to be equally loud when the reference tone is adjusted to 86 dB at the observer location. This makes the loudness level of the machine noise 86 phons.

At 1000 Hz the decibel and phon levels are numerically identical, by definition. However, at other frequencies sinewave tones may have numerically quite different sound- and loudness-levels, as seen in Fig. 20.1.3. The dashed contour curves show the decibel level at each frequency corresponding to the loudness level identifying the curve at 1000 Hz. For example, a tone at 80 Hz and 70 dB lies on the contour marked 60 phons. Its sound level must be 70 dB for it to be as loud as a 60-dB tone at 1000 Hz. Such differences at low frequencies, especially at low sound levels, are a characteristic of the sense of hearing. The fluctuations above 1000 Hz are caused by sound-wave diffraction around the head of the listener and resonances in his ear canal. This illustrates how human physiological and psychological characteristics complicate the application of purely physical concepts.

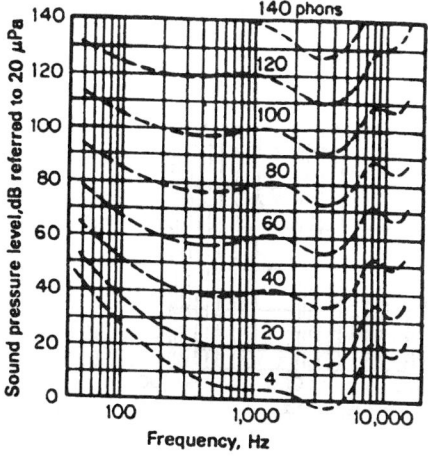

FIGURE 20.1.3 Equal-loudness-level contours.

Since loudness level is related to 1000-Hz tones defined physically in magnitude, the loudness-level scale is not really psychologically based. Consequently, although one can say that 70 phons is louder than 60 phons, one cannot say *how much* louder.

Loudness

By using the phon scale to overcome the effects of frequency, psychophysicists have developed a true loudness scale based on numerous experimental procedures involving relative-loudness judgments. *Loudness*, measured in *sones*, has a direct relation to loudness level in phons, which is approximated in Fig. 20.1.4. (Below 30 phons the relation changes slope. Since few practical problems require that range, it is omitted for simplicity.) A loudness of 1 sone has been defined equivalent to a loudness level of 40 phons. It is evident in Fig. 20.1.4 that a 10-phon change doubles the loudness in sones, which means *twice as loud*. Thus a 20-phon change in loudness level quadruples the loudness.

Another advantage of the sone scale is that the loudness of components of a complex sound are additive on the sone scale as long as they are well separated on the frequency scale. For example (using Fig. 20.1.4), two tonal components at 100 and 4000 Hz having loudness levels of 70 and 60 phons, respectively, would have individual loudnesses of 8 and 4 sones, respectively, and a total loudness of 12 sones.

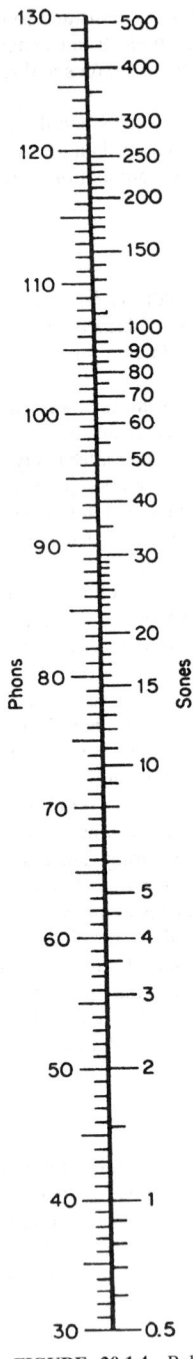

FIGURE 20.1.4 Relation between loudness in sones and loudness level in phons.[3]

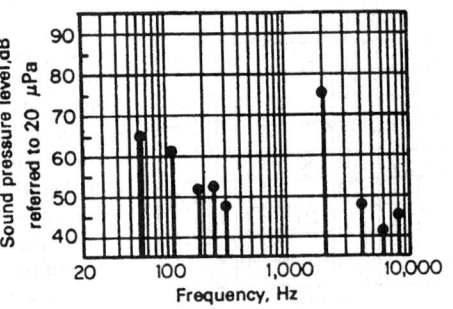

FIGURE 20.1.5   Typical line spectrum.[1]

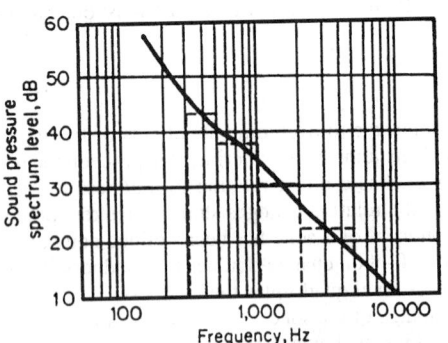

FIGURE 20.1.6   Continuous-level spectrum curve for a motor and blower.[1]

Detailed loudness computation procedures have been developed for highly complex sounds and noises, deriving the loudness in sones directly from a complete knowledge of the decibel levels for individual discrete components or noise bands. The procedures continue to be refined.

TYPICAL FORMATS FOR SOUND DATA

Sound and audio electronic data are frequently plotted as functions of frequency, time, direction, distance, or room volume. Frequency characteristics are the most common, in which the ordinate may be sound pressure, sound power, output-input ratio, percent distortion, or their logarithmic-scale (level) equivalents.

Sound Spectra

The frequency spectrum of a sound is a description of its resolution into components of different frequency and amplitude. Often the abscissa is a logarithmic frequency scale or a scale of octave (or fractional-octave) bands with each point plotted at the geometric mean of its band-limiting frequencies. Usually the ordinate scale is sound-pressure level. Phase differences are often ignored (except as they affect sound level) because they vary so greatly with measurement location, especially in reflective environments.

Line spectra are bar graphs for sounds dominated by discrete frequency components. Figure 20.1.5 is an example.

Continuous spectra are curves showing the distribution of sound-pressure level within a frequency range densely packed with components. Figure 20.1.6 is an example. Unless stated otherwise, the ordinate of a continuous-spectrum curve, called *spectrum level*, is assumed to represent sound-pressure level for a band of 1-Hz width. Usually level measurements L_{band} are made in wider bands, then converted to spectrum level L_{ps} by the bandwidth correction

$$L_{ps} = L_{\text{band}} - 10 \log(f_2 - f_1) \text{ dB} \qquad (2)$$

in which f_1 and f_2 are the lower- and upper-frequency limits of the band.

When a continuous-spectrum curve is plotted automatically by a level recorder synchronized with a heterodyning filter or with a sequentially switched set of narrow-bandpass filters, any effect of *changing* bandwidth on curve slope must be considered.

Combination spectra are appropriate for many sounds in which strong line components are superimposed over more diffuse continuous spectral backgrounds. Bowed or blown musical tones and motor-driven fan noises are examples.

Octave spectra, in which the ordinate is the sound-pressure level for bands one octave wide, are very convenient for measurements and for specifications but lack fine spectrum detail.

Third-octave spectra provide more detail and are widely used. One-third of an octave and one-tenth of a decade are so nearly identical that substituting the latter for the former is a practical convenience, providing a 10-band pattern that repeats every decade. Placing third-octave band zero at 1 Hz has conveniently made the band numbers equal 10 times the logarithm (base 10) of the band-center frequency; e.g., band 20 is at 100 Hz and band 30 at 1000 Hz.

Visual proportions of spectra (and other frequency characteristics) depend on the ratio of ordinate and abscissa scales. There is no universal or fully standard practice, but for ease of visual comparison of data and of specifications, it has become rather common practice in the United States for 30 dB of ordinate scale to equal (or slightly exceed) 1 decade of logarithmic frequency on the abscissa. Available audio and acoustical graph papers and automatic level-recorder charts have reinforced this practice. When the entire 120-dB range of auditory area is to be included in the graph, the ordinate is often compressed 2:1.

Response and Distortion Characteristics

Output-input ratios versus frequency are the most common data format in audio-electronics engineering. The audio-frequency scale (20 Hz to 20 kHz) is usually logarithmic. The ordinate may be sound- or electric-output level in decibels as the frequency changes with a constant electric or sound input; or it may be a ratio of the output to input (expressed in decibels) as long as they are linearly related within the range of measurement.

When the response-frequency characteristic is measured with the input frequency filtered from the output, a distortion-frequency characteristic is the result. It can be further filtered to obtain curves for each harmonic if desired.

Directional Characteristics

Sound sources radiate almost equally in all directions when the wavelength is large compared to source dimensions. At higher frequencies, where the wavelength is smaller than the source, the radiation becomes quite directional.

Time Characteristics

Any sound property can vary with time. It can build up, decay, or vary in magnitude periodically or randomly. A reverberant sound field decays rather logarithmically. Consequently the sound level in decibels falls linearly when the time scale is linear. The rate of decay in this example is 33 dB/s.

REFERENCES

1. Harris, C. M. (ed.), "Handbook of Acoustical Measurements and Noise Control," Chaps. 1 and 2, McGraw-Hill, 1991.

2. Acoustical Terminology (Including Mechanical Shock and Vibration), S1.1–1994, Acoustical Society of America, 1994.

3. ANSI Standard S3.4-1980 (R1986).

CHAPTER 20.2
SPEECH AND MUSICAL SOUNDS

Daniel W. Martin, Ronald M. Aarts

SPEECH SOUNDS

Speech Level and Spectrum

Both the sound-pressure level and the spectrum of speech sounds vary continuously and rapidly during connected discourse. Although speech may be arbitrarily segmented into elements called phonemes, each with a characteristic spectrum and level, actually one phoneme blends into another.

Different talkers speak somewhat differently, and they sound different. Their speech characteristics vary from one time or mood to another. Yet in spite of all these differences and variations, statistical studies of speech have established a typical "idealized" speech spectrum. The spectrum level rises about 5 dB from 100 to 600 Hz, then falls about 6, 9, 12, and 15 dB in succeeding higher octaves.

Overall sound-pressure levels, averaged over time and measured at a distance of 1 m from a talker on or near the speech axis, lie in the range of 65 and 75 dB when the talkers are instructed to speak in a "normal" tone of voice. Along this axis the speech sound level follows the inverse-square law closely to within about 10 cm of the lips, where the level is about 90 dB. At the lips, where communication microphones are often used, the overall speech sound level typically averages over 100 dB.

The peak levels of speech sounds greatly exceed the long-time average level. Figure 20.2.1 shows the difference between short peak levels and average levels at different frequencies in the speech spectrum. The difference is greater at high frequencies, where the sibilant sounds of relatively short duration have spectrum peaks.

Speech Directional Characteristics

Speech sounds are very directional at high frequencies. Figure 20.2.2 shows clearly why speech is poorly received behind a talker, especially in nonreflective environments. Above 4000 Hz the directional loss in level is 20 dB or more, which particularly affects the sibilant sound levels so important to speech intelligibility.

Vowel Spectra

Different vowel sounds are formed from approximately the same basic laryngeal tone spectrum by shaping the vocal tract (throat, back of mouth, mouth, and lips) to have different acoustical resonance-frequency combinations. Figure 20.2.3 illustrates the spectrum filtering process. The spectral peaks are called *formants*, and their frequencies are known as formant frequencies.

The shapes of the vocal tract, simplified models, and the acoustical results for three vowel sounds are shown in Fig 20.2.4. A convenient graphical method for describing the combined formant patterns is shown in Fig 20.2.5. Traveling around this vowel loop involves progressive motion of the jaws, tongue, and lips.

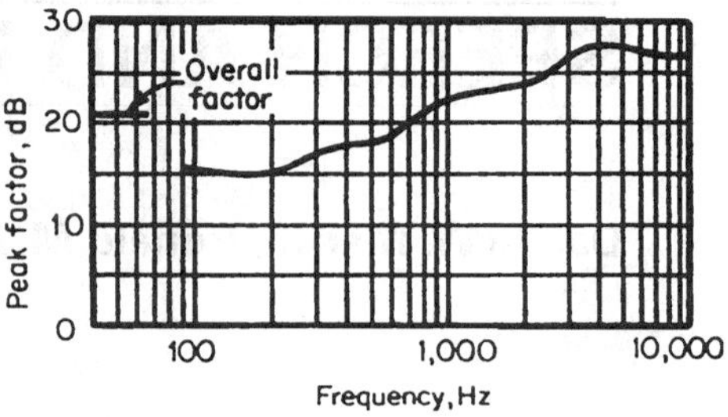

FIGURE 20.2.1 Difference in decibels between peak pressures of speech measured in short ($\frac{1}{8}$-s) intervals and rms pressure averaged over a long (75-s) interval.

Speech Intelligibility

More intelligibility is contained in the central part of the speech spectrum than near the ends. Figure 20.2.6 shows the effect on articulation (the percent of syllables correctly heard) when low- and high-pass filters of various cutoff frequencies are used. From this information a special frequency scale has been developed in which each of 20 frequency bands contributes 5 percent to a total *articulation index* of 100 percent. This distorted frequency scale is used in Fig. 20.2.7. Also shown are the spectrum curves for speech peaks and for speech minima, lying approximately 12 and 18 dB, respectively, above and below the average-speech-spectrum curve. When all the shaded area (30-dB range between the maximum and minimum curves) lies above threshold and below overload, in the absence of noise, the articulation index is 100 percent.

If a noise-spectrum curve were added to Fig 20.2.7, the figure would become an articulation-index computation chart for predicting communication capability. For example, if the ambient-noise spectrum coincided with the average-speech-spectrum curve, i.e., the signal-to-noise ratio is 1, only twelve-thirtieths of the shaded area would lie above the noise. The articulation index would be reduced accordingly to 40 percent.

Figure 20.2.8 relates monosyllabic word articulation and sentence intelligibility to articulation index. In the example above, for an articulation index of 0.40 approximately 70 percent of monosyllabic words and 96 percent of sentences would be correctly received.

However, if the signal-to-noise ratio were kept at unity and the frequency range were reduced to 1000 to 3000 Hz, half the bands would be lost. Articulation index would drop to 0.20, word articulation to 0.30, and sentence intelligibility to 70 percent. This shows the necessity for wide frequency range in a communication system when the signal-to-noise ratio is marginal. Conversely a good signal-to-noise ratio is required when the frequency range is limited.

The articulation-index method is particularly valuable in complex intercommunication-system designs involving noise disturbance at both the transmitting and receiving stations. Simpler effective methods have also been developed, such as the *rapid speech transmission index* (RASTI).

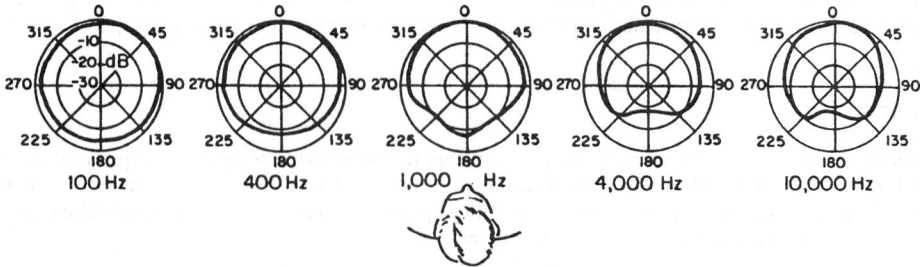

FIGURE 20.2.2 The directional characteristics of the human voice in a horizontal plane passing through the mouth.

Speech Peak Clipping

Speech waves are often affected inadvertently by electronic-circuit performance deficiencies or limitations. Figure 20.2.9 illustrates two types of amplitude distortion, center clipping and peak clipping. Center clipping, often caused by improper balancing or biasing of a push-pull amplifier circuit, can greatly interfere with speech quality and intelligibility. In a normal speech spectrum the consonant sounds are higher in frequency and lower in level than the vowel sounds. Center clipping tends to remove the important consonants.

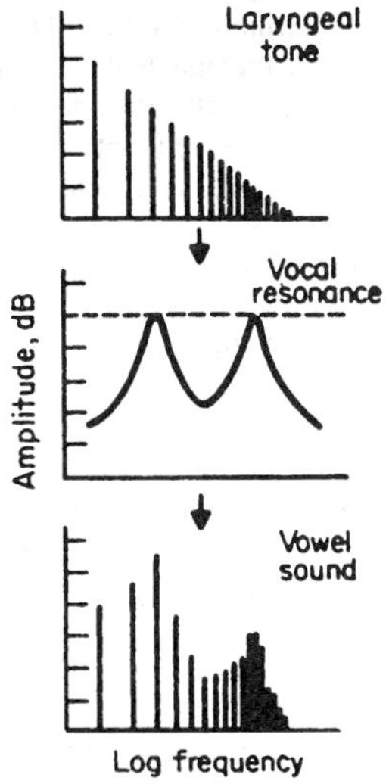

By contrast peak clipping has little effect on speech intelligibility as long as ambient noise at the talker and system electronic noise are relatively low in level compared with the speech.

Peak clipping is frequently used intentionally in speech-communication systems to raise the average transmitted speech level above ambient noise at the listener or to increase the range of a radio transmitter of limited power. This can be done simply by overloading an amplifier stage. However, it is safer for the circuits and it produces less intermodulation distortion when back-to-back diodes are used for clipping ahead of the overload point in the amplifier or transmitter. Figure 20.2.10 shows intelligibility improvement from speech peak clipping when the talker is in quiet and listeners are in noise. Figure 20.2.11 shows that caution is necessary when the talker is in noise, unless the microphone is shielded or is a noise-canceling type.

Tilting the speech spectrum by differentiation and flattening it by equalization are effective preemphasis treatments before peak clipping. Both methods put the consonant and vowel sounds into a more balanced relationship before the intermodulation effects of clipping affect voiced consonants.

Caution must be used in combining different forms of speech-wave distortion, which individually have innocuous effects on intelligibility but can be devastating when they are combined.

FIGURE 20.2.3 Effects on the spectrum of the laryngeal tone produced by the resonances of the vocal tract.[5]

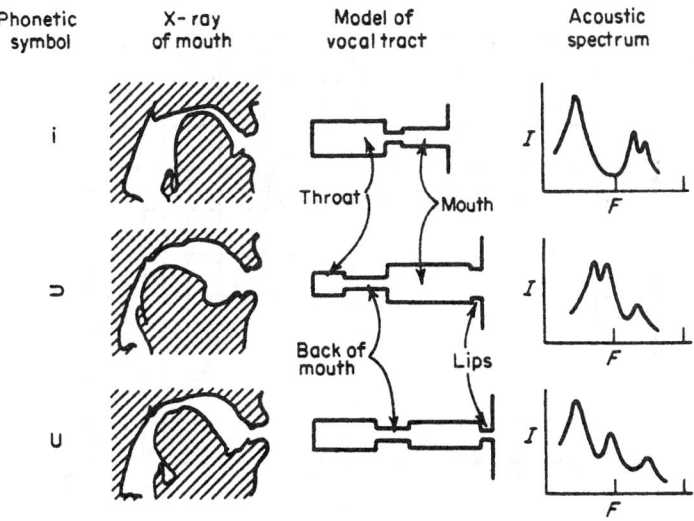

FIGURE 20.2.4 Phonetic symbols, shapes of vocal tract, models, and acoustic spectra for three vowels[6].

MUSICAL SOUNDS

Musical Frequencies

The accuracy of both absolute and relative frequencies is usually much more important for musical sounds than for speech sounds and noise. The international frequency standard for music is defined at 440.00 Hz for A_4, the A above C_4 (middle C) on the musical keyboard. In sound recording and reproduction, the disc-rotation and tape-transport speeds must be held correct within 0.2 or 0.3 percent error (including both recording and playback mechanisms) to be fully satisfactory to musicians.

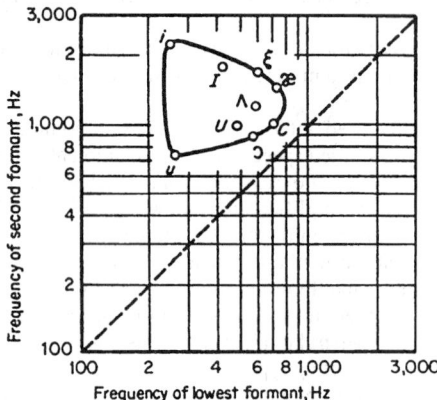

FIGURE 20.2.5   The center frequencies of the first two formants for the sustained English vowels plotted to show the characteristic differences.[7]

The mathematical musical scale is based on an exact octave ratio of 2:1. The subjective octave slightly exceeds this, and piano tuning sounds better when the scale is stretched very slightly.

The equally tempered scale of 12 equal ratios within each octave is an excellent compromise between the different historical scales based on harmonic ratios. It has become the standard of reference, even for individual musical performances, which may deviate from it for artistic or other reasons.

Different musical instruments play over different ranges of *fundamental* frequency, shown in Fig.20.2.12. However, most musical sounds have many harmonics that are audibly significant to their tone spectra. Consequently high-fidelity recording and reproduction need a much wider frequency range.

Sound Levels of Musical Instruments

The sound level from a musical instrument varies with the type of instrument, the distance from it, which note in the scale is being played, the dynamic marking in the printed music, the player's ability, and (on polyphonic instruments) the number of notes (and stops) played at the same time.

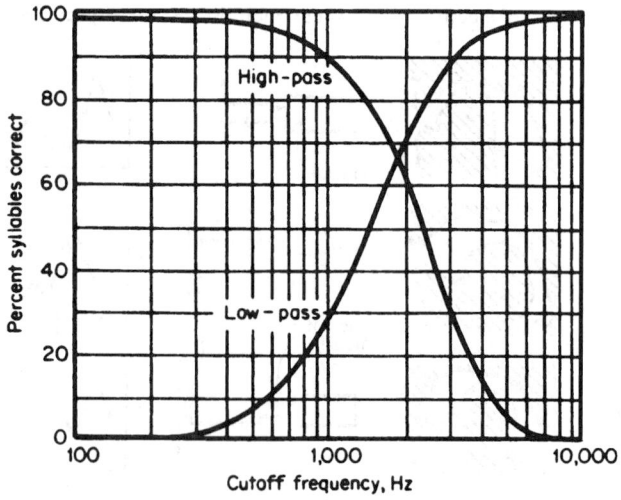

FIGURE 20.2.6   Syllable articulation score vs. low- or high-pass cutoff frequency.[8]

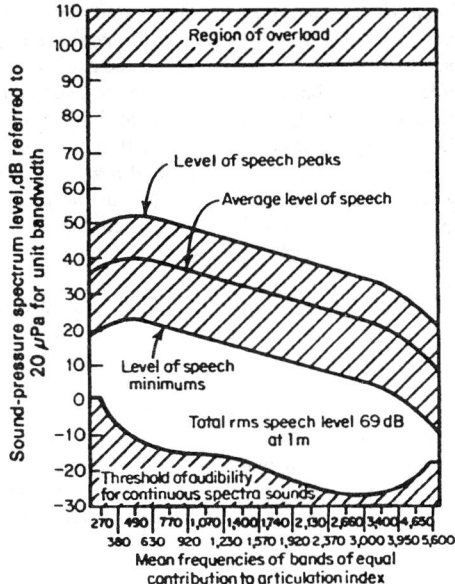

FIGURE 20.2.7 Speech area, bounded by speech peak and minimum spectrum-level curves, plotted on an articulation-index calculation chart.[9]

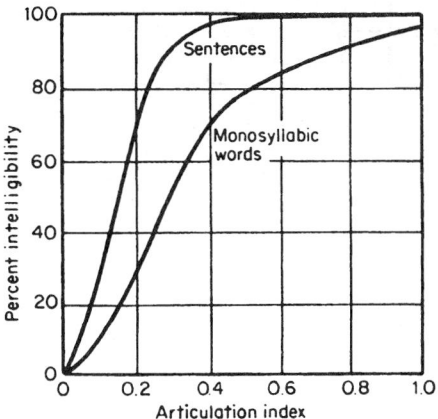

FIGURE 20.2.8 Sentence- and word-intelligibility prediction from calculated articulation index.[10]

Orchestral Instruments. The following sound levels are typical at a distance of 10 ft in a nonreverberant room. Soft (pianissimo) playing of a weaker orchestral instrument, e.g., violin, flute, bassoon, produces a typical sound level of 55 to 60 dB. Fortissimo playing on the same instrument raises the level to about 70 to 75 dB. Louder instruments, e.g., trumpet or tuba, range from 75 dB at pianissimo to about 90 dB at fortissimo.

Certain instruments have exceptional differences in sound level of low and high notes. A flute may change from 42 dB on a soft low note to 77 dB on a loud high note, a range of 35 dB. The French horn ranges from 43 dB (soft and low) to 93 dB (loud and high).

Sound levels are about 10 dB higher at 3 ft (inverse-square law) and 20 dB higher at 1 ft. The louder instruments, e.g., brass, at closer distances may overload some microphones and preamplifiers.

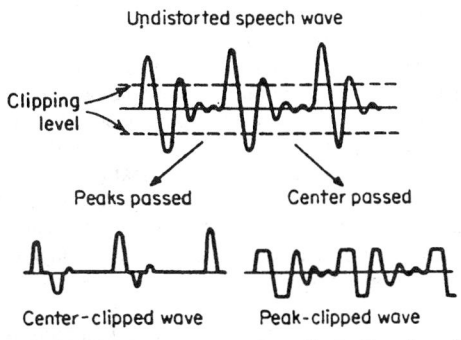

FIGURE 20.2.9 Two types of amplitude distortion of speech waveform.[5]

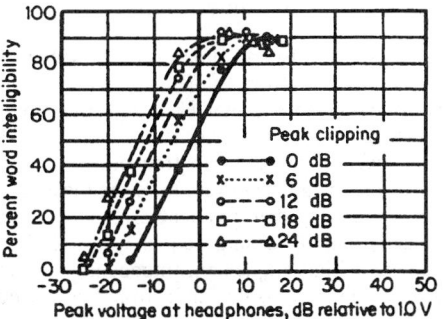

FIGURE 20.2.10 Advantages of peak clipping of noise-free speech waves, heard by listeners in ambient aircraft noise.[10]

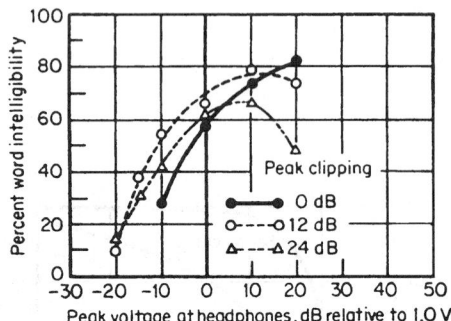

FIGURE 20.2.11   Effects of speech clipping with both the talker and the listener in simulated aircraft noise. Note that excessive peak clipping is detrimental.[10]

Percussive Instruments.   The sound levels of shock-excited tones are more difficult to specify because they vary so much during decay and can be excited over a very wide range. A bass drum may average over 100 dB during a loud passage with peaks (at 10 ft) approaching 120 dB. By contrast a triangle will average only 70 dB with 80-dB peaks. A single tone of a grand piano played forte will initially exceed 90 dB near the piano rim, 80 dB at the pianist, and 70 dB at the conductor 10 to 15 ft away. Large chords and rapid arpeggios will raise the level about 10 dB.

Instrumental Groups.   Orchestras, bands, and polyphonic instruments produce higher sound levels since many notes and instruments (or stops) are played together. Their sound levels are specified at larger distances than 10 ft because the sound sources occupy a large area; 20 ft from the front of a 75-piece orchestra the sound level will average about 85 to 90 dB with peaks of 105 to 110 dB. A full concert band will go higher. At a similar distance from the sound sources of an organ (pipe or electronic) the full-organ (or crescendo-pedal) condition will produce a level of 95 to 100 dB. By contrast the softest stop with expression shutters closed may be 45 dB or less.

Growth and Decay of Musical Sounds

These characteristics are quite different for different instruments. Piano or guitar tones quickly rise to an initial maximum, then gradually diminish until the strings are damped mechanically. Piano tones have a more rapid decay initially than later in the sustained tone. Orchestral instruments can start suddenly or smoothly, depending on the musician's technique, and they damp rather quickly when playing ceases. Room reverberation affects both growth and decay rates when the time constants of the room are greater than those of the instrument vibrators. This is an important factor in organ music, which is typically played in a reverberant environment.

Many types of musical tone have characteristic transients which influence timbre greatly. In the "chiff" of organ tone the transients are of different fundamental frequency. They appear and decay before steady state is reached. In percussive tones the initial transient is the cause of the tone (often a percussive noise), and the final transient is the result.

These transient effects should be considered in the design of audio electronics such as "squelch," automatic gain control, compressor, and background-noise reduction circuits.

Spectra of Musical Instrument Tones

Figure 20.2.13 displays time-averaged spectra for a 75-piece orchestra, a theater pipe organ, a piano, and a variety of orchestral instruments, including members of the brass, woodwind, and percussion families. These vary from one note to another in the scale, from one instant to another within a single tone or chord, and from one instrument or performer to another. For example, a concert organ voiced in a baroque style would have lower spectrum levels at low frequencies and higher at high frequencies than the theater organ shown.

The organ and bass drum have the most prominent low-frequency output. The cymbal and snare drum are strongest at very high frequencies. The orchestra and most of the instruments have spectra which diminish gradually with increasing frequency, especially above 1000 Hz. This is what has made it practical to preemphasize the high-frequency components, relative to those at low frequencies, in both disc and tape recording. However, instruments that differ from this spectral tendency, e.g., coloratura sopranos, piccolos, cymbals, create problems of intermodulation distortion, and overload.

Spectral peaks occurring only occasionally, for example, 1 percent of the time, are often more important to sound recording and reproduction than the peaks in the average spectra of Fig. 20.2.13. The frequency

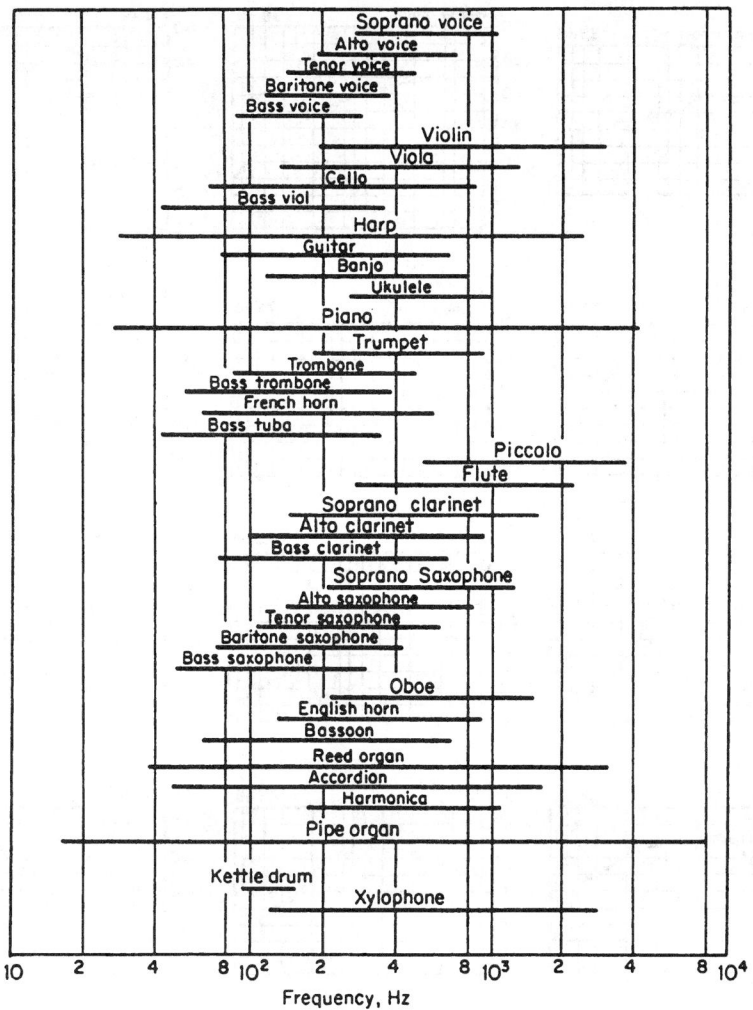

FIGURE 20.2.12   Range of the fundamental frequencies of voices and various musical instruments. (Ref. 8).

ranges shown in Table 20.2.1 have been found to have relatively large instantaneous peaks for the instruments listed.

Directional Characteristics of Musical Instruments

Most musical instruments are somewhat directional. Some are highly so, with well-defined symmetry, e.g., around the axis of a horn bell. Other instruments are less directional because the sound source is smaller than the wavelength, e.g., clarinet, flute. The mechanical vibrating system of bowed string instruments is complex, operating differently in different frequency ranges, and resulting in extremely variable directivity. This is significant for orchestral seating arrangements both in concert halls and recording studios.

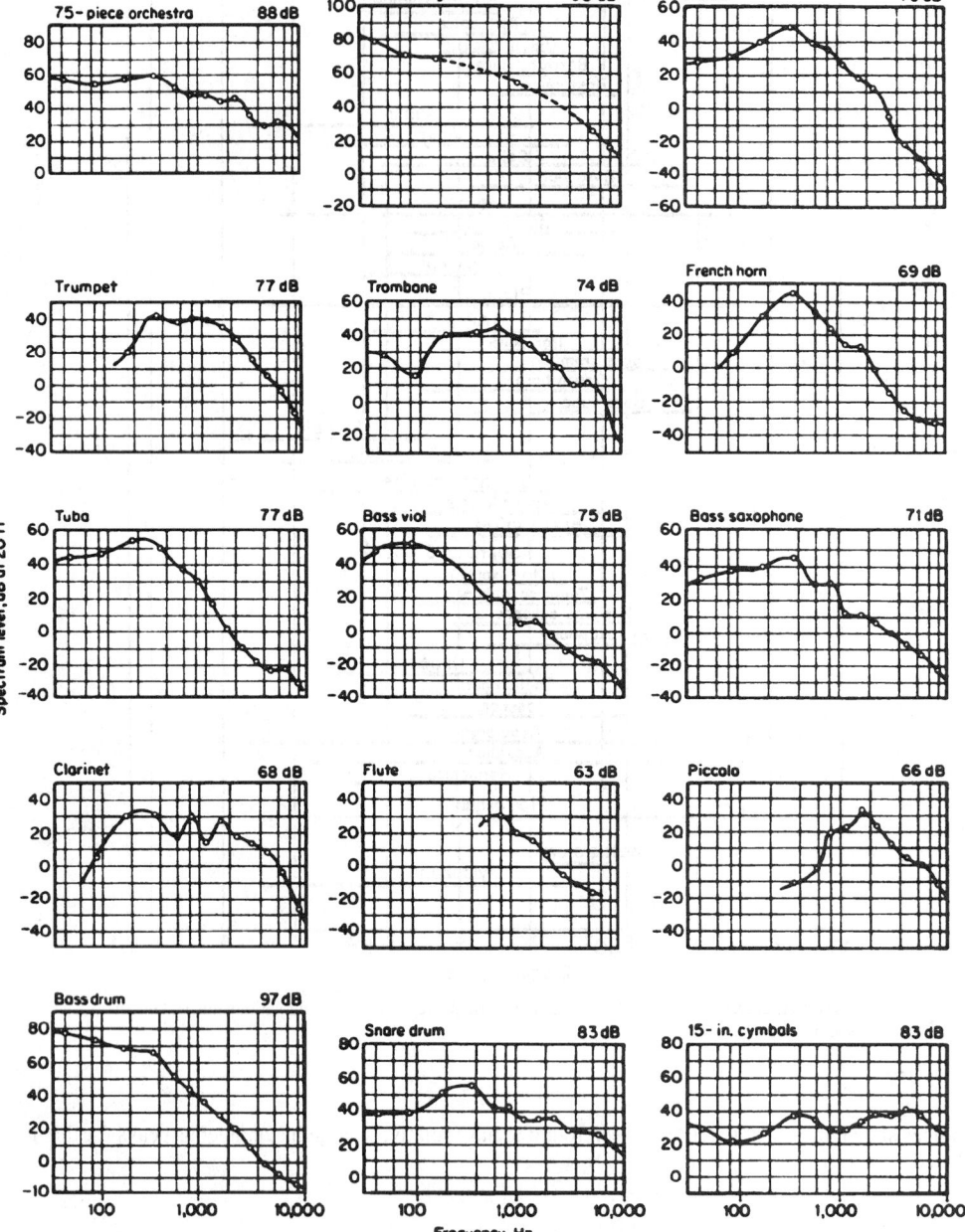

FIGURE 20.2.13 Time-averaged spectra of musical instruments.

TABLE 20.2.1 Frequency Band Containing Instantaneous Spectral Peaks

Band limits, Hz	Instruments
20–60	Theater organ
60–125	Bass drum, bass viol
125–250	Small bass drum
250–500	Snare drum, tuba, bass saxophone, French horn, clarinet, piano
500–1,000	Trumpet, flute
2,000–3,000	Trombone, piccolo
5,000–8,000	Triangle
8,000–12,000	Cymbal

Audible Distortions of Musical Sounds

The quality of musical sounds is more sensitive to distortion than the intelligibility of speech. A chief cause is that typical music contains several simultaneous tones of different fundamental frequency in contrast to typical speech sound of one voice at a time. Musical chords subjected to nonlinear amplification or transduction generate intermodulation components that appear elsewhere in the frequency spectrum.

Difference tones are more easily heard than summation tones because the summation tones are often hidden by harmonics that were already present in the undistorted spectrum and because auditory masking of a high-frequency pure tone by a lower-frequency pure tone is much greater than vice versa.

When a critical listener controls the sounds heard (an organist playing an electronic organ on a high-quality amplification system) and has unlimited opportunity and time to listen, even lower distortion (0.2 percent, for example) can be perceived.

REFERENCES

1. Miller, G. A. "Language and Communication," McGraw-Hill, 1951.

2. Dunn, H. K. *J. Acoust. Soc. Am.*, 1950, Vol. 22, p. 740.

3. Potter, R. K., and G. E. Peterson *J. Acoust. Soc. Am.*, 1948, Vol. 20, p. 528.

4. French, N. R., and J. C. Steinberg *J. Acoust. Soc. Am.*, 1947, Vol. 19, p. 90.

5. Beranek, L. L. "Acoustics,"Acoustical Society of America, 1986.

6. Hawley, M. E., and K. D. Kryter Effects of Noise on Speech, Chap. 9 in C. M. Harris (ed.), "Handbook of Noise Control," McGraw-Hill, 1957.

7. Olson, H. F. "Musical Engineering," McGraw-Hill, 1952.

8. Olson, H. F. "Elements of Acoustical Engineering," Van Nostrand, 1947.

9. Sivian, L. J., H. K. Dunn, and S. D. White *IRE Trans. Audio*, 1959, Vol. AU-7, p. 47; revision of paper in *J. Acoust. Soc. Am.*, 1931, Vol. 2, p. 33.

10. Hawley, M. E., and K. D. Kryter Effects of Noise on Speech, "Handbook of Noise Control," McGraw-Hill, 1957.

CHAPTER 20.3
MICROPHONES, LOUDSPEAKERS, AND EARPHONES

Daniel W. Martin

MICROPHONES

Sound-Responsive Elements

The sound-responsive element in a microphone may have many forms (Fig. 20.3.1). It may be a stretched membrane (*a*), a clamped diaphragm (*b*), or a magnetic diaphragm held in place by magnetic attraction (*c*). In these the moving element is either an electric or magnetic conductor, and the motion of the element creates the electric or magnetic equivalent of the sound directly.

Other sound-responsive elements are straight (*d*) or curves (*e*) conical diaphragms with various shapes of annular compliance rings, as shown. The motion of these diaphragms is transmitted by a drive rod from the conical tip to a mechanical transducer below.

Other widely used elements are a circular piston (*f*) bearing a circular voice coil of smaller diameter and a corrugated-ribbon conductor (*g*) of extremely low mass and stiffness suspended in a magnetic field.

Transduction Methods

Microphones have a great variety of transduction methods shown in Fig. 20.3.2.

The loose-contact transducer (Fig. 20.3.2*a*) was the first achieved by Bell in magnetic form and later made practical by Edison's use of carbonized hard-coal particles. It is widely used in telephones. Its chief advantage is its self-amplifying function, in which diaphragm amplitude variations directly produce electric resistance and current variations. Disadvantages include noise, distortion, and instability.

Moving-iron transducers have great variety, ranging from the historic pivoted armature (Fig. 20.3.2*b*) to the modern ring armature driven by a nonmagnetic diaphragm (Fig. 20.3.2*h*). In all these types a coil surrounds some portion of the magnetic circuit. The reluctance of the magnetic circuit is varied by motion of the sound-responsive element, which is either moving iron itself (Fig. 20.3.2*c* and *d*) or is coupled mechanically to the moving iron (Fig. 20.3.2*e–h*). In some of the magnetic circuits that portion of the armature surrounded by the coil carries very little steady flux, operating on differential magnetic flux only. Output voltage is proportional to moving-iron velocity.

Electrostatic transducers (Fig. 20.3.2*i*) use a polarizing potential and depend on capacitance variation between the moving diaphragm and a fixed electrode for generation of a corresponding potential difference. The *electret microphone* is a special type of electrostatic microphone that holds polarization indefinitely without continued application of a polarizing potential, an important practical advantage for many applications.

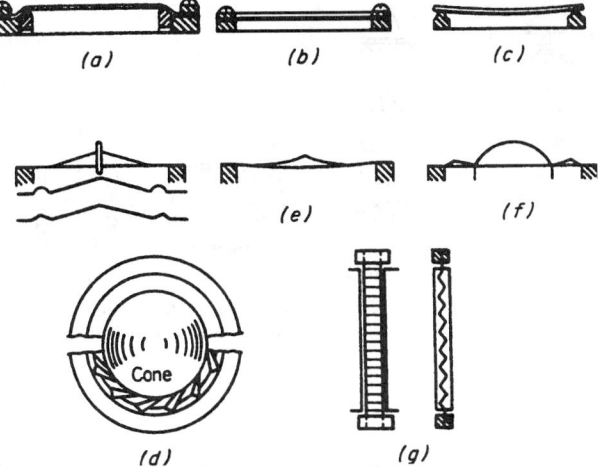

FIGURE 20.3.1   Sound-responsive elements in microphones.[1]

Piezoelectric transducers (Fig. 20.3.2*j*) create an alternating potential through the flexing of crystalline elements which, when deformed, generate a charge difference proportional to the deformation on opposite surfaces. Because of climatic effects and high electric impedance the rochelle salt commonly used for many years has been superseded by polycrystalline ceramic elements and by piezoelectric polymer.

Moving-coil transducers (Fig. 20.3.2*k*) generate potential by oscillation of the coil within a uniform magnetic field. The output potential is proportional to coil velocity.

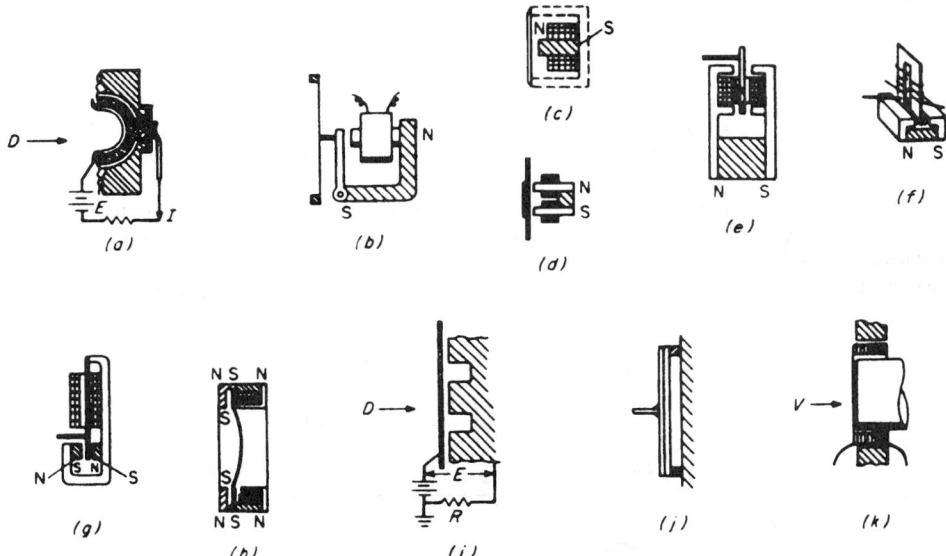

FIGURE 20.3.2   Microphone transduction methods.[1]

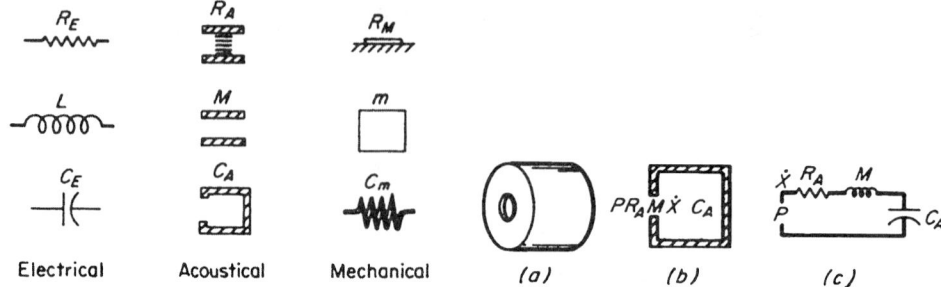

FIGURE 20.3.3   Equivalent basic elements in electrical, acoustical, and mechanical systems.[2]

FIGURE 20.3.4   Helmholtz resonator in (*a*) perspective and (*b*) in section and (*c*) equivalent electric circuit.[2]

Equivalent Circuits

Electronics engineers understand electroacoustic and electromechanical design better with the help of equivalent or analogous electric circuits. Microphone design provides an ideal base for introduction of equivalent circuits because microphone dimensions are small compared with acoustical wavelengths over most of the audio-frequency range. This allows the assumption of lumped circuit constants.

Figure 20.3.3 shows equivalent symbols for the three basic elements of electrical, acoustical, and mechanical systems. In acoustical circuits the resistance is air friction or viscosity, which occurs in porous materials or narrow slots. Radiation resistance is another form of acoustical damping. Mechanical resistance is friction. Mass in the mechanical system is analogous to electric inductance. The acoustical equivalent is the mass of air in an opening or constriction divided by the square of its cross-sectional area. The acoustical analog of electric capacitance and mechanical-spring compliance is acoustical capacitance. It is the inverse of the stiffness of an enclosed volume of air under pistonlike action. Acoustical capacitance is proportional to the volume enclosed.

Figure 20.3.4 is an equivalent electric circuit for a Helmholtz resonator. Sound-pressure and air-volume current are analogous to electric potential and current, respectively. Other analog systems have been proposed. One frequently used has advantages for mechanical systems.

Microphone Types and Equivalent Circuits

Different types of microphone respond to different properties of the acoustical *input* wave. Moreover, the electric *output* can be proportional to different internal mechanical variables.

Pressure Type, Displacement Response.   Figure 20.3.5 shows a microphone responsive to the sound-pressure wave acting through a resonant acoustical circuit upon a resonant diaphragm coupled to a piezoelectric element responsive to displacement. (The absence of sound ports in the case or in the diaphragm keeps the microphone pressure responsive.) In the equivalent circuit the sound pressure is the generator. L_a and R_a represent

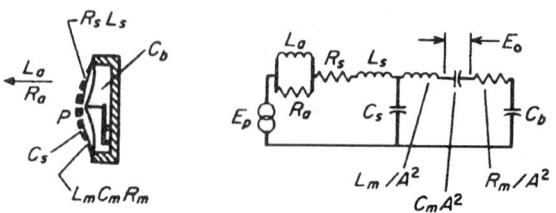

FIGURE 20.3.5   Pressure microphone, displacement response.[1]

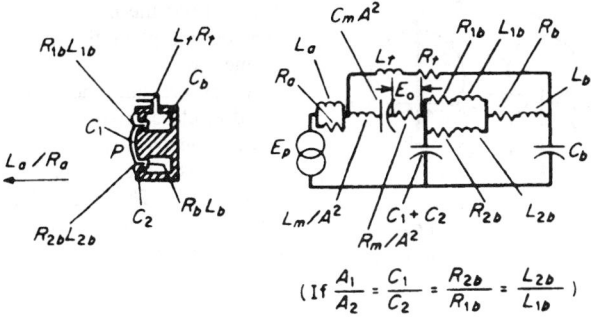

FIGURE 20.3.6   Pressure microphone, velocity response.[1]

the radiation impedance. L_s and R_s are the inertance and acoustical resistance of the holes; C_s is the capacitance of the volume in front of the diaphragm; L_m, C_m, and R_m are the mass, compliance, and resistance of the piezoelectric element and diaphragm lumped together; and C_b is the capacitance of the entrapped back volume of air. The electric output is the potential differential across the piezoelectric element. It is shown across the capacitance in the equivalent circuit because microphones of this type are designed to be stiffness-controlled throughout most of their operating range.

Pressure Type, Velocity Response.   Figure 20.3.6 shows a moving-coil pressure microphone, which is a velocity-responsive transducer. In this microphone three acoustical circuits lie behind the diaphragm. One is behind the dome and another behind the annular rings. The third acoustical circuit lies beyond the acoustical resistance at the back of the voice-coil gap and includes a leak from the back chamber to the outside. This microphone is resistance-controlled throughout most of the range, but at low frequencies its response is extended by the resonance of the third acoustical circuit. Output potential is proportional to the velocity of voice-coil motion.

Pressure-Gradient Type, Velocity Response.   When both sides of the sound-responsive element are open to the sound wave, the response is proportional to the *gradient* of the pressure wave. Figure 20.3.7 shows a rib-bon conductor in a magnetic field with both sides of the ribbon open to the air. In the equivalent circuit there are two generators, one for sound pressure on each side. Radiation resistance and reactance are in series with each generator and the circuit constants of the ribbon. Usually the ribbon resonates at a very low frequency, making its mechanical response mass-controlled throughout the audio-frequency range. The electric output is proportional to the conductor velocity in the magnetic field. Gradient microphones respond differently to distant and close sound sources.

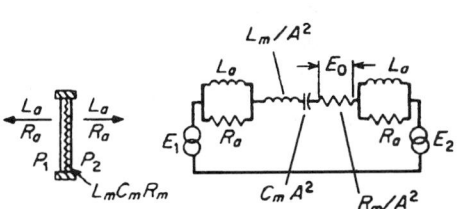

FIGURE 20.3.7   Gradient microphone, velocity response.[1]

Directional Patterns and Combination Microphones

Because of diffraction, a pressure microphone is equally responsive to sound from all directions as long as the wavelength is larger than microphone dimensions (see Fig. 20.3.8*a*). (At high frequencies it is somewhat directional along the forward axis of diaphragm or ribbon motion.)

By contrast a pressure-gradient microphone has a figure-eight directional pattern (Fig. 20.3.8*b*), which rotates about the axis of ribbon or diaphragm motion. A sound wave approaching a gradient microphone at 90° from the axis produces balanced pressure on the two sides of the ribbon and consequently no response.

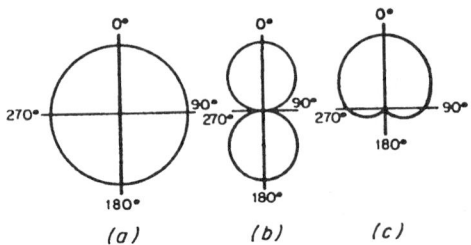

FIGURE 20.3.8 Directional patterns of microphones: (*a*) nondirectional; (*b*) bidirectional; (*c*) unidirectional.[2]

This defines the *null plane* of a gradient microphone. Outside this plane the microphone response follows a cosine law.

If the pressure and gradient microphones are combined in close proximity (see Fig. 20.3.9) and are connected electrically to add in equal (half-and-half) proportions, a heart-shaped cardioid pattern (Fig. 20.3.8*c*) is obtained. (The back of the ribbon in the pressure microphone is loaded by an acoustical resistance line.) By combining the two outputs in other proportions other limacon directional patterns can be obtained.

Phase-Shift Directional Microphones

Directional characteristics similar to those of the combination microphones can also be obtained with a single moving element by means of equivalent circuit analysis using acoustical phase-shift networks. Figure 20.3.10 shows a moving-coil, phase-shift microphone and its simplified equivalent circuit. The phase-shift network is composed of the rear-port resistance R_2 and inertance L_2, the capacitance of the volume under the diaphragm and within the magnet, and the impedance of the interconnecting screen. The microphone has a cardioid directional pattern.

Special-Purpose Microphones

Special-purpose microphones include two types that are superdirectional, two that overcome noise, and one without cables.

Line microphones use an approximate line of equally spaced pickup points connected through acoustically damped tubes to a common microphone diaphragm. The phase relationships at these points for an incident plane wave combine to give a sharply directional pattern along the axis if the line segment is at least one wavelength.

Parabolic microphones face a pressure microphone unit toward a parabolic reflector at its focal point, where sounds from distant sources along the axis of the parabola converge. They are effective for all wavelengths smaller than the diameter of the reflector.

Noise-canceling microphones are gradient microphones in which the mechanical system is designed to be stiffness-controlled rather than mass-controlled. For distant sound sources the resulting response is greatly attenuated at low frequencies. However, for a very close sound source, the response-frequency characteristic is uniform because the *gradient* of the pressure wave near a point source decreases with increasing frequency. Such a microphone provides considerable advantage for nearby speech over distant noise on the axis of the microphone.

FIGURE 20.3.9 Combination unidirectional microphone.[1]

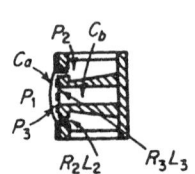

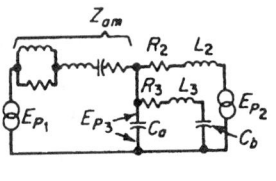

FIGURE 20.3.10 Phase-shift unidirectional microphone.[1]

Contact microphones are used on string and percussion musical instruments, on seismic-vibration detectors, and for pickup of body vibrations including speech. The throat microphone was noted for its convenience and its rejection of airborne noise. Most types of throat microphone are inertia-operated, the case receiving vibration from the throat walls actuated by speech sound pressure in the throat. The disadvantage is a deficiency of speech sibilant sounds received back in the throat from the mouth.

Wireless microphones have obvious operational advantages over those with microphone cords. A wireless microphone contains a small, low-power radio transmitter with a nearby receiver connected to an audio communication system. Any of the microphone types can be so equipped. The potential disadvantage is in rf interference and field effects.

Microphone Use in Recordings

The choice of microphone type and placement greatly affects the sound of a recording. For speech and dialogue recordings pressure microphones are usually placed near the speakers in order to minimize ambient-noise pickup and room reverberation. Remote pressure microphones are also used when a maximum room effect is desired.

In the playback of monophonic recordings room effects are more noticeable than they would have been to a listener standing at the recording microphone position because single-microphone pickup is similar to single-ear (monaural) listening, in which the directional clues of localization are lost. Therefore microphones generally need to be closer in a monophonic recording than in a stereophonic recording.

In television pickup of speech, where a boom microphone should be outside the camera angle, unidirectional microphones are often used because of their greater ratio of direct to generally reflected sound response.

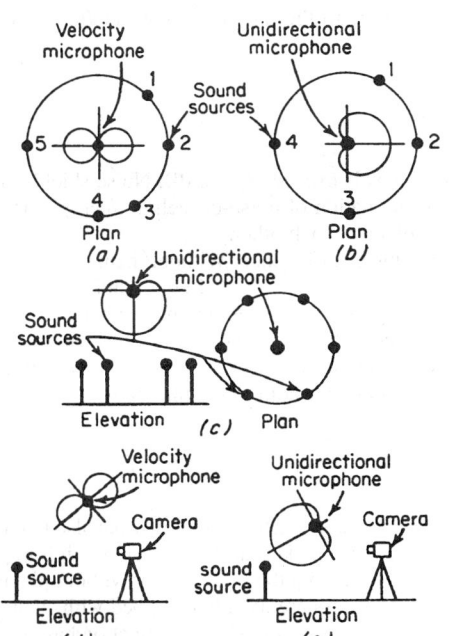

FIGURE 20.3.11 Use of directional microphones.[2]

Both velocity (gradient) microphones and unidirectional microphones can be used to advantage in broadcasting and recording. Figure 20.3.11a shows how instruments may be placed around a figure-eight directivity pattern to balance weaker instruments 2 and 5 against stronger instruments 1 and 3 with a potential noise source at point 4. In Fig. 20.3.11b source 2 is favored, with sources 1 and 3 somewhat reduced and source 4 highly discriminated against by the cardioid directional pattern. In Fig. 20.3.11c an elevated unidirectional microphone aimed downward responds uniformly to sources on a circle around the axis while discriminating against mechanical noises at ceiling level. Figure 20.3.11d places the camera noise in the null plane of a figure-eight pattern, and Fig. 20.3.11e shows a similar use for the unidirectional microphone. Camera position is less critical for the cardioid microphone than for the gradient microphone.

Early classical stereo recordings used variations of two basic microphone arrangements. In one scheme two unidirectional microphones were mounted close together with their axes angled toward opposite ends of the sound field to be recorded. This retained approximately the same arrival time and phase at both microphones, depending chiefly on the directivity patterns to create the sound difference in the two channels.

In the second scheme the two microphones (not necessarily directional) were separated by distances of 5 to 25 ft, depending on the size of the sound field to be recorded. Microphone axes (if directional) were again directed toward the ends of the sound field or group of sound sources. In this arrangement the time of arrival and phase differences were more important, and the effect of directivity was lessened. Each approach had its advantages and disadvantages.

With the arrival of tape recorders having many channels a trend has developed toward the use of more microphones and closer microphone placement. This offers much greater flexibility in mixing and rerecording, and it largely removes the effect of room reverberation from the recording. This may be either an advantage or a disadvantage depending on the viewpoint. Reverberation can be added later.

In sound-reinforcement systems for dramatic productions and orchestras the use of many microphones again offers operating flexibility. However, it also increases the probability of operating error, increased system noise, and acoustical feedback, making expert monitoring and mixing of the microphone outputs necessary.

An attractive alternative for multimicrophone audio systems is the use of independent voice-operated electronic control switches in each microphone channel amplifier, in combination with an automatic temporary reduction of overall system gain as more channels switch on, in order to prevent acoustical feedback. Automatic mixers have been devised to minimize speech signal dropouts, and to prevent the inadvertent operation of channel control switches by background noises.

Microphone Mounting

On podiums and lecterns microphones are typically mounted on fixed stands with adjustable arms. On stages they are mounted on adjustable floor stands. In mobile communication and in other situations where microphone use is occasional, handheld microphones are used during communication and are stowed on hangers at other times. For television and film recording, where the microphone must be out of camera sight, the microphones are usually mounted on booms overhead and are moved about during the action to obtain the best speech-to-noise ratio possible at the time. In two-way communication situations which require the talker to move about or to turn his head frequently, the microphone can be mounted on a boom fastened to his headset. This provides a fixed close-talking microphone position relative to the mouth, a considerable advantage in high-ambient-noise levels.

Microphone Accessories

Noise shields are needed for microphones in ambient noise levels exceeding 110 dB. Noise shields are quite effective at high frequencies, where the random-noise discrimination of noise-canceling microphones diminishes. Noise shields and noise-canceling microphones complement each other.

Windscreens are available for microphone use in airstreams or turbulence. Without them aerodynamically induced noise is produced by turbulence at the microphone grille or openings. Large windscreens are more effective than small ones because they move the turbulence region farther from the microphone.

Special sponge-rubber mountings for the microphone and cable to reduce extraneous vibration of the microphone are often used. Many microphone stands and booms have optional suspension mounting accessories to reduce shock and vibration transmitted through the stand or boom to the microphone.

Special Properties of Microphones

The source impedance of a microphone is important not only to the associated preamplifier but also to the allowable length of microphone cable and the type and amount of noise picked up by the cable. High-impedance microphones (10 kΩ or more) cannot be used more than a few feet from the preamplifier without pickup from stray fields. Microphones having an impedance of a few ohms or less are usually equipped with stepup transformers to provide a line impedance in the range of 30 to 600 Ω, which extensive investigation has established as the most noise-free line-impedance range.

The microphone unit itself can be responsive to hum fields at power-line frequencies unless special design precautions are taken. Most microphones have a hum-level rating based on measurement in a standard alternating magnetic field.

For minimum electrical noise balanced and shielded microphone lines are used, with the shield grounded only at the amplifier end of the line.

Microphone linearity should be considered when the sound level exceeds 100 dB, a frequent occurrence for loud musical instruments and even for close speech. Close-talking microphones, especially of the gradient type, are particularly susceptible to noise from breath and plosive consonants.

Specifications

Microphone specifications typically include many of the following items: type or mode of operation, directivity pattern, frequency range, uniformity of response within the range, output level at one or more impedances for a standard sound-pressure input (for example, 1 Pa or 10 dyn/cm^2), recommended load impedance, hum output level for a standard magnetic field (for example, 10^{-3} G), dimensions, weight, finish, mounting, power supply (if necessary), and accessories.

LOUDSPEAKERS

Introduction

A loudspeaker is an electroacoustic transducer intended to radiate acoustic power into the air, with the acoustic waveform equivalent to the electrical input waveform. An earphone is an electroacoustic transducer intended to be closely coupled acoustically to the ear. Both the loudspeaker and earphone are receivers of audio-electronic signals. The principal distinction between them is the acoustical loading. An earphone delivers sound to air in the ear. A loudspeaker delivers sound indirectly to the ear through the air.

The transduction methods of loudspeakers and earphones are historically similar and are treated together. An overview of loudspeaker developments of the closing 50 years of the last millennium is given by Gander.[3] However, since loudspeakers operate primarily into radiation resistance and earphones into acoustical capacitance, the design, measurement, and use of the two types of electroacoustic transducers will be discussed separately.

Transduction Methods

Early transducers for sound reproduction were of the mechanoacoustic type. Vibrations received by a stylus in the undulating groove of a record were transmitted to a diaphragm, placed at the throat of a horn for better acoustical impedance matching to the air, all without the aid of electronics. Electro-acoustics and electronics introduced many advantages and a variety of transduction methods including moving-coil, moving-iron, electrostatic, magnetostrictive, and piezoelectric (Fig. 20.3.12).

Most loudspeakers are moving-coil type today, although moving-iron transducers were once widely used. Electrostatic loudspeakers are used chiefly in the upper range of audio frequencies, where amplitudes are small. Magnetostrictive and piezoelectric loudspeakers are used for underwater sound. All the transducer types are used in earphones except magnetostrictive.

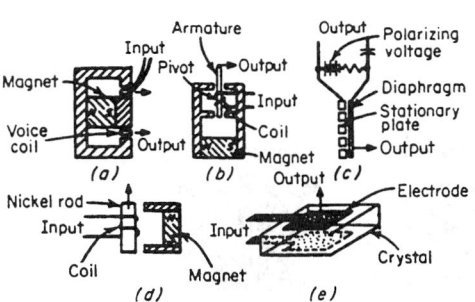

FIGURE 20.3.12 Loudspeaker (and earphone) transduction methods: (*a*) moving-coil; (*b*) moving-iron; (*c*) electrostatic; (*d*) magnetostrictive; (*e*) piezoelectric.[4]

Moving-Coil. The mechanical force on the moving coil of Fig. 20.3.12*a* is developed by the interaction of the current in the coil and the transverse magnetic field disposed radially across the gap between the magnet cap and the iron housing, which completes the magnetic circuit. The output force along the axis of the circular coil is applied to a sound radiator.

Moving-iron transducers reverse the mechanical roles of the coil and the iron. The iron armature surrounded by the stationary coil is moved by mechanical forces developed within the magnetic circuit. Moving-iron magnetic circuits have many forms. As an example in the balanced armature system (Fig. 20.3.12*b*) the direct magnetic flux passes only transversely through the ends of the armature centered within the two magnetic gaps. Coil current polarizes the armature ends oppositely, creating a force moment about the pivot point. The output force is applied from the tip of the armature to an attached sound radiator. In a balanced-diaphragm loud-speaker the armature is the radiator.

Electrostatic.   In the electrostatic transducer (Fig. 20.3.12*e*) there is a dc potential difference between the con-ductive diaphragm and the stationary perforated plate nearby. Audio signals applied through a blocking capac-itor superimpose an alternating potential, resulting in a force upon the diaphragm, which radiates sound directly.

Magnetostrictive transducers (Fig. 20.3.12*d*) depend on length fluctuations of a nickel rod caused by vari-ations in the magnetic field. The output motion may be radiated directly from the end of the rod or transmit-ted into the attached mechanical structure.

Piezoelectric transducers are of many forms using crystals or polycrystalline ceramic materials. In simple form (Fig. 20.3.12*e*) an expansion-contraction force develops along the axis joining the electrodes through alternation of the potential difference between them.

Sound Radiators

The purpose of a sound radiator is to create small, audible air-pressure variations. Whether they are produced within a closed space by an earphone or in open air by a loudspeaker, the pressure variations require air motion or current.

Pistons, Cones, Ports.   Expansion and contraction of a sphere is the classical configuration but most practi-cal examples involve rectilinear motion of a piston, cone, or diaphragm. In addition to the primary direct radi-ation from moving surfaces, there is also indirect or secondary radiation from enclosure ports or horns to which the direct radiators are acoustically coupled.

Attempts have been made to develop other forms of sound radiation such as oscillating airstreams and other aerodynamic configurations with incidental use, if any, of moving mechanical members.

Directivity.   Figure 20.3.13 shows the directional characteristics of a rigid circular piston for different ratios of piston diameter and wavelength of sound. (In three dimensions these curves are symmetrical around the axis of piston motion.) For a diameter of one-quarter wavelength the amplitude decreases 10 percent (approximately

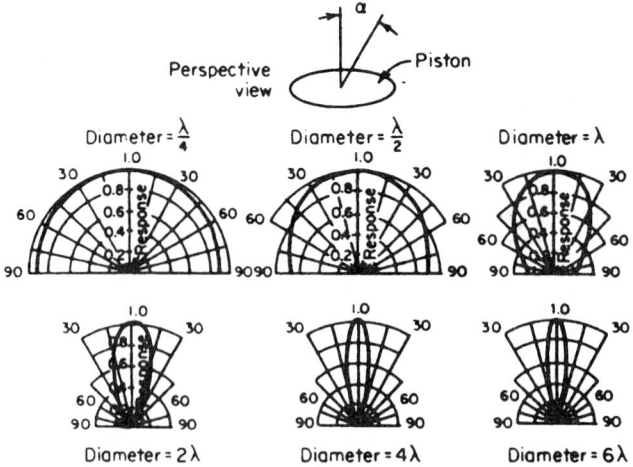

FIGURE 20.3.13   Directional characteristics of rigid circular pistons of differ-ent diameters or at different sound wavelengths.[2]

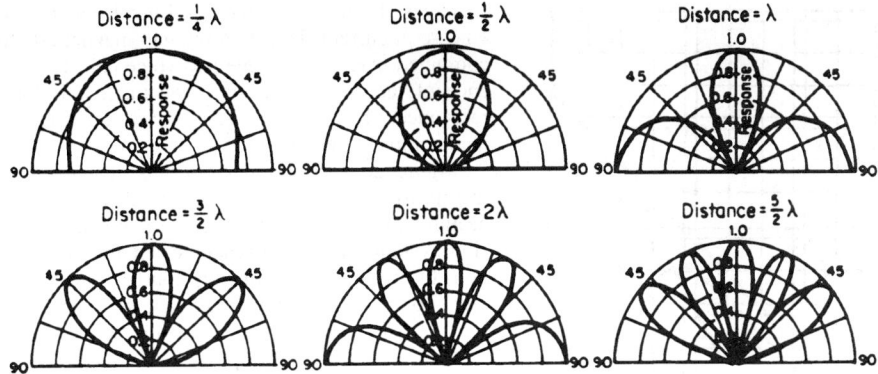

FIGURE 20.3.14 Directional characteristics of two equal small in-phase sound sources separated by different distances or different sound wavelengths.[2]

1 dB sound level) at 90° off axis. For a four-wavelength diameter the same drop occurs in only 5°. (The beam of an actual loudspeaker cone is less sharp than this at high frequencies, where the cone is not rigid.) Note that all the polar curves are smooth when the single-source piston vibrates as a whole.

Radiator Arrays. When two separate, identical small-sound sources vibrate in phase, the directional pattern becomes narrower than for one source. Figure 20.3.14 shows that for a separation of one-quarter wavelength the two-source beam is only one-half as wide as for a single piston. At high frequencies the directional pattern becomes very complex. (In three dimensions these curves become surfaces of revolution about the axis joining the two sources.)

Arrays of larger numbers of sound radiators in close proximity are increasingly directional. Circular-area arrays have narrow beams which are symmetrical about an axis through the center of the circle. Line arrays, e.g., column loudspeakers, are narrowly directional in planes containing the line and broadly directional in planes perpendicular to the line.

Direct-Radiator Loudspeakers

Most direct-radiator loudspeakers are of the moving-coil type because of simplicity, compactness, and inherently uniform response-frequency trend. The uniformity results from the combination of two simple physical principles: (1) the radiation resistance increases with the square of the frequency, and hence the radiated sound power increases similarly for constant velocity amplitude of the piston or cone; (2) for a constant applied force (voice-coil current) the mass-controlled (above resonance) piston has a velocity amplitude which decreases with the square of the frequency. Consequently a loudspeaker designed to resonate at a low frequency combines decreasing velocity with increasing radiation resistance to yield a uniform response within the frequency range where the assumptions hold.

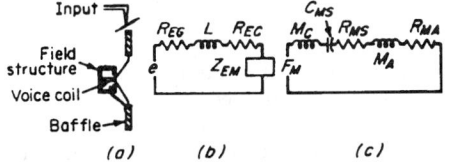

FIGURE 20.3.15 (*a*) Structure, (*b*) electric circuit, and (*c*) equivalent mechanical circuit for a direct-radiator moving-coil loudspeaker in a baffle.[4]

Equivalent Electric Circuits. Figure 20.3.15 shows a cross-sectional view of a direct-radiator loudspeaker mounted in a baffle, the electric voice-coil circuit, and the equivalent electric circuit of the mechanoacoustic system. In the voice-coil circuit e is the emf and R_{EG} the resistance of the generator, e.g., power-amplifier output, L and R_{EC} are the inductance and resistance of the voice coil. Z_{EM} is the motional electric impedance from the mechanoacoustic system.

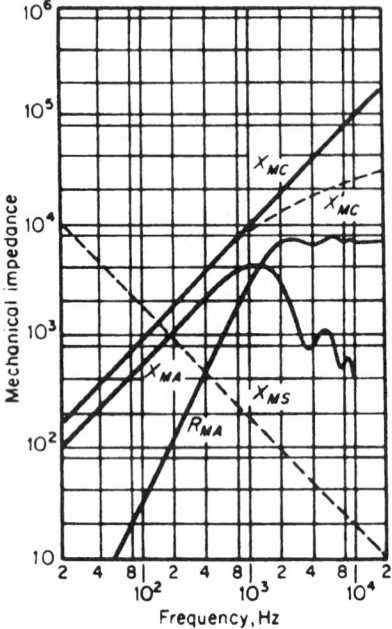

FIGURE 20.3.16 Components of a mechanical impedance of a typical 4-in loudspeaker.[4]

F_M is the driving force resulting from interaction of the voice-coil current field with the gap magnetic field. M_C is the combined mass of the cone and voice coil. C_{MS} is the compliance of the cone-suspension system. R_{MS} is the mechanical resistance. The mass M_A and radiation resistance R_{MA} of the air load complete the circuit.

Figure 20.3.16 summarizes these mechanical impedance factors for a 4-in direct-radiator loudspeaker of conventional design. Above resonance (where the reactance of the suspension system equals the reactance of the cone-coil combination) the impedance-frequency characteristic is dominated by M_C. From the resonance frequency of about 150 Hz to about 1500 Hz the conditions for uniform response hold.

Efficiency.   Since R_{MA} is small compared to the magnitudes of the reactive components, the efficiency of the loudspeaker in this frequency range can be expressed as

$$\text{Efficiency} = \frac{100(Bl)^2 R_{MA}}{R_{EC}(X_{MA} + X_{MC})^2} \text{ percent} \qquad (1)$$

where B = gap flux density (G)
  l = voice-coil conductor length (cm)
  R_{EC} = voice-coil electric resistance (abohms)

Since R_{MA} is proportional to the square of the frequency and both X_{MA} and X_{MC} increase with frequency, the efficiency is theoretically uniform.

All this has assumed that the cone moves as a whole. Actually wave motion occurs in the cone. Consequently at high frequencies the mass reactance is somewhat reduced (as shown in the dashed curve of Fig. 20.3.16), tending to improve efficiency beyond the frequency where radiation resistance becomes uniform.

Magnetic Circuit.   Most magnets now are a high-flux, high-coercive permanent type, either an alloy of aluminum, cobalt, nickel, and iron, or a ferrite of iron, cobalt, barium, and nickel. The magnet may be located in the core of the structure or in the ring, or both. However, magnetization is difficult when magnets are oppositely polarized in the core and ring.

Air-gap flux density varies widely in commercial designs from approximately 3000 to 20,000 G. Since most of the reluctance in the magnetic circuit resides in the air gap, the minimum practical voice-coil clearance in the gap compromises the maximum flux density. Pole pieces of heat-treated soft nickel-iron alloys, dimensionally tapered near the gap, are used for maximum flux density.

Voice Coils.   The voice coil is a cylindrical multilayer coil of aluminum or copper wire or ribbon. Aluminum is used in high-frequency loudspeakers for minimum mass and maximum efficiency. Voice-coil impedance varies from 1 to 100 Ω with 4, 8, and 16 Ω standard. For maximum efficiency the voice-coil and cone masses are equal. However, in large loudspeakers the cone mass usually exceeds the voice-coil mass. Typically the voice-coil mass ranges from tenths of a gram to 5 g or more.

Cones.   Cone diameters range from 1 to 18 in. Cone mass varies from tenths of a gram to 100 g or more. Cones are made of a variety of materials. The most common is paper deposited from pulp on a wire-screen form in a felting process. For high-humidity environment cones are molded from plastic materials, sometimes with a cloth or fiber-glass base. Some low-frequency loudspeaker cones are molded from low-density plastic foam to achieve greater rigidity with low density.

So far piston action has been assumed in which the cone moves as a whole. Actually at high frequencies the cone no longer vibrates as a single unit. Typically there is a major dip in response resulting from quarter-wave

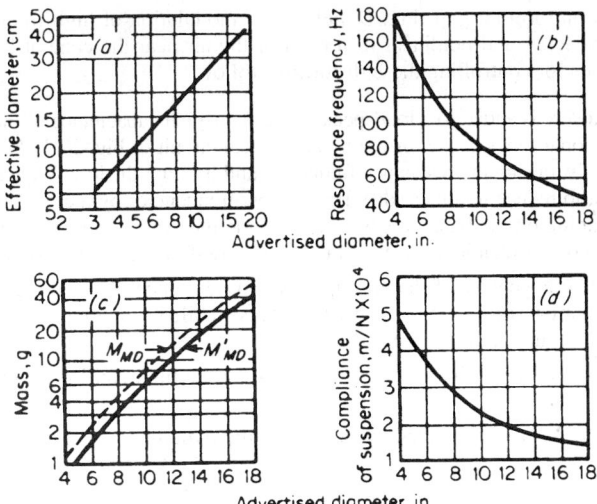

FIGURE 20.3.17   Typical cone and coil design values.[5]

reflection from the circular rim of the cone back to the voice coil. For loudspeaker cones in the range of 8 to 15 in. diameter this dip usually occurs in the range of 1 to 2 kHz.

Typical Commercial Design Values.   Figure 20.3.17 shows typical values for several cone and voice-coil design parameters for a range of loudspeaker diameters. These do not apply to extreme cases, such as high-compliance loudspeakers or high-efficiency horn drivers. The effective piston diameter (Fig. 20.3.17a) is less than the loudspeaker cone diameter because the amplitude falls off toward the edges. A range of resonance frequencies is available for any cone diameter, but Fig. 20.3.17b shows typical values. In Fig. 20.3.17c typical cone mass is M including the voice coil and M' excluding the voice coil. Figure 20.3.17d shows typical cone-suspension compliance.

Impedance.   A major peak results from motional impedance at primary mechanical resonance. Impedance is usually uniform above this peak until voice-coil inductance becomes dominant over resistance.

Power Ratings.   Different types of power rating are needed to express the performance capabilities of loudspeakers. The large range of typical loudspeaker efficiency makes the acoustical power-delivering capacity quite important. The electrical power-receiving capacity (without overload or damage) determines the choice of power amplifier.

Loudspeaker efficiencies are seldom measured but are often compared by measuring the sound-pressure level at 4 ft on the loudspeaker axis for 1-W audio input. High-efficiency direct radiators provide 95 to 100 dB. Horn loudspeakers are typically higher by 10 dB or more, being both more efficient and more directional.

Loudspeakers are also rated by the maximum rms power output of amplifiers which will not damage the loudspeaker or drive it into serious distortion on peaks. Such ratings usually assume that the amplifier will seldom be driven to full power. For example, a 30-W amplifier will seldom be required to deliver more than 10 W rms of music program material. Otherwise music peaks would be clipped and sound distorted.

However, in speech systems for high-ambient-noise levels the speech peaks may be clipped intentionally, causing the loudspeaker to receive the full 30 W much of the transmission time. Then the loudspeaker must handle large excursions without mechanical damage to the cone suspension and without destroying the cemented coil or charring the form.

Distortion.   Nonlinear distortion in a loudspeaker is inherently low in the mass-controlled range of frequencies. However, distortion is produced by nonlinear cone suspension at low frequencies, voice-coil motion beyond the limits of uniform air-gap flux, Doppler shift modulation of high-frequency sound by large cone velocity at low frequencies, and nonlinear distortion of the air near the cone at high powers (particularly in horn drivers). Methods for controlling these distortions follow.

1. When a back enclosure is added to a loudspeaker, the acoustical capacitance of the enclosed volume is represented by an additional series capacitor in the mechanical circuit of Fig. 20.3.15. Insufficient volume stiffens the cone acoustically, raising the resonance frequency and limiting the low-frequency range of the loudspeaker. It is convenient to reduce nonlinear distortion at low frequencies by increasing the cone-suspension compliance and depending on the back enclosure to provide the system stiffness. Since an enclosed volume is more linear than most mechanical springs, this lowers low-frequency distortion.

2. Distortion from inhomogeneity of the air-gap flux can be reduced by making the voice-coil length either considerably smaller or larger than the gap width. This stabilizes the total number of lines passing through the coil, but it also reduces loudspeaker efficiency.

3. Doppler distortion can be eliminated only by separating the high and low frequencies in a multiple loudspeaker system.

4. Air-overload distortion can be avoided by increasing the radiating area.

Loudspeaker Mountings and Enclosures

Figure 20.3.18 shows a variety of mountings and enclosures. An unbaffled loudspeaker is an acoustic doublet for wavelengths greater than the rim diameter. In this frequency range the acoustical power output for constant cone velocity is proportional to the fourth power of the frequency.

Baffles.   In order to improve efficiency at low frequencies it is necessary to separate the front and back waves. Figure 20.3.18a is the simplest form of baffle. The effect of different baffle sizes is given in Fig. 20.3.19. Response dips occurring when the acoustic path from front to back is a wavelength are eliminated by irregular baffle shape or off-center mounting.

Enclosures.   The widely used open-back cabinet (Fig. 20.3.18b) is noted for a large response peak produced by open-pipe acoustical resonance. A closed cabinet (Fig. 20.3.18c) adds acoustical stiffness at low frequencies where the wavelength is larger than the enclosure. At higher frequencies the internal acoustical resonances create response irregularities requiring internal acoustical absorption.

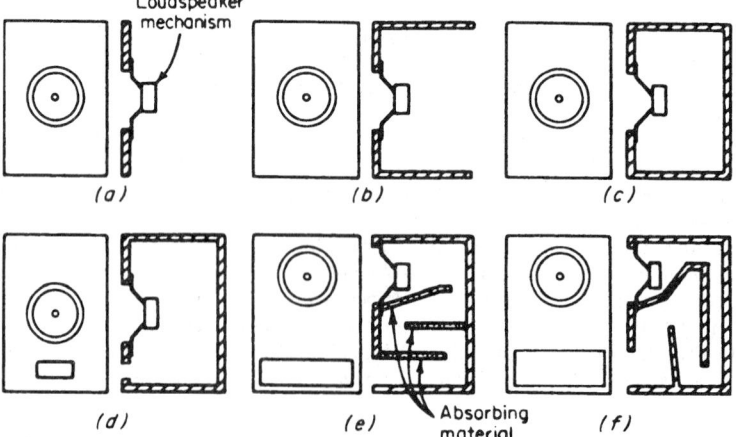

FIGURE 20.3.18   Mountings and enclosures for direct-radiator loudspeaker: (*a*) flat baffle; (*b*) open-back cabinet; (*c*) closed cabinet; (*d*) ported closed cabinet; (*e*) labyrinth; (*f*) folded horn.[4]

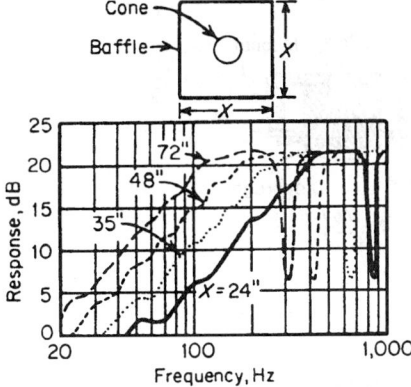

FIGURE 20.3.19   Response frequency for loud-speaker in 2-, 3-, 4-, and 6-ft square baffles.[6]

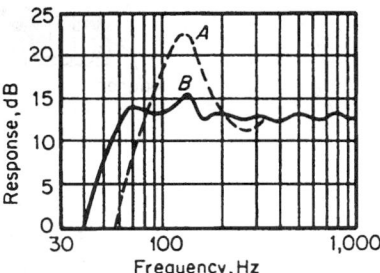

FIGURE 20.3.20   Response frequency for loud-speaker in closed (*A*) and ported (*B*) cabinets.[6]

Ported Enclosures (Fig. 20.3.18d).   Enclosure volume can be minimized without sacrificing low-frequency range by providing an appropriate port in the enclosure wall. Acoustical inertance of the port should resonate with the enclosure capacitance at a frequency about an octave below cone-resonance frequency. *B*, Fig. 20.3.20, shows that this extends the low-frequency range. This is most effective when the port area equals the cone-piston area. Port inertance can be increased by using a duct. An extreme example of ducting is the acoustical labyrinth (Fig. 20.3.18*e*). When duct work is shaped to increase cross section gradually, the labyrinth becomes a low-frequency horn (Fig. 20.3.18*f*).

Direct-radiator loudspeaker efficiency is typically 1 to 5 percent. Small, highly damped types with miniature enclosures may be only 0.1 percent. Transistor amplifiers easily provide the audio power for domestic loudspeakers. However, in auditorium, outdoor, industrial, and military applications much higher efficiency is required.

Horn Loudspeakers

Higher efficiency is obtained with an acoustic horn, which is a tube of varying cross section having different terminal areas to provide a change of acoustic impedance. Horns match the high impedance of dense diaphragm material to the low air impedance. Horn shape or taper affects the acoustical transformer response. Conical, exponential, and hyperbolic tapers have been widely used. The potential low-frequency cutoff of a horn depends on its taper rate. Impedance transforming action is controlled by the ratio of mouth to throat diameter.

Horn Drivers.   Figure 20.3.21 shows horn-driving mechanisms and straight and folded horns of large- and small-throat types. A large-throat driver (Fig. 20.3.21*a*) resembles a direct-radiator loudspeaker with a voice-coil diameter of 2 to 3 in. and a flux density around 15,000 G. A small-throat driver (Fig. 20.3.21*b*) resembles a moving-coil microphone structure. Radiation is taken from the back of the diaphragm into the horn throat through passages which deliver in-phase sound from all diaphragm areas. Diaphragm diameters are 1 to 4 in. with throat diameters of $^1/_4$ to 1 in. Flux density is approximately 20,000 G.

Large-Throat Horns.   These are used for low-frequency loudspeaker systems. A folded horn (Fig. 20.3.21*c*) is preferred over a straight horn (Fig. 20.3.21*d*) for compactness.

Small-Throat Horns.   A folded horn (Fig. 20.3.21*e*) with sufficient length and gradual taper can operate efficiently over a wide frequency range. This horn is useful for outdoor music reproduction in a range of 100 to 5000 Hz. Response smoothness is often compromised by segment resonances. Extended high-frequency range requires a straight-axis horn (Fig. 20.3.21*f*).

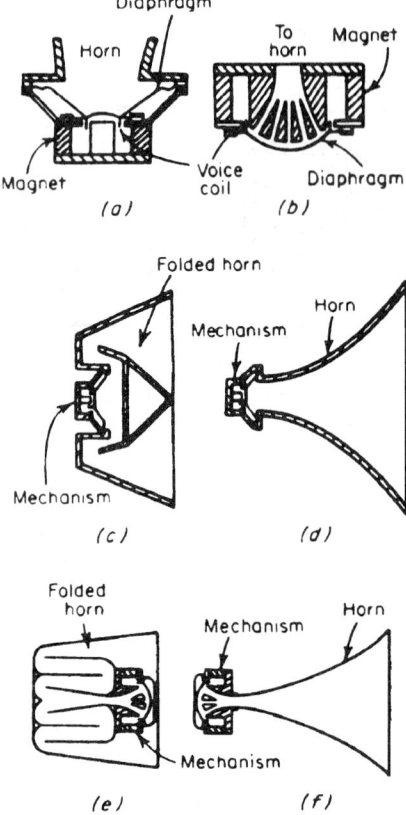

FIGURE 20.3.21   Horns and horn drivers: (*a*) large-throat driver; (*b*) small-throat driver; (*c*) folded large-throat horn; (*d*) straight large-throat horn; (*e*) folded small-throat horn; (*f*) straight small-throat horn.[4]

Horn Directivity.   Large-mouth horns of simple exponential design produce high-directivity radiation that tends to narrow with increasing frequency (as in Fig. 20.3.13). In applications requiring controlled directivity over a broad angle and a wide frequency range, a horn array (shown in Fig. 20.3.22*a*) can be used, with numerous small horn mouths spread over a spherical surface and throats converging together. Figure 20.3.22*b* shows the directional characteristics. Single sectoral horns with radial symmetry can provide cylindrical wavefronts with smoother directional characteristics which are controlled in one plane. Recent rectangular or square-mouth "quadric" horns, designed by computer to have different conical expansion rates in horizontal and vertical planes, provide controlled directivity in both planes over a wide frequency range.

Special Loudspeakers

Special types of loudspeakers for limited applications include the following.

Electrostatic high-frequency units have an effective spacing of about 0.001 in. between a thin metalized coating on plastic and a perforated metal backplate. This spacing is necessary for sensitivity comparable to moving-coil loudspeakers, but it limits the amplitude and the frequency range. Extension of useful response to the lower frequencies can be obtained with larger spacing, for example, $1/16$ in., with a polarizing potential of several thousand volts. This type of unit employs push-pull operation.

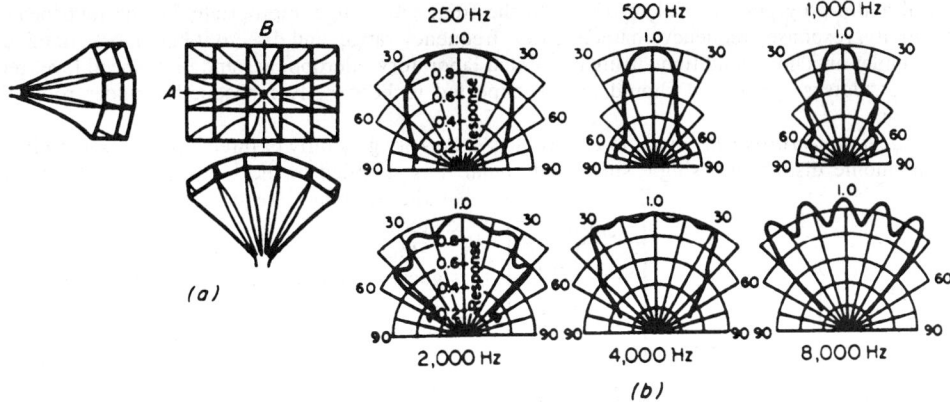

FIGURE 20.3.22 Horn array (cellular) and directional characteristics: (*a*) array; (*b*) horizontal directional curves.[6]

Modulated-airflow loudspeakers have an electromechanical mechanism for modulating the airstream from a high-pressure pneumatic source into a horn. Low audio power controls large acoustical power in this system. A compressor is also needed. Nonlinear distortion in the air and reduced speech intelligibility have been limitations of this high-power system.

Loudspeaker Specifications and Measurements

Typical loudspeaker specifications are shown in Table 20.3.1 for a variety of loudspeaker types.

Loudspeaker impedance is proportional to the voltage across the voice coil when driven by a high-impedance constant-current source. Continuous power ratings are obtained from sustained life tests with

TABLE 20.3.1   Characteristics of a Variety of Loudspeaker Types

Company	Altec	Altec	Bozak	RCA
Model no.	775C	1505B horn 290D driver	CM-109–23	LC1B
Type	Direct radiator	Cellular horn (3 × 5)	Three-way column	Duo-cone
Sensitivity (at 4 ft for 1 W), dB	95	110	106	95
Frequency range, Hz	40–15,000	300–8,000	65–13,000	25–16,000 (± 4 dB)
Impedance, Ω	8	4	8	15
Power rating, W	15	100	200	20
Distribution angle, deg	90	105 horizontal 60 vertical	90 horizontal 30 vertical	120
Voice-coil diameter, in.	2	2.8	(3 sizes)	(2 cones)
Cone resonance, Hz	52	…	(3 sizes)	22
Crossover frequency, Hz	…	500	800, 2,500	1,600
Diameter, in.	$8^{3}/_{8}$	$18^{1}/_{2}$ high $30^{1}/_{2}$ wide	57 in. high $22^{3}/_{4}$ wide	17
Depth, in.	$2^{1}/_{4}$	30	$15^{3}/_{4}$	$7^{1}/_{2}$
Weight, lb	$3^{3}/_{4}$	43	250	21

typical audio-program material restricted to the frequency range appropriate for the loudspeaker type. Sensitivity, response-frequency characteristics, frequency range, and directivity are most effectively measured under anechoic conditions using calibrated laboratory microphones and high-speed level recorders. However, data so measured should not be expected to be exactly reproducible under room-listening conditions.

Distortion measurements in audio-electronic systems are generally of three types shown in Fig. 20.3.23. For harmonic distortion a single sinusoidal signal A is supplied to the loudspeaker and wave analysis at the harmonic frequencies determines the percent distortion.

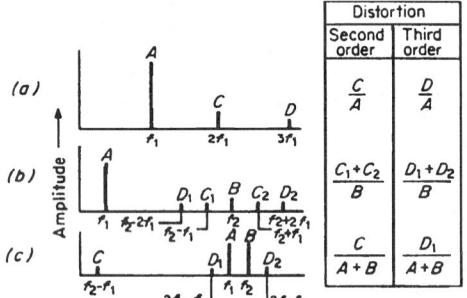

FIGURE 20.3.23 Methods of measuring nonlinear distortion: (*a*) harmonic; (*b*) intermodulation method of SMPTE; (*c*) intermodulation method of CCIF.[7]

Both intermodulation methods supply two sinusoidal signals of different frequency to the loudspeaker. In the older Society of Motion Picture and Television Engineers (SMPTE) method the frequencies are widely separated, and the distortion is expressed in terms of sum and difference frequencies around the higher test frequency. This method is meaningful for wide-range loudspeaker systems.

The CCIF (International Telephone Consultative Committee) method is more applicable to narrow-range systems and loudspeakers receiving input at high frequencies. It supplies two high frequencies to the loudspeaker and checks the low difference frequency.

Transient intermodulation distortion, resulting from nonlinear response to steep wavefronts, is measured by adding square-wave (3.18-kHz) and sine-wave (15-kHz) inputs, with a 4:1 amplitude ratio, and observing the multiple sum- and difference-frequency components added to the output spectrum.

EARPHONES

The transduction methods are the same as for loudspeakers. Telephone and hearing aid receivers are usually moving-iron. Most military headsets are now moving-coil. Piezoelectric, moving-coil, and electrostatic types are used for listening to recorded music.

Equivalent Electric Circuits

Figure 20.3.24 shows a cross section of a moving-coil earphone and the equivalent electric circuit. The voice-coil force drives the voice coil and diaphragm. (Mechanical resonance of earphone diaphragms occurs at a high audio frequency in contrast to loudspeakers.) Diaphragm motion creates sound pressure in several spaces behind the diaphragm and the voice coil and between the diaphragm and the earcap. Inertance and resistance of the connecting holes and clearances combine with the capacitance of the spaces to add acoustical resonances. Z is the acoustical impedance of the ear.

Idealized Ear Loading

The ear is approximately an acoustical capacitance. However, acoustical leakage adds a parallel resistance-inertance path affecting low-frequency response. At high frequencies the ear canal-length resonance is a factor.

Since the ear is a capacitance, the goal of earphone design is a constant diaphragm amplitude throughout the frequency range. This requires a stiffness-controlled system or a high-resonance frequency. The potential across the ear is analogous to sound pressure within the ear cavity. This sound pressure is proportional to

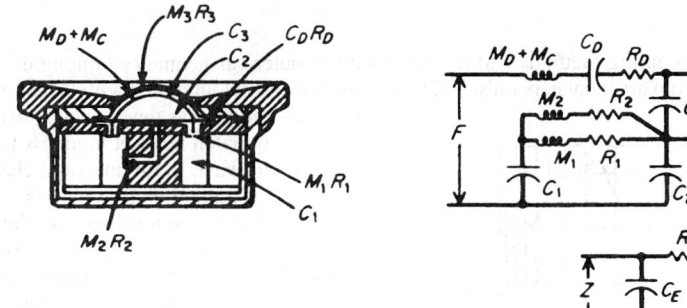

FIGURE 20.3.24 Moving-coil earphone cross section and equivalent electric circuit.[8]

diaphragm area and inversely proportional to enclosed volume. Earphone loading conditions are extremely varied for different types of earphone mountings.

Earphone Mountings

The most widely used earphone is the single receiver unit on a telephone handset. It is intended to be held against the ear but is often tilted away, leaving considerable leakage.

Headsets provide better communication than handsets because they supply sound to both ears and shield them.

A remote earphone can drive the ear canal through a small acoustic tube. The length may be an inch or two for hearing aids and several feet for music listening on aircraft.

Efficiency, Impedance, and Driving Circuits

Moving-iron earphones and microphones can be made efficient enough to operate as sound-powered (batteryless) telephones. Efficient magnet structures, minimum mechanical and acoustical damping, and minimum volume of acoustical coupling are required for this purpose. In some earphone applications overall efficiency is less critical, and wearer comfort is important.

Insert earphones need less efficiency than external earphones because the enclosed volume is much smaller; however, they require moderate efficiency to save the amplifier batteries.

Circumaural earphones are frequently driven by amplifiers otherwise used for loudspeakers. Here efficiency is less important than power-delivering capacity.

Typically 1 mW of audio power to an earphone will produce 100 to 110 dB in a standard 6-cm^3 coupler. The same earphone will produce less sound level in an earmuff than in an ear cushion and more when coupled to an ear insert.

The shape of the enclosed volume also affects response. The farther the driver is from the eardrum the lower the frequency of standing-wave resonance. Small tube diameters produce high-frequency attenuation.

The response-frequency characteristic of moving-iron or piezoelectric earphones is quite dependent on source impedance. A moving-iron earphone having uniform response when driven at constant power will have a rising response (with increasing frequency) at constant current and a falling response at constant voltage (Fig. 20.3.25).

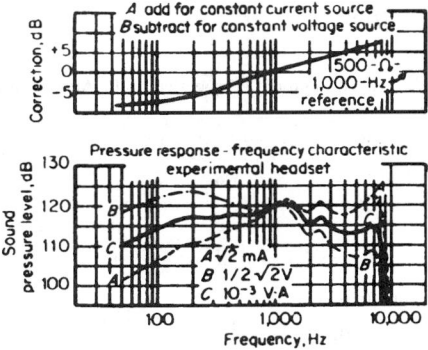

FIGURE 20.3.25 Effect of source impedance upon earphone response curve: (*a*) constant current; (*b*) constant voltage; (*c*) constant power.[9]

Real-Ear Response

The variety of earphone-coupling methods and the variability of outer-ear geometry (among different listeners) make response data from artificial ears only indicative, not definitive. Out of necessity a real-ear response-measuring technique was developed. A listener adjusts headset input to match headset loudness to an external calibrated sound wave in an anechoic chamber. From matching data at numerous frequencies an equivalent free-field sound-pressure level can be plotted for constant input to the earphone. This curve usually differs from a sound-level curve on a simple earphone coupler. The reason is that probe measurements of sound at the eardrum and outside the ear in a free field differ because of ear amplification and diffraction about the head (Fig. 20.3.26).

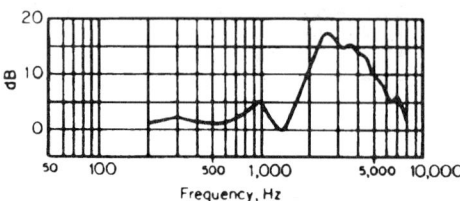

FIGURE 20.3.26 Relative level of sound pressures at the listener's eardrum and in the free sound field.[10]

Acoustic attenuation by earphones can be measured either by threshold shift or by matching the loudness of tones heard from an external loudspeaker, with and without the headset on. The sound-level difference is plotted as attenuation in decibels.

Monaural, Diotic, and Binaural Listening

A handset earphone provides monaural listening. Diotic listening with the same audio signal in both earphones localizes sound within the head. This is not unpleasant and may actually be an aid to concentration. In natural-binaural listening the ears receive sound differently from the same source unless it is directly on the listening axis. Usually there are differences in phase, arrival time, and spectrum (because of diffraction about the head).

Recordings provide true binaural effects only if the two recording microphones are on an artificial head. Stereophonic microphones are usually separated much farther, so that headset listening gives an exaggerated effect. For some listeners this is an enhancement.

REFERENCES

1. Bauer, B. B. *Proc. IRE*, 1962, Vol. 50, 50th Anniversary Issue, p. 719.

2. Olson, H. F. "Musical Engineering," McGraw-Hill, 1952.

3. Gander, M. R. Fifty Years of Loudspeaker Developments as Viewed Through the Perspective of the Audio Engineering Society, *J. Audio Eng. Soc.*, 1998, Vol. 46, No. 1/2, pp. 43–58.

4. Olson, H. F. *Proc. IRE*, 1962, Vol. 50, 50th Anniversary Issue, p. 730.

5. Beranek, L. L. "Acoustics," Acoustical Society of America, 1986.

6. Olson, H. F. "Elements of Acoustical Engineering," Van Nostrand, 1947.

7. Beranek, L. L. *Proc. IRE*, 1962, Vol. 50, p. 767.

8. Anderson, L. J. *J. Soc. Motion Pict. Eng.*, 1941, Vol. 37, p. 319.

9. Martin, D. W., and L. J. Anderson *J. Acoust. Soc. Am.*, 1947, Vol. 19, p. 63.

10. Wiener, F. M., and D. A. Ross *J. Acoust. Soc. Am.*, 1946, Vol. 18, p. 401.

DIGITAL AUDIO RECORDING AND REPRODUCTION

Daniel W. Martin, Ronald M. Aarts

INTRODUCTION

A digital revolution has occurred in audio recording and reproduction that has made some previous techniques only of historical interest. Although analog recording and reproduction systems have been greatly improved (Fig. 20.4.1), their capabilities are still short of ideal. For example, they could not provide the dynamic range of orchestral instrument sounds (e.g., from 42 dB on a soft low flute note to 120 dB for a bass drum peak), plus a reasonable ratio of weakest signal to background noise. Mechanical analog records are still limited by inherent nonlinear distortions as well as surface noise, and magnetic analog recording is limited by inherent modulation noise.

Digital audio signal transmission, recording, and playback have numerous potential advantages, which, with appropriate specifications and quality control, can now be realized as will now be shown.

DIGITAL ENCODING AND DECODING

There is much more to digital audio than encoding the analog signal and decoding the digital signal, but this is basic. The rest would be largely irrelevant if it were not both advantageous and practically feasible to convert analog audio signals to digital for transmission, storage, and eventual retrieval.

A digital audio signal is a discrete-time, discrete-amplitude representation of the original analog audio signal. Figure 20.4.2 is a simple encoding example using only 4 bits. The amplitude of the continuous analog audio signal wavetrain A is sampled at each narrow pulse in the clock-driven pulse train B, yielding for each discrete abscissa (time) value a discrete ordinate (voltage) value represented by a dot on or near the analog curve. The vertical scale is subdivided (in this example) into 16 possible voltage values, each represented by a binary number or "word." The first eight words can be read out either in parallel

$$1000, 1010, 1011, 1011, 1010, 1000, 0110, 0101, \ldots$$

on four channels, or in sequence

$$1000101010111011101010000110 0101 \ldots$$

on a single channel for transmission, optional recording and playback, and decoding into an approximation of the original wavetrain. Unless intervening noise approaches the amplitude of the digit 1, the transmitted or played-back digital information matches the original digital information.

The degree to which digitization approximates the analog curve is determined by the number of digits chosen and the number of samplings per second. Both numbers are a matter of choice, but the present specifications for digital audio systems generally use 16 bits for uniform quantization (65,536 identifiable values),

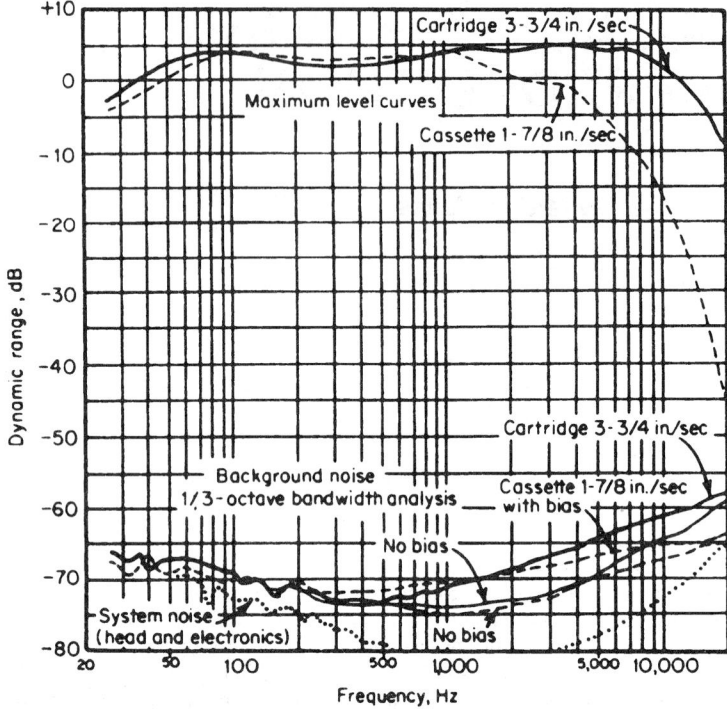

FIGURE 20.4.1 Dynamic range of analog tape cartridges and cassettes. (*After Ref.* 1)

corresponding to a theoretical dynamic range of 16(6 dB) = 96 dB. The sampling frequency, according to the Nyquist criterion, must be at least twice the highest audio frequency to be transmitted or recorded. Three different sampling frequencies are being used, 48 kHz "for origination, processing, and interchange of program material"; 44.1 kHz "for certain consumer applications"; and 32 kHz "for transmission-related applications."

Figure 20.4.3 shows the main electronic blocks of a 5-bit digital system for encoding and decoding audio signals for various transmitting and receiving purposes. The digital audio signal may be transmitted and

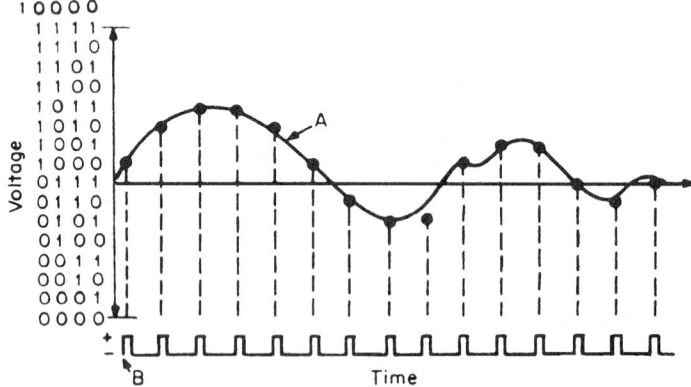

FIGURE 20.4.2 Digital encoding of an analog waveform: (*a*) continuous analog signal wavetrain; (*b*) clock-driven pulse train. At equal time intervals, sample values are encoded into nearest digital word.

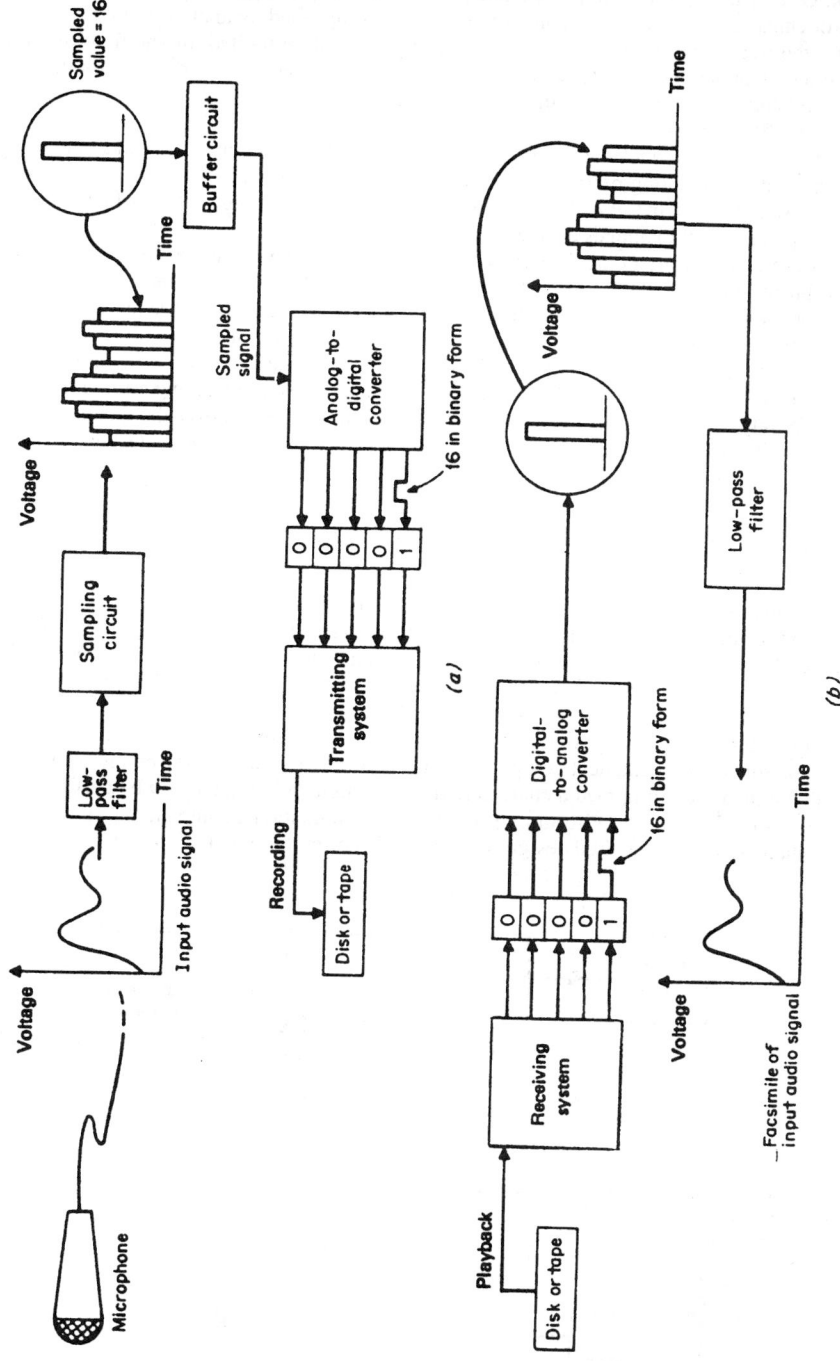

FIGURE 20.4.3 The basic electronic system components for encoding and decoding digital audio signals for (*a*) transmitting (or recording) and (*b*) receiving (or play-back). (Ref. 2)

received conductively or electromagnetically. Alternatively, it may be stored and later retrieved by recording and playback. Or transmission may simply be into and out of a digital signal processing system, which purposely alters or enhances the signal in a manner not easily accomplished by analog means.

In any case the frequency range of the analog input signal must be limited, by the first low-pass filter of Fig. 20.4.3, to less than one-half the sampling frequency. For 16-bit accuracy in digitization this filter probably needs a stop-band attenuation greater than 60 dB, a signal-to-noise ratio of 100 dB, bandpass ripple less than 0.2 dB, and differential nonlinearity less than 0.0075 percent.

The next block is a sample-and-hold analog circuit which tracks the input voltage and samples it during a very short portion of the sampling period; it then holds that value during the remainder of the sampling period until the next sampling begins. Possible faults in sampling include timing "jitter," which adds modulation noise, and "droop" in the held voltages during digitization.

The analog-to-digital converter quantizes each of the succession of held voltages shown and turns them into a sequence of binary numbers, the first of which (1000, corresponding to 16) is shown at the outputs of the converter. For practical reasons the parallel output information from the converter is put into sequential form in the transmitting system by multiplexing, for example, before transmission or recording occurs.

Demultiplexing in the receiving system puts the data back into parallel form for digital-to-analog conversion. Possible faults in the conversion include gain errors, which increase quantizing error, and nonlinearity or relative nonuniformity, which cause distortion. The second low-pass filter removes the scanning frequency and its harmonics, which, although inaudible themselves, can create audible distortion in the analog output system.

TRANSMISSION AND RECEPTION OF THE DIGITAL AUDIO SIGNAL

As previously stated, other digital functions and controls are required to assist in the encoding and decoding. Figure 20.4.4 complements Fig. 20.4.3 by showing in a block diagram that analog-to-digital (A/D) conversion and transmission (or storage) have intervening digital processing and that all three are synchronized under digital clock control. In reception (or playback), equivalent digital control is required for reception, digital reprocessing, and digital-to-analog (D/A) conversion. Examples of these functions and controls are multiplexing of the A/D output, digital processing to introduce redundancy for subsequent error detection and correction, servo system control when mechanical components are involved, and digital processing to overcome inherent transmission line or recording media characteristics. The digitization itself may be performed in any of a number of ways: straightforward, uniform by successive approximations, companding, or differential methods such as delta modulation. Detailed design of much of this circuitry is in the domain of digital-circuit and

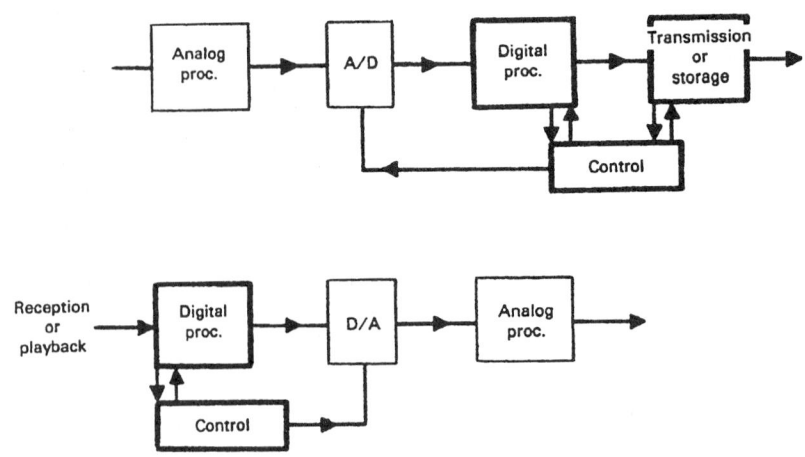

FIGURE 20.4.4   Block diagram of the basic functions in a digital audio system.

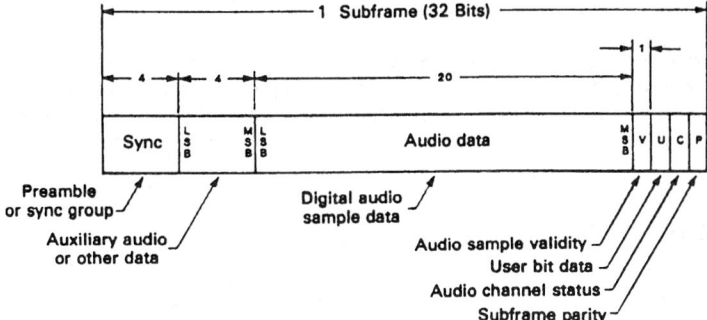

FIGURE 20.4.5 Subframe format recommended for serial transmission of linearly represented digital audio data.

integrated-circuit engineering, beyond the scope of this chapter. However, the audio system engineer is responsible for the selection, specification, and (ultimately) standardization of sampling rate, filter cutoff frequency and rate, number of digits, digitization method, and code error-correction method, in consultation with broadcast, video, and transmission engineers with whose systems compatibility is necessary.

Compatibility should be facilitated by following the "Serial Transmission Format for Linearly Represented Digital Audio Data" recommended by the Audio Engineering Society, in which digital audio sample data within a subframe (Fig. 20.4.5) is accompanied by other data bits containing auxiliary information needed for functions and controls such as those listed above. Two 32-bit subframes in sequence, one for each channel (of stereo, for example), comprise a frame transmitted in any one period of the sampling frequency. A channel (or modulation) code of the biphase mark, self-clocking type is applied to the data prior to transmission, in order to embed a data-rate clock signal which enables correct operation of the receiver. In this code all information is contained in the transitions, which simplifies clock extraction and channel decoder synchronization.

The audio signal data may occupy either 20 or 24 bits of the subframe, preceded by 4 bits of synchronizing and identifying preamble for designating the start of a frame and block, or the start of the first or the second subframe. If the full 24 bits are not needed for the audio sample, the first four can be auxiliary audio data.

Following the audio data are four single bits that indicate (V) whether the previous audio sample data bits are valid; (U) any information added for assisting the user of the data; (C) information about system parameters; and (P) parity for detection of transmission errors for monitoring channel reliability.

Within the audio field it is the Audio Engineering Society (AES) that has determined many standards. Among these there are a few digital interconnect standards [http://www.aes.org/standards/]. The AES30-1985 document has formed the basis for the international standards documents concerning a two-channel digital audio interface. The society has been instrumental in coordinating professional equipment manufacturers' views on interface standards although it has tended to ignore consumer applications to some extent, and this is perhaps one of the principal roots of confusion in the field.[3] The consumer interface was initially developed in 1984 by Sony and Philips for the CD system and is usually called Sony-Philips digital interface (SPDIF). The interface is serial and self-clocking. The two audio channels are carried in a multiplexed fashion over the same channel and the data are combined with a clock signal in such a way that the clock may be extracted at the receiver side.

A further standard was devised, originally called multichannel audio digital interface (MADI), which is based on the AES3 data format and has been standardized as AES10-1991. It is a professional interface that can accommodate up to 56 audio channels.

Bluetooth

Bluetooth is a low-cost, low-power, short-range radio technology, originally developed as a cable replacement to connect devices.[4] An application of Bluetooth is as a carrier of audio information. This functionality allows to build devices such as wireless headsets, microphones, headphones, and cellular phones. The audio quality provided by Bluetooth is the same as one would expect from a cellular telephone.

IEEE1394

IEEE1394 is a standard defining a high-speed serial bus. This bus is also named FireWire or i.Link. It is a serial bus similar in principle to UBS, but runs at speeds of up to 400 Mbit/s, and is not centered around a PC (i.e., there may be none or multiple PCs on the same bus). It has a mode of transmission that guarantees bandwidth that makes it ideal for audio transmission digital video cameras and similar devices.

DIGITAL AUDIO TAPE RECORDING AND PLAYBACK

The availability of adaptable types of magnetic tape video recorders accelerated digital-audio-recording development in the tape medium. Nippon Columbia had developed a video-type recorder into a PCM tape recorder for eight channels of audio information with each channel sampled at 47.25 kHz. Now numerous manufacturers produce audio tape recorders for professional recording studios, and some large recording companies have developed digital master tape recording systems including digital editors and mixers.

An inherent disadvantage of digital recording and playback, especially in the tape medium, has been dropout caused by voids or scratches in the tape. Some dropouts are inevitable, so protective or corrective means are used such as interlacing the encoded signal with redundancy, or reserving and using bits for error-detection schemes, e.g., recording sums of words, for comparison with sums simultaneously calculated from playback of the words. Such error detection can trigger the substitution of adjacent data, for example, into the dropout gap.

Digital audio tape recorders are of two different types, helical-scan and multitrack using rotary and stationary heads, respectively. Helical-scan systems already had the needed bandwidth, but improved recording densities and multitrack head stacks allowed multitrack systems to become competitive. A variety of tape formats has been developed. Table 20.4.1 shows part of the specifications for a multitrack professional digital recorder, the Sony PCM-3324.

Two new modes of recording on magnetic tape have permitted large increases in lineal density of recording and signal-to-noise ratio, both great advantages for digital magnetic recording. Perpendicular (or vertical) recording (see Fig. 20.4.6a) uses a magnetic film (e.g., CoCr crystallites), which has a preferred anisotropy normal to the surface. In contrast to conventional longitudinal magnetic recording, demagnetization is weak at short wavelengths, increasing the signal amplitude at high frequencies. Another advantage is that sharp transitions between binary states is possible. Vector field recording (Fig. 20.4.6b) with isotropic particles and microgap heads has also led to higher bit densities.

TABLE 20.4.1   Specifications for the PCM-3324

Number of channels (one track per channel):
  digital audio 24, analog audio 2, time code 1, control 1; total 28
Tape speed, sampling rate:

70.01 cm, 44.1 kHz
76.20 cm/s, 48.0 kHz $\Big\}$ with ±12.5% vernier

(selectable at recording, automatic switching in playback)
Tape: 0.5-in. (12.7-mm) digital audio tape
Quantization: 16-bit linear per channel
Dynamic range: more than 90 dB
Frequency response: 20 Hz to 20 kHz, +0.5, −1.0 dB
Total harmonic distortion: less than 0.05%
Wow and flutter: undetectable
Emphasis: 50 μms/15 μs (EIAJ format and compact disc compatible)
Format: DASH-F (fast)
Channel coding: HDM-1
Error control: cross-interleave code

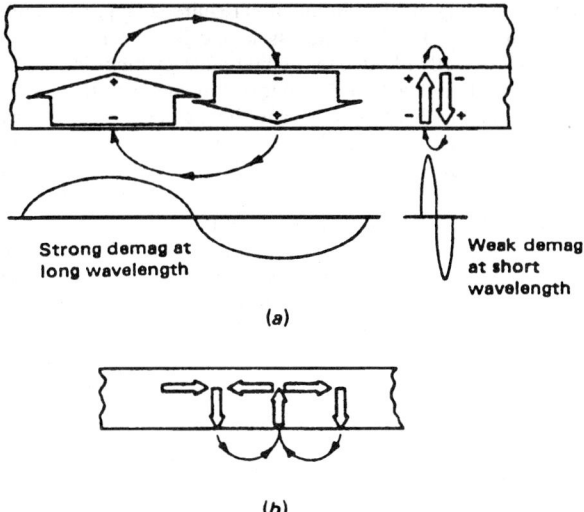

FIGURE 20.4.6 New recording modes: (*a*) perpendicular recording (CoCr); adjacent dipole fields aid; (*b*) vector field recording (isotropic medium): longitudinal and perpendicular fields aid at short wavelength.

Digital Compact Cassette (DCC)

After intensive research on digital audio tape (DAT), Philips built on this research to develop the digital compact cassette (DCC) recorder (Fig. 20.4.7). To make DCC mechanically (and dimensionally) compatible with analogue cassettes, and their tape mechanisms, the same tape speed of 4.76 cm/s ($1^{7}/_8$ in./s) was adopted. At

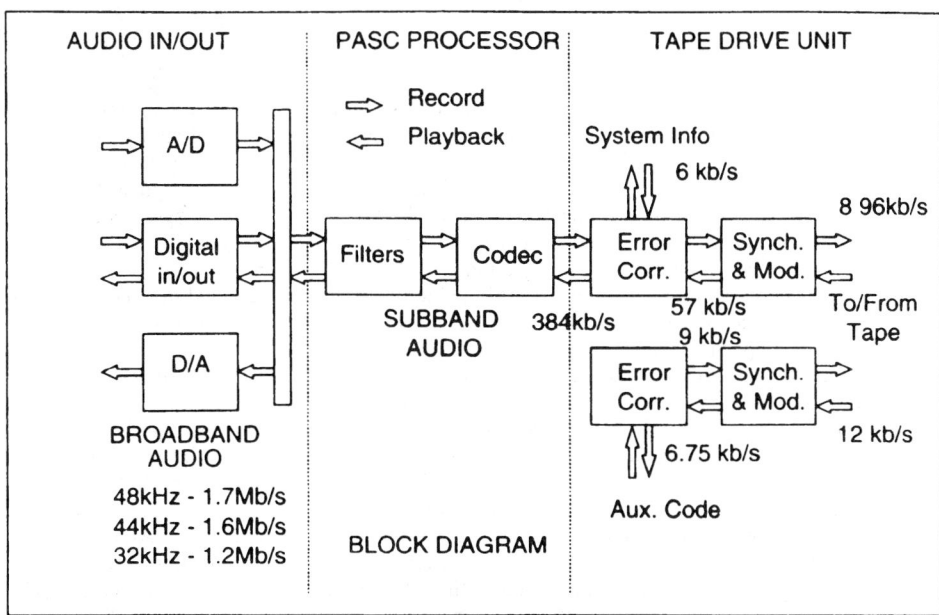

FIGURE 20.4.7 DCC recorder block diagram (Ref. 5).

that tape speed, matching the quality of CD reproduction required compression of the digital audio data stream (1.4 million bits/s) by at least 4 to 1. DCC's coding system, called precision adaptive sub-band coding (PASC) achieves the compression after digitally filtering the audio frequency range into 32 sub-bands of equal bandwidth. The PASC signal processor adapts its action dynamically in each sub-band (a) by omitting data for sounds lying below the hearing threshold at that frequency and time, and (b) by disregarding data for weak sounds in any sub-band that would be masked (rendered inaudible) at that time by the presence of stronger sounds in adjacent or nearby sub-bands. Bits are reallocated to sub-bands when needed for accuracy, from sub-bands where not needed at the time, to optimize coding accuracy overall. The PASC sequential data, together with error correction codes and other system information are multiplexed into eight channels for recording on eight 185μ m-wide tracks. The 3.78 mm wide tape accommodates two sets of eight tracks, one forward and one reverse, alongside auxiliary code data on a separate track providing track and index numbers, time codes, and so forth. In playback the tape output signals are amplified, equalized, demodulated, with error detection and correction. The PASC processor reconstructs the input data that are fed to the digital-to-analogue converter. PASC is compatible with all three existing sampling frequencies, 32, 44.1, and 48 kHz with sub-band widths of 500, 690, and 750 Hz, respectively, and corresponding frame periods of 12, 8.7, and 8 msec.

The 18-channel thin-film playback head has magnetoresistive sensors having resistance that varies with the angle between the sensor's electrical bias current vector and the magnetization vector. Although the recorded digital track is 185 μm wide, the playback sensor is only 70 μm wide, allowing considerable azimuth tolerance. When playing analog audio cassette tapes each audio track uses more than one sensor, as shown, to improve S/N ratio.

DIGITAL AUDIO DISC RECORDING AND PLAYBACK

As in video discs, two general types of digital audio discs were developed. One type recorded binary information mechanically or electrically along a spiral groove that provides guidance during playback for a lightly contacting pickup. The second type used optical laser recording of the digital information in a spiral pattern and optical playback means which track the pattern without contacting the disc directly. The optical type now appears to be dominant.

Optical Digital Discs

The compact disc optical digital storage and reproduction system, a milestone in consumer electronics, was made possible by the confluence of significant progress in each of a number of different related areas of technology. Optical media capable of high storage density had long been available at high cost, but more durable optical surfaces of lower costs, integrated solid-state lasers, and mass-producible optical light pens were all required to permit economical optical recording and playback. Mechanical drive systems of higher accuracy were needed under servocontrol by digital signals. Advanced digital signal processing algorithms, complex electronic circuitry, and very large-scale integration (VLSI) implementation were part of the overall system development. Many research organizations contributed to the state of the art, and in 1980 two of the leaders, Philips and Sony, agreed on standardization of their compact disc optical systems which had been developing along similar but independent paths.

On the reflective surface of the compact optical disc is a spiral track of successive shallow depressions or pits. The encoded digital information is stored in the length of the pits and of the gaps between them, with the transitions from pit to gap (or vice versa) playing a key role. The disc angular rotation is controlled for constant linear velocity of track readout on the order of 1.3 m/s. A beam from a solid-state laser, focused on the disc, is reflected, after modulation by the disc track information, to a photodiode that supplies input to the digital processing circuitry. Focusing of the laser spot on the spiral track is servocontrolled.

In the compact disc system, as in most storage or transmission of digital data, the A/D conversion data are transformed to cope with the characteristics of the storage medium. Such transformation, called modulation, involves (1) the addition of redundant information to the data, and (2) modulation of the combined data to compensate for medium characteristics (e.g., high-frequency losses). The modulation method for the compact disc system, called eight-to-fourteen modulation (EFM), is an 8-data-bit to 14-channel-bit conversion block code

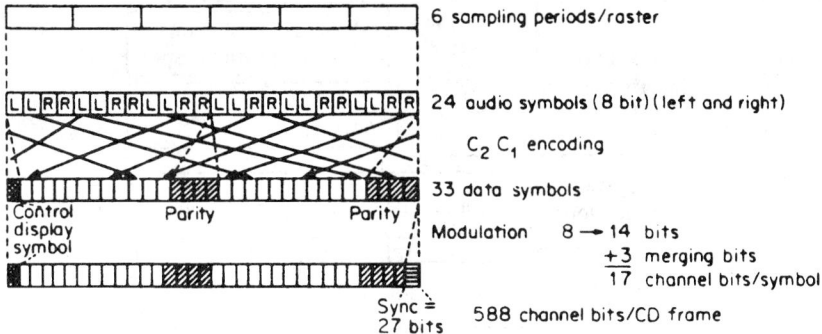

FIGURE 20.4.8   Formats in the compact disc encoding system.

with a space of 3 channel bits (called merging bits) for every converted 14 channel bits for connecting the blocks. Figure 20.4.8 shows the format in the compact disc encoding, and Table 20.4.2 the disc specification.

The purpose of the redundant information is to be able to detect and correct errors that occur because of storage medium imperfections. It is important to minimize the probability of occurrence of such imperfections. The use of optical noncontacting readout from a signal surface protected by a plastic layer allows most of the signal errors at the surface to be reduced to random errors of several bits or larger burst errors. The error-correcting code, the cross-interleave Reed-Solomon code (CIRC), adopted in the standardization provides highly efficient detection and correction for errors of these types. It happens that the EFM modulation method and the CIRC error-correction method used in the compact disc system are well matched. This combination is credited with much of the system's success.

Between tape mastering and replication lies a complex and sophisticated disc mastering process which gets the information into the CD standard format and onto the surface of the CD disc master. Optical disc preparation, recording, development, electroplating, stamping, molding, and protection film coating are the major steps in the highly technological production process.

MiniDisc (MD) System

For optical digital discs to compete more favorably with digital audio tape, a recordable, erasable medium was needed. Magneto-optical discs combined the erasability of magnetic storage with the large capacity and long life of optical storage. An optical disc, with its digital data stream recordable in a tight spiral pattern, provides rapid track access for selective playback or re-recording.

TABLE 20.4.2   Specifications for a Compact Disc

Playing time: 75 min
Rotating speed: 1.2–1.4 m/s (constant linear velocity)
Track pitch: 1.6 μm
Disc diameter: 120 mm
Disc thickness: 1.2 mm
Center hole: 15 mm
Signal surface: 50–116ϕ mm (signal starts from inside)
Channel number: 2
Quantization: 16-bit linear per channel
Sampling rate: 44.1 kHz
Data rate: 2.0338 Mb/s
Channel bit rate: 4.3218 Mb/s
Error protection: CIRC (cross-interleave Reed-Solomon code), redundancy 25% ($^4/_3$)
Modulation: EFM (eight-to-fourteen modulation)

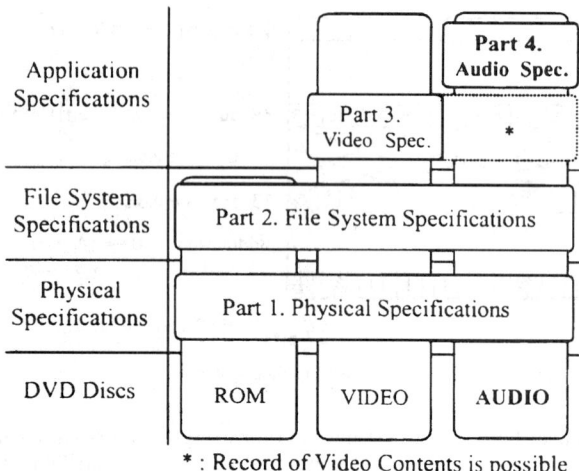

FIGURE 20.4.9 DVD specifications for DVD-ROM, DVD-video, and DVD-audio read only disks, parts 1 to 4.

On the blank disc is a thin film of magneto-optic material embedded within a protective layer, with all of the magnetic domains pointing north pole down (a digital zero). The magnetic field needed for reversal of polarity (to convert from zero to one) is very temperature dependent. At room temperature reversal requires a very strong magnetic field. However, at about 150°C only a small coercive force, provided by a dc magnetic bias field, is needed. During recording, a high-power laser beam, modulated by the digital data stream, heats microscopic spots on the rotating magneto-optic surface (within nanoseconds) to temperatures that allow the dc magnetic bias field to convert zeroes to ones. When the laser beam is off, the spots on the medium cool very rapidly, leaving the desired pattern of magnetic polarity. Erasure can be effected by repeating the procedure with the dc bias reversed.

Playback uses a low-power laser beam, which, because of the Kerr magneto-optic effect, has its plane of polarization rotated one way or the other depending on the magnetic polarity of the recorded bit. An opto-electronic playback head senses the polarization and delivers the digital playback signal.

The Sony magneto-optic MiniDisc is 6.4 cm (2½ in.) in diameter, half that of a CD. To compensate for reduced recording area, a digital audio compression technique called *adaptive transform acoustic coding* (ATRAC) is used. The analog signal is digitized at 44.1 kHz sampling frequency with 16-bit quantization. Waveform segments of about 20 ms and 1000 samples are converted to frequency components that are analyzed for magnitude by the encoder and compressed. Threshold and masking effects are used as criteria for disregarding enough data for an overall reduction of about 5 to 1. During playback, the ATRAC decoder regenerates an analog signal by combining the frequency components recorded on the magneto-optic disc. An added feature of the compression circuit is storage of 3 s of playback time when potential interruptions could occur owing to system shock or vibration.

Digital Versatile Disc-Audio (DVD-A)

DVD-Audio is a HiFi music format based on the same DVD technology as the DVD-Video discs and DVD-ROM computer discs, see Fig. 20.4.9. The disc structure is basically the same.

Recorded with current CD recording methods (PCM), DVD-Audio has a theoretical sampling rate of 192 kHz with 24-bit processing. Like Super Audio CD (SACD) and normal DVD-Video and DVD-Data formats, DVD-Audio discs can store 4.7-GB with a choice of 2-channel and 6-channel audio tracks or a mix of both (see Table 20.4.3).

Like SACD, information such as track names, artists' biographies, and still images can be stored. The format is supported by DVD-Video players made after about November 2000. Manufacturers are making audio machines compatible with playing both types of disc. Titles are available in both Dolby digital mix (so they are compatible on all DVD-Video players) and specific DVD-Audio (requiring the separate player).

TABLE 20.4.3 Specification of DVD-A

Audio combination	Configuration	Playing time in minutes (single layer)		Playing time in minutes (dual layer)	
		PCM	MLP	PCM	MLP
2 channels	48 kHz, 24 bits, 2 ch	258	409	469	740
2 channels	192 kHz, 24 bits, 2 ch	64	119	117	215
6 channels	96 kHz, 16 bits, 6 ch	64	201	117	364
5 channels	96 kHz, 20 bits, 5 ch	61	137	112	248
2 channels & 5 channels	96 kHz, 24 bits, 2 ch + 96 kHz, 24 bits, 3 ch & 48 kHz, 24 bits, 2 ch	43 each	79 each	78 each	144 each

Note: MLP is an acronym for Meridian Lossless Packing, a lossless coding scheme (see Lossless Coding section).

Super Audio CD

Super Audio CD is a new format. It uses direct stream digital (DSD) and a 4.7-GB disc with 2.8 MHz sampling frequency (i.e., 64 times the 44.1 kHz used in CD) enabling a very high quality audio format. Technical comparison between conventional CD and SACD is detailed in Table 20.4.4.

The main idea of the hybrid disc format (see Fig. 20.4.10) is to combine both well-known technologies, CD and DVD, respectively, to keep compatibility with the CD players in the market, and to use the existing DVD-video process tools to make a two-layer disc, i.e., to add a high-density layer to a CD reflective layer. As shown in Table 20.4.4, the storage capacity of the high-density layer is 6.9 times higher than the storage capacity of a conventional CD.

Direct Stream Digital

The solution came in the form of the DSD signal processing technique. Originally developed for the digital archiving of priceless analog master tapes, DSD is based on a 1-bit sigma-delta modulation together with a fifth-order noise-shaping filter and operates with a sampling frequency of 2.8224 MHz (i.e., 64 times the 44.1 kHz used in CD), resulting in an ultrahigh signal-to-noise ratio in the audio band.

TABLE 20.4.4 Comparison Between Conventional CD and SACD

	Conventional compact disc	Super Audio CD
Diameter	120 mm (4–3/4 in.)	120 mm (4–3/4 in.)
Thickness	1.2 mm (1/20 in.)	2 × 0.69 mm = 1.2 mm (1/20 in.)
Max. substrate thickness error	+/–100 μm	+/–30 μm
Signal sides	1	1
Signal layers	1	2: CD-density reflective layer + high-density semitransmissive layer
Data capacity		
Reflective layer	680 MB	680 MB
Semitransmissive layer	—	4.7 GB
Audio coding		
Standard CD audio	16-bit/44.1 kHz	16-bit/44.1 kHz
Super Audio	—	1-bit DSD/2.8224 MHz
Multichannel	—	6 channels of DSD
Frequency response	5–20,000 Hz	DC(0)–100,000 Hz (DSD)
Dynamic range	96 dB across the audio bandwidth	120 dB across the audio bandwidth (DSD)
Playback time	74 min	74 min
Enhanced capability	CD text	Text, graphics, and video

Hybrid Disc Content

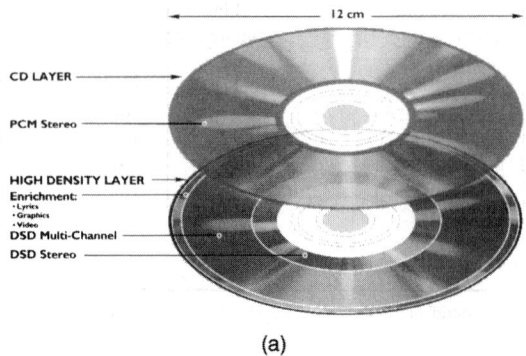

CD LAYER

PCM Stereo

HIGH DENSITY LAYER
Enrichment:
· Lyrics
· Graphics
· Video
DSD Multi-Channel
DSD Stereo

(a)

Hybrid Disc Construction

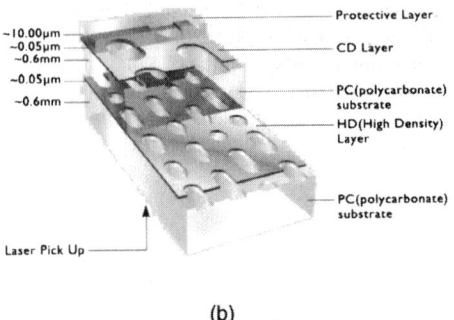

~10.00μm
~0.05μm
~0.6mm

~0.05μm
~0.6mm

Protective Layer

CD Layer

PC(polycarbonate) substrate

HD(High Density) Layer

PC(polycarbonate) substrate

Laser Pick Up

(b)

Hybrid Disc Signal Reading

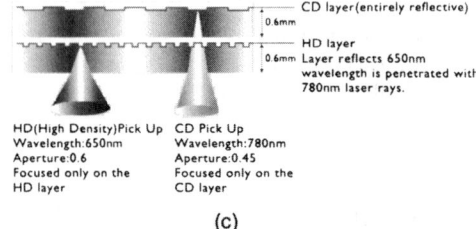

CD layer(entirely reflective)

0.6mm

HD layer

0.6mm Layer reflects 650nm wavelength is penetrated with 780nm laser rays.

HD(High Density)Pick Up
Wavelength:650nm
Aperture:0.6
Focused only on the
HD layer

CD Pick Up
Wavelength:780nm
Aperture:0.45
Focused only on the
CD layer

(c)

FIGURE 20.4.10 Hybrid disc content of the Super Audio CD (*a*), hybrid disc construction (*b*), and hybrid disc signal reading (*c*).

The Three Type of Super Audio CD

The SACD standard, published by Philips and Sony in March 1999, defines three possible disc types (see Fig. 20.4.10). The first two types are discs containing only DSD data; the single layer disc can contain 4.7 GB of data, while the dual layer disc contains slightly less than 9 GB. The third version—the SACD Hybrid—combines a single 4.7 GB layer with a conventional CD that can be played back on standard CD players. (For more information see http://www.sacd.philips.com/)

RDAT

Rotary head digital audio tape (RDAT) is a semiprofessional recording format, an instrumentation recorder, and a computer data recorder.[6] Mandatory specification are:

- 2 channels (optional more)
- 48 or 44.1 kHz sampling rate
- 16 bits quantization
- 8.15 mm/s tape speed
- 2 h playing time (13 μm tape)
- The cassette has a standardized format of $73 \times 54 \times 10.5$ mm, which is rather smaller than the compact cassette.

OTHER APPLICATIONS OF DIGITAL SIGNAL PROCESSING

The main applications of audio DSP are high-quality audio coding and the digital generation and manipulation of music signals. They share common research topics including perceptual measurement techniques and knowledge and various analysis and synthesis methods.[7] Chen[8] gives a review of the history of research in audio and electroacoustics, including electroacoustic devices, noise control, echo cancellation, and psychoacoustics.

Reverberation. For some years digital processing of audio signals has been used for special purposes (e.g., echo and reverberation effects) in systems that were otherwise analog in nature. The possibility was suggested by computer-generated "colorless" artificial reverberation experiments. When high-quality A/D and D/A conversion became economical, digital time-delay and reverberation units followed. Figure 20.4.11a is a block diagram of a digital audio reverberation system in which the complete musical impulse sound reaching a listener (Fig. 20.4.11b) consists of slightly delayed direct signal, followed by a group of simulated early reflections from a tapped digital delay line and a "reverberant tail" added to its envelope by a reverberation processor using multiple recursive structures to produce a high time density of simulated reflections.

Dither is used to prevent perceptually annoying errors like quantizers. It is a random "noise" process added to a signal prior to its (re)quantization in order to control the statistical properties of the quantization error.[9,10] A common stage to perform dithering is after the various digital signal processing stages just ahead of the quantization before storing the signal or sending it to a digital-to-analog converter (DAC).

A special topic in signal processing for sound reproduction is overcoming the limitations of the reproduction set-up, e.g., reproduction of bass frequencies through small loudspeakers.[11] Another limitation is the distance between the two loudspeakers of a stereophonic setup. If one likes to increase the apparent distance, frequency dependent cross talk between the channels can be applied.[12]

Lossless Coding. Lossless compression is a technique to recode digital data in such a way that the data occupy fewer bits than before. In the PC world these programs are widely used and known under various names such as PkZip. For digital audio these programs are not very well suited, since they are optimized for text data and programs. Figure 20.4.12 shows a block diagram representing the basic operations in most lossless compression algorithms involved in compressing a single audio channel.[13]

All of the techniques studied are based on the principle of first removing redundancy from the signal and then coding the resulting signal with an efficient coding scheme. First the data are divided into independent frames of equal time duration in the range of 13 to 26 ms, which results in a frame of 576 to 1152 samples if a sampling rate of 44.1 kHz is used. Then the bits in each frame are decorrelated by some prediction algorithm as shown in Fig. 20.4.13.

The value of a sample $x[n]$ is predicted using the preceding samples $x[n-1]$, $x[n-2]$, ..., by using the filters A, B and quantizer Q. The error signal $e(n)$ that remains after prediction is in general smaller than x, and will therefore require fewer bits for its exact digital representation. The coefficients of the filters A and B are transmitted as well, which makes an exact reconstruction of $x[n]$ possible.

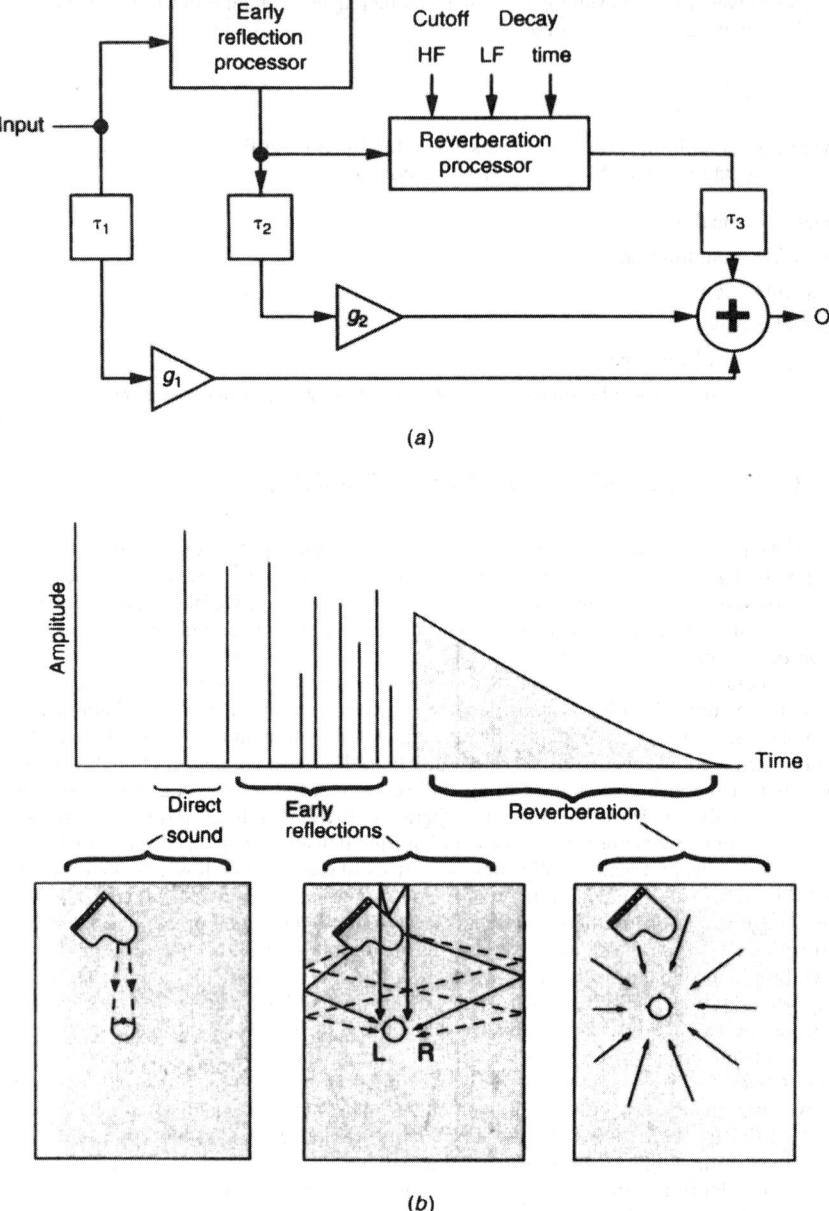

FIGURE 20.4.11 The basic operations in most lossless compression algorithms.

The third stage is an entropy coder, which removes further redundancy from the residual signal $e[n]$, and again in this process no information is lost. Most coding schemes use one of these three algorithms:

- Huffman, run length, and Rice coding, see Ref. 13 for more details
- Meridian Lossless Packing (MLP) for DVD-A
- Direct Stream Transfer for SACD

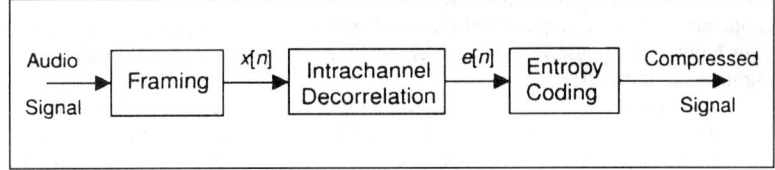

FIGURE 20.4.12   The basic operations in most lossless compression algorithms.

Watermarking.   The advantages of digital processing and distribution of multimedia, such as noise-free transmission and the possibility of digital signal processing on these media, are obvious. The disadvantage, from the viewpoint of media producers and content providers, can be the possibility of unlimited coping of digital data without loss of quality. Digital copy protection is a way to overcome these problems. Another method is the embedding of digital watermarks into the multimedia.[14]

The watermark is an unremovable digital code, robustly and imperceptibly embedded in the host data and typically contains information about the origin, status, and/or destination of the data. While copyright protection is the most prominent application of watermarking techniques, other methods exist, including data authentication by means of fragile watermarks that are impaired or destroyed by manipulations, embedded transmission of value-added services, and embedded data labeling for other purposes than copyright protection such as monitoring and tracking.

Multimedia Content Analysis

Multimedia content analysis refers to the computerized understanding of semantic meanings of multimedia documents such as a video sequence with an accompanying audio track. There are many features that can be used to characterize audio signals. Usually audio features are extracted in two levels: short-term frame level and long-term clip level, where a frame is about 10 to 40 ms. To reveal the semantic meaning of an audio signal, analysis over a much longer period is necessary, usually from 1 to 10 s.[15]

Special Effects.   If a single variably delayed echo signal ($\tau > 40$ ms) is added to direct signal at a low frequency (< 1 Hz), a sweeping comb filter sound effect is produced called *flanging*. When multiple channels of lesser delay (e.g., 10 to 25 ms) are used, a "chorus" effect is obtained from a single input voice or tone.

Time-Scale Modification.   Minor adjustment of the duration of prerecorded programs to fit available program time can be accomplished digitally by loading a random-access memory with a sampled digital input signal and then outputting the signal with waveform sections of the memory repeated or skipped as needed under

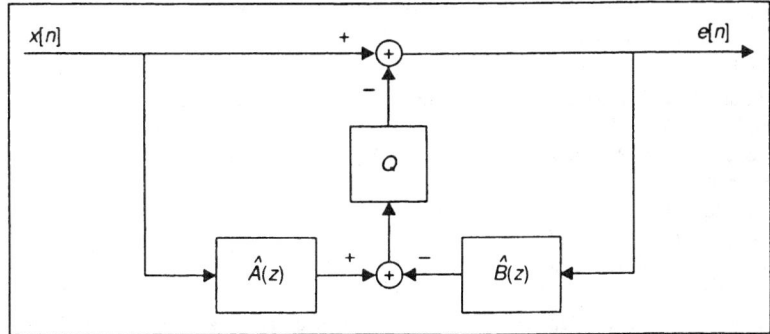

FIGURE 20.4.13   General structure for prediction.

computer control, in order to approximate a continuous output signal of different duration. A related need, to change the fundamental (pitch) frequency of recorded speech or music without changing duration, involves the use of different input and output clock frequencies, along with repeating or skipping waveform segments as needed to retain constant duration.

Other digital audio components that offer advantages, or indeed are essential, once the system goes digital, include filters, equalizers, level controllers, background-noise reducers, mixers, and editors.

The compact disc, developed especially for audio uses, provides a multimegabyte storage technique, which is very attractive in many other applications for read-only memories. Conversely, in the evolution of telecommunication networks, new techniques for signal decomposition and reconstruction, and for echo cancellations, suggest further audio improvements in conference pickup and transmission, for example. The interchange between digital audio and other branches of digital communication continues.

MPEG Audio Coding General

Moving Picture Experts Group (MPEG) is well known for its developments of a series of standards for the coding of audiovisual content [http://www.cselt.it/mpeg/]. Initially targeted at the storage of audiovisual content on compact disc media, the MPEG-1 standard was finalized in 1992 and included the first generic standard for low-bit-rate audio within the audio part. Then the MPEG-2 standard was completed and extended MPEG-1 technology toward the needs of digital video broadcast. On the audio side, these extensions enabled coder operation at lower sampling rates (for multimedia applications) and coding of multichannel audio. In 1997 the standard of an enhanced multichannel coding system (MPEG-2 Advanced Audio Coding, AAC) was defined. The so-called MP3 is the popular name for MPEG-1 Layer III. Then the MPEG-4 standard was developed, with new functionalities such as object-based representation, content-based interactivity, and scalability; the MPEG-4 standard was developed in several steps (called versions), adding extensions to the basic technology for audio. Reference 16 describes in some detail the key technologies and main features of MPEG-1 and MPEG-2 audio coders. In 1996 the effort behind MPEG-7 was started. MPEG-7 defines a universal standardized mechanism for exchanging descriptive data that are able to characterize many aspects of multimedia content with a worldwide interoperability,[17] or as the official name says, a "multimedia content description interface." Work on the new standard MPEG-21 "Multimedia Framework" was started in June 2000. The vision for MPEG-21 is to define a multimedia framework to enable transparent and augmented use of multimedia resources across a wide range of networks and devices used by different communities.

REFERENCES

1. Gravereaux, D. W., A. J. Gust, and B. B. Bauer *J. Audio Eng. Soc.*, 1970, Vol. 18, p. 530.

2. Bernhard, R. *IEEE Spectrum*, December, 1979, p. 28.

3. Rumsey, F., and J. Watkinson "The Digital Interface Handbook," Focal Press, 1995.

4. Bray, J., and C. F. Sturman "Bluetooth Connect Without Cables," Prentice Hall PTR, 2001.

5. Lokhoff, G. C. P. "dcc—Digital Compact Cassette," *IEEE Trans. Consumer Electron.*, 1991, Vol. 37, p. 702.

6. Watkinson, J. "RDAT," Focal Press, 1991.

7. Kahrs, M., and K. Brandenburg, eds. "Applications of Digital Signal Processing to Audio and Acoustics," Kluwer Academic Publishers, 1998.

8. Chen, T., ed., "The Past, Present, and Future of Audio Signal Processing," *IEEE Signal Process. Magazine*, September 1997, Vol. 14, No. 5, pp. 30–57.

9. Wannamaker, R., S. Lipshitz, J. Vanderkooy, and J. N. Wright "A Theory of Nonsubstractive Dither," *IEEE Trans. Signal Process.*, Feb. 2000, Vol. 48, No. 2, pp. 499–516.

10. Norsworthy, S. R., R. Schreier, and G. C. Temes, eds. "Delta-Sigma Data Converters: Theory, Design, and Simulation, IEEE Press, 1996.

11. Larsen, E., and R. M. Aarts "Reproducing Low Pitched Signals Through Small Loudspeakers," *J. Audio Eng. Soc.*, March 2002, Vol. 50, No. 3, pp. 147–164.

12. Aarts, R. M. "Phantom Sources Applied to Stereo-Base Widening," *J. Audio Eng. Soc.*, March 2000, Vol. 48, No. 3, pp. 181–189.

13. Hans, M., and R. W. Schafer, "Lossless Compression of Digital Audio," *IEEE Signal Process. Magazine*, July 2001, Vol. 18, No. 4, pp. 21–32.

14. Special Issue on Identification and Protection of *Multimedia Information, Proc. IEEE*, 1999, Vol. 87, No. 7, pp. 1059–1276.

15. Wannamaker, R., S. Lipshitz, J. Vanderkooy, and J. N. Wright "A Theory of Nonsubstractive Dither," *IEEE Trans. Signal Process*, February 2000, Vol. 48, No. 2, pp. 499–516.

16. Noll, P. "MPEG Digital Audio Coding," *IEEE Signal Process. Magazine*, Sept. 1997, Vol. 14, No. 5, pp. 59–81.

17. Lindsay, A. T., and J. Herre "MPEG-7 and MPEG-7 Audio—An Overview," *J. Audio Eng. Soc.*, July/August 2001, Vol. 49, Nos. 7/8, pp. 589–594.

SECTION 21

VIDEO AND FACSIMILE SYSTEMS

Although much of this section describes basic video and facsimile technologies that have not changed over the years, newer material is also included. For example, international agreement was reached recently on the use of 1920×1080 as a common image format for high-definition (HD) production and program exchange. The 1920×1080 format has its roots in the CCIR (Consultative Committee in International Radio) sampling standard and brings international compatibility to a new level.

Set-top boxes and high-definition or digital-ready TV sets will be the mechanism that brings digital technology to the consumer for the next several years as the transition from analog to digital takes place. In the United States, three modulation techniques have become "standards" in a particular application: vestigial sideband (VSB) for terrestrial, quadrature amplitude modulation (QAM) for cable, and quaternary phase-shift keying (QPSK) for direct-to-home satellite.

With Internet facsimile, store-and-forward facsimile occurs when the sending and receiving terminals are not in direct communication with one another. The transmission and reception takes place via the store-and-forward mode on the Internet using Internet e-mail. In this mode, the facsimile protocol "stops" at the gateway to the Internet. It is reestablished at the gateway leaving the Internet. Real-time facsimile is covered by Recommendation T.38 approved by the International Telecommunications Union, Telecommunications (ITU-T) in 2002.   R.J.

In This Section:

On the CD-ROM:

The following is reproduced from the 4th edition of this handbook:

"Television Cameras," by Laurence J. Thorpe.

CHAPTER 21.1
TELEVISION FUNDAMENTALS AND STANDARDS

James J. Gibson, Glenn Reitmeier

INTRODUCTION

This chapter summarizes analog and digital television signal standards and the principles on which they are based.

The technical standards for color television developed in 1953 for the United States by the National Television System Committee (NTSC) are described on a few pages in the rules of the Federal Communications Commission (FCC Rule Part 73). The rules only specify the radiated signal in sufficient detail for a receiver manufacturer to produce receivers, which convert this signal into a television picture with sound. This traditional approach to formulating standards leaves implementation to competitive forces. Since 1953 many international standards and recommended practices have evolved. A similar philosophy was used in the FCC's adoption of the Advanced Television System Committee (ATSC) digital television standards in 1996.

All color television standards are based on the same principles:

- The psychophysics of the human visual system (HVS).

- Picture-signal conversion by sampling/display, at field rate, of three primary colors in a flat rectangular dynamic picture on a raster of horizontal scan lines, scanned from left to right and top to bottom.

- The signals are conveyed as three components: one luminance signal, which essentially provides brightness information, and two chrominance signals which essentially provide hue and color saturation information.

- For radio frequency transmission these three signals and audio signals are multiplexed to form a single r.f. signal, which occupies a channel in the frequency spectrum.

Some of these principles are illustrated in Fig. 21.1.1, which shows a block diagram of a standard analog television system for terrestrial broadcasting. The figure shows that the video and audio signals are multiplexed separately to form *composite video and audio signals*, which are subsequently delivered to separate picture and sound transmitters generating signals, which are diplexed to form a radiated signal that occupies a 6, 7, or 8 MHz band in the radio frequency spectrum from 40 to 900 MHz. This is the usual practice in broadcasting of analog television signals in the NTSC, PAL (Phase Alternation Line), and SECAM (Séquentiel à mémoire) systems. These systems, which are compatible with black and white reception, use frequency division multiplex: The chrominance signals are bandlimited and modulate one or two subcarriers that are "inconspicuously" added to (multiplexed with) the luminance signal. Besides analog television terrestrial broadcast standards, there are many standards for analog television production, storage, and distribution (terrestrial, satellite, and cable) developed by several organizations. In analog television there are two picture scanning standards specified as N/F_v where N = total number of scanning lines and F_v = number of picture fields per second. These standards are 625/50 and 525/60 (including

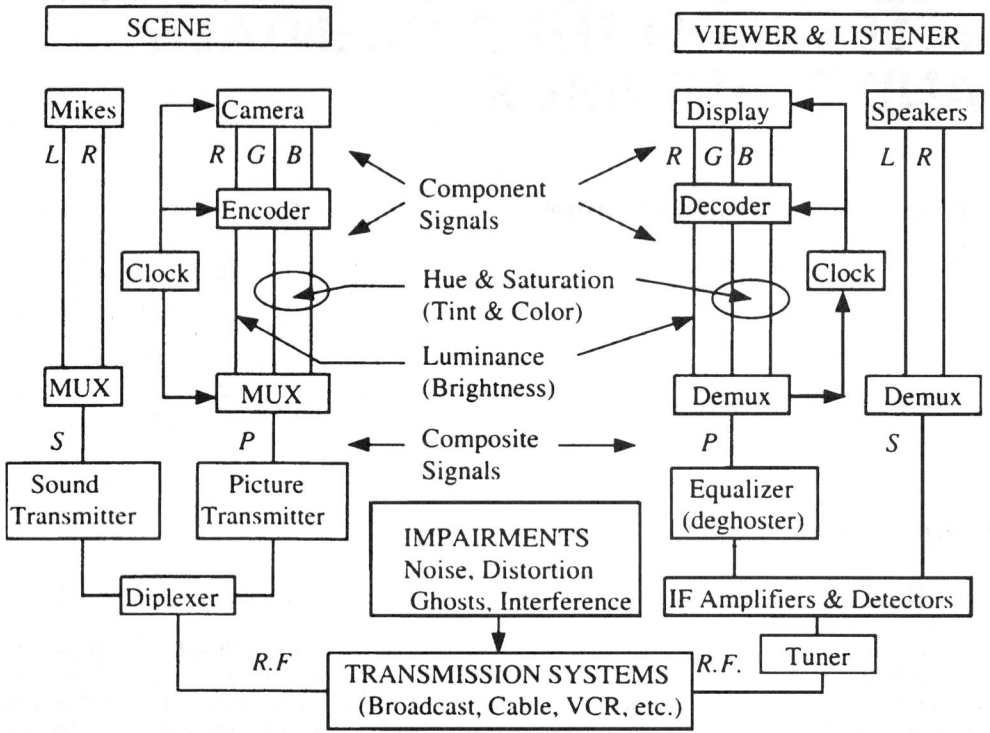

FIGURE 21.1.1 Functional block diagram of a standard analog television broadcast system. Some systems may multiplex sound and picture before delivery to a transmitter.

525/59.94). There are three basic *composite color video signal standards* that carry all the color information in one signal: NTSC, PAL, and SECAM. In addition there are nine standards (with variations) describing the radiated signal carrying the composite color video signals with various types of r.f. modulation and bandwidths and with various standards for audio signals and audio signal modulation. These nine standards are referred to as B, G, H, I, K, K1, L, M, N. Only system M is of type 525/60.

In digital television (DTV), the three component signals and audio are sampled, digitized, and data compressed to eliminate redundant and psychophysically irrelevant data. Digital signals use available spectrum more effectively than analog signals. Digital television standards have been developed with consideration for flexibility, extensibility (for future applications), and interoperability with other systems for information production and distribution (e.g., computers).

In 1982 the Radio Consultative Committee of the International Telecommunications Union (ITU-R), formerly called the International Radio Consultative Committee (CCIR) adopted an international digital television component standard, ITU-R Recommendation 601. In this chapter this important standard is referred to by its popular old designation: CCIR601. This standard was primarily intended for production and for tape recording (SMPTE format D-1), but is now used in many applications, including DTV transmission.

In digital television (DTV), including high-definition television (HDTV), digital techniques are used for video compression, data transport, multiplexing, and r.f. transmission. DTV promises to be more than television, in the sense that it delivers to homes a digital channel with high data rate which may carry more than high quality pictures and sound. Of particular importance are the standards developed by the Moving Picture Expert

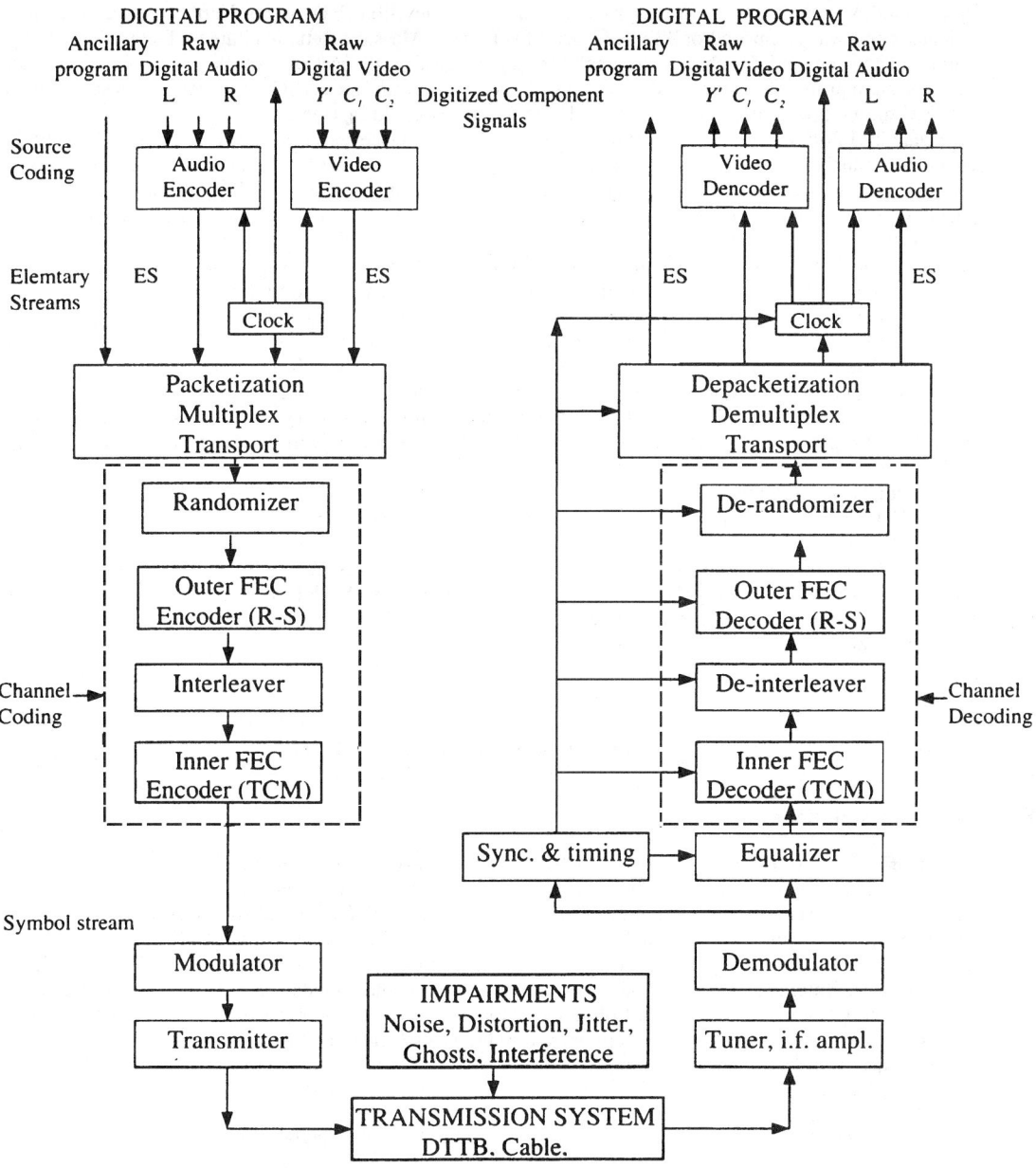

FIGURE 21.1.2 Functional block diagram of a standard television broadcast system. Different modulation techniques and signal bandwidths are used for different transmission media.

Group (MPEG) of ITU. MPEG standards are quite flexible, but include specific approaches to television data compression and packetization. MPEG standards have been adopted by the International Standards Organization (ISO) and the International Electro-technical Committee (IEC). Based on the MPEG-2 format, the FCC Advisory Committee on Advanced Television Systems (ACATS) provided oversight for the development of a "Grand Alliance," a system for high-definition digital television (HDTV) for North America. The Grand Alliance itself was comprised of AT&T, General Instrument, Massachusetts Institute of Technology, Philips, Sarnoff, Thomson, and Zenith, which joined forces to forge a single best-of-the-best system from four competing system proposals. The Grand Alliance system is the basis of the ATSC digital television standard and the FCC digital broadcast standards for the United States adopted in 1996.

Figure 21.1.2 shows basic functional blocks of the "Grand Alliance" HDTV system. The basic principles listed above for analog systems still apply to this digital system. The video and audio components are basically the same as in the analog system shown in Fig. 21.1.1, but source coding (removal of irrelevant and redundant data), ancillary signals, multiplexing, transport, packetization, channel coding (for error management), and modems are entirely different.

In both analog and digital systems, the relation between the picture, as observed by a viewer, and the signals is not a simple one. Signal quality measures are not easily related to the subjective quality of pictures. In fact, there are very few objective measures of picture quality, while there are many objective measures of signal quality and signal tolerances. The selection of television standards, including recent HDTV standards, are based on subjective picture quality evaluations. Relations between signal quality and picture quality in NTSC, PAL, and SECAM are fairly well known. Numerous measurement and monitoring techniques using test signals have been developed for these signals. Relations between objective measures of signal quality and picture and sound quality are not as well correlated in digital television and HDTV.

Ongoing work on digital TV standardization is carried out by numerous committees. ATSC continues to develop its terrestrial transmission standard. The Society of Cable Telecommunications Engineers (SCTE) develops standards for cable transmission. The Digital Video Broadcast (DVB) group also continues to develop its standards for terrestrial (DVB-T), cable (DVB-C), and satellite (DVB-S) transmission. The Society of Motion Picture and Television Engineers (SMPTE) is involved in standards for related professional production equipment. The Consumer Electronics Association (CEA) establishes industry standards for consumer equipment.

PHOTOMETRY, COLORIMETRY, AND THE HUMAN VISUAL SYSTEM

Radiance and Luminance

The HVS is sensitive to radiation over a 2:1 bandwidth of wavelengths extending from 380 to 760 nm, i.e., from extreme blue to extreme red. When adapted to daylight it is most sensitive to green light at 555 nm. As the wavelength of monochromatic light departs from 555 nm, the *radiance* (radiation from a surface element in a given direction defined by the angle Θ from the normal and measured in watts/steradian per unit projected area = the actual surface of the radiating element times $\cos\Theta$) must be increased for constant perception of brightness. The International Commission on Illumination (CIE) has standardized the response versus wavelength of the HVS $\bar{y}(\lambda)$ of a standard observer adapted to daylight vision (photopic vision). Figure 21.1.3 shows $\bar{y}(\lambda)$ versus λ. Luminance, sometimes referred to as brightness and measured in cd/m², *Candelas per projected area in m^2*, is defined as

$$Y = 680 \int_0^\infty E(\lambda)\bar{y}(\lambda)d\lambda \quad cd/m^2 \qquad E(\lambda) = \text{spectral density of radiance in } (W/nm)/m^2 \qquad (1)$$

where λ is the wavelength in nanometers. Older units for luminance are: a footlambert (ft.L) = 3.42626 cd/m² and a millilambert (mL) = 3.18310 cd/m².

A surface is *diffuse* at a wavelength λ if $E(\lambda)$ is independent of the direction Θ. The face plate of a TV tube is essentially diffuse for all wavelengths of visible light, but this may not be the case for a projection screen. Thus the luminance of the face-plate of a picture tube is roughly independent of the direction from where it is seen. Typically the peak luminance levels of picture tubes is about 120 cd/m², but bright displays may have luminance levels exceeding 250 cd/m². The luminance of bright outdoor scenes may well exceed 10,000 cd/m²,

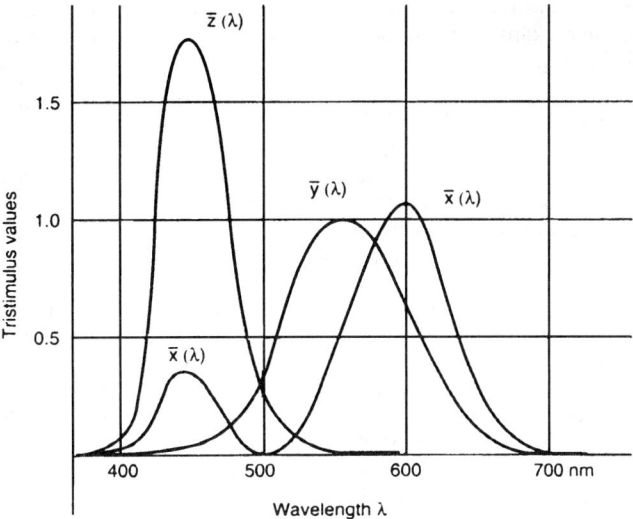

FIGURE 21.1.3 Tristimulus values of the CIE nonphysical primaries, $\bar{x}(\lambda)$, $\bar{y}(\lambda)$, and $\bar{z}(\lambda)$. Note that $\bar{y}(\lambda)$ is the CIE standard luminosity function.

while a motion picture screen may be 30 cd/m². *Equal Energy Radiation*, E(λ) = constant, appears to have the color "white." Corresponding radiometric and photometric units are:

Radiant flux	Watt	Luminous flux	Lumens
Irradiance	Watt/m²	Illuminance	Lux = lumens/m²
Radiant intensity	Watt/steradian	Luminous intensity	Candela = lumens/steradian
Radiance	(Watt/steradian)/m²	Luminance	Candela/m²

At $\lambda = 555$ nm there are 680 lm/w. Without a color filter, the illuminance I of a target in a camera or of the retina of an eye is proportional to the luminance Y of the scene (theoretically $I \cong \pi Y/4$(f-number of the lens)²).

Contrast and Resolution

The HVS is sensitive to relative changes in luminance levels. The just noticeable difference (JND) between the luminance $Y + \Delta Y$ of a small area and the luminance Y of a surrounding large area can be expressed by the ratio

$$F = \Delta Y/Y \tag{2}$$

where F is the *Fechner ratio*.

The Fechner ratio is remarkably constant over the large range of the high luminance levels used in TV displays. It ranges from 1 percent for Y > 10 cd/m² up to 2 percent at Y = 1 cd/m². Assuming a constant Fechner ratio, the number of distinguishable small area gray levels between a "black" level Y_b and a highlight "white" level Y_w is

$$n \cong (\ln Y_w/Y_b)/F \qquad \text{or} \qquad Y_w/Y_b \cong \exp(nF) \cong (1 + F)^n = \text{contrast} \tag{3}$$

For example, for n = 255 levels (8-bit quantization of log luminance) and F = 1.5 percent, the contrast is 45. With 9 bits (511 levels @ 1 percent) the contrast is 165. Contrast rarely exceeds 50:1 in TV displays owing to ambient light and light scattered within the picture tube. Thus, due to the contrast available in consumer displays, DTV systems use 8-bit luminance (and chrominance) signals.

More relevant test patterns for measuring the contrast sensitivity of the HVS are luminance sinewave gratings added to a uniform background luminance level. The contrast ratio of a sinewave grating, also referred to as *modulation*, is defined as:

$$m = (Y_{maz} - Y_{min})/(Y_{max} + Y_{min}) = \text{(luminance amplitude/average luminance)} \qquad (4)$$

The just perceptible modulation (JPM) of a stationary sinewave grating is sensitively dependent on a large number of factors: the spatial frequency v of the grating expressed in cycles per degree (cpd), the average luminance level, the orientation of the wavefront, the adaptation of the pupil of the eye, and ambient factors. If the line from the eye to the observed point is perpendicular to the picture, v cycles per degree can be converted into k cycles per picture height (cph) by multiplication with 57.3/D, where D is the viewing distance in picture height. The most sensitive m is 0.3 percent at about 3.5 cpd = 200/D cph and increases to 100 percent at about 35 cpd = 2000/D cph (500 cph @ D = 4). This upper limit can be taken as an estimate of the resolution limit of the HVS. TV systems have been designed with HVS characteristics in mind. Analog and standard definition (SD) digital systems assume D = 6; therefore, the HVS resolution limit is approximately 383 cph. HDTV systems assume a larger display or closer viewing, D = 2; therefore, the HVS resolution limit is approximately 1000 cph.

The ratio of the perceived to the displayed modulation is the modulation transfer function (MTF) of the HVS. It rolls off with spatial frequency depending on a number of parameters including the adaptation of the pupil of the eye. It is down about 6 dB at 750/D cph for a luminance of 140 cd/m². One performance measure of a TV system is its MTF = ratio of displayed m(k) to input m(k) versus spatial frequency for various orientations of the wavefront of the grating.

The eye has high resolution only within a sustained angle of about 2°. For D = 6 this is 2.5 percent of a TV display area. The remaining area is viewed with low resolution rod vision which is sensitive to flicker.

Gamma "Pre-Correction"

Since the HVS is sensitive to contrast, it can tolerate more noise and larger luminance errors in bright areas than in dark areas. Since noise and errors are added to the luminance signal during transmission, the visibility of the added noise is reduced if the luminance signal is compressed at the source and expanded in the receiver. A constant Fechner ratio suggests logarithmic compression and exponential expansion of the luminance, a proposal that was recommended by the NTSC in 1941 and is still allowed in the FCC rules for black and white TV. In the early days of black and white TV, it was found, however, that the picture tube itself acted as a good and inexpensive expander. The displayed luminance Y on the face plate of a picture tube in response to an applied signal voltage V is approximately equal to:

$$Y = \text{const.} \times V^{\gamma} \qquad (5)$$

where γ is *gamma* ranges from 2 to 3. Assuming that the output signal Y from a TV camera is proportional to the illuminance of the target (a fair assumption for modern cameras), the luminance signal Y delivered by such a linear camera is compressed right at the output of the camera to yield a *gamma "corrected" luminance signal*, which, in black and white TV, is approximately proportional to $Y^{1/\gamma}$. In DTV systems, the same practice continues to be employed, both for reducing the visibility of compression-related artifacts as well as maintaining compatibility with legacy analog systems, which is an economic consideration in dual digital/analog receivers. Table 21.1.1 shows standards for gamma and modifications of Eq. (5) in recent standards.

Flicker

The human visual system is quite sensitive to large area luminance fluctuations at frequencies below 100 Hz (flicker). The critical flicker frequency f_f Hz (threshold of perception) is determined by on-off modulation of the luminance of a large area. The critical frequency increases, according to the "Ferry-Porter law," in proportion to the logarithm of the highlight Y cd/m² of the luminance:

$$f_f = 30 + 12.5 \log_{10}(Y \text{ cd/m}^2) \qquad \text{Hz} \qquad (6)$$

TABLE 21.1.1   Standards for Electro-Optical Transfer Characteristic: $V = F(L)$, $L = F^{-1}(V)$ Example: $R' = F(R = R)$ in "gamma corrector," $R = F^{-1}(R')$ in display

SYSTEM	γ (gamma)	V_O (bias)	k(slope)	L^*	V^*
NTSC	2.2	0	0	0	0
PAL/SECAM	2.8	0	0	0	0
TU R-709, SMPTE170M, SMPTE274M	1/0.45	0.099	4.5	0.018	0.081
SMPTE 240M	1/0.45	0.1115	4	0.0228	0.0912

V = *normalized electric signal level* and L = normalized optical stimulus level (at Maximum level $V = L = 1$ by definition):
$V = F(L)$: $V = kL$ for $0 \leq L < L^*$ and for $L^* \leq L \leq 1$ $V = (1 + V_o)L^{1/\gamma} - V$
$L = F^1(V)$: $L = V/k$ for $0 \leq V < V^* = kL^*$ and for $V^* \leq V \leq 1$ $L = [(V + V_o)/(1 + V_o)]^\gamma$

The result varies with individuals and with the adaptation of the eye. It shows that f_f increases by about 12.5 Hz when Y increases by a factor of 10. For example, f_f is 48, 50, and 60 Hz for a peak luminance Y = 25.4, 36.6, and 227.6 cd/m². The HVS is more sensitive to flicker for light entering the eye at an angle (rod vision). In motion pictures the picture frame rate is 24 pictures per second. To avoid flicker at acceptable luminance levels, every picture is shown at least twice, when the film is projected (this is referred to as double-shuttering). In television the frame rate is 25 Hz in most of the world and 30 Hz in the United States, Japan, and some other countries. In all analog TV systems, large area flicker rates are doubled by using two fields with interlaced scan lines to complete a frame.

DTV systems decouple the display frame rate and the transmitted frame rate. DTV systems generally support the legacy analog frame rates and interlaced scanning formats, and additionally provide new capabilities for progressive scan formats and film-related frame rates.

Color Space, Color Primaries, and CIE Chromaticities

It is a fact that a wide range of colors can be reproduced with only three sources of light, e.g., red, green, and blue. The basic laws of *colorimetry*, the science devoted to color vision are:

• The HVS can only perceive three attributes of color: *brightness, hue,* and *saturation,* more precisely defined as *luminance, dominant wavelength*, and *purity.*

• Colors can be represented as vectors in a three-dimensional linear vector space referred to as *color space*. Colors add and subtract as vectors, but a distance in color space is not a measure of perceived difference in colors. In color space the luminance component Y $\geq$ 0 traditionally points up from a horizontal chrominance plane Y = 0.

Figure 21.1.4 shows a color vector $\mathbf{A} = Y\mathbf{W} + \mathbf{C}$ as the sum of a white vector $Y\mathbf{W}$ with the same luminance (brightness) Y as $\mathbf{A}$ and a chrominance vector $\mathbf{C}$ in a constant luminance plane with an orientation related to hue (tint) and a magnitude related to saturation (color).

Based on experiments with hundreds of observers matching colors composed of monochromatic light of different wavelengths, the CIE standardized X- and Z-axes at right angles in the chrominance plane, such that all color vectors have nonnegative components X, Y, Z, and that X = Y = Z for equal energy light. Thus, basis-vectors $\mathbf{X, Y, Z}$ with unit length along the *XYZ-axes* are *artificial or nonphysical primary vectors.* The X, Y, Z components of monochromatic colors all having the same radiance are $\bar{x}(\lambda)$, $\bar{y}(\lambda)$, and $\bar{z}(\lambda)$ (Fig. 21.1.3). Given a color with a radiance having a spectral power density $E(\lambda)$ (radiance per nm), the relative X, Y, Z components are

$$X = \int_0^\infty E(\lambda)\bar{x}(\lambda)d\lambda \qquad Y = \int_0^\infty E(\lambda)\bar{y}(\lambda)d\lambda \qquad Z = \int_0^\infty E(\lambda)\bar{z}(\lambda)d\lambda \qquad (7)$$

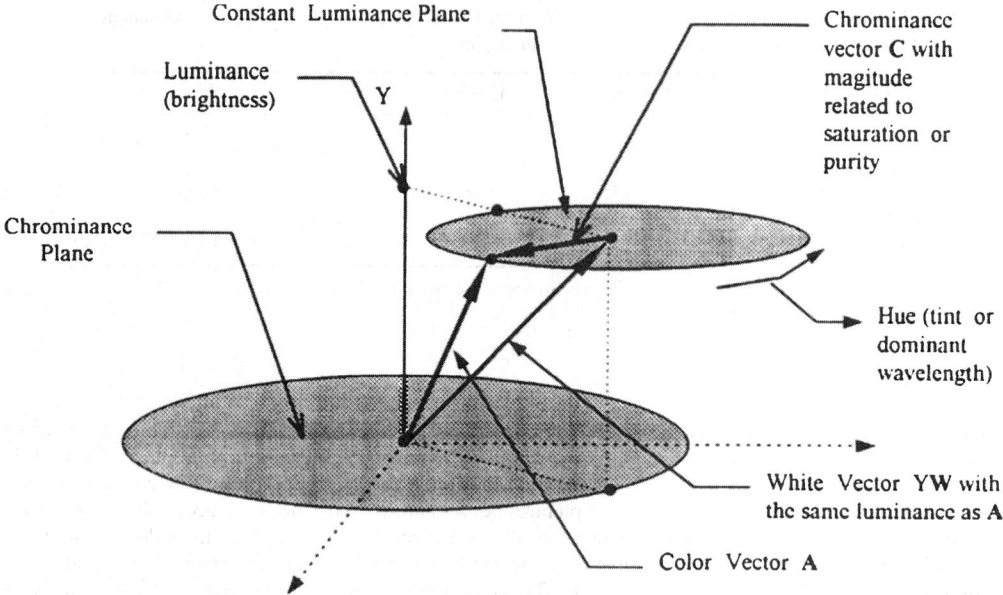

FIGURE 21.1.4 Representation of a color as a vector **A** with three attributes luminance (brightness), hue (tint), and saturation. **A** has a vertical luminance component Y. **A** is the sum of a white vector **YW**, which has the same luminance as **A**, and a chrominance vector **C**, which is in the constant luminance plane.

Since $X = Y = Z$ for equal energy white (E = const.), the areas under the curves $\bar{x}(\lambda)$, $\bar{y}(\lambda)$, and $\bar{z}(\lambda)$ are the same. Clearly lights with different spectral content $E(\lambda)$ can have the same color. The direction of a color vector is specified by the point [x, y, z] where the vector or its extension penetrates the unit plane $X + Y + Z = 1$.

$$[x, y, z] = [X, Y, Z]/(X + Y + Z) \tag{8}$$

thus $x + y + z = 1$ and $Y/y = X/x = Z/z = X + Y + Z =$ "gain."

The coordinates x, y, z are referred to as the *CIE chromaticities* of the color. Clearly $z = 1 - x - y$ is redundant. It is common and practical to specify a color vector **A** by chromaticities x and y and luminance Y as

$$\mathbf{A} = [X, Y, Z] = (Y/y)[x, y, 1 - x - y)] \tag{9}$$

Figure 21.1.5 shows the CIE color space and the CIE chromaticity plane penetrated by color vectors or their extensions. Figure 21.1.6 shows the horseshoe-shaped locus of the xy-chromaticities of monochromatic light $[x(\lambda), y(\lambda)] = [\bar{x}(\lambda), \bar{y}(\lambda)] / (\bar{x}(\lambda) + \bar{y}(\lambda) + \bar{z}(\lambda))$. All realizable chromaticities are within the horseshoe. The magenta colors along the straight line from extreme blue to extreme red are not monochromatic, but are obtained as red-blue mixtures. Areas in the CIE chromaticity chart of just noticeable differences in chromaticities (JNDs) are shaped like ellipses which are larger in the green area than in the red area, which in turn are larger than in the blue area. The number of chromaticities (color vector directions) required for perfect color reproduction has been estimated to range from 8000 (13 bits) to 130,000 (17 bits). The number of chromaticity JNDs within the color gamuts of current display devices has been estimated to be about 5000. In television standards all color vectors are defined in CIE color space.

The CIE components, X, Y, Z are not practical. Television standards specify realizable red, green, and blue *primary colors* **R, G, B**, which are not in the same plane. In terms of these primaries a color is represented as a vector

$$\mathbf{A} = [X, Y, Z] = R\mathbf{R} + G\mathbf{G} + B\mathbf{B} = Y\mathbf{W} + \mathbf{C} \tag{10}$$

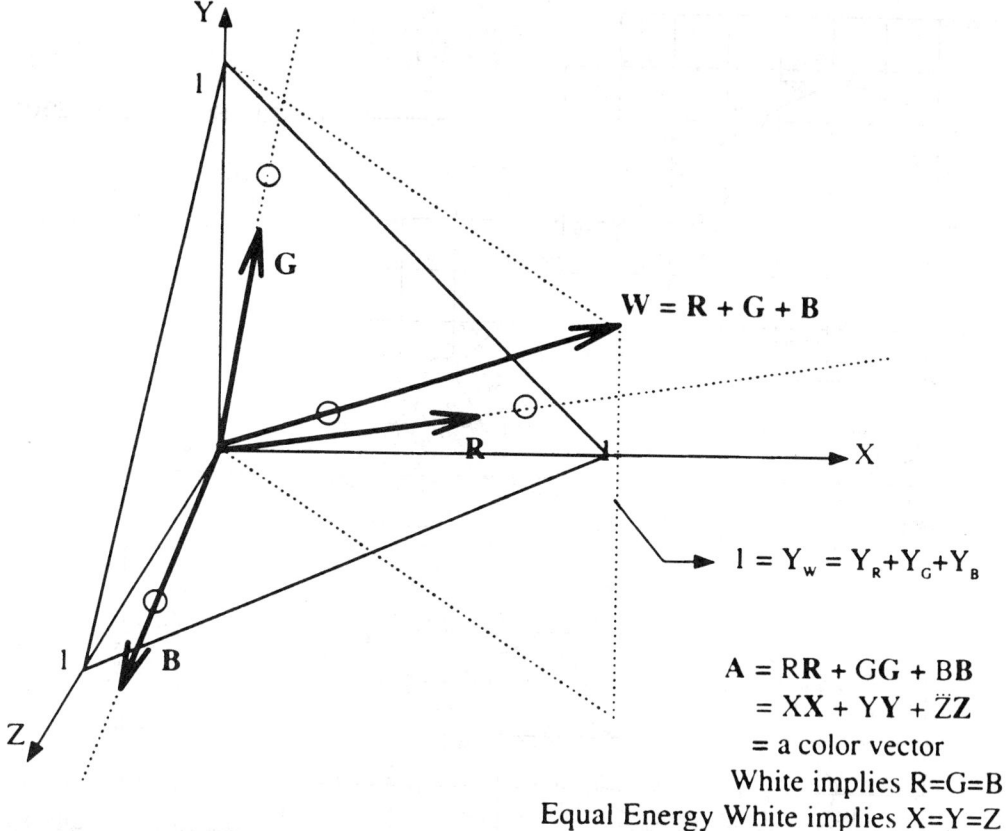

FIGURE 21.1.5   CIE color space with CIE chromaticity plane $X + Y + Z = 1$. The CIE chromaticities of a color vector are the coordinates x, y, and $z = 1 - x - y$, where the vector or its extension penetrates the CIE chromaticity plane. These points are shown for a set of **R**, **G**, **B** basis vectors and the associated white vector **W**, which by definition has a luminance of unity. **X**, **Y**, **Z** are unit vectors along the X, Y, Z axes.

where R, G, B are *tristimulus values* = quantity of each primary color in the mixture. The luminance Y of **A** multiplies a "reference *white vector*" **W** = **R** + **G** + **B**, which, by definition, has a luminance component normalized to unity ($Y_w = 1$). Thus *chrominance vector* **C** is in a *constant luminance plane* ($Y_c = 0$). The luminance components of **R**, **G**, and **B** are Y_r, Y_g, and Y_b, respectively. Consequently:

$$Y = Y_r R + Y_g G + Y_b B = \text{the luminance component of } \mathbf{A} \tag{11a}$$

$$1 = Y_r + Y_g + Y_b = \text{normalized luminance component of the white vector } \mathbf{W} \tag{11b}$$

$$\mathbf{W} = \mathbf{R} + \mathbf{G} + \mathbf{B} = \text{white vector with luminance component } Y_w = 1 \tag{11c}$$

$$C = M(R - Y) + N(B - Y) = \text{chrominance vector with luminance component } Y_c = 0 \tag{12a}$$

$$\mathbf{M} = \mathbf{R} - (Y_r/Y_g)\mathbf{G} = \text{the R} - \text{Y basis vector in the chrominance plane} \tag{12b}$$

$$\mathbf{N} = \mathbf{B} - (Y_b/Y_g)\mathbf{G} = \text{the B} - \text{Y basis vector in the chrominance plane} \tag{12c}$$

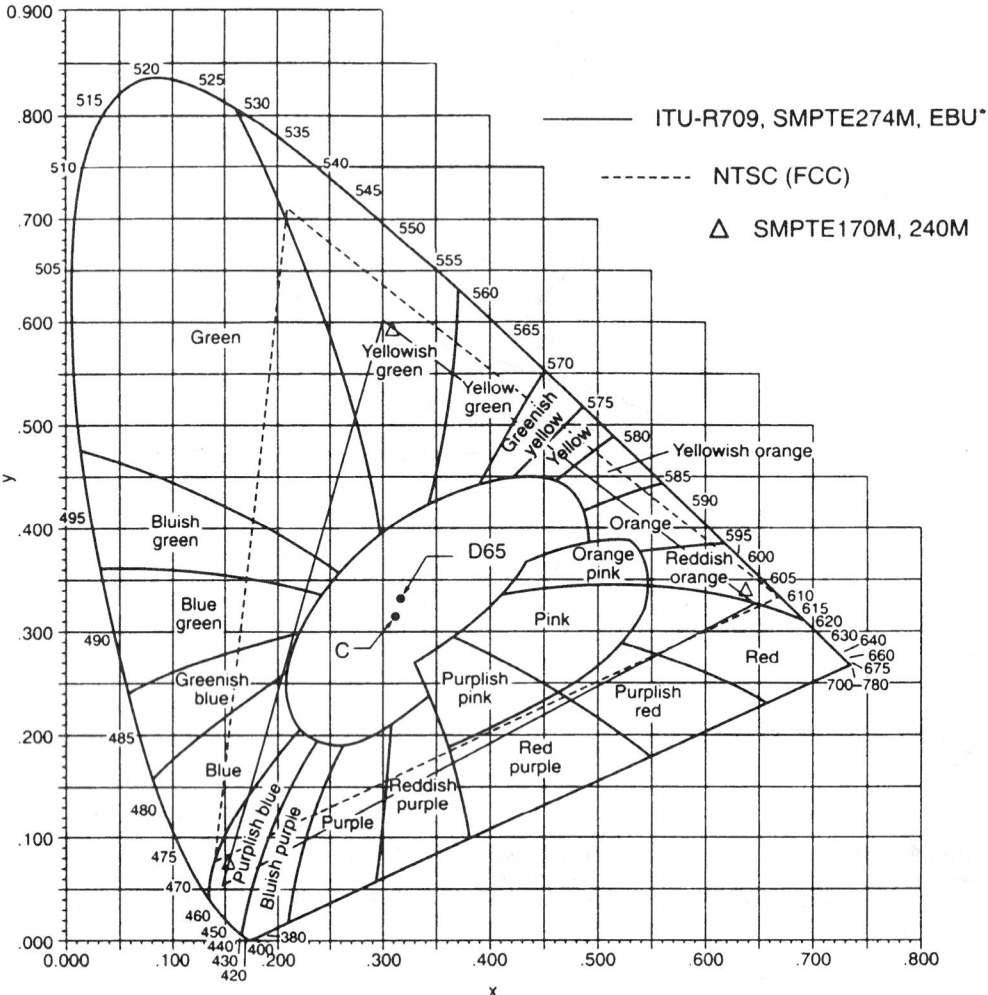

FIGURE 21.1.6 The *xy*-chromaticity diagram of the CIE system. Also shown are television standard rgb chromaticities and white illuminants. *The EBU *x*-coordinate for green is 0.29.

A white vector implies $C = 0$ and $R = G = B = Y$. Equation (12) is derived from Eqs. (10) and (11). Figure 21.1.5 shows vectors **R, G, B**, and **W**. Figure 21.1.7 shows the vectors **M** and **N**. Given the X, Y, Z components of the primaries **R, G, B** the R, G, B tristimuli can be related to the CIE tristimuli X, Y, Z by a matrix P as:

$$
\begin{vmatrix} X \\ Y \\ Z \end{vmatrix} = P \begin{vmatrix} R \\ G \\ B \end{vmatrix} = \begin{vmatrix} X_r & X_g & X_b \\ Y_r & Y_g & Y_b \\ X_r & Z_g & Z_b \end{vmatrix} \begin{vmatrix} R \\ G \\ B \end{vmatrix}
\tag{13}
$$

RGB components of two different basis systems (e.g., NTSC and PAL) specified by matrices P_1 and P_2 are related by a matrix $P_1 P_2^{-1}$ where P^{-1} is the inverse of P. In television standards the **R, G, B** primaries

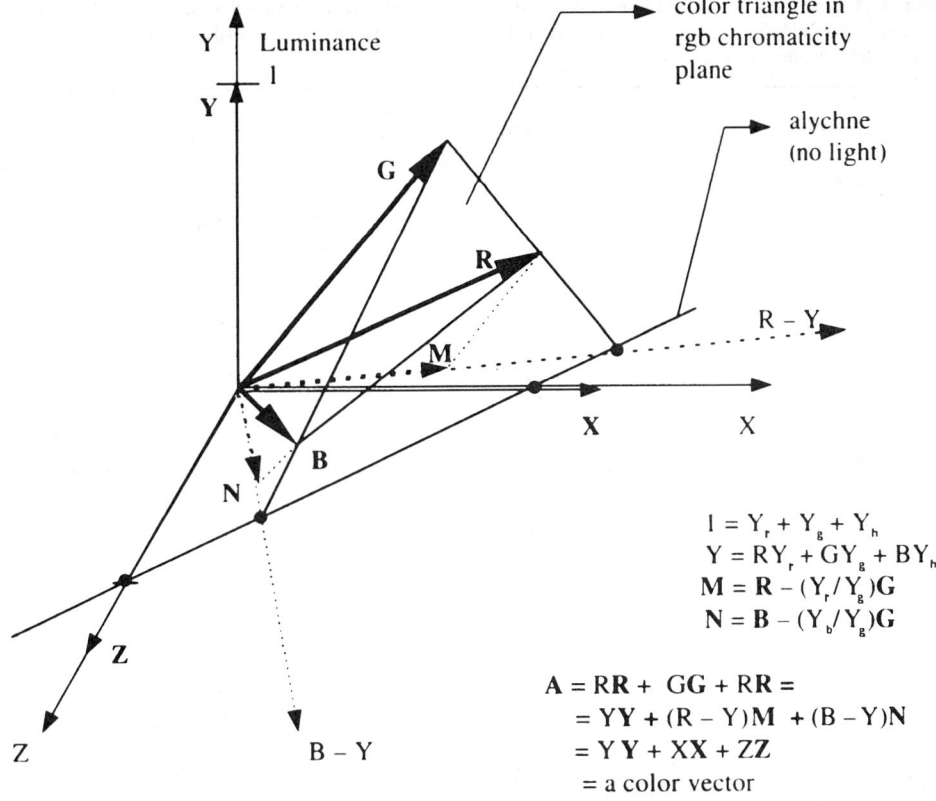

$$1 = Y_r + Y_g + Y_h$$
$$Y = RY_r + GY_g + BY_h$$
$$M = R - (Y_r/Y_g)G$$
$$N = B - (Y_b/Y_g)G$$

$$A = RR + GG + RR =$$
$$= YY + (R - Y)M + (B - Y)N$$
$$= YY + XX + ZZ$$
$$= \text{a color vector}$$

FIGURE 21.1.7   The rgb chromaticity plane goes through the tips of the basis vectors **R, G,** and **B**. It intersects the chrominance plane along the alychne. Also shown are the CIE basis vectors **X, Y,** and **Z** and the basis vectors used in television **Y, M,** and **N**. The chrominance values defined by the chrominance basis vectors **M** and **N** are the "color difference" values $M = R - Y$ and $N = B - Y$.

are specified by their CIE chromaticities and by the chromaticity of **W** (illuminant). The luminance values of the primaries Y_r, Y_g, Y_b can be derived from these eight standardized chromaticities by inverting the matrix **P** for the white vector $R = G = B = Y_w = 1$ and noting that $X/Y = x/y$ and $Z/Y = z/y$ and that $Y_r + Y_g + Y_b = 1$:

$$
\begin{vmatrix} X_w \\ Y_w \\ Z_w \end{vmatrix} = \begin{vmatrix} x_w/y_w \\ 1 \\ z_w/y_w \end{vmatrix} = \begin{vmatrix} x_r/y_r & x_g/y_g & x_b/y_b \\ 1 & 1 & 1 \\ z_r/y_r & z_g/y_g & z_b/y_b \end{vmatrix} \begin{vmatrix} Y_r \\ Y_g \\ Y_b \end{vmatrix}
\tag{14}
$$

Given the Y values of the primaries, the X and Z values can be determined by Eq. (9). Table 21.1.2 lists the xy-chromaticities and the luminance values of the primaries in standard TV systems. Figure 21.1.6 shows the NTSC, ITU-R709 (SMPTE274M), EBU, and the SMPTE170M (SMPTE240M) chromaticities in the CIE diagram. Television programs are produced with the assumption that display devices have (phosphors with) these chromaticities. The chromaticities of the primaries of a display device can be marked as corners of a triangle in Fig. 21.1.6. Since the tristimuli R, G, B are nonnegative (see Figs. 21.1.5 and 21.1.7, only colors with chromaticities within the triangle can be displayed.

TABLE 21.1.2   Chromaticities and Luminance Components in Standard Television Systems. White Illuminant (last row) is C in NTSC, Otherwise D_{65}. Note that ITU-R709 is Almost the Same as EBU, and That NTSC Accommodates More Green Colors

	NTSC (FCC) System M			ITU-R709 & SMPTE274 EBU* (PAL* & SECAM*)			SMPTE 170M & 240M		
	x	y	Y	x	y	Y	x	y	Y
Red	0.67	0.33	0.299	0.64	0.33	0.2125 0.222*	0.630	0.340	0.212
Green	0.21	0.71	0.587	0.30 0.29*	0.60	0.7154 0.707*	0.310	0.595	0.701
Blue	0.14	0.08	0.114	0.15	0.06	0.0721 0.071*	0.155	0.070	0.087
White	0.3101	0.3162	1	0.3127	0.3291	1	0.3127	0.3291	1

In television the key components are $R - Y$, Y, and $B - Y$ of the **M, W, N** primaries. They are related to the tristimuli R, G, B by the matrix H:

$$\begin{vmatrix} R - Y \\ Y \\ B - Y \end{vmatrix} = H \begin{vmatrix} R \\ G \\ B \end{vmatrix} = \begin{vmatrix} 1 - Y_r & -Y_g & -Y_b \\ Y_r & Y_g & Y_b \\ -Y_r & -Y_g & 1 - Y_b \end{vmatrix} \begin{vmatrix} R \\ G \\ B \end{vmatrix}, \quad \begin{vmatrix} R \\ G \\ B \end{vmatrix} = \begin{vmatrix} 1 & 1 & 0 \\ -Y_r + Y_g & 1 & -Y_b - Y_g \\ 0 & 1 & 1 \end{vmatrix} \begin{vmatrix} R - Y \\ Y \\ B - Y \end{vmatrix} \quad (15)$$

Two colors are said to be *complementary* if some weighted mixture of them results in a white color **WW**. Saturated complementary colors to the primaries **R, G, B** are Cyan **CY** $= $ **G** $+$ **B**, Magenta **MA** $=$ **B** $+$ **R** and Yellow **YE** $=$ **R** $+$ **G**. If a primary has components M, Y, N, the saturated complementary color has components −M, 1 − Y, −N. The RGB and MYN values of these basic colors in NTSC, ITU-R709 & SMPTE274M, and SMPTE240M are listed in Table 21.1.3. The EBU colors in PAL & SECAM differ only slightly (third decimal) from ITU-R709. These saturated colors can be seen on TV as color bars.

The rgb-Chromaticities and the Color Triangle

Figure 21.1.7 shows a plane connecting the tips of the primary vectors **R, G, B**. This plane is referred to as the rgb-chromaticity plane, not to be confused with the CIE xyz-chromaticity plane. A color vector **A** $=$ R**R** $+$ G**G** $+$ B**B** $= (R + G + B)(r$**R** $+ g$**G** $+ b$**B**$)$ has rgb-chromaticities $= [r,g,b] = [R,G,B]/(R + G + B)$. Clearly $g = 1 - r - b$ is redundant. An rgb-chromaticity vector is conveniently represented as **G** $+ r($**R** $-$ **G**$) + b($**B** $-$ **G**$)$.

TABLE 21.1.3   Values of Y, M $= R - Y$ and N $= B - Y$ for Saturated Colors

		NTSC			ITU-R709 & SMPTE 274M			SMPTE 240M & 170M		
COLOR	R G B	Y	R − Y	B − Y	Y	R − Y	B − Y	Y	R − Y	B − Y
White	1 1 1	1	0	0	1	0	0	1	0	0
Yellow	1 1 0	0.886	+0.114	−0.886	0.928	+0.072	−0.928	0.913	+0.087	−0.913
Cyan	0 1 1	0.701	−0.701	+0.299	0.787	−0.787	+0.213	0.788	−0.788	+0.212
Green	0 1 0	0.587	−0.587	−0.587	0.715	−0.715	−0.715	0.701	−0.701	−0.701
Megenta	1 0 1	0.413	+0.587	−0.587	0.285	+0.715	+0.715	0.299	+0.701	+0.701
Red	1 0 0	0.299	+0.701	−0.299	0.213	+0.787	−0.213	0.212	+0.788	−0.212
Blue	0 0 1	0.114	−0.114	+0.886	0.072	−0.072	+0.928	0.087	−0.087	+0.913

The rgb-chromaticity of the white vector $\mathbf{W}$ is r = g = b = 1/3. Figure 21.1.7 also shows that the **GR** and **GB** planes intersect the chrominance plane along the M = R − Y and N = B − Y axes. The xyz chromaticities are nonlinearly related to the rgb chromaticities. The relations are easily determined with Eq. 13, given [R,G,B] = [r,g,b] *or* [X,Y,Z] = [x,y,z]. The figure also shows a line, the *alychne* (no light), which is the intersection of the rgb-chromaticity plane and the chrominance plane Y = 0.

The tips of the **R, G,** and **B** primary vectors are the corners of the ***color triangle***. The rgb-chromaticities of the saturated complementary colors, cyan, magenta, and yellow, are midpoints on the sides of the triangle. The color triangle can be mapped to take more convenient shapes, for example, in a right angle or coordinate system. Figure 21.1.6 shows color triangles for some television standards first polar-projected onto the CIE chromaticity plane and subsequently parallel projected onto CIE xy-chromaticity diagram. Display devices using the standard chromaticities can only display colors with chromaticities inside the color triangle (nonnegative R, G, B tristimuli).

Relation Between Chromaticities and Chrominance-to-Luminance Ratios

Since R − Y, Y, B − Y are related to signals in TV transmission, it is of interest to relate the ratios R/Y and B/Y to the xyz (CIE) chromaticities. In the CIE chromaticity diagram (Fig. 21.1.6) all lines R/Y = constant go through the intersection of the projection of the green-blue side (R/Y = 0) of the color triangle with the x-axis and all B/Y = constant lines go through the intersection of the projection of the green-red side (B/Y = 0) of the color triangle with the x-axis. Linear scales of R/Y and B/Y can be made on any line parallel with the x-axis. The scales are easily calibrated since lines R/Y = 1 and B/Y = 1 for lines going through the illuminant white point. The intersection between the pivoting R/Y and B/Y lines determines the chromaticity. Errors in the (R − Y)/Y and (B − Y)/Y ratios cause errors in the slope of these pivoting lines and may require the display device to produce negative light if they intersect outside the color triangle. Chroma noise and incorrect chroma gain ("color" on TV sets) may cause such errors. Errors in the chrominance cause large chromaticity errors when Y is small, which it is in the blue-magenta-red areas, especially in dark areas.

Implications of Chromaticity Standards

Television standards specify the chromaticities that are expected to be used in the display devices at the receivers. Receiver manufacturers do not have to meet these standards, and in fact, in the United States, more efficient phosphors are used, which do not have the standard FCC chromaticities. The SMPTE has recommended other chromaticities to be used at the source which better match the more efficient phosphors now used in most receivers. Various chromaticity standards are shown in Table 21.1.2. SMPTE 170M is proposed as NTSC broadcast studio standards and ITU-R709 (equal to SMPTE 274M) and 240M for DTV standard. White illuminant D_{65} is becoming universal. Figure 21.1.6 shows that the "old" NTSC gamut of colors (triangle) is substantially larger in the green-cyan area than the gamuts of the newer chromaticities. However, the gamuts of the new phosphors cover almost the gamut of paints, pigments, and dies, and there are few complaints about the green color. Future displays may provide larger gamuts and force a change in *de facto* standards.

The implications of chromaticity standards or of *de facto* standards are that in production of television programs it is expected that most displays in the receivers have luminous phosphors and a reference white illuminant with chromaticities close to the standards. Cameras usually have sensors with nonstandard chromaticities. The output signals from a camera can be matrixed to conform with a large gamut of chromaticity standards. A cameraman can adjust this matrix and process the output signals (linearly as well as nonlinearly) to produce an artistically desirable picture on the faceplate of displays with standard or *de facto* standard primaries. The chromaticities listed in television standards are not enforceable, but are taken as guidelines for designing receivers and displays.

Progress in Psychophysics of the Human Visual and Aural Systems

Current television standards are based on an understanding of psychophysics which dates back a number of decades. Although current standard analog television systems are clever, economical, and robust, it is clear that they make inefficient use of channel capacity. In a 6-MHz transmission channel, over 90 percent of the energy of the visual information occupies a bandwidth of a few hundred kHz. With the advent of digital television, methods for effective *reduction of irrelevancy and redundancy* in communication of moving pictures and audio benefit from a new technology basis in digital compression.

PICTURE SIZE, SCANNING PROCESS, AND RESOLUTION LIMITS

Picture Size and Scanning Raster

In analog TV systems, the transmitted signal was designed to be directly related to the image representation of the camera and the display. The white rectangular area in Fig. 21.1.8 shows the *active area* of the picture, i.e., the area containing pictorial information. The active area is assumed to be flat. The height of the active area is the length unit ph (picture height) and the width A ph of the active area is the *aspect ratio*. In current standard television systems the aspect ratio is 4/3 and in HDTV it is 16/9. The viewing distance D is commonly assumed to be 6 to 8 ph in standard TV and 3 to 4 ph in HDTV.

Also shown in Fig. 21.1.8 is an imaginary *total picture area* that includes inactive areas representing durations used in the scanning process for synchronization and retrace blanking. These durations are referred to as the *vertical blanking interval VBI* and the *horizontal blanking interval HBI*. The *total vertical height* including the VBI is $1/\eta_V$ and the *total horizontal width* including the HBI is A/η_H where η_V and η_H are vertical and horizontal scanning efficiencies. They are shown in Table 21.1.4 for analog TV systems and their standard definition DTV equivalents.

In all television systems the total picture is scanned along horizontal *lines* from left to right while the scan lines progress downward from the top of the picture. The duration of a horizontal scan line including the duration of the HBI is $H = 1/F_H$, where F_H = line rate in Hz. The duration of a *field* including the VBI is $V = 1/F_V$, where F_V = field rate in Hz. In television standards a scanning system is specified as N/F_V where $\eta_V N = N_0$ = number of nonoverlapping lines displayed in the visible picture area (active lines). In *progressive scan* ("proscan" or "1:1 scan") displayed lines in successive fields are co-located (overlap) and consequently the total number of scan lines is $N = F_H/F_V = V/H$. In 2:1 *interlaced* scan, displayed lines in even numbered fields are interlaced with the lines in odd-numbered fields and consequently $N = 2F_H/F_V = 2V/H$. When referring to "interlaced scan" it is generally assumed to be 2:1 interlaced. Multiple interlace without storage causes visible

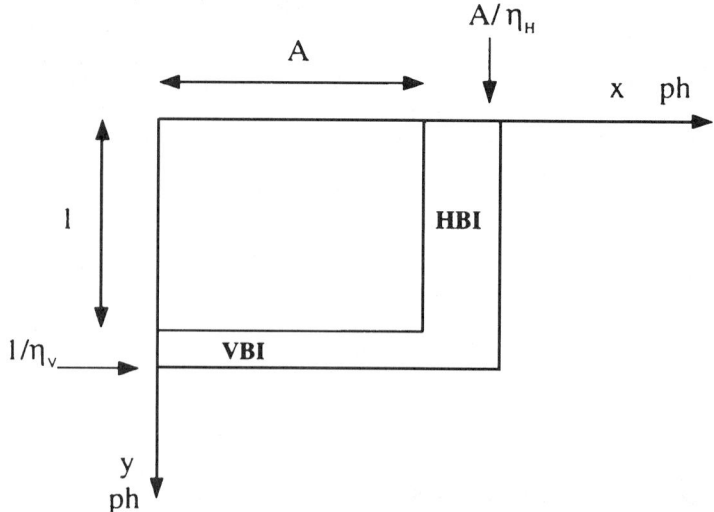

FIGURE 21.1.8   Active picture area and the total area that includes the horizontal and vertical blanking intervals. η_v and η_H are scanning efficiencies.

TABLE 21.1.4 Scanning Parameters and Scanning Efficiencies in Interlaced Analog
TV Systems and Their Digital Equivalents

	Analog		Digital CCIR601	
	525/60	625/50	525/60	625/50
A	4/3	4/3	4/3	4/3
N	525	625	525	625
F_V Hz	60/1.001	50	60/1.001	50
F_H Hz	15,750/1.001	15,625	15,750/1.001	15,625
η_v	0.92	0.92	480/525	576/625
η_H	0.8285	0.8125	720/858	720/864
$N_o = N\eta_v$	483	575	480	576
$V_v = F_v/\eta_v$ ph/s	65.152	54.348	65.559	54.253
$V_H = F_H A/\eta_H$ ph/s	25,321	25,641	25,000	25,000
$V_T = F_v/N_o$ ph/s	0.1241	0.0870	0.1249	0.0868

line-crawl. In interlaced scan *one frame comprises an odd-numbered field followed by an even-numbered field*. The frame rate is $F_v/2$ Hz. In proscan a frame and a field is the same thing. Automatic interlace can be obtained if N = number of lines per frame is odd, as shown in Fig. 21.1.9. Interlaced scan can be achieved with an even number of lines per frame and proscan can be achieved with an odd number of half lines per field. The PAL/SECAM systems are 625/50(2:1) systems and the NTSC system is a 525/59.94(2:1) system or simply as a 525/60(2:1) system (although the field rate is exactly 60/1.001 Hz). Figure 21.1.9 shows the scanning raster for the interlaced 525/60 system. The scanning raster shown in Fig. 21.1.9 covers the total area including the blanking intervals. Interlaced scan has been used in all analog television systems to achieve good vertical resolution and little large area flicker, given available transmission bandwidth.

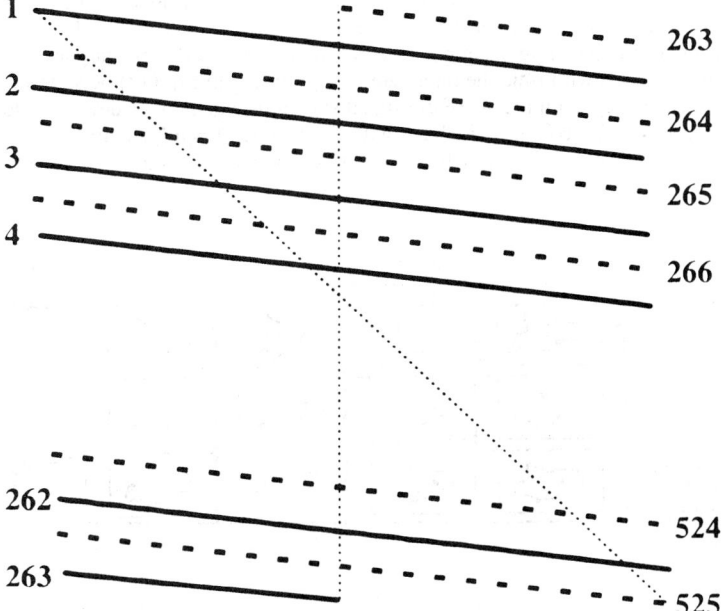

FIGURE 21.1.9 Interlaced scanning raster covering the total picture (including blanking intervals) for a 525/60 system (NTSC). Retrace during horizontal blanking is not shown.

Table 21.1.4 shows scanning parameters for various raster formats including the number of visible scan lines $N_o = \eta_v N$, the *vertical scanning velocity* $v_V = F_V/\eta_v$ ph/s, and *the horizontal scanning velocity* $v_H = F_H A/\eta_H$ ph/s in interlaced scan (v_H must be doubled for proscan). Velocities of moving objects in the picture are conveniently expressed relative to the reference velocity $v_T = F_V/N_o$ ph/s (= one line-spacing per field). In digitized systems η_H = ratio of active samples per line to total samples per line.

The Video Signal Spectrum.   The dynamic image on the target of a TV camera is a three-dimensional message: two spatial and one temporal. In the scanning process this message is converted into a one-dimensional signal. Consider first a stationary image having a property, e.g., a response to exposure, which can be specified as a function $g(x,y)$ defined over the total picture area shown in Fig. 21.1.8. This is a rectangular area of width $A/\eta_H = v_H/F_H$ and a height $1/\eta_V = v_V/F_V$. The scanning beam can be represented as progressing with time along a straight line $x = v_H t$ and $y = v_V t$ over the periodically repeated image as shown in Fig. 21.1.10. The figure shows an interlaced scan with five lines per frame. A two-dimensional Fourier series component of the periodically repeated function $g(x,y)$ is sampled by the scanning beam to yield a one-dimensional signal:

$$C_{mn}\cos 2\pi[(mF_H/v_H)x + (nF_V/v_V)y + c_{mn}] = C_{mn}\cos 2\pi[(mF_H + nF_V)t + c_{mn}] \qquad (16)$$

where $n = 0, \pm 1, \pm 2$, when $m = 1,2,3,\ldots$, and $n = 0,1,2,3,\ldots$, when $m = 0$. A Fourier component in the picture is a sinewave grating with a constant phase wavefront perpendicular to the spatial frequency vector $\mathbf{k}_{mn} = [(mF_H/v_H), (nF_V/v_V)]$ = *cycles per picture height* (cph). In the scanning process this grating generates a spectral component in the signal at the discrete frequency $mF_H + nF_V$ Hz. A grating that cannot be expressed with m and n as integers has "kinks" at the borders of the total pictures shown in Fig. 21.1.10 and consists of several gratings of the type shown in Eq. (16).

Figure 21.1.11 shows a part of the spectrum of an interlaced system with $N = 9$ lines. Spectral components for $n > 0$ interleave with components for $n < 0$. The figure shows that a frequency component can be generated in more than one way (*aliasing*) if $|n| > N/2$. Thus, as expected, the highest vertical spatial frequency that can be conveyed without aliasing is $N_o/2$ cph. Aliasing shows in certain pictures: venetian blinds, striped shirts, resolution test-patterns. If the horizontal scan frequency is doubled to $2F_H$ (progressive scan with *N* lines per field), the spectral components shown as dotted lines in Fig. 21.1.11 disappear. The components are spaced F_V in proscan and $F_V/2$ interlaced scan, but aliasing occurs in both systems if $|n| > N/2$. Optical defocusing and vertical defocusing of the scanning beam reduces aliasing at the expense of sharpness.

As the picture changes with time the parameters C_{mn} and c_{mn} in Eq. (14) become functions of time. As a consequence the frequency components of the signal shown in Fig. 21.1.11 develop sidebands just as amplitude and phase modulated carriers. A sideband with a frequency $f = mF_H + nF_V + f_{mn}$ could have been generated by the sinewave grating defined by Eq. (16) with C_{mn} = constant and $c_{mn} = f_{mn}t$ + constant. This moving

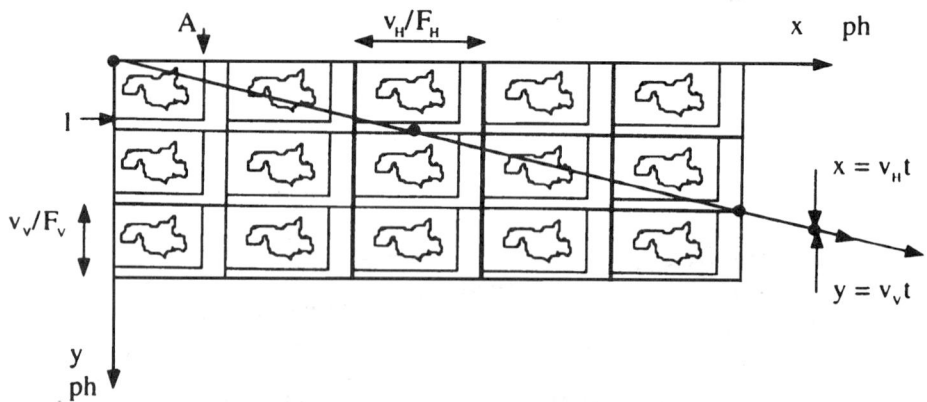

FIGURE 21.1.10   Interlaced scan of a stationary picture with $N = 5$ lines per frame.

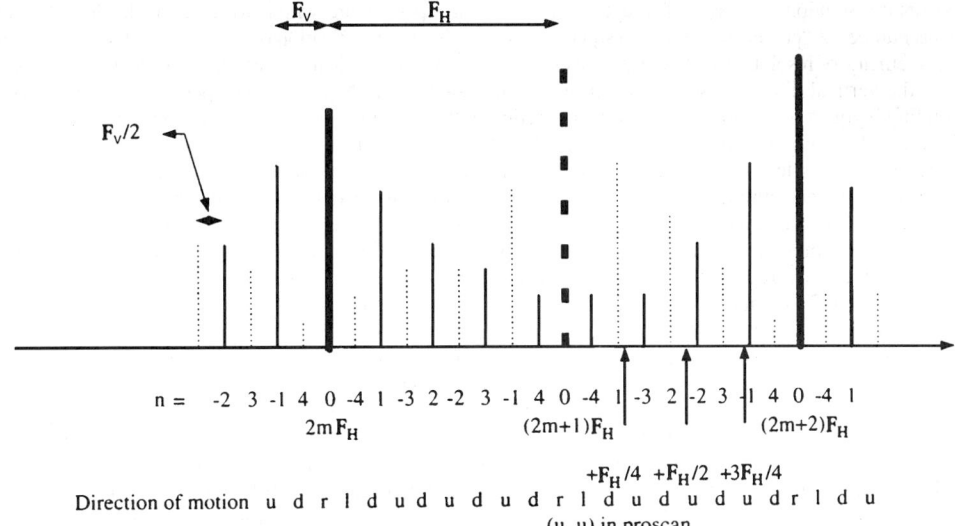

FIGURE 21.1.11 Part of the frequency spectrum generated by interlaced scan of a stationary picture with $N = (4p + 1) = 9$ lines per frame. In proscan at line-rate $2F_H$, dotted line components disappear. Motion directions up, down, left, right are for sidebands associated with "carriers" at the endpoints of the $F_V/2$ intervals.

grating causes the amplitude at every active point x, y in the picture to oscillate with a frequency f_{mn}. The constant phase front of the grating moves in *a direction opposite to* $\mathbf{k}_{mn}$ *with a phase velocity*

$$v_{mn} = f_{mn}/|\mathbf{k}_{mn}| = f_{mn}\,\lambda_{mn} \quad \text{ph/s} \quad \lambda_{mn} = \text{wavelength in ph} \tag{17}$$

A component at a frequency f in the video signal can, however, be generated by a number of moving sinewave gratings with different velocities and spatial frequencies, and it is up to the viewer to interpret and choose between $f = mF_H + nF_V + f_{mn}$ and $f = pF_H + qF_V + f_{pq}$. More often than not this ambiguity causes little confusion, because a moving object generates sidebands to a large number of carriers, including low-frequency carriers, with the consequence that the viewer will choose a good fit to his or her expectation of the motion. This psychophysical phenomenon is also taken advantage of in a 24 frame per second motion picture (aliasing: wagon wheels appearing as rotating the wrong way). In television, however, the confusion is accentuated by the scanning raster.

Figure 21.1.11 indicates perceived direction of motion (up, down, left, right) of a spectral component, under the assumption that the viewer interprets it as a sideband to either the nearest carrier to the left or the nearest carrier to the right. The motion designation (u, d, l, r) of these frequency intervals of width $F_V/2$ remain in proscan. Viewers tend to choose a carrier with the lowest spatial resolution, i.e., with the lowest value of $|n|$. As the velocity of a grating increases and $|f_{mn}|$ goes through $F_V/2$ the grating may appear to reverse direction of motion. This may cause visible aliasing effects, e.g., when busy pictures are panned. However, confusion is not likely if $|\mathbf{k}_{mn}| < (dF_V/2)/|v_{mn}|$ ph/s, where d = 1 in proscan and $d = 1 - 2|n|/(N - 1)$ in interlaced scan. Thus, for a given velocity, the highest nonaliasing spatial frequency depends only on d times the frame rate, which is the same for HDTV and standard TV. For example, for d = 1 and v = 1 ph/s, the maximum nonaliasing spatial frequency is 30 cph in N/60 systems and 25 cph in N/50 systems. Motion is handled better in progressive scan (d = 1) than in interlaced scan (d < 1). If C_{mn} varies with time but c_{mn} is constant there is level variation but no motion.

Resolution

Resolution is expressed in terms of the highest spatial frequencies k_{max} in *cycles per (picture) height* (cph), which can be conveyed by the television system and displayed given a scanning process and a transmission bandwidth. Sometimes resolution is still expressed in the old unit "TV lines," which should mean $2k_{max}$ cph. Sometimes horizontal resolution is expressed in cycles or in TV lines per picture width, which is misleading. Sometimes it means number of "perceived" lines of a square wave test signal. Specification of resolution in terms of TV lines and measurements of resolution with square waves should be avoided. Square waves may have lost some harmonics.

In the vertical direction, stationary information is sampled at N_0 active lines per picture height. The corresponding Nyquist bandwidth $N_0/2$ cph is theoretically the maximum *vertical resolution* for stationary pictures. The vertical resolution in the NTSC system is $k_v = N_0/2 = 241.5$ cph. The "perceived" vertical resolution of stationary pictures is less than $N_0/2$ because of many factors: aliasing, source processing, scanning spot sizes, contrast, brightness, ambient conditions, test signals, viewing distance, line flicker, luminance-chrominance cross talk, and confusion caused by the display of the scanning raster (display of high spatial frequency repeat spectra). The combination of all these factors on perceived vertical resolution is sometimes expressed in terms of some perceived vertical resolution $KN_0/2$, where K = *Kell factor* < 1. The Kell factor, which has been quoted to range from 0.7 to 1. must be used with many qualifications and a great deal of caution. It is not a basic system parameter. It should preferably be avoided. In interlaced scan *line flicker* can be very disturbing in busy slow-moving pictures. In some receivers line flicker is eliminated by converting interlaced scan into progressive scan ("line doubling") by motion-adaptive interpolation.

The *horizontal resolution* of stationary pictures is $k_H = B/v_H$ cph, where B is the highest frequency in Hz of the information, which can theoretically be conveyed and which modulates "the scanning beam." For example, in the NTSC system the highest frequency of the luminance signal is B = 4,200,000 Hz. Consequently the horizontal resolution of luminance in NTSC is $k_H = 4,200,000/25,321 = 166$ cph. The "perceived" horizontal resolution is less than k_H depending on many factors: video filter transfer functions, luminance-chrominance cross talk, type of test signal, contrast, and so forth. The horizontal resolution of the NTSC chrominance signals are $k_I = 52$ and $K_Q = 24$ cph, which are much less than the vertical chrominance resolution (241.5 cph).

When the NTSC system was developed it was observed that horizontal and vertical *chrominance resolution* can be reduced to about half the luminance resolution without significantly reducing perceived quality. This observation was crucial for the success of NTSC.

Table 21.1.5 shows the theoretical maximum bandwidth and resolution of luminance and chrominance in various systems. In the analog TV systems chrominance resolution is significantly reduced in the horizontal

TABLE 21.1.5   Potential Resolutions in cph in the Interlaced Systems Listed in Table 21.1.2 and the HDTV Common Usage Format (The Digital CCIR System is a 4:2:2 System with a Luminance Sample Rate 13.5 MHz and Chrominance Sample Rate 6.75 MHz. In the 1125/60 System, Luminance is Sampled at 74.25 MHz and in the 1250/50 System Luminance is Sampled at 72 MHz Nyquist Maximum Frequencies and Resolutions Are Shown for all Sampled Systems)

		Analog (A = 4/3)		Digital CCIR 601[4]		Digital HDTV (A = 16/9)		
		NTSC	PAL	525/60	625/50	750/60[6]	1125/60	1250/50
M_0/N_0		443/483	520/575	720/480	720/576	1280/720	1920/1080	2048/1152
Max.Hor.Y Freq. in MHz		4.2	5[3]	6.75[4]	6.75[4]	37.125	37.125	36
Max.Hor R − Y Freq. in MHz		1.3[1]	1.3	3.375[4]	3.375[4]	18.5625	18.5625	18
Max.Hor B − Y Freq. in MHz		.6[2]	1.3	3.375	3.375	18.5625	18.5625	18
Y′	Hor.	166	195[3]	270[4]	270[4]	360	540	576
Y	Vert.	241.5	288	240	288	360	540	576
R − Y	Hor.	51.5[1]	50.5	135[4]	135[4]	180	270	288
B − Y	Hor.	24[2]	50.5	135[4]	135[4]	180	270	288
R − Y, B − Y	Vert.	241.5[1,2]	288[*]	240	288	180	270[5]	288[5]

(1) Applies to I = (R − Y)cos33° − (B − Y)sin33° (2) Applies to Q = (R − Y)sin 30° + (B − Y)cos33° (3) 5.5 MHz in PAL/I and 6 MHz in SECAM. The corresponding max resolutions of Y are 214.5 and 234 cph, respectively. [*]144 in SECAM. (4) *Standard CCIR601* luminance and chrominance bandwidths are 5.75 and 2.75 MHz respectively and A = 4/3. Corresponding resolutions are *230 and 115 cph*. (5) In the HDTV systems the resolution of R − Y and B − Y is half the luminance resolution horizontally as well as vertically. (6) Progressive scan.

direction, but it is not reduced in the vertical direction. In digital TV with rectangular pixels the maximum resolution is the Nyquist bandwidth in cph determined by the horizontal and vertical distances between neighboring samples. The meaning of 4:2:2 is that for every four samples of the luminance signal Y along a horizontal line there are two samples of $B - Y$ and $R - Y$. In 4:2:2 systems the chrominance resolution is half the luminance resolution in the horizontal direction. In 4:2:0 systems the chrominance resolution is half the luminance resolution both horizontally and vertically. Tables 21.1.4 and 21.1.5 show a 4:2:2 version of CCIR601, while HDTV systems are 4:2:0 versions. The resolutions shown in Table 21.1.5 can be related to the luminance resolution of the HVS, which is at most 250 cph and 500 cph at viewing distances of 8 and 4 ph, respectively. It should be emphasized that the resolutions shown in Table 21.1.5 are theoretical maxima. The resolutions of the pictures displayed by a receiver depend on many factors and are usually much lower than the maximum resolutions shown in Table 21.1.5. One reason is that the signals conveyed by the system are not proportional to the tristimuli Y, $R - Y$, $B - Y$. In all systems the tristimuli are nonlinearly compressed before matrixed and bandlimited to form video signals Y', $B' - Y'$, and $R' - Y'$. Whatever the resulting resolutions are in cph, the displayed resolution in cycles per visible height is usually lower owing to the overscan needed to allow for deflection tolerances.

In summary, the critical resolutions in cph, given the number of visible scan lines N_o, a maximum horizontal frequency B Hz, a grating phase velocity v, and a viewing distance D ph are:

$$k_{max} \text{ (hor.)} = B/v_H, \qquad k_{max} \text{ (vert.)} = N_o/2, \qquad k_{max}(\text{move}) = dF_v/2v, \qquad k_{max}(\text{HVS}) \cong 2000/D \qquad (18)$$

where d = 1 in proscan and $1 - 2|k_y|/N_z$ in interlaced scan (k_y = vertical spatial frequency of grating in cph).

Table 21.1.5 summarizes the static resolutions of the interlaced systems listed in Table 21.1.4 and the ATSC (Grand Alliance) HDTV system formats. The IQ chrominance axes in NTSC were chosen because the HVS can better resolve chrominance detail in the I direction (approximately red-cyan) than in the Q direction (approximately blue-yellow).

Standard Frequencies in Color Television Systems

Figure 21.1.12 shows that the key frequencies used in worldwide color television systems are related to a common frequency = 2.25 MHz. Multiples of this frequency are used as international standards for sample rates in HDTV and in the digital CCIR601 standard used in production and in the professional tape recording in the D1 format. The horizontal line rate in NTSC is $F_H = 2.25/143$ MHz and in PAL and SECAM it is $F_h = 2.25/144$ MHz. The field rate in NTSC is 60/1.001 Hz and in PAL and SECAM it is 50 Hz. ATSC Grand Alliance HDTV formats when operating with 60 Hz field rates have frequencies related to 2.25 MHz, but when operating with 60/1.001 they are related 2.25/1.001 MHz.

In the NTSC, PAL, and SECAM systems, which are *compatible* with black and white reception, most of the chrominance information $R - Y$ and $B - Y$ is conveyed by modulating one or two subcarriers to form a *chrominance subchannel*, which is added to (multiplexed with) the signal that carries most of the luminance information Y. The resulting signal is a *composite video signal*. In the FCC rules the NTSC color subcarrier frequency is specified to be $f_c = 315/88$ MHz = 227.5 F_H = 3.58…MHz. In PAL the color subcarrier is specified to be $f_c = 283.75F_H + F_v/2 = 4,433,618,75$ MHz. SECAM operates with two frequency modulated subcarriers: $f_{ob} = 4.25$ MHz and $f_{or} = 4.40625$ MHz. The subcarriers are chosen to provide acceptable compatibility with black and white reception and to minimize visible cross talk between luminance and chrominance in color reception.

The peculiar frequencies in NTSC resulted from a slight modification of the monochrome standards (60 Hz and 15.75 kHz) for the purpose of reducing the visibility of a beat between the color subcarrier and the 4.5 MHz sound carrier. F_H was chosen to be 4.5/286 MHz to make the beat fall at an odd multiple of $F_H/2$ ((286 – 227.5)F_H = 920 kHz). This is how the 2.25 MHz frequency came about. In digital processing of NTSC and PAL signals, including professional recording in the D2 format as well as in signal processing in consumer products, $4f_c$ is often used as the sample rate. This sample rate exceeds the Nyquist rate somewhat in most applications, but it is readily available and phase-locked to F_H and F_v. It is particularly well suited for sampling composite video signals because it performs the function of a synchronous detector separating the in-phase and quadrature chrominance component which modulate the color subcarrier.

Some frequencies used or proposed for use in television are not well related to the frequencies shown in Fig. 21.1.12. One is 24 frames per second used in motion pictures. In NTSC broadcasting the movie frame rate

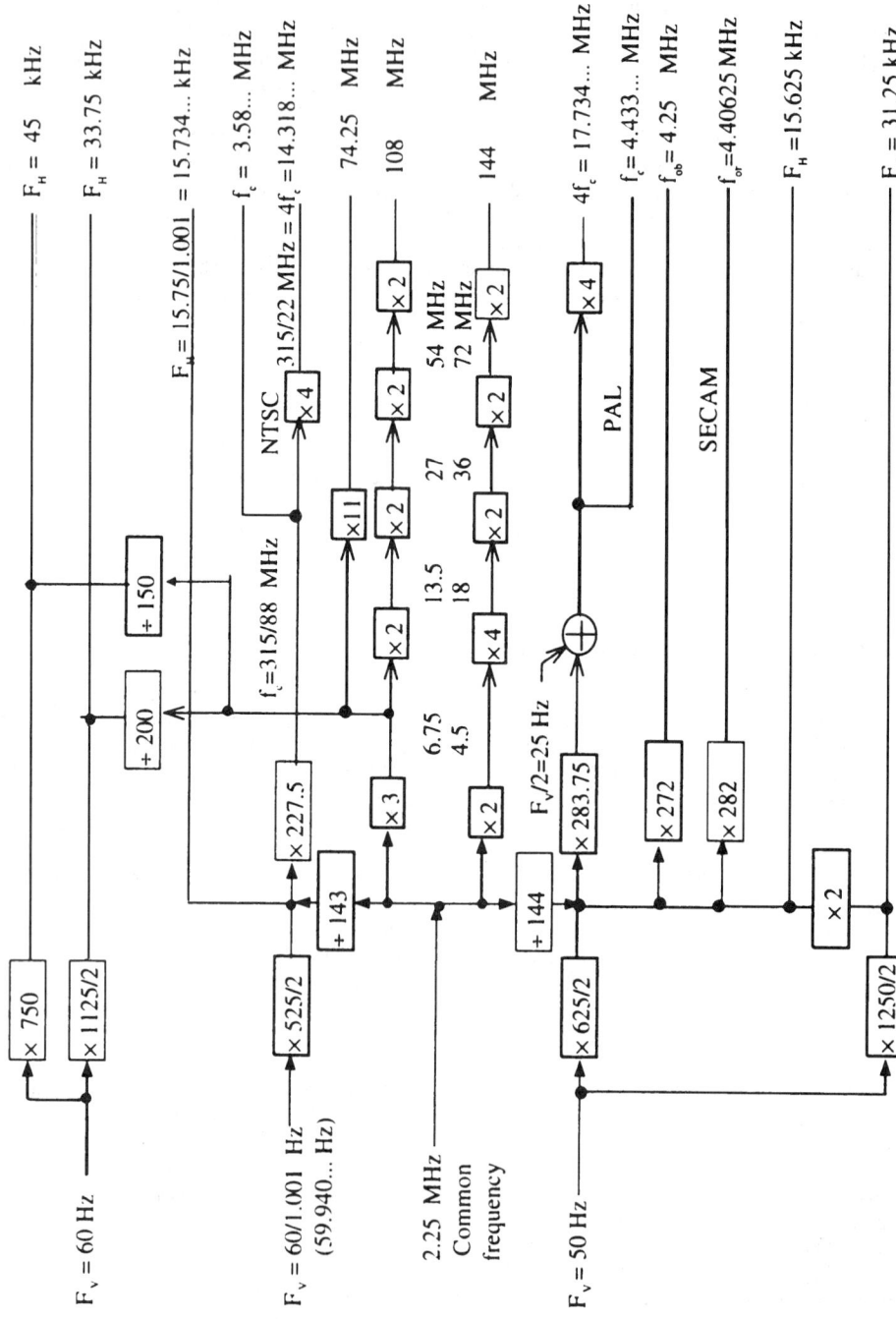

FIGURE 21.1.12 Relations between frequencies in standard analog and digital television systems.

is first reduced to 24/1.001 Hz. This is followed by a so-called 3-2 pull-down used to fit four movie frames to 10 NTSC fields, alternating three fields and two fields per movie frame. In Europe it has been common to speed up the movie by a ratio 25/24, but this increases the pitch of the sound unless digital pitch-preserving nonlinear time base techniques are used to accomplish the audio speedup. With digital techniques, better frame conversions can now be made. Other frequencies that are not well related to 2.25 MHz are: digital audio sample rates 48 and 44.1 kHz (8/375 and 49/2500 times 2.25 MHz).

ANALOG VIDEO SIGNALS

Gamma Pre-Correction

Three signals *R*, *G*, and *B* emerge from the camera amplifiers and associated signal processing circuits (aperture correction, peaking, correction for camera nonlinearity, and so forth). These signals are assumed to be proportional to the tristimulus values R, G, B defined by standard or *de facto* standard chromaticities and the illuminant of receiver display devices (see Photometry, Colorimetry, and the Human Visual System in this chapter). These signals and the tristimuli are all normalized so that

$$R = \text{R} \quad G = \text{G} \quad B = \text{B} \quad \text{and} \quad R = G = B = 1 \text{ for reference white highlight} \tag{19}$$

In television standards the signals *R*, *G*, *B* are usually denoted E_r, E_g, E_b to distinguish them from tristimulus values. Simplified notations are used in what follows: italics for signals and romans for optical tristimulus values.

True luminance and chrominance signals are:

$$Y = \text{Y}_r R + \text{Y}_g G + \text{Y}_b B \quad M = R - Y \quad N = B - Y \tag{20}$$

$$R = G = B = Y \text{ for white with luminance Y} = Y \tag{21}$$

The signals *M, Y, N* are related to the signals *R, G, B* by the matrix H [Eq. (5) and Table 21.1.3]. It is noted that only the luminance values Y_r, Y_g, Y_b appear in this matrix. Tristimuli components X and Z as well as chromaticities, while implicit in the significance of *M, Y,* and *N* in electro-optical conversion in the display are of no concern in video signal analysis.

The original intent of the NTSC was to convey the "true" signals *M, Y, N* over separate channels. This was referred to as the *constant luminance principle*, because all the luminance would be conveyed by the luminance signal. Reasons for this approach are:

- compatibility with black and white receivers
- luminance information requires more picture detail than chrominance
- noise is more visible in luminance than in chrominance. To minimize the visibility of noise added in transmission, the luminance signal was companded in black and white television using the nonlinear transfer characteristic of the picture tube as expander in the receivers. The idea of using the picture tube as an expander has also been adopted in color television

To display a red color with a tristimulus value R, a nonlinear display must be driven by a "*gamma corrected*" signal *R'* which is a function of R. It is assumed that green and blue have the same electro-optical transfer functions and must be driven by "gamma corrected" signals

$$R' = \text{F(R)} \quad G' = \text{F(G)} \quad B' = \text{F(B)} \quad \text{where} \quad \text{F}(0) = 0 \quad \text{and} \quad \text{F}(1) = 1 \tag{22}$$

assuming display systems with inverse electro-optical transfer functions

$$\text{R} = \text{F}^{-1}(R') \quad \text{G} = \text{F}^{(1}(G') \quad \text{B} = \text{F}^{-1}(B') \quad \text{where} \quad \text{F}^{-1}(0) = 0 \quad \text{and} \quad \text{F}^{-1}(1) = 1 \tag{23}$$

In NTSC and PAL/SECAM standards, these functions are as in black and white television:

$$R' = R^{l/\gamma}, \quad G' = G^{l/\gamma} \quad \text{and} \quad B' = B^{l/\gamma} \tag{24}$$

In NTSC, $\gamma = 2.2$ ("but not enforced"). In PAL and SECAM, the recommendation is $\gamma = 2.8$. Recent standards are listed as electro-optical transfer characteristics in Table 21.1.1.

The nonlinear transfer characteristics of picture tubes usually differ from the production standards $L = F^{-1}(V)$ listed in Table 21.1.1. It is often assumed that they are close to $L = V^{2.5}$.

The Luminance Signal

In all current analog and digital standard television systems the gamma-corrected primary signals R', G', B' are matrixed in a matrix T to form signals for transmission which are proportional to:

$$Y' = T_R R' + T_G G' + T_B B' \qquad M' = R' - Y' \qquad N' = B' - Y' \qquad (25a)$$

$T_R + T_G + T_B = 1$. When the color is gray $R' = G' = B' = Y'$ and $M' = N' = 0$. Conversely

$$G' = Y' - (T_R/T_G)M' - (T_B/T_G)N' \qquad R' = Y' + M' \qquad B' = Y' + N' \qquad (25b)$$

Y' is called the *luminance signal*, while signals proportional to M' and N' are called *chrominance signals*. In standard definition television (SDTV), including NTSC, PAL, SECAM, CCIR601, and SMPTE170M, the transmission coefficients T_R, T_G, T_B, are equal to the luminance components Y_r, Y_g, Y_b of the NTSC primaries (Table 21.1.2), i.e.,

$$Y' = 0.299R' + 0.587G' + 0.114B' \qquad (25c)$$

The coefficient should *not* be rounded to two decimals as they unfortunately are in the official FCC standards. In HDTV, the transmission coefficients are equal to the luminance components of the primaries associated with the system (Table 21.1.2), i.e.,

$$Y' = 0.2125R' + 0.7154G' + 0.0721B' \ldots \text{SMPTE274M and ITU-4 709} \qquad (25d)$$

$$Y' = 0.212R' + 0.701G' + 0.087B' \ldots \text{SMPTE240M} \qquad (26e)$$

While the signal Y' is traditionally referred to as the luminance signal or just "the luminance" it is not a function of true luminance Y only. This happens only when the color is gray. For all other colors, part of the true luminance information is conveyed by the chrominance signals. When the chrominance (color) in a color receiver is turned off, the displayed luminance in the resulting black-and-white picture is less than in the color picture in all but originally gray areas. The ratio of the displayed luminance when the chrominance signals are turned off (compatible monochrome TV) to the displayed luminance when they are turned on can be taken as a viewer's perception of how much true luminance is conveyed by luminance signal Y'. This ratio is $G(Y')/Y$, where $G(V)$ is the transfer characteristic of the display. The ratio can be calculated given $G(V)$, Eqs. (20) and (25), and parameters for various systems given in Table 21.1.2. The ratio becomes smaller with increased saturation of the colors, and becomes exceptionally small for saturated blue. For a display with $G(V) = V^{2.5}$ the ratio for saturated blue is 3.8 percent in NTSC, 6.2 percent in PAL/SECAM, and 2 percent in ITU-R709. For $G(V) = F^{-1}(V)$ according to Table 21.1.1, corresponding numbers are 7.4, 3.2, and 22.2 percent. While saturated blue is an extreme case and the ratios are sensitive to the transfer function of the display at low signal levels, it is clear that a significant amount of luminance information is conveyed by the chrominance channels. Since the chrominance channels have less bandwidth, and in HDTV also less vertical resolution than the luminance channel, luminance resolution is lost, particularly in saturated colors. Another consequence is that noise (including various coding and transmission errors and defects) in the chrominance channels is displayed as luminance noise. The HVS is more sensitive to luminance noise than to chrominance noise. These effects are consequences of the violation of the constant luminance principle caused by pre-gamma correction of the R, G, B primaries.

The Chrominance Signals

The chrominance signals, also referred to as the *color difference signals*, are proportional to $M' = R' - Y'$ and $N' = B' - Y'$ and defined in Eq. 25a, b, c, d, and e. The N' blue-yellow range $\pm(1 - T_B)$ is larger than the M'

TABLE 21.1.6 Chrominance Scale (gain) Factors K_R and K_B Multiplying the Basic Chrominance Signals $M' = R' - Y'$ and $N' = B' - Y'$ to Yield Transmitted Chrominance Signals

SYSTEM	$K_R(R' - Y')$	$K_B(B' - Y')$	K_R	K_B
NTSC, PAL, SMPTE170M	V	U	0.877 (1/1.14)	0.493 (1/2.03)
SECAM[*]	D_R	D_B	-1.902	1.505
CCIR601[†]	C_R	C_B	0.713 (.5/.701)	0.564 (.5/.886)
ITU-R709, SMPTE274M	P'_R	P'_B	0.6349 (.5/.7875)	0.5389 (.5/.9279)
SMPTE240M	E'_{PR}	E'_{PB}	0.6345 (.5/.788)	0.5476 (.5/.913)

[*]Normalized frequency deviations for FM modulation.
[†]Includes all SDTV systems when luminance and chrominance components are digitized. See MPEG standards Rec. ITU-T H262.

red-cyan range $\pm(1 - T_R)$. In transmission, the chrominance signals $M' = R' - Y'$ and $N' = B' - Y'$ are multiplied by scale (gain) factors K_R and K_B. Scale factors of digitized chrominance signals are $K_R = (1 - T_R)/2$ and $K_B = (1 - T_B)/2$ to yield scaled chrominance signal ranges of ± 0.5, equal the unity range of the luminance signal Y'. Standard *chrominance scale factors* are shown in Table 21.1.6.

In NTSC and PAL the chrominance signals can also be represented by a vector

$$C = [U, V] = [0.493(B' - Y'), 0.877(R' - Y')] = [\,Ccos(c), Csin(c)] \tag{26}$$

$$C = (U^2 + V^2)^{1/2} \text{ amplitude} \qquad \text{and} \qquad c° = \arctan(V/U) = \text{phase in degrees}$$

The chrominance vector C is shown in Fig. 21.1.15. The vertical V-axis is in a "reddish" direction and the horizontal U-axis, in a "bluish" direction, is the reference direction of zero phase. The chrominance vector is displayed in an instrument called a *vector scope* as illustrated at the bottom of Fig. 21.1.16. Table 21.1.7 shows Y', M', N', V, U, C, and $c°$ for white and for the saturated colors yellow, cyan, green, magenta, red, and blue. These are the colors shown in the ubiquitous color bars. While the color bar signals are the same in NTSC, PAL, and SECAM standards, the colors vary from system to system as well as within each system depending on the chromaticities of the display and adjustments of primary levels (gains).

In PAL the chrominance signals $V = .877(R' - Y')$ and $U = .493(B' - Y')$ are both bandlimited to 1.3 MHz. Because of lack of available transmission bandwidth in NTSC the V and U signals are further matrixed in a rotational network (33°) yielding the transmitted signals $I = .839V - .545U$ and $Q = .545V + .839U$. The I and Q axes are shown in Figs. 21.1.15 and 21.1.16. The NTSC standards specify the maximum I-frequency to be 1.3 MHz and the maximum Q-frequency to be 0.6 MHz. The reason for this choice is that the HVS has better resolution for colors along the I-axis (red-cyan) than for colors along the Q-axis (blue-yellow). More often than not I and Q both roll off toward a max. frequency < 0.6 MHz.

TABLE 21.1.7 Luminance and Color Difference Signals for Saturated Colors (Color Bar Signal) in SDTV. The Sum of Complementary Colors (180° Apart) is White

COLOR	R' G' B'	Y'	$R' - Y'$	$B' - Y'$	V	U	C	$c°$
White	1 1 1	1.000	0	0	0	0	0	0
YEllow	1 1 0	0.886	+0.114	−0.886	+0.100	−0.437	0.448	+167.11
CYan	0 1 1	0.701	−0.701	+0.299	−0.615	+0.147	0.632	−76.56
Green	0 1 0	0.587	−0.587	−0.587	−0.515	−0.289	0.590	−119.30
MAgenta	1 0 1	0.413	+0.587	+0.587	+0.515	+0.289	0.590	+60.70
Red	1 0 0	0.299	+0.701	−0.299	+0.615	−0.147	0.632	+103.44
Blue	0 0 1	0.114	−0.114	+0.886	−0.100	+0.437	0.488	−12.89

Video Component Signals

While the primary signals *R, G, B* and the precorrected signals *R', G', B'* are video component signals, what is usually meant by *video component signals* is the set consisting of the luminance signal *Y'* and the chrominance signals $K_R(R'-Y')$ and $K_b(B'-Y')$. Figure 21.1.13 illustrates an NTSC transmission system of *video component*

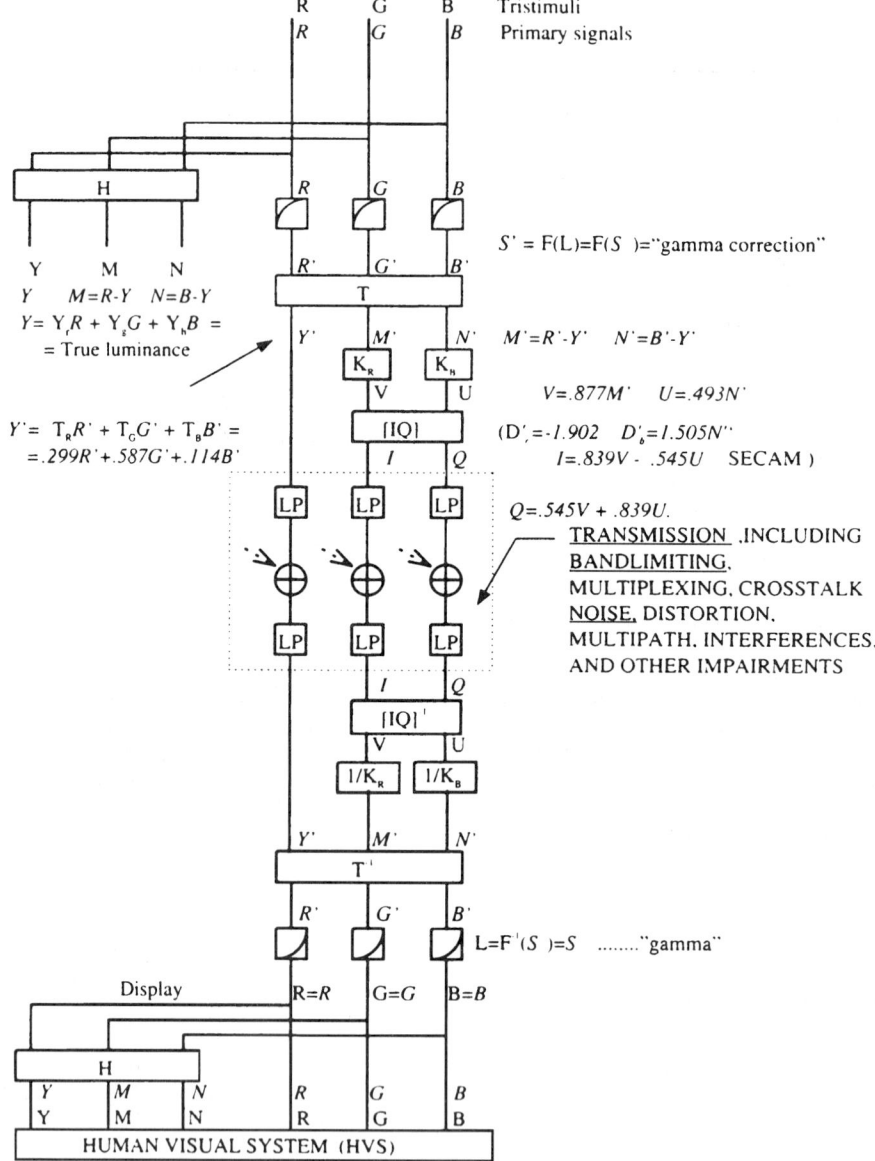

FIGURE 21.1.13 NTSC video transmission system. In PAL and SECAM drop the IQ matrices. In SECAM replace *V* by D'_r and *U* by D'_b. In digital systems, K_R and K_B are different. In HDTV the T matrices are different but, as in NTSC, the H and T matrices are identical.

signals. By dropping the IQ and IQ^{-1} matrices and ignoring numerical illustrations of K_R, K_B, T_R, T_G, and T_B it is valid for all television systems. All SDTV systems use the same T-matrix (NTSC). The T and H matrices are identical in NTSC as well as in HDTV systems. The H matrices, which are not in the transmission path, only show conversion from R, G, B to true luminance and chrominance Y, M, N. The block with dotted borders shown in Figure 21.1.13 represents bandlimiting and impairments of the component signals in transmission. In digital television the block includes data compression. A/D and D/A conversions, contributing with quantization and compression errors, are usually also inside this block. If this block is bypassed, the transmission is perfect if the processes at the transmitter have corresponding inverse processes at the receivers. Many proposals have been made to "correct" for luminance lost in the narrowband chrominance channels. Since the early 1950s there have also been proposals to replace Y' [see Eq. (25)] with the compressed luminance signal F(Y) [see Eqs. (20) and (22)], thereby conveying all the luminance in the luminance channel. No such *constant luminance* standards have been adopted, partly because it would not be compatible with current receivers and production standards, and partly because of added receiver costs for nonlinear expansion.

The NTSC Composite Video Signal, the HBI, and the Color Bar Signal

In NTSC, PAL, and SECAM the video component signals are multiplexed to form a single signal. In NTSC and PAL the chrominance signals modulate *a color subcarrier* which is "inconspicuously" added to the luminance signal to form a *composite video signal*. SECAM uses two frequency modulated subcarriers. In NTSC there is only one *color subcarrier* at a frequency $f_c = 227.5 \times F_H = 315/88$ MHz $\cong 3.58$ MHz. It is amplitude modulated in-phase and in quadrature by the chrominance signals (QAM) and added to the luminance signal to form the *composite NTSC signal*:

$$S = Y' + U\sin(\omega_c t) + V\cos(\omega_c t) = Y' + Q\sin(\omega_c t + 33°) + I\cos(\omega_c t + 33°) = Y' + C\sin(\omega_c t + c°)$$

$$V = .877(R' - Y') \qquad U = .493(B' - Y') \qquad I = .839V - .545U \qquad Q = .545V + .839U$$

$$C = (U^2 + V^2)^{1/2} = (Q^2 + I^2)^{1/2} \qquad C° = \arctan(V/U) = 33° + \arctan(I/Q) \qquad (27)$$

The vector diagram in Fig. 21.1.15 illustrates the instantaneous level of the color subcarrier, given the chroma vector [U,V]. A reference phase ($c = 180°$, "yellow") for synchronous detection is transmitted during the horizontal blanking interval by a short *burst* (8 to 11 cycles) of the color subcarrier (see Figs. 21.1.14, 21.1.16, and 21.1.17). The subcarrier components are [0, V] when $\omega_c t = 0$, [U,0] when $\omega_c t = 90°$; and (−0.4,0) for the burst at $\omega_c t = 270°$. As time progresses the instantaneous NTSC color subcarrier amplitude progresses in the order V, U, −V, −U or I, Q, −I, −Q, as illustrated in Fig. 21.1.15.

The signal *S* occupies active timeslots along horizontal scan lines and is bordered by horizontal and vertical blanking signals as shown in Figs. 21.1.14, 21.1.16, and 21.1.17 which also show the timing signals: horizontal and vertical sync pulses as well as the color subcarrier burst. Video signal levels in NTSC are specified in IRE units. The peak of the sync pulses is at −40 IRE, the blanking level at 0 IRE, and the reference white level at 100 IRE. The black level "setup" is at +7.5 IRE. Figure 21.1.14 includes a table with specification of durations, levels, and tolerances for NTSC and most PAL standards. Figure 21.1.14 also shows that the composite signal "inversely" modulates the main carrier of a broadcast transmitter with zero carrier (0 percent) at the whiter-than-white level of 120 IRE and 75 percent at blanking level and 100 percent at − 40 IRE (peak of sync).

Table 21.1.7 shows that the peak composite signal $Y' + C$ for saturated yellow, cyan, and green would exceed reference white level and overmodulate a broadcast transmitter. Similarly the lower level $Y' − C$ would drop significantly below blanking level, which may cause sync problems. To transmit the composite signal *S* within tolerable limits it must be reduced to $g_s S$, where g_s is a gain factor <1:

$$\text{NTSC video signal} = 7.5 + g_s \times 92.5 \times [Y' + C \sin(\omega_c t + c)] \qquad \text{IRE units.} \qquad (28)$$

In NTSC and PAL/BG the color bar signal as broadcast is at 75 percent amplitude ($g_s = 0.75$).

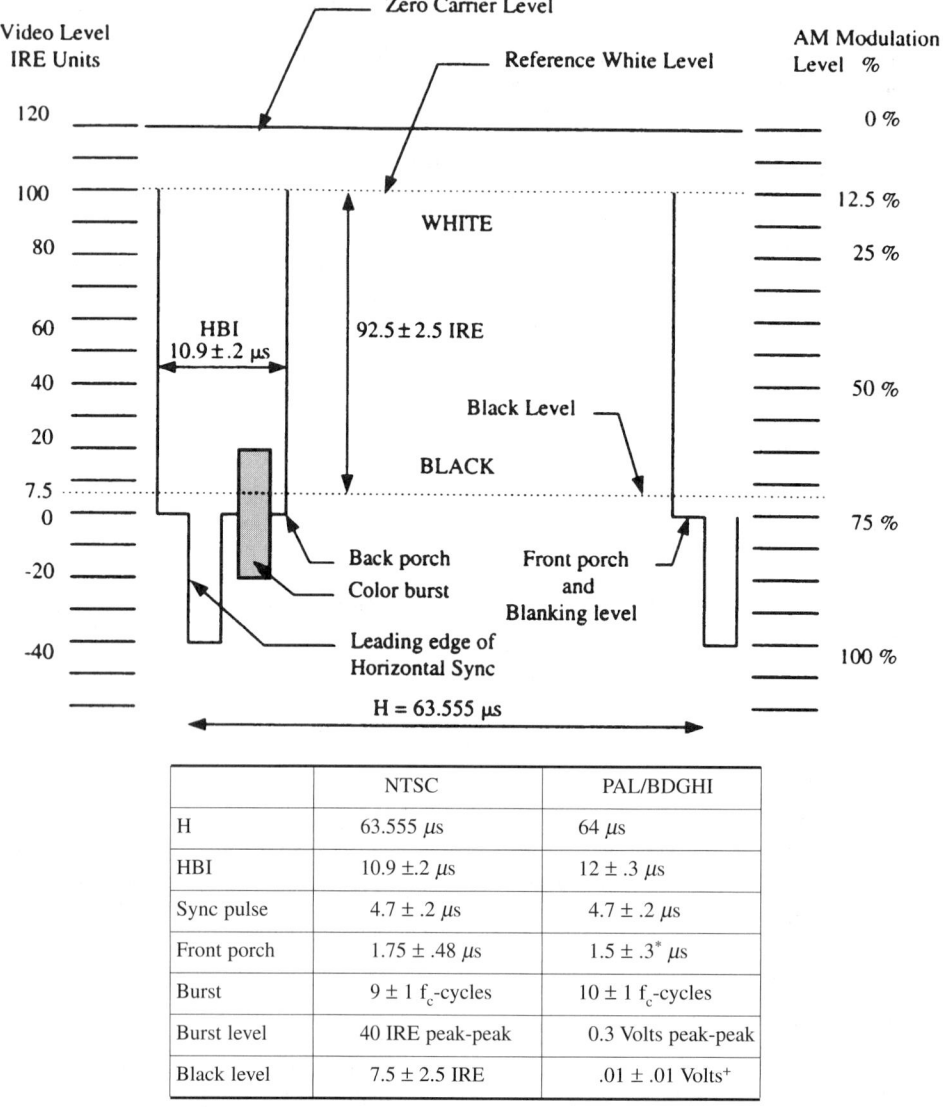

FIGURE 21.1.14 Signal levels along a horizontal line in NTSC. Levels are shown in IRE video level units and in r.f. modulation levels. NTSC and PAL specifications are shown in the table. Modulation is negative in the sense that the white reference level is at a low carrier level and peak of sync is at peak r.f. power level.

	NTSC	PAL/BDGHI
H	63.555 μs	64 μs
HBI	10.9 ±.2 μs	12 ± .3 μs
Sync pulse	4.7 ± .2 μs	4.7 ± .2 μs
Front porch	1.75 ± .48 μs	1.5 ± .3* μs
Burst	9 ± 1 f_c-cycles	10 ± 1 f_c-cycles
Burst level	40 IRE peak-peak	0.3 Volts peak-peak
Black level	7.5 ± 2.5 IRE	.01 ± .01 Volts+

*1.65 ± 0.1 in PAL/I.
+ In PAL 100 IRE = 0.7 Volts. Sync pulse is 0.3 Volts and modulation levels at white reference level are 10 percent in BDG and 20 percent in I.

Figure 21.1.16 shows a horizontal line signal in NTSC with color bars at 75 percent. The peak signal just about touches the 100 IRE white reference level, but drops to −16 IRE in blue and red (peak of sync is at −40 IRE). The figure also shows the vector diagram (vector scope) of the chrominance components and the burst. In PAL the color bar signals have ratio C/Y' as in NTSC but there is no 7.5 IRE black level setup. In PAL the phase is $+c°$ and $−c°$ on alternate lines. In PAL/BG the EBU-color bars at 75 percent are

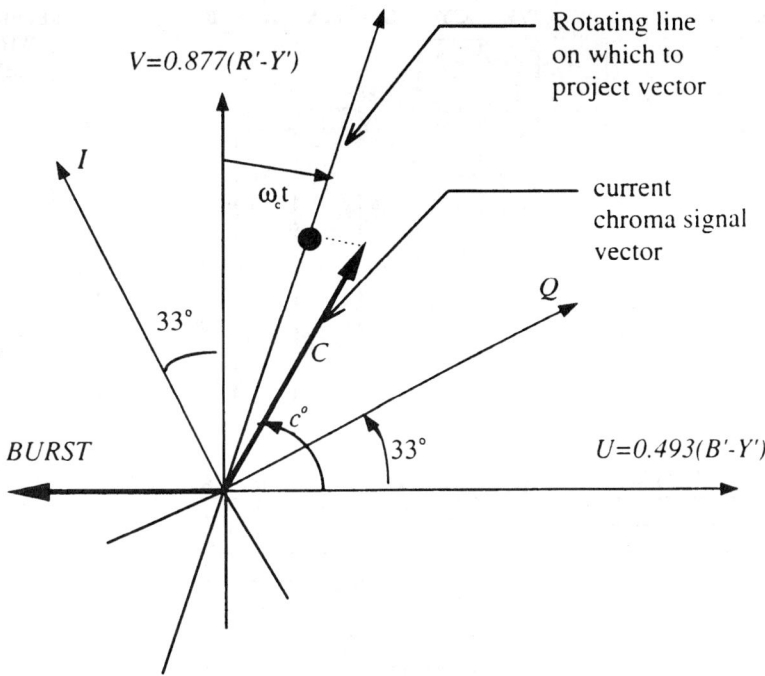

$$\text{NTSC Color subcarrier signal} = U \sin(\omega_c t) + V \cos(\omega_c t) = Q \sin(\omega_c t + 33°) + I \cos(\omega_c t + 33°)$$

$$= C \sin(\omega_c t + c)$$

FIGURE 21.1.15   Vector diagram showing quadrature modulation of the NTSC color subcarrier. With time the signal progresses in order $I, Q - I, - Q$.

$0.3 + 0.7 \times 0.75 \times S$ volts, except that white is shown at 100 percent, i.e., at 1 V. In PAL/I they are at 100 percent amplitude and 95 percent saturation (V and U multiplied by 0.95). If saturated colors are avoided in programs, the gain g_s could be larger than 0.75, but this presents a risk for overmodulation, which causes noise in the received audio. In compatible black-and-white reception this gain reduction results in a 2.5-dB loss in signal-to-noise ratio.

The PAL Composite Video Signal

In NTSC a phase error with respect to the burst reference phase used to be a problem with early tape recorders. Adding a phase error ϕ to $\omega_c t$ in Eq. (24) causes the synchronously detected U and V outputs to be:

$$U(\phi) = U\cos\phi - V\sin\phi, \qquad V(\phi) = V\cos\phi + U\sin\phi \tag{29}$$

This UV cross talk changes the tint because of the $\sin\phi$ contributions. To abate this effect, the PAL system alternates the polarity of V and $V(\phi)$ on successive lines in a field (by switching during the HBI) and on the same line after two frames by choosing a subcarrier frequency that has an odd number of cycles in four frames = a PAL color cycle. Equation (29) shows that this is equivalent to altering the sign of the $\sin\phi$ contribution to the error. These chroma errors may then be averaged and "canceled" in the HVS. However,

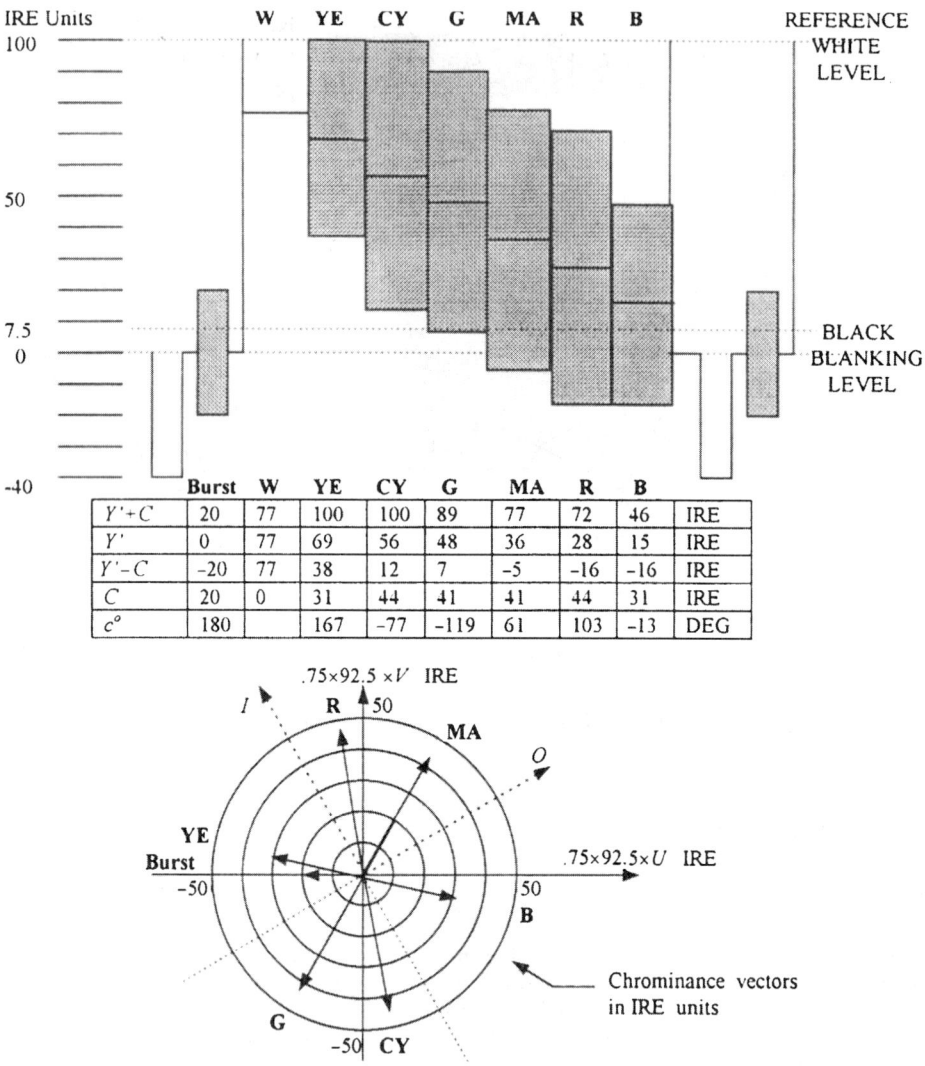

	Burst	W	YE	CY	G	MA	R	B	
$Y'+C$	20	77	100	100	89	77	72	46	IRE
Y'	0	77	69	56	48	36	28	15	IRE
$Y'-C$	-20	77	38	12	7	-5	-16	-16	IRE
C	20	0	31	44	41	41	44	31	IRE
$c°$	180		167	-77	-119	61	103	-13	DEG

FIGURE 21.1.16   NTSC Color bars at 75 percent amplitude (EIA standard RS 189-A). See Table 21.1.8 for U, V, $B'-Y'$, $R'-Y'$, Y', R', G', B'.

the alternating errors tend to be visible as crawling ("Hanover") bars, partly due to the nonlinearity of the display. This problem is solved in "standard PAL" by canceling the $\sin\phi$ error with the use of a comb filter that averages luminance and chrominance signals on successive lines. The price is reduced vertical resolution. In PAL the phase $c°$ of the chroma signal vector **C** shown in Fig. 21.1.15 alternates sign on every other line. To identify A-lines with chroma vector (U, V) and B-lines with chroma vector (U, V) the burst swings from a phase $c = 135°$ on A-lines to $-135°$ on B-lines. The reference phase $c = 180°$ is derived by averaging burst on A and B lines. The color subcarrier progresses with time in the order V, U, $-V$, $-U$ on A-lines and in the order U, V, $-U$, $-V$ on B-lines.

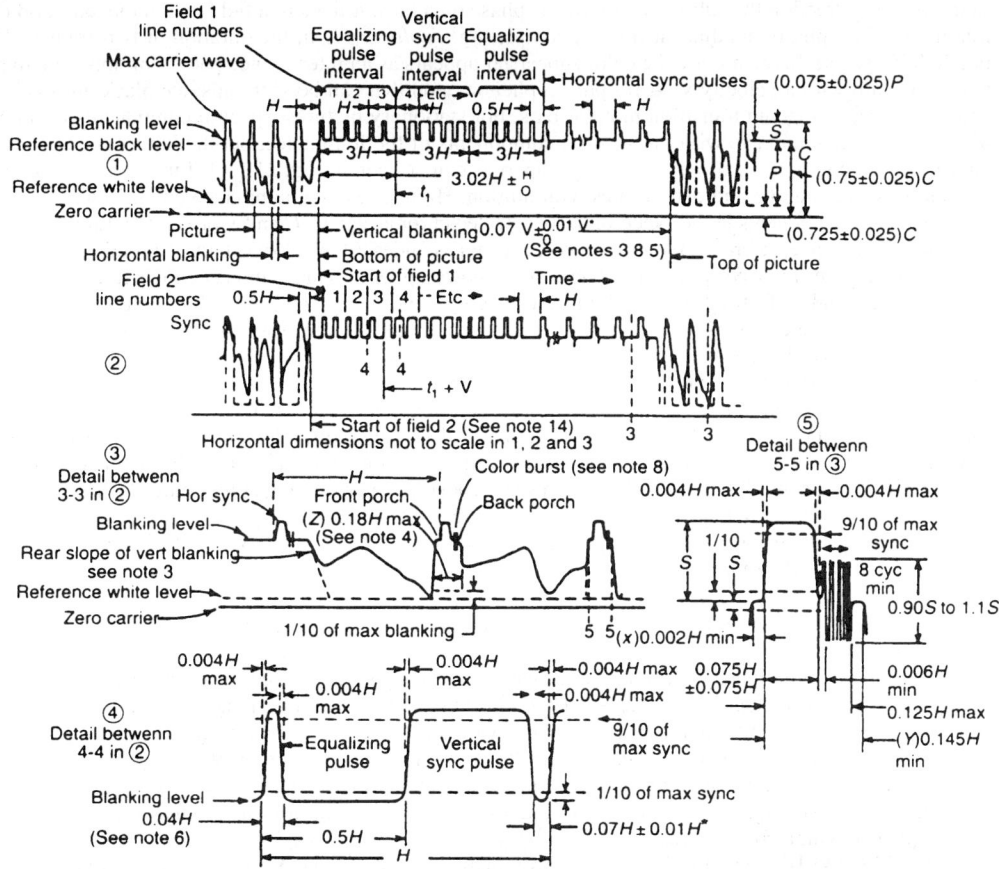

FIGURE 21.1.17 National Television System Committee color synchronizing-signal waveform (FCC Standard).

Notes: (1) H = time from start of one line to start of next line. (2) V = time from start of one field to start of next field. (3) Leading and trailing edges of vertical blanking should be complete in less than $0.1H$. (4) Leading and trailing slopes of horizontal blanking must be steep enough to preserve minimum and maximum values of $(x + y)$ and (z) under all conditions of picture content. *(5) Dimensions marked with an asterisk indicate that tolerances are permitted only for long time variations and not for successive cycles. (6) Equalizing pulse area shall be between 0.45 and 0.5 of the area of a horizontal synchronizing pulse. (7) Color burst follows each horizontal pulse but is omitted following the equalizing pulses and during the broad vertical pulses. (8) Color burst to be omitted during monochrome transmission. (9) The burst frequency shall be 3.579545 MHz. The tolerance on the frequency shall be ±10 Hz with a maximum rate of change not to exceed 0.1 Hz/s. (10) The horizontal scanning frequency shall be $2/455$ times the burst frequency. (11) The dimensions specified for the burst determine the times of starting and stopping the burst but not its phase. The color burst consists of amplitude modulation of a continuous sine wave. (12) Dimension P represents the peak excursion of the luminance signal from blanking level but does not include the chrominance signal. Dimension S is the synchronizing amplitude above blanking level. Dimension C is the peak carrier amplitude. (13) Start of field 1 is defined by a whole line between the first equalizing pulse and the preceding H sync pulses. (14) Start of field 2 is defined by a half line between the first equalizing pulse and the preceding H sync pulses. (15) Field 1 line numbers start with the first equalizing pulse in field 1. (16) Field 2 line numbers start with the second equalizing pulse in field 2.

Composite Signal Spectra: NTSC, PAL, and SECAM III

In NTSC the color subcarrier frequency is $f_c = 455 \times (F_H/2) \cong 3.58$ MHz. FCC rules specify f_c as $(63/88) \times 5$ MHz ± 10 Hz. This frequency is located exactly between two horizontal frequency components and exactly between vertical components $n = \pm (N - 1)/4 = \pm 131$ marked as $+ F_H/2$ in the spectrum shown in Fig. 21.1.11.

In the vertical direction the subcarrier alternates phase on adjacent lines in a field and in the temporal direction it alternates phase on adjacent frames, completing a *color-cycle* in the duration of two frames (1/15 s or 238,875 cycles). In the picture the color subcarrier appears as an interference pattern of dots seen to move slowly upward with a velocity $v_T \cong 1/8$ ph/s, a phenomenon that can be seen on some black-and-white sets and in the green-magenta transition in color bars. For stationary pictures, spectral components of the modulated subcarrier interleave with the spectral components of the luminance, but the luminance and chrominance spectral components are now only $F_V/4$ apart as can be seen in Fig. 21.1.11. Luminance-chrominance cross talk is unavoidable and deteriorates with motion. High-frequency luminance components show up as low-frequency components in the synchronously detected chroma signals U and V or I and Q. Luminance cross talk into chrominance, referred to as *cross-color*, appears as undesirable color patterns. Chrominance cross talk into the luminance, referred to as *cross-luminance*, shows up as moving luminance high frequency patterns (dots, gratings, and so forth). Since the luminance energy is primarily concentrated at even multiples of $F_H/2$ and the chrominance energy is concentrated at odd multiples of $F_H/2$, the cross talk can be reduced by passing the frequency band with the shared spectrum through a 1H-delay comb filter. The comb filter partially separates luminance from chrominance, but also averages luminance and chrominance on successively transmitted lines, which are two lines apart in interlaced scan. The price is "tolerable" reduction of diagonal luminance resolution, tolerable reduction of vertical chrominance resolution, and *hanging dots*, which occurs when the chrominance information changes significantly from one line to the next, and some other artifacts. The picture quality is usually improved by a comb filter, particularly with some cooperative processing at the source. Chroma-luma separation can be significantly improved with more sophisticated techniques.

In PAL the color subcarrier has a frequency $f_c = (1135/4)F_H + F_V/2 = 4,433,618.75$ Hz (709,379 cycles in four frames). The PAL composite signal is

$$S = Y' + U\sin(\omega_c t) \pm V\cos(\omega_c t) = Y' + C\sin(\omega_c t \pm c) \qquad (30)$$

The *V*-subcarrier is multiplied by a square wave with a frequency $F_H/2$. As a consequence, V modulates a number of subcarriers at frequencies $f_c \pm (2p + 1)(F_H/2)$ having amplitudes diminishing in proportion to $1/(2p + 1)$. The PAL color subcarriers appear as gratings which would move up slowly if it were not for the added 25 Hz. Shifting them all up in frequency by 25 Hz turn them into less visible gratings which appear to move three times faster downwards. A complete color cycle takes a duration of four frames = 0.16 s (709,379 cycles of the color subcarrier). An important feature of the PAL system is that the chrominance signals U and V can be partially separated with a comb filter.

In the SECAM III system (625/50) the chrominance signals U and V are conveyed on alternate lines (line sequential chroma). A 1H-delay line is used to make the U and V signals available simultaneously. Since there are an odd number of lines per frame, a line carrying U on one frame carries V in the next frame. The pre-emphasized U signal frequency modulates a subcarrier at a frequency $f_{ob} = 272\ F_H = 4.25$ MHz and the pre-emphasized V signal frequency modulates a carrier at a frequency $f_{or} = 282\ F_H = 4.40625$ MHz. The sidebands of the carriers are also pre-emphasized, and the polarity of the subcarriers (not the polarity of U and V) are altered in a sequence chosen to minimize their visibility. Flag signals are transmitted during vertical and horizontal blanking intervals to inform the receiver about parameters needed for proper detection. Specifications of the PAL and SECAM III systems are summarized in Fig. 21.1.18.

The Vertical Blanking Interval (VBI)

Figure 21.1.17 shows specifications for the vertical blanking interval in NTSC with the FCC designations of line numbers. The VBI occupies 21 lines on field 1 and 22 on filed 2. Three lines of equalizing pulses are followed by three lines of vertical sync pulses and another three lines of equalizing pulses, all needed for reliable interlaced scan in adverse reception conditions. Synchronization pulses are followed by line 10 through line 22. Most of these lines can be used compatibly (without unblanking receivers) for transmission of auxiliary information including test signals, data transmission (teletext, ceefax, antiope), and reference signals to improve reception. Each line is worth the equivalent of continuous transmission with 13-kHz bandwidth covering large areas with excellent signal-to-noise ratios. Most lines have been allocated for

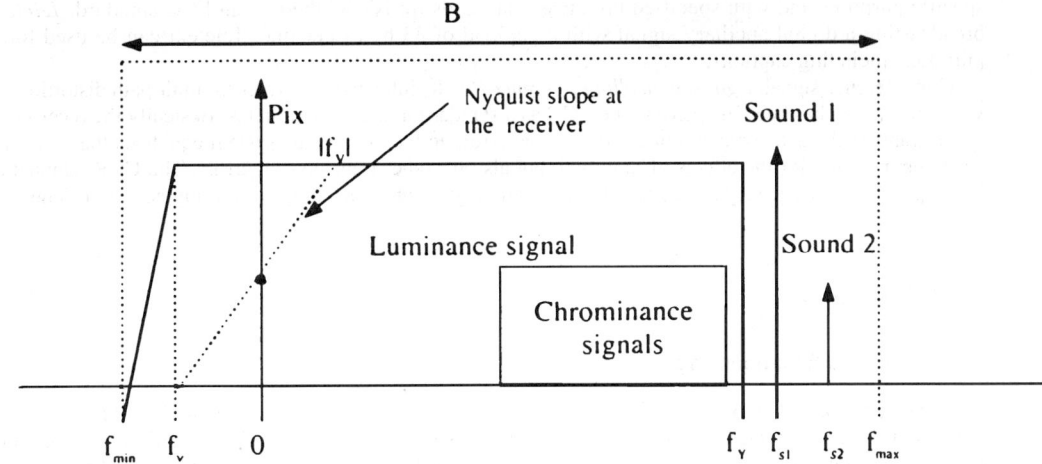

	NTSC	PAL/BGH	PAL/DHK	PAL/I	SECAM III/L
B MHz	6	7(B) 8(G)	8	8	8
f(min) MHz	−1.25	−1.25	−1.25	−1.25(4)	−1.25
f(max) MHz	4.75	5.75	6.75	6.75	6.75
f(Y) MHz	4.2	5	5	5.5	6
f(v) MHz	−0.75	−0.75 (−1.25 H)	−0.75	−1.25	−1.25
f($s1$) MHz	4.5	5.5	5.5	6	6.5
f($s2$) MHz		5.7421875		6.552	
Luminance modulation	White Level 12.5 ± 2.5%	White Level 10 to 12.5 %	White Level 10 to 12.5 %	White Level 20 ± 2 %	Peak of sync < 6%
Audio 1 level rel max.pix dB	BTSC-Stereo FM, Δf = 0.073 −10 to −7 (3)	Mono FM, Δf = 0.050 −13 to −10	L + R (Mono) FM, Δf = 0.050 −13	(Mono) FM, Δf = .050 −10 to −7	Mono AM −10
Audio 2 level rel max pix dB			R FM, Δf = 0.050 −20	NICAM728 DPSK Stereo −20	
Chroma 1	0.5 MHz Q inphase on f_c = 3.58 MHz	0.5 MHz U inphase on f_c = 4.43 MHz	1.3 MHz U inphase on f_c = 4.43 MHz	1.3 MHz U inphase on f_c = 4.43 MHz	1.3 MHz U FM on f_c = 4.25 MHz
Chroma 2	1.3 MHz I in in quadrature on f_c = 3.58 MHz*	1.3 MHz ± V in quadrature on f_c = 4.43 MHz	1.3 MHz ± V in quadrature on f_c = 4.43 MHz	1.3 MHz ±V in quadrature on f_c = 4.43 MHz	1.3 MHz V FM, on f_c = 4.40 MHz (1)
Picture i.f. Freq. MHz	45.75 (2)	38.9 (2)	38.9 (2)	38.9 (2)	32.7 (2)

*DSBAM up to 0.5 MHz and SSB-Lower Sideband to 1.3 MHz. (1) For detailed standards of SECAM preemphasis, deviations, and level control see Ref. 1. (2) Usual practice. I. F.-spectrum is flipped, i.e., sound is below pix. (3) Cable TV practice in the United States is −15 dB. (4) Roll-off actually spills over into lower adjacent channel.

FIGURE 21.1.18   Some r.f. transmission standards for NTSC, PAL, and SECAM.

specific purposes and with specified broadcast standards. In NTSC there is an FCC standard, *Teletext*, for broadcasting a digital ancillary signal with a payload of 33 bytes per line. Teletext can be used for many purposes, including captions.

One reference signal, a *ghost cancelling reference* (GCR) intended to correct for multipath distortion in television reception, has been adopted by the FCC as a standard in the United States. Basically the receiver adapts its r.f. transfer characteristics to correct detectable errors in the GCR. This system equalizes the transfer characteristic, not only for the effects of multipath but also for other transmission errors. The GCR adopted by the FCC appears as a frequency sweep (chirp) with carefully designed leading and trailing edge transients.

THE AUDIO SIGNALS

Monophonic Sound Broadcasting

In NTSC, PAL, and SECAM broadcasting, a "sound carrier" is modulated by an audio signal and added to the "picture carrier," which is modulated by the composite video signals. The summation is usually done at a high power level in a passive linear hybrid circuit called a diplexer (Fig. 21.1.1). In NTSC and PAL the sound carrier is frequency modulated, in SECAM/L it is amplitude modulated. In NTSC the ratio of the peak picture carrier to the sound carrier is 7 to 10 dB (15 dB in CATV). As in FM radio broadcasting, the audio signals are pre-emphasized (75 μs in NTSC, 50 μs in all other systems) and limited to a bandwidth from 40 to 15,000 Hz. The frequency deviation in the NTSC monophonic standard is 25 kHz, ensuring excellent audio quality even when the picture is quite degraded by interference.

Stereophonic Sound Broadcasting

In the United States a standard (BTSC) was adopted by the FCC in 1984 for multiplexing two analog stereo signals, L + R and L − R, to form a composite "stereo baseband signal." The system is also referred to as MTS (multichannel television sound) and is described in the chapter on broadcasting.

In Germany (PAL/B,G) an additional r.f. carrier, 7 dB weaker than the main sound carrier, has been added at a frequency $15.5F_H \cong 242$ kHz above the main sound carrier, to transmit the right signal R. In this dual-sound-carrier system the compatible monophonic signal $M = L + R$ is transmitted on the main audio carrier.

In terrestrial broadcasting in Great Britain (PAL/I) an additional r.f. carrier, 10 dB weaker than the main sound carrier has been added at a frequency about 552 kHz above the main sound carrier, to convey digital stereophonic sound referred to as NICAM728 (*near instantaneous companding audio multiplex @ 728 kb/s*). To provide 15 kHz stereo the L and R signals are sampled at a 32-kHz rate with 14 bits per sample. Companding reduces the required number of bits per sample to be conveyed to 10. The 14 bits are restored in the receiver. In terrestrial broadcasting, bit-pairs, one bit from each channel, modulate the r.f. carrier in *differential quadrature phase shift keying* (DPSK) by shifting the phase of the carrier from the current phase by a multiple (0, 1, 2, 3) of 90°. NICAM was also the preferred digital format for D-MAC satellite broadcasting in Great Britain. Some SDTV television audio transmission standards are summarized in Fig. 21.1.18.

R.F. TRANSMISSION SIGNALS

Basics and Status

Video signals are transported to consumers by carriers having frequencies in various bands in the r.f. spectrum. A modulated carrier has the form

$$z(t) = x(t) \cos(\omega_c t) - y(t) \sin(\omega_c t) = E(t) \cos(\omega_c t + p(t)) \tag{31}$$

where x = in-phase signal

y = quadrature signal

$E = (x^2 + y^2)^{1/2}$ = envelope

$p = \arctan (y/x)$ = phase

$f = (1/2\pi)(dp/dt)$ = frequency deviation

In double sideband amplitude modulation (DSBAM) x is linearly related to the video signal while $y = 0$. In digital quadrature amplitude modulation (QAM) both x and y take a number of discrete levels. In phase shift keying (PSK), p takes discrete levels. In frequency modulation (FM) f carries the video information. In PSK and FM, E carries no information and is approximately constant. In orthogonal frequency division multiplex (OFDM), a vast number of carriers transport a video signal. In fiber optic transmission, the carrier is light, not always with a single frequency.

Television is broadcast to a large number of viewers by three transmission technologies: terrestrial broadcasting, cable, and satellite. Terrestrial broadcasting includes a microwave multipoint distribution system (wireless cable). Cable broadcasting includes all technologies that can deliver television into homes via some conduit. In satellite communication FM is used for analog video transmission and usually PSK (or differential PSK) for digital video transmission. While satellite broadcasting of digital television is rapidly becoming popular, open (i.e., not proprietary) signal standards are still under development. Detailed specifications of signals used in digital television satellite broadcast systems and in digital video disks (DVD) are proprietary at the time of this writing. A major effort to develop an international digital video broadcast standard is pursued by several standards organizations, in particular by the DVB group of the EBU. It is beyond this section to discuss various standards for point-to-point transmission of video, e.g., satellite and broadband digital networks (BISDN). In what follows the discussion is limited to standards for analog and digital terrestrial broadcasting in the VHF and UHF television broadcast bands (54 to 890 MHz in the United States), where the channel bandwidth constraints (6, 7, or 8 MHz) have had a major impact on video signal standards. Cable TV systems operate with essentially the same r.f. signals.

R.F. Signals in Analog Terrestrial and Cable TV Broadcasting

All high-power television stations radiate a signal consisting of a picture carrier that has a frequency close to the lower band limit of the channel and one or two sound carriers with frequencies close to the upper band limit of the channel as shown in Fig. 21.1.18. The picture carrier is amplitude modulated by the composite video signal, but only a part of the lower sideband is transmitted. This is referred to as a *vestigial sideband amplitude modulation*. VSBAM is usually generated by passing a DSBAM signal through a vestigial sideband filter. Figure 21.1.18 illustrates the bandpass characteristic of the radiated VSBAM picture signal. In the i.f. amplifiers of receivers, the signal is transferred through a "Nyquist" filter, which in theory is phase-linear and has an antisymmetric amplitude (not necessarily linear) response around the picture carrier frequency (Fig. 21.1.18). The output from the Nyquist filter consists of an in-phase signal x amplitude modulated by the desired video and a quadrature signal y amplitude modulated by a video frequency component delayed 90° (Hilbert transform of video). With a perfect Nyquist filter synchronous detection yields undistorted video, while envelope detection contains some quadrature distortion. Ghosts also contain the quadrature signal. For these reasons the Nyquist slope must not be too steep. VSBAM wastes almost half the transmitter power to reduce radiated bandwidth. In the United States some low-power TV stations are allowed to operate with DSBAM.

Figure 21.1.18 shows the radiated spectrum of the NTSC, PAL, and SECAM signals including the sound carrier(s). Also shown are a number of r.f. signal standards. The NTSC sound carrier frequency is 4.5 MHz (286 F_H) above the picture carrier frequency and is frequency modulated by the BTSC (MTS) stereo baseband signal. The power of the NTSC sound carrier is 7 to 10 dB weaker (in Cable TV 15 dB weaker) than the power of the picture carrier at peak of sync (peak power). The radiated power of an analog television signal is always specified by the peak power of the picture carrier. Modulation levels at peak of sync, blanking, and white level are shown in Fig. 21.1.14. Broadcast technology, recommended practices, and FCC rules are included in the section on broadcast transmitters.

Transmission Impairments of Analog Television Signals

Major impairments are:

- Random noise and man-made noise (including impulse noise)
- Linear distortion including multipath distortion (ghosts)
- Interference from other broadcast stations (including television co-channel interference, adjacent channel interference, and at UHF, interference from taboo channels)
- Jittery local oscillators
- Nonlinear distortion (including crossmodulation, triple beats)

Random noise ("snow"), caused primarily by receiver noise, is the principal impairment limiting the area, which can be covered by a broadcast station or the number of homes that can be reached from a cable television headend. Reception impaired by random noise is specified by the carrier-to-noise ratio (CNR) defined to be the ratio in dB of carrier peak power to total noise power in an r.f. bandwidth = B MHz. In NTSC terrestrial broadcasting the practice is to specify B = 6 MHz (TASO definition) and in cable television the practice is to specify B = 4 MHz (NCTA definition). Thus

$$\text{CNR(NCTA)} - \text{CNR(TASO)} = 1.76 \quad \text{dB} \tag{32}$$

Video signal-to-noise ratio, unweighted for subjective perception of noise, is defined as the ratio in dB of the white-minus blanking level to rms noise in the luminance bandwidth. In NTSC the unweighted SNR is related to the CNR as

$$\text{SNRU} = \text{CNR(TASO)} - 5.3 = \text{CNR(NCTA)} - 7.06 \text{ dB} \tag{33}$$

With noise weighted for subjective perception (CCIR), the NTSC weighted SNR is

$$\text{SNRW} = \text{SNRU} + 7.6 \quad \text{dB} \tag{34}$$

CNR(TASO) is related to the field strength E dBμV/m (dBu) at the location of the receiving antenna:

$$\text{CNR(TASO)} = E + G - F - 20 \log_{10}(f \text{ MHz}) + 31.1 \text{ dB} \tag{35}$$

where G is the antenna gain minus cable losses and F is the system noise figure, both in dB.

The FCC has based determination of "grade B" noise-limited coverage on a just barely acceptable CNR(TASO) = 28.2 dB to be exceeded 90 percent of the time at 50 percent of the locations. To meet this performance with FCC "planning factors" (G, F, and a 90 to 50 percent time factor), the field strength on the grade B contour must exceed E(50/50) shown in Table 21.1.8 50 percent of the time at 50 percent of the locations.

Determination of E(50/50) and the 90 to 50 percent time factor (50/10) − E(50/50) is discussed in the section on broadcast transmitters.

The effects of an r.f. impulse (*impulse noise*) in the received composite video is a pulse with a shape, amplitude, and ringing which is determined by the video bandwidth of the receiver and the phase of the carrier at the instant of the impulse. Impulse responses are colorful, and chroma streaks can be long because of the narrow chrominance bandwidth.

Co-channel interference shows up as crawling venetian blinds. Their visibility for a given relative level of the interfering channel can vary by as much as 20 dB depending on the difference between the frequencies of

TABLE 21.1.8 Grade-B Field Strength at 50 Percent of the Locations 50 Percent of the Time 9 m Above Ground

	Low VHF (54–88 MHz)	High VHF (174–216 MHz)	UHF (470–806 MHz)
Grade-B E(50/50) dBμV/m	47	56	64

the co-channels. A precise visual carrier frequency offset of ±10010 Hz is cooperatively used by many NTSC broadcasters to minimize the visibility of co-channel interference.

Multipath transmission has two effects: (1) it changes the amplitude of the carrier which carries the desired signal and consequently the CNR and (2) it causes linear distortion of the video signal which can be characterized by an r.f. to video transfer function. Because VSBAM transmission also transmits the Hilbert transform of the video signal on a quadrature carrier, part of the ghost as it appears in the picture may consist of this quadrature signal. The transfer function varies rapidly with the relative carrier-phase of the ghosts. This is noticeable if a ghost is a reflection from an airplane (*airplane flutter*). The transfer characteristic can be equalized with a *deghoster* operating with a ghost cancelling reference signal (GCR) in the VBI. The standard GCR for NTSC is discussed in the sections on television transmitters and receivers. The carrier level, and consequently CNR, can be significantly affected by ghosts, not the least caused by reflections from nearby clutter, i.e., buildings, hills, trees, and so forth. A deghoster can straighten out the transfer characteristic, but it cannot improve degradation of CNR by ghosts.

There are numerous other impairments of the video signal, some caused in the r.f. path, some in the transmitter, and some in the receiver. Distortion, distortion specifications, measurement techniques, performance standards, and vertical interval test signals (VITS) for measurements of impairments are discussed in the chapter on broadcast transmitters.

Cable TV introduces impairments resulting from nonlinear distortion in broadband amplifiers. The most significant consequences of intermodulation between the many r.f. signals passed through a cascade of broadband amplifiers, are beats between three channels (triple beats), which primarily cause streaks in the picture, and cross-modulation, whereby a carrier is modulated by video conveyed by other carriers. Progress in amplifier design has made possible the distribution of the sum of more than 50 standard VSBAM television signals on a cable with imperceptible cross-modulation and triple beat interference levels in the video below −55 dB relative to 100 IRE. For a detailed discussion see the section on Cable TV.

DIGITAL VIDEO SIGNALS

Sampling and Quantizing Video Signals

In theory, a dynamic picture is sampled at lattice points in a three-dimensional space, two spatial and one temporal, and at each point (pixel or pel), three digital values specifying a color are sampled. In practice this process is often accomplished by sampling and quantizing analog component video signals. Conventional practice is to align samples along vertical lines ("rectangular" pels) in identical locations on successive frames.

Of key importance is where in the video transmission path, shown in Fig. 21.1.13, the A/D conversion takes place, what signals are sampled, and how noisy they are. Other key parameters are: sample rate, number of quantization levels, and anti-alias filtering preceding A/D conversion. A camera with CCD or CMOS sensors may be the first device to sample the image along a line. The number of reliably distinguishable levels is limited to a few hundreds by camera noise. Aliasing in CCD cameras in the spatial directions can be partially controlled by sensor size and defocusing, and in the temporal direction by the integration time. The data sampled on the faceplate of a CCD camera are, however, converted to analog signals, which are subject to very sophisticated signal processing. At the other extreme, digitization of the composite signal occurs far down the transmission path. While the sampled composite signal is free of horizontal aliasing, it is oversampled and impaired by luminance-chrominance cross talk, bandwidth limitations, and other artifacts that cannot be corrected. Composite digital signals may be useful and inexpensive for some NTSC and PAL applications, but repeated A/D and D/A conversion cause significant degradation of the signals.

Composite NTSC and PAL Signals

NTSC or PAL composite signals are almost always sampled at a sample rate equal to four times the color subcarrier frequency ($4 \times f_c$). Figure 21.1.15 shows that if the samples are taken at nulls of the in-phase and quadrature components of the color subcarrier, the sequence of sampled values in NTSC are $Y' + I$, $Y' + Q$, $Y' - I$, $Y' - Q$, ... In PAL they are $Y' + V$, $Y' + U$, $Y' - V$, $Y' - U$, ... on A-lines and $Y' - V$, $Y' + U$, $Y' + V$,

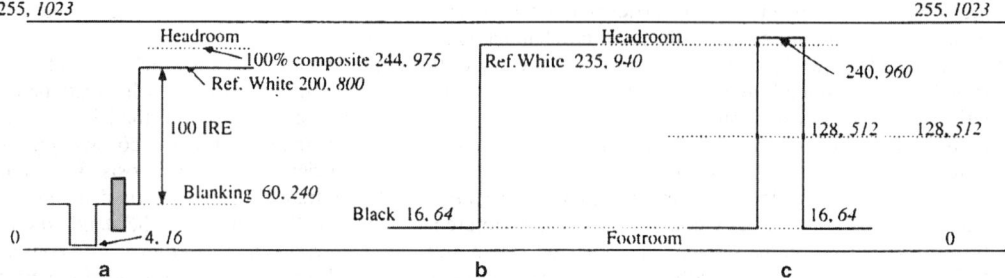

FIGURE 21.1.19   Digitized levels for composite NTSC and for components in CCIR601. Standard is for 8- and 10-bit quantization. 10-bit quantization levels are shown in italics.

  a. Composite NTSC
  b. Luminance component Y' in CCIR601
  c. Chrominance components in CCIR601

$Y' - U, \ldots$ on B-lines. There are 910 samples on a NTSC line and 1135 on a PAL line. Chrominance and luminance can be separated in digital circuits with inputs from adjacent lines in a field.

A typical NTSC scale for 256 levels (8 bits) and 1024 levels (10 bits) is shown in Fig. 21.1.19*a* (SMPTE259M). According to this scale some levels are reserved for headroom and sync. In the 8-bit version, only a little more than 7 bits are available for 100 IRE units (blanking to white). Quantization causes distortion of the signal which sometimes appears in the picture as *quantization contours or puddles*. One way to reduce the visibility of contours is to "dither" the least significant bit with random noise. In busy picture quantization may have noiselike appearance. These artifacts are visible in 8-bit NTSC, so 10 bits per sample is commonly used in professional applications, implying a data rate of 143 Mb/s requiring a transmission bandwidth of about 80 MHz.

The adoption in 1982 of an international standard, ITU-R Recommendation 601 (CCIR601), for component digital television, was a major milestone in the development of digital television. The D1 tape format and numerous other applications are based on the CCIR601 standard. Television studio equipment is now almost all digital, and digital interfaces between equipments are almost always based on CCIR601.

TABLE 21.1.9   ATSC Format Interoperability Considerations

	Format	Aspect Ratio			Frame Rate		
	1920 x 1080 (square pixels)	16:9			60 I	30 P	24 P
	1280 x 720 (square pixels)	16:9		60 P		30 P	24 P
TV	704 x 480 (CCIR 601)	16:9	4:3	60 P	60 I	30 P	24 P
	640 x 480 (square pixels)		4:3	60 P	60 I	30 P	24 P
		Film	TV		TV		Film
	Computer			Computer		Computer	

Supported frame rates include both 60.0 and 59.94 Hz related rates

COMPONENT DIGITAL VIDEO SIGNALS

Digitization of composite analog signals is still common, e.g., in the D2 and D3 tape recording formats, in various production systems, e.g., for time base correction, frame synchronization, editing, as well as for video processing in consumer products. Advantages of digital technology: perfect multigeneration reproducibility; precise time and level controls, digital storage, and signal processing; data compression; easy manipulation by computers, multimedia; incorporation in digital data transport packets for packet switching (ATM), broadband digital communication (BISDN); error control; and so forth. Recently, international agreement has been reached on the use of 1920 × 1080 as a common image format for high-definition production and program exchange. The 1920 × 1080 format has its roots in the CCIR sampling standard, and brings international compatibility to a new level. Table 21.1.9 shows the picture sensing characteristics of the 1920 × 1080 format for a variety of frame rates.

HIGH-DEFINITION TELEVISION (HDTV)

Fundamentals and Picture Quality

The fundamentals of HDTV are the same as the fundamentals of television in general, which have been reviewed in the previous sections of this chapter. From the viewers point of view, HDTV differs from "standard" definition TV (SDTV) by a significantly improved viewing experience in many respects:

1. Sharper, higher resolution images. By design, HDTV is intended to be as good as the resolution of the human visual system at a viewing distance of four picture heights (D = 4 ph). A large screen and a viewing distance of D < 4 stimulates viewer involvement in the scene. Compared to standard definition TV, HDTV increases horizontal luminance resolution by a factor of $\cong$ 3, vertical luminance resolution by a factor of $\cong$ 2, and horizontal chrominance resolution by a factor of $\cong$ 5.

2. A wider, cinemalike picture with an aspect ratio of 16/9 instead of 4/3.

3. High-quality surround sound audio that produces an immersive, theaterlike experience.

4. A substantial reduction in picture artifacts, including the cross-luminance and cross-chrominance artifacts of analog color systems. In addition, digital HDTV eliminates the analog transmission impairments such as noise, ghosts, distortion, and interference from other stations.

Also, a digital HDTV channel has all the potentials of digital TV in general and can offer many information services to the consumer. A packetized data format with headers and descriptors provide flexibility, interoperability, and extensibility.

HDTV, however, suffers from some imperfections:

1. Although HDTV supports multiple raster formats, each one has certain trade-offs among spatial and temporal resolution. Interlaced scan formats at 30, 30/1.001, or 25 Hz and progressive scan formats at the film rates of 25 or 24 Hz continue to have the same motion-related artifacts as today's analog TV systems and film. Also, in interlaced scan formats, line flicker may be visible at viewing distances D < 3.3 ph. While HDTV progressive scan formats at 60, 60/1.001, and 50 Hz improve motion rendition, they do so at the expense of horizontal and vertical spatial resolution.

2. Large area flicker may be perceived at viewing distances < 4 ph at typical TV brightness. At D = 4 ph the width of the picture is sustained by a 25° viewing angle, which means that a part of the picture is seen by flicker-sensitive rod vision. While a 60-Hz flicker may be acceptable, 50 Hz may not be acceptable and may require frame doubling as in motion pictures.

3. In digital HDTV over narrowband channels, there may be some visible video compression and quantization artifacts.

4. In DTTB of HDTV, there is the "cliff effect": Either a virtually perfect picture is received or no picture is received.

HDTV Production Standards

Efforts to establish national and international technical standards for production, distribution, and transmission of HDTV have been going on ever since Japan Broadcasting Corporation (NHK) started development of the analog 1125/60 HDTV system (MUSE) in the 1970s. In the mid-1980s the EBU proceeded with a project, referred to as "Eureka," to develop European 1250/50/2:1 and 1250/50/1:1 production standards. In the United States, the Society of Motion Pictures and Television Engineers (SMPTE) has proposed production standards for an 1125/60 high-definition production system, referred to as the SMPTE274M and SMPTE296M, which have been adopted by ATSC, while SMPTE240M is used in Japan.

Analog HDTV Broadcast Standards

The first operational system was the analog 1125/60 MUSE system ("multiple sub-Nyquist sampling encoding"), which has been adopted in Japan as a standard for HDTV satellite direct broadcasting (DBS). In the MUSE system, video bandwidth requirements are reduced from 21 to 8.1 MHz by a technique of horizontal and temporal subsampling, whereby diagonal and motion resolution is traded off to improve horizontal luminance resolution. Four fields are required to complete a picture with all details. Some motion compensation is used. Chrominance is also subsampled and the chrominance signals C_r and C_b are time compressed by factors of 4 and 5, respectively, and transmitted on alternate lines in the horizontal blanking interval which occupies about 20 percent of the time. This "time compressed integrated" (TCI) signal is converted into an analog signal that modulates the satellite FM transmitter.

In Europe an analog DBS system, HD-MAC, was developed based on a 1250/50/2:1 production standard. HD-MAC has been abandoned in favor of digital television systems. HD-DIVINE developed by a Scandinavian consortium for DTTB was demonstrated in 1992.

In the United States and Canada, the Advanced Television Systems Committee (ATSC) was formed in 1982 by the television industry to coordinate standards for advanced television systems. In 1987 a number of committees were formed to study and tests proposed systems for DTTB of HDTV. Originally over a dozen analog systems were proposed and many transmission methods were considered. But in June of 1990, a landslide move toward digital television terrestrial broadcasting began when General Instrument Corporation announced its DigiCipher all digital system. Others soon proposed all digital systems and in 1992 and 1993, four digital systems and Narrow MUSE, were tested at the ACATS' Advanced Television Test Center and by Cable Television Laboratories. Subjective tests were conducted in Canada by the Advanced Television Evaluation Laboratory (ATEL). The tests showed many advantages of the digital systems, which all performed better than Narrow-MUSE. In May 1993, the four proponents of digital systems formed a *Grand Alliance* which would develop a system combining the best features of each system. In 1994 and 1995, the Grand Alliance system was a subject of extensive tests (objective, subjective, and field tests), that validates its performance as a "best-of-the-best" digital system. The Grand Alliance system formed the basis for the ATSC standard (which added standard-definition formats) and the FCC transmission standard (which requires the use of the ATSC standard, with the exception that no picture format requirements whatsoever are specified by the FCC).

A better, but more expensive choice in terms of data rate, is A/D conversion of component video signals components. Usually the components are the pre-gamma corrected luminance signal Y' and the scaled chrominance signals $C_r = K_r(R' - Y')$ and $C_b = K_b(B' - Y')$ where $K_r = 0.5/(1 - T_R)$ and $K_b = 0.5/(1 - T_b)$. However, the choice can also be the pre-gamma corrected primaries R', G', and B' or the uncorrected primary signals R, G, and B. In the latter case the quantization must be at least 10 bits because of the high gain of the gamma-correcting circuit at low signal levels and the sensitivity of the HVS to contrast ratios. A quantum step at low luminance levels is more perceptible than at high luminance levels.

CCIR601 is a family of compatible international standards for component television. Figures 21.1.19b, 21.1.19c, and 21.1.20 summarize some key aspects of CCIR601. The common ground of 525/59.94 and 625/50 systems is that the duration of a TV line is about the same and that at a 13.5-MHz sample rate there is an integral number of sample points per line in all standard analog systems: $6 \times 143 = 858$ in 525/59.94 systems and $6 \times 144 = 864$ in 625/50 systems, as shown in Fig. 21.1.20. In both systems there are *720 active samplepoint locations* per line. Along a line the active samples are preceded and followed by sequences of digital timing

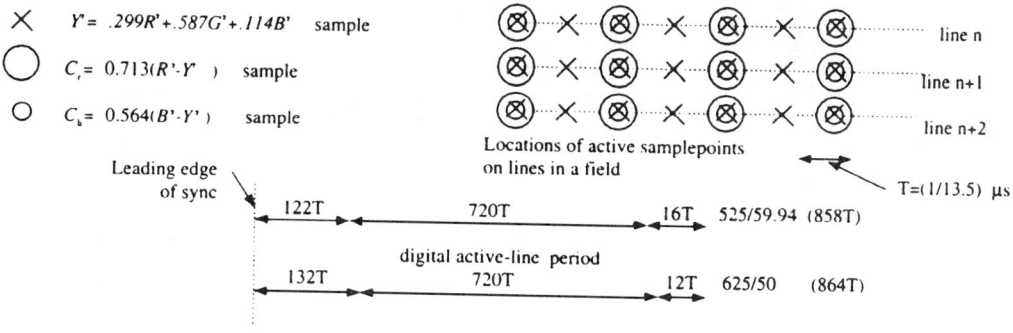

FIGURE 21.1.20 Sample point locations in 4:2:2 CCIR601 (ITU-R Rec. 601) for component SDTV digital television systems 525/59.94 and 625/50. Total samples per second = 27,000,000. In 4:2:0 the density of chrominance samples along a vertical line is reduced by a factor of 2 (see text). In 4:4:4 there are chrominance samples at every samplepoint location.

and ancillary data. This inactive period coincides with the horizontal blanking interval in all analog standard systems.

The CCIR601 4:2:2 format (family member) is summarized in Fig. 21.1.20. In a sequence of four Y' samples along a horizontal line there are two $C_r = K_R(R' - Y')$ samples and two $C_b = K_B(B' - Y')$ samples which are both co-located with every other Y' sample. The total number of active sampled values per line is thus 1440. The Nyquist bandwidths are 6.75 MHz for luminance and 3.875 MHz for chrominance. The CCIR601 standard specifies a 5.75 MHz low pass cutoff for luminance and a 2.75 MHz cutoff for chrominance. CCIR601 4:2:2 standard specifies either 8 or 10 bits per sampled value. The luminance and chrominance signal level as well as headroom and footroom are shown in Figs. 21.1.19b and 21.1.19c. The scale factors K_R and K_B (Table 21.1.6) are chosen to fit the chrominance signals C_r and C_b at 100 percent amplitude in the range 128 ± 112 (512 ± 448), which is about the same as the range of the luminance level.

In the CCIR601 4:4:4 format chrominance samples are co-located with every luminance sample. In the CCIR601 4:2:0 format the density of chrominance samples along a vertical line is reduced by a factor of 2 usually by taking some weighted average of 4:2:2: chrominance samples. In interlaced scan the so-derived 4:2:0 chroma samples appear to be uniformly interlaced in successive fields. The vertical chroma resolution in 4:2:0 is a factor of 2 less than in 4:2:2 or 4:4:4. The "raw" data rate of digitized video is high: 115 Mb/s for composite NTSC, up to 270 Mb/s for 4:2:2 CCIR601, and about 1 Gb/s for HDTV. By removing *statistical redundancy* and taking advantage of psychophysics to remove *irrelevant information*, the data rate of high quality HDTV has been reduced by a factor of 50 to about 20 Mb/s, which can be conveyed over a 6 MHz television broadcast channel.

With digitization of audio and video signals, television is becoming an integral part of a much broader field of digital information production and distribution. Digital standards strive to develop formats which provide high-quality data-compressed television and audio with considerations to *flexibility* of use and of receiver standards, *extensibility* to various formats, and *interoperability* with other applications, e.g., computers and data communications. In order to meet these objectives, a layered digital system architecture approach has been adopted as the basis for all DTV standards. This approach is loosely analogous to the ISO Open System Interconnect layered model for data communications. An international video compression standard, MPEG-2, provides the flexibility required to implement the three uppermost layers of the DTV system architecture. This standard was developed by the Motion Picture Expert Group of ITU and adopted by the International Standards Organization (ISO) and the International Electrotechnic Committee (IEC) in 1993. This accomplishment is the second major breakthrough toward international standards for digital television.

MPEG has a standard syntax (language) and protocol (rules) allowing for a great variety of inputs and for compatibility with many applications, i.e., not just one television standard or ancillary program. For example, there are "flags" specifying aspect ratio, samplepoint format, scanning standard, chromaticities, electro-optical transfer characteristics, sample rates, and so on. *Digital Television Standard* is referred to as the *ATSC*

TABLE 21.1.10 Picture Scanning Characteristics of the 1920 × 1080 High-Definition Common Image Format (ITU-R BT. 709-3)

Item	Parameter	System values									
		60/P	60/I	30/PsF	30/P	50/P	50/I	25/PsF	25/P	24/PsF	24/P
6.1	Order of sample presentation in a scanned system	Left to right, top to bottom. For interlace and segmented frame systems, 1st active line of field 1 at top of picture									
6.2	Total number of lines	1125									
6.3	Field/frame/segment frequency (Hz)	60 (60/1.001)	60 (60/1.001)	30 (30/1.001)	30 (30/1.001)	50	50	25	25	48 (48/1.001)	24 (24/1.001)
6.4	Interlace ratio	1:1	2:1	1:1	1:1	1:1	2:1	1:1	1:1	1:1	1:1
6.5	Picture rate (Hz)	60 (60/1.001)	60 (60/1.001)	30 (30/1.001)	30 (30/1.001)	50	50	25	25	24 (24/1.001)	24 (24/1.001)
6.6	Line frequency* (Hz)	67500 (67500/1.001)	67500 (67500/1.001)	33750 (33750/1.001)	33750 (33750/1.001)	56250	56250	28125	28125	27000 (27000/1.001)	27000 (27000/1.001)
6.7	Samples per full line –R, G, B, Y / –C_B, C_R	2200 / 1100	2200 / 1100	2200 / 1100	2200 / 1100	2640 / 1320	2640 / 1320	2640 / 1320	2640 / 1320	2750 / 1375	2750 / 1375
6.8	Nominal analog signal bandwidths† (MHz)	60	30	30	30	60	30	30	30	30	30
6.9	Sampling frequency –R, G, B, Y (MHz)	148.5 (148.5/1.001)	74.25 (74.25/1.001)	74.25 (74.25/1.001)	74.25 (74.25/1.001)	148.5	74.25	74.25	74.25	74.25 (74.25/1.001)	74.25 (74.25/1.001)
6.10	Sampling frequency‡ –C_B, C_R (MHz)	74.25 (74.25/1.001)	37.125 (37.125/1.001)	37.125 (37.125/1.001)	37.125 (37.125/1.001)	74.25	37.125	37.125	37.125	37.125 (37.125/1.001)	37.125 (37.125/1.001)

*The tolerance on frequencies is ± 0.001 percent.
†Bandwidth is for all components.
‡C_B, C_R sampling frequency is half of luminance sampling frequency.

TABLE 21.1.11 Some Key Transmission Parameters of the ATSC System for DTTB of HDTV

Transmission parameters	Terrestrial mode
Modulation method	VSB
Channel bandwidth MHz	6
Symbol rate megasymbols/second	10.76
Excess bandwidth	11.5%
Bits per symbol	3
Reed-Solomon FEC	T = 10 (208, 188)
Trellis coding	2/3 rate
Segment length symbols	836
Sync symbols per segment	4
Frame sync	1 per 313 segments
Payload data rate Mb/s	19.3
CNR threshold dB	14.9
Transport parameters	
Multiplex technique	MPEG-2 systems layer
Packet size bytes	188
Packet header bytes	4 including sync

system for HDTV. The ATSC standards accommodate one interlaced and one proscan HDTV format, shown in Table 21.1.10. It accommodates a number of frame rates: 60, 30, and 24 frames per second as well as these rates divided by 1.001. The total samples per line shown in Table 21.1.10 are for 60 frames per second proscan (format 1) and for 30 frames per second interlaced (format 2) and imply a 74.25 MHz (74.25/1.001 MHz) sample rate in accord with the SMPTE274M and SMPTE296M production standards. The ATSC standard also accommodates two SDTV formats with 480V/704-H and 480V/640-H active luminance samples per frame and a standard 13.5 MHz sample rate for 60/1.001 field per second, as in CCIR601. The ATSC has adopted the Dolby AC-3 digital audio system, while European standards for digital TV may lean toward MPEG2 for audio. The ATSC system uses VSB for transmission.

Some of the key transmission parameters for the ATSC system are shown in Table 21.1.11. MPEG-2 has been rapidly and enthusiastically adopted for many applications besides DTTB, e.g., satellite broadcasting, digital video disc (DVD). Digital satellite systems (DSS) are in operation worldwide. In the United States DIRECTV, for home reception with an 18 in. dish, provides 150 television channels.

MPEG-2 Format for Video Compression

Source coding or compression of audiovisual data has two objectives: removal of redundancy (*entropy coding*) and reduction of irrelevancy (*perceptual coding*). The first objective is to take advantage of statistical correlation in the bit stream to increase the information (entropy) per transmitted bit. There is no redundancy in a picture if all samples are statistically independent as they are in random noise, while there is a lot of redundancy in a patch of blue sky. Statistical coding requires buffering to absorb a variable information rate. The second objective is to eliminate information which cannot be perceived by human observers. While a significant amount of information which is irrelevant to human observers has been removed in standard analog television (number of lines and fields, relative bandwidth of components, and so forth.), further elimination of imperceptible information is possible with digital techniques. Removal of redundancy and reduction of irrelevancy, while having different objectives, are not independent of each other or of the constraints imposed by the transmission channel.

Data compression obviously increases the information carried by remaining bits, which must be conveyed with a very low error rate. If at most one error per minute can be tolerated in a 20 Mb/s HDTV signal, the error rate must be less than one in a billion. A channel coder encodes the source coded bit-stream to minimize transmission errors given the characteristics of the channel.

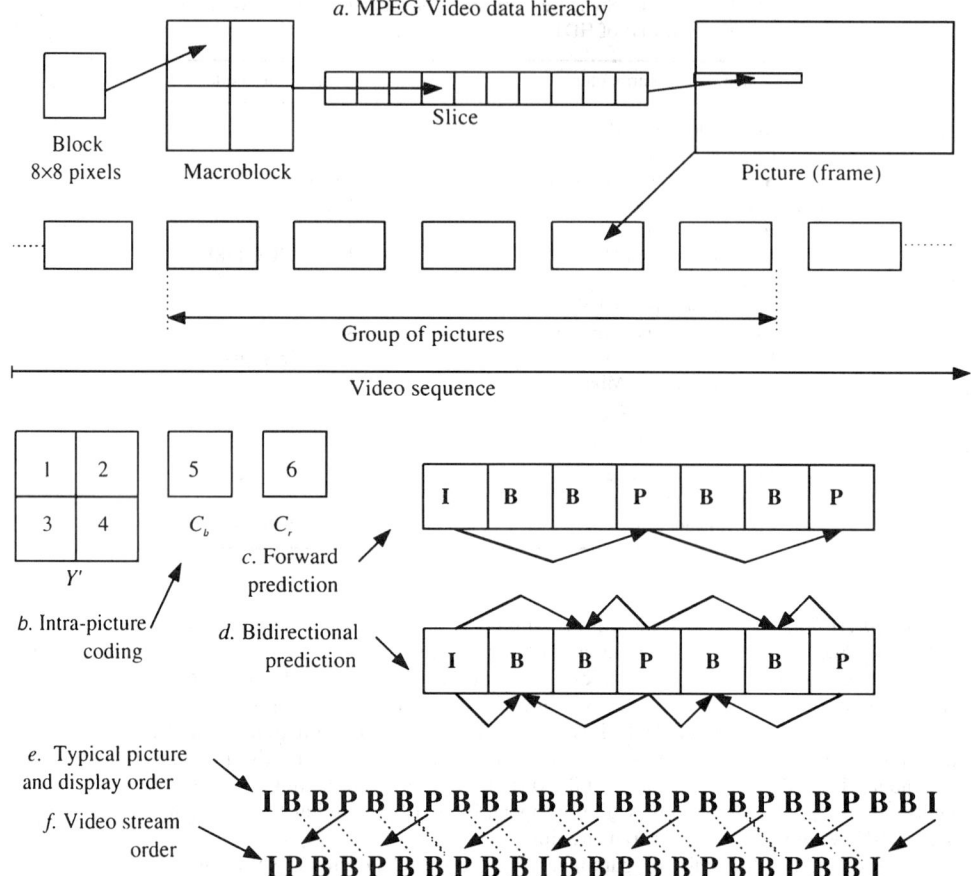

FIGURE 21.1.21 Basic principles of video compression coding. I frames are intraframes coded only with DCT (see Fig. 21.1.23). In forward predicted frames *P* and bidirectionally predicted frames B, the difference between actual data and motion predicted data is DCT coded. Data in DCT "frequency" space are adaptively quantized and zig-zag scanned with run-length coding. From S. N. Baron and W. R. Wilson (1994).

Figure 21.1.21 illustrates the basic elements of the MPEG-2 compression format. A video sequence consists of a series of pictures. A picture is usually a progressively scanned or interlaced scanned frame, but it may also be a single interlaced field. The pictures are organized into groups of pictures (GoPs), which are a basic unit for compression (Fig. 21.1.22a). Within a GoP, the first picture is coded entirely with spatial, or intraframe coding, and is referred to as an I-picture (I-frame). Subsequent pictures in the GoP are predicted using block-based motion estimation techniques. P-pictures are predicted from preceding I- or P-pictures using forward motion compensation. B-pictures are bidirectionally predicted from adjacent I- and/or P-pictures.

All pictures are composed of Y, U, and V pixels in a 4:2:0 sampling grid organized as 8×8 blocks, as shown in Fig. 21.1.22b. A macroblock consists of the four luminance blocks and two chrominance blocks covering the same spatial area of the picture, and macroblocks are organized into slices. A macroblock is the basic unit for motion compensation, and each macroblock in a P- or B-picture may have motion vector information indicating its prediction in $1/2$ pixel components from the adjacent reference frames.

Within a *Video Sequence*...

a *Group of Pictures* (GOP) is composed of I- P- and B- pictures (frames)

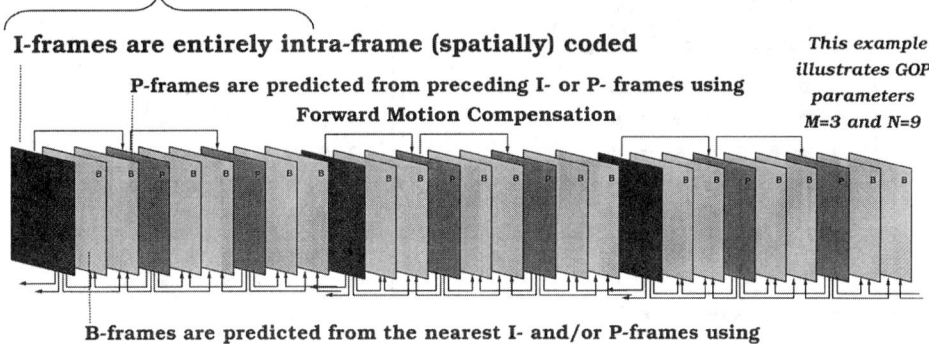

I-frames are entirely intra-frame (spatially) coded

This example illustrates GOP parameters M=3 and N=9

P-frames are predicted from preceding I- or P- frames using Forward Motion Compensation

B-frames are predicted from the nearest I- and/or P-frames using Bidirectional Motion Compensation

Time

FIGURE 21.1.22   (*a*) MPEG Group of Pictures.

Each 8 × 8 block is encoded with the DCT. DCT coefficients are quantized and scanned in zigzag order.
Two methods of data compression are used in MPEG-2: (1) intra-picture transform coding using discrete cosine transform (DCT), adaptive quantization, and variable length coding (VLC) and (2) motion compensation for prediction of P and B pictures. Motion vectors together with the intra-frame transform coded difference between predicted and actual picture is transmitted.

Pictures are composed of Slices

a *Slice* is a collection of adjacent Macroblocks

Luminance Picture   Cr Picture   Cb Picture

a *Macroblock* is a set of co-located luminance and chrominance blocks

(4 blocks of Y
and 1 block
each of
Cr and Cb)

a *Block* is an 8 x 8 array of pixels

FIGURE 21.1.22   (*b*) Slices, Macroblocks, and Blocks.

Block of 8×8 samples

8×8 DCT transform space ("frequency" space)
to be adaptively quantized, zig-zag scanned,
and run-length coded

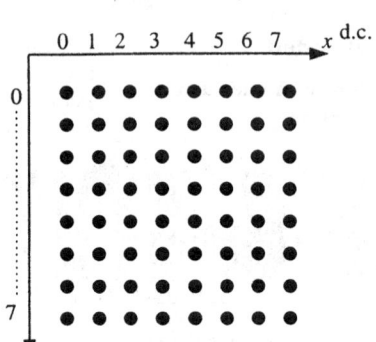

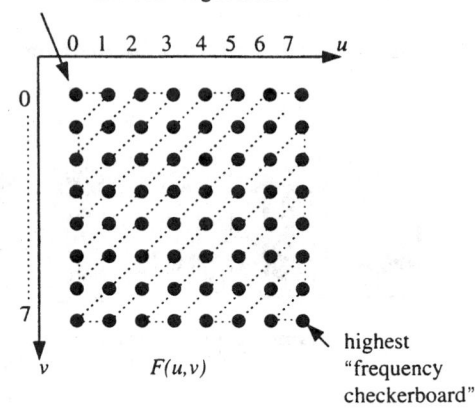

The $N \times N$ two-dimensional discrete Fourier transform (DCT) is defined as:

$$F(u,v) = \frac{2}{N} C(u)C(v) \sum_{x=0}^{N-1} \sum_{y=0}^{N-1} f(x,y) \cos\frac{(2x+1)u\pi}{2N} \cos\frac{(2y+1)v\pi}{2N}$$

where $x, y = 0, 1, 2, \ldots, N-1$ are discrete spatial coordinates in the pixel domain

and $u, v = 0, 1, 2, \ldots, N-1$ are discrete coordinates in the transform domain

and
$$C(u), C(v) = \begin{cases} \dfrac{1}{\sqrt{2}} & \text{for } u,v = 0 \\ 1 & \text{otherwise} \end{cases}$$

The inverse discrete Fourier transform (IDCT) is defined as:

$$f(x,y) = \frac{2}{N} \sum_{u=0}^{N-1} \sum_{v=0}^{N-1} C(u)C(v)F(u,v) \cos\frac{(2x+1)u\pi}{2N} \cos\frac{(2y+1)v\pi}{2N}$$

FIGURE 21.1.23   The discrete cosine transform (DCT). Many transform components have quantization level = zero. If x and y were continuous (which they are not) each transform component would be a product of a vertical and a horizontal cosine grating. From ISO/IEC 1.3818-2 Rec. H262.

The DCT transform is applied to an 8×8 block of samples $f(x,y)$ as shown in Fig. 21.1.23. In the transform domain the information is represented by an 8×8 array of real "frequency" components $F(u,v)$. A component $F(u,v)$ can be considered as the amplitude of an interference pattern consisting of the product of a horizontal and a vertical cosine wave (grating) forming a "checker-board" with soft-edged rectangular "squares." The pattern with the smallest "squares" is represented by the component in the lower right corner in the frequency domain shown in Fig. 21.1.23, and the biggest "square" is represented by the dc component in the upper-left corner. The pels can be considered as samples of a superposition of these interference patterns. It must be kept in mind that the pixels $f(x,y)$ are *not* meant to represent a continuous function periodically mirrored in the x and y axis and band limited to the Nyquist bandwidth of the samplepoint lattice, and that a "frequency" component $F(u,v)$ does *not* specify a sine-wave grating. The DCT is strictly a transformation of an $N \times N$ array of numbers into another $N \times N$ array of numbers using Fourier techniques.

The DCT is remarkably effective in removing redundancy in typical pictures. When the pels $f(x,y)$ in a block are highly correlated, a large number of coefficients $F(u,v)$ in the transform domain are zero or very small, particularly the high-frequency components. The components in the transform domain are quantized. While the HVS is sensitive to quantization of low-frequency components, it tolerates large steps for high frequencies. An 8×8 *quant matrix* specifies the quantization steps to be associated with the array of $F(u,v)$ components. With quantum steps increasing with frequency, a large number of high-frequency components are quantized to level zero. MPEG allows for adapting the quant matrix for each block for an optimum trade-off between quality and available buffer memory. The quant matrix is specified in a header. The components are scanned in a zigzag pattern as shown in Fig. 21.1.23, generating a stream of symbols with bursts of zeros. The scanned data are subsequently variable length coded in run-amplitude pairs, specifying a nonzero component and an associated run length of zero coefficients. As in the Morse code, shorter codes are used for common pairs and longer for less common pairs. These symbols are delivered for transmission. DCT is the only technique used for I-picture data compression.

Motion compensation is used for P- and B-picture data compression. Before intra-frame coding (DCT), samples in blocks of P or B pictures are subtracted from predicted samples. The intra-frame encoded data of these "differential" pictures is transmitted together with data, *motion vectors*. Motion vectors are determined by matching a macroblock in the actual picture with macroblocks in a nearby area in the reference pictures (the reference picture for a P picture is the previous P or I picture and the reference pictures for B pictures are nearby I and P pictures). MPEG has standardized the vector range and how to specify macroblock motion vectors, but not how they are determined. After motion compensation, the pels of the P and B frames are highly decorrelated. This makes the subsequent DCT and quantization processes extremely efficient. In the atypical case where a predicted macroblock has higher entropy than the original image macroblock, intra-coding may still be used in P + B pictures. Given the capacity of the channel and the buffer memory, increased range and more accurate block prediction can reduce the call for quant matrices with course quantization. The consequence is compatible improvement of picture quality. Buffering the data for adaptive quantization is a sophisticated process. The outputs of video and audio data compression encoders are called program elementary streams (ES). MPEG classifies video ESs compression types, called *profiles*, intended for various applications. The profiles are called *simple, main, snr, spatial, and high.* MPEG also identifies various *levels* of video ESs characterized by the maximum number of active luminance samples per second. They are

Low:	352H × 288V × 30Hz	Main:	720H × 576V × 30Hz
High-1440:	1440H × 1152V × 60Hz	High:	1920H × 1152V × 60Hz

The two most important are main profile at main level (MP@ML) used in standard definition DTV system such as DVB and main profile at high level (MP@HL) and in high definition DN systems such as the ATSC standard.

Digital Audio for DTV

Just as analog video signals are digitized by sampling and quantization, analog audio signals are similarly converted into digital audio. A *de facto* international standard (AES/EBU) for high-quality digital audio is 48,000 samples per second and 16 bits per sample. Other standard sample rates are 44.1 kHz (compact disc) and 32 kHz. And just as digital video pels are compressed to discard redundant and imperceptible visual information, audio can also be compressed. By taking advantage of the psychophysics of the Human Aural System the data rate per audio channel can be reduced for transmission from $48 \times 16 = 768$ kb/s to substantially less than 128 kb/s with virtually imperceptible degradation. While there is an MPEG-2 audio standard, the Dolby AC-3 system, used in cinema theaters, has been adopted for the ATSC system for HDTV in North America. Both systems use *psychoacoustic coders* and deliver high-quality audio. The ATSC "Dolby Digital" system delivers five 20 kHz audio channels and a low-frequency (3 to 120 Hz) enhancement channel (LFE). This package of "5.1 channels" requires a maximum digital channel capacity of 384 kb/s.

Strings of digitized samples are converted from the time domain to the frequency domain, where spectra in 24 "critical bands" are processed. The basis for data compression is that in the vicinity of strong components, relatively weak components can be ignored without perceptible degradation (masking). The frequency components are coded and packetized. There are six levels of services including surround sound (complete main = CM, music and effects = CE), sound for visually impaired (VI) or hearing impaired (HI), as well as dialogue (D), emergency (E), and voice over (VO). Principles of *psychoacoustic coders* are discussed in the chapter on audio.

The ES data are put as payload in *packets*, like merchandise in freight cars to be assembled to form a train and referred to as *Packetized Elementary Streams* (*PES*). A packet has a header including a label (packet ID, *PID*) and timing information. The packet also includes an *adaptation field*, including program clock, reference addresses, and descriptors for how the payload is to be used. In the MPEG-2 transport system a packet contains 188 bytes. Just as MPEG-2 compression is highly flexible, MPEG-s transport contains flags for local program insertion points, encryption for controlled access (CA), multimedia, interactive data, and many other descriptors.

It is one of the most important attributes of digital television that the information can be packetized and transmitted in a data communications format which is interoperable with other applications, and can be distributed not only by broadcasting (terrestrial, satellite, cable), but over various data communication networks, for example, as *Asynchronous Transfer Mode* (ATM) using communication over BISDN. BISDN includes fiber optics transmission. ATM uses fixed length packets (53 bytes) with a header containing descriptors as well as addresses for switching. The convergence of broadcasting, multimedia, telecommunications, data communications, and computer applications presents complex compatibility and standardization problems which at present are addressed by a number of international standardization committees. Digital television is now part of the broader industry of information production and distribution. MPEG is part of this activity, and the various adaptation layers have been designed to provide interoperability, so that MPEG-2 transport streams may be easily layered on top of other digital communication systems.

Digital Television Terrestrial Broadcasting (DTTB)

Considerations driving the selection of digital modulation approaches for DTTB include striking an appropriate balance among *quality, coverage, and compatibility* as determined by public policy priorities of various countries:

- Reception of high-quality HDTV and audio
- Good coverage of population and area with acceptable reliability
- Acceptable levels of DTTB interference into reception of existing services, particularly reception of standard analog television signals (NTSC, PAL, SECAM) (Fig. 21.1.18)

In the United States the FCC has only made available the existing TV broadcast bands for DTTB of HDTV, allocating an additional 6-MHz TV channel for DTV to every existing broadcast station to be radiated from an antenna which is close to the NTSC antenna of the associated station. The consequence is that current rules for station allocations, including so-called UHF taboos, must be changed to accommodate about 1700 new DTTB channels in the TV broadcast bands, primarily at UHF, with tolerable interference from NTSC transmitters and acceptable DTTB interference into existing NTSC reception. The FCC's channel allocation plan is based on extensive measurements, field tests, and computer modelling. Thus DTV coverage is primarily limited by interference into existing NTSC service, so achieving the required data rate (approximately 20 Mb/s) at the lowest possible power is paramount.

CNR in DTTB is defined as the ratio in dB of *average power* to total noise power in the channel. With the USB modulation used in the ATSC standard, the threshold of visibility of errors caused by random noise occurs at CNR $\cong$ 15 dB. At about 1 dB below this threshold the picture is not useable. Random noise is only one of a number of possible transmission impairments (see previous section). In contrast with broadcasting of analog video, impairments in proposed DTTB systems add up toward a threshold at which reception abruptly changes from perfect to useless. This abrupt coverage limit in DTTB is referred to as the "cliff effect," and is similar to the effect in FM radios.

The radiated DTTB signal is noiselike and has a peak-to-average power ratio of 8 to 10 dB. Co-channel DTTB interference into NTSC appears essentially as random noise. Comparing DTTB coverage with NTSC grade-B coverage on the same channel and with the same receiving antenna, transmission line and receiver noise figure as in NTSC reception, the required effective radiated *peak power* in DTTB of HDTV could be 3 to 5 dB less than NTSC *peak power* radiating from the same antenna. While HDTV quality of DTTB is far better than NTSC grade-B quality, statistics of field strength variability with time and location is critical considering the "cliff effect" in DTTB. DTTB reception reliability can, however, be improved by as much as 10 dB with a better receiving antenna and an antenna amplifier.

Modulation Methods for Digital TV

Four modulation methods (see Eq. 31) widely used for r.f. transmission of digital television:

- PSK: Phase shift keying may be preferred in digital TV satellite broadcasting
- QAM: Quadrature amplitude modulation in digital cable television system
- VSB: Vestigial sideband is used in the ATSC standard for terrestrial DTV broadcasting
- COFDM: Coded orthogonal frequency division multiplex is used in the DVB-T standard for terrestrial DTV broadcasting

In QAM, a carrier in the middle of the channel is amplitude modulated double sideband in-phase by one sequence of digital symbols and in quadrature by another sequence of symbols. In a 2.5-MHz Nyquist bandwidth 5 megasymbols per second can be transmitted in the I channel as well as in the Q channel. Thus, a 5-MHz channel can convey 10 megasymbols per second. To reduce extended ringing and intersymbol interference, the spectrum is rolled off antisymmetrically around the Nyquist band edges. This calls for some extra bandwidth. Figure 21.1.24 illustrates typical roll-off for transmission of about 10 megasymbols per second over a 6-MHz television channel.

The quantized symbols in the I and Q channels are coincident, and consequently the amplitude and phase of the carrier at symbol detection time can be represented by a point in a constellation of points in the *xy-plane* representing all possible transmitted vectors as shown in Fig. 21.1.24. If the number of constellation points is K, the signal is referred to as K-QAM (K = 64 for 3 bits/symbol and 16 for 2 bits/symbol). Figure 21.1.24 shows that each constellation point is associated with a "decision" cell. An error occurs if transmission impairments at decision time moves a signal vector into the wrong cell. The distance from a constellation point to the border of its cell represents the allowed "budget" for transmission impairments. Figure 21.1.24 shows an approximate relation between the error rate and the rms level σ of Gaussian noise (expressed in terms of the distance from a constellation point to the border of its box). For a typical error rate of 10^{-3}, $\sigma =$ 0.35 or −11 dB. The figure also shows the carrier level for $(2N)^2$-QAM. For 16QAM the required CNR for an error rate of 10^{-3} is seen to be about 18 dB. Channel coding (see below) can improve this relation to an error rate of 10^{-9} out of the channel decoder for a CNR of 15 dB at the input of the demodulator. However, other impairments leave only a part of the error budget to random noise. Phase jitters are particularly damaging to the peripheral vectors. Carrier phase and symbol phase recovery in the demodulator are also critical and complex functions.

VSB is quite similar to *QAM*. In fact if *I* and *Q* symbols in *QAM* are interleaved in time rather than coincident, they can be detected with a synchronous carrier at the Nyquist band-edges as well as in the center of the band. In VSB detection, the symbols arrive in a sequence I,Q, −I, −Q, I, Q, . . . , i.e., at twice the QAM rate but with half the number of bits per symbol. Thus, for a given data rate, the required CNR for a given error rate is about the same as in QAM. As in standard analog television VSB transmits a quadrature signal that eliminates a sideband but doubles the average power. Peaks in the transmitted signal are caused by the quadrature signal. The peak-to-rms ratio is, however, about the same as in QAM (8 to 10 dB). The antisymmetric roll-off around the carrier reduces peaks, ringing, and intersymbol interference. A residual carrier pilot tone is also used. ATSC standard has to improve robustness and reduce lock-in time.

While VSBAM is an extreme time division multiplex system, OFDM is an extreme frequency division multiplex system. OFDM is an entirely different approach to over-the-air transmission of digital data. OFDM uses hundreds of narrowband channels equally spaced along the frequency axis. Each carrier is modulated (QAM)

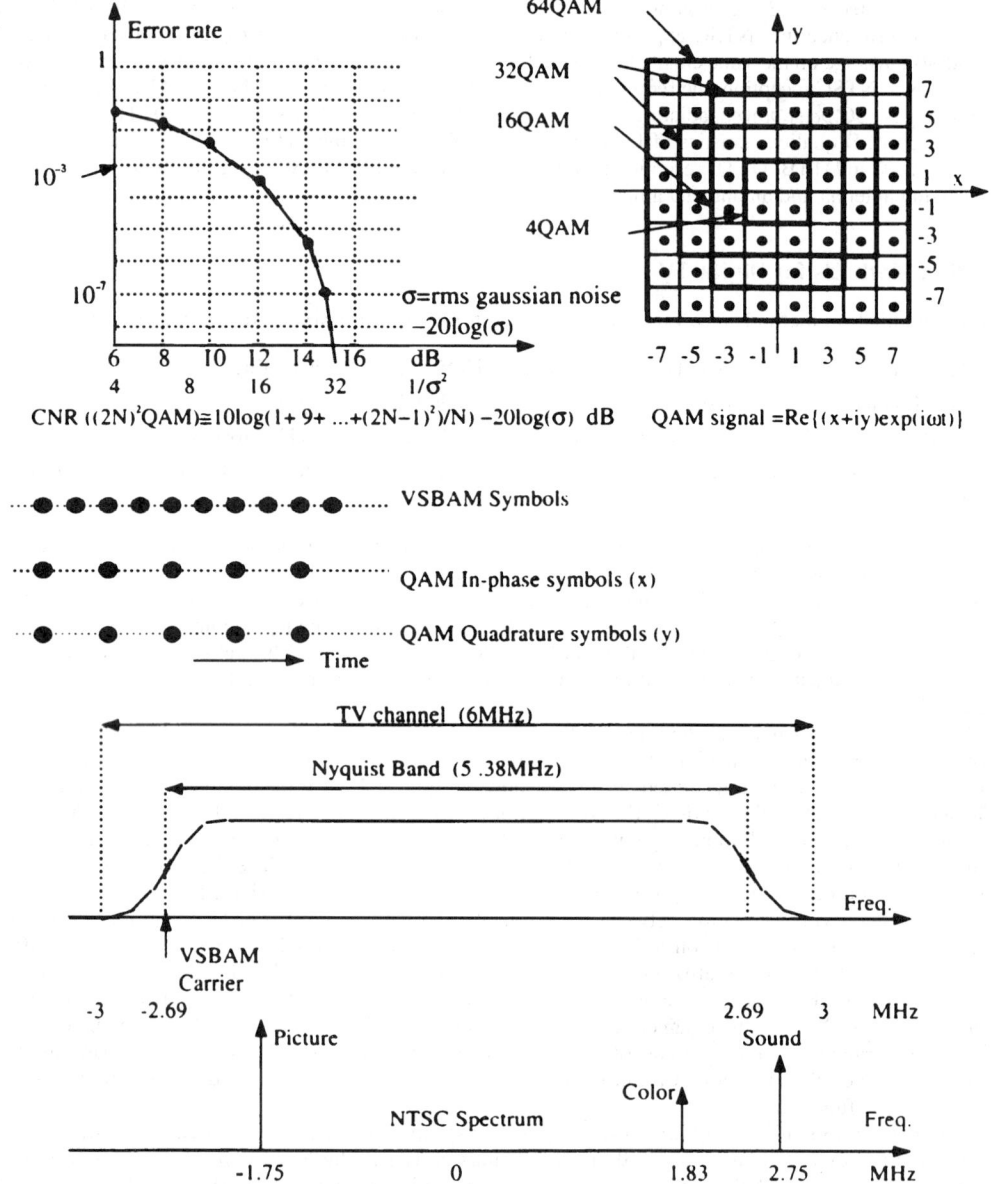

FIGURE 21.1.24 QAM signal vector constellation points and CNR. Spectra VSBAM, DTTB broadcasting in the United States in the presence of co-channel NTSC radiation.

by digital symbols in such a way that the spectrum of the channel has a sinx/x response with nulls at the carrier frequencies of the other channels, as illustrated in Fig. 21.1.25. As a consequence these channels are "orthogonal" along the frequency axis and do not interfere with each other. Figure 21.1.25 shows that each QAM symbol occupies a long period of time T. In the frequency domain the symbols are spaced F = 1/T Hz. There is a period D between symbols in the time domain to guard against intersymbol interference due to ghosts.

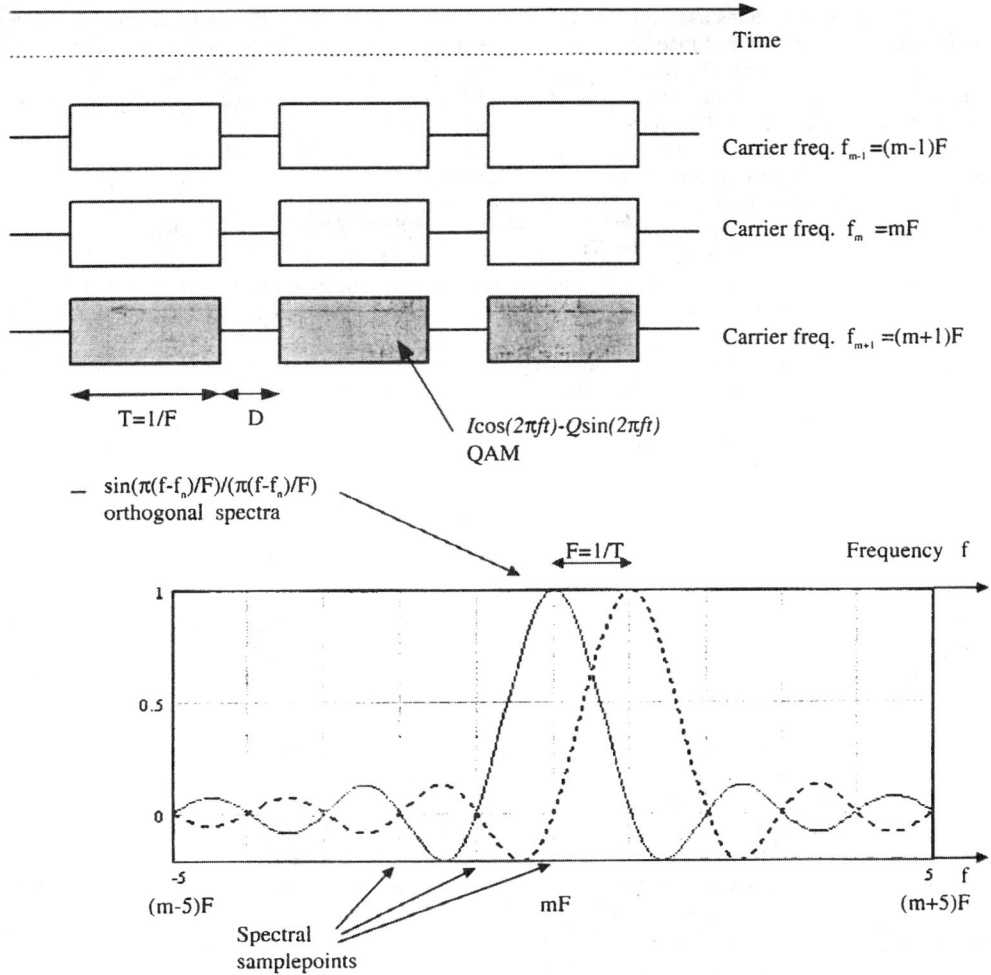

FIGURE 21.1.25 OFDM (orthogonal frequency division multiplex) time-frequency representations. Temporal guardband D is chosen to minimize intersymbol interference due to multipath.

Each symbol is not distorted by a ghost, but the level of the symbol is affected by ghosts. With an equalizer, the CNR of different channels may be different. The terrestrial DVB standard (D1 3-T) includes two OFDM formats: 2k with 1705 carriers and 8k with 6817 carriers. In the 8k format the guard interval D is scalable.

With coded OFDM (*COFDM*) lost information in different channels can be managed. While OFDM is tolerant of impulse noise, a disturbance occupying a part of the spectrum may wipe out a number of channels. For broadcasting in the presence of strong interference from standard analog television transmitters, it has been proposed that OFDM channels coinciding with bands around the picture, sound, and color subcarrier of an analog TV co-channel signal be left vacant (Fig. 21.1.24). Other methods using error correction designed for this type of interference are also possible. COFDM can be used for same channel cellular digital broadcasting of identical information. Co-channel interference is then equivalent to a ghost. The time guard D has a significant impact on the geographic spacing of the cell. In detection of OFDM, the symbols transmitted during a period T are derived by a spectral analysis of a fast Fourier transform of the temporal signal received during the period.

All transmission systems used for DTTB take the packetized bitstream through a channel coder before it is delivered to a modulator. Since the source coders have eliminated a large amount of redundancy in the video and audio programs, the bitstream feeding *the channel coder* is extremely intolerant to errors. The channel coder adds redundancy in the form of *forward error correction* (FEC) and possibly by *trellis coding* as illustrated in Fig. 21.1.2. The purpose of FEC is to correct transmission errors at the receiver. The Reed-Solomon block FEC can very effectively correct random errors. In the Grand Alliance system 20 parity bytes are added to the 188 byte packet to correct for up to 10 errors (t = 10 R-S code). The price is a 10 percent increase in required channel capacity (code rate R = 188/208). However, since errors induced by RF interference are generally burst errors, the data interleaving technique is used to regroup data, so that a burst error in transmission is dispersed to appear as random errors spread across many Reed-Solomon FEC blocks, each of which can correct the random error. The purpose of trellis-coding is to prevent errors from occurring, particularly errors caused by random noise, or conversely, to reduce the required CNR for a given error rate. In trellis-coded transmission, fewer symbols are used than the number of symbols that the modulator can generate, i.e., more bits per symbols are available than are transmitted. In the decoder, hard decisions are not made on a symbol by symbol basis, but detected levels of a number of symbols in a sequence form a vector (soft data) which is compared with vectors which possibly could have been transmitted (allowed vectors). Using "soft" data from several symbols and cleverly selected allowed vectors, the required CNR for a given error rate can be reduced.

BIBLIOGRAPHY

"CCIR and FCC TV Standards," a condensed pamphlet of worldwide SDTV standards, Rohde & Schwarz, P.O. 801469, D-8000 München 80, Germany.

Hutson, G. H., and P. J. Shepderd, "Colour Television," 2nd ed., McGraw-Hill, 1990. Emphasis on European standards.

Jordan, E. C. (ed.) "Reference Data for Engineers: Radio, Electronics, Computer, and Communications," 8th ed., Howard W. Sams (McMillan), 1993. Includes numerous references to TV standards and FCC F(50/50) charts. *Journal of the SMPTE* publishes many excellent papers on Television Fundamentals and Standards. The papers include numerous references. Since 1994 the journal regularly publishes tutorials. There are a number of tutorials on digital television by S. Baron and from 3/95 to 3/96 by D. Strachan.

Pearson, D. E. "Transmission and Display of Pictorial Information," Pentech Press, 1975.

Poynton, C. "A Technical Introduction to Digital Video," John Wiley & Sons, 1996.

Pritchard, D. H., and J. J. Gibson, "Worldwide Color Television Standards—Similarities and Differences, *SMPTE J.*, 89: 111–120, February 1980. Also in Rzeszewski, *Color Television.*

Rzeszewski, T. "Color Television," IEEE Press, 1993.

Symes, P. D., "Video Compression," McGraw-Hill, 1998.

TV Technology publishes many timely papers, including informative tutorials on advanced TV by M. Weiss, C. Rhodes, and others.

Whitaker, J. C. "DTV: The Revolution in Digital Video," 2nd ed., McGraw-Hill, 1999.

Whitaker, J. C., and K. B. Benson (eds.) "Standard Handbook of Video and Television Engineering," 3rd ed. (formerly the Television Engineering Handbook), McGraw-Hill, 2000.

Standards

DVB Standards are Published by ETSI, at http://www.etsi.org in the "Publications Download Area."

FCC Rules, Part 73, Broadcast Services, Federal Communications Commission, Washington, D.C.

ITU-R Volume XI. Part 1. *Broadcasting Service-Television.* Report 624-3: *Characteristics of Television Systems.* Report 476-2: *Colorimetric Standards in Colour Television.* Recommendation ITU-R 601-1: *Encoding Parameters of Digital Television for Studios,* also in *SMPTE J.* 90–10, October 1981, *SMPTE* 103–12, December 1994 (SMPTE125M) and *EBU Review* #187 June 1981.

MPEG-2, Recommendation ITU-T H.222.0, (ISO/IEC 13818-1) Generic Coding of Moving Pictures and Associated Audio: SYSTEMS.

MPEG-2, Recommendation ITU-T H.262, (ISO/IEC 13818-2) Generic Coding of Moving Pictures and Associated Audio: VIDEO.

MPEG-2. *ISO/IEC*. Generic Coding of Moving Pictires and Associated Audio" *International Standard 13818-2, Recommendation H.262, (1995E)*. ISO/IEC Copywrite Office, Case Postale 56, CH1211 Geneva 20, Switzerland. ATSC Digital Television Standard and Guide to the Use of ATSC Digital Television Standard, documents A/53 and A/54 published by ATSC in 1995. ATSC, 1700 K Street NW, Suite 800, Washington, D.C.

Recommendation ITU-R BT.470-6, Conventional Television Systems.

Recommendation ITU-R BT.601-5, Studio Encoding Parameters of Digital Television for Standard 4:3 and Wide-Screen 16:9 Aspect Ratios.

Recommendation ITU-R BT. 656-4, Interfaces for Digital Component Video Signals in 525-line and 625-line Television Systems Operating at the 4:2:2 Level of Recommendation ITU-R BT. 601 (Part A).

Recommendation ITU-R BT.709-3, Parameter Values for the HDTV Standards for Production and International Programme Exchange.

Recommendation ITU-R BT.1381, SDI-Based Transport Interface for Compressed Television Signals in Networked Television Production Based on Recommendations ITU-R BT.656 and ITU-R BT.1302.

SMPTE 170M, Composite Analog Video Signal—NTSC for Studio Applications.

SMPTE 310M, Synchronous Serial Interface for MPEG-2 Digital Transport Streams.

SMPTE 314M, Data Structure for DV Based Audio, Data and Compressed Video—25 Mb/s and 50 Mb/s.

SMPTE proposed Standards are published in the *Journal of the SMPTE*. SMPTE 240M-1988. *Standard for-1125/60 high-definition production.* September 1989. SMPTE274M *Standard* for *1020 × 1080 Scanning*, October 1994. SMPTE296M *Standard for 1280 × 720 Scanning*, May 1996.

Audio

Gibson, J. J., *J. Audio Engineering Society*, September 1986. p. 647–660. Reviews the FCC-BTSC Standard (FCC Bulletin OET60) for and the performance of Television Multichannel Sound (MTS). Includes FCC specifications in FCC Bulletin OET-60.

Digital Television and HDTV

Advanced Digital HDTV System Description, Sarnoff, Thomson, Philips and NBC submission to the Advisory Committee on Advanced Television Service, 1992.

Anaslassiou, D. (ed.) "Special Issue on HDTV," *IEEE Communications Magazine*, Vol. 34, No. 6, June 1996.

ATVA-Progressive System Description, MIT and General Instruments submission to the Advisory Committee on Advanced Television Service, 1992.

Baron, S. N. "An overview of the DTTB model," *J. SMPTE*, Vols. 103–105, May 1994.

Baron, S. N. "Digital television scalability and interoperability, *J. SMPTE*, Vol. 103, No. 10, October 1994, p. 673.

Baron, S. N. and W. R. Wilson, "MPEG Overview," *J. SMPTE*, Vol. 103 No. 6. June 1994.

Beyers, B. W. "Digital Television: Opportunities for Change," *IEEE Transactions on Consumer Electronics*, Vol. 38, No. 1, February 1992.

Brandenburg K. and G. Stoll. "ISO-MPEG-1 Audio: Ageneric Standard for Coding of High Quality Digital Audio," *J. Audio Engineering Society*, Vol. 42, No. 10, October 1994.

Bretl, W., et al. "VSB Modern Subsystem Design for Grand Alliance Digital Television Receiver," *IEEE Transactions on Consumer Electronics*, Vol. 41, No. 3, August 1995.

Bryan, D. "QAM for Terrestrial and Cable Television," *IEEE Transactions on Consumer Electronics*, Vol. 41, No. 3, August 1995.

Chopra A., et al. "The HDTV Broadcast Standard: Subjective Evaluation of Proponent Systems," *SMPTE J.*, July 1998.

Cugnini A., et al. "MPEG-2 Decoder for the Digital HDTV Grand Alliance System," *IEEE Transactions on Consumer Electronics*, Vol. 41, No. 3, August 1995.

DigiCipher System Description, General Instruments submission to the Advisory Committee on Advanced Television Service, 1992.

Digital HDTV Grand Alliance System Description, Grand Alliance (AT&T, General Instruments, Sarnoff, Thomson, Philips, MIT and Zenith) submission to the Advisory Committee on Advanced Television Service, 1994.

"Digital Television in the U.S.," *IEEE Spectrum*, Vol. 4, No. 5, March 1999.

Digital Spectrum Compatible HDTV System Description, Zenith and AT&T submission to the Advisory Committee on Advanced Television Service, 1992.

Hopkins, R. "Digital Terrestrial HDTV for North America: The Grand Alliance HDTV System," *IEEE Transactions on Consumer Electronics*, Vol. 40, No. 3, August 1994.

Joseph, K., et al. "Prioritization and Transport in the ADTV Digital Simulcast System," *IEEE Transactions on Consumer Electronics*, Vol. 38, No. 3, August 1992.

Kraus, J. A. "Source Coding, Channel Coding and Modulation Techniques Used in the DigiCipher System," *IEEE Transactions on Broadcasting, Special Issue on HDTV Broadcasting*, Vol. 37, No. 4, December 1991.

Kraus, J. A. "Source Coding, Channel Coding and Modulation Techniques Used in the ATVA-Progressive System," *IEEE Transactions on Broadcasting, Special Issue on HDTV Broadcasting*, Vol. 37, No. 4, December 1991.

Krivocheev, M. edited by S. N. Baron," The First Twenty Years of HDTV: 1972–1992," *J. SMPTE*, Vol. 102, No. 10, October 1993. (Includes many references to international standards committees).

Luplow, W. "Digital Spectrum Compatible HDTV," *HD World Review* (ISSN 1055-6931) Vol. 2, No. 4, Fall 1991.

Luplow, W., et al. "Source Coding, Channel Coding and Modulation Techniques Used in the Digital Spectrum-Compatible HDTV System," *IEEE Transactions on Broadcasting, Special Issue on HDTV Broadcasting*, Vol. 37, No. 4, December 1991.

Lyons, P. "Grand Alliance Prototype Transport Stream Encoder Design and Implementation," *IEEE Transactions on Consumer Electronics*, Vol. 41, No. 3, August 1995.

Mailhot J., et al. "The Grand Alliance HDTV Video Encoder," *IEEE Transactions on Consumer Electronics*, Vol. 41, No. 3, August 1995.

Miller, H. N. "A Market Perspective on the HDTV Interoperability Standards Question," *IEEE Transactions on Consumer Electronics*, Vol. 38, No. 1, February 1992.

Netravali, A. B. Haskell, "Digital Pictures—Representation and Compression," Plenum Press, 1988.

Reitmeier, G. "The U.S. Digital television Standard and Its Impact on VLSI," *J. VLSI SIGNAL PROCESSING SYSTEMS for Signal, Image and Video Technology*, Vol. 17, No. 2/3, pp. 281–290, November 1997.

Reitmeier, G., and C. Basile "Advanced Digital Television," *HD World Review* (ISSN 1055-6931) Vol. 2, No. 4, Fall 1991.

Reitmeier, G., et al. "Source Coding, Channel Coding and Modulation Techniques Used in the ADTV System," *IEEE Transactions on Broadcasting, Special Issue on HDTV Broadcasting*, Vol. 37, No. 4, December 1991.

Sablatash, M. "Transmission of all Digital Advanced Television: State of the Art and Future Reflections," *IEEE Transactions on Broadcasting*, Vol. 40, No 2, June 1994.

Sgrignoli, G. "Field Test Results of the ATSC VSB Transmission System," *J. SMPTE*, Vol. 106, p. 601, September 1997.

Sgrignoli G., et al. "VSB Modulation Used for Terrestrial and Cable Broadcasts," *IEEE Transactions on Consumer Electronics*, Vol. 41, No. 3, August 1995.

"Special issue on consumer electronics," *Proc. IEEE*, Vol. 82–84, April 1994.

"Special issue on digital television," *IEEE Communications Magazine*, 35–5, May 1994.

"Special Issue on Digital Television," *IEEE Spectrum*, Vol. 32, No. 4, April 1995.

"Special Issues on Digital Television," *Proc. IEEE*, Vol. 83, No. 6, June 1995; Vol. 83, No. 7, July 1995.

Wu, Y., et al. "Orthogonal Frequency Division Multiplex: A Multi-Carrier Modulation Scheme," *IEEE Transactions on Consumer Electronics*, Vol. 41, No. 3, August 1995.

CHAPTER 21.2
VIDEO SIGNAL SYNCHRONIZING SYSTEMS

Richard A. Kupnicki

TYPES OF SYNCHRONIZATION DEVICES

Video synchronization is the process of aligning or matching two or more dissimilar signals with respect to their timing relationship. In order to fulfill this function, it is first necessary to have a reference video signal such as that produced by a master or a slave sync generator. Then it is necessary to perform the actual signal synchronization using a frame synchronizer.

The three types of synchronization devices which will be discussed herein are the master sync generator, the slave sync generator, and the frame synchronizer. Each fulfills a specific, independent function within the television facility. For each, there are often subgroups which relate to the different video signal formats.

The master sync generator is usually free running and is used as a lock source for all other devices within the facility. The master generator is further characterized by its own high-stability internal oscillator. Frequently two master sync generators will be used together in conjunction with an automatic change over unit in order to minimize the impact of equipment failure. A master sync generator does not need to lock a reference input signal. Sometimes a high stability frequency source may be used as a reference in critical applications.

A slave sync generator will have all the attributes of the master with the addition of circuits to allow the generator to lock to an incoming reference sync signal (generated by a master). Some design savings are generally achieved in the slave generator by relaxing the specifications of the local oscillator circuits. The main purpose of the slave generator is to produce a local sync reference, with timing capability, thus permitting timing offsets with respect to the master generator. Note that as a good practice, color black should be used as the reference video signal due to its constant average picture level (APL) in order to minimize jitter introduced into the phase lock loop.

A video frame synchronizer is used whenever it is necessary to bring an unrelated video source back into timing relationship within a facility. A typical example of this requirement would be a satellite feed or other external source. As well as providing the necessary timing controls, the majority of frame synchronizers will also provide some measure of video conditioning (processing amplifier functions) such as video level control and hue control.

A modern television studio processes audio along with video. Use of frame synchronizers in the signal paths adds considerable delay to the video signal, to the point where video gets delayed more than audio. A human finds it objectionable when audio is more than one video frame ahead or two frames behind its associated video signal. To avoid this problem, audio must be delayed to the same degree as video. To aid in audio synchronization process a suitable reference signal, digital audio reference signal (DARS) needs to be generated by the master or slave sync generators. SMPTE time code is also generated to aid in editing of audio and video streams.

MULTI-STANDARD SLAVE SYNCHRONIZING GENERATOR

Slave sync generators are generally used to provide a local color black reference signal which is locked to a master sync generator system, either directly or via a distribution system.

The use of slave sync generators is an integral part of any television or related facility. They provide a means for timing local devices or groups of devices to a common master reference. Modern slave sync generators will generally always provide an infinite phasing capability, thus allowing the timing outputs of the slave generator to be located anywhere in time relative to the master generator. *Infinite phasing* refers to the capability to adjust vertical phase anywhere within a color frame. "H" phasing refers to adjustments within a line and "fine" phasing refers to adjustment within a subcarrier cycle.

A complete block functional diagram of a slave synchronizing generator (Fig. 21.2.1) should be used as a reference for the following description. The figure shows a multi-standard generator capable of locking to and generating reference NTSC and PAL signals. The implementation differences are summarized in Table 21.2.1, as it is practical to build one piece of hardware with programmable registers to satisfy the requirements of both television systems. In general practice, analog NTSC is used as a reference for all 29.97-frame-per-second-based systems and analog PAL as reference for 25-frame-per-second-based systems.

The reference signal's color burst component is separated using a bandpass filter. The stripped color burst signal is then fed to the complex multiplier. The subcarrier generator block generates two other signals referred to as Sin and Cos since they have a quadrate relationship. The subcarrier generator is composed of an accumulator feeding a set of Sin and Cos ROM look up tables. The ROM outputs are converted to analog via a DAC before feeding the multipliers. The multiplier output is low-pass filtered to retain the difference frequencies, i.e., the frequency error between the crystal oscillator and the input signal. In PAL applications,

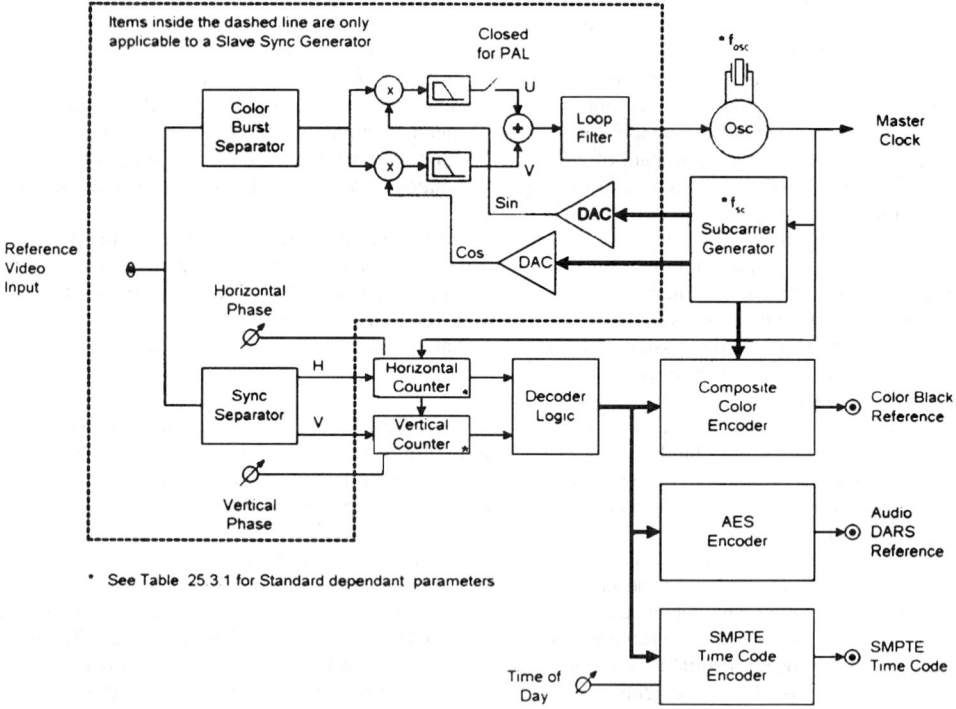

FIGURE 21.2.1   Multi-Standard Sync generator.

TABLE 21.2.1 Standard Related Parameters

Reference video standard	f_{osc}(MHz)	f_{sc}(MHz)	Horizontal counter counts	Vertical counter counts
NTSC	13.5	3.579545	858	525
PAL	13.5	4.43361875	864	625

both filtered axis (U and V) are summed together in order to average the swinging color burst. The resultant error is further filtered by a loop filter before it is applied to the voltage controlled crystal oscillator (VCXO). The oscillator output drives the subcarrier generator as previously described, thus closing the loop. Fine phasing can be implemented by placing an adder between the accumulator and ROM tables in the subcarrier generator. Adding a constant has the effect of statically rotating the sinusoids, thus achieving a phase offset between the input and output and a phasing resolution in the subnanosecond range.

The generation of the blanking interval pulses is performed by two very similar circuits running at horizontal (H) and vertical (V) rates, respectively. The H logic is composed of 858 counter counts for NTSC and 864 counter counts for PAL, which is clocked by the local crystal oscillator running at 13.5 MHz. The output of the counter function is a 10-bit address to a look up table. The H counter is itself synchronized by using the horizontal sync from the reference video input and using the leading edge to preload the counter. The instantaneous load value at the time of initiation of the load defines the relative H timing output to input.

Generation of the vertical pulses is almost identical to the horizontal except that the counter length is 525 for NTSC and 625 for PAL and the clock used is twice the horizontal rate (31.5 kHz) derived from the horizontal counter and look up table decoder.

The respective outputs of both the horizontal and vertical decoders are gated together in a decoder logic block to produce the required final composite outputs of sync, burst flag, and blanking.

Having derived the ingredient signals, the remaining task is to generate the actual output reference signals. In the past, up to the mid-1990s, a reference synchronization generator would generate the basic drive signals (color subcarrier, sync, burst flag, composite blanking), since the process of encoding and decoding was expensive and cumbersome. At the same time one had to deal with routing and timing of multiple signals. With the advancements in silicon, encoding of the drive signals into one composite signal became the norm. When encoding video it is important to remember to transfer the color frame relationship of the input reference signal to the generated reference signal. In addition, generation of audio DARS reference and SMPTE time code are considered to be important functions performed by a sync generator.

The following standards best described the parameters associated with the reference signals mentioned previously.

Reference signal	Associated standard
Video	SMPTE 170M and SMPTE 318M
Audio (DARS)	AES 3, AES 5, and AES 11
Time code	SMPTE 12M

MULTI-STANDARD MASTER SYNCHRONIZING GENERATOR

Master sync generators are generally used in pairs, via an automatic change over device, as the primary reference for a facility. The design of a master generator is in some ways much simpler than a slave since sections of circuitry are simply removed, as can be seen from Fig. 21.2.1. However, the remaining circuits are usually upgraded to meet the tight tolerances required for a master generator, particularly in the case of the crystal oscillator.

Referring to Fig. 21.2.1, it can be seen that the phase lock loop and associated circuit are deleted and the oscillators left to free run. In television plants where interchange of program material extends beyond the plant boundaries and the production is a live event (e.g., the Olympics), it is desirable to lock the oscillator to an external high stability reference, like an atomic standard or GPS. In this case the internal oscillator is placed in a PLL configuration with an external input. The dividers need to be selected so the comparator sees the same frequency from the forward the feedback paths.

The internal crystal oscillator of a master sync generator is generally of very high quality to maintain the required specification of 3.579545 MHz ± 10 Hz maximum variation over the operating temperature range and in conjunction with the aging of the crystal.

Two methods are used to maintain a consistent environment for the internal crystal oscillator. The first method has the crystal enclosed in an oven, which is maintained at a constant temperature, typically 66°C (high enough such that the ambient temperature is not a factor). This has the associated drawbacks of considerable power consumption and being able to achieve this temperature instantaneously upon power up. The second method uses a temperature compensation network where the temperature is sensed and a control voltage for the VCO is offset.

There is an interesting derivative of the oven approach, which relies on operation of the crystal closer to the ambient room temperature. In this scenario, if the ambient temperature is above the set crystal temperature, cooling is required and vice versa. This is accomplished using a device based on the Peltier Effect for both heating and cooling under closed loop control.

VIDEO FRAME SYNCHRONIZER

A frame synchronizer is a device that accepts video information at one rate and provides an output that is "in time" with respect to a reference. Synchronization is achieved by writing the incoming video into memory at one rate, and then reading it out at a rate locked to the facility reference (master sync generator). The memory, therefore, acts as a buffer, allowing the data to accumulate prior to discharge.

A mechanical example to illustrate this buffering action of memory would be a stream flowing downhill into a reservoir contained by a dam. At the bottom of the dam the water is discharged at a constant rate. The volume of water the stream is supplying can increase or decrease, but the water discharged at the bottom of the dam will remain constant due to the buffering action of the accumulating water in the reservoir.

A complete functional block diagram of a typical frame synchronizer (Fig. 21.2.2) should be used as a reference for the following description.

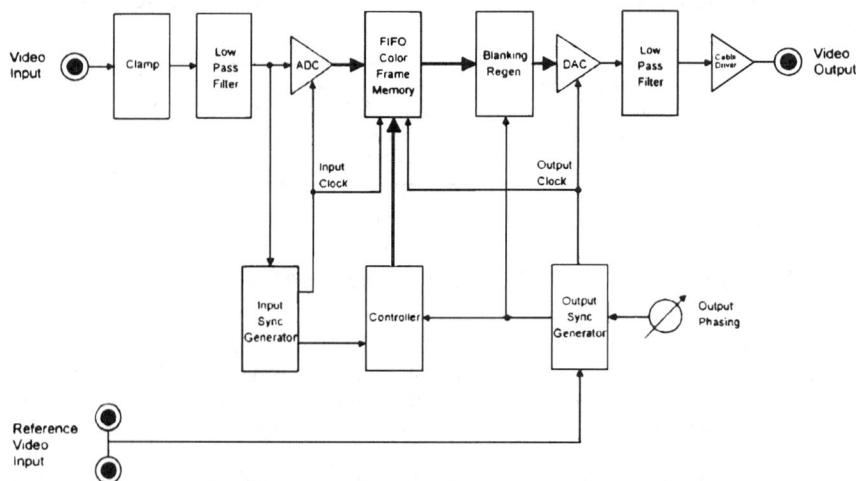

FIGURE 21.2.2   Typical video frame synchronizer.

The input video is first clamped in order to establish a known dc level, which is within the window of the analog-to-digital converter. This is followed by a low-pass filter that prevents any harmonics above one-half the sampling frequency (14.31818 MHz in the case of NTSC) from passing through, thus satisfying the Nyquist sampling criteria.

The signal is then digitized using either an 8- or 10-bit converter. The choice of 8 or 10 bits is the determining factor in frame synchronizer classes. Higher end applications will use 10-bit quantization. However, this is a cost versus performance function.

The digital video data are then stored in a FIFO memory where, after a period of time, it is read back out. New blanking is inserted into the signal and the information is then converted back into the analog domain and low-pass filtered.

Information is required for both the input and output processing sections for the purposes of timing. This is produced by the input and output sync generator sections, as illustrated in Fig. 21.2.2.

The input sync generator samples the incoming video signal and locks a local oscillator to the incoming subcarrier at a 4 fsc rate (fsc = the frequency of the subcarrier). The horizontal and vertical portions of the incoming signal are decoded to provide the information required by the controller and subsequent memory.

The output sync generator performs a similar function except that it samples an external video reference in order to establish a timing reference and clock for the output video. The output sync generator also provides phasing capability which permits the positioning of the output video signal anywhere in time with respect to the reference video signal.

MEMORY

The quantity of memory used must be enough to store one complete color frame of video. This is the smallest unit of video, which can be deleted or repeated while still maintaining the proper color relationship. This means four fields of video in the case of NTSC and eight fields in the case of PAL video signals. It is worthwhile to note that some video frame synchronizers store the complete video line, blanking included, while others store only the active portion of the line.

The most common type of memory used is DRAM where the controller provides read and write addressing. The latest trend is to use FRAM (Field RAM) which employs a FIFO structure and only read and write pointer resets need be provided by the controller.

It is necessary to provide a mechanism such that memory addressing is synchronized to the sync structure of the video signal. For example, if pixel 1 of line 1 is written into location X of the memory, at the time that location X is read back, that information must be placed back on line 1, pixel 1 of the output sync structure. In the case where FRAM is used, this is accomplished by generating a write/reset pulse coincident with the leading edge of sync of the first full line in the vertical interval of field 1 in the color frame sequence based on the input sync generator. Similarly, the read/reset is generated in the same manner, except it is based on the output sync generator. In the case of DRAM memory, the controller will contain read and write address counters that are reset with pulses as described in the FRAM implementation.

FRAME SYNCHRONIZERS WITH DIGITAL I/O

The emergence of digital video standards has increased the importance of frame synchronizers because of the increased processing throughout delays associated with digital equipment. The structure of the frame synchronizer is easily adapted to digital interface in either the parallel or serial video domains.

The analog-to-digital and digital-to-analog converter sections are replaced with the appropriate digital interface, and the input sampling clock is derived directly from the incoming video signal rather than having to extract this information (these codes are embedded directly as part of the digital signal's structure). The output sync generator remains the same since the majority of house reference signals are still analog color black. The change over to a digital house reference will likely happen in the future. Once this does happen, the output sync generator will shift in appearance to that of the input sync generator just described.

BIBLIOGRAPHY

Application Note: Field and Frame Synchronizers, Leitch Technology Int.

AES3-1992 AES recommended practice for digital audio engineering—serial transmission format for linearly represented digital audio data. Audio Engineering Society, 1992.

AES5-1984 AES recommended practice for professional digital audio applications employing pulse code modulation Preferred Sampling Frequencies. Audio Engineering Society, 1984.

AESII-1997 AES recommended practice for digital audio engineering—Synchronization of digital audio equipment in studio operations.

EBU Tech. 6267 EBU Interfaces for 625-line Digital Video Signals at the 4:2:2 Level of CCIR Recommendation 601 (2nd ed., 1992).

ITU-R BT.470-6 Conventional Television Systems.

Kano, K., et al. Television Frame Synchronizer, *SMPTE J.*, March 1975.

Kupnicki, R., et al. Digital Processing in the DPS-I, *SMPTE, Digital Video*, Vol. 2, March 1979.

Kupnicki, R., Software Based Digital Signal Processing, *SMPTE, Digital Video*, Vol. 3, June 1980.

SMPTE 12M-1999 For Television, Audio and Film—Time and Control Code.

SMPTE 170M-1999 Composite Analog Video Signal-NTSC for Studio Applications.

SMPTE 259M-1997 10-Bit 4:2:2 Component and 4fsc Composite Digital Signal Serial Digital Interfaces.

SMPTE 318M-1999 For Television and Audio—Reference Signals for the Synchronization of 59.94- or 50-Hz Related Video and Audio Systems in Analog and Digital Areas.

CHAPTER 21.3
DIGITAL VIDEO RECORDING SYSTEMS

Peter H. N. de With, M. van der Schaar

INTRODUCTION TO MAGNETIC RECORDING

Historical Development

Recording of video signals has established itself as a continuously growing and indispensable technique in video communication services and the video entertainment area. The currently available video recorders for professional and consumer applications have evolved from the early experiments in magnetic tape recording, which were focused on analog voice and audio recording. In the 50s, a number of companies experimented and succeeded to record video signals for broadcasting purposes, however, using rather diverging technical approaches, as the optimal concept for video recording was still being explored. The discrepancies in the applied techniques were due to a new technical hurdle that had to be taken: the required video bandwidth is roughly 300 times the bandwidth of audio signals. It is therefore not surprising that a number of system proposals were based on extending the traditional longitudinal audio recording technique by using a plurality of longitudinal tracks that are registered simultaneously, and, in order to further cope with the demanding bandwidth, a very high tape speed.

In 1956, Ampex announced a system (Quadruplex, or simply Quad) that enabled a substantially lower tape speed, by applying a rotary drum containing a head wheel that turns around at a much higher speed than the tape is transported. Another technique, which had a major impact on further developments in video recording, was the frequency modulation of the video signal prior to storage. This principle, in which a carrier having a relatively low frequency, e.g., 5 MHz, is frequency modulated by the wideband video signal, is still being applied in the current analog video recorders for registration of the luminance signal. The major drawback of the Quad system was its segmented storage of television fields on tape caused by relatively short tracks, which demanded accurate time-base correction for picture reconstruction. The short tracks resulted from the applied transversal recording technique, in which the tracks were written perpendicular to the tape length, so that several rotations, and thus tracks, were required to cover one complete television field.

The segmentation problem was solved a few years later (1959) by other companies, which realized a helical-scan video tape recorder (VTR) that formed the basis of the currently available VHS format. In the 1960s, emphasis was put on the implementation of consumer video cassette recorders (VCRs)—a logical successor of audio cassette recording—that allowed a much simpler tape loading mechanism. One of the first VCRs for professional use is the U-matic system, introduced by Sony in 1970. Since then, many systems have been realized and passed through the consumer's horizon,[1] among which a VCR by Philips in 1970, called the N-1500 (the first consumer VCR), Betamax by Sony in 1975, the VHS system by JVC in 1976, V-2000 again by Philips in 1979, and 8 mm by a company consortium in 1984.

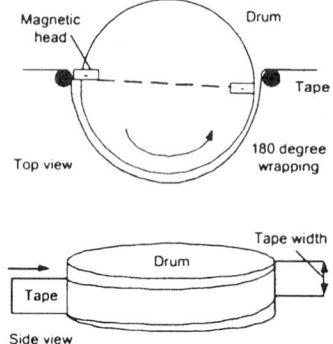

FIGURE 21.3.1 Helical-scan tape recorder with two heads.

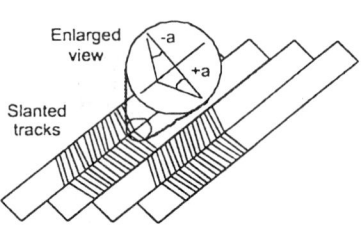

FIGURE 21.3.2 Azimuth recording track pattern and reading with a magnetic head having corresponding azimuth.

As a result of the fast progress, video recording has even started to influence audio recording by means of the introduction in 1987 of the digital R-DAT format, which is fundamentally a miniaturized video recorder.[2] The generation of video recorders from the past two decades is based on helical-scan recording, which is depicted in Fig. 21.3.1. In such a system,[3] the magnetic heads are mounted on a rotary head wheel inside a cylindrical drum, which has the tape wrapped helically around it. The tracks written on tape have a shallow angle with respect to the tape direction and are gradually crossing the tape width, so that a single track can be long enough to store an entire television field.

The continuous drive for higher quality and the demand for more compact mechanics have been satisfied by an increased recording density with each new generation of video recorders. The enormous augmentation in information density is mainly a result of better recording media and heads.[4] The linear-density improvement resulted from better magnetic tapes, offering higher signal-to-noise ratio (SNR) and more high-frequency output, thereby enabling more flux reversals per unit of length. This is expressed by the decrease in the minimum wavelength.[5] The track density improves with the higher SNR and better tracking systems, enabling a further reduction of the track pitch, i.e., the distance between the centers of two adjacent tracks.

Another technique, introduced by Sony in the Betamax system (1975), which significantly contributed to a higher track density, is azimuth recording, in which the tracks are recorded without guardband (the empty space between tracks) and with alternating azimuth (Fig. 21.3.2). Azimuth is the angle between the actual gapline of the magnetic head and the direction perpendicular to the track length. Azimuth recording, which reduces the effects of "sidereading" on adjacent tracks, is now a universally applied technique in consumer video recorders.

The areal recording density improves considerably from higher track densities than from higher linear densities. Although for the consumer the volume density is the most important—a parameter in which magnetic recording clearly outperforms optical disk recording, one of its key parameters, the tape thickness, has not decreased very rapidly. In the future, it is therefore expected that tape manufacturers will work toward tapes with thinner base films, which further contributes to a higher volume density. In recent years, the mass production of optical disks, which was initiated by the success of the compact disc for audio and the CD-ROM for data, has become the leading technology for video recording systems. This has resulted in the DVD recorder and playback-only systems, which are available for various home and personal computer systems.

Analog Consumer Video Recording

From the systems mentioned above, two analog recorders are still in use in the home today: VHS and the 8 mm system. From both standards, camcorders are available as well. The 8 mm cassette is smaller than that of the VHS system, therefore a special compact VHS-C cassette was later introduced for camcorder use only.

The small cassette needs as adaptor for playback in the desktop VHS recorder. With respect to the tape format, the following characteristics apply.

Video. Video fields originating from the odd and even lines of complete pictures (frames) are recorded line by line in individual tracks, thus one field per track. This holds both for most professional systems and the consumer VHS and 8 mm system. This recording allows access to each field individually, for trick modes, such as still picture or fast search (also called picture shuttle).

Audio. VHS has two audio tracks in the direction of the tape travel, while 8 mm uses audio recording inside the slanted tracks.

Mechanics. All home recorders use 180° wrap angle with two video heads diametrically opposed, such as in Fig. 21.3.3. The heads are mounted on a wheel inside a rotating drum. The scan speed, i.e., the actual speed of reading or writing over the tape surface varies between 25 m/s for professional recorders and about

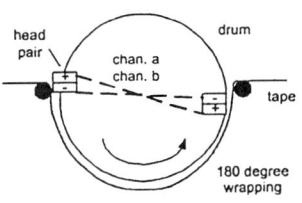

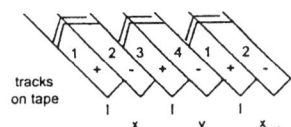

4 m/s for consumer recorders. The tape travel speed is only a fraction of this. Furthermore, the systems are usually cassette based. Some professional recorders use an omega wrap that covers nearly a full circle.

Recording density. The consumer recorders apply relatively narrow tracks of, say, 20 μm. The consumer recorders apply azimuth recording for high recording density, whereas professional recorders use guard bands between two succeeding slanted tracks. The tracks in professional systems are less narrow.

Tape format. Figure 21.3.4 shows a typical footprint of a tape format. VHS uses longitudinal tracks for the audio signal. The video tracks are slanted. The 8 mm system uses an audio slanted track with an edit gap between the video and the audio part, thereby enabling separate editing of video and audio. The control track at the bottom allows the insertion of markers and time code on tape for programming and fast searching to a specific point.The place and location of control and audio tracks may vary from standard to standard, especially for professional analog video recorders such as the B- and C-type recorders and the Betacam and M-machines (see, e.g., Ref. 6).

FIGURE 21.3.3 Headwheel with two head pairs and 180° wrap angle used in consumer recorders.

Comparison of VHS and 8 mm Systems

Table 21.3.1 depicts the key parameters of VHS and 8 mm systems. The numbers of the Betamax system have been added for comparison. Table 21.3.1 shows that Betamax and VHS are comparable and are nearly equal systems. The picture quality of Betamax was slightly better, because the drum was larger and more tape was consumed, allowing a broader recording frequency band. For home recorders, VHS finally won the battle because more prerecorded movies were available in the market, and it offered a longer playing time. The 8 mm system was introduced later and offers a higher picture quality than VHS, merely as a result of using newer, thinner tapes with an improved magnetic layer. Consequently, the system size could be reduced (drum, cassette), combined

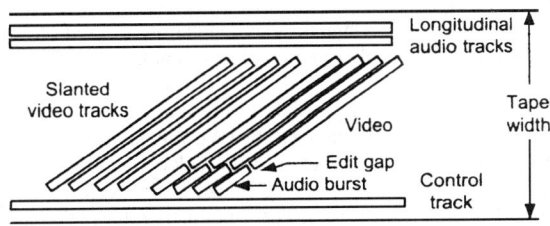

FIGURE 21.3.4   Tape format as used in consumer recorders.

TABLE 21.3.1   Key Parameters of Betamax, VHS, and 8 mm Helical-Scan Cassette Recorders

	Betamax	VHS	8 mm
Tape width	12.65 mm (0.5 in.)	12.65 mm (0.5 in.)	8 mm
Tape thickness	20/16.5 μm	20/16.5 μm	13 (10 Hi-8) μm
Tape material		CoγFe oxide	Metal particle
Head drum diameter	75 mm	62 mm	40 mm
Tape wrap angle	180°	180°	180°
Head-tape speed	7.0 m/s	5.8 m/s	3.76 m/s
Linear tape speed	4.0/2.0/1.33 cm/s	3.34/1.67/1.12 cm/s	1.43 cm/s
Video track width	58.6/29.2/19.5 μm	58.6/29.2/19.5 μm	20.5 μm
Audio track width	1.05 mm	1.0 mm	0.65 mm
Track length		97 mm	62.6 mm
Track angle		6°	4° 53′
Azimuth angle		+/−6°	+/−10°
Video modulation	FM composite	FM composite	FM composite
Color modulation	color under	color under	color under
FM carrier, sync tip	3.5 MHz	3.4 MHz	4.2 MHz
P.White in spectrum	4.8 MHz	4.4 MHz	5.4 MHz
		7 MHz (SVHS)	7.7 MHz (Hi-8)
Chrominance carrier	688 kHz	629 kHz	743 kHz
Cassette size	156 × 96 × 25 mm	162 × 104 × 25 mm	95 × 62.5 × 15 mm
Playing time	60/120/180/240 min	120/240/360/480 min	90 min

with a smaller track width. The carrier frequencies were raised to a higher level. Because of its smaller size, the 8 mm camcorder products became an almost immediate success and they have gradually pushed VHS camcorders out of the consumer market.

Nowadays, the VHS and VHS-C camcorders are being used less often, and camcorders are typically based on either the analog 8 mm format or the digital DV system discussed later. The success of the DV system has also resulted in the reintroduction of the 8 mm format, but now based on digital recording using the video compression technique that was developed for the DV system. This product was introduced at the end of the 1990s by Sony and is called "Digital 8." Camcorders are basically video tape recorders in small form with a lens, optical pickup device based on a CCD sensor (resolution 250,000 pixels or higher), a viewfinder using a color liquid crystal display (LCD), and a microphone for capturing audio signals.

Color Modulation in Analog Recorders

The video signal recording on VHS and 8 mm is such that the color is modulated with a specific technique, called "color under." This will be explained now, starting from the limitation of the magnetic recording channel properties. The television signal contains about 5 MHz bandwidth. The magnetic recording channel is bandlimited, that is, it cannot record either zero frequency or very high frequencies. Analog professional and consumer recorders therefore modulate the signal to a bandpass signal. The modulation is based on frequency modulation (FM).

Professional analog recorders such as the type C machines apply direct recording, in which the frequency carrier is modulated by the composite video signal. The composite video signal contains the baseband luminance (Y) signal with a 5-MHz spectrum onto which the chrominance or color (C) signal is added with amplitude modulation using a specific carrier frequency. For PAL this is 4.43 MHz and for NTSC it is 3.58 MHz. With direct recording, the sync tip (lowest voltage) of the video signal has a carrier frequency of 7.06 MHz, the blanking signal (zero video) is modulated with 7.8 MHz and the peak white signal (full video) with 9.3 and 10 MHz, for PAL/Secam and NTSC, respectively. These numbers refer to the analog type C recorder. This type

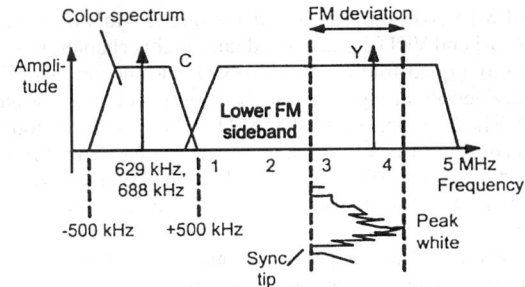

FIGURE 21.3.5 Spectrum of a full color video signal after applying the "color under" system.

of modulation requires a total bandwidth of 15 MHz, which is too expensive for consumer systems. The advantage of direct recording is the excellent quality.

Consumer analog recorders have a different system for adding the color signal to the modulated spectrum. In order to save bandwidth, the luminance signal is modulated with a lower carrier frequency. The color signal is then modulated with a carrier frequency residing at the low-frequency part of the luminance spectrum, so that the color signal is located close to zero frequency and below the luminance signal. This approach is therefore called the "color under" system. The principle is visualized in Fig. 21.3.5 and Fig. 21.3.6. In analog consumer recorders such as VHS, the modulation spectrum looks typically like Fig. 21.3.5. Figure 21.3.5 shows the resulting spectrum with the modulation carrier frequencies and Fig. 21.3.6 depicts an example of an encoder block diagram required for modulation. The required bandwidth for the modulated video signal is well below 10 MHz. However, owing to the bandwidth restrictions, the color bandwidth is in practice limited to about 300 kHz for the VHS system.

Digital Video Recording and the Growth of Bit-Rate Reduction

The advent of digital signal processing techniques and its subsequent rapid development has resulted in a reconsideration of the current concepts for video recording. Digital processing and recording not only outperforms analog recording with respect to time stability, robust picture reconstruction, and (flawless) reproduction after multiple copying, but also allows for advanced image manipulation techniques that were impossible with the conventional analog technology. In the past decade, we have witnessed a transition from analog recording to digital recording systems. As always, at the introduction stage of a new technology where costs are still critical, the first systems are targeted for professional (broadcast) applications. The first full-digital video recorder brought to the market was the D-1,[7] which was introduced around 1985. It was a YUV-component video recorder, capable of recording a CCIR-601 video signal. Soon thereafter, a digital recorder, referred to as D-2, was agreed upon, for composite NTSC[8] and PAL video signals. From 1995 onwards, the gradual introduction of a digital format for consumer applications took place, called the DV system. The DV system established itself as a compact camcorder format, initially for semiprofessional and ENG applications but recently also for the consumer. Principal manufacturers are Sony, JVC, and Matsushita. From a technological viewpoint, a small-sized digital video recorder became feasible and the IEEE-1394 digital interface to computers for off-line digital editing has proven its value.

In addition to the developments in consumer magnetic recording, video recording has further broadened its scope of application by new video (image) storage equipment based on optical disc recording. One of the first systems was the compact disc interactive (CD-i) system of Philips[9] from about 1990, of which the video-coding technique was standardized for multimedia applications MPEG-1,[10,11] electronic photography using magnetic and optical storage equipment, or even solid-state memory. The CD-i was soon succeeded by the

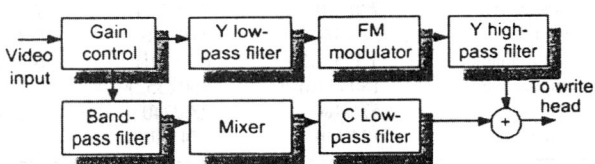

FIGURE 21.3.6 Modulation block diagram for the "color under" system.

video-CD (VCD) format, offering straightforward MPEG-1 compressed video[10] on 1.2 Mb/s on a compact disc. The system is still being used in Asia. Both CD-i and VCD are discussed later in this chapter. In the past years, the optical recording density was improved to a quad-fold of the audio CD, and the new video compression standard MPEG-2[12,13] was used to store video in strongly compressed form. The disk format was standardized as the Digital Versatile Disc (DVD). The high density of the DVD and the compression factor of 25 of MPEG-2 for broadcast video enables recording of a full movie on a single disk. This format has resulted in a number of recordable DVD formats[14] of which DVD − RW and DVD + RW are most known. Other DVD formats are DVD-ROM and DVD-RAM. Details of the above-cited systems are disclosed in the subsequent sections.

The importance of video bit-rate reduction techniques (data compression) cannot be underestimated. The implementation of this technology in various forms combined with the implementation in cost-efficient silicon chips has fueled the realization of many new products in consumer electronics as described above. Video compression has been widely studied, which is proven by the excellent introductory textbooks on digital image processing and associated video coding,[15–17] which are available for the interested reader. More details about the algorithms video data compression for recording can be found in reviews, see e.g. Ref. 18. From this literature, it can be deduced that transform coding[15] has proven its merits for efficient coding of video signals. During the 90s, advanced video compression has become an emerging and enabling technology for new systems in digital transmission and storage. An example of successful video compression supporting digital video broadcasting in the MPEG standard,[10–13] which will be discussed in a succeeding section.

From Composite to YUV-Component Video

Except for the video cassette recorders such as VHS and 8 mm systems, which store the video signal in composite form, all modern recording systems are based on YUV- or YCrCb-component recording. Composite recording means that the video signal is basically one signal with the color signal modulated into it around the color subcarrier frequency. With component recording, the input video signal, which is mostly generated by a camera in three color components—red, green, and blue (RGB), is also recorded in three separate signals, mostly YCrCb. The three RGB components are converted to YCrCb components at the input of the recorder (e.g., with camcorders) or the signal is already ordered in YCrCb format (with, e.g., DVD playback). The conversion is a 3×3 matrix multiplication with a RGB sample triplet at the input and the new YCrCb triplet at the output. The Y-signal refers to the luminance portion (black and white) of the full-color signal, the UV or CrCb to the color-difference signals, i.e., B-Y and R-Y. The CrCb components are shifted and centered around the value 128 in an 8-bit format between 0 and 255, in accordance with the Y component, whereas U and V signals are centered around zero. Since the color-difference signals contain less energy than the Y-component, they are usually sampled with a lower frequency than the Y sample frequency, e.g., with half or a quarter of the Y sample frequency.

TABLE 21.3.2 Key Parameters of Typical YCrCb Sampling Standards Used in Digital Video Products.

Sampling standard	4:1:1	4:2:0	4:2:2
Sample frequency Y	13.5 MHz	13.5 MHz	13.5 MHz
Y samples/line active	720 (704)	720 (704)	720
Y lines active (50/60)	576/480	576/480	576/480
Cr/Cb sample frequency	3.375 MHz	6.75 MHz	6.75 MHz
C samples/line active	180 (176)	360 (352)	360
C lines active	576/480	288/240	576/480
Active bit rate (8-bit)	124 Mb/s	124 Mb/s	166 Mb/s

Note: When interlacing is used, the number of lines per picture splits into two groups (fields) of odd and even lines.

Table 21.3.2 summarizes the most important properties of common sampling standards. The 4:2:2 sampling standard is mostly found in professional recorders, such as the D-1. The DV format applies 4:2:0 in Europe and 4:1:1 in the United States and Japan. The DVD standards are all based on 4:2:0 sampling in both 50 and 60 Hz countries. Older products such as VCD use 4:2:0, albeit at halved horizontal and vertical resolution of the input signal, also known as SIF or CIF resolution (standard or common intermediate format). The halved resolution at the input is required to obtain the low bit rate on CD. It should be noted that the conversion from RGB to YUV or YCrCb signals reduces the quality of the full-color signal, because the difference signals are sampled at reduced frequencies. The advantage is a smaller input bit rate (see Table 21.3.2) at the expense of some loss of sharpness in color transients.

PROFESSIONAL DIGITAL VIDEO RECORDING

The Lossless D-1, D-2, and D-3 Systems

The gradual introduction of digital signal processing in the studio and an increased use of computer graphics in video signals resulted in the introduction of various digital video tape recorders for professional use. The advantages of digital signal storage in a studio environment are, apart from conventional issues such as increased robustness, that multiple editing generations (copy of copy) are allowed. The deterioration in quality is much less than that with analog recorders. In the early 1990s, both composite (D-2, D-3) and component digital studio recorders (D-1) were used. D-1 is the oldest one (introduced in the mid-1980s), but due to its robustness for editing it is still being used actively. All systems are cassette recorders, based on helical-scan recording and use 8 bits for video sampling.[6,19] The recording mechanics are constructed in such a way that three different types of cassettes can be inserted without adaptors, leading to various playing times. Another common property is that they do not use video compression in order to offer maximum quality for studio applications. For further comparison, see Table 21.3.3.

When uncompressed video is used, the bit rates after sampling with sufficiently high frequency are too high (e.g., about 100 Mb/s at minimum) for using a single recording head. Therefore, a head *pair* is used for recording two tracks simultaneously. Professional recorders sometimes use a plurality of heads to offer high picture quality during special modes such as "still picture," slow motion, and fast forward or backward search. The picture quality of digital VTRs in such modes is further improved by *shuffling* the pixels in a special way prior to recording on tape.[7] The shuffling balances and smoothes the effect on the picture quality of crossing various tracks that inevitably lead to data discontinuities.[18] D-1 can search up to 64 times the normal speed and

TABLE 21.3.3 Key Parameters of Digital Professional Video Recorders (f_{sc} = Color Subcarrier Frequency, 3.58 MHz for NTSC and 4.43 MHz for PAL)

	D-1	D-2	D-3
Video recording type	Component	Composite	Composite
Tape width	19 mm (3/4 in.)	19 mm (3/4 in.)	12.65 mm (0.5 in.)
Tape magnetic layer	Fe oxide	Metal	Metal particle
Tape speed	28.6 cm/s	13.2 cm/s	8.4 cm/s
Track width	40 μm	39 μm	20 μm
Bit length	0.45 μm	0.43 μm	0.38 μm
Recording density	55 kb/mm^2	59 kb/mm^2	130 kb/mm^2
Video sample freq. (Y/C)	13.5/6.75 MHz	$4f_{sc}$	$4f_{sc}$
Video sampling	8 bits	8 bits	8 bits
Audio sampling	20 bits	20 bits	20 bits
Audio recording	4 tracks	4 tracks	4 tracks
Recording bit rate	216 Mb/s	105 Mb/s	105 Mb/s
Playing time	76/34/11 min	208/94/32 min	240/125 min

still show a complete—though noisy—picture. The noise is caused by a picture that is constituted from data blocks of pixels originating from various pictures with different time origin. The data shuffling redistributes the pixels over the image leading to an effect perceived as pixel "noise." The noise increases with the search speed.

The recording density continuously improves because of the introduction of better magnetic tapes, using special magnetic materials for higher density. The D-3 that was introduced in 1990 by Panasonic uses a higher recording density than the D-1 and D-2 machines. It is based on the small size of Betacam of Sony and a format called M-II of Panasonic, but now designed for digital recording. Another difference is that the D-3 cassettes contain half-inch tapes, whereas D-1 and D-2 use $^3/_4$-in. tapes. The high recording density reduces the robustness against channel errors, such as scratches and dirt, but this is combated by *error correction codes*. The aforementioned recorders apply Reed-Solomon (R-S) block codes for error protection. A group of pixels is mapped into a data block onto which a string of parity check symbols is appended. These codes can reduce a fluctuating Byte Error Rate (BER) in the order of 10^{-4} 10^{-5} to as low as 10^{-10} 10^{-12}. This extra coding step is also an aid in achieving sufficient robustness for multiple editing iterations to, e.g., 100 times.

D-5 Recorder

The D-5 of Panasonic is an extension of the D-3 machine and is backward compatible to D-3. The unique feature of D-5 is the recording of 16:9 aspect ratio television signals, using three video components of 10-bit resolution. For wide screen, an 18-MHz sampling frequency is used. Additionally, an adapter can be used with 4:1 video compression for recording HDTV signals of 1.2 Gb/s.

Professional Recorders with Compression

The DV format that is being discussed later in the chapter has resulted in three different formats:

- **DVCAM** (Sony) is very similar to DV, using 25 Mb/s video with DV compression, 8-bit sampling on 4:1:1, and the 6.35-mm DV tape with a metal evaporated layer. However, 15 μm tracks are used, giving higher robustness. It is DV backwards compatible. Audio is done with 2-track recording, 16-bit sampling, and 738 kb/s subcode can be added.
- **DVCPRO** (Panasonic) is also compatible with DV and even DVCAM. The major differences are the use of two longitudinal tracks for improved editing (cue and control). The track pitch is 18 μm for robustness. This system was later extended to 50 Mb/s recording, called DVCPRO50, offering 4:2:2 broadcast quality recording with less compression (slightly more than 3:1) than DV (5:1).
- **Digital-S** (JVC) uses 0.5-in. tape based on S-VHS recording mechanics and a twin DV compression system for high-quality 4:2:2 broadcast recording. This system also records with a compression of 3:1:1. The 50 Mb/s machines have 4-track audio and the double subcode capacity of DVCAM, i.e., 1.47 Mb/s.

The Betacam SX recorder has also 8-bit video with compression using a 0.5-in. tape. The compression is higher—about 10:1—than DV, since it applies the MPEG 4:2:2 Studio Profile. This was chosen for compatibility reasons with emerging MPEG-based DVB standards. The video data rate is 18 Mb/s.

Finally, a number of digital high-definition recorders were introduced with the first experimental transmissions of digitally compressed HDTV in the United States. Examples are HDCAM, HDD-1000, and HDD-2700 machines. Typical resolutions are 1440 through 1920 samples by 720 progressive, 1035 or 1080 interlaced lines, using 10-bit video sampling. Audio is performed with 20-bit sampling and recorded with 4 or 8 tracks.

D-VHS

D-VHS (or Data-VHS) is the latest development of the VHS video format developed by JVC that enables digital recording of high-definition broadcasts and multichannel broadcasts. Three different D-VHS modes have been standardized: the STD (standard) mode, the HS (high speed) mode which is compatible with high-definition (HD) digital images and multichannel broadcasts, and the LS (low speed) mode which enables a maximum recording time of 49 h (Fig. 21.3.7 and Table 21.3.4). The technical format specifications of D-VHS were standardized in April 1996. D-VHS is compatible with all types of digital broadcasts in the world including new digital high-definition broadcasts.

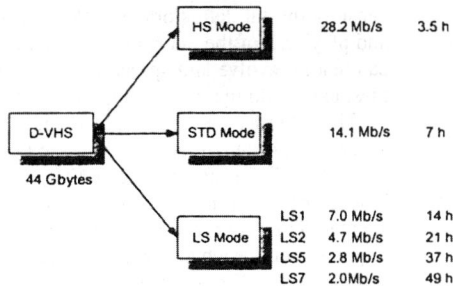

FIGURE 21.3.7   Various D-VHS formats.

Main Characteristics of D-VHS

D-VHS[39] enables bit-stream recording of digital broadcasts, in addition to analog recording of conventional broadcasts. D-VHS takes advantage of the characteristics of tape as a medium to provide extremely large

TABLE 21.3.4   Basic Specification of the Three D-VHS Modes

			LS Mode			
	HS Mode	STD	LS2	LS3	LS5	
Cassette	D-VHS cassette					
Head Configuration	Dual × 2	Single × 2	Single × 1 / Dual × 1	Single × 2		
Mechanics	Based on conventional VHS system					
Recording time						
Using DF-420	3.5 h	7 h	14 h	21 h	35 h	49 h
Using DF-300	2.5 h	5 h	10 h	15 h	25 h	35 h
Track Composition — Tape speed (mm/s)	33.35	16.67	8.33	5.55	3.33	2.38
Head azimuth	$+/-30°$					
Drum rotation	1,800 rpm					
Tracking system	CTL track system					
Main data input rate	28.2 Mbit/s	14.1 Mbit/s	7 Mbit/s	4.7 Mbit/s	2.8 Mbit/s	2 Mbit/s
Recording Specifications — Sub-data input rate	0.292 Mbit/s	0.146 Mbit/s	73.0 kbit/s	48.7 kbit/s	29.3 kbit/s	20.9 kbit/s
Recording rate	38.28 Mbit/s	19.14 Mbit/s				
Track structure	1 sector					
Sync block length	112 Bytes					
Inner correction code	RS code					
Outer correction code	RS code					
Code word shuffling	6 track					
Modulation system	SI-NRZI					
Interface	IEEE-1394 digital interface					

capacity digital data storage. D-VHS maintains compatibility with conventional VHS by allowing recording and playback of the enormous amount of VHS software available as well as stored around the world. D-VHS can readily evolve and spread as a household device owing to the possibility of using current VHS technologies, parts, and manufacturing facilities.

The data that are recorded from digital video broadcasts are actually an MPEG-compressed video and audio program, as standardized by, e.g., the DVB standard. This standard is based on MPEG-2 compression and so-called MPEG-2 transport stream (TS) packets of 188 bytes. The DVB standard has adopted most of the MPEG-2 standard as the basis for transmission of digital programs. The MPEG standard is explained later in this chapter. The recording of digital data on a basically analog machine is not without problems because the required media such as magnetic heads and tape for digital data differ from heads and tape that are optimal for analog signals. Fortunately, the size of the VHS headwheel (62 mm) is large enough to contain additional heads and transformers for constituting a compatible recording system.

DV RECORDING SYSTEM

Introduction

The design of the DV format has shown that for digital TV recording (and HDTV), small-sized recording mechanics giving sufficient playing time can only be achieved if the recording bit rate is reduced substantially. The initial bandwidth expansion due to the digitization of the video signal, combined with binary recording instead of the analog multilevel recording, must be compensated to a large extent by bit-rate reduction.[43] The relation between recording mechanics and video compression has played an important role in setting the DV standard so that both will be addressed.

For consumer recording,[19] the trend is toward small recording mechanics, especially for portable applications, and digital recording of compressed digital video signals. These aspects have been combined into the DV system, which was first announced in 1993[20] and which has been further developed by a substantial group of companies. The main focus of the standard was to introduce attractive small camcorder products in the mid-90s. Experimental digital video recording systems were initiated at the end of the 80s.[21–25] Since September 1995, a number of companies have introduced DV camcorders into the market. Based on the DV technology employing also new tapes,[26–30] the professional systems DVCPRO and DVCAM have emerged, which are increasingly popular.

Despite the acceptance of the MPEG and JPEG standards in numerous applications, the development of the DV video compression standard was largely guided and justified by the targeted product and cassette size, power consumption, and the consumer price level. Irrespective of the previous issues, a number of recording system constraints have to be satisfied in any case and are rather typical, namely:

- Editing, preferably on picture basis
- Robust for repeated (de-)compression, resulting from analog video dubbing
- Multitrack format, using a set of tracks for one frame (see Fig. 21.3.9)
- Forward and backward search on tape, from small data bursts pictures have to be recovered (see Fig. 21.3.9)
- Very high robustness
- High picture quality

DV System Architecture

Figure 21.3.8 from Ref. 31 portrays the block diagram of the recording system. Signal processing is divided into two main classes:

- Source coding employing video compression and source data formatting
- Channel coding using error-correcting codes and modulation codes for spectrum shaping of the bit stream

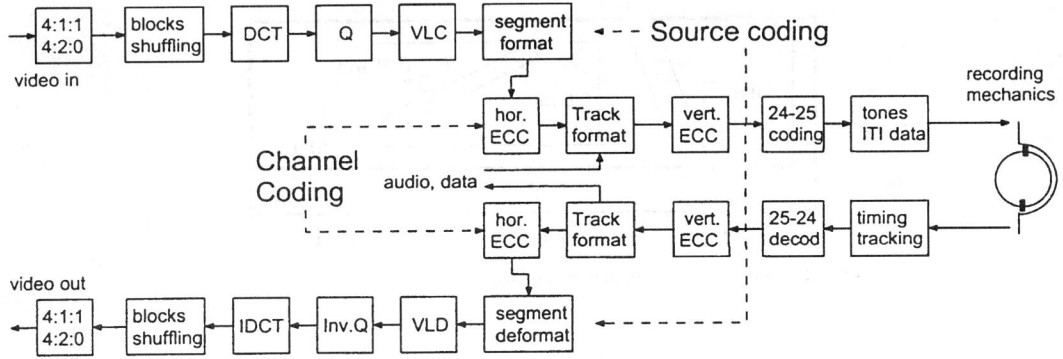

FIGURE 21.3.8 Block diagram of complete processing of consumer DV recording system.

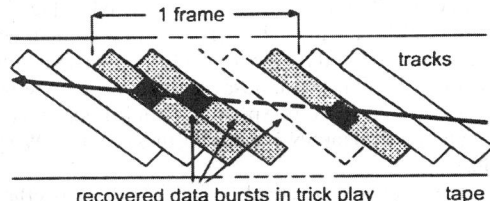

FIGURE 21.3.9 Multitrack format and trick play.

The video compression is based on the discrete cosine transform (DCT), and subsequent quantization and variable-length coding (VLC) of the transformed video data. Since the operations are performed on blocks of 8 × 8 samples, the video input signal is stored for line-to-block conversion. Shuffling of data blocks is performed for improving the picture quality under normal and trick play. The sampling of the video signal is derived from the CCIR-601 standard (4:2:2) with 13.5 MHz sampling frequency for the luminance (Y) signal. The color-difference signals (Cr and Cb) are subsampled either horizontally with a factor 2 (4:1:1) for 60 Hz or vertically with a factor 2 (4:2:0) for 50 Hz systems.

For error-correction coding (ECC), the DV recording format employs the obligatory Reed-Solomon (R-S) product codes, of which the error-correcting properties are defined within a track. After compression, the data are organized in small packets, called sync blocks. Each sync block gets some parity bytes (HECC) assigned to it for random byte errors and for error detection of burst errors. Subsequently, the complete video data block (or audio) is extended with extra sync blocks containing vertical parity bytes (VECC). The performance of these codes is very good for removing random errors and general burst errors. As a second step in channel coding,[40] a high-efficiency DC-free 24-25 channel code was adopted, described in Refs. 32, 33, to constitute pilot tones for tracking, which are embedded in the data.

The recording mechanism is based on a small drum of 21.7 mm, which rotates at 9000 rpm. The basic recording mode (2 × 1 head) has a capacity of 41.85 Mb/s (equivalent to 25 Mb/s video), but the system can be upgraded mechanically and electronically to 83.7 Mbit/s (50 Mbit/s net video rate) for HDTV recording.[31] Possibilities to go to 12.5 Mbit/s video rate or lower, required for recording of MPEG signals (MP@ML SDTV), have been indicated earlier.[32] Table 21.3.5 portrays the key parameters of the experimental recording system.

TABLE 21.3.5 Key Parameters of Digital Consumer DV Recorder

Scanner	2 heads
Scanner diameter	21.7 mm
Scanner speed	9000 rpm
Tape width	6.35 mm
Tape thickness	7 μm
Track pitch	10 μm
Recording rate	41.85 Mb/s
Tracking	Embedded tracking tones
Channel modulation	24-25 code
Error correction code	R-S product code
Video bit rate	25 Mb/s
Video coding	8 × 8 intraframe DCT
Trick mode implementation	Video macroblock shuffling

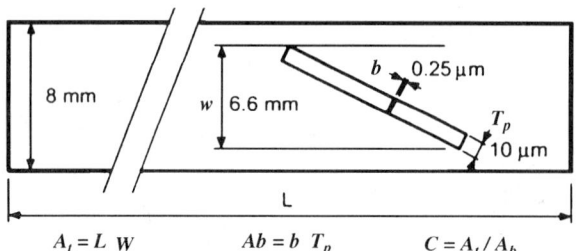

FIGURE 21.3.10 Important 8-mm tape parameters.

DV Cassette Size and Capacity

Cassette Size Related to the 8 mm System. The well-known analog 8 mm cassette was the reference and the starting point for establishing the cassette parameters in relation to the playing time. The largest tape length in the 8 mm system is about 125 m. Some relevant parameters of the 8 mm tape format are depicted in Fig. 21.3.10. The effective width w of the tape area for analog video and PCM audio (221 ± wrap) equals 6.6 mm, which is somewhat lower than the tape width of 8 mm. The edges cannot be used for the helical tracks because of the limited head-tape contact. The product of L and w results in 0.83 m² maximum usable tape area A_t. The cassette capacity C in bits is the ratio of A_t to the bit area A_b: It was estimated from experiments on ME tape that a track width of 10 μm with a bit length of 0.25 μm would result in sufficient SNR and robustness for the DV system. The gross 8 mm cassette capacity is therefore 330 Gb.

To derive the net capacity for compressed video, the estimated practical contributions for overhead are tabulated in the left column of Table 21.3.6. The estimations are educated guesses of the expected overhead, assuming a sync block size of around 1000 bits and a not usable track margin of ±5 − 6. For clarification, the real values of the DV system have been added in the right columns of Table 21.3.6. The overhead for channel modulation and digital audio is now briefly clarified. A well-known block code such as 8-10 modulation needs an overhead of 20 percent. Some form of channel modulation is a must, primarily for tracking and robustness reasons, but the overhead should be limited. The sophisticated 24-25 modulation from Ref. 33 with only 4 percent overhead has proven to be an attractive candidate. The overhead for digital audio is in principle more an absolute value in bit rate and less a relative overhead. However, we have assumed that the

TABLE 21.3.6 Contributions to Recording Rate

Topic	Estim. (%)	DV (kbit)	DV fraction (%)
Channel modulation	4	5.4	3
Tracking information	0	1.8	1
Synchronization + identification	4	6.5	5
Two-dim. error correction	15	20.3	15
Digital audio	6	5.2	4
Auxiliary data	2	2.2	2
Subcode information	2	1.2	1
Margins at track begin/end	3	(1.2)	
Edit gaps and pre/post-amble	3	9.1	7
Video	61	83.2	62
Total one track	100	134.9	100

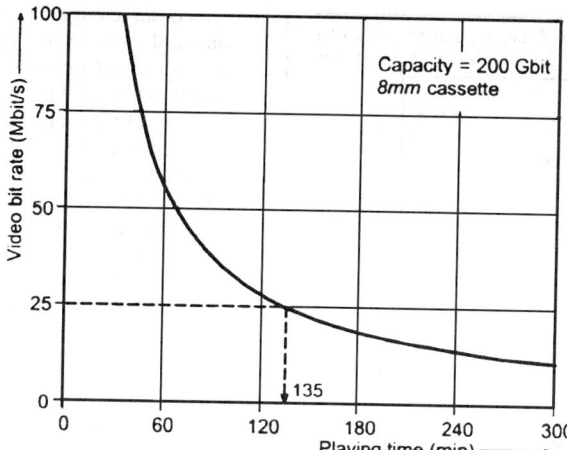

FIGURE 21.3.11   Video bit rate and playing time.

bits needed for digital audio are a reasonable 10 percent fraction of the bits for compressed video, hence 6 percent of the total.

From the previously discussed overhead inventory it is clear that, even with an economical solution for channel modulation, only around 60 percent of the gross capacity is available for compressed video. The net cassette capacity for the 8 mm system is therefore 200 Gb. From this number a simple relation is obtained between the compressed video bit rate and the playing time, which is graphically presented in Fig. 21.3.11. This curve explains why the video bit rate of 25 Mb/s was chosen.

Requirements of the DV System.   Prior to discussing the principal DV system parameters, we list the basic aims and assumptions for defining the recording system:

- Pocketable camcorders, which are substantially more compact than the analog 8 mm system. Playing time is less important; 60 min is regarded as sufficient. Picture quality and editability are of utmost importance.
- Home-use recording of HDTV signals with a minimum playing time of 135 min needed to record almost all movies in one piece. At the time of standardization, HDTV systems such as MUSE in Japan, HDMAC in Europe, and ATV in the United States were seen as very important for the near future.
- Home recording of SDTV signals, with the same compression as for the camcorder. Ratio of HDTV to SDTV bit rate of 2:1, resulting in a minimum playing time of 270 min for SDTV.
- Preferably, a one-cassette system has to be adopted.

It will be clear that the last point can be conflicting with the first two, owing to the significantly higher bit rate and playing time demands of the HDTV system, when compared to the camcorder application. This will become more apparent from a further evaluation of Fig. 21.3.11. At a playing time of 135 min for HDTV, the available video bit rate for an 8 mm cassette equals 25 Mb/s, and consequently 12.5 Mb/s for SDTV. It is elucidated later that this bit rate is considered too low for an economical, well editable, and high-quality compression scheme for the camcorder. In fact, the double of this bit rate was regarded as realistic. This inevitably leads to a two-cassette system with a larger cassette for home use and a compact cassette with less playing time for the camcorder. Currently, the camcorder cassette has proven to be the main cassette in practical use.

Cassette Sizes for DV System.   The net cassette capacity for home use was planned to be 400 Gb, requiring a substantially larger cassette compared to the analog 8 mm system. The required capacity of the small

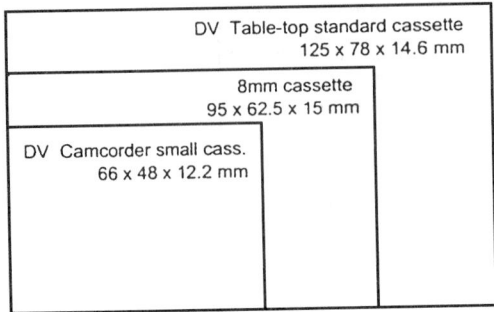

FIGURE 21.3.12   Preferably, a one-cassette system has to be adopted.

camcorder cassette is roughly one quarter of the standard cassette, owing to the playing time ratio, and has therefore half its size. This is substantially smaller than the 8 mm cassette. The relative sizes of the final DV cassettes are depicted in Fig. 21.3.12.

Thus far the thickness of the cassettes was assumed identical to the 8 mm cassette. This is suitable for the larger standard cassette, but a small camcorder cassette with this thickness is not very elegant. Additionally, the compactness of the camcorder is influenced by this parameter, which was therefore reduced as far as possible to the well-known quarter inch (6.35 mm) owing to the camcorder application. The final thicknesses for the standard and small cassettes are 14.6 and 12.2 mm, respectively.

Drum Diameter and Speed

For a camcorder the focus is on an economical drum for the camcorder, which is the well-known drum concept having two heads of different azimuth on opposite sides and using a wrap angle of 180°. This not only leads to the lowest drum complexity, but also minimizes the required read/write electronics and therefore power dissipation, because only single recording channel electronics are needed. Alternative drum configurations do exist and their relations to other applications will be elucidated briefly.

The relation between the drum and video signals asks for a single worldwide recording system with identical rates and drum speeds everywhere. Furthermore, an integer number of tracks/frame was realized to support editing, combined with a small drum size for compactness. The results are given in Table 21.3.7. A comparison between Table 21.3.7 and the wish list at the beginning of this section shows that 10/12 tracks/frame is a well-balanced choice, taking into account a few additional considerations. First, an even number of tracks/frame is preferred for both systems, owing to azimuth recording. Second, higher drum speeds generally yield a higher power dissipation of the drum motor, but too small drum diameters pose problems with mounting internal electronics. Fine-tuning of the tape format resulted in 41.85 Mb/s channel rate, 34.024 mm track length, 21.7 mm drum size, and 9.1668° track angle for the DV system. The effective track length is 32.890 mm, which corresponds to an effective wrap of 174°.

Alternative drum concepts are important for possible extensions toward HDTV, using a doubled recording rate. This is at best achieved with a drum containing two head pairs that are mounted diametrically. The second double head requires continuous two-channel recording, allowing for an HD mode with two times the bit rate of SD. Full compatibility to the SD and even longplay (LP) modes is realized by using only one head pair. If the second double head is mounted with a slightly modified height position, it can also be used for read after write in generic data recording applications (see Ref. 31 for further details).

DV Video Blocks and Sync Block Format

In the DV system, video data is mapped onto the tracks, where each track consists of data packets, called sync blocks. Video data in the form of images is divided into macroblocks, i.e., a 16×16 pixel area from an image.

TABLE 21.3.7   Various Drum Configurations

Tracks/ frame (30/25)	Track rate (tr/s)	Drum speed (rpm)	Track length (mm)	Drum diam. (mm)	Track angle (deg.)
5/6	150	4500	67	42	5
10/12	300	9000	33	21	10
15/18	450	13500	22	14	15
20/24	600	18000	17	11	20

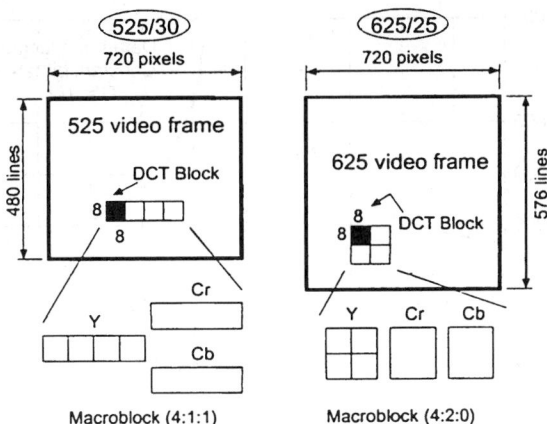

FIGURE 21.3.13 Macroblocks (MBs) taken from 525/30 and 625/25 frames and corresponding MB structure.

This section addresses the mapping of those macroblocks (MBs) on the sync blocks in the tracks, under the assumptions of 10/12 tracks/frame and 25 Mbit/s for compressed video. After sampling of the analog signal, the effective area of the video frame consists of 720 pixels horizontally by 480 or 576 lines vertically for the 525/30 and 625/25 systems, respectively. This is depicted in Fig. 21.3.13. The pixels are grouped in DCT blocks of 8×8 pixels, resulting in 5400 or 6480 luminance DCT blocks/frame. With 10/12 tracks/frame, this leads to an identical number of 540 luminance DCT blocks/track for both systems. Although the subsampling of the chrominance signals is different for 525/30 and 625/25, the ratio of luminance to chrominance is 4:2 in both cases. Macroblocks are formed, consisting of four luminance and two chrominance DCT blocks, resulting in 135 MBs per track.

The remaining restrictions on sync block (SB) mapping. For reasons of convenience in hardware implementation, robustness, and trick-play performance, a fixed mapping of a limited number of MBs (a unit) onto an integer number of SBs has been adopted. Moreover, a multiple of such mapping units should exactly fit into one track for obvious reasons. As a result, the number of MBs per unit is a divider of 135. Possibilities are 1, 3, 5, 9, 15, . . . Similarly, the same units, in the sequel called segments, will be used as independent coding units for the video compression. The video segment size was fixed to 5 MBs, because it was found from simulations that no essential improvement in picture quality could be obtained for larger segments (see Fig. 21.3.20) in the subsection on DV video compression. This results in 27 video segments per track.

The remaining key parameter in the mapping is the number of SBs per segment. It is clear that this number will vary the total amount of video SBs in a track, which, as a consequence of the previous result, should be a multiple of 27.

For determining the last parameter of the format, the highest possible trick-play speed was a decisive factor (thereby adopting a constant bit-rate concept). Assuming a reasonable positioning of the head for reading (mapped into a "quality factor" $Q = 1/2$) and relative tape speed $n > 1$, the recovered data burst length D during fast reading can be approximated by $D = T_L/n$. The track length is denoted T_L. Naturally, as the speed increases, the amount of data read becomes smaller. Figure 21.3.14 portrays the data recovered as a function from the search speed (relative to the reference playback speed). Because the position of the track crossings with respect to the SBs is arbitrary at high speeds, the data burst must be at least two SBs in order to retrieve one correctable SB from tape. The maximum trick-play speed $n_{\max}$ is then given by

$$n_{\max} = \frac{T_L}{D_{\min}} = \frac{T_L}{2} \tag{1}$$

with T_L expressed in SBs being the total length of a track for 180° wrap. The total overhead in a 180° track is assumed to be 40 percent, of which 15 percent is SB-based. The remaining 25 percent of the overhead can be expressed as SBs on top of the video SBs. Therefore, the total number of SBs in a track (T_L) is 4/3 times the

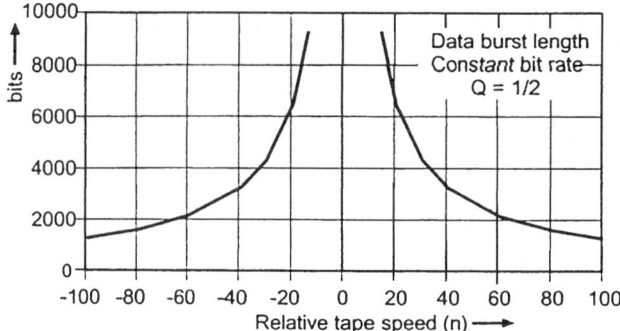

FIGURE 21.3.14 Data burst length vs. search speed.

number of video SBs. Results for T_L and n_{max} are given in Table 21.3.8. In practice, the value of n_{max} will be somewhat lower.

Table 21.3.8 shows why the 5-5 mapping has been chosen in the DV standard. First, this mapping leads to the highest possible trick-play speed. Second, note that with this mapping, in contrast to the other possible mappings, one MB is stored in one SB, thereby enabling a fixed allocation of the most important data. For trick play, this enables the decoding on SB basis, which was assumed implicitly for the calculation of n_{max}. Third, as a bonus, the 5-5 mapping results in the shortest SB length. This property, combined with the fixed data allocation, proves to be very beneficial for the robustness of the system. With the chosen 5-5 mapping, the size of an SB can be determined. With 83 kbit and 135 MBs in a track for video, a 77-byte data area is required to store one macroblock. The addition of a 2-byte Sync pattern, a 3-byte ID, and 8-byte parity for the inner (horizontal) ECC results in a total SB length of 90 bytes. For completeness, it is mentioned that this number must be a multiple of 3 bytes, because of the 24-25 channel coding. The SB structure of the DV system is given in Fig. 21.3.15.

DV Video Compression

It has been explained that video compression is required to establish a low bit rate and sufficient playing time. For this reason, an intraframe coding system was adopted. Intraframe coding means that each picture is compressed independent of other pictures. However, to support the high-speed search the system goes one step further and concentrates on coding segments of a picture as independent units. This will be elaborated in this section. We consider a feedforward-buffered bit-rate reduction scheme, based on DCT coding, in which the pictorial data are analyzed prior to coding. The aim is to define a compression system with fixed-length coding of a relatively small group of MBs because this is beneficial for recording applications such as trick play and robustness. Given the system constraints from previous sections, the target system is based on fame-based DCT coding with VLC. Such a system operates well using compression factors 5 to 8. This results globally in a bit rate after compression of 20 to 25 Mb/s. Numerous subjective quality experiments during development

TABLE 21.3.8 Various Mappings of MBs to SBs

MB's per segment	SB's per segment	Video SB's per track	Total SB's per track	n_{max}
5	3	81	108	54
5	4	108	144	72
5	5	135	180	90

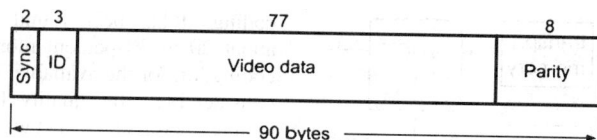

FIGURE 21.3.15   Format of a single sync block (SB).

of the DV standard provided evidence that about 25 Mb/s yielded the desired picture quality, which is well above analog consumer recorders such as VHS and 8 mm.

DV Feedforward Coding Concept.   In most transform coding systems using VLC techniques, the variable-rate output is buffered and monitored by feedback quantizer control to obtain—on the average—a constant bit rate, although the bit rate is locally varying. With feedback compression systems having a locally varying output bit rate, the relation between the recovered data bits and the location of the data in the picture is lost. The major advantage of the feedforward coding system is that relatively small groups of DCT blocks, henceforth termed segments, are coded independently and, in contrast with a feedback system, as a fixed entity. This property makes the segments independently accessible on the tape, while the fixed code length ensures a unique relation between the segment location on tape and its location in the reconstructed picture. The latter property, in combination with the 1 macroblock-per-SB data allocation (see previous section on mapping), is exploited for optimizing the picture quality during trick modes (see later).

In the feedforward coding system (see Fig. 21.3.16), video is first organized into blocks and subsequently in groups of blocks, called segments. Each segment is then compressed with DCT coding techniques into a fixed code length (bit cost), despite the application of VLCs. Fixed-length coding of a small group of DCT blocks (several tenths only) can only be realized if the transformed data are analyzed prior to coding, requiring temporal data storage. During the storage of a segment, several coding strategies are carried out simultaneously, from which only one is chosen for final quantization and coding. This "analysis of the limited future" explains the term feedforward buffering. Feedforward coding has two important advantages: a fixed relation between the data on tape and the reconstructed image, and a high-intrinsical robustness as the channel error propagation is principally limited within a video segment.

DV Motion-Adaptive DCT.   The DCT has become the most popular transform in picture compression, since it has proven to be the most efficient transform for energy compaction, whereas the implementation has limited complexity. The definition of the DCT used in the DV system is

$$F(u,v) = C(u)C(v) \sum_{i=0}^{N-1} \sum_{i=0}^{N-1} f(i,j) \cos \frac{(2i+1)u\pi}{2N} \cos \frac{(2j+1)v\pi}{2N} \tag{2}$$

where a block of samples $f(i,j)$ has size $N \times N$. The two constants $C(u)$ and $C(v)$ are defined by $C(w) = \frac{1}{2}$, for $w = 0$ and $C(0) = \frac{1}{2}\sqrt{2}$. The DV standard applies a block size of 8×8 samples because it provides the best compromise between compression efficiency and complexity and robustness.

One of the first main parameters for coding efficiency is to choose between intrafield and intraframe coding. In the latter system, the odd and even fields are first combined into a complete image frame prior to block

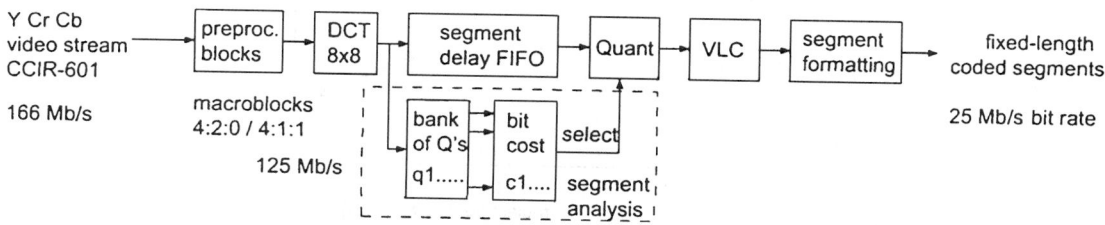

FIGURE 21.3.16   Architecture of feedforward video compression system.

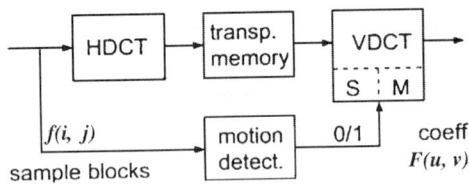

FIGURE 21.3.17 Architecture of a motion-adaptive DCT transformer.

coding. It has been found that intraframe coding is about 20 to 30 percent more efficient than intrafield coding, or, for the available 25 Mb/s bit rate, it offers a considerable better quality. For this reason, intraframe coding was adopted in the standard. However, it was found in earlier investigations that local motion in sample blocks leads to complicated data structures after DCT transformation, which usually are particularly difficult to code. The solution for this problem is to split the vertical transform into two field-based transforms of length $N/2$. Hence, first an N-point horizontal transform (HDCT) is performed yielding intermediate data $F_h(I, v)$, and subsequently, two vertical ($N/2$)-point transforms (VDCT), specified by

$$F(u,v) = C(u)C(v) \sum_{i=0}^{N/2-1} \sum_{j=0}^{N-1} g_s(i,j) \cos \frac{(2i+1)u\pi}{N} \cos \frac{(2j+1)v\pi}{2N}$$

$$F(u+4,v) = C(u)C(v) \sum_{i=0}^{N/2-1} \sum_{j=0}^{N-1} g_d(i,j) \cos \frac{(2i+1)u\pi}{N} \cos \frac{(2j+1)v\pi}{2N}$$

(3)

$$g_s(i,j) = [f(2i,j) + f(2i+1,j)]$$

$$g_d(i,j) = [f(2i,j) - f(2i+1,j)]$$

(4)

Note that in the first field-based output coefficient block, the sum of the two fields is taken as an input, while in the second coefficient block the difference of the two fields is considered. Hence, this corresponds to a two-point transform in the temporal domain.

The required DCT processor architecture is depicted in Fig. 21.3.17.

DV Adaptive Quantization

The primary elements for quantization of the coefficients $F(u, v)$ are frequency-dependent weighting, adaptivity to local image statistics, and global bit-rate control. These elements are individually addressed briefly. The weighting is based on the decaying sensitivity of the transfer function H of the human visual system (HVS). As an example, as HVS model from the literature has been plotted in Fig. 21.3.18, in which the transfer function of the HVS has been normalized.

When using that model the image is divided in 8×8 blocks, in Fig. 21.3.18 we find that the HVS decreases exponentially, is multiplicative in nature, and hence $F_W(u, v) = W(u, v)F(u, v)$. The weighting function can be simplified using special multiplying factors (see Table 21.3.10). For simplicity, the matrix of factors $W(u, v)$

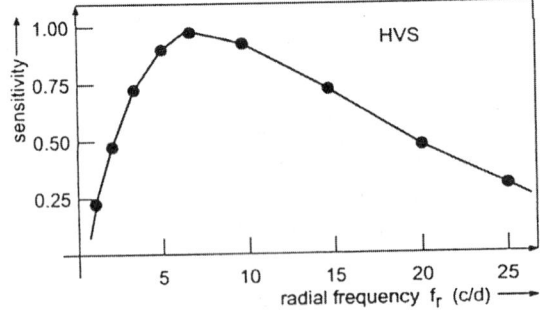

FIGURE 21.3.18 HVS function of radial frequency f_r.

TABLE 21.3.9 Area Numbers of Weighting for Static 8 × 8 Block (left) and Moving 2 × (4 × 8) Block

```
W(u, v)      v ⟶              W(u, v)      v ⟶
u  X 0 0 | 1 1 1 | 2 2         u  X 0 | 1 1 1 | 2 2 | 3      ⎤
↓  0 0 | 1 1 1 | 2 2 2        ↓  0 | 1 1 | 2 2 2 | 3 3      ⎟ sum
   0 | 1 1 1 | 2 2 2 | 3         1 1 | 2 2 2 | 3 3 3        ⎦
   1 1 1 | 2 2 2 | 3 3           1 | 2 2 2 | 3 3 3 3
   1 1 | 2 2 2 | 3 3 3           0 0 | 1 1 | 2 2 2 | 3      ⎤
   1 | 2 2 2 | 3 3 3 3           0 | 1 1 | 2 2 2 | 3 3      ⎟ difference
   2 2 2 | 3 3 3 3 3             1 1 | 2 2 2 | 3 3 3        ⎦
   2 2 | 3 3 3 3 3 3             1 | 2 2 | 3 3 3 3 3
```

are not different for each (u, v) combination. Instead, groups of (u, v) combinations apply the same weighting factor, according to Table 21.3.9.

The second element of quantization is the adaptivity to local image statistics. The adaptivity is to be worked out by measuring the contents of the DCT coefficient block. One of the most well-known metrics for local activity is the "ac energy" of the block, $\Sigma_{u,v} F(u, v)^2$. However, simpler metrics, such as the maximum of all $F(u; v)$ within a block, perform satisfactorily as well. The DV system allows that any metric in the encoder is acceptable: the decoder simply follows the two decision bits reserved for the quantizer classification. This freedom also allows different quantization of luminance (Y) and color (Cr, Cb) blocks. Generally, more activity or information content results in more coarse quantization. The third element in the quantizer is a final block quantization by a linear division with a step size S. The advantage of this approach is its simplicity and it leads to uniform quantization. The variable S defines the accuracy of the global block quantization. When taking into account the elements previously discussed, the overall quantization is specified by $F_Q(u, v) = W(u, v)F(u, v)/S$. For camcording, low implementation cost is of utmost importance. A particular simple system is obtained by taking $W(u, v) = S = 2^{-p}$ with p being an integer that is controlled by all three elements discussed in this subsection. The final quantization table is shown in Table 21.3.10. The weighting area numbers in Table 21.3.10 refer to Table 21.3.9.

TABLE 21.3.10 Table of Step Sizes Using Area Indication for Weighting and Strategy Number for Global Uniform Quantization

Q strategy	Q class 0	Q class 1	Q class 2	Q class 3*	W area 0	W area 1	W area 2	W area 3
15								
14								
13								
12		15						
11		14						
10		13		15	1	1	1	1
9		12	15	14	1	1	1	1
8		11	14	13	1	1	1	2
7		10	13	12	1	1	2	2
6		9	12	11	1	1	2	2
5		8	11	10	1	2	2	4
4		7	10	9	1	2	2	4
3		6	9	8	2	2	4	4
2		5	8	7	2	2	4	4
1		4	7	6	2	4	4	8
0		3	6	5	2	4	4	8
		2	5	4	4	4	8	8
		1	4	3	4	4	8	8
		0	3	2	4	8	8	16
			2	1	4	8	8	16
			1	0	8	8	16	16
			0		8	8	16	16

*If class 3 occurs, all step sizes are multiplied by 2.

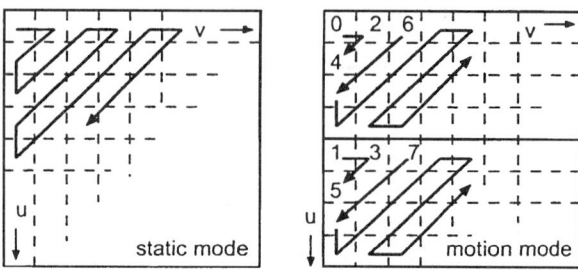

FIGURE 21.3.19   Diagonal zigzag scanning in 8×8 and $2 \times (4 \times 8)$ mode for clustering of zeros.

DV Variable-Length Coding

A bit-assignment using one coding table only was chosen for simplicity. The coding algorithm is fundamentally the same as in MPEG compression and is based on runlength counting of zeros only and the use of an end-of-block (EOB) codeword.

The principle of the algorithm is that first the block of quantized coefficients is scanned using diagonal zigzag scanning in order to create a one-dimensional stream of numbers. The scanning is adapted to motion. The purpose of the scanning is to cluster zero coefficients (see Fig. 21.3.19), so that they can be coded efficiently. Second, from the start of the string, zeros are counted until a nonzero coefficient is noticed. The magnitude of this nonzero coefficient is combined with the preceding length of zeros into a single event (run, amplitude), which is jointly coded with a single codeword. The sign of the coefficient is appended at the end of the codeword. An excerpt of the encoding table showing the variable wordlengths and codewords is given in Table 21.3.11. The coding table is optimized to prevent large codewords and low implementation cost.

Total DV Video Performance

The control and overall performance of the compression system is now indicated. The system optimizes the picture quality by testing a number of quantizer settings in parallel. The optimal quantization strategy m_{opt} is the quantizer setting (strategy) that yields a bit rate just below or equal to the desired bit rate. The choice of m_{opt} can vary between 0 and $M - 1$ when M different quantization strategies are used. The picture quality of the compression system for various segment sizes K and number of quantization strategies M was measured. The resulting picture quality is expressed in SNR (dB), which refers to the mean squared error (MSE) with the original picture compared to squared peak white (255^2). The results of the measurements are shown in Fig. 21.3.20.

The optimal choice of the segment size K and the number of strategies M can be derived from Fig. 21.3.20. Evidently, the recording system designer opts for the smallest possible value of K (small segments), because it yields a high robustness and it enables higher search speeds. However, Fig. 21.3.20 shows that if the size becomes too small, i.e., $K < 30$, the picture quality deteriorates rapidly. For $K = 30$ to 60 DCT blocks, the resulting SNR remains nearly constant. Therefore, a segment size of $K = 30$ was adopted as being the best compromise. The 30 DCT blocks are not arbitrarily chosen from the picture, but clustered in groups, called MBs. An MB (see Fig. 21.3.13) is a group consisting of 2×2 DCT blocks of 8×8 Y samples each and the two corresponding 8×8 color blocks Cr and Cb. In order to improve the picture quality of the compression system, MBs are selected from different areas of the picture, so that the influence of local image statistics is smoothed. The result is a stable and high picture quality for a large class of images. However, after compression, the MBs are redistributed in order to improve the visual performance during search (see later in this DV section). A second conclusion is that $M = 16$ gives a substantial improvement in picture quality, compared to $M = 8$. The quality improvement can be fully explained by a more efficient use of the available bit rate, which becomes particularly important for small segment sizes.

TABLE 21.3.11   Table of Wordlengths (a) and Codewords (b) of DV System

Run	Amplitude (abs.) → 0	1	2	3	4	5	6	7	8	9	10	11	12	13	14	15	16
0	11	2	3	4	4	5	5	6	6	7	7	7	8	8	8	8	8
1	11	4	5	7	7	8	8	8	9	10	10	10	11	11	11	12	12
2	12	5	7	8	9	9	10	12	12	12	12	12					
3	12	6	8	9	10	10	11	12									
4	12	6	8	9	11	12							coeff. sign not incl.				
5	12	7	9	10									EOB = 4				
6	13	7	9	11													
7	13	8	12	12													
8	13	8	...	...													

(a)

Event	Codeword		Event	Codeword
(0, 1)	00s		(4, 1)	110001s
(0, 2)	010s		(0, 7)	110010s
EOB	0110s		(0, 8)	110011s
(1, 1)	0111s		(5, 1)	1101000s
(0, 3)	1000s		(6, 1)	1101001s
(0, 4)	1001s		(2, 2)	1101010s
(2, 1)	10100s		(1, 3)	1101011s
(1, 2)	10101s		(1, 4)	1101100s
(0, 5)	10110s		(0, 9)	1101101s
(0, 6)	10111s		(0, 10)	1101110s
(3, 1)	110000s		(3, 11)	1101110s

(b)

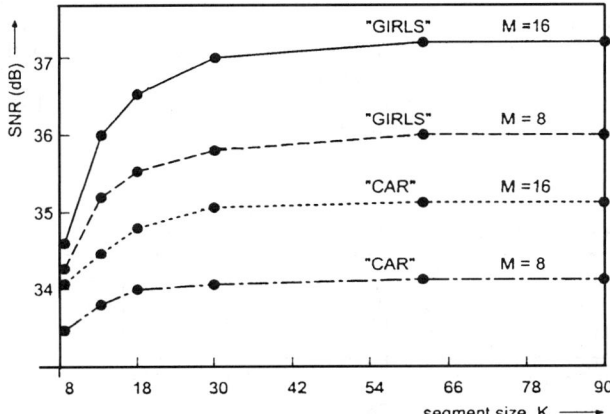

FIGURE 21.3.20   SNR of DV compression system for various segment sizes *K* and number of strategies *M*(CCIR-601 images, 4:2:0 sampling, 25 Mb/s).

The subjective picture quality of the system is known as excellent and regarded as very close to levels for professional use. For regular input pictures, the SNR approaches 40 dB, and the resulting subjective image quality of the system comes rather close to the quality of existing professional recording systems. For complex and detailed imagery, the SNR is a few dBs lower.

Macroblock-Based SB Format

The DV format uses a special compressed data format inside a sync block (SB) to improve robustness for high-speed search, where a part of the error-correction coding (ECC) cannot be applied, because only data portions of a track are recovered. This considerably multiplies the chance of having errors in the signal. Second, at higher tape speeds, the head-to-tape contact is reduced and less stable, which also leads to a lower robustness. The special data format enables the compression decoder to cope with residual errors in the video data. In order to construct a robust format, it is absolutely essential to limit the propagation of errors that emerge from erroneous variable-length decoding. The propagation of errors is limited in three ways, which are briefly discussed.

First, *fixed-length coding of segments* is realized by the choice of a feedforward coding scheme. Every segment is compressed into a fixed bit cost, so that the decoder should periodically reset itself at a segment border. A proper numbering of SBs allows to identify the start of a new segment without using the compressed data. Error propagation from segment to segment is therefore impossible. Second, *identification of individual macroblocks* is enabled. A segment consists of five full-color MBs as indicated in the previous section. For robustness, a single MB is put into a single SB. Note that the MB is sometimes smaller than a SB and sometimes larger. Furthermore, every SB has a fixed unique location on tape. As a result, each MB—at least the low-frequency information—can be addressed. Third, *identification of individual DCT blocks* is possible. Within an SB (thus one MB) six DCT blocks are located. Each compressed DCT block is of variable length. Similarly, to the MBs, by putting the low-frequency data of each DCT block on a fixed position, each DCT block can be addressed and partially decoded, and error propagation is limited to high-frequency components of DCT blocks only.

The internal SB format is depicted in Fig. 21.3.21. A group of five SBs forms a fixed-length segment, preventing error propagation. As a bonus, the fixed-length segment compression allows replacement of individual segments for post editing or error concealment in the compressed domain, without decoding the full picture. The individual segments can also easily be transmitted over a digital interface (IEEE-1394) to other equipment. Figure 21.3.21 also shows the fixed positions of the start of each DCT block.

DV Data Shuffling for Trick Play

The DV standard applies special data shuffling techniques to optimize the picture quality both for normal playback and for high-speed search on tape. It was elucidated in preceding subsections that a coding unit, called video segment, consists of five MBs only. When considering MBs for shuffling, the following general statement applies. For smoothing statistics, a regular distribution over the picture area of the MBs for one segment results in the highest average picture quality for all practical video material. This subsection describes MB shuffling in more detail, taking the 625/25 system as an example.

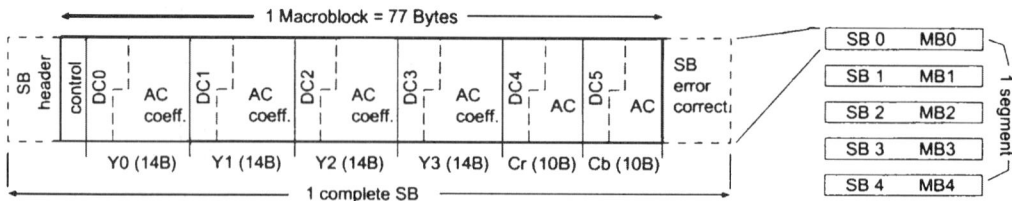

FIGURE 21.3.21   Inner SB data format showing the fixed predetermined positions of low-frequency DCT coefficients of each block and the construction of segments using these sync blocks.

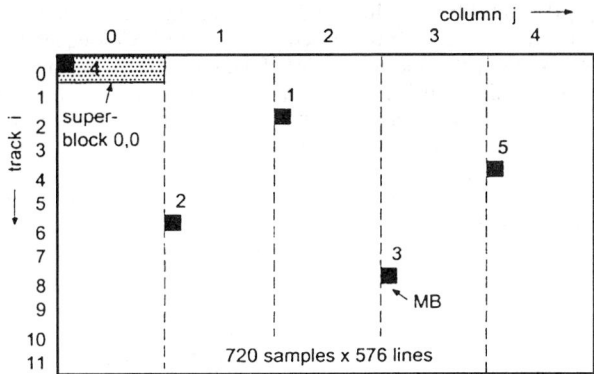

FIGURE 21.3.22 Selection of MBs for segment construction and assignment of picture areas on tracks.

Data Ordering for Optimal Picture Quality at Normal Speed. As depicted in Fig. 21.3.13, a picture consists of 480/576 lines of 720 pixels wide. With 10/12 tracks per frame, it is clear that the data of 48 lines or 135 MBs is stored in one track. In the case of the 625/25 video system, the picture of 36 by 45 MBs is divided into 12 horizontal rows of three MBs high and 45 MBs wide, where each row corresponds to one track. On the other hand, one segment consists of five MBs, which should originate from distributed parts of the picture, preferably with the highest distance. A maximum horizontal distance is achieved when they are distributed regularly, leading to a horizontal pitch of nine MBs. Consequently, the picture is divided into five columns of nine MBs wide. The row-column structure is depicted in Fig. 21.3.22. Despite the difference in video lines, the 525/30 picture is divided similarly in 10 rows and five columns. In the vertical direction, a regular distribution with maximum distance is also the optimal for a good picture quality. Taking into account the 10/12 rows for both systems and a universal algorithm, a distance of two rows is the best option.

A unit of 3 by 9 MBs is called a superblock. A picture consists of superblocks $S_{i,j}$ with i, j being the row and column number, respectively. The numbering of the MBs within the superblock can be found in Fig. 21.3.23. The construction of a superblock for the 525/30 system is somewhat different because of modified MB dimensions (see Fig. 21.3.13). Since the distances are known now, a suitable algorithm has to be defined such that the five MBs forming one segment are spread out over the picture area. The following algorithm has been adopted:

$$V_{i,k} = \sum_{p=0}^{4} MB[(i+2p) \bmod n, 2p \bmod 5, k] \tag{5}$$

with $0 \leq i < n - 1$, $0 \leq k \leq 26$, and $n = 10$ or 12. Furthermore, $M[i, j, k]$ denotes the macroblock number k in superblock $S_{i,j}$. The MBs forming segment $V_{0,0}$ are indicated in Fig. 21.3.22.

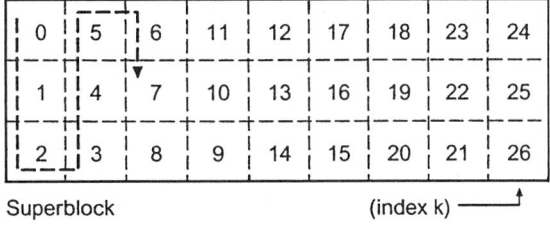

Superblock   (index k) ⟶

FIGURE 21.3.23 Ordering of macroblocks in repetitive clusters called superblocks.

Macroblock Mapping for Trick Play.   Editing and picture reconstruction in normal play are performed on a picture basis, so that the mapping can be optimized for trick play. Consequently, the data of one picture is recorded in an alternative order within the boundaries of the set of tracks assigned to the complete picture. The following aspects are used in the mapping: *Coherency and speed adaptivity.* Coherency means that the best subjective trick-play picture quality is obtained if the largest possible coherent picture parts are refreshed in one piece.18 The picture quality should be adaptive to the speed, meaning that the highest quality is achieved for the lowest search speeds. At the introduction of the MBs, it was shown that the recovered data burst length from tape increases for lower speeds (see Fig. 21.3.14). Consequently, neighboring pictorial information should be recorded side by side in a track.

The previous insight contradicts with the MB shuffling for efficient video encoding at normal speed requiring that the pictorial information should be spread out over the picture. As indicated in Fig. 21.3.21, the majority of the information for one MB is stored in one SB. This is an attractive property for trick play. It enables decoding on a SB rather than on a segment basis, resulting in a slightly reduced but still adequate picture quality for trick play. This feature allows for a block mapping on MB basis instead of segments, thereby solving the paradox between MB shuffling and actual mapping on tape.

The selection of the optimal mapping for a large set of trick-play speeds is a difficult issue, especially if the possibilities for alternative scanners have to be considered simultaneously. The highest flexibility in all situations is achieved by a more or less direct projection of the picture on the tape. This is realized by recording the superblocks in one row of Fig. 21.3.22 one after the other in a track, with the row number equal to the track number. MBs within a superblock are stored in the order indicated in Fig. 21.3.23. A schematic representation of the final mapping is given in Fig. 21.3.24. On top of this mapping, trick-play speeds are chosen carefully at special fractional noninteger values, leading to sufficient refresh of data over time. More details can be found in Ref. 18. In general, it can be concluded that the search speed should be such that all other picture parts are refreshed between the updates of a specific part.

Conclusion on DV System

It has been clarified that the adopted cassette size and tape thickness have determined the required compression factor and the resulting bit rate of 25 Mb/s after compression. At this bit rate, sufficient playing time can be obtained, both for an ultrasmall camcorder cassette and the larger desktop cassette. For desktop video recording, the trend is clearly to disk-based recording, which is discussed later in this chapter. The compression system has been optimized for low-cost intraframe coding because of the use in portable equipment. The feedforward coding architecture is different from MPEG-based recording in DVD, but it offers high robustness for high-speed search and error concealment. The compression in fixed-length segments offers besides robustness also easy packet transmission over digital links. The system yields a high picture quality, even in the case of motion, owing to a special motion adaptivity.

The independent compression and coding of segments, based on five macroblocks only, allows for high search speeds during trick play and provides a very robust tape format. In normal play, the powerful R-S product ECC ensures an error-free and flawless reproduction of the recorded images. A special macroblock format mapped

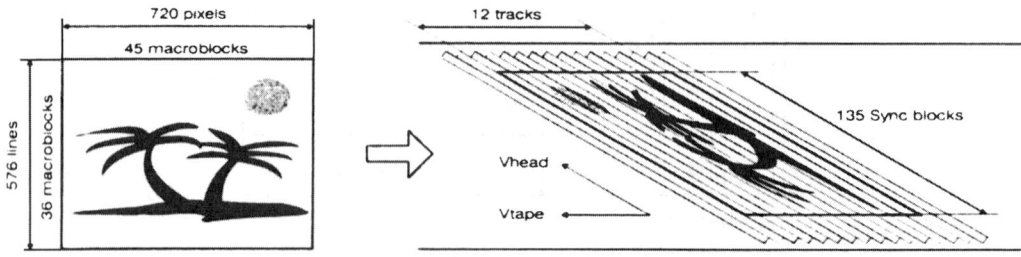

FIGURE 21.3.24   Reorganization of macroblocks on tape for trick play so that the picture is stored in coherent form on tape.

onto single channel sync blocks enables data recovery at very high search speeds. Even at high search speeds, the format limits error propagation severely and shows only errors in high-frequency information. Moreover, the special data shuffling scheme enables a relatively high perceptual performance during trick-play modes.

CD-I AND VIDEO CD

Introduction to Optical Recording

Optical recording can support much higher packing densities than magnetic recording. A typical optical recorder has a 1.6 μm track spacing, and a minimum recorded wavelength of 1.2 μm. A single side of a 30-cm optical disc can record as much as 60 min of video. Several systems are in use since the early 90s.

The first system is designed for mass replication. An expensive master produces numerous very cheap copies, in a similar manner to phonograph discs. However, the process is more refined because of the smaller dimensions involved. This system is highlighted subsequently in more detail. The second system uses an irreversible change produced by laser heating selected spots in the medium. The change may be melting, bubble formation, a dye change, or a chemical change. The systems in which writing is irreversible are termed *write once* or WORM (write once, read many times) systems. A typical example is the Panasonic 20-cm optical recorder that stores nearly 14 min of an NTSC signal. The disc rotates at 1800 rpm and the system can either record a single spiral track, or 24,000 concentric tracks.

The third method uses laser heating to define the small recorded area, but the recording is magnetic and therefore erasable. The system is similar to WORM systems except that the disc is coated with a thin magnetic layer. Like any magnetic recording, the signal can be erased or it can be overwritten. Such a system using analog recording can store over 30 min of NTSC television per side of a 30-cm disc. If digital recording is used, 50 s on each side of the 13-cm disc can be recorded.

We describe herein briefly a type of system that is suited for mass replication. There have been several approaches to the home optical disc system, including one pioneered by Philips, named Laservision. The Philips video disc consists of a transparent plastic material with a diameter of 30 cm and a thickness of 2.5 mm for a double-sided recording. The audio disc has grooves, whose walls move a transducer to reproduce the audio signals. The video disc must meet a requirement of much higher information density. It has no grooves and has tracks with a much finer structure. The track spacing is about one-sixtieth of that of an audio disc. The rotational speed is synchronous with the vertical frame rate to enable recording of one complete frame per revolution (1800 rpm for NTSC, 1500 rpm for PAL and SECAM). A longer-playing version of the Philips system maintains a constant track-to-pickup velocity, allowing approximately 1 h per side of recording.

Let us now briefly look into the basic optical readout technology. In both Laservision and the CD-based system, the signal is recorded in the form of a long spiral track, consisting of pits of about 0.5 μm width (see Fig. 21.3.25 (*a*)). The pits have been pressed into a plastic type of substrate

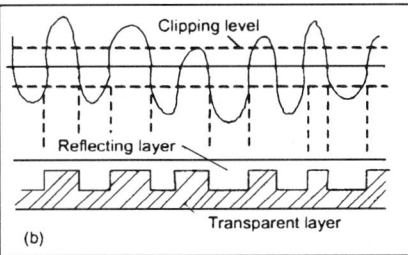

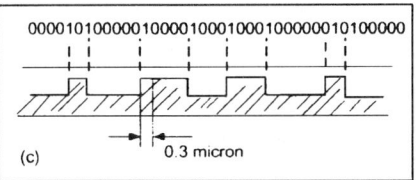

FIGURE 21.3.25 (*a*) The pits in an optical disk of Laservision or CD with tracks of about 0.5 μm width, (*b*) the analog waveform in Laservision is clipped and the transitions defining pits and walls on the disc, (*c*) the conversion of a binary string into pits and walls in reflective layer on the compact disc where each transition corresponds with a "1" in the binary signal. Between each binary 1, every flat distance of 0.3 μm represents a binary 0. The reflective layer is the backside of the optical disc.

with a special stamp. The stamp originates from a parent disc that was generated by a real laser-burned disc. The recorded layer is protected by a transparent coating. Since the coating is much thicker than the recorded layer, scratches and dirt on the surface of the disc are outside the focal plane of the optical pickup that reads out the data.

An essential difference between Laservision and all CD successors is that Laservision registers analog and CD holds digital information. Figure 21.3.25(*b*) depicts how the analog signal is recorded on a Laservision disc. The video signal is frequency modulated (FM) and an analog sound signal is superimposed on the FM-modulated video signal. The resulting signal is clipped for not exceeding maximum positive and negative signal values. The transitions of the signal coincide with the pits and walls on the disc. A predecessor of the video-CD was a system called *CD-Video* of Philips that recorded a digital audio signal in special modulation area on top of the FM-video signal.

Figure 21.3.25(*c*) portrays the recording of digital data on the CD. The length of each pit or wall is 0.3 *μ*m or a multitude of this length with a certain maximum. With a special coding algorithm, called *eight-to-fourteen modulation* (EFM), the number of zeros between two ones is kept between a pre-defined minimum and maximum value. Such a code is often referred to as a *run-length-limited* (RLL) code. The bounding of zeros helps in robustness by avoiding DC energy and it helps with synchronization. As can be noticed, each transition corresponds with the bit value 1 and the intermediate bits get the bit value 0. These bits are called *channel bits* and they originate via a coding procedure from the data bits. The data bits for digital audio (CD-DA) are resulting from 16-bit audio samples that represent a stream of binary numbers, called *pulse code modulation* (PCM). An illustration of the focus motor and detection optics is illustrated in Fig. 21.3.26. To track the line of very narrow pits by a closed-circuit servo like the one depicted in the illustration, two auxiliary light beams are formed, which are slightly displaced from the centerline of the track, one on each side of the correct position. Two photodiodes are introduced into the optical paths on either side of the quadrant detectors, as portrayed in Fig. 21.3.26. The error signal generated by the difference in output of these diodes, after appropriate filtering, is used to deflect a galvanometer mirror and thus move the beam laterally to correct the error. Similar techniques can be used for optical tracking, where the signal envelope of recovered waveform after opto-electrical transformation by receiving photodiodes can be kept as high as possible.

Figure 21.3.27 depicts an outline of the signal path of the compact disc. Assuming correct servo control and focus of the laser beam, the electrical signal coming from photodiodes is amplitude controlled by a gain amplifier.

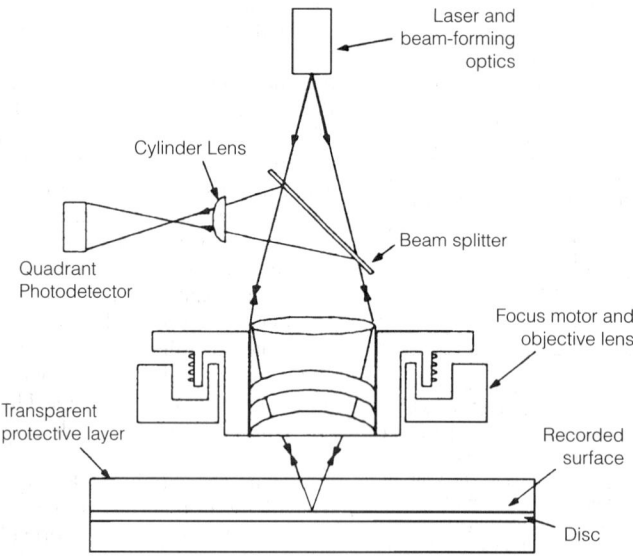

FIGURE 21.3.26   Example construction for focusing the laser beam on the optical disc.

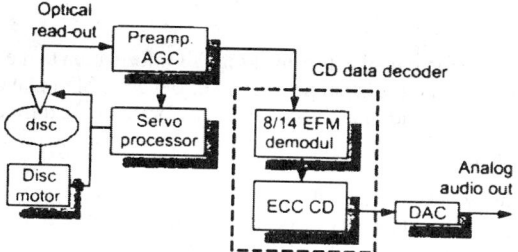

FIGURE 21.3.27 Block diagram of signal path in compact disc.

The output in the form of a digital string of channel bits is buffered and demodulated by the EFM demodulator. This involves a mapping of groups of 14 channel bits and three merging bits into consecutive bytes. Subsequently, these bytes are buffered in the error-correction decoder (ECC). This ECC decoder recomputes data packets and corrects possible errors. The decoder consists of a Reed-Solomon block decoder combined with a so-called *cyclic interleaved redundancy check* (CIRC). The latter step involves data interleaving to spread out errors and cut the length of long burst errors into shorter pieces that can be handled by the Reed-Solomon decoder. After the complete ECC decoding, the audio samples come available as PCM data and are available for digital-to-analog conversion (DAC).

History and CD-i Concept

The CD format was invented in 1980 and was introduced in Europe and Japan in 1982 and in the United States in 1983. This originally introduced format was CD-audio (CD-A, CD-DA, or "Red Book" CD, as specified in the International Electrotechnical Commission IEC-908 standard available from the American National Standards Institute). Additional CD formats were subsequently introduced: CD-ROM (1984), CD-i (1986), CD-WO [CD-R] (1988), Video-CD (1994), and CD-RW (1996).

The primary focus in this section is on the compact disc interactive (CD-i). Even if CD-i is not very much used anymore, its importance to the evolution of the digital multimedia recording systems aimed at applications such as home entertainment, education, and training has been fundamental. It should be noted that the strong development of the CD-based formats was the fundament for the recent success of the DVD,[5] which is discussed later in this chapter. Furthermore, the CD-i format resembles very much the video-CD format that is currently very popular in Asia. CD-i combines audio, video, text, graphics, and animation, while providing features such as interactivity.

At the time of its introduction, CD-i provided several advantages over PC-based interactive systems, some of which still hold today:

- *Cost* compared with that of building a PC with the same audiovisual performance and functionality.
- *Compatibility*, since each CD-i disc is compatible with every CD-i player. There are no "system requirements," such as a type of display adapter, sound card, version of the operating system, screen resolution, CD-ROM drive speeds, drivers, hardware conflicts, and so forth, like in the PC-based scenario.
- *Ease-of-use.* A CD-i player and software are very easy to use and do not require software setup, adjusting the hardware or other complex installation procedures. Moreover, CD-i can be connected to a variety of devices, such as TV sets and stereosystems. As an additional advantage, the user interfaces resemble those of CE devices, making it a far more comfortable system for many people to use over a PC.
- *Worldwide standard.* CD-i is a worldwide standard, crossing the borders of various manufacturers and TV systems. Every disc is compatible with every player, regardless of its manufacturer or the TV system (PAL, NTSC, or SECAM) that is being used.

CD-i System Specification

The CD-i players are based on a 68000 CPU running at least at 15 MHz, with at least 1 Mbyte of RAM, a single speed CD-drive, dedicated audio and video decoding chips, at least 8 kbyte of nonvolatile storage memory, and a dedicated operating system called CD-RTOS, which stands for *compact disc real-time operating system.* CD-RTOS is based on version 2.4 of Microware's OS-9/68K operating system that is very similar to the Unix operating system, and supports multitasking and multiuser operation. The operating system as well as other player-specific software such as the player's start-up shell are hard coded in a ROM of at least 512 kbyte.

CD-i Sector Format

CD-i is able to retrieve in real time the audiovisual data stored on the disc and send this information to the appropriate decoder ICs, without putting a heavy load on the overall system performance. Hence, a CD-i player does not need much RAM or processing power, since all audio and video decoding is performed in real time without storing large amounts in RAM for later decoding. To enable the simultaneous retrieval of both audio and video information, data are interleaved on the CD-i disc.

Since a CD-i disc is read at a constant continuing speed, the designer needs to be aware of the audio or video bit-stream quality. For instance, when a lower audio quality is used, fewer sectors will be occupied than with a higher quality. Alternatively, it is also possible to read only the sectors belonging to one audio channel at a time, and then move back to the beginning of the disc and read the sectors of another audio channel. Since a CD-i disc lasts for 74 min and the lowest audio quality only uses one out of every 16 sectors, the audio can be digitally recorded for (16×74 min) over 19 h.

Because of this real-time reading of sectors, every CD-i player reads data at the same speed, sometimes referred to as normal speed or single speed. It would be unnecessary to make a CD-i player running at a higher speed CD-drive, since data are to be read in real-time according to the specifications (thus single speed) and audio, video, and animation would be out of sync when being read at a higher speed. Special attention has been paid to the development of encoding techniques that enable high-quality audio and video within the single data speed and hence resulting in a longer playing time, instead of using a high-speed drive and by such reducing the playing time.

For CD-ROM, the mode 1 sector format is defined that allows for 2048 bytes of user data in every sector, with an accompanying 280 bytes of error correction information in each sector. When data are read at 75 sectors per second (the normal CD speed), this results in a data rate of 150 kbytes per second. For CD-i, it is not always necessary to have error correction in each sector. For example, audio and video need a much lower degree of correction than data. Instead, the 280 bytes used for error correction in mode 1 could be added to the 2048 of user bytes, resulting in 2324 bytes of user data per sector. This larger sector size results in an improved data rate of about 170 kbyte per second, which is referred to as mode 2. Within mode 2, two forms were defined: form 1 does incorporate the original error correction and is used for data and form 2 which lacks the error correction is used for multimedia. Mode 2 added an additional header to the header of mode 1, which holds information about the type of data that are contained in a sector (audio, video, data, and so forth), the way it is coded (for example, which audio level is used), and an indication of the used sector form. This header is interpreted by the CD-i system for each sector, which is then processed by the appropriate decoders. Both forms of mode 2 sectors can be interleaved, so that program data and audio and video can be read instantaneously from the disc.

Note that when all sectors are form 1, the disc holds 648 Mbyte. When all sectors are form 2, the capacity is 744 Mbyte. CD-i's disc capacity can hence be between 648 and 744 Mbyte. Although a CD-i disc consists of only mode 2 sectors, a CD-i system must be able to read mode 1 sectors on CD-ROM discs, and of course the audio sectors that are defined for CD audio.

Physical Dimensions and Specifications

A CD disc is 120 mm in diameter (60 mm radius), with a hole 15 mm diameter (7.5 mm radius) and a thickness of 1.2 mm. Starting at the hole edge at 7.5 mm radius, there is a clamping area extending from 7.5 to 23 mm radius (this area is partly clear and partly metalized, and may include a visible inscription stamped by the manufacturer), then a 2 mm wide lead-in area extending from radius 23 to 25 mm (containing information used to control the player), then the 33 or 33.5 mm wide data area (program area) extending from radius 25 to a maximum of c. 58 mm, a lead-out area (which contains digital silence or zero data) of width 0.5 to 1 mm from radius starting maximally at c. 58 mm, and finally at c. 1 mm unused area extending to the outer edge.

CD-i Disc Structure

A CD-i disc is divided into tracks. A CD-i disc contains at least one CD-i track, and may also optionally contain additional CD-Audio tracks that may also be played on a standard CD-Audio player. The first 166 sectors of the CD-i track are message sectors, followed by the disc label. Subsequently, an additional 2250 message sectors follow that contain a spoken message in CD-Audio format, which informs users who put the disc in a regular CD-Audio player about the possible damage to equipment or speakers when the disc is not taken out immediately. Usually, a modern CD-Audio player will recognize the CD-i track as a data track and will not play it, so you won't hear the message. The disc label contains some specified fields that offer a lot of information about the disc, such as the title and creator, but also the name of the CD-i application file that needs to be run at start-up. Furthermore, the disc label contains the file structure volume descriptor, which is loaded into RAM at start-up. This allows the system to find a certain file on a CD-i disc in only one stroke. After these message sectors and disc label, the actual CD-i data start.

CD-i Audio

A minimal configuration (denominated also "Base Case") of a CD-i player should be able to decode standard PCM audio as specified for CD-Audio, as well as a dedicated audio coding scheme called *adaptive delta pulse code modulation* (ADPCM). The difference with PCM is that audio is not stored individually per time segment, but that only the difference (delta) to the previous sample is recorded. Owing to the existing correlation between adjacent samples, a significant decrease in the used storage space on the disc can be achieved, and hence in the bit stream being read from the disc. When normal PCM CD-Audio would be used (which occupies all successive sectors), this would not leave room for video or animations to be read without interrupting the audio playback.

CD-i provides three levels of ADPCM audio, all of which can be used either in mono or stereo, as shown in Table 21.3.12. Note that level A provides Hifi quality, Level B gives FM radio quality, while Level C is for voice quality. Level C mono can be used for up to 16 voice channels, e.g., in different languages. The sector structure facilitates switching between languages on the fly, as the sectors are interleaved on the disc.

Thus, when ADPCM Level C is used, only 1 out of every 16 sectors needs to be used for audio, leaving all other sectors for other data such as video or animation. It is also possible to record different audio channels at once, allowing for the seamless switching between, e.g., various languages. The disc may also be read from the beginning while decoding a different audio channel, allowing for increased audio-playing times, as indicated in Table 21.3.12.

A CD-i player equipped with a digital video cartridge is also able to decode MPEG-1 Layer I and II audio. MPEG is far more efficient in coding audio, resulting in an even longer playing time while providing a highly increased audio quality when compared to ADPCM. This is because of the fact that MPEG audio is based on precision adaptive subband coding (PASC),[35] which uses perceptual coding to only store those audio signals that are audible, while filtering out other signals. Note that CD-i offers a very flexible way of using MPEG audio (for example, at various bit rates and quality levels), but cannot decode MPEG-1 Level III, or MP3 files.

CD-i Video

The video image of a CD-i player consist of four "planes," which are overlaid on top of each other. The first plane is used by a cursor and its size is limited to 16 × 16 pixels. The second and third planes are shown

TABLE 21.3.12   Different ADPCM Formats

Format	Frequency (kHz)	Numbers of	Used sectors	Recording time
CD-audio PCM	44.1	bits per sample 16	all sectors	up to 74 min
ADPCM Level A stereo	37.8	8	1 in 2 sectors	up to 2.4 h
ADPCM Level A mono	37.8	8	1 in 4 sectors	up to 4.8 h
ADPCM Level B stereo	37.8	4	1 in 4 sectors	up to 4.8 h
ADPCM Level B mono	37.8	4	1 in 8 sectors	up to 9.6 h
ADPCM Level C stereo	18.9	4	1 in 8 sectors	up to 9.6 h
ADPCM Level C mono	18.9	4	1 in 16 sectors	up to 19.2 h

underneath the cursor and are used for full screen images. The fourth plane is used for a single-colored background or for MPEG full motion video (or to display video from an external source on some players). Parts of an image on one of the middle two planes can be transparent, so that the underlying plane becomes visible. This can be used, for example, to show subtitles or menus on an image. Both planes can also be used for blending and fading effects.

There are various encoding techniques for video that can be used in CD-i, which are indicated below.

DYUV. DYUV or Delta YUV is used for the encoding of high-quality photographs and other natural images. It is based on the fact that the human eye is more sensible to differences in brightness than to differences in color. Therefore, it stores one color for a set of pixels, and a brightness value for each pixel. The result is an image of slightly more than 100 kbyte. Owing to the complexity of a DYUV image, the storage on the disc must take place in advance, and it cannot be created nor modified in the player. DYUV is used mostly in CD-i titles because of its high quality and efficient storage.

RGB555. RGB555 is a compression format that allows only 5 bits per *R*, *G*, and *B* value, resulting in a picture with a maximum of over 32,000 colors. Since RGB555 uses both planes to display the image, it cannot be used in combination with other graphics. An RGB555 image is roughly 200 kbyte in size. The image can be altered by the player at run time. RGB555 is actually never used in regular CD-i titles because of its inefficiency and limitations in usage.

CLUT. CLUT, or color look-up table, is a compression method aimed at coding simple graphics. The colors used in a certain picture are stored in a CLUT-table, which reduces the size of the image dramatically, because color values refer to the appropriate CLUT-entry instead of indicating, for example, a 24-bit color value. In CD-i, a CLUT image can have an 8-bit (256 colors), 7-bit (128 colors), 4-bit (16 colors), or 3-bit (8 colors) resolution.

Run Length Encoding (RLE). RLE is a variation of the CLUT compression method that besides storing the CLUT-color table in an image, further reduces the image size by storing certain "run lengths" of repeating horizontal pixels with the same color. The results are usually pictures between 10 and 30 kbyte in size. This makes RLE ideal for compressing animations that contain large continuous areas with similar colors.

QHY. QHY, or quantized high Y, is an encoding technique that combines DYUV and RLE, resulting in a very sharp high-quality natural image, that is displayed in CD-i's high-resolution mode. A QHY image is usually about 130 kbyte in size. Since it consists of a DYUV component, it cannot be modified by the player. QHY is, for example, used to display the images of a photo-CD in high resolution on a CD-i player.

CD-i can display both main planes in either normal, double, or high resolution, which are 384 × 280, 768 times a DYUV image is always standard resolutions. It is possible for the images on each of the planes to be displayed 280 and 768 × 560 pixels, respectively. Some encoding techniques are limited to a single resolution, for example at once, even if they are in different resolutions. For example, a double-resolution CLUT4 menu bar can be overlayed on a standard resolution DYUV image. CD-i highest resolution (768 × 560 pixels), used for QHY images, is the highest resolution that can be made visible on a normal TV set.

To enable audiovisual data to be coded at CD-i bit-rates, MPEG-1 compression has been employed. This standard allows CD-i to display 384 × 280 progressive video sequences, at a video quality roughly comparable to standard VHS.

Full Motion CD-i Players

The quality of the "Base Case" CD-i players can be further extended to "Full Motion" audiovisual quality.[9] For this, two coding formats are defined for video sectors: "MPEG video" and "MPEG still picture" for storage of MPEG-coded progressive video and interlaced still picture data. One new coding format is defined for audio sectors: "MPEG audio" for storage of coded MPEG audio data. All these video, still picture, and audio sectors are a form 2 sector with a usable datafield of 2324 bytes, while the Full Motion application program is stored in a normal data sector.

To code audio, CD-i applies MPEG layer I and layer II. Audio can be coded in stereo and in mono, at varying bit rates. For instance, for speech and other audio not requiring a high quality, the lowest bit rate for mono of 32 kbit/s can be used. For high-quality audio, a bit rate of 192 kbit/s in stereo is necessary to achieve similar quality to the compact disc. Audio can be coded in stereo, joint stereo (i.e., intensity stereo), dual channel, or single channel. All bit rates allowed by MPEG for these layers are supported. The applied audio sampling frequency is 44.1 kHz. The bit rates can vary from 32 to 448 kbit/s for layer I and from 32 to 384 kbit/s for layer II. For more information, see Ref. 9.

The Full Motion system in CD-i[41] supports the video parameters defined in the video part of the MPEG standard, but some parameters have a larger range in the CD-i system. For instance, while in the MPEG-1 specification the bit rate should not exceed 1.856 Mb/s; in the Full Motion system a maximum bit rate of about 5 Mb/s is allowed. Also, in the Full Motion system, video coding with a variable bit rate is allowed, leading to a higher visual quality. The maximum picture width is 768 pixels and the maximum height is 576 lines. The supported picture rates are 23.976, 24, 25, 29.97, and 30 Hz. The maximum size of the VBV buffer is 40 kbyte. More information on the specific video parameters employed for CD-i Full Motion video can be found in Ref. 9. Digital video is displayed on the background plane and can be overlayed with images coded in CLUT or RLE format.

The system part of the MPEG standard applies a multiplex structure consisting of packs and packets. In each MPEG video sector, still picture sector, and audio sector, one pack is stored. Each pack consists of a pack header and one or more packets containing a packet header followed by data from one audiovisual stream. More information can be found in Ref. 9.

The authoring system provided by the CD-i[41] allows to optimize the audiovisual compression parameters to the requirements of the application. On a CD-i disc multiple MPEG audio and video streams can be recorded in parallel, e.g., for applications requiring audio in different languages. Moreover, during the authoring process, trade-offs can be made between the number of streams, the bit rate, the audio or picture quality of each stream, and, in the case of video, the picture size.

Compatibility to Other Formats

A CD-i player only plays discs that incorporate a dedicated application program that was designed for CD-i's operating system and hardware components. For some disc types, such as video-CDs, photo-CDs, and CD-BGM, this CD-i application is a mandatory part of the disc's specification. Next to this, the CD-i standard requires the player to be able to play back standard CD-Audio discs.

Differences between Video-CD and CD-i.   A video-CD is a compact disc with up to 75 min of VHS quality video with accompanying sound in CD quality. Audio and video are coded according to the MPEG-1 standard and the disc layout (see Fig. 21.3.28) is based on the CD-i Bridge (see Fig. 21.3.29) specification to allow for the playback on a variety of plackback devices such as CD-i players and dedicated video-CD players.

Video-CD became very popular mainly in Asia, while elsewhere, it is mainly used as a prototype tool. Although video-CD compatibility is not required for DVD-video players, it is very likely that video-CD playback

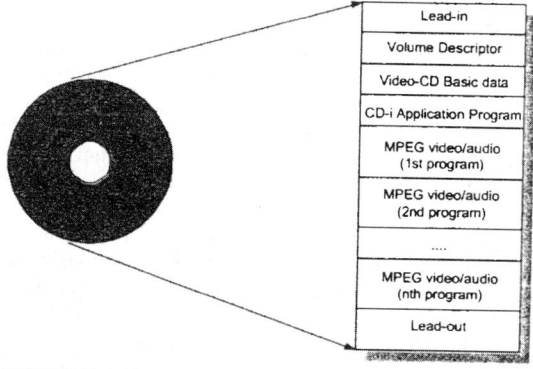

FIGURE 21.3.28   Video CD—disc data allocation.

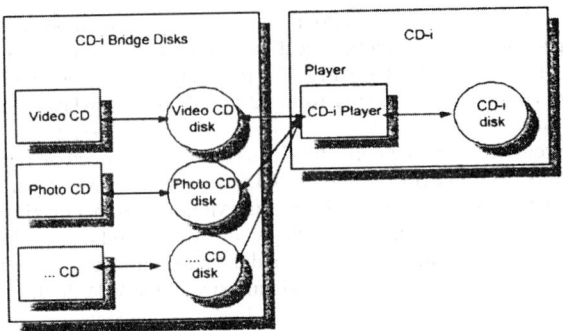

FIGURE 21.3.29 CD-i bridge.

functionality is included since every DVD-video player must be able to decode MPEG-1 as well. Another difference is that the resolution of the MPEG video on a CD-i movie disc is slightly higher than the defined resolution of a video-CD disc (384 × 280 for the CD-i "Base Case" players instead of 352 × 240 for video-CD). This also prevents extracting the video from a CD-i disc in order to burn a video-CD. To do this, reencoding of the video according to the White Book (video-CD) specification is necessary, leading to a decreased picture quality.

BRIEF DESCRIPTION OF MPEG VIDEO CODING STANDARD

The Meaning of I, P, and B Pictures

MPEG video coding is described here because it applies both to VCD (and CD-i) and the DVD system that will be described next. Many good overviews of the MPEG video coding standards can be found in a variety of books[13] and articles.[11] The MPEG video compression algorithm relies on two basic techniques: block-based motion compensation for the reduction of the temporal redundancy and DCT-based compression for reducing spatial correlations. The DCT-based compression has been described in detail in the section on DV recording. The intraframe DCT coder of DV comes close to the MPEG video processing with respect to DCT, quantization, and VLC coding. Therefore, the focus is in this section on exploiting the temporal redundancy.

In MPEG, three picture types are considered: intra pictures (I), predicted pictures (P), and bidirectional prediction pictures (B). Intra pictures are intraframe coded, i.e., temporal correlation is not exploited for the compression of these frames. I pictures provide access points for random access but achieve only moderate compression, because as their name indicates, compression is limited within the same picture. Predicted pictures, or P pictures, are coded with reference to a past I or P picture and will in general be used as a reference for a future frame. Note that only I and P pictures can be used as a reference for temporal prediction. Bidirectional pictures provide the highest level of compression by temporally predicting the current frame with respect to both a previous and current reference frame. B pictures can achieve higher compression since they can handle effectively uncovered areas, since an area just uncovered cannot be predicted from the past reference, but can be properly predicted from the future reference. They also have the additional benefit that they decouple prediction and coding (no error propagation is incurred from the prediction errors of B pictures).

In all cases, when a picture is coded with respect to a reference, motion compensation is applied to improve the prediction and the resulting coding efficiency. The relationship with respect to the three picture types is illustrated in Fig. 21.3.30. The I pictures and the B and P pictures predicted based on it form a group-of pictures (GOP). The GOP forms an information layer in MPEG video and is from the data point of view buildup as shown at the lower side of Fig. 21.3.30. First, the I picture is coded at the start and kept in a memory in the encoder. These memories are indicated at the bottom in Fig. 21.3.31. Second, the next

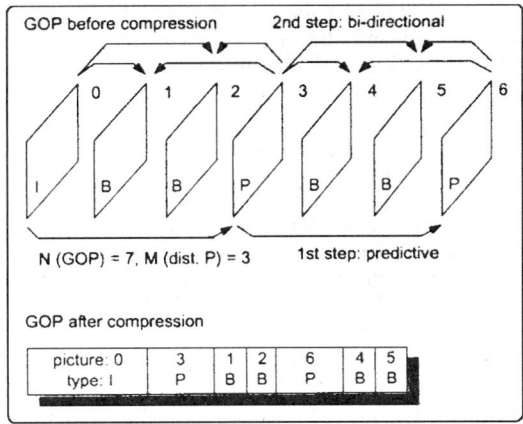

FIGURE 21.3.30 A group-of-pictures (GOP) divided in I, P, and B pictures.

P picture is selected and coded with reference to the I picture. The pictorial result after local decoding is also stored in a second memory. Third, the two B pictures in between are processed in sequential order. Each B picture is coded depending on the past I picture and the near P picture in future. This explains the term *bidirectional*. When the B pictures in between have been processed, the process repeats itself and the next P picture is first coded and stored as reference, and so on. Note that MPEG video coding as a result changes the transmission order of pictures over the channel (see bottom of Fig. 21.3.30 for the modified order).

MPEG Coding and Motion Compensation

Coding and processing with reference to a past and future picture involves motion compensation (MC). This means that the motion of objects is taken into account and predicted and only the difference between

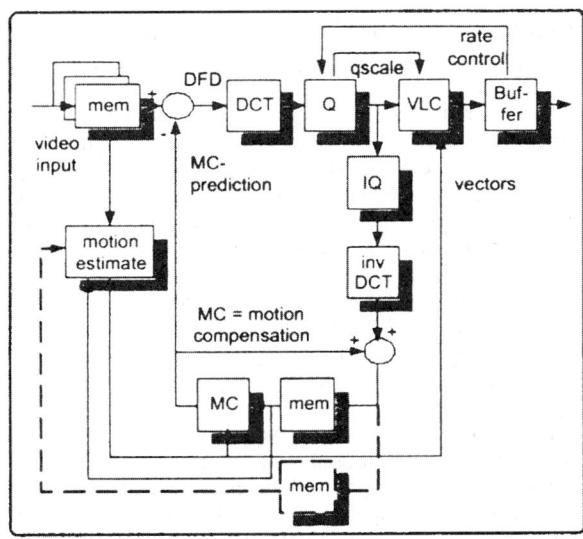

FIGURE 21.3.31 MPEG video encoder block diagram.

the temporal prediction and the actual video data is coded (see Fig. 21.3.31). In order to predict the motion of video data, it is first measured in an initial step, called *motion estimation* (ME). Motion compensation and estimation in MPEG are based on block-based processing. Hence, the blocks of video samples in actual and reference picture are compared. The typical block size for comparison in MPEG is a macroblock of 16×16 pixels.

There is a trade-off between the coding gain due to motion compensation and the cost of coding the necessary motion information (prediction type, motion vectors relative to the reference pictures, and so forth). Hence, in MPEG, the choice of 16×16 macroblocks for the motion compensation unit is the result of such a trade-off. The motion information is coded differentially with respect to the motion information of the previous adjacent macroblock. The differential motion information is likely to be small, except at object boundaries, and it is coded using a variable-length code to provide greater efficiency.

To reduce the spatial correlation in I, P, and B pictures, DCT-based coding is employed. After the computation of the transform coefficients, they are quantized using a visually weighted quantization matrix. Subsequently, after quantization the coefficients are zigzag scanned and grouped into (run, amplitude) pairs. To further improve the coding efficiency, a Huffman-like table for the DCT coefficients is used to code (run, amplitude) pairs. Only those pairs with a high probability of occurrence are coded with a VLC, while less likely pairs are coded with an escape symbol followed by fixed-length codes, in order to avoid extremely long codewords and reduce the cost of implementation. All these steps are indicated in Fig. 21.3.31.

The quality of video compressed with the MPEG-1 algorithm at rates about 1.2 Mb/s is comparable to VHS quality recording. The quality of MPEG-2 video is higher, but it is also operated at a higher bit rate between 3 to 5 Mb/s. This is discussed further in the next section about DVD.

THE DVD SYSTEM

Introduction and History

The DVD system was introduced in 1996[14] and has experienced a strong growth of interest and sales in the consumer area. The key factors of its success are the very high quality of the reproduced video and audio signals and the fact that the storage is based on a high-capacity optical disc medium. The latter factor builds clearly on the success of the compact disc technology, albeit with strongly improved densities to boost the total storage capacity of the disc.

The success of DVD results also from the perfect fit with the trends that are taking place in the consumer and computer industry. In the 90s, computer applications have been expanding continuously toward high-quality audio and video recording and playback functions. For example, video compression of MPEG-1[10] and MPEG-2[12] have been adopted gradually and the CD audio playback function evolved to CD-ROM and later CD-R/CD-RW recording subsystems in the multimedia computers. Nowadays, DVD-ROM has become an integral part of the modern multimedia computer, thereby solving (temporarily) the ever increasing request for more storage capacity on a removable medium. DVD-video players share most of the same physical and mechanical components with DVD-ROM drives in the computer. In the coming years, DVD recording (R/RW), which is already introduced, will occupy a significant part of consumer and general computing applications.

A DVD holds 4.4 to 15.9 GB (gigabytes), which is the same as 4.7 to 17 billion bytes. The capacity improvement of the typical DVD of 4.7 GB compared to CD (of 0.65 GB) results from different factors, which are shown in Table 21.3.13. The expansion to larger capacities results from extra bit layers, which will be elucidated in the next subsection. The DVD capacity was tuned to satisfy one of the main applications: storing a full-length and full-quality movie on a single disc. Over 95 percent of the Hollywood movies are shorter than 2 h 15 min, so that 135 min was taken as a guideline for the design of a digital video disc system. Uncompressed movies, e.g., with 4:2:2 10-bit sampling, run at 270 Mb/s and consume more than 32 Mbyte/s, requiring more than 250 gigabytes storage capacity. With resampling on a somewhat lower resolution to 124 Mb/s and MPEG-2 video coding offering a compression factor of 30, the average bit rate is reduced to 3.5 Mb/s for the video signal. Additionally, on the average, three tracks of audio consuming 384 kb/s each are added to come to 4.4 Mb/s average bit rate.

TABLE 21.3.13   Key Parameters of CD and DVD Standard Where Improvement Factors in Both Density and Channel Coding Lead to an Improvement of a Factor of 7 in Capacity

	CD	DVD	Factor
Pit length	0.83 μm	0.40 μm	2.08
Track pitch	1.6 μm	0.74 μm	2.16
Data area surface	8605 mm^2	8759 mm^2	1.02
Channel modulation ratio	8/17 EFM	8/16 EFM+	1.06
Error correction ratio	1064/3124 (34%)	308/2366 (13%)	1.32
Sector overhead ratio	278/3390 (8.2%)	62/2418 (2.6%)	1.06

Unlike the DV format, almost everything in the DVD standard is variable. The previous discussion on playing time was entirely based on average numbers. Let us consider some variations on the application. If the movie producer skips two audio tracks, the playing time increases to a near 160 min. The difference can, e.g., be used to increase the picture quality. The maximum bit rate can be as high as 9.8 Mb/s for the video and the total bit rate including audio is limited to 10.08 Mb/s. This maximum rate can be sustained and would lead to a playing time of 62 min. On the other side of the spectrum, if a movie is recorded in MPEG-1 at about 1.2 Mb/s, the playing time becomes more than 9 h. This discussion applies to DVD-video. When considering DVD-ROM, the flexibility in usage and the recorded data is infinite. If a new video coding technique is found in the future, the disc may be used to store more than 3 h of high-quality film.

Bit Layers in DVD

The increase in storage capacity from 4.7 to 15.9 gigabytes results from a new technology that was applied in the DVD system, where information bits can be stored in more than one layer. The laser that reads the disc can be focused at two different levels. Hence it can read the top layer or it can look through the top layer and read the layer beneath. This process is enabled by a special coating on the top layer, which is semireflective, thereby allowing the laser wavelengths to pass through. The reading process of a disc starts at the inside edge and gradually moves toward the outer edge, following a spiral curve. The length of this curve is about 11.8 km. If the laser reaches the end of the first layer, it quickly refocuses on the second layer and starts reading backwards. The switch to the second layer takes place in the order of 100 ms, which is fast enough to match this electronically with data buffering in order to prevent hiccups in presentation of the video or audio data. The reading process of bits from the disc is faster than the decoding of it to compensate for time losses from refocusing. Possibilities for various forms of reading and layering are shown in Fig. 21.3.32. First, DVDs can be single-sided or double-sided. Second, each side can have one or two information layers, indicated with the tiny square waves in Figs. 21.3.32(a) through 21.3.32 (e). The information layers are in the middle of the disc for protection and enabling the double-sided options. A double-sided disc is constructed by binding two

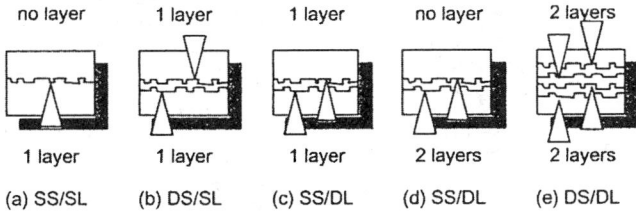

FIGURE 21.3.32   Various configurations of bit layers and sides: SS = single side, DS = double side, SL = single layer, DL = double layer.

TABLE 21.3.14 DVD Capacities Resulting from the Various Combinations of Layers, Sides, and Disc Sizes. The Triangles Represent the Focused Laser Beam. CD-ROM is Added for Reference (Assuming Four Times the Speed)

Type	Diam.	Sides	Layers	Gigabytes	Pl.-time/(h)
DVD-5	12 cm	SS	SL	4.70	2.25
DVD-9	12 cm	SS	DL	8.54	4.0
DVD-10	12 cm	DS	SL	9.40	4.50
DVD-18	12 cm	DS	DL	17.08	8.0
	8 cm	SS	SL	1.46	0.75
	8 cm	SS	DL	2.66	1.25
	8 cm	DS	SL	2.92	1.5
	8 cm	DS	DL	5.32	2.5
CD-ROM	12 cm	SS	SL	0.68	0.25

stamped substrates back to back. Single-sided discs have a blank substrate on one side. There are even more possibilities, since DVDs can also vary in size, i.e., by using 8 cm or 12 cm diameter for the disc. The most important options are shown in Table 21.3.14. Physically, each substrate has a thickness of 0.6 mm. The thickness of a DVD is thus 1.2 mm so that it equals the thickness of a CD. The glued construction of the DVD makes them more rigid than a CD, thereby enabling more accurate tracking by the laser and thus a higher density.

A special topic is backward compatibility to the CD-ROM format. Although this is not required, the key manufacturers see this point of vital importance, because of large market presence of CD formats. One of the problems is that with CD, the data are within a stamped metal layer deep inside the disc (virtually at the backside), whereas in DVD it is in the middle. The focusing is therefore different. This leads to a player with three focusing levels. Another problem is that CD discs do not well reflect the 635 to 650 nm wavelength transmitted by a DVD player. This sometimes leads to extra lasers or optical devices. A last aspect is the compatibility with data formats (VCD, CD-i, photo CD, and so forth) requiring extra electronics inside data decoding circuitry. In practice, most DVD players have backward compatibility with VCD and sometimes with one or more CD-R/RW formats.

The DVD Disc

The disc is read from the bottom, starting with data layer 0. The second layer, called layer 1, is further from the readout surface. The layers are actually very closed, spaced about 55 μm from each other. The backside coating of the deepest layer is made from metal (such as aluminium) and is thus fully reflective, while it has a semitranspar-ent coating on top. Layer 1 has a semireflective coating to enable reading of layer 0. The first layer is about 20 percent reflective and the second layer 70 percent. The substrate is transparent. The laser light has a wave-length of 650 or 635 nm (red laser). The lens has a numerical aperture of 0.60 and the optical spot diame-ter is 0.58 to 0.60 μm.

Figure 21.3.33 portrays the key areas of the disc. The burst cutting area is for unique identification of a disc via a 188-byte code and can be used for automat-ic machines with multiple disc storage capacity (e.g., a juke box). Table 21.3.15 shows relevant parameters for recovering bits from a DVD-ROM or DVD-video. What strikes the eye here is the robustness: a scratch of 6 mm can be covered by the ECCs, which clearly outperforms the CD channel coding.

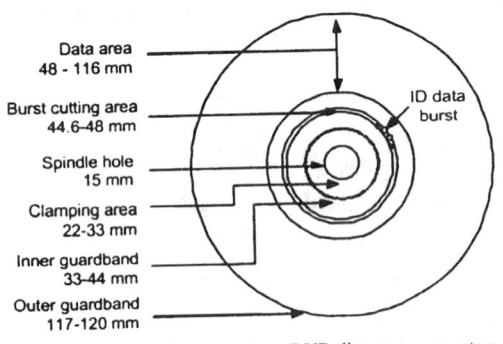

Data area
48 - 116 mm

Burst cutting area
44.6–48 mm

Spindle hole
15 mm

Clamping area
22-33 mm

Inner guardband
33–44 mm

Outer guardband
117–120 mm

ID data burst

FIGURE 21.3.33 Physical 12 cm DVD disc area parameters.

TABLE 21.3.15 Key Physical Bit Reading Parameters for DVD-ROM

Pit length SL	0.40–1.87 μm
Pit length DL	0.44–2.05 μm
Track pitch	0.74 μm
Average data bit length	0.267 (SL)/0.293 μm (DL)
Average channel bit length	0.133 (SL)/0.147 μm (DL)
Velocity	570–1630 rpm (574–1528 rpm data)
Scanning speed	3.49 m/s (SL), 3.84 m/s (DL)

DVD Data Format

The logical data format of DVD is indicated in Table 21.3.16. A key parameter is the easy to use sector size of user data of 2048 bytes that nicely fits with computer applications. The modulation code is called EFMPlus, a more efficient successor of the EFM code that was applied for the CD. This code enables a high density of bits on the disc, by limiting the number of zeros between two transitions. In this case, the code bounds the number of zeros between 2 and 8, between each group of ones. The modulation also helps keep DC energy low and provides synchronization in the channel. The ECC is a powerful R-S product code having considerably less overhead than in the CD. By using row-column processing of the rectangular data sector block of 192 by 172 bytes, it is well suited for removing burst errors resulting from scratches and dirt and so on. Note that the gross number of user data here is 11.08 Mb/s, instead of the 10 Mb/s earlier. The extra 1 Mb/s is available for extra insertion of user data such as subpictures and the like. The following figures explain the data sector construction in more detail.

The construction of a DVD data sector is portrayed by Fig. 21.3.34. The sector consists of 12 rows of 172 bytes each. The first sector starts with identification bytes (sector number with error detection on it) and copy protection bytes and the last sector ends with four EDC bytes for a payload ECC. Figure 21.3.35 depicts the conversion from data sectors into ECC blocks. All data sectors together then form a block of 192 rows of 172 bytes each. The rows of the ECC block are interleaved to spread out possible burst errors, thereby enhancing the removal of those errors. Looking vertically in Fig. 21.3.35 each of the 172 columns of the ECC block gets an outer parity check data of 16 bytes assigned to it. This extra data forms the outer parity block. Similarly, for each of the 208 rows (192 + 16 bytes), a 10-byte inner parity check is computed and appended. The power of the R-S product code is such that a (random) error rate of 2 percent is reduced to less than one error out of 10^{15} bits (million-billion). The product code works on relatively small blocks of data, resulting in a maximum correctable burst error of approximately 2800 bytes, which corresponds to about 6 mm scratch length protection.

To come to recording sectors, each group of 12 rows of the complete ECC block gets one row of parity added to it, leading to a spread of parity codes as well. Thus, a recording sector consists of 13 (12 + 1) rows

TABLE 21.3.16 Data Format Parameters of DVD

Data sector (user data)	2048 bytes
Logical data sector size	2064 bytes (2048 + 12 header + 4 EDC)
Recording sector size	2366 bytes (2064 + 302 ECC bytes)
Unmodulated physical sector	2418 bytes (2366 + 52 sync bytes)
Modulated physical sector	4836 bytes (2418 × 2 bytes)
Modulation code	8–16 (EFMPlus)
Error correction code	R-S product code (208,192,17) × (182,172,11)
ECC block size	16 sectors (32,768 bytes user data)
ECC overhead	15% (13% of recording sector)
Format overhead	16% (37,856 bytes for 32,768 user B)
User/Channel data rate	11.08/26.16 Mb/s

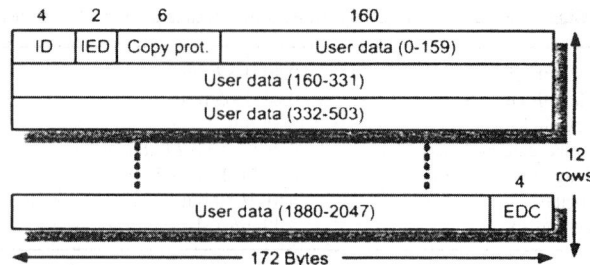

FIGURE 21.3.34 Construction of a DVD data sector.

of 182 (172 + 10) bytes. A recording sector is split in the middle and 2 bytes are inserted for synchronization and 2 sync bytes are added at the beginning of the first half. This leads to two parts of 2 + 91 bytes (total 186 bytes) and 13 rows, which form 2418 bytes together. This block is modulated to 4836 bytes by the EFMPLus code.

DVD-Video

The DVD-video system is an application of the DVD-ROM standard and aims at playback of high-quality movies together with multichannel audio. Presently, this is one of the mostly accepted standards of the DVD products. DVD-video consists of one stream of MPEG-2 video coded at variable rate, up to eight channels of Dolby digital multichannel audio or MPEG-2-coded multichannel audio or linear PCM audio, and up to 32 streams of subpicture graphics with navigation menus, stills pictures, and control for jumping forward and backward through the video program. With respect to video resolution, the format intends to serve and output both analog standard resolution TV as well as high-quality digital TV and possibly in future digital HDTV. Many multimedia computers can playback DVD-video as well.

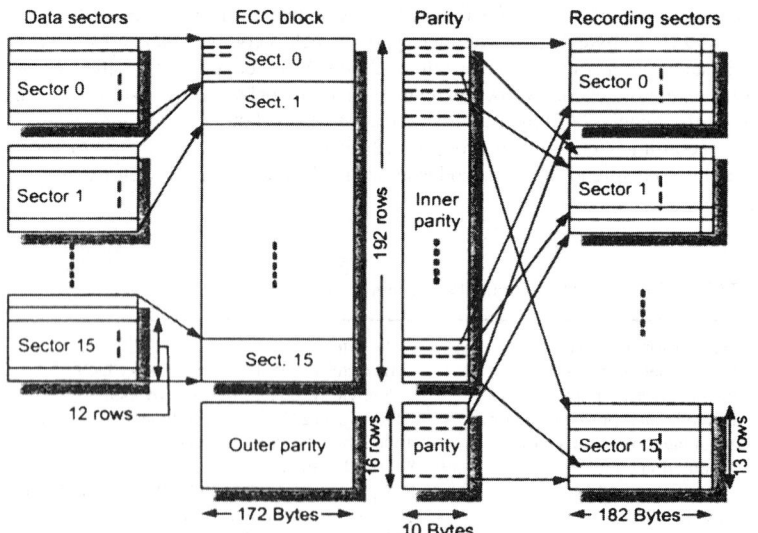

FIGURE 21.3.35 Data sectors mapped into DVD ECC blocks that are subsequently converted into recording sectors.

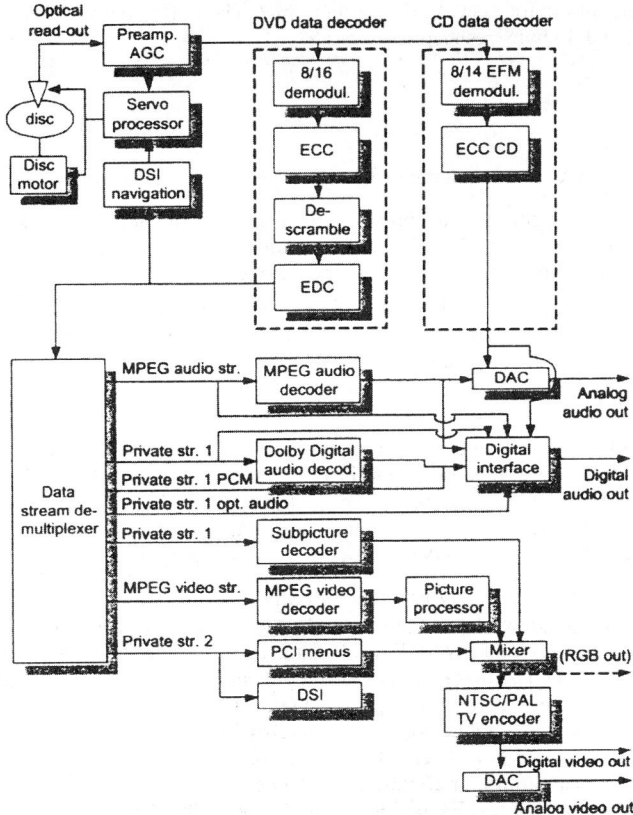

FIGURE 21.3.36 Block diagram of DVD player with CD compatibility.

The system diagram of DVD-video is depicted in Fig. 21.3.36. The diagram depicts audio data reading on disc with either CD audio data decoding or via the DVD demultiplexing of audio streams. The video results from demultiplexing the DVD data into the various data streams. Audio is decoded with MPEG or Dolby digital decoders. The amount of postprocessing is growing continuously and may involve noise reduction, sharpness improvement, and even 100 Hz conversion (Europe). The NTSC/PAL encoder ensures seamless connection to existing TV receivers.

The data flow and buffering operates as follows. The channel bits from disc are read in a constant 26.16 Mb/s rate. The EFMPlus 8-16 demodulator halves this to 13.08 Mb/s and this is reduced after ECC to a constant user data stream of 11.08 Mb/s. Data search information (DSI) is copied from this stream prior to writing this in a track buffer. A so-called MPEG program stream of variable rate is recovered at 10.08 Mb/s, which contains five different packetized elementary streams (PES): video, audio, subpicture, presentation control information (PCI), and the data search information (DSI). The latter two streams are system overhead data. The 10.08 Mb/s data rate is the maximum rate of the video and audio stream together. Video may be coded as MPEG-1 or MPEG-2 streams and has the form of a PES stream; audio, if coded as MPEG, is also a PES stream. The PCI and DSI streams are mapped as MPEG private streams. Another private stream contains the subpicture data and alternative audio data. Since audio consumes data rate, the video is limited to either 9.8 Mb/s for MPEG-2 and 1.85 Mb/s for MPEG-1. The pulse code modulation (PCM) audio stream is maximally 6.14 Mb/s, which can consist of eight channels audio sampled at 16 to 48 kHz. Note that each AV decoder has its own buffers for keeping data at the input. The buffer sizes for video, audio, subpicture, and PCI are 232, 8, 52, and 3 kbyte, respectively.

DVD Audio and Data

Audio streams in DVD are based on three different possibilities as follows:

- *PCM audio.* This is a straightforward stream of digital samples of audio without compression, referred to as LPCM. The minimum bit rate is 768 kb/s and for multiple channels, it can be maximally 6.144 Mb/s. The typical bit-rate is 1.536 Mb/s for 48 kHz sampling frequency.

- *Dolby digital.* This is based on the Dolby digital coding scheme, having a minimum, typical, and maximum bit rate of 64, 384, and 448 kb/s, respectively. The coding scheme is also known as AC-3. There can be one, two, three, four, or five channels with an optional subwoofer channel (all together indicated as Dolby 5.1). All Dolby digital decoders are required to downmix 5.1 channels to two channels stereo PCM and analog output.

- *MPEG audio.* MPEG-1 audio layer II is based on coding the audio spectrum in individual subbands. It is limited to 384 kb/s bit rate, the typical rate is 192 kb/s. MPEG-2 audio can extend this from 64 kb/s to maximally 912 kb/s. The typical rate is 384 kb/s. At maximum, seven channels can be used (7.1 with subwoofer). The advanced audio coding (AAC) mode is optimized for perception and is part of MPEG-2 audio coding. The primary audio track of DVD is always MPEG-1 compatible.

The subpicture data of 10 kb/s may grow to 3.36 Mb/s maximally, whereas the DTS data can have a rate up to 1.536 Mb/s. Typically, the rates for this type of data are rather limited in capacity. Various coding types and parameters can be used in the above options. Coding algorithms can be besides the MPEG equal to LPCM, AC-3, DTS, and SDDS. The last two modes involve optional multichannel formats (e.g., 5.1) using compression on PCM-based channels. Audio sampling can be performed at 16, 20, or 24-bit resolution. Sampling rate is 48 or 96 kHz and the dynamic range compensation (DRC) of the audio signals can be switched on and off. Along with certain applications, a code can be used, such as surround (multichannel HQ audio), karaoke (for singing with supporting subtitling), or an unspecified code.

The data are organized in files. Each side of the disk contains one logical volume, using one partition and one set of files. Each file is a contiguous area of data bytes to support easy transfer and decoding. The complete list of constraints applies to the normal DVD-video application that are used at playback. However, the format allows to record special data files that do not satisfy the above constraints. This is allowed if they are formatted into the DVD-video "Data" stream, behind the regular data that are required for playback. Then a specially designed player can access the extra data. This option allows many special applications in the future.

DVD Conclusions

DVD is an evolutionary product that successfully builds further on the widely accepted and distributed optical recording technology of the CD. The principal advantage is the excellent video and audio quality recorded on a single disc. Table 21.3.17 provides the main system parameters of DVD compared with DV and D-VHS. The table clearly shows that VHS is lagging behind in picture quality and that the digital formats of DV and DVD are comparable with each other.

The compression factor of DVD is high but the quality is approaching the DV quality in spatial resolution. The MPEG compression relies heavily on the motion compensation, which is not used in DV. Thus the temporal quality of DV clearly outperforms that of DVD and this explains why there are semiprofessional DV systems (e.g., DVCPRO) for portable studio applications.

The coming years will show further penetration of recordable DVD formats in the market (DVD-RW, DVD + RW, DVD-R[36] with the above-mentioned resolution and quality. Meanwhile, even higher capacity optical discs are being developed to pave the way for HDTV and to stay comparable with computer hard disks.

TABLE 21.3.17   DVD Compared to Other Existing Recording Systems

	DV	D-VHS	DVD
Video	Component digital	Composite analog or digital bit stream	Component digital
Playing time	0.5–1 h (mini) 4.5 h (standard)	2–6 h (VHS) 4–40 h (D-VHS)	2.25 h (1 layer) 4.5 h (2 layers)
Data capacity	5–50 Gbytes	32–40 Gbytes (300–400 m tape)	4.7–17 Gbytes
Compression	Intraframe DV factor 5	External (MPEG-2)	MPEG-1/MPEG-2 factor 30
Data rate	25.146 Mb/s video 1.536 Mb/s audio	28.2 Mb/s HD, 14.1 Mb/s STD, 2–7 Mb/s LP	Up to 9.8 Mb/s combined VBR or CBR V + A
Error correction	R-S product code + track interleave	R-S (inner/outer)	R-S product code
Channel modulation	24-25 code	NA	EFMPlus (8–16 block)
TV video systems	525/60/2:1 29.97 Hz 625/50/2:1 25 Hz	NA	525/60, 1:1 24 Hz/ 2:129.97 Hz 625/50/2:1 25 Hz 1:1/24 Hz
V 525 resolution	720 × 480 (346 kpixels)	320 × 480 (154 kpixels) VHS	720 × 480 (346 kpixels)
V 625 resolution	720 × 576 (415 kpixels)	320 × 580 (187 kpixels) VHS	720 × 480 (415 kpixels)
Audio	2 tracks of 2 channels (32 kHz, 12-bit nonlinear PCM) or 1 track of channels (32/44.1/48 kHz 16-bit PCM)	Analog VHS stereo HiFi or digital bit stream (MPEG/Dolby like)	8 tracks of up to 8 channels each (LPCM/ Dolby digital/MPEG-2)
Audio SNR	72 dB (12-bit), 96 dB (16-bit)	40 dB(mono), 90 dB HiFi	96–144 dB
Trick play, searching	Up to 70–90 times, with complete picture	Up to 10–20 times, with distorted picture (VHS)	Jumping to I pictures several speeds, stills

NETWORKED AND COMPUTER-BASED RECORDING SYSTEMS

Personal Video Recorder (PVR)

The final part of the chapter is devoted to new emerging technologies and products in recording of which most are related to further applying the computer technology into consumer systems. The personal video recorder (PVR) is a TV recorder that records on a computer hard disk as opposed to tape. The PVR is also called personal digital recorder, digital video recorder, digital network recorder, smart TV, video recording computer, time-shifted television, hard disk recorder, personal television receiver, television portal, or on-demand TV. PVR has evolved as a direct consequence of digital recording and low-cost hard disk storage. The advantages of hard-disk-based recording are:

- Allowing random access to content
- Metadata can be stored along with the content (allowing easier searching, selection and management of content, segmentation of content, richer interaction, and so forth
- Easier transfer and redistribution of content

Some of the features provided by the PVR justify its popularity in first product trials.

Simultaneous record and replay, allowing "trick" modes such as pausing and rewinding live TV. This allows pausing of a live TV program when the phone rings, later viewing of a program that has already started but is not yet finished, "fast-forward" through parts of programs when watching a time-shifted program, and so forth.

Sophisticated interactive multimedia information services, regular program and news updates that are instantly available to consumers. For instance, downloads of electronic program guides (EPGs) with the TV signal or via the phone network simplify program selection, browsing (by time, by channel, and so forth), and recording.

Automatic indexing of recorded programs allowing easy retrieval and manipulation of content.

"Intelligent" agents that can automatically record or recommend TV programs that they think the viewer might want to watch. For instance, the EPG can also include metadata describing the programs (e.g., the actors, the genre) or a brief text summary, that can be used by the PVR to manage the recording process. Consequently, the PVR can use this metadata to automatically record or recommend similar types of programs the users have recorded in the past or preferences they have entered as part of the set-up procedure.

Nonlinear viewing of programs that were recorded in a linear manner. The viewing is performed by, e.g., skipping certain segments.

Several brands of PVRs exist—TiVo, Replay TV, Ultimate TV, Axcent, and so on. Subsequently, we will explain some of the technology and issues behind the PVR, by giving some implementation examples.

Content-Format I/O and Storage.   The various PVR brands use different formats for the I/O and for the storage format. This choice is very important since it influences the overall PVR architecture, the necessity for content protection as well as the resulting quality of the stored content. For instance, TiVo's I/O interfaces are analog, even for the digital services, thereby eliminating issues such as conditional access and scrambling. For storage, the analog signals are converted to digital and then compressed using MPEG-2 coding. This implies that in the case digitally broadcasted content is stored using TiVo, the audiovisual quality can potentially be decreased by the cascaded encoding/decoding process. To solve this problem, other PVR configurations store the original broadcast digital signal in scrambled form on its hard disk, giving the potential for better audiovisual quality and the possibility for the conditional access operator to control the replay of stored programs. On the positive side, however, because TiVo compresses the content, easy trade-offs between quality and storage capacity can be made. For example, if a TiVo has 40 Gigabytes of hard disk capacity, this represents about 40 h storage at "basic" quality or about 12 h at "best" quality. It should be noted, however, that the "best" quality is limited by the incoming audiovisual content quality.

Event- Versus Time-Driven Programming.   To better understand the differences between these types of programming, let us consider two different implementations. For the TiVo PVR, data are downloaded daily via the telephone network, and depending on the availability of the data, TiVo can have program details for up to two weeks in advance. The TiVo data are entirely time-driven, such that the program is stored based on the a priori available time information. Alternatively, other PVRs are event-driven, i.e., detect the program junctions. For instance, Axcent PVR uses a 64 kb/s channel on the satellite transponder to which the PVR tunes by default to keep recording when a program runs late.

Software Downloading to Existing PVRs for Enabling New Services.   Different services can be enabled using the PVR functionality. For instance, "trailer selection" services are expected to appear enabling the viewer to simply press a button when a trailer for a program is presented, and the PVR will automatically record the program whenever this is broadcast. This can be done using the program metadata (see next paragraph). Furthermore, other e-commerce services can also be enabled via similar mechanisms, as described below.

Metadata.   Since the introduction of the PVR, it was argued that the scheduling of the programs at attractive hours will lose its power to attract viewers, because the PVR can present desired video programs that were already stored on its hard disk on-demand. If scheduling will indeed lose its power to attract the viewers, its place will probably be taken by the metadata and services that support PVR functions. Consequently, the battle for the attention of the viewers will be won by the provider that can describe programs and services most fully and attractively using metadata. Moreover, with a PVR it is possible for the viewer to skip all advertisements. Nevertheless, the content still has to be paid for, such that new mechanisms are necessary to encourage the viewers to watch commercials. To enable this, ubiquitous metadata can be employed to target advertisements at specific interest groups. Furthermore, PVR's combination of storage, processing power, and supporting metadata offers potential for new kinds of programs (e.g., educational), where the viewer interactively navigates around the stored program.

Note that standards are necessary for describing multimedia content, i.e., metadata, which is required by these services. MPEG-7 provides such a standard[37] (see the "MPEG-7" section). Furthermore, while many PVRs are already existing on the market, proprietary solutions restrict the user to a single service or content provider and lock together the broadcaster, service provider, and the PVR manufacturer. Consequently, to enable PVR proliferation, standards are necessary for metadata to choose programs, for finding the programs in a multi-channel, web-connected broadcasting environment, for management of the rights that must be paid for in order to support the services, and so forth. These standards are generated in a new worldwide body, the TV-Anytime forum (see "TV-Anytime" section).

Brief MPEG-7 Overview

As mentioned in the PVR section, digital audiovisual recorded material is increasingly available to users in a variety of formats. For instance, persistent large-volume storage that allows nonlinear access to audiovisual content, such as hard-disk storage in powerful PC platforms and personal video recorders (PVR), is becoming available in many consumer devices. Consequently, there is a need for rapid navigation and browsing capabilities to enable users to efficiently discover and consume the contents of the stored audiovisual programs. Users will also benefit from having nonlinear access to different views of a particular program, adapted to the user's personal preferences, interests, or usage conditions such as the amount of time the user wants to spend in consuming the content or the resources available to the user's terminal. Such adaptability will enhance the value provided by the multimedia content. The MPEG-7 standard,[38] formally named *Multimedia Content Description Interface*, provides a rich set of standardized tools to describe such multimedia content (see Fig. 21.3.37). A good description of the MPEG-7 standard can be found in Refs. 37 and 38. In this section, we will summarize only those MPEG-7 components that can be useful in finding, retrieving, accessing, filtering, and managing digitally recorded audiovisual content.

The main elements of the MPEG-7 standard[37,38] are presented below.

- *Descriptor* (*D*) is a representation of a feature. A descriptor defines the syntax and the semantics of the feature representation.
- *Description Scheme* (*DS*) specifies the structure and semantics of the relationships between its components, which may be both Ds and DSs.

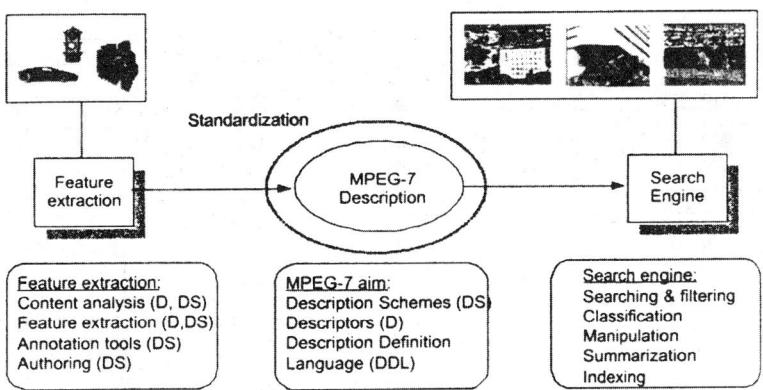

FIGURE 21.3.37   MPEG-7 standardization role.

- *Description Definition Language* (*DDL*) is a language to specify description schemes. It also allows the extension and modification of existing DS. MPEG-7 decided to adopt XML Schema Language as the MPEG-7 DDL. However, DDL requires some specific extensions to XML Schema Language to satisfy all MPEG-7 requirements.

- *Binary representation* provides one or more ways (e.g., textual, binary) to encode descriptions. A coded description is a description that has been encoded to fulfill relevant requirements such as compression efficiency, error resilience, random access, and so forth.

MPEG-7 offers a comprehensive set of audiovisual description tools that can be used for effective and efficient access (search, filtering, and browsing) to multimedia content. For instance, MPEG-7 description tools allows the creation of descriptions of content that may include:[37]

- Information describing the creation and production processes of the content (director, title, short feature movie)
- Information related to the usage of the content (copyright pointers, usage history, broadcast schedule)
- Information of the storage features of the content (storage format, encoding)
- Structural information on spatial, temporal, or spatio-temporal components of the content (scene cuts, segmentation in regions, region motion tracking)
- Information about low-level features in the content (colors, textures, sound timbres, melody description)
- Conceptual information of the reality captured by the content (objects and events, interactions among objects)
- Information about how to browse the content in an efficient way (summaries, variations, spatial and frequency subbands, and so forth)
- Information about collections of objects
- Information about the interaction of the user with the content (user preferences, usage history)

The process of generating metadata for content description that can later be used for retrieving, accessing, filtering, and managing digitally recorded audiovisual content is portrayed in Fig. 21.3.38. Note that the MPEG-7 metadata can be either physically co-located on the same storage system with the associated audiovisual content or could also be stored elsewhere. In the latter case, mechanisms that link multimedia and its MPEG-7 descriptions are needed.

Example: MPEG-7 Usage in PVR Applications

For a better understanding of the MPEG-7 importance to the field of digital recording, let us consider its benefits in a PVR application. In this application, content descriptions could be generated by a service provider,

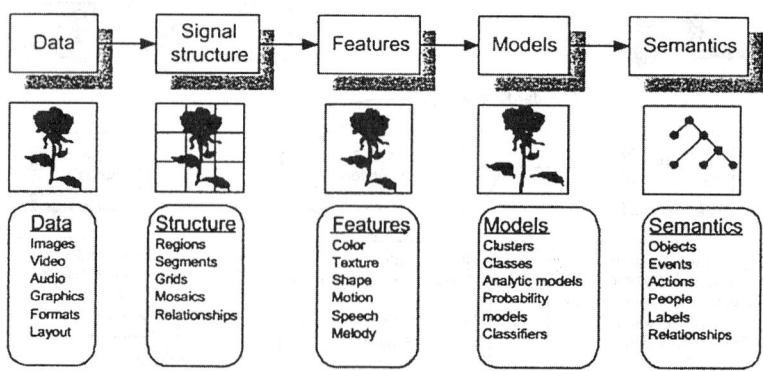

FIGURE 21.3.38   Generating metadata for content description.

separate from the original content provider or broadcaster. Certain high-level content descriptors, such as the program name and channel name, can be downloaded by the PVR in advance to provide an on-screen electronic program guide (EPG). This EPG enables the user to efficiently navigate at the program level and record programs easily. Moreover, summary descriptions can be made available in advance as well (e.g., in the case of movies), or downloaded at the end of a program (e.g., in the case of sports events). Furthermore, low-level descriptions that describes the various features of the content can also be provided. These descriptions can include information for video transcoding (i.e., transcoding hints), that can be used by the PVRs or transcoding proxies whenever transcoding from the "source" (service provider) content quality to a lower quality that can be used for local storage is desired.

Additionally, other descriptions may be generated by the user, e.g., by marking highly entertaining segments for later review. Such a feature can be simply provided by the PVR by copying the XML-fragment associated with the selected segment, including its name and locators and storing this element separately, along with some high-level elements that will allow its easy identification at a later time. Such fragments can be exchanged with friends or relatives. MPEG-7 is also developing description schemes that can capture the user's preferences with respect to a specific content, and store them on the PVR under the user's control. These schemes support personalized navigation and browsing, by allowing the user to indicate the preferred type of view or browsing, and automatic filtering of content based on the user's preferences.

TV Anytime

The *TV Anytime Forum*[42] is an international consortium of companies dedicated to producing standards for PVRs. TV Anytime aims at developing a generic framework that incorporates standards, tools, and technologies for an integrated system providing a multitude of services such as movies on demand, broadcast recording, broadcast searching and filtering, retrieving associated information from web pages, home banking, e-commerce, home shopping, and remote education. To enable this vision, TV Anytime will define specifications

- That will enable applications to exploit local persistent storage in consumer electronics platforms
- That are network independent with regard to the means for content delivery to consumer electronics equipment, including various delivery mechanisms (e.g., ATSC, DVB, DBS, and others) and the Internet and enhanced TV
- For interoperable and integrated systems, from content creators/providers, through service providers, to the consumers
- That provide the necessary security structures to protect the interests of all parties involved

Two important components in the TV Anytime framework are the digital storage and recording and the metadata, because they allow consumers to access the content they want, whenever they want it and how they

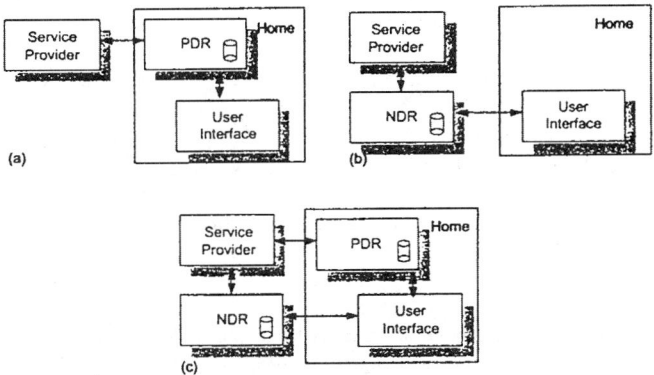

FIGURE 21.3.39 Various digital storage models: (*a*) in-home consumer storage—personal digital recorder (PDR), (*b*) remote consumer storage—network digital recorder (NDR), (*c*) PDR + NDR combination.

want it (i.e., presented and tailored according to the user preferences/requests). Metadata can be easily stored along with the content to enable searching, selection, and management of the content in a much easier fashion as compared to the current analog VCRs, and also allow a richer interaction with the stored content. Other benefits of local digital storage are that the TV viewers can order a program to be recorded using a single button during a trailer, making available intelligent agents that based on the stored user preferences can record TV programs that they think a viewer might want to watch and consuming a program in a nonlinear rather than linear manner (e.g., a news program). Note also that the digital recording process can be performed within the TV Anytime framework locally, remotely, or in a combined manner (see Fig. 21.3.39).

Hence, digital recording of content on local storage and digital broadcasting together with the framework provided by TV Anytime for content referencing and location resolution, metadata and rights management and protection, are providing significant benefits as compared with alternative forms of content delivery such as analog broadcasting, Internet, and broadband networks.

REFERENCES

1. Y. Shiraishi, "History of Home Videotape Recorder Development," *SMPTE J.*, pp. 1257–1263, December 1985.

2. K. Sadashige, "Transition to Digital Recording: An Emerging Trend Influencing All Analog Signal Recording Applications," *SMPTE J.*, pp. 1073–1078, November 1987.

3. H. Sugaya, and K. Yokoyama, Chapter 2 from C. D. Mee and E. D. Daniel, *Magnetic Recording*, Vol. 3, McGraw-Hill, 1988.

4. S. B. Luitjens, "Magnetic Recording Trends: Media Developments and Future (video) Recording Systems," *IEEE Trans. Magnetics*, Vol. 26, No. 1, pp. 6–11, January 1990.

5. M. Umemoto, Y. Eto, and T. Fukinuki, "Digital Video Recording," *Proc. IEEE*, Vol. 83, No. 7, pp. 1044–1054, July 1995.

6. J. C. Whitaker, and K. Blair Benson, Standard Handbook of Video and Television Engineering, 3rd ed., Chapter 8, McGraw-Hill, 2000.

7. S. Gregory, *Introduction to the 4:2:2 Digital Video Tape Recorder*, Pentech Press, 1988.

8. R. Brush, "Design Considerations for the D-2 NTSC Composite DVTR," *SMPTE J.*, pp. 182–193, March 1988.

9. J. van der Meer, "The Full Motion System for CD-I," *IEEE Trans. Consum. Electron.*, Vol. 38, No. 4, pp. 910–920, November 1992.

10. *Coding of Moving Pictures and Associated Audio*, Committee Draft of Standard ISO 11172, ISO MPEG 90/176, December 1990.

11. D. LeGall, "MPEG: A Video Compression Standard for Multimedia Applications," *Commun. ACM*, Vol. 34, No. 4, pp. 46–58, April 1991.

12. *Coding of Moving Pictures and Associated Audio*, International Standard, ISO/IEC 13818, November 1994.

13. J. L. Mitchell, W. B. Pennebaker, C. Fogg, and D. LeGall, *MPEG Video Compression Standard*, Chapman and Hall, 1996.

14. J. Taylor, *DVD Demystified: The Guide Book for DVD-ROM and DVD-Video*, McGraw-Hill, ISBN 0-07-064841-7, 1998.

15. R. J. Clarke, *Transform Coding of Images*, Academic Press, 1985.

16. N. S. Jayant, and P. Noll, *Digital Coding of Waveforms: Principles and Applications to Speech and Video*, Prentice Hall, 1984.

17. M. Rabbani, and P. W. Jones, *Digital Image Compression Techniques*, SPIE Tutorial Texts, Vol. TT 7, SPIE Opt. Engineering Press, Bellingham, 1991.

18. P. H. N. de With, *Data Compression Techniques for Digital Video Recording*, Ph.D. Thesis, University of Technology Delft, June 1992.

19. J. Watkinson, The Art of Digital Video, Focal Press, Chapter 6, pp. 226–274, 1990.

20. *Outline of Basic Specifications for Consumer-Use Digital VCR*, Matsushita, Philips, Sony, Thomson, July 1993.

21. S. M. C. Borgers, W. A. L. Heijnemans, E. de Niet, and P. H. N. de With, "An Experimental Digital VCR with 40 mm Drum and DCT-Based Bit-Rate Reduction," *IEEE Trans. Consum. Electron.*, Vol. CE-34, No. 3, pp. 597–605, August 1988.

22. N. Doi, H. Hanyu, M. Izumita, and S. Mita, "Adaptive DCT Coding for Home Digital VTR," *IEEE Proc. Global Telecomm. Conf., Hollywood (USA)*, Vol. 2, pp. 1073–1079, November 1988.

23. T. Kondo, N. Shirota, K. Kanota, Y. Fujimori, J. Yonemitsu, and M. Nagai, "Adaptive Dynamic Range Coding Scheme for Future Consumer Digital VTR," *IERE Proc. 7th Int. Conf. Video, Audio & Data Recording*, York (U.K.), Publ. No. 79, pp. 219–226, March 1988.

24. C. Yamamitsu, A. Ide, A. Iketani, T. Juri, S. Kadono, C. Matsumi, K. Matsushita, and H. Mizuki, "An Experimental Study for Home-Use Digital VTR," *IEEE Trans. Consum. Electron.*, Vol. CE-35, No. 3, pp. 450–457, August 1989.

25. H.-J. Platte, W. Keesen, and D. Uhde, "Matrix Scan Recording, a New Alternative to Helical Scan Recording on Videotape," *IEEE Trans. Consum. Electron.*, Vol. CE-34, No. 3, pp. 606–611, August 1988.

26. K. Kanota, H. Inoue, A. Uetake, M. Kawaguchi, K. Chiba, and Y. Kubota, "A High Density Recording Technology for Digital VCR's," *IEEE Trans. Consum. Electron.*, Vol. 36, No. 3, pp. 540–547, August 1990.

27. M. Kobayashi, H. Ohta, and A. Murata, "Optimization of Azimuth Angle for Some Kinds of Media on Digital VCR's," *IEEE Trans. Magnetics*, Vol. 27, No. 6, pp. 4526–4531, November 1991.

28. Y. Eto, "Signal Processing for Future Home-Use Digital VTR's," *IEEE J. Sel. Areas Commun.*, Vol. 10, No. 1, pp. 73–79, January 1992.

29. M. Shimotashiro, M. Tokunaga, K. Hashimoto, S. Ogata, and Y. Kurosawa, "A Study of the Recording and Reproducing System of Digital VCR Using Metal Evaporated Tape," *IEEE Trans. Consum. Electron.*, Vol. 41, No. 3, pp. 679–686, August 1995.

30. C. Yamamitsu, A. Iketani, J. Ohta, and N. Echigo, "An Experimental Digital VCR for Consumer Use," *IEEE Trans. Magnetics*, Vol. 31, No. 2, pp. 1037–1043, March 1995.

31. P. H. N. de With, and A. M. A. Rijckaert, "Design Considerations of the Video Compression System of the New DV Camcorder Standard," *IEEE Trans. Consum. Electron.*, Vol 43, No. 4, pp. 1160–1179, November 1997.

32. R. W. J. J. Saeijs, P. H. N. de With, A. M. A. Rijckaert, and C. Wong, "An Experimental Digital Consumer Recorder for MPEG-coded Video Signals," *IEEE Trans. Consum. Electron.*, Vol. 41, No. 3, pp. 651–661, August 1995.

33. J. A. H. Kahlman, and K. A. S. Immink, "Channel Code with Embedded Pilot Tracking Tones for DVCR," *IEEE Trans. Consum. Electron*, Vol. 41, No. 1, pp. 180–185, February 1995.

34. P. H. N. de With, A. M. A. Rijckaert, H.-W. Keessen, J. Kaaden, and C. Opelt, "An Experimental Digital Consumer HDTV Recorder Using MC-DCT Video Compression," *IEEE Trans. Consum. Electron.*, Vol. 39, No. 4, pp. 711–722, November 1993.

35. G. C. P. Lokhoff, "Precision Adaptive Subband Coding (PASC) for the Digital Compact Cassette (DCC)," *IEEE Trans. Consum. Electron.*, Vol. 38, No. 4, pp. 784–789, November 1992.

36. S. G. Stan and H. Spruit, "DVD + R—A write-once optical recording system for video and data applications," *Int. Conf. Consum. Electron.*, Los Angeles, 2002, Digest of Techn. Papers, pp. 256–257, June 2002.

37. S. F. Chang, T. Sikora, and A. Puri, "Overview of the MPEG-7 standard," *IEEE Trans. Circuits Syst. Video Technol.*, Vol. 11, No. 6, pp. 688–695, June 2001.

38. B. S. Manjunath, P. Salembier, and T. Sikora, *Introduction to MPEG-7: Multimedia Content Description Interface*, Wiley, April 2002.

39. D-VHS: A First Look at a New Format, http://www.thedigitalbits.com/articles/dvhs.

40. K. A. S. Immink, *Codes for Mass Data Storage Systems*, Shannon Found. Publishers, ISBN 90-74249-23-X, Venlo, The Netherlands, November 1999.

41. F. Sijstermans, and J. van der Meer, "CD-i Full-Motion Video on a Parallel Computer," *Commun. ACM.* Vol. 34, No. 4, pp. 82–91, April 1991.

42. http://www.tv-anytime.org

43. P. H. N. de With, M. Breeuwer, and P. A. M. van Grinsven, "Data Compression Systems for Home-Use Digital Video Recording," *IEEE J. Sel. Areas Commun.*, Vol. 10, No. 1, pp. 97–121, January 1992.

CHAPTER 21.4
TELEVISION BROADCAST RECEIVERS

Lee H. Hoke, Jr.

GENERAL CONSIDERATIONS

Television receivers are designed to receive signals in two VHF bands and one UHF band, and optionally a complement of the cable-TV channels, according to the United States and Canadian standards. The lower VHF band (channels 2 to 6) extends from 54 to 88 MHz in 6-MHz channels, with the exception of a gap between 72 and 76 MHz. The higher VHF band (channels 7 to 13) extends from 174 to 216 MHz in 6-MHz channels. The UHF channels are spaced 254 MHz above the highest VHF channels, comprising 56 6-MHz channels extending from 470 to 806 MHz. Cable channels extend continuously from 54 to approximately 1000 MHz, also with 6-MHz spacing. Figure 21.4.1[1] shows the CATV channelization plan adopted jointly by the NCTA and EIA in 1983 and revised in 1994. The television tuner is thus required to cover a frequency range of more than 15:1. TV tuners of past generations use separate units to cover the UHF and VHF bands. Current design practice includes the circuitry to receive all bands in a single unit.

The signal coverage of TV transmitters is generally limited to line-of-sight propagation, with coverage extending from 30 to 100 mi depending on antenna height and radiated power. The coverage area is divided into two classes of service, depending on the signal level. The service area labeled class A is intended to provide essentially noise-free service and specifies the signal levels shown.

Channels	Peak signal level, μV/m	Peak open-circuit antenna voltage, μV
Class A service		
2–6	2500	3500
7–13	3500	1800
14–69	5000	800
Class B service		
2–6	225	300
7–13	630	300
14–69	1600	250

For the limiting area of fringe service the signal levels are defined as shown for class B service. The typical level of the sound signal is from 3 to 20 dB below the picture level, due to radiated sound power and antenna gain. The block diagram of a monochrome TV receiver for analog signal reception is shown in Fig. 21.4.2.

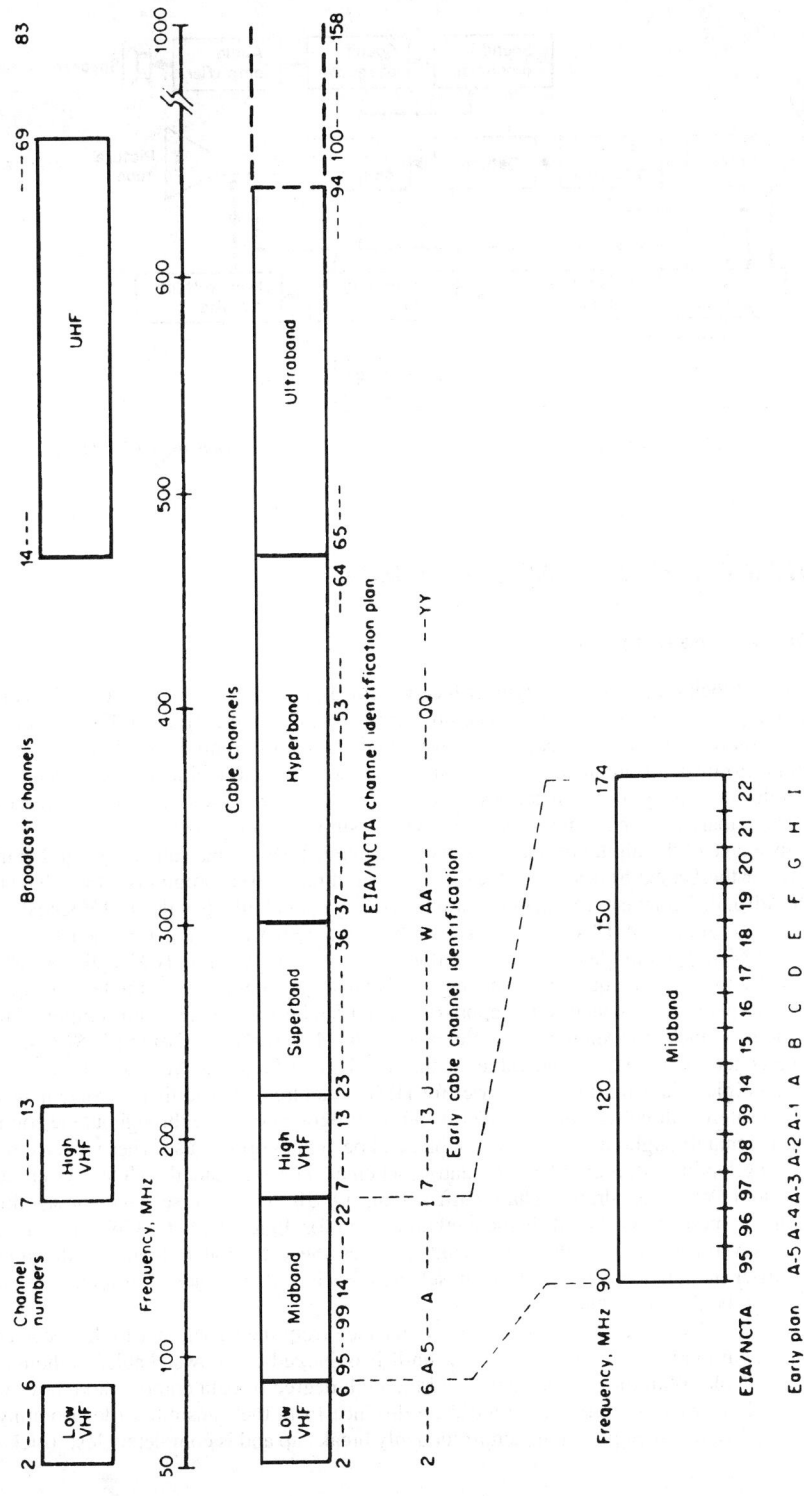

FIGURE 21.4.1 Frequency spectrum and channel designation for broadcast and cable television. *(From EIA IS-6. Revised Aug./1992)*

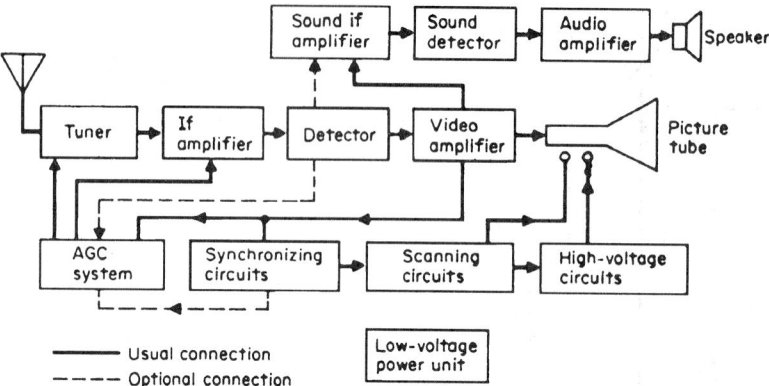

FIGURE 21.4.2   Fundamental block diagram, monochrome receiver.   (*From Benson & Whitaker, Ref. 2*)

RECEIVERS FOR DIGITAL TELEVISION TRANSMISSIONS

From Analog to Digital—the Contrast

When comparing a block diagram of the typical receiver for digital transmission to that of a current analog receiver, the first impression is the difference in complexity. A digital receiver (Fig. 21.4.3) contains more functional blocks, several of which differ greatly and have a high degree of complexity while still containing others that are similar to those of an analog set. The digital set contains more silicon ICs, as both memory and signal processor devices, many of which are custom designs at this date, a factor that may increase the set cost by $200 to $300 compared to a baseline NTSC set having a similar display size.

One basic advantage of the digital format is the increase in information that can be transmitted in a standard channel (6 MHz for the United States). This leads to packing three to four programs into a channel, or in the case of the U.S. Advanced Television System Committee (ATSC) Standard, a single HDTV signal or a group of up to four standard definition (SD) TV programs. Complexity here involves comparing the nearly 150,000 picture elements in an NTSC picture display to the 2 million picture elements of an HDTV display, an increase of 13 to 1. This equates to an RGB studio signal having 3×1080 active lines $\times$ 1920 samples per line $\times$ 8 bits per sample $\times$ 30 pictures per second, which equals approximately 1.5 Gb/s.[3] By using video compression, especially MPEG-2, this can be reduced to a more reasonable value of 20 Mb/s for transmission as a TV signal.

Many private concerns (broadcast and cable) within the United States as well as other countries are not interested in using digital transmission as a carrier for HDTV pictures. Instead their involvement with digital transmission is for satellite direct-to-home broadcast and for program coverage throughout the country, including a more robust signal throughout all areas for mobile and personal portable use. These diverse requirements have led to differing, optimized, digital RF modulation schemes. For example, the ATSC system uses a vestigial sideband system (VSB), quadrature phase-shift keying (QPSK) has been selected for satellite-to-home, quadrature-amplitude modulation (QAM) is the method chosen for digital cable (D-CATV), and for terrestrial area coverage in several countries outside the United States the orthogonal frequency division multiplex (OFDM) multicarrier modulation scheme has been selected. Each of these leads to differences in the receiver configuration as will be described later.

With current analog TV signal transmission, as the distance from the transmitter to the receiver increases, the picture becomes progressively noisier (snowy), until it is judged as "unwatchable," although the sound might still be acceptable. With the digital signals and the high degree of data compression, however, the picture remains crystal clear and noisefree until a certain distance from the transmitter (deteriorating signal-to-noise ratio by 1 to 2 dB) at which point the picture suddenly breaks up and is completely lost (brick wall effect,

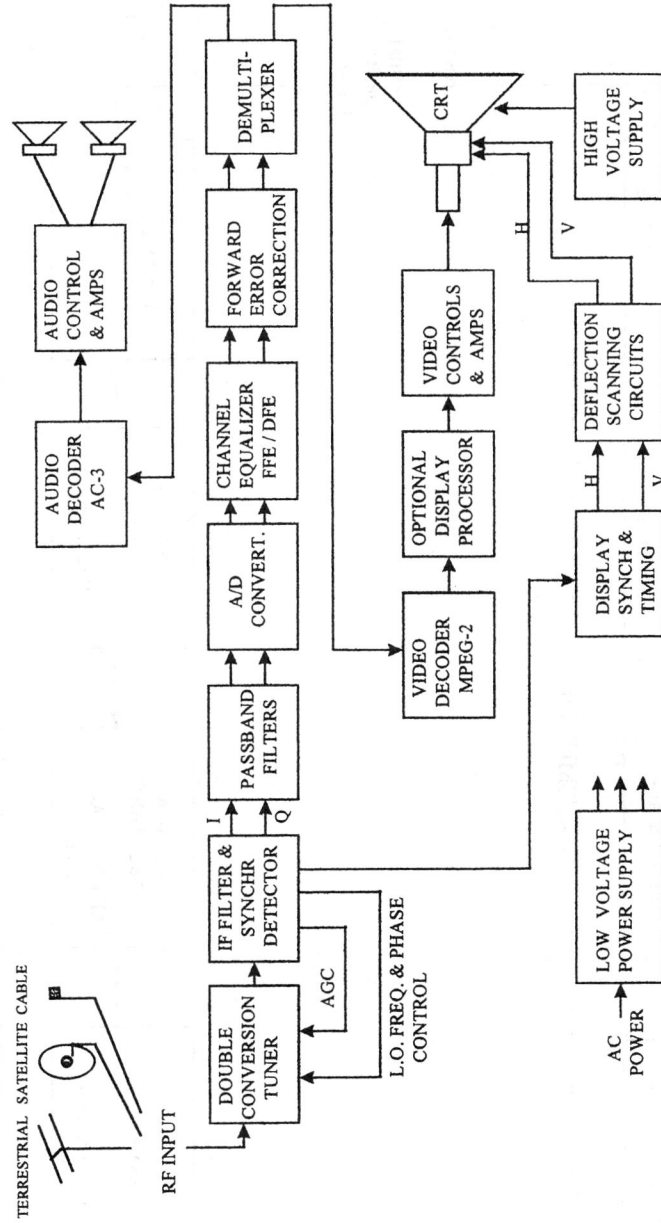

FIGURE 21.4.3 Block diagram of typical receiver for digital TV transmissions.

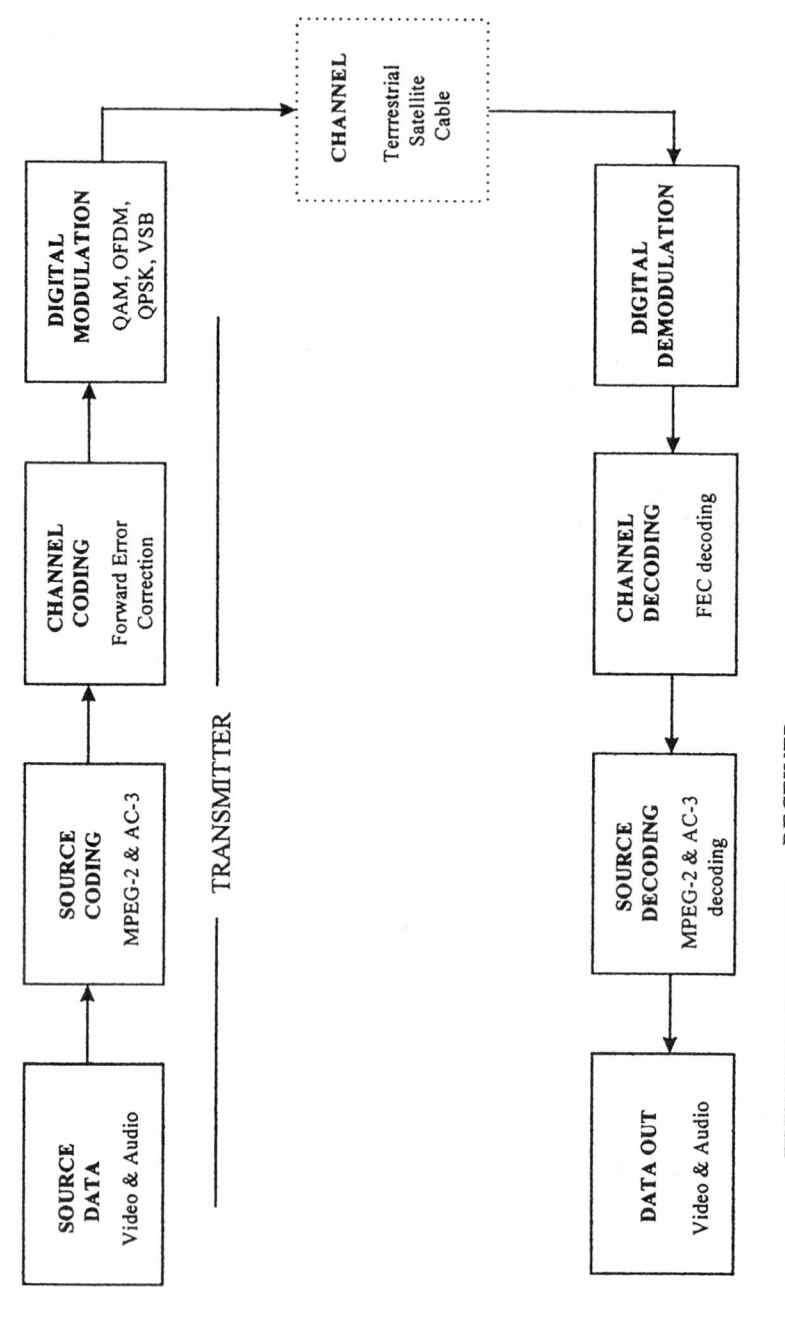

FIGURE 21.4.4 Basic elements of a digital TV system.

sometimes referred to as the waterfall effect) as compared to the gradual deterioration of an analog TV signal. A similar phenomenon can be observed in a satellite-to-home system. A storm cloud or leaves on trees can reduce the signal level just below the threshold at which point the received picture is lost completely, again a change of only 1 to 2 dB. In an effort to overcome the deficiencies of the transmission channel such as ghosts, noise, fades, and co-channel interference, the digital signal is encoded with forward error correction (FEC), often called channel coding, at the transmitter end. This necessitates complementary decoding at the receiving end, as shown in Fig. 21.4.4. The various types of encoding used with current digital TV systems are Reed-Solomon, Viterbi, trellis, interleaving, and convolutional, or a combination of these (concatenated coding system), depending on the robustness of the channel (cable, satellite or terrestrial).[4] For a cable system, this additional signal processing ensures high-quality service throughout the entire system with no signal deterioration at the extremities.

The final difference to be discussed is the mapping of the demodulated video signal onto the display means (CRT, LCD, and so forth). Typically, this will be one-to-one, that is, the display system will be designed to match the specification (scan lines and pixels or samples per line) of the decoded video. This is true except for the ATSC system where five distinctively different video formats have been allowed for transmission. In the receiver, these must be converted to the natural parameters of the display device (display field rate, interlaced or progressive display, lines per field, and samples per line). Details will be covered in a later section.

Set-top boxes and high-definition (HD) or digital-ready TV sets will be the mechanism that brings digital technology to the consumer for the next several years as the transition from analog to digital takes place. Currently within the United States three of the modulation techniques to be discussed later have become "standards" in a particular application, i.e., VSB for terrestrial, QAM for cable, and QPSK for direct-to-home satellite. Although the ability to design a TV set that can accommodate all three exists, the cost to the consumer would be prohibitive. To include a range of sets in the retail product line to handle each individual application might be cost effective, but would be prohibitive to the retailer as well as very confusing to the customer. In order to achieve flexibility needed by customers who have changed their TV delivery service frequently over the years, a set-top box that is unique to the service has become the answer. Not only does this solve the input signal demodulation problem in an economic way, the output of each box connects to a standard, existing NTSC receiver, thereby allowing the customer to use his old set with the new service. Typically, for cable and satellite, these boxes are rented from the local cable provider or the retail outlet that sells and installs the satellite system hardware. Recent set-top boxes from at least two manufacturers have included the dual function of being a terrestrial HDTV decoder as well as a satellite decoder for the DirecTV system. This eases the "Which box should I buy?" decision for the consumer. Signals available from these boxes include composite video (CVBS), S-video (Y/C) for standard TV receivers, and high-resolution component video (Y, P_b, P_r) and RGB to drive HDTV-ready monitor receivers.

Although new fully integrated digital HDTV sets are available on the market, their price is still considerably above the standard market level for a TV set. The most popular designs currently available are the "HD Ready TVs," which have upgraded display capability consisting of wider bandwidth video, up to 30 MHz, progressive scan, a higher-resolution CRT or projection display mechanism, and input connectors and wide-band circuitry which accepts progressive ($2f_H$) video signals or interlaced signals ($1f_H$) and convert them to progressive. A block diagram of an HD-ready receiver design is shown in Fig. 21.4.5.

QAM DIGITAL MODULATION

QAM digital modulation has been found to be advantageous for cable systems because of its simplicity, robustness, and its ability to deliver high-quality video signals in a 6-MHz channel over a hybrid fiber/coaxial network to subscriber homes where the signals are received via set-top boxes. The QAM signal can be considered to be a double-sideband suppressed carrier amplitude-modulated scheme. The input data bit stream is split into two independent data streams with one modulating the in-phase carrier while the other modulates the quadrature carrier component. Higher-order systems (M), e.g., $M = 16, 64,\ldots$, contain additional sets of carriers that are evenly phase spaced from the others.[5] Figure 21.4.6 shows a block diagram of a QAM receiver that can decode either the usual 64-QAM (20 Mbit/s) or the higher information density 256-QAM signal (40 Mbit/s).[6]

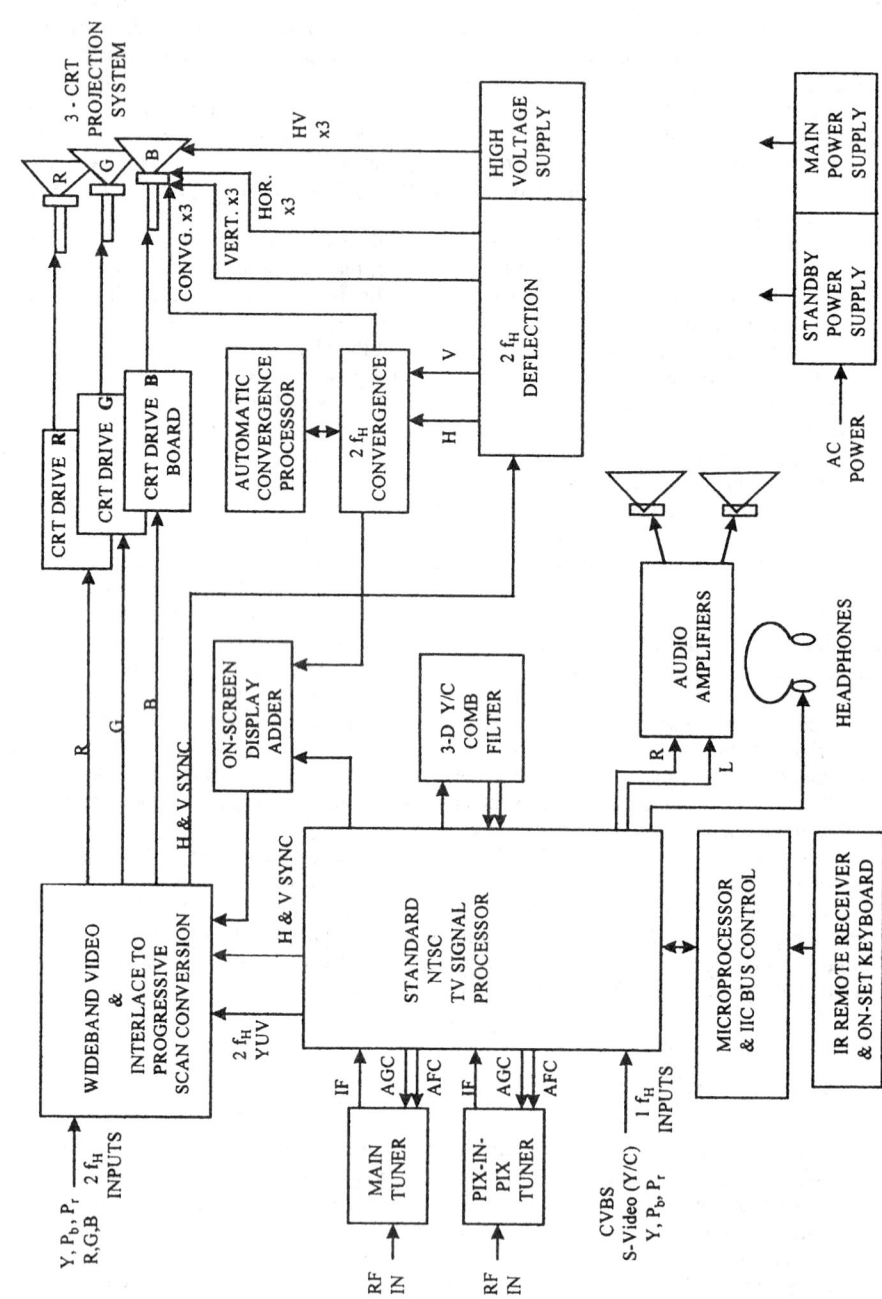

FIGURE 21.4.5 HD-ready TV receiver block diagram.

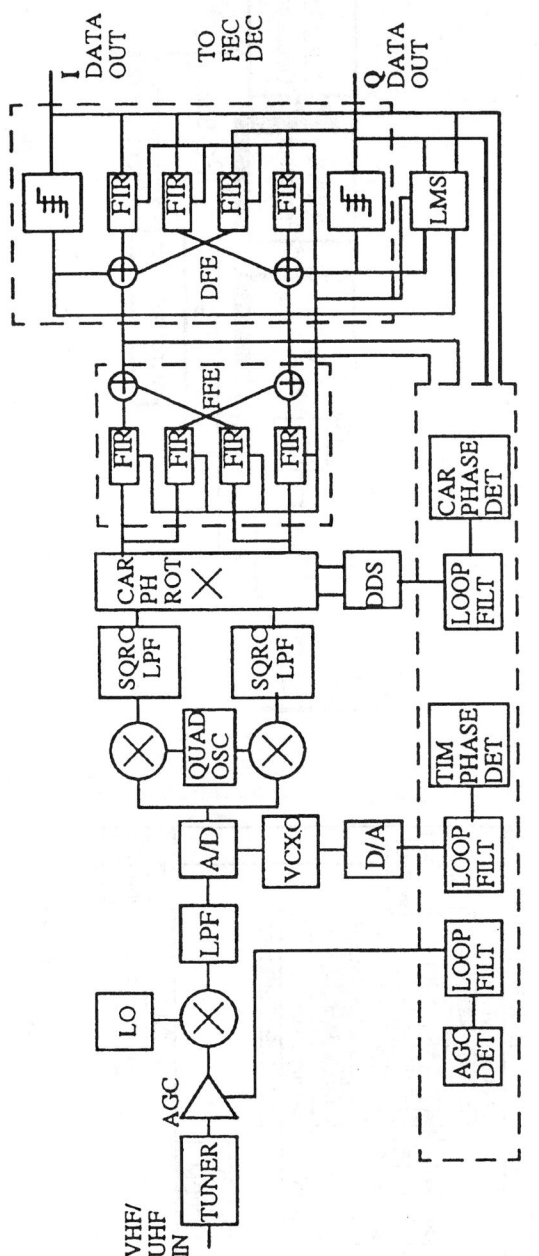

FIGURE 21.4.6 Block diagram of a QAM receiver. *(From Ref. 6)*

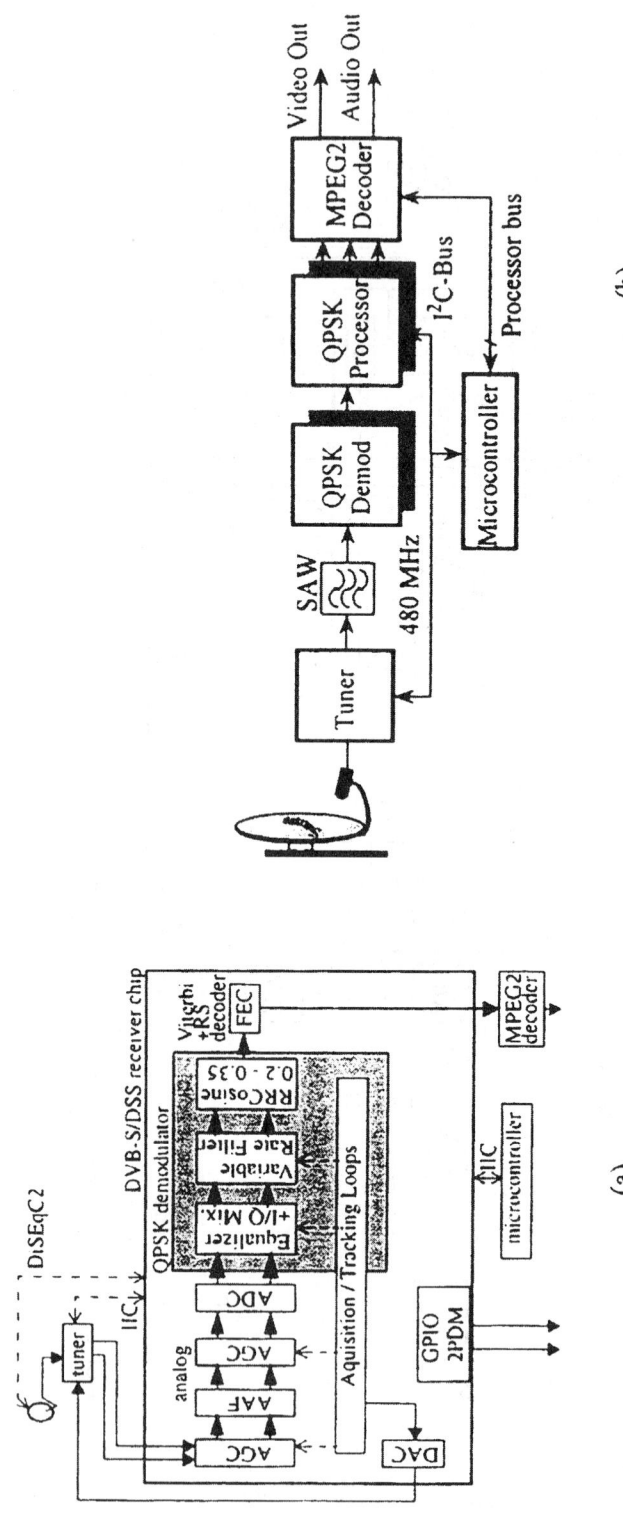

FIGURE 21.4.7 Receivers for QPSK transmission: (a) demodulation after A/D conversion; (b) demodulation prior to A/D conversion. *(From Refs. 9 and 10)*

The tuner is of conventional TV tuner design with somewhat tighter specs for local oscillator phase noise and hum modulation. An AGC amplifier maintains a constant signal level to the second mixer that transposes the IF signal to baseband where it is low-pass filtered to remove all frequencies that would cause aliasing in the analog-to-digital A/D converter. The signal is then demodulated into in-phase I and quadrature Q components and again low-pass filtered. Because free-running oscillators were used in earlier stages, a carrier phase rotation corrector stage is needed to realign the amplitudes of the I and Q components. This stage is part of a feedback loop that uses equalized I and Q as inputs. An adaptive equalizer consisting of a feed-forward section and a decision-feedback section removes amplitude and phase distortion caused by reflections and imperfections in the previous filters and the transmitter upconverter. These equalizer stages are made up of sections of programmable-tap FIR filter sections. Following equalization and symbol quantization, the forward error correction takes place. The data are then fed to MPEG-2 and AC-3 source decoders (not shown) for final processing. Two somewhat modified implementations of QAM demodulators for cable TV application are given in recent literature.[7,8] In the first, the QAM waveform is sampled at a low IF instead of baseband. This has led to simplified hardware implementation, whereby QAM decoding, including equalization, is achieved on one IC chip. The second design, also intended for use in a set-top cable box, describes two ICs; one performing as downconverter, containing the IF amplifier, local oscillator, quadrature demodulator, and AGC. The second IC contains antialias filters, A/D converters, digital I/Q demodulators, frequency and time domain equalizers, symbol-to-byte mapping, and Reed-Solomon forward error correction.

QPSK QUADRATURE-PHASE-SHIFT KEYING

QPSK quadrature-phase-shift keying is the accepted digital modulation for satellite-to-home application. This technique is used in direct broadcast satellite (DBS) systems in the United States (DirecTV and others), the European direct video broadcast (DVB) Eureka system, and the Japanese 8-PSK system. *M*-PSK is similar to *M*-QAM in that multiple phases of the carrier ($M = 4$ for QPSK, 8 for 8-PSK) are modulated by split bit streams. The modulation, however, is only phase, thereby leading to a constant amplitude RF signal. At the receiver, the process is similar to that for QAM, except that the decision leading to reconstruction of the transmitted bit stream is made only on phase information.[5] A receiver block diagram and circuitry, therefore, is very similar to that shown earlier for the QAM case. Circuits that convert the signal to a baseband digital signal before demodulating the QPSK signal as well as the alternate process of demodulating QPSK as an analog signal at IF, then doing the remaining signal processing, have been built.[9,10] In the former, extensive analog circuitry is used after the tuner to accomplish antialias band filtering, AGC and A/D conversion. A block diagram is shown in Fig. 21.4.7*a*. The block diagram of the second approach, Fig. 21.4.7*b*, is similar to that shown for a QAM receiver in Ref. 7, except that the QPSK demodulator consisting of a quadrature demodulator with $4f_{if}$ local oscillator is located in the first IC with outputs to the A/D converters being baseband analog I and Q signals. In both cases, the QPSK demodulation is followed by channel decoding, which includes filtering, deinterleaving and FEC, usually Reed-Solomon type. Synchronizing either to the carrier or to the recovered I and Q signals, Fig. 21.4.8, is also an important feedback loop that is contained in most receiver designs of this type.[11]

ORTHOGONAL FREQUENCY DIVISION MULTIPLEX

Orthogonal frequency division multiplex (*OFDM*) can be thought of as a multiple carrier version of QAM in which the individual carriers are equally spaced in frequency across the channel bandwidth. The input data stream is split into parallel blocks of symbols, each of which modulates a separate carrier. The carriers are then summed and transmitted. Owing the orthogonality of the carriers, the sampled output is effectively the inverse discrete Fourier transform of the input sequence. This parallel transmission or multiple carrier modulation (MCM) technique avoids several problems such as fading and impulse noise, which affect single carrier modulation (SCM) systems. At the receiver end, the signal is down-converted and sampled at the appropriate frequency, locked to the transmitted signal, then passed to the discrete Fourier transform demodulator where the

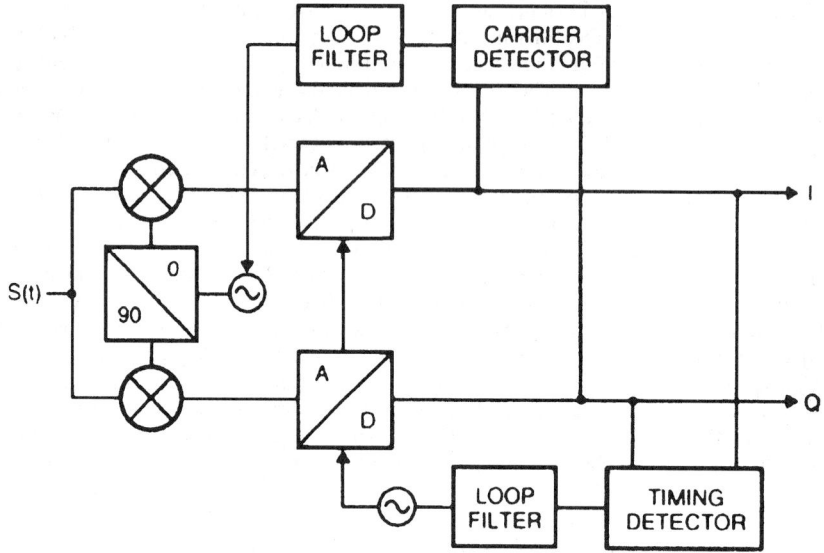

FIGURE 21.4.8   Demodulator concept having both carrier and clock recovery.   (*From Ref. 11*)

symbols are recovered.[5] Figure 21.4.9 shows a block diagram of the classical OFDM system for television, including transmitter and receiver. At the transmitter the processes that occur prior to the IFFT are done in the frequency domain, while those after the IFFT are in the time domain. At the receiver, the process is complementary with those processes ahead of the FFT being in the time domain and those after the FFT being in the frequency domain.[12]

OFDM modulation has been selected for digital terrestrial TV broadcast (dTTb) in Europe not only for fixed location reception but also for mobile and portable applications. In the United States, some factions have vigorously pushed for OFDM as opposed to the Grand Alliance vestigial-sideband system, which will be covered later in this section.

An OFDM receiver that follows the block diagram shown previously is shown in Fig. 21.4.10. This design features a single IC chip that contains both analog and digital circuits for implementing much of the OFDM demodulation. The analog part contains an antialiasing filter, AGC stage, and A/D converter, which outputs 8 bit resolution to the digital part of the IC. The digital part of the chip contains a signal detection unit that aids in start-up, channel change and provides AGC; an I/Q mixer for down-converting the signal to baseband; and an I/Q resampler where carrier and sampling clock frequency adjustments are made. The FFT unit can perform either 2k or 8k demodulation. The parallel symbols are then converted to serial bit stream and sent through the Viterbi plus Reed-Solomon error decoding. The output of the chip then consists of a bit stream that is sent to the MPEG-2 decoder for source decoding.

VESTIGIAL SIDE-BAND MODULATION

Eight-level vestigial side-band modulation (8-VSB) was proposed by the Grand Alliance in 1993. A testing phase was completed and the FCC accepted the system in 1995 as the standard for terrestrial digital television including both high and standard definition for the United States. The tests showed that the proposed system was robust to not only in-channel noise, ghosts, or reflections, but also the coexistent NTSC stations which would be broadcasting on the same and adjacent channels during the years of transition from analog to digital. In extensive field tests made in the Charlotte, N.C. area in 1994, the VSB system using 12 dB lower radiated signal than

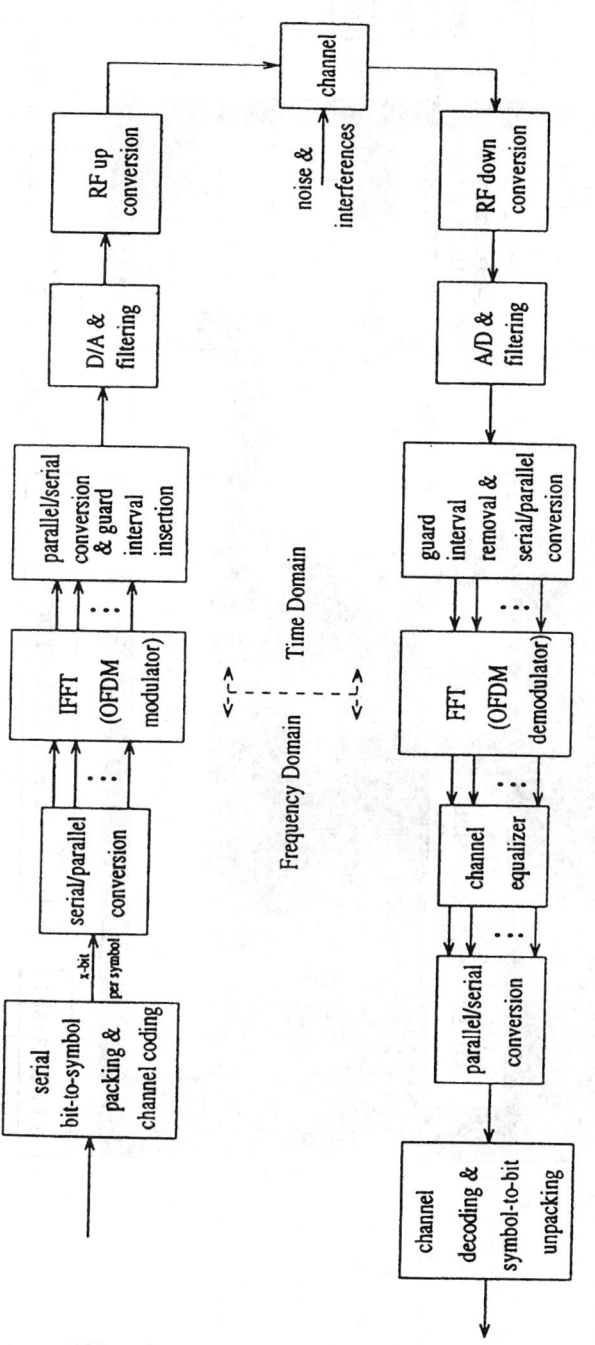

FIGURE 21.4.9 OFDM system diagram. *(From Ref. 12)*

21.119

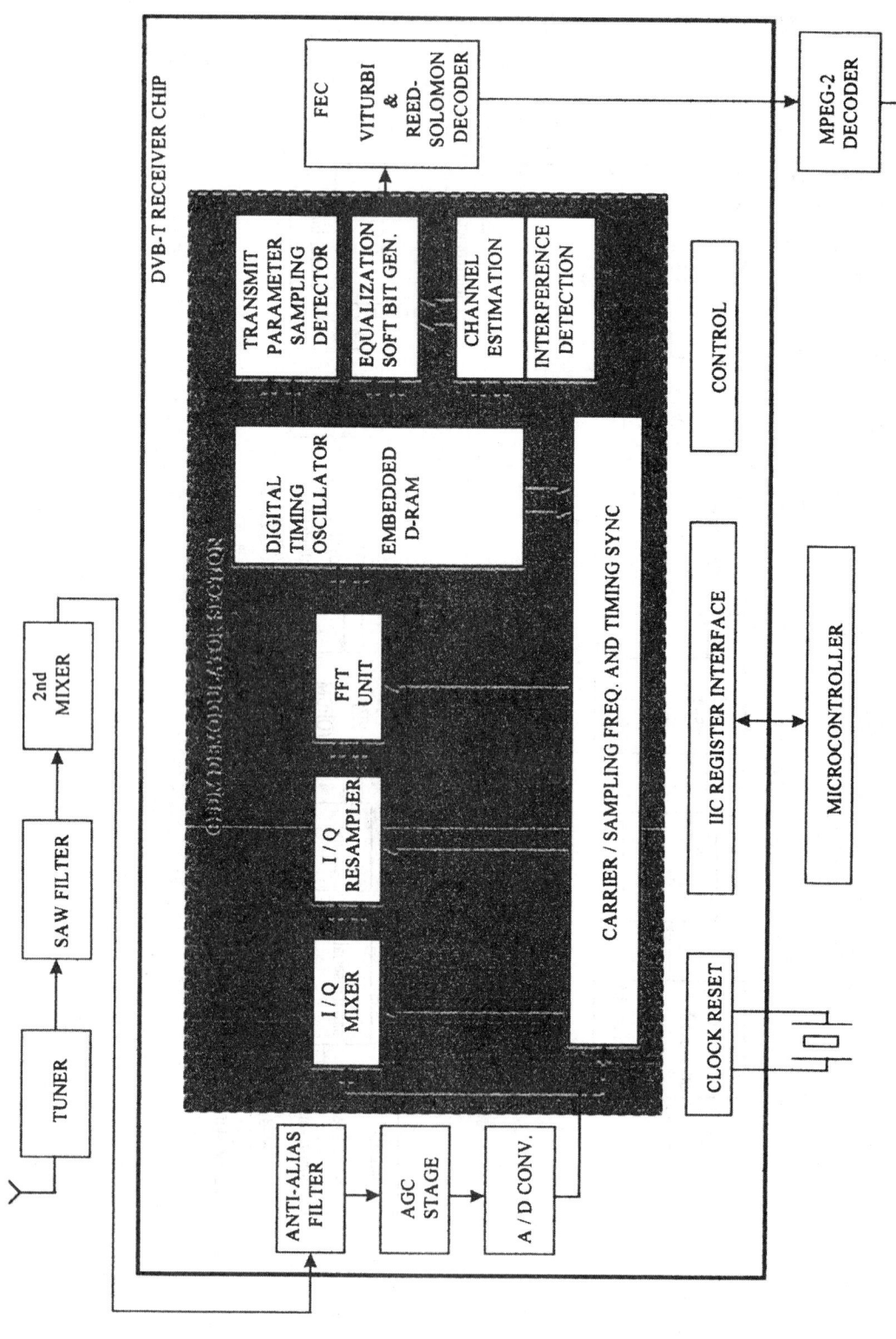

FIGURE 21.4.10 DVB-T OFDM receiver block diagram. *(Redrawn from Ref. 13)*

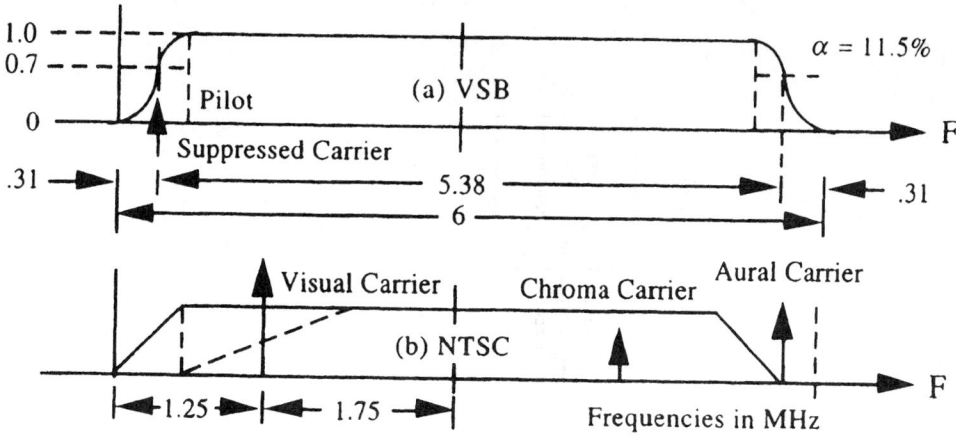

FIGURE 21.4.11   VSB and NTSC RF spectra.   (*From Ref. 14*)

the NTSC broadcast outperformed NTSC by a significant margin.[14] The actual calculated and measured margin at which picture or sound deterioration takes place in the presence of white noise favors 8-VSB over NTSC by 19 dB. The spectrum of VSB and NTSC are similar in that both completely fill the 6-MHz channel and that both use a carrier that is located near the lower channel edge. In VSB, however, the spectrum of the modulation within the channel is nearly flat and uniform, as compared to NTSC's chroma subcarrier and sidebands and the FM modulated sound carrier at the upper end of the channel. A comparison is shown in Fig. 21.4.11. Much testing has also been done on cable systems of a 16-VSB modulation system that has a throughout of nearly 40 Mbit/s, double that of the 8-VSB system.[15] The discussions between the Advanced Television Systems Committee and the cable industry to set one standard for HDTV transmission appear to be nearing the consensus stage.

To date, most TV manufacturers have developed and marketed digital TV sets having capability to decode the full HDTV 8-VSB standard. The blueprint for the system and prototype hardware has been reported in numerous technical publications.[14–17] A simple block diagram of the receiver is shown in Fig. 21.4.12.

A similarity can be seen between these major blocks and those shown for the digital TV receivers described earlier, especially the OFDM receiver. Trellis decoding by Viterbi means and Reed-Solomon FEC is a dominant part of each of the receivers. Each has some means to equalize the channel to correct for ghosts and bursts. The VSB system, however, uses only the I component for data recovery and therefore needs only a single A/D converter and channel equalizer instead of the two matched units used in other systems. While the other systems synchronize by using the demodulated data symbols and therefore need quadrature correction circuitry, the VSB system has three transmitted mechanisms for synchronizing the receiver to transmitter. The first is the pilot carrier located 0.31 MHz in from the lower band edge. A frequency/phase-locked loop (FPLL) in the receiver, Fig. 21.4.13, establishes synchronization to this carrier. A noncoherent AGC feedback adjusts the gain of the IF amplifier and tuner to bring the signal into range of the A/D converter. Repetitive data segment syncs consisting of four symbols per segment provide the second synchronizing means. These sync pulses are detected from among the synchronously detected random data by a narrow bandwidth filter. A feedback loop, Fig. 21.4.14*a*, then creates a properly phased 10.76 MHz symbol clock along with a coherent AGC control voltage that locks the proper demodulated signal level to the A/D converter. A third mechanism is to compare the received data field segment with a set of ideal segments (field 1 and field 2) contained in a frame sync recovery circuit within the receiver. This circuit is shown in Fig. 21.4.14*b*.

Following this, a circuit determines if there is a co-channel NTSC signal. If so, the NTSC interference rejection comb filter, a one-tap linear feed-forward filter that has nulls nearly corresponding to the frequencies of the NTSC video carrier, chroma subcarrier, and aural carrier, is switched in Fig. 21.4.15. This filter degrades white noise performance by 3 dB and will not be included in receivers after all NTSC transmitters are silent.

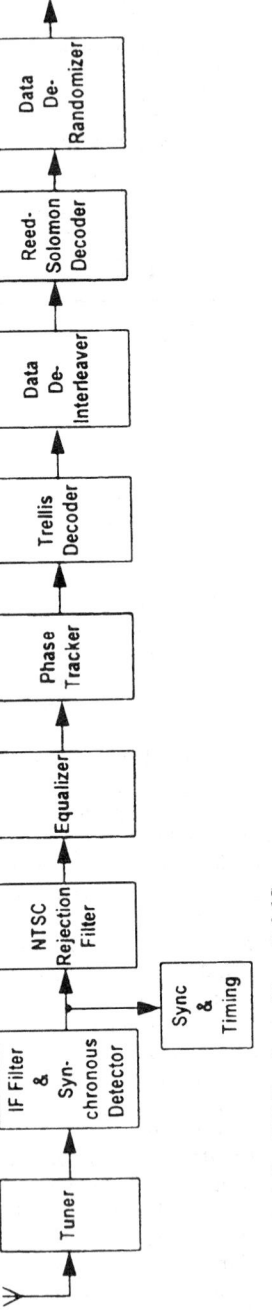

FIGURE 21.4.12 VSB Receiver. *(From Ref. 15)*

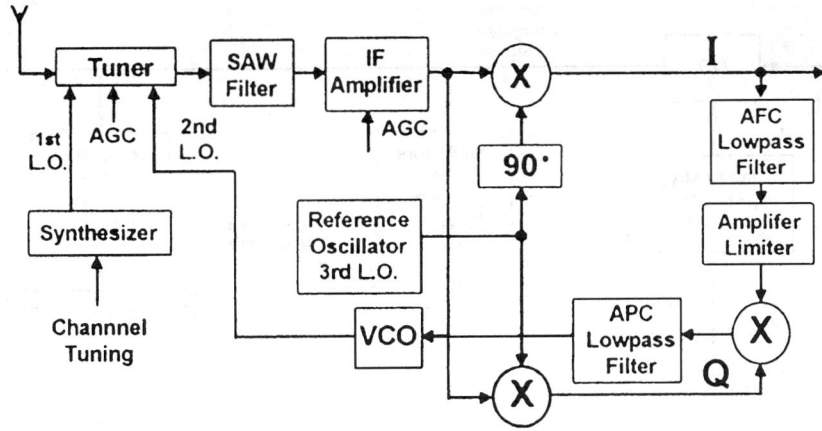

FIGURE 21.4.13   Tuner—IF frequency/phase lock loop.   (*From Ref. 15*)

Following the NTSC interference rejection filter is a channel equalizer that compensates for linear channel distortions such as tilt and ghosts. The prototype system used a 64 tap feed-forward transversal filter followed by a 192 tap decision feedback filter. A least-mean-square algorithm was used to compare the transmitted training signal, pseudo-noise sequences which are a part of the data filed sync, to a stored image of the training signal with the error being feed-back to set the tap coefficients. Once equalization is achieved at this level, the circuit can lock on to either data symbols throughout the frame or the data itself for further fine-tuning of the ghost canceling. Airplane flutter is usually too rapid for a full tap evaluation and is therefore handled by the latter technique. A block diagram of the equalizer is shown in Fig. 21.4.16.

The next block in the receiver chain is a phase tracking loop, Fig. 21.4.17, which tracks out phase noise that had not been removed by the Tuner-IF PLL operating on the pilot carrier. This circuit consists of a digital filter that constructs a Q signal from the existing I signal. These signals are then used to control a complex multiplier or phase derotator. It has been reported that the 8-VSB receiver system consisting of the front-end FPLL and the phase-tracking circuit can compensate for phase errors up to –77 dBc/Hz at a 20 kHz offset from the carrier.

The next block provides deinterleaving of the 12 symbol intersegment code interleaving which was applied in the transmitter. At the same time, trellis decoding takes place in a structure shown in Fig. 21.4.18. Here, one trellis decoder is provided for each branch although in more recent designs, a single trellis decoder is used in a time-multiplexed fashion to reduce IC complexity.[18,19] Following trellis decoding, Reed-Solomon decoding takes place. At this point, channel decoding is complete and the data are ready to be split into the appropriate audio and video packets and sent to the source-decoding circuitry.

The receiver-to-transmitter lock-up and signal-decoding process takes place in the following sequence:[16]

1. Tuner first local oscillator synthesizer acquisition

2. Noncoherent AGC reduces unlocked signal to within A/D range

3. Carrier acquisition (FPLL)

4. Data segment sync and clock acquisition

5. Coherent AGC of signal (IF and RF gains properly set)

6. Data field sync acquisition

7. NTSC rejection filter insertion decision made

8. Equalizer completes tap adjustment algorithm

9. Trellis and RS data decoding begins

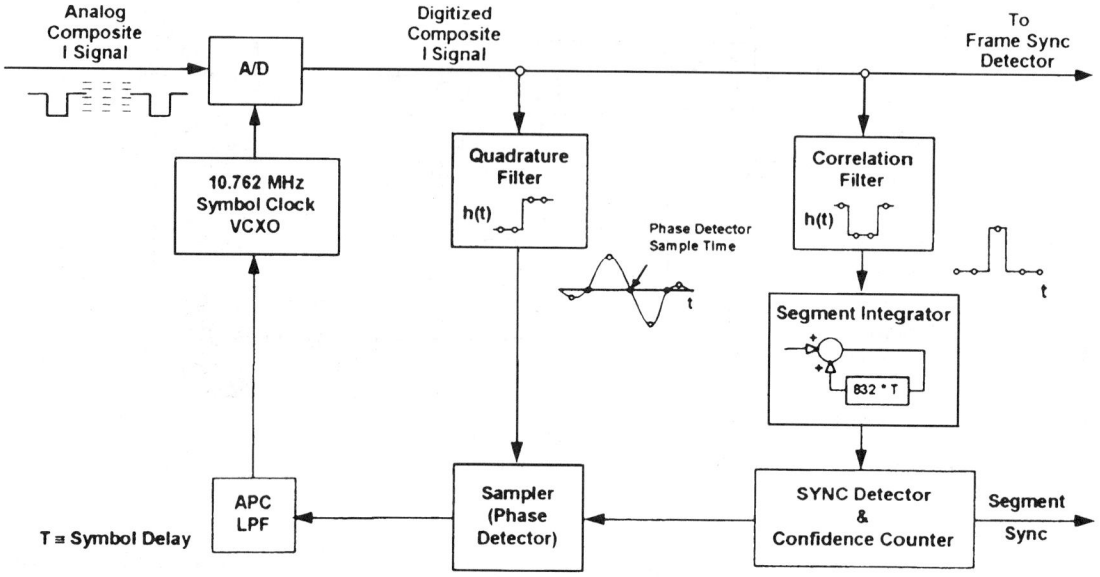

a. Segment sync and symbol clock recovery.

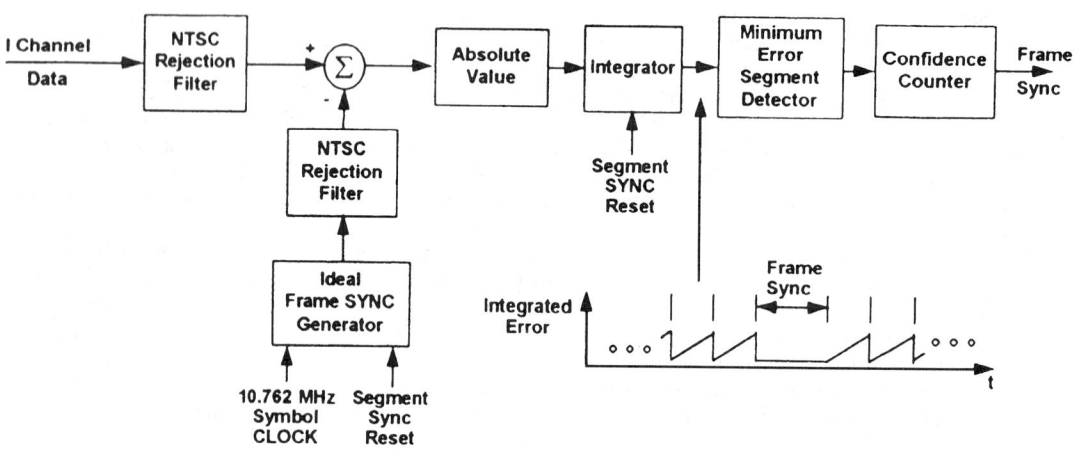

b. Data frame sync detection.

FIGURE 21.4.14   (a) Segment sync and symbol clock recovery; (b) data frame sync recovery.   (*From Ref. 14*)

SOURCE DECODING

Source decoding consists of decoding the AC-3 audio and the MPEG-2 video which had been encoded at the transmitter using the main profile at high level (MP@HL) specification.[3,16,20] A block diagram of an MPEG-2 decoder is shown in Fig. 21.4.19. Here video frames are created from the compressed packet data and stored

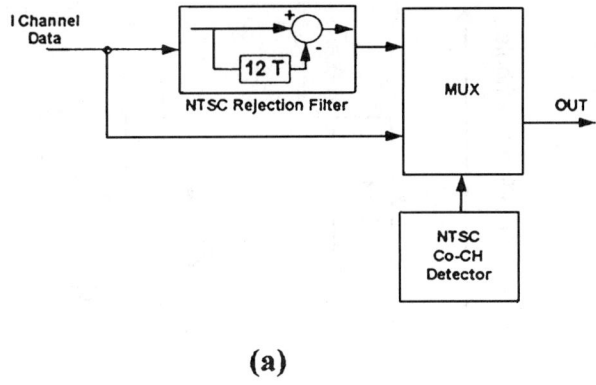

(a)

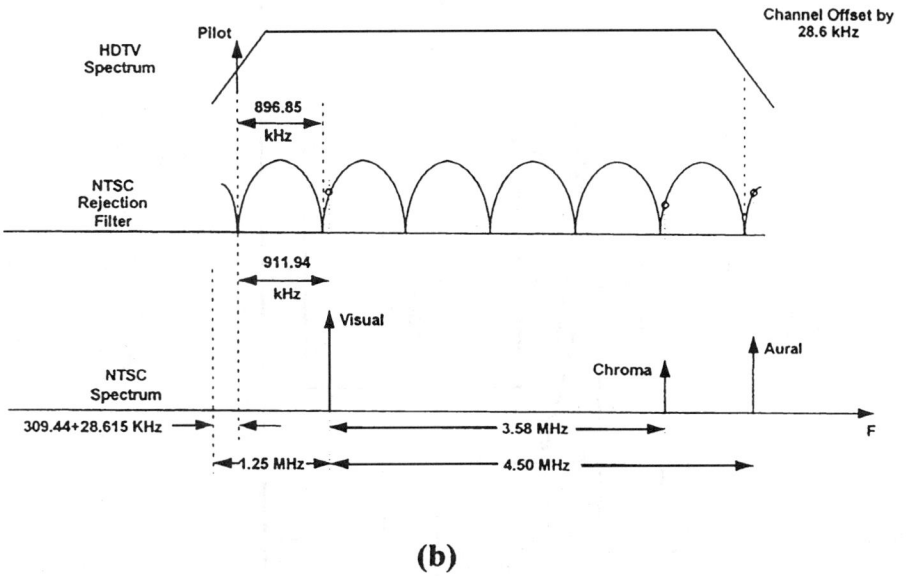

(b)

FIGURE 21.4.15   (a) NTSC interference rejection filter; (b) comb filter spectrum.   (*From Ref. 14*)

in frame memory. The decoded video is then read out and passed on to the display circuitry of the receiver in whatever format is required by that display.

The final piece of the TV receiver system is the display. The drive requirements of the display do not necessarily match the format of the decoded video signal. In fact, the ATSC Standard permits the transmission of any one of numerous video formats (Table 21.4.1).[16]

The requirement for this section of the receiver, therefore, is to scan convert or format convert the decoded video from the MPEG-2 decoder into the form needed by the display. Typically, this is accomplished by loading the video into RAM-type memory (frame buffer) and clocking the video out at a rate and in a format needed for the display (pixels per line, lines per frame/field, progressive or interlaced fields).[21] In the case where there is not a 1:1 match between the total number of video pixels in a frame and the display pixels, e.g., displaying an HDTV

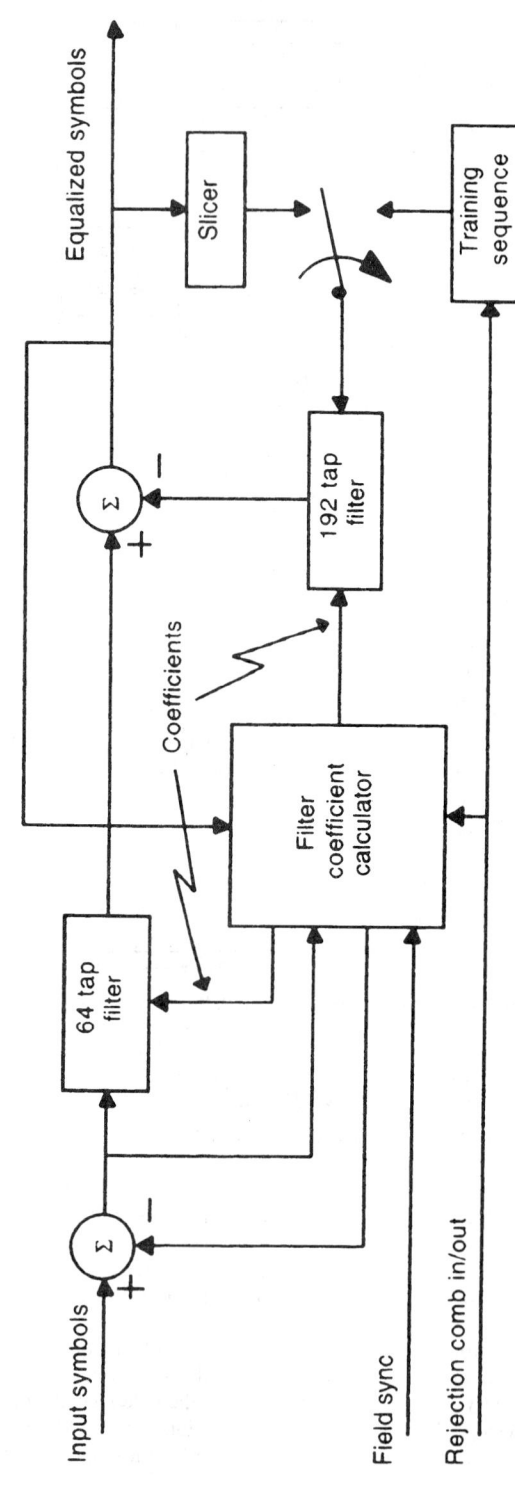

FIGURE 21.4.16 VSB receiver equalizer. *(From Ref. 3)*

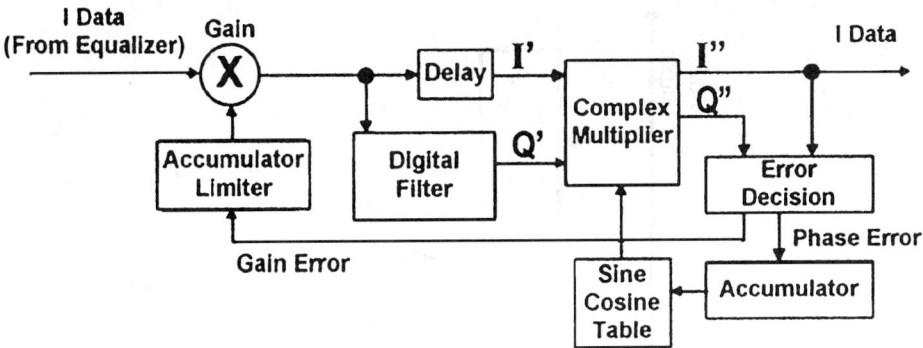

FIGURE 21.4.17   Phase-tracking loop block diagram.   (*From Ref. 15*)

signal on a standard NTSC-type SDTV display, an intermediate step of data interpolation, frame storage, and smoothing is needed (Fig. 21.4.20). Often, noise reduction and motion compensation is included in this step. In some newer designs a considerable saving in memory requirements can be achieved when the downconversion of the video information is accomplished within the MPEG-2 decoding process[22,23] (Fig. 21.4.21).

The video signals in the display section are usually in the Y/C format, then converted to Y, P_b, P_r format and finally to analog R, G, B format, especially if driving a direct-view CRT or CRT projection display. The parameters of the final video signals for several of the more popular ATSC display formats of Table 21.4.1 is shown in Table 21.4.2.

Figure 21.4.22 gives a comparison of bandwidth requirements for various values of horizontal picture resolution. It is interesting to note that although the video channel bandwidth requirements are identical for the

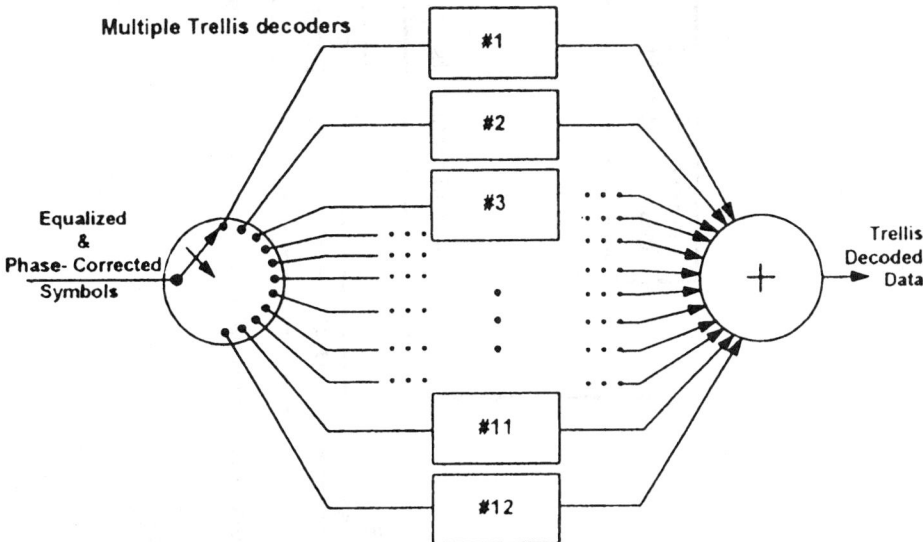

FIGURE 21.4.18   Intersymbol code de-interleaver and trellis decoders.   (*From Ref. 15*)

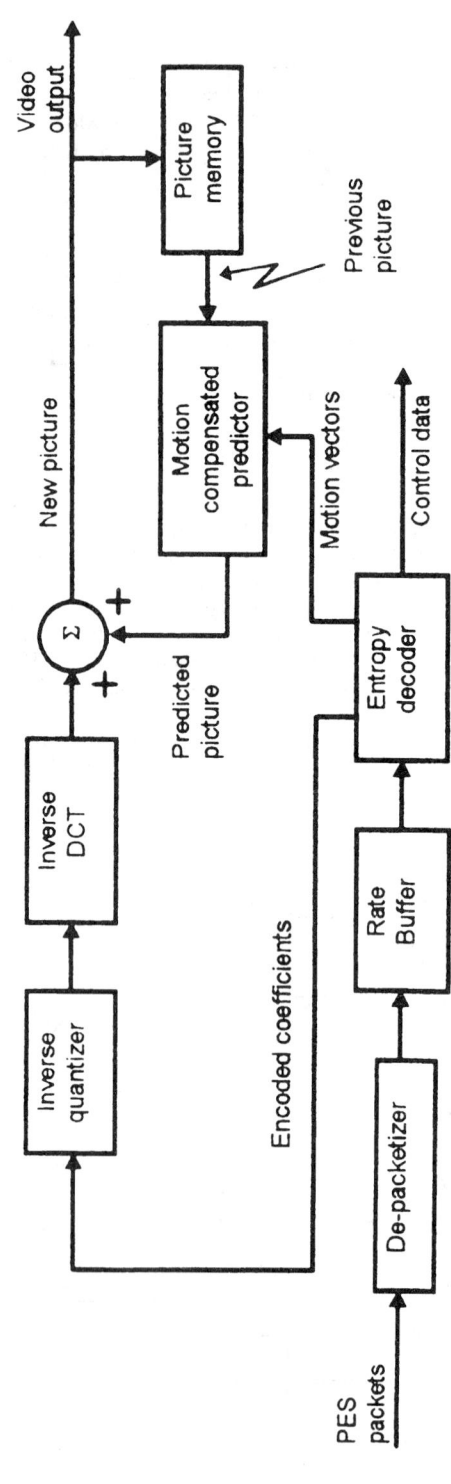

FIGURE 21.4.19 MPEG-2 video decoder block diagram. *(From Ref. 3)*

TABLE 21.4.1   ATSC Digital Television Standard Video Formats
(From Ref. 16)

Vertical lines	Pixels	Aspect ratio	Picture rate
1080	1920	16:9	60I, 30P, 24P
720	1280	16:9	60P, 30P, 24P
480	704	16:9 and 4:3	60P, 60I, 30P, 24P
480	640	4:3	60P, 60I, 30P, 24P

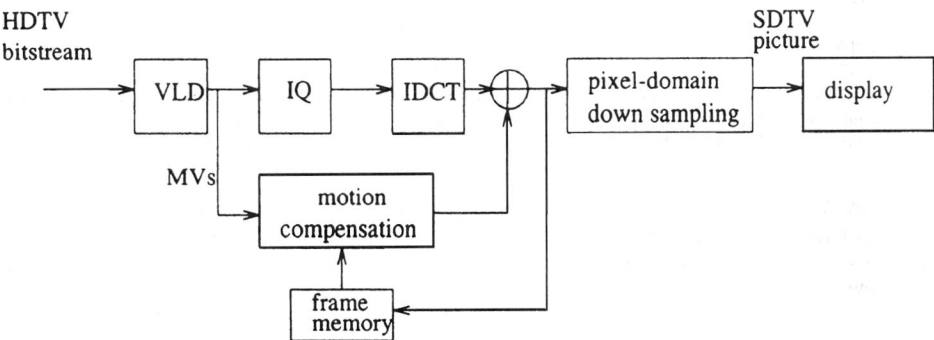

FIGURE 21.4.20   The traditional down-conversion method in the pixel domain.   (*From Ref. 22*)

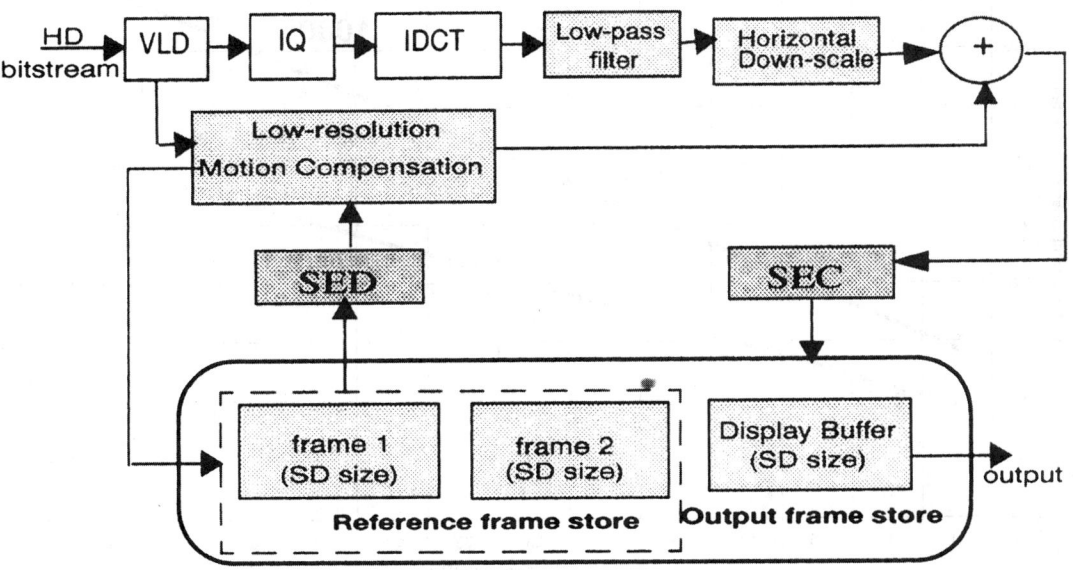

FIGURE 21.4.21   HD to SD low cost decoder.   (*From Ref. 23*)

TABLE 21.4.2   Picture Parameters, Video Bandwidth Requirements and Scan Frequencies for Several Display Systems

System	Type	Active picture elements		Total elements	Rate per second	Video bandwidth[*] (MHz)	Horizontal frequency (kHz)
HDTV	1080i	H pixels	1920	2200			
		V lines	1080	1125	30 frames (60 fields)	37.125	33.57
HDTV	720p	H pixels	1280	1650			
		V lines	720	750	60 frames	37.125	45
NTSC (16:9)	525i	H pixels	704	858			
(SDTV)		V lines	480	525	30 frames (60 fields)	13.5	15.75
VGA (16:9)[†]	480p	H pixels	704	858			
		V lines	480	525	60 frames	27	31.5

[*]Similar to ITU-R BT.601-4 which uses only 704 of 720 pixels and only 480 of 483 lines.
[†]Video bandwidth values are for the Nyquist criterion.

1080i and 720p systems, the picture resolution of the two systems differs by a factor of 1.5. Discussion of how much resolution is really necessary to sell HDTV to the public has been going on for the past 10 years, and may continue. Figure 21.4.23 shows the results of a study made on several integrated HD and HD-Ready TV sets now available in the marketplace. These sets were driven at the video inputs with a 1080i signal and each

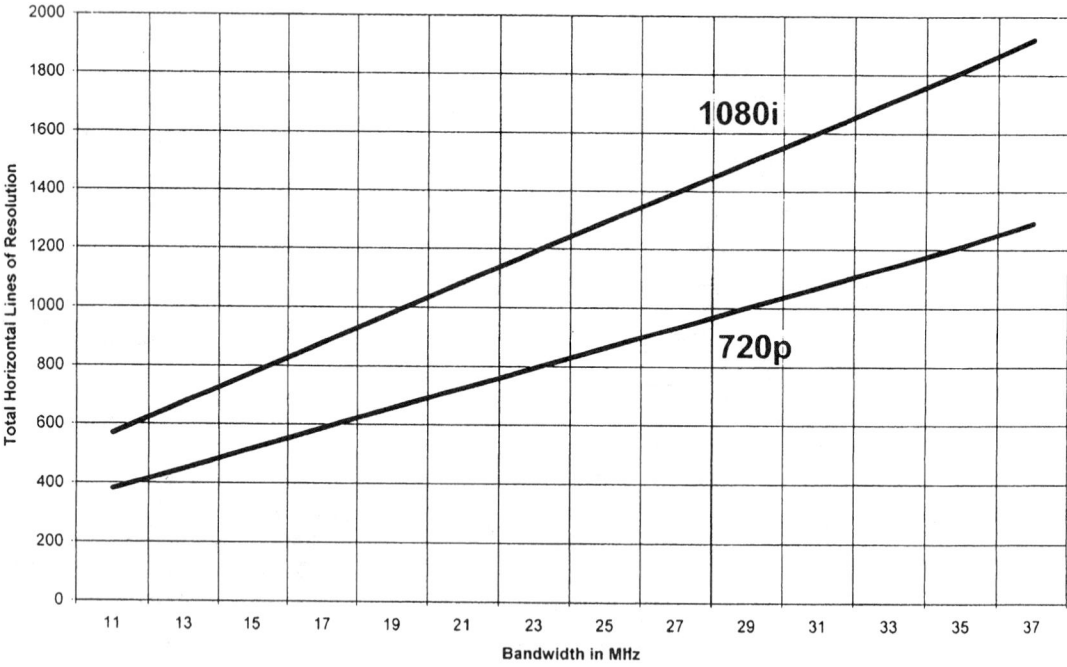

FIGURE 21.4.22   High-definition picture resolution vs. video bandwidth.

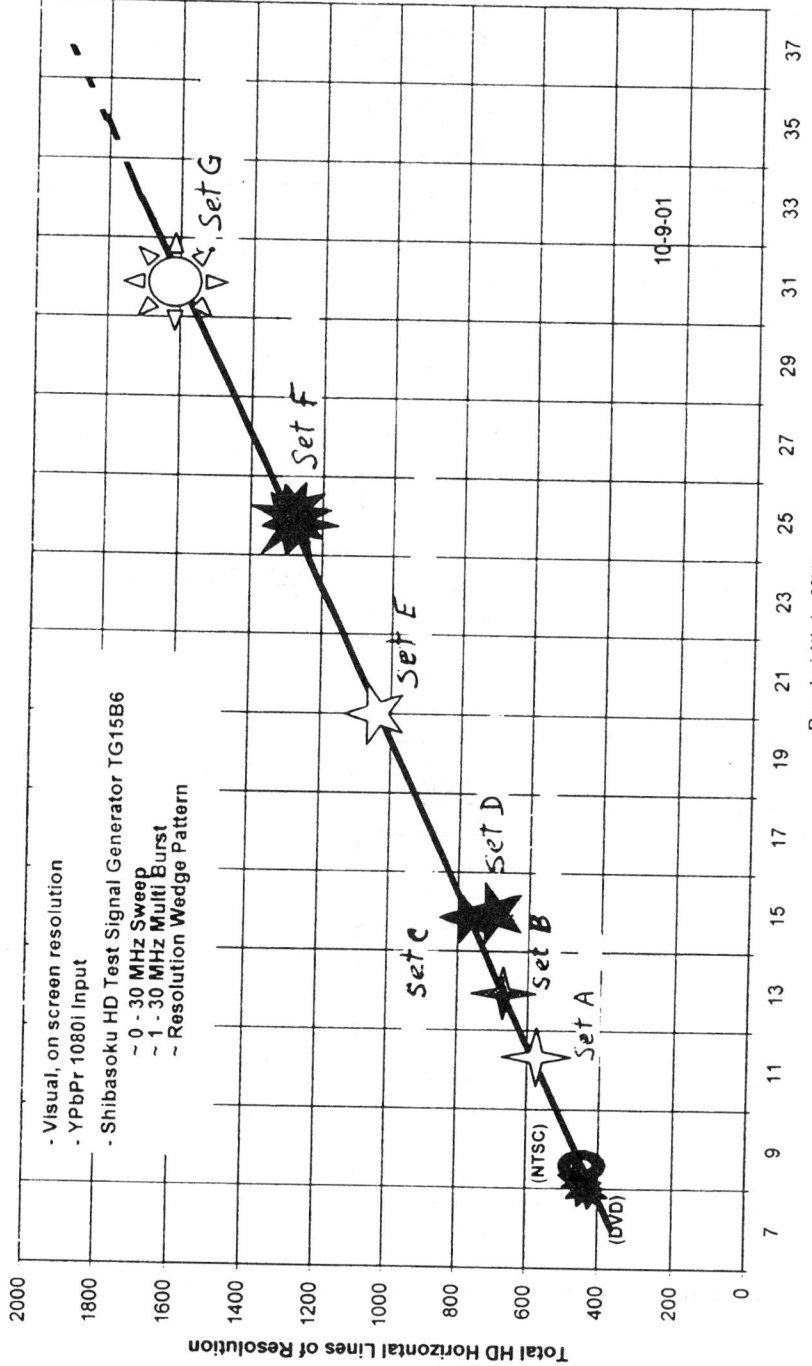

FIGURE 21.4.23 Picture resolution performance of several HD and HD-Ready TV sets.

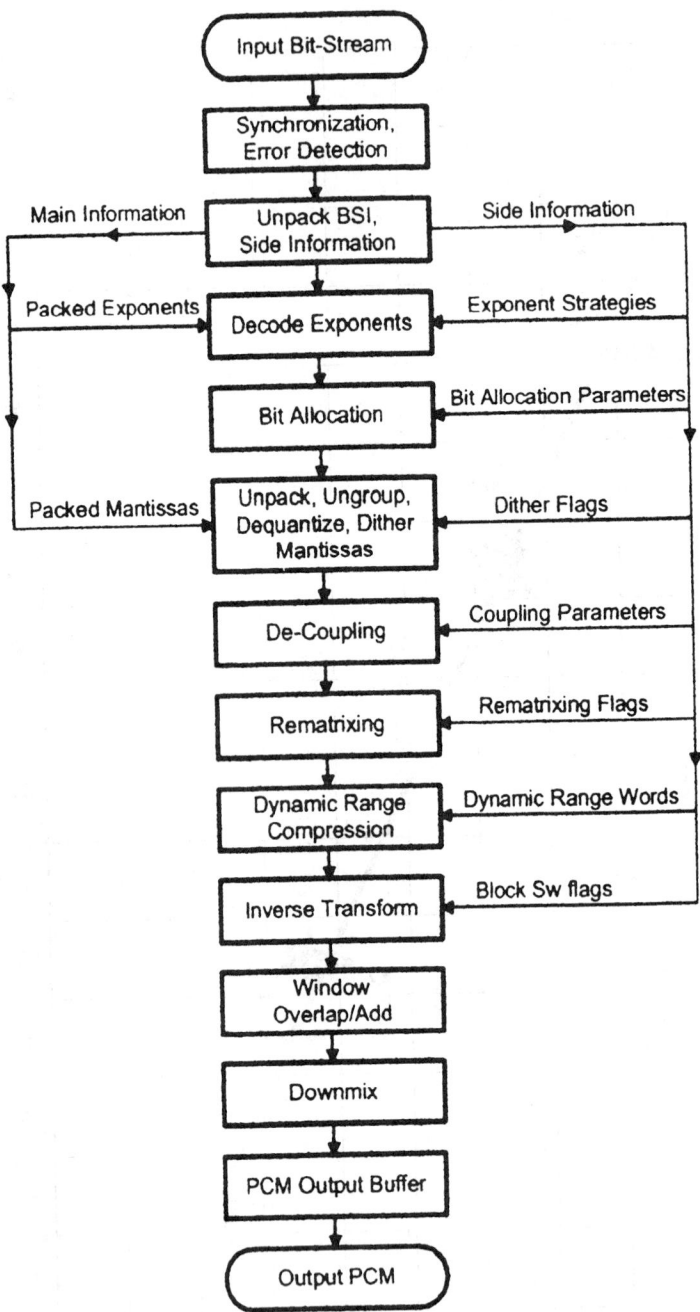

FIGURE 21.4.24   Flow diagram of the AC-3 decoding process.   (*From Ref. 25*)

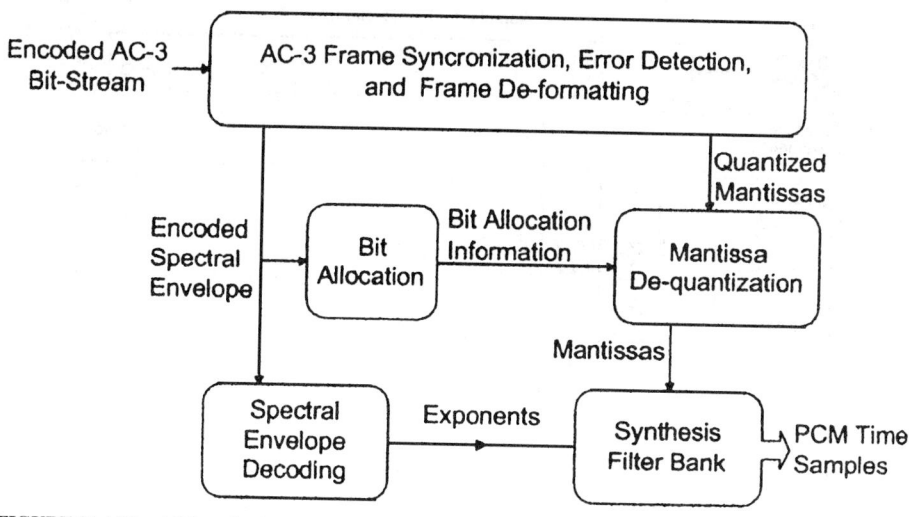

FIGURE 21.4.25   AC-3 audio decoder.   *(From Ref. 25)*

either displayed full resolution or down-converted to the native scan/video capability of the receiver. The major diagonal line on the chart represents the 1080i locus with full HD resolution at one end and NTSC at the other. In the case of several receivers, severe aliasing of the multiburst pattern caused the observed picture to be judged to have a lower frequency value than the design intent. Second generation designs will most likely correct these deficiencies.

The audio that had been encoded at the transmitter using the AC-3 specifications is decoded into one to six audio channels, all at line level (left, center, right; left surround, right surround, and low-frequency enhancement sub woofer[25]). Since the low-frequency enhancement channel has an upper bandwidth limit of 120 Hz, it is usually not counted as a complete channel, but only as 0.1 channel, leading to the designation of AC-3 as having 5.1 channels. It is not necessary for a receiver to decode all the available channels. A monophonic receiver will need to provide only one output audio signal. In this case, the receiver's decoder will down-mix the six channels into one. A popular decoder design will provide a down-mix of six into two audio outputs for use in lower cost stereo TV sets.

The AC-3 bit stream is composed of frames each containing sync, encoding/decoding information, and six blocks of audio data. These frames are decoded using the process shown in Figs. 21.4.24 and 21.4.25. Each frame can be decoded as a series of nested loops in which each channel can be handled independently. This process can be accomplished in an audio DSP IC. One such implementation for the full 5.1 channel output uses 6.6K RAM, 5.4K ROM, and 27.3 MIPS. A lower-cost 2 channel implementation requires the same amount of ROM and nearly the same MIPS, but only 3.1K RAM.[24] PCM to analog (D/A) converters, amplifiers, level and balance controls, and speakers complete the audio system.

DISPLAYS

Liquid-Crystal Displays (LCDs)

Use of both monochrome and color LCDs has become popular, especially in small personal portable television receivers. The operation of these devices is not limited by the high-voltage requirements of conventional CRTs. Instead, the picture raster is constructed of a rectangular MOS switching matrix of from 240 to 600 horizontal elements and from 200 to 400 vertical elements.[26] The gates of all the thin-film transistors (TFTs) in a given

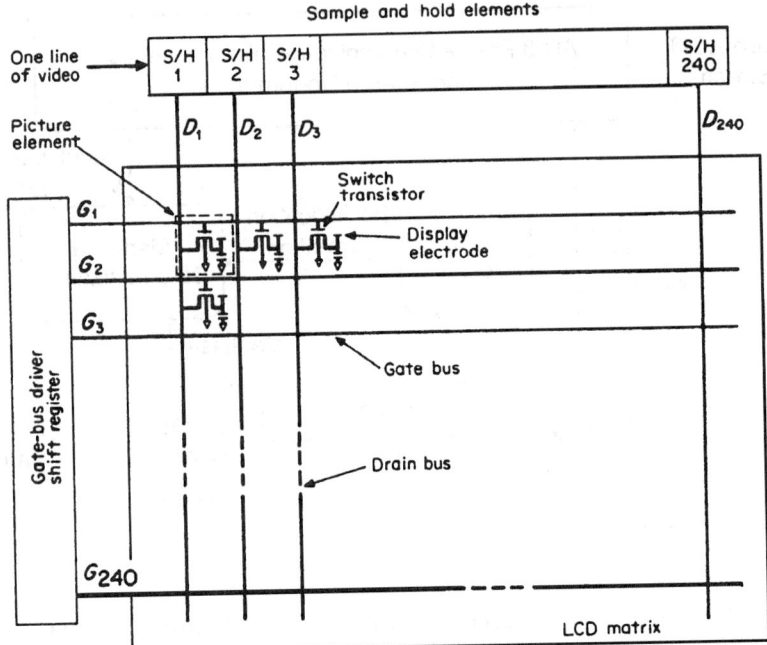

FIGURE 21.4.26 LCD television picture display. (*From Benson & Whitaker, Ref. 2*)

horizontal row are connected to a common bus (Fig. 21.4.26). Likewise the drains of all transistors in a vertical column are connected to a common bus. Vertical scan (row addressing) is produced by sequentially driving the gate buses from the shift register. Horizontal scan, which contains the video information (column addressing), is somewhat more difficult because of the stray capacitance and cross-under resistance associated with drain bus. A given line of video is broken into the same number of pieces as there are pixels in the horizontal row, and stored in the sample-and-hold (S/H) stages, which all drive their respective drain bus lines simultaneously, thus creating a line sequential display. The information on a drain is, therefore, changed only once for each horizontal period (63.5 μs).

A color LCD contains a repeating sequence of red, green, and blue filters covering adjacent pixels of a horizontal row.[27] The sequence is offset by one pixel in adjacent rows. The video and chroma signals are decoded and matrixed in the conventional manner. The *R-G-B* signals are then clocked into the line S/H stages in the appropriate sequence.

LARGE-SCREEN PROJECTION SYSTEMS

The display for picture sizes of up to 36 in. diagonal usually consists of a direct-view CRT. For pictures above 36 in., a relatively new technology has emerged and a number of projection technologies have become popular for domestic use.

Plasma display panel (PDP) systems are a relatively new type of direct view display. Typical sizes have been 40 in. with 4:3 length-to-height ratio, 42 and 60 in. diagonal with a 16:9 length-to-height ratio. The major advantage to a PDP is that it has a depth of only a few inches. This product has been touted as the "picture on the wall." The structure consists of two pieces of glass separated by an insulating structure in the form of small pockets, cells, or stripes. Each cell is filled with an ionizing gas such as neon and xenon. On the facing sides of the glass plates are metal electrodes, vertical (column) on one and horizontal (row) on the other (Fig. 21.4.27).

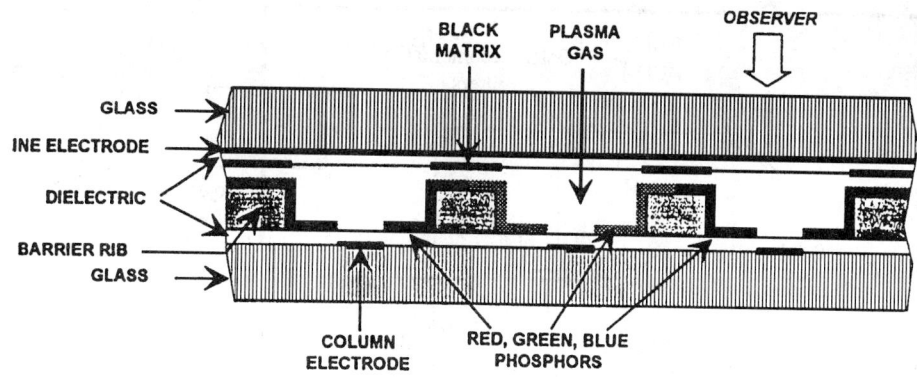

FIGURE 24.4.27   Cross-section of one pixel of an AC plasma display panel.   (From Ref. 28)

When a voltage of several hundred volts is applied to a given row and column electrodes, the gas in the appropriate cell ionizes giving off ultraviolet light, thus exciting the color phosphor deposited on the glass plate. Since a cell is either on or off, pulse modulation is used to obtain a shade of gray or a desaturated color. This is accomplished in the driving circuitry by slicing each video field into 8 to 10 subfields and then driving all cells at the subfield rate. The 60-in. unit possesses a resolution of 1280×720 (HDTV quality) and has a 500-to -1 contrast ratio in a dark room. The 42-in. units have 852×480 or 1024×1024 pixel counts. Light output over 500 cd/m^2 has been measured. Although this technology has been shown to produce bright, outstanding pictures, the consumer price is still somewhat higher than any of the other display systems at this time.

   CRT projection systems consisting of three cathode-ray tubes, with a typical raster diagonal of 3 to 5 in., produce three rasters in red, green, and blue, being driven by the respective *R*, *G*, and *B* drive signals. These images are projected through three wide-aperture lenses to either a highly reflective screen having a diagonal dimension of 50 in. to 8 or 10 ft (front projection) or to the back of a diffused translucent screen having diagonal dimension of 40 to 70 in. (rear projection) (Fig. 21.4.28). By careful adjustment of the deflection and orientation of the CRTs and lenses, the rasters are brought into precise registry and geometric congruence.[29,30] Various surfaces (typically two or four) of the rear projection screen are impressed with patterns of fine-pitch grooves that form lens elements to "focus" or control the direction of the light as it leaves the screen. Medium screen gains (three to six) are preferred and can be designed for more uniform image brightness over a wider viewing angle. Currently, brightness levels of 600 cd/m^2 can be achieved with rear projection systems having a picture size of 35 to 50 in. measured diagonally. Since the projection system has no shadow mask in its electrooptical path, the system can achieve better resolution than a conventional large-screen direct-view color CRT. At this time, CRT projection is the preferred display system for large screen HDTV.

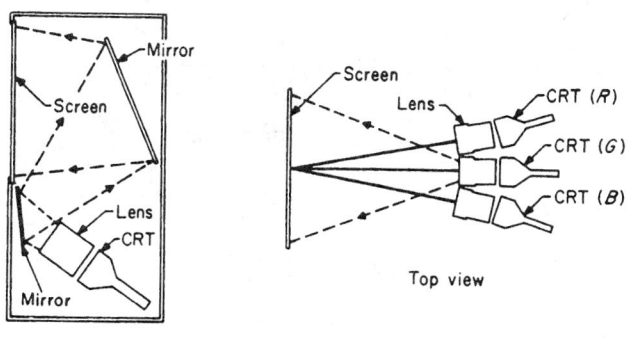

FIGURE 21.4.28   Mechanical arrangement of large-screen rear-projection receiver.

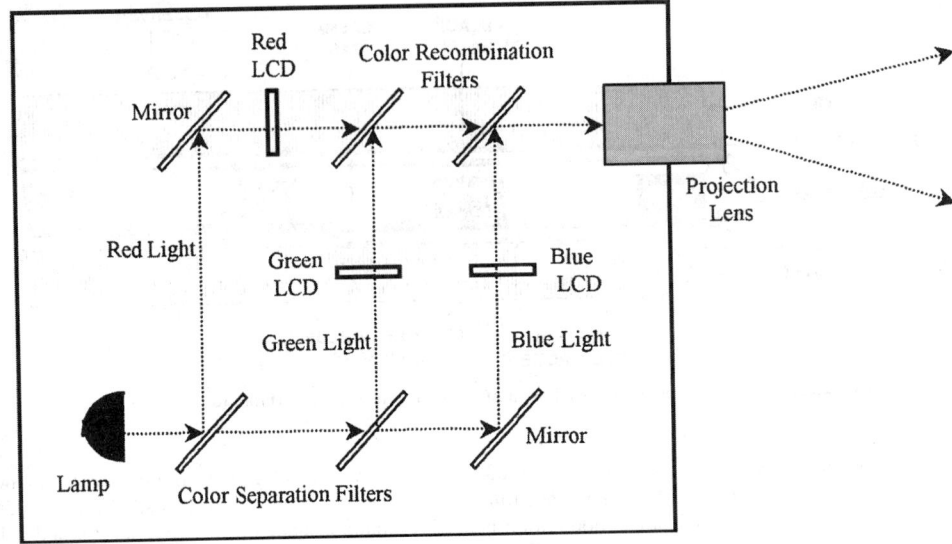

FIGURE 21.4.29 Typical LCD projection system.

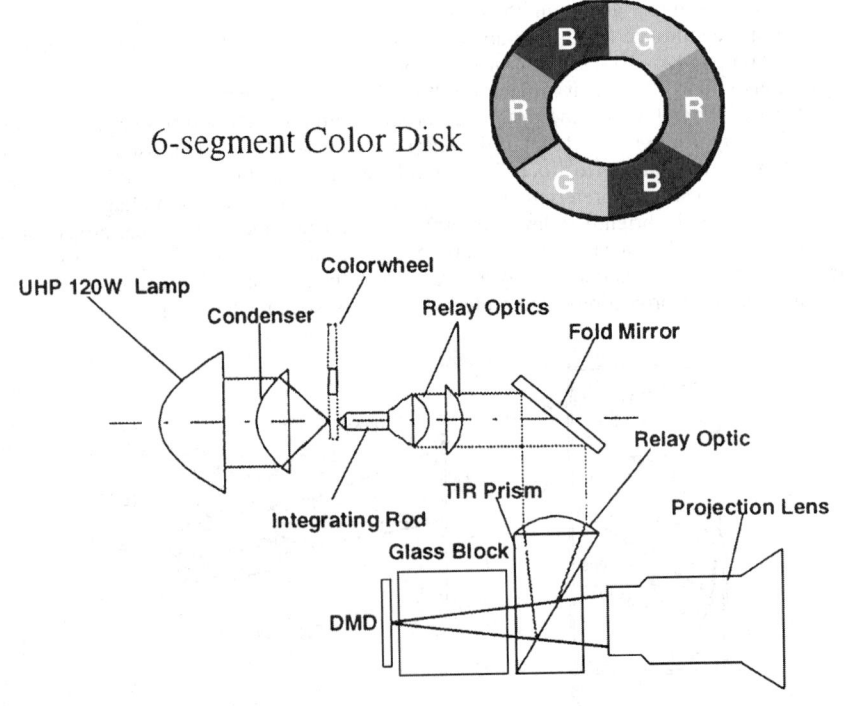

FIGURE 21.4.30 Optical system using a digital micromirror device. (From Ref. 33)

LCD projection systems use three LCD panels each measuring 1 to 2 in. diagonal. A single high-intensity light bulb illuminates a series of color-selective dichroic mirrors that split the light into three paths. Each light path passes through its respective LCD panel (red, green, and blue). The three light paths are then combined into one by another series of dichroic mirrors and passed through a lens that projects the image onto either a front or a rear screen (Fig. 21.4.29). Light output of the unit can be improved by adding a collimating element consisting of several microlenses just in front of the illuminating lamp.[31] As in the case of CRT projection, the precise alignment of LCD panels and mirrors is essential to register the three color images. Another novel design that has not yet been universally adopted uses only two LC panels, one black and white (B/W) and one three-color unit. This simplifies the mechanical design, uses fewer components, and has simpler convergence alignment. The B/W panel supplies the brightness to make up for the low transparency of the color panel. Video signals to the two panels are matrixed in a unique manner to yield correct color performance.[32]

LCD projectors currently exist in home theater size (front projection), rear projection self-contained large-screen TV sets, and small portable units having weight of 5 to 10 lb and light output of 500 lm to greater than 1000 lm for use in classroom.

Digital micromirror device (*DMD*TM) *display* systems also called digital light processing (DLPTM) use a semiconductor IC that contains an array of small mirror elements mounted on its surface, one for each pixel. When an electrical charge is applied to the substrate under a mirror, it will tilt by +10° (ON) and −10° (OFF). In the +10° configuration, the light from a high-intensity light is reflected through a lens and projected onto the viewing surface. In order to obtain shades of gray, the electrical charge to the mirrors is pulse modulated. Two configurations of this basic architecture have been developed. In the first, three DMD ICs, one for each color, are used. The light splitting, combining, and convergence of the three images is similar to that described for the LCD projector. The second configuration, now becoming more popular, uses a small six-segmented color wheel in the light path (Fig. 21.4.30).[33] The video signal sent to the DMD is that of a field sequential format in which the DMD is activated to the red picture image when the light is passing through the red portion of the color filter wheel; likewise for blue and green. Currently DMDs having picture definition of 1280×720 pixels are being used in HDTV applications.[34,35] Because of the light weight of the mechanism, DLP projectors are also becoming popular as portable units for classroom and traveling use. (DMD and DLP are trademarks of Texas Instruments.)

REFERENCES

1. Farmer, J. The Joint EIA/NCTA Band Plan for Cable Television, *IEEE Trans. Consum. Electr.*, August 1994, pp. 503–513.

2. Benson, K. B., and J. Whitaker (eds.) "Television Engineering Handbook," McGraw-Hill, 1992.

3. Hopkins, R. Chapter 13, HDTV Broadcasting and Reception, "Digital Consumer Electronics Handbook," McGraw-Hill, 1997, p. 13.7; Also Digital Terrestrial HDTV for North America: The Grand Alliance HDTV System, *IEEE Trans. Consum. Electr.*, August 1994, pp. 185–198.

4. Ghosh, M. Error Correction Schemes for Digital Television Broadcasting, *IEEE Trans. Consum. Electr.*, August 1995, pp. 400–404.

5. Shi, Q. Chapter 5, Digital Modulation Techniques, "Digital Consumer Electronics Handbook," McGraw-Hill, 1997, pp. 5.1–5.79.

6. Bryan, D. QAM for Terrestrial and Cable Transmission, *IEEE Trans. Consum. Electr.*, August 1995, pp. 383–391.

7. Lane, F., et al. A Single Chip Demodulator for 64/256 QAM, *IEEE Trans. Consum. Electr.*, November 1996, pp. 1003–1010.

8. Haas, M., et al. Flexible Two IC Chipset for DVB on Cable Reception, *IEEE Trans. Consum. Electr.*, August 1996, pp. 335–339.

9. Menkhoff, A., et al. Performance of an Advanced Receiver Chip for DVB-S and DSS, *IEEE Trans. Consum. Electr.*, August 1999, pp. 965–969.

10. Haas, M., et al. Advanced Two IC Chipset for DVB on Satellite Reception, *IEEE Trans. Consum. Electr.*, August 1996, pp. 341–345.

11. van der Wal, R., and L. Montreuil QPSK and BPSK Demodulator Chip-set for Satellite Applications, *IEEE Trans. Consum. Electr.*, February 1995, p. 34.

12. Wu, Y., and W. Zou Orthogonal Frequency Division Multiplexing: A Multi-Carrier Modulation Scheme, *IEEE Trans. Consum. Electr.*, August 1995, pp. 392–399.

13. Fetchel, S., et al. Advanced Receiver Chip for Terrestrial Digital Video Broadcasting: Architecture and Performance, *IEEE Trans. Consum. Electr.*, August 1998, pp. 1012–1018.

14. Sgrignoli, G., et al. VSB Modulation used for Terrestrial and Cable Broadcasts, *IEEE Trans. Consum. Electr.*, August 1995, pp. 367–382.

15. Bretl, W., et al. VSB Modem Subsystem Design for Grand Alliance Digital Television Receivers, *IEEE Trans. Consum. Electr.*, August 1995, pp. 773–786.

16. Advanced Television Systems Committee, Doc A/54 Guide to the Use of the ATSC Digital Television Standard, October 1995.

17 Tsunashima, K., et al. An Integrated DTV Receiver for ATSC Digital Television Standard, *IEEE Trans. Consum. Electr.*, August 1998, pp. 667–671.

18. Bryan, D., et al. A Digital Vestigial-Sideband (VSB) Channel Decoder IC for Digital TV, *IEEE Trans. Consum. Electr.*, August 1998, pp. 811–816.

19. Lin, W., et al. A Trellis Decoder for HDTV, *IEEE Trans. Consum. Electr.*, August 1999, pp. 571–576.

20. Shi, Q. Chapter 8, MPEG Standards, "Digital Consumer Electronics Handbook," McGraw-Hill, 1997, pp. 8.55–8.68.

21. Bhatt, B., et al. Grand Alliance HDTV Multi-format Scan Converter, *IEEE Trans. Consum. Electr.*, November 1995, pp. 1020–1031.

22. Zhu, W., et al. A Fast and Memory Efficient Algorithm for Down-Conversion of an HDTV Bitstream to an SDTV Signal. *IEEE Trans. Consum. Electr.*, February 1999, pp. 57–61.

23. Peng, S., and K. Challapali Low-Cost HD to SD Decoding, *IEEE Trans. Consum. Electr.*, August 1999, pp. 874–878.

24. Vernon, S. Design and Implementation of AC-3 Coders, *IEEE Trans. Consum. Electr.*, August 1995, pp. 754–759.

25. Advanced Television Systems Committee, Doc A/52A Digital Audio Compression Standard (AC-3), August 2001.

26. Kokado, N., et al. A Pocketable Liquid-Crystal Television Receiver, *IEEE Trans. Consum. Electr.*, August 1981, Vol. CE-27, No. 3, p. 462.

27. Yomamo, M., et al. The 5-Inch Size Full Color Liquid Crystal Television Addressed by Amorphous Silicon Thin Film Transistors, *IEEE Trans. Consum. Electr.*, February 1985, Vol. CE-31, No. 1, pp. 39–46.

28. Mercier, B., and E. Benoit A New Video Storage Architecture for Plasma Display Panels, *IEEE Trans. Consum. Electr.*, February 1996, pp. 121–127.

29. Howe, R., and B. Welham Development in Plastic Optics for Projection Television Systems, *IEEE Trans. Consum. Electr.*, February 1980, Vol. CE-26, No. 1, pp. 44–53.

30. Yamazaki, E., and K. Ando CRT Projection, *Proceedings—Projection Display Technology, Systems and Applications*, SPIE, Vol. 1081, January 19–20, 1989, pp. 30–37.

31. Ohuchi, S., et al. Ultra Portable LC Projector with High-Brightness Optical System, *IEEE Trans. Consum. Electr.*, February 2000, pp. 221–226.

32. Lee, M.-H., et al. Hybrid LCD Panel System and Its Color Coding Algorithm, *IEEE Trans. Consum. Electr.*, February 1997, pp. 9–16.

33. Ohara, K., and A. Kunzman Video Processing Technique for Multimedia HDTV with Digital Micro-Mirror Array, *IEEE Trans. Consum. Electr.*, August 1999, p. 604.

34. Hutchison, D., et al. Application of Second Generation Advanced Multi-Media Display Processor (AMDP2) in a Digital Micro-Mirror Array based HDTV, *IEEE Trans. Consum. Electr.*, August 2001, pp. 585–592.

35. Suzuki, Y., et al. Signal Processing for Rear Projection TV Using Digital Micro-Mirror Array, *IEEE Trans. Consum. Electr.*, August 2001, pp. 579–584.

CHAPTER 21.5
FACSIMILE SYSTEMS

Stephen J. Urban

INTRODUCTION

Facsimile is one of the original electrical arts, having been invented by Alexander Bain in 1842. In its long history it has prompted or helped accelerate the development of a variety of devices and methods, including the photocell, linear phase filters, adaptive equalizers, image compression, television, and the application of transform theory of signals and images. Facsimile systems have been used for a variety of services.[1] The *Wall Street Journal*, *USA Today*, and other newspapers have used facsimile to distribute their newspapers electronically, so that they can be printed locally. Weathermaps and satellite images are transmitted by other facsimile systems. Associated Press uses a form of facsimile to send photographs to newspapers. The most well-known facsimile system is the familiar one used in businesses and homes called simply "fax"—short for the Group 3 facsimile machine. The following discussion concentrates on Group 3, since it embodies techniques found in most other systems.

GROUP 3 FACSIMILE STANDARDS

Facsimile standards have been developed with international cooperation by the Telecommunication Standardization Sector of the International Telecommunication Union (ITU-T, formerly the CCITT). The ITU-T is made up of a number of study groups, each working on a particular aspect of telecommunications standardization. Study Group 16 (SG16) develops facsimile standards (called "Recommendations" by the ITU-T). National bodies may have their own slightly different version of the international standard. In the United States, the Telecommunications Industry Association TR-29 Committee was responsible for the development of standards for facsimile terminals and facsimile systems. TR-29, organized in the early 1960s, was a major contributor to the 1980 Group 3 Recommendation. Facsimile and modem standardization efforts have now been combined in the TIA-30 committee.

The ITU-T produced the first international facsimile standard in 1968, Recommendation T.2 for Group 1 facsimile. In North America a facsimile standard was used that was similar to Rec. T.2, but with enough of a difference so that North American machines were not interoperable with the rest of the world. The Group 1 standard provided for a 6-min transmission of a nominal 210 mm by 297 mm page at a scanning density of 3.85 lines per mm. In 1976 the first truly international facsimile standard, T.3 for Group 2 facsimile, was published. A Group 2 facsimile machine transmitted a page in half the time of a Group 1 machine with about the same quality. Both of these machines were analog, and neither used image compression.

In 1980, the Group 3 standards were published. Group 3 provided significantly better quality and shorter transmission time than Group 2, accomplished primarily by digital image compression. The Group 4 standard followed in 1984, with the intent to provide higher quality (twice the resolution of Group 3), higher speed (via digital networks), and more functionality. Since 1984 work has continued on both the Group 3 and Group 4 standards, with most of the advances being made in Group 3. Although Group 4 was intended to be the next generation facsimile terminal, it is clear that Group 3 has equaled or surpassed Group 4 in terms of performance and functionality. Group 3 can now use the same compression algorithm as Group 4, can provide the same image quality, and can operate on digital networks, thereby matching the short transmission times of Group 4. Actually Group 3 may be a little faster, because its protocol overhead is less.

Both Group 1 and Group 2 facsimile are now obsolete; neither are included in most new facsimile terminal implementations. The following discussion concentrates on Group 3 facsimile, which represents the vast majority of facsimile machines in current use.

The Group 3 facsimile recommendations, ITU-T Recommendations T.4[2] and T.30,[3] were first published in 1980. Recommendation T.4 covered the compression algorithm and modulation specifications, and Recommendation T.30 described the protocol. Group 3 originally was designed to operate on the general switched telephone network (GSTN) in half-duplex mode. The protocol specifies a standard basic operation that all Group 3 terminals must provide, plus a number of standard options. The various parameters and options that may be used are negotiated at 300 b/s, before the compressed image is transmitted at higher speed. *All* Group 3 machines must provide the following minimum set of capabilities: a pel density of 204 pels/in. horizontally and 98 pels/in. vertically; one-dimensional compression using the modified Huffman code; a transmission speed of 4800 b/s with a fallback to 2400 b/s; and a minimum time per coded scan line of 20 ms (to allow real time printer operation without extensive buffering). This requirement ensures that even the newest Group 3 facsimile machines can communicate with those designed to the 1980 Recommendations.

The standards options defined in the 1980 Recommendations are the following: use of the modified READ code to achieve two-dimensional compression; a vertical pel density of 196 pels/in. to provide higher image quality; a higher transmission speed of 9600 b/s with a fall back to 7200 b/s; and a minimum coded scan line time of zero to 40 ms. Many new options have been added over time. These are described in the following pages.

RESOLUTION AND PICTURE ELEMENT (PEL) DENSITY

The resolution of the scanners and printers used in facsimile apparatus (and the associated transmitted pel density) has a direct effect on the resulting output image quality. The highest pel density specified for the original Group 3 terminal was approximately 204 by 196 pels per 25.4 mm. The actual specification is 1728 pels per 215 mm by 7.7 lines/mm. This is referred to as "metric based" pel density. The Group 4 recommendations support the following standard and optional pel densities: 200×200, 240×240, 300×300, and 400×400 pels per 25.4 mm (referred to as "inch based" pel density). Note that the pel densities specified for Group 3 are "unsquare," that is, not equal horizontally and vertically. Group 4 pel densities are "square." This difference causes a compatibility problem. If the Group 3 pel densities were to be extended in multiples of their current pel densities, for example to 408×392, then a distortion of approximately 2 percent horizontally and vertically occurs when communicating with a 400×400 "square" machine. The ITU-T has decided to accept this distortion and encourage a gradual migration to the square pel densities. Accordingly, Group 3 has been

TABLE 21.5.1 Metric Based Pel Densities

Pel density (approximate) (pels/25.4 mm)	Tolerance	Number of picture elements along a scan line		
		ISO A4	ISO B4	ISO A3
Horizontal 204 Vertical 98	±1%	1728/215 mm	2048/225 mm	2432/303 mm
Horizontal 204 Vertical 196	±1%	1728/215 mm	2048/225 mm	2432/303 mm
Horizontal 408 Vertical 392	±1%	3456/215 mm	4096/225 mm	4864/303 mm

TABLE 21.5.2 Inch Based Pel Densities

Pel density (pels/25.4 mm)	Tolerance	Number of picture elements along a scan line		
		ISO A4	ISO B4	ISO A3
Horizontal 200 Vertical 100	± 1%	1728/219.45 mm	2048/260.10 mm	2432/308.86 mm
Horizontal 200 Vertical 200	± 1%	1728/219.45 mm	2048/260.10 mm	2432/308.86 mm
Horizontal 300 Vertical 300	± 1%	2592/219.45 mm	3072/260.10 mm	3648/308.86 mm
Horizontal 400 Vertical 400	± 1%	3456/219.45 mm	4096/260.10 mm	4864/308.86 mm

enhanced to include higher pel densities, including both multiples of the original Group 3 pel densities and "square pel densities." Specifically, optional Group 3 pel densities of 408 × 196, 408 × 392, 200 × 200, 300 × 300, and 400 × 400 pels per 25.4 mm have been added. These are summarized in Tables 21.5.1 and 21.5.2.

The metric-based pel densities and their picture elements are given in Table 21.5.1. Specific values for the number of pels per line are given in Table 21.5.1 for all the Group 3 pel densities for ISO A4, ISO B4, and ISO A3.

The optional inch-based pel densities and their picture elements are shown in Table 21.5.2. Specific values for the number of pels per line are given in Table 21.5.2 for all the Group 3 pel densities for ISO A4, ISO B4, and ISO A3.

NOTE: An alternative standard pel density of 200 pels/25.4 mm horizontally × 100 lines/25.4 mm vertically may be implemented provided that one or more of 200 × 200 pels/25.4 mm, 300 × 300 pels/25.4 mm, and 400 × 400 pels/25.4 mm are included.

PROTOCOL

A facsimile call is made up of five phases, as shown in Fig. 21.5.1. In phase A (call establishment) the telephone call is placed, either manually or automatically by the facsimile terminal(s). In phase B (premessage procedure) the call station responds with signals identifying its capabilities. The calling station then sends signals to indicate the mode of operation to be used during the call (e.g., transmission speed, resolution, page size,

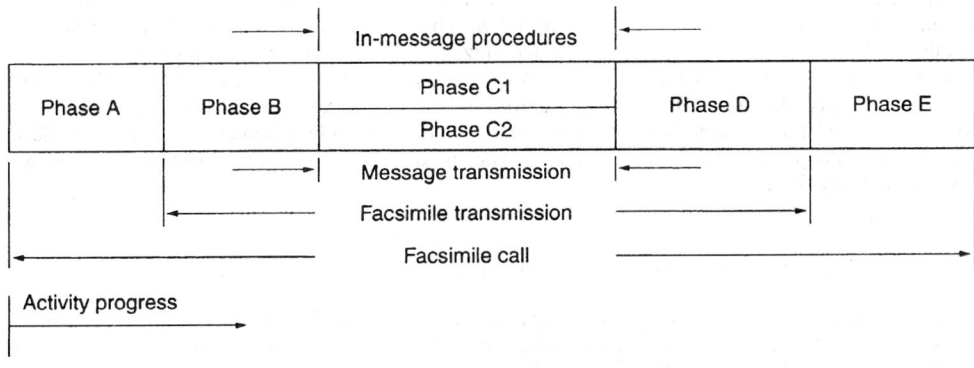

FIGURE 21.5.1 The five phases of a facsimile call.

image coding method). The option selection must be consistent with the capability set received from the called station. Phase C (in-message procedure and message transmission) includes both procedural signaling and the transmission of the facsimile image. Procedural signaling includes, for example, a modem training sequence from the sending station that allows the receiving station to adapt its modem to the telephone line characteristics. If the receiving station trains successfully and is ready, it responds with a confirmation to receive (CFR) signal. Upon receipt of the CFR, the sending station sends the facsimile image at the negotiated higher speed. As in phases B and D, all signaling is communicated at 300 b/s using high-level data link control (HDLC), before the compressed image is transmitted at higher speed. Phase D (postmessage procedure) includes information of end-of-message, confirmation of the reception of the message, and transmission of further message information. In phase E the call is disconnected.

DIGITAL IMAGE COMPRESSION

One-Dimensional Coding Scheme—Modified Huffman Code (MHC)

A digital image to be transmitted by facsimile is formed by scanning a page from left to right and top to bottom, producing a bit map of picture elements. A scan line is made up of runs of black and white pels. Instead of sending bits corresponding to black and white pels, coding efficiency can be gained by sending codes corresponding to the lengths of the black and white runs. A Huffman procedure[4] uses variable length codes to represent the run lengths; the shortest codes are assigned to those run lengths that occur most frequently. Run-length frequencies are tabulated from a number of "typical" documents, and are then used to construct the code tables. True Huffman coding would require twice 1729 code words to cover a scan line of 1728 pels. To shorten the table, the Huffman technique was modified for Group 3 facsimile to include two sets of code words, one for lengths of 0 to 63 (terminating code table), and one for multiples of 64 (make-up code table). Run lengths in the range of 0 to 63 pels use terminating codes. Run lengths of 64 pels or greater are coded first by the appropriate make-up code word specifying the multiple of 64 less than or equal to the run length, followed by a terminating code representing the difference. For example, a 1728-pel white line would be encoded with a make-up code of length 9 representing a run of length 1728, plus a terminating code of length 8 representing a run of length zero, resulting in a total code length of 17 bits (without the synchronizing code). When the code tables for Group 3 were constructed, images containing halftones were deliberately *excluded*, so as not to skew the code tables and degrade the compression performance for character-based documents.

The modified Huffman code is mandatory for all Group 3 machines, providing a basis for interoperability. It is relatively simple to implement, and produces acceptable results on noisy telephone lines. In order to ensure that the receiver maintains color synchronization, all coded lines begin with a white run length code word. If the actual scan line begins with a black run, a white run length of zero is sent. Black or white run lengths, up to a maximum length of one scan line (1728 picture elements or pels) are defined by the code word in Tables 21.5.3 and 21.5.4. Note that there is a different list of code words for black and white run lengths.

Each coded line begins with an End-of-Line (EOL) code. It is a unique code word that can never be found within a valid coded scan line; therefore, resynchronization after an error burst is possible.

A pause may be placed in the message flow by transmitting FILL. FILL is a variable length string of 0s that can only be placed after a coded line and just before an EOL, but never within a coded line. Fill must be added to ensure that the transmission time of the total coded scan line is not less than the minimum transmission time established in the premessage control procedure. The end of a document transmission is indicated by sending six consecutive EOLs.

Two-Dimensional Coding Scheme—Modified READ (MR)

The MR coding method makes use of the vertical correlation between black (or white) runs from one scan line to the next (called vertical mode). In vertical mode, transitions from white to black (or black to white) are coded relative to the line above. If the transition is directly under (zero offset), the code is only one bit. Only fixed

TABLE 21.5.3 Terminating Codes

White run length	Code word	Black run length	Code word
0	00110101	0	0000110111
1	000111	1	010
2	0111	2	11
3	1000	3	10
4	1011	4	011
5	1100	5	0011
6	1110	6	0010
7	1111	7	00011
8	10011	8	000101
9	10100	9	000100
10	00111	10	0000100
11	01000	11	0000101
12	001000	12	0000111
13	000011	13	00000100
14	110100	14	00000111
15	110101	15	000011000
16	101010	16	0000010111
17	101011	17	0000011000
18	0100111	18	0000001000
19	0001100	19	00001100111
20	0001000	20	00001101000
21	0010111	21	00001101100
22	0000011	22	00000110111
23	0000100	23	00000101000
24	0101000	24	00000010111
25	0101011	25	00000011000
26	0010011	26	000011001010
27	0100100	27	000011001011
28	0011000	28	000011001100
29	00000010	29	000011001101
30	00000011	30	000001101000
31	00011010	31	000001101001
32	00011011	32	000001101010
33	00010010	33	000001101011
34	00010011	34	000011010010
35	00010100	35	000011010011
36	00010101	36	000011010100
37	00010110	37	000011010101
38	00010111	38	000011010110
39	00101000	39	000011010111
40	00101001	40	000001101100
41	00101010	41	000001101101
42	00101011	42	000011011010
43	00101100	43	000011011011
44	00101101	44	000001010100
45	00000100	45	000001010101
46	00000101	46	000001010110
47	00001010	47	000001010111
48	00001011	48	000001100100
49	01010010	49	000001100101
50	01010011	50	000001010010

(Continued)

TABLE 21.5.3 Terminating Codes (*Continued*)

White run length	Code word	Black run length	Code word
51	01010100	51	000001010011
52	01010101	52	000000100100
53	00100100	53	000000110111
54	00100101	54	000000111000
55	01011000	55	000000100111
56	01011001	56	000000101000
57	01011010	57	000001011000
58	01011011	58	000001011001
59	01001010	59	000000101011
60	01001011	60	000000101100
61	00110010	61	000001011010
62	00110011	62	000001100110
63	00110100	63	000001100111

offsets of zero and ±1, ±2, and ±3 are allowed. If vertical mode is not possible (for example, when a nonwhite line follows an all-white line) then horizontal mode is used. Horizontal mode is simply an extension of MHC; that is, two consecutive runs are coded by MHC and preceded by a code indicating horizontal mode. To avoid the vertical propagation of transmission errors to the end of the page, a one-dimensional (MHC) line is sent every Kth line. The factor K is typically set to 2 or 4, depending on whether the vertical scanning density is 100 or 200 lines per inch. The K-factor is resettable, which means that a one-dimensional line may be sent more frequently (than 2 or 4) when considered necessary by the transmitter. The synchronization code (EOL) consists of eleven 0s followed by a 1, followed by a tag bit to indicate whether the following line is coded one-dimensionally or two-dimensionally.

The two-dimensional coding scheme is an optional extension of the one-dimensional coding scheme. It is defined in terms of changing picture elements (see Fig. 21.5.2).

A changing element is defined as an element whose color (i.e., black or white) is different from that of the previous element along the same scan line.

a_0 The reference or starting changing element on the coding line. At the start of the coding line a_0 is set on an imaginary white changing element situated just before the first element on the line. During the coding of the coding line, the position of a_0 is defined by the previous coding mode.

a_1 The next changing element to the right of a_0 on the coding line.

a_2 The next changing element to the right of a_1 on the coding line.

b_1 The first changing element on the reference line to the right of a_0 and of opposite color to a_0

b_2 The next changing element to the right of b_1 on the reference line.

Code words for the two-dimensional coding scheme are given in Table 21.5.5.

Extended Two-Dimensional Coding Scheme—Modified Modified READ (MMR)

The basic facsimile coding scheme specified for Group 4 facsimile (MMR) may be used as an option in Group 3 facsimile. This coding scheme must be used with the Error Correction Mode (ECM) option, described below. This coding scheme, described in ITU-T Recommendation T.6,[5] is very similar to that of Group 3 (Rec. T.4). The same modified READ code is used, but only two-dimensional lines are transmitted with no EOL codes for synchronization. An error-free communication link makes this possible. A white line is assumed before the first actual line in the image. No fill is used; adequate buffering is assumed to provide a memory to memory transfer.

TABLE 21.5.4 Make Up Codes

White run lengths	Code word	Black run lengths	Code word
64	11011	64	0000001111
128	10010	128	000011001000
192	010111	192	000011001001
256	0110111	256	000001011011
320	00110110	320	000000110011
384	00110111	384	000000110100
448	01100100	448	000000110101
512	01100101	512	0000001101100
576	01101000	576	0000001101101
640	01100111	640	0000001001010
704	011001100	704	0000001001011
768	011001101	768	0000001001100
832	011010010	832	0000001001101
896	011010011	896	0000001110010
960	011010100	960	0000001110011
1024	011010101	1024	0000001110100
1088	011010110	1088	0000001110101
1152	011010111	1152	0000001110110
1216	011011000	1216	0000001110111
1280	011011001	1280	0000001010010
1344	011011010	1344	0000001010011
1408	011011011	1408	0000001010100
1472	010011000	1472	0000001010101
1536	010011001	1536	0000001011010
1600	010011010	1600	0000001011011
1664	011000	1664	0000001100100
1728	010011011	1728	0000001100101
EOL	000000000001	EOL	000000000001

Note: For those machines that choose to accommodate larger paper widths or higher pel densities, the following Make Up Code Set may be used:

Run length (black and white)	Make up codes
1792	00000001000
1856	00000001100
1920	00000001101
1984	000000010010
2048	000000010011
2112	000000010100
2176	000000010101
2240	000000010110
2304	000000010111
2368	000000011100
2432	000000011101
2496	000000011110
2560	000000011111

Note: Run lengths in the range of lengths longer than or equal to 2624 pels are coded first by the make-up code of 2560. If the remaining part of the run (after the first make-up code of 2560) is 2560 pels or greater, additional make-up codes(s) of 2560 are issued until the remaining part of the run becomes less than 2560 pels. Then the remaining part of the run is encoded by terminating code or by make-up code plus terminating code according to the range as mentioned above.

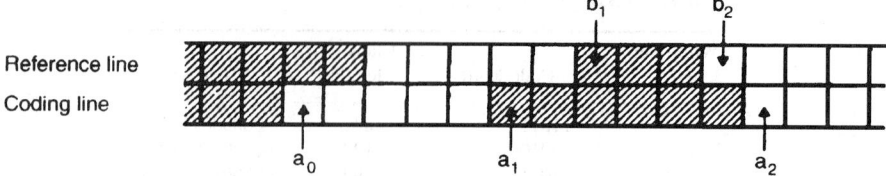

FIGURE 21.5.2 Changing picture elements.

The objective of each of the image compression schemes is to reduce the number of bits transmitted and thus the transmission time (and cost). The more aggressive schemes provide the most compression, at a cost of increased implementation complexity. In general, T.6 coding outperforms T.4 two-dimensional coding which outperforms T.4 one-dimensional coding. The differences tend to be smaller on "busy" images and greater on images containing more white space. On dithered images and halftones the compression is very poor (or even negative), and one-dimensional coding outperforms the other methods in some cases.

MODULATION AND DEMODULATION METHODS (MODEMS)

Every Group 3 facsimile machine must be able to operate at the standard modem speeds of 4.8 and 2.4 kb/s according to ITU-T Recommendation V.27ter. The receive modem has an automatic equalizer to compensate for the telephone line amplitude distortion and envelope-delay distortion, improving the accuracy of the delivered digital facsimile signal. Optional modems are V.29 (9.6 and 7.2 kb/s), and V.17 with Trellis coding for improved error immunity (14.4, 12, 9.6, and 7.2 kb/s) and V.34 for speeds up to 33.6 kb/s.

Group 3 facsimile modems adapt to the transmission characteristics of the telephone connection. Before sending fax data, the modem sends a standard training signal. The receiver uses this signal to "adapt" to the electrical characteristics of the connection. The highest rate available in both facsimile machines is tried first. If this speed would give too many errors, the next lower speed is tried by the transmitter. If this fails, the modem rate again steps down to the next lower speed. This system assures transmission at the highest rate consistent with the quality of the telephone line connection.

Standard Operation—V.27ter (4.8 and 2.4 kb/s)

At 4.8 kb/s, the modulation rate (or baud rate) is 1600 baud or 1.6 kB. The data stream to be transmitted is divided into groups of three consecutive bits (tribits). Each is encoded as a phase change relative to the phase

TABLE 21.5.5 Two-Dimensional Code Table

Mode	Elements to be coded		Notation	Code word
PASS	b_1, b_2		P	0001
HORIZONTAL	$a_0 a_1, a_1 a_2$		H	$001 + M(a_0 a_1) + M(a_1 a_2)_{\text{Note}}$
VERTICAL	a_1 just under b_1	$a_1 b_1 = 0$	$V(0)$	1
	a_1 to the	$a_1 b_1 = 1$	$V_R(1)$	011
	right of	$a_1 b_2 = 2$	$V_R(2)$	000011
	b_1	$a_1 b_1 = 3$	$V_R(3)$	0000011
	a_1 to the	$a_1 b_1 = 1$	$V_L(1)$	010
	left of	$a_1 b_1 = 2$	$V_L(2)$	000010
	b_1	$a_1 b_1 = 3$	$V_L(3)$	0000010

Note: Code M() of Horizontal mode represents the code words in Tables 21.5.3 and 21.5.4.

TABLE 21.5.6 Tribit Values and Phase Changes

Tribit values	Phase change
0 0 1	0°
0 0 0	45°
0 1 0	90°
0 1 1	135°
1 1 1	180°
1 1 0	225°
1 0 0	270°
1 0 1	315°

TABLE 21.5.7 Dibit Values and Phase Changes

Dibit values	Phase change
00	0°
01	90°
11	180°
10	270°

of the preceding signal tribit element (see Table 21.5.6). At the receiver, the tribits are decoded and the bits are reassembled in correct order.

At 2.4 kb/s, the modulation rate is 1200 baud or 1.2 kB. The data stream to be transmitted is divided into groups of two bits (dibits). Each dibit is encoded as a phase change relative to the phase of the immediately preceding signal element (see Table 21.5.7). At the receiver the dibits are decoded and reassembled in the correct order.

Optional Operation A—V.29 (9.6 and 7.2 kb/s)

For optional higher speed operation, such as may be possible on high-quality circuits, the optional mode may be used.

At all speeds, the modulation rate is 2.4 kB. At 14.4 kb/s the scrambled data stream to be transmitted is divided into groups of six consecutive data bits and mapped onto a signal space of 128 elements. At 12 kb/s the scrambled data stream is divided into groups of five consecutive data bits and mapped onto a signal space of 64 elements. At 9.6 kb/s the data are divided into groups of four consecutive data bits and mapped onto a signal space of 32 elements. At 7.2 kb/s the data are divided into groups of three consecutive data bits and mapped onto a signal space of 16 elements. The V.17 modem is more robust than V.29, due to its trellis coding method; e.g., V.17 can operate at 9.6 kb/s on telephone lines where V.29 cannot.

Higher data rates are theoretically possible, and are highly desirable for transmission of continuous tone color. The V.34 modem standard approved in 1994 provides speeds to 33 kb/s.

The use of this modem is described in Annex F of Recommendation T.30. This modem also uses trellis coding and a number of symbol rates from 2400 to 3429 symbols/per second. It requires the use of ECM and runs in half-duplex for facsimile. An unusual feature of this modem is the use of Recommendation V.8 for the modem start-up procedures. V.8 provides the means to determine the best mode of operation *before* the initiation of the handshake.

INTERNET FACSIMILE

The availability of the Internet for international communication provides the potential for using this transmission medium in the transfer of Group 3 facsimile messages between terminals. Since the characteristics of IP networks differ from those provided by the PSTN, some additional provisions need to be standardized to maintain successful facsimile operation. Service requirements for Internet facsimile are defined in Recommendation F.185. Two methods have been defined to meet these service requirements: a store-and-forward method akin to email and a real-time method that is largely transparent as far as the endpoints are concerned.

Store-and-Forward Facsimile

Store-and-forward Internet facsimile occurs when the sending and receiving terminal are not in direct communication with each other. The transmission and reception takes place via the store-and-forward mode on the Internet using Internet email. In store-and-forward mode the facsimile protocol "stops" at the gateway to the Internet. It is reestablished at the gateway leaving the Internet.

Two modes of store-and-forward facsimile are defined. In "Simple" mode store-and forward only the coded image is transmitted. In "Full" mode store-and-forward facsimile over the Internet three requirements must be satisfied:

1. The capabilities of the terminals are exchanged.

2. An acknowledgement of receipt is exchanged between gateways and may be transferred from the receiving terminal to the sending terminal.

3. The contents of standard messages used by the transmitting terminal are preserved.

ITU-T Recommendation T.37, *procedures for the transfer of facsimile data via store-and-forward on the Internet*, approved in 1998, standardized the Simple mode. It references a set of IETF documents called "Request for Comment" documents (RFCs) that define the procedures for facsimile communication over the Internet using email. Full mode was added in 1999 with Addendum 1. The intention is that Full mode should support, to the greatest extent possible, the standard and optional features of Group 3 facsimile. This would include, among others, delivery notification, capabilities exchange, color and other optional coding mechanisms, and file transfer.

Real-Time Internet Facsimile

Recommendation T.38, *procedures for real-time Group 3 facsimile communication between terminals over IP networks*, was approved by the ITU-T in 2002. This standard describes how facsimile transmission takes place between Internet gateways. Communication between the facsimile machine and the gateway is by means of T.30. The T.38 procedures work in conjunction with Rec. H.323 and Rec. H.225.

The transmission method in T.38 may be described as a demodulation/remodulation (demod/remod) procedure. The facsimile transmission arrives at the gateway as a data stream modulated by the facsimile modem into an analog signal. At the gateway the signal is demodulated. It is then packaged into data packets to be conveyed over the Internet by the emitting gateway. The receiving gateway reassembles the packets and modulates them into a modem signal that continues on to the receiving facsimile terminal. Neither facsimile terminal is aware that part of the transmission was over the PSTN and part over the Internet. Once the PSTN calls are established on both ends, the two Group 3 terminals are virtually linked. All T.30 session establishment and capabilities negotiation is carried out between the terminals. An alternate scenario would be a connection to a facsimile-enabled device (for example, a PC), which is connected directly to an IP network. In this case, there is a virtual-receiving gateway as part of the device's facsimile-enabling software and/or hardware.

Transmission Protocols.   There are two types of transmission protocols over the Internet: TCP and UDP. The TCP protocol is error free (via retransmission) so that errors translate into delay. UDP has no error control. Errors manifest themselves as lost packets. T.30 is very sensitive to delay, since there are a number of timers active in the protocol, which, if they expire, would cause transmission to fail. Lost packets could have serious consequences. If a packet is lost in the image transmission phase, it could cause lost scan lines in the image, unless ECM error control was in effect.

Modem Rate Alignment.   When the TCP protocol method is selected, each gateway independently provides rate negotiation with the facsimile modems and training is carried out independently. It would therefore be possible to have each end of the connection negotiate to a different data rate, which would be a problem. To prevent this from occurring, a set of messages is provided to align the modem speeds. Because of the lower delays experienced with using UDP, the modems can negotiate data rates without relying on gateways. The V.34 modem is not used with Internet facsimile because of the added complexity.

Error Correction Mode

The optional error correction mode applies to one-dimensional and two-dimensional coding, and provides true error correction. The primary objective of ECM is to perform well against burst errors. Additional objectives included backwards compatibility with existing facsimile machines, and minimizing the transmission overhead in channels with low error rates.

The error-correction scheme is known as page selective repeat ARQ (automatic repeat request). The compressed image data are embedded in high-level data link control (HDLC) frames of length 256 octets or 64 octets, and transmitted in blocks of 256 frames. The communication link operates in a half-duplex mode; that is, the transmission of image data and the acknowledgment of the data are not sent at the same time. The technique can be thought of as an extension to the Group 3 protocol. The protocol information is also embedded in HDLC frames, but does not use selective repeat for error control. Every Group 3 facsimile machine must have the mechanism to transmit and receive the basic HDLC frame structure, including flags, address, control, and frame check sequence. Thus the use of an extended HDLC scheme helped to minimize changes to existing facsimile designs.

The transmitting terminal divides the compressed image data into 256-octet or 64-octet frames and sends a 256-octet block of frames to the receiving terminal. (The receiving terminal must be able to receive both frame sizes). Each transmitted frame has a unique frame number. The receiver requests retransmission of bad frames by frame number. The transmitter retransmits the requested frames. After four requests for retransmission for the same block, the transmitter may stop or continue, with optional modem speed fall back.

The page selective repeat ARQ is a good compromise[6] that balances complexity and throughput. A continuous selective repeat ARQ provides slightly higher throughput, but requires a modem back channel. Forward error correction (FEC) schemes typically have higher overhead on good connections, can be more complex, and may break down in the presence of burst errors. In addition to providing the capability of higher throughput on noisy lines, the error correction mode option supports an error-free environment that has enabled many new features. Examples of these are T.6 encoding, color fax, and secure operation.

COLOR FACSIMILE

Group 3 facsimile is capable of transmitting color, including color or gray scale photographs and colored text and line art. Much research has been devoted to the development of a "universal" compression technique that would apply to both photographs and text, but the best approach to achieve high compression ratios and retain quality is to compress the different image types according to their individual attributes. Photographs are compressed using an approach that puts a high emphasis on maintaining the smoothness and accuracy of the colors. Text and line art are compressed with an approach that puts high emphasis on maintaining the detail and structure of the input.

The method to transmit continuous tone color was added to Group 3 facsimile in 1994, with the approval of Recommendation T.42. This method is based on the popular Joint Photographic Experts Group (JPEG) standard. Although this was a good first step toward a color facsimile capability, it was not an efficient solution for text or line art. To correct this problem, Recommendation T.43, based on JBIG, was developed to accommodate text and line art with some limited color capability. The Mixed Raster Content (MRC) option was added to correct a weakness of the color facsimile options that use JPEG and JBIG; that is, before MRC there was no standard way of efficiently coding a page that contained *both* text and color photographic content. The MRC option, as implemented by ITU-T Recommendation T.44, is a way of describing documents with both bi-level and multilevel data within a page. These methods are outlined in the following paragraphs.

Coding Continuous Tone Color and Gray Scale Images

The facsimile standards include an option for the transmission of continuous-tone color and gray-scale images, based on an ISO/ITU-T standard[7] commonly known as JPEG. As a result of diverse requirements, the JPEG compression algorithm is not a single algorithm but a collection of techniques that are often referred to as a *toolkit*. The intent is that applications, such as facsimile, will have a "customized" subset of the JPEG components. A detailed description of JPEG with a comprehensive bibliography is given in Ref. 8.

JPEG specifies two classes of coding processes: lossy and lossless processes. The lossy processes are all based on the discrete cosine transform (DCT), and the lossless are based on a predictive technique. There are four modes of operation under which the various processes are defined: the sequential DCT-based mode, the progressive DCT-based mode, the sequential lossless mode, and the hierarchical mode.

In the sequential DCT-based mode, 8×8 blocks of pixels are transformed, the resulting coefficients are quantized and then entropy coded (losslessly) by Huffman or arithmetic coding. The pixel blocks are typically formed by scanning the image (or image component) from left to right, and then block row by block row from top to bottom. The allowed sample precisions are 8 and 12 bits per component sample. All decoders that include any DCT-based mode of operation must provide a default decoding capability, referred to as the baseline sequential DCT process. This is a restricted form of the sequential DCT-based mode, using Huffman coding and 8 bits per sample precision for the source image. The application of JPEG to facsimile is based on the baseline sequential DCT process.

Continuous tone color was added to Group 3 facsimile with the approval of Recommendation T.42. In order to represent continuous tone color data accurately and uniquely, a device-independent interchange color space is required. This color image space must be able to encode the range of hard copy image data when viewed under specified conditions. In addition to the basic color space, the reference white point, illuminant type, and gamut range must also be specified. The image pixel values are represented in the CIE 1976 ($L^* a^* b^*$) color space, often referred to as CIELAB. This color space, defined by the CIE (Commission Internationale de l'Eclairage), has approximately equal visually perceptible difference between equispaced points throughout the space. The three components are L^*, or luminance, and a^* and b^* in chrominance. The luminance-chromaticity spaces offer gray-scale compatibility and higher DCT-based compression than other spaces. The human eye is much more sensitive to luminance than chrominance; thus it is easier to optimize the quantization matrix where luminance and chrominance are separate components. Subsampling the chrominance components provides further compression.

The basic resolution is 200 pels/25.4 mm. Allowed values include 200, 300, and 400 pels/25.4 mm, with square (or equivalent) pels. At 200×200 pels/25.4 mm, a color photograph (A4 paper size) compressed to 1 bit per pixel and transmitted at 64 kb/s would require about 1 min to send. The selection of 8-bit or 12-bit data precision can also have an effect on the data compression. Subsampling and data precision, as well as the ability to send color, are negotiated in Phase B.

Lossless Coding of Color and Gray Scale Images

Recommendation T.43 was developed to accommodate text and line art with some limited color capability. Recommendation T.43 defines a lossless color data representation method using Recommendation T.82, which was prepared by the Joint Bi-level Image Experts Group (JBIG). In this recommendation, three types of images are treated: the first type is a one-bit-per-color CMY(K) or RGB image; the second type is a palletized color image in which palette tables are specified with the CIELAB color space defined in Recommendation T.42; and the last type is a continuous-tone color and gray-scale image also specified with the CIELAB color space. Images can be created in a variety of ways, including conventional scanning, computer generation, or by image processing techniques such as one of the dither methods. Recommendation T.43 was approved in July 1997. The following section presents an overview of JBIG compression, followed by a description of the three image types treated by Recommendation T.43.

JBIG overview (ITU-T T.82/ISO11544). The progressive bi-level coding technique consists of repeatedly reducing the resolution of a bi-level image, creating subimages each having one-half the number of pels per line and one-half the number of lines of the previous image. The lowest-resolution image, called the base layer, is transmitted losslessly (free of distortion) by binary arithmetic coding. The next higher resolution image is transmitted losslessly, using its own pels and previously transmitted (causal) pels as predictors in an attempt to predict the next pel to be transmitted. If prediction is possible (both transmitter and receiver are equipped with rules to tell whether this is the case), the predicted pel value is not transmitted. This progressive build-up is repeated until the final image has been losslessly transmitted (the process stopped at the receiver's request). A sequential mode of transmission also exists. It consists of performing the entire progressive transmission on successive horizontal stripes of the original image. The algorithm performs image reduction, typical prediction, deterministic prediction, and binary arithmetic encoding and decoding.

Recommendation T.43.   The one-bit-per-color mode was intended to represent images with primary colors using the CMY(K) or RGB color space. Each bit-plane indicates the existence of one of the primary colors. The image is encoded with JBIG and transmitted. The receiver represents the image on a CRT (soft-copy) or a printed (hard copy) using its own primary colors. The colors of the document may not be represented accurately at the receiver in the one-bit-per-color mode since neither RGB nor CMY(K) are device-independent color spaces. A typical application of the mode is to transmit business correspondence containing a colored logo.

The palletized color mode expands the possible number of colors that may be used to characterize an image, and in addition provides the capability for accurate color reproduction. Both of these features are achieved by using color palette table data as specified by the device-independent interchange color space (CIELAB) defined in Recommendation T.42. The price for this added capability is coding (and transmission) efficiency, that is, the compression is less so the facsimile transmission takes longer.

The continuous-tone color mode provides the highest color accuracy of the three modes in Recommendation T.43 and the lowest compression efficiency. In this mode, an original continuous-tone color or gray-scale image is generated in the color space defined in Recommendation T.42 (CIELAB). This mode provides lossless encoding using JBIG bit-plane compression. Gray-code conversion on the bit planes is used to improve the compression efficiency.

Coding Images with Mixed Raster Content

The MRC option, as implemented by ITU-T Recommendation T.44, is a way of describing raster-oriented (scanned or synthetic) documents with both bi-level and multilevel, continuous-tone or palettized colors (contone) usually associated with naturally occurring images: bi-level detail associated with text and line art and multilevel colors associated with the text and line art. The goal of MRC is to make the exchange of raster-oriented mixed content color documents possible with higher speed, higher image quality, and modest computing resources. This efficiency is realized through segmentation of the image into multiple layers (planes) as determined by image type, and applying image specific encoding and spatial and color resolution processing. The MRC method defines no new image compression methods, but does require that all previously defined compression methods used in Group 3 facsimile be supported.

SECURE FACSIMILE

Group 3 secure facsimile, including new Annexes to T.30 and a new Recommendation T.36, was approved by the ITU-T in July 1997. The new and amended recommendations accommodate two methods—one based on a public-key cryptosystem and one based on a secret-key cryptosystem. The public-key management system is based on the method devised by Ron Rivest, Adi Shamir, and Leonard Adleman. It is called RSA after the initials of its inventors. The secret-key method is based on the Hawthorne Key Management (HKM) system, the Hawthorne Facsimile Cipher (HFX40), and the HFX40-I message integrity system, hereafter referred to as the HKM/HFX cryptosystem. Both systems are incorporated into the Group 3 facsimile protocol, and either may be used independently.

Procedures Using the RSA Security System

The procedures for the RSA security system are defined in Annex H of ITU-T Recommendation T.30. The RSA security system uses one pair of keys (encipherment public key and encipherment secret key) for document encryption.

The registration mode permits the sender and receiver to register and store the public keys of the other party prior to secure facsimile transmission. Two parties wishing to communicate can register their public keys with another user in two steps. First, the sender and receiver each hash their identities and public key(s) and the hash results are exchanged out of band (directly, by mail, by phone, and so forth) and stored in the terminals. Then the identities and public keys of the two parties are exchanged and stored in the registration mode of the T.30 protocol. The validity of the identity and public key(s) of the other party is assessed by hashing these values and comparing them with the hash result that was exchanged out of band.

An optional security page follows the last page of the transmitted document. The security page contains the following parameters: security page indicator, identity of the sender, public key of the sender, identity of the recipient, random number created by the recipient, time stamp, length of the document, digital signature of the entity in brackets, certificate of the public key of the sender, and security-page-type identification.

One other significant issue is the key length. Because of varying regulations of different governments, it was agreed to limit the session key length for the RSA secure facsimile system to 40 bits. The amendment also adds a redundancy mechanism that repeats the 40-bit session key to fit the length of the various encipherment algorithms whose key lengths are required to be longer than 40 bits.

Procedures Using the HKM/HFX Cryptosystem

The secret-key method consists of the Hawthorne Key Management system, the HFX40 carrier cipher (encryption algorithm), and the HFX40-I message integrity system (hashing algorithm). The HKM key management algorithm includes a registration procedure, and a secure transmission of a secret key that enables subsequent transmission to be provided securely. These procedures are defined as ITU-T Recommendation T.36.

In the registration mode, the two terminals exchange information that enables them to uniquely identify each other. This is based on the agreement between the users of a secret one-time key that must be exchanged securely (not defined by the Recommendations). Each terminal stores a 16-digit number that is uniquely associated with the terminal with which it has carried out registration. This number is used together with a challenge procedure to provide mutual authentication and document confirmation. The procedure is also used to transmit the session key to be used for document encryption and hashing.

An override mode is also provided that bypasses the exchange of security signals between two terminals. Just as in the classic secret-key cryptosystem approach, this mode depends on the secure exchange of a secret key outside the system. This key is used by the transmitting terminal to encrypt the document and by the receiving terminal to decrypt the document.

REFERENCES

1. K. R. McConnell, D. Bodson, and S. Urban, *FAX*, 3rd ed., Artech House.
2. ITU-T Recommendation T.4, Standardization of Group 3 Facsimile Apparatus for Document Transmission.
3. ITU-T Recommendation T.30, Procedures for Document Facsimile Transmission in the General Switched Telephone Network.
4. D. A. Huffman, "A Method for the Construction of Minimum Redundancy Codes," *Proc. IRE*, Vol. 40, pp. 1098–1101, September 1952.
5. ITU-T Recommendation T.6, Facsimile Coding Schemes and Coding Control Functions for Group 4 Facsimile Apparatus, Vol. VII-Facsimile VII.3, pp. 48–57.
6. "Error Control Option for Group 3 Facsimile Equipment," National Communications System Technical Information Bulletin, 87-4, January 1987.
7. ISO DIS 10918-1/ITU-T T.81, "Digital Compression of Continuous-Tone Still Images, Part I: Requirements and Guidelines."
8. J. L. Mitchell, and W. B. Pennebaker, JPEG Still Image Data Compression Standard, Van Nostrand Reinhold, 1993.

SECTION 22

BROADCAST AND CABLE SYSTEMS

Coded orthogonal frequency multiplex (COFDM) is used in new audio transmission systems so that numerous low-power broadcast transmitters spaced over a wide area can broadcast the identical signal on the same identical frequency. As an automobile travels around the region, CD quality reception is obtained continuously with no interference or receiver retuning required.

Radio broadcast data systems have digital subcarriers superimposed or added to the existing channel spectrum. These subcarriers transmit such information to the receivers as station call letters, road conditions, program or music names, and other data believed to be useful in a moving automobile.

One of digital technology's main advantages is the applicability of computer processing power and features so that the challenges of scrambling and signal security can be approached with new techniques. Unfortunately, with cable television systems, this advanced digital technology also provides new means of attack for the would-be cable television pirate. To counter this problem, it must be possible to respond to a security attack by replacing the breached security element. If the cable operator owns the set-top boxes, they can be replaced. If the set-top boxes were sold to the subscriber, a plug-in card or module, called the point of deployment module, is used. It remains the property of the cable operator and can be disabled electronically. The subscriber would then be given a new POD device based on a system not yet breached.   R.J.

In This Section:

On the CD-ROM:

Frequency Channel Assignments for FM Broadcasting

Numeral Designation of Television Channels

Zone Designations for Television Broadcasting

Minimum Distance Separation Requirements for FM Broadcast Transmitters in KM (MI)

Medium Wave AM Standard Broadcasting Definitions

Frequency Modulation Broadcasting Definitions

Analog Television (NTSC) Definitions

ATSC Digital Transmission System Definitions

Short Wave Broadcasting Definitions

Projections for Digital Cable

CHAPTER 22.1
BROADCAST TRANSMISSION PRACTICE

Earl F. Arbuckle, III

INTRODUCTION

Broadcasting refers to those wireless services, such as standard broadcast, FM, television, and short-wave, which are intended for free reception by the general public. Point-to-point communications links (i.e., two-way radio) and point-to-multipoint private or subscription-based transmission systems (i.e., MMDS, ITFS, and DBS satellite) are not included in this definition and are therefore not discussed in this section.

In the United States, broadcasting activities are regulated by the Federal Communications Commission (FCC). The FCC is empowered by the Communications Act of 1934 (as amended) to administratively control the use of radio frequency (RF) spectrum by means of licensing users within an allocations framework developed through international consensus at various World Administrative Radio Conferences (WARC). Potential licensees must demonstrate by formal application and engineering studies that a license grant would conform to spectrum allocation, technical performance, and public service criteria spelled out in the FCC Rules. In the following paragraphs, reference is made, when appropriate, to the specific section of the FCC Rules relevant to the paragraph. Because technology is developing rapidly and the FCC Rules are dynamic, the reader is warned to consult them directly whenever specific compliance must be determined. For the same reason, this handbook will state only the fundamental tenets of those rules which are likely to remain immutable.

ALLOCATIONS

Standard Broadcast (AM)

An amplitude-modulated (AM) standard broadcast channel is that band of frequencies occupied by the carrier and the upper and lower sidebands with the carrier frequency at the center. Channels are designated by their assigned carrier frequencies. The AM broadcast band consists of 117 carrier frequencies which begin at 540 kHz and progress in 10 kHz steps to 1700 kHz. (FCC Rules 73.14)

Classes of AM broadcast channels and stations are:

Clear channel. One on which stations are assigned to serve wide areas. These stations are protected from objectionable interference within their primary service areas and, depending on the class of station, their secondary service areas. Stations operating on these channels are classified as follows:

Class A station. An unlimited time station that operates on a clear channel and is designed to render primary and secondary service over an extended area and at relatively long distances from its transmitter. Its primary service area is protected from objectionable interference from other stations on the same and adjacent channels, and its secondary service area is protected from interference from other stations on the same channel. The operating power shall not be less than 10 kW nor more than 50 kW.

Class B station. An unlimited time station, which is designed to render service only over a primary service area. Class B stations are authorized to operate with a minimum power of 0.25 kW (or, if less than 0.25 kW, an equivalent of RMS antenna field of at least 141 mV/m at 1 km) and a maximum power of 50 kW, or 10 kW for stations that are authorized to operate in the 1605–1705 kHz band.

Class D station. Operates either daytime, limited time, or unlimited time with nighttime power less than 0.25 kW and an equivalent RMS antenna field of less than 141 mV/m at 1 km. Class D stations shall operate with daytime powers not less than 0.25 kW nor more than 50 kW. Nighttime operations of Class D stations are not afforded protection and must protect all Class A and Class B operations during nighttime hours. New Class D stations that had not been previously licensed as Class B will not be authorized.

Regional channel. One on which Class B and Class D stations may operate and serve primarily a principal center of population and the rural area contiguous thereto.

Local channel. One on which stations operate unlimited time and serve primarily a community and the suburban and rural areas immediately contiguous thereto.

Class C station. Operates on a local channel that is designed to render services only over a primary service area that may be reduced as a consequence of interference. The power shall not be less than 0.25 kW, nor more than 1 kW. Class C stations that are licensed to operate with 0.1 kW may continue to do so. (FCC Rules 73.21)

Frequency Modulation Broadcast (FM)

The FM broadcast band consists of that portion of the radio frequency spectrum between 88 and 108 MHz. It is divided into 100 channels of 200 kHz each. For convenience, the frequencies available for FM broadcasting (including those assigned to noncommercial educational broadcasting) are given numerical designations that are shown in the accompanying CD-ROM.

Different classes of FM station are provided so as to best serve the intended coverage area. Class A stations, for example, are authorized to serve a single community of limited geographic size. Class B stations have increased capability, which is suitable for metropolitan area coverage. Class C stations, on the other hand, are considered regional services and are granted increased power and antenna height accordingly. The maximum effective radiated power (ERP) in any direction, reference height above average terrain (HAAT), and distance to the class contour for each FM station class are listed in Table 22.1.1.

TABLE 22.1.1   Maximum ERP, HAAT, and Distance to the Class Contour for Each FM station Class

Station class	Maximum ERP	Reference HAAT in meters (ft)	Class contour distance in kilometers (mi)
A	6 kW (7.8 dBk)	100 (328)	28 (17)
B1	25 kW (14.0 dBk)	100 (328)	39 (24)
B	50 kW (17.0 dBk)	150 (492)	52 (32)
C3	25 kW (14.0 dBk)	100 (328)	39 (24)
C2	50 kW (17.0 dBk)	150 (492)	52 (32)
C1	100 kW (20.0 dBk)	299 (981)	72 (45)
C	100 kW (20.0 dBk)	600 (1968)	92 (57)
D*	0.01 kW (−20.0 dBk)	N/A	N/A

*Secondary

Analog NTSC Television Broadcast (TV)

A television channel is a band of frequencies 6 MHz wide in one of the television broadcast bands (VHF or UHF) and designated either by number or by the extreme lower and upper frequencies. Numerical designation of channels is given on the CD-ROM. Channels 2 through 13 are considered very high frequency (VHF), while channels 14 through 69 are considered ultrahigh frequency (UHF). The FCC Rules provide specific channel allocations to cities and towns so as to maximize spectrum reuse and efficiency.

Unlike radio stations, television stations are not actually described by a class designation, but rather by channel number and service level; namely, a station would be considered low-band VHF if assigned to a channel from 2 through 6, high-band VHF if assigned to a channel from 7 through 13, or UHF if assigned to a channel from 14 through 69. A station is considered either full power or low power. The following shows the allowable effective radiated power and antenna height for full power stations, which also vary by geographic zone (see Zone Descriptions on the CD-ROM).

Minimum power requirements require at least −10 dBk (100 W) horizontally polarized visual effective radiated power in any horizontal direction. No minimum antenna height above average terrain is specified. Maximum power may not exceed the permitted boundaries specified in the following formulas:

1. Channels 2 to 6 in Zone I:

   $ERPMax = 102.57 - 33.24^* \, Log10 \, (HAAT)$, and $-10 \, dBk \leq ERPMax \leq 20 \, dBk$

2. Channels 2 to 6 in Zones II and III:

   $ERPMax = 67.57 - 17.08^* \, Log10 \, (HAAT)$ and $10 \, dBk \leq ERPMax \leq 20 \, dBk$

3. Channels 7 to 13 in Zone I:

   $ERPMax = 107.57 - 33.24^* \, Log10 \, (HAAT)$, and $-4.0 \, dBk \leq ERPMax \leq 25 \, dBk$

4. Channels 7 to 13 in Zones II and III:

   $ERPMax = 72.57 - 17.08^* \, Log10 \, (HAAT)$, and $15 \, dBk \leq ERPMax \leq 25 \, dBk$

5. Channels 14 to 69 in Zones I, II, and III:

   $ERPMax = 84.57 - 17.08^* \, Log10 \, (HAAT)$, and $27 \, dBk \leq ERPMax \leq 37 \, dBk$

   *Where ERPMax = maximum effective radiated power in decibels above 1 kW (dBk) and HAAT = height above average terrain in meters.

The boundaries specified are to be used to determine the maximum possible combination of antenna height and ERPdBk. When specifying an ERP less than that permitted by the lower boundary, any antenna HAAT can be used. Also, for values of antenna HAAT greater than 2300 m the maximum ERP is the lower limit specified for each equation.

Service Levels and Coverage Prediction

In general, AM broadcast service is determined by field strength and interference protection levels. Interference protection is necessary since AM medium wave frequencies have the potential to travel great distances. FM and TV broadcast service, on the other hand, is determined mainly by field strength, since interference protection is typically not provided in the FCC Rules. FM and TV operate on frequencies that can be characterized as having "line-of-sight" propagation.

Standard broadcast (AM) service levels are as follows:

Mainland U.S. Class A stations operate with a power of 50 kW. These stations are afforded protection as follows (stations in Alaska are subject to slightly different standards). In the daytime, protection is provided to the 0.1 mV/m groundwave contour from stations on the same channel, and to the 0.5 mV/m groundwave contour from stations on adjacent channels. In the nighttime, such protection extends to the 0.5 mV/m—50 percent skywave contour from stations on the same channel.

Class B stations are stations that operate on clear and regional channels with powers not less than 0.25 kW nor more than 50 kW. Class B stations should be located so that the interference received from other stations will not limit the service area to a groundwave contour value greater than 2.0 mV/m nighttime and to the 0.5 mV/m groundwave contour daytime, which are the values for the mutual protection between this class of stations and other stations of the same class.

Class C stations operate on local channels with powers not less than 0.25 kW, nor more than 1 kW. Such stations are normally protected to the daytime 0.5 mV/m contour. On local channels the separation required for the daytime protection shall also determine the nighttime separation.

Class D stations operate on clear and regional channels with daytime powers of not less than 0.25 kW (or equivalent RMS field of 141 mV/m at 1 km if less than 0.25 kW) and not more than 50 kW. Class D stations that have previously received nighttime authority operate with powers of less than 0.25 kW (or equivalent RMS fields of less than 141 mV/m at 1 km) are not required to provide nighttime coverage and are not protected from interference during nighttime hours.

When a station is already limited by interference from other stations to a contour value greater than that normally protected for its class, the individual received limits shall be the established standard for such station with respect to interference from each other station.

The four classes of AM broadcast stations have in general three types of service area, i.e., primary, secondary, and intermittent. Class A stations render service to all three areas. Class B stations render service to a primary area but the secondary and intermittent service areas may be materially limited or destroyed due to interference from other stations, depending on the station assignments involved. Class C and Class D stations usually have only primary service areas. Interference from other stations may limit intermittent service areas and generally prevents any secondary service to those stations which operate at night. Consistent intermittent service may still be obtained in many cases depending on the station assignments involved.

The groundwave signal strength required to render primary service is 2 mV/m for communities with populations of 2500 or more and 0.5 mV/m for communities with populations of less than 2500. (See FCC Rules Part 73.184 for curves showing distance to various groundwave field strength contours for different frequencies and ground conductivities.)

A Class C station may be authorized to operate with a directional antenna during daytime hours provided the power is at least 0.25 kW. In computing the degrees of protection which such antenna will afford, the radiation produced by the directional antenna system must be assumed to be no less, in any direction, than that which would result from nondirectional operation using a single element of the directional array, with 0.25 kW.

All classes of AM broadcast stations have primary service areas subject to limitation by fading and noise, and interference from other stations to the contours set out for each class of station.

Secondary service is provided during nighttime hours in areas where the skywave field strength, 50 percent or more of the time, is 0.5 mV/m or greater. Satisfactory secondary service to cities is not considered possible unless the field strength of the skywave signal approaches or exceeds the value of the groundwave field strength that is required for primary service. Secondary service is subject to some interference and extensive fading whereas the primary service area of a station is subject to no objectionable interference or fading. Only Class A stations are assigned on the basis of rendering secondary service. (Standards have not been established for objectionable fading because of the relationship to receiver characteristics. Selective fading causes audio distortion and signal strength reduction below the noise level, objectionable characteristics inherent in many modern receivers. The automatic volume control circuits in the better designed receivers generally maintain the audio output at a sufficiently constant level to permit satisfactory reception during most fading conditions.)

Intermittent service is rendered by the groundwave and begins at the outer boundary of the primary service area and extends to a distance where the signal strength decreases to a value that is too low to provide any service. This may be as low as a few μV/m in certain areas and as high as several millivolts per meter in other areas of high noise level, interference from other stations, or objectionable fading at night. The intermittent service area may vary widely from day to night and generally varies over shorter intervals of time. Only Class A stations are protected from interference from other stations to the intermittent service area.

Broadcast stations are licensed to operate unlimited time, limited time, daytime, share time, and specified hours. New stations may be licensed only for unlimited time operation. Unlimited time stations may operate 24 h per day. Limited time stations (Class B only) may operate only during daytime hours until local sunset or sunset at the nearest Class A station on the same frequency, whichever is easternmost. Daytime stations may operate only between local sunrise and local sunset. Share time stations, of which there are very few, operate on a frequency that is shared by another station in the same service area in accordance with a mutually agreed upon schedule which the FCC incorporates into the stations' licenses. Stations authorized to operate during specified hours will have those hours enumerated in their FCC license.

FM Broadcast Service Levels

FM broadcast service levels are determined by the class of FCC authorization. Within any class, coverage will be a function of effective radiated power (ERP) and antenna height above average terrain (HAAT). In general, higher ERP and higher HAAT result in larger coverage areas. The FCC table of allotments takes these factors into consideration to minimize the potential for interference or coverage overlap. Thus, FM broadcast stations are not generally protected from interference caused by the operation of other FM stations, so long as those stations operate in compliance with the FCC Rules. Class D (secondary) stations may not cause interference to primary stations.

The primary protection against interference is based on a minimum allowable spacing between FM broadcast transmitter sites (see accompanying CD-ROM).

Determination of predicted field strengths for FM broadcast stations is made using the FCC F(50/50) graphs. (Refer to FCC Rules 73.699 for the current graphs.) The 50 percent field strength is defined as that value exceeded for 50 percent of the time. The procedure to be used is as follows:

The F(50/50) chart gives the estimated 50 percent field strengths exceeded at 50 percent of the locations in dB above 1 μV/m. The chart is based on an effective power radiated from a half-wave dipole antenna in free space, that produces an unattenuated field strength at 1 km of about 107 dB above 1 μV/m (221.4 mV/m).

To use the chart for other ERP values, convert the ordinate scale by the appropriate adjustment in dB. For example, the ordinate scale for an ERP of 50 kW (17 dBk) should be adjusted by 17 dB and, therefore, a field strength of 40 dBu would be converted to 57 dBu. When predicting the distance to field strength contours, use the maximum ERP of the main radiated lobe in the pertinent azimuthal direction. When predicting field strengths over areas not in the plane of the maximum main lobe, use the ERP in the direction of such areas, determined by considering the appropriate vertical radiation pattern.

The antenna height to be used with this chart is the height of the radiation center of the antenna above the average terrain along the radial in question. In determining the average elevation of the terrain, the elevations between 3 and 16 km from the antenna site are used.

Profile graphs are drawn for eight radials beginning at the antenna site and extending out 16 km. The radials should be drawn for each 45° of azimuth starting with True North. At least one radial must include the principal community to be served even though it may be more than 16 km from the antenna site. However, in the event none of the evenly spaced radials include the principal community to be served, and one or more such radials are drawn in addition, these radials must not be used in computing the antenna height above average terrain.

The profile graph for each radial is then plotted by contour intervals from 12 to 30 m, and, where the data permit, at least 50 points of elevation (generally uniformly spaced) should be used for each radial. In instances of very rugged terrain where the use of contour intervals of 30 m would result in several points in a short distance, 60- or 12-m contour intervals may be used for such distances. On the other hand, where the terrain is uniform or gently sloping, the smallest contour interval indicated on the topographic map should be used, although only relatively few points may be available. The profile graph should indicate the topography accurately for each radial, and the graphs should be plotted with the distance in kilometers as the abscissa and the elevation in meters above mean sea level as the ordinate. The profile graphs should indicate the source of the topographical data used. The graph should also show the elevation of the center of the radiating system. The graph may be plotted either on rectangular coordinate paper or on special paper that shows the curvature of the earth (commonly called 4/3 radius paper). It is not necessary to take the curvature of the earth into consideration in this procedure, as this factor is taken care of in the charts showing signal strengths. The average elevation of the 13 km distance between 3 and 16 km from the antenna site should then be determined from the profile graph for each radial. This may be obtained by averaging a large number of equally spaced points, by using a planimeter, or by obtaining the median elevation (that exceeded for 50 percent of the distance) in sectors and averaging those values. In the event that any of the radials encompass large amounts of water or non-U.S. territory, slightly different rules are applied.

Here is an example calculation of HAAT where all of the radials are over land within the United States and the heights above average terrain on the eight radials are as follows:

Radial	Meters
0°	120
45°	255
90°	185
135°	90
180°	−10
225°	−85
270°	40
315°	85

The antenna height above terrain is computed as follows:

$$(120 + 255 + 185 + 90 - 10 - 85 + 40 + 85)/8 = 85 \text{ m}$$

In cases where the terrain in one or more directions from the antenna site departs widely from the average elevation of the 3- to 16-km sector, the prediction method may indicate contour distances that are different from what may be expected in practice. For example, a mountain ridge may indicate the practical limit of service although the prediction method may indicate otherwise. In such cases, the prediction method should be followed, but a supplemental showing may be made concerning the contour distances as determined by other means. Such supplemental showings should describe the procedure used and should include sample calculations. Maps of predicted coverage should include both the coverage as predicted by the regular method and as predicted by a supplemental method. When measurements of an area are required, these should include the area obtained by the regular prediction method and the area obtained by the supplemental method. In directions where the terrain is such that antenna heights less than 30 m for the 3- to 16-km sector are obtained, an assumed height of 30 m should be used for the prediction of coverage. However, where the actual contour distances are critical factors, a supplemental showing of expected coverage must be included together with a description of the method used in predicting such coverage. In special cases, the FCC may require additional information as to terrain and coverage.

The effect of terrain roughness on the predicted field strength of a signal at points distant from an FM transmitting antenna is assumed to depend on the magnitude of a terrain roughness factor which is a measure of the number of points along a profile segment where the elevations are exceeded by all points on the profile for 10 and 90 percent, respectively, of the length of the profile segment. This factor is simply added to the profile average height. No terrain roughness correction need be applied when all field strength values of interest are predicted to occur 10 km or less from the transmitting antenna. The FCC F(50/50) field strength charts were developed assuming a terrain roughness factor of 50 m, which is considered to be representative of average terrain in the United States. Where the roughness factor for a particular propagation path is found to depart appreciably from this value, a terrain roughness correction should be applied.

Television Broadcast Service Levels

In the authorization of TV stations, two field strength contours are represented in the FCC F(50,50) coverage chart. These are specified as Grade A and Grade B and indicate the approximate extent of coverage over average terrain in the absence of interference from other television stations. Under actual conditions, the true coverage may vary greatly from these estimates because the terrain over any specific path is likely to be different from the average terrain on which the field strength charts were based. The required field strength in decibels above 1 μV/m (dBu) for the Grade A and Grade B contours is as follows:

	Grade A (dBu)	Grade B (dBu)
Channels 2–6	68	47
Channels 7–13	71	56
Channels 14–69	74	64

It should be realized that the F(50/50) curves when used for Channels 14 to 69 are not based on measured data at distances beyond about 48.3 km (30 mi). Theory would indicate that the field strengths for Channels 14 to 69 should decrease more rapidly with distance beyond the horizon than for Channels 2 to 6. For this reason, the curves should be used with appreciation of their limitations in estimating levels of field strength. Further, the actual extent of service will usually be less than indicated by these estimates due to interference from other stations. Because of these factors, the predicted field strength contours give no assurance of service to any specific percentage of receiver locations within the distances indicated.

As in the case of FM broadcast stations, television stations are not guaranteed any protection from interference except for that arising out of minimum spacing requirements. In the special case of channel 6 television stations, additional protection is afforded by restrictions on the operation of noncommercial educational stations on the adjacent lower end of the FM broadcast band (FCC Rules 73.525).

Technical Standards

In the United States, the Federal Communications Commission (FCC) establishes the technical standards with which all broadcast facilities must comply. Such standards have the goal of providing maximum service to the public and, therefore, address issues such as transmitter emissions purity, frequency, power, and modulation. These standards help ensure that each broadcaster radiates a signal that is compatible with the intended receivers without causing undo interference to other stations.

Medium wave AM standard broadcasting definitions are included on the accompanying CD-ROM.

Amplitude Modulation Broadcasting

The emissions of AM broadcast stations must not cause harmful interference. Such emissions may be measured using a properly operated and suitable swept-frequency RF spectrum analyzer using a peak hold duration of 10 min, no video filtering, and a 300-Hz resolution bandwidth, except that a wider resolution bandwidth may be employed above 11.5 kHz to detect transient emissions. Alternatively, other specialized receivers or monitors with appropriate characteristics may be used to determine compliance with the FCC Rules. Measurements made of the emissions of an operating station are to be made at ground level approximately 1 km from the center of the antenna system.

FCC Rules require that emissions 10.2 to 20 kHz removed from the carrier must be attenuated at least 25 dB below the unmodulated carrier level, emissions 20 to 30 kHz removed from the carrier must be attenuated at least 35 dB below the unmodulated carrier level, emissions 30 to 60 kHz removed from the carrier must be attenuated at least [5 + 1 dB/kHz] below the unmodulated carrier level, and emissions between 60 and 75 kHz of the carrier frequency must be attenuated at least 65 dB below the unmodulated carrier level. Emissions removed by more than 75 kHz must be attenuated at least 43 + 10 Log (Power in watts) or 80 dB below the unmodulated carrier level, whichever is the lesser attenuation, except for transmitters having power less than 158 W, where the attenuation must be at least 65 dB below carrier level.

The percentage of modulation is to be maintained at as high as is consistent with good quality of transmission and good broadcast service. Generally, the modulation should not be less than 85 percent on peaks of frequency recurrence, but where lower modulation levels may be required to avoid objectionable loudness or to maintain the dynamic range of the program material, the degree of modulation may be reduced to whatever level is necessary for this purpose, even though under such circumstances, the level may be substantially less than the level that produces peaks of frequency recurrence at a level of 85 percent.

Maximum modulation levels for AM stations must not exceed 100 percent on negative peaks of frequent recurrence, or 125 percent on positive peaks at any time. AM stations transmitting stereophonic programs must not exceed the AM maximum stereophonic transmission signal modulation specifications of the stereophonic system in use. For AM stations transmitting telemetry signals for remote control or automatic transmission system operations, the amplitude of modulation of the carrier by the use of subaudible tones must not be higher than necessary to effect reliable and accurate data transmission and may not, in any case, exceed 6 percent. If a limiting or compression amplifier is employed to maintain modulation levels, precaution must be taken so as not to substantially alter the dynamic characteristics of programs.

AM Stereophonic Transmission

The FCC has never issued a technical standard for AM stereo transmission, but rather chose to adapt a market-driven approach wherein several competing systems were allowed to coexist. Only two systems survived this competition, Motorola's C-Quam and Kahn's ISB (independent sideband). The systems are incompatible, but both are mono compatible.

C-Quam uses a quadrature phase modulation scheme wherein the main amplitude modulated carrier contains the left plus right channel information, thus insuring monaural receiver compatibility. The left minus right stereo information is transmitted on two subcarriers, phase modulated in quadrature, which are recovered in the C-Quam receiver and combined with the main carrier modulation in a matrix, resulting in separate left and right channels that are then amplified and fed to speakers.

The ISB system essentially uses the redundant characteristics of normal full carrier, double sideband amplitude modulation to provide left and right channels. All of the information being transmitted by such a system can be recovered, using a suitable receiver, from either the upper or the lower sideband, with or without the carrier. So the ISB system simply transmits the left channel on the lower sideband and the right channel on the upper sideband. The carrier is not suppressed, so as to ensure compatibility with monaural receivers, which recover both channels additively to produce monaural audio. ISB receivers recover the sidebands independently, routing left and right channels to the corresponding amplifiers. However, AM stereo failed in the marketplace.

Frequency Modulation Broadcasting

 Definitions are given on the accompanying CD-ROM.

FM broadcast stations must maintain the bandwidth occupied by their emissions in accordance with the FCC specification detailed below. If harmful interference to other authorized stations occurs, the problem must be corrected promptly or the station may be forced to cease operation.

1. Any emission appearing on a frequency removed from the carrier by between 120 kHz and 240 kHz inclusive must be attenuated at least 25 dB below the level of the unmodulated carrier. Compliance with this requirement will be deemed to show the occupied bandwidth to be 240 kHz or less.

2. Any emission appearing on a frequency removed from the carrier by more than 240 kHz and up to and including 600 kHz must be attenuated at least 35 dB below the level of the unmodulated carrier.

3. Any emission appearing on a frequency removed from the carrier by more than 600 kHz must be attenuated at least 43 + 10 Log 10 (power, in watts) dB below the level of the unmodulated carrier, or 80 dB, whichever is the lesser attenuation.

Preemphasis shall not be greater than the impedance-frequency characteristics of a series inductance resistance network having a time constant of 75 μs.

The percentage of modulation is to be maintained at as high a level as is consistent with good quality of transmission and good broadcast service. Generally, the modulation should not be less than 85 percent on peaks of frequent recurrence, but where lower modulation levels may be required to avoid objectionable loudness or to maintain the dynamic range of the program material, the degree of modulation may be reduced to whatever level is necessary for this purpose, even though under such circumstances the level may be substantially less than the level that produces peaks of frequent recurrence at a level of 85 percent.

Maximum modulation levels must not exceed 100 percent on peaks of frequent recurrence referenced to 75-kHz deviation. However, stations providing subsidiary communications services using subcarriers concurrently with the broadcasting of stereophonic or monophonic programs may increase the peak modulation deviation as follows:

1. The total peak modulation may be increased 0.5 percent for each 1.0 percent subcarrier injection modulation.

2. In no event may the modulation of the carrier exceed 110 percent (92.5-kHz peak deviation).

If a limiting or compression amplifier is employed to maintain modulation levels, precaution must be taken so as not to substantially alter the dynamic characteristics of programs. Modern audio processing, often using digital circuitry, is able to significantly increase the density of modulation without adverse impairment to the audio.

Multiplex Stereo

FM broadcast stations may transmit stereophonic programming via a compatible multiplex system. This system uses a matrix to derive sum (left + right) and difference (left − right) signals from the applied audio. The monaural and stereophonic frequency response is 50–15,000 Hz.

The sum signal is used to frequency modulate the main carrier. A monaural receiver will demodulate this main carrier into an audio signal containing both left and right signals, so no information will be lost. The difference signal is used to modulate a double-sideband stereo subcarrier operating at 38 kHz. This subcarrier, which must be suppressed to a level of less than 1 percent of the main carrier, is developed as a phase-locked second harmonic of the 19,000 ± 2 Hz stereo "pilot" signal, which frequency modulates the main carrier between the limits of 8 and 10 percent. This pilot signal is used by stereophonic receivers to enable stereo demodulating circuits and, almost always, some kind of stereo mode indicator. The stereo demodulator will provide a difference signal that can then be combined with the sum signal to regenerate the original left and right audio signals at the receiver output.

Because of the "triangular" shape of the noise spectrum, the effective noise floor of the 38 kHz difference subcarrier is higher than that of the baseband audio frequencies. The result is that signal-to-noise ratios are somewhat poorer for a stereo transmission system than for a monaural system. (This is the reason for the use of a companded difference channel in the BTSC television stereo system.)

Subsidiary Communications Authorization (SCA)

The FCC may issue a subsidiary communications authorization to an FM broadcast station so that station may provide limited types of subsidiary services on a multiplex basis. Permissible uses fall within the following categories:

1. Transmission of programs or data which are of a broadcast nature but which are of interest primarily to limited segments of the public wishing to subscribe thereto. Examples include background music, storecasting, detailed weather forecasting, real-time stock quotations, special time signals, and other material of a broadcast nature expressly designed and intended for business, professional, educational, religious, trade, labor, agricultural, or other groups.

2. Transmission of signals which are directly related to the operation of FM broadcast stations, i.e., relaying broadcast material to other FM and standard broadcast stations, remote cueing and order circuits, remote control telemetry, and similar uses. SCA operations may be conducted without time restriction so long as the main channel is programmed simultaneously.

SCA subcarriers must be frequency modulated and are restricted to the range of 20 to 75 kHz, unless the station is also broadcasting stereophonically in which case the restriction is 53 to 75 kHz.

Digital Audio Radio Service (DARS)

In 1997, the FCC finally awarded spectrum after a long rule-making process for DARS to two winners in an auction. XM Satellite Radio and Sirius Radio each inaugurated pay-to-listen subscription satellite-delivered services in 2001 with no free-to-air component, so they do not fit the definition of broadcasting and will not be discussed here.

Currently there is no free digital radio broadcasting available in the United States. The FCC continues to consider various approaches to an analog-to-digital transition for the incumbent terrestrial radio broadcasters. However, no system has yet been deployed as of 2002.

Analog NTSC Television Broadcasting

 The accompanying CD-ROM includes definitions.

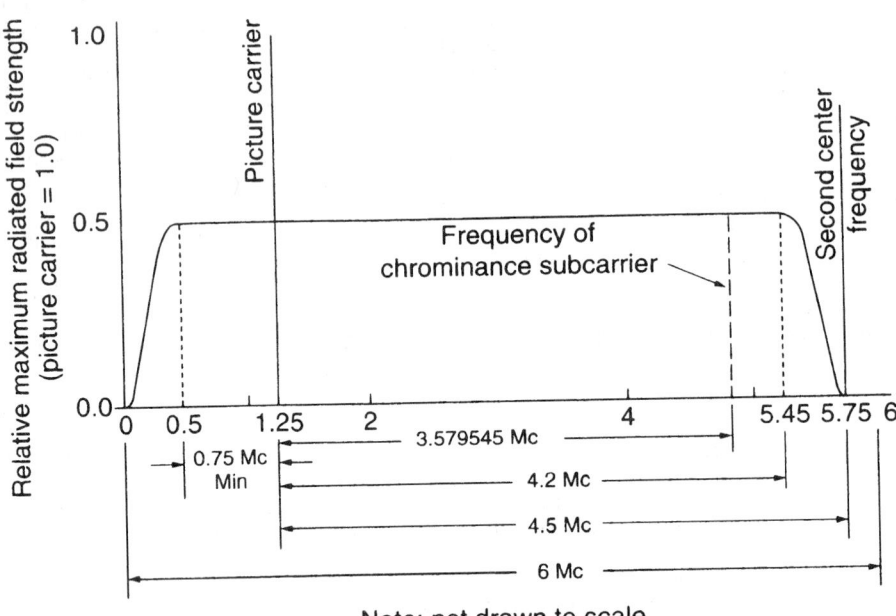

FIGURE 22.1.1 Visual amplitude characteristic.

Visual Transmission Systems

Transmission standards are:

1. The width of the television broadcast channel is 6 MHz.

2. The visual carrier frequency is nominally 1.25 MHz above the lower boundary of the channel.

3. The aural center frequency is 4.5 MHz higher than the visual carrier frequency.

4. The visual transmission amplitude characteristic is generally required to conform to that shown in Fig. 22.1.1.

5. The chrominance subcarrier frequency is 63/88 times precisely 5 MHz (3.57954545 . . . MHz). The tolerance is ±10 Hz, and the rate of frequency drift must not exceed 0.1 Hz per second (cycles per second squared).

6. For monochrome and color transmissions the number of scanning lines per frame is 525, interlaced two to one in successive fields. The horizontal scanning frequency is 2/455 times the chrominance subcarrier frequency; this corresponds nominally to 15,750 Hz (with an actual value of 15,734.264 ± 0.044 Hz). The vertical scanning frequency is 2/525 times the horizontal scanning frequency; this corresponds nominally to 60 Hz (the actual value is 59.95 Hz). For monochrome transmissions only, the nominal values of line and field frequencies may be used.

7. The aspect ratio of the transmitter television picture is four units horizontally to three units vertically.

8. During active scanning intervals, the televised scene is scanned from left to right horizontally and from top to bottom vertically, at uniform velocities.

9. A carrier is modulated within a single television channel for both picture and synchronizing signals.

10. A decrease in initial light intensity will cause an increase in radiated power (negative transmission).

11. The reference black level is always represented by a definite carrier level.

12. The blanking level is transmitted at 75 ± 2.5 percent of the peak carrier level.

13. The reference white level of the luminance signal is 12.5 ± 2.5 percent of the peak carrier level.

14. It is customary to employ horizontal antenna polarization. However, circular or elliptical polarization may be employed if desired, in which case clockwise (right-hand) rotation, as defined in the IEEE Standard Definition. 42A65-3E2, and transmission of the horizontal and vertical components in time and space quadrature shall be used. For either omnidirectional or directional antennas the licensed effective radiated power of the vertically polarized component may not exceed the licensed effective radiated power of the horizontally polarized component. For directional antennas, the maximum effective radiated power of the vertically polarized component must not exceed the maximum effective radiated power of the horizontally polarized component in any specified horizontal or vertical direction.

15. The effective radiated power of the aural transmitter must not exceed 22 percent of the peak radiated power of the visual transmitter, but may be less. Typically, 10 percent aural power is used.

16. The peak-to-peak variation of transmitter output within one frame of video signal due to all causes including hum, noise, and low-frequency response, measured at both scanning synchronizing peak and blanking level, must not exceed 5 percent of the average scanning synchronizing peak signal amplitude.

17. The reference black level must be separated from the blanking level by setup interval, which shall be 7.5 ± 2.5 percent of the video range from blanking level to the reference white level.

18. For monochrome transmission, the transmitter output should vary in substantially inverse logarithmic relation to the brightness of the subject.

19. The color picture signal consists of a luminance component transmitted as amplitude modulation of the picture carrier and a simultaneous pair of chrominance components transmitted as the amplitude modulation sidebands of a pair of suppressed subcarriers in quadrature.

20. The radiated chrominance subcarrier must vanish on the reference white of a transmitted scene.

Visual transmitter performance:

1. The field strength or voltage of the lower sideband must not be greater than 20 dB for a modulating frequency of 1.25 MHz or greater and in addition, for color, shall not be greater than -42 dB for a modulating frequency of 3.579545 MHz (the color subcarrier frequency). For both monochrome and color, the field strength or voltage of the upper sideband shall not be greater than -20 dB for a modulating frequency of 4.75 MHz or greater. For stations operating on Channels 15 to 69 and employing a transmitter delivering maximum peak visual power output of 1 kW or less, the field strength or voltage of the upper and lower sidebands may depart from the visual amplitude characteristic by no more than the following amounts:

   • 2 dB at 0.5 MHz below visual carrier frequency
   • 2 dB at 0.5 MHz above visual carrier frequency
   • 2 dB at 1.25 MHz above visual carrier frequency
   • 3 dB at 2.0 MHz above visual carrier frequency
   • 6 dB at 3.0 MHz above visual carrier frequency
   • 12 dB at 3.5 MHz above visual carrier frequency
   • 8 dB at 3.58 MHz above visual carrier frequency (for color transmission only)

   The field strength or voltage of the upper and lower sidebands must not exceed a level of -20 dB for a modulating frequency of 4.75 MHz or greater. If interference to the reception of other stations is caused by out-of-channel lower sideband emission, the technical requirements applicable to stations operating on Channels 2 to 13 must be met.

2. The attenuation characteristics of the visual transmitter must be measured by application of a modulating signal to the transmitter input terminals in place of the normal composite television video signal. The signal applied will ordinarily be a composite signal composed of a synchronizing signal to establish peak output voltage plus a variable frequency sine wave voltage occupying the interval between synchronizing pulses. (The "synchronizing signal" referred to in this section means either a standard synchronizing wave form or

any pulse that will properly set the peak.) The axis of the sine wave in the composite signal observed in the output monitor should be maintained at an amplitude 0.5 of the voltage at synchronizing peaks. The amplitude of the sine wave input must be held at a constant value. This constant value should be such that at no modulating frequency does the maximum excursion of the sine wave, observed in the composite output signal monitor, exceed the value 0.75 of peak output voltage. The amplitude of the 200-kHz sideband should be measured and designated 0 dB as a basis for comparison. The modulation signal frequency is then varied over the desired range and the field strength or signal voltage of the corresponding sidebands measured. As an alternate method of measuring in those cases in which the automatic d-c insertion can be replaced by manual control, the above characteristic may be taken by the use of a video sweep generator and without the use of pedestal synchronizing pulses. The d-c level should be set for midcharacteristic operation.

3. A sine wave, introduced at those terminals of the transmitter which are normally fed the composite color picture signal, should produce a radiated signal having an envelope delay, relative to the average envelope delay between 0.05 and 0.20 MHz, of 0 microsecond (μs) up to a frequency of 3.0 MHz; and then linearly decreasing to 4.18 MHz so as to be equal to 0.17 μs at 3.58 MHz. The tolerance on the envelope delay is ±0.05 μs at 3.58 MHz. The tolerance increases linearly to ± 0.1 μs down to 2.1 MHz, and remains at ± 0.1 μs down to 0.2 MHz. (Tolerances for the interval of 0.0 to 0.2 MHz are not specified at the present time.) The tolerance should also increase linearly to ± 0.1 μs at 4.18 MHz.

4. The change rate of the frequency of recurrence of the leading edges of the horizontal synchronizing signals should not be greater than 0.15 percent per second, the frequency to be determined by an averaging process carried out over a period of not less than 20, nor more than 100 lines, such lines not to include any portion of the blanking interval.

Requirements applicable to both visual and aural transmitters are:

1. Automatic means must be provided in the visual transmitter to maintain the carrier frequency within ±1 kHz of the authorized frequency; automatic means must also be provided in the aural transmitter to maintain the carrier frequency 4.5 MHz above the actual visual carrier frequency within ±1 kHz.

2. The transmitters should be equipped with suitable indicating instruments for the determination of operating power and with other instruments necessary for proper adjustment, operation, and maintenance of the equipment.

3. Adequate provision must be made for varying the output power of the transmitters to compensate for excessive variations in line voltage or for other factors affecting the output power.

4. Adequate provisions should be provided in all component parts to avoid overheating at the rated maximum output powers.

5. The construction, installation, and operation of broadcast equipment is expected to conform with all applicable local, state, and federally imposed safety regulations and standards, enforcement of which is the responsibility of the issuing regulatory agency.

6. Spurious emissions, including radio frequency harmonics, must be maintained at as low a level as the state of the art permits. As measured at the output terminals of the transmitter (including harmonic filters, if required), all emissions removed in frequency in excess of a 3 MHz above or below the respective channel edge shall be attenuated no less than 60 dB below the visual transmitter power.

7. If a limiting or compression amplifier is used in conjunction with the aural transmitter, due care should be exercised because of the preemphasis in the transmitting system.

8. TV broadcast stations operating on Channel 14 and Channel 69 must take special precautions to avoid interference to adjacent spectrum land mobile radio service facilities.

Vertical Interval Signals

The interval beginning with line 17 and continuing through line 20 of the vertical blanking interval of each field may be used for the transmission of test signals, cue and control signals, and identification signals. Test signals may include signals designed to check the performance of the overall transmission system or its individual components. Cue and control signals shall be related to the operation of the TV broadcast station. Identification signals may be transmitted to identify the broadcast material for its source, and the date and time of its origination.

Modulation of the television transmitter by such signals must be confined to the area between the reference white level and the blanking level, except where test signals include chrominance subcarrier frequencies, in which case positive excursions of chrominance components may exceed reference white, and negative excursions may extend into the synchronizing area. In no case may the modulation excursions produced by test signals extend beyond peak-of-sync, or to zero carrier level.

The use of such signals must not result in significant degradation of the program transmission of the television broadcast station, nor produce emission outside of the frequency band occupied for normal program transmissions.

Vertical interval signals may not be transmitted during the portion of each line devoted to horizontal blanking.

Closed Captioning

Line 21, in each field, may be used for the transmission of a program related data signal, which, when decoded, provides a visual depiction of information simultaneously being presented on the aural channel (closed captions). Such data signal shall conform to the format described in Fig. 22.1.2 and may be transmitted during all periods of regular operation. On a space available basis, Line 21 in field 2 may also be used for text-mode data and extended data service information.

The signals on fields 1 and 2 would ordinarily be distinct data streams, for example, to supply captions in different languages or at different reading levels. The data signal shall be coded using a non-return-to-zero (NRZ) format and shall employ standard ASCII 7 bit plus parity character codes. (Note: For more information on data formats and specific data packets, see EIA-608, "Line 21 Data Services for NTSC" available from the Electronics Industries Association.)

At times when Line 21 is not being used to transmit a program-related data signal, data signals that are not program related may be transmitted, provided the same data format is used and the information to be displayed is of a broadcast nature.

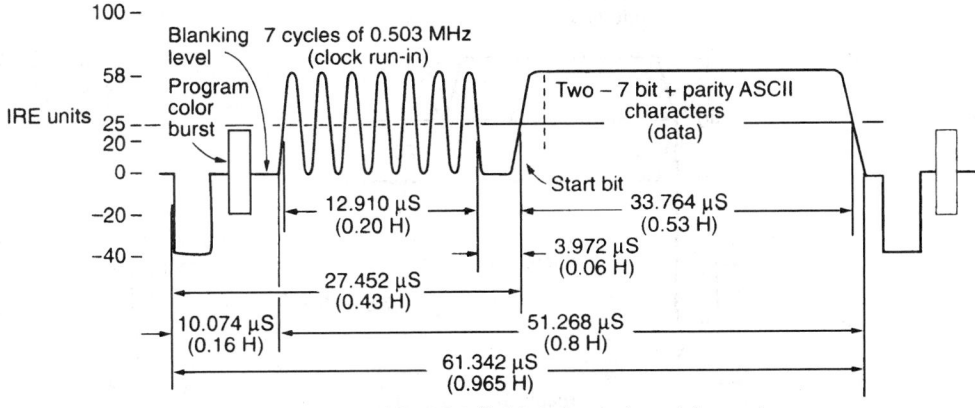

Line 21 field 1 data signal format

Horizontal dimensions not to scale

1. Data "1" = 50 IRE units, data "0" = 0.
2. Data pulse rise time = 2 T bar rise time.
3. Data time base = 32 f_H (0.50349650 MHz.)
4. Data bit interval = H/32 (1.986 μS.)
5. Negative going zero crossings of clock are coherent with data transitions.
6. Data and clock run-in coherent with H.

FIGURE 22.1.2   Line 21 (closed captioning) data format.

Test Signals

Vertical interval test signals include the following:

Composite: Includes white bar to detect the line time distortions (such as over- or underdamped response) and 12.5 "T" pulse, which enables measurement of group (envelope) delay.

Multiburst: Contains packets of various frequencies, which give an indication of amplitude versus frequency response.

Modulated stairstep: Used to determine differential gain and phase response of a transmission system.

Line 19, in each field, may be used only for the transmission of the Ghost-Cancelling Reference (GCR) signal described in OET Bulletin No. 68. The vertical interval reference (VIR) signal formerly permitted on Line 19 and described in Fig. 22.1.3, may be transmitted on any of Lines 10 to 16.

The GCR signal, shown in Fig. 22.1.4, serves as a reference for adaptive filter delay networks in receivers and demodulators (see also Fig. 22.1.5). By correcting the network response to reconstitute a proper GCR in the receiver, active picture area is also corrected to reduce the effects of multipath reception and "ghosts."

Aural Transmission Systems

Transmission Standards are as follows:

1. The modulating signal for the main channel shall consist of the sum of the stereophonic (biphonic, quadraphonic, and the like) input signals.

2. The instantaneous frequency of the baseband stereophonic subcarrier must at all times be within the range of 15 to 120 kHz. Either amplitude or frequency modulation of the stereophonic subcarrier may be used.

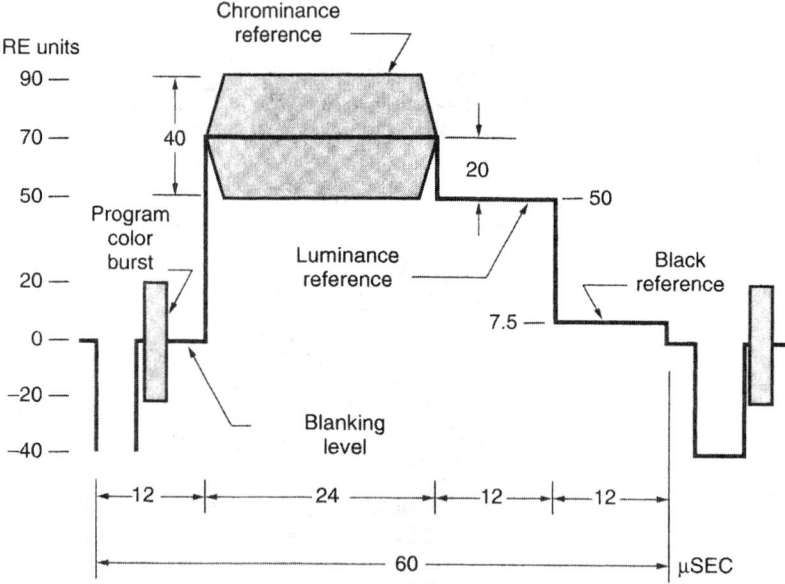

Note: the chrominance reference and the program color burst have the same phase.

FIGURE 22.1.3 Vertical interval reference signal.

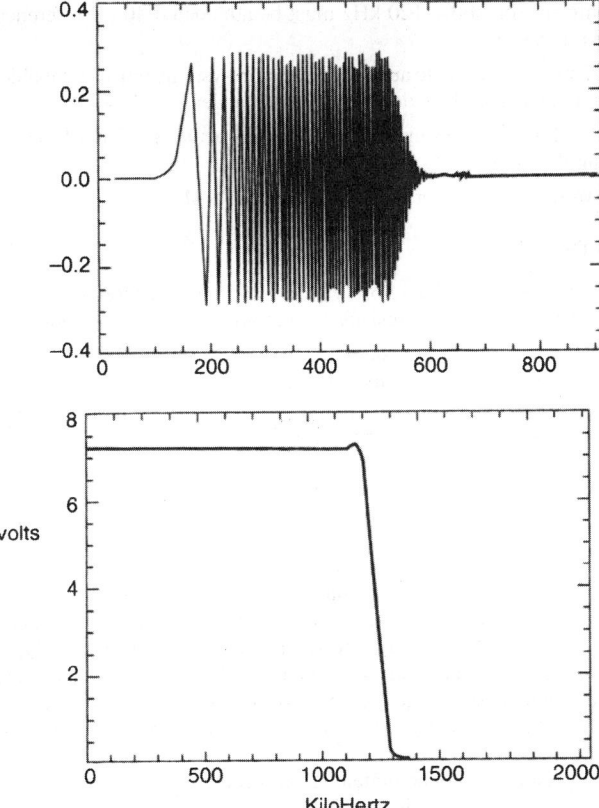

FIGURE 22.1.4   The 525-line system ghost canceling reference signal (A).
The spectrum of the 525-line system GCR (B).

3. One or more pilot subcarriers between 16 and 120 kHz may be used to switch a TV receiver between the stereophonic and monophonic reception modes or to activate a stereophonic audio indicator light, and one or more subcarriers between 15 and 120 kHz may be used for any other authorized purpose; except that stations employing the BTSC system of stereophonic sound transmission and audio processing may transmit a pilot subcarrier at 15,734 Hz ±2 Hz. Other methods of multiplex subcarrier or stereophonic aural transmission systems must limit energy at 15,734 Hz ±20 Hz, to no more than ±0.125 kHz aural carrier deviation.

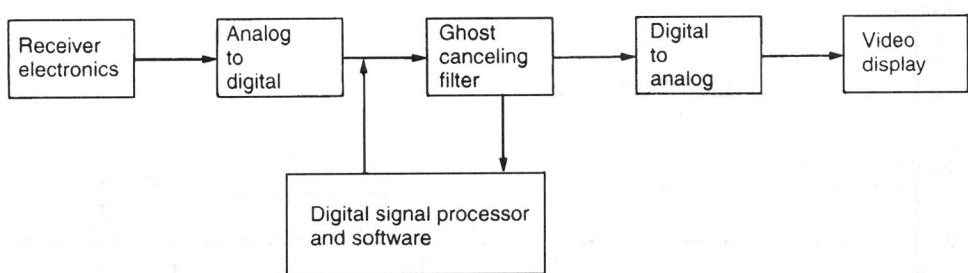

FIGURE 22.1.5   Ghost canceling reference block diagram.

4. Aural baseband information above 120 kHz must be attenuated 40 dB referenced to 25 kHz main channel deviation of the aural carrier.

5. Multiplex subcarrier or stereophonic aural transmission systems must be capable of producing and must not exceed ±25 kHz main channel deviation of the aural carrier.

6. The arithmetic sum of nonmultiphonic baseband signals between 15 and 120 kHz must not exceed ±50 kHz deviation of the aural carrier.

7. Total modulation of the aural carrier must not exceed ±75 kHz.

Aural transmitter performance:

1. Preemphasis should be employed as closely as practicable in accordance with the impedance-frequency characteristics of a series inductance-resistance network having a time constant of 75 μs.

2. If a limiting or compression amplifier is employed, care should be exercised in its connection in the circuit due to the use of preemphasis in the transmitting system.

3. For the aural transmitter of TV broadcast stations, a frequency deviation of ±25 kHz is defined as 100 percent modulation (except in the case of BTSC MTS).

BTSC MTS (Stereo)

The Broadcast Television Systems Committee (BTSC) Multichannel Television Sound (MTS) system was developed by the broadcast industry to provide a standardized approach to the addition of aural subcarriers to the television broadcast sound carrier for the purpose of transmitting multiple sound channels. The system provides for stereo, second audio program (SAP), and a professional cueing (PRO) channel. While the FCC has not enacted MTS as an absolute standard, it has promulgated regulations that protect the technical parameters so that the manufacturers, broadcasters, and the public can expect a consistent multichannel television sound service.

The objectives of the BTSC system, which was adopted in 1984, are as follows:

1. Compatibility with existing monophonic television receivers

2. Simultaneous stereo and SAP capability

3. Provision for other professional or station use subcarriers

4. Noise reduction in both the stereo and SAP channels

5. Increased aural carrier deviation to accommodate the new subcarriers

The BTSC MTS baseband spectrum is shown in Fig. 22.1.6.

The BTSC monophonic (L + R sum) channel is identical to the normal non-BTSC baseband aural signal, ensuring compatibility with existing nonstereo television receivers. The stereo subcarrier is very similar to that

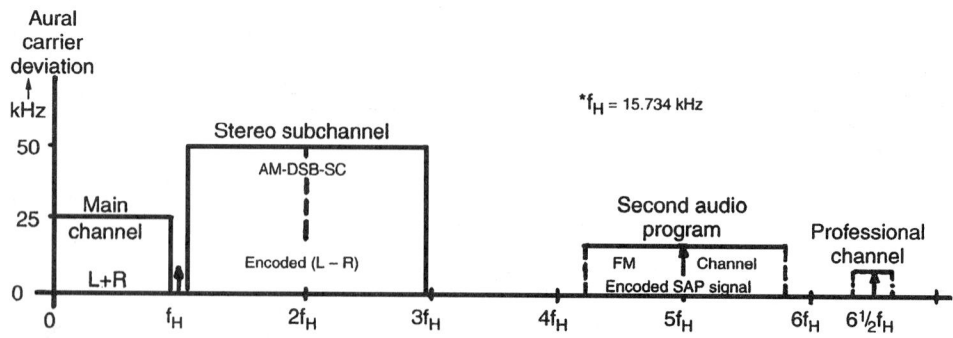

FIGURE 22.1.6   BTSC baseband spectrum.

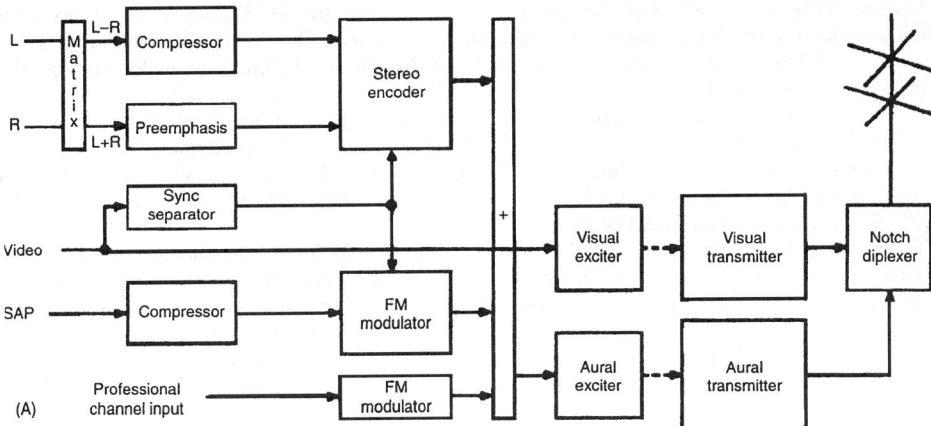

FIGURE 22.1.7 BTSC transmission block diagram.

used in FM broadcasting, except that it operates at a frequency of 31,468 Hz (twice the NTSC horizontal scanning rate of 15,734 Hz, or "H"). This subcarrier is a double-sideband, suppressed-carrier, amplitude modulated signal. An unmodulated pilot signal at 15,734 Hz, derived from the associated video horizontal synchronizing (H), is also transmitted, which is used by the receiver to reinsert, at its second harmonic (2H), the reference carrier for the stereo subcarrier. Special provisions are made in transmitting and receiving equipment to ensure that the H components from the visual signal do not interfere with the H sound pilot. In addition, the BTSC stereo subcarrier is modulated with a difference signal (L-R) that is companded, unlike FM broadcast, to provide improved signal-to-noise performance amidst the noise, multipath distortion, and intercarrier "buzz" often generated by television transmitting and receiving systems. The compander circuits employ a combination of fixed preemphasis, "spectral" compression, and amplitude compression. Expander circuits found in BTSC-compatible receivers restore the proper dynamic range and stereo image while reducing undesirable artifacts of the transmission process.

The Second Audio Program (SAP) channel is located at 5H (78,670 Hz) in the aural baseband. It is frequency modulated with audio, which may or may not be related to the main program. SAP audio is companded using the same system as the stereo subcarrier, which results in fairly good fidelity at the receiver. Typical uses include the simulcasting of a second language version of the main program, aural "captioning" of the main program to benefit the visually impaired, or transmission of other services, such as local weather information.

The remaining BTSC subcarrier is called the Professional (PRO) channel and is centered at 6.5 H (102,271 Hz). It is frequency modulated with a very narrow deviation of only ±3 kHz. It may be used to transmit voice or data.

A typical BTSC transmission block diagram is shown in Fig. 22.1.7 and a typical BTSC receiver-decoder is shown in Fig. 22.1.8.

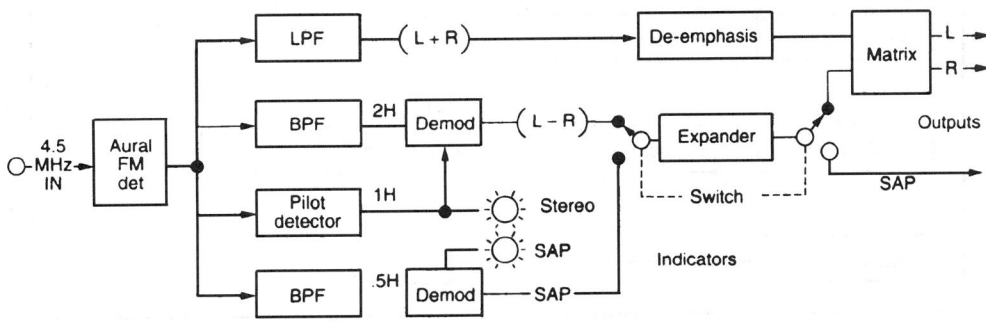

FIGURE 22.1.8 BTSC receiver-decoder block diagram.

Because BTSC specifications are protected, but not mandated, by the FCC, other subcarriers are permitted to be transmitted in the aural baseband between 16 and 120 kHz. The only restriction is that they must not interfere with BTSC or monophonic receiver operation. Amplitude or frequency modulation carrying analog or digital signals may be used.

BTSC stereo has several significant advantages over the simpler FM broadcast stereo system in that there is no "stereo noise penalty," owing to the use of companding, and there is no required reduction in the allowable deviation of the monophonic signal. The FCC does not require stations transmitting BTSC MTS to meet any particular audio performance standards. Rather, a market-driven approach has been taken. Nevertheless, BTSC MTS is capable of excellent performance.

The most significant impairments to BTSC performance result from crosstalk and incidental carrier pulse modulation (ICPM) in the transmission system. Crosstalk can occur due to nonlinearities in the TV aural exciter, high-power RF amplifier stages, combining networks, or the antenna system. ICPM results from phase modulation of the visual carrier which is independent of the aural carrier. Since most receivers employ intercarrier detection of the sound carrier, ICPM will cause a "buzz" in the sound unless kept to levels below 3° to 5°. Excessive ICPM will show up in the demodulated video as differential phase shift.

ATSC Digital Television Systems

 ATSC digital transmission systems definitions are included on the accompanying CD-ROM.

The ATSC Digital Television Standard is intended to provide a system capable of transmitting high-quality video and audio and ancillary data over a single 6-MHz channel. The system is specified to deliver 19.39 Mbps of throughput in a 6-MHz terrestrial broadcasting channel. This means that encoding a high-resolution video source containing about four times as much detail as that of conventional television (NTSC) resolution requires a bit-rate reduction by a factor of up to 50. To achieve this bit-rate reduction, the system is designed to exploit complex video and audio compression technology so that the resultant payload may be transmitted in a standard 6-MHz-wide television channel. The video encoding relies on MPEG-2 compression and audio is encoded using multichannel Dolby AC-3. Figure 22.1.9 depicts a typical ATSC encoder. The resultant transport stream is then randomized and transmitted using 8VSB (8 level, vestigial sideband) modulation. The transport specifications are shown in Table 22.1.2. Figure 22.1.10 shows the basic arrangement of an ATSC decoder. Figures 22.1.11 and 22.1.12 depict block diagrams of an 8VSB transmitter and receiver, respectively.

The objective is to minimize the amount of data required to represent the video image sequence and its associated audio. The goal is to represent the video, audio, and data sources with as few bits as possible while preserving sufficient quality to recreate those sources as accurately as required for the particular application. While the transmission subsystems described in the ATSC Standard are designed specifically for terrestrial transmission, the objective is that the video, audio, and service multiplex/transport subsystems be useful in

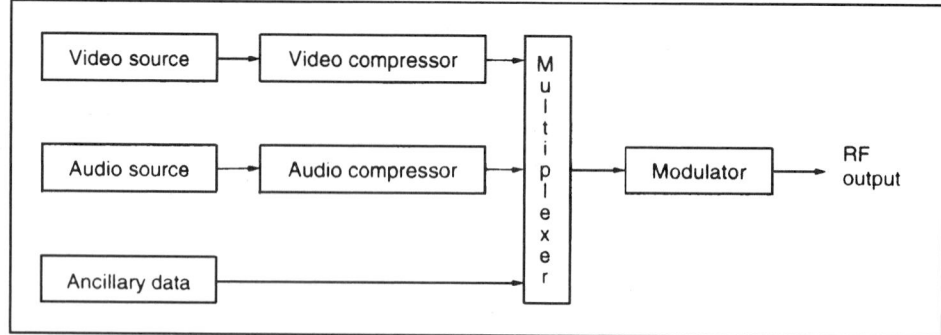

FIGURE 22.1.9   Grand Alliance encoder.

TABLE 22.1.2   ATSC Transmission Specifications

Transmission parameter	Terrestrial mode
Channel bandwidth	6 MHz
Excess bandwidth	11.5 percent
Symbol rate	10.76 Msymbols/s
Bits per symbol	3
Trellis FEC	2/3 rate
Reed-Solomon FEC	(208,188) T = 10
Segmented length	836 symbols
Segmented sync	4 symbols per segment
Frame sync	1 per 313 segments
Payload data rate	19.3 Mbit/s
NTSC co-channel rejection	NTSC rejection filter in receiver
Pilot power contribution	0.3 dB
C/N threshold	14.9 dB

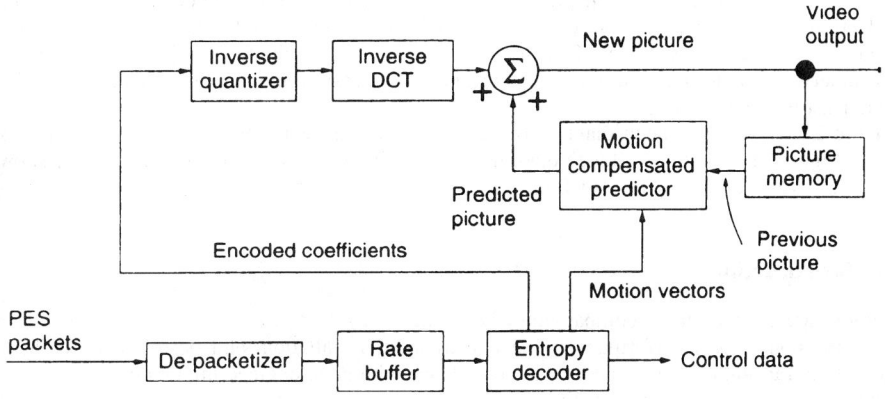

FIGURE 22.1.10   ATSC decoder.

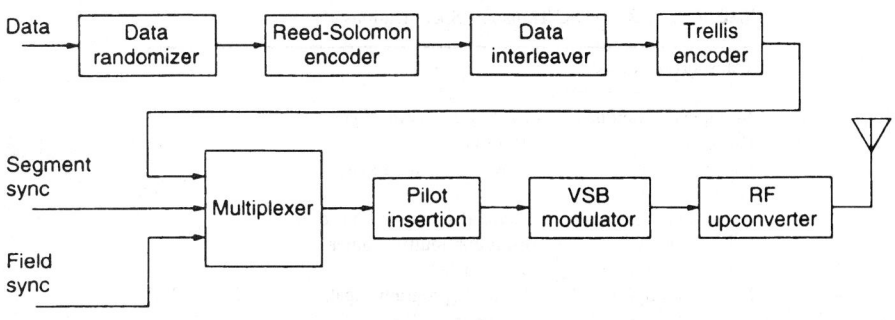

FIGURE 22.1.11   ATSC VSB transmitter.

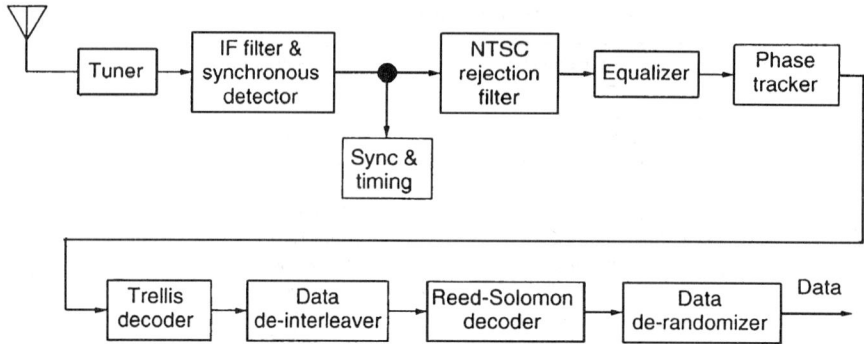

FIGURE 22.1.12   ATSC VSB receiver.

other applications, including those not presently identified. Conditional access provisions are included in the standard, to allow future subscription and data services.

The entire process is so efficient that once the conversion of broadcast television from NTSC analog to ATSC digital is complete, part of the present television spectrum will be reallocated away from broadcast service. UHF television channels 52 to 69 will ultimately be assigned to other nonbroadcast services. While the FCC anticipates that this will occur in 2006, the roll out of ATSC digital television in the United States currently underway is not progressing as fast as originally forecast. Some pundits predict the transition will continue at least through 2015.

Figure 22.1.9 shows a block diagram of the video encoder, while Fig. 22.1.10 shows the corresponding decoder. Figure 22.1.11 shows a block diagram of an ATSC VSB transmitter. Figure 22.1.12 shows a diagram of the ATSC VSB receiver. (See also Table 22.1.3.)

Short Wave Broadcasting

 Definitions are given on the accompanying CD-ROM.

Transmission system performance: The construction, installation, and performance of an international broadcasting transmitter system shall be in accordance with good engineering practice.

Spurious emissions must be effectively limited.

1. Any emission appearing on a frequency removed from the carrier frequency by between 6.4 kHz and 10 kHz inclusively, must be attenuated at least 25 dB below the level of the unmodulated carrier.

TABLE 22.1.3   ATSC Transport Specifications

Transport parameter	
Multiplex technique	MPEG-2 systems layer
Packet size	188 bytes
Packet header	4 bytes including sync
Number of services	
Conditional access	Payload scrambled on service basis
Error handling	4-bit continuity counter
Prioritization	1 bit/packet
System multiplex	Multiple program capability described in PSI stream

2. Any emission appearing on a frequency removed from the carrier frequency by more than 10 kHz and up to and including 25 kHz must be attenuated at least 35 dB below the level of the unmodulated carrier.

3. Any emission appearing on a frequency removed from the carrier frequency by more than 25 kHz must be attenuated at least 80 dB below the level of the unmodulated carrier.

4. In the event spurious emissions from an international shortwave broadcasting transmitter cause harmful interference to other stations or services, additional steps may be required to eliminate the interference.

The transmitter must be equipped with automatic frequency control apparatus so designed and constructed that it is capable of maintaining the operating frequency within 0.0015 percent of the assigned frequency.

No international broadcast station will be authorized to install or be licensed for operation of transmitter equipment with a rated carrier power of less than 50 kW.

The percentage of modulation should be maintained as high as possible consistent with good quality of transmission and good broadcast practice. In no case should it exceed 100 percent on positive or negative peaks of frequent recurrence. It also should not be less than 85 percent on peaks of frequency recurrence. The highest allowable modulating frequency is 5 kHz.

BROADCASTING EQUIPMENT

In the following sections, general characteristics and typical examples of some current broadcasting equipment are presented. A brief theory of operation and a block diagram are provided.

Medium Wave AM Transmission Systems

Medium wave transmitters are now generally solid state up to about 10 kW. Higher power transmitters may still employ one or more vacuum tube stages to achieve power levels of 100 kW or more. The maximum power level allowed in the United States is 50 kW, but other jurisdictions may authorize power levels exceeding 1000 kW. Such transmitters are usually fixed-tuned to operate on only one frequency.

Figure 22.1.13 shows a block diagram of a Continental Electronics 419F 250 kW power tube transmitter. This transmitter employs low-level solid-state audio and RF stages which drive an unmodulated intermediate-level tube RF power amplifier (IPA). The output of this IPA then drives a pair of large power amplifier tubes, arranged in the efficient "Doherty" configuration where modulation is applied with a 90° phase shift to the screens. One tube is considered to be the "carrier" tube and contributes most of the unmodulated power output. The other tube is called the "peak" tube and produces no power in the absence of modulation. On 100 percent negative modulation peaks, the carrier tube cuts off, reducing the transmitter output to zero. On 100 percent (or greater) positive peaks, the carrier tube and the peak tube both contribute approximately equal amounts of power. Efficiency of this design exceeds 60 percent.

Figure 22.1.14 shows a block diagram for the Nautel AMPFET 5 W and AMPFET 10 kW solid-state transmitters. They employ a high-efficiency MOSFET pulse width modulation system which collector-modulates MOSFET RF amplifiers operating in class D to achieve an overall transmitter efficiency exceeding 72 percent. The transmitters differ mainly in the number of RF amplifier modules necessary to achieve rated power.

Modern medium wave broadcast stations typically employ one or more vertical radiators operating over a ground system (counterpoise) comprised of multiple copper radials to provide either nondirectional or directional coverage patterns over the service area of interest. Directional arrays, consisting of one or more individual towers, are normally specified to protect the service areas of other stations by producing nulls in the directions to be protected. Antenna towers are normally at least one quarter wavelength in height, though shorter towers may be electrically lengthened by means of capacitive or inductive loading. Towers may be self-supporting or guyed. Each tower normally has an antenna tuning unit at its base, which serves to match the tower's feed point impedance to that of the transmission line. Directional antenna systems will also have a phasing network that delivers power with the appropriate amplitude and phase shift to each tower so as to achieve the desired coverage pattern (Fig. 22.1.15).

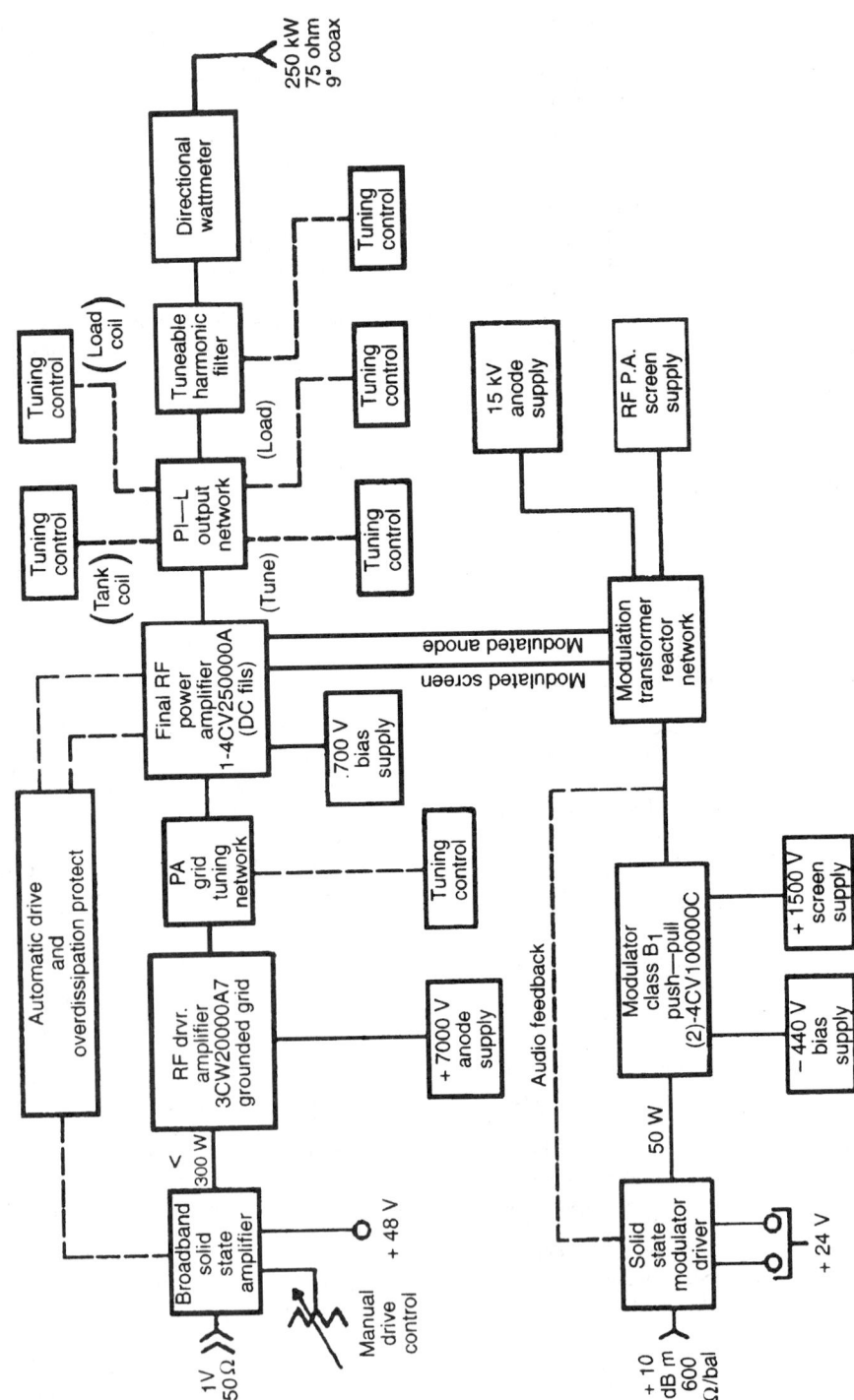

FIGURE 22.1.13 Continental Electronics Model 419F 250-kW shortwave transmitter.

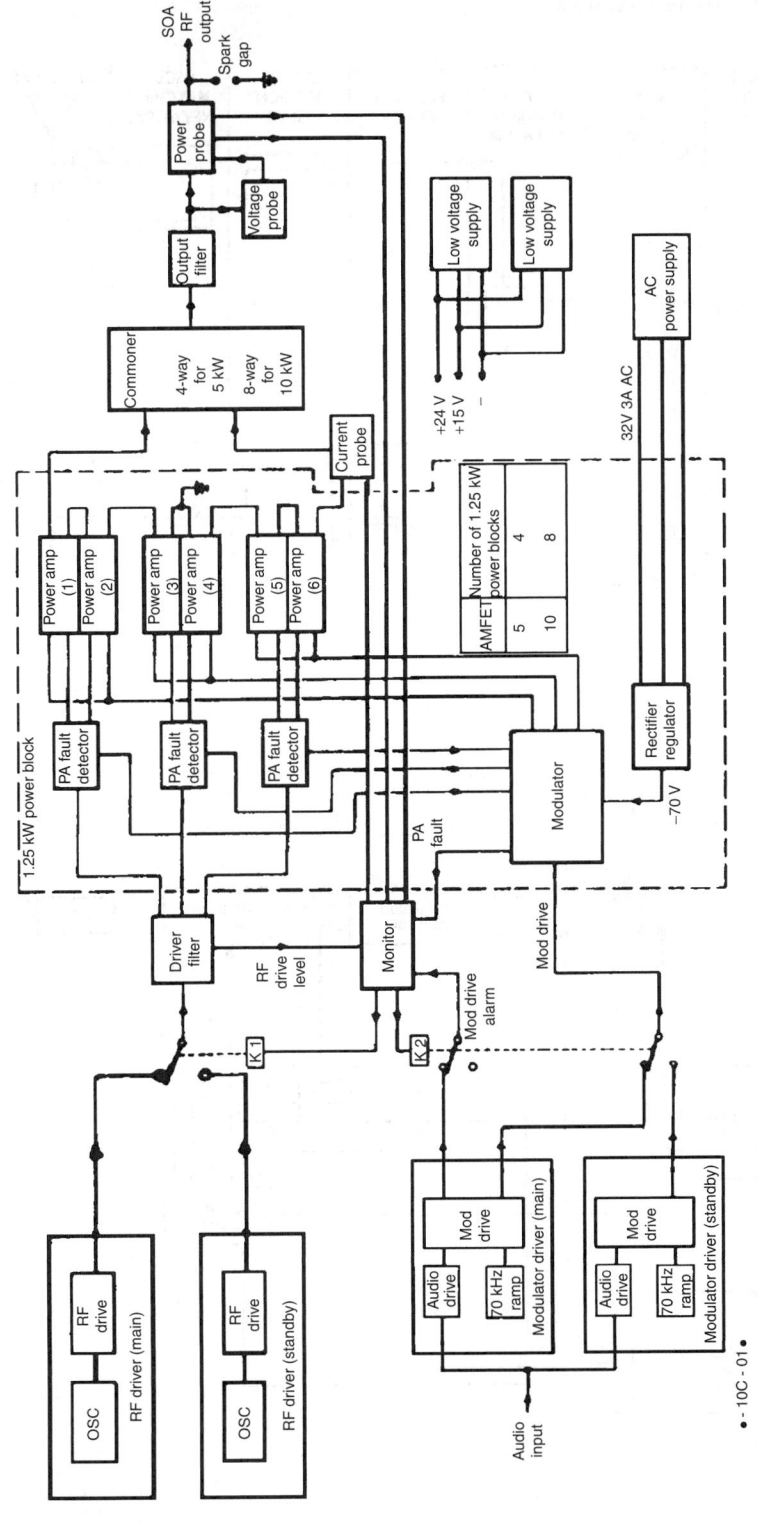

FIGURE 22.1.14 Nautel AMPFET solid-state AM broadcast transmitter.

22.25

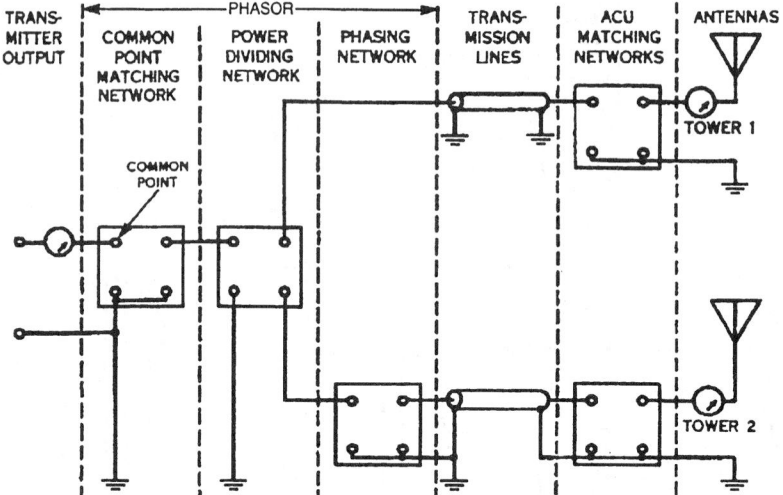

FIGURE 22.1.15 Directional antenna system block diagram.

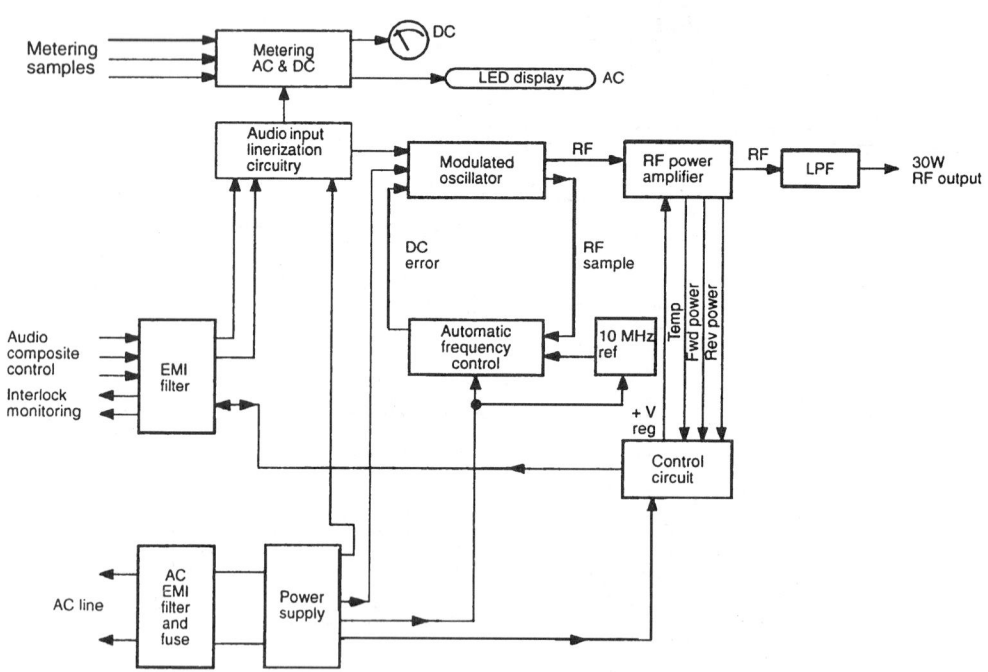

FIGURE 22.1.16 FM exciter.

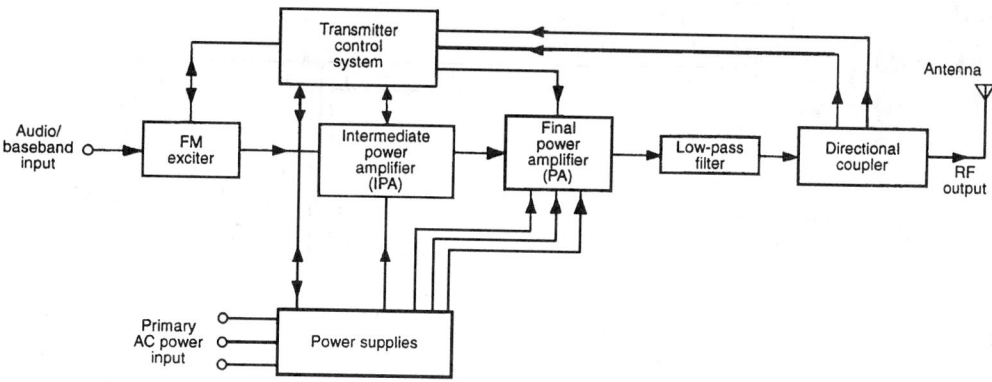

FIGURE 22.1.17   FM amplifier and control circuits.

Wire antennas, such as the flat tops and dipoles used in the early days of medium wave broadcasting, are not favored today, principally because of their high angles of radiation and poor ground wave efficiency. Such characteristics result in inferior local coverage and increased interference to distant stations, especially at night.

Frequency Modulation Transmission Systems

FM transmitters consist of several main components, namely:

1. An exciter, Fig. 22.1.16, which converts the audio baseband to frequency-modulated RF. The exciter determines the key qualities of the transmitted signal. All modern exciters employ direct modulation.
2. Power amplifier stages, Fig. 22.1.17, which boost the output of the exciter to the full rated output of the transmitter. Such stages usually operate class C, a nonlinear though highly efficient mode.
3. Power supplies, which provide the necessary ac and dc voltages to operate the transmitter.
4. Control circuits, Fig. 22.1.17, which allow the transmitter to be turned on and off, the power to be adjusted, and protect the transmitter against damage from overload.
5. Low-pass filtering to suppress undesired harmonic frequencies.

FM broadcast stations may use antennas with either horizontal or circular polarization. The advantage of circular polarization is that twice as much power may be used as compared to the horizontal polarization case, since the FCC allows the vertical component of radiated power to be equal to or less than the horizontal component. Most FM antennas are based on dipole or loop designs. Two or more elements are usually stacked to obtain increased gain at the expense of vertical bandwidth. Since it is possible to make very wide-band antennas relative to the width of a single FM channel, it is common in some areas to combine (or multiplex) several FM stations onto one physical antenna. This is done using large networks of three- or four-port combiners.

Analog NTSC Television Transmission Systems

Modern VHF Television transmitters are solid state at all power levels to 60 kW and more. Reliability and performance have been enhanced by the elimination of heat-producing tubes and high voltage in the amplifier stages of the transmitter. Instead, large amounts of RF power are produced by combining many

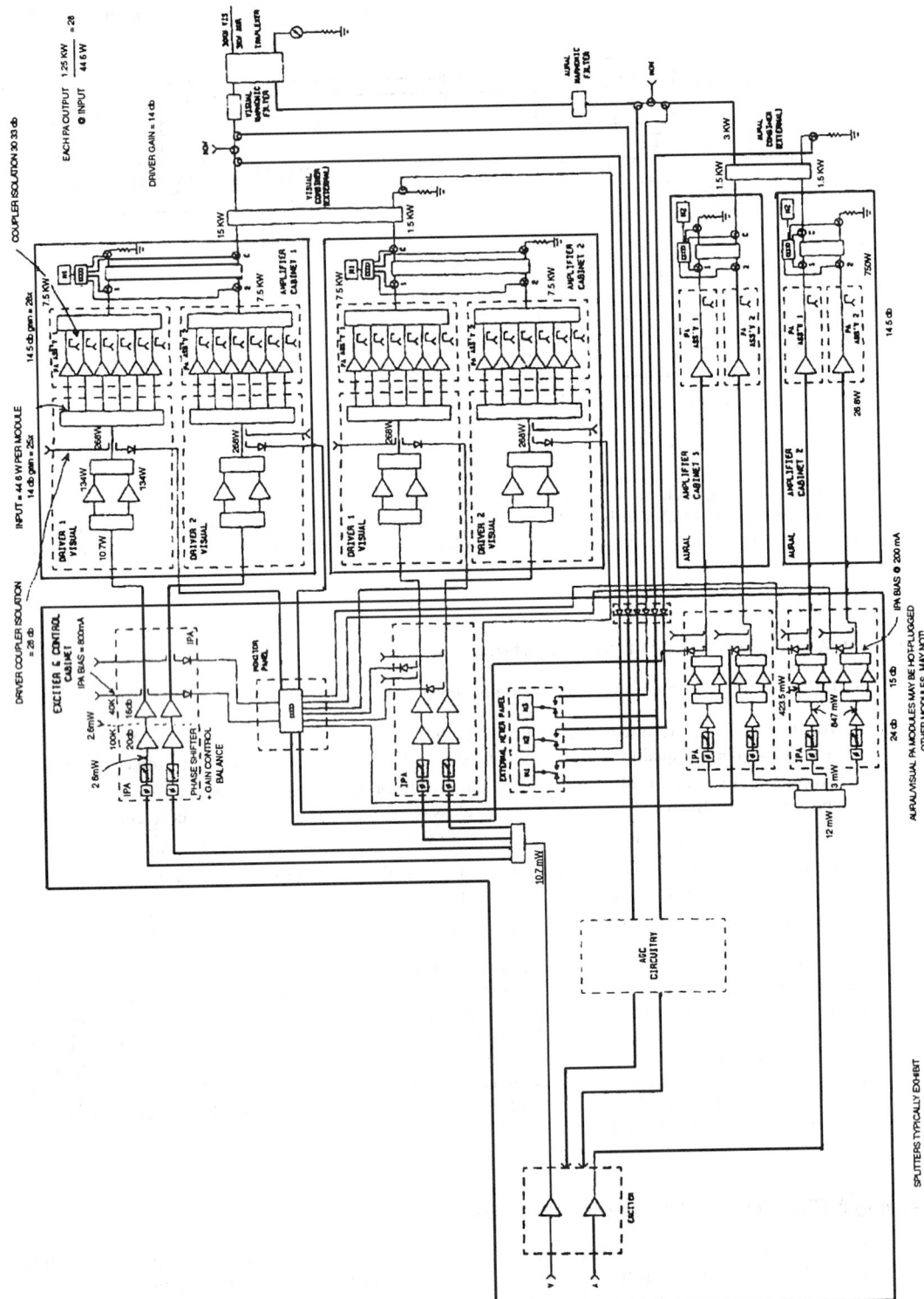

FIGURE 22.1.18 Larcan 30-kW solid-state VHF TV transmitter.

TABLE 22.1.4   Typical Solid-State VHF Transmitter Specifications

VISUAL

Type of emission	5M45C3F
Frequency range (channels 2 to 6)	54 to 88 MHz
(channels 7 to13)	174 to 216 MHz
Rated power output	250 W peak sync
Output impedance	50 ohms
Input impedance: video	75 ohms
Input level	0.5 to 2 V p-p
Regulation of output: black to white picture	1%
Variation of output: over one frame	2%
Modulation capability	1.0%
LF linearity	4%
Variation in frequency response with brightness[1]	±0.3 dB
Carrier frequency stability[2]	±200 Hz
Differential phase: reference burst	
(12.5 to 75 percent modulation, subcarrier modulation 10%)[3]	±1°
Differential gain:	
(12.5 to 75 percent modulation, subcarrier modulation 10%)	3%
K factor: 2T pulse	1%
Incidental carrier phase modulation	±1°
Signal-to-noise ratio: RMS below sync level	−55 dB
Harmonic radiation	−80 dB
Spurious radiation	−60 dB
Envelope delay:	
0.2 to 2.0 MHz	±40 ns
at 3.58 MHz	±30 ns
at 4.18 MHz	±60 ns
Sideband response:	
Carrier +200 kHz	0 dB ref
Carrier −0.5 MHz to + 4.18 MHz	+0.5, −1 dB
Carrier +4.75 MHz and higher	−35 dB or better
Carrier −1.25 MHz and lower	−20 dB or better
Carrier −3.58 MHz	−42 dB or better
Carrier −4.5 MHz	−30 dB or better
Blanking level variation: black to white picture	1.5%
Intermodulation distortion	−52 dB

Visual footnotes:
[1] With respect to response at midcharacteristic. Measured at 85 percent and 20 percent modulation.
[2] Maximum variation over 30 days, at an ambient temperature of 0 to 45°C.
[3] Maximum variation with respect to burst.

smaller stages, each of which operates on low voltage and comparatively high current. The Larcan 30-kW transmitter shown in Fig. 22.1.18 is typical. The transmitter has been designed to operate conservatively at 30 kW peak sync visual power and 3 kW average aural power.

Typical solid-state VHF transmitter specifications are given in Table 22.1.4.

Ruggedly constructed, this transmitter is modular in format. Many of the modules are standard, thereby providing a very high degree of commonality in systems using transmitters of various power ratings and frequencies. The simplicity of design and use of standard, readily available components enhances serviceability.

All important parameters are monitored and can be displayed on the meters built into the exciter and amplifier. This equipment is suitable for automatic or remote-control operation.

The exciter-modulator provides a fully processed, precorrected, on-channel signal. Low-level IF modulation and surface acoustic wave (SAW) filter technology are employed to ensure superior color performance and simplicity of operation.

The exciter (model TEC-3V) is modular in construction and BTSC stereo compatible. In addition to the normal audio input, the exciter also has two wideband inputs to accept the signals from stereo, separate audio program (SAP), and pro-channel generators. A built-in incidental carrier phase modulation (ICPM) corrector ensures optimum performance when transmitting multichannel sound or teletext. All routine on-air adjustments are screwdriver adjustable from the front panel, with an extender module provided for servicing and more extensive adjustments. The precorrection circuits, aural preemphasis, and SAW filters can be switched out of circuit by front panel locking-type toggle switches. These circuits are all unity-gain, thus no level adjustments are required when switching in and out of circuit. Each IF and RF module has a BNC test point located on the front panel. These test points are fed from a directional coupler at the module output, and have no effect on the module output when connected to a 50-Ω load.

The visual output of the exciter is fed to four broadband solid-state 9 kW amplifiers that require no tuning or adjustment. All of the modules in each amplifier are operated well below their maximum ratings.

It can be seen from the RF flow diagram that there are four stages of amplification. The preamplification stages are high gain, broadband, thin-film integrated circuit amplifiers operating class A. The IPA stages also operate class A and consist of two transistors whose outputs are combined by 3-dB quadrature couplers. The driver stage consists of two amplifiers operating class A, again with their outputs combined by a 3-dB quadrature coupler. Each device is a pair of push–pull FETs in a single case.

The four visual final stages each consist of six modules whose outputs are combined in a six-way stripline combiner. This method of combining provides excellent isolation between modules and the failure of one or more modules will not affect the performance or reliability of the others. To combine the four final stages, 3-dB couplers are used with the output fed to an external diplexer. Each output module has AGC control, VSWR protection, and a test jack for monitoring purposes. The aural chain of the transmitter operates in a similar manner.

The control circuitry in this transmitter is extremely simple because no interlocking or sequential starting of power supplies is required. The transmitter is basically built as two parallel halves; i.e., amplifiers 1-1A and 1-1B, complete with their own power supplies and protection circuits, can be operated independently or together with amplifiers 1-2A and 1-2B. The transmitter control panel, situated below the visual and aural IPA units in cabinet 2, enables single or joint control to be made.

All control, status, and telemetry signals are brought out on a 25-pin D connector.

For UHF broadcasters, transmitters have been the subject of ongoing development work ever since the inception of UHF stations in the 1950s. When the UHF band was initially allocated for television service, most stations signed on-the-air with lower power tetrode and traveling wave tube transmitters. Later, klystrons were developed that allowed generation of much higher power at UHF frequencies. Klystrons have the advantage of no delicate grid or large anode structure, while providing extremely high gain. The problem was, as stations increased power with klystrons, very complex and expensive transmitters were required. Power efficiency, while better than a tetrode, was fairly low. In the late 1970s, as oil prices rose and cost of energy became an important factor, stations began to employ klystron pulsers, which boosted efficiency somewhat, to cut their electrical usage. At the same time, many development projects were undertaken to improve the klystron's fundamental efficiency. The MSDC (multistage, depressed collector) klystron was one result. By using multiple collectors, operating at graduated voltages, more electrons could be collected and efficiency increased. Sometime later, tube manufacturer Eimac developed a new type of gridded klystron, which they called a *Klystrode inductive output tube* (IOT). The IOT achieved even better efficiency than the MSDC klystron without the complication of multiple high-voltage power supplies. Almost all modern high-power UHF transmitters employ IOT power amplifiers.

Figure 22.1.19 shows the block diagram of a typical UHF transmitter. Table 22.1.5 gives typical UHF TV transmitter characteristics.

Antenna designs vary depending on whether the intended use is VHF or UHF television. It is generally not possible to use one antenna to cover both UHF and VHF bands, nor even to cover the entirety of any one band with a single antenna. Antennas for television broadcast usually provide an omnidirectional antenna pattern, but directional antennas are sometimes employed at UHF stations.

VHF antennas will use designs similar to those shown in Fig. 22.1.20. Antennas similar to those shown in Fig. 22.1.21 are often used in UHF applications.

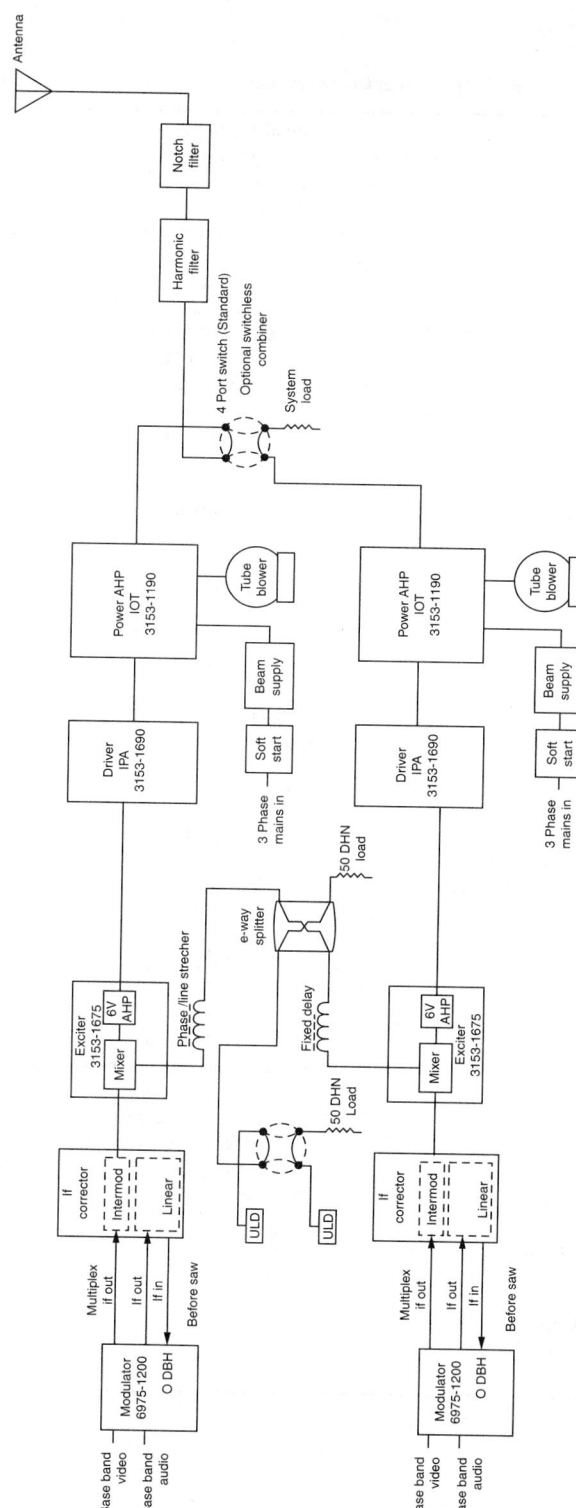

FIGURE 22.1.19 Larcan/TTC redundant UHF transmitter.

TABLE 22.1.5   Typical UHF TV Transmitter Characteristics

Vision performance	Sound performance

Frequency range:
 470–806 MHz
Output impedance:
 50 or 75 Ω
Video output:
 75 Ω, 36 dB return loss
Sideband response:
   −3.58 MHz   −42 dB or better
   −1.25 MHz and below −20 dB or better
   Carrier to −0.5 MHz +0.5 dB, −1.0 dB
   Carrier to +4.0 mHz   ±0.5 dB
   +4.00 to +4.18 MHz +0.5 dB. −2.0 dB
   +4.75 MHz and above −40 dB or better
Variation in response with brightness:
 ±0.75 dB
Frequency stability:
 ±250 Hz (maximum—30 days)
Modulation capability:
 100%
Differential gain:
 0.5 dB
Differential phase:
 3°
Low frequency nonlinearity:
 1.0 dB
Envelope delay:
   0.2–2.0 MHz   ±40 nS
   3.58 MHz   ±30 nS
   4.18 MHz   ±60 nS
Regulation of output:
 3%
Variation of output:
 2%
AM noise (RMS):
 −55 dB
K-factors:
 2T—2%
 12.5T—3%
Spurious and harmonics:
 −60 dB
Incidental carrier phase modulation:
 ±1.5°
Common amplification applications in
 band intermodulation:
 −60 dB
Stereo pilot carrier protection:
 Full compliance with FCC specification
73.682(c)(3)
	Intercarrier (+4.5 MHz)
	 Frequency stability:
	   ±15 Hz. phase locked to video line
	Monaural input:
	 600 Ω, balanced
	Broadband input (2):
	 75 Ω, unbalanced
	Monaural performance
	 (±25 kHz deviation)
	 Frequency response:
	   ±0.5 dB, 50 Hz to 15 kHz
	 Preemphasis:
	   75 μS or flat
	 Distortion (with deemphasis):
	   0.5%
	 FM noise:
	   −60 dB
	 AM noise:
	   −55 dB
	 AM synchronous noise:
	   −40 dB
	Stereo performance
	 Frequency response:
	   ±0.5 dB, 50 Hz to 120 KHz
	 FM noise:
	   −70 dB, ±75 KHz reference
	 Distortion (THD):
	   0.5%
	 Distortion (IMD):
	   0.5%
	 Stereo separation
	   (equivalent mode):
	   40 dB
	 BTSC pilot protection:
	   −45 dB or better
	(*Note*: Patent pending noise reduction circuit
	required in common amplification systems.)
	General performance
	Operating ambient temperature:
	 0 to 50°C
	Altitude:
	 Sea level to 7,500 FT: Consult factor for
	 other altitudes.
	Relative humidity:
	 0 to 95% noncondensing

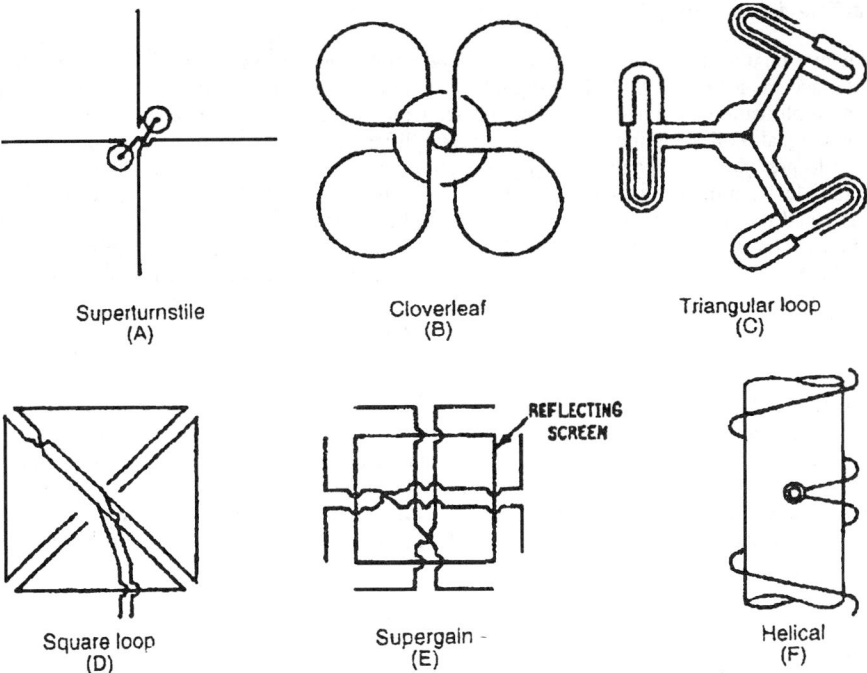

Superturnstile
(A)

Cloverleaf
(B)

Triangular loop
(C)

Square loop
(D)

Supergain -
(E)

Helical
(F)

FIGURE 22.1.20   Omnidirectional antenna configurations.

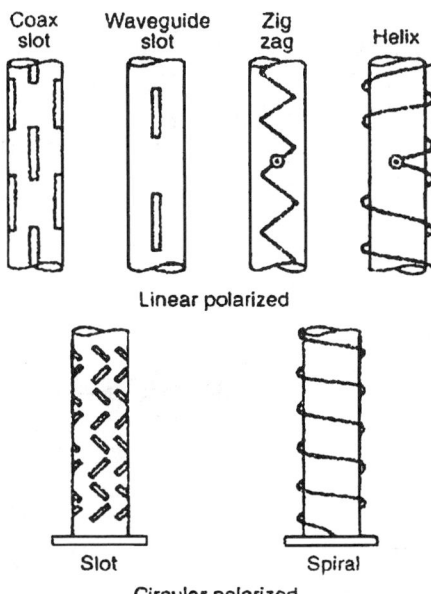

Coax slot

Waveguide slot

Zig zag

Helix

Linear polarized

Slot

Spiral

Circular polarized

FIGURE 22.1.21   Traveling wave antennas.

ATSC Digital Television Transmission Systems

Digital television transmitters are similar to combined-amplification analog transmitters, except that the modulator/exciter accepts a single 19.39 Mbps digital bit stream, usually via a SMPTE 310 interface. The following stages of amplification are tuned to pass the 6-MHz wide, broadband noise-like spectrum of the ATSC 8VSB-modulated signal. Following the final stage(s) of amplification, a constant-impedance mask filter is normally employed to provide the stringent band-edge rejection levels required by the FCC Rules.

Digital television transmitters are typically rated by average power, instead of the peak power ratings normally ascribed to analog transmitters. Solid-state ATSC transmitters on either VHF or UHF channels may achieve power levels of up to 30 kW. UHF digital stations may require higher transmitter output power levels, up to 100 kW or more, which may currently only be economically produced by IOT transmitters, in the same manner as is done for high-power NTSC UHF transmitters. Such transmitters are typically assembled from multiple IOT amplifiers, each producing up to 25 kW of average digital power, which are electrically connected by "magic tee" combiners. Figure 22.1.22 shows a block diagram of the Harris CD3200P2 two-IOT ATSC digital transmitter, capable of 42 kW average power.

The broadband nature of the ATSC signal means that the venerable klystron, still used in some analog UHF television transmitters, is not suitable because of its relatively narrow bandwidth capability. However, that same broadband characteristic has enabled tube manufacturers to apply the multistage depressed-collector (MSDC) technology, originally developed around the klystron, to the IOT. This technology, which consists of multiple electron collectors operating at progressively lower beam voltages, results in a substantial increase in the efficiency of the conventional IOT. Figure 22.1.23 shows the internal structure of a typical MSDC IOT manufactured by L-3 communications.

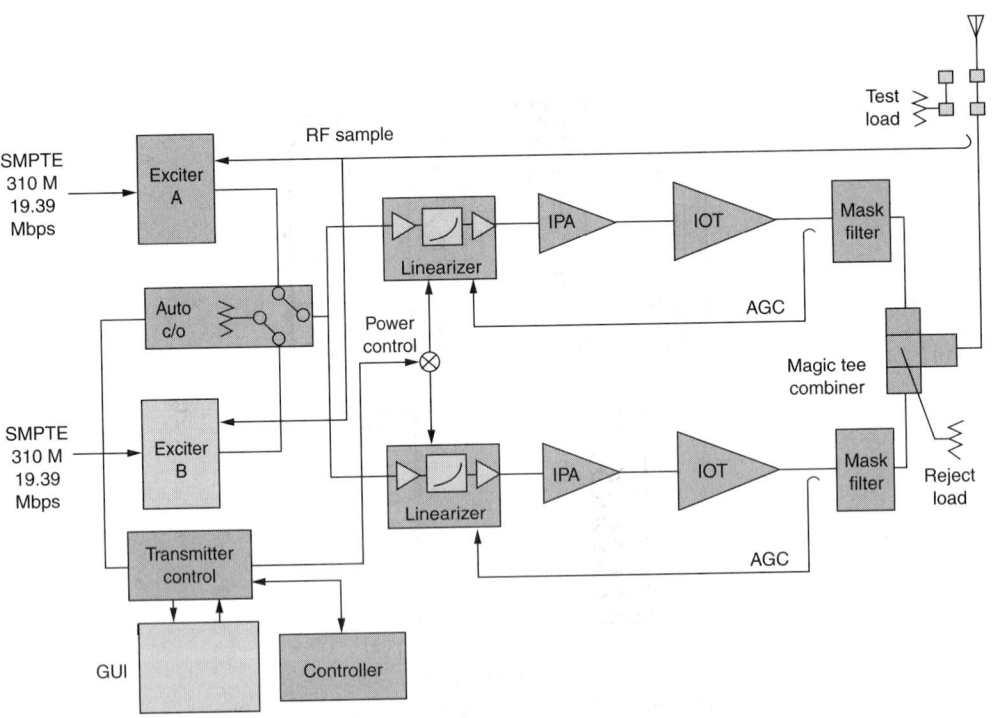

FIGURE 22.1.22 Harris Sigma ATSC transmitter block diagram.

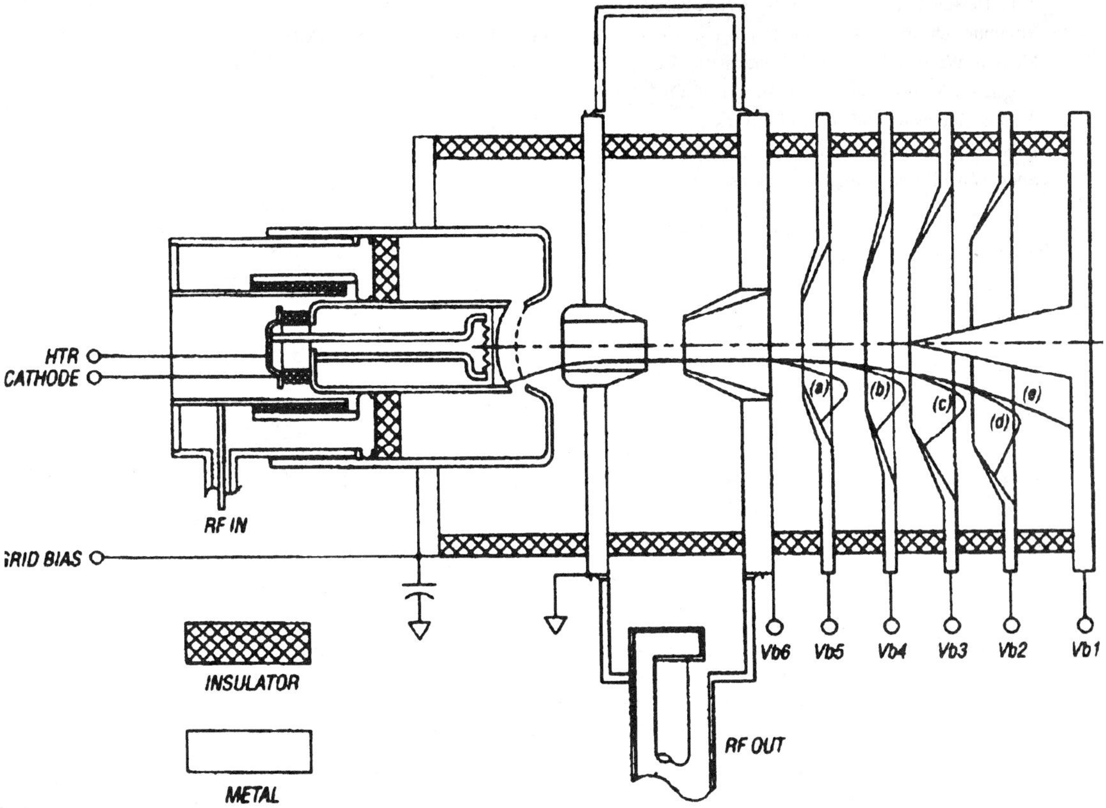

FIGURE 22.1.23   MSDC IOT structure, typical.

BIBLIOGRAPHY

Advanced Television Systems Committee, Washington, D.C., *ATSC Standard: Digital Television Standard, Revision B*, 7 August 2001.

Advanced Television Systems Committee, Washington, D.C., *Guide to the Use of the ATSC Digital Television Standard*, 4 October 1995.

Federal Communications Commission, *Fourth Report and Order Re: Advanced Television Systems and Their Impact upon the Existing Television Broadcast Service* (Adopted), December 24, 1996.

Public Broadcasting Service, Washington, D.C., *Advanced Television Transmission*, 1995.

Silbergleid, M., and M. Pescatore (eds.), *Guide to Digital Television*, 2nd ed., 1999.

National Association of Broadcasters Washington, D.C., *Engineering Handbook*, 9th ed., 1999.

Television Engineering Handbook, revised edition, McGraw-Hill, 1992.

ON THE CD-ROM:

 Frequency Channel Assignments for FM Broadcasting
Numeral Designation of Television Channels

Zone Designations for Television Broadcasting
Minimum Distance Separation Requirements for FM Broadcast Transmitters in KM (MI)
Medium Wave AM Standard Broadcasting Definitions
Frequency Modulation Broadcasting Definitions
Analog Television (NTSC) Definitions
ATSC Digital Transmission System Definitions
Short Wave Broadcasting Definitions

CHAPTER 22.2
AM AND FM BROADCAST RECEIVERS

Lee H. Hoke, Jr.

AM RECEIVERS: GENERAL CONSIDERATIONS

AM broadcast receivers are designed to receive amplitude-modulated signals between 530 and 1700 kHz (566 to 177 m wavelength), with channel assignments spaced 10 kHz. To enhance ground-wave propagation the radiated signals are transmitted with the electric field vertically polarized.

AM broadcast transmitters are classified, according to the input power supplied to the power amplifier, from a few hundred watts up to 50 kW. The operating range of the ground-wave signal, in areas where the ground conductivity is high, is up to 200 mi for 50-kW transmitters. During the day the operating range is limited to the ground-wave coverage. At night, refraction of the radiated waves by the ionosphere causes the waves to be channeled between the ionosphere and the earth, resulting in sporadic coverage over many thousands of miles. The nighttime interference levels thus produced impose a restriction on the number of operating channels that can be used at night.

The signal-selection system is required to have a constant bandwidth (approximately 10 kHz), continuously adjustable over a 3:1 range of carrier frequencies. The difficulty of designing cascaded tuned rf amplifiers of this type has resulted in the universal use of the superheterodyne principle in broadcast receivers.

A block diagram of a typical design is shown in Fig. 22.2.1. In this figure the signal is supplied by a vertical monopole (whip) antenna in automobile radio receivers or by a ferrite-rod loop antenna in portable and console receivers. An rf amplifier is used in most automobile designs but not in small portable models.

In some receivers the local oscillator is combined with the mixer, which simplifies the rf portion of the receiver. An intermediate frequency of 455 kHz is used in portable and console receivers, while 262.5 kHz is common in automobile radio designs. Diode detectors have been used in discrete component designs for detection of the i.f. signal. Push-pull class B audio-power amplifiers are used to minimize current drain. A moving coil permanent-magnet speaker is used as the output transducer.

A potential major improvement in AM radio sound quality occurred in 1990 with the adoption of the *Amax* voluntary standard by the National Radio Systems Committee (NRSC)[1] and implementation by most radio stations. This standard established a receiver audio frequency response of not less than 50 to 7500 Hz, with limits of +1.5 dB, –3.0 dB referenced to 0 dB at 400 Hz. The receiver shall have less than 2 percent total harmonic distortion plus noise (THD+N) and shall have a deemphasis curve that complements the preemphasis now added to the AM broadcast modulation. Attenuation at the 10-kHz adjacent frequencies shall be at least 20 dB. In addition, the receiver must be equipped with a noise blanker circuit (impulse noise suppressor) and have means for connecting an external antenna. To many listeners, AM and AM-stereo receivers designed to these specifications gave performance rivaling FM receivers of equivalent class in tests conducted at the 1992 NAB convention.

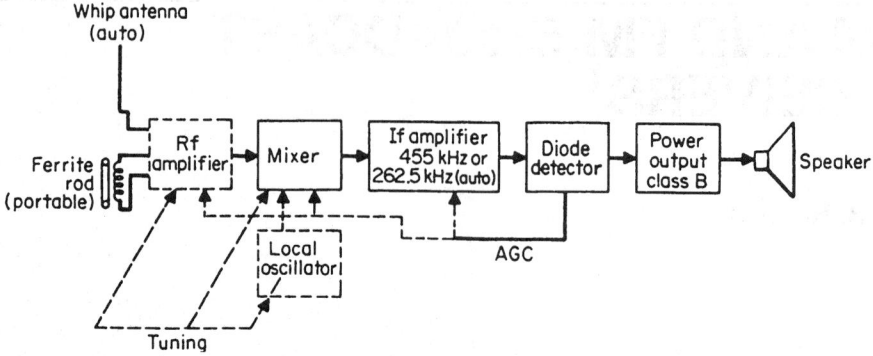

FIGURE 22.2.1 Block diagram typical of AM receivers.

Design Categories

AM receiver designs currently fall into three categories:

Portable Battery-Powered Receivers without External Power Supply.   These units vary in size from small pocket radios operating on penlite cells to larger hand-carried units using D cells for power. The power output in the larger portable units is about 250 mW.

Console and Component Type AM Receivers Powered by the Power Line.   These units are usually a part of an AM-FM receiver, with high audio-power output capability. The power output ranges from several watts to more than 100 W. Most audio systems use push-pull class B operation. Most such systems are equipped with two amplifier systems for FM stereo operation.

Automotive Receivers Operated on the 12-V Battery-Generator System of the Automobile.   The primary current used in transistorized receivers usually does not exceed 1 A. Because of operation in the high ambient noise of the vehicle, the power output is relatively high (2 to 20 W).

Sensitivity and Service Areas

The required sensitivity is governed by the expected operating field strengths. Typical field strengths for primary (ground-wave) and secondary (sky-wave) service are as follows:

	Field strength, mV/m	
Primary service:		
Central urban areas (factory areas)	10–50	
Residential urban areas	2–10	
Rural areas	0.1–1.0	
Secondary service		Areas where sky-wave signals > 0.5 mV/m at least 50 percent of the time

Co-channel protection is provided for signals exceeding 0.5 mV/m. The receiver sensitivity and antenna system should be adjusted to provide usable outputs with signals of the order of 0.1 mV/m if the receiver is to be used over the maximum coverage area of the transmitter.

The required circuit sensitivity is controlled by the efficiency of the antenna system. A car-radio receiver vertical antenna is adjustable to about 1 m in length. Since the shortest wavelengths are of the order of 200 m, the antenna can be treated as a nonresonant short-capacitive antenna. The open-circuit voltage of such a short monopole antenna is

$$E_a = 0.5 l_{\text{eff}} E_f \quad \text{mV}$$

where l_{eff} is the effective length of antenna (m) and E_f is the field strength (mV/m).

The radiation resistance of the short monopole is small compared with the circuit resistance of the receiver input circuit, but the antenna is not matched to the input impedance since matching is not critical, adequate antenna voltage being available at the minimum field strength (0.1 mV/m) needed to override noise. The car-radio antenna is coupled to the receiver by shielded cable. This, with the receiver input capacitance, forms a capacitive divider, reducing the antenna voltage applied to the receiver. To ensure adequate operation the receiver should offer 20 dB signal-to-noise ratio when 10 to 15 μV is applied to the input terminals.

Portable and console receivers use much shorter built-in antennas, usually horizontally polarized coils wound on ferrite rods. The magnetic antenna can be shielded from electric field interference. Although the effective length of a ferrite rod is shorter than that of a whip antenna, the higher Q of the ferrite rod and coil could provide approximately the same voltage to the receiver. The unloaded Q of a typical ferrite-rod antenna coil is of the order of 200. The voltage at the terminals of the antenna coils is QE_a.

Selectivity

Channels are assigned in the broadcast band at intervals of 10 kHz, but adjacent channels are not assigned in the same service area. In superheterodyne receivers, selectivity is required not only against interference from adjacent channels but also to protect against image and direct i.f. signal interference.

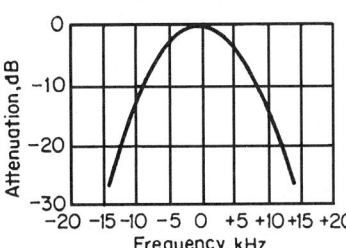

FIGURE 22.2.2 Typical selectivity curve of an AM receiver using a ferrite-rod antenna without rf amplifier.

The primary adjacent-channel selectivity is provided by the i.f. stages, whereas image and direct i.f. selectivity must be provided by the rf circuits. In receivers using a ferrite-rod antenna and no rf stage, the rf selectivity is provided by the antenna. High Q in the antenna coil thus not only provides adequate signal to override the mixer noise but also protects against image and i.f. interference. With a Q of 200 the image rejection at 1400 kHz is about 40 dB, while the direct i.f. rejection at 600 kHz is about 24 dB. With an rf stage added, the image rejection is about 50 dB and the i.f. rejection about 46 dB.

Since car-radio receivers are subjected to an extreme dynamic range of signal levels, the selectivity must be slightly greater to accommodate strong signal conditions. The image rejection at 1400 kHz is typically 58 dB in spite of the lower i.f. frequency; the i.f. rejection is typically 50 dB, and the adjacent-channel selectivity is about 20 dB.

Figure 22.2.2 shows the overall response of a typical portable receiver using a ferrite rod without an rf amplifier.

High-Signal Interference

When strong signals are present, the distortion in the rf and i.f. amplification stages can generate additional interfering signals. The transfer characteristic of an amplifier system can be expressed in a power series as

$$E_{\text{out}} = G_1 E_{\text{in}} + G_2 (E_{\text{in}})^2 + G_3 (E_{\text{in}})^3 + \cdots + G_n (E_{\text{in}})^n$$

where E_{out} = output voltage
E_{in} = input voltage (same units)
$G_1, G_2, \ldots, G_n$ = voltage gains of successive amplifier stages

When the input signal consists of two or more modulated rf signals, E_{in} becomes

$$E_{in} = e_1 \cos \omega_1 t + e_2 \cos \omega_2 t$$

where

$$e_1 = E_1[1 + Ea_1(t)/E_1], \qquad e_2 = E_2[1 + Ea_2(t)/E_2]$$

where E_1 = signal 1 carrier
E_2 = signal 2 carrier
$Ea_1(t)$ = audio modulation first signal
$Ea_2(t)$ = audio modulation second signal

I.F. Beat.   When two strong signals are applied to the amplifier with carrier frequencies separated by a difference equal to the intermediate frequency, a difference-frequency signal appears, which is independent of the local oscillator frequency. Because of the wide frequency spacing, interference of this kind can take place only in the mixer or rf amplifier and only with strong signals (signal strengths of several volts per meter). These signals are derived from the G_2 term:

$$E_{if.beat} = G_2 e_1 e_2 \cos (\omega_1 - \omega_2)t$$

where $e_1 e_2$ and $\omega_1 \omega_2$ are the respective amplitude and angular frequencies.

Cross Modulation.   When a strong interfering signal is present, the modulation on the interfering signal can be transferred to the desired signal by third-order distortion components. This type of interference does not occur at a critical frequency of the interfering signal, provided that it is close enough to the desired signal frequency not to be attenuated by the selectivity of the receiver. This type of distortion can take place in the rf, mixer, or i.f. stages of the receiver. These signals are derived from the G_3 terms:

$$E_{crossmod} = G_3(e_2)^2 e_1/4 \cos \omega_1 t$$

The cross modulation is proportional to the square of the strength e_2 of the interfering signal.

Intermodulation.   Another type of interference due to third-order distortion is caused by two interfering carriers. When these signals are so spaced from the desired signal that the first signal is arbitrarily spaced by Δf and the second signal is $2\Delta f$, third-order distortion can create a signal on the desired carrier frequency. These signals are derived from the G_3 terms:

$$E_{intermod} = G_3 e_1^2 e_2/4 \cos (2\omega_1 - \omega_2)t$$

The interference is proportional to the square of the amplitude of the closer carrier times the amplitude of the farther carrier. Intermodulation is sometimes masked by cross modulation; it occurs only when the e_2 signal is stronger than the desired carrier after attenuation by the rf selectivity of the receiver.

Harmonic Distortion.   Harmonic-distortion interference usually arises from harmonics generated by the detector following the i.f. amplifier, which radiated back to the tuner input.

Choice of Intermediate Frequency

Two intermediate frequencies are used in broadcast receivers: 455 kHz and 262.5 kHz. The 455-kHz i.f. has an image at 1450 kHz when the receiver is tuned to 540 kHz, thus allowing good image rejection with simple selective circuits in the rf stage. At 540 kHz the selectivity must be sufficient to prevent i.f. feedthrough since the receiver is particularly sensitive at i.f. frequencies because converter circuits typically have higher gain at i.f. than at the converted carrier frequency. The choice of the higher i.f. frequency also makes i.f. beat interference less likely. The second harmonic falls at 910 kHz and the third harmonic at 1365 kHz.

The 262.5 kHz i.f. has a lower limit of image frequency at 1065 kHz, which requires somewhat more rf selectivity than is needed when the higher i.f. is used. The second harmonic of the 262.5-kHz frequency falls below the broadcast band (at 525 kHz) and hence does not interfere. On the other hand, there are more higher-order responses in the passband (787.5, 1050, 1312.5, and 1575 kHz). Sensitivity to i.f. feedthrough when the receiver is tuned to 540 kHz is greatly reduced by the use of the lower i.f. frequency.

Circuit Implementation

Although hobbyists still build AM broadcast receivers using multiple stages of discrete components as described in previous editions of this handbook, all present-day commercially built receivers are designed with ICs which require only a handful of additional components to complete their function. Figure 22.2.3 shows one such implementation with its associated peripheral components.

This circuit is intended for shirt-pocket use and drives audio earphones.[2,3] In other applications, a simple mono audio amplifier IC can be added to provide table-top or clock-radio application. The design follows the classic superheterodyne block diagram of mixer, oscillator, intermediate frequency amplifier, detector, and audio amplifier. In designs intended for automotive radio, an RF amplifier stage precedes the mixer and a much higher power amplifier is included. Station selection is provided by an RF L-C tank circuit and an oscillator L-C tank circuit.

Differential amplifiers, in single or multiple dc-coupled stages, are generally used for rf and i.f. amplification. This configuration gives excellent high-frequency stability, extremely low distortion (low generation of both cross-modulation and intermodulation spurious responses), and has wide AGC range capability (typically 30-dB gain reduction for a single stage, 55 dB for a dual stage).

Double-balanced mixers are typically used to achieve a high degree of isolation between the rf and local oscillator and between the local oscillator and i.f. In some applications the rf stage can be dc-coupled to the mixer with no need for an interstage frequency-selection network. When compared to classical single transistor mixer, the spurious response level of a double-balanced mixer is considerably lower, thereby reducing the stop-band filter requirement.

The *local oscillator* typically employs internal control circuitry to stabilize the amplitude of the waveform and maintain low distortion. This is particularly important in varactor-diode tuning applications where a change in amplitude will tend to shift the bias voltage of the oscillator tank circuit, thereby causing mistracking with the other varactor-tuned circuits.

Selectivity is achieved more often by the use of the block filter-block gain configuration rather than the conventional distributed filter, cascade stage design approach. Mechanically vibrating elements can be designed with Q's much higher than electric circuits. Piezoelectric vibrator plates made of quartz or (for low and medium frequencies) barium titanate form the equivalent of a high-Q coil-and-capacitor combination which can be used singly or in combination to form i.f. filters with steep band-edge selectivity characteristics. The piezoelectric vibrator converts energy from electric to mechanical and back to electric with relatively little loss.

Barium titanate resonators have been used in some broadcast receiver designs. While they have the advantage of small size (at 455 kHz), the disadvantage is the numerous spurious responses caused by multiple modes of vibration. Hence those resonators must be used in combination with coils and capacitors to suppress the spurious responses. Three configurations of block filter networks are shown in Fig. 22.2.4. The main frequency selection element is the piezoelectric ceramic filter.[4]

Both single-ended and full-wave balanced *envelope detectors* are used, depending on the intended application. The full-wave detector produces a lower level of distortion in the recovered audio over a very wide range, including lower absolute level, of input carrier levels.

An *AGC amplifier* provides a control voltage to the i.f. and rf stages. This voltage is proportional to the carrier amplitude. Second-order filtering is often incorporated to reduce audio distortion, even at low audio frequencies, and to give fast settling time which is advantageous with electronic tuning systems when operated in the "station search" mode.

Some IC designs contain an *audio preamplifier* that provides additional filtering to remove residual carrier from the audio signal. Preamplifier output levels are typically 0.25 to 0.3 V for a 30 percent modulated medium-level carrier input to the i.f.

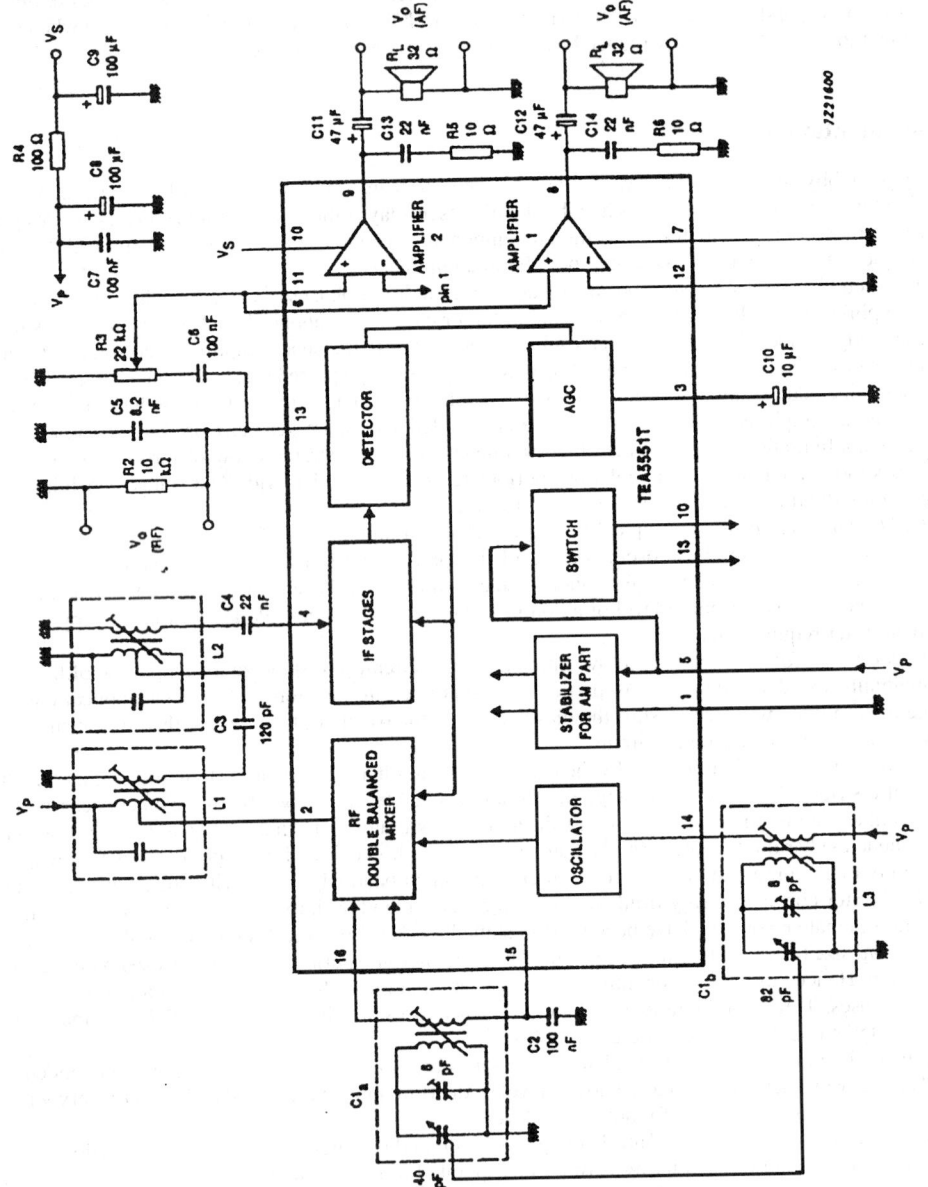

FIGURE 22.2.3 Typical AM receiver integrated circuit. *(From Philips Semiconductor)*

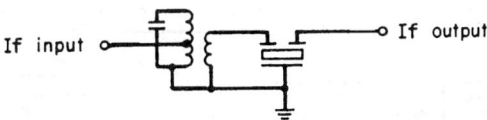

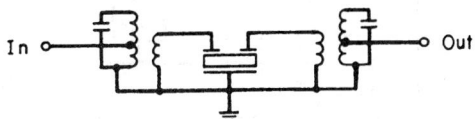

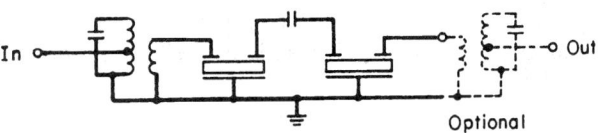

FIGURE 22.2.4   Three configurations of i.f. block filter networks.

IC *audio power amplifiers* fall into four major categories. Both monaural and two-channel (stereo) versions are available in each category.

1. Fifty to 500 mW, 1.5 to 9 V supply, intended for personal applications with 1.5- to 2-in. loudspeaker or earphone.
2. One-half to 2 W for table model, power-line-connected applications.
3. Two to 20 W, 12 to 16 V supply, for car radios.
4. Ten to 100 W for high-fidelity systems. The higher-powered designs in this category are typically a hybrid of IC's mounted on a ceramic thick-film substrate.

AM Stereo

In March 1982, after many months of testing and evaluation by the National AM Stereo Radio Committee, the FCC ruled that it could not select a single system for AM stereo and therefore made a "free marketplace" ruling. Of the five systems under consideration at that time, three were put into operation throughout the United States. These were

1. Compatible quadrature amplitude modulation (C-QUAM)
2. Amplitude and phase modulation (AM/PM)
3. Independent sideband modulation (ISB)

By 1993, the C-QUAM system was recognized as the overwhelming marketplace leader and was given full endorsement by the FCC.

QUAM can be thought of as the addition of two carriers having the same frequency in phase quadrature (separated by 90°), one being modulated by the left audio signal (L), the other by the right audio signal (R). In practice, this is achieved by amplitude-modulating a single carrier with the sum of L and R, and deviating the carrier phase according to a function made up of the sum of quadrature sidebands of $L + R$ and $L − R$. At the transmitter a correction factor is added to have the envelope truly represent $L + R$ and therefore maintain compatibility (low distortion in recovered audio) with existing monophonic AM receivers (C-QUAM).[5] Stereo identification is provided by a 25-Hz tone phase-modulating the carrier.

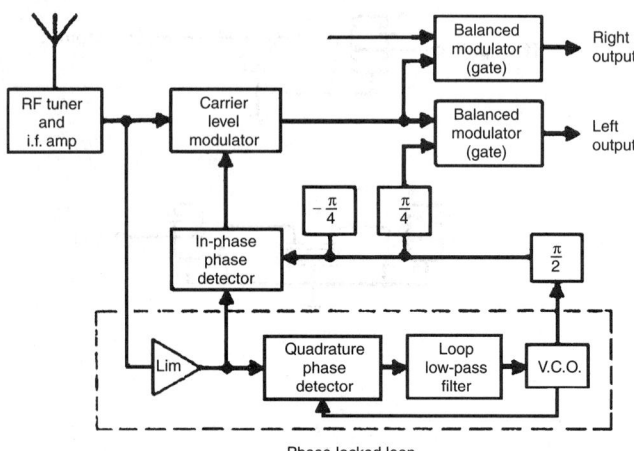

FIGURE 22.2.5   Block diagram of a C-QUAM AM-stereo receiver.   (*From Parker et al., Ref. 6*)

The block diagram of a stereo receiver designed to receive C-QUAM is shown in Fig. 22.2.5. The PLL is locked to the carrier frequency and supplies the cw reference for demodulation. The complementary correction to transform the C-QUAM signal to QUAM is performed by multiplying the carrier in the receiver by a function of the angle between the transmitted sidebands. Recovery of the *L* and *R* audio signals is accomplished by conventional quadrature demodulation. The receiver front end must be designed to a higher-quality level than that of a conventional monophonic AM receiver. Since phase components of the signal are now important, the local oscillator noise and tuning system flutter must be kept to a minimum.[6] Incidental carrier phase modulation (ICPM) caused by tuned circuit misalignment must also be minimized.

FM BROADCAST RECEIVERS: GENERAL CONSIDERATIONS

Broadcast FM receivers are designed to receive signals between 88 and 108 MHz (3.5 to 2.8 m wavelength). The broadcast carrier is frequency-modulated with audio signals up to 15 kHz, and the channel assignments are spaced 200 kHz. The FM band is primarily intended to provide a relatively noise-free radio service with wide-range audio capability for the transmission of high-quality music and speech. The service range in the FM band is generally less than that obtainable in the AM band, especially when sky-wave signals are relied on for extending the AM coverage area.

VHF signals are limited to usable service ranges of less than 70 mi. Since sky-wave signals do not materially affect the transmission of FM signals, there is no equivalent night effect and licenses are not limited to daylight hours only, as with many AM operations. In the past, all FM signals in the United States were horizontally polarized. However, the rules have been changed to allow maximum power to be radiated in both the horizontal and vertical planes.

Unlike AM broadcasting, where the station power is measured by the power supplied to the highest-power rf stage, FM transmitters are rated in *effective radiated power* (ERP), i.e., the power radiated in the direction of maximum power gain. This method of power measurement is used since the transmitting antenna has significant power gain, resulting in an ERP many times the input power supplied to the rf output stage of the transmitter.

Although FM receivers have been principally used in high-fidelity audio installations, more recently small pocket-sized FM receivers have been designed with limited audio-output capabilities. Also, an increasing number of FM receivers are being included in automobile installations.

In 1960 the FCC amended the broadcast rules to allow the transmission of stereophonic signals on FM stations. This transmission is equivalent to the transmission of two audio signals, each having 15 kHz bandwidth, transmitted on the same carrier as used for monophonic FM signals. Since the FM signal was initially designed to have sufficient additional bandwidth to achieve improved signal-to-noise ratio at the receiver, there is room to multiplex the second component of the stereophonic signal with no increase in the radiated bandwidth. However, the signal-to-noise ratio is reduced when the multiplexing technique is employed.

Sensitivity

The field strength for satisfactory FM reception in urban and factory areas is about 1 mV/m. For rural areas, 50 μV/m is adequate signal strength. These signal levels are considerably lower than the equivalent levels for AM reception in the standard broadcast bands. These effects make satisfactory reception with these lower signal levels possible: (1) the effects of lightning and atmospheric interference (static) are negligible at 100 MHz, compared with the interference levels typical of the standard broadcast band; (2) the antenna system at 100 MHz can be matched to the radio-receiver input impedance, providing more efficient coupling between the signal power incident on the antenna and receiver, and (3) the use of the wide-band FM method of modulation reduces the effects of noise and interference on the audio output of the receiver.

The open-circuit voltage of a dipole for any length up to one-half wavelength is given by

$$E_{oc} = E_f (5.59 \sqrt{R_r})/F_s \ \text{mV}$$

where E_{oc} = open-circuit antenna voltage
E_f = field strength at antenna (mV/m)
F_s = received signal frequency (MHz)
R_r = antenna radiation resistance (Ω)

For a half-wave dipole $R_r = 72 \ \Omega$; for a folded dipole R_r is 300 Ω. For antennas substantially shorter than one-half wavelength, $R_r = 8.75 l^2 F_s^2 \times 10^{-3}$, where l is the total length of the dipole in meters. For a folded dipole one-half wavelength long operating at 100 MHz, the open-circuit voltage is $E_{oc} = 0.97 E_f$. The voltage delivered to a matched transmission and receiver input is one-half of this value, $0.48 E_f$.

The noise in the receiver output is caused by the antenna noise plus the equivalent thermal noise at the receiver input. The input impedance generates an excess of noise compared with the noise generated by an equivalent resistor at room temperature. The noise generated is given by

$$E_{nr} = E_n \sqrt{2NF - 1} \ \text{V}$$

where E_{nr} = equivalent noise generated in receiver input
E_n = equivalent thermal noise (V) = ($\sqrt{4R_{in}kT\Delta_f f}$)
R_{in} = receiver input resistance
k = Boltzmann's constant = 1.38×10^{-23} J/K
T = absolute temperature = 290 K
Δf = half-power bandwidth of receiver response taken at discriminator
NF = receiver noise figure. (If the noise figure is given in decibels, $NF_{dB} = 10 \log NF$.)

The generator noise and receiver noise add as the square root of the sum of the squares. Figure 22.2.6 shows that the equivalent receiver input noise is $0.707 E_n \sqrt{NF}$. For a receiver with 300-Ω input resistance and 200-kHz noise bandwidth, $E_n = 0.984 \ \mu$V. A typical noise factor for a well-designed receiver is 3 dB, or two times power ($\sqrt{2}$ voltage) increase, giving an equivalent noise input of 1.39 μV.

In an AM receiver the signal-to-noise (S/N) ratio at the receiver input is a direct measure of the S/N ratio to be expected in the audio output. In an FM receiver using frequency deviation greater than the audio frequencies transmitted, the S/N ratio in the output may greatly exceed that at the rf input. Figure 22.2.7 shows typical output S/N ratios obtained with receiver bandwidths adjusted to accommodate transmitted signals with modulation indexes of 1.6 and 5.0, compared with the audio S/N ratio when AM modulation is used and the bandwidth of the receiver is adjusted to accommodate the AM sidebands only. As shown in this figure, the S/N

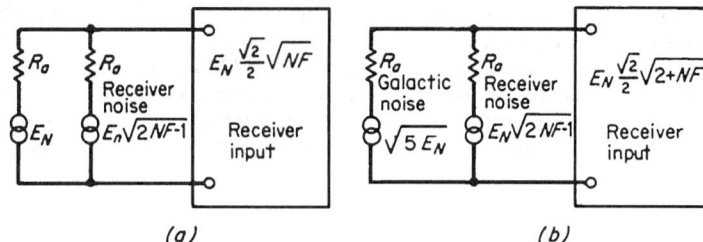

FIGURE 22.2.6 Equivalent sources of receiver noise; (*a*) with resistive generator; (*b*) with galactic noise source included.

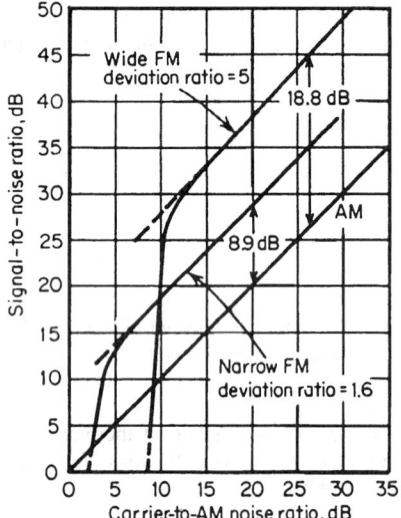

FIGURE 22.2.7 Typical signal-to-noise ratios for FM and AM for deviation ratios of 1.6 and 5.

ratio for a properly designed receiver operating with a modulation index of 5 is 18.8 dB higher than that of an AM receiver with the same rf S/N ratio.

For rf S/N ratios in FM higher than 12 dB, the audio S/N ratio increases 1 dB for each 1-dB increase in rf S/N ratio. For FM S/N ratios lower than 12 dB, the S/N ratio in the audio drops rapidly and falls below the AM S/N ratio at about 9 dB. The point at which the ratio begins to fall rapidly is called the *threshold signal level*. It occurs where the carrier level at the discriminator is equal to the noise level. The threshold level increases directly as the square root of the receiver bandwidth, i.e., approximately the square root of the modulation index. The equation for S/N improvement using FM is

$$\frac{(S/N)_{FM}}{(S/N)_{AM}} = \sqrt{3}\Delta$$

where Δ is the deviation ratio. Since broadcast standards in the United States for FM call for a modulation index of 5 for the highest audio frequency, the direct S/N improvement factor is 18.8 dB for rf S/N ratios exceeding 12 dB.

In the FM system a second source of noise improvement is provided by pre-emphasis of the high frequencies at the transmitter and corresponding de-emphasis at the receiver. The pre-emphasis network raises the audio level at a rate of 6 dB/octave above a critical frequency, and a complementary circuit at the receiver decreases the audio output at 6 dB/octave, thus producing a flat overall audio response. Figure 22.2.8 shows simple *RC* networks for pre-emphasis and de-emphasis.

The additional S/N improvement using de-emphasis in an FM receiver is

$$\frac{(S/N)_{out}}{(S/N)_{in}} = \frac{f_a^3}{3[f_a f_0^2 - f_0^3 \tan^{-1}(f_a/f_0)]}$$

where $(S/N)_{out}$ = signal-to-noise at de-emphasis output
$(S/N)_{in}$ = signal-to-noise ratio at de-emphasis input
f_a = maximum audio frequency
f_0 = frequency at which de-emphasis network response is down 3 dB

For f_a = 15 kHz and a 75-μs time constant (f_0 = 2.125 kHz), the S/N improvement is 13.2 dB. The total S/N improvement over AM is 32 dB when the carrier is high enough to override noise plus 12 dB for the 75-kHz deviation used in U.S. broadcast stations. The minimum coherent S/N ratio is therefore 44 dB.

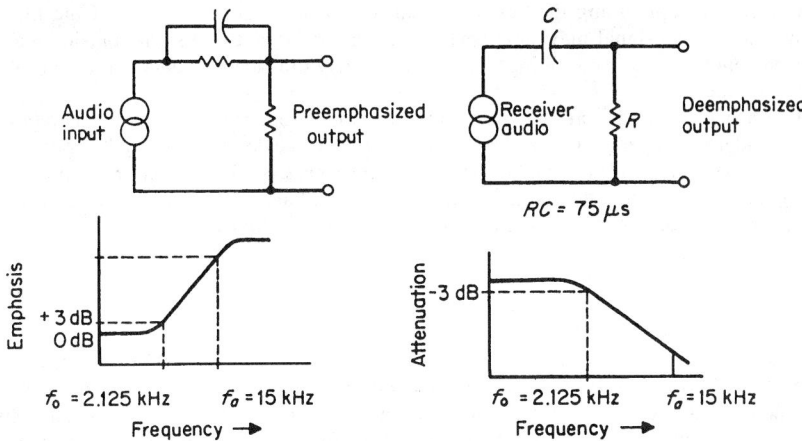

FIGURE 22.2.8   Pre-emphasis and de-emphasis circuits and characteristics.

When a dipole receiving antenna is used, an additional noise component is produced by *galactic noise*, because the dipole pattern does not discriminate against sky signals. The ratio of signal to galactic noise can be improved by using an antenna array with gain in the horizontal direction, i.e., one that discriminates against sky-wave signals. The additional noise is shown in Fig. 22.2.6. Using the calculated value of E_n and assuming a 3-dB noise factor in the receiver gives an equivalent noise input to the receiver of 1.39 μV. The required field strength to produce a 12-dB S/N ratio at the receiver and a 44-dB S/N ratio at the audio output when using a half-wide dipole is 11.5 μV/m.

Selectivity

When the FM system uses a high-modulation index, the system is not only capable of improving the S/N ratio but will reject an interfering co-channel signal. The FM signal modulation involves very wide phase excursions, and since the phase excursion that can be imparted to the carrier by an interfering signal is less than 1 rad, the effect of the interference is markedly reduced. The co-channel-interference-suppression effect requires that the interfering signal be smaller than the desired signal, since the larger signal acts as the desired carrier, suppressing the modulation of the smaller signal. This phenomenon is called the *capture* effect since the larger signal takes over the audio output of the receiver.

The capture effect produces well-defined service areas, since signal-level differences of less than 20 dB provide adequate signal separation. Although it is useful in suppressing undesired signals of a level less than the desired signal, the capture effect can also produce an annoying tendency for the receiver to jump between co-channel signals when fading, for example, caused by airplanes makes the desired signal drop below the interfering signal by only a few decibels. This effect also occurs in FM radios used in automobiles when motion of the antenna causes the relative signal levels to change.

Tuners

The FM tuner is matched to the antenna input. An rf amplifier is used to override the mixer noise. The mixer provides a 10.7-MHz i.f. signal. Most FM tuners contain a single stage of rf amplification. The mixer may be self-oscillating or employ a separate oscillator. The rf stage must have a low noise figure to reach the minimum threshold-signal level, but its most important requirement is to provide the mixer and i.f. amplifier with signals free of distortion.

When the rf amplifier is overloaded, the signal supplied to the i.f. amplifier may be distorted or suppressed. For single interfering signals there are three significant sources of difficulty: (1) image signals may capture the receiver, suppressing the desired signal; (2) strong signals at one-half the i.f. frequency (5.35 MHz) above the desired signal may capture the receiver; or (3) a strong signal outside the range of the i.f. beat but strong enough to cause limiting in the rf stage may drastically reduce the output of the rf amplifier at the carrier frequency.

When two strong signals are present, three conditions may produce unsatisfactory operation: (1) cross modulation of two adjacent upper- or lower-channel signals may produce an on-channel carrier; (2) two strong signals spaced 10.7 MHz in the rf passband may produce an i.f. beat, or (3) submultiple i.f. beats may be produced by strong signals spaced at i.f. submultiple spacings in the rf band. To minimize the effects of distortion and provide a low noise figure, many FM tuners employ an FET type transistor rf stage.

I. F. Amplifiers

To provide sufficient image separation, a higher intermediate frequency (10.7 MHz) is used in FM than in standard broadcast AM. The i.f. amplifier must provide sufficient gain for the noise generated by the rf amplifier to saturate the limiting stages fully if the benefits of wide-band FM are to be obtained at low signal levels. The high gain should be supplied with a low noise figure in the first i.f. amplifier stage, so that the noise introduced by the i.f. is small compared with the noise from the rf amplifier.

One of the most important characteristics of the i.f. amplifier is phase linearity, since envelope-delay distortion in the passband is a principal cause of distortion in FM receivers. Care must also be taken to avoid regeneration since this would cause phase distortion and hence audio distortion in the detected signal.

Although AGC is theoretically unnecessary, it is sometimes used in the rf stage to avoid overload. Such overload, coming before sufficient selectivity is present, could produce cross modulation, causing capture by an out-of-band signal. In the classical cascade amplifier design the requirements of high gain and good phase linearity are generally met by using amplifiers with double-tuned circuits adjusted to operate at critical coupling.

Limiters

The design of the limiter is critical in determining the operating characteristics of an FM receiver. The limiter should provide complete limiting with constant-amplitude signal output on the lowest signal levels that override the noise. In addition, the limiting should be symmetrical, so that the carrier is never lost at low signal levels. This is essential if the receiver is to capture on the strongest signal when there is little difference in signal strengths between the weaker and the stronger signals. Finally, the bandwidth in the output must be wide enough to pass all the significant sideband terms associated with the carrier, to prevent spurious amplitude modulation due to the lack of sufficient bandwidth to provide the constant-amplitude FM signals. The differential amplifier with dc coupling can be made to provide highly symmetrical limiting.

FM Detectors

There are five well-known types of FM detectors: (1) the balanced-diode discriminator (Foster-Seeley circuit); (2) the ratio detector using balanced diodes; (3) the slope-detector-pulse-counter circuit using a single diode with an RC network to convert FM to AM; (4) the locked-oscillator or PLL circuit, which uses a variation in current as the frequency is varied to convert the output of an oscillator (locked to the carrier frequency) to a voltage varying with the modulation; and (5) the quadrature detector circuit that produces the electrical product of the two voltages, the first derived from the limiter, the second from a tuned circuit that converts frequency variations into phase variations. The output of the product device is a voltage that varies directly with modulation.

Circuit Implementation

Integrated circuits designed for use in audio broadcast receivers in many parts of the world contain functional combinations of AM/FM and AM/FM stereo circuitry.

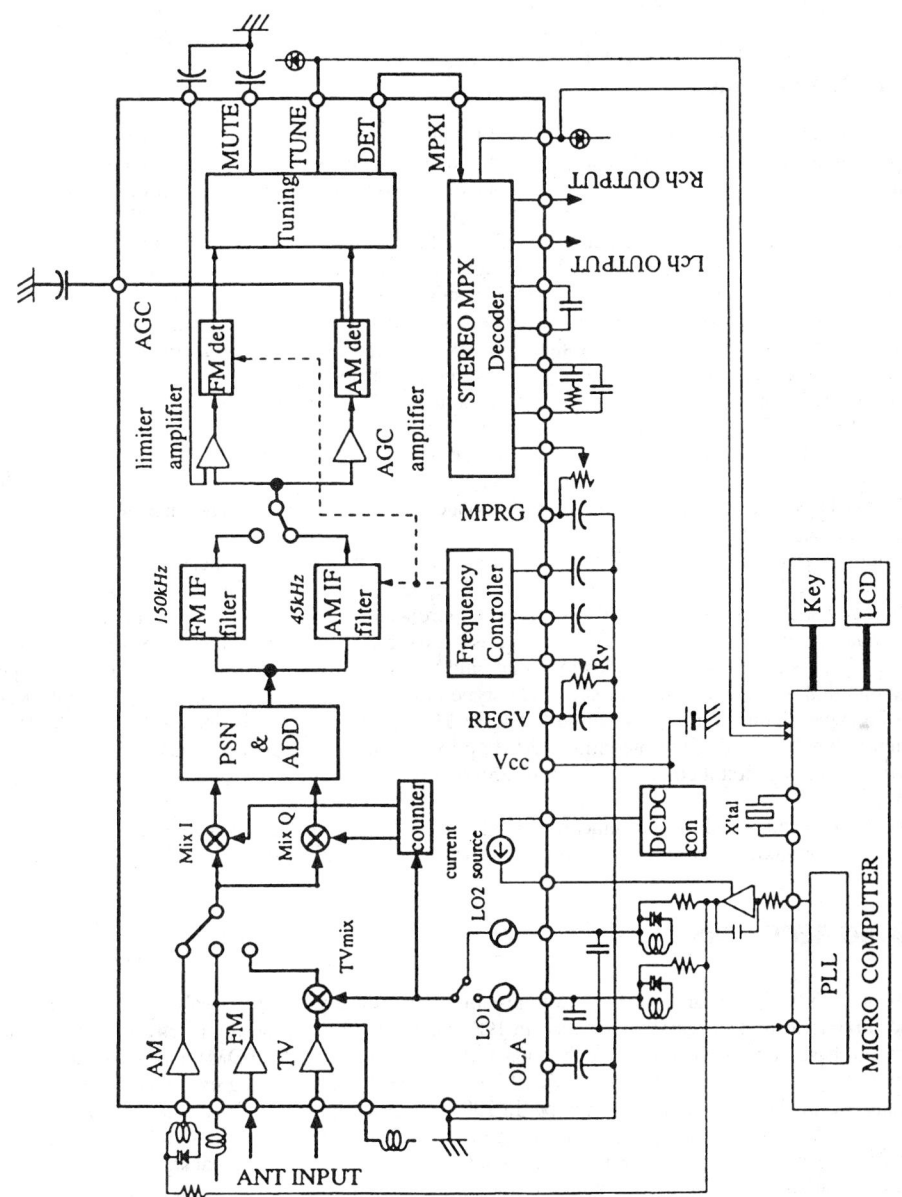

FIGURE 22.2.9 Block diagram of AM/FM/TV sound receiver having a digital tuning system. *(From Ref. 9)*

Differing classes of these ICs find use in battery operated shirt-pocket radios, portable boom-boxes, line-connected table radios, component stereo audio systems, and automobile radio receivers. In recent years, designs have been developed to provide broadcast audio reception in personal computers.[7] Most ICs contain all the functions from antenna through low-level audio for both AM and FM reception.[8–10] A block diagram of one such unit is shown in Fig. 22.2.9.

FM Stereo and SCA Systems

Since FM broadcasting uses a bandwidth of 200 kHz and a modulation index of 5 for the highest audio frequency transmitted, it is possible by using lower modulation indexes to transmit information in the frequency range above the audio. This method of using a supersonic carrier to carry additional information is used in FM broadcasting for a number of purposes, most notably in FM stereo broadcasting.

In broadcasting stereo the main-channel signal must be compatible with monophonic broadcasting. This is accomplished by placing the sum of the left- and right-hand signals $(L + R)$ on the main channel and their difference $(L - R)$ on the subcarrier. The subcarrier is a suppressed-carrier AM signal carrying the $L - R$ signal.

The suppressed-carrier method causes the carrier to disappear when L and R vary simultaneously in identical fashion. This occurs when the $L - R$ signal goes to zero and allows the peak deviation in the monophonic channel to be unaffected by the presence of the stereo subcarrier.

In the receiver it is necessary to restore the subcarrier. In the U.S. Standards, the technique used provides a pilot signal at 19 kHz, one-half the suppressed carrier. The frequency subcarrier is restored by doubling the pilot-signal frequency and using the resulting 38-kHz signal to demodulate the suppressed-carrier $(L - R)$ signal. The suppressed carrier has a peak deviation of less than 2, and the subcarrier is amplitude-modulated. The composite stereo signal thus has a S/N ratio 23 dB below that of monophonic FM broadcasting. The main $(L + R)$ channel is not affected by the stereo subcarrier $(L - R)$ signal.

The stereo signal can be decoded in two different ways. In the first, the subcarrier is separately demodulated to obtain the $L - R$ signal. The $L + R$ and $L - R$ signals are then combined in a matrix circuit to obtain the individual L and R signals. In the second method, more widely used, the composite signal is gated to obtain the individual L and R signals directly. The circuit uses a doubler circuit to obtain the 38-kHz reference signal. The latter signal is added to the composite signal, and the signal is decoded in a pair of full-wave peak rectifiers to obtain the L and R signals directly. This type of demodulation is also performed in present-day ICs by use of a synchronous demodulator circuit. The 38-kHz reference signal is derived from a 76-kHz oscillator, which is synchronized to the incoming 19-kHz pilot via a PLL and the necessary dividers. This configuration yields a simple circuit consisting of an IC and relatively few simple, i.e., no coils, resistor and capacitor networks.[11]

In the SCA system an additional subcarrier is placed well above the stereo sidebands at 67 kHz. This subcarrier is used for auxiliary communication services (SCA—subsidiary communications authorization).

DIGITAL RADIO RECEIVERS

Since about 1985 the European Broadcast Commission has been evaluating proposals for a digital audio service to include both terrestrial and satellite delivery. In 1987 the Eureka 147 Consortium was established to develop a novel system based on coded orthogonal frequency division multiplex (COFDM) approach and the advanced audio compression scheme now known as MUSICAM. In late 1994, following extensive testing, this proposal became the accepted digital method for Europe. In 1995, it was accepted as the Canadian standard and was well on its way to being a global standard. The advantage of COFDM modulation is that it allows for numerous low-power broadcast transmitters spaced over a wide area to each broadcast the identical signal on the same identical frequency. As an automobile travels around the region, the reception will continue to have CD quality with no interference or receiver retuning required. A more detailed description of the Eureka 147 system is contained in Ref. 12.

A typical DAB (Eureka 147) receiver block diagram is shown in Fig. 22.2.10. Several IC manufacturers make chip sets for this type receiver. Typically two large-scale ICs are used in the design, one for channel decoding and the other for audio source decoding.

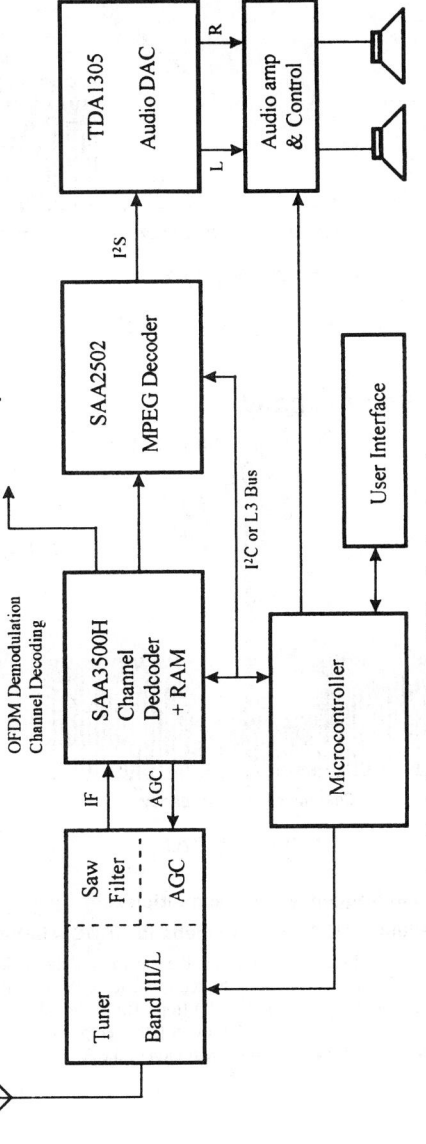

FIGURE 22.2.10 DAB receiver block diagram. *(From Philips Semiconductor)*

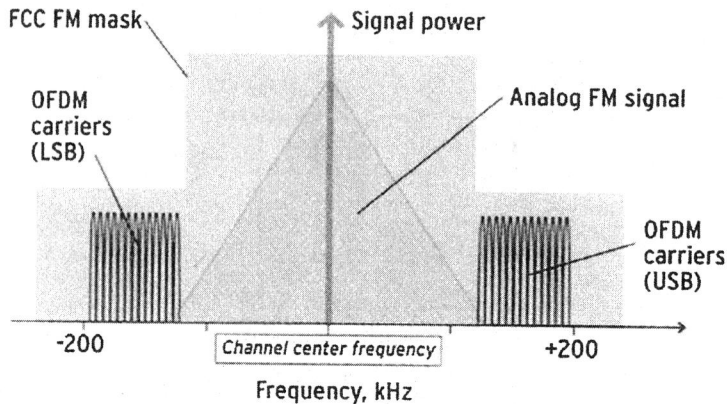

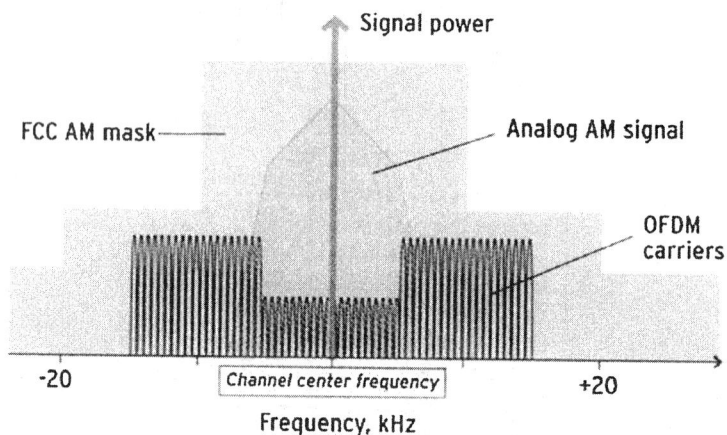

OFDM = Orthogonal frequency division multiplexing

LSB = lower sideband   USB = upper sideband   IBOC = in-band/on-channel

FIGURE 22.2.11 The digital signal is included in the spectrum of the analog channel. The digital radio signals fall within the frequencies allocated to stations, the so-called FCC FM or AM masks, indicated by the cross-hatched area. For FM [top], IBOC digital sidebands are placed on either side of the ananlog host signal. For AM [bottom], the setup is similar except that, in addition, some digital carriers are also co-located in the region of the AM host itself.   (*From Ref. 13*)

U.S. IBOC Digital Broadcast Systems for AM and FM

In the United States the Eureka 147 system has not been accepted because of the wide bandwidth, 1.5 MHz, needed for each transmission system and the large number of independent broadcasters who would need this bandwidth for each station. There exists no available unused spectrum in the United States for additional services.

A current proposal, which occupies the same channel space as present AM and FM stations, is being field test-ed by the Digital Audio Broadcast Subcommittee of the National Radio Systems Committee (NRSC). This sys-tem and the staffing belongs to iBiquity Digital Corporation, which was formed as a merger of several other proposals and proponents over the period from the early 1990s through 2001.[13] The system provides a digital signal on the upper and lower edges of the main analog (AM or FM) channel from which derives the name in-band-on-channel (IBOC) (Fig. 22.2.11). The digital signal will have the same identical program material (simulcast) as the analog transmission. A time delay will exist between the analog and the digital signals in order to provide diversity. In the receiver, this will lead to a more robust reception quality. It is anticipated that sometime in the future as the transition from analog to digital occurs that the analog signal will disappear and the digital signal will occupy the entire channel.

This technology will enhance both the AM and FM band's audio fidelity. AM will sound like FM does today and FM will have compact-disc-like audio. Multipath, noise, and interfering signals that cause the static, hiss, pops, and fades heard on today's analog radios will be virtually eliminated, thus ensuring crystal clear recep-tion. Additionally, the technologies will allow for new wireless data services to be received from AM and FM stations, such as station information, artist and song identification, traffic, weather, and news and sports scores.[14]

A receiver design for the IBOC service would have a similar block diagram as present AM or FM receivers with the addition of two blocks: (1) an OFDM decoder IC and time delay memory and control to handle the diversity provided by the two carriers and (2) a decoder IC to restore the audio perceptual audio coder (PAC) compression provided at the transmitter. Several major broadcast manufacturers and radio receiver manufac-turers have already signed license agreements with the technology owner. At least 20 broadcast stations situ-ated in major market areas of the United States were involved in the field testing phase. The service was launched in 2003 and receivers are now available.

Radio Broadcast Data Systems

Unlike the two systems described earlier (DAB and IBOC), which were "totally digital," this section will deal with the existing analog FM systems that have digital subcarriers superimposed or added into the existing channel spectrum. These subcarriers would transmit to the receiver information such as station call letters, road conditions, program or music name, and other data believed to be useful to a moving automobile. In Europe the system has been called *radio data system* (RDS) and has been in operation for several years, whereas in the United States it is known as *radio broadcast data system* (RBDS). In Japan a similar system called *data radio channel* (DARC) is being implemented. Currently in the United States numerous FM radio stations are broad-casting RBDS signals and receivers which make some use of RBDS number in the hundreds of thousands.[15]

Superimposed subcarriers in the United States system exist at 57 kHz (third harmonic of the 19 kHz stereo pilot tone) with injection level of 2 to 3 percent. ICs to decode this signal from the main FM stereo multiplex have been developed by several companies and have been used in receiver design for several years for both RDS and RBDS service.

The Japanese DARC system employs a 76 kHz level-controlled minimum shift keying (L-MSK) modula-tion. The level being controlled by the level of the $R - L$ signal to reduce interference in the $L - R$ channel and therefore reduce noise and interference in the recovered stereo signals. Most of the circuitry added to an FM receiver to handle this service is contained in one LSI-IC.[16]

Satellite Radio

Numerous satellite digital audio radio service (SDARS) proposals have been on the table both internation-ally and in the United States during the 1990s. At the present time, only four appear to be viable. The WorldSpace system covers Africa (AfriStar launched in October 1999) and Asia (AsiaStar launched in March 2000). A third satellite will cover Central and South America.[17] Development of the Japanese Communication Satellite System operated by Mobile Broadcasting Corporation dates from the early 1990s.[18,19] Two U.S. services are the XM system (launched in 2001) and the Sirius system (launched in 2002). Both are aimed at auto radio application and are subscription services (XM at $9.95 per month and Sirius at $12.95 per month).

Receivers for the WorldSpace system are available from several major radio receiver manufacturers and are on sale in the intended countries. Over 50 channels of low noise, interference-free audio and multimedia programming are available in the 1467–1492 MHz (L–band) spectrum. The system is primarily intended for direct-to-home broadcast. RF modulation is digital QPSK combined with convolutional and Reed-Solomon forward error correction. Ninety-six channels of 16 kbit/s each are combined using time-division multiplex (TDM). High-quality audio is ensured by encoding the audio using the MPEG 2.5 layer 3 compression scheme. The basic diagram of a typical receiver using two ICs is shown in Fig. 22.2.12.[20]

Several other semiconductor manufacturers have either two IC or three IC solutions for WorldSpace radios. Functional block diagrams for all are very similar.

The Japanese system receivers are being developed to handle both satellite and terrestrial signals with automatic switching and antenna diversity. Reception of the signals in the 2.6 GHz band will be by an antenna assembly consisting of two microstrip antennas, i.e., a high gain with a low-noise amplifier for satellite reception and a low-gain unit for the terrestrial pickup. Multimedia signals containing up to 50 programs will be available from the satellite. These include MPEG-4 video, CD-quality audio, and digital information services to mobile receivers.[18]

The two U.S. SDARS differ from each other in several ways. XM employs two geostationary satellites, while Sirius has three satellites covering the western hemisphere in an elliptical pattern. A technical comparison of the two U.S. satellite radio services is shown in Table 22.2.1.[13] Both systems employ three degrees of diversity in order to guarantee uninterrupted service to the receiver. These three are:

1. Spatial. Use of two widely separated satellites and numerous terrestrial repeaters to fill in the holes created by building shadows, tunnels, and other obstructions to the satellite signal in metropolitan areas.

2. Frequency. The two satellites will transmit the same signal but in different frequency bands.

3. Time. A time delay of several seconds will exist between the transmissions of the same signal from each satellite.

At the receiver, these diversity situations can be resolved with the strongest and cleanest signal at the moment taking command of the circuitry. The receiving antenna will be of a simple omnidirectional design mounted on the roof or side window of the car.

The XM receiver uses two custom-integrated circuits that are fabricated by using CMOS and proprietary high-speed bipolar technology.[21] A block diagram of the receiver is shown in Fig. 22.2.13.

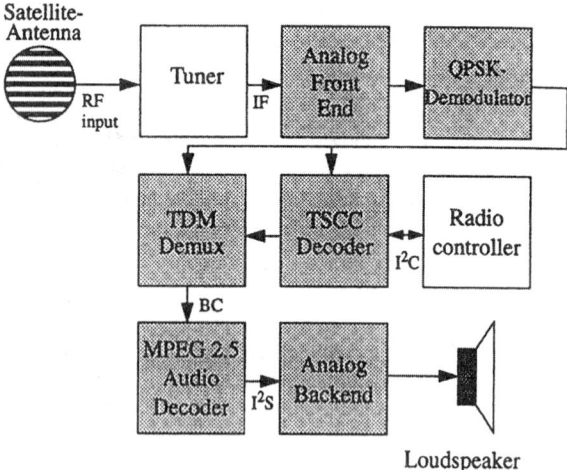

▧ **integrated in the chip set**

FIGURE 22.2.12 Basic WorldSpace receiver. (*From Ref. 20*)

TABLE 22.2.1.   The U.S. Satellite Digital Audio Radio Services (SDARS)

Parameter	Satellite radio		Comments
	Sirius	XM	
IN ORBIT			
No. of satellites (longitude)	3 (nominal 100° W)	2 (85° and 115° W)	Sirius satellites are in a highly elliptical orbit, rising and setting every 16 h: XM types are geostationary
Uplink frequencies	7060–7072.5 MHz	7050–7075 MHz	Sirius also uses Ku-band (14–12 GHz) uplink to feed repeaters
Downlink frequencies	2320.0–2324.0 MHz and 2328.5–2332.5 MHz	2332.5–2336.5 MHz and 2341.0–2345.0	Redundant downlink signals are for spatial/frequency/time diversity
Satellite elevation angle	60°	45°	Typical
ON LAND			
Location	New York City	Washington, D.C	
No. of studios	75	82	In main facility
No. of terrestrial repeaters	105 (46 markets)	1500 (70 markets)	Approximate numbers
Repeater EIRP	Up to 40 kW	90% are 2 kW	
OTHER CHARACTERISTICS			
No. of CD-quality (64-kb/s) channels	50		Lower-quality services use 0.5–64 k/s
No. of news-talk-sports channels	50		System is reconfigurable on the fly
Satellite modulation	TDM-QPSK		Each carrier is about 4 MHz wide
Terrestrial repeater modulation	TDM-COFDM		Carrier ensemble is about 4 MHz wide
Channel coding scheme	Concatenated		Error-correcting Reed-Solomon outer code and rate 1/2 convolutional inner code
Source coding scheme	Lucent PAC		Nominal rate for top-quality music: 64 k/s
Transmission rate	4.4 M/s	4.0 M/s	Before channel coding

COFDM = coded orthogonal frequency-division multiplexing; EIRP = equivalent isotropic radiated power;
PAC = Perceptual Audio Coder; OPSK = quadrature phase shift keying; TDM = time-division multiplexing.
Source: Sirius Satellite Radio, XM Satellite Radio.

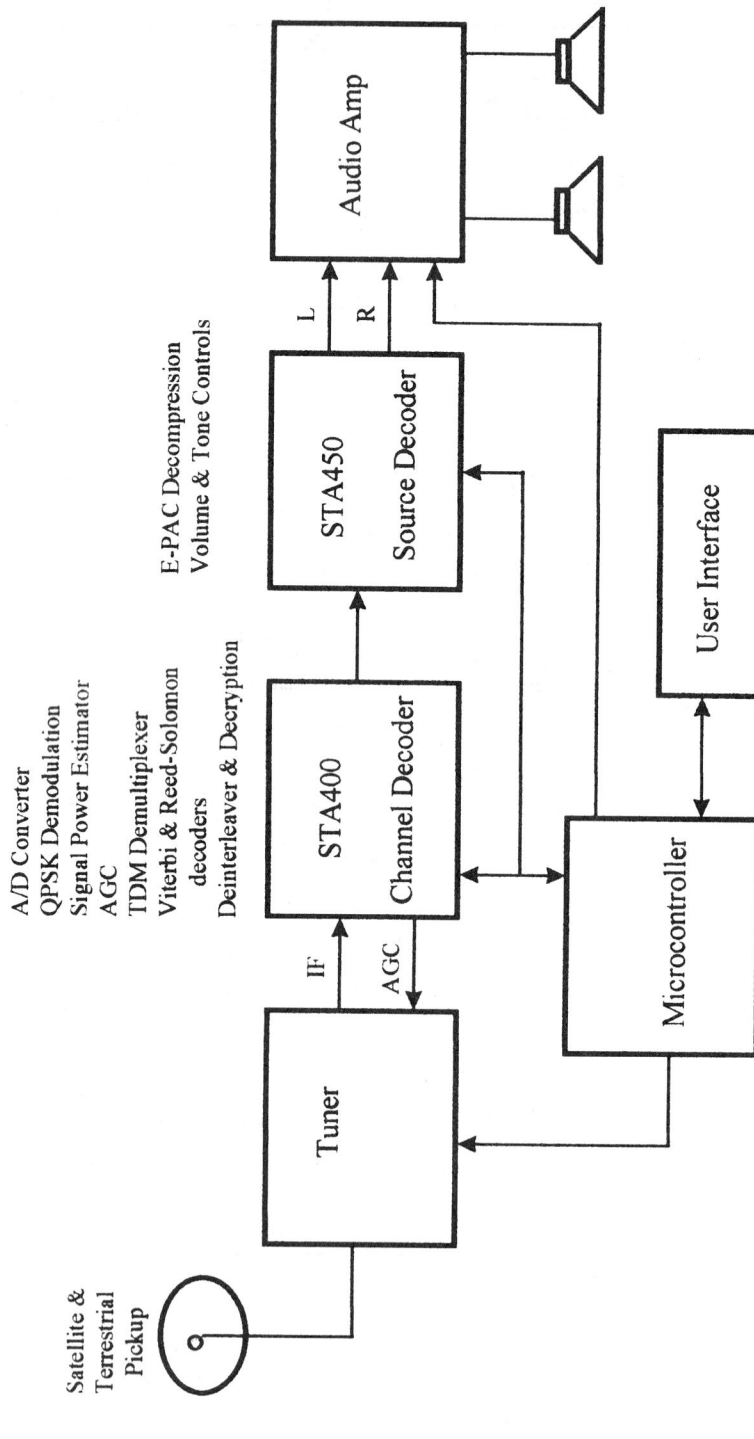

FIGURE 22.2.13   Block diagram of XM satellite receiver.

REFERENCES

1. *Audio Bandwidth and Distortion Recommendations for AM Broadcast Receivers.* National Radio Systems Committee, June 1990. Also published as EIA/IS-80, Electronic Industries Association, March 1991.

2. TEA555IT Product Specification, Philips Semiconductors, October 1990.

3. "Portable and Home Hi-Fi/Radio Designers Guide," Philips Semiconductors, June 1996.

4. "Ceramic Filter Applications Manual," Murata Manufacturing Co., 1982.

5. Parker, N., F. Hilbert, and Y. Sakaie A Compatible Quadrature System for AM Stereo, *IEEE Trans. Consum. Electron,* November 1977, Vol. CE-23, No. 4, pp. 454–460.

6. Ecklund, L., and O. Richards A New Tuning System for Use in AM Stereo Receivers, *IEEE Trans. Consum. Electron.,* August 1986, Vol. CE-32, No. 3, pp. 497–500.

7. Brekelmans, H., et al. A Novel Multistandard TV/FM Front-end for Multimedia Applications, *IEEE Trans. Consum. Electron.,* May 1998, Vol. 44, No. 2, pp. 280–288.

8. Sato, A., et al. Development of an Adjustment-free Audio Tuner IC, *IEEE Trans. Consum. Electron.,* August 1996, Vol. 42, No. 3, pp. 328–334.

9. Okanobu T., et al. An AM/TV/FM Radio IC Including Filters for DTS, *IEEE Trans. Consum. Electron.,* August 1997, Vol. 43, No. 3, pp. 655–661.

10. Yamazaki, D., et al. A Complete Single Chip AM Stereo/FM Stereo Radio IC, *IEEE Trans. Consum. Electron.,* August 1994, Vol. 40, No. 3, pp. 563–569.

11. Nolde, W., et al. An AM-Stereo and FM-Stereo Receiver Concept for Car Radio and Hi-Fi, *IEEE Trans. Consum. Electron.,* May 1981, Vol. CE-27, No. 2, pp. 135–143.

12. Kozamernik, F. Digital Audio Broadcasting, Chapter 14, in: R. Jurgen (ed.), *Digital Consumer Electronics Handbook,* McGraw-Hill, 1997.

13. Layer, D. Digital Radio Takes to the Road, *IEEE Spectrum,* July 2001, Vol. 38, No. 7, pp. 40–46.

14. "What is iBiquity Digital," www.ibiquity.com, iBiquity Digital Corp., 2000.

15. Clegg, A. The Radio Broadcast Data System, Chapter 15, in: R. Jurgen (ed.), *Digital Consumer Electronics Handbook,* McGraw-Hill, 1997.

16. Suke, M., et al. Development of DARC Decoding LSI for High-Speed FM Subcarrier System, *IEEE Trans. Consum. Electron.* August 1994, Vol. 40, No. 3, pp. 570–579.

17. Bonsor, K. How Satellite Radio Works, *Marshall Brain's HowStuffWorks,* 2000, available at:www.howstuffworks.com/satellite-radio

18. Press Release—TDK Semiconductor Corp. and Kenwood Corp., Jan. 21, 1999, available at:www.tdk.co.jp/teaah01/aah38000.htm

19. "Loral to build digital satellite for Mobile Broadcasting Corp." *Loral Space & Communications News Release,* August 20, 2001, available at: www.spaceflightnow.com/news/n0108/20mbsat

20. Bock, C., et al. Receiver IC-Set for Digital Satellite System, *IEEE Trans. Consum. Electron.,* November 1997, Vol. 43, No. 4, pp. 1305–1311.

21. Press Release "XM Announces Delivery of Production Design Chipsets to Radio Manufacturers" March 30, 2001, ST Digital Radio Solutions, STMicroelectronics, available at:http://us.st.com/stonline/prodpres/dedicate/digradio/digradio.htm

CHAPTER 22.3
CABLE TELEVISION SYSTEMS

Walter S. Ciciora

INTRODUCTION

Cable television service is enjoyed by almost 70 million U.S. households. This is a market penetration of over 68 percent. Cable systems have expanded capacity and added digital signal capability. High-speed cable modems and telephony are important growth areas. Video on demand (VOD) is offered to ever more subscribers. Interactive television is beginning to gain acceptance. It is expected that these growth trends will continue.

HISTORICAL PERSPECTIVE

The original purpose for cable television was to deliver broadcast signals in areas where they were not received in an acceptable manner with an antenna. These systems were called Community Antenna Television, or CATV. In 1948, Ed Parson of Astoria Oregon, built the first CATV system consisting of twin-lead transmission wire strung from housetop to housetop. In 1950, Bob Tarlton built a system in Landsford, Pennsylvania, using coaxial cable on utility poles under a franchise from the city.

In the mid-1970s, an embryonic technology breathed new life into cable television. This technology was satellite delivery of signals to cable systems, which added more channels than were available from terrestrial broadcasters. While satellites and earth stations were very expensive investments, these programming pioneers understood that the costs could be spread over many cable operators who, in turn, serve many subscribers.

Subscribers are offered a variety of video services. The foundation service taken by most subscribers is called *basic*. Off-air channels, some distant channels, and some satellite-delivered programs are included. The satellite programs include the super stations and some of the specialty channels. *Pay television* constitutes premium channels, usually with movies and some special events, which are offered as optional channels for an extra monthly fee. Some cable systems offer *Pay-Per-View* (PPV) programming which is marketed on a program-by-program basis. Recent movies and special sport events are the mainstay of PPV programming. *Impulse Pay Per View* (IPPV) allows the subscribers to order the program spontaneously, even after it has begun. The ordering mechanism usually involves an automated telephone link or, occasionally, two-way cable.

Ways of providing conditional access to allow for a limited selection of service packages at differing price points are often included in the cable system. Simple filters remove the unsubscribed channels in some systems, while elaborate video and audio scrambling mechanisms involving encryption are used in other cable systems.

Since cable television systems must use the public right-of-way to install their cables, they, like power, telephone, and gas companies, must obtain a franchise from the local governmental authorities. This is a nonexclusive

Note: This overview is based on a CableLabs publication, "Cable Television in the United States."

franchise. However, experience with multiple cable systems has shown that the economics of the business generally only support one system per community.

The decade of the 1990s introduced major changes into the cable industry. From a technical perspective, the addition of optical fiber improved signal quality, increased bandwidth, enhanced reliability, and reduced operating costs while making two-way cable practical. The second major technical advance was the introduction of digital television signals. Digital video multiplied the program capacity dramatically. These two technical advances enabled significant new service opportunities. Near video on demand (NVOD) took advantage of increased signal capacity. True VOD was made possible by the dramatic performance increases and cost reductions on hard drive storage systems. High-speed cable modem services and telephony became possible as cable operators learned to send digital signals reliably. Cable systems changed from entertainment-only systems into general-purpose, comprehensive communications facilities.

SPECTRUM REUSE

For NTSC, each television channel consumes 6 MHz because of vestigial side-band amplitude modulation, VSB-AM. Compared with double side-band amplitude modulation's need for 8.4 MHz, VSB-AM transmits one complete side-band and only a vestige of the other. At the time the standard was created, the design requirements of practical filters determined the amount of side-band included. The consumer's receiver selects the channel to be watched by tuning a 6-MHz portion of the assigned spectrum. In the terrestrial broadcast environment, channels must be carefully assigned to prevent interference with each other. The result of this process is that most of the terrestrial broadcast television spectrum is vacant. Better television antennas and better television circuits allow more of the spectrum to be used. However, with nearly 300 million receivers and more than 100 million VCRs in consumers' hands, the changeover process to upgraded systems would be difficult, costly, and require something like 20 years.

The rest of the terrestrial spectrum that is not assigned to broadcast has other important uses. These include aircraft navigation and communications, emergency communication, and commercial and military applications. The terrestrial spectrum is too limited to supply the video needs of the U.S. viewer.

Cable television is made possible by the technology of coaxial cable. Rigid coaxial cable has a solid aluminum outer tube and a center conductor of copper-clad aluminum. Flexible coaxial cable's outer conductor is a combination of metal foil and braided wire, with a copper-clad steel center conductor. The characteristic impedance of the coaxial cable used in cable television practice is 75 Ω. The well-known principles of transmission line theory apply fully to cable television technology.

The most important characteristic of coaxial cable is its ability to contain a separate frequency spectrum and respect the properties of that separate spectrum so that it behaves like over-the-air spectrum. This means that a television receiver connected to a cable signal will behave as it does when connected to an antenna. A television set owner can become a cable subscriber without an additional expenditure on consumer electronics equipment. The subscriber can also cancel the subscription and not be left with useless hardware. This ease of entry and exit from an optional video service is a fundamental part of cable's appeal to subscribers.

Since the cable spectrum is tightly sealed inside an aluminum environment (the coax cable), a properly installed and maintained cable system can use frequencies assigned for other purposes in the over-the-air environment. This usage takes place without causing interference to these other applications or without having them cause interference to the cable service. New spectrum is "created" inside the cable by the "reuse" of spectrum. In some cable systems, dual cables bring two of these sealed spectra into the subscriber's home, with each cable containing different signals.

The principal negative of coaxial cable is its relatively high loss. Coaxial cable signal loss is a function of its diameter, dielectric construction, temperature, and operating frequency. A ballpark figure is 1 dB of loss per 100 ft.

CABLE NETWORK DESIGN

Since cable television is not a general-purpose communication mechanism, but rather a specialized system for transmitting numerous television channels in a sealed spectrum, the topology or layout of the network can be customized for maximum efficiency. The topology that has evolved over the years is called tree-and-branch architecture.

There are five major parts to a cable system: (1) the headend, (2) the trunk cable, (3) the distribution (or feeder) cable in the neighborhood, (4) the drop cable to the home and in-house wiring, and (5) the terminal equipment (consumer electronics).

Flexible coaxial cable is used to bring the signal to the terminal equipment in the home. In the simplest cases, the terminal equipment is the television set or VCR. If the TV or VCR does not tune all the channels of interest because it is not "cable compatible," a converter is placed between the cable and the TV or VCR tuner.

The home is connected to the cable system by the flexible drop cable, typically 150 ft long.

The distribution cable in the neighborhood runs past the homes of subscribers. This cable is tapped so that flexible drop cable can be connected to it and routed to the residence. The distribution cable interfaces with the trunk cable through an amplifier called a bridge amplifier, which increases the signal level for delivery to multiple homes. One or two specialized amplifiers called line extenders are included in each distribution cable. Approximately 40 percent of the system's cable footage is in the distribution portion of the plant and 45 percent is in the flexible drops to the home.

The trunk part of the cable system transports the signals to the neighborhood. Its primary goal is to cover distance while preserving the quality of the signal in a cost-effective manner. Broadband amplifiers are required about every 2000 ft depending on the bandwidth of the system. The maximum number of amplifiers that can be placed in a run or *cascade* is limited by the buildup of noise and distortion. Twenty or 30 amplifiers may be cascaded in relatively high-bandwidth applications. Older cable systems with fewer channels may have as many as 50 or 60 amplifiers in cascade. Approximately 14 percent of a cable system's footage is in the trunk part of the system.

The headend is the origination point for signals in the cable system.

When the whole picture is assembled, the tree shape of the topology is evident. The trunk and its branches become visible.

The cable industry had an early interest in fiber optics. Since signal losses in optical fiber are orders of magnitude lower than signal losses in coaxial cable, power-consuming amplifiers in the trunk became unnecessary.

Initially, laser technology was the most significant challenge to implementing fiber optics in cable systems. The laser had to have extreme linearity in order to avoid distortions of the broadband analog signal. This degree of linearity is not necessary in a digital transmission system. Once the process of producing sufficiently linear lasers were mastered, the rapid roll out of optical fiber in cable systems began.

Coaxial cable design had become an optimized, well-understood process prior to the introduction of fiber. Optical fiber offered a wide variety of options that caused cable system design to become much more complex. Many approaches were tried and as the cost of the various components came down, new approaches became possible. The technique had generally become known as *hybrid fiber coax* (HFC). Optical fiber is relatively inexpensive. It is supplied in a plastic protective sheath that is a bit more expensive on a per foot basis. But the most expensive element is the cost of installation. As a result, when optical fiber is installed, multiple strands of unused fiber are included in anticipation of future needs. These fibers are called "dark fiber" since they have not yet had lasers installed. As future growth in subscribers and services leads to demands for more bandwidth, these fibers are activated.

The point where the optical fiber ends and the coaxial distribution begins is called a "node." The node serves a group of subscribers. The size of the group depends on the demographics and the philosophy of the system designer. The size ranges from a few hundred to a few thousand homes. "Node splitting" is the process of activating dark fibers to create additional smaller nodes out of a larger node. This is done when additional upstream signal return capacity is required or when additional downstream frequency reuse is needed.

Downstream frequency reuse is an important concept, giving rise to one of cable's most sustainable competitive advantages against other communications systems. Since cable trunks or fibers extend from the cable head end into neighborhoods, different programming can be placed on the same frequencies. This technique greatly expands the reach of cable's spectrum. There are services that dedicate signals to groups of subscribers or individual subscribers. These services include VOD high-speed cable modem services and telephony. These services do not comply with the broadcast model and require directed signals to individual locations. Nodalization, the process of splitting larger nodes into smaller groups, makes this possible with limited spectrum capacity. As these services increase in penetration, they burden the assigned frequencies that are shared among the subscribers connected to a node. Two choices are possible: assign more channels to the service or split the node. So nodalization is a means of creating additional capacity when more spectra are unavailable.

SIGNAL QUALITY

The ultimate goal of the cable system is to deliver pictures of adequate quality at an acceptable price while satisfying stockholders, investors, and holders of the debt generated by the huge capital expenses of building the cable system plant. This is a difficult balancing act. It would be a simple matter to deliver very high-quality video if cost were not a consideration. Experience teaches that subscriber satisfaction is a complex function of a variety of factors led by program quality, and variety, reliability of service, video and sound quality, and the amount of the subscriber's cable bill.

The principal picture impairments can be divided into two categories—coherent and noncoherent. Coherent impairments result in a recognizable interfering pattern or picture. They tend to be more objectionable than noncoherent impairments of equal strength.

The principal noncoherent picture impairment is noise. Random noise behavior is a well-understood part of general communication theory. The familiar Boltzmann relationship, noise figure concepts, and the like apply fully to cable television technology. Random noise is the consequence of the statistical nature of the movement of electric charges in conductors. This creates a signal of its own. This noise is inescapable. If the intended analog signal is ever allowed to become weak enough to be comparable to the noise signal, it will be polluted by it yielding a snowy pattern in pictures and a seashore sounding background to audio.

Noise levels are expressed in cable system practice as ratios of the video carrier to the noise in a television channel. This measure is called the carrier-to-noise ratio (CNR) and is given in decibels (dB). The target value for CNR is 45 to 46 dB. Noise in the picture, called snow, is just visible when CNR is 42 to 44 dB. Snow is objectionable at CNRs of 40 to 41 dB.

Coherent interference includes ingress of video signals into the cable system, reflections of the signal from transmission-line impedance discontinuities, cross modulation of video, and cross modulation of the carriers in the video signal. This latter phenomenon gives rise to patterns on the screen, which are called *beats*. These patterns often look like moving diagonal bars or herringbones.

The evaluation of signal quality takes place on two planes, objective and subjective. In the objective arena, measurements of electrical parameters are used. These measurements are repeatable. Standardized, automated test equipment has been developed and accepted by the video industries. Table 22.3.1 lists the parameters usually considered important and the values of good, current practice.

TABLE 22.3.1 Signal Quality Parameters

Parameter	Symbol	Value
Carrier/noise (CNR)	C/N	46 dB
Composite second-order	CSO	–53 dB
Composite triple beat	CTB	–53 dB
Signal level at TV		0 to +3 dBmV

Signal Quality Target Values

The ultimate performance evaluation involves the subjective reaction of viewers. One example of the difficulties experienced is the fact that different frequencies of noise have differing levels of irritation. High-frequency noise tends to become invisible, while low-frequency noise creates large moving blobs, which are highly objectionable. Subjective reactions to these phenomena are influenced by such factors as the age, gender, health, and attitude of the viewer. The nature of the video program, the characteristics of the viewing equipment, and the viewing conditions also impact the result.

As time progresses, the level of performance of consumer electronics will continue to increase. As ATV and HDTV are introduced, still more demands will be made on cable system performance. The trend to larger screen sizes also makes video impairments more evident.

DIGITAL VIDEO

In the 1980s, the over-the-air broadcast industry attempted to upgrade its services by proposing high-definition television (HDTV).

The HDTV baseband analog signal coming from a television camera has substantially greater bandwidth than the NTSC baseband signal. Digitizing the signal results in even greater bandwidth requirements. This made complying with the FCC's 6-MHz channel requirement impossible without further signal processing. Fortunately, the digital nature of the signal enabled massive data compressing. The principle feature of digital signal processing, which enables these data compressions, is inexpensive large-scale memory. Such memory does not exist in analog implementations. Since video consists of a series of 30 complete still pictures every second, which usually have only small portions of the image change from picture to picture, the unchanging portions can be stored in digital memory. Only the differences need to be transmitted. This saves a tremendous amount of bandwidth. For example, if the image is a "talking head" against a stationary background, once the background is transmitted, it can be stored and further retransmission minimized or avoided. Only the relatively small fraction of the picture that changes needs to be transmitted. Also, the motion of much of the images consists of the horizontal or vertical translation of elements of the picture. So rather than resending those elements, an instruction to relocate them will suffice. Even more compression is possible by taking advantage of the human visual perception mechanism. Humans are more sensitive to luminance detail than color detail. Perception of detail diminishes if the object is in motion. Such factors enable still more compression of the visual data. The overall consequence is that only about 2 percent of the data generated by digitizing the analog signal is required for excellent reconstruction of the video at the receiver.

The observation was made that if an HDTV signal could be compressed with digital means to fit into a 6-MHz spectral slot, multiple standard-definition TV (SDTV) signals could be placed in that same spectral slot. This development has become commercially much more important than HDTV. The multiplication of capacity by digital compression made direct broadcast satellite (DBS) practical. DBS with a few dozen analog channels simply didn't have enough variety to be a realistic competitor to 50, 70, or 100 cable channels. However, digital compression enabled DBS to have several hundred channels, much more than cable in analog format. DBS could offer NVOD. The stagger starting of movies on multiple channels reduced the waiting time for a movie. A 2-h movie assigned eight program streams would require a maximum wait time of just 15 min, averaging to 7.5. Several tens of movie titles are offered at various stagger starts yielded a compelling service. The need to launch the digital DBS service in a timely manner prevented waiting for the complete resolution of the details of the broadcast digital television standard. Consequently, the digital broadcast compression standard and the DBS compression system are very similar, but not completely identical.

The competitive threat of DBS forced a response from cable that had to be implemented before the broadcast standard was complete. This resulted in yet a third digital approach that was mostly, but not completely, the same as the broadcast standard. The broadcast signal environment is more difficult than that found in cable and both are very different from the satellite system. Thus there are three different, incompatible modulation techniques. The satellite transmitter is a saturation device that is either on or off and transmits only one digital bit per symbol, i.e., time element. The broadcast signal must deal with multipath and low signal strengths at great distances from the transmitter. The broadcast modulation scheme is called 8-VSB (vestigial side band) and transmits three digital bits per symbol. Cable's signal environment has essentially no multipath and only modest signal strength differences between channels. Consequently, the cable signal can transmit many more digital bits per symbol. Two modulation schemes are common: 64 QAM (quadrature amplitude modulation) carrying five digital bits per symbol and 256 QAM conveying seven digital bits per symbol. QAM is very bandwidth efficient. All of these methods include extensive error detection and error correction methods with the broadcast signal the most protected. These powerful techniques allow for the recovery of transmission errors.

An important trade-off exists between image quality and program quantity. The more of the data capacity in a 6-MHz channel that is assigned to one program, the better its image quality. Five or six Mb/s (million bits per second) yields essentially perfect video under nearly all circumstances. But some types of video provide acceptable results with just 1.5 to 2 Mb/s. The decoder at the receiver that converts the bits to the analog signals necessary to drive the display device can accommodate any of these bit rates. So the choice is made at the point of origination of the video. A 256 QAM system has a payload of data of about 38 Mb/s. Thus six program streams of 6 Mb/s can be carried or double that for 3 Mb/s. The bit rate does not have to be the same for all streams or for all times. The data capacity can be allocated based on the type of programming and changed as needed. A further efficiency technique called *statistical multiplexing* can be employed. Statistical multiplexing averages out the bit capacity assignments depending on the needs of the multiple program streams sent in a 6-MHz channel. Thus when one program has high detail or high action or both, it is temporarily assigned the bits it needs while other less demanding channels make a contribution. This allotment changes from moment to moment depending on the needs of the program streams. Statistically, better quality and perhaps an additional program stream can be accommodated in this manner. The same techniques can be applied in satellite and broadcast. Because of the more difficult signal conditions in broadcast, only four to six program streams can be accommodated in the broadcast 6 MHz.

The video compression method used by broadcast, cable, and satellite is called MPEG after the committee that standardized it, the Moving Pictures Expert Group. MPEG is actually a collection of standards that can be applied to fit a variety of applications. The standard is unsymmetrical in that the equipment necessary to encode the signal is much more complex than the hardware needed to decode it. This makes very good economic sense since there will be many more receiving devices than transmitters. A further important feature is that improvements in the encoder result in better video reception without the need to modify the decoder. So as more processing power becomes affordable in encoders and as the techniques evolve, improved results are obtained in the system with previously installed receivers. In some cases, the bit rate can be reduced as well.

While the United States has selected 8-VSB as its broadcast modulation scheme, much of the rest of the world has selected versions of QAM. Frustratingly, the opportunity for much more signal commonality was not achieved. Signal differences prevent a worldwide receiver that can function on cable, broadcast, and satellite anywhere in the world. Multiple subsystems are required. Fortunately, the digital portion of the signal allows relatively easy interconversion between standards.

Techniques for delivering 3 to 6 Mb/s of data in the analog television signal without significant interference to the analog television signal have been developed. These developments might have made the original FCC's hope of a compatible system possible, at least in cable. But the standards are now too entrenched. However, these methods can be used to add either data or encoded video to an analog signal.

The auxiliary service of transmitting data in either an analog or a digital television signal has been called datacasting. A wide variety of services are proposed for this technique including data to computers, additional video signals, downloading to hard drives, and other services.

CABLE SYSTEM TRADE-OFFS

The experienced cable system designer has learned how to balance noise, nonlinear distortions, and cost to find a near optimal balance.

Signals in cable systems are measured in decibels relative to 1 mV across 75 Ω. This measure is called dBmV. Applying the well-known Boltzmann noise equation to 75-Ω cable systems yields an open-circuit voltage of 2.2 μV in 4 MHz at room temperature. When terminated in a matched load, the result is 1.1 μV. Expressed in dBmV, the minimum room-temperature noise in a perfect cable system is –59.17 dBmV.

Starting at the home, the objective is to deliver at least 0 dBmV, but no more than 10 dBmV to the terminal on the television receiver. Lower numbers produce snowy pictures and higher numbers overload the television receiver's tuner, resulting in cross modulation of the channels. If a converter or descrambler is used, its noise figure must be taken into account. There are two reasons for staying toward the low side of the signal range: cost and the minimization of interference in the event of a signal leak caused by a faulty connector, damaged piece of cable, or defect in the television receiver. Low signal levels may cause poor pictures for the subscriber who insists on splitting in the home to serve multiple receivers. Working our way back up the plant, we need a signal level of 10 dBmV to 15 dBmV at the tap to compensate for losses in the drop cable.

The design objectives of the distribution part of the cable system involve an adequate level of power not only to support the attenuation characteristics of the cable but to allow energy to be diverted to subscribers' premises. Energy diverted to the subscriber is lost from the distribution cable. This loss is called *flat loss* because it is independent of frequency. Loss in the cable itself is a square-root function of frequency and is therefore contrasted to flat loss. Because of flat losses, relatively high power levels are required in the distribution part of the plant, typically 48 dBmV at the input to the distribution plant. These levels force the amplifiers in the distribution part of the plant to reach into regions of their transfer characteristics that are slightly nonlinear. As a result, only one or two amplifiers, called line extenders, can be cascaded in the distribution part of the plant. These amplifiers are spaced 300 to 900 ft apart depending on the number of taps required by the density of homes.

Because the distribution part of the plant is operated at higher power levels, nonlinear effects become important. The television signal has three principal carriers—the video carrier, the audio carrier, and the color subcarrier. These concentrations of energy in the frequency domain give rise to a wide range of "beats" when passed through nonlinearities. To minimize these effects, the audio carrier is attenuated about 15 dB below the video carrier.

When cable systems only carried the 12 VHF channels, second-order distortions created spectrum products that fell out of the frequency band of interest. As channels were added to fill the spectrum from 54 MHz to as much as 650 MHz, second-order effects were minimized through the use of balanced, *push–pull* output circuits in amplifiers. The third-order component of the transfer characteristic dominates in many of these designs. The total effect of all the carriers beating against each other gives rise to an interference called *composite triple beat* (CTB). In an older 35-channel cable system, about 10,000 beat products are created. More channels create more beats. Channel 11 suffers the most with 350 of these products falling in its video. Third-order distortions increase about 6 dB for each doubling of the number of amplifiers in cascade. A 1-dB reduction in amplifier output level will generally improve CTB by 2 dB. If these products build to visible levels, diagonal lines will be seen moving through the picture. When these components fall in the part of the spectrum that conveys color information, spurious rainbows appear.

The design objective of the trunk part of the cable system is to move the signal over substantial distances with minimal degradation. Because distances are significant, lower-loss cables are used. One inch and 0.75 in. diameter cable is common in the trunk, while 0.5 in. cable is found in the distribution. Signal levels in the trunk at an amplifier's output are 30 to 32 dBmV, depending on the equipment used. Cable trunk is rapidly being replaced by fiber. No new cable systems are being built with coax in the trunk.

It has been determined through analysis and confirmed through experience that optimum noise performance is obtained when the signal is not allowed to be attenuated more than about 20 to 22 dB before being amplified again. Amplifiers are said to be "spaced" by 20 dB. The actual distance in feet is a function of maximum frequency carried and the cable's attenuation characteristic. Modern high-bandwidth cable systems have their amplifiers fewer feet apart than older systems with fewer channels. Since attenuation varies with frequency, the spectrum in coaxial cable develops a *slope*. This is compensated with equalization networks in the amplifier housings.

Since the signal is not repeatedly tapped off in the trunk part of the system, high-power levels are not required to feed splitting losses. As a result, signal levels are lower than in the distribution portion of the plant. Typical levels are about 30 dBmV. For the most part, the amplifiers of the trunk are operated within their linear regions. The principal challenge of trunk design is keeping noise under control. Each doubling of the number of amplifiers in the cascade results in a 3-dB decrease in the CNR at the end of the cascade and a 6-dB increase in the amount of CTB.

If the noise at the end of the cascade is unacceptable, the choices are to employ lower noise amplifiers, shorter cascades, or a different technology such as microwave links of fiber optic links.

TECHNICAL DETAIL

The balance of this chapter concentrates in more detail on issues relating to the technical performance of cable systems. Some generalizations have been made in order to group explanations. This section is intended to serve as a briefing, so selective trade-offs were made on the amount of detail given. There are always exceptions, for cable systems do not neatly fall into clear types. A reading list is provided. The following topics will be covered: channel capacities and system layouts, FCC spectrum regulation, means of increasing channel capacity, scrambling methods, the interface between the cable and the customer's equipment, and fiber in cable television practice.

Channel Carriage Capacity

Channel carriage capacity is based on radio frequency (RF) bandwidth. It is a useful characteristic for classifying cable systems. There are three types of system (Table 22.3.2). Systems are categorized by their highest operating frequency. Downstream signals are transmitted to the customers' homes.

A cable system configuration consists of: (1) the headend (the signal reception, origination, and modulation point), (2) main trunk (or tree) cable, which is coaxial in older systems and fiber in new and upgraded systems and runs through the central streets in communities, (3) coaxial distribution (branch) cable to the customer's neighborhood, including distribution taps, (4) subscriber drops to the house, and (5) subscriber terminal equipment (television sets, converter/descramblers, VCRs, and so forth). Distribution plant is sometimes called *feeder* plant. Programming comes to the headend by satellite signals, off-air signals from

TABLE 22.3.2 Classification of Cable Systems

	Bandwidth (MHz)	Operating frequencies (MHz)	Number of channels
Small	170	50–220	12–22 (single coax)
Medium	220	50–270	30 (single coax)
	280	50–330	40 (single coax)
	350	50–400	52 (single coax)/104 (dual coax)
Large	400	50–450	60 (single coax)/120 (dual coax)
	500	50–550	80 (single coax)
	700	50–750	116 (single coax)
	950	50–1 GHz	158 (single coax)

broadcast stations, and signals imported via terrestrial microwave. Signals originating from the headend are from a co-located studio facility, VCRs, character generators, or commercial insertion equipment.

Plant mileage is calculated using the combined miles of strand that support the coaxial cables in the air and the footage of trenches where cables are installed in the ground. There are about a million miles of plant in more than 11,000 U.S. cable systems.

Extension cables, or drops, interconnect main coaxial plant lines to customers' homes. 220 MHz systems built 15 or more years ago are found in rural areas or in areas with clusters of smaller established communities. Some of these systems operate trunk lines running over 20 mi with 50 or more amplifiers in cascade. Total plant mileage for an average 220 MHz systems extends from 50 to 500 mi and services up to 15,000 cable customers. New construction of 220 MHz systems occur only where there are small numbers of potential customers (no more than 300) and where plant mileage does not exceed 10 mi.

Medium capacity cable systems operate with upper frequencies at 270 and 300 MHz, and total bandwidths of 220 and 280 MHz, respectively. 270 MHz systems deliver 30 channels, while 330 MHz systems deliver 40. Although new cable systems are seldom built with 40-channel capacity, plant extensions to existing 270-, 300-, and 330-MHz systems occur. Electronic upgrade is frequently employed to increase 270 MHz systems to 330 MHz. Some 220 MHz systems are upgrading to 300 MHz.

Medium capacity systems account for about 75 percent of total plant mileage. They serve a wide range of communities, from rural areas (population of 5000 to 50,000) to some of the largest systems built in the late '70s.

Large capacity cable systems achieve high channel capacities through extended operating frequencies and through the installation of dual, co-located, coaxial cable. Single coaxial cable systems range from 54-channel 400 MHz to 80-channel 550 MHz. With dual cable, it is not unusual to find 120 channels.

Large capacity cable systems account for about 15 percent of total cable plant miles. They are primarily high-tech systems designed for large urban areas previously not cabled. Recently, cable systems extending to 750 MHz have been built. Three cable systems with 1 GHz bandwidths have been constructed to date, the first in Queens, New York. That system, called Quantum, introduced NVOD for the first time to U.S. subscribers. Many cable systems are now constructed so that they can easily be upgraded to 750 MHz or 1 GHz. It is not uncommon to use GHz-rated passive components and amplifier housings to facilitate a later upgrade.

Large capacity systems are designed, and some operate, as two-way cable plant. In addition to the downstream signals to the customers (50 MHz to upper band edge), upstream signals are carried from customers to the cable central headend, or hub node. They are transmitted using frequencies between 5 and 30 MHz.

Channelization

There are three channelization plans to standardize the frequencies of channels. The first plan has evolved from the frequency assignments that the Federal Communications Commission (FCC) issued to VHF television broadcast stations. This plan is called the standard assignment plan.

The second channelization plan is achieved by phase locking the television channel carriers. It is called the incrementally related carriers (IRC) plan. The IRC plan was developed to minimize the effects of third-order distortions generated by repeated amplification of the television signals as they pass through the cable plant. As channel capacities increased beyond 36 channels, composite, third-order distortions became the limiting distortion.

The third channelization type is the harmonically related carriers (HRC) plan. It differs from the standard and IRC plan by lowering carrier frequencies by 1.25 MHz. With HRC, carriers are phase locked and fall on integer multiples of 6 MHz starting with channel 2 at 54 MHz. This plan was created to further reduce the visible impact of amplifier distortions.

FM radio services are carried at an amplitude that is 15 to 17 dB below Channel 6's video carrier level. The services are carried on cable in the FM band slot of 88 to 108 MHz. In an IRC channel plan, Channel 6's aural carrier falls at 89.75 MHz, which reduces the available FM band to 90 to 108 MHz.

Low-speed data carriers are transmitted in the FM band or in the guard band between Channels 4 and 5 in a standard frequency plan. The amplitude of these carriers in at least 15 dB below the closest video carrier level.

Frequencies Under Regulation

FCC rules and regulations govern the downstream cable frequencies that overlap with the over-the-air frequencies used by the Federal Aviation Administration (FAA). The frequencies are from 108 to 137 MHz and from 225 to 400 MHz. They are used by the FAA for aeronautical voice communications and navigational information. Since cable plant is not a perfectly sealed system, the FCC and FAA want to maintain a frequency separation between signals carried on cable and frequencies used by airports near the cable system boundaries. In 400-MHz systems, over 30 channels are affected by the FCC rules on frequency offset and related operating conditions.

Effects of the FCC Rules

The maximum, unregulated, carrier power level rule has been reassessed and changed. The previous limit of 1×10^{-5} W (28.75 dBmV) has been raised to 1×10^{-4} W (38.75 dBmV). Carriers with power levels below 38.75 dBmV are not required to follow the frequency separation and stability criteria. Carriers within ±50 kHz of 156.8 MHz, ±50 kHz of 243 MHz, or ±100 kHz of 121.5 MHz, which are emergency distress frequencies, must be operated at levels no greater than 28.75 dBmV at any point in the cable system.

Increasing Channel Capacity

There are several ways to increase channel capacity. If the actual cable is in good condition, channel capacity is upgraded by modifying or replacing the trunk and distribution amplifiers. If the cable has seriously deteriorated, the cable plant is completely rebuilt.

Upgrades (Retrofitting) and Rebuilds

An upgrade is defined as a plant rehabilitation process that results in the exchange or modification of amplifiers and passive devices (such as line splitters, directional couplers, and customer multitaps). A simple upgrade requires new amplifier circuit units called hybrids. A full upgrade replaces all devices in the system. In an upgrade project, most of the cable is retained. Goals of an upgrade project include increasing the plant's channel capacity and system expansion to outlying geographic areas. New amplifier technology such as feedforward and/or power doubling circuitry and advances in amplifier performance have greatly enhanced the technical and financial viability of upgrades. Upgrades are often the least expensive solution to providing expanded service.

In a rebuild, the outside plant is replaced.

System Distortion and System Maintenance

Constraints on the design and implementation of cable systems are imposed by each device used to transport or otherwise process the television signal. Each active device adds small distortions and noise to the signal.

Even passive devices contribute noise. The distortions and noise compound so that with each additional device the signal becomes less perfect.

Any nonlinear device, even bimetallic junctions, cause distortions. The primary contributors are the slight nonlinearities of amplifiers. Because the amplifiers are connected in cascade, the slight damage to the signal accumulates.

Noise in any electronic system can come from many sources. The major source is the random thermal movement of electrons in resistive components. For a cable system at 20°C or 68°F, the thermal noise voltage in a 4-MHz bandwidth will be 1.1 μV or −59.1 dBmV. This is the minimum noise level, or noise floor. Noise contributions from amplifiers add on a power basis, with the noise level increasing 3 dB for each doubling of the number of identical amplifiers in cascade. Eventually, the noise will increase to objectionable levels. The difference between the RF peak level and the noise level is measured to quantify the degree of interference of the noise power. The power levels in watts are compared as a ratio. This is called the signal-to-noise ratio, or SNR. In a cable system, the apparent effect of noise is its interference with the video portion of the TV channel. This level is compared to the video carrier and called the carrier-to-noise ratio (CNR).

As the CNR value decreases, the interference of noise with the signal becomes visible as a random fuzziness, called *snow*, that can overwhelm the picture resolution and contrast. The point where the picture becomes objectionably noisy to viewers is at a CNR = 40 dB. In well-designed systems, the CNR is maintained at 46 dB. While an increase in signal level would improve the CNR, unfortunately, there can be no level increase without increases in distortions.

The distortion products of solid-state devices used in cable amplifiers are a function of the output levels and bandwidths. The higher the signal level, the greater the distortion products produced. Modern amplifiers use balanced configurations, which almost completely cancel the distortion caused by the squared term of the amplifier's transfer characteristic. The dominant remaining distortions are called triple beats. They are caused by the cubed term. Because distortion products add on a voltage basis, the composite triple beat (CTB) to carrier ratio changes by 6 dB for each doubling of the number of amplifiers in cascade, whereas the CNR decreases by 3 dB for each doubling.

As signal levels are increased in the distribution sections, additional allowances must be made in the system design. As a rule of thumb, CNR is determined primarily by the conditions of trunk operation and signal-to-distortion ratio (SDR) primarily by the conditions of distribution operation.

Two other factors limit the geography of a cable system. Cable attenuation rises with increasing frequency. More equal gain amplifiers are required to transmit the signal a given distance. But noise limits the maximum number of amplifiers used. The second factor is that amplifier distortion is a function of channel loading. The more channels carried, the greater the distortions.

System Reflections

Signal reflections occur throughout the cable plant and are called *microreflections*. They are caused by the individual slight errors in impedance match. The severity of the mismatch is measured by the magnitude of the return-loss ratio. The larger the return loss, the better. Perfection is infinite. Mismatches include connectors, splices, and even damage to the cable itself.

Microreflections are likely to be a serious problem for ATV transmission because of their digital nature.

Phase Noise

Phase noise is added to the original signal through modulation and frequency conversion processes. A significant amount of phase noise must be added to the video carrier before generated impairments become perceptible. Narrowband phase noise (measured 20 kHz from the video carrier) in a TV channel produces variations in the luminance and chrominance levels that appear as an extremely grainy pattern within the picture. The perceptibility level of phase noise on the video carrier is 53 dB below the carrier at 20 kHz. If the frequency conversion or modulation processes are operating close to specification, phase noise impairments should not be perceptible on the customer's TV unless the converter/descrambler is malfunctioning or is of poor quality.

Amplifier Distortions and Their Effects

New amplifier technology based on feedforward and power-doubling techniques increases power levels with fewer distortions. However, additional sources of minutely delayed signals have been created. The signal delays produced in these amplifiers have similar end results in picture degradation as the delayed signals generated by reflected signals in the cable plant. But they are caused by a different mechanism. These amplifiers use parallel amplification technology. The signals are split, separately amplified, and then recombined.

With a feedforward amplifier, the signals are purposely processed with delay lines. If the propagation time is not identical through each of the amplifiers' parallel circuits, signals will be recombined that are delayed by different amounts of time. In most circumstances, the amount of differential delay is small and will not produce a visible ghost, but it may cause loss of picture crispness. Since the hybrids used in these amplifiers are normally provided in matched pairs or in a single hybrid package, these delays are only a problem when the hybrids are not replaced as a matched set.

In systems that carry more than 30 channels, CTB is the limiting distortion. However, cross modulation (X-MOD) distortions, which is often the limiting factor in systems with less than 30 channels, can reappear as the controlling factor in dictating system design. The HRC and the IRC channelization plans discussed in the first section were developed to minimize the visible degradation in picture quality that is caused by CTB.

X-MOD is one of the easiest distortions to identify visually. Moderate X-MOD appears as horizontal and vertical synchronizing bars that move across the screen. In severe cases, the video of multiple channels is visible in the background.

Moderate CTB is the most misleading distortion since it appears as slightly noisy pictures. Most technicians conclude that there are low signal levels and CNR problems. CTB becomes visible as amplifier operating levels exceed the design parameters. Once CTB reaches a severe stage, it becomes more readily identifiable because it causes considerable streaking in the picture.

Composite second order (CSO) beats can become a limiting factor in systems that carry 60 or more channels and use the HRC or IRC channelization plans. This distortion appears as a fuzzy herringbone pattern on the television screen. The CSO beats fall approximately 0.75 and 1.25 MHz above the video carrier in a television channel. An IRC channelization will frequency-lock these beats together while increasing their amplitude relative to the carrier level.

Hum modulation caused by the 60-Hz amplifier powering is identified by its characteristic horizontal bar that rolls through the picture. If the hum modulation is caused by the lack of ripple filtering in the amplifier power supply, it will appear as two equally spaced horizontal bars that roll through the picture.

Frequency Bands Affected by Radio Frequency Interference

Discrete beat products can be difficult to identify by the displayed picture impairment. Radio frequency interference that leaks into the cable system from nearby RF transmitters causes spurious carriers to fall in the cable spectrum. Common sources of signal leakage are cracked cables and poor quality connections. When either of these situations happen, strong off-air television and FM radio broadcast signals interfere.

If television stations are carried at the same frequency on cable as broadcast and the headend channel processing equipment is phase locked to the off-air signal, the effects of this interference will be ghosting. The ghost appears before (to the left of) the cable signal since propagation time through the air is less than through cable. If the signals are not phase locked together, lines and beats appear in the picture.

Often there is interference from off-air signals due to consumer electronics hardware design. If the internal shielding of the equipment is inadequate, the internal circuits will directly pick up the signal. This phenomenon is called DPU for *direct pick-up interference*. This is the original motivation for cable converters. Those early set-top boxes tuned no more channels than the TV set, but they protected against DPU by incorporating superior shielding and connecting to the TV set through a channel not occupied off-air.

DPU can be misleading. When the subscriber switches to an antenna, he might receive better pictures than from the cable connection. He concludes that his TV receiver is operating correctly and the cable system is faulty. The only convincing argument is a demonstration with a receiver, which does not suffer from DPU. Viacom Cable has measured off-air field intensities of eight volts per meter. The German specification for immunity to DPU is 4 v/m. VCR tuners are generally inferior to TV tuners because the VCR market is even more price competitive. The Electronic Industries Association (EIA) and NCTA Joint Engineering Committee are studying this issue.

The second most likely source of radio frequency interference is created by business band radios, paging systems, and amateur radio operators. These signals leak into the cable system and interfere with cable Channels 18 through 22 and Channels 23 and 24 (145 to 175 and 220 to 225 MHz). It is easy to determine that these signals are caused by an RF transmitter because of the duration of the interference and, sometimes, by hearing the broadcast audio. Since the signals are broadcast intermittently, it is almost impossible to determine the exact location(s) of ingress. Cable systems that operate above 450 MHz may find severe forms of interference. They are subjected to high-power UHF television stations, mobile radio units and repeaters, as well as a group of amateur radio operators signals in the top 10 to 12 channels. The extreme variation of shortwave signals in time and intensity makes location of the point(s) of infiltration of these signals difficult.

The upstream 5 to 30-MHz spectrum is a problem for operators who have two-way cable systems. There are many sources of interference and these signals accumulate upstream. In a two-way plant, a single leak in the system can make that portion of the upstream spectrum unusable throughout the entire plant; whereas in the downstream spectrum a leak may only affect a single customer's reception.

Signal Security Systems

Means of securing services from unauthorized viewership of individual channels range from simple filtering schemes to remote controlled converter/descramblers. The filtering method is the commonly used method of signal security and is the least expensive.

Trapping Systems

There are two types of filtering or trapping schemes: positive trapping and negative trapping. In the positive trapping method, an interfering jamming carrier(s) is inserted into the video channel at the headend. If the customer subscribes to the secured service, a positive trap is installed at the customer's house to remove the interfering carrier. The positive trapping scheme is the least expensive means of securing a channel where less than half the customers subscribe.

A drawback to positive trap technology is its defeatability by customers who obtain their own filters through theft, illegal purchase, or construction. Another drawback is the loss of resolution in the secured channel's video caused by the filter's effect in the center of the video passband. Pre-emphasis is added at the headend to correct for the filter's response, but loss of picture content in the 2- to 3-MHz region of the baseband video signal remains.

Negative trapping removes signals from the cable drop to the customer's home. The trap is needed for customers who do not subscribe. This is the least expensive means of securing a channel when over half the customers subscribe. The negative trap is ideal. There is no picture degradation of the secured channel because the trap is not in the line for customers who take the service. A drawback occurs for customers who do not subscribe to the secured service but want to view adjacent channels. These customers may find a slightly degraded picture on the adjacent channels due to the filter trapping out more than just the secured channel. This problem becomes more significant at higher frequencies, owing to the higher Q (efficiency) required of the filter circuitry. From a security standpoint, it is necessary for the customer to remove the negative trap from the line to receive an unauthorized service. Maintaining signal security in negative trapped systems depends on ensuring that the traps remain in the drop lines.

Scrambling and Addressability

There are two classes of scrambling technologies: (1) RF synchronization suppression systems and (2) baseband scrambling systems.

The concept of addressability should be considered separately from the scrambling method. Non-addressable converter/descramblers are programmed via internal jumpers or a semiconductor memory chip called a programmable read only memory (PROM) to decode the authorized channels. These boxes' authorization must be physically changed by the cable operator. Addressable converters are controlled by a computer-generated data signal originating at the headend either in the vertical blanking interval (VBI) or by an RF carrier. This signal remotely configures the viewing capabilities of the converter. Impulse-pay-per-view (IPPV) technology is supported by addressable converter/descrambler systems.

RF Synchronization Suppression Systems.   Converter-based scrambling systems that perform encoding and decoding of a secured channel in an RF format comprise the commonly used scrambling technology. The more common is known as *gated* or *pulsed* synchronization suppression. With this method, the horizontal synchronizing pulses (and with some manufacturers, the vertical synchronization pulses) are suppressed by 6 dB and/or 10 MHz dB. This is done in the channel's video modulator at the IF frequency. The descrambling process in the converter/descrambler occurs at its channel output frequency. This is accomplished by restoring the RF carrier level to its original point during the horizontal synchronization period. Variations of this method pseudorandomly change the depth of suppression from 6 to 10 dB or only randomly perform suppression.

A phase-modulated RF scrambling technique based on precision matching of SAW filters constructed on the same substrate has been introduced. This low-cost system is extending operators' interest in RF scrambling techniques for use within addressable plants.

Baseband Scrambling Systems.   Baseband converter/descrambler technology provides a more secure scrambling technology for delivering video services. The encoding format is a combination of random or pseudorandom synchronization suppression and/or video inversion. Because the encoding and decoding are performed at baseband, these converter/descramblers are complex and expensive.

Maintenance of the system's video quality is an ongoing issue. The encoders are modified video processing amplifiers. They provide controls to uniquely adjust different facets of the video signal. The potential for setup error in the encoder, in addition to the tight tolerances that must be maintained in the decoders, has presented challenges to the cable operator.

Off-Premises Systems

The off-premises approach is compatible with recent industry trends to become more consumer electronics friendly and to remove security-sensitive electronics from the customer's house. This method controls the signals at the pole rather than at a decoder in the home. This increases consumer electronics compatibility since authorized signals are present in a descrambled format on the customer's drop. Customers with cable-compatible equipment can connect directly to the cable drop without the need for converter/descramblers. This allows the use of all VCR and TV features.

Interdiction technology involves a scheme similar to that of positive trap technology. In this format, the pay television channels to be secured are transported through the cable plant in the clear (not scrambled). The security is generated on the pole at the subscriber module by adding interference carrier(s) to the unauthorized channels. An electronic switch is incorporated allowing all signals to be turned off. While this method of signal access control has generated a lot of interest, practical problems and its incompatibility with advanced services and digital signals has precluded any wide-spread application.

Digital Encryption

One of digital technology's main advantages is the applicability of computer-processing power and features. As a result, the old challenges of scrambling and signal security can be approached with new techniques.

However, the advanced digital technology that makes better security possible also provides new means of attack for the would-be pirate. The microprocessors that are the heart of personal computers are available in ever-faster speeds. More memory, more hard disk speed and capacity, and high-speed cable modems make these more potent means of attack. Additionally, the high-speed cable modem makes it possible to combine the processing power of multiple personal computers. Consequently, it must be possible to respond to a security attack by replacing the breached security element. This can be done in two ways. If the cable operator owns the set-top boxes, the cable operator can replace the set-top boxes when it is judged that the security breach is excessive. This occurs when the breach is relatively accessible to a significant fraction of the subscriber base, not just a handful of experimenters. However, if the set-top boxes were sold to subscribers, this option is not readily available. In the case of set-top boxes sold to subscribers, the security element must be replaceable in the event of significant breach. A plug-in card or module called the point of deployment (POD) module is used. The POD remains the property of the cable operator and can be electronically disabled. The subscriber would then be given a new POD device based on a system not yet breached.

Signal Splitting at the Customer's Premises

The common devices at the cable drop to the customer's home are grounding safety devices called *ground blocks* and a two-way signal splitter that sometimes has a built-in grounding terminal.

Splitters or ground blocks should have little effect on picture quality provided there is adequate signal to handle the splitter's loss. The signal strength may be below specifications because of an excessively long drop or more activated cable outlets in the house than the cable design anticipated. To compensate, some systems use an AC-powered drop amplifier. These amplifiers can create problems—a reduced CNR or increased distortions.

Consumer electronics switching devices, designed to allow convenient control and routing of signals between customer products and cable systems' converters/descramblers, have built-in amplification stages to overcome the losses associated with the internal splitters. These amplifiers add distortions or noise. When cable systems were designed, consumer electronics switching devices were not taken into account because they did not exist.

Signal splitting in VCRs can be a problem. To compensate for recording SNR deficiencies, inexpensive VCRs sometimes split the signal unequally between the by-pass route and the VCR tuner. This gives the VCR a stronger signal than the TV receiver to improve VCR performance. In addition, this strategy reduces the quality of the signal routed to the TV. When it is compared with VCR playback, the disparity in performance is reduced.

Consumer Electronics Compatibility

Cable systems are becoming more consumer friendly by trapping popular secured services. More cable-ready television sets are appearing in the customer's home. With VCRs that are also cable ready, some of the interface issues are becoming simpler. Some television sets have built-in A/B selector switches and video input ports that allow signal source switchings to be performed through the television's remote control. Up to three A/B switches, two splitters, and two converter/descramblers have been wired into configuration allowing the consumer to watch and record the programming desired. Even with this configuration, consumers lose the ability to preprogram the VCR to record more than one channel.

Hopefully, the days of complex interfaces will soon be over. A positive step in this direction is the EIA Decoder Interface Interim Standard, IS-105. This connection system is oriented toward supporting external, low cost, descramblers. If a descrambler is connected to the TV set or VCR via this technique, the user regains use of the advanced features precluded by converters.

PROJECTIONS FOR DIGITAL CABLE

The advent of digital television has created new complexities for the interface between cable and consumer electronics, as well as exciting new opportunities. What can or may transpire is governed not only by technical developments but by rules and regulations imposed by Congress and the FCC, as well as by consumer preferences and the marketing objectives of the cable industry, broadcasters, and consumer electronics manufacturers. Many of these factors are related and are discussed at length on the accompanying CD-ROM, in the section on Projections for Digital Cable. Also discussed therein are Digital Must Carry, Cable Modems and Telephone, and Interactive Television.

BIBLIOGRAPHY

Adams, M., "Open Cable Architecture, The Path to Cable Compatibility and Retail Availability in Digital Television," ISBN 1-57870-135-X.

Bartlett, E. R., "Cable Television Technology & Operations," McGraw-Hill, 1990, ISBN 0-07-003957-7.

Brinkley, J., "Defining Vision, The Battle for the Future of Television, How Cunning, Conceit, and Creative Genius Collided in the Race to Invent Digital, High-Definition TV," ISBN 0-15-100087-5.

Ciciora, W., J. Farmer, and D. Large, "Modern Cable Television Technology: Video, Voice, and Data Communications," ISBN 1-55860-416-2.

Farnsworth, E. G., "Distant Vision, Philo T. Farnsworth, Inventor of Television," ISBN 0-9623276-0-3.

Grant, W. O., "Cable Television," 3rd ed., GWG Associate, 1994, Library of Congress Cataloging-in-Publication Data, Application number: TXu 661-678.

Hodge, W. W., "Interactive Television," McGraw-Hill, 1995, ISBN 0-07-029151-9.

Laubach, M., D. Farber, and S. Dukes, "Delivering Internet Connections over Cable," ISBN 0-471-38950-1.

National Cable Television Association, Technical Papers, NCTA Science & Technology Dept., NW, 20036, 1996, ISBN 0-940272-24-5.

NCTA Recommended Practices, NCTA Science & Technology Dept. NW, 20036, 1st ed., 1983, ISBN 0-940272-09-1.

Rzeszewski, T. S. (ed.), "Color Television," IEEE Press, 1983, ISBN 0-87942-168-1.

Rzeszewski, T. S. (ed.), "Digital Video, Concepts and Applications Across Industries," IEEE Press, 1983, ISBN 0-7803-1099-3.

Rzeszewski, T. S. (ed.), "Television Technology Today," IEEE Press, 1985, ISBN 0-87942-187-8.

Schwartz, M., "Information, Transmission, Modulation, and Noise," 2nd ed., McGraw-Hill, 1970, ISBN 07-055761-6.

Society of Motion Picture and Television Engineers, *SMPTE J.*, January 1985 to present, ISSN: 0036-1682.

Southwick, T., "Distant Signals, How Cable TV Changed the World of Telecommunications," ISBN 0-87288-702-2.

Standage, T., "The Victorian Internet, The Remarkable Story of the Telegraph and the Nineteenth Century's On-Line Pioneers," ISBN 0-8027-1342-4.

Taylor, A. S., "History between their Ears, Recollections of Pioneer CATV Engineers," ISBN 1-89182-101-6.

Thomas, J. L., "Cable Television Proof-of-Performance," Prentice Hall, 1995, ISBN 0-13-306382-8.

Weinstein, S. B., "Getting the picture," IEEE Press, 1986, ISBN 0-87942-197-5.

ON THE CD-ROM

Ciciora, W. Projections for Digital Cable, including Digital Must Carry, Cable Modems and Telephony, and Interactive Television.

Useful cable television URLs for further study.

SECTION 23

NAVIGATION AND DETECTION SYSTEMS

Electromagnetic wave theory is fundamental to all navigation and detection systems. Global positioning system (GPS) applications have brought an added dimension to this area and in many applications works with radar systems in more sophisticated applications. The basic operation of the radar system has not changed in years; however, computers and digital technology have significantly enhanced the way data are processed.

Underwater sound systems are nothing more than radar systems at lower frequencies. We can work with underwater systems like we do for the higher-frequency electromagnetic systems to communicate, navigate, detect, track, classify, and so on. We are just working with substantially longer wavelengths.   C.A.

In This Section:

CHAPTER 23.1
RADAR PRINCIPLES

David K. Barton

THE RADAR SYSTEM AND ITS ENVIRONMENT

Radar is an acronym for radio detection and ranging, and is defined[7] as a device for transmitting electromagnetic signals and receiving echoes from objects of interest (targets) within its volume of coverage. The signals may be in the frequency range from the high-frequency radio band (3 to 30 MHz) to light (10^{15} Hz), although most systems operate between 300 MHz and 40 GHz. The target is a passive reflecting object in primary radar, while in secondary radar a beacon (transponder) is used to reinforce and identify the echo signal.

A radar system in its environment is shown in Fig. 23.1.1. Transmission is typically through a directional antenna, whose beam can either scan a coverage volume (search radar) or follow the echo from a previously detected target (tracking radar). Transmission is over a line-of-sight path through the atmosphere or space, except for over-the-horizon radar using ionospheric bounce paths. The target can be a man-made object (e.g., an aircraft or missile) or a natural surface or atmospheric feature (land formation, ocean wave structure, or precipitation cloud). The target echo is often received with the same antenna used for transmission, from which it is routed to the receiver by a duplexer. Environmental factors that influence radar performance include the characteristics of the propagation path (attenuation, refraction, and reflection) and reception of noise and interference from objects both within and beyond the radar beam (thermal radiation, accidental or international radio interference, and echoes from unwanted target objects, called clutter). The radar information may be used locally by an operator, or may be transmitted as analog or digital data to a remote site or network.

Radar Frequencies

Radar can be operated at any frequency at which electromagnetic energy can be generated and radiated, but the primary bands used are identified in Table 23.1.1. The band letter designations provide the radar engineer with a convenient way of specifying radar frequency with sufficient accuracy to indicate the environmental and antenna problems, but without disclosing potentially secret tuning limitations. The International Telecommunications Union (ITU) defines no specific service for radar, and the assignments listed are derived from those radio services which use radar: radiolocation, radionavigation, meteorological aids, earth exploration satellite, and space research. Where the ITU defines UHF as extending from 300 to 3000 MHz, radar engineers use L and S bands to refer to frequencies above 1000 MHz.

The applications of the several frequency bands are discussed in the following section.

Radar Functions and Applications

The basic functions performed by the radar are implicit in the definition: target detection, and measurement of target position (not only range but angles and often radial velocity). Other measured data may include target amplitude, size and shape, and rate of rotation. The major fields of applications are listed in Table 23.1.2.

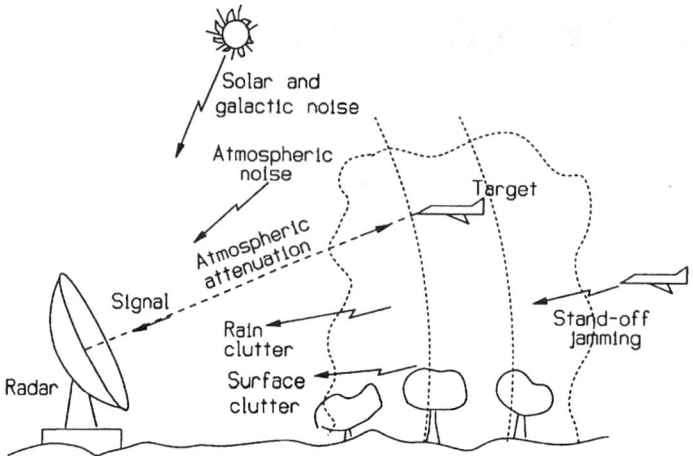

FIGURE 23.1.1   Radar system in its environment.

Radar Subsystems.   A block diagram of a simple, coherent pulsed radar is shown in Fig. 23.1.2. The dashed lines divide the equipment into seven subsystems, according to the technology used in implementation (see the following chapter on radar technology). In this radar, operation is controlled by a synchronizer, which initiates the pulsing of the transmitter and the gates and triggers that control the receiver, signal processor, and display. The synchronizer may also serve as the frequency reference used in the exciter to generate the r.f. drive signal, at carrier frequency f_0, to the transmitter and the local oscillator signals used for downconversion of the signal to intermediate frequency f_c in the superheterodyne receiver. The system is termed coherent because the exciter maintains the transmission (carrier) frequency at a consistent phase with respect to the local oscillator signals.

The transmitted pulse is generated by amplification of the exciter r.f. drive, during the pulse supplied by the modulator and triggered by the synchronizer. This pulse is passed to the antenna through the duplexer, an r.f. switch that connects the transmitter to the antenna, with low loss, during the pulse transmission. During transmission, the duplexer protects the sensitive receiver from damage by establishing a short circuit across the receiver input

TABLE 23.1.1   Radar Frequency Bands (IEEE Standard 521-1984)

Band designation	Nominal frequency range	Specific frequency range for radar based on ITU assignments for regin 2 (N. and S. America)
HF	3–30 MHz	No specific assignment for radar
VHF	30–300 MHz	138–144 and 216–225 MHz
UHF	300–1000 MHz	420–450 and 890–942 MHz
L	1–2 GHz	1.215–1.4 GHz
S	2–4 GHz	2.3–2.5 and 2.7–3.7 GHz
C	4–8 GHz	5.25–5.925 GHz
X	8–12 GHz	8.5–10.68 GHz
K_u	12–18 GHz	13.4–14 and 15.7–17.7 GHz
K	18–27 GHz	24.05–24.25 GHz
K_a	27–40 GHz	33.4–36 GHz
V	40–75 GHz	59–64 GHz
W	75–110 GHz	76–81 and 92–100 GHz
mm	110–300 GHz	126–142, 144–149, 231–235, and 238–248 GHz

TABLE 23.1.2 Radar Applications

Type of application	Specific applications	Usual bands
Air surveillance	Long-range early warning	UHF, L
	Ground-controlled intercept, air-route surveillance	L
	Acquisition for weapon system, height finding and three-dimensional radar, air collision avoidance	S
Space and missile surveillance	Ballistic missile early warning, missile acquisition, satellite surveillance	VHF, UHF
Surface search and battlefield surveillance	Sea search, navigation and collision avoidance, ground mapping	X, K_u, K_a
	Mortar and artillery location	C, X
	Airport taxiway control, intrusion detection, land vehicle collision avoidance	K_u, K_a
Weather radar	Observation and prediction, weather avoidance (aircraft), cloud-visibility indicators	S, C
Tracking and guidance	Antiaircraft fire control, surface fire control, missile guidance, range instrumentation, satellite instrumentation, precision approach and landing	C, X, K_u
	Smart weapons, projectiles, bombs	K_a, V, W
Astronomy and geodesy	Planetary observation, earth survey, ionospheric sounding	VHF, UHF, L

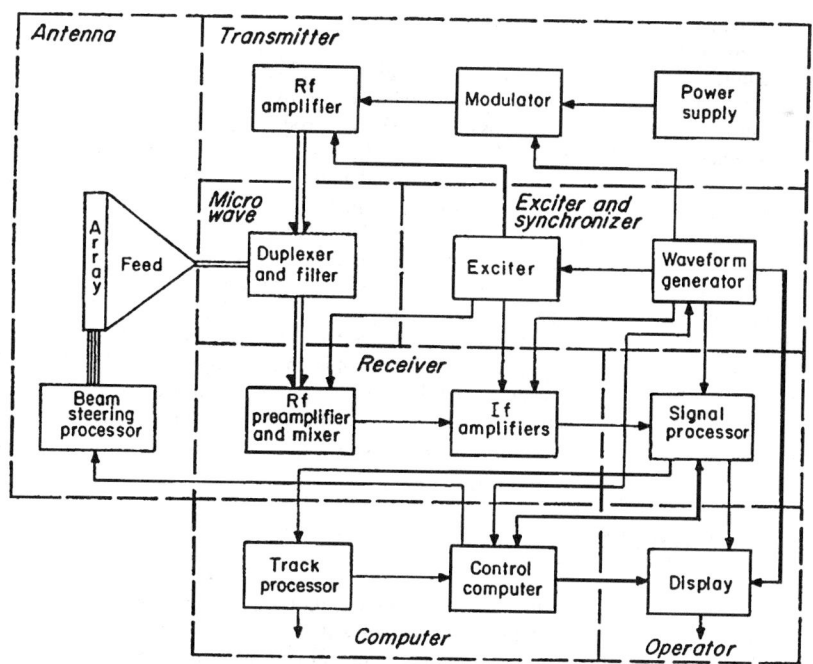

FIGURE 23.1.2 Block diagram of simple, coherent pulsed radar.

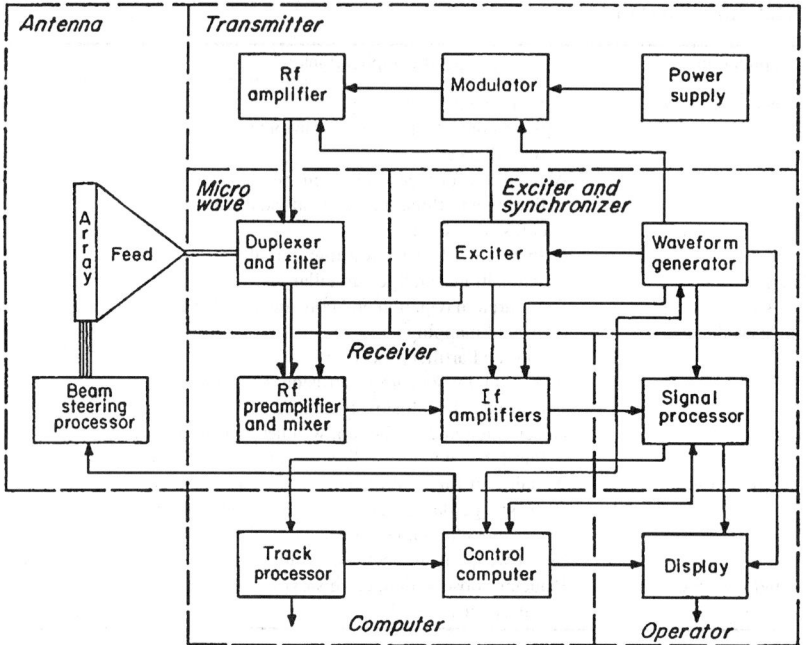

FIGURE 23.1.3   Block diagram of phased-array radar (Ref. 2).

terminal. The antenna shown is a parabolic reflector, steered mechanically in one or two axes by servomechanisms. The received signal passes from the antenna to the receiver through the duplexer, which has disconnected the transmitter and established a low-loss path to the receiver terminal. A low-noise r.f. amplifier is used at the receiver input, followed by a mixer for conversion to intermediate frequency (i.f.). Following amplification of several tens of decibels, the signal passes to the signal processor, which is designed to optimize the ratio of signal to noise and clutter. Outputs to the display will consist, ideally, of target echoes, appearing at locations on the display corresponding to the target range and angles. In a tracking radar, the signal outputs are fed back to control the antenna steering and the position of the range gate in the receiver.

A more advanced radar, using a computer-controlled phased-array antenna, is shown in Fig. 23.1.3. An eighth subsystem has been added, consisting of the control computer and a track processor, both implemented digitally. Beam steering is also controlled digitally by the beam steering processor. The simple synchronizer has been replaced by a digitally controlled waveform generator. Digital control is also applied to the receiver, signal processor and display. The phased-array radar may be programmed to perform a variety of functions in rapid sequence, including search, target tracking, and tracking and guidance of interceptor missiles. Such a radar is known as a multifunction array radar (MFAR), favored in modern U.S. weapon systems. Phased-array radar may be used solely for search (usually as a three-dimensional, or 3D radar, scanning in elevation as well as in azimuth), or solely for tracking, in which case multiple-target tracking is possible through rapid sequencing among several targets.

RADAR-RANGE EQUATIONS

A radar range equation is a relationship by which the detection range expected on a given target by a given radar may be calculated. Different equations may be derived, depending on whether the radar and its mode of operation can be specified in detail or only in a more general way, and on the complexity of the environment to which the calculation is applicable.

Basic Radar Range Equation

The basic radar range equation applies to a radar for which the known parameters include observation time (the time during which the beam lies on the target), and which operates in a benign environment (thermal noise only). The peak signal power S received by the radar antenna over a free-space path is calculated from the transmitted peak power P_t and a series of factors that describe, basically, the geometry of the radar beam relative to the target:

$$S = \frac{P_t G_t A_r \sigma}{(4\pi)^2 R^4}$$

(1)

where G_t = transmit antenna gain
A_r = effective receive antenna aperture area
σ = target cross section
R = target range

Since the receive aperture and gain are related by

$$A_r = \frac{G_r \lambda^2}{4\pi}$$

(2)

where λ is the carrier wavelength, the signal power can also be given as

$$S = \frac{P_t G_t G_r \lambda^2 \sigma}{(4\pi)^3 R^4}$$

(3)

Thermal noise in a radar receiver is a combination of receiver-generated and environmental noise, extending over the entire radar frequency band with a power density N_0 given by

$$N_0 = kT_s$$

(4)

where k is the Boltzmann's constant = 1.38×10^{-23} W/(Hz · K) and T_s is the system noise temperature referred to the antenna terminal. The system noise temperature[8] is calculated from the receiver noise factor F_n, line losses L_r, and antenna temperature T_a:

$$T_s = T_a + T_r + L_r T_e$$

(5)

$$T_a = \frac{0.88 T_a' - 254}{L_a} + 290$$

(6)

$$T_r = T_{tr}(L_r - 1)$$

(7)

$$T_e = T_0(F_n - 1)$$

(8)

where T_r = temperature contribution of loss L_r
T_a' = sky temperature from Fig. 23.1.4
L_a = antenna ohmic loss
T_{tr} = physical temperature of the receiving line
T_e = temperature contribution of the receiver
T_0 = standard temperature (290 K) used in measuring noise factor

It can be seen that low noise factor and low losses can lead to $T_s < 290$ K when the antenna is looking upward into space.

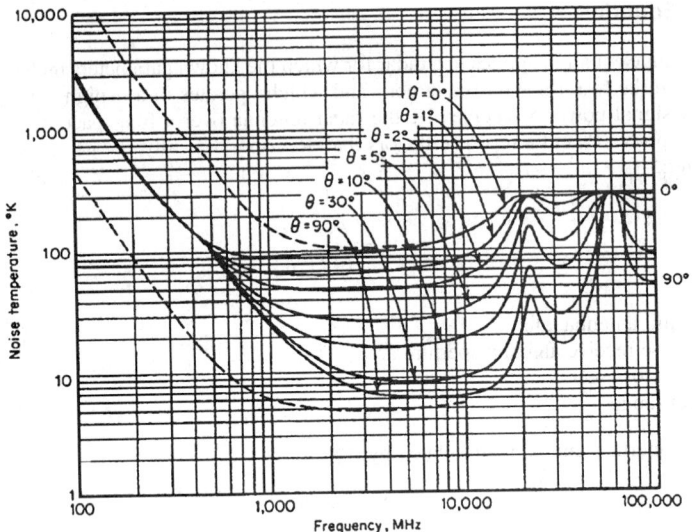

FIGURE 23.1.4 Sky temperature for an idealized antenna (lossless, no earth-directed sidelobes) located at the earth's surface, as a function of frequency, for a number of beam elevation angles. Solid curves are for geometric-mean galactic temperature, sun noise 10 times quiet level, sun in unity-gain side lobe, cool-temperature-zone troposphere, 2.7-K cosmic black-body radiation, zero ground noise. Upper dashed curve is for maximum galactic noise (center of galaxy, narrow-beam antenna), sun noise 100 times quiet level, zero elevation angle, other factors the same as the solid curves. Lower dashed curve is for the minimum galactic noise, zero sun noise, 90° elevation angle.[8]

The noise power at the i.f. output of the receiver will depend on receiver bandwidth and gain, but this noise power is equivalent to an input power at the antenna terminal of

$$N = N_0 B_n = kT_s B_n \tag{9}$$

where B_n is the noise bandwidth of the i.f. filter. For a wideband filter and a simple pulse ($B_n \gg 1/\tau$, where $\tau =$ pulse width) the signal peak is not affected by the filter, and $S/N = S/kT_s B_n$. In general, however, the SNR at the receiver i.f. output is calculated from the ratio of received pulse energy $S\tau$ to noise density:

Ideal energy ratio for single pulse:

$$\frac{E_1}{N_0} = \frac{P_t \tau G_t G_r \lambda^2 \sigma}{(4\pi)^3 R^4 kT_s} \tag{10}$$

Intermediate-frequency power ratio for single pulse:

$$\frac{S}{N} = \frac{E_1}{N_0 L_m} = \frac{P_t \tau G_t G_r \lambda^2 \sigma}{(4\pi)^3 R^4 kT_s L_m} \tag{11}$$

where $L_m =$ i.f. filter matching loss. For a rectangular pulse, this loss is shown in Fig. 23.1.5a, as a function of the product $B_n \tau$ for different filter shapes. For a system using linear-fm pulse compression (chirp), the loss will be a function of the weighting applied to achieve the desired time-sidelobe level, as shown in Fig. 23.1.5b.

The expression of i.f. output SNR in terms of transmitted pulse energy $P_t \tau$ permits the range equation to be used with any type of pulse modulation, without in-depth knowledge of that modulation and its processing

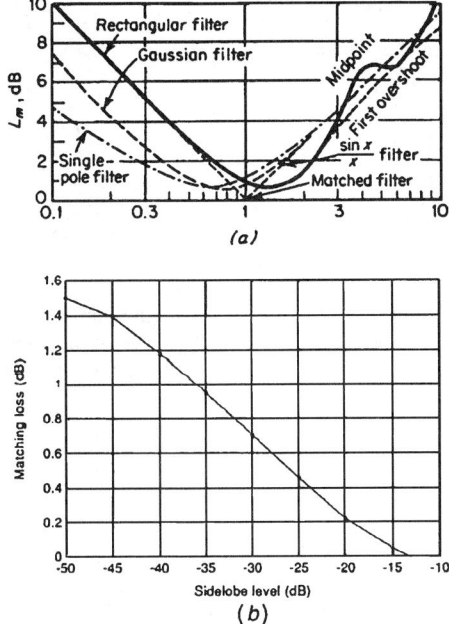

(a)

(b)

FIGURE 23.1.5 Intermediate-frequency filter matching loss: (a) rectangular pulse with different filter shapes; (b) linear-fm pulse compression with weighting for different sidelobe levels.

in the receiving system. Most systems will be characterized by i.f. matching loss L_m between 0.7 and 1.5 dB, and little error will result from assuming $L_m = 1$ dB even if the actual modulation and processing are completely unknown.

When the radar maintains coherence and processes the target echo over a coherent processing interval t_f, the output of the i.f. processor will be

$$\frac{S}{N} = \frac{E}{N_0 L_m} = \frac{P_{av} t_f G_t G_r \lambda^2 \sigma}{(4\pi)^3 R^4 k T_s L_m} \tag{12}$$

where P_{av} = average transmitter power = $P_t \tau f_r$ for pulsed radar operating at pulse repetition frequency f_r, and L_m = matching loss of the filter to the entire waveform received over t_f s. In this form, the equation can be applied directly to a radar using any type of waveform (and processing): pulsed (noncoherent processing), by setting $t_f = t_r$, the pulse repetition interval; pulsed (coherent doppler processing) or CW, where $t_f = 1/B_f$ is the integration time of the doppler filter with bandwidth B_f. In the limit, for systems with coherent processing, $t_f = t_o$, the total observation time during which the target lies within the radar beam, and there will be only one output sample from the doppler filter. In most cases, however, $t_f < t_o$, and there will be $n' = t_o/t_f$ independent output samples available for subsequent noncoherent integration.

RF Loss Factors

Equations (10) to (12) consider free-space transmission conditions and idealized radar operation (except for a possible matching loss). In practice, a number of other factors must be included in expressing the energy ratio of the received signal.

a. *Signal attenuation prior to receiver:* Transmission line loss L_t; antenna losses (to the extent they are not included in G_t, G_r, T_s); receiving line and circuit losses at r.f. (to the extent they are not included in T_s); atmospheric attenuation L_α (see Figs. 23.1.20 to 23.1.22); and atmospheric noise (included in T_s through T'_a, Fig. 23.1.4, for clear air, and using Eq. (54) when precipitation is present).

b. *Surface reflection and diffraction effects:* Pattern-propagation factor F, calculated from Eq. (57) with data from Figs. 23.1.24 and 23.1.25, appears as F^4 in the numerator of the radar equation ($F > 1$ implies extra gain from the surface interaction).

c. *Antenna pattern and scanning:* The gains G_t and G_r are defined on the beam axis, and apply directly for tracking radar, with signal energy calculated for arbitrarily chosen observation time t_o; for one-coordinate scan at ω rad/s, the observation time is calculated using $t_o = \theta_3/\omega$, where θ_3 is the one-way half-power beamwidth (in radians) in the scan coordinate, and signal energy is calculated using gains G_t and G_r, evaluated at the point of the scan nearest the target; for two-coordinate scan, the observation time is $t_o = t_s \theta_a \theta_e/\psi_s$, where θ_a and θ_e are azimuth and elevation beamwidths, and ψ_s is the solid angle searched in time t_s, and signal energy is calculated using maximum gains. The variation in actual antenna gains brought to bear on the target, as a function of target position in the beam, is included as a beamshape loss in calculation of the *deductibility factor*, which is the energy ratio required to achieve given detection performance.

d. *Signal-processing losses:* A number of losses resulting from nonideal signal processing are discussed in Radar Functions and Applications, and included in the deductibility factor.

When the r.f. losses are included, the equations for energy ratio become

$$\frac{E_1}{N_0} = \frac{P_t \tau G_t G_r \lambda^2 \sigma F^4}{(4\pi)^3 R^4 k T_s L_t L_\alpha} \tag{13}$$

$$\frac{E}{N_0} = \frac{P_{av} t_f G_t G_r \lambda^2 \sigma F^4}{(4\pi)^3 R^4 k T_s L_t L_\alpha} \tag{14}$$

Calculation of Detection Range in Noise

The deductibility factor for noncoherent integration is denoted by $D_x(n, n_e)$, where n is the number of noncoherently integrated pulses and $n_e \leq n$ is the number of independent target samples included in the integration. By setting $E_1/N_0 = D_x(n, n_e)$, we may solve for the maximum range at which the given detection performance is achieved, for a noncoherent system:

$$R_m^4 = \frac{P_t \tau G_t G_r \lambda^2 \overline{\sigma} F^4}{(4\pi)^3 k T_s D_x(n, n_e) L_t L_\alpha} \tag{15}$$

The average target cross section $\overline{\sigma}$ is used in this equation, and any target fluctuation effects are included in the deductibility factor. When coherent integration is used, the deductibility factor becomes $D_x(n', n_e)$ where n' is the number of independent noise samples available from a doppler filter. By setting $E/N_0 = D_x(n', n_e)$, we may solve for the maximum range at which the given detection performance is achieved, for any system (including the noncoherent system, for which $t_f = t_r$, $P_{av} t_f = P_t \tau$, and $n' = n$):

$$R_m^4 = \frac{P_{av} t_f G_t G_r \lambda^2 \overline{\sigma} F^4}{(4\pi)^3 k T_s D_x(n', n_e) L_t L_\alpha} \tag{16}$$

It should be noted that Eqs. (15) and (16) are transcendental, since the atmospheric loss L_α depends on range R_m. It may be necessary to use an iterative calculation, in which an initial estimate of R_{m1} is made with $L_{\alpha 1} = 1$, followed by one or two refinements in which $L_{\alpha 2}$ is evaluated at R_{m1} to calculate R_{m2}, and $L_{\alpha 3}$ is evaluated at R_{m2} to calculate the final R_m. Another method is to apply Eq. (13) or (14) repeatedly with varying range until the resulting energy ratio equals the required D_x, or to perform such calculations at fixed range intervals and interpolate to find the range giving the required D_x. If the factor F^4 shows oscillatory behavior with target range, as it may when surface reflections are present, there may be several values of range at which the equation is satisfied, corresponding to the beginnings and endings of detection regions.

The deductibility factor D_x used in the range equations will be found from the theoretical value derived in Radar Functions and Applications for the various fluctuating target cases, increased by the filter matching loss L_m (Fig. 23.1.5), the beamshape loss L_p, and the miscellaneous signal processing loss L_x.

Search-Radar Equation

The potential performance of a search radar, or of a tracking radar during acquisition scan, can be determined from its average power, receiving aperture, and system temperature, without regard to its frequency or waveform. The steps in deriving optimum search performance from Eq. (16) are as follows:

a. Group all r.f. and other losses into a combined search loss given by

$$L_s = \frac{L_m L_t L_\alpha L_p^2 L_i L_f L_c L_x L_n}{F^4} \tag{17}$$

where L_p^2 = two-coordinate beamshape loss
L_i = integration loss
L_c = collapsing loss
L_f = fluctuation loss
L_x = miscellaneous signal processing loss
L_n = a beamwidth factor

b. *Assume uniform search, without overlap,* of an assigned solid angle ψ_s in a time t_s using a rectangular beam whose solid angle is

$$\psi_b = \theta_a \theta_e = \frac{4\pi}{G_t L_n} \ll \psi_s = A_m(\sin\theta_m - \sin\theta_0) \tag{18}$$

where θ_a and θ_e = 3-dB beamwidths
L_n = beamwidth factor relating gain to solid angle of the beam
A_m = azimuth sector searched
θ_m and θ_0 = upper and lower elevation search limits

c. *Express the observation time t_o for a target as*

$$t_o = \frac{t_s \psi_b}{\psi_s} = \frac{4\pi t_s}{G_t \psi_s L_n} \tag{19}$$

and assume that all signal energy reaching A_r during t_o is integrated for one detection decision. The definition of two-coordinate beamshape loss, which is included in L_s, is consistent with Eq. (19).

d. *Substitute Eqs. (2), (17), (18), and (19) into (16) to obtain the search radar equation:*

$$R_m^4 = \frac{P_{av} A_r t_s \bar{\sigma}}{4\pi \psi_s kT_s D_0(1)L_s} \tag{20}$$

Neither wavelength nor waveform appears directly in Eq. (20) although wavelength and aperture size must permit Eq. (18) to be satisfied in such a way as to concentrate energy within ψ_s, and the loss terms may vary with wavelength, waveform, and scan procedure. The new loss term L_n appearing in Eq. (19) is an antenna beam loss that accounts for energy outside the main lobe and not available for integration. The often used gain expression $G = 25,000/\theta_a\theta_e$, with θ in degrees, corresponds to a 40 percent loss of useful energy, or $L_n = 2.0$ dB. Reduced directivity caused by illumination taper does not appear in L_n but enters the search radar equation through A_r. Practical minimum values for L_s are between 11 and 13 dB for $P_d = 0.9$ with optimum diversity and scan procedure.

Beacon-Range Equations

One-way transmission from a radar to a beacon gives the interrogation power level S_b at the beacon receiver:

$$S_b = \frac{P_t G_t G_r \lambda^2}{(4\pi)^2 R^2 L_t L_{\alpha 1} L_b} \tag{21}$$

where G_b = beacon antenna gain in the radar direction
$L_{\alpha 1}$ = one-way atmospheric loss
L_b = loss between the beacon antenna and receiver

On the return link, a beacon peak power P_b determines the signal power S at the radar antenna terminal:

$$S = \frac{P_b G_b G_r \lambda^2}{(4\pi)^2 R^2 kT_s L_m L_{\alpha 1}} \qquad (22)$$

DETECTION

The term *detection* in radar can refer to either of two steps in signal processing. First, it may refer to the process by which the signal envelope is extracted from the sinusoidal i.f. waveform (i.e., envelope *detection*). Second, it may refer to the determination that a target signal is present at the output of the processor (i.e., threshold *detection*). Both processes are discussed in this section. The meaning will usually be apparent from the context.

Signal and Noise Statistics

Most radar signals are sinusoids with narrowband modulation (in amplitude or phase) superimposed by the transmitter, antenna, and target. Random noise is also constrained approximately to the signal bandwidth when it reaches the detection circuits after passage through the receiver. The instantaneous i.f. output noise voltage can be described in terms of in-phase and quadrature components

$$V = V_i \cos(2\pi f_c t) + V_q \sin(2\pi f_c t) \qquad (23)$$

Either component of V can be described by a gaussian probability distribution

$$dP_v = \frac{1}{\sqrt{2\pi N}} \exp\left(\frac{-V^2}{2N}\right) dV \qquad (24)$$

where dP_v = probability that voltage lies between V and $V + dV$, and N = mean-square noise voltage (average noise power). The video noise voltage E_n out of a linear envelope detector, corresponding to $E_n = \sqrt{V_i^2 + V_q^2}$ has a Rayleigh distribution

$$dP_e = \frac{E_n}{N} \exp\left(\frac{-E_n^2}{2N}\right) dE_n \qquad (25)$$

These noise distributions are shown in Fig. 23.1.6. The probability P_f that the noise envelope will exceed a given threshold level E_t is the shaded area to the right of E_t in Fig. 23.1.6b.

$$P_f = \int_{Et}^{\infty} \frac{E_n}{N} \exp\left(\frac{-E_n^2}{2N}\right) E_n = \exp\left(\frac{-E_t^2}{2N}\right) \qquad (26)$$

If a sinusoidal signal of peak amplitude E_s is present with the noise, the distribution of envelope voltage E_s is Rician

$$dP_s = \frac{E_n}{N} \exp\left(\frac{-E_n^2 + E_2^2}{2N}\right) I_0\left(\frac{E_n E_s}{N}\right) dE_n \qquad (27)$$

where I_0 is the Bessel function with imaginary argument. The probability of detection P_d for a sample taken at the peak of the signal is the area under this curve and above the threshold E_t, as shown in Fig. 23.1.7 for different signal-to-noise power ratios $S/N = E_s^2/2N$.

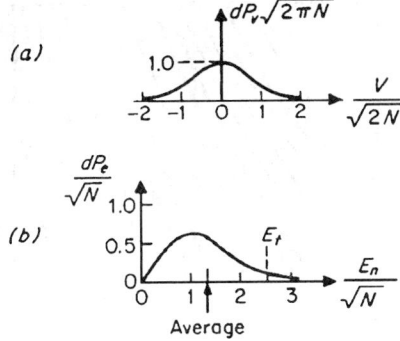

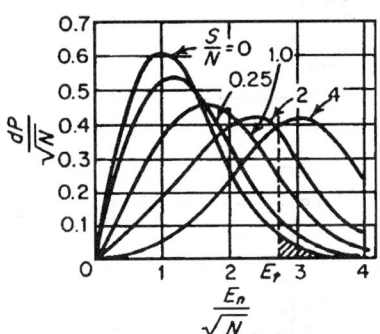

FIGURE 23.1.6 Probability distributions of noise: (*a*) Gaussian distribution of I and Q components of i.f. noise voltage; (*b*) Rayleigh distribution of detected envelope.[2]

FIGURE 23.1.7 Probability distributions of envelope of signal plus noise.[2]

Single-Sample Detection Probability

When a single sample is available for detection, the threshold is set to give the desired false-alarm probability according to Eq. (26), and the detection probability follows from the integral of Eq. (27). The results are plotted in Fig. 23.1.8, where each curve gives P_d versus S/N for the threshold setting corresponding to the P_f value shown. The value of S/N required to achieve given P_d with fixed P_f is denoted by $D_0(1)$, the single-pulse deductibility factor for the steady-signal case. For example, for $P_d = 0.90$, $P_f = 10^{-6}$, we find $D_0(1) = +13.2$ dB. Although steady signals are seldom encountered in radar, data from this figure are used as the basis for establishing deductibility factors for fluctuating targets in systems using single or multiple pulses.

Filters and Signal Spectra

Detection performance for a single sample has been shown to depend on the S/N ratio at the i.f. output. Noise power at this point is a function of the receiver gain, the bandwidth, and the noise spectral density in the early stages of the receiver, before bandwidth is established. It is customary in detection analysis to assume an ideal (noise-free) receiver of unity gain and to replace all actual noise sources with an equivalent broadband source of spectral density N_0, given by Eq. (4), added to the input signal. The output noise power will then be $N = N_0 B_n$, where B_n is the equivalent noise bandwidth of the receiver. The signal output poser will also depend on the receiver bandpass characteristics, and on the input signal energy.

The maximum possible output S/N will be

$$(S/N)_{max} = E/N_0 \qquad (28)$$

where E is the total signal energy and $(S/N)_{max}$ is measured at the maximum of the signal envelope out of a matched filter. Practical filters can approach, but never exceed, the performance of a matched filter.

The matched filter for a single, uncoded pulse of duration τ and energy E_1 can be approximated by a conventional bandpass filter whose bandwidth is $B \approx 1/\tau$. A matching loss is defined for this case as

$$L_m = \frac{E_1/N_0}{S/N} \qquad (29)$$

(see Fig. 23.1.5). Pulses with internal coding or modulation require more complex filters.

A train of pulses transmitted coherently (with controlled phase) produces a spectrum consisting of many separate lines (Fig. 23.1.9), and requires a *comb filter* with a matched series of response bands properly phased

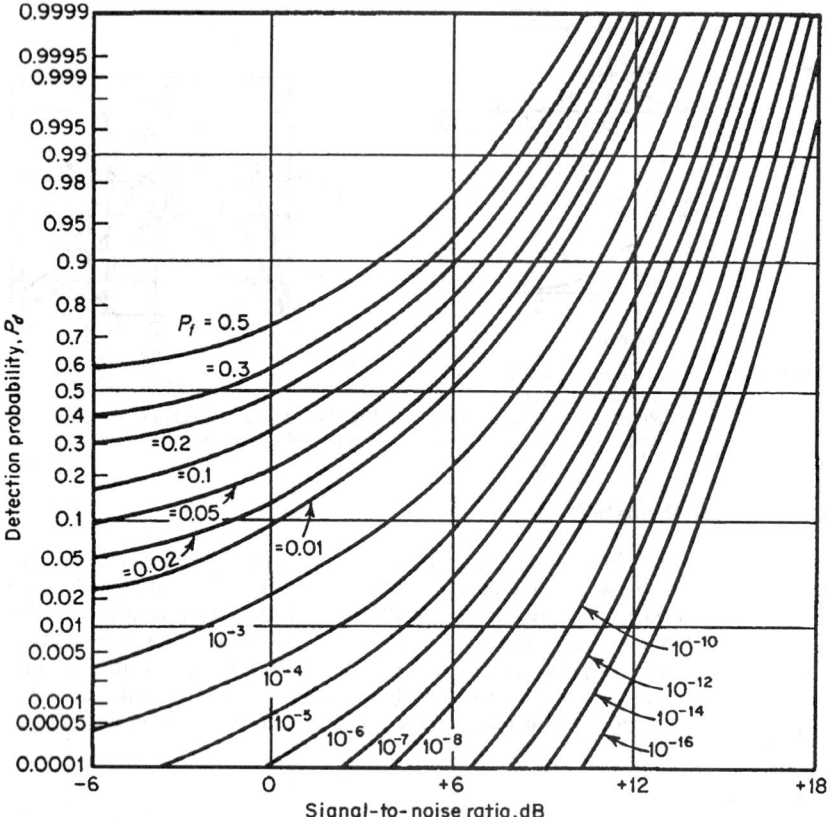

FIGURE 23.1.8 Detection probability versus signal-to-noise ratio of steady signal.

to add all signal components into one output. Although such a filter is readily synthesized for a selected point in the time-frequency domain, using a range gate and narrowband filter, it is difficult to match to signals of unknown time delay and doppler shift. Hence the integration of pulse trains is often carried out at video (after phase has been removed by an envelope detector). The IF filter is matched approximately to a single pulse, giving $S/N = (E_1/N_0)L_m$ at the envelope detector. Successive video pulses are added (integrated) on the radar display or in some other range-ordered storage device, prior to threshold detection.

Video Integration

Pulse train integration at video reduces the required energy of each pulse but introduces a loss in performance relative to predetection (matched-filter) integration in a comb filter. This integration loss $L_i(n)$ is defined as the increase in total signal energy required for a given P_d and P_f, relative to that needed with predetection integration. The deductibility factor $D_0(n)$ for n-pulse video integration as defined by Blake[8] and the IEEE[7] is

$$D_0(n) = \frac{L_i D_0(1)}{n} \qquad (30)$$

where $D_0(1)$ is the detectability factor for the single-sample case, from Fig. 23.1.8. Curves of integration loss L_i versus n are shown in Fig. 23.1.10, with $D_0(1)$ as a parameter. From these two figures, the steady-target data originally derived by Marcum[14] may be reproduced with an accuracy of about 0.1 dB. For example, if $P_d = 0.90$ and $P_f = 10^{-6}$ are needed, $D_0(1) = +13.2$ dB from Fig. 23.1.8, and $L_i = 5.7$ dB for $n = 100$ pulses, giving a total energy ratio requirement $100 \times D_0(100) = +18.9$ dB and a single-pulse requirement $D_0(100) = -1.1$ dB.

False-Alarm Time

The D_0 curves of Fig. 23.1.8 correspond to a given probability P_f of a false alarm from each independent sample applied to the threshold. If the detection circuits are operative for $t_g < t_r$ seconds in each repetition interval t_r, there will be $\eta = t_g B_n$ independent samples applied to the threshold during the integration interval nt_r, producing an average of ηP_f false alarms. The ratio of total time to average number of alarms is the false-alarm time t_{fa}. In some discussions, following Marcum's usage,[6] the false-alarm time t_{fa} is defined as the interval in which the probability of at least one false alarm is 0.5, and a false-alarm number n' is defined as the number of independent samples applied to the threshold during t_{fa}. Since $t_{fa} = 0.69\, t_{fa}$, Marcum's false-alarm number is related to false-alarm probability by

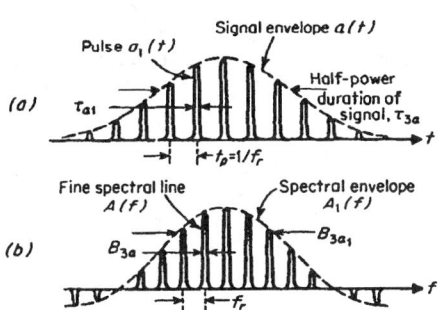

FIGURE 23.1.9 Waveform and spectrum of coherent pulse train.[5]

$$n' = \frac{0.69 B_n t_g \overline{t_{fa}}}{nt_r} = \frac{0.69}{P_f} \tag{31}$$

Collapsing Loss

Practical radars seldom preserve their full r.f. signal resolution through the integration and thresholding processes, where use of insufficient video bandwidth or broadened range gates may prove economical. The n video samples of signal plus noise are then integrated along with m extra samples of noise, a process described in terms of collapsing ratio

$$\rho = 1 + \frac{m}{n} \tag{32}$$

The effect, when using a square-law detector, is the same as if the signal energy were redistributed over $\rho n = m + n$ pulses, leading to a larger L_i value (Fig. 23.1.10) associated with the integration of ρn pulses. The additional loss is referred to as the collapsing loss L_c:

$$L_c(\rho,n) = \frac{L_i(\rho n)}{L_i(n)} \tag{33}$$

Formulas for ρ in different cases are given in Table 23.1.3. In cases for which the number of independent threshold samples varies inversely with ρ, P_f may be permitted to increase, and the energy added to overcome collapsing will not be as great as in other cases.

Detection of Fluctuating Signals

Swerling Case 1 (Rayleigh Target Fluctuation).   Essentially all actual radar targets fluctuate in amplitude, as a function of aspect angle and hence of time, following the Rayleigh distribution. The correlation time t_c of this fluctuation can be estimated, for a target of width L_x rotating at rate ω_a relative to the radar line of sight, from

$$t_c = \frac{\lambda}{2\omega_a L_x} \tag{34}$$

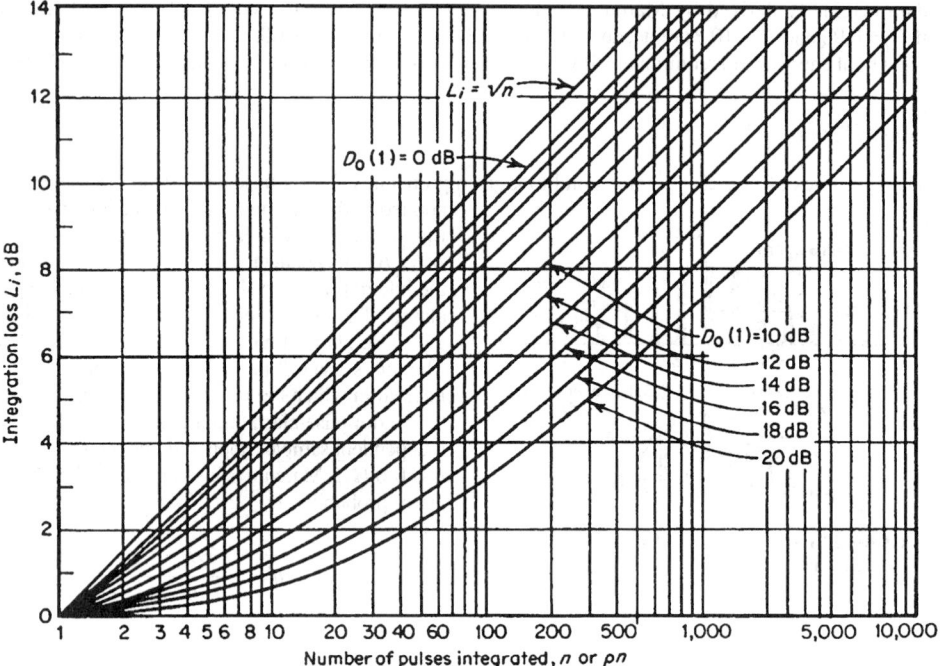

FIGURE 23.1.10 Integration loss versus number of pulses (Ref. 10).

Since this time is generally much greater than the observation time t_o, the target amplitude remains correlated over the n pulses are integrated, meeting the definition of the Swerling Case 1 (slowly fluctuating) target.[11] In order to ensure high probability of detection, it is necessary to increase the average signal power to provide a margin for fading, and this increase is called fluctuation loss. This loss is primarily a function of required detection probability, as shown in Fig. 23.1.11. The Rayleigh distribution is a chi-square distribution with two

TABLE 23.1.3 Equations for Collapsing Ratio $\rho = (m + n)/n$

Cases for which P_{fd}/ρ remains constant	
1. Restricted CRT sweep speed s, where d = spot diameter and τ = pulse width	$\rho = 1 + \dfrac{d}{s\tau}$
2. Restricted video bandwidth B_v, where $B = 1/\tau$ = i.f. signal bandwidth	$\rho = 1 + \dfrac{1}{2B_v\tau} = 1 + \dfrac{B}{2B_v}$
3. Collapsing of coordinates onto the display, where $2\Delta_r/c$ = time-delay interval displayed per display cell, $\omega_e t_v$, ω_d/t_v = elevation and azimuth scans during integration time t_v, and θ_e and θ_a = beamwidths	$\rho = \dfrac{2\Delta_r}{c\tau}$ or $\pi = \dfrac{\omega_e t_v}{\theta_e}$ or $\pi = \dfrac{\omega_a t_v}{\theta_a}$
Cases for which P_{fa} remains constant	
4. Excessive i.f. bandwidth $B_n > 1/\tau$, followed by matched video $2B_v = 1/\tau$	$\rho = 1 + B_n\tau$ (use L_c in place of L_m)
5. Receiver outputs mixed at video, where M = number of receivers	$\rho = M$
6. i.f. filter followed by gate of width τ_g and by video integration	$\rho = B_n\tau\left(1 + \dfrac{\tau_g}{\tau}\right)$

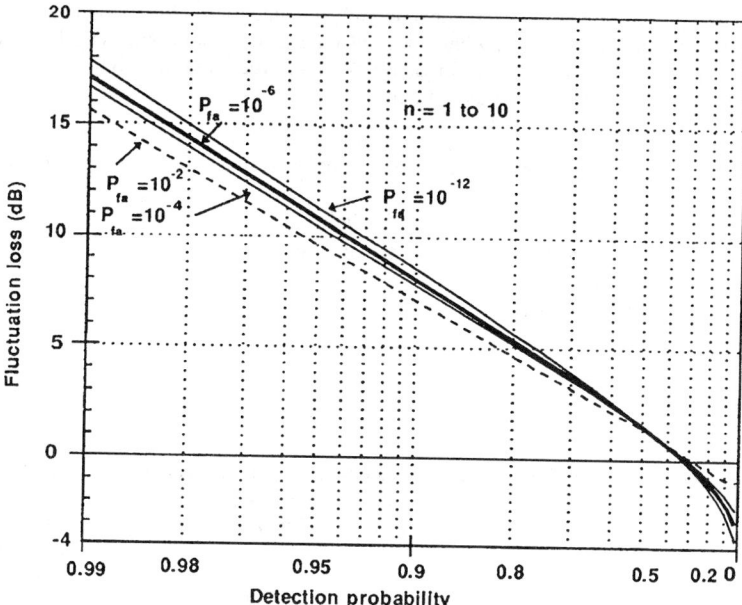

FIGURE 23.1.11 Fluctuation loss versus detection probability for Rayleigh fluctuating target (Swerling Case 1).

degrees of freedom, corresponding to the I and Q components of the reflected signal, each of which has a gaussian distribution. The figure gives data for $1 < n < 10$, and the loss will be a few tenths of a decibel higher for integration of tens or hundreds of pulses.

Having calculated the detectability factor for a steady target, we can now adjust this to find the detectability factor $D_1(n)$ for the Rayleigh fluctuating target:

$$D_1(n) = D_0(n)L_f(1) = \frac{D_0(1)L_i(n)L_f(1)}{n} \tag{35}$$

For example, for $P_d = 0.90$, $P_f = 10^{-6}$, $n = 100$, we can estimate $L_f = 8.6$ dB, from which the total energy requirement will be $100 \times D_1(100) = +27.5$ dB and the single-pulse requirement $D_1(100) = +7.5$ dB.

Reduction of Fluctuation Loss with Diversity

As in communications, the application of diversity can reduce the required fade margin (fluctuation loss). When the number of diversity samples (independent samples of target amplitude) available for integration is n_e, the fluctuation loss becomes[10]

$$L_f(n_e) = [L_f(1)]^{1/n_e} \tag{36}$$

or, in decibels,

$$[L_f(n_e)]_{dB} = \frac{1}{n_e}[L_f(1)]_{dB} \tag{37}$$

Since diversity samples must be integrated noncoherently, the integration loss will increase as n_e increases. The optimum number of diversity samples, when coherent integration would otherwise be available, depends on the required P_d, and for $P_d = 0.90$ there is a broad optimum between four and eight samples. Diversity is available in time, frequency, polarization, or aspect angle.

Time Diversity.   The number of independent target samples available in an observation time t_o is

$$n_e = 1 + t_o/t_c \leq n \tag{38}$$

In the limit, when $t_c < t_r$, $n_e = n$, the detectability factor is that given by Swerling for the Case 2 (rapidly fluctuating) target. Coherent integration cannot be carried out on the Case 2 target, but such rapid fluctuation is not normally seen for stable targets.

Frequency Diversity.   The number of independent target samples available when the n pulses are distributed uniformly over a frequency interval Δf is

$$n_e = 1 + \Delta f/f_c \tag{39}$$

where $f_c = c/2L_r$ and L_r is the radial dimension of the target (along the radar line of sight). For example, a target of length $L_r = 15$ m will provide an independent target sample for each 10 MHz of frequency shift. A dual-diversity radar system, with two fixed frequencies separated by any amount greater than f_c, will provide $n_e = 2$, regardless of the number of pulse transmitted. There are two degrees of freedom for each channel, and hence the signal available for integration at the combined output of the two channels will have four degrees of freedom, corresponding to the statistics of the Swerling Case 3 model. Use of pulse-to-pulse frequency agility with adequate total bandwidth can provide $n_e = n$, giving Swerling Case 2 statistics.

Polarization Diversity.   Use of two orthogonal polarizations can also provide $n_e = 2$. A system operating with pulse-to-pulse frequency agility on two orthogonal polarizations can provide Swerling Case 4 statistics, $n_e = 2n$. It is not necessary to have separate L_f curves for each Swerling case, since Eq. (.37) permits all data to be calculated with sufficient accuracy from Fig. 23.1.11.

Aspect Angle (Space) Diversity.   Radar systems using more than one site can observe a target over spatially diverse paths, obtaining independent samples on each path. If the echo signals are then combined for integration at a central point, diversity gain can be achieved. This mode of operation leads to complexity and high cost, because of the need to duplicate equipment and supporting facilities and the need to compensate for differing signal delays and doppler shifts before combining the signals. Hence, it is not usually a practical option for the system designer.

Detectability Factor with Diversity.   The final value of theoretical detectability factor with diversity, $D_e(n,n_e)$, will be found using $L_f(n_e)$ in place of $L_f(1)$ in Eq. (35). The value $D_x(n,n_e)$ used in the radar equation will be this theoretical value increased by filter mismatch, beamshape, and miscellaneous signal processing losses.

Cumulative Probability of Detection

It is not always possible to combine successive signals from a given target through integration. For example, if the signals are obtained from more than one scan across the target position, or if the repetition interval is long and the target velocity high, the target may have moved in range by an amount greater than the radar resolution, and successive signals will not remain in the same integration cell. The conventional means of using all the signal information in such cases is to accumulate the probabilities of detection resulting from each observation, rather than the energy of the observations. If the probability of detection on each observation (or scan) is P_d, the probability of obtaining at least one detection on k observations is the cumulative probability of detection,

$$P_c = 1 - (1 - P_d)^k \tag{40}$$

The cumulative probability of detection will build up quite rapidly over several scans, even when the single-scan probability is below 50 percent. Large fluctuation loss can be avoided by scanning the search region several times with low P_d. However, such a process is less efficient than integration of energy from all scans with adequate diversity.

TARGETS AND CLUTTER

Target Cross Section

The primary parameter used to describe a radar target is its radar cross section, defined as 4π times the ratio of the reflected power per unit solid angle in the direction of the source to the power per unit area of the incident wave. A large sphere (whose radius $a \gg \lambda$) captures power from an area πa^2, scattering it uniformly in solid angle, and hence has a radar cross section $\sigma = \pi a^2$ equal to its projected area.

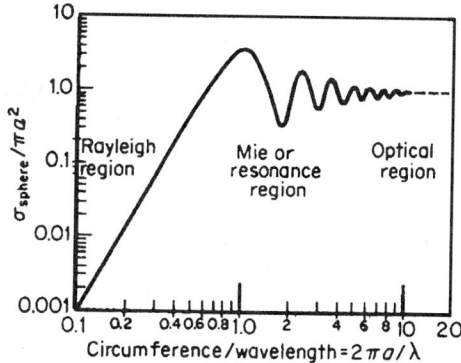

FIGURE 23.1.12 Normalized cross section of a sphere (Ref. 1).

The variation of sphere cross section with wavelength (Fig. 23.1.12) illustrates the division into three regions of the spectrum:

a. *The optical region, $a \gg \lambda$*, where cross section is essentially constant with wavelength.

b. *The resonant region, $a \approx \lambda/2\pi$*, where the cross section oscillates about its optical value due to interference of the direct reflection with a creeping wave, propagated around the circumference of the object.

c. *The Rayleigh region, $a \ll \lambda/2\pi$*, where the cross section drops rapidly below its optical value, varying as $1/\lambda^4$.

Although the sphere cross section varies with wavelength, it is constant over all aspect angles, and hence can serve as a reference for radar testing and evaluation. The cross section of all other objects vary with aspect angle, requiring more complex descriptions: amplitude probability distributions, fluctuation frequency spectra (or time correlation functions), and radar frequency correlation functions. For a few simple shapes, it is possible to write expressions for maximum cross section as a function of size and wavelength:

$$\sigma = 0.88\lambda^2 \text{ resonant dipole viewed normal to axis}$$

$$\sigma(0) = 2\pi a L^2/\lambda \text{ cylinder viewed normal to axis}$$

where a is the radius and L is the length, both assumed $\gg \lambda$,

$$\sigma(0) = 4\pi A^2/\lambda^2 \text{ flat plate viewed normal to surface}$$

where $A = wL$ is the plate area, assumed $\gg \lambda^2$. The cross section of a cylinder or a rectangular plate varies with aspect angle, with a pattern in the plane that includes dimension L given by

$$\sigma(\theta) = \sigma(0) \left[\frac{\sin\left(\frac{2\pi L}{\lambda}\sin\theta\right)}{\frac{2\pi L}{\lambda}\sin\theta} \cos\theta \right]^2 \tag{41}$$

For most other shapes, it is necessary to calculate cross section from complex equations or computer codes, or to measure it with a calibrated radar.

Amplitude Distributions

The cross section of a complex object is best described statistically by its probability density function, examples of which are shown in Fig. 23.1.13. These functions represent the Swerling fluctuation models

$$\text{Cases 1 and 2: } dP = \frac{1}{\bar{\sigma}} \exp\left(\frac{-\sigma}{\bar{\sigma}}\right) d\sigma \quad \sigma \geq 0 \tag{42}$$

$$\text{Cases 3 and 4: } dP = \frac{4\sigma}{\bar{\sigma}^2} \exp\left(\frac{-2\sigma}{\bar{\sigma}}\right) d\sigma \quad \sigma \geq 0 \tag{43}$$

where $\bar{\sigma}$ is arithmetic mean of the distribution. The median value σ_{50} is used as the center of each plot.

Spectra and Correlation Intervals. Swerling Cases 1 and 3 describe slowly fluctuating targets, for which the correlation time t_c is such that all pulses integrated in the time t_o within a single scan are correlated but successive

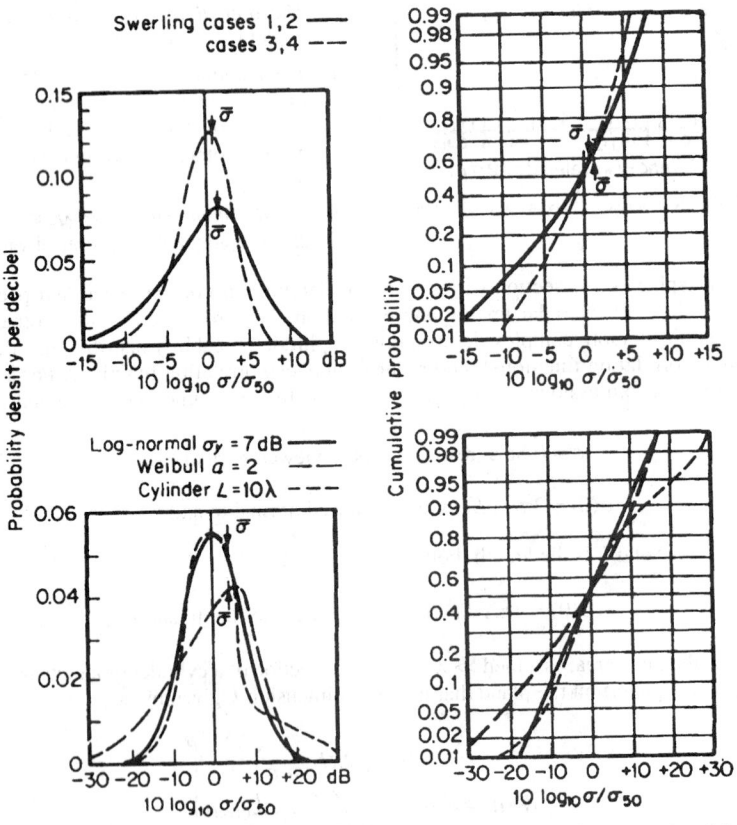

FIGURE 23.1.13 Amplitude distributions of cross section. Upper plots: Cases 1 and 2; lower plots: Cases 3 and 4.

scans separated by t_s give uncorrelated values:

$$t_o \ll t_c < t_s \qquad (44)$$

Cases 2 and 4 describe fast (pulse-to-pulse) fluctuations, $t_c < t_r$.

Target Glint and Scintillation. Targets composed of multiple scattering elements whose phase variations cause fluctuations in signal amplitude are subject to errors in radar position measurement. The apparent source of the composite echo signal wanders back and forth across the target, and at times the signal appears to originate from points well beyond the physical spread of the target itself. In principle, the variance in position estimate is infinite, for a measuring system with unlimited dynamic range and measurement bandwidth. However, for practical reasons this *glint error* is closely approximated by a gaussian distribution with standard deviation $\sigma_r = 0.35L_r$ (in range), or $\sigma_\theta = 0.35L_x/R$ rad (in angle), where L_r is the spread of the target scattering elements along the radar beam and L_x that normal to the beam. On typical aircraft targets, the distribution of significant scattering elements produces values of L between one-third and two-thirds of the maximum target dimensions, leading to glint errors from 0.1 to 0.25 the aircraft dimensions.

Measurement systems using sequential samples of target echo amplitude (conical-scanning and sector-scanning systems) are subject to additional *scintillation errors* caused by interaction of the target amplitude fluctuations with the measurement process. Typical values of this error are about $\sigma_\theta = 0.03\theta_3$ for conical-scanning trackers and $\sigma_\theta = 0.1\theta_3$ for sector scanning search radar.

Clutter Echoes

Unwanted radar echoes (clutter) may originate from land or sea surfaces (characterized by a dimensionless surface reflectivity σ^0), from precipitation or chaff occupying a volume of space (with a volume reflectivity η_v), or from discrete objects such as birds (described by cross sections in square meters). The cross section of distributed clutter can be found by multiplying the area or volume of the radar resolution cell by the appropriate reflectivity parameter:

For surface clutter,

$$\sigma_c = A_c \sigma^0 = \left(\frac{R\theta_a}{L_p} \right) \left(\frac{\tau_n c}{2} \sec \psi \right) \sigma^0 \qquad (45)$$

For volume clutter,

$$\sigma_c = V_c \eta_v = \left(\frac{R\theta_a}{L_p} \right) \left(\frac{R\theta_e}{L_p} \right) \left(\frac{\tau_n c}{2} \right) \eta_v \qquad (46)$$

where τ_n is the processed pulse width, $L_p = 1.33$ is beamshape loss, and ψ is the grazing angle between the radar beam and the surface.

Models for surface clutter reflectivity can be very complex, taking into account grazing angle, terrain type and vegetation or sea state, wind speed and direction, polarization, and other parameters. However, a simple approximate model uses a parameter γ to characterize the surface, and accounts for a broad region in grazing angle using

$$\sigma^0 = \gamma \sin \psi, \quad \psi_c < \psi < \pi/3 \qquad (47)$$

$$\psi_c = \frac{\lambda}{4\pi\sigma_h} \qquad (48)$$

where the surface roughness parameter σ_h is the rms deviation of surface heights from the average. For grazing angles below ψ_c, interference or reflected and direct rays reduces the illumination of the surface scatterers,

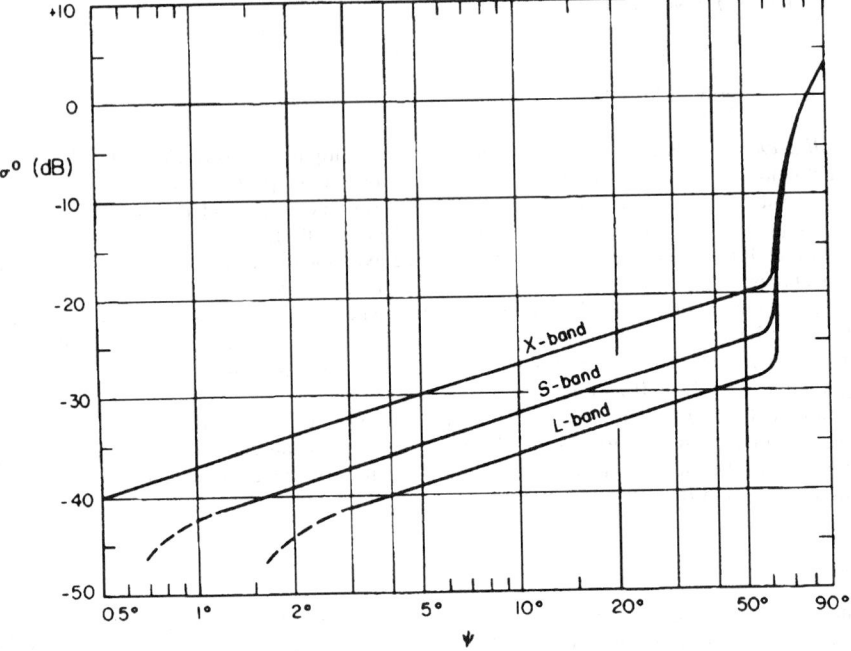

FIGURE 23.1.14   Sea clutter reflectivity versus grazing angle.[2]

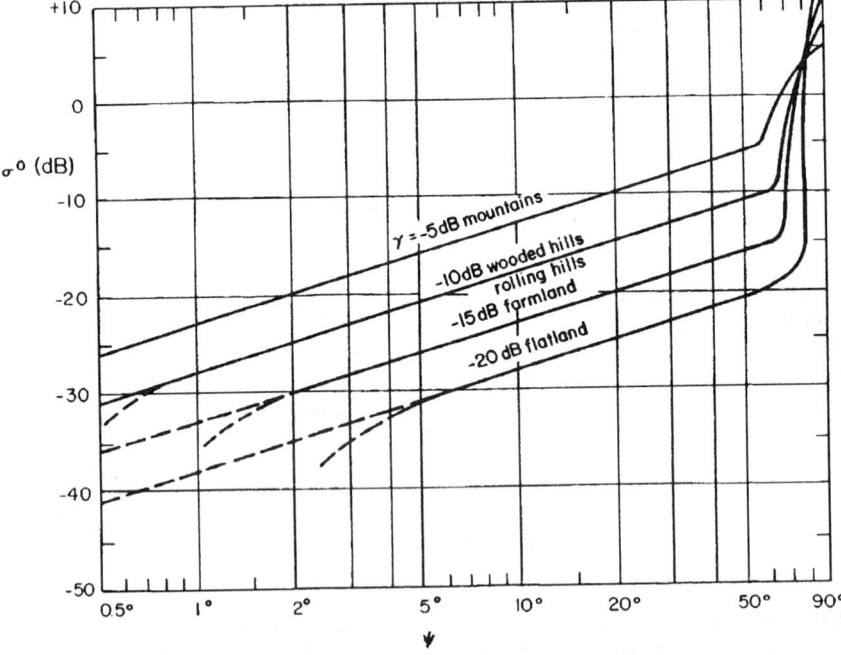

FIGURE 23.1.15   Land clutter reflectivity versus grazing angle.[2]

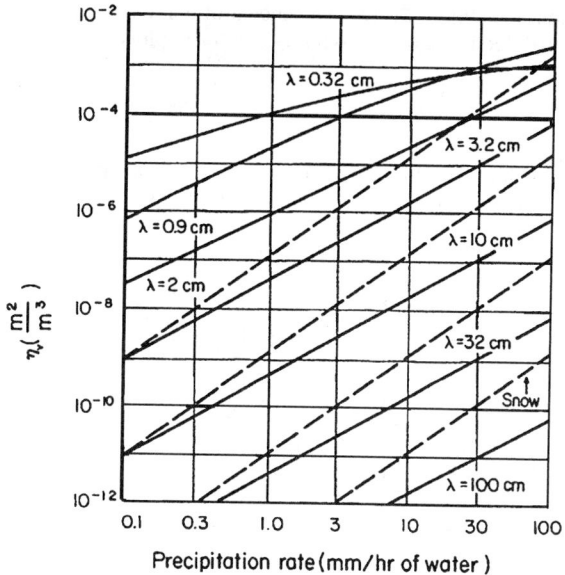

FIGURE 23.1.16 Radar reflectivity of rain and snow.[2]

causing the clutter propagation factor to drop below unity: $F_c = \psi/\psi_c$. The effective surface reflectivity, seen by the radar, becomes $\sigma^0 F_c^4$. At angles near vertical incidence, σ^0 increases steeply as a result of quasispecular reflection. Typical variation in the reflectivity is shown in Fig. 23.1.14 for the medium-rough sea and Fig. 23.1.15 for land surfaces. Sea clutter has an inherent frequency sensitivity, but for land clutter the effect of frequency is primarily in the propagation factor.

Reflectivity of precipitation clutter depends on the rate r in the mm/h of water content, and on wavelength:

$$\text{Rain:} \quad \eta_v = 5.7 \times 10^{-14} r^{1.6}/\lambda^4, \quad \lambda > 0.02 \text{ m} \tag{49}$$

$$\text{Snow:} \quad \eta_v = 1.2 \times 10^{-13} r^2/\lambda^4 \tag{50}$$

These values are plotted in Fig. 23.2.16, with rain values extended into the millimeter-wave bands, based on measurements.[12]

RESOLUTION

Definition of Resolution

A target is said to be resolved if its signal is separated by the radar from those of other targets in at least one of the coordinates used to describe it. For example, a tracking radar may describe a target by two angles, time delay, and frequency (or doppler shift). A second target signal from the same angle and at the same frequency, but with different time delay, may be resolved if the separation is greater than the delay resolution (processed pulse width) of the radar.

Resolution, then, is determined by the relative response of the radar to targets separated from the target to which the radar is matched. The antenna and receiver are configured to match a target signal at a particular angle, delay, and frequency. The radar will respond with reduced gain to targets at other angles, delays, and frequencies. This *response function* can be expressed as a surface in a five-dimensional coordinate system, the fifth coordinate representing amplitude of response. Because five-dimensional surfaces are impossible to plot, and because angle response is almost always independent of delay-frequency response, these pairs of coordinates are usually separated, requiring only two three-coordinate plots.

Antenna Resolution

In angle, the response function $\chi(\theta,\phi)$ is simply the antenna voltage pattern. It is found by measuring the system response as a function of angles from the beam center. It has a main lobe in the direction to which the

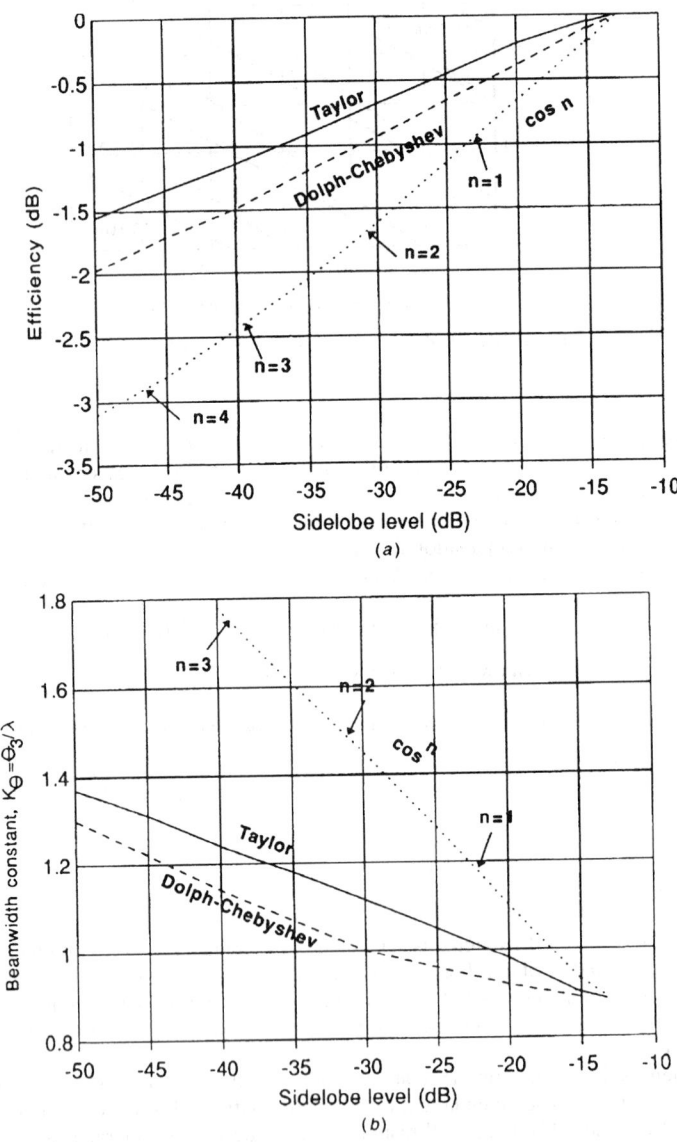

FIGURE 23.1.17 Efficiency and beamwidth constant for antennas with tapered illumination: (a) efficiency versus sidelobe level; (b) beamwidth constant versus sidelobe level.

antenna is scanned, and sidelobes extending over all visible space. Angular resolution, i.e., the main-lobe width in the θ and ϕ coordinates, is generally taken to be the distance between the -3-dB points of the pattern. The width, amplitude, and location of the lobes are determined by the aperture illumination (weighting) functions in the two coordinates across the aperture.

Because the matched antenna is uniformly illuminated, its response has relatively high sidelobes which are objectionable in most radar applications. To avoid these, the antenna illumination may be mismatched (tapered), with resulting loss in gain and broadening of the main lobe. Figure 23.1.17 shows the effects of tapering for sidelobe control on the gain and beamwidth of a rectangular antenna. As the sidelobe level is reduced, aperture efficiency η (the ratio of gain to that of the uniformly illuminated antenna having the same dimensions) falls below unity. At the same time, the beamwidth (which would have been $\theta_3 = 0.886\lambda/w$ for uniform illumination) increases.

Waveform Resolution

Time delay and frequency can also be viewed as if they were two angular coordinates, i.e., as a two-dimensional plane above which the response can be plotted to describe the filter response to a given signal as a function of the time delay t_d and the frequency shift f_d of the signal relative to some reference point to which the radar is matched. Points on the surface are found by recording the receiver output voltage while varying these two target coordinates. The response function $\chi(t_d, f_d)$ is given, for any filter and signal, by

or

$$\chi(t_d, f_d) = \int_{-\infty}^{\infty} H(f)A(f - f_d)\exp(j2\pi f t_d)df \tag{51}$$

$$\chi(t_d, f_d) = \int_{-\infty}^{\infty} h(t_d - t)a(t)\exp(j2\pi f_d t)dt \tag{52}$$

where the functions $A(f)$ and $a(t)$, $H(f)$ and $h(t)$, are Fourier transform pairs describing the signal and filter, respectively.

The transform relationship, Eqs. (50) and (51), governing the time-frequency response function are similar to those which relate the far-field antenna patterns to aperture illumination functions. Hence, data derived for waveforms can be applied to antennas, and vice versa, by interchanging analogous quantities between the two cases. There is a significant difference between waveform and antenna response functions and constraints, however, because the two waveform functions (in time delay and frequency) are dependent on each other through the Fourier transform. The two antenna patterns (in θ and ϕ coordinates) are essentially independent of each other, depending on aperture illuminations in the two aperture coordinates x and y. Further differences arise from the two-way pattern and gain functions applicable to the antenna case.

Ambiguity Function of a Single Pulse

When the filter response is matched to the waveform, $H(f) = A^*(f)$, $h(t) = a^*(t_d - t)$, the magnitude of the squared response function $|\chi(t_d, f_d)|^2$ is called the *ambiguity function* of the waveform. Figure 23.1.18 shows the square root of this function (the voltage response of the matched filter) for three pulses with different modulations. For a simple pulse with constant carrier frequency and phase (Fig. 23.1.18a), there is a single main lobe whose time response is a triangle extending over $\pm\tau$ in time with zero amplitude outside that region. In frequency, the function has the $(\sin^2 x)/x^2$ shape, with sidelobes extending over infinite bandwidth.

Introduction of phase modulation during the transmission of the pulse broadens the frequency spread of the response and narrows the response along the time axis. This is the principle of pulse compression, of which the most common form is linear fm (chirp), shown in Fig. 23.1.18b. With the linear-fm function, very low sidelobes can be obtained in region on both sides of the main, diagonal response ridge. Along this ridge, however, the response falls very slowly from its central value, and targets separated by almost one transmitted pulse width will be detected if they are offset in frequency by the correct amount. Pseudorandom phase coding can generate a single narrow spike in the center of the ambiguity surface (Fig. 23.1.18c), at the expense of large-amplitude sidelobes elsewhere on the ambiguity surface.

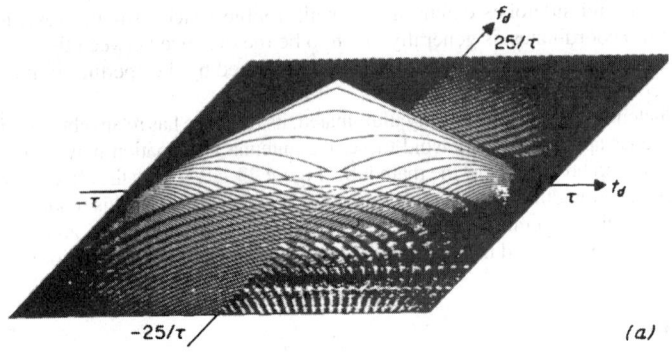

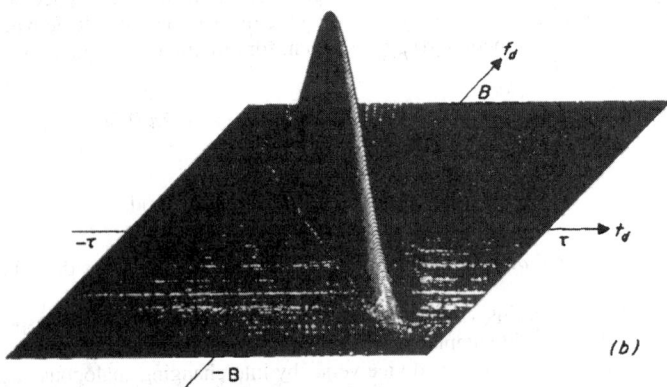

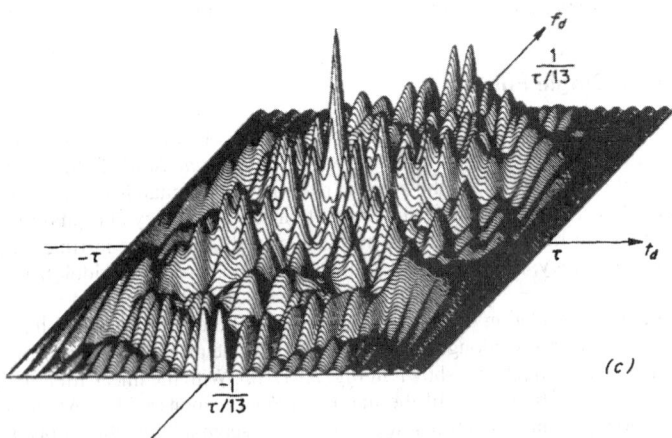

FIGURE 23.1.18 Ambiguity functions for single pulses: (*a*) response for a constant carrier pulse with rectangular envelope; (*b*) chirp response for Hamming weighting on transmit and receive; (*c*) response for the 13-element Barker code (Ref. 4).

The plots of Fig. 23.1.18 illustrate an important property of the ambiguity function: the total volume under the surface is equal to the signal energy (or to unity, when normalized) for all waveforms. Compression of the response along the time axis must be accompanied by an increase in response elsewhere in the time-frequency plane, either along a well-defined ambiguity ridge (for linear fm) or in numerous random lobes (for random-phase codes). For mismatched filters, the response function (cross-ambiguity function) has similar properties, although the central peak will be reduced in amplitude.

Mismatched filters are often used to reduce sidelobe levels in linear-fm pulse compression. The matching efficiency and pulse broadening factors have the same relationship to sidelobe levels as were illustrated in Fig. 23.1.17 for aperture efficiency and beam broadening for tapered aperture illuminations.

Ambiguity Functions of Pulse Trains

The principle of constant volume under the ambiguity function is also applicable to pulse trains. A train of coherent pulses of width τ, with pulse repetition interval $t_r = 1/f_r \gg \tau$, merely generates a repeating set of surfaces similar to Fig. 23.1.18a at intervals t_r in time. The added volume equals the energy of the additional pulses, and if the energy is normalized to unity (by division among the several pulses), the amplitude of each response peak is reduced proportionately. The location of peaks and associated sidelobes is shown in Fig. 23.1.19a. The result is to form a series of responses spaced in range by the *unambiguous range interval* R_u. When the signal is coherent over an observation time $t_o = nt_r$, the time response of the matched filter stretches the ambiguity function to $\pm t_o$ along the time axis. At the same time, there is formed a series of ambiguous responses spaced in radial velocity by the *blind speed* v_b, as shown in Fig. 23.1.19b, c.

Within the central lobes, this response is concentrated in spectral lines separated by f_r and approximately $1/t_o = f_r/n$ wide in frequency (Fig. 23.1.19b). Near the ends of the ambiguity function, where the matched-filter impulse response overlaps $n' < n$ pulses of the received train, the lines broaden to width f_r/n', and at the end of the ambiguity function, where $n' = 1$, no line structure remains. If the repetition rate is increased, holding constant the pulse width and number of pulses in the train, t_o is decreased and the ambiguity volume is redistributed into a smaller number of broader lines. A decrease in pulse width, such that t_o is restored to its original value and n is increased, leads to a broader overall ambiguity function, with the original number of lines in frequency but with narrower and more numerous response bands along the time axis (Fig. 23.1.19c).

The coherent pulse train is thus characterized by a pattern of ambiguities in range and velocity, where the *unambiguous range* is $R_u = c/2f_r$ and the *unambiguous velocity* (or blind speed interval) is $v_b = \lambda f_r/2$. The product of these two quantities (the unambiguous area in the range-velocity plane) depends only on wavelength: $R_u v_b = \lambda c/4$.

Resolution of Targets in Clutter

The choice of the waveform is often dictated by the need to resolve small targets (aircraft, buoys, or projectiles) from surrounding clutter. The clutter power at the filter output is found by integrating the response function over the clutter region, with appropriate factors for variable clutter density, antenna response, and the inverse fourth-power range dependence included in the integrand. Signal-to-clutter ratio S/C for a target on the peak of the radar response is then given by

$$\frac{S}{C} = \frac{\sigma G_t(0)G_r(0)|\chi(0,0)|^2}{\int_v \eta_v(\theta,\phi,f_d t_d)G_t(\theta,\phi)G_r(\theta,\phi)\chi(f_d,t_d)^2 \ (R/R_c)^4 \, dv} \tag{53}$$

where σ = target cross section
$\quad \eta_v$ = clutter reflectivity
$\quad R/R_c$ = target-to-clutter range ratio
$\quad v$ = four-dimensional volume containing clutter

The usual equations for S/C ratio are simplifications of Eq. (53) for various special cases (e.g., surface clutter, homogeneous clutter filling the beam, and so forth). Clearly, the S/C ratio is improved by choosing a waveform

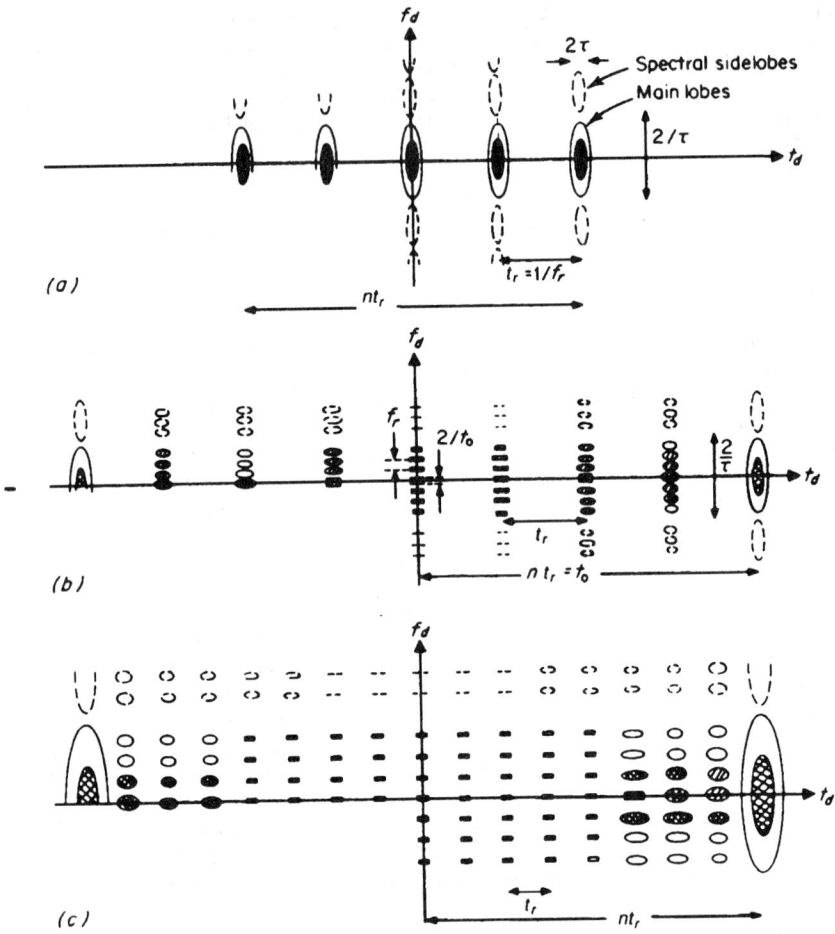

FIGURE 23.1.19 Ambiguity functions of uniform pulse trains: (*a*) noncoherent pulse train; (*b*) coherent pulse train; (*c*) coherent pulse train with reduced pulsewidth τ, increased f_r.

and filter such that $\chi(f_d, t_d)$ is minimized in clutter regions while maintaining a high value $\chi(0,0)$ for all potential target positions. In a search radar, a two-dimensional bank of filters and range gates would be constructed to cover the intervals in doppler and delay occupied by targets, and the clutter power for each of these filters would then be evaluated using Eq. (53).

RADAR PROPAGATION

In the radar equation (15), the echo signal power is seen to be proportional to pattern-propagation factor F^4, which is the product of F_t^2 for the transmit path and F_r^2 for the receive path, and inversely proportional to L_α, the two-way path attenuation. The attenuation depends on the length of the path in which molecules of the atmosphere and of clouds or precipitation are encountered, and the wavelength of the radar wave. The pattern-propagation factor depends on the interaction of the wave with the underlying surface, and the antenna gains in the directions of the target and the reflected wave from the surface.

Apart from the issue of echo signal strength, propagation will affect the accuracy of target position measurements. Errors will be introduced by refraction of the wave as it passes from the target to the radar, and by the multiple signal components which may reach the radar from the underlying surface as a result of reflection and diffraction.

Atmospheric Attenuation.   The frequency bands used for radar were selected to minimize the effects of the atmosphere while achieving adequate bandwidth, antenna gain, and angular resolution. Attenuation is introduced by air and water vapor, by rain and snow, by clouds and fog, and (at some frequencies) by electronics in the ionosphere.

Clear-Air Attenuation.   Attenuation in the clear atmosphere is seldom a serious problem at frequencies below 16 GHz (Fig. 23.1.20). The initial slope of the curves shows the sea-level attenuation coefficient k_α in decibels per kilometer, and this coefficient is reduced as the path reaches higher altitude. Above 16 GHz, atmospheric attenuation is a major factor in system design (Fig. 23.1.21). The absorption lines of water vapor (at 22 GHz) and oxygen (near 60 GHz) are broad enough to restrict radar operations above 16 GHz in the lower troposphere to relatively short range, even under clear-sky conditions. Attenuation versus frequency for two-way paths through the entire atmosphere is shown in Fig. 23.1.22.

Precipitation, Cloud, and Fog Effects.   Above 2 GHz, rain causes significant attenuation, with k_α roughly proportional to rainfall rate r and to the 2.5 power of frequency. The classical data on rain attenuation[13,14] were based on drop-size distributions given by Ryde and Ryde,[15] which gave generally accurate results, except for a 40 percent underestimate of the loss between 8 and 16 GHz, at low rainfall rates. Later data were derived by Wexler and Atlas[16] from a modified Marshall-Palmer distribution.[17] At high rates (100 mm/h) the loss coefficient k_α/r is doubled between 8 and 16 GHz, giving better agreement with measurements and matching the measurements of Medhurst[18] above 16 GHz. The Wexler and Atlas data provide the most satisfactory estimates for general use, and these were used in preparing Fig. 23.1.23.

Very small water droplets, suspended as clouds or fog, can also cause serious attenuation, especially since the affected portion of the transmission path can be tens or hundreds of kilometers. Attenuation is greatest at 0°C (Fig. 23.1.24). Transmissions below 2 GHz are affected more seriously by heavy clouds and fog than by rain of downpour intensity.

Water films that form on antenna components and radomes are also sources of loss. However, such surfaces can be specially treated to prevent the formation of continuous films.[19,20]

Apparent Sky Temperature.   Associated with the atmospheric loss is a temperature term, which must be added to the radar receiver input temperature. Figure 23.1.4 showed this loss temperature T'_a as a function of frequency, for clear-air conditions. When precipitation is present, there will be additional loss along the atmospheric path, generating additional loss temperature into the antenna. For this situation, the sky temperature is calculated from total atmospheric loss L_α using

$$T'_a = 290\left(1 - \frac{1}{\sqrt{L_\alpha}}\right) + T_g \tag{54}$$

where T_g is the galactic background noise, a significant component for $f \le 1$ GHz.

Ionospheric Attenuation.   In the lowest radar bands, the daytime ionosphere may introduce noticeable attenuation.[21] However, above the 100 MHz, this attenuation seldom exceeds 1 dB.

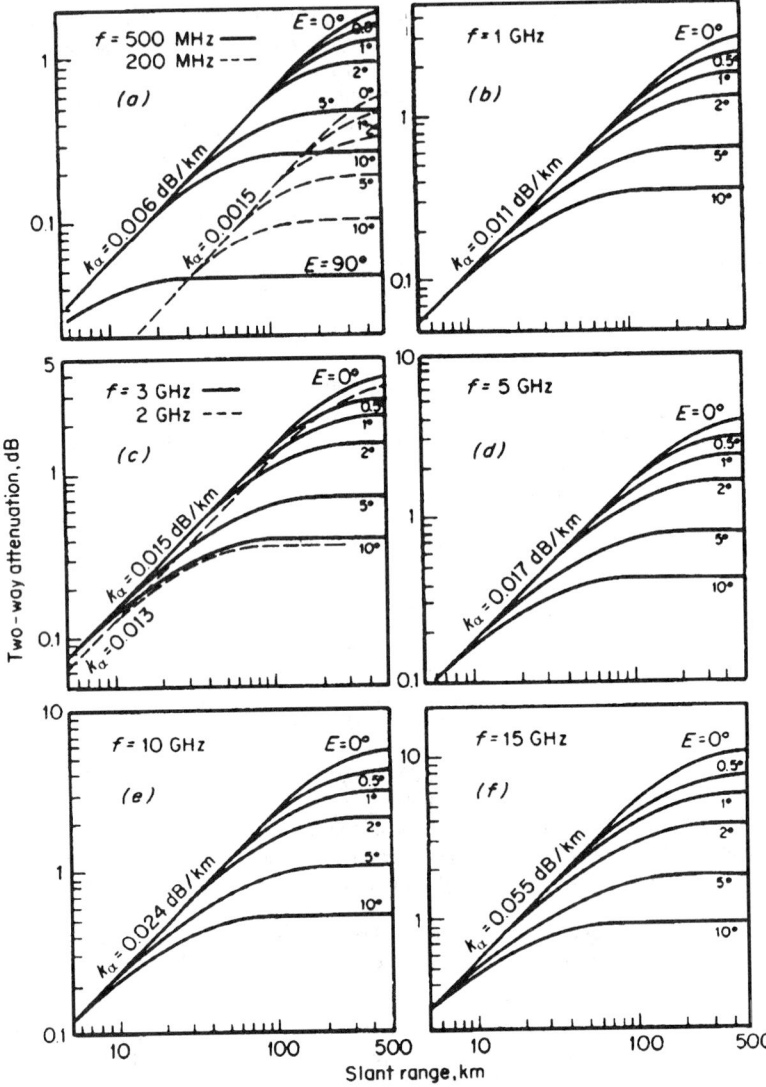

FIGURE 23.1.20   Atmospheric attenuation (0.2 to 15 GHz) versus range and elevation angle E. (*Data from Ref. 8*)

Surface Reflections

The radar target scatters power in all directions, and some of this power arrives at the radar antenna via reflection from the surface. If the radar receiving antenna pattern has significant response in the direction from which these reflections arrive, the receiver will be presented with a composite (multipath) signal, in which the reflected components interfere with the direct signal, alternately adding to and subtracting from the direct signal magnitude. This will affect the detectability of the signal, and will introduce multipath error in the measurement of target position. On the transmit path, the same phenomenon will modulate the illumination of the target, affecting the magnitude of the echo but not its arrival angle.

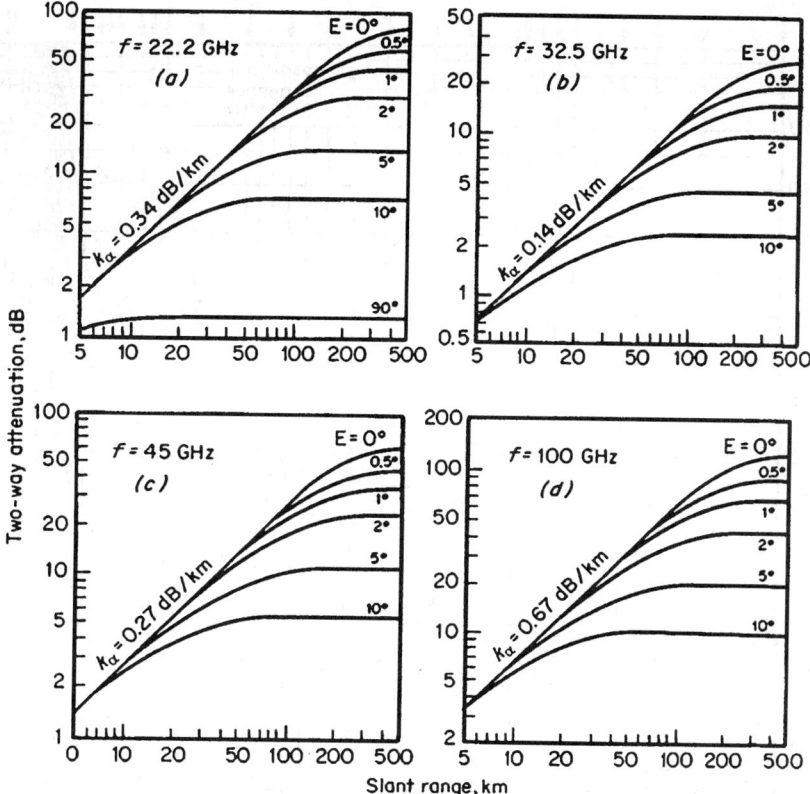

FIGURE 23.1.21 Atmospheric attenuation (20 to 100 GHz) versus range and elevation angle *E*. (*Data from Ref. 8*)

Specular Reflection. The simple model for surface reflection applies to a flat, smooth surface (Fig. 23.1.25). Ignoring curvature of the earth, the specular reflection from this surface arrives at the radar from a negative elevation angle θ_r, approximately equal to the positive elevation angle θ_t, of the direct ray from the target:

$$\text{Target elevation} = \theta_t = \sin^{-1}\left(\frac{h_t - h_r}{R}\right) \approx \frac{h_t - h_r}{R} \tag{55}$$

$$\text{Depression angle of reflection} = \theta_r = \psi = \sin^{-1}\left(\frac{h_t + h_r}{R}\right) \approx \frac{h_t + h_r}{R} \tag{56}$$

The depression angle from the radar is equal to the grazing angle ψ at the surface. The extra pathlength for the reflected ray will be

$$\delta_0 = R\left(\cos\theta_r - \cos\theta_t\right) \approx \frac{2h_t h_r}{R} = 2h_r\theta_t \tag{57}$$

For a radar beam pointed at elevation angle θ_b, the voltage gain for the direct ray will be $f(\theta_t - \theta_b)$, and for the reflected ray will be $f(\theta_t + \theta_b)$. The reflected ray will arrive at the antenna with a relative amplitude equal

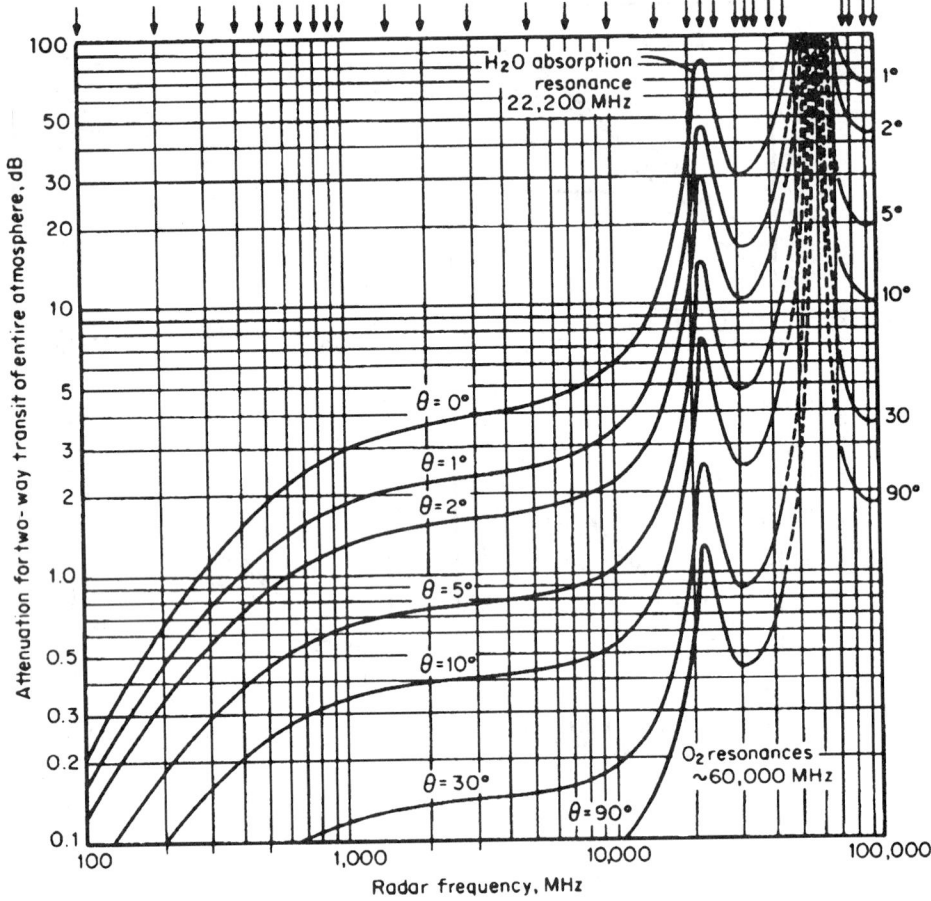

FIGURE 23.1.22 Absorption loss for two-way transit of the entire troposphere, at various elevation angles (Ref. 8).

to the surface reflection coefficient ρ, resulting in an apparent antenna gain for the composite signal given by

$$f_r'(\theta_t) = f_r(\theta_t - \theta_b) + \rho f_r(-\theta_r - \theta_b) \exp\left[-j\left(\frac{2\pi\delta_0}{\lambda} + \phi\right)\right] = F_r(\theta_t)f(0) \tag{58}$$

where ϕ is the phase angle of the reflection coefficient and F_r is the pattern-propagation factor for the receive path. A similar expression involving the transmit pattern $f_t(\theta)$ will give the transmit pattern-propagation factor F_t. A two-way pattern-propagation factor $F^4 = F_t^2 F_r^2$ is used in the radar equation, and the detection range will be directly proportional to F.

As a result of the reflected signal, the coverage pattern of a search radar that has a broad elevation pattern will appear as in Fig. 23.1.25b. Near the horizon, where $f(\theta_t - \theta_b) \approx f(-\theta_r - \theta_b)$, $\rho \approx 1$, and $\varphi \approx \pi$, the reflection lobes extend the coverage to twice the free-space value, while the nulls give zero signal. The nulls will appear at angles such that

$$\sin\theta_n = i\left(\frac{\lambda}{4h_r}\right) \approx \theta_n, \quad i = 0, 1, 2, \ldots \tag{59}$$

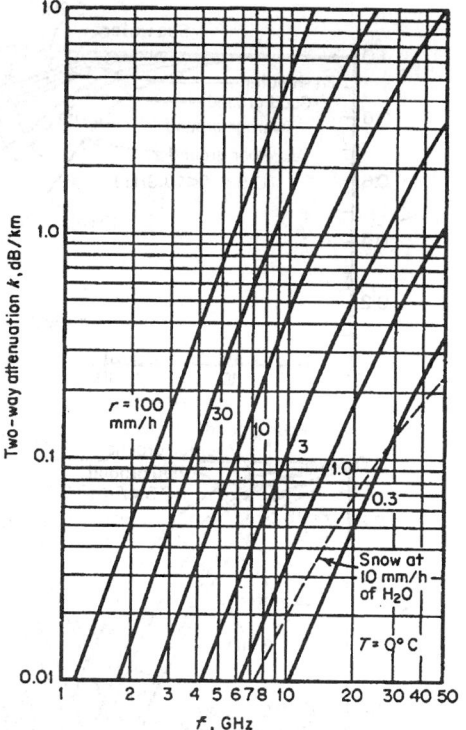

FIGURE 23.1.23   Attenuation in rain.

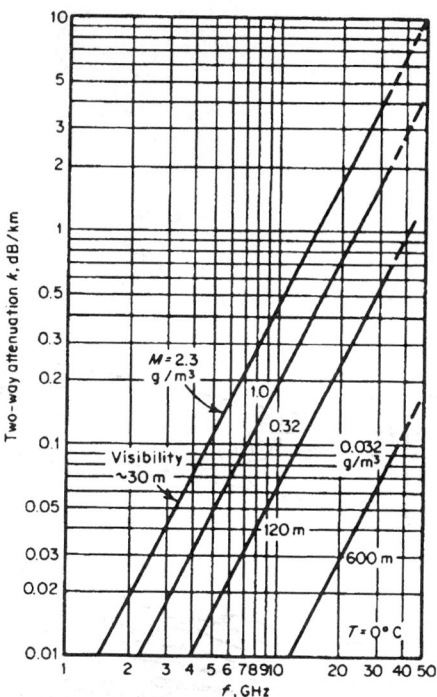

FIGURE 23.1.24   Attenuation in clouds or fog (values of k_a are halved at $T = 18°C$).

For elevated radars or long-range targets, curvature of the earth cannot be ignored, and equations that correct for curvature must be used [Ref. 3, p. 553]. However, flat earth approximations are adequate for radars up to 100 m from the surface viewing targets at ranges of a few tens of km.

The Fresnel reflection coefficient ρ_0 as a function of grazing angle ψ, for various surface materials, is plotted in Fig. 23.1.26. For horizontal polarization, the reflection coefficient remains near -1 for grazing angles below about 10°. For vertical polarization, the coefficient goes from -1 at low angles, through a minimum near zero amplitude, and to a positive value at high angles. The angle at which the real part of the coefficient goes through zero is known as the Brewster angle, and at this angle most of the power is absorbed by the surface.

Reflection from Rough Surface.   Actual land and water surfaces are irregular, reducing the magnitude of the specular reflection to a value $\rho_0\rho_s$, where the specular scattering factor ρ_s is a function of the rms surface height deviation σ_h, the wavelength, and the grazing angle:

$$\rho_s = \exp\left[-2\left(\frac{2\pi\sigma_h \sin\psi}{\lambda}\right)^2\right] \tag{60}$$

The specular scattering factor is plotted in Fig. 23.1.27 as a function of normalized surface roughness.

As the specular scattering coefficient decreases, diffuse reflections appear, containing the power which has been lost to the specular component. These diffuse components arrive from a broad elevation region surrounding the point of specular reflection, and much of their power may fall outside the beamwidth of the antenna. They have random fluctuations in amplitude and phase, and produce little effect on detection range. However, they are important sources of tracking error.

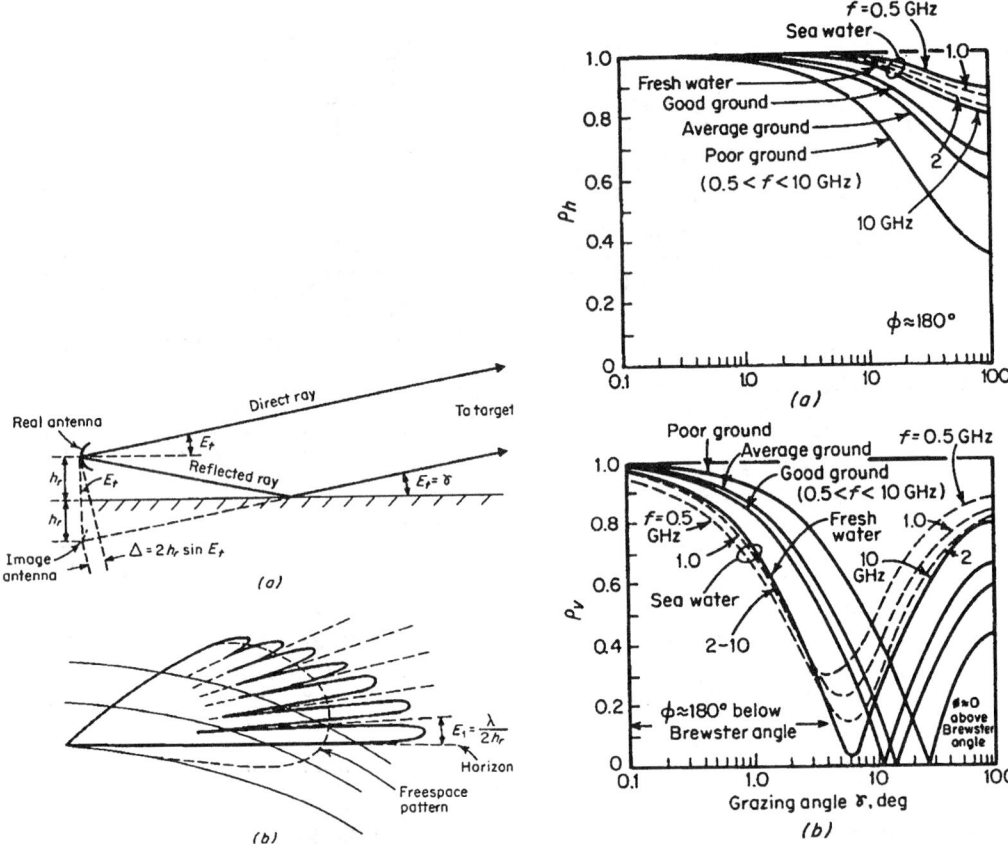

FIGURE 23.1.25   Effect of surface reflections: (*a*) geometry of specular reflection; (*b*) lobing pattern produced by reflections (Ref. 2).

FIGURE 23.1.26   Fresnel reflection coefficient, as a function of grazing angle, for different surfaces: (*a*) horizontal polarization; (*b*) vertical polarization.

Diffraction at the Surface.   Rays that pass close to the curved earth surface, or to an obstacle rising from the surface, are affected by diffraction. Smooth-sphere diffraction modifies the pattern-propagation factor at elevation angles less than $\theta_n/2$, where the pathlength difference between direct and reflected rays is less than $\lambda/4$. The calculation of F for such paths is described in [Ref. 3, pp. 297–302] and in Sec. 12 of this handbook.

Diffraction over obstacles such as trees, ridges, or fences will follow the knife-edge diffraction theory. Figure 23.1.28 shows curves for smooth-sphere and knife-edge diffraction.

Tropospheric Refraction

The refraction index of the troposphere, for all radar frequencies, can be expressed in terms of a deviation from unity in parts per million, or *refractivity*

$$N \equiv (n-1) \times 10^6 = \frac{77.6}{T}\left(P + \frac{4810p}{T}\right) \tag{61}$$

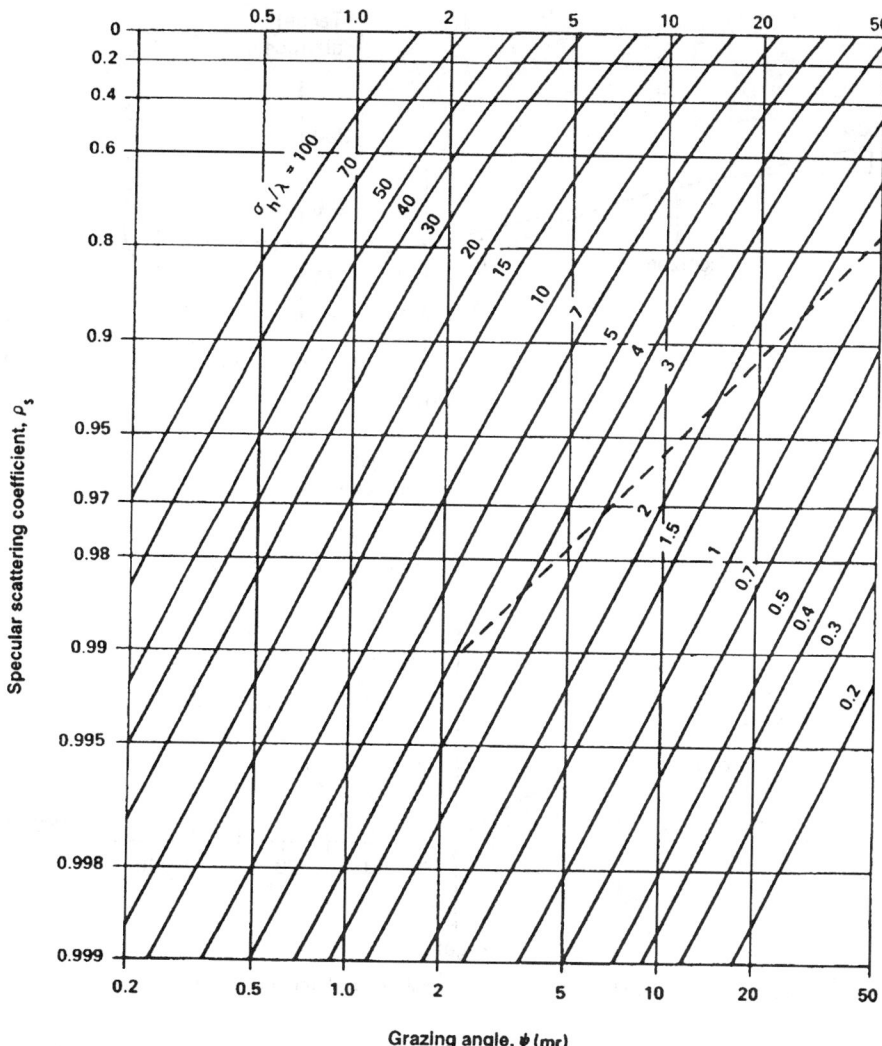

FIGURE 23.1.27   Specular scattering factor versus normalized surface roughness.

where T = temperature in kelvins
   P = total pressure in millibars
   p = partial pressure of water-vapor component
   n = refractive index

Dry air at sea level can have a value as low as $N = 270$, but normal values lie between 300 and 320.

The Central Radio Propagation Laboratory (CRPL) of the National Bureau of Standards (now the National Oceanic and Atmospheric Administration) established a family of exponential approximations to the variation in refractivity with altitude for the normal atmosphere, in which the average U.S. conditions are represented by

$$N(h) = 313.0 \exp(-0.14386h) \tag{62}$$

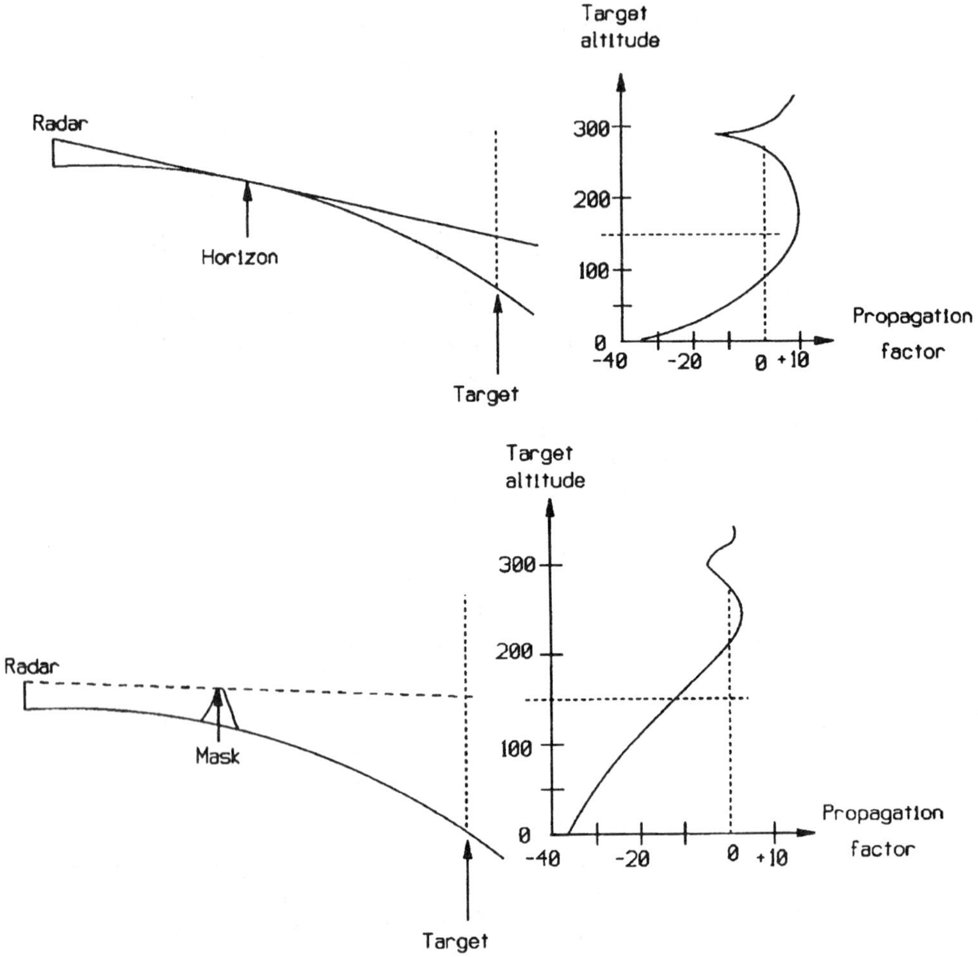

FIGURE 23.1.28 Diffraction for paths near surface: (*a*) smooth-sphere diffraction; (*b*) diffraction over obstacle.

where h is the altitude in km above sea level. The velocity of wave propagation is $1/n$ times the vacuum velocity, including an extra time delay in radar ranging and causing radar rays to bend downward relative to the angle at which they are transmitted. Figure 23.1.29 shows, on an exaggerated scale, the geometry of tropospheric refraction.

For surveillance radars, the effects of refraction are adequately expressed by plotting ray paths as straight lines above a curved earth whose radius k_a is 4/3 times the true earth's radius: $a = 6.5 \times 10^6$ m, $ka = 8.5 \times 10^6$ m.

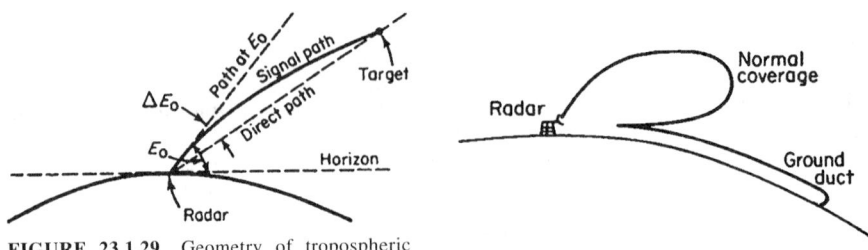

FIGURE 23.1.29 Geometry of tropospheric refraction.

FIGURE 23.1.30 Low-angle ducting effect (Ref. 2).

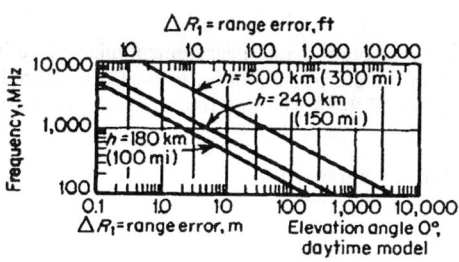 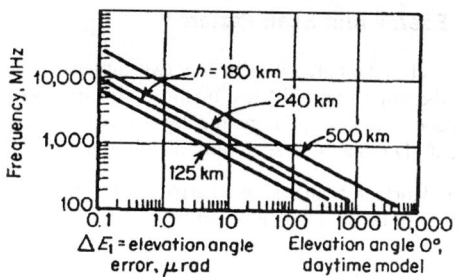

FIGURE 23.1.31 Ionospheric errors in range and elevation angle, for different target altitudes: (*a*) range error versus frequency; (*b*) elevation angle error versus frequency.

A special problem can arise when the ray is transmitted at an elevation below 0.5° into an atmosphere whose refractivity has a rapid drop, several times greater than the standard 45 *N*-units per km. Under those conditions, the ray can be trapped in a surface duct (Fig. 23.1.30) or in an elevated duct bounded by layers of rapidly decreasing *N*. The result is a great increase in radar detection range for targets within the duct (and for clutter) at the expense of coverage just above the ducting layer. Although there is some leakage of energy through the top of the duct, increasing at lower frequencies, the duct will usually trap all radar frequencies sufficiently to create a gap just above the horizon, through which targets can pass undetected.

Ionospheric Refraction

The refractivity of the ionospheric at radar frequencies is given by

$$N_i = (n-1) \times 10^6 = -\frac{40N_e}{f^2} \times 10^6 = \frac{-1}{2}\left(\frac{f_c}{f}\right)^2 \times 10^6 \tag{63}$$

where N_e is the electron density per m³ and f_c is the critical frequency in hertz ($f_c \approx 9\sqrt{N_e}$) Since f_c seldom exceeds 14 MHz, the refractivity at 100 MHz is less than 10^4 *N*-units, and above 1 GHz it does not exceed 100 *N*-units. Figure 23.1.31 plots the errors in range end elevation angle for normal ionospheric conditions for targets at different altitudes. Ionospheric errors are not generally significant for radars operating in the gigahertz region, but can dominate the analysis of tracking error in the 200- and 400-MHz bands.

SEARCH-RADAR TECHNIQUES

A *search radar* is one that is used primarily for the detection of targets in a particular volume of interest. A *surveillance radar* is used to maintain cognizance of selected traffic within a selected area, such as an airport terminal area or air route.[7] The difference is terminology implies that the surveillance radar provides for the maintenance of track files on the selected traffic, while the search radar output may be simply a warning or one-time designation of each target for acquisition by a tracker. There is no significant difference in radar characteristics between the two uses, but the requirements for detection probability and search frame time may lead to different operating modes, scan patterns, and maximum ranges. The following discussions of search radar are equally applicable to surveillance systems.

Search Beams and Scan Patterns

Search radars are described as two-dimensional (2D) when resolution and measurement takes place in range and azimuth coordinates, or three-dimensional (3D) when the elevation coordinate is also measured. Beam shapes and scan patterns for the two types are shown in Fig. 23.1.32. The first four parts of this figure represent 2D radars that scan only in azimuth:

(*a*) Horizon scan, using a narrow beam, fixed in elevation at the horizon
(*b*) Fan beam, using a broad beam whose lower edge is at the horizon

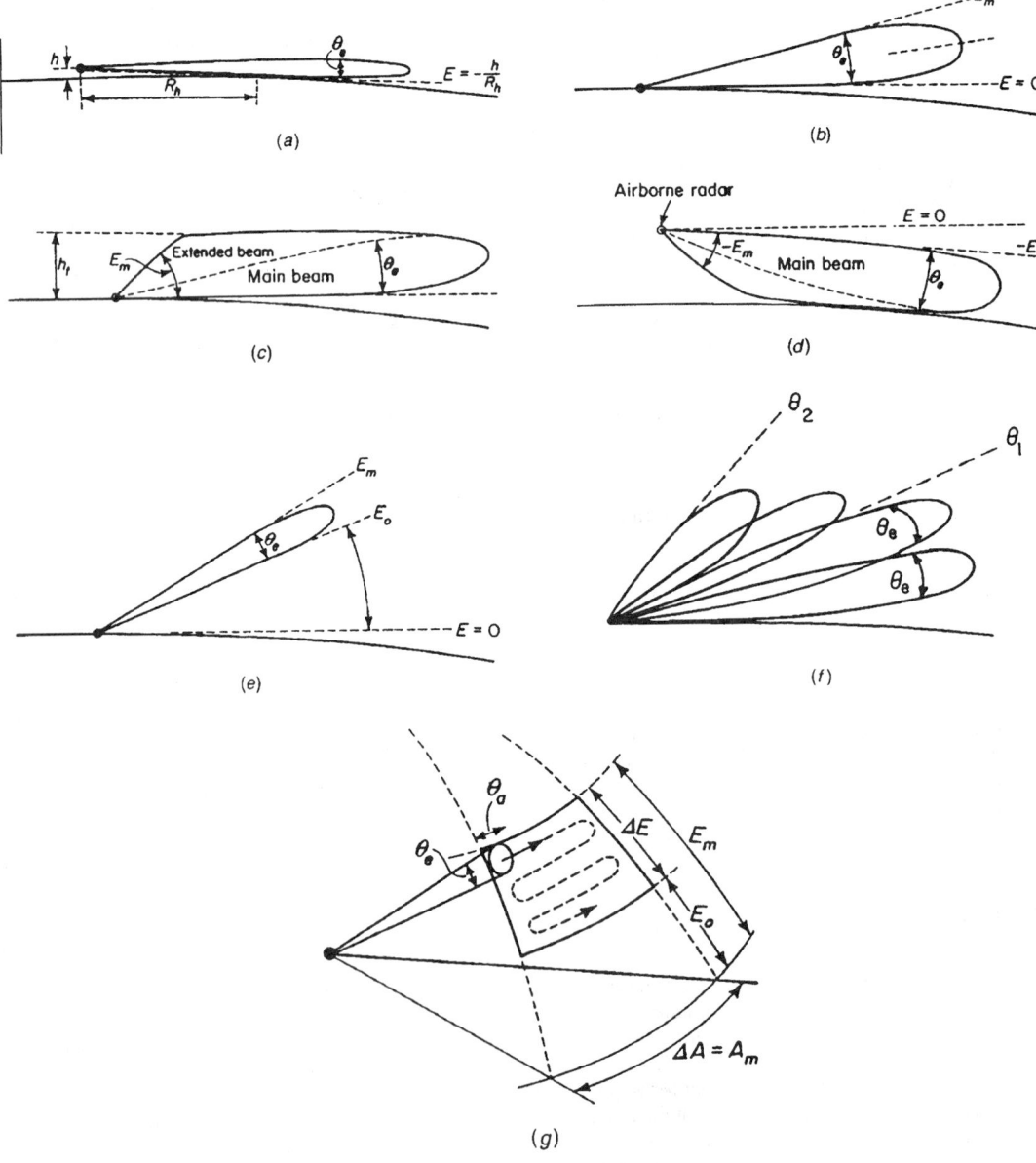

FIGURE 23.1.32 Basic types of search beams and scans.

(*c*) Cosecant-squared beam, similar to the fan beam but with extended coverage in elevation above the main beam, at reduced range

(*d*) Inverted csc^2 beam, for coverage to the surface from an airborne radar

The last three parts of the figure represent 3D radars:

(*e*) Elevation-scanning pencil beam, used at fixed azimuth for height-finding radar and combined with azimuth scan for volume 3D coverage;

(*f*) Stacked beams in elevation, scanning in azimuth for volume 3D coverage;

(*g*) Raster scan of a sector in both azimuth and elevation, often used in the search mode of a multifunction array radar.

Whatever the type of scan, the pulsed search radar will resolve and measure in range, for each beam position, and may also use doppler processing to resolve in radial velocity. The CW search radar may or may not resolve in range, as well as in radial velocity, depending on the modulation applied to the carrier.

Search-Radar Detection Process

Early search radars depended entirely on cathode-ray-tube displays with human operations for target detection, and this process remains one of the most efficient and adaptable. The curves for detection probability (Fig. 23.1.8) may be applied to the human operator if a suitable miscellaneous signal processing loss component for the operator, $L_{xo} \approx 4$ dB, is included in calculation of D_x for use in the radar equation. When the operator is fatigued or distracted, even larger losses will be encountered.

Electronic and digital integrators and automatic target detector circuits are included in many modern search radars to ensure more consistent performance than can be provided by human operators. The performance of such systems is not necessarily better than that of an alert operator, especially in the presence of clutter and interference or jamming. In order to hold the number of false alarms below the level which would overload the subsequent data processing and tracking, it is essential that the automatic detector system include constant-false-alarm-rate (CFAR) circuits to adapt the threshold to actual interference levels at the receiver-processor output. Typical CFAR techniques are shown in Fig. 23.1.33. The circuits (*a*) and (*b*) provide averaging of surrounding resolution cells in the frequency-domain; (*c*) averages in the time domain, over range cells surrounding the detection cell; (*d*) averages in the angle domain, over sectors each side of the antenna main lobe.

Use of CFAR detection processing leads to a loss, which must be included as a component of miscellaneous signal processing loss L_x. This loss may be estimated from Fig. 23.1.34. The CFAR ratio is defined as the ratio of the negative exponent x of false-alarm probability (e.g., $x = 6$ for $Pf = 10^{-6}$) to the number of independent interference samples, m_e, averaged in setting the threshold.

Moving-Target Indication

A moving-target indicator (MTI) is a device that limits the display of radar information primarily to moving targets. The sensitivity to target motion is usually provided through their doppler shifts, although *area MTI* systems have been built, which cancel targets on the basis of overlap of their signal envelopes in both range and angle.

In the usual pulse-amplifier coherent MTI system (Fig. 23.1.35), two cw oscillators in the radar are used to produce a phase and frequency reference for both transmitting and receiving, so that echoes from fixed targets have a constant phase at the detector, over the train of pulses received during a scan. These echoes will be canceled, leaving at the output only those signals whose phase varies from pulse to pulse as a result of target motion.

Coherent MTI can also be implemented with a pulsed oscillator transmitter (Fig. 23.1.36), in which the coherent oscillator at i.f. is locked in phase to a downconverted sample of each transmitted pulse. Although the transmitted and received signals are noncoherent, the phase-detected output is coherent and clutter components can be canceled. Both systems attenuate targets in a band centered on zero radial velocity, the depth and width of the rejection notch depending on the design of the canceler and the stability of the received signals.

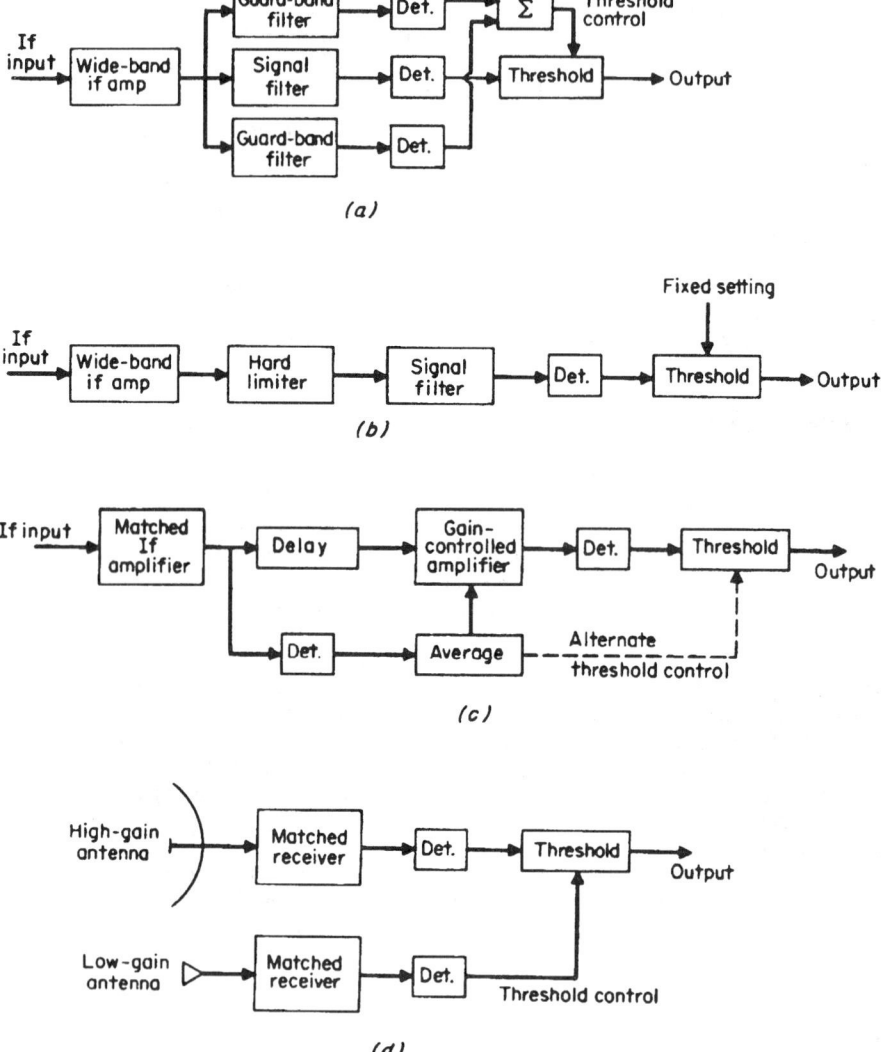

FIGURE 23.1.33 CFAR techniques: (*a*) guard-band system; (*b*) Dicke fix system; (*c*) range-averaged AGC; (*d*) sidelobe blanker.

Two variations on the coherent MTI are available for rejection of clutter with nonzero radial velocity. In the clutter-locked MTI, the average doppler shift of a given volume of clutter is measured and used to control an offset frequency oscillator in the receiver, shifting the received clutter components into the rejection notch. Short- or long-term averages may be used to obtain rapid adaptation to varying clutter velocity (as in weather clutter) or better rejection of selected parts of a complex clutter background. The alternative is *non-coherent MTI*, in which the clutter surrounding a target provides the phase reference with which the target signal is mixed to produce a baseband signal having target doppler shift. Although simpler to implement,

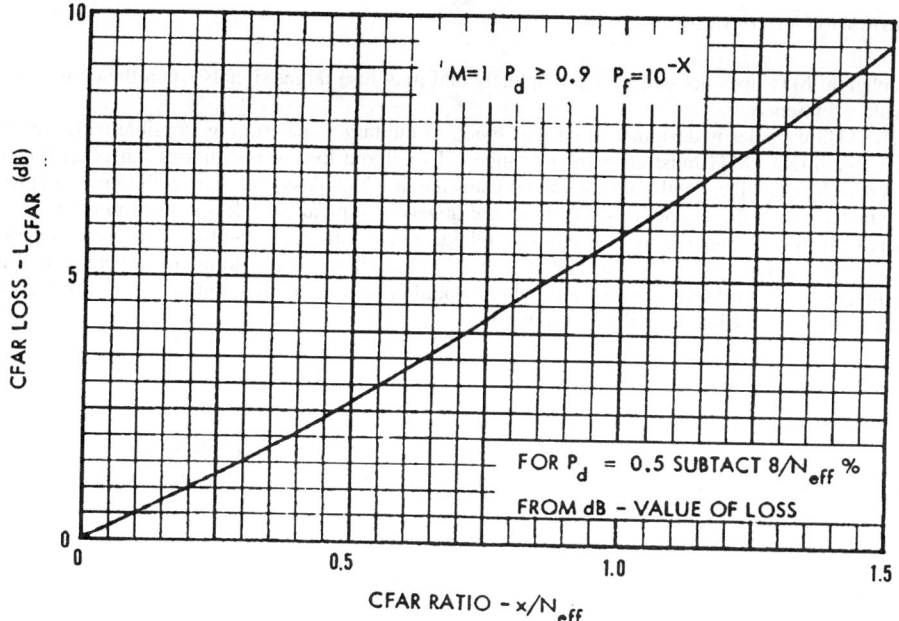

FIGURE 23.1.34 Universal curve for CFAR loss (Ref. 23).

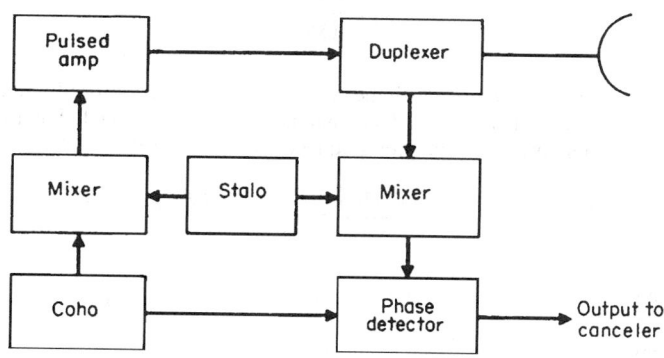

FIGURE 23.1.35 Coherent MTI system.

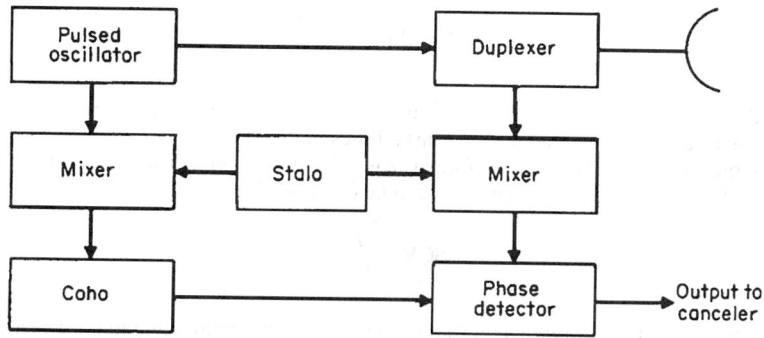

FIGURE 23.1.36 Coherent-on-receive MTI system.

noncoherent MTI does not cancel as completely and may lose target signals when the clutter is too small to provide a reference.

The MTI canceler is designed to pass as much of the target spectrum as possible while rejecting clutter. Since search-radar MTI must cover many range cells without loss of resolution, canceling filters are implemented with delay lines, with multiple range gates feeding bandpass filters, or with range-sampled digital filters which perform these functions. The response of several typical cancelers is shown in Fig. 23.1.37. A wide variety of response shapes is available through the use of feedback and multiple, staggered repetition rates.[24] In particular, through proper use of stagger (Fig. 23.1.37c), it is possible to maintain detection of most targets with nonzero radial velocities, even those which would fall in one of the blind speeds v_{bj} (ambiguous rejection notches) of an MTI with a single repetition rate:

$$v_{bj} = j\frac{\lambda f_r}{2} \quad j = \pm 0, 1, 2, \ldots \tag{64}$$

The MTI radar system must transmit, in each beam position, at least two pulses (for a single-delay canceler), and when feedback is used the pulses in each beam must extend for the duration of significant impulse response of the filter.

Performance of MTI

The basic measure of MTI performance is the *MTI improvement factor I*, defined by

$$I = \frac{(S/N)_{out}}{(S/N)_{in}}\bigg|_{all\ v_t}$$

This is equal to the *clutter attenuation* when the targets are distributed uniformly over one or more blind-speed intervals. The basic relationships between I and parameters of the radar and clutter can be expressed in terms of the ratio of rms clutter spread to repetition frequency or blind speed:

$$z = \frac{2\pi\sigma_f}{f_r} = \frac{2\pi\sigma_v}{v_{bl}} = \frac{2\pi\sigma_v}{\lambda f_r} \tag{65}$$

where σ_f = standard deviation of clutter power spectrum in hertz
f_r = repetition rate
σ_v = standard deviation in m/s
v_{bl} = first blind speed from Eq. (64)

For a scanning beam viewing motionless clutter, motion of the beam induces a velocity spread such that

$$z = 2\sqrt{\ln 2}\ \frac{\omega}{f_r\theta_3}\ \frac{1.665}{n} \tag{66}$$

In general, the rms spread must be calculated using the rms sum of components due to scanning, internal motion of clutter, radar platform motion, and instabilities in the radar.

Another term describing MTI radar performance is *subclutter visibility* (SCV), which is the maximum input clutter-to-signal ratio at which target detection can be obtained:

$$SCV = \frac{I}{D_{xc}} \tag{67}$$

where D_{xc} is the clutter detectability factor (output signal-to-clutter ratio required for detection). Depending on the integration gain that follows the MTI, D_{xc} may be as large as 100 (or 20 dB), or as small as unity (0 dB).

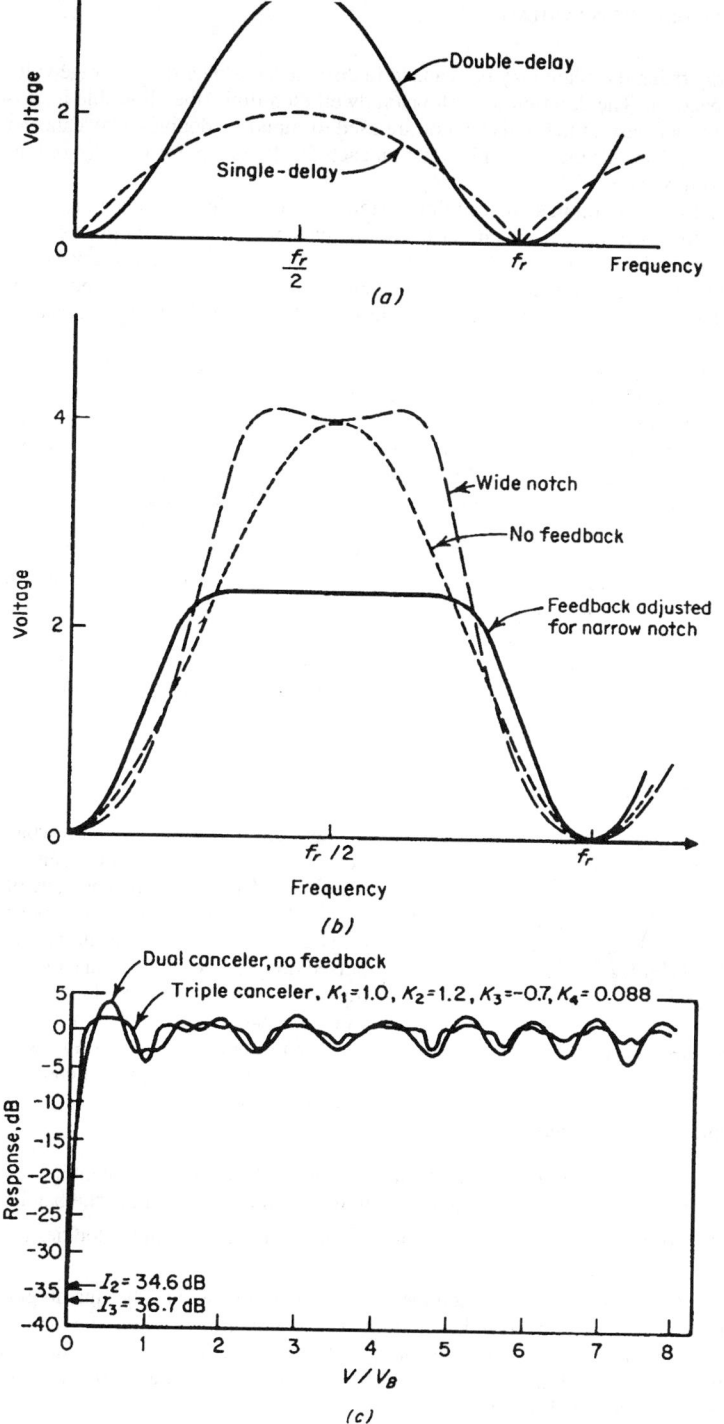

FIGURE 23.1.37 Frequency response of MTI filters: (*a*) single- and double-delay without feedback; (*b*) double-delay with feedback; (*c*) double- and triple-delay with staggered prf and feedback (Ref. 24).

In phased array radar, the beam may be scanned in discrete steps over the coverage volume without spreading the clutter spectrum. The duration of each beam dwell then limits the allowable impulse response time of the filter. In many such cases, three-pulse bursts are used to support a double-delay canceler without feedback. Stagger to avoid blind speeds may be applied within each dwell, on a pulse-to-pulse or burst-to-burst basis, or on a dwell-to-dwell basis.

An important consideration in step-scanning systems is the possible presence of clutter beyond the unambiguous range of the waveform. Such clutter cannot be canceled unless extra pulses (*fill pulses*) are included in the dwell to produce identical clutter inputs on those pulse repetition intervals processed in the MTI filter. Pulse-to-pulse stagger may not be used when clutter beyond the unambiguous range is to be canceled, and burst-to-burst stagger waveforms must include fill pulses in each burst, increasing the required dwell time.

When an average is taken over all target velocities, the MTI system does not change the signal-to-noise ratio (SNR). However, for given average SNR, the probability of detection may be severely degraded by several factors. First, the number of independent noise samples available for integration is reduced because successive outputs on successive are not independent at the canceler output. The effective number of samples integrated is $(2/3)n$ for a single-delay canceler, and $(1/2)n$ for double-delay, where n is the number of pulses per beamwidth. The number is further reduced by the number of fill pulses and the necessity to gate the canceler output after each step in scanning and each change in PRF, when burst-to-burst stagger is used. If the quadrature canceler channel is not implemented, the number of target samples is reduced by a factor of 2. For a Case 1 target viewed over a short dwell, the Rayleigh distribution is reduced to a single-sided gaussian distribution, with twice the decibel value of fluctuation loss. Finally, the clutter rejection notches in the velocity response curve present some targets from being detected, making it difficult to achieve high probabilities of detection regardless of SNR. These losses in information, expressed in terms of required increases in SNR to maintain a given detection performance, constitute MTI loss components of miscellaneous signal processing loss, often totaling 6 to 12 dB.

Pulsed Doppler Radar Systems

A *pulsed doppler radar* is a system in which targets are selected in one or more narrowband filters. For search radar, these filters must cover the velocity band occupied by targets, omitting or desensitizing filters containing clutter. As a result, the envelope of filter response will have a shape similar to that of an optimized MTI canceler (Fig. 23.1.38). To generate the narrow filter responses, a pulsed doppler radar must transmit, receive, and process a train of pulses considerably longer than required for MTI. Coherent integration is inherent in the narrow doppler filters used in pulsed doppler systems, and hence these systems may operate with greater energy efficiency than MTI radars.

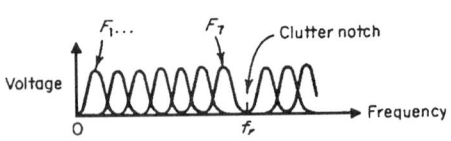

FIGURE 23.1.38 Frequency response of filter bank in pulsed doppler radar.

Pulsed doppler radar can operate in any of three modes:

1. *Low-PRF mode*, in which target detections are intended only within the unambiguous range of the waveform ($R_{t\max} < R_u$). In most cases, this mode has multiple ambiguities within the band of target velocities.

2. *Medium-PRF mode*, in which target detections are required at ranges and velocities beyond the unambiguous values ($R_{\max} > R_u$, $v_{\max} > v_b$).

3. *High-PRF mode*, in which target detections are intended only within the unambiguous velocity interval ($v_{\max} < v_b$). In most cases, this mode has multiple range ambiguities within the range interval occupied by targets. When the radar platform is moving at velocity v_p, high-PRF operation requires ($v_{t\max} + 2v_p) < v_b$, to avoid aliasing of sidelobe clutter into the target filter. Thus, airborne radars operating at X-band must typically use PRFs in excess of 100 kHz.

An example of low-PRF pulsed doppler processing is the *moving target detector* (MTD), shown in Fig. 23.1.39. Designed for application to conventional airport surveillance radars, this processor accepts digitized baseband signals from in-phase and quadrature phase detectors. The radar transmits 24 pulses per beam position, in two 12-pulse

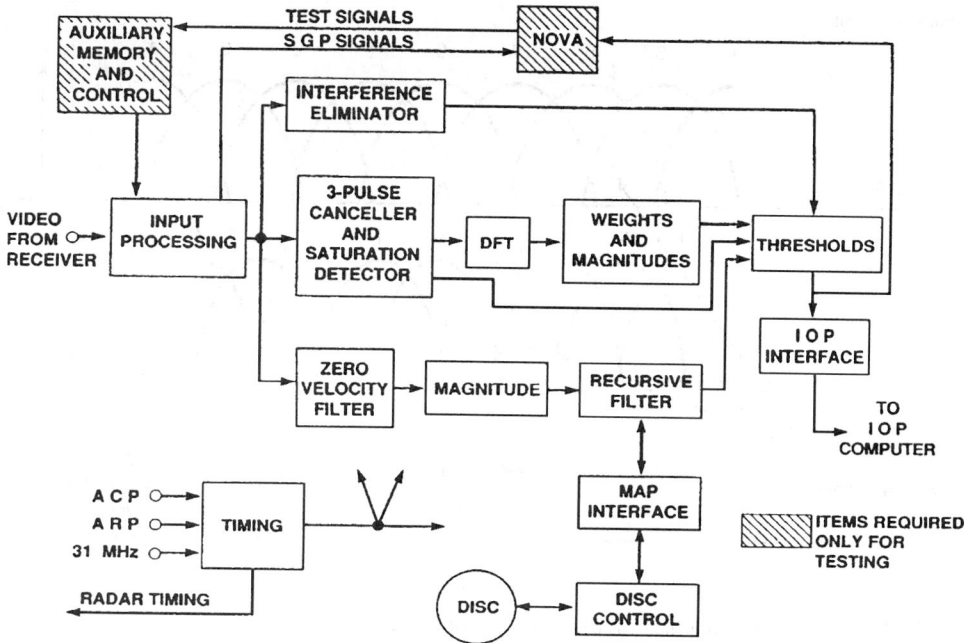

FIGURE 23.1.39 Block diagram of moving target detector system (Ref. 25).

coherent processing intervals (CPIs). In each CPI, after two fill pulses, a 10-pulse burst is applied to the triple-delay cancelers, producing an eight-pulse burst for subsequent doppler filtering in the discrete Fourier transform. A parallel zero-velocity channel feeds a clutter map through a recursive filter, averaging over several scans of the antenna. The eight filter channels are passed to range-cell averaging CFAR thresholds that desensitize those filters having range-extensive clutter.

Blind speeds can be eliminated by the burst-to-burst PRF diversity used in the MTD, provided that the clutter spectrum is not excessively wide. The two sets of filter response curves are shown in Fig. 23.1.40, along with a typical rain clutter spectrum. In this example, the rain clutter will desensitize filters 6, 7, and 0 at both PRFs. The target aliased into filters 6 and 7 at one PRF, appears in filter 5 at the other PRF, ensuring detection. If the clutter spectrum expands to cover more than about 20 percent of the blind speed interval, loss of targets may be expected.

An airborne high-PRF doppler processor is shown in Fig. 23.1.41. The mainlobe clutter in this example appears just below the platform velocity, corresponding to a beam position displaced from the velocity vector. Sidelobe clutter extends from $+v_p$ to $-v_p$, with enhancement at zero velocity where the clutter is viewed at vertical incidence (the *altitude line*). To reduce the dynamic range of digital processing, analog filters pass only velocities between v_p and $v_b - v_p$. Within the passband of the analog filters, many doppler filters are formed, usually through FFT processing.

Because the high-PRF radar normally has only a few range cells within the pulse repetition interval, CFAR will be based on doppler filter averaging. If a medium-PRF mode is included in the airborne radar, a combination of doppler and range cell averaging may be used.

The requirement for fill pulses is the same in pulsed doppler radar as in MTI, to ensure that the clutter input has reached steady state before coherent processing begins. However, as PRF increases into the medium- or high-PRF region, the number of such pulses may become very large. The basic requirements is that the time during which the initial transient must be gated out of the processor, after each change in PRF or beam position, is the delay time for the most distant significant clutter sources.

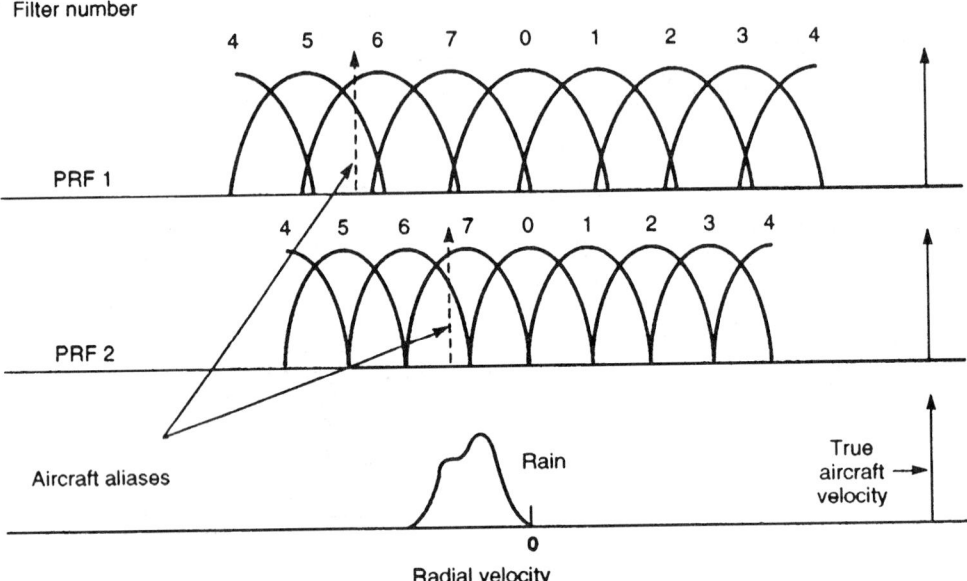

FIGURE 23.1.40 MTD filter responses and clutter spectrum (Ref. 25).

CW Radar

A cw transmissions has no velocity ambiguity, and so cw radar equipment can be designed with a broad clutter notch, providing up to 130 dB rejection of fixed and moving clutter. Coherent integration of target signals in selected doppler bands is also provided by a single set of narrowband filters, rather than the multiple sets required

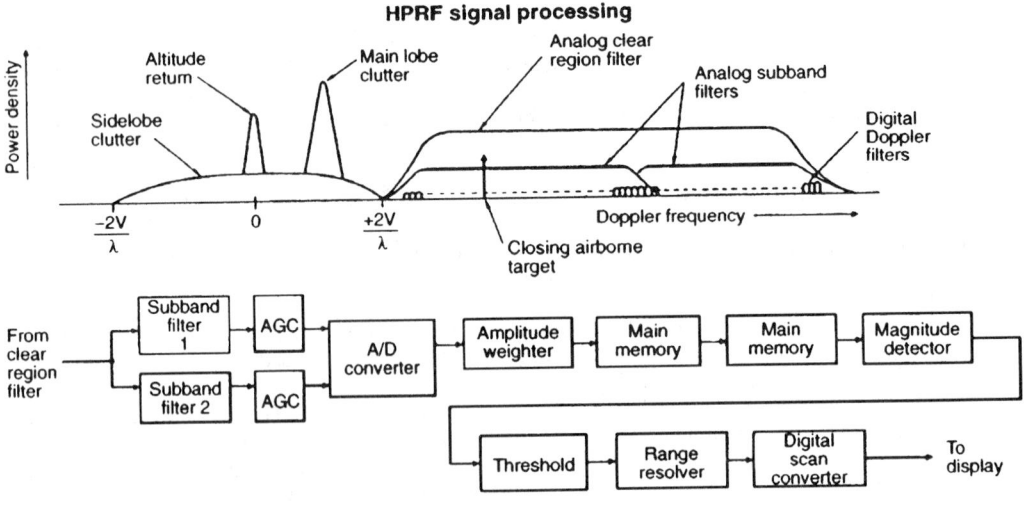

FIGURE 23.1.41 High-PRF airborne radar spectrum and processor (Ref. 26).

for range-gated operation in pulsed systems. Three problems, however, restrict the usefulness of cw radar for search:

1. *Isolation of receiver from transmitter.* Direct feed-through of transmitter power to the receiver must be minimized, requiring separate antennas in high-power systems and careful design in all systems to avoid receiver saturation.

2. *Magnitude of short-range clutter echo.* The echo power received from clutter in a cw radar is proportional to the integrated product of the reflectivity, antenna gain factors, and (range)$^{-4}$ over the common volume of the transmitting and receiving beams. In both volume and surface clutter, the echo power is dominated by the clutter at the shortest range in the common volume, and the effective clutter cross section is a function of beamwidth and range R_c to the point where the beams substantially overlap. The required clutter improvement for a cw radar may therefore be very high because of the $(R/R_c)^4$ term, where R is target range.

3. *Transmitter noise.* Both the direct feed-through from transmitter to receiver and the echoes from short-range clutter will contain random noise components from the transmitter. Special circuits may be designed to cancel the direct feed-through and low-frequency components of reflected noise, but the higher-frequency components will appear with phase shift from the range delay and cannot be canceled completely. Subclutter visibility in cw systems is generally controlled by these noise components.

TRACKING AND MEASUREMENT

Detection of a target within a given resolution cell implies at least coarse measurement of target coordinates, to the value describing the center of the cell in each coordinate. However, measurements can be made with much greater accuracy by interpolating target position within the resolution cell. This process is normally carried out in both search and tracking radars, with emphasis on high accuracy in the tracking case.

The Basic Process of Measurement

The basic process by which target position is interpolated within a resolution cell involves the formation of two offset response channels in the coordinate of interest (Fig. 23.1.42). The difference Δ and sum Σ of these two channels are formed, and the ratio Δ/Σ provides a measure of the target displacement from the equisignal axis.

The two channels may be formed either simultaneously or sequentially. Simultaneous channel formation is usual in range and doppler measurement, as well as in monopulse angle measurement. Sequential channel formation is often used for angle measurement, where a second antenna beam and receiving channel would be relatively expensive. The two outputs are stored over one or more switching cycles, permitting the Δ and Σ channels to be generated (Fig. 23.1.43a). A convenient implementation of the ratio process is to apply slow automatic gain control (AGC) to the common receiving channel (Fig. 23.1.43b). In the case of simultaneous channel formation (Fig. 23.1.43c), the AGC operates as a closed loop in the Σ channel, and open-loop in the Δ channel. Whichever method is used, the normalized output ratio Δ/Σ can be calibrated in terms of target position and used either to correct the output data relative to the axis position or to close a tracking loop.

Measurement Accuracy

The rms error in measurement σ_z, caused by thermal noise, can be expressed in terms of the half-power width z_3 of the resolution cell, the slope constant k_z for the measurement system, and the signal-to-noise energy ratio E/N_0 in the Σ channel of the system:

$$\sigma_\theta = \frac{\theta_3}{k_m \sqrt{2E/N_0}} \tag{68}$$

When other types of interference are present, the ratio E/N_0 may be replaced by $E/I_{0\Delta}$, where $I_{0\Delta}$ is the spectral density of interference in the Δ channel.

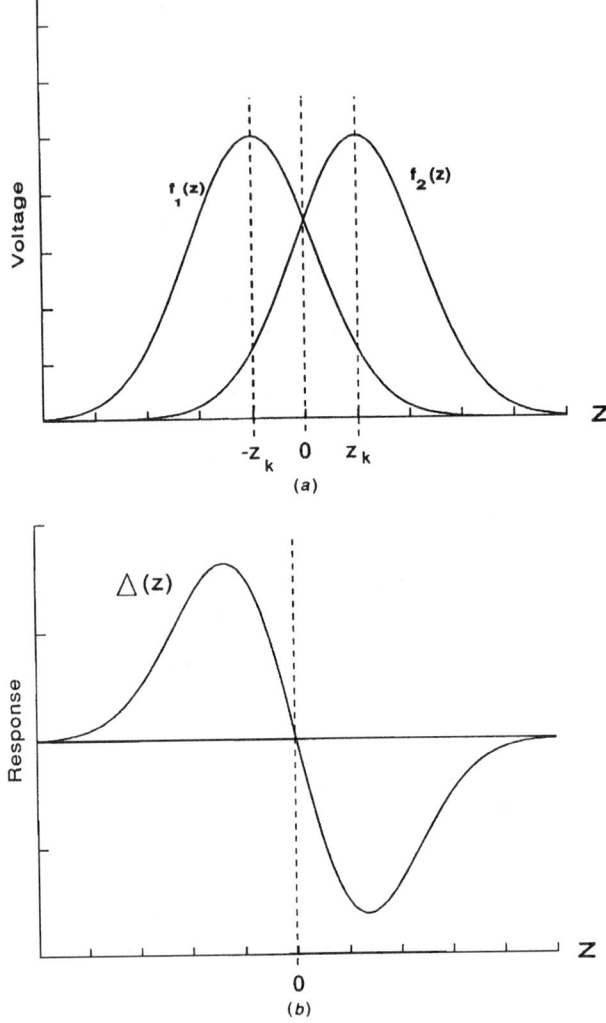

FIGURE 23.1.42 Basic process of measurement: (*a*) two displaced response channels in coordinate *z*; (*b*) difference Δ of two channels; (*c*) sum Σ of two channels; (*d*) normalized ratio Δ/Σ. (*Continued*)

Ideal estimators can be characterized[8] in terms of their slope constants:

Angle measurement: $k_m = L\theta_3/\lambda = 1.61$ (for uniform illumination of aperture)

Time delay measurement: $k_t = \beta\tau_3 = 1.61$ (for rectangular spectrum)

Frequency measurement: $k_f = \alpha B_3 = 1.61$ (for rectangular pulse)

Coherent frequency measurement: $k_f = \alpha_f B_{3f} = 1.61$

Here L = rms aperture with = $(\pi/\sqrt{3})w$ for rectangular aperture of width w, β = rms spectral width = $(\pi/\sqrt{3})B$ for rectangular spectrum of width B, α = rms time duration = $(\pi/\sqrt{3})\tau$ for rectangular pulse of width τ and α_f = rms time duration = $(\pi/\sqrt{3})t_f$ for coherent pulse train received with uniform amplitude over time t_f. The half-power resolution cell widths are θ_3 in angle, τ_3 in time, and B_3 or B_{3f} in frequency.

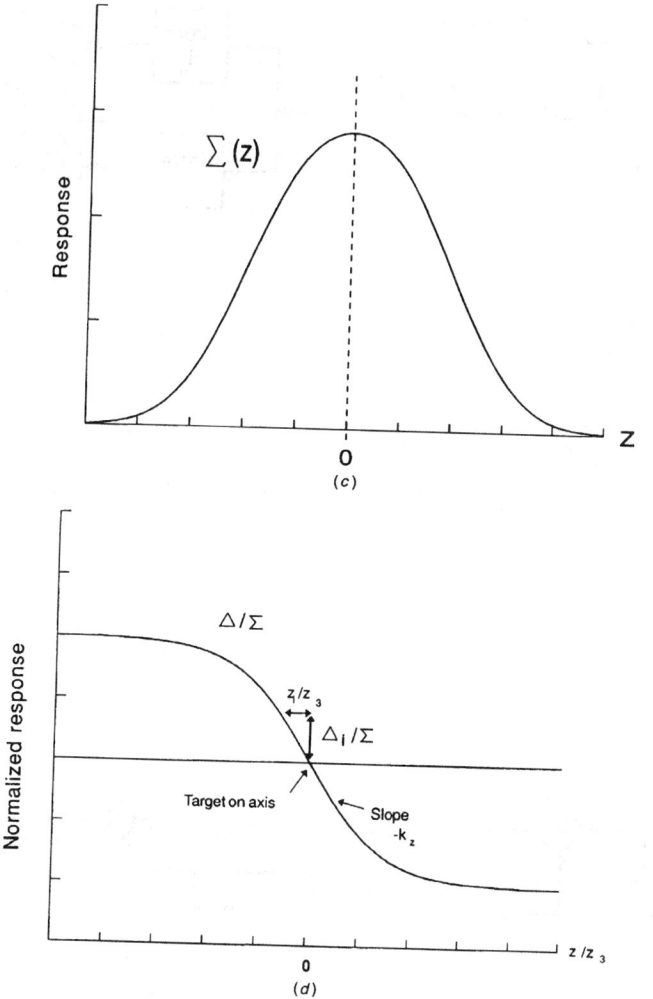

FIGURE 23.1.42   (*Continued*)

The ideal estimators are implemented using a Σ channel derived from uniform aperture illumination, uniform spectral weighting, or uniform weighting in time, giving $(\sin x)/x$ response (sidelobe levels -13.5 dB). This is combined with a Δ channel derived from linear-odd illumination or weighting. The linear-odd function generates a response, which is the derivative of the Σ response (sidelobe levels -11 dB relative to Σ mainlobe).

Practical Estimators.   In most practical cases, the need for sidelobe control dictates use of tapered or weighted functions for aperture illumination, signal spectrum, and pulse train amplitude. The weighting tends to increase the half-power width of the resolution cell and decrease the rms width in the transform coordinate, leaving the slope constant near 1.6.

In a tracker that observes the target continuously, the received signal energy E will increase without limit. For purposes of evaluating tracking error, it must be assumed that the tracker averages over a number of pulses $n = f_r t_o = f_r / 2\beta_n$, where t_o is the time constant of the tracking loop and β_n is its (one-sided) noise bandwidth. For multiple-target phased arrays, t_o is the dwell time on the target of interest.

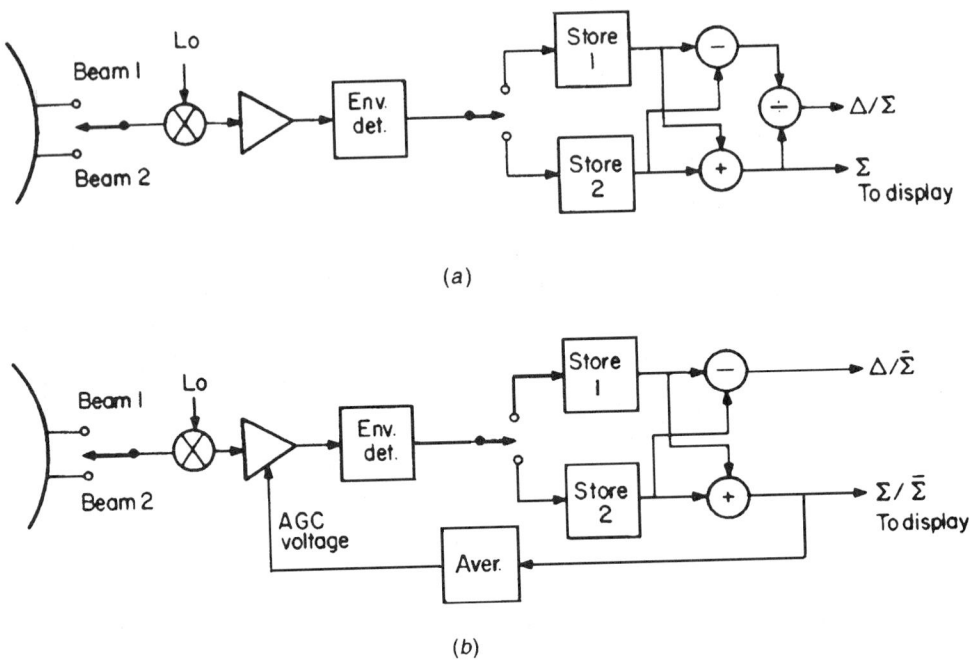

(a)

(b)

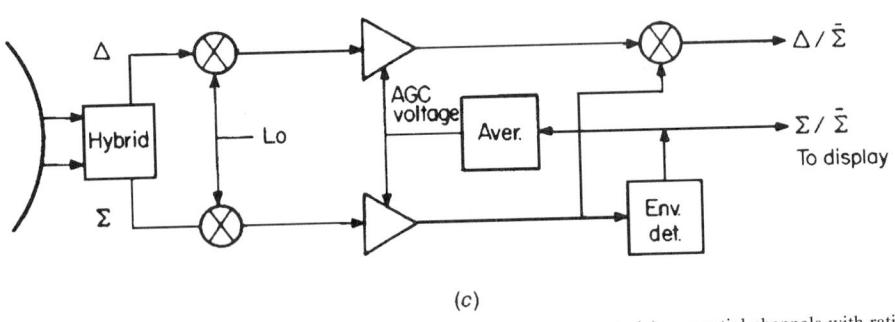

(c)

FIGURE 23.1.43 Sequential and simultaneous channels for measurement: (*a*) sequential channels with ratio circuit; (*b*) sequential channels with AGC; (*c*) simultaneous channels with AGC.

Angle Tracking

In a tracking radar, the antenna pattern is usually a narrow, circular beam, directed at the selected target either continuously, with a mechanically steered antenna, or with a time-shared phased array beam. The electro-mechanical or electronic servo loop is controlled to minimize the angle errors, as measured by an error-sensing antenna and receiver system. In early tracking radars, the beam was scanned about the tracking axis in a narrow cone, producing amplitude modulation on signals from targets displaced from the axis. The conical-scanning radar has been largely replaced by monopulse radar, which forms the sum and difference beams simultaneous-ly. The normalized error is formed on a single pulse (or a few pulses received within the AGC time constant),

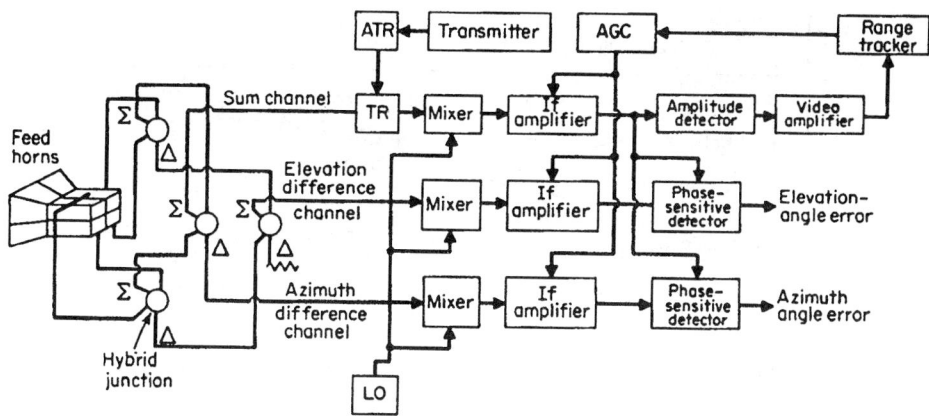

FIGURE 23.1.44   Block diagram of conventional monopulse tracking radar (Ref. 6).

eliminating errors caused in conical scanning radar by target amplitude scintillation. A block diagram of a typical monopulse tracker is shown in Fig. 23.1.44.

The conventional system illustrated uses three identical receiver channels to process the Σ and two Δ signals. The Σ channel is used for transmitting, and on receiving its output provides target detection, ranging, and AGC. It also serves as a phase reference for detection of the Δ signals. These detected error signals, appearing as bipolar video pulses at the phase detector output, are smoothed to DC and applied as inputs to the angle servo channels, causing the antenna to follow the target in azimuth and elevation.

Monopulse Antennas.   The patterns required from a monopulse antenna consist of the Σ pattern, generally a circular pencil beam with sidelobes controlled by tapering of the aperture illumination, and a pair of Δ patterns. A typical Δ pattern may approximate that of Fig. 23.1.45a, calculated for a cosine illumination taper. The azimuth Δ pattern will have two azimuth lobes of opposite polarity ($0°$ and $180°$ phase), approximating that of

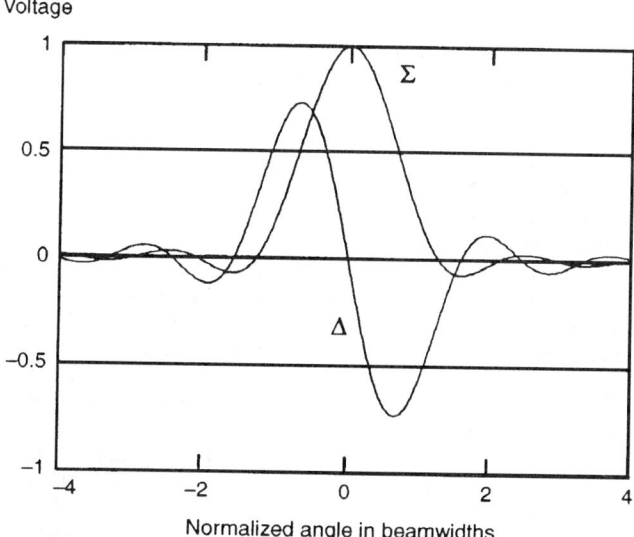

FIGURE 23.1.45   Typical monopulse antenna patterns, using cosine taper.

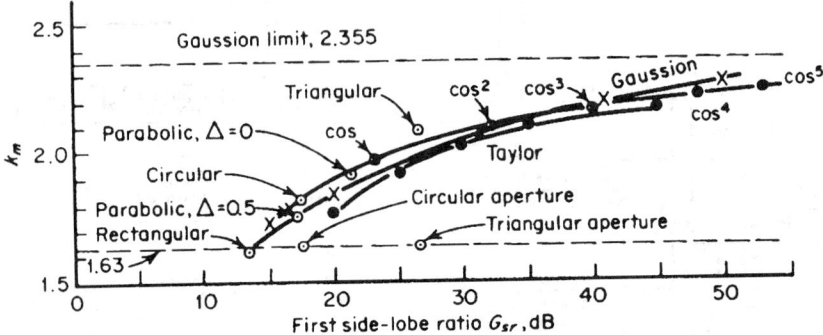

FIGURE 23.1.46 Normalized measurement slope versus sidelobe ratio for monopulse antenna in which Δ pattern is derivative of Σ pattern.

Fig. 23.1.45b. This pattern was calculated for an illumination given by a linear-odd function multiplied by the cosine taper. The azimuth Δ pattern in the elevation coordinate will reproduce the Σ elevation pattern. The elevation Δ pattern will have the two lobes in elevation, and will reproduce the Σ pattern in azimuth.

For an antenna with a Δ pattern following the derivative of the Σ pattern, the beam broadening is more rapid than the decrease in slope, and the normalized slope k_m actually increases as sidelobe levels are reduced (Fig. 23.1.46). In most practical antennas, however, departure from the derivative Δ pattern and various losses restrict the normalized slope to values near 1.6.

The rms error of the angle estimate in the presence of thermal noise will be given by

$$\sigma_\theta = \frac{\theta_3}{k_m\sqrt{2E/N_0}} \tag{69}$$

where θ_3 is the half-power beamwidth of the Σ pattern, k_m is the normalized slope constant ≈ 1.6, and E/N_0 is assumed large. This slope is the derivative of the normalized Δ pattern, as shown in Fig. 23.1.45:

$$k_m \equiv \left[\frac{d(\Delta/\Sigma)}{d(\theta/\theta_3)} \right]_{\theta=0} \tag{70}$$

When the Σ signal-to-noise ratio presented to the phase-sensitive detectors of Fig. 23.1.44 is not high ($S/N < 4$), estimation error will be increased by detector loss, and monopulse tracking performance will be degraded. Detailed analysis of this effect is given in Ref. 3, pp. 467–472.

When a reflector or lens is the radiating aperture, monopulse patterns are generated by a cluster of horns in the focal plane. The early four-horn feed clusters have given way to more advanced horn structures, using additional horns or multiple modes to generate efficient, low-sidelobe illumination functions.[27] Table 23.1.4 shows the performance of five types of horn cluster, in terms of sum-channel aperture efficiency η_a, measurement slopes k_m in both coordinates, and sidelobe ratios G_s (mainlobe-to-sidelobe power) in sum and difference patterns. These values were derived by Hannan[27] for rectangular apertures, but the absolute levels of performance are unchanged when an elliptical aperture having the same maximum dimensions is used. The efficiency will be higher for the elliptical aperture because it is referred to the smaller area of the ellipse.

Monopulse array systems using constrained feeds can be designed to have arbitrary Σ and Δ illumination functions (e.g., Taylor functions for Σ and Bayliss functions[28] for Δ), independently controlled in the feed networks. Such systems will have their efficiencies reduced by feed and phase shifter losses, as well as by the selected taper functions. Phased array systems using horn-fed reflectors or lenses can be described by the parameters of Table 23.1.4, with efficiencies reduced by losses in the phase shifter. Introduction of more complex

TABLE 23.1.4 Monopulse Feed Horn Performance

Type of horn	η_a	H-plane		E-plane		G_{sr}, dB	G_{se}, dB	Feed shape
		$K_r\sqrt{\eta_y}$	K_m	$K_r\sqrt{\eta_x}$	K_m			
Simple four-horn	0.58	0.52	1.2	0.48	1.2	19	10	
Two-horn dual-mode	0.75	0.68	1.6	0.55	1.2	19	10	
Two-horn triple-mode	0.75	0.81	1.6	0.55	1.2	19	10	
Twelve-horn	0.56	0.71	1.7	0.67	1.6	19	19	
Four hour triple-mode	0.75	0.81	1.6	0.75	1.6	19	19	

Source: D. K. Barton, and H. R. Ward, "Handbook of Radar Measurement," copyright 1984. Reprinted by permission of Artech House, Norwood, MA.

horn structures having additional modes can produce illumination functions having low sidelobes and spillover, with gain and slope approximating those of constrained-feed arrays but with higher efficiencies due to decreased feed losses.

Range Tracking

Measurement of target range is carried out by estimating the time delay t_d between transmission and reception of each pulse, and calculating range as

$$R = \frac{t_d}{c} \tag{71}$$

where c is the velocity of light ($c = 2.997925 \times 10^8$ m/s in vacuum, somewhat less in the atmosphere). The time delay is estimated by counting clock pulse between the times of transmission and reception of the centroid of the pulse, and correcting for any fixed delays within the radar components (transmission line, filters, and the like). The accuracy of the measurement is dependent on the accuracy with which the centroid of the received pulse, contaminated by noise and other interference, can be determined.

The ideal estimation process is to identify the centroid of the received pulse by passing the signal through a matched filter, differentiating with respect to time, and stopping the counter when the derivative passes

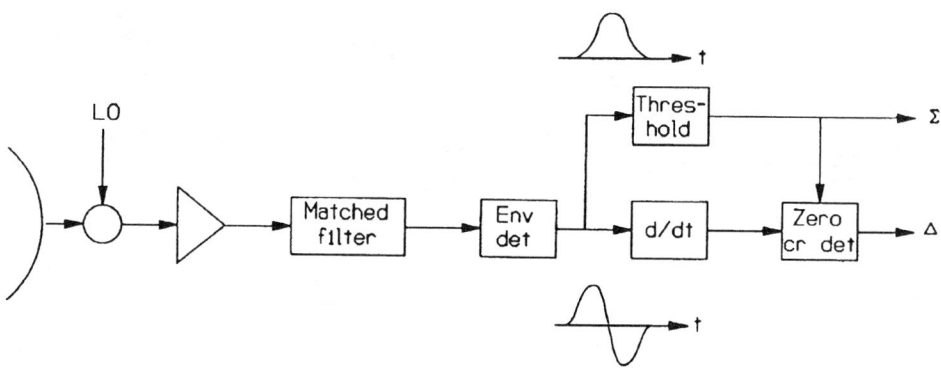

FIGURE 23.1.47   Block diagram of optimum pulse centroid estimator.

through zero. A block diagram of this process is shown in Fig. 23.1.47. The resulting time-delay accuracy is then given by Eq. (68) with parameters appropriate to the time coordinate:

$$\sigma_1 = \frac{1}{\beta\sqrt{2E/N_0}} = \frac{\tau_3}{k_t\sqrt{2E/N_0}} \tag{72}$$

where β = rms signal bandwidth
τ_3 = half-power width of processed pulse
k_t = slope constant ≈ 1.61

In order for the circuit of Fig. 23.1.47 to produce centroid estimates without gross error from false zero crossings on noise, the signal energy ratio E/N_0 must be high enough to avoid the occurrence of false alarms at the Σ threshold prior to reception of the target pulse.

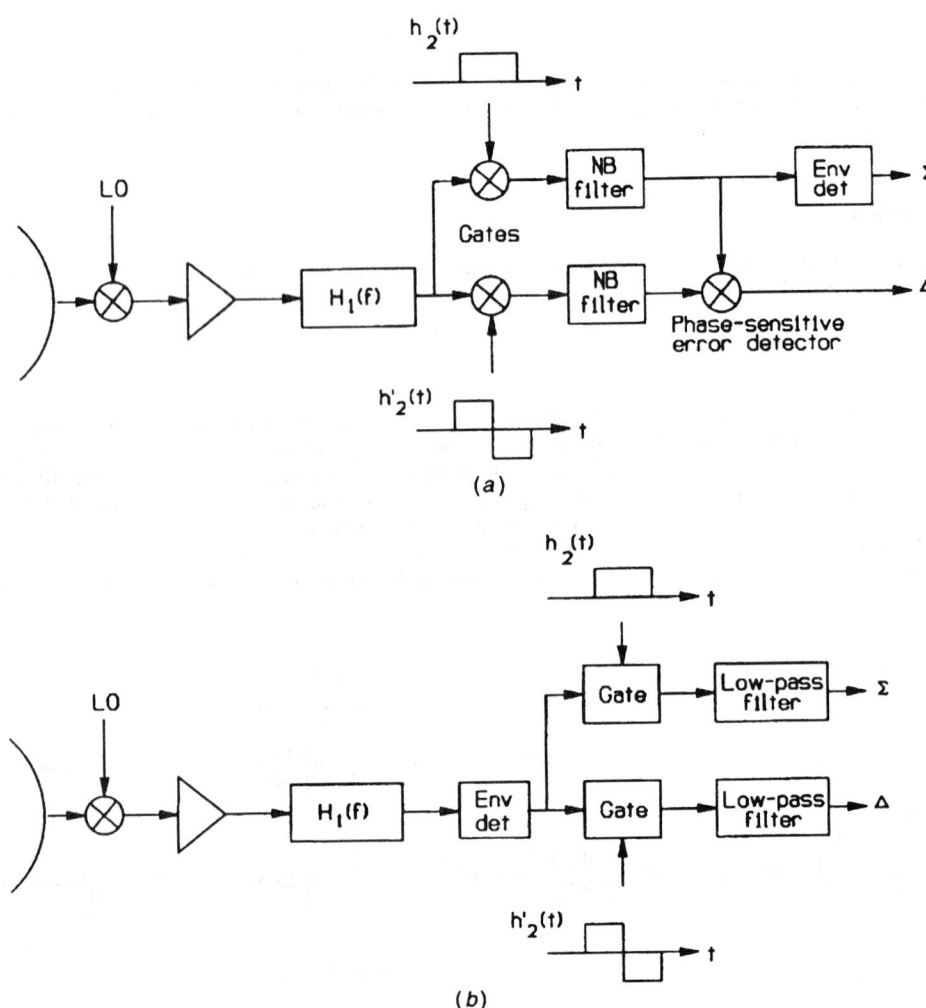

FIGURE 23.1.48 Centroid estimators for pulses in a train: (*a*) coherent processing; (*b*) noncoherent processing.

When the energy ratio of the individual pulse in a pulse train is low, time-delay estimation will normally be carried out using a tracking correlator, as shown in Fig. 23.1.48. Reception of a coherent pulse train permits the correlation to be performed prior to envelope detection (Fig. 23.1.48a). The i.f. filter $H_1(f)$ has a broad bandwidth passing the complete spectrum of the individual pulse, providing pulse compression if that pulse contains phase modulation. The locally generated reference signals labeled $h_2(t)$ and $h_2'(t)$ are approximately matched to the pulse width, and often take the form of rectangular gate and split-gate functions as shown. The Δ output of this system is a DC voltage proportional to the time difference between the center of the gate function and the received pulse centroid. This voltage is then normalized to the Σ amplitude and used as input to the variable time-delay generator that initiates the gates. The noise error in coherent processing is given by Eq. (72), using the energy ratio of the pulse train as integrated in the narrowband filters.

When the pulse train is not coherent, the individual pulses must be passed through envelope detectors prior to narrowband filtering, as shown in Fig. 23.1.48b. The performance of the noncoherent processor, when the energy ratio of the individual pulse is low, is reduced by loss (small-signal suppression) in the envelope detector, in a way similar to that of the angle tracker.

Doppler Tracking

Doppler tracking is used in coherent radars for two purposes: (1) It permits the signal to pass a narrowband filter, providing coherent integration and rejecting clutter; and (2) It provides accurate radial velocity data on the target. A transmission at frequency f_0, reflected from a target moving with radial velocity $v_t = \dot{R}$, will be received at $f_0 + f_d$. The change in frequency f_d is known as the doppler shift:

$$f_d = f_0 \left(\frac{c - v_t}{c + v_t} - 1 \right) = \frac{-2 f_0 v_t}{c} \left(1 - \frac{v_t}{c} + \frac{v_t^2}{c^2} - \cdots \right) \approx \frac{2 f_0 v_t}{c} = -\frac{2 v_t}{\lambda} \tag{73}$$

Target velocity is calculated from the measured doppler shift as

$$v_t = -\frac{f_d c}{2 f_0} \left(1 - \frac{f_d}{2 f_0} + \frac{f_d^2}{4 f_0^4} - \cdots \right) \approx -\frac{f_d c}{2 f_0} = -\frac{f_d \lambda}{2} \tag{74}$$

In most cases, the signal bandwidth is small enough relative to f_0 so that the doppler shift can be regarded as a simple displacement of the spectrum relative to that transmitted, and the target velocity will be small enough relative to c that the approximations in Eqs. (73) and (74) are valid.

The frequency accuracy of an ideal estimator is given by Eq. (68) with parameters appropriate to the frequency coordinate:

$$\sigma_f = \frac{1}{\alpha \sqrt{2E/N_0}} = \frac{B_3}{k_f \sqrt{2E/N_0}} \tag{75}$$

When a coherent pulse train is measured, the applicable bandwidth B_3 in Eq. (75) is the bandwidth of each spectral line shown in Fig. 23.1.9. Figure 23.1.49 shows a block diagram of a range-gated doppler centroid estimator for use with a coherent pulse train. Two narrowband filters, displaced each side of the tracking frequency f_c, are used to form Σ and Δ channels in the frequency coordinate. The Σ i.f. signal is used as a reference to phase detect the Δ signal, providing a DC error value that can be used to control an oscillator shifting the LO to center the signal at the frequency f_c.

Before measurements can be made with the accuracy implied by Eq. (75), the central line of the received spectrum must be placed in the narrowband i.f. filter-discriminator of Fig. 23.1.50. When the waveform duty factor is small, or when pulse compression is used, there will be many spectral lines present ($2B/f_r$ lines within

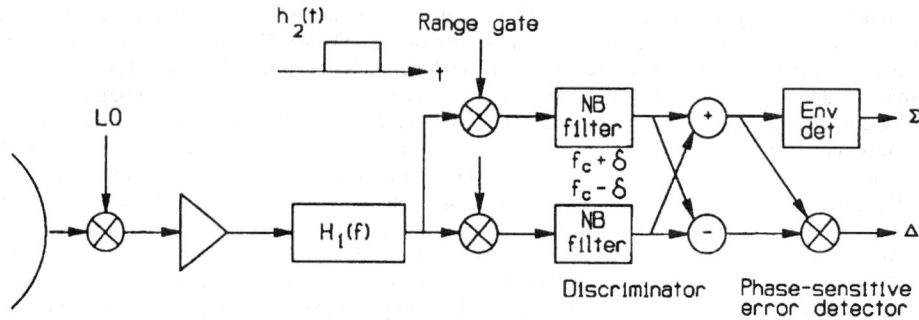

FIGURE 23.1.49 Block diagram of frequency centroid estimator for use with coherent pulse train.

the main lobe of the spectrum, for a signal bandwidth B at repetition frequency f_r). The frequency ambiguity can be resolved in one of several ways:

1. A discriminator operating at i.f. on the spectral envelope may obtain coarse frequency data adequate to place the central line in the fine discriminator bandwidth, if duty factor is not too small;
2. For systems with adequate range resolution, range data may be differentiated to obtain an estimate of v_t and hence of f_d;
3. Observation of the target at two or more PRFs may resolve the ambiguity.

Search-Radar Measurements

As the search radar beams scans (usually in azimuth), measurements can be made of target range and angle (and sometimes of radial velocity). The equations given above for tracking radar can be applied to the range and velocity measurements, using the energy ratio E/N_0L_p obtained during the scan across the target. Measurements of azimuth angle are made by estimating the time (and hence the antenna angle) of the centroid of the pulse train envelope received by the radar. This time measurement may be performed with a split-gate tracker similar to that used in ranging, but operating on a time-scale matched to observation time t_o rather than pulse width. The target to be measured is selected by a range gate or equivalent channel of the detection processor. Alternatively, the signal may be integrated, differentiated, and the centroid estimated as the point where the derivative passes through zero.

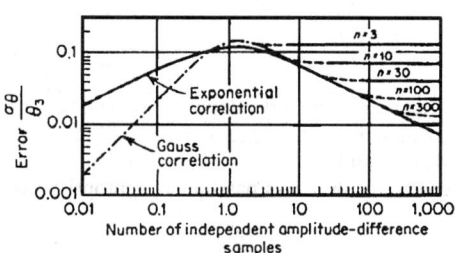

FIGURE 23.1.50 Scintillation error in a scanning radar.[5]

The azimuth angle error due to thermal noise will be

$$\sigma_\theta = \frac{\theta_3}{2\sqrt{E/N_0}} \tag{76}$$

where the constant 2 in the denominator includes the slope constant, the doubling of energy ratio normally appearing in error equations, and the beamshape loss effect. This factor is essentially constant for all beam shapes and for one-way and two-way beam patterns,[8] where θ_3 is the one-way, half-power beamwidth.

Thermal noise and other interference are not the only factors affecting search radar angle accuracy. Fluctuating targets will induce a *scintillation error* in the estimate, independent of SNR. This error depends on the relationship between the target amplitude correlation time t_c and the observation time t_o of the beam. The number of independent samples obtained is

$$n_e = 1 + \frac{t_o}{t_c} \tag{77}$$

Figure 23.1.50 shows the dependence of scintillation error on the quantity $(n_e - 1) = t_o/t_c$. When this ratio is between 0.1 and 10, the error will exceed $0.05\theta_3$.

Error Analysis

The total error in estimating a target coordinate is found by evaluating each of several error components, and forming the rss sum

$$\sigma = \sqrt{\sigma_1^2 + \sigma_2^2 + \cdots} \tag{78}$$

The individual error components include the thermal noise error from Eqs. (68) to (76), errors from other interference components, the scintillation error for a scanning radar, and other components due to dynamic lags in tracking, target glint, atmospheric refraction, multipath reflections, and imperfections in the radar components and circuits.

Random Interference Components.   In addition to thermal noise, the following interference components will cause random errors, which are estimated by substitution of E/I_Δ for E/N_0 in Eqs. (68) to (76):

1. Clutter, for which I_Δ is estimated from the power spectral density of clutter within $\pm\beta_n$ of the target spectral lines; additional error may result when the clutter in the Σ and Δ channels is correlated (Ref. 3, pp. 531–533);

2. Jamming, in the case of stand-off and escort jammers, using the jamming spectral density J_Δ; in the case of a self-screening noise jammer, the jamming constitutes a beacon signal and J_0/N_0 is the energy ratio;

3. Multipath reflections, for which the energy ratio can be equated with the power ratio S/M_Δ, determined by integration of the reflected power over the Δ pattern (Ref. 2, pp. 512–531);

4. Cross-polarized signal, using the ratio $(\sigma/\sigma_c)(\Delta_c/\Sigma)^2$, where σ_c is the cross-polarized target cross section and Δ_c is the cross-polarized difference channel voltage on the tracking axis (Ref. 3, pp. 415–416).

Target-Induced Error Components.   The error components induced by target characteristics are glint, dynamic lag, and scintillation errors (the latter in angle estimates of scanning radars only, as discussed above).

Target glint results from interaction of the several scatterers of the target, changing the phase pattern of the echo in space and time. Where the scatterers are distributed uniformly over a target span L, the rms error will be approximately $L/3$ in that coordinate (producing an angle error $L/3R$ radians).

Dynamic lag results primarily from target accelerations, which can be followed only imperfectly by the tracking loops. The error for an acceleration a_t in any coordinate will be

$$\epsilon_a = \frac{a_t}{K_a} = \frac{a_t}{2.5\beta_n^2} \tag{79}$$

where K_a is the acceleration error constant of the tracking loop and β_n is (one-sided) loop bandwidth. It is the requirement for adequate bandwidth to follow the accelerations that prevents the tracking radar from operating with heavy smoothing of random errors.

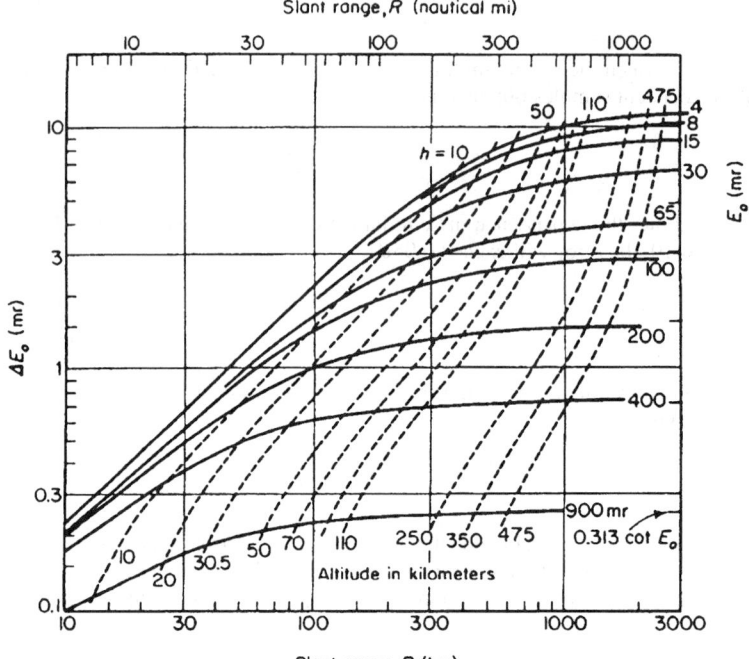

FIGURE 23.1.51 Elevation angle error versus target range, for different elevation angles and altitudes.

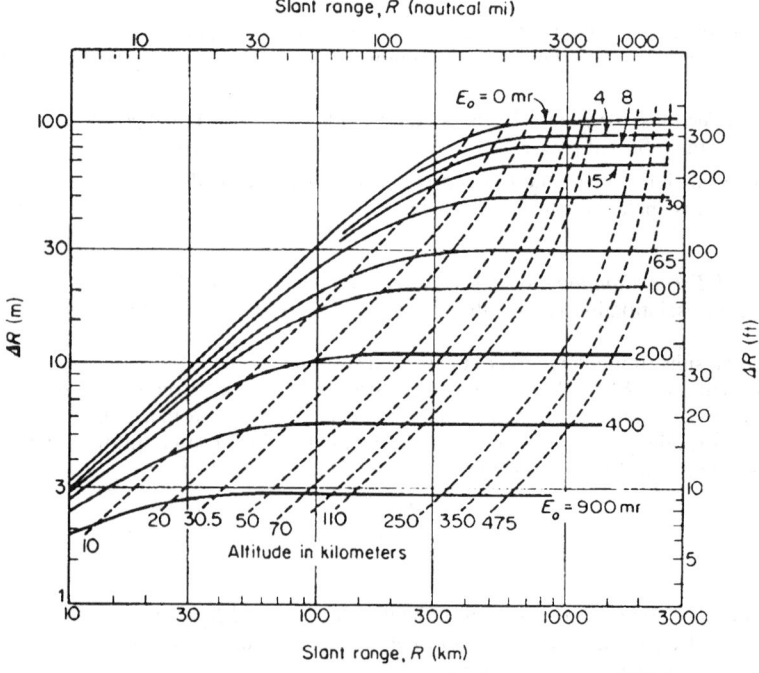

FIGURE 23.1.52 Range error versus target range, for different elevation angles and altitudes.

Atmospheric Refraction.   Major errors in elevation angle and range result from the change in refractive index of the atmosphere with altitude. The beam tends to bend downwards as it leaves the earth, causing the angle of arrival measured by the radar to be higher than the actual target elevation. The reduced velocity of propagation in the atmosphere also makes the time delay somewhat greater than would be measured in a vacuum. These bias errors in elevation and range, for a radar at sea level, are plotted in Figs. 23.1.51 and 23.1.52 as a function of target range R and elevation angle E_o.

Other Sources of Error.   The remaining error components that must be evaluated in order to estimate radar accuracy consist of instrumental errors in the design and construction of the radar. These are both mechanical and electrical in origin, and must be evaluated for each particular radar using detailed data on design parameter and construction or circuit tolerances.

REFERENCES

Reference Books on Radar

1. Skolnik, M. I. "Introduction to Radar Systems," 2nd ed., McGraw-Hill, 1980.

2. Barton, D. K. "Modern Radar System Analysis," Artech, 1988.

3. Nathanson, F. E. "Radar Design Principles: Signal Design and the Environment," 2nd ed., McGraw-Hill, 1991.

4. Rihaczek, A. "Principles of High Resolution Radar," McGraw-Hill, 1969.

5. Barton, D. K., and H. R. Ward "Handbook of Radar Measurement," Artech, 1984.

6. Skolnik, M. I. (ed). "Radar Handbook," 2nd ed. McGraw-Hill, 1990.

Other References on Radar Principles

7. IEEE Standard Dictionary of Electrical and Electronics Terms, ANSI/IEEE Std. 100–1988, Institute of Electrical and Electronics Engineers, 1988.

8. Blake, L. V. Prediction of Radar Range, Chap. 2 in Ref. 2.

9. Marcum, J. I. A Statistical Theory of Target Detection by Pulsed Radar, *IRE Trans.,* April 1960, Vol. IT-6, No. 2, pp. 59–267.

10. Barton, D. K. Simple Procedures for Radar Detection Calculations, *IEEE Trans.,* September 1969, Vol. AES-5, No. 5, pp. 837–846.

11. Swerling, P. Probability of Detection for Fluctuating Targets, *IRE Trans.,* April 1960, Vol. IT-6, No. 2, pp. 145–268.

12. Crane, R. K. Microwave Scattering Parameters for New England Rain, *MIT Lincoln Laboratory Tech. Report No. 426,* October 3, 1966.

13. Goldstein, H. Attenuation by Condensed Water, Sec. 8.6 in D. E. Kerr (ed.), "Propagation of Short Radio Waves," McGraw-Hill, 1951.

14. Gunn, K. L. S., and T. W. R. East The Microwave Properties of Precipitation Particles, *Q. J. R. Meteorolog. Soc.* October 1954, Vol. 80, pp. 522–545.

15. Ryde, J. W., and D. Ryde Attenuation of Centimeter and Millimeter Waves by Rain, Hail, Fogs and Clouds, *General Electric Co., Report 8670,* 1945.

16. Wexler, R., and D. Atlas Radar Reflectivity and Attenuation of Rain, *J. Appl. Meteorol.,* April 1963, Vol. 2, pp. 276–280.

17. Marshall J. S., and W. M. Palmer The Distribution of Raindrops with Size, *J. Meteorol.,* August 1948, Vol. 5, pp. 165–166.

18. Medhurst, R. G. Rainfall Attenuation of Centimeter Waves: Comparison of Theory and Measurement, *IEEE Trans.,* July 1965, Vol. AP-13, No. 4, pp. 550–564.

19. Blevis, B. C. Losses Due to Rain on Radomes and Antenna Reflecting Surfaces, *IEEE Trans.,* January 1965, Vol. AP-13, No. 1, pp. 175–176.

20. Ruze, J. More on Wet Randomes, *IEEE Trans.* September 1965, Vol. AP-13, No. 5, pp. 823–824.

21. Millman, G. H. Atmospheric Effects on VHF and UHF Propagation, *Proc. IRE,* August 1958, Vol. 46, No. 8, pp. 1492–1501.

22. Rihaczek, A. W. "Principles of High-Resolution Radar," McGraw-Hill, 1969.

23. Gregers-Hansen, V. Constant False Alarm Rate Processing in Search Radars, *Radar-73, Radar—Present and Future, IEEE Publ. No. 105*, October 1973, pp. 325–332.

24. Shrader, W. W. MTI Radar, Chap. 15 in Ref. 9.

25. Cartledge, L., and R. M. O'Donnell. Description and Performance Evaluation of the Moving Target Detector, *MIT Lincoln Laboratory Project Report ATC-69*, March 8, 1977.

26. Schleher, D. C. "MTI and Pulsed Doppler Radar," Artech House, 1991.

27. Hannan, P. W. Optimum Feeds for All Three Modes of a Monopulse Antenna, *IRE Trans.,* September 1961, Vol. AP-9, No. 5, pp. 444–461.

28. Bayliss, E. T. Design of Monopulse Antenna Difference Patterns with Low Side-lobes, *Bell System Tech. J.*, May–June 1968, Vol. 47, No. 5, pp. 623–650.

CHAPTER 23.2
RADAR TECHNOLOGY

Harold R. Ward

Radar development, since its beginning during World War II, has been paced by component technology. As better rf tubes and solid-state devices have been developed, radar technology has advanced in all its applications. This subsection presents a brief overview of radar technology. It emphasizes the components used in radar systems that have been developed specifically for the radar application. Since there are far too many devices for us to mention, we have selected only the most fundamental to illustrate our discussion.

In this subsection, the sequence in which each subsystem is discussed parallels the block diagram of a radar system. Pictures of various radar components give an appreciation for their physical size, while block diagrams and tabular data describe their characteristics. The material for this subsection was taken by permission largely from Ref. 1, to which we refer the reader for more detail and references.

RADAR TRANSMITTERS

The requirements of radar transmitters have led to the development of a technology quite different from that of communication systems. Pulse radar transmitters must generate very high power, pulsed with a relatively low duty ratio.

The recent development of high-power rf transistors capable of producing a few hundred watts of peak output power at S band has made solid-state radar transmitters feasible. See Ref. 2a. A corporate combiner is needed to sum the outputs of many devices to obtain the power levels required of medium- and long-range search radars. Such solid-state transmitters offer the following advantages over tube transmitters: low-voltage operation (typically 36 V), no modulator required, and reliable operation through the redundant architecture. While the solid-state transmitter still costs more than its tube equivalent, the solid-state technology is developing more rapidly.

Tube-type power oscillators and power-amplifier stages consist of three basic components: a power supply, a modulator, and a tube. The power supply converts the line voltage into dc voltage of from a few hundred to a few thousand volts. The modulator supplies power to the tube during the time the rf pulse is being generated. Although the modulation function can be applied in many different ways, it must be designed to avoid wasting power in the time between pulses. The third component, the rf tube, converts the dc voltage and current to rf power. The devices and techniques used in the three transmitter components are discussed in the following paragraphs.

RF Tubes

The tubes used in radar transmitters are classified as *crossed-field, linear-beam,* or *gridded* (see Sec. 7). The crossed-field and linear-beam tubes are of primary interest because they are capable of higher peak

TABLE 23.2.1 Comparison of Modulators

Modulator	Fig.	Flexibility		Pulse-length capability		Pulse flatness	Crowbar required		Modulator voltage level
		Duty cycle	Mixed pulse lengths	Long	Short		Load arc	Switch arc	
Line-type: Thyratron/SCR	25–50	Limited by charging circuit	No	Large PFN	Good	Ripples	No		Medium/Low
Magnetic modulator	25–52	Limited by reset and charging time	No	Large C's and PFN	Good	Ripples	No		Low
Hybrid SCR-magnetic modulator	···	Limited by reset and charging time	No	Large C's and PFN	Good	Ripples	No		Low
Active switch: Series switch	25–51a	No limit	Yes	Excellent; large capacitor bank	Good	Good	Maybe	Yes	High
Capacitor-coupled	25–51b	Limited	Yes	Large coupling capacitor	Good	Good	Maybe	Yes	High
Transformer-coupled	25–51c	Limited	Yes	Difficult; XF gets big; large capacitor bank	Good	Fair	Maybe	Yes	Medium-high
Modulator anode	···	No limit	Yes	Excellent; large capacitor bank	OK, but efficiency low[*]	Excellent	Yes	Yes	High
Grid	···	No limit	Yes	Excellent; large capacitor bank	Excellent	Excellent	Yes	···	Low

[*]Unless ON and OFF tubes carry very high peak current or unless modulator anode has high mu. After Weil, Ref. 2.

powers at microwave frequencies. Gridded tubes such as triodes and tetrodes are sometimes used at UHF and below. Since these applications are relatively few, gridded tubes will not be described here (see Sec. 7).

Modulators

If a pulsed radar transmitter is to obtain high efficiency, the current in the output tube must be turned off between pulses. The modulator performs this function by acting as a switch, usually in series with the anode current path. Some rf tubes have control electrodes that can also be used to provide the modulation function. There are three kinds of modulators in common use today: the line-type modulator, magnetic modulator, and active-switch modulator. Their characteristics are compared in Table 23.2.1.

The *line-type modulator* is the most common and is often used to pulse a magnetron transmitter. A typical circuit including the high-voltage power supply and magnetron is shown in Fig. 23.2.1. Between pulses, the pulse-forming network (PFN) is charged. A trigger fires the thyratron V_1, shorting the input to the PFN, which causes a voltage pulse to appear at the transformer T_1. The PFN is designed to produce a rectangular pulse at the magnetron cathode, with the proper voltage and current to cause the magnetron to oscillate. The line-type modulator is relatively simple but has an inflexible pulse width.

Active-switch modulators are capable of varying their pulse width within the limitation of the energy stored in the high-voltage power supply. A variety of active-switch cathode pulse modulators is shown in Fig. 23.2.2.

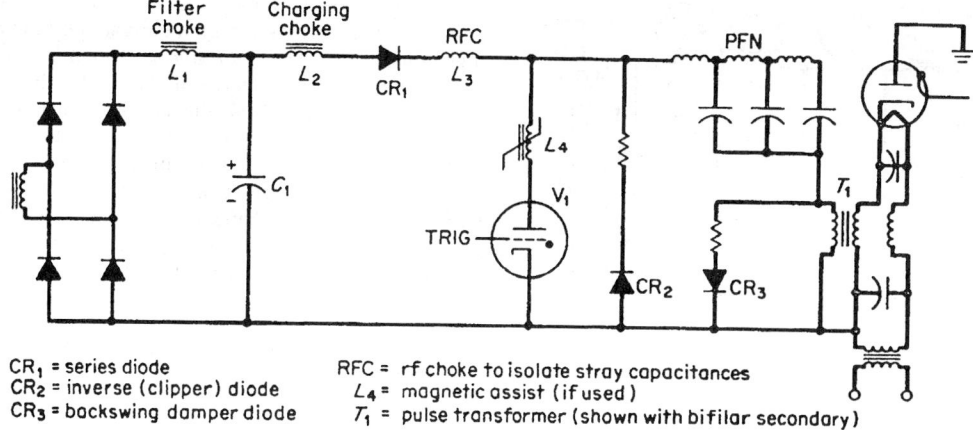

CR$_1$ = series diode
CR$_2$ = inverse (clipper) diode
CR$_3$ = backswing damper diode

RFC = rf choke to isolate stray capacitances
L_4 = magnetic assist (if used)
T_1 = pulse transformer (shown with bifilar secondary)

FIGURE 23.2.1   Line-type modulator.[12]

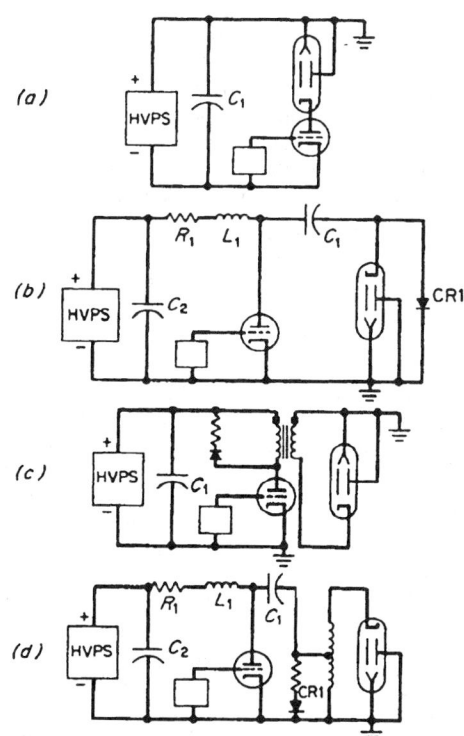

FIGURE 23.2.2   Active-switch cathode pulsers: (*a*) direct-coupled; (*b*) capacitor coupled; (*c*) transformer coupled; (*d*) capacitor- and transformer-coupled.[2]

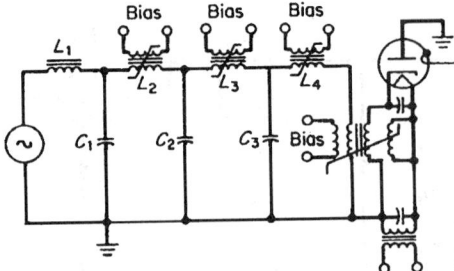

FIGURE 23.2.3   Magnetic modulator.[2]

Active-switch modulators using a vacuum tube free of gas but capable of passing high current and holding off high voltage are called *hard-tube modulators*.

The *magnetic modulator*, a third type of cathode pulse modulator (Fig. 23.2.3) has the advantage that no thyratron or switching device is required. Its operation is based on the saturation characteristics of inductors L_2, L_3, and L_4. A long, low-amplitude pulse is applied to L_1 to charge C_1. When C_1 is nearly charged, L_2 saturates, and the energy in C_1 is transferred resonantly to C_2. The process is continued to the next stage, where the transfer time is about one-tenth that of the stage before. The energy in the pulse is nearly maintained so that at the end of the chain a short-duration high-amplitude pulse is generated.

Power Supplies and Regulators

The power supply converts prime power from the ac line to dc power, usually at a high voltage. The dc power must be regulated to remove the effects of line-voltage and load variation.

Protective circuitry is usually included with the high-voltage power supply to prevent the rf tube from being damaged in the event of a fault. Improper triggers and tube arcs are detected and used to trigger a crowbar circuit that discharges the energy stored in the high-voltage power supply. The crowbar is a triggered spark gap capable of dissipating the full energy of the power supply. Thyratrons, ignitrons, ball gaps, and triggered vacuum gaps are used.

Stability.   Radar systems with moving-target-indicator (MTI) place unusually tight stability requirements on their transmitters. Small changes in the amplitude, phase, or frequency from one pulse to the next can degrade MTI performance. In the transmitter, the MTI requirements appear as constraints on voltage, current, and timing variations from pulse to pulse. The relation between voltage variations and variation in amplitude and phase shift differs with the tube type used. Table 23.2.2 lists stability factors for the various tube types used with a high-voltage power supply (HVPS).

RADAR ANTENNAS

The great variety of radar applications has produced an equally great variety of radar antennas. These vary in size from less than a foot to hundreds of feet in diameter. Since it is not feasible even to mention each of the types here, we shall discuss the three basic antenna categories, search antennas, track antennas, and multifunction antennas, after first reviewing some basic antenna principles.

A radar antenna directs the radiated power and receiver sensitivity to the azimuth and elevation coordinates of the target. The ability of an antenna to direct the radiated power is described by its antenna pattern. A typical antenna pattern is shown in Fig. 23.2.4. It is a plot of radiated field intensity measured in the far field (a distance greater than twice the diameter squared divided by the wavelength from the antenna) and is plotted as a function of azimuth and elevation angle. Single cuts through the two-dimensional pattern, as shown in Fig. 23.2.5, are more often used to describe the pattern. The principle of reciprocity assures that the antenna pattern describes its gain as a receiver as well as transmitter. The gain is defined relative to an isotropic radiator.

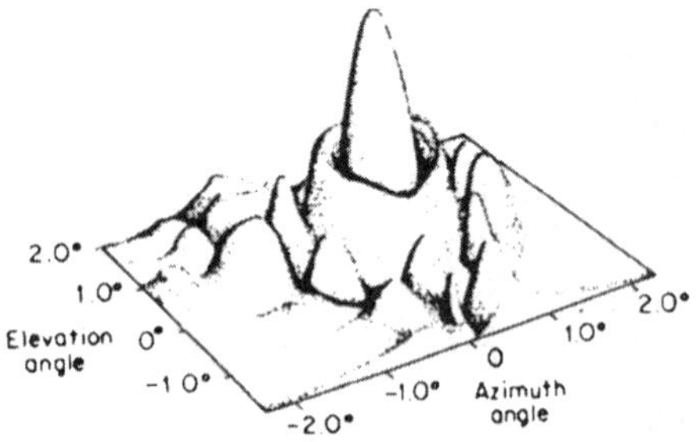

FIGURE 23.2.4   Three-dimensional pencil-beam pattern of the AN/FPQ-6 radar antenna. (*Courtesy of D. D. Howard, Naval Research Laboratory*)

TABLE 23.2.2   Stability Factors

	Frequency- or phase-modulation sensitivity	Impedance, ratio Dynamic static	Current or voltage change for 1% change in HVPS voltage	
			Line-type modulator	Low-impedance* hard-tube modulator, or dc operation
Magnetron	$\dfrac{\Delta f}{f} = \begin{pmatrix} 0.001 \\ \text{to} \\ 0.003 \end{pmatrix} \dfrac{\Delta I}{I}$	0.05–0.1	$\Delta I = 2\%$	$\Delta I = 10\text{–}20\%$
Stabilotron or stabilized magnetron	$\dfrac{\Delta f}{f} = \begin{pmatrix} 0.002 \\ \text{to} \\ 0.0005 \end{pmatrix} \dfrac{\Delta I}{I}$	0.05–0.1	$\Delta I = 2\%$	$\Delta I = 10\text{–}20\%$
Backward-wave CFA	$\Delta \phi = 0.4$ to $1°$ for 1% $\Delta I/I$	0.05–0.1	$\Delta I = 2\%$	$\Delta I = 10\text{–}20\%$
Forward-wave CFA	$\Delta \phi = 1.0$ to $3.0°$ for 1% $\Delta I/I$	0.1–0.2	$\Delta I = 2\%$	$\Delta I = 5\text{–}10\%$
Klystron	$\dfrac{\Delta \phi}{\phi} = \dfrac{1}{2}\dfrac{\Delta E}{E} \quad \phi \approx 5\lambda$ $\Delta \phi \approx 10°$ for 1%$\Delta E/E$	0.67	$\Delta E = 0.8\%$	$\Delta E = 1\%$
TWT	$\dfrac{\Delta \phi}{\phi} \approx \dfrac{1}{3}\dfrac{\Delta E}{E} \quad \phi \approx 15\lambda$ $\Delta \phi \approx 20\%$ for 1%$\Delta E/E$	0.67	$\Delta E = 0.8\%$	$\Delta E = 1\%$
Triode or tetrode	$\Delta \phi = 0$ to $0.5°$ for 1% $\Delta I/I$	1.0	$\Delta I = 1\%$	$\Delta I = 1\%$

*A high-impedance modulator is not listed because its output would (ideally) be independent of HVPS voltage.
Source: Weil, Ref. 2.

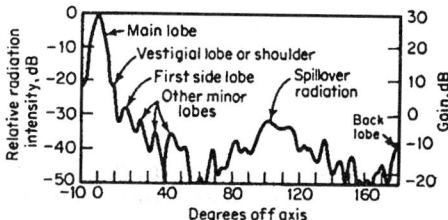

FIGURE 23.2.5   Radiation pattern for a particular paraboloid reflector antenna illustrating the main-lobe and side-lobe radiation.[3]

The gain used as a defining parameter is the gain at the peak of the beam or main lobe (see Fig. 23.2.5). This is the one-way power gain of the antenna

$$G_P = 4\pi A \eta_a / \lambda^2$$

where A = area of antenna aperture (reflector area for a horn-fed reflector antenna)
λ = radar wavelength in units of A
η_a = aperture efficiency, which accounts for all losses inherent in the process of illuminating the aperture

Tapered-aperture-illumination functions designed to produce low side lobes also result in lower aperture efficiency η_a and larger beamwidth θ_3, as shown in Fig. 23.2.6. A second gain definition sometimes used is directive gain. This is defined as a maximum radiation intensity, in watts per square meter, divided by the average radiation intensity, where the average is taken over all azimuth and elevation angles.

Directive gain can be inferred from the product of the main-lobe widths in azimuth and elevation, over a wide range of tapers (including uniform). For example, an array antenna, with no spillover of illumination

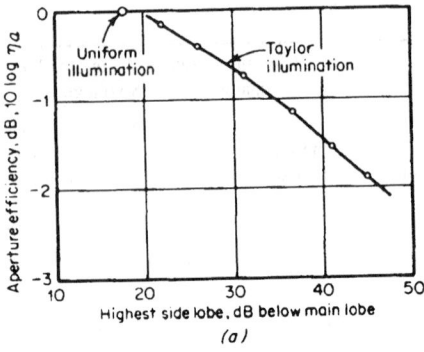
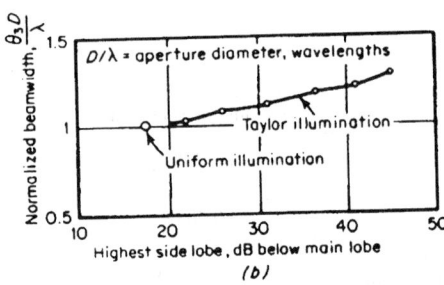

FIGURE 23.2.6 Aperture efficiency and beamwidth as a function of highest side-lobe level for a circular aperture: (a) aperture efficiency; (b) normalized beamwidth.[5]

power, gives

$$G_d = 36,000/\theta_{3a}\theta_{3e} \tag{80a}$$

where θ_{3a} = 3-dB width of the main lobe in the azimuth coordinate (degrees) and θ_{3e} = 3-dB main-lobe width in the elevation coordinate (degrees). For horn-fed antennas the constant in Eq. (80a) is about 25,000.

Search Antennas

Conventional surface and airborne search radars generally use mechanically scanned horn-fed reflectors for their antennas. The horn radiates a spherical wavefront that illuminates the reflector. The shape of the reflector is designed to cause the reflected wave to be in phase at any point on a plane in front of the reflector. This focuses the radiated energy at infinity. Mechanically scanning search radars generally have fan-shaped beams that are narrow in azimuth and wide in the elevation coordinate. In a typical surface-based air-search radar the upper edge of the beam is shaped to follow a cosecant-squared function. This provides coverage up to a fixed altitude. Figure 23.2.7 illustrates the effect of cosecant-squared beam shaping on the coverage diagram as well as on the antenna pattern. In the horn-fed reflector the shaping can be achieved by either the reflector or the feed, and the gain constant in Eq. (80a) is reduced to about 20,000.

Tracking-Radar Antennas

The primary function of a tracking radar is to make accurate range and angle measurements of a selected target's position. Generally, only a single target position is measured at a time, as the antenna is directed to follow the target by tracking servos. These servos smooth the errors measured from beam center to make pointing corrections. The measured errors, along with the measured position of the antenna, provide the target-angle information.

Tracking antennas like that of the AN/FPS-16 use circular apertures to form a pencil beam about 1° wide in each coordinate. The higher radar frequencies (S, C, and X band) are preferred because they allow a smaller aperture for the same beamwidth. The physically smaller antenna can be more accurately pointed. In this section we discuss aperture configurations, feeds, and pedestals.

One of the simplest methods of producing an equiphase wavefront in front of a circular aperture uses a parabolic reflector. A feed located at the focus directs its energy to illuminate the reflector. The reflected energy is then directed into space focused at infinity. The antenna is inherently broadband because the electrical path length from the feed to the reflector to the plane wavefront is the same for all points on the wavefront.

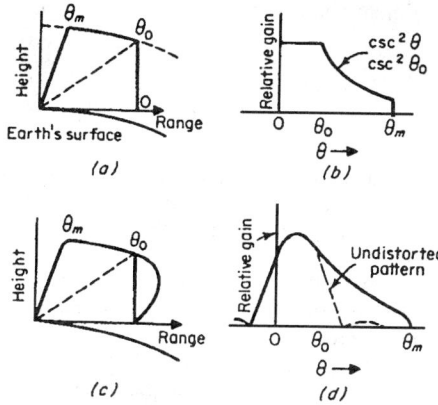

FIGURE 23.2.7 Elevation coverage of a cosecant-squared antenna: (*a*) desired elevation coverage; (*b*) corresponding antenna pattern desired; (*c*) realizable elevation coverage with pattern shown in (*d*); (*d*) actual cosecant-squared antenna pattern.[8]

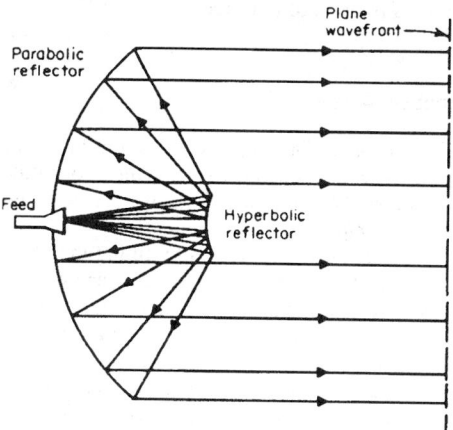

FIGURE 23.2.8 Schematic diagram of a Cassegrain reflector antenna.[9]

   Locating the feed in front of the aperture is sometimes inconvenient mechanically. It also produces spillover lobes where the feed pattern misses the reflector. The Cassegrain antenna shown in Fig. 23.2.8 avoids these difficulties by placing a hyperbolic subreflector between the parabolic reflector and its focus. The feed now illuminates the subreflector, which in turn illuminates the parabola and produces a plane wavefront in front of the aperture.

   Lenses can also convert the spherical wavefront emanating from the feed to a plane wavefront over a larger aperture. As the electromagnetic energy passes through the lens, it is focused at infinity (see Fig. 23.2.9).

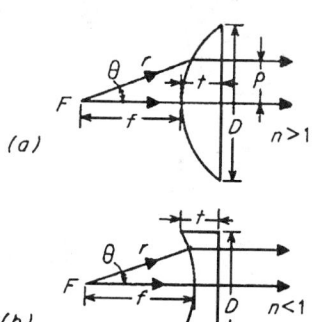

FIGURE 23.2.9 Geometry of simple converging lenses: (*a*) $n > 1$; (*b*) $n < 1$.[9]

Depending on the index of refraction n_g of the lens, a concave or convex lens may be required. Lenses are typically heavier than reflectors, but they avoid the blockage caused by the feed or subreflector.

   A single feed providing a single beam is unable to supply the angle-error information necessary for tracking. To obtain azimuth and elevation-error information, feeds have been developed that scan the beam in a small circle about the target (conical scan) or that form multiple beams about the target (monopulse). Conical scanning may be caused by rotating a reflector behind a dipole feed or rotating the feed itself. It has the advantage compared with monopulse that less hardware is required in the receiver, but at the expense of somewhat less accuracy.

   Modern trackers more often use a monopulse feed with multiple receivers. Early monopulse feeds used four separate horns to produce four contiguous beams that were combined to form a reference beam and azimuth and elevation-difference beams. More recently, multimode feeds have been developed to perform this function more efficiently with fewer components (see Fig. 23.1.44).

   Shaft angle encoders quantize radar pointing angles through mechanical connections to azimuth and elevation axes. The output indicates the angular position of the mechanical bore-site axis relative to a fixed angular coordinate system. Because these encoders make an absolute measurement, their outputs contain 10 to 20 bits of information. A variety of techniques is used, the complexity increasing with the accuracy required. Atmospheric errors ultimately limit the number of useful bits to about 20, or 0.006 mrad. In less precise tracking applications, synchros attached to the azimuth and elevation axes indicate angular position within a fraction of a degree.

Multifunction Arrays

Array antennas form a plane wavefront in front of the antenna aperture. These points are individual radiating elements which, when driven together, constitute the array. The elements are usually spaced about 0.6 wavelength apart. Most applications use planar arrays, although arrays conformal to cylinders and other surfaces have been built.

Phased arrays are steered by tilting the phase front independently in two orthogonal directions called the *array coordinates.* Scanning in either array coordinate causes the beam to move along a cone whose center is at the center of the array. The paths the beam follows when steered in the array coordinates are illustrated in Fig. 23.2.10, where the z axis is normal to the array. As the beam is steered away from the array normal, the projected aperture in the beam's direction varies, causing the beamwidth to vary proportionately.

Arrays can be classified as either active or passive. *Active arrays* contain duplexers and amplifiers behind every element or group of elements; *passive arrays* are driven from a single feed point. Only the active arrays are capable of higher power than conventional antennas.

Both passive and active arrays must divide the signal from a single transmission line among all the elements of the array. This can be done by an optical feed, a corporate feed, or a multiple-beam-forming network. The *optical feed* is illustrated in Fig. 23.2.11. A single feed, usually a monopulse horn, illuminates the array with a spherical phase front. Power collected by the rear elements of the array is transmitted through the phase shifters that produce a planar front and steer the array. The energy may then be radiated from the other side of the array, as in the lens, or be reflected and reradiated through the collecting elements, where the array acts as a steerable reflector.

Corporate feeds can take many forms, as illustrated by the series-feed networks shown in Fig. 23.2.12 and the parallel-feed networks shown in Fig. 23.2.13. All use transmission-line components to divide the signal among the elements. Phase shifters can be located at the elements or within the dividing network.

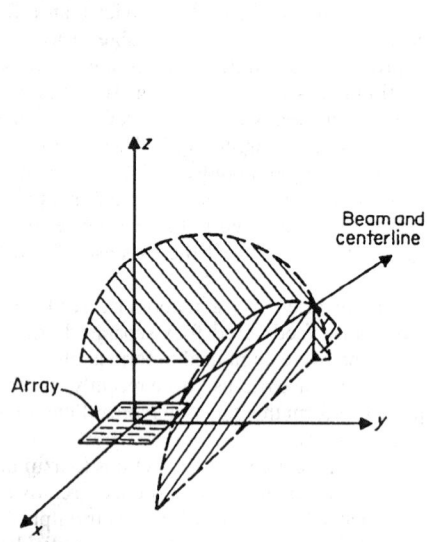

FIGURE 23.2.10 Beam-steering contours for a planar array.

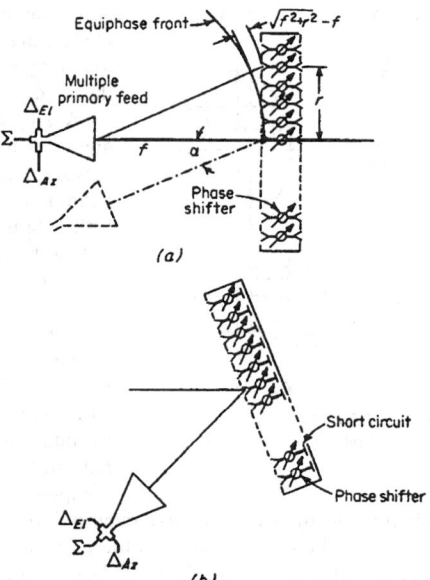

FIGURE 23.2.11 Optical-feed systems: (*a*) lens; (*b*) reflector.[11]

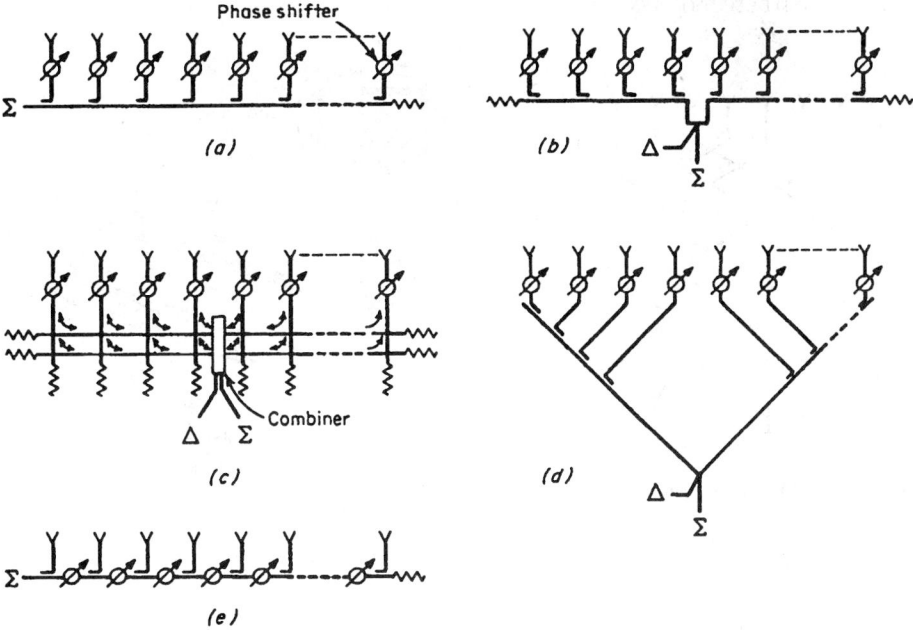

FIGURE 23.2.12 Series-feed networks: (a) end feed; (b) center feed; (c) separate optimization; (d) equal path length; (e) series phase shifters.[11]

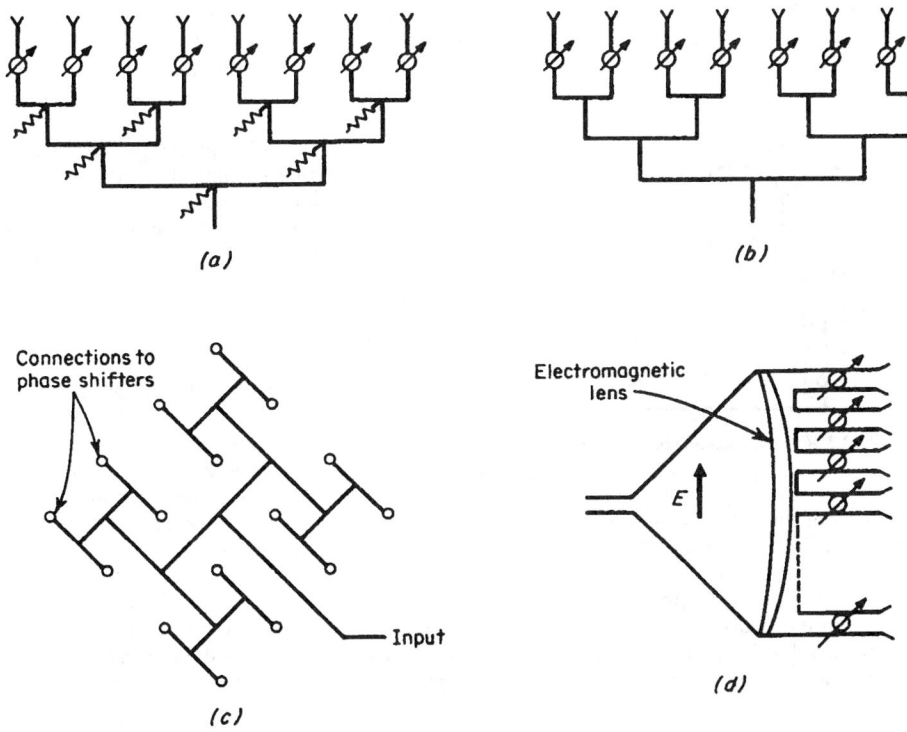

FIGURE 23.2.13 Parallel-feed networks: (a) matched corporate feed; (b) reactive corporate feed; (c) reactive stripline; (d) multiple reactive divider.[11]

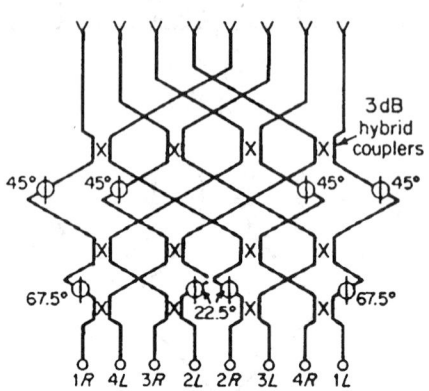

FIGURE 23.2.14   Butler beam-forming network.[11]

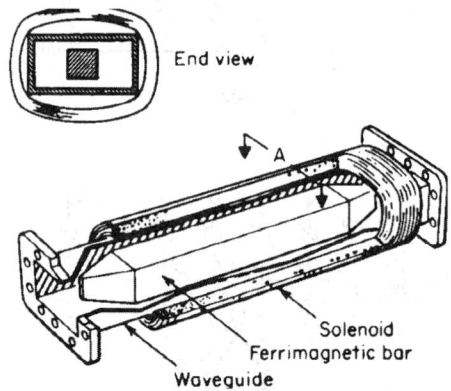

FIGURE 23.2.15   Typical Reggia-Spencer phase shifter.[12]

Multiple-beam networks are capable of forming simultaneous beams with the array. The Butler matrix shown in Fig. 23.2.14 is one such technique. It connects the N elements of a linear army to N feed points corresponding to N beam outputs. It can be applied to two-dimensional arrays by dividing the array into rows and columns.

The phase shifter is one of the most critical components of the array. It produces controllable phase shift over the operating band of the array. Digital and analog phase shifters have been developed using both ferrites and *pin* diodes. Phase shifter designs always strive for a low-cost, low-loss, and high-power-handling capability.

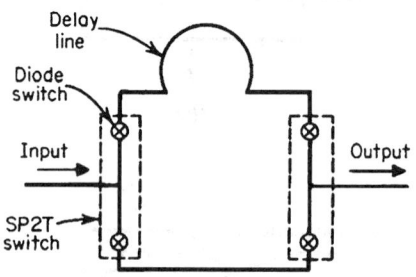

FIGURE 23.2.16   Switched-line phase bit.[12]

The Reggia-Spencer phase shifter consists of a ferrite inside a waveguide, as illustrated in Fig. 23.2.15. It delays the rf signal passing through the waveguide. The amount of phase shift can be controlled by the current in the solenoid, through its effect on the permeability of the ferrite. This is a reciprocal phase shifter that has the same phase shift for signals passing in either direction. Nonreciprocal phase shifters (where phase-shift polarity reverses with the direction of propagation) are also available. Either reciprocal or nonreciprocal phase shifters can be locked or latched in many states by using the permanent magnetism of the ferrite.

Phase shifters have also been developed using *pin* diodes in transmission-line networks. One configuration, shown in Fig. 23.2.16, uses diodes as switches to change the signal path length of the network. A second type uses *pin* diodes as switches to connect reactive loads across a transmission line. When equal loads are connected with a quarter-wave separation, a pure phase shift results.

When digital phase shifters are used, a phase error occurs at every element due to phase quantization. The error in turn causes reduced gain, higher side lobes, and greater pointing errors. Gain reduction is tabulated in Table 23.2.3 for typical quantizations. Figure 23.2.17 shows the rms side-lobe levels caused by phase quantization in an array of N elements. The rms pointing; error relative to 3-dB beamwidth is given by

$$\sigma_\theta / \theta_3 \approx 1.12 / 2^m \sqrt{N}$$

TABLE 23.2.3 Gain Loss in a Phased
Array with m-Bit Digital Phase Shifters

Number of Bits, m	Gain Loss, dB
3	0.228
4	0.057
5	0.0142
6	0.00356
7	0.0089
8	0.0022

where m is the number of bits of phase quantization and N is the number of elements in array.[5]

Frequency scan is a simple array-scanning technique that does not require phase shifters, drivers, or beam-steering computers. Element signals are coupled from points along a transmission line as shown in Fig. 23.2.18. The electrical path length between elements is much longer than the physical separation, so that a small frequency change will cause a phase change between elements large enough to steer the beam. The technique can be applied only to one array coordinate, so that in two-dimensional arrays, phase shifters are usually required to scan the other coordinate.

MICROWAVE COMPONENTS

The radar transmitter, antenna, and receiver are all connected through rf transmission lines to a duplexer. The duplexer acts as a switch connecting the transmitter to the antenna while radiating and the receiver to the antenna while listening for echoes. Filters, receiver protectors, and rotary joints may also be located in the various paths. See Section 7 for a description of microwave devices and transmission lines.

A variety of other transmission-line components are used in a typical radar. Waveguide bends, flexible waveguide, and rotary joints are generally necessary to route the path to the feed of a rotating antenna. Waveguide windows provide a boundary for pressurization while allowing the microwave energy to pass through. Directional couplers sample forward and reverse power for monitoring, test, and alignment of the radar system.

Duplexers

The duplexer acts as a switch connecting the antenna and transmitter during transmission and the antenna and receiver during reception. Various circuits are used that depend on gas tubes, ferrite circulators, or *pin* diodes as the basic switching element. The duplexers using gas tubes are most common. A typical gas-filled TR tube is shown in Fig. 23.2.19. Low-power rf signals pass through the tube with very little attenuation. Higher power causes the gas to ionize and present a short circuit to the rf energy.

Figure 23.2.20 shows a balanced duplexer using hybrid junctions and TR tubes. When the transmitter is on, the TR tubes fire and reflect the rf power to the antenna port of the input hybrid. On reception, signals received by the antenna are passed through the TR tubes and to the receiver port of the output hybrid.

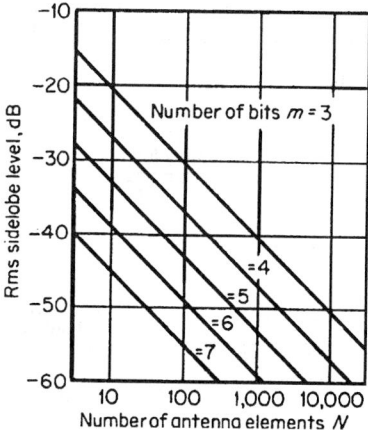

FIGURE 23.2.17 RMS side lobes due to phase quantization.[11]

FIGURE 23.2.18 Simple types of frequency-scanned antennas: (*a*) broad-wall coupling to dipole radiators; (*b*) narrow-wall coupling with slot radiators.[14]

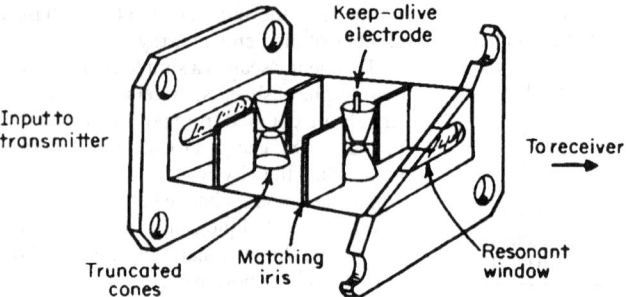

FIGURE 23.2.19   Typical TR tube.[3]

Circulators and Diode Duplexers

Newer radars often use a ferrite circulator as the duplexer. A TR tube is required in the receiver line to protect the receiver from the transmitter power reflected by the antenna due to an imperfect match. A four-port circulator is generally used with a load between the transmitter and receiver ports so that the power reflected by the TR tube is properly terminated.

In place of the TR tube *pin* diode switches have been used in duplexers. These are more easily applied in coaxial circuitry and at lower microwave frequencies. Multiple diodes are used when a single diode cannot withstand the required voltage or current.

Receiver Protectors.   TR tubes with a lower power rating are usually required in the receive line to prevent transmitter leakage from damaging mixer diodes or rf amplifiers in the receiver. A keep-alive ensures rapid ionization, minimizing spike leakage. The keep-alive may be either a probe in the TR tube maintained at a high dc potential or a piece of radioactive material. Diode limiters are also used after TR tubes to further reduce the leakage.

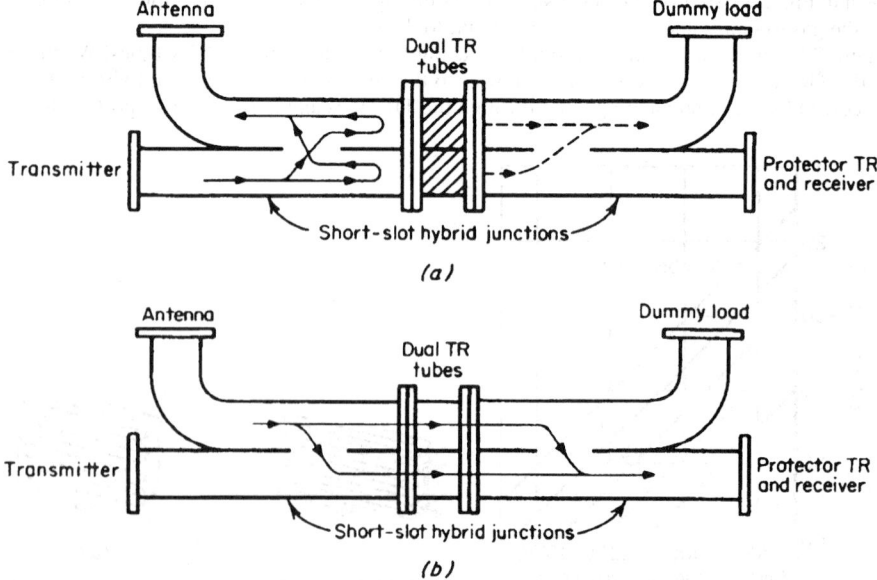

FIGURE 23.2.20   Balanced duplexer using dual TR tubes and two short-slot hybrid junctions: (*a*) transmit condition; (*b*) receive condition.[3]

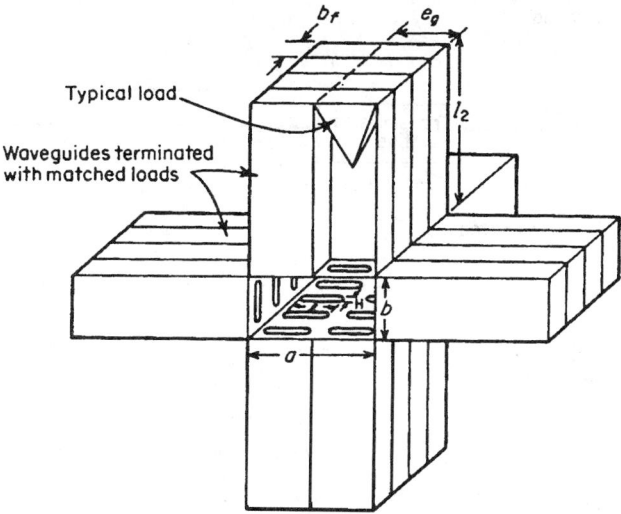

FIGURE 23.2.21   Typical construction of a dissipative waveguide filter.[15]

Filters

Microwave filters are sometimes used in the transmit path to suppress spurious radiation or in the receiver signal path to suppress spurious interference. Because the transmit filters must handle high power, they are larger and more difficult to design.

Narrow-band filters in the receive path, often called *preselectors*, are built using mechanically tuned cavity resonators or electrically tuned YIG resonators. Preselectors can provide up to 80 dB suppression of signals from other radar transmitters in the same rf band but at a different operating frequency.

Harmonic filters are the most common transmitting filter. They absorb the harmonic energy to prevent it from being radiated or reflected. Since the transmission path may provide a high standing-wave ratio at the harmonic frequencies, the presence of harmonics can increase the voltage gradient in the transmission line and cause breakdown. Figure 23.2.21 shows a harmonic filter where the harmonic energy is coupled out through holes in the walls of the waveguide to matched loads.

RADAR RECEIVERS

The radar receiver amplifies weak target returns so that they can be detected and displayed. The input amplifier must add little noise to the received signal, for this noise competes with the smallest target return that can be detected. A mixer in the receiver converts the received signal to an intermediate frequency where filtering and signal decoding can be accomplished. Finally, the signals are detected for processing and display.

Low-Noise Amplifiers

Because long-range radars require large transmitters and antennas, these radars can also afford the expense of a low-noise receiver. Considerable effort has been expended to develop more sensitive receivers. Some of the devices in use will be described here after a brief review of noise-figure and noise-temperature definitions.

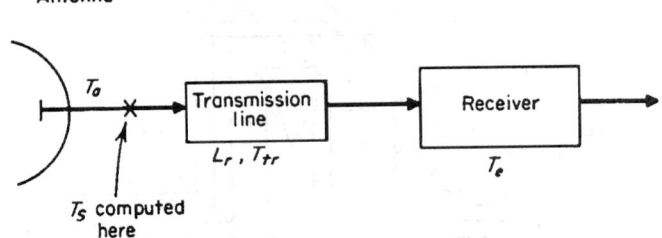

FIGURE 23.2.22   Contribution to system noise temperature.

Noise figure and noise temperature measure the quality of a sensitive receiver. Noise figure is the older of the two conventions and is defined as

$$F_n = \frac{S/N \text{ at output}}{S/N \text{ at input}}$$

where S is the signal power, N is the noise power, and the receiver input terminates is at room temperature. Before low-noise amplifiers were available, a radar's noise figure was determined by the first mixer, which would be typically 5 to 10 dB. For these values of F_n it was approximately correct to add the loss of the waveguide to the noise figure when calculating signal-to-noise ratio. As better receivers were developed with lower noise figures, these approximations were no longer accurate, and the noise-temperature convention was developed.

Noise temperature is proportional to noise-power spectral density through the relation

$$T = N/kB$$

where k is Boltzmann's constant and B is the bandwidth in which the noise power is measured. The noise temperature of an rf amplifier is defined as the noise temperature added at the input of the amplifier required to account for the increase in noise due to the amplifier. It is related to noise figure through the equation

$$T = T_0(F_n - 1)$$

where T_0 = standard room temperature = 290 K.

The receiver is only one of the noise sources in the radar system. Figure 23.2.22 shows the receiver in its relation to the other important noise sources. Losses, whether in the rf transmission line, antenna, or the atmosphere, reduce the signal level and also generate thermal noise. The load presented to the rf transmission line by the antenna is its radiation resistance. The temperature of this resistance T_a depends on where the antenna beam is pointed. When the beam is pointed into space, this temperature may be as low as 50 K. However, when the beam is pointed toward the sun or a radio star, the temperature can be much higher. All these sources can be combined to find the system noise temperature T_s, according to the equation

$$T_s = T_a + (L_r - 1)T_{tr} + T_e L_r$$

where T_a = temperature of the antenna
L_r = transmission-line loss defined as ratio of power in the to power out
T_{tr} = temperature of transmission line
T_e = receiver noise temperature

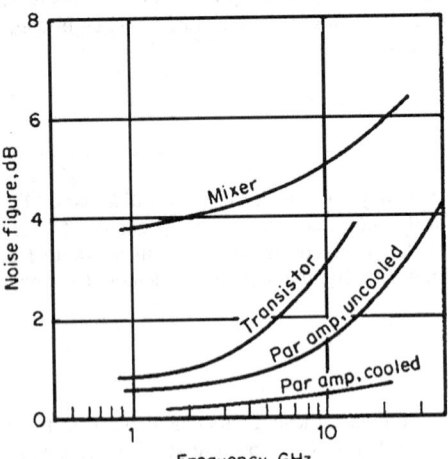

FIGURE 23.2.23   Noise characteristics of radar front ends.

Figure 23.2.23 shows the noise temperature as a function of frequency for radar-receiver front ends. All are noisier at higher frequencies. Transistor amplifiers and uncooled

TABLE 23.2.4 Approximations to Matched Filters

Pulse shape	Filter	Optimum bandwidth-time product			Mismatch loss, dB
		6 dB	3 dB	Energy	
Gaussian	Gaussian bandpass	0.92	0.44	0.50	0
Gaussian	Rectangular bandpass	1.04	0.72	0.77	0.49
Rectangular	Gaussian bandpass	1.04	0.72	0.77	0.49
Rectangular	5 synchronously tuned stages	0.97	0.67	0.76	0.50
Rectangular	2 synchronously tuned stages	0.95	0.61	0.75	0.56
Rectangular	Single-pole filter	0.70	0.40	0.63	0.88
Rectangular	Rectangular bandpass	1.37	1.37	1.37	0.85
Rectangular chirp	Gaussian	1.04×6 dB width of equivalent $(\sin x)/x$ pulse, $(0.86 \times$ width of spectrum$)$			0.6

Source: Taylor and Mattern, Ref. 16.

parametric amplifiers are finding increased use in radar receivers. *Transistor amplifiers* have been improved steadily, with emphasis on increased operating frequency. Although the transistor amplifier is a much simpler circuit than the parametric amplifier, it does not achieve the parametric amplifier's low noise temperature.

A balanced mixer is often used to convert from rf to i.f. Balanced operation affords about 20 dB immunity to amplitude noise on the local-oscillator signal. Intermediate frequencies of 30 and 60 MHz are typical, as are 1.5 to 2 dB intermediate-frequency noise figures for the i.f. preamplifier. Double conversion is sometimes used with a first i.f. at a few hundred megahertz. This gives better image and spurious suppression.

The *matched filter* is usually instrumented at the second i.f. frequency. This filter is an approximation to the matched filter and therefore does not achieve the highest possible signal-to-noise ratio. This deficiency is expressed as mismatch loss. Table 23.2.4 lists the mismatch loss for various signal-filter combinations when the optimum bandwidth is used.

Pulse Compression

Pulse compression is a technique in which a rectangular pulse containing phase modulation is transmitted. When the echo is received, the matched-filter output is a pulse of much shorter duration. This duration approximately equals the reciprocal of the bandwidth of the phase modulation. The compression ratio (ratio of transmitted to compressed pulse lengths) equals the product of the time duration and bandwidth of the transmitted pulse. The technique is used when greater pulse energy or range resolution are required than can be achieved with a simple uncoded pulse.

Linear FM (chirp) is the phase modulation that has received the widest application. The carrier frequency is swept linearly during the transmitted pulse. The wide application has both caused and resulted from the development of a variety of dispersive analog delay lines. Delay-lines techniques covering a wide range of bandwidths and time durations are available. Table 23.2.5 lists the characteristics of a number of these dispersive delay lines.

Range lobes are a property of pulse-compression systems not found in radar using simple cw pulses. These are responses leading and trailing the principal response and resembling antenna side lobes; hence the name range lobes. These lobes can be reduced by carefully designing the phase modulation or by slightly mismatching the compression network. The mismatch can be described as a weighting function applied to the spectrum.

Detectors

Although bandpass signals on an i.f. carrier are easily amplified and filtered, they must be detected before they can be displayed, recorded, or processed. When only the signal amplitude is desired, square-law characteristics may be obtained with a semiconductor diode detector, and this provides the best sensitivity for detecting pulses in noise when integrating the signals returned from a fluctuating target. Larger i.f. signal amplitudes derive

TABLE 23.2.5 Characteristics of Passive Linear-fm Devices

	B, MHz	T, μs	BT	f_0, MHz	Typical loss, dB	Typical spurious, dB
Aluminum strip delay line	1	500	200	5	15	−60
Steel strip delay line	20	350	500	45	70	−55
All-pass network	40	1000	300	25	25	−40
Perpendicular diffraction delay line	40	75	1000	100	30	−45
Surface-wave delay line	40	50	1000	100	20	−50
Wedge-type delay line	250	65	1000	500	50	−50
Folded-type meander line	1000	1.5	1000	2000	25	−40
Waveguide operated near cutoff	1000	3	1000	5000	60	−25
YIG crystal	1000	10	2000	2000	70	−20

Source: Farnett et al., Ref. 17.

the diode detector into the linear range, providing a linear detector. The linear detector has a greater dynamic range with somewhat less sensitivity. When still greater dynamic range (up to 80 dB) is required, log detectors are often used. Figure 23.2.24 shows the functional diagram of a log detector. The detected outputs of cascaded amplifiers are summed. As the signal level increases, stages saturate, reducing the rate of increase of the output voltage.

Some signal-processing techniques require detecting both phase and amplitude to obtain the complete information available in the i.f. signal. The phase detector requires an i.f. reference signal. A phase detector can be constructed by passing the signal through an amplitude limiter and then to a product detector, where it is combined with the reference signal, as shown in Fig. 23.2.25. An alternative to detecting the amplitude and phase is to detect the in-phase and quadrature components of the i.f. signal. The product detector shown in Fig. 23.2.25 can also provide this function when the input signal is not amplitude-limited. Quadrature detector circuits differ only in that the reference signal is shifted by 90° in one detector relative to the other.

Analog-to-Digital Converters

Digital signal processors require that the detected i.f. signals be encoded by an analog-to-digital converter. A typical converter may sample the detected signal at a 1-MHz rate and encode the sampled value into a 12-bit

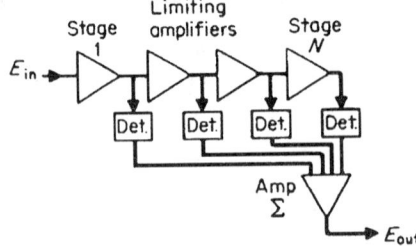

FIGURE 23.2.24 Logarithmic detector.[16]

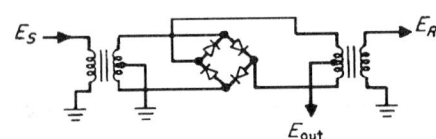

FIGURE 23.2.25 Balanced-diode detector.[16]

binary word. Encoders operating at higher rates have been built, but with fewer bits in their output. Encoders typically have errors that about equal the least significant bit.

EXCITERS AND SYNCHRONIZERS

Exciters

Two necessary parts of any radar system are an *exciter* to generate rf and local-oscillator frequencies and a *synchronizer* to generate the necessary triggers and timing pulses.

The components used in exciters are oscillators, frequency multipliers, and mixers. These can be arranged in various ways to provide the cw signals needed in the radar. The signals required depend on whether the transmitter is an oscillator or a power amplifier.

Transmitters using power oscillators such as magnetrons determine the rf frequency by the magnetron tuning. In a noncoherent radar, the only other frequency required is that of the local oscillator. It differs from the magnetron frequency by the i.f. frequency, and this difference is usually maintained with an automatic frequency control (AFC) loop. Figure 23.2.26 shows the circuit of a simple magnetron radar, illustrating the two alternative methods of tuning the magnetron to follow the *stable local oscillator* (*stalo*) or tuning the stalo to follow the magnetron.

If the radar must use coherent detection (as in MTI or pulse doppler applications), a second oscillator, called a *coherent oscillator* (*coho*), is required. This operates at the i.f. frequency and provides the reference for the product detector. Because an oscillator transmitter starts with random phase on every pulse, it is necessary to quench the coho and lock its phase with that of the transmitter on each pulse. This is accomplished by the circuit shown in Fig. 23.2.27.

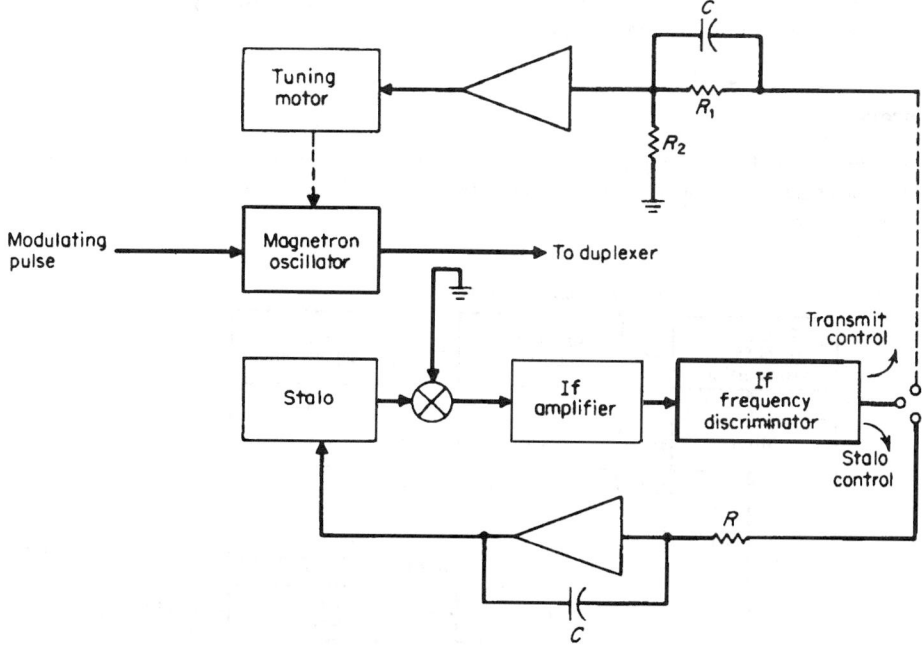

FIGURE 23.2.26   Alternative methods for AFC control.[16]

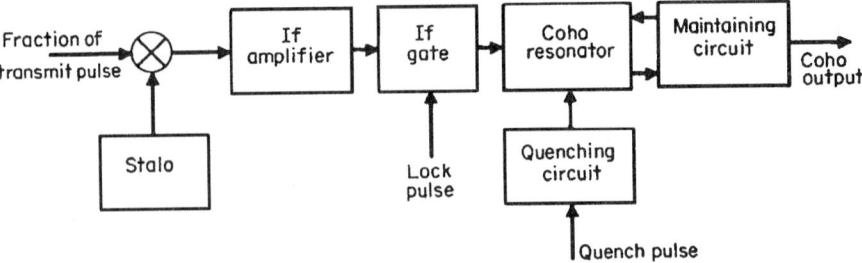

FIGURE 23.2.27 Keyed coho.[16]

When an amplifier transmitter is used, coho locking is not required. The transmit frequency can be obtained by mixing the stalo and coho frequencies, as shown in Fig. 23.2.28. The stalo and coho are not always oscillators operating at their output frequency. Figure 23.2.29 shows an exciter using crystal oscillators and multipliers to produce the rf and local-oscillator frequencies. Crystals may be changed to select the rf frequency without changing the i.f. frequencies.

The stability required of the frequencies produced by the exciter depends on the radar application. In a simple noncoherent radar a stalo frequency error shifts the signal spectrum in the i.f. passband, and an error which is a fraction of the i.f. bandwidth can be allowed. In MTI or pulse doppler radars, phase changes from pulse to pulse must be less than a few degrees. This requirement can be met with crystal oscillators driving frequency multipliers or fundamental oscillators with high-Q cavities when sufficiently isolated from electrical and mechanical perturbation. Instability is often expressed in terms of the phase spectrum about the center frequency.

Crystal oscillators driving frequency multipliers are finding increased use as stalos. A typical multiplier might multiply a 90-MHz crystal oscillator frequency by 32 to obtain an S-band signal. This source has the long-term stability of the crystal oscillators, but with degraded short-term stability. This is because the multiplier increases the phase modulation on the oscillator signal in proportion to the multiplication factor; i.e., each dobbler stage raises the oscillator sideband 6 dB. Frequency may be varied by tuning the crystal oscillator (about 0.25 percent) or by changing crystals.

Synchronizers

The synchronizer delivers timing pulses to the various radar subsystems. In a simple marine radar this may consist of a single multivibrator that triggers the transmitter, while in a larger radar 20 to 30 timing pulses may

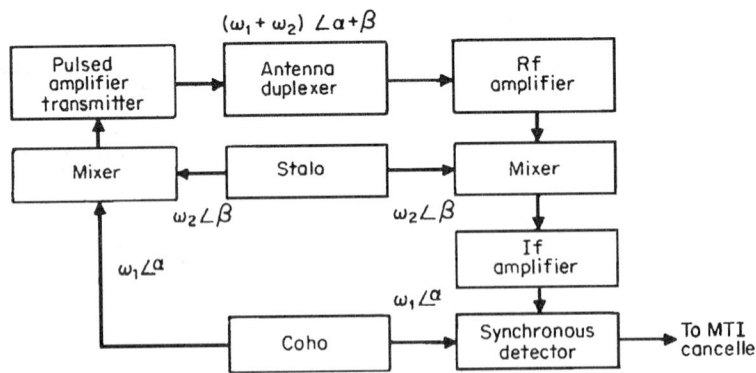

FIGURE 23.2.28 Coherent radar.[16]

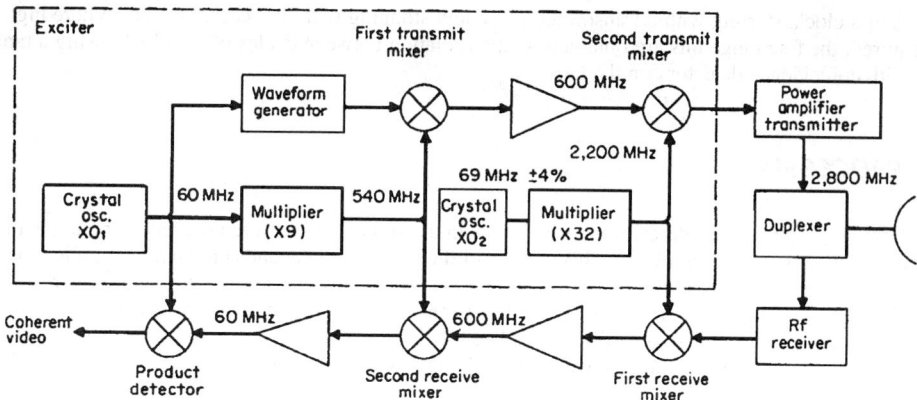

FIGURE 23.2.29   Coherent-radar exciter.

be needed. These may turn on and off the beam current in various transmitter stages; start and stop the rf pulse time attenuators; start display sweeps; etc.

Timing pulses or triggers are often generated by delaying a pretrigger with delays that may be either analog or digital. New radars are tending toward digital techniques, with the synchronizer incorporated into a digital signal processor. A diagram of the delay structure in a digital synchronizer is shown in Fig. 23.2.30. A 10-MHz clock moves the initial timing pulse through shift registers. The number of stages in each register is determined by the delay required. Additional analog delays provide a fine delay adjustment to any point in the 100-ns interval between clock pulses.

The synchronizer will also contain a range encoder, in radars where accurate range tracking is required or where range data will be processed or transmitted in digital form. Range is usually quantized by counting

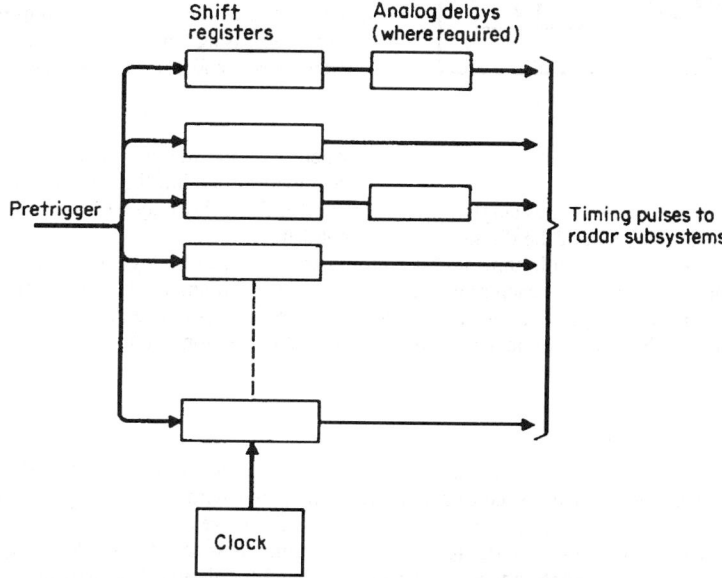

FIGURE 23.2.30   Digital synchronizer.

cycles of a clock, starting with a transmitted pulse and stopping with the received echo. Where high precision is required, the fine range bits are obtained by interpolating between cycles of the clock, using a tapped delay line with coincidence detectors on the taps.

SIGNAL PROCESSING

The term *signal processing* describes those circuits in the signal path between the receiver and the display. The extent to which processing is done in this portion of the signal path depends on the radar application. In search radars, postdetection integration, clutter rejection, and sometimes pulse compression are instrumented in the signal processor. The trend in modern radar has been to use digital techniques to perform these functions, although many analog devices are still in use. The following paragraphs outline the current technology trends in postdetection integration, clutter rejection and digital pulse compression.

Postdetection Integration

Scanning-search radars transmit a number of pulses toward a target as the beam scans past. For best detectability, these returns must be combined before the detection decision is made. In many search radars the returns are displayed on a plan-position indicator (PPI), where the operator, by recognizing target patterns, performs the postdetecton integration. When automatic detectors are used, the returns must be combined electrically. Many circuits have been used, but the two most common are the video sweep integrator and binary integrator.

The simplest video integrator uses a single delay line long enough to store all the returns from a single pulse. When the returns from the next pulse are received, they are added to the attenuated delay-line output. Figure 23.2.31 shows two forms of this circuit. The second (Fig. 23.2.31*b*) is preferred because the gain factor K is less critical to adjust. The circuit weights past returns with an exponentially decreasing amplitude where the time constant is determined by K. For optimum enhancement

$$K = 1 - 1.56/N$$

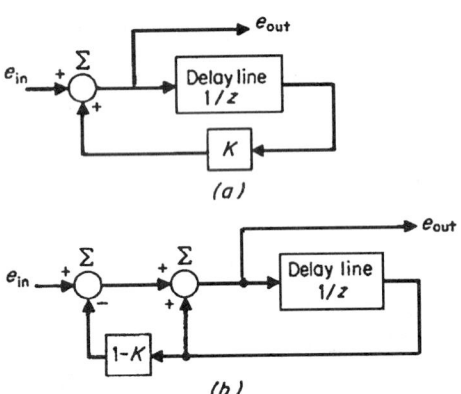

FIGURE 23.2.31   Two forms of sweep integrator.[18]

where N is the number of hits per one-way half-power beamwidth. By limiting the video amplitude into the integrator, the integrator eliminates single-pulse interference. The delay may be analog or digital, but the trend is to digital because the gain of the digital loop does not drift.

The binary integrator or double-threshold detector is another type of integrator used in automatic detectors. With this integrator, the return in each range cell is encoded to 1 bit and the last N samples are stored for each range cell. If M or more of the N stored samples are 1s, a detection is indicated. Figure 23.2.32 shows a functional diagram of a binary integrator. This integrator is also highly immune to single-pulse interference.

Clutter Rejection

The returns from land, sea, and weather are regarded as clutter in an air-search radar. They can be suppressed in the signal processor when the spectrum is narrow compared with the radar's pulse repetition rate (prf). Filters that combine two or more returns from a single-range cell are able to discriminate between the desired targets and clutter. This allows the radar to detect targets with cross section smaller than that of the clutter. It also provides a means of preventing the clutter from causing false alarms. The two classes of clutter filters are moving target indicator (MTI) and pulse doppler.

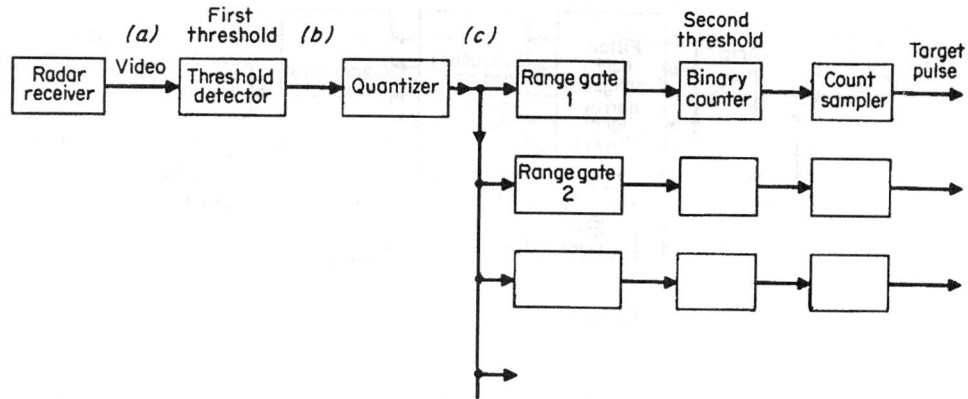

FIGURE 23.2.32 Binary integrator.[3]

MTI combines a few pulse returns, usually two or three, in a way that causes the clutter returns to cancel. Figure 23.2.33 shows a functional diagram of a digital vector canceler. The in-phase and quadrature components of the i.f. signal vector are detected and encoded. Stationary returns in each signal component are canceled before the components are rectified and combined. The digital canceler may consist of a shift register memory and a subtractor to take the difference of the succeeding returns. Often only one component of the vector canceler is instrumented, thereby saving about half the hardware, but at the expense of signal detectability in noise.

A pulse doppler processor is another class of clutter filter where the returns in each range resolution cell are gated and put into a bank of doppler filters. The number of filters in the bank approximately equals the number of pulse returns combined. Each filter is tuned to a different frequency, and the passbands contiguously positioned between zero frequency and prf. Figure 23.2.34 shows a functional diagram of a pulse doppler processor. The pulse doppler technique is most often used in either airborne or land-based target-tracking radars, where a high ambiguous prf can be used, thus providing an unambiguous range of doppler frequencies. The filter bank may be instrumented digitally by a special-purpose computer wired according to the fast Fourier transform algorithm.

Digital Pulse Compression

Digital pulse compression performs the same function as the analog pulse compression devices described earlier, except that it is instrumented using digital technology. IF samples of a phase-coded echo are processed to produce samples of a compressed pulse with much shorter duration. Now that the digital technology is comparative in cost with the analog dispersive delay lines, its freedom from temperature variation causes it to be preferred for

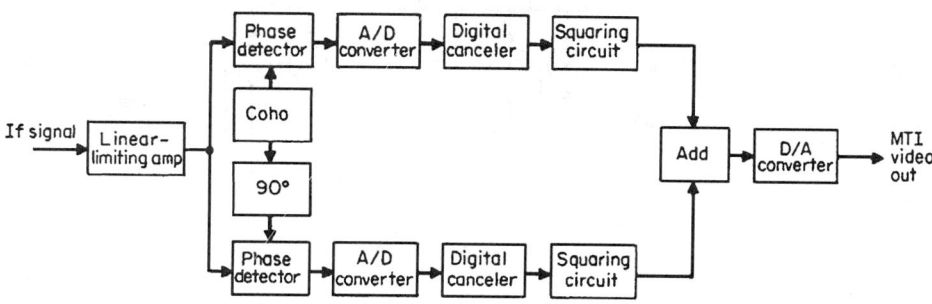

FIGURE 23.2.33 Digital vector canceler.[18]

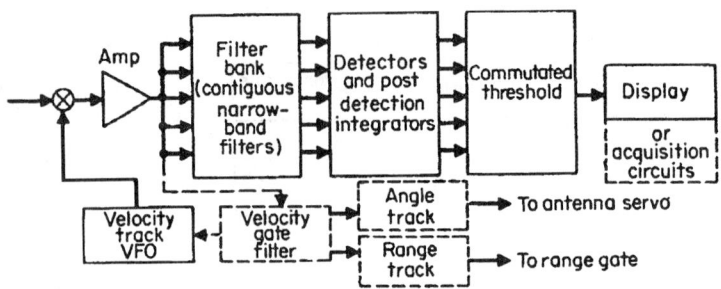

FIGURE 23.2.34 Typical pulse doppler processor.[19]

new designs. A typical functional implementation is illustrated in Fig. 23.2.35. Quadrature samples of the IF signal are digitized and stored for an interval at least as long as the transmitted pulse. The stored samples are then correlated with a set of complex weights that represent the time-inverted transmit waveform. The output data stream represents time samples of the compressed IF pulse.

A second important advantage of digital pulse compression is the increased dynamic range compared with analog compression techniques. Since the analog-to-digital converter is usually the limitation on dynamic range, the gain in peak signal provided by pulse compression adds to the dynamic range available to the subsequent digital signal processing.

DISPLAYS

Radar indicators are the coupling link between radar information and a human operator. Radar information is generally a time-dependent function of range or distance. Thus the display format generally uses displacement of signal proportional to time or range. The signal may be displayed as an orthogonal displacement (such as an oscilloscope) or as an intensity modulation of brightening of the display. Most radar signal presentations have the intensity-modulated type of display where additional information such as azimuth, elevation angle, or height can be presented.

A common format is a polar-coordinate, or plan-position, indicator (PPI), which results in a maplike display. Radar azimuth is presented on the PPI as an angle, and range as radial distance. In a cartesian coordinate display, one coordinate may represent range and the other may represent azimuth elevation, or height.

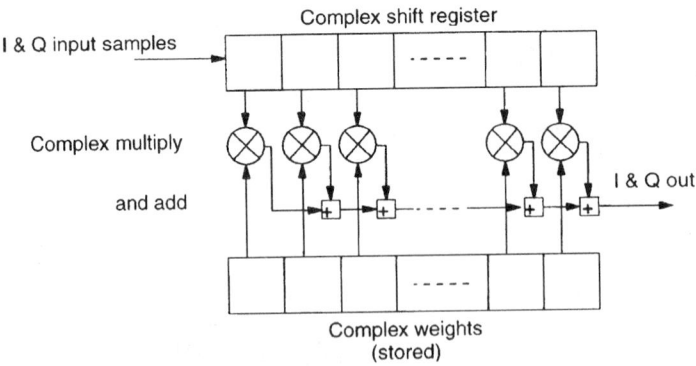

FIGURE 23.2.35 Digital pulse compressor that correlates samples of the receive waveform with a stored replica.

Variations may use one coordinate for azimuth and the other coordinate for elevation and gate a selected time period or range to the display. The increasing use of processed radar data can provide further variations. In each case, the display technique is optimized to improve the information transfer to the operator.

Marker signals may be inserted on the displays as operator aids. These can include fixed and variable range marks, strobes, and cursors as constant-angle or elevation traces. Alphanumeric data, tags, or symbols may be incorporated for operator information or designation of data.

Cathode-Ray Tubes

The cathode-ray rube is the most common display device used for radar indicators. The cathode-ray tube is used most because of its flexibility of performance, resolution, dynamic range, and simplicity of hardware relative to other display techniques. Also, the cathode-ray tube has varied forms, and parameters can be optimized for specific display requirements (see Secs. 5 and 9).

Cathode-ray tubes using charge-storage surfaces are used for specialized displays (see Sec. 5). The direct-view storage tube is a high-brightness display tube. Other charge-storage tubes use electrical write-in and read-out. Such tubes may be used for signal integration, for scan conversion so as to provide simpler multiple displays, for combining multiple sensors on a single display, for increasing viewing brightness on an output display, or for a combination of these functions.

REFERENCES

1. Skolnik, M. I. (ed.) "Radar Handbook," McGraw-Hill, 1970.

1a. Skolnik, M. I. (ed.) "Radar Handbook," 2nd ed., McGraw-Hill, 1990.

2. Weil, T. A. Transmitters, Chap. 7 in Ref. 1.

2a. Borkowski, M. T. Solid-State Transmitters, Chap. 5 in Ref. 1a.

3. Skolnik, M. I. "Introduction to Radar Systems," McGraw-Hill, 1980.

4. Sherman, J. W. Aperture-Antenna Analysis, Chap. 9 in Ref. 1.

5. Barton, D. K., and H. R. Ward "Handbook of Radar Measurement," Prentice Hall, 1969.

6. Ashley, A., and J. S. Perry Beacons, Chap. 38 in Ref. 1.

7. Croney, J. Civil Marine Radar, Chap. 31 in Ref. 1.

8. Freedman, J. Radar Chap. 14 in "System Engineering Handbook," McGraw-Hill, 1965.

9. Sengupta, D. L., and R. E. Hiatt Reflectors and Lenses, Chap. 10 in Ref. 1.

10. Dunn. J. H., D. D. Howard, and K. B. Pendleton Tracking Radar, Chap. 21 in Ref. 1.

11. Cheston, T. C., and J. Frank Array Antennas, Chap. 11 in Ref. 1.

12. Stark, L., R. W. Burns, and W. P. Clark Phase Shifters for Arrays, Chap. 12 in Ref. 1.

13. Kefalas, G. P., and J. C. Wiltse Transmission Lines, Components, and Devices, Chap. 8 in Ref. 1.

14. Hammer, I. W. Frequency-Scanned Arrays, Chap. 13 in Ref. 1.

15. Matthaei, G. L., L. Young, and E. M. T. Jones "Microwave Filters, Impedance Matching Networks, and Coupling Structures," McGraw-Hill, 1964.

16. Taylor, J. W., and J. Mattern Receivers, Chap. 5 in Ref. 1.

17. Farnett, E. C., T. B. Howard, and G. H. Stevens Pulse-Compression Radar, Chap. 20 in Ref. 1.

18. Shrader, W. W. MTI Radar, Chap. 17 in Ref. 51.

19. Mooney, D. H., and W. A. Skillman, Pulse-Doppler Radar, Chap. 19 in Ref. 1.

20. Nathanson, F. "Radar Design Principles: Signal Processing and the Environment," McGraw-Hill, 1969.

21. Berg, A. A. Radar Indicators and Displays, Chap. 6 in Ref. 1.

21a. Brookner, E. "Radar Technology," Artech House, 1986.

21b. DiFranco, J. V., and W. L. Rubin "Radar Detection,"Artech House, 1986.

21c. Currie, N. C., and C. E. Brown (eds.) "Principles and Applications of Millimeter-Wave Radar," Artech House, 1987.

CHAPTER 23.3
ELECTRONIC NAVIGATION SYSTEMS

Walter R. Fried

Some form of electronic navigation system is used in virtually all types of vehicles, including commercial and military aircraft, ships, land vehicles, and space vehicles. In recent years, they have also found application in automobiles and trucks and by individual personnel, both civil and military. Electronic navigation and positioning systems are also used for surveying and for mineral, oil, or other resource applications. Typical parameters measured by electronic navigation systems are the distance and bearing from a vehicle to a known point or the present position of a vehicle in a particular coordinate system. From the knowledge of present position, the course and distance to a destination can then be computed.

Most modern electronic navigation systems are based on the use of electromagnetic (radio) waves. The primary exceptions are systems using gyroscopes and accelerometers; those using optical celestial observations and those using pressure transducers. The use of radio waves has been found attractive because of their nearly constant and known speed of propagation, namely, the speed of light, which is about 3×10^8 m/s. With the knowledge of that velocity of propagation, if the time of travel of the radio signal between two points is accurately measured, the distance (range) between the points can be accurately determined. This is expressed by the equation $d = ct$, where d is the distance between the points, c is the speed of light, and t is the time of travel of the signal between the points. Also, measurement of the phase of the signal can be used for the determination of distance between the points, as well as relative bearing. In addition, the capability of high-frequency electromagnetic systems to generate narrow radiation beams can be useful for the measurement of the relative bearing from a vehicle to another vehicle or from a known point to a vehicle.

TYPES OF SYSTEMS

Electronic navigation systems can be classified in a number of ways, both from an electronics viewpoint and from a navigational viewpoint. From an electronics viewpoint, they can be categorized as *cooperative* or *self-contained*. The cooperative systems, in turn, are divided into *point source* systems and *multiple source* systems. Finally, the multiple source systems can be categorized as hyperbolic systems, pseudoranging systems, one-way synchronous (direct) ranging systems, and two-way (round-trip) ranging systems.

From a navigational viewpoint, systems are frequently classified as *positioning* systems and *dead-reckoning* systems. Most positioning systems are cooperative systems, while most dead-reckoning systems are self-contained. In dead-reckoning systems, the initial position of the vehicle must be known, and the system determines the distance and direction traveled from the departure point by continuous mathematical integration of velocity or acceleration. In this handbook, the system technologies are described primarily from the electronics viewpoint. In many modern electronic vehicle navigation systems, the data from cooperative and self-contained sensors are combined, typically using Kalman filters, in order to obtain a more accurate solution for position. Such systems are called *multisensor* or *hybrid* systems.

Cooperative Systems

The two general categories of cooperative electronic navigation systems are *point source* systems and *multiple source* systems. The point source systems typically determine the vehicle's present position by measuring the distance and bearing (azimuth) to a known source. They may determine only distance or only bearing, if these are the only desired parameters. Direction finders are examples of a bearing-only measurement. Multiple source systems determine vehicle position and in some systems also vehicle velocity, either in an absolute or a relative coordinate system. This may be accomplished by multiple ranging (multilateration), differential ranging, multiple angle measurements, multiple pseudorange measurements, or by a combination of these. These methods are also categorized as (1) range and bearing (rho-theta), (2) true ranges (rho-rho or rho-rho-rho), (3) angle-angle (theta-theta), (4) differential ranging, and (5) pseudoranging. The first one results in a combination of circular and radial lines of position (LOPs), the second in circular LOPs, the third in radial LOPs, and the fourth in hyperbolic LOPs. The fifth method exhibits spherical LOPs, but they do not cross at the correct position unless the user's time offset is determined. Therefore, its geometric (GDOP) behavior is not exactly equivalent to that of a rho-rho type system discussed later.

Point Source Radio Systems.   Perhaps the earliest form of a point source radio system is the *direction finder.* Its principle of operation is based on the use of a single transmitter source whose signal is received at two known points or elements. The direction from the vehicle to the source is determined by the measurement of the differential phase of the signals at the two points or elements. For operational convenience (notably size), it is frequently desirable to have the two receiving points close together and to use common circuitry at both measuring points. A loop antenna fulfills both of these requirements. Typically, a square loop is physically rotated until the currents in the two vertical arms of the loop are equal in amplitude and phase so that the output of the receiver is zero. The transmitter source is then located 90° from the plane of the loop. In simple loops there would be a 180° ambiguity, but this is resolved by temporarily connecting an omnidirectional antenna to the receiver input. Such direction finders are used for backup navigation and emergency homing to a station. Only a single transmitter source (beacon) and a simple rotating antenna and receiver on the vehicle are needed for operation. Lateral navigational error decreases as the transmitter source is approached, which is a property common to all point source angle-measuring systems.

Another class of point source angle-measuring systems are based on the use of scanning antenna beams at the transmitter source and reception of the transmitted signal by the user vehicle receiver. For example, if a ground transmitter generates a rotating cardiod amplitude pattern at a fixed rate plus an omnidirectional reference signal, the user receiver can measure the relative phase difference between these two signals and can thereby determine the bearing to the transmitter source. The operation of the VHF omnidirectional range (VOR), which is used worldwide for short-distance navigation is based on this principle.

The most common type of point-source system for *range* determination is based on the *two-way* (round-trip) ranging principle, which is illustrated in Fig. 23.3.1. The interrogator, which may be located on the navigating vehicle or at a reference site, transmits a signal, typically a pulse (or pair of pulses), at a known time, the transmission time being stored in the equipment. The signal is received at a transponder and, after a fixed known delay, it is retransmitted toward the interrogator, is received by the interrogator's receiver, and input to a ranging circuit. The latter measures the time difference between the original transmission time and the time of reception (less the known fixed delay), which is a direct measure of the two-way distance when multiplied by the speed of light. An important advantage of this technique is that the signal returns to the point of initial generation for the time difference measurement process. Therefore, the interrogator's clock oscillator does not need to be highly stable, since the resulting time error due to any clock instability is only a function of the round-trip time multiplied by the clock drift, the round-trip time being very short, inasmuch as the signal travels at the speed of light. If the transponder in Fig. 23.3.1 is replaced by a passive reflector, for example an aircraft, the principle of operation illustrated is that used in primary surveillance radars, such as those used for air traffic control, as well as military ground-based and airborne radars.

A second fundamental technique of range determination is a *one-way* (versus two-way) signal delay (time of arrival) measurement, between a transmitter source at a known location and a user receiver (Fig. 23.3.2). In this case, an accurate measurement is possible only if the transmitter oscillator (clock) and the user receiver oscillator (clock) are precisely time synchronized. If such time synchronization is not maintained, *true* range cannot be determined, since the exact time of transmission is not known with respect to the user's clock time.

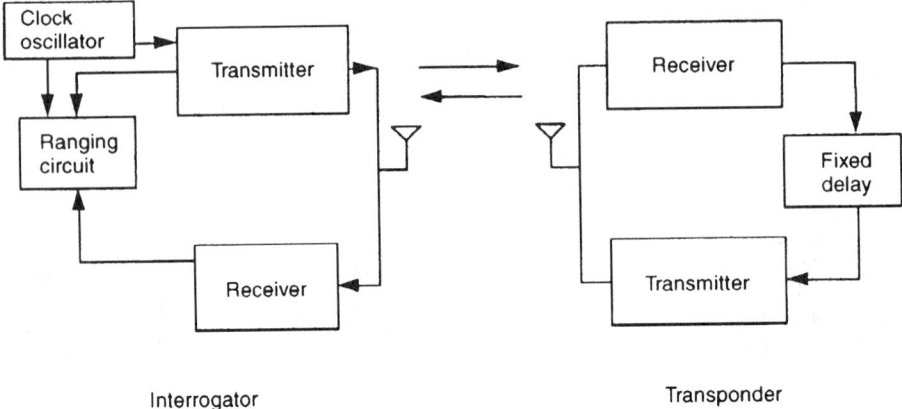

FIGURE 23.3.1   Two-way (round-trip) ranging.

Such precise time synchronization between two individual pices of equipment is frequently not possible at reasonable equipment cost. Therefore, no practical *point-source* distance measurement system based on this technique has been developed to date; however, several modern multiple source systems (e.g., PLRS, JTIDS-RelNav, and some hyperbolic systems) have modes that use *one-way* synchronous ranging after independent time synchronization has first been accomplished.

Multiple Source Radio Systems.   Many systems containing multiple transmitter sources have been developed for the determination of vehicle position. Normally, the user vehicle equipment includes a receiver or a receiver/transmitter. There are five categories of such systems (with some implementations using combinations of these), namely (1) hyperbolic systems, (2) inverse hyperbolic systems, (3) pseudoranging systems, (4) one-way synchronous ranging (direct ranging) systems, and (5) two-way (round trip) ranging systems.

The hyperbolic systems were the first to be developed and are currently in widespread use. The principle of operation is based on the use of three or more transmitter sources, which continuously or periodically transmit time synchronized signals. These may be continuous wave or pulsed signals. The minimum size *chain*, a triad,

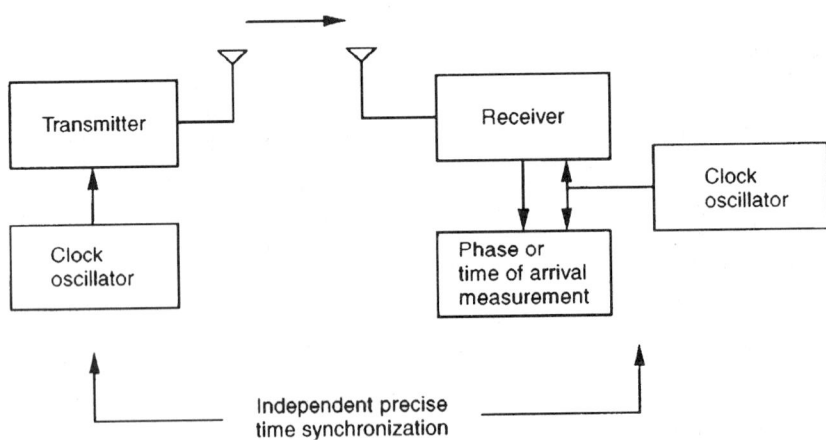

FIGURE 23.3.2   One-way synchronous ranging.

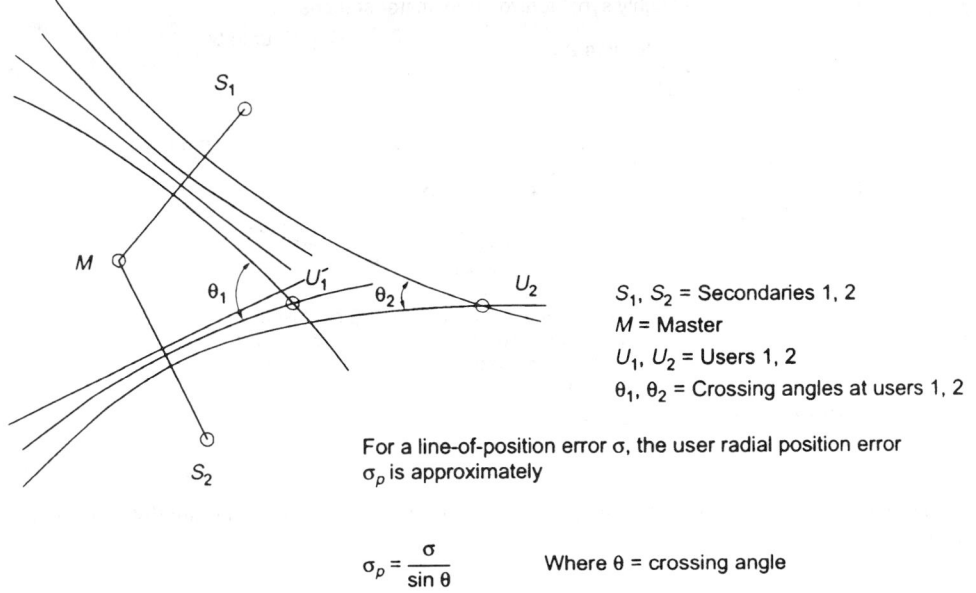

S_1, S_2 = Secondaries 1, 2
M = Master
U_1, U_2 = Users 1, 2
θ_1, θ_2 = Crossing angles at users 1, 2

For a line-of-position error σ, the user radial position error σ_p is approximately

$$\sigma_p = \frac{\sigma}{\sin \theta}$$ Where θ = crossing angle

FIGURE 23.3.3 Hyperbolic Navigation System; geometric effects.

usually consists of one Master and two Secondary (slave) stations (Fig. 23.3.3). The user receiver measures the time differences of arrival (TDs or TDOAs) of signals from pairs of stations. Loci of constant time differences of arrival, or (equivalently) constant differences in distance, from two stations form hyperbolic lines of position (LOPs). The point where two lines of position cross is the position of the user vehicle (Fig. 23.3.3). One major advantage of this technique is that the user needs only a receiver and the receiver does not need a high-quality time synchronized clock oscillator, since only *time differences* are used. Theoretically, three pairs of sources are needed for a completely unique horizontal position fix, but in practice two pairs will suffice. The "differences in distance" can be measured either in terms of differences in times of arrival (for example of pulses), differences in electrical carrier phase, or both. As depicted in Fig. 23.3.3, achievable accuracy is very much a function of the relative geometry between the sources and the user, i.e., the crossing angles of the LOPs. The smaller the crossing angle, the larger the position error. This accuracy degradation is called geometric dilution of precision (GDOP). The GDOP is essentially a multiplier on the basic time difference of LOP measurement error.

A related concept, used primarily for external position location, could be called *inverse hyperbolic* system. In this system the vehicle carries a transmitter, which periodically transmits a signal to several receiving stations, whose positions are precisely known. The time differences of arrival (TDOAs) of the signal from the vehicle at pairs of stations is measured and the loci of constant difference in time of arrival form hyperbolic lines of positions. The point where two such lines of positions cross is the position of the vehicle. This technique has been applied to automatic vehicle location (AVL) systems.

The third multiple source radio system concept, which has recently become very important, is called *pseudoranging*. In this concept, several transmitter sources, whose positions are made known to the user, transmit highly time synchronized signals on established *system time* epochs (Fig. 23.3.4). With these time epochs known to the user, the user measures the time of arrival (TOA) of each signal with respect to its own clock, which normally has some time offset from system time. The resulting range measurement (by multiplying by the speed of light) is called *pseudorange* (PR), since it differs from the *true range,* as a result of the user's time offset. From several successive or simultaneous such TOA (pseudorange) measurements from four (or more) sources, the user receiver then calculates the three-dimensional position coordinates and its own time offset (from system time). This is accomplished by solving four simultaneous quadratic equations, involving the three known position coordinates of the sources and the four unknowns, namely, the three user position coordinates and the user's time offset.

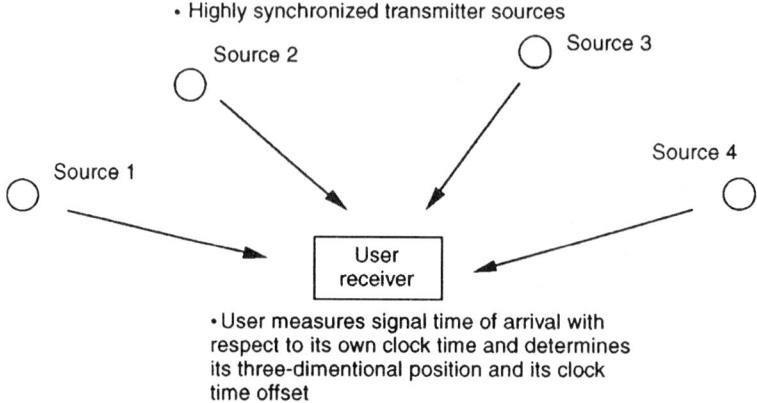

FIGURE 23.3.4   Pseudoranging system.

The basic solution equations for the position of a user in a pseudoranging multiple source system are given by

$$R_{0i} = c.\text{TOA}_i = [(x_u - x_i)^2 + (y_u - y_i)^2 + (z_u - z_i)^2]^{1/2} - c. \Delta T_u \tag{81}$$

where R_i = pseudorange from the user to source i
  c = speed of light
  TOA_i = time of arrival of the signal from the ith source
  x_u, y_u, z_u = unknown three-dimensional components of the user position
  x_i, y_i, z_i = known three-dimensional components of the source i's position
  ΔT_u = user's clock oscillator's time offset
  $i = 1, 2, 3, 4$

With x_i, y_i, z_i known and transmitted to the user and four TOAs from four sources measured, the user position coordinates x_u, y_u, z_u and the user's time offset ΔT_u can be determined by solving these four equations with four unknowns. In some implementations, more than four sources are used and the solution is then overdetermined. Thus, this system not only accurately determines the user's three-dimensional position, but also system time, which can be related (in a message from the sources) to standard *universal time coordinated* (UTC). The operation of the satellite-based U.S. Global Positioning System (GPS) and the Russian GLONASS are based on this technique. Also, certain modes of some terrestrial systems, e.g., JTIDS-RelNav use this technique. In addition, by properly combining the known velocity of the sources (from transmitted data) and the measured Doppler shift of the signals received from the sources, the three user velocity coordinates and the user clock frequency offset (drift) can also be determined. Specifically, if the set of Eqs. (81) are differentiated, the basic solution equations for the determination of the user velocity in a pseudoranging multisource system can be obtained. In many practical implementations, the delta pseudorange rate or integrated Doppler measurement is used via a Doppler count over a short time interval. Since Eq. (81) is nonlinear, the solution is normally obtained via linearization about an assumed position and time.

In order to provide high-accuracy TOA measurement capability, these pseudoranging navigation systems use wide bandwidth, spread spectrum modulation methods, notably *direct sequence spreading*. Propagation delay corrections are mode as required.

The fourth multiple source radio system concept is based on one-way synchronous ranging or direct ranging (see Fig. 23.3.2) and the earlier discussion. The concept is used in some military systems (e.g. JTIDS-Relnav and PLRS) and in a second mode of some hyperbolic systems, e.g., Loran-C and Omega. When applied to the latter, the concept is called *direct ranging*. Two implementations are possible, namely, the *rho-rho* method or the *rho-rho-rho* method. The rho-rho method requires only two source transmitters, but it also requires a highly stable user receiver oscillator (clock) and precise knowledge of the time of transmission from the source station. A *direct range* is then developed to each station. The disadvantage of the rho-rho method is that a very high stability oscillator is required. The rho-rho-rho scheme is an extension of the rho-rho scheme requiring the use of at least three stations. Using three LOPs permits clock oscillator *self-calibration*, somewhat similar to the previously discussed pseudoranging concept and therefore leads to a less stringent clock oscillator requirement. The LOPs of both schemes are circles rather than hyperbolas, with the intersection of the circles representing the user position, thereby leading to more favorable geometry conditions.

The fifth multiple source system for position determination is based on multiple two-way (round trip) *true range* measurements. It is therefore a direct extension of the concept depicted in Fig. 23.3.1. To obtain a completely unambiguous horizontal position fix, three two-way ranges are required; however, in most practical cases, two are sufficient. Since the LOPs are circular, the geometric accuracy behavior is generally more favorable than for hyperbolic systems.

Self-Contained Systems

These electronic navigation systems are called self-contained because the navigation equipment required is located totally on the vehicle, i.e., operation of the system is not dependent on any outside sources or stations. These systems can be classified as radiating or nonradiating. The radiating systems described below are (1) the radar altimeter, (2) the airborne mapping radar (3) map matching systems, and (4) Doppler navigation radars. Nonradiating systems offer essentially complete protection against radio interference or jamming. The two systems in this category described below are inertial navigation systems and celestial navigation systems.

Radar Altimeter (Ref. 1, Chapter 10).   The device (also known as *radio altimeter*) is a small radar with an antenna on the bottom of an aircraft generating a beam toward the earth's surface. The signal is back-scattered by the surface and received and processed by the radar, which measures the two-way (round-trip) delay experienced by the signal, thereby generating altitude above the surface. The modulation may be frequency-modulated continuous wave (FM-CW) or pulse modulation. For civil aviation use, the systems typically have a range of 0 to 2500 ft (0 to 750 m), are used for landing guidance, and are of the FM-CW type. In FM-CW systems, the carrier is frequency modulated in a triangular pattern, a portion of the transmitted signal is mixed with the received signal and the resulting beat frequency is directly proportional to altitude above terrain. In pulse radar altimeters, primarily used in military aircraft, the carrier is modulated by narrow pulses and the time of reception of the leading edge of the return pulses are measured with respect to their time of transmission. The time difference is a direct measure of the two-way distance to the ground straight below the aircraft. The frequency band of operation in both types of altimeters is 4200 to 4400 MHz. Typical accuracy performance is 2 ft or 2 percent. Military aircraft and helicopters use these altimeters for terrain avoidance.

Mapping Radars (Ref. 1, Chapter 11).   These radars scan the ground using specially shaped beams, which effectively map the terrain. They are used by pilots for navigation by recognizing certain terrain features, for example, rivers and bridges, and by manually or semiautomatically position fixing their navigation systems through designation on the known ground mapped objects. Synthetic aperture radars (SARs) provide particularly high resolution of the mapped terrain, making these position update functions very accurate.

Terrain (Map) Matching Systems (Ref. 1 Chapter 2).   The output of an airborne mapping radar or a radar altimeter can be used to generate a digital map or terrain profile which is then compared with on-board stored digital maps or terrain profiles, in order to allow a military aircraft or missile to automatically fly a prescribed track.

Doppler Radar Navigation (Ref. 1, Chapter 10).   Radio waves that are transmitted from a moving aircraft toward the ground and back-scattered by the ground to the aircraft experience a change of frequency, or Doppler shift, which is directly proportional to the ground speed of the aircraft and the cosine of the angle between the radiated beam center line and the velocity vector. A Doppler navigation radar consists of a transmitter, a receiver, and a frequency tracker which measures and tracks the Doppler shift of the signals in each of three or four antenna beams directed at steep angles toward the earth's surface. Modern systems operate at the 13.325-GHz frequency. The transmission may be pulse modulated, frequency-modulated continuous-wave (FM-CW), or continuous wave (CW). Because the earlier, pulse modulated systems were less efficient, current systems are either of the FM-CW or CW type. The FM-CW systems transmit sinusoidally modulated signals which are mixed with the back-scattered signal in the receiver. The Doppler shift of one of the (Bessel function) sidebands of the mixed signals is measured and tracked. In the pure CW systems, the beat frequency Doppler between the transmitted and received signals are measured and tracked. Typical transmitted antenna patterns consists of four narrow beams, directed at steep angles toward the ground and are generated by planar array antennas. For the determination of three-dimensional velocity components, a minimum of three such beams are required. However, in modern systems, four beams are used because they are easily generated by planar array antennas and the redundant information can provide higher accuracy and a self-test function. If the antenna is fixed to the airframe the ground velocity components are computed by resolving the beam Doppler shifts through pitch and roll angles obtained from a vertical gyro or inertial navigation system. In order to determine vehicle present position, the velocity components in aircraft coordinates are converted to earth-referenced coordinates by resolving them about true heading obtained from a heading reference, such as a magnetic compass corrected for variation, an attitude-heading reference or an inertial navigation system. The velocity components are then integrated into distance traveled north and east from a previously known position (dead reckoning). The position accuracy of a Doppler navigation system is therefore a function of the accuracy of the heading reference. The basic velocity accuracy of modern lightweight Doppler radars is 0.2 percent or ± 0.2 knot. These systems have been used by military aircraft and transoceanic airliners. Currently, Doppler radar navigation systems are widely used on military helicopters for navigation and hovering. Doppler radars have also been used for achieving lunar and planetary landings. The Doppler velocity measurement concept has been incorporated into modern airborne search and mapping radars (Ref. 1, Chapter 11) and is also used in sonar systems for measuring ship's velocity.

Inertial Navigation Systems (Ref. 1, Chapter 7).   Inertial navigation equipment which is based on Newton's second law can determine aircraft acceleration and direction of travel. From Newton's second law, it is known that the force acting on a body is proportional to its mass and acceleration. In one implementation, acceleration may be determined by measuring the deflection of a spring attached to a known mass. If the acceleration is doubly integrated, the distance traveled can be determined. An inertial navigation system (INS), consisting of accelerometers, gyros, and processors, continuously determines position, velocity, acceleration, direction (with respect to north), and attitude (pitch and roll) of a vehicle. Since the position is obtained from doubly integrating acceleration of the vehicle from a known position, an inertial navigation system is inherently a *dead reckoning* system. In a good quality inertial navigation system, the accelerometers must be capable of measuring acceleration with high accuracy; e.g., a 10^{-4} g accelerometer bias error causes a 0.3-nmi-peak position error, where g is the magnitude of the gravitational acceleration (32.2 ft/s^2). Misalignment of an accelerometer with respect to vertical causes it to read a component of the gravity vector g. Therefore, in some systems the three accelerometers (for sensing acceleration in three dimensions) are mounted on a gimbaled platform, which is stabilized by gyroscopes in order to keep the accelerometers in a horizontal plane. The drift rate of the gyroscopes must be low, since a 0.017° drift rate gives rise to a 1 nmi/h average position error. In many modern systems, the inertial sensors are strapped to the vehicle, e.g. the aircraft, and data from the gyros are used to correct the data from the accelerometers. These are called *strapdown* systems. The basic inertial navigation system error equations for motion over the earth give rise to a sinusoidal oscillatory behavior at the so-called Schuler frequency. The latter is derived from the square root of the ratio of the magnitude of the gravitational acceleration g and the radius of the earth. Its period is 84.4 min. As a result, the inertial position and velocity errors are oscillatory at that frequency, although increasing linearly with time, with the Schuler oscillations superimposed. Inertial navigation systems must be aligned initially as accurately as possible, and their accuracy degrades with time, even when stationary (due to gyro drift and accelerometer bias). Alignment is degraded at high latitudes (above 75°) and in a moving vehicle. The gyroscopes used in earlier systems were precision mechanical devices. Recently, ring laser gyros (RLGs) have been developed and are used in many modern inertial systems. Their operation is based on the use of an optical laser that generates two counter rotating light beams within a triangular structure. When that structure is physically rotated around the axis normal to the plane of the structure, the beat or difference frequency of the light beams is proportional to the angular rate and is sensed as an output at one corner of the structure (Sagnac effect). This results in the device being a rate gyro. These gyros are less expensive than mechanical gyros. Another gyro based on a similar optical concept, which has been developed and shows great promise, is the fiberoptic gyro (FOG). Typical modern inertial navigation systems have position accuracies between 0.5 and 1 nmi/h.

Celestial Systems, (Ref. 1, Chapter 12).   These systems operate by making optical observations on celestial bodies and solving the astronomical triangle. Accurate knowledge of the position of the vehicle with respect to the local horizontal (pitch and roll) is required to obtain optimum performance with these systems. Fortunately, this information is readily available from inertial instruments or inertial navigation systems. These systems are particularly useful in bounding the long-term errors of inertial navigation systems. Previously they were used extensively on all types of ships and commercial airliners for position fixing. Currently, they are used primarily on military aircraft.

GEOMETRIC DILUTION OF PRECISION

In hyperbolic and pseudoranging navigation systems, such as Loran-C and GPS, there is a significant effect on accuracy due to the relative geometric location between the user vehicle and the transmitter sources. This effect is represented by a nondimensional term called *geometric dilution of precision* (GDOP). GDOP is essentially a factor by which the basic range difference or ranging error is multiplied to obtain the resulting position error.

Horizontal Hyperbolic System GDOP

In a two-dimensional hyperbolic navigation system such as Loran-C, using a chain of three stations (triad), the GDOP factor is given by the following equation:

$$\text{GDOP} = \frac{1}{\sin(\phi_1 + \phi_2)} \sqrt{\frac{1}{\sin^2 \phi_1} + \frac{1}{\sin^2 \phi_2} + \frac{2r\cos(\phi_1 + \phi_2)}{\sin\phi_1 \sin\phi_2}} \tag{82}$$

where ϕ_1, ϕ_2 = half-angles subtended by the two station pairs
r = correlation coefficient between two lines of position (LOPs) (typically assumed to be 0.5)
$\phi_1 + \phi_2$ = crossing angle of the LOPs

The radial standard deviation horizontal position error is the product of GDOP and the standard deviation range difference (LOP) measurement error.

Three-Dimensional Pseudoranging System GDOP

For pseudoranging, satellite-based systems, such as GPS and GLONASS, which provide information on the three-dimensional components of position of the vehicle and time, the GDOP factor is given by the following equation:

$$\text{GDOP} = [\text{Trace } (H^T H)^{-1}]^{1/2} \tag{83}$$

where H is the measurement matrix relating the pseudorange measurements to the three user position components and the user time offset (bias) from system time, i.e.,

$$HX_u = PR$$

H is defined such that the ith row (for satellite i) of $H(h_i)$ is given by $h_i = (1_{xi}\ 1_{yi}\ 1_{zi} - 1)$ and the elements 1_{xi}, 1_{yi}, 1_{zi} are the unit vectors defining the direction cosines from the user position to the satellite positions; (H has at least four rows), X_u is the vector of the user xyz position coordinates (m), and the time offset expressed in meters, i.e., $X_u = (x_u, y_u, z_u, c\Delta T_u)^T$, ΔT_u = user time offset (seconds), c = speed of light (3×10^8 m/s). Trace ($\bullet$) indicates the sum of the diagonal elements of ($\bullet$). PR = vector of measured pseudoranges to the satellites, including errors. x, y, z can be the orthogonal right-handed cartesian Earth Centered Earth Fixed (ECEF) coordinates, with z from the center of the earth to the North Pole, x through the Greenwich meridian, and y orthogonal to x in the plane of the equator.

The solution for x_u, y_u, z_u, ΔT_u involves four (or more) pseudoranges and requires linearization about an estimated position. This is typically done using a minimum variance, Kalman filter, or least squares estimator and leads to pseudorange residuals, such that:

$$\delta PR = H(\delta X_u^T c \cdot \delta \Delta T_u) \tag{84}$$

where δ_{PR} = four element pseudorange error vector
δX_u = user position error vector
$\delta \Delta T_u$ = user time offset error

If the four (or more) pseudorange errors are statistically uncorrelated with equal, zero mean one-sigma values of σ_{PR}, the one-sigma position and time error is given by

$$\sigma_{x,t} = \text{GDOP}.\sigma_{PR} = (\sigma_x^2 + \sigma_y^2 + \sigma_z^2 + c^2\sigma_{\Delta t}^2)^{1/2} \tag{85}$$

The position dilution of precision (PDOP) is computed by deleting the fourth diagonal element in the trace of the GDOP equation. The time dilution of precision (TDOP) is the square root of the fourth element of the GDOP equation. The first two diagonal elements in the trace are used to compute HDOP, and only the third diagonal element is used to compute VDOP. The ECEF coordinate residuals must be transformed to local tangent plane coordinates, in order to compute HDOP and VDOP (Ref. 1, Chapter 5).

INTERNATIONALLY STANDARDIZED SYSTEMS

The systems described next are used by hundreds of thousands of vehicles throughout the world. Standardization is therefore desirable and refers principally to the radiated signal characteristics and the performance of the system. Each manufacturer and country may decide on the equipment design that best satisfies the operational requirements.

Omega (Ref. 1, Chapter 4)

This is a hyperbolic navigation system that provides worldwide service with a high enough data rate to be used by aircraft. It is a hyperbolic phase-comparison system using eight stations located in Norway, Liberia, Hawaii, North Dakota, La Reunion Island, Argentina, Australia, and Japan. The system operates in the 10 to 14-KHz, very low frequency (VLF) band, using a fixed transmission pattern. The overall signal format is shown in Fig. 23.3.5. Each station transmits on four common frequencies and one station-unique frequency. The signal frequencies are time-shared among the stations, so that a given frequency is transmitted by only one station at a given time. Each station transmits on one frequency at a time, for about 1 s, the cycle being repeated every 10 s. At the transmitted VLF frequencies, lane ambiguity occurs approximately every 8 n.mi., i.e., every half-wavelength. However, by using the beats between the basic frequencies, these ambiguities can be extended, e.g., to 24 n.mi for two frequencies and to 72 n.mi for three frequencies. Most modern receivers use at least three frequencies. In addition, continuous lane counting is used from a starting point, in order to avoid any lane ambiguity problems. Omega signal propagation basically takes place in the wave guide formed by the surface of the earth and the ionosphere. Most airborne receivers also use the system in the rho-rho-rho mode, backed up by 10 very low-frequency (VLF) communication transmitter stations in the 16 to 24 kHz band.

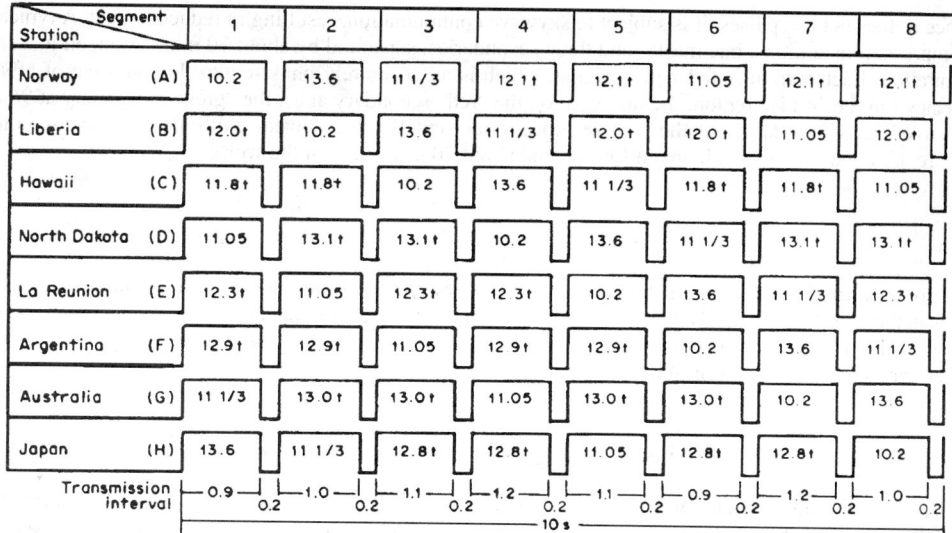

FIGURE 23.2.5 Omega System Signal Transmission Format (frequencies in kilohertz). Frequencies marked † are the unique frequencies for the respective stations.

There is a marked difference in propagation time between day and night (diurnal variation), but this is predictable and amenable to computer storage, and can be applied to the solution in the receiver, as a function of time, data, and approximate user position. Also, an earth's conductivity map can be stored in the computer. The position accuracy of Omega is 2 to 4 n. mi (3.7 to 7.4 km), 2 DRMS.

Loran C (Ref. 1, Chapter 4)

This is a hyperbolic system, with coverage in the United States, the North Atlantic, the Mediterranean, and the North Pacific. The combination of low-frequency transmission and pulses provides long range and eliminates sky waves. The system uses a carrier frequency of 100 kHz and matches the individual rf cycles within each pulse, thereby gaining added resolution and accuracy. Since all stations operate on the same frequency, discrimination between chains is by the pulse-repetition frequency. A typical chain comprises a master and two secondaries about 600 n.mi from the master. At the receiver, the first three rf cycles are used to measure the signal time of arrival. At this point the pulse is about half amplitude. The remainder of the pulse is ignored, since it may be contaminated by skywave interference. To obtain greater average power at the receiver without resorting to higher peak powers, the master transmits groups of nine pulses 1000 μs apart and the secondaries transmit eight pulses, also 1000 μs apart. These groups are repeated at rates ranging from 10 to 25/s. Within each pulse, the rf phase can be varied in a code for communication purposes. At the receiver, phase-locked loops track the master and secondary signals and generate their time differences. A digital computer can provide direct readouts in latitude and longitude and steering information. There has been a tremendous growth in the use of Loran by general aviation. Position accuracy is 0.25 n.mi (0.4 km) 2 DRMS.

Decca (Ref. 1, Chapter 4)

Unlike all other systems described herein, Decca is a proprietary system, with most of the stations owned by the Decca company and the receivers leased by it to the users, who are primarily in Europe. It is a continuous-wave hyperbolic system operating in the 70 to 130-kHz band. It uses the same frequency band as Loran-C, but since it does not use pulses, it is subject to skywave contamination, resulting in reduced range. A typical chain comprises four stations, one master and three secondaries, separated by about 70 n.mi, arranged in a star configuration. Each station is fed with a signal, which is an accurately phase-controlled multiple of a base frequency f in the 14 kHz region, the master at $6f$, the "red" secondary at $8f$, the "green" secondary at $9f$, and the "purple" secondary at $5f$. At the receiver, these four signals are received, multiplied, and phase-compared. There are about 25 Decca chains in Europe and about 20 elsewhere in the world.

Beacons

As sources for shipboard and airborne direction-finders, these are the oldest and most numerous navigation aids in the world. Since the frequency bands of beacons are adjacent to the amplitude-modulation (AM) broadcast band, receivers are easily designed to serve a dual purpose and they are consequently popular with boat operators and with aircraft of all types. Direction-finding accuracy can be as good as $\pm 3°$.

Very High Frequency Omnidirectional Range (VOR) (Ref. 1, Chapter. 4)

This aviation system uses the VHF band and is thus free of atmospherics and skywave contamination. It places the directional burden on the ground, rather than in the aircraft, where more extensive means can be employed to alleviate site errors. Line-of-sight limits its service area to about 200 n.mi for high-flying aircraft, and some stations are intended for only 25 n.mi service to low-flying aircraft. There are more than 1000 stations in the United States and about an equal number in the rest of the world. There are two variations, i.e., conventional VOR and Doppler VOR, with the latter providing increased site error reduction.

Conventional VOR. It operates on 40 channels, 100 kHz apart, between 108 and 112 MHz, interleaved between ILS Localizer channels, and on 120 channels, spaced 50 kHz, between 112 and 118 MHz. The airborne receiver is frequently common with the airborne Localizer receiver and may use the same airborne antenna. Power output from the ground transmitter varies from 25 to 200 W, depending on antenna design and on the desired service area. The ground-antenna pattern forms a cardioid in the horizontal plane which is rotated 30 times per second. The CW transmission is amplitude-modulated by a 9960-Hz tone which is frequency modulated ± 480 Hz at a rate of 30 Hz. This latter, 30-Hz "reference" tone, when extracted in the airborne receiver, is compared with the 30-Hz amplitude modulation provided by the rotating antenna. The phase angle between these two 30-Hz tones is the bearing of the aircraft with respect to north.

VOR is the internationally standardized en route navigation system, widely implemented through-out the world which meets the FAA en route integrity requirements for a warning of within 10 s if there is a failure of ground equipment. The basic VOR instrumental error is $\pm 1.4°$ 2-sigma. Site errors can degrade this error; this lead to the development of the Doppler VOR, described below.

Doppler VOR reduces site error about tenfold by using a large-diameter antenna array at the ground station. This array consists of a 44-ft diameter circle of antennas. Each antenna is sequentially connected to the transmitter in a manner so as to simulate the rotation of a single antenna around the 44-ft diameter circle at 30 rps. The receiver sees an apparent Doppler shift in the received rf of 480 Hz at a 30 rps rate, and at a phase angle proportional to the receiver's bearing with respect to north. This signal is therefore identical with the conventional VOR reference tone. It remains to transmit a 30-Hz AM tone, separated 9960 Hz as a reference, in order to radiate an identical signal to the conventional one, receivable in an identical receiver but benefiting from a manifold increase in ground-antenna aperture.

Distance-Measuring Equipment (DME) (Ref. 1, Chapter 4)

DME is an interrogator-transponder *two-way* ranging system (Fig. 23.3.1). About 2000 ground stations and 70,000 pieces of airborne equipment are in use worldwide.

The airborne interrogator transmits 1kw pulses of 3.5 μs duration, 30 times a second, on one of 126 channels 1 MHz apart, in the band 1025 to 1150 MHz. The ground transponder replies with similar pulses on another channel 63 MHz above or below the interrogating channel. (This allows use of the transmitter frequency as the receiver's local oscillator frequency if the intermediate frequency is 63 MHz). A single antenna is used at both ends of the link. In order to reduce interference from other pulse systems, paired pulses are used in both directions, their spacing being 12, 30, or 36 μs. The fixed delay in the ground transponder is 50 μs.

In the airborne unit, the received signal is compared with the transmitted signal, their time difference derived, and the distance is determined and displayed. Ground transponders are arranged to handle interrogation from up to 100 aircraft simultaneously, each aircraft recognizing the replies to its own interrogation by virtue of the pulse repetition frequency being identical with that of the interrogation. Analog models require about 20 s for identity to be initially established (after which a continuous display is provided); modern digital models perform the search function in less than 1 s.

The DME is nearly always associated with a VOR, the two systems forming the basis for a rho-theta area navigation (RNAV) system. Some use is being made of DME in the rho-rho mode, particularly by airlines. In this mode, the airborne interrogator scans a number of DME channels, automatically selecting two or more having the best signal and thus providing a fix of greater accuracy than would be obtained from VOR/DME. This is known as DME/DME and is thus a way of achieving a two-way ranging *multiple* source navigation system. DME is an international standard which, together with VOR, has been the basis for the most widely used line-of-sight rho-theta aircraft navigation system. Standard airborne DME equipment provides a two-sigma accuracy of ±0.1 n. mi (±185 m).

Tacan (Ref. 1, Chapter 4)

This is a military modification of DME, using the same channels, and adding a bearing capability in the same frequency band. This results in a small ground antenna system, a property useful for ships. It is a military system used widely on aircraft carriers. The DME antenna is replaced by a rotating directional antenna generating two superimposed patterns. One of these is a cardiod rotating at 15 rps. The other is a nine-lobe pattern, also rotating at 15 rps. The squitter pulses and replies are amplitude-modulated as the antenna rotates. Reference pulses are transmitted at 15 and 135 Hz. In the aircraft, a coarse phase comparison can then be made at 15 Hz which is supplemented by a fine comparison at 135 Hz. The overall instrumental accuracy of TACAN is 0.2° two-sigma (bearing) and 0.1 n.mi. (185 m) two-sigma (distance).

Vortac (Ref. 1, Chapter 4)

In countries having a common air traffic control system for civil and military users (e.g., the United States and Germany), the civil rho-theta system is implemented by the use of Tacan rather than DME for distance measurement. Tacan transponders are colocated with VOR stations, and civil aircraft get their DME service from the Tacan stations. In the United States, over 700 VORs have colocated Tacan transponders.

Instrument Landing Systems (ILS) (Ref. 1, Chapter 13)

This is the internationally standardization aircraft approach and landing system, which provides a fixed electronic path to almost touchdown at major runways throughout the world. The ground equipment is made up of three separate elements: the *localizer*, giving left-right guidance; the *glide slope*, giving up-down guidance; and the *marker beacons*, which define progress along the approach course. Using VHF, ILS is free of atmospheric and sky-wave effects, but is subject to site effects.

The *localizer* operates on 40 channels spaced 50 KHz apart, in the 108 to 112 MHz band, radiating two antenna patterns which give an equisignal course on the centerline of the runway, the transmitter being located at the far end of the runway. The left-hand pattern is amplitude-modulated by 90 Hz, the right-hand pattern by 150 Hz (Fig. 23.3.6a). The airborne receiver detects these tones, rectifies them, and presents a left-right display in the cockpit. The accuracy is better than ±0.1°. Minimum performance calls for the airborne meter to remain hard left or hard right to a minimum of ± 35° from the centerline, i.e., there must be no ambiguous or "false" courses within this region. More sophisticated systems exist in which a usable "back course" (with reverse

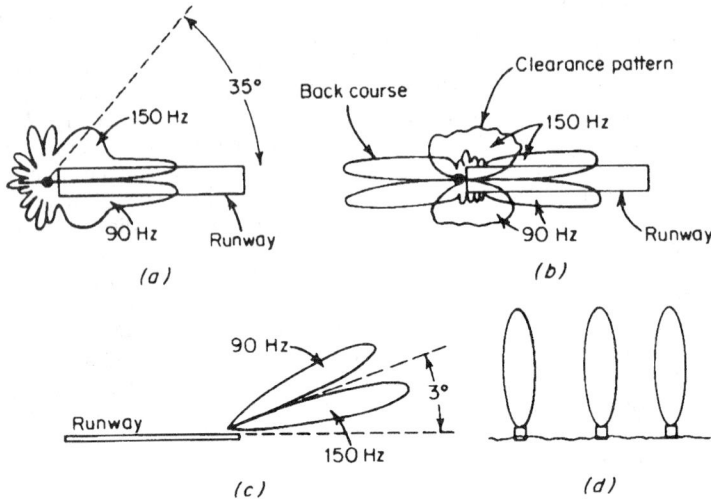

FIGURE 23.3.6   Instrument Landing System: (*a*) minimum ICAO localizer pattern; (*b*) localizer pattern with back course and clearance; (*c*) glide slope pattern; (*d*) marker beacons   (*From* S. H. Dodington, *Electronic Engineers' Handbook*, McGraw Hill, 3d ed.).

sense) is obtained, and a separate transmitter, offset by about 10 kHz, provides "clearance," so that no ambiguities exist throughout ±180° (Fig. 23.3.6*b*).

The *glide-slope* transmitter, of about 7W power, is located at the approach end of the runway and up to about 500 ft to the side (Fig. 23.3.6*c*). It operates in the 329 to 335-MHz band, each channel being paired with a localized channel. In the airborne receiver, both channels are selected by the same control. Two antenna patterns are radiated, giving an equisignal course about 3° above the horizontal. The lower pattern is modulated with 150 Hz, and the upper pattern with 90 Hz. The airborne receiver filters these tones, rectifies them, and presents the output on a horizontal zero-centered meter mounted in the same instrument case as the localizer display, the two together being called a *cross-pointer display*. The accuracy is better than ±0.1°. Required accuracies for ILS and MLS are sometimes specified in distance (meters) at the decision height for the three landing categories described below.[9] The glide slope suffers from course bends because of the terrain in front of the array, and is generally not depended on below 50 ft of altitude. In this phase of the landing maneuver, either visual observation or a radar altimeter are frequently used.

Marker beacons operate at a fixed frequency of 75 MHz, and radiate about 2 W upward toward the sky with a fan-beam antenna pattern whose major axis is across the direction of flight (Fig. 23.3.6*d*). There is an "outer" marker about 5 n.mi from touchdown, and a "middle" marker about 3500 ft from touchdown. There can also be an "inner" marker at about 1000 ft from the touchdown threshold. Each type is modulated by audio tones which are easily recognized as the aircraft passes through their antenna pattern. Alternatively, differently colored lamps are set to light in the cockpit as each marker is passed.

Categories of ILS performance have been established for different visibility conditions and the quality of the installation. These place the following minimum limits on how an aircraft may approach the touchdown point:

Category I: 200-ft ceiling and $\frac{1}{2}$-mi visibility

Category II: 100-ft ceiling and $\frac{1}{4}$-mi visibility

Category III: zero ceiling and 700-ft visibility

ILS meets the FAA requirements for approach mode integrity; that is, failure of the ground equipment is evident to the pilot within 2 s.

Microwave Landing System (MLS) (Ref. 1, Chapter 13)

While the ILS described above has served well worldwide for over 30 years, it has been thought that requirements for the future will necessitate more channels, more flexible (curved) approach paths, and greater freedom from site effects. These can be obtained at microwave frequencies where greater antenna directivity and a wider frequency spectrum are available. Since a range of only 20 n.mi or so is needed, line-of-sight limitations pose no problem. Angular guidance is obtained from fan-shaped beams that scan the airspace, using 200 radio frequencies between 5.00 and 5.25 GHz. The perceived angle in the aircraft is proportional to the time it takes the beam to pass through the aircraft, first in one direction and then in the other. The scanning rate is 50 μs/degree. The system is thus known as a time-referenced scanning beam (TRSB) system. Use of the high microwave frequencies and the nature of the ground-based antenna patterns, as well as the possible inclusion of a high-precision version of DME, can provide higher accuracy performance than that of the ILS in some sites and a capability for curved approaches. However, in view of the great potential for using GPS for approach and landing, the U.S. FAA has decided to curtail development efforts of the MLS. In Europe, the civil aviation community, however, has shown continued interest in using MLS, in view of the unique requirements there. Also, the U.S. military services have had continued interest in microwave landing systems. The official, basic specified accuracy performance of MLS is the same as that of ILS.[9]

JTIDS-RelNav (Ref. 1, Chapter 6)

The Joint Tactical Information Distribution System (JTIDS) is a decentralized, military, spread spectrum data communication and navigation system, using wide bandwidth phase coding, frequency hopping and time division multiple access. It operates in the 960 to 1215-MHz band. It includes a *Relative Navigation* (RelNav) function that permits all members of a JTIDS network, such as aircraft and ships, to accurately determine their position in both absolute and relative grid coordinates. Its operation is based on highly precise TOA measurement of signals received by cooperating units. The system includes a means for independent precise time synchronization of each unit with *system time*. The system includes two modes of operation, namely, one-way synchronous ranging and pseudoranging.

Position Location Reporting System (PLRS) (Ref. 1, Chapter 6)

The Position Location Reporting System (PLRS) is a centralized, military, spread spectrum *position location and navigation system* for military aircraft, land vehicles, and personnel. It uses wide bandwidth phase coding, frequency hopping and time division multiple access and also incorporates data communications capability. It operates in the 420 to 450 MHz frequency band. It provides military commanders with information on the location of all of his elements, as well as own position and relative guidance information to each unit, in the absolute Military Grid Reference System (MGRS). Its operation is based on multilateration using highly precise TOA measurements of signals exchanged between units. Multiple relays are used to combat line-of-sight problems. The system includes two modes of operation, namely, two-way (round-trip) ranging and one-way synchronous ranging.

IFF

To distinguish friend from foe, radars employ an interrogator-transponder system operating at a different set of frequencies from those of the basic radar. The "friend" is assumed to be transponders-equipped and the "foe" is not. The interrogator is pulsed at about the same time as the radar, and the transponder produces coded replies shortly after the direct radar reply from the aircraft skin. In theory, even if the foe used the same transponder equipment, he would not know the code of the day. This "identification of friend or foe" system became known as *IFF*. Interrogation take place at 1030 MHz and replies at 1090 MHz. Typical pulse powers are 500 W, with a 1-μs length for interrogation and 0.5-μs length for reply.

SSR (Ref. 1, Chapter 14)

Secondary surveillance radar (SSR) is an outgrowth of the military identification of friend or foe (IFF) system, using the same frequencies. It is the principal means by which air traffic controllers identify and track civil and military aircraft worldwide. Secondary surveillance radars are the primary components of the FAA's Air Traffic Control Radar Beacon System (ATCRBS). The *secondary surveillance radar (beacon)* units are frequently co-located with, and mounted on top of, conventional *primary* radars. The beacon ground stations interrogate airborne transponders at 1030 MHz and receive replies at 1090 MHz, in order to measure distance using two-way (round-trip) ranging and bearing to an aircraft using the SSR's narrow antenna beam. Paired pulses are used for interrogation, and a third pulse between the two is radiated omnidirectionally to reduce triggering by side lobes. The airborne transponder replies only when the directional pulses are stronger than the omnidirectional pulse. The reply pulses comprise a train of up to 14 pulses, lasting 21 μs, which are currently combined into 4094 codes. These can be used to identify the aircraft or to communicate its altitude to the ground controller. The major problem of SSRs is the interference (garbling) which occurs when two or more aircraft are at about the same azimuth and distance from the interrogator. To alleviate this effect, the U.S. FAA has developed a new mode of interrogation coding (Mode S) which will allow each aircraft to be addressed by a discrete code, and thus only "speak when spoken to." This system will be compatible with the present SSRs, to allow an orderly transition. In the current system, the replies are pulse-coded with identity (mode A) and altitude (mode C). Mode S provides higher angular accuracy through monopulse techniques, discrete addressing of each aircraft, more efficient code modulation and a much higher capacity date link capability.

ATCRBS

The ATCRBS is the ICAO standard ground-based civil aircraft surveillance system based on the use of primary radars and secondary radars (SSRs) as sensors, as well as on extensive data processing, display, and communication systems. It is used by the air traffic control authorities to track aircraft.

TCAS

The Traffic Alert and Collision Avoidance System (TCAS) is based on the use of SSR technology. In operation, a TCAS-equipped aircraft interrogates all aircraft in the vicinity; the interrogated aircraft responding by means of the SSR transponder and the response is received by the interrogating aircraft. The latter can thus determine the relative altitudes and positions of the two aircraft. Computation of the relative rate of change (closing velocity) of these aircraft provides an indication of whether the two aircraft are on a collision threat. Thus, unmodified SSR transponders are used at the standard SSR frequencies (1030 and 1090 MHz). Three versions of TCAS have been or are under development, i.e., TCAS I, II, and III. TCAS I is intended for general aviation. It tracks proximate aircraft in range and altitude and displays traffic advisories (TAs) for intruders, thereby aiding the flight crews in visually acquiring threat aircraft. The most widely implemented version is TCAS II, currently used by the commercial airlines. In this version, not only TAs are displayed, but, if a threat is calculated, a resolution advisory (RA) is displayed showing a vertical escape maneuver (e.g., "climb") for avoiding collision. The two TCAS aircraft must exchange escape maneuver intentions in order to assure that they are complementary. The SSR Mode S data link can be used for that purpose. The U.S. Congress has passed a law making TCAS II mandatory on most airline type aircraft. In 1996, improved versions of TCAS were under development, notably those directed at a capability for horizontal-twin escape maneuvers. Also, techniques for the broadcast by each aircraft of on-board derived position via data link is under investigation, called ADS-B and TCAS IV.

Transit (Ref. 6)

This satellite-based system was originally developed for use by the U.S. Navy. However, after its release for civil use, the marine industry and also land mobile users saw its advantage of worldwide service and high accuracy. Each satellite operates in a polar orbit at 600-mi altitude and radiates two CW frequencies near 150 and 400 MHz. As it passes over an observer on the surface of the earth, these frequencies undergo a Doppler shift. The user receiver records the Doppler frequency history of the signal. Since the satellite orbit is accurately known as a function of time, the time of zero Doppler shift and the slope at zero Doppler are sufficient to determine the user's position on the earth. The satellite positions are continually determined by a ground based tracking network and are broadcast by the satellites to the user. The user computes its position from the Doppler history. Six to seven satellites in polar orbit are used to provide global coverage. While a single radiated frequency would lead to a position accuracy of about 500 m 2 DRMS, two frequencies allow errors due to ionospheric propagation effects to be greatly reduced and the resulting accuracy with a two-frequency receiver is 25 m 2DRMS. Most receivers use two transmitted frequencies. The system is not suitable for aircraft because of its low update rate. For six satellites in orbit, the satellites pass a given point, on the average, 90 min apart. The system is in operation in 1996, but the 1992 U.S. Federal Radio Navigation Plan (Ref. 9) lists Transit as a candidate for phase-out in the late 1990s, based on the operational status of GPS.

Global Positioning System (GPS) (Ref. 1, Chapter 5)

This statellite-based radio system was developed by the U.S. Air Force to provide worldwide coverage, high-accuracy three-dimensional position, velocity and time, and permitting completely passive (receive-only) operation by all types of military users. The system has now found wide acceptance by both military and civil users, e.g., military aircraft, ships, land vehicles, and foot soldiers, and a large variety of civil users, such as commercial and general aviation aircraft, commercial ships and pleasure boats, automobiles and trucks, and operators of surveying systems. Two services are available: the precise positioning service (PPS) for authorized (military) users provides horizontal position accuracy of 21 m 2 DRMS, a two-sigma vertical accuracy of 29 m and a one-sigma time accuracy of 100 ns; the standard positioning service (SPS) for all other users provides a two-DRMS, 95 percent probability, horizontal position accuracy of 100 m, a two-sigma vertical accuracy of 140 m, and a one-sigma time accuracy of 170 ns. The orbital configuration of the satellites is designed to provide a GDOP of normally near 2.3, and is intended to always be lower than 6. The GPS consists of 24 satellites, including three operational spares. The satellites are in 12-h orbits with four satellites in six orbit planes inclined at 55°, all at an orbital altitude of 10,900 n.mi. The satellites transmit highly synchronized, pseudonoise coded, wide bandwidth signals, which include data on their ephemerides and clock errors. A ground-based master control station (MCS) and five monitor stations track the satellites and periodically determine satellite ephemerides and clock errors and uplink these to the satellites via three uplink stations. The overall system configuration of GPS is shown in Fig. 23.3.7. The system operates on two frequencies, i.e., 1575.42 MHz (L1) and 1227.5 MHz (L2), to permit compensation for ionospheric propagation delays. The satellites transmit two codes, the 1.023 Mbps C/A code and the 10.23 Mbps P code. The latter code is encrypted into a Y code for military users. Each satellite has a unique code. System data (e.g., satellite ephemeris) is modulo-2 added to both codes at 50 bps.

GPS is a pseudoranging system (see Fig. 23.3.3). The user receiver determines at least four *pseudoranges* by TOA measurements with respect to its own clock time, and can also determine four pseudorange rates or *delta pseudoranges* via Doppler measurements with respect to its own clock frequency. From these measurements, the user receiver computes its own three-dimensional position coordinates and its clock time offset, as well as (in some receivers) its three-dimensional velocity coordinates and its clock frequency offset. The basic functional block diagram of a generic GPS receiver is shown in Fig. 23.3.8. GPS receiver processing includes both code and carrier tracking functions, which aid each other in two ways. The carrier (Doppler) tracking function sends Doppler velocity estimates to the code tracking information so that the resulting tracking loop bandwidth can be made very narrow. The code tracking function sends the *prompt* (on-time versus the early or late) estimate of the tracked code to the carrier tracking function, so that the code can be properly removed to allow tracking of the Doppler frequency.

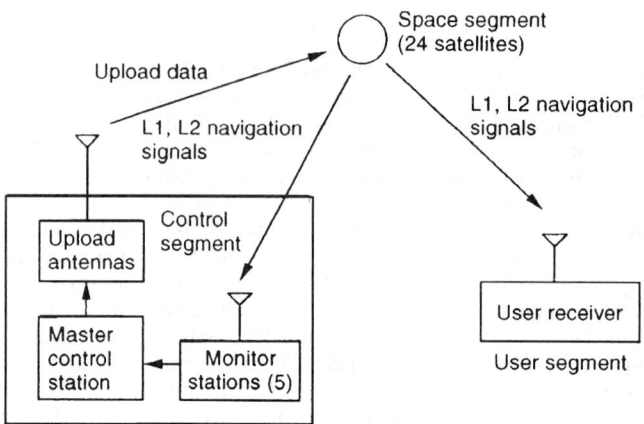

FIGURE 23.3.7   GPS system configuration.

Since the civil aviation community requires a very high integrity, i.e., warning of a failure, a number of system augmentations have been under development for this purpose. One is the receiver autonomous integrity monitoring (RAIM) function within the user receiver, which automatically detects and isolates a satellite failure. The other is the ground integrity broadcast (GIB) which consists of a network of ground-based monitoring stations, which monitor all satellite signals, and transmit certain error data to a central computer which computes integrity data and sends these to a satellite earth station. The latter then transmits appropriate integrity information via geostationary satellites to all GPS user receivers which are equipped to receive these signals. In the United States, this system has been named *Wide Area Augmentation System (WAAS)* and includes additional functions.

Some applications, for example aircraft landing and airport surveillance, require higher accuracies than those available from the basic GPS service. For these, the differential GPS (DGPS) concept has been implemented (Fig. 23.3.9). This concept employs a GPS reference station, whose position is precisely known. The reference station compares the predicted pseudoranges (from its known position) to the actually measured pseudoranges, computes the differences (corrections) for all satellites in its view and broadcasts these over a separate data link to the vehicles in the general vicinity, say at a 100 n.mi (160 km) radius. The user receiver then applies these corrections to its own pseudorange measurements for computation of its position. By means of this process, ionospheric and tropospheric propagation and satellite position errors common to the

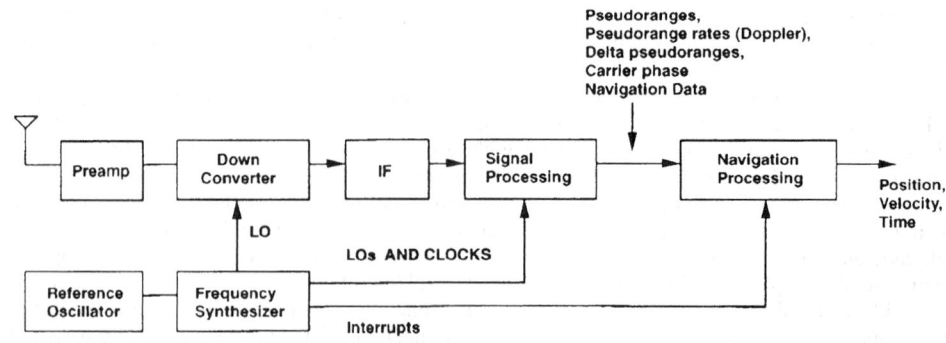

FIGURE 23.3.8   Generic GPS receiver block diagram.

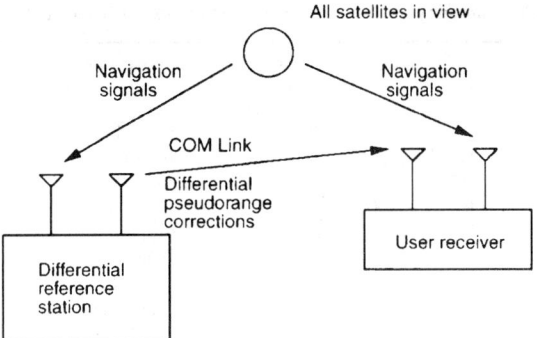

FIGURE 23.3.9 Differential GPS concept.

user and the reference station can be eliminated. With the DGPS technique, user position accuracies of 2 to 5 m can be obtained. A further refinement of this technique is the use of pseudosatellites (*pseudolites*) located on the ground and transmitting GPS-like signals to the user. This technique tends to reduce the GDOP of the system, notably in the vertical dimension. For certain very high accuracy performance applications, such as surveying and category II/III landing, the carrier phase of the GPS signal, rather than only the code phase (pseudorange), is measured to determine user position. In order to do this successfully, the carrier cycle ambiguity (of 19 cm) must be resolved. Using GPS carrier phase measurements on a postprocessing basis, centimeter accuracies have been obtained for surveying applications. In the late 1990s GPS is likely to be in widespread use throughout the world by a very large variety of users. A U.S. Department of Commerce report predicts 1 million commercial users by the year 2000.

Global Navigation Orbiting Satellite System (GLONASS) (Ref. 1, Chapter 5)

The GLONASS is a worldwide satellite radio navigation system being developed by the Russian government, whose concept is very similar to that of the U.S. GPS. It uses 24 satellites in 25,500-km orbits, with eight satellites in three orbital planes, at an inclination of 64.8° and an 8-day ground track repeat. This inclination is somewhat higher than that of GPS, thus providing somewhat better polar coverage. Each GLONASS satellite operates on two codes, one wide band and the other narrow band, and in two frequency bands. Unlike GPS, all satellites use the same two codes, but each satellite operates on slightly different frequencies. The higher frequency band currently goes from 1602 to 1615.5 MHz. The lower frequency band goes from 1246.4 to 1256.5 MHz. The higher GLONASS frequency assignments are likely to be changed in the future, in order to avoid interference with certain communication satellite frequencies. Also, it is currently planned in the future to use the same frequency on satellites in antipodal positions (i.e., halfway around the globe) in order to conserve the frequency spectrum. The two GLONASS code clock frequencies are 0.5 and 5.0 MHz. The accuracy performance of the standard (nonmilitary) GLONASS service is a horizontal position accuracy of 100 m, a vertical accuracy of 150 m, a velocity accuracy of 15 cm/s, and a time accuracy of 1 μs. Integrated receivers are under development, which can process both GPS and GLONASS signals, thereby providing the use of a total of 48 satellites, increasing system availability.

SUMMARY OF ELECTRONIC NAVIGATION SYSTEM CHARACTERISTICS

Table 23.3.1 presents a summary of the technical and performance characteristics of the major electronic navigation and position location systems currently in operational use worldwide. In the case of the accuracy values,

TABLE 23.3.1 Summary of Electronic Navigation System Characteristics

System	Type[a]	Frequency (MHz)	Accuracy (absolute)	Number of stations
Omega/VLF	H	0.010–0.014	3.7–7.4 km	8
Loran C	H	0.100	460 m	50
Decca	H	0.070–0.130	<300 ft	150
Beacons	R	0.200–1.6	±3 to ±10°	5000
VOR	R	108–118	±1.4°	2000
DME	C	960–1215	185m	1000
Tacan	R, C	960–1215	±1.0°, 185 m	1500
ILS	R	108–112,	±4.6° (az) Cat2	2000
		329–335	±1.4° (el) Cat2	2000
MLS	R	5000–5250	±4.6° (az) Cat2	117
		5000–5250	±1.4° (el) Cat2	
		960–1220 (DME)	100 ft (30 m)	
JTIDS-RelNav	C, PR	960–1215	variable	variable
PLRS	C	420–450	15–25 m	variable
SSR/IFF	R, C	1030, 1090	0.16°, 124 ft	800
		1030, 1090	0.04°, 24 ft[b]	
TCAS	R, C	1030, 1090	variable	
Transit	H	150, 400	25 m[d], 500 m[e]	7 satellites
GPS	PR	1575, 1227	21 m[f], 100 m[g]	24 satellites
GLONASS	PR	1602–1616[c]	100 m[h]	24 satellites
		1246–1257[e]		
Altimeter	S	4200–4400	2 ft or 2%	
Mapping radar	S	various	variable	
Map matching	S	various	variable	
Doppler radar	S	13,325	0.2% or 2 kn	
Inertial	S		0.5–1 nmi/h	
Celestial	S	optical	variable	

[a]H = hyperbolic; R = radial, C = circular, lines of position; PR = pseudoranging; S = self-contained
[b]Mode-S with monopulse
[c]May be changed in the future
[d]Dual-frequency receivers
[e]Single-frequency receivers
[f]PPS receivers (2DRMS)
[g]SPS receivers (2DRMS)
[h]Civil version

in most cases the values cited are the horizontal, 2DRMS, 95 percent probability, officially specified values. (For a definition of 2DRMS, see Ref. 1, Chapter 2 and Ref. 9.)

REFERENCES

1. Kayton, M. and W. R. Fried "Avionics Navigation Systems," 2nd ed., John Wiley & Sons, 1997.

2. Skolnik, M. I. "Introduction to Radar Systems," 2nd ed., McGraw-Hill, 1980.

3. Global Positioning System, Vols 1–4, Compendium of Papers Published in *Navigation, Journal of the Institute of Navigation*, 1980, 1984, 1986, 1993.

4. *Proceedings of the International Technical Meetings of the Satellite Division of the Institute of Navigation*, Annually from 1988.

5. *Proceedings of the IEEE Position Location and Naviagation Symposium* (PLANS), published by the IEEE, biannually from 1976.

6. Stansell, T. A. The TRANSIT Navigation Satellite System, Status, Theory, Performance, Applications, Magnavox Government and Industrial Electronics Company (currently Leica, Inc.), 1978.

7. Daly, P. "Review of Glonass System Characteristics," *Proceedings of the Third Intern. Technical Meeting of the Satellite Division of the Institute of Navigation*, September 1990, pp. 267–275.

8. Parkinson B. W. and James J. Spilker, Jr. "Global Positionings System: Theory and Applications," Vols. 1 & 2, AIAA, 1996.

9. Federal Radionavigation Plan, Published biannually, in even years, by the U.S. Department of Transportation and Department of Defense; National Technical Information Service.

CHAPTER 23.4
UNDERWATER SOUND SYSTEMS

**James F. Bartram, Stanley L. Ehrlich, Donald A. Fredenburg,
Jack H. Heimann, Joseph A. Kuzneski, Paul Skitzki,
Joseph E. Blue, A. L. Van Buren, R. Y. Ting**

PRINCIPLES, FUNCTIONS, AND APPLICATIONS

Sound energy travels in water as a result of particle motion initiated by the application of physical forces to the particles from a vibrating diaphragm, collapsing air bubbles, or other energy sources with sufficient mechanoacoustical coupling for the transfer of the energy. It can be controlled, directed, and transmitted for many useful purposes.[1-3]

Water is an excellent medium in which to transmit compressional sound waves. Liquids have higher specific acoustic impedances by several orders of magnitude than gases. The high acoustic impedance of water (1.5 MN·s/m^3 for seawater) makes it possible to design transducers whose internal mechanical impedance approaches the radiation load impedance, with conversion efficiency on the order of 50 percent over a band of an octave or over 80 percent over a narrow band.

The transmission and reception of underwater sound can be controlled and directed to perform the functions of communications, navigation, detection, tracking, classification, and so forth, which in aerospace are accomplished with electromagnetic energy. The wavelengths of underwater sound systems and radar systems are of the same order of magnitude, since the frequencies employed differ by the ratio of the speed of sound to the speed of electromagnetic waves. The term *sonar*, derived from sound navigation and ranging, is used synonymously with *underwater sound* and *underwater acoustics*.

The applications of underwater sound for defense purposes, both pro- and antisubmarine, have advanced with development of the nuclear submarine and other platforms. In military applications, underwater sound is used for depth sounding; navigation; ship and submarine detection, ranging, and tracking (passively and actively); underwater communications; mine detection; and or guidance and control of torpedoes and other weapons. Most systems are monostatic, but bistatic systems are also employed.

Civilian applications of underwater sound are numerous and are containing to increase as attention is focused on the hydrosphere, the ocean bottom, and the subbottom. These applications include depth sounding, bottom topographic mapping, object location, underwater beacons (pingers), wave-height measurement, doppler navigation, fish finding, subbottom profiling, underwater imaging for inspection purposes, buried-pipeline location, underwater telemetry and control, diver communications, ship handling and docking aid, antistranding alert for ships, current flow measurement, and vessel velocity measurement.

PROPAGATION

Propagation of sound in water can be represented by the sound pressure, the sound particle velocity, and/or the sound intensity as a function of position, time, and frequency. Because the sound pressure can usually be measured more nearly directly, it is the preferred parameter for most experimental data.[4] The

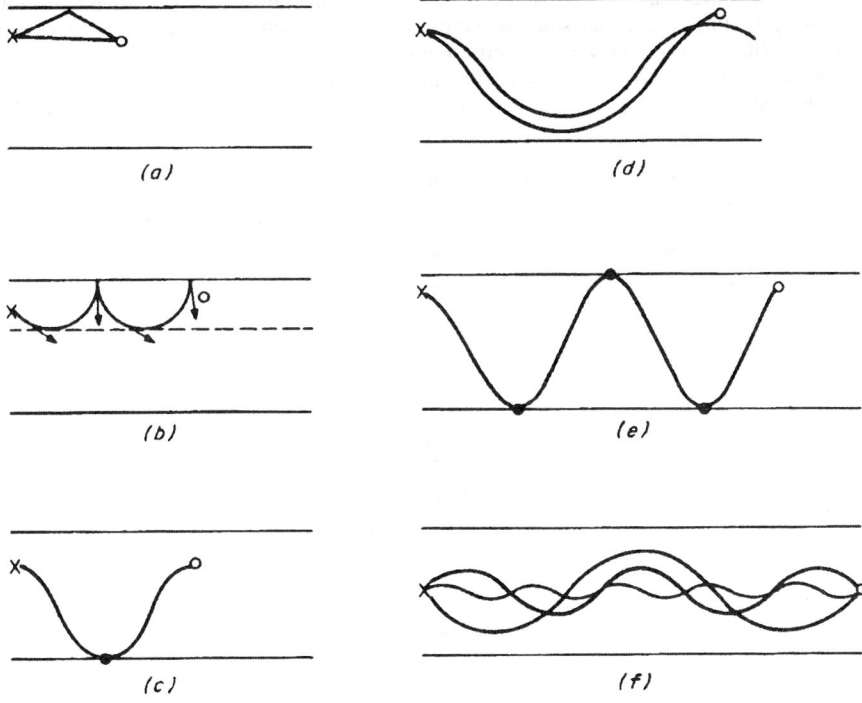

FIGURE 23.4.1 Propagation paths.

sound-pressure amplitude p in water is expressed in pascals.[*] The logarithmic unit of a *sound-pressure level* L_p is expressed in decibels with respect to the reference sound pressure amplitude, that is, 1 μPa = p_o, where $L_p = 20\log(p/p_o)$. The phase of the sound pressure is expressed in degrees or in radians with respect to a specified reference.

The difference between the sound-pressure level at the reference position and the sound-pressure level at a point in the sound field is called the *propagation loss* N_w for that point. For a small sound source, the reference position may be at a standard distance of 1 m in the direction of the maximum response. For a larger source, far-field data may be extrapolated back to the reference distance r_0.

The propagation loss may be considered[1,3,5-7] to consist of two basic components, one due to the spreading of sound energy with increasing radial distance r from the sound source N_{spr} and the other to attenuation of sound as it propagates through the medium N_{att}

$$N_w = N_{spr} + N_{att}$$

The *spreading loss* is given by $10n \log (r/r_0)$, where n is dependent on the spreading law and is equal to 2 for the theoretical case of *spherical spreading*. The *attenuation loss* is given by $10^{-3} \alpha r$, where α, the *attenuation coefficient*, is as discussed next.

Common propagation paths are illustrated[3] in the ray diagrams of Fig. 23.4.1. The paths of the direct ray and a ray with a single surface reflection in water with constant sound speed, are shown in Fig. 23.4.1a. In *b* a surface-layer sound channel confines the ray near the surface, with leakage rays due to diffraction and reflected waves from a rough surface. A ray that experiences a single bottom reflection is shown in *c*, while two bottom reflections with an intermediate surface reflection are shown in *e*. A pair of rays that diverge and return to

---

[*]1 pascal = 1 N/m^2 = 10 dyn/cm^2 = 10 μbars.

a crossover in the convergence zone is shown in *d*. The three diverging rays, trapped in a deep sound channel and crossing each other several times before converging at the receiver, are shown in *f*. The reliable acoustic path, RAP (not shown), exists between a source at moderate depth and a surface receiver.

The theoretical treatment of sound propagation in water depends on assumptions used to simplify the mathematical formulation. Typical assumptions used by various authors include combinations of one or more of the following.[1,3,8]

- One sound source with constant frequency and spherically isotropic radiation
- Propagation medium with linear transmission characteristics and sound speed dependent only on depth
- An ideal horizontal plane sea surface with zero acoustic impedance
- An ideal horizontal plane sea bottom with infinite acoustic impedance
- A sound receiver with zero rise time, flat frequency response, and spherically isotropic reception

Speed of Sound

Essentially, all seawater can be represented by the following conditions:

$$\text{Temperature } T = -3 \text{ to } +30°C$$
$$\text{Depth } d = 0 \text{ to } 10{,}000 \text{ m}$$
$$\text{Salinity } S = 33 \text{ to } 37 \text{ parts per thousand}$$

with atmospheric pressure (absolute pressure) 0.1013 MPa at zero depth at sea level. For these conditions, the speed of sound *c* in seawater is given by an empirical formula due to Wilson,[9–11] here simplified to

$$c = 1449.3 + 4.572T - 0.0445T^2 + 0.0165d + 1.398(S - 35) \quad \text{m/s}$$

At zero depth the accuracy of *c* is about 3 m/s, or 0.2 percent; for extreme conditions with comparable accuracy a more complete equation, including higher-order and cross-product terms, becomes necessary. At $T = 0°C$, $d = 0$ m, and $S = 35$ parts per thousand, $c = 1449.3$ m/s. The nominal value $c = 1500$ m/s, corresponding to about $T = 13°C$, is convenient for engineering calculations.

Attenuation of Sound

The attenuation of sound in seawater has been studied by many investigators to determine its variation in different frequency bands as well as its dependence on temperature, salinity, and depth. At frequencies greater than 1 MHz, the attenuation mechanism is generally attributed to shear and dilatational viscous losses.[1] In the frequency band between 10 and 40 kHz, the increased attenuation is almost solely because of a relaxation-type mechanism in the $MgSO_4$ salts dissolved in the seawater.[12] Recent work indicates that between 0.1 and 1.0 kHz another relaxation-type mechanism, not yet identified, dominates the attenuation.[13] Below 50 Hz other attenuation mechanisms are of greater importance.

An expression for the attenuation coefficient of seawater from 0.1 kHz to 100 MHz includes three components resulting from the three main attenuation mechanisms,[1,3,13,14] multiplied by a depth-dependent term:[15]

$$\alpha = \left[\frac{0.11f^2}{1 + f^2} + \frac{0.70f_T f^2 (S/35)}{f_T^2 + f^2} + \frac{0.03f^2}{f_T} \right] (1 - 65 \times 10^{-6} d) \quad \text{dB/km}$$

where f = frequency (kHz)
f_T = relaxation frequency[14] (kHz) = $21.9 \times 10^{6-1.520/(T+273)}$
T = temperature (°C),
S = salinity (parts per thousand)
d = depth below air-water boundary surface (m)

At $T = 4°C$ ($f_T \approx 71$ kHz), $d = 0$ m, and $S = 35$ parts per thousand, the equation simplifies to

$$\alpha = \frac{0.11f^2}{1+f^2} + \frac{50f^2}{5000+f^2} + 0.0004f^2 \quad \text{dB/km}$$

Further simplification is possible for the above listed conditions, at frequencies below about 20 kHz, to approximately

$$\alpha = \frac{0.11f^2}{1+f^2} + 0.010f^2 \quad \text{dB/km}$$

and above 200 kHz to approximately

$$\alpha = 50 + 0.004f^2 \quad \text{dB/km}$$

Measured values of attenuation in seawater include absorption losses, scattering losses due to random internal inhomogeneities, and interaction losses with the bottom boundary, the subbottom, and upper surface boundary.

Reflection and Refraction

Reflection and refraction of sound are normally in accordance with Snell's law, which states that $(\cos \theta_i)/c_i$ is a constant, where θ_i is the angle between the direction of propagation and the horizontal plane and c_i is the sound velocity at the point i of the ray. At a boundary where the sound velocity is discontinuous, the angle θ_i is the angle of incidence with respect to the plane tangent to the boundary, which is not necessarily horizontal.

A special case of interest occurs in a region with a constant *sound speed gradient* ∇c. This results in a circular sound ray path with radius equal to $c_i/(\nabla c \cos \theta_i)$. The center of the circle corresponds to a position where c_i becomes zero. In a surface layer, the resultant upward refraction leads to shadow zones which theoretically contain no propagated sound energy.

A second special case occurs at a boundary where the incident sound speed is lower than the refracted sound speed c_r, such as a plane interface between water and air. The well-known phenomenon of total internal reflection results when θ_i is less than a critical angle θ_{crit}, at which $c_i/(\cos\theta_{crit})$ is equal to c_r.

Another important case occurs when the sound in water is propagated by a direct path and also by a single reflection from a boundary of slower sound speed, such as air. Because of a 180° phase shift in the sound pressure at such a reflecting boundary, the resultant sound pressure exhibits maxima and minima as a function of the position of the receiver. This is called the *Lloyd mirror*, an image interference effect, since the condition can be represented by an additional sound source of equal amplitude and opposite phase as a virtual image that provides the constructive and destructive interference with the real sound source.

Other important cases include a multiple-layered medium where the sound field at each discontinuity must be accounted for; irregular and nonstationary boundaries that tend to randomize the reflection and refraction properties; moving boundaries that tend to modify the frequency of the sound energy because of the doppler effect; and intentional discontinuities introduced in transducer and array designs.

Reverberation

Reverberation of sound in water produces energy usually unwanted at the receiver. It is caused by scattering, i.e., reflection and refraction of sound from discontinuities other than those of primary interest. When the sound source and sound receiver are at the same location, reverberation is produced primarily by backscattering. When the scatterers are boundaries of the medium, the effect is called *surface reverberation*, which may be subdivided into *sea-surface* and *sea-bottom reverberation*. When the scatterers are contained within the medium, the effect is called *volume reverberation*, which may be due to fine particles, fish, or other inhomogeneities, including the structure of the sea.

NOISE

Background Noise

Underwater sound systems operate in a medium that has a very low acoustic-noise level under quiet conditions. Stimulation from natural and human causes, however, can generate and propagate acoustic noise at various levels and frequencies. The acoustic-noise background consists of an *ambient-acoustic-noise level* and a *self-noise level* caused by the sonar-platform presence and its movement.[16] Passive sonar systems detect target-generated noise, while active sonar systems detect target echoes in the presence of the noise background.

Ambient Noise

The main contributions to ambient noise are shown in Fig. 23.4.2. Tides, waves, and seismic disturbances predominate at the very low frequencies, in the region of 1 Hz. At frequencies used for sonar systems, the main sources are sea surface agitation due to meteorological effects, noise from marine life, and human noises from shipping and other activities.[3] Of these, surface agitation due to wind and wave actions is most significant. The acoustic spectrum levels for various sea states are shown in Fig. 23.4.3.

At the high frequencies (above 100 kHz), thermal molecular motion of the water is the principal noise contributor but does not limit sonar performance at lower frequencies. It has a spectrum level that rises with frequency at a rate of 6 dB/octave. The mean-square sound pressure in a 1-Hz band is

$$p_T^2 = 4\pi kT\rho c/\lambda^2$$

where k = Boltzmann's constant
T = temperature of water (K)
pc = specific acoustic impedance
λ = wavelength

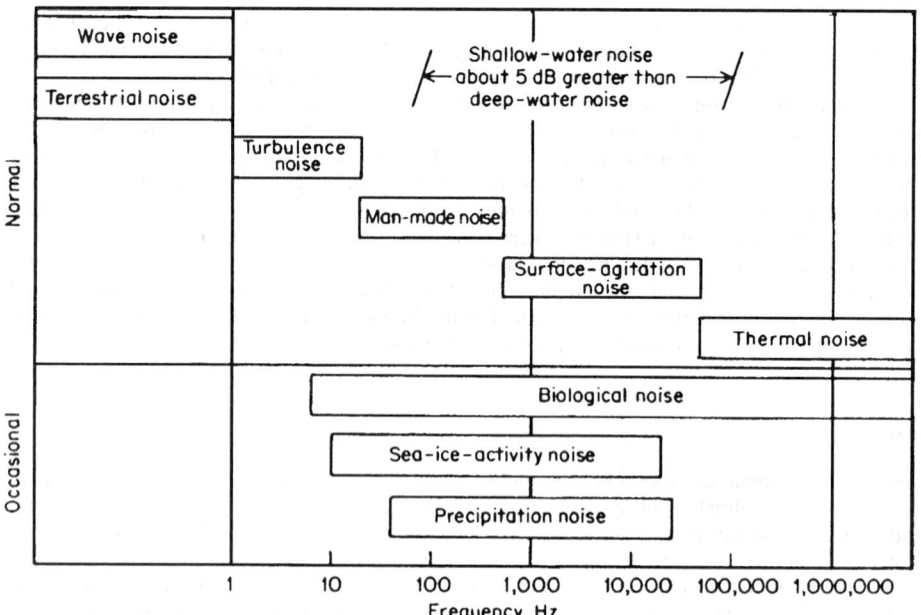

FIGURE 23.4.2 Principal ambient-noise sources.

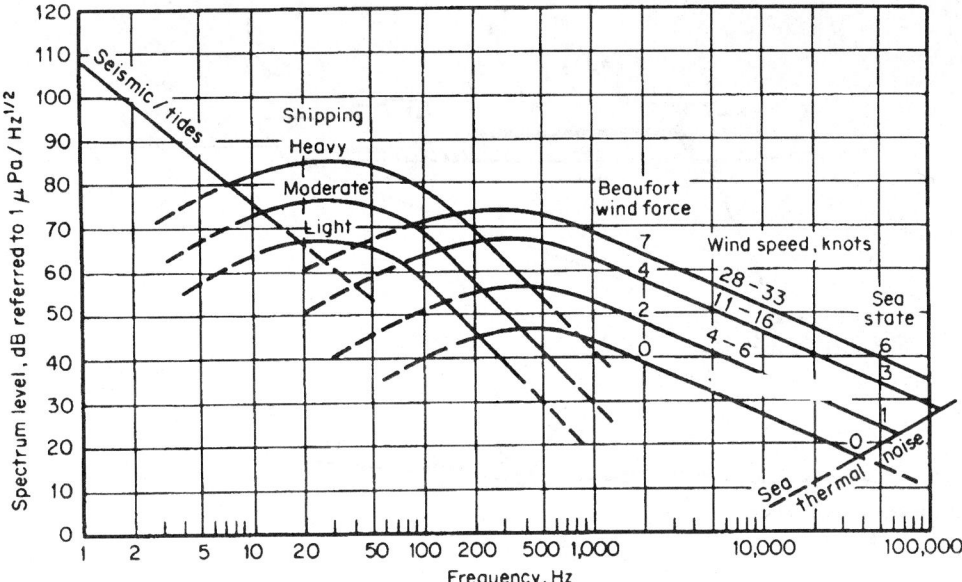

FIGURE 23.4.3   Average deep water noise.

Acoustic thermal noise is consistent with its more familiar electrical counterpart. It is far below sea state levels at frequencies below 10 kHz. At higher frequencies than 100 kHz, it sets a threshold on the minimum detectable pressure levels in the medium. The ambient-noise level, as a sonar parameter, is the intensity of the noise background as measured with a nondirectional hydrophone, referred to the intensity of a plane wave having an rms pressure of 1 μPa.

Platform Noise

Platform noise is a degrading factor in the performance of underwater sound systems, particularly in the case of mobile platforms such as ships, submarines, aircraft, torpedoes, and other sonar-carrying vehicles. Fixed-position platforms, including sonobuoys, moored structures, bottom-mounted structures, mines, and so forth, are also plagued with platform-noise problems, primarily induced by hydrodynamic flow or motion. Good sonar performance with mobile platforms at any speed above approximately 10 knots is achievable only by most careful attention to platform-noise reduction and by design of the sonar system for minimum susceptibility to local noise.

Platform noise may enter the sonar system via radiation in the medium or by conduction through the platform structure. Generally, conducted noise can be reduced below the level arriving at the array via the medium. The techniques involve sensor design for minimum response to acceleration forces, isolation of the sensor elements from the mounting structure, and isolation and location of the array away from hull-borne and structural vibrations. The noise radiated into the medium may reach the array directly or via reflected paths, as shown in Fig. 23.4.4. The ship's propellers and machinery are the dominant noise sources. Since the noise level is highest on stern bearings, baffles are generally employed behind the sonar array to minimize stern noise, even though this results in loss of sonar performance over a portion of the azimuth.

Radiated Noise

Platform-generated noise, radiated into the medium, produces an acoustic signature that can be detected by passive sonar systems. Machinery noise is defined as noise caused by propulsion and auxiliary machinery on

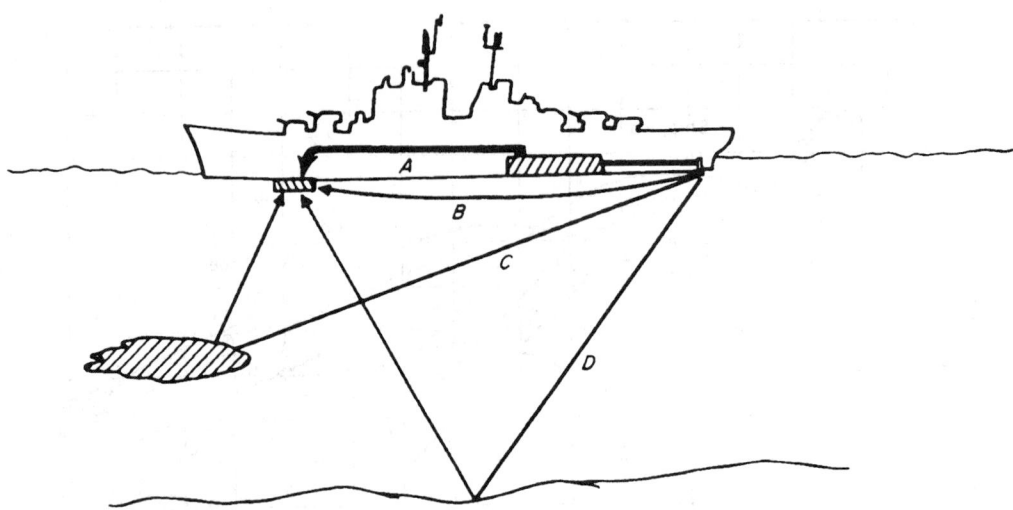

FIGURE 23.4.4 Paths of self-noise.

the vessel. The noise produced by the various machines, generators, pumps, actuators, and so forth, travels by diverse paths to the hull structure, where it is introduced into the medium. Propeller noise originates outside the hull and is mainly due to cavitation at the propeller blades. Cavitation-produced bubbles generate acoustic noise, the acoustic spectrum of which differs from machinery noise and varies with speed and depth. In addition to the cavitation noise, propellers produce amplitude-modulated noise modulated at a frequency equal to the shaft rotation speed times the number of blades. Such *propeller beats* are most pronounced just beyond the onset of cavitation and are swamped by cavitation noise at higher speeds. Propellers may also produce a "singing" noise due to vibrational resonance of the blades.[16]

Hydrodynamic noise results from the flow of fluid past the moving platform. It increases with hydrodynamic structural irregularities of the platform and the fluid flow rate. Breaking bow and stern waves can excite the hull or structural members. Hydrodynamic noise is a minor contributor to the platform-radiated noise, and is usually swamped by machinery and propeller noise. However, it is an important element in consideration of self-noise for underwater sound systems associated with the platform.

Radiated Noise Levels

Radiated noise consists of broadband noise and tonal noise (line components). Measurements of radiated noise are made at some distance, say 200 yd, from the vessel and reduced to source spectrum level values by correction for the test distances and the measurement bandwidths. Tonals are determined by fine-grain spectral analysis. Figure 23.4.5 illustrates broadband and tonal noise from a submarine at two speeds. In *a* the broadband noise from cavitation at the propeller begins to appear, and the tonals from machinery noise are predominant. In Fig. 23.4.5*b* the broadband noise has increased as a result of higher speed, and many tonals are masked by the broadband noise while tonals are changed in amplitude or frequency.[3] Figure 23.4.6 shows average radiated-noise spectra for several classes of ships, and Fig. 23.4.7 illustrates noise from running torpedoes.

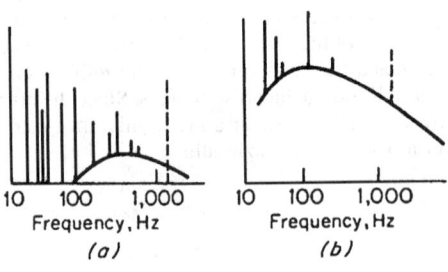

FIGURE 23.4.5 Diagrammatic spectra of submarine noise: (*a*) low speed; (*b*) high speed.

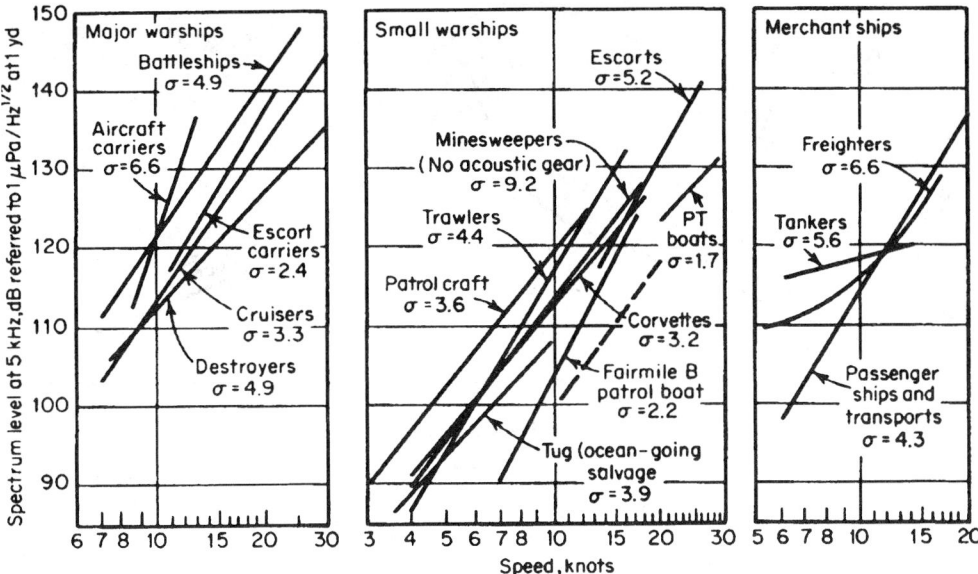

FIGURE 23.4.6   Average radiated spectrum levels of ships.

TRANSDUCERS AND ARRAYS

Introduction

Transducers for underwater sound applications perform the functions of generating a sound wave in the medium or detecting the existence of a sound wave and its properties (e.g., amplitude and phase) in the medium. In the generating case, the transducer is commonly referred to as a *source*, or *projector*, and in the detecting case, as a *hydrophone*. Often the transducer is required to perform both functions. Single transducers or arrays of transducers may be designed to control the directional properties (i.e., directivity or beam pattern) of the generated acoustic energy, and to discriminate against the noise in the receiving case.

The function of a projecting transducer is to convert the input energy (usually electric) to acoustic energy in a manner that is efficient and compatible with the other components of the transmitting system, e.g., amplifiers. In the hydrophone, linear conversation of the acoustic signal to an electric signal is the basic function, and compatibility must be maintained with the other components of the receiving subsystem.

The conversion of energy is accomplished by any of a variety of physical phenomena, e.g., piezoelectricity and electrostriction, piezomagnetism and magnetostriction, electrodynamics and magnetodynamics, and chemical transformations and hydrodynamics.

The selection of the transduction mechanism and the design of the transducer are based on the following considerations: operating frequency; bandwidth; power (acoustical and electric); directional properties; the characteristics of available energy-converting materials; the characteristics of, and the materials used for, packaging (such factors as stability and static pressure, temperature, and

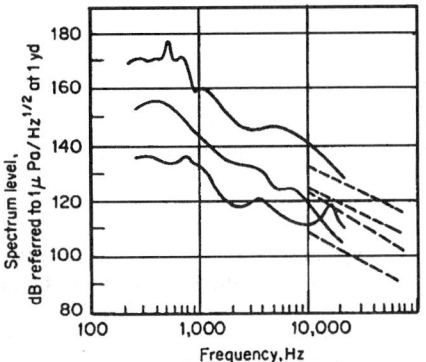

FIGURE 23.4.7   Noise spectra of torpedoes.

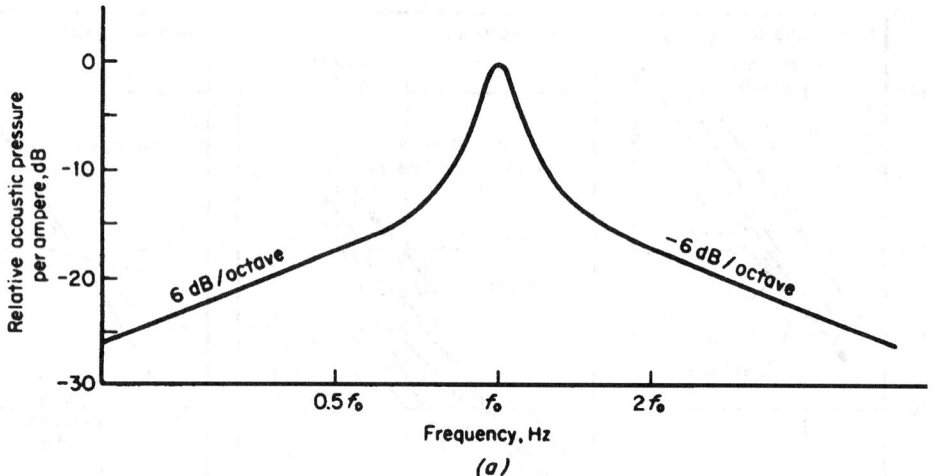

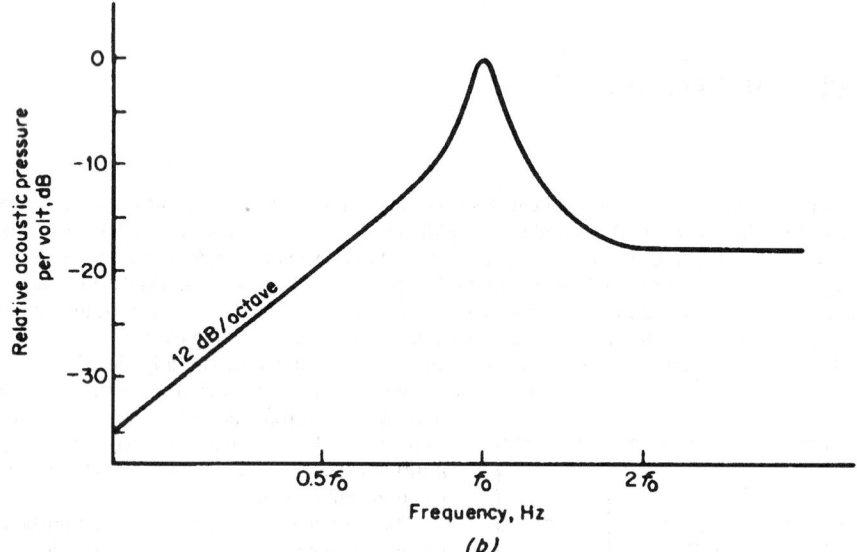

FIGURE 23.4.8 Idealized transmitting current and voltage responses for a piezoelectric transducer: (*a*) current response; (*b*) voltage response.

time and resistance to corrosive effects of the medium); cavitation and other nonlinear effects of the medium; and the effect of static pressures encountered at great depths on the overall transducer design and its operating characteristics.

Significant developments have been achieved in the calibration and performance testing of transducers and arrays both in the laboratory and in the ocean environment, and standard techniques have been established.[17]

Types of Projectors

There are two main transducer types used as underwater sound sources: those operating with a continuous-wave or modulated (amplitude, frequency) input and those operating as impulse sources. The former are used for most military and many commercial applications;[18–20] the latter are used mainly for oceanographic and geophysical applications.[3]

CW and Modulated Sources

Transducers designed for cw or modulated-input underwater applications use piezoelectric, electrostrictive, or magnetostrictive energy-conversion materials. Piezoelectric crystals, e.g., quartz, have a linear relationship between strain and electric field. However, their application is limited by low dielectric constant, low electro-mechanical coupling coefficient (the ratio of the converted energy to the total input energy in a transducer), narrow bandwidth, low power-handling capability, and limited availability of geometrical shapes. They can yield very high conversion efficiencies (quartz transducers having efficiencies in excess of 90 percent have been built).[18]

Ferroelectric crystals, in either single-crystal or polycrystalline ceramic form, are often electrostrictive and have higher-order nonlinear properties in their natural state. With the application of a polarizing electric field, the electromechanical processes in these materials can be linearized over a wide range of operating conditions. These materials have the advantage of high dielectric constant (which results in low impedance), high electromechanical coupling coefficients, broad bandwidth, and high power-handling capability when properly used, and are available in a wide variety of shapes (plates, cylinders, rings, spherical zone sections). Efficiencies in excess of 70 percent can be achieved. Operating frequencies from less than 1 Hz to more than 10 MHz can be achieved.

The most widely used materials in the ferroelectric class are modified lead zirconate titanate and modified barium titanate. Idealized transmitting current and voltage responses of a transducer that uses these materials are shown in Fig. 23.4.8*a* and *b*, where f_0 is the frequency of mechanical resonance of the transducer. Figure 23.4.9 shows a typical impedance characteristic.

Magnetostrictive transducers depend on the interchange of energy between magnetic and mechanical forms. In an unpolarized state, such transducers are nonlinear, frequency-doubling devices. However, in a polarized state (achieved by the use of permanent magnets, direct current, or operation at remanence) they are linear (i.e., piezomagnetic). Commonly used materials include various nickel alloys and ferrites.

Properly designed magnetostrictive transducers can achieve radiated acoustic powers up to several kilowatts with efficiencies in excess of 50 percent. Operating frequency is usually limited to frequencies below 100 kHz. Figure 23.4.10 shows a typical idealized transmitting current characteristic.

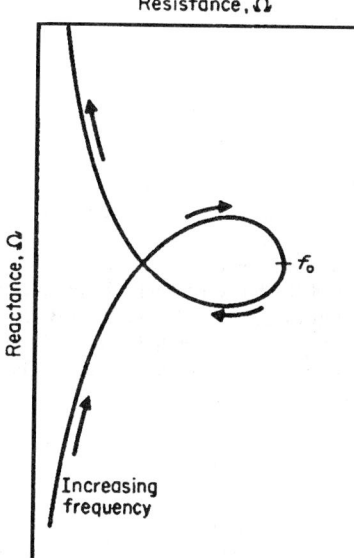

Resistance, Ω

Reactance, Ω

f_0

Increasing frequency

FIGURE 23.4.9 Idealized impedance locus for a piezoelectric transducer.

Electrodynamic sources have been used in underwater acoustics for low-frequency applications. One noteworthy application is the low-frequency *standard projector.*[21]

Impulse Sources

Impulse sound sources produce a short-time duration, high-amplitude transition of more or less regular waveform. For example, explosives (TNT or other high-burning-rate chemical) with provision for hydrostatic, electrical, or fused detonation produce in the ocean medium a pressure wave that is initially steep-fronted, displays approximately exponential decay, and is followed by a sequence of bubble pulses. The shock wave is usually

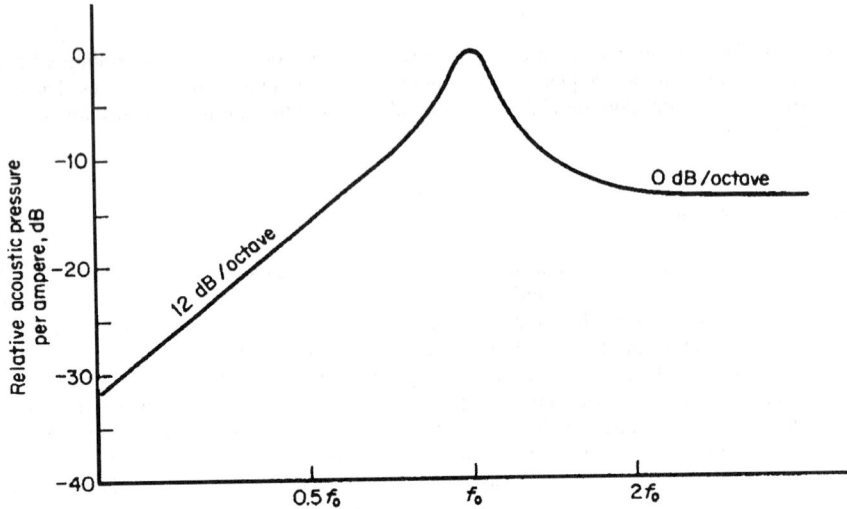

FIGURE 23.4.10 Idealized transmitting current response for a magnetostrictive transducer.

so intense that the resulting finite-amplitude effects are appreciable. Figure 23.4.11 shows a typical pressure-time characteristic.

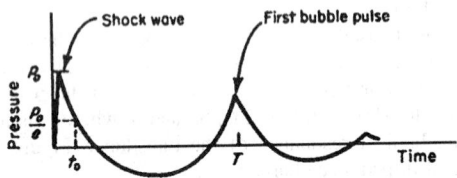

FIGURE 23.4.11 Pressure-time characteristic for an explosive source.[3]

Empirical formulas yield $p_0 = 5.16 \times 10^7 (w^{1/3}/r)^{1.13}$, $t_0 = 92 \times 10^{-6} w^{1/3}(r/w^{1/3})^{0.22}$, and $T = 5.67\ w^{1/3}/(3.28d + 33)^{5/6}$, where d is depth (m), r is range (m), and w is equivalent yield of TNT (kg).

Charges ranging in weight from 1 oz to 60 lb are in common use. A 4-lb charge of TNT will produce a peak pressure at 1 km of 26 kPa (that is 208 dB re 1 μPa). Other impulse type sources include implosive devices, spark-gap generators, and pneumatic- and mechanical-impact mechanisms.

Hydrophones

The receiver in a sonar system employs a hydrophone or hydrophone array coupled to an amplifier. Hydrophone elements generally use piezoelectric energy-conversion materials, although magnetostrictive and electrodynamic mechanisms are sometimes used. Typical hydrophone sensitivities are on the order of −180 to −200 dB re 1 V/μPa. Proper impedance termination and suitable amplification are necessary to obtain useful electrical levels. Figure 23.4.12 shows a typical idealized open-circuit receiving response for a piezoelectric hydrophone.

PRODUCTION OF SOUND FIELDS

Acoustic Principles

The production of a sound field in the medium involves an electrical input that is converted by the transducing mechanism into the motion of a surface in contact with the medium. The motion initiated by the moving surface of the transducer is communicated to the adjacent water principles, and a sound wave is propagated from its surface.

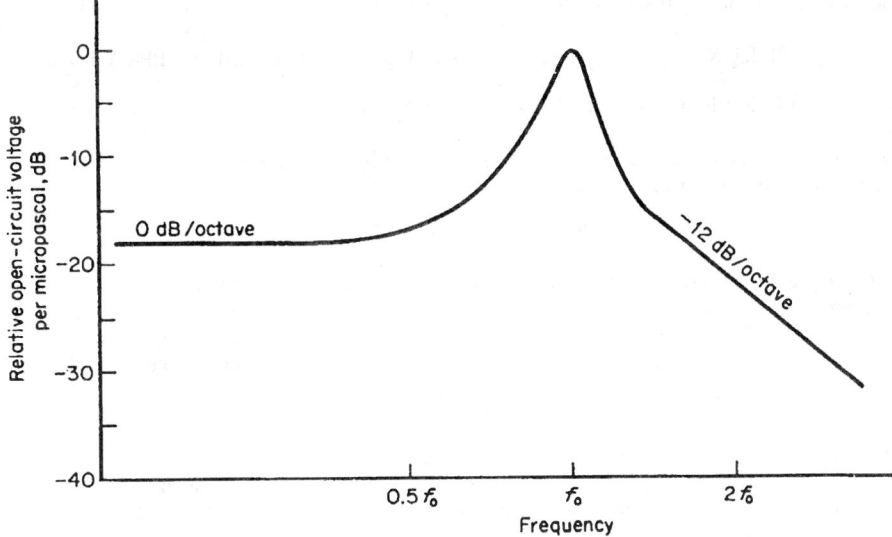

FIGURE 23.4.12   Idealized open-circuit receiving response for a piezoelectric transducer.

The *far-field sound pressure* produced by the source can be described in terms of the radiated acoustic power P_a by

$$p^2(r,\phi,\theta) = \rho c P_a D_i R(\phi,\theta)/4\pi r^2 \tag{86}$$

where $p^2(r,\phi,\theta)$ = mean-square acoustic pressure (Pa): r,ϕ,θ = spherical coordinates, r in meters: ρc = product of density and speed of sound of medium, i.e., specific acoustic impedance of medium ($\mathrm{N \cdot s/m^3}$); $R(\phi,\theta)$ = normalized pattern function; and D_i = directivity factor of source.

The directivity factor is defined as

$$\frac{1}{D_i} = \frac{1}{4\pi r_0^2} \int\int \frac{p^2(r,\phi,\theta)}{p_0^2} \, dS \tag{87}$$

where p_0 is the pressure at distance r_0 in direction of maximum response and dS is the element of surface area on sphere having radius r_0.

An example of a normalized beam pattern is shown in Fig. 23.4.13. This figure shows plot of $10 \log p^2(\theta,\phi)$ versus θ for a particular value of $\phi = \phi_d$. Dividing Eq. (86) by the square of the input current to the transducer I^2 and rearranging gives

$$\frac{p^2(r,\phi,\theta)}{I^2} = \frac{\rho c D_r R(\phi,\theta) R_e}{4\pi r^2} \frac{P_a}{P_e} \tag{88}$$

where P_e is the real part of the electrical input power to transducer (W) and R_e is the electrical input resistance of transducer (Ω).

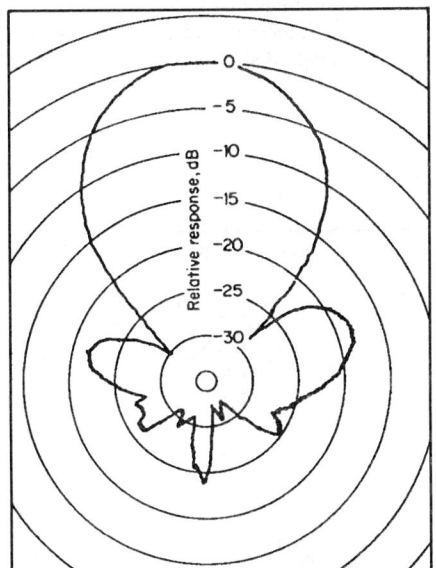

FIGURE 23.4.13   Typical beam pattern of an underwater sound transducer.

The current transmitting response ($20 \log S_0$) is given by

$$20 \log S_0 = 10 \log R_e + 10 \log D_i + 10 \log \eta_{ea} + 170.8 \quad \text{dB re 1 } \mu\text{Pa/A at 1 m} \tag{89}$$

The term $10 \log D_i$ is referred to as the *directivity index N_{DI}*, or gain, of the transducer, $\eta_{ca} = P_o/P_e$, or electrical efficiency.

The free-field open-circuit receiving response M_0 in volts per micropascal is related to the current transmitting response for a reciprocal transducer by

$$20 \log M_0 = 20 \log S_0 - 20 \log f - 294 \quad \text{dB re 1 V/}\mu\text{Pa} \tag{90}$$

where f is the frequency (Hz) and nominal conditions in the water are assumed.

In determining the reaction of the medium on the moving surface of the transducer, it is assumed that the vibrating surface of the source has a velocity u, and that the surface exerts a force F_r on the water, and the force exerted by the water on the moving surface of the source is $-F_r$. The radiation impedance Z_r is expressed as[22]

$$Z_r = -F_r/u = R_r + jX_r \tag{91}$$

where R_r is the radiation resistance and X_r is the radiation reactance.

In a linear system consisting of a continuous source, the value of Z_r is frequency-dependent, but is a constant at constant frequency. If the radiation impedance is known, calculating the acoustic power P_a of the source is greatly simplified, since

$$P_a = 1/2 \, u^2{}_{\text{peak}} R_r = u^2_{\text{rms}} R_r \tag{92}$$

Table 23.4.1 lists the radiation impedances for pistons and spheres, and Fig. 23.4.14 shows plots of radiation impedance for typical surfaces.

Nonlinear Acoustic Principles

The equations governing the propagation of sound in water are in fact nonlinear. The conventional linear relationships are really approximations, strictly valid only in the limit of infinitesimal sound amplitudes. Although the observable effects of this fundamental nonlinearity when relatively intense sound amplitudes are employed have normally been viewed as a source of performance degradation, e.g., distortion and loss, beneficial practical applications have been found.

TABLE 23.4.1 Radiation Impedance for Simple Geometries

Type of radiator		Radiation impedance
Rigid circular piston in infinite baffle		$Z = \pi a^2 \rho c \left[1 - \dfrac{J_1(2ka)}{ka} + \dfrac{S_1(2ka)}{ka} \right]$ where J = Bessel function S = Struve function
Sphere	Pulsating	$Z = \dfrac{4}{3}\pi a^2 \rho c \dfrac{(ka)^2 + jka}{1 + (ka)^2}$
	Oscillating	$Z = \dfrac{4}{3}\pi a^2 \rho c \dfrac{(ka)^4 + jka(1 + k^2 a^2)}{4 + (ka)^4}$

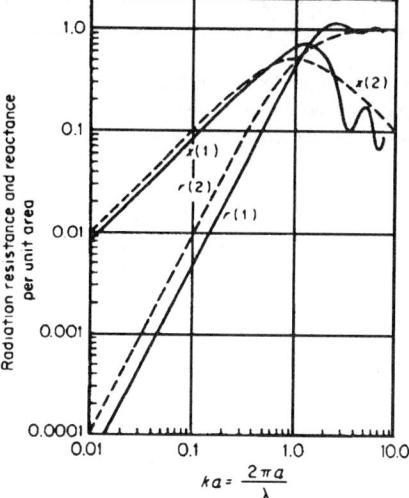

FIGURE 23.4.14 Radiation resistance and reactance per unit area divided by ρc as a function of a ka (a = radius) for (1) a circular position in a rigid baffle; (2) a pulsating sphere.

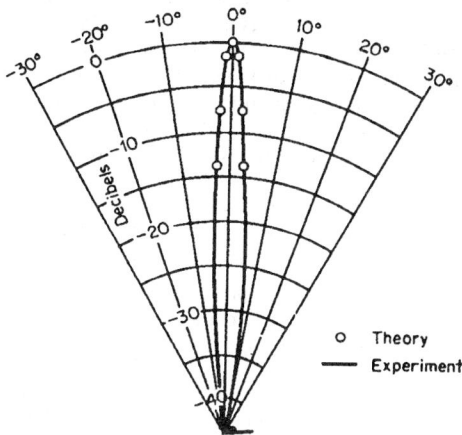

FIGURE 23.4.15 Difference-frequency beam pattern.

In particular, if two relatively intense sound waves of nearly equal frequency are caused to propagate through water in the same direction, the nonlinearity causes them to interact with each other to form waves having the sum and difference frequencies. The original primary waves are eventually absorbed in propagating through the water, and the sum-frequency wave, being at a higher frequency, is absorbed even sooner. The difference-frequency wave, being at a lower frequency, persists and has an independent existence.

The far-field sound pressure (all pressures expressed in pascals) of the difference-frequency wave P_d can be described in terms of the radiated sound pressures, p_1, p_2 of the two collimated primary waves by the following basic equation,[23] which is subject to various refinements discussed below:

$$p_d = \frac{(1 + 1/2\ B/A)\ \omega_2^2 p_1 p_2 S}{8\pi r \rho c^4} \frac{1}{j\alpha + k_d \sin^2(\theta/2)} \tag{92a}$$

where B/A = parameter of nonlinearity (5.25 in 20°C seawater[24])
ω_d = frequency of difference-frequency wave (rad/s)
S = cross-sectional area of collimated primary beams (m²)
r = distance to measuring point (m)
ρ = density of medium (kg/m³)
c = speed of sound (m/s)
α = attenuation coefficient (np/m)
k_d = wave number of difference-frequency wave (m⁻¹)
θ = angle to measuring point with respect to axis of beam

Figure 23.4.15 is an example of the difference-frequency wave beam pattern. It will be noticed that this beam has no side lobes and is quite narrow. θ_d, the value of θ for which the intensity is down 3 dB, is that which causes the real and imaginary parts of the denominator of Eq. (92a) to be equal. Because this value of θ is small, it is approximately

$$\theta_d = 2\sqrt{\alpha/k_s} \tag{92b}$$

The beamwidth, which is twice this angle, is narrower than that achievable when a wave of the same (difference) frequency is generated directly using a transducer of the same size. Also, this beamwidth does not very significantly over a wide frequency range.

Reference 25 provides a nomogram for finding the far-field source level and beamwidth of the difference-frequency wave when the medium is seawater. As an example, assume primary waves with frequencies of 95 and 105 kHz. The mean primary frequency is 100 kHz, and the difference frequency is 10 kHz, resulting in a half beamwidth of 1.1°. A mean primary source level minus directivity index of 170 dB re 1 μPa results in a difference-frequency source level of 143 dB at 1 m re 1 μPa at 1 m.

It will be observed that there is an adverse conversion loss between primary and difference-frequency source levels, particularly if the primary directivity index is taken into account. This is the price paid for the benefits gained in terms of the narrow bandwidth.

Several practical factors serve to modify the above relationships:

1. Equation (29.92a) assumes collimated plane primary waves and essentially zero primary beamwidth. More realistic nonzero primary beamwidths require a modification. Reference 26 provides useful curves for taking the shape and dimensions of the transducer into account.

2. As the intensity of the primary waves increases, a point is reached where a shock wave is formed, at which point a different set of relationships govern the behavior. This phenomenon is still a subject of research; a good current reference is Ref. 27.

In addition to the transmitting-array application described above, it is also possible to create a highly directive receiving array by placing high-frequency high-intensity acoustic source, called a *pump,* at a relatively great distance from the receiver, in line with a relatively low-frequency plane wave arriving from a distant source. Again waves at the sum and difference frequencies are generated by the nonlinear interaction in the medium. One or both of these sidebands are separated from the pump frequency by filtering. This application and the governing relations are presented in Ref. 28.

TRANSDUCER MATERIALS

Piezoelectric Materials

Since 1950, synthetic piezoelectric ceramics as transducing materials have reached maturity by achieving reproducibility for a given composition and by diversification.[29]

The electromechanical nature of these materials in a polarized state is described by linear equations. With stress and electric field as the independent variables,

$$S = s^E T + dE \quad D = dT + \epsilon^T E \tag{93a}$$

With stress and charge density (electric displacement) as the independent variables,

$$S = s^D T + gD \quad E = - gT + \beta^T D \tag{93b}$$

where S = strain
T = stress
E = electric field
D = electric displacement
s^E = elastic compliance at constant electric field
s^D = elastic compliance at constant electric displacement
ϵ^T = dielectric constant at constant stress
β^T = dielectric impermeability at constant stress

d and g are piezoelectric constants defined as

$$d = (\partial S/\partial E)_T = (\partial D/\partial T)_E \quad g = (-\partial E/\partial T)_D = (\partial S/\partial D)_T \tag{94}$$

The electromechanical coupling coefficient is an important measure of the effectiveness of the energy conversion mechanism and is defined as

$$k = U_m/U_e U_d = d/\epsilon^T s^E \tag{95}$$

where U_m = mutual elastic and dielectric energy density
U_e = elastic self-energy density
U_d = dielectric self-energy density. This is a quasi-static parameter that can be related to the fundamental material constants for one-dimensional transducers. The definition is not necessarily applicable to all geometries.

A more complete treatment of the piezoelectric equations and the measurement of the various constants can be found in Refs. 29–35.

Although piezoelectric ceramics can be operated in various modes, two types are of major importance to underwater sound transducers. In the parallel, or *33-mode* type, the stress, strain, and electric field are in the same direction. In the transverse, or *31-mode* types, the stress and strain are the same direction but orthogonal to the electric field. Each of these mode types is characterized by its associated constants, for example, d_{31} or d_{33}, g_{31} or g_{33}, k_{31} or k_{33}, respectively. The dielectric constant of interest is in the direction of the electric field for both cases and is denoted ϵ_{33}. For the above constants, the direction of the electric vectors are denoted by the three directions, and the direction of the mechanical variable, if different, is the other subscript. An additional electromechanical coupling coefficient k_p is useful for some applications. It is related to k by

$$k_p = \sqrt{2/(1-\sigma^E)}\ k_{31} \tag{96}$$

where σ^E is Poisson's ratio at constant electric field.

The most important piezoelectric ceramic materials for underwater sound transducers are the modified lead zirconate titanate (PZT) compositions and to a lesser extent modified barium titanate and lead titanate compositions. Table 23.4.2 lists some more important properties of these materials. A more comprehensive table of properties is available in Ref. 29, Chapter 3. These materials are characterized as being "very hard" lead zirconate titanate, "hard" lead zirconate titanate, and "soft" lead zirconate titanate. Progress has been made in specifying and classifying the various ceramic compositions.[36] The properties of the piezoelectric ceramics vary as functions of time, static stress, stress cycling, and electric field strength.[29]

When the transducer design calls for operation in a hydrostatic-mode as opposed to the conventional 33- or 31-mode, the commonly known PZT or barium titanate compositions are not useful because of their low hydrostatic piezoelectric coefficients such as g_h, which is $g_h = g_{33} + g_{32} + g_{31}$. Low g_h values are characteristic of bulk ceramics due to the fact that in these materials $g_{32} = g_{31}$, both have an opposite sign to g_{33} and are nearly half the magnitude of g_{33}. In order to improve the hydrostatic piezoelectric properties, multiphase composite materials with different connectivity patterns have been developed.[37] The basic concept is to reduce the g_{31} contribution by introducing inert polymeric phases and to maximize the g_{33} effect of the piezoelectric phase. Composite properties are optimized by varying the configurational patterns of the active ceramic component and the inactive polymeric components in the composite. This is achieved by manipulating the connectivity, which describes how one phase connects itself in one or several dimensions. The more popular piezocomposites include the 0-3 and the 1-3 types. The former consists of ceramic particles of lead titanate composition dispersed in a neoprene rubber matrix, whereas the 1-3 piezocomposite is formed by aligning ceramic rods inn the poled direction and imbedded them in a polymeric matrix. Because these piezocomposites contain a polymer phase, they are lighter in weight than bulk ceramics, and can be made flexible and/or comfortable to curved surfaces. In general, they can also be processed to form large sheets for large area coverage, as is the case for piezoelectric polyvinyledene fluoride polymers (PVDF). PVDF also exhibits good g_h property, but is more difficult to pole, and its dielectric constant is very low. Mostly formed by using an extrusion process from polymer melts, PVDF also shows higher g_{31} in the extrusion direction than the transverse g_{32}. On the other hand, the inert polymer phase in a piezocomposite can potentially be designed to greatly decouple the g_{31} effect of the ceramic, resulting in a dramatic improvement in the hydrostatic-mode properties. Table 23.4.3 shows the hydrostatic-mode properties of these new piezoelectric materials in comparison with those of well-known ceramic compositions. Hydroacoustic behaviors of these piezocomposites have been investigated.[38] Transducers fabricated by using the composite materials showed a constant broadband sensitivity, which in many cases is also independent of temperature and pressure.

TABLE 23.4.2 Characteristics of Commonly Used Piezoelectric Ceramics, Low-Signal Properties at 75°C

Quantity	PZT-4*	PZT-5*	PZT-8*	95% wt BaTiO$_3$, 5% wt CaTiO$_3$
k_p	0.58	0.60	0.51	0.36
k_{31}	0.334	0.344	0.30	0.212
k_{33}	0.70	0.705	0.64	0.50
$\varepsilon_{33}^T/\epsilon_0$ (T = at constant stress)	1,300	1,700	1,290	1,700
$\varepsilon_{33}^S/\epsilon_0$ (S = at constant strain)	635	830	580	1,260
tan δ	0.004	0.02	0.004	0.006
d_{33}, pC/N	289	374	225	149
d_{31}	−123	−171	−97	−58
g_{33}, mV·m/N	26.1	24.8	25.4	14.1
g_{31}	−11.1	−11.4	−10.9	−5.5
s_{11}^E, pm^2/N (E = at constant electric field)	12.3	16.4	13.5	8.6
s_{33}^E	15.5	18.8	11.5	9.1
s_{11}^D (D = at constant displacement)	10.9	14.4	10.4	8.3
s_{33}^D	7.90	9.46	8.0	7.0
Q_M	500	75	1,000	400
ρ, 10^3 kg/m^3	7.5	7.75	7.6	5.55
N_1, Hz·m[†]	1,650	1,400	1,700	2,290
N_3[‡]	2,000	1,770	2,070	2,740
Curie point, °C	328	365	300	115
Heat capacity, J/Kg·°C	420	420	420	500
Thermal conductivity, W/m·°C	2.1	2.1	2.1	3.5
Static tensile strength, lb/in^2	13,000	13,000	12,000	12,000
Rated dynamic tensile strength, lb/in^2	6,000	4,000	7,000	7,500

*Trademark, Vernitron Piezoelectric Division.
[†]N_1 = frequency constant of a thin bar with electric field perpendicular to length, f_l.
[‡]N_3 = frequency constant of a thin plate with electric field parallel to thickness, f_t.

Magnetostrictive Materials

Magnetostrictive materials offer certain advantages for some underwater sound transducer applications. One example is that of a large low-frequency source that is submerged to a great depth and must operate unattended for long periods of time. Two forms of magnetostrictive materials are used, metal alloys and ceramic compositions.

The physical quantities regarding the magnetic and mechanical state of a polarized material are the magnetic field strength H, the magnetic flux density B, the mechanical stress T, and the mechanical strain S. For a sinusoidal variation, they are related by

$$S = s^H T + dH \quad B = dT + \mu^T H \qquad (97a)$$

or

$$T = c^B S - hB \quad H = -hS + v^s B \qquad (97b)$$

where s^H = elastic compliance at constant magnetic field strength
c^B = elastic stiffness at constant magnetic flux density
μ^T = permeability at constant stress
v^s = reluctivity at constant strain

d and h are the piezomagnetic constants, which are defined as

TABLE 23.4.3 Piezoelectric Properties of Composite Materials in Hydrostatic Mode

Composite type	K^* (1 kHz)	d_h (pC/N)	g_h (mV-m/N)
3–3	200	92	53
3–3	478	200	48
0–3	54	22	47
0–3	38	27	81
0–3	37	41	124
0–3	51	69	153
1–3	45	28	71
1–3	200	40	18
PVDF	10	10	112
PVDF	8	18	270
PZT4	1300	43	4
PZT5	1700	21	2
PbTiO$_3$	210	48	26

*Dielectric constant (measured at 1 kHz).

$$d = (\partial S/\partial H)_T = (\partial B/\partial T)_H \quad h = -(\partial T/\partial B)_S = -(\partial H/\partial S)_B \tag{98}$$

More detailed information regarding the magnetostrictive equations and constants can be found in Refs. 29, 39–41.

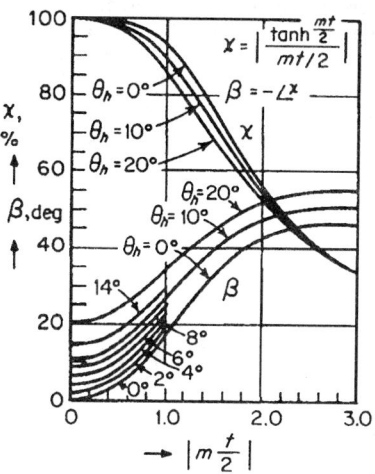

FIGURE 23.4.16 Eddy-current loss factor, magnitude and phase angle, of magnetic materials; $\theta_h < \mu(0)$.[41]

The electromechanical coupling factor k of magnetostrictive materials has the same physical meaning as for piezoelectric materials, with the same limitations.

Eddy currents can play an important role in the efficiency of magnetostrictive materials. For this reason, magnetostrictive assemblies are often constructed from thin laminations cemented together, usually in an annealed state. Eddy current losses can be taken into account by multiplying the permeability by a complex eddy current factor x. This factor depends on the geometry. With a modification of the analysis in Ref. 41, a skin-effect parameter m^2 can be defined as

$$m^2 = j\omega\mu(0)/\rho_e \tag{99}$$

where $\mu(0)$ = permeability with no eddy current effect
ω = angular frequency
ρ_e = resistivity

The apparent permeability is given by

$$\mu = \mu(0)\frac{\tanh(mt/2)}{mt/2} \tag{100}$$

where t is the thickness of sheet of material.

Figure 23.4.16 shows a plot of the magnitude and phase angle of the complex correction factor X versus $mt/2$.

Some important properties of a number of magnetostrictive materials can be found in Ref. 29.

TRANSDUCER ARRAYS

Beam Formation

Transducers and hydrophones can be arranged individually or in arrays to possess omnidirectional or directional characteristics, depending on effective aperture dimensions, geometrical shape, and vibrational modes used. At high frequencies, since the wavelengths are short, highly directional individual units can be designed. At lower frequencies, multiple transducers or hydrophones are used in arrays of planar, cylindrical, spherical, or volumeric configuration.

Directionality is highly desirable in underwater sound detection systems because it makes both directional transmission and the determination of the direction of arrival of a signal possible. As in directional radar or communications antennas, this reduces the noise relative to the signal from other directions. Arrays can be steered mechanically by physical rotation or electrically by phasing or time delay networks. The direction of maximum sensitivity of a plane array of elements can be rotated into a direction lying at angle θ_0 to a reference direction by delaying differentially the output of each element. In this way, an irregular array can be effectively converted into a line array.

In their simplest form, arrays are arranged with elements along a line or distributed along a plane. The acoustic axis of such line or plane arrays, when unsteered, lies at right angles to the line or plane. The beam pattern of a line array may be visualized as a doughnut-shaped figure having supernumerary attached doughnuts formed by the side lobes of the pattern.

Lines of Equally Spaced Elements[3]

The beam pattern of a line of equally spaced, equally phased (i.e., unsteered) elements is derived as follows. Let a plane sinusoidal sound wave of unit pressure be incident at an angle θ to a line of n such elements, each spaced from the next a distance d. The output of the mth element relative to that of the zeroth element is delayed by the time necessary for sound to travel the distance $l_m = md \sin\theta$. The corresponding phase delay for sound of wavelength λ, at frequency $\omega = 2\pi f$, is $u_m = mu$ where the phase delay between adjacent elements, in radians, is $u = (2\pi d/\lambda) \sin\theta$. The output voltage of the mth element of voltage response R_m is $V_m = R_m \cos(\omega t + mu)$ and the array voltage is the sum of such terms:

$$V = R_0 \cos\omega t + R_1 \cos(\omega t + u) + \cdots + R_m \cos(\omega t + mu) + \cdots + R_{n-1} \cos[\omega t + (n-1)u]$$

In the complex notation, the array voltage will be

$$V = (R_0 + R_1 e^{iu} + R_2 e^{2iu} + \cdots + R_n e^{(n-1)iu})e^{i\omega t}$$

If the array elements all have unit response $(R = 1)$,

$$V = (1 + e^{iu} + e^{2iu} + \cdots + e^{(n-1)iu})e^{i\omega t}$$

Multiplying by e^{iu} and subtracting gives

$$V = [(e^{inu} - 1)/(e^{iu} - 1)]e^{i\omega t}$$

Neglecting the time dependence, this becomes

$$V = [\sin(nu/2)]/[\sin(u/2)]$$

Finally, expressing u in terms of θ, the beam pattern, the square of the function-normalized to unity at $\theta = 0$ is

$$b(\theta) = \left(\frac{V}{n}\right)^2 = \left\{ \frac{\sin(n\pi d \sin\theta/\lambda)}{n \sin(\pi d \sin\theta/\lambda)} \right\}^2 \tag{101}$$

Continuous-Line and Plane Circular Arrays[3]

When the array elements are so close together that they may be regarded as adjacent, the array becomes a continuous-line transducer and the beam pattern is found by integration rather than by summation. For this case, let the line transducer be of length L and have a response per unit length of R/L. The contribution to the total voltage output produced by a small element of line length dx located at a distance x from the center is (neglecting the time dependence)

$$dv = (R/L)e^{(i2\pi/\lambda)\sin\theta} \, dx$$

The beam pattern, the square of V normalized so that $b(\theta) = 1$, is

$$b(\theta) = \left(\frac{V}{R}\right)^2 = \left\{ \frac{\sin[(\pi L/\lambda)\sin\theta]}{(\pi L/\lambda)\sin\theta} \right\}^2 \tag{102}$$

In a similar manner, the beam pattern of a circular-plane array of diameter D of closely spaced elements can be shown to be

$$b(\theta) = \left\{ \frac{2J_1[(\pi D/\lambda)\sin\theta]}{(\pi D/\lambda)\sin\theta} \right\}^2 \tag{103}$$

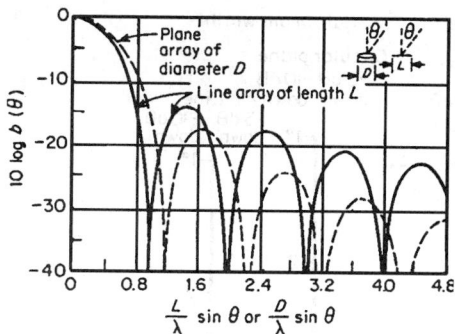

FIGURE 23.4.17 Beam patterns of a line array of length L, and a circular-plane of diameter D.

where $J_1[(\pi D/\lambda) \sin\theta]$ is the first-order Bessel function of argument $(\pi D/\lambda) \sin\theta$.

Generalized beam patterns for continuous-line and circular-plane arrays are drawn in Fig. 23.4.17 in terms of the quantities $(L/\lambda) \sin\theta$ and $(D/\lambda) \sin\theta$.

Figure 23.4.18 is a nomogram for finding the angular width between the axis and the −3-dB and −10-dB points of the beam pattern for continuous-line and circular-plane arrays. The dashed lines indicate how the nomogram is to be used. Thus a circular-plane array of 500 mm diameter at a wavelength of 100 mm (corresponding to a frequency of 15 kHz at a sound speed of 1500 m/s) has a beam pattern 6° wide between the axis of the pattern and the −3-dB points.

PASSIVE SONAR SYSTEMS

Passive sonar systems, also referred to as *listening sonar systems*, are designed to respond to acoustic energy radiated by sources in the band of the sonar receiver. These systems are designed to accentuate the response to wanted signals while suppressing unwanted background noise. Passive systems are designed to maximize the signal-to-noise ratio to the degree that the characteristics of the signal and noise are known or can be predicted.[42–44]

Passive Sonar Equations

The fundamental relation of passive sonar can be written in terms of the signal-to-interfering-background ratio. Since in sonar the usual practice is to write equations in the decibel notation, the ratio is defined as the *signal differential* $\Delta L_{S/N}$, in decibels, between the equivalent plane-wave levels of signals and interfering noise at the passive sonar receiving array:

$$\Delta L_{S/N} = (L_{sf} - N_W + N'_{BW}) - [(L_{Nf} - N_{DI} + N_{BW}) + N_\delta] \quad (104)$$

where L_{sf} = target-radiated-noise spectrum level versus frequency at 1 m from effective center of radiating source (dB re 1 μPa/Hz$^{1/2}$), N_W = one-way acoustic propagation loss between radiating source and passive sonar array (dB), $N'_{BW} = 10 \log$ (signal bandwidth) = signal-bandwidth-level correction, L_{Nf} = equivalent-plane-wave interfering-noise spectrum level versus frequency at passive sonar array resulting from summation of noise from all sources (dB re 1 μPa/Hz$^{1/2}$), $N_{DI} = 10 \log D$, where D = effective directivity of the passive sonar array and beam former against isotropic noise (dB) (if the noise background has directional components, the effectiveness of the array in decreased background noise is modified), $N_{BW} = 10 \log$ (noise bandwidth) = (noise-bandwidth-level correction, and N_δ = receiving deviation loss (dB).

The value of the signal difference required at the array output varies with the application, e.g., detection, classification, or localization.

The signal differential can be considered to be the sum of two terms: the detection threshold N_{DT}, defined as that value of the signal differential required to just detect the signal, and the signal excess N_{SE}, which is the amount in decibels by which $\Delta L_{S/N}$ exceeds N_{DT}. A detection threshold N_{DT} adjustment in the system is usually set to a value consistent with a sufficiently low false-alarm rate on the display in use. Substituting this sum into Eq. (92) gives

$$N_{SE} = (L_{Sf} - N_W + N'_{BW}) - [(L_{NF} - N_{DI} + N_{BW}) + N_\delta + N_{DT}] \quad (105)$$

Equation (105) is arranged to show that signal excess in decibels is the differential between a set of signal terms and a set of effective-noise terms, i.e.,

$$N_{SE} = L_I - L_{MD} \quad (106)$$

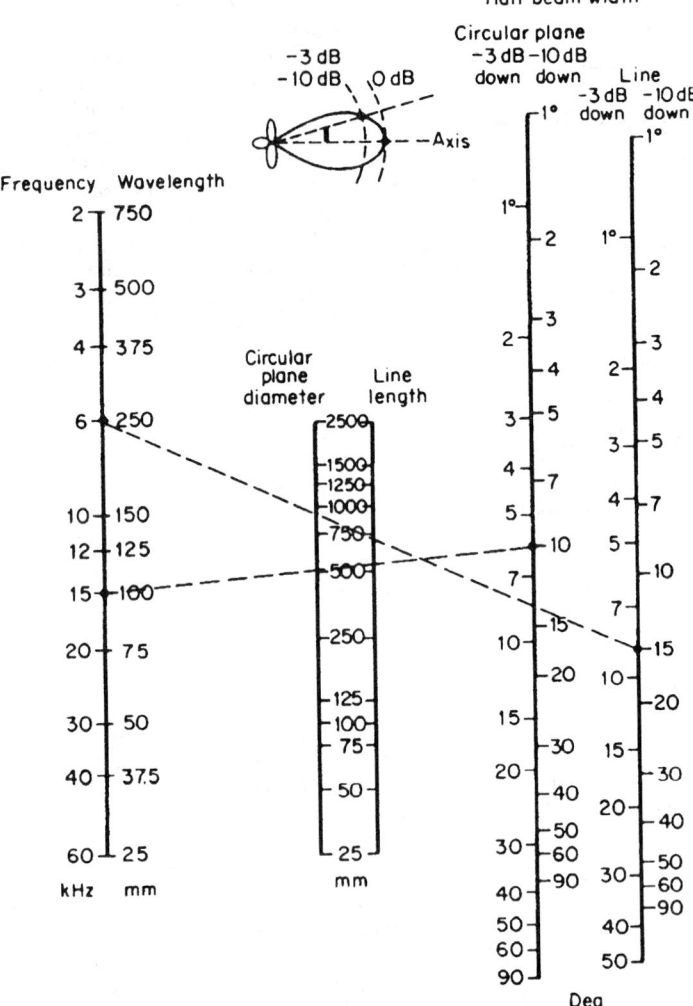

FIGURE 23.4.18   Nomogram for finding width of beam pattern.

where L_I is the incident signal level and L_{MD} is the minimum detectable signal level, given by

$$L_{MD} = L_{Nf} - N_{DI} + N_{BW} + N_{\delta} + N_{DT} \tag{107}$$

Another useful measure is the figure of merit N_{FM}, defined as the maximum allowable one-way propagation loss under the condition of zero signal excess. From Eq. (105),

$$N_{FM} = (L_{Sf} + N'_{BW}) - [(L_{Nf} - N_{DI} + N_{BW}) + N_{\delta} + N_{DT}] \tag{108}$$

and from Eq. (107), the figure of merit can also be written

$$N_{FM} = L_{Sf} + N'_{BW} - L_{MD} \tag{109}$$

Sonar Parameters

The radiated-noise level of a target is usually composed of broadband and narrowband noise from the propeller, machinery, and possibly echo-ranging pings from the target, as well as from other ships.[3,16] Since this radiation can vary considerably in frequency and transmitted intensity as a function of time, the signal excess, and consequently the maximum detection range, can also vary over a wide range.

Acoustic propagation loss, also called *transmission loss*, varies according to an applicable spreading law modified by absorption, refraction, and reflection. In deep water, spherical spreading applies, and the propagation loss N_W in decibels is given by

$$N_W = 20 \log R + \alpha R + 60 \tag{110}$$

where R = range (km); α = absorption (dB/km), which varies with frequency f, and 60 represents the conversion from meters to kilometers. For shallow sources in deep oceans when a surface layer is present, spherical spreading applies for the first kilometer and cylindrical spreading thereafter, and the propagation loss then becomes

$$N_W = 10 \log R + \alpha R + 60 \tag{111}$$

The interfering background noise places a limit on detectability, but depending on the character of the noise, on the array interelement correlation, and on the signal processing used, the threshold of detection can be lowered considerably below the rms noise band level. Background noise is composed of self-noise, ambient noise, and, when present, reverberation. The noise components in these categories are listed in Table 23.4.4.

Self-noise is generated by all those sources associated with the listening vessel and its interaction with the surrounding water. *Ambient noise* is caused by both natural and man-made sources, while *reverberation* results from reflections of sonar pings by the ocean surface, bottom, volume, and scattering layers.

Self-noise has many directional characteristics (near-field effects); ambient noise generally has an omnidirectional distribution, with exceptions because of man-made and precipitation components; and reverberation is largely directional in nature.

The *effective directivity index* is an indicator of the degree to which the receiving array and beam former discriminates against background noise, assuming that this noise is isotropic in character and that the array geometry results in low correlation from element to element in the operating frequency band.

Examination of Eq. (108) shows that it is possible to distinguish three groups of parameters that determine figure of merit. First, associated with own ship, its sonar, background noise, and the operator are N_{DI}, N_{BW}, N_δ, N_{DT}, and these components of L_N associated with own ship and the surrounding sea. The second group of parameters is associated with the acoustic properties of the sea (or freshwater) medium and its boundaries, and it consists of N_W and the ambient components of L_N, including surface noise and reflections of noise from the sea surface. The third group of parameters, summarized in the terms L_S and N'_{BW}, is concerned with the radiating acoustic source or target, i.e., the description of the characteristics of the target in terms of radiated-noise spectrum level as a function of frequency, including both broadband and narrowband spectra, and the bandwidth of the radiated noise.

TABLE 23.4.4   Sources of Background Noise

A. Self-noise	B. Ambient noise	3. Man-made
1. Hydrodynamic	1. Sea	*a.* Shipping
a. Flow-excited	*a.* Bubbles	*b.* Shore-radiated
b. Turbulence	*b.* Seismic	4. Precipitation
c. Cavitation	*c.* Waves	C. Reverberation
d. Bubbles	*d.* Boundary turbulence	1. Bottom
e. Splash	*e.* Ice motion	2. Surface
2. Machinery	2. Biological	3. Volume
a. Turbines	*a.* Shrimp	4. Scattering layers
b. Pumps	*b.* Whales	
c. Blowers, etc.	*c.* Carp, etc.	
3. Propeller		

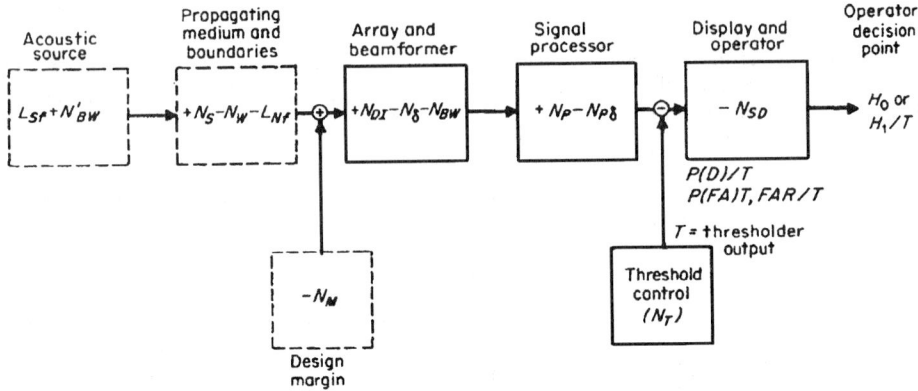

At operator decision point, N_{SE} is given in Eq. (112)

FIGURE 23.4.19 Passive sonar system detection model.

System Configuration and Parameters

A generalized passive sonar detection model, shown in Fig. 23.4.19, depicts all the parameters associated with a passive sonar system and its operator. To account for system hardware and operator gains and losses, terms must be added to Eqs. (104), (105), (107), and (108). While these equations are applicable at the output of the idealized beam former, the modified equations apply at the operator decision point indicated in Fig. 23.4.19. Specifically, at the decision point the signal excess is

$$N'_{SE} = (L_{Sf} + N_S - N_W + N'_{BW} + N_P) - [(L_{Nf} - N_{DI} + N_{BW}) + N_M + N_\delta + N_T + N_{SD})] \qquad (112)$$

The added terms are N_S, to account for more than one ray path from the acoustic source to the sonar N_P, for signal processing gain; N_M, for hardware design margin (loss); N_δ, for signal processing loss, as in a clipping processor; N_T for the effect of the threshold level, and N_{SD}, to allow for the signal differential required for signal recognition on the particular display used.

The sonar detection system is placed between the hydrophone array and the decision point. It consists of a receiver, a visual and/or aural display, and an operator. The detection threshold N_{DT} is defined[3] as the ratio in decibels of the signal power S in the receiver band to the noise power N_0 in a 1-Hz band, measured at the receiver terminals; i.e.

$$N_{DT} \equiv 10 \log (S/N_0) \qquad (113)$$

The signal and signal-plus-noise values are taken in the receiver band, and N_{DT} is computed for use in the sonar equation. Noise backgrounds are expressed as power spectrum levels, i.e., as powers in 1-Hz bands.

Detection decisions are binary in nature; i.e., signal is present or signal is absent. But since a signal can actually be present or actually absent at the receiver input, there are four possible situations, summarized in Table 23.4.5. The correct decisions are shown on the diagonal of the matrix, namely, the detection and null decisions, with probabilities $P(D)$ and $1 - P(FA)$, respectively, where $P(D)$ is the probability of detection and $P(FA)$ is the probability of false alarm. False-alarm and miss decisions can occur with respective probabilities $P(FA)$ and $1 - P(D)$.

To implement the detection threshold criterion, it is necessary to apply a threshold voltage to a comparison circuit, at the receiver output. The threshold voltage is either fixed at a level corresponding to the desired N_{DT} or it can be controlled and calibrated in terms of N_{DT}. Whenever the level of the waveform at the comparison point exceeds the threshold voltage, the decision of signal present is made, unless a rule is adopted or circuit implemented, which, for example, counts the number of times the threshold is exceeded during a fixed time interval and bases the detection decision on this count.

The effects of setting the threshold voltage at various levels are shown in Fig. 23.4.20, at three target signals in a noise background, with two possible threshold voltages. High settings like T_1 allow only strong targets to

TABLE 23.4.5 Binary Decision Matrix

When at the input:	And the decision is	
	Signal present	Signal absent
Signal present	Correct decision: detection with probability $P(D)$	Incorrect decision: miss with probability $1 - P(D)$
Signal absent	Incorrect decision: false alarm with probability $P(FA)$	Correct decision: null with probability $1 - P(FA)$

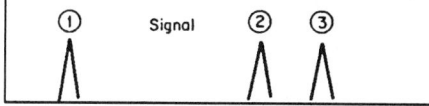

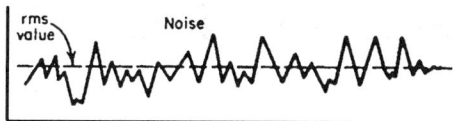

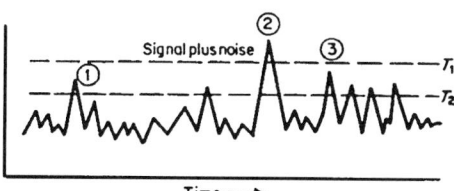

FIGURE 23.4.20 Signal and noise at two threshold settings.

be detected and extremely few, if any, false alarms; consequently, both $P(D)$ and $P(FA)$ are low. Low settings like T_2 allow many more possible signals to be detected and also many false alarms to occur, consequently both $P(D)$ and $P(FA)$ are high. Both these threshold settings are unsatisfactory. A threshold setting should be found that produces a display with a uniformly distributed (gray) background in the presence of noise only and which prevents the buildup of the number of noise markings as a function of time and the decay rate of the display storage system in use; i.e., the false-alarm rate must be bounded.

Detection decisions depend upon the independent probabilities $P(D)$ and $P(FA)$, and consequently on the distributions of noise and signal-plus-noise at the receiver output. Figure 23.4.21 shows the probability density of noise alone and signal-plus-noise plotted as a function of amplitude[3] a. The curves are assumed to have a gaussian distribution with variance σ^2, the first with the mean noise amplitude $M(N)$ and the second with mean signal-plus-noise amplitude $M(S + N)$. Then the parameter detection index d is defined as

$$d \equiv (M_{S+N} - M_N)^2/\sigma^2 \tag{114}$$

which is equivalent to the signal-to-noise ratio of the envelope of the receiver output effectively at the terminals where the threshold voltage T is established. The area (integral) under the probability density curve of signal-plus-noise to the right of T in Fig. 23.4.21 is the probability that an amplitude in excess of T is due to signal-plus-noise and is equal to the probability of detection $P(D)$. Similarly, the area under the probability curve of noise alone to the right of T is equal to the probability of false alarm $P(FA)$. Since $P(D)$ and $P(FA)$ vary as the threshold T is changed, their values depend on the parameter d.

The probabilities associated with various threshold voltage levels can be plotted, as shown in Fig. 23.4.22, with the detection index d as a parameter. Thus, for high-T settings, for example, the corresponding low $P(D)$, $P(FA)$ points are plotted in the lower left portion of the figure above the diagonal. These are known as receiver operating characteristic (ROC) curves.

If the curves of Fig. 23.4.21 are imagined to refer to the receiver input, the likelihood ratio for an input sample of amplitude a is

$$L = B/A \tag{115}$$

Thus, the ROC curves are related to the signal-to-noise ratio at the receiver input required for detection.

Following the detector, a postdetection averaging filter is used to smooth out fluctuations due to detector output noise. The optimum integration time T of this filter is the time equal to the signal duration t. In passive sonar systems t can be as long as it is operationally useful to integrate the incoming signal in the noise and/or reverberation background. Here, factors such as stabilization of own ship's receiving beams (for rotating or preformed beams), rotating receiving beam RPM (which determines the time on target during each rotation), receiving beamwidth, possible range of the target, and possible speed of the target must be considered.

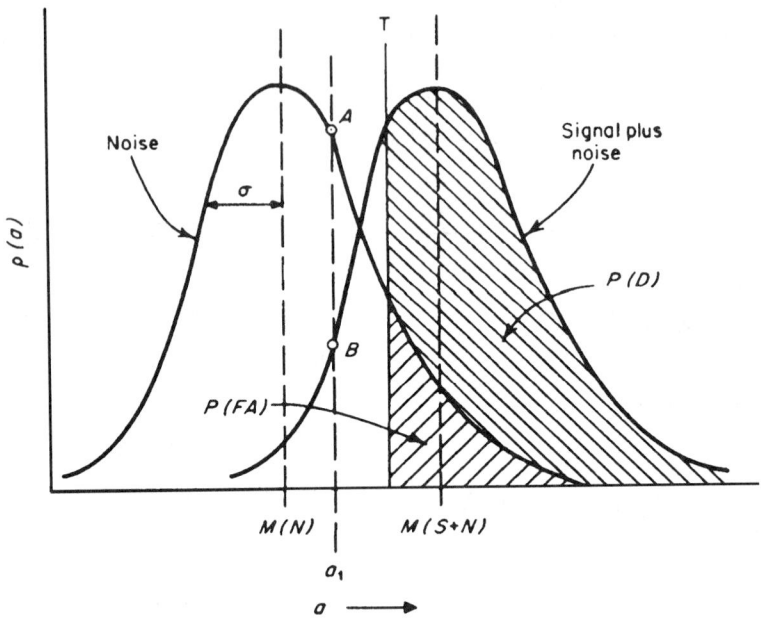

FIGURE 23.4.21 Probability density distributions of noise.

ACTIVE SONAR SYSTEMS

Introduction

Active sonar systems, like radar, make use of *reflected energy reception*, referred to as *echo ranging*. Active sonar systems are used primarily to determine target range and bearing. In addition, active systems may be used for determining target depth, aspect angle, course, and speed. Target motion may be computed from successive echo returns or measurement of target-generated doppler on each ping.

A wide variety of signal waveforms, pulsed or continuous, may be used in echo-ranging operations. Pulsed continuous waves of 1 ms to several seconds duration are commonly used, short pulses providing maximum range resolution and long pulses providing maximum energy return from the target, hence greater detection range. Pulsed FM waves with up or down sweeps and pseudo-random noise waves are used to obtain better target discrimination in a reverberant background. Combinations of the foregoing waveforms may be used to reduce mutual interference.[42–44]

Continuous transmission systems (cw or FM) are employed in short-range work, such as navigation sonars. Continuous transmission is not used in long-range detection systems because of the difficulty of acoustic isolation between the receiving and transmitting arrays.

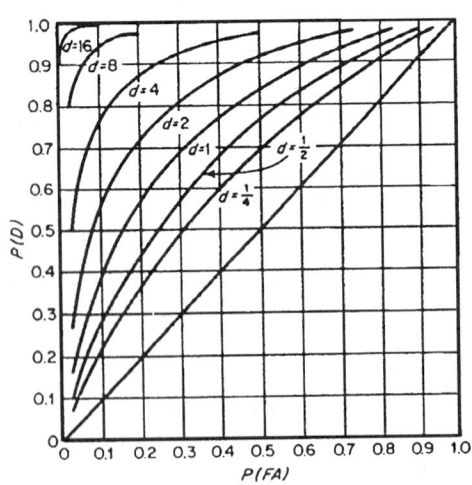

FIGURE 23.4.22 Receiver operating characteristic (ROC) curves.

Active Sonar Equation

The active sonar equation, corresponding to the model shown in Fig. 23.4.23, is used for performance prediction. The following equation is commonly used for applications with two-way propagation loss:

$$N_{SE} = L_S - (N_{\delta T} + N_{\delta R}) + N_{TS} - 2N_W - L_B - N_{DT} \qquad (116)$$

where N_{SE} = signal excess (dB)
$\quad L_S$ = source level (dB re 1 μPa at 1 m) on transmit beam axis
$N_{\delta T}, N_{\delta R}$ = transmit and receive deviation losses due to target ray being off axis of transmit and receive beams, respectively (dB)
$\quad N_{TS}$ = target strength (dB)
$\quad N_W$ = one-way propagation loss (dB)
$\quad L_B$ = noise level in receiver band, including ambient noise, self-noise, and reverberations (dB ar 1 μPa)
$\quad N_{DT}$ = detection threshold in receiver band and at receiver input to produce desired display statistic (dB)

Equation (104) also applies to applications with one-way propagation losses, such as communications or telemetry, with only one propagation loss taken into account and background noise corresponding to the receive platform only.

The figure of merit (FOM) is another commonly used criterion for evaluating sonar performance. The FOM is equal to the allowable propagation loss when the signal excess is zero.

Source Level

The radiated acoustic powers of shipboard sonars[3] range from a few hundred watts to tens of kilowatts, with transmitting directivity indexes between 10 and 30 dB. The equation for the value of the source level L_S is

$$L_S = 170.8 + 10 \log P_A + N_{DI} - N_D \qquad (117)$$

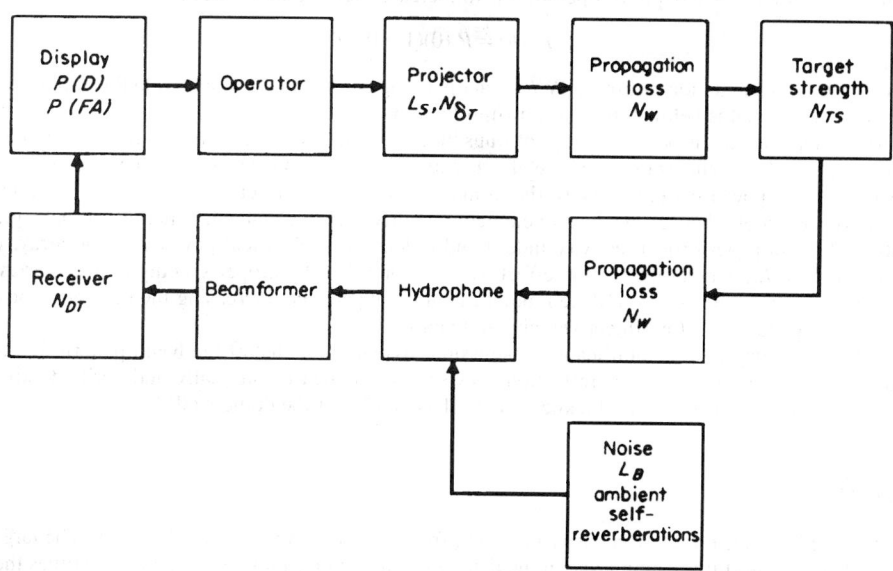

FIGURE 23.4.23   Echo-ranging block diagram.

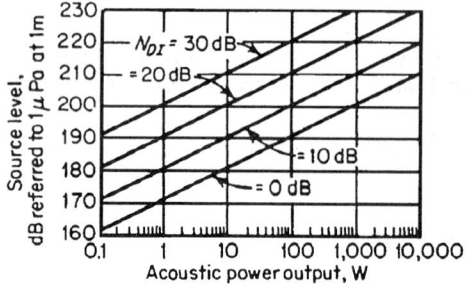

FIGURE 23.4.24 Plot of Eq. (117)

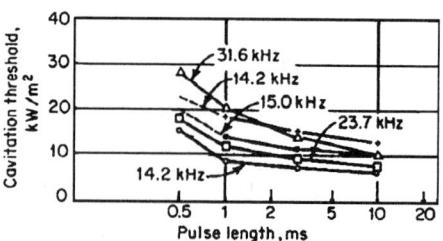

FIGURE 23.4.25 Cavitation threshold vs. pulse duration.

where P_A = acoustic radiated power (W)
   N_{DI} = directivity index (dB)
   N_D = dome loss (dB) where applicable

A plot of L_S versus P_A is shown in Fig. 23.4.24 with N_{DI} as a parameter. The efficiency of converting electric to acoustic power varies with the transducer design, the match between transmitter source and transducer load impedance, and array-mutual-interaction effects.

There are two major acoustic limitations that limit achievable source level: cavitation and array mutual interaction. Cavitation limits occur[3] when power applied to a transducer is such that bubbles begin to form on the face and just in front of transducer. These bubbles are a manifestation of cavitation and are caused by the rupture of water through negative pressure of the generated sound field. Cavitation causes a deterioration of transducer performance and life by erosion of the transducer surface; loss of radiated acoustic power in absorption and scattering by cavitation bubbles; deteriorization in beam pattern; and reduction of the acoustic impedance into which the transducer must radiate. Cavitation threshold occurs at the point of departure from linearity of the input-output power curve and is a function of frequency, pulse length, and operating depth.

The effect of increased depth of operation is to increase the cavitation threshold

$$P_C(h) = P_C(0)(1 + 0.1h)^2 \qquad (118)$$

where $P_C(h)$ is the cavitation threshold (W/m^2) at operating depth h m. The cavitation threshold increases with decreasing pulse lengths below 5 ms, as shown in Fig. 23.4.25.

Unequal loading effects occur in arrays of transducer elements because of mutual radiation impedances[45–47] between independent sound sources. These effects cause hot spots in the array (i.e., one element receives many times the average power per element), so that it may be driven to destruction. Some elements can also absorb the acoustic output of others even when all elements are driven in like manner. Other effects are a gradual deterioration of beam pattern (or directivity index) and reduction of electrical power into the array, with corresponding loss of L_S. Mutual-interaction effects can be controlled by proper spacing between array elements, tuning the transducer element to reduce mutual-radiation impedance, increasing the self-radiation impedance and controlling individual transducer velocity and phase.

Sonar domes may have a significant transmission loss and may distort the beam pattern. Expressions for the transmission loss and specular reflections have been obtained theoretically and verified experimentally. Both increase with frequency and thickness, as well as density of the dome wall.

Target Strength

Target strength relates the echo intensity returning from a target to the incident intensity. The target strength, a concept equally useful for submarine, mine detection, and fish location, is defined as 10 times the logarithm to the base 10 of the ratio of the intensity of the sound returned by the target, at a distance of 1 m from its acoustic center to the incident intensity from a distant source. The theoretical target strength of a number of geometric shapes and forms is shown in Table 23.4.6. Nominal values of measured target strength for typical underwater targets are tabulated in Table 23.4.7. Target strength is a function of aspect angle, frequency, and pulse length.

Propagation Loss

The *propagation loss* is a highly variable function, and only average values can be predicted. Average propagation-loss data have been published as a function of ocean model, propagation path, wind state, and season.[3] In the absence of empirical data, several theoretical models may be used: spherical or cylindrical spreading loss as a function of range, sound absorption loss as a function of range and frequency, and sound wave or ray propagation.

The propagation from the target to the transducer is the same as that from the transducer to the target but may include several propagation paths. Propagation paths of interest for active sonar include in-layer direct path, across-layer direct path, bottom bounce, and convergence zone. Separate elevation angles may be required for the different propagation paths, depending on the ocean model.

Background Noise

Background noise in an active sonar system[3,16] is the random addition of ambient and self-noise as well as reverberations. Ambient and self-noise are assumed to be stationary during the ping cycle, while reverberations vary with detection range. Ambient and self-noise are usually assumed to be isotropic in azimuth and elevation for performance predictions (i.e., in the active sonar equation), so that the ambient-noise band level L_B at the receiver input is

$$L_B = L_{Nf} + N_{BW} - N_{DI} \quad \text{dB} \tag{119}$$

where L_{Nf} = equivalent isotropic spectrum level of ambient or self-noise (dB re 1 μPa) in 1-Hz band
$\quad N_{BW}$ = receiver input band level (dB re 1 Hz)
$\quad N_{DI}$ = directivity index (dB)

Neither the ambient nor the self-noise is in fact isotropic, so that performance predictions based on equivalent isotropic noise levels predict average conditions. Ambient noise varies with elevation angle, as shown in Fig. 23.4.26, and varies in azimuth as well as with the maximum noise arriving from the direction of the wave swells. In addition to being anisotropic, ambient noise varies with the ocean configuration, wind, current, diurnal heating and cooling cycles, and widely varying velocity structures.

Self-noise is also highly directional, and, as expected, the highest self-noise on ships and submarines originates at the propeller. Self-noise can be separated into platform and equipment noise. Sonar systems are usually designed so that equipment noise is small compared with ambient or platform noise.

The receiving transducer may be the same as the transmitting transducer or can be completely independent. When the same transducer is used, a transmit/receive switch connects the transducer to either the transmitter or beam former as required. The receiver transducer (or set of transducers) accepts the echo plus noise and furnishes these to a beam former. Beam formers reject noise and have either an adaptive or fixed configuration. Adaptive beam formers are designed to reject variable directional noise sources such as self-noise and reverberations by monitorizing the noise field and adjusting internal parameters to minimize the background noise. Fixed beam formers are designed to reject isotropic noise and take advantage of the directional property of the echo.

Reverberation[3]

Reverberation often limits performance on modern high-power sonars. The boundaries of the ocean at the surface and bottom and inhomogeneities in the ocean, such as schools of fish, layers of air bubbles near the surface, the deep scattering layer, as well as pinnacles and seamounts on the sea bed, all reradiate a portion of

TABLE 23.4.6 Target Strength of Simple Forms

Form	t Target strength = $10 \log t$	Symbols	Direction of incidence	Conditions
Any convex surface	$\dfrac{a_1 a_2}{4}$	$a_1\, a_2$ = principal radii of curvature r = range $k = 2\pi/$wavelength	Normal of surface	$ka_1, ka_2 \gg 1$ $r > a$
Sphere Large	$\dfrac{a^2}{4}$	a = radius of sphere	Any	$ka \gg 1$ $r > a$
Small	$(61.7)\dfrac{V^2}{\lambda^4}$	V = volume of sphere λ = wavelength	Any	$ka \ll 1$ $kr \gg 1$
Cylinder Infinitely long Thick	$\dfrac{ar}{2}$	a = radius of cylinder	Normal to axis of cylinder	$ka \gg 1$ $r > a$
Thin	$\dfrac{9\pi^4 a^4}{\lambda^2}\,r$	a = radius of cylinder	Normal to axis of cylinder	$ka \ll 1$
Finite	$\dfrac{aL^2}{2\lambda}$	L = length of cylinder a = radius of cylinder	Normal to axis of cylinder	$ka \gg 1$ $r > L^2/\lambda$
	$\dfrac{aL^2}{2\lambda}\left(\dfrac{\sin\beta}{\beta}\right)^2 \cos^2\theta$	a = radius of cylinder $\beta = kL\sin\theta$	At angle θ with normal	
Plate Infinite (plane surface)	$\dfrac{r^2}{4}$		Normal to plane	
Finite Any shape	$\left(\dfrac{A}{\lambda}\right)^2$	A = area of plate L = greatest linear dimension of plate l = smallest linear dimension of plate	Normal to plate	$r > \dfrac{L^2}{\lambda}$ $kl \gg 1$
Rectangular	$\left(\dfrac{ab}{\lambda}\right)^2 \left(\dfrac{\sin\beta}{\beta}\right)^2 \cos^2\theta$	a, b = side of rectangle $\beta = ka\sin\theta$	At angle θ to normal in plane containing side a	$r > \dfrac{a^2}{\lambda}$ $kb \gg 1$ $a > b$
Circular	$\left(\dfrac{\pi a^2}{\lambda}\right)^2 \left(\dfrac{2J_1(\beta)}{\beta}\right)^2 \cos^2\theta$	a = radius of plate $\beta = 2\,ka\sin\theta$	At angle θ to normal	$r > \dfrac{a^2}{\lambda}$ $ka \gg 1$
Ellipsoid	$\left(\dfrac{bc}{2a}\right)^2$	a, b, c = semimajor axes of ellipsoid	Parallel to axis of a	$ka, kb, kc \gg 1$ $r \gg a, b, c$

Conical tip	$\left(\dfrac{\lambda}{8\pi}\right)^2 \tan^4\psi\left(1-\dfrac{\sin^2\theta}{\cos^2\psi}\right)^{-3}$	At angle θ with axis of cone	ψ = half angle of cone	$\theta < \psi$
Average overall aspects Circular disk	$\dfrac{a^2}{8}$	Average overall directions	a = radius of disk	$ka \gg 1$ $r > \dfrac{(2a)^2}{\lambda}$
Any smooth convex object	$\dfrac{S}{16\pi}$	Average overall directions	S = total surface area of object	All dimensions and radii of curvature large compared with λ
Triangular corner reflection	$\dfrac{L^4}{3\lambda^2}(1-0.000760\theta^2)$	At angle θ to axis of symmetry	L = length of edge of reflector	Dimensions large compared with λ

Source: Urick, Ref. 3, Par. 9.5.

TABLE 23.4.7 Nominal Values of Target Strength

Target	Aspect	Target strength, dB
Submarines	Beam	+25
	Bow-stern	+10
	Intermediate	+15
Surface ships	Beam	+25 (highly uncertain)
	Off-beam	+15 (highly uncertain)
Mines	Beam	+10
	Off-beam	+10 to −25
Torpedoes	Bow	−20
Fish of length L, in . . .	Dorsal view	19 log L − 54 (approx.)

Source: Urick, Ref. 3, Par. 9.15.

acoustic energy incident on them. This reradiation of sound is called *scattering*, and the sum-total acoustic energy at the receiving transducer from all the scatterers is called *reverberation*. There are many different types of reverberation, and it is important to visualize the kinds of reverberation occurring during the ping cycle and what azimuth and elevation angle these will be coming from.

Certain characteristics of reverberation can be used to good advantage by the sonar designer. Knowledge of the directions of reverberation may be used to reduce their effect on sonar operations. For instance, with a direct propagation path, a narrow vertical transmit-and-receive beam pattern will first decrease the incident acoustic energy on the bottom, and second, reject reverberation coming from the bottom outside the main lobe of the receive beam pattern.

A knowledge of the reverberation spectrum can also be used to minimize the effect. The sonar system them becomes self-noise or ambient noise limited with a significant drop in the background-noise level and corresponding improvement in system performance. Long cw pulses have reverberation spectrums that are small compared with the receiver input bandwidth and lend themselves readily to this scheme. Own-ship doppler nullification (ODN) circuits compensate for platform velocity-induced shifts in center frequency of the reverberation spectrums. The notch filter needed for rejection of the reverberation has to be wide enough for transmitted pulse spectrum convolved with the scatter motion. Homogeneous scatterer motion causes center-frequency shifts, and random scatterer motion causes frequency spreading. For echo-ranging applications, it is usually possible to reject the reverberations with a notch filter with a bandwidth corresponding to a ± 0.25 m/s random scatterer motion.

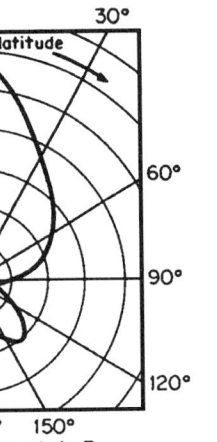

FIGURE 23.4.26 Vertical directivity of ambient sea noise.

The doppler shift of the center frequency of the reverberation from that of the transmitted pulse spectrum can be used to measure ship speed with respect to the immediate surrounding water.

Detection Threshold

The detection threshold N_{DT} is the signal-to-noise ratio S/N required in the receiver input band at the receiver input to produce the desired display statistics.[3] The term N_{DT} is usually applied to surveillance displays and refers to the desired single-ping probability of detection and false alarm. For other sonar modes of operation, such as fine-grained measurement of target range, doppler range rate, bearing, and elevation angle, the target has already been detected, and the required S/N at the receiver input is related to the desired standard deviation of the estimator.

The desired display statistics for a surveillance display are usually a 50 percent probability of detection and an acceptably low number of false alarms on the display. A typical surveillance display is a B scan with linear presentation of range along the vertical axis and linear presentation of bearing along the horizontal axis. The number of independent range cells is usually taken as the echo-ranging time required for the display range gate, in seconds, times the receiver output bandwidth in hertz. The number of independent bearing cells is usually the number of beams required to give continuous bearing coverage for the desired azimuthal sector. The false-alarm probability of any one of the independent range-bearing cells is just the number of allowed false alarms on the display divided by the total number of such range-bearing cells.

Once the desired probability of detection $P(D)$ and probability of false alarm $P(FA)$ have been established, the required S/N at the display input can be derived through the use of receiver operating curve (ROC) as described earlier.

The N_{DT} at the receiver input can be established through the use of S/N transfer functions for a given signal processor.[3,48–49] It should be noted that the desired $P(FA)$ of independent range-bearing cells is usually also a function of the signal processor because of different receiver output bandwidths for different signal processors.

A sonar receiver for surveillance has to work over a wide range of conditions, which, because of the nature of the targets, are generally unknown. A *robust receiver* (i.e., one that works well over all the expected variations of input parameters) or a combination of several receivers to cover such variations is usually desired. The "ideal" N_{DT} for any particular receiver degrades with target doppler, multipath arrivals, and nonstationary noise background.

The selection of the receiver for a particular application cannot be based on the lowest N_{DT} alone, but has to take many other receiver characteristics into account, such as cost, reliability, maintainability, logistics, power consumption, cooling requirement, operator training, weight, space, plus additional items peculiar to the application.

The optimum receiver for a signal known exactly, with white, stationary, gaussian noise, is a matched filter. The signal processing gain for this type of receiver is 10 times the logarithm to the base 10 of the time-bandwidth product of the signal, and the resulting N_{DT} is the lower limit. Cross-correlators and comb-filter receivers are matched-filter-type receivers.

Incoherent receivers are optimum for unknown signals, with white, stationary, gaussian noise. The signal processing gain for this type of receiver is five times the logarithm to the base 10 of the time-bandwidth product. Energy and envelope detectors, as well as autocorrelators and spatial correlators, are examples of this type of receiver.

Since the echo for an active sonar is neither known exactly nor completely unknown, semicoherent receivers are sometimes used, which have matched-filter features for those echo characteristics which are relatively well known and incoherent receiver characteristics for those echo characteristics which are likely to change significantly. The postdetection pulse-compression receiver is an example of this kind of receiver. The signal processing gain of this type of receiver is between that of the matched-filter type and the incoherent receiver.

Surveillance Receivers

A typical active surveillance display is shown in Fig. 23.4.27. This is a rectangular B-scan format with linear presentation of range along the vertical axis and linear presentation of bearing along the horizontal axis. The center bearing of the B-scan display can usually be selected to be true north or ship's bearing by the operator. It is driven by a thresholded output of a multichannel surveillance receiver. The signal excess over the threshold controls the display intensity in each bearing-range cell. The ability to threshold relies on noise-power normalization, so that changes in background-noise levels caused by variations in ambient noise and/or self-noise, as well as reverberation, do not cause an excessive number of false alarms.

A number of different surveillance receivers are available to drive the display: linear or clipped correlators, autocorrelator, spatial correlator, comb-filter receivers (with or without reverberation suppression), energy or envelope detectors, and pulse compression receivers. Performing prediction for the various types of receivers is covered in the literature on signal processing techniques. It is important to remember that these receivers should be compared on the basis of equal display statistics and nonstationary reverberations, self-noise, and ambient background noise, as well as "ideal" conditions.

Noise-power variations may be reduced by clipping, nonlinear amplification such as logarithmic amplifiers, or variable linear amplification such as time variable gain (TVG), automatic gain control (AGC), or step gain control (SGC) ahead of the receiver. Clipping is inexpensive but has the disadvantage of possible capture of the clipper by strong interfering noises. Logarithmic or other nonlinear amplifications can hold

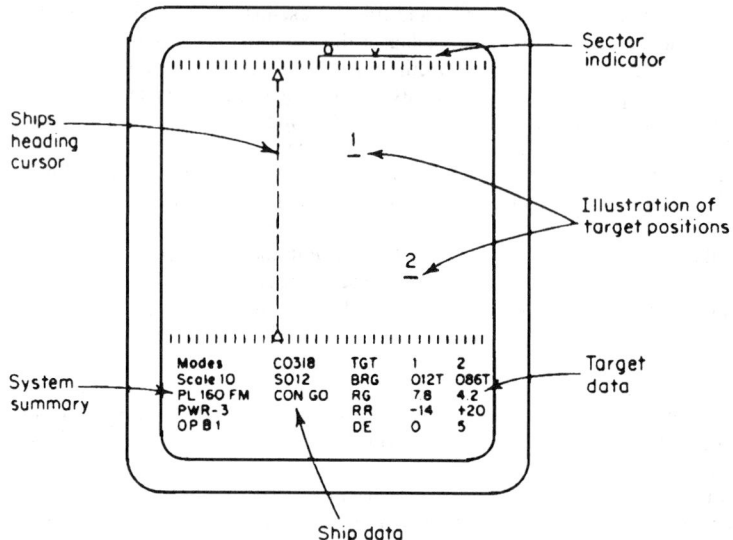

FIGURE 23.4.27   B-scan surveillance display.

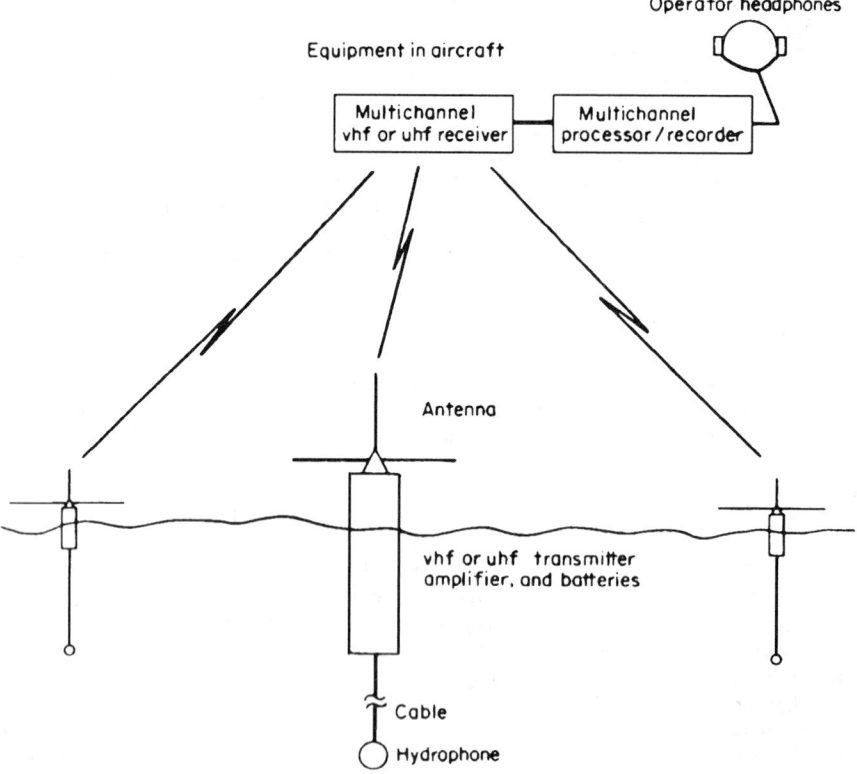

FIGURE 23.4.28   Passive sonobuoy system.

the noise power constant but cause frequency distortion and spreading. TVG is generally used clearly in the ping cycle to prevent system saturation right after transmission and then to increase the gain versus time in a predetermined manner as a function of the transmitted source level and pulse length. AGC is closed-loop gain control where the noise power is estimated and the gain inversely controlled to maintain constant noise power. SGC is like AGC except that the gain is controlled in steps, so that inverse proportional gain control is not needed. Both AGC and SGC are sometimes referred to as reverberation gain control (RGC).

Another type of active surveillance display is the plan position indicator (PPI) display. The PPI format has the ship at the display center, with bearing linearly presented as display angle and the ship's heading presented vertically upward. Range is linearly presented as the radius from the display center.

Track and localization displays are usually presented in A-scan format, with a range gate linearly presented along the horizontal axis and either azimuth or elevation angle gate or doppler range rate linearly presented along the vertical axis. Localization and track displays and receivers are generally rated according to the resolution in range, bearing, elevation, and doppler range rate available to the operator as a function of S/N.

Sonobuoy Systems

Sonobuoy systems are miniature passive and/or active sonar systems designed for deployment from aircraft, making use of VHF or UHF radio channels for transmission of information from the buoy to the aircraft. In its simplest configuration, a passive sonobuoy consists of a cable-connected omnidirectional hydrophone, which is released from the buoy after water entry and which sinks to a predetermined operating depth, an amplifier to raise the level of the hydrophone output, a VHF or UHF transmitter, antenna system, and batteries for power. These items are often contained in a canister less than 4 ft long and 5 in. in diameter, fitted with a rotochute or other device for control of descent after release from the aircraft. The buoy is designed to float so as to maintain its antenna system above the water surface. On the aircraft the sonobuoy signal is extracted from the radio channel and applied to a processor-recorder and operator's headphones. Multichannel receiving and processing equipment is used to monitor the outputs of several sonobuoys simultaneously (Fig. 23.4.28).

More complex sonobuoy systems result when directional arrays are used in lieu of the simple omnidirectional hydrophone, necessitating the transmission of array orientation and target-bearing information via the radio link. Greater complexity, posing a severe challenge to the sonobuoy designer, results from combination of passive and active systems and control of the sonobuoy functions from the aircraft via a command radio link. Because of the advances made in micro-miniaturization, integrated circuits, multiplexing, miniature acoustic arrays, digital signal processing, and high-density packaging, it is now possible to design sophisticated multifunction sonobuoy systems.

The design of such systems includes heavy emphasis on sonobuoy performance, cost, size, weight, life, and reliability factors—in view of the end use of the buoy as a expendable item. System partitioning, in which the functions to be performed in the buoy and in the aircraft equipment are defined, is an important element of the performance per cost consideration, particularly for the more sophisticated sonobuoy systems. Sonobuoy system design includes the application of passive and active sonar system design techniques described in this section of the handbook, plus aerodynamic and hydrodynamic technologies.

BIBLIOGRAPHY

1. Horton, J. W. "Fundamentals of Sonar," U.S. Naval Institute, 1959.

2. Gray, D. E. "American Institute of Physics Handbook," 2d ed., McGraw-Hill, 1963.

3. Urick, R. J. "Principles of Underwater Sound," McGraw-Hill, 1975.

4. Bobber, R. J. "Underwater Electroacoustic Measurements," Peninsula Publishing, 1988.

5. Tolstoy, I., and C. S. Clay "Ocean Acoustics," McGraw-Hill, 1966.

6. Albers, W. M. "Underwater Acoustics Handbook," Penn State University Press, 1960.

7. Clay, C. S., and H. Medwin "Acoustical Oceanography," Wiley, 1977.

8. Stephens, R. W. B. (ed.), "Underwater Acoustics," Wiley-Interscience, 1970.

9. Wilson, W. D. *J. Acoust. Soc. Am.*, 1960, Vol. 32, p. 1357.

10. Ibid., p. 641, 1960.

11. Ibid, p. 866, 1962.

12. Liebermann, L. N. *Phys. Rev.*, 1949, Vol. 76, p. 1520.

13. Thorp, W. H. *J. Acoust. Soc. Am.*, 1967, Vol. 42, p. 270.

14. Schulkin, M., and H. W. Marsh *J. Acoust. Soc. Am.*, 1962, Vol. 34, p. 864.

15. Ross, D. "Mechanics of Underwater Noise," Peninsula Publishing, 1987.

16. Special Issue on Detection Theory and Its Applications, *Proc. IEEE*, May 1970.

17. American National Standards Institute ANSI SI.20, *Procedures for Calibration of Underwater Electroacoustic Transducers*, 1988.

18. Hueter, T. F., and R. H. Bolt "Sonics," Wiley, 1955.

19. Stansfield, D. "Underwater Electroacoustic Transducers," Bath University Press, 1991.

20. Wilson, O. B. "An Introduction to the Theory and Design of Sonar Transducers," U.S. Government Printing Office, 1985.

21. Sims, C. C. High-Fidelity Underwater Sound Transducers, *Proc. IRE*, 1959, Vol. 47, p. 866.

22. Kinsler, L. E., Frey, A. R., Coppens, A. B., and J. V. Sanders "Fundamentals of Acoustics," Wiley, 1982.

23. Westervelt, P. J. *J. Acoust. Soc. Am.*, 1963, Vol. 35, p. 535.

24. Beyer, R. T. "Nonlinear Acoustics," Naval Sea Systems Command, Department of the Navy, 1974.

25. Lockwood, J. C. Nomographs for Parametric Array Calculations, *Univ. Tex. Tech. Appl. Res. Lab. Mem.* 73-3, 1973.

26. Berktay, H. O., and D. J. Leahy *J. Acoust. Soc. Am.,* 1974, Vol. 55, p. 539.

27. Moffett, M. B., and R. H. Mellen *J. Acoust. Soc. Am.*, 1977, Vol. 61, p. 325.

28. Barnard, G. R., J. G. Willette, J. J. Truchard, and J. A. Shooter *J. Acoust. Soc. Am.*, 1972, Vol. 52, p. 1437.

29. Berlincourt, D. A., D. R. Curran, and H. Jaffee "Piezoelectric and Piezomagnetic Materials and Their Function in Transducers," in W. P. Mason (ed.), *Physical Acoustics*, Vol. I, Pt. A, Academic, 1964.

30. Cady, W. G. "Piezoelectricity," Vols. 1 and 2, Dover, 1964.

31. Mason, W. P. "Piezoelectricity Crystals and Their Application to Ultrasonics," Van Nostrand, 1950.

32. Piezoelectric Crystals, *IEEE Standard 176*, 1949.

33. Definitions and Methods of Measurements of Piezoelectric Vibrators, *IEEE Standard 177*, 1966.

34. Piezoelectric Crystals: Determination of the Elastic, Piezoelectric, and Dielectric Constants; The Electromechanical Coupling Factor, *IEEE Standard 178*, 1958.

35. Measurement of Piezoelectric Ceramics, *IEEE Standard 179*, 1961.

36. Piezoelectric Ceramic for Sonar Transducers, DOD-Std 1376 (SH), 1984.

37. Newnham, R. E. *Ann. Rev. Mater. Sci.*, 1986, Vol. 16, p. 47.

38. Ting, R. Y. *Ferroelectrics*, 1990, Vol. 102, p. 215.

39. Kikuchi, Y. "Ultrasonic Transducers," Corona, 1969.

40. Magnetostrictive Materials: Piezomagnetic Nomenclature, *IEEE Standard 319*, 1971.

41. Kikuchi, Y. Magnetostrictive Metals and Piezomagnetic Ceramics as Transducer Materials, in O. E. Mattiat (ed.) ,"Ultrasonic Transducer Materials," Plenum, 1971.

42. Hassad, J. C. "Underwater Signal and Data Processing," CRC Press, 1989.

43. Scharf, L. L. "Statistical Signal Processing: Detection, Estimation, and Time Series Analysis," Addison-Wesley, 1991.

44. Widrow, B., and S. D. Stearns, "Adaptive Signal Processing," Prentice Hall, 1985.

45. Sherman, C. H. and D. F. Kass Radiation Impedances, Radiation Patterns, and Efficiency for Large Array on a Sphere, *U.S. Navy Underwater Sound Lab Res. Rep.* 429, July 17, 1959.

46. Carson, D. L. Diagnosis and Cure of Errant Velocity Distributions in Sonar Projector Arrays, *J. Acoust. Soc. Am.,* September 1962, Vol. 34, No. 9.

47. Hansh, S. "A Treatise on Acoustic Radiation, Vol. IV—Mutual Radiation Impedance and Other Special Topics," Naval Research Laboratory, 1987.

48. Skolnick, M. J. "Introduction to Radar Systems," McGraw-Hill, 1962.

49. Davenport, W. B., and W. L. Root "Introduction to Random Signals and Noise," McGraw-Hill, 1958.

SECTION 24

AUTOMOTIVE ELECTRONICS

The importance of sensors and actuators in automotive electronics cannot be overemphasized. The ability to sense and measure automotive parameters accurately is key to any automotive electronics application, as is the ability to initiate control actions accurately in response to commands. Chapter 24.1 contains a wealth of information on sensors and actuators and their applications in cars.

Major automotive on-board systems are discussed in Chap. 24.2. These systems include engine, transmission, cruise, braking, traction, stability, suspension, and steering control systems. Also presented is information on object detection, collision warning, collision avoidance, and adaptive cruise control. Intelligent transportation systems are discussed in Chap. 24.3.

It should be noted that the information presented in this section is based principally on condensations of chapters in the book, *Automotive Electronics*, 2nd ed., Ronald K. Jurgen (ed.), McGraw-Hill, 1999. Credit to individuals who provided the original material for that book is given where appropriate in the following three chapters.   R.J.

In This Section:

CHAPTER 24.1
AUTOMOTIVE SENSORS AND ACTUATORS

Ronald K. Jurgen

INTRODUCTION

Automotive sensors and actuators are key to all automotive electronic control systems. Sensors, which send electric signals proportional to the parameters they measure to a car's control systems, are grouped herein by the parameters they sense (pressure, force, position, flow, and the like). Actuators, which take action upon receiving signals from the car's control systems, are classified as either electromechanical or electrical.

PRESSURE SENSORS*

Various pressure measurements are required to optimize performance, determine safe operation, assure conformance to government regulations, and to advise the driver. Those measurements can also be used to provide data log information for diagnostic purposes.

Different units for indicating pressure are used, depending on the particular application, e.g., tire pressure in psi and manifold pressure in kPa. Table 24.1.1 lists the various pressure units together with their conversion constants. The type of pressure measurement that is made can be divided into five basic areas: gage, absolute, differential, liquid level, and pressure switch.

Figure 24.1.1 illustrates the basic differences in gage pressure, absolute pressure, and differential pressure measurements. For gage pressure measurements, the pressure is applied to the top of a silicon diaphragm creating a positive output. The back side of the diaphragm is exposed to atmospheric pressure. Absolute pressure measurement is made with respect to a fixed (usually a vacuum) reference sealed within the sensor. Differential or Delta-P measurements are made with the higher pressure applied to the top of the diaphragm and the lower pressure (possibly a reference) to the opposite side.

Technologies for Sensing Pressure

Among the technologies used for on-vehicle measurements of static and dynamic pressure are diaphragm-potentiometer, linear variable differential transformer (LVDT), aneroid, silicon or ceramic capacitive,

---

*Frank, R., Chapter 2, "Pressure Sensors," in *Automotive Electronics Handbook*, 2nd ed., Ronald K. Jurgen (ed.), McGraw-Hill, 1999, pp. 2.3–2.25.

TABLE 24.1.1 Pressure Unit Conversion Constants

(Most commonly used per international conventions)

	psi[*]	In H$_2$O[†]	In Hg[‡]	k Pascal	Millibar	cm H$_2$O[§]	mm Hg[¶]
psi[*]	1.000	27.680	2.036	6.8947	68.947	70.308	51.715
In H$_2$O[†]	3.6127×10^{-2}	1.000	7.3554×10^{-2}	0.2491	2.491	2.5400	1.8683
In Hg[‡]	0.4912	13.596	1.000	3.3864	33.864	34.532	25.400
k Pascal	0.14504	4.0147	0.2953	1.000	10.000	10.1973	7.5006
Millibar	0.01450	0.40147	0.02953	0.100	1.000	1.01973	0.75006
cm H$_2$O[§]	1.4223×10^{-2}	0.3937	2.8958×10^{-2}	0.09806	0.9806	1.000	0.7355
mm Hg[¶]	1.9337×10^{-2}	0.53525	3.9370×10^{-2}	0.13332	1.3332	1.3595	1.000

[*]psi = pounds per square inch
[†]At 39°F
[‡]At 32°F
[§]At 4°C
[¶]At 0°C

piezoresistive strain gage, piezoelectric ceramic or film, and optical phase shift (combustion pressure). Common mechanical devices include the Bourdon tube, bellows, diaphragms, deadweight gage, and manometer.

Diaphragm-Potentiometer Pressure Sensors. These sensors are based on the principle that the distance that a rubber diaphragm travels when pressure is applied is a mechanical indication of pressure. The movement of the diaphragm also moves the sliding element or wiper of a potentiometer to provide an electric signal that can be applied to a remote gage such as an oil pressure gage.

Linear Variable Differential Transformer. In an LVDT pressure sensor, Fig. 24.1.2, the moving core is attached to a force collector that provides differential coupling from the primary to the secondaries resulting in a position output that is proportional to pressure. A pressure applied to a Bourdon tube or diaphragm causes the core to be displaced from its null position and the coupling between the primary and each of the

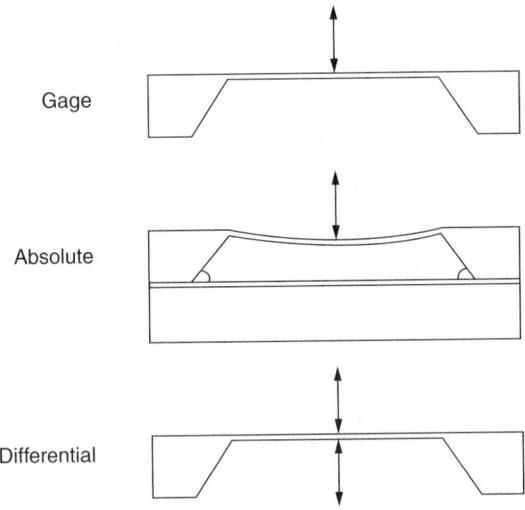

FIGURE 24.1.1 Types of pressure measurements: (*a*) gage, (*b*) absolute, and (*c*) differential.

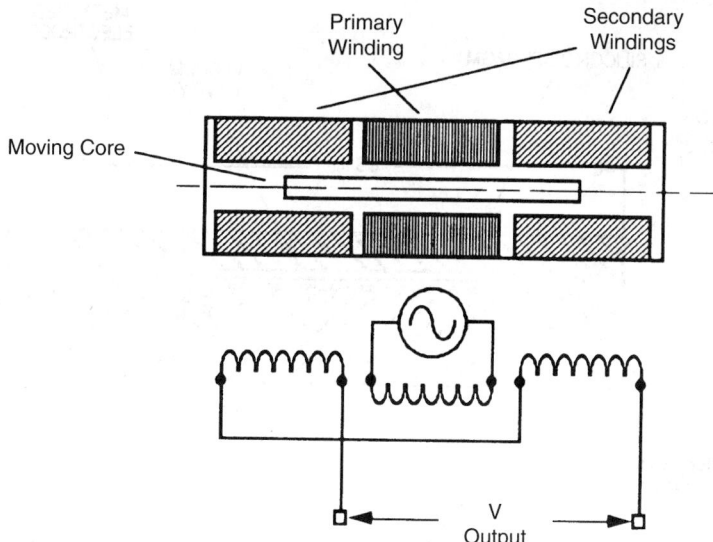

FIGURE 24.1.2 LVDT pressure sensor.

secondaries is no longer equal. The resulting output varies linearly within the design range of the transducer and has a phase change of 180° from one side of the null position to the other.

Dual-sealed aneroid diaphragms are bonded and sealed with a vacuum inside each unit to a metallized conductive ring on opposite sides of a ceramic substrate. The substrate serves as the fixed plates of two separate capacitors. Manifold vacuum is applied to one chamber and the second chamber serves as a reference for compensating and signal conditioning that minimizes common mode errors due to vibration and shock.

Capacitive Pressure Sensors. These sensors have one plate connected to a force collector (usually a diaphragm) and the distance between the two plates of the capacitor varies as a result of the applied pressure. A silicon capacitive absolute pressure (SCAP) sensing element is shown in Fig. 24.1.3. The value of the capacitor changes linearly from approximately 32 to 39 pF with applied pressure from 17 to 105 kPa. The output of the sensor is typically signal-conditioned to provide a frequency variation with applied pressure for easy interface to microcontrollers.

Strain Gage Pressure Sensors. These sensors convert the change in resistance in four (sometimes in two) arms of a Wheatstone bridge. The change in the resistance of a material when it is mechanically stressed is called piezoresistivity. Different approaches to piezoresistive strain gages range from traditional bonded and unbonded to integrated silicon pressure sensors. A bonded sensor consists of a filament-wire or foil, metallic or semiconductor, and bulk material or deposited film bonded to the surface of a diaphragm. Pressure applied to the diaphragm produces a change in the dimension of the diaphragm and, consequently, in the length of the gage, and therefore a change in its resistance. When a pressure sensor is used for measuring an applied force it is called a local cell.

With integrated silicon pressure sensors, the sensor signal can be provided from a single piezoresistive element located at a 45° angle in the center of the edge of a square diaphragm that provides an extremely linear measurement. In addition to the basic sensing element, an interactively laser-trimmed four-stage network has also been integrated into a single monolithic structure (Fig. 24.1.4). A polysilicon thin-film sensor that consists of a thin film of silicon that is doped with boron and vapor-deposited over a stainless steel diaphragm is shown in Fig. 24.1.5.

Piezoelectric Pressure Sensors. Piezoelectric pressure sensors produce a change in electric charge when a force is applied across the face of a crystal or piezoelectric film. Transducers are constructed with rigid multiple plates

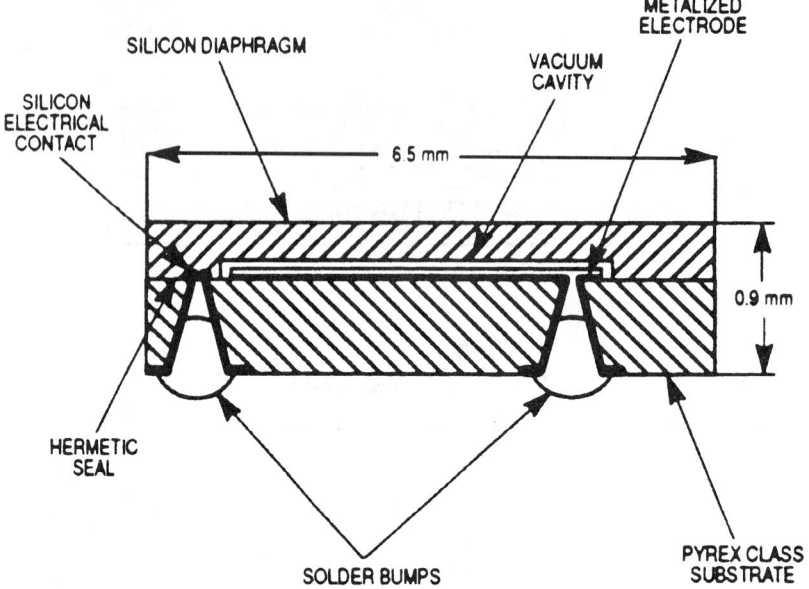

FIGURE 24.1.3 Silicon capacitive absolute pressure (SCAP) sensor.

and a cultured-quartz-sensing element that contains an integral accelerometer to minimize vibration sensitivity and suppress resonances. A typical unit also contains a built-in microelectronic amplifier to transform the high-impedance electrostatic charge output from the crystals into a low-impedance voltage signal (Fig. 24.1.6). Surface micromachining techniques have been combined with piezoelectric thin-film materials, such as zinc oxide, to produce a semiconductor piezoelectric pressure sensor.

Fiber Optic Combustion Pressure Sensor. For extremely high pressure (0 to 1000 psi normal, 3000 psi over-range) or pressure measurements at high temperatures (up to 550°C), a fiber optic combustion pressure sensor (Fig. 24.1.7) can be used. The amount of light that is returned to the sensor fiber after it is reflected from the diaphragm depends on the gap D between the fiber and the diaphragm. The diaphragm can be sized to allow the sensor to be integrated into a spark plug for easy access to the combustion pressure of each cylinder.

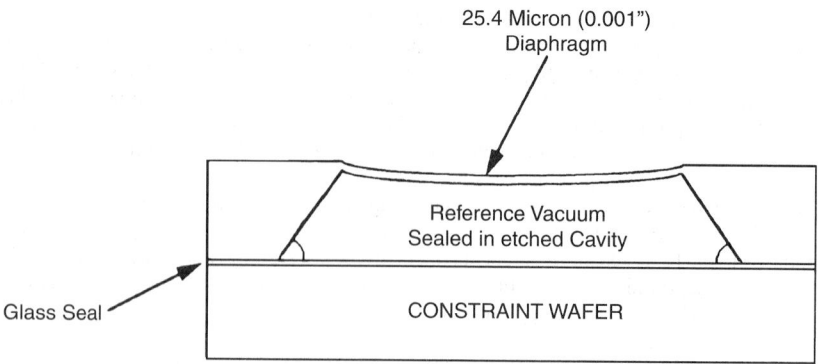

FIGURE 24.1.4 Cross section of piezoresistive silicon pressure sensor for measuring absolute pressure.

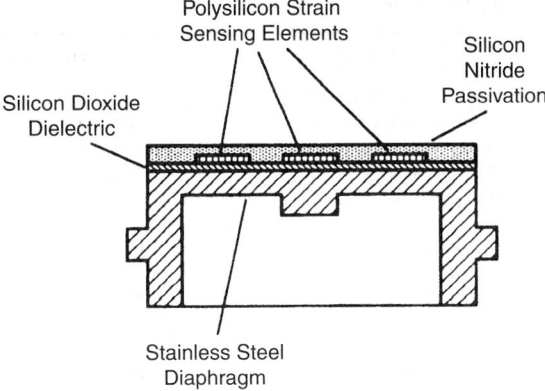

FIGURE 24.1.5 Polysilicon pressure sensor on stainless steel diaphragm.

Pressure Switch.   In a pressure switch, sufficient motion of the force-collecting diaphragm allows a spring to be compressed and a set of contacts to be closed in a traditional mechanical switch.

Automotive Applications for Pressure Sensors

Table 24.1.2 lists a number of potential pressure measurements versus vehicle systems and provides an indication of the pressure range and type of measurement.

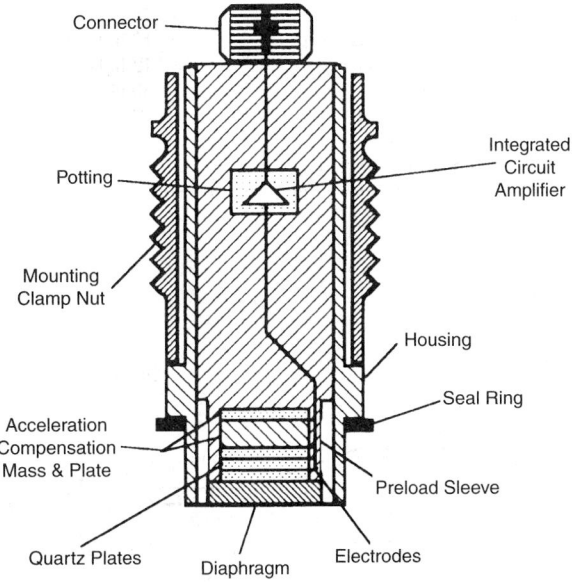

FIGURE 24.1.6 Piezoelectric pressure sensor.

TABLE 24.1.2 Pressure Sensing Requirements for Various Vehicle Systems

System	Parameter	Pressure range	Type
Engine control	Manifold absolute pressure	100 kPa	Absolute
	Turbo boost pressure	200 kPa	Absolute
	Barometric pressure (altitude)	100 kPa	Absolute
	EGR pressure	7.5 psi	Gage
	Fuel pressure	15 psi—450 kPa (13.5 MPa peak)	Gage
	Fuel vapor pressure	15 in H_2O	Gage
	Mass air flow	—	Differential
	Combustion pressure	100 Bar (16.7 MPa peak)	Differential
	Exhaust gas pressure	100 kPa	Gage
	Secondary air pressure	100 kPa	Gage
Elect transmission	Transmission oil pressure	80 psi	Gage
(continuously variable transmission)	Vacuum modulation	100 kPa	Absolute
Idle speed control	AC clutch sensor/switch	300–500 psi	Absolute[*]
	Power steering pressure	500 psi	Absolute[*]
Elect power steering (also elect assisted)	Hydraulic pressure	500 psi	Absolute[*]
Antiskid brakes/ traction control	Brake pressure	500 psi	Absolute[*]
	Fluid level	12 in H_2O	Gage
Air bags	Bag pressure	7.5 psi	Gage
Suspension	Pneumatic spring pressure	1 MPa	Absolute[*]
Security/keyless entry	Passenger compartment pressure	100 kPa	Absolute
HVAC (climate control)	Air flow (PC) Compressor pressure	300–500 psi	Absolute[*]
Driver information	Oil pressure	80 psi	Gage
	Fuel level	15 in H_2O	Gage
	Oil level	15 in H_2O	Gage
	Coolant pressure	200 kPa	Gage
	Coolant level	24 in H_2O	Gage
	Windshield washer level	12 in H_2O	Gage
	Transmission oil level	12 in H_2O	Gage
	Tire pressure	50 psi	Gage/absolute
	Battery fluid level	1-2 in below	Optical
Memory seat	Lumbar pressure	7.5 psi	Gage
Multiplex/diagnostics	Multiple usage of sensors	—	

[*]Gage measurement but absolute sensors used for failsafe.

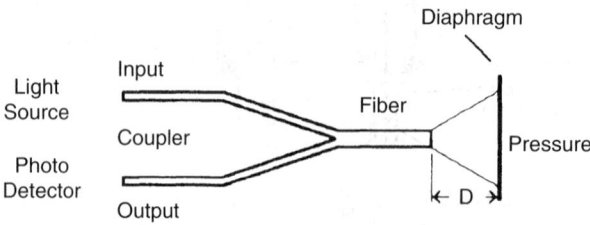

FIGURE 24.1.7 Fiber optic combustion pressure sensor.

TABLE 24.1.3 Potentiometer Specifications

Parameter	Minimum	Maximum
Electrical travel	90°, 10 mm	360°, 3000 mm
Nominal resistance	500 Ω	20 kΩ
Resistance tolerance	10%	20%
Resistance temperature coefficient (TC)		500 ppm/°C
TC of V_{out} in voltage divider mode		5 ppm/°C
Linearity error	0.01%	1%

LINEAR AND ANGLE POSITION SENSORS*

Position sensors are electromechanical devices that translate position information into electric signals and can be grouped into two basic categories: incremental and absolute. Incremental sensors measure position as the distance from an arbitrary index or zero and must rely on some method of pulse counting. One pulse in the sequence is designed to be wider or of opposite polarity than the others so that it may be used as a nominal zero. Absolute position sensors produce an unambiguous output at power-up. Each position or angle has a unique value. The output may be a voltage or frequency or other analog of the input position.

Position sensors may be directly coupled by some shaft or linkage or by some noncontact or proximity means. The most common forms of proximity sensors are based on various methods of magnetic field detection.

Position Sensor Technologies

Microswitches. The simplest form of contact sensor is a switch and may be as simple as a microswitch that operates anything from brake lights to courtesy lights in an automobile. Many applications of microswitches in position sensing are as limit switches, usually wired to limit or warn of the extent of travel of a mechanical component by disconnecting power to an electric motor or by operating an indicator lamp.

Optical Angle Encoders. These encoders for incremental shaft angular position measurement are constructed of a disk with a series of transparent- and opaque-spaced sectors. The disk is illuminated on one side and light sensors on the other side detect the passage of light and dark sectors as the disk is rotated. Linear incremental optical encoders allow direct measurement of linear motion.

Potentiometers. Potentiometers are widely used as position sensors in automotive applications such as throttle and accelerator pedal position measurement. Potentiometers have been developed that are capable of operational life far in excess of the life of the average car and in some cases capable of continuous rotational speeds of above 1000 rpm for more than 1000 h.

All potentiometers are ratiometric sensors. That is, the wiper voltage at a given position is some fraction of the reference voltage applied across the resistive track. If the reference voltage is varied, the potentiometer output remains in the same ratio to the reference voltage. Typical potentiometer specifications are shown in Table 24.1.3.

Magnetic. The largest category of position sensors relies on electromagnetic induction principles. Variable reluctance devices operate by sensing changes in the reluctance within a magnetic circuit and are used to detect the position and speed of rotating toothed or slotted wheels in crank-, cam-, and wheel-monitoring applications.

*Nickson, P., Chapter 3, "Linear and Angle Position Sensors," in *Automotive Electronics Handbook*, 2nd ed., Ronald K. Jurgen (ed.), McGraw-Hill, 1999, pp. 3.1–3.21.

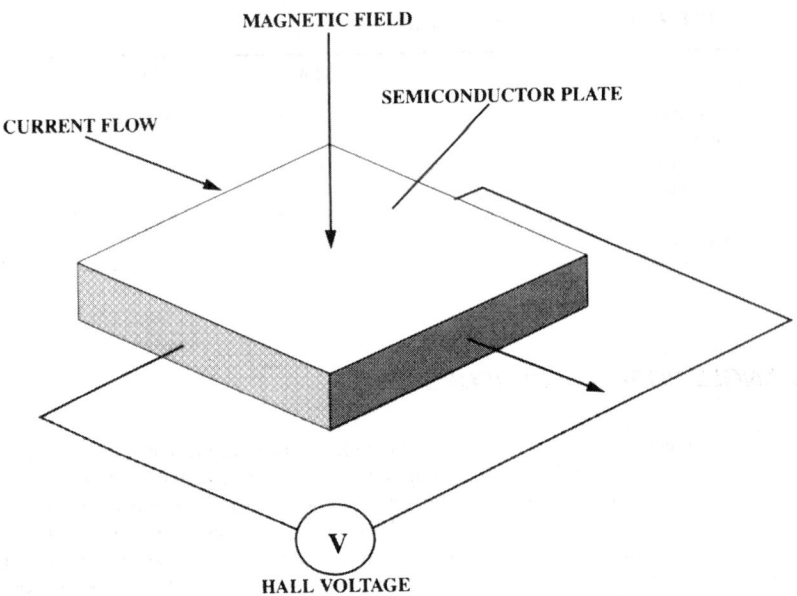

FIGURE 24.1.8   Hall-effect device.

Hall-effect devices rely on a galvanomagnetic effect in which a transverse direct voltage is developed across a current-carrying conductor inserted in a magnetic field so that its direction is perpendicular to the direction of current flow in the conductor. Devices can be constructed using semiconductor materials that can use the Hall effect to detect the strength of magnetic fields. Figure 24.1.8 illustrates the construction and operation of a silicon Hall-effect device. Hall-effect-integrated circuits are best categorized in terms of their output characteristics. Digital devices are commonly used for speed or incremental position sensing. Analog output devices produce a voltage proportional to magnetic field and can be used as components in sensor assemblies that are designed to replace contacting potentiometric sensors. A Hall probe gear position sensor is shown in Fig. 24.1.9. A Hall probe rotary position sensor is shown in Fig. 24.1.10.

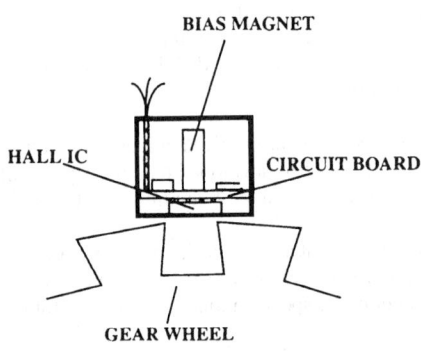

FIGURE 24.1.9   Hall probe gear position sensor.

Synchro resolvers, or simply resolvers, are absolute angle transducers. Resolvers are basically rotating transformers. The construction of a typical device is shown in Fig. 24.1.11.

Shading ring or short-circuit ring sensors (Fig. 24.1.12) are absolute displacement sensors consisting of an E-shaped core with a winding on the central leg of the E. The winding is excited with high-frequency alternating current. An electrically conducting ring of Al or Cu is allowed to slide, maintaining an air gap along the central leg. The ring is attached to the mechanical component, whose position is to be measured.

Another form of absolute linear displacement sensor is the LVDT. LVDTs are used, for example, mounted inside hydraulic cylinders in suspension control systems.

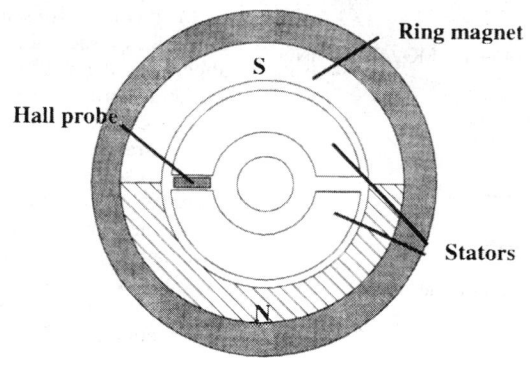

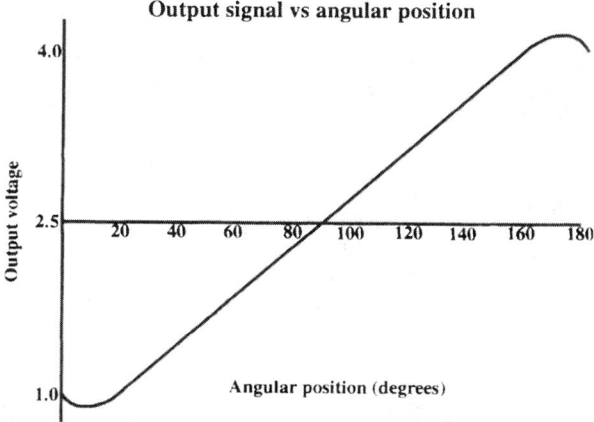

FIGURE 24.1.10   Rotary position sensor.

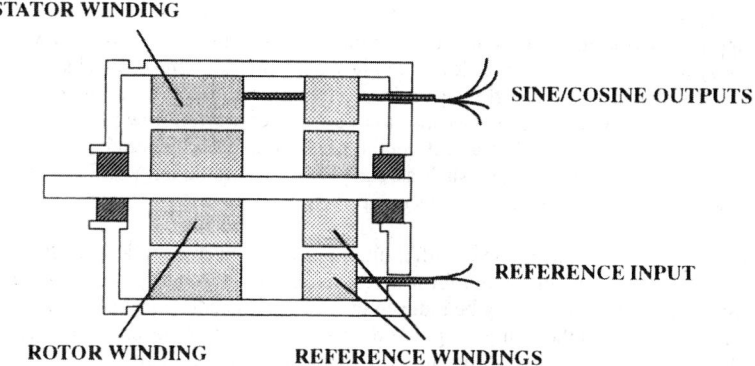

FIGURE 24.1.11   Resolver construction.

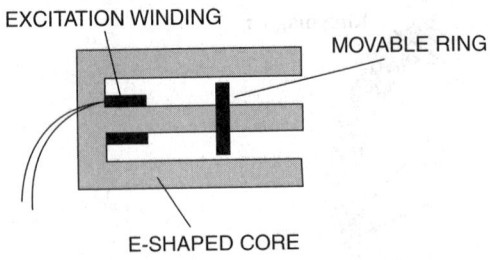

FIGURE 24.1.12   SCR sensor construction.

Magnetoresistive sensors use the property of some FeNi alloys such that their resistivity is strongly affected by the presence of a magnetic field. Despite the nonlinearity of the phenomenon, accurate linear sensors can be constructed by using the sensor in a bridge arrangement together with flux nulling means, such as a servo-driven coil surrounding the sensor to effectively operate the sensor at constant resistance.

Magnetostrictive linear displacement sensors take advantage of the phenomenon that the magnetostriction is a property of materials that respond to a change of magnetic flux by developing an elastic deformation of their crystal structure.

FLOW SENSORS*

Measurement of flow rate is important to optimize the performance of several key engine control subsystems. Mass air flow sensors are replacing the indirect calculation of intake mass air flow for improved performance driveability, and economy. New requirements for on-board diagnostics are opening new applications for flow sensing in the automobile.

If the parameter to be measured is a gaseous mass flow as opposed to a volume flow, this further focuses the sensing technology selection since only a few technologies inherently measure mass flow. For liquid flow, either a mass flow or volumetric flow approach may be used since the density of a liquid changes only a small amount with atmospheric pressure and temperature.

Flowing fluid possesses both potential and kinetic energy. One approach to flow measurement extracts energy from the flow. Alternatively, energy may be added to the flow and its effect observed.

Flow Measurement Technologies

Gaseous Flow.   A vane or paddle located in the flow duct is restrained by a spring so that it blocks the duct with no flow. The deflection of the vane is thus proportional to flow. This deflection is read out by a potentiometer. A preferred approach is a thermal one. A variety of designs exist, from the straightforward hot-wire air flow sensor to more complex schemes. The basic idea is to heat a fine wire, and then as the gas flows past the wire, convection removes heat. The amount of heat removed can be measured with an electronic circuit and is proportional to mass air flow rate. A hot-wire air flow sensor and its control circuit is shown in Fig. 24.1.13. Control circuits typically either supply constant power to the heated element or operate it at a constant temperature delta above the ambient temperature. Micromachined air flow sensors are beginning to be used in automotive applications. Table 24.1.4 compares the main types of thermal mass air flow sensors.

A simple way to measure volumetric flow is to place an obstruction in a flow channel and measure the differential pressure drop across it. The flow is proportional to the square root of the differential pressure. This only works well for narrow flow ranges because to operate over a given flow range, the pressure sensor must operate over the square of that range. The differential pressure approach could be suitable for some low accuracy or low dynamic range applications such as exhaust gas recirculation (EGR) valve flow. Venturis, flow restrictions, and pitot tubes all operate on the same principle.

Liquid Flow.   The previous discussion on differential pressure sensing applies equally to liquid flows. A turbine blade placed in a flow channel can offer low restriction to flow, and by counting the speed of rotation, it can also measure flow. Oscillations may be induced in a fluid by placing an obstruction in the flow stream. The oscillations may be measured thermally, by pressure changes or using ultrasonics. This approach doesn't work well at low flow rates because of instability in the vortex shedding mechanism.

---

*Bicking, R. E., "Flow Sensors," in *Automotive Electronics Handbook*, 2nd ed., Ronald K. Jurgen (ed.), McGraw-Hill, 1999, pp. 4.1–4.9.

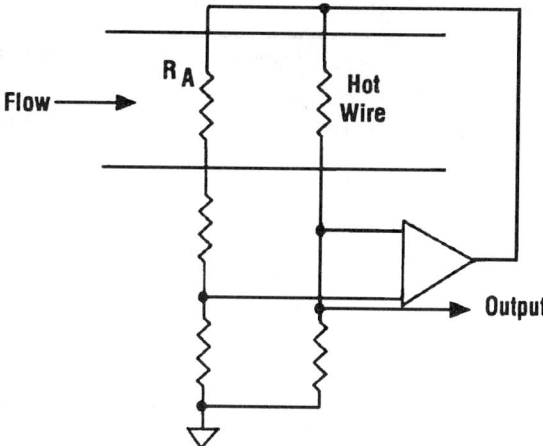

FIGURE 24.1.13   Hot-wire air flow sensor.

Automotive Applications of Flow Sensors

Intake Air Mass Flow.   To do electronic fuel injection, the mass flow rate of air going into the engine must be determined. The speed density approach calculates the mass flow rate by measuring engine speed, intake air temperature, and intake manifold pressure.

Fuel Flow for Gas Mileage Measurement.   Fuel flow is calculated from air flow and these data are already in the engine control computer. By simply summing the injector "on" time over a revolution or multiple revolutions, the fuel flow is known. Another approach is to measure the difference between the fuel coming into the fuel rail and that being returned.

Exhaust Gas Recirculation Flow.   Exhaust gas recirculation is performed to reduce the emission of nitrous oxides by cooling the combustion process. Measurement of the flow is one way to diagnose a faulty EGR valve.

Secondary Air Pump Flow.   The secondary air pump is used to reduce the emissions of carbon monoxide and hydrocarbons. Measurement of its flow rate is an approach to verify that it is operating properly and is doing its part to reduce emissions.

Fuel Flow for Fuel-Air Ratio Feedback Control.   By measuring the fuel flow, the accuracy of fuel injection could be improved. Table 24.1.4 summarizes the performance requirements by application.

TABLE 24.1.4   Performance Requirements for Fuel-Air Ratio Feedback Control

Application	Measurement type	Range, kg/h	Accuracy, %
Intake air	Mass	10–1000	±4
Fuel flow	Mass/vol.	1–66	±10
EGR flow	Mass	30–100	±10
Air pump flow	Vol.	50	±20
Fuel flow	Mass/vol.	1–66	±4

TABLE 24.1.5 Heat Sources in Vehicles

General category	Example	Max. temperature °C
Engine	Combustion/ignition process	>1000
Catalytic converter	Chemical reaction	>1000
Road/tire friction	Tires	<100
Brakes	Disk/drum	250
Mechanical motion (gears, bearings)	Transmission/rear axle/air pump/ power steering pump	200
Heat exchangers	Radiator (coolant, transmission), heater, heatsink	>175
Electric heaters	Windshield, backlight, seats, mirrors	Ta + 25
Electric windings	Motors, alternator, solenoids	<180 (Class H)
Resistors	Ballast resistor	150
Lamps	Headlamps, tail lamps, dash lamps	125
Power transistors	Ignition driver, voltage regulator	Up to 200
Electric vehicle battery	Sodium sulfur	300

TEMPERATURE, HEAT, AND HUMIDITY SENSORS*

Temperature sensing and taking into account the effect of temperature on the performance and reliability of automotive components, and therefore the systems that contain these components, is one of the more important aspects of vehicle design. The heat sources that are present on modern automobiles are given in Table 24.1.5. Temperatures at various vehicle locations are given in Table 24.1.6.

Humidity affects the comfort of the drivers and the passengers as well as the performance of the engine and impacts the reliability of vehicle components. High humidity and low dew point can lead to increased moisture content that can affect the operation of vehicle fluids such as fuel and brake fluid and impact solid surfaces such as the windshield or brake linings.

TABLE 24.1.6 Temperatures at Various Vehicle Locations

Location	Temperature, °C Low	High	Slew rate	Humidity High	Low	Frost
Underhood—Engine						
Exhaust manifold	−40	+649	−7°C/min	95% at 38°C	0	Yes
Intake manifold	−40	+121	−7°C/min	95% at 38°C	0	Yes
Underhood—Dash Panel						
Normal	−40	+121	Open	95% at 38°C		
Extreme	−40	+141	Open	80% at 66°C		
Chassis						
Isolated	−40	+85	NA	98% at 38°C	0	Yes
Near heat source	−40	+121	NA	80% at 66°C	0	Yes
At drivetrain temperature	−40	+177	NA	80% at 66°C	0	Yes

*Frank, R., "Temperature, Heat, and Humidity Sensors," in *Automotive Electronics Handbook*, 2nd ed., Ronald K. Jurgen (ed.), McGraw-Hill, 1999, pp. 5.1–5.24.

TABLE 24.1.7 Sensing Requirements Versus Vehicle Systems

System	Parameter
Engine control	Air charge (intake air) temperature
	Engine coolant temperature
	Intake air humidity
	Ambient temperature
	Fuel temperature
HVAC	Humidity in passenger compartment (PC)
(climate control)	Temperature (PC)
	Outside air temperature
Electronic transmission	Transmission oil temperature
(continuously variable transmission)	
Driver information	Coolant temperature
(body computer inputs)	Ambient air temperature
	Temperature (PC)
	Tire surface temperature
	Rain sensor
	Windshield moisture
	Sun sensor
	Battery temperature
Multiplex/diagnostics	Multiple usage of sensors
Cruise control	
Idle speed control	
Memory seat	
Navigation	
Antiskid brakes	Brake fluid temperature
Traction control/ABS	Brake moisture sensor
Air bags	
Suspension	
Electronic power steering	Power steering fluid temperature
	(also electric assisted)
Four-wheel steering	
Security and keyless entry	

Automotive Temperature and Humidity Measurements

A list of the potential temperature and humidity measurements that can be made and may be incorporated in future systems is shown in Table 24.1.7. The choice in technology for production sensors depends on performance, reliability, and cost.

Measuring Liquid Temperature.   Mounting location, fluid contacting the sensor, and packaging are critical concerns for liquid temperature sensors. Upper temperature limits are typically around 150 to 200°C and operation down to −40°C is usually required. Battery temperature measurements in vehicle development use glass thermometers or glass-shielded thermocouple probes to protect the sensor from the corrosive electrolyte.

The charging voltage of a lead-acid battery must be modified for temperature. Compensation circuitry in the voltage regulator in the charging system is designed to provide a voltage that is within an acceptable range over the entire vehicle-operating temperature range. Semiconductor sensing techniques are used in this approach.

Air Temperature Measurements. Air is measured for ambient temperature, passenger compartment temperature, and inlet air on production vehicles. Upper temperature ranges are lower than liquid measurements, with 85 or 125°C being common upper limits.

Catalyst Temperature. Measuring catalyst temperature during vehicle development has been performed since the earliest implementation of these devices. The converter must be at a minimum temperature, usually above 350°C, to be effective. A linear temperature sensor has been developed to measure the temperature across the catalyst.

Tire Temperature Sensing. Sensing the temperature within a tire is part of the pressure measurement and fault detection provided by tire pressure measuring and automatic adjustment systems. In a system such as the one developed by Michelin (Fig. 24.1.14), both a pressure and a temperature sensor are located on each wheel. A circular antenna and a transceiver allow these signals to be sent to an electronic processing module. An air compressor attempts to maintain desired tire pressure under normal conditions.

Humidity Sensing. Correlation of a number of vehicle tests requires that humidity be monitored. Variations in humidity can explain variations that occur in test results. Significant, repeatable, and controllable results

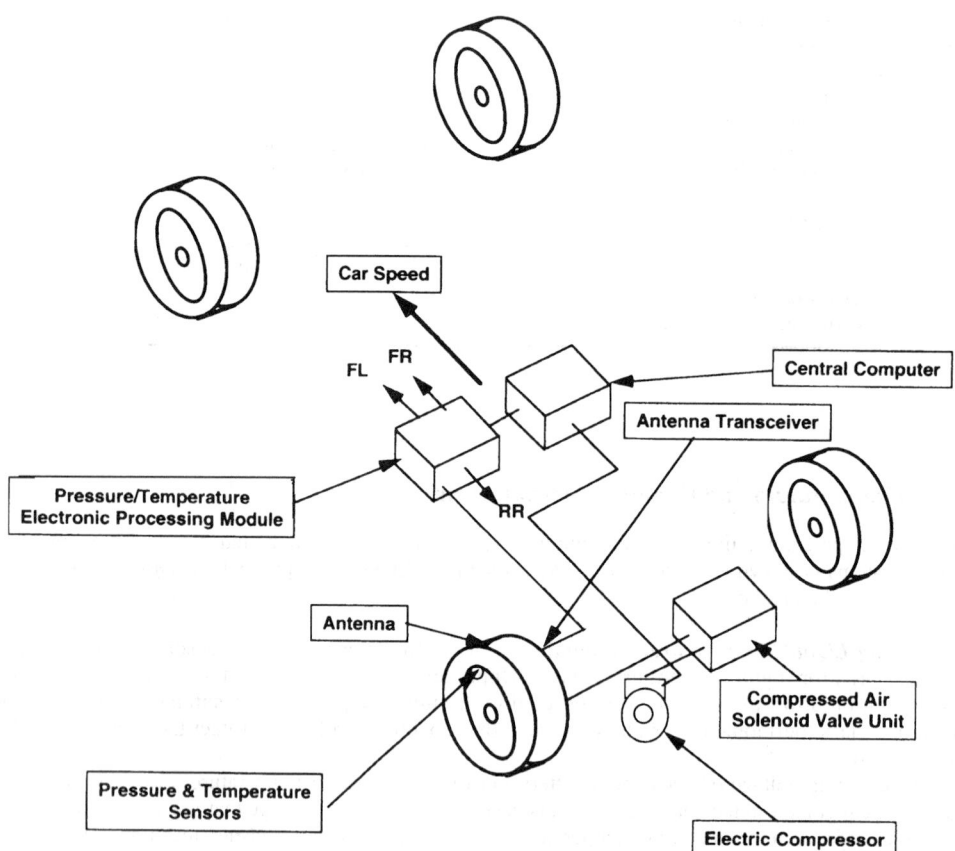

FIGURE 24.1.14 Temperature sensing in tire pressure control.

TABLE 24.1.8 Temperature Sensing Techniques versus Operating Range

Sensing technique	Temperature range, °C	Usage: production (P) development (D) future (F)
Thermistor	0 to 500	P
Thermocouple	−200 to +3000	D
Bimetallic switch	−50 to +400 (650)	P
Potentiometer temperature sensor	−40 to +125	P
Platinum wire resistor (RTD)	−200 to +850 (−40 to + 200)	P
Semiconductor (junction)	−40 to +200	P
Thermostat (pressure spring)	−50 to +500	P
Fiber optic temperature sensor	+1800	D/F
Temperature indicator	+40 to +1350	D
Infrared	>Ta	D
Liquid thermometer	−200 to +1000	D

may lead to humidity sensing being part of future engine control systems and passenger compartment comfort control systems.

Temperature Sensors

Several different sensing techniques are used in production vehicles and during vehicle development to provide the needed temperature measurements. Table 24.1.8 shows a list of these techniques and the temperature range that is typical for each approach. Table 24.1.9 shows common types of thermocouples and their operating characteristics.

Humidity Sensors

Techniques to measure humidity are listed in Table 24.1.10. Most of these are actually laboratory instruments. Three sensing techniques that have potential for future vehicle use are capacitive, resistive, and oxidized porous silicon.

TABLE 24.1.9 Common Thermocouples and Application Factors

ISA code	Conductor Positive	Conductor Negative	Temperature range, °C	Standard error limit, °C	Seebeck coefficient, μV/°C
E*	Chromel	Constantan	0 to +316	±2	62
J*	Iron	Constantan	0 to +277	±2	51
T*	Copper	Constantan	−59 to +93	±1	40
K*	Chromel	Alumel	0 to +277	±2	40
N*	Nicrosil	Nisil	0 to +277	±2	38
S*	Platinum†	Platinum	0 to +538	±3	7
R*	Platinum‡	Platinum	0 to +538	±3	7

*Other temperature ranges and error limits are available.
†10 percent Rhodium.
‡13 percent Rhodium.

TABLE 24.1.10 Techniques for Measuring Humidity/Moisture

Principle	Type measurement
Gravimetric hygrometer	Instrument
Pressure-humidity generator	Instrument
Wet bulb/dry bulb (psychrometer)	Instrument
Hair element	Instrument
Electric conductivity	Instrument
Dew cell	Instrument
Chilled mirror	Instrument
Karl Fisher titration	Instrument
Electrolytic	Instrument
Lithium chloride	Instrument
Capacitance hygropolymer	Production sensor
Bulk polymer (resistance)	Production sensor
Thin-film polymer (capacitance)	Production sensor
Gold/aluminum oxide	Production sensor
Oxidized porous silicon (OPS)	Experimental sensor

EXHAUST GAS SENSORS*

From the model year 1994 onward, on-board devices capable of monitoring the operation of all emission-relevant vehicle components have been required in the United States. Pollutant emission sensors to measure levels of CO, NO_x, and HC emissions downstream from the catalytic converter represent an ideal means of monitoring the performance of both converter and oxygen sensor. There also exist concepts that employ a second oxygen sensor behind the catalytic converter to detect aging in the converter and/or the lambda sensor. This concept is currently used in almost all On-Board Diagnosis II applications worldwide.

Lambda Sensors

Lambda = 1 Sensor: Nernst Type (ZrO_2). The lambda sensor operates as a solid-electrolyte galvanic oxygen-concentration cell (Fig. 24.1.15). A ceramic element consisting of zirconium oxide and yttrium oxide is employed as a gas-impermeable solid electrolyte designed to separate the exhaust gas from the reference atmosphere.

Lambda = 1 Sensor: Semiconductor Type. Oxidic semiconductors such as TiO_2 and $Sr\ TiO_3$ rapidly achieve equilibrium with the oxygen partial pressure in the surrounding gas phase at relatively low temperatures. Changes in the partial pressure of the adjacent oxygen produce variations in the oxygen vacancy concentration of the material, thereby modifying its volume conductivity.

Lean A/F Sensor: Nernst Type. It is always possible to employ a potentiometric lambda = 1 sensor as a lean A/F sensor by using the flat part of the Nerst voltage curve to derive the values at lambda >1.

Lean A/F Sensor: Limiting Current Type. An external electrical voltage is applied to two electrodes on a heated ZrO_2 electrolyte to pump O_2 ions from the cathode to the anode. When a diffusion barrier impedes the flow of O_2 molecules from the exhaust gas to the cathode, the result is current saturation beyond a certain

*Weidenmann, H. M., Hotzel, G., Neumann, H., Riegel, J., Stanglmeier, F., and Weyl, H., "Exhaust Gas Sensors," in *Automotive Electronics Handbook*, 2nd ed., Ronald K. Jurgen (ed.), McGraw-Hill, 1999, pp. 6.1–6.25.

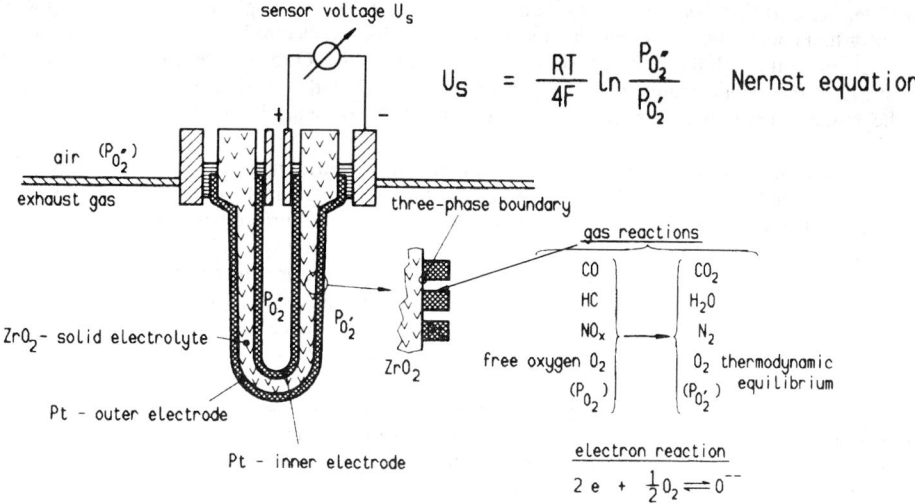

FIGURE 24.1.15   Diagram illustrating design and operation of a lambda = 1 sensor.

pumping voltage threshold. The resulting limiting current is roughly proportional to the exhaust gas's oxygen concentration.

Wide-Range A/F Sensor: Single-Cell.   When the anode of a limiting current sensor is exposed to reference air instead of to the exhaust gas, the total voltage at the probe will be the sum of the effective pumping voltage and a superimposed Nernst voltage. In operation, holding the total voltage to, as an example, 500 mV will produce a positive pumping voltage in lean exhaust gases. The diffusion limits the rate at which O_2 is pumped from the cathode to the anode. At lambda = 1, the pumping voltage and, with it, the pumping current drop toward zero.

Wide-Range A/F Sensor: Dual-Cell.   Skillful combination of a limiting-current sensor with a Nernst concentration cell on a single substrate will produce a dual-cell, wide-range A/F sensor. An electronic circuit regulates the current applied to the pumping cell to maintain a constant gas composition in the measurement gap. If the exhaust gas is lean, the pumping cell drives the oxygen outward from the measurement gap. If the exhaust gas is rich, the flow direction is reversed and oxygen from the surrounding exhaust gas is pumped into the measurement gap.

Other Exhaust Gas Components

Sensors capable of monitoring the levels of the regulated toxic exhaust substances—CO, NO_x, and HC—would be desirable as elements of the On-Board Diagnosis Systems specified by the California Air Resources Board and EURO III/IV, especially for Gasoline Direct Injection systems. For the more stringent emissions standards, the need for selective exhaust gas sensors have to be developed.

Mixed-Potential Sensors.   If reduced catalytic activity prevents gas equilibrium from being achieved at the electrode of a galvanic ZrO_2 cell, competing reactions can occur. These, in turn, prevent a state of reduction/oxidation equilibrium in the oxygen, and lead to the formation of a mixed potential. Mixed-potential exhaust gas sensors are available for NO_x, HC, CO, and NH_3 to study new diagnostic concepts.

Dual Pumping Cell Gas Sensors.   These sensors are based on a zirconia planar type similar to the wide-range A/F sensor and contain two different chambers with gas-selective electrodes. The exhaust gas penetrates into the first chamber through the first diffusion barrier. By a selective O_2 electrode, oxygen is removed from the exhaust gas. In the second chamber, NO_x is decomposed into N_2 and O_2 and the O_2 is pumped out of the chamber. The associated pumping current is a measure of the NO_x concentration in the exhaust gas.

Semiconductor Gas Sensors.   On the surface of nonstoichiometric metal oxides such as SnO_2, TiO_2, InO_3, and Fe_2O_3, oxygen is absorbed and dissociated in air at high temperatures and is bonded to the crystal lattice. This leads to the formation of a thin depletion layer at the crystallite surface with a resultant arch in the potential curve. This phenomenon produces reductions in surface conductivity and higher intercrystalline resistance at the boundaries between the two crystallites; this is the major factor determining the total resistance of the polycrystalline metal oxide. Oxidation gases such as CO, H_2, and C_xH_y, which react with surface oxygen, increase the density of charge carriers in the boundary layer and reduce the potential barrier. Reduction gases such as NO_x and SO_x raise the potential barrier and, with it, the surface/intercrystalline resistance.

Catalytic Gas Sensors.   The catalytic gas sensor is essentially a temperature sensor featuring a catalytically active surface. An exothermic reaction at the catalytically active surface causes the temperature of the sensor to rise. The increase in temperature is proportional to the concentration of an oxidation gas in an excess-oxygen atmosphere.

SPEED AND ACCELERATION SENSORS*

Speed sensing can be divided into rotational and linear applications. Rotation speed sensing has two major application areas: engine speed monitoring to enhance engine control and performance and antilock braking and traction control systems for improved road handling and safety. Linear sensing can be used for ground-speed monitoring for vehicle control, obstacle detection, and crash avoidance. Acceleration sensors are used in air bag deployment, ride control, antilock brake, traction, and inertial navigation systems.

Speed-Sensing Devices

Variable Reluctance Devices.   These devices are essentially small ac generators with an output voltage proportional to speed, but they are limited in applications where zero speed sensing is required. These devices have a linear output voltage with frequency and need an A/D converter to generate a digital signal for compatibility with the master control unit.

Hall-Effect Devices.   A Hall-effect device can be used for zero speed sensing (it can give an output when there is no rotation). Hall devices give a frequency output that is proportional to speed, making them compatible with microcontrollers.

Magnetoresistive Devices.   The magnetoresistive effect is the property of a current-carrying ferromagnetic material to change its resistivity in the presence of an external magnetic field. The resistivity value rises to a maximum when the direction of current and magnetic field are coincident, and is a minimum when the fields are perpendicular to each other. These devices give an output frequency that is proportional to speed, making for ease of interfacing with an MCU.

Ultrasonic Devices.   Ultrasonic devices can be used to measure distance and ground speed. They can also be used as proximity detectors. To give direction and beam shape, the signals are transmitted and received via specially

---

*Dunn, W. C., "Speed and Acceleration Sensors," in *Automotive Electronics Handbook*, 2nd ed., Ronald K. Jurgen (ed.), McGraw-Hill, 1999, pp. 7.1–7.29.

configured horns and orifices. To measure speed, the distance variation with time can be measured and the velocity calculated. A more common method is to use the Doppler effect, which is a change in the transmitted frequency as detected by the receiver due to motion of the target (or, in this case, the motion of the transmitter and receiver).

Optical and Radio Frequency Devices.   Optical devices are still being used for rotational speed sensing. They normally consist of light-emitting diodes (LEDs) with optical sensors. An optical sensor detects light from an LED through a series of slits cut into the rotating disc, so that the output from the sensor is a pulse train whose frequency is equal to the rpm of the disc multiplied by the number of slits. Radio frequency devices use gallium arsenide on Gunn devices to obtain the power and high frequency (about 100 GHz) required in the transmitter.

Automotive Applications for Speed Sensing

There are several applications for speed sensing. First it is necessary to monitor engine speed. This information is used for transmission control, engine control, cruise control, and possibly for a tachometer. Wheel speed sensing is required for use in transmissions, cruise control, speedometers, antilock brake systems, traction control, variable ratio power steering assist, four-wheel steering, and possibly in inertial navigation and air bag deployment applications. Linear speed sensing is used to measure the ground speed and could also be used in antilock brake systems, traction control, and inertial navigation.

Acceleration Sensing Devices

Acceleration sensors vary widely in their construction and operation. In applications such as crash sensors for air bag deployment, mechanical devices (simple mechanical switches) have been developed and are in use. Solid-state analog accelerometers have also been designed for air bag applications.

Mechanical Sensing Devices.   Mechanical switches are simple make-break devices. Figure 24.1.16 shows the cross section of a Breed-type of switch or sensor.

Piezoelectric Sensing Devices.   Piezoelectric devices (Fig. 24.1.17) consist of a layer of piezoelectric material (such as quartz) sandwiched between a mounting plate and a seismic mass. Electric connections are made to both sides of the piezoelectric material.

Piezoresistive Sensing Devices.   The property of some materials to change their resistivity when exposed to stress is called the piezoresistive effect. Piezoresistive sensing can be used with bulk micromachined accelerometers (Fig. 24.1.18).

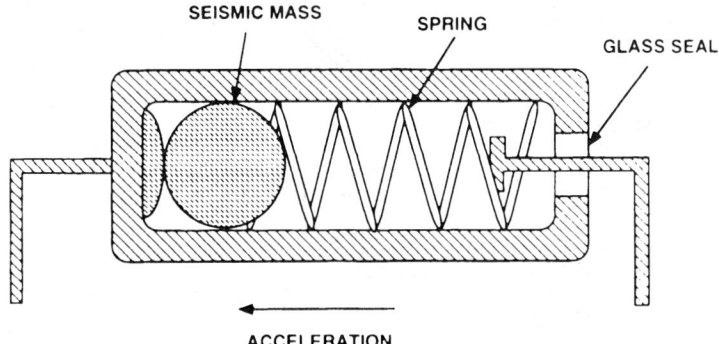

FIGURE 24.1.16   Cross section of a mechanical sensor.

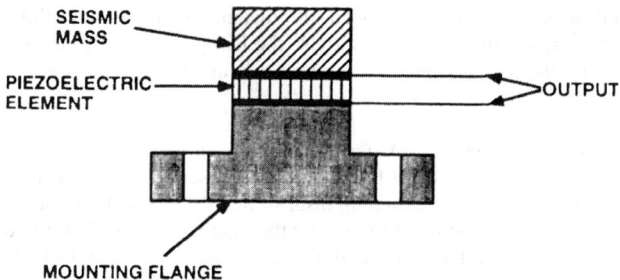

FIGURE 24.1.17 Cross section of a piezoelectric accelerometer.

Capacitive Sensing Devices.   Differential capacitive sensing has a number of attractive features when compared to other methods of sensing: easily implement self-test, temperature insensitivity, and smaller size. Capacitive sensing has the advantages of dc and low-frequency operation and well-controlled damping.

Open-loop signal-conditioning circuits amplify and convert the capacitance changes into a voltage. Such a CMOS circuit using switched capacitor techniques is shown in Fig. 24.1.19. A closed-loop system (Fig. 24.1.20) can be configured to give an analog or digital output.

Automotive Applications for Accelerometers

Accelerometers have a wide variety of uses in the automobile. The initial application is as a crash sensor for air bag deployment. An extension of this application is the use of accelerometers for the detection of side

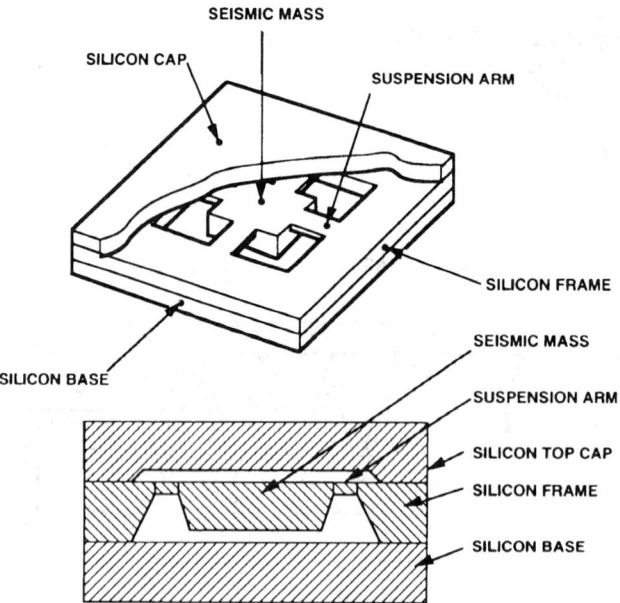

FIGURE 24.1.18   Bulk micromachined accelerometer.

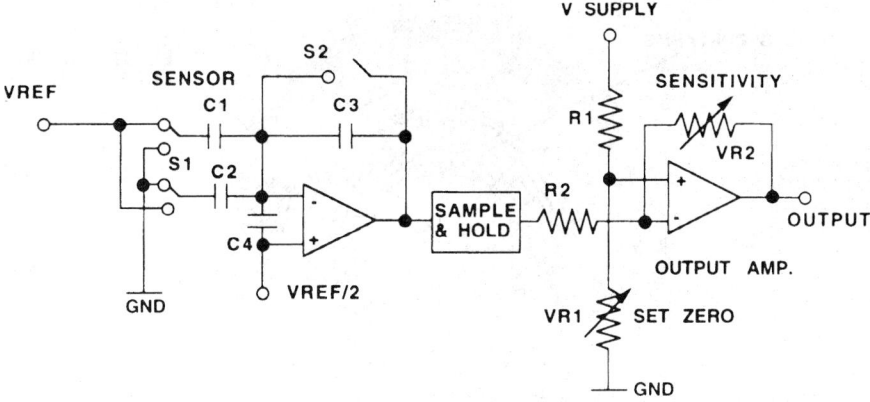

FIGURE 24.1.19   Capacitive sensing integrator circuit.

impact or rollover. Low g sensors are being developed for ride control, antilock braking, traction, and inertial navigation requirements.

New Sensing Devices

New cost-effective sensors are continually being developed. For rotational speed sensing, a number of new devices are being investigated to detect magnetic fields: flux gate, Weigand effect, magnetic transistor, and magnetic diode. For linear speed-sensing, ultrasonics, infrared, laser, and microwaves (radar) can be used in the detection of objects behind vehicles and in the blind areas. Recent developments in solid-state technology

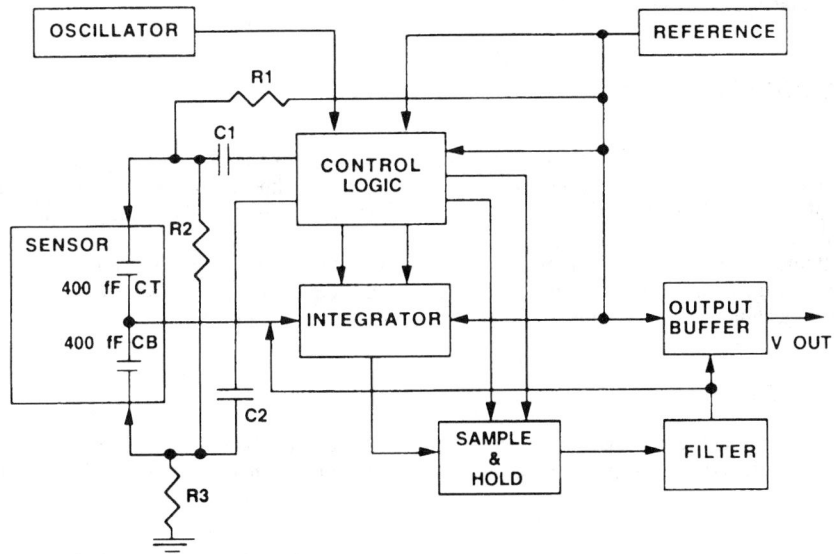

FIGURE 24.1.20   Block diagram of analog closed-loop control.

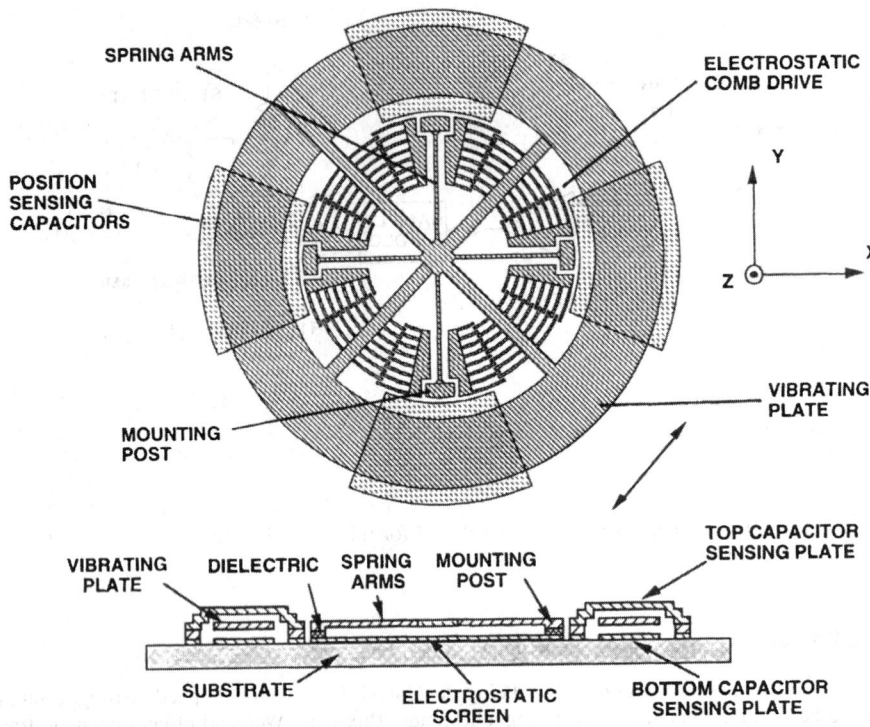

FIGURE 24.1.21 Solid-state gyroscope.

have made possible very small cost-effective devices to sense angular rotation. The implementation of one such gyroscopic device is shown in Fig. 24.1.21.

ENGINE KNOCK SENSORS*

Knock is a phenomenon characterized by undesirable structural vibration and noise generation and is peculiar to spark-ignited engines. The terms *ping* (light, barely observable knock) and *predetonation* (knock caused by ignition of the charge slightly before the full ignition of the flame front by the spark plug) are also commonly used in the industry.

An attempt to measure the cause of the phenomenon leads one to the difficult problem of observing pressure waves in the cylinder. In fact, over the years these difficulties led the industry to devise an experimental comparison technique that measured the octane rating of the fuel, not of the engine.

Technologies for Sensing Knock

A number of different technologies have been selected for measuring knock in real time on board, and sensors have been developed for this purpose. These sensors measure the magnitude of a consequential parameter

*Wolber, W. G., "Engine Knock Sensors," in *Automotive Electronics Handbook*, 2nd ed., Ronald K. Jurgen (ed.), McGraw-Hill, 1999, pp. 8.1–8.10.

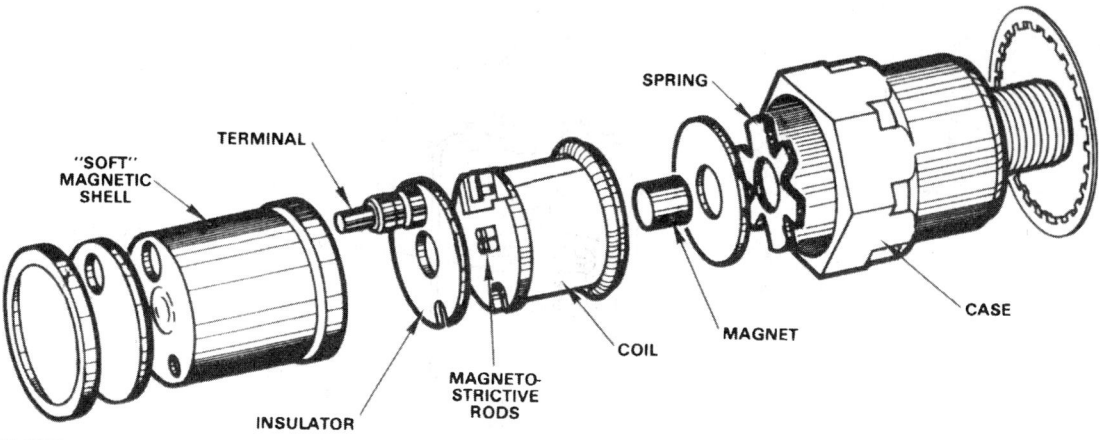

FIGURE 24.1.22 An exploded view of a jerk sensor.

driven by the knock, rather than the knock phenomenon itself. The overall effectiveness of the control is determined not only by the intrinsic performance and stability of the sensor and control, but also by how robust the chain of causality is between the knock phenomenon and the parameter measured. Knock can be controlled either by retarding spark timing or by opening a manifold boost wastegate valve.

Jerk Sensor.   The first turbocharged engine knock control used a jerk sensor, a productionized version of the kind of sensor used in the laboratory (Fig. 24.1.22). When the sensor is fully assembled, the spider spring is preloaded, and all parts of the sensor except the coil cover are in compression. The vibrations picked up and transmitted from the engine block through the mounting study appear in the nickel alloy rods. The waves present in the rods linearly modulate the magnetic reluctance of the magnetic circuit. The many-turn coil wound around the magnetostrictive rods generates a voltage proportional to the rate of change of the magnetic reluctance of the rods. Since the vibrations picked up are already due to accelerations from the knock reverberations transmitted though the engine block, the voltage from the coil represents the third-time derivative of displacement, or jerk.

Accelerometer Sensors.   The jerk sensor has too many parts to be a low-cost solution to measuring knock on board. A more economical approach is to use the second-time derivative of displacement—acceleration. Piezoelectric, piezoceramic, and silicon accelerometers can be used.

Other Sensors.   An instantaneous cylinder pressure sensor permits the extraction of the pressure reverberation signal, which is the direct cause of knock, but it has not been implemented in on-board knock control systems for several reasons:

- Either the same cylinder must always be the one to experience knock first and most severely or one must have a sensor on every cylinder.
- The cylinder pressure wave is complex, and knock is only one of many signature elements present. Deriving a unique knock signal requires considerable signal processing in real time.
- While cylinder pressure sensors exist, which are suitable for test purposes, a durable, low-cost, mass-producible on-board pressure sensor is not yet available.

Piezoceramic Accelerometers.   These accelerometers (Fig. 24.1.23) lend themselves to low-cost mass production. As a result they have become pretty much the knock sensor of choice for the automobile industry.

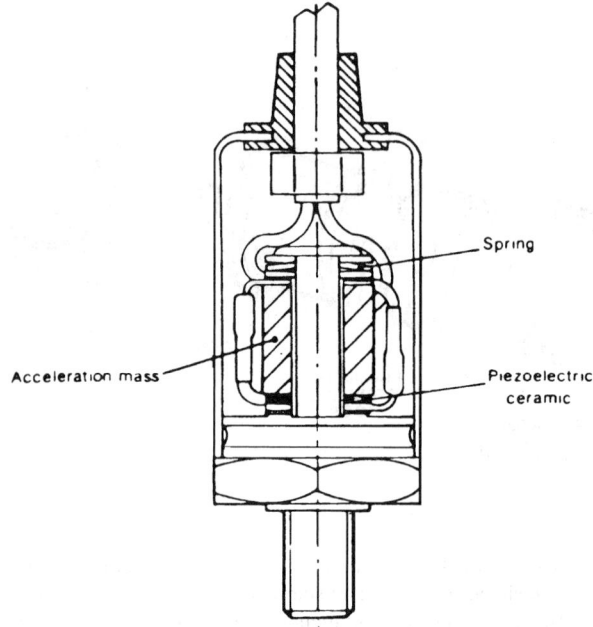

FIGURE 24.1.23 A piezoceramic accelerometer knock sensor.

ENGINE TORQUE SENSORS*

Torque can be defined as the moment produced by the engine crankshaft tending to cause the output driveline to turn and thereby deliver power to the load.

Direct Torque Sensors

Magnetic Vector Sensors. A strain-gaged torsional Hooke's law sensor has been used in off-board torque measurements. A more practical approach to an on-board torque sensor is a noncontacting design called a magnetic vector sensor. It operates on the principle that the magnetic domains in a ferromagnetic shaft delivering torque are randomly distributed when no torque is being delivered, but that each domain, on average, is slightly rotated in the tangential direction when torque is delivered by the shaft and twists it. If an ac-driven coil is placed near the shaft and surrounded by four sensing coils arranged mechanically and electrically as a bridge, the amplitude of the bridge offset is proportional to the magnetic vector tangential component, and therefore to twist and torque.

Optical Twist Measurement. Research has been reported on a sensor that changes the duty factor of light pulses sensed after passing through sets of slits at both ends of a shaft. The principle of its operation is shown in Fig. 24.1.24.

---

*Wolber, W. G., "Engine Torque Sensors," in *Automotive Electronic Handbook*, 2nd ed., Ronald K. Jurgen (ed.), McGraw-Hill, 1999, pp. 9.1–9.14.

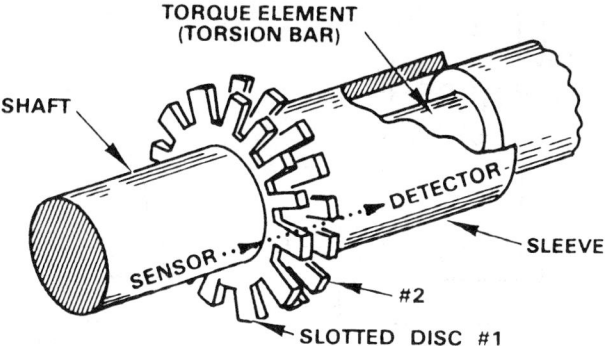

FIGURE 24.1.24 Optical torque meter. (*Courtesy of The Bendix Corp.*)

Capacitive Twist Sensor. An electrode pattern can be made using interdigitated electrodes spaced one or two degrees apart on two facing discs. One of the discs is stationary; the other rotates with the crankshaft. Two such pairs of electrodes can be operated with phase detection measurement to provide a virtually instantaneous signal proportional to the twist of the shaft. The rotating halves of the electrode pairs are attached to the ends of the Hooke's law torsional spring.

Inferred Torque Measurement

Instantaneous Cylinder Pressure Sensors. Much work continues on development of a mass-producible on-board cylinder pressure sensor. The signals from cylinder pressure sensors need considerable real-time data processing to produce inferred "torque" signals.

Digital Period Analysis (DPA). When an engine is run at low speed and heavy load, the instantaneous angular velocity of its output shaft on the engine side of the flywheel varies at the fundamental frequency of the cylinders, since the compression stroke of each cylinder abstracts torque and the power stroke adds a larger amount. The signal-to-noise ratio of the measurement of instantaneous angular velocity (or rather of its reciprocal, instantaneous period) degrades with increasing engine speed and lighter load, but is a useful way to infer torque-like measures of engine performance.

ACTUATORS *

Actuators in automobiles respond to position commands from the electronic control unit to regulate energy, mass, and volume flows. Basic actuator elements are shown in Fig. 24.1.25. Either the control unit or the actuator itself will feature an integral electronic output amplifier.

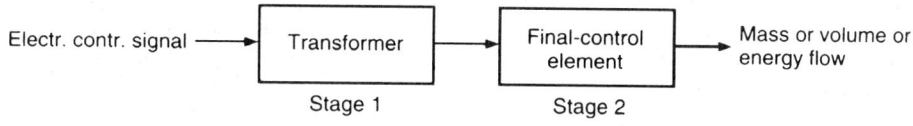

FIGURE 24.1.25 Basic actuator elements.

---

*Müller, K., "Actuators," in *Automotive Electronics Handbook*, 2nd ed., Ronald K. Jurgen (ed.), McGraw-Hill, 1999, pp. 10.1–10.35.

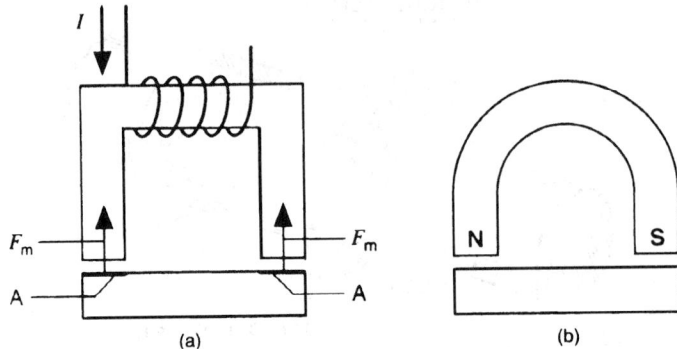

FIGURE 24.1.26   Flat-armature solenoid featuring field excitation (a) via coil, (b) via permanent magnet.

Types of Electromechanical Actuators

Magnetic Actuators.   In order to operate, these actuators depend on the forces found at the interfaces in a coil-generated magnetic field when current passes through it. With a flat armature solenoid (Fig. 24.1.26), a particular solenoid force is specified for each technical application. The pole face area, the magnetic circuit, and the coil are then determined for this force.

Torque Motors.   The torque motor consists of a stator and an armature—both made of soft magnetic material—and a permanent magnet. The pivoting armature can be equipped with either one or two coils.

Electrostatic Step Motors.   Electrostatic step motors are drive elements in which a special design operates in conjunction with pulse-shaped control signals to carry out rotary or linear-stepped movements. The step motor is capable of transforming digital control signals directly into discontinuous rotary motion. In principle, the step motor is essentially a combination of dc solenoids. Depending on the configuration of the magnetic circuit, a distinction is made between three types of step motors: the variable-reluctance step motor (neutral magnetic circuit), heteropolar units (polarized magnetic circuit), and hybrid devices.

Moving Coils.   The moving coil is an electrodynamic device (force is applied to a current-saturated conductor in a magnetic field). A spring-mounted coil is located in the ring gap of a magnetic circuit featuring a permanent magnet. When current flows through the coil, a force is exerted against it. The direction of this force is determined by the flow direction of the current itself.

Electrical Actuators

Piezoelectric Actuators.   When mechanical compression and tension are brought to bear on a piezoelectric body, they produce an asymmetrical displacement in the crystal structure and in the charge centers of the affected crystal ions. The result is charge separation. An electric voltage proportional to the mechanical pressure can be measured at the metallic electrodes. If electric voltage is applied to the electrodes on this same body, it will respond with a change in shape; the volume remains constant. This piezoelectric effect can be exploited to produce actuators.

Electrostatic Actuators.   Microactuator technology makes it possible to apply small electrical field forces in mechanical drive devices. They combine high switching speeds with much smaller energy loss than that found in electromagnetic actuators. The disadvantages are the force-travel limitations and the high operating voltages.

Electrorheological Fluids.   The electrorheological effect is based on polarization processes in minute particles suspended in a fluid medium. These particles modify the fluid's viscosity according to the orientation in the electrical field. This effect is exploited in controlled transfer and damping elements.

Thermal Actuators

Temperature-Sensitive Bimetallic Elements.   The temperature-sensitive bimetallic element is composed of at least two bonded components of varying thermal-expansion coefficients. When heat is applied, the components expand at different rates, causing the bimetallic element to bend. When electrically generated heat is applied, these devices become actuators.

Memory Alloys.   Memory alloys are metallic materials that exhibit a substantial degree of "shape memory." If the element is heated to beyond the transformation temperature (from martenistic to austenitic), it returns to its former shape. If the component is then reshaped after cooling, the entire process can be repeated.

Automotive Actuators

Actuators for Braking Intervention.   Braking pressure is regulated normally via 2/2 solenoid valves, that is, valves with two ports and two switch positions. When no current is applied, the inlet valve remains open and the outlet valve is closed, allowing unrestricted wheel braking.

Electronic Throttle Control (ETC) with Throttle-Aperture Adjustment.   Either of two standard methods can be employed for throttle regulation. ETC systems feature an actuator (servo motor) mounted directly at the throttle valve. On systems incorporating a traction-control actuator, the servo motor is installed in the throttle cable. Another approach to engine intervention is embodied in a design in which the linkage spring is integrated within the throttle body.

Fuel Injection for Spark-Ignition Engines.   Electronically controlled fuel injection systems meter fuel with the assistance of electromagnetic injectors. The injector's opening time determines the discharge of the correct amount of fuel.

Fuel Injection for Diesel Engines.   Distributor-type fuel injection pumps contain rotary solenoid actuators for injection quantity, and two position valves for engine operation and shutoff.

Passenger-Safety Actuators.   Pyrotechnical actuators are used for passenger-restraint systems such as the air bag and the automatic seal belt tensioner. When an accident occurs, the actuators inflate the air bag (or tension the seat belt) at precisely the right instant.

Electronic Transmission Control.   Continuous operation actuators are used to modulate pressure, while switching actuators function as supply and discharge valves for shift-point control. Types of actuators used included on/off solenoid valves, variable-pressure switches, and pulse-width-modulated solenoid valves.

Headlight Vertical Aim Control.   Headlight-adjustment actuators are dc or step motors producing a rotary motion that gear-drive units then convert to linear movement.

Future Actuators

The motivation to develop new actuators is created by the potential advantages of new manufacturing and driving techniques in combination with new materials. Representative new fields of innovation include micromechanical valves and positive-engagement friction drive (ultrasonic motor).

CHAPTER 24.2
AUTOMOTIVE ON-BOARD SYSTEMS

Ronald K. Jurgen

INTRODUCTION

In the previous chapter in this section, the significance of sensors at the input of electronic automotive systems and actuators at the output of such systems was stressed. Equally important, of course, are the automotive microcontrollers that accept the sensor signals, process them, and then send command signals to the actuators. The combination of sensors, microcontrollers, and actuators is what makes possible the myriad of automotive control systems now in use in production cars.

This chapter begins with a brief discussion of microcontrollers and then proceeds with descriptions of most major automotive control systems.

MICROCONTROLLERS*

A microcontroller can be found at the heart of almost any automotive electronic control module or electronic control unit (ECU) in production today. Automotive systems such as antilock braking control (ABS), engine control, navigation, and vehicle dynamics all incorporate at least one microcontroller within their ECU to perform necessary control functions.

A microcontroller can essentially be thought of as a single-chip computer system and is often referred to as a single-chip microcomputer. It detects and processes input signals, and responds by asserting output signals to the rest of the ECU. Fabricated on a highly integrated, single piece of silicon are all of the features necessary to perform embedded control functions. Microcontrollers are fabricated by many manufacturers and are offered in just about any imaginable mix of memory, input/output (I/O), and peripheral sets. The user customizes the operation of the microcontroller by programming it with his or her own unique program. The program configures the microcontroller to detect external events, manipulate the collected data, and respond with appropriate output. The user's program is commonly referred to as code and typically resides on-chip in either read-only memory (ROM) or erasable programmable read-only memory (EPROM). In some cases where an excessive amount of code space is required, memory may exist off-chip on a separate piece of silicon. After power-up, a microcontroller executes the user's code and performs the desired embedded control function.

---

*Boehmer, D. S., "Automotive Microcontrollers," in *Automotive Electronics Handbook*, 2nd ed., Ronald K. Jurgen (ed.), McGraw-Hill, 1999, pp. 11.3–11.55.

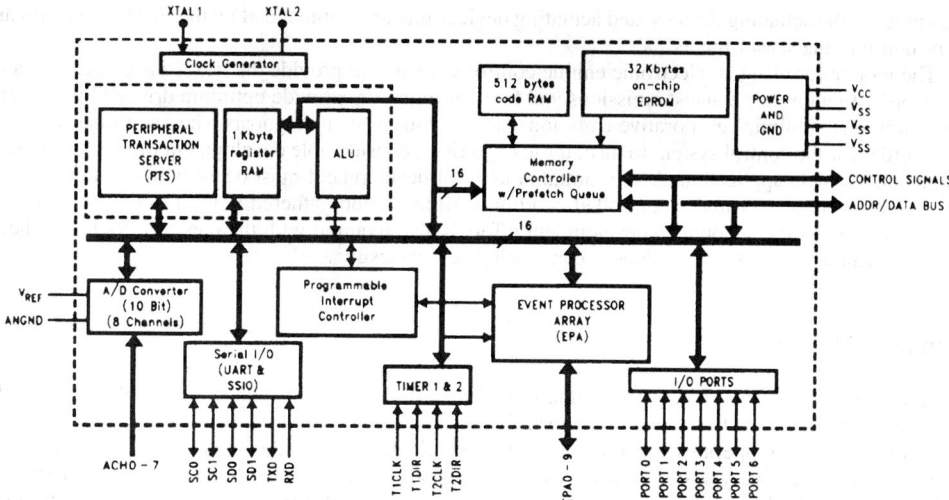

FIGURE 24.2.1   Microcontroller block diagram.

Microcontrollers differ from microprocessors in several ways. A microcontroller can be thought of as a complete microcomputer on a chip that integrates a central processing unit (CPU) with memory and various peripherals such as analog-to-digital (A/D) converters, serial communication units, high-speed input and output units, timer/counter units, and standard low-speed I/O ports. Microcontrollers are designed to be embedded within event-driven control applications and generally have all necessary peripherals integrated onto the same piece of silicon.

Microprocessors, on the other hand, typically require external peripheral devices to perform their intended function and are not suited to be used in single-chip designs. Microprocessors basically consist of a CPU with register arrays and interrupt handlers. Peripherals such as A/D are rarely integrated onto microprocessor silicon. Microprocessors are designed to process large quantities of data and have the capability to handle large amounts of external memory.

Choosing a microcontroller for an application is a process that takes careful investigation and thought. Items such as memory size, frequency, bus size, I/O requirements, and temperature range are all basic requirements that must be considered when choosing a microcontroller. The microcontroller family must possess the performance capability necessary to successfully accomplish the intended task. The family should also provide a memory, I/O, and frequency growth path that allows easy upgradability to meet market demands. Additionally, the microcontroller must meet the application's thermal requirements in order to guarantee functionality over the intended operating temperature range. Items such as these must all be considered when choosing a microcontroller for an automotive application.

A typical block diagram of a microcontroller is shown in Fig. 24.2.1.

ENGINE CONTROL*

An electronic engine control system consists of sensing devices that continuously measure the operating conditions of the engine, an ECU that evaluates the sensor inputs using data tables and calculations and determines

---

*Hirschlieb, G. C., Schiller, G., and Stottler, S., "Engine Control," in *Automotive Electronics Handbook*, 2nd ed., McGraw-Hill, 1999, pp. 12.1–12.36.

the output to the actuating devices, and actuating devices that are commanded by the ECU to perform an action in response to the sensor inputs.

The motive for using an electronic engine control system is to provide the needed accuracy and adaptability in order to minimize exhaust emissions and fuel consumption, provide optimum driveability for all operating conditions, minimize evaporative emissions, and provide system diagnosis when malfunctions occur.

In order for the control system to meet these objectives, considerable development time is required for each engine and vehicle application. A substantial amount of development must occur with an engine installed on an engine dynamometer under controlled conditions. Information gathered is used to develop the ECU data tables. A considerable amount of development effort is also required with the engine installed in the vehicle. Final determination of the data tables occurs during vehicle testing.

Spark Ignition (SI) Engines

Fuel Delivery Systems.   Fuel management in the spark ignition system consists of metering the fuel, formation of the air/fuel mixture, transportation of the air/fuel mixture, and distribution of the air/fuel mixture. The driver operates the throttle valve, which determines the quantity of air inducted by the engine. The fuel delivery system must provide the proper quantity of fuel to create a combustible mixture in the engine cylinder. In general, two fuel delivery system configurations exist: single-point and multipoint (Fig. 24.2.2).

For single-point systems such as carburetors or single-point fuel injection (Fig. 24.2.3), the fuel is metered in the vicinity of the throttle valve. Mixture formation occurs in the intake manifold. Some of the fuel droplets evaporate to form fuel vapor (desirable) while others condense to form a film on the intake manifold walls (undesirable). Mixture transport and distribution is a function of intake manifold design. Uniform distribution under all operating conditions is difficult to achieve in a single-point system.

For multipoint systems, the fuel is injected near the intake valve. Mixture formation is supplemented by the evaporation of the fuel on the back of the hot intake valve. Mixture transport and distribution occurs only in

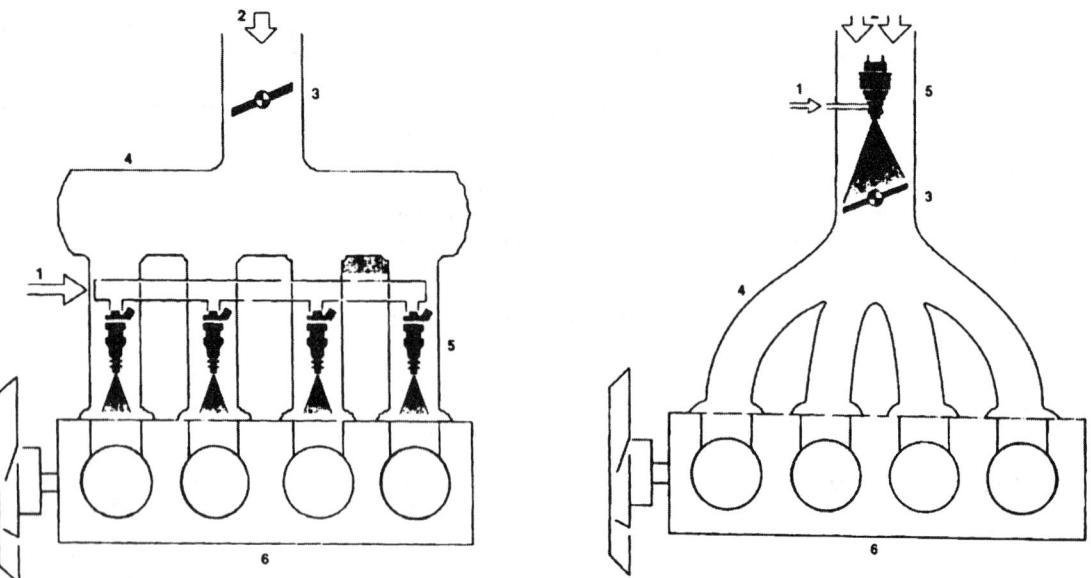

FIGURE 24.2.2   Air-fuel mixture preparation: right, single-point fuel injection; left, multipoint fuel injection with fuel (1), air (2), throttle valve (3), intake manifold (4), injector(s) (5), and engine (6).

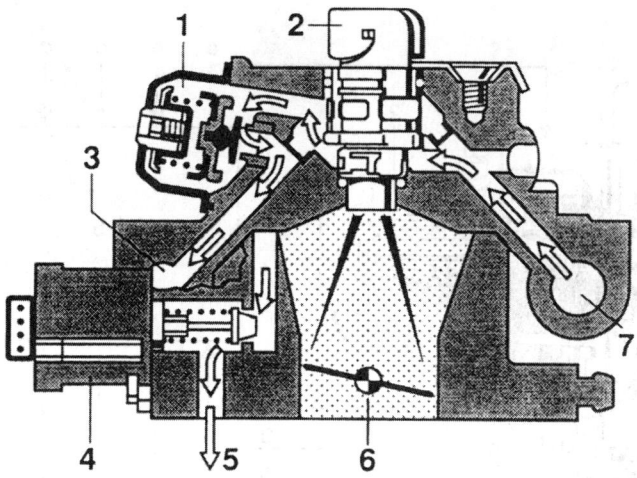

FIGURE 24.2.3   Single-point injection unit: pressure regulator (1), injector (2), fuel return (3), stepper motor for idle speed control (4), to intake manifold (5), throttle valve (6), and fuel inlet (7).

the vicinity of the intake valve. The influence of the intake manifold design on uniform mixture distribution is minimized. Since mixture transport and distribution is not an issue, the intake manifold design can be optimized for air flow.

Ignition Systems.   The general configuration of an ignition system consists of the following components: energy storage device, ignition timing mechanism, ignition triggering mechanism, spark distribution system, and spark plugs and high-tension wires. Table 24.2.1 summarizes the various ignition systems used on SI engines.

Compression Ignition Engines

Electronic engine controls are now being used on compression ignition (diesel) engines. These controls offer greater precision and control of fuel injection quantity and timing, engine speed, exhaust gas recirculation (EGR), turbocharger boost pressure, and auxiliary starting devices. The following inputs are used to provide the ECU with information on current engine operating conditions: engine speed; accelerator position; engine coolant, fuel, and inlet air temperatures; turbocharger boost pressure; vehicle speed; control rack or control collar position

TABLE 24.2.1   Overview of Various Ignition Systems

Ignition function	Ignition designation				
	Coil system	Transistorized coil system	Capacitor discharge system	Electronic system with distributor	Electronic distributorless system
Ignition triggering	Mechanical	Electronic	Electronic	Electronic	Electronic
Ignition timing	Mechanical	Mechanical	Electronic	Electronic	Electronic
High-voltage generation	Inductive	Inductive	Capacitive	Inductive	Inductive
Spark distribution to appropriate cylinder	Mechanical	Mechanical	Mechanical	Mechanical	Electronic

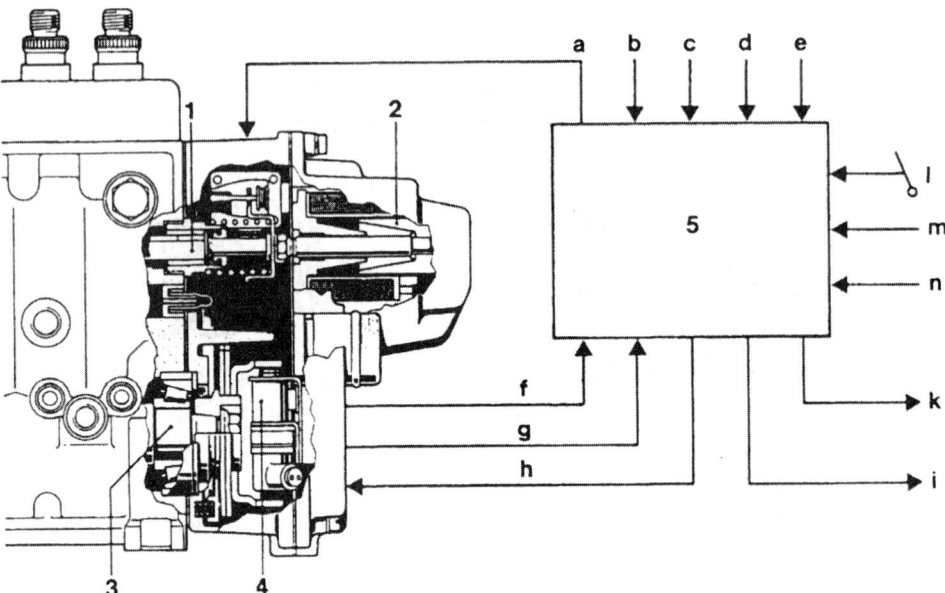

FIGURE 24.2.4   Electronic engine control system for an in-line injection pump: control rack (1), actuator (2), camshaft (3), engine speed sensor (4), ECU (5), input/output: redundant fuel shutoff (a), boost pressure (b), vehicle speed (c), temperature—water, air, fuel (d), intervention in injection fuel quantity (e), speed (f), control rack position (g), solenoid position (h), fuel consumption and engine speed display (i), system diagnosis information (k), accelerator position (l), preset speed (m), and clutch, brakes, engine brake (n).

(for control of fuel quantity; and atmospheric pressure). Figure 24.2.4 is a schematic of an electronic engine control system on an in-line diesel fuel injection pump application.

Engine Control Modes

Engine Crank and Start.   During engine cranking, the goal is to get the engine started with the minimal amount of delay. To accomplish this, fuel must be delivered that meets the requirements for starting for any combination of engine coolant and ambient temperatures. For a cold engine, an increase in the commanded air-fuel ratio is required due to poor fuel vaporization and "wall wetting," which decreases the amount of usable fuel. Wall wetting is the condensation of some of the vaporized fuel on the cold metal surfaces in the intake port and combustion chamber. It is critical that the fuel does not wet the spark plugs, which reduces the effectiveness of the spark plug and prevent the plug from firing. Should plug wetting occur, it may be impossible to start the engine.

Engine Warm-Up.   During the warm-up phase, there are three conflicting objectives: keep the engine operating smoothly (i.e., no stalls or driveability problems), increase exhaust temperature to quickly achieve operational temperature for the catalyst and lambda sensor so that closed-loop fuel control can begin operating, and keep exhaust emissions and fuel consumption to a minimum. The best method for achieving these objectives is very dependent on the specific engine application.

Transient Compensation.   During transitions such as acceleration or deceleration, the objective of the engine control system is to provide a smooth transition from one engine operating condition to another (i.e., no hesitations, stalls, bumps, or other objectionable driveability concerns), and keep exhaust emissions and fuel consumption to a minimum.

Full Load.   Under steady-state full-load conditions, such as for climbing a grade, it is desirable to control the air/fuel mixture and ignition timing to obtain maximum power and to also limit engine and exhaust temperatures.

Idle Speed Control.   The objectives of the engine control system during idle are to provide a balance between the engine torque produced and the changing engine loads, thus achieving a constant idle speed even with various load changes due to accessories (i.e., air conditioning, power steering, and electrical loads) being turned on and off and during engagement of the automatic transmission. In addition the idle speed control must be able to compensate for long-term changes in engine load, such as the reduction in engine friction that occurs with engine break-in. The idle speed control must also provide the lowest idle speed that allows smooth running to achieve the lowest exhaust emissions and fuel consumption (up to 30 percent of a vehicle's fuel consumption in city driving occurs during idling).

To control the idle speed, the ECU uses inputs from the throttle position sensor, air conditioning, automatic transmission, power steering, charging system, engine RPM, and vehicle speed.

TRANSMISSION CONTROL*

Since the introduction of electronic transmission control units in the early 1980s, the acceptance of the automatic transmission (AT) rose sharply. The market for ATs is divided into stepped and continuously variable transmissions (CVTs). For both types the driver gets many advantages. In stepped transmissions, the smooth shifts can be optimized by the reduction of engine torque during gear shift, combined with the correctly matched oil pressure for the friction elements (clutches, brake bands). The reduction of shift shocks to a very low or even to an unnoticeable level has allowed the design of five-speed ATs where a slightly higher number of gear shifts occur.

With the CVT, one of the biggest obstacles to the potential reduction in fuel consumption by operating the engine at its optimal point is the power loss from the transmission's oil pump. Only with electronic control is it possible to achieve the required yield by matching the oil mass-stream and oil pressure for the pulleys to the actual working conditions.

To guarantee the overall economic solution for an electronically controlled transmission, either stepped or CVT, the availability of precision electrohydraulic actuators is imperative.

System Components

Transmission.   The greatest share of electronically controlled transmissions currently on the market consists of four- or five-speed units with a torque converter lockup clutch, commanded by the control unit. In a conventional pure hydraulic AT, the gear shifts are carried out by mechanical and hydraulic components. These are controlled by a centrifugal governor that detects the vehicle speed and a wire cable connected to the throttle plate lever. With an electronic shift point control, on the other hand, an electronic control unit detects and controls the relevant components.

Present electronically controlled ATs usually have an electronically commanded torque converter clutch, which can lock up the torque converter between the engine output and the transmission input. The torque converter clutch is activated under certain driving conditions by a solenoid controlled by the electronic TCU. Locking up the torque converter eliminates the slip of the converter, and the efficiency of the transmission system is increased. This results in an even lower fuel consumption for cars equipped with ATs.

*Neuffer, K., and Engelsdorf, K., "Transmission Control," in *Automotive Electronics Handbook*, 2nd ed., Ronald K. Jurgen (Ed.), McGraw-Hill, 1999, pp. 13.1–13.23.

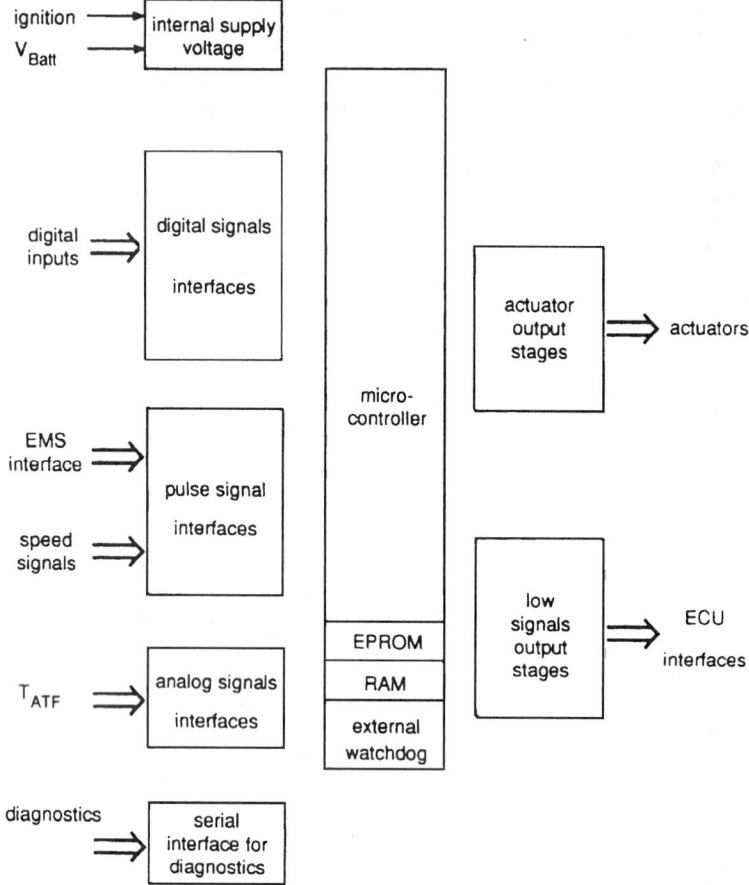

FIGURE 24.2.5   Overview of hardware parts.

Electronic Control Unit.   Another important component in electronic transmission control is the ECU, which is designed according to the requirements of the transmission and the car environments. The ECU can be divided into two main parts: the hardware and the corresponding software.

The hardware consists of the housing, the plug, the carrier for the electronic devices, and the devices themselves. The housing, according to the requirements, is available as an unsealed design for applications inside the passenger compartment or within the luggage compartment. It is also possible to have sealed variants for mounting conditions inside the engine compartment or at the bulkhead. An overview of hardware parts is shown in Fig. 24.2.5.

The software within the electronic transmission control system is gaining increasing importance because of the increasing number of functions, which, in turn, requires increasing software volume. The software for the control unit can be divided into two parts: the program and the data. The program structure is defined by the functions. The data are specific for the relevant program parts and have to be fixed during the calibration stage. The most difficult software requirements result from the real-time conditions coming from the transmission design. This is also the main criterion for the selection of the microcontroller (Fig. 24.2.6).

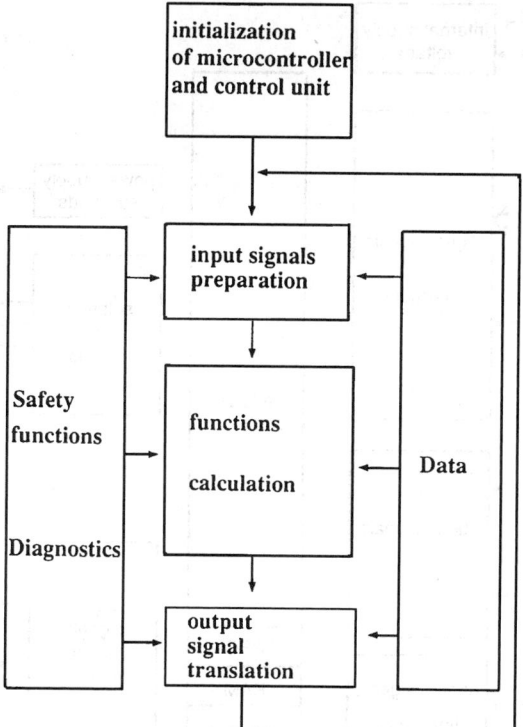

FIGURE 24.2.6   Software structure overview.

Actuators.   Electrohydraulic actuators are important components of the electronic transmission control systems. Continuously operating actuators are used to modulate pressure, while switching actuators functions as supply and discharge valves for shift-point control.

System Functions

Functions can be designated as system functions if the individual components of the total electronic transmission control system cooperate efficiently to provide a desired behavior of the transmission and the vehicle.

Basic Functions.   The basic functions of the transmission control are the shift point control, the lockup control, engine torque control during shifting, related safety functions, and diagnostic functions for vehicle service. The pressure control in transmission systems with electrical operating possibilities for the pressure during and outside shifting can also be considered as a basic function. Figure 24.2.7 shows the necessary inputs and outputs as well as the block diagram of an electronic TCU suitable for the basic functions.

Shift Point Control.   The basic shift point control uses shift maps that are defined in data in the unit memory. These shift maps are selectable over a wide range. The shift point limitations are made, on the one hand, by the highest admissible engine speed for each application and, on the other hand, by the lowest engine speed that is practical for driving comfort and noise emission. The inputs of the shift point determination

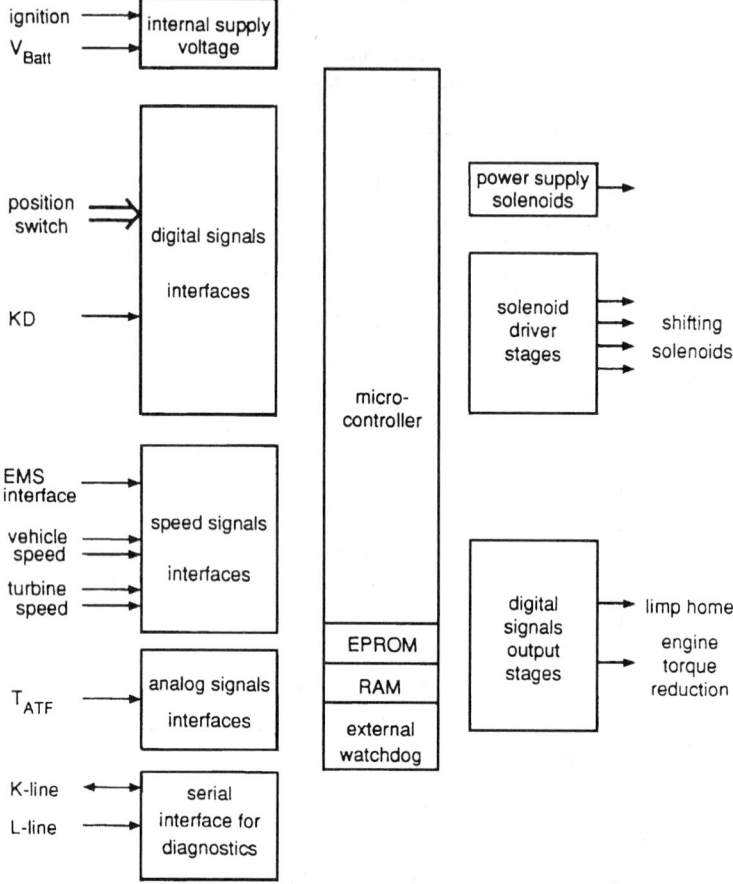

FIGURE 24.2.7 Structure of a basic transmission electronic control unit.

are the throttle position, the accelerator pedal position, and the vehicle speed (determined by the transmission output speed).

Lockup Control/Torque Converter Clutch. The torque converter clutch connects both functional components of the hydraulic converter, the pump and the turbine. The lockup of the clutch reduces the power losses coming from the torque converter slip. To increase the efficiency of the lockup, it is necessary to close the clutch as often as possible but the activation of the lockup is a compromise between low fuel consumption and high driving comfort.

Engine Torque Control During Shifting. The engine torque control requires an interface to an electronic engine management system. The target of the engine torque control, torque reduction during shifting, is to support the synchronization of the transmission and to prevent shift shocks.

Pressure Control. The timing and absolute values of the pressure, which is responsible for the torque translation of the friction elements is aside from the engine torque reduction, the most important influence to shift comfort. The electronic TCU offers additional possibilities for better function than a conventional hydraulic system. The pressure values during and outside shifting can be calculated by different algorithms or can be determined by characteristic maps. The inputs for a pressure calculation are engine torque, transmission input speed, turbine torque, throttle position, and so on.

Safety and Diagnostic Functions.   The functions, which are usually known as diagnostic functions of the electronic TCU, can be divided into real safety functions to prevent critical driving conditions and diagnostic functions that affect an increasing availability of the car and a better failure detection for servicing.

Improvement of Shift Control

Shift Point Control.   This basic function can be improved significantly by adding a software function, the so-called adaptive shift point control. This function requires only signals that are available in an electronic TCU with basic functions. The adaptive shift point control is able to prevent an often-criticized attribute, the tendency for shift hunting especially when hill climbing and under heavy load conditions.

Lockup Control.   There are some additional functions that can improve considerably the shift comfort of the lockup. In a first step, it is possible to replace the on/off control of the lockup actuator by a pulse control during opening and closing. This can be achieved using conventional hardware only by a software modification. In a further step, the on/off solenoid is replaced by a pressure regulator or a PWM solenoid.

By coordinating intelligent control strategies and corresponding output stages within the electronic TCU, a considerable improvement of the shift behavior of the lockup results.

Engine Torque Reduction During Gear Shifting.   By an improved interface design to the engine management system, it is possible to extend the engine torque reduction function. It is necessary to use a PWM signal with related fixed values or a bus interface. The engine torque reduction is controlled directly by the TCU.

Pressure Control.   The pressure control can be improved in a similar way as the shift point control with an adaptive software strategy. The required inputs for the adaptive pressure control are calculated from available signals in the transmission control. The main reasons for the implementation of the adaptive pressure control are the variations of the attributes of the transmission components such as clutch surfaces and oil quality as well as the changing engine output torque over the lifetime of the car.

Adaptation to Driver's Behavior and Traffic Situations

In certain driving conditions, some disadvantages of the conventional AT can be prevented by using self-learning strategies. The intention of the self-learning functions is to provide the appropriate shift characteristic suitable to the driver under all driving conditions. Additionally, the behavior of the car under special conditions can be improved by suitable functions. The core of the adaptive strategies is the driver's style detection, which can be detected by monitoring the accelerator pedal movements.

Future Developments

Independent of the type of automatic transmission, the shift or ratio control of the future will be part of a coordinated powertrain management, which, in turn, will be included in an overall control architecture encompassing all electronic systems of a vehicle. In the next years, development work on transmission control will concentrate on improving production costs, reliability, and size and weight, which are relevant to the product success in the market. A fundamental step toward these targets is higher integration of the components of the control system (i.e., combining mechanical, hydraulic, and electronic elements in mechatronic modules).

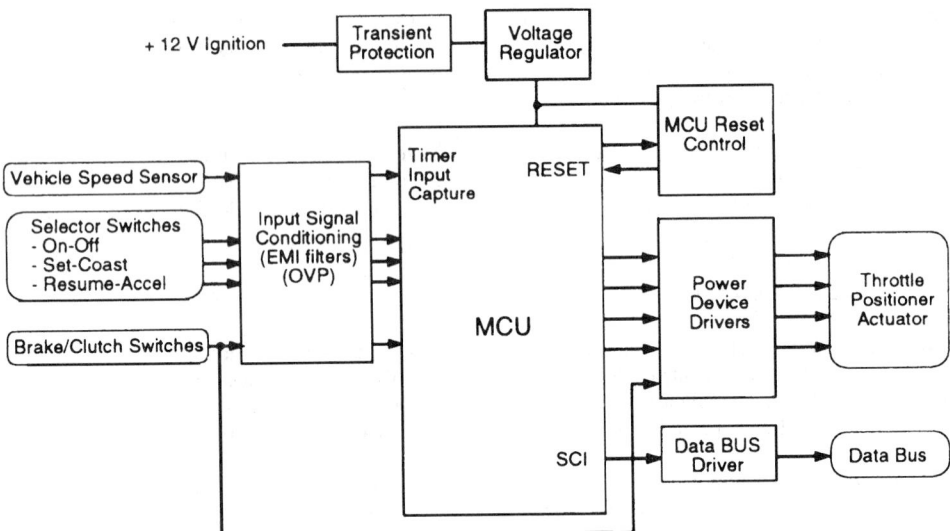

FIGURE 24.2.8   Cruise control system.

CRUISE CONTROL*

A vehicle speed control system can range from a simple throttle-latching device to a sophisticated digital controller that constantly maintains a set speed under varying driving conditions. The next generation of electronic speed control systems will probably still use a separate module (black box), the same as present-day systems, but will share data from the engine, ABS, and transmission control systems. Futuristic cruise control systems that include radar sensors to measure the rate of closure to other vehicles and adjust the speed to maintain a constant distance are possible but need significant cost reductions for widespread private vehicle use.

The objective of an automatic vehicle cruise control is to sustain a steady speed under varying road conditions, thus allowing the vehicle operator to relax from constant foot throttle manipulation. In some cases, the cruise control system may actually improve the vehicle's fuel efficiency value by limiting throttle excursions to small steps. By using the power and speed of a microcontroller device and fuzzy logic software design, an excellent cruise control system can be designed.

The cruise control system is a close-loop speed control as shown in Fig. 24.2.8. The key input signals are the driver's speed setpoint and the vehicle's actual speed. Other important inputs are the faster-accel/slower-coast driver adjustments, resume, on/off, brake switch, and engine control messages. The key output signals are the throttle control servo actuator values. Additional output signals include cruise ON and service indicators, plus messages to the engine and/or transmission control system and possibly data for diagnostics.

The ideal cruise system features would include the following specifications:

- Speed performance: ±0.5 percent m/h control at less than 5 percent grade, and ±1 m/h control or vehicle limit over 5 percent grade.

- Reliability: Circuit designed to withstand overvoltage transients, reverse voltages, and power dissipation of components kept to minimum.

- Application options: By changing EEPROM via a simple serial data interface or over the multiplexing network, the cruise software can be upgraded and optimized for specific vehicle types. These provisions allow for various sensors, servos, and speed ranges.

*Valentine, R., "Cruise Control," in *Automotive Electronics Handbook*, 2nd ed., Ronald K. Jurgen (ed.), McGraw-Hill, 1999, pp. 14.1–14.10.

- Driver adaptability: The response time of the cruise control can be adjusted to match the driver's preferences within the constraints of the vehicle's performance.
- Favorable price-to-performance ratio: The use of integrated actuator drivers and a high-functionality MCU reduce component counts, increase reliability, and decrease the cruise control module's footprint.

  The MCU for cruise control applications requires high functionality. The MCU would include the following: a precise internal timebase for the speed measurement calculations, A/D inputs, PWM outputs, timer input capture, timer output compares, serial data port (MUX), internal watchdog, EEPROM, and low-power CMOS technology.

Insofar as cruise control software is concerned, the cruise error calculation algorithm can be designed around traditional math models such as PI or fuzzy logic. Fuzzy logic allows somewhat easier implementation of the speed error calculation because its design syntax uses simple linguistics. For example, *if* speed difference is negative and small, *then* increase throttle slightly. The output is then adjusted to slightly increase the throttle. The throttle position update rate is determined by another fuzzy program that looks for the driver's cruise performance request (slow, medium, or fast reaction), the application type (small, medium, or large engine size), and other cruise system factor preset parameters.

ADAPTIVE CRUISE CONTROL*

Adaptive cruise control (ACC) introduces a function to automobiles, which relieves the driver of a significant amount of the task of driving, in a comfortable manner. ACC differs from other vehicle control functions, especially in that the function is performed by several ECUs. While conventional control systems consist of a sensor and actuator environment around a central ECU, ACC adds functions to existing systems. A truly new component is the sensor for measuring the distance, relative speed, and lateral position of potential target vehicles, using laser optics or millimeter waves. This component often contains the logic for controlling vehicle movement. The latter is effected by commands to ECUs for engine and brake control electronics. Existing components are also used or adapted for operator use and display. The example in Fig 24.2.9 shows all basic components of an ACC system as a broad overview.

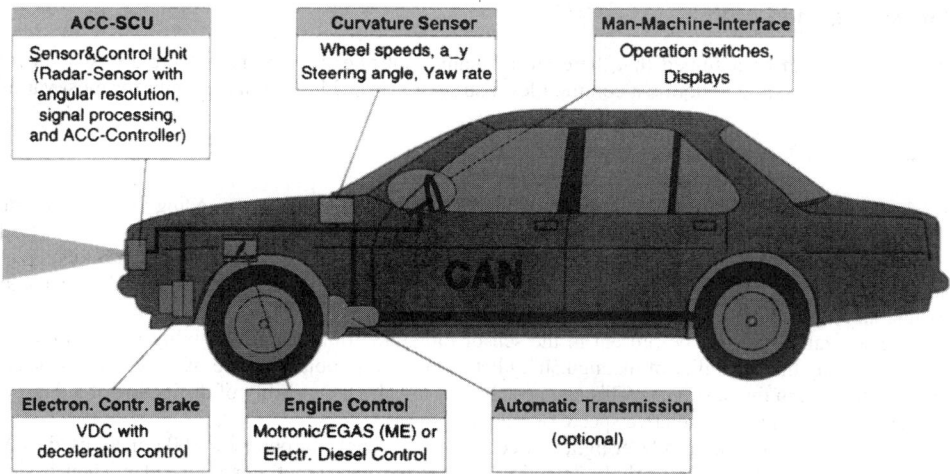

FIGURE 24.2.9 Basic components for typical adaptive cruise control.

---

*Winner, H., "Adaptive Cruise Control," in *Automotive Electronics Handbook*, Ronald K. Jurgen, (ed.), McGraw-Hill, 1999, pp. 30.1–30.30.

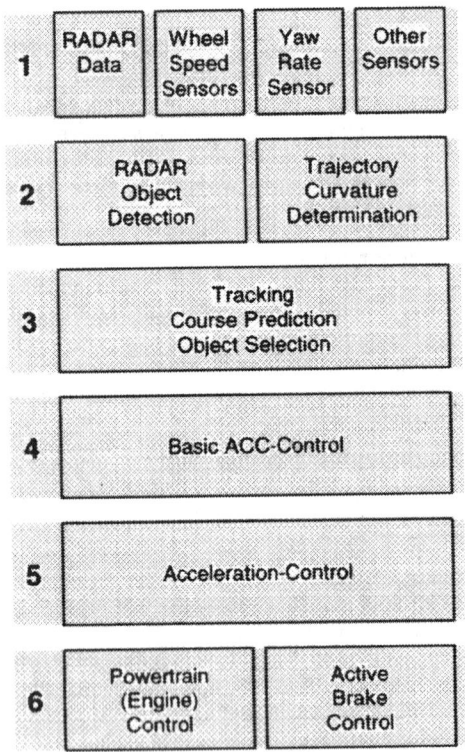

FIGURE 24.2.10 Levels of ACC signal processing and control.

The significant new function of ACC, compared to cruise control, consists of following a preceding vehicle that is being driven slower than the desired speed set by the driver of the ACC vehicle. If the preceding vehicle is being driven at a constant speed, the ACC vehicle follows at the same speed at an approximately constant distance. In general, this distance depends on the speed.

If the sensor's field of view in sharp curves is not sufficient to maintain detection with the target vehicle while following it at an adequate distance, then resumption of set-speed control can be delayed. An additional reason for reducing the acceleration and possibly even the current speed can be to achieve a comfortable reduction in the lateral acceleration while in the curve.

The ACC system, with its ACC sensor for monitoring the environment, and its access to vehicle dynamics variables and driver information systems, provides a good basis for simple collision warning functions.

Currently known ACC designs target use on freeways and other high-speed roads. Extension to urban traffic and to automatic traffic congestion following is desirable, but has not yet been satisfactorily achieved with the technology currently available.

An especially severe limitation of the ACC function is the discrimination of stationary targets. The result of this is that ACC only considers objects for which a minimum speed has been measured (typical values are 20 km/h or 20 percent of the speed of the ACC vehicle). This limitation is necessary not only because of the sensor's limited capabilities, but also because of the control definition.

Signal Processing and Control

Even though it can be assumed that there are as many controller variations as there are ACC development teams, it is still possible to define a common level structure (Fig. 24.2.10) that is valid for most variations.

ACC Sensor

The most elementary task of the ACC sensor is to measure the distance to preceding vehicles in a range that extends from about 2 to 150 m. A maximum allowable error of about 1 m or 5 percent is generally adequate.

The relative speed toward the target vehicle is of special importance for the ACC control cycle, since this variable is much more strongly weighted in the controller. It should be possible to detect all relevant speeds of possible target objects.

Since several targets can be present in the sensor range, multiple target capability is very important. This means especially the capability of distinguishing between relevant objects in the ACC vehicle's lane and irrelevant objects (e.g., in the next lane). This can be achieved by a high capability of distinction of at least one measurement variable (distance, relative speed, or lateral position).

Especially in Japan, infrared laser light sources with wavelengths around $\lambda = 800$ nm are used as ACC sensors. They are often called Lidar (light detection and ranging) or sometimes laser-radar. Their basic element (Fig. 24.2.11) is a laser diode, which is stimulated by a pulse controller to emit short light pulses. Another type of ACC sensor is a millimeter-wave radar sensor operating in the frequency band from 76 to 77 GHz. The use of the Doppler shift for direct determination of the relative speed is a prominent feature of millimeter-wave

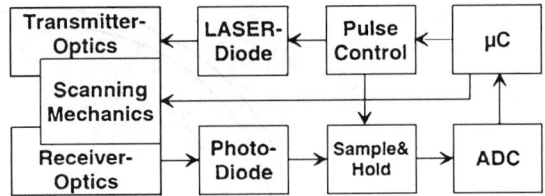

FIGURE 24.2.11   Lidar block diagram.

technology. It allows a high separation capability of objects with this variable and delivers the primary control variable for ACC at high quality.

ACC Controller

The basic structure of an ACC controller is shown in Fig. 24.2.12. The first step is to select the relevant preceding vehicle, which is done by comparing the object data with the geometry of the predicted course. Once the target vehicle is selected (several vehicles may be in the range of the predicted course), the required acceleration is calculated based on the distance and the relative speed.

BRAKING CONTROL*

The braking force generated at each wheel of a vehicle during a braking maneuver is a function of the normal force on the wheel and the coefficient of friction between the tire and the road, which is not a constant. It is a function of factors, most prominent being type of road surface and the relative longitudinal slip between the tire

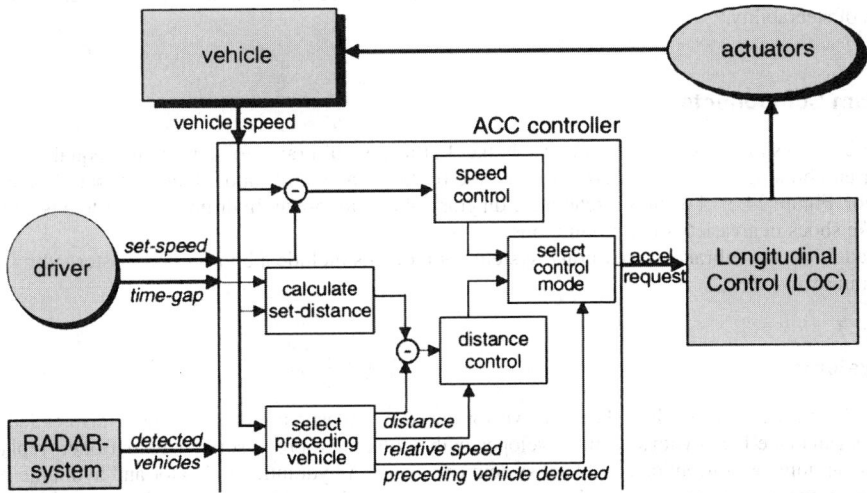

FIGURE 24.2.12   Basic structure of the ACC controller.

---

*Cage, J. L., "Braking Control," in *Automotive Electronics Handbook*, 2nd ed., Ronald K. Jurgen, (ed.), McGraw-Hill, 1999, pp. 15.1–15.17.

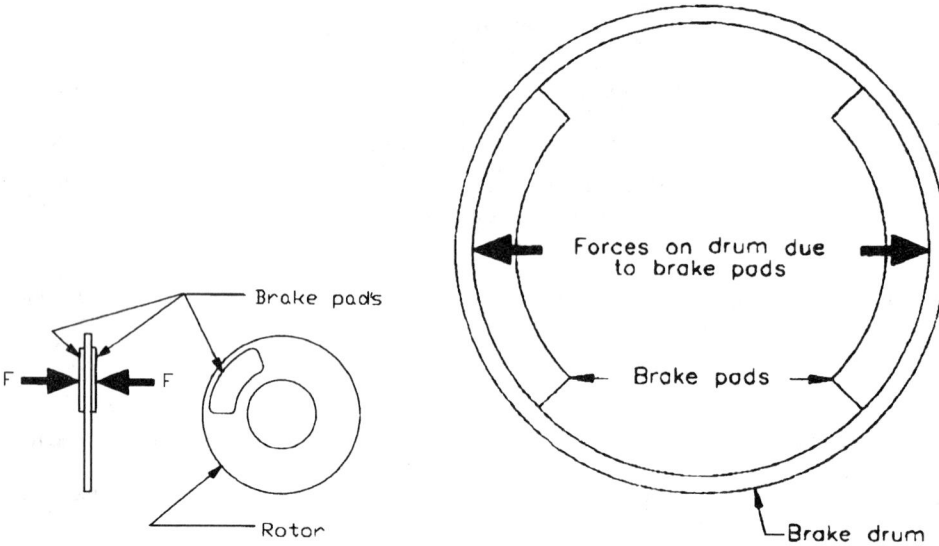

FIGURE 24.2.13 Disk brake schematic.   **FIGURE 24.2.14** Drum brake schematic.

and the road. Another characteristic of automotive tires important in braking is lateral force versus slip. Lateral force is the force keeping a tire from sliding in a direction normal to the direction of the vehicle. The lateral coefficient of friction drops off quickly once a wheel begins to slip longitudinally, as can happen during braking. Excessive wheel slip at the rear wheels of a vehicle and the resulting loss of lateral frictional force will contribute to instability as the rear of the vehicle tends to slide sideways with relatively small lateral forces on the vehicle. Excessive wheel slip and the resulting loss of lateral friction force on the front wheels of a vehicle will contribute to loss of steerability.

Brake System Components

Figure 24.2.13 shows a schematic of a disc brake. In this type of brake, force is applied equally to both sides of a rotor and braking action is achieved through the frictional action of inboard and outboard brake pads against the rotor. Figure 24.2.14 depicts a schematic diagram of a drum brake. In drum brakes, force is applied to a pair of brake shoes in a variety of configurations.

In addition to the brakes, other brake system components include a booster and master cylinder, and a proportioning valve.

Antilock Systems

Although antilock concepts have been known for decades, widespread use of antilock (also called antiskid and ABS) began in the 1980s with systems developed with digital microprocessors/microcontrollers replacing the earlier analog units. A conventional antilock system consists of a hydraulic modulator and hydraulic power source that may or may not be integrated with the system master cylinder and booster, wheel speed sensors, and an ECU. The fundamental function of an antilock system is to prohibit wheel lock by sensing impending wheel lock and taking action through the hydraulic modulator to reduce the brake pressure in the wheel sufficiently to bring the wheel speed back to the slip level range necessary for near-optimum braking performance.

Antilock components include wheel speed sensors, hydraulic modulators, electric motor/pump, and an ECU (Fig. 24.2.15).

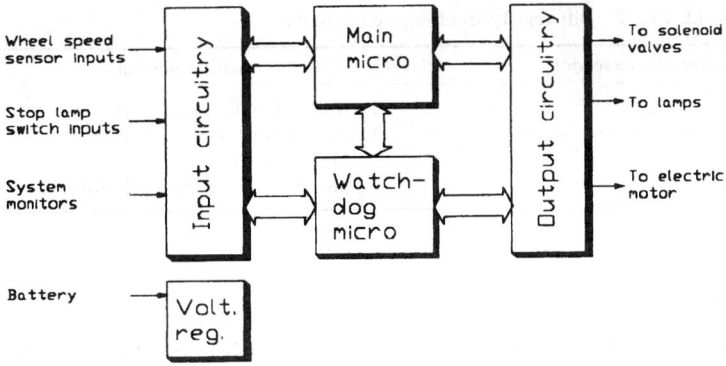

FIGURE 24.2.15   Electronic control unit block diagram.

Future Vehicle Braking Systems

A number of trends are developing relative to future vehicle braking systems. Some of the prominent ones are as follows:

- The expanding use of multiplexing
- Vehicle dynamics control during nonbraking as well as braking maneuvers
- Collision-avodiance systems
- Regenerative braking in electric vehicles
- Brake-assist systems
- Electrohydraulic brake by wire
- Electric actuators

TRACTION CONTROL*

Traction control systems or ASRs designed to prevent the drive wheels from spinning in response to application of excess throttle have been on the market since 1987. Vehicles with powerful engines are particularly susceptible to drive-wheel slip under acceleration from standstill and/or on low-traction road surfaces. The results include attenuated steering response on front-wheel-drive (FWD) vehicles, diminished vehicle stability on rear-wheel-drive (RWD) cars, and both on four-wheel-drive (4WD) cars—depending on their concept.

One technique for optimizing traction is to apply braking force to the spinning wheel. A second option is the application of fixed, variable, or controlled differential-slip limitation mechanisms. These provide fixed coupling to ensure equal slippage rates at the drive wheels, thereby allowing them to develop maximum accelerative force.

System Arrangements

The demand on the reaction time from the intervention up to the effect at the driven wheels depends on the drive concept. Different philosophies of the optimization of driving stability, traction, and comfort (and also the costs of the setting element) bring about the existence of different ASR system arrangements. Table 24.2.2 shows different types of engine interventions and their setting elements. Figure 24.2.16 shows an ASR system overview.

---

*Sauter, T., "Traction Control," in *Automotive Electronics Handbook*, 2nd ed., Ronald K. Jurgen (ed.), McGraw-Hill, 1999, pp. 16.1–16.19.

TABLE 24.2.2 Different Types of Engine Intervention

Type of intervention	Setting element
Airflow control	Electronic throttle control (ETC)
Fuel-injection suppression	Ignition and fuel-injection unit (MOTRONIC)
Spark retard	Ignition and fuel-injection unit (MOTRONIC)

Control Algorithm

The wheel velocities are determined by the speed sensors in the ABS/ASR control unit. If the drive wheels spin as a result of too high engine torque, the ASR determines a transferable moment, which corresponds to the friction coefficient and which is transferred by a data bus to the engine control unit. Meanwhile the spinning wheels are decelerated by the ASR hydraulic unit. By the reduction of engine torque, which can either be effected by closing the throttle valve or by injection suppression and spark retard, wheel slip is limited to such a degree that the lateral forces can be transferred again and the vehicle gets stabilized.

By the influence of the transmission control with vehicles with automatic transmissions, additional safety can be obtained, for example, by early shifting into a higher gear or by preventing downshifting when cornering. Figure 24.2.17 demonstrates the interaction of different ECUs and setting elements of a traction control system.

STABILITY CONTROL*

Driving a car at the physical limit of adhesion between the tires and the road is an extremely difficult task. Most drivers with average skills cannot handle those situations and will lose control of the vehicle. Several solutions for the control of vehicle maneuverability in physical-limit situations have been published.

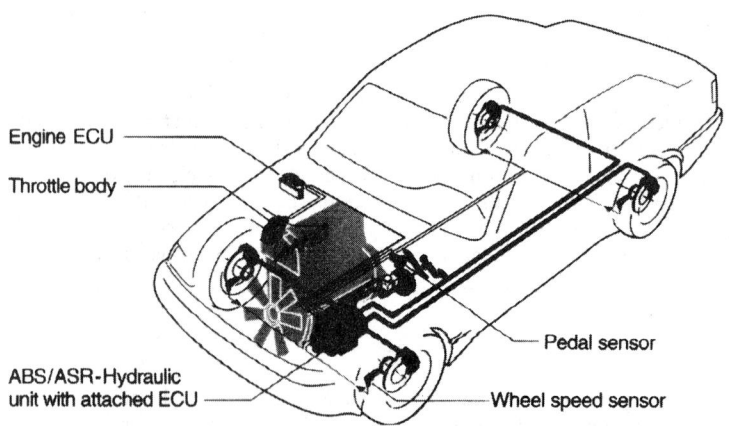

FIGURE 24.2.16 ASR system overview.

*Van Zanten, A., Erhardt, R. Landesfeind, K., and Pfaff, G., "Stability Control," in *Automotive Electronics Handbook*, 2nd ed., Ronald K. Jurgen (ed.), McGraw-Hill, 1999, pp. 17.1–17.33.

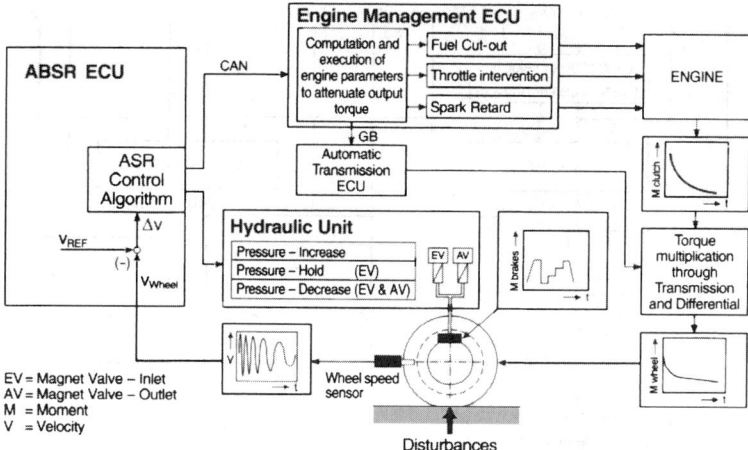

FIGURE 24.2.17 ASR control loop.

Physical Concept

Since the nominal trajectory desired by the driver is unknown, the driver's inputs are taken to obtain nominal state variables that describe the desired vehicle motion instead. These inputs are the steering wheel angle, the engine drive torque as derived from the accelerator pedal position, and the brake pressure. To determine which state variables describe the desired motion best, a special situation is taken (Fig. 24.2.18). In this illustration a vehicle is shown negotiating a turn after a step input in the steering angle. The lower curve shows the path the vehicle will follow if its lateral acceleration is smaller than the coefficient of friction of the road for the given tires. In this case the vehicle follows the nominal motion. If the road is slippery, with a coefficient of friction smaller than the nominal lateral acceleration, the vehicle will not follow the nominal value and the radius of the turn will become larger than that of the nominal motion.

One of the basic state variables that describe the lateral motion of the vehicle is its yaw rate. It therefore seems reasonable to design a control system that makes the yaw rate of the vehicle equal to the yaw rate of the nominal motion (yaw rate control). If this control is used on the slippery road, the lateral acceleration and the yaw rate will not correspond to each other as they do during the nominal motion. The slip angle of the vehicle increases rapidly as is shown by the upper curve in Fig. 24.2.18. Therefore both the yaw rate and the slip angle of the vehicle must be limited to values that correspond to the coefficient of friction of the road. For this reason in the Bosch vehicle dynamics control (VDC) system, both the yaw rate and the vehicle slip angle are taken as the nominal state variables and thus as the controlled variables. The result is shown by the curve in the middle of Fig. 24.2.18. It requires the installation of a yaw rate sensor and a lateral acceleration sensor.

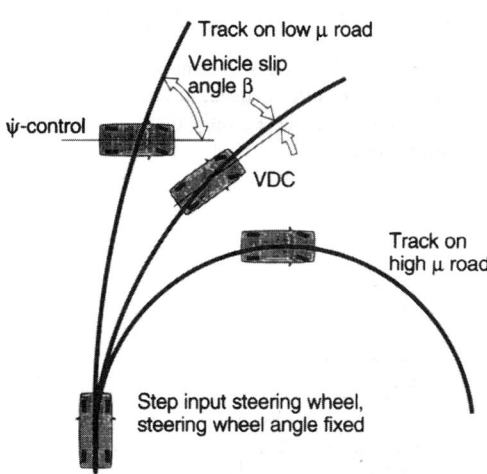

FIGURE 24.2.18 VDC versus yaw rate control.

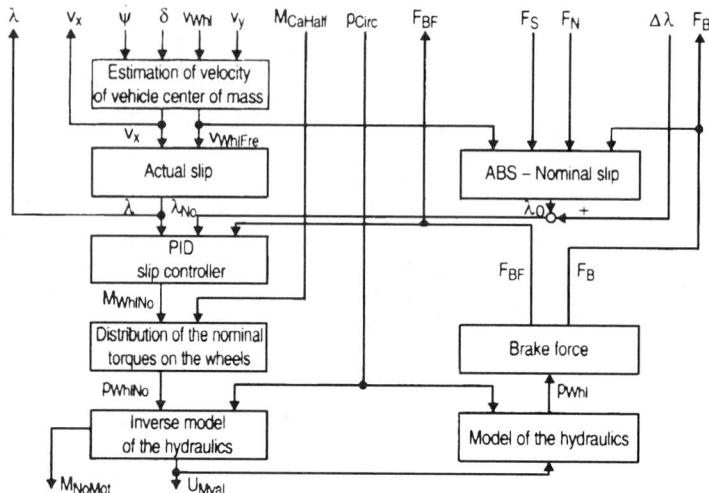

FIGURE 24.2.19 Brake slip controller.

Control Concept

From the driver's inputs, measured by a steering wheel angle sensor, a throttle position sensor, and a brake pressure sensor, the nominal behavior as described by the nominal values of the controlled variables must be determined. From the signal values of the wheel speed sensors, the yaw rate sensor, and the lateral acceleration sensor, the actual behavior of the vehicle as described by the actual values of its controlled variables is determined. The difference between the nominal and the actual behavior is then used as the set of actuating signals of the VDC.

The Slip Controller. The slip controller consists of two parts—the brake slip controller (Fig. 24.2.19) and the drive slip controller (Fig. 24.2.20). Depending on which slip controller is used, different nominal variables are passed by the VDC to the slip controller. The drive slip controller is used only for the slip control of the driven wheels during driving. Otherwise the brake slip controller is used.

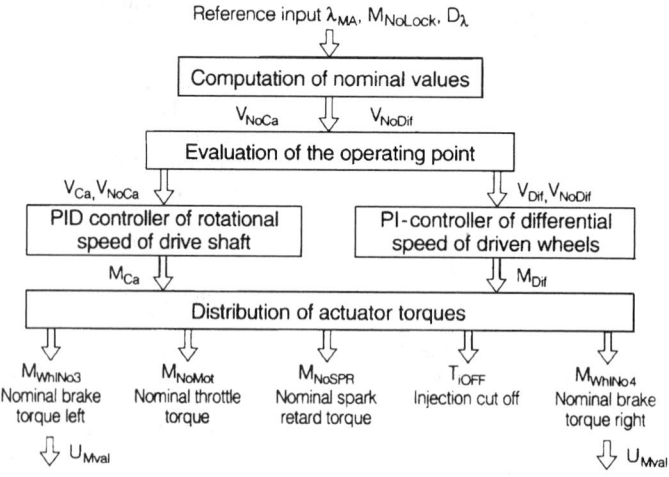

FIGURE 24.2.20 Drive slip controller.

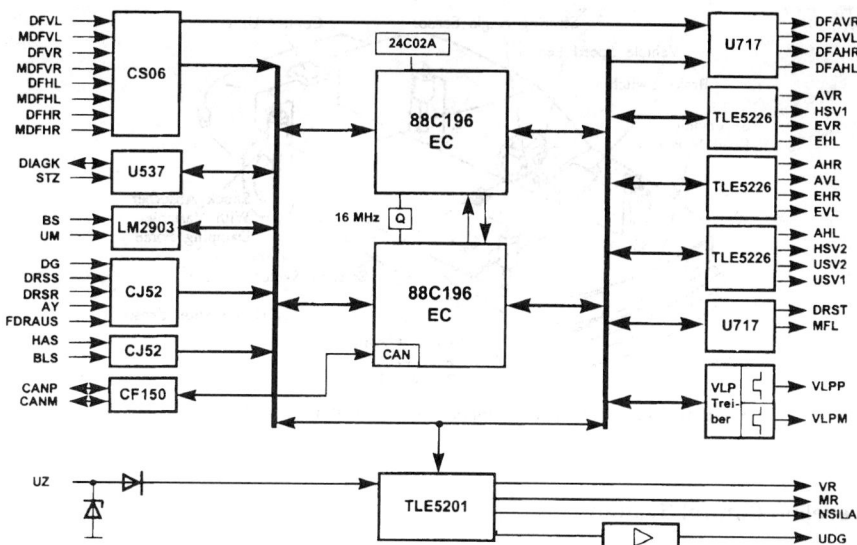

FIGURE 24.2.21   Block diagram of a typical ECU for VDC.

Electronic Control Units

There are two different types of ECUs. One is attached to the hydraulic unit and the other is separated from it. Figure 24.2.21 shows a typical separated ECU.

*SUSPENSION CONTROL**

The function of a suspension system in an automobile is to improve ride comfort and stability. An important consideration in suspension design is how to obtain both improved ride comfort and stability, since they are normally in conflict. Advances in electronic control technology, applied to the automobile, can resolve this conflict.

Shock Absorber Control System

There are three main parts of a damping control system: a damping control device (actuator), sensors, and software (control stategy). Optimum damping forces should be set for various running conditions in order to improve ride comfort and handling stability. One damping control system configuration is shown in Fig. 24.2.22. It uses five sensors to detect running conditions: vehicle speed sensor, steering angle sensor, acceleration and deceleration sensor, a brake sensor, and a supersonic sensor to detect road conditions. Control signals are sent to adjust the damping force of the variable shock absorbers to optimum values.

Hydropneumatic Suspension Control System

A hermetically sealed quantity of gas is used in the hydropneumatic suspension control system. The gas and hydraulic oil are separated by a rubber diaphragm (Fig. 24.2.23). The mechanical springs are replaced by gas. The shock absorber damping mechanism is achieved by the orifice fitted with valves.

---

*Yohsuke, A., "Suspension Control," in *Automotive Electronics Handbook*, 2nd ed., Ronald K. Jurgen (ed.), McGraw-Hill, 1999, pp. 18.1–18.19.

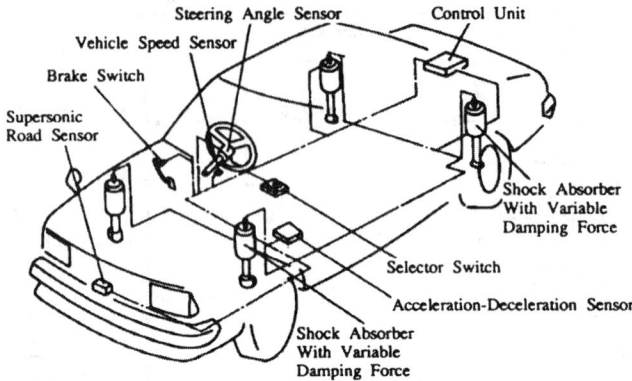

FIGURE 24.2.22 Principal components of a damping control system.

Electronic Leveling Control System

An electronic leveling control system (Fig. 24.2.24) permits a low spring rate to achieve good ride comfort independent of load conditions, an increase in vehicle body height on rough road surfaces, and a changing spring rate and damping force in accordance with driving conditions and road surfaces.

Active Suspension

An active suspension (Fig. 24.2.25) is one where energy is supplied constantly to the suspension and the force generated by that energy is continuously controlled. The suspension incorporates various types of sensors and a unit for processing their signals that generate forces that are a function of the signal outputs.

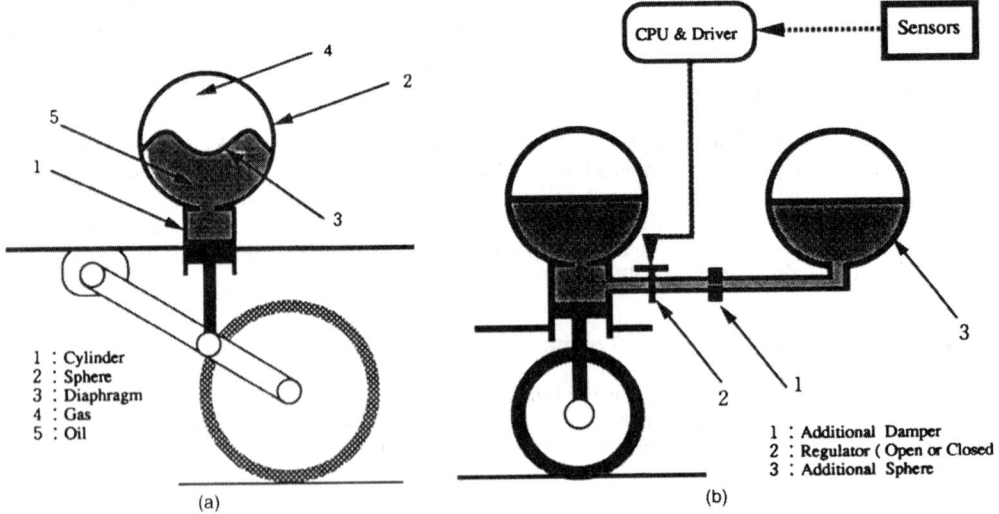

FIGURE 24.2.23 (a) Hydropneumatic suspension system; (b) Controllable hydropneumatic suspension system.

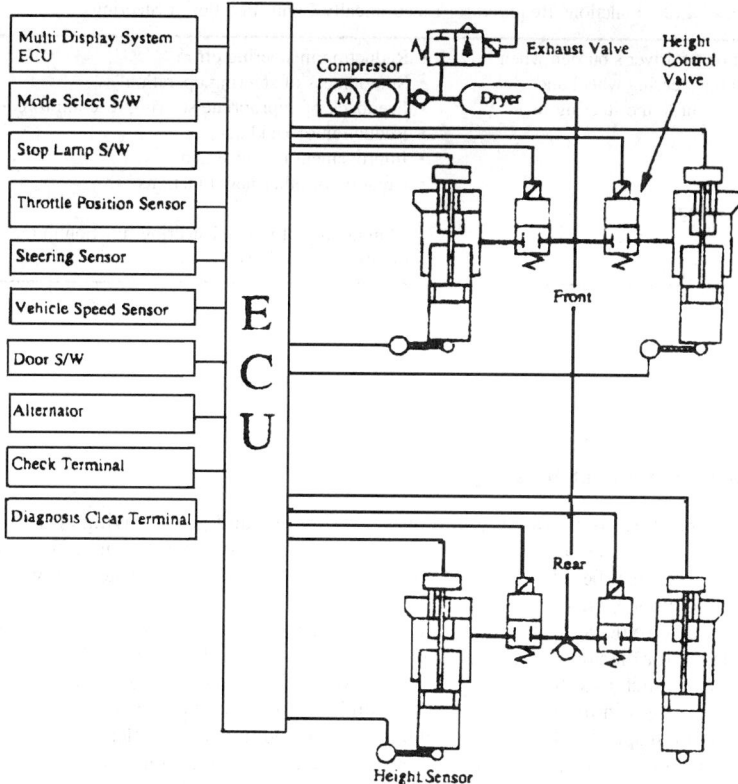

FIGURE 24.2.24 Principal components of an air-suspension system.

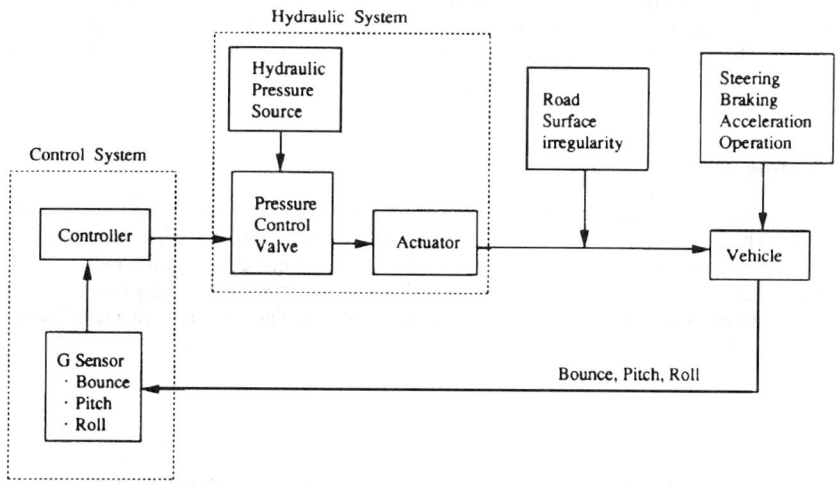

FIGURE 24.2.25 Hydraulic and control system for an active suspension.

TABLE 24.2.3   Functions Required for Electronically Controlled Power Steering

Reduction of driver's burden when turning the steering wheel and improvement in the steering feel	• Reduction in steering effort • Smoothness of steering operation • Feedback of appropriate steering reaction forces • Reduction of kickback[1] • Improvement in convergence[2] • Creation of other new functions
Power saving	
Failsafe	• Maintaining of manual steering function in the event of any malfunctions

STEERING CONTROL*

Electronically Controlled Power Steering

Electronically controlled power steering improves steering feel and power-saving effectiveness and increases steering performance. It does so with control mechanisms that reduce the steering effort. An electronic control system, for example, may be added to the hydraulic booster or the whole system may be composed of electronic and electric components.

The intent of electronic controls, initially, was to reduce the steering effort when driving at low speeds and to supply feedback for the appropriate steering reaction force when driving at high speeds. In order to achieve these goals, devices such as vehicle speed sensors were used to detect vehicle speed in order to make smooth and continuous changes in the steering assist rate under conditions ranging from steering maneuvers at zero speed to those at high speeds. However, as vehicles became equipped with electrohydraulic systems and fully electronic and electric systems, the emphasis for these systems started to include reduction in power requirements and higher performance.

The main functions required for electronically controlled power steering are listed in Table 24.2.3 and the various types of electronically controlled power systems are given in Table 24.2.4.

Electric power steering is a fully electric system, which reduces the amount of steering effort by directly applying the output from an electric motor to the steering system. This system consists of vehicle speed sensors, a steering sensor (torque, angular velocity), an ECU, a drive unit, and a motor (Fig. 24.2.26).

Hybrid systems use a flow control method in which the hydraulic power steering pump is driven by an electric motor. The steering effort is controlled by controlling the rotating speed of the pump (discharge flow).

Four-Wheel Steering Systems (4WS)

For vehicles with extremely long wheel bases and vehicles that need to be operated in narrow places, the concept of a four-wheel steering system is attractive. In such systems, the rear wheels are turned in the opposite direction to the steering direction of the front wheels in order to make the turning radius as small as possible and to improve the handling ability. Four-wheel steering systems that are currently being implemented in vehicles are classified according to their functions and mechanisms. The aims and characteristics of each system are briefly explained in Tables 24.2.5 and 24.2.6.

*Sato, Makoto, "Steering Control," in *Automotive Electronics Handbook*, 2nd ed., Ronald K. Jurgen (ed.), McGraw-Hill, 1999, pp. 19.1–19.33.

TABLE 24.2.4 Classification of Electronically Controlled Power Steering System

Basic structure	Control method	Control objects	Sensors				Actuator	Major effects	
			Vehicle speed	Steering torque	Angular velocity	Current		Steering force responsive to vehicle speed	Power saving
Electronically controlled hydraulic system	Flow	Flow supply to power cylinder	○			○	Solenoid	○	○
	Cylinder bypass	Effective actuation pressure given to cylinder	○			○	Solenoid	○	
	Valve characteristics	Pressure generated at control value	○			○	Solenoid	○	
	Hydraulic reaction force control	Pressure acting on the hydraulic reaction force mechanism	○			○	Solenoid	○	
Hybrid system	Flow	Flow supply to power cylinder	○	○	○	○	Motor	○	○
Full electric system	Current	Motor torque	○	○		○	Motor	○	○
	Voltage	Motor power	○	○	○	○	Motor	○	○

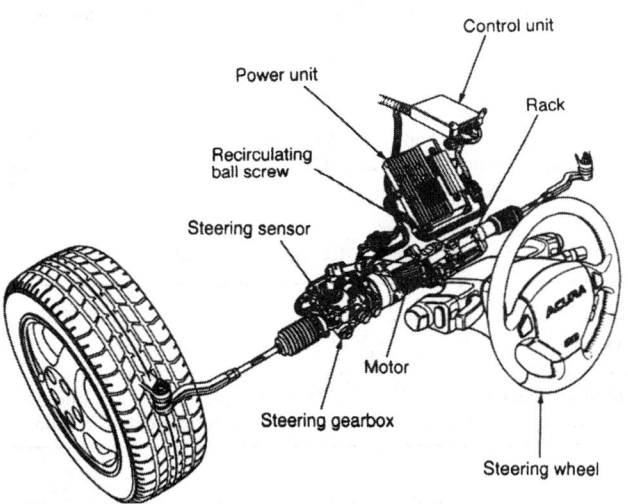

FIGURE 24.2.26 Structure of an electric power steering system (rack assist-type ball screw drive).

TABLE 24.2.5 Four-Wheel Steering Systems: Functions and Aims

Classification by functions	Aims
Small range of rear steer angle only controlled electronically	Improvement of steering response and vehicle stability in medium to high speed
Not only small range in medium to high speed but large range in low speed of rear steering angle are controlled electronically	In addition to the above, making the minimum turning radius small

OBJECT DETECTION, COLLISION WARNING, COLLISION AVOIDANCE*

Although object detection and collision-avoidance systems may still be regarded as being in their infancy, their perceived value in enhancing safety and reducing accidents is high. Two categories of systems are currently available or under development: passive collision-warning systems and active collision-avoidance systems. A passive system will detect a hazard and alert the driver to risks, whereas an active system will detect the hazard and then take preventive action to avoid a collision. Both types require object detection. The only difference between them is in how a collision-diverting event is actuated following object detection—by the driver or automatically.

Development engineers are proceeding cautiously with active-collision work. Any systems that take control of the brakes (and, in the future, steering) from the driver are potential sources of litigation, particularly in North America. Standards and federal regulations will emerge during the next few years that will cover such systems.

Active and Passive Systems

A passive collision-warning system (Fig. 24.2.27) includes a visual and/or audible warning signaled to the driver, but there is no intervention to avoid a collision. An active collision-warning system (Fig. 24.2.28) interacts with the powertrain, braking, and even the steering systems with the objective of sensing objects that present a collision risk with the host vehicle, then taking preventative measures to avoid an accident.

The most essential element in both passive and active systems is the object detection system. A key difference between the object detection systems used in active and passive systems is that the active system will require more accurate object recognition, so as to prevent collision-avoidance maneuvers against objects such as road signs.

Vehicular Systems

There are three types of warning systems: front, rear, and side. Frontal systems, both active and passive, operate on the same principles of object detection. There are different techniques used for obstacle detection.

TABLE 24.2.6 Four-Wheel Steering Systems: Mechanisms and Features

Classification by mechanism	Feature
Full mechanical system	Simple mechanism
Electronic-hydraulic system	High degree for control freedom (compact actuator)
Electronic-mechanical-hydraulic system	High degree of freedom (mechanism is not simple)
Full electric system (electronic-electric system)	High degree of control freedom simple mechanism

*Bannatyne, R., "Object Detection, Collision Warning, Collision Avoidance," in *Automotive Electronics Handbook*, 2nd ed., Ronald K. Jurgen (ed.), McGraw-Hill, 1999, pp. 29.3–29.22.

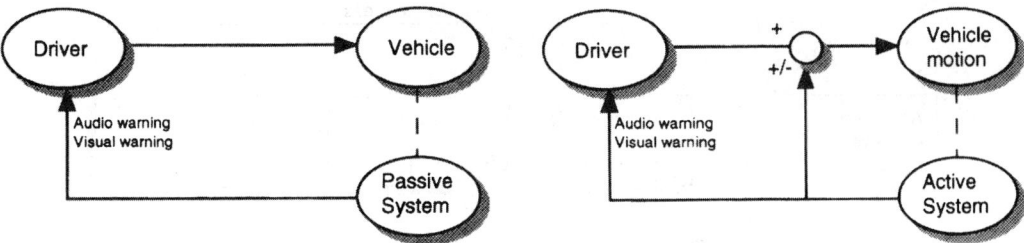

FIGURE 24.2.27 Passive collision-warning system.   **FIGURE 24.2.28** Active collision-warning system.

The main approaches are a scanning laser radar sensor, a frequency modulated constant wavelength (FMCW), or a video camera used in conjunction with an algorithm that will detect hazardous objects. This detection system is usually mounted at the front of the host vehicle to detect objects in the vehicle's forward path. Other techniques may involve a combination of different sensors, including backup sensors.

To be a hazardous obstacle, the target must lie in the corridor or the intended path for the host vehicle (Fig. 24.2.29).

Rear warning systems can use shorter-range, often nonscanning sensors to provide close-range sensing for parking assist capability, or scanning radar for more advanced sensing capability. Some systems use a combination of both types of sensors.

Side-warning systems use radar sensors to detect objects in the traditional blind spots that are often responsible for causing accidents. The sensors are mounted in the rear quarter area of the vehicle and detect objects in adjacent lanes.

An active collision-avoidance system interacts with other vehicle systems, as well as providing a warning (usually via the audio system and perhaps a head-up display). The most critical system with which it interacts is the braking system. A simplified algorithm that senses hazards and adjusts brake pressure accordingly is shown in Fig. 24.2.30.

Future Systems

A fully equipped, high-end system, such as that shown in Fig. 24.2.31, would support an advanced system such as the automated highway. The collision-avoidance electronic control unit is at the center of the diagram. The frontal vehicular detection system features two sensors: a 77-GHz FMCW and a camera system. Data from both of these sources are fused to map out a reliable picture of the potential hazards.

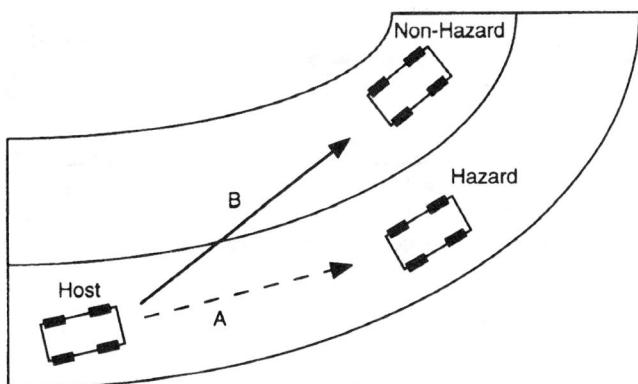

FIGURE 24.2.29 Trajectory corridor.

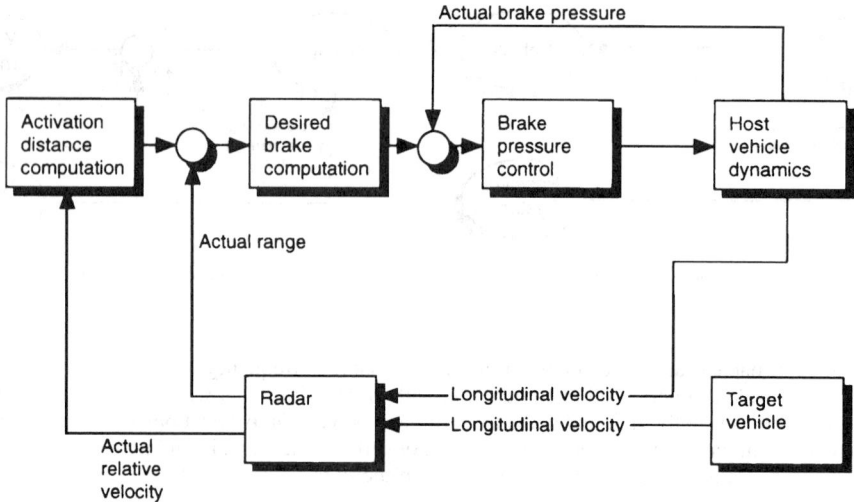

FIGURE 24.2.30 Object detection system with braking system interaction.

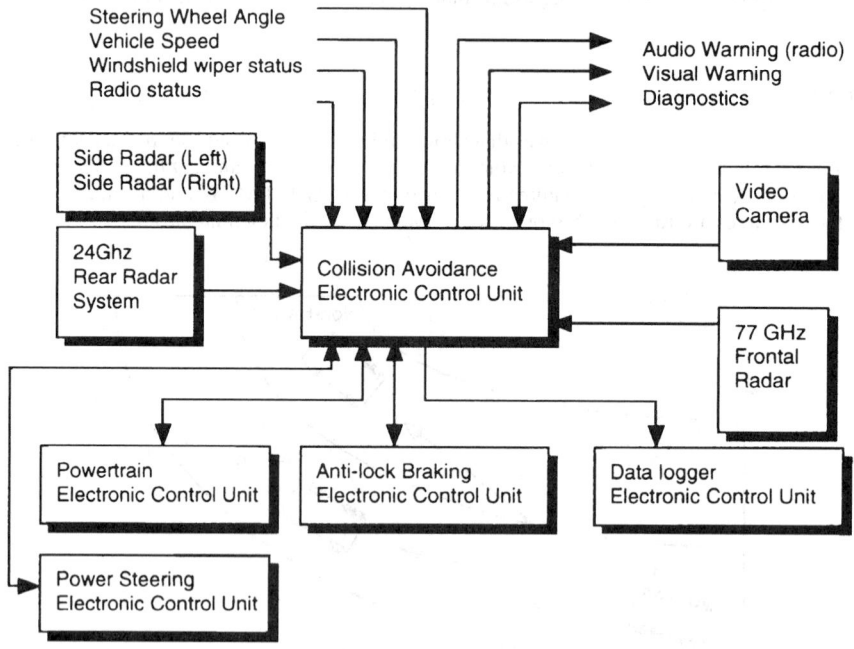

FIGURE 24.2.31 Complete "high-end" collision-avoidance system.

*NAVIGATION AIDS AND DRIVER INFORMATION SYSTEMS**

Navigation aids and driver information systems provide various combinations of communicated, stored, and derived information to aid drivers in effectively planning and executing trips. Navigation aids and driver information systems are a major subset of advanced traveler information systems, which, in turn, are a subset of intelligent transportation systems (ITS) as discussed in Chap. 25.3.

Automobile Navigation Technologies

Positioning technologies are fundamental requirements of both vehicle navigation systems and vehicle tracking systems. Almost all current vehicular navigation systems include a global positioning system (GPS) receiver, and most also include dead reckoning with map matching.

Map matching as well as route guidance must be supported by digital road maps. Another important supporting technology is mobile data communications for traffic and other traveler information.

Radiopositioning. Radiopositioning is based on processing special signals from one or more radio transmitters at known locations to determine the position of the receiving equipment. Although a number of radiopositioning technologies are potentially applicable, satellite-based GPS is by far the most popular. GPS, which is operated by the U.S. Department of Defense, includes 24 satellites spaced in orbits such that a receiver can determine its position by simultaneously analyzing the travel time of signals from at least four satellites.

Proximity beacons provide another form of radiopositioning, which is used in some vehicle navigation systems, particularly those that also use proximity beacons for communications purposes. Proximity beacons are installed at key intersections and other strategic roadside locations that communicate their location and/or other information to receivers in passing vehicles via very short-range radio, microwave, or infrared signals. The reception of a proximity beacon signal means that the receiving vehicle is within 50 m or so of the beacon, and provides an occasional basis for confirming vehicle position.

Dead Reckoning. Dead reckoning, the process of calculating location by integrating measured increments of distance and direction of travel relative to a known location, is used in virtually all vehicle navigation systems. Dead reckoning gives a vehicle's coordinates relative to earlier coordinates. Distance measurements are usually made with an electronic version of the odometer. Electronic odometers provide discrete signals from a rotating shaft or wheel, and a conversion factor is applied to obtain the incremental distance associated with each signal. Vehicle heading may be measured directly with a magnetic compass, or indirectly by keeping track of heading relative to an initial heading by accumulating incremental changes in heading.

Digital Road Maps. The two basic approaches to digitizing maps are matrix encoding and vector encoding. A matrix-encoded map is essentially a digitized image in which each image element or pixel, as determined by an *X-Y* grid with arbitrary spacing, is defined by digital data-giving characteristics such as shade or color. The vector-encoding approach applies mathematical modeling concepts to represent geometrical features such as roadways and boundaries in abstract form with a minimum of data. By considering each road or street as a series of straight lines and each intersection as a node, a map may be viewed as a set of interrelated nodes, lines, and enclosed areas.

Map Matching. Map matching is a type of artificial intelligence process used in virtually all vehicle navigation and route guidance systems that recognize a vehicle's location by matching the pattern of its apparent

---

*French, R. L., and Krakiwsky, E. J., "Navigation Aids and Driver Information Systems," in *Automotive Electronics Handbook*, Ronald K. Jurgen (ed.), McGraw-Hill, 1999, pp. 31.1–31.15.

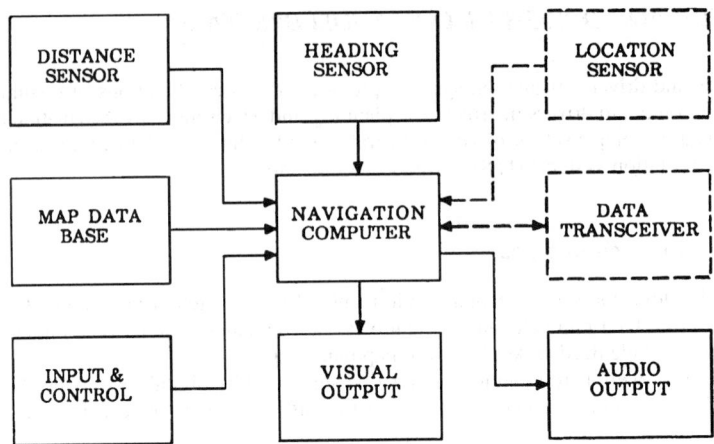

FIGURE 24.2.32   Typical components and subsystems of vehicle navigation system.

path (as approximated by dead reckoning and/or radiopositioning) with the road patterns of digital maps stored in computer memory. Most map-matching software may be classified as either semideterministic or probabilistic.

Navigation Systems

Figure 24.2.32 is a block diagram showing the major elements of a typical automobile navigation system. Distance and heading (or heading change) sensors are almost invariably included for dead-reckoning calculations, which, in combination with map matching, form the basic platform for keeping track of vehicle location. However, dead reckoning with map matching has the drawback of occasionally failing due to dead-reckoning anomalies, extensive travel off mapped roads, ferry crossings, and so forth.

The "location sensor" indicated by dashed lines in Fig. 24.2.32 is an optional means of providing absolute location to avoid occasional manual reinitialization when dead reckoning with map matching fails. Although proximity beacons serve to update vehicle location in a few systems, almost all state-of-the-art systems used GPS receivers instead.

CHAPTER 24.3
INTELLIGENT TRANSPORTATION SYSTEMS*

Ronald K. Jurgen

INTRODUCTION

Intelligent transportation systems (ITS) is an umbrella term that covers the application of a wide variety of communication, positioning, sensing, control, and other information-related technologies to improve the efficiency, safety, and environmental aspects of surface transportation. Major categories of ITS include advanced traffic management systems, advanced public transportation systems, advanced traveler information systems, advanced vehicle safety systems, and commercial vehicle operations.

The U.S. National ITS Architecture was developed under a three-year program funded by the U.S. Department of Transportation (DOT) that included a parallel consensus-building effort with industry and local governments. The architecture provides a consistent basis for establishing standards and assuring ITS interoperability throughout the country. To this end, ITS installations by state and local governments using federal funding are required by legislation to be consistent with the National Architecture.

Table 24.3.1 gives a comprehensive list of ITS user services that are candidates for an emerging ISO standard based in part on program information from many countries around the world. Although the level of emphasis for individual user services may vary greatly from country to country, generally similar ITS user services are defined in the United States, Europe, and Asia.

Table 24.3.2 gives a brief statement of the function of each of the 29 user services originally defined by the National ITS Program Plan developed jointly by DOT and ITS America.

SYSTEM ARCHITECTURE

The system architecture identifies necessary subsystems, defines the functions to be performed by each subsystem, and identifies the data that must flow among them, thus providing a consistent basis for system design. In particular, it defines the input, output, and functional requirements of several hundred individual processes required for implementing ITS. More important, the architecture defines the intricate interconnections required among the myriad subsystems in order to synergistically integrate ITS services ranging from advanced traffic management, which has already been implanted extensively, to future automated highway systems. It is important to understand,

---

*French, R. L., and Chen, K., "Intelligent Transportation Systems (ITS)," in *Automotive Electronics Handbook*, 2nd ed., Ronald K. Jurgen (ed.), McGraw-Hill, 1999, pp. 32.1–32.11.

TABLE 24.3.1 ITS User Services

Service category	Service no.	Service name
Traveler information (ATIS)	1	Pretrip information
	2	On-trip driver information
	3	On-trip public transport information
	4	Personal information services
	5	Route guidance and navigation
Traffic management (ATMS)	6	Transportation planning support
	7	Traffic control
	8	Incident management
	9	Demand management
	10	Policing/enforcing traffic regulations
	11	Infrastructure maintenance management
Vehicle (AVCS)	12	Vision enhancement
	13	Automated vehicle operation
	14	Longitudinal collision avoidance
	15	Lateral collision avoidance
	16	Safety readiness
	17	Precrash restraint deployment
Commercial vehicle (CVO)	18	Commercial vehicle preclearance
	19	Commercial vehicle administrative processes
	20	Automated roadside safety inspection
	21	Commercial vehicle onboard safety monitoring
	22	Commercial vehicle fleet management
Public transport (APTS)	23	Public transport management
	24	Demand-responsive transport management
	25	Shared transport management
Emergency (EM)	26	Emergency notification and personal security
	27	Emergency vehicle management
	28	Hazardous materials and incident notification
Electronic payment	29	Electronic financial transactions
Safety	30	Public travel security
	31	Safety enhancement for vulnerable road users
	32	Intelligent junctions

however, that the architecture is neither a design nor a set of physical specifications. Instead the architecture provides a basis for developing standards and protocols required for exchanging data among subsystems.

Logical Architecture

The logical architecture presents a functional interpretation of the ITS user services, which is divorced from likely implementations and physical interface requirements. In particular, it defines all functions (called process specifications) necessary to perform the user services and indicates the data flows required among these functions.

Physical Architecture

The physical architecture identifies the physical subsystems and the architecture flows between subsystems that implement the processes and support the data flows of the logical architecture. It assigns processes from the logical architecture to each of the subsystems and defines the data flows between the subsystems based on the data exchange implied by the process specifications. The physical architecture consists of the four top-level categories of subsystems shown in Fig. 24.3.1.

TABLE 24.3.2 Functions of ITS User Services

User service	Function
En route driver information	Provides driver advisories and in-vehicle signing for convenience and safety
Route guidance	Provides travelers with simple instructions on how to best reach their destinations
Traveler information services	Provides a business directory of "yellow pages" containing service information
Traffic control	Manages the movement of traffic on streets and highways
Incident management	Helps public and private organizations identify incidents quickly and implement a response to minimize their effects on traffic
Emissions testing and mitigation	Provides information for monitoring air quality and developing air quality improvement strategies
Demand management and operations	Supports policies and regulations designed to mitigate the environmental and social impacts of traffic congestion
Pretrip travel information	Provides information for selecting the best transportation mode, departure time, and route
Ride matching and reservation	Makes ride sharing easier and more convenient
Public transportation	Automates operations, planning, and management functions of public transit systems
En route transit information	Provides information to travelers using public transportation after they begin their trips
Personalized public transit	Provides flexibility-routed transit vehicles to offer more convenient customer service
Public travel security	Creates a secure environment for public transportation patrons and operators
Electronic payment services	Allows travelers to pay for transportation services electronically
Commercial vehicle electronic clearance	Facilitates domestic and international border clearance, minimizing stops
Automated roadside safety inspections	Facilitates roadside inspections
Onboard safety monitoring	Senses the safety status of a commercial vehicle, cargo, and driver
Commercial vehicle administration	Provides electronic purchasing of credentials and automated mileage and fuel reporting and auditing
Hazardous material incident response	Provides immediate description of hazardous materials to emergency responders
Freight mobility	Provides communication between drivers, dispatchers, and intermodal transportation providers
Emergency notification	Provides immediate notification of an incident and an immediate request for assistance
Emergency vehicle management	Reduces the time needed for emergency vehicles to respond to an incident
Longitudinal collision avoidance	Helps prevent head-on, rear-end, or backing collisions between vehicles, or between vehicles and objects or pedestrians
Lateral collision avoidance	Helps prevent collisions when vehicles leave their lane of travel
Intersection collision avoidance	Helps prevent collisions at intersections
Vision enhancement for crash avoidance	Improves the driver's ability to see the roadway and objects that are on or along the roadway
Safety readiness	Provides warnings about the conditions of the driver, the vehicle, and the roadway
Precrash restraint deployment	Anticipates an imminent collision and activates passenger safety systems before the collision occurs, or much earlier in the crash event than would otherwise be feasible
Automated highway systems	Provide a fully automated, hands-off operating environment

FUTURE DIRECTIONS

A new DOT initiative in 1998 is the Intelligent Vehicle Initiative (IVI). The objective of the IVI is to improve significantly the safety and efficiency of motor vehicle operations by reducing the probability of crashes. To accomplish this, the IVI will accelerate the development, availability, and use of driving-assistance and control-intervention systems to reduce deaths, injuries, property damage, and the societal loss that results from motor vehicle crashes. These systems include provisions for warning drivers, recommending control

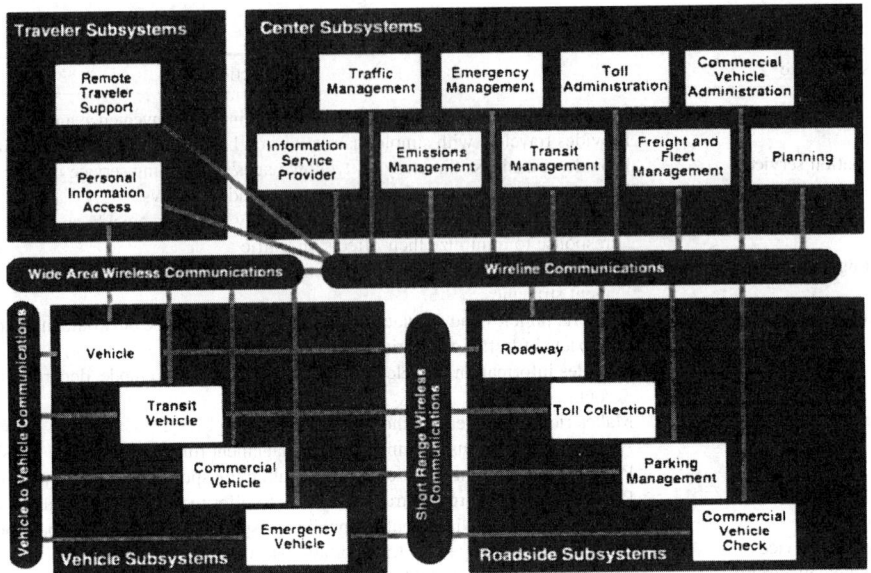

FIGURE 24.3.1 National ITS architecture for the United States.

actions, intervening with driver control, and introducing temporary or partial automated control of the vehicle in hazardous situations.

The improve safety category includes nine candidate IVI services: rear-end collision avoidance, road departure collision avoidance, lane-change and merge collision avoidance, intersection collision avoidance, railroad crossing collision avoidance, vision enhancement, location-specific alert and warning, automatic collision notification, and smart restraints and occupant protection systems.

The safety impacting category includes four services: navigation/routing, real-time traffic and traveler information, driver comfort and convenience, and vehicle stability warning and assistance.

SECTION 25

INSTRUMENTATION AND TEST SYSTEMS

The increasing importance of instrumentation in all phases of design, development, manufacture, deployment, and maintenance of electronic systems has prompted us to add this entirely new section to this edition of the handbook. It complements Sec. 15, Measurement Circuits, but deals with the instruments themselves and how they are used.

The first two chapters are devoted to instruments for measuring basic parameters: current, voltage, frequency, and time. Chapter 25.3 covers signal sources, and Chap. 25.4 logic analyzers.

Arguably the most powerful instrument, the oscilloscope, is treated in some detail in Chap. 25.5. During its 60-year history, it has been constantly improved in performance and sophistication.

Chapter 25.6 deals with reconfigurable instruments. One example would be configuring a portable ground support test system for military avionics from VXI modules configured in a dual-rack structure. Finally, the embedment of sensor- and computer-based instrumentation into complex systems can enable the implementation of self-repair concepts.

The editors wish to thank Stephen E. Grossman, a consultant to Agilent Technologies, for his invaluable assistance in organizing this section.   C.A.

In This Section:

CHAPTER 25.1
INSTRUMENTS FOR MEASURING FREQUENCY AND TIME*

Rex Chappell

INTRODUCTION

For measurements of frequency, time interval, phase, event counting, and many other related signal parameters, the ideal instrument to use is an electronic counter or its cousin, the frequency and time-interval analyzer. These instruments offer high precision and analysis for research and development applications, high throughput for manufacturing applications, and low cost and portability for service applications.

Electronic counters come in a variety of forms (Fig. 25.1.1).

FREQUENCY COUNTERS

The earliest *electronic counters* were used to count such things as atomic events. Before counters were invented, frequency measurement was accomplished with a frequency meter, a tuned device with low accuracy. *Frequency counters* were among the first instruments to digitally measure a signal parameter, yielding a more precise measurement. Today, they are still among the most precise instruments. Characterizing transmitters and receivers is the most common application for a frequency counter. The transmitter's frequency must be verified and calibrated to comply with government regulations. The frequency counter can measure the output frequency as well as key internal points, such as the local oscillator, to be sure that the radio meets specifications.

Other applications for the frequency counter can be found in the computer world where high-performance clocks are used in data communications, microprocessors, and displays. Lower-performance applications include measuring electromechanical events and switching power-supply frequencies.

UNIVERSAL COUNTERS

Electronic counters can offer more than just a frequency measurement. When it offers a few simple additional functions, such as period (the reciprocal of frequency), the instrument is sometimes known as a *multifunction*

---

*Adapted from Chapter 19, "Electronic Instrument Handbook," 3rd ed., Clyde Coombs Jr. (ed.), McGraw-Hill, 1999.

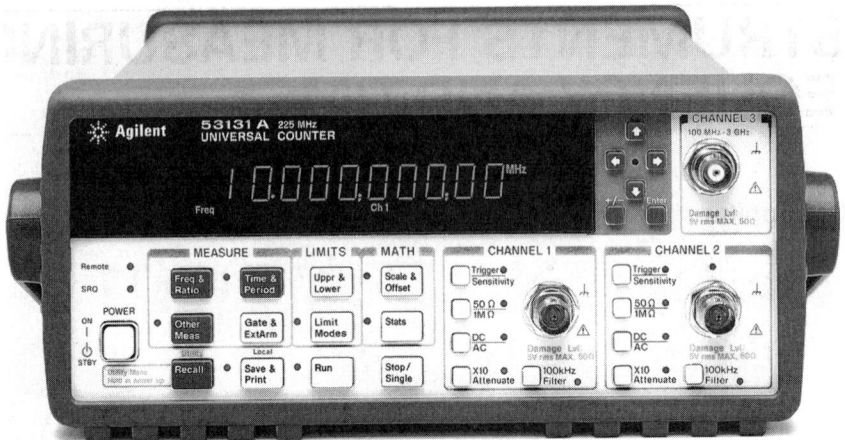

FIGURE 25.1.1 Frequency counters have become the most commonly employed instrument for measuring frequencies. The counter shown here is the Agilent Technologies 53131A with 10-digit resolution, 20-mV sensitivity, and frequency coverage up to 225 MHz.

counter. When two-channel functions, such as time interval, are provided, the instrument is usually called a *universal counter.* This name reflects this instrument's wide variety of applications. Several measurement functions provided by universal counters are shown in Fig. 25.1.2.

Time interval measures the elapsed time between a start signal and a stop signal. The start signal is usually fed into one channel (A), while the stop signal is fed into a second channel (B). The function is often called time interval A to B. The resolution of the measurement is usually 100 ns or better, depending on the *time interpolators* employed. Applications range from the measurement of propagation delays in logic circuits to the measurement of the speed of golf balls.

Variations of time interval that are particularly useful for digital circuit measurements are pulse width, rise time, and fall time. And, if the trigger points are known for rise and fall time, a calculation of the slew rate (volts-per-second) can be displayed. In all of these measurements the trigger levels are set automatically by the counter.

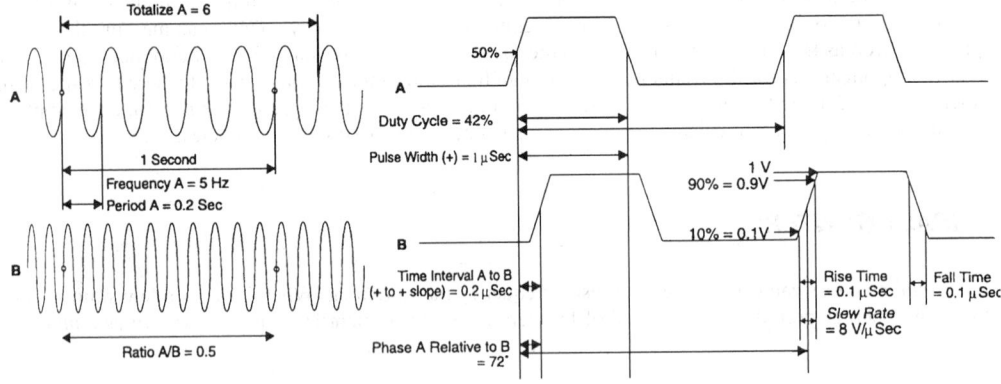

FIGURE 25.1.2 Several measurements that a universal counter can make. The A and B refer to the channels of the arriving signals. Frequency counters often have just one input channel, but universal counters need two for performing timing and signal comparison measurements.

Time interval average is a function that yields more resolution and can be used to filter out jitter in signals. *Totalize* is the simple counting of events. It is useful for counting electronic or physical events, or where a digitized answer is needed for an automated test application. The counter accumulates and displays event counts while the gate is open. In some instruments, this function is called *count*.

Frequency ratio A/B compares the ratio between two frequencies. It can be used to test the performance of a frequency doubler or a prescaler (frequency divider).

Phase A relative to B compares the phase delay between two signals with similar frequencies. The answer is usually displayed in degrees. Some instruments display only positive phase, but it is more convenient to allow a display range of ±180°, or 0° to 360°.

CW MICROWAVE COUNTERS

Continuous wave (CW) microwave counters are employed for frequency measurements in the microwave bands. Applications include calibrating the local oscillator and the transmitting frequencies in a microwave communication link. These counters can measure 20 GHz and higher. They offer relatively good resolution, down to 1 Hz, in a short measurement time. This makes them popular for manufacturing test applications. Their low cost and high accuracy, relative to a spectrum analyzer, make them popular for service applications. Some microwave counters also include the ability to measure power. This is a typical measurement parameter in service applications.

FREQUENCY AND TIME-INTERVAL ANALYZERS

Modern applications, particularly those driven by digital architectures, often have more complex measurement requirements than those satisfied by a counter. These measurements are often of the same basic parameters of frequency, time, and phase—but with the added element of being variable over time. For example, digital radios often modulate either frequency or phase. In *frequency shift keying* (FSK), one frequency represents a logic "0" and another frequency represents a "1." It is sometimes necessary to view frequency shifts in a frequency-versus-time graph. A class of instruments called *frequency and time-interval analyzers* (FTIA) can be used for this measurement. An FTIA provides a display that is similar to the output of a frequency discriminator displayed on an oscilloscope, but with much greater precision and accuracy (Fig. 25.1.3). This type of instrument is also referred to as a *modulation domain analyzer*. Other applications to which an FTIA can be applied include voltage-controlled oscillator (VCO) analysis, phase lock loop (PLL) analysis, and the tracking of frequency agile or hopping systems.

A variation of the frequency and time-interval analyzer simply measures time. These are called *time-interval analyzers* (TIA), and they tend to be focused on data and clock jitter applications. These are generally a subset of an FTIA, although some may offer special arming and analysis features tuned to particular applications, such as testing the read/write channel timing in hard-disk drives.

SPECIFICATIONS AND SIGNIFICANCE

Product data sheets and manuals often provide fairly detailed specifications. This section covers how to interpret some of the more important specifications.

Universal Counter Specifications

Many of the specifications found in a universal counter data sheet will also be found in other types of counters and analyzers.

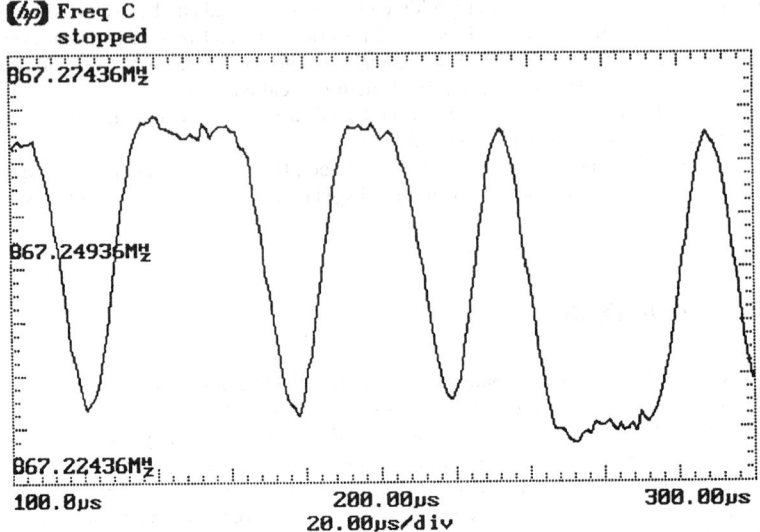

FIGURE 25.1.3 The display of a frequency-shift-keyed digital radio signal on a frequency and time-interval analyzer. Although it resembles an oscilloscope display, it is actually showing frequency vs. time, not voltage vs. time. The two frequencies, representing 0 and 1, are about 33 kHz apart. Each bit is about 15 μs long.

Sensitivity and Bandwidth. *Sensitivity* refers to how small a signal the instrument can measure. It is usually given in mV rms and peak-to-peak (p-p) values. In RF and microwave counters, it is given in dBm. Because of the element of noise and the broadband nature of a counter's front end, the sensitivity is rarely better than 10 mV rms or 30 mV p-p.

For frequency measurements, the sensitivity can be important if measuring a signal off the air with an antenna. Under most situations, however, there is sufficient signal strength to measure, and sensitivity is not an issue.

The counter's front-end *bandwidth* is not always the same as the counter's ability to measure frequency. If the bandwidth is lower than the top frequency range, the sensitivity at the higher frequencies will be reduced. This is not necessarily a problem if there is enough signal strength at the frequencies being measured. For timing measurements, however, bandwidth can affect accuracy. An input channel with a low front-end bandwidth can restrict the minimum pulse width of the signal it can measure, and, for very precise measurements, it can distort the signal. So, if precise timing measurements are being made, it is best to get a counter with as high a bandwidth as possible.

Resolution. Counters are commonly used because they can provide reasonably good resolution in a quick, affordable way. This makes resolution an important specification to understand. As will be seen, frequency measurement resolution is one of the specifications most dependent on the instrument's architecture.

The older *basic counter* architecture is still used in very low-cost counters and in some DVMs, oscilloscopes, and spectrum analyzers that include a frequency or time function. This architecture has the advantages of a simple low-cost design and rapid measurement, but has lower resolution and reduced measurement flexibility. For example, frequency resolution is limited, particularly for low frequencies, where resolution greater than 1 Hz is desired (assuming measurement times greater than 1 s are impractical).

The basic counter was invented well before digital logic was able to economically perform an arithmetic division. Division is needed because a general way of calculating frequency is to divide the number of periods (called *events*) counted by the time it took to count them (*events/time = frequency*). The basic counter accomplishes the division by restricting the denominator (time) to decade values. After it became practical to perform a full division, the instrument no longer had to be restricted to decade gate times. More importantly, the time base could be synchronized to the input signal and not the other way around. This means that the resolution of the measurement can be tied to the time base (usually time-interpolated) and not to the input frequency. This particularly improves the measurement of low frequencies (see Fig. 25.1.4). This class of instrument is called a *reciprocal counter*.

Microwave Counter Specifications

A few specifications are unique to microwave counters, and some of the most common are covered here.

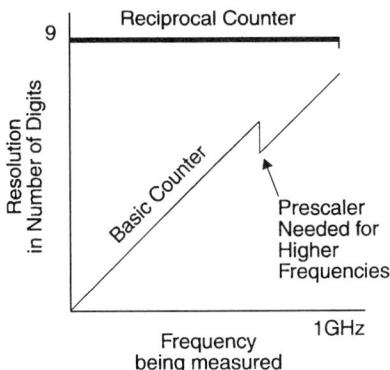

FIGURE 25.1.4 Frequency resolution of a reciprocal counter with 1-ns interpolators vs. a basic counter, given a 1-s gate time. The reciprocal counter has the advantage of a constant number of digits of display, no matter what the input frequency.

Acquisition Time. Because of the way that most microwave counters operate, using a harmonic sampler technique or a YIG preselector, a time lapse is required to determine where in the spectrum the signal is. This is called *acquisition time*. This is usually in the 50 to 250 ms range. YIG-based counters generally exhibit longer acquisition times. If there is a general idea of where the signal already is, some counters work in a manual acquisition mode. The counter then skips the acquisition phase of the measurement and makes the measurement after properly positioning the LO to the manually specified area, taking less time (< 20 ms).

FM Tolerance. This is largely determined by the IF bandwidth of the counter or the bandwidth of the preselector. If the input signal's FM *deviation* is such that the signal occasionally falls outside of these bandwidths, the measurement will be in error. Also, because the IF or preselector is seldom exactly centered on the signal, a margin is accounted for that narrows the FM tolerance of the instrument.

Counters that use prescaler technology do not have a problem with FM tolerance—or with acquisition time—for that matter. The trade-off is that some loss of resolution compared with a down-converter-type counter is realized, but this may not be a problem if advanced interpolators are used. Prescaling to 12.4 GHz is available in some counters today.

Power. Some microwave counters have the ability to measure the signal's power. This is useful for checking to see if a transmitter is within specifications or for finding faults in the cabling or connectors. Two main methods are used—true rms and detectors. The former produces much better accuracy, down to 0.1 dB, and the latter is less expensive, but has an accuracy of around 1.5 dB.

Frequency and Time Interval Analyzer Specifications

Although many specifications are similar to those found in a counter, two are unique and need some explanation.

Sample Rate. Also called *measurement rate*, this is how rapidly the instrument can make measurements. The speed needed depends on the phenomenon being measured. For example, to measure the 5-MHz jitter bandwidth of a clock circuit, the instrument must sample at 10 MHz (or faster). Sample rates in the 1-MHz range are good for moderate data-rate communications, VCO testing, and electromechanical measurements. Sample rates of 10 MHz (and higher) are better for high data-rate communications, data storage, and radar applications.

Memory Depth. Memory depth for storing the digitized data is the other important instrument parameter. Some applications only need a short amount of memory to take a look at transient signals, such as the step response of a VCO. Other applications need a very deep memory, where probably the most extreme is the surveillance of unknown signals.

INSTRUMENTS FOR MEASURING CURRENT, VOLTAGE, AND RESISTANCE

Scott Stever

INTRODUCTION

Voltage, and current, both ac and dc, and resistance are quantities commonly measured by electronic instruments. In the simplest case, each measurement type is performed by an individual instrument—a voltmeter measures voltage, an ammeter measures current, and an ohmmeter measures resistance. These instruments have many elements in common. The more classical, *electromechanical meters* are the easiest-to-use instrument for performing these measurements. A *multimeter* combines these instruments, and sometimes others, together into a single, general-purpose multifunction instrument.

Categories of Meters

There are two primary types of meters—*general purpose* and *specialty*. General-purpose meters measure several types of electrical parameters such as voltage, resistance, and current. A *digital multimeter* (DMM) is an example of a general-purpose meter (Fig. 25.2.1). Specialty meters are generally optimized to measure a single parameter very well, emphasizing either measurement accuracy, bandwidth, or sensitivity. Listed in Table 25.2.1 are various types of meters and their measuring capabilities.

The general-purpose multimeter is a flexible cost-effective solution. It is most suitable for most common measurements. Although DMMs can achieve performance rivaling the range and sensitivity of specialty meters while delivering superior flexibility and value, the presence of many display digits on a digital meter does not automatically mean that the meter has high accuracy. Meters often display significantly more digits of resolution than their accuracy specifications support. This can be very misleading to the uninformed user.

General Instrument Block Diagram

The function of a meter is to convert an analog signal into a human- or machine-readable equivalent. Analog signals may be quantities such as a voltage or current, ac or dc, or a resistance. Shown in Fig. 25.2.2 is a block diagram of a typical signal-conditioning process used with meters.

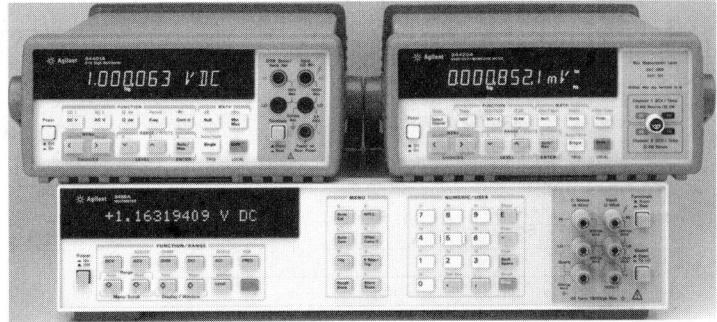

FIGURE 25.2.1 Multimeters are obtainable with varying degrees of resolutions and accuracies. Shown here are the Agilent Technologies 34401A, 34420A, and 3458A Digital Multimeters. They provide resolutions of $6^{1}/_{2}$, $7^{1}/_{2}$, and $8^{1}/_{2}$ digits, respectively, along with corresponding increases in accuracy on both their voltage and resistance measurement ranges.

Signal Conditioning, Ranging, and Amplification

The input signal must first pass through some type of signal conditioner that typically comprises switching, ranging, and amplification circuits, as shown in Fig. 25.2.2. If the input signal to be measured is a dc voltage, the signal conditioner may be composed of an attenuator for the higher voltage ranges and a dc amplifier for the lower ranges, whereas if the signal is an ac voltage, a converter is employed to change the ac signal to its equivalent dc value. Resistance measurements are performed by supplying a known dc current to an unknown resistance, thereby converting the unknown resistance value to an easily measurable dc voltage. In nearly all cases, the input signal switching and ranging circuits, along with the amplifier circuits, convert the unknown quantity to a dc voltage that falls within the measuring range of the analog-to-digital converter (ADC).

TABLE 25.2.1 Meters—Types and Features

Type of meter	Multi-function	Measuring range	Frequency range	Speed, max readings/second	Best accuracy	Digits
General Purpose						
Handheld DMM	Y	10 μV–1000 V; 1 nA–10 A; 10 mΩ–50 MΩ	20 Hz–20 kHz	2	0.1%	$3^{1}/_{2}$–$4^{1}/_{2}$
Bench DMM	Y	10 μV–1000 V; 1 nA–10 A; 10 mΩ–50 MΩ	20 Hz–100 kHz	10	0.01%	$3^{1}/_{2}$–$4^{1}/_{2}$
System DMM	Y	10 nV–1000 V; 1 pA–1 A; 10 μΩ–1 GΩ	1 Hz–10 MHz	50–100,000	0.0001%	$4^{1}/_{2}$–$8^{1}/_{2}$
Specialty						
ac Voltmeter	N	100 μV–300 V	20 Hz–20 MHz	1–10	0.1%	$3^{1}/_{2}$–$4^{1}/_{2}$
Nanovoltmeter	N	1 nV–100 V		1–100	0.005%	$3^{1}/_{2}$–$7^{1}/_{2}$
Picoammeter	N	10 fA–10 mA		1–100	0.05%	$3^{1}/_{2}$–$5^{1}/_{2}$
Electrometer	pA, high Ω	1 Ω–100 MΩ				
Microohmmeter	N			1–100	0.05%	$3^{1}/_{2}$–$4^{1}/_{2}$
High resistance	N	>10 TΩ		1–10	0.05%	$3^{1}/_{2}$–$4^{1}/_{2}$

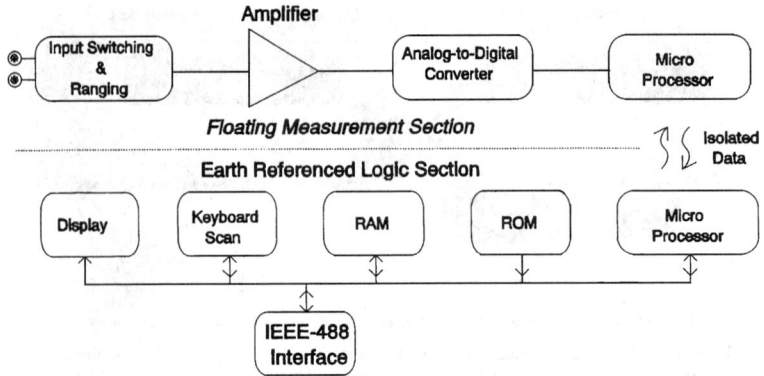

FIGURE 25.2.2   Generalized block diagram of most modern meters.

Analog-to-Digital Conversion

The role of the ADC is to transform a prescaled dc voltage into digits. For example, the ADC for a $6\frac{1}{2}$ digit resolution (21-bit) instrument is capable of producing over 2.4-million unique reading values. You can think of this as a bar chart with 2.4 million vertical bars with each bar increasing in size from the previous bar by an identical amount. Converting the essentially infinite resolution of the analog input signal to a single bar in our chart is the sole function of the ADC. However, the continuum of analog input values is partitioned— quantized—into 2.4-million discrete values in our example.

The ADC used in a meter governs some of its most basic characteristics. These include its measurement resolution, its speed, and in some cases, its ability to reject spurious noise. The various methods used for analog-to-digital conversion can be divided into two groups—*integrating* and *nonintegrating*. Integrating techniques measure the average input value over a relatively long interval, while nonintegrating techniques sample the instantaneous value of the input—plus noise—during a very short interval.

ADCs are designed strictly for dc voltage inputs. They are single-range devices with some exhibiting a 3-V full-scale input, while others have a 12-V full-scale input. For this reason, the input switching and ranging circuits must attenuate higher voltages and amplify lower voltages to enable the meter to provide a selection of ranges.

Managing the Flow of Information

The first three blocks in Fig. 25.2.2 combine to produce a meter's overall analog performance characteristics— measuring functions, ranges, sensitivity, reading resolution, and reading speed. The two microprocessor blocks manage the flow of information within the instrument, ensuring that the various subsystems are properly configured and that internal computations are performed in a systematic and repeatable manner. Convenience features such as automatic range selection are managed by these microprocessors. Electrical isolation is provided between the earth-referenced outside world and sensitive measuring circuits. The earth-referenced microprocessor also acts as a communications interpreter, receiving keyboard and programming instructions and managing the outward flow of data to the display, or the IEEE-488 computer interface.

DC Voltage Measurement Techniques

Signal conditioning and analog-to-digital conversion circuits have the greatest influence on the characteristics of a dc meter. The ADC measures over only a single range of dc voltage and it usually exhibits a relatively low input resistance. To configure a useful dc meter, a *front end* is required to condition the input before the analog-to-digital conversion. Signal conditioning increases the input resistance, amplifies small signals, and attenuates large signals to produce a selection of measuring ranges.

Signal Conditioning for DC Measurements

Input signal conditioning for dc voltage measurements includes both amplification and attenuation. Shown in Fig. 25.2.3 is a typical configuration for a dc input switching and ranging section of a meter. The input signal is applied directly to the amplifier input through switches K_1 and S_1 for lower voltage inputs—generally those less than 12 V dc. For higher voltages, the input signal is connected through relay K_2 to a precision 100:1 divider network formed by resistors R_4 and R_5. The low voltage output of the divider is switched to the amplifier input through switch S_2.

The gain of amplifier A_1 is set to scale the input voltage to the full-scale range of the ADC, generally 0 ± 12 V dc. If the nominal full-scale input to the ADC is 10 V dc, the dc input attenuator and amplifier would be configured to amplify the 100-mV range by 100 times and to amplify the 1-V range by 10 times. The input amplifier would be configured for unity gain, $X1$, for the 10-V measuring range. For the upper ranges, the input voltage is first divided by 100, and then gain is applied to scale the input back to 10 V for the ADC—inside the meter, for instance, 100 V dc is reduced to 1 V dc for the amplifier, whereas 1000 V dc is divided down to become 10 V dc.

For the lower voltage measuring ranges, the meter's input resistance is essentially that of amplifier A_1. The input amplifier usually employs a low bias current—typically less than 50 pA. It is often an FET input stage providing an input resistance greater than 10 GΩ. The meter's input resistance is determined by the total resistance of the 100:1 divider for the upper voltage ranges. Most meters provide a 10-MΩ input resistance for these ranges.

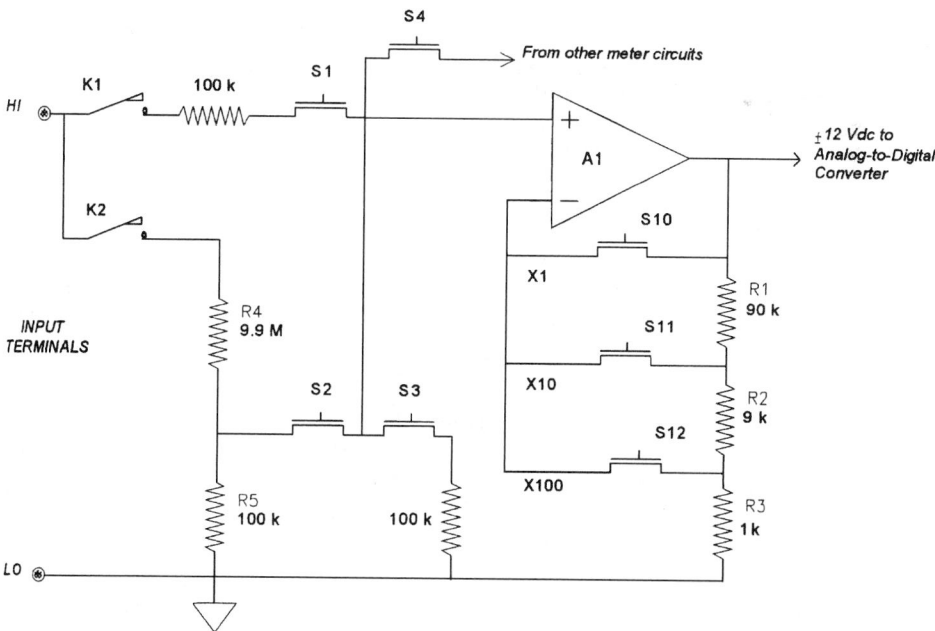

FIGURE 25.2.3 Simplified schematic of the input switching, measuring range selection, and amplifier for a dc voltage meter.

Amplifier Offset Elimination (Autozero)

The main performance limitation of the dc signal-conditioning section is usually its offset voltage. This affects the meter's ability to read zero volts with a short applied. Most meters employ some method for automatically zeroing out amplifier offsets. Switch S_3 in Fig. 25.2.3 is used to periodically *short* the amplifier input to ground to measure the amplifier offset voltage. The measured offset is stored and then subtracted from the input signal measurement to remove amplifier offset errors. Switches S_1 and S_2 are opened simultaneously during the offset measurement to avoid shorting the meter's input terminals together.

In a multifunction instrument, all measurements are eventually converted into a dc voltage, which is measured by an ADC. Other dc signals are often routed to the ADC through a dc voltage measuring front end. Switch S_4 in Fig. 25.2.3 could be used to measure the dc output of an ac voltage function or a dc current measuring section.

AC VOLTAGE MEASUREMENT TECHNIQUES

The main purpose of an ac front end is to change an incoming ac voltage into a dc voltage that can be measured by the meter's ADC. The type of ac voltage to dc voltage converter employed in a meter is quite critical. There are vast differences in behavior between rms, average-responding, and peak-responding converters.

Signal Conditioning for AC Measurements

The input signal conditioning for ac voltage measurements includes both attenuation and amplification, similar to the dc voltage front end already discussed. Shown in Fig. 25.2.4 are typical input switching and ranging circuits for an ac-voltage instrument. Input-coupling-capacitor C_1 blocks the dc portion of the input signal so

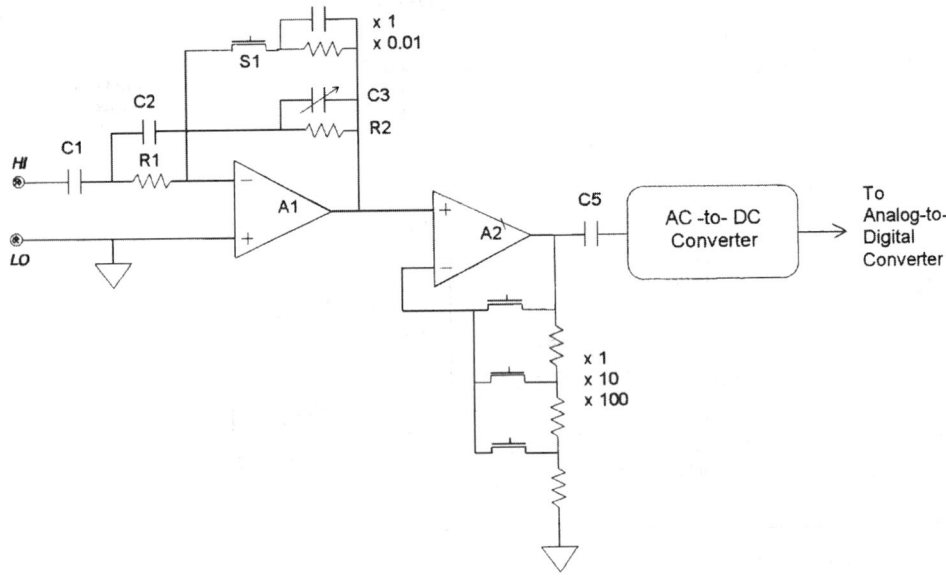

FIGURE 25.2.4 The input switching and ranging sections of a typical ac voltage measurement section—simplified schematic.

that only the ac component is measured by the meter. Ranging is accomplished by combining signal attenuation from first-stage amplifier A_1 and gain from second-stage amplifier A_2.

The first stage implements a high input impedance—typically 1-MΩ switchable compensated attenuator. The value of capacitor C_3 is adjusted so that the R_2C_3 time constant precisely matches the R_1C_2 time constant, yielding a compensated attenuator in which the division ratio does not vary with frequency. Switch S_1 is used to select greater attenuation for the higher input voltage ranges. The second stage provides variable-gain, wide-bandwidth signal amplification to scale the input to the ac converter to the full-scale level. The output of the second stage is connected to the ac converter circuit. Residual dc offset from the attenuator and amplifier stages is blocked by capacitor C_5.

An ac-voltage front end, similar to the one discussed above, is also used in ac-current-measuring instruments. Shunt resistors are used to convert the ac current into a measurable ac voltage. Current shunts are switched to provide selectable ac current ranges.

Amplifier bandwidth and ac converter limitations are the main differences between various ac front ends. As mentioned earlier, the type of ac-to-dc converter circuit has a profound effect on overall measurement accuracy and repeatability. True rms converters are superior to both *average responding* and *peak-responding* ac converters in almost every application.

CURRENT MEASUREMENT TECHNIQUES

An ammeter measures the current flowing through it, while approximating a short circuit between its terminals. A conventional ammeter is connected in series with the circuit or device being measured so that current flows through both the meter and the test circuit. There are two basic techniques for making current measurements: *in-circuit methods* and *magnetic field sensing methods*.

In-Circuit Methods

In-circuit current sensing meters employ either a *current shunt* or *virtual ground amplifier* technique, similar to those shown in Fig. 25.2.5a and b. Shunt-type meters are very simple. A resistor R_S, shown in Fig. 25.2.5a, is connected across the input terminals so that a voltage drop proportional to the input current is developed. The value of R_S is kept as low as possible to minimize the instrument's *burden voltage*, or IR drop. This voltage drop is sensed by an internal voltmeter and scaled to the proper current value.

Virtual-ground-type meters are generally better suited for measuring smaller current values—usually 100 mA to below 1 pA. These meters rely on low-noise, low-bias-current operational amplifiers to convert the input current to a measurable voltage, as illustrated in Fig. 25.2.5b. Negligible input current flows into the negative input terminal of the amplifier. Therefore the input current is forced to flow through the amplifier's feedback resistor R_f, causing the amplifier output voltage to vary by IR_f. The meter burden voltage, which is the voltage drop from input to LO, is maintained near 0 V by the high-gain amplifier forming a *virtual ground*. Since the amplifier must source or sink the input current, the virtual ground technique is generally limited to low current measurements.

Magnetic Field Sensing Methods

Current measurements employing magnetic field sensing techniques are extremely convenient. The reason is that measurements can be performed without interrupting the circuit or producing significant loading errors. Since there is no direct contact with the circuit being measured, complete dc isolation is also ensured. These meters use a transducer—usually a *current transformer* or *solid-state Hall-effect sensor*—to convert the magnetic field surrounding a current-carrying conductor into an ac or dc value that is proportional. Sensitivity can be very good, since simply placing several loops of the current-carrying conductor through the probe aperture will increase the measured signal level by the same factor as the number of turns.

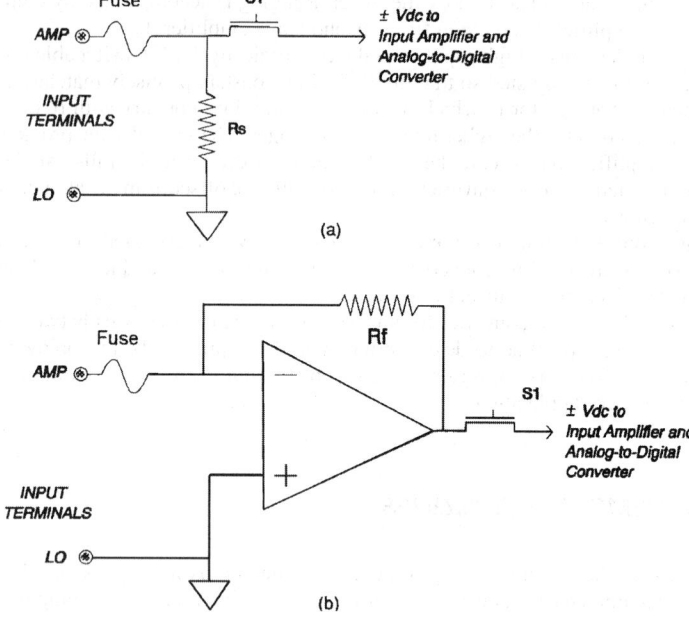

FIGURE 25.2.5   Two common methods for in-circuit current measurements. (a) Shunt resistor R_s is connected across the input terminals, developing a voltage proportional to the input current. (b) The input current is forced to flow through R_f, while the meter burden voltage is limited to the drop across the fuse and the amplifier offset voltage.

Alternating Current

Alternating current measurements are very similar to dc measurements. However, ac measurements employ *shunt-type current-to-voltage converters* almost exclusively. The output of the current-to-voltage sensor is measured by an ac voltmeter. Signal and ac converter issues discussed in "AC Voltage Measurement Techniques" are relevant to ac measurements as well. The input terminals of in-circuit ac meters are always direct coupled—ac + dc coupled—to the shunt so that the meter maintains dc continuity in the test circuit. The meter's internal ac voltmeter section can be either ac coupled or ac + dc coupled to the current-to-voltage converter.

RESISTANCE MEASUREMENT TECHNIQUES

An ohmmeter measures the dc resistance of a device or circuit connected to its input. Resistance measurements are performed by supplying a known dc to an unknown resistance, thereby converting the resistance value to an easily measured dc voltage. Most meters use *an ohms converter* technique similar to the current source or voltage ratio types shown in Fig. 25.2.6.

Signal Conditioning for Resistance Measurements

The current-source method shown in Fig. 25.2.6a employs a known current source value I that flows through the unknown resistor when it is connected to the meter's input. This produces a dc voltage proportional to the unknown resistor value: by Ohm's law, $E = IR$. Thus dc voltmeter input-ranging and signal-conditioning circuits

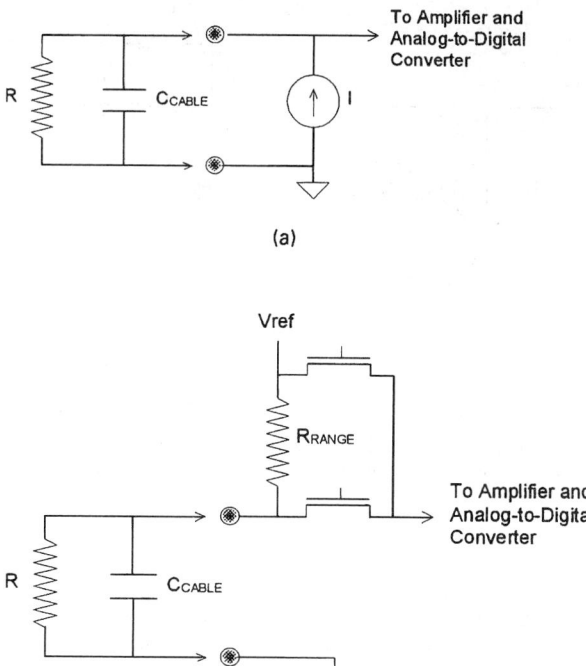

FIGURE 25.2.6   Ohms converter circuits used in meters. (a) The current-source ohms converter employs a constant current source, forcing current I through unknown resistance R, developing a voltage to be measured by a dc voltage front end. (b) The voltage-ratio type ohms converter calculates the unknown resistor value R from dc voltage measurements in a voltage divider circuit.

are used to measure the voltage developed across the resistor. The result is scaled to read directly in ohms. Shown in Fig. 25.2.6b is the *voltage-ratio-type* ohms converter technique.

This method uses a known voltage source V_{ref} and a known *range* resistor R_{range} to compute the unknown resistor value. The range resistor and the unknown resistor form a simple voltage divider circuit. The meter measures the dc voltage developed across the unknown resistor. This voltage, along with the values of the internal voltage source and range resistor, is used to calculate the unknown resistor value. In practice, meters have a variety of resistance-measuring ranges. To achieve this, the ohms test current—or range resistor—is varied to scale the resulting dc voltage to a convenient internal level, usually between 0.1 and 10 V dc. This measurement is relatively insensitive to lead resistances R_l in the high-impedance input of the dc voltmeter. Voltage drops in the current source leads do not affect the voltmeter measurement. However, they can affect the accuracy of the current source itself.

Two-Wire Sensing

The ohms, converters discussed above use *two-wire sensing*. When the same meter terminals are used to measure the voltage dropped across the unknown resistor as are used to supply the test current, a meter is said to use a two-wire ohms technique. With two-wire sensing, the lead resistances R_l shown in Fig. 25.2.7a, are indistinguishable from the unknown resistor value, causing potentially large measurement errors for lower-value resistance measurements.

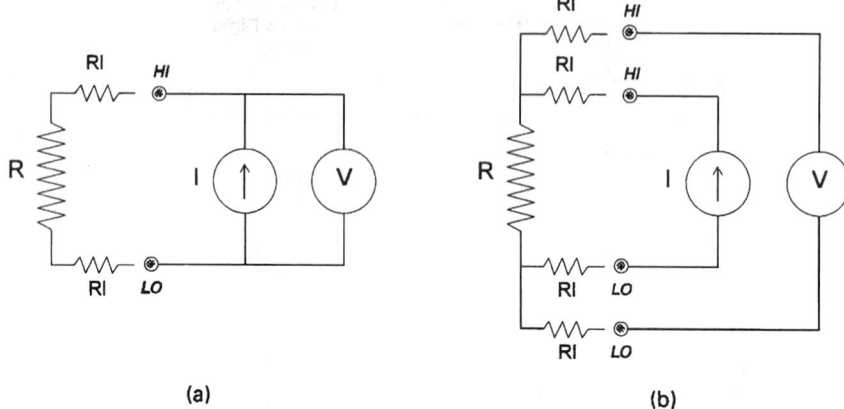

FIGURE 25.2.7 (a) Simplified schematic for two-wire sensing. Lead resistances R_l are inseparable from the unknown resistance measurement. (b) Simplified schematic for four-wire sensing.

The two-wire technique is widely used by all types of ohmmeters because of its simplicity. Often, meters provide a *relative* or *null* math function to allow lead resistances to be measured and subtracted from voltage, current, and resistance-measuring instruments' subsequent resistance measurements. This works well unless the lead resistances vary due to temperature or connection integrity. The *four-wire sensing* technique, or *Kelvin sensing*, is designed to eliminate lead resistance errors.

Four-Wire Sensing

The four-wire sensed-resistance technique is the most accurate way to measure small resistances. Lead resistances and contact resistances are virtually eliminated using this technique. A four-wire converter senses the voltage drop across only the unknown resistor. The voltage drops across the lead resistances are excluded from measurement, as shown in Fig. 25.2.7*b*.

The four-wire converter employs two separate pairs of connections to the unknown resistor. One connection pair, often referred to as the *source leads*, supplies the test current that flows through the unknown resistor, similar to the two-wire measurement case. Voltage drops are still developed across the source lead resistances.

A second connection pair, referred to as the *sense leads*, connects directly across the unknown resistor. These leads connect to the input of a dc voltmeter. The dc voltmeter section is designed to exhibit an extremely large input resistance so that virtually no current flows in the sense input leads. This enables the meter to measure only the voltage drop across the unknown resistor. This scheme thereby removes from the measured voltage, the drops in both the source leads and the sense leads.

Generally, lead resistances are limited by the meter manufacturer. There are two main reasons for this. First, the total voltage drop in the source leads will be limited by the design of the meter and is usually limited to a fraction of the meter measuring range being used. Second, the sense lead resistances will introduce additional measurement noise if they are allowed to become too large. Sense leads less than 1 kΩ usually will contribute negligible additional error.

The four-wire technique is widely used in situations where lead resistances can become quite large and variable. It is used almost exclusively for measuring lower resistor values in any application, especially for measuring values of 10 Ω or less. It is also used in automated test applications where cable lengths can be quite long and numerous connections or switches may exist between the meter and the device under test. In a multichannel system, the four-wire method has the obvious disadvantage of requiring twice as many switches and twice as many wires as the two-wire technique.

CHAPTER 25.3
SIGNAL SOURCES

Tomas Fetter

INTRODUCTION

The simplest useful definition for a *signal* is an electrical voltage (or current) that varies with time. To characterize a signal, an intuitive yet accurate concept is to define the signal's waveform. A waveform is easy to visualize by imagining the picture a pen, moving up and down in proportion to the signal voltage, would draw on a strip of paper being steadily pulled at right angles to the pen's movement. Shown in Fig. 25.3.1 is a typical periodic waveform and its dimensions. A *signal source*, or *signal generator*, is an electronic instrument that generates a signal according to the user's commands with regard to its waveform (Fig. 25.3.2). Signal sources serve the frequent need in engineering and scientific work for energizing a circuit or system with a signal whose characteristics are known.

KINDS OF SIGNAL WAVEFORMS

Most signals fall into one of two broad categories—*periodic* and *nonperiodic*. Signal source instruments generate one or the other, and sometimes both. A periodic signal has a waveshape, which is repetitive: the pen, after drawing one period of the signal waveform, is in the same vertical position where it began, and then it repeats exactly the same drawing. A sine wave is the best-known periodic signal. By contrast, a nonperiodic signal has a nonrepetitive waveform. The best-known nonperiodic signal is random noise.

The familiar sinusoid, illustrated in Fig. 25.3.3, is the workhorse signal of electricity. The simple mathematical representation of a sine wave can be examined to determine the properties that characterize it:

$$s(t) = A \sin(2\pi f t)$$

where s = represents the signal, a function of time
t = time, s
A = peak amplitude of the signal, V or A
f = signal frequency, cycles/second (Hz)

HOW PERIODIC SIGNALS ARE GENERATED

Introduction

Periodic signals are generated by *oscillators*. Some signal generators use the waveform produced directly by an oscillator. However, many signal generators combine a number of different techniques to increase the

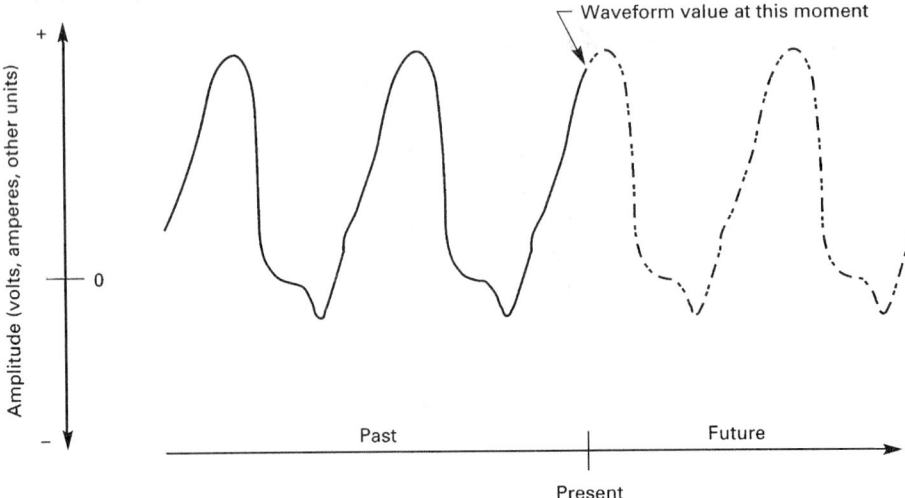

FIGURE 25.3.1 Waveform of an active typical period signal.

generator's performance and capabilities. Key performance attributes include frequency range, frequency accuracy, amplitude accuracy, frequency switching speed, types of waveforms produced, and various modes of signal modulation.

Oscillators

Electronic oscillators are the basic building blocks of signal generators. Any oscillator circuit fits into one of these broad categories:

- AC amplifier with filtered feedback
- Threshold decision circuit

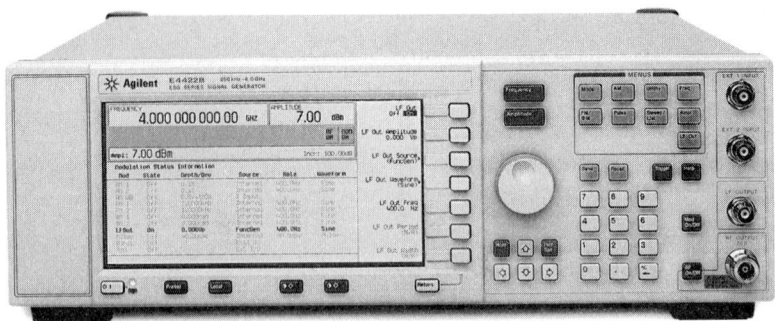

FIGURE 25.3.2 Today's signal generators must meet demanding specifications, including accuracy, spectral purity, and low phase noise. This Agilent Technologies 4422B Signal Generator covers 225 kHz to 4 GHz and can be tailored to meet changing requirements as technologies evolve.

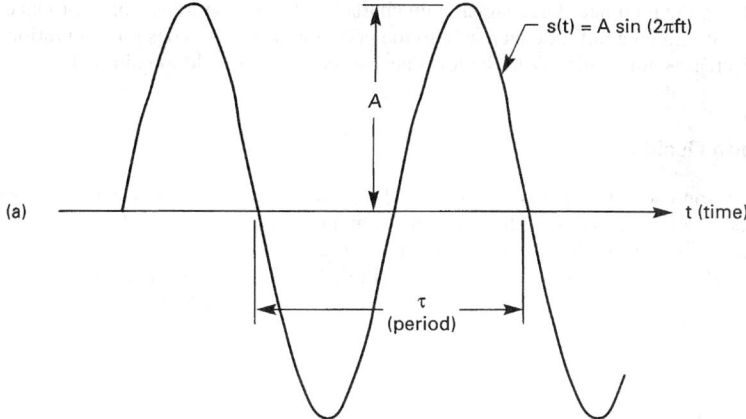

FIGURE 25.3.3   A typical sine wave.

Feedback Oscillators

The feedback technique is historically the original, and still the most common form of oscillator circuit. Shown in Fig. 25.3.4 are the bare essentials needed for the feedback oscillator. The output from the amplifier is applied to a frequency-sensitive filter network. The output of the network is then connected to the input of the amplifier. Under certain conditions, the amplifier output signal, passing through the filter network, emerges as a signal, which, if supplied to the amplifier input produces the output signal. Because of the feedback connection, the circuit is capable of sustaining a particular output signal indefinitely. This is, in fact, an oscillator. The circuit combination of the amplifier and the filter is called a feedback loop. To understand how the combination can oscillate, mentally break open the loop at the input to the amplifier; this is called the open-loop condition. The open loop begins at the amplifier input and ends at the filter output. Here are the particular criteria that the open loop must satisfy in order that the closed loop will generate a sustained signal at some frequency f_0:

1. The power gain through the open loop (amplifier power gain times filter power loss) must be unity at f_0.

2. The total open-loop phase shift at f_0 must be 0° (or 360°, 720°, …).

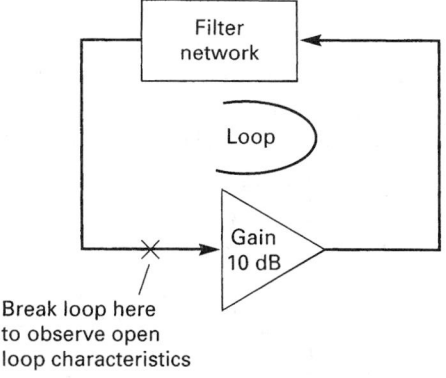

FIGURE 25.3.4   Oscillator, using an amplifier and a filter to form a feedback loop.

Both criteria are formal statements of what was said previously: the loop must produce just the signal at the input of the amplifier to maintain the amplifier output. Criterion 1 specifies the amplitude and criterion 2 the phase of the requisite signal at the input.

When the above criteria are met, the poles of the oscillator are on the $j\omega$, or imaginary, axis when plotted on the complex s-plane, which corresponded to a stable amplitude oscillation. The loop gain actually must be greater than unity for the oscillation to start and build to the final stable amplitude, which corresponds to poles being in the right half plane. Once the final amplitude is reached, something must reduce the loop gain to exactly 1. In a gain controlled oscillator, some form of level detection and a variable gain element are used as a part of the oscillator circuit feedback loop to precisely sustain the conditions of oscillation, keeping the poles on the imaginary axis. Self-limiting oscillators, such as the single transistor Colpitts oscillator, rely on the nonlinear large signal gain characteristics of the transistor to produce a stable amplitude oscillation. As the signal amplitude increases to the point the transistor goes into saturation or cutoff, the large signal gain decreases automatically to the level necessary to ensure a loop gain of 1.

Threshold Decision Oscillators

Threshold decision oscillators generally consist of an integrator or capacitor, which is driven from a current source switched by level detectors. The integrator signal ramps up to a preset high level. The current source is then reversed, and the integrator integrates down to a preset low level, where the process is started over. The result is a triangle wave. The triangle wave can be filtered or shaped to form a sine wave, or run though a high gain-limiting amplifier to produce a square wave. Threshold decision oscillators are generally used at frequencies below RF.

The Colpitts oscillator

The Colpitts oscillator (Fig. 25.3.5) and circuits derived from it that operate on the same basic principle are the most commonly used configurations in transistor oscillator design. The inductor L and the capacitors C, C_1, and C_2 form a parallel resonant circuit. The output voltage is fed back to the input in the proper phase to sustain oscillation via the voltage divider formed by C_1 and C_2, parts of which may be internal to the transistor itself. Typically bipolar *silicon* (Si) transistors are used up to 10 or 12 GHz and *gallium arsenide* (GaAs) FETs are usually selected for coverage above this range, though bipolar Si devices have been used successfully to 20 GHz. Bipolar Si devices generally have been favored for lower phase noise, but advances in GaAs FET design have narrowed the gap, and their superior frequency coverage has made them the primary choice for many designs.

Electrically Tuned Oscillators

The use of an electrically tuned capacitor, such as C_R in Fig. 25.3.6, enables the frequency of oscillation to be varied and to be phase-locked to a stable reference. A common technique for obtaining a voltage-variable capacitor is to employ a *variable-capacitance diode*, or *varactor*. This device consists of a reverse-biased junction diode with a structure optimized to provide a large range of depletion-layer thickness variation with voltage as well as low losses (resistance) to ensure high Q. Major advantages of varactor oscillators are the potential for obtaining high tuning speed and the fact that a reverse-biased varactor diode does not dissipate dc power, as does a magnetically biased oscillator. Typical tuning rates in the microsecond range are realized without great difficulty.

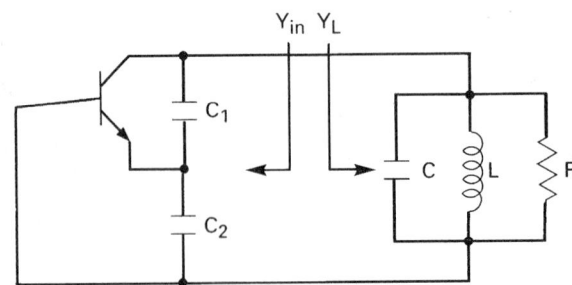

FIGURE 25.3.5 Colpitts oscillator circuit.

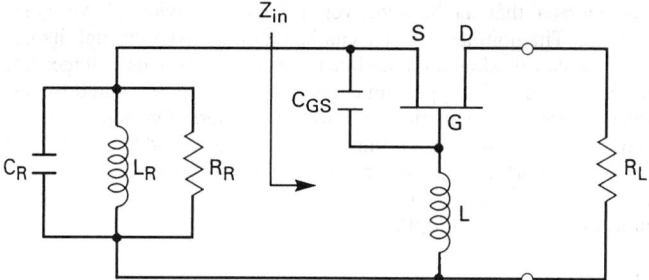

FIGURE 25.3.6   Negative-resistance oscillator. C_R, L_R, and R_R represent a resonator. The inductance L is required to achieve a negative resistance in Z_{in}.

YIG-Tuned Oscillators

High Q resonant circuits suitable for tuning oscillators over very broad frequency ranges can be realized with polished, single-crystal spheres of *yttrium-iron-garnet* (YIG). When placed in a dc magnetic field, ferromagnetic resonance is attained at a frequency that is a linear function of the field. The microwave signal is usually coupled into the sphere (typically about 0.5 mm in diameter) via a loop, as shown in Fig. 25.3.7. The equivalent circuit presented to the transistor is a shunt-resonant tank that can be tuned linearly over several octaves in the microwave range. Various rare earth *dopings* of the YIG material have been added to extend performance to lower frequency ranges in terms of spurious resonances (other modes) and nonlinearities at high power, but most ultra-wideband oscillators have been built above 2 GHz. Frequencies as high as 40 GHz have been achieved using pure YIG, and other materials, such as hexagonal ferrites, have been used to extend frequencies well into the millimeter range.

Frequency Multiplication

Another approach to signal generation involves the use of frequency multiplication to extend lower-frequency sources up into the microwave range. By driving a nonlinear device at sufficient power levels, harmonics of

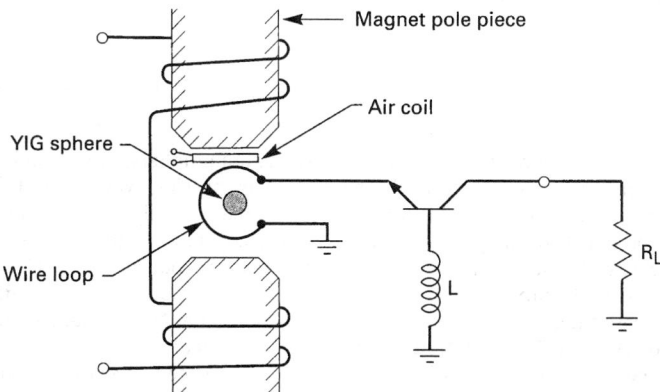

FIGURE 25.3.7   YIG-tuned oscillator.

the fundamental are generated that can be selectively filtered to provide a lower-cost, less complex alternative to a microwave oscillator. The nonlinear device can be a diode driven through its nonlinear *i* versus *v* characteristic, or it can be a varactor diode with a nonlinear capacitance versus voltage. Another type of device consists of a *pin* structure (*p*-type and *n*-type semiconductor materials separated by an intrinsic layer) in which charge is stored during forward conduction as minority carriers. On application of the drive signal in the reverse direction, conductivity remains high until all the charge is suddenly depleted, at which point the current drops to zero in a very short interval. When this current is made to flow through a small drive inductance, a voltage impulse is generated once each drive cycle which is very rich in harmonics. Such *step-recovery diodes* are efficient as higher-order multipliers.

Crystal Oscillator

A crystal oscillator can be made by using a quartz crystal resonator as the frequency selective element in a transistor amplifier-based oscillator. The equivalent circuit of a crystal is a series resonant LC circuit, so the oscillator must be designed around a series resonator. A quartz crystal is a very high Q resonator that delivers two major benefits. First, the accuracy and stability of the frequency of the resonance of the crystal makes the crystal oscillator a very good frequency reference for a signal generator. Second, the very high Q, or narrow bandwidth, of the resonator allows very little of the broadband noise of the transistor, and other components, to modulate the oscillator signal. Noise that is present on an oscillator and affects the phase stability is called phase noise. Low phase noise is important if a signal is to be modulated onto a carrier signal because the phase noise will limit the dynamic range of the modulated signal.

Frequency Synthesis

The methods by which frequencies can be generated using addition, subtraction, multiplication, and division of frequencies derived from a single reference standard are called *frequency synthesis techniques*. The accuracy of each of the frequencies generated becomes equal to the accuracy of the reference. Normally, an oscillator that covers a wide range of frequencies cannot, by itself, be tuned to an absolute frequency accurately. And an oscillator that is very accurate in absolute frequency, such as a crystal oscillator, cannot be tuned over a wide range of frequencies, due to its high Q. Even if it could be tuned, the crystal oscillator would then lose its absolute frequency accuracy because some element in addition to the crystal would be determining the frequency of oscillation. However, if that wide ranging oscillator is *locked* to the accurate fixed frequency reference oscillator through synthesis techniques, then the absolute frequency accuracy of the reference oscillator can be extended to the full frequency range of the wide tuning oscillator. Three classifications are commonly referred to: *indirect synthesis, direct synthesis*, and *direct digital synthesis* (DDS).

Direct Synthesis

By assembling a circuit assortment of frequency dividers and multipliers, mixers and bandpass filters, an output *m/n* times the reference can be generated. There are many possible ways to do this, and the configuration actually used is chosen primarily to avoid strong spurious signals, which are low-level, nonharmonically related sinusoids. The principal advantage of the direct synthesis technique is the speed with which the output frequency may be changed. Shown in Fig. 25.3.8 is one way to produce a 13-MHz output from a 10-MHz reference. The inputs to the mixer are 10 and 3 MHz, the mixer produces sum and difference frequency outputs, and the bandpass filter on the output selects the 13-MHz sum. Notice that another bandpass filter could have been used to select the 7-MHz difference, if that were wanted.

A key advantage of direct synthesis is frequency switching speed. Because no control loops are directly involved in the signal generation that have relatively slow time constants, switching speeds in the microsecond range are easily achievable.

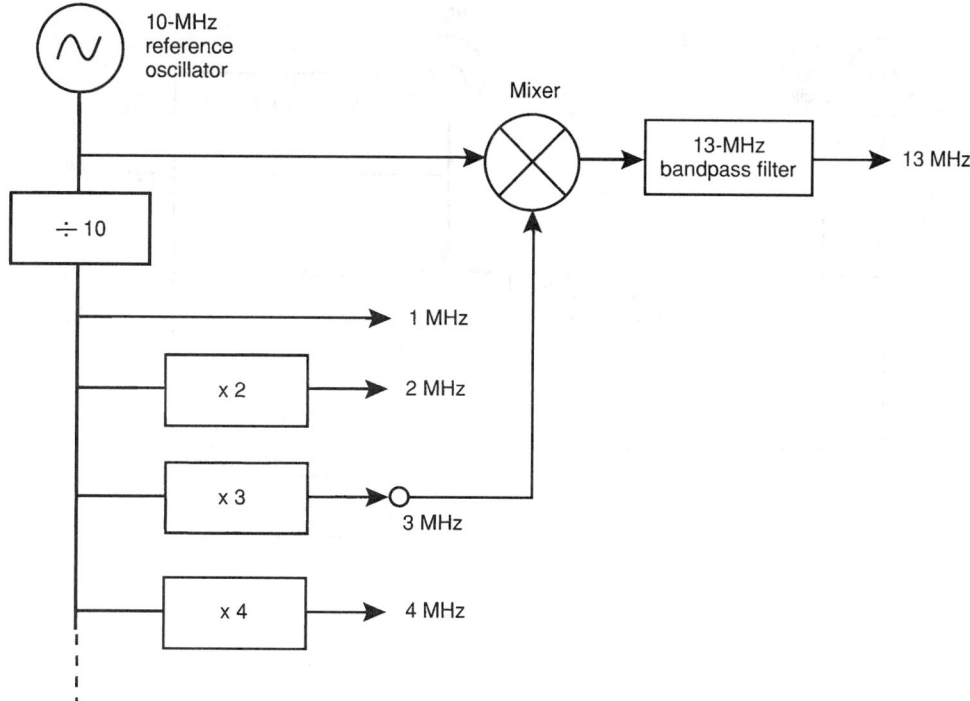

FIGURE 25.3.8 Direct frequency synthesis.

Indirect Synthesis

This name derives from the use of an oscillator other than the reference to generate the output. This technique involves placing the oscillator in a *phase-locked loop* (PLL). The phase-locked loop starts with a tunable oscillator, or voltage-controlled oscillator (VCO). The output of the oscillator is then divided down in frequency by a frequency divider with a divide number m. The divided down signal is compared to a reference oscillator signal in a device called a phase detector. A phase detector produces a voltage output that is proportional to the difference in phase between the two input signals. The output voltage of the phase detector, or error signal, is then fed into an integrator/loop filter, which produces the voltage that is necessary to tune the tunable oscillator to minimize the error voltage, or phase difference between the reference oscillator and divided down tunable oscillator. If the phase difference between the two signals is minimized, then the two signals must also be at the same frequency. But the tunable oscillator is actually running at m times the reference oscillator due to the divider. When this occurs, the two oscillators are said to be *phase-locked* together. If the loop filter is a true integrator, then the steady-state frequency error must be zero.

The finest frequency resolution of a phase-locked loop is called the step size of the synthesizer. For a simple phase-locked loop with a single divider in the feedback path between the tunable oscillator and the phase detector, output frequency is m times the frequency at the phase detector, and the step size is the frequency of the reference signal. For higher frequency resolution, or smaller step size, more sophisticated phase-locked loop topologies are required. Shown in Fig. 25.3.9 is a slightly more complicated phase-locked loop synthesizer in which the reference oscillator is also divided down by a divide number n. The tunable oscillator is then locked to m/n times the reference frequency. If the dividers are programmable, then a wide range of

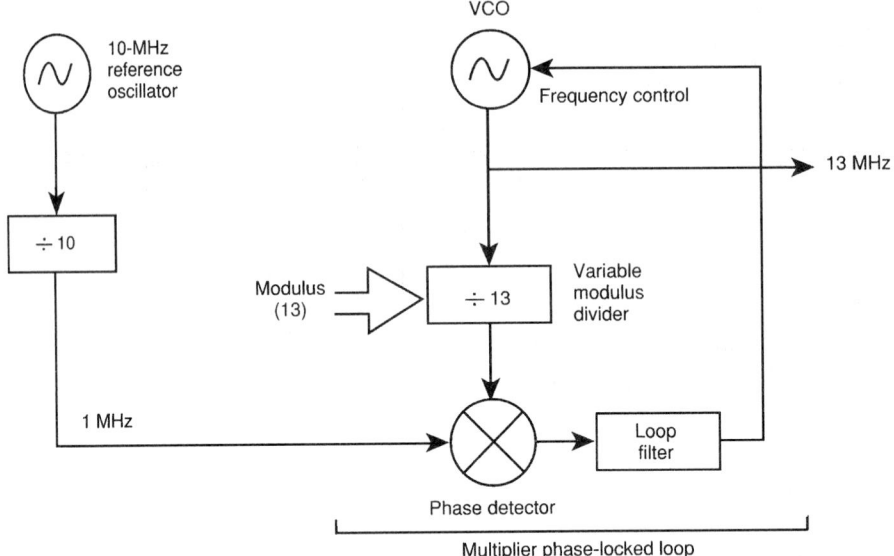

FIGURE 25.3.9   Indirect frequency synthesis.

output frequencies that are m/n times the reference can be generated by this technique. As the divide number n is increased, the resolution of the phase-locked loop increases because the step size becomes smaller. As n is increased to reduce the step size of the loop, m must be increased to maintain the same range of output frequencies.

A practical limitation to resolution is reached due to the noise performance of the phase-locked loop. Any additive noise generated in the phase detector sees a gain of m to the output of the tunable oscillator within the bandwidth of the phase-locked loop. So effectively, as the resolution of the loop increases, so does the phase noise. This limitation is overcome either by summing multiple synthesis loops together in a way to reduce the maximum divide number and thus the noise gain of each while preserving the resolution and frequency range, or by using sophisticated frequency interpolation techniques known as fractional frequency dividers, or *fractional-n*. When the tunable oscillator is divided by a divider that has both an integer and fractional part, the resolution can be extended without increasing the noise gain simply by increasing the number of digits of fractional resolution. Fractional-n dividers work by switching between two adjacent whole divide numbers at a rate that time averages to the correct fraction between those two numbers. The output frequency averages to exactly the correct frequency, but it also has a large frequency modulation due to the divider changing back and forth between two divide numbers. Several techniques can be used to remove the FM component and leave a spectrally pure output signal. A key advantage of indirect synthesis is the simplicity of the design required to produce a wide range of output frequencies that are locked to a fixed reference frequency by simply changing the divider numbers.

Arbitrary Waveform Synthesizers

In this technique, the complete period of some desired waveshape is defined as a sequence of numbers representing sample values of the waveform, uniformly spaced in time. These numbers are stored in digital memory and then, paced by the reference, repetitively read out in order. The sequence of numbers is then converted into a sequence of voltage levels using a *digital-to-analog converter* (DAC). The output of the DAC is then filtered to produce an accurate replica of the numerically defined signal. Any *arbitrary* waveform can be produced with this technique with frequency components from DC up to half the sample frequency.

TYPES OF SIGNAL GENERATORS

Introduction

The various types of signal generators are categorized by frequency range and functional capability. Many of the historical boundaries between generator types have changed over the years as the technology for generating electrical signals has changed. For example, the digitally based arbitrary waveform generator has mostly replaced the traditional function generator for producing complex, sinusoidal, and nonsinusoidal baseband waveforms.

In addition to generating a signal at a specific frequency, a signal generator accurately controls the amplitude of the signal, sometimes over a wide range of amplitudes through use of precision level control circuitry and precision step attenuators. Absolute amplitudes accurate to within several dB over the full frequency range of a frequency generator are possible.

Audio Oscillators

At audio frequencies (~1 Hz to 100 KHz), there is more emphasis on purity of waveform than at RF, so it is natural to use an oscillator circuit that generates a sine wave as its characteristic waveform. This is, in effect, an amplifier with filtered feedback. For audio frequencies, however, the elements of an LC resonant circuit for the filter become expensive and large, with iron or ferrite core construction necessary to realize the large values of inductance required. Another handicap is that ferrocore inductors exhibit some nonlinear characteristics which increase harmonic output when they are used in a resonant filter circuit. Using resistors and capacitors, it is possible to establish a voltage transfer function that resembles that of a resonant circuit in both its phase and amplitude response. The four-element RC network in the dashed portion of Fig. 25.3.10 is

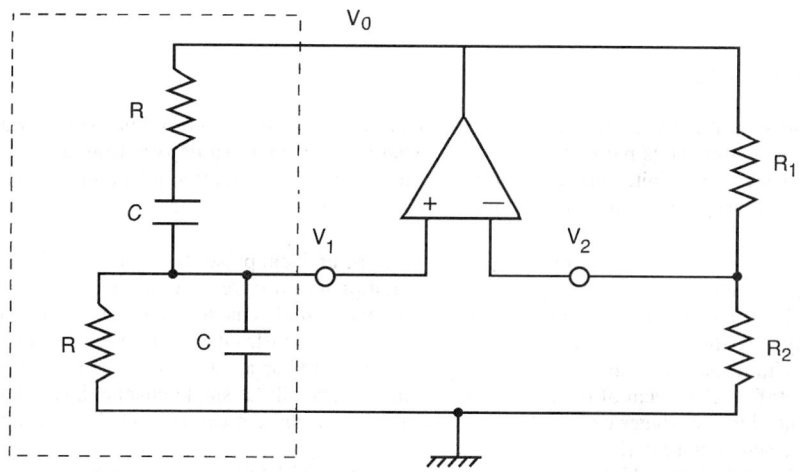

FIGURE 25.3.10   Wien bridge oscillator.

characterized by an input-output voltage ratio V_1/V_0 (called a transfer function) that varies with frequency. This transfer function has an amplitude peak of $(1/3)$ V_0 at the *resonant frequency* $(1/2)\pi RC$ and a phase shift passing through zero at this frequency. Very low distortion audio oscillators use some form of level detection and variable gain as a part of the oscillator circuit feedback loop to precisely sustain the conditions of oscillation, keeping the poles on the imaginary axis, without adding distortion. Self-limiting oscillators, such as the single transistor Colpitts, rely on the nonlinear large signal gain characteristics of the transistor to produce a stable amplitude oscillation at the expense of harmonic distortion performance. Because the gain of the active device in a self-limiting oscillator changes throughout the cycle of the sinusoidal oscillation to sustain an average loop gain of 1, the waveform is distorted, and harmonics of the frequency of oscillation are created. The effects of the distortion can be minimized by taking advantage of the filtering action of the resonator, but a highly linear gain-controlled oscillator can have much lower distortion.

Function Generators

The term *function generator* describes a class of oscillator-based signal sources in which the emphasis is on versatility. Primarily, this means providing a choice of output waveforms. It also connotes continuous tuning over wide bands with max-min frequency ratios, sub-Hz to MHz frequencies, flat output amplitude, and sometimes modulation capabilities—*frequency sweeping*, *frequency modulation* (FM), and *amplitude modulation* (AM). With regard to their frequency accuracy and stability, function generators are inferior to sine-wave oscillators, but their performance is quite adequate for many applications. Some frequency synthesizers are also called *precision function generators* by their manufacturers. What this means is that circuitry has been added to the synthesizer to produce other waveforms in addition to sine waves.

Early function generators were based on integrators driven from current sources switched by level detectors to form triangle wave generators, which is a form of threshold decision circuit. The triangle waves were then either put through high gain-limiting circuits to produce square waves, or shaped with diode-shaping circuits to produce sine waves. Because the signal was based on the RC time constant of an integrating amplifier, traditional function generators are not synthesized. Modern function generators are based on digital arbitrary waveform generators and are capable of generating an infinite variety of waveforms. And because arbitrary waveform generators are based on a single clock frequency, they are technically synthesizers. Function generators, and now mostly arbitrary waveform generators, are often included in RF and microwave signal generators that offer signal modulation capabilities. Because these function generators create the baseband signal that contains the information that is then modulated onto the RF or microwave carrier signal, these function generators are often called *baseband generators*.

Pulse Generators

A *pulse generator* is an instrument that can provide a voltage or current output whose waveform may be described as a continuous pulse stream. A *pulse stream* is a signal that departs from an initial level to some other single level for a finite duration, then returns to the original level for a finite duration. This type of waveform is sometimes referred to as a *rectangular pulse*.

Pulse Nomenclature.   The common terms that describe an ideal pulse stream are identified in Fig. 25.3.11. However, in order to describe real pulses, some additional terms are needed. These terms are explained in Fig. 25.3.12. If a pulse generator provides more than just a single channel output, the multiple outputs do not necessarily have to be time synchronous. The majority of these instruments have the capability of delaying the channels with respect to each other. This delay may be entered as an absolute value (e.g., 1 ns) or as a phase shift (e.g., 90° = 25 percent of the period). Sometimes, especially in single-channel instruments, these values are referenced to the trigger output. The maximum delay or phase range is limited and in most cases dependent on the actual pulse period.

All pulse generators offer variable levels (HIL, LOL, AMP, OFS), variable pulse period (PER, FREQ), and variable pulse width (PWID, NWID, DCYC). Some instruments also offer variable transition times

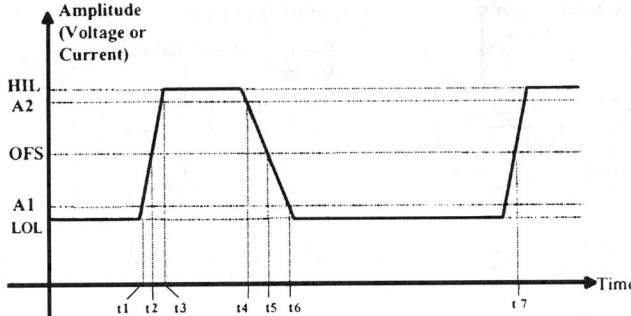

FIGURE 25.3.11   Ideal pulse nomenclature. HIL = high level (voltage or current); LOL = low level (voltage or current); AMP = HIL − LOL = amplitude (voltage or current); OFS = (HIL + LOL)/2 = offset (voltage or current); $A1$ = LOL + X*AMP, $A2$ = HIL − X*AMP, where X is defined as 0.1 (normally) or as 0.2 (for ECL devices). PER = $t7 - t2$ = pulse period; FEWQ = 1/PER = pulse frequency; PWID = $t5 - t2$ = positive pulse width; NWID = $t7 - t5$ = negative pulse width; DCYC = PWID/PER = duty cycle; TRISE = $t3 - t1$ = rise time; TFALL = $t6 - t4$ = fall time; $t1$, $t7$ rising edge crosses $A1$; $t2$ = rising edge crosses $A1$; $t4$ = rising edge crosses OFS; $t3$ = rising edge crosses $A2$; $t2$ = falling edge crosses $A2$; $t5$ = falling edge crosses OFS; $t6$ = falling edge crosses $A1$. Remarks: ECL stands for emitter-coupled logic. Rise and fall times are sometimes also referred to as transition times or edge rates.

(TRISE, TFALL). The minimum and maximum values and the resolution with which these parameters can be varied are described in pulse-generator data sheets.

Special Pulse Generators

This section describes pulse generators with special capabilities and distinguishes them from instruments called data generators.

Programmable Data.   Some pulse generators can generate not only repetitive pulse but also serial data streams. These instruments have a memory with a certain depth and an address counter in the signal path of

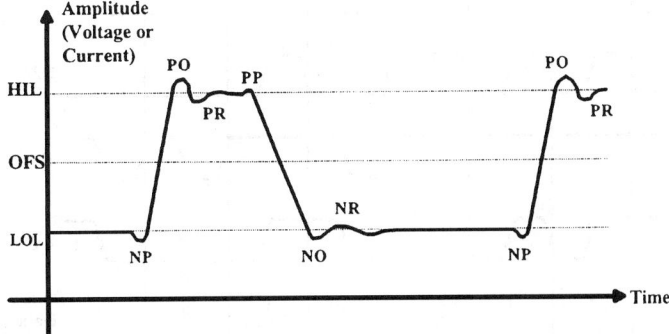

FIGURE 25.3.12   Real pulse generator nomenclature. NP = negative preshoot; PO = positive overshoot; PR = positive ringing; PP = positive preshoot; NO = negative overshoot; NR = negative ringing. Remark: Preshoot, overshoot, and ringing are sometimes also referred to as pulse distortions.

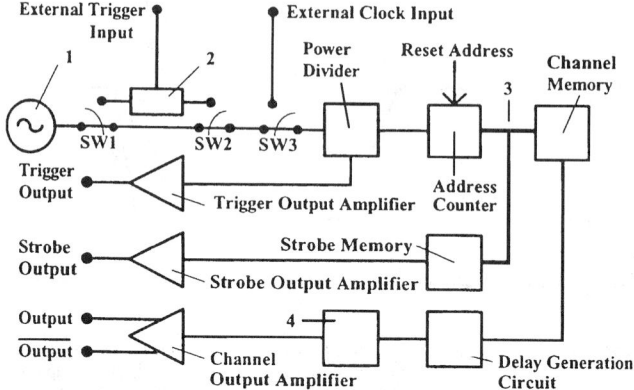

FIGURE 25.3.13   Block diagram of a single-channel pulse generator with data capabilities. 1: Internal clock generation circuit; 2: start/stop generation circuit; 3: memory address bus (n lines); 4: width generation circuit (NRZ, RZ with variable duty cycle). Data stream length (selected by the user): m bit ($m \le 2^n$); memory depth: 2^n; reset address: 2^{n-m}. Remark: When the address counter value reaches 2^{n-1}, the counter resets itself to the reset address (calculated and programmed by the microprocessor).

each channel, as shown in Fig. 25.3.13. Thus, single-shot or repetitive data streams of a programmable length can be generated. The maximum length of such a data stream is limited by the memory depth. The strobe output provided by these instruments can be used to generate a trigger signal that is synchronous to one specific bit of this data stream, which is useful when this bit—or the corresponding reaction of the Device Under Test (DUT)—has to be observed with an oscilloscope.

There are several different *data formats*, as shown in Fig. 25.3.14. The most popular one is Non-Return-to-Zero modulation (NRZ), because the bandwidth of a data stream with that format and a data rate of, for instance, 100 Mbit/s is only 50 MHz (100 MHz if the format is RZ, R1, or RC). RC is used if the average duty

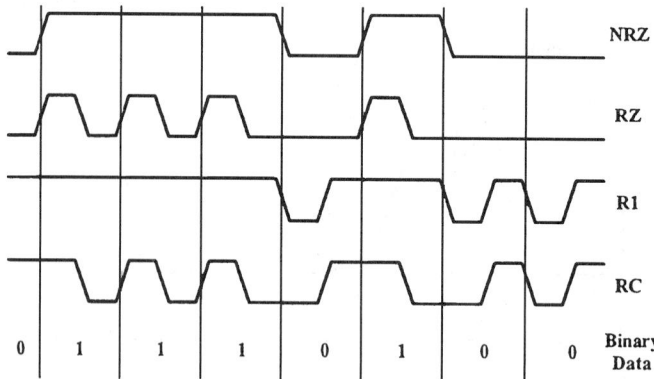

FIGURE 25.3.14   Different data formats. NRZ: nonreturn to zero (zero = logic "0" = –LOL); RZ: return to zero; R1: return to "1" (Logic "1" = HIL); RC: return to complement ("0" "1"; "1" "0").

cycle (see Fig. 25.3.11) of the data stream has to be 50 percent, no matter what the data that are being transmitted look like.

For instance, if the data link that has to be tested cannot transmit a dc signal, an offset (or threshold) of 0 V (or 0 A, respectively) and the data format RC has to be chosen. Sometimes, when selecting RZ, users can even select the duty cycle of the signal. If they choose 25 percent, for instance, a logic "1" will return to "0" after 25 percent of the pulse period. In Fig. 25.3.14, a duty cycle of 50 percent was chosen.

Random Data.   Some pulse generators with data capabilities can generate random data streams, which are also referred to as *pseudorandom binary sequences* or *pseudorandom bit sequences* (PRBS). These sequences are often generated according to an international standard.

Difference between Data/Word Generators and Pulse Generators.   In contrast to pulse generators, most *data generators* are modular systems that can provide up to 50 channels, or more. Data generators usually exhibit a much higher memory depth and more functionality in terms of sequencing—*looping* or *branching*. Some data generators even can provide asynchronous data streams (generated with several independent clock sources). This means that a pulse generator with data capabilities cannot replace a data generator, especially if more than one or two channels are needed. However, having some data capabilities in a pulse generator can be very beneficial.

Frequency Synthesizers

Starting in the early 1960s, frequency synthesis techniques have grown steadily in instrument applications and have become the dominant technology in signal sources. Now, almost all signal generators, whether baseband, RF, or microwave, are based on synthesis technology and offer very good frequency resolution, stability, and absolute frequency accuracy. Many also offer the option of locking to an external reference frequency, often 10 MHz. If all the instruments in a system are locked to the same external reference, then there will be no relative frequency error among the instruments. If the external reference has very good absolute frequency stability and absolute accuracy such as a rubidium or cesium atomic clock frequency standard, then the entire system will have the same stability and absolute frequency accuracy.

Radio-Frequency (RF) Signal Generators

Frequencies from 100 kHz to 1 GHz.   This important class of instrument was developed in the late 1920s when the need was recognized for producing radio receiver test signals with accurate frequencies (1 percent) and amplitudes (1 or 2 dB) over wide ranges. The basic block diagram of these instruments remained nearly unchanged until about 1970, when frequency synthesizer circuits began to usurp *free-running oscillators* in the waveform generator section of the instruments. Shown in Fig. 25.3.15 is a simplified form of a typical RF sine-wave oscillator used in a signal generator. An amplifier producing an output current proportional to its input voltage (a transconductance amplifier) drives a tunable filter circuit consisting of a parallel LC resonant circuit with a magnetically coupled output tap. The tap is connected to the amplifier input with the proper polarity to provide positive feedback at the resonant frequency of the filter. To assure oscillation, it is necessary to design the circuit so that the gain around the loop—the amplifier gain times the filter transfer function—remains greater than unity over the desired frequency tuning range. But since the output signal amplitude remains constant only when the loop gain is exactly unity, the design must also include level-sensitive control of the loop gain.

Microwave Signal Generators

Frequencies from 1 to 30 GHz are usually designated as being in the microwave range. The lower boundary corresponds approximately to the frequency above which lumped-element modeling is no longer adequate for

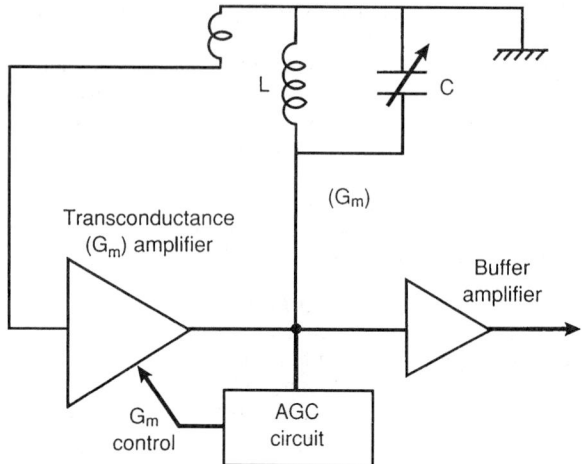

FIGURE 25.3.15 Radio-frequency sine-wave oscillator.

most designs. The range above is commonly referred to as the *millimeter range* because wavelengths are less than 1 cm. It extends up to frequencies where the small wavelengths compared with practically achievable physical dimensions require quasi-optical techniques to be used for transmission and for component design.

Previous generations of microwave sources were designed around vacuum tubes such as the klystron and the backward-wave oscillator. These designs were bulky, often requiring very high voltages and currents, and were subject to drift with environmental variations. More recently, compact solid-state oscillators employing field-effect transistors (FET) or bipolar transistors and tuned by electrically or magnetically variable resonators have been employed with additional benefits in ruggedness, reliability, and stability. Frequency synthesis techniques are now used to provide accurate, programmable sources with excellent frequency stability, and low phase noise. Several varieties of microwave signal generators, each optimized for certain ranges of applications, are described in this section.

CW Signal Generators

For certain applications, a *continuous-wave* (CW) signal generator without modulation capability may be all that is required, providing a cost savings over more sophisticated models. Output power is of importance in typical applications such as where the signal generator serves as a local oscillator (LO) driving a mixer in an up- or down-converter. The signal level needs to be high enough to saturate the mixer to assure good amplitude stability and low noise. If the phase noise of the converted signal is not to be degraded, the phase noise of the signal generator must be sufficiently low.

Other applications for CW sources include exciters, in transmitter testing, sources driving amplifiers, and modulators. High-level accuracy, low spurious and harmonic signal levels, and good frequency resolution may all be important specifications in these and other applications.

Swept Signal Generators

Frequency *swept signal generators* are used for the test and characterization of components and subsystems and for general-purpose applications. They can be used with scalar and vector network analyzers or with power meters or detectors. Sweep can be continuous across a span, or in precise discrete steps. Techniques exist for phase locking throughout a continuous sweep, allowing for high accuracy in frequency. Step sweep techniques, in which the source is phase-locked at each discrete frequency throughout the sweep, are more compatible with some measurement systems.

Signal Generators with Modulation

Microwave signal generators designed to test increasingly sophisticated receivers are being called upon to provide a growing variety of modulations at accurately calibrated signal levels over a wide dynamic range, without generating undesired spurious signals and harmonics. Application-oriented signal generators are available, which provide combinations of modulation formats. These range from simple amplitude and frequency modulation to those that employ a wide variety of digital modulation formats in which discrete symbols are transmitted as combinations of phase and amplitude of the carrier.

For applications where simulations of complex scenarios involving multiple targets or emitters need to be carried to test radar or Electronic Warfare (EW) receivers, or to perform certain tests on satellite and communications receivers, higher-performance signal generators featuring very fast frequency switching and/or sophisticated modulation capabilities are employed. These sources may also feature a variety of software interfaces that can provide suitable *personalities* for various receiver test applications, allowing entry of parameters in familiar form and enabling the creation of complex scenarios involving lengthy sequences of a number of signals.

Types of Modulation

Some of the more commonly required modulation formats found in signal generators are described below, along with some common applications.

Pulse Modulation.   Pulse modulation is used to simulate target returns to a radar receiver, to simulate active radar transmitters for testing EW surveillance or threat warning receivers, or to simulate pulse code modulation for certain types of communications or telemetry receivers. Microwave signal generators can have inputs for externally applied pulses from a system under test or from a pulse generator.

Some microwave signal generators have built-in pulse sources, which may be free-running at selectable rates or can be triggered externally with selectable pulse widths and delays—the latter being used to simulate a variety of distances to a radar target.

Amplitude Modulation (AM).   In addition to the simulation of microwave signals having AM and for AM to PM (amplitude to phase modulation) conversion measurements, amplitude modulation of a microwave signal generator may be needed for simulation of signals received from remote transmitters in the presence of fading phenomena in the propagation path, or from rotating radar antennas. AM should be externally applicable or internally available over a broad range of modulation frequencies without accompanying undesired (incidental) variations in phase.

Frequency Modulation (FM).   For applications where signals with FM need to be provided, signal generators with an external input and/or an internal source are available. The *modulation index* β is defined as

$$\beta = \Delta f/f_m$$

where Δf is the peak frequency deviation and f_m is the modulation frequency.

I/Q (Vector) Modulation.   Digital modulation techniques have essentially supplanted analog modulation methods for communications and broadcasting applications. Modulation is said to be digital if the signal is allowed to assume only one of a set of discrete states (or symbols) during a particular interval when it is to be read. Data are transmitted sequentially at a rate of n bits per symbol, requiring 2^n discrete states per symbol. By representing the unmodulated microwave carrier as a vector with unity amplitude and zero phase, as shown in Fig. 25.3.16, we can display various modulation formats on a set of orthogonal axes commonly labeled I (*in-phase*) and Q (*quadrature phase*).

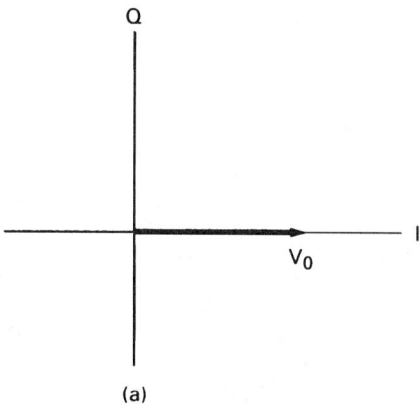

(a)

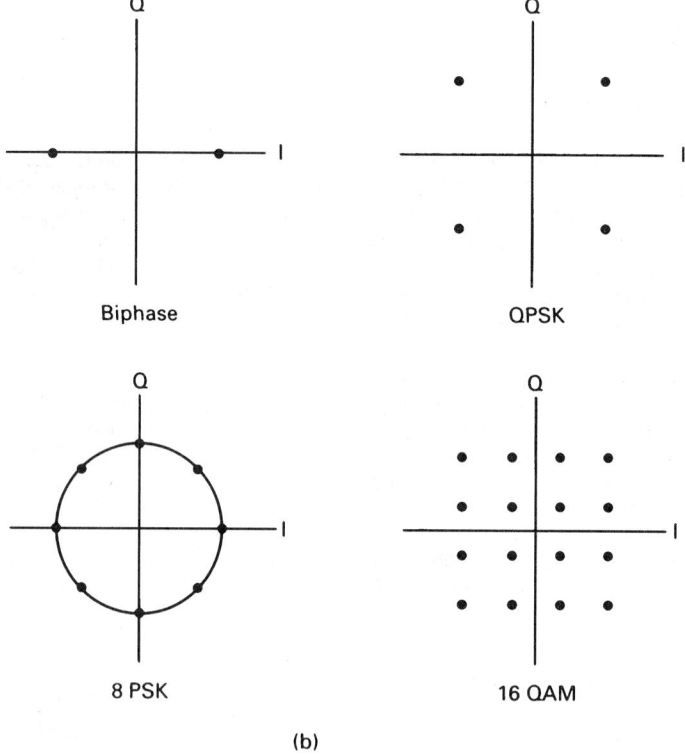

(b)

FIGURE 25.3.16 (a) Unmodulated carrier displayed as a vector with zero phase. (b) Some examples of digital modulation formats. In biphase modulation, there are two states characterized by a carrier at 0° or 18° of relative phase. QPSK (quadrature phase shift keying) has the four states shown and 8-psk (eight-phase shift keying) has the eight states of equal amplitude. The last example illustrates 16QAM (quadrature amplitude modulated) where there are three different amplitudes.

CHAPTER 25.4
LOGIC AND PROTOCOL ANALYZERS

Steven B. Warntjes, Steve Witt

LOGIC ANALYZERS

Steven B. Warntjes

Introduction

The advent of digital circuits dramatically changed the concerns of engineers and technicians working with electronic circuits. Ignoring for a moment digital signal quality or signal integrity, the issues switched from the world of bias points and frequency response to the world of logic ones, zeros, and logic states (see Fig. 25.4.1). This world has been called the *data domain*. Using off-the-shelf components virtually guarantees correct values of voltage and current if clocks are kept to moderate speeds—less than 50 MHz—and fan-in/fan-out rules are observed. The objective for circuit verification and testing focuses on questions of proper function and timing. Although parametric considerations are simplified, the functional complexity and sheer number of circuit nodes are increased tremendously. Measurements to address these questions and to manage the increased complexity are the forte of the *logic analyzer* (Fig. 25.4.2). Logic analyzers collect and display information in the format and language of digital circuits. Microprocessors and microcontrollers are the most common logic-state machines. Software written in either high-level languages, such as C, or in the unique form of a processor's instruction set—*assembly language*—provide the direction for these state machines that populate every level of electronic products.

Most logic analyzers can be configured to format their output as a sequence of assembly processor instructions or as high-level language source code. This makes them very useful for debugging software. For real-time or time-crucial embedded controllers, a logic analyzer is a superb tool to trace program flow and to measure event timing. Because logic analyzers do not affect the behavior of processors, they are excellent tools for system performance analysis and verification of real-time interactions.

Data-stream analysis is also an excellent application for logic analyzers. A stream of data from a digital signal processor or digital communications channel can be easily captured, analyzed, or uploaded to a computer.

Basic Operation

The basic operation of logic analysis consists of acquiring data in two modes of operation: *asynchronous timing mode* and *synchronous state mode*. Logic analysis also consists of emulation solutions used to control the processor in the embedded system under development and the ability to time correlate high-level language source with the captured logic analyzer information.

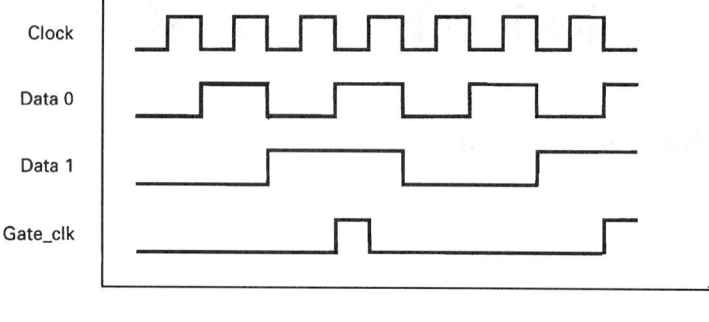

(a)

(b)

FIGURE 25.4.1   (a) Logic timing diagram. Logic values versus time is shown for four signals. (b) Logic state diagram. Inputs *I* and *S* control transitions from state to state. *O* and *E* are outputs set to new values on entry to each state.

Asynchronous Mode.   On screen, the asynchronous mode looks very much like an oscilloscope display. Waveforms are shown, but in contrast to an oscilloscope's two or four channels, there are a large number of channels—eight to several hundred. The signals being probed are recorded either as a 1 or a 0. Voltage variation—other than being above or below the specified logic threshold—is ignored, just as the physical logic elements would do. In Fig. 25.4.3, an analog waveform is compared with its digital equivalent. A logical view of signal timing is captured. As with an oscilloscope, the logic analyzer in the timing mode provides the time base that determines when data values are clocked into instrument storage. This time base is referred to as the

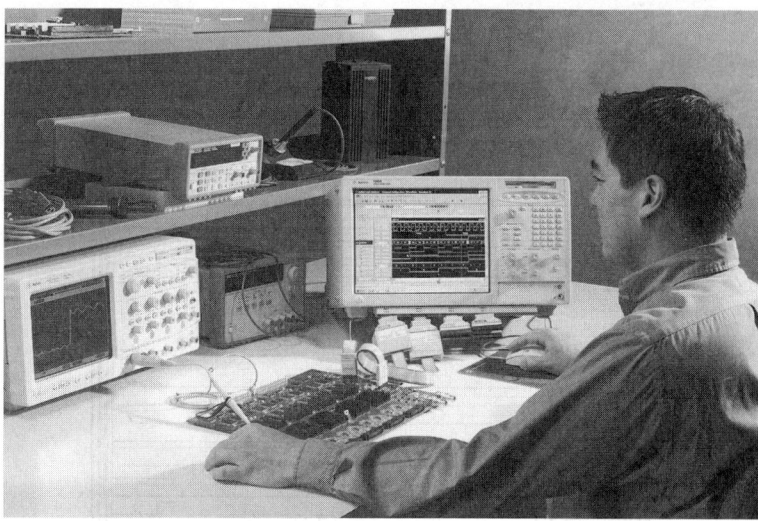

FIGURE 25.4.2   Logic analyzers are supplied as PC-hosted or benchtop versions. Shown here at right is the Agilent Technologies 1680 Benchtop Protocol Analyzer. Note the pods that connect to the device under test in the foreground.

internal clock. A sample logic analyzer display, showing waveforms captured in timing mode, appears in Fig. 25.4.4.

Synchronous Mode.   The synchronous state mode samples the signal values into memory on a clock edge supplied by the system under test. This signal is referred to as the *external clock*. Just as a flip-flop takes on data values only when clocked, the logic analyzer samples new data values or states only when directed by the clock signal. Groupings of these signals can represent state variables. The logic analyzer displays the

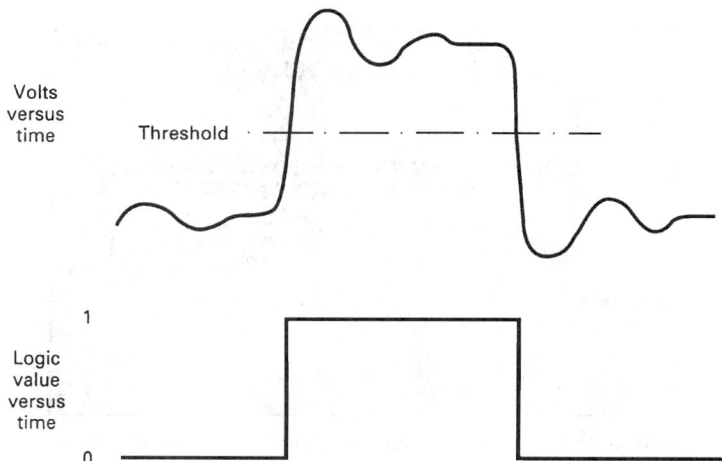

FIGURE 25.4.3   Analog versus digital representations of a signal.

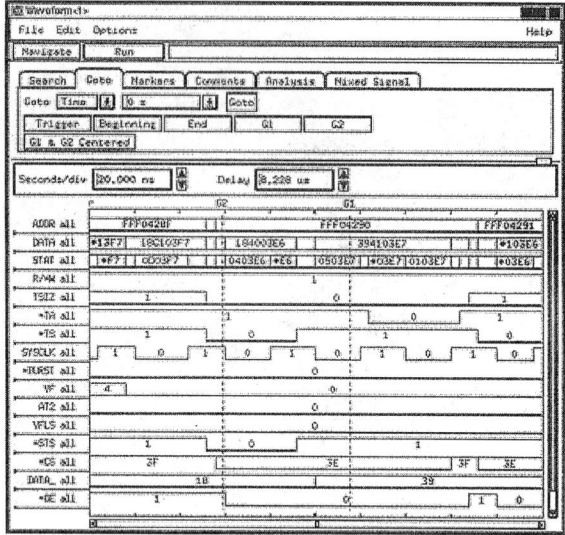

FIGURE 25.4.4 Timing mode display. Display and acquisition controls are at the top. Waveforms are displayed on the bottom two-thirds of the display. Note the multibit values shown for "A8-15."

progression of states represented by these variables. A sample logic analyzer display showing a trace listing of a microprocessor's bus cycles, in *state mode*, is shown in Fig. 25.4.5.

Block Diagram

An understanding of how logic analyzers work can be gleaned from the block diagram in Fig. 25.4.6. Logic analyzers have six key functions: the probes, the high-speed memory, the trigger block, the clock generator, the storage qualifier, and the user interface.

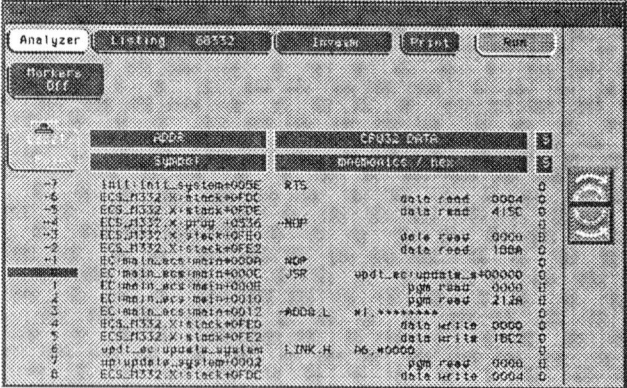

FIGURE 25.4.5 State mode display. Listing shows inverse assembly of microprocessor bus cycles.

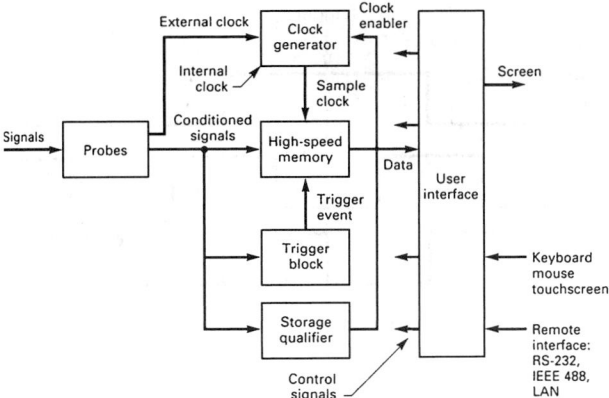

FIGURE 25.4.6 Logic analyzer block diagram.

Probes.   The first function block is the probes. The function of the probes is to establish physical connections with the target circuit under test. To maintain proper operation of the target circuit, it is vital that the probes do not unduly load down the logic signal of interest or disturb its timing. It is common for these probes to operate as voltage dividers. By dividing down the input signal, voltage comparators in the probe function are presented with the lowest possible voltage slew rate. Higher-speed signals can be captured with this approach. The voltage comparators transform the input signals into logic values. Different logic families, such as TTL, ECL, and CMOS, have different voltage thresholds, so the comparators must have adjustable thresholds.

High-Speed Memory.   The second function is high-speed memory, which stores the sampled logic values. The memory address for a given sample is supplied internally. The typical memory depth is hundreds of thousands of samples. Some analyzers can store several megasamples. Usually the analyzer user is interested in observing the logic signals around some event. This event is called the *measurement trigger*. It is described in the next functional block. Samples have a timing or sequence relationship with the trigger event, but are arbitrarily placed in the sample memory, depending on the instantaneous value of the internally supplied address. The memory appears to the user as a continuously looping storage system.

Trigger Block.   The third functional block is the trigger block. Trigger events are a user-specified pattern of logical ones and zeros on selected input signals. Shown in Fig. 25.4.7 is how a sample trigger pattern corresponds with timing and state data streams. Some form of logic comparator is used to recognize the pattern of interest. Once the trigger event occurs, the storage memory continues to store a selected number of post-trigger samples. Once the post-trigger store is complete, the measurement is stopped. Because the storage memory operates as a loop, samples before the trigger event are captured, representing time before the event. Sometimes this pre-trigger capture is referred to as *negative time capture*. When searching for the causes of a malfunctioning logic circuit, the ability to view events leading up to the problem—*the trigger event*—makes the logic analyzer extremely useful.

Clock Generator.   The fourth block is the clock generator. Depending on which of the two operating modes is selected, state or timing, sample clocks are either user supplied or instrument supplied. In the state mode, the analyzer clocks in a sample based on a rising or falling pulse edge of an input signal. The clock generator function increases the usability of the instrument by forming a clock from several input signals. It forms the clocking signal by *ORing* or *ANDing* input signals together. The user could create a composite clock using

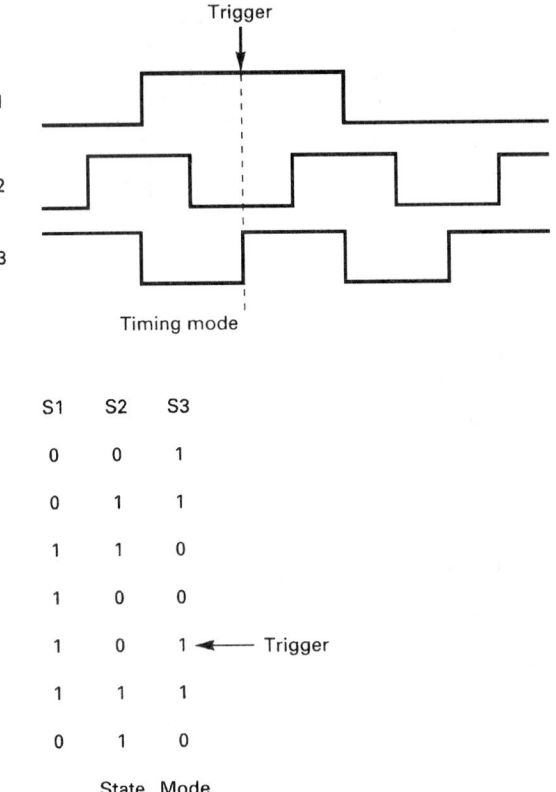

FIGURE 25.4.7 Example of trigger pattern showing match found with timing mode data and then state mode data. Trigger pattern is "1 0 1" for input signals S_1, S_2, and S_3.

logic elements in the circuit under test, but it is usually more convenient to let the analyzer's clock generator function do it. In timing mode, two different approaches are used to generate the sample clock. Some instruments offer both approaches, so understanding the two methods will help obtain more from the instrument. The first approach, or *continuous storage mode*, simply generates a sample clock at the selected rate. Regardless of the activity occurring on the input signals, the logic values at the time of the internal clock are entered in memory (see Fig. 25.4.8).

The second approach is called *transitional timing mode*. The input signals are again sampled at a selected rate. The clock-generator-function clocks the input signal values into memory only if one or more signals change their value. Measurements use memory more efficiently because storage locations are used only if the inputs change. For each sample, a time stamp is recorded. Additional memory is required to store the time stamp. The advantage of this approach over continuous storage is that long-time records of infrequent activity or bursts of finely timed events can be recorded, as shown in Fig. 25.4.9.

Storage Qualifier. The fifth function is the storage qualifier. It also has a role in determining which data samples are clocked into memory. As samples are clocked, either externally or internally, the storage qualifier function looks at the sampled data and tests them against a criterion. Like the trigger event, the qualifying

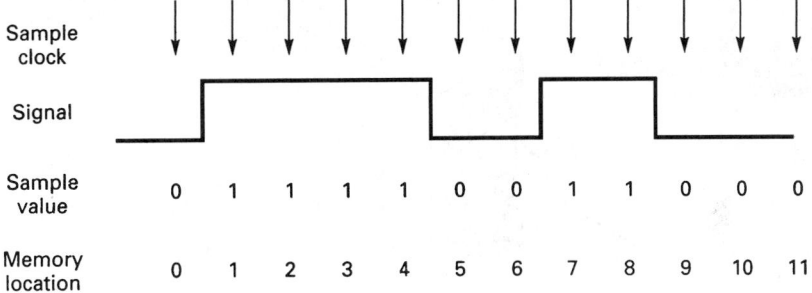

FIGURE 25.4.8 Continuous storage mode. A sample value is captured at each sample clock and stored in memory.

criterion is usually a one-zero pattern of the incoming signal. If the criterion is met, then the clocked sample is stored in memory. If the circuit under test is a microprocessor bus, this function can be used to separate bus cycles, steering them to a specific input/output (I/O) port—distinct from instruction cycles or cycles steered to other ports.

User Interface.   The sixth function, the user interface, allows the user to set up and observe the outcome of measurements. Benchtop analyzers typically use a dedicated keyboard and either a *cathode-ray tube* (CRT) or a *liquid crystal display* (LCD). Many products use *graphic user interfaces* similar to those available on personal computers. *Pull-down menus, dialog boxes, touch screens*, and *mouse pointing devices* are available.

   Since logic analyzers are used sporadically in the debug process, careful attention to a user interface that is easy to learn and use is advised when purchasing. Not all users operate the instrument from the built-in keyboard and screen. Some operate from a personal computer or workstation. In this case, the *user interface* is the remote interface: IEEE-488 or local area network (LAN). Likewise, the remote interface could be the user's Web browser of choice, if the logic analyzer is Web enabled.

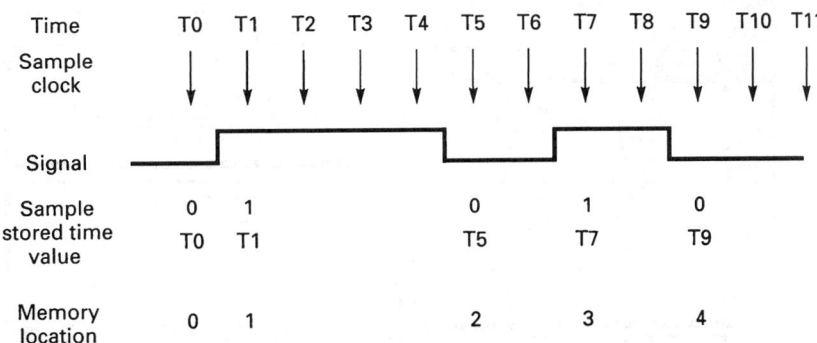

FIGURE 25.4.9 Transitional storage mode. The input signal is captured at each sample clock, but is stored into memory only when the data changes. A time value is stored at each change so that the waveform can be reconstructed properly.

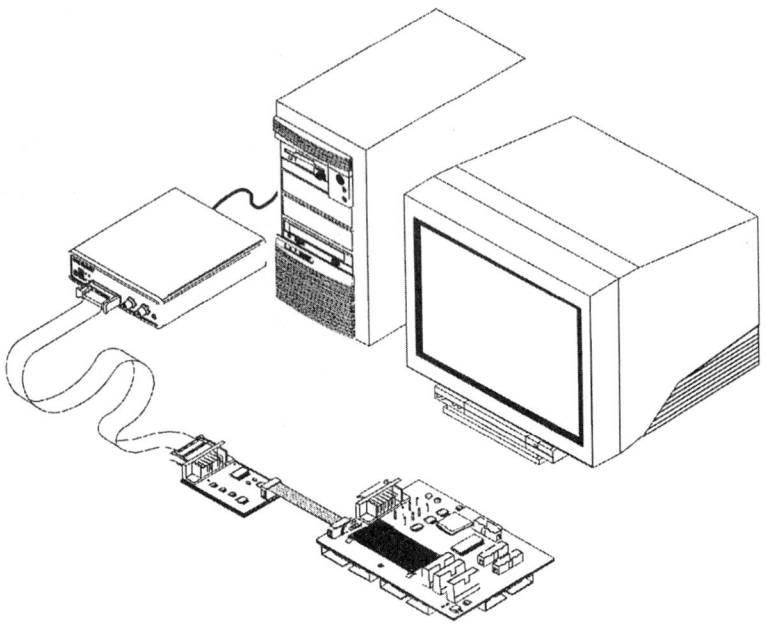

FIGURE 25.4.10 Example of an emulation solution with a PC connected to a customer target to control the processor under development.

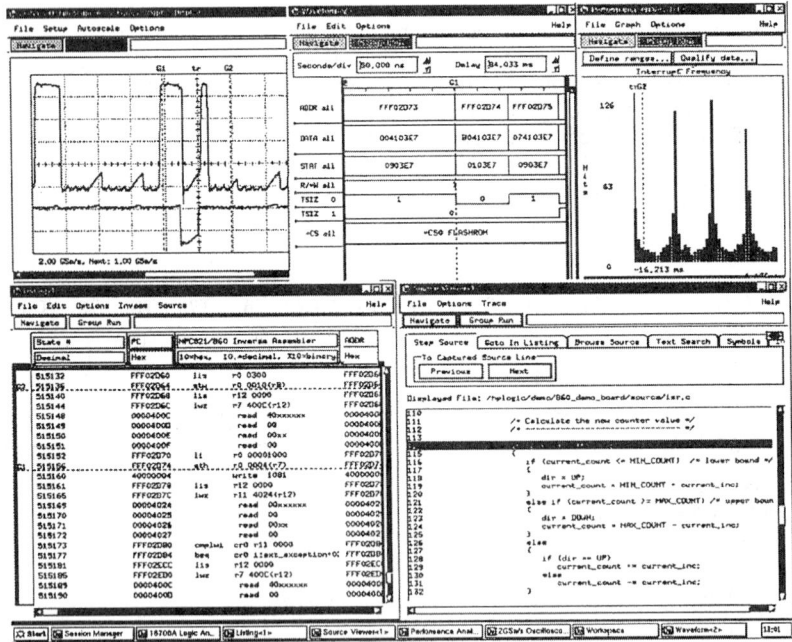

FIGURE 25.4.11 Correlated high-level language display. Lower right is the high-level source code; lower left, the assembly code listing. Upper left is a time-correlated oscilloscope display; upper middle, a timing waveform display; upper right, a performance analysis snapshot.

Emulation Solutions

An emulation solution is a connection to the processor in the embedded system used to control the program execution. The connection is usually on the order of five to 10 processor signals. These signals control the program execution, including the state of the processor, such as *running, reset,* or *single stepping,* and also the ability to quickly download the processor's executable code and examine/modify processor memory and registers. This processor control is usually accomplished through dedicated, on-processor debug resources. An emulation solution enable the user to perform software and hardware debugging as well as time correlate software and hardware activity (see Fig. 25.4.10).

High-Level Language Source Correlation

High-level language source correlation provides a real-time trace or acquisition of processor address, data, and status information linked to a software high-level source language view. This information is then *time correlated* to activity captured by the rest of the logic analyzer's acquisition modules, such as oscilloscope channels. Symbols from the user's software program can also be used to specify trigger conditions and are listed in the analyzer display (Fig. 25.4.11). This feature uses the information provided in the object file from the customer's compiler to build a database of source files, line numbers, and symbolic information. The HLL source correlation is a nonintrusive tool that typically does not require any major changes in the software compilation process.

PROTOCOL ANALYZERS

Steve Witt

Introduction

Computer communication networks are made up of many different computer systems, applications, and network topologies. The capital investment in cabling and transmission infrastructure is massive. The number of users demanding access to computer networks is ever increasing and these users are demanding more bandwidth, increased performance, and new applications. There is a constant stream of new equipment and services being introduced in the marketplace. In this complex environment, computer networking is made possible only by equipment and services vendors adhering to standards covering protocols, physical connectors, electrical interfaces, topologies, and data formats. Protocol analysis is used to ensure that the products implemented according to these standards behave as per specifications.

The term protocol analyzer describes a class of instruments that are dedicated to performing protocol analysis. The instruments are special-purpose computer systems that act as a node on the network, but, unlike a typical node on the network, monitor and capture all of the network traffic for analysis and testing.

Protocol Definition. Generally speaking, a protocol is a code or a set of rules specifying the correct procedure for a diplomatic exchange. In terms of computer networks, a protocol is a specific set of rules, procedures, and conventions defining the format and timing of data transmission between devices connected to a computer network. Protocols are defined so that devices communicating on a computer network can exchange information in a useful and efficient manner. Protocols handle synchronization, addressing, error correction, header and control information, data transfer, routing, fragmentation and reassembly, encapsulation, and flow control. Protocols provide a means for exchanging information so that computer systems can provide services and applications to end users. The wide range of protocols in use is a result of the wide range of transmission medium in use, the wide range of applications and services available to end users, and the numerous independent organizations and vendors creating protocol standards.

Communication via a computer network is possible only when there is an agreed-upon format for the exchange of information and when there is a common understanding of the content of the information being exchanged. Therefore, protocols must be defined by both semantic and syntactic rules. Semantics refers to the meaning of the information in the frame of data including control information for coordination and error handling. An example of the semantic information in a frame is a request to establish a connection, initiated by one computer and sent to another computer. Syntax refers to the structure, arrangement, and order of the protocol including data format and signal levels. An example of the syntax of a protocol is the relative position of a protocol field in the frame, such as the network address. Protocol analysis is concerned with both the syntax and the semantics of the protocols.

Protocol Standards. Communication between devices connected to computer networks is controlled by transmission and protocol standards and recommendations. These standards are necessary for different vendors to offer equipment and services that interoperate with one another. While standards can be defined and implemented in the private sector by computer and network component vendors such as Cisco Systems, Hewlett-Packard, and IBM, most standards and recommendations are created by organizations including but not limited to ANSI (American National Standards Institute), CCITT (Consultative Committee on International Telegraphy and Telephony), ETSI (European Telecommunications Standards Institute), IEEE (Institute of Electrical and Electronic Engineers), ISO (International Standards Organization), and the ITU (International Telecommunications Union). The ATM Forum and the IETF (Internet Engineering Task Force) are technical working bodies that develop standards for networking products.

The OSI Reference Model. The ISO, located in Geneva, Switzerland, is responsible for many of the international standards in computer networking. The ISO defined a model for computer communications networking called the Open Systems Interconnection Reference Model. This model, commonly called the OSI model, defines an open framework for two computer systems to communicate with one another via a communications network. The OSI model (see Fig. 25.4.12) defines a structured, hierarchical network architecture. The OSI model consists of the communications subnet, protocol layers 1 to 3, and the services that interface to the applications executing in the host computer systems (protocol layers 4 to 7). The combined set of protocol layers 1 to 7 is often referred to as a protocol stack. Layer 7, the applications layer, is the interface to the user application executing in the host computer system. Each layer in the protocol stack has a software interface to the application below it and above it. The protocol stack executes in a host computer system. The only actual physical connection between devices on the network is at the physical layer where the interface hardware connects to the physical media. However, there is a logical connection between the two corresponding layers in communicating protocol stacks. For example, the two network layers (layer 3 in the OSI reference model) in two protocol stacks operate as if they were communicating directly with one another, when in actuality they are communicating by exchanging information through their respective data link layers (layer 2 in the OSI reference model) and physical layers (layer 1 in the OSI reference model).

Current network architectures are hierarchical, structured, and based in some manner on the OSI reference model. The functionality described in the OSI reference model is embodied in current network architectures, albeit at different layers or combined into multiple layers.

Network Troubleshooting Tools. There are two broad categories of products used to implement and manage computer networks—those that test the transmission network and those that test the protocol information transferred over the transmission network. Testing the protocols is commonly referred to as protocol analysis and it can be accomplished with several different types of products including:

Network Management Systems—comprehensive, integrated networkwide systems for managing and administrating systems and networks. Protocol analysis is one of many applications performed by network management systems. Network troubleshooting is performed by acquiring network data from devices on the network and from instrument probes distributed through the network.

Distributed Monitoring Systems—performance monitoring and troubleshooting applications that are implemented with instrument probes or protocol analyzers that are distributed throughout the network. The probes and analyzers are controlled with a management application running on a workstation or PC.

THE OSI REFERENCE MODEL

The OSI Reference Model defines an internationally accepted standard for computer network communications.
Host system applications exchange information in a hierarchical and structured manner.

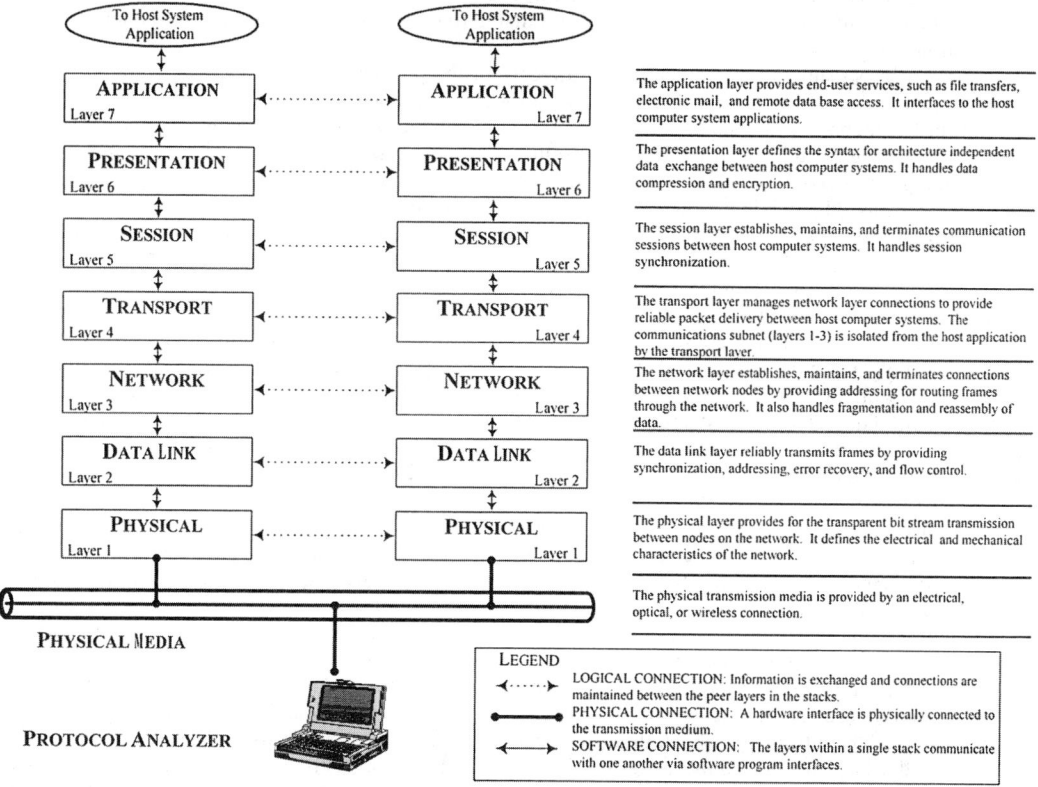

FIGURE 25.4.12   OSI model.

Protocol Analyzers—specialized instrumentation dedicated to protocol analysis. Protocol analyzers are used to troubleshoot network problems and to monitor the performance of networks.

Handheld Test Tools—special purpose tools that are very small, lightweight, and usually battery operated. They perform a variety of measurements such as continuity tests, transmission tests, and simple protocol analysis measurements such as simple statistics and connectivity tests.

The Need for Protocol Analysis.   In order for two applications running on two different computer systems to communicate with one another (e.g., a database application executing on a server and a client application performing database queries), meaningful information must be continually, efficiently, and correctly exchanged. This requires a physical connection to exist, either twisted pair copper wire, coaxial cable, optical fiber, or wireless (e.g., radio transmission). The physical characteristics and specifications of the transmission media must be standardized so that different computer systems can be electrically connected to one another. The bit streams exchanged over the physical media must be encoded such that the analog stream of information can be converted to digital signals. In order for two people to effectively communicate they must speak the same language. Similarly, for two computer systems to communicate they must speak the same "language." Therefore, the bit stream must conform to a standard that defines the encoding scheme, the bit order (least significant bit first or most significant bit first), and the bit sense (a high value defined either as a 1 or a 0). Any errors in the transmission must be detected and recovered and, if necessary, the data must be retransmitted. A protocol analyzer is used to examine the bit stream and ensure that it conforms to the protocol standards that define the encoding schemes, bit sequences, and error conditions.

Once a bit stream can be transmitted and received, the physical communication is established and the exchange of information can be accomplished. Information is exchanged in logical units of information. A protocol frame, packet, message, or cell (in this chapter, the term frame will be used to mean any or all of these) is the logical unit of information transmitted on the physical infrastructure of a computer network. Depending on the type of network and protocols, these frames of data are either fixed in size, such as the 53 byte cells used by ATM networks, or they can be variable in size, such as the 64 to 1518 byte frames used by Ethernet networks. The most fundamental aspect of protocol analysis is the collection and analysis of these frames.

Networks usually have more than one path connecting different devices. Therefore, the frames in which the data are contained must be addressed properly so that they can traverse single or multiple routes through the network. Fragmentation and reassembly issues must be properly handled—frames are often disassembled and reassembled so that they can be of a proper size and can be encapsulated with the proper header information to ensure that the intermediate and end devices in the network can properly manipulate them. The network must also handle error conditions such as nodes that stop responding on the network, nodes that transmit error frames or signals, and nodes that use excessive bandwidth. A protocol analyzer is used to examine the addresses of the frames, check fragmentation and reassembly, and investigate errors.

Connections are established so that communication is efficient. This prevents the communication channel from being redundantly set up each time a frame is transmitted. This is similar to keeping a voice line open for an entire telephone conversation between two people, rather than making a phone call for each sentence that is exchanged. To ensure this efficiency, connections or conversations are established by devices on the network so that the formal handshaking doesn't have to be repeated for each information exchange. Protocol analysis includes scrutinizing protocol conversations for efficiency and errors.

The data that are transferred to the host system application in the frames must conform to an agreed-upon format, and if necessary it must be converted to an architecture-independent format so that both computer systems can read the information. Each time a user enters a command, downloads a file, starts an application, or queries a database, the preceding sequence of processes is repeated. A computer network is continuously performing these operations in order to execute an end user's application. Protocol analysis involves critically examining the formatted data that are exchanged between host system applications.

Protocol Analyzers

A protocol analyzer is a dedicated, special purpose computer system that acts as a node on the network but, unlike a typical node on the network, it monitors and captures *all* of the network traffic for analysis and testing (Fig. 25.4.13). Protocol analyzers provide a "window into the network" to allow users to see the type of traffic on the network. Many problems can be solved quickly by determining what type of traffic is or is not present on the network and if protocol errors are occurring. Without a "window into the network" network managers must rely on indirect measures to determine the network behavior, such as observing the nodes attached to the network and the problems the users are experiencing.

Protocol analyzers provide capabilities to compare data frames with the protocol standards (protocol decodes), load and stress networks with traffic generation, and monitor network performance with statistical analysis. The measurement capabilities have expanded to focus on a broad set of applications such as network troubleshooting, network performance monitoring, network planning, network security, protocol conformance testing, and network equipment development. These new applications go far beyond simply examining network traffic with protocol decodes.

The term protocol analyzer most commonly refers to the portable instruments that are dispatched to trouble sites on the network, i.e., the critical links or segments that are experiencing problems. The emphasis in these products is

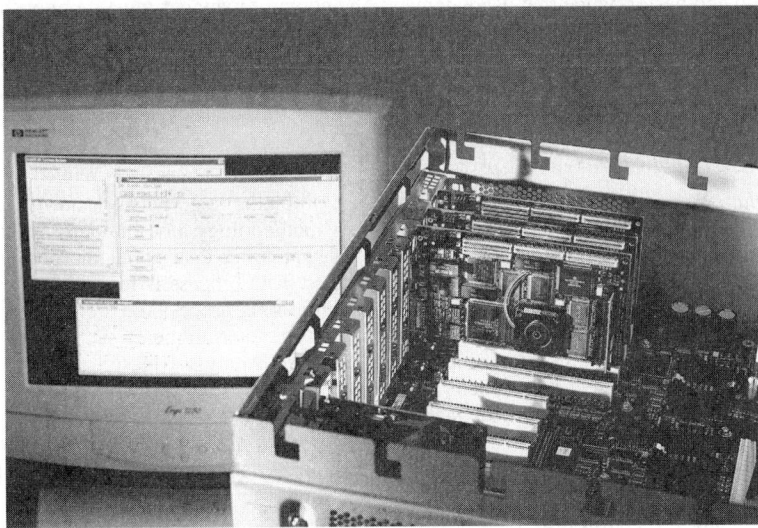

FIGURE 25.4.13 Protocol analyzers come in a number of physical configurations. Shown here are 3 Agilent Technologies E2920 PCI/PCI-X Exerciser and Analyzer cards plugged into a server undergoing production testing.

portability—products that are light-weight, rugged, and include the maximum amount of functionality. Network troubleshooters require a product that connects to any point in an internetwork, regardless of the physical implementation of the network. Therefore, most protocol analyzers are able to accommodate multiple network interface modules. To avoid multiple trips back to the office to get additional equipment, network troubleshooters require a "one handle solution," a product that integrates as much test capability as possible into one portable product.

Portable protocol analyzers are most commonly used for network troubleshooting in installation and maintenance applications. Such products focus on installing networks and troubleshooting network problems. Installing networks requires the network engineer to stress the network or network devices using scripts that emulate setting up logical connections, placing calls, and creating high traffic scenarios. Troubleshooting network problems requires that the network engineer have extensive protocol decodes available to examine whatever frames are present on the network. Protocol statistics and expert analysis are used to identify network errors.

Protocol Analysis

A network fault is any degradation in the expected service of the network. Examples of service degradation include an excessively high level of bit errors on the transmission medium, a single user monopolizing the network bandwidth, a misconfigured device, or a software defect in a device on the network. Regardless of the cause, the network manager's fundamental responsibility is to ensure that such problems are fixed and that the expected level of service is restored. To accomplish this, the network manager must troubleshoot the network and isolate faults when the inevitable problems occur. Many of the difficult to detect, intermittent problems can only be recreated if the network is stressed by sending frames. Many of the network problems are caused by the constantly changing configurations of the devices attached to the network, so the network manager must manage the configuration of the network devices. In order to ensure the bandwidth and performance that users demand, the network manager must monitor the performance of the network and plan accordingly for future growth. Network managers are also concerned with the security of their networks and use protocol analysis tools to monitor for illegal or unauthorized frames on network segments.

Protocol Analysis Applications

Fault isolation and troubleshooting. Most network problems (e.g., network down time) are solved by following a rigorous troubleshooting methodology. This methodology, not unique to network troubleshooting, consists of observing that a problem has occurred, gathering data about the problem, formulating a hypothesis, and then proving or disproving the hypothesis. This process is repeated until the problems are resolved. Protocol analysis is used in the network troubleshooting process for first observing that a problem has occurred, and next gathering data (using protocol analysis measurements such as protocol decodes and protocol statistics, as described in the section entitled "Protocol Analysis Measurements"). The user then formulates a hypothesis and uses the protocol analysis measurements to confirm cause of the problem, ultimately leading to a solution. The protocol analyzers can then be used to confirm that the problem has indeed been repaired.

Performance monitoring. Determining the current utilization of the network, the protocols in use, the errors occurring, the applications executing, and the users on the network is critical to understanding if the network is functioning properly or if problems such as insufficient capacity exist. Monitoring performance can be used over short time periods to troubleshoot problems or it can be used over long time periods to determine traffic profiles and optimize the configuration and topology of the network.

Network baselining. Every network is unique—different applications, distinct traffic profiles, products from numerous vendors, and varying topologies. Therefore, network managers must determine what is normal operation for their particular network. A network baseline is performed to determine the profile of a particular network over time. A profile is made up of statistical data including a network map, the number of users, protocols in use, error information, and traffic levels. This information is recorded on a regular basis (typically daily or weekly) and compared to previously recorded results. The baselining information is used to generate reports describing the network topology, performance, and operation. It is used to evaluate network operation, isolate traffic-related problems, access the impact of hardware and software changes, and plan for future growth.

Security. Networks are interconnected on a global scale; therefore, it is possible for networks to be illegally accessed. Illegal access can be knowingly performed by someone with criminal intent or it can be the result of an error configuration of a device on the network. Protocol analysis tools, with their powerful filtering, triggering, and decoding capabilities can detect security violations.

Stress testing. Many errors on networks are intermittent and can only be recreated by generating traffic to stress network traffic levels, error levels, or by creating specific frame sequences and capturing all of the data. By stress testing a network and observing the results with protocol statistics and decodes many difficult problems can be detected.

Network mapping. Networks continually grow and change, so one of the big challenges facing network engineers and managers is determining the current topology and configuration of the network. Protocol analysis tools are used to provide automatic node lists of all users connected to the network as well as graphical maps of the nodes and the internetworks. This information is used to facilitate the troubleshooting process by quickly being able to locate users. It is also used as a reference to know the number and location of network users to plan for network growth.

Connectivity testing. Many network problems are the result of not being able to establish a connection between two devices on the network. A protocol analyzer can become a node on the network and send frames, such as a PING, to a device on the network and determine if a response was sent and the response time. A more sophisticated test can determine, in the case of multiple paths through the network, which paths were taken. In many WANs connectivity can be verified by executing a call placement sequence that establishes a call connection to enable a data transfer.

Conformance testing. Conformance testing is used to test data communications devices for conformance to specific standards. These conformance tests consist of a set of test suites (or scenarios) that exercise data communications equipment fully and identify procedural violations that will cause problems. These conformance tests are used by developers of data communications equipment and by carriers to prevent procedural errors before connection to the network is allowed. Conformance tests are based on the applicable protocol standard.

Protocol Analysis Measurements.   Protocol analyzer functionality varies depending on the network technology (e.g., LAN vs. WAN), the targeted user (e.g., R&D vs. installation), and the specific application (e.g., fault isolation vs. performance monitoring). Protocol analysis includes the entire set of measurements that allow a user to analyze the information on a computer communications network. These measurements include:

- Protocol decodes
- Protocol statistics
- Expert analysis
- Traffic generation
- Bit error rate tests
- Stimulus/response testing
- Simulation

Because protocol analysis requires associating the results of different measurements, the above measurements are typically made in combination. Thus, protocol analyzers include different sets or combination of these measurements. A good user interface combines pertinent information for the user and integrates the results of the different measurements.

Protocol decodes.   Protocol decodes, also referred to as packet traces, interpret the bit streams being transmitted on the physical media. A protocol decode actually *decodes* the transmitted bit stream. The bit stream is identified and broken into fields of information. The decoded fields are compared to the expected values in the protocol standards, and information is displayed as values, symbols, and text. If unexpected values are encountered, then an error is flagged on the decode display. Protocol decodes follow the individual conversations and point out the significance of the frames on the network by matching replies with requests, monitoring packet sequencing activity, and flagging errors. Protocol decodes let the user analyze data frames in detail by presenting the frames in a variety of formats.

Figure 25.4.14*a* shows a summary of protocol packets that have been captured on the network. Figure 25.4.14*b* shows a detailed view of one particular packet, in this case, the packet with reference number 4264. It displays all of the detailed fields of an http packet.

Protocol statistics.   An average high-speed computer network such as a 10 Mbit/s Ethernet handles thousands of frames per second. Protocol statistics reduce the volumes of data captured into meaningful statistical information providing valuable insight for determining the performance of a network, pinpointing bottlenecks, isolating nodes or stations with errors, and identifying stations on the network. Protocol analyzers keep track of hundreds of different statistics. The statistical information is typically displayed as histograms, tables, line graphs, and matrices.

Expert analysis.   Troubleshooting computer networks is made complicated by the wide range of network architectures, protocols, and applications that are simultaneously in use on a typical network. Expert analysis reduces thousands of frames to a handful of significant events by examining the individual frames and the protocol conversations for indications of network problems. It watches continuously for router and bridge misconfigurations, slow file transfers, inefficient window sizes, connection resets, and many other problems. Thus, data are transformed into meaningful diagnostic information.

Traffic generation.   Many network problems are very difficult to diagnose because they occur intermittently, often showing up only under peak load. Traffic generators provide the capability to simulate network problems by creating a network load that stresses the network or a particular device sending frame sequences on to the network.

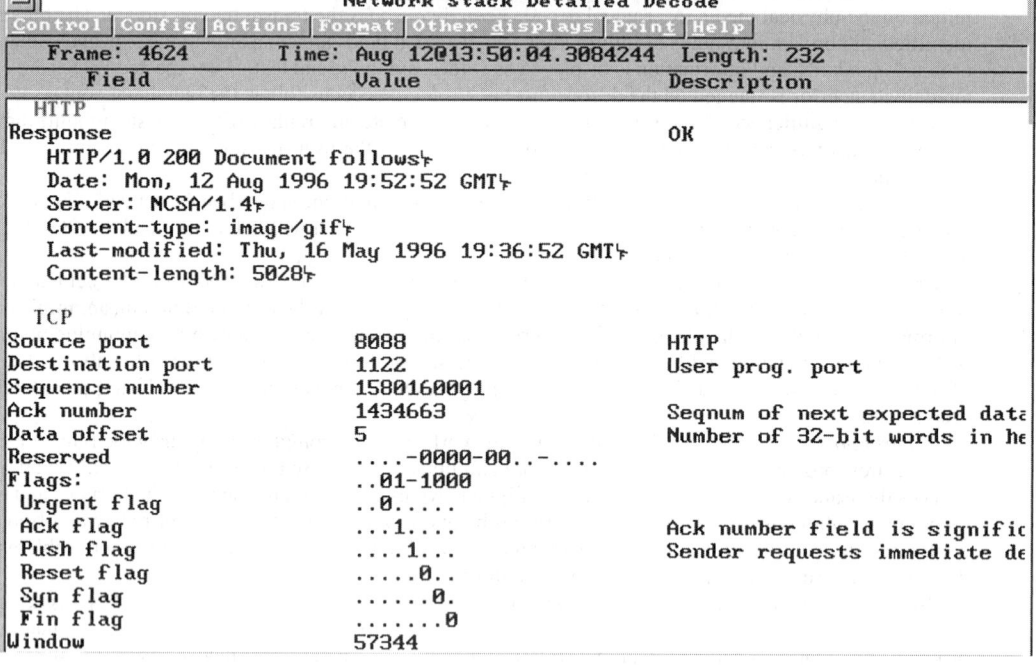

FIGURE 25.4.14 (*a*) Summary view of protocol packets as captured on a network. (*b*) Detailed view of a particular packet, with reference number 4264.

Bit error rate tests.   Bit error rate (BER) tests are transmission tests used to determine the error rate of the transmission media or the end-to-end network. While advanced BER measurements reside in the domain of sophisticated transmission test sets, protocol analysis, particularly in a WAN or ATM environment, often requires a verification of the media. BER tests are performed by transmitting a known bit pattern onto the network, looping it back at a point on the network, and receiving the sequence. The bit error rate is calculated as a percentage of the bits in error compared to the total number of bits received.

Stimulus/response testing.   While many networking problems can be quickly solved with decodes and statistics, many of the more difficult problems cannot be solved in a nonintrusive manner. In this case, it is necessary to actively communicate with the devices on the network in order to recreate the problem or obtain necessary pieces of information to further isolate a problem. So, the user can actively query or stress the network and observe the results with decodes and statistics.

Simulation.   In the context of protocol analysis, simulation can take two forms: protocol simulation and protocol emulation. Protocol simulation allows the user to send strings of data containing selected protocol headers along with the encapsulated data. In this way, the operation of a network device can be simulated for the purpose of testing a suspected problem or for establishing a link to confirm operation. Protocol emulators are software that controls the operation of the protocol analyzer automatically.

Protocol Analysis Measurement Functions.   Protocol analysis consists of using measurements such as those described in the previous section to isolate network problems or monitor network performance. The protocol analysis measurement options described in this section are a set of orthogonal capabilities to the measurements described in the previous section. For example, a protocol analyzer can gather statistics on the network traffic, but a more effective troubleshooting approach is to set a capture filter and run the statistics so only the data between a file server and a router is analyzed.

Data capture.   The most fundamental attribute of protocol analysis is the ability to capture the traffic from a live network and to store these data into a capture buffer. The captured data can then be analyzed and reanalyzed at the user's discretion. Once data are captured in the capture buffer they can be repeatedly examined for problems or events of significance. Search criteria, such as filter criteria, are based on address, error, protocol, and bit patterns.

Data logging.   Many network troubleshooting sessions can be spread over hours and days. The capture buffer on a typical protocol analyzer is filled in seconds on a high-speed network. Therefore, data logging capabilities are crucial for setting up long troubleshooting sessions. The user can specify a file name and a time interval for recording critical information. This information is then regularly stored to hard disk and can be examined by the user at a later time. Information that is typically stored to disk includes frames matching a user specified filter, statistics results, or the results of a programmed test.

Filtering.   The key to successfully troubleshooting network problems is based on eliminating the unnecessary information and focusing on the critical information that is essential to solving the problem. Computer networks process thousands of frames per second. A protocol analyzer can quickly fill a capture buffer with frames, and a user can sift through protocol decodes searching for the errors. But this is a time-consuming and tedious task. The most powerful function of protocol analysis is the ability to filter the data on the network in order to isolate problems. The function of a filter is very similar to that of a trigger (sometimes called a trap). Specific filter patterns are set by the user of a protocol analyzer, and these filters are then compared with the data from the network. Filters range from simple bit pattern matching to sophisticated combinations of address and protocol characteristics. Fundamentally, there are two types of filters—capture filters and display filters.

Capture filters are used to either include or exclude data from being stored in a protocol analyzer's capture buffer. Capture filters make it possible to only collect the frames of interest by eliminating extraneous frames. This effectively increases the usage of the capture buffer. Rather than a capture buffer that only contains six error frames out of 40,000 captured frames in the buffer, the data are filtered so that only error frames are in the buffer. More frames of interest can be captured, and they can be located more quickly. A disadvantage of capture filters is that it is necessary for the user to know what to filter, i.e., to have some idea of what problem to investigate. A second disadvantage is that the frames that were filtered out may contain the sequence of events leading up to the error frame.

In many situations the source of a network problem is not known; therefore it is necessary to capture all of the frames on the network and use display filters to repeatedly filter the frames. Because all of the frames are stored in the capture buffer, the frames can be played back through the display filters. Display filters act on the frames once they have been captured. Frames can be selected for display by measurements such as protocol decodes.

Filter conditions can be combined to form more powerful filter criteria. Typically, as the troubleshooting process progresses, the user discovers more and more information about the network problem. As each new fact is discovered, it can be added to the filter criteria until finally the problem is identified. For example, to isolate a faulty Ethernet network interface card it is necessary to filter on the MAC address of the suspicious node and bad frame check sequence (FCS) simultaneously.

Triggers and actions.   In order to troubleshoot network problems, it is often necessary to identify specific frames or fields in frames. Triggers are used to detect events of significance to the user and then initiate some action. Triggers and filters operate the same way in terms of recognizing conditions on the network. The parameters for setting trigger criteria are the same as the "filter types." The trigger is a key capability of protocol analyzers, since it allows the automatic search of a data stream for an event of significance, resulting in some action to be taken. Possible trigger actions include:

- Visual alarm on the screen
- Audible alarm
- Start capturing data in the capture buffer continuously
- Start capturing data, fill the capture buffer and stop
- Position the trigger in the capture buffer and stop capturing data
- End the data capture
- Increment a counter
- Start a timer
- Stop a timer
- Make an entry in the event log
- Start a specific measurement
- Send an SNMP trap
- Log data to disk

Protocol Analyzer Block Diagram

There are three main components to any protocol analyzer:

- Computing platform
- Analysis and acquisition system
- Line interface

The functions of a protocol analyzer are described in Fig. 25.4.15.

Computing Platform.   The computing platform is a general-purpose processing system—typically a PC or a UNIX workstation. The computing platform executes the user interface for the product and controls the measurements that are executed in the analysis and acquisition systems. It is very common for other applications to be run in conjunction with a protocol analyzer. These include spreadsheet applications for analyzing data and measurement results, network management software, and terminal emulation applications that log into computer systems on the network. Therefore, computing platforms are usually based on an industry standard system that allows a user to openly interact with the application environment.

Analysis and Acquisition System.   Fundamentally a protocol analyzer *acquires* data from the network and then *analyzes* the data. Thus, the analysis and acquisition system is the core of a protocol analyzer. This system

Protocol Analyzer Block Diagram

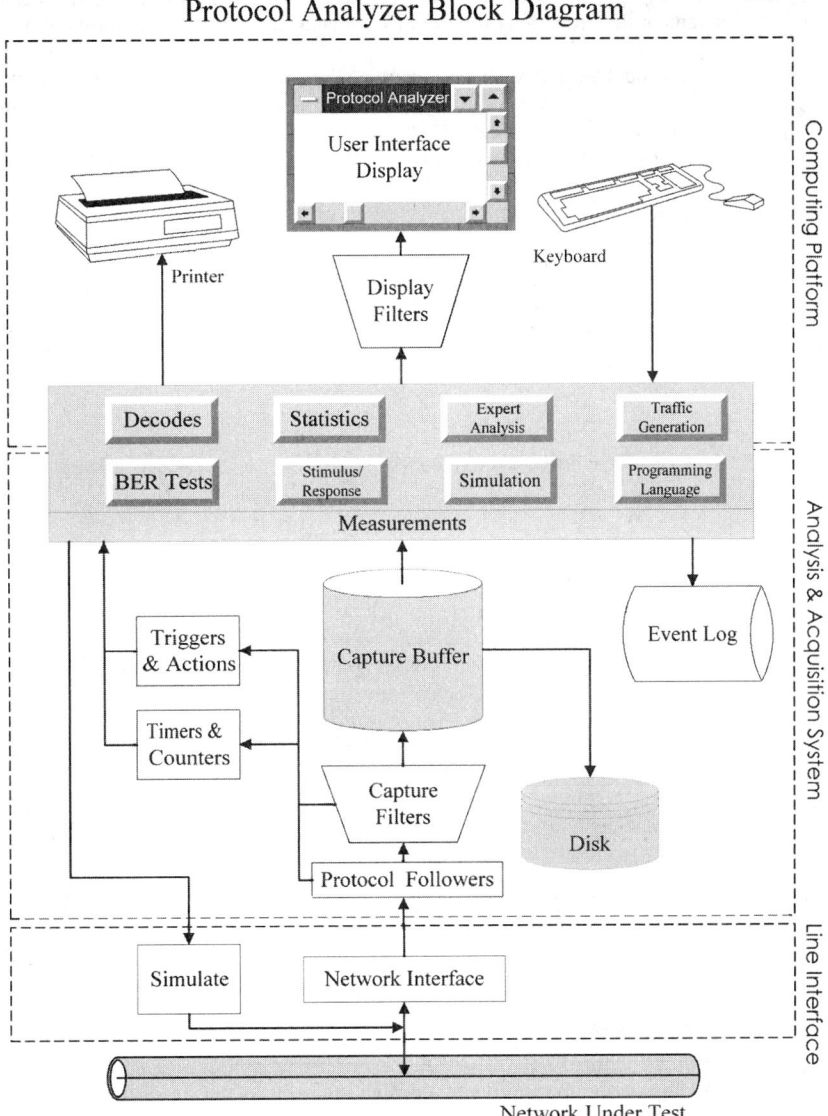

FIGURE 25.4.15   A protocol analyzer block diagram.

is essentially responsible for transferring data from the line interface to the capture buffer, ensuring that all of the error conditions, the protocol state information, and the protocol data are correctly stored and time stamped. During real-time and in postprocess mode the triggers and actions, the timers and counters, and the protocol followers are executed in the analysis and acquisition system. Additionally the measurements are typically executed in a distributed fashion between the computing platform and the analysis and acquisition system. In low-cost software-based analyzers the analysis and acquisition system functions are performed by the computing platform. In high-performance protocol analyzers a dedicated processor and special-purpose hardware are used to implement the processing required by the analysis and acquisition functions.

Line Interface.   The physical hardware and firmware necessary to actually attach to the network under test are implemented in the line interface. Additionally the line interface includes the necessary transmit circuitry to implement simulation functions for intrusive testing. The function of the line interface is to implement the physical layer of the OSI reference model and provide framed data to the analysis and acquisition system.

CHAPTER 25.5
OSCILLOSCOPES

Jay A. Alexander

INTRODUCTION

The word *oscilloscope* has evolved to describe any of a variety of electronic instruments used to observe, measure, and record transient physical phenomena and present the results in graphic form (Fig. 25.5.1). Perhaps the popularity and usefulness of the oscilloscope spring from its exploitation of the relationship between vision and understanding. In any event, several generations of technical workers have found it to be an important tool in a wide variety of settings.

Basic Functions

The prototypical oscilloscope produces a two-dimensional graph with the voltage applied at the input plotted on the *vertical axis* and time plotted on the *horizontal axis* (Fig. 25.5.2). Usually the image appears as an illuminated trace on the screen of a cathode-ray tube (CRT) or liquid-crystal display (LCD) and is used to construct a model or representation of how the instantaneous magnitude of some quantity varies during a particular time interval. The quantity measured is often a changing voltage in an electronic circuit. However, it could be something else, such as electric current, acceleration, or light intensity, which has been changed into a voltage by a suitable transducer. The time interval over which the phenomenon is viewed may vary over many orders of magnitude, allowing measurements of events that proceed too quickly to be observed directly with the human senses. Instruments currently available measure events occurring over intervals as short as picoseconds (10^{-12} s) and up to tens of seconds.

The measured quantities can be uniformly repeating or essentially nonrecurring. The most useful oscilloscopes have multiple input channels so that simultaneous observation of multiple phenomena is possible, enabling the measurement of the time relationships among events.

GENERAL OSCILLOSCOPE CONCEPTS

General-purpose oscilloscopes are classified as *analog oscilloscopes* or *digital oscilloscopes*. Newly produced models are almost exclusively of the digital variety, although many lower-bandwidth (< 100 MHz) analog units are still being used in various industrial and educational settings. Digital oscilloscopes are often called *digital storage oscilloscopes* (DSOs), for reasons that will become apparent below.

Analog and Digital Oscilloscope Basics

The classic oscilloscope is the analog form, characterized by the use of a CRT as a direct display device. A beam of electrons, *cathode rays,* is formed, accelerated, and focused in an electron gun and strikes a phosphor

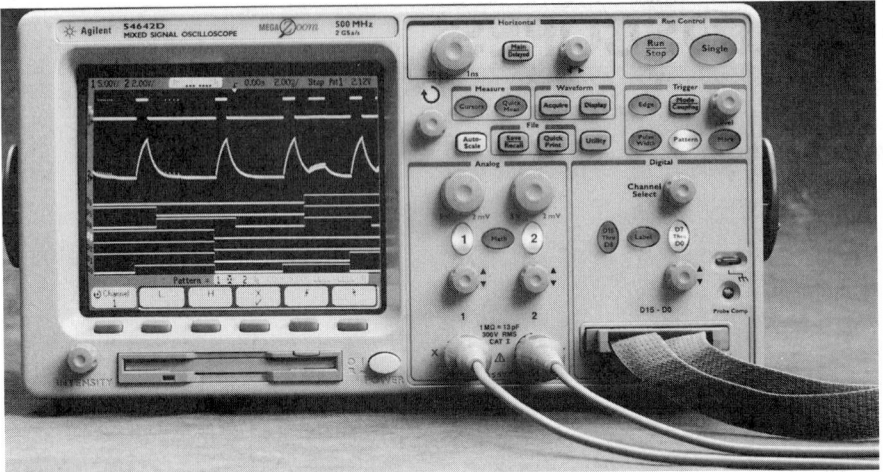

FIGURE 25.5.1 Epitomizing recent developments in oscilloscopes is the mixed analog and digital signal oscilloscope, recently introduced by several manufacturers . The version shown here is the Agilent Technologies 54642D. Mixed signal oscilloscopes are discussed later.

screen, causing visible light to be emitted from the point of impact (Fig. 25.5.3). The voltage signals to be displayed are amplified and applied directly to vertical deflection plates inside the CRT, resulting in an angular deflection of the electron beam in the vertical direction. This amplifier system is referred to as the *vertical amplifier*. The linear vertical deflection of the point at which the electron beam strikes the screen is thus proportional to the instantaneous amplitude of the input voltage signal. Another voltage signal, generated inside the oscilloscope and increasing at a uniform rate, is applied directly to the horizontal deflection plates of the CRT, resulting in a simultaneous, uniform, left-to-right horizontal motion of the point at which the electron beam strikes the phosphor screen. The operator of the oscilloscope may specify the rate of this signal using the *time-per-division* or horizontal scale control. For example, with the control set to 100 μs/div on a typical oscilloscope with 10 horizontal divisions, the entire horizontal extent of the display will represent a time span of 1 ms.

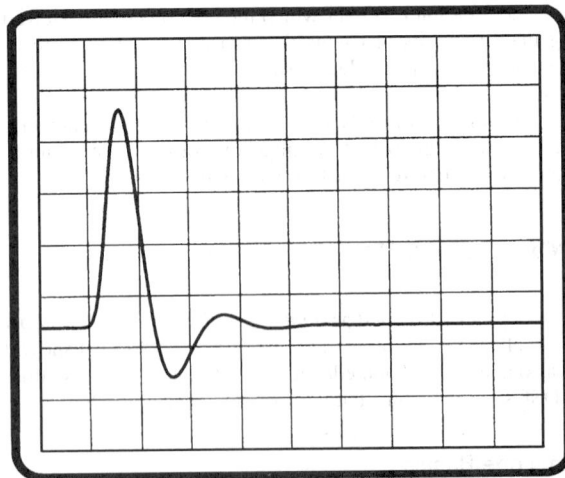

FIGURE 25.5.2 Voltage is plotted on the vertical axis and time horizontally on the classic oscilloscope display.

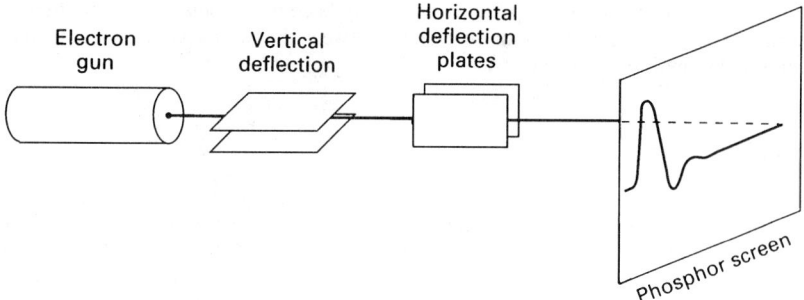

FIGURE 25.5.3   Analog oscilloscope cathode-ray tube.

The electronic module that generates the signals that sweep the beam horizontally and controls the rate and synchronization of those signals is called the *time base*. The point on the phosphor screen illuminated by the electron beam thus moves in response to the vertical and horizontal voltages, and the glowing phosphor traces out the desired graph of voltage versus time.

The digital oscilloscope has been made practical and useful by advances in the state of the art of digitizing devices called *analog-to-digital converters* (ADCs). For the purposes of this discussion, an ADC is a device, which at suitable regular intervals measures or *samples* the instantaneous value of the voltage at the oscilloscope input and converts it into a digital value (a number), representing that instantaneous value (Fig. 25.5.4). The oscilloscope function of recording a voltage signal is achieved by storing in a digital memory a series of samples taken by the ADC. At a later time, the series of numbers can be retrieved from memory and the desired graph of volts versus time can be constructed. The graphing or display process, since it is distinct from the recording process, can be performed in several different ways. The display device can be a CRT employing direct beam deflection methods. More commonly, a *raster-scan display*, similar to that used in a conventional television receiver or a computer monitor, can be used. The samples may also be plotted on paper using a printer with graphics capability.

The digital oscilloscope is usually configured to resemble the traditional analog instrument in the arrangement and labeling of its controls, the features included in the vertical amplifier, and the labeling and presentation of the display. In addition, the circuits that control the sample rate and timing of the data-acquisition cycle are configured to emulate the functions of the time base in the analog instrument. This has allowed users who

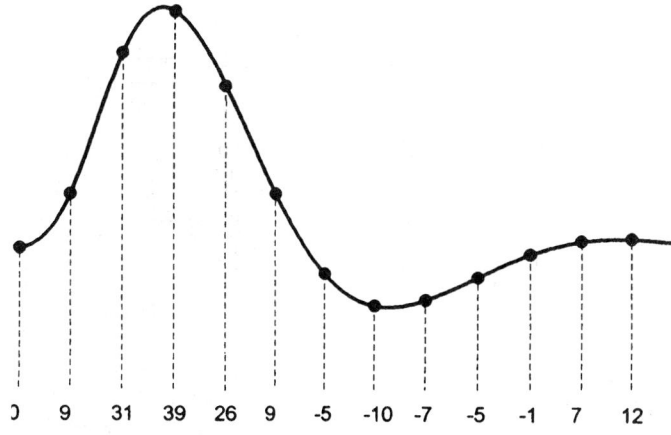

FIGURE 25.5.4   Sampling in a digital oscilloscope.

are familiar with analog oscilloscope operation to quickly become proficient with the digital version. Indeed, while there are fundamental and important differences between the two measurement technologies, many common elements and shared requirements exist.

Oscilloscope Probing

Most oscilloscope measurements are made with *probes* of some type. Probes connect the oscilloscope to the signals being measured and form an important part of the overall measurement system. They must both load the circuit minimally and transmit an accurate version of the signal to the oscilloscope. Many varieties of probes are available, and they may be classified in several ways, such as passive versus active, single-ended versus differential, and voltage versus current. Passive probes are the most common. They are relatively inexpensive and are effective for many measurements below 500 MHz. Passive probes typically feature high input resistance, which minimizes loading at low frequencies. To effectively measure signals with frequency content above 1 GHz, active probes are usually employed. Active probes contain amplifiers in the probes themselves, and they present lower capacitive loading to the circuit being measured, which allows higher frequencies to be measured accurately. Because they are more complex, active probes are significantly more expensive than passive probes. Newer active probes are often differential in nature; this reflects the increased use of differential signals in high-speed digital systems. An increasingly important requirement for oscilloscope probes is small physical size. This is driven by the fine-pitch geometry of modern *surface mount technology* (SMT) components and is particularly important for active probes, which tend to be larger because of their amplifier circuits.

THE ANALOG OSCILLOSCOPE

A complete block diagram for a basic two-channel analog oscilloscope is shown in Fig. 25.5.5.

Vertical System

The vertical preamps and associated circuitry allow the operator of the oscilloscope to make useful measurements on a variety of input signals. The preamps provide gain so that very small input signals may be measured,

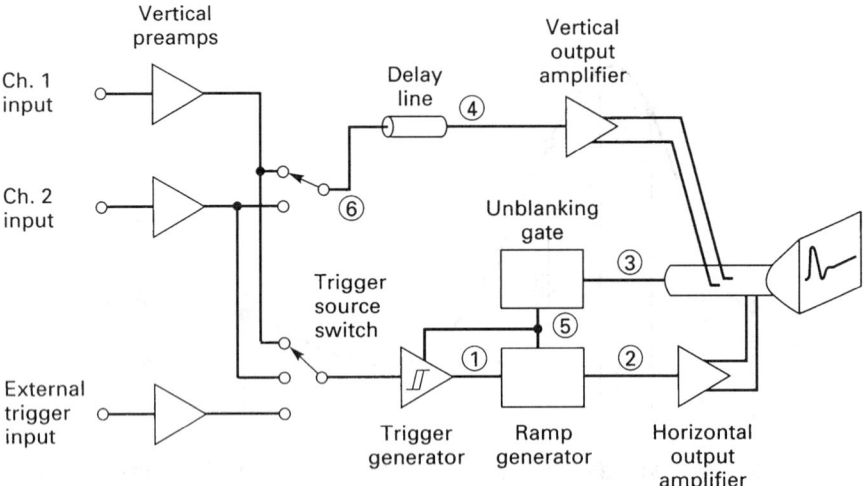

FIGURE 25.5.5 Analog oscilloscope block diagram.

and one or more *attenuators*, usually implemented with switches or relays, are available for reducing the amplitude of large signals. Together, the preamp and attenuator settings typically manifest as a vertical sensitivity or *scale* control on the oscilloscope's control panel. The scale control is specified in terms of *volts per division* (volts/div), where a division corresponds to a fixed fraction (typically 1/8th) of the vertical extent of the display. Thus when the scale control is set to 100 mV/div, for example, a signal with a peak-to-peak amplitude of 800 mV will occupy the full vertical extent of the display. A provision for injecting dc-shift or *offset* is also provided; this aids in measuring signals that are not symmetric about zero volts. Circuits and controls for changing the input *coupling* and *impedance* are often included as well. Coupling options usually consist of ac and dc, and sometimes GROUND, which is useful for quickly viewing a 0 V reference trace on the display. Impedance selections are generally 50 Ω and 1 MΩ. The 50 Ω selection is useful for making measurements where input sources having 50 Ω output impedance are connected to the oscilloscope with 50 Ω cables. In this situation the oscilloscope preserves the impedance of the entire system and does not introduce undesirable effects such as signal reflections. The 50-Ω selection is also used for many active probes, whose output amplifiers are commonly designed to drive a 50-Ω load. The 1-MΩ selection is appropriate for most other measurements, including those employing passive probes.

Trigger

In Fig. 25.5.6, the signals occurring at the indicated nodes in Fig. 25.5.5 are shown for a single acquisition cycle of a signal connected to the channel 1 input. The trigger source is set to the *internal, channel 1* position, and a positive slope-edge trigger is selected, so a trigger pulse is generated as the input signal crosses the indicated trigger level. In response to the trigger pulse, the ramp signal initiates the motion of the spot position on the display from left to right, and the unblanking gate signal reduces the CRT grid-to-cathode negative bias, causing the spot to become visible on the display screen. The ramp increases at a precisely controlled rate,

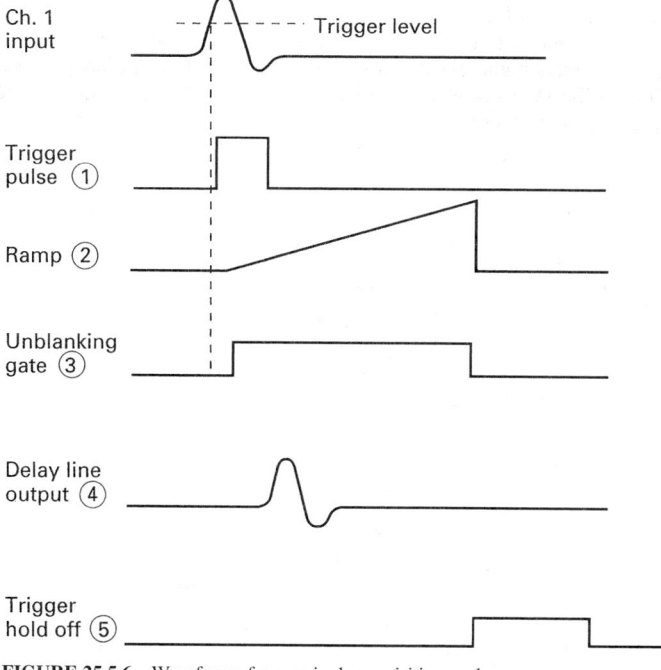

FIGURE 25.5.6   Waveforms from a single acquisition cycle.

causing the spot to progress across the display screen at the horizontal rate determined by the current time-per-division setting. When the spot has moved across the horizontal extent of the display, the unblanking gate switches negative, the trigger holdoff period begins, and the ramp retraces the spot to the starting point. At the end of trigger holdoff, the system is ready to recognize the next trigger and begin the next signal acquisition.

Delay Line

Some time is necessary after initiation of the trigger pulse for the sweep to attain a linear rate and for the display spot to reach full brightness. Since it is desirable to be able to view the voltage transient that has caused the trigger, a means of storing the input signal during the startup delay is needed. This is accomplished by placing a delay line in the vertical path. The triggering signal at the output of the delay line is visible on the display screen while the unblanking gate signal is at its most positive level (Fig. 25.5.6). A total delay time of between 25 and 200 ns is required depending on the oscilloscope model, with higher bandwidth units requiring shorter delays.

Dual-Trace Operation

Oscilloscopes are equipped with two or more channels because the most important measurements compare time and amplitude relationships on multiple signals within the same circuit. However, a conventional analog oscilloscope CRT has only one write beam and thus is inherently capable of displaying only one signal. Thus the channel switch (see Fig. 25.5.5) is used to timeshare, or *multiplex*, the single display channel among the multiple inputs. The electronically controlled channel switch can be set manually first to channel 1 and then later switched by the user to channel 2. However, a better emulation of simultaneous display is attained by configuring the oscilloscope to automatically and rapidly switch between the channels. Two different switching modes are implemented, called *alternate* and *chop*. In *alternate mode*, the channel switch changes position at the end of each sweep during retrace while the write beam is blanked. This method works best at relatively fast sweep speeds and signal repetition rates. At slow sweep speeds the alternating action becomes apparent, and the illusion of simultaneity is lost. In *chop mode*, the channel switch is switched rapidly between positions at a rate that is not synchronized with the input signals or the sweep. This method is effective at slower sweep speeds but requires a relatively higher bandwidth output amplifier to accurately process the combined chopped signal. Many analog oscilloscopes provide a control for the user to select between alternate and chop modes as appropriate to their measurement situation.

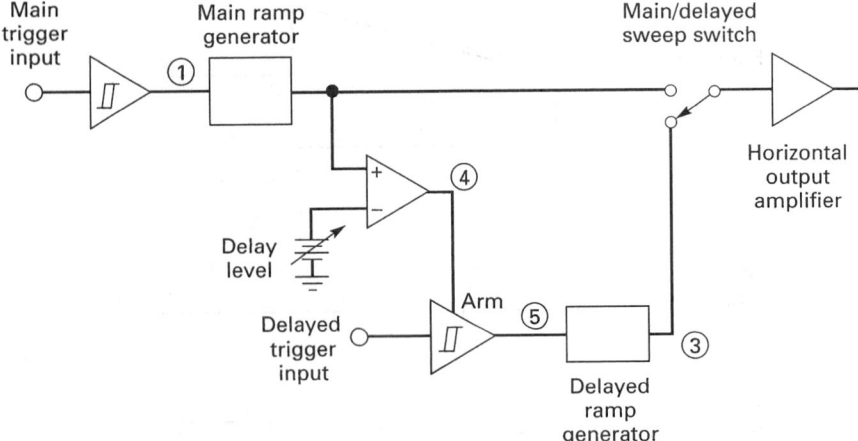

FIGURE 25.5.7   Main and delayed sweep generator block diagram.

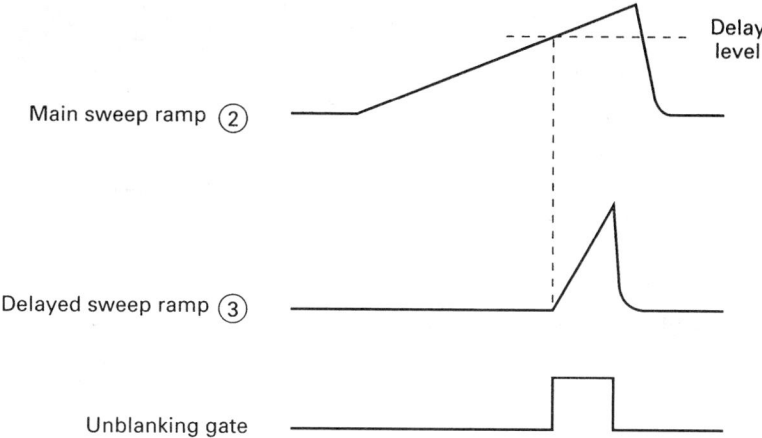

FIGURE 25.5.8 Delayed sweep starting when the main sweep ramp exceeds the delay level.

Delayed Sweep

Some analog oscilloscope models include a second ramp generator, called a *delayed sweep*, and a second trigger generator, called a *delayed trigger* (Fig. 25.5.7), providing an additional method for controlling the placement of the *window* in time relative to the main trigger event.

The horizontal amplifier can still be connected to the output of the main ramp generator, in which case the operation is identical to that of the standard oscilloscope configuration described earlier. A comparator circuit is added whose output (4) switches when the main ramp signal exceeds a voltage called the *delay level* (Fig. 25.5.8). This dc voltage can be adjusted by the oscilloscope operator using a calibrated front panel control. The main ramp is initiated by the main trigger pulse and increases at a precisely controlled rate (volts per second) determined by the sweep-speed setting. Therefore, the time elapsed between the trigger pulse and comparator output state change is the reading on the delay control (in divisions) multiplied by the sweep-speed setting (in seconds per division). The delayed sweep ramp is initiated after the delay period in one of two ways: The *normal* method immediately starts the delayed sweep when the delay comparator switches (see Fig. 25.5.8). The *delayed trigger* mode uses the delay comparator output to arm the delayed trigger circuit. Then when the delayed trigger condition is met, the delayed sweep starts (Fig. 25.5.9). The delayed sweep rate always has a shorter time-per-division setting than the main sweep.

THE DIGITAL OSCILLOSCOPE

A block diagram of a two-channel digital oscilloscope is shown in Fig. 25.5.10. Signal acquisition is by means of an ADC. The ADC samples the input signal at regular time intervals and stores the resulting digital values in memory. Once the specified trigger condition is satisfied, the sampling process is interrupted, the stored samples are read from the acquisition memory, and the volts-versus-time waveform is constructed and graphed on the display screen. In some advanced models, the storage and readout of the samples is managed using two or more memory blocks, so that the oscilloscope can continue to acquire new data associated with the next trigger event while the display operation is proceeding. This keeps the display update rate high and helps to reduce the *dead time* during which the oscilloscope is unable to acquire data. The time interval between samples is called the *sample time* and is the reciprocal of the *sample rate*. The signal that regulates the sampling process in the ADCs is called the *sample clock*, and it is generated and controlled by the time base circuit. A crystal

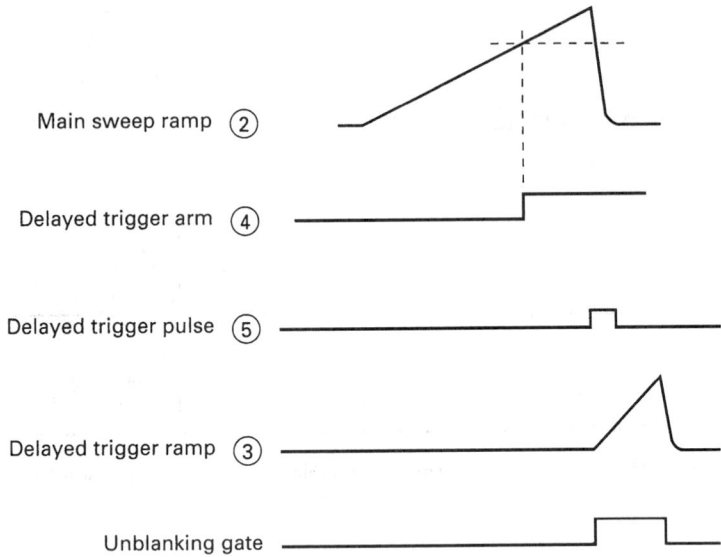

Main sweep ramp ②

Delayed trigger arm ④

Delayed trigger pulse ⑤

Delayed trigger ramp ③

Unblanking gate

FIGURE 25.5.9   The delayed trigger starts the delayed sweep.

oscillator is used as a reference for the time base to ensure the accuracy of the sample interval and ultimately of the time measurements made using the digital oscilloscope.

Sampling Methods

The exact process by which the digital oscilloscope samples the input signal in order to present a displayed waveform is called the *sampling method*. Three primary methods are employed. In *real-time* or "single-shot" sampling, a complete memory record of the input signal is captured on every trigger event. This is the most

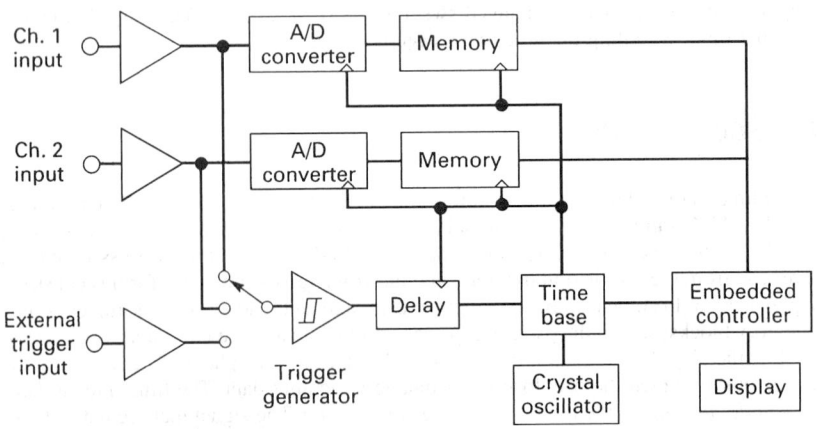

FIGURE 25.5.10   Digital oscilloscope block diagram.

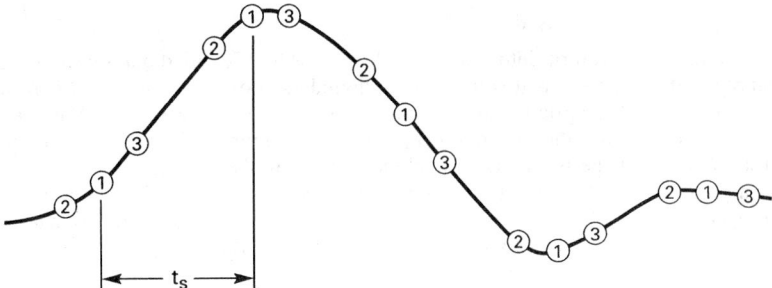

FIGURE 25.5.11 Random repetitive sampling builds up the waveform from multiple data acquisitions. Dots from the same acquisition bear the same number.

conceptually straightforward type of sampling and its main advantage is that it can capture truly transient or nonrecurring events, provided the sample rate is sufficiently fast. Another advantage is the ability to capture "negative time," or time before the trigger event. This is especially valuable in fault analysis, where the oscilloscope is used to trigger on the fault and the operator may look "backward in time" to ascertain the cause. The main disadvantage of real-time sampling is that the effective bandwidth of the oscilloscope can be no higher than 50 percent of the maximum sample rate of the ADC. When higher bandwidths are desired, *random repetitive sampling* may be used. This method is also called *equivalent-time sampling*. In this method, a complete waveform for display is built up from multiple trigger events. The ADC samples at a slower rate, and the samples are taken with a random time offset from the trigger event in order to ensure adequate coverage of all time regions of the input signal. This process is illustrated in Fig. 25.5.11. The major disadvantage of random repetitive sampling is that a repetitive input signal and stable trigger event are required. Many digital oscilloscopes are capable of sampling in either real-time or repetitive mode, with the selection performed either by the operator or automatically by the oscilloscope based on the sweep speed.

The final type of sampling is referred to as *sequential sampling*. In this method, only one point is acquired per trigger event, and successive samplings take place farther away from the trigger point. This is illustrated in Fig. 25.5.12. Sequential sampling is used to achieve even higher bandwidths (20 GHz and above) than are possible with random repetitive sampling due to the nature of the ADCs required. Oscilloscopes that use sequential sampling are not capable of capturing negative time and typically contain very limited trigger functionality.

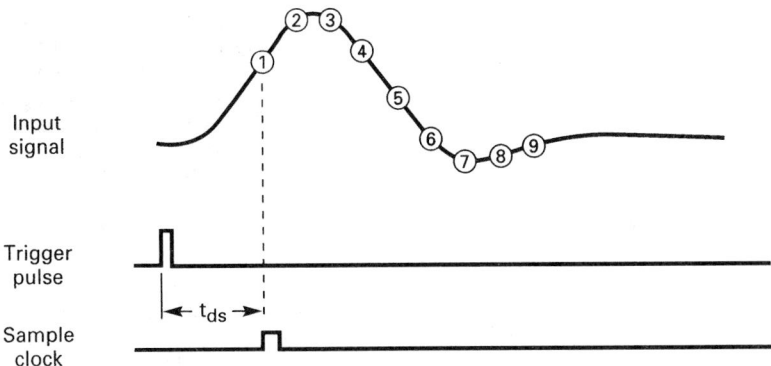

FIGURE 25.5.12 Sequential sampling captures one sample per trigger and increases the delay t_{ds} after each trigger.

Quantization

Converting a continuous waveform into a series of discrete values is called *quantization* and several practical limitations apply to this process, as it is used in the digital oscilloscope. The signal is resolved into discrete levels only if it is within a specific range of voltages (Fig. 25.5.13). If the input signal is outside this range when a sample is taken, either the maximum code or the minimum code will be the output from the ADC. The limited *window* in voltage is similar to that encountered in the analog oscilloscope CRT. A sufficiently large signal amplitude causes the trace to disappear off the edge of the display screen. As in the analog instrument, the vertical scale and offset controls are used to adjust the input waveform so that the desired range of voltages on the waveform will fall within the digitizer voltage window.

Resolution

Voltage resolution is determined by the total number of individual codes that can be produced. A larger number permits a smoother and more accurate reproduction of the input waveform, but increases both the cost of the oscilloscope and the difficulty in achieving a high sample rate. ADCs are usually designed to produce a total code count that is an integer power of 2, and a unit capable of 2^n levels of resolution is called an *n-bit digitizer*. Digital oscilloscopes are available in resolutions from 6 to 12 bits, with the resolution varying generally in an inverse relationship to the maximum sample rate. Eight bits is the most frequently used resolution. The best possible intrinsic voltage resolution, expressed as a fraction of the full-scale range, is 2^{-n}, e.g., 0.4 percent for an 8-bit ADC. Many digital oscilloscopes provide averaging and other types of filtering modes that can be used to increase the effective resolution of the data.

Acquisition Memory

Each sample code produced must be stored immediately in the acquisition memory and so that memory must be capable of accepting data from the digitizer continuously at the oscilloscope's sample rate. For example, the memory in each channel of an 8-bit, 2 GSa/s oscilloscope must store data at the rate of 2×10^9 bytes/s. The memory is arranged in a serially addressed, conceptually circular array (Fig. 25.5.14). Each storage location is written to in order, progressing around the array until every cell has been filled. Then each subsequent sample overwrites what has just become the oldest sample contained anywhere in the memory, the progression continues, and a memory with n_m storage locations thereafter always contains the most recent n_m samples. The n_m samples captured in memory represent a total time of waveform capture of n times the sample time interval. For example, an oscilloscope operating at 2 GSa/s with a 32,000-sample acquisition memory captures a 16-µs segment of the input signal.

If sampling and writing to acquisition memory stop immediately when the trigger pulse occurs (delay = 0 in Fig. 25.5.15), then the captured signal entirely precedes the trigger point (Fig. 25.5.15a). Setting delay greater than zero enables sampling to continue for a predetermined number of samples after the trigger point. Thus a signal acquisition could capture part of the record before and part after the trigger (Fig. 25.5.15b), or capture information after the trigger (Fig. 25.5.15c).

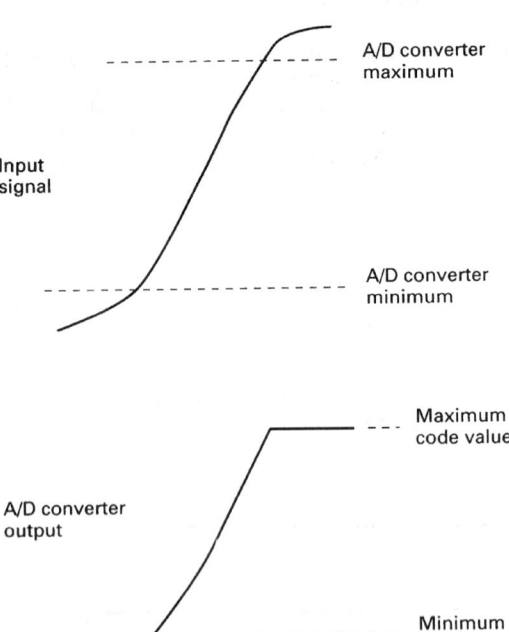

FIGURE 25.5.13   An ADC quantizes a continuous signal into discrete values having a specific range and spacing.

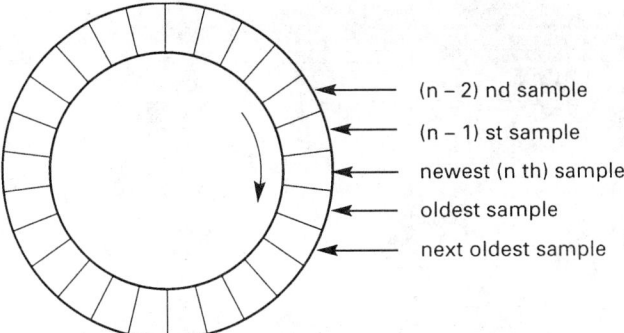

FIGURE 25.5.14   Acquisition memory is arranged in a circular array so that the next sample always overwrites the oldest sample in memory.

In recent years, deep memory (>1 M sample) digital oscilloscopes have become more prevalent. Deep memory is valuable because it allows the oscilloscope to capture longer periods of time while maintaining a high sample rate. To see why this is so, consider an oscilloscope with a maximum sample rate of 5 GSa/s and 10k acquisition memory. If the operator sets the horizontal scale control to 100 us/div, the oscilloscope must capture 1 ms of time in order to present a full screen of information. To do this, it must reduce the sample rate to 10 MSa/s (10k samples/1 ms) or lower. The oscilloscope will exhaust its memory if it samples faster than 10 MSa/s. An oscilloscope with a maximum sample rate of 2 GSa/s but 8M of memory, on the other hand,

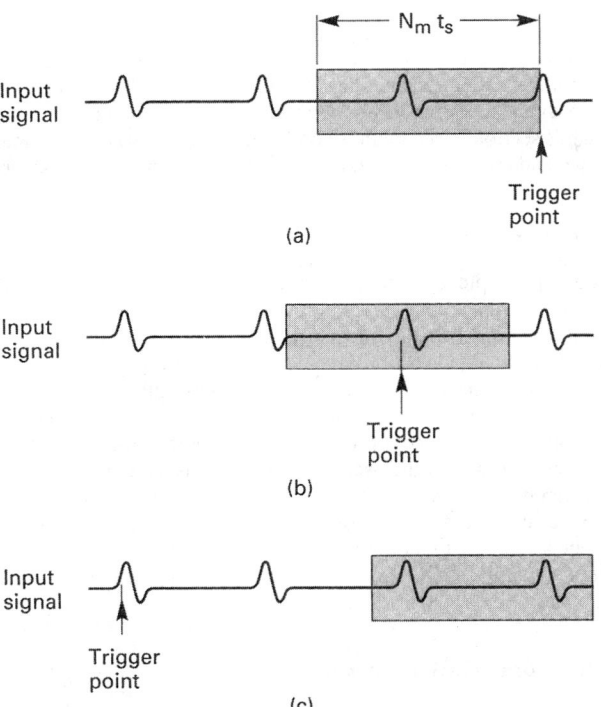

FIGURE 25.5.15   Trigger point can be placed anywhere within or preceding the captured record. (a) Delay = 0; (b) delay = $n_m/2$; (c) delay = n_m.

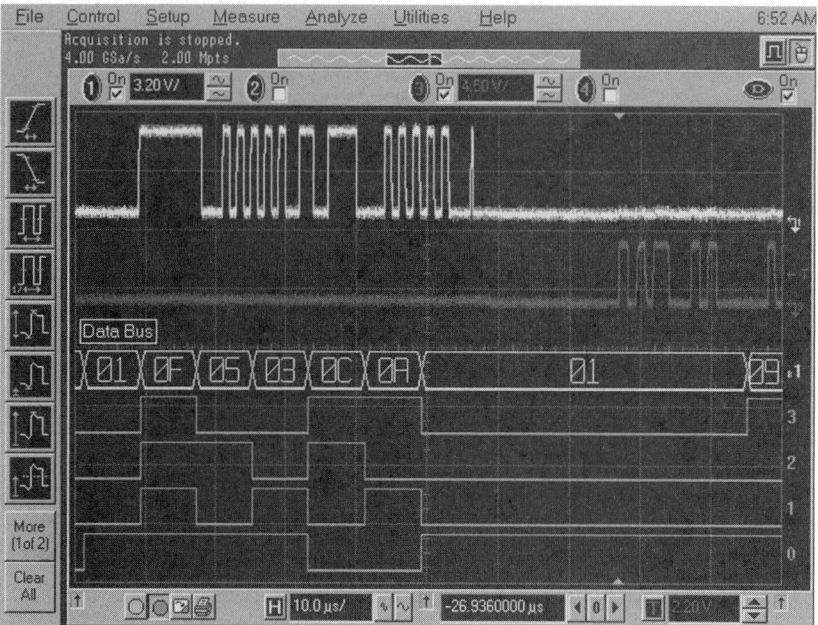

FIGURE 25.5.16 The relationship between memory depth and sample rate.

is able to maintain its maximum sample rate even at the 100 us/div setting. Maintaining sample rate in a digital oscilloscope is important because it reduces the likelihood of *aliasing* (undersampling) the high-frequency components of the input signal, and ensures that all the signal details will be captured, even at slower sweep speeds. Figure 25.5.16 shows the relationship between sweep speed and actual sample rate for the two oscilloscopes discussed in this section. Deep memory both delays the onset of the sample rate reduction as the oscilloscope is slowed down and reduces the extent of the reduction once it does occur.

Advanced Digital Oscilloscope Features

Modern digital oscilloscopes typically contain many features that never appeared in their analog counterparts. These include automatic signal scaling, automatic measurements, extensive waveform math functions, signal mask testing, saving and recalling of setup information, waveform data, and display images to internal or external storage media, and integrated calibration and self-test capabilities. Many of these features are made possible by the fact that these instruments are controlled by increasingly powerful microprocessors. Indeed, most digital oscilloscopes now contain more than one microprocessor, with each one focused on a different aspect of the overall product operation. In recent products, one of the processors is often a complete personal computer (PC) that is an integrated part of the oscilloscope. This has led to still more features in the areas of communications and connectivity. For example, PC-based oscilloscopes are currently available with capabilities such as networking, voice control, and web control. Advances in integrated circuit (IC) capabilities have also led to new features such as units with more than two channels; units with advanced trigger modes like pattern, state, sequence, risetime, and pulse width; and units with segmented memory, for storing separate acquisitions from different trigger events for later recall and analysis.

Mixed Analog and Digital Signal Oscilloscopes

In response to the growing digital signal content in electronic circuits, several manufacturers have developed *mixed analog and digital signal oscilloscopes* (abbreviated as MSO) that contain 16 digital inputs in addition to the two or four standard oscilloscope channels. These digital inputs pass through comparators (1-bit ADCs) and function similar to the inputs on a logic timing analyzer. They are displayed along with the standard channels, and may also be used for triggering and automatic measurements. Measuring up to 20 channels with these oscilloscopes provides the user with more information at one time about his or her circuit. An example of a display from an MSO is shown in Fig. 25.5.1.

THE FUTURE OF OSCILLOSCOPES

Oscilloscopes have existed in various forms for over 60 years. They are at once basic general-purpose tools and advanced powerful measurement systems. As with other technology-based products, they will continue to evolve, in response to both emerging user needs and the availability of new technologies. Bandwidths, sample rates, and memory depths will continue to increase, and the trend toward more postacquisition analysis will continue. An area of increased attention is customization and application-specific measurement sets. Another is the combination of classic oscilloscope capabilities with those of other test and measurement products such as logic analyzers and signal generators. Mixed signal oscilloscopes are an example of such a combination. While the primary value of the oscilloscope, showing the user a picture of his or her signals versus time, will remain unchanged, many other attributes will indeed change, likely in as big a way as the digital oscilloscope has changed from its analog predecessor.

CHAPTER 25.6
STANDARDS-BASED MODULAR INSTRUMENTS

William M. Hayes

INTRODUCTION

Modular instruments employ a frame (Fig. 25.6.1) that serves as a host. These frames allow multiple switch, measurement, and source cards to share a common backplane. This makes it possible to configure instruments that can accommodate a range of input/output (I/O) channels. It also makes possible tailoring measurement capabilities to the specific applications being addressed.

Modular Standards

The modular standards described below are industry standards:

- VME standard
- VXI standard
- Personal computer (PC) plug-ins
- CompactPCI standard.

PC plug-ins are not part of a formal instrument standard. However, the ubiquity of the personal computer has made the PC motherboard I/O bus a de facto standard for instruments. Although all these standards are used for instrument systems, only VXI and a derivative of CompactPCI, called PXI, were developed expressly for instrumentation. For general-purpose instrumentation, VXI has the most products. PXI is emerging in the market, and generally offers the same features as VXI.

Open standards-based modular instruments are compatible with and can therefore accept products from many different vendors, as well as user-defined and constructed modules. Modular instruments generally employ a computer user interface instead of displays and controls embedded in the instrument's frame or package. By sharing a computer display, modular instruments save the expense of multiple front-panel interfaces. Without the traditional front panel, the most common approach to using modular instruments involves writing a test program that configures the instruments, conducts the measurements, and reports results. For this reason, modular instruments typically are supplied with programmatic software interfaces, called *drivers*, to ease the task of communicating with an instrument module from a programming language.

FIGURE 25.6.1 A typical modular instrument: VXI modules have been configured in this dual-rack structure, set on castors, to serve as a portable ground support test system for military avionics.   (*Photo courtesy of Agilent Technologies*)

Advantages of Modular Instruments

Modular instruments are an excellent choice for high-channel-count measurements, automated test systems, applications where space is at a premium, and complex measurements where several instruments need to be coordinated.

ELEMENTS OF MODULAR INSTRUMENTS

Figure 25.6.2 shows the key elements of modular, standard-based instruments including a frame that contains a backplane with measurement and switching modules—all working with the *device under test* (DUT). Control is by system software, running on the computer, including the programming language and drivers that make up the test program.

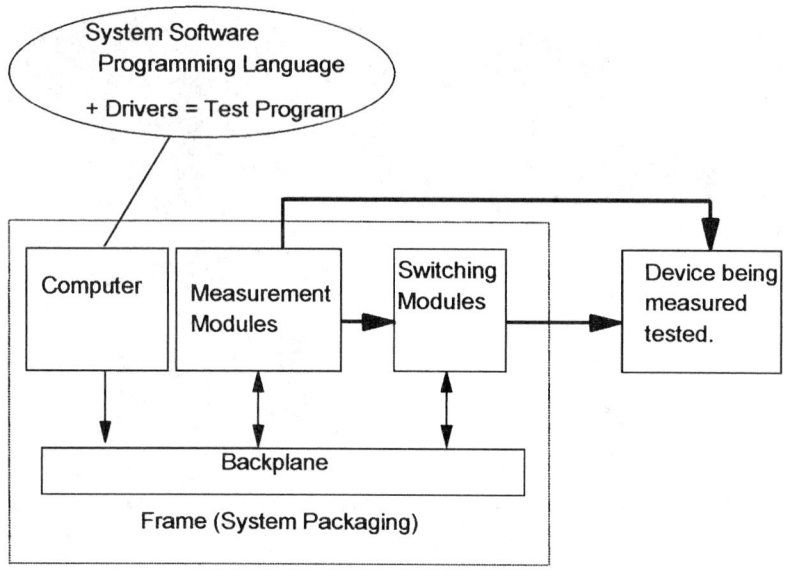

FIGURE 25.6.2 Key elements of modular standards-based instruments.

THE SYSTEM BACKPLANE

At the heart of a modular instrument is the system backplane, or set of buses connecting system modules together. These buses are high-speed parallel buses. All of the modular standards have a data-transfer bus, a master/slave arbitration bus—except ISA PC plug-ins—an interrupt bus and a bus for special functions. The VME (32-bit) backplane is one example.

FORM FACTORS

One place to begin understanding the similarities and differences between standards is board size.

Board Size

To ensure interchangeability among vendors, standardization of the instrument module's board size and spacing is important. With the exception of the PC plug-ins, all the other modular forms use Eurocard board sizes (see Table 25.6.1). These board sizes were standardized as IEEE 1101.1, ANSI 310-C, and IEC 297.

VME and VXI Standards

The VME standard uses the first two sizes, referring to them as single-height and double-height boards. The VXI standard uses all four sizes and refers to them as sizes A, B, C, and D, respectively. Most VXI manufacturers have adopted the B and C sizes. Compact PCI uses sizes 1 and 2, referring to the Eurocard 3U and 6U sizes.

TABLE 25.6.1 Eurocard Sizes

Eurocard Size	VME name	VXI name	CompactPCI name
10×16 cm	Single height	A	3U
23×16 cm	Double height	B	6U
23×34 cm	—	C	—
36×34 cm	—	D	—

PC Plug-in Modules

PC plug-in modules use the full- and half-card sizes adopted in the original IBM PC. The board size is approximately 12.2 cm × 33.5 cm for the full-size card and 12.2 cm × 18.3 cm for the half-size card.

VMEBUS (VME STANDS FOR VERSAMODULE EUROCARD)

The VMEbus specification was developed for microcomputer systems using single or multiple microprocessors. The specification was not originally intended for instrumentation, but the concept of computer functions integrated with measurement and control functions on a single backplane was a factor in its creation. The VMEbus International Trade Association (VITA) released the specification in August 1982. It was approved as IEEE 1014-87 in March 1987. In 1994, the VMEbus specification was amended to include 64-bit buses and became known as *VME64*. Additional backplane connectors are involved with VME64, so care in choosing system components is required. VMEbus products are used in a wide variety of applications including industrial controls and telecommunications.

VME System Hardware Components

A VME system comprises one or more frames, an embedded controller module, optionally various computer storage and I/O modules (e.g., LAN), and various measurement and switch modules. There is not a common programming model or instrument driver standard. As with other devices on a computer bus, programs must read/write device registers. In some cases, vendors supply software functions or subroutines to help with the programming.

VME Frames.   *Frames* refer to the backplane, power supply as well as the packaging that encloses a system (Fig. 25.6.3). System builders can choose a solution at many different levels from the component level (i.e., a backplane) to complete powered desktop enclosures. Backplanes, subracks, and enclosures are available from many manufacturers, in single-, double-, and mixed-height sizes. Typically a 19-inch rack will accommodate 21 cards. Power supplies are available from 150 to 1500 W.

VME-Embedded Computers.   VME is well suited for building single- and multiple-processor computer systems. A large number of embedded processors are available.

VME Switching.   Many measurement systems require switching between measurement channel and signal paths from the device being tested. Typically these switches are multiplexers or matrix switches. There are several simple relay cards available in VME, but few multiplexers or matrix switches.

VME Software.   Many VME systems are developed with real-time operating systems. Software is available for VxWorks, pSOS+, OS-9, and various other computer operating systems. Source code for the C programming language is the most common. There is no standardized driver software.

FIGURE 25.6.3   VME Backplanes and subracks are available in many different configurations. (*Photo courtesy of APW*)

VXI (VMEBUS EXTENSIONS FOR INSTRUMENTATION)

The VXI standard is a derivative of the VME standard. It was driven by a U.S. Air Force program in the 1980s to design a single *Instrument-on-A-Card* (IAC) standard with the objective to substantially reduce the size of electronic test and operational systems. In April 1987, a number of companies started discussions on creating an IAC standard that would benefit both the commercial and military test communities. The VXIbus consortium was formed and the VXIbus system specification was released in July of 1987. The IEEE adopted it as IEEE 1155 in 1992. After the basic VXIbus specification was adopted, it was recognized that the software aspect of test system development should be addressed. In 1993, the VXIplug&play Systems Alliance was formed. Several system *frameworks* were defined, based on established industry-standard operating systems. The Alliance released Revision 1.0 of the VXIplug&play specification in 1994; the latest revision, Revision 2, was released in 1998.

Common VXI Functions and Applications.   VXI is used for most instrument and switching applications, including:

- *Automated test systems*. With its aerospace/defense roots, some of the specific applications have been for weapons and command/control systems, and as the foundation for operational systems, such as an artillery firing control system.

- *Manufacturing test systems*. Examples include cellular phone testing and testers for automotive engine-control modules.

- *Data acquisition*. In particular, DAC applications include physical measurements such as temperature and strain. Examples include complex and sophisticated measurements, such as satellite thermal vacuum testing or aircraft stress analysis.

VXII Standard

The VXI standard including VXIplug&play is a comprehensive system-level specification. It defines back-plane functionality, electrical specifications, mechanical specifications, power management, electromagnetic compatibility (EMC), system cooling, and programming model. The backplane functionality has features intended specifically for instrumentation.

VXI System Hardware Components

VXI systems can be configured with either an external computer connected to a VXI mainframe, or with an embedded computer. External computers are the more common configuration because the cost is low and they allow the user to choose the latest and fastest PC available. A VXI system configured with an embedded computer is formed with an VXI computer module, a mainframe, various measurement and switching modules, and the supporting software including VXI Plug-and-Play drivers.

VXI Mainframes.   VXI system builders have a choice of backplanes or complete mainframes. Mainframes are the most common choice (Fig. 25.6.4). They include backplane, power supply, and enclosure. B-size mainframes are typically available with 9 or 20 slots and supply up to 300 W. Mainframes for C-size modules are available with 4, 5, 6, or 13 slots. D-size mainframes are available with 5 or 13 slots. Power capabilities range from 550 to 1500 W. Backplanes are available with basically the same slot choices as mainframes.

VXI External Connection to Computers.   A VXI module is required to connect and communicate with an external computer. Two choices are available: *Slot 0 controllers* and *bus extenders*. The most common bus extender is called MXIbus. Slot 0 controllers and bus extenders must be placed in a unique slot of the mainframe. They receive commands from the external computer, interpret them and then direct them to the various measurement and switch modules. Both provide special VXI functions that include the *Resource Manager*. Bus extenders translate a personal computer's I/O bus to the VXI backplane using a unique high-speed parallel bus. Transactions over a bus extender have a register-to-register bit level flavor. The advantage of bus extenders is high speed. Cable lengths must be relatively short.

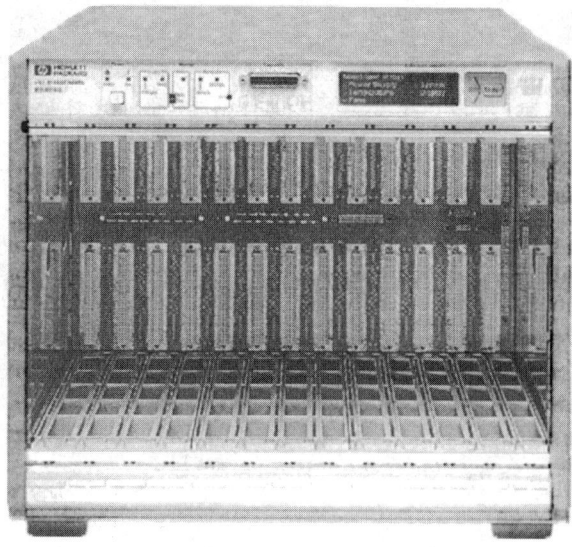

FIGURE 25.6.4   A 13-slot C-size VXI mainframe. Notice the system monitoring displays, which are unique, relative to other modular instrument types.   (*Photo courtesy of Agilent Technologies*)

Slot 0 controllers support these register level transactions but also support message-based transactions. These transactions are ASCII commands following the standard commands for programmable instrumentation (*SCPI*) standard, accepted by the instrument industry in 1990. SCPI commands are very readable, although slower to interpret. The advantages of message-based transactions is easy and fast program development. The introduction of VXI plug-&-play drivers has superseded SCPI for easier, quicker programming.

Slot 0 controllers require a relatively high-speed communication interface to communicate with an external computer. *GP-IB*, known as *IEEE-488*, and *Firewire*, known as *IEEE-1394*, are the standard interfaces for Slot 0 controllers. The external computer must have a matching communication interface to the Slot 0 controller.

Firewire interfaces are a computer-industry standard and are often built into PCs as standard equipment. Basic Firewire software-driver support was included in Windows 98 from Microsoft. GP-IB has been an instrument communication standard since 1975. Because it is not a computer standard, a plug-in interface board

FIGURE 25.6.5 VXI switch module and termination blocks for wiring, C-size. (*Photo courtesy of Agilent Technologies*)

must be added to the external PC for connection to the Slot 0 controller. GP-IB interfaces are about five times slower than Firewire.

VXI Embedded Computers. Embedded controllers are available for C-size main-frames. Products with Intel-architecture processors running Windows 3.11/ 95/98/NT are available from several vendors.

VXI Measurement Modules. VXI offers the broadest and most complete line of instrumentation modules (see Fig. 25.6.5 for an example of a VXI measurement module). Like the other modular standards, VXI offers a full line of digital I/O modules and modules containing ADCs and DACs. These functions are commonly used for data-acquisition and industrial-control applications.

Switching Modules. VXI also has the broadest line of switching modules among the modular instrument standards (see Fig. 25.6.5 for an example of a switching module). Switching can be used in many different ways and include simple relays, multiplexers and matrices, RF/microwave switches, and analog/FET switches.

VXI Software

The VXI standard has emphasized system software as an important element to improve the interoperability of modules from multiple vendors and to provide system developers a head start in developing their software. The VXIplug&play Systems Alliance developed a series of system frameworks encompassing both hardware and software. Six frameworks have been defined:

- Windows 3.1
- Windows 95
- Windows NT
- HP-UX
- Sun
- GWIN

In common with all of the frameworks is a specification for basic instrument communication called *virtual instrument software architecture* (VISA). It is a common communications interface regardless of whether the physical interconnect is GP-IB, Ethernet, or Firewire. Also in common among the frameworks is a specification for instrument drivers. *Instrument drivers* are functions, which are called from programming languages to control instruments. VXIplug&play drivers must include four features:

- C function library files
- Interactive soft front panel
- Knowledge base file
- Help file

The C function library files must include a dynamic link library (.DLL or .SL), ANSI C source code, and a function panel file (.FP). The function must use the VISA I/O library for all I/O functions. The interactive soft-front panel is a graphical user interface for directly interacting with a VXI instrument module. Some front panels closely resemble the front panel of a traditional box instrument. The *knowledge base file* is an ASCII description of all the instrument module's specifications. The *help file* provides information on the C function library, on the soft front panel, and on the instrument itself.

VXI Standard Specifications

Because the intent of the VXI standard was to create a standard specifically for instruments, several extensions were made to the VME specification. These extensions include additions to the backplane bus, and power, cooling, and RFI specifications.

Unique VXI Signals

The A-board size has only one connector (P1) and has no backplane extensions over VME. The B-, C-, and D-size boards have at least a second connector (P2). On P2 are the following additions:

- Additional supply voltages to support analog circuits: −5.2 V, −2 V, +24 V, and −24 V.
- Additional pins were also added for an increase in the maximum current capacity of the +5 V supply.
- 10-MHz differential ECL clock for synchronizing several modules.
- Two parallel ECL trigger lines.
- Eight parallel TTL trigger lines.
- A module identification signal.
- A 12-line local bus that connects adjacent modules. The manufacturer of the module defines the functionality of these lines.
- An analog summing bus terminated in 50 Ω.

Trigger Lines

The TTL and ECL trigger lines are open collector lines used between modules for *trigger, handshake, clock,* and *logic state* communications. Several standard protocols have been defined for these lines, including *synchronous* (SYNC), *asynchronous* (ASYNC), and *start/stop* (STST).

Module Power

The VXIbus specification has set standards for mainframe and module power, mainframe cooling, and electro-magnetic compatibility between modules. This ensures that products from multiple vendors will operate together.

Cooling Specification

The mainframe and module-cooling specification focuses on the test method for determining whether proper cooling will be available in a system.

PERSONAL COMPUTER PLUG-INS (PCPIS)

Since the IBM PC was introduced in the mid-1970s, a number of companies have designed products to plug into the open slots of the PC motherboard. Three standards have defined those open slots. They are the ISA (Industry Standard Architecture), EISA (Extended Industry Standard Architecture), and PCI (Peripheral Component Interconnect) standards. EISA is an extension of ISA. All three were defined by the computer industry and none have support for instrumentation. In 1994, a group of industrial computer vendors formed a consortium to develop specifications for systems and boards used in industrial computing applications. They called themselves PICMG (PCI Industrial Computer Manufacturers Group). Today, it includes more than 350 vendors. The group's first specification defined passive backplane computers. PICMG 1.0 *PCI-ISA Passive Backplane* was adopted in October 1994. A second specification, PICMG 1.1 *PCI-PCI Bridge Board,* was adopted in May 1995.

PCPI Common Functions AND Applications

The most common applications are data-acquisition, industrial-control, and custom electronic-control systems.

FIGURE 25.6.6   PC plug-in: Multifunction data acquisition module, 16 analog inputs, 2 analog outputs, 24 digital I/O. (*Photo courtesy of ComputerBoards, Inc*)

PCPI System Components

Two different configurations are common with PC plug-in measurement systems. The first configuration consists simply of a personal computer and a few measurement modules. The second configuration comprises a passive backplane, a single-board computer, and several measurement modules. This latter approach can also be referred to as an *industrialized PC*. In both cases, the measurement modules are the same (Fig. 25.6.6).

PCPI Frames.   In the simplest and most common configuration, the personal computer comprises the backplane, power supply, and packaging for the measurement system. System builders can choose a variety of PC form factors—from desktops to server towers. PC backplanes usually provide seven or eight slots. After the installation of standard PC peripherals, only a couple of slots are free for instrumentation. Extender frames are available, but not common. This type of PC plug-in system tends to be small, where only a few instrument functions are needed.

PCPI Measurement Modules.   PC plug-in instrument functions are primarily digital I/O modules and modules containing ADCs and DACs. In addition, there is a small number of basic instruments including oscilloscopes, digital multimeters, and pulse/function generators.

COMPACTPCI

CompactPCI is a derivative of the Peripheral Component Interconnect (PCI) specification from the personal-computer industry. CompactPCI was developed for industrial and embedded applications, including real-time data acquisition and instrumentation. The specification is driven and controlled by PICMG, the same group covered in PC plug-ins. The CompactPCI specification was released in November 1995. To support the needs of general-purpose instrumentation, a PICMG subgroup developed a CompactPCI instrumentation specification, called PXI (PCI *eXtended for Instrumentation*). The first public revision of PXI was released in August 1997. The PXI specification continues to be developed and maintained by the PXI Systems Alliance (PXISA), a separate industry group.

CompactPCI Specification

Similar to VME, the CompactPCI specification defines backplane functionality, electrical specifications, and mechanical specifications. The functionality is the same as the PCI bus, as defined by the PCI local bus specification, but with some additions. These additions provide pushbutton reset, power-supply status, system-slot identification, and legacy IDE interrupt features. A unique feature of CompactPCI is *Hot Swap*, the ability to insert or remove modules while power is applied. This is an extension to the core specification.

CompactPCI System Components

A CompactPCI measurement system usually consists of one or more frames, an embedded computer module, and various measurement modules.

Frames. Frames refer to the backplane, power supply, and packaging that encloses a system. System builders can choose a solution at many different levels from the component level (i.e, a backplane to complete powered desktop enclosures). CompactPCI backplanes are available with two, four, six, or eight slots. It is possible to go beyond eight slots using a bridge chip on the frame or a bridge card.

Embedded Computers. The PCI bus is commonly used in many computers, from personal computers to high-end workstations. For that reason, a large number of CompactPCI embedded processors are available.

Measurement Modules. Instrument functions in CompactPCI are primarily digital I/O modules and modules containing ADCs and DACs. In addition, there is a small but growing number of traditional instruments, including oscilloscopes, digital multimeters, and serial data analyzers.

Switching Modules CompactPCI, specifically PXI, includes a wide range of switching products, including simple relays, multiplexers and matrices, RF switches, and FET switches.

Compact PCI Software. PXI adopted as part of its specification many of the features of VXI plug&play software. It adopted software frameworks for Windows 95 and Windows NT. These frameworks are required to support the VISA I/O standard.

EMBEDDED COMPUTERS IN ELECTRONIC INSTRUMENTS

Tim Mikkelsen

INTRODUCTION

All but the simplest electronic instruments have some form of embedded computer system. More and more computing is being embedded in the instrument because of reductions of computing cost and size. This is increasing computing power as well as increasing the number of roles for computing in the instrument domain. Consequently, systems that were previously a computer or PC and an instrument are now just an instrument.

This transition is happening because of the demand for functionality, performance, and flexibility in instruments and also because of the low cost of microprocessors. Embedded computers are almost always built from microprocessors or microcontrollers. In fact, the cost of microprocessors is sufficiently low and their value sufficiently high, hence most instruments have more than one embedded computer.

Embedded Computer Model

The instrument and its embedded computer normally interact with four areas of the world: the measurement, the user, peripherals, and external computers. The instrument needs to receive measurement input and/or send out source output. A *source* is defined as an instrument that generates or synthesizes signal output. An *analyzer* is an instrument that analyzes or measures input signals. These signals can consist of analog and/or digital signals. The *front end* of the instrument is the portion of the instrument that conditions, shapes, and modifies the signal to make it suitable for acquisition by the analog-to-digital converter.

The instrument normally interacts with the user of the measurement. The instrument also generally interacts with an external computer that is usually connected for control or data-connectivity purposes. Finally, the instrument is often connected to local peripherals, primarily for printing and storage. Figure 25.7.1 is an example of an embedded computer. A generalized block diagram for the embedded computers in an instrumentation environment appears in Fig. 25.7.2.

The embedded computer is typically involved in the control and transfer of data via external interfaces. This enables the connection of the instrument to external PCs, networks, and peripherals. Examples include local area networks (LAN), *IEEE 488* (also known as *GPIB* or *HP-IB*), *RS-232* (serial), *Centronics* (parallel), *Universal Serial Bus* (USB), and *IEEE 1394* (Firewire).

The embedded computer is also typically involved in the display to and input from the user. Examples include *keyboards*, *switches*, *rotary pulse generators* (RPGs, i.e, knobs), *LEDs* (single or alpha-numeric displays), *LCDs*, *CRTs*, and *touch screens*.

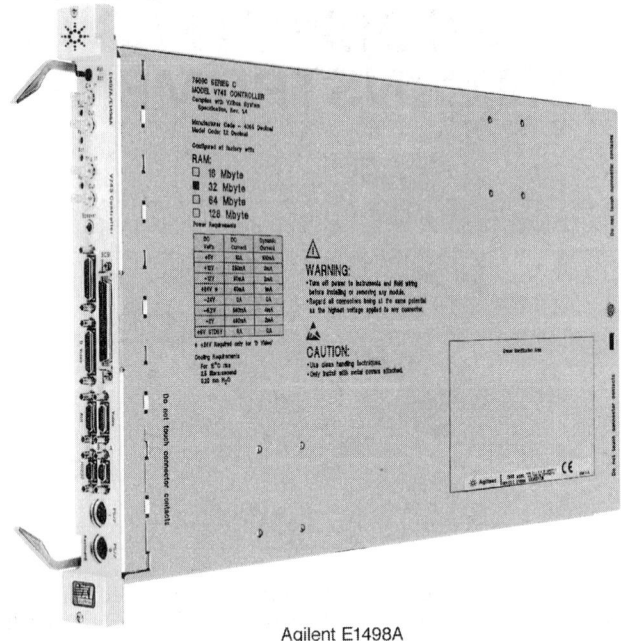

Agilent E1498A

FIGURE 25.7.1 Typical of the embedded computers on the marketplace is the Agilent Technologies E1498A Embedded Controller. This single-slot, C-sized message-based computer was developed specifically for VXI.

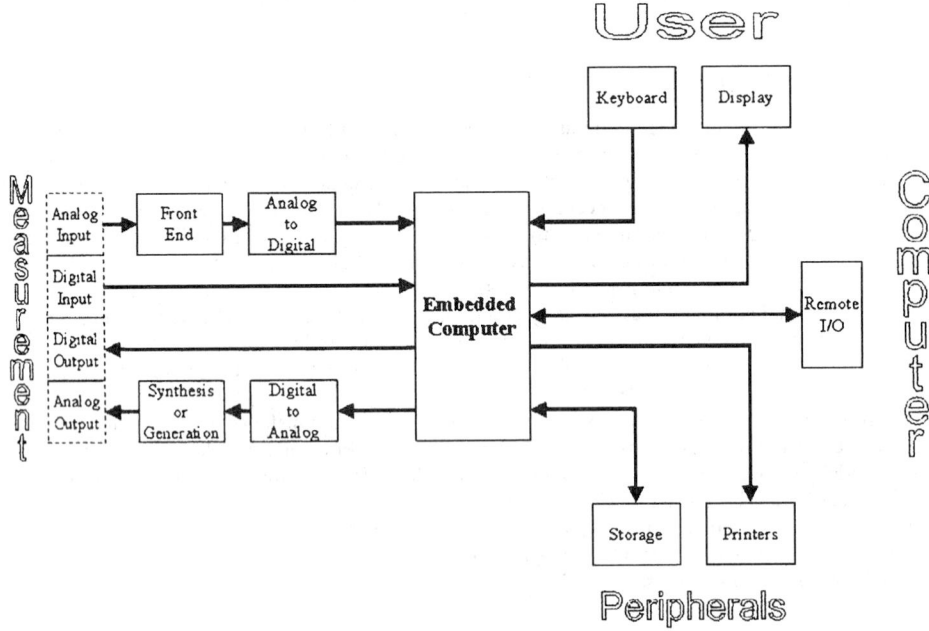

FIGURE 25.7.2 An embedded computer generalized block diagram.

Many instruments often have a large amount of configuration information because of their advanced capabilities. The embedded computer enables saving and recalling of the instrument state. Also, the embedded computer is used sometimes for user customization of the instrument. This can range from simple configuration modifications to complete instrument programmability.

With very powerful embedded computers available in instruments, it is often unnecessary to connect the instrument to an external computer for more-advanced tasks. Examples include *go/no-go*, also known as *pass/fail* testing and data logging.

The embedded computer almost always performs calculations, ranging from very simple to very complex, which convert the raw measurement data to the target instrument information for measurement or vice versa for source instruments. The embedded computer generally controls the actual measurement process. This can include control of a range of functions such as analog-to-digital conversion, switching, filtering, detection, and shaping. The embedded computer is almost always used to perform at least a small amount of self-testing. Most instruments use embedded computers for more extensive calibration tests.

BENEFITS OF EMBEDDED COMPUTERS IN INSTRUMENTS

In addition to the direct uses of embedded computers, it is instructive to think about the value of an embedded computer inside an instrument. The benefits occur throughout the full life cycle of an instrument, from development through maintenance.

One of the biggest advantages of embedding computers within an instrument is that they allow several aspects of the hardware design to be simplified. In many instruments, the embedded computer participates in acquisition of the measurement data by servicing the measurement hardware. Embedded computers also simplify the digital design by providing mathematical and logical manipulations, which would otherwise be done in hardware. They also provide calibration both through numerical manipulation of data and by controlling calibration hardware. This is the classic transition of function from hardware to software.

The embedded computer allows for lower manufacturing costs through effective automated testing of the instrument. They are also a benefit because they allow for easier and lower-cost defect fixes and upgrades (with a ROM or program change).

When used as a stand-alone instrument, embedded computers can make the setup much easier by providing online help or setup menus. This also includes automatic or user-assisted calibration. Although many are stand-alone instruments, a large number are part of a larger system. The embedded computers often make it easier to connect an instrument to a computer system by providing multiple interfaces and simplified or automatic setup of interface characteristics.

Support Circuitry

Although requirements vary, most microprocessors require a certain amount of support circuitry. This includes the generation of a system clock, initialization hardware, and bus management. In a conventional design, this often requires two or three external integrated circuits (ICs) and five to ten discrete components. The detail of the design at this level depends heavily on the microprocessor used. In complex or high-volume designs, an *application-specific integrated circuit* (ASIC) can be used to provide much of this circuitry.

Memory

The microprocessor requires memory both for program and data store. Embedded computer systems usually employ both ROM and RAM.

Read-only memory (ROM) retain their contents if power is removed.

Random access memory (RAM) is a historical but inadequate term that really refers to read/write memory—memory whose contents can be changed. RAM memory is volatile; it will lose its contents when power is no longer applied. RAM is normally implemented as either static or dynamic devices.

Static memory is a type of electrical circuit that will retain its data, with or without access, as long as power is supplied. It is normally built with latches or flip-flops.

Dynamic memory is built out of a special type of circuit that requires periodic memory access (every few milliseconds) to refresh and maintain the memory state. It uses a switched capacitor for the storage element. This is handled by memory controllers and requires no special attention by the developer.

The advantage of the dynamic memory RAM is that it consumes much less power and space. ROM is used for program storage because the program does not usually change after power is supplied to the instrument. A variety of technologies are used for ROM in embedded applications.

Nonvolatile memory. Some instruments are designed with special nonvolatile RAM that retains its contents after power has been removed. This is necessary for storing such information as calibration and configuration data. This can be implemented with regular RAM memory that has a battery backup. It can also be provided by special nonvolatile memory components, most commonly, flash memory devices.

Flash memory is a special type of EEPROM that uses block transfers—instead of individual bytes—and has a fairly slow, in computer terms, write time. So, it is not useful as a general read/write memory device, but is perfect for nonvolatile memory purposes. Also, a limited number of writes are allowed (on the order of 10,000). All embedded systems have either a ROM/RAM or a flash/RAM memory set so that the system will be able to operate the next time the power is turned on.

INSTRUMENT HARDWARE

Given that the role of these microprocessors is instrumentation (measurement, analysis, synthesis, switches, and the like), the microprocessor needs to have access to the actual hardware of the instrument. This instrument hardware is normally accessed by the microprocessor like other peripheral components such as registers or memory locations. Microprocessors frequently interface with the instruments' analog circuits using *analog-to-digital converters* (ADCs) and *digital-to-analog converters* (DACs).

In an analog instrument, the ADC bridges the gap between the analog domain and the digital domain. In many cases, substantial processing is performed after the input has been digitized. Increases in the capabilities of ADCs enable the analog input to be digitized closer to the front end of the instrument, allowing a greater portion of the measurement functions to occur in the embedded computer system. This has the advantages of providing greater flexibility and eliminating errors introduced by analog components.

Just as ADCs are crucial to analog measuring instruments, DACs play an important role in the design of source instruments such as signal generators. They are also very powerful when used together. For example, instruments can have automatic calibration procedures where the embedded computer adjusts an analog circuit with a DAC and measures the analog response with an ADC.

PHYSICAL FORM OF THE EMBEDDED COMPUTER

Embedded computers in instruments take one of three different forms: a separate circuit board, a portion of a circuit board, or a single chip. In the case of a separate circuit board, the embedded computer is a board-level computer that is a circuit board separate from the rest of the measurement function. An embedded computer that is a portion of a circuit board contains a microprocessor, its associated support circuitry and some portion of the measurement functions on the same circuit board. A single-chip embedded computer can be a microcontroller, digital signal processor, or microprocessor core with almost all of the support circuitry built into the chip.

Digital Signal Processor (DSP)

A DSP is a special type of microcontroller that includes special instructions for digital signal processing, allowing it to perform certain types of mathematical operations very efficiently. These math operations are primarily *multiply and accumulate* (MAC) functions, which are used in filter algorithms.

Microprocessor Cores

These are custom microprocessor, IC segments, or elements that are used within custom-designed ICs. Assume an instrument designer has a portion of an ASIC that is the CPU core. The designer can then integrate much of the rest of the system, including some analog electronics, on the ASIC, creating a custom microcontroller. This approach minimizes size and power. In very high volumes, the cost can be very low. However, these chips are intended for very specific applications and are generally difficult to develop.

Architecture of the Embedded Computer Instrument

Just as an embedded computer can take a variety of physical forms, there are a number of ways to configure an embedded computer instrument. The architecture of the embedded computer has significant impact on many aspects of the instrument including cost, performance, ease of development, and expandability. The range of choices include:

- Peripheral-style instruments (externally attached to a PC)
- PC plug-in instruments (circuit boards inside a PC)
- Single-processor instruments
- Multiple-processor instruments
- Embedded PC-based instruments (where the embedded computer is a PC)
- Embedded workstation-based instruments.

EMBEDDED COMPUTER SYSTEM SOFTWARE

As stated earlier, the embedded computer in an instrument requires both hardware and software components. Embedded computer system software includes:

- *Operating system*—the software environment that the instrument applications run within
- *Instrument application*—the software program that performs the instrument functions on the hardware
- *Support and utility software*—additional software the user of the instrument requires to configure, operate, or maintain the instrument such as reloading or updating system software, saving and restoring configurations.

USER INTERFACES

Originally, instruments used only panel-mounted, *direct controls* that were connected directly to the analog and digital circuits. As embedded computers became common, instruments began employing *menu* or *keypad-driven* systems, in which the user input was read by the computer, which then modified the circuit operation. Today, designs have progressed to the use of *graphical user interfaces* (GUIs). Although some instruments are intended for automated use or are *faceless* (have no user interface), most need some way for the user to interact with the measurement or instrument. All of these user interface devices can be mixed with direct control devices such as meters, switches, and potentiometers/knobs. There are a variety of design challenges in developing effective user interfaces in instruments.

EXTERNAL INTERFACES

Most instruments include external interfaces to a *peripheral*, another instrument device, or to an external computer. An interface to a peripheral allows the instrument to use the external peripheral, normally printing or plotting the measurement data. An interface to another device allows the instrument to communicate with or control another measurement device. The computer interface provides a communication channel between the embedded computer and the external computer. This allows the user to:

- Log (capture and store) measurement results
- Create complex automatic tests
- Combine instruments into systems
- Coordinate stimulus and response between instruments

The external computer accomplishes these tasks by transferring data, control, setup, and/or timing information to the embedded computer. At the core of each of these interfaces is a mechanism to send and receive a stream of data bytes.

Hardware Interface Characteristics.   Each interface that has been used and developed has a variety of interesting characteristics, quirks, and trade-offs. However, external interfaces have some common characteristics to understand and consider:

- *Parallel or serial*—How is information sent (a bit at a time or a byte at a time)?
- *Point to point or bus/network*—How many devices are connected via the external interface?
- *Synchronous or asynchronous*—How are the data clocked between the devices?
- *Speed*—What is the data rate?

Probably the most fundamental characteristic of hardware interfaces is whether they send the data stream one bit at a time, *serial*, or all together, *parallel*. Most interfaces are serial. The advantage of serial is that it limits the number of wires down to a minimum of two lines (data and ground). However, even with a serial interface, additional lines are often used (transmitted data, received data, ground, power, request to send, clear to send, and so forth). Parallel interfaces are normally 8 bit or 16 bit. Some older instruments had custom *binary-coded decimal* (BCD) interfaces, which usually had six sets of 4-bit BCD data lines. Parallel interfaces use additional lines for handshaking-explicit indications of *data ready* from the sender and *ready for data* from the receiver.

TABLE 25.7.1   Common Instrument Software Protocols

Software protocol	Description
The IEEE 488.2	The *IEEE 488.2, Codes, Formats, Protocols and Common Commands for Use with IEEE 488.1* is a specification that defines: 39 common commands and queries for instruments, the syntax for new commands and queries, and a set of protocols for how a computer and the instrument interact in various situations. Although this is a companion to the IEEE 488.1 interface, it is independent of the actual interface, but it does depend on certain interface characteristics.
SCPI	The *Standard Commands for Programmable Instruments* (SCPI) is a specification of common syntax and commands so that similar instruments from different vendors can be sent the same commands for common operations. It also specifies how to add new commands that are not currently covered in the standard.
TCP/IP	The *Transmission Control Protocol/Internet Protocol* (TCP/~P) specification is the underlying protocol used to connect devices over network hardware interfaces. The network hardware interface can be a LAN or a WAN.
FTP	The *File Transfer Protocol* (FTP) specification is a protocol used to request and transfer files between devices over network hardware interfaces.
HTTP	The *HyperText Transfer Protocol* (HTTP) specification is a protocol used to request and transfer Web (HyperText Markup Language, HTML) pages over network hard\~are interfaces.
VXI-11	The *VXI-11 plug-and-play specification* is a protocol for communicating with instruments that use the VXI-bus (an instrument adaptation of the VME bus), GPIB/HP-IB, or a network hardware interface.

SOFTWARE PROTOCOL STANDARDS

The software protocol standards listed in Table 25.7.1 provide the hardware interface between devices and computers. The *physical layer* is necessary, but not sufficient. To actually exchange information, the devices (and/or computers) require defined ways to communicate, called *protocols*. If a designer is defining and building two or more devices that communicate, it is possible to define special protocols (simple or complex). However, most devices need to communicate with standard peripherals and computers that have predefined protocols that need to be supported. The protocols can be very complex and layered (one protocol built on top of another). This is especially true of networked or bus devices.

USING EMBEDDED COMPUTERS

Instruments normally operate in an analog world. It has characteristics, such as noise and nonlinearity of components, that introduce inaccuracies in the instrument. Instruments generally deal with these incorrect values and try to correct them by using software in the embedded computer to provide calibration (adjusting for errors inside the instrument) and correction (adjusting for errors outside of the instrument).

Using Instruments That Contain Embedded Computers

In the process of selecting or using an instrument with an embedded computer, a variety of common characteristics and challenges arise. This section covers some of the common aspects to consider:

- *Instrument customization*—What level of instrument modification or customization is needed?
- *User access to the embedded computer*—How much user access to the embedded computer as a general-purpose computer is needed?
- *Environmental considerations*—What is the instrument's physical environment?
- *Longevity of instruments*—How long will the instrument be in service?

User Access to an Embedded Computer

As described earlier, many instruments now use an embedded PC or workstation as an integral part of an instrumentation system. However, the question arises: "Is the PC or workstation visible to the user?" This is also related to the ability to customize or extend the system. Manufacturers realize benefits from an embedded PC because there is less software to write and it is easy to extend the system (both hardware and software). Manufacturers realize these benefits—even if the PC is not visible to the end user. If the PC is visible, users often prefer an embedded PC because it is easy to extend the system, the extensions (hardware and software) are less expensive, and they don't require a separate PC. However, there are problems in having a visible embedded PC. For the manufacturer, making it visible exposes the internal architecture. This can be a problem because competitors can more easily examine their technologies. Also users can modify and customize the system. This can translate into the user overwriting all or part of the system and application software. This is a serious problem for the user, but is also a support problem for the manufacturer. Many instrument manufacturers who have faced this choice have chosen to keep the system closed, and not visible to the user, because of the severity of the support implications.

The user or purchaser of an instrument has a choice between an instrument that contains a visible embedded PC and an instrument that is just an instrument, independent of whether it contains an embedded PC.

It is worth considering how desirable access to the embedded PC is to the actual user of the instrument. The specific tasks that the user needs to perform using the embedded PC should be considered carefully. Generally, the instrument with a visible embedded PC is somewhat more expensive.

INDEX

ABOUT THE EDITORS

DONALD CHRISTIANSEN, former staff director for the Institute of Electrical and Electronics Engineers, is a publishing consultant and president of Informatica. He is a Fellow of the IEEE, an Eminent Member of Eta Kappa Nu, and Editor Emeritus of *IEEE Spectrum*, where for many years he held the post of Editor and Publisher. Mr. Christiansen received his electrical engineering degree from Cornell University and worked for Philco Corporation and CBS Electronics, where he was an engineering group leader. He is a registered professional engineer. His interests include the history of science and technology, engineering education, and ethics in engineering and engineering management. He is a Fellow of the Radio Club of America and a member of the Society for the History of Technology. He served as a member of the National Research Council Commission on the Education and Utilization of the Engineer, and initiated the *IEEE Spectrum* Precollege Math/Science Education Award program to recognize innovative projects that stimulate young people's interest in technical careers. He was elected to membership in the New York Academy of Sciences and the Royal Institution (London), and is a Fellow of the World Academy of Art and Science. He was a member of the IEEE Publications Board for several years and a member of the magazine policy committee of the American Institute of Physics. He is the editor of *Engineering Excellence*, published by the IEEE Press.

CHARLES K. ALEXANDER is Professor of Electrical and Computer Engineering and Dean of Fenn College of Engineering at Cleveland State University. A Fellow of the Institute of Electrical and Electronics Engineers, he is active in IEEE affairs and served as IEEE president in 1997. Dr. Alexander received his BSEE from Ohio Northern University, and his MSEE and Ph.D. from Ohio University. He has served as a consultant to numerous companies and governmental organizations, including the Navy and the Air Force. He is Director of the Center for Research in Electronics and Aerospace Technology at CSU. His research and development projects have ranged from solar energy to software engineering. He was named Distinguished Professor at Youngstown State University in recognition of his outstanding teaching and research, and was awarded the Distinguished Engineering Education Award and the Distinguished Engineering Education Leadership Award, both by the San Fernando Engineering Council. He is co-author of *Fundamentals of Electric Circuits*, published by McGraw-Hill, and has written numerous technical papers.

RONALD K. JURGEN is the editor of the *Automotive Electronics Handbook,* 2nd edition (McGraw-Hill, 1999) and the *Digital Consumer Electronics Handbook* (McGraw-Hill, 1997). Mr. Jurgen received his electrical engineering degree from Rensselaer Polytechnic Institute. He is the editor of the Society of Automotive Engineers'. He is a Life Senior Member of the Institute of Electrical and Electronics Engineers and former Senior Editor of *IEEE Spectrum*.